The HARPER
ENCYCLOPEDIA of
MILITARY
HISTORY

Books by T. N. Dupuy

THE BATTLE OF AUSTERLITZ

REVOLUTIONARY WAR LAND BATTLES
(with Gay M. Hammerman)

REVOLUTIONARY WAR NAVAL BATTLES
(with Grace P. Hayes)

MODERN LIBRARIES FOR MODERN COLLEGES

COLLEGE LIBRARIES IN FERMENT

MILITARY HISTORY OF THE CHINESE CIVIL WAR

MILITARY HISTORY OF WORLD WAR I

MILITARY HISTORY OF WORLD WAR II

HOLIDAYS

CIVIL WAR NAVAL ACTIONS

CIVIL WAR LAND BATTLES

CAMPAIGNS OF THE FRENCH REVOLUTION AND OF NAPOLEON

FAITHFUL AND TRUE

MILITARY LIVES—ALEXANDER THE GREAT TO WINSTON CHURCHILL

ALMANAC OF WORLD MILITARY POWER
(ed., with John A. C. Andrews and Grace P. Hayes)

A DOCUMENTARY HISTORY OF ARMS CONTROL AND DISARMAMENT
(ed., with Gay M. Hammerman)

PEOPLE AND EVENTS OF THE AMERICAN REVOLUTION
(with Gay M. Hammerman)

MONGOLIA

A GENIUS FOR WAR: THE GERMAN ARMY AND GENERAL STAFF, 1807–1945

NUMBERS, PREDICTIONS, AND WAR

ELUSIVE VICTORY: THE ARAB-ISRAELI WARS, 1947–1974

THE EVOLUTION OF WEAPONS AND WARFARE

GREAT BATTLES OF THE EASTERN FRONT
(with Paul Martell)

OPTIONS OF COMMAND

FLAWED VICTORY: THE ARAB-ISRAELI CONFLICT AND THE 1982 WAR IN LEBANON
(with Paul Martell)

UNDERSTANDING WAR: HISTORY AND THEORY OF COMBAT

DICTIONARY OF MILITARY TERMS
(with Curt Johnson and Grace P. Hayes)

UNDERSTANDING DEFEAT

ATTRITION: FORECASTING BATTLE CASUALTIES AND EQUIPMENT LOSSES IN MODERN WAR

IF WAR COMES, HOW TO DEFEAT SADDAM HUSSEIN
(with Curt Johnson, David L. Bongard, and Arnold C. Dupuy)

FUTURE WARS

THE ENCYCLOPEDIA OF MILITARY BIOGRAPHY
(with Curt Johnson and David L. Bongard)

Consultants and Contributors

RICHARD C. ANDERSON, JR., Senior Historian, The Dupuy Institute

DAVID I. BONGARD, Senior Historian, The Dupuy Institute

D.G.E. HALL, Professor Emeritus of History, London University

JOHN D. HAYES, Rear Admiral, United States Navy, Rtd.

HAROLD C. HINTON, Professor of Political Science, George Washington University

JAMES D. HITTLE, Brigadier General, United States Marine Corps, Rtd.

C. CURTISS JOHNSON, Vice President, Research, HERO-TNDA, Inc.

W. BARTON LEACH, Brigadier General, United States Air Force Reserve, Rtd., Story Professor of Law, Harvard University

LOUIS MORTON, Professor of History, Dartmouth University

JOHN F. SLOAN, Lieutenant Colonel, United States Army, Rtd.

CHESTER G. STARR, Professor of History, University of Illinois

THOMAS D. STAMPS, Brigadier General, United States Army, Rtd., Professor Emeritus of Military Art and Engineering, United States Military Academy

CHARLES H. TAYLOR, H.C. Lea Professor of Medieval History, Harvard University

FREDERICK TODD, Curator Emeritus, United States Military Academy Museum

The HARPER ENCYCLOPEDIA *of* MILITARY HISTORY

From 3500 B.C. *to the* Present

R. Ernest Dupuy
and
Trevor N. Dupuy

FOURTH EDITION

HarperCollins*Publishers*

To the memory of

THEODORE AYRAULT DODGE *and* JOHN FREDERICK CHARLES FULLER,

who pointed the way

FIRST EDITION

HarperCollins books may be purchased for educational, business, or sales promotional use. For information, please write: Special Markets Department, HarperCollins Publishers, Inc., 10 E. 53rd St., New York, NY 10022.

Library of Congress Cataloging-in-Publication Data

Dupuy, Trevor Nevitt, 1916–
   The Harper encyclopedia of military history : from 3500 B.C. to the present /
Trevor N. Dupuy and Ernest R. Dupuy. —
4th ed.
      p.   cm.
   Includes bibliographical references and index.
   ISBN 0-06-270056-1
   1. Military history—Dictionaries.   2. Military art and science—History—
Dictionaries.   I. Dupuy, Ernest R. (Richard Ernest), 1887–1975.   II. Title.
D25.D86 1993
355'.009—dc20   92-17853

95 96 97 PS/RRD 10 9 8 7 6 5 4 3

CONTENTS

MAPS

ix

PREFACE TO THE FOURTH EDITION

It is now more than thirty years since my father and I began to work on *The Encyclopedia of Military History,* and almost twenty-two years since we completed the first edition. The acceptance of that work as a standard reference by the scholarly community was gratifying to both of us, particularly since we were so acutely aware of its shortcomings. So, we had barely finished that first edition before we were at work on a revised edition, attempting to correct those errors and fill those gaps of which we were aware, and those also brought to our attention by constructive critics. Sadly, my father died before that revised edition appeared.

But the world has not stood still. Wars and other events of military significance have occurred with the passage of years—as they always have. And I have found more of the gaps and errors inevitable in a fact-filled work as massive as this. So to keep up with these events, it became necessary to produce a Second Revised (or Third) Edition.

The impetus for going through this process once more—and more comprehensively—came from a visit to China in 1988, and was reinforced by another visit to that country in 1989. When visiting the excavation site where not long ago Chinese archaeologists found the famous life-size terra-cotta soldiers of Emperor Shih Huang Ti, founder of China's Ch'in Dynasty in 222 B.C., I was suddenly struck by something that I knew but had never previously thought about: Chinese armies were making effective use of sophisticated cross-bows a millenium before these weapons were significant in Western warfare. That led me to do some research, which in turn caused me to change substantial portions of the early chapters of this book. I sent the draft of these changes to scholars at the Chinese Academy of Military Science, just outside Beijing, for comment. I received the benefit of their comments in an all-day session at the Academy in July 1989. (Particularly significant in that meeting was that the Chinese historians convinced me that recent archaeological work had finally confirmed the true identity of the famed military philosopher Sun Tse.) The result was further changes and additions to the revisions I had made the previous year.

Soon after I returned from China, I informed Mr. M. S. ("Buz") Wyeth, executive editor at Harper & Row (now HarperCollins) of these changes, as well as others that I felt should be made. (Buz has shepherded this project, from the publisher's standpoint, from its inception: a conversation I had with him and the legendary Cass Canfield.) Buz and I agreed that the time had come for a truly comprehensive revision of the encyclopedia. And so, with the assistance of several colleagues, I set about that task, which is now finished—even though I recognize that it will never be complete. I am particularly indebted to David L. Bongard, not only for his thorough research and excellent writing, but also for overall coordination of the efforts of all of us working on the revision project. Also making major contributions to the revisions were C. Curtiss Johnson, John F. Sloan, Richard C. Anderson, Jr., Peter A. Frandsen, and Lawrence D. Higgins.

TREVOR N. DUPUY

MCLEAN, VIRGINIA
NOVEMBER 1991

PREFACE TO THE
SECOND REVISED EDITION

It is hard for me to realize that it has been fourteen years since the publication of the first edition of this book. Those years—including the eight since the publication of the revised edition—have gone by more quickly than a mortal likes to contemplate. Passage of time has another effect upon the author of an encyclopedia, and unfortunately it seems particularly to affect a work devoted to the recording of wars and military affairs: the work is once more outdated.

This edition, therefore, attempts to catch up with time and to make this *Encyclopedia of Military History* current as of the end of 1983 and even early 1984. It also seeks to correct those inaccuracies or omissions in the earlier editions that my colleagues at the Historical Evaluation and Research Organization and I have discovered, or have had brought to our attention. In this process of updating I have been assisted particularly by my colleague, Dr. Charles R. Smith, and also by the publisher of HERO Books, Guy P. Clifton.

TREVOR N. DUPUY

FAIRFAX, VIRGINIA
SEPTEMBER 1985

PREFACE TO THE
REVISED EDITION

The six years since the original Preface was written have seen the conclusion of the war in Vietnam, the beginning and end of a short but important war in the Middle East, and a series of brief revolutions, *coups d'état,* and politico-military developments that warrant the publication of a new edition.

Another reason for revision was the confirmation of our comment in the original Preface that such a work cannot be completely accurate. Accordingly, we have made minor corrections, and a few substantial revisions, throughout the book where inaccuracies or omissions have been brought to our attention. We welcome comments from our readers, and always seek to improve the book.

The revision was begun jointly by both of the authors, but time claimed one of us during the course of this work. I know, however, that my father would have wanted to join me in expressing our thanks for assistance not only to those who have been kind enough to send us their comments, but also to our colleagues of the Historical Evaluation and Research Organization who have given so much help in this revision, and particularly to Mrs. Grace P. Hayes.

TREVOR N. DUPUY

DUNN LORING, VIRGINIA
June 1976

PREFACE TO THE FIRST EDITION

In writing this book, the authors have had two general purposes in mind: to present to both scholar and general reader a comprehensive survey of the history of war and of military affairs in the world throughout recorded human experience; and to provide a reliable, relatively complete, and authoritative reference work covering the entire sweep of world military history. As military historians, we were acutely aware of the lack of any work or works which even attempted to accomplish either of these purposes.

War has been a concomitant of man's existence ever since the first rival cavemen started trying to beat one another to death with club and stone. Through the ages the means and techniques of conflict have become increasingly sophisticated and complex, but there has been no change in the fundamental human objectives of war. Not even the recent advent of weapons possessing the potentiality of utter extermination has yet substantially modified man's willingness to resort to war to force his will upon others.

It is partly because this present situation demands the most thorough possible understanding of war, and partly because of the historian's belief that knowledge of the past is worthwhile in itself, that we have been impelled to accomplish our first purpose. Much has been written about war, and about the broader implications of military affairs within societies, but we know of no previous attempt to present the entire cause of the world's military history in orderly, readable form. As professional soldiers, as well as military historians, we have attempted not only to present the major facts of the world's military history, but also to interpret them—although with no intent that this should require the thoughtful reader to accept our interpretation.

As to our second purpose, our own research has made us acutely aware of the lack of any over-all reference work on military history to which other historians and other researchers could refer with reasonable confidence. We also recognize the danger of attempting to fill this void. No work which deals with the activities of men of all nations and all parts of the world since the dawn of history can hope to be either completely accurate or totally comprehensive. In part this is unavoidable; even if the authors were not subject to human error and human bias, they would be forced to rely upon sources which have all too many human shortcomings; the absolute truth can never be known about any past event. In many instances, we have been forced to choose among amazingly diverse and conflicting accounts of a single event, guided by common sense and the most objective possible evaluation of the sources. Furthermore, in such a vast undertaking, we have undoubtedly missed sources which, had we been aware of them, might have modified our presentation of some of the facts. Finally, no historical work can include everything about anything.

This question of what to include and what to exclude is a particularly difficult matter in a single-volume reference work, even a very large one. With a subject as vast as ours, we had to be arbitrary. We have tried to include everything of significance in world military history. Unfortunately, significance is not only subjective; it can change from time to time and from circumstance to circumstance. So some readers may wonder why we have included some material which they consider insignificant, and omitted other facts which they believe essential. This, however, is their opinion; what is included herein reflects the authors' joint opinion.

We recognize the fallibility of all authors, even within the parameters that they have set for themselves. To reduce such fallibility in this book to limits acceptable to other scholars—and to ourselves—we have had the good fortune to obtain advice and commentary from the distinguished consultants listed elsewhere in these pages. We are vastly indebted to them for their interest and for their assistance.

To two of our professional colleagues of the Historical Evaluation and Research Organization—Mrs. Gay M. Hammerman and Mrs. Grace P. Hayes—we owe a particular debt of gratitude. While indexing the book—in itself a stupendous task, as can be readily seen—they also read it thoroughly and gave us the benefit of their historical knowledge and experience by drawing our attention to discrepancies, errors, and questionable interpretations. They added substantially to such scholarly merit as this work may possess. Their work was greatly facilitated by research assistance from Mrs. Edith Kilroy, Mrs. Bonnie Marsh, Miss Karen Rice, and Mrs. Jonna Dupuy.

Messrs. Cass Canfield and M. S. Wyeth, Jr., of HarperCollins, have been most patient and solicitous in their editorial support and guidance, while the expert copy-editing of Mr. James Fergus McRee has gone far to improve the text and to preserve the authors from being unduly influenced by their personal biases.

However, the authors alone stand responsible for the final determination of what is to be included in this book, of how it is to be presented, and for the military assessments made and opinions expressed.

We must also express our gratitude to the ladies who have spent so much time and effort, over so many years, in typing the drafts and redrafts of thousands upon thousands of manuscript pages. In particular, we are indebted to Mrs. Jean D. Brennan, Mrs. Judith B. Mitchell, and Mrs. Billie P. Davis, who have never lost either interest or good humor under the pressure of this trying task. Finally, we want to acknowledge the inspirational encouragement which we received during the ten years this book was in preparation from Laura N. Dupuy, wife of one of the authors and mother of the other.

R. ERNEST DUPUY
TREVOR N. DUPUY

MCLEAN, VIRGINIA
JANUARY 1970

ORGANIZATION OF THIS BOOK AND ITS USE

The schema of this encyclopedia consists of a series of chronologically and geographically organized narratives of wars, warfare, and military affairs. The course of recorded history is arbitrarily divided into twenty-two specific time periods, and one chapter is devoted to each period. Each chapter is prefaced by an introductory essay presenting a professional assessment of the principal military trends of the period, including its outstanding leaders, and the general progression of the military art in tactics, strategy, weaponry, organization, and the like. The chronological surveys of the principal wars of the period which follow are treated separately from the regional or national presentations because they possess a significance or geographical scope that extends beyond the confines of a single region. The remainder of each chapter is divided into the major geographical regions known during the era. Within these regional sections brief subsections present chronologically the principal military events affecting the affairs of each state, or nation, or people within the region.

An exhaustive General Index includes all names and events mentioned in the text, thoroughly cross-referenced. There are separate index sections for Battles and Sieges and for Wars. The indexes include the major abstract, conceptual, and topical terms associated with military affairs where it is believed that they would facilitate index search. Thus a reader in search of some specific incident, or battle, or war, or person, or type of military activity should be able, once he has located it, to relate it immediately to the local and world situation at the time, as well as to the current practice of warfare.

It is also possible to pursue the general course of the military history of a region, or nation, by following the appropriate sections from one chapter to another.

In addition to the essays on military trends at the beginning of each chapter, essays on the principal military systems of ancient and medieval history are included where appropriate. Such descriptions of national military systems are omitted in the chapters on modern military history, because these chapters cover much briefer periods, and the various military systems (such as Napoleonic, Prussian, etc.) are believed to be covered adequately in the introductory essays and in the sectional narratives.

In a work of this sort it is neither practical nor particularly useful to try to document any of the individual facts by reference citations. At the end of the book, however, there is an extensive bibliography, which includes general references as well as those of specific importance for each individual chapter.

The HARPER
ENCYCLOPEDIA of
MILITARY
HISTORY

I

THE DAWN
OF MILITARY HISTORY:

---

to 600 B.C.

MILITARY TRENDS

WARFARE BEGINS

Primitive clashes of force first occurred when groups of Paleolithic men, armed with crude stone implements, fought with other groups for food, women, or land. Somewhere along the prehistoric road other drives—such as sport, the urge for dominance, or the desire for independence—became further causes for armed conflicts. Archaeology tells us—by dating fortifications at Jericho to 6000 B.C. and at Catal Huyuk (Anatolia) to 7000 B.C.—that Neolithic men were waging organized warfare centuries before the invention of writing or the discovery of how to work metal.

The dawn of history and the growing sophistication of organized warfare went hand in hand. Most primitive societies learned the use of metals at the same time that they developed a system of writing. This phenomenon appeared almost simultaneously, and apparently quite independently, in Mesopotamia and Egypt sometime between 3500 and 3000 B.C., when the use of copper for weapons, household implements, and decorations began. Several hundred more years elapsed before men mastered the secret of hardening copper into bronze by mixing tin with it. Comparable development of Bronze Age culture occurred in the Indus Valley sometime before 2500 B.C., and in the Yellow River Valley of China about the same time. Iron metallurgy began to replace bronze in the Middle East shortly before 1000 B.C., and in Europe soon thereafter. It was a few centuries later when the Iron Age appeared in India and China.

The ancient history that began with Bronze Age cultures is known to us largely in the terms of military history. The record is almost entirely devoted to migrations, wars, and conquests. Not until about 1500 B.C., however, are we able to visualize the actual course of any of the constant wars of the Middle East, or dimly to perceive primitive military organization and methods of combat. By the 6th century B.C. relatively comprehensive and more or less continuous records

1

of wars become available. These records reveal that three of the five great military societies of antiquity were flourishing prior to 600 B.C., and the origins of the other two were evident.

GENERAL DEVELOPMENTS

Despite the embellishments of myth and legend, we can discern four broad general trends in warfare: (1) the introduction of military transport—on land and on water; (2) the introduction and then the relatively early decline of the chariot; (3) increasing ascendancy of the horseman—whether barbarians or the elite in society—on the battlefield over the inchoate masses of plebeian foot soldiers; and (4) the all-important introduction of iron and steel, replacing bronze in the manufacture of weapons.

By the year 600 B.C. our knowledge is sufficient to reveal that the art of war had become highly developed in the major centers of civilization. Unquestionably there were gifted military leaders in these early centuries, but with the possible exception of **Thutmosis III** of Egypt, we can hardly more than surmise the reasons for tactical success on the battlefield.

WEAPONS

Primitive Weapons

Weapons fall into two major categories: shock and missile. The original shock weapon was the prehistoric man's club; the first missile weapon was the rock that he hurled at the hunted prey or human enemy. The next important development was the leather sling for hurling small, smooth rocks with greater force for longer distances than was possible by arm power alone. In some regions the rock gradually was displaced by a light club or throwing stick, which in turn evolved into darts, javelins, and the boomerang. The club was modified in a number of other ways. The shock-action counterpart of the javelin was the heavy pike, or thrusting spear. The basic club itself took on a variety of forms, of which the American Indian tomahawk is an axlike example, while clubs with sharpened edges became Stone Age prototypes of the sword. The bow, developed late in the Stone Age, was also invaluable to the early fighting man and to his successors over many centuries.

The most important form of protective armor devised by primitive man was the shield, held almost invariably in the left hand, or on the left arm, leaving the right arm free to wield a weapon. Shields most often were simple wooden frameworks, covered with leather hide, though some were made entirely of wood, and in Asia wicker shields were common.

Other types of protective covering for head, torso, and legs appeared before the Bronze Age. These were of leather, wicker, padded or quilted cloth, or wood.

Historic Weapons

The most important weapons improvement during the early historic period was the adoption of metal for the points, edges, or smashing surfaces in the Bronze and Iron Ages.

The first new weapons of the metallic age were the axe and the mace, the dagger, and then the sword. The long thin blade that characterizes the sword could not have been created until metallurgy had sufficiently developed to permit the working of hard malleable metal. This occurred in the Bronze Age sometime before 2000 B.C., and the sword was probably introduced into warfare by the Assyrians.

Protective armor was also greatly improved during the Bronze Age. Although leather remained the basic and most common material, this was often reinforced with metal; some helmets, breastplates, and greaves were made entirely in metal—at first bronze, later iron.

TACTICS

Sometime before 1500 B.C., weapons, missions, and relative mobility began to dictate the composition of armies. Their bulk consisted of large, tight masses of infantry, wearing little or no armor, probably carrying spears or axes and shields. This infantry component was made up of men from the poorer classes of society, and its purpose was almost solely to provide a solid and stable base around which the more important and better-armed groups could operate. There were additional foot troops with missile weapons, either slingers or archers.

Until about 700 B.C. the elite striking force of this army of antiquity was usually a contingent of chariots. These were small carts, usually light-weight, generally two-wheeled, sometimes armored, and sometimes with sharp blades projecting from the whirling axles of the wheels, and drawn by armored horses. Great nobles and members of the royal family rode to battle in chariots and at times fought from them, though occasionally they dismounted for the actual hand-to-hand fighting. Chariots were apparently first introduced in Sumer about 2500 B.C. They were particularly dominant in warfare from about 1700 to about 1200 B.C. The Egyptian chariot was a mobile firing platform for well-trained archers. The Hittites also fought from the chariot, but with spearmen rather than archers. The Assyrians developed chariot warfare to its greatest sophistication in Western Asia, with light chariots for archers and heavier chariots carrying as many as four spearmen. Chinese chariot warfare was similarly sophisticated.

Chariot

Cavalry, when it appeared soon after 1000 B.C., was often composed of the lesser nobles, who possessed enough wealth to own horses and to supply themselves with good weapons and armor. In other instances cavalry contingents were recruited from neighboring barbarian tribes. In some regions the principal weapon of the horseman was the bow; in others it was the javelin or spear. The Assyrians were the first military power to use cavalry in large numbers.

There was little organization when such an army went to war. The sole objective was to reach a suitable place of battle in order to overwhelm the enemy before he was able to prevail. As time went on and the results of a number of such conflicts brought several towns, or entire regions, under the dominance of one ruler, geographical horizons widened and wars became series of battles, or even campaigns, rather than a single encounter between the forces of two small towns.

A campaign was a huge raid, in which large regions were overrun, defeated armies slaughtered, cities destroyed, and entire peoples enslaved. Men were compelled to fight through both fear and the prospect of loot and booty. There was perhaps some effort to weld units with discipline and to prepare them for battle by training. When armies met, the infantry spearmen stayed together in large groups. The nobles in their chariots or on their horses took positions in the front and on the flanks, and the swarms of lightly armed archers and slingers were out in front.

Maneuver in battle was generally accidental. As one or both armies advanced, the archers and slingers maintained harassing fire until the chariots or horsemen started to charge; the light troops then drifted to flank and rear positions through intervals in the heavy infantry masses. Sometimes the initial charge of chariots and horsemen would strike terror into the opposite side, in which case the battle quickly became a chase, with only the fleetest men of the pursued army escaping the slaughter. More often, the two masses simply converged to carry on the butchery in earnest. This horrible process could last for an hour or more, with the lines swaying back and forth over the growing numbers of dead and wounded, until one side suddenly sensed defeat. This quickly communicated itself through mass hysteria to all the soldiers of that side, and again only a small proportion of the defeated army would escape.

After 1000 B.C. more order was introduced into warfare by the Egyptians, who certainly understood and employed maneuver by well-organized and disciplined units. Soon afterward the Assyrians contributed even more order, organization, and discipline into military affairs, both on and off the battlefield.

MEDITERRANEAN–MIDDLE EAST*

EGYPT, 3100–600 B.C.

c. 3100†–c. 2600. Early Egypt. The first identifiable figure in history was **Menes** (Narmer), the warrior ruler who established a unified kingdom of Egypt. For the next 1300 years

---

* "Middle East" is a modern, and rather imprecise, term. In this text it is used to designate that region of southwestern Asia and eastern North Africa lying roughly between 24° and 60° east longitude, i.e., the area covered by modern Iran, Asiatic Turkey, Iraq, Arabia, the Levantine states, and Egypt.

† There are substantial differences in the dates shown by different respected authorities for the period up to 600 B.C. (and to some extent later). In the case of Menes, for instance, dates ascribed vary from about 3400 to 2900 B.C. In this and all other instances, dates shown in this text will be those that we believe are most reliable. In general we shall avoid indicating either a possible range of dates or the questionable nature of any particular date. The use of the identifying abbreviations B.C. and A.D. will not be used further in the text (as opposed to the headings) save in some instances in the latter part of the 1st century B.C. or in the early part of the 1st century A.D.

Egyptian civilization flourished along the banks of the Nile River, relatively isolated from the rest of the world.

c. 2600–c. 2000. The Old Kingdom. Under the III to VII Dynasties the capital was at Memphis. There were many military expeditions to the neighboring regions of Palestine and Nubia, with Egyptian influence reaching far up the Nile into what is now the Sudan. Internal disorders were frequent, with kings (pharaohs) often embroiled in civil wars against unruly provincial nobles. Consequently there were wide fluctuations in the power exercised by a central authority, the rhythm of change being occasionally punctuated by the violent collapse of dynasties. Standing armies were at first unknown, contingents of provincial militia being assembled into active armies only temporarily when the pharaohs embarked on foreign adventures, or needed forces for internal security or to protect the frontiers.

c. 2000–c. 1600. The Middle Kingdom. Beginning with the XII Dynasty the pharaohs maintained several, well-trained, professional standing armies, usually including Nubian auxiliary units. These became the nuclei for larger forces when required for defense against invaders, or for expeditions up the Nile or across the Sinai. The military strategy was essentially defensive, with fortifications blocking the Isthmus of Suez and the southern frontier at the First Cataract of the Nile.

c. 1800–c. 1600. Hyksos Invasion. Historic Egypt suffered its first foreign invasion during a period of internal weakness. Semitic **Hyksos** (usually translated as "shepherd kings") introduced the horse and the horse-drawn chariot in Egypt. Gradually the invaders expanded southward until (c. 1700) they completely overran Egypt. For another hundred years they maintained their sway, ruling from the Delta through local princes, and ruthlessly repressing frequent Egyptian uprisings.

1600. Revolt of Thebes. The native nobles drove the Hyksos from Upper Egypt.

1580. Emergence of the New Kingdom. Borrowing much from the Hyksos, the pharaohs developed a sophisticated and formidable military machine, distinguished by the introduction of chariots, new weapons, and archery.

1580–1557. Reign of Amosis. He drove the Hyksos into Palestine, again uniting Egypt under native rule. He established a strong central government, greatly reducing the former autonomy of the provincial nobles. He then reconquered Nubia, which had fallen away from Egyptian control during the Hyksos occupation. He created the first permanent army in Egyptian history and placed his main reliance upon comparatively well-drilled, disciplined archers; he adopted the dreaded Hyksos war chariot.

1546–1507. Reigns of Amenophis I and Thutmosis I. They extended Egyptian rule westward into Libya, further south into Nubia, and northeastward into Palestine and Syria, Egyptian armies actually reaching the banks of the Euphrates. Under more peaceful successors, Egyptian control of outlying regions was relaxed.

1525–1512. Reign of Thutmosis I. A strong ruler, he invaded Syria as far as the Euphrates River, conquered northern Nubia, and reduced the power of local princes.

1491–1449. Reign of Thutmosis III. For the first 20 years of his reign he was merely nominal co-ruler with his aunt, **Hatshepsut.** Upon her death (1472) the Hyksos King of Kadesh, in northern Palestine, led a highly organized revolt of the tribes of Palestine and Syria against the supposedly weak young Pharaoh.

1469. Battle of Megiddo. (First recorded battle of history.) Thutmosis led an Egyptian army (possibly 20,000 men) on a rapid and unexpected march into central Palestine. The rebellious chieftains assembled an army at Megiddo, north of Mount Carmel, sending outposts to hold the three passes leading from the south. But Thutmosis pushed through the Megiddo Pass, scattering the defenders in a bold attack he himself led. In the valley beyond, the rebel army, under the King of Kadesh, was drawn up on high ground near the fortress of Megiddo. Thutmosis' army was aligned in a concave formation. While the southern wing engaged the rebels in a holding attack, Thutmosis personally

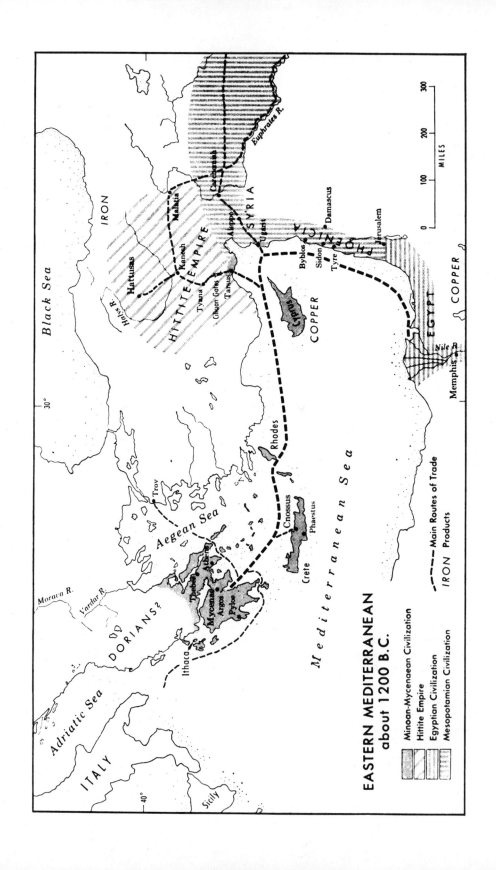

EASTERN MEDITERRANEAN
about 1200 B.C.

Minoan-Mycenaean Civilization
Hittite Empire
Egyptian Civilization
Mesopotamian Civilization

Main Routes of Trade
IRON Products

Black Sea

Adriatic Sea

ITALY

Sicily

Morava R.

Vardar R.

DORIANS?

Ithaca

Pylos
Argos
Mycenae
Thebes
Athens

Aegean Sea

Trov

Rhodes

Crete
Phaestus
Cnossus

Mediterranean Sea

IRON

Halys R.

Hattusas

HITTITE EMPIRE

Tyana
Kanesh
Cilican Gates
Tarsus

Malatia

Carchemish

Euphrates R.

Aleppo

SYRIA

Ugarit

Damascus

Byblos
Sidon
Tyre

PALESTINE

Jerusalem

Cyprus

COPPER

EGYPT

COPPER

Nile R.

Memphis

MILES

0   100   200   300

30°

40°

led the north "horn" in an attack that seems to have driven between the rebel flank and the fortress. The result was envelopment of the rebel flank, and overwhelming victory for the Egyptians.

1470–1450. Height of Egyptian Power. After some 17 campaigns, Thutmosis had not only subdued the rebellious rulers of Palestine and Syria, he had pushed Egyptian rule to the edge of the Hittite Empire in Asia Minor and had expanded into northwestern Mesopotamia. His fleet controlled the eastern Mediterranean.

1380–1365. Reign of Ikhnaton. Serious internal religious disputes weakened Egyptian hold over outlying regions. The warlike Hittites seized Syria, and with local allies overran much of Palestine.

1352–1319. Reign of Haremhab. Just as the Egyptian Empire seemed on the verge of dissolution, **Haremhab,** a general, seized the throne to establish the XIX Dynasty. He restored internal order and halted the erosion of the frontiers by firm defense and by sending offensive expeditions beyond the borders.

1317–1299. Reign of Seti I. A reorganized Egyptian army reconquered Palestine, but was unable to shake Hittite control of Syria.

1299–1232. Reign of Ramses II (son of Seti). He was partly successful in efforts to restore the empire to its old boundaries.

1294. Battle of Kadesh. Ramses led an army of about 20,000 men, including 2,500 chariots, composed largely of Numidian mercenaries against the Hittite stronghold, Kadesh, on the Orontes River. In his haste to capture Kadesh before the main Hittite army could arrive, he and his advance guard were for a while cut off and surrounded by a surprise Hittite attack. Holding out until reinforcements arrived, Ramses repulsed the Hittites. He was unable to capture Kadesh, however, and eventually made peace, with the Hittites controlling most of Syria. Ramses' superior leadership was offset by the fact that many of his enemies were evidently armed with new iron weapons, while his mercenaries were still using bronze weapons.

c. 1200. Invasions by the "Peoples of the Sea." These were seaborne raiders from Mediterranean islands and southern Europe. Most of these raids were repulsed, but some of the invaders, like the Philistines in Palestine, succeeded in establishing themselves along the coast (see pp. 9, 13).

1198–1167. Reign of Ramses III. When he ascended the throne, the Philistines were advancing from Palestine toward Egypt, the Libyans were approaching from the west, and the Peoples of the Sea were again harassing the Delta coast. Ramses decisively defeated all these threats, and reestablished Egyptian control over Palestine. He was the last great pharaoh; in following centuries the power and influence of Egypt declined steadily; the country was frequently overrun by foreign conquerors.

c. 730. Ethiopian Conquest of Egypt. The invaders were led by King **Piankhi,** whose capital was at Napata near the Fourth Cataract of the Nile.

671–661. Assyrian Conquest. Assyrians drove out the Ethiopians (see p. 12).

661–626. Period of Turmoil. Constant Egyptian revolts finally culminated in ejection of the Assyrians (see p. 12).

609–593. Brief Egyptian Resurgence. This was under Pharaoh **Necho,** who led an Egyptian invasion of Palestine and Syria.

609. Second Battle of Megiddo (or Armageddon). Necho easily defeated a Jewish army under **Josiah** and pushed northward to the Euphrates.

605. Battle of Carchemish. Necho was disastrously defeated by **Nebuchadnezzar** of Babylonia; the Egyptians were driven completely from Syria and Palestine.

MESOPOTAMIA

The importance of the Mesopotamian valley as a prize for military activity lay not only in the relatively rich local agriculture and craft production, but also in its preeminence as the center of the most significant trade routes in the ancient world between the Persian Gulf and the

Mediterranean. The early military history of the region is composed of two interrelated processes. The first is the internal cycle created by the efforts of cities in the valley to gain supremacy over the entire region and its trade routes, versus the countervailing efforts of the other cities to retain their independence. The other process was the continual external pressure exerted by ''barbarian'' tribes from the southeastern deserts and the eastern and northern mountains to seize part or all of the fertile valley or occupy strategic points on the trade routes. In fact, most of the military activity was the result of efforts by the inhabitants to keep strategic trade centers out of the hands of interlopers. Strategic success by one state frequently was the result of seizing one end of a trade route rather than direct attack on the enemy. Military power gave a state the capability to control the commerce in critical strategic minerals and other resources, while control of this commerce contributed to military superiority. Gradually the process expanded in scope as the contending parties became larger political entities with greater available military resources. The constant ebb and flow gave the region a much more dynamic and changing character than that of Egypt, which lay virtually unchanged for centuries behind its natural barriers. Finally, these processes culminated in the conquest of the entire region and its immediate environs by one superpower, Persia.

SUMER, AKKAD AND BABYLONIA, 3500–1200 B.C.

c. 3500. Emergence of Sumer. A people of undetermined racial origin migrated southward through Asia Minor, or through the Caucasus Mountains, and settled in southern Mesopotamia. Although these Sumerians developed a civilization contemporaneously with the Egyptians, they never created a stable, unified kingdom. Sumer was divided among a number of independent, constantly warring city-states.

c. 2400. Reign of Lugalzaggisi of Erech (Umma). He created a temporary Sumerian Empire, and may have controlled all of Mesopotamia and part of Syria and Asia Minor, his realm reaching from the Persian Gulf to the Mediterranean, though there is some doubt that his armies ever reached the Mediterranean.

2371–2316. Reign of Sargon of Akkad. He led a Semitic people to conquer Sumer. Sargon extended his empire northwestward into Asia Minor and the Mediterranean coast. This empire lasted nearly 200 years.

c. 2200. Turmoil in Mesopotamia. Akkad collapsed under pressure from a new wave of migrations. During the confusion, the Sumerians reasserted their supremacy in southern Mesopotamia for approximately two centuries.

2006. Fall of Sumer. Elamites, invading from the eastern mountains, destroyed the Sumerian Empire.

c. 2000. Establishment of the First (Old) Babylonian Empire. A new Semitic people, the Amorites, probably from Syria, became dominant in Mesopotamia, with Babylon their capital.

1792–1750. Reign of Hammurabi. This able warrior and enlightened king of Babylonia, ancient history's first famous law-giver, extended his rule over all Mesopotamia.

c. 1700–c. 1300. Decline of the Old Babylonian Empire. In the confusion of the destructive Hittite raids, the Kassites, an obscure barbarian mountain people from east of Babylonia, overran southern Mesopotamia, adopted the civil justice of the conquered area, and established a kingdom that lasted over 4 centuries.

HITTITE KINGDOM, 2000–1200 B.C.

c. 2000. Rise of the Hittites. An Indo-European people who apparently originated northeast of the Caucasus, the Hittites became dominant in northern and central Asia Minor. They maintained steady pressure against neighbors to the east and south. Hittite pressure probably pushed the Hyksos into Egypt (see p. 5).

c. 1590. Reign of Murshilish I. He raided extensively in Mesopotamia, overrunning the Old Babylonian Empire, bringing it to the verge of collapse. He also captured Aleppo, expanding his kingdom's southern boundaries deep into Syria. For the next two centuries the Hittites were occupied with internal disorders, as well as almost constant warfare with the Mitanni of northwestern Mesopotamia.

c. 1460. Defeat by Thutmosis III of Egypt. (See p. 5.) The Egyptian conquerer drove the Hittites out of most of Syria, and the weakened Hittite kingdom paid tribute to Egypt.

1375–1335. Reign of Shubbiluliu. The Hittites revived to re-establish control over most of Anatolia and to conquer the Mitanni. For the next century the Hittites and Egyptians struggled for control of Syria and Palestine (see p. 7).

1281–1260. Reign of Hattushilish III. He made a treaty of peace and alliance with Ramses II (1271), accepting Egyptian sovereignty over Palestine in return for recognition of Hittite control of Syria. In his era the Hittites introduced weapons made of iron. In subsequent years the Hittite kingdom, shaken by internal disorders, declined rapidly. The great Aegean migrations of the "Peoples of the Sea" (see pp. 7, 13) also began to threaten Hittite control of western Anatolia, while a powerful new Mesopotamian kingdom was pushing from the east.

c. 1200. Disintegration of the Hittite Kingdom.

ASSYRIA, 3000–612 B.C.

Early Assyria, 3000–727 B.C.

c. 3000. Emergence of Assyria. The Assyrian people appeared in the upland plains of northeastern Mesopotamia, along the upper reaches of the Tigris River. Flat Assyria, with no natural frontiers, was constantly threatened by neighbors on all sides, particularly the Hittites to the northwest and the Sumerian–Babylonians to the southeast.

c. 2000–c. 1200. Military Development. The Assyrians, engaged in a never-ending struggle to maintain freedom, became the most warlike people of the Middle East (c. 1400). Initially they relied upon an informal militia system, though constant campaigning gave exceptional military proficiency to these part-time soldiers. But the Assyrian economy was severely strained by the long absence of militiamen from fields and workshops. After growing in size, wealth, and power, Assyria temporarily declined (1230–1116).

1116–1093. Reign of Tiglath-Pileser I. Assyria became the leading power of the Middle East, a position she was to maintain almost continuously for five centuries. He expanded Assyrian power into the heart of Anatolia and across northern Syria to the Mediterranean.

c. 1050. Period of Retrenchment. Another wave of migrations—this time Aramean nomads—swept across Mesopotamia. The hard-pressed Assyrians finally repelled, or absorbed, the migrating tribes, and reestablished control over all the main routes of the Middle East.

883–824. Reigns of Ashurnasirpal II and Shalmaneser III. They carried fire and sword across Mesopotamia, into the Kurdish mountains, and deep into Syria. Then came a brief lull in Assyrian expansion, as weak successors were unable to retain the northern conquests against vengeful foes. The Aramean tribes in Mesopotamia also became restive and unruly.

745–727. Reign of Tiglath-Pileser III. He firmly reestablished internal order throughout Mesopotamia, then undertook a systematic series of military expeditions around the periphery of Assyria's borders, reestablishing Assyria's frontiers on the Armenian highlands north of Lake Van and Mount Ararat, then conquering Syria, Palestine, and the lands east of the Jordan. In later years he campaigned repeatedly along the new borders he had established, maintaining order by inspiring fear. His last important operation was to invade Babylonia, reasserting vigorously the hitherto nominal Assyrian sovereignty.

ASSYRIAN MILITARY ORGANIZATION, C. 700 B.C.

Tiglath-Pileser III established the most efficient military, financial, and administrative system the world had yet seen. The army was its heart. He abolished the militia organization and built the state around a standing regular army. The principal business of the nation became war; its wealth and prosperity were sustained by booty and by supervision of trade and finance. A semimilitary bureaucracy carried out the functions of government at home and in the conquered regions, setting the first pattern of centralized imperial control over far-flung provincial territories.

This was the first truly military society of history. No effort was spared that would contribute to the efficiency of the army or assure continued Assyrian supremacy over all possible foes. The Assyrians were the first to recognize fully the advantage of iron over bronze. As early as 1100 their militia armies had been completely equipped with weapons, chariots, and armor made of iron. Tiglath-Pileser I saw to it that this technical superiority was maintained by constant and systematic improvement of weapons, and by the careful training of the soldiers in the use of their arms.

The bulk of the army was comprised of large masses of spearmen, slow-moving and cumbersome, but relatively more maneuverable than similar infantry formations of other peoples of the time. Their irresistible advance was the culminating phase of a typical Assyrian battle plan.

In the Assyrian Army the archers were more highly organized than their counterparts elsewhere and evidently had stronger bows, from which they fired iron-tipped arrows with deadly accuracy. They created confusion in the enemy ranks in preparation for a closely coordinated chariot and cavalry charge.

The main striking force of the Assyrian Army was the corps of horse-drawn, two-wheeled chariots. Their mission, after a preliminary attack by Assyrian infantry, was to smash their way through the ranks of the shaken enemy infantry. Like their contemporaries, the Assyrians used chariots in simple, brute force, but employed them in larger numbers, with more determination, and in closer coordination with archers, spearmen, and cavalry.

The cavalry was the smallest element of the army, but probably the best trained and equipped. The horsemen—some noblemen, but most Scythian mercenaries—fought with a combination of discipline, skill, and ingenuity not possible in the other elements of the army.

Assyrian mounted archer

Only the cavalry could be employed in the occasional maneuvers attempted in battle. By the time of the Assyrian revival under Sargon II (see below), the cavalry had increased in proportional strength and had largely replaced the chariots.

The art of fortification had been well developed in the Middle East before 1000 B.C. The great walls of the large cities were almost invulnerable to the means of attack available within the limited technology of the times. The Assyrians greatly improved the techniques of siegecraft and attack of fortifications. Accompanying their armies were siege trains and various forms of specialized equipment, including materials for building large movable wooden towers (protected from the flaming arrows of defenders by dampened leather hides) and heavy battering rams. From the tops of the wooden towers, skilled archers would sweep the walls of the defenders, to prevent interference with the work of demolition, while nearby other archers, sheltered by the shields of spearmen, would fire arrows—some of them flaming—in a high trajectory over the walls, to harass the defenders and to terrify the population. The methods used by the Assyrians did not originate with them but were apparently borrowed from the Sumerians. Yet it was the skill and organization of employment which brought success to Assyrian siegecraft.

The high degree of organization of the Assyrian army is clearly evidenced by its ability to fight successfully over all kinds of terrain. The organizational details have not been preserved in the fragmentary records available to us, but their field armies may occasionally have approached a strength of 100,000 men. Forces of such size would have required large supply trains for desert or mountain operations, and could have functioned only with smoothly operating staff and logistical systems.

Terror was another factor contributing greatly to Assyrian success. Their exceptional cruelty and ferocity were possibly reflections of callousness developed over centuries of defense of their homeland against savage enemies. But theirs was also a calculated policy of terror—possibly the earliest example of organized psychological warfare. It was not unusual for them to kill every man, woman, and child in captured cities. Sometimes they would carry away entire populations into captivity.

The policies and procedures of Tiglath-Pileser III were employed with vigor and ferocity by his successors and proved invaluable in maintaining security.

Assyria, 722–612 B.C.

722–705. Reign of Sargon II. He was faced by a powerful alliance of the northern provinces, combined with the neighboring tribes and nations of Armenia, the Caucasus, and Media. In a series of campaigns he reconquered the rebellious provinces and extended his rule further north, as well as into central and southern Anatolia. He then returned to Mesopotamia to suppress brutally another Babylonian uprising.

705–681. Reign of Sennacherib. He was faced with comparable insurrections in Palestine, Syria, and Babylonia; among these major setbacks was his repulse at Jerusalem (701, or possibly in a later campaign, 684; see 2 Kings xviii and xix). This repulse was probably the result of a pestilence that ravaged his army. However, he regained the lost provinces, and his successes culminated in the capture and destruction of Babylon (689).

681–668. Reign of Essarhaddon. He was able to maintain better internal order than his immediate predecessors. After repelling incursions of the Cimmerians, an Indo-European people inhabiting south Russia and the Caucasus, Essarhaddon conquered Egypt (671). Three years later he died while suppressing a revolt in that country (see p. 7).

688–625. Reign of Ashurbanipal (Essarhaddon's son). He put down Egyptian revolts (668 and 661) as well as undertaking a number of successful campaigns along the northern frontier. Babylonia rebelled once more under the leadership of his half-brother, **Shamash-Shu-mukin** (652). In a bitter four-year struggle Ashurbanipal put down the revolt with typical Assyrian barbarity. Meanwhile, Egypt had risen again and driven out the Assyrian garrisons, while Arabs and Elamites took advantage of Assyria's troubles to attack from the north, west, and east. Ashurbanipal subdued the Arabs, then turned east to crush and practically exterminate the Elamites. Despite his successes, the desperate struggles had exhausted the country, almost wiping out the sturdy Assyrian peasantry, the backbone of the army. Assyria, having reached the zenith of her power and magnificence, was forced now to rely largely on mercenaries, mostly from the wild Scythian tribes who had replaced the Cimmerians along the northern frontier. Upon the death of Ashurbanipal their hordes poured across the eastern frontiers, roaming almost at will across the disintegrating empire.

626. Babylonian Revolt. The rebel leader, the satrap, **Nabopolassar,** formed an alliance with **Cyaxares** of Media, also rebelling against Assyria (see p. 13).

616–612. Fall of Assyria. The Median and Babylonian allies (their armies including many Scythians) invaded Assyria. Nineveh was captured and destroyed; the fall of the capital was the end of Assyria (612), although some resistance persisted in the northwest (612–610).

PALESTINE AND SYRIA, 1200–700 B.C.

c. 1200–c. 800. Warring States. Between the decline of the Egyptian and Hittite empires, and before the height of Assyrian power, the various tribes of Palestine and Syria coalesced into a number of petty, independent, constantly warring states. Outstanding among these were the Jewish nation, the Philistines of southwestern Palestine, the Phoenician cities of northern and western Syria, and the Aramean kingdoms of eastern Syria, of which Damascus was the most important.

c. 1100. Gideon. This most famous of the early Jewish warriors temporarily united most of the independent Israelite tribes in repelling the incursions of the Midianites, an Arabic people living east of the Jordan.

1080–1025. Rise of the Philistines. Israel was invaded and dominated by the Philistines (see p. 7).

1028–1013. Reign of Saul. The Jews rose against their oppressors. Despite many successes, internal squabbles prevented Saul from driving the Philistines completely out of Israel. He was killed by them in the **Battle of Mount Gilboa** (1013).

1010–973. Reign of David. He checked then destroyed resurgent Philistine power. He was successful in recruiting defeated Philistine soldiers, who thenceforth made up a major portion of his army. He reunited the Jews, conquered all Palestine, and apparently dominated most of Syria. He defeated all the external enemies of the Jews, but the later years of his reign were marred by several bloody internal insurrections, one led by his son **Absalom.**

973–933. Reign of Solomon. A period of peace and prosperity. After his death, the Jewish kingdom split into two parts, the kingdoms of Israel and Judah. For two centuries Jewish history was a succession of wars in which these two rival kingdoms were either pitted against each other or against their many small neighbors.

854. Battle of Qarqar. The temporary alliance of **Ahab** of Israel and **Ben Hadad II** of Damascus postponed Assyrian conquest by a victory over Shalmaneser III.

c. 750. Assyrian Conquest. Palestine and Syria remained under foreign control for the next 27 centuries.

724–722. Revolt in Israel. Assyrian King **Shalmaneser V** heavily besieged **Samaria.** Upon his death, **Sargon II** stormed the city and suppressed the revolt.

CHALDEA, 1500–600 B.C.

c. 1500–c. 700. Appearance of the Chaldeans. During the time of Assyrian supremacy, the Chaldeans, a Semitic desert people, infiltrated into southern Mesopotamia. They provided much of the vitality evidenced by Babylonia's frequent efforts to throw off the Assyrian yoke.

612. The New Babylonian, or Chaldean, Empire. Following the conquest of Assyria, the Chaldeans took all of Assyria west of the Tigris, while the Medes' share was the former Assyrian provinces east of the river.

612–605. Reign of Nabopolassar. He had no difficulty in establishing his authority in Mesopotamia, but the Egyptian pharaoh, Necho, challenged his assertion of dominion over Syria and Palestine. Nabopolassar sent an army into Syria under his son **Nebuchadnezzar,** who defeated Necho at Carchemish (see p. 7).

605–561. Reign of Nebuchadnezzar. He campaigned in Syria, Palestine, and Phoenicia on several occasions, to subdue sporadic uprisings. He failed to take **Tyre,** which resisted a 13-year siege (585–573). Otherwise his long reign was relatively peaceful, and Babylonia reached the pinnacle of ancient Oriental culture.

MEDIA, 800–600 B.C.

c. 800–c. 625. Appearance of the Medes. These semibarbaric descendants of the Asiatic Scythians and Indo-European Iranians occupied what is now northwestern Iran, Kurdistan, and Azerbaijan. They were more or less under Assyrian control (700–625).

625–585. Reign of Cyaxares. He joined Nabopolassar of Babylonia to throw off Assyrian rule (see p. 12). His south-western frontier secure as a result of a cordial alliance with Chaldea, Cyaxares expanded his empire rapidly to the west as far as Lydia and to the east almost as far as the Indus. This was the largest empire the world had yet seen, but it lacked the administrative machinery of Assyria, Egypt, or Chaldea.

GREECE, 1600–600 B.C.

The Early Greeks

c. 1600. The Minoans of Crete. Their highly developed and artistic civilization had spread to southern Greece and to most of the Aegean Islands.

c. 1400. The Fall of Crete. The island was overrun by invaders from the mainland, probably part of the Achaean (Indo-European) migrations from central Europe.

c. 1400–c. 1200. The Achaeans Take to the Sea. Under pressure from succeeding migration waves, the Achaeans, in company with other Mediterranean peoples, stimulated and took part in the "Peoples of the Sea" movement which so seriously affected the Mediterranean coast of the Middle East (see pp. 7, 9).

c. 1184. The Siege of Troy. The half-legendary story of this war, as passed on to posterity by **Homer,** can be considered as the beginning of Greek history.

1100–600. Coalescence of Greece. The different peoples who had migrated into Greece, the Aegean Islands, and the west coast of Asia Minor gradually became the relatively homogeneous Greek people known to history. But for all of their cultural homogeneity, the mountainous, insular, and peninsular geography of Greece divided them politically into many tiny, independent, energetic states. Much Greek energy was consumed in a great colonization effort, in a sense merely a continuation of the migratory urge that had brought them originally into Greece. To this basic wanderlust were added the impulses of trade and the pressures of population. This colonization had important military consequences: (1) the Greeks became a seafaring people; (2) the adventurous found ample opportunity for maintaining a high standard of combat proficiency beyond the seas, and (3) the Greeks learned, from combat and observation, much about the strengths and weaknesses of the barbarian and civilized nations of the Mediterranean and Middle East.

Sparta, 1000–600 B.C.

c. 1000. The Founding of Sparta. The early military development of this small town in the middle of the Peloponnesian Peninsula was indistinguishable from that of other inland Greek towns.

c. 700. The Legacy of Lycurgus. Under this semilegendary leader, Sparta became, and remained, a completely military society, always maintained on a war footing. From his earliest years the Spartan citizen had only one mission in life: military service. The state was the army and the army was the state. The result was the development of the best individual soldiers in Greece and the creation of what was, for its size and time, possibly the best small army in the history of the world. The Spartan army was not significantly different from those of other Greek city-states in composition, armament, or tactics; essentially an infantry force of armored spearmen, it was composed primarily of the free-born citizens of the upper and middle classes. The principal distinguishing characteristics were the more thoroughly developed individual military skills, greatly superior organization, higher order of unit maneuverability, and the iron discipline for which the Spartans became renowned throughout Greece.

c. 700–c. 680. The First Messenian War. Sparta conquered the rich Messenian Plain to become the dominant state in southern Peloponnesus.

c. 640–620. The Second Messenian War. After a prolonged struggle Sparta again subdued Messenia and enslaved the conquered survivors.

ITALY AND ROME, 2000–600 B.C.

c. 2000. Arrival of the Ancestors of the Latins. An Indo-European people, closely related to the Greeks, migrated across the Alps from central Europe, bringing the Bronze Age to Italy.

c. 1000. Introduction of Iron. The development came with another migration.

c. 900. Arrival of the Etruscans. These were people of different racial stock who arrived in northwestern Italy by sea from the east. At the same time Greek traders and colonizers were gaining considerable influence in Sicily and southern Italy.

c. 700.* Founding of Rome. From its beginning Rome was important militarily; archaeology confirms legend. This was inevitable from its location on the border between the Indo-European settlements of Latium and the Etruscan city-states farther north. Possibly the town was established as a Latin outpost to hold off the expanding Etruscans.

c. 700–c. 500. Continuous Wars with the Etruscans. Rome in its early history may have fallen to the status of an Etruscan colony; it was for a time ruled by kings of Etruscan ancestry. A distinctive military system began to evolve during this formative period of constant combat.

* It is generally agreed that Rome was founded later than the traditional date of 750 B.C.

SOUTH ASIA

INDIA, 2000–600 B.C.

1800–1000 B.C. The Aryan Migrations. The military history of South Asia is generally considered to begin with the arrival of Aryan invaders in the Indus Valley. They were an Indo-European people, closely related by language, religion, and customs to the Persians or Iranians. We know nothing of the details of the conflicts between the invaders and the more cultured, dark-skinned early inhabitants, the Dravidians. The synthesis of the Aryan and earlier cultures gradually produced a new **Hindu** civilization in the Ganges Valley after 1400 B.C. This was the background against which the

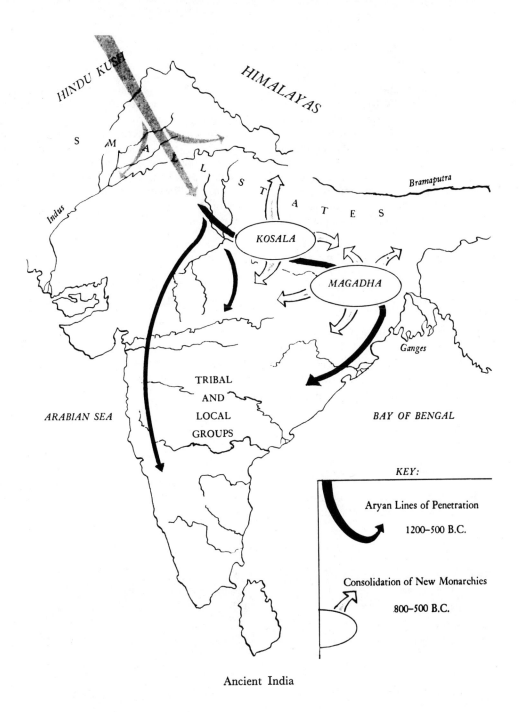

HINDU KUSH

HIMALAYAS

S   M   A   L   L   S   T   A   T   E   S

Bramaputra

Indus

KOSALA

MAGADHA

Ganges

TRIBAL
AND
LOCAL
GROUPS

ARABIAN SEA

BAY OF BENGAL

KEY:

Aryan Lines of Penetration

1200–500 B.C.

Consolidation of New Monarchies

800–500 B.C.

Ancient India

great Hindu epic, the Rigveda, was composed, and thus is often called the Vedic period.

1000–600 B.C. Later Vedic Culture and Hindu Consolidation. The predominantly Aryan Hindus firmly established themselves as rulers of northern and central India, although their influence in the South and on Ceylon was limited. Numerous small states arose and engaged in incessant struggles for dominance.

c. 900. Mongoloid Invaders. These brief raids presaged the future. The north Indian plain was vulnerable to invasion from the northwest. Though the fertile Ganges and Brahmaputra valleys must have been at least equally attractive to northeastern neighbors, it is of some significance that there is no record of a great invasion from that direction. Formidable though they were, the Hindu Kush and Iranian highlands were fairly easily traversable by substantial military forces; the Himalayan-Tibetan complex of mountains and desolate plateaux were not.

Early Hindu Military Organization

Our knowledge of warfare and military practices during the period prior to 600 B.C. has been gleaned from the earliest classical literature of India, particularly, the **Rigveda** and the **Mahabharata.**

The armies were made up almost entirely of footmen. The bow was their principal weapon. There was apparently no cavalry; horses were scarce and therefore were reserved for pulling the two-man war chariots of the kings and nobles. The warriors were the most honored and leading class of society. Iron weapons did not appear in India until about the 5th century B.C., which would indicate that military techniques were probably less advanced than in the Middle East.

EAST ASIA

China, 1600–600 B.C.

c. 1600. Emergence of Historical China. This was in the Yellow River Valley. It is difficult to distinguish between legend and history until sometime after 900 B.C. Archaeology, however, has supported most traditional legend.

Ancient China

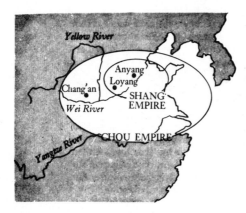

Shang and Chou empires

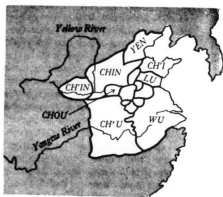

The warring states

c. 1523. The Battle of Ming Chiao. (Near modern Shang Chu, in Honan.) The semilegendary Shao Dynasty was defeated and eliminated by the Shang Dynasty.

c. 1523–1027. The Shang Dynasty. The first clearly identifiable ancestors of the modern Chinese were a highly civilized people known to history by the name of their ruling dynasty. The Shang ruled over a relatively limited inland area around their capital, Anyang, in what is now the northern tip of Honan province. Their history is a record of wars, expansion, and internal troubles not unlike that of Egypt at about the same time. During the 12th century B.C., a semibarbaric people called **Chou** began to press eastward from the region near the junction of the Wei and Yellow rivers.

c. 1057 (possibly 1027). Battle of Mu Yu. (Southern Honan.) **Wu Wang,** the "Martial King" of the Chou, decisively defeated the Shang.

c. 1027. Establishment of the Chou Dynasty. Wu Wang established Chou authority over the former Shang domains. He died soon after, **Chou Kung,** became the regent for 7 years, during the minority of the Martial King's son. Chou Kung firmly repressed a Shang uprising and seems to have established a ruling organization comparable to that of Thutmosis III. It was largely due to this organization that the Chou Dynasty lasted for about 800 years, despite the weakness and impotence of the kings during the latter two-thirds of that period.

c. 1000–c. 900. Chou Expansion. They drove eastward, to the sea, and pressed north to the vicinity of the present Manchurian border. Later they moved south into the lower Yangtze Valley and the coastal regions in between.

c. 800–600. Decline of Chou Power. As royal control became nominal, power gravitated to the provincial nobles, and for several centuries China's history became one of constant fighting among approximately 140 autonomous warlords. Of these, some 7 were important: Ch'i, Chin, Ch'in, Wu, Yueh, Sung, and Ch'u. By the end of the period the economically-strapped Chou exercised authority over only a small area immediately outlying Loyang.

EARLY CHINESE MILITARY ORGANIZATION

The art of warfare in China by about 600 B.C. was apparently as well advanced as that of the Middle East. Chinese weapons, however, were apparently not so advanced as those of the Assyrians. While their bronze metallurgy seems to have been the equivalent of that of the Middle East, and their bronze workmanship was probably superior, iron metallurgy seems to have lagged behind by several centuries.

From earliest times the Chinese appear to have relied upon the bow as their principal weapon. Apparently the bow was always of the reflex variety, constructed of wood, horn, and sinew, and longer and more powerful than those normally found in the West. Arrows were probably made of bamboo, with metal tips appearing in historical times.

Bronze helmets appeared in China during the time of the Shang, and armor development, despite a slower start than in the Middle East, had probably caught up by about 600 B.C.

Cavalry would not appear in China for several centuries. Chariots, however, were introduced sometime between 2100 and 1600 B.C., and became increasingly important. When the Shang defeated the Shao at the Battle of Ming Chiao (see above), they probably had about 70 chariots. Five hundred years later, when the Chou defeated the Shang at the Battle of Mu Yu (see above), they had about 300 chariots. Two centuries after that, King Hsuan of Western Ch'u had about 3,000 chariots when he inflicted a defeat on the Chou (823 B.C.).

Initially, only the principal leaders rode and fought in chariots, which seem to have been used primarily as mobile command posts by kings and generals, usually accompanied by a drummersignaller, as well as by a driver and one or more archers. But as time wore on, greater advantage was taken of the shock-action capabilities of the chariot, and it became the primary instrument of war. The strength of a state came to be estimated in terms of the number of chariots it could mobilize for war. As the use of chariots became more common, each became the central feature of a small combined-arms combat team, roughly the equivalent of a modern platoon, called a *cheng*. Each chariot carried three to five people, at least two of whom were archers, and was accompanied by about 20 infantrymen, usually armed with spears.

The Western Ch'u developed a well-organized military system, adopted by most of the other warlords. This system was characterized by the equivalents of platoons, companies, battalions, brigades, divisions, corps, and armies.

During the 8th and 7th centuries B.C., warfare became somewhat less violent and more ritualistic. It would not long retain this relatively benign character.

II

WAR BECOMES AN ART:

600–400 B.C.

MILITARY TRENDS

Significant, in this era of change and progress in all human activity, is the transition from semilegendary chronicles to serious, reliable histories. The principal military trends are thus more clear-cut. Two were of significance: (1) Within the limits of technology sound concepts evolved for the employment of weapons; and (2) as a consequence, theories of tactics and military doctrine emerged. By 400 B.C. war had assumed the major characteristics which it retained at least into the dawn of the nuclear age.

WEAPONS

Weapons themselves were essentially unchanged. There would, in fact, be relatively little modification in weapons and related implements of warfare for nearly 2,000 more years. Fundamental changes would have to await the appearance of sources of power transcending brute strength and mechanical attempts to harness the forces of gravity and of the wind.

TACTICS AND DOCTRINE ON LAND

Substantial advances were made, however, in the use of existing weapons. During the middle of the 6th century, **Cyrus the Great** made conscious effort to instill concepts of discipline and training into his army. Though these concepts had been recognized from early Mesopotamian times, they were gradually assuming greater importance among all peoples geared to military action. The principal result was to increase the value and importance of the infantry, whose unwieldy masses in earlier centuries had rarely been capable of maneuver.

Battle formations, prebattle rituals, and religious rites, however, were stereotyped, and had their origins in the unrecorded past. Astrologers and soothsayers were consulted; offerings were made to the gods; and omens foretelling victory were anxiously sought. After haranguing the troops drawn up for battle, the general would order the advance to combat. This was usually done to the accompaniment of rousing military bands.

The armies approached each other in parallel lines, with the infantry in the center, chariots and cavalry generally on the flanks (sometimes in front), and with swarms of irregular slingers and bowmen usually screening the advance until the main bodies were within a hundred yards of one another. Variations in battle orders were rare, save for relative locations of cavalry and chariots, either in front or on the flanks. Just before the clash of the main lines of the opposing armies, the light troops would slip away around the flanks, or back through intervals left in the lines for this purpose. Sometimes one side would stand fast on the defensive, but more often both sides would stride purposefully toward each other, their shouts and clash of arms creating a terrible noise.

Experience had shown that last-minute maneuvers were likely to create dangerous gaps in the lines or expose a marching flank to missile and shock attack. Therefore, tactical ingenuity was not often attempted beyond the point where an enemy would be forced to enter battle on unfavorable ground or with only a portion of his available forces. The usual objective in battle was to outflank the enemy, since only the flanks and rear of well-armed infantry—10 to 30 ranks deep—were sensitive and vulnerable. Though we shall note a few examples of successful deviation from the parallel order of battle, such deviations more often led to failure.

Cavalry still played a great role, particularly on the wide plains of Central and Southwest Asia. Except in India and Persia, the chariot had lost much of its terror for disciplined, maneuverable footmen, and was no longer the main weapon of battle.

Cyrus of Persia won his earliest successes with foot troops—particularly expert archers— much more alert and resourceful than those in other Asiatic countries. But he discovered that he needed cavalry of his own to neutralize the effectiveness of the horsemen on whom many of his foes still mainly depended. His early conquest of Media, however, enabled him to recruit excellent horsemen. Following this, the Persians quickly adapted themselves to the horse, and soon the Persian heavy cavalry and mounted archers were by far the best in the world.

As the mountains of Greece were unfavorable to cavalry movements, the Greeks in general neglected that arm, except in the northern, flatter regions of Thessaly and Macedonia. Elsewhere in the Greek peninsula, the Aegean Islands, and along the Ionian coast, the steadily improving infantry **phalanx** was relied on chiefly. This disciplined body of heavy infantry formed itself for battle in long lines, which varied in depth from 8 to 16 men. The individual soldier of the phalanx was called a **hoplite**—a well-trained, disciplined soldier, kept in excellent physical condition by sport or combat. His major weapon was a pike, 8 to 10 feet long. His short sword was usually sheathed while he was in the phalanx formation. He also wore a helmet, breastplate, and greaves, and carried a round shield. In battle the hoplites in the front ranks pointed their spears toward the foe; those in the rear rested theirs on the shoulders of the men in front, forming a sort of hedge to break up flights of enemy arrows.

The phalanx and its individual units were capable of limited maneuvers in combat formation. In battle the invariable formation was a long, solid line, with narrow intervals through which light troops could pass. Battle was waged on the flattest ground available, since movement over rough ground created gaps which could be fatal.

The hoplites came from the upper and middle classes of the free citizens of the Greek states. The **psiloi,** or light troops, generally poorly armed, were neither so well trained nor disciplined. For the most part these came from the lower classes of society, but many of them were mercenaries. Along with the generally inferior cavalry, they protected the flanks of the phalanx on the march and in battle. Some of the archers and slingers, such as those from Crete and Rhodes, were quite effective. In addition to rigorous training and excellent physical condition, the Greek hoplite possessed the military advantages and disadvantages of alert, intelligent, literate free citizens of proud and independent countries. These qualities made the Greeks suspicious of regimentation, even though intellect clearly accepted the need for tactical discipline.

One aspect of the military art in which this period failed to approach the limits imposed by existing technology was in the area of engineering. Neither Persians, Greeks, nor Chinese achieved any marked improvement over the engineering techniques which had been developed by the Assyrians. Fortification had, in fact, progressed about as far as available means would permit; the art of siegecraft had failed to keep pace. Save for a few exceptional instances of surprise, ruse, or betrayal, walled cities or fortresses were impervious to everything but starvation.

Greek hoplites

NAVAL TACTICS AND DOCTRINE

The use of ships for warlike purposes had long been a common practice of seafaring peoples living along the shores of the Mediterranean and Aegean seas. Prior to the 7th century B.C., however, this had been largely limited to employing merchant ships as troop and supply transports. These short, broad-beamed craft, combining sails and oars, were essentially adjuncts of land power.

About 700 B.C. the Phoenicians introduced the first vessels designed essentially for fighting. These were speedy, oar-propelled galleys—called biremes because they had two banks of oars—longer and narrower than the typical merchant ship.

The Greeks, particularly the Athenians, improved the Phoenician galleys, and brought to naval warfare a skill and perfection in technique hitherto unknown in fighting on land or sea. The Athenian **trireme** was long, low, and narrow, deriving its name from the fact that its oars were ranged in groups of three along one bank of oars, with oarsmen on each side of the vessel. Seaworthiness, comfort, cargo capacity, and range were deliberately sacrificed to achieve speed, power, and maneuverability. In addition to its oars, the trireme carried sails on its two masts as a means of auxiliary power; in battle, however, it was propelled exclusively by its 150 oarsmen.

A war fleet could not carry food and water for a long voyage; it had to be accompanied by a flotilla of supply vessels and transports. Vulnerable to storms, trireme fleets endeavored to keep near to sheltering coasts; long voyages far from land were avoided if possible.

The principal weapon of the trireme was a metal beak projecting some 10 feet in front of the prow at the water line. When this beak was rammed into the side of another vessel, the results were deadly. The difficulties of accomplishing this, however, were such that most of the Greeks usually preferred to rely upon the older tactics of pulling up alongside a foe and boarding.

Athenian sailors, however, relied upon superior seamanship, speed, and maneuverability to bring victory. When there was not an immediate opportunity to smash directly into the side of an opponent, Athenian vessels would swerve unexpectedly beside their foes, shipping their oars at the last moment, and breaking those of the surprised enemy. The disabled foe was then literally a sitting duck, to be rammed at leisure by one of the Athenian vessels. Athen's foes were never able to match this superiority in seamanship.

THEORY OF WARFARE

Economic and logistical considerations played a particularly important part in the major wars of the 5th century B.C. For Persia, the great land power, the problem was lines of communications thousands of miles long, vulnerable to harassment and interruption by sea and by land. For the smaller Greek states, particularly in their wars against Persia and among themselves, there were two main problems: (1) Their relatively complex societies were not self-sufficient, and in many instances were dependent upon distant, overwater sources of supply to maintain both peacetime and wartime economies. (2) The military security of several Greek states was based upon an extremely expensive and relatively sophisticated weapon system (the trireme fleet), which could be maintained and operated only at great cost in treasure, and in highly trained, skilled manpower.

Not the least remarkable development of this age was the serious study of wars and warfare, which is suddenly discernible in the 5th century B.C. The first known histories—those of **Herodotus** and **Thucydides**—were not conscious military histories, but inevitably they dealt mainly with military events. And about the same time, in China, **Sun Tzu** was composing his treatise, *The Art of War,* revealing an understanding of the practical and philosophical fundamentals of war and of military leadership so sound and enlightened as to warrant serious study by scholars and soldiers today.

MEDITERRANEAN—MIDDLE EAST

EGYPT, 600–525 B.C.

c. 590. Operations in Lower Nubia. The campaigns of **Psammetichus II** were inconclusive.

c. 586–568. Reign of Apries. His vain efforts to prevent the Chaldean conquest of Syria and Palestine culminated in his defeat at Jerusalem by Nebuchadnezzar (c. 580), who then consolidated control over Palestine (see p. 13). Turning west, Apries was repulsed in efforts to conquer the Greek colony of Cyrene (c. 570). He was overthrown by the revolt of **Ahmose II.**

567. Invasion of Palestine. Ahmose was repulsed by the aged Chaldean emperor, Nebuchadnezzar.

547–546. Alliance with Croesus of Lydia. Ahmose sent a large contingent of Egyptian heavy infantry to join Croesus against Cyrus of Persia. At the **Battle of Thymbra** the Egyptians stood firm in the rout of the Lydians (see below). Cyrus made separate terms with the Egyptians, who returned home with honor.

525. Persian Conquest. (See p. 24). Egypt was to remain under foreign rule for more than 24 centuries.

PERSIA, 600–400 B.C.

The Decline of Media, 600–559 B.C.

600–585. War with Lydia. A long, inconclusive war in Asia Minor ended with the Halys River accepted as the boundary.

585–559. Uneasy Balance of Power in the Middle East. This was shared between Media, Chaldean Babylonia, Egypt, and Lydia.

The Early Persian Empire, 559–400 B.C.

559. Independence of Persia. Led by their prince, **Cyrus,** the Persians, an Aryan people

closely related to the Medes, revolted against **Astyages** of Media, who was deposed by Cyrus.

559–530. Reign of Cyrus the Great. His first task was the consolidation of his conquest of Media (559–550).

547. Lydian Invasion. Under their king, Croesus, they crossed the Halys into Cappadocia, a province of Persia-Media, either for the purpose of restoring Croesus' brother-in-law Astyages to the throne of Media or to try to forestall a Persian invasion of Lydia. Croesus had organized an alliance against Persia with Chaldea, Egypt, and tiny, but militarily potent, Sparta.

547 (546?). Battle of Pteria. Cyrus marched to meet Croesus, fighting a savage but indecisive winter battle. Croesus withdrew across the Halys and prepared for a new campaign. From Sardis, his capital, he sent messages to his allies, suggesting an advance into Persia when the weather improved. Cyrus invaded Lydia and (early 546) approached Sardis with a large army (but certainly no more than one-quarter of the 200,000 men reported in Xenophon's *Cyropaedia*).

546. Battle of Thymbra. Croesus hastily reassembled an even larger allied army and marched to meet Cyrus on the nearby Plain of Thymbra. Badly outnumbered, Cyrus deployed his troops with flanks refused in a great square formation, the first recorded deviation from the normal parallel order of combat. He organized most of his army in depth, in 5 relatively short lines. The flanks were covered by chariots, cavalry, his best infantry, and a newly improvised camel corps, facing outward, perpendicular to the front. As Cyrus expected, the wings of the Lydian army wheeled inward to envelop this novel formation. As the Lydian flanks swung in, gaps appeared at the hinges of the wheeling wings. Disorder was increased by effective overhead fire of Persian archers and dart throwers, stationed within the square. Cyrus then gave the order to attack. His flank units smashed Croesus' disorganized wings; shortly afterward the Persian cavalry slashed through the gaps at the hinges. In a short time the Lydian army was routed. Cyrus pursued and captured Sardis by storm. He treated the captured with magnanimity rare for the age.

545–539. Eastward Expansion. Cyrus now turned his attention to the arid plateaux to the east, which had owed nominal allegiance to Media, but which now were attempting to reestablish independence. In a few years he reconquered most of Parthia, Sogdiana, Bactria, and Arachosia.

539–538. Conquest of Chaldea. Cyrus next invaded Babylonia, which had joined Croesus' anti-Persian alliance, defeating King **Nabonidus** and investing Babylon. For nearly 2 years the tremendous walls of the city defied Cyrus. He finally diverted the waters of the Euphrates, and his troops dashed in through the lowered stream bed, catching the defenders by surprise. The Chaldean Empire was quickly annexed to Persia.

537–530. Expedition in the East. Cyrus decided to round out his eastern dominions be-

fore dealing with Egypt and Sparta. He conquered much of the region west of the Indus River, and campaigned north as far as the Jaxartes. Here he was killed in battle against the Massagetae.

COMMENT. **Cyrus** *was the first great captain of recorded history. His conquests were more extensive than those of any earlier conqueror and proved to be more permanent. This was largely due to his administrative genius and his ability to win the confidence of the conquered peoples.*

530–521. Reign of Cambyses (son of Cyrus). An able warrior, he was an inadequate ruler. He carried out his father's ambition to conquer Egypt, defeating **Psammetichus III** at **Pelusium** (525), and also seized the Greek colony of Cyrene. He was unsuccessful, however, in an expedition up the Nile against Nubia and Ethiopia. He died while marching back to Persia to deal with an imposter who had seized the throne.

522–521. Civil War. Cambyses' cousin, **Darius,** led a successful revolt against the imposter who had usurped the throne.

521–486. Reign of Darius the Great. The wise and brilliant rule of this organizational genius

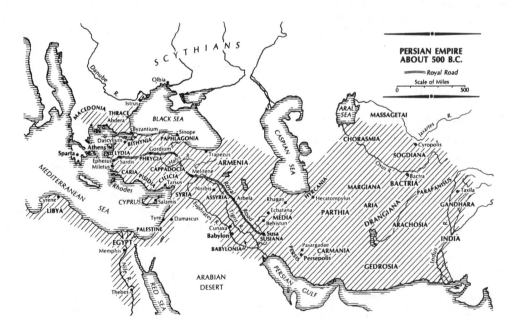

assured the stability of the Persian Empire. Many subject peoples had taken advantage of Persia's internal turmoil to try to regain independence. Darius promptly and efficiently put down the revolts (521–519), then spent the remainder of his reign in consolidation. He did not personally participate in many campaigns after his authority had been firmly established. He did, however, supervise some operations along the Indus River, north into the Pamirs of Central Asia, and against the Scythians in the steppes east of the Caspian Sea. His generals subjugated eastern Asia Minor and Armenia and established the northern frontier of the empire along the crest of the Caucasus Mountains.

511. Invasion of Southeastern Europe. Darius personally led this expedition, which had three objectives: (1) to establish a base in Thrace for subsequent absorption of the Greek states; (2) to protect the long lines of communications leading to and from Thrace; and (3) to strike the rear of the Scythian tribes of the steppes region. (He did not realize that the Scythians of the Danube area were some 2,000 miles west of their brethren south of the Aral Sea.) The expedition, carefully prepared, utilized both land and sea forces drawn from all parts of the empire and included some Greek mercenaries. (The total strength was probably one-tenth the 700,000 attributed by Herodotus.) The navy, after building and maintaining a floating bridge over the Bosporus, patrolled the western shores of the Black Sea. Darius marched north to the Danube, where the navy constructed another floating bridge. Leaving a strong force to protect the bridge, Darius continued northward for several hundred miles, living off the country. The Scythian horsemen refused to stand and fight, but continuously harassed the Persian army. After two or more months of frustrating and costly marches, Darius returned to the Danube with the bulk of his army intact. The Scythians, however, had been sufficiently awed by Darius' armed might so that they made no move south of the Danube during subsequent Persian wars with Greece; Thrace and Macedonia were firmly annexed to the Persian Empire.

499–448. Graeco-Persian Wars. (See pp. 26–32.)

486–465. Reign of Xerxes. This was the beginning of a slow decline of Persia.

401–400. Revolt of Cyrus the Younger. This revolt against his brother, **Artaxerxes II,** led to the **Battle of Cunaxa** and the *Anabasis* of Xenophon (see p. 37).

PERSIAN MILITARY SYSTEM, C. 500 B.C.

How much of the military and political organization of the Persian Empire was due to the genius of Cyrus and how much to the innovations of Darius is unclear. Cyrus, the more imaginative, probably established the system that Darius then perfected.

The basis of the system was the spirit, skill, and resourcefulness of the Persians. An important weapon was the bow, used effectively by both cavalry and infantry. Insofar as possible the Persians avoided close-quarters infantry combat until their foes had been thoroughly disorganized by swarms of foot archers from the front and the daring onrushes of horse archers against flanks and rear. The Persians were versatile in adapting their methods of warfare to all conditions of terrain. They respected the shock action of the Lydian cavalry lancers and incorporated this concept into their mounted tactics.

Subject peoples were required to render military service. The garrisons scattered throughout the empire were principally composed of units from other regions (including many Greek mercenaries) but always included a Persian contingent. Imperial expeditionary forces were also multinational. The Persians received a surprisingly high standard of loyalty from these diverse groups, due largely to their policies of leniency toward the conquered and a carefully supervised but decentralized administration.

The empire was divided into about 20 provinces, or satrapies, each governed by a trusted and able official. The principal military garrison in each satrapy was under the command of a general directly responsible to the emperor, which prevented dangerous accumulation of power in any region. In the emperor's court an inspector general (the "eye of the king") was responsible for the supervision of all provincial activities. This complex, but not cumbersome, system of control and of checks and balances was facilitated by a mounted-messenger system utilizing an excellent network of roads.

GREECE, 600–494 B.C.

Peninsular and Aegean Greece

c. 600. Ascendancy of Sparta. Her military prowess particularly dominated the Peloponnesus. This was the so-called "Age of Tyrants," and many of the Greek states were torn by civil wars between the forces of democracy and oligarchy. The tyrants were as frequently leaders of the democratic elements as they were of the oligarchy, and the word "tyrant" in those days simply meant authoritative rule by a single individual. Sparta, though far from a democratic state herself, was consistently anti-tyrant, and sometimes intervened in the internal struggles of the other states. Though none of the other Greek states felt strong enough to challenge Sparta's ascendancy by itself, four of these (Argos, Athens, Corinth, and possibly Thebes), in shifting alliances among themselves and with smaller states, were able to preserve a balance of power.

c. 560–c. 520. Rise of Athens. Her growing commerce, population, and wealth might have put Athens in a position to challenge Sparta, had it not been for recurrent civil strife between the forces of oligarchy and democracy.

519–507. War of Athens and Thebes. This desultory conflict resulted from Theban efforts to take advantage of Athens' difficulties and to force little Plataea, an ally of Athens, into a Boeotian League.

510–507. Spartan Intervention in Athens. The oligarchs overthrew the popular Athenian tyrant **Hippias** with the help of **Cleomenes,** King of Sparta. Hippias fled to Persia, where he soon found favor with Darius. When the internal struggle in Athens continued, Cleomenes again intervened and captured Athens (507). The Athenians, rallying under the democratic leader **Cleisthenes,** expelled the Spartans. Cleomenes could not obtain support in Sparta for another invasion of Attica to avenge this repulse.

494. Battle of Sepeia. Cleomenes overwhelmed Argos, then established himself as virtual tyrant in Sparta. However, in an ensuing civil war he was captured and deposed.

Ionian Greece

c. 550. Lydian Control. Some of the Greek cities on the coast of Asia Minor, with the principal exception of Miletus, were annexed to Lydia by Croesus.

546. Persian Conquest. The general **Harpagus** subdued the Ionian cities, which had tried to reassert their independence after the fall of Croesus (see p. 23).

c. 512. Unrest in Ionia. This was largely inspired by the free Greek states farther west. As a result Darius decided to conquer European Greece.

510. Ionian Revolt. Misled by false reports that Darius had been defeated by the Scythians, some of the northern Ionian cities revolted, but were quickly and firmly brought back under Persian rule. Unrest seethed in Ionia.

THE GRAECO–PERSIAN WARS, 499–448 B.C.

Ionian Revolt, 499–493 B.C.

499. Outbreak of the Revolt. This was led by the city of Miletus, which requested assistance from Greece. Sparta refused, but Athens and Eretria (on Euboea) sent small land and naval contingents.

498. Rebel Setbacks. The rebels captured Sardis, capital of the satrapy of Lydia, but **Artaphernes,** the satrap, quickly recaptured his capital and drove the Greeks back to the sea, defeating them in the **Battle of Ephesus.** The rebellious cities were unable to maintain a united front against the Persians and most were reconquered.

494. Siege of Miletus; Battle of Lade. To cut the city off from its contact with European Greece, Darius assembled a large fleet, which defeated the Ionians in a great battle off the tiny island of Lade, near Miletus. The city soon surrendered and the revolt collapsed.

493. Darius Determines to Conquer Greece. In particular he wanted to punish Athens for having supported the Ionian revolt.

492. Preparations for a Land-Sea Expedition. The Persian general **Mardonius** consolidated control over restive Thrace and Macedonia, in preparation for an invasion of Greece. His fleet, however, was wrecked during a storm off the rocky promontory of Mount Athos, so he withdrew to Asia Minor.

491. Amphibious Preparations. Darius now decided to send an amphibious force directly across the Aegean to attack Athens and Eretria. In command were Artaphernes and the Median general **Datis.** The expedition contained the cream of the Persian Army and Navy and probably numbered nearly 50,000 men (not the 100,000 averred by Herodotus). The former Athenian tyrant Hippias was with the expedition.

CAMPAIGN AND BATTLE OF MARATHON, 490 B.C.

The Athenians first learned of the expedition when Eretria was attacked. A message asking Sparta for help was immediately sent, carried by the famed runner **Pheidippedes,** and the Athenians prepared themselves for battle. Sparta sent word that it would help but would be delayed for nearly two weeks by a religious festival. About the same time word reached Athens that the Persians were landing near Marathon, some 26 miles away. The Athenians, about 9,000 hoplites and a smaller number of light troops, immediately marched to high ground overlooking the Persian debarkation. Here they could block the route leading from the narrow coastal plain toward Athens. They were soon joined by a small force from Plataea. **Callimachus** commanded the Athenians, and under him were 10 other generals, of whom the most respected and most experienced was **Miltiades.**

The Persians apparently knew that many Athenians, fearing defeat and the destruction of their city, were ready to surrender. They had landed at Marathon for the express purpose of drawing the Athenian troops away from the city. Having accomplished this, Artaphernes embarked with half the Persian army, to sail around Attica for Athens, while Datis and the remainder (possibly 20,000 men) stayed ashore to hold the Athenian army immobilized at Marathon.

Miltiades guessed the Persian plan and urged an immediate attack. After a heated council of war, Callimachus voted in support of Miltiades' bold plan and entrusted him with command of the battle.

Immediately the Athenians and Plataeans marched down the slopes to form up facing the Persian outposts, the Plataeans on the left. Miltiades had lengthened the Greek line so that the flanks rested on two small streams flowing to the sea. This thinned the center of the line substantially below the 12-man depth then favored for the phalanx, making it vulnerable to

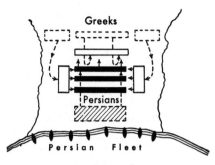

Battle of Marathon

penetration by Persian cavalry charges. But Miltiades kept his wings at full phalanx depth. The result was a formation providing a powerful striking force on each flank, connected by a very thin line in the center.

The Greeks advanced across the narrow plain toward the Persian camp and beach, until within bowshot range (less than 200 yards) of the Persian archers, and then charged; the opposing archers could do no more than fire a few hastily aimed arrows before seeking safety behind the main Persian formation.

It is probable that the Greek center advanced somewhat less rapidly than the flanks, either by design or because they were exposed to the heaviest fire from the Persian archers. As the two lines met in the shock of combat, the Persians were able to throw back the thin center with relative ease. The Greek line almost immediately became concave, as the two heavy phalangial wings rapidly drove back the flanks of the lightly armored Persians. The Greek wings now began to wheel inward, compressing the Persians in a perfectly executed double envelopment. (Authorities differ as to whether this had been planned or was accidental. In any event, Miltiades had displayed his understanding of the capabilities and limitations of both armies, and of the fundamental military principles of concentration and of economy of force.)

The Persian flanks, followed by the center, took flight back to the shore and to the transports drawn up along the beach. Datis seems to have organized some sort of a rear guard to cover the panicky embarkation of his defeated troops. This is the only explanation of his ability to get away with most of his fleet and with relatively little loss of men and transports. It was in the final confused and desperate fighting at the shoreline that the Greeks lost most of their 192 killed, among these Callimachus. The Persians are reputed to have lost 6,400 killed.

Miltiades now promptly set his tired but jubilant men marching back toward Athens. In advance, in hopes that tidings of the victory would strengthen the wavering citizens sufficiently for them to hold out until the army arrived, he sent word back by a runner, reputedly Pheidippedes, on the first Marathon run. As the Athenian army arrived, the Persian fleet was only beginning to approach the shore for a landing. Realizing he was too late, Artaphernes withdrew.

That evening the Spartans arrived, to learn to their chagrin that they had missed the fight.

Persian Preparations for Invasion, 490–480 B.C.

490. Darius' Plans for Revenge. Infuriated, Darius began elaborate preparations for the complete subjugation of Greece, this time by a combined land-sea expedition.

486–484. Revolt in Egypt. This forced a temporary diversion of Persian military strength. The revolt was subdued, but meanwhile Darius had died (486).

484–481. Xerxes Resumes Preparations. Within three years Xerxes had gathered at Sardis a force of about 200,000 men—probably the largest army ever assembled up to that time. Two long floating bridges were built across the Hellespont, over which the army could march in two parallel columns. To prevent the Greek states from receiving any assistance from the powerful Greek colonies in Sicily, Xerxes made a treaty with Carthage, which agreed to attack Sicily when he began his invasion of Greece. These preparations reveal a remarkable Persian capacity for diplomacy and for strategic and administrative planning. Despite Marathon, the Greeks still feared and respected the military might of Persia, and were alarmed by reports of Xerxes' preparations. Most Athenians and most of the Peloponnesian states, led by Sparta, manfully determined to resist. Most of the remaining Greek states, convinced that Persian power was overwhelming, either endeavored to stay neutral or supported Persia.

484–483. Military Policy Debate in Athens. A lingering naval war with rival Aegina caused many citizens, led by **Themistocles,** to urge an increasing emphasis on sea power—particularly since they saw no possibility of matching Persian land power. The other party, under **Aristides,** pointed to the vulnerability of Athens to overland invasion, insisting that the largest navy in the world could not protect the city from the Persian army. The issue was resolved by a popular vote; Aristides was defeated, and Themistocles immediately began a tremendous trireme-building program (483).

481–480. Strategic Debate between Sparta and Athens. The patriotic states now disagreed on the strategy to meet the expected invasion. The Peloponnesians urged the abandonment of all of Greece north of the Isthmus of Corinth; they felt this 4½-mile corridor could easily be defended. The Athenians, however, refused to abandon their city. Themistocles pointed out the vulnerability of the Peloponnesus to Persian sea power, and insisted that the Persian advance could be successfully disputed on land and on sea much farther north. The Spartans, recognizing the value of the Athenian navy, reluctantly agreed to Themistocles' strategy.

The Campaigns of Thermopylae and Salamis, 480 B.C.

480, Spring. The Persian Advance. The Persian host crossed the Hellespont and marched westward along the Thracian and Macedonian coasts, then south into Thessaly. In direct command, under Xerxes, was **Mardonius.** Just offshore the great Persian fleet kept pace. According to Herodotus, the fleet consisted of approximately 1,500 warships and 3,000 transports.

Greek Defensive Measures. Northern Greece was abandoned without a blow because holding the passes south of Mount Olympus required too many men. The next suitable defensive position was the defile of Thermopylae. At the West and Middle Gates of the defile, the Ledge, probably not more than 14 feet wide, provided perfect defensive positions where a few determined hoplites could indefinitely hold off any number of the more lightly armed Persians. To Thermopylae went Spartan King **Leonidas,** with about 7,000 hoplites and some archers. Save for Leonidas' bodyguard of 300 men, few of these were Spartans. The failure of the Peloponnesian states to send more troops to hold Thermopylae is evidence of their halfhearted interest in carrying out any defense north of Corinth. To prevent the Persian fleet from attacking or bypassing the sea flank of the troops at Thermopylae, the Greek fleet of about 330 triremes was stationed off Artemisium, on the northeastern coast of Euboea. In nominal command was the Spartan **Eurybiades,** though Themistocles, with nearly two-thirds of the total Greek naval strength, exercised a major voice in the councils of war. As the Persian fleet approached, the forces of nature took a hand; severe storms inflicted great damage on the Persian fleet, which lost nearly half of its fighting strength. Apparently the Greeks did not suffer so seriously.

480, August (?). Battle of Artemisium. An indecisive naval conflict took place off Artemisium in two cautious engagements on successive days, with few losses on either side. The Greeks were prepared to continue the battle more decisively on the third day, but on hearing the news from Thermopylae they sailed for Athens.

480, August (?). Battle of Thermopylae. Leonidas had carefully and soundly prepared for defense. With his main body, about 6,000 strong, he held the Middle Gate. He had posted a force of 1,000 men high on the mountains to his left, to cover the one forest track which led around the defile. As expected, the Persians tried to force their way through the pass, but the Greek hoplites repulsed them. For three days the Persians vainly tried to break through; then a Greek traitor told Xerxes of the forest track across the mountain behind Thermopylae. Xerxes promptly dispatched along this trail the "Immortals" of his bodyguard, who quickly overwhelmed the Greek flank guard in a surprise attack. Though Leonidas sent about 4,500 men to block the Persian envelopment, they were too late, and were crushed by the Immortals. The Thebans, and perhaps some of the other Greeks with Leonidas, now surrendered. But the Spartan king and his bodyguard fought on courageously till all were killed.

480, August–September. Persian Advance on Athens. All the Peloponnesians retired behind the fortifications of the Isthmus of Corinth. Themistocles, however, refused to withdraw his fleet as the Spartans requested, but instead used the vessels to ferry the population of Athens to the nearby island of Salamis. The remainder of the Greek fleet reluctantly agreed to stay and fight.

480, September. Persian Occupation of Athens. Xerxes' army, which had suffered few casualties, had been augmented by contingents from Thebes and other northern Greek states. The Persian fleet probably still numbered more than 700 fighting vessels—about double the number of Greek triremes. Themistocles feared a Persian blockade of the Greek fleet, while a powerful Persian army contingent was landed behind the defenses of the isthmus. He therefore sent a secret message to Xerxes, saying that if the Persian fleet attacked, the Athenians would join the Persians and the rest of the Greek fleet would flee. Xerxes ordered the Persian fleet to move out that very night. While the Egyptian contingent blocked the western exit south of Salamis, the main fleet, at least 500 strong, formed in line of battle opposite the eastern entrance of the strait. Before dawn a force of Persian infantry landed on the islet of Psyttaleia, at the entrance to the channel. On the mainland, overlooking the strait from a hilltop, Xerxes sat on his throne to observe the battle which would win Greece for his empire.

480, September 23 (?). Battle of Salamis. Part of the Greek fleet was sent to defend the narrow western strait, and the rest of the triremes were drawn up in a line, behind a bend in the eastern strait, waiting for the main Persian fleet. Where the Persians came around the bend, the channel narrowed somewhat, forcing the ships to crowd together, with resultant confusion. At this moment the Greeks attacked. Maneuver was now impossible, superior numbers to no avail. Advantage lay with the heavier, more solidly built Greek triremes, carrying the whole Athenian army of at least 6,000 men. Literally hundreds of small land battles took place across the decks of the jammed vessels. Man for man the Greek hoplite was far superior to his foes. The battle lasted for 7 or more hours. Half the Persian fleet was sunk or captured; the Greeks lost only 40 ships. The remaining Persians broke off the fight and fled back to Phalerum Bay. A contingent of Greeks—mostly Athenians—under Aristides, now landed on Psyttaleia, overwhelming the Persians who had been isolated there.

480, September–October. Persian Retreat. Xerxes' army, largely dependent upon supply by sea, could no longer hold Athens. With about half of his army and the remnants of his fleet, he marched back to the Hellespont, leaving Mardonius with the remainder in northern Greece.

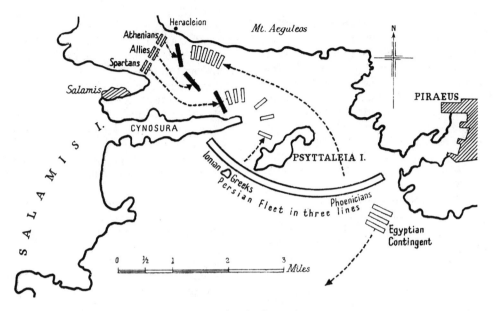

Battle of Salamis

Campaigns of 480–479 B.C.

480–479. Operations of Mardonius. He restored Persian influence and prestige among the northern states, particularly Thebes, through combinations of threats and promises. In the spring, with perhaps 100,000 men, he marched south and captured Athens again. But upon the approach of the Spartan King **Pausanias,** with the main Greek army from Corinth, Mardonius withdrew northward to Thebes, after destroying Athens (June). Pausanias followed cautiously, with less than 80,000 men, of whom about half were hoplites.

479, July (?). Campaign of Plataea. The Greeks found the Persian army holding the line of the Asopus River, about 5 miles south of Thebes. After a brief skirmish with the Persian cavalry, they advanced to a ridge running just south of the Asopus–Spartans on the right,

Athenians on the left, and allied contingents in the center. Both sides held their positions for 8 days, each waiting for the other to attack. Finally Mardonius sent his cavalry raiding behind the Greek positions, destroying supply trains coming over the mountain passes from Athens and polluting the springs from which the Greeks obtained water. Pausanias decided to withdraw that night to a new position at the base of the mountains, just east of Plataea. Here he could cover the 3 passes from Attica and have an assured supply of water.

Battle of Plataea. During the Greek withdrawal some units lost their way in the darkness. At dawn Mardonius discovered the Greeks stretched out in 3 uncoordinated groups. He ordered an immediate attack. The brunt of the first Persian cavalry and archery blows fell upon the Spartans. Thinking the Spartans were about to collapse, Mardonius led the Persian infantry in a charge. But he underestimated Spartan staunchness and discipline. They repulsed the charge, then counterattacked. A terrible struggle ensued, with neither side prevailing. Meanwhile the Athenians, to the left front of the Spartans, were engaged heavily by a mixed force of Persians, Thebans, and other pro-Persian Greeks. The allied contingents, formerly the Greek center, who had already reached Plataea, now marched promptly back to assist the Spartans and Athenians. Those going to help the Athenians were attacked by Persian and Theban cavalry. Though unable to make any progress, these Greeks indirectly saved the Athenians from an envelopment that would probably have been decisive. As a result the Athenians, no longer harassed by cavalry, began to gain the upper hand, and pushed the Thebans and Persians back toward the Asopus. Since Mardonius had been killed in the struggle against the Spartans, the Persians began to lose heart. As allied reinforcements arrived, the Spartans redoubled their efforts, and soon the Persians were fleeing back toward the river. The Greeks followed, driving them across the stream, inflicting terrible losses. Thebes was invested, and within a month surrendered. Greek losses were few, but probably more than the 1,360 reported by Plutarch. The Persians lost over 50,000. Victory was not due to superior Greek leadership; Mardonius appears to have outgeneraled Pausanias throughout the campaign. The battle was won by technical military superiority, in the first clear-cut example of the value of superior discipline and training.

479, August (?). Battle of Mycale. Meanwhile a Greek fleet under Spartan **Leotychidas** was operating off the Ionian coast. The Persian fleet withdrew to Mycale, near Samos, where Xerxes had left a strong army. Unable to entice the Persian fleet into a naval battle, Leotychidas landed his troops and attacked the Persian army—a foolhardy move, since the Greeks were greatly outnumbered. However, as the battle began, the contingent of Ionian Greeks in the Persian army changed sides. The Greeks won a complete victory, capturing Mycale and the Persian fleet.

Concluding Campaigns of the Persian War, 479-448 B.C.

479. Greek Capture of Cyprus and Byzantium. A fleet and army under Pausanias captured Cyprus, then returned through the Aegean and the Hellespont to seize Byzantium.

478-470. Athens Continues the War. Sparta and most of the other states of European Greece withdrew from the war, but Athens, becoming ever more dependent upon overseas commerce, particularly upon grain supplies from the Black Sea regions, continued to assist the Greek cities of Asia Minor to break away from Persia. This alliance was called the **Delian League.** Soon the alliance became a façade for virtual Athenian sovereignty over all member states.

466. Battle of Eurymedon. Cimon of Athens won a great naval battle against the Persians off the Eurymedon River in Asia Minor, ending the war in the Aegean area, though desultory fighting continued elsewhere.

460-454. Operations in Egypt. In response to an Egyptian appeal, a strong Athenian fleet sailed to Egypt and helped rebels capture **Memphis,** the capital. The Persian garrison held out in the citadel, however, for nearly 4 years, until a new army from Persia drove off the Athenians (456). Athenians and rebels were now besieged for 2 years on an island in the Nile, to be finally annihilated when the Persians diverted the course of the river and attacked.

450. Battle of Salamis (in Cyprus). Cimon's Athenian fleet thoroughly defeated the Persians.

448. The Peace of Callias. The Greco-Persian War came to an end.

GREECE, 480-400 B.C.

Background of the Peloponnesian Wars, 480-460 B.C.

480-479. Themistocles Rebuilds Athens. The Spartans, secretly pleased by Mardonius' destruction of rival Athens (479), unsuccessfully opposed Themistocles' plans to reconstruct the city's walls. Themistocles also improved and fortified the harbor of the Piraeus, strengthening Athens' links with the sea, and in general stimulated the city's commercial greatness.

478–420. Growing Rivalry of Athens and Sparta. Sparta was jealous of Athens' growing prosperity and power. Like other Greeks, the Spartans also abhorred Athens' increasingly autocratic leadership of the Delian League. Athenian distaste for the military regimentation of Spartan society, and for ruthless Spartan suppression of the Messenian helots, was equally strong. Thus the paradox: a democratic state suppressing the freedom of its allies, while a militaristic oligarchy became the champion of self-determination.

The First Peloponnesian War, 460–445 B.C.

460. Athens' War with Corinth. Despite involvement in Egypt (see p. 32), economic ri-

valries caused Athens to go to war with Corinth and other Peloponnesian states. **Aegina** joined Corinth, but her fleet was soon overwhelmed by Athens (458), and the city was besieged and captured (457).

457. Sparta Joins the War. She denounced her alliance with Athens to enter the war, and a Spartan-led army defeated the Athenians at the **Battle of Tanagra,** near Thebes. After this, Spartan participation was only halfhearted, and soon afterward the Athenians crushed Thebes at the **Battle of Oenophyta.**

457–447. Athenian Successes. Despite some defeats, the Athenians under the military and political leadership of **Pericles** were generally successful in fighting on land and sea.

446. Change of the Tide. Athens was driven by Sparta and her allies from her conquests in mainland Greece, and threatened with revolts within her empire. The Athenian fleet was barely able to retain control of the seas.

445. Thirty Years' Peace. Concluded on the initiative of Pericles.

The Golden Age of Pericles, 445–432 B.C.

The fallacy of a policy of expansion in Greece proper was now obvious to Pericles; Athens lacked manpower and wealth to maintain both a large fleet and a large army. Like Themistocles he concluded that the glory and prosperity of Athens must be in trade and colonization overseas. The **Long Walls** he built connecting Athens with her seaport—the Piraeus—symbolized his strategic concept of a self-contained, invulnerable metropolis that could exist indefinitely, isolated from the rest of mainland Greece, so long as she retained command of overseas supply routes. Pericles hoped that his defensive policy on the mainland would eliminate the former causes of war between Athens and her neighbors. But Sparta was still jealous.

435. Naval War between Corinth and Corcyra. Corinth was a member of the Peloponnesian League; Corcyra (Corfu) was an ally of Athens.

433. Athenian Intervention. Athens threw her influence on the side of Corcyra and began economic reprisals against Corinth and her allies.

432. Sparta Declares War. She charged Athens with breach of the Thirty Years' Peace. The Pel-

oponnesian and Boeotian Leagues joined Sparta.

The Second Peloponnesian War, 432–404 B.C.

431-430. Pericles' Strategy of Attrition. Athens defended on land, while taking the offensive at sea. Spartan armies invaded and ravaged Attica but were stopped by the Long Walls; command of the sea assured uninterrupted supplies to Athens. Athenian fleets ravaged the Peloponnesian coast.

430–429. Plague Strikes Athens. Brought by ship, the disease decimated the overcrowded populace. Pericles died (429); his place was taken by **Cleon,** who prosecuted the war with equal determination but less skill.

429. Battles of Chalcis and Naupactus (Lepanto). Athenian admiral **Phormio** won two great victories over superior Peloponnesian fleets to establish and maintain a blockade of the Gulf of Corinth. His death soon after was a blow to Athens no less severe than the loss of Pericles.

429–427. Siege of Plataea. A combined Spartan-Theban force invested Plataea, Athens' faithful ally. This siege saw the first known instance of besiegers establishing both lines of contravallation (facing inward, toward the city) and circumvallation (facing outward, to protect the besiegers from outside attack). When Plataea fell, the defenders were massacred and the city destroyed.

427. Revolts on Corcyra and Lesbos. Athens concentrated on suppression of the revolts. A Spartan expedition to help the rebels at Mitylene, capital of Lesbos, hastily withdrew upon the appearance of an Athenian fleet. The Athenians landed and speedily captured the city, suppressing the revolt.

426. Cleon Takes the Offensive on Land. He sent his most able general, **Demosthenes,** to Aetolia, preliminary to a projected two-front attack to crush Thebes and Boeotia. Demosthenes, due to insufficient force, was repulsed.

426. Battle of Tanagra. At the same time another Athenian general, **Nicias,** made a half-hearted invasion of Boeotia from the east, retiring to Attica after an inconclusive victory.

426. Battle of Olpae. Demosthenes retrieved his reputation by a brilliant victory, enticing the numerically superior Spartans into an ambush.

425. Battle of Pylos (Navarino). Demosthenes landed on the west coast of the Peloponnesus and established a fortified base at Pylos. The Spartans reacted violently, but Demosthenes repulsed the land attacks and captured an entire Spartan fleet, leaving a Spartan contingent isolated on the island of Sphacteria, just off the coast.

425. Battle of Sphacteria. Cleon brought reinforcements, and the two Athenian generals overwhelmed the Spartans, capturing 292 survivors. Sparta, anxious to secure release of the prisoners, sued for peace, but Cleon foolishly refused.

424. Battle of Delium. The Athenians again planned convergent land attacks against Boeotia, but were frustrated when Theban general **Pagondas** defeated the main Athenian army under **Hippocrates,** who was killed.

424–423. Brasidas' Invasion of Thrace. Sparta's greatest general, **Brasidas,** probably following directives of the Spartan ephors (the body of magistrates responsible for Spartan policy and strategy), undertook a daring strategic diversion. The main base protecting the Athenian supply route to the Black Sea was a complex of colonies around the Chalcidice Peninsula. Brasidas, with less than 2,000 hoplites, plus auxiliary troops, marched north to threaten this base, defeating two Athenian armies en route (424). He captured Amphipolis, the most important Athenian colony in the Chalcidice. He was prevented from seizing the nearby port of Eion only by the timely arrival of the Athenian admiral **Thucydides** (historian of the war, who soon after was accused of negligence by Cleon and exiled for 20 years).

423. Truce. Largely because of Brasidas' threat to her supply route, Athens concluded a year's truce with Sparta. Brasidas, however, ignored the truce and continued to capture Athenian colonies in Thrace and Chalcidice.

422. Cleon and Nicias Rushed to Thrace. They forced Brasidas back to Amphipolis.

422. Battle of Amphipolis. Cleon advanced toward the city with inadequate security. Brasidas made a surprise attack, completely defeating the numerically superior Athenians. Cleon and Brasidas were both killed.

Peace of Nicias, 421–415 B.C.

421. Fifty Years' Peace. The treaty was intended to last for 50 years. However, Sparta's allies were generally dissatisfied with the terms, which restored the conquests of each side. This led to a shifting of alliances and the outbreak of numerous minor conflicts throughout Greece. Soon Athens' new allies—Argos, Mantinea, and Elis—were at war with Sparta.

418. Battle of Mantinea. Spartan King **Agis** invaded Argos and Mantinea. Athens sent a small force to assist her allies, but the allied army was decisively beaten by Agis. This, the largest land action of the Peloponnesian War, took place when the major antagonists were nominally at peace. The battle restored Sparta to unquestioned hegemony of the Peloponnesus.

417–416. Athenian Troops Assist Argos. Athens' relations with Sparta were strained, but both refrained from war.

416. Rise of Alcibiades. This brilliant young Athenian soldier and politician, portrayed by many historians as an evil, opportunistic villain, by others as a victimized hero, played the leading role in the last 11 years of the war. He turned his countrymen's attention toward Sicily, considered by many Greeks to be a land of opportunity and the gateway to Greek expansion. Syracuse, richest city of the island, was involved in a local quarrel with an Athenian ally. Alcibiades suggested that this was an opportunity to crush Syracuse, capture Sicily, and establish for Athens indisputable leadership of the Greek world.

The Sicilian Expedition, 415–413 B.C.

415. Athenians at Syracuse; Alcibiades Flees to Sparta. Against the advice of Nicias, the Athenian Assembly voted to send an expedition against Syracuse under the joint leadership of Nicias, Alcibiades, and the able general **Lamachus.** There were significant strategic reasons for the expedition. Sicily was a major source of grain for Greece. Furthermore, Syracuse was a major trading partner of Corinth, the chief maritime rival of Athens. The expedition consisted of 136 triremes and an equal number of transports, carrying 5,000 hoplites and a somewhat smaller force of light troops. Before they reached Sicily, Alcibiades was recalled to stand trial on charges of religious sacrilege (possibly trumped up by political enemies). On the way back to Athens and certain execution, he escaped and fled to Sparta. Meanwhile Lamachus urged an immediate attack on Syracuse, before the city was alerted to danger. Cautious Nicias, however, so delayed the approach that the Syracusans were given ample notice. The Athenians won a battle outside the walls, but the strengthened defense of Syracuse defied them, and Lamachus was killed. Accordingly the Athenians encamped and began to build siege lines around the land side of the city.

414. Spartan Assistance to Syracuse. Upon the advice of Alcibiades, the Spartans sent their general **Gylippus** to assist the defenders. Under Gylippus' leadership the Syracusans prevented the Athenians from completing their planned wall of contravallation.

413. Athenian Disaster. The Athenian fleet, trying to operate in narrow waters close to Syracuse, was defeated by a combined Corinthian–Syracusan fleet. Soon after this, Demosthenes arrived from Athens with land and sea reinforcements. Realizing that Athenian morale and health were at low ebb, he urged immediate withdrawal. Nicias procrastinated as usual. While the two generals debated, the Syracusans and their allies blockaded and then annihilated the Athenian fleet. Nicias and Demosthenes now tried to escape overland, abandoning their sick and wounded. The victorious Syracusans pursued and captured the remnants of the Athenian army. Nicias and Demosthenes were executed; the survivors became slaves.

Final Phase, 414–404 B.C.

414. Sparta Declares War. She established a virtually permanent siege of Athens.

412. Naval Struggle for Ionia and the Aegean. A Spartan fleet sailed to Ionia, to lead a revolt against Athens. Alcibiades, now a trusted representative of Sparta, negotiated a treaty between Sparta and the Persian satrap of Sardis, **Tissaphernes,** recognizing Persian sovereignty over Athens' dependencies in Asia Minor. In return the Persians agreed to provide funds to support the Peloponnesian fleet. Amazingly, Athens replaced the fleet lost at Syracuse and was soon challenging the Peloponnesians in the Aegean Islands and off the coast of Asia Minor.

411. Alcibiades Rejoins Athens. Tissaphernes failed to provide the promised funds, making it difficult for Sparta to maintain her fleet. Alcibiades was secretly angling to rejoin Athens, and had suggested that Persia should stand aloof from the struggle. When an upstart Athenian government began to negotiate secretly with Sparta, patriotic Athenian citizens, and the strongly democratic fleet, overthrew the oligarchs, calling Alcibiades back to command the fleet. He at once sailed northward to counter Spartan efforts to incite rebellion in the Thracian colonies and along the Hellespont, threatening Athens' vital grain route from the Black Sea. Off **Cynossema** a victory was won over the Peloponnesian fleet, increasing the confidence of the newly raised Athenian crews and augmenting Alcibiades' prestige in Athens.

410. Battle of Cyzicus. Alcibiades won an overwhelming victory over the Peloponnesian fleet and a Persian army in a combined land-sea operation in the Sea of Marmora. Sparta now offered to make peace on the basis of the *status quo,* but the demagogue **Cleophon,** who had just seized power in Athens, rejected the offer.

408. Alcibiades Recaptures Byzantium. Athens regained undisputed control of the Bosporus.

408–407. Cooperation of Sparta and Persia. They marshaled their forces in a major effort to humble resurgent Athens. A new Spartan fleet was built at Ephesus with funds and materials provided by Persia. Supervising the construction and training the crews was the Spartan general **Lysander.** The new satrap, **Cyrus,** lent him every support.

406. Battle of Ephesus. Alcibiades tried vainly to entice Lysander into combat. But when Alcibiades sailed off with part of his fleet to collect supplies, Lysander attacked and defeated the blockading Athenian squadron. Alcibiades rushed back, but Lysander again refused to fight. Word of this setback provided an opportunity for Alcibiades' personal enemies in Athens to persuade the amazingly fickle people to relieve him of command. **Conon** was appointed to command the Athenian fleet. Because Spartan law permitted an admiral to command a fleet for only one year, Lysander was replaced by **Callicratidas.**

406. Blockade of Mitylene. Callicratidas outmaneuvered Conon and blockaded the Athenian fleet in Mitylene Harbor. In a desperate and again amazing effort, impoverished Athens raised a new fleet, which was dispatched to raise the blockade of Mitylene.

406, August. Battle of Arginusae. The Athenians gained a great victory; Callicratidas was drowned. Once more Sparta offered to make peace, and once more Cleophon incredibly rejected the offer.

405. Return of Lysander to Command. When Cyrus demanded that Lysander be restored to command of the Persian–Peloponnesian fleet, he was permitted by Sparta to accompany the fleet, and in fact, if not in name, he commanded the fleet and the Persian army as well. Cautiously he sailed northward to the Hellespont, skillfully avoiding the Athenian fleet. Conon immediately sailed for the Hellespont to counter this new threat to Athens' grain supply route. Unable to entice Lysander into battle, Conon established a base at Aegospotami. In the following days he tried to force a battle, but Lysander refused. Soon, despite warnings from Alcibiades, Conon and his men relaxed their watchfulness.

405. Battle of Aegospotami. Lysander struck suddenly, while the Athenian fleet was moored for the night, with most of the crews ashore. This was the decisive action of the Peloponnesian War. The Athenian fleet of nearly 200 vessels was completely destroyed; most of the crews were captured and slaughtered. Lysander now sailed to blockade the Piraeus, while King **Pausanias** invested Athens by land.

404, April. Surrender of Athens. Forced by starvation after a six months' siege. Sparta was supreme in Greece.

The Anabasis, 401–400 B.C.

401. Revolt of Cyrus. Persian satrap of Lydia, Cyrus planned to overthrow his elder brother, Emperor Artaxerxes II. Cyrus marched on Susa with an army probably 50,000 strong, including some 13,000 Greek mercenaries, veterans from both sides of the Peloponnesian War, commanded by a Spartan general named **Clearchus**.

401. Battle of Cunaxa. At Cunaxa, near Babylon, Cyrus' army was met by Artaxerxes, with a Persian host numbering perhaps 100,000. The Greeks, on the right of Cyrus' army, utterly defeated the left wing of Artaxerxes' army. On the other side of the field, the struggle was prolonged. Then Cyrus was killed, and all the rebel army—except for the Greeks—fled. Changing fronts, Clearchus' phalanx now advanced against the victorious right wing of Artaxerxes' army and drove it from the field. This amazing battle was won by the Greeks—so Xenophon tells us—at the cost of only one hoplite wounded. Tissaphernes subsequently invited Clearchus and his senior officers to a feast, at which all the Greek leaders were treacherously seized. Clearchus was murdered and the others were sent to Artaxerxes, who had them beheaded.

401–400. "March of the 10,000." The younger officers—mostly Spartans and Athenians— took over, marching to the nearest friendly haven: the Greek Black Sea colony of Trapezus, over 1,000 miles away, across the wild mountains of Armenia. They lived off the country, continuously fighting off the barbarian hill peoples, many subsidized by Persian agents and all resentful of the more than 12,000 plus men who requisitioned all available food. The supe-

rior energy and ability of **Xenophon,** a young Athenian officer, was soon evident.

400. Arrival at the Sea. After 5 months of fighting and marching, 6,000 survivors reached Trapezus. Xenophon was primarily responsible for the accomplishment. His later *Anabasis* ("Upcountry March") is one of the outstanding military histories of all time. In this book were fateful words: "Persia belongs to the man who has the courage to attack it."

ROME, 600–400 B.C.

The Kingdom, c. 600–509 B.C.

c. 600–509. Rome under the Tarquins. The city seems to have prospered under this line of Etruscan kings, achieving hegemony over most of Latium. The history of the period is most unreliable.

578–534. Reign of Servius Tullius. He was a man of plebeian birth who married into the royal family, then seized the throne. To him are attributed a number of the early laws and customs on which the subsequent military might of Rome was based. He is, however, a legendary figure, and it is possible that his supposed reforms were actually instituted about a century later, during the Republic.

The Republic, 509–400 B.C.

509. Republican Revolt. For several years the Tarquins, supported by their Etruscan relatives, endeavored unsuccessfully to regain the throne. It was during one of these struggles that legendary **Horatius Cocles** is reputed to have held the bridge over the Tiber River against an Etruscan host under **Lars Porsena** (508).

496. Battle of Lake Regillus. Another legendary encounter, this may have been fought against Latin foes rather than Etruscans. During the century there were frequent conflicts with neighboring Latin, Etruscan, or Sabine states, or with the Aequian and Volsci hill people to the northeast.

458. Dictatorship of Cincinnatus. With the Aequi threatening to overwhelm Rome, the

Roman Senate called the respected warrior **L. Quinctius Cincinnatus** from his farm to take over the dictatorship (see below). Cincinnatus decisively defeated the Aequi, gave up the dictatorship before it was due to end, and returned to his plow.

439. Cincinnatus Again Called to Dictatorship. This time he defeated the Volsci.

438–425. War with Veii. This Etruscan state was about 10 miles northwest of Rome, on the far bank of the Tiber. Rome was victorious in a drawn-out siege.

431. Defeat of the Aequians. By victories of dictator **A. Postumius Tubertus.**

405–396. Renewed War with Veii. The Romans besieged Veii for 9 years. Since an army was kept constantly in the field during this time, Rome began regular payments to the troops, inaugurating the concept of a regular career service. When Veii fell (396), it was destroyed and its territory and people absorbed by Rome. The Roman Republic was now unquestionably the leading state of central Italy (see also p. 64).

ROMAN MILITARY SYSTEM, C. 400 B.C.

Servius Tullius—or some later organizer—divided the population for military purposes into six groups, in accordance with wealth, since soldiers furnished their own arms and equipment. The first class provided the cavalry and the best-armed heavy infantry, similar to the Greek hoplite. The next two classes provided slightly less elaborately equipped heavy infantry. The fourth and fifth classes furnished javelin men, slingers, and other unarmored auxiliaries. In the sixth class were those exempted from military service for physical, religious, or other reasons. Like the Greeks, Roman citizens served only when called upon, but frequent wars, a well-developed martial spirit, an inherent sense of discipline, and constant peace-time exercise and maneuvers made these part-time soldiers the terror of their enemies.

Each class was organized in units of 100 men, called **centuries.** In combat formation the heavy infantry was arranged like a phalanx, the better-armed men in front. The light troops functioned ahead of the main body and covered the flanks.

Under the Republic, executive and military authority were exercised by two coequal **consuls,** elected to office each year by popular vote. This system effectively ensured a balance of political power, but had serious military drawbacks. The consuls shared responsibility for combat operations, usually exercising command on alternate days. Recognizing the military inefficiency of such an arrangement, Roman law provided that in time of emergency one individual—termed a **dictator**—could be called upon to exercise complete authority over the state and the armed forces, though only for a limited time, usually 6 months.

CARTHAGE AND SICILY, 800–400 B.C.

c. 800. Carthage Founded. This was the most important colonial outpost established in the central Mediterranean by the Phoenicians. At this time, Greek seafarers also began to penetrate into the same region, establishing colonies mainly in Cyrenaica, eastern Sicily, and southern Italy. Friction between Greeks and Phoenicians, and between their colonies, began early in the 8th century and continued for some 500 years. At the same time there was incessant conflict between colonizers and the barbarian peoples inhabiting the littoral.

c. 650–c. 500. Expansion of Carthage; Conflicts with Greek Colonies. Phoenician colonization dwindled as Phoenicia was overrun by foreign conquerors. Carthage soon assumed the role of protector of the other Phoenician colonies, while aggressively founding colonies of her own in North Africa, Iberia, western Sicily, and other Mediterranean islands. With equal vigor, though less centralization, Greek colonization continued along the northern shores of the Mediterranean. By the 6th century there had been a number of violent clashes between Greeks and Carthaginians in Sicily, Corsica, Sardinia, southern Gaul, and eastern Iberia. The Carthaginians were successful in Corsica and Sardinia, while late in the 6th century the Greek colony of Massilia (Marseille) drove the Carthaginians from the coast of Gaul. Though the Greeks absorbed much of Sicily, Carthage established a firm foothold in the west.

c. 500. Rise of Syracuse. Under the wise rule of the tyrant **Gelon,** Syracuse began to rival Carthage, not only in Sicily but as a power in the central Mediterranean.

481. War of Carthage and Syracuse. The Carthaginians were easily persuaded by envoys from Xerxes of Persia to attack Syracuse (see p. 29). As a consequence, Gelon was forced to refuse an appeal for assistance from Greece, to devote all his attention to dealing with the Carthaginians.

480. Battle of Himera. Gelon, heading an alliance of Greek Sicilian states, decisively defeated a large Carthaginian army under **Hamilcar,** who was killed. The Carthaginians, forced to pay an indemnity, temporarily abandoned their ambitions in Sicily.

480–410. Carthaginian Expansion Westward. They greatly augmented their sea power, gaining complete control of the western Mediterranean and contiguous regions farther west beyond the Straits of Gibraltar, with outposts down the African coast and possibly on the Madeira and Canary islands.

474. Battle of Cumae. Etruscan expansion into Campania was halted by the Syracusan fleet under **Hiero.**

409. Carthaginian Return to Sicily. An early **Hannibal*** gained revenge for the defeat of his grandfather, Hamilcar, at Himera by capturing that and other cities of northern and western Sicily. The Carthaginians pressed eastward, taking Agrigentum (406) and threatening to overwhelm Syracuse and the other Greek strongholds on the island (400).

* Note that the names Hamilcar, Hannibal, Mago, etc., were common among Carthaginians.

SOUTH ASIA

INDIA, 600–400 B.C.

c. 600 B.C. The Mahajanapandas. These were 16 kingdoms and oligarchic republics in the Ganges Plain, coalescing from a patchwork of feuding tribal states. In the next century (c. 500 B.C.), 4 of them, all in the eastern Ganges Valley, had gained ascendancy: the Kingdoms of Kosala, Kasi, and Magadha, and the Republic of Vrjji.

c. 600–c. 500. Rise of Kosala. This Hindu kingdom became the leading power of northern India by defeating and absorbing Kasi.

c. 543–491. Rise of Magadha. Under King **Bimbisara,** this Hindu kingdom (modern Bihar) expanded greatly.

c. 537. Cyrus of Persia Reaches India. He campaigned across Bactria and Arachosia (modern Afghanistan and southern Soviet Turkestan) and turned south to the vicinity of modern Peshawar (see p. 24). He may have reached the Indus.

517–509. Darius Annexes the West Bank of the Indus. He also conquered part of the northwestern Punjab, east of the river. He sent the Greek admiral **Skylax** to explore the Indus as far as the Arabian Sea. These regions remained under Persian control until the beginning of the 4th century.

c. 490–c. 350. Magadha Predominant in Northwest India. Under **Ajatusatra,** Magadha began a series of successful wars with Kosala.

HINDU MILITARY SYSTEMS, C. 500 B.C.

The invasions of Cyrus and Darius seem to have stimulated the Hindus to develop cavalry, hitherto neglected in India. Hampered, however, by the difficulties of breeding good horses in the Indian climate, they continued to place primary reliance on the chariot, which had become the decisive element in Indian warfare. The best horses, therefore, were reserved for chariots. It was probably about the 6th century, also, that war elephants began to appear on Indian battlefields. By 400 B.C., it appears that the Hindu princes considered elephants equally important instruments of war as chariots.

The bow, the primary Hindu weapon, remained unchanged for approximately 2,200 years. It was usually 4 to 5 feet long. Though other materials, including metal, were tried, bamboo remained the preferred material. The arrows, made of bamboo or of cane, were 2 to 3 feet long, usually tipped with metal. The effective range of the bow was 100 to 120 yards, or slightly less when a heavy, iron anti-elephant arrow was used. Fire arrows were also used against elephants.

Bowmen did not usually carry shields but were protected by a front rank of shield-bearing javelin throwers. Though hand-to-hand combat was avoided if possible, both archers and javelin men were also armed with a fairly long, broad-bladed sword. If a decision could not be reached by long-range fire, recourse was made to the sword in a confused melee.

CEYLON, 500–400 B.C.

A band of some 700 Aryan Hindu adventurers, under **Vijaya** (an exiled prince of Sinhapura), landed near Puttalam on Ceylon (c. 483 B.C.). Vijaya and his followers established a kingdom in the northeast, and his grandson (the third Vijayan monarch) reputedly founded Anuradhpura, for centuries the capital and chief city. Ceylon's early history is colored by frequent struggles for power and by periodic incursions by South Indian Tamils.

EAST ASIA

CHINA, 600–400 B.C.

c. 600–c. 500. Decline of the Chou. Warfare in China lost its ritualistic character and became more violent as the nominal vassals of Chou struggled for supremacy in the region including the Yangtze and Yellow River valleys. By this time the contest had narrowed to three states: (1) Ch'in, to the northwest, in the Wei Valley that had produced the Chou; (2) Ch'u, dominating the area from the Yangtze River almost as far north as the Yellow River; and (3) Wu, holding the mouth of the Yangtze and the seacoast region almost up to Shantung.

519–506. War between Wu and Ch'u. This culminated half a century of conflict between these rivals. It began with a Wu attack on the Ch'u city of Zholai (modern Fenglai in Anhwei). A decade of inconclusive warfare followed. King **He Lu** of Wu (513–494) seems to have raised the first peasant conscript army for the latter part of this war. The final campaign began with General **Sun Wu (Sun Tzu,** see below) taking an army 300 kilometers by boat up the Hwai River to its confluence with the Hanshwei River (506). Two Ch'u armies converged on Sun Wu, who withdrew overland.

He ambushed and decisively defeated his pursuers in the **Battle of Bai ju** (modern Ma-Zhang, Hubei). Sun Wu pursued the fleeing survivors to Ying (modern Jiang-ling, Hubei), the Ch'u capital. The Wu army occupied, looted, and destroyed Ying, then withdrew, taking great wealth back to Wu. Meanwhile, King **Shao** of Ch'u fled to Chin, where he sought assistance. A Chin army occupied the ruins of Ying, but elected not to continue the war.

c. 511. Sun Tzu. Modern Chinese historians seem to have resolved the scholarly controversy over both the identity of the author of the first great military classic, *The Art of War,* and when it was written. Sun Wu, a renowned general of Ch'i, was enticed to Wu by King He Lu to assume leadership of the ongoing war with Ch'u (c. 511, see p. 40). He is sometimes referred to as Wu Sun Wu, or Sun Tzu (Honorable Sun) of Wu. The book is distinguished by Chinese scholars from other works of the same or similar titles as "the book of thirteen chapters." Until recently, many scholars confused **Sun Ping** of Ch'i (also often called Sun Tzu), who lived about 150 years later, with the original Sun Tzu. There are two reasons for the confusion. Sun Wu had also originally come from Ch'i, while Sun Ping also wrote a book on the art of war, apparently identical in title. That book—more contemporary and tactical, and less conceptual than Sun Wu's book—disappeared in A.D. 7th century but was rediscovered by archaeologists excavating a Han tomb in 1972. (From 5th century B.C. to A.D. 18th century there have been more than 200 Chinese books on the art of war; these have not been properly researched, owing in part to the difficulties of translating ancient Chinese.) Sun Wu's (or Sun Tzu's) book reveals a profound understanding of the philosophy of human armed conflict, providing valuable lessons for military men, even today. The earliest known translation was to Japanese in A.D. 8th century; the first European translation was in French about 1780; a Russian translation appeared in 1860. The first English edition (translated from the Japanese) was in 1905; a German version was published in 1910.

c. 475–221. Era of the Warring States. This was merely an intensification of the pre-existing anarchy. Different scholars give different beginning dates, for instance 8th century; 519; 453; 403. During this period armored infantry formations—consisting of spearmen, halbardiers, swordsmen, and crossbowmen—dominated China's battlefields. Cavalry appeared toward the end of the period, while the chariot was increasingly relegated to the role of mobile command post.

c. 473. Wu Overthrown by Yueh. Yueh inherited both the coastal region and Wu's growing maritime tradition.

453. Overthrow of Chin. Chin was defeated and split among Chao, Han, and Wei. (The independence of these three states was acknowledged 50 years later (403) by Emperor Kan of Chou.)

c. 400. Rise of Ch'in and Ch'u. Although other states continued to participate in the multi-lateral struggle, these two emerged as the leading military powers of China.

III

THE ERA OF THE GIANTS:

400–200 B.C.

MILITARY TRENDS

The emergence of strategy and of the tactical application of the combined arms—horse, foot, and artillery—highlighted this period, together with a dazzling display of military leadership. **Alexander the Great** and **Hannibal** were geniuses. But great generals in their own right also were **Philip of Macedon**—Alexander's father—and Alexander's successors, the **Diadochi.** To these we must add Hannibal's father—**Hamilcar**—and his pupils, the Romans **Scipio, Marcellus,** and **Nero;** also **Pyrrhus** of Epirus; **Epaminondas, Xanthippus,** and **Philopoemen** of Greece; **Dionysius** and **Agathocles** of Syracuse; the Indian **Chandragupta;** the Chinese **Shih Huang Ti;** and many others. No other comparable period of history produced more capable military leaders.

THEORY OF WARFARE

Hannibal has been called the "father of strategy," although Alexander was no less aware of strategic fundamentals. Few other important leaders of the period demonstrated comparable strategic grasp, until the lessons learned from Hannibal were utilized to some extent by Nero in the Metaurus campaign, and by Scipio in Spain and Africa.

On the other hand, there was a general awareness of—and frequently an emphasis on—economic warfare. In consequence, save for the wars of Alexander, Hannibal, and the offensive-minded Romans, pitched battles were relatively infrequent. The opposing generals concentrated on raiding one another's resources while at the same time blockading towns and fortresses.

MILITARY ORGANIZATION

Another new development was the integration of basic military components into a combined fighting team. Asian leaders, particularly Persian, had understood how to employ cavalry, but they had never been able to coordinate this arm effectively with their infantry. In Europe, Epaminondas at Leuctra showed how horsemen could be used for tactical screening and delaying purposes in coordination with his infantry. Dionysius of Syracuse was also apparently successful, about this same time, in creating a combined fighting force.

But the first scientifically organized military force of history was that of Philip II of Macedon. His concept of the use of heavy missile engines in coordination with the field operations of infantry and cavalry was the genesis of field artillery. His son, Alexander, developed this concept further.

Philip also developed a new type of light infantry—his **hypaspists**—combining the discipline of the trained heavy **hoplite** with the speed and flexibility of the irregular **psiloi.** This development grew out of the Athenian **Iphicrates'** introduction of the lightly armed, disciplined **peltast**—usually a mercenary—into Greek warfare about half a century earlier.

WEAPONS

The **catapult** and **ballista** came into their own during the period. The catapult,* probably appearing first in Sicily under Dionysis I of Syracuse about 400 B.C., was essentially a large, crew-served bow using tension or torsion from twisted animal sinew to propel large arrows—and later large rocks—up to 500 yards, with an accurate range of over 200 yards. The ballista,* similar in concept, was generally larger, and used torsion to hurl large rocks—and later, arrows—for comparable distances. These heavy and cumbersome engines were usually used in siege operations—as well as in the defense of fortifications—but both Philip and Alexander also used light catapults and ballistae in field operations. The **onager,** a later development (c. 200 B.C.), was named after the wild ass because of its vicious kick when the firing arm slammed into the crossbeam.

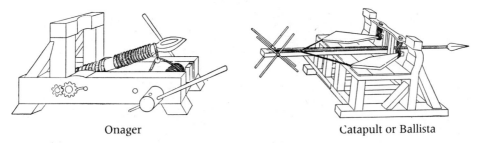

Onager   Catapult or Ballista

There were no important developments in small arms and armor, although there were some significant refinements: the **sarissa,** the long Macedonian pike; and the Roman **gladius** (short sword) and **pilum** (javelin). Also significant was the introduction of lightweight armor and equipment for Greek **peltasts** and Macedonian **hypaspists.**

SIEGE WARFARE

The basic siege weapons remained the battering ram and the movable tower. **Mantelets**—great wicker or wooden shields, sometimes mounted on wheels—were used to shelter outpost

---

* There are some contradictions in nomenclature and identification of catapult and ballista among modern authorities.

guards and operators of siege engines within range of weapons on the city walls. **Diades,** Alexander's engineer, invented a mural hook, or **crow,** consisting of a long, heavy bar or lever, suspended from a high vertical frame, to knock down the upper parapets of a wall. He also invented the **telenon,** a box or basket large enough to contain a number of armed men, slung from a boom. This boom was in turn suspended from a tall mast or vertical frame, on which it could be raised or lowered by tackle. By this elevator a group of infantrymen could be hoisted above parapet height, swung over any intervening obstacle, such as a moat, and deposited directly upon the enemy's battlements.

   Archimedes, the great mathematician and scientist of his day, apparently created some other special engines for use in the defense of Syracuse (213–211) against the Romans. Unfortunately, no designs of these have come down to modern times. From the meager descriptions extant, these would appear to have been refinements on weapons already known and used (such as those of Diades). Archimedes was partial to the use of huge grappling devices, or tongs, to be used against battering rams or to seize hostile warships approaching the sea wall of Syracuse.

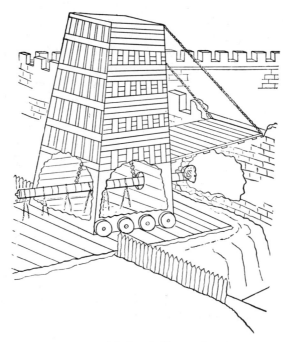

Tower with drawbridge and ram

Tactics on Land

There were two major developments: the introduction of tactical maneuver by Epaminondas of Thebes and of tactical flexibility by the Romans.

   In earlier periods there were crude, tentative—and usually fortuitous—attempts at tactical maneuver, the most notable being at Marathon. But at Leuctra, Epaminondas deliberately introduced the concepts of mass—or concentration—and of economy of force, in his oblique order of battle. His additional contribution was cavalry-infantry coordination in his plan of maneuver—a sharp departure from past reliance solely on cavalry shock action.

Tactical flexibility was introduced by the Romans in the cellular battle order of their legion, a drastic deviation from the solid mass of the Greek phalanx. The Roman formation was originally adopted to permit easier movement in combat over uneven terrain. It took the Carthaginian Hannibal and the Greek Philopoemen to teach the Romans how this flexible formation could further tactical maneuver. Scipio, at Ilipa, in turn showed how well the Romans could take this lesson to heart.

The tactical use of elephants, originating earlier in India, was brought forcibly to the Western military mind at the Hydaspes, where both the potentialities and the limitations of the war elephant were demonstrated. Alexander's horses refused to face the beasts, yet his disciplined phalanx, despite initial surprise and dismay, eventually turned the elephants back in panic-stricken flight. **Seleucus'** impressions, however, were sufficiently favorable for him—20 years later—to cede substantial territory to Chandragupta in exchange for 500 elephants, which he proceeded to use to advantage in his victory at Ipsus. After that time the use of the war elephant spread rapidly to Greece and Carthage. That the beast was a valuable weapon is clear from its use by such objective warriors as Hannibal and Pyrrhus (the latter, in fact, owed to his elephants his hard-won victories over the Romans).

As proven at Beneventum and Heraclea, elephants were most successful when used against troops unacquainted with them. Disciplined and resourceful opponents, however, could stam-

Mantelets   Telenon and mural hook

pede the elephants, which then became more dangerous to friend than foe. For this reason, the war elephant's mahout (driver) carried a steel spike to hammer into the beast's brain should he stampede.

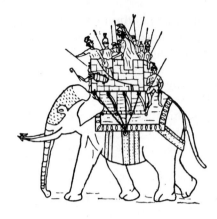

War elephant

Anti-elephant measures included the use of fire arrows, by the Indians, while some unidentified Greek genius evolved the prototype of an antitank mine field: iron spikes, chained and anchored in place, to rip the tender feet of the pachyderms. This last device was used effectively by **Ptolemy** at Gaza.

NAVAL WARFARE

Though navies declined in size during the 4th century, they increased greatly in the 3rd. During the First and Second Punic Wars the Romans and Carthaginians each frequently had at sea up to 500 war vessels, manned by as many as 150,000 seamen and marines. Individual fleets ran as high as 350 ships. To control these vast maritime forces, Rome and Carthage had administrative organizations comparable to a modern admiralty or navy department.

The trireme was generally displaced by larger vessels during this period. Dionysius of Syracuse was apparently the first to build such vessels. By the time of the Punic Wars, the quinquireme had become the standard warship. The complement of these decked galleys consisted of up to 300 rowers and seamen, plus as many as 100 seagoing soldiers—equivalent of latter-day marines.

Naval tactics were unchanged until the middle of the 3rd century, when the Romans, recognizing both the inferiority of their relatively slower and clumsier ships and the superiority of Carthaginian seamanship, introduced a new concept: their combined grappling device and boarding bridge called the **corvus.** In order to come close enough to ram or to break the oars of the clumsier Roman ships, the Carthaginians had to risk being caught by the corvus, which was followed by an irresistible charge of Roman legionaries.

ADMINISTRATIVE AND LOGISTICAL SERVICES

We know little of the manner in which supply operations were organized in this period, and in most instances these were undoubtedly quite haphazard. There is, however, clear evidence of much advance planning and systematic organization of the supply service of the long-range (both in time and distance) operations of both Alexander and Hannibal.

Macedonians, Carthaginians, and Romans had also systematized their baggage trains with the emphasis on austerity in impedimenta. Essential equipment, weapons and the like were carried on pack animals, though wagons and hand carts were also used.

EURASIA—MIDDLE EAST

PERSIA, 400–338 B.C.

c. 400. Persistence of Persian Power. Largely by default, Persia remained the major power of the Middle East, the Greeks having exhausted themselves in the Peloponnesian War. Although royal authority had greatly declined, and outlying provinces were drifting away, most of the empire had peace, prosperity, and order.

386–358. Continuing Decline in Central Authority. The satraps of Asia Minor, in particular, became virtually independent.

358–338. Reign of Artaxerxes III. In large part due to the Greek general **Mentor,** imperial authority was temporarily reasserted over the satraps. Egypt was reconquered (342).

GREECE AND MACEDONIA, 400–336 B.C.

Spartan Period, 400–371 B.C.

400–371. Spartan Hegemony. By virtue of victory in the Peloponnesian War, Sparta was supreme in Greece. This hegemony was briefly maintained by Lysander and Kings Agis and **Agesilaus,** while simultaneously conducting successful operations against Persia in Asia Minor (see below).

400–387. War of Sparta and Persia. Sparta came to the aid of the Ionian cities when Tissaphernes, satrap of Lydia and Caria, began to punish them for their support of Cyrus' revolt against Artaxerxes II (see p. 37). After desultory operations and extended truces, Agesilaus II campaigned aggressively and victoriously across western Asia Minor (396–

394) until recalled to Greece where war had spread.

395–387. Corinthian War. Resentful of Spartan arrogance, and taking advantage of the war in Asia Minor, Athens, Thebes, Corinth, Argos, and some smaller states allied themselves with Persia against Sparta. On land Lysander won some initial successes before he was killed in an unsuccessful attack on the town of **Haliartus** (395). Agesilaus, returning from Asia Minor, avenged the defeat at **Coronea** (394).

394. Battle of Cnidus. The former Athenian admiral **Conon,** commanding the Athenian-Persian fleet, destroyed the Spartan fleet (commanded by Peisander, Lysander's brother-in-law, who was killed) off Cnidus, near Rhodes, ending Sparta's brief term as a maritime power, though the Spartan admiral **Antalcidas** had some later success against the Persians and the Athenians (388).

394. Siege of Corinth. Agesilaus besieged Corinth, but soon the war became a stalemate (393). Athens, seizing the opportunity to recover some of the strength lost in the Peloponnesian War, rebuilt the Long Walls, and began to reestablish control over some old colonies. An Athenian army under Iphicrates relieved Corinth (390). Persia, alarmed, gave surreptitious assistance to its enemy (Sparta) against its ally (Athens).

387. The Peace of Antalcidas (or **King's Peace).** A compromise settlement. Athens' partial recovery was acknowledged; Sparta's hegemony, though shaken, continued. The Greek states agreed to nominal Persian suzerainty over Greece in hopes this would reduce jealousies.

387–379. Spartan Supremacy Continues.
Sparta vigorously crushed several challenges, maintaining garrisons and puppet rulers in a number of Greek cities, including Thebes.

379–371. War of Independence. Thebes revolted against Sparta. Athens soon joined her, and although the allies were unable to win a decisive victory, they maneuvered the Spartans out of central Greece. At the same time Athens had the best of war at sea, **Chabrias** winning a great naval victory off **Naxos** (376). Sparta, however, taking advantage of a split between Athens and Thebes, recovered some initial land losses (375–372).

371. Persia's Efforts Bring Peace Settlement. Sparta, however, refused to allow Thebes to represent the other members of the Boeotian League, upon which the Theban leader, **Epaminondas,** withdrew from the negotiations. Spartan King **Cleombrotus** immediately invaded southern Boeotia with an army of 11,000 men. Epaminondas could muster only 6,000.

371, July. Battle of Leuctra. The Spartans drew up for battle in the conventional phalangial line, the best troops on the right, a few

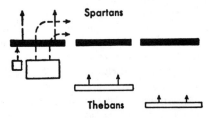

Battle of Leuctra

cavalrymen and light troops covering the flanks. They expected the Thebans to form in similar fashion. In such a battle the Spartans, superior both in numbers and in fighting quality, would unquestionably have been victorious. Epaminondas, however, refused to fight on Spartan terms. He quadrupled the depth of his left wing, forming a column 48 men deep and 32 wide. The remainder of his army, covered by a cavalry screen, was echeloned to his right rear in thin lines facing the left and center of the Spartan army. This is the first known

example in history of the deep **column of attack** and of a **refused flank,** prototypes of the main effort and holding attack of more modern times. Epaminondas personally led his left-wing column in a vigorous charge against the Spartan right, while his cavalry and the infantry of the refused center and left advanced slowly, occupying the attention of the Spartans to their front, but without engaging them. The Spartans were hopelessly confused by these novel tactics. The weight of the Theban column soon crushed the Spartan right. Epaminondas completed the victory by wheeling against the exposed flank of the remaining Spartans, who promptly fled when simultaneously engaged by the Theban center and right. The Spartans lost over 2,000 men; Theban casualties were negligible. Spartan military prestige was shattered forever.

Theban Period, 371–355 B.C.

371–362. Theban Hegemony. The most startling event of a decade of desultory warfare was an alliance of Sparta and Athens to circumvent Theban supremacy (369). But the military and political genius of Epaminondas prevailed on land and sea. Thebes' most dangerous enemy was **Alexander** of Pherae, ruler of Thessaly, who at **Cynoscephalae** fought a drawn battle with Theban general **Pelopidas,** who was killed (364). The following year Epaminondas defeated Alexander.

362. Battle of Mantinea. The Arcadian League of the Peloponnesus—founded by Thebes (370) as a counterweight to Sparta—broke apart, a number of members joining Sparta and Athens. Epaminondas marched to the Peloponnesus, where his army was joined by some faithful allies. Near Mantinea he met an army composed of troops of Mantinea, Sparta, Athens, and dissident former members of the Arcadian League. Each army numbered about 25,000 men. By a combination of unexpected maneuvers and deception, Epaminondas completely surprised his enemies, then overwhelmed them with an oblique attack almost identical to that he had introduced at Leuctra. He, however, was killed in the moment of victory, and his followers were unable to take advantage of the success. Theban supremacy collapsed. In the words of Xenophon, Greece fell into "even greater confusion and indecision."

359–355. Rise of Macedon. Philip II began a new phase in the history of Greece. After securing his throne from a pretender supported by Athens, Philip began to reorganize his army, stabilizing the political and social conditions in his backward kingdom and expanding Macedonia's frontiers in all directions. He was undoubtedly inspired by the example of Epaminondas, whom he had known as a youth, while a hostage at Thebes. After successful campaigns in Illyria and against barbarian tribes between Macedonia and the Danube, Philip took advantage of Athens' disastrous involvement in the **Social War** (358–355) with her allies and colonies, to seize some Athenian possessions in Thrace.

Macedonian Period, 355–336 B.C.

355–346. Third Sacred War. The Amphictyonic Council of central Greece, supported by Thebes, declared war against Phocis (which had profaned the temple at Delphi) and against Phocian allies: Sparta, Athens, and Pherae (in Thessaly). Philip offered to aid the Amphictyonic Council, and seized Thessaly after a bitter two-year struggle against Thessalians, Phocians, and Athenians (355–353), climaxed by a victory at **Volo,** in which the Phocian general **Onomarchus** was killed. Blocked at Thermopylae by a combined Phocian, Spartan, Athenian, and Achaean force, Philip returned north to carry out a systematic conquest of Athenian colonies in Thrace and Chalcidice (352–346). When Athens sued for peace, Philip's terms were generous (346). He now turned against Phocis. Bribing his way through the pass at Thermopylae, he crushed the Phocians in a brief campaign and was elected chairman of the Amphictyonic Council.

345–339. Philip's Consolidation of Northern Conquests. Moving westward, Philip subdued all remaining opposition in Epirus, Thessaly, and southern Illyria (344–343). Next, moving north to the Danube, he brought all the wild tribes of that region under his sway. The following years were devoted to extending his domains eastward in Thrace, as far as the Black Sea. Here he met the only serious military failures of his career: After he had conquered Propontis, Athens supported uprisings in Perinthus and Byzantium. Philip was repulsed in efforts to capture these two fortified seaports (339).

339–338. Fourth Sacred War. Demosthenes of Athens, whose famed Philippic orations (351) had warned Greece to unite against the growing power of Macedon, again stirred Athens and Thebes to war against Philip and the Amphictyonic Council.

338. Battle of Chaeronea. Philip, with 32,000 men, crushed an Athenian-Theban army of 50,000, which included the best mercenaries obtainable from other Greek states. Philip's young son, Alexander, distinguished himself in command of the cavalry of the Macedonian left flank. Greek casualties were about 20,000; those of the Macedonians are unrecorded, but must have been severe. Philip was now the unquestioned master of Greece.

337–336. The Hellenic League. Philip called a congress of Greek states at Corinth; all—except Sparta—participated, creating a **Hellenic League,** in perpetual alliance with Macedonia. Philip's plan for a war against Persia—for the ostensible purpose of freeing the Greek cities of Asia Minor—was approved; he was appointed chairman of the League for the duration of the war, and at the same time commanding general of the combined Graeco-Macedonian army. His trusted general **Parmenio** was then sent to Asia with an advance body, to carry out a reconnaissance in force (336).

336. Assassination of Philip. The murder was probably instigated by his divorced wife, **Olympias,** mother of Alexander. Historians agree that Alexander was not implicated in the plot.

MACEDONIAN MILITARY SYSTEM, 350–320 B.C.

Philip, as soon as he came to power, completely reorganized the Macedonian army (359). The result was the finest fighting force the world had yet seen: a national army, combining the disciplined skill of Greek mercenaries with the patriotic devotion of Greek citizen soldiers. For the first time in history, scientific design—based on exhaustive analysis of the capabilities and limitations of the men, weapons, and equipment of the time—evolved into a clear concept of the coordinated tactical action of the combined arms. Careful organization and training programs welded the mass into a military machine, which under the personal command of Philip (or later Alexander) probably could have been successful against any other army raised during the next 18 centuries—in other words, until gunpowder weapons became predominant.

The backbone of the army was its infantry. The Macedonian phalanx was based on the Greek model, but 16 men deep, instead of 8 to 12, and with a small interval between men, instead of the shoulder-to-shoulder mass of the Greek phalanx. There were two types of hoplites: **pezetaeri** and **hypaspists.** The more numerous pezetaeri carried **sarissas,** or spears more than 13 feet long.* For training purposes, a heavier, longer sarissa was used.) In addition, each man carried, slung over his shoulder, a shield large enough to cover his body when kneeling, with a short sword worn on a belt, plus helmet, breastplate, and greaves. The sarissa was held 3 to 6 feet from its butt, so that the points of the first 4 or 5 ranks protruded in front of the phalanx line in battle. Despite the heavier armament, constant training made pezetaeri units more maneuverable than the normal Greek phalanx. They were capable of performing a variety of movements and maneuvers in perfect formation.

More adaptable to any form of combat, however, was the hypaspist, cream of the Macedonian infantry. He was distinguished from the pezetaeri only by his shorter pike, probably 8 to 10 feet in length, and possibly by slightly lighter armor. Formations and evolutions of the hypaspist phalangial units were identical to those of the pezetaeri. The hypaspists were, if possible, better trained, more highly motivated, faster, and more agile. Since Alexander usually used an oblique order of battle, echeloned back from the right-flank cavalry spearhead, the hypaspists were usually on the right flank of the phalanx, to provide a flexible hinge between the fast-moving cavalry and the relatively slow pezetaeri.

Although Philip designed this heavy infantry formation as a base of maneuver for the shock action of his cavalry, the phalanx was a highly mobile base, which, completing a perfectly aligned charge at a dead run, would add its powerful impact upon an enemy not yet recovered from a cavalry blow. To exploit these tactics, Philip and Alexander tried to choose flat battlefields; but the concept was applicable, and was applied, on rough terrain.

The diagram (p. 51) of a simple phalanx is a graphic representation of the "idealized phalanx" of Asclepiodotus, dating from the 1st century B.C., but differs little from the formation employed by Philip and Alexander.

To protect the flanks and rear of the phalanx, and to maintain contact with the cavalry on the battlefield, the Macedonian army of Philip and Alexander included light infantry. Generally referred to as **peltasts,** a term borrowed from contemporary Greek usage, these troops were

---

* Some authorities assert that the war sarissa was 21 feet long, the training sarissa 24 feet long. This is not totally unreasonable (as other authorities insist) since medieval Swiss pikemen wielded spears of comparable length.

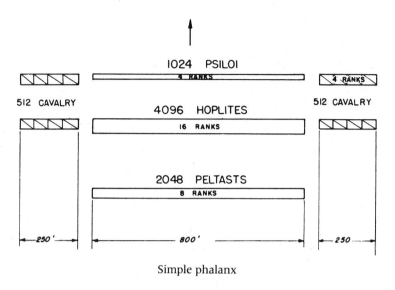

Simple phalanx

unarmored or lightly armored, and equipped with bows and arrows, slings, or javelins. The peltasts would also cover the advance of the phalanx, and retire to the flanks or rear before the moment of impact. Additionally, armed servants and camp followers, called **psiloi,** usually guarded the camp and baggage trains. Sometimes they also served as foragers and scouts.

The organization of the hoplites, or pezetaeri, under Philip (apparently retained by Alexander, and probably by the Successors too) was based on the file (**dekas**) of 16 men. Four files composed a **tetrarchy** of 64 men, and 4 tetrarchies comprised a **syntagma** or **speira** of 256 men. The largest standard unit, the **taxis** of about 1,500 men, was a territorially based unit containing 6 syntagmata. There were apparently 12 taxeis. When he invaded Asia, Alexander probably took 8 taxeis with him and left the other 4 behind in Macedon. The taxeis apparently did not long survive Alexander's death, and the armies of the Successors (c. 320–c. 200 B.C.) apparently contained **chiliarchies** (a Greek term approximating the modern "regiment") of 4 syntagmata, totaling about 1,000 men in place of the earlier taxis.

Like the modern division, the simple phalanx was a self-contained fighting unit of combined arms; in addition to the heavy infantry, it included (at theoretical full strength) 2,048 peltasts, 1,024 psiloi, and a cavalry regiment (**epihipparchy**) of 1,024, for a total of 8,192 men. The grand phalanx, composed of four simple pahalanxes, could be likened to a small modern field army, and had a strength of about 32,000 men.

Cavalry was a decisive arm of the Macedonian army, as well trained and as well equipped as the infantry. The elite were the Macedonian aristocrats of the Companion cavalry, so called because Philip, and later Alexander, habitually led them personally in battle. Hardly less skilled, and also relying upon shock action, were the mercenary Thessalian horsemen. The Companions usually were on the right of the infantry phalanx, the Thessalians on the left. The principal weapon of these heavy cavalrymen was a pike, about 10 feet long, light enough to be thrown, heavy enough to be used as a lance to unhorse an opposing cavalryman or to skewer an infantry foe. They were equally adept at using the short swords carried at their belts. They wore a scale-

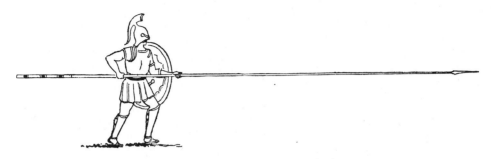

Pezetaerus, with sarissa couched

armor breastplate, plus shield, helmet, and greaves. Their horses also had scale-armor head-pieces and breastplates.

There were other, intermediate, cavalry formations; some organized as lancers, others—prototypes of dragoons—capable of fighting on horse or on foot; both varieties carried lighter weapons and armor. Finally, there were the light cavalrymen, mounted equivalents of the psiloi, who carried a variety of weapons: javelins, lances, bows. These light horsemen rarely wore armor, save for a helmet. Their functions were screening, reconnaissance, and flank protection.

The stirrup had not yet been invented; the horseman was seated on a pad, or saddle blanket of some sort (though the light cavalryman sometimes rode bareback), with bridle and headstall comparable to those of our own times. To become effective in combat, long training and practice were essential for both men and beasts.

The Macedonian army was the first to use prototypes of field artillery. Philip devised light-weight catapults and ballistae to accompany his siege train; it is not clear whether he actually used them in field operations. Alexander, however, habitually used these weapons in battle, particularly in mountain and river-crossing operations. Philip designed these engines so that the essential parts could be carried on a mule or pack horse; the bulky wooden elements would be hewn on the spot from tree trunks. This, of course, would delay their employment in field operations, so Alexander carried a number of the assembled weapons in wagons.

As noted earlier, Philip, Alexander, and their engineers introduced several innovations in siege warfare, and were far more successful in their sieges than their Greek predecessors. The highly organized Macedonian corps of engineers was responsible not only for the siege train but also for a bridge train for river crossing. As in the case of the artillery, the essential manufactured components of the specialized equipment were packed on animals or in wagons; these were then assembled with lumber hewn on the spot.

The details of the Macedonian staff system are not clear, though obviously well developed. Command was exercised by voice, by trumpet, and by spear movements. Long-range communication was accomplished by smoke signals in the daytime, by fire beacons at night. For battlefield messages Alexander used his seven aides-de-camp, or one of a more numerous corps of youthful pages. This latter corps was an officer-training unit, with programs of instruction and development comparable to those of modern military academies.

The most thorough administrative and logistical organization yet seen was developed by Philip of Macedon. Surgeons were attached to the Macedonian army, and there is even some evidence of something like a medical field hospital service. There was also an efficient engineer corps, whose major function was to perform the technical tasks of siege operations and river crossings.

This was the compact, competent, smoothly organized, scientific instrument which Philip bequeathed to his son, Alexander the Great.

Macedonian cavalryman

CONQUESTS OF ALEXANDER, 336–323 B.C.

336–323. Reign of Alexander III (b. 356). The accession of this youth appeared to many Greeks to offer an opportunity to throw off Macedonian domination. Alexander, however, quickly marched to Greece at the head of an army; opposition disappeared. At Corinth he was elected captain general of the Hellenic League, in place of his father, for the projected operations against Persia (336).

335. Campaigns of Consolidation. The death of Philip had also caused the northern barbarian tribes to become restive. Alexander marched across the Danube, punishing recalcitrants firmly and reestablishing unquestioned Macedonian suzerainty over the region. Moving westward, he put down another uprising in southern Illyria, where he received word that Thebes and Athens had risen against Macedonian leadership and that most of Greece was wavering. **Darius III** of Persia had apparently been instrumental in arousing Greece against

Macedonia. By forced marches Alexander arrived quickly in central Greece, captured Thebes by surprise, and virtually destroyed it. The lesson was salutary; Athens surrendered (and was treated generously); opposition ceased.

334. Invasion of Persia. Having assured the security of the Hellenic base, Alexander crossed the Hellespont into Asia with an army of 30,000 infantry and 5,000 cavalry. He had expanded his father's objective. He was determined to conquer Persia. He left **Antipater,** one of his most trusted generals, with an army of slightly more than 10,000 to hold Macedonia and Greece.

334, May. Battle of the Granicus. Alexander was met in western Asia Minor by a Persian army of about 40,000 men—about half Greek mercenaries. Alexander led an assault across the Granicus River, and was victorious in a short, sharp battle. He quickly liberated the Greek coastal cities of Asia Minor from Persian control, meeting little opposition, save at **Miletus,** which he captured after a brief siege.

334, July. Alexander's Strategy. Alexander now made the basic strategic decision for his subsequent campaigns against Persia. His only line of communications with Macedonia and Greece was overland, but across the Hellespont. The Persian fleet dominated the Aegean and eastern Mediterranean; not only could it cut his line of communications at the Straits, it could support dissident uprisings in Greece, which could ruin his entire plan. He decided that before he could hope to conquer Persia, he must destroy Persian sea power. Lacking an adequate fleet, he would seize the entire Mediterranean seacoast of the Persian Empire; not only would this assure the security of his base, it would force the surrender of the Persian fleet. Then he could advance into Persia without fear for his communications.

334–333. Alexander conquered the coastal regions of Asia Minor, encountering difficulty only at **Halicarnassus,** captured after a hard siege. He went on to secure all important Persian strongholds in the interior, including Gordium, where occurred the famous incident of the Gordian knot.

333, October. Arrival of the Persian Army. Darius III, learning that Alexander was moving southward into Syria, hastened to place himself behind him and across the Macedonian line of communications, near Alexandretta (Iskendurun), with an army estimated at more than 100,000 men. Alexander, with some 30,000 men, turned to meet the threat.

333. Battle of Issus. Alexander found the Persians drawn up in a very deep formation on the narrow coastal plain, just north of the Pinarus River. Because of the tremendous discrepancy in numbers, Alexander decided to adopt the tactics of Epaminondas at Leuctra. With the Companion cavalry and the hypaspists he planned to attack the Persian left, with the remainder of the phalanx echeloned to his left rear. The Thessalian cavalry guarded the left flank of the phalanx from the dangerous Persian cavalry. In a preliminary action, he drove back a strong Persian covering force in the foothills, south of the Pinarus. Then, leading the Companions, he put his plan into effect. The leading echelons of the phalanx were briefly in trouble when the Persian center counterattacked while they were crossing the stream. Meanwhile Alexander's cavalry assault had smashed the Persian left; with the hypaspists he wheeled westward, into the exposed center of the Persian army. With this support, the Macedonian center recovered and renewed its attack. The Persian cavalry on Darius' right flank had meanwhile crossed the Pinarus, to be repulsed by the left of the phalanx and the Thessalians. As his center crumbled under the combined Macedonian cavalry-infantry attack, Darius fled the field, amidst the panic-stricken survivors of his army. Persian losses were tremendous, probably more than 50,000. Macedonian casualties were 450 killed. The Macedonians captured, among others, the family of Darius, including his queen, children, and mother. Alexander pursued briefly, then returned to his original plan of securing the seacoast as a base.

332, January–August. Siege of Tyre. The principal Phoenician seaport of Tyre (Sur) was situated on an island less than half a mile off the mainland. The main base of the Persian navy, Tyre's capture was essential to Alexander's plan. To get at the city, Alexander built a mole 200 feet wide from the mainland out to the island. Tyrian opposition was vigorous. Using fire ships, they several times interrupted the work, burning down part of the mole and the wooden besieging towers on it. Redoubling his efforts by land, Alexander also scraped up a naval force from other captured Phoenician towns. After winning a tough sea fight, he cooped up the Tyrian ships in their harbors. Finally, as the mole approached the island city's walls, a breach was made by shipborne engines, and the city was stormed. As an example to other towns on his route, Alexander treated the survivors harshly; the city was practically destroyed and most of the inhabitants scattered as slaves.

332. Peace Overtures from Darius. The Persian ruler offered Alexander 10,000 gold talents ($300 million), all of the Persian Empire west of the Euphrates, and his daughter in marriage.

Alexander refused, replying that he intended to take all of Persia, and that he could already marry Darius' captive daughter, if he wished, without Darius' consent.

332, September–November. Siege of Gaza. While besieging Tyre, Alexander had sent troops to seize the rest of Syria and Palestine. Gaza alone resisted. Alexander immediately marched there and began another siege. The most memorable feature of the operation was Alexander's construction of a great earthen mound, 250 feet high and a quarter of a mile in circumference at its base, on which he mounted catapults and ballistae with which to bombard the defenders. After two months the city was stormed and sacked.

332–331, December–March. Occupation of Egypt. There was no significant opposition. Alexander's seacoast base was now secure. As in Asia Minor and Syria, Alexander established firm control with military garrisons in the chief cities. While in Egypt he founded the city of Alexandria—one of many of that name. He also made a journey to the Temple of Zeus Ammon, at the Siwa Oasis, some 200 miles west of Memphis in the Lybian desert, to be hailed by the priests as the son of Zeus.

332–331. Spartan Revolt. King **Agis II** of Sparta, with Persian financial support, had roused several Greek states to revolt against Macedonia while Alexander was away on his expedition. He was joined by most of the southern Greek states, and besieged **Megalopolis.** Antipater marched south and defeated the rebels outside Megalopolis (331). After this victory he sent a substantial reinforcement of infantry and cavalry to Alexander, which joined the king in Egypt.

331, April–September. Learning that Darius was assembling a vast army in Mesopotamia, Alexander rapidly marched to Tyre, then turned eastward to cross the Euphrates and Tigris rivers without opposition. He located the Persian host—probably about 200,000 strong—drawn up for battle on the Plain of Gaugamela, near ancient Nineveh, and about 70 miles west of Arbela (Erbil). Alexander, who now had about 47,000 men, halted seven miles from the Persian camp to reconnoiter and rest his troops. Another peace offer from Darius was refused; this time the Persian offered 30,000 gold talents ($900 million), half his kingdom, and his daughter's hand.

BATTLE OF ARBELA (OR GAUGAMELA), OCTOBER 1, 331 B.C.

Darius, whose best foot troops, the Greek mercenaries, had been almost destroyed at Issus, was relying mainly on his cavalry, chariots, and elephants. The Persians were in two long, deep lines, with cavalry on each flank. In the center were some remaining Greek mercenaries and the Persian Royal Guard cavalry. Numerous chariots lined the front of the entire army, with a clump of elephants in front of the center. Darius, expecting a night attack, had kept his tired troops in position all night.

As at Issus, Alexander advanced in echelon from the right, where his cavalry Companions, screened by light infantry, were to strike the hammer blow. Next came the hypaspists, with the main phalanx in the center of his line. The left was composed of the Greek and Thessalian

horse, commanded by Parmenio. Behind each flank of the Macedonian line moved a column of light horse and foot, prepared to protect the flank from envelopment by the long Persian line. Behind the center, and covering the camp, was a thin phalanx of Thessalian infantry. These last three elements comprised what was probably the first recorded battlefield use of a tactical reserve.

Apparently the Macedonian advance "drifted" obliquely to its right; the cumbersome Persian host endeavored to shift correspondingly to its left, the movement creating some gaps in the Persian line. As Alexander led his Companions forward in a charge, the Persian wings swept in to envelop the Macedonian flanks but were met and repulsed by the flank reserve columns Alexander had disposed for that purpose. Alexander noticed a gap near the left-center of the Persian line, and led his charge there. Creating a giant wedge with his cavalry and the hypaspists, he smashed through completely; Darius, in the path of the onslaught, fled. Panic spread throughout the Persian center and left, and they crumpled and gave way. Alexander was forced to turn back to rectify conditions on his own left flank, which had been driven back by a determined Persian cavalry charge in great force. But the reserve had held, and Alexander's drive into the rear of the Persian attackers ended the threat, and con-

cluded the battle as well. He now led his entire army in a vigorous pursuit of the fleeing Persians, scattering the defeated foe hopelessly. Alexander lost 500 men killed, and probably about 3,000 wounded. Persian casualties are unknown, but there were at least 50,000 slain.

331–330, October–July. Pursuit. Alexander marched into Babylon, which surrendered without a fight. Subduing wild mountain tribes as he advanced through the heart of the Persian Empire, he destroyed Persepolis, the ancient Persian capital, in somewhat delayed retribution for the burning of Athens in 480. Then he turned northward to Ecbatana (Hamadan),

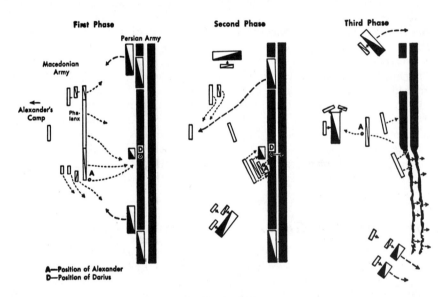

Battle of Arbela

where Darius had taken refuge; but on the approach of the Macedonians, Darius fled eastward through the Caspian Gates. Choosing 500 of his strongest men, Alexander dashed ahead of his army and, after marching 400 miles in eleven days, caught up with the fleeing Persians. As Alexander approached, only 60 men still with him, the Persian nobles led by **Bessus,** satrap of Bactria, murdered Darius, then scattered. Alexander now was unquestioned ruler of the Persian Empire.

329. Consolidation and Advance into Central Asia. Alexander, pursuing the murderers of Darius, and at the same time consolidating his hold on the empire, marched eastward through Parthia into Bactria. Capturing and executing Bessus, he turned northward across the Oxus into Sogdiana in central Asia. He was forced to fight a number of bitter battles against the wild Scythian tribesmen in the mountain passes south and west of the Jaxartes, and was wounded—once seriously—in two of these.

His crossing of the Jaxartes was brilliant. Rafts for the infantry were improvised by sewing up tents stuffed with hay. Scythian bowmen, lined up along the far river bank, were driven back by heavy missile fire covering the crossing. After inflicting a crushing defeat on the Scythians, he returned to Sogdiana, which had risen in revolt, under **Spitamenes,** former satrap of the province.

328–327. Advance to India. After a long, hard campaign against the rebels, Alexander subdued Sogdiana, Spitamenes being murdered by his own adherents. Alexander married **Roxana,** daughter of **Oxyartes,** one of the Sogdianan chieftains who had fought most valiantly against him and who now became viceroy for the Macedonian emperor. He next

prepared for a campaign into India, encouraged in this by the King of Taxala—at war with **Porus,** leading monarch of the Punjab—who thought Alexander would be a good ally. Despite bitter opposition from the natives of the Hindu Kush region, Alexander fought his way through passes north of the Kabul Valley (part of his army went through the Khyber Pass) to the Indus River (327). Crossing into India, he was welcomed by the King of Taxala (near Rawalpindi), who again asked Macedonian assistance against Porus.

326, March–May. March to the Hydaspes. Pleased to have an excuse to invade central India, Alexander marched eastward, until he was stopped at the unfordable Hydaspes (Jhelum) River, torrential and swollen from

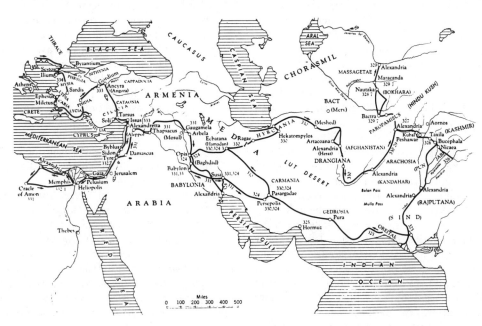

Alexander's conquests

rains. On the far bank lay the army of Porus, about 35,000 men. Alexander had about 20,000 with him. A crossing of the swollen river against opposition was out of the question; Alexander established a camp on the river bank, and endeavored to convince Porus that he

would not attempt to cross before the river fell. To confuse the Indians, however, he undertook a series of feints up and down the river near the two opposing camps. Kept constantly on edge by this ceaseless activity, Indian reactions to the feints and alarms became perfunctory.

326, May. Battle of the Hydaspes. Sensing decreased Indian vigilance, Alexander moved rapidly. He had reconnoitered a crossing place 16 miles upstream from his camp. With about half his army, he marched to the selected point during a stormy night, leaving the remainder to continue the feints along the bank near his camp. Careful preparations had been made, boats were ready, and the crossing was completed shortly after dawn. Porus, bewildered by reports that Alexander was on his side of the river, drew up his army near his camp, with about 100 elephants lined up in front; he knew that the horses of Alexander's cavalry would not face the elephants. Alexander arrived in front of this formidable array with about 6,000 cavalry and 5,000 infantry. He promptly sent his general **Coenus** with half of his cavalry in a wide encirclement of the Indian right flank. The remainder of his small army he drew up beside the river, with his left refused, to prevent Porus from enveloping his open flank. His light infantry began to harass the elephants to his front. A number of the maddened beasts turned and dashed through their own lines, putting Porus' ranks into confusion. Just as the Indian right wing was advancing to envelop the open Macedonian flank, it was struck in the rear by Coenus, who then swept down the rear of the entire Indian line, adding to the existing confusion. At this moment Alexander led his Companions in a charge along the river bank, while the small phalanx smashed into the Indian left wing. For a while the Indians fought stoutly and casualties were severe on both sides. But, assailed from front, flank, and rear, Porus' men finally gave way and took to flight. Porus, badly wounded, was captured by Alexander.

326, July. Mutiny. Alexander now decided to continue into north-central India and planned to continue to the Ganges. He got as far as the Hyphasis (Beas) River when his exhausted, homesick Macedonians simply refused to go further. It was respectful but determined mutiny. Sadly, Alexander gave in to their demands and began the return.

326–324. Return to Persia. Alexander marched south down the Indus, meeting considerable resistance. In a battle with the Mallians near modern Multan he was seriously wounded. He recovered, and reached the mouth of the Indus, exploring the surrounding countryside. He built a fleet and sent it westward under **Nearchus** across the Arabian Sea to the Persian Gulf, exploring a hitherto unknown sea route. A portion of his army under **Craterus** he sent back to Persepolis via the Bolan Pass and Kandahar. With the remainder Alexander marched across the mountains and deserts of Gedrosia, in one of the most grueling and difficult marches of military history. At two points en route he made contact with the fleet of Nearchus. Though we have no details, the administrative arrangement for this march must have been superb—evidence of a major factor in Alexander's invariable success.

324–323. Planned Union of East and West. Arriving back in Persia and Mesopotamia, Alexander discovered that his empire had become shaky while he was away campaigning. Promptly he restored order and set in motion a grandiose plan for a melding of the best features of the cultures of Greece and Persia. How much further he would have changed the course of history can only be speculative; he died in Babylon of a fever (probably malaria).

COMMENT. *Of particular interest was Alexander's ability to adjust his tactics and tactical formations to fit conditions of the moment, as shown in his central Asian operations (329–327). Against guerrilla resistance he reorganized his army into light, mobile columns moving independently but in coordination. Much use was made of light cavalry bowmen. Exhaustive terrain reconnaissance assisted his supply needs. Military colonies, established at important road junctions, not only protected his communications, but these colonies, becoming cities, brought civilization to a large area of the Middle East. His accomplishments in mountain warfare and against irregular forces have never been equaled. No man in history has surpassed his intellectual, military, and administrative accomplishments; not more than two or three are worthy of comparison.*

THE DIADOCHI—SUCCESSORS OF ALEXANDER, 323–200 B.C.

Wars of the Diadochi, 323–281 B.C.

Upon the death of Alexander his leading generals (the Diadochi, or successors) immediately fell to wrangling over his empire in a multilateral conflict that lasted for more than 40 years.

The principal contestants were **Perdiccas** (d. 321), Alexander's prime minister or chief of staff, whom the dying king made his regent; **Antipater** (398?–319), able regent in Greece and Macedonia from 334 to 323; **Eumenes** (360?–316), staff secretary; **Ptolemy** (367?–283) and **Lysimachus** (361?–281), personal aides and principal staff officers; **Seleucus** (358–280), commanding the officers' training corps of pages; **Craterus** (d. 321, son-in-law of Antipater), **Polysperchon** (d. 310?), **Antigonus** (382–301), and **Cassander** (350–297, son of Antipater), all the equivalent of infantry division commanders; and **Demetrius** (d. 283, son of Antigonus), who had commanded a squadron of Companion cavalry.

These were the men Alexander had selected as his principal subordinates on the basis of combat performance; in addition to intrinsic toughness and ability, all had been trained by an unexcelled master of war. None of them, however, really understood Alexander's system or possessed his spark of genius. They were skilled professionals, but not great captains. Much of the struggle was waged by guile, treachery, and bribery. Their armies were completely mercenary and would turn against their leaders if offered more money by the enemy.

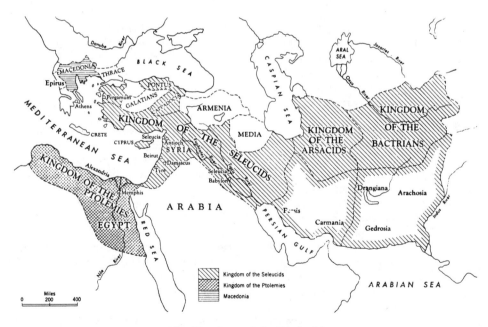

The kingdoms of the Diadochi

Between them, these men set Alexander's Macedonian Empire aflame from the Balkans to the Indus River and from the Caucasus to Egypt's southern border. The conflict consumed all of them and for the remainder of the 3rd century partitioned the empire loosely into three warring segments ruled by the descendents of Antigonus, Seleucus, and Ptolemy. A detailed recital of the military events of the Diadochian Wars is of no military interest. However, a few highlights merit attention.

322. Lamian War. Revolt by Athens and most of the Greek cities was crushed by Antipater and Craterus at the **Battle of Crannon.** The Macedonian fleet obliterated the Athenian navy forever at **Amorgos.** Demosthenes, leader of the revolt, took poison as the Macedonians occupied Athens.

321–319. Deaths of Perdiccas, Craterus, and Antipater. Perdiccas was killed by mutineers bribed by Ptolemy (321); Craterus, invading Cappadocia, was killed by Eumenes (320); Antipater died (319).

317. Battle of Paraetakena. This engagement, between Antigonus and Eumenes, in Iran, was indecisive. The next year, after another drawn battle, Eumenes was killed by his own men, who had been bribed by Antigonus.

311. Truce. Cassander was to hold Macedonia until Roxana's son, **Alexander IV,** came of age; Lysimachus was to keep Thrace and the Chersonese; Ptolemy held Egypt, Palestine and Cyprus; Antigonus retained Asia Minor and Greece; Seleucus kept the vast region east of the Euphrates as far as India.

310. Cassander's Assassination of Roxana and Alexander IV. The line of Alexander was extinguished. Again the surviving Diadochi clashed with one another (309).

308. Battle of Salamis (Cyprus). There was a decisive naval victory by Demetrius over Ptolemy's brother, **Menelaus.** As a result Antigonus regained Cyprus (306).

307–306. Demetrius' Invasion of Greece and Palestine. He captured Athens and much of Greece, then marched to reconquer Palestine, but was repelled from Egypt by Ptolemy (305).

305–304. Siege of Rhodes. Demetrius attacked Ptolemy's garrison there. A two-year siege ensued. All the devices known to the times were tried by both sides: rams, attacking towers, liquid fire, mines and countermines, all the engines of attack and defense, raids and assaults

in both directions. Ptolemy's sea power provided adequate logistical support to the garrison, causing Demetrius to withdraw and return to Greece.

301. Battle of Ipsus. In Asia Minor, the aged Antigonus and Demetrius were defeated by Seleucus and Lysimachus, allies of Cassander, who made excellent use of war elephants, and who were aided by deserters from Antigonus' own forces. Antigonus was killed, but Demetrius escaped and established control over western Asia Minor. Cassander was recognized as king of Macedonia, but died soon after (300).

294. Demetrius Seizes Macedonia. He seized the throne after murdering the son of Cassander.

290–245. Rise of the Aetolian and Achaean Leagues in Greece. The Aetolian League, first organized in the 4th century in western Greece, became more active in this period, including central Greece. The Achaean League (280) covered most of the Peloponnesus. Athens and Sparta, generally independent, frequently became temporarily allied with one or the other of these leagues.

286. Downfall of Demetrius. King **Pyrrhus** of Epirus, allied with Ptolemy and Lysimachus, drove Demetrius from Macedonia. He retreated to Asia Minor, but his troops deserted him and he surrendered to Seleucus, dying in prison three years later (283).

283–280. End of the Diadochi. Ptolemy died (283). Seleucus and Lysimachus, the last two surviving Diadochi, clashed. Aged Seleucus, aided by **Ptolemy Keraunos** (disinherited son of Ptolemy) defeated and killed Lysimachus in hand-to-hand combat at the **Battle of Corus** (Corupedion) (281). Seleucus, now ruling all of Alexander's Macedonian Empire except Egypt, was himself murdered (280) by Ptolemy Keraunos while en route to Macedonia.

Antigonid Macedonia, Seleucid Persian, Ptolemaic Egypt, 281–200 B.C.

279–275. Celtic Invasion. A migratory wave of Celts invaded Macedonia, Greece, and Thrace, then crossed to Asia Minor, where they established the kingdom of Galatia. Ptolemy Keraunos was killed in battle with the invaders (277). **Antigonus Gonatus** (son of Demetrius) then regained control of Macedonia, driving out the Celts (276). **Antiochus I** (son of Seleucus) finally subdued the Galatian-Celts (275).

280–279. Damascene War. The first of a series of wars between the Ptolemys and the Seleucids for the control of Syria and Palestine. **Ptolemy II,** second son of Ptolemy I, defeated Antiochus I.

276–272. First Syrian War. Ptolemy II again defeated Antiochus I, at the same time occupying Antiochus' ally, Antigonus, by subsidizing Pyrrhus' invasion of Macedonia (274–273).

266–255. Alliance of Antiochus I and Antigonus against Ptolemy II. Ptolemy subsidized invasions of Macedonia by Athens, Sparta, and Epirus. Sparta and Athens, led by Chremonides of Athens, were defeated by Antigonus in the **Chremonidean War.** Sparta was eliminated by a defeat near **Corinth** (265); Athens was then invested, and surrendered after a siege of two years (262). **Alexander** of Epirus (son of Pyrrhus) was more successful, capturing most of Macedonia (263), but was finally driven out by Antigonus (255). To keep Antiochus busy in Asia Minor, Ptolemy subsidized **Eumenes I,** who successfully defended the independence of Pergamum from the Seleucid Empire (263). Antiochus invaded Syria, to initiate the **Second Syrian War** (260). Antigonus defeated Ptolemy in a naval battle off **Cos** (258). Ptolemy, admitting failure, made peace (255).

252–215. Turmoil in Greece. The principal events: **War of Demetrius** (238–229), in which **Demetrius II** of Macedon fought an inconclusive war against the Achaean and Aetolian leagues. **Cleomenes III** of Sparta defeated the Achaeans (228–227); **Antigonus III** of Macedon defeated Cleomenes at the **Battle of Sellasia** (222). In the **Social War** (219-217), **Philip V** of Macedonia crushed the Aetolians.

250–227. Decline of the Seleucid Empire. Bactria became an independent Macedonian kingdom (250) under **Diodotus I** (see p. 86). **Arsaces I,** Scythian nomad leader, controlled the province of Parthia (235), which he expanded at the expense of eastern portions of the Seleucid Empire. These losses were largely due to the preoccupation of **Seleucus II** (246–227) with the invasion of Syria by Ptolemy III (see below) and a civil war in Asia Minor (against his brother, **Antiochus Hierax**), in which the Galatians and Pergamum were also involved. Antiochus Hierax, aided by the Galatians, defeated Seleucus at **Ancyra** (Ankara; 236), but Antiochus and the Galations were soon thereafter defeated in turn by **Attalus I** of Pergamum (230–229), leaving Seleucus secure on his throne, but with Pergamum dominant in western Asia Minor.

246–241. Third Syrian War. Ptolemy III conquered Syria and much of southern Asia Minor from Seleucus II, despite a defeat by Antigonus in the naval **Battle of Andros** (245).

224–221. War between Pergamum and the Seleucids. Attalus was at first successful, but then was defeated by the new Seleucid emperor, **Antiochus III (the Great),** who regained most of central Asia Minor from Pergamum.

223–200. Revival of the Seleucid Empire. After his victory over Pergamum, and suppressing a revolt in Mesopotamia (221), Antiochus III became involved in the inconclusive **Fourth Syrian War** with **Ptolemy IV** (221–217); Antiochus was defeated at **Raphia** (Rafa; 217) after earlier victories and left Palestine in Egyptian hands. He next subdued a serious revolt in Asia Minor (216–213). He then devoted his efforts, generally successfully, to restoring the vigor and domains of the empire. He defeated Armenia (212–211), which was forced to acknowledge Seleucid sovereignty. After reestablishing control over Media (210), he invaded Parthia (209) and quickly forced **Arsaces III** to become his vassal, as a result of a great victory at the **Battle of the Arius.** He continued on to Bactria (see p. 87), where, after a hard-fought campaign (208–206), he obtained the qualified submission of the Greek ruler, **Euthydemus.** He next marched down the Kabul River as far as the Indus and possibly into the Punjab (see p. 87). Returning to his capital, Seleucia, he then conducted a successful amphibious expedition down the Arabian coast of the Persian Gulf, to capture Gerha (modern Bahrain, 205–204).

215–205. First Macedonian War. Philip V of Macedon made a treaty with **Hannibal** of Carthage against Rome and threatened to invade Italy. Rome formed an alliance against Macedonia including the Aetolian League, Pergamum, Elis, Mantinea, and Sparta. Macedonia was joined by the Achaean League, under the inspired leadership of **Philopoemen,** last of the great Greek generals. Rome, fighting for her life against Hannibal, was able to contribute few ground forces but aided with a sizable fleet, which gave the allies naval superiority. Generally inconclusive, the most notable feature of this war was the brilliant leadership of Philopoemen, whose victory at

Mantinea (207) crushed Sparta, whose able king and general, **Machanidas,** was killed.

203–195. Fifth Syrian War. Upon the accession of the infant **Ptolemy V** to the Egyptian throne (205), Antiochus III and Philip V made a secret agreement to strip the Ptolemaic Empire of all its holdings in Palestine, Syria, Asia Minor, and the Aegean Islands. In Palestine and Syria, Antiochus was succesful, the crowning victory being that of the **Battle of Panium** (198). He soon occupied all Palestine and other Ptolemaic possessions in Syria and southeast Asia Minor, except for Cyprus. Philip was less successful. In Asia Minor he was repulsed by Pergamum (under aged but still able Attalus I) and Rhodes, allies of Egypt, and in the Aegean he was defeated in the naval **Battle of Chios** (201). He was unable to redress these losses because he soon became engaged in the disastrous **Second Macedonian War** with Rome (200-196; see p. 95).

202–201. Rise of the Achaean League. Philopoemen defeated **Nabis,** tyrant of Sparta, at **Messene** and again in a naval battle off **Tegea.** By this time the military efficiency and skill of Philopoemen had brought the Achaean League to preeminence in Greece proper.

CENTRAL MEDITERRANEAN

CARTHAGE, 400–200 B.C.

400–264. Expansion Westward. Though Carthage was preoccupied with Sicily in the 4th and 3rd centuries B.C., she was simultaneously engaged in land operations and overseas expeditions along the Mediterranean coasts of Africa and Spain, and also along the Atlantic shores beyond the Pillars of Hercules (Straits of Gibraltar).

398–397. First War with Dionysius of Syracuse. A Carthaginian army under **Himilco** besieged Syracuse, but was repulsed with great losses. Carthage was forced to abandon its outposts in the eastern and central portion of Sicily.

392. Second War with Dionysius. Again the Syracusans were successful, and the war ended with Dionysius in control of most of the island; Carthage retained only a few footholds in the west.

385–376. Third War with Dionysius. The Carthaginians were successful and substantially increased their holdings in western and central Sicily.

368–367. Fourth War with Dionysius. This was inconclusive, and ended with the death of the Syracusan ruler.

347. Treaty with Rome. This, apparently, was a reaffirmation of an earlier treaty (509), in which Rome's trade was limited to Italy, while Carthage was to keep out of Italy entirely.

344–339. War with Timoleon of Syracuse. Initially successful, a Carthaginian army again besieged Syracuse, and actually occupied all the city except the citadel. But dissension and plague in Carthage weakened the armies in the field. **Timoleon** drove out the invaders, and then defeated them decisively at the **Battle of the Crimissus** (340). Peace terms were unfavorable to Carthage.

323–312. Expansion in Sicily. Again Carthage was able to profit from internal troubles in Syracuse to regain control of most of Sicily.

311–306. War with Agathocles. Hamilcar defeated Agathocles of Syracuse at **Himera** (311), and then laid seige to Syracuse (311–310). Agathocles led an army to Africa, and in turn besieged Carthage (310–307). The Carthaginians defeated the invaders outside the walls of their city, and Agathocles was forced to flee to Sicily, where his son had recently been defeated by a Carthaginian army. Despite these setbacks, peace terms were not unfavorable to Agathocles.

306. Treaty with Rome. Carthage and Rome limited their respective areas of Mediterranean commerce.

278–276. War with Pyrrhus. On the outbreak of another war, Syracuse called for the assistance of **Pyrrhus** of Epirus, who was in control of southern Italy after having defeated Rome (see p. 66). Pyrrhus drove the Carthaginnans from their investment of Syracuse. He was generally victorious, but was unable to drive the Carthaginians from their strongholds in western and central Sicily. Carthage concluded a defensive-offensive alliance with Rome against Pyrrhus (277). Pyrrhus, who probably could have conquered Sicily, returned to Italy to meet a Roman threat (see p. 66).

264–241. First Punic War. (See p. 66.) Carthage lost Sicily to Rome.

241–237. Rise of Hamilcar. Political rivalry between Hamilcar Barca, hero of the First Punic War, and **Hanno,** renowned for victories against Numidians and Mauretanians in Africa. A revolt of unpaid mercenaries (inspired in part by Hanno's inept leadership) broke out under **Matho.** The rebels, 25,000 strong, laid siege to Carthage (238). Hamilcar was called to command the Carthaginian army. By a brilliant stratagem he got 10,000 men out of the city and defeated the rebels at the **Battle of Utica.** Soon afterward, he ambushed and defeated another rebel army near **Tunes** (Tunis). This ended the mutiny. Hamilcar was now the acknowledged leader of Carthage.

238. Loss of Sardinia to Rome. (See p. 68.)

237–228. Conquest of Spain. To offset the loss of Sicily and Sardinia and to provide a base for renewed war with Rome, Hamilcar decided to conquer Spain. Raising an army largely with his own money, he expanded Carthaginian footholds in Iberia, conquering most of the peninsula below the Tagus and Ebro rivers before his death. Hanno, meanwhile, regained a position of political ascendancy in Carthage.

228–221. Hasdrubal Barca in Spain. Son-in-law of Hamilcar, Hasdrubal consolidated conquests and concluded a treaty with Rome, which recognized Carthaginian sovereignty of all territories south of the Ebro.

221. Assassination of Hasdrubal. He was succeeded by **Hannibal Barca** (247–183), son of Hamilcar.

221–219. Hannibal Consolidates. After conducting two successful campaigns against barbarian tribes in outlying regions, and advancing his control northward to the Durius River, Hannibal attacked the seaport of Saguntum (Sagunto), an ally of Rome, the only territory south of the Ebro that rejected Carthaginian sovereignty. This precipitated the **Second Punic War** (see pp. 68–79).

MAGNA GRAECIA (SICILY AND SOUTH ITALY), 400–264 B.C.

405–367. Reign of Dionysius, Tyrant of Syracuse. After gaining power, Dionysius concluded a war with Carthage (404), then expanded Syracusan control over neighboring regions (403–400).

398–367. Wars with Carthage. (See p. 62.) Dionysius on balance was successful in these wars, and extended his control over most of Sicily.

390–379. Conquest of South Italy by Dionysius. Syracuse became the leading power of Magna Graecia, and in fact of the entire central Mediterranean area. Dionysius crushed the Italiote League by his decisive victory at the **Battle of the Elleporus** (389).

366–344. Turmoil. For more than 20 years after the death of Dionysius there was constant turmoil in Syracuse and in the Greek states of Sicily. Carthage regained much of the areas lost to Dionysius. Order was reestablished in Syracuse by the rise of another strong man, **Timoleon.**

344–339. Timoleon's War with Carthage. (See p. 63.)

338–330. Greek Intervention in Italy. War between the Italian Greeks and native Italians in southern Italy (Samnites, Lucanians, Umbrians) had been going on sporadically since the beginning of the century. The Greeks asked for and received the assistance of **Archimadus** of Sparta (338) and, after his death, that of Alexander of Epirus.

334–330. Campaigns of Alexander of Epirus. He came to Italy to assist the Tarentines against the Lucanians, Bruttians, and Samnites. He entered into an alliance with Rome against the Samnites, but was defeated and killed at the **Battle of Pandosia** (331). Alexander of Epirus was the uncle of Alexander of Macedon; before his death it is reported that when he received glowing reports of his nephew's victories against the Persians, he replied that his nephew fought women while he was fighting men.

317. Rise of Agathocles. He brought order to Syracuse, which had been in decline since the death of Timoleon (323).

311–306. War of Agathocles with Carthage. (See p. 63.)

302. Agathocles Invades Southern Italy. An expedition into southern Italy against the Italians, undertaken at the request of Tarentum, was indecisive.

282–275. Pyrrhus in Southern Italy and Sicily. (See above, p. 63, and below, p. 66.)

c. 275. The Mamertines. After the death of Agathocles, a group of Italian mercenaries revolted against his successor, **Hiero II** of Syracuse. These former mercenaries, called Mamertines, soon established themselves in Messana (Messina), which became a base for wholesale brigandage by land and sea. Carrying on more or less constant war with Syracuse, they eventually became hard-pressed by Hiero, and dissension broke out in Messana. One faction of Mamertines called on Rome for help; another asked for Carthaginian aid (265). Carthage immediately sent a garrison, which seized and held Messina; Rome had an expedition on the way. This was the basis of the First Punic War (see p. 66).

ROME, 400–200 B.C.

Conquest of Italy, 405–265 B.C.

405–396. War with Veii. The first of Rome's life-or-death contests, **Marcus Furius Camillus,** Rome's first great general, was appointed dictator (first of five times) after the besieging Roman army was defeated outside the walls of Veii. Camillus had a tunnel dug under the walls of Veii; he sent a party of picked men through to strike the defenders, while he attracted their attention by an external assault (see also p. 38). Complete victory ended the 9-year siege.

390. Sack of Rome by the Gauls. Celts, or Gauls, had migrated into the Po Valley (c. 400), spreading down the Adriatic coast as far south as the Aesis River. Invading central Italy (391), they laid waste to much of Etruria, then decisively defeated a Roman army at the **Battle of the Allia** (390). They then swept into Rome itself, seizing all the city save for the citadel, on Capitoline Hill. Furius Camillus, again appointed dictator, raised an army in outlying districts, but got rid of the Gauls by paying a large tribute. Sporadic Celtic raids into central Italy continued for another 50 years.

389–343. Expansion in Latium. Quickly recovering from the Celtic raid, Rome expanded steadily in all directions in Latium and southern Etruria. Furius Camillus finally ended the threat of the Aequians and Volsci by defeating both tribes completely (389). Many conquered cities were accepted as allies, in a Latin confederacy under Rome's leadership.

367. Second Celtic Invasion. Furius Camillus, again called as dictator, drove the Gauls away. He reorganized the legion (see p. 79).

362–345. Latin Uprisings. Rome retained her leadership with great difficulty.

343–341. First Samnite War. The cities of Campania called on Rome for assistance against the warlike Samnite hill tribes. **Marcus Valerius Corvus** won a major victory at **Mount Gaurus** (342), but was unable to conquer the Samnites. Rome did, however, establish a virtual protectorate over Campania.

340–338. Latin War. An uprising of the Latin allies and colonies, joined by the Campanians, for a time threatened to overthrow Rome. Fortunately for her, the Samnites were busy against the Italian Greeks (see p. 64). At the **Battle of Vesuvius** (339) the left wing of the Roman army, commanded by Consul **Publius Decius Mus,** was shattered; Decius, to permit the withdrawal of the remainder of the army, under his colleague **T. Manlius Torquatus,** deliberately sacrificed himself in a forlorn hope attack against the Latins; the result was a drawn battle. Manlius soon thereafter won the decisive **Battle of Trifanum** (338), crushing the revolt. Rome treated the defeated Latins with a mixture of firmness and leniency, which thereafter assured their steadfast loyalty. Roman consolidation and slow expansion continued.

327–304. Second Samnite War. Initially successful (327–322), Rome was several times subsequently close to disaster. A Roman army, under the consuls **Spurius Postumius** and **T. Veturius Calvinus,** was decisively defeated by **Gavius Pontius** and forced to surrender at the **Battle of the Caudine Forks** (321). Rome agreed to a temporary peace or armistice under very unfavorable terms. A complete reorganization of the Roman military system resulted (see p. 79). Fighting broke out soon

again, with Roman successes at first. Then they were defeated at **Lautulae** (316), but won an important battle at **Ciuna** (315). The construction of the Via Appia (312) gave Rome a logistical advantage, which enabled her to drive the Samnites from Campania. This was soon offset by the entrance of the northern Etruscans into the war against Rome (311). The Etruscans were defeated at the **Battle of Lake Vadimo** (310) by **Q. Fabius Rullianus** and forced to make peace (308). Meanwhile, **L. Papirius Cursor** had won a great victory over the Samnites in the mountains to the south (309). Then the Umbrians, Picentini, and Marsians (all Italian peoples inhabiting the southeast slopes of the Apennines) joined the Samnites (308). Rome sent land and sea expeditions against these peoples—the first use of naval force by the Romans in the Adriatic. The consuls **M. Fulvius** and **L. Postumius** defeated the Samnites in the decisive battle of the war at **Bovianum** (305). All of Rome's enemies were soon forced to make peace (304).

298–290. Third Samnite War. An early Samnite success at **Camerinum,** against **Lucius Scipio,** inspired the Etruscans to join another alliance against Rome. The Gauls and Umbrians also joined the Samnites. In the decisive **Battle of Sentinum** (295) the Romans under Fabius Rullianus and **Publius Decius Mus** (son of the hero of the Latin War) defeated a combined army of Etruscans, Gauls, and Samnites. Decius, like his father, deliberately sacrificed his life when the battle began to go against Rome; his men rallied to win an overwhelming victory. The Gauls, Umbrians, and Etruscans made peace, but the Samnites continued the war for a few years until crushed by **Manius Curius Dentatus** at the **Battle of Aquilonia** (293). In recognition of their gallantry, the Samnites were permitted to enter the Roman confederation as allies rather than as subjects of Rome.

285–282. Revolt of Etruscans and Gauls. A Roman army under **Lucius Caecilius** was annihilated by the Gauls at **Arretium** (285), but a combined invasion by Etruscans and Gauls was smashed by **P. Cornelius Dolabella** north of

Rome at **Lake Vadimo** (283). Final Etruscan resistance was crushed at **Populonia** (282).

281-272. War with Pyrrhus. Roman expansion into southern Italy alarmed Tarentum, which declared war and asked **Pyrrhus,** king of Epirus, to come to her assistance. Pyrrhus arrived in southern Italy with an army of about 20,000 infantry, organized as a Macedonian-type phalanx, and more than 3,000 Thessalian and Epirote cavalry. He established himself as the virtual master of the Greek cities of southern Italy.

280. Battle of Heraclea (or of the Siris River). The Romans, under **Publius Valerius Laevinus,** numbered about 35,000; Pyrrhus had about 30,000. After a terrible struggle, Pyrrhus routed the Roman cavalry by judicious use of his elephants, which the Romans had never before encountered. He then drove the Romans infantry back across the Siris in great disorder. The Romans lost 7,000 to 15,000 killed; Pyrrhus had 4,000 to 11,000 dead. The Epirote king is reputed to have said: "One more such victory and I am lost." After advancing toward Rome, Pyrrhus gained further evidence of the toughness of his new foes and the loyalty of their allies when he learned that a Roman-allied army was marching to meet him. He retired to southern Italy, where he recruited a large army of about 70,000, including substantial contingents of Samnites and other southern Italians, as well as Greeks.

279. Battle of Asculum. The Roman consuls **Caius Fabricius** and **Quintus Aemilius,** with an allied-Roman army about the same size as that of Pyrrhus, marched south to Apulia, meeting Pyrrhus near Asculum (Ascoli). The first day of a two-day battle was indecisive. On the second day of fierce fighting Pyrrhus, badly wounded, again won by use of his elephants against the Roman cavalry. The Roman army, however, withdrew in good order. The losses on both sides were about 11,000 men; Pyrrhus lost particularly heavily in the contingents he had brought from Greece, and again was dissatisfied with the hard-won victory. From Heraclea and Asculum are derived the term "Pyrrhic victory."

278–276. Pyrrhus in Sicily. (See p. 63.)

275. Battle of Beneventum. Shortly after his return from Sicily, Pyrrhus met another Roman army, commanded by M. Curius Dentatus. In a seesaw battle, the Romans were again near defeat through Pyrrhus' judicious use of elephants. Driven back to the walls of their camp, they were joined by the camp garrison, and turned the elephants back into the phalanx, creating great confusion. A prompt Roman counterattack following, Pyrrhus was decisively defeated with great loss. Soon afterward he returned to Greece, reporting to have said on his departure: "What a fine field of battle I leave here for Rome and Carthage." Tarentum fell to the Romans the same year that Pyrrhus was killed in a street fight in Argos (272).

272–265. Consolidation of Italy. The capture of Rhegium (270) brought all southern Italy under Roman control. After defeating a Samnite uprising (269) Rome was unchallenged mistress of all Italy south of the Arnus River.

First Punic War, 264-241 B.C.

265. War Breaks Out at Messana. The leaders of the Mamertines of Messana, engaged in war with **Hiero II** of Syracuse, sent for assistance from Carthage; another faction sent to Rome for help (see p. 64). The Carthaginians arrived first and seized Messana. Driven out by the Romans, the Carthaginians, in alliance with Hiero, then besieged Messana, but were repelled (264) by a Roman army under Consul **Appius Claudius Caudex,** who then unsuccessfully besieged **Syracuse.** However, Roman successes under Consul **Marcus Valerius Maximus** in eastern Sicily caused Hiero to make peace with Rome and to join an alliance against Carthage (263). The Romans then invaded western Sicily and laid siege to the Carthaginian stronghold of Agrigentum.

262. Siege and Battle of Agrigentum (Girgenti). The defenders under **Hannibal Gisco** put up a stout defense, and Carthage sent a relieving army under **Hanno** (one of several leaders of that name). In the ensuing battle Hanno was defeated, but Hannibal escaped the city with his army. This victory gave Rome control of most of Sicily, save for a few scattered Carthaginian fortresses along the western coast.

260. Battle of the Lipara Islands. A Roman naval squadron under **C. Cornelius Scipio** was defeated.

260. Battle of Mylae. C. Duillius commanded the Roman fleet, first revealing the new Roman methods of naval warfare (see p. 82) and winning a decisive victory. Control of the sea was wrested from the Carthaginians. Roman expeditionary forces invaded Corsica and Sardinia.

256. Battle of Cape Ecnomus. A fleet of 330 vessels under the consuls **M. Atilius Regulus** and **L. Manlius Volso** sailed from Sicily with a total force of about 150,000 soldiers and sailors to invade Africa. A Carthaginian fleet of 350 ships, commanded by **Hamilcar** and **Hanno** (common Carthaginian names), met the Romans off the coast of Sicily. In a hard-fought battle, the new Roman tactics again prevailed; 30 Carthaginian ships were sunk and 64 captured, while only 24 Roman ships were lost.

256. Invasion of Africa. The Roman fleet continued to the coast of Africa near Carthage. An army of 20,000, under Regulus, was landed. Regulus won the decisive **Battle of Adys;** the Carthaginians sued for peace. The terms set by Regulus were so severe that the Carthaginians decided to continue the war; in desperation they asked for assistance from the Spartan soldier of fortune **Xanthippus,** who arrived at Carthage with a force of Greek mercenaries.

255. Battle of Tunes. Xanthippus, having reorganized, trained, and inspired the Carthaginian army, took the field against Regulus. The two armies were approximately equal in size— slightly less than 20,000 men each. Xanthippus, making good use of his cavalry, elephants, and his phalanx of Greek mercenaries, defeated Regulus, capturing him and about half of the Roman army. The remainder of the Romans—probably less than 5,000 men—entrenched themselves on an inaccessible promontory, and were later rescued by a Roman

fleet. The fleet, however, was caught in a storm between Africa and Sicily, losing 284 ships out of 364—a disaster which cost Rome nearly 100,000 of her best soldiers, sailors, and marines.

254. Carthaginian Recovery. The loss of the Roman fleet, and elimination of the threat to their capital, permitted the Carthaginians to reinforce their besieged strongholds in Sicily and to recapture Agrigentum. The Romans responded with a successful amphibious assault on Panormus. A stalemate followed.

251. Battle of Panormus. In northwest Sicily the Carthaginian general **Hasdrubal** was defeated by Consul **L. Caecilius Metellus;** both armies were about 25,000 strong. The Carthaginians again made peace overtures. They sent the captured Roman general, Regulus, on parole, to Rome with the mission of at least negotiating an exchange of prisoners, if satisfactory peace terms could not be achieved. According to legend, Regulus advised his countrymen to reject both Carthaginian proposals; he then honored his parole by returning voluntarily to Carthage, where he was tortured to death.

249. Battle of Drepanum (Trapani). The Roman fleet blockading Lilybaeum, commanded by the Consul **P. Claudius Pulcher,** engaged a Carthaginian fleet under the admiral **Adherbal;** each fleet comprised about 200 warships. Just before the battle Claudius invoked the blessing of the gods by placing sacred chickens on the deck, where they were supposed to eat grain spread out before them and thus provide a favorable omen. The chickens refused to eat, so Claudius ordered them to be thrown overboard, with the words, "Then let them drink." The Romans were defeated with a loss of 93 ships; 8,000 Romans were killed, 20,000 captured. No Carthaginian ships were lost. Claudius survived the battle, was called back to Rome, and fined heavily by the Senate when he refused to appoint a suitable successor as dictator. This was a bad year for Rome. Land and sea attacks against a Carthaginian force commanded by **Hamilcar Barca** were repulsed at **Eryx.** Soon after this most of the remainder of the Roman fleet was lost at sea in a storm. This was the fourth time the Romans had suffered such a disaster, having lost more than 700 ships and 200,000 men in storms. Accepting this as a warning from the gods, for several years they refrained from major efforts at sea.

247–242. Hamilcar Barca Commands in Western Sicily. For five years he repulsed all Roman efforts to recapture the Carthaginian strongholds. The superiority of the legion was offset by the skill and ingenuity of Hamilcar. Without serious Roman opposition at sea, he sent his warships to harry the coast of Italy.

242. Conquest of Lilybaeum and Drepanum. Rebuilding their navy, the Romans sent a fleet of about 200 ships to western Sicily, under the command of **L. Lutatius Catulus.** By combined land and sea operations they captured the Carthaginian strongholds.

241. Battle of the Aegates Islands. The Carthaginians sent Hanno with a battle fleet of 200 ships to Sicilian waters. Catulus won a complete victory, 70 Carthaginian ships being captured, 50 more sunk. This disaster forced Carthage to make peace—after crucifying Hanno. They agreed to evacuate Sicily and to pay Rome 3,200 talents (about $95 million*) over ten years. Rome permitted Syracuse to retain control over eastern Sicily, but organized western Sicily as her first overseas province.

Between the Wars, 241–219 B.C.

238. Roman Seizure of Sardinia. Turmoil in Sardinia due to a mutiny of Carthaginian mercenaries (see p. 63) brought Roman intervention. For several years Roman troops were engaged in pacifying the wild tribes of Sardinia.

235. Peace. For the first time in recorded history the doors of the Temple of Janus—always open when Rome was at war—were closed.

229–228. First Illyrian War. The Greek states asked the assistance of Rome in suppressing Illyrian piracy in the Adriatic and Ionian seas.

* Assuming these were gold talents; the usually quoted equivalent of $4 million seems relatively insignificant.

Roman ambassadors sent to the court of Queen **Teuta** of Illyria were murdered. A Roman army soon crushed the Illyrians.

225–222. Gallic Invasion of Central Italy. The Gauls won a victory at **Faesulae** (225). Then Roman armies under **Aemilius Papus** and **Caius Atilius Regulus** defeated them at the **Battle of Telamon,** though Regulus was killed (224). Forty thousand Gauls were slain, 10,000 captured, and 20,000 got away. The Romans pursued the remnants northward to the Po Valley, where **M. Claudius Marcellus** won the **Battle of Clastidium** (222).

219. Second Illyrian War. King **Scerdilaidas** provoked the Romans into another punitive expedition, which smashed the Illyrians.

Second Punic War, 219–202 B.C.

219. Siege of Saguntum. Hannibal, son of Hamilcar Barca, demanded the submission of Saguntum, a Greek city and Roman ally, the sole spot in Spain south of the Ebro not under Carthaginian control. When Saguntum refused, Hannibal immediately invested it, realizing this would probably provoke war with Rome. He was following his father's plan of gaining revenge for the First Punic War. Rome demanded that Carthage halt the siege of Saguntum and surrender Hannibal. Carthage refused; Rome declared war. After a siege of 8 months, Hannibal captured Saguntum by storm. His Iberian base now secure, he was ready to carry out his carefully planned strategy.

218. Hannibal's Plan: To circumvent Roman control of the seas, he planned to take a large army overland from Spain, through southern Gaul, across the Alps to the Po Valley. He had already sent agents to secure allies in Transalpine and Cisalpine Gaul, thus assuring a line of communications back to Spain and a secure advanced base in northern Italy. He planned to recruit reinforcements from among the warlike Celtic tribes who hated Rome. He had also opened communications with **Philip V** of Macedon with a view to forcing Rome into a two-front war. He planned to leave about

HANNIBAL'S THEATRE OF OPERATIONS
(Showing his path of invasion into Italy)

20,000 men under his brother, **Hasdrubal,** to hold Spain.

218. Roman Plans: Consul Titus Sempronius, with about 30,000 men, and a fleet of 60 warships were to invade Africa and attack Carthage; Consul **Publius Cornelius Scipio,** with his brother **Gnaeus Cornelius Scipio,** would invade Spain with an army of about 26,000 men and a fleet of 60 warships; Praetor **Lucius Manlius,** with about 22,000 men, would hold Cisalpine Gaul to keep the restless Celts in check while the consular armies were engaged with the Carthaginians. The Romans had no inkling of Hannibal's planned invasion.

218, March–June. Across the Pyrenees. Crossing the Ebro with about 90,000 men, Hannibal subdued the country south of the Pyrenees. He left a strong garrison in this region and eliminated from his army all men unfit for the field. He entered Gaul with less than 50,000

infantry, 9,000 cavalry, and about 80 elephants.

218, July–October. Through Gaul. Though he met some opposition on the march—notably when crossing the Rhone River—the passage through Gaul was generally quick and easy, due to Hannibal's advance preparations. Scipio, learning of this movement, landed with his army at Massilia (Marseille) in hope of cutting the Carthaginians off. But Hannibal, to avoid interference, had already turned north up the Rhone Valley, planning to cross the Alps well inland—possibly at the Traversette. Scipio, despairing of catching the Carthaginians, hastened along the coast to north Italy with a small force; he sent the bulk of his army to Spain under his brother.

218, October. Passage of the Alps. The Alpine passes were already heavy with snow, but Hannibal pressed on. Many of his men and animals perished in the wintry climate; many others were killed in smashing through unexpectedly fierce resistance from mountain tribes. He reached the Po Valley with 20,000 infantry, 6,000 cavalry, and a few elephants.

218, November. Battle of the Ticinus. Hannibal was as surprised by the presence of Scipio as the Roman consul was by the rapidity of the Carthaginian advance. Scipio had taken command of Manlius' army—shaken from a recent defeat by the Gauls—and rushed to meet Hannibal at the Ticinus River. In an engagement mostly confined to cavalry, the Romans were defeated and Scipio wounded.

218, December. Battle of the Trebia. Having learned of Hannibal's arrival, Sempronius took most of his army from Sicily by sea through the Adriatic to the Po Valley, to join Scipio. Hannibal, who had increased his army to over 30,000 by recruiting Gauls, enticed Sempronius to attack across the Trebia River (against Scipio's advice). While Hannibal counter-attacked the shivering Romans, a small force of infantry and cavalry under his brother **Mago,** concealed in a ravine upstream, struck the Roman flank and rear. Of the Roman army of 40,000, only 10,000 escaped, cutting their way through the Carthaginian center; the remainder were slaughtered. Hannibal's loss probably exceeded 5,000.

218. Spain. Meanwhile, Gnaeus Scipio had landed north of the Ebro River in Spain, and had defeated and captured **Hanno,*** gaining control of the region between the Ebro and Pyrenees.

217, January–March. Winter Quarters in the Po Valley. Hannibal rested his men and recruited Gauls, while collecting information from an efficient spy network in Italy. He learned that the two new consuls, who took office March 15, were **Gaius Flaminius,** who

---

* Presumably this was Hannibal's third brother, who apparently later escaped or was released or exchanged. The name Hanno (like Hamilcar, Hasdrubal, and Hannibal) was very common among the Carthaginians, and it is difficult to distinguish between several individuals bearing the same name.

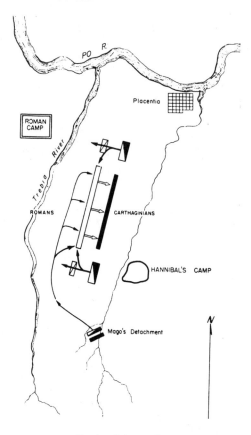

Battle of the Trebia

had about 40,000 men at Arretium (Arezzo), and **Gnaeus Servilius,** who had about 20,000 at Arminium (Rimini). The consular armies blocked the two main roads leading toward central Italy and Rome.

217, March–April. Advance into Central Italy. In the first conscious turning movement of history, Hannibal with about 40,000 men made a surprise crossing of the snowy Apennine passes north of Genoa, marched south along the seacoast, and in 4 days struggled through the treacherous Arnus marshes—supposedly impassable during the spring floods. Pressing on, he soon reached the Rome-Arretium road, near Clusium (Chiusi), thus placing himself between the Roman armies and their capital. (During this arduous march an eye infection caused Hannibal to lose the sight of one eye.)

217, April. Battle of Lake Trasimene. Headstrong Flaminius, realizing too late that his line of communications had been cut, marched

Battle of Lake Trasimene

south rapidly to seek battle; security was sacrificed for speed. Hannibal, acquainted both with Roman practice and with the nature of his antagonist, set up an ambush by his entire army where the main road passed Lake Trasimene in a narrow defile under overhanging cliffs. His light infantry was posted under cover on the mountainside, the cavalry concealed behind them. At the southern end of the defile he posted his heavy infantry, blocking the road. Upon reaching the southern end of the defile, the head of the Roman column was halted by Hannibal's infantry. Gradually the entire Roman army closed up into the 4-mile-long defile.

Hannibal ordered his cavalry to close the northern end of the defile, then struck the east flank of the Roman column with his light infantry. The result was surprise, panic, and slaughter. About 30,000 Romans—including Flaminius—were killed or captured; another 10,000, in scattered groups, fled through the mountains to notify Rome of the terrible defeat. Hannibal then continued south, seeking a base in southern Italy, where he expected to be joined by the cities and tribes which were nominally allies (but really vassals) of Rome.

217, May–October. Quintus Fabius Appointed Dictator by the Senate. Recognizing that he could not cope with Hannibal on the battlefield, Fabius wisely chose to avoid formal combat while conducting a campaign of delays and harassment. These "Fabian tactics" soon earned for Fabius the nickname of **Cunctator,** or "Delayer." Many Romans soon grew impatient; they knew only the tradition of offensive warfare. **M. Minucius Rufus,** his senior lieutenant, expressing scorn for Fabian tactics, was rewarded by the Senate with a command status coequal with the dictator. Hannibal had been doing everything he could to entice the Romans into combat, and suddenly his efforts were rewarded at **Geronium,** when Minucius accepted the challenge. Hannibal attacked at once. Minucius, on the verge of defeat, was saved by the timely arrival of Fabius, who threatened the Carthaginian flank. Hannibal prudently withdrew. Manfully, Minucius acknowledged his error and thereafter gave Fabius loyal support.

217–211. Spain and Africa. Meanwhile, Publius Scipio had joined his brother in Spain, with 8,000 reinforcements. In the following years the two Scipios were generally successful, forcing Hasdrubal and Mago to withdraw from the Ebro line. They also induced **Syphax,** King of Numidia, to revolt against Carthage (213) However, Hasdrubal returned to Africa, and with **Massinissa,** a Numidian prince, defeated Syphax. Hasdrubal then returned to Spain with reinforcements, including Massinissa's Numidian cavalry (212). At this time the Scipios recaptured Saguntum.

216, April–July. Thanks to the time gained by Fabius, Rome gathered an army of 8 Roman and 8 allied legions—80,000 infantry plus 7,000 cavalry—sending them south to Apulia under the two new consuls, **Aemilius Paulus** and **Terentius Varro,** to seek battle with Hannibal. Hannibal, who had about 40,000 infantry and 10,000 cavalry, sought favorable conditions for battle. Paulus, a cool, cautious leader, was careful to avoid presenting such an opportunity, and for a while was able to prevail upon his impetuous colleague, Varro, to follow the same policy. The two consuls alternated in command each day. In an effort to force the issue, Hannibal made a night march to Cannae, capturing a Roman supply depot, and gaining possession of the grain country of southern Apulia. The Roman army followed; both forces established fortified camps six miles apart on the south bank of the Aufidus River.

BATTLE OF CANNAE, AUGUST 2, 216 B.C.

Early in the morning on a day when he knew Varro would be in command, Hannibal drew up for battle near the Roman camp, with one flank on the stream, thus secured from envelopment by the more numerous Romans. He left about 8,000 troops to hold his camp. His center was composed of Spanish and Gallic infantry, spread out in a thin line. The wings each consisted of a deep phalanx of heavy, reliable African foot. On the left of his line were his 8,000 heavy Spanish and Gallic cavalrymen under Hasdrubal; on the right he posted his 2,000 Numidian light cavalry.

Varro accepted the challenge with most of his army; he sent 11,000 men to attack the Carthaginian camp. Perceiving that he could not envelop the well-protected flanks of the Carthaginian army, Varro decided to crush his opponent by weight of numbers. He doubled the depth of each maniple, the intervals also being greatly reduced so that his infantry front, comprising about 65,000 men, corresponded with that of Hannibal's 32,000. Varro placed 2,400 Roman cavalry on his right flank; on his left were 4,800 allied horsemen.

Under the cover of the preliminary skirmishing of light troops, Hannibal personally advanced the thin central portion of his line until it formed a salient toward the Romans; his heavy infantry wings stood fast.

The battle was opened on the left by a charge of the heavy Spanish and Gallic horse, who crushed the Roman cavalry, then swung completely around the rear of the Roman army to smash the rear of the allied cavalry, engaged in indecisive combat with the Numidians. The allied horsemen were driven off the field, pursued by the Numidians. The heavy Carthaginian cavalry now turned to strike the rear of the Roman infantry.

The infantry combat, meanwhile, had gone according to Hannibal's plan. His central salient slowly withdrew under fierce Roman pressure. Varro sent the maniples of his second line into the intervals of his hastati; then ordered his triarii and even the velites to add their weight, in order to drive the Carthaginians into the river. Hannibal's line had now become concave, but was still intact. The Romans, in a dense phalangial mass, pressed ahead.

Suddenly Hannibal gave the signal to the commanders of his African wings, thus far hardly engaged. They advanced, wheeling inward against the Romans, who were already raising shouts of victory. At this time the Carthaginian cavalry struck the rear of the Roman line. Cries of victory turned to screams of consternation. The Romans became a herd of panic-stricken individuals, all cohesion and unity lost. There was only slight resistance offered by this hemmed-in mob during the following hour of grim butchery, though one contingent of about 10,000 fought free. At the close of the day about 60,000 Romans—including Paulus—lay dead on the field; with another 2,000 lost when they were repulsed from Hannibal's well-guarded camp. Ingloriously, Varro was among the handful of fugitives. Hannibal's losses were at least 6,000 men.

The Battle of Cannae was the high-water mark of Hannibal's career; it has also provided military theorists with a symbol of tactical perfection.

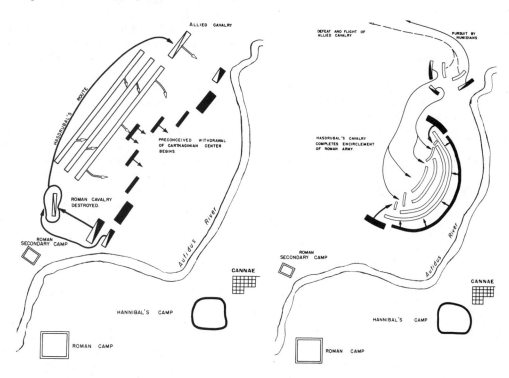

BATTLE OF CANNAE (Opening Phase)   BATTLE OF CANNAE (Final Phase)

216, August–December. Response of Rome.

Never before nor since has a state survived after suffering such crushing defeats in close succession as those of Rome at the Trebia, Lake Trasimene, and Cannae. When the news of Cannae reached Rome, there were a few faint hearts, but as a group the Romans had but one thought: perseverance to victory. **M. Junius Pera** was appointed dictator by the Senate. All able-bodied men, regardless of age or occupation, were mobilized; the principal field commander was Proconsul **Marcus Claudius Marcellus,** who immediately marched south from Rome with two legions to maintain successfully the confidence of Rome's allies in ultimate victory. If her allies had deserted her, Roman valor and determination could never have prevailed against the genius of Hannibal. But the majority remained loyal. Hannibal, without a siege train, was unable to seize Naples from a Roman garrison rushed there by Marcellus. Capua, the second largest city in Italy, and a few other small towns of Campania joined Hannibal, as did some Samnites and

Lucanians. But wavering Italian towns were impressed when Marcellus, taking advantage of the walls of Nola, repulsed the great Carthaginian in the **First Battle of Nola.** A few reinforcements from Carthage arrived late in the year, but luke-warm support from the Carthaginian Senate, then dominated by Hanno, his father's old political opponent, plus Roman naval superiority, prevented arrival of reinforcements in bulk, which might have tempted Hannibal to risk an attack on Rome itself. He has been criticized for not marching on Rome right after Cannae. But Hannibal rightly knew that without a siege train his own motley army had no chance of capturing the powerful fortress, garrisoned by 40,000 men. Accordingly, he concentrated upon the task of building a base in southern Italy, and in this he was relatively successful, despite the solidarity of the Italian cities with Rome.

215. Stalemate in Campania. Hannibal captured a number of towns and fortresses, but made no real gains. Rome had about 140,000 soldiers under arms, including detachments in Spain, Gaul, and Sicily; of these about 80,000 were concentrated against Hannibal's 40,000 to 50,000. The Romans, however, avoided battle, under the new policy established by the Senate. Marcellus, seizing another favorable opportunity, again repulsed Hannibal in the **Second Battle of Nola.**

215–205. First Macedonian War. Although Hannibal had succeeded in getting Philip of Macedon to join him in an alliance against Rome, the results were disappointing to him (see p. 62).

214–213. Inconclusive Campaigning. Rome now had more than 200,000 soldiers under arms, with 85,000 to 90,000 of these cautiously observing Hannibal, now able to maintain his army at 40,000 only by recruiting halfhearted Italians. He fought the indecisive **Third Battle of Nola** with Marcellus, then marched into Apulia, hoping to capture Tarentum as a seaport. His brother Hanno, with 18,000 men, was severely defeated at **Beneventum** (Benevento) by **Tiberius Gracchus** with 20,000. Marcellus went to Sicily, where he won some

successes against the Syracusans (who had declared for Carthage) and Carthaginians. The following year Hannibal devoted himself to operations against Tarentum, while Hanno defeated **Tiberius Gracchus** in Bruttium (213).

213–211. Siege of Syracuse. For a year the assaults of Marcellus were frustrated by a number of brilliantly conceived defensive engines designed by Archimedes. The defense was skillfully conducted by the Syracusan general, **Hippocrates.** Finally (212) Marcellus was able to force his way into the outer city, timing his assault with celebration of a local festival. Archimedes was killed. For 8 more months operations continued in Syracuse, as Marcellus chipped away at the fortifications of the inner city and citadel, finally overwhelming the garrison by assault.

212. Tarentum and Capua. Hannibal captured Tarentum, but a Roman garrison held out in the citadel. Meanwhile, the Roman consuls **Quintus Fulvius Flaccus** and **Appius Claudius** invested Capua, which was already short of food. In response to Capuan pleas for help, Hannibal sent Hanno to relieve the city. Hanno gathered large quantities of food in a fortified camp at Beneventum, and then enticed the Roman armies away from Capua. He got some food to the beleaguered city, but the Capuans were slow in responding to Hanno's skillful arrangements, and while he was away on a foraging expedition, Fulvius made a successful night attack on Hanno's camp, capturing several thousand Capuan wagons and great quantities of supplies; 6,000 Carthaginians were killed and 7,000 captured. Hanno escaped back to Bruttium. The Romans resumed the siege of Capua. Hannibal now advanced from Tarentum with about 20,000 men, and though the Romans had more than 80,000 men in southern Italy, they were unable or unwilling to prevent him from marching into Capua.

212. The First Battle of Capua. Hannibal defeated the consuls in an indecisive battle outside the city. To lure him away from Capua, they marched off in different directions, threatening his strongholds in Campania and Lucania. Hannibal followed Appius into Lucania, but was unable to catch him. He did, however, meet and destroy the army of Praetor **M. Centenius Penula** in northwest Lucania, probably near the **Silarus River.** Centenius had about 16,000 men, Hannibal about 20,000; Centenius was killed; only 1,000 of his men escaped slaughter or capture. A few days later Hannibal met the army of Praetor **Gnaeus Fulvius** (not the consul) and also destroyed it in another decisive victory at **Herdonia;** out of 18,000 only 2,000 Romans (including Fulvius) escaped. Meanwhile, the consuls had reestablished the siege of Capua, but since the city was now well supplied, Hannibal returned to the south coast, where he was repulsed in an effort to capture Brundisium (Brindisi).

211. Spain. Hasdrubal's reinforced Carthaginian armies defeated the Scipio brothers in two separate battles in the **Upper Baetis** Valley, both Roman leaders being killed. Carthage again controlled all Spain south of the Ebro.

211. Siege and Second Battle of Capua. During the winter the Romans completed lines of circumvallation and contravallation around Capua. The new consuls, **Publius Sulpicius Galba** and **Gnaeus Fulvius Centumalus,** with more than 50,000 men, guarded against approach by Hannibal from the south, while the proconsuls Fulvius and Appius continued the siege with 60,000 more. Upon another appeal from Capua, Hannibal with 30,000 men marched north and again either eluded or overawed the Roman forces supposed to bar the way. While the garrison of Capua made a sortie, he attacked the Roman lines from the outside. He was repulsed by Fulvius, while Appius drove the Capuans back into their city.

211. The March on Rome. Hannibal decided to march on Rome itself, in hopes that the threat would cause all Roman forces to rush to the defense of their capital and abandon the siege of Capua. In fact, the two consuls hastened to Rome, and Fulvius took a small force from Capua, while Appius continued the siege. This must have brought the garrison of Rome to more than 50,000 men. Hannibal merely demonstrated, then marched slowly south again, harassed by the consuls, while Fulvius returned to command outside Capua. Now close to starvation, the city surrendered; the worst blow Hannibal had yet suffered in Italy.

210. Roman Offensives. Though still trying to avoid anything like an equal battle with Hannibal personally, the Romans now decided to attempt to destroy his base and sources of supply. Hannibal, however, destroyed the army of Proconsul **Fulvius Centumalus,** at the **Second Battle of Herdonia,** Centumalus being killed. Soon after this he defeated Marcellus in the **Battle of Numistro.**

210–209. Spain. After the death of Publius Scipio the Roman Senate sent his 25-year-old son **Publius Cornelius Scipio**—known to history as "Africanus"—to take command in Spain. He quickly reestablished Roman control north of the Ebro. Then with 27,500 men he made a surprise march to the Carthaginian capital of New Carthage, while his fleet blockaded the town from the sea (209). He captured the town in a surprise assault.

209–208. Tarentum. Although Rome was near bankruptcy, and the people of Italy were close to starvation due to lack of men to till the fields, she again had 200,000 troops in the field. Hannibal could muster barely 40,000, mostly Italians, and save for a few veterans the quality of his army was much inferior to the legions. He was now holding on, awaiting reinforcements from his brother, Hasdrubal, in Spain. The Roman objective was Tarentum, which had become Hannibal's main base in Italy. Surprisingly the Roman garrison of the citadel was still holding out, supplied by sea. In a hard-fought two-day battle, Hannibal defeated Marcellus at **Asculum,** but again was unable to gain a decisive victory over his most persistent foe. Meanwhile, Fabius Cunctator (consul for the fifth time) recaptured Tarentum through the treachery of Hannibal's Italian allies. Amazingly, despite the loss of his base at Tarentum, Hannibal was able to carry on and hold at bay the far more numerous and far more efficient Roman armies (208). Yet the Romans—and particularly Marcellus—were no longer afraid to try battle with him. Marcellus, however, was killed in an ambush this year.

208. Spain; Battle of Baecula. After extensive maneuvering and skirmishing, Scipio defeated Hasdrubal in an inconclusive battle near modern Cordova. Hasdrubal, whose losses had not been serious, now responded to Hannibal's orders to bring reinforcements to Italy, even though it might mean the abandonment of Spain to Scipio. He marched to Gaul, where he spent the winter, resting his men and recruiting reinforcements.

207. Hasdrubal in Italy. Early in the year Hasdrubal pushed over the Alps, arriving in the Po Valley with about 50,000 men, more than half Gauls. He sent a message to his brother, reporting his arrival, and began to advance slowly toward central Italy. Meanwhile, Hannibal had been maneuvering against the efficient consul **Caius Claudius Nero.** At the **Battle of Grumentum,** Nero, with 42,000 men, gained a slight advantage over Hannibal (who probably had about 30,000), but was unable to prevent the Carthaginian from marching north to Canusium (Canosa di Puglia) to await word from his brother. But Hasdrubal's messengers missed him and were captured by Nero. The Roman consul now conceived a brilliant plan. Leaving most of his army facing Hannibal, he took 6,000 infantry and 1,000 cavalry—the best troops of his army—and marched north as quickly as possible to join his colleague, **M. Livius Salinator,** who was facing Hasdrubal in northeast Italy. He marched 250 miles in 7 days, joining Livius secretly south of the Metaurus River.

BATTLE OF THE METAURUS, 207 B.C.

Hasdrubal's patrols reported the arrival of Roman reinforcements, and he decided to make a night withdrawal north of the Metaurus to more favorable ground. He was deserted by his Italian guides, however, and his army got lost during the darkness. The Roman consuls caught up with him after dawn, just south of the river. Hasdrubal hastily prepared for battle, placing his least reliable troops on his left flank behind a deep ravine. His right was soon engaged heavily by Livius, but Nero on the Roman right was prevented by the ravine from reaching the Gauls to his front. Reasoning that the obstacle would be just as formidable to the Carthaginians, Nero pulled his troops out of line and quickly marched behind the rest of the Roman army, swinging in behind the right rear of the Spanish infantry. The surprise rear attack completely demoralized the Spaniards, and despite valiant efforts of Hasdrubal, panic spread through his army. Hasdrubal, seeing that all was lost, deliberately rode into a Roman cohort, to die fighting. The Carthaginian army was hopelessly smashed; more than 10,000 men were killed and the rest scattered; the Romans lost 2,000. Immediately after the battle Nero marched back to south Italy in 6 days. According to legend, the first news Hannibal received of his brother's arrival in Italy was when Hasdrubal's head was catapulted into the Carthaginian camp. Sorrowfully he withdrew to Bruttium.

207–206. Spain. Despite determined opposition from Mago and Hasdrubal Gisco, Scipio rapidly spread his control over most of Spain. The climax came in the **Battle of Ilipa** (or Silpia), where Scipio with 48,000 men decisively defeated 70,000 Carthaginians in a battle of brilliant maneuver (206). Spreading out the center of his army in a manner somewhat reminiscent of Hannibal's formation at Cannae, Scipio employed it in an entirely different manner. The center was refused, while the Roman general undertook a successful double envelopment with his wings. This ended Carthaginian rule in Spain. Soon after this Scipio made a bold trip to North Africa, where he concluded an alliance with Massinissa, rival of Syphax for the throne of Numidia.

206–204. Hannibal at Bay. Incredibly, Hannibal maintained himself against tremendous odds in Bruttium, despite the inferior quality of

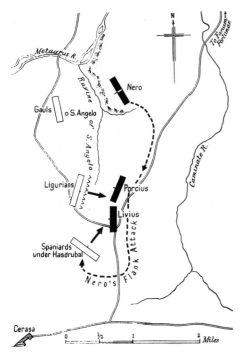

Battle of the Metaurus

redeem their honor. He landed near Utica, which he invested. Hannibal's brother Hanno was apparently killed in the preliminary skirmishing here. The arrival of a large Carthaginian army under Hasdrubal Gisco and Syphax forced Scipio to give up the siege and to establish a fortified camp near the coast. An armistice was arranged and the two armies went into winter quarters.

203. Battles Near Utica. Suddenly ending the armistice, Scipio made a surprise night attack on the Carthaginian and Numidian camps, setting them on fire and destroying the entire army. He immediately renewed the siege of Utica. Hasdrubal and Syphax soon raised a new army, and met Scipio at the **Battle of Bagbrades,** near Utica. Scipio won an overwhelming victory; Syphax was captured.

203. Return of Hannibal. The Carthaginian Senate, in desperation, sent for Hannibal and Mago; at the same time they sued for peace. During the ensuing armistice, Hannibal sailed from Italy with about 18,000 men, mostly Italians who remained loyal to their foreign leader. Mago, who had been defeated in Liguria, also returned with a few thousand men, but he died of wounds en route. Upon the return of Hannibal, the Carthaginian Senate broke off negotiations and helped Hannibal raise a fresh army around his nucleus of Italian veterans.

202. March to Zama. Hannibal marched inland from Carthage with an army of about 45,000 infantry and 3,000 cavalry. Apparently this was to draw Scipio away from the area around Carthage, which the Romans were systematically devastating. Scipio followed, with 34,000 infantry and 9,000 cavalry, having been joined by Massinissa and Numidian reinforcements.

his troops. Although there were many skirmishes, the only important action was the drawn **Battle of Crotona** (204) against **Sempronius.** That same year his brother Mago landed in Liguria with a small army. Scipio, meanwhile, had been elected consul (205), and was in Sicily preparing an army to invade Africa.

204. Invasion of Africa. Scipio, as proconsul, sailed from Lilybaeum with a magnificently trained and equipped army of about 30,000 men. Many of these were surviving veterans of Cannae, as was Scipio, and were anxious to

BATTLE OF ZAMA, 202 B.C.

With the two armies drawn up in battle formation, Hannibal may have met Scipio in an indecisive parley; then the battle began. Scipio's army was in the usual three lines, but with the distances between lines increased, and the maniples in column, rather than in the usual checkerboard formation. This was to create lanes through which the Carthaginian elephants could be herded. Hannibal's infantry was also in three lines—as early as Cannae he had begun to adopt much of the Roman formations and tactical system. But more than half of his infantry were raw recruits, the remainder being his Italian veterans and a few Ligurians and Gauls that had come back with Mago. He was particularly weak in cavalry, the arm on which he had depended in so many of his great victories. Consequently, he was unable to employ his favorite maneuvers.

   The Carthaginian elephants attacked the legions, but were handled as Scipio had planned. At the same time the Roman and Numidian cavalry were driving Hannibal's horsemen off the field. The main infantry lines then clashed, and the Romans quickly disposed of the first two lines of Carthaginian infantry. The triarii then advanced against Hannibal's reserve, but under his inspiring leadership the veterans stood fast. Just as the remainder of the Roman infantry joined the attack, Massinissa and the Numidians returned from their pursuit of the Carthaginian cavalry and struck the rear of Hannibal's line. This ended the battle. Hannibal and a few survivors fled to Carthage. Left on the field were 20,000 Carthaginian dead and 15,000 prisoners. The Romans lost about 1,500 dead and probably another 4,000 wounded.

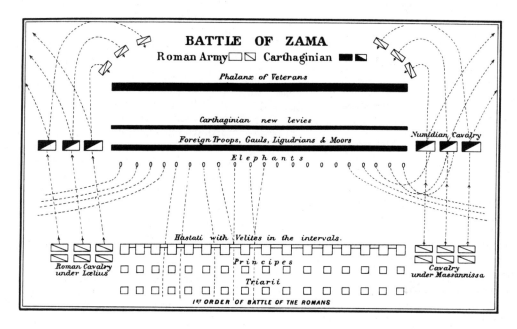

202. Peace. Carthage, suing for peace, accepted Scipio's terms: she handed over all warships and elephants; she agreed not to make war without the permission of Rome; Massinissa was reinstated as King of Numidia in place of Syphax; Carthage agreed to pay Rome 10,000 talents ($300 million) over the next 50 years.

202–183. The Tragedy of Hannibal. In the years immediately after the war, Hannibal was so successful in reviving Carthaginian prosper-

ity that the Romans accused him of planning to break the peace. He fled Carthage (196) and joined Antiochus III, but was forced to flee again when Antiochus was defeated by the Romans (see p. 96). Pursued by the Romans to Bithynia, he committed suicide (183).

COMMENT. *No other general in history faced such adversity or such formidable odds as Hannibal. His inspiring leadership, his consummate tactical and strategic skill, and his accomplishments with inferior*

material against the most dynamic and militarily effi-
cient nation in the world have prompted many histo-
rians and military theorists to rank him as the greatest
general of history. Objective assessment makes it im-

possible to rank him ahead of Alexander, Genghis
Khan, or Napoleon; equally, it is impossible to rank
them significantly ahead of him.

ROMAN MILITARY SYSTEM, C. 220 B.C.

Army Organization and Tactics

The security of Rome depended upon a semi-militia citizen army that was essentially professional. All able-bodied male Romans between 17 and 60 were obliged to serve (men over 47 were employed in garrisons only). In 220 B.C. the military manpower of Rome was about 750,000 men out of a total population of 3,750,000. Rome was frequently at war, so there was always a leaven of veterans. There were four major factors in the strength and success of Roman arms: (1) the moral strength of an army composed of free, intensely patriotic citizens; (2) the development of the legion—a new type of military organization superior to any previously seen on the battlefield; (3) maintenance of a high order of military competence, to some extent attributable to frequent combat experience, but resulting in particular from insistence on constant training and enforcement of severe discipline; (4) a traditional, intense, but intelligent reliance upon bold, aggressive doctrine, even in adversity. The wisdom of Rome's political system of confederation and colonization in Italy, including generous and magnanimous treatment of defeated Italian foes, was also an important element of Roman strength.

The evolution of the legion had begun with the legendary reforms of **Servius Tullius** (see p. 37). Many modifications were introduced by **Marcus Furius Camillus,** including an organizational breakdown based primarily upon age and experience, rather than on wealth and quality of personally owned weapons. The individual was still expected to supply his own weapons, but frequently these were purchased from the state.

The lessons of mountain combat during the Samnite wars, particularly the disaster of the Caudine Forks (see p. 65), affected Roman doctrine. By about 300 the Roman legion had attained the cellular type of organization described below. It was this flexible, disciplined legion which acquitted itself so well against the Macedonian-Epirote phalanx of **Pyrrhus,** and which was employed also in the First and Second Punic Wars. During the last half of the 3rd century B.C. the Roman legion probably achieved its highest development and greatest competency.

There were four classes of soldiers. The youngest, most agile, and least trained men were the **velites,** or light infantry. Next in age and experience came the **hastati,** who comprised the first line of the legion heavy infantry. The **principes** were veterans, averaging about 30 years of age; the backbone of the army, mature, tough, and experienced, they made up the second line of the legion. The oldest group, the **triarii,** who contributed steadiness to offset the vigor of the more youthful classifications, comprised the third line of heavy infantry.

The basic tactical organization was the **maniple,** roughly the equivalent of a modern company. Each maniple was composed of two **centuries,** or platoons, of 60 to 80 men each, except that the maniple of the **triarii** was one century only. The **cohort,** comparable to a modern battalion, consisted of 450 to 570 men (120 to 160 **velites,** the same number of **hastati** and **principes,** 60 to 80 **triarii,** and a **turma** of 30 cavalrymen). The cavalry component of the cohort rarely fought with it; the horsemen were usually gathered together in larger cavalry formations.

The legion itself—the equivalent of a modern division—comprised some 4,500 to 5,000 men, including 300 cavalrymen. For each Roman legion, there was one allied legion, organized identically, except that its cavalry component was usually 600 men. (Some authorities suggest that allied contingents were not organized in this formal manner, but that it was merely Roman policy to support each legion with an approximately equal number of allied troops, whose largest formal organization was the cohort.) A Roman legion, with its allied counterpart, was the equivalent of a modern army corps, a force of some 9,000 to 10,000 men, of whom about 900 were cavalry. Two Roman and two allied legions comprised a field army, known as a consular army, commanded by one of Rome's two consuls.

A consular army was usually 18,000 to 20,000 men, with a combat front of about one and a half miles. Often the two regular consular armies would be joined together, in which case the consuls would alternate in command, usually on a 24-hour basis. In times of war or great danger, however, Rome might have more than the 8 standard legions (4 Roman, 4 allied) under arms. In such cases, if a dictator had been appointed, he would directly command the largest field force, exercising overall control over the others as best he could under the circumstances. Whether or not there was a dictator, additional armies were usually commanded by proconsuls (former consuls, appointed by the Senate), or by praetors, elected officials.

Since consuls were elected civil officials, as well as field commanders, both military and political power lay in their hands; rarely were Roman commanders harassed by directives from home. On the other hand, this system often resulted in mediocre top military leadership. Another drawback to this system was that consuls changed each year; yet in a long drawn-out war, such as against Hannibal, Roman generals had to keep the field for years on end.

Under the consul or proconsul was a staff of senior officers or quaestors, who took care of administrative and planning tasks delegated by the army commander. The senior officers of the legion were the 6 tribunes—2 for each combat line. In a peculiar arrangement, the 6 tribunes rotated in command of the legion, though later a legate was frequently appointed over the tribunes as legion commander. Below the tribunes were 60 centurions, 2 for each maniple.

The flexibility of the legion lay in the tactical relationship of the maniples within each line, and between the lines of heavy infantry. Each maniple was like a tiny phalanx, with a front of about 20 men, 6 deep, but with the space between men somewhat greater than in the phalanx. Each man occupied space 5 feet square. Between the maniples in each line were intervals of the same frontage as that of a maniple, about 20 yards. The maniples in each line were staggered, with those of the second and third lines each covering intervals in the line to their front. There were approximately 100 yards between each line of heavy infantry.

This cellular, checkerboard type *quincunx* formation had a number of inherent advantages over the phalanx: it could maneuver more easily in rough country, without fear of losing alignment, and without need for concern about gaps appearing in the line—the gaps were built in. If desired, the first line could withdraw through the second, or the second could advance through the first. With its triarii line, the legion had an organic reserve, whether or not the commander consciously used it as such. The intervals were, of course, a potential source of danger, but one that was kept limited by the stationing of other troops immediately behind those of the first two lines. In battle it appears that the lines would close up to form a virtual

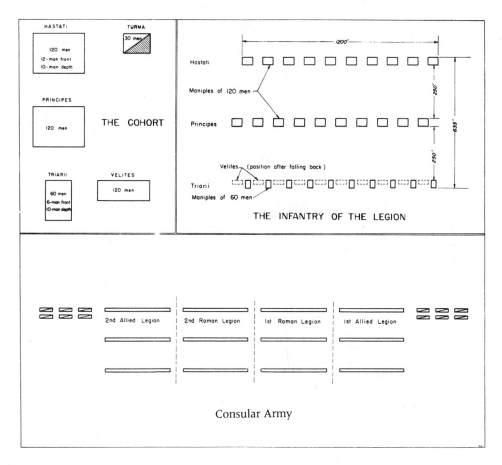

phalanx, but these could quickly resume their flexible relationship when maneuver became necessary once more.

The hastati and principes were each armed with two sturdy javelins, about 7 feet long, and with a broad-bladed short sword, about 2 feet long. The javelins were usually thrown at the enemy just before contact, with the sword (**gladius**) being wielded at close quarters. The tactical concept would be comparable to modern bayonet attacks preceded by rifle fire. The triarii each carried a 12-foot pike, as well as the gladius. The velites were armed with javelins and darts. To obtain greater diversity in range and effective missile weapons, the Romans sometimes employed foreign mercenaries, such as Balearic slingers and Aegean bowmen.

The Roman javelin, or **pilum,** was an ingeniously constructed weapon. Its head of soft iron was connected to the shaft by a slender neck. Once it had struck, the point became bent, and the head usually broke off from the shaft, thus making the weapon useless to the enemy. In addition, the soft iron head would frequently remain imbedded in the shields and armor of the enemy, adding to their discomfiture.

One of the great Roman military innovations was the practice of castrametation, or camp

Roman legionary

building. Every night during a campaign the Romans would build a fortified camp. Thus, no matter how far they might be from Rome, or from other friendly forces, the Roman troops always had a secure base, and the commander always had a choice of offensive or defensive combat. The construction of the camp was a relatively quick process, in which every man had a specific job, with which he was thoroughly familiar, due to constant practice. To build a palisade, each man carried two long stakes as part of his march equipment. A ditch was dug completely around the camp, with the earth thrown against the palisade to add thickness and sturdiness. Inside, the camp was laid out in a regular street pattern, each unit always occupying the same relative position.

Although the stakes and other equipment, as well as armor and weapons, gave each man a weight of 75 to 80 pounds on the march, the Romans marched quickly, and the legion was highly mobile. Surprisingly, there was no fixed march organization, which sometimes led to carelessness in reconnaissance and march security. This deficiency was corrected, however, during the Second Punic War, as a result of some costly lessons from Hannibal.

The other principal Roman military deficiency in this period was in siegecraft. They were far behind the Macedonians, for instance, and their sieges were usually drawn-out affairs of attrition. Again lessons of the Second Punic War brought improvements.

Naval Organization and Tactics

The Romans were not outstanding seamen. They relied largely upon allied and subject peoples—particularly the Greeks of southern Italy—to provide ships and seamen. Nevertheless, when necessary, the Romans applied characteristic efficiency and logic to maritime affairs.

Major developments in the Roman navy came during the First Punic War. The Romans soon realized that neither they nor their allies possessed ships as maneuverable as those of the Carthaginians; they were also hopelessly outclassed in seamanship. Methodically, they attacked the problem. First, they began building vessels copied from the Carthaginian **quinquireme.**

Rightly confident of the superiority of the Roman legionary over the Carthaginian soldier and sailor in hand-to-hand fighting, the Romans then introduced two major modifications, permitting them to create conditions of land warfare at sea. The most important was the **corvus,** or "raven," which was a combined grappling device and gangway. This was, in fact, a narrow

Roman corvus

bridge, some 18 feet long. Mounted on a pivot near the bow of the galley, it was held in a near-vertical position by ropes and pulleys and could be swung outboard in any direction. When an enemy ship approached, the corvus was let down with a crash on the opposing deck, where it was held fast by a large spike attached under its outer end. Immediately a swarm of Roman soldiers dashed across this bridge to fight a land battle on the foe's decks. The Romans also installed fighting turrets fore and aft on their ships, from which additional soldiers supported the boarding party with missile weapons and discouraged any enemy attempt at boarding.

These modified Roman warships were one of the first truly "secret" weapons of history. Unlike many later developers of technological innovations, the Romans waited until they had built a significant number of these new ships, then surprised the Carthaginians at the decisive Battle of Mylae (see p. 67).

Rome adapted castrametation to its naval tactics. As usual in ancient times, war fleets were beached on the shore every night. The Romans always entrenched the camp of rowers and marines, thus protecting men and vessels.

SOUTH ASIA

INDIA, 325–200 B.C.

The history of India remained virtually blank until about 325 B.C. It would appear that by the middle of the 4th century the Indian domains of the Persian Empire had achieved independence.

327-325. Alexander's Invasion of India. (See p. 57.)

c. 325. Magadha Dominant in Ganges Valley. Magadha power grew under the wealthy but low caste Nanda Dynasty (479–323 B.C.).

323–297. Reign of Chandragupta. The Magadhan general Chandragupta overthrew the Nandas and assumed the Magadhan throne. As a young man, Chandragupta had met Alexander, and had apparently spent some time with him while the Macedonian was in northern India, learning much of the art of war. Chandragupta expanded his domains in all directions. He ejected the Macedonians from northwestern India during the early wars of the Diadochi.

305. Invasion of Seleucus Nicator. (See p. 62.) Details of Seleucus' campaign east of the Indus in an effort to reestablish Macedonian rule in India are unknown. It is doubtful if any great battles were fought with Chandragupta. Whatever the nature of the operations, Chandragupta evidently had the best of it. A

treaty of alliance was drawn up in which Seleucus ceded to Chandragupta not only the
provinces east of the Indus, but also substantial
areas of Arachosia and Gedrosia west of the
river, in return for 500 war elephants—which
Seleucus used to advantage at Ipsus (see p. 60).
Possibly the demands of the continuing Diadochian Wars caused Seleucus to leave India
without attempting a decisive encounter with
Chandragupta. Whatever the reasons, the result was to assure stability and to increase the
power of India's first great dynasty, known as
the Maurya Empire.

**297–274. Reign of Bindusara (Chandragup-
ta's Son).** He apparently expanded the empire
considerably to the southward.

**274–232. Reign of Asoka (Grandson of
Chandragupta).** He brought the Maurya Empire to the height of its power and grandeur. He
apparently had to fight several claimants for the
throne. After coming to power he undertook

only one major campaign, in which he conquered the east coast of India between the
Ganges and the Deccan. He was so saddened by
the slaughter and misery of warfare that he
forsook conquest, became a Buddhist, and devoted the rest of his life to good works and to
the orderly, efficient administration of his vast
empire. These domains included all India (save
the southern tip and Assam) plus Nepal and a
substantial portion of Afghanistan. Most historians consider him to be the greatest and no-
blest ruler of Indian history.

232–180. Decline of the Mauryas. The later
emperors lacked the zeal, energy, and organizing ability of the first three. Within 50 years the
great empire had disappeared.

206. Antiochus III in the Kabul Valley. He
possibly invaded the Punjab. He received the
nominal vassalage of the Indian ruler **Sopha-
gasenous,** apparently without fighting.

THE MILITARY SYSTEM OF THE MAURYAS

To rule his large dominions effectively, Chandragupta established at his capital of Pataliputra
(Patna) an efficient governmental machinery, which provided for complete control over national economy as well as over military affairs and administration. He maintained a large
permanent military establishment, which probably comprised one-quarter or one-third of the
vast armies that he had raised for his victorious campaigns: 600,000 infantry, 30,000 cavalry,
and 9,000 elephants. This possibly exaggerated report on the strength of Chandragupta's army
comes from fragments remaining of the reports of the Seleucid Greek ambassador, **Megas-
thenes;** it is not clear whether he omits mention of chariots because of his contempt for such
instruments of warfare or whether Chandragupta had learned from Alexander the uselessness
of chariots.

It is interesting to note that in his highly organized military administration Chandragupta had
an efficient secret service and also a naval affairs office. His fleets operated on the Indus and
Ganges and probably along the coasts as well.

It was probably during the Maurya period that one of the most significant documents of
Indian history—and of military history—was written. This was the *Arthasastra (Manual of
Politics),* which has been attributed to **Kautilya,** close friend and principal administrative
assistant to Chandragupta. Whether or not the able and powerful Kautilya was the actual
author or whether—more likely—the manual was prepared by an unknown writer a century
or more later, it unquestionably reflects the ideas of Chandragupta and Kautilya, and is a
remarkably clear portrayal of political, military, and social conditions during the 3rd century
B.C. and later.

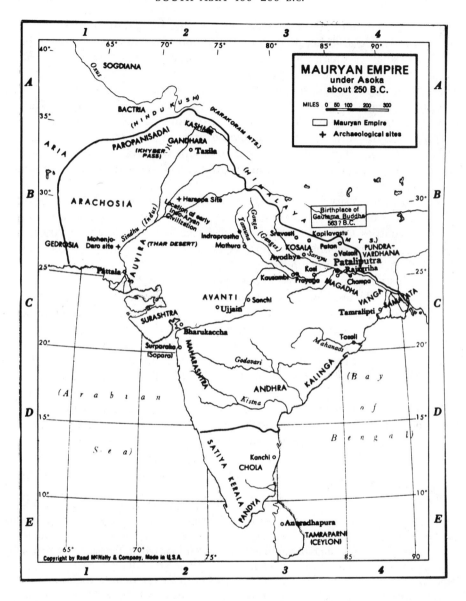

MAURYAN EMPIRE
under Asoka
about 250 B.C.

MILES 0 50 100 200 300

☐ Mauryan Empire
+ Archaeological sites

Essentially the *Arthasastra* was an exhaustive, thorough summation of early Hindu concepts of government, law, and war. It has been compared with the writings of Machiavelli, while at the same time its military maxims have much in common with the earlier ideas of Sun Tzu. Its military sections include information on the composition of armies, the functions of the branches of service, duties of officers, rules of field operations, siegecraft, and fortification, plus a list of tactical and strategical maxims.

Standing armies had become normal in India by the 4th century, and a substantial portion of national revenues went to maintain such armies. In major campaigns, however, rulers usually found it necessary to expand their forces. One of the principal sources was from the so-called guild levies. These were members of the various trades or crafts, trained as soldiers, and mobilizable in time of war as a sort of militia. In addition to these part-time soldiers, there were a number of military guilds whose business was war; essentially military companies such as were seen in Europe in the Middle Ages, they sold their services to the various kings of India. The rulers apparently had some well-grounded suspicions of the guild levies, who had been known on occasion to usurp power during their term of military service.

It is clear that the importance of training and discipline was well known to the Hindus. They understood the theory, but apparently found it difficult to impose military fundamentals on their troops in practice.

It was much the same story with elephants and chariots. The unreliability of the one and the unwieldiness of the other, in the face of competent, trained infantry and cavalry, had been obvious at the Hydaspes. Chariots continued to be used in India for another 1,000 years, though their importance steadily declined. The post-Vedic chariots of India were usually rather large, drawn by 4 horses, and carrying 4 to 6 men.

Despite the constant danger that wounded elephants might do more harm to friends than to foe, they continued to be used by Indian warriors right up to the 19th century. In Maurya times they usually carried 3 or 4 men, including the driver. The principal weapon employed with elephants was the bow. The Hindus were able to make effective use of elephants in special circumstances. For instance, they were most useful in battering the gates of besieged forts or cities. And on some occasions they were used to bridge unfordable streams, infantrymen apparently making a precarious crossing on the backs of elephants lined side by side, facing the current.

From the *Arthasastra* we learn that the Maurya armies used entrenched camps, though apparently this was only when staying for a long period in one location. Such camps are clearly described by Kautilya, as well as the measures for defense, readiness, and alertness.

BACTRIA AND PARTHIA, 323–200 B.C.

323–311. Decline of Macedonian Influence. The eastern provinces of Alexander's old empire were under the successive nominal rule of **Perdiccas, Antipater, Eumenes,** and **Antigonus** (see pp. 59–60). In fact, however, these regions were gradually falling away from the empire under the rule of local princes and tribal leaders. At the same time, the Scythians began to move in from central Asia.

321–302. Reconquest by Seleucus. He carried out a systematic reconquest of the eastern provinces of the empire.

255. Revolt of Diodotus. Seleucid satrap of Bactria, he rebelled against **Antiochus II** and established an independent kingdom. He conquered Sogdiana and expanded also eastward toward the Indus and westward into Parthia.

250. Arsaces in Parthia. This Scythian tribal leader was driven by Diodotus into western Parthia, where he established an independent kingdom (249).

239–238. Seleucid Setback. Diodotus and **Arsaces II** (247–212) joined to repulse attempts of **Seleucus II** to reconquer the region. After initial success, Seleucus was forced to abandon his expedition in order to deal with a revolt by his brother, Antiochus Hierax, in Asia Minor (see p. 61).

230. Rise of Euthydemus. This Graeco-Macedonian overthrew and slew Diodotus and made himself King of Bactria.

209–208. Invasion of Antiochus III. He defeated Arsaces III in Parthia (212–171). Arsaces appealed for help to Euthydemus, but Antiochus defeated the combined Bactrian-Parthian forces at the **Battle of the Arius** (208). Arsaces made peace and acknowledged the suzerainty of the Seleucids.

208–206. War of Antiochus III and Euthydemus. Although Antiochus won a number of minor successes, he was unable to conquer Bactria and Sogdiana. Finally, he offered generous terms to Euthydemus, who acknowledged nominal Seleucid suzerainty.

EAST ASIA

CHINA, 400–200 B.C.

475–221. Era of the Warring States. (See p. 41.) At the beginning of this period (c. 400), the principal antagonists were Wei, Ch'i, Ch'u, Yen, Ch'ao, Han, and Ch'in. Of these, at the outset, Wei was the strongest. Ch'in, a backward, semibarbarian state in the northwest region of the Yellow River Valley, was the weakest.

c. 354–353. War between Wei and Han. A Wei army invaded Han, which asked Ch'i for aid. In response, Ch'i sent an army commanded by General **T'ien Chi** to invade Wei, threatening the capital. General T'ien's principal adviser was a noted strategist, Sun Ping (one of those to whom authorship of Sun Tzu's *The Art of War* has been attributed, see p. 41). The Wei army, commanded by General **P'ang Chuan,** a former colleague of Sun Ping, rushed back from Han to defend its capital. Following Sun Ping's advice, the Ch'i army retreated, leaving evidence of disorder and panic in its abandoned camp. General P'ang pursued.

c. 353. Batttle of Ma Ling. Sun Ping had placed 10,000 crossbowmen in ambush. The Wei army rushed into the trap and was almost totally annihilated. (Some sources show this battle as fought in 341 B.C..)

c. 350. Rise of Ch'in. This was the result of political and military reforms instituted by **Shang Yang,** a refugee from Wei.

c. 342–341. War between Wei and Ch'ao. Having recovered from the disaster at Ma Ling, Wei invaded the neighboring state of Ch'ao and besieged the capital. Ch'ao asked Ch'i for aid, much as Han had done about 12 years earlier. As before, Ch'i invaded Wei and threatened its

capital, and Sun Ping was again principal advisor to the Ch'i general.

c. 341. Battle of Guai Ling. Once more Wei withdrew its invading army to rush home to relieve the capital. This time Sun Ping ambushed them en route. The ambush tactics used by Sun Ping at Ma Ling and Guai Ling are known in Chinese as "surrounding Wei," and were later praised and used by Mao Zedong.

334–286. Expansion of Ch'u. By conquereing first Yueh, along the coast, and then Sung (in modern Anhwei), Ch'u dominated the Yangtze Valley from the Gorges to the sea.

330–316. Expansion of Ch'in. Simultaneously, Ch'in was consolidating its control to the north and west. By conquereing Shu and Pa in modern Szechuan, Ch'in established itself in the western Yangtze Valley, posing a direct threat to Ch'u (c. 316). The rivalry between these two states intensified, but shifting alliances among less powerful states preserved a balance of power.

c. 320. Cavalry Appears in China. Horsemen had been used in war in China as early as the 5th century B.C. Their employment as a major, formal arm of an army appears to have been initiated by King **Wu Ling** of Ch'ao, one of the "Warring States." This led to changes in military uniforms. The Chinese had traditionally worn long robes. They now changed to short robes and trousers, like the more primitive, horse-oriented nomads further north.

c. 315–223. The Struggle of Ch'u and Ch'in. Slowly Ch'in became ascendant, until finally King **Cheng** (see p. 88) defeated and conquered Ch'u.

c. 280. Ch'in Defeats Wei. Ch'in's conquest of

Wei set the stage for a more intensive confrontation between Ch'in and Ch'u.

260. Battle of Ch'an-P'ing. Modern Gao Ping in Shansi. Ch'in defeated Ch'ao in a fierce, bloody, and decisive battle.

249. End of Chou Dynasty. The last vestiges of Chou power were eliminated by King **Chao Zhao** of Ch'in.

247–210. Reign of Cheng (Zheng), Ruler of Ch'in (Qin).* Ying Cheng, presumed son of Prince **Zi Ch'u** of Ch'in, and grandson of King Chao, was probably the son of **Lin Bu Wei,** a native of Ch'ao, who became prime minister of Ch'in. Cheng was one of the great figures of Chinese history. He undertook a systematic conquest of the other feudal states. He was not only a consummate politician, but an accomplished general. The backbone of his army was the powerful Ch'in cavalry, developed through extended combat with the warlike Hsiung-nu and Yueh Chih nomads of Mongolia. During the final struggle with Ch'u, Cheng also conquered the five other remaining independent states: Ch'ao, Han, Wei, Yen, and Ch'i (228–221).

222–210. Reign of Shih Huang Ti. Having overcome the last feudal opposition, Cheng declared himself Emperor of China, with title of Shih Huang Ti. This marked the beginning of the short-lived but memorable **Ch'in Dynasty,** from which the name "China" is derived. Shih Huang Ti now devoted his exceptional talents as a planner and organizer to the creation of the system of government that was to last in China for more than 2,100 years. In many respects this organization was comparable to that created by Cyrus and Darius in Persia, the most obvious similarities being in the creation of military satrapies and the construction of an excellent system of roads. After

* The old system of romanization, or transliteration, from Chinese to English (Wade-Giles) is used in this book. Since 1949, however, the government of the People's Republic of China has introduced a new system of romanization, called "pin-yin." Thus in recent translations from Chinese the old form "Cheng" (hard ch) is replaced by "Zheng," the old "Ch'in" (soft ch) is replaced by "Qin." Similarly T'ai (soft t) has been replaced by Tai, while Teng (hard t) has been replaced by Deng.

defeating the Hsiung-nu (later known to Europe as the Huns), Shih Huang Ti built the Great Wall along the northern border of his domains as a barrier to prevent further nomadic inroads. Several of the northern states had independently over many years built walls to protect their respective domains from the nomadic raiders, and Shih Huang Ti essentially joined them into a single wall, 3,000 kilometers long. Later dynasties would further expand it to its ultimate 6,000-kilometer length. Internally he disarmed all the feudal armies; the confiscated weapons were melted down and used for statues and farm implements. His stern, ruthless, autocratic rule was efficient but was resented by many Chinese.

221–214. Expansion of China. Shih Huang Ti undertook a series of great military expeditions south of the Yangtze, conquering the regions later known as Fukien, Kwantung, Kwangsi, and Tongking. He also conquered north Korea. Apparently most of these operations were under the immediate command of his generals **Ming T'ien** and **Chao T'o.**

210–207. Reign of Hu Hai. The weak son of Shih Huang Ti was soon beset by a combination of palace intrigue, a resurgence of the feudal lords, and widespread popular uprisings.

207–202. Anarchy. Hu Hai was killed by his courtiers. **Hsiang Yu,** a powerful and able general of the Ch'u principality, and **Liu Pang,** a popular revolutionary leader, joined to overthrow the Ch'in Dynasty. A great struggle between these two followed, with Liu Pang eventually triumphing. North Korea broke away from Chinese control during the struggle.

202. The Han Dynasty. Liu Pang, taking the imperial name of **Kao Tsu,** was the first emperor of the dynasty. His position was very shaky in the early years of his reign, due to the internal unrest which followed the violent collapse of the Ch'in Dynasty, and due also to inroads of the Hsiung-nu, who had gained supremacy in Mongolia by defeating the Yueh Chih tribes (see p. 132).

201. Invasion of the Hsiung-nu. The various Hsiung-nu tribes having been unified by **Mo Du,** the barbarian king invaded northwest China with an army of nearly 300,000 men.

200. Defeat by the Hsiung-nu. Kao Tsu took the field against the invaders. He was surrounded for several days in a fortified border town and finally defeated. He was forced to conclude a humiliating treaty with Mo Du, who retained control of the occupied border region, and was given the Chinese emperor's daughter in marriage.

THE MILITARY SYSTEM OF THE CH'IN DYNASTY

Although the Ch'in Dynasty established by Emperor Shih Huang Ti (see p. 88) was short-lived, the political and military systems he established were enduring. He replaced a semifeudal governmental system with a centralized, autocratic government, the essential features of which lasted for more than 2,000 years until the collapse of the Ch'ing (Manchu) Dynasty in 1911. According to historian Sidney Shapiro, Shih Huang Ti "promulgated a uniform code of law and standardized currency, weights and measures, the written language and the axle length of wagons and chariots."

The military system established by the Ch'in emperor was not the product of a revolution. It had a background of hundreds of years of development. The army of Shih Huang Ti was clearly the culmination of an evolutionary process, symbolically capped by his introduction of iron weapons. This system survived with little change for more than 1,500 years, until it was smashed, and then absorbed, by the Mongols in the 13th century. The discovery near Sian in 1974 of a vast underground army of terra-cotta warriors guarding the tomb of the emperor provides an exceptionally complete picture of the early manifestation of that system.

The typical field army of the Ch'in and Han periods was a combined arms force of infantry, cavalry, chariots, and crossbowmen. The principal element had been heavy armored infantry, but increasing reliance was placed upon cavalry as time went by. Shih Huang Ti did not introduce the crossbow into Chinese armies, since we know that these weapons were in extensive use as early as the Battle of Ma Ling (353 B.C., see p. 87). He seems, however, to have relied upon crossbowmen more heavily than his predecessors and may have been responsible for establishing a substantial contingent of mounted crossbowmen in his army. He also coordinated the employment of the reflex longbow with the crossbow, but (unlike the Mongols) does not seem to have had mounted longbowmen.

The combined arms concept seems to have been adopted for units as small as a 1,000-man equivalent of a modern regiment. Thus the Chinese appear to have been able to deploy units capable of decentralized, independent action, as well as to combine them into large, massed, but articulated armies, in which the major units were brigades of 2 or 3 regiments. Heavy armored infantry predominated. Light unarmored infantry—archers, crossbowmen, and spearmen—functioned as skirmishers and provided security by screening flanks and rear. Ch'in cavalry was also armored and performed reconnaissance and security duties, such as pursuit of a defeated enemy.

The bulk of the soldiers, infantry and cavalry alike, had bronze-tipped—or iron-tipped—spears as their primary weapons. The secondary weapon for most soldiers, archers or spearmen, mounted or dismounted, was a single-edged sword nearly three feet long, suspended in a scabbard from a waist belt. All, except apparently for lightly-armed skirmishers, wore armor made up of small metal (bronze) plates attached by a form of rivet to a quilted fabric base. Some protection seems to have been provided even those without armor by a heavy quilted robe. The Chinese apparently relied entirely upon their armor for passive protection and did not carry shields.

Although it is clear from the recent archaeological finds near Sian that there were chariots in the Ch'in army, these were being phased out as major weapons and may have disappeared altogether by the beginning of the Christian era. The principal use of chariots in the terra-cotta army near Shih Huang Ti's mausoleum seems to have been as elevated, mobile command post platforms from which senior officers could control their units.

The most remarkable aspect of the military system of the Ch'ins, and later of the Hans and their successors, was the reliance on a sophisticated crossbow, in general use in China more than a thousand years before that weapon was extensively employed in the West. The bronze mechanisms of the terra-cotta army's crossbows are precisely fashioned, as if they came from a machinist's lathe. There does not appear to have been a windlass, or comparable mechanism such as those found in heavier Western crossbows in the 10th to 14th centuries. However, in drills in which they seem to have used both their feet and their waists to pull the bowstring tight, the Chinese appear to have been able to achieve sufficient tension to enable sharp-tipped bronze bolts to penetrate metal armor.

It was the Ch'in-Han crossbow that seems to have assured victory to the Chinese in their somewhat legendary encounter with the soldiers of Rome at the Battle of Sogdiana, 36 B.C. (see p. 133).

While the Chinese in the 2nd and 3rd centuries B.C. seem to have been far ahead of their Western contemporaries in weapons design, they clearly lagged in metallurgy, to the extent that under Shih Huang Ti they were only beginning to introduce iron weapons and were still largely relying upon bronze. This was long after military men of the Mediterranean world had graduated to iron. On the other hand, the bronze weapons of the Chinese seem to have been every bit as sturdy and tough as comparable contemporary iron weapons in the West. According to Shapiro, "Chemical analysis has revealed that the swords and arrowheads are primarily bronze and tin with traces of rare metals, and that their surfaces were treated with chromium."

While the nobility still filled the key command positions in the army, junior officers were being chosen from the ranks for ability, not heredity, and they appear to have been able to advance to high rank despite lowly birth.

Manpower for the army was obtained by conscription of peasants into a semiregular militia force, liable to call-up at any time. Men between the ages of 17 and 60 had to serve for many years as soldiers or laborers. At one time Shih Huang Ti apparently had any army of 1 million men under arms, out of a population of about 12 million. He marshalled an army of 600,000 men for his final campaign against Ch'u.

Administratively there was a complex, sophisticated, organization of 20 pay grades in the army, partially related to rank and partially to years of service. Advancement from one grade to the next was partly on the basis of seniority, partly on merit. The Emperor was the commander in chief, but rarely commanded in battle. Discipline was strict and punishment for infractions was severe.

A well-developed logistics system was essential to maintain such large armies. Armories, magazines, and granaries were established across the country, with priority emphasis on the capital, next on the northern regions where barbarian depradation could be expected, and next on the lines of communications. These installations were carefully guarded with strong permanent garrisons, it being recognized that they would be primary targets in the event of rebellion.

IV

THE RISE OF GREAT EMPIRES
IN EAST AND WEST:

200–1 B.C.

MILITARY TRENDS

The most significant phenomenon of this period was the parallel growth of two great military powers: the Han Empire of China and the Roman Empire of the Mediterranean world. Standards of military leadership were generally lower than in the preceding period. **Caesar** alone is worthy of consideration with Alexander and Hannibal; even his illustrious predecessors, **Scipio Aemilianus, Sulla,** and **Lucullus,** were hardly up to the caliber of the Roman heroes of the Second Punic War or to the competent professionals of the Diadochi.

MILITARY THEORY AND TACTICS

There were no major developments in theory, strategy, tactics, or weapons, but there were some important refinements, innovations, and modifications, particularly in Roman military organization. In Europe infantry remained the supreme arbiter of battles. In Asia cavalry predominated. The Macedonian phalanx and Roman legion met in two major battles— Cynoscephalae and Pydna; the legion was victorious.

The Romans employed field fortifications in battle to a degree hitherto unknown; a logical development of their earlier practice of using fortified camps as offensive and defensive bases. Following the initiative of Sulla at the Battle of Chaeronea, the Romans on numerous occasions were able to wield their shovels and axes in such a way as to integrate field fortifications into aggressive, offensive battle plans. They began also to use light missile engines as field artillery, in conjunction with their field fortifications.

CAVALRY

By the beginning of the Christian era the horse archer dominated warfare in Asia, save only in India. Though cavalry was the major arm in northern India, the principal weapons were lance and short sword; in central and southern India cavalry was less important than infantry, since geography and climate made it impossible to raise and support enough good horses for large cavalry units. Most of the good horses of southern India were used for chariots. (Chariots were still used elsewhere—notably by **Mithridates** of Pontus—but their importance had generally declined.)

The horse archer had been introduced into warfare by the nomadic barbarians of central Asia. Alexander encountered them as the major components of the Scythian tribes he defeated in Sogdiana and north of the Jaxartes. The Parthians, descendants of these Scythians, brought the horse archer to dramatic prominence in southwestern Asia in this period. Horse archers were the only type of warrior in the hordes of the fierce Mongoloid and Indo-European peoples who then dominated Mongolia and Turkestan: the Hsiung-nu and the Yueh Chih. Their depredations forced the Chinese to make cavalry their major arm.

One of the most significant portents of future military developments was the Battle of Carrhae, where the horse archers of the Parthian leader **Surenas** gained an overwhelming victory over the infantry legions of Roman **Crassus.** Alexander had had little trouble with similar foes, but no leader of genius comparable to Alexander appeared for centuries who could meet the horse archer with truly balanced forces of well-trained infantry and cavalry, working together as an integrated team.

The Parthians had little success, however, in their raids and invasions west of the Euphrates in the decades after Carrhae. They rightly were more fearful and respectful of the military power of Rome than vice versa. Nevertheless, Carrhae pointed to a trend. A few centuries later the horse archer would replace the legionary as the principal guardian of the eastern frontiers of Rome and Byzantium.

India was behind other civilized regions of the world in military developments, with one exception: the stirrup apparently was first used by Indian lancers as early as the 1st century B.C.

SIEGE WARFARE

Next to Alexander, Caesar was the outstanding director of siege operations of the ancient world. He, like other logical, methodical Romans, brought systematic procedure to siege operations. While the sequence of the details of operational and engineering actions naturally varied as required by local circumstances and the reactions of the besieged, an outline of a typical implementation of the system does give a clear picture of ancient siege operations:*

    a. Reconnaissance of the fortifications and of the surrounding region to note local resources in lumber, stones, animals, food, fodder.

    b. Establishment of a fortified camp.

    c. Collection of materials needed for construction and for siege engines.

    d. Manufacture of mantelets and movable gallery sections. (These latter were like roofed huts, which, when placed together and the roofs covered with dampened skins, provided

* With modifications, this is based upon the discussion in T. A. Dodge, *Julius Caesar* (Boston: Houghton Mifflin), 1892, 387–399.

fireproof and missile-proof galleries through which troops and workmen could walk to the most advanced works and entrenchments.)

e. Building of redoubts around the circumference of the fortified place, then connecting these with lines of contravallation. (This would usually be begun at the same time as the two previous steps.) Sometimes a wall of circumvallation would also be built; Caesar almost invariably built double walls.

f. By use of mantelets, galleries, and entrenchments, the preparation of covered ways toward the enemy walls; these would lead to mine heads, subterranean passages, and the advanced siege engines, which by now were beginning to harass the besieged troops and civilian population. Use of heavy and light machines became constant on both sides.

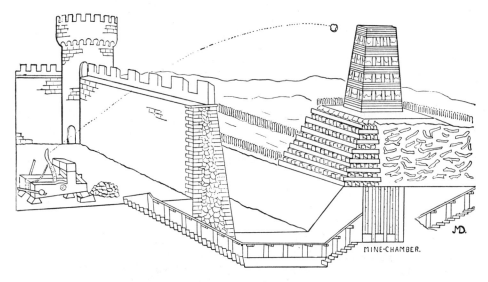

MINE-CHAMBER.

Fort, tower, mound, mantelets

g. Building a terraced mound, raised one level at a time, and advanced gradually toward the walls, under the cover of a rampart of mantelets along its forward crest.

h. Erection of towers (usually on the terrace); these would be placed on great logs, then rolled gradually forward toward the besieged walls. The fronts of these towers were covered with dampened skins to prevent burning by incendiary arrows; their bases were well protected by infantry units, to prevent a destructive sortie by the defenders.

i. If the town or fort was surrounded by a moat, this would usually be filled in ahead of the mound and at any other point where a breach was desired. Breaches were achieved in two principal ways: (1) by use of rams, under protected galleries; or (2) by mine galleries under the walls, which when collapsed would cause the wall to crumple. If an alert defender built new interior walls, this breaching operation might have to be repeated several times. Alert defenders could also interfere with mining operations by construction of countermines.

j. The final assault; usually by charging through the breach made in the defending wall. Sometimes an assault would be attempted without a breach, the attackers rushing onto the

Penthouse and ram picked up by tongs

ramparts from movable towers, up scaling ladders, or by telenons (see p. 40). Sometimes an advance party would be sent secretly into the interior of the defended place through a mine shaft; thence to open the city gates and/or attack the defenders from the rear. One typically Roman innovation in the assault phase was the advance of a cohort to the walls under the protection of a **testudo** made by raising and interlocking their shields over their heads.

Testudo

NAVAL WARFARE

War at sea changed little. Triremes and quinquiremes remained the major combat vessels, though there were many other types, including special flat-bottomed craft (**pontones**), which the Romans used for their river flotillas. In battle the main objectives were still to run down the opponents, or to sink them by ramming, or to break their oars, or to board and capture, or to set them on fire. Cruise procedures, such as camping along the shore at night whenever possible, were also unchanged.

The major innovation was introduced by **Octavian**'s admiral, **Agrippa: the harpax** or **harpago.** This was a pole, with a hook on its end, shot from a catapult into the side of an opposing ship, where the hook would hold it fast. Attached to the end of the harpago was a rope, which could be winched in to bring the two vessels together and facilitate boarding. This, the precursor of the whaling harpoon, was first used to good effect by Agrippa at the Battle of Naulochos and again at Actium (see p. 126).

EUROPE–MEDITERRANEAN

ROME, MACEDONIA, GREECE, AND PERGAMUM, 200–150 B.C.

200–191. Resubjugation of the Po Valley. After the defeat of **Hannibal,** Rome began the pacification of Cisalpine Gaul. Major Roman victories were won at **Cremona** (200) and **Mutina** (Modena, 194).

Second Macedonian War, 200–196 B.C.

Philip V of Macedon, in alliance with **Antiochus III** of Seleucid Syria, after the indecisive First Macedonian War (see pp. 62, 74), tried to dominate Greece, Thrace, and the Aegean coast of Asia Minor. Rome received appeals for help from Pergamum, Rhodes, and Athens.

200. Rome Declares War. Nabis, tyrant of Sparta, lined up with Philip. Two years of inconclusive minor operations followed.

198. Battle of the Aous. Titus Quinctius Flamininus, newly appointed Roman consul, drove Philip from a strong defensive position, while Roman diplomatic moves undermined his position in Greece. The Macedonian, taking the initiative, moved south from Larissa (197) as Flamininus advanced into Thessaly. Both armies were approximately equal in strength— about 26,000 each.

197. Battle of Cynoscephalae. They met unexpectedly in hill country, in a fog. Philip, encouraged by initial successes, brought on a general engagement on terrain unfavorable to the phalanx. His right wing drove back the Roman left;

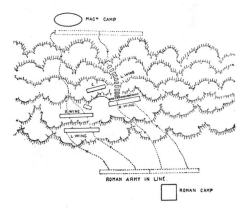

Battle of Cynoscephalae

but while the Macedonian left was deploying from march column on uneven ground, it was struck by the Roman right, led by Flamininus, and routed. Part of the advancing Roman right now swung around—apparently without orders from Flamininus—hitting the Macedonian right and driving it from the field in confusion. Macedonian losses were about 13,000; Roman, a few hundred. This was the first battle in the open between the flexible Roman legion and the Macedonian phalanx.

196. Liberation of Greece. Philip was forced to give up all claim to Greece, as well as his possessions in Thrace, Asia Minor, and the Aegean.

195. Revolt in Spain. This was quelled by Consul **Marcus Porcius Cato.** Unrest was endemic on the peninsula until 155.

195–194. War between the Achaean League and Sparta. With some assistance from Flamininus, **Philopoemen** defeated Nabis, gaining an overwhelming victory at the **Battle of Gythium.** Flamininus, however, prevented Philopoemen from seizing Sparta.

194. Rome Declares Greece to Be Independent. Roman troops were withdrawn.

War with Antiochus III of Syria, 192–188 B.C.

Tensions between Rome and Seleucid Syria had mounted. Conciliation failed.

192. Rome Declares War. Antiochus, on invitation of the Aetolian League, invaded Greece, as a Roman army under **M. Acilius Glabrio** and Cato arrived in Epirus from Italy (early 191). Antiochus was driven from **Thermopylae** by the Romans. He evacuated Greece and sailed to Ephesus. Soon after, a Roman fleet commanded by **Gaius Livius** defeated a Syrian fleet between **Ionia** and **Chios** (191). Next year Roman admiral **Lucius Aemilius Regillus,** assisted by the Rhodians, defeated a Syrian fleet under Hannibal (taking part in his first and last naval battle) at **Eurymedon,** and then defeated a second Syrian fleet at **Myonnessus** (190). Meanwhile, a Roman army under **Lucius Cornelius Scipio** crossed the Hellespont to invade Asia. Lucius, a man of moderate ability, was accompanied by his bril-

liant brother, **Publius Cornelius Scipio Africanus,** victor of Zama, who actually planned the operations. But as the two armies approached each other, about 40 miles east of Smyrna, Scipio Africanus became ill and was unable to accompany the Roman army; tactical direction was assumed by **Cnaeus Domitius.** The Romans had been joined by **Eumenes II** of Pergamum, with a small army of about 10,000, mostly cavalry. The combined Roman-Pergamenian army was about 40,000 strong. Antiochus had about 75,000.

190, December. Battle of Magnesia. The Romans took the initiative, crossing the Hermus River to attack the Syrians, waiting behind a formidable line of war elephants. Eumenes, with the Pergamenian cavalry, was on the Roman right; Roman and allied cavalry were on the left of the legions and allied infantry. Antiochus had divided his cavalry into two elements, one on each flank of his phalanx; he commanded the right wing. At the outset the right flank cavalry contingents on each side won quick successes. Antiochus drove the Roman horsemen from the field, pursuing them across the river to the Roman camp, where he was barely repulsed by the Roman security detachment. Eumenes, however, halted his pursuit of the Syrian left wing cavalry and turned to strike the exposed flank of the Syrian phalanx. Meanwhile, the legions had repulsed the elephants, a number of the maddened beasts being driven back through their own phalanx. While still in confusion, the Syrian infantry's left flank was struck by Eumenes and his cavalry. At the same time the legions pressed a vigorous frontal assault. The Syrians broke and fled. The Romans—once again under the strategic and political direction of Scipio Africanus—followed up their victory to gain almost complete control over western Asia Minor (189).

188. Peace of Apameax. Antiochus was forced to give up his possessions in Greece and in Anatolia west of the Taurus Mountains. These lands were divided between Rome's allies, Rhodes and Pergamum, who were strengthened to offset Macedonia and Syria. Rome kept only the islands of Cephalonia and Zacynthus.

190. War between the Achaean League and Sparta. Complete victory of Philopoemen over Nabis and the suppression of Nabis' social reforms. Philopoemen's ruthlessness caused him to be censured by Rome, which established a virtual protectorate over Sparta.

189. Galatian Invasion of Pergamum. A Roman army under **G. Manlius Volso** helped Eumenes to defeat and repel the Celts of Galatia.

189–179. Macedonia Prepares for War. Philip V of Macedon, who had assisted Rome in the war against Antiochus, was embittered by Roman failure to reward his services. After his death (179), his son **Perseus** continued quiet and efficient preparations for war.

186–179. Expansion of Pergamum. Eumenes greatly increased the power and influence of Pergamum in Asia Minor by victories over **Prusias I** of Bithynia (186) and **Pharnaces I** of Pontus (183–179). Rome was suspicious of his ambitions.

184. Death of Philopoemen. During a revolt of Messene against the Achaean League, Philopoemen was captured and executed by the rebels.

183. Deaths of Hannibal and Scipio. For the death of Hannibal in Bithynia, by his own hand, see page 78. Scipio died at about the same time, in self-imposed exile in southern Italy, where he had retired after undergoing political attack, abuse, and trial in Rome on charges (probably false) of embezzlement.

Third Macedonian War, 172–167 B.C.

An unsuccessful attempt to murder Eumenes II of Pergamum, instigated by Perseus of Macedonia, led to war between those states.

172. Rome Enters the War. Almost equally suspicious of Perseus and Eumenes, Rome decided to support her old ally, sending an army against Macedonia, via Illyria, under **P. Licinius Crassus** (171). Perseus defeated Crassus at the **Battle of Callicinus** (near Larissa, 171). Perseus then repulsed **A. Hostilius Mancinus,** commanding another Roman army, in Thessaly (170). Still another Roman invasion attempt, under **Q. Marcius Philippus,** failed the following year. Perseus tried, unsuccessfully, to bribe Eumenes and the Rhodians—both Roman allies—to join him. **Genthius** of Illyria and **Clondicus,** chief of the Gallic tribes north of Macedonia, accepted Perseus' bribes, but their ardor waned when he failed to make promised payments.

168. Lucius Aemilius Paulus to Command in Greece. Son of the Roman consul at Cannae (see p. 72), Paulus arrived with reinforcements. At the same time Praetor **Lucius Anicus** took a small army to Illyria and quickly overwhelmed Genthius. When Paulus joined the dispirited Roman army it was encamped on the south bank of the Enipeus River, beside the shore of the Gulf of Thessalonika; Perseus' army was camped to the northwest, on the far bank of the river. Paulus used stern measures to restore traditional Roman discipline (April–May).

168. June. Paulus Advances. He attempted a wide envelopment of the Macedonian army by sending a portion of his army around Mount Olympus. Perseus learned of the movement in time to withdraw to a position behind the Aeson River.

168, June 22. Battle of Pydna.* The action was started by accident during the afternoon while both sides were watering horses on opposite banks of the river. Perseus, seizing the initiative, formed his phalanx and attacked across the Aeson. Despite the efforts of Paulus to organize and rally his men, the phalanx swept forward irresistibly on the flat terrain near the river, but it was unable to keep alignment in the rolling ground farther south. Paulus counterattacked, taking advantage of gaps in the phalanx. Once the Romans penetrated, the phalanx fell apart; 20,000 Macedonians were killed, 11,000 captured; only a handful—including Perseus—escaped. Roman losses

---

* The date is known from an eclipse of the moon the night before the battle, an omen the Romans considered favorable; it caused consternation among the Macedonians.

were less than 1,000. Perseus later surrendered, and died in captivity in Italy.

167. Rome Controls the Eastern Mediterranean. Macedonia was divided into 4 separate republics under Roman protection. A virtual protectorate was established over Greece; **Antiochus IV,** who had conquered Egypt, had accepted Rome's demands that he withdraw and restore Egyptian sovereignty (168), thus acknowledging virtual Roman suzerainty. A protectorate was established over Anatolia; the power of Pergamum was cut sharply, since it was no longer needed by Rome to offset the power of Macedonia and Syria.

SELEUCID SYRIA AND PTOLEMAIC EGYPT, 200–50 B.C.

192–188. Seleucid War with Rome. (See p. 96) Ptolemy V allied Egypt with Rome. A consequence of Antiochus' defeat was the defection of Armenia and Bactria from weakened Seleucid rule.

187. Death of Antiochus. This was in Luristan, while he was attempting to recover the lost eastern provinces.

171–168. Seleucid War with Egypt. Antiochus IV twice successfully invaded Egypt (171–170 and 168). His conquest of Egypt was almost complete when, while investing Alexandria, he was warned by Rome to restore Egypt to the Ptolemies. During his ignominious withdrawal from Egypt, he occupied Jerusalem, destroyed the city walls, and decreed the abolition of Judaism, causing a protracted Jewish revolt (168).

168–143. Wars of the Maccabees. (See below)

166–163. Seleucid Campaigns in the East. Antiochus IV recovered Armenia and other eastern provinces.

162–143. Dynastic Strife within the Seleucid Empire. Demetrius I and his son, **Demetrius II,** were opposed by **Alexander Balas.** Ptolemy VI of Egypt supported the Demetrii, while Eumenes of Pergamum supported Balas. Crucial conflict of this struggle was the victory of Ptolemy over Alexander

Balas at the **Battle of the Oenoparus** (145). Ptolemy died of wounds; Balas was soon thereafter murdered and succeeded by Demetrius II.

c. 161–159. Revolt of Timarchus. The governor of Babylonia, **Timarchus** of Miletus, threw off Seleucid rule and conquered Media, taking the title of "Great King." Demetrius I defeated Timarchus, recovering Babylonia and Media for the empire (159).

150. Parthian Conquest of Media. Mithridates I took advantage of chaos in the Seleucid Empire.

141–139. Seleucid War with Parthia. Mithridates invaded Babylonia and captured Babylon (141). Demetrius II then drove the Parthians from Mesopotamia (130). Next year, however, he was captured through treachery.

145–51. Decline of Ptolemaic Egypt. Constant unrest and civil strife tore Egypt in an ugly succession of treachery, murder, and incest among the decaying Ptolemies. Frequent Roman intervention failed to bring lasting peace. **Cleopatra VII** and her brother **Ptolemy XII** jointly ascended the throne (51).

130–127. Renewed Seleucid War with Parthia. Antiochus VIII recovered Babylonia, forcing **Phraates II** of Parthia to release his brother, Demetrius II, from captivity. Continuing on into Media, Antiochus was defeated and killed at the **Battle of Ecbatana** by Phraates (129). This defeat cost the Seleucid Empire all of the region east of the Euphrates River.

125–64. Decline of Seleucid Syria. The dwindling Seleucid Empire was racked by dynastic disorders, while neighboring Parthia continued its encroachments in the east, and Armenia from the north. Finally, Rome moved in to restore order and to prevent further Parthian and Armenian expansion (64); Syria, conquered by Pompey, became a province of Rome (see p. 105).

JUDEA, 168–66 B.C.

168. Revolt of the Maccabees. Suppression of Judaism by Antiochus IV (see above) ignited a Jewish revolt led by a priest, **Mattathias,** and his five sons, of whom the most prominent and most able was **Judas Maccabeus.**

166–165. Victories of Judas. Upon the death of Mattathias, Judas became leader of the rebellious Jews. In brilliant guerrilla actions, he defeated a succession of Syrian generals, his most renowned victories being at **Beth Horon** (166), **Emmaus** (166), and **Beth Zur,** near Hebron (165). After this latter victory he captured Jerusalem, liberating the temple, though a Seleucid garrison continued to hold out in the citadel.

165–164. Struggle for Jerusalem. Judas extended his control over much of Judea, while keeping a tight siege of Jerusalem's citadel. Antiochus being away on his successful campaign to the east, the Syrian regent **Lysias** led an invading army into Judea in an effort to recapture Jerusalem. He defeated the Jews at **Beth Zachariah,** but returned to Syria to suppress a revolt (164).

163–161. The Last Campaigns of Judas. Syrian general **Bacchides,** now commanding Seleucid forces in Judea, defeated Judas and drove him from Jerusalem (162). Judas quickly resumed the offensive, defeating the Syrian general **Nicanor** at **Adasa,** near the site of his earlier victory of Beth Horon (161); Nicanor was killed. Later that same year, however, Judas was defeated and killed in battle by Bacchides at **Elasa.**

161–143. Leadership of Jonathan (Judas' brother). He continued generally successful guerrilla warfare against the Syrians. Recognized as *de facto* ruler of Judea, he made his headquarters in Jerusalem (152). At **Ptolemais** (modern Acre) he was ambushed, captured, and later killed by the Syrians and dissident Jews (143).

143–66. Independence of Judea. Simon, another brother of Judas Maccabeus, was recognized as King of Judea by the Seleucids (143). The following decades were marked by internal violence and frequent invasions by the Seleucids and neighboring Arab chieftains. The turmoil was brought to a close by Rome after Pompey captured Jerusalem (see p. 105); Judea was annexed by Rome, to round out

her control over the eastern Mediterranean (64).

ROME AND THE MEDITERRANEAN WORLD, 150–60 B.C.

152–146. Uprisings in Macedonia. (Sometimes called Fourth Macedonian War.) **Andriscus,** supposed son of Perseus, led an uprising which temporarily united Macedonia. Declaring himself king, Andriscus repelled initial Roman efforts to regain control. He was crushed by **Q. Caecilius Metellus** (148), who then suppressed subsequent outbreaks (148–146). Macedonia became a Roman province (146).

Third Punic War, 149–146 B.C.

BACKGROUND. Troubles long brewing between Carthage and aged **Massinissa** of Numidia broke out into open war (150). Carthage-haters in Rome—led by Marcus Porcius Cato—now acted against Carthage. Upon Roman demand Carthage ceased operations against Numidia, gave up 300 hostages to Rome, surrendered most of its weapons, and dismantled the battlements of the city. Carthage, however, refused the final crushing Roman demand to abandon the existing city and to move the populace inland.

149. Rome Declares War. Initial Roman land and sea efforts against Carthage were foiled by the vigorous defense of the hastily fortified city (148).

147. Scipio Aemilianus to Command. Son of L. Aemilius Paulus, victor of Pydna, and adopted grandson of Scipio Africanus, **Publius Scipio Aemilianus** bore his two names worthily. Upon arrival in Africa, he vigorously pressed the land and sea blockade of Carthage, whose population suffered terribly from starvation and disease.

146. Fall of Carthage. A determined assault by Scipio Aemilianus was followed by house-to-house conflict through the city. When it was over, nine-tenths of the population had perished by starvation, disease, or battle. By order of the Roman Senate, and despite Scipio's protests, the city was completely destroyed and the survivors sold as slaves.

149–139. Lusitanian and Celtiberian Wars. Earlier disorders in Spain had been suppressed by **M. Claudius Marcellus,** who defeated the Lusitanians and Celtiberians (154–151). The Lusitanians and Iberians took advantage of the Third Punic War, revolting under the leadership of **Viriathus,** who held Lusitania successfully against repeated Roman efforts. After Viriathus was assassinated by a traitor in Roman pay, the Lusitanians collapsed (139). The Celtiberians had already been crushed by Metellus, conqueror of Macedonia (144).

146. Achaean War. Hoping to take advantage of Roman preoccupation with Carthage, the Achaean League attacked Sparta, still under Roman protection. Consul **Lucius Mummius** led an army to Greece and defeated the Achaean general **Critolaus** near Corinth. Mummius then captured and destroyed Corinth. The Achaean League was dissolved, and Rome subjugated all Greece.

137–133. Numantian War. Numantia, city of the upper Durius (Douro) River in Spain, became the center of revived Celtiberian independence efforts. The consuls **Quintus Pompeius** and **M. Papilius Laenas** were defeated and disgraced by the Numantians (137–132). Scipio Aemilianus, victor over Carthage, was appointed to command in Spain (134). He quickly reorganized the Roman armies, then defeated and captured the city.

c. 135. Growing Internal Disorders in Rome. During the quarter-century beginning about 160 there had been a decline in Roman military capabilities reflecting serious political, economic, and social unrest. The Roman Senate was unequal to the task of governing the new empire; administration was inefficient, corruption rife. This deterioration had begun with the devastation and dislocation created by Hannibal and the Roman efforts to defeat him. The growth of large estates, operating with slave labor procured in overseas conquests, hastened the disappearance of the sturdy farm peasantry who had formed the backbone of the Roman militia army. These now pauperized peasants became the mobs of the city or the permanent professional soldiery of the army. These soldiers lacked traditional Roman discipline. They owed allegiance only to generals and were inspired by loot rather than patriotic ardor. The near-collapse of law and order was demonstrated in repeated uprisings of slaves, in the domination of the Mediterranean by pirates, and in recurrent civil strife. This state of affairs, gradually worsening, continued for a century. Yet, surprisingly, the essential vigor of Rome was demonstrated by a few military leaders of ability, like Scipio Aemilianus and Metellus, who inspired the degenerating army to fight worthily and to hold secure the far-flung Roman borders.

135–132. First Servile War. A slave uprising in Sicily, long defying repeated Roman efforts at suppression, was finally quelled by **Publius Rupilius.**

133. Assassination of Tiberius Gracchus. The violent death of the democratic tribune marked the beginning of endemic riot and bloodshed in the streets of Rome itself.

133–129. Conquest of Pergamum. Attalus **III** of Pergamum bequeathed his kingdom to Rome on his death. **Aristonicus,** pretender to the throne, defied Roman efforts to take over. He was finally defeated by Proconsul **P. Licinius Crassus,** with the assistance of Cappadocian forces. Pergamum became the Roman province of Asia.

125–121. Expansion into Transalpine Gaul. Consul **Marcus Fulvius Flaccus** commenced conquest of the region between the Alps, the Rhone, and the Mediterranean (125). Consul **Q. Fabius Maximus,** allied with the Aedui, defeated the Arverni and Allobroges tribes near the confluence of the Rhone and Isere rivers (121). The province of Transalpine Gaul (known as Provincia, or *the* Province) was established.

124. Revolt of Fregellae. Unrest among the Italian allies—partly due to their unsatisfied demands for the franchise, partly due to the extension of the democratic-conservative struggle in Rome—was evidenced by the revolt of Fregellae, in Latium, second largest city of Italy. Captured by ruse, the city was destroyed by the Romans.

115. Emergence of Caius Marius (155–86). As a praetor he subdued Further Spain.

113. Appearance of the Cimbri. The Teutonic tribe of the Cimbri reached the Carnic Alps, defeated a Roman army under **G. Papirius Carbo** in the Drava Valley, then repulsed subsequent Roman punitive attempts.

Jugurthine War, 112–106 B.C.

BACKGROUND. A dynastic struggle in Numidia between descendants of Massinissa resulted in the victory of **Jugurtha** over **Adherbal** (119). Rome, however, divided the kingdom between the two contestants. Jugurtha refused to accept the Roman verdict, and intermittent fighting continued, marked by drawn-out negotiations and shameless bribery of leading Romans by the Numidian princes.

112. Jugurtha Declares War. After he won a few minor successes, a truce was reached (110), but fighting soon broke out again. Jugurtha began a systematic campaign to sweep the Romans from Numidia (109). **Caecilius Metellus** (nephew of Metellus Macedonicus) was then sent to take command in Africa. Reorganizing the shattered Roman forces, he invaded Numidia and defeated Jugurtha in the **Battle of the Muthul** (108). He then occupied most of the settled regions of Numidia. Jugurtha took refuge in the desert and began a successful guerrilla campaign against the Romans. Uncouth and boorish Marius was a subordinate of Metellus in these operations.

106. Marius to Command. Elected consul, he superseded his former commander. His principal subordinate was a young quaestor, **Lucius Cornelius Sulla.**

106. Capture of Jugurtha. By a combination of his own energy and Numidian treachery, Sulla captured Jugurtha, bringing the war to a conclusion (106). Differing versions as to the respective roles of Marius and Sulla in this campaign gave rise to jealousy between them, heightening the bitterness of their later struggle

as leaders respectively of the democratic and aristocratic factions of Rome.

109–104. Migrations of the Cimbri and Teutones. The Cimbri, accompanied by the related tribe of Teutones, migrated through what is now Switzerland to southern Gaul; near the Rhone River they defeated the army of **M. Junius Silanus.** After several futile efforts to subdue the barbarians, Consul **Mallius Maximus** led an army of 80,000 against them. The Roman army was defeated and virtually annihilated at the **Battle of Arausio** (Orange); 40,000 Roman noncombatants were also killed (105). This, one of the worst disasters ever to befall Roman arms, created consternation in Rome. The barbarians then moved toward Spain, but were repulsed in the Pyreneean passes by the Celtiberians. Returning to central Gaul, they clashed with the Belgae, then retired to southern Gaul.

105. Military Reforms in Rome. Marius, who had been reelected consul, initiated sweeping reforms of the Roman military system, assisted by **P. Rutilius Rufus** (see p. 107).

104–101. War with the Cimbri and Teutones. Following the disaster of Arausio, Marius took command in Roman Gaul. At first he avoided battle, devoting efforts to training, rebuilding the discipline and confidence of his demoralized troops, and to the reorganization of the logistical system of the province. The Cimbri and Teutones now determined to invade Italy. Most of the Cimbri marched northeastward, through Switzerland, heading for the Brenner Pass. The Teutones, with some of the Cimbri, advanced toward the Little Saint Bernard Pass. Marius built a powerful fortified camp at the junction of the Rhone and Isère rivers and repulsed repeated barbarian assaults (102). The Teutones then marched down the Rhone, heading for the passes over the Maritime Alps. Marius followed cautiously.

102. Battle of Aquae Sextae (Aix-en-Provence). Marius enticed the barbarians to attack him in a carefully selected hill position. At the height of the battle a small force of Romans in ambush attacked the rear of the Teutones and threw them into confusion. In the ensuing slaughter 90,000 barbarians were killed, 20,000 captured.

102–101. Advance of the Cimbri. Traversing the Brenner Pass, the barbarians defeated Consul **Q. Lutatius Catulus** in the **Adige** Valley (102). They then wintered in the Po Valley. Marius hastened back to Italy and joined forces with Catulus.

101. Battle of Vercellae. Marius literally annihilated the Cimbri; 140,000 barbarians (men, women, and children) were killed, 60,000 captured.

104–99. Second Servile War. The major uprising was in Sicily. After most of the island had been overrun by the slaves, the rebellion was suppressed by Consul **Manius Aquillius** (101–99).

100. Marius Elected Consul for the Sixth Time. This precipitated violence and bloodshed in a pitched battle in the streets of Rome. Marius, a good soldier, proved incompetent as a politician and statesman; the struggle between democrats and aristocrats was intensified by his misrule.

93–92. War with Tigranes of Armenia. Expanding the power of his kingdom, **Tigranes** invaded the Roman protectorate of Cappadocia. Sulla, then praetor of the Asian provinces, concluded an alliance against Tigranes with **Mithridates** of Parthia, but repulsed Tigranes without Parthian help.

91–88. Social War. Most of Rome's Italian allies rose in revolt because Rome refused to grant them citizenship. A new Italian republic was established, with capital at Corfinium. At first the rebels were generally successful. Rome made concessions by granting citizenship to those allies who remained loyal (mostly Latins, Etruscans, and Umbrians), and then offering it to rebels who laid down their arms to acknowledge Roman sovereignty. Consul **Lucius Porcius Cato** was defeated and killed at **Fucine Lake** (89), but his colleague **C. Pompeius Strabo** won a decisive victory at **Asculum** (89). Meanwhile, Sulla had besieged and captured **Pompeii**. Most of the rebels now accepted the Roman offer, and the revolt died out. The generally poor showing of Roman forces in the early stages of the war was due to a shocking deterioration of discipline in the Roman army.

First Mithridatic War, 89–84 B.C.

BACKGROUND. **Mithridates VI** of Pontus, who had greatly expanded the power and prestige of his kingdom in Asia Minor and on the eastern and northern shores of the Black Sea, had had a long-standing dispute with **Nicomedes III** of Bithynia over the province of Cappadocia. Rome had previously intervened twice (95 and 92) and warned him to keep hands off Bithynia (89).

89. Mithridates Invades Bithynia and Cappadocia. Completely successful, he overran the Roman provinces in Asia Minor, then invaded Greece, where his successes fanned flickering flames of revolt against Rome. Sulla, placed in command of operations in the east, was en route to Greece (88), when he was diverted by the outbreak of civil war in Rome (see p. 103).

87. Arrival of Sulla. Having temporarily settled affairs in Rome, Sulla returned to Greece, where he promptly drove the two Mithridatic-Greek armies (commanded by **Archelaus** and **Aristion**) into the fortifications of Athens and the Piraeus, which he then invested. Sulla captured Athens by storm (86), and Archelaus escaped from the Piraeus by sea, landing in Boeotia. Meanwhile, Sulla's young subordinate, Lucius Licinius Lucullus, raised a fleet, and decisively defeated the fleet of Mithridates in a battle off **Tenedos** (86).

86. Battle of Chaeronea. Sulla, with about 30,000 men, moved to Boeotia, seeking battle with Archelaus, who had assembled an army of 110,000 men and 90 chariots. In the first known offensive use of field fortifications, Sulla built entrenchments to protect his flanks against envelopment by the Mithridatic-Greek cavalry and erected palisades along the front of his position to provide protection against the chariots. The battle opened with a charge by the Mithridatic cavalry, some of whom were able to avoid the entrenchments and the palisades. Sulla, his legions formed into squares, easily repulsed the charge. The chariot attack was handled according to plan; the maddened

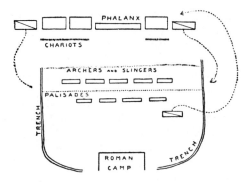

Battle of Chaeronea

horses that survived the Roman arrows and javelins dashed back through the phalanx, throwing it into confusion. Sulla immediately launched a combined infantry and cavalry counterattack and swept the foe from the field.

85. Battle of Orchomenus. Archelaus, who had received reinforcements from Mithridates and from his Greek allies, again outnumbered Sulla by about the same margin as at Chaeronea. Sulla, contemptuous of his adversaries, but exercising judicious care in his plans and preparations, again used field fortifications to assist him in advancing against the sluggish foes, defeating them by a decisive envelopment. Sulla now prepared to invade Asia.

85–84. Arrival of a New Roman Army. At this time the new democratic government in Rome under Marius and Cinna (see below), antagonistic to Sulla, sent another army to the east under **L. Valerius Flaccus,** who was to supplant Sulla. Sulla refused to acknowledge the authority of Flaccus and continued his plans, assisted by Lucullus and his fleet. Flaccus was murdered by **Gaius Flavius Fimbria,** who then assumed command of the democratic Roman army in the east; apparently he tacitly supported Sulla's operations against Mithridates. The Pontine king, his territory threatened by two Roman armies, now made peace. When Sulla persuaded the army of Fimbria to join him, Fimbria committed suicide. Leaving this army in Asia under Lucullus, Sulla returned to Italy to intervene in the civil war there.

Roman Civil War, 88–82 B.C.

88. Democratic Uprising. Rebellious democrats, led by tribune **P. Sulpicius Rufus,** with the support of Marius, were crushed by conservative Sulla, who was called back while en route to Greece (see p. 102). Marius and other rebels fled to Africa.

87. Renewed Uprising. After Sulla again left for Greece, the democrats rose again, now led by **Lucius Cornelius Cinna.** With the democrats in power, Marius returned to Rome and instituted a reign of terror, carried on by Cinna after Marius' death (86). Cinna was killed by a mutiny of his troops (84), but the despotic regime carried on for 2 more years.

83. Return of Sulla. After landing at Brundisium, Sulla defeated Consul **Caius Norbanus** at **Mount Tifata** near Capua. He spent the winter in Capua. The 5-year reign of terror in Rome ended when Sulla, with his 40,000 veterans, marched up from southern Italy (82). Defeating the allied forces of the democrats and revolting Samnites in the **Battle of the Colline Gate,** Sulla seized Rome, made himself dictator, and restored law and order after slaying his political opponents. He sent **Cnaeus Pompeius** (Pompey), one of his subordinates, to Sicily and then to Africa to stamp out the remnants of democratic dissension.

82–79. Sulla Dictator. He reformed the government and restored the authority of the Senate. He then permitted free elections and retired, dying the next year (78).

83–81. Second Mithridatic War. Essentially a local and apparently accidental clash between Mithridates and the Roman governor of Asia, **L. Licinius Murena.** Peace was reestablished on the basis of the *status quo.*

80–72. Sertorian War. Quintus Sertorius, a democratic supporter of Marius, established an independent regime in Lusitania (Portugal and western Spain) after Sulla gained control of Rome. He raised an army of Lusitanians and defeated the legal governor, **Lucius Fufidias,** in the **Battle of the Baetis** (Guadalquivir) River (80). **Quintus Metellus Pius,** principal subordinate of Sulla, took the field, but

Sertorius had the best of inconclusive maneuvering and skirmishing (80–78). Pompey arrived with an army from Italy, via the Pyrenees, to join Metellus (77), but Sertorius kept the two Sullan leaders at bay in a brilliant series of guerrilla and conventional campaigns (76–73). Sertorius was assassinated in a plot directed by his principal subordinate, **Marcus Perperna,** who was then immediately defeated by Pompey, bringing the war to a close (72).

79–68. Operations against the Pirates. Sporadic punitive efforts to subdue or to limit the depredations of the Mediterranean pirates were generally unsuccessful.

78–77. Revolt of Lepidus. Marcus Aemilius Lepidus, leader of the democratic party, became consul and attempted to overthrow the constitution of Sulla (78). Defeated in a battle outside Rome by **Q. Lutatius Catulus,** Lepidus and his supporters fled into Etruria, where they were wiped out by Pompey (77).

Third Mithridatic War, 75–65 B.C.

BACKGROUND. Nicomedes III of Bithynia, to thwart the ambitions of Mithridates, bequeathed his kingdom to Rome on his death (75). Roman troops occupied Bithynia.

75. Mithridates Declares War. With 120,000 men, he invaded Cappadocia, Bithynia, and Paphlagonia, at the same time encouraging revolt in the Roman provinces. The two consuls, **M. Aurelius Cotta** and L. L. Lucullus (former colleague of Sulla), led armies of about 30,000 each to Asia. While Lucullus reestablished control in Roman possessions, Cotta and the fleet sailed to Chalcedon on the Bosporus. Mithridates moved against Cotta, defeated him outside Chalcedon, and drove him back into the city. The Mithridatic fleet at the same time defeated and destroyed the Roman fleet, cutting off Cotta.

74. Battle of Cyzicus. Lucullus marched to assist his colleague, defeating Mithridates' lieutenant **Marius** near Brusa. Mithridates, bottled up on the peninsula of Cyzicus between the two Roman armies, escaped by sea, while his army cut its way out overland, with terrible losses. Lucullus immediately pursued into Pontus.

72. Battle of Cabira (Sivas). Lucullus defeated Mithridates completely, after which he overran the entire kingdom. Mithridates fled to join his son-in-law and ally, Tigranes of Armenia, now the most powerful ruler of the Middle East. Tigranes refused Lucullus' demand to surrender Mithridates.

70–67. Lucullus Invades Armenia. With seeming reckless self-confidence, Lucullus, who had 10,000 men, attacked and defeated Tigranes, who had about 100,000, at the **Battle of Tigranocerta** (69). Advancing into northeastern Armenia and winning another victory at **Artaxata** (68), he returned to the Euphrates Valley when his worn-out troops refused to go farther (68–67). Mithridates meanwhile had invaded Pontus.

67. Pompey Given Command in the East. He reaped the fruits of the amazing victories of Lucullus. After being ambushed and utterly defeated by Pompey in the **Battle of the Lycus** (66), Mithridates escaped to the Crimea, where he committed suicide (64). Tigranes, defeated and captured, was forced to give up all his previous conquests (65; see p. 128).

73–71. Third Servile War. Led by the gladiator **Spartacus,** rebellious slaves established a base on the slopes of Mount Vesuvius, terrorizing southern Italy. Defeating the praetor **Varinius,** Spartacus with 40,000 men exercised virtual control over most of Campania. He defeated both Roman consuls, then ranged almost at will over most of Italy (72). Finally he was defeated by **M. Licinius Crassus** (71), and the revolt was stamped out completely by Pompey, just returned from the Sertorian War in Spain.

67. Pompey's War against the Pirates. The Senate now gave to Pompey (at his suggestion) unlimited authority over the Mediterranean Sea and its littoral for 50 miles inland. In three months Pompey completely defeated the pirates of both western and eastern Mediterranean, conquering their bases and restoring relative tranquillity to that sea for the first time in more than half a century. Following this brilliant success, the Senate gave Pompey dictatorial powers in the east in order to bring to a conclusion the war against Mithridates (see above, p. 104).

65–61. Pompey in the East. Following his victories over Mithridates and Tigranes, Pompey swept through the Middle East. Reaching the Caspian Sea in Armenia (65), he annexed Syria (64), and when Palestine refused to accept Roman sovereignty, he captured Jerusalem, annexing Palestine as well (64). Then completely reorganizing the system of Roman provinces and protectorates in the east, he returned to Rome in triumph (61).

63–62. Insurrection of Catiline. A democratic conspiracy led by **Lucius Sergius Catiline,** to kill the consuls and seize power in Rome, was discovered and exposed by Consul **Marcus Tullius Cicero** in his famous orations. Some violence in Rome itself was quickly suppressed by the government, but Catiline fled to Etruria, where his adherents had raised an army. Pursued by Consul **Gaius Antonius,** Catiline was defeated and killed in a violent battle near **Pistoria** in Etruria (January 62).

61–60. Caesar in Spain. Gaius Julius Caesar (102–44), a democratic politician of aristocratic birth, first gained military prominence as propraetor and governor of Further Spain. Prior to this he had been a staff officer under Sulla in the east (82–78) and had been a quaestor in Spain (69). Most of his adult life, however, had been devoted to politics. Now, exercising military command for the first time, he suppressed an uprising of unruly barbarians in his province, then conquered all of Lusitania for Rome.

THE FIRST TRIUMVIRATE, 60–50 B.C.

60. Establishment of the Triumvirate. An informal association of Pompey, Crassus (victor over Spartacus), and Caesar established political ascendancy in Rome. This was a period of violence and upheaval in Rome itself, with the triumvirs playing confused and intricate roles as each attempted to use the internal discord to his own advantage, while at the same time they divided responsibility for control of the colonial areas beyond Italy.

58–50. Caesar's Gallic Wars. (See p. 113.)

55–38. War with Parthia. (See p. 129.)

54–53. Campaign of Carrhae. In an effort to gain military renown comparable to his rivals and colleagues, Crassus had himself appointed proconsul for Syria. He then marched against the Parthians. He was defeated at Carrhae by the Parthian general **Surenas,** and was killed during the ensuring retreat (see p. 129).

52–50. Pompey Seizes Power. Partly because of anarchy in Rome, and partly because he was increasingly jealous of the growing fame of Caesar in Gaul, Pompey had himself elected sole consul and became virtual dictator. Relations between the two men grew more strained; the Senate, supporting Pompey, passed laws that would cause Caesar's political and military power to lapse on March 1, 49.

ROMAN MILITARY SYSTEM, c. 50 B.C.

In the century and a half following the Second Punic War the upheavals and disorders of the Roman state were mirrored in its armed forces. Yet despite much individual incompetence and a decline in the civic and military virtues of Rome, its military system remained founded on the basic principles which had brought it to preeminence: regularity, discipline, training, flexibility, and unbounded faith in the efficacy of offensive action.

The decline in numbers and vigor of the sturdy Italian peasantry, and the tremendous expansion of year-round Roman military commitments due to the steady growth of imperial dominion, had seriously compromised the Roman militia concept of annual levies. Rome was now, in effect, maintaining a standing army of professional soldiers whose trade was fighting.

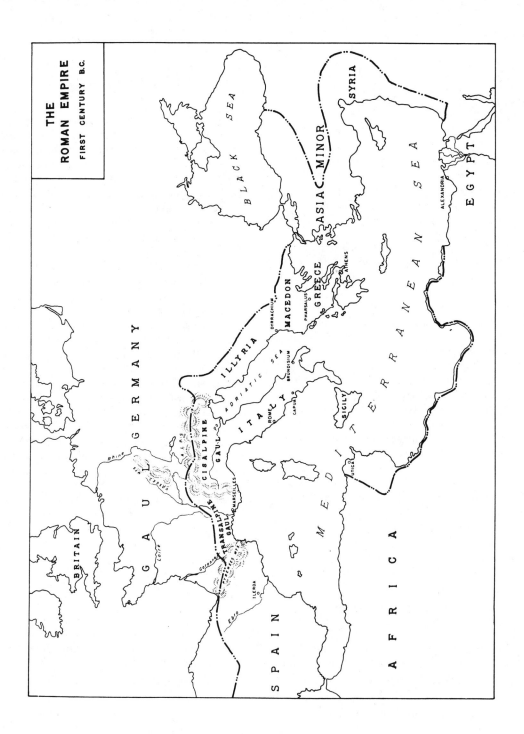

THE
ROMAN EMPIRE
FIRST CENTURY B.C.

BLACK SEA

ASIA MINOR

SYRIA

MEDITERRANEAN SEA

EGYPT

ALEXANDRIA

GERMANY

GREECE

MACEDON

PHARSALUS

ATHENS

ILLYRIA

DYRRACHIUM

ADRIATIC SEA

BRITAIN

GAUL

CISALPINE GAUL

Po

ALPS

Rhine

Ebro

TRANSALPINE GAUL

MARSEILLES

ITALY

ROME

CAPUA

BRUNDISIUM

SICILY

Loire

Garonne

ILERDA

UTICA

SPAIN

MEDITERRANEAN SEA

AFRICA

106

This was particularly true in the overseas provinces, since the government could not afford to send out new armies each year.

The civil and military administrations of the Roman state remained essentially identical, and while this continued to have a number of advantages in the ability of the nation to prosecute war, the decline of the militia system had a serious impact on the level of competence and military leadership. Professional politicians, eschewing the hardships of campaign as legionaries and junior officers, were frequently thrust into positions of military command for which they had little background, experience, or inclination. The general lack of trustworthiness of troops recruited mostly from the less reliable elements of the society reduced discipline and training; increasing lack of confidence between commander and troops created a tendency to reduce the intervals between the maniples of the legion, which began to approach the old Greek phalanx in battle order. This in turn decreased the inherent superiority of the Roman formation over those of its enemies, and contributed to a number of Roman defeats.

The Reforms of Marius

The disaster of Arausio (see p. 101) was the death knell of the old militia system. The efforts to raise new armies caused the state to confirm the professionalization of the army by enlisting men for terms of up to 16 years. It also caused one tough-minded Roman, ignoring sentimental and theoretical attachment to old virtues, to adapt the military system to the realities of the time.

During his terms as consul, Marius established a new system of organization that would continue to be effective through the early years of the Christian era. Though Caesar, too, introduced a number of refinements and adaptations, the armies with which he fought were essentially modeled on those created by Marius about the time of Caesar's birth. Regardless of his political shortcomings, Marius' military reforms entitle him to a place of honor in Roman history.

The old aristocratic distinctions between militia classes were eliminated, as were also the distinctions of age and experience which had led to the creation of hastati, principes, and triarii (though these terms continued to be used). This permitted interchange of units and individual soldiers, greater operational flexibility and maneuverability, increased efficiency in recruitment and replacement.

A complete and revised manual of drill regulations was produced by Publius Rufus (105), a colleague of Marius. Though later refined, particularly by Sulla, these were the regulations in effect in Caesar's time. Thus in this respect, as in others, the trend to professionalism in the army tended to offset the decline in martial spirit and in civic responsibility to the state.

The Marian Legion

Accepting the trend toward a phalangial formation, Marius made the cohort his major tactical organization. The maniple remained merely an administrative element within the cohort. Ten cohorts, 400 to 500 men each, continued to comprise a legion.

The cohort formed for battle in a line of 10 or 8 ranks, with a frontage of about 50 men. In close order, which was used for maneuvering and for massed javelin launching—but rarely for hand-to-hand combat—there was an interval of about 3 feet between men. This did not leave adequate room for wielding a sword, so the open formation, with 6 feet between men, was used for close combat. To permit rapid extension from close to open formation, it was necessary to keep an interval of one cohort's width between cohorts prior to actual engagement. Thus, with a legion formed in 2 or 3 lines, Marius was able (1) to retain the traditional flexibility and maneuverability of the legion by a cellular, checkerboard arrangement of cohorts (rather than maniples), (2) to keep the traditional sword-length interval between legionaries engaged in combat, and yet (3) at the same time to adapt this flexibility to the natural phalangial tendency by permitting a continuous front when engaged in close combat. It was a simple, brilliant, practical development, perpetuating the inherent virtues of the old legion.

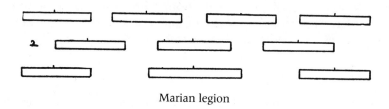

Marian legion

The cohort formed battle line from marching column in 4s or 5s simply by closing up to massed double columns, then facing right or left. The marching evolutions to achieve this and various changes of front and direction were comparable to those of modern close-order drill.

The usual formation of the legion was 3 lines, with 4 cohorts in the first line, and 3 each in the second and third lines, alternatively covering the intervals of the lines to the front in the traditional **quincunx,** or checkerboard, concept. In 2-line formation, obviously, there were 5 cohorts to the line. On rare occasions the legion would be drawn up in one line and even more rarely in 4. The front of a cohort, about 120 to 150 feet, with an equal interval between cohorts, meant that in the normal 3-line formation the legion covered a front of about 1,000 feet. The distance between lines was usually about 150 feet, giving the legion in normal formation a depth of about 350 feet.

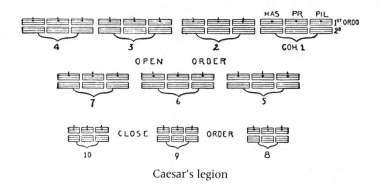

Caesar's legion

An army of 8 legions, then, with an average strength of 4,500 men per legion, would in the normal 3-line formation take up a front of about a mile and a half. The Marian legion, with some 13 men per yard of front, had about half the density of the Macedonian phalanx, which had had about 25 per yard.

The major defensive formations of the legion were the line, square, and circle. The line was usually a single line of 10 cohorts when formed behind fortifications or entrenchments. The square was formed from a normal 3-line formation by simple facing movements of 7 of the cohorts, leaving 3 facing front, while 3 faced the rear, and 2 to each flank. This, or its modification, the circle, was employed in defense against cavalry. Usually, however, if the flanks were protected by friendly horsemen or light auxiliary troops, the legion preferred to face the cavalry in its normal line formation; the combination of **pilum** (javelin), **scutum** (shield), and **gladius** (short sword) was usually too much for even the shock of the most desperate cavalry charge.

The standard of the legion was a silver replica of an eagle, wings outstretched, perhaps a foot in height, mounted on the top of a staff. The legion's eagle was revered perhaps even more than the colors of a modern military unit. Marius apparently regularized the system of legion insignia, which previously had included various other kinds of emblems. Each cohort had its own ensign, usually a device or a medallion of metal or wood, perhaps 6 inches in diameter, also carried on a staff or lance. Each maniple also had an ensign, to provide a rallying point like the modern company guidon. These were always a life-sized human fist, of wood or bronze, mounted on top of a lance, with other distinguishing symbols below it.

Light Infantry

A small but important component of the legion was its contingent of 10 scouts—**speculatores.** These formed, in effect, a kind of reconnaissance squad. The speculatores of several legions could be grouped for army reconnaissance missions.

Though not combined into legions, the light troops or auxiliaries of the Roman army were similarly organized in cohorts. Like the velites of old, these auxiliaries could operate in regular formation like the cohorts of the legion, as well as in their normal irregular skirmishing role to front and flank. Traditionally the best light infantry came from Liguria, in Cisalpine Gaul. As in centuries past, the best slingers came from the Balearic Islands, bowmen from Crete and other Aegean islands.

Cavalry

Under Marius the old Roman cavalry, made up of the **equites** or nobility, completely disappeared. The importance of cavalry, and its relative proportionate strength, increased however. Even more than formerly, therefore, the Romans relied upon allies and mercenaries to provide cavalry. In the time of Marius these came mostly from Thrace and Africa, and to a lesser extent from Spain. Caesar relied almost entirely upon Gallic and German mercenaries and allies to provide him with horsemen.

Naturally, therefore, the organization and discipline of the cavalry of the Roman army were less formal and rigid than in the legion. The most forceful Roman commanders, however, were able to impose a substantial degree of regularity upon their horsemen. A **turma,** or troop, of cavalry consisted of 32 troopers under a **decurion** (sergeant), and was formed for battle in 4 ranks. Twelve turmae formed an **ala** (wing), the equivalent of a squadron, apparently commanded by an officer of rank equivalent to that of a tribune. The ala formed up in 2 or 3 lines, with intervals between the turmae, in a checkerboard formation similar to the cohorts of the legion.

The Legionary

The average soldier of the legion was an Italian peasant or lower-class city dweller. By Caesar's time all Italians were Roman citizens. The spread of the franchise throughout Italy was given much of its initial impetus by Marius, who, because of their conduct in action, gave Roman citizenship to allied cohorts at Vercellae (see p. 102), justifying this action to the Senate by saying that in the din of battle he could not distinguish the voice of the laws. But though the majority of the legionaries were still Italians in Caesar's day, there were growing numbers of other subject peoples and barbarians, some in separate legions, some intermingled with the Italians.

By the time of Marius the decline of the Roman military system had caused the loyalty of the soldier to be transferred from the state—as it had been in the days of the patriotic citizen militia—to the commanding general. The soldier swore allegiance to the general, who provided his daily pay—about 11 cents, the average wage of a day laborer in Rome. It was the general who gave the soldier opportunities for loot and plunder, and who obtained from the Senate, sometimes grudgingly, awards and retirement benefits (usually a plot of land) after the soldier had done with the campaigns. The average legionary was a tough, hard-bitten man, with values and interests—including a rough, heavy-handed sense of humor—comparable to those always found among professional private soldiers. Individually rarely more than 5'6" in height, robust and well muscled, the Italian legionary had a healthy respect for his huskier barbarian foes. In fact, until the time of Caesar, the almost unreasoning Roman fear of the Gauls and Germans—reinforced by the disaster of Arausio—was reflected in the individual emotions of even veteran soldiers. Yet they realized that regular formations and discipline made them militarily superior to the barbarians and, despite personal fears, under good leadership fought staunchly against Gauls and Germans.

Lacking the patriotic ardor of his militia predecessor, the professional Roman legionary under a leader like Marius, Sulla, or Caesar was at least equally tough in combat, and probably even more skillful in the essentials of drill and field campaigning.

Command and Administration

In the old militia army the centurions, tribunes, and staff officers were appointed to their positions each time a levy was made. A centurion one year might theoretically be a simple private the next time he was called to the ranks. With the professionalization of the army, a professional officer corps developed, divided into two main classes. The centurions, who still rose from the ranks, retained a permanent status as officers once they had proven their worth. But they rarely rose above the modern equivalent of company officer rank, though sometimes the senior centurion of the legion—the **primipilus**—would exercise command in battle. The primipilus normally carried the legion's revered eagle.

Officers of field rank—tribune and above—came from the aristocracy. The relationship between centurion and tribune was similar to that between sergeant and lieutenant in modern times.

Theoretically the command of the legion was still rotated among the 6 traditional tribunes, while each cohort was commanded by its senior centurion. It became common practice, however, to assign one officer—a legate—to command of a legion, with the tribunes acting as staff officers and commanding detached cohorts or task groups. Caesar made permanent assignments of legates to his legions.

As before, the general (**imperator**) was assisted by a small staff of quaestors, whose functions were primarily logistical and administrative. In addition, he was served by a group of volunteer aides—**comites praetori**—usually young aristocrats. To protect the general and his headquarters—the **praetorium**—there was now a special guard detachment, composed usually of veteran, trusted legionaries, called the **cohors praetorians.** Scipio Aemilianus had first created such a guard in the Numantian campaign (see p. 100). This was the origin of the famed Praetorian Guards of Imperial Rome.

The interrelationship of civil and military responsibilities exercised by the general and his staff facilitated administrative arrangements. The regimented, military nature of the early Republic had created in the Romans an almost inherent efficiency, which was perpetuated, even if degraded, during the decline of the Republic. The combined civil-military organization of the outlying provinces greatly facilitated supply, logistics, and military administration in general. The ingrained efficiency of the Romans assured a smoothly functioning system of reporting, financial control, and the like.

Every army had an engineering detachment, skilled in the construction of bridges and the specialized structures of siege operations. They carried with them, on a special baggage train, tools and equipment needed for their missions, though their major reliance was on materials and lumber found on the scene of operations.

The Legion on Campaign

On the march a Roman army formed with advance, rear, and flank guards similar to those of modern armies. Each legion was usually accompanied by its baggage train of 500–550 mules. On these were carried skin tents—1 per 10 men—rations, the assigned ballistae and catapults, and miscellaneous equipment. In dangerous country the legion would often march in a square, with the train in the center. In flat, open country all the baggage trains could be assembled, with the entire army forming a large square as it advanced.

If action was likely, the soldier naturally wore his armor. To keep the size of baggage trains to a minimum, Marius insisted that the legionary carry his armor even on administrative marches. To make this easier, and also to help carry the normal 50-pound load of personal equipment and 15 days' rations, each man was given a forked stick—nicknamed "Marius' mule"—to permit him to hoist the load on his shoulder.

The practice of **castrametation**—preparing a fortified camp at the close of each day's march—was continued, and further developed by Caesar. Normally in a square or rectangle—with rounded corners for easier defense—the shape of the camp could vary with the details of terrain. A location next to a convenient source of water was important. It took 3 to 4 hours for the troops to dig the ditch, erect the rampart and palisade, lay out the streets, and pitch tents.

The time was longer, of course, in hostile territory, when as much as a third or half of each legion mounted guard while the remainder made camp. If the encampment was prolonged, towers were usually built, the ditch deepened, and ramparts raised in the days subsequent to arrival. The only difference between the normal field camp and the camps for winter quarters was in the substitution of huts for the skin tents.

Traditionally, the camp was not only a measure of local security; it provided a Roman army with a base for offensive and defensive action wherever it might be, and was virtually a means for multiplying the combat value of the Roman soldier.

Battle Tactics

Whenever possible the Romans, like their enemies, tried to obtain the important advantage of being on higher ground than the foe. This added to the range of missiles, increased the shock effect of a charge while reducing the physical effort in making it, and even made it slightly easier to wield sword and spear. Usually—but not always—Caesar had his best cohorts in the first line, to get the maximum results from the initial shock of battle.

After the skirmishing and missile harassment by light troops had come to an end, the main battle lines approached each other. The legion deliberately advanced, or awaited the enemy, until the lines were about 20 yards apart. Then the first two ranks of the front lines hurled their javelins. Usually by this time the legion had adopted the open-order, semiphalangial formation, though sometimes this maneuver would be delayed until the javelins had been thrown.

Even on the defensive, for moral and physical effect, the legion almost always charged just before the actual hand-to-hand contact of the main battle lines. The first line—all 8 or 10 ranks—dashed violently against the foe, with the first 2 ranks only being able to employ their swords. The ranks behind would then throw their javelins over the heads of the melee. After a few minutes, the second set of 2 ranks would move forward to relieve the men already engaged, and so on, for as long as the fight lasted. Meanwhile, the rear-rank men would be resupplied with javelins by the light troops, who—in addition to protecting rear and flanks—had the mission of salvaging all usable javelins or darts they could find on the field.

Sometimes the initial clash of conflict would be delayed while all ranks of the first line threw their javelins, thus permitting 4 or 5 heavy volleys before the actual charge of the swordsmen.

If the first line was unable to prevail, or was hard-pressed, the second line would advance through the 6-foot intervals in the first line, and the first line would fall back to recuperate and reorganize. Finally, the third line was available, as the commander's reserve. Throughout the battle, therefore, there was incessant movement by ranks within lines, and between 2 or 3 main lines themselves. The discipline and organization which made this movement and replacement possible gave to the Romans a tremendous advantage over barbarian enemies, and is the main explanation why small Roman forces, under good leadership, were able consistently to defeat vastly larger aggregations of barbarians.

Even in drawn-out battles the casualties of a victorious army in antiquity were usually relatively light, while the losses of a defeated army were frequently catastrophic. This was particularly true of Roman battles. The large scutum (shaped like a segment of a cylinder) was probably the most efficient shield of antiquity, and its dexterous employment combined with helmet, breastplate, and greave (right leg only), gave the legionary excellent protection. But when ranks were broken, or an assault was received from flank or rear, the massed ranks of ancient armies were very vulnerable. It was rare that an army could be rallied after sustaining such a blow. Those who could escape, fled. The others were either slaughtered or captured. In a victorious army there were usually 3 to 10 times as many wounded as there were killed. In a defeated army few of the wounded survived.

There was growing use of small missile engines. By the time of Caesar each legion apparently had a complement of 30 small catapults and ballistae, each served by 10 men. These were primarily used in sieges, for defense of field fortifications, and to cover river crossings. Apparently they were also used on some open battlefields, during the preliminary phase and before the actual shock of heavy infantry lines.

THE GALLIC WARS, 58–51 B.C.

58. Caesar to Gaul. By agreement of the triumvirate, after he had served as consul, Caesar was appointed governor of the Roman provinces of Gaul, as proconsul. His area of responsibility included Istria, Illyricum, Cisalpine Gaul (roughly the Po Valley of northern Italy), and the Province of Transalpine Gaul (roughly the French provinces of Provence, Dauphiné, and Languedoc).

58. Migration of the Helvetians. When Caesar took over his provinces at the beginning of the year, he discovered that the entire Helvetian people, a Gallic tribe inhabiting modern Switzerland, numbering 386,000, of whom more than 100,000 were warriors, were heading south for Gaul. Moving through the northern portion of the Province, they planned to concentrate on the Rhone by summer. Building a number of fortifications to block the main route of march of the Helvetians down the Rhone Valley, Caesar collected the scattered regular forces of the Province, and also accepted contingents from a number of Gallic tribes fearful of being overrun by the Helvetians (March–May). Finding their way blocked, the Helvetians continued their move westward, but across the wild Jura country north of the Rhone.

58, June. Battle of the Arar (Saône). Caesar with about 34,000 men caught the Helvetians in the process of crossing the Arar River. In a surprise attack, after a long night march, he overwhelmed and annihilated the 30,000-odd Helvetian warriors still on the east bank. The remainder of the horde continued west toward the Liger (Loire) River; Caesar followed cautiously.

58, July. Battle of Bibracte (Mount Beuvray.) Turning on Caesar, the Helvetians attacked. They still had about 70,000 warriors. Caesar had about 30,000 legionaries, about 20,000 Gallic auxiliaries, and 4,000 Gallic cavalry. Driving the Helvetians back to their camp, the Romans found the enemy's ranks swollen by women and children. In the violent struggle 130,000 Helvetians of all ages and sexes fell; Roman losses were heavy but are not known precisely. The remaining Helvetians submitted, and returned to their homes east of the Jura as demanded by Caesar.

58, August–September. Campaign against Ariovistus. A Germanic tribe, under the chieftain **Ariovistus,** had been terrorizing the Aedui, Sequani, and Arverni (Gallic tribes) in the region later comprising Alsace and Franche-Comté in France. The Gauls asked Caesar for assistance, and though the Romans were even more fearful and respectful of the Germans than of the Gauls, Caesar answered the call. Caesar and Ariovistus maneuvered cautiously in the region east of Vesontio (Besançon). Caesar had about 50,000 men; Ariovistus probably 75,000. Near modern Belfort, Mulhaus, or Cernay, Caesar found a favorable opportunity to attack (September 10). The Germans were completely routed; the remnants fled back across the Rhine, closely pursued by Caesar. Most of central Gaul now acknowledged

Roman supremacy. Caesar went into winter quarters near Vesontio.

57. Campaign against the Belgae. The Belgae, collective name for the tough, Gallic-Germanic people of northeastern Gaul, were alarmed by Caesar's two successful campaigns along the southern fringes of their domains. They formed a coalition against the Romans, planning to march south with about 300,000 warriors. Caesar, learning of this, struck before they were ready. With about 40,000 legionaries and 20,000 Gallic auxiliaries, he invaded Belgica (probably April).

57, April–May. Battle of the Axona (Aisne). Hastily gathering a force of about 75,000–100,000, the Belgae, under **Galba,** King of Suessiones (Soissons), attempted to stop Caesar at the Axona. He defeated them, then pressed farther north into Belgica. A number of tribes submitted. But others, led by the Nervii, prepared for further conflict.

57, July. Battle of the Sabis (Sambre). Advancing with inadequate reconnaissance, Caesar was preparing to make camp on the banks of the Sabis when ambushed by an army of 75,000 Nervii. Fortunately the legions did not panic. The battle was desperate, Caesar going from legion to legion to fight in the front ranks and inspire his hard-pressed men. The Romans beat off the attacks, then seized the initiative. The Nervii suffered about 60,000 dead; Roman loss was also heavy.

57, September. Siege of Aduatuca (Tongres). Continuing into the country of the Aduatuci, Caesar besieged and captured their capital of Aduatuca. The Aduatuci treacherously attacked the Romans as they marched in the town. Caesar repelled the attacks, then overwhelmed the barbarians. Most of Belgica now submitted to the Romans. Caesar took up winter quarters along the line of the Loire, personally returning to Cisalpine Gaul to look after his political interests, as he did almost every winter he was in Gaul.

56. Campaign against the Veneti. During the winter the Veneti, inhabitants of Armorica (Brittany), seized some Roman ambassadors. At the same time scattered outbreaks against the Romans occurred throughout Gaul. Early in the spring Caesar, with three legions, advanced into Armorica, just north of the Loire. Another legion, under **Decimus Brutus,** manned a fleet hastily constructed near the mouth of the river. At the same time **Publius Crassus** (son of Caesar's triumvir partner), with a force slightly larger than a legion, invaded Aquitania (southwestern Gaul), which was becoming hostile to the Romans, while a small force under **Titus Labienus** patrolled the region near the Rhine, and another under **Q. Titurius Sabinus** was in modern Normandy. The campaign against the Veneti progressed slowly; a series of protracted sieges of small fortified towns ensued. The decisive action was a naval battle in Quiberon Bay (or **Gulf of Morbihan**), under the eyes of Caesar and the Roman army. The light Roman galleys had difficulty coping with the powerful sailing vessels of the Veneti, but discovered they could disable the Gallic vessels by slashing the rigging with sickles tied on the ends of long poles. Suppression of the Veneti now proceeded rapidly, Caesar punishing the people severely for having mistreated his ambassadors. Meanwhile, Sabinus and Crassus had been successful against serious opposition in their respective areas of operations.

56, Fall. Campaign against Morini and Menapii. Caesar now moved to suppress dissident tribes in northwestern Belgica. He was frustrated by the Morini and Menapii tribes, most of whom sought refuge in the trackless seacoast marshes of the Low Countries; a few others escaped from the Romans in the wilderness of the Ardennes. Aside from these small areas of Belgica, all of Gaul was now under Roman domination.

55. Campaign against the Germans. During the winter two Germanic tribes (Usipetes and Tencteri) crossed the Rhine into Gaul, establishing themselves on the lower Meuse, near modern Maastricht. Totaling about 430,000 people in all, they had more than 100,000 warriors. Caesar marched to the Meuse and entered into negotiations with the Germans, with a view to persuading them to return to Germany

(May). Discovering that they planned a treacherous attack under cover of the negotiations, Caesar decided to make an example of the invaders, to dissuade further German inroads. Using guile himself, he made a surprise attack during the negotiations, somewhere between the Meuse and Rhine rivers. He annihilated the Germanic armies, then massacred the women and children. There were no survivors. In Rome, Caesar's political enemies professed indignation at this cold-blooded act. Caesar, however, insisted that it had been necessary to prevent further Germanic inroads. He marched to the Rhine.

55, June. Crossing the Rhine. Near the site of modern Bonn, Caesar built a great bridge across the Rhine, in a memorable feat of engineering. He then marched into Germany, to further intimidate the German tribes. After receiving the submission of several tribes, he returned to Gaul, destroying his bridge.

55, August. First Invasion of Britain. With two legions, Caesar landed near **Dubra** (Dover), where he encountered serious opposition from the Britons on the beach; the landing was covered by catapults mounted on the ships. After some hard fighting, followed by a truce, Caesar returned to Gaul, having spent three weeks in Britain.

54, July. Second Invasion of Britain. With 5 legions and 2,000 cavalry (about 22,000 men), Caesar returned in a fleet of 800 small craft. He landed unopposed northeast of Dubra. After the debarkation a severe storm destroyed a number of vessels and damaged others. A large force of Britons quickly gathered under a chieftain named **Cassivellaunus.** Caesar marched inland, sweeping aside harassing Britons, crossing the Thames somewhere west of modern London. Cassivellaunus was unable to face the Romans in a major engagement. His diversionary attacks against the entrenched camp of the Roman fleet repulsed, he asked for peace near Verulamium (St. Albans). Caesar was apparently happy to halt the campaign and return to Gaul after receiving the nominal submission of the Britons.

54–53. Uprisings in Gaul. Unrest was seething throughout Gaul. The warlike tribes, who could probably muster more than 1 million

Caesar's Rhine bridge

warriors, began to realize that they had been conquered by an army that rarely exceeded 50,000 men. The major uprising was led by **Ambiorix** of the Nervii. He seized the opportunity presented by the dispositions of small Roman detachments in 8 winter camps scattered across northern Gaul. Ambiorix first attacked Sabinus, near Aduatuca. He then offered Sabinus a safe-conduct to rejoin the other legions. When the Romans were on the march, they were attacked and annihilated. Ambiorix then attacked the fortified camp of **Quintus Cicero** near modern Binche. Repulsed, he tried to entice Cicero into the open, as he had with Sabinus; Cicero refused, so the siege continued. A messenger from Cicero reached Caesar in north-central Gaul. Gathering the nearest forces, totaling only 7,000 men, Caesar hastened to the rescue. Ambiorix with 60,000 men marched to meet him near the **Sabis,** leaving a strong force still besieging Cicero. Caesar, feigning indecision, enticed Ambiorix to undertake a careless attack, then counter-attacked vigorously, driving the Gauls from the field. This exemplified Caesar's audacity and inspirational qualities. His never-failing luck was also working for him, but it was luck partly created by a restless energy and amazing vitality that always enabled him to seize the initiative from his foes, regardless of odds or circumstances. Marching on, Caesar relieved the hard-pressed forces of Cicero. Meanwhile, Labienus, who had also been under attack, had repelled his assailants and joined Caesar, who now consolidated his forces into more secure winter quarters.

53. Suppression of the Belgae. Caesar, who now had 10 legions, seized the initiative early in the spring, to crush the rebellions. In a systematic campaign, in which there was little fighting but considerable marching and pursuit, Caesar completely subdued Belgica. Since Ambiorix had been assisted by some Germans, Caesar again built a bridge and crossed the Rhine, to repeat his demonstrations in Germany. By the end of the summer the insurrection had been completely suppressed.

53–52. Revolt in Central Gaul. The conquered regions which Caesar considered most secure now revolted, under the leadership of Arverni chieftain **Vercingetorix,** by far the most able of Caesar's Gallic opponents. Raising an army in central Gaul, Vercingetorix trained and disciplined it in a manner hitherto unknown among the barbarians. Most of Caesar's legions were in northern Gaul, while Caesar himself was in Italy. Learning of the uprising, he hastened back to Gaul (January), to arrange for the protection of the Province, his main base in southern Gaul. Then, with a small force, he made a rapid, difficult march through the snowcovered Cevennes Mountains, evading Vercingetorix (February). Having reunited his forces north of the Loire, Caesar recaptured **Cenabum** (Orléans), where the rebellion originated, then turned to fight his way south to the heart of the rebellious territory, leaving Labienus to hold northern Gaul. As Caesar advanced, capturing town after town, Vercingetorix retreated in front of him, fighting a partisan war, destroying all food and supplies useful to the Romans.

52, March. Siege of Avaricum (Bourges). The Gallic defense was determined and skillful. Short of food, the Roman troops were in a serious situation. Vercingetorix, harassing the Romans constantly, tried to relieve the defenders. But Caesar, a master of the art of siegecraft, soon captured the town. He then marched rapidly south, hoping to capture Gergovia (Gergovie, near Puy-de-Dôme), capital of the Arverni, before it could be prepared for defense.

52, April–May. Siege of Gergovia. Vercingetorix, taking advantage of the natural strength of Gergovia, on a high, steep hill, was ready. Plentiful supplies had been gathered in the town, but the nearby country was denuded. The Roman situation became further complicated by the revolt of the remaining tribes of central and northern Gaul, including some who had been faithful Roman allies. Caesar sent for Labienus, before the Gauls could isolate and destroy his detachment. Endeavoring to obtain a quick solution at Gergovia, he tried an assault and was repulsed with heavy losses. Short of food, he now had no choice but to withdraw. He marched northward to meet Labienus, who had meanwhile won a battle at **Lutetia** (Paris), enabling him to march south without interference.

52, June (?). Retreat to the Province. Having united his army south of the Seine, and realizing that all his Gallic conquests were temporarily lost, Caesar now headed back to the Province, his main base, to refit and to get a new start. Furthermore, the seeds of rebellion were beginning to sprout even in that Romanized region of Gaul. Vercingetorix, with an army of 80,000 infantry and 15,000 cavalry, probably the best Gallic army ever assembled, now attempted to intercept and to destroy Caesar in central Gaul. He took a position in the hills along the Vingeanne (a small tributary of the upper Saône) where he could block all possible routes by which Caesar could move from the Seine to the Saône Valley.

52, July (?). Battle of the Vingeanne. Caesar suddenly discovered the army of Vercingetorix in position on his line of march, but was alert and ready. After an indecisive cavalry skirmish, Vercingetorix retired without attempting to bring on a major engagement. Typically, Caesar grasped the initiative. He pursued.

52, July–October (?). Battle and Siege of Alesia. Vercingetorix retired to the strongly fortified mountaintop town of Alesia (Alise-Ste.-Reine, on Mount Auxois, near the source of the Seine River). He had over 90,000 men, Caesar had about 55,000, of whom approximately 40,000 were his legionaries, the remainder being auxiliaries, plus 5,000 faithful German-Gallic cavalry. Attacking vigorously, Caesar drove the Gauls inside the walls of Alesia. He built walls of contravallation and circumvallation, each about 14 miles in circumference. His wisdom in undertaking this formidable engineering feat was soon demonstrated. Responding to messages from Vercingetorix, a tremendous Gallic relief army, more than 240,000 in all, gathered around Alesia,

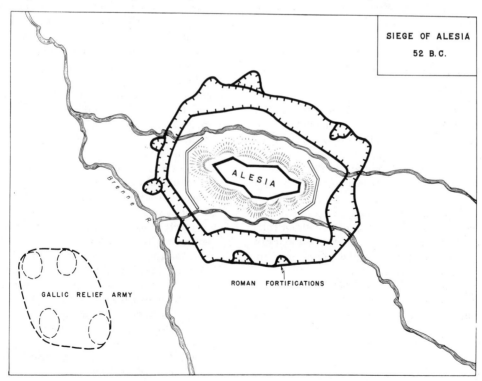

SIEGE OF ALESIA
52 B.C.

ALESIA

Brenne R.

GALLIC RELIEF ARMY

ROMAN FORTIFICATIONS

besieging the besiegers. Caesar had collected large quantities of food and had an assured water supply, so he calmly continued with his siege approaches to Alesia. He repulsed three relief attempts with heavy losses. To delay starvation, Vercingetorix tried to send out the women and children from Alesia, but Caesar refused to let them through his lines. The situation in Alesia was now hopeless. To save his people from further disasters Vercingetorix surrendered. (He was later taken to Rome for Caesar's triumph, then executed.) This defeat broke the back of the Gallic insurrection; most of the Gauls hastened to renew their fealty to Rome.

51. Final Pacification of Gaul. A few remaining embers of insurrection were squelched promptly and effectively. Caesar traversed the entire country, impressing indelibly on Gallic minds the power and glory of Rome and his own skill and force of character. The Gallic Wars were over, and Gaul would be an integral part of the Roman Empire for half a millennium to come.

THE GREAT ROMAN CIVIL WAR, 50–44 B.C.

50. Caesar Ordered to Return to Rome. Meanwhile, Pompey, illegally appointed sole consul by the Senate (52), urged that body to order Caesar to give up his provinces, to disband his army, and to return to Rome, or else be declared a traitor. At this time Caesar, with one legion, was at Ravenna, in Cisalpine Gaul. Under Roman law, a general was forbidden to bring his forces into Italy proper without consent of the Senate. The southern boundary of Cisalpine Gaul on the Adriatic coast was the tiny Rubicon River, south of Ravenna.

49, January 11. Crossing of the Rubicon. Upon receipt of the Senate's order, Caesar immediately marched to the Rubicon. He made a night crossing, uttering his famous "The die is cast!" In addition to the one legion with him, he had 8 legions back in Transalpine Gaul, a total force of perhaps 40,000 first-rate veterans,

plus about 20,000 auxiliaries and cavalry. Available to Pompey and the Senate were two legions in Italy, seven in Spain, eight more being raised in Italy, and at least nominal control of all Roman forces in Asia, Africa, and Greece, totaling perhaps ten or more additional legions, plus an even larger number of auxiliary troops. Caesar hoped to make up for this great discrepancy in forces and resources by the energy which had brought him victory in Gaul. He marched rapidly southward along the Adriatic coast, collecting recruits and reinforcements en route. Pompey and most of the Senate abandoned Rome, seeking safety and adherents in the south. The only unfavorable news reaching Caesar was that his most able and trusted subordinate, Labienus, former lieutenant of Pompey, had defected to his old leader. All Caesar's other legates and all his legions remained steadfastly loyal.

49, January–February. Flight of Pompey. Not trusting the legions in Italy, and having been too slow in raising the forces he had ordered mobilized, Pompey fled to Epirus from Brundisium (Brindisi), taking 25,000 troops and much of the Senate with him. He retained control of the Roman navy. Caesar, in Rome, thereupon consolidated his position in Italy and wrestled with his difficult strategic problem. Lacking control of the sea, he would have to go overland through Illyricum to get at Pompey in Greece. To do this would leave Gaul and his communications through Cisalpine Gaul exposed to the powerful Pompeian force in Spain. Boldly, he counted on Pompey's lethargy in Greece to give him time to dispose of the threat in Spain.

49, March 9. March to Spain. Caesar said, "I set forth to fight an army without a leader, so as later to fight a leader without an army." He left **Marcus Aemilius Lepidus** as prefect of Rome; the remainder of Italy was under **Marcus Antonius** (Mark Antony), Illyria under Gaius Antonius, Cisalpine Gaul under **Licinius Crassus; Gaius Curio** was sent to gain control of Sicily and Africa.

49, March–September. Siege and Battle of Massilia (Marseille). Lucius Domitius

Ahenobarbus, a supporter of Pompey, arrived by sea at Massilia with a small force of troops and persuaded the city to declare for Pompey. Caesar, who had sent most of his army ahead to seize the passes over the Pyrenees, quickly invested Massilia with three legions (April 19). Then he hastened off to Spain, leaving **Gaius Trebonius** to prosecute the siege. He also left Decimus Brutus to raise a naval force to blockade Massilia from the sea. The siege continued throughout Caesar's campaign in Spain and was marked by Brutus' victory over the Pompeian–Massilian naval forces in a battle off the city.

49, June. Arrival in Spain. Caesar's troops meanwhile had seized the Pyrenees passes barely in time to forestall a Pompeian army of about 65,000 men under **L. Afranius** and **M. Petreius.** Frustrated in their plan to block the passes, Afranius and Petreius waited for Caesar at Ilerda (Lerida). Two more Pompeian legions, plus about 45,000 auxiliaries, held the rest of Spain, under **Vibellius Rufus** and **M. Varro.** Caesar advanced into northern Spain with about 37,000 men, confronting the Pompeian force at Ilerda.

49, July–August. Ilerda Campaign. Both sides were anxious to avoid battle; Caesar because of the preponderance of force against him, Afranius and Petreius because of their respect for Caesar's reputation. Gaining and keeping the initiative by maneuvering and skirmishing, Caesar decided to try to capture rather than destroy the Pompeian army. Thus he would not only avoid a major bloodletting among Romans, but also might be able to add recruits to his own army. Discouraged by Caesar's energy, Afranius and Petreius decided to withdraw, only to have their retreat cut off by Caesar's rapid movements. They returned to Ilerda, where Caesar surrounded them and cut off their water supply (July 30).

49, August 2. Surrender at Ilerda. Afranius and Petreius surrendered; their legions were disbanded. Caesar, gaining some recruits, as he hoped, immediately marched south as far as Gades (Cádiz) to impress his authority on Spain. Then, leaving a small force to complete domination of the unruly province, he hastened back to Massilia.

49, August 24. Defeat of Curio in Africa. Caesar's legate, Gaius Curio, had established Caesar's authority in Sicily without trouble. In Africa, however, he was opposed by a Pompeian force under **Attius Varus,** in alliance with **Juba,** King of Numidia. After defeating the allies near **Utica,** Curio was defeated at the **Bagradas River,** thus assuring Pompeian hold over Africa. Curio killed himself rather than surrender.

49, September 6. Surrender of Massilia. When Caesar arrived from Spain, Massilia surrendered, Domitius escaping by sea. Caesar hastened to Rome to discover that his small fleet in the Adriatic had been defeated near **Curicta** (Krk).

49, October (?). Caesar Appointed Dictator. That portion of the Senate remaining in Rome appointed him dictator. From that time on Caesar was a virtual monarch. This was the end of the Roman Republic. Finding Italy calm and peaceful, he prepared to move to Greece after Pompey. Despite Pompey's control of the sea, Caesar decided to risk the crossing over the Adriatic with 12 understrength legions.

48, January 4. Caesar Sails from Brundisium (Brindisi). Using every available vessel, Caesar was able to take with him only 7 legions and a few cavalry, about 25,000 men in all. His fabulous luck stayed with him, and he avoided Pompey's fleet, landing south of Pompey's base at Dyrrhachium (Durazzo). Caesar sent his vessels back to Brundisium to pick up Mark Antony, whom he had left in command of the 20,000 men remaining there. But Pompey's fleet, now alerted, blockaded Antony in Brundisium.

48, January–February. Maneuvering around Dyrrhachium. After Caesar landed, Pompey moved from eastern Epirus to Dyrrhachium, forestalling Caesar's plan to seize the town. By this time Pompey had raised an army of about 100,000 men, including a number of veterans, but the quality of his troops was inferior to Caesar's. Strangely, Pompey made no effort to take advantage of his 4-to-1 numerical superiority by forcing Caesar to battle. Caesar, in fact, seized the initiative in a series of bold but careful maneuvers south of Dyrrhachium.

48, March. Arrival of Antony. Sneaking out of Brundisium, Antony landed north of Dyrrhachium with the remainder of Caesar's army. Pompey moved promptly eastward, intending to defeat Caesar's divided forces in detail. But Caesar was even more prompt. He linked forces with Antony at Tirana, then cut Pompey off from his base at Dyrrhachium by a clever march through the mountains. This did not cause Pompey serious difficulty, since he still controlled the sea and was able to maintain contact with his base on the coast a few miles south of Dyrrhachium.

48, April–July. Siege of Dyrrhachium. Pompey, realizing that the countryside was denuded of food, and having access by sea to plentiful supplies in Dyrrhachium, decided to avoid battle and to let Caesar's army starve. Caesar, however, was able to keep his army fed, and began a bold and amazing investment of an army more than twice the size of his own, building a chain of redoubts and ramparts around Pompey's beachhead. Pompey imme-diately built a similar line of fortifications, while skirmishing was conducted all along the line.

48, July 10. Battle of Dyrrhachium. Pompey, who had been cut off from water and fodder for his horses by Caesar's investment, now mounted simultaneous attacks on both ends of Caesar's wall of contravallation. With his great numerical superiority, and with the assistance of his fleet, he had no trouble breaking through. Faced with apparent disaster, Caesar rallied his men, collected his army, and withdrew successfully into Thessaly. His losses were more than 1,000 killed; Pompey's were much less. Pompey, failing to take advantage of his victory, followed cautiously, leaving a strong garrison in Dyrrhachium.

48, July–August. Caesar, obtaining supplies, regrouped in Thessaly. He now had about 30,000 infantry in his 12 thin legions, plus 1,000 cavalry. Pompey had about 60,000 infantry and 7,000 cavalry. The armies camped on opposite sides of the plain of Pharsalus. Caesar, typically, attempted to entice his enemy to battle. Pompey, typically, sat still.

BATTLE OF PHARSALUS, AUGUST 9, 48 B.C.

Finally Pompey decided to try to overwhelm Caesar's numerically inferior army. He formed a line of battle on the plain between the two opposing camps. Caesar at once advanced to fight. With his left flank resting securely on the steep bank of the Enipeus River, he realized that the chief danger lay on his right, where his cavalry was outnumbered 7 to 1 by Pompey's horsemen. He formed his legions in the customary 3 lines, but held out 6 cohorts—about 2,000 men—to cover his right rear. He extended the intervals between the cohorts of his main body, to match the frontage of Pompey's army, drawn up in normal formation. The 6 cohorts Caesar had held out were posted behind his right flank to support his cavalry (see diagram). The third line of his main body was, as usual, held out as a reserve to the first 2 lines. Caesar took his own post initially with the 6 cohorts—the so-called "Fourth Line"—on the right rear.

His dispositions completed, Caesar ordered his first 2 lines of infantry to attack the motionless army of Pompey. At the moment of impact, Pompey launched his cavalry, supported by archers and slingers, against Caesar's small contingent of horse. These fought stubbornly, but were forced back by weight of numbers. At the decisive moment, Caesar personally led his selected force of 6 cohorts against the flank of Pompey's advancing horsemen. Scattering the surprised cavalry, the small contingent pushed on, slaughtered the archers and slingers, then turned against the left flank of Pompey's main body. Caesar then galloped to join his third line, which he led through the intervals of the first 2 to smash into the front of the Pompeian legions. This

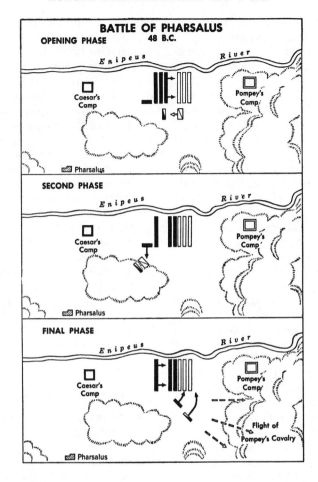

BATTLE OF PHARSALUS
48 B.C.
OPENING PHASE

Enipeus River

Caesar's Camp

Pompey's Camp

Pharsalus

SECOND PHASE

Enipeus River

Caesar's Camp

Pompey's Camp

Pharsalus

FINAL PHASE

Enipeus River

Caesar's Camp

Pompey's Camp

Flight of
Pompey's Cavalry

Pharsalus

charge, combined with the surprise envelopment by the 6 cohorts, broke Pompeian resistance. Pompey and his army fled to their camp. Without pause Caesar followed, stormed the camp, and, without letting his men stop to plunder, pursued the fugitives. Pompey fled in disguise; reaching the coast with only 30 horsemen, he embarked on a vessel and sailed to Egypt. In this decisive battle Caesar lost 230 killed and perhaps 2,000 wounded; Pompey lost 15,000 killed and wounded and 24,000 prisoners. The Roman provinces and protectorates of Greece and Asia immediately declared for Caesar. Only Juba and the Pompeians in Africa continued to defy him, while Egypt and seething Spain remained uncertain.

Operations in Egypt, 48–47 B.C.

48, August–September. Pursuit of Pompey.
Caesar, with only 4,000 men, pursued Pompey to Egypt. At Alexandria he learned that Pompey had been assassinated by his associates (September 28). These former Pompeians, however, succeeded in arousing young **Ptolemy XII,** co-ruler with his sister **Cleopatra,** to defy Caesar. With his handful of men, Caesar was besieged in a corner of

Alexandria by an army of 20,000, led by the young king and his Roman advisers.

48, August–47, January. Siege of Alexandria. Caesar held only a portion of Alexandria and part of the eastern harbor. Disdaining flight, he sent for help while vigorously defending himself. Help was slow to arrive, however, since the former Pompeians still controlled the sea. Nevertheless, land and naval reinforcements trickled in, and Caesar was able to prevent a sea blockade by narrow victories in two desperate naval engagements just outside the harbor. Endeavoring to extend his control over the entire harbor, he was repulsed in a battle on the mole leading out to the harbor entrance. Then his ships were defeated in a third naval engagement nearby.

47, January. With his future bleak, Caesar learned of the arrival at the Nile River of a small army which had been led overland from Asia Minor by his ally, **Mithridates** of Pergamum (not to be confused with either Mithridates of Pontus or Mithridates of Parthia). Leaving a small garrison to hold his positions in Alexandria, Caesar slipped out of the city to join Mithridates. Ptolemy and the Roman-Egyptian army followed. The sizes of the opposing forces are unknown, but each probably totaled about 20,000 men.

47, February. Battle of the Nile. Caesar and Mithridates completely defeated Ptolemy, who was killed. Caesar then relieved his beleaguered forces in Alexandria.

47, February–March. Caesar established complete control over Egypt, placing on the throne with Cleopatra her still younger brother, **Ptolemy XIII.** For nearly two months Caesar lingered in Egypt, engaged in amorous dalliance with Cleopatra.

Pontic Campaign, 47 B.C.

BACKGROUND. Meanwhile in Asia Minor, **Pharnaces,** King of Bosporus Cimmerius (Crimean region), son of Mithridates of Pontus, had taken advantage of the Roman civil war to recreate his father's kingdom of Pontus. He extended his domains along the northern coast of Asia Minor and into Cappadocia, having defeated **Domitius Calvinus,** a subordinate of Caesar, in the **Battle of Nicopolis** (Nikopol) (October 48).

47, April–May. Caesar Leaves for Pontus. Sailing from Alexandria to Syria with a portion of his army, Caesar collected reinforcements from the Roman garrison of Syria, then rapidly marched north through Asia Minor.

47, May. Battle of Zela. Caesar, met in Pontus by Pharnaces, in greater strength, was victorious. He sent to Rome his famous message: **"Veni, vidi, vici."** ("I came, I saw, I conquered.") He then reorganized the eastern dominions, giving to his ally Mithridates of Pergamum nominal rule over the kingdom of Pharnaces.

47, August. Mutiny. Caesar, back in Rome, was soon faced by mutiny of many of his veterans, who felt they should be discharged and rewarded for their efforts. He subdued the mutineers by personal leadership, then enlisted most to accompany him to Africa.

Operations in Africa, 47–46 B.C.

47, October. Invasion of Africa. Having concentrated an army of 25,000 in Sicily, Caesar sailed for Africa. There stood the remnants of the Pompeian forces defeated in Spain and Greece. In command was **Metellus Scipio,** assisted by—among others—Caesar's former lieutenant, Labienus. This army of more than 50,000 men was combined with a Numidian force of nearly equal size under Juba, and was supported by the formidable Pompeian fleet.

47, October–November. Operations around Ruspina. Though Caesar evaded the Pompeian fleet, his own squadron was scattered by a storm as it approached the eastern Tunisian coast, and he landed at Ruspina (Monastir) with a handful of troops. His opponents failed to seize this opportunity. By the time Labienus approached with a Roman–Numidian army of 60,000 men, most of Caesar's scattered units had joined him. Somewhat recklessly, Caesar with 12,000 men let himself be cut off from his base at Ruspina and completely surrounded by Labienus. His former lieutenant, however, was reluctant to press an all-out attack with light troops. Caesar, in a series of intricate maneuvers, followed by an attack against one portion of the encircling line, broke out and made his way back to Ruspina, where he was soon blockaded by the entire army of Scipio and Juba, more than 100,000 strong.

47, December–46, January. Maneuvers around Utica. Having collected all his scattered army, and having received reinforcements, Caesar again marched inland with 40,000 men, seizing the initiative in typical energetic maneuvers between Ruspina, Utica, and Thapsus. He laid siege to Thapsus, inviting an attack.

46, February. Battle of Thapsus. Desertions and illness had reduced the Pompeian–Numidian army to about 60,000. However, they attacked Caesar's positions outside of Thapsus and were utterly defeated. Caesar's casualties were less than 1,000; Scipio and Juba lost more than 10,000 dead; at least as many more were wounded and captured. The remainder scattered, a few fleeing to Spain, where the Pompeian banner had been raised once more by Pompey's young sons. Having subdued Africa, Caesar returned to Rome (May).

Operations in Spain, 46–45 B.C.

46, December. Return to Spain. Caesar sailed to Spain with a small contingent of veterans. Upon arrival there he took command of the forces he had left after the Ilerda campaign. His total strength was about 40,000.

45, January–March. Corduba Campaign. Discovering that young **Gnaeus Pompey,** with 50,000 to 60,000 men, was in the neighborhood of Corduba (Cordova), Caesar marched there and at once commenced his typical maneuvering and skirmishing. Apparently Labienus exercised field command of the Pompeian army, which retreated southward to the support of its fleet on the seacoast, meanwhile

hoping to entice Caesar to battle on favorable terrain.

45, March 17. Battle of Munda. (Exact modern location unknown, probably the modern village of Montilla, north of the Singulis [Aguilar] River.) Finding Pompey and Labienus drawn up for battle on a formidable hill position, Caesar reluctantly decided to attack, impelled by his desire to prevent Pompey's escape, and by the aggressive confidence of his legions. Having halted Caesar's initial uphill attack, Pompey counterattacked and came very close to success. Caesar suppressed panic only by personally rushing into the center of the fight, first with one legion, then with another. The protracted struggle was perhaps the most bitterly contested of Caesar's battles. Finally, under his personal inspiration, his men began to forge ahead up the hill, breaking through the center of the Pompeian line. Resistance then collapsed, and the battle became a massacre, 30,000 Pompeians being killed, among them Labienus. Gnaeus Pompey was captured and executed; his younger brother, **Sextus,** escaped, to join the remnants of the Pompeian fleet, with which he conducted piratical operations for several years. Caesar lost more than 1,000 killed and at least 5,000 wounded.

45, March–July. From Spain to Rome. Caesar again marched through Spain, resubjugating the province, then returned to Rome, where he busied himself with the political, economic, and legal affairs of the Roman state, of which he was now the uncrowned but undisputed monarch.

44, March 15. Assassination of Caesar. Caesar was assassinated in the Senate by a group of conspirators including sincere democrats alarmed by his autocratic despotism, as well as former Pompeians and a number of his own disgruntled adherents. Among these latter were two of his favored subordinates, **Marcus Junius Brutus** and Caesar's distant kinsman, the naval leader, Decimus Brutus. (If he uttered his famous *"Et tu Brute!"* it was probably to the latter.)

COMMENT. *Probably lacking the superlative, balanced military genius of Alexander or Hannibal,*

Caesar was, nonetheless, one of the greatest generals of world history. His energy and audacity have never been excelled, and his charismatic leadership inspired the devotion of his soldiers to a degree matched by few other great leaders. His one serious military weakness was in carrying audacity to the extreme of recklessness, as at Dyrrhachium, Alexandria, and Ruspina. No general has ever been luckier, and this of course was because, to a great extent, he made his own luck by seizing and maintaining the initiative. No other man has ever matched his unique combination of talents: genius as a politician, statesman, lawgiver, and classic author, in addition to being a great captain.

THE STRUGGLE FOR POWER, 44–43 B.C.

44, March–October. Confusion and Apathy. Despite severe denunciations by Mark Antony, the surviving consul and Caesar's most trusted subordinate, no punitive action was taken against the conspirators who had assassinated Caesar. Decimus Brutus went to Cisalpine Gaul to take over the governorship of the province, with Senatorial approval; Marcus Brutus became governor of Macedonia; **Gaius Cassius Longinus,** the ringleader of the conspirators and an experienced soldier, took over the governorship of Syria. These conspirators, with some connivance from friends in the Senate, planned to raise sufficient forces in their provinces to be able to return to Rome to reestablish the Republic. During this period 18-year-old **Gaius Julius Caesar Octavianus** (or **Octavian**), nephew and civil heir of Julius Caesar, appeared on the scene. Mark Antony considered himself Caesar's successor and refused to accept Octavian as Caesar's political heir. Antony, Octavian, and the Republicans maneuvered to gain political power and to raise forces for the violent struggle which was inevitable. Octavian increased his importance by raising a substantial force from among Caesar's veterans, whom the great general had settled in Campania.

44, November. Octavian Takes the Field. Though he despised Decimus Brutus as one of his uncle's assassins, young Octavian had apparently formulated a long-range plan to ally himself temporarily with Brutus in order to eliminate Antony and achieve complete control over Rome. With the force he had raised, he marched north to Cisalpine Gaul to join Brutus. Though he was only a mediocre soldier, he soon proved himself one of the ablest politicians of history.

44, December–43, April. Siege of Mutina. Antony marched north to Cisalpine Gaul and besieged Brutus in Mutina (Modena). The new consuls, **Aulus Hirtius** and **C. Vibius Pansa,** Republicans, came to Cisalpine Gaul to support Brutus and Octavian. Joining forces with Octavian, the two consuls advanced toward Mutina.

43, April 14. Battle of Forum Gallorum. Leaving his brother **Lucius** to continue the siege of Mutina, Antony marched to meet the threat. A few miles east of Mutina he defeated Pansa, who was killed. But, while Antony's men were celebrating their victory, Hirtius came unexpectedly on the scene and routed them. Antony rallied the fugitives and withdrew to Mutina.

43, April 21. Battle of Mutina. Antony had the worst of another battle against Hirtius, though the consul was killed. Antony withdrew westward, crossing the Apennines into Liguria, and then going to the Province of Transalpine Gaul, to join Aemilius Lepidus, one of Caesar's loyal adherents. Soon after this Decimus Brutus was killed by brigands.

43, August. Octavian Seizes Power. Returning to Rome, Octavian forced the Senate to declare him consul. He then made the Senate outlaw Caesar's assassins and acknowledge him as Caesar's heir under the terms of Caesar's will.

43, November. The Second Triumvirate. Octavian marched north again to Cisalpine Gaul with the intention of reaching an accommodation with Antony. At Bononia (Bologna) he reached an agreement with Antony and Lepidus to establish joint rule over the empire and to punish the assassins of Caesar.

WARS OF THE SECOND TRIUMVIRATE, 43–34 B.C.

43–42. Brutus and Cassius in the East. Marcus Brutus and Cassius, establishing control over their provinces of Macedonia and Syria, seized the wealth of subject cities and protectorates to raise funds to support their armies. They met at Sardis (July 42), their combined forces totaling about 80,000 infantry and 20,000 cavalry. They then returned across the Hellespont to Thrace (September).

42, September. Antony and Octavian to Greece. Antony and Octavian, with about 85,000 infantry and 13,000 cavalry, moved to Epirus from Brundisium. When their advance elements, moving northeastward, made contact with the Republicans near Philippi, Antony hastened ahead with most of the army, leaving ailing Octavian to follow more slowly. Throughout this entire campaign Antony was the principal triumvirate leader.

42, October 3. First Battle of Philippi. Discovering the Republicans in two fortified camps west of Philippi, Antony carefully planned a surprise attack through a swamp against the camp of Cassius, south of Brutus'. The attack was successful, but at the same time Brutus unexpectedly advanced against the left wing of the triumvirate army, smashing it, temporarily seizing its camp, and forcing Octavian to flee. Thus the honors and casualties were about even in this strange battle. But Cassius, the most able Republican leader, had committed suicide when he thought the Republican army had been defeated.

42, October 23. Second Battle of Philippi. The relative positions of the opposing armies were unchanged. Antony, hoping to cut the communications of the Republican army, again advanced secretly through the swamp, to envelop Brutus' left flank. At the same time Octavian, with the remainder of the army, attracted Brutus' attention to the front. In the ensuing battle south of Philippi, Antony routed the Republicans. Brutus escaped with about 4 legions, but committed suicide soon after, bringing the war to a close. The triumvirs now returned to their agreed areas of responsibility: Octavian and Lepidus in the west, Antony to the east. In Cilicia Antony met Cleopatra, Queen of Egypt, and followed her to Egypt, to begin one of the most famous and most fateful love affairs of history.

41. Perusian War. Lucius Antonius (brother of Mark Antony), who had become consul, clashed with Octavian. Civil war flared; Lucius was supported by **Fulvia,** wife of Mark Antony. Both were defeated and captured by Octavian at **Perusia** (Perugia). Fulvia soon died.

40–36. Octavian's War with Sextus Pompey. Sextus, younger son of Pompey the Great (see pp. 104 ff.), subjugated Sardinia, Sicily, Corsica, and the Peloponnese, threatening the vital grain route to Rome from Africa. This caused desultory hostilities between him and Octavian.

40. Treaty of Brundisium. Shortly after the Battles of Philippi, the conflict between Octavian and Antony's brother and wife brought the leaders to the brink of war. Allying himself briefly with Sextus Pompey, Antony landed near Brundisium with a small army. Octavian, however, negotiated an agreement whereby Antony agreed to support him against Pompey, in return for assistance in a proposed invasion of Parthia. A revised division of the empire was also agreed upon: Octavian to have Italy, Dalmatia, Sardinia, Spain, and Gaul; Lepidus to retain only Africa; Antony to have everything to the east. The agreement was cemented by Antony's marriage to **Octavia,** sister of Octavian.

38. Uprisings in Gaul and Germany. M. Vipsanius Agrippa, sent by Octavian, successfully put down disorders along the Rhine.

37. Treaty of Tarentum. The agreement of Brundisium (see above) was reaffirmed. Antony loaned 130 warships to Octavian to help against Sextus Pompey; Octavian in turn loaned Antony 1,000 men and promised 4 legions for the invasion of Parthia.

36, June–October. Antony's Invasion of Parthia. With an army of about 60,000 infantry and 10,000 cavalry, Antony moved via the Euphrates into Armenia, past Erzerum and

Mount Ararat through Tabriz into Media Atropatene (Azerbaijan) southward as far as Phraaspa. There he encountered **Phraates IV** of Parthia. In an action reminiscent of Carrhae, Antony lost his siege train and suffered heavy casualties, but repulsed the Parthian attacks and extricated his army—losing 30,000 men in the process.

36, September 3. Battle of Naulochus (or Mylae). Octavian's fleet, commanded by his loyal lieutenant, **M. V. Agrippa,** defeated the fleet of Sextus Pompey, ending the war. In this victory the revolutionary new harpax played a major role (see p. 95). Meanwhile, Lepidus had brought an army to Sicily, ostensibly to support Octavian but actually intending to seize the island himself. His army mutinied and surrendered to Octavian, who kept Lepidus in luxurious captivity in Rome until his death 23 years later.

36. Battle of Sogdiana. (See p. 133.)

34. Border Expeditions of Octavian and Antony. Octavian successfully pacified Dalmatia, Illyricum, and Pannonia. At the same time Antony, to punish the Parthians for repeated invasions of Syria, led another invasion. Though unable to win a clear-cut victory, he succeeded in regaining Armenia.

War of Octavian Against Antony, 33–30 b.c.

33. Rupture between Antony and Octavian. Antony's repudiation of his marriage to Octavia, and his subsequent marriage to Cleopatra—who had already borne him three children—hastened the inevitable break. Octavian aroused the people of Rome and of Italy against Antony and Cleopatra by a vicious propaganda attack against Cleopatra, pictured as seeking to gain dominion over Rome and its empire.

32. Declaration of War. The Senate declared war against Cleopatra and divested Antony of his triumviral title.

32, April–May. Antony and Cleopatra to Greece. Octavian convinced the Romans that this movement of an army and a powerful fleet to Greece was preparatory to an invasion of Italy. It seems clear, however, that Antony's objective was merely to discourage an Octavian invasion of the east. His army consisted of about 73,000 infantry and 12,000 cavalry. The fleet comprised about 480 vessels, with aggregate crews of nearly 150,000 men. The couple kept their fleet and army in winter quarters near Actium (Punta) on the west coast of Greece (32–31).

32–31. Octavian Prepares for Combat. He assembled a powerful army of 80,000 infantry and 12,000 cavalry at Brundisium. His fleet of more than 400 vessels, under Agrippa, was at Tarentum.

31. Octavian Crosses to Greece. Early in the year, while Agrippa demonstrated against Antony's and Cleopatra's supply line along the west coast of Greece, Octavian crossed the Adriatic to Illyricum and Epirus. With most of his army he marched south to seize and fortify a strong position five miles north of Actium. Meanwhile, by seizing islands and key points along the Greek coast, Agrippa had broken the Antonine supply line to Egypt and Asia (June). Antony and Cleopatra apparently decided that their best hope would be a naval battle, though doubtful of the loyalty of a substantial portion of their fleet. Nevertheless, they felt that in the event of defeat they could outsail Agrippa's ships, leaving the army under **P. Crassus Canidus** to try to fight its way overland to Syria and Egypt.

31, September 2. Battle of Actium. The 2 fleets were almost equal in strength, each having more than 400 vessels. Agrippa's fleet was formed in two wings and a center, each about equal in strength. These were opposed by three similar, but smaller elements under Antony; Cleopatra commanded a reserve squadron of more than 60 vessels. Both Agrippa and Antony planned to envelop each other's north flank. The result was an immediate and violent struggle on that flank, in which Antony at first had slightly the better of the engagement. But almost at once the unfaithful center and left wings of the Antonine fleet either fled back into

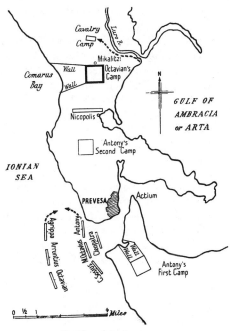

Battle of Actium

Actium or surrendered. Seeing that the situation was hopeless, Antony signaled to Cleopatra to escape, then tried to fight his own way clear to the open sea. Despite the odds, the fight was desperate. Antony's flagship being held by a harpax, he changed to another vessel. With a few of his ships he succeeded in breaking away, joining Cleopatra in flight to Egypt. On land Canidus tried to break out, according to plan, but almost all of his army mutinied and surrendered to Octavian, whose victory was complete and decisive.

30, July. Octavian Invades Egypt. Antony, in the depths of despondency since Actium, aroused himself to repulse Octavian's initial advance on Alexandria. Then, being misinformed that Cleopatra had committed suicide, he killed himself. Cleopatra surrendered, but upon learning that she would be led as a captive through the streets of Rome in Octavian's triumph, she too killed herself, probably by permitting herself to be bitten by a snake. Aged 39, she had proven herself a courageous, resourceful, and able national leader, as well as a profli-

gate, fascinating woman. A measure of her stature is the fact that of all Rome's enemies only Hannibal had ever been more feared and hated in the city on the Tiber.

THE BEGINNINGS OF IMPERIAL ROME AND THE PAX ROMANA, 30–1 B.C.

29. Octavian's Triumphal Return to Rome. Granted the title of imperator which had been held by his uncle, Octavian dominated Rome. For the third time in Roman history the doors of the Temple of Janus were closed; Rome was at peace. Octavian began immediately to build up the imperial system of Rome, being careful, however, like Caesar before him, to retain the forms of republican democracy.

27. Octavian Becomes Augustus. The Senate conferred on Octavian the semigodly name of **Augustus,** by which he is known to history as the first emperor of Rome.

27–1 B.C. Military Reforms. Augustus reduced the army from an overall strength of about 501,000 to about 300,000—25 legions, of 6,000 men each, and an approximately equal number of auxiliaries. Almost all of these were stationed to guard the frontiers of outlying provinces. In addition he created the Praetorian Guard—10 cohorts of 1,000 each—to give himself a private army to control Rome and Italy, without appearing to station regular troops near the capital. The soldiers of this new army were enlisted for terms of 20 years and were promised bounties of land upon discharge. Noncitizens of the *auxilia* were also granted automatic citizenship for themselves and their families upon discharge.

20. Treaty with Parthia. By a policy of mixed conciliation and firmness, Augustus made a favorable treaty with Parthia. Phraates IV recognized Roman sovereignty over Armenia and Osroene (upper Mesopotamia), and also returned the eagles and other standards captured from Crassus and Antony.

20–1 B.C. Operations along the Northern Frontiers. Peace and prosperity settled over the Roman Empire and existed along most of its frontiers. There were a few military operations, however. **Marcus Lollius,** legate on the Rhine, was defeated by a horde of German invaders (16). Augustus himself went to Gaul, while sending his stepsons **Tiberius** and **Drusus** on successful punitive expeditions into Raetia and Pannonia. Drusus, after a hard-fought battle against great odds at the **Battle of the Lupia** (Lippe) **River** (11), continued to push Roman control toward the Elbe, until his death (9). Meanwhile, Tiberius and Drusus had also suppressed another revolt in Pannonia (12–9). Tiberius then completed the methodical advance to the Elbe (9–7).

SOUTHWEST ASIA

PARTHIA AND ARMENIA

188. Independence of Armenia. Following the defeat of Antiochus III by Rome (see p. 96), Armenia, nominally subject to the Seleucids, asserted her independence.

c. 175. Expansion of Parthia. Phraates I conquered the region along the south shores of the Caspian Sea, hitherto nominally owing allegiance to the Seleucid emperors.

c. 170–c. 160. War between Parthia and Bactria. A long and inconclusive war between Mithridates I of Parthia (171–138) and Eucratides of Bactria finally ended with the Parthians gaining control of some border regions of Turania.

166–163. Antiochus IV Reconquers Armenia. (See p. 98.)

163–150. Parthian Expansion. Mithridates I conquered Media. He continued to press west and southwest against the declining Seleucid Empire, as well as eastward against waning Bactrian power.

141–139. Struggle for Babylonia. (See p. 98.)

130–124. Scythian and Seleucid Alliance against Parthia. While **Phraates II** was engaged in the west against the Seleucids, the Tochari tribe of the Scythians mounted a major invasion from the northeast, defeating and killing Phraates in battle. Following this the Scythians overran and devastated most of the Parthian Empire. They defeated and killed **Artabanus I,** successor of Phraates (124).

123–88. Mithridates II Restores the Power of Parthia. In a series of successful campaigns in the east, Mithridates drove the Scythians from Parthian territory. He then turned westward to meet Armenian threats to the northwestern frontiers of the Parthian Empire, defeating **Artavasdes** of Armenia, who was forced to acknowledge Parthian suzerainty (c. 100). Less successful in dealing with Tigranes, successor of Artavasdes, Mithridates made a treaty with Rome against Armenia (92; see p. 102, and below).

95–70. Rise of Armenia under Tigranes. Early in his reign, Tigranes was repulsed from an invasion of Cappadocia (92) by Sulla (see p. 102). However, after the death of Mithridates II of Parthia (88), Tigranes conquered most of Media and northern Mesopotamia and received the submission of the rulers of Atropatene, Gordyene, Adiabene (Assyria), and Osroene. He then invaded Syria, overrunning the vestiges of the Seleucid Empire (83). After the death of Sulla he again invaded Cappadocia and annexed it (78). In the following years, he consolidated his control over his conquests and was generally acknowledged the most powerful ruler of southwest Asia.

89–70. Decline of Parthia. While Tigranes was conquering western Parthia, the Scythians were expanding again into Parthia from the east, and at one time actually dominated the country sufficiently to place a puppet ruler on the throne (77).

70–65. Tigranes Overthrown by Rome. (See p. 104.)

69–60. Initial Contacts between Parthia and Rome. Phraates III (70–57) of Parthia, endeavoring to restore the power and prestige of his empire, concluded an alliance with Lucullus against Tigranes (69). Later he supported **Tigranes the Younger** in a revolt against his father (65). Pompey defeated and captured the younger Tigranes; he refused to permit Parthia to annex the states of Gordyene and Osroene, instead annexing them to Rome. (See p. 105.) Fearful of Roman strength, Phraates grudgingly made peace with Pompey. Phraates was deposed and murdered by his sons, **Mithridates III** and **Orodes I** (57). Orodes then forced Mithridates to flee to Syria.

First War with Rome, 55–38 B.C.

BACKGROUND. **Aulus Gabinius,** Roman governor of Syria, supported the refugee Mithridates against Orodes, providing him with forces to invade Mesopotamia. **Surenas,** the great general of Orodes, met and defeated Mithridates at the **Battle of Seleucia** (55). Mithridates was then besieged in Babylon, where he was captured and killed (54).

54–53. Campaign of Carrhae. The Roman involvement in this campaign gave an excuse to Crassus for a campaign into Parthia (see p. 105). Shortly after crossing the Euphrates, Crassus and his army of 39,000 were surprised near Carrhae (Haran) by Surenas, with a cavalry army of unknown size. The Romans, surrounded in the semidesert plains, promptly formed a square against the Parthian cavalry. But Surenas made no effort to close, contenting himself with firepower—long-range harassment by a hail of arrows. As the bowmen of each contingent depleted their quivers, they were replaced by a new unit, then fell back to replenish their ammunition supply from a camel train loaded with arrows. The Romans, unable to come to grips with their foes, and suffering from the sun and lack of water, sustained great losses. Crassus sent his son, **Publius Crassus,** with a picked force of 6,000 legionaries, cavalry, and auxiliary bowmen, in an attack designed to pin down the elusive tormentors. The Parthian cavalry fell back, enticing the small column away from the main body, then, cutting it off, surrounded and anni-

hilated the entire detachment. Meanwhile, the harassment of the main body continued unabated. At nightfall Crassus withdrew westward, leaving 4,000 wounded to be massacred by the Parthians, but was brought to bay by Surenas again the next day. In negotiations following that fight, Crassus was treacherously killed. In subsequent days the Roman retreat continued and so did the Parthian harassment. Less than 5,000 Romans returned from the disastrous campaign; 10,000 were captured and enslaved by the Parthians; the remainder perished.

COMMENT. *Carrhae had both political and military significance. Politically, the Roman invasion unnecessarily aroused the undying enmity of the Parthians. For the short term it gave Parthia control of Mesopotamia and Armenia. Militarily, it was a cavalry victory over the hitherto invincible infantry of Rome. An example of the superiority of mobility and firepower over shock action, it was the harbinger of eventual dominance of the Middle East by horse archers.*

53–38. Sporadic and Inconclusive Warfare. Several subsequent Parthian invasion attempts into Syria were easily repulsed by the Romans. In the last of these, the able Parthian general **Pacorus** was killed at the **Battle of Gandarus** in northern Syria (38).

36–34. Antony's Invasions of Parthia. (See p. 125.)

32–31. Parthian Reconquest of Atropatene. Taking advantage of the civil war between Antony and Octavian, Phraates IV reconquered the region.

31–26. Dynastic War in Parthia. A revolt by **Tiridates** was put down with difficulty by Phraates IV.

20. Peace between Parthia and Rome. (See p. 127.)

BACTRIA AND THE HELLENIC STATES OF THE EAST

c. 200–c. 175. Bactrian–Hellenic Invasion of India. Euthydemus, now virtually independent of the Seleucid Empire (see p. 98), began to expand his dominions southeastward into Gandhara (northeastern Afghanistan) and the

Punjab. After his death (c. 195), his son **Demetrius** apparently conquered at least the northern half of the Indus Valley, and then began to probe eastward. This was the height of Hellenic-Bactrian power and influence.

c. 175–c. 162. Revolt of Eucratides. A general of Demetrius, Eucratides, took advantage of his master's adventures in India to seize control of Bactria proper. A violent civil war ensued, in which Eucratides conquered most of Gandhara and the western Punjab from Demetrius and his successors, who retained that portion of the Punjab east of the Jhelum River. At the same time Eucratides was engaged in a prolonged struggle with Mithridates of Parthia (see p. 128).

c. 162–c. 150. Continuing Civil Strife. Assassination of Eucratides caused renewal of struggle for control of the Bactrian kingdom. One of Demetrius' descendants, **Menander,** seems to have been victorious, though the descendants of Eucratides retained some lands in the western Punjab and Kabul Valley.

c. 160. Scythian and Parthian Invasions of Bactria. The Scythians (mixed Indo-European and Mongoloid nomads) of central Asia, closely related to the Parthians, now pressed into Bactria from the north. (A major reason for this migration was pressure from the Indo-European Yueh Chih, who had been driven from Mongolia by the Hsiung-nu; see p. 132.) At the same time the Parthians, under Mithridates I, also began to expand into Bactria from the west and northwest. Thus, by the time Menander gained supremacy among the Hellenic Bactrians, their control was limited to southern Bactria, Gandhara, and parts of Arachosia and the Punjab.

c. 150–c. 140. Expansion of Menander's Kingdom. Though unable to stop the progressive erosion of his Bactrian dominions, Menander appears to have been able to expand his control over north India as far as Pataliputra— former capital of the Mauryas. Known to Indian history as **Milinda,** he appears to have had more influence on India than any other Greek, save possibly Alexander himself.

c. 140–40. Decline of the Hellenic Kingdoms. Even during the time of Menander, the Greek kingdoms were breaking up from internal and external pressures. Crushed between the Hindus and the barbarians, they soon disappeared from India (c. 100). Their last strongholds in Gandhara were finally overwhelmed by the Scythians (c. 40).

140–100. Three-Way Struggle over Bactria. The situation in Bactria and neighboring regions of central and southwest Asia was extremely confused. The Scythians, under pressure from the Yueh Chih, tried to press southwestward into Parthia as well as southeastward into Gandhara. Temporarily repulsed by the Greeks from Gandhara, they found themselves engaged in a violent struggle with the Parthians, who were continuing their efforts to expand eastward. At the same time the first waves of the Yueh Chih appeared in northern Bactria.

100–1. Yueh Chih Absorption of Bactria. Gradually, Yueh Chih domination over Bactria checked Parthian expansion efforts, driving the Scythians farther south and southeast into Arachosia, Baluchistan, Gandhara, and the Punjab.

SOUTH ASIA

NORTH INDIA AND THE DECCAN

200–180. Decay of the Maurya Empire. As so often before and since, the decline of a strong regime in north India was an invitation to invasion from the northwest. This time it was begun by the Greek rulers of Bactria, who again brought Hellenic influence to India (see above).

c. 190–50. Confusion in North India. Three new dynasties, two of them short-lived, rose and struggled for supremacy during this era: the Kalinga, Sunga, and Andhra. Most of the petty Greek enclaves were absorbed in the struggle. By the end of the period Andhra had prevailed, only to face a new flux of invasion.

c. 80–c. 40. Scythians in the Punjab. Pushed by Yueh Chih pressure from Central Asia, and pulled by the attractions of India, the Scythians under a leader named **Maues** had meanwhile obtained a foothold in the Indus Valley (c. 80), finally overwhelming the last descendant of Eucratides at Kabul, ending the Greek episode in Southern and Central Asia (40). By the middle of the 1st century B.C., the Scythians (known to Indian history as Sakas) were in complete control of the Punjab. Successive waves of Saka nomads now poured through the Bolan and Khyber Passes, precipitating a sanguinary internecine struggle for the newly won land. These latest arrivals had become thoroughly mixed with the Parthians, and Indian history is unable to make a clear distinction between the Sakas and the Pahlavas (or Parthians). This influx of invaders into the Punjab caused the Saka-Pahlavas to push farther south and east, where they found resurgent Andhra prepared to dispute control of northern India.

50–1. Struggle of Andhra and the Sakas. The invaders were at first successful, but a semi-legendary Andhra king named **Vikramaditya** halted the nomads and drove them back to the Punjab sometime during this period. By the dawn of the Christian era, Andhra had firmly established its supremacy in north-central and central India, while the Sakas retained their grasp over the Indus Valley. To the northwest, however, Yueh Chih pressure on the Sakas was increasing ominously.

26–20. Indian Embassy to Rome. An embassy from India arrived at the court of **Augustus,** apparently from Andhra.

SOUTH INDIA

Though Andhra apparently conducted a number of expeditions into the Tamil regions at the southern tip of the peninsula, they were either unwilling or unable to exert the effort required to overcome the three fiercely independent kingdoms of Chola (Cola) in the southeast, Pandya at India's southern tip, and Cera (Kerala) in the southwest. The fortunes varied among these three Tamil kingdoms, almost constantly at war with one another. A rough balance seems to have been maintained among them for many centuries, each retaining its independent integrity.

CEYLON (SRI LANKA)

c. 200–c. 160. Constant Turmoil.

c. 160–c. 140. Tamil Conquest. The Chola leader **Elara** (Elala) seems to have brought temporary peace to the island through brief conquest. He and his adherents were finally defeated and ejected, however, by the Sinhalese national hero, Prince **Duttha-Gamani** (Dutegemunu) Abhaya, who reigned for 24 years (161–137).

c. 130–44. Anarchy. After the death of Duttha-Gamani, anarchy again pervaded Ceylon, in a welter of dynastic feuds and tribal wars.

44–29. New Tamil Invasion. The invaders from India, at first successful, were ejected by native uprisings (29). Chaos again ruled in Ceylon.

EAST ASIA*

CHINA

202–195. Rise of the Han Dynasty. Defeats at the hands of the Hsiung-nu and the loss of Korea (see p. 88) were not the only difficulties which Liu Pang (or **Kao Tsu;** first emperor of the Han Dynasty encountered at the commencement of his reign. Other outlying regions had slipped away from central control during

* The earliest historical information about Korea and Vietnam begins to emerge in this period. This information is so spotty and incomplete as not to warrant separate mention. See chapter VIII.

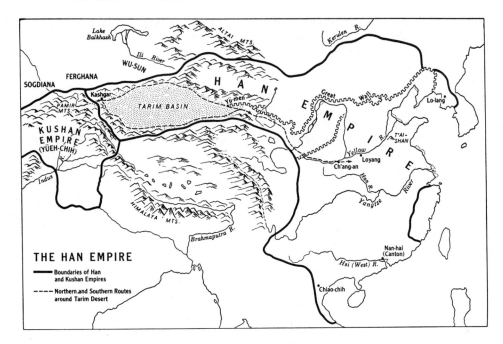

THE HAN EMPIRE

▬▬ Boundaries of Han
and Kushan Empires

▬ ▬ ▬ Northern and Southern Routes
around Tarim Desert

the collapse of the Ch'in Dynasty and in the ensuing civil war. Before his death (195), however, the new emperor established order and centralized control over most of the remainder of the former Ch'in Empire, save in the southeastern coastal region of Yueh (modern Chekiang and south to Tonkin).

196–181. Rise of the Kingdom of Yueh. Chao T'o, former general of Shih Huang Ti, was recognized by Kao Tsu as the autonomous king of Yueh (196). A subsequent imperial invasion was repulsed by Chao T'o (181).

c. 176–166. Supremacy of the Hsiung-nu in Mongolia. The Mongoloid Hsiung-nu people decisively defeated the Indo-European Yueh Chih tribes in the area of modern Kansu and western Mongolia, driving the survivors into Central Asia (see pp. 88, 130). The victors then stepped up their depredations against the border regions of northwest China. One raid reached the vicinity of Loyang, the Han capital (166).

154. Feudal Revolt against the Han Dynasty. Seven feudal princes revolted against Emperor **Ching,** in an effort to overthrow the

dynasty. Suppressing the rebellion with great difficulty, Ching began systematic measures to eradicate feudalism, a policy continued by his successors.

140–87. Expansion of China under Wu Ti. Known to history as "the Martial Emperor," Wu Ti ascended the throne at the age of 16. His reign was one of the great periods of Chinese military glory. The young Emperor considered that the Hsiung-nu constituted the greatest threat to his realms and undertook detailed, long-range plans to end this barbarian menace. He paid particular attention to improving the cavalry forces of his army, and to breeding good horses, so as to excel the frontier peoples at their forte.

138–126. Embassy of Chang Ch'ien. This envoy of the emperor undertook a dangerous mission to Central Asia to seek an alliance with the Yueh Chih against the Hsiung-nu. En route Chang was captured by the Hsiung-nu and imprisoned for several years. Escaping, he reached Central Asia and went as far as Bactria, the new center of Yueh Chih power. Remembering their disastrous defeats at the hands of the Hsiung-nu half a century earlier, and satisfied with their newly conquered lands in Central Asia, the Yueh Chih declined to join any such alliance.

133–119. War against the Hsiung-nu. After long preparations, Wu Ti sent his general **Wei Ch'ing** to invade Hsiung-nu territory between the Great Wall and the northern bend of the Yellow River. At first the Chinese were repulsed, but Wei Ch'ing persisted. He won a great victory over the barbarians and recovered most of the territory lost by Kao Tsu at the end of the previous century.

121. Invasion of Kansu. Wei Ch'ing sent his nephew, **Ho Ch'u Pang,** barely 20 years old, to invade Hsiung-nu territory west of the Yellow River with an army of 100,000 horsemen. In a great battle at **He Si,** Ho defeated the Hsiung-nu, capturing their king and about 40,000 of his warriors. He then consolidated the region, putting it firmly under Han control.

119. Defeat of the Hsiung-nu. Generals Wei and Ho, advancing northward in two main parallel thrusts, drove the Hsiung-nu north of the Gobi Desert. Pursuing, the Chinese advanced as far north as modern Ulaan Bataar (Ulan Bator), where they decisively defeated the Hsiung-nu in the **Battle of Mo Bei,** 70,000 Hsiung-nu reputedly being killed. This broke the back of the Hsiung-nu, who never again threatened China.

117. Death of Ho Ch'u Pang. This brilliant young general was barely 22 years old at the time of his death. One major result of his brief career was the opening up of invasion and trade routes to the West.

111–109. Wu Ti's Conquests in the South. Having consolidated the regions conquered from the Hsiung-nu, Wu Ti next turned his attention southward, to the upstart kingdom of Yueh (which now included Tonkin and Annam). He quickly defeated and reannexed Yueh to the empire. The chieftains of the mountain tribes of Yunnan also acknowledged Chinese sovereignty.

108. Expansion into Manchuria. The kingdom of Ch'ao Hsien—southern Manchuria and northern Korea—was next defeated and annexed.

105–102. Chinese Penetration into Central Asia. Li Kuang Li led a Chinese army into modern Sinkiang and across the mountains into Ferghana, in the Jaxartes Valley. After winning a number of successes against the independent tribes of Central Asia, he was defeated by a coalition of nomadic tribesmen in Ferghana. Withdrawing into Sinkiang, he reorganized and again invaded Ferghana, this time successfully. Ferghana acknowledged Chinese suzerainty, but the cost had been high. Barely half of his army of 60,000 troops reached Ferghana and only 10,000 ever returned to China. Li returned, however, with 3,000 prize horses for breeding purposes.

100–80. Consolidation of the Conquests of Wu Ti. The final years of Wu Ti's reign were devoted to consolidation and to undertaking economic measures to restore the imperial fortunes, which had been impoverished in constant wars.

80 B.C.–1 B.C. Pax Sinica. This was a period of substantial peace and prosperity in China, marked by a number of generally successful expeditions against the Hsiung-nu in Outer Mongolia.

73. Hsiung-nu Invasion of Turkestan. This was repelled by the Chinese and the Indo-European Wu Sun tribe (related to the Yueh Chih peoples) inhabiting the area northwest of the Jaxartes River.

54. Renewed Hsiung-nu Invasion of Turkestan. Again the invaders were repulsed by joint Chinese–Wu Sun efforts.

36. Battle of Sogdiana. A Han expedition into central Asia, west of the Jaxartes River, apparently encountered and defeated a contingent of Roman legionaries. The Romans may have been a part of Antony's army invading Parthia (see p. 126). Sogdiana (modern Bukhara), east of the Oxus River, on the Polytimetus River, was apparently the most easterly penetration ever made by Roman forces into Asia. The margin of Chinese victory appears to have been their crossbows, whose bolts or darts seem easily to have penetrated Roman shields and armor.

THE MILITARY SYSTEM OF THE HAN DYNASTY

The Han Dynasty inherited the government and military institutions of the Ch'in Dynasty. The basis of Han military power was the militiaman. Han law required males between the ages of 23 and 56 to undergo one month of military training each year at provincial training centers. Each man was also required to serve a 1-year tour with the Imperial Guards army in the capital and a 3-year tour at a frontier post. The milita was also called up during local emergencies and for foreign campaigns, such as those of Wu Ti against the Hsiung-nu.

V

THE PAX ROMANA:

A.D. 1–200

MILITARY TRENDS

This, the era of the **Pax Romana,** was perhaps the least eventful period of military history. Not that there was any less strife outside of the relatively calm provinces of the Roman Empire. It was simply that the stability and overwhelming military strength of that empire were too solidly based to be seriously challenged either by external threats or by occasional internal discord. And, with possibly two exceptions, there were no military leaders of exceptional ability who could stand out above the level of their warrior contemporaries in the welter of conflicts across the continent of Asia. Possibly **Pan Ch'ao** of China and the dim figure of **Kanishka,** the Kushan emperor, can be compared with the four great soldier-emperors of Rome in this period: **Tiberius, Trajan, Marcus Aurelius,** and **Septimus Severus.** None of these was a military genius, but all were exceptionally energetic, sound, competent military professionals.

THE PAX ROMANA

The **Pax Romana** is generally considered to have begun in 29 B.C., when Octavian—Augustus—returned to Rome after his victory over Antony. It ended during the reign of Marcus Aurelius; probably the year 162 (beginning of the Eastern War) should be selected as the close of this golden age, even though the energy and skill of Marcus Aurelius were able to protect the interior of the empire from growing external challenges.

In 13 B.C. Augustus reduced the army of the Roman Empire to 25 legions. At this time total Roman armed strength, including auxiliaries and the Praetorian Guard, was probably not over 300,000. Gradually, increasing barbarian pressures along the Rhine and Danube frontiers had raised this to somewhere between 350,000 and 400,000 by the time of Marcus Aurelius. The bases of the superiority of this relatively small army over the millions of the warrior races along the frontiers were the same as they had been for centuries: superior training, discipline, and organization.

Augustus showed no greater wisdom than in the policies he set to provide for retired veterans. He established a permanent fund—the **aerarium militare**—from which assured

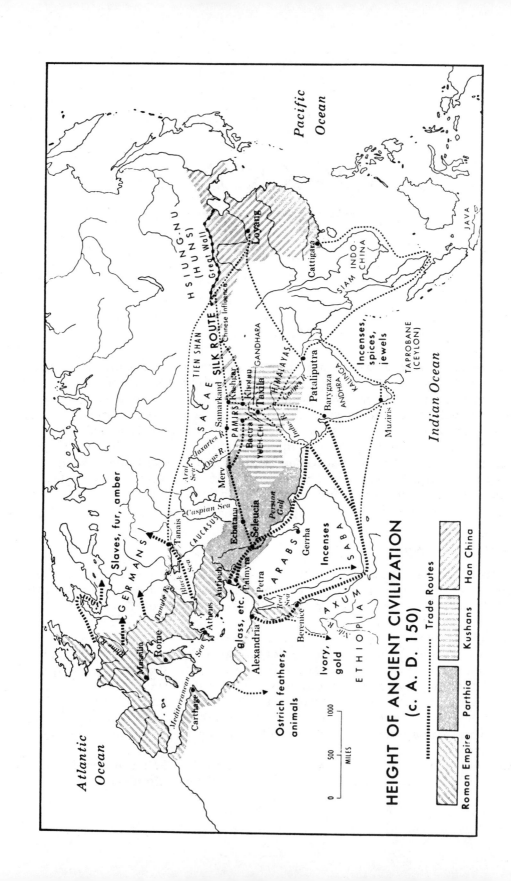

HEIGHT OF ANCIENT CIVILIZATION (c. A.D. 150)

Atlantic Ocean

Pacific Ocean

Indian Ocean

Slaves, fur, amber

Ostrich feathers, animals

Ivory, gold

Incenses

Incenses, spices, jewels

GERMANS

Rome

Massilia

Carthage

Mediterranean Sea

Athens

Danube R.

Black Sea

Tanais

CAUCASUS

Caspian Sea

Aral Sea

Antioch

Palmyra

Petra

Alexandria

Berenice

Red Sea

Nile R.

ETHIOPIA

AXUM

SABA

ARABS

Gerha

Ecbatana

Seleucia

Persian Gulf

glass, etc.

Merv

Oxus R.

Jaxartes R.

SACAE

TIEN SHAN

SILK ROUTE

Chinese Influence

Great Wall

HSIUNG-NU (HUNS)

Loyang

Cattigara

SIAM INDO-CHINA

JAVA

TAPROBANE (CEYLON)

Muziris

BARYGAZA ANDHRA KALINGA

Pataliputra

Ganges R.

Indus R.

HIMALAYAS

GANDHARA

Taxila

Khotan

Kashgar

PAMIRS

Bactra

YUEH-CHI

Samarkand

MILES
0   500   1000

Trade Routes

Han China

Kushans

Parthia

Roman Empire

retirement benefits were to be paid (A.D. 6). In addition, he apparently encouraged retired soldiers to settle in the frontier provinces, although this may have merely been the result of veterans settling down spontaneously near their old garrisons. In any event, as a result, in times of invasion or border squabbles substantial numbers of trained, steady soldiers were available to fight. In time these settlements became virtually a part of the over all military defense system composed of the permanent camps of the legions and the small fortresses, or blockhouses (**castella**), strung along the length of the frontiers. From these settlements grew a sturdy race of frontier colonists—raised in tradition of military service, and loyal to a grateful and benevolent government—from whom more young soldiers could be continually recruited. Finally, this resulted in the development of a flourishing agricultural economy in these formerly barbarian regions; a matter of great importance in view of the steady deterioration of the farming population in southern Italy and other central regions of the empire. Thus the administration of veterans' affairs would become a vital aspect of the economic as well as the colonial and military policies of the empire.

During this period the Roman navy seems to have deteriorated to a police or coast guard status, utilized primarily for the suppression of piracy and smuggling rather than a fleet in being.

The least satisfactory aspect of Roman political and military policy during this period was in the control and status of the Praetorian Guard. There was no basis, constitutional, moral, or physical, for assuring the subserviency of the Guard to the state. They had little to occupy them, and so they were frequently idle, profligate, and vicious, at the same time possessing the means to influence policies and the succession to the throne. After the terrible events of A.D. 69—"the Year of the Four Emperors"—the Praetorians were brought under firm imperial control. A century later, however, the floodgates were again opened by the example of degeneracy given by **Commodus.** Again the Praetorians began to make and break emperors, and thereby ushered in the era of permanent military despotism which began with the reign of Septimus Severus.

WEAPONS, TACTICS, AND TECHNIQUES

There were no startling or significant developments in the design or employment of weapons, either on land or sea. The Romans, with unchallengeable superiority in all aspects of warfare, had no incentive to modify or improve, though they were solicitous in maintaining the elements of their military superiority. In fact, for all practical purposes, the scientific knowledge and technology of the times had not advanced sufficiently in this period to provide a real basis for major changes.

EUROPE–MEDITERRANEAN

THE ROMAN WORLD

Early in his reign Augustus decided to establish the frontiers of Rome along clearly defined, easily defensible barriers. He created border provinces as military outposts and buffer regions to prevent incursions into the heart of the empire by the barbarian tribes of Eurasia. In the northwest and west, the English Channel and Atlantic Ocean simplified the defensive problem, as did the Sahara and Arabian deserts to the south. On the east the Euphrates and Lycus rivers—flowing through rugged mountains—and the Black Sea were natural frontiers, the potential threat from Parthia having diminished due to his diplomacy (see p. 127) and the internal weakness of the nomadic Parthian kingdom. The regions north and east of the Alps, however, posed military problems. These lands were populated by fierce, restless Teutonic tribes, one of which had penetrated into Gaul and Italy itself during the early years of Augustus' reign (see p. 128). Save for the Roman outposts on the Danube in Raetia and Pannonia, there was no clear barrier to barbarian migrations, no buffer region to provide maneuvering room for the legions protecting the frontier. The Rhine, Augustus felt, was too close to the rich province of Gaul; particularly dangerous was the deep salient between the upper Rhine and upper Danube, extending southwestward as far as modern Basel. Before the beginning of the Christian era Augustus had determined to establish the northern frontier of the empire along the Elbe and Danube rivers, and the campaigns of his adopted sons **Drusus** and **Tiberius** in Germany and Raetia (see p. 128) were the initial steps of this project.

A.D. 1–5. Revolts in Germany. Sporadic revolts (which had begun in 1 B.C.) spread throughout Germany, for a time threatening the earlier conquests. Augustus recalled his adopted son Tiberius from the east and sent him to Germany (A.D. 4). In two campaigns Tiberius skillfully suppressed the disorders; the second of these was a combined land and naval operation, with a flotilla operating along the north coast and up the rivers, in coordination with an overland expedition which swept as far as the Elbe (A.D. 5).

6–9. Revolts in Pannonia. Severe uprisings in Pannonia and Illyricum caused Tiberius to hasten to the Danube, leaving the consolidation of Germany to the legate **P. Quintilius Varus.** In a series of workmanlike, unspectacular operations, Tiberius pacified Pannonia. Moesia and upper Pannonia became frontier provinces (6 and 9).

9. Arminius and Varus in Germany. Varus commanded five legions in Germany, plus a number of auxiliaries. One auxiliary unit of the Cherusci tribe was commanded by **Arminius,** a young German chieftain, who had recently served under Tiberius in Pannonia. Varus, with 3 legions and Arminius' auxiliaries, was encamped in a summer garrison in central Germany east of the Visurgis (Weser) River, near modern Minden. As the end of the summer approached, Varus prepared to march back to his winter quarters camp at Aliso (Haltern?) on the Lupia (Lippe) River. Arminius now set in motion well-laid plans for an uprising against Rome. At his command a small insurrection broke out in the region between the Visurgis and Aliso. Varus, warned of the conspiracy by friendly Germans, refused to believe that Arminius was unfaithful; he decided to suppress the uprising during the march back to Aliso.

9, September (October?). Battle of the Teutoberg Forest. With his 3 legions, plus the auxiliaries, Varus had perhaps 20,000 men. The force was accompanied by the families of the soldiers—at least 10,000 noncombatants—and a long baggage train. After crossing the Visurgis, the column entered the difficult, wooded, mountain region of the Teutoberger Wald. Heavy rains contributed to the difficulties the Romans began to encounter from harassing German guerrillas. Suddenly Arminius and his contingent deserted (not far from modern Detmold), annihilating a Roman detachment and creating havoc in the unsuspecting Roman column. Varus, rallying his troops, tried to press on to Aliso. Learning that his base had been invested by hordes of aroused Germans, he apparently decided to try to march northward along the Ems Valley to establish a base at one of the Roman outposts on the North

Sea coast. But his advance was slowed by the encumbrance of his noncombatants, the baggage train, the terrible weather, and the lack of adequate forest trails. Constantly harassed by the barbarians, his men tried to hew new trails through the woods. Finally, after several days of a bitter, running fight, the Germans broke through the legions and began to cut them to pieces. Varus and other surviving officers, all wounded, killed themselves. The few surviving legionaries, and nearly all of the women and children, were massacred. The exact location of the fight is unknown but was somewhere between modern Detmold and Münster. Meanwhile the garrison of Aliso, better handled than Varus' legions, fought its way out to the Rhine. All of central Germany was lost by the Romans.

COMMENT. *The disaster of the Teutoberg Forest had far-reaching consequences. Augustus, grief-stricken, according to legend, during the remaining years of his life frequently burst into tears, crying: "Varus, give me back my legions!" Although he sent punitive expeditions into Germany during these years, the disaster of the Teutoberg Forest caused him to abandon his plans for the conquest and colonization of Germany; he settled for a boundary along the Rhine and the Danube. This decision, accepted by his successors, was of the utmost significance to the future of Europe. For this reason the victory of Arminius is generally considered one of the decisive battles of world history.*

9–13. Tiberius and Germanicus in Germany. Augustus sent Tiberius to avenge the Teutoberg disaster, and the Roman prince did his usual skillful job. Satisfied with this progress, Augustus again sent Tiberius to the east, replacing him with **Germanicus,** young son of Drusus (and adopted son of Tiberius), who led an expedition to the Elbe River (13).

14–37. Reign of Tiberius. Upon the death of Augustus, his selected heir, Tiberius, was invested by the Senate with the powers held by his adopted father. A forbidding and controversial figure, Tiberius was as able an administrator as he was a soldier, and followed the policies of his illustrious predecessor. The empire continued to prosper. Aside from continuing operations in Germany, there were few events of military significance in his reign. A brief revolt in Gaul (21) was promptly suppressed by **Gaius Silius.** There were nationalistic and religious disorders in Judea attendant upon the crucifixion of **Jesus Christ** by Jewish leaders, with the authorization of **Pontius Pilate,** Roman governor (30). Tiberius foiled a conspiracy led by his chief minister and commander of the Praetorian Guard, **Lucius Aelius Sejanus,** who was executed (31). His troops defeated **Artabanus** of Parthia (35–36; see p. 144).

14–16. Operations against Arminius. Germanicus suppressed a mutiny by his Panonian legions, then continued his successful campaigns in Germany. He fought a drawn battle with Arminius in the Teutoberg Forest (15). He finally defeated Arminius east of the Weser near **Minden** (16). The eagles of Varus' legions were recovered. But despite this, and despite the ability of the legions to march across Germany almost at will, there was no further effort to establish colonies or bases in Germany. Germanicus, whose relations with Tiberius became strained, was sent to the east, where he died under suspicious circumstances (19).

17–200. Internal Strife in Germany. The relaxation of Roman control led to widespread tribal warfare in Germany. **Marboduus,** leader of the Marcomanni tribe inhabiting modern Bohemia, began a fierce struggle with Arminius and other German leaders for the control of central Germany. Marboduus was eventually defeated, and forced to seek refuge in the empire (19). Violent internal disorders continued in Germany for several centuries.

21. Assassination of Arminius. This occurred during tribal warfare.

37–54. Reigns of Caligula and Claudius. Gaius Caesar Caligula, youngest son of Germanicus and declared heir of Tiberius, soon became insane. He was assassinated by the Praetorian Guard, who replaced him with his uncle, **Tiberius Claudius Drusus Nero Germanicus,** son of Drusus (41). Roman forces patrolled successfully beyond the frontiers in Germany and Syria; Mauretania was occupied and incorporated into the empire (42), and the conquest of Britain began (see below).

43–60. Conquest of Britain. The invading forces—4 legions and auxiliaries, some 50,000 in all—were under the command of **Aulus Plautius.** The landing was made on the Kentish coast. Claudius personally led reinforcements, including some elephants (44). British chieftain **Caractacus,** of the Catuvellauni tribe, was finally defeated by Plautius (47) and driven into south Wales, from whence he led frequent forays into Roman-held territory to the east. He was defeated at **Caer Caradock** (in modern Shropshire) and captured by **Ostorius Scapula,** governor of Britain, and sent to captivity in Rome (50).

54–68. Reign of Nero. Claudius, poisoned by his wife **Agrippina,** was succeeded by his adopted stepson and Agrippina's son, **Nero Claudius Caesar Drusus Germanicus.** Despite his vicious, cruel, murderous nature, the empire continued to prosper.

56–63. War with Parthia. Vologases of Parthia invaded the Roman protectorate of Armenia, drove out the Roman-supported ruler, and placed his brother, **Tiridates,** on the throne. **Gnaeus Domitius Corbulo** was sent out to command in the east, where years of inaction had lowered the quality of Roman forces. After reorganizing the legions in Syria, Corbulo advanced into Armenia, capturing Artaxata (58). He invaded Mesopotamia, defeated the Parthians, and captured Tigranocerta (59). He then set up a new ruler, **Tigranes,** in Armenia. Vologases thereupon invaded Armenia, and in an inconclusive campaign forced Corbulo to withdraw (61). An armistice was negotiated; both sides agreed to evacuate Armenia. Nero then sent out **L. Caesennius Paetus** to replace Corbulo (who was reduced to command of a garrison of Syria). Paetus was defeated at the **Battle of Rhandeia** (62), by mixed Parthian-Armenian forces under Tiridates. Corbulo, restored to command, invaded Armenia again and defeated Tiridates, who accepted Roman sovereignty over Armenia (63). Parthia now withdrew from the war. Nero, jealous of the popularity and ability of Corbulo,

accused him of treason and forced him to commit suicide (67).

61. Revolt in Britain. Boudicca (Boadicea), queen of the Iceni tribe of Britons in modern Norfolk and Suffolk, led a revolt against the Romans after **Suetonius Paulinus,** Roman governor, conquered the Druid center in Anglesea. After some initial success, she was joined by the **Trinovantes** of modern Essex. Paulinus suppressed the revolt in a victory near modern **Towcester.** Boadicea committed suicide.

68. Overthrow of Nero. Widespread disgust at the excesses of Nero led to a popular uprising. The Senate declared the emperor to be a public enemy. Nero, with some assistance from an eager slave, committed suicide.

66–73. Revolt in Judea. (The Roman garrison was driven from Jerusalem with heavy losses. **Titus Flavius Vespasianus (Vespasian),** who had an excellent military and administrative record in Germany, Britain, and Africa, was sent to reconquer Judea (67). In a deliberate campaign, highlighted by the desperate defense of **Jotapata** by the Jews under historian **Josephus** (68), Vespasian overran the country and laid siege to Jerusalem (early 69). After Vespasian became emperor, the siege was brought to a successful conclusion by his son, **Titus Flavius Sabinus Vespasianus (Titus),** an able soldier. Roman forces consisted of 4 legions and perhaps 25,000 auxiliaries, plus a siege train of 340 catapults. The last Jewish stronghold, **Masada,** was beseiged by the Romans (72–73); when it fell **Eleazar ben Yair** and 900 men, women, and children killed themselves rather than be Roman captives.

69. Year of the Four Emperors. Upon the death of Nero, **Servius Sulpicius Galba,** legate in Spain, was saluted as emperor by his legions, and then recognized by the Praetorian Guards and the Senate. In Germany, however, the legate **Aulus Vitellius** claimed the throne with the backing of his troops, and marched on Rome. Meanwhile, the fickle Praetorian Guards shifted their support to **Marcus Salvus Otho,** who had Galba murdered and seized the throne. Otho then marched north to meet

Vitellius, but was defeated (April) in the **First Battle of Bedriacum** (near Cremona). Otho committed suicide, and Vitellius marched to Rome, where he was recognized as emperor by the Senate. Meanwhile, **Antonius Primus,** legate of Pannonia, joined the legates of Egypt and Syria in nominating Vespasian as emperor. Antonius invaded Italy and defeated Vitellius in the **Second Battle of Bedriacum** (October). Vitellius withdrew to Rome, where he was slain in street fighting (December). The Senate thereupon recognized Vespasian as emperor; he arrived in Rome about four months later.

69–71. Revolt in Batavia. Uprising of Batavian auxiliaries spread to legions and auxiliary units in northeastern Gaul and the Roman areas of Germany, under the leadership of **Claudius Civilis.** It was crushed near modern **Treves** by **Petillius Cerialis** (who, though one of the most popular generals of the army, had refused to take part in the earlier scramble for the throne). Vespasian disbanded 4 disloyal legions.

70–79. Reign of Vespasian. The new emperor was an enlightened and able ruler, who brought peace and prosperity back to the empire. The only important military operations were the gradual expansion of Roman control in Britain (72–84) and in the Agri Decumates region of Germany south of the Main River (73–74). In Britain the earlier conquests were consolidated, and Roman rule expanded into southern Scotland and Wales under the able leadership of **Gnaeus Julius Agricola** (77–84). Britain was peaceful for 3 centuries after his victory over the Caledonians at **Mons Graupius** (Mt. Kathecrankie?; 84).

79–81. Reign of Titus. Save for operations in Britain, there were no significant military events in this short reign. Titus died under suspicious circumstances.

81–96. Reign of Domitian. Titus Flavius Domitianus, younger son of Vespasian, succeeded his brother. He frequently took the field as emperor. To subdue the unruly Chatti tribe, he crossed the Rhine at Mainz and continued the conquest and colonization of the Agri Decumates (83). He began the construction of a line of fortifications (the **Limes**) from the Rhine to the Danube, to protect this frontier region. He led Roman forces in repulsing a Dacian invasion of Moesia (85) under the Dacian King **Decebalus.** In a subsequent campaign north of the Danube in the area of modern Hungary, against the Dacians, Marcomanni and Quadi tribes, Domitian was defeated (89) and was forced to buy a humiliating peace from Decebalus (90). Jealous of the successes of Agricola in Britain, he recalled him, and probably was responsible for his death by poison (85). A revolt of **Antonius Saturninus,** legate of upper Germany (88–89), was crushed, and as a result Domitian established a policy of keeping legions permanently quartered in separate camps, thus losing the military mobility and flexibility envisaged by Augustus in establishing his policy of frontier outpost provinces.

96–98. Reign of Nerva. Marcus Cocceius Nerva, a senator, was elected emperor upon the assassination of Domitian. His reign was military uneventful, save for a threatened military mutiny of the Praetorian Guard and other army units, which he averted by adopting the successful general, **Marcus Ulpius Traianus (Trajan)** as his successor (97).

98–117. Reign of Trajan. The first Roman emperor born outside of Italy (a native of Spain), Trajan was one of the most gifted militarily, and was an able, benevolent ruler.

101–107. Dacian Wars. The first of these (101–102 reestablished clear Roman sovereignty over the arrogant Decebalus, nominally a vassal of Rome. When Decebalus rebelled, Trajan took the field, conquered Dacia, and defeated Decebalus, who was killed (103–107). Rome's frontier was advanced to the Carpathians and the Dniester.

107. Annexation of Arabia Petrea. Forces sent by Trajan rounded out the southeastern frontiers.

113–117. Eastern War. Osroes of Parthia violated the old treaty with Rome by installing a puppet ruler on the throne of Armenia (113). Trajan (then age 60) immediately marched east, repeatedly defeated the Parthians, and overran Armenia and northern Mesopotamia

(114). He then invaded Assyria and southern Mesopotamia, capturing the Parthian capital at Ctesiphon and reaching the Persian Gulf. Roman fleets based on Egypt explored and raided along the Red Sea and Persian Gulf coasts of Arabia. Osroes raised a new army and attacked the Roman forces scattered in the consolidation of Mesopotamia and Assyria; Trajan was cut off in southern Mesopotamia (115). Trajan united his scattered forces and, despite a setback at Hatra, reconquered most of Mesopotamia and Assyria (116). He subdued a Judean revolt (117) and planned to continue the operations in Mesopotamia the following year but, falling seriously ill, he started for Rome, leaving his able kinsman, **Publius Aelius Hadrianus (Hadrian),** in command in Syria and Mesopotamia; he died on the journey.

117–138. Reign of Hadrian. Knowing that he could not simultaneously conduct a major foreign war and establish control of the empire, Hadrian made peace with Parthia, abandoning Trajan's conquests east of the Euphrates, but retaining the nominal vassalage of Armenia. He then returned to Rome (via Dacia, where he suppressed a conspiracy of discontented generals). He spent most of his reign traveling to every corner of the empire, which, on the whole, he ruled wisely and well. On a visit to Britain (122), he supervised the construction of the great northern wall which bears his name. After personally suppressing an insurrection in Mauretania, he hastened to the east, where new war with Parthia threatened; this he averted by a personal meeting with Osroes (123). After other travels, he went to Judea to suppress another revolt under Jewish leader **Bar Kochba** (132–135). Apparently Hadrian assumed personal command of the operations, which culminated in the crushing of the Jewish people and their dispersal throughout the world.

138–161. Reign of Antoninus Pius. The only important military events of this quiet reign were the suppression of a revolt of the Brigantes tribe of modern Yorkshire by **Q. Lollius Urbicus** (142–143) and a very brief, inconclusive border war with Parthia (155). Minor up-

risings in Mauretania (152) and in Egypt (153) were easily suppressed.

161–180. Reign of Marcus Aurelius Antoninus. Last, and perhaps best, of the five "good" emperors, this scholar and philosopher was forced to devote most of his attention to repulsing almost incessant external threats to the empire.

162–165. Eastern War. Vologases III of Parthia invaded Syria and declared sovereignty over Armenia; Marcus Aurelius sent **Lucius Verus** (who was in effect "assistant emperor") to command Roman armies in the east. Verus remained in Antioch while **Avidius Cassius** led a successful counteroffensive against the Parthians, capturing Artaxata, Seleucia, and Ctesiphon, and occupying Armenia and Mesopotamia. The Parthians sued for peace. Troops returning from the campaign were infected with the plague; the result was a violent epidemic (166–167), depopulating vast areas of the empire.

166–179. Wars on the Danube Frontier. The Marcomanni, Langobardi, and Quadi tribes broke across the Danube in Pannonia and Noricum (Austria); some of the Teutons crossed the Alps into Italy to reach Verona, before being repulsed. Marcus Aurelius, accompanied by Verus, after severe fighting forced the Marcomanni to make peace (168). The following year the Marcomanni again burst across the Danube from Bohemia, and 3 years of bitter warfare followed, Marcus Aurelius remaining constantly in the field. He finally defeated the Marcomanni but permitted many of them to stay inside the empire and to settle on lands depopulated by the pestilence. The Quadi and other barbarians were finally crushed by Marcus (174). At about this same time the Sarmatians crossed the lower Danube into Moesia, while further unrest spread along the upper Danube in Germany. Sending subordinates to deal with the Sarmatians, Marcus marched to Germany, and then to Syria to suppress a revolt of the legate Avidus Cassius (175). After a brief respite, Marcus again took the field on the Danube frontier, accompanied by his son, **Lucius Aelius Aurelius Commodus** (179).

Again success crowned the brilliant leadership of the emperor who hated war. Early the following year he became ill and died, probably at Vindobona (Vienna).

180–192. Reign of Commodus. A murderous nightmare, comparable to the terrible days of Caligula and Nero. Commodus was finally assassinated by the joint efforts of his favorite concubine and the leader of the Praetorian Guards. During this time Roman garrisons in Moesia repelled invasions by Scythians and Sarmatians.

193. Struggle for the Throne. Commodus was succeeded by elderly, noble **Publius Helvius Pertinax,** whose austere measures to restore the shattered economy of the empire led to his murder by the pleasure-loving Praetorians, after a reign of 3 months. The Praetorians then offered the throne to the highest bidder, profligate **M. Didius Severus Julianus.** Three rival claimants immediately arose: **D. Clodius Septimus Albinus,** legate of Britain; **Lucius Septimus Severus,** legate of Pannonia; and **C. Pescennius Niger Justus,** legate of Syria. Severus marched quickly to Rome (800 miles in 40 days), where he found that the Senate, in anticipation of his arrival, had executed Julianus. Severus disbanded the Praetorians, establishing a new guard from elements of the frontier armies. Gaining the support of one of his rivals, Albinus, by recognizing him as his successor, he then marched east to deal with the other rival, Niger.

193–211. Reign of Septimus Severus. In three battles—**Cyzicus, Nicaea,** and **Issus** (193–194)—Severus defeated Niger; after Issus, Niger tried to flee but was overtaken and killed outside the walls of Antioch. Severus then devoted himself to pacifying the frontier regions and to restoring order and imperial authority in the east. His only serious opposition was in the defiance of Byzantium, which he besieged, captured, and sacked (196). Returning to Rome, Severus, alleging that Albinus had revolted, marched to Gaul and defeated his rival (197) in the exceptionally bitter and hard-fought **Battle of Lugdunum** (Lyon). Returning to Rome, he established what was, in effect, a military government over the entire empire, ignoring and humiliating the Senate. A mean and petty man in many ways, he was vigorous and energetic, perhaps the most able Roman soldier since Caesar.

195–202. Parthian War. Mesopotamia, under nominal Roman rule since the time of Marcus Aurelius, was invaded by Parthian ruler **Vologases IV.** Severus reconquered Mesopotamia (197), then invaded Parthia in an inconclusive and unremunerative punitive expedition.

208–211. Campaigns in Britain. Revolts in northern Britain caused Severus to take personal command there. After mixed successes, he abandoned the region north of Hadrian's Wall. He died at Eboracum (York).

SOUTHWEST ASIA

PARTHIAN EMPIRE

BACKGROUND. This was a period of constant, violent internal strife in southwest Asia. The nomadic Parthians were never able to establish unchallenged control over the region; they clashed with other nomadic peoples to the north and northeast; they were frequently hard-pressed to retain control over their restless subjects; their royal princes were almost always intriguing or fighting against one another in fierce dynastic struggles. Despite their one great victory over the Romans at Carrhae (53 B.C.; see p. 129), and a few other scattered successes, they realized their military inferiority to Rome, and Parthian rulers generally acknowledged the supremacy of the Roman emperors. However, as a civilized and martial kingdom, they shared with Rome and China the responsibility of defending the heartland of civilization from the barbarians to the north.

2 B.C.–A.D. 35. Dynastic Struggles. Phraates V was succeeded by **Orodes II** (5), who, in turn, was overthrown by **Vonones I** (8), with Roman support. Vonones almost immediately became engaged in a prolonged and disastrous war against **Artabanus II** (11). Vonones fled to Armenia, then to Roman Syria, where he hoped to receive support from Augustus, his original sponsor. But Augustus died (14), and Tiberius refused to become embroiled in the internal Parthian struggle; he sent his heir and nephew, Germanicus, to make a treaty with Artabanus, in which the Parthian ruler recognized Roman sovereignty over Armenia (18). Though his control was shaky, intransigent Artabanus remained in power for nearly twenty years thereafter.

27–43. Revolt of Seleucia. Taking advantage of the internal breakdown of the Parthian Empire, the Hellenic city of Seleucia (on the Tigris) revolted and maintained its independence until overthrown and destroyed by the Parthian emperor **Vardanes I.**

35–37. Invasion of Armenia. Artabanus' challenge to the Roman protectorate over Armenia caused Tiberius to sponsor a rival, **Tiridates III,** for the throne of Parthia, and to send a Roman army under **L. Vitellius** to Armenia. Artabanus was defeated (36), though he later temporarily deposed Tiridates (38–39).

38–42. Struggle of Vardanes and Gotarzes. In a three-way struggle for the throne, Var-danes deposed **Artabanus II** (39), but had more difficulty in dealing with his major rival, **Gotarzes.** Finally victorious (42), Vardanes drove Gotarzes into Hyrcania, then restored some of the lost power and prestige of the empire. He was assassinated, and Gotarzes returned to the throne (47).

51–77. Reign of Vologases I. Vologases was defeated by Corbulo in a war with Rome (see p. 140). However, his brother Tiridates became King of Armenia, as a vassal of Rome. At the outset of this war Vologases suppressed a revolt of his son, **Vardanes II** (54–55). At the same time he was hard-pressed by incursions of Scythians in the east. In Vespasian's war with Vitellius (see p. 140), Vologases offered Vespasian an army of 40,000 Parthian horse archers. Soon after, when the Alanis invaded Media and Armenia, Vologases vainly asked for help from Vespasian.

77–147. Chaos in Parthia. During this period of anarchy the only leader of importance was **Vologases II,** who maintained some control over major portions of the empire against a series of rival claimants (111–147).

148–192. Reign of Vologases III. Early in his reign Vologases reunited the troubled Parthian Empire. He was defeated in a disastrous war with Rome (see p. 142). He was overthrown by the successful revolt (191–192) of his son **Vologases IV.**

195–202. War with Rome. (See p. 143.)

SOUTH ASIA

NORTHERN INDIA

c. A.D. 1–50. Rise of the Kushans. The Kushans, apparently an offshoot of the Yueh Chih people who had conquered central Asia in the preceding century, began to follow the Sakas and Pahlavas they had driven into India. The Kushans soon held most of modern Afghanistan (25) and completed the conquest of the Kabul Valley (50). Soon after this, under the leadership of **Vima Kadphises,** they reached the Punjab.

c. 78–c. 103. Reign of Kanishka. The greatest of the Kushan rulers, **Kanishka,** extended his rule to Pataliputra on the Ganges and southward to include all of Rajputana. He established his capital on the site of modern Peshawar, from whence he vigorously controlled his vast empire, which included Bactria, at least a portion of Parthia, and most of modern Russian Turkestan. Much of his reign was taken up with a series of inconclusive wars with China for the control of central Asia. Dates are vague, but apparently Kanishka, or his subordinate general, **Hsieh,** were the principal opponents of the brilliant Chinese general **Pan Ch'ao** (see p. 146).

c. 103–200. The Successors of Kanishka. The direct successors of Kanishka allowed much of their authority to be usurped by satraps and feudal lords. One of the most renowned of these Saka satraps was **Rudradaman** (c. 150), who ruled much of northeastern India, was successful in subduing rebellious subjects, as well as in punishing wild tribes along the northern mountain frontiers. He also inflicted severe defeats on the Andhra kingdom to the south. The Hindus found it hard to distinguish between the Kushans and their Saka vassals; Kanishka's successors, as well as the surviving Saka chieftains, are all lumped together in Hindu history as "Sakas." The distinctions, in fact, soon disappeared, as both of the invader races rapidly became assimilated.

CENTRAL AND SOUTHERN INDIA

Barely maintaining its domains against the pressure of Sakas and Kushans, the Andhra kingdom rapidly declined in the years following its defeats at the hands of Rudradaman (see above). By the end of the period it had dwindled to its original area in the eastern Deccan.

The decline of Andhra encouraged the northward expansion of the Tamil states of Cera (Kerala) and Chola. The interminable wars between these two states and neighboring Pandya continued, punctuated by occasional overseas adventures against Ceylon.

TAMIL AND HINDU WARFARE, C. A.D. 200

The conflicts of the Dravidian Tamils were generally more ferocious than those of the Aryan Hindus farther north. Hindu warfare was conducted in accordance with formal, elaborate rules—not always adhered to, but nonetheless honored, even in the breach. War in the north was a sport of kings and rarely took the form of a national struggle for existence—save when a new invader from the northwest rudely smashed the rules. Wars were usually limited in objectives and conducted for the most part with far less savagery than elsewhere in the world. Provinces would change hands, but dynasties were rarely overthrown. Defeated foes were treated with chivalrous generosity; rarely did the Hindus indulge in mass slaughter after a victorious battle. This chivalrous, ritualistic conduct of war facilitated conquest by less punctilious invaders. It is interesting to note, however, that the invaders soon adopted the martial customs as well as the civilian culture of India, while Hindu kings and generals who violated the rules were looked upon contemptuously by their fellows.

No such inhibitions bothered the Tamils, whose conflicts were marked by copious bloodshed, violence, and treachery.

CEYLON (SRI LANKA)

The history of Ceylon was marked by endemic internal warfare and by numerous raids and invasions from the neighboring Tamil kingdoms of the mainland.

EAST ASIA*

A.D. 1–23. Rule of Wang Mang. First and only emperor of the Hsin (New) Dynasty, he seized the Chinese throne after first acting as a regent for child emperors of the decadent Han

* For initial entries on Japan, Korea, and Southeast Asia, see chapter VIII (A.D. 600–800).

Dynasty. He invaded and announced the annexation of the Hsiung-nu lands (Mongolia and modern Turkestan). Meanwhile the so-called "Red Eyebrow" rising, a peasant revolt at home, reached such proportions as to sap Wang's military strength, and the invasion of Hsiung-nu territory failed. While he was occupied with successive revolts at home, China lost much of Turkestan. Wang Mang was killed during a revolt.

24–220. Reestablishment of the Han Dynasty. After a brief period of anarchy, **Kuang Wu Ti,** a warrior member of the former imperial family, seized control, restored internal order, and reasserted Chinese authority over most of the border regions.

40–43. Expansion in the South. Kuang Wu Ti sent his general **Ma Yuan** to crush a revolt in Tongking. Ma, an outstanding cavalry leader, also conquered Annam and Hainan.

c. 50–60. Operations against the Hsiung-nu. Renewed border raiding by the Hsiung-nu was met firmly and sternly, the nomads being driven from Kansu into the Altai Mountains and the area that is now northeastern Sinkiang (Chinese Turkestan).

73–102. Central Asian Campaigns of Pan Ch'ao. Soldier member of an exceptional and gifted family, **Pan Ch'ao** (32–102) was one of the great generals of Chinese history. After playing a subordinate role in a successful expedition against the Hsiung-nu, he was sent southwestward, in command of a small army. He conquered the Tarim Basin (eastern Turkestan). He next crossed the Tien Shan Mountains into western Turkestan, defeating the various nomadic tribes of the region between the Hindu Kush Mountains and the Aral Sea, and forcing them to accept Chinese sovereignty. Even the powerful Kushans (see p. 144) were forced to send tribute to China (c. 90). Possibly Pan Ch'ao's conquests extended as far as the Caspian Sea; Chinese reconnaissance elements under his command certainly reached the eastern shores of that sea.

89–91. Smashing the Hsiung-nu. A punitive expedition against the Hsiung-nu was led by **Tou Shien** (apparently a lieutenant of Pan Ch'ao), who overwhelmed the nomads and drove most of them westward. It was the final Chinese victory of this campaign—probably in

the Kirghiz steppes—which apparently set in motion the great Hun migrations that swept into Europe a few centuries later. Replacing those Hsiung-nu who had departed (some remained in the northwestern fringes of Mongolia), the Mongol tribe of Hsien Pi moved into the desert-mountain region north and northwest of Kansu. In a few years (by 101), the Hsien Pi were raiding the Chinese frontiers just as their predecessors had done.

94–97. Expedition to Persia. Kan Ying, another lieutenant of Pan Ch'ao, led an expedition into Persia, reaching the Persian Gulf. This was apparently a form of ambassadorial mission rather than a raid or invasion; it would never have been possible, however, had the Parthians not been fearful and respectful of Chinese power.

100–200. Decline of the Han Dynasty. All of the areas north and west of Kansu gradually fell away from Chinese control. The Hsien Pi, the revitalized Hsiung-nu, and the Ch'iang people of Tibet all proved particularly troublesome. The process was halted only temporarily by the victories of Chinese general **Chao Chung** over the Ch'iangs (141–144).

190–200. China under the Control of Military Dictators. Control of the empire became centered in the hands of powerful general **Tung Cho,** ruling in the name of a puppet Han emperor. The assassination of Tung Cho (192), however, was followed by a breakdown of central authority; regional warlords became practically autonomous. A civil war broke out between two rival generals, **Ts'ao Ts'ao** and **Liu Pei** (194). Ts'ao Ts'ao was successful, and became the new dictator of the empire (196). Though a weak Han emperor was still nominally on the throne at the close of the century, the dynasty was on the verge of collapse. The Hans, however, had brought a cultural and territorial unity to China which has persisted; to this day the Chinese refer to themselves as "sons of Han."

VI

THE DECLINE OF ROME AND THE RISE OF CAVALRY:

200–400

MILITARY TRENDS

In the Battle of Adrianople (378) we see the two most important trends of this period: (1) the impending collapse of history's greatest empire at the hands of Germanic barbarians, and (2) one of the momentous tactical revolutions of military history—the eclipse of the infantry by the cavalry. As in the era immediately preceding, there were no really great captains of the first rank, though there was perhaps a slightly higher overall standard of professional competence— at least among the Roman leaders. The Roman emperors **Claudius II, Aurelian Probus,** and **Julian** were all capable, but all died before they had a chance to prove themselves great captains. Other generals of exceptional ability were the Romans **Carus, Constantine,** and **Theodosius;** the Romanized barbarian **Stilicho;** the Arab **Odenathus;** the Chinese **Ssu Ma Yen;** the Persians **Ardashir** and **Shapur I;** and possibly the Indian **Samudragupta.**

THE TEUTONIC BARBARIANS

The uncontrolled fury and bravery of the German tribes during most of this period could not compete successfully with Roman skill, discipline, and organization. Not until the middle of the 3rd century did they become a threat, when internal Roman disorders almost gave them the empire by default. But from that time on the barbarians became increasingly Romanized and were able to compete with the Romans with increasing confidence and some success. Even so, and despite their victory at Adrianople, the Teutons could not take cities from the Roman legions nor could they eject the Romans from imperial territory—save only in Dacia and (later) Britain, where Roman problems were more administrative than military. The Germans did not overthrow the Roman Empire; they merely inherited it—or its western portion—when the old Roman virtues finally died out.

There was no common pattern of barbarian military methods or tactics. The **Franks** of

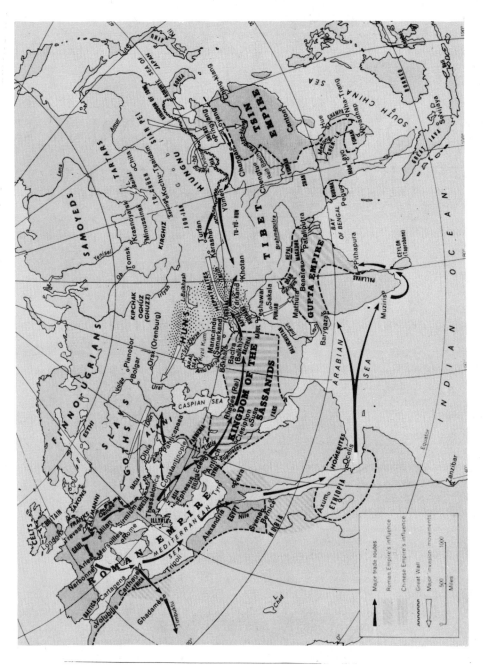

The ancient world, c. 400

northwestern Germany were essentially foot soldiers, though they did make some use of cavalry. The **Alemanni** of southern Germany, like the **Quadi** farther east, were primarily horsemen, but they mixed light infantry in with their mounted formations. The **Sarmatians** were also horsemen, more on the Asian model; apparently they were a mixed Scythian-Germanic people, closely related to the Turklike **Alans** farther east.

The **Goths** seem also to have been of mixed Scythian and German stock, though less Asian than the Sarmatians. They were divided into two main groupings: the **Ostrogoths,** or East Goths, of the Dnieper-Don steppes, were primarily horsemen; the military formations of the **Visigoths,** or West Goths, of the Carpathian-Transylvania region, were mixed horse and foot, with primary reliance on infantry. Like their cousins the **Heruli,** the Goths also became a seafaring people, and their most destructive raids into the Roman Empire were by water across the Black and Aegean seas. The **Saxons** of north Germany were also sea rovers, who frequently raided the coasts of Britain and Gaul.

The most interesting military innovation of the Germans was probably the Gothic wagon fort. When migrating, or when on campaigns, they moved in great wagon convoys. Every night they assembled the wagons in circular laagers, creating crude but effective forts. Not only were these useful for defense in hostile territory; they also provided ready-made bases for forage and plunder. The Germans were even known to bring their wagons together while on the march to create moving forts. Whether this was an original idea, or was inherited from Asia, or was an adaptation of the Roman system of castrametation is not clear.

As time went on, the barbarians played an ever-increasing role in the Roman army. **Constantine** made more use of barbarians in positions of authority and responsibility than had his predecessors. This trend continued in the reign of **Julian,** and reached its climax—so far as the unified empire was concerned—under **Theodosius,** when the barbarians were clearly the predominant element of the army, and most imperial military leaders were also barbarian in origin.

THE RISE OF CAVALRY

Horsemen had long dominated warfare in Asia, save only in India, but Asian cavalry had never been able consistently to defeat the disciplined infantry of Greece, Macedon, and Rome. However, in this era a number of factors brought cavalry to a position of gradually increasing importance in Europe until, at Adrianople, it became suddenly evident that horsemen were supreme in war in Europe as well as in Asia.

The Romans, never having had the balance of arms that had brought success to Alexander of Macedon, found that they needed greater mobility, speed, and maneuverability when operating over the great distances and flat spaces of eastern deserts and East European steppes. Hence, they sought the capability to match and offset the principal and most effective methods of cavalry warfare of their Parthian, Persian, and Teutonic enemies. At the same time, the increased use of missile weapons created a tendency to extend and thin out the formations of the infantry, making them more vulnerable to cavalry charges, while at the same time reducing the occasions on which decisions were reached in hand-to-hand infantry fighting. Also, there was the natural human desire to win wars and campaigns by movement and maneuver, avoiding the risks and losses of pitched battle as much as possible. The Parthians had given the

Romans a lesson in this kind of warfare by their tactics at Carrhae. Finally, the slow but perceptible weakening of Roman discipline made it more difficult for legionaries to stand up against the terror of a cavalry charge. Adrianople, then, was the culmination of a long, gradual evolution which had begun at least as early as Carrhae.

A great impetus to the employment of cavalry, particularly for shock action, came through Asian developments. First, the invention of the saddle, with stirrups, gave to the horse soldier a firm base from which a stout lance could brutally apply the force resulting from the speed of the horse multiplied by the weight of horse and rider. Second, in Persia and on the steppes of Central Asia new breeds of heavy horses appeared, particularly suitable for such shock action. These were soon adopted by the Romans, who—like the Persians—covered man and horse with coats of chain mail to make them relatively invulnerable to small missiles and light hand weapons.

The Romans and their enemies discovered that these heavy lancers did not displace the light and heavy archers which the Parthian, Chinese, and Central Asian peoples had long used so effectively. These two major types of horsemen complemented one another: the horse archers preparing a foe for the charge of the lancers, while the threat of the lancers forced an enemy to remain in close order, thus becoming most vulnerable to the archers. Interestingly, in the Arabian and Nubian deserts, the Romans showed their adaptability by using light cavalry on the Arabian model, as well as a camel corps.

Rightly or wrongly, so far as the Romans were concerned, the lesson of Adrianople meant that the legion was finished as an offensive instrument. It was replaced by heavy cavalry—horse archers (cataphracts) and lancers—as the main reliance of the army. The heavy infantry was relegated to a purely secondary, passive, defensive role, in which it provided a base for maneuver by cavalry and light infantry. The phalanx had returned again, with all its vulnerability. The Romans, recognizing this vulnerability, generally kept the new phalangeal legion stationary.

Military Theory

During the latter part of this period, a shadowy aristocrat named **Flavius Vegetius Renatus** prepared a compilation of Roman military theory in a book called *De Re Militari (On Military Affairs,* commonly known as *The Military Institutions of the Romans).* Vegetius evidently wrote his work during the reign of **Valentinian II,** after the Battle of Adrianople, between 383 and 392.

Vegetius was not a great military leader; his work reveals that he had little practical military experience. But he was a student of military history, and he believed that Rome's past greatness could be restored by reviving the tactical and operational principles of the early Romans. His proposals for a return to the organization and doctrines of the early legion were impractical in the political, social, and cultural environment of the late 4th century, and were based upon an inadequate understanding of the effects of improved missile weapons and particularly of improvements in the armament and equipment of cavalry.

Vegetius' writings, therefore, did not greatly influence the practical Roman and East Roman soldiers of his own times and of the centuries immediately following. During the Middle Ages, however, his book was the principal reference work of military men, partly because it was the only compendium of sound military concepts available to educated soldiers of feudal Western

Europe, and partly because his discussions of the developing cavalry tactics of his time were easily applicable to the crude military concepts of medieval Europeans. Vegetius became even more useful almost a millennium after he had written, when the combination of the crossbow (and longbow) plus heavier armor for cavalrymen reduced cavalry's shock capability, and led to the revival of the infantry. By the fourteenth century, *De Re Militare* had truly become the military bible of the Western world.

OTHER DEVELOPMENTS

Weapons

Weapons remained essentially unchanged. The Romans did refine and lighten ballistae and catapults and used them in increasing numbers in their formations.

Fortifications and Siegecraft

There were no major innovations. Only the Chinese and the Romans had a real facility for siegecraft; while both made some refinements over methods of previous centuries, nothing startling appeared.

Naval Warfare

Here, too, there was nothing really new. In addition to river patrols, the Romans had standing fleets in the Adriatic and Tyrrhenian seas and on the northern and western coasts of Gaul. But their functions were to prevent depredations by pirates, not to dispute command of the sea with a foreign power. The only important use of sea power in this period was in the campaign of **Constantine** and **Licinius,** in combination with their armies, during the struggle for control of the straits between Europe and Asia Minor. Like the Romans, the Chinese also employed naval power on the large rivers and maintained fleets to protect their coasts. Apparently they did use sea power to further overseas trade to some extent, but little is known of this.

EUROPE–MEDITERRANEAN AREA

ROMAN EMPIRE, 200–235

193–211. Reign of Septimus Severus. (see p. 143.)

211–217. Reign of Caracalla (Bassianus Marcus Aurelius Antonius, son of Severus). He and his brother **Geta** inherited the empire; Caracalla then murdered his brother, to become sole Augustus. Personally a vicious monster, he was a generally successful military leader and popular with the troops because he greatly increased their pay. He repelled an invasion of the Alemanni in southern Germany (213); the following year he drove invading Goths back across the lower Danube. He reconquered Armenia, Osrhoene, and Mesopotamia from Parthia (216). While preparing to invade Parthia he was murdered at Carrhae by his officers, under the leadership of **Marcus Opelius Macrinus,** who became his successor (217).

217–218. Reign of Macrinus. He soon lost to Parthia most of the territory gained by Caracalla. An insurrection against his weak rule was led by Caracalla's young cousin **Varius Avitus,** who defeated Macrinus at the **Battle of Antioch** (218). Macrinus fled and was soon slain.

218–222. Reign of Elagabalus. Varius, recognized as emperor by the army and the Senate, assumed the name of **Elagabalus,** the Syrian sun god. A worthless ruler, he was murdered by the Praetorian Guards, who elevated his cousin **Alexander Severus** to the throne (222).

222–235. Reign of Alexander Severus. Potentially a good soldier and ruler, Alexander's greatest mistake was to reject the advice and support of the able Roman soldier-historian **Dio Cassius** (229). Alexander personally led a Roman army into Mesopotamia and Armenia (231–233) to repel an invasion by **Ardashir I,** founder of the new Sassanid Empire of Persia. Renewed Alemanni incursions across the upper Rhine brought Alexander to Gaul and Germany, where he combined a successful campaign with astute diplomacy. During this campaign he was murdered by a group of military conspirators, led by fierce tribune **C. Julius Verus Maximinus,** a gigantic, untutored Thracian peasant.

CHAOS IN THE EMPIRE, 235–268

This was a period of fierce but tiresome strife between numerous rival claimants to the title of "Augustus." That the empire did not collapse is attributable to the fact that residual Roman military qualities—even though terribly debased and abused—remained superior to those of more vigorous, but less skillful, neighbors. The empire suffered severely during this period. As Gibbon wrote:

> The discipline of the legions, which alone, after the extinction of every other virtue, had propped the greatness of the state, was corrupted by the ambition, or relaxed by the weakness, of the emperors. The strength of the frontiers, which had always consisted in arms rather than in fortifications, was insensibly undermined; and the fairest provinces were left exposed to the rapaciousness or ambition of the barbarians, who soon discovered the decline of the Roman empire.

Internal Turmoil

235–244. Maximinus and the Gordiani. The troops in Germany elected the crude Maximinus emperor. The Senate nominated two of its own members—**Maximus** and **Balbinus.** In Africa the legions proclaimed 80-year-old **Marcus Antonius Gordianus I** as emperor (238). A few weeks later his son, **Marcus A. Gordianus II** was defeated and killed in the **Battle of Carthage** by adherents of Maximinus, and the aged imperial aspirant committed suicide. His troops immediately proclaimed his grandson, **M. A. Gordianus III,** emperor. Maximinus meanwhile was killed by his own troops during the siege of Aquileia; the two Senatorial emperors were soon killed by the Praetorian Guards, and Gordianus became sole emperor (238). While engaged in war with the Persians (see p. 174), he was murdered by one of his officers, **M. Julius Philippus Arabus,** who seized the throne (244).

244–268. Philip, Decius, Gallus, Aemilianus, Valerian, and Gallienus. A period of incessant internal strife, intrigue, and murder as general displaced general as emperor. During the reign of Gallienus the competition became so widespread that it was called the "Age of Thirty Tyrants," though Gibbon comments that there were really only 19 (259–268).

259–274. Separation of Gaul from the Empire. M. Cassianus Postumus (one of the "Thirty Tyrants") first established control over Gaul, then over Britain and Spain. Momentarily attaining greater stability than the parent state, his provincial empire soon collapsed (274; see p. 155).

Wars with Persia; the Rise of Palmyra

241–244. First Persian Invasion. (See p. 174.)

258–261. Shapur and Valerian. At **Edessa** the Persian king defeated and captured the Roman emperor, then plundered Syria, Cilicia, and Cappadocia (see p. 174).

259–261. The Rise of Odaenathus of Palmyra. Septimus Odaenathus, prince of Palmyra, was a Romanized Arab. Apparently he preferred to accept Roman authority rather than Persian. He may have tried to obtain Shapur's good will after the capture of Valerian; either his efforts were rebuffed or he was merely gaining time while raising a new Roman-Arab army to dispute Shapur's control of the Roman dominions of the East. The threat of Odaenathus' small army seems to have caused Shapur to withdraw eastward from Cappadocia (261). West of the **Euphrates River,** Odaenathus and his small army surprised and routed the Persians, who were carrying great quantities of booty from Antioch and Asia Minor. Abandoning most of their loot, the Persians fled across the river, harassed by Odaenathus' light cavalry.

262. Odaenathus and Gallienus. Odaenathus then attacked, defeated, and executed **Quietus,** one of the "Thirty Tyrants," or usurpers, who was endeavoring to establish his rule in Syria. As a reward the then reigning emperor, Gallienus, appointed Odaenathus as virtual coruler in the East, conferring on him the title of "Dux Orientis."

262–264. Odaenathus Invades Persia. Having been substantially reinforced by Gallienus, Odaenathus invaded the lost Roman provinces east of the Euphrates with a small army composed mainly of light foot archers, heavy cataphracts and lancers, and irregular light Arabian cavalry. He drove off a Persian army investing Edessa, and recaptured Nisibis and Carrhae (262). In the two following years he harassed Armenia and raided deep into Mesopotamia,

consistently defeating Shapur and his lieutenants, and twice capturing Ctesiphon, the Sassanid capital. Apparently Odaenathus was accompanied and assisted on his campaigns by his beautiful and able wife, **Zenobia.** Shapur sued for peace (264).

266. Odaenathus against the Goths. Odaenathus undertook a successful punitive expedition against the Goths, then ravaging Asia Minor. Soon after this he was murdered, and was succeeded as Prince of Palmyra by his son, **Vaballathus.** However, the virtual ruler of Palmyra—and thus of Rome's eastern dominions—was **Zenobia,** Odaenathus' widow.

267. Independence of Zenobia. Not sure of Zenobia's loyalty, Gallienus sent an army to the East to reassert imperial control. Zenobia, assisted by her general, **Zobdas,** defeated the Romans, then confirmed her virtual independence by conquering Egypt.

Incursions of the Barbarians

The Germanic barbarians of central and southeastern Europe were not long in taking advantage of the internal wars of the Augusti. Prominent among the tribes taking part in the almost incessant raids across the frontiers were the Quadi, Sarmatians, Heruli, and (mostly by sea) Saxons. Three Germanic peoples, however, played the greatest roles in weakening the foundations of the empire at this time: Franks, Alemanni, and Goths.

236–258. Depredations of the Franks. Their raiding parties crossed the lower Rhine and ranged almost at will over Gaul; some penetrated the Pyrenees and plundered Spain, where many seized vessels, sailed across the Mediterranean, and continued their destructive activities in Mauretania. **Valerian** was defeated in Gaul in an unsuccessful attempt to subdue and punish the Franks (256). After the revolt of **Postumus** (see p. 152), who had established a Gallic empire (259–274), the Franks were limited to relatively minor border incursions (259–268).

236–268. Incursions of the Alemanni. Like the Franks to their north, the Alemanni penetrated Gaul with little difficulty. They also frequently crossed the Alps to raid Italy. On one of these raids they defeated Valerian north of the Po, overran Milan, and reached Ravenna (257). For years the interior cities of the empire had allowed their walls to decay; now old walls were hastily repaired and new walls built in northern Italy, Gaul, and Illyricum.

238–268. Raids of the Goths. Most devastating of all were the Goths, who began a reign of terror in Moesia and Thrace (238). However, when **Gaius Trajanus Decius** was legate of Dacia, he won a number of notable successes against the barbarians on both sides of the Danube (245–249).

250–252. First Gothic War. Taking advantage of Roman internal troubles, **Cuiva,** King of the Goths, crossed the Danube in great force and defeated a Roman army at the **Battle of Philippopolis** (250). He continued on, penetrating as far as northern Greece. Decius, now emperor, marched against the invaders, and in two campaigns drove them back to the marshes south of the Danube mouth.

251. Battle of Forum Terebronii. Backed into a corner, the barbarians fought desperately. Early in the battle a son of Decius was killed, when one of the emperor's generals— **C. V. Tribonianus Gallus**—failed to push home an attack that could have assured Roman victory. The Goths counterattacked. Most of the Roman army—save for Gallus' legions—was shattered; Decius was killed while trying to rally his troops. Gallus, became emperor, then concluded a shameful peace, permitting the Goths to keep their booty and to withdraw peacefully across the Danube, while at the same time promising an annual tribute in return for Gothic agreement not to repeat the invasion (252).

252–268. Aemilianus and Claudius on the Danube. Although the Goths almost immediately broke this promise, they were promptly and decisively defeated by **Aemilianus,** the new legate of Moesia (252). When he left to seize the throne (see p. 152), his successor,

Marcus Aurelius Claudius, retained command on the Danube frontier under emperors Valerian and Gallienus. His energy and skill discouraged the Goths from further attempts to cross the river.

253–268. Raids of Gothic Sea Rovers. The Goths now took to the Black Sea as an easy route to reach the defenseless coasts of Moesia, Thrace, and northern Asia Minor. Valerian endeavored to halt these depredations, but had generally the worst of several inconclusive encounters along the Black Sea littoral (257–258). The Goths—joined by the kindred Heruli—expanded the scope of their raids and penetrated still farther inland. Capturing bases along the coast, they ranged over Asia Minor, Caucasia, and Georgia, then forced their way through the Bosporus and Hellespont to reach the Aegean. They sacked the Ionian city of Ephesus, destroying the famed Temple of Diana, one of the Seven Wonders of the World (262). Pressing on to Greece, they captured and sacked Athens, Corinth, Sparta, Argos, and other cities (265–267). The only serious checks they encountered were at the hands of Odaenathus (see p. 153) and in Greece, where Athenian general-historian **Publius Herennius Dexippus** drove them northward from central Greece (267). By the end of the reign of Gallienus, the Goths and Heruli were in effective control of most of the Aegean area, save for Greece (268).

REVIVAL UNDER THE ILLYRIAN EMPERORS, 268–305

268–270. Reign of Claudius II. Having loyally and effectively held the Danube frontier under three emperors, M. A. Claudius now came to the Roman throne by unanimous choice of army and people. He secured his position by defeating and executing his rival, **Aureolis** of Milan (268).

268. Battle of Lacus Benacus (Lake Garda). Almost immediately Claudius was faced with a new Alemanni incursion of Italy. This he apparently routed in the Alpine foothills, by the shores of Lake Garda. He then marched to

Thrace to deal with the Gothic threat. Upon the approach of the emperor's army, the barbarians attempted to evade him and to march to Italy.

269. Battle of Naissus (Nish). Claudius pursued and decisively defeated the Goths in a major battle in the Morava Valley of the Balkan Mountains. He then marched south to capture and destroy the main Goth fleet at Thessalonika. These victories brought relative peace to southeastern Europe for the first time in three decades, and won for Claudius the appellation of "Gothicus." He died in an outbreak of the plague (270).

270–275. Reign of Aurelian. Another able Illyrian general and trusted subordinate of Claudius, **Lucius Domitius Aurelianus,** was proclaimed emperor by the army.

270. Renewed War with the Goths. Learning of the death of Claudius, the Goths turned back across the Danube. Aurelian marched against them and drove them completely out of Moesia, and back across the Danube. He decided to abandon Dacia, which had been overrun by the barbarians for several decades. He resettled most of the surviving Roman colonists of Dacia in Moesia.

271. War with the Alemanni. The Alemanni made another raid across the upper Danube to ravage Pannonia and northern Italy. Aurelian defeated them just south of the Danube, only to have them escape due to the laxity of a legate. They headed for Italy, pursued by the emperor. Catching up, he was severely defeated at the **Battle of Placentia,** withdrawing in considerable disorder. The Alemanni then headed for Rome. Aurelian rallied, regrouped his troops, and continued the pursuit. He caught up with the barbarians at the Metaurus River and defeated them at the **Battle of Fano.** The Alemanni now retreated northward, closely pursued by Aurelian, who practically annihilated them at the **Battle of Pavia.**

271–276. Reconstruction of the Walls of Rome. Aurelian wished to make sure that the heart of the empire would not again be defenseless against barbarian raids.

271–273. War against Zenobia. Most of the empire's eastern dominions—Syria, Egypt, Mesopotamia, and most of Asia Minor—were still ruled by the beautiful widow of **Odaenathus** (see p. 153). Aurelian now marched east, and in the very bitterly contested **Battle of Immae,** near Antioch, defeated **Zenobia** and her general, **Zobdas** (271). The Palmyrans withdrew in good order, but Aurelian pursued and defeated them again in the decisive **Battle of Emesa** (272). Zenobia fled to her desert capital, Palmyra, where she was besieged by Aurelian. Despite harassment from nomadic guerrillas, and some interference from the Persians, Aurelian prosecuted the siege vigorously, feeding and supplying his army by means of excellent administrative arrangements. When Zenobia surrendered, Aurelian forgave her and left her in control of Palmyra and its vicinity (272). But soon after the departure of the emperor and his army, Zenobia again declared her independence. Aurelian promptly returned, renewed the siege, and captured and sacked Palmyra. Zenobia, attempting to flee, was captured and taken back to Rome as a prisoner, to be led through the streets in chains during the emperor's triumph (274).

273. Revolt in Egypt. During the second siege of Palmyra, a pretender named **Firmus** proclaimed himself emperor in Egypt. Aurelian marched to the Nile Valley, suppressed the revolt, and executed Firmus.

273–274. War with Tetricus. Aurelian now turned to reunite with the empire the one remaining area of dissidence—the provincial empire of Gaul, Britain, and Spain, now ruled by **Tetricus.** He defeated his rival—whose resistance was only half-hearted—at **Châlons** (late 273), and within a few months had obtained the submission of all three provinces, thus restoring the unity of the empire.

274–275. Plans for War against Persia. The Sassanid Empire had given some assistance to Zenobia, and had taken advantage of Aurelian's internal preoccupations to annex Roman Mesopotamia and to strengthen its hold on Armenia. The emperor, preparing for war, began to march eastward. While in Thrace, near Byzantium, a cabal of officers, resenting his stern discipline, assassinated him. Aurelian was a vigorous, able general, a wise and respected ruler, and a man of exemplary personal conduct. He well deserved his title "Restorer of the World."

275–276. Reign of Tacitus. An elderly, respected Italian statesman and soldier, **Marcus Claudius Tacitus,** was chosen by Senate and army to succeed Aurelian. He marched to Asia Minor to deal with continuing depredations of the Goths and the Alans—a Scythian-type people. Despite his advanced years, he conducted a vigorous campaign and defeated the barbarians in Cilicia (276). He became ill and died (or was assassinated) after a reign of 6 months.

276–281. Reign of Probus. Another Illyrian general, **Marcus Aurelius Probus,** proclaimed emperor by the army upon the death of Tacitus, promptly disposed of **Florianus,** younger brother of Tacitus, who also claimed the throne. The death of Aurelian had encouraged the barbarians to renew their incursions into the empire. Probus, in a series of campaigns distinguished equally by his brilliant generalship and his legendary personal valor, inflicted crushing defeats on them. He expelled the Franks, Burgundians, and Lygians from Gaul (276); he then invaded Germany, marching without serious opposition as far as the Elbe (277). He rebuilt the old—and long-abandoned—**Limes,** with a permanent masonry wall from the Rhine to the Danube. He vigorously punished other barbarian raids across the Danube farther south, into Illyricum (278). His subordinate, **Saturninus,** commanding all Roman forces in the East while Probus attended to the Germans, revolted and declared himself emperor. Probus promptly marched to Asia Minor and defeated Saturninus, who was killed (279). It was possibly about this time that Probus crushed a long-standing revolt of the fierce Isaurian mountaineers in central Asia Minor. A new revolt in Gaul, however, called the warrior-emperor back to that province. He promptly defeated the leaders of the revolt—his legates **Bonosus** and **Proculus**—bringing complete peace to the empire for the first time in more than a century (280). To keep his soldiers busy, he put them to work on public projects: building roads, draining swamps, and the like. Resentful of this, and of his strict discipline, a group of soldiers killed him while he was inspecting their work (281).

281–283. Reign of Carus. The commander of the Praetorians, **Marcus Aurelius Carus** (another Illyrian), has sometimes been suspected (probably wrongly) of having instigated the death of Probus. Selected emperor by the army, he decided to invade Persia, to regain Mesopotamia and Armenia. Leaving his elder son, **Carinus,** to administer the western portions of the empire, Carus marched east via Illyria, defeating invading Quadi and Sarmatian tribes en route.

282–283. War with Persia. Carus defeated **Bahram I** of Persia in Mesopotamia. He then marched on to defeat the Persians again near their capital, Ctesiphon, which he captured. He then marched east of the Tigris—possibly intending to overrun the entire Sassanid Empire. Soon after this he died under mysterious circumstances, supposedly killed by lightning, but possibly murdered. The Romans, now led by Carus' younger son, **Numerianus,** withdrew behind the Tigris.

283–284. Carinus and Numerianus as Co-emperors. Near Chalcedon, while the army was marching back to Europe, Numerianus was murdered, evidently by **Arius Aper,** commander of the Praetorians (284). Meanwhile, in the west, Carinus assumed the throne on hearing of the death of his father.

284. Diocletian Chosen Emperor. Upon the death of Numerianus, the Illyrian commander of the imperial bodyguard, **Gaius Aurelius Valerius Diocletianus,** was chosen emperor by the army. Immediately he personally slew Aper, presumed assassin of Numerianus.

284–285. Civil War between Carinus and Diocletian. Carinus took the field against Diocletian. The armies met in Moesia, and Carinus had the better of the initial engagements. The climax came in the decisive **Battle of the Margus** (Morava). Again Carinus was having the better of the fight, when he was apparently murdered by one of his own officers. Diocletian became sole ruler of the empire (285).

284–305. Reign of Diocletian. Never known as an outstanding general, Diocletian is worthy of comparison with Augustus as an administrator and politician. His reign, bringing to fruition the military successes of his Illyrian predecessors Claudius, Aurelian, Probus, and Carus, was unquestionably the most important single factor in the survival of the Roman Empire for almost two more centuries. He instituted a number of major political, administrative, and military reforms. He completely separated military organization from governmental machinery; for the first time local military commanders were subordinate to regional officials of government. He reorganized the army completely, drastically reducing the size of the legion. At the same time he greatly strengthened the frontier defensive system, which had suffered seriously during the recent decades of barbarian invasions. By the end of his reign the barbarians found it practically impossible to penetrate the chain of forts, camps, and walled frontier colonies.

286–292. Reorganization of Imperial Administration. Diocletian believed that the size of the empire, combined with the threats of external enemies, constituted an administrative and military burden beyond the capacities of any single man. Accordingly he divided the empire into 2 major divisions—East and West—ruled by coequal emperors, or **Augusti.** Each of these, in turn, was assisted by a carefully selected imperial prince, or **Caesar,** who directly controlled about half of that portion of the empire ruled by his superior **Augustus.** This meant that the empire was divided into 4 major administrative regions and had 4 principal armies to cope with external threats and to maintain internal order. The careful selection of the junior associates of the Augusti, furthermore, was intended to assure orderly and peaceful succession to imperial authority, in contrast to the violence and death which had characterized the rapid changes in rulers for the previous centuries. Diocletian put his new system into effect gradually, first choosing able general **M.A. Valerius Maximianus** (or Maximian) as co-Augustus, to rule

over the West; Diocletian retained the East, with his capital at Nicomedia, in Asia Minor (286). Then, a few years later, he and Maximian selected two more generals, **Gaius Galerius Valerius** and **Flavius Valerius Constantius,** to be their assistant rulers, or Caesars (292). Each commanded one of the empire's four major armies. Diocletian was in direct control of Thrace, Egypt, and Asia; Galerius was administrator of Illyria and the Danubian frontiers from a capital at Sirmium (Mitrovica in Yugoslavia); Maximian directly controlled Italy and Africa from Mediolanum (Milan); Constantius ruled Gaul, Britain, and Spain from Augusta Trevirorum (Trier). By avoiding use of Rome as a capital, Diocletian reduced the importance of both the Praetorians and the Senate. Each of the four princes was practically sovereign in his own region, though sharing jointly overall responsibility for the empire. Diocletian was acknowledged by his associates as first among equals; so long as he reigned there was no jealousy in this system of corporate, decentralized control. But, though it solved some of the empire's most pressing problems, the system had two major dangers: (1) Some discord would be inevitable among such a group of rulers. (2) The cost of four imperial courts would be likely to create a staggering economic burden. These dangers soon became facts.

286. Peasant Uprising in Gaul. This was suppressed promptly by **Marcus Aurelius Carausius,** commander of the Roman fleet at Gessariacum (Boulogne).

287–293. Revolt of Carausius. Carausius enriched himself by plundering the shores of Germany, when he was presumably protecting the northern coasts of the empire against Germanic pirates. Suddenly assuming the title of Augustus, he defied Maximian and Diocletian and established firm control over northern Gaul and Roman Britain. The revolt was quelled by Constantius (294) after 5 years of indecisive fighting, and shortly after Carausius' murder by a subordinate, **Allectus.**

294–296. Disorders in Egypt. A usurper named **Achilleus** established himself as emperor in Alexandria, controlling northern Egypt. Diocletian himself took the field, captured Alexandria after a siege of 8 months, and executed Achilleus (296). Southern, or Upper, Egypt was still in turmoil, however, due to incursions of a wild barbarian tribe—the Blemmyes—who apparently originated south of Nubia, and who had been terrorizing Upper Egypt for many years. Diocletian assisted the Upper Egyptians and Nubians in repelling the marauders and in establishing effective defenses.

295–297. Pacification of Mauretania (Morocco). The coastal regions of Mauretania were ravaged by an invasion of wild desert and mountain tribes of the interior. Maximian went to northwest Africa and drove the invaders back into the desert and to their mountain strongholds. He then systematically pacified the mountain regions, capturing their strongholds, removing all their inhabitants, and resettling them in other parts of the empire.

298. Invasion of Gaul by the Alemanni. This major barbarian invasion was met promptly by Constantius, who defeated the Alemanni decisively at **Lingones** (Langres) and at **Vindonissa** (Windisch, Switzerland).

295–297. Persian War. (See p. 175.)

305. Abdication of Diocletian and Maximian. With the Roman Empire reorganized, law and order pervading its interior provinces, and the frontiers respected by chastened external foes, Diocletian abdicated, to facilitate the peaceful succession system he had planned. He persuaded his somewhat reluctant colleague, Maximian, to do likewise. Diocletian spent the last 8 years of his life as a gentleman-farmer at Salona (near modern Split), on the Dalmatian coast of the Adriatic Sea; he lived long enough to realize that his hopes of orderly succession had been unduly optimistic.

ROMAN MILITARY SYSTEM, C. 300

From its earliest days the power and glory of Rome were derived from a superb military system unmatched in skill or efficiency by any potential enemy. There were many changes in the details of the system in the first $3\frac{1}{2}$ centuries of the Roman Empire: some were normal and evolutionary; others were the result of internal stresses and strains; some were inspired by the example of enemies. The supremacy of the Roman military system—and thus the continued existence of the empire—was in large part due to the continued pragmatic, logical approach of the Romans to practical problems. They respected tradition, but they were not slaves to it, and were extremely flexible in adapting themselves to military change. The fact that there were no more fundamental changes in the Roman military system from about 50 B.C. to A.D. 300, therefore, reflects both the lack of technological change in these centuries and the thoroughness with which the Republican Romans had adapted existing technology to the art and science of warfare.

The Military Policy of Augustus

Augustus established certain fundamentals of military policy which could not be improved by his successors. The most basic of these was relating the security of the empire to economic soundness as well as to military excellence and adequacy. The Roman Empire had become about as large as could be managed by one man or one single administrative system. Armed forces were required only for the defense of the frontiers and to maintain domestic tranquillity.

Augustus wished to keep the armies as small as possible, in order to place the least possible strain upon the economic fabric of the empire. By organization and skill, a small, efficient army could perform the essential defensive missions. The force of 300,000 men under arms was small, in the light of the size and the population of the empire (some 50 million) and considering the number of warlike foes around the frontiers. Yet it was adequate.

Augustus well understood that aspect of economy of forces now often called "cost effectiveness." His successors, almost without exception, adhered to this principle also. Even when the empire's frontiers seemed most seriously threatened, they remained rightly confident that Roman military abilities would assure victory over the greatest numerical odds without the need for raising vast levies of expensive, and relatively inefficient, mass armies. As external pressures increased, so too did the size of the standing army. But even in the middle of the 4th century the total size of the Roman army would have been considered small by the standards of any other age, before or since.

This policy of military economy permitted the empire to avoid an expensive central reserve. Save for the 10,000 Praetorians—who were intended by Augustus more to maintain internal tranquillity than as a personal bodyguard—all the armed forces of the empire were scattered along the frontiers. A threat in any region would be met by dispatching detachments from other portions of the frontiers. The superb road network behind the frontiers, and connecting the provinces, was deliberately built up as a substitute for a military reserve.

The Augustan Legion

Augustus standardized the size of the legion, setting it at 6,000 men, composed of the traditional 10 cohorts. Apparently the first and tenth cohorts were 1,000 each in strength; the other 8 were 500 strong. The reason for this is not clear, and it is possible that the most common organization was 10 cohorts of 600 men each. Command arrangements were much as in Caesar's time. Augustus had 25 such legions deployed along the frontiers of the empire.

With some exceptions, soldiers were enlisted for 20 years. There was apparently no difficulty in obtaining recruits; most of them were the sons of veterans. Although Augustus forbade his soldiers to marry, he did not intend that this should prevent them from raising families. There were always arrangements for taking care of the families of soldiers in or near the main camps of the legions, and the liaisons of soldiers and camp followers were always legalized upon the veteran's discharge. Part of the veteran's pension was usually a plot of farmland in the vicinity of his frontier post, and the veteran's son more likely than not joined his father's old unit.

The morale of the legions was maintained by *esprit de corps,* by discipline, and by rigorous training. It was no small contribution to unit *esprit* for the individuals to know that they belonged to an organization which had been in existence for centuries, with a proud record of almost uninterrupted victories, and whose heroes more often than not were their own family ancestors.

As in the past, training and disciplinary measures were harsh, intensive, and effective. Standards fluctuated over the centuries, and in later years unquestionably declined. Nevertheless, even at their lowest, these disciplinary standards were never approached by Rome's foes, and they provided, to a greater degree than any other factor, the usually wide margin of Roman military superiority over all enemies.

Auxiliaries of the Imperial Army

To support his 25 legions scattered along the frontier, Augustus maintained an approximately equal force of auxiliaries—about 150,000 in all: archers, slingers, light infantry, and cavalry; most were recruited from the barbarian or semibarbarian tribes outside the empire.

The auxiliaries were paid less than the legionaries. Their terms of service were usually for 25 years. They could then look forward to automatic citizenship and—sometimes—to veterans' compensation in land and money comparable to that of the legionary veteran.

Initially most auxiliary units were permitted to retain their original tribal organization and leadership. Save for special troops—such as archers and slingers, who usually came from the eastern regions of the empire—the auxiliaries at first served in the general area in which they were recruited. This sometimes led to revolts or mutinies in which the auxiliaries joined forces with local fellow tribesmen. Thus, after the time of Augustus, barbarian auxiliary units were usually shifted from their homelands to other frontier regions, where their homogeneous tribal organization was deliberately diluted by reinforcements from other tribes.

By the time of Trajan this policy of discouraging revolt had been carried even further; tribes were completely mixed up in auxiliary units and officers were no longer tribal chiefs. Consequently tribal enthusiasm had to be replaced by unit *esprit de corps.* This, in turn, led logically to the creation of certain regular standing units of auxiliaries, comparable to the cohorts of the legions. These permanent units were given numerical identifications and became known as **numeri.** This increased permanence and regularity in auxiliary units lessened still further the diminishing distinctions between the Romanized barbarians of the auxiliary units and the increasingly barbarian composition of the legions during the 3rd and 4th centuries.

As early as the time of Augustus the training and equipment provided the barbarian auxiliaries by the Romans were sometimes used against Rome. Discharged auxiliaries and deserters served in the barbarian ranks during raids across the frontier. At the same time the barbarians learned much by experience from their battles against Roman formations. Certainly this continual improvement in barbarian methods of war contributed to the final overthrow of Rome. That this did not happen earlier is a tribute to Roman political skill, and to the organizational and leadership abilities of the outstanding Roman generals.

Innovations of Hadrian, Marcus Aurelius, and Septimus Severus

Despite the essentially defensive military policy of Augustus and his successors, the empire grew slightly in the century after his death. Hadrian, giving up some of the eastern conquests of his predecessor, Trajan, decided that this expansion must stop; the empire was already too vast to be effectively administered politically or controlled militarily. He modified the mobile defensive concept of Augustus to one of rigid frontier defense. He had no intention that this should change the inherent mobility of the legion or its tactical flexibility. Rather, he wanted to establish man-made obstacles which would supplement natural barriers—rivers and mountains. This would make it harder for barbarian raiders to cross the frontiers, and easier to cope with them if they did, without the necessity for shifting reinforcements from one frontier region to the other, or for conducting large-scale campaigns within or without the frontiers.

The **Limes** in Germany and Hadrian's Wall in Britain were high mounds of earth, topped by wooden palisades. These were not manned permanently by Roman troops—there were not enough soldiers in the army to do anything like that. Rather, the barriers provided protection and concealment for Roman border patrols and made it more difficult for barbarian raiders to cross the frontier secretly. More important, they were obstacles to the easy escape of barbarian raiding parties, many of whom were caught and slaughtered under the walls by pursuing imperial troops.

Hadrian made increasing use of riverboat patrols along the Danube and Rhine to make invasion more difficult. He also cultivated and expanded the already-existing Roman intelligence network in barbarian areas beyond the imperial frontier.

Marcus Aurelius, exemplifying Roman imperial generalship and the intelligent application of policy to strategy at their best, made few changes in organization. He did, however, add 2 legions to the standing army, and an even greater proportion of auxiliaries. In his time the Roman standing army probably numbered more than 350,000.

Septimus Severus added 3 more legions, one of these being stationed in Italy to provide the first mobile reserve of the Imperial Roman Army. The total strength of the army probably reached 400,000 men in his time. Severus also improved the lot and comfort of the soldiers in a number of ways, including permission for legal marriage. He made some changes in command organization, the most important being to split the major regional frontier commands, reducing the temptation and opportunity for regional commanders to revolt against central authority. He also changed Hadrian's Wall in northern Britain from a wooden palisade to a masonry wall, 16' high and 8' thick.

It was about this time that cavalry began to assume a role of importance in the Roman army. Related to this was the appearance of mounted infantry to provide greater mobility to foot troops and to assure rapid, distant movements of combined arms.

Reforms of Diocletian and Constantine

During the chaotic times of the mid-3rd century, it became obvious that the administrative, political, and military policies of Augustus, though modified and adapted by his successors, were in need of drastic revision. A frontier defensive system without a central reserve was not readily adaptable to a situation in which the empire had to face increasingly dangerous raids by improved, Romanized barbarian forces, as well as the threats of a strengthened, militant Persia. Simply increasing the size of the forces, without changing their organization and operational concepts, was not sufficient.

This, however, was the only recourse of the early Illyrian emperors, who had no time for reorganization and who had to make do with the system and concepts they had inherited, beating back multiplying threats in whatever way they could. The overall strength of the army grew to about 500,000 men. By stern and harsh measures Aurelian, Probus, and Carus restored training and discipline to standards at least approximating those of the early empire. With makeshift forces, they ejected the Germanic barbarians and chastened the Persians.

One of the major causes of disorganization in the Roman army had been the old system of sending units from one portion of the frontiers to reinforce armies engaged in major wars. Naturally, it was rarely possible to take a full legion, or even a major portion of a legion, from one area, since this would have left a dangerous gap in the frontier defenses. The logical solution was to take detachments from different cohorts, legions, and numeri, and to form these into temporary task forces, called **vexillations,** which then operated as tactical units attached to the army in the threatened area. It was discovered that the most manageable vexillations were units of about 1,000 infantry or 500 cavalry.

The vexillation system had initially been satisfactory. As soon as the threat was taken care of, the task forces were dissolved, and the detachments returned to their parent organizations. But during the turbulent period from 235 to about 290, detachments and vexillations were shifted so rapidly from one frontier to another that units became hopelessly mixed up. The tradition of the legion, and its impact on *esprit de corps,* almost disappeared.

This was only one of the circumstances which caused Diocletian to make sweeping reforms in military policy and organization, reforms later carried on and brought to completion by Constantine.

To provide a mobile reserve, the army was divided into two major portions: frontier troops (known as **limitanei** or **riparienses**) and mobile field forces (composed of more mobile units known variously as **palatini** and **comitatenses**). Approximately two-thirds of the army strength was in the frontier forces. The remainder was in the mobile units, which the emperors (Augusti and Caesars) kept centrally located in their respective domains. The mobile forces received slightly higher rates of pay than the frontier troops, later a cause of trouble.

Experience with the system of vexillations caused Diocletian to reduce the size of the legions of the field forces to about 1,000 men. This assured greater strategic and tactical flexibility, without need for detachments. The legions of the frontier forces were kept at 6,000-man strength. Auxiliary units were usually 1,000 each, in both mobile and frontier forces.

Diocletian also eliminated the position of Praetorian prefect, a post which combined in one individual duties approximating those of an imperial chief of staff, with direct command of the Praetorian Guard. The power of these prefects had too often been used to overthrow an emperor or to win the throne. Instead, each Augustus and each Caesar had two major military subordinates: a master of foot and a master of cavalry. Not only did this divide military power and thus reduce political danger, it also indicated the increasingly important position of cavalry in the Roman army.

Constantine later abolished the Praetorian Guard; in its place each emperor had a personal bodyguard of about 4,000 men.

The New Formations

Save for the obvious change in size, there was little superficial change in the tactical organizations of the legions, or in their soldiers, as a result of the reforms of Diocletian and Constantine. On the whole these changes were good and contributed to the extension of the life of the Roman Empire. There were, however, some unfortunate practical and psychological effects— unforeseen and probably unavoidable.

The infantry of the mobile field forces were generally more lightly equipped than the frontier units. This, combined with their greater pay, created jealousy in the frontier units. Since there was no longer much difference in the background or nature of the men in the legions and the auxiliary soldiers, service in the frontier regions became less popular than in the less heavily encumbered and less strictly disciplined auxiliaries, or in the better-paid field forces. The result was a decline in morale affecting much of the army. It also resulted in changes in equipment and training programs, reducing still further the distinction between legionaries and auxiliary soldiers.

The soldiers—legionaries and auxiliaries, field forces and frontier forces—were now mostly barbarians. The result of this, by 375, was to cause most barbarian warriors to be essentially Romanized in weapons and tactics. For the most part, barbarians in Roman service thought of themselves as Romans and generally maintained the full loyalty of professional soldiers to their units and to their commanders. Nevertheless, though barbarians seemed to have no objection to fighting their fellows when ordered to do so by their Roman commanders (who also were more often than not barbarians themselves), this did create a number of opportunities for collusion, mutiny, and mass desertion.

Two new types of soldier also appeared in the Roman army, contributing to the slowly declining standards of training and discipline.

First, the increasing numbers of barbarian tribes which had been permitted to settle within the empire resulted in a reversal of Trajan's trend away from tribal auxiliaries. Tribal units of auxiliaries, under their own chiefs and retaining their own weapons and methods of warfare, were incorporated into the army in increasing numbers. These were called **federati.** Due to the overall Romanization of the barbarian warriors, this had little tactical impact. But it did create new opportunities for unrest and mutiny.

Second, a form of conscription appeared in some portions of the empire, when it was not possible to fill the ranks by volunteers. The large landholders were then required to provide new recruits, on a *pro rata* or on a rotational basis. Terms of service of these conscripts were less onerous than for the regulars. This did not happen often, but it was a step in decreasing the professionalization of the army. It was also a step toward later medieval feudalism.

Tactics of the Later Empire

Between the time of Julius Caesar and Julian, no leading soldier wrote about his experiences. Nor did any other first-class historian bother to describe the combat of his times. As a result, the tactical details of operations in the 4-century period from 50 B.C. to A.D. 350 are quite obscure.

So far as the basic formations of the legion are concerned, apparently there was increasing use of the 2-line formation, with 5 cohorts in each line. There are some indications that the formations became more dense and phalangeal as a result of the increasing importance of cavalry and the concurrent decline in disciplinary standards.

On the other hand, there are also indications that the fundamental facts of human combat which impelled Marius to adapt the ancient quincunx formation to the cohorts of his legion were as well known to later Romans as to the Republicans. It would appear from the records of Constantine's tactics at the Battle of Turin that well-trained, well-disciplined infantry of the later empire could in the traditional close formation face cavalry charges with as much confidence as Caesar's legions, and that they could extend their front and otherwise maneuver and fight in cellular formations with the same speed, flexibility, and decisiveness.

Nevertheless, granting that there had been no change in the basic fundamentals of close-order and open-order infantry tactics, the increasing role of cavalry in warfare undoubtedly had its effect. When standards of discipline were high, the formations probably were similar to the ancient quincunx; when discipline was low, generals probably kept their legions closed up in something resembling the even-more-ancient phalanx.

One indication of the effect of cavalry on the declining legion was the adaptation of the old pilum into a "throwing spear." Apparently the change was adopted reluctantly, since the new pike was light enough to be hurled, and this was always done, in order to get rid of the encumbrance, before the legion came to hand-to-hand combat with other infantry. Presumably, in order to obtain room to use the gladius, the front was extended to gain the old open formation of the early legion. Yet the very change in the pilum was indicative of the trend toward lessened flexibility and reduced offensive capability, and in turn showed that discipline was declining.

By the beginning of the 4th century, cavalry made up about one-fourth of the strength of the average Roman army; the percentage was much higher in the eastern deserts in combat with the Persians and Arabians. Cavalry had become the arm of decision.

Missile weapons, too, increased in importance in the Roman army. These combined developments posed a tactical dilemma. To face cavalry charges the infantry had to rely increasingly on close formations; yet the problems of using, and being subjected to, missile engines of war created a tendency toward a more extended order. Recognizing that this same dilemma faced their foes, the Romans simultaneously increased their proportion of missile weapons and added more and more horsemen to their armies. Quite naturally, they endeavored to retain their flexibility so that they could adapt themselves to whatever danger posed the greatest threat at any time.

By the beginning of the 4th century the Romans provided more than one ballista, catapult, or onager to every 100 men of the legion. In addition, possibly half of the foot auxiliaries were archers or slingers, and many of the cavalry were horse archers on the Asian model. One result of this great reliance on missiles of one sort or another was to reduce significantly the occasions when the legion engaged in hand-to-hand combat with the gladius.

ROME AND THE BARBARIANS, 305–400

305–306. Confusion in the Imperial Succession. Upon the abdication of Diocletian and Maximian, they were succeeded as Augusti by Constantius in the west and Galerius in the east. Two new Caesars appointed were **Flavius Valerius Severus,** who took over control of Africa and Italy, and **Galerius Valerius Maximinus Daia.** Ignored were two young men who had expected the honor: **Flavius Valerius Aurelius Constantinus** (Constantine), estranged son of Constantius, and **Marcus Aurelius Velerius Maxentius,** son of Maximian. When Constantius died, later in the year, the legions of Britain and Gaul declared Constantine his successor as Augustus. Galerius reluctantly accepted Constantine, but only as Caesar, insisting that Severus should become Augustus. At about this same time, in Rome, the Praetorian Guard acclaimed Maxentius as Augustus. Maxentius, however, would accept only the title of Caesar, calling his willing father Maximian, back from retirement to resume his old title of Augustus.

306–307. Civil War; Severus against Maximian and Maxentius. A year of strife ended with the execution of Severus, the appointment of Constantine as junior Augustus, and the recognition of Maximian as the senior. Galerius, however, refused to accept either Maximian or Constantine as a fellow Augustus.

307. Invasion of Italy by Galerius. Maximian and Galerius now maneuvered cautiously against each other in a stalemate. Galerius appointed **Valerius Licinianus Licinius** as Augustus in place of Severus; then when his nephew Daia complained, because he felt that as a Caesar he should have had precedence over Licinius to the title of Augustus, Galerius conferred the title on him as well (308). Thus there were five Augusti. Meanwhile, Maxentius, with the support of the Praetorians, drove his father, Maximian, from the throne, and himself assumed the title of Augustus. Maximian fled to join Constantine in Gaul, refusing to relinquish his title.

308. Congress of Carnuntum (Hainburg in Modern Austria). Arbitration by Diocletian brought an accommodation among 5 of the 6 bickering Augusti. Maximian abdicated again (he remained at the court of Constantine). Maxentius was ignored in the Carnuntum agreements, but his control of Italy was not disputed by Licinius, who was content with Illyricum.

310. Revolt of Maximian. Taking advantage of the absence of Constantine on an expedition against the Franks, the old emperor again tried to seize active power. Constantine returned promptly, drove Maximian from Arelate (Arles) to Marseilles, where he captured his elderly rival and father-in-law, whom he permitted to commit suicide. Galerius died the following year (311). This left four Augusti: Constantine, Maxentius, Licinius, and Daia.

311–312. Civil War; Maxentius against Constantine. Maxentius, with the secret support of Daia, prepared to invade Gaul through Raetia with an army of 170,000. Constantine had about 100,000, of whom more than half were required to hold the Rhine and Caledonian frontiers. Learning of Maxentius' plans, he decided to seize the initiative. Before the snow had completely melted, Constantine marched over the Mount Cenis Pass with 40,000 men (312). He won successes at **Susa, Turin,** and **Milan** against superior forces under subordinates of Maxentius. Of these, the **Battle of Turin** was a major engagement, in which Constantine displayed great tactical skill. He was next met by **Ruricius Pompeianus,** the principal general of Maxentius. Constantine defeated him in the hard-fought battles of **Brescia** and **Verona.**

312. Battle of the Milvian Bridge. Gathering reinforcements as he went, Constantine marched rapidly south. By the time he reached the vicinity of Rome, he probably had about 50,000 men. Maxentius had about 75,000 to protect the capital. Here it was, legend has it, that Constantine saw a cross in the sky and heard the words: *"In hoc signo vinces!"* Vowing to become a Christian if he won, Constantine plunged into battle, gaining a decisive victory on the banks of the Tiber near the Milvian Bridge. Maxentius was drowned trying to escape. Constantine, entering the capital, first disbanded the remnants of the Praetorian Guard, then announced his conversion to Christianity. He recognized Licinius as Augustus of the East.

313. Civil War; Daia against Licinius. Daia, endeavoring to gain control of the entire eastern portion of the empire, marched from Syria, planning to invade Europe with an army of 70,000 men. Licinius, with 30,000 veterans of Danubian campaigns against the barbarians, met him near Heraclea Pontica, in western Asia Minor. In the **Battle of Tzirallum,** Licinius won a complete victory. Daia fled, but soon died. Licinius was now unchallenged ruler of the East.

313. Frank Incursion into Gaul. Constantine, marching from Italy, defeated the barbarians and ejected them from Gaul.

314. Civil War; Licinius against Constantine. Discovering that Licinius was inciting a conspiracy against him, Constantine suddenly marched with 20,000 men to invade the eastern part of the empire. Licinius, hastily gathering 35,000, met the invaders in southeastern Pannonia.

314. Battle of Cibalae (Vinkovci?). Constantine, in a good defensive position, repulsed Licinius' attacks, then counterattacked. Licinius withdrew after losing more than half of his army as casualties in the violent, indecisive struggle. Constantine pursued into Thrace.

314. Battle of Mardia. Again the issue was hotly disputed. Constantine gained the victory by a surprise turning movement, but Licinius again withdrew in fair order. He sued for peace, and in the subsequent negotiations agreed to give up Illyricum and Greece to Constantine, retaining Asia, Egypt, and Thrace.

315. Operations against the Goths. Taking advantage of the civil war, the Goths crossed the Danube into Constantine's new dominions. He repelled them, then crossed the Danube to punish the barbarians in the old Roman province of Dacia.

323. Renewed War between Licinius and Constantine. After a smoldering dispute, war broke out again between the two emperors. Each assembled great armies and large fleets.

323, July 3. First Battle of Adrianople. Each army comprised 120,000 to 150,000 men. Again Constantine made use of a turning movement to win a great victory, which was distinguished by his own personal valor in leading the frontal attack. Licinius lost 35,000 to 50,000 casualties during and after the battle, when his fortified camp was taken by storm. He fled to Byzantium, which was besieged by Constantine.

323, July (?). Battle of the Hellespont. **Crispus,** elder son of Constantine, led his father's fleet of 200 vessels in a 2-day victorious naval battle against the 350 warships of Licinius, destroying 130 of the defending fleet and scattering the remainder. His line of retreat across the Bosporus thus threatened, Licinius fled to Chalcedon. Constantine followed with about 60,000 men, leaving the remainder to press the siege of Byzantium.

323, September 18. Battle of Chrysopolis (Scutari). Licinius had been able to gather an army of about 60,000. Again the fight was prolonged and bloody, but Constantine finally won a crushing victory. Licinius, fleeing, later surrendered and was executed (324).

324–337. Constantine as Sole Emperor. Most of the declining years of this controversial, flamboyant man were spent in the construction of a new imperial capital—Constantinople—on the site of Byzantium.

332–334. Intervention in War of Goths and Sarmatians. Constantine took the field to assist the Sarmatians (a Scythian-Germanic peo-

ple). He decisively defeated **Araric,** King of the Goths, who had crossed the Danube to invade Moesia (332). When the ungrateful Sarmatians began their own raids into the empire, Constantine encouraged the Goths to resume the war, and stood aside while **Geberic,** new Gothic ruler, crushed **Wisumar,** King of the Sarmatians. Constantine then let the remnants of the Sarmatian tribe—some 300,000 people—settle in the empire (334).

337–340. The Successors of Constantine. Constantine had divided the empire among his three surviving sons, though he retained supreme authority until his death (337). **Constantine II** held Britain, Gaul, and Spain; **Constans** had Illyricum, Italy, and Africa; **Constantius** ruled Thrace, Greece, and the East. Almost immediately Constantine II invaded Italy. He was killed in an ambush near Aquileia (340), and his domains immediately seized by Constans. Constantius, who had become engaged in a war with Persia, was not involved in this struggle.

337–350. Persian War of Constantius. (See p. 175)

343. Barbarian Raids in Britain. Incursions of Picts and Scots forced Constans to go to northern Britain to suppress the disorders. About this same time he was also concerned with the depredations of Saxon sea rovers along the coasts of Britain and Gaul.

c. 350–376. Rise of Ostrogothic Kingdom of Ermanaric. The aged Ostrogothic ruler **Ermanaric** (reputedly about 80 years old in A.D. 350) united the Ostrogoths and Visigoths in a powerful kingdom, and by conquest extended his rule from the Dnieper Valley to a vast empire extending from the Black Sea to the Baltic, an area comprising roughly modern Ukraine, Poland, White Russia, Great Russia, Rumania, Slovakia, and Lithuania.

350. Revolt of Magnentius. In Gaul the general Magnentius led a successful revolt; Constans, attempting to flee to Spain, was murdered, and Magnentius declared himself emperor.

350–351. War of Magnentius and Constantius. Patching up a temporary peace with Shapur of Persia, Constantius marched west to deal with the usurper who had murdered his brother. He discovered that in Illyricum the aged general **Vetranio** had been crowned Augustus by **Constantina** (sister of Constantius) and had joined an alliance with Magnentius. In a dramatic confrontation between their two armies, Constantius persuaded Vetranio to resign and added the Illyrian army to his own. Meanwhile, Magnentius gathered a large army and marched to meet Constantius in lower Pannonia. The armies were about equal in size, nearly 100,000 each. Magnentius had somewhat the best of the subsequent cautious maneuvering.

351. Battle of Mursa (Osijek, Yugoslavia). In a hard-fought battle, Constantius' more maneuverable army, with a superiority in auxiliary archers and in light and heavy cavalry, enveloped Magnentius' left flank; then his heavy cavalry drove through the harassed Gallic legions in a climatic charge. Both sides suffered enormous losses; Magnentius apparently lost about 30,000 dead, and Constantius nearly as many. Magnentius withdrew to Italy.

351. Battle of Pavia. Magnentius defeated the imprudently pursuing forces of Constantius. But the Italian population rose in favor of Constantius, and Magnentius retreated to Gaul. Here, too, the populace rose against him, as did his own soldiers. Magnentius committed suicide, leaving Constantius sole emperor (351).

351–355. Internal and Frontier Disorders. German barbarians were making serious inroads into Gaul, apparently bribed by Constantius himself during his civil war with Magnentius. The barbarians found the experience so pleasant that they were unwilling to stop when the civil war ended. The situation in Gaul was further complicated by the brief revolt of the general **Sylvanus** (355). During this same period Constantius was forced to deal with the gross misrule and conspiracy of his cousin **Gallus,** who had been left in command in the East. Gallus was relieved and executed.

355–358. Constantius along the Danube. Invasions of the Quadi and Sarmatians now occupied most of Constantius' attention, though he was able briefly to operate with **Julian** along the Rhine (358; see below). The emperor attacked the barbarian invaders along the upper and middle Danube, drove them back across the river, then conducted a successful punitive expedition into their homeland (357). Following this, he moved southward along the Danube, subduing other barbarian disorders.

355. Julian as Caesar. Somewhat reluctantly, Constantius appointed his scholarly cousin, **Flavius Claudius Julianus** (Julian), the only other surviving male member of the Constantine family, as Caesar of the West in Gaul. To the surprise—and alarm—of the emperor, the 24-year-old Julian proved to be an outstanding military leader.

356. Julian's First Campaign. This began inauspiciously with a defeat at the hands of the Alemanni at **Reims.** He recovered, defeated the raiders, and advanced to the Rhine at Cologne. Here he worked in cooperation with Constantius, who advanced from Raetia up the right bank of the Rhine, while Julian operated to the west. There were no large-scale battles, and Julian had an excellent opportunity to learn something about war from his enemies and from his experienced cousin. Late in the year, while in winter quarters at **Sens,** he survived and repulsed a violent Alemanni surprise attack.

357. Julian Takes the Offensive. A joint campaign had again been planned, with Constantius' general **Barbatio** operating from a base in Italy and Raetia. As Julian closed in on the Alemanni King **Chnodomar** in the area of modern Alsace, Barbatio approached from the East. But instead of assisting, Barbatio suddenly withdrew from the campaign, leaving Julian with 13,000 men, opposed to Chnodomar's 35,000, near Argentorate (Strasbourg).

357. Battle of Argentorate. Julian's heavy cavalry stampeded early in the battle, apparently because of the Alemanni tactic of sending lightly armed men to crawl under the armored horses and to stab their unprotected bellies. The legions stood firm, and Julian rallied the cavalry. After heavy losses, the Alemanni finally fled; Chnodomar was captured. Julian chased the survivors across the Rhine and raided briefly across the river himself. He then moved northward against the Franks, and before winter had completely cleared the left bank of the Rhine of all barbarians.

357–359. Julian Consolidates Gaul. After repulsing one final Frank effort to cross the Rhine, Julian devoted himself to rebuilding frontier defenses and to reestablishing Roman outposts east of the Rhine. Making his headquarters at Lutetia (Paris), he also encouraged the reconstruction of ravaged cities and towns, sending troops to assist whenever possible.

358–363. Renewed War with Shapur of Persia. A Persian invasion of Roman Mesopotamia forced Constantius to turn once more to the east (see pp. 176).

360–361. Civil War; Julian against Constantius. Constantius, jealous of the successes of Julian in Gaul, ordered him to send most of his best legions to the East for the war against Persia. Julian protested, but prepared to comply with the order. His legions, however, refused to leave their brilliant young leader and proclaimed him emperor. Julian's attempt to reach an accommodation with Constantius was rejected. In a surprise march of amazing rapidity, Julian took his army eastward through the Black Forest and Raetia, arriving in Illyricum so unexpectedly that he captured Constantius' legate, **Lucilian,** at Sirmium without a struggle. As he continued toward Constantinople, he received word that Constantius, marching back to oppose him, had died en route in Asia Minor (November 361). Julian, meanwhile, had renounced Christianity and returned to paganism; thus he is known to history as "Julian the Apostate."

362–363. Julian Prepares for War with Persia. After a few months in Constantinople, where he cleaned up the dissolute court and established a regime of austerity, Julian marched to Antioch. There, during the fall and early winter, he collected an army of 95,000 men, the largest expeditionary force Rome had ever assembled in the East.

363. Julian's Invasion of Persia. Early in the year Julian advanced to Carrhae, where he detached a force of 30,000 men under generals **Procopius** and **Sebastian** to march northwestward into Armenia. In accordance with promises of King **Tiranus** of Armenia, Julian expected his two lieutenants would be joined by an Armenian army of 24,000, and that the combined force would then move down the east bank of the Tigris River, toward Ctesiphon, the Persian capital. Julian, with the remander of his army, accompanied by a fleet of 1,100 river supply ships and 50 armed galleys, marched down the east bank of the Euphrates at a pace of 20 miles per day. Reaching the point where the two rivers most closely approach each other, he besieged and captured two strong forts protecting the approach to Ctesiphon. He then set his men to widening the canals between the Euphrates and Tigris, so as to be able to bring his fleet to Ctesiphon.

363. Battle of Ctesiphon. Discovering that Shapur's army held the far bank of the Tigris in force, Julian, with the help of his fleet, undertook a successful river crossing and defeated Shapur in a violent battle under the walls of the capital. Shapur and part of his army retreated westward; the remainder took refuge behind the city walls. Julian was now disappointed by the nonarrival of Procopius, Sebastian, and Tiranus. (He did not know that the Armenian king would not cooperate as he had promised, and that the two Roman generals were unable to agree on a plan of action.) Julian decided not to try to besiege and capture Ctesiphon, but rather to pursue Shapur into the heart of Persia.

363. Julian's March from Ctesiphon. Destroying his fleet, and all supplies that could not be carried by his men and a small baggage train, Julian marched east. The Persians, however, had conducted a "scorched earth" policy, and Julian, deceived by the treachery of a Persian nobleman, was lost for several days. Realizing that his army was running dangerously short of supplies, the emperor decided to withdraw north, up the Tigris, to establish a new base in Armenia. As soon as the Romans changed direction, they were harassed by swarms of Persian light horsemen. Though supplies were getting low, the army moved rapidly, and Julian expected to reach Armenia in about two weeks. During a night attack on his camp, Julian led a counterattack, without waiting to put on his armor. He was mortally wounded and died a few hours later.

363–364. Disaster under Jovian. The army elected the general **Flavius Claudius Jovian** to succeed Julian. Alarmed by the shortage of supplies, weak and indecisive Jovian accepted Shapur's offer to negotiate, halting his army for several days. Finally, all food gone, Jovian had no choice but to accept harsh Persian terms: loss of all provinces east of the Tigris and of Nisibis and other Roman fortified towns in Mesopotamia, abandonment of suzerainty over Armenia and other Caucasian regions. With but the emaciated remnants of the great army led eastward by Julian, Jovian returned to Antioch. Early the next year, en route to Constantinople, he died under mysterious circumstances.

364–378. Reigns of Valentinian and Valens. An able soldier, **Flavius Valentinianus I,** was chosen by the army to replace Jovian. He immediately selected his brother **Valens** as coemperor in the East. Both Augusti were soon occupied with renewed, intensive barbarian incursions, and with some internal disorders.

365–367. War with the Alemanni. The violence of the barbarian onslaughts caused Valentinian to take the field in person in Gaul. Repulsing them at **Châlons,** he drove them back across the Rhine and won a great victory at **Solicinium** (Sulz) on the Neckar (367). He then devoted himself to strengthening the defenses along the Rhine.

366. Revolt of Procopius in the East. Gaining control of Constantinople and much surrounding territory, Procopius declared himself emperor. He was soon defeated, however, by Valens' generals **Arbetio** and **Lupicinus.** The usurper was captured and executed.

367–369. Gothic War. Valens and his generals imprisoned a force of Gothic mercenaries who had aided Procopius in the civil war. Ermanaric, ancient King of the Goths, protested and sent a Visigothic army, under **Athanaric,** across the Danube. During a period of confused fighting and negotiations, Valens released the prisoners. Finally repelling Athanaric, Valens and his generals **Victor** and **Arintheus** invaded Visigothic territory north of the Danube. Peace was restored by a treaty recognizing the Danube as the boundary between the Gothic and Roman empires.

368–369. Chaos in Britain. The inroads of Saxon sea rovers, combined with Scottish and Pict uprisings in the north, kept Britain in turmoil. Valentinian's general **Theodosius** restored order in 2 well-conducted campaigns.

371–372. Revolt of Firmus. This Moorish leader revolted against Roman rule, nearly driving the Romans from Mauretania. Theodosius, however, arrived and quickly defeated Firmus, who committed suicide to avoid capture.

372–374. Appearance of the Huns. The Hsiung-nu tribes driven from Mongolia by the Chinese 2 centuries earlier (see p. 146) were apparently the ancestors of a fierce Mongoloid people—the Huns—who now entered European history. They invaded the lands of the Scythian-Germanic Alans (cousins of the Sarmatians) in the region between the Volga and the Don. The Huns won a great victory at the **Battle of the Tanais River** (373?); in less than 2 years the kingdom of the Alans was overwhelmed. Some of the survivors were absorbed by the Huns; other refugees wandered through the lands of the Goths, and some reached the Roman Empire, where they joined the imperial cavalry.

373–377. War with Persia. (See p. 175.)

374–375. Wars with the Quadi and Sarmatians. While trying to settle a border dispute peacefully, **Gabinus,** King of the Quadi, was treacherously killed by the Roman general **Marcellinus.** The infuriated Quadi immediately invaded and laid waste to Pannonia; all the upper and central Danube region flamed. While young **Theodosius** (son of Firmus' conqueror) held Moesia against the Sarmatians, Valentinian marched from Gaul, defeating the Quadi and driving them back over the Danube. He died of apoplexy while planning a punitive expedition north of the river (375). He was succeeded as emperor in the West by his son,

Flavius Gratianus (Gratian), a youth of sixteen.

376. Huns Invade Gothic Empire. While vainly attempting to repulse a major Hunnish invasion over the Dnieper, the ancient Ermanaric was killed, or committed suicide. His successor, **Withimer,** was soon after this also defeated and killed. The Ostrogoths—men, women, and children—now led by **Alatheus** and **Saphrax,** began to stream across the Dniester, seeking refuge from the Huns. Athanaric, leader of the Visigoths, planned to stand and fight the invaders, but most of his people, infected by the panic of their Ostrogothic cousins, also began to migrate en masse toward the Danube, under the leadership of **Fritigern** and **Alavius.** There were between 700,000 and 1,000,000 refugees, of whom more than 200,000 were warriors. Athanaric and the remainder of his people then sought refuge in the Carpathian and Transylvanian forests.

376. Goths at the Danube. The panic-stricken Visigoths appealed to Valens to grant them refuge and protection. Valens reluctantly agreed, on condition that the warriors give up their arms, and that all male children under military age be surrendered as hostages. The frantic Goths agreed to the terms and began to cross the Danube. Most of the boys were surrendered and were scattered through Asia Minor, but the Visigoths were slower to give up their weapons, bribing venal Roman officers with gold and other treasures, including the favors of their wives and daughters. Meanwhile, the remnants of the Ostrogoths reached the Danube and appealed for refuge in the empire. When this was refused, they crossed the river anyway, since the Romans were too busy trying to look after the Visigoths—and their women—to pay much attention to the new arrivals.

377. Outbreak of War between Goths and Romans. Roman officials in Thrace, not knowing what to do with the great influx of barbarians, took advantage of every opportunity to exploit and mistreat them. Fritigern and Alavius apparently tried to cooperate with their presumed protectors, but soon lost patience and began to negotiate with the Os-

trogoths to present a united front against the Romans. At about this time the Romans treacherously attacked the Visigothic leaders at a parley. Alavius was killed, but Fritigern escaped and immediately led his men in a successful attack against the forces of Lupicinus at **Marianopolis** (Shumla, eastern Bulgaria). He then joined forces with Alatheus and Saphrax in the region between the lower Danube and Black Sea (modern Dobruja).

377. Battle of the Salices (Willows). Valens, making a hasty peace with the Persians, sent strong reinforcements to Thrace. His generals **Saturninus, Trajan,** and **Profuturus** drove the Goths northward, blockading them between sea and river in the marshy region just south of the Danube mouth. Here the Goths made a stand behind the protection of their wagon forts. A bloody, indecisive battle ensued. While the Romans were reorganizing to make another attack, Fritigern made a successful secret move through the marshes and escaped. Once free of the blockade, the Goths streamed through Thrace and Moesia, joined in their mad, destructive rampage by raiding parties of Sarmatian, Alan, and Hun horsemen. Valens rushed back to Thrace with reinforcements from the East, sending an appeal for help to Gratian, his youthful nephew and coemperor.

378. Alliance among the Barbarians. Either by coincidence or—more likely—some sort of informal alliance among the Germanic tribes, the entire European border of the empire now erupted from the mouth of the Rhine to the lower Danube.

378. Gratian Defeats the Alemanni. Gratian, gathering forces to assist Valens, was forced instead to march to Gaul to meet serious Frank and Alemanni incursions. He defeated and killed **Prianus,** ruler of the Alemanni, in the **Battle of Argentaria** (Colmar), nearly annihilating the 40,000 invaders. Boldly crossing the Rhine, Gratian seriously punished the barbarians, pacifying the northern frontier. He then marched southeastward to help Valens.

378, July–August. Campaign in Thrace. Meanwhile, Valens' lieutenant, **Sebastian,**

had slowly gained the upper hand in Thrace. Relying primarily on light, mobile task forces of well-trained infantry and cavalry, he inflicted a series of defeats on the Goths and their allies, of which the most important occurred at the **Maritza River.** By early August the bulk of the combined Visigoth-Ostrogoth forces of Fritigern, Alatheus, and Saphrax were brought to bay in an immense wagon-camp fort, or series of wagon forts, on a hill in a valley some 8 to 12 miles from Adrianople. Here they were joined by a force of Gothic mercenaries deserting from the Roman army. The exact number of barbarian warriors present near Adrianople is not known, probably somewhere between 100,000 and 200,000, of which about half were infantry, mostly Visigoths, and about half were horsemen—Ostrogoths, Sarmatians, Alans, and possibly some Huns. Fritigern seems to have been in overall command, but Alatheus and Saphrax apparently exercised considerable independent authority. The Goths, having ravaged Thrace for several months, were now having serious difficulties obtaining food for themselves and their families (numbering at least 200,000 women and children). While the Visigothic infantry held the wagon camp, the horsemen spent most of their time foraging and raiding in central Thrace. Fritigern, realizing the danger of his situation, tried to negotiate with Valens, who was marching from Constantinople to reinforce Sebastian. Valens, however, jealous of Sebastian's and Gratian's successes, saw an opportunity for a great victory before Gratian, approaching from the north, could arrive to share the glory.

SECOND BATTLE OF ADRIANOPLE, AUGUST 9, 378

As soon as he reached Adrianople, Valens moved on the Gothic camp with his combined army, some 60,000 men, of which about two-thirds were infantry, the remainder heavy and light cavalry. Fritigern's scouts learned of the advance, and he sent for Alatheus and Saphrax, who—as usual—were out with their horsemen on a foraging expedition. In order to gain time for the Gothic cavalry to return, he also sent a message to Valens, offering to negotiate. Valens, whose troops were tired and sluggish after a long morning's march in the midsummer sun, ostensibly entered into negotiations, but in fact used the time to deploy his exhausted troops for an attack on the Gothic camp. He apparently neglected proper security patrols to the flank and rear.

While this was going on, battle began prematurely, apparently by Roman auxiliaries opening fire on a Visigothic negotiating party. Though the legions were still only partially deployed from their march column, the Roman cavalry was ready on the flanks; so Valens ordered a general attack. Just at this moment, Alatheus and Saphrax arrived on high ground overlooking the valley, where the battle was just beginning. The Gothic horsemen fell like a thunderbolt on the Roman right-wing cavalry just as they were reaching the wagon camp and swept them from the field. Some of the Gothic horsemen then streamed through the camp; others swept around behind the Roman army to attack the Roman left-flank cavalry, in coordination with Visigothic counterattacks from behind the wagon ramparts. The Roman cavalry was routed, leaving only the infantry on the field, still not completely deployed, and without maneuvering room. The Gothic cavalry now swarmed around the flanks and rear of the legions, while the Visigoths charged down from their camp on foot against the Roman front lines. The battle became a slaughter. Valens was soon wounded. He, Sebastian, Trajan, and 40,000 other Romans perished in this climactic defeat of the Roman legions. The Goths, unable to capture Adrianople or Constantinople, swarmed unchecked through Thrace.

379–395. Reign of Theodosius in the East. Gratian now called upon Theodosius, son of his father's great general and an able leader in his own right, to be Emperor of the East. The new emperor established himself at Thessalonika, built up the defenses of the principal cities of Greece and Thrace, and restored the morale and discipline of the dispirited army. With cooperation from Gratian, Theodosius sent small, mobile forces to punish Gothic detachments. Gradually (379–381) he built up the confidence of his troops, and in two campaigns reestablished a measure of order in portions of Thrace. He then felt strong enough to conduct two major campaigns against the Goths. He defeated Fritigern in central Thrace; the Gothic leader apparently died or was killed in the operations (382–383). At the same time Theodosius' general, **Promotus,** cleared the region south of the Danube by a series of successes against the Ostrogoths under Alatheus and Saphrax. Combining skill, prudence, daring, and conciliation, Theodosius finally pacified Thrace and Moesia; large numbers of the Goths were driven north across the Danube and even more were permitted to settle as peaceful citizens of the empire.

383. Revolt of Maximus. In Britain the general **Magnus Clemens Maximus** assumed the title of emperor, then invaded Gaul. Gratian moved to meet the threat, but after some minor setbacks was murdered by adherents of Maximus at Lugdunum (Lyon). Maximus then gained control of Gaul and Spain, while the young brother of Gratian, **Valentinian II,** became the ruler in Italy under the regency of his mother.

387–388. Civil War; Maximus against Theodosius. Invading Italy, Maximus drove out Valentinian II. Theodosius, marching to the assistance of his young coemperor, met the usurper at the **Save,** in Illyricum, winning by a bold and unexpected river crossing (388). Maximus fled to Italy, closely pursued by Theodosius, who besieged the usurper at Aquileia (near Venice). Maximus was murdered by his followers, who then surrendered to Theodosius; Valentinian II was reinstated as Augustus of the West.

387. Renewed Gothic Inroads. Alatheus led a new invasion over the Danube. He was defeated and killed by Promotus. Theodosius then imposed easy terms on the Goths.

390. First Appearance of Alaric. The young Visigothic chieftain **Alaric** raided across the Danube into Thrace, but was soon subdued by Theodosius. Alaric was then permitted to join the Roman army with most of his men, in a typical instance of Theodosius' mixture of firmness and conciliation.

392–394. Revolt of Arbogast and Eugenius. After the defeat of Maximus, the Frankish general **Arbogast** pacified Gaul at the behest of Theodosius (388–389). Having gained almost complete authority in Gaul, he resented the interference of the youthful Valentinian, and evidently instigated his murder (392). Arbogast then appointed his protégé **Eugenius** as emperor, and consolidated his position in Gaul by two successful campaigns against Frankish invaders along the Rhine. Theodosius, however, refused to accept the usurpers and, with an army composed mainly of Goths (including Alaric), marched to avenge the death of Valentinian II. Arbogast and Eugenius met Theodosius in northeastern Italy.

394, September 5–6. Battle of Aquileia or of the Frigidus (Vipacco, a tributary of the Isonzo). Theodosius recklessly attacked a position carefully chosen by Arbogast and was repulsed with heavy losses. Rallying his troops, Theodosius spent the night alternately in prayer and in the reorganization of his shattered formations. Next day he attacked again, and apparently was aided by the effects of a violent windstorm. By evening of the second day of the battle, Theodosius, thanks largely to the efforts of his brilliant Vandal general, **Stilicho,** had won a complete victory. Eugenius was killed; Arbogast committed suicide two days later. The Empire was reunited.

395–400. Renewed Chaos in the Empire. The death of Theodosius was a signal for a new rising of the barbarians against his relatively inept sons: **Arcadius** in the East and **Honorius** in the West.

395–396. Alaric against Stilicho. Alaric, marching and pillaging practically without opposition through Thrace and Greece, was only temporarily deterred by the genius of Stilicho, who had promptly marched with his army from Italy to Greece (396). But Arcadius, more jealous of his brother and his great Vandal general than he was fearful of the Goths, insisted that Stilicho leave the Eastern Empire. Reluctantly Stilicho obeyed the order, and Alaric resumed his plunders unhindered. At the same time two other rebellious Gothic leaders—an Ostrogoth, **Tribigild,** and **Gainas,** a former Visigothic general of the Roman army—were terrorizing Asia Minor and the region around Constantinople and forced Arcadius to pardon them (399). Gainas, made master general of the armies, was soon overthrown by trickery, and

was defeated by the loyal Gothic general **Fravitta** while trying to cross the Hellespont with a small army. With the remnants he fled north and was defeated and killed in Thrace by **Uldin,** King of the Huns (400).

395–400. The Rise of Stilicho. Only the strong character and ability of Stilicho kept the empire from collapse. In addition to his brief campaign against Alaric in Greece, Stilicho generally maintained the authority of Honorius in the West. One of his most brilliant victories was gained in Africa subduing a rebellion by the Moorish leader **Gildo** (396). By the end of the century Stilicho was the uncrowned ruler of the West.

SOUTHWEST ASIA

DECLINE OF PARTHIA, 200–226

Torn by a series of violent civil wars, the Parthian Empire's collapse was hastened through defeats by Roman emperors **Septimus Severus** (see p. 143) and **Caracalla** (see p. 151). **Artabanus V** rallied from this latter disaster to defeat weak and vacillating Roman emperor **Macrinus** at **Nisibis** (217), regaining all of the territory captured by Caracalla. Meanwhile, Artabanus was engaged in a debilitating civil war with his brother, **Vologases V,** which he finally brought to a conclusion by defeating (and probably killing) Vologases in southern Babylonia (222). But the drain on Parthian strength caused by Artabanus' efforts to gain these victories facilitated the rise of an even greater threat to his throne.

SASSANID PERSIA, 226–400

c. 208–226. Rise of Ardashir of Sassan (Central Persia). Subduing his unruly neighbors in the region around Persepolis, Ardashir (or Artaxerxes), the Sassanid ruler, challenged the weakened authority of Artabanus (c. 220). A series of campaigns was climaxed by the bloody **Battle of Ormuz** (southern Iran, on the Per-

sian Gulf), in which Artabanus was killed (226).

226. Establishment of the Sassanid Empire. Ardashir seized Ctesiphon, the Parthian capital, and declared himself the successor of the Achaemenid Dynasty of Cyrus and Darius. In a series of whirlwind campaigns he reasserted Persian supremacy over most of the former dominions of Darius. He defeated the Massagetae and other Scythians who had been harassing the northern boundaries of Parthia, to gain uncontested control over Hyrcania (south Caspian shore), Khorasan, and Kharesan (Oxus Valley). He also conquered Kushan dominions between the Oxus and Jaxartes rivers, as well as in the mountains of modern Afghanistan and Baluchistan. Only in the west were his ambitions balked by the wavering, but still formidable, power of Rome.

230–233. Ardashir's War with Rome. Ardashir demanded that Rome withdraw from all of her Asiatic provinces. When this demand was ignored, he invaded Syria and Armenia (ruled by **Chosroes,** Parthian-Arsacid vassal and ally of Rome). Some Persian scouting and raiding parties reached the Mediterranean near Antioch and pushed into the mountains of Cappadocia in Asia Minor (230–231). Alexander Severus responded to this threat by gathering a large army at Antioch, then marching eastward on a broad front. One Roman column went to the assistance of Chosroes in Armenia, another was sent toward Babylon, along the Euphrates River; the main force, under Alexander himself, reconquered Roman Mesopotamia, marching by way of Carrhae and Nisibis. Apparently, according to incomplete records, the coordination between the Roman columns was poor; the Romans seem to have won all of the major engagements, but lacked the ability to exploit success. Having reached the Tigris, but also having sustained heavy losses, Alexander withdrew to Roman Mesopotamia, and the war simply stopped by mutual consent (233). Desultory fighting continued for several years in Armenia, where Chosroes, with Roman assistance, repulsed sporadic Sassanid invasions.

241–244. Shapur's First War with Rome. Shortly before his death, Ardashir raised his son, **Shapur,** to be coruler. The Persians decided to exploit Rome's internal difficulties (see p. 152), and Shapur was leading an army into Mesopotamia when his father died. The new emperor continued his advance, capturing the Roman outposts of Nisibis (Nusaybin, southeastern Turkey) and Carrhae (Haran), then penetrated into Syria (242). Gordianus III, however, assisted by his father-in-law, **G. F. Sabinus Aquila Timesitheus,** inflicted a decisive defeat on Shapur at the **Battle of Resaena,** on the upper Araxes (Araks) River (243). Gordianus was planning to exploit this victory when he was murdered and succeeded by Philippus Arabus, who hastily concluded a peace generally favorable to Shapur, leaving the Roman-Sassanid boundaries essentially unchanged.

c. 250. War with the Kushans. Shapur defeated the Kushan ruler, **Vasuveda,** somewhere in Bactria or Gandhara, ending Kushan influence in Central or Southwestern Asia (see p. 177).

258–261. Shapur's Second War with Rome. Again Rome's internal decay invited attack. Chosroes of Armenia was murdered at the in-

stigation of Shapur, who then overran and conquered Armenia (258). Continuing into Syria, he captured and plundered Antioch (258). Soon after this, however, **Valerian** arrived from the west and apparently defeated Shapur in a number of minor engagements in central and eastern Syria (259). He drove the Persians from Antioch and back across the Euphrates. Shapur, however, defeated Valerian at the **Battle of Edessa** (Urfa) and succeeded in blockading the Romans (260). His army surrounded and in dire straits, Valerian opened negotiations with Shapur, who treacherously captured the Roman emperor. The Roman army surrendered, and Valerian was sent to Persia, where he died in captivity. Shapur promptly marched back into Roman territory, devastating Syria, Cilicia, and Cappadocia. Delayed for a long time by the tenacious defense of Caesarea in Cappadocia by the general **Demosthenes,** Shapur's final capture and sack of that city appeared to assure the collapse of all Rome's dominions in Asia (261). But a new threat to his lines of communication caused Shapur to withdraw from Asia Minor.

261–266. Shapur's Wars with Odaenathus of Palmyra. The Persians were driven from Rome's Asiatic provinces (see p. 153).

264–288. Persian Occupation of Armenia. Weakened by his defeats by Odenathus, and occupied with troubles elsewhere in his vast dominions, Shapur maintained only a tenuous hold over Armenia. He appointed as his satrap there the somewhat mysterious Chinese refugee prince, **Mamgo** (see p. 178), who with his army of Chinese cavalry for several years maintained nominal Persian control over the unruly Armenian nobility (c. 275–288).

282–283. Carus' Invasion of Mesopotamia. (See p. 156.) This was facilitated by a dynastic war between **Bahram I,** son of Shapur, and his brother, **Hormizd.**

288–314. Reign of Tiridates III of Armenia. Diocletian determined to reestablish Roman suzerainty over Armenia. **Tiridates,** son of Chosroes, had taken refuge in the Roman Empire. Diocletian sent the Armenian prince back to his homeland with the support of a small

Roman force. At first opposed by Mamgo, Tiridates slowly gained ground as the Armenian nobility rose against the Persians and their Chinese mercenaries. Finding himself unsupported by the Persians, Mamgo changed sides and helped Tiridates regain control of Armenia and expel the remaining Persians. Tiridates now invaded Assyria (293).

294–295. Persian Conquest of Armenia. Having overthrown his nephew, **Bahram III,** the new Persian emperor, **Narses,** turned to punish Tiridates. Driving the Armenians from Assyria, Narses soon reconquered all of Armenia; Tiridates fled to refuge in the Roman Empire.

295–297. Persian War with Rome. The Persian invasion of Armenia and the expulsion of his protégé, Tiridates, caused Diocletian to declare war. He placed his colleague **Galerius** in charge of the campaign. Tiridates joined Galerius with a contingent of Armenian exiles. Narses, meanwhile, invaded Roman Mesopotamia. With a relatively small army, Galerius marched into Mesopotamia, gaining a number of minor successes against the wary Persians (296).

296. Battle of Callinicum. Near Carrhae, site of Crassus' defeat, Galerius undertook a rash piecemeal attack against Narses and the main Persian army. The Romans were decisively defeated; Galerius and Tiridates escaped with only a remnant of their forces. Diocletian publicly rebuked Galerius at Antioch, but did not relieve him from command. With a reinforced, reorganized army, Galerius marched back against the Persians, while Diocletian sent for more troops from the Danube frontier.

297. Victory of Galerius over Narses. Marching by way of southern Armenia, Galerius with about 25,000 men encountered Narses with 100,000 at an unidentified site, probably along the Tigris River between Amida and Nisibis. Galerius routed the Persians in a surprise attack, capturing much booty and many prisoners—including the family and harem of Narses. The Persian emperor sued for peace.

297. Peace of Nisibis. Diocletian and Galerius, their joint armies now assembled at Nisibis,

dictated stiff terms. The Persians gave up their claim to Roman Mesopotamia; they ceded five provinces northeast of the Tigris River to Rome; they recognized Roman suzerainty over the entire Caucasus region, and recognized Tiridates as ruler of the Roman protectorate of Armenia. In return Narses got back his wives and concubines. This peace lasted for about 40 years, though intermittent bickering continued along the eastern borders of Armenia, largely inspired by religious differences between Christian Armenians (Tiridates III had been converted to Christianity about 300) and Zoroastrian Persians.

c. 320–328. Arab Wars. Thair, a king of Arabia (or Yemen), led a successful expedition into Persia. Eight years later **Shapur II** (then about 18) led a punitive expedition into Arabia, defeating Thair (328).

337–350. Wars of Constantius and Shapur II. Shapur II, probably provoked by religious differences, suddenly invaded Roman Mesopotamia (337). Constantine prepared for war, but died before he could take the field. His son, **Constantius,** continued the preparations, and marched to meet Shapur in Armenia. This was the beginning of a drawn-out war, very inadequately recorded. Apparently 9 major battles were fought. The most renowned was the inconclusive **Battle of Singara** (Sinjar, in Iraq), in which Constantius was at first successful, capturing the Persian camp, only to be driven out by a surprise night attack after Shapur rallied his troops (344—or 348?). Gibbon asserts that Constantius was invariably defeated by Shapur, but there is reason to believe that the honors were fairly evenly divided between two capable antagonists. (Since Singara was on the Persian side of the Mesopotamian frontier, this alone is evidence that the Romans had not seriously lost ground in the war up to that time.) The most notable feature of this war was the consistently successful defense of the Roman fortress of Nisibis in Mesopotamia. Shapur besieged the fortress 3 times (337, 344?, and 349), and was repulsed each time by Roman general **Lucilianus.** The war was concluded—or at least recessed—by a hasty truce patched up between the two emperors, who simultaneously were faced with severe threats in other regions of their respective domains (350).

349–358. Scythian Raids on Northeastern Persia. Destructive invasions by the Scythian Massagetae and other Central Asian tribes forced Shapur to hasten eastward to restore the northeastern borders of his empire. Most able and most persistent of his opponents was **Grumbates,** ruler of the Chionites (353–358). Finally defeating Grumbates, Shapur enlisted the barbarian leader and his light cavalrymen into the Persian army.

358–363. Renewed War with Rome. Having restored order in his eastern dominions, Shapur returned to his struggle with Rome. With an army that included the Chionites of Grumbates, the Persian emperor invaded southern Armenia, but was held up by the valiant Roman defense of the fortress of Amida (Diyarbekir, in Turkey), which finally surrendered after a 73-day siege in which the Persian army suffered great losses (358). The delay forced Shapur to halt operations for the winter. Early the following year he continued his operations against the Roman fortresses, capturing Singara and Bezabde. (The fact that the Romans were still in possession of this town in northeastern Assyria, east of the Tigris, is further evidence that Constantius had not been defeated in his earlier war with Shapur.) Constantius arrived from the west at this time, and unsuccessfully tried to recapture Bezabde. Finding his forces too weak to undertake a major counteroffensive, he sent for his cousin Julian to help. This led to civil war between Constantius and Julian, which was resolved (361) by the death of Constantius (see p. 168).

362–363. Julian's Invasion of Persia. (See p. 168.)

363. Shapur's Victory over Jovian. (See p. 169.)

364–373. Struggle over Armenia. By treachery, Shapur captured and imprisoned **Arsaces III** of Armenia, who soon committed suicide. He then endeavored to establish Persian control over the country. The Armenian nobles, however, led by **Olympias,** widow of Arsaces, resisted bitterly. Roman Emperor Valens then lent support to the Armenian cause, giving clandestine assistance to **Para** (or Pap), son of Arsaces, who established a "government-in-exile" in Roman Pontus.

373–377. Renewed War between Persia and Rome. Valens' support to Para, in violation of Jovian's treaty of 363, caused Shapur to declare war. While Valens remained in Antioch, his generals defeated the Persians in southern Armenia and again near Nisibis. The Roman-Armenian alliance was marked by intrigue and perfidy, however, and the Roman general **Trajan** had Para assassinated (374). Gradually Shapur gained the upper hand in Armenia. Valens now being forced to concentrate much of his attention on affairs along the Danube, the war trailed out inconclusively. Shapur, realizing he could never gain a major victory over the Romans, and satisfied with his conquest of Armenia, was happy to make peace when Valens had to hasten back to meet the Gothic threat in Thrace (377). Unrest and violence continued in Armenia, however, the Armenian nobles constantly endeavoring—with some success—to obtain Roman assistance against Persia.

390. Partition of Armenia. The prolonged wars of the Roman and Sassanid empires were brought to a conclusion by a treaty between **Theodosius** of Rome and **Bahram IV** of Persia, which divided Armenia between the two empires.

ARABIA AND ABYSSINIA 200–400

220–400. Rise of the Kingdom of Aksum. Centered at Aksum (in modern Tigré province), this state grew rich from the trade flowing through its ports, and controlled most of modern Ethiopia, eastern Sudan, and northern Somalia, as well as the Minaean and Sabaean kingdoms in Yemen, east of the Red Sea.

c. 325. Ezana. This monarch gained the throne of Aksum between 320 and 325. He annexed the remnants of the ancient Cushite Kingdom of Meroe on the Nile, solidified Aksumite control in Yemen, and helped Christianize the kingdom. He died sometime after 345.

c. 350. Aksum Destroys Meroe. A major expedition from Aksum sacked Meroe, destroying the Cushite Kingdom, which had been waning for several centuries as trade routes shifted from the Nile to the Red Sea.

c. 375. Sabaean Revolt. A revolt among the Jewish Sabaeans in northeastern Yemen restored that area's independence and weakened Aksumite control in southwestern Arabia.

SOUTH ASIA

The 3rd century A.D. is another frustrating period of obscurity in the history of India, perhaps due to the fact that no major power appeared on the scene during that time.

200–250. Decline of the Kushans. The Kushan, or "Saka," Dynasty established by Kanishka lingered on into midcentury, to be crushed when the Sassanid Emperor **Shapur I** completely defeated the Kushan King **Vasuveda,** probably in Bactria or Gandhara (c. 250). However, Persian internal disorders, and the obsession of Shapur and his successors with the struggle against Rome, kept the Sassanids from seriously following up this success. Some Sassanid expeditions apparently reached the Indus River region in the latter quarter of the century. As a result of the expulsion of the Kushans from Central and Western Asia, Hindu India lost even nominal ties with those regions. Otherwise this defeat had little effect on affairs in Hindustan; the Saka kings and princes still dominated the Punjab, Rajputana, and Sind; the smaller Hindu states to the east and southeast continued their decentralized, warring existence.

c. 230. Fall of Andhra. Farther south, the old ruling dynasty of Andhra fell, to be replaced by the vigorous Pallava family. The new Pallava-Andhra regime began a tentative expansion of its influence in the eastern Deccan, and north toward the Ganges, as well as pressing south against the Tamil Cholas.

c. 300. Revival of Magadha. A new ruler of ancient Pataliputra began to raise Magahda once more to a position of ascendancy in the Ganges Valley comparable to that which it had held in the time of the Mauryas. **Chandra-gupta** (Chandra Gupta, or Candra Gupta) of Pataliputra claimed to be descended from his great namesake, founder of the Maurya Dynasty. After a number of victories that brought most of the central Ganges region under his control, he had himself crowned "Chandragupta I, King of Kings," establishing the Gupta Dynasty (320).

330–375. Reign of Samudragupta (son of Chandragupta). He was the greatest military leader of the Gupta line. Like his father, his aim was to reestablish the power, glory, and geographical area of the old Maurya Empire. He came close to his goal, conquering Rajputana, most of the northern Deccan, and the east coast region southwest of the Ganges mouths. In the process he subjugated the Pallava kings of Andhra. He also gained a more tenuous authority over Nepal, Assam, the Punjab, and neighboring Gandhara. He made at least one successful raid deep into south India. The area nominally under his control was not much smaller than that governed by Asoka; his actual authority, however, was substantially less extensive and less clear-cut, in part because Gupta administration was not centralized or well-controlled, and was significantly feudal in nature.

c.350. Temporary Eclipse of Pallava-Andhra Influence. Defeat at the hands of the Guptas seriously weakened the growing power of Pallava. As a result the local principalities of the Deccan grew in strength. One of these—the warlike Vakataka Dynasty—which dominated the region south of Nagpur, appears to have been primarily responsible for limiting Gupta expansion farther south.

375–413. Reign of Chandragupta II. The son

of Samudragupta, **Chandragupta II,** brought the Gupta Empire to its zenith. He defeated the Saka princes of the Punjab, establishing in fact the purely nominal sovereignty his father had exercised over that region. Symbolic of this victory was the addition to his name of that of **Vikramaditya,** legendary Andhra conqueror of the early Sakas. He also annexed Malwa, Saurashtra, and Gujarat. Under his wise and able administration, Hindustan experienced its golden age of art, literature, and learning.

EAST ASIA

CHINA

196–220. Last Years of the Han Dynasty. The actual ruler of China during the last quarter-century of nominal Han ascendancy was the warlord **Ts'ao Ts'ao** (see p. 146). Most of the regions south of the Yangtze River, however, refused to acknowledge his authority. In an effort to reestablish central control farther south, he sent a naval expedition up the river. This was defeated by the fleet of a coalition of southern warlords in the **Battle of the Red Cliff** (208).

220–280. The Era of the Three Kingdoms. Upon the death of Ts'ao Ts'ao, his son, **Ts'ao P'ei,** deposed the last of the Han emperors and established himself as the first ruler of the Wei Dynasty (220). Since the regions south of the Yangtze still refused to recognize the central authority of the Ts'ao Ts'ao family, the former Han Empire now fell apart into 3 major regions: Wei north of the Yangtze, Shu to the southwest, and Wu to the southeast. The constant strife among the trio offered increasing opportunities for barbarian raids along the outer fringes of Chinese civilization. This was the beginning of a period of anarchy in China that would last—with rare interludes of peace—for almost 4 centuries.

220–234. Rise of Shu. The Shu, or Shu Han, Dynasty of southwest China was established by **Liu Pei,** member of a collateral line of the Han family. His capital was at Ch'engtu. The period of greatest strength and prosperity of the Shu kingdom was under the leadership of its great general and prime minister, **Chu Ko Liang** (223–234).

222–280. The Kingdom of Wu. This kingdom, and dynasty, founded by **Sun Ch'uan,** was the first of the so-called ''Six Dynasties'' whose capital was Nanking (220–589). The kingdom extended along the seacoast from the lower Yangtze Valley to below Tongking. During its period of greatest power (c. 250), this maritime kingdom sent a naval-commercial reconnaissance mission into the Indian Ocean.

234–264. Decline of Shu. After the death of Chu Ko Liang, the Shu kingdom steadily lost strength. It was conquered by Wei general **Ssu Ma Yen.** After this defeat, the remnants of the Shu army—apparently accompanied by their families—fled westward through Turkestan to Persia under the leadership of a Han prince, known to Persian history as Mamgo. He and his followers offered their services to Shapur I, who welcomed them. Some years later Ssu Ma Yen—by this time emperor of China—sent a demand to Shapur to surrender Mamgo and his adherents or else suffer a Chinese invasion. Shapur, facing war with Rome, wished to retain good relations with China; he simply sent Mamgo and his people to Armenia, then informed Ssu Ma Yen that they had been banished to ''certain death'' at ''the ends of the earth.'' This apparently satisfied the Chinese emperor (see p. 174).

265. Appearance of the Ch'in Dynasty. Seizing the throne of Wei, Ssu Ma Yen established the Ch'in Dynasty, with himself as the first emperor, under the title of **Wu Ti.** He then conquered the kingdom of Wu, to reunite China temporarily and to end the Era of the Three Kingdoms (280).

290–316. Return to Chaos. The death of Wu Ti signaled a new outbreak of civil war. In the southwest, Szechwan established its indepen-

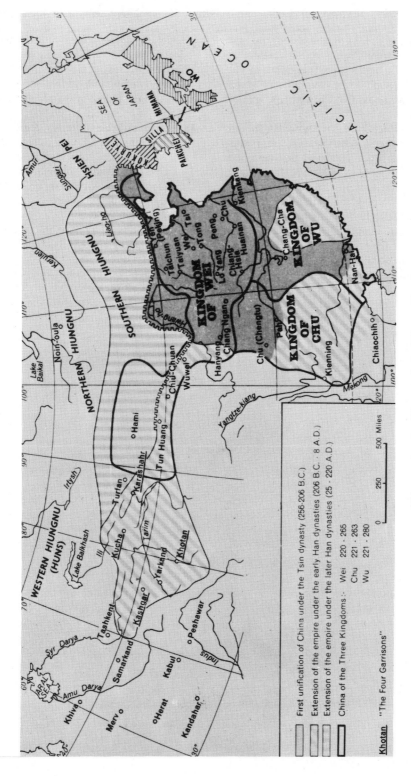

China under the Hans and the Three Kingdoms

dence (304). In the north, the Hsiung-nu and Hsien Pi (Hsien Pei) barbarians raided unopposed, and soon held most of the country north of the Yangtze Valley. The Hsiung-nu captured and killed two successive Chin emperors (311 and 316).

317–420. The Eastern Chin Dynasty. Driven from their northern capitals, the Chin emperors abandoned north China and reestablished themselves south of the Yangtze at Nanking, to continue as the Eastern Chin Dynasty, with their authority limited to South China.

317–589. Anarchy in North China. Northern China was now completely overrun and dominated by barbarians—or their direct descendants—for more than 2½ centuries. It was a period of utmost confusion: ceaseless fighting among the barbarians in the region, with an almost constant influx of new waves of barbarians, punctuated by frequent raids to the south, into the remaining dominions of the Chins. Initially the Hsiung-nu predominated.

Later the Ch'iang tribe of Tibet became predominant (c. 350).

370. Reconquest of Szechwan. The beginning of a very slow and gradual rise of the Eastern Chins was marked by the reconquest of Szechwan by General **Huan Wen.**

383. Defeat of the Ch'iangs. A great invasion of Chin territory by Ch'iang ruler **Fu Chien** was decisively defeated by the Chinese at the **Battle of the Fei River.** Following this, the other barbarians in North China revolted against the Ch'iangs, bringing about the immediate collapse of their regime.

386–534. Rise of the To Pa. A branch of the Mongolian or Turkish Hsien Pi people, the To Pa (or T'u Pa, or Toba) tribe now established themselves as the dominant power in North China. Calling themselves the Northern Wei Dynasty, they held most of the region north of the Yangtze, including much of Mongolia and Turkestan.

VII

THE OPENING OF
THE MIDDLE AGES:

400—600

MILITARY TRENDS

This period witnessed one of the great turning points of world history: the dissolution of the decaying Western elements of the divided Roman Empire and the consequent emergence of the Empire of the East as the sole guardian of Western civilization. The decline of the old Empire was accompanied by a parallel decline in military leadership, ingenuity, and systematic study of war as an instrument of national policy. The crumbling of the West ushered in a new military era; armed force was applied vigorously and violently, but with little inspirational spark.

The Eastern Roman Empire was a partial exception, yet even there—with only slight modification during the reign of **Justinian**—national policy and military strategy were fundamentally defensive and passive in nature. (See chapter VIII, p. 233, for discussion of the Byzantine military system.)

As might be expected under these circumstances, there were new military developments. The principal trend of the preceding era—predominance of cavalry—was accentuated in this, the opening phase of the so-called Middle Ages.

The one truly outstanding general during this early medieval period, **Belisarius,** was an exception to the trend and may even deserve consideration as one of the great captains of history. **Narses,** who had comparable skill in employment of combined arms, unquestionably learned the art of war from Belisarius (despite the parodox that he was nearly 30 years older). **Stilicho,** and possibly **Aeitus,** were also professionals in the Roman tradition, far excelling their contemporaries and immediate opponents in skill, ingenuity, leadership, and the ability to integrate the efforts of infantry and cavalry in their battles. These four leaders were distinctly superior to **Attila,** who was an able tactician and an inspiring leader of light cavalry, but apparently unworthy of serious comparison with later outstanding Mongol, Turk, Tartar, or Arab leaders of light cavalry. **Alaric, Odoacer, Gaiseric, Theodoric,** and **Clovis** evidently owed their victories simply to superior vigor, craftiness, and charismatic inspirational ability in

181

comparison to their foes. **Vitiges** and **Totila** were probably as able as any of these; they simply had the misfortune to be opposed by Belisarius and Narses.

THE BARBARIAN MERCENARIES

To an ever-increasing degree the East and West Roman empires made use of barbarian mercenaries in the early 5th century. These were enlisted in the dwindling number of regular units of the army, or more frequently were incorporated as **federati** in their respective national or tribal organizations under their own barbarian tribal leaders. Quite naturally, with the growing importance of cavalry, imperial generals favored the tribes who were natural horsemen. Thus the tribes of Asian origin—Huns, Alans, Avars, and Bulgars—were enlisted as light-cavalry bowmen. The German tribes inhabiting the plains between the Danube and Black Sea—mainly Goths, Heruli, Vandals, Gepidae, and Lombards—provided heavier cavalry who relied upon shock action with lance or pike.

The East Roman army got many of its footmen from within its own provinces, particularly in Anatolia, though some barbarian foot soldiers—Germans and Slavs—were also enlisted. In the West, only a relatively small percentage of the armies were from the Roman provinces, and these mainly from Gaul. Thus foot soldiers as well as cavalrymen were obtained from the barbarian tribes, with the infantry coming mainly from the Franks and Burgundians.

The example of the fall of Rome undoubtedly influenced the East Roman Emperor **Leo I** and his successor **Zeno** to reduce their reliance upon federati. They vigorously endeavored to raise units from within the confines of their borders. The Isaurians, in particular, began to replace the hitherto preponderant foreign barbarians in the East Roman army. By the time of Justinian the imperial armies were almost equally divided between federati and professional citizen troops.

CAVALRY TACTICS

In the early part of the 5th century, as the Romans were endeavoring to adjust their military system to the catastrophic lessons of the Battle of Adrianople, they particularly admired the effectiveness of Hun horse archers. This, combined with experience against the Persians, influenced the development of Roman cavalry. At first this new type of Roman cavalry was light and distinguished from the barbarian federati only by better organization and discipline. Gradually it became more heavily armored, and carried lance, sword, and shield as well as bow. Thus emerged (by the beginning of the 6th century) the **cataphract,** who, as the mainstay of the Byzantine army of the future centuries, would be the most reliable soldier of the Middle Ages. This heavy Roman horse archer combined fire power, discipline, mobility, and shock-action capability. He was the true descendant of the Roman legionary.

In his battles with Persians, Vandals, and Goths, Belisarius relied primarily upon his cavalry, which consisted of three main types: the heavy Roman horse archer, the light Hun federati horse archer, and the less important and less reliable Goth, Heruli, or Lombard heavy pike cavalrymen.

Only when combined with steadier organizations, and incorporated into a regular army organization with adequate logistical arrangements, could the Asiatic light cavalry conduct sustained operations. This was done successfully by the Romans and the Chinese. Otherwise, they were severely limited in their effectiveness. They could move only where they could find pasturage for their horses and (since they usually moved as tribes in great wagon trains) for the flocks of domestic animals that accompanied them. They were unable to campaign effectively in winter due to the lack of forage for horses. So, since they could only live off the country, and since they had so many animals foraging, they could never stay long in one spot, regardless of the requirements of the military situation.

In battle the Asian light cavalryman always avoided shock action against an organized foe. If long-range harassment with arrows was sufficiently effective, the Hun and his cousins would occasionally close to annihilate the enemy with sword and lasso. But if the foe could sustain punishment and return it, or if he was able to withdraw in good order, the Hun could not reach decisive conclusions and had to be satisfied with whatever raiding and looting he could accomplish.

INFANTRY TACTICS

By the end of the 5th century the legion had disappeared from the armies of the West and from the East Roman Empire. The frantic Roman search for a new answer to the combined threats of missile weapons and cavalry shock action had been relatively successful in the cavalry arm, but had failed utterly in the infantry. There was no infantry organization anywhere in the world that combined both the strength and the flexibility of the old legion. It was one extreme or the other—a relatively immobile mass of heavy infantry or swarms of lightly armed and armored missile-throwing skirmishers.

In consequence, with one significant but transitory exception, infantry became completely subsidiary to cavalry. The ponderous masses (sometimes combined with missile engines such as ballistae) could provide a base of maneuver for mobile cavalry. The light skirmishers—bowmen, javelin throwers, or both—could confuse, distract, and soften up a foe for the climactic shock of a cavalry charge. But (save for the East Romans and probably the Chinese) the military leaders of the age found it difficult to coordinate such combinations of missile and shock, steadiness and mobility. The result was stereotyped tactics and a general reversion to methods of warfare antedating the Battle of Marathon.

The one important exception to reliance upon cavalry was among the Franks. They had crudely combined infantry mass with mobility, missile power, and shock action. Lightly armored—or almost completely unarmored before the 6th century—the Franks rushed into action in dense, disorganized masses, comparable to the tactics their ancestors used against the early Romans. Just before contact with their enemy they would hurl a heavy throwing ax (**francisca** it was called by the Visigoths) or a javelin, then dash in with sword to take advantage of the confusion thus created. The fearless barbarians awaited cavalry charges in their dense masses, then swarmed around and under the stalled horsemen, cutting down mount and rider.

Oman* suggests that the Franks had learned nothing from their centuries of contact with the Romans so far as weapons, discipline, or tactics were concerned, and that their successes were due simply to their extraordinary vitality, and to the degeneration of the military art among

* Charles Oman, *A History of the Art of War in the Middle Ages,* London, 1898, Book II, Chapter I.

their enemies. There is undoubtedly much to this; yet it cannot be the complete answer. There is evidence that Clovis was able to instill some discipline in his fierce warriors, and that he was an admirer of the Roman military system. His armies were small, and apparently he was outnumbered by his foes in many of his campaigns—particularly against the heavy cavalry of the Visigoths. He could not, then, have been victorious without a greater degree of organization and tactical control than Oman would suggest.

Frankish experiences against Ostrogoths, Visigoths, Lombards, Avars, and East Romans gave them a healthy respect for cavalry, and by the end of the 6th century they too had come to rely upon the shock action of heavy, lance-carrying horsemen.

The Angles, Saxons, and Jutes, in their invasion of Britain, employed the same kind of disorganized infantry tactics as the Franks. Infantry remained predominant in Britain simply because there was no challenge from an enemy possessing good cavalry.

NAVAL DEVELOPMENTS

If anything, war at sea had declined to an even greater degree than on land. The East Roman navy showed flashes of effectiveness and retained a measure of control over the Black Sea and eastern Mediterranean. Yet Teutonic sea raiders consistently flouted the naval might of Constantinople, and the sailors of the East were not even able to prevent mass crossings of the Hellespont. **Gaiseric** the Vandal displayed considerable ingenuity and energy in organizing his long-range marauding expeditions and in dealing with threats to his own African coasts. But it seems doubtful that the Vandal fleets—or indeed those of any of their contemporaries—could have stood up to the navy of Themistocles or the Roman and Carthaginian fleets of the Punic Wars.

EUROPE–MEDITERRANEAN

ROME AND THE BARBARIANS, 400–450

The West

401. Alaric's First Invasion of Italy. With his Gothic-Roman army, **Alaric** marched from Thessalonika through Pannonia and across the Julian Alps (October). He besieged and captured Aquileia (at the head of the Adriatic), then overran Istria and Venetia. **Stilicho,** Patrician of the Empire and virtual ruler of the West for the incompetent Emperor **Honorius** (see p. 173), was taken by surprise. However, by harassing and delaying actions, his outnumbered forces so hindered the Gothic advance that Alaric wintered in north Italy.

401–402. Stilicho Raises an Army. Stilicho personally crossed the snow-covered Alpine passes in midwinter to raise forces among German federati in Raetia and southern Germany. At the same time he ordered most of the Rhine garrisons and units in Gaul to join him in Italy.

402. February (?). Siege of Milan. Stilicho returned (probably via St. Gotthard or Splügen) with his German levies, an exceptional logistic feat in winter, particularly for his contingent of Alan cavalry. Joined by units from Gaul, he advanced rapidly toward Milan, which Alaric had invested. Alaric raised the siege and pressed southwestward after Honorius, who had fled to Asta (Asti). He apparently hoped to defeat the Romans in detail and to capture the emperor before Stilicho could interfere. (The exact sequence of events before April 6 is obscure.)

402, March (?). Battle of Asta. Stilicho pursued Alaric and seems to have had the better of an inconclusive battle outside Asta. Alaric withdrew up the Tanarus (Tanaro) River to the vicinity of Pollentia (Bra). Stilicho again pursued.

402, April 6. Battle of Pollentia. Early Easter morning, Stilicho attacked, surprising Alaric. The Goths rallied and Alaric's disciplined Gothic cavalry finally drove Stilicho's Alani federati off the field. Meanwhile, however, Stilicho's infantry had thrown back the Gothic foot troops and captured Alaric's camp. Alaric thereupon withdrew across the Apennines toward Tuscany, while opening negotiations with Stilicho and Honorius. A treaty followed: Alaric agreed to leave Italy, and retired to Istria for the winter.

403, June. Battle of Verona. Alaric, however, had no intention of giving up his invasion. He planned to advance through the Brenner Pass to Raetia as soon as the passes were clear of snow, thence to Gaul, now weakly garrisoned. Stilicho learned of the Gothic plan. Somewhere north of Verona in the narrow valley of the Athesis (Adige) River, the Goths were stopped by part of Stilicho's alerted army. Stilicho then attacked from the rear, inflicting a crushing defeat. Apparently Alaric was able to rally his troops sufficiently to withdraw eastward in good order—possibly via modern Trent or Bassano. Or, possibly, he made a deal with Stilicho. Honorius now moved his capital from Milan to heavily fortified Ravenna, inaccessible behind lagoons and marshes, save by two narrow, easily defensible spits of land.

404. Reconciliation of Alaric and Stilicho. Alaric renounced his adherence to the Eastern Emperor Arcadius, was appointed by Honorius (really by Stilicho) as master general of Illyricum, and was subsidized to hold that province—which he had previously held for Constantinople—for Rome against barbarians and against the Eastern Empire.

405. Invasion of Radagaisus. Sweeping down from the Baltic area of modern Poland,

Lithuania, and East Prussia came what was possibly the largest of all the barbarian migrations. The mixed Germanic peoples in that region, finding Hunnish pressures intolerable, banded together under a chieftain named **Radagaisus.** The exact numbers in this movement are unknown, probably nearly 500,000, of whom about one-third were warriors. Vandals, Suevi, and Burgundians predominated; there were also numbers of Goths, Alans, and other smaller tribes. The invaders apparently crossed the Alps late in the year and spent the winter in the Po Valley, observed by Stilicho, who lacked sufficient strength to attack.

406. Radagaisus Moves South. With half to one-third of his warriors—approximately 70,000—Radagaisus advanced into central Italy, leaving the remainder of his army, and most of his noncombatants, in the north. Stilicho, who now had about 45,000 men—30,000 legionaries plus a sizable force of Huns under Uldin, as well as detachments of Alan and Gothic cavalry—moved to meet the barbarians.

406. Siege of Florence. Apparently while the poorly disciplined barbarians were settling themselves into an investment of Florence, Stilicho sent a heavily guarded convoy of provisions into the town. He then blockaded the barbarians, constructing a line of blockhouses, connected by trenches, after the fashion of Caesar. The barbarians were soon starved into submission. Radagaisus was executed; his surviving warriors were sold into slavery.

406–410. Barbarian Migration to Gaul. The remainder of Radagaisus' host now withdrew back over the Alps into south Germany, moving as separate tribes, of which the largest were the Germanic Suevi, Vandals, and Burgundians, and the Asian Alans. A portion of the Vandals were intercepted and seriously defeated by the Franks under **Marcomir;** the Vandal King **Godigisclus** was killed. In the following two or three years, all of the tribes, including the remnants of the Vandals, crossed the upper Rhine into Gaul to initiate the climactic Germanic invasion of that province.

407–408. Usurpation of Constantine. In Britain a simple soldier named **Constantine** had been elected emperor by his fellows (406?). To extend his dominion over Gaul and Spain, he took practically all the Roman garrisons from Britain and crossed the Channel. While he established tenuous authority over Gaul, sharing control with marauding barbarians, his son **Constans** and the general **Gerontius** conquered most of Spain in his name.

407–450. Roman Abandonment of Britain. Constantine's withdrawal meant the virtual abandonment of Britain, which became semi-independent under its local regional British-Roman leaders, who raised their own levies for defense against Saxon sea rovers.

408. Murder of Stilicho. At the instigation of Honorius, Stilicho was murdered. Mass murders of his adherents followed. Stilicho's barbarian auxiliaries thereupon turned against the emperor, urging Alaric to return to Italy to protect them and to avenge their murdered leader.

Alaric's Operations Against Rome, 409–410

409. Alaric's Second Invasion of Italy. Crossing the Julian Alps by way of Aquileia again, Alaric marched into northern Italy, crossed the Po at Cremona, and advanced to invest Rome (409). Honorius, safe in inaccessible Ravenna, did nothing to aid the mother city of his empire, soon reduced to desperation by famine and pestilence. The Roman Senate agreed to pay Alaric a heavy tribute to raise the siege. The Goths then wintered in Tuscany, from whence Alaric opened negotiations with Honorius. Failing to obtain any satisfaction from the emperor, Alaric briefly blockaded Rome again, set up a puppet emperor, then marched on Ravenna. Unable to capture the city, he returned to the outskirts of Rome.

410, August 24. Capture and Sack of Rome. After a short siege Rome was betrayed to Alaric, who turned the city over to his Gothic soldiers for 6 days of controlled looting. Rome had never before been captured by foreign invaders. He then marched through Campania to southern Italy. He was preparing to invade Sicily and

Africa when he died (December). He was succeeded by his brother-in-law, **Athaulf** (Adolphus).

409–411. Chaos in Gaul and Spain. Meanwhile Honorius, unmindful of Italy's sufferings, set about recovering Gaul from Constantine and Constans. A three-cornered free-for-all followed between the usurper, his own general Gerontius—who had rebelled against him—and Honorius' general **Constantius.** After Gerontius defeated and killed Constans at Vienne (411), he lost most of his army by desertion to Constantius. With the remainder he returned to Spain, where he was halted by invading barbarians. Meanwhile, Constantine was besieged at Arelate (Arles) by Constantius. After Constantius defeated a relief army under **Edobic,** Constantine surrendered and was put to death (411).

412–414. Visigothic Invasion of Gaul. Honorius' chicanery now resulted in another free-for-all, with four principal contenders again battling in Gaul: Athaulf and his Visigoths; **Jovinus,** a self-styled emperor supported by the Burgundians; Constantius; and the Roman-barbarian general **Bonifacius** (Count Boniface). The prize was control of Gaul. Also a prize was **Galla Placidia,** half-sister to Honorius, already promised by him to Constantius, but now promised to Athaulf if he would rid Gaul of the other barbarians. Jovinus was killed by Athaulf (412). The Visigoths then overran all southern Gaul save Massilia (Marseille), which was successfully defended in Honorius' name by Bonifacius (413). Constantius then defeated Athaulf, but was ordered back to Italy. Athaulf then announced Gaul had been returned to the empire, and married Placidia (414).

415–419. Visigothic Invasion of Spain. On Honorius' invitation Athaulf crossed the Pyrenees to reconquer Spain for the Empire (415), but was murdered (416). His successor, **Wallia,** defeating the Alans, Suevi, and Vandals, drove them into Galicia (northwestern Spain). Having reestablished imperial control over Spain, Wallia was rewarded by Honorius with the region of Aquitaine, or Toulouse, the first barbarian kingdom within the old empire (419). Upon Wallia's death, **Theodoric I,** son of Alaric, became King of Toulouse (419–451). Meanwhile, Placidia, Athaulf's widow, married Constantius (419).

420–428. Vandal Resurgence in Spain. Under their King **Gunderic,** the Vandals in Galicia quickly recovered from their defeat by the Visigoths. They defeated the Suevi, then defeated **Castinus,** the new Roman master general in eastern Spain (421), making themselves preeminent in the Iberian peninsula. Upon Gunderic's death (428), he was succeeded by his half-brother **Gaiseric** (Genseric). The Suevi under **Hermanric** now made one more effort to throw off the Vandal yoke, but Gaiseric crushed them on the Anas (Guadiana) River at the **Battle of Mérida** (428).

423–425. Usurpation of John. Upon the death of Constantius (421), his widow Placidia became estranged from her brother Honorius and, with her son **Valentinian,** sought refuge in Constantinople at the court of **Theodosius,** emperor of the East. Upon the death of Honorius, the West Roman throne at Ravenna was usurped by his prime minister, **John** (Johannes). Theodosius sent forces under the command of father-and-son generals, **Ardaburius** and **Aspar,** to depose the usurper. Ardaburius' fleet was scattered by a storm, and he was captured. Aspar, however, marched overland by way of Aquileia to the vicinity of Ravenna. Aided by confederates within the city, he captured it, deposing and killing John (425). Valentinian was enthroned, but Placidia became the virtual ruler of the Western empire. Illyricum was given to Theodosius in return for his aid.

424–430. Rise of Aetius. The usurper John had been supported by the Roman-barbarian general **Aetius** (390–454), who recruited a barbarian army (composed principally of Huns supplied by Aetius' friend **Ruas,** King of the Huns) which he brought to Italy from Pannonia (424). Arriving too late to rescue John, Aetius promptly made peace with Placidia and Valentinian and was placed in command of Gaul. He defeated Theodoric, King of Toulouse, at **Arles**

(425), foiling a Visigothic attempt to conquer Provence. After making peace with Theodoric, Aetius in a series of campaigns subdued the Franks and other Germanic invaders of Gaul, reestablishing Roman control over all of Gaul save Visigothic Aquitaine (430).

428. Revolt of Bonifacius. During the usurpation of John, Bonifacius (see p. 186) remained loyal to Valentinian and Placidia, and was rewarded with the governorship of Africa. Feeling that his influence with Placidia had been undermined by Aetius, and disgusted with lack of imperial appreciation for his past services to the throne, Bonifacius revolted, calling for assistance from **Gaiseric,** the Vandal.

429–435. Vandal Invasion of Africa. Responding to Bonifacius' appeal, Gaiseric led an army (perhaps 50,000 men) of Vandals and Alans into Africa. Bonifacius, having meanwhile been reconciled to Placidia, attempted to call off the Vandal movement, but to no avail, and found himself at war with the barbarians he had invited into his province. Gaiseric twice defeated him (430) near **Hippo** (Bône, Algeria), then captured the city after a 14-month siege (431). (**St. Augustine,** Bishop of Hippo, died during the siege.) Taking advantage of internal religious dissensions, the Vandals soon held all northwest Africa except eastern Numidia (modern Tunisia).

432. Battle of Ravenna. Placidia, fearful of the growing power of Aetius, called Bonifacius back from Africa during the siege of Hippo. Aetius, marching into Italy from Gaul, was decisively defeated by the imperial forces under Bonifacius. Aetius fled to Pannonia and refuge with his old friends the Huns. Bonifacius was mortally wounded in the battle (possibly by Aetius personally), and soon died.

433–450. Aetius Returns to Power. When Aetius returned from Pannonia with a large army of Huns, he was restored to favor and given the title of Patrician by Placidia. He became virtual ruler of the West. He made a treaty with Gaiseric, confirming Vandal control of northwestern Africa save the environs of Carthage (435). He spent most of his time in restoring and maintaining order in Gaul with an army composed mainly of Huns and Alans. He defeated a Burgundian uprising (435). He repulsed Theodoric at **Arles** in a new Visigothic effort to capture Provence, then defeated the Visigoths again at **Narbonne** (436). After desultory war, Aetius and Theodoric signed a treaty (442). To assure control of the Loire Valley from further Gothic encroachments, Aetius placed a colony of Alans at Orléans. For several years he campaigned against the Salian Franks under **Chlodian** (Chlodio). He defeated them repeatedly, but finally permitted the persistent barbarians to settle in the region north of the Somme River (c. 445). Aetius also subdued a number of peasant revolts, one of which briefly threatened his control of Gaul (437).

435–450. Expansion of Vandal Power under Gaiseric. After consolidating his hold on Mauretania and western Numidia, and building up a powerful fleet of sea rovers, Gaiseric seized Carthage and eastern Numidia (October 439). Next year he raided Sicily, to begin a Vandal piratical career that would terrorize the Mediterranean for a century. Aetius, heavily involved in Gaul, asked help from the East Roman Empire. Theodosius sent a fleet to Sicily, intending an invasion of Africa, but Gaiseric, whose sea raiding had given him wealth, bribed his ally, Attila, ruler of the Huns, to attack Illyricum and Thrace; the East Roman fleet was ordered back to Contantinople (see below).

439–450. Resurgence of the Suevi in Spain. The departure of the Vandals provided the Suevi, under their king **Rechila,** with an opportunity to overthrow weak Roman rule in Spain. He captured Mérida (439) and Seville (441), and had overrun all of Spain save Tarraconensis (modern Catalonia) before his death (447). His successor, **Rechiari,** endeavored to conquer northeastern Spain, but was repulsed when Aetius sent reinforcements.

The Eastern Empire and the Huns, 408–450

408. Death of Arcadius. Theodosius II was only 7 at the time of his father's death; the reins of government were assumed by his elder sister, **Pulcheria,** and the able Praetorian prefect **Anthemius.**

409. Hun Invasion of Thrace. Uldin, king of the Huns, led his fierce horsemen into Thrace, but was defeated by Anthemius and forced to fall back across the Danube.

421–422. War with Persia. Persian persecution of Christians under **Bahram V** (see p. 208) led to war, that was quickly ended by a treaty after the East Roman general Ardaburius and his son Aspar won several minor successes in Mesopotamia. Persia agreed to allow freedom of worship to Christians, while Constantinople granted similar freedom to Zoroastrianism.

424–425. Expedition to Italy against John. (See p. 187.)

431. Expedition to Africa. At the request of Placidia, Theodosius and Pulcheria sent Aspar with a strong land-naval force to assist Bonifacius in the defense of Africa (see p. 187). Arriving at Hippo, the combined Roman-East Roman force, under the command of Bonifacius, marched out and was defeated by Gaiseric, who renewed the siege of Hippo. Presumably disgusted by the situation in Africa, Aspar and his force soon returned to Constantinople.

432. Treaty with Ruas, King of the Huns. The growing importance of the Huns in central Europe led Theodosius to pay tribute to Ruas and to make him a general in the Roman army. This in effect recognized the gradual extension of Hun sovereignty over Pannonia.

433–441. Conquests of Attila in the East. Upon the death of Ruas, his nephews Attila and **Bleda** became joint rulers of the Huns. Renewing the treaty with Constantinople, the new Hun leaders undertook extensive conquests in Scythia (south Russia), Media, and Persia.

441. War with Persia. Another short religious war followed renewed persecution of Christians by Persian ruler **Yazdegird II.** Again, Aspar was successful in a few minor conflicts, and peace was quickly restored on the basis of Persian promises to adhere to the earlier treaty (see p. 208).

441. Expedition against the Vandals. Upon the request of Aetius, Theodosius sent a large fleet and army to Sicily, with the intention of invading Africa. The East Romans were successful in several naval encounters with the Vandals, but the force was recalled to Constantinople when Attila invaded the East Roman Empire (see p. 188 and below).

441–443. Attila's First Invasion of the Eastern Empire. Bribed and encouraged by Gaiseric, Attila invaded Illyricum. A truce with the imperial court lasted for less than a year. Attila led his Huns into Moesia and Thrace to the very walls of Constantinople. He drove the main imperial army, under Aspar, into the **Chersonese** peninsula, where he practically annihilated it, only Aspar and a few survivors escaping by sea. Continuing to range over the Balkan peninsula at will, Attila suffered only one setback, being repulsed with heavy losses from the town of **Asemus** (or Azimus, modern Osma, 20 miles south of Sistova). Finally Theodosius made peace, promising an increased tribute (August 443).

445. Murder of Bleda by Attila. Attila now became the sole ruler of a vast empire of undetermined extent, reaching roughly from southern Germany on the west to the Volga or Ural River in the east, and from the Baltic in the north to the Danube, Black Sea, and Caucasus in the south. He was a bold, fierce leader of light cavalry, an excellent tactician, and had some rudimentary strategic ability.

447. Attila's Second Invasion of the Eastern Empire. As the Huns advanced toward Thrace, panic broke out in Constantinople, where the walls had just been shattered by an earthquake. But the southward drive of Attila was checked briefly by the East Roman army in the indecisive **Battle of the Utus** (Vid). Though forced to withdraw, the imperial forces deflected the invaders from Constantinople toward Greece, where the Huns were finally halted by the fortifications at Thermopylae. Theodosius again sought terms. This time the annual tribute was trebled, and he was forced to cede the whole right bank of the Danube to Attila, from Singidunum (Belgrade) to Novae (Svistov, Bulgaria) for a depth of about 50 miles.

450. Death of Theodosius. Pulcheria, who had exercised virtual rule over the empire during the reign of Theodosius, married **Marcian,** who became the new Eastern emperor, and immediately stopped the tribute to the Huns. Attila was furious, but he had already made plans to invade the West, so decided to deal with Marcian later.

ROME AND THE BARBARIANS, 450–476

The West

450. Attila Decides to Invade the Western Empire. Because of a quarrel with Gaiseric, Theodoric of Toulouse prepared to invade North Africa. Gaiseric again called upon his friend Attila to make a diversion, suggesting that opportunities for rapine and loot were much greater in Gaul and Spain than in the devastated Balkan provinces of the East. At the same time a quarrel had broken out between the two sons of **Chlodian,** King of the Franks. One of these, **Meroveus,** asked Aetius for help; the other called on Attila. Another stimulus to Attila was his earlier rebuff when he called upon Valentinian, emperor of the West, for the hand of his sister **Honoria** and for half of the Western Empire.

451. Attila Crosses the Rhine. The Hun emperor led a great host which has been reported at 500,000 warriors (it was probably closer to 100,000), accompanied by a substantial wagon train of supplies and Hun families. He crossed the Rhine north of Moguntiacum (Mainz) in the territory of his Frank allies. The bulk of the army was Hunnish light cavalry, but there were substantial detachments of Ostrogoths, Gepidae, Sciri, Rugi, Ripuarian Franks, Thuringi, and Bavarians. They advanced on a front of more than 100 miles, north and west of the Mosella (Moselle) River. Most of the towns of northern Gaul were sacked. Paris was saved, according to legend, by divine intervention besought by St. Geneviève. While advancing, Attila sent messages to Theodoric urging him to join in a campaign against Roman dominion in Gaul.

451. Aetius Raises an Army. Aetius raised a large army, his Gallo-Roman legions and Roman heavy cavalry forming the core, with substantial contingents of Franks under Meroveus, plus Burgundians and other German federati and Alan cavalry. Even with the unreliable and wavering Alans (kinsmen of the Huns), Aetius had a force probably no more than half the strength of Attila's. His personal appeal, however, persuaded Theodoric that the security of his Visigothic kingdom depended upon joining Aetius against the Huns.

451, May–June. Siege of Orléans. With more than half of his army Attila advanced through Metz (April 7) to the Loire Valley, where he laid siege to Orléans. The remainder of his forces devastated northern France. Starving Orléans was on the verge of surrender when the combined forces of Aetius and Theodoric approached. After some inconclusive skirmishing, Attila retreated precipitously, sending for the rest of his army. Aetius pursued closely. With his entire host assembled, the Hun king chose a position suitable for his cavalry army somewhere between Troyes and Châlons, and probably near Méry-sur-Seine. His wagon train was formed in a great laager behind the army, and was probably reinforced with entrenchments.

451, Mid-June. Battle of Châlons. Finding Attila thus prepared for battle, Aetius advanced cautiously, to be joined by a substantial number of Frank deserters from the Hun army. Attila drew his army up in three major divisions: he commanded the Huns in the center, the Ostrogoths were on his left, and most of his other German allies were on the right. Aetius placed Theodoric and the Visigoths (mostly heavy cavalry) on his right, and personally commanded the left, which consisted mainly of his legions, heavy cavalry, and the Frank infantry. The untrustworthy Alans were in the center, probably supported by a contingent of heavy Roman infantry or cavalry to dissuade them from deserting or changing sides. With Aetius was **Thorismund** (son of Theodoric) and a Visigothic contingent, possibly in the role of hostages. The battle was apparently opened when

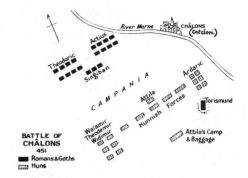

BATTLE OF
CHÂLONS
451
◼ Romans & Goths
▨ Huns

Aetius sent Thorismund's contingent to seize a commanding height overlooking the Huns' right flank. Attila replied with a general counterattack, which penetrated the allied center, as the Alans either deserted or fled. The Franks and Romans on the left and the main body of Visigoths on the right held firm, however, while fierce Hun attacks were unable to dislodge Thorismund from his isolated position. Theodoric now counterattacked the Ostrogoths to his front. The aged ruler was struck down in the confused fighting, but despite the death of their leader, the enraged Visigoths slowly forced their Ostrogothic kinsmen back. Meanwhile, on the Allied left, Aetius regained contact with Thorismund. Now, apparently in compliance with Aetius' preconceived battle plan, the Huns were threatened by a double envelopment. Realizing this, some of Attila's allies, and even his own Huns, began to waver. As darkness descended over the wild scene, Thorismund on the allied left, and the main Visigothic contingent on the right, apparently had routed the opponents to their immediate front; Thorismund himself actually reached the fortified Hun camp early in the evening, but was repulsed by Hun cavalry sent back by Attila. Realizing that the day had been lost, Attila had already ordered a general retirement. Fierce, confused fighting continued throughout the night as scattered Hun contingents tried to regain their camp. Aetius ordered his own wing of the army to stand fast under arms all night, while he tried to reorganize the shattered center. Apparently he was cut off by some of the retreating Huns and barely escaped to the bivouac fires of the main body of Visigoths, where he spent the night.

451. Aftermath of Châlons. Attila, expecting Aetius would attack his camp, prepared to fight to the end. Aetius, however, did not attack. It is not clear why he did not at least blockade Attila and starve the Huns to submission. Possibly he feared that complete annihilation of the common enemy would permit the Visigoths to take over all of Gaul. It has even been suggested that he entered into secret negotiations with Attila, promising no retribution if the Huns withdrew immediately from Gaul. In any event, he encouraged Thorismund, new King of the Visigoths, to return to Toulouse with most of his army to assure the security of his new crown. Attila quietly and quickly returned eastward across the Rhine. The casualties at Châlons are unknown; losses were apparently frightful, particularly among the Huns and their allies.

COMMENT: *Châlons is generally considered to have been one of the decisive battles of world history, since a victory by Attila would have meant the complete collapse of the remaining Roman civilization and Christian religion in Western Europe, and could even have meant domination of Europe by an Asian people.*

452. Attila Invades Italy. Having returned to Pannonia, Attila again demanded the hand of Honoria; when this was refused, he crossed the Julian Alps into northeastern Italy. Aquileia, traditional doormat of barbarian invaders, was completely destroyed. As the Hunnish horde advanced, the inhabitants of Venetia withdrew to the islands off the coast, which resulted, according to tradition, in the founding of Venice, though fisher villages already existed on these islands. Destroying Padua, Attila advanced to the Mincio. Aetius, whose main army was still in Gaul, had rushed back to Italy, and apparently held the principal crossings over the Po, cautiously observing the Huns with a small force. At this time Attila apparently learned that one of his lieutenants had been defeated by an East Roman army in northeastern Illyricum. Famine and pestilence were raging in Italy, making it difficult for Attila to collect supplies for his men and horses, and already causing sickness in his army. At this time a Roman mission, led by Pope **Leo I,** visited Attila's camp. Whether they offered tribute to the Hunnish leader if he withdrew (which is likely), or whether he was miraculously awed by the demeanor of the Pontiff (according to tradition), or whether Attila was simply fearful for the security of his army and its line of communications (also likely), does not matter. He did withdraw, and Leo has received most of the credit in history.

453. Death of Attila. The vast Hun empire immediately fell apart as Attila's sons fought for the vacant throne and the Ostrogoths, Gepidae, and other German subject tribes revolted. **Ardaric,** King of the Gepidae, who had been Attila's right-hand man at Châlons, defeated and killed **Ellac,** son of Attila, at the **Battle of the Netad** in Pannonia (454). Dacia was occupied by the Gepidae, Pannonia by the Ostrogoths. The remnants of the Huns held on to their territory north and east of the Danube for a while under **Dengisich,** another son of Attila, but their dominions continued to dwindle as German tribes revolted and Dengisich was defeated and killed in south Germany (469). One contingent, under Attila's youngest son, **Irnac,** withdrew to the Volga-Ural region, but was soon overwhelmed and absorbed by the Avars. The Huns disappeared from European history.

454. Death of Aetius. Under circumstances remarkably similar to the end of Stilicho (see p. 186), Aetius was murdered personally by his jealous sovereign, Valentinian.

455. Death of Valentinian. Petronius Maximus, a protégé of Aetius, murdered the emperor, assumed the throne, and forced **Eudoxia,** widow of Valentinian, to marry him. Eudoxia thereupon appealed to Gaiseric, the Vandal, for aid.

455, June 2–16. Sack of Rome. Responding promptly to Eudoxia's plea, Gaiseric led a Vandal fleet to the mouth of the Tiber. Maximus was killed by his own people as he fled; the Vandals occupied and sacked Rome for two weeks. They then sailed back to Carthage, taking Eudoxia with them as a hostage.

456. Rise of Ricimer. Upon the death of

Maximus, the master general of Gaul, **Avitus,** briefly held the throne with the support of **Theodoric II** of Toulouse. But Duke **Ricimer** (of mixed Swabian and Visigothic royal blood) now became the major figure in Italy. He defeated the Vandals at sea near Corsica, then expelled them from a foothold in Sicily (456). He next overthrew Avitus at the **Battle of Piacenza** (October 456). At about the same time the general **Marjorianus** (Marjorian), a former subordinate of Aetius, defeated an invading army of Alemanni at the **Battle of the Campi Cannini** (Valley of Bellinzona, southern Switzerland).

456. Visigothic Invasion of Spain. With the approval of Avitus, Theodoric II invaded Spain. He defeated Rechiari, King of the Suevi (see p. 188) at the **Battle of the Urbicas** (in Galicia), ending Suevi supremacy in Spain (456).

457–461. Reign of Marjorian. Set on the throne by Ricimer, Marjorian refused to play a puppet role. Defeating a Vandal force raiding near the mouth of the **Liris** (Garigliano) River, he decided that Vandal depredations could be ended only by destruction of the seat of Vandal power in Africa. First he was determined to reunite the Western Empire. He led an invading army into Gaul (early 458) over snow-clogged Alpine passes. Marching westward near Toulouse, he met and defeated Theodoric II, who had hurried back from Spain to protect his capital. Marjorian granted magnanimous peace terms to Theodoric, facilitating his consolidation of renewed Roman control over the remainder of Gaul and Spain (458–460). He now bent every effort toward his main objective: invasion of Gaiseric's kingdom in Africa. He built up a large fleet in Cartagena. With Marjorian's preparations approaching completion, Gaiseric now began to sow treason among the Romans in Cartagena by bribery. Local treachery then permitted a Vandal fleet to surprise and destroy Marjorian's fleet shortly before he was ready to embark (461). Undaunted by this disaster, he began new preparations, but Ricimer now revolted in Italy. Refusing to rule an ungrateful nation, Marjorian abdicated, and shortly afterward was murdered at the instiga-

tion of Ricimer. Thus ended the last brief burst of glory of the Western Roman Empire.

461–467. Ricimer Uncrowned Ruler of Italy. Ricimer established dictatorial control over north Italy, while dominating the entire peninsula through a series of puppet emperors. A rival, **Marcellinus** (another former subordinate of Aetius), established himself as ruler of Dalmatia, and built up a fleet that controlled the Adriatic. Another rival, the general **Aegidus,** became virtual ruler of Gaul, and was accepted as such by both Visigoths and Franks. Meanwhile, continuing Vandal depredations along the coast of Italy caused Ricimer to request aid from Leo I, the Eastern emperor. Marcellinus joined in the alliance against Gaiseric, driving the Vandals from Sardinia (468). East Roman expeditions against Africa were repulsed by Gaiseric (see p. 193), and Marcellinus withdrew from Africa to Sicily, where he was murdered by an agent of Ricimer (468). **Anthemius,** who had been installed as emperor by Ricimer (467), later quarreled with the kingmaker. Ricimer, his army reinforced by Suevi and Burgundian contingents, marched on Rome, which he captured by storm after a three-month siege; Anthemius was killed. Ricimer himself died soon afterward, leaving Italy in chaos.

461–476. Visigothic Expansion. After the death of Marjorian, Theodoric II of Toulouse rejected his alliance with Rome and quickly conquered Narbonne, thus bringing his dominions to the Mediterranean. He campaigned repeatedly, and generally successfully, against the Suevi in central and northwestern Spain. He invaded east and central Gaul, but was repulsed by Aegidus at **Arles,** and again at **Orléans.** Theodoric was assassinated soon after this and succeeded by his brother **Euric** (466). Euric continued Visigothic expansion in both Gaul and Spain. He defeated **Remismund,** King of the Suevi (468), extending control over all of Spain, with the Suevi, as vassals, confined to Galicia. In Gaul, Euric successfully carried the Visigothic frontiers to the Loire and Rhone rivers.

461–477. Continuing Success of Gaiseric. All

coastal regions of the Western Mediterranean suffered from destructive Vandal raids, which extended as far east as Thrace, Egypt, Greece, and Asia Minor. Gaiseric's greatest success was in repelling an invasion (468) by the combined forces of the Eastern and Western empires (see below). Following this, he reconquered Tripoli and Sardinia (temporarily wrested from him by the allies; see p. 192 and below), then conquered Sicily. At home in Africa, Gaiseric was plagued by constant religious unrest, and during most of his reign he and his Arian warriors cruelly persecuted the Catholic population. At the time of the death of Gaiseric (January 477), Vandal sea power was virtually unchallenged in the Mediterranean.

475. Orestes Seizes Power in Italy. The master general in Italy, **Orestes,** seized power and deposed the emperor, **Julius Nepos.** He placed his own son, **Romulus Augustus** (or Augustulus), on the throne.

476. Uprising of Odoacer. A barbarian (part Hun) general named **Odoacer,** leading a mutiny by most of the barbarian mercenaries in Orestes' army, defeated him in the **Battle of Pavia** (August 23). After a brief siege Orestes was captured and killed.

476, September 4. Fall of the West Roman Empire. Odoacer deposed Romulus Augustulus and seized control of Italy. He did not assume the title of emperor. This date is usually accepted as the end of the Roman state established more than a millennium earlier.

The East

450–457. Reign of Marcian. After refusing to continue tribute to the Huns (see p. 189), Marcian evidently personally commanded expeditions to repel nomadic attacks on Syria and Egypt (452), and in Armenia (456). After the Ostrogoths threw off the Hun yoke, Marcian accepted them under their leaders **Walamir, Theodemir,** and **Widemir** (three brothers who had led the Ostrogoth contingent under Attila at Châlons) as subsidized federati in the East Roman army (454). At about the same

time a substantial contingent of Ostrogoths under **Theodoric Strabo** enlisted directly in the army under the master general Aspar (also of Gothic ancestry).

457–474. Reign of Leo I. Leo, who owed his throne to the support of Aspar, soon became estranged from his general. To reduce Aspar's influence, the emperor changed his bodyguard from Goths to Isaurians. Ending the subsidy to the Ostrogoths (461), Leo was forced to renew it the following year when Theodemir (now sole ruler of the Ostrogoths) ravaged Illyricum. An incursion of Huns over the Danube from Dacia was repelled by Leo's general Anthemius (466); a second Hun invasion was turned back by **Anagastus.** Meanwhile, following an agreement between Leo and Ricimer, in return for East Roman support against the Vandals, Anthemius had become emperor of the West (see p. 192).

468. War against the Vandals. Having entered an alliance against Gaiseric with Anthemius (and Ricimer) and with Marcellinus, Leo dispatched two invasion forces against Vandal territory. The smaller of these, under **Heraclius,** captured Tripoli and then prepared to march overland toward Carthage to join the main army. This, under **Basiliscus** (Leo's brother-in-law), was 100,000 strong (the figure probably includes the fleet strength). Sailing from Constantinople, Basiliscus, joined by Marcellinus, landed near Cape Bon. Gaiseric, after slight skirmishing, requested a truce, which Basiliscus unwisely granted. While this was in force, Gaiseric successfully attacked the allied fleet with fire ships and all of the fighting vessels he had; more than half of the allied fleet was destroyed. Basiliscus, embarking with only about half of his army, fled back to Constantinople with the remnants of his expedition. Heraclius was able to withdraw successfully to Egypt, across the desert. Leo used this disaster as a pretext for charging Aspar with treason; the general and his son **Ardaburius** were both executed (471).

471–474. Ostrogothic Dispute. The death of Aspar caused great unrest among the Gothic soldiers of the East Roman army; Leo attempted to placate his Gothic troops by recognizing the claim of their leader—Theodoric Strabo—as King of the Goths. This in turn infuriated Theodemir, who began raiding into Illyricum and Thrace. This enmity between the rival Gothic houses of Strabo and Amal continued when the death of Theodemir brought his son **Theodoric** to the throne of the Goths in Pannonia (474). Leo died the same year and was succeeded by **Zeno,** who endeavored to continue the partially successful policy of playing the two rival Theodorics off against each other. Intermittent raiding and fighting continued.

The Barbarian Kingdoms, 476–600

Italy and the Ostrogoths, 476–553

476–493. Reign of Odoacer. After the fall of Rome, ex-Emperor **Nepos** established himself as ruler of Dalmatia. After Nepos' assassination (480), Odoacer sent forces to Dalmatia to punish the assassins and to annex the area (481). A few years later he led an army over the Alps into Noricum (Bavaria and Austria south of the Danube) to defeat and capture **Fava** (Feletheus), King of the Rugi (487). The Rugi war was brought to a close the following year by the victory of **Onulf** (Odoacer's brother) over **Frederick,** son of Fava. Odoacer annexed Noricum (488).

476–488. Rise of Theodoric the Ostrogoth. The struggle between Theodoric Strabo and Theodoric Amal for recognition as King of the Ostrogoths was settled by the death of Strabo (481) and the treacherous murder of his successor by Theodoric Amal (484). The successful Theorodic then quarreled with Zeno, whereupon he led his people and army into Thrace to ravage that province again (486). To get rid of the troublesome Ostrogoths, and at the same time to reduce the dangerously growing power of Odoacer, Zeno appointed Theodoric as Patrician of Italy, and sent him off to try to seize Odoacer's kingdom.

488–489. Theodoric Invades Italy. Marching from Novae (Sistova), Theodoric took his entire kingdom, some 150,000–200,000 souls, of whom 50,000–75,000 were probably warriors. The Gepidae tried to halt the advance, but Theodoric defeated them at the **Battle of Sirmium,** then continued on to cross the Julian Alps into Italy (August 489).

489–493. War between Theodoric and Odoacer. Odoacer marched to meet the invaders, but was defeated (August 28, 489) at the **Battle of the Sontius** (Isonzo). Falling back, he was defeated again at the **Battle of Verona** (September 30). Odoacer withdrew to the impregnable defenses of Ravenna, where Theodoric blockaded him. Reinforcements arriving from the south, Odoacer sortied from Ravenna to defeat Theodoric at the **Battle of Faenza** (490). Theodoric fell back to Pavia, where he constructed a fortified camp, which was blockaded by Odoacer. At this time a force of Visigoths and Burgundians invaded Liguria, causing Odoacer to divide his forces. Taking advantage of this situation, Theodoric defeated Odoacer at the **Battle of the Adda** (August 11, 490), and Odoacer was again forced to flee to Ravenna.

490–493. Siege of Ravenna. For 3½ years Theodoric besieged Ravenna, until a naval blockade caused Odoacer to sue for peace. Theodoric, still uncertain of success, agreed to share the rule of Italy with Odoacer, bringing the war and the siege to an end (February 27, 493). Two weeks later Theodoric treacherously murdered his rival at a banquet, becoming undisputed ruler of Italy (March 15, 493).

493–526. Reign of Theodoric. He campaigned extensively beyond the borders of Italy, conquering Raetia, Noricum, and Dalmatia. After the defeat of the Visigoths by Clovis (see p. 197), Theodoric intervened in Gaul, seizing Provence, defeating the Burgundians, and blocking Frank expansion to the Mediterranean (507). Becoming involved in a war with the Eastern Empire, Theodoric defeated the general **Sabinian** at the **Battle of the Margus** (508). After some minor East Roman raids against the coasts of Italy, peace was concluded, confirming Ostrogothic control of Pannonia as far south as Sirmium. Soon after this, the declining Visigothic kingdom virtually recognized the sovereignty of Theodoric, who now came close to restoring the boundaries of the Western Empire in one great Gothic kingdom. This linking of the Gothic kingdoms, however, ended with his death (526).

534–553. Gothic War with Justinian. The power of the Ostrogoths was shattered by **Belisarius** and finally—after an exhausting war of nearly twenty years—destroyed completely by Narses at the **Battle of Monte Lacteria** (see pp. 202–205).

Britain; Angles, Saxons, and Jutes, 450–600

c. 450–500. Invasion by Angles, Saxons, and Jutes. The Saxons were the predominant element of the three related peoples gaining footholds in eastern Britain during this period. They had been raiding the coast of Britain for at least two centuries prior to actual invasion. Whether they actually settled on their own initiative or were invited by the Roman-Briton leader **Vortigern** to help repel the devastating raids of Picts and Scots is not clear. The most important early arrivals were the Saxon (or Jute) brothers **Hengist** and **Horsa,** who soon established themselves in Kent despite bitter resistance by Vortigern and later his son **Vortimer.** Horsa was slain at the **Battle of Aiglaesthrep** (c. 455). By the time of the death of Hengist, the barbarians were in complete control of Kent (488). Meanwhile, other landings had taken place along the coast: the Saxons held the littoral around the mouth of the Thames and southeast Britain from Kent to the Isle of Wight; the Jutes held part of the Isle of Wight and the neighboring coastal regions; the Angles the coastal strip of modern Anglia and as far north as the mouth of the Humber.

500–534. Anglo-Saxon Conquest of Eastern Britain. In a century of almost constant, fierce warfare, the Germanic invaders gradually extended their hold over southeastern and central Britain. One outstanding early Saxon leader was **Cerdic,** who landed near Southampton (495) and was engaged in a series of violent but inconclusive conflicts by the Roman-Briton leader **Natanleod.**

c. 516–c. 537. Reign of Arthur. This semilegendary leader of the Roman-Briton peoples of southern Britain was the principal opponent of Cerdic and other Saxon invaders. For more than 20 years Arthur seems to have repulsed all or most of the Saxon efforts to penetrate central Britain. He may even have reconquered some of the territories seized by the Saxons to the south and the east. He probably commanded the Britons in a victory over Cerdic at **Mount Badon** (Badbury, Dorset, c. 517) and in another about 10 years later. Nevertheless, by Cerdic's death (534), the Saxons controlled most of Hampshire and (with the Jutes) had conquered the Isle of Wight.

534–600. Saxon Expansion in South Central Britain. Cynric, son of Cerdic, penetrated into Wiltshire, against fierce resistance, and **Ceawlin,** son of Cynric, advanced as far as the Severn, winning a major battle at **Dearham** (577). Attempting to push into Wales, however, Ceawlin was decisively defeated by the Britons west of the upper Severn at **Faddiley** (near Nantwich, 583). Further Saxon progress stopped, not only because of the increasingly effective resistance of the Britons but also due to internal strife among the petty Saxon kingdoms.

The Visigoths, 476–600

476–485. Continued Expansion under Euric. (See p. 192.) Euric consolidated Visigothic control over Spain, whose Catholic inhabitants were restive under their barbarian Arian conquerors. He also subdued the sturdy Gallo-Roman mountaineers of the upper Loire River. Defeating the Burgundians, he captured Arles, Marseille, and western Provence. This was the height of the power of the Kingdom of Toulouse.

506–507. War with the Franks. Alaric II, who had been an ally of **Clovis,** King of the Salian

Franks, later went to war and was defeated by Clovis (see p. 197). That the Visigothic kingdom survived at all, and that it retained this foothold in Gaul, was due to the intervention of Theodoric the Ostrogoth, father-in-law of Alaric (see p. 194).

526–600. Amalgamation of Visigoths and Romans in Spain. Subsequent, relatively ineffectual Visigothic rulers continued intermittent bickerings with the Suevi of northwestern Spain. A new Roman conquest under the resurgent East Roman Empire threatened temporarily (534–554), when Justinian sent small forces which—with the enthusiastic support of the populace—briefly conquered southern and southeastern Spain (see p. 206). The Byzantines made no real effort to retain their conquests, however, and the Visigoths soon retook all of the territory save a few coastal cities, which remained in East Roman hands at the end of the century. The Suevi were finally destroyed by Visigothic King **Leovigild** (585). Even more dangerous, and more persistent, was the problem of subduing recalcitrant Catholic, Romanized Iberians, particularly in the mountainous regions. The constant small wars were only brought to a conclusion by the conversion of the Visigoths from Arianism to Catholicism (589).

Franks and Burgundians, 407–600

407–500. Consolidation of Burgundians in Rhone and Saône Valleys. After the expedition of Radagaisus (see p. 185), the Burgundians entered Gaul across the upper Rhine (407–413) and under their leader **Gundicar** established themselves in control of an indeterminate region astride the upper Rhine. They were pushed out of southwestern Germany, and Gundicar was killed by the Huns (436). His son, **Gunderic,** carved out a new kingdom in the Rhone and Saône valleys, and was involved in frequent wars with Visigoths and Romans. After a defeat at the hands of Aetius, however, Gunderic remained a fairly constant, though highly autonomous, ally of decaying Rome.

The Burgundians became truly independent, and reached the height of their power under **Gundobad** (473–516), grandson of Gunderic, who for some time shared the kingship with his brother **Godegesil.**

c. 450–481. Consolidation of the Franks in Rhine Valley and North Gaul. One cause of Attila's invasion of the west had been a dynastic dispute among the Salian Franks (from the Somme River to the lower Rhine). The Battle of Châlons (see p. 190) virtually established the Merovingian dynasty of the Franks, under **Meroveus,** who had fought with Aetius. He and his son **Childeric** were faithful allies of the Romans, but slowly expanded the regions under their direct control. At the same time the Ripuarian Franks held the central Rhineland, striving to extend their influence southward against the Alemanni, who held the entire upper Rhine and southwest Germany. At the same time the Ripuarians were pushing up the Moselle Valley beyond Metz. Clashes were frequent between these two branches of the Frankish peoples, with the Meuse River the boundary between them.

481–511. Reign of Clovis. Upon the death of Childeric, his son **Clovis** (b. 466) became King of the Salian Franks. A gifted soldier, Clovis greatly admired the Roman military system and instilled a greater degree of order and discipline in his army than was common among the Franks. He soon came into conflict with **Syagrius,** a Roman general who had established himself as ruler of north Gaul west of the Somme and Meuse rivers, and north of the Loire. Clovis defeated Syagrius at the **Battle of Nogent** (or Soissons, 486). Syagrius fled to take refuge with Alaric II of Toulouse, but was surrendered to Clovis upon demand. Clovis gradually annexed the region north of the Loire, encountering considerable resistance from the local Gallo-Romans. He formed an alliance with Gundobar and Godegesil of Burgundy, marrying their niece **Clotilda.** At this time the Franks were still pagan, but the Burgundians had become Catholic Christians. Clovis noted that the relations between Burgundians and Gallo-Romans were eased by their common religion, while the Visigoths, Arian Christians, had constant religious trouble with their peoples. This undoubtedly facilitated his conversion to Catholic Christianity by Clotilda, a matter of great military and political importance, since it gave him popular support and facilitated his later conquests in Gaul.

496. War with the Alemanni. The Ripuarian Franks called upon Clovis for help against an Alemanni invasion of the Kingdom of Cologne. He broke the power of the Alemanni west of the Rhine by his decisive victory at the **Battle of Tolbiac** (Zulpich, southwest of Cologne). His actual conversion to Christianity is by legend attributable to his successfully calling for aid from Clotilda's God at a critical moment in this hard-fought conflict. This victory extended his domains to the upper Rhine and made him the virtual leader of the Ripuarian as well as the Salian Franks.

500. War of the Franks and Burgundians. A dispute between the brothers Gundobar and Godegesil caused Clovis to intervene on the latter's behalf. Clovis defeated Gundobar at the **Ouche,** and advanced down the Saône and Rhone valleys as far as Avignon. The continuing vigorous resistance of Gundobar, and his skillful, prolonged defense of besieged Avignon, caused Clovis to grant peace without conquest; the Burgundians became nominal vassals of the Franks. Almost immediately Gundobar defeated and killed Godegesil in a successful assault of the stronghold of **Vienne.** Clovis did not intervene, possibly because he was again engaged against the Alemanni, eventually breaking their power east of the Rhine (506).

507. War with the Visigoths. Despite the mediation of Theodoric, the Ostrogoth, war broke out between Clovis and Alaric II of Toulouse, ostensibly on religious grounds. The more numerous Visigoths lacked the enthusiasm, discipline, training, and leadership of the Franks. By his victory at **Vouillé** (see p. 195), where he defeated and personally killed Alaric, Clovis extended the Frankish kingdom to the Pyrenees, although the Visigothic foothold in Septimania, and that of the Ostrogoths in Provence, blocked

him from the Mediterranean. This victory assured the triumph of Catholicism over Arianism in Western Europe. Soon thereafter Clovis was elected King of the Ripuarian Franks, which gave him personal rule over all northern Gaul and most of western Germany. His reign is generally accepted as the beginning of the French nation.

511—600. Expansion of the Franks. The amazing vigor and military prowess of the relatively few Franks, combined with the support of their fellow-Catholic Gallo-Romans, resulted in continual Frankish expansion in Gaul and Germany. It should be noted that these operations, unlike the migrations of most of the other barbarian peoples, were basically military expeditions, unencumbered by families and household possessions. Much energy was consumed, however, by violent dynastic wars, engendered by the Frankish custom of dividing the realm amongst all of the male survivors of a king. Thus, on the death of Clovis, his kingdom was shared by four sons. Nevertheless, Burgundy and Provence were conquered by the Franks (523—532), and Frankish armies frequently invaded Germany, Italy, and Spain. **Clotaire I,** longest-lived son of Clovis, briefly reunited the Frankish empire (558—561). In a series of great battles in central Germany (562), his sons halted the western migration of the Avars (see p. 199).

Vandals, 476—534

477. Death of Gaiseric. The great Vandal leader was succeeded by his son **Hunneric** (477—484), who aroused animosity among the Catholic Afro-Roman population by his persecutions. This continued under his Arian successors **Gunthamund** (484—496) and **Thrasamund** (496—523). During this time Vandal possessions in Africa and the Mediterranean islands were somewhat extended and consolidated; piracy continued, but not so intensively as under Gaiseric.

531—532. Hilderic and Gelimer. Hilderic, son of Hunneric and the former empress Eudoxia (see p. 191), who was secretly a Catholic, ceased the persecutions of the Afro-Romans, but aroused the resentment of his Vandal people. His younger cousin, **Gelimer,** overthrew and imprisoned him (531) and began persecution of Catholics again. Imprisonment of the son of Eudoxia and the persecution led Justinian to send an expedition under Belisarius to invade Africa.

533–534. War with the Eastern Empire. The Vandal kingdom was destroyed (see p. 202).

Lombards, 500–600

c. 500–565. Lombard Domination of the Central Danube Valley. After the Rugi had been overwhelmed by Odoacer (see p. 194), the Lombards (or Langobardi, related to the Suevi) moved into the region just north of Pannonia and Noricum and began a struggle for control with the Heruli. **Tato** of the Lombards finally defeated **Rodulf** of the Heruli (508). In subsequent years the Lombards increased their dominions north of the Danube and began a series of violent wars with the Gepidae, as both peoples moved into Pannonia and Noricum after the collapse of the Ostrogothic kingdom in Italy. Playing the two tribes off against each other, Justinian gave nominal approval to Lombard expansion south of the Danube. The Lombard King **Audoin** (546–565) also fought under Narses in the Gothic War in Italy (see p. 204).

565–572. Reign of Alboin. The son of Audoin, **Alboin** precipitated a war with **Cunimund,** King of the Gepidae. At first defeated, he formed an alliance with the Avars (see p. 199) and again invaded Gepidae territory, defeating and killing Cunimund, and marrying the dead king's reluctant daughter, **Rosamond** (567). Lombards and Avars then completely destroyed the Gepidae nation. Alboin, apparently somewhat fearful of the numbers and ferocity of his new allies, made an agreement with **Baian,** King of the Avars. He abandoned Pannonia and Noricum to the Avars in return for assistance in starting a migration southward (568).

568. Lombard Invasion of Italy. The entire Lombard nation (which probably included not more than 50,000 warriors), accompanied by 20,000 Saxons, invaded Italy by the traditional route across the Julian Alps. Alboin defeated the imperial general **Longinus,** Exarch of Ravenna in northeast Italy (569), and quickly overran the Po Valley. Milan was soon captured; Pavia fell after a three-year siege (572). Settling in northern Italy (thereafter known as Lombardy), with their capital at Pavia, the Lombards soon drove imperial forces from most of Italy, save for a number of large coastal cities, which remained under Constantinople. Alboin was murdered by his vengeful wife (573), but the conquest of Italy continued unabated despite a lack of strong central authority. The Frankish King **Childebert** intervened briefly in the drawn-out war between Lombards and imperial forces in Italy (585). Subsidized by Constantinople, Childebert crossed the Alps three times to gain mixed success against **Autharis,** King of the Lombards. Failing to receive any cooperation from imperial forces, Childebert then withdrew. Anarchy pervaded Italy, with the semi-independent Lombard dukes constantly at war with each other.

Bulgars and Slavs, c. 530–600

c. 530. Appearance of Bulgars and Slavs. Two new barbarian peoples began to cross the Danube in great force. One group—the Bulgars—comprised fierce, nomadic Asian Turanian horsemen, related to Huns and Avars. The other group—the Slavs—were a blond Indo-European people, similar in appearance to the Germans. Avar pressure had driven the Bulgars into southeastern Europe (see p. 199). The Slavs, who seemingly lacked the inherent ferocity or aggressiveness of either the German or Mongoloid barbarians, had been slowly pushed southward from their homeland north of the Carpathians by successive tides of Asian invasians. Stolid foot soldiers, particularly reliable and effective on the defense, the Slavs were frequently incorporated into the fighting forces of the nomadic armies which swept through their lands. It was thus that they began to appear south of the Danube, first with the Huns, then with the Bulgars.

540–600. Bulgar and Slav Raids into the Balkans. Following time-honored invasion routes across the Danube, both Bulgars and Slavs—sometimes together, more often separately—ravaged the Balkans, and on occasion continued across the straits into Asia. One such invasion was stopped (550) by **Germanius,** nephew of Justinian, who won a great victory over the Slavs at **Sardica** (Sofia). A dangerous raid of Bulgars brought Belisarius out of retirement for his last victory under the walls of Constantinople (559; see p. 207). A few years later the Slav people began to settle in large numbers in depopulated regions of Thrace and Greece (c. 578).

Avars, 550–600

c. 555. Avars Driven into Europe. A Mongoloid people, akin to Tartars and Huns, the Avars of European history were probably basically a part of the Uighur people inhabiting the steppes lying generally between the Volga, Kama, and Ural rivers (c. 460–c. 555). When the Avar Empire of central Asia was overthrown by the Turks (see p. 214), the Avar-Uigurs were driven south into the northern Caucasus (c. 555). Obtaining a subsidy from Justinian, they appear to have migrated to the Danube Valley to assist the East Romans against Bulgars and Slavs. Like their kinsmen the Huns, Bulgars, and Turks, the Avars were essentially light cavalry, their principal weapon the bow.

562–600. Rise of the Avar Empire. Under their great chief, **Baian,** the Avars soon carved out a large empire. Invading Germany, they were repulsed in Thuringia by the Franks (562). After a raid into the Eastern Empire (564), the Avars allied themselves with the Lombards to destroy the power of the Gepidae (see p. 198) and to seize their lands in modern Transylvania (567). After the migration of the

Lombards, the Avars moved into Pannonia and Noricum. Aside from the defeat by the Franks, Baian was for most of his reign universally victorious in his campaigns in central and southern Europe. He repeatedly invaded the Eastern Empire, capturing Sirmium (Sirmione, 580) and sweeping as far south as the Aegean (591 and 597). His successes, however, were finally halted by the emperor **Maurice** and his general **Priscus** in a series of victorious campaigns from the Black Sea to the Theiss River (595–601), culminating on the south bank of the Danube at the Battle of **Viminacium** (601). By the end of the century, nevertheless, Baian was nominal suzerain of an empire extending from the Julian Alps to the Volga, and from the Baltic to the Danube.

East Roman Empire, 474–600

The Empire on the Defensive, 474–524

474–491. Reign of Zeno. A successful Isaurian general, **Zeno,** followed Leo I. A revolt by **Verina,** widow of Leo, and her brother **Basiliscus** (475) compelled Zeno to take refuge in Isauria, but within a year he had regained his throne and exiled the rebels (476). He handled the Ostrogothic threat by playing off Theodoric Strabo and Theodoric Amal against each other, and when the latter was successful, persuaded him to invade Italy (see p. 194).

491–518. Reign of Anastasius. A former civil official, **Anastasius** was an able and vigorous administrator. He soon faced a civil war started by **Longinus,** brother of Zeno.

492–496. Isaurian War. Though nominally an integral part of the empire, and though Isaurians comprised an important element in the East Roman army, the mountain fastnesses of Isauria had long been semi-independent. Anastasius directed war vigorously against Longinus, with a view to subduing the wild Isaurians. His essentially Gothic army defeated them at the **Battle of Cotyaeum** (Kutahya, western Anatolia) and drove them back into their mountains in south central Anatolia (493). Continuing guerrilla resistance by the

Isaurians was stamped out by systematic capture of their fortified towns.

502–506. War with Persia. Though essentially a war to prevent Persian expansion on the Black Sea through Colchis (or Lazica), most of the fighting took place along the Mesopotamian and Armenian frontiers of the two nations. The Persians succeeded in capturing Theodosiopolis (Erzerum) in Roman Armenia and—after a three-month siege—Amida in Roman Mesopotamia. They were repulsed from Edessa, however, and the remainder of the war consisted mainly of long-range raids by each side. After the East Romans recovered Amida, peace was restored on the basis of the *status quo ante.*

507–512. Anastasius' Fortifications. To offset the Persian fortress at Nisibis, Anastasius built a powerful fort at Daras (Dara), just west of Nisibis (507). Soon after this, alarmed by the depredations of Slavs and Bulgars in Thrace, he built the "Anastasian Wall" from the Black Sea to Propontis (Sea of Marmara), across the narrow peninsula a few miles from Constantinople (512).

514–527. Reign of Justin. A former general in the Isaurian and Persian wars, and probably of mixed Gothic-Slav ancestry, **Justin's** reign is particularly noteworthy because of the important administrative role played by his nephew, **Justinian.** Prior to Justin's death, another war broke out with Persia, partly on religious grounds and partly due to a dispute over the responsibility for protecting the Caucasus passes against barbarian inroads into both empires. In an earlier agreement the Romans had agreed to contribute to the support of Persian garrisons and fortifications at the Caspian (Albanian) Gate and the Caucasus (Iberian) Gate. Roman failure to keep up the payments had irritated the Persian Emperor **Kavadh I.**

The Wars of Justinian, 527–565

527–565. Reign of Justinian (nephew of Justin). He was the most illustrious ruler of the East Roman or Byzantine Empire.

JUSTINIAN'S FIRST PERSIAN WAR, 524–532

524–528. Persian Successes. At the outset Kavadh, assisted by his Arabian vassal, **al Mondhir** of Hira, had the better of fighting in Mesopotamia. He was held up, however, by the successful defense of the powerful Roman frontier fortresses. The Romans also repulsed Persian attacks into Lazica (Colchis or western Georgia).

529–531. Rise of Belisarius. The later years of the war were distinguished by the exploits of a young Thracian general, **Belisarius** (505–565). He led a successful long-range raid into Persian Armenia (529). As commander in the East, with 25,000 men he defeated a combined Persian-Arab army of 40,000 at **Dara** (530) by entrenching his infantry in a refused position in the center of his line, then carrying out a cavalry envelopment to culminate a classic defensive-offensive battle plan. The following year he repulsed several Persian efforts to invade Syria, but was defeated by greatly superior forces at **Callinicum.** Withdrawing to the islands of the Euphrates, at **Sura** he then repulsed all Persian efforts to overwhelm his shattered army, and so gained considerable glory despite the earlier defeat. The Persians, discouraged, withdrew and peace was soon made. As a result of this inconclusive war the East Romans strengthened their previously tenuous hold on Lazica, but agreed to make regular payments to subsidize the Persians' forts in the Caucasus passes.

532. Nika Uprising in Constantinople. Belisarius, withdrawn before the conclusion of the Persian War to command an expedition to Africa, was in Constantinople when an uprising almost overthrew Justinian. The firmness of the empress **Theodora** and the prompt military action of Belisarius and the generals **Narses** and **Mundas,** with 3,000 troops, saved the vacillating Justinian. Some 30,000 of the mob were killed and order restored.

THE VANDAL WAR IN AFRICA, 533–534

533. Belisarius' Expedition. For religious and political reasons, Justinian decided to recon-

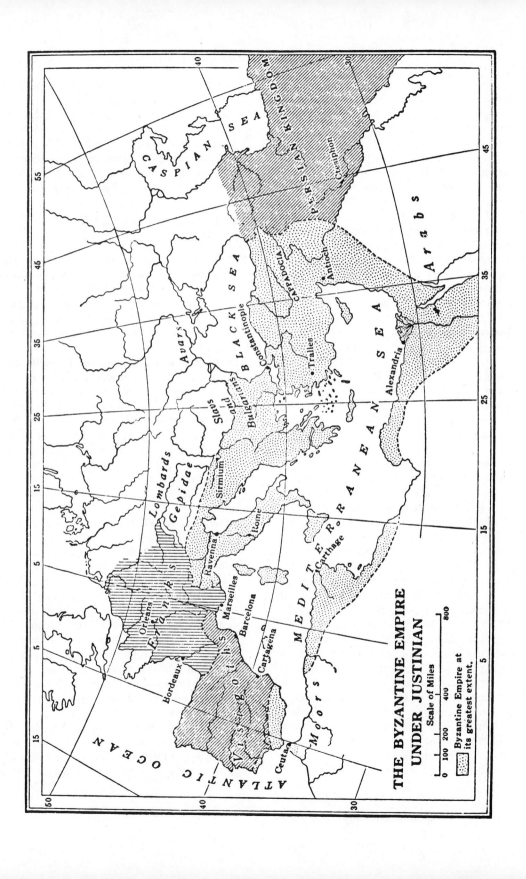

THE BYZANTINE EMPIRE
UNDER JUSTINIAN

Scale of Miles

0 100 200   400   800

☐ Byzantine Empire at
   its greatest extent.

CASPIAN SEA

PERSIAN KINGDOM

Ctesiphon

BLACK SEA

CAPPADOCIA

Antioch

Arabs

Avars

Constantinople

Tralles

Alexandria

MEDITERRANEAN SEA

Slavs
and
Bulgarians

Sirmium

Lombards

Gepidae

Ravenna

Rhone

Carthage

Franks

Marseilles

Barcelona

Moors

Orleans

Visigoths

Cartagena

Bordeaux

Ceuta

ATLANTIC OCEAN

quer Africa from the Vandals (see p. 198). Belisarius took a joint army-navy task force, consisting of 10,000 infantry and 5,000 cavalry, in 500 transports escorted by 92 war vessels manned by 20,000 seamen. He stopped at Sicily, which he used as a staging area, with the permission of **Amalasuntha,** daughter of Theodoric and regent of Italy. At this time **Tzazon** (Zano), brother of the Vandal King **Gelimer,** led a force of 5,000 Vandal soldiers on an expedition to suppress an uprising in Sardinia that had been fomented by Justinian.

533, September 13. Battle of Ad Decimum. Belisarius landed (early September) at Cape Vada and immediately marched north toward Carthage. Gelimer attempted an ambush in a defile at the 10th milestone from Carthage, but due to inadequate coordination and the alertness of Belisarius and his men, the attack was repulsed with heavy losses and the Vandals scattered into the desert. Belisarius promptly seized Carthage, which fell without further contest (September 15). Gelimer collected his shattered forces at Bulla Regia (Hamman Daradji) and sent urgently for Tzazon, who arrived a few weeks later.

533, December. Battle of Tricameron. With the arrival of Tzazon, Gelimer had assembled an army of about 50,000 men, mostly cavalry. He now advanced toward Carthage. Belisarius moved out to meet the Vandals. Leading the way with his cavalry, he discovered the Vandals, not fully prepared for battle, on the far side of a shallow stream. Without waiting for his infantry to come up, Belisarius charged, despite odds of almost 10-to-1, and threw the Vandals back in confusion. As soon as the infantry arrived, he captured the Vandal camp by storm. Tzazon was killed, the Vandal army shattered, and Gelimer forced to flee. He finally surrendered (March 534), bringing the Vandal kingdom to an end; he and large numbers of his captured soldiers were carried back to Constantinople with Belisarius—who was recalled because Justinian was jealous of his successes (early 534). Belisarius left a small force in Africa under the general **Solomon** to continue the subjugation of the province. Initially suc-

cessful in subduing the Vandals and pacifying the Moors and Numidians, Solomon was then plagued by a mutiny (536), which he suppressed only with the help of Belisarius (see below), and soon after was defeated and killed in a battle with the Moors. Justinian, however, sent reinforcements, and the gradual pacification of Africa continued (536–548).

The Italian (or Gothic) War, 534–554

534. Justinian Declares War on the Ostrogoths. The Ostrogothic Queen Amalasuntha (who had acknowledged the sovereignty of Justinian) was murdered by her coruler, **Theodatus,** providing Justinian an excuse for intervention in Italy so as to reunite it with the Empire.

535. Expedition of Belisarius to Sicily and Italy. Justinian sent Belisarius with an expeditionary force of only 8,000 soldiers (about half were heavy East Roman cavalry) to begin the reconquest of Italy. Landing first in Sicily, Belisarius met little opposition, save for the Gothic garrison of Palermo. Laying siege to the citadel, he brought his ships into the harbor, then hoisted archers to the mastheads, from which they could shoot down over the walls at the garrison. The Goths quickly surrendered.

536. Mutiny in Africa. While preparing to invade Italy itself, Belisarius was forced to rush to Africa with 1,000 men to suppress a mutiny. With the addition of 2,000 loyal troops, he promptly defeated 8,000 mutineers, then hurried back to Sicily. About this time Theodatus offered to surrender Italy to imperial control.

536. Battle of Salona (near Split). Meanwhile, the imperial general **Mundas,** who had invaded Dalmatia with only 4,000 men, was defeated by the far more numerous Gothic defenders of Dalmatia. This so encouraged vacillating Theodatus that he withdrew his offer of peace and prepared to defend Italy.

536. Belisarius Invades Italy. Crossing the Strait of Messina, Belisarius marched to Naples, which he captured after a month's siege by sending troops into the city through an abandoned aqueduct. Meanwhile a Frankish army, allied to Justinian, prepared to cross the Alps. The Goths, disgusted by the vacillation and inaction of Theodatus, deposed and killed him, electing the warrior **Vitiges** in his place. Vitiges immediately bought off the Franks by ceding Provence to them.

536, December 10. Capture of Rome. Upon the approach of Belisarius, the Gothic garrison of 4,000 fled Rome, permitting Belisarius to enter the city without a struggle. Since Vitiges was collecting a powerful army in north Italy, Belisarius decided not to risk further advance without more troops. He sent an urgent request for reinforcements to Justinian, meanwhile preparing Rome for a siege by bringing in great quantities of food and other supplies and repairing the neglected walls of the city. With him he had only 5,000 men, of whom half were his personal bodyguard, probably the finest troops in the world at that time. He stretched chains across the Tiber to hold both parts of the city, and he recruited approximately 20,000 young Romans to help man the long city walls.

537, March 2–538, March 12. Siege of Rome. Vitiges arrived with an army probably about one-third the strength of 150,000 attributed to it by the chroniclers, but at least ten times the size of Belisarius' field force. Conducting a delaying action outside the **Flaminian Gate,** Belisarius was almost cut off, but fought his way back to the city, inflicting numerous casualties on the Goths. Vitiges attempted an assault, but was repulsed with heavy losses from ballistae, catapults, and Belisarius' veteran archers (March 21). The Goths' ranks were so depleted that they were unable to keep a continuous ring around the city, Belisarius being able to send and receive messages, receive reinforcements, and even obtain supplies. One reinforcing body of 2,000 men was sent by Belisarius on a raid to the east coast of Italy under the command of **John "the Sanguinary."** When he realized that the people and garrison of Rome were not starving, Vitiges attempted another assault, but was repulsed again (early 538). The arrival of an East Roman fleet in the Tiber and of another 5,000 reinforcements forced Vitiges to raise the siege. He marched east in pursuit of John the Sanguinary, whom he besieged in Rimini.

538–539. Siege of Ravenna. The arrival of an army under Narses and pursuit by Belisarius across the Apennines forced Vitiges to raise the siege of Rimini and to retreat to Ravenna. There he was besieged by Belisarius, who was having difficulty in coordinating the operations of the numerous semi-independent imperial commands in Italy. Meanwhile, **Theodebert,** Frankish King of Austrasia, sent a small force across the Alps to operate with Gothic forces against imperial troops near Milan (538), and the next spring led a much larger army into north Italy. Taking advantage of the confused fighting in the Po Valley, the Franks indiscriminately attacked both Goths and imperial troops, and ravaged north Italy. Belisarius, still besieging Ravenna, negotiated a treaty with Theodebert (whose forces were suffering from pestilence) and the Franks withdrew. Justinian now became alarmed by renewed barbarian (Bulgar and Slav) incursions across the Danube and the threat of renewed war with Chosroes of Persia (who was negotiating with Vitiges). Justinian offered to make peace with Vitiges, but Belisarius refused to transmit the message. The Goths then offered to support Belisarius as emperor of the West. The general pretended to agree, but when Vitiges surrendered Ravenna (late 539), Belisarius—ever faithful to Justinian—captured the Gothic king and sent him to Constantinople as a prisoner.

540–541. Belisarius Consolidates in Italy. The East Roman general now began mopping-up operations all over Italy. One by one the remaining Gothic fortifications were captured. Finally the only remaining Gothic strongholds of any importance were the besieged cities of Pavia and Verona. At this point Justinian, still jealous of Belisarius, and fearful that he might belatedly accept the proffered crown of emperor in the West, called his great general back to Constantinople (541). Imperial operations in Italy were now being directed, without central coordination or supervision, by eleven different imperial generals.

541–543. Resurgence of the Goths. Upon the departure of Belisarius, the Goths, under their new King **Ildibad,** broke out of Verona and defeated the uncoordinated imperialists at the **Battle of Treviso,** regaining control over all of the Po Valley (541). After some internal disputes and the death of Ildibad, **Totila** (Baduila), nephew of Vitiges, became King of the Goths. He immediately carried the Gothic offensive into central Italy, defeating the imperialists decisively at the **Battle of Faenza,** then at **Mugello** (542). He next marched south to besiege and capture Naples and to reconquer most of Italy, save for a handful of cities remaining under imperial control.

544–549. Return and Departure of Belisarius. Once more Justinian sent Belisarius—as usual with totally inadequate forces (4,000 men)—to retrieve the situation. After a year's siege, which Belisarius' harassment was unable to disrupt, Totila captured Rome (545). But when he moved to Lucania to attack imperial forces there, Belisarius promptly recaptured Rome. Totila hurried back, but was repulsed in three costly assaults (546). The four following years were reminiscent of the operations of Hannibal in Italy. The Goths were unable to oppose Belisarius in the field; he was able to march over the country at will, but he was unable to obtain sufficient forces from the jealous emperor to permit him to accomplish anything. Finally he was recalled to Constantinople (549). Totila, immediately besieging and recapturing Rome, not only reestablished firm Gothic control over most of Italy but was able to reconquer Sicily, Sardinia, and Corsica as well. He tried to capture Ravenna by a combined land and sea blockade, but his naval force in the Adriatic was disastrously defeated and he was forced to give up the attempt (551). Soon afterward an imperial expedition under **Artaban** again wrested Sicily from the Goths.

551. Narses Commands Imperial Forces in Italy. Justinian now finally realized that he could not succeed in Italy without a major effort, in which an able general would have to be placed in command of adequate forces. Still jealous of Belisarius, he first selected his nephew **Germanius,** who had distinguished himself as a subordinate of Belisarius in Persia and who had recently won a substantial victory over the Slavs (see p. 199). Germanius, however, died near Sardica, so Justinian selected the aged eunuch **Narses** (478–573) for command in Italy. Narses refused to accept the post without adequate forces. When he marched north from Salona, later that year, he probably had a total force of 20,000–30,000 men. Arriving in Venetia, he discovered that a powerful Gothic-Frank army at least 50,000 strong, under the Goth leader **Teias** and the Frank King **Theudibald,** blocked the principal route to the Po Valley. Not wishing to engage such a formidable force and confident that the Franks would soon tire of their alliance with the Goths, as they had in the past, Narses cleverly skirted the lagoons along the Adriatic shore by using his vessels to leapfrog his army from point to point along the coast, some going by ship, some marching, in a manner similar to a modern truck and foot march. In this way he arrived at Ravenna without encountering any opposition. Near Ravenna he attacked and crushed a small Gothic force at **Rimini.**

552. Battle of Taginae (near Modern Gubbio). Narses now began an advance on Rome. Crossing the Apennines with nearly 20,000 men, he was met by Totila, who probably did not have more than 15,000. In a narrow mountain valley suitable for the shock tactics of his heavy-cavalry lancers, Totila had chosen a position which Narses could not bypass. The imperial general immediately deployed his army in a concave formation. He dismounted his Lombard and Heruli cavalry mercenaries, placing them as a phalanx in the center. His heavy Roman cavalry cataphracts were on each flank, reinforced with all his infantry—who were foot archers. On his left he sent out a mixed force of foot and horse archers to seize a dominant height. Totila's army was in two lines: the heavy cavalry lancers in the front, with his archers and a line of spear and axe-wielding infantry behind. The Goths opened the battle with a determined cavalry charge. As they swept down the valley they first came under the fire of the advanced force on Narses' left, then rode into the cross fire of the archers in his concave wings. Halted by the devastating fire, the attackers were then thrown back in confusion on the infantry advancing behind them. Efforts of the Gothic archers to support their cavalry were foiled by the more aggressive, more mobile imperial horse archers on the flanks. Covered by continued fire from the foot archers, these heavy cataphracts then swept into the milling mass of Goths in a double envelopment. More than 6,000 Goths, including Totila, were killed. The remnants fled. Narses then continued on to Rome, which he captured after a brief siege.

553. Battle of Monte Lacteria (or of the Samus, near Cumae). Teias was now elected King of the Goths. He had reassembled most of the remnants of Totila's army in the Po Valley, while his brother, **Aligern,** still retained some strongholds in Campania. In a secret, well-conducted march, Teias joined forces with his brother in Campania. Narses immediately followed, blockading the combined Gothic force west of Naples. The Goths were crushed in a hopeless last-ditch fight, and Teias was killed. Aligern escaped, but surrendered a few months later. Narses now divided his forces to besiege the remaining Gothic strongholds and fortresses scattered throughout Italy.

553–554. Frankish Invasion. Two Frankish-Alemanni dukes—the brothers **Lothaire** and **Buccelin**—now crossed the Alps from Germany with a force of 75,000 men, mostly Frankish infantry. In the Po Valley, they won an easy victory over a much smaller imperial force at **Parma,** and were then joined by remnants of the Gothic armies, bringing the total strength of the invaders to about 90,000. They began to march south. Narses, gathering his forces as quickly as possible, marched north to harass the Franks, but did not yet feel strong enough to engage them in battle. In Samnium (south-central Italy) the brothers divided their forces; Lothaire went down the east coast, then returned to the north, to winter in the Po Valley. Buccelin followed the west coast to the very toe of the boot, where he spent the

winter—his army being seriously wasted by high living and disease.

554. Battle of Casilinum. In the spring Buccelin marched north, his army reduced to about 30,000 men. Narses with 18,000 men (including a contingent of Goths under Aligern) marched south to meet him at Casilinum (on the banks of the Volturno, near Capua). Outmaneuvering Buccelin, Narses forced the Franks to battle on ground of his own choosing. He took up a concave formation, similar to that he had used at Taginae. The Frankish infantry, in a wedgelike phalanx, advanced against the imperial center, much as the Romans had done at Cannae. The result was almost identical. The imperial foot archers engaged the flanks of the Frank phalanx, while the heavy Roman horse archers swung around behind them. Pausing, the imperial horse and foot archers softened up the milling mass of Franks with volley after volley of arrows—then they charged. The Frankish army was annihilated. Meanwhile, in the north, Lothaire and his men had been struck by an epidemic which killed most of them. The Italian, or Gothic, War was over; Italy had been rewon for the Empire. But it was a prostrate, exhausted, depopulated Italy which, after a few years of corrupt imperial misrule, was ready to accept another barbarian migration (see p. 198).

JUSTINIAN'S SECOND PERSIAN WAR, 539–562

539. Chosroes Declares War. Partly because of unsatisfactory Roman performance in helping to pay for the garrisons in the Caucasus passes, partly for religious reasons, and partly because he feared imperial strength was growing too much from the victories in Africa and Italy, Chosroes declared war on the Empire.

540. Persian Invasion of Syria. Advancing up the Euphrates, Chosroes swept into the heart of Syria, capturing Antioch, plundering the region, and taking back with him vast numbers of captives.

541–542. Chosroes in Lazica (Colchis, on coastal Armenia). The Persian ruler now shifted his attention to the northern front, where he was generally successful. He captured Petra (near Phasis) from the Romans (541), and in two campaigns succeeded in establishing firm Persian control over the country.

542–544. Belisarius in Mesopotamia. Recalled from Italy (see p. 203), Belisarius was sent by Justinian to take command in Syria and Mesopotamia. Driving the Persians out of the positions they had captured early in the war, Belisarius forced Chosroes to return from Lazica (543). In a campaign of maneuver, with no major battles, Belisarius recovered all of Roman Mesopotamia and raided deep into Persia. Persian efforts to take Dara and Edessa were repulsed, as was a Roman attempt to invade Armenia. Chosroes and Justinian then agreed to an armistice (545). Belisarius was thereupon sent back to Italy (see p. 204).

549. Resumption of War in Lazica. Upon appeals from the Christian population of Lazica, being cruelly oppressed by the Persians, Justinian sent an expedition of 8,000 troops under General **Dagisteus,** who laid siege to Petra. A Persian relieving force drove back the Romans, but reinforcements under **Besas** enabled imperial forces to renew the siege, which lasted for more than a year. After a heroic defense, the Persian garrison was finally overwhelmed (551). The war in Lazica dragged on for several years, but the climactic action was the **Battle of Phasis,** in which the imperial forces decisively defeated the Persian general **Nacoragan.**

562. Peace Restored. With both nations engaged in other wars, and both close to bankruptcy, the final years of the war had been desultory. The *status quo ante* was restored, with Lazica firmly in imperial hands, but with Justinian again agreeing to help pay for the Caucasus fortifications by an increased subsidy.

JUSTINIAN AND BELISARIUS

Justinian combined pettiness with great vision, parsimony with soaring ambition, timidity with vigor, brilliant organizational ability with inept administration, bellicosity with appeasement. It has been suggested that his reputation as the greatest of all East Roman emperors was due to three people: his steel-willed ex-prostitute wife, Theodora; the noble general, Belisarius; the crafty old eunuch-soldier, Narses. Yet though his debt to each of these was great, no man could stamp his imprint so forcibly on his own times, and on the future history of his nation, without exceptional natural ability. One explanation of the paradoxes of his nature and of his accomplishments is the fact that he had an unusual ability to recognize his own weaknesses, and relied upon able assistants to make amends for his deficiencies.

In addition to four major wars, Justinian's overextended, underpaid, and inadequate military forces were engaged in widespread operations in Spain, the Balkan-Danubian region, and elsewhere along the entire Mediterranean littoral. He built several tremendous systems of fortifications to protect vulnerable frontiers and to control unsettled and disputed regions—along the Danube; throughout Thrace, Epirus, Thessaly, and Macedonia; at the defiles of Thermopylae, Corinth, the Chersonese (Gallipoli) Peninsula, and the Tauric Chersonese (Crimean) Peninsula; along the eastern frontier with Persia—and he extended and reinforced the wall Anastasius had built across the peninsula approach to Constantinople. At the same time he subsidized and encouraged his Ghassanid and Abyssinian allies in Arabia to harass the southern fringes of the Persian Empire.

Justinian recognized the danger that incessant wars, and the expenses of widespread fortifications, could ruin the economy of the nation, and this recognition undoubtedly had much to do with his notorious and frequently shortsighted parsimony, reflected by the tiny forces he provided his commanders and by the delays in payment of troops that sometimes caused entire armies to melt away in desertion or to turn against the empire in mutiny. The regular forces of his army never exceeded 150,000 men, scattered in Spain, Italy, Africa, Egypt, and along the Danube and Persian frontiers. Unquestionably, even his budgetary caution could not prevent serious straining of the economic fabric of the entire empire, and his parsimony contributed to the total collapse of the economy of Italy through unnecessary protraction of the war in that unfortunate country.

Yet on the whole, despite the strain on its economy, it is probable that the wars and conquests of Justinian halted the decline of the East Roman Empire. Even though most of these conquests were partial or temporary, the victories of the empire restored the prestige of Roman arms, and provided new and justified confidence to the rulers of Constantinople in the tremendous residual superiority of even their debased Roman military system over the military capabilities of their civilized and barbarian neighbors.

Most reprehensible—and probably most shortsighted—of all of Justinian's controversial conduct was his jealousy and mistreatment of Belisarius. No prince has ever been better served by a loyal subject than was the emperor by his great general. To prevent Belisarius (or any other general) from becoming too powerful, Justinian frequently hampered him by imposing divided command, and by sending him on vast projects with totally inadequate forces; more often than not Belisarius succeeded anyway. The inevitable result of such success was relief and insults at the hands of Justinian. Yet, in renewed crises, the noble soldier responded loyally and effectively to the desperate appeals of his emperor.

Having been relieved for the second time from his command in Italy (see p. 204), Belisarius lived quietly in Constantinople until Justinian called him out of retirement to complete the consolidation of reconquered regions of southern Spain (554). Again he was thrust back into retirement and obscurity. But the emperor had no hesitation in calling on Belisarius, and the soldier had no hesitation in responding, when a combined Bulgar-Slav invasion of Moesia and Thrace, under the Bulgarian leader **Zabergan,** reached to the outer defenses of Constantinople (559). All of the regular forces of the empire were scattered in fortifications or in campaigns against Persians or barbarians. With a force of 300 of his veteran cavalry, plus a few thousand hastily raised levies, Belisarius repulsed a Bulgarian attack near **Melanthius,** just outside the walls, then counterattacked to drive the barbarians away. Having saved Constantinople by his last victory, Belisarius, with little thanks from the emperor, returned to his retirement.

Soon after this the jealous emperor accused him of treason, disgraced and imprisoned him (562). Remorse and better judgment caused Justinian a year later to release Belisarius, to restore to him the properties and honors which he had stripped from him, and to permit the general to live in relative honor, if in obscurity, until his death shortly before that of the emperor himself (565).

After Justinian, 565–600

565–574. Reign of Justin II (nephew of Justinian). He began his reign by refusing further subsidies to the Avars, who thereupon conducted several large-scale raids through the Balkan Peninsula. He responded favorably to an embassy from the Turks, who were at war with Persia in central Asia (see p. 209), and as a result of this, and the continuing Persian persecution of Christians in Armenia, embarked on another war with Persia (572).

572–596. War with Persia. (See p. 209.)

574–582. Reign of Tiberius. Due to recurring seizures of insanity, Justin abdicated in favor of the general **Tiberius.** The war with Persia continued without important developments (see p. 210), while the Avars continued to raid at will over the European areas of the Empire (see p. 199).

582–602. Reign of Mauricius (Maurice). Having distinguished himself in the war with Persia, this general was selected as emperor by the dying Tiberius. Prior to ascending the throne, Maurice had written an encyclopedic work on the science of war (the *Strategikon*), which was to exercise a major influence on the military system of the Byzantine Empire for centuries to come (see p. 233). His personal attention to the military affairs of the empire resulted in solid imperial victories over both the Persians (591; see p. 209) and the Avars (598–601; see p. 199). His insistence on strict discipline as the basis of the old military virtues of the Romans, however, resulted in a mutiny in the Danube army, which spread to the populace of Constantinople. In an effort to restore peace, Maurice abdicated, only to be cruelly murdered by his successor, **Phocas** (602).

SOUTHWEST ASIA

SASSANID EMPIRE OF PERSIA

399–438. Reigns of Yazdegird I and Bahram V. Following the relatively peaceful, prosperous, and benevolent reign of **Yazdegird** (399–421), his son **Bahram V** provoked a brief, unsuccessful war with Rome (421–422) as a result of his violent persecutions of Christians (see p. 188). Bahram, too, was the first of the Sassanid kings to be subjected to the depredations of the White Huns, or Ephthalites—apparently related to the Kushans, and like them descended from the Aryan-Mongolian people known to the Chinese as the Yueh Chih (see p. 130). Bahram drove them back over the Oxus River.

438–459. Reigns of Yazdegird II and Hormizd III. Like his father, Bahram, Yazdegird II (438–457) provoked a brief, unsuccessful war with Rome (441; see p. 188). He also repulsed Ephthalite inroads with the help of subsidies from Constantinople (443–451). In a later war against the White Huns he was defeated near the Oxus (457). **Hormizd III** (457–459) succeeded his father Yazdegird to the throne, but was forced to fight continuously against his brothers, as well as the White Huns.

459–488. Reign of Peroz (Firuz) and Balash. Peroz (459–484), with the assistance of the White Huns, overthrew and killed his younger brother, Hormizd III. He was soon engaged in a series of disastrous wars with his former allies, who overran Bactria and other eastern regions of the Sassanid Empire (464–484). Seriously defeated by the Ephthalites, or their cousins the Kushans, who lived east of the Caspian (481), he prepared a major expedition to recover his lost provinces. Advancing eastward with a large army, apparently he became lost in the desert and fell an easy prey to the White Huns; he and all his army perished (484). He was succeeded by another brother, **Balash** (484–488), who ended a revolt in Armenia (481–484) by agreeing to end persecutions of Christians. Balash spent most of his reign in a civil war with still another brother, **Zareh.** Because of his ineptitude and failure to deal with the dangerous Ephthalite threat to the kingdom, he was deposed.

488–531. Reign of Kavadh (Nephew of Balash). Initially he had no more success against the Ephthalites than his father or uncles. He was deposed briefly in favor of his brother **Jamasp** (496–499), and took refuge with the Ephthalites. When he returned to his throne, Kavadh was angered when the East Roman Emperor Anastasius refused his request for subsidies, so contracted an alliance with the Ephthalites and went to war with Rome (502–506; see p. 200): This inconclusive struggle was brought to an end by the invasion of the western Huns through the Caucasus (505), prompting the two empires to make common cause against the barbarians and to agree on joint responsibility for defense of the Caucasus passes.

524–531. Kavadh's Second War with Rome. (See p. 200).

531–570. Reign of Chosroes I (Anushirvan). He concluded "eternal peace" with Justinian, but was forced to fight a bitter dynastic war before establishing firm control over the Sassanid Empire, of which he became the most illustrious ruler.

539–562. Chosroes' First War with Rome. (See p. 203.)

c. 554–c. 560. Persian and Turk Alliance against the Ephthalites. Chosroes entered an alliance with the newly risen Turkish people in Central Asia (see p. 214). The Ephthalites were crushed. Later Chosroes conducted a successful campaign in the steppes north of the Caucasus against the Khazars.

572–579. Chosroes' Second War with Rome. Justin, allied with the Turks, started a new war by supporting an uprising of the Christian Armenians (see p. 207). Chosroes, now an old man, immediately dispatched an expeditionary force into Roman Mesopotamia, while holding off the Turks in the east. His grandson, also named **Chosroes,** successfully besieged **Dara,** while the general **Adarman** raided effectively into Syria as far as Antioch. The Romans, dismayed by this unexpected result of a war they had initiated, sought a truce, which lasted three years (573–575). Both sides used the lull to prepare for a renewal of the struggle. Chosroes, again seizing the initiative, sent his grandson once more into Roman territory. The Persians were driven from Cappodocia (575) by the young general Mauricius (Maurice). The following year the Roman general **Justinian** (son of Germanius; see p. 204) defeated young Chosroes in the hard-fought **Battle of Melitene,** just west of the upper Euphrates, and drove the Persians westward. Justinian then invaded Persian Armenia, reaching the Caspian, where he established a base and constructed a fleet (577). The following year he invaded As-

syria, causing the elder Chosroes to seek peace (see p. 208), which was being negotiated when he died (589).

579–590. Reign of Hormizd IV; Continued War with Rome and the Turks. The new king, son of Chosroes, refused to give up any territory, and so the war with Rome and the Turks continued, generally to the disadvantage of Persia. The East Romans cleared the Persians from much of Armenia and the Caucasus, capturing a number of fortresses along the Mesopotamian frontier. At the same time the Turks overran most of Khorasan, and reached Hyrcania on the shores of the Caspian. Meanwhile, Hormizd's cruelty and misrule stimulated widespread rebellions within Persia and alienated allies and vassals. With the Sassanid Empire on the verge of collapse, an able general, **Bahram Chobin,** appeared on the scene. With a small army he ambushed the Turks to win a great victory at the **Battle of the Hyrcanian Rock** (588). Pursuing vigorously, he regained much of the lost areas south of the Oxus. He then turned against the Romans (589), winning an initial victory at **Martyropolis** (near modern Mus, Turkey). Maurice defeated Bahram at **Nisibis,** and drove him back into Armenia. Attempting to invade Lazica, Bahram was defeated again at the **Battle of the Araxes** (589). Jealous, Hormizd seized this opportunity to dismiss Bahram, but the general refused to be dismissed, and revolted. Hormizd was overthrown by other dissidents, who placed young Chosroes (son of Hormizd) on the throne, but Bahram also refused to accept him, defeating and deposing the new ruler and seizing the throne himself (590). Chosroes took refuge with Roman forces at Circesium (on the Euphrates).

590–596. Roman Intervention. Maurice, seeing an opportunity to end the prolonged war to the advantage of Constantinople, immediately supported Chosroes against Bahram. A Roman army, probably about 30,000 strong, under the general **Narses** (apparently unrelated to Justinian's great general) was sent into Persia through Assyria to restore Chosroes to the throne. At the same time Roman forces in Armenia advanced into Media. Bahram rushed west with an army of about 40,000 and attempted to halt the invasion at the Zab River. Narses, whose total strength, including Persians defecting to Chosroes, was about 60,000, decisively defeated Bahram at the **Battle of the Zab** and pursued him into Media, where the Romans won another victory, smashing Bahram's army (591). Bahram then fled east to seek refuge with the Turks, but soon died or was killed. Chosroes, returned to the throne with Roman assistance, promptly made peace. Maurice, hoping to establish perpetual peace through magnanimity, merely insisted upon Persian recognition of the traditional frontiers and the cessation of subsidies for the Caucasus forts; Roman troops were withdrawn from the positions they held in Armenia.

591–628. Reign of Chosroes II. The only important military events in the early years of this reign were operations in Media to suppress a revolt by Prince **Bistam** (591–596).

Arabia and Abyssinia, 400–600

c. 400. Rise of Hira. Located at the northeastern edge of the Arabian desert on the fringe of Mesopotamia, and about 100 miles south of the Persian capital of Ctesiphon, the principality of Hira had become prosperous as a vassal of Persia. The most renowned ruler of Hira was **al Mondhir,** an able warrior, whose support was decisive in Bahram V's gaining the throne of Persia (421) in the dynastic disputes following the death of Yazdegird I (see p. 208). Another **al Mondhir** was a leading Persian general under Chosroes I (see p. 200).

c. 400. Rise of Ghassan. A Christian tribe, apparently originating in Yemen, the Ghassanids established themselves in northwestern Arabia, just east of the Jordan (c. 200). Vassals of Rome, they gradually extended their control over most of the region between Palmyra and Petra (east of the Jordan River). They clashed frequently with Hira.

c. 400–525. Struggle for Yemen. One of the three fertile regions of Arabia, Yemen had long figured in struggles between native Arabs and the Abyssinians of Aksum (see p. 176). After the Sabaean revolt, Aksum retained only a few

coastal enclaves, but continued the struggle in Yemen with varying fortune. King **Ella-Abeha** (Caleb) of Aksum appointed **Abrahia** viceroy of Himyar (southwest Yemen) about 545. Abraha worked vigorously to expand Aksumite power, and built a cathedral at San'a between campaigns against the Sabeaans and Minaeans. By the late 560s, most of Yemen was under Abraha's control, but his activities were increasingly affected by the Sassanid Persian–Byzantine struggle to the north.

c. 500–583. Wars of Ghassan and Hira. On the fringes of the two great empires, the two Arab kingdoms waged almost constant war on each other, largely at the behest of their suzerains (East Roman Empire for Ghassan, Persia for Hira). The Ghassanids had generally the best of these encounters, but the close proximity of the center of Persian power prevented them from fully exploiting their victories. The most outstanding warrior of the Ghassanids was **Harith Ibn Jabala,** who inflicted a crushing defeat on Hira (528). Harith led the highly effective Bedouin cavalry contingents that fought under Belisarius in his Mesopotamian campaigns (541–544). Dynastic troubles, combined with Persian intervention in northern Arabia, soon after brought about the collapse of Ghassan (583).

543–575. Christianization of Sudan. The kingdoms of Nobatia (around Faras), Maqurrah (around Dunqulah), and 'Alwah (around Khartoum) all adopted Christianity from Egyptian missionaries during this period.

570. Abraha's Invasion of Hejaz. His operations were at first successful, but an attack on Mecca, which had become a major commercial rival of San'a, was repulsed. Abraha died soon after, and was succeeded by his son **Taksoum.**

572–585. Persian Intervention. The Sassanid Persian Emperor, Chosroes II, infuriated with Abraha's intrigues and expansion, launched a naval expedition, which landed at Aden in 572. Within a few years the Persians and their allies had driven the Abyssinians out of most of Himyar and had conquered Sabaea and Minaea. By 585, despite assistance from the Byzantine emperors Justin II and Tiberius, Aksum retained only a few coastal enclaves, and the ancient kingdoms of "Arabia Felix" were destroyed.

SOUTH ASIA

NORTH INDIA

c. 400–c. 450. Ephthalite (Huna) Expansion into Gandhara. In the years following the reign of Chandragupta Vikramaditya (see p. 178), northwest India began to feel the pressure of another people migrating from central Asia. These were the Ephthalites, or White Huns (see p. 208), a people of mixed Indo-European and Mongoloid blood, evidently descended from the Hsiung-nu people of Mongolia, and closely related to the Yueh Chih and their offshoot, the Kushans. Pushing into the Sassanid Empire, the Ephthalites (known to Indian history as Hunas) became firmly established in Bactria (c. 420) and dominated all of modern Russian Turkestan southeast of the Aral Sea. From here they began to conduct raids into Gandhara and as far as the Punjab, while they gradually pressed their bases southward into the mountains.

c. 450–500. Clash of the Hunas and the Gupta Empire. At first successful in minor incursions into the Punjab in the reign of **Kumaragupta** (413–455), the Ephthalites began a full-scale invasion of the Gupta Empire. They were decisively defeated (457) by **Skandagupta** (455–467) and made no further inroads during his lifetime. However, his successors were helpless against the invaders, and the sagging Gupta Empire was further weakened (470–500).

500–530. Rise and Fall of the Huna Kingdom. As the Guptas were driven back, the Hunas established their own kingdom in the Punjab and Rajputana. Their ruler **Toramana** (500–510) conquered Malwa. His son, **Mihiragula** (510–530), continued the expansion southward and eastward, only to be defeated (530) by a coalition of Hindu princes, of whom the most important were **Balditya** of Magadha and **Yasodharman** of Ujjain. The Huna kingdom dwindled in the wake of this defeat and disappeared c. 550. It is probable that several late 6th-century Rajput and *ksatriya* (warrior caste) dynasties sprang from Huna roots. Northern India reverted to a patchwork of warring Hindu states.

SOUTH INDIA AND CEYLON

c. 400–575. Resurgence of the Pallava. The Pallava dynasty of the eastern Deccan steadily gained strength as the Gupta Empire waned. Pallava power was based on the fertile and densely populated coastal plains, which also gave access to the lucrative trade with Southeast Asia. Having gained control over much of the Deccan, Pallavan King **Simhavishnu** drove the Chola kingdom south of the Kaveri River valley (575).

432–459. Pandya Conquest of Ceylon. An invading army from Pandya overthrew the Lambakanna dynasty (who had ruled since 65 A.D.), but they were driven out by **Dhatusena** in 459. He ruled until 477, restoring Sinhalese rule and founding the Moriya Dynasty.

550–600. Rise of Chalukya. A new power now appeared in the western Deccan, near Bijapur: the Chalukya Dynasty, established by **Pulakesin I** (550–566). Toward the close of the century, the Chalukya defeated Pallava in a series of violent wars, and established themselves in control of the central Deccan. Pallava was hard-pressed to hold the core lands on the east coast against the sons of Pulakesin: **Kirtivarman I** (566–597) and **Mangalesa** (597–608).

575–600. Resurgence of Pandya. The defeats suffered at the hands of the Pallava caused the Chola Dynasty to go into decline, and to lose most of its power in the southern tip of India. Its influence was replaced by that of the Pandya Dynasty.

HINDU MILITARY SYSTEM (C. 500)

It was sometime during the 5th or 6th centuries that one of the great military classics of India was written—or collected—by an unknown author or editor, possibly named **Siva,** or **Sadasiva.** This was the *Siva-Dhanur-veda.* Like other Dhanur-vedas that preceded and followed this, its central theme was archery and the employment of the bow as the main weapon of Hindu warfare. This preeminence of the bow, in fact, caused the Hindus to use the term Dhanur-veda—or archery manual—to apply to all writings on the art of warfare, and the *Siva-Dhanur-veda* dealt with the employment of other weapons, and with military science in general. Significantly, it repeated much that had appeared several centuries earlier in the *Arthasastra* of Kautilya (see p. 84).

During this period the only significant change in the art of warfare in south Asia had been the gradual disappearance of the chariot—which Kautilya had not much favored anyway. The growing importance of cavalry had posed a great challenge to the Hindus, who still had difficulty in raising horses of requisite quality and stamina. The Deccan, in particular, was unfavorable to horse raising, and the warrior dynasties of the south customarily imported horses from the north. As in the time of Kautilya, the Hindu princes placed much reliance upon their powerful hill forts to defend their lands from the raids of neighbors, and to provide bases for their own frequent raids.

Weapons and armor changed little. Mail armor had become the common raiment of the horsemen, whose principal weapons were the lance and the mace. Despite their reliance upon infantry bowmen, the Hindus never showed much interest or competence in use of the bow on horseback, even though mounted archers were usually the main components of the armies that had invaded from central Asia. The fact that the invaders soon adapted themselves to the Hindu form of cavalry employment would seem to indicate that Indian horses had neither the stamina nor agility necessary for mounted archery tactics.

There had been no increase in interest in naval warfare. Rulers along the seacoast near the mouths of the Indus and Ganges usually had small naval forces, whose primary missions were to provide merchant shipping and coastal towns with some protection from the depredations of the pirates who based themselves in the river deltas. But the employment of warships as instruments of national power was practically unknown, save in the Tamil kingdoms of the extreme south, whose naval expeditions were almost exclusively raids against the island of Ceylon.

EAST AND CENTRAL ASIA

CHINA

400–589. Continuing Anarchy in China. This was a continuation of the period known as that of the "Six Dynasties," which had begun with the collapse of the Han Empire (see p. 178). China remained divided into two major portions, one generally north of the Yangtze River, the other to the south. This was China's "Dark Age," a time of incessant warfare between these two regions, and of almost constant internal strife within both.

South China, 400–589

420–549. Appearance of the Third, Fourth, and Fifth Dynasties. The Eastern Chin Dynasty was overthrown by general **Liu Yu,** who established the Liu Sung, or "Former" Sung Dynasty (420). This, in turn, was overthrown by general **Hsiao Tao-ch'eng,** who killed the last two Liu Sung emperors to establish himself as the first ruler of the Southern Ch'i Dynasty (479). This disappeared when general **Hsiao Yen** seized the throne to establish the Liang Dynasty (502). He took the imperial title of **Liang Wu Ti,** and during his reign of 47 years south China had a modicum of internal peace, regained some strength, and demonstrated some vigor. A revolt in Annam (541–547) was suppressed (see p. 294), and Liang Wu Ti then attempted an invasion north of the Yangtze. This quickly bogged down, however, through the protracted siege of Hsiangyang. Soon after the failure of this invasion, Liang Wu Ti died, and his dynasty disappeared with him.

557–589. Last of the Six Dynasties. After eight years of violent and confused civil war, the Ch'en Dynasty was established by **Ch'en Pa Hsien.** This disappeared under the impact of invasion from the north.

North China

400–500. Ascendancy of the Northern Wei Dynasty. The T'upa, or Northern Wei, Dynasty maintained its supremacy in the north. Though apparently originally a branch of the Turklike Hsien Pi people, the Northern Wei had become thoroughly Sinicized, but relied militarily almost completely on cavalry. One reason for the relative longevity of this dynasty was the attention lavished on horse units, particularly by utilizing much of the best of the northern farmlands for pasturage—an estimated 7 acres per horse. The Northern Wei probably introduced the stirrup into China (c. 477), possibly having adopted this valuable aid to horsemanship from the Hsiung-nu. Having achieved complete control of north China, the Northern Wei expanded their empire to include part of Sinkiang (c. 450).

500–534. Decline of the Northern Wei. A principal cause of the dynasty's collapse was a disastrous defeat suffered in an attempted invasion of south China (507). Violent civil war broke out in the north (529–534), leading to the split of the Wei Empire into two warring parts: Western Wei and Eastern Wei.

534–557. Internal and External Disorders. While Eastern and Western Wei kingdoms fought each other bitterly, both were troubled from within, and from barbarian pressures in the north. The Eastern Wei reconstructed damaged portions of the Great Wall in the face of Avar pressure (543), before the kingdom was overthrown in civil war and replaced by the Northern Ch'i Dynasty (550). The Turks soon replaced the Avars in the north (see p. 214), and again the Great Wall was strengthened (556). That same year the Western Wei Dynasty was overthrown and replaced by the Northern Chou Dynasty. Warfare between the two portions of north China continued, despite the changes in dynasties, with Northern Chou soon defeating and annexing the dominions of Northern Ch'i (557).

581–600. Rise of the Sui Dynasty. North China was unified by **Yang Chien,** who established a new and vigorous dynasty. Contemplating an invasion of the south, with a view to reuniting China, Yang renewed the work on the Great Wall so as to protect his base area from the barbarians. At the same time he completed plans for his expedition. The Sui invasion of the south was a complete success (589). The Ch'en Dynasty was overthrown, and China was reunited for the first time in almost four centuries.

INNER ASIA*

Throughout most of history the land which we here call Inner Asia has been inhabited by racially and linguistically varied, nomadic, warlike, restless peoples, usually at war with each other and with the more civilized peoples to the west, south, and east. These incessant conflicts and other migratory influences have stimulated almost constant centrifugal population pressures, which have posed equally constant threats to the more civilized neighbors around the periphery. Since the dawn of history the tides of migration from Inner Asia have had periodically earth-shaking effects upon other regions of Europe and Asia.

c. 400–546. Ascendancy of the Avars. A Mongol people (also called the Juan Juan or Gougen), the Avars, under a leader named **Toulun,** had defeated the Hsiung-nu, or Huns, driving them south and west (c. 380). Toulun then established a powerful nomadic empire spreading out to the north of that of the Northern Wei. He took the title of Khan or Cagan (c. 400). The Avars then became engaged in almost constant border warfare with the Northern Wei.

546–553. Rise of the Turks. A peasant people, of mixed Indo-European and Mongol stock, the Turks were subject to, and serfs of, the Avars, most of them working in the iron mines of the Altai Mountains. Revolting, under a chieftain named **Tuman** (or **Tumere**) they soon overthrew the Avar Empire, to establish themselves as the most powerful force in northern and central Asia, and to begin a career of military conquest which would have a profound effect upon history for more than a thousand years. The Avars were completely shattered. Some of them slipped south into China, where they were absorbed. Most migrated west, reaching the Urals (558) and (in combination with other nomads, who adopted the Avar name; see p. 199) continuing on to the Danube (c. 565).

553–582. Expansion of the Turkish Empire. After consolidating the former Avar Empire, the Turks began to press southwestward. They defeated the White Huns in Transoxiana, then continued into Khorasan, conquering most of the territory between the Aral Sea and the Hindu Kush Mountains from the Sassanid Empire (see p. 209). They raided frequently and successfully into China. They also pushed after the Avars into the south Russian steppes, apparently reaching as far as the Black Sea. Tuman and his immediate successors, however, lacked the organization and ability to administer the vast empire which had been conquered so rapidly by their fierce mounted bowmen, and the first Turkish Empire soon collapsed into a number of squabbling principalities.

KOREA

c. 400. The Three Kingdoms. In prehistoric times (traditionally 2300 B.C.) Korea was occupied by a Mongoloid, or Tartar people; little is known of the country or people prior to the 1st century B.C. Off and on during the five centuries preceding A.D. 400, various Chinese dynasties had conquered north Korea, but had never been able to maintain a permanent foothold, or to penetrate substantially south of the Han

---

* For the purposes of this book, Inner Asia is that vast region of mountains, deserts, and steppes lying generally north and west of China, north of the Himalayas, north and east of Persia, east of the Volga River and Ural Mountains, and bounded on the north by Arctic Siberia. (For all practical purposes, it is the area which Mackinder has called the "Heartland.") Although this region has affected and been affected by the regions to the west, southwest, and south, its history has been more closely linked with that of China and East Asia. Accordingly, in this book it will normally be included in general sections dealing with East and Central Asia.

River. Chinese influence and culture, however, pervaded the entire peninsula. During most of this time Korea was divided between three warring kingdoms: Kokuryo in northeast Korea (which also usually included northwest Korea, when that area was not under Chinese control), Paekche in the southwest, and Silla (oldest of the three) in the southeast. During this time there had been some contact with the Japanese, who apparently had colonies on the south coast, between Silla and Paekche, for much of the 4th century. Silla, fearful of the Japanese, traditionally sought an alliance with China. Paekche usually maintained connections with the Japanese.

c. 500–600. Expansion of Silla. Silla gradually achieved predominance in Korea, mainly at the expense of Paekche. Expanding to the west coast (554), Silla eliminated the last Japanese footholds in the south (562). The ascendancy of Silla was also favored by the rise of the strong Sui Dynasty in China (see p. 213), whose suzerainty was accepted by Kokuryo (589).

JAPAN

c. 400. The Clan Period. Like the Koreans, the original Japanese people probably migrated from northern Asia before 2000 B.C. They soon expelled the aboriginal Ainus from most of the three southern islands, establishing themselves in a number of cohesive tribes or clans. Japanese history traditionally begins with the accession of the first emperor, **Jimmu Tenno,** a legendary warrior from Kyushu, who supposedly united the country (600 B.C.). The clan structure of society continued, however, and unity was nominal, though the ascendancy of the imperial clan was recognized by the others. Fierce wars between the clans were frequent, as were overseas expeditions (some joint, some by clan) to Korea. A major invasion of Korea resulted in the establishment of colonies on the south coast of that peninsula (c. 360).

c. 500–600. Rise of the Soga Clan. By a combination of intrigue, martial prowess, and assassination, the leaders of the Soga Clan gradually achieved ascendancy in the imperial government. This resulted in relegation of the emperors to positions of nominal or symbolic power.

VIII

THE RISE AND
CONTAINMENT OF ISLAM:

600–800

MILITARY TRENDS

This was a period of dynamic change marked by the explosive rise of Islam and its sweep of conquest across nearly half of the civilized world. Equally remarkable was the resilience of the Byzantine Empire in sustaining and repelling the direct shock of the violent Moslem assault. This was far more significant than the eventual success of the Frankish and Chinese empires in snubbing the westward and eastern expansion of Mohammedanism, thousands of miles farther away from Mecca.

One general of the period is worthy of consideration as a great captain of history: the Byzantine Emperor **Heraclius.** There were a number of other capable, first-rate soldiers in this period: the Chinese **Li Shih-min** (Emperor **T'ai Tsung**), the Indians **Harsha** and **Pulakesin II,** the Moslems **Khalid ibn al-Walid** and **Harun al-Rashid,** the Byzantines **Leo III** and **Constantine V,** and the Frankish Emperor **Charlemagne.**

Of these Charlemagne had the greatest impact on the history of his times and the greatest influence on future events. His military and political successes were due more to sound, deliberate organizational skill than to imaginative genius. He imparted to chaotic Western Europe an order, prosperity, and stability which had been missing since the collapse of the Roman system. Though much that he accomplished died with him, enough survived to create a pattern for European development for several centuries to come.

THE IMPACT OF ISLAM ON WARFARE

The major cause of the meteoric expansion of Islam was the fanaticism engendered by the charismatic leadership of **Mohammed,** and by specific tenets of his teachings which promised everlasting pleasure in heaven to those who died in holy wars against the infidel. No other religion has ever been able to inspire so many men, so consistently and so enthusiastically, to be completely heedless of death and of personal danger in battle.

Thus it was energy more than skill, religious fanaticism rather than a superior military system, and missionary zeal instead of an organized scheme of recruitment which accounted for Moslem victories. It should be noted, furthermore, that these successes might never have been possible had it not been for unusual circumstances which existed in Southwest Asia at the moment Islam appeared as a major religious, political, and military force in the region. The Byzantine and Persian empires were both exhausted by prolonged wars. Both were plagued by serious internal political and religious unrest, leaving their outlying provinces rebellious and ripe for plucking. Without the impetus derived from their early, relatively easy victories over Byzantines and Persians, it is doubtful if the Moslems would have been able to expand so fast or so far.

Once their initial headlong rush had run its course, the Moslems began to realize that even their own religious fervor could not afford the appalling loss of life resulting from heedless light-cavalry charges—almost entirely by unarmored men wielding sword and lance—against the skilled bowmen of China and Byzantium or the solid masses of the Franks. Having by this time come into contact with practically every important military system in the world, the Mohammedans sensibly adopted many Byzantine military practices (see p. 233). They were never so well disciplined as the East Romans nor so well organized. They relied primarily upon tribal levies rather than upon a standing military force. Their original fanaticism, nevertheless, combined with astute adoption of Byzantine tactics and strategic methods, made them still the most formidable offensive force in the world at the close of this period.

By that time, however, dynastic and sectarian differences had caused the Moslems to turn more of their energies inward. By the end of the 8th century the perimeter of Islam was generally stabilized, but endemic warfare persisted—and would persist for centuries—along three flaming frontiers: the mountains of Andalusia, the mountains of Anatolia, and the mountains and deserts of central Asia.

WEAPONS

Only one major new weapon appeared during this period: Greek fire, decisive and history-making, introduced to warfare by the Byzantines during the first Moslem siege of Constantinople (see p. 242). The exact composition of this explosive-inflammatory material, a closely guarded secret, has not survived to modern times. But this prototype of the modern flame thrower apparently was based upon a mixture of sulphur, naphtha, and quicklime that burst explosively into flames when wetted.

The combustible mixture was evidently packed into brass-bound wooden tubes, or siphons. Water was then pumped from a hose at high pressure into the tube. The material burst into flames and was projected a considerable distance by its own explosion as well as by the force of water pressure. The deadly effect of this weapon upon wooden ships, and upon the flesh of opposing soldiers, can well be imagined. Greek fire retained Byzantine maritime supremacy against a strong Moslem challenge. It also helped to keep the walls of Constantinople inviolate.

There were the usual modifications and improvements of existing weapons, of which the development of the cavalry lance was perhaps the most important. This was becoming the principal weapon of Western Europe, and was an important secondary weapon of the Byzantine cataphract (see p. 235).

On a worldwide basis, however, the bow was by far the predominant weapon. Its almost universal use had caused the more highly developed armies (such as the Byzantine) to discard, for field operations, the heavier and clumsier catapult and ballista used so effectively by the Romans (though these weapons were used extensively by the Byzantines in the siege and defense of fortifications). The general effectiveness and utility of the bow were such that it was being adopted extensively by the Moslems at the end of the period, and Charlemagne was making a determined but vain effort to introduce it to the armies of Western Europe (see p. 226).

TACTICS

The most important tactical developments of the period are discussed in some detail in connection with the military systems of Charlemagne and of the Byzantines (see pp. 225 and 233).

In general the most significant feature of tactics was the worldwide supremacy of cavalry. This, in combination with the almost equally prevalent reliance upon the bow as a missile weapon, meant that warfare throughout most of the world was one of fire and movement, with shock action relegated to a secondary role and used normally to confirm a decision already achieved with missile weapons. There were two important exceptions: the Moslems and the Western Europeans.

Only among the Byzantines did there seem to be any real effort or ability to establish coordination between the actions of infantry and cavalry. This required a high state of training, discipline, and control—conditions rarely found save among the East Romans.

WESTERN EUROPE

VISIGOTHIC SPAIN, 600–712

The Visigothic monarchy in Spain declined steadily during the 7th century. Internal dissension between the nobility and the inept kings, and between Visigoths and the Ibero-Roman population, had brought the kingdom to the verge of disintegration when **Roderick** ascended the throne (710). The suddenness and completeness of the collapse in the face of the Moslem invasion (711–712) was less a measure of the military capability of the Saracens than proof of the utter incompetence of a regime on the verge of collapse. Important events of the period were:

616–624. Elimination of the Last East Roman Footholds in Southern Spain. This occurred while the Byzantine Empire was fighting for its life against the Persians and Avars (see p. 228 ff). During this period Ceuta (on the Moroccan shore of the Strait of Gibraltar) was also conquered from the Byzantines (618).

710. Moslems Reach the Strait of Gibraltar. Count **Julian** repulsed a Moslem attack on Ceuta led by **Musa ibn Nusair.** Julian, disloyal to King Roderick, then made an alliance with Musa, promising assistance in an Arab invasion of Spain. With his assistance a Moslem raiding party led by **Abu Zora Tarif**

made a reconnaissance in force in the Algeciras area of southern Spain (July).

711. Moslem Invasion of Spain. Musa sent a force of 7,000 under **Tarik ibn Ziyad** across the strait. Landing at Gibraltar (whose name commemorates the incident: Gebel el Tarik) he proceeded northwest, seeking the Visigoths, finding them between the Barbate and Guadalete rivers.

711, July 19. Battle of the Guadalete (near Medina Sidonia). Roderick, with 90,000, was completely defeated by Tarik, now reinforced to about 12,000. The Moslem victory was facilitated by Visigothic dissension, treachery, and wholesale defections during the battle. Roderick fled while the issue was still in doubt, and was drowned while trying to cross the river. Tarik swept through southern Spain to win another victory at the **Battle of Ecija,** then captured the Visigothic capital at Toledo without opposition.

712. Arrival of Musa. Taking the command from Tarik, Musa completed the conquest of Spain save for isolated regions in the Asturias Mountains.

LOMBARDY, 600–774

The Lombards never achieved unity among themselves; as a result they never succeeded in conquering all of Italy. Until overwhelmed by **Charlemagne,** the Lombard kings were constantly at war with the Byzantine Empire in vain efforts to absorb the remnants of the East Roman Empire in Italy. At the same time they had to contend not only with their own unruly nobility but with frequent inroads of Franks, Avars, and Slavs into north Italy. Kings **Liutprand** (712–744) and **Aistulf** (749–756) were the strongest of the Lombard rulers. Liutprand broke the power of the southern dukes of Benevento and Spoleto, and briefly captured Ravenna, capital of Byzantine holdings in Italy. Building on his accomplishments, Aistulf again captured Ravenna and twice came close to capturing Rome. These successes caused Pope **Stephen II** to appeal to **Pepin,** King of the Franks, for assistance. Pepin twice marched into Italy to defeat Aistulf, and by the "donation of Pepin" ceded to the Pope territory formerly belonging to the Byzantine emperors and to the Lombard kings, thus establishing the temporal powers of the papacy. Renewed Lombard threats against the new papal possessions caused Pope **Adrian I** to appeal for assistance to Charlemagne, son of Pepin. In two campaigns Charlemagne defeated and captured **Desiderius,** last King of the Lombards, took the Lombard crown himself, and annexed all of Lombardy to his empire. Important events of the period were:

728. Capture of Ravenna by Liutprand. Soon after, it was retaken by the Byzantine Exarch of Ravenna.

752. Capture of Ravenna by Aistulf.

754. Invasion of Italy and Recapture of Ravenna by Pepin. Aistulf sued for peace, agreeing to end attacks against Rome.

756. Pepin's Second Invasion. Upon renewed attacks against Rome by Aistulf, Pepin again invaded Italy, defeated Aistulf near **Ravenna,** and granted lands to Pope **Stephen II.**

773–774. Charlemagne Defeats Desiderius. After besieging and capturing Desiderius at Pavia, Charlemagne took the crown of Lombardy for himself.

BRITAIN

Early in the 7th century the Britons were eliminated from England proper—save for Cornwall, where they retained a foothold until the 9th century—and were driven into Wales. Subsequent violent struggles among the seven Anglo-Saxon kingdoms of England were not decisive; supremacy shifted from one to the other, with Northumbria and Mercia most often predominant. Nowhere in the Western world did the art of war sink lower than in Britain during this period. Savage ardor had overcome all vestiges of Roman system and discipline. Retaining little direct contact with continental affairs, Anglo-Saxon methods of warfare had probably regressed since the time of Hengist and Horsa. Strategy and discipline were unknown; tactics simply consisted of the disorderly alignment of opposing warriors in roughly parallel orders of battle followed by dull, uninspired butchery until one side or the other fled. Fortifications were crude and rudimentary palisades; armor was scarce and poor. In consequence warfare was, as Oman says, "spasmodic and inconsequent." Near the end of the 8th century, Norse raiders began to appear along the coasts of Britain and Ireland, to find lands ripe for conquest. The important events of the period were:

597–616. Supremacy of Kent. King **Aethelbert** was ruler of Kent.

603. Scottish Invasion. This was repelled by King **Aethelfrith** of Northumbria, who defeated King **Aidan** of the Dalriad Scots at the **Battle of Daegsastan.**

615. Battle of Chester. Aethelfrith defeated the Britons, pushing on to reach the Irish Channel, separating the Britons in Wales from those of northwest England.

616. Battle of the Idle. Raedwald, King of East Anglia, defeated and killed Aethelfrith to help **Edwin** become the first Christian King of Northumbria.

616. Supremacy of Northumbria. Edwin defeated the Britons in north Wales and Anglesea. His principal rival was the Briton, **Cadwallon,** a king in north Wales. The supremacy of Northumbria continued under **Oswald,** successor of Edwin (632–641)

632. Battle of Hatfield Chase. Edwin was defeated and killed by the alliance of the Christian Briton Cadwallon and the heathen Angle, **Penda** of Mercia.

633. Battle of Hefenfelth (Rowley Water). Oswald defeated and killed Cadwallon to drive the Britons completely from northwest England.

641. Battle of Maserfeld (Oswestry, Shrop- shire). Penda defeated and killed Oswald to establish Mercia temporarily as the leading Anglo-Saxon kingdom. He was defeated and killed at the **Battle of Winwaed** (655) by **Oswy** (641–670), younger brother of Oswald, who reestablished Northumbria in a position of preeminence.

685. Battle of Dunnichen Moss (north of Tay River in Forfar, Scotland). Egfrith, successor of Oswy, was defeated and killed by the Scots after he had overrun most of southern Scotland. This battle assured the independence of Scotland from Anglo-Saxon England.

716–796. Supremacy of Mercia. This was under Kings **Aethelbald** (716–757) and **Offa** (757–769). Aethelbald successfully invaded Wessex (733) and Northumbria (744). His cousin Offa succeeded to the throne after a violent civil war. He reestablished Mercian hegemony in a series of hard-fought campaigns, which resulted in the first real unification of England. He was treated as an equal by his contemporary, Charlemagne.

789–795. Arrival of the Vikings. The first recorded Viking raid on England was near Dorchester (789). The Danes raided Lindisfarne Island off the Northumbrian coast (793). The first recorded Viking raids against Scotland (794) and Ireland (795) soon followed.

KINGDOM OF THE FRANKS, 600–814
Decline of the Merovingians, 600–731

The vigor and bellicosity of the Franks were undiminished despite the decline and degeneracy of the ruling Merovingian family. A remarkable noble family seized the reins of leadership. **Pepin I** became mayor of the palace (prime minister) of Austrasia early in the 7th century. After an anarchic interval and a series of violent civil wars, he was succeeded by his son **Pepin II,** who expanded his authority over Neustria as well. He was succeeded as mayor of the palace early in the 8th century by his illegitimate son **Charles Martel** (Charles the Hammer).

In Aquitaine, Count **Eudo** (Eudes, or Odo) was emerging as Charles Martel's principal rival for Frankish supremacy. Most of Eudo's attention, however, was directed to repelling Moslem incursions across the Pyrenees, which began in 712 and which waxed in frequency and intensity in the following decade. At first successful, by 731 Eudo found himself unable to cope with nearly simultaneous invasions of his realm from the northeast by Charles Martel and from the south by the Moslems. The principal events were:

613. Temporary Reunification of the Principal Frankish Kingdoms. Austrasia, Neustria, and Burgundy were united under Merovingian King **Lothair II.**

623–629. Pepin I, Mayor of the Palace. This was during the minority of **Dagobert** (last strong Merovingian king). After coming of age (628), Dagobert exiled Pepin. After the death of Dagobert, Pepin again became mayor of the palace briefly before his own death (638–639).

c. 675–678. Rise of Pepin II. He led Austrasian nobles in civil war to overthrow **Ebroin,** whom he replaced as mayor of the palace.

687. Battle of Tertry (near St. Quentin). Pepin conquered Neustria. In subsequent campaigns against Frisians, Alemanni, and Burgundians, he reunited the old Frankish kingdom of Clovis, save for Aquitaine.

712. Arrival of the Moslems. First raid north of the Pyrenees.

714. Rise of Charles Martel. The death of Pepin caused temporary anarchy, resolved in Austrasia by victories of Charles Martel over his rivals in the battles of **Amblève** (near Liège, 716) and **Vincy** (near Cambrai, 717).

c. 715. Rise of Eudo. He consolidated his control over Aquitaine. He soon made an alliance with **Ragenfrid,** mayor of the palace of Neustria and rival of Charles.

717–719. Moslem Raids into Aquitaine and Southern France. The Moslem leader **Hurr** captured Narbonne (719) to obtain a base in western Septimania (coastal strip along the Mediterranean between the Pyrenees and the Rhone River).

719. Battle of Soissons. Charles gained control over Neustria by victory over Ragenfrid and Eudo. Eudo made peace temporarily by surrendering Ragenfrid to Charles.

721. Battle of Toulouse. Eudo decisively defeated and killed **Samah,** Moslem governor of Spain, who had invaded Aquitaine and invested Toulouse; remnants of the Moslem army were driven back into Spain and Septimania.

721–725. Appearance of Abd er-Rahman. The governor of Moslem Spain, he led repeated raids into Aquitaine and southern France.

725–726. Moslem Invasion of Southern France. Under the leadership of **Anbaca** (Anabasa), the Moslems captured Carcassonne and Nîmes and conquered all Septimania. The expedition culminated in a raid up the Rhone Valley as far as Besançon and the Vosges.

726–732. Expeditions of Charles Martel into Germany. These resulted in the conquest of Bavaria, expansion of the kingdom into Thuringia and Frisia, and limited success against the Saxons.

731. Charles Martel Invades Northeastern Aquitaine. He annexed Berry (region around Bourges).

Campaign and Battle of Tours, 732

Abd er-Rahman led a Moslem army of unknown strength (probably about 50,000) into Aquitaine, slipping past the western flank of the Pyrenees at Irun. This army, almost entirely cavalry, was comprised mostly of Berbers and other Moors, with a leavening of Arab leadership. Abd defeated Eudo at the **Battle of Bordeaux.** With the remnants of his defeated army, Eudo fled to Austrasia to make peace and swear fealty to Charles Martel, who had hastily returned from a campaign along the upper Danube upon hearing of the Moslem invasion.

The Moslems, meanwhile, had been halted temporarily by the fortified city of Poitiers. Leaving a part of his army to invest the city, Abd advanced to the Loire, near Tours, plundering en route. The Moslems had just laid siege to Tours when they became aware of the secret and rapid approach of Charles and Eudo from the east, south of the Loire, threatening their lines of communication. Abd hastily dispatched his great train of booty to the south, following in a slow withdrawal toward Poitiers.

The Frank army evidently made contact with the foe somewhere south of Tours early in October, 732. From scanty records it appears that for the next six days Abd er-Rahman endeavored to cover the retreat of his train of booty in a classical delaying action marked by frequent but indecisive skirmishes. It would also seem that Charles maintained strong pressure on the Moslems and forced them back steadily toward Poitiers. Accordingly, Abd er-Rahman decided to fight a major battle somewhere between Tours and Poitiers, probably near Cenon, on the Vienne River. The subsequent action, known usually as the Battle of Tours, is sometimes (and more appropriately) called the Battle of Poitiers.

Though the respective strengths are unknown, the Frank army was probably larger than that of the Moslems. Charles had both infantry and cavalry, probably in nearly equal proportions, consistent with the increasing trend toward cavalry. His rapid and secret march into Touraine, and the nature of the skirmishing during the week prior to the battle, support the assumption that Charles had a substantial cavalry contingent, which he used skillfully against the Moslem horsemen.

The Franks had engaged in more or less constant warfare against the Moslems for nearly two decades. Thus Charles was undoubtedly aware of the respective strengths and weaknesses of his own and Abd er-Rahman's forces. He evidently realized that the heavy Frank cavalry was undisciplined, sluggish in comparison with the mobile light Moslem cavalry, and extremely difficult to control in mounted combat. He also realized that the Moslems were effective only in attacking, that they were deadly in taking advantage of a gap in a battle line, but that they had no defensive staying power and that they lacked the weight to deliver an effective blow by shock action against a strong defensive force. These must have been the considerations which led him to dismount his cavalry when he saw the Moslems preparing for a decisive encounter. Apparently he formed his army into a solid phalanx of footmen, presumably on the most commanding terrain available in the rolling country of west-central France.

The classic and frequently quoted description of the battle is that of Isodorus Pacensis, as contained in Bouquet's *Recueil des Historiens des Gaules et de la France* and translated freely and dramatically by Oman:

> The men of the north stood as motionless as a wall; they were like a belt of ice frozen together, and not to be dissolved, as they slew the Arabs with the sword. The Austrasians, vast of limb, and iron of

hand, hewed on bravely in the thick of the fight; it was they who found and cut down the Saracen king.

This and other accounts, including those of the Moslems, indicate that repeated and violent cavalry attacks were repulsed in desperate fighting that lasted till nightfall. One report implies that the Frank right wing, under Eudo, enveloped the Moslem left and forced them to withdraw to protect their threatened camp. At any rate, when night fell, the disheartened and exhausted Moslems had withdrawn to their camp, when they discovered that Abd er-Rahman had been killed during the fight. They then apparently panicked; abandoning their train, they fled south after nightfall.

At dawn the next morning Charles formed his army again to meet a renewed assault. When cautious reconnaissance revealed the flight of the enemy, he rightly refused to pursue. In pursuit his own undisciplined troops were at their weakest, and not amenable to control. It was a favorite Arab tactic to entice the cumbersome Frankish cavalry to such pursuit, and then to turn and slaughter them when they were spread out. Furthermore, he did not wish to leave the great quantities of booty which he had recaptured. Finally, he possibly did not wish to eliminate completely Moslem opposition to his rival, Eudo.

The Battle of Tours is considered one of the decisive battles of history, complementing the even more decisive victory won by Leo III of Byzantium fourteen years earlier at Constantinople (see p. 242). Together these two victories stemmed the hitherto irresistible tide of Moslem expansion and assured Christian Europe of several centuries of growth and development before being faced with comparable pagan threats in the 13th and 15th centuries.

The Battle of Tours cemented the authority of Charles Martel (whose sobriquet was won there) over the Franks, laying the basis for the establishment of the Holy Roman Empire by his grandson Charlemagne.

Rise of the Carolingians, 732–814

Following the Battle of Tours, Charles and Eudo drove the Moslems from all of their conquests in France except for Septimania, where they held on for nearly three decades. There were a few more Arab raids to the north, but the Franks dealt with these readily. They aggressively whittled at Septimania, and **Pepin III,** son and successor of Charles Martel, finally drove the Moslems back over the Pyrenees.

After serving for ten years as the actual ruler of France in the pose of Mayor of the Palace, Pepin ended the mockery of the Merovingian Dynasty by deposing **Childeric III** and, with the approval of Pope **Zacharias,** having himself crowned King of the Franks. In subsequent years (751–768), in addition to wars against the Moslems, Pepin campaigned aggressively in Germany, in Aquitaine, and in Italy. His campaigns in Germany expanded Frankish influence over Bavaria, but he was unable to subdue the stubborn Saxons. He defeated **Waifer,** Duke of Aquitaine and successor to Eudo, bringing that province firmly under central control. In Italy, at the request of Pope **Stephen** (see p. 219), he defeated the Lombards and established the temporal power of the papacy.

Upon the death of Pepin the kingdom was divided equally between his two sons, **Charles** (the elder) and **Carloman,** according to the Frankish custom of succession. Relations between the brothers were cool, but on the death of Carloman (771), Charles became sole ruler of the

THE GROWTH of the CAROLINGIAN EMPIRE

Kingdom of Clovis
Conquests of Clovis' Sons
Conquests of Charles Martel and Pepin I
Conquests of Charlemagne
Byzantine Empire

Franks, to commence the brilliant reign which caused him to be known to history as "the Great," or **Charlemagne.**

Charlemagne's principal attributes were vigor, brilliance, efficiency, and intellectual curiosity. He stimulated and encouraged revivals of learning, art, and literature. Under his firm, benevolent rule most of the Christian nations of Western Europe enjoyed peace and prosperity to an extent unknown since the Antonine emperors of Rome. His cultural, economic, political, and judicial accomplishments were based upon a remarkably efficient military system which he devised in a series of aggressive campaigns around the frontiers of his realm. He kept his external enemies so busy that they had neither the time nor inclination to think of invading or harassing his territories. The principal events of the period were:

735. Rebellion in Aquitaine. After Eudo's death his sons rebelled against Charles Martel, who invaded Aquitaine and forced the rebels to pay homage.

734–739. Charles Martel's Rhône Valley Campaigns against the Moslems. During this time the Moslems attempted only two major raids into Frankish territory, which Charles repulsed after the Moslems reached Valence and Lyon (737 and 739).

741–747. Joint Rule of Carloman and Pepin III (Pepin the Short). Upon the death of

Charles Martel, his sons ruled jointly as Mayors of the Palace.

743–759. Moslems Driven from France. Pepin drove the Moslems across the Pyrenees from Septimania, which he annexed to the Frankish kingdom.

747. Resignation of Carloman. Pepin became sole ruler of the Franks.

751. Coronation of Pepin.

754–756. Campaigns of Pepin against the Lombards. (See p. 219.)

757–758. Pepin's Campaigns in Germany.

He defeated **Tassilo III,** Duke of Bavaria, who had tried to assert his independence (757), and defeated the Saxons, requiring them to pay tribute (758).

760–768. Campaigns in Aquitaine. Pepin finally defeated **Waifer,** rebellious Duke of Aquitaine.

763. Revolt of Tassilo of Bavaria. Frankish control was eliminated from Bavaria for fifteen years.

768–771. Joint Rule of Charlemagne and Carloman. After the death of Pepin, his two sons divided the kingdom. During this period Charlemagne suppressed another insurrection in Aquitaine (769).

771–814. Reign of Charlemagne. The kingdom was reunited after the death of Carloman.

772–799. Charlemagne's Campaigns against the Saxons. Generally successful in these eighteen campaigns, Charlemagne was frequently frustrated by Saxon uprisings as soon as his main army withdrew. Minor setbacks were suffered by the Franks when the Saxons captured **Eresburg** by treachery (776) and took **Karlstadt** by storm (778). The back of Saxon resistance was broken by Charlemagne's vigorous winter campaign (785–786), although sporadic resistance continued for 13 more years.

773–774. Charlemagne Defeats Desiderius. Charlemagne besieged and captured Desiderius, last King of the Lombards, in **Pavia,** and took the Lombard crown himself (see p. 219). The following year he suppressed a Lombard revolt (775).

777–801. Frankish Conquest of Northern Spain. The first expedition against the Emir of Cordova was successful, but ended in a disaster when the Christian Basques of Pamplona joined the Moslems in ambushing and destroying the rearguard of the Frankish army under Charlemagne's nephew, **Roland,** in the **Pass of Roncesvalles** (778). (One result was the classic epic poem, *The Song of Roland.*) In subsequent campaigns Charlemagne subdued the Basques and drove the Moslems south of the Ebro River. These campaigns culminated in the successful siege of Barcelona (800–801).

780 and 787. Campaigns in South Italy. Charlemagne's successful punitive expeditions against the Duke of Benevento.

787–788. Reconquest of Bavaria. Charlemagne defeated Tassilo.

789. Occupation of Istria. Charlemagne ignored Byzantine objections.

791–796. Defeat and Conquest of the Avars. During these campaigns Frank armies under Charlemagne and his son **Pepin** operated in the central Danube Valley, and penetrated east of the Theiss River. The Franks also conquered parts of Croatia and Slovenia from the Slavs. Subsequent Avar revolts were suppressed (799, 803).

800, December 25. Charlemagne, Holy Roman Emperor. He was crowned by Pope **Leo III**

803–810. Operations against Byzantines. These desultory campaigns were fought on land in Dalmatia and at sea on the Adriatic. Though generally successful, Charlemagne made a conciliatory peace with the Byzantines, surrendering most of his conquests in Dalmatia, but retaining Istria. In return Byzantine Emperor **Nicephorus I** recognized Charlemagne as Emperor of the West (see p. 284).

809–812. Arrival of the Norsemen. Charlemagne and his subordinates repelled Danish raids up the Elbe. His fleet, based in Boulogne, prevented harassment of the sea coast of the Frank Empire.

THE MILITARY SYSTEM OF CHARLEMAGNE, C. 800

Charlemagne's military system was crude in comparison with the organization of the Macedonians, Romans, and Byzantines. It was, however, a revolutionary departure from the military anarchy which had prevailed in Western Europe for four centuries, and which returned after his death. Furthermore, it provided a basis for the development of feudalism and the chivalry of the Middle Ages.

Prior to Charlemagne the principal military characteristics of the Franks were exceptional vigor and exceptional indiscipline. Even under such able warriors as Charles Martel and Pepin, Frank armies were unreliable, Frank conquests impermanent. The principal reason why opponents of Charlemagne were consistently outclassed was because he was able to harness Frankish vigor in a disciplined, efficient organization, while at the same time providing a high order of personal leadership.

One demonstration of the working of Charlemagne's military system was its adaptation to the Lombard heavy cavalry. He had relatively little trouble in defeating the Lombards in two brief campaigns, but recognized an intrinsic superiority of the Lombard cavalry to his own Frankish horsemen. While working successfully to improve his Frank cavalry, he henceforth made much use of the Lombards, who provided the major component of the Frank armies which defeated the Avars. Prior to their reorganization and disciplining under Charlemagne, these same Lombards had been relatively helpless against Avar raids into north Italy.

As in every successful military system, discipline was the prime ingredient. Cavalry and infantry were responsive to the commands of superior authority in a way unknown to the Teutons since they had overthrown the Roman legions. By the end of his reign, Charlemagne established a system of calling men to service through his noble vassals, which enabled him to maintain standing armies in the field indefinitely without placing an undue strain on the economy, without being forced to employ unreliable riffraff, and without denuding the provinces of local resources for preserving law and order.

One reason why the predecessors of Charlemagne had been unable to keep armies in the field for any length of time had been lack of logistical organization. Frankish armies had subsisted on foraging and plunder. In nominally friendly areas this soon antagonized the inhabitants and contributed to internal unrest. In hostile territory the dispersal of forces on plundering missions often led to disaster at the hands of an alert, concentrated foe. Supply shortages almost always caused the dissolution of Frankish armies after a few weeks in the field. Charlemagne established a logistical organization, including supply trains with food and equipment sufficient to maintain his troops for several weeks. Replenishment of supplies was done on an orderly basis, both by systematic foraging and by convoying additional supply trains to the armies in the field. This permitted Charlemagne to carry war a thousand miles from the heart of France and to maintain armies in the field on campaign, or in sieges, throughout the winter—something unknown in Western Europe since the time of the Romans.

Charlemagne also revived the Roman and Macedonian practice of maintaining a siege train. Thus he was never embarrassed by inability to deal with hostile fortifications. The supply and siege trains slowed down the advance of his main armies, but assured reliable progress. Furthermore, by increasing reliance upon cavalry, accompanied by mule pack trains, he was still able to project his strength quickly and forcefully.

A key element of Charlemagne's military system was his use of frontier fortified posts, or burgs. These were built along the frontiers of every conquered province, and were connected with each other by a road along the frontier. In addition, another road led directly back to the old frontier from each burg. Stocked with supplies, these forts became bases for the maneuver of the disciplined Frank cavalry, either to maintain order in the conquered territory or to project Frankish power forward in further operations.

Charlemagne also brought the bow back into the arsenals of Western Europe, probably as a result of experience against the Avars and Byzantines. For reasons that are not clear, this weapon was soon again discarded by most European armies after his death, although it remained popular as a hunting weapon.

Another evidence of the efficacy of the system was uniform tactical doctrine. Indoctrination and training of subordinate commanders is only possible in a highly efficient military organization. He had an excellent intelligence system. The completeness and directness of Charlemagne's orders—some of which have been preserved—prove not only his own professional competence but also that this competence had created an effective staff system.

The main elements of Charlemagne's system are contained in a series of five imperial ordinances, issued between 803 and 813, which were a form of field-service regulations. In these he prescribed the duties of vassals in preparing forces to be raised on call; the property basis for military call-ups of individual soldiers; the organization of units; the weapons, armor, and equipment to be carried by each man and to be brought with each unit; lists of punishments for common offenses, and the like.

MOSLEM AND CHRISTIAN SPAIN, 712–800

Under Musa ibn Nusair the Moslem conquest of Spain was completed quickly, save for scattered resistance in the narrow strip of mountains of the northwest. He was delayed only by the fortifications of Seville and Mérida, which he took after protracted sieges. The Ibero-Roman inhabitants were glad to be rid of the Visigoths, and the Moslems were lenient in allowing Christians and Jews to continue the practice of their faiths. Moslem leadership had been Arab initially, and the core of the army had been Arab-Syrian. Most of the fighting men, however, had been Berber, and as more Berbers arrived from Morocco, the flavor of the occupation became essentially Moorish. The internal dissensions of Islam were reflected in the Moslem community in Spain, and additional discord was created by friction among Arabs, Syrians, and Berbers. The establishment of the independent Omayyad Dynasty in Spain by refugee Prince **Abd-er-Rahman ibn Mu'awiya** (see below) eventually led to relative peace and centralized control. Meanwhile, the dissension among the Moslems had given the Visigothic refugees in the Asturias Mountains a breathing spell to organize their defenses and to begin tentative efforts to expand their tiny enclave of Christian freedom. By the end of the 8th century, Moslem-Christian warfare, which would be endemic in Spain for more than seven centuries, had begun. Principal events were:

712–720. Consolidation of the Moslem Conquest of Iberia. Raids were begun across the Pyrenees (see p. 221).

c. 718. Establishment of the Christian Kingdom of the Asturias. Visigothic Prince **Pelayo** (718–737) made his capital at Cangas de Onís. He repulsed the Moslems at the **Battle of Covadonga** (718).

732. Battle of Tours. (See p. 222.)

740. Berber Revolt. This created chaos in Morocco and Spain, and permitted the tiny kingdom of the Asturias to expand to the south and west, culminating in the Christian conquest of Galicia (750).

755–756. Establishment of the Omayyad Emirate of Cordova. Abd-er-Rahman ibn Mu'awiya, refugee from the Abbassid massacre of the Omayyads (see p. 252), founded the emirate.

756–786. Moslem Consolidation. Abd-er-Rahman suppressed internal disorders in Moslem Spain and halted Christian resurgence in the northwest.

777–801. Invasion of Spain by Charlemagne. (See p. 225.) He captured Barcelona.

SCANDINAVIA

Beginning in the 7th century, bands of Scandinavian warrior-adventurers, called Vikings, began to strike into parts of northern Europe. The first areas subjected to their raids were southwestern Finland and the Baltic coast from the Vistula north to the Gulf of Bothnia. Most of the Vikings who operated in those areas were from Sweden, but others from Denmark and Norway also took part. By the late 8th century, Vikings from Norway and Denmark began to strike at the British Isles and the western European coast, while raiders and traders from Sweden voyaged deep into Russia, eventually reaching the Black Sea, and thence Constantinople.

At this time, none of the Scandinavian countries was unified, and political power rested with local lords and chieftains. Among the major developments were:

c. 650–c. 800. Swedes Subjugate Sweden. Sweden was settled by a number of regional tribes. The Swedes, who lived around Uppsala in the northeast, gradually subjugated their neighbors, like the Goths of Västergötland and Östergötland, and by the early 9th century dominated the country.

c. 700. Jarls in Norway. By the early 8th century, regional chieftains called *jarls* ("earls") came to prominence in Norway, ruling over small states based on clans or tribes.

EURASIA AND THE MEDITERRANEAN

BYZANTINE EMPIRE, 602–642

Disaster under Phocas, 602–610

602. Overthrow and Murder of Maurice. Phocas, a centurion of the Byzantine army on the Danube, fomented a mutiny, displaced Priscus (see p. 199), and led the mutinous army back to Constantinople, forcing the abdication of Maurice, whom Phocas then cruelly murdered.

603–604. Resurgence of the Avars. Aided by the mutiny, Avar raiders swept through imperial lands south and west of the Danube. Phocas brought temporary peace by paying a large tribute (604).

603–610. War with Persia. Chosroes II of Persia, who owed his throne to Maurice (see p. 210), declared war on the murderer of his benefactor. Persian armies were victorious in Mesopotamia and Syria, capturing the fortress towns of Dara, Amida Haran, Edessa, Hierapolis, and Aleppo, though they were repulsed from Antioch and Damascus. They then overran Byzantine Armenia and raided deep into Anatolia through the provinces of Cappadocia, Phrygia, Galatia, and Bithynia. Byzantine resistance collapsed; cowardly, inept Phocas did nothing to halt the invasion. A Persian army penetrated as far as the Bosporus (608).

610. Revolt of Heraclius (the Elder). The Exarch of Africa, **Heraclius,** a loyal subordinate of Maurice, had refused to accept the usurpation of Phocas. He now sent a fleet from Carthage to Constantinople, under the command of his son, also named **Heraclius.** Assisted by an uprising in the city, Heraclius overthrew Phocas, who was killed by the mob. Young Heraclius then accepted the throne thrust upon him by popular demand.

Heraclius and the Last Persian War, 610–628

610. Continuation of the War with Persia. The downfall of Phocas should have satisfied

Chosroes, but the Persian ruler now had visions of reestablishing the empire of Darius, and intensified his war effort. The Byzantine army, ruined by defeat, corruption, and misadministration, offered only halfhearted opposition.

611–620. Persian Victories in Syria and Anatolia. Antioch and most of the remaining Byzantine fortresses in Syria and Mesopotamia and Armenia were captured (611). After long sieges the invaders took Damascus (613) and Jerusalem (614). Chosroes then began a determined invasion of Anatolia (615). Persian forces under General **Shahen** captured Chalcedon on the Bosporus after a long siege (616). Here the Persians remained, within one mile of Constantinople, for more than 10 years. Meanwhile, they captured Ancyra and Rhodes (620); remaining Byzantine fortresses in Armenia were captured; the Persian occupation cut off a principal Byzantine recruiting ground.

616–619. Persian Conquest of Egypt. After defeating Byzantine garrisons in the Nile Valley, Chosroes marched across the Libyan desert as far as Cyrene. These victories cut off the usual grain supplies from Egypt to Constantinople.

617–619. Renewed Avar Invasions. Avar raiders swept the Balkan provinces, reaching the walls of Constantinople.

619–621. Heraclius and Sergius. Heraclius vainly attempted to reorganize the army. In despair, he prepared to leave Constantinople and to return to Africa. At this point **Sergius,** Patriarch of Constantinople, sparked a popular resurgence of patriotism in Constantinople. By entreaty and reproach he obtained from Heraclius an oath that he would never abandon the capital. Sergius promised to make available all of the resources of the Church. With renewed energy and confidence, Heraclius turned to the task of reorganization. With the somewhat reluctant approval of Sergius, he emptied the overcrowded monasteries to recruit monks into the army, and seized much of the wealth of the churches of Constantinople. He made peace with the Avar chagan (chieftain) just outside the walls of Constantinople by paying a large indemnity. Meanwhile, negotiating with Shahen, he pretended to consider unacceptable Persian terms (which would have recognized all of Chosroes' conquests), while preparing to take the field.

622, April. Heraclius Sails from Constantinople. Taking advantage of continuing Byzantine command of the sea, Heraclius sailed from Constantinople with an army of unknown size (probably less than 50,000). He landed a few days later at the junction of Cilicia and Syria, near Alexandretta and ancient Issus.

622–623. The Campaign of Issus.* The coastal plain of Issus, ringed by mountains, is accessible only through three passes, all of which Heraclius promptly seized. A much larger Persian army under the general **Shahr Baraz** probed cautiously at the passes, but was repulsed. To restore martial prowess in his army, Heraclius combined tough drill-field training with carefully controlled engagements against the frustrated Persians encamped outside his enclave. Finally ready (probably October), Heraclius apparently pretended to march through the Amanic Gates as though striking toward Armenia. He enticed the Persians into battle on the heights overlooking Issus. Pretending flight, he apparently drew the Persians back toward the plain, while he sent the main body of his army eastward through another pass. Counterattacking with his rear guard, he halted the Persian advance in the pass, then hurried after the main body, leaving Shahr Baraz and his army far behind. The **Battle of Issus** was a minor engagement, but the emperor and his army were elated by the ease with which they had outmaneuvered and outwitted their enemy. After several days of indecision and countermarching, Shahr Baraz followed Heraclius into Pontus. At the end of the year the two armies were again facing each other, this time in the Anti-Taurus Mountains along the upper Halys River. After several weeks of skirmishing, Shahr Baraz attempted an ambush, but Byzan-

* The details of this and the subsequent operations of Heraclius are inadequately recorded. What he accomplished is known; how he did it is often vague.

tine reconnaissance discovered the Persian dispositions (January 623). Heraclius pretended to fall into the trap while setting one of his own. As the concealed Persians burst out of their hiding place, Byzantine troops converged on them from the surrounding mountains. The entire Persian enveloping force was annihilated. Heraclius immediately turned to attack Shahr Baraz' main body, but the remaining Persians were so shaken by the unexpected turn of events that they soon fled in disorder. Following this decisive **Battle of the Halys,** Heraclius put his army into winter quarters, then personally returned to Constantinople to supervise defensive measures being taken against the continuing Avar threat.

623. Invasion of Media. Leaving his son **Constantine** to command the capital against the Avars and the Persians still at Chalcedon, Heraclius sailed with 5,000 reinforcements to join his army at Trapezus (Trebizond). Raising additional forces in Pontus, Heraclius struck quickly through the mountains of Armenia to northern Media, heading for the capital, Tauris (Tabriz). The strength of his army is not known, but was probably more than 50,000. Chosroes, dashing to protect Media with about 40,000 men, declined battle and abandoned Tauris to Heraclius. The Byzantine emperor then advanced to Thebarma (Urmia, or Rizaiyeh), where he captured the shrine at the supposed birthplace of Zoroaster. He then withdrew northeastward from the mountains to winter in the fertile Araxes Valley of Albania (northern Azerbaijan).

624. Invasion of Central Persia. Heraclius apparently led his army southeastward along the shores of the Caspian, then turned south through the Hyrcanian (Elburz) Mountains into the heart of Persia, advancing as far as Ispahan—the farthest penetration into Persia ever made by a Roman or Byzantine army. Chosroes, withdrawing most of his troops from Chalcedon, assembled three great armies, converging on Heraclius in central Persia. Heraclius seems to have operated successfully on interior lines between the Persian armies, then withdrew to spend the winter in northwest Persia or Media. When the Persians went into winter

quarters nearby, Heraclius made a surprise attack on Shahr Baraz, routing his army and successfully storming a walled city in which the Persian general had taken refuge. Shahr Baraz escaped, and the Byzantine army spent the winter in and around the captured Persian fortifications.

625, Spring. Campaigns in Corduene and Mesopotamia. Heraclius marched west through the mountains of Corduene (Kurdistan) and across the Tigris, accompanied by a great train of booty. The speed of his march seems to have surprised the Persians, and he recaptured the fortified towns of Amida and Martyropolis (Maipercat), apparently without a struggle. The Persian army in northern Mesopotamia withdrew westward across the Euphrates. Heraclius pursued into Cilicia, where he found the reinforced Persians, under his old foe Shahr Baraz, holding the line of the Sarus River.

625. Battle of the Sarus. Heraclius was victorious in an assault river crossing. Shahr Baraz withdrew northwestward, pursued by Heraclius. Clearing the Persians from Cappadocia and Pontus, Heraclius returned to spend the winter at Trapezus.

625–626. Alliance of Persia and the Avars. Learning of this alliance, Heraclius evidently made a sea voyage to Constantinople to organize the defense of the city against the double threat. Presumably while he was there, he made arrangements for a new army under his brother **Theodore** to operate against the Persians in western Anatolia while he returned to his own army in Pontus.

626. Persian Plans. During the winter Chosroes planned an all-out effort against Constantinople. He returned to Anatolia with three armies—of unknown size, presumably more than 50,000 men each. One of these (possibly commanded by Chosroes himself) was to contain Heraclius in Pontus; another, under Shahen, was to prevent any junction between the armies of Heraclius and Theodore; the third, evidently under Shahr Baraz, was to cooperate with the Avars in an attack against Constantinople.

626, June 29–August 10. The Defense of Constantinople. An Avar army about 100,000 strong (including large contingents of Slavs, Germans, and Bulgars) stormed the walls protecting the approach to the city and advanced to the city wall itself. Rejecting Byzantine bribes, the Avars closely invested the city. Heraclius, himself hard-pressed in Pontus, rushed 12,000 of his best veterans by sea to bolster the defenders of the city, commanded by his son, Constantine. After careful coordination of plans between the Avar chagan and Shahr Baraz, the allies mounted a furious offensive (August 1). While the Avars stormed at the walls, the Persians attempted to cross the Bosporus by small boats and rafts above Constantinople, aided by a swarm of Slavic small craft. While Constantine and his soldiers repulsed every Avar assault, the Byzantine Navy smashed the Slavic and Persian flotillas on the Bosporus. The action continued incessantly for ten days until the allies admitted defeat (August 10). The Persians had been unable to effect a landing either at Constantinople itself or behind the Avar lines. The Avars, having suffered terrible losses, running short of food and supplies, were forced to withdraw. This was one of the great moments in the long history of Constantinople.

626, Summer. Operations in Anatolia. Meanwhile, Heraclius, his army reduced by campaigning and detachments to less than 30,000 men, was on the defensive in Pontus. Apparently he left a strong garrison in Trapezus and withdrew slowly northeastward along the Black Sea into Colchis (Lazica), where he halted the Persians by aggressive defensive-offensive operations along the Phasis River. By attracting the main Persian army in Anatolia, he provided his brother Theodore with the opportunity to defeat the one remaining Persian army somewhere in western Anatolia. By the end of the summer Theodore's army was probably in central Anatolia, threatening the communication of both the Persian armies at Chalcedon and in Colchis. The defensive strategy of Heraclius—with the able assistance of his son and his brother—had proven as successful as his offensive operations.

626–627. Preparations for a Renewed Offensive. During the winter Heraclius made an alliance with **Ziebel,** Khan of the Khazars, for a joint invasion of Persia the following spring. The Khazars were to advance across the Oxus into eastern Persia, and to raid from the Caucasus into Armenia and Media; the Byzantines would meanwhile drive through Syria, Mesopotamia, and Assyria toward the Persian capital of Ctesiphon. In preparation Heraclius (reinforced by troops from Constantinople) marched southward to Edessa, either during the winter or (more likely) in the early spring.

627, Spring. Opening the Campaign. With 70,000 men Heraclius swept through Syria, Mesopotamia, and southern Armenia, recapturing most of the Byzantine fortresses lost to the Persians 10 and 15 years earlier. The army of Shahr Baraz, still at Chalcedon, was now cut off completely. The Persian general then learned that Chosroes, dissatisfied with his failure to capture Constantinople, was planning to have him executed. Accordingly, Shahr Baraz surrendered to Heraclius, although refusing to join the Byzantine army against his ungrateful sovereign. Anatolia was now free of the invaders.

627. Invasion of Assyria. Heraclius now marched from southern Armenia to the upper Tigris, then continued southward into Assyria. The hoped-for cooperation of the Khazars was not forthcoming; Ziebel had died unexpectedly, and his successor did not see fit to continue the alliance. Heraclius, however, was determined to carry the war into Persia. A Persian army, under the general **Rhazates,** hovered to the eastward, but refused battle. The aged Chosroes ordered Rhazates to stand and fight near ancient Nineveh.

627, December. The Battle of Nineveh. On the same Assyrian plain where Alexander had won the climactic victory of Arbela, the decisive battle of the last war between the East Roman and Sassanid empires took place. The two armies were apparently about evenly matched; Rhazates probably had a substantial numerical superiority, but Heraclius had the moral ascendancy. The battle was hard-fought, and

Heraclius—as usual—was in the thick of the fray. Although wounded, he refused to leave the field, and in a final charge personally killed the Persian general. By nightfall the Persian army was completely smashed.

627. Pursuit. No greater tribute to the leadership of Heraclius can be paid than by noting that, despite his wounds and his army's severe losses at Nineveh, he immediately pursued the remnants of the Persian army. Within 48 hours of his great victory he had marched at least 40 miles to seize the bridges over the Great and Lesser Zab rivers. He continued on into Assyria, Chosroes fleeing ahead of him to Ctesiphon. Arriving in front of the Persian capital, Heraclius evidently toyed with the idea of attempting an assault. But his army had suffered severely and he had been unable to bring a siege train with him in his rapid pursuit. Apparently remembering the fate of the army of Julian (see p. 168), Heraclius, after a brief demonstration, turned northeastward into the Persian highlands again, to rest his army and to recuperate in the familiar country around Tauris. He now sent an offer of peace to Chosroes.

628. The End of the Persian Wars. When stubborn Chosroes refused Heraclius' generous terms, he was overthrown by the war-weary Persians, who installed his son **Kavadh II**

(Siroes) on the throne. Kavadh immediately accepted Heraclius' terms. These are not precisely known, but Heraclius apparently had no desire "to enlarge the weakness of the empire" (in Gibbon's words) and asked for no territorial cessions. The Persians were forced to relinquish all of their conquests and to give up all of the trophies they had captured, including the relic of the True Cross. Evidently there was also a large financial indemnity. Having dictated peace on his own terms in the capital of ancient Media, Heraclius returned in triumph to Constantinople.

COMMENT. *In six campaigns Heraclius had led his army from total disaster to glorious victory. Had he died at this moment, his name would be ranked in military history with those of Alexander, Hannibal, and Caesar—as it deserves to be. Later events have unfairly beclouded the reputation of this truly great captain. The fact remains that he inherited an empire on the verge of destruction, which unquestionably would have fallen early in the 7th century had it not been for his perseverance and military skill. At the end of his reign, admittedly, some of the frontier provinces he reconquered from Persia had been lost forever to a new enemy. But he had recovered, rebuilt, and bequeathed to his successors a cohesive core of empire which could and did survive this shock and which would shield Christendom for eight more centuries.*

HERACLIUS AND THE IMPACT OF ISLAM, 629–641

The long war with Persia had prostrated the Byzantine Empire. Some 200,000 men had been killed and untold riches destroyed or squandered. Hardly had Heraclius set himself to the task of recovery when a new foe appeared on the eastern frontiers of the empire (629). Islam—the religion and empire founded by Mohammed (see p. 247)—swept out of Arabia to strike the Persian and Byzantine empires when both were totally exhausted. Overwhelming Persia, the Moslems quickly conquered the eastern provinces of the Byzantine Empire (633–642), aided by the passive neutrality of the Syrian and Egyptian peoples, who had little loyalty to the Byzantines (see p. 249). Heraclius, ill and infirm, unpopular with his church and his people as a result of marriage to his niece, was unable to take the field himself. His subordinates were unable to cope with the fanatic onslaughts of the Moslems. When Heraclius died of the dropsy (edema) (641), all of the provinces east of the Taurus had been lost—most of them forever. But, thanks to his earlier accomplishments, the Anatolian and Thracian heart of the empire survived.

BYZANTINE MILITARY SYSTEM, C. 600–1071

The Theoretical Basis

The remarkable longevity of the Byzantine Empire was due primarily to the fact that, as Oman says, its army "was in its day the most efficient military body in the world." The system that produced this army was the result of superior discipline, organization, armament, and tactical methods, which in turn combined with treasured Roman traditions to generate an unsurpassed *esprit de corps.* The Byzantines were able to achieve and maintain these superiorities by their emphasis on analysis: analysis of themselves, of their enemies, and of the geophysical factors of combat.

The results of this analysis are evident in a number of military treatises, of which three are outstanding. The first of these was the *Strategikon* of Maurice, written about 580, just before he became emperor. (See p. 208.) The second is the *Tactica* of **Leo the Wise,** written about 900. The third is a small manual, apparently inspired by the warrior emperor **Nicephorus Phocas,** and probably written by a staff officer, about 980. These three documents reveal that the fundamentals of the system changed very little in the half-millennium after Belisarius, despite many evolutionary changes in detail which would inevitably be found in a system based upon frequent, objective analysis.

The organization and doctrine herein discussed emerged in the century and more following the reign of Heraclius as his successors struggled to adapt their system to the challenge of Islam. During those years—as in the century preceding—the system was not universally successful. But despite some defeats, and occasional disaster, the superior bases on which the system rested reasserted themselves under thoughtful and energetic leadership. For five centuries such leadership always appeared in time to reestablish Byzantine military supremacy over its neighbors; and for almost four centuries after that, the vestiges of the system helped to postpone the final demise of the empire.

During most of this period the basic military textbook was the *Strategikon.* This was a comprehensive manual on all aspects of warfare and military leadership, not unlike the field-service regulations of modern armies. It covered training, tactical operations, administration, logistics, and discussions of the major military problems to be encountered in operations against any of the many foes of the empire.

Organization

The basic administrative and tactical unit of the Byzantine army, for infantry as well as cavalry, was the **numerus,** or **banda,** of 300–400 men; the equivalent of a modern battalion. The numerus was commanded by a tribune, or count, or (later) a **drungarios**—the equivalent of a colonel. Five to eight numeri were combined to form a **turma**—or division—under a **turmarch,** or duke. Two or three turmae comprised a **thema**—or corps—under a **strategos.** A deliberate effort was made to keep units different in size, so as to make it difficult for opponents to estimate the exact strength of any Byzantine army.

All officers of the rank of drungarios and higher were appointed by the emperor and owed their allegiance to him. This method of encouraging allegiance to central national authority, rather than to army commanders as was the Roman practice, was apparently introduced by Maurice.

The field organization of the Byzantine army, as outlined and largely introduced by Maurice, was integrated into a geographical, military district organization during the latter part of the reign of Heraclius and the reigns of his successors **Constans II** and **Constantine IV.** The inroads of Persians and Arabs had completely obliterated the old provincial organization of the empire. Thus, as the Anatolian provinces were reoccupied, local governmental authority was perforce exercised by the military commanders responsible for defense of each region. In the face of the continuing threat of Saracen raids, the emperors adopted this civil-military administrative system permanently. Each district was called a **theme,** under a strategos; his thema was the garrison of the district, which was broken down into smaller administrative districts under the turmarchs and drungarios. Some of the most critical frontier regions, especially around the key Taurus passes, were organized into smaller districts called **clissuras,** each under a **clissurarch,** where the garrisons were maintained in an especially high state of readiness against attack.

By the end of the 7th century there were 13 themes: 7 in Anatolia, 3 in the Balkans, and 3 in island and coastal possessions in the Mediterranean and Aegean seas. By the 10th century the number of themes had grown to about 30. The army had not grown comparably in those 300 years; there were simply fewer forces in each theme. During most of this period the standing army of the empire numbered about 120,000–150,000 men, proportionately half and half in horse and foot.

Apparently themes closer to the frontier had larger standing forces than those in the interior. On the average each strategos could take the field on short notice with two to four turmae of heavy cavalry. He also had available approximately an equal force of infantry. Depending upon the situation (the nature of the foe, the area of expected operations, etc.), he could leave some or all of his foot soldiers for local garrisons.

Man Power and Recruitment

Theoretically maintained by universal military service, in practice the standing forces of each theme were kept up by selective recruitment from the most promising of the local inhabitants. No longer was the empire dependent upon the barbarians for its soldiers, though some barbarian units were usually maintained in the army. For the most part the empire raised its own soldiers, and of these the best came from Armenia, Cappadocia, Isauria, and Thrace. The Greeks were deemed the least suitable military material.

The theme system included a militia concept of home-guard local defense. This was satisfactory where the local inhabitants were hardy and warlike, but was useless in regions like Greece. Where this concept worked, the guerrilla tactics of the local inhabitants greatly assisted the regular forces of the empire in repelling or destroying invading forces.

Byzantine Strategic Concepts

During most of its existence the Byzantine Empire had no incentive for conquest or aggression, once its people became reconciled to the losses of their extensive peripheral possessions. Living standards were high; the nation was the most prosperous in the world. Foreign conquests were expensive in lives and treasure in the first place, and merely resulted in increased costs for administration and defense. At the same time, they recognized that their wealth was a constant attraction to predatory barbarian neighbors.

The essentially defensive Byzantine military policy is thus easily understood. The objective was the preservation of territory and resources. Byzantine strategy was essentially a sophisticated medieval concept of deterrence, and was based upon the desire to avoid war if possible—but when necessary to fight, to do so by repelling, punishing, and harassing aggressors with the minimum possible expenditure of wealth and manpower. Their method was usually that of elastic defensive-offensive, in which the Byzantines would endeavor to throw the invaders back against defended mountain passes or river crossings, then to destroy them in a coordinated, concentric drive of two or more themas.

Economic, political, and psychological warfare assisted—and often obviated—the use of brute force. Dissension among troublesome neighbors was craftily fomented. Alliances were contracted from time to time to reduce danger from formidable foes. Subsidies to allies and to semi-independent barbarian chieftains along disputed frontiers also helped to reduce the burden on the armed forces. In all of this, imperial action was facilitated by an efficient, widespread intelligence network comprised mainly of merchants and of trusted, well-paid agents in key positions in hostile and friendly courts.

The emperors were not above using religion for temporal ends. Sincerely anxious to spread Christianity to heathen peoples, they also found that missionaries could exert subtle and helpful influence at the courts of converted rulers—and that common adherence to the Christian faith automatically created a bond against pagan and Moslem.

Cavalry: Weapons, Equipment, and Uniforms

The basic military strength of the empire lay in its disciplined heavy cavalry. The cataphract of the Byzantine Empire symbolized the power of Constantinople in the same way that the legionary had represented the might of Rome.

The individual horseman wore a casque or conical helmet, topped with a colored tuft of horsehair. His chain-mail shirt covered him from neck to thighs. On his feet were steel shoes, usually topped with leather boots or greaves to protect the lower leg. Hands and wrists were protected by gauntlets. He carried a relatively small round shield strapped to his left arm, thus leaving both hands free to control the horse's reins and to use his weapons, yet available to provide protection to his vulnerable left side in hand-to-hand fighting. Over his armored shirt he wore a lightweight cotton cloak or surcoat, dyed a distinctive color for each unit; helmet tuft and shield also were of this same color for uniform purposes. A heavy cloak for cool weather, which served also as a blanket, was strapped to the saddle. Some horses—those normally deployed in front rank positions—wore armor on their heads, necks, and chests.

The cataphracts' weapons usually included bow, quiver of arrows, long lance, broadsword, dagger, and sometimes an ax strapped to the saddle. Apparently a proportion of the heavy cavalry were lancers only, but most seem to have carried both bow and lance. Presumably, when he was using the bow, the soldier's lance rested in a stirrup or saddle boot, like the carbine of more modern cavalrymen. In turn, the bow was evidently slung from the saddle when he

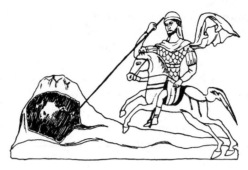

Byzantine cataphract

was using lance, sword, or ax. Attached to the lance was a pennon of the same distinguishing color as helmet tuft, surcoat, and shield.

Men and horses were superbly trained and capable of complex evolutions on drillfield and battlefield. There was great emphasis on archery marksmanship and on constant practice in the use of other weapons.

Infantry: Weapons, Equipment, and Uniforms

The infantry was almost equally divided into two classes: heavy and light. The heavy infantry-men, known as **scutati** from the round shields they carried, were equipped much like the cataphracts. They wore helmet, mail shirts, gauntlets, greaves (or knee-length boots), and surcoats. They carried lance, shield, sword, and (sometimes) ax. Uniform appearance was achieved by color of surcoat, helmet tuft, and shield.

Most of the light infantrymen were archers, though some were javelin throwers. To permit maximum mobility they carried little in the way of armor or additional weapons, though apparently there was some leeway allowed to individual desires. Most wore leather jackets, some may have worn helmets, and apparently they usually carried a short sword in addition to bow and quiver (or javelins).

Battlefield Tactics

Although it was not at all unusual for Byzantine armies to be composed entirely of cavalry, more frequently the two arms were combined on campaign in about equal proportions. The infantry, in turn, was usually divided equally between heavy and light.

Byzantine tactics were based upon offensive, or defensive-offensive, action, and envisaged a number of successive coordinated blows against the enemy. Their normal tactical formation, which could be varied greatly depending upon the circumstances, comprised five major elements. There were (1) a central front line; (2) a central second line; (3) a reserve/rear security, usually in two groups behind each flank; (4) close-in envelopment/security flank units; (5) distant envelopment/screening units. In a force of combined arms, with infantry and cavalry present in about equal numbers, the first two of these elements were infantry, with the scutati in the center, the light troops to their flanks; the latter three were always cavalry. If the infantry contingent was small, it might compose only the central second line or be placed as an additional reserve behind two cavalry central lines.

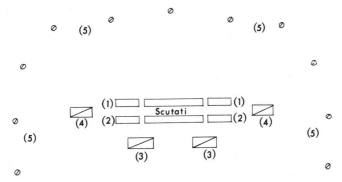

Standard Byzantine battle formation

When the opposing army was mainly cavalry, and their own army included substantial infantry, the infantry front line would await enemy attack. Byzantine scutati, confident of flank and rear protection by their cavalry, were as effective against horsemen as Roman legionaries had been. The enemy's first attack would be struck on the flanks by the close-in envelopment/ security flank units. Soon thereafter would come a heavier blow to the hostile flank and rear from the distant envelopment/screening units. If these counterpunching tactics failed to achieve their objective, and if the Byzantine front line should be forced to fall back, it could do so through the intervals in the second line—left there in the traditional Roman manner for this purpose. The enveloping units would withdraw, regroup, and return to the attack. Finally, if the second Byzantine line should fail and the former front line had not yet had time to rally, the day could still be saved by a smashing coordinated counterattack by the fresh reserve units, almost always conducted as a double envelopment rather than a frontal attack.

Obviously there was room for many variants in such a set-piece battle, and many different combinations possible against different enemies and different kinds of forces. The important thing to note here is the existence of a standard tactical doctrine: the emphasis on envelopment, on coordinated action (including coordination between missiles and shock action, between the arms, and between all elements of the force), and on retaining a fresh reserve with which ultimately to gain the day in a hard-fought action.

Though subsidiary to the cavalry, Byzantine infantry doctrine was far from passive. Whenever opposed by infantry—either in a combined-arms battle or in essentially infantry operations in rough terrain—the scutati, in close coordination with archers and missile throwers, were wont to seize the initiative and to carry the attack to the foe. The normal formation of the scutati was 16 deep, and separate numeri were capable of individual evolutions, extending and closing the ranks like the old Roman cohort. In the attack they would rush on the foe, throwing their lances just before contact, again like the cohort of the legion. Thus the numerus of the scutati combined the attributes of legion and phalanx, though lacking the moral feeling of superiority which contributed so much to the success of the infantry of Alexander and Caesar.

The numeri of the cavalry usually formed in lines 8 to 10 horsemen in depth. The Byzantines acknowledged that this was perhaps more cumbersome than an optimum formation, but were willing to accept a slight decrease in flexibility in exchange for the greater feeling of security the men derived from the deep formation.

Adaptation to Hostile Characteristics

Byzantine military theoreticians spent as much time in study of the traits of their foe as they did in elaboration of their own tactical formations. Combining these different studies and analyses was an important aspect of their consistent superiority.

Maurice, for instance, suggested that whenever possible campaigns should be undertaken during seasons in which the various and diverse neighbors were least prepared to fight. The Huns and Scythians of Eastern Europe, for example, were in poorest condition in February and March, when their horses were suffering most from lack of forage. A little earlier—midwinter— was best against the Slavic marsh dwellers, since the Byzantine troops could cross the ice to their hideouts and the defenders would be unable to take refuge in water and reeds. Fall, winter, and spring were good against mountain tribes, since the snow would reveal their tracks and the lack of foliage would reduce their concealment. Any cold or rainy weather was good for campaigning against the Persians or Arabs, since they were depressed and less effective at such times.

Later, Leo the Wise had much similar advice for the officers of his time—then the main enemies were European Franks and wily Arab Moslems. Of the Franks, Leo had this to say (adapted from Oman's translation):

> The Franks (and Lombards) are bold and daring to excess; they regard the smallest movement to the rear a disgrace, and they will fight whenever you offer them battle. So formidable is the charge of the Frankish chivalry with their broadsword, lance and shield, that it is best to decline a pitched battle with them till you have put all the chances on your own side. You should take advantage of their indiscipline and disorder; whether fighting on foot or on horseback, they charge in dense, unwieldy masses, which cannot maneuver because they have neither organization nor discipline. . . . Hence they readily fall into confusion if suddenly attacked in flank and rear—a thing easy to accomplish, as they are utterly careless and neglect the use of outposts and reconnaissance. They camp in confusion, without fortifying themselves, so that they can be easily cut up by a night attack. Nothing succeeds better against them than a feigned flight, which draws them into an ambush; for they follow hastily and invariably fall into the snare. But perhaps the best tactics of all are to protract the campaign, and lead them into hills and desolate regions, for they take no care of their supply, and when their stores run low their vigor melts away, and after a few days of hunger and thirst desert their standards and steal home as best they can. For they have no respect for authority—each noble thinks himself as good as the other—and will deliberately disobey when they are discontented. Nor are their chiefs above the temptation of taking bribes; a moderate outlay of cash will frustrate one of their expeditions. On the whole, therefore, it is easier and less costly to wear out a Frankish army by skirmishes, protracted operations in desolate districts, and the cutting off of its supplies, than to attempt to destroy it at a single blow.

Leo had much more respect for the Arab Moslems. As quoted by Oman, he says of the Saracens: "Of all the barbarous nations they are the best advised and most prudent in their military operations. . . . They have copied the [Byzantines] in most of their military practices, both in arms and strategy." Yet he noted that for all of their imitation the Arabs had failed to absorb the organizational and disciplinary basis of Byzantine success. Despite the superior numbers of the Moslems, their fanatic courage, and their willingness to learn from experience, Leo believed that Byzantine skill, discipline, and organization should prevail—and usually they did.

Administration on Campaign

Although each soldier carried his weapons, elementary necessities of life, and food for several days, each army was always accompanied by a supply and baggage train with sufficient additional supplies and equipment to permit sustained operations and to undertake siegework if necessary. This baggage train was composed partly of carts and partly of pack animals. Accompanying the baggage train were the camp followers: servants and foragers for the officers and men. Apparently female camp followers were discouraged.

Basic equipment included picks and shovels, necessary for the practice of castrametation, observed by the Byzantines as faithfully as by the early Romans. A camp site was selected and marked out in advance by the army's engineer unit. While part of the army deployed to provide security, the remainder took the picks and shovels from pack animals and quickly prepared entrenchment and palisade in the old Roman fashion.

Attached to each numerus was a medical detachment, usually consisting of a doctor and a surgeon, plus 8 to 10 stretcherbearers or medical aid men. As an incentive to their efficient performance of duty, aid men were given a substantial cash reward for each wounded soldier they rescued during battle.

A highly developed signal service existed. In addition to the expected corps of messengers, the Byzantines had developed a system of beacon signal fires whereby warning of attack could be sent from the frontiers to Constantinople in a matter of minutes.

Chaplains were always present with the army. As in Western Europe, priests and monks were able to take their place in the battle line, although the Orthodox Church was more strict than that of Rome in tempering crusading zeal by adherence to the Biblical injunction, "Thou shalt not kill."

Staff and Command

The training of an officer began when a youth—usually of noble family—enlisted in a cadet corps. The peacetime training of these youths was probably not much different in concept from a modern officer training program: emphasis on the basic tasks of the soldier, mastery of weapons and horses, study of the writings of military experts of the past and present, exercises in which the theoretical knowledge was put to practice. In addition, during wartime the cadet corps served on the staffs of the various strategi, acting as clerks and messengers—undoubtedly occasionally getting opportunities to assist staff officers in the simpler aspects of writing orders and preparing plans.

The advancement of the young officer through staff and command duties was apparently arranged to give him experience and to give his superiors an opportunity to observe him in action. The most promising turmarchs were given opportunity for independent command as clissurarchs; if successful, they could be assured of promotion to strategos. During these formative years conscious emphasis seems to have been given to encouragement of objective analysis, since the Byzantines were rightly convinced that this provided the basis of their success and that it was essential to the development of good commanders and staff men.

Evidently the strategi were rotated from one theme to another. This kept them from getting either too entrenched politically or too settled from the standpoint of personal attitude toward the rigors of combat. Furthermore, the senior strategi were assigned to the largest and most important themes—usually those on the frontiers, save for the Anatolian Theme of central Asia Minor. The army's senior strategos was usually the commander of the Anatolian Theme and (in the absence of the emperor) exercised overall field command.

The Byzantine Navy

Only occasionally after the Punic Wars had the Romans—and later the Byzantines—given much thought to the concept of naval power and control of the sea. They took for granted their mastery of the Mediterranean, even when it was challenged by pirates, Gothic raiders, or Vandal sea rovers. But the development of Moslem sea power in the latter half of the 7th century stirred Constantinople to energetic reaction. By the early years of the 8th century, Byzantine supremacy had been reestablished—as was demonstrated by Leo III in the siege of Constantinople. Though occasionally shaken or threatened, this supremacy was generally maintained for the next four centuries.

The principal base of Byzantine sea power was the Theme of Cibyrrhaeots, on the south coast of Anatolia, where a hardy race of sailors had descended from the pirates of Roman republican days. Ships and sailors were also supplied by the Aegean islands and other maritime regions of Anatolia. Cibyrrhaeots, however, was the one theme which supplied almost no soldiers to the Byzantine army, yet maintained a standing fleet of nearly 100 vessels and more than 20,000 sailors, about half of the manpower of the Byzantine navy.

The navy was made up of relatively small, light, fast galleys, most with two banks of oars (usually 30–40 on each side), two masts, and two lateen sails. In addition to the oarsmen— trained to fight hand-to-hand if necessary—there was a small force of marines on each vessel, which had a total complement of 200–300 men. The larger warships had revolving turrets, mounting war engines. But the most deadly weapons on these ships—beginning about 670— were the bow tubes from which the dreaded and deadly Greek fire was hurled with explosive force (see p. 242).

Byzantine Honor in War

Guile and fraud were admired by the Byzantines and used whenever possible. Scorning the often hypocritical honor of Western chivalry, their objective was to win with minimum losses and the least possible expenditure of resources—if possible without fighting. Bribery and trickery were common, and considered respectable. They were masters at forms of psychological warfare which caused dissension in hostile ranks; they did not hesitate to use false propaganda to raise the morale of their own men.

In the light of some modern practices of warfare one cannot be too critical of the Byzantines.

They probably merit admiration for a practical, no-nonsense, alert attitude toward the basic issues of national survival. Furthermore, they *did* have a moral code of conduct for war. Signed treaties were inviolate. Ambassadors and negotiators were scrupulously protected. Captured male and female noncombatants were not mistreated. A brave defeated foe was treated with generosity and respect.

BYZANTINE EMPIRE, 642–800

The Successors of Heraclius, 642–717

The successors of Heraclius devoted most of their attention to defense against the Arabs, who had completely displaced the Persians on the empire's eastern boundaries. Frequently close to disaster, and subject to numerous Moslem raids deep into Anatolia, the Byzantines nevertheless maintained the frontier generally along the line of the Taurus Mountains. To the southwest, however, the Arabs soon conquered all Byzantine possessions on the shores of north Africa, aided by religious dissension between the provinces and Constantinople. They then began seriously to challenge Byzantine control of the sea, defeating the fleets of the emperor **Constans II** and capturing some islands in the eastern Mediterranean. They next undertook a determined effort to overthrow the empire, and for six years their fleets and armies literally hammered at the gates and walls of Constantinople in a protracted siege. Thanks to Greek fire, and to the efficiency of the Byzantine military system, **Constantine IV** repulsed the Moslems so violently that they sued for peace and paid tribute to the empire for several years.

Meanwhile, the emperors also had to cope with nearly constant threats to their European dominions. The Avars and Slavs continued depredations south and west of the Danube. By the end of the 7th century a new Bulgar kingdom had established itself permanently in the eastern Balkans (see p. 245). Lombard pressure against Byzantine possessions in Italy was constant (see p. 219).

Though these external threats were unremitting, the inherent toughness and vigor of the empire and its military system were demonstrated by the repulse of all major invasions. Then, at the turn of the century, internal discord once again wracked Byzantium. Seizing the opportunity, the Arabs began a climactic effort to overthrow the empire. The major events were:

641–668. Reign of Constans II. This grandson of Heraclius established a new civil-military defensive organization based upon geographical military districts (themes).

645–659. Continued War with the Arabs. The Arabs repulsed a halfhearted Byzantine effort to recapture Alexandria (645). Pushing westward, they began their invasion of the province of Africa (647). The Arabs also temporarily conquered Cyprus (647–653) and part of Byzantine Armenia (653). At sea they plundered Rhodes (654) and defeated a Byzantine fleet, commanded personally by Constans, in a great naval battle off the coast of Lycia (655). Because of other pressures, a temporary peace was negotiated between the Byzantine Empire and the Caliphate (659).

662–668. Operations in Italy. Constans personally campaigned against the Lombards in Italy with mixed success. He was killed during a mutiny in Italy, which was vigorously suppressed by his son, **Constantine IV.**

668–679. Renewed War with the Caliphate. An Arab invasion of Anatolia reached Chalcedon; the Moslems crossed the Bosporus to attack Constantinople, but were repulsed

(669). The Arab army was then virtually destroyed at **Armorium.** An Arab naval attack on Constantinople was repulsed; the Arab fleet was destroyed in the **Battle of Cyzicus** (in the Sea of Marmara); Greek fire, apparently first used, contributed greatly to the victory (672). The Arabs returned to establish a land and sea blockade and intermittent siege of Constantinople (673–677). Greek fire continued to play a great part in destroying Moslem warships and in repulsing assaults against the city walls. The climax came when the Byzantines destroyed the Arab fleet at the **Battle of Syllaeum** (southern Asia Minor). These disasters led Caliph **Mu'awiya** to sue for peace; he agreed to evacuate Cyprus, to keep the peace for 30 years, and to pay an annual tribute to Constantinople.

675–681. Operations in the Balkans. Repeated Slav assaults on Thessalonika were repulsed. Constantine recognized the virtual independence of Bulgaria (680; see p. 245).

685–695. First Reign of Justinian II. He was defeated by the Bulgars, but gained a victory over the Slavs in Macedonia (689).

690–692. Arab War. A generally unsuccessful war with the Arabs was culminated by the **Battle of Sebastopolis** (Phasis), which resulted in Moslem conquest of Armenia, Iberia, and Colchis, last remaining Byzantine footholds east of the Taurus. Justinian was also forced to agree to joint Byzantine-Moslem control of Cyprus.

695–698. Usurpation by Leontius. Justinian was overthrown and banished to the Crimea.

697–698. Arab Conquest of Carthage. This eliminated Byzantine influence from north Africa.

698–705. Usurpation by Tiberius Absimarus. Overthrowing Leontius, he assumed the title of Tiberius III.

705. Restoration of Justinian. Assisted by Bulgar King **Terbelis,** Justinian captured Constantinople in a surprise attack.

705–711. Second Reign of Justinian II. He executed all of his former opponents (including Leontius and Tiberius) and their adherents in a 6-year reign of terror. At the same time he was defeated in intermittent wars with the Arabs and with Terbelis of Bulgaria.

711. Overthrow of Justinian. He sent an expedition to the Crimea to punish his subjects who had mistreated him during his exile. The troops mutinied under the leadership of **Bardanes Philippicus,** with the assistance of the Khazars. The mutineers returned to Constantinople; Philippicus defeated and killed Justinian in a battle in northwestern Anatolia and seized the throne.

711–717. Disaster and Anarchy. The frontiers collapsed during the reign of incompetent Philippicus (711–713). The Bulgars reached the walls of Constantinople (712). The Arabs overran Cilicia, then invaded Pontus to capture Amasia (Amasya). The Byzantine Army mutinied, overthrew Philippicus, and installed **Anastasius II** in his place (713). Anastasius restored internal order and began to rebuild the army. His harsh reforms were resented, however; part of the army mutinied and proclaimed **Theodosius III** as emperor (715). Theodosius captured Constantinople after a 6-month siege and banished Anastasius to a monastery. Meanwhile, a major Moslem invasion reached Pergamum (716). Failure of Theodosius to take adequate action against the threat caused his leading general **Leo ("the Isaurian")** to overthrow the emperor and assume the throne himself as Leo III (717).

The Siege of Constantinople, 717–718

717, June. The Moslems Cross the Hellespont. The Moslem general **Maslama** led his army of 80,000 men from Pergamum to Abydos, where he crossed the Hellespont. To prevent interference by the Bulgars, or by any Byzantine forces in Thrace, he sent part of his army to a covering position near Adrianople; with his main body, Maslama laid siege to Constantinople (July).

717, August. The First Moslem Assault. Leo had been hurriedly preparing the city for siege. Maslama's assault was repulsed with heavy loss.

717, September. Arrival of Moslem Reinforcements. The Moslem general **Suleiman** led a great armada of 1,800 vessels and 80,000 additional troops through the Hellespont. After landing most of his men, Suleiman tried to take his fleet past the city to blockade the Bosporus. Leo led his own fleet in a successful attack, winning an overwhelming victory and driving the surviving Moslem vessels back into the Sea of Marmara. Greek fire played a great part in this victory, which assured a supply line from Constantinople through the Bosporus to Byzantine provinces on the Black Sea. The Moslems made no more direct attacks on the city during the remainder of the year, being mostly occupied in keeping themselves warm and alive during an exceptionally cold winter. Leo made a treaty with Terbelis, King of Bulgaria, who promised to join the war against the Arabs.

718, Spring. Renewed Moslem Pressure. Taking advantage of night and of stormy weather, Suleiman got part of his rebuilt fleet up the Bosporus to tighten the blockade of the city. At the same time approximately 50,000 more reinforcements arrived to swell the besieging forces on both sides of the straits. Leo repulsed several assaults.

718, June (?). Leo Counterattacks. Leo's fleet surprised and annihilated the Moslem blockading squadron in the Bosporus. He then landed an army on the far (Asiatic) shore of the Bosporus and defeated the Arab blockading army near Chalcedon.

718, July (?). Battle of Adrianople. Terbelis, King of Bulgaria, now honored his alliance by leading an army into Thrace. He attacked and defeated Maslama south of Adrianople.

718, August. Moslem Withdrawal. The appearance of the Bulgars, combined with the terrible losses which Leo had inflicted, caused Maslama and Suleiman to abandon the siege of Constantinople. Part of their army marched back through Anatolia, harassed by Leo's army. The remainder of the Moslem army embarked in Suleiman's fleet, which was destroyed in a storm; only 5 vessels are said to have survived. Of more than 200,000 Moslems committed to the siege of Constantinople, only 30,000 eventually returned across the Taurus Mountains.

COMMENT. *The successful defense of Constantinople saved Christian Europe from Moslem invasion. Its importance probably transcends even that of the later Battle of Tours (see p. 222), Islam's high-water mark in the West.*

Byzantine Empire, 717–802

Leo III consolidated his victory by sweeping reforms: administrative, economic, legal, social, military, and religious. Most controversial were his religious reforms—pertinent to this text because of direct military consequences. About 726 Leo began to issue a series of imperial edicts against both the veneration and worship of religious images, causing him to be known as the "Iconoclast." A major motive in his iconoclasm was evidently a belief that the elimination of frills from Christian worship would restore to it a vigor that would enable it to compete more successfully with the simple and austere tenets of Islam. Most of the clergy—particularly in Italy and Greece—were violently opposed to these edicts, insisting that veneration was not image worship. The result was turbulence during much of his reign. Pope Gregory inspired the Italian provinces to revolt against Leo, and became the virtual temporal ruler of most Byzantine possessions in Italy. Although the religious disorders probably weakened imperial control over some possessions, Leo's iconoclasm seems eventually to have solidified popular support in Anatolia; his other reforms were unquestionably beneficial. Two Arab invasions later in his reign were easily repulsed; long-term Byzantine stability was unquestionably enhanced.

Leo's policies were vigorously continued during the long and successful reign of his son,

Constantine V. The empire regained the initiative in its struggles with the Arabs, Slavs, and Bulgars, and reoccupied some of the lands lost in Europe and Asia in the previous century. Internal religious disputes over the issue of the distinction of veneration and image worship continued to rage, however. These contributed to the only important military and territorial losses of the reign, in Italy, where the Lombards were able to take advantage of the disorders to conquer most of the Exarchate of Ravenna and to threaten Rome. The popes, seriously estranged from the emperor on religious grounds, then turned to the Franks for assistance, leading to the end of Byzantine control over Rome and the surrounding territories (see p. 209).

In the last quarter of the 7th century there was renewed dynastic and political disorder in the empire; again the Moslems were able to exploit these troubles by military victories. The dominant personage of these years was **Irene,** wife of feeble **Leo IV** and mother of **Constantine VI.** Young Constantine finally tried to throw off his mother's able but dictatorial dominance, but through intrigue she soon returned to power, had her son deposed and blinded, and seized the throne for herself. She contributed to internal peace by ending the iconoclastic policies of the empire. She bought peace temporarily from the Moslems. The major events were:

717–741. Reign of Leo III.

721. Revolt of Anastasius. This was suppressed by Leo, who had Anastasius executed.

726. Arab Invasion. The Arabs were repulsed in eastern Anatolia.

726–727. Revolt in Greece. Defiance of the first imperial edict against image worship was crushed by Leo after his fleet destroyed a rebel fleet sailing against Constantinople.

726–731. Revolt in Ravenna against Leo's Iconoclasm. This resulted in the virtual independence of the exarchate, after part of an imperial invasion fleet was lost in an Adriatic storm and the remainder of the invasion force repulsed from the city walls (731). The Lombards had earlier attempted to take advantage of the dispute and had temporarily captured part of the city, but were then repulsed (728; see p. 209).

739. Arab Invasion. After initial success, the invaders were repulsed by Leo at the **Battle of Akroinon** (Afyon Karahisar).

741–775. Reign of Constantine V.

741–752. Arab War. Constantine invaded Syria, but had to withdraw to deal with civil war at home (see below). Having reestablished control of his empire, he again invaded Syria, reconquering a few small border regions (745). His fleet won a victory over the Arabs near Cyprus; the Moslems were eliminated from Cy-

prus (746). The war came to an indeterminate conclusion after Constantine defeated the Arabs in Armenia, regaining part of the province (751–752).

741–743. Civil War. While Constantine was campaigning against the Arabs, immediately after his accession, his brother-in-law, **Artavasdus,** led a religious revolt with the support of the image worshipers. Constantine successfully fought on two fronts and overcame the rebels.

752–754. Final Loss of Ravenna. First to Lombards, then to Franks (see p. 209).

755–772. Intermittent Campaigns in Thrace. Constantine fought wars against the Bulgars and the Slavs in Thrace. In the west he overwhelmed the Slavs (758). He defeated the Bulgars in eastern Thrace at the battles of **Marcellae** (759) and **Anchialus** (763). After a lull of several years, he again defeated the Bulgars (772).

775–780. Reign of Leo IV (Son of Constantine V).

778–783. Arab War. An invasion of Anatolia was repulsed by the Byzantine victory at the **Battle of Germanicopolis** (Gangra, 778). Byzantine general **Michael** captured Mas'ash (in Cilicia, 778).

780–797. Reign of Constantine VI. Irene was regent for ten years (780–790). During this time an Arab invasion reached the Bosporus, where it was bought off by tribute (783). Wars with the Bulgars were intermittent. Byzantine general **Staurakios** defeated the Slavs in Macedonia and Greece (783). An army mutiny forced the retirement of Irene (790). Irene soon again became virtual coruler (792).

797–802. Usurpation of Irene. She had her son blinded. She was deposed by a revolt of Byzantine nobility.

797–798. Renewed Arab Invasion of Anatolia. Caliph Harun al-Rashid defeated Byzantine general **Nicephorus** at the **Battle of Heraclea.** The Moslems were again bought off by tribute.

SLAVS, AVARS, BULGARS, AND KHAZARS

By the beginning of the 7th century the Slavs had spread over much of Eastern Europe. There was no unity among their many tribes; those north of the Danube and Carpathians were involved in more or less constant conflict among themselves and with German neighbors, but otherwise played little part in the development of military history. The southern Slav tribes maintained a semi-independent status under the suzerainty of Avars or of the Eastern Empire. They soon drove out the remnants of the former inhabitants of the inland regions of Illyricum, Macedonia, and Greece; their frequent raids into Thrace and the coastal regions still under Byzantine control presented a constant military problem to the empire. As their strength grew, they eliminated Avar influence south of the Danube, thus ending the frequent wars between Avars and Byzantines.

By the middle of the 8th century, both Slavs and Avars found themselves under simultaneous pressure from Bulgars to the northeast and Franks to the northwest. By the end of the century, the Franks had destroyed the Avar Empire (see p. 225), but their further advance into the Balkans ceased as a result of treaties between Franks and Byzantines (see p. 225).

The Bulgars were a Turanian people, related to the Avars and Huns, inhabiting the steppes north of the Caspian in the 5th and 6th centuries. Early in the 7th century, under a leader named **Kubrat** (or Kurt), they threw off Avar domination and established an independent kingdom in the area roughly bounded by the Volga and Don rivers and the Caucasus Mountains. Under pressure from the Avars and Khazars, they migrated southwestward in the middle of the 7th century, to reach the lower Danube Valley. Gaining ascendancy over the Slavs in this region, their leader **Isperich** (Asparukh) established an independent kingdom, centered between the Balkan Mountains and the lower Danube, but also extending north of that river into Wallachia, Moldavia, and Bessarabia. A slow process of intermingling and amalgamation with the Slavs began; a few centuries later Bulgaria was essentially Slavonic. During most of the 8th century, the Bulgars were at war with Byzantium, though **Terbelis** (Tervel) did form an alliance with Leo III against the Arabs during the great siege of Constantinople (see p. 243).

The Khazars were another Turanian people, possibly descended from the Huns, who became predominant in the steppes north of the Black and Caspian seas in the latter 6th century. They were probably part of the great Turkish empire which then flourished in Central Asia (see p. 258). With the collapse of central Turkish control, the Khazars created a powerful and extensive empire of their own, extending westward to frontiers with the Avars and Bulgars in the Carpathians and on the lower Danube, southward to the Caucasus, and eastward beyond the Caspian Sea. During much of the 7th and 8th centuries, they were engaged in fierce warfare with the expanding Moslem Caliphate on both sides of the Caspian. They were usually allied with the Byzantines, but had some wars with them. The principal events were:

601–602. Byzantine-Avar War. Byzantine general **Priscus** defeated the Avars south of the Danube (see p. 199).

603. Independence of the Slavs in Moravia. Avar rule was thrown off.

617–620. Avar Raids to the Walls of Constantinople. (See p. 229).

619. Byzantine-Bulgar Alliance. Heraclius and Kubrat against the Avars.

626. Avar Siege of Constantinople. In alliance with the Persians (see p. 231).

627. Byzantine-Khazar Alliance. Heraclius and **Ziebel,** Khan of the Khazars, against the Persians (see p. 231).

c. 634. Independence of Bulgaria in the Volga Region.

640. Independence of the Slavonian Croats. Avar rule was thrown off.

643–701. Reign of Isperich. Successor to Kubrat as leader of the Bulgars. He led a migration from the Volga to the Danube (c. 650–c. 670). Later he led the Bulgars across the Danube in force, defeated the Byzantines, and forced Constantine to cede the province of Moesia (679–680; see p. 242).

661–790. Recurring Wars between Khazars and Moslems. Started by an Arab invasion north of the Caucasus, which drove the Khazars from Derbent on the Caspian (661). Slowly the initiative passed to the Khazars, who invaded Armenia, Georgia, Albania, and Azerbaijan (685–722). The tide changed again with an Arab invasion of Khazar country, penetrating the Caucasus to capture Balanjar (near Daryal Pass, 727–731). In turn the Khazars struck south, reaching into Mesopotamia before being driven back to the Caucasus (731–733). Sporadic border warfare continued; the Khazars raided Arab-held Georgia (790).

701–718. Reign of Terbelis (Tervel) of the Bulgars. He defeated the Byzantines at the **Battle of Anchialus** (708). Following this, he raided through Thrace to the walls of Constantinople (712).

717–718. Alliance of Terbelis with Leo III. (See p. 243).

755–772. Byzantine-Bulgar Wars. Marked by victories of Constantine V over **Telets** (see p. 244).

777–802. Reign of Kardam of the Bulgars. His frequent invasions of Thrace forced the Byzantines to pay tribute.

791–796. Overthrow of the Avars by Charlemagne. (See p. 225.)

SOUTHWEST ASIA

SASSANID PERSIA, 600–650

This was a half-century of violent vicissitudes in the ever-fluctuating fortunes of the Sassanids. Under **Chosroes II** the Persians virtually eliminated the Byzantines from all their Asiatic and Egyptian provinces, expanding Sassanid dominions practically to the extent of the empire of Darius. Then in six brilliant campaigns Heraclius gained complete victory over Persia (see pp. 229–232).

As the exhausted Byzantine and Persian empires endeavored to recover from this disastrous and costly quarter-century war, both were struck simultaneously by a new, vigorous, and fanatic power: Islam. The Sassanids, lacking the political and military stability of the East Roman Empire, completely succumbed to the Arab invaders in less than 20 years (see p. 249). The principal events were:

603–628. Persian War with Byzantine Empire. (See pp. 229–232).

610. Battle of Dhu-Qar. An army of raiding Arab tribes defeated a small Persian force south of the Euphrates before being driven back into the Arabian desert.

628–629. Reign of Kavadh II (Siroes). Son of Chosroes II, who was overthrown and killed by an army revolt under the leadership of the general **Gurdenaspa.** Kavadh made peace with Heraclius (see p. 232). His death (of plague) was followed by chaos in Persia.

633. Arab Invasion. Taking advantage of the disorders in Persia, the Arabs, under **Khalid ibn al-Walid,** invaded, capturing Hira and Oballa. They were checked temporarily at the **Battle of the Bridge** (see p. 249).

634–642. Reign of Yazdegird III (grandson of Chosroes II).

637–650. Arab Conquest of Persia. (see p. 249.)

640–651. Flight of Yazdegird. Driven from Persia to Media and to Balkh, Yazdegird appealed in vain to the emperor of China for assistance. He was later murdered near Merv (651).

THE RISE OF ISLAM, 622–800

The Early Caliphate, 622–750

Before Mohammed's death (632) and even before the complete consolidation of his control over Arabia, the new religious tide of Islam began to sweep northwestward and northeastward against the exhausted Byzantine and Persian empires. The eastern provinces of the Byzantine Empire—particularly Syria and Egypt—were estranged from Constantinople by a sectarian Christian dispute. So bitter was this that the Syrians and Egyptians for the most part welcomed the Moslems as deliverers from tyranny; they gave no support to the imperial armies, and in some instances actually aided the invaders. At the same time Persia, prostrated by defeat, had been thrown into anarchy by the untimely death of Kavadh II. In a little more than a decade the Byzantines were ejected from Syria and Egypt, to be thrown back across the Taurus Mountains of Anatolia. Simultaneously, Sassanid power was completely destroyed, and the vast Persian Empire fell to the Arabs.

Internal disputes now caused a brief pause in Islamic expansion. But these were soon—though only temporarily—settled, and the amazing vitality of Islam was demonstrated by a centrifugal push in all directions. The main thrust was against the Byzantines in Anatolia, but simultaneous advances were made westward along the North African coast, eastward toward India, northward against the Khazars through the Caucasus, and northeastward across the Oxus into Central Asia.

Twice repulsed at Constantinople, the Arab caliphs finally were forced to recognize that the Byzantines had a vitality (if not an aggressiveness) to match their own. They were also stopped along the line of the Caucasus Mountains by the fierce resistance of the Khazars. But they continued a slow advance into South and Central Asia, while sweeping across North Africa and through Spain.

After more spectacular successes, the initial Moslem tide was halted early in the 8th century by several factors. Most important of these was the resilience of the Byzantine Empire. Next, perhaps, was the vitality of the Franks, demonstrated by Charles Martel at the Battle of Tours (see p. 222). Overextension unquestionably drained Moslem resources, despite remarkable ability to inspire converts to a religious zeal matching that of the original Arab disciples of

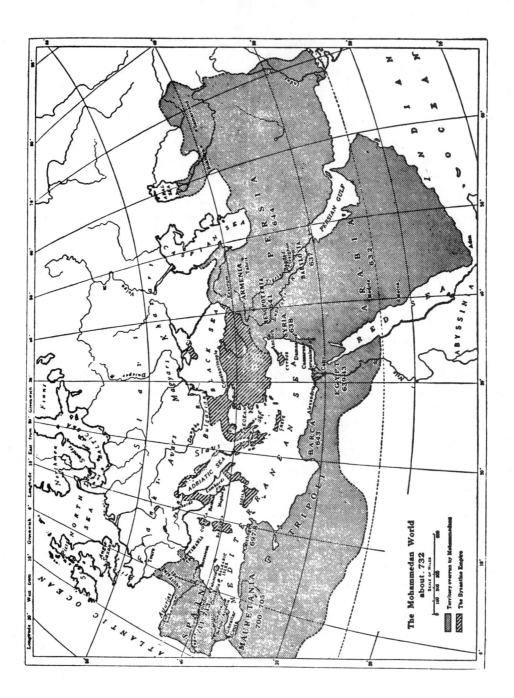

The Mohammedan World
about. 732

SCALE OF MILES
100 200 300 . . . 600

▨ Territory overrun by Mohammedans
▧ The Byzantine Empire

Mohammed. Finally, renewal of internal disputes, in even more violent form, split the Caliphate and caused the Moslems to turn their still-fiery energy upon each other. The principal events were:

622. The Hegira. Mohammed's flight from Mecca to Medina.

624–630. War between the Arabs of Medina and Mecca. Mohammed was victorious at the **Battle of Badr** (624), defeated at the **Battle of Ohod** (625), repulsed a Meccan attack on Medina (627), and captured Mecca by assault (630). During this period he had also begun the conquest of nearby Arab tribes; this continued rapidly after the capture of Mecca.

629. Battle of Muta. Repulse of the first Moslem raid into Byzantine Palestine.

632–634. Reign of Abu Bekr. This loyal disciple of the prophet became the first caliph upon the death of Mohammed. He vigorously suppressed the revolts of "false prophets" **Tulayha** and **Musaylima;** the actual military operations against the rebels were undertaken (632–633) by **Khalid ibn al-Walid,** the first and one of the greatest Moslem generals. The final victory was won over Musaylima at the **Battle of Akraba** (633).

633. First Expansion of Islam. Khalid invaded Persian Mesopotamia (see p. 247). At the same time another Arab force under **Amr ibn al-As** invaded Palestine and Syria.

634–644. Reign of the Caliph Omar.

634–636. Operations in Syria and Palestine. A Byzantine counteroffensive threatened the forces of Amr and other Arab leaders in Palestine. Khalid made a forced march from Mesopotamia to join them and to defeat Theodore between Jerusalem and Gaza at the **Battle of Ajnadain** (July 634). Pursuing relentlessly, Khalid defeated the Byzantines again at the **Battle of Fihl** (Pella or Gilead, near Baisan; January 635). Continuing north, he defeated the Byzantine General **Baanes** at the **Battle of Marjal-Saffar** (near Damascus), then captured Damascus and Emesa (Homs). He abandoned these cities when threatened by a new Byzantine army under Theodore, and retired to the Yarmuk River (636). Here he repulsed a

diversionary attack by the Sassanid allies of the Byzantines. Taking advantage of a mutiny in Theodore's army, Khalid attacked. After a bitter struggle, decided only when the Byzantines exhausted their supply of arrows, Khalid won a decisive victory in the **Battle of the Yarmuk** (August 636). He then recaptured Damascus and Emesa.

634–637. Operations in Persia. After Khalid left Mesopotamia for Palestine, the Persian general **Mihran** defeated the remaining Arabs at the **Battle of the Bridge,** on the Euphrates River, and drove them back to Hira (634). Arab reinforcements under **Muthanna** halted the Persian pursuit by a victory at the **Battle of Buwayb** (south of Kufa, 635). After Khalid's victory at the Yarmuk, Omar sent a new Arab army into Mesopotamia under the command of **Sa'd ibn abi-Waqqas.** With 30,000 men, Sa'd defeated more than 50,000 under the Persian chancellor, **Rustam,** in the 3-day **Battle of the Qadisiya** (June 637), then captured Ctesiphon (September ?). Pursuing, Sa'd defeated the Persians again at the **Battle of Jalula** (50 miles north of Madain, December 637).

637–645. Completion of the Conquest of Syria. This included capture of Jerusalem and Antioch (638), Aleppo (639), Caesarea and Gaza (640), Ascalon (644), and Tripoli (645). Most of the fortified places, stoutly defended, were taken only after long sieges.

639–641. Completion of the Conquest of Byzantine Mesopotamia.

639–642. Invasion of Egypt. Amr ibn al-As defeated the Byzantines at the **Battle of Babylon** (near Heliopolis, July 640). He next captured the fortress city of Babylon (April 641), then Alexandria (September 642), both after long sieges.

640–650. Invasion of Persian Highlands. The Arabs won decisive victories at the **Battles of Ram Hormuz** (near Shushtar, 640) and **Nahavend** (641). Organized resistance ended in Persia; consolidation took about a decade. The Oxus became the boundary between the Arabs and the Turks.

642–643. Expansion in North Africa. Cyrene and Tripoli were captured by **Abdulla ibn Zubayr,** who raided within 100 miles of Carthage.

644–656. Reign of the Caliph Othman.

645. Revolt in Alexandria. A Byzantine invasion fleet was repulsed and the uprising suppressed by Amr, who recaptured the city by assault.

649–654. Expansion at Sea. The growing Moslem navy demonstrated its prowess by capturing Cyprus (649), then the island of Aradus (off the coast of Syria); Moslem raiders next pillaged Sicily (652) and captured Rhodes (654).

652–664. Recurring Moslem Invasions of Afghanistan. Temporary capture of Kabul (664).

655. Naval Battle of Dhat al-Sawari (off Phoenix, Lycian Coast). Victory by Moslems over Byzantine fleet commanded by Emperor Constans.

656–661. Reign of Ali. Cousin, son-in-law, and adopted son of Mohammed, he succeeded Othman, who was murdered.

656–657. Revolt of Talha and Zubayr. The rebels were supported by **Ayesah,** widow of Mohammed. They were defeated by Ali in Mesopotamia near Basra in **"The Battle of the Camel."**

657–661. Civil War. This was precipitated by the revolt of **Mu'awiya,** cousin of Othman, governor of Syria. Ali invaded Syria; he and Mu'awiya fought the drawn **Battle of Siffin** (657). After prolonged and inconclusive negotiations, Mu'awiya was proclaimed Caliph in Jerusalem (660). Ali was murdered soon thereafter; his first son, **Hassan,** after brief resistance, abdicated (661). This civil war led to the division of Islam into two major sects: the Shi'ites, or supporters of Ali, and the more orthodox Sunnites, who abhorred him.

659. Peace with Byzantine Empire. To permit him to devote full attention to the civil war, Mu'awiya concluded peace and agreed to pay an annual tribute to Constantinople.

660–680. Reign of Mu'awiya. First Omayyad caliph.

661–663. Arabs Reach India. Raid into Sind and lower Indus Valley by **Ziyad ibn Abihi.**

668–679. War with Byzantines. The Arabs were defeated (see p. 241).

674–676. Moslem Invasion of Transoxiana. Temporary capture of Bokhara (674) and Samarkand (676).

680–683. Reign of Yazid I (son of Mu'awiya). **Hussain,** second son of Ali, revolted, was defeated at the **Battle of Kerbela** (on the Euphrates), and was cruelly murdered with most of his family (680). Another revolt in Arabia was led by **Abdulla ibn Zubayr** (son of the conqueror of Tripoli; see above), who successfully conducted the defense of Mecca (682–683) against Yazid's army, which raised the siege upon news of the death of Yazid. Zubayr was recognized as caliph in Arabia, Iraq, and Egypt.

681–683. Invasion of Morocco. An Arab army from Egypt, led by **Okba ibn Nafi,** reached the Atlantic, but was then driven back to Cyrene by the Berbers, in alliance with the Byzantines based in Carthage. Okba was killed in the retreat.

683–684. Continued Civil War. Marwan ibn Hakam of the Omayyad family defeated adherents of Zubayr at the **Battle of Marj Rahit** (near Damascus) to affirm his claim to the Caliphate (684).

684–685. Reign of Marwan. He immediately reconquered Egypt for the Omayyad Caliphate. He died before he could attack Arabia or Iraq.

685–705. Reign of Abd ul-Malik (son of Marwan). Complicated religious strife continued throughout the Moslem world during the early years of this reign. At the same time Abd was occupied putting down Byzantine-inspired revolts in Syria (685–690).

690. Reconquest of Iraq. Abd ul-Malik defeated **Mus'ab ibn-Zubayr** (brother of Abdulla) on the Tigris River, near Basra.

691–692. Reconquest of Arabia. Al-Hajjaj ibn Yusuf, general of Abd ul-Malik, captured Medina (691) and Mecca (692). Abdulla was killed in al-Hajjaj's successful assault on Mecca, following a 6-month siege.

691–698. Continued Disorders in Iraq. The dissident Kharijite sect was suppressed by the generals Muhallab and al-Hajjaj. This left Abd ul-Malik the undisputed ruler of the Moslem world.

693–698. Conquest of Tunisia. Byzantine influence in North Africa was eliminated with the capture of Carthage (698).

699–701. Revolt in Afghanistan. An Arab army, commanded by **Ibn al-Ash'ath,** revolted against al-Hajjaj, overall governor of the Eastern Moslem domains. Ibn al Ash'ath marched back to Iraq, where he occupied Basra, and marched against al-Hajjaj at Kufa. Al-Hajjaj was forced to withdraw into Kufa after losing the indecisive **Battle of Dair al-Jamajim.** Receiving reinforcements, al-Hajjaj finally defeated Ibn at the **Battle of Maskin** (on the Dujail River, 701) to suppress the revolt.

703. Repulse in Algeria. The Berbers defeated an Arab army under **Hassan ibn No'man** near Mount Aurasius (Aures Mountains). For reasons not quite clear the Berbers then entered into an alliance with the Arabs, facilitating their subsequent conquest of all North Africa (705).

705–715. Reign of Caliph Al-Walid (son of Abd ul-Malik). Zenith of the Omayyad Dynasty, and greatest extent of a single Moslem empire. Continued expansion in the east under the overall supervision of al-Hajjaj. **Qutayba ibn Muslim** reconquered Bokhara and Samarkand; he conquered Khwarizm, Ferghana, and Tashkent. He then raided into Sinkiang as far as Kashgar (713). After capturing Kabul (708), **Mohammed ibn al-Kassim** invaded and conquered Sind (708–712), after defeating the Indian King **Dahar** and capturing Multan after a long siege. He then raided the Punjab.

708–711. Conquest and Pacification of Northwest Africa. Musa ibn Nusair led the Arab forces.

710–714. Arab Invasion of Anatolia. The Byzantine province of Cilicia was conquered (711); the Arabs gained partial control of Galatia (714).

711–712. Conquest of Spain. (See p. 219.)

715–717. Reign of Caliph Sulaiman. Another son of Abd ul-Malik.

716–719. Invasions of Southern France. (See p. 221.)

716. Invasion of Transcaspian Region. The area between the Oxus and the Caspian Sea was conquered by Yemenite general **Yazid ibn Mohallib.**

717–718. Siege of Constantinople. (See p. 242.)

717–720. Reign of Omar II. There were no major military operations, other than frontier raids in France and Central Asia.

720–724. Reign of Caliph Yazid II. Another son of Abd ul-Malik. A revolt by Yazid ibn Mohallib was suppressed by Maslama at the **Battle of Akra** (on the Euphrates, 721). Internal disorders continued, however, with the southern Arabian, or Yemenite (or Kalb), faction generally opposed to the Caliphate, which was supported by the Qais (or Maadite, or northern Arabian) faction.

724–743. Reign of Caliph Hisham. Another son of Abd ul-Malik. He reorganized the administrative and military organizations of the empire. Recognizing that much greater expansion was undesirable, he established defensive regions to help stabilize the frontiers.

727–733. War with the Khazars in the Caucasus. Initially successful, the Moslems established a foothold north of the Daryal Pass, then were defeated and thrown back to Mesopotamia by the Khazars (see p. 244). Counterattacking, the Moslems reconquered Georgia, establishing their northern frontier on the Caucasus, with an outpost at Derbent.

730–737. War with the Turks in Transoxiana. After suffering a disastrous defeat at the hands of Chinese-led Turks near Samarkand (730), and again near Kashgar (736), the Arabs under **Nasr ibn Sayyar** rallied to defeat the Turks near Balkh (737). The Arab struggle against the Chinese and Turks continued without a decision for many more years (see pp. 252 and 260).

732. Battle of Tours. (See p. 222.)

739. Unsuccessful Invasion of Anatolia. (See p. 244.)

741-742. Revolt of Kharijites and Berbers. Omayyad troops were driven out of Morocco; they barely retained their control over their other North African provinces.

743-750. Turmoil in the Caliphate. Dynastic struggles, regional revolts, and religious disputes were interrelated in confusing abundance. The inept caliphs **Al-Walid II** (743-744) and **Yazid III** (744) were followed by the more able **Marwan II** (744-750). By hard fighting, Marwan restored order in Syria, Iraq, Arabia, and much of Persia (744-748).

Commencement of the Abbasid Caliphate, 750-800

The Abbasids—descendants of the Prophet's companion and first cousin, **Al-Abbas**—led a revolt which flamed from Khorasan through Persia into Mesopotamia. The violence of the fratricidal warfare impeded the expansion of Islam, providing a breathing spell for its enemies. By the end of the period, Abbasid caliphs were firmly on the throne in the new city of Baghdad, while in Morocco and Spain an independent Omayyad state had been established. The principal events of the period were:

747-749. Outbreak of Abbasid Rebellion in Khorasan. Leader was **Abu Muslim,** henchman of **Ibrahim** and **Abu'l Abbas,** grandsons of Al-Abbas. Despite the opposition of the governor of Khorasan, **Nasr ibn Sayyar,** Abu Muslim captured Merv (748). Ibrahim's general **Kahtaba** next defeated Nasr at the **Battle of Nishapur,** and again at the **Battles of Jurjan, Nehawand,** and **Kerbela.** A general rising in Persia and Mesopotamia permitted Abu'l Abbas (Ibrahim being dead) to proclaim himself caliph at Kufa (749).

750, January. Battle of the Greater Zab. Kahtaba decisively defeated Marwan, who fled to Egypt, where he was killed. This was followed by the systematic murder of most of the Omayyad family.

750-754. Reign of Abu'l-Abbas, First Abbasid Caliph. Recurring Omayyad revolts in Syria and Mesopotamia. Taking advantage of these disorders, the Byzantines raided deep into Moslem territory.

751. Battle of Talas. The Moslems finally drove the Chinese from Transoxiana (see p. 260).

754-775. Reign of Al-Mansur (brother of Abu'l Abbas). Al-Mansur's succession was challenged by other Abbasids. These revolts were suppressed with the assistance of Abu Muslim, who was now governor of Khorasan. Al-Mansur then had Abu Muslim assassinated, causing renewed revolts in Khorasan and elsewhere throughout the Caliphate. These and continuing Omayyad revolts were finally suppressed, but efforts to restore Abbasid control in Spain failed (see p. 227). Having firmly established the Abbasid Dynasty, Al-Mansur established a new capital at Baghdad (762-766).

762. Shi'ite Rebellion. This was suppressed at the **Battle of Bakhamra** (48 miles from Kufa).

775-785. Reign of Al-Mahdi. Son of Al-Mansur. After suppressing several revolts, he renewed the war with the Byzantines (778; see p. 244).

779-783. Intensification of War with Byzantine Empire. After a number of successful penetrations of Anatolia by **Harun al-Rashid** (son of Al-Mahdi), Byzantine Empress Irene made peace and agreed to pay tribute (783; see p. 245).

786-809. Reign of Harun al-Rashid. The zenith of the power and prosperity of the Abbasid Caliphate. There were a number of continuing disorders in the empire, but the Moslems were generally successful in recurring wars with the Byzantine Empire (see pp. 245 and 284).

AFRICA

EAST AFRICA

Aksum continued to be a major, wealthy power, despite the loss of its Arabian provinces. Aksum also retained control of a few coastal enclaves in Yemen, including Zabid. Its ships were active in the Red Sea, and successfully raided the Arabian port of Jiddah (702), although relations with the Caliphate were generally peaceful.

Other Christian kingdoms in the region included the Nile states of Nabotia, Maqurrah, and 'Alwah (see p. 210). In the early 7th century Maqurrah absorbed Nabotia, and thereafter endured constant raiding by Moslem Arabs based in Egypt until a peace treaty was concluded (c. 675) with **'Abd Allah ibn Sa'd,** whose army had sacked Dongola (Dunqalah). On the Red Sea coast, from Ethiopia to Egypt, lay the small kingdoms of Qata, Jarin, Baza, Baqlin, and Naqis (on the Egyptian frontier). The first two were vassals of Aksum. Farther south, the Bantu kingtom of Zanj held much of what is now Kenya and southern Somalia.

SOUTH ASIA, 600–800

Two leading figures of Hindu history dominated the subcontinent and vied for supremacy during the first half of the 7th century. The north was conquered by **Harsha,** who started his career as the ruler of the obscure kingdom of Thaneswar in the eastern Punjab. The Deccan was dominated by the greatest of the Chalukyas, **Pulakesin II.**

NORTH INDIA

Harsha was a conqueror in the tradition of the Chandraguptas; his empire approximated that of the earlier Maurya and Gupta dynasties. He subjugated most of north India in the first 15 years of his long reign. Turning southward, he invaded Chalukya territory, but was repulsed by Pulakesin. They concluded a treaty establishing a boundary along the Narbada River. During the remainder of his reign, Harsha devoted his energy and administrative skill to the consolidation of his empire and to the inspiration of a cultural revival comparable to the Golden Age of the Guptas.

Harsha's empire fell apart on his death. One of his ministers, an adventurer named **Arjuna,** seized the eastern provinces—modern Bihar and Bengal. He then made the mistake of molesting an embassy that the Chinese emperor had sent to Harsha. The amazing Chinese ambassador raised a small Nepalese and Tibetan army, then overthrew Arjuna.

The history of Hindustan in the centuries after Harsha is a drab and confusing story of endemic warfare between numerous rival dynasties. As in earlier times, these wars were punctuated by the arrival of new invaders from the north and west. The first of these was an obscure group of nomads—the Gurjaras—from central Asia who arrived in Rajputana about the beginning of the 7th century. By the middle of the century, they became the leading power of northwestern India. They were followed by the Arabs, whose centrifugal drive reached

Baluchistan late in the 7th century. A few years later the Arabs conquered Sind, but their further advance was checked by the fierce resistance of the Gurjaras.

Farther north the brief resurgence of Kanauj was shattered by the rising power of Kashmir, which, by the close of the 8th century, was engaged in a four-way struggle for supremacy in north India with the Gurjaras, the new Pala Dynasty in Bengal, and the Rashtrakuta Dynasty which was pressing northward from the Deccan. The principal events were:

606–648. Reign of Harsha.

648–649. War between Arjuna and Wang Hsuan-Tsu. When his embassy was attacked in north India by Arjuna, Wang—evidently a man of spirit and ability—withdrew to Nepal where he collected about 7,000 soldiers from the King of Nepal and another 1,200 from the King of Tibet (both vassals of the Chinese emperor). Wang then marched back to the Ganges, defeated Arjuna, and carried him away to captivity in China. (If these strength figures are correct, one must conclude that Wang was an exceptionally gifted general, or that he was extremely lucky, or that the strength figures usually given for Indian armies were grossly exaggerated—or possibly all three.)

c. 650. Arrival of the Gurjaras. They established themselves in Rajputana.

c. 650–707. Arrival of the Moslems. Arab conquest of Baluchistan.

708–712. Arab Conquest of Sind. Arab general **Mohammed ibn Kasim** seized the province. Subsequent Moslem raids into Rajputana and Gujarat were repulsed by the Gurjaras.

720–740. Resurgence and Fall of Kanauj. Yasovarman, a collateral descendant of Harsha, conquered Bengal (c. 730), bringing most of the Ganges Valley under his control. While attempting to expand northward, he was defeated in the Punjab by **Lalitaditya,** King of Kashmir; the new Kanauj Empire collapsed (740).

730–750. Rise of Kashmir. Under Lalitaditya, Kashmir became the leading power of the Punjab. He campaigned against the Gauda kings of Western Bengal, and evidently defeated an invading Tibetan army in the awesome mountain regions near the headwaters of the Indus. Like his Kashmiri predecessors, Lalitaditya maintained close relations with China and acknowledged the suzerainty of the Chinese emperor.

740–750. Brief Rise of the Gauda Tribe. They dominated western Bengal following the collapse of Yasovarman's empire.

c. 740–c. 780. Rise of the Pratihara Dynasty. Gurjara King **Nagabhata** decisively repulsed Moslem invasions of Rajputana and Gujarat (740–760). In subsequent years the Hinduized Gurjaras extended and consolidated their control over all Rajputana. Toward the end of the century they were being pressed from the south by Rashtrakuta (see p. 255).

750–770. Rise of the Pala Dynasty. Gopala established the dynasty in eastern Bengal, then extended his rule over all Bengal.

770–810. Reign of Dharmapala. This son of Gopala continued expansion of Pala dominions, reaching the edge of the Punjab by conquest of Kanauj (c. 800).

SOUTH INDIA AND CEYLON

Early in his reign, Pulakesin II defeated the Pallavas and extended Chalukya rule to the east coast of the Deccan. This began an almost incessant series of wars between Pallava and Chalukya. For several years the great Pallava ruler **Mahendravarman** fought Pulakesin on almost equal terms. But after Mahendravarman's death the Chalukya king steadily compressed the Pallava kingdom to the coastal area near modern Madras. Before this he had repulsed Harsha's invasion of the Deccan (see p. 253). He also defeated the Chola, Pandya, and Kerala kingdoms. Having eliminated practically all rivals in south India, Pulakesin's power was jolted by a revolt led by his brother. Pallava seized this opportunity to rise in revolt, and Pulakesin was defeated and killed by Pallava King **Narasimharvarman I.**

The Chalukya-Pallava feud continued with varying fortunes until the middle of the 8th century. By that time Chalukya had again achieved ascendancy, when it was suddenly faced with a new state led by the Rashtrakutas, former Chalukya vassals who had violently discarded Chalukya sovereignty. The Rashtrakuta established a military state, comparable in concept and organization to that of ancient Assyria, which steadily expanded in all directions. Its principal thrust, however, was toward the Ganges Valley, at the expense of the Pratihara Dynasty of Rajputana. By the end of the century, under **Govinda III,** the greatest Rashtrakuta, it had not only consolidated supremacy in the Deccan but had become one of the four major powers contending for hegemony in Hindustan. The principal events were:

600–625. Reign of Mahendravarman of Pallava.

608–642. Reign of Pulakesin II of Chalukya.

608–625. Wars between Pulakesin and Mahendravarman. Initially successful, Pulakesin's expansion in the northeast Deccan was halted after he pushed to the Bay of Bengal (609). Soon after this he undertook his victorious campaigns in the far south.

620. Pulakesin's War with Harsha. Harsha's attempt to invade the Deccan was repulsed.

625–630. Pulakesin's War with Pallava. This established Chalukya supremacy in south India.

630. Revolt of Kubja. The brother of Pulakesin, Viceroy of Vengi, declared his independence and established the Eastern Chalukya Dynasty.

630–642. Resurgence of Pallava. Narasimharvarman I defeated and killed Pulakesin in a battle outside Vatapi, capital of Chalukya (642). Pallava then pillaged Vatapi.

655. Chalukya Victory over Chola and Pandya. **Vikramaditya,** son of Pulakesin II, defeated Chola and Pandya to conclude a drawn-out war.

674. Chalukya Victory over Pallava. Vikramaditya destroyed the Pallava capital of Kanchi, revenging his father's defeat and death.

684. Pallavan Expedition to Ceylon. A Pallavan naval expedition with Tamil mercenaries placed **Manavamma** on the throne, thereby ending the Moriya Dynasty and inaugurating the second Lambakanna Dynasty.

c. 730–740. Chalukya War with Pallava. **Vikramaditya II,** grandson of Vikramaditya I, smashed Pallava power, capturing Kanchi three times.

753. Advent of the Rashtrakuta Dynasty. **Dantidurga,** founder of the Dynasty, threw off Chalukya rule and defeated King **Kirtivarman II,** beginning his new kingdom's rapid territorial expansion.

793–814. Reign of Govinda III of Rashtrakuta. He conquered Malwa and Gujarat, as well as defeating the Pallava, who began to recover power as the Chalukya state shrank to the status of a minor kingdom.

EAST AND CENTRAL ASIA

CHINA, THE TURKS, TIBET, AND NANCHAO

The resurgence of China during this period caused the military history of the Turks, Tibetans, and T'ais to be almost completely intermingled with that of their more powerful and more civilized neighbor.

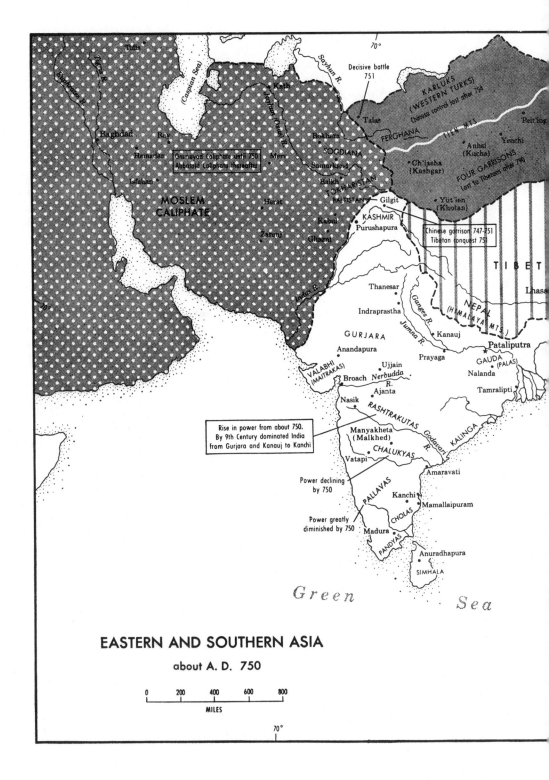

Tiflis

(Caspian Sea)

Kath

Sayhun R.

70°

Decisive battle
751

KARLUKS
(WESTERN TURKS)
Chinese control lost after 751

Peit'ing

Baghdad   Ray

Talas

FERGHANA

Bokhara

SOGDIANA

TIEN SHAN

Anhsi
(Kucha)

Yenchi

Hamadan   Merv

Samarkand

Ch'iasha
(Kashgar)

FOUR GARRISONS
lost to Tibetans after 790

Isfahan

Balkh

Ommayad Caliphate until 750.
Abbasid Caliphate thereafter.

TOKHARISTAN

Herat

MOSLEM
CALIPHATE

BALTISTAN

Gilgit

Yüt'ien
(Khotan)

Kabul

Ghazni

KASHMIR
Purushapura

Chinese garrison 747-751
Tibetan conquest 751

Zaranj

Indus R.

Thanesar

Ganges R.

Jumna R.

NEPAL
(HIMALAYA MTS)

TIBET

Lhasa

Indraprastha

Kanauj

Pataliputra

GURJARA
Anandapura

Prayaga

GAUDA
(PALAS)

Ujjain

VALABHI
(MAITRAKAS)

Broach   Nerbudda
R.

Nalanda

Tamralipti

Nasik

Ajanta

RASHTRAKUTAS

Godavari R.

KALINGA

Rise in power from about 750.
By 9th Century dominated India
from Gurjara and Kanauj to Kanchi

Manyakheta
(Malkhed)

CHALUKYAS

Amaravati

Vatapi

Power declining
by 750

PALLAVAS

Kanchi

Mamallaipuram

CHOLAS

Power greatly
diminished by 750

Madura

PANDYAS

Anuradhapura

SIMHALA

Green   Sea

EASTERN AND SOUTHERN ASIA

about A. D. 750

0   200   400   600   800

MILES

70°

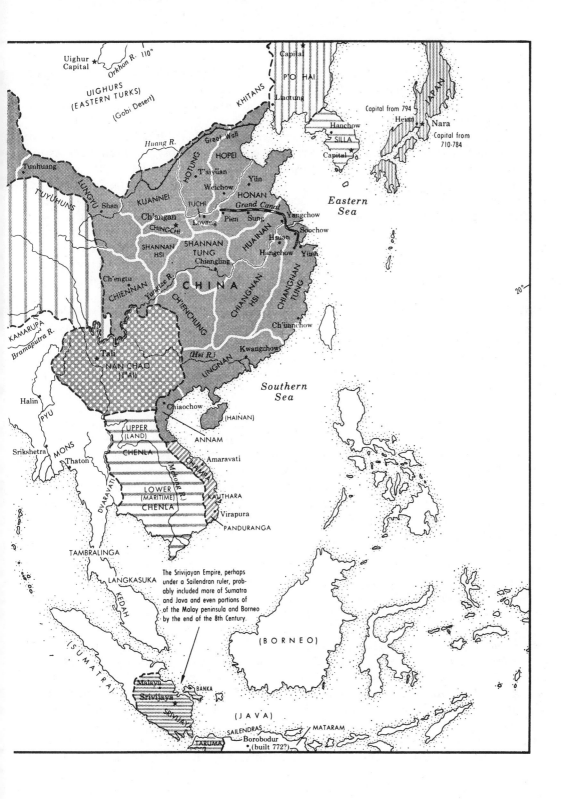

The Srivijayan Empire, perhaps under a Sailendran ruler, probably included more of Sumatra and Java and even portions of of the Malay peninsula and Borneo by the end of the 8th Century.

At the beginning of the 7th century, China was just emerging from four dismal centuries of chaos, to enter one of its great periods. The revival began under the short-lived Sui Dynasty. The T'ang Dynasty, which followed, made China the largest, most powerful, most prosperous, and (possibly) most cultured nation in the world at the time.

The man most responsible for this was a military genius named Li Shih-min, better known to history as the Emperor **T'ai Tsung,** whose reign combined almost universally successful external conquests with internal peace and prosperity. China's resurgence had begun when T'ai Tsung inherited a still-shaky kingdom from his father—the first of the T'ang Dynasty. The growing power of the Eastern Turks was menacing the northern and northwestern frontiers. Based upon extensive and successful combat experience against the Turks in his youth, the new emperor decided to reorganize Chinese military policy and doctrine. Concluding that the only sure way to prevent nomad invasions was to defeat them on their own grounds, he refused to strengthen the Great Wall, as his counselors advised. Instead, he fought delaying actions on the frontiers, while building up a new army—primarily cavalry—which he trained partly on the battlefield and partly in the interior.

As the basis for his new army, T'ai Tsung revived the Ch'in-Han militia system. He divided the country into 10 circuits (*tao*) and 634 prefectures (*fu*). Within each prefecture the peasants were organized into militia units (*fuping*) and received regular, periodic training. These militia units took turns serving with the Imperial Guards Division in the capital. Otherwise, they were called to active duty only when needed—either locally within the prefecture or elsewhere in the country, or in foreign expeditions.

When ready, T'ai Tsung began a series of punitive expeditions against the Eastern Turks. He destroyed that khanate in about 10 years of systematic campaigning. Simultaneously, his diplomacy weaned the Uighur Turks from their allegiance to the khanate of the Western Turks. At about this time the Tufans (Tibetans) invaded western China. T'ai Tsung inflicted a crushing defeat on them and forced the Tibetan ruler to acknowledge Chinese suzerainty. With Uighur assistance, he then defeated and conquered the Western Turks, annexing all of modern Chinese Turkestan and most of the area between the Altai, the Pamirs, and the Aral Sea. His sovereignty was acknowledged by the tribes inhabiting most of modern Afghanistan; the kings of Kashmir were his vassals.

By the middle of the 7th century, China had no rival in northern, eastern, or central Asia. T'ai Tsung's death, however, was followed by a sharp decline, as many of the conquered peoples tried to throw off Chinese rule. But by the end of the century a revival of Chinese power took place under the Empress **Wu,** one of the great women of history. An able, unscrupulous monarch, she dispatched armies to the frontier regions that soon restored order and Chinese control.

The power of the T'ang Dynasty peaked just before the middle of the 8th century under **Hsuan Tsung,** who extended and consolidated Chinese control of the Oxus and Jaxartes valleys. He threw back Arab penetration into Central Asia—mostly by playing off the local rulers against the Moslems. At the same time, with Chinese blessing, the Uighurs extended their semi-independent dominions northeastward from the Altai over the remnants of the Eastern Turks in Mongolia.

The T'ang Dynasty declined in the last half of the 8th century. This was precipitated by Arab victories, which expelled the Chinese from Transoxiana. Only the assistance of the loyal

Uighurs enabled the Chinese to retain Sinkiang. The Khitan people of eastern Mongolia and Manchuria began raiding deep into north China; the Tufans of Tibet threw off Chinese rule, as did Korea and the T'ai Kingdom of Nanchao in Yunnan. At this time the general **An Lu-shan** started a revolt, which grew into one of the more violent civil wars of Chinese history. The rebellion was finally quelled—with Uighur assistance—but its ravages hastened still further the political and military decline of the T'ang Dynasty. Another reason for the decline was the historical inability of the Chinese to develop horse-breeding competence, or to match the desert nomads either in horsemanship or in cavalry tactics. Also contributing to the decline were the effects of corruption upon the formerly high standards of the militia system.

This decline facilitated the rise of the powerful new T'ai Nanchao Kingdom in Yunnan. Having thrown off Chinese rule, Nanchao expanded steadily during the last half of the 8th century, to gain control over the northern Irrawaddy Valley in Burma. Nanchao raiding expeditions undoubtedly reached the Gulf of Martaban, and possibly even crossed over the mountains into India. The principal events were:

589–605. Reign of Yang Chien. Consolidation of the reunited Chinese empire under the first of the Sui Dynasty.

c. 600–603. Temporary Reunification of the Turk Empire. Under **Tardu,** a Western Turk chieftain, the Turks briefly threatened Ch'ang-an, the Chinese capital (601), but this menace disappeared with the collapse of Tardu's empire (603).

c. 600. Rise of the Tufan People of Tibet. Related to the earlier Ch'iang tribe, the Tufans controlled parts of Kansu, Sinkiang, and north India, under a northern Indian ruling family.

602–605. Chinese Reconquest of Tongking and Annam. General **Lui Fang** subdued the rebellious provinces. He defeated the neighboring Chams, forcing them to pay tribute. Liu possibly marched across Cambodia to the Gulf of Siam; at any rate, the Khmer states of Cambodia began to pay tribute to China.

605–618. Reign of Emperor Yang Ti. He undertook aggressive expansion in all directions.

607–609. Chinese Advance into Sinkiang. General **P'ei Chu** defeated the Western Turks, driving them into Tibet and Sinkiang from footholds in Kansu and Kokonor. He occupied the oasis of Hami (Khamil, 608) and penetrated into Sinkiang (609). At the same time Yang Ti undertook a major reconstruction program on the Great Wall, to consolidate his northern conquests and to protect the left flank of a proposed invasion of Manchuria-Korea.

607–610. Conquest of Yunnan. Chinese forces subdued the barbarian T'ai inhabitants.

610. Conquest of Formosa. An overseas expedition under General **Ch'en Leng** conquered at least part of Formosa.

611–614. Operations in Manchuria and Korea. Yang Ti undertook a series of major campaigns against the Kokuryo State of southern Manchuria and northwestern Korea. These costly operations were generally unsuccessful and greatly weakened a nation already overextended militarily and economically.

613–618. Revolts against Yang Ti. Economic hardships and the losses of the disastrous Korean campaigns stimulated a rash of uprisings. The Eastern Turks resumed their raids against the northern frontiers. Yang Ti took the field personally; he was defeated and surrounded in the fortified town of Yenmen (615). He was rescued by a daring attack led by young Li Shih-min, then about 20 years old. Three years later Yang Ti was murdered during internal disorders.

618–626. Reign of Kao Tsu. Li Yuan, a Wei nobleman, aided by his brilliant soldier-son, Li Shih-min, seized the throne and restored order. Li Yuan took the imperial name **Kao Tsu,** starting the T'ang Dynasty.

c. 620–650. Reign of Song-tsan Gampo of Tibet. He conducted many forays into neighboring countries, possibly including a brief conquest of Upper Burma. He established his capital at Lhasa.

c. 622. Rise of the Western Turks. They conquered the Oxus region and parts of Khorasan from Persia, cooperating with Byzantine Emperor Heraclius.

624–627. Eastern Turk Raids into China. These were repulsed at the gates of Ch'ang-an (Sian), the capital, by Li Shih-min, who combined force, threat, and diplomacy to establish a temporary peace with the Turks.

626–649. Reign of T'ai Tsung. Li Shih-min, under the title of T'ai Tsung, succeeded his father after a brief struggle with rival claimants.

629–641. Wars with the Eastern Turks. T'ai Tsung's punitive expeditions destroyed Eastern Turk power. By diplomacy he kept peace with the Western Turks and established an alliance with the semi-independent Uighur Turks. During this period the Khitan Mongols, east of the Gobi Desert, submitted to T'ai Tsung (c. 630).

641. War with Tibet. T'ai Tsung inflicted a crushing defeat on invading Tibetans under Song-tsan Gampo. To avoid a multifront war against barbarians, he concluded an easy peace, establishing an alliance which was sealed by giving his niece as wife to the Tibetan king, who acknowledged Chinese sovereignty. Nepal also began sending tribute about this time.

641–648. Wars with the Western Turks. T'ai Tsung defeated and conquered the Western Turks with the assistance of the Uighurs. He reestablished Chinese sovereignty over Sinkiang, and received tribute from tribes and principalities west of the Pamirs.

645, 647. Chinese Expeditions into Korea. These had limited success, but failed to subdue the area.

646. War of Wang Hsuan-tsu and Arjuna in India. (See p. 254.)

649–683. Reign of Kao Tsung. The expansion of T'ang power continued in the early years of his reign, but declined near the end.

c. 650. Emergence of Nanchao. This T'ai state in western Yunnan was nominally subject to China.

657–659. Revolt of the Western Turks. This was crushed, the Khan being captured. Chinese control over the Oxus Valley was strengthened.

660–668. Conquest of Korea. This was accomplished in alliance with the Korean vassal state of Silla.

663–683. Military Disasters. The Tibetans revolted and seized portions of the Tarim Basin, cutting off China from its dominions farther west, which thereupon asserted their independence. Nanchao, Korea, and the Western and Eastern Turks all threw off the Chinese yoke.

684–704. Reign of Empress Wu. Chinese power revived. The Tibetans were driven from Sinkiang, and portions of the other outlying regions were recovered.

712–756. Reign of Hsuan Tsung. By a combination of force and diplomacy, Chinese control over the Oxus and Jaxartes valleys was reestablished. Arab invaders were defeated in a series of hard-fought campaigns (730–737; see p. 251). The Tibetans were driven from the passes of the Nan Shan and Pamir mountains and forced again to acknowledge Chinese suzerainty. The Koreans resumed payment of tribute. A series of campaigns in Yunnan punished Nanchao and forced the T'ais to reaccept Chinese sovereignty (730).

c. 745. Expansion of the Uighurs in Northern Mongolia. They conquered the remnants of the Eastern Turks, creating an empire—nominally subject to China—from Lake Balkash to Lake Baikal.

747–751. War with Arabs and Tibetans. After conclusion of an Arab-Tibetan alliance, General **Kao Hsien-chih** (of Korean descent) surprised and defeated both allies, following an amazing march across the Pamirs and the Hindu Kush (747). Kao then engaged in some questionable, and probably treasonable, intrigues in the Oxus region, whereupon the Prince of Tashkent called on the Arabs for assistance. Kao was decisively defeated (751) by the Arabs in the **Battle of Talas** (in Kirghiz SSR). This ended Chinese control in the areas west of the Pamirs and Tien Shan mountains.

Further losses in Turkestan were prevented with the assistance of the Uighurs.

751–755. More Military Disasters. The Khitan Mongols (descendants of Hsien Pi) invaded North China, where they were barely held through the efforts of general **An Lu-shan,** a Turkish adventurer in the Chinese Army. Nanchao again rebelled, and decisively repulsed two Chinese invasion attempts (751, 754). Tibet and Korea again became independent.

755–763. Revolt of An Lu-shan. Declaring himself emperor, he quickly overran the Yellow River Valley, but was held up by the loyal garrison of Sui Yang. This delay permitted the T'ang army, reinforced by Uighur allies, to reorganize, and to defeat him just after he captured Sui Yang (757). He was then killed by his son, who continued the revolt, which was not crushed until an Uighur force captured Loyang, the rebel capital (763).

755–797. Ascendancy of Tibet. The height of Tibetan power under **Khrisong Detsen.** Taking advantage of the civil war in China, he captured and sacked the Chinese capital, Ch'ang-an (763).

c. 760–800. Expansion of Nanchao. King **Kolofeng** (748–779) conquered the upper Irrawaddy Valley. Nanchao expansion was continued by **I-mou-hsun,** his grandson and successor.

760–800. Progressive Weakening of T'ang Authority. The seacoast was raided constantly by pirates, mostly Indonesian, but evidently including some Arab and Persian Moslems, who had gained control of the Indian Ocean, and who had established themselves on the trade routes between the Indian Ocean and the China Sea.

KOREA, 600–800

The unceasing struggle between the three kingdoms of Kokuryo, Paekche, and Silla continued into the early years of the 7th century. At the same time Kokuryo was being subjected to increasing pressure from the expanding Chinese Empire of the Sui and T'ang Dynasties. Soon after the middle of the century, Silla, in alliance with the Chinese, overcame both of its rivals and accepted Chinese sovereignty. Apparently Paekche or Kokuryo received some assistance from Japanese allies, but to no avail. T'ang influence in Korea soon declined, but was reestablished in the early 8th century. Important events of the period were:

611–647. Chinese Invasions. Kokuryo repulsed five major invasion efforts (611–614, 645, 647).

660–663. War of Paekche with Silla and China. The allies overthrew Paekche, defeating also a combined Japanese army and fleet sent to the assistance of Paekche (see p. 262). The enlarged Kingdom of Silla accepted Chinese suzerainty.

663–668. War of Kokuryo with Silla and China. The allies overthrew Kokuryo. All of Korea was united under one rule, subject to Chinese suzerainty.

c. 670–c. 740. Autonomy of Silla. During the decline of the T'angs, the Korean Kingdom of Silla became virtually independent. The Koreans began paying tribute again, however, with T'ang resurgence.

JAPAN, 600–800

The Soga clan, which had long dominated Japan, was eclipsed during the early years of the 7th century by brilliant Prince **Shotoku,** nephew and heir to the Empress **Suiko** (593–628).

Acting as regent for the empress until his death (621), Shotoku introduced a constitution and created the basis for a centralized Japanese state. Like his successors, he was engaged in constant frontier war with the Ainu tribes in northern Honshu. Apparently he also planned an invasion of Korea, but abandoned the idea to meet pressing internal problems. After his death the Soga clan reestablished its ascendancy over the imperial family, but was soon overthrown in a brief struggle by an imperial prince who took the imperial name **Kotoku.** He was aided by **Nakatomi-no-Kamatari,** who became the founder of the Fujiwara clan. These two instituted the Taika Reforms, carrying further the centralization of the government that had been begun by Shotoku. After one disastrous expedition to Korea, the Japanese devoted their attention to internal affairs and to the continuing war with the Ainu. By the end of the 8th century these wars were drawing to a close, with the Japanese consolidating the newly conquered regions in northern Honshu. The important events were:

593–621. Regency of Prince Shotoku. Promulgation of a constitution established the basis of a centralized state (604).

621–645. Resurgence of the Soga Clan. This renewed ascendancy was abruptly ended by the victory of Kotoku.

662–663. Japanese Land and Sea Expedition to Korea. (See p. 261.)

710–781. Period of Internal Unrest. This was marked by intrigue and frequent *coups d'état.* A decline in central authority resulted in several defeats at the hands of the Ainu.

781–806. Reign of the Emperor Kammu. He vigorously reestablished a strong central government and revitalized the army. By the end of the century his general **Sakanoue Tamuramaro** had crushed most Ainu resistance in northern Honshu.

IX

THE DARK AGES—
BATTLE-AX AND MACE:

800–1000

MILITARY TRENDS

This, the darkest period of the Middle Ages, was characterized by aimless and anarchic strife. Any effort to catalogue the multitudinous wars would be hopelessly confusing, dull—and relatively unimportant. Presented here are only those events which are most significant, as well as representative of the period as a whole.

There were no great captains. Among the Byzantines, however, there were a number of outstanding soldiers, including **Basil I, Nicephorus Phocas** (probably the best), **John Zimisces,** and **Basil II.**

In the welter of chaotic conflict, four significant historical and military trends stand out:

First, the rise of feudalism in Western Europe, in the Moslem world, and in Japan.

Second, the continued superiority of the Byzantine military system, permitting that empire to stand like a mighty rock amidst swirling tides and waves which buffeted it from all directions. The manner in which the Byzantine Empire withstood the first two centuries of Moslem assaults, and then was able to expand its power and influence in Asia and Europe during these two centuries, provides a useful object lesson for those impatient or fearful of the tide of modern events.

Third, the growing force of the Turkish migration, moving slowly but steadily from central Asia against the Moslem eastward current. This Turkish migratory process had two facets. There was, of course, the actual westward and southwestward movement of tribes and peoples from the area known today as Turkestan. At the same time, the martial prowess of the Turks was utilized increasingly by the Abbasid Caliphate in military units comprised of Turkish mercenaries or slaves. The most important of these was the large imperial guard, the only standing, professional force in the Caliphate. The Turkish generals commanding these units began to appear first as governors, then as independent princes, in provinces as far to the west as Egypt.

Fourth was the appearance of a number of states clearly identifiable as the precursors of modern nations. (Save for China, there had been little connection between the kingdoms and peoples of earlier periods and states of our modern world.) In some instances the link over the intervening millennium is tentative—as for Russia and Burma, for instance. In others, such as France and Japan, the relationship is clear and direct.

FEUDALISM

Western Europe

The trend toward feudalism in Western Europe had been evident in the tumultuous period following the fall of Rome, as barbarian tribal chieftains became the landowning nobles of the Teutonic kingdoms, providing protection to their people and receiving services and goods in return. This trend had been suspended by the stability of the centralized empire of Charlemagne. Yet Charlemagne's policies, demanding higher military standards of the contingents which the nobles furnished to his armies, provided a basis for subsequent acceleration of the feudalizing process.

The immediate stimuli for this acceleration in Western Europe were the Viking and Magyar invasions. Kings and nobles took frantic measures to protect their resources—people, livestock, and commercial centers. The chaotic dynastic disputes among the successors of Charlemagne precluded centralized effort against the devastations of the raiders; there was no leader with the ability to re-create his centralized military and administrative machinery. Consequently, defensive and protective measures had to be local, and largely uncoordinated. These measures took two principal forms: the construction of fortifications to protect rural populations as well as commercial and communications centers, and creation by each landowner of permanent military forces to man his fortifications and to harass the raiders whenever possible.

It was in this latter respect that the military standards of Charlemagne contributed to the establishment of excellent, even though small, professional units under the standards of the nobles. The trend toward cavalry continued; these standing forces were entirely mounted men—knights and men-at-arms. The nobles would, on rare occasions, call up levies of all their able-bodied men who had had some training as foot soldiers, but who were generally inadequately armed, protected, and organized; the role of such infantry was always passive and defensive.

From these developments emerged feudal society, based on the mounted knight and fortified castle, in which the strong protected the weak—and the weak had to pay a price. The independent middle class disappeared. Freemen of some wealth and property became vassals of the neighboring lord and, in return for his promise of protection, pledged themselves (and retainers, if any) to serve him as cavalry soldiers under certain clearly defined conditions. The poorer freemen simply became serfs of the gentry or nobility. Though liable to call-up for military service, they rarely were mobilized, save possibly to assist in the defense of the lord's castle. In return for tilling the soil or other menial service they, too, were given protection by their betters—and they found this to be a much more satisfactory arrangement than being left at the mercy of Vikings or Magyars.

Feudalism was based upon a military concept concerned primarily with local defense. Each great lord held his lands from the king. In return he was to be prepared to take his men out of his own local district on operations at the call of the king for a given period each year—usually 40 days. His primary responsibility, however, was for the local security of his own lands, 365 days a year. The military result of this system was that when royal armies were assembled and employed for offensive operations, they lacked homogeneity; there was no common loyalty to king and nation, they had no cohesive discipline based upon a common organization and integrated training, and there was no effective unity of command.

The one common social force was sincere devotion to Christendom. This provided a basis for the essentially moral concept of knightly honor which, in turn, was the principal ingredient of the chivalry of the Middle Ages, which created the romantic aura that tints our distant view of that essentially barbarous grim age and society.

The Viking raids provided the same impetus toward feudalism in Britain as on the Continent, but the standing forces of the Anglo-Saxons—and later the Danes—were almost entirely infantry. Otherwise, the feudalizing process was parallel on both sides of the English Channel, although the art of fortification was much less advanced in Britain, where simple ditches and wooden palisades substituted for the stone castles of Europe. The feudal lords and military leaders in England were known as thegns.

The local defensive measures which soon slowed, then halted, the Norse and Magyar depredations were found by the nobles to be equally useful in permitting defiance of the central authority of kings and emperors with relative impunity. The remainder of the medieval period in European history, therefore, was to be largely a struggle between central authority and the jealously guarded local power and privileges of nobles and of walled cities.

Islam

The rise of feudalism in Moslem lands had a somewhat different basis, at least initially. This was the pressure of the violent internal centrifugal religious and political forces which tore at the early Caliphate and its successors, forces which led to the rise of numerous, largely independent, great and small Moslem principalities and heretical religious communities. The rivalries among these independent Moslem groups, and between them and central authority, combined with raids by Byzantines, Khazars, Turks, and Spanish Christians, led to the same kind of local defensive and protective measures which appeared about the same time in Western Europe. As the local defensive capabilities were enhanced, this feudalizing process of course compounded the fragmentation of Islam.

Japan

Though feudalism appeared in Japan also at about this same time, it had a slightly different origin, and followed a somewhat different course than in the West. This is discussed elsewhere in this chapter (see p. 299).

WEAPONS AND ARMOR

There were no important innovations in weapons during this period, though there were some modifications, mainly in Europe. Swords became somewhat heavier and longer; no longer suitable for thrusting, they were used mainly as cutting weapons. The double-edged ax became more popular in Europe. These were indications of the greater emphasis on ponderous, brute force—rather than nimble skill—in West European warfare in the Middle Ages.

Charlemagne's efforts to introduce the bow into western Europe had failed completely. Aside from the Byzantines, the bow was used in Europe only by the Norsemen and the Turko–Scythian invaders from Asia (Bulgars, Magyars, and Pechenegs). One mark of their Norse background was the occasional use of the bow by the Normans—but usually for hunting rather than warfare.

Defensive armor became more common and more effective among Europeans and Moslems. For the most part Byzantine example was followed, and in some respects improved on. The ancient crested helmet disappeared, to be replaced by an iron conical headpiece, to which Europeans began to attach a nosepiece—precursor of the visor. The mail shirt was universally the basic item of armor, and was increased in length so that its flaps would cover the knees of a mounted man. One of the most important innovations of the Western Europeans was the evolution of the kite-shaped shield, which may have been the result of a conscious effort to combine the best features of the ancient Roman scutum and the more common round targe. This was a much more sensible item of equipment for mounted men, providing more protection, with less bulk, than a round shield. Another useful innovation was the hauberk, to protect the neck between helmet and mail shirt.

LAND TACTICS

General

There were no really important tactical innovations. Cavalry remained supreme throughout most of the world. Byzantine practices were much as they had been; their enemies tried to copy Byzantine tactics, but lacked the discipline, training, and organization to do so with full effectiveness.

The great revival of fortification throughout most of the world naturally put a premium on siegecraft. Again Byzantine example was copied, but neither they nor their enemies had improved upon the techniques of Julius Caesar—with the sole exception of Byzantine introduction of Greek fire (see p. 242). In Western Europe, both weapons and techniques were crude in comparison with those the Romans had employed a thousand years earlier. On the other hand, all of these techniques and weapons were well known and applied—even by the Vikings, as was demonstrated in the well-documented siege of Paris.

In siegecraft and in the defense of fortified places—and indeed in tactics in general—the Western Europeans tried to follow Roman example as best they could understand and adapt it to their own form of ponderous, heavy shock cavalry. The only important military manual to be found in the West was *De Re Militari* of Vegetius (see p. 150), which became required reading for those among the gentry who could read.

Viking Tactics

The Vikings were essentially raiders, more interested in plunder and the spoils of victory than in any kind of permanent conquest. On the other hand, they had a fierce love of combat, and though they rarely sought battle—and would wisely avoid it against odds—under most circumstances they were never averse to a good fight.

The Vikings, skillful warriors, initially had a higher standard of discipline—based on loyalty to their immediate chieftain—than was to be found in Western Europe. They were foot soldiers, usually armed with spears, swords, and axes—sometimes also carrying bows. Defensive armor consisted of helmet, round shield, and leather jacket. Later many adopted the mail shirt.

As European opposition to their inroads became more effective, the individual marauding Viking bands of 100–200 men would join together to form armies that sometimes were quite numerous. The Viking force besieging Paris in 885–886 must have been close to 30,000 men.

When fighting Western Europeans, the Vikings found defensive-offensive tactics to be effective against the more numerous, but poorly armed, poorly trained, poorly led militia levies with whom they first had to deal. The same tactics were obviously the best against the new professional class of cavalrymen developed in Europe to meet the Viking threat.

The Europeans, from their standpoint, found that this cavalry professionalization (combined with fortification) was an effective answer to the Norse raids. The heavy cavalrymen had greater shock power, and could fight the Vikings on equal terms or better if not outnumbered. And if the Vikings were too strong, then fast-moving, professional cavalrymen could operate from a secure, fortified base, keep up with the Viking foot columns, and frequently harass them effectively. They could also concentrate rapidly with other contingents to force battle on the raiders.

This led the Norsemen to two countermeasures. First, wherever they went ashore—on the coast, or on the river banks in inland waters—they would seize all horses in the vicinity, mounting as many men as possible to permit rapid movement. At first they used such horses only for transportation. Later, as their cavalry opposition became more formidable, they maintained large, permanent, well-defended bases on coastal or river peninsulas and islands, and developed their own cavalry units. To the end, however, the bulk of the Viking forces were infantry.

Magyar Tactics

The Magyars fought and raided as Scythians had since the dawn of history. They were light horsemen, usually unarmored, whose principal weapon was the bow and whose most important characteristic was mobility. They could not stand up and fight the heavy West European cavalry, and avoided hand-to-hand combat if it was at all possible. They fought as the Parthians had against the Romans at Carrhae (see p. 129). Using their superior mobility and exploiting their missile weapon—the bow—they would try to circle their more ponderous foes, harassing them for hours until the combination of casualties, exhaustion, and frustration led to gaps in the European formations. They would exploit such gaps—usually by attacking from the rear—endeavoring to cut off and overwhelm any isolated groups.

On their long raids the Magyars relied mainly on speed and rapid changes of direction to avoid large concentrations of West European cavalry. But though they were always more mobile than their principal enemies, the steady improvement in effectiveness and mobility of European heavy cavalry, combined with the growing number of fortifications, gradually reduced the returns from their raids.

NAVAL WARFARE

There were no important improvements in naval warfare. There was only one really effective navy: that of the Byzantine Empire, though its fleets were allowed to decline during the early part of this period. The principal employment of sea power—if it can be termed that—was by raiders and pirates: the Vikings against the coast of Western Europe, Moslem corsairs in the Mediterranean, Varangian (Scandinavian Russian) raiders in the Black Sea.

Thus naval engagements were rare, since a battle at sea was the last thing a raider wanted. In the 10th century, however, the revitalized Byzantine navy systematically hunted down the Moslem corsair and pirate fleets, with the result that there were several spectacular sea battles in the Mediterranean.

Viking ships were considerably different than those earlier used for warlike purposes in the Mediterranean. Generally less than 100 feet in length, they had 10 to 16 oars to a side, plus a mast carrying a square sail when favored by a following wind. These vessels at first carried only 60–100 men. But during the latter years of the 9th century, some larger ships carried as many as 200. Viking seamanship was as admirable as their fighting abilities on land.

Much has been made of the fact that **Alfred the Great** of England built a navy with which to defend his coasts against the Viking raiders. This is a fact of considerable military and historical significance, as was the subsequent development of that navy by his successors **Edward** and **Aethelstan** into an instrument of offensive warfare against the Danes and Scots. It is a mistake, however, to consider Alfred as the father of the Royal Navy of today, since the Anglo-Saxon fleets of the 9th and 10th centuries soon thereafter disappeared completely, and new beginnings for Britain's sea power were made centuries later.

In the combined naval and land operations of Nicephorus Phocas against Crete (see p. 286), the Byzantines demonstrated a high order of skill and inventiveness in amphibious warfare. His transport vessels were equipped with bridges, or ramps, whereby his mounted cavalrymen could charge ashore directly onto the beaches in opposed landings. These were the prototypes of the modern LST (landing ship tank).

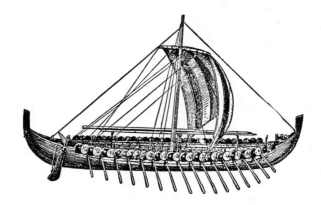

Viking ship

WESTERN EUROPE

The Viking Invasions and Scandinavia

The history of Scandinavia emerged from obscurity at the beginning of the 9th century, about the same time that Scandinavian raiders—known variously as Vikings, Norsemen, and Northmen—began their fierce raids against Western Europe and the British Isles. The three principal Scandinavian countries of Denmark, Norway, and Sweden were already separate entities, with Denmark the most advanced.

The reasons for the relatively sudden commencement of the great Viking raids of the 9th century have never been clear. With few exceptions, these were neither nationalistic nor even tribal invasions or migrations, but rather were independent forays for loot and plunder, led by Scandinavian nobles and adventurers.

For the most part the Norwegian and Danish Vikings sailed west and southwest, Swedes confining themselves mostly, but not exclusively, to the Baltic area. The Viking raids began late in the 8th century; within 50 years they achieved an amazing level of intensity and destructiveness, and did not begin to taper off until the early 10th century.

The most important result of the raids was the impetus given to the development of feudalism in Western Europe (see p. 264). At the same time, the course of events in France, the British Isles, and Russia was profoundly affected, with indirect results later evident in Southern Europe, the Mediterranean, and the Near East.

The Vikings in Ireland, 800–1000

Norse raids grew in intensity during the early years of the 9th century, during which they established numerous permanent posts in northern and eastern Ireland. In the vicissitudes of constant conflict over the next century and a half, the Norse were driven from their positions in the north, but established themselves firmly in eastern and central Ireland, Dublin, Waterford, and Limerick being their main centers of strength. They played a dominant role in Ireland during most of the 10th century, but by the end of the period their decline was hastened by repeated defeats at the hands of **Brian Boru,** King of Munster. The principal events were:

807–832. First Serious Viking Raids. Subsequently the raiders occupied islands off the coast and in the Irish Sea. The Norse chieftain **Thorgest** made the first deep overland penetration into the interior of Ireland (832).

841. Vikings Seize and Fortify Dublin and Annagassan. Thorgest sailed up the Bann River into Lough Neagh; later up the Shannon to Lough Ree. By this time he was the ruler of half of Ireland. He was killed in battle in the interior of Ireland by **Mael Sechnaill,** King of Mide (845).

853. Olaf, the First Norse High King of Ireland.

862–879. Irish Resurgence in the North. Aed Findliath, Irish high king, drove the Vikings from northern Ireland, but was unable to prevent their expansion on the east coast.

914–920. Norse Expansion. The Vikings seized and fortified Waterford (914) and Limerick (920). **Niall Glundub,** Irish high king, was defeated and killed in an attack on Norse-held Dublin (919).

c. 965–975. Reign of Mathgamain, King of Munster. He defeated the Vikings at Limerick, capturing the city (968). He was defeated and killed by **Mael Muaid,** a rival Irish prince who seized the throne of Munster (975).

976. Rise of Brian Boru. The brother of Mathgamain, he defeated, killed, and succeeded Mael Muaid. He steadily increased his power and domains at the expense of Norse and Irish rivals.

999. Battle of Glen Mama. Brian defeated the Irish King of Leinster (who was allied with the Norse ruler of Dublin), forcing Leinster and Dublin to acknowledge his suzerainty. (For subsequent events in the career of Brian, see p. 315.)

The Vikings in Britain, 800–914

Beginning shortly after their inroads into Ireland, Norse raiders ranged along the coasts of Scotland and—somewhat later—England. By the middle of the 9th century they had seized many Scottish islands and had footholds in Scotland proper. After establishing bases on islands and peninsulas along the English coast, Danish rovers invaded and completely destroyed the kingdoms of Northumbria, Mercia, and East Anglia. In a few years they had seized and colonized all of Anglo-Saxon Britain except for Wessex. Alfred (the Great) of Wessex barely maintained his independence, and later drove the Vikings from Southern England. His victories over the Danes were due to a high order of battlefield leadership and administrative ability. He and his successors encouraged the development of the feudal warrior class of thegns (see p. 265). They effectively utilized crude fortifications for defense and as bases for invasions of the Danish lands of northern England; they built up an effective navy to meet and repel the Vikings at sea. During the 10th century the Anglo-Saxon struggle with the Danes was no longer a matter of Viking raiders against local inhabitants, but rather a more or less constant war between Southern and Northern England. The principal events of the Viking invasions of Britain were:

802–835. Norse Depredations Begin. Frequent raids against the coasts of Scotland and England, seizure of many Scottish islands. Viking settlements were established on the Scottish mainland (c. 835).

838. Battle of Hingston Down. Egbert of Wessex defeated a combined Viking-Welsh invasion force.

850. Viking Raids in Southeast England. The raiders sacked London and Canterbury, but were defeated by **Aethelwulf** of Wessex at **Ockley.**

851. First Viking Settlement in England. This was on the island of Thanet (near Margate). Soon afterward they established a permanent base on Sheppy Island at the mouth of the Thames (853).

865–874. Danish Conquest of Northumbria, Mercia, and East Anglia. The Danes decisively defeated the Northumbrians at **York** (867). Under their leader **Halfdan,** most of the Norsemen settled in the conquered regions. They repulsed **Aethelred** of Wessex and his brother Alfred in an effort to reconquer Mercia (868). **Edmund** (Saint) of East Anglia was defeated and killed at **Hoxne** (870).

870–871. Danish Invasion of Wessex. The English under Aethelred and Alfred won the first engagement at **Englefield** (December 870), were badly defeated at **Reading** (January 871), were victorious at **Ashdown** (January), defeated at **Basing** (January), fought two indecisive engagements (February), and were defeated again at **Marton** (March). After the death of Aethelred, Alfred was defeated at **Wilton** when retreating Danes turned upon disorganized English pursuers. A 5-year peace ensued.

876–878. Renewed Invasion of Wessex. The Danes captured Wareham (876) and Exeter (877). Blockaded in Exeter by Alfred, they sought terms and withdrew to Mercia. Soon after, **Guthrum,** the Danish leader, made a surprise attack on Alfred at **Chippenham** (January 878), smashing his army and forcing him to flee with a few survivors to a fort at Athelney, where he gathered a new force while using guerrilla tactics to harass the Danes. Then Alfred attacked, winning a decisive victory at the **Battle of Edington** (May 878). Guthrum sued for peace. The Danes agreed to abandon their efforts to conquer Wessex, but retained control of Northern England. It was during this war that Alfred began to build a navy.

884–885. Renewed War between Alfred and Guthrum. This was started by a Danish raid on Kent, which was repulsed. Alfred captured London (885) and in the ensuing peace expanded his holdings north of the Thames.

893–896. Danish Invasions of Kent and Wessex. A large Danish expedition was blockaded by Alfred immediately after landing. Danes from North England simultaneously invaded Wessex by land. Part of the shipborne raiding force broke out from Alfred's blockade. In confused fighting that ranged all over southern and western England, Alfred, assisted by his son **Edward** and by his great vassals, defeated both invading forces (893, 894). The following year the Danes sailed up the Thames and Lea rivers, but Alfred cut off their escape by building two forts at the mouth of the Lea and blocking the stream by a great boom. The Danes abandoned their ships and marched north to join their countrymen in Danish England.

905. Renewed Danish Invasion. An overland expedition reached the Thames, where it was repulsed by Edward (who had succeeded Alfred). Edward retaliated by a successful and destructive raid through much of Danish England, culminating in a major victory at **Tetlenhall** (910).

914. Viking Invasion from Brittany. The invading fleet sailed up the Severn, but was blockaded by Edward and forced to withdraw. Though much border fighting continued against the Danes in northern and eastern England, the initiative now passed to the Anglo-Saxons. Final victory came when Edward captured **Tempsford,** killing **Guthrum II,** Danish King of East Anglia (918). The Viking era in Britain was ended—though the Danish era was not.

The Vikings in Western Europe, 800–929

During the reign of Charlemagne the few Viking raids on the coast of Western Europe were dealt with effectively on land and sea (see p. 225). The depredations increased in intensity during the reign of **Louis the Pious** (son of Charlemagne), but burst like a storm over the Continent when the Vikings discovered that they could take advantage of the internal confusion caused by the dynastic struggles of the later Carolingians. For half a century they ranged over most of France, the Low Countries, and northern Germany, plundering, killing, and destroying, meeting little effective resistance save when the kings and great nobles temporarily halted their civil and dynastic wars to deal with the raiders. And though the Vikings preferred plunder to battle against organized opposition, they frequently and successfully engaged the best Frankish armies during the third quarter of the 9th century.

The height of Viking power and destructiveness came during the reign of inept Emperor **Charles the Fat,** during the darkest of the Dark Ages, and was marked by their vigorous and partially successful siege of Paris (885–886). But the forces they had set in motion among their victims—the development of effective feudal cavalry armies and the fortification of important places—now began to thwart the Norse raiders. In the latter years of the 9th century the cost of plundering often exceeded the rewards; Viking armies were as often defeated as they were victorious.

The last gasp of the Vikings was the invasion of northern France by **Rollo,** or Rolf, in the early years of the 10th century. Though generally successful, he found the opposition so strong that he was happy to accept an offer from **Charles the Simple,** King of France, to settle in the region around Rouen in return for his promise to be a faithful vassal. Thus was Normandy created, with Rollo as its first duke. Only once, after the death of Charles the Simple, did the Normans return to their old ways, when they made an unsuccessful raid through western France. By the end of the 10th century they had become completely assimilated; though unruly, they were the most faithful and most vigorous vassals of the kings of France. The principal events were:

810. Invasion of Frisia. Danish King **Godfred** was repulsed by forces of Charlemagne, who later made a treaty with Godfred or his successor (811).

811–840. Increasing Viking Raids. Against outlying islands and coastal regions of Western Europe. Their most dramatic exploits were the plundering of Dorstadt and Utrecht (834).

c. 840–c. 890. Viking Conquest of Frisia. This included most of modern Holland, from Walcheren Island to the mouth of the Weser River.

843. First Viking Settlement at the Mouth of the Loire River. Viking raiders also ravaged the Mediterranean coasts of Spain and France.

845. First Viking Raid on Paris.

851, 880. Plundering of Hamburg. In both cases the raiders sailed up the Elbe River.

852–862. The Vikings against Charles the Bald. They first defeated him at **Givald's Foss** (exact location unknown, 852). A few years later he unsuccessfully besieged the Viking base on the island of Oiselle (in the Seine River, 10 miles above Rouen, 858). Charles was again repulsed in a similar attack three years later (861). A peasant uprising in the Loire region failed in a vain effort to drive out the Vikings (859). A Viking raid through Paris and up the Marne River was frustrated by Charles, who blockaded the mouth of the Marne River, forcing the raiders to abandon their boats and to retreat overland to the sea (862).

863. Vikings Raid Deep up the Rhine Valley.

882. Battle of Ashloh (exact location unknown). After a setback at Saucourt (881) at the hands of **Louis III,** the Vikings defeated **Charles the Fat.**

885–886. Siege of Paris. The Viking army of about 30,000 was led by **Siegfried** and **Sinric.** This is generally considered the high point of Norse power in Western Europe. Using the fortifications begun by Charles the Bald, Count **Odo** (Eudes) of Paris, assisted by Bishop **Gozelin,** successfully defended the island city and its fortified bridgeheads for 11 months against fierce and skillful Viking assaults and siege operations. Unable to maintain a complete land and water investment of the city, which received a trickle of reinforcements and provisions by river and from relieving armies, the Vikings, however, were able to drive away relief armies led by Duke **Henry of Saxony** and by Charles the Fat. Although they were defeated near **Montfaucon** by Odo, they were still vigorously prosecuting the siege in September 886, when Charles the Fat bought them off by paying a large ransom and permitting them to ravage Burgundy (which did not acknowledge his authority) without interference.

886–887. Six Months' Siege of Sens. The Vikings under Sinric were again unsuccessful.

891. Battle of Louvain. Arnulf, King of the East Franks (or Germans), stormed the Viking fortified base and killed many of the defenders. There were no further deep Viking raids into Germany.

c. 896–911. Raids of Rollo in Northern France. These culminated in the Treaty of St. Clair-sur-Epte; Rollo accepted the region around Rouen as a fief from Charles the Simple to establish the Duchy of Normandy (911).

929. Norman Raid. The expedition through west-central France toward Aquitaine was defeated by King **Rudolph** at the **Battle of Limoges.**

The Vikings in Russia, c. 850–900

Origins of Norse influence in Russia are not clear. Apparently raiders from Sweden had penetrated far inland from the Baltic Sea by the middle of the 9th century. Soon after this, **Rurik,** a semilegendary leader of the Rus, or Varangian, tribe of Sweden, established himself as the ruler of Novgorod (c. 862), and a few years later Norse rule extended as far as Kiev (c. 865). By the end of the century (c. 880) Kiev was the capital of an extensive Varangian empire extending from the Gulf of Finland to the northern Ukraine and the Carpathians. Varangian raiders had already reached the Black Sea, and had made their first effort against Constantinople (865).

Denmark

c. 800. Danish War with Charlemagne. The Danish Kings **Gudfred** (d. 810) and **Hemming** (d. 812) won a series of clashes with Charlemagne's Frankish empire. The unified state they had created did not survive their deaths.

c. 930. Gorm Gains Danish Throne. This chieftain, bynamed "the Old," made himself King of Jutland, with his seat at Jellig.

c. 940–c. 985. Harald Blåtand (Bluetooth). Harald, Gorm's son, succeeded his father as king, and largely completed the work of unification. He was baptized, thereby introducing Christianity (c. 960). Harald meddled continually in Norway, opposing both **Haakon I** and **Harald II Gråfell** in his efforts to control that country.

c. 985–1014. Sweyn I Tveskaeg (Forkbeard). Sweyn, Harald's son, revolted against his father and slew him in battle, thereby gaining the kingdom. Sweyn led several attacks on England (see p. 276), leading to the outright conquest of that country (see p. 309). Sweyn feuded with **Olaf Tryggvason,** and after Olaf's defeat at the naval **Battle of Svolder** (1000) became *de facto* ruler of Norway.

Norway

c. 825. Rise of Regional States. Contemporary with the appearance of the first *things* (*tings*), or regional assemblies, was the rise of four regional states: Gulatingslag in western Norway, Frostatinglag around Trondheim Fjord (and farther north), one around Oslo Fjord, and another around Lake Mjosa.

c. 886. Battle of Hafrs Fjord. Sometime between 872 and 900, **Harald I Harfager** (Fairhair), the son of **Halfdan the Black,** in alliance with *jarls* from Gulatingslag and Frostatinglag, defeated his opponents near Stavanger and proclaimed himself king, reigning until his death (c. 930).

c. 930–935. Reign of Erik Bloodax. Erik, a son of Harald Harfager, received his nickname because he murdered or killed in battle seven of his eight brothers. He was deposed by rebellious nobles, and his sole living brother Haakon took the throne. Erik, having moved to England, later ruled in Northumbria (950, 952–954), and died at York (954).

c. 935–c. 961. Reign of Haakon I Den Gode (the Good). Raised in the court of English King Athelstan, Haakon returned to Norway and helped depose his brother Erik. His efforts to Christianize Norway met with limited success, but he was able to establish a national navy, and created law codes and effective administration throughout the kingdom. Erik's sons, sheltering in Denmark, launched repeated raids on Norway with the aid of King Harald I Blatand, and Haakon was killed in battle at the island of **Fitjar** in southwestern Norway (c. 961).

c. 961–c. 970. Reign of Harald II Gràfell (Greycloak). The eldest son of Erik Bloodax, he seized the throne after Haakon I fell at Fitjar. His rule was harsh, and his efforts toward forced Christianization aroused opposition. He killed **Haakon,** *jarl* of Lade, and 2 sub-kings of the Oslo region, to suppress opposition. Harald was killed in battle by **Haakon,** son of *jarl* Haakon (c. 970).

c. 970–c. 995. Reign of Haakon Siggurdsson den Store (the Great). Returning from exile in Denmark, he defeated Harald Grafell and became principal ruler of Norway. He supported Harald I Blatand against German Emperor Otto II, but revolted against Harald's Christianizing (c. 974). His rule grew harsh after 990, and Haakon's own followers murdered him soon after **Olaf Tryggvasson** invaded (995).

995–1000. Reign of Olaf I Tryggvasson. The great-grandson of Harald I Harfager, Olaf was the first Christian king of Norway. He was defeated and killed at the naval **Battle of Svolder** or Svalde (1000) by Eric, *jarl* of Lade, King Sweyn I of Denmark, and **Olaf Skötkonung** of Sweden, a tale oft retold in medieval Scandinavian sagas.

Sweden

c. 990–c. 1022. Reign of Olaf Skötkonung (the Taxgatherer). Olaf was the first national Swedish ruler, although his efforts at Christianization encountered resolute opposition from the pagans of Uppsala. Olaf also took part in the coalition against Olaf Tryggvason of Norway (see above). The allied victory of Svolder enabled Olaf to annex much of southeastern Norway.

Iceland

c. 870. First Norwegian Settlement. Although Iceland was earlier visited by Irish monks, the 1st permanent settlements were founded by Norwegian immigrants arriving between 870 and 930.

c. 930. Icelandic Commonwealth. About 40 priest-chieftains, called *godar,* formed the ruling class and exercised their authority through the annual midsummer assembly called the *Althing,* a combined parliament and law court, held at Thingvellir.

c. 999. In either 999 or 1000, according to Icelandic records, the *Althing* decided that all Icelanders would become Christians.

THE MAGYAR INVASIONS, c. 850–955

The Magyar tribe (related racially and linguistically to both Finns and Turks) inhabited the lower Don Basin in the early 9th century, where they were vassals of the Khazar Turks. Militarily they were typical Scythian light cavalry, with the bow their principal weapon. Soon after the middle of the century, the first recorded Magyar raid reached Frankish outposts in the middle Danube Valley. At this time the Magyars were under increasing pressure from the Pechenegs (or Patzinaks, a Turkish people living between the Volga and the Urals) who gradually drove them across the Dnieper and Dniester rivers to the lower Danube Valley. About 890, Magyar contingents were serving under Byzantine emperors in wars against the Bulgars, and under German King Arnulf against the Slavs of Moravia. After being badly defeated by a Pecheneg–Bulgar alliance, the Magyars, under their chieftain **Arpad,** migrated across the Carpathians into the middle Danube and Theiss valleys, driving out the Slavic and Avar inhabitants and establishing the Hungarian nation which has continued to the present.

From their new home the Magyars first began to raid their Slavic neighbors, then pushed farther into Germany and north Italy. As with the Vikings a few decades earlier, the Magyars raids were facilitated by the anarchy then existing in the Holy Roman Empire. The raids gave further impetus in eastern Germany to the process of feudalism already progressing rapidly farther west as a result of the Viking inroads.

During the early 10th century the Magyars ranged unchecked across Germany, Italy, and France. Their first setback came at the hands of German King **Henry the Fowler.** Their raids continued, but opposition became more effective because of the fortifications built by Henry and his son **Otto I,** and because of the steadily improving quality of the Holy Roman Empire's feudal cavalry. The last, most destructive, and most extensive Magyar raid came shortly after the middle of the 10th century, a national Hungarian effort to take advantage of civil war in Germany. Attempting to repeat this performance the following year, the Magyars were crushed by Otto in a fierce battle near Augsburg. From then on they were on the defensive against the Germans and never again threatened western Europe. The principal events were:

862. First Magyar Raid West of the Danube. Into the Frankish Ostmark.

862–889. Magyars Driven from Don Basin. They were pushed into Moldavia by the Pechenegs.

c. 895. Magyar Defeat by Pechenegs and Bulgars. Arpad led the Magyars and other future Hungarian tribes into the Middle Danube Valley (c. 896).

896–906. Establishment of Magyar Hungary. The Magyars drove the Slavs from the Danube and Theiss Valleys, destroying the Kingdom of Moravia (906).

899. First Magyar Raids into North Italy.

900. First Magyar Raids into Bavaria. These were followed by increasingly bold and destructive inroads (900–933).

910. Battle of Augsburg. The Magyars defeated the Germans (under the nominal leadership of King **Louis III,** The Child) by ambush after pretending to flee.

924. Raid through Germany and France. The Magyars swept through Bavaria, Swabia, Alsace, Lorraine, Champagne, back across the Rhine through Franconia to the Danube Valley.

926. Raid into North Italy and South France. The Magyars marched through Ventia and Lombardy, and were repulsed at the Pennine Alps (probably one of the St. Bernard passes) by **Rudolph** of Burgundy and **Hugh** of Vienne. Turning south, they crossed the Maritime Alps, raided across Provence and Sep-

timania to the Pyrenees, then returned, after further inconclusive engagements in the Rhone Valley against Rodolf and Hugh.

933. Battle of Riade (or Merseberg). Henry the Fowler defeated the Magyars near Erfurt. The Magyars did not seriously contest the field, apparently being surprised by the determination and vigor of their foes.

933–954. Declining Intensity of Raids into Germany. The result of fortifications in East Germany built by Henry and Otto. During this period the Bavarians began to raid back into Hungary (950).

942. Raid to Constantinople. The Magyars were bought off by the Byzantines.

954. The Great Magyar Raid. A force between 50,000 and 100,000 swept through Bavaria and into Franconia. **Conrad** of Lorraine, then in revolt against Otto I, made a treaty with them, helped them to cross the Rhine at Worms, and facilitated their movement into Lorraine. They then crossed the Meuse to devastate northeastern France, through Rheims and Châlons into Burgundy, then to Italy via the Great St. Bernard Pass, through Lombardy, then across the Carnic Alps to the Drava and Danube valleys.

955, August. Battle of Lechfeld. A Magyar force of about 50,000 men invaded Bavaria. It was besieging Augsburg when Otto I arrived with an army of about 10,000. Raising the siege, the Magyars accepted battle, making a surprise turning movement, which captured the German camp and drove one-third of Otto's army off the field. Assisted by his former enemy, Conrad of Lorraine, Otto repulsed the enveloping force, then charged the Magyar main body. He drove them off the field with heavy losses, captured their camp and its booty, then pursued vigorously for three days. Conrad was killed at the moment of victory. This ended Magyar depredations into Germany.

BRITAIN

During the early years of the 9th century, Wessex became the predominant kingdom of Anglo-Saxon England, under King **Egbert.** Shortly after the middle of the century Britain was struck by the Viking invasions (see p. 270). During the 10th century the successors of Alfred the Great slowly regained central and northern England from the Danes, and conquered Scotland by using their new navy. The end of the century, however, saw a new Danish invasion under King Sweyn I Tveskaeg (Forkbeard) (see p. 273). This was a Danish national effort, and thus not properly classifiable as a part of the Viking era.

802–839. Egbert of Wessex. He established supremacy in England following a crucial victory over **Beornwulf** of Mercia at **Ellandun** (825).

856–914. The Viking Invasions. (See p. 270.)

871–899. Reign of Alfred the Great. He established the basis of Anglo-Saxon feudalism.

899–924. Reign of Edward (the Elder, son of Alfred). At the outset Edward was forced to fight a sanguinary civil war with his cousin, **Aethelwold,** who allied himself with the Danes (c. 900–905). After repelling the final Viking invasions of southern England (see p. 271), Edward devoted his reign to the conquest of Danish England, which was more than half completed at the time of his death.

924–939. Reign of Aethelstan (son of Edward). He completed the conquest of remaining Danish holdings in northern England, and also received the homage of the Britons of Wales and of **Constantine III** of Scotland, thus unifying Britain under one ruler for the first time.

934–937. Invasion of Scotland. By land and sea Aethelstan subdued recalcitrant Constantine, who had formed an alliance with some resurgent Welsh leaders and with Norse chieftains based in Ireland. After three years of hard fighting, Aethelstan won a decisive victory over the allies at the **Battle of Brunanburgh** (southeast Dumfrieshire) to reestablish his control over all Britain (937).

954. Battle of Stanmore. Eric Bloodax (exiled King of Norway; see p. 273) was defeated and killed in an effort to seize northern England.

978–1016. Reign of Ethelred the Redeless (the Uncounseled, or Unready). In the early years of his reign, a revival of minor Norse raids (c. 980) soon led to full-scale Danish invasion under King Sweyn I Tveskaeg (Forkbeard) (991). One of a number of Danish victories that year was the **Battle of Maldon** in Essex. Repeated Danish invasions resulted in the defeat and death of Aethelred and the conquest of England by Sweyn and his son Canute (see pp. 273 and 309).

CAROLINGIAN EMPIRE, 814–887

The decline of the Carolingian Empire began immediately after the death of Charlemagne. His son, Louis the Pious, was ineffectual as a ruler and as a soldier. Viking raiders became increasingly bold, and while Louis was alive his sons were engaged in civil war as they endeavored to stake out their inheritance claims. Immediately after his death the civil war became more violent as the two younger sons—**Louis the German** and **Charles the Bald**—combined to fight a bloody battle at **Fontenay** against their elder brother, **Lothair I.** A temporary settlement was achieved through the subsequent **Treaty of Verdun,** in which Lothair was recognized as emperor and as ruler of Italy, Burgundy, and Lotharingia (a strip of land extending north from the Jura Mountains through the Rhine and Meuse valleys to the North Sea), while Louis obtained the German lands to the east, and Charles held the future French kingdom to the west.

For nearly 50 years, near anarchy pervaded the Carolingian domains, resulting from the combination of continuing dynastic civil wars and the depredations of the Vikings. Lothair soon disappeared from the scene, and was succeeded as emperor first by his son **Louis II,** and then (briefly) by his brother Charles the Bald. Following Charles's death, anarchy became absolute, and the situation was hardly improved by the temporary unification of most of the Carolingian domains under **Charles the Fat**—who was deposed because of his ineptitude in dealing with internal problems and his timidity in the face of the Vikings at Paris (see p. 272).

For all practical purposes this ended the Carolingian Empire, though Carolingian descendants continued to reign for more than two decades in Germany and (nominally) for a century in France. As central authority declined, the power of the nobles increased. In Germany, in particular, the old tribal organizations became virtually independent duchies. Moslem raiders from North Africa took advantage of the situation to ravage and to occupy extensive regions of Sicily and south and central Italy. Rome was sacked (846; see p. 290). The principal events were:

814–840. Reign of Louis I the Pious.

840–855. Reign of Lothair I (as Emperor). A period of continual strife between Lothair and his brothers Louis and Charles.

841. Battle of Fontenay (Fontenat, near Sens). Louis and Charles defeated Lothair, who was forced to sue for peace.

843. Treaty of Verdun. Charlemagne's empire was divided among his three grandsons.

843–876. Reign of Lewis the German (over the East Frankish Kingdom, Germany). During most of this period he was at war with his brothers (Lothair and Charles the Bald), with Lothair's sons (Louis II, Lothair, and Charles) and later with his own sons (Carloman, Louis, and Charles). He was possibly the most able ruler and soldier-descendant of Charlemagne.

843–877. Reign of Charles the Bald (over the West Frankish Kingdom, France). By his policy of building extensive fortifications, Charles laid the foundation for the eventual neutralization of the Vikings in France (see p. 272). In shifting alliances, Charles was almost constantly at war with his brothers and nephews. By clever diplomacy he outmaneuvered Carloman (son of his brother Louis) to become Emperor (875), in return for which the elderly Louis invaded and ravaged France. After Louis' death the next year, Charles endeavored to annex Germany, but was decisively defeated by his brother's son (also named Louis) at the **Battle of Andernach** (October 876).

855–875. Reign of Louis II (as Emperor). Son of Lothair, he participated in the confused Carolingian family wars with and against his uncles, brothers, and cousins. He was unable to recover Lotharingia, which his uncles Louis and Charles had seized and shared after the death of his father. He was successful in campaigns against the Moslem invaders of southern Italy (866–875). He made an alliance with Byzantine Emperor **Basil I** for joint land and sea operations against Bari (871–875; see p. 285).

870. Treaty of Mersen. Lotharingia was divided between Louis the German and Charles the

Bald, thus bringing about a temporary settlement of their violent disputes. Central Lotharingia (or Lorraine) continued to be disputed between them, however, until the death of Louis.

879. Independence of Lower Burgundy (Provence). This was under **Boso** during the anarchic period which followed the death of Charles the Bald.

884–887. Reign of Charles the Fat (as Emperor). All of Charlemagne's empire save Burgundy was briefly united under him. He was deposed by an assembly of nobles after his shameful deal with the Vikings at Paris (see p. 272).

France, 888–1000

After the deposition of Charles the Fat, Count **Odo** (defender of Paris against the Vikings; see p. 272) was elected King of the West Franks. During the last five years of his reign he was engaged in a generally unsuccessful war against Charles III, "The Simple," nephew of Charles the Fat, who claimed the throne. Charles III finally became sole ruler on the death of Odo. A group of nobles later rebelled against Charles, electing Count **Robert** of Paris, brother of Odo, as king. Charles was defeated and deposed. Robert, killed during the war, was replaced by **Rudolph,** Duke of Burgundy (a Carolingian, not to be confused with Rodolph, founder of Upper Burgundy; see p. 279). Following the reign of Rudolph, the Carolingians declined in authority, while the great vassals became practically independent.

After the death of the last Carolingian ruler of France, the nobles elected **Hugh Capet,** Count of Paris, as king. By the time of his death, he had established the basis for a strong monarchy. The important events were:

887–898. Reign of Odo.

893–923. Reign of Charles III. One of the most important events was the **Treaty of St. Clair-sur-Epte** (911) between Charles and Rollo, the Viking, which resulted in establishing the Duchy of Normandy (see p. 272).

921–923. Robert's Revolt against Charles. Charles was defeated and deposed, but Robert, Count of Paris, was killed in the climactic **Battle of Soissons** (923).

923–936. Reign of Rudolph. Marked by his victory over the Normans and Aquitanians at the **Battle of Limoges** (see p. 273).

942. Franco–German War. Between **Louis IV** of France and **Otto I** of Germany (see p. 279).

947–948. Civil War. This was between Louis IV and **Hugh the Great,** Count of Paris. Initially successful, Hugh was forced by Otto I of Germany to make peace and to restore Louis to the throne (see p. 279).

987–996. Reign of Hugh Capet. It began with civil war between Hugh and his rival, **Charles of Lorraine,** in which Hugh was victorious.

Holy Roman Empire (Germany, Italy, Burgundy), 887–1000

Arnulf, Carolingian King of Bavaria and Lombardy, ringleader of the uprising which deposed Charles the Fat, was elected King of the East Franks. His victory over the Vikings at the **Battle of Louvain** (see p. 272) signaled the decline of the Norsemen in Germany. For much of his reign Arnulf was engaged in a struggle with the Slavs of Moravia and Bohemia. He twice came to the assistance of the Pope against unruly Italian nobles, and in return was crowned Emperor. Central authority in Germany declined during the reigns of **Louis the Child** and **Conrad I,** but revived markedly under **Henry the Fowler,** able soldier, diplomat, and administrator. His son, **Otto I, the Great,** became virtual creator of a new Holy Roman Empire. The most important events were:

887–899. Reign of Arnulf.

888. Independence of Upper Burgundy. Established by **Rodolph,** who repulsed Arnulf's attempts to reestablish German control.

890–975. Moslem Footholds in Southern France. Raiders seized and held the southern coast of Provence (Lower Burgundy).

891. Battle of Louvain. Arnulf's victory over the Vikings (see p. 272).

892–893. Slavic Invasion of Bavaria. With Magyar assistance, Arnulf repulsed invaders under **Sviatopluk** of Moravia.

894–896. Arnulf's Successful Expedition to Italy. He was crowned Emperor (896).

899–911. Reign of Louis III, the Child.

900. First Magyar Raids into Germany. (See p. 275)

911–918. Reign of Conrad I. (The former Duke of Franconia.) Plagued by widespread civil war through Germany and Italy.

919–936. Reign of Henry I, the Fowler. (The former Duke of Saxony.) Henry waged successful war against **Arnulf,** Duke of Bavaria, claimant to the German throne (920–921).

924. Truce with the Magyars. This was in return for payment of tribute for 10 years. During that time Henry built up the defenses of Eastern Germany (see p. 275).

933. Battle of Riade. Victory of Henry over the Magyars (see p. 275).

936–973. Reign of Otto I, the Great. During the early years he successfully fought two civil wars to assure his succession: The first was against his half-brother **Thankmar** (938–939). The second was against a group of rebellious nobles (including his brother **Henry**), supported by **Louis IV** of France (939–941). Otto won the battles of **Xanten** (940) and of **Andernach** (941). His victories brought Lorraine definitely under German control and increased his central authority.

942. Invasion of France. Otto's purpose was to punish Louis IV for the support given the rebels. Peace was quickly made, and Otto withdrew.

944–947. Bavarian Revolt. Otto was defeated by **Bertold,** Duke of Bavaria, at the **Battle of Wels.** Later, combining diplomacy with military action, Otto gained control over Bavaria (947).

948. Invasion of France. Otto was supporting his brother-in-law, Louis IV, imprisoned by Hugh the Great, Count of Paris. Otto captured Rheims; Hugh submitted and restored Louis to the throne.

950. Invasion of Bohemia. Otto defeated Duke **Boleslav** and forced him to accept his suzerainty.

951–952. Otto's First Expedition to Italy. Upon request of **Adelaide,** widow of King **Lothair** of Italy, Otto defeated rival claimants, crowned himself King of the Lombards, and married Adelaide.

953–955. Rebellion in Germany. The revolt was led by Otto's son, **Ludolf,** assisted by **Conrad,** Duke of Lorraine, and **Frederick,** Bishop of Mainz. After initial defeat and capture by the rebels, Otto escaped, then defeated the rebels, his victory culminating with the capture of the rebel stronghold of Regensburg.

955. Victory over the Magyars at the Battle of Lechfeld. (See p. 275.)

955. Battle of the Recknitz. Otto defeated the Slavic Wends.

961–964. Otto's Second Expedition to Italy. Upon the appeal of Pope **John XII,** Otto led his second expedition to Italy against **Berengar II,** King of Italy. He defeated Berengar and forced him to vassalage (961). Otto was crowned as Holy Roman Emperor by Pope **John XII** (962). Soon after this he deposed John and appointed a new Pope, **Leo VIII** (963).

966–972. Otto's Third Expedition to Italy. He suppressed a revolt in Rome, where **Benedict V** had deposed Leo VIII. Since Leo had died, Otto appointed another new Pope, **John XIII.** He then marched to South Italy and operated successfully against the Saracens and Byzantines.

973–983. Reign of Otto II. This began with five years of violent civil wars, in which **Henry the Wrangler** of Bavaria and Boleslav of Bohemia were the main rebels (973–978).

978–980. War with France. A dispute over Lorraine led to an invasion of Germany by **Lothair** of France. The French were repulsed after briefly occupying Aachen. Otto invaded France and besieged Paris. Forced to withdraw due to an epidemic in his army, he was harassed severely by the French during the withdrawal. Lothair abandoned his claim to Lorraine.

981–983. Expedition to Italy. Otto was disastrously defeated by a Moslem-Byzantine alliance, being repulsed from **Crotona** while his fleet was crushed nearby at the **Battle of Stilo** (982), and barely escaped with his life.

983–1002. Reign of Otto III (age 3). During his minority, his vigorous mother, **Theophano,** suppressed a number of revolts led by Henry the Wrangler and others.

996. Expedition to Italy. Otto appointed a new Pope, **Gregory V,** and was crowned emperor.

998. Expedition to Italy. Otto suppressed an antipapal revolt by **John Crescentius,** who was captured and killed in the castle of **St. Angelo** at Rome. Otto died unexpectedly, 4 years later, while attempting to suppress still another revolt in Rome (1002).

SPAIN

During the 9th century the Christian rulers of the Kingdom of the Asturias expanded their control over all Galicia (Northwest Spain). For most of the following century, however, the pendulum swung the other way, under pressure from the resurgent Omayyad caliphs of Cordova. Frontier warfare between Christians and Moslems was literally continuous. The principal events were:

796–822. Reign of al-Hakam I, Emir of Cordova. There were recurrent revolts, the most serious being in Cordova (805, 817) and Toledo (814).

792–842. Reign of Alfonso II of the Asturias. He began the advance of the Christians from the northern mountains toward the central plains of Spain, until stopped by **Abd-er-Rahman II** (see below).

801. Capture of Barcelona by Charlemagne. (See p. 227.)

822–852. Reign of Abd-er-Rahman II. He defeated Alfonso II and drove the Franks back in Catalonia. He suppressed a revolt of Christians and Jews in Toledo (837).

852–886. Reign of Mohammed I. He conducted successful expeditions into Galicia and Navarre.

866–910. Reign of Alfonso III. The greatest of the kings of Asturias. Before his death Alfonso had extended the southern frontier of Christian Spain to the Duero (Douro) River. His successor, **Garcia** (910–914), moved the capital to León, and the kingdom became known by that name.

903. Moslem Capture of the Balearic Islands. The Franks were defeated by **Isam-al-Khamlani.**

912–961. Reign of Abd-er-Rahman III. The greatest of the Omayyads of Spain, he reunited Moslem Spain and took the title of Caliph of Cordova (929). He was generally successful in wars with León and Navarre, forcing the Christian kings to pay tribute.

914–924. Reign of Ordono II of León. His occasional military successes prevented threatened Moslem conquest of the kingdom.

924–950. Reign of Ramiro II of León. He regained the initiative for the Christians by great victories over Abd-er-Rahman at **Simancas** (934) and at **Zamora** (939), where he drove off a besieging army. However, during the latter part of his reign the eastern provinces of the kingdom, known as Castile, under the Counts of Burgos, became virtually independent.

961–976. Reign of al-Hakam II (son of Abd-er-Rahman). This able ruler continued to enjoy success over the Christians (thanks largely to his excellent general **Ghalib**). Morocco was conquered (973).

976–1013. Nominal Reign of Hisham II of Cordova. His weakness resulted in the establishment of a military dictatorship under the powerful palace chamberlain, **Ibn Abi Amir al-Mansour** (981–1002), who seized power by defeating and killing his father-in-law, Ghalib, in a brief civil war. Al-Mansour inflicted decisive defeats on León, Navarre, and Barcelona, capturing and sacking León (988) and ravaging most of the kingdom. He also suppressed rebellions in Morocco.

EASTERN EUROPE

BULGARIA

During the 9th century and first half of the 10th, Bulgarian power steadily increased, despite some setbacks at the hands of the Byzantines, and despite the loss of holdings north of the Danube to the Magyars and Pechenegs. After the middle of the 10th century, the Bulgars suffered crushing defeats by the Byzantines and Russians. Briefly resurgent under **Samuel,** the Bulgars were engaged in a disastrous war with Byzantium at the close of the century (see p. 327). The principal events were:

808–814. Reign of Krum. He fought successfully against the Byzantines (808–813; see p. 000).

817. Battle of Mesembria. Victory of **Leo V** over **Omortag,** son of Krum (see p. 285).

827–829. Bulgar Raids into Croatia and Pannonia.

852–889. Reign of Boris I. In the early years he failed in attempts to expand into Serbia and Croatia. Under Byzantine duress, he accepted Christianity (866). Bulgaria prospered under his rule. He abdicated to enter a monastery (889).

893. Return of Boris. Boris came out of retirement to subdue a revolt against his inept son, **Vladimir.** He deposed Vladimir in favor of his second son, **Symeon,** then returned to his monastery.

893–927. Reign of Symeon. Repeated and generally successful wars with the Byzantine Empire (894–897, 913–924). Preoccupied with ambitions to capture Constantinople, Symeon lost the outlying semi-autonomous provinces of Transylvania and Pannonia to the Magyars, and Wallachia to the Pechenegs. Offsetting these losses were his conquest of Macedonia, Thessaly, and Albania from the Byzantines (c. 914) and of Serbia (918–926). The Byzantines paid him tribute.

927–969. Reign of Peter (son of Symeon). He had to meet frequent Magyar and Pecheneg raids.

967–969. Byzantine and Russian Invasion. Nicephorus Phocas led a Byzantine army overland in an alliance with Prince **Sviatoslav,** who came by sea. The Russians captured the new Bulgarian King **Boris II.** Bulgaria was overrun.

969–972. Russian–Byzantine War over Bulgaria. The result of a dispute over division of

Bulgaria (see p. 286). The victorious Byzantine Empire annexed all of Eastern Bulgaria up to the Danube.

976–1014. Reign of Samuel. A Bulgarian noble in the West, he reestablished the Bulgarian state (976).

981–996. Resurgence of Bulgaria. Samuel defeated a Byzantine invading army under **Basil II** near Sofia (981). After periodic raids to the south, he invaded Thessaly, to capture Larissa and Dyrrhachium (986–989). Taking advantage of internal discord in the Byzantine Empire, also involved in war with the Fatimids, Samuel began to extend his control over Eastern Bulgaria (989–996) and also northwestward into Serbia.

996–1014. Conquest of Bulgaria by Basil II. (See p. 302.)

SERBIA

The Serb tribes coalesced as a kingdom toward the end of the 8th century. From its earliest years Serbia suffered constant internal turmoil, while at the same time fighting off incessant encroachments by the neighboring Bulgars. The Serbs allied themselves with the Byzantines against Bulgaria and acknowledged the suzerainty of Constantinople (c. 870). Despite this, practically all of Serbia was conquered by Symeon of Bulgaria (918–926; see p. 281). Part of Serbia regained its independence under Prince **Chaslav** (931), but he was unable to establish a truly unified state. Most of Serbia was again conquered by the Bulgars, under Tsar Samuel (c. 989; see above), and remained under Bulgarian control until early in the 11th century.

CROATIA, C. 900–1000

The various Croat and Slovene tribes had not yet coalesced into a state when they were conquered by Charlemagne, late in the 8th century (see p. 225). During the 9th and early 10th centuries Croatia was buffeted by tides of warfare and migration, and was at different times dominated by Franks, Moravians, Byzantines, Bulgars, and Magyars. A true Croatian state, under the nominal suzerainty of the Byzantine Empire, finally appeared under the leadership of King **Tomislav** (c. 925). **Drzislav** (969–997) assisted Byzantine Basil **II** in his wars against the Bulgars, and in return was granted recognition as an independent monarch.

BOHEMIA, C. 900–1000

Late in the 9th century the wild Slavonic tribes of Bohemia became Christianized, and coalesced into a state, initially subject to Moravia. After the destruction of Moravia by the Magyars, Bohemia became independent. After several decades of constant warfare, during the reign of **Boleslav I** (929–967) Bohemia was forced by Otto I to accept German suzerainty (950). Boleslav led a contingent in the German Army which defeated the Magyars at the **Battle of Lechfeld** (955; see p. 275). At the very end of the century Bohemia lost some of its territory to Poland (999–1000; see p. 283).

POLAND, C. 960–1000

Poland was a unified, pagan Slavonic state when it first came into contact with the Germans (963). Its first historical ruler was **Mieszko I** (c. 960–992), who expanded his frontiers westward as far as the Oder River, thus arousing German alarm. After a decade of warfare, Mieszko voluntarily accepted German suzerainty (973) in order to accelerate the civilization of his country. His son **Boleslav the Brave** (992–1025) continued the process of organizing and modernizing the Polish state, and had ambitions of uniting all of the western Slavs. One of the leading soldiers of his time, Boleslav conquered east Pomerania (994), then defeated the Bohemians to conquer Silesia, Moravia, and Cracow (999). His military success continued in the following century (see p. 325).

RUSSIA, C. 900–1000

Late in the 9th or early in the 10th century, the two Varangian states of Kiev and Novgorod were united into a single Russian state by Prince **Oleg,** who had probably led the original Norse expedition to Kiev. He obtained a commercial treaty with the Byzantine Empire (911), renewed by his son **Igor.** Igor's son, **Sviatoslav,** had ambitions which led him to foreign adventures that eventually cost his life. The son of Sviatoslav, **Vladimir the Saint,** was the last of the Vikings and the first of the Christian Russians. His early years were spent in civil and foreign wars; after his conversion to Christianity he appears to have become truly a man of peace—but he retained both the vigor and the will to protect his people from foreign invaders. The principal events were:

c. 880–912. Reign of Prince Oleg. Creator of the Russian state.

912–914. Reign of Prince Igor.

964–972. Reign of Prince Sviatoslav. He brought greater unity to the Russian state, and defeated the Khazars on the lower Volga (965).

967–972. Sviatoslav's Wars in Bulgaria. (See pp. 281 and 286.)

967–968. Pecheneg Invasion of Russia. The threat to Kiev forced Sviatoslav to return from Bulgaria. After repulsing the Pechenegs (968), he immediately returned to Bulgaria, where he fought successfully against the Bulgarians, but was disastrously defeated by the Byzantines.

972. Death of Sviatoslav. He was defeated and killed by the Pechenegs while returning to Kiev from Bulgaria.

972–980. Dynastic Struggle. The three sons of Sviatoslav—**Yaropolk, Oleg,** and **Vladimir**—vied for control. Vladimir went back to Scandinavia to attract Vikings to his cause (977–978). Returning, he subdued the principality of Polotsk, and regained control of Novgorod (978–979). He then defeated and killed his brother Yaropolk, captured Kiev, and established himself as the unquestioned ruler of all Russia (980).

981–985. Foreign Wars of Vladimir. He conquered Chervensk (modern Galicia, 981), then overran the heathen tribes between Poland and Lithuania (modern Byelorus, 983). He conquered all or part of the Bulgar state on the Kama River, east of the Volga (985).

988. Conversion of Vladimir to Christianity.

EURASIA AND THE MIDDLE EAST

THE BYZANTINE EMPIRE

There was an amazing revival in the strength, stability, and prosperity of the Byzantine Empire. Byzantine generals were generally successful in repelling the frequent foreign inroads—the Caliphate remaining the most important and most dangerous foe. Border fighting was incessant

Byzantine themes, c. 950

on the eastern frontier. Toward the end of the period a series of outstanding soldier-emperors, and their able subordinates, deliberately expanded the imperial frontiers in the Balkans, Asia Minor, Syria, Armenia, Italy, and the Mediterranean islands. This was not a reversal of the traditionally defensive strategic policy of the empire, but was rather a logical implementation of that policy, in consonance with the growing strength of the empire in contrast to the gradual weakening of the Moslem threat. Expansion was limited to areas which, in previous centuries, had been traditionally Byzantine. The new frontiers were always carefully selected with two main considerations in mind: for their natural defensive strength (mountains, rivers, or deserts), and to protect by coverage and depth the vital regions of the empire and the traditional invasion routes.

Decline of the Byzantine Navy during the latter 8th and early 9th centuries was corrected by the energetic and gifted emperor **Basil I,** who also revitalized the army and began the trend of imperial expansion on land. A slight recession in the progress of imperial power about the beginning of the 10th century was reversed dramatically by the military brilliance of the emperors **Nicephorus Phocas, John Zimisces,** and **Basil II.** The principal events were:

802–811. Reign of Nicephorus I. He was engaged in almost constant war with the Caliph **Harun al-Rashid** (see p. 252). Able Nicephorus was handicapped in the prosecution of this war by various conspiracies and a revolt led by the general **Bardanes.** He made peace with Charlemagne—whom he recognized as Emperor of the West—by treaties delineating the frontiers between the two empires in Italy and Illyria (803, 810). After Harun's army captured **Heraclea Pontica** (Eregli) by assault (806), Nicephorus sued for peace.

809–817. War with Krum of Bulgaria. Nicephorus personally led his army into Bulgaria, with consistent success. After capturing Krum's capital, Pliska, he was defeated and killed in a surprise Bulgarian night attack in a nearby mountain pass (811).

811–820. Reigns of Stauracius, Michael I, and Leo V. The Bulgarians were generally successful in subsequent operations against weak Stauracius (811) and Michael I (811–813). Disgusted by the ineptitude of Michael, his general, **Leo the Armenian,** refused to support him in the **Battle of Versinikia** (813), won by the Bulgars. Krum took Adrianople and advanced to the walls of Constantinople. Leo revolted against Michael, usurped the throne as Leo V, and then was consistently victorious. He defeated **Omortag,** son of Krum, at the **Battle**

of Mesembria (817), forcing the Bulgarians to agree to a 30-year peace.

820–829. Reign of Michael II. Called "The Stammerer," he was a general who had conspired against Leo, and who was made emperor after Leo was assassinated. His fellow general **Thomas** led a revolt which collapsed after two unsuccessful attempts to capture Constantinople (822–824). During his reign the Arabs took advantage of Byzantine naval decline by conquering Crete (825–826) and invading Sicily (827).

829–842. Reign of Theophilus. This was marked by intensified but inconclusive war against the Caliphate. Byzantine armies five times invaded the Abbasid dominions, but the Empire was also frequently ravaged by Moslem raids and invasions. In his most successful invasion, Theophilus reached the Euphrates in northeastern Syria, where he captured and sacked the towns of Samosata and Aibatra (837). Next year, however, the Caliph **al-Mu'tasim** led a great force into Anatolia, defeating Theophilus in the **Battle of Dasymon** (or Anzen) on the Halys River. The Moslems then captured and sacked Amorium (after a long siege) and Ancyra before retiring to Syria (838).

842–867. Reign of Michael III (age 3). Under the influence of his uncle, **Bardas,** who became virtual dictator, Michael successfully renewed the war against the Caliphate (851–863; see p. 287).

c. 850–1000. Periodic Paulician Revolts. This heretical Christian sect in northeastern Anatolia was sternly persecuted by the Orthodox Byzantine emperors.

865. Arrival of the Varangians (Russians). A pillaging expedition reached the Bosporus and threatened Constantinople.

866. Invasions of Bulgaria. Michael and Bardas forced the conversion of the Bulgarian King **Boris I** to Christianity.

867–886. Reign of Basil I. Known as "the Macedonian," he was a peasant by birth. He overthrew Michael to establish the most powerful dynasty of Byzantine history.

871–879. Renewed War with the Caliphate.

Basil was generally successful; he moved the frontiers eastward to the Euphrates after a decisive victory at **Samosata** (Sam-sat; 873).

875–885. Byzantine Expansion in South Italy. The Moslems were driven from **Bari** (875) and **Tarentum** (880), and Calabria was reconquered (885). Basil's simultaneous efforts to drive the Moslems from Sicily, however, were not successful. There was a rare instance of collaboration between the Byzantine and Holy Roman empires in the siege of Bari (871–875); the Byzantine fleet blockaded the port, while a land army, under Emperor Louis II, conducted the siege by land (see p. 277).

880–881. Eastern Mediterranean Secured. Basil's reorganized navy smashed Moslem pirates.

886–912. Reign of Leo VI. Known as "the Wise," he was author of the *Tactica* (see p. 233). He continued Basil's wars against the Moslems in Sicily and South Italy.

889–897. Bulgarian Wars. Despite early success and an alliance with the Magyars, the Byzantines were generally unsuccessful (see p. 281).

912–959. Reign of Constantine VII (age 7). Real authority was exercised by his father-in-law, the admiral **Romanus,** at first regent, later coemperor (919–944).

913–924. Renewed Wars with Bulgaria. Symeon repeatedly threatened Constantinople, temporarily captured Adrianople (914), and gained a major victory at the **Battle of Anchialus** (near Bargas, 917).

915. Battle of the Garigliano. The Byzantines crushed a Moslem threat to their hold on south Italy.

920–942. Northeastern Campaigns of John Kurkuas. This general captured Theodosiopolis (modern Erzerum), east of the upper Euphrates (928) and Melitene (Malatya, 934). He brought Byzantine authority once again to the upper Tigris.

924. Battle of Lemnos. The Byzantine Navy crushed the power of Moslem pirate **Leo** of Tripoli, who had threatened control of the eastern Mediterranean for 20 years.

941. Russian Raid to the Bosporus. Defeated by the Byzantine navy.

956–963. Eastern Campaigns of Nicephorus Phocas. This general drove Moslem invaders from eastern Anatolia, then raided deep into Syria. He invaded and reconquered Crete (960–961). Following this he invaded Cilicia and Syria, temporarily capturing Aleppo, while his brother, **Leo Phocas,** repelled a Moslem invasion of Anatolia (963; see p. 288).

963–969. Reign of Nicephorus Phocas (Nicephorus II). He was coemperor with the infant sons of **Romanus II** (959–963): **Basil II** and **Constantine.**

964–969. Eastern Conquests. Nicephorus invaded Cilicia, captured **Adana** (964) and **Tarsus** (965), completing the conquest of the province the following year. Meanwhile, his general **Nicetas** invaded and conquered **Cyprus** (965). Nicephorus then invaded Upper Mesopotamia and Syria, capturing **Antioch** and **Aleppo,** forcing the Caliphate to sue for peace (969).

966–969. War with Bulgaria. (See p. 281)

966–967. Failure in Sicily. Nicetas was repulsed by the Fatimid Moslems.

967–969. War against Otto. Making peace with the Fatimids, Nicephorus entered an alliance with them against Otto I, who had been encroaching on Byzantine and Moslem territory in south Italy. Otto reduced both Moslem and Byzantine holdings in south Italy (see p. 280).

969–976. Reign of John Zimisces. He assassinated his uncle, Nicephorus, to become coemperor with Basil and Constantine, and maintained control by suppressing the insurrection of **Bardas Phocas** (971).

969–972. War with Russian Prince Sviatoslav in Bulgaria. John defeated the Russians near Adrianople at the **Battle of Arcadiopolis** (970). While the Byzantine navy drove the Russian fleet away and blockaded the mouth of the Danube, John pursued Sviatoslav to the Danube, defeated him again in the **Battle of Dorostalon** (971), then invested the Russians on the land and river sides of Dristra with his army and navy. After a siege of two months Sviatoslav was forced to submit and to abandon Bulgaria (972; see p. 282).

973–976. War with the Moslems. John eliminated all Moslem holdings west of the middle Euphrates River, and extended Byzantine control south and east in Syria. After capturing **Damascus,** he advanced as far south as Jerusalem, where he was halted by Fatimid resistance (976). He died before he could resume the offensive.

976–1025. Reign of Basil II. Initially under the regent, **Basil Paracoemomenus,** until Basil seized full power (985).

976–979. Revolt of Bardas Skleros. This was suppressed mainly because of the victory of loyal general Bardas Phocas at the **Battle of Pankalia.**

981. Unsuccessful Invasion of Bulgaria. (See p. 282.)

987–989. Rebellion of Bardas Phocas and Bardas Skleros. The rebels briefly threatened Constantinople (988). Phocas was then defeated by Basil at the **Battle of Abydos** (989). Phocas died shortly thereafter, and the revolt collapsed.

995–996. War in Syria. The Fatimid Caliphate in Egypt began encroaching on Byzantine Syria. In two campaigns Basil brought all Syria under control.

996–1018. War with Bulgaria. (See p. 282.)

MOSLEM DOMINIONS

The Abbasid Caliphate and Syria

The Abbasid Caliphate reached the height of its glory in the first half of the 9th century. But by the end of the period the Abbasids had lost almost all of their temporal power through internal dynastic struggles, constant conflicts between the rival Moslem principalities, and innumerable foreign wars along the northern and eastern frontiers of Islam. They retained considerable religious prestige as the lineal successors of Mohammed—though even this was disputed by rival Caliphates at Cairo and Cordova. The principal events were:

786–809. Reign of Harun al-Rashid. (See p. 252.)

803–809. Renewed Byzantine War. Nicephorus I broke the treaty between al-Rashid and Irene (see p. 252). Al-Rashid led invading armies across the Taurus three times (803, 804, 805–807), defeating Nicephorus in several engagements, of which the most decisive was the **Battle of Crasus** (in Phrygia, west-central Anatolia, 805). While Harun was capturing numerous cities in Anatolia, his fleets ravaged Rhodes and captured Cyprus (805–807). Nicephorus, however, rallied, seized the offensive, and had recovered much lost territory when (in 808) Harun was forced to go to Khorasan (see below).

806–809. Rebellion in Khorasan. Harun died while leading his armies there.

809–813. Civil War. Between Harun's sons: **al-Amin,** the new caliph, and **al-Ma'mun.** The Khorasanian rebels, led by the general **Tahir ibn Husain,** sided with al-Ma'mun. Tahir defeated al-Amin's troops, marched to Mesopotamia, capturing Baghdad (811), after a siege of nearly two years. Al-Amin was killed while attempting to escape.

813–833. The Reign of Al-Ma'mun. Despite internal discord and revolt (814–819), al-Ma'mun restored relative stability. The remainder of his reign was one of the most splendid of the Abbasid Caliphate.

816–837. Khurramite Revolt. This sect in Azerbaijan, under their leader **Babek,** received considerable Byzantine support; they raided constantly into Persia and Mesopotamia. Babek was finally defeated by Abbasid general **Afshin** (835–837).

821. Autonomy of Khorasan. Under Tahir, who had been appointed governor.

825–826. Conquest of Crete. By Arab refugees expelled from Spain by the Omayyads. It became a Moslem pirate base. (See p. 285.)

829–833. Byzantine War. Al-Ma'mun repulsed Byzantine invasions of Syria, and in turn led successful expeditions into Anatolia (see p. 285). During this time he also crushed continuing disorders in Egypt (832).

833–842. Reign of Abu Ishak al-Mu'tasim. The last strong Abbasid caliph, he created a new imperial bodyguard composed mainly of Turkish slaves and mercenaries. He suppressed a revolt in southern Mesopotamia (834).

837–842. Renewal of the Byzantine War. (See p. 285.) Al-Mu'tasim's plan to besiege Constantinople was frustrated by the destruction of his fleet in a storm (839).

842–847. Reign of Harun Al-Wathik. He sent the imperial guard, under Turkish general **Bogha,** to subdue an uprising in Arabia (844–847).

847–861. Reign of Ja-far Al-Mutawakkil. He was usually influenced by the Turkish guards, who eventually murdered him.

851–863. Renewal of the Byzantine War. A Byzantine amphibious force sacked **Damietta** in Egypt (853). The Moslems, augmented by Paulician refugees, gained a great victory over Michael III on the Euphrates in North Syria (860), following which there was a truce and exchange of prisoners. A powerful Moslem army under Abbasid general **Omar** invaded Anatolia (863). After sacking Amisus (Samsun) and ravaging Paphlagonia and Galatia, Omar withdrew eastward toward the Anti-Taurus Mountains, pursued by **Petronas,** uncle of Michael III. Petronas virtually annihilated Omar's army (863).

863–870. Anarchy in the Caliphate. The Turkish guards made and murdered caliphs. Central authority disappeared from the outlying provinces.

870–892. Reign of Ahmad Al-Mu'tamid. The real ruler was his soldier-brother, **Abu Ahmad al-Muwaffak,** who regained control of the Turkish guards and reestablished order in Mesopotamia and Iraq (870–883). He also repulsed an attack on Baghdad (876) by the Saffarids of Khorasan (see p. 291). However, he lost most of Syria to the Egyptian Tulunids (see p. 289).

892–902. Reign of Ahmad Al-Mu'tadid. The son of al-Muwaffak, he somewhat strengthened central authority and regained partial control of Egypt, Syria, and Khorasan.

899–903. Carmathian Revolt. This Shi'ite sect was led by **Abu Sa'id al-Jannabi** in the Arabia-Mesopotamia border region south and west of the Euphrates. After overrunning Syria, al-Jannabi withdrew into northeastern Arabia to establish an independent state on the shores of the Persian Gulf. The remaining Carmathians in Syria were defeated by Abbasid general **Mohammed ibn Sulaiman** (901–903).

902–932. Decline of the Caliphate. There were further Carmathian uprisings in Syria and Arabia, including the sacking of **Basra** (923), **Kufa** (925), and **Mecca** (929). The Caliphate survived mainly because of the ability of the general **Mu'nis,** who held the resurgent Byzantines at bay, repulsed early Fatimid attempts to conquer Egypt (915–921; see p. 289), and was able to limit (but not stop) the depredations of the Carmathians.

c. 929. Rise of the Hamdanid Dynasty. Autonomous princes in northeastern Syria and Kurdistan.

932–946. Impotence of the Caliphate. Mu'nis was executed by the ungrateful Caliph **al-Kahir** (932–934). The Buyid dynasty of Persia (see p. 291) captured Baghdad (946). Succeeding caliphs were merely Buyid puppets.

936–944. Anarchy in Syria. Control was seized by an adventurer, **Ibn Raik,** who repulsed feeble Abbasid attempts to reconquer the country. In the drawn **Battle of Lojun,** he also repulsed **Ibn Tughj,** Ikhshid ruler) of Egypt (see p. 289), who attempted to conquer Syria. After the death of Ibn Raik, the Ikhshidites quickly overran Syria (941). The Hamdanid prince, **Saif al-Dawla,** entered the struggle and captured **Aleppo** and much of northern Syria from the Ikhshidites (944).

945–948. Byzantine–Hamdanid–Ikhshidite Struggle for Syria. At first Saif al-Dawla defeated Ikhshid general **Kafur,** but then had to deal with Byzantine inroads. Kafur then regained southern and central Syria. Al-Dawla, retaining northern Syria, held the Byzantines at bay by skillful defensive tactics and counter-raids into Anatolia.

956–969. Campaigns of Nicephorus Phocas in Syria. (See p. 286.)

963. Hamdanid Invasion of Anatolia. Saif al-Dawla was defeated by **Leo Phocas,** brother of Nicephorus, in the **Battle of Maghar-Alcohl,** a pass in the Anti-Taurus Mountains of Charisiana, near the head-waters of the Halys River.

969. Battle of Ramleh. Fatimid general **Ja'far** defeated the Ikhshidite forces in Syria and Palestine between Jerusalem and Jaffa.

970–971. Carmathian Invasions of Palestine and Egypt. After defeating and killing Ja'far (970), they were repulsed from Fatimid Egypt by **Jauhar** (see p. 289).

974–975. Renewed Carmathian Invasions. They were repulsed from Egypt by Fatimid Caliph **Mu'izz** (see p. 289), who then drove them out of Palestine and Syria.

974–977. Turmoil in Syria. This was largely stirred up by Byzantine intrigue and invasions. Hamdanid general **Aftakin** (a Turk), in alliance with the Carmathians, drove the Fatimids under Jauhar from central and southern Syria (976). Reinforced from Egypt, Jauhar and Fatimid Caliph **Aziz Billah** defeated Aftakin and the Carmathians at another **Battle of Ramleh** (977). Aziz, however, was unable to recover all of the lost territory since the Byzantines and Hamdanids had moved in.

980–1000. Fatimid-Byzantine Struggle for Syria. (See p. 286.)

NORTH AFRICA

Egypt and Cyrenaica

Eastern North Africa was the scene of almost constant war and revolt until the last quarter of the 10th century. Intermittently and sporadically Egypt was under the direct control of the Abbasid caliphs at Baghdad. More frequently, it was ruled by independent princes, only nominally vassals of the caliph. Most important among these were the Tulinids, during the latter 9th century, and the Ikhshidite dynasty, during the middle of the 10th. The Ikhshidites were then overthrown by the Fatimids from Tunisia (see below). By the end of the 10th century the Fatimid caliphs, who had moved their capital to Cairo, controlled all of North Africa and much of Arabia, and were engaged in a struggle for Syria. The principal events were:

817–827. Piratical Refugees from Spain Occupy Alexandria. Recaptured for the Abbasids by general **Abdullah ibn Tahir** (827).

828–832. Uprising of Arabs and Christian Copts in Egypt. Cruelly suppressed by Caliph al-Ma'mun after his general Afshin defeated the rebels at the **Battle of Basharud** in the Nile Delta.

868–876. Rise of the Tulunid Dynasty. Ahmed ibn Tulun, Turkish governor of Egypt, became virtually independent. He later conquered most of Syria (878–884; see p. 287).

904–905. Abbasid Authority Reestablished. Abbasid general **Mohammed ibn Sulaiman** defeated and overthrew the Tulinids in a combined land and sea invasion of Egypt.

914–915. Fatimid Invasion of Egypt. Abu'l Kasim al-Kaim (son of 'Obaidallah; see below) briefly occupied Alexandria, but was repulsed by Abbasid general Mu'nis.

919–921. Second Fatimid Invasion of Egypt. Al-Kaim again occupied Alexandria, but once more Mu'nis repulsed the Fatimids, on sea as well as on land.

922–935. Chaos in Egypt. Finally ended by Abbasid general **Mohammed ibn Tughj,** appointed by the Caliph to be semi-independent governor with title of **Ikhshid** (ruler), beginning the Ikhshidite Dynasty (935), which controlled Egypt for 30 years.

969. Fatimid Conquest of Egypt. The general Jauhur conquered Egypt and overthrew the Ikhshidites at the **Battle of Gizeh,** then sent part of his army to Syria to defeat Ikhshidite forces there (see p. 288).

970–971. Carmathian Invasions. (See p. 287.) Jauhur, on the verge of disaster, rallied and defeated the invaders at the **Battle of Cairo** (971). He then continued construction of the new city of Cairo (meaning "Victory") near the site of ancient Memphis. This became the capital of his master, Fatimid Caliph **Mu'izz li-Din allah** (973).

974–975. Renewed Carmathian Invasion. Fatimid Caliph Mu'izz was briefly besieged in Cairo, but rallied to crush the invaders, then reconquered Syria. This was the height of the power of the Fatimid Caliphate, which was recognized in all North Africa, Syria, western Arabia, and much of the western Mediterranean.

982. Byzantine Invasion of Sicily. Repulsed by the Fatimids.

Tunisia, Tripolitania, Eastern Algeria

At the outset of the 9th century, central-north Africa came under the control of the vigorous Aghlabid family, who devoted most of their attention to invasions of Sicily and south Italy, and to extensive piratical raids of western Mediterranean islands and coasts. Shortly after the beginning of the 10th century the Aghlabids were overthrown by **'Obaidallah al-Mahdi,** who claimed descent from the union of Fatima, a daughter of Mohammed, and the controversial Caliph **Ali** (see p. 250). 'Obaidallah founded the Shi'ite Fatimid Dynasty. He and his successors continued Moslem incursions into Southern Europe. The important events were:

801. Rise of Aghlabid Dynasty. Ibrahim ibn Aghlab, Abbasid governor of Africa, established a virtually independent emirate at Kairouan (Tunisia).

827–831. Aghlabid Invasion of Sicily. The Byzantines were driven from all of the island except Syracuse and Taormina.

836–909. Aghlabid Intrusions in Southern

and **Central Italy.** Extensive areas were seized. Part of Rome was sacked (846; see p. 277).

878. Aghlabid Capture of Syracuse.

902. Aghlabid Capture of Taormina. This ended the last Byzantine foothold in Sicily.

902–909. Revolt of Fatimid Shi'ites. Abu Abdullah al-Husain (known as al-Shi'i), overthrew the Aghlabids. 'Obaidallah al-Mahdi was declared Caliph, to establish the Fatimid Dynasty at Kairouan.

934. Fatimid Capture of Genoa. A Moslem fleet seized and sacked the city.

934–947. Kharijite Rebellion. A religious sect under **Abu Yazid Makhlad** was finally overcome by Fatimid Caliph **al-Mansur.**

965. Byzantine Recapture of Taormina.

975. Fatimid Conquest of Egypt. (See p. 289.)

Morocco and Western Algeria

In the extreme northwest of Africa, **Idris ibn Abdulla,** a descendant of Mohammed, had established an independent emirate late in the 8th century, and his successors reigned for most of the two following centuries, rarely acknowledging even the nominal sovereignty of the Caliphs of Baghdad. During the 10th century, however, the Idrisids were forced to submit first to the Fatimids, then to the Omayyads of Spain, then again to the Fatimids, before finally succumbing to a native Berber tribe. The principal events were:

c. 800–922. Idrisid Independence. Morocco and Western Algeria were controlled by the Arab–Berber dynasty.

922. Fatimid Conquest of Morocco. 'Obaidallah defeated the Idrisids, who accepted Fatimid suzerainty.

973. Omayyad Conquest of Morocco. By Caliph **al-Hakam II** of Spain (see p. 281). The Fatimids temporarily regained control two years later (975).

975–1000. Turmoil in Morocco. Authority and control were disputed between Omayyads of Spain, the Fatimids, the decaying Idrisids, and various Berber factions.

PERSIA AND SOUTH-CENTRAL ASIA, 821–1000

When Abbasid Caliph al-Ma'mun appointed Tahir as viceroy of Khorasan and the eastern provinces early in the 9th century (see p. 287), Persia became virtually independent. The Tahirid Dynasty retained control of most of Persia and much of Transoxiana for more than half a century, but minorities refused to acknowledge either the authority of the Caliph or the Tahirids; principal among these was a heretical Shi'ite community in the region around Dailam in the Elburz Mountains.

Shortly after the middle of the century a revolt broke out among the frontier warriors of Seistan (southwestern Afghanistan and Baluchistan). Known as the Saffarids, they were led by **Yakub ibn Laith.** They soon conquered most of Persia and Khorasan from the Tahirids, and even made a vain attempt to capture Baghdad. Meanwhile, the Persian Samanid family had long reigned as governors of Transoxiana under the caliphs and the Tahirids. In the inevitable clash between Saffarids and Samanids, at the beginning of the 10th century, Samanid leader **Isma'il** was successful. The Samanids held Transoxiana and most of eastern Persia until the end of the century.

Western expansion of the Samanids was blocked by the Shi'ites of Dailam. During the third decade of the 10th century, under the leadership of **Merdawj ibn Ziyar,** these Shi'ites had just begun to expand southward into the plains of Persia when three brothers of the Buyid family suddenly seized power in Dailam. The Ziyarids were forced back into the Caucasus Mountains. The Buyid Shi'ites spread over southwestern Persia and shortly before the middle of the century they occupied Baghdad (see p. 288).

Late in the 10th century, Samanid control over Transoxiana and eastern Persia was challenged by the rising Turkish Ghaznevids (of modern Afghanistan) and by the Ilak Turks, pressing down from central Asia. By the end of the century the Ghaznevids had occupied much of Khorasan. At the same time the Ilak khans had begun to move into Transoxiana. The principal events were:

821. Tahir Appointed Viceroy of Khorasan and the East.

c. 860. Growing Importance of Dailam. It became the Persian center of Shi'ism.

866–900. Rise of the Saffarids. Under **Yakub ibn Laith,** who overthrew the Tahirids (872).

876. Saffarid Attempt to Capture Baghdad. Repulsed by al-Muwaffak (see p. 287).

903. Overthrow of the Saffarids. Samanid leader **Isma'il** defeated **Amron ibn Laith,** brother and successor of Yakub.

c. 925. Shi'ite Expansion. From Dailam, under the Ziyarids.

932. Emergence of the Buyid Dynasty of Dailam. The brothers **Imad al-Dawla, Rukn al-Dawla,** and **Mu'izz al-Dawla** succeeded to Shi'ite leadership, driving the Ziyarids into the Caucasus.

946. Conquest of Baghdad. Buyid Mu'izz al-Dawla captured Baghdad and the caliph, who became a Buyid puppet.

c. 962. Beginning of Ghaznevid Dynasty. Turkish slave **Alptagin,** fleeing from the Samanid court at Bokhara, established a semi-independent state in the mountains of Ghazni.

977–997. Reign of Sabuktagin of Ghazni. The son-in-law of Alptagin, he gained control of much of Khorasan, and began to expand into Northern India (see p. 292).

c. 990–999. The Ilak Turks Conquer Transoxiana. The Samanids were expelled.

1000. Mahmud of Ghazni Conquers Khorasan from the Samanids. He left control of the frontier north of the Oxus to a vassal Turkish tribe under the leadership of **Seljuk** (see pp. 336 and 347).

SOUTH ASIA

NORTH INDIA

During the 9th century there was a drawn-out but fierce three-way struggle between the Pratihara (or Gurjara) dynasty of Rajputana, Pala of Bengal, and the Rashtrakuta kings from the Deccan. Rashtrakuta, under King **Govinda III,** drove Pratihara from the Malwa plateau and southwestern Ganges Valley, while the Pratihara King **Nagabhata II** was ejecting Pala from the recently conquered province of Kanauj. Pratihara then moved the capital to the city of Kanauj from whence, by the end of the century, the dynasty gradually gained ascendancy in north India under Kings **Bhoja I** and **Mahendrapala.** They pressed Rashtrakuta back from the Ganges Valley, and drove Pala back into native Bengal. In the early years of the 10th century, Pratihara held all of north India save for Moslem Sind, Pala Bengal, and a Kashmiri foothold in the northern Punjab. The unsuccessful Pratihara attempts to eject the Kashmiris, combined with persistent Rashtrakuta attacks from the south, contributed to a rapid decline in Pratihara power. By the middle of the 10th century, north India reverted to its usual condition as a cockpit of war between minor dynasties. Rashtrakuta, whose unremitting efforts had created this situation, was unable to take advantage of it because of interference from two resurgent powers in the south (see below).

But a new force was ready to exploit the near anarchy existing in the Indus and Ganges valleys. Early in the 10th century the Turks from Central Asia, like so many nomads before them, had begun to move into modern Afghanistan. This became the Moslem emirate of Ghazni, feudatory to the Samanids of Bokhara (see p. 290). The Ghazni Emir **Sabuktagin** began to raid into the northern Punjab late in the century. Finding relatively slight resistance, he seized Peshawar, in a repetition of the old familiar pattern of invasion of India from the northwest. The principal events were:

793–814. Reign of Govinda III of Rashtrakuta. (See p. 255).

793–833. Reign of Nagabhata II of Pratibara. He reorganized Pratihara power, and later conquered Kanauj from Pala (c. 830).

890–910. Reign of Mahendrapala of Pratihara. Pratihara became predominant in North India.

c. 916. Temporary Occupation of Kanauj by Rashtrakuta.

c. 962. Establishment of Ghazni by Alptagin. (See p. 291.)

977–997. Reign of Sabuktagin of Ghazni. He began expansion into India by the seizure of Peshawar (c. 990).

SOUTH INDIA AND CEYLON (SRI LANKA)

After the collapse of the Chalukya Dynasty in the middle of the 8th century, its old rival, Pallava, recovered some of its former power on the east coast of the Deccan. During most of the 9th century resurgent Pallava engaged in more or less constant warfare with its immediate neighbors: to the north the offshoot Chalukya Dynasty of Vengi and the Ganga Dynasty of Kalinga; to the south Chola and Pandya. During most of this time Pallava was favored by the benevolent neutrality—or outright assistance—of more powerful Rashtrakuta.

Farther south, by the middle of the 9th century, Chola under King **Vijayala** (836–870) had become the dominant Tamil state, ruling over the Coromandel coast and much of the Eastern Deccan. Late in the century, in alliance with Chalukya of Vengi, Chola King **Aditya I** (c. 870–906) defeated and annexed the Pallava kingdom of Kanchi (Conjeeveram, 888). Early in the 10th century, Aditya's son, **Parantaka I** (906–953), extended Chola control southward to Cape Comorin, at the expense of the Pandya kings of Madura. But growing Chola power now attracted the jealous attention of **Krishna III** (939–968) of Rashtrakuta, who consequently shifted his interest from north to south India. In a series of campaigns around the middle of the century, Rashtrakuta overwhelmed Chola. King Parantaka was killed in battle (953).

The Rashtrakuta's old enemy, the Chalukyas—who ruled several small principalities after their empire fell—suddenly rose under **Taila II** (973–997), overthrowing and completely destroying the Rashtrakuta Dynasty (973). Taila and his successors are known as the Later Chalukya Dynasty, or as Chalukya of Kalyani.

The disappearance of Rashtrakuta permitted Chola to reestablish itself as a major power of southern India under **Rajaraja I** (985–1014). This led to a struggle between resurgent Chola and resurgent Chalukya. Though Chola had somewhat the best of these conflicts, Rajaraja was unable to conquer Chalukya. However, he did conquer Pandya, Kerala (or Chera), Vengi, and Kalinga on the mainland, while overseas his fleets and armies conquered most of Ceylon (993) and overran the Maldive Islands.

SOUTHEAST ASIA (TO A.D. 1000)

BACKGROUND OF MILITARY HISTORY OF SOUTHEAST ASIA

The early history of Southeast Asia is vague and spotty, the origins of its people indefinite. The entire region has from the earliest recorded times been strongly influenced by its two powerful neighboring civilizations, India and China.

Chinese mercantile and military contacts came earliest, particularly along the east coast. But Indian cultural influence began to affect the region beginning about the 1st century A.D., becoming significant about the 5th century. This Indian influence was expressed in literature, art and architecture, court ceremonials, religions and other customs, and is particularly noticeable in the names and titles of rulers and military leaders.

CHAMPA AND VIETNAM

The area of Tonkin and northern Annam had been strongly influenced by China as early as the 9th century B.C. The people were of mixed Mongoloid-Indonesian stock. Late in the 3rd century B.C. the area was conquered by Chinese general **Chao T'o** and combined by him in a rebellious regime he established in South China (c. 208) at the time of the fall of the Ch'in Dynasty (see p. 88). The name Nam-viet, which was at first applied to Chao T'o's entire Canton kingdom, soon was limited to the Tonkin and northern Annam provinces. When the Canton kingdom was overthrown and reannexed by the powerful Han emperor, Wu Ti (111 B.C.), he also annexed the provinces of Nam-viet, which remained under Chinese rule until A.D. 939.

Sometime after the Chinese conquest of Nam-viet, in southern Annam, south of modern Hué, an Indonesian people, known to history as the Chams, had established an independent kingdom (c. A.D. 200). In due course the Chams adopted a form of Indian culture. Military contacts with China were sporadic, the semibarbaric Chams raiding by land and sea into the civilized area to the north, and being subjected to occasional Chinese punitive expeditions.

During the 5th and early 6th centuries, a period of weakness and anarchy in China, the Chams made numerous raids into Nam-viet. One of the leading Cham warriors and rulers who conducted such expeditions was King **Bhadravarman** (c. 400). A Chinese naval punitive expedition against Champa was repulsed (431). At this time the Cham King **Yang Mah** unsuccessfully sought the aid of Funan (see p. 294) in a major expedition into Tonkin. An overland Chinese punitive expedition some years later led by **T'an Ho-ch'u,** governor of Tonkin, was successful and sacked the Cham capital (446). This apparently greatly reduced Cham interest in raiding into Nam-viet. About a century later, however, Cham King **Rudravarman** sought to take advantage of local disorders in Nam-viet (now also known as Vietnam) and led an unsuccessful raiding expedition to the north.

Under the leadership of **Li-Bon,** the Vietnamese rebelled and temporarily expelled the Chinese (541). Li-Bon then repulsed a raiding expedition from Champa (543) under King Rudravarman (see above). The Chinese soon defeated Li-Bon and suppressed the revolt (547). Internal weaknesses in China encouraged Vietnamese patriots to rise and again temporarily throw off the Chinese yoke. The Sui Emperor **Yang Chien,** however, soon sent forces which re-established Chinese control over the rebellious provinces of Tonkin and Annam. Farther south the independent Chams remained relatively peaceful, having been severely punished (605) by an aggressively conducted Chinese expedition under General Liu Fang (see p. 259). During the latter half of the 18th century the coasts of Champa and of Chinese Vietnam were constantly harassed by Indonesian raiders—probably mostly from Srivijaya (see p. 295).

In the early years of the 9th century, King **Harivarman I** renewed Cham harassment of Chinese Annam, and apparently also did some raiding against the Khmer Kingdom of Cambodia. But from the middle of that century until the middle of the next (when King **Indravarman III** repulsed a Khmer invasion; see p. 295), Champa was relatively peaceful. Chinese Vietnam, meanwhile, had sustained an invasion by Nanchao (862–863; see p. 296). Some time after this the Annamese took advantage of the chaos in China after the fall of the T'ang Dynasty (907; see p. 298) to begin a struggle for independence that was finally successful (939).

The latter years of the 10th century saw almost incessant war between Champa and newly independent Annam. The conflict was begun by an invasion of Annam by Cham King **Paramesvaravarman** (979). He was repulsed, and the Annamese King **Le Hoan** led a destructive raid into Champa that resulted in the sacking of the Cham capital and the death of the Cham King (982). At this time a revolt in Annam created chaos in that country from which a new dynasty, founded by King **Harivarman II,** emerged (989). The war between Champa and Annam was soon renewed, and continued far into the next century (p. 350).

FUNAN AND CHENLA (CAMBODIA)

During the 1st and 2nd centuries A.D. an Indonesian people in the lower Mekong Valley coalesced into a kingdom of Funan. The first ruler of importance was a conqueror known as **Fan Shih-man,** who extended his domains to the sea in modern Cochin-China, and who apparently also controlled all of modern Cambodia and possibly much of modern Thailand (c. 200). The Kingdom of Funan prospered, continued to grow, and had extensive contacts with both China and India. One of its most illustrious early rulers was **Chandan,** who may have been an exile from India, of Hindu-Kushan descent (c. 350).

Under King **Jayavarman** (c. 480–514), the Empire of Funan reached its greatest extent and the height of its prosperity. It is possible that military expeditions from Funan actually reached and conquered parts of the lower Irrawaddy Valley. During the reign of **Rudravarman,** an uprising (apparently starting as a dynastic dispute) broke out in the province of Chenla (northern Cambodia, along the middle Mekong from the Mun River to Bassak), led by the brothers **Bhavavarman** and **Chitrasena** (c. 550). Rudravarman was deposed and driven into southern Cambodia, where a remnant of the Funan Kingdom continued to exist for a few more decades. Bhavavarman assumed the throne of Chenla and, with the help of his warrior brother, absorbed much of the former Funan Empire (c. 500–c. 600).

After the death of Bhavavarman his brother, Chitrasena, took the throne with the royal name of **Mahendravarman** (c. 600–c. 611). This great conqueror was followed by his son, **Isan-avarman I** (c. 611–635), who completed the conquest of Funan (c. 627), and who extended the boundaries of Chenla steadily westward to include most of modern Cambodia. This expansion was continued under **Jayavarman I** (c. 655–695), who overran southern and central Laos to establish a common frontier with Nanchao. Apparently Chenla became overextended during this reign, since there was much unrest, and the empire was crumbling at Jayavarman's death. A few years later (c. 706) Chenla broke apart into regions known as Upper and Lower Chenla. We know little of what happened in Upper Chenla, save for some fighting against the Chinese in Tonkin. Lower Chenla apparently was in turmoil during the remainder of the period. This was intensified in the latter half of the 8th century by frequent raids from the Indonesian islands of Sumatra and (possibly) Java. For part of the time Lower Chenla was apparently tributary to the Sumatran Kingdom of Srivijaya (see below).

During the first half of the 9th century the Khmer Kingdom of Lower Chenla revived under **Jayavarman II** (802–850). Throwing off Indonesian suzerainty (see below), he in effect was the founder of the magnificent Angkor monarchy. The actual construction of the temples and palaces of Angkor was begun under his successor, **Indravarman I** (877–889). The Angkor Kingdom expanded steadily, and reached its first peak under King **Yasovarman** (889–900), who ruled an empire which probably included Cambodia, southwestern Vietnam, most of Laos and Thailand, and Southern Burma. There was a decline in the Khmer Kingdom under the usurper **Jayavarman IV** (928–942) and his son, **Harshavarman II** (942–944), who moved the capital from Angkor. Harshavarman was overthrown by a revolt which placed **Ravendravarman II** on the throne, and which brought the capital back to Angkor. Though Ravendravarman was repulsed in an attack on Champa (945–946), the fortunes of the Khmer Kingdom again rose under him and his successors.

MALAYSIA (INDONESIA AND MALAYA)

In the 2nd and 1st centuries B.C. the islands of Indonesia and the Malay Peninsula had some contacts with Chinese seafarers and traders. However, Indian influence became more significant in the early centuries of the Christian era. By the beginning of the 5th century there were a number of independent, relatively civilized principalities, Hindu in nature, on the peninsula as well as in the neighboring large islands. The records are very sketchy. It is clear, however, that by the end of the 7th century the powerful Kingdom of Srivijaya had grown up in southern Sumatra with its capital at Palembang. This kingdom conquered at least the western portion of Java, and raided frequently along the mainland coasts to the north, exacting tribute. Srivijaya dominated western Indonesia and the South China Sea for several centuries.

Though the power of Srivijaya appears to have declined somewhat by the early 9th century, it retained its ascendancy in Indonesia throughout this period. Its position in Java was challenged by the states of Sailendra and Sanjaya, which in turn were engaged in a bitter struggle for supremacy in central Java. This was won by Sanjaya after a final clash around the Sailendra fortified stronghold on the Ratubaka Plain (856). **Balaputra,** the last Sailendra to rule in Java, married a Srivijayan princess and eventually succeeded to the throne. Since Balaputra brought with him his claim to Central Java, Srivijayan pressure apparently caused the removal of the Sanjaya capital to east Java (early 10th century). Javanese tradition has it that during subsequent warfare, **Dharmauamsa** (c. 958–c. 1000), King of East Java, attacked Sumatra but was repelled by Srivijaya, which counterattacked (1016–1017) and destroyed the East Javanese capital.

THE THAIS AND SHANS

The early inhabitants of the lower Menam River Valley of modern Thailand were the Mons, whose principalities also controlled Lower Burma. By the 8th century, extensive migrations from the northwest had brought a Mongoloid people, related to the Chinese, into the mountainous region between the upper Salween and Mekong Rivers in modern Yunnan. These people, known as the Nanchao, were the ancestors of the modern Shans (of Burma), Thais, and Laotians. For several centuries, however, the history of Nanchao was inextricably bound with that of China.

Late in the 8th century a powerful independent Nanchao Kingdom was established by **Kolofeng** in Yunnan, Laos, and northeast Burma (see p. 261). Nanchao prospered and expanded under his grandson and successor, **I-mou-hsun,** early in the 9th century. Central and southern Burma were overrun (832) and the Pyu Kingdom destroyed (see below). The Nanchao then invaded south China and Tonkin (858–863), and sacked Hanoi (863). They were finally repelled from Tonkin by Chinese General **Kao P'ien.** Turning to the northwest, they invaded Szechuan and laid siege to the Chinese city of Ch'engtu, where once again they were repulsed by Kao P'ien (874). A long period of relative peace seems to have followed; though Tai mercenary contingents played an active part in other wars in Southeast Asia. Late in the 10th century the Chinese Sungs made tentative attempts to reconquer Yunnan, but evidently gave it up in the face of determined Nanchao resistance. (All during this period the Thai, or Shan, people seem to have been slowly shifting southward through Yunnan into the regions they had conquered in eastern Burma, northern Siam, and Laos.)

BURMA (OR MAYNMAR)

As early as the 2nd century B.C., Chinese traders were using an overland route from the area of modern Yunnan, through Upper Burma, and into India through Assam. Despite the ferocity of wild tribes in this region, the route was apparently used extensively for about 500 years, and was traversed by embassies between China and India, as well as by embassies from Rome and Constantinople to the courts of China. Apparently the use of this route was discontinued for several centuries, sometime around the middle of the 4th century A.D. Meanwhile in the south, in the Irrawaddy Valley, a people known as Pyus had established a civilized society strongly influenced by Hindu culture.

By the beginning of the 9th century the Kingdom of Nanchao held the northern portion of what is now modern Burma or Myanmar (see above), while a Pyu kingdom (evidently subject to Nanchao) flourished in central Burma, with its capital at Srikshetra (Hmawza, near Prome). Southern Burma and central Thailand contained a number of small states inhabited by the Mon (Talaing) people. The principal center of Mon culture was in Siam, where their Buddhist Kingdom of Dvaravati flourished from the 6th century onward. Nanchao armies destroyed the Pyu Kingdom, and two of the Mon principalities in Lower Burma also fell. However, the Mons were finally able to repulse the invaders (935). Some time after this, the Mons appear to have expanded northward into the remnants of the old Pyu Kingdom and, in collaboration with Pyu refugees, established a new capital in central Burma at Pagan (c. 950).

Meanwhile, shortly after the middle of the 9th century a new people, related to the Pyus, had begun to migrate into central Burma around the region of Kyaukse, where Mons and Pyus had established a highly developed irrigation system. These new people were evidently an offshoot of the old Ch'iang Tibetan tribes (see p. 146) and over a period of centuries had moved slowly, under pressure from Chinese and Nanchao, from Kansu to eastern Tibet, to western Yunnan, to central Burma. These warlike invaders, ancestors of the modern Burmans, began to drive the Mons southward. Spreading out over the central Irrawaddy Valley, by the end of the 10th century they firmly held the region from Shwebo south to Prome.

AFRICA

EAST AFRICA

The power and wealth of Aksum waned between 800 and 1000. Maqurrah and more prosperous 'Alwah survived. Most of the Red Sea coastal states were absorbed by Moslems from Egypt and Arabia. Coastal Somalia was settled by Arab and Persian traders; by 1000 there was a string of Arab trading settlements at least as far south as Pemba.

In addition to Arab-Moslem expansion, Aksum was threatened by the Damot Kingdom (apparently of Jewish origins and possibly identical to the Demdem) in modern southwestern Ethiopia. In 976, the Damot Queen Esato (Judith) led a serious invasion of Aksum, laying waste the countryside nearly to the gates of Aksum itself.

WEST AFRICA

In the early 9th century A.D., the Empire of Ghana, centered on its capital at Kumbi Saleh (200 miles north of modern Bamako, Mali), was first mentioned by Arab geographers. The empire, known to Europeans by the title of its ruler (*ghana*) and called Wagadu by the natives, predates 800, and may have existed as early as the mid-4th century. Ghana's wealth and power depended on its control of the gold trade from fields along the upper Niger. During the 900s, Ghana conquered the Sanhajah Berber tribes to their north, and so gained control of the highly lucrative salt trade.

Meanwhile, in the mid-9th century the kingdom of Kanem arose around Lake Chad, where it gained wealth from trans-Sahara caravan trade. Its capital was at Njimi, northeast of Lake Chad.

EAST AND CENTRAL ASIA

CHINA

The T'ang Dynasty gradually declined in the 9th century. As central authority lessened, the power of semiautonomous warlords increased. During much of this time the T'ang were engaged in inconclusive war with their nominal vassals, the Nanchao kings of Yunnan. During the growing lawlessness of the latter part of the century, a popular leader named **Huang Ch'ao** set himself up as a sort of Robin Hood, taking goods of the wealthy and government officials to distribute to starving peasants. As Huang's strength grew, so did his ambition; he started a revolt, conquered much of the country, and declared himself emperor. Eventually, however, the T'ang defeated him, and he committed suicide. This prolonged civil war exhausted the country and accelerated the T'ang decline.

Early in the 10th century the T'ang Dynasty was overthrown by a warlord, and there followed a period of anarchy called the Era of Five Dynasties. Toward the end of the century, however, most of the country was reunited under a general of one of these dynasties—**Chao K'uang-yin**—who inaugurated the stable Sung Dynasty. He reduced the power of the warlords, centralizing military control. The northern areas of China had been absorbed by barbarians, however, and resisted Chao's unification efforts. These regions included the best horse-breeding and grazing grounds of the Far East, thus embarrassing Chao and later Sung emperors, who were unable to create worthwhile cavalry. This contributed largely to the fact that the Sung were constantly on the defensive against the barbarians of the north for two and a half centuries. The most important events were:

829–874. Intermittent War between the T'ang and Nanchao. The T'ang repulsed three major Nanchao invasions against Ch'engtu (829 and 874) and Hanoi (863).

848. Reconquest of Kansu. The Tibetans—who had gradually worked their way back into Kansu—were driven out by T'ang forces.

875–884. Revolt of Huang Ch'ao. He conquered the capital cities (Loyang and Ch'ang-an) and declared himself emperor (880). T'ang Emperor **Hsi Tsung** fled to the southwest and raised a new army, which he put under the command of General **Li K'o-yung** (of Turkish stock). Li defeated Huang (882), who fled and later committed suicide (884).

c. 908. The Introduction of Gunpowder. The initial uses of this explosive were for demolitions, or as noisemakers.

907. End of the T'ang. General **Chu Wen** seized control of China, murdering the last T'ang emperor.

907–959. The Era of Five Dynasties. An anarchic period during which five successive families sought to maintain imperial authority.

960–976. Establishment of Sung Dynasty. Chao K'uang-yin (general of the Later Chou Dynasty) established the Sung Dynasty, after being declared emperor by his soldiers, taking the name **T'ai Tsung.**

976–997. Reign of T'ai Tsung. He completed the reunion of China (essentially modern China proper), less Kansu, Inner Mongolia, and northeast China.

979–1004. War with the Khitans. T'ai Tsung was repulsed in efforts to capture Peking (986). The Khitans then began a slow advance into north China.

INNER ASIA

Toward the middle of the 9th century the Uighur Turk Empire was overthrown by the Kirghiz and Karluk Turks—apparently with some assistance from the Chinese T'ang, who had been frequently embarrassed by the power of their nominal vassals and allies (see p. 260). The Uighurs migrated southward into the Tarim Basin, where they established a new kingdom and eliminated the last remaining traces of earlier Indo-European influence (c. 846). Early in the 10th century, however, they were forced to submit to the khans of the Ilak Turks, who began to dominate Central Asia.

Early in the 10th century, **Ye-lu A-pao-chi,** chieftain of the Khitan Mongol people of eastern Mongolia, conquered all of Inner Mongolia, southern Manchuria, and much of north China (907–926). He received tribute from the Uighurs of Sinkiang. His successor, **Ye-lu Te-kuang** (927–947), strengthened the Khitan Empire, which he designated as the Liao Dynasty. Playing an important role in Chinese affairs during the anarchic Five Dynasties period, Te-kuang helped to establish the Later Chin Dynasty (936), only to overthrow it later (946).

Meanwhile the Tangut tribe—a Tibetan people—again established themselves in Kansu (c. 990), where they created the Western Hsia Kingdom, with Ninghsia as their capital. They were almost constantly at war with Uighurs, Khitans, and Chinese.

KOREA

Most of the 9th century in Korea was peaceful. Silla, under strong influence from the Chinese T'ang, was prosperous; the country became homogeneous. Toward the end of the century, however, there were several revolts. Early in the 10th century the general **Wanggun** rebelled to establish the new Kingdom of Koryo in West-Central Korea, with the capital at Kaesong (918). Soon after this the last king of Silla abdicated, and the Wang Dynasty of Koryo peacefully absorbed Silla. Paekche, however, attempted to reestablish its independence (935). The Wang immediately invaded Paekche, destroying it and annexing it to a reunified Korea (936).

Earlier, when the T'ang Dynasty of China had collapsed (907), the kings of Silla had voluntarily submitted to the nominal suzerainty of the Khitan Mongols of North China and Manchuria (see above). The Wang Dynasty, however, ignored this Khitan suzerainty and established an alliance with the Sung emperors (c. 960). The Khitan Liao Dynasty resented this and late in the century forced the Wang to renounce their alliance with the Sung (985), and to submit once more to Khitan suzerainty (996).

JAPAN

Aside from sporadic warfare with the Ainu in the north and some inroads by Korean pirates (c. 869), the 9th century was generally peaceful in Japan. The period saw two developments of great significance to the future history of the country. The first was the growing power and influence of the Fujiwara clan, which gained virtual control over the country by its complete domination of the imperial family; **Fujiwara Yoshifusa** was the first to achieve this power (858) and soon thereafter he took the title of regent.

The second development was the trend toward feudalism. This was the result of the growing professionalism of the retainers of the great nobles—partly due to the military reforms of **Sakanoue Tamuramaro** (see p. 262), and partly to the accumulation of wealth and power by the nobles in a prospering nation. This trend to feudalism obviously had something in common with the feudal trend in Europe, but had one unique aspect. This was the appearance of two warrior leagues, each affiliated with a great family of imperial origins: the Taira and Minamato. Evidently, in addition to their fealty to their own feudal lords, all Japanese warriors affiliated themselves with one or the other of these warrior clans. There seems to have been no particular system or pattern of such relationships.

Though there was some growing internal unrest, stimulated partly by clan rivalries, the 10th century was almost as peaceful as the 9th—save for one short period of intense civil strife (935–941). The warrior leagues were involved in this, but not in a formal or unified way. One of the most serious uprisings was led by **Taira Masakado** (of the warrior league family) in eastern Japan, and was finally suppressed (940). Another, led by **Fujiwara Sumitomo** (of the ruling clan) established piratical control over the Inland Sea, but was crushed with the assistance of the Taira League (941).

X

THE RETURN OF SKILL
TO WARFARE:

1000–1200

MILITARY TRENDS

Savage strife continued to rock all corners of the so-called civilized world, but political aims—some vague, others more definite—were emerging. At the same time there was a noticeable rise in the level of military skill, particularly in Western Europe and among the Moslem Turks.

Although no true military genius emerged during this period, there were a few outstanding soldiers: **Mahmud** of Ghazni; **Richard the Lionhearted** of England; **Alexius** of Byzantium; the Seljuk Turk, **Alp Arslan; Saladin** of Egypt and Syria; and the Sicilian Norman, **Robert Guiscard.** All six demonstrated characteristics of leadership higher than heretofore seen outside the Byzantine Empire, and all six played a part in one or more of the four major historical developments of this period.

The first of these was a continuation of an earlier trend: the westward progression of the Turks north and south of the Black and Caspian seas. The northern current—that of the Pechenegs and Cumans—would be halted, after a bitter struggle, by the combined forces of the Byzantine Empire and its Slavic neighbors. But the southern surge—probably because of the interaction of Turkish vigor and Moslem fanaticism—was more significant. It swept away most of Byzantium's Asian holdings, pervaded the entire Middle East, and, although checked by the Crusades, was continuing as the period ended.

The second trend was in part contributory to, and in part a result of, the first. This was the decline of the Byzantine Empire. Despite erosion of energy and of military competence, Constantinople was still unsurpassed in theoretical mastery of the art of war, and this, together with amazing resiliency, would prolong the life of the empire for several more centuries. However, as the Battle of Adrianople had clearly foreshadowed the fall of Rome, so the Battle of Manzikert sounded the knell for Byzantium.

The third trend was the growing centralization of power in the kingdoms of Europe. Despite the inhibiting effects of the still flourishing feudal system, a kind of nationalistic coagulation was in process. This was particularly true in England and France; less so in Germany, Spain, and Poland.

The fourth and most momentous of these trends, from the standpoint of military, religious, and cultural history, was that of the Crusades.

WEAPONS AND ARMOR

One important new weapon appeared: the crossbow. This was literally a hand ballista, somewhat more clumsy and with a slower rate of fire than the traditional bows of military history. But the metal bolts or arrows fired from these machines flew farther, faster, and generally more accurately than the lighter arrows of conventional archers. Most important, they could penetrate armor impervious to other missiles. Apparently the Chinese and the Romans had experimented with hand ballistae, but the concept was forgotten in the West after the fall of Rome. It was revived, however, in the 11th century; crossbows made their first unmistakable appearance in action in Europe with the Norman Army at the Battle of Hastings. Significantly, however, the crossbow had flourished throughout these intervening centuries in China (see p. 90).

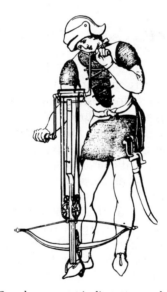

Crossbowman winding up crossbow

The crossbow was the West European answer to Turkish horse archery. Usually European crossbowmen were on foot. The Crusaders experimented with mounted crossbowmen, but discovered that the added mobility was more than offset by the decrease in accuracy and in rate of fire. Interestingly, the Chams of Southeast Asia apparently used mounted crossbowmen about this same time (see p. 351).

Another new weapon in West Europe was the infantry halberd. This modified the pike by the addition of an axhead near the point to permit the weapon to be used for cutting or smashing as well as thrusting. More important than the weapon itself was the fact that West Europeans were devoting attention and ingenuity to improving the fighting capabilities of the once-despised foot soldier. Thus the halberd, like the crossbow, was an early indication of the revival of infantry in Europe.

One other example of weapons improvement was the perfection of the Moslem scimitar. The significance of this curve-bladed sword lay in quality of metallurgy rather than in any radical change of design. The craftsmen of Damascus and Toledo, particularly, became known for the magnificent steel blades they created: amazingly supple, yet tough and sturdy and capable of being honed to razor keenness.

In Europe, defensive armor continued to improve. It also became increasingly heavy. The mail shirt was shortened, its long skirts replaced by mail breeches. Sleeves were lengthened to the wrists and a coif, or mailed hood, was often added, replacing a helmet. Such a suit of chain mail weighed between 30 and 50 pounds. To enhance protection and to prevent bruises from blows against its hard surface, mail armor was worn over a coat of heavy leather or felt. Such leather or felt jackets were usually the only body protection of the foot soldier. But even this was enough to stop most arrows; Turkish chroniclers describe unharmed Crusader infantrymen in battle as often looking like pincushions.

The fit of the helmet was improved. The nosepiece was lengthened and strengthened. In fact, complete facial coverage was provided by many armorers, who produced flat-topped casques covering the entire head and neck, with slits in the front for vision and breathing. Helmets of this sort, however, were so heavy and so suffocating that they were usually carried on the saddle pommel until action was about to begin. The most common pot helmets of the period weighed 15 or 20 pounds. By the end of the period armorers were experimenting with pointed helmet fronts to deflect frontal blows and to reduce the unpleasant possibility of having the helmet smashed back into the wearer's face.

These improvements in armor kept casualties low in European battles. They also made for great disproportion in the losses of Crusaders and Moslems in their battles in the East. When the Crusaders won, their casualties were always relatively light; when they were defeated, however, they suffered heavily in the final phases of the battle, since they were unable to escape from their more mobile foes.

FORTIFICATION AND SIEGECRAFT

There were no new developments in fortification. However, the Crusaders learned lessons from the Byzantines that completely changed West European concepts of fortification and of the defense of cities (see p. 346). But in one important respect the Europeans differed from the Byzantines in their application of these lessons. To the Byzantines, fortresses were essentially bases for defensive-offensive operations in the field, and thus they were usually located on commanding, but accessible, ground. The Europeans, more defensive-minded, and still limited by their feudal concepts of short-time, small-scale military operations, sited their new forts or castles in the most inaccessible spots possible. Thus it was extremely difficult for an attacker to reach and to assault such forts. But it was almost as difficult for the defenders to debouch rapidly, and so they had little opportunity to seize the initiative from a besieging or blockading force.

In siege operations, one new weapon was introduced. This was the **trebuchet,** or **mangonel,** a missile-hurling machine for battering fortifications, or for throwing rocks or other

weight
(up to
10 tons)   sling
                            with
                            projectile

Trebuchet

projectiles over walls. Unlike the ballista and catapult, which obtained their power from tension or torsion (see p. 43), the propelling force of the trebuchet was provided by a counterweight.

Siege methods were unchanged. Human ingenuity had long before devised practically every possible way of battering down, climbing over, or tunneling under walls. European medieval siegecraft, however, rarely visualized employment of the favorite, and effective, siege technique of the Macedonians and Romans. This was the **agger,** or mound, which could be built up to dominate the defending fortifications. In the first place, manpower shortages (discussed elsewhere) kept European armies relatively small, and the construction of a great siege mound required tremendous manpower. Second, the usual location of European castles on inaccessible rocky promontories precluded employment of mounds.

Thus, to a greater extent than ever, the principal siege weapon was starvation. Since it was very difficult and expensive for feudal monarchs and nobles to keep armies in the field indefinitely, a well-provisioned castle had a very good chance to survive siege.

TACTICS OF LAND WARFARE

General

Cavalry remained the dominant arm. But infantry was reviving in importance. Even prior to the Crusades the Western Europeans had discovered that—all things being equal—an army which included a reliable foot element had an advantage over a completely cavalry force. This was because the infantry provided a base of maneuver for the cavalry, and because the infantry could seize and hold commanding or vital ground. For this reason, many European leaders would habitually dismount a portion of their knights and men-at-arms, to obtain a reliable nucleus for the less trustworthy footmen of the feudal levies. Sometimes the only foot element of an army would be its dismounted knights. The reason for this was that the knights, with their heavier armor, better weapons, and a code of honor which gave them a kind of discipline, could stand firm against a cavalry attack that would scatter the fyrd or militia. Because this was obviously an uneconomical use of expensive cavalry, a kind of medieval intuitive cost effectiveness led to development of standing forces of well-equipped, disciplined infantry.

This phenomenon was accelerated by Crusade experience. In fighting the mobile Moslems, the Crusaders found it essential to maintain a solid infantry base from which to launch their overwhelming cavalry charges. The significance of this was soon realized by the Moslems, who then made it one of their important objectives to separate the Crusader cavalry from the infantry, then to defeat each in detail. This is how Saladin won his decisive victory at Hattin (see p. 343). This Moslem tactic, in turn, taught the Crusaders to devote more attention to close coordination of the actions and movements of both infantry and cavalry, leading to some really effective operations of the combined arms, as in Richard's great victory over Saladin at Arsouf (see p. 345).

Part of this coordination of foot and horse elements was centered around the related concept of fire and movement. Reliance upon crossbows was greatly accelerated by the Crusaders' experience, and by their realization that they needed firepower to offset that of the Turkish horse archers. Whenever possible, the Crusaders would launch their battle-winning heavy cavalry charges immediately after a cross-bow volley had shaken the opposing force.

The Turks, in turn, found that they needed combined arms against the formidable Crusaders. Saladin was apparently the first who effectively combined fanatic Arab and Egyptian foot soldiers with Mameluke (Turkish slaves) horse archers. But in such a contest the more lightly armed Moslems had no hope of success against well-coordinated European combined arms.

Infantry Tactics

Even by the end of this period, the foot soldier was still decidedly inferior to the cavalryman in prestige and effectiveness. There was little infantry maneuver in battle. Its purpose was to stand firm as the base for battle-winning cavalry maneuvers. An additional purpose, as discussed above, was to provide firepower which would enhance the chances of success of cavalry charges.

Cavalry Tactics

There were three distinct cavalry types during this period. First was the horse archer of the Byzantine and Turkish armies, with the Byzantine being far better disciplined, more heavily armored, and capable of functioning also as the second type: heavy shock-action cavalry. The West Europeans were supreme in this type; no other military force in the world could stand up against equal numbers of the mailed knights and men-at-arms of the Franks (as Moslems and Byzantines called all West Europeans). The third type was light cavalry, usually lightly armored and equipped with lance and sword. Only the Arabs, Egyptians, and North Africans attempted shock tactics with such horsemen; this accounts for the ability of the Crusaders—prior to the time of Saladin—to operate so successfully against much more numerous Egyptians.

The Crusaders, however, learning from Byzantines and Moslems, did make use of light cavalry themselves for screening and reconnaissance purposes. They also used light horse archers. Later, in addition to using Moslem mercenaries in these light-cavalry roles, they also had units of what were known as **turcopoles:** lightly armed European horse bowmen who were usually second-generation Europeans born in Syria.* The Christian Spanish also developed a similar light cavalry—known as **genitours**—armed with light lance or javelin, for reconnaissance and screening, and for skirmishing with their Moorish opponents.

* J. F. C. Fuller, although a great historian, is in error when he applies the term turcopole to the bastard offspring of Crusaders and Syrian women, *Military History of the Western World,* vol I, p. 424.

Efforts to introduce horse archers in Western Europe were unsuccessful. However, Hungarian or Turkish mercenary horse archers sometimes appeared on European battlefields.

THE SOLDIERY OF MEDIEVAL EUROPE

As we have seen, heavy armored cavalry, designed for shock action, was the central feature of the feudal military forces of Western Europe. Because the weapons and armor were expensive, at first only the wealthy members of the aristocracy could afford to equip themselves. Only to them was the old Roman term *miles* (soldier) originally applied, and thus the elite of the heavy cavalry were also the knights of medieval chivalry.

Two factors caused a change in this situation. First, the nobles and kings found that their offensive or defensive needs required more numerous forces than they could raise among the feudal gentry who owed them allegiance. Furthermore, the short terms of service which a vassal owed his liege lord for offensive military operations seriously inhibited the military ambitions of the many feudal lords. An obvious solution was to provide the horses and heavy equipment necessary to arm and train more promising commoners, who would then ride and fight beside the knights, and who would be available for operations all year round—if the liege lord had the wherewithal to pay and feed them. These were called men-at-arms, and the term *miles* gradually began to be applied to them as to the knights. To help support these standing forces, the king or noble would accept money instead of service from his feudal vassals, an arrangement satisfactory to all.

We have seen how the unreliability of the medieval militia foot levies first led commanders to dismount their knights and men-at-arms, which in turn led to the more economical measure of raising dependable standing formations of foot soldiers, well-armed and well-equipped, though not so lavishly as the mounted soldiers. Most of these were pikemen, whose main role in battle was to stand solidly in a heavy phalanx as a firm base for the firepower of crossbowmen and the ponderous maneuvers of the heavy cavalry. It was such footmen as these who repulsed the German knights and provided the base for the Milanese cavalry charge that defeated Frederick Barbarossa at the Battle of Legnano (see p. 320). Perhaps the best pike units of the 12th century were raised in the Netherlands.

Earlier, in England, the unreliability of the fyrd, or militia, had caused Canute to create a large standing force of bodyguards, or **housecarls** (*huscarls*), who became the nucleus of his feudal armies. The housecarls remained the mainstay of English military forces until they were annihilated at the Battle of Hastings (see p. 310).

Some noble adventurers hit upon the scheme of raising mercenary forces, which they would then hire out to other nobles or kings who could not afford to maintain full-time standing armies, or who had special requirements which could not be met by their regular standing forces. Such mercenary units made up a substantial portion of the army of Duke **William** of Normandy when he invaded England.

NAVAL WARFARE

The most important naval development was the rise of the maritime states of Italy, particularly Venice, Pisa, and Genoa. The Norman Kingdom of Sicily also developed a substantial naval capability. By the close of the 12th century the navies of these four states dominated most of the Mediterranean. The Byzantine navy declined markedly during this period, in large part due to the devastation of the maritime regions of Anatolia after the Battle of Manzikert.

Sea power was a decisive element in the Crusades. Thus, during the 12th century the maritime states of Italy and Sicily were able to make healthy profits by ferrying Crusader armies to the Holy Land, and then by providing them with logistical support. Had the Moslems not been driven from the Mediterranean during the early 11th century, even a partial Christian success would have been impossible.

The principal fighting vessels of the era were—as they had been for centuries—long, low galleys, whose oarsmen were usually slaves. Methods of naval warfare were unchanged. The objective of a naval fight was to ram the opposing vessel, or to capture it by boarding.

There is evidence that during the 11th century in China the navy of the Northern Sung Dynasty began to make use of the first crude compass.

THE CHURCH AND MILITARY AFFAIRS

The Western Church

In theory the Catholic Church was consistently opposed to the warlike brutality of the age in Western Europe. Yet in practice the Church realistically accepted its bellicose environment, and not infrequently leading churchmen—popes and bishops—employed force to obtain both spiritual and temporal objectives of the Church. This was particularly true during the prolonged struggle between the popes and the German emperors.

One of the paradoxes of the time, in fact, was the personal participation of some priests in the actual business of battle. Many of these were hardly a credit to their calling; they were generally members of the nobility who had entered the Church to enjoy its temporal benefits, rather than because of any religious vocation. On the other hand, many warrior priests were devout men of God, meticulous in eschewing combat for any cause that they could not sincerely reconcile with the teachings of the Church. Such men as Bishop **Adhemar du Puy,** for instance, believed that fighting the infidel Moslems was a direct contribution to Christianity and to the glory of God.

The first attempts to control or to limit war were made by the Church during these centuries. There were at least three manifestations of this.

First was the "Peace of God." Late in the 10th century (c. 990), the Church in France had begun to propagate the understandable maxim that priests, monks, and nuns must not be harmed during military operations. This was later extended to cover shepherds, schoolchildren, merchants, and travelers. A further extension applied the Peace of God to churches, and to people going to or from church on Sundays. From this grew almost universal acceptance of church buildings as sanctuaries. The provisions of the Peace of God were fairly faithfully adhered to in most of Europe, possibly because they did not interfere greatly with actual military operations.

More sweeping—and thus less effective—was the "Truce of God." This Church precept, which apparently first appeared in Aquitaine early in the 11th century, forbade fighting on Sundays. At first the teaching was observed; but as bishops extended the Truce, first to cover the entire week end and later to last from Vespers on Wednesday to sunrise on Monday, as well as all of Lent, Advent, and Holy Days, it was generally ignored.

The third Church ruling affecting the warfare of the period was a direct measure of arms control. It was also an indirect testimonial to the effectiveness of the crossbow. This weapon was outlawed by a Vatican edict in 1139 as being too barbarous for use in warfare between Christians; its employment against Moslems, or other infidels was, however, deemed perfectly appropriate.

The Eastern Church

The Orthodox Church of the Byzantine Empire was somewhat less compromising in its attitude toward war and slaughter than was the Roman Church—as might be expected in a more cultured and less aggressive society. But Orthodox religious leaders recognized and granted absolution for killing in self-defense—both individual and that of the state.

WESTERN EUROPE

ENGLAND

Dynastic Struggles of Saxons, Danes, and Normans, 1000–1066

Soon after the beginning of the century, able Danish King **Sweyn Forkbeard** took advantage of the ineptitude of the Saxon King **Aethelred** ("the Unready") to begin the conquest of England, completed by his brilliant son, **Canute** (after the death of Aethelred's eldest son, **Edmund Ironside**). England enjoyed a period of unprecedented peace and prosperity under Canute, who at the same time extended his empire by conquest of Norway, Scotland, and part of Sweden.

The basis for a complicated and momentous series of dynastic wars was Canute's marriage to Aethelred's widow, **Emma,** daughter of **Richard I,** Duke of Normandy. By an earlier liaison Canute had a son of questionable legitimacy, **Harold Harefoot.** By Emma he had a son, **Harthacanute.** Meanwhile, **Edward** and **Alfred,** two sons of Aethelred and Emma (brothers of Edmund Ironside), were residing in exile at the court of their kinsmen, the Dukes of Normandy. The election of Harold to succeed Canute resulted in a three-way war between Harold, Harthacanute, and Alfred. Harold captured and killed Alfred, and repulsed the efforts of Harthacanute (supported by his mother, Emma) to gain the throne. Upon the death of Harold, Harthacanute finally achieved his objective, but died after reigning two years. He selected his elder half-brother, Edward, to be his successor.

Edward, a deeply religious man, known to history as "the Confessor," created dissension in England by bringing many Norman advisers with him. This led to an abortive civil war in which **Godwine,** Earl of Wessex, and his son, **Harold,** tried to end Norman influence. After a brief exile, Godwine became the closest and most trusted adviser of Edward, a relationship continued by Harold after Godwine's death. This ended Norman influence at Edward's court. Apparently Edward had earlier selected his cousin **William,** Duke of Normandy, as his heir. At the time of his death, however, Edward chose Earl Harold (son of Godwine) as his successor, and Harold was elected king by the Witan (council of nobles). William of Normandy, considering himself the rightful heir to the throne, immediately prepared to invade England. But before

William's fleet could sail from Normandy, Harold's brother, **Tostig,** with **Harold Hardraade,** King of Norway, led another invasion to Northumbria. Harold defeated and killed these two invaders at the **Battle of Stamford Bridge,** just before William landed on the south coast of England. The principal events were:

968–1016. Reign of Aethelred the Unready.

1003–1013. Danish Conquest of England. By Sweyn, with the assistance of his son, Canute. Aethelred took refuge in Normandy.

1014–1016. War between Aethelred and Canute. Aethelred, unsuccessful in an invasion attempt, died in London while Canute was approaching with a victorious army.

1016. War between Canute and Edmund (son of Aethelred). Edmund was victorious at the **Battle of Pen** in Somersetshire. The **Battle of Sherston** in Wiltshire was indecisive. Edmund drove Canute from his siege of London, but was badly defeated by Canute at the **Battle of Assandun** (Ashington) in Essex, thanks in part to treachery in the Saxon army. Canute and Edmund then agreed to divide England, Canute reigning in the north, Edmund in the south. Edmund died soon afterward, and Canute was elected king of the entire country by the Witan.

1016–1035. Reign of Canute (See p. 316).

1035–1040. Reign of Harold Harefoot.

1040–1042. Reign of Harthacanute.

1042–1066. Reign of Edward the Confessor.

1051. Revolt of Godwine and Harold. They were restored to favor the next year.

1053. Death of Godwine. Harold became Earl of Wessex.

1055–1057. Civil War. Harold's influence was challenged by **Aelfgar,** Earl of Mercia and East Anglia. Aelfgar was assisted in this inconclusive struggle by **Gruffyd,** Prince of North Wales.

1063. Invasion of North Wales. Harold, assisted by his brother **Tostig,** Earl of Northumbria, defeated Gruffyd.

1064. Harold's Oath. Shipwrecked off the coast of Normandy, Harold was forced by Duke William to swear to support his claim to the throne of England as heir of Edward the Confessor.

1065. Revolt in Northumbria. Tostig was ejected in favor of **Morkere,** son of Aelfgar, and brother of **Edwin,** Aelfgar's successor as Earl of Mercia. Harold recognized the new earl, gaining the undying enmity of his brother, Tostig.

1066, January 6. Death of Edward. Harold was elected king by the Witan. William of Normandy prepared to invade South England. During the spring and summer Harold repelled several raids of Norse adventurers under his brother Tostig.

1066, September. Norwegian Invasion. Harald Hardraade, with Tostig, brought an army by ship up the Humber River. Harald (of Norway) agreed that Tostig would be made Earl of Northumbria if he helped the Norwegians to conquer the rest of England.

1066, September 20. Battle of Fulford. Harald and Tostig defeated Edwin and Morkere, Earls of Mercia and Northumbria, then occupied York. Meanwhile, Harold (of England) had collected an army and started north from London (September 16).

1066, September 25. Battle of Stamford Bridge. King Harold defeated the invaders in an exceedingly hard-fought and bloody battle. Fighting raged for several hours before the Norwegian army gave way. Only 24 ships were needed to take the remnants back to Norway, while 300 ships had sailed for England. Both Tostig and Harald Hardraade were killed. Casualties were heavy among King Harald's housecarls, the only regular standing force in England. The Northumbrian and Mercian levies were also depleted.

1066, September 28. Norman Invasion. Landing of William at Pevensey in Sussex.

1066, October 2. Harold Marches South. As soon as he learned of William's landing, Harold started marching south with his housecarls.

THE BATTLE OF HASTINGS (OR SENLAC), OCTOBER 14, 1066

As soon as Harold ascended the throne of England, William of Normandy began to raise an army to fight for what he considered his rightful inheritance. He could not use his feudal levies for such a large-scale protracted operation outside of their own localities. Accordingly, most of his troops were mercenaries or feudal contingents attracted to his standard by the promise of lands and wealth in England. The exact size of his army is unknown. Estimates of reliable military historians vary from 7,000 to 50,000. The lower figure is probably closer to the truth. Oman suggests that his total strength might have been as much as 12,000 cavalry and 20,000 infantry.

William was ready to sail for England in a great armada by midsummer, but was long delayed by unfavorable winds. Finally (September 27) the wind changed; the next day the Norman army began to debark at Pevensey. William certainly knew of the invasion of Tostig and Harold Hardraade. (They may even have been secret allies.) He decided to let the Norse and Anglo-Saxon armies punish each other, and remained on the defensive near the south coast. Building a powerful wooden fortification as a base near the coast at Pevensey, he sent his cavalry ravaging the Sussex countryside to gather supplies and to force Harold to act.

Harold covered the 200 miles between York and London in less than 5 days. He waited a few days at London, from October 6 to 11, to gather up all available militia and to rest his exhausted housecarls briefly. He then marched to Senlac, arriving the afternoon of the 13th—about 56 miles in 48 hours. Choosing a ridge about 8 miles northwest of Hastings, Harold organized his army for defense since he felt certain that William would attack as soon as possible. As with the Norman army, we have no accurate figures for Harold's strength. But on the basis of the description of the battle, and the known frontage of the Anglo-Saxon army, it would appear that he had about 9,000 men, including some 3,000 housecarls. Larger numbers have been suggested, but are unlikely in view of the limited battle area. It has been suggested that if Harold had waited a few days, he would have been joined by the Northumbrian and Mercian levies, and that he would have been able to gather more south Englishmen as well. There is some doubt whether the northern levies ever started out, or indeed if they were in any condition to. Whether he could have raised more men in the south or not, Harold apparently felt his political and military situations were sufficiently insecure to make it essential to seek a decision as soon as possible.

Believing (probably correctly) that he was outnumbered, and that—save for his depleted housecarls—his troops were neither so well trained nor so well armed as the Norman mercenaries, Harold decided to fight on the defensive. He dismounted his cavalry housecarls, and these, with the infantry housecarls, formed the center of his line at the highest point on the ridge. The remainder of his army—men of the fyrd, or militia—were stretched out along the ridge on each side, some 300–400 yards, in a dense infantry mass, probably about 20 men deep. Harold's army awaited the Norman attack early on October 14. Possibly during the afternoon of the 13th, the Anglo-Saxons hastily erected a rough hedge, or abatis, in front of their line; there is dispute about this among scholars.

The Normans advanced soon after dawn in three lines. In the front were William's archers, including a number of crossbowmen (first recorded appearance of the medieval crossbow in battle). In the second line were William's infantry pikemen. The third line contained the cavalry knights.

The battle was opened by the Norman archers, firing from less than 100 yards. But since they

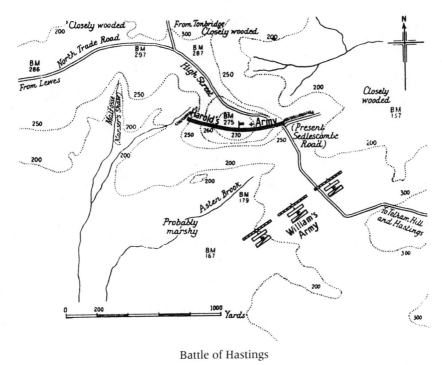

Battle of Hastings

had to fire uphill, most of the arrows and bolts were either short, or deflected by Anglo-Saxon shields or flew harmlessly overhead. The archers then evidently fell back through intervals in the line of Norman pikemen. The pikemen rushed up the slope, but were met by a shower of javelins and rocks—some thrown by hand, some hurled from slings—and were thrown back by the Anglo-Saxons, who were wielding swords, pikes, and great two-handed, double-bladed battleaxes.

With his infantry discomfited, William led his cavalry in a charge up the slope, with no better results. The Norman left wing was so roughly handled, in fact, that it fell back in panic. The elated Anglo-Saxon militiamen on the right immediately dashed down the hill in pursuit. Panic began to spread through the rest of the Norman army as the rumor of William's death passed like wildfire through the ranks.

Throwing off his helmet so the men could see his face, William galloped along the line of his retreating center, which quickly rallied. He then led a charge against the Anglo-Saxon right wing, which had become scattered in premature, heedless pursuit. The Norman cavalry quickly overwhelmed the surprised pursuers.

William now led his cavalry once more against the Anglo-Saxon center, and again was repulsed. Hoping to lure more of the English to leave their solid defensive position, William ordered his men to feign flight. Despite Harold's efforts to hold his men fast, a number of them fell into the Norman trap, and were cut down in the open at the base of the hill when William led another cavalry counterattack. But the main Anglo-Saxon line still stood firm and impervious to repeated assaults.

For several hours the Normans alternated harassment by bowmen with renewed infantry and

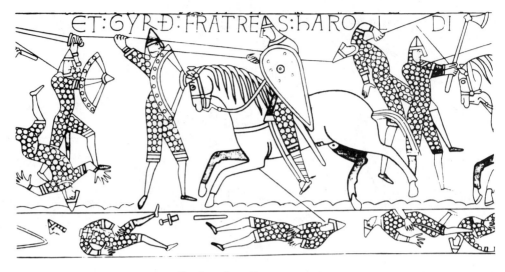

ET·GYRÐ·FRATRE·AS·hARO L DI

Hastings from Bayeux tapestry

cavalry assaults. William ordered his archers to use high-angle fire, so that the arrows and bolts would plunge down on the Anglo-Saxon line. Though many were killed this way, Harold's men were still unshaken by midafternoon, but, having had no respite from alternating Norman fire and movement, were close to exhaustion.

At this time a chance arrow struck Harold in the eye, mortally wounding him. Encouraged, the Normans renewed their assaults, and now the leaderless English line began to collapse. The fyrd gave way, and soon only the housecarls were left, at the top of the ridge, around the body of their dying king. But their fight was now hopeless; completely surrounded, their line began to break. By dark the Normans held the crest. William led a pursuit into the woods behind the ridge, and was almost killed when a few housecarls rallied to renew the fight. But these brave men were soon overwhelmed. The Battle of Hastings was over.

No battle was ever more hard-fought than Hastings; no battle has had more momentous results. In one respect this was merely the conclusion of a dynastic war establishing the succession to the crown of a small island kingdom. But in fact this battle was a turning point in history: the initiation of a series of events which would lead a revitalized Anglo-Saxon-Norman people to a world leadership more extensive even than that of ancient Rome.

Following the battle, William seized Dover, then advanced on London. At first the capital rejected his demand for its surrender. William thereupon began to devastate the surrounding countryside, and London quickly capitulated. William's claim to the throne was acknowledged, and on Christmas Day, 1066, in Westminster Abbey, he was crowned William I, King of England.

The Early Norman Kingdom, 1067–1200

Within four years William had completely consolidated his control over England. He broke the power of the great Anglo-Saxon earldoms. His was still a feudal kingdom, but with a far greater degree of centralized authority than would exist in any other European nation for several centuries. For instance, no baron could build a stone castle without royal authority; when feudal levies were called to his service, they had to swear allegiance to the king.

Under subsequent rulers in the following century, there was some decline in the centralized authority of the king; but despite the feudal pressures of the time, and despite a number of civil wars, the nation remained unified.

During the remainder of this period, the rulers of England were also the greatest landowners of France. Thus the dynastic succession to the Duchy of Normandy was as important to them as the crown of England. Their French possessions involved them in frequent armed conflicts with other feudal lords in France, and not infrequently with the French kings.

The last king of England during this period was **Richard I,** known as "the Lion-hearted." Possibly the greatest European soldier of the Middle Ages (though opinions differ widely), he spent most of his time in foreign wars, and was a poor king. Ironically, the campaigns in which he won his military reputation (the Third Crusade, see p. 344) were failures, politically and strategically. The principal events were:

1067–1071. Consolidation of England by William I. The most serious of his several campaigns of pacification was in suppressing the rebellion led by **Hereward the Wake,** the "great uprising of the north" (1069–1071). The rebels, aided by a Danish force under **Jarl Osbiorn** sent by King **Sweyn Estrithson,** captured York. William defeated the joint force, recaptured York, and drove the Danes back to their ships. The revolt ended with William's successful assault on Hereward's island stronghold of Ely.

1072. Invasion of Scotland. King **Malcolm** was forced to pay homage.

1073. William's Reconquest of Maine. This French province had revolted from his feudal control while he was busy in England (1069).

1075. Revolt of the Earls of Hereford and Norfolk. Suppressed by William.

1076. Invasion of Brittany. The Duke of Brittany was sheltering the fugitive Earl of Norfolk. William abandoned the campaign under pressure from **Philip I,** King of France.

1077–1082. Dynastic Unrest. This was marked in Normandy by sporadic revolts of **Robert,** eldest son of William.

1087. War with Philip of France. During this campaign William died as a result of an accident.

1087–1100. Reign of William II (second son of William I). He was almost constantly at war with his elder brother Robert, who had become Duke of Normandy, over succession to the throne of England and over control of family holdings in France.

1100–1135. Reign of Henry I (fourth son of William I). He continued the conflict with Robert of Normandy, culminating in the **Battle of Tinchebrai.** Henry defeated and captured Robert, and declared himself Duke of Normandy as well as King of England. A series of wars with **Louis VI** of France, who assisted Norman rebels, ended with the defeat of Louis at the **Battle of Bermule** (1119).

1135–1154. Reign of Stephen (nephew of Henry). For 18 years (1135–1153) Stephen was engaged in a seesaw, unimaginative, exhausting, and inconclusive dynastic war with his cousin **Matilda** (daughter of Henry) and her son, also named **Henry.** Stephen's principal opponent in this warfare was **Robert,** Earl of Gloucester, half-brother of Matilda. The "Anarchy" was ended by a truce concluded after the drawn **Battle of Wallingford** (1154).

1154–1189. Reign of Henry II (son of Matilda). He founded the Plantagenet dynasty. An able ruler and general, Henry restored general peace and prosperity to England. His marriage (1152) to **Eleanor** of Aquitaine, whose marriage to **Louis VII** of France had been annulled, made him the Lord of Poitou, Guienne, and Gascony, and thus the largest landholder in France, laying the basis for the later Hundred Years' War (see p. 382). Henry was sporadically at war with his suzerain, Louis (1157–1180). He invaded and subdued Wales in a series of campaigns (1158–1165). He sent his vassals to invade Ireland (1169–1175; see p. 315), participating there briefly himself (1171). He vigorously suppressed the last important feudal revolt of the Anglo-Saxons (1173–1174). The murder of Saint **Thomas à Becket,** Archbishop of Canterbury and enemy of Henry, by Norman knights (December, 1070) may have led to that revolt, which was also sponsored in part by at least two of Henry's sons. This was the beginning of sporadic intrigue and frequent open warfare in England and in his French domains between Henry and his sons, **Henry, Richard, Geoffrey** and **John,** with constantly shifting alliances be-tween sons and father (1173–1189). At the time of his death Henry was engaged in an unsuccessful war against Richard and King **Philip II Augustus** of France, the victorious allies having just been joined by Henry's youngest son, John.

1189–1199. Reign of Richard I (son of Henry II). He was the very epitome of the chivalrous knight, performing well-documented prodigies of valor and strength. During the Third Crusade he also proved himself one of the few brilliant generals of the Middle Ages (see p. 344). He was a poor ruler, spending very little time in England because of his occupation with the Crusaders (1190–1191), a period of captivity in Austria (1192–1194), and his protracted war with Philip of France (1194–1199; see p. 318). Richard won the only important battle of this war of sieges at **Gisors** (near Paris, 1197). He died of a wound received at the siege of **Châlus** near Limoges. He was succeeded by his brother **John** (1199–1216).

SCOTLAND

At the outset of the 11th century, Scotland comprised four principal communities, or kingdoms: Scots, Picts, Britons, and Angles, upon whom were superimposed a number of independent Norse settlements, particularly in the islands off the coast. These separate entities coalesced into a single kingdom of Scotland under Scottish Kings **Malcolm II** (1005–1034) and **Duncan** (1034–1040), though the islands still remained Norse. But despite this apparent unification, and the general acceptance of the suzerainty of a single Scottish king, the nation was in a state of anarchy during this period. In addition to almost constant internal wars and feuds, the Scots raided frequently into the northern provinces of England, and in turn were subjected to a number of English punitive expeditions. The principal events were:

1009–1010. Danish Invasions. Malcolm was defeated by Sweyn Forkbeard at **Nairn** (1009), but the Danes withdrew. Returning the following summer, Sweyn was repulsed by Malcolm in a desperate battle at **Mortlack.**

1018. Battle of Carham. Malcolm invaded Northumbria, defeated the Northumbrians on the Tweed River and annexed the region between the Tweed and the Firth of Forth.

1040–1057. Reign of Macbeth. A usurper who seized the throne, probably after killing Duncan. He was defeated by Duncan's son **Malcolm** at the **Battle of Dunsinane** (1054), which led to his overthrow and death three years later.

1057–1093. Reign of Malcolm III. He engaged in sporadic border wars with the English. William the Conqueror invaded Scotland and forced Malcolm to acknowledge his suzerainty (1072). Full-scale war with the Anglo-Normans broke out again (1077–1080), and Malcolm finally was killed in his last invasion of England at the **First Battle of Alnwick** (1093).

1124–1153. Reign of David I. Supporting Matilda (mother of Henry II) in her wars against Stephen of England (see p. 313), David conquered most of the English provinces of Northumberland and Cumberland, despite a defeat by the local English nobles and militia at the **Battle of the Standard,** or of Northallerton (1138). He was later repulsed by Stephen when making further inroads into England (1149).

1153–1165. Reign of Malcolm IV (grandson of David). He was forced to surrender Cumberland and Northumberland to Henry II of England.

1165–1214. Reign of William the Lion (brother of Malcolm IV). He supported the feudal rebels and sons of Henry II in the Great Rebellion (1173–1174); he was defeated and captured by Henry at the **Second Battle of Alnwick,** and forced to pay ransom to the English king (1174).

IRELAND

As the 11th century opened, **Brian Boru** was consolidating his position as the leading king of the several Irish kingdoms (see p. 270). In the following decade, he drove out the Danes to gain unchallenged sovereignty over the entire island. One further Danish effort to reestablish themselves in Ireland was smashed by Brian at the **Battle of Clontarf** (just north of Dublin), but Brian was killed during the battle. There was no successor who could hold the turbulent Irish in check; the following century and a half were chaotic. Early in the 12th century **Magnus,** King of Norway, led the final Norse invasion of Ireland, but was killed in Ulster, and his men withdrew. Just after the middle of the century, a number of Norman-English adventurers invaded Ireland with the approval of Henry II and under the leadership of **Richard of Clare.** After they established themselves on the east coast of Ireland, Henry himself came with a substantial army and with papal authority to rule Ireland. There was little further fighting. The Irish, awed by the Norman strength and respectful of Henry's church support, submitted. The island was relatively peaceful for the rest of the century. The principal events were:

1002. Brian Boru High King. He defeated **Mael Sechnaill II,** former high king, who recognized Brian as his successor.

1014. Danish Invasion. Danes from the Orkney Islands were joined by an uprising of Danes around Dublin. Brian defeated the invaders and suppressed the insurrection by his last great victory at **Clontarf.**

1014–1167. Chaotic Internal Squabbling.

1167–1171. Norman Conquest.

SCANDINAVIA

Denmark

This period opened and closed with Denmark one of the leading powers of Northern and Western Europe. But for much of the intervening time the country had been relatively weak, and at one point was on the verge of disintegration. The important events were:

985–1014. Reign of Sweyn I, Forkbeard. He made Denmark predominant in Scandinavia, and conquered most of England (see pp. 273, 308). He left his English dominions to his son

Canute and his Danish kingdom to another son, **Harald** (1014–1018). When Harald died, Canute also inherited the Danish throne.

1019–1035. Reign of Canute II. He secured his position in Denmark, with the aid of English troops, after the death of his brother Harald (1018).

1026–1030. War with Norway and Sweden. Canute attacked Norway, and **Olaf II Haraldsson** called on **Anund Jakub** of Sweden for aid. Canute was defeated at the indecisive naval **Battle of Stangebjerg** (1026), but his continued control of trade routes encouraged opposition to Olaf. Olaf fled to Russia after Canute decisively won the **Battle of Helgeaa** (1028). With Anund Jakub's help, Olaf returned in 1030 to try to regain his throne, but was defeated and killed at the **Battle of Stiklestad** by a much larger Norwegian—Danish army (August 31, 1030). Following this Canute overran Norway and established control over most of the south coast of the Baltic.

1035. Death of Canute. The empire was divided between his sons **Harold Harefoot** (England), **Harthacanute** (Denmark), and **Sweyn** (Norway).

1035–1042. Reign of Harthacanute. He was more interested in England than in Denmark (see p. 309).

1042–1047. Reign of Magnus I Olafsson of Norway. (See p. 317.) Magnus repulsed Wendish (Slav) attacks on Jutland, winning a victory at the **Battle of Lysborg** (1043). Magnus also faced a revolt by **Sweyn Estridsen,** Canute II's nephew (c. 1045–1047) which, despite repeated victories, he was unable to quash.

1047–1074. Reign of Sweyn II Estridsen. He seized the throne after Magnus I died at Skibby (October 25, 1047), but had to fight off the efforts of Norwegian **Harald III Hardraade** (or Hardrade) to conquer Denmark (1047–1064). Sweyn later raided England in support of an Anglo-Saxon rising against William I (1069). Sweyn founded the Valdemar dynasty, and the throne peacefully passed in turn to each of his five sons (1074–1134).

1131–1157. Dynastic Civil War. A quarrel arose between **Magnus the Strong,** son of King **Niels** (1104–1134), and **Knud (Canute) Lavard,** son of former king **Erik Ejegod** (c. 1095–1103). Magnus's murder of Knud sparked a war which flared off and on until Knud's son **Valdemar** gained the throne (1157).

1157–1182. Reign of Valdemar I, the Great. Assisted by the remarkable soldier-statesman-bishop, **Absalon,** Valdemar restored internal order and expanded Danish influence in the Baltic at the expense of Swedes, Wends, and Germans.

1160–1169. War with the Wend Pirates. Absalon eventually conquered the Wend island stronghold of Rügen.

1170–1182. Danish Expansion. Absalon extended Danish control over the coast of Mecklenberg, with expeditions establishing outposts in Estonia.

1182–1202. Reign of Knud (Canute) IV. Danish power and prestige continued to expand, with Absalon continuing as the principal royal adviser. He repulsed a German invasion (1182), and smashed a Pomeranian effort to challenge Danish supremacy on the Baltic in an overwhelming victory at the naval **Battle of Strela** (Stralsund, 1184).

Norway

Norway, conquered by Sweyn Forkbeard at the end of the 10th century, was briefly resurgent about 20 years later under **Olaf II Haraldson,** while Canute was engaged in completing the conquest of England. A few years later, however, Canute reconquered the country (see above). After Canute's death Norway regained its independence. The principal events were:

1015–1028. Reign of Olaf II Haraldson. Expelled by Canute (1028), he tried to come back, and was defeated and killed at the **Battle of Stiklestad** (see above).

1035. Reign of Sweyn (son of Canute). He was soon expelled in favor of **Magnus,** son of Olaf II.

1035–1047. Reign of Magnus I the Good. The Norwegian chieftains, after deposing Sweyn, acclaimed 11-year-old Magnus king. Magnus also became king of Denmark after Hardecanute (Harde-knut or Harthacanute) died (1042). After Harald Sigurdsson returned, Magnus shared the kingdom with him (1045), but fell ill and died (October 1047).

1045–1066. Reign of Harald III Hardraade (the Ruthless). At age fifteen, Harald fled Norway after **Stiklestad** (see p. 316), and served in the Byzantine Empire's Varangian Guard (c. 1034–1042). He became sole king of Norway after Magnus died. His cunning and battlefield prowess won him the byname Hardraade (literally "hard-counseled," or ruthless), also attested by his battle-standard, "Land-Waster." He waged war against Sweyn II of Denmark, but after his victory at the naval **Battle of the Nissa** (Niz) (August 9, 1062) made peace with Sweyn (1064). Harald invaded England (1066), smashing local English forces at the **Battle of Fulford,** but was defeated and killed at the **Battle of Stamford Bridge** (see p. 309).

1066–1093. Reign of Olaf III Kyrre (the Quiet). He ruled jointly with his older brother Magnus (both sons of Harald III), becoming sole king after Magnus died (1069). Olaf waged no warfare during his reign.

1093–1103. Reign of Magnus III Barfot (Bareleg). Bynamed for his wearing of Scottish kilts, he led several expeditions to the Orkneys and Hebrides to reestablish Norse control. He died on an expedition to Ireland (August 1103).

1103–1130. Reigns of the the Sons of Magnus III. Of these the most renowned was **Sigurd I,** the Crusader (see p. 342).

1130–1217. Civil Wars and Church–Crown Struggles. (See p. 400.)

Sweden

These two centuries in Sweden, less civilized and less Christianized than either Norway or Denmark, can be divided into three equal subperiods. In the first, Sweden was involved continuously in the wars of Norway and Denmark. In the second, the country was in a state of anarchy. In the third, peace was restored under a strong central monarchy. The principal events were:

994–1021. Reign of Olaf I Skötkonung. He was driven from the Norwegian provinces captured in 1000 (see p. 274) by **Olaf II Haraldsson** (1015–1019).

1021–1050. Reign of Anund Jakub. Anund and Olaf II of Norway formed an alliance against Canute of Denmark, but were decisively defeated (1026–1030; see p. 316).

1050–1060. Reign of Edmund. He was the younger brother of Anund Jakub.

1060–1066. Reign of Stenkil (Steinkel). When Edmund died heirless, the throne passed to his son-in-law, Stenkil. He was defeated by Harald Hardraade of Norway and Sweyn II of Denmark (c. 1063). His death without heirs marked the start of a long period of civil war.

1066–1134. Civil Wars. These combined repeated succession struggles with the final efforts of the pagans to resist Christianization.

c. 1130–1156. Reign of Sverker. This Östergötland magnate reestablished a measure of central government and ended the civil wars.

1156–1160. Reign of Erik III Eriksson. He led a crusade to Finland (1157; see p. 318), and was assassinated (1160). He was later canonized as the patron saint of Sweden.

1160–1210. Further Civil Wars. These were waged between the descendants of Erik III and Sverker. During this period the Archbishopric of Uppsala was created (1164), with five bishoprics in Sweden, and later one (Åbo) in Finland.

Finland

The Swedes continued their efforts at Christianization and colonization of the southwestern coast of Finland, a process which had begun in the 8th century. Among the major events were:

1157. Swedish Crusade into Finland. King Erik III Eriksson launched a crusade to subdue and convert the pagan Finns. Accompanying him was an English bishop named **Henry,** who was later murdered and became Finland's patron saint.

c. 1172. Papal Bull. Pope Alexander III issued a bull encouraging the Swedes to force the Finns to submit and become Christians.

FRANCE

A period of gradual expansion of the authority and the domains of the kings of France, the successors of Hugh Capet (see p. 278). Royal authority and the very existence of the kingdom were seriously threatened when **Henry of Anjou** married **Eleanor of Aquitaine** (former wife of Louis VII of France) to become feudal lord of all of southwestern France in addition to the Norman dominions of northern and northwestern France, followed by his coronation as Henry II of England. But Louis VII and his son and successor, **Philip Augustus,** encouraged and profited from the revolts of Henry's sons, thus preventing the English king from taking over all of France. During the last two decades of the 12th century, Philip Augustus began military and diplomatic successes which brought him unquestioned supremacy. The principal events were:

1031–1060. Reign of Henry I. He defeated a rebellion of his brother **Robert** (1032). He then was engaged in a series of wars against the counts of Blois, of Champagne, and of other great fiefs of northern France (1033–1043). In the opening campaigns he was greatly assisted by **Robert,** Duke of Normandy.

1035. Accession of William I as Duke of Normandy. Because of his illegitimate birth, many Norman nobles revolted against him (1035–1047). The revolt was finally suppressed, with the assistance of Henry I of France, by victory at the **Battle of Val-des-Dunes,** near Caen. Two years later William and Henry were at war, and the Normans repulsed a French invasion (1049). They fought another prolonged, inconclusive war (1035–1058).

1060–1108. Reign of Philip I. He accomplished little save to gain a victory over William the Conqueror (1079).

1108–1137. Reign of Louis VI. He fought two inconclusive wars with Henry I of England (1109–1112, 1116–1120). He repulsed a German invasion (1124), and generally expanded the power of the monarchy at the expense of the great nobles.

1137–1180. Reign of Louis VII. He inspired and participated in the Second Crusade (see p. 343). The most important events of his reign were the annulment of his marriage to Eleanor of Aquitaine and her subsequent marriage to Henry II of England. As a result, the power and prestige of the French monarchy were greatly diminished.

1180–1223. Reign of Philip II, Augustus. Early in his reign he subdued the great lords of the north and northeast (1180–1186). He allied himself with the rebellious sons of Henry II of England in order to reduce Plantagenet power in France (1187–1190). Save for a brief and stormy alliance in the Third Crusade (see p. 344), Philip was almost constantly at war with Richard I of England (see p. 314). By the end of the century he had gotten off to a good start in his efforts to restore and enhance the power and prestige of the French monarchy.

GERMANY

In the early 11th century there was a gradual rise in imperial power and prestige under three able emperors. Then complicated and intermingled political and religious strife rocked Germany. The power of the emperors (Holy Roman Empire) declined steadily, despite some restoration of imperial prestige under **Frederick I** in the latter part of the 12th century. Yet the basic vigor and vitality of the German people were demonstrated by steady expansion eastward, at the expense of neighboring Slavs.

Germany's internal troubles stemmed mostly from a prolonged struggle between the emperors and the popes in both spiritual and temporal affairs. A reform movement in the Church caused a line of vigorous 11th-century popes (**Gregory VII** being outstanding) to insist upon the right of appointing bishops and eliminating the graft and corruption which had stemmed largely from imperial control over the clergy in Germany. German nobles took advantage of the dispute to increase their power at the expense of the emperors, while the popes also exploited Germany's internal political troubles to gain their ends. For their part, the German emperors (**Henry IV** and Frederick I being most renowned) conducted frequent military expeditions into Italy for the purpose of controlling the papacy.

Early in the 11th century, the Hohenstaufen family of Swabia and the so-called Welf family of Saxony and Bavaria became engaged in a bitter struggle for the imperial succession. The resultant civil wars became involved with the concurrent church–state disputes. Because of the Hohenstaufen estates near Waiblingen, this civil war became known as that of the Welfs and Waiblingens. In Italy these same names, with some linguistic modifications, were applied to the participants in the church–state struggle: Guelphs for adherents of the papal party, Ghibellines for supporters of the emperor. The principal events were:

1002–1024. Reign of Henry II. During the early years he conducted a series of wars against **Ardoin,** King of Lombardy, whom he finally defeated and deposed (1002–1014). He was less successful in his wars against **Boleslav** of Poland, who conquered Lusatia and Silesia from the Germans (1003–1017). He suppressed a revolt by **Baldwin,** Count of Flanders (1006–1007).

1024–1039. Reign of Conrad II. A vigorous, capable soldier and administrator, he asserted his authority over the unruly German nobles and in Italy as well. He was repulsed in an attempted invasion of Hungary by (Saint) **Stephen I** (1030). He defeated the Poles and temporarily recovered Lusatia for Germany (1031).

1039–1056. Reign of Henry III. He centralized authority and expanded the empire by foreign conquests. He defeated **Bratislav** of Bohemia (1041). He was involved in a number of conflicts with his vassals, **Baldwin V** of Flanders and **Godfrey** of Lorraine, suppressing their most serious combined revolt (1047). He was repulsed in repeated invasions of Hungary by **Andrew I** (1049–1052).

1056–1106. Reign of Henry IV (age 6). When he came of age, Henry was forced to exert great efforts to recover central power lost during a weak regency. He suppressed a particularly violent Saxon revolt (1073–1075). His struggle with the papacy (1073–1077) ruined his reign and seriously weakened the empire. The real victor in this struggle was Bishop **Hildebrand,** later Pope Gregory VII, before whom Henry finally abased himself at Canossa.

1077–1106. Civil War in Germany. Rebellious

nobles elected **Rudolph** of Swabia to replace Henry. Though the emperor soon defeated and killed Rudolph (1080), the revolt continued. Henry finally defeated another rival, **Herman** of Luxembourg (1086–1088). The revolt continued sporadically to the end of the reign.

1081–1085, 1090–1095. Henry's Expeditions to Italy. He briefly captured Rome (1083), but Gregory took refuge with the Norman duke, **Robert Guiscard,** of southern Italy, who recaptured the city (1084).

1093–1106. Revolt of Henry's Sons. Henry (later **Henry V**) and **Conrad** joined German and Italian rebels. The emperor was captured, but escaped (1105), and died while raising a new army (1106).

1106–1125. Reign of Henry V. The political-religious-military struggle between emperor and popes continued. Much of this reign was occupied with domestic and foreign wars: Henry campaigned successfully in Bohemia (1107–1110), but was repulsed in invasions of Hungary (1108) and Poland (1109). An expedition to Italy (1110–1111) forced the temporary submission of Pope **Paschal.** He was only partially successful in dealing with a series of revolts in Lorraine and other German provinces (1112–1115). He campaigned repeatedly and with mixed success in Holland (1120–1124). He was repulsed by Louis VI in an invasion of France (1124; see p. 318).

1125–1137. Reign of Lothair II. He was elected by the German nobles over a rival, **Frederick Hohenstaufen,** Duke of Swabia. The rivalry erupted into civil war (1125–1135) to initiate the struggle of Welf (Guelph) and Waiblingen (Ghibelline). After defeating the Hohenstaufens, Lothair led an unsuccessful expedition against **Roger II,** King of Sicily (1136–1137).

1138–1152. Reign of Conrad III (first of Hohenstaufen Dynasty). The Welfs, under Dukes of Saxony **Henry the Proud** and his son, **Henry the Lion,** rebelled. Henry the Lion (1142–1180) became a virtually independent king, expanding his dominions eastward against the Slavs, while successfully defying the emperor. Conrad, accompanied by his able warrior nephew, **Frederick,** participated in the disastrous Second Crusade (1146–1148; see p. 343).

1152–1190. Reign of Frederick I, "Barbarossa." The greatest of the Hohenstaufens. Sporadic war against Henry the Lion continued, but Frederick eventually defeated and deposed the Welf duke (1182). He was victorious in a number of campaigns in Poland, Bohemia, and Hungary (in intervals between his campaigns in Italy, 1156–1173).

1154–1186. Frederick's Six Expeditions to Italy. Frederick had mixed success in the imperial struggle with the papacy. Though he captured Rome on his fourth expedition (1166–1168), he was forced to leave Italy by a pestilence that destroyed his army. His fifth (1174–1177) ended in disaster at the **Battle of Legnano** (May 29, 1176), when, with a force consisting only of cavalry, he rashly accepted battle with a much larger infantry and cavalry army of the Lombard League. The steady Italian pikemen repulsed Frederick's attacks; the Lombard cavalry then routed the Germans by an envelopment. (This battle has sometimes been incorrectly cited as the first medieval victory of infantry over cavalry; the Lombard victory was due to the coordinated use of both arms.)

1189–1190. The Third Crusade. Frederick was drowned in Cilicia (1190; see p. 344).

1190–1197. Reign of Henry VI. Welf revolt erupted again under aged, but still vigorous, Henry the Lion, resolved by the **Peace of Fulda** (1190). The emperor campaigned inconclusively in southern Italy against **Tancred** of Sicily (1191–1193); he returned to conquer Sicily and be crowned king (1194–1195). He died in south Italy while suppressing a rebellion (1197). This premature death caused chaos in Germany, where two antikings (**Rudolph** of Swabia and **Otto** of Saxony) were elected as rivals to each other and to Henry's young son, **Frederick II.**

THE MARITIME STATES OF NORTH ITALY

Genoa

At the beginning of the 11th century, Genoa, which had suffered severely from Moslem piratical raids in previous centuries, had begun to gain wealth and power as an independent maritime city-state. In conjunction with the Pisans, Genoese warships ended the naval dominance of the Moslems in the northern Mediterranean, a process climaxed by a joint naval expedition which drove the Moslem pirate **Mogahid** from his bases in Sardinia (1005–1016). The two republics also collaborated in a number of raiding expeditions against Moslem cities in North Africa. The joint capture of Mahadia (1087) gave the Genoese and Pisans effective control of the western Mediterranean. Genoese power and prosperity waxed, and early in the 12th century Genoa challenged Pisan preeminence in the northern Mediterranean and on the islands of Corsica and Sardinia (1118–1132). That conflict was a standoff, but warfare became endemic between the two city republics. Genoa gained its first actual foothold on Corsica by the capture of Bonifacio (1195).

Pisa

Despite the persistent challenges of Genoa, Pisa was the predominant maritime power of western Italy during this period. After defeating Lucca (1003), Pisa was sacked in the last major Moslem raid against Italy (1011). Recovering rapidly, Pisa joined Genoa in operations against the Moslems (see above). One noteworthy success was the capture and sack of Palermo by a Pisan fleet (1062). With papal approval, Pisa obtained virtual sovereignty over Corsica (1077). Pisan warships, joining the First Crusade, provided decisive logistical support in the operations around Antioch (1098) and in the advance on Jerusalem (1099; see p. 340). Pisan expeditions in the western Mediterranean were marked by two particularly successful and profitable raids on the Balearic Islands (1113, 1115). Despite almost continuous warfare with Genoa, Pisa was at the height of its prosperity and power.

Venice

The independence of Venice was recognized by the Byzantines in 584. The Venetians successfully defied combined land and sea attacks by Pepin against their island stronghold (774). In the following centuries Venice became a wealthy trading center. Depredations of Dalmatian pirates, who dominated the Adriatic in the late 10th century, forced the Venetians to build a fleet to protect their commerce. Under Doge **Pietro Orseolo II,** the Venetian fleet defeated the pirates and captured their strongholds of **Curzola** and **Lagosta,** assuring Venetian predominance in the Adriatic. Soon after this the Venetians captured Bari from the Moslems (1002). Later Venice lost most of the Dalmatian coast in a war with King **Calomar I** of Hungary (1097–1102). King **Bela III** of Hungary repulsed Venetian efforts to reconquer Dalmatia (1172–1196). During most of this period Venice was allied with Byzantium, and Venetian fleets played an important role in the struggles between the Byzantines and the Normans. A dispute with Byzantium over trading concessions, however, led to war (1171), in which an attempted Venetian invasion was repulsed and the Venetians were defeated at sea (see p. 331).

Norman States in South Italy and Sicily, 1016–1200

Norman involvement in south Italy and Sicily was apparently accidental, resulting from the presence of a group of Norman pilgrims, returning from the Holy Land, who helped defend Salerno from a raiding Moslem fleet (1016). Soon more adventurers arrived from Normandy, and some of them established a permanent stronghold at Aversa, near Naples. In subsequent years Norman control spread over south Italy. The most important Norman leader, flourishing in the middle and latter portion of the 11th century, was **Robert Guiscard,** who expelled the Byzantines and Moslems from southern Italy and conquered most of Sicily from the Moslems. He then attempted to invade the Byzantine Empire via Epirus, Thessaly, and Macedonia, but was eventually repulsed. During the next century the Normans consolidated their positions in southern Italy and Sicily, and played an important role in the struggles between the German emperors and the popes, while continuing to be involved in Mediterranean affairs and in the Crusades. The principal events were:

1027. First Permanent Norman Settlement in Italy. Established by **Rainulf** in the fortress of Aversa.

1030–1059. Extension of Norman Possessions in Apulia and Calabria.

1038. Byzantine Invasion of Eastern Sicily. The Byzantine army, under **George Maniakes,** included many Normans. When Maniakes was recalled to Constantinople, the Moslems rallied and drove the Byzantines out.

1053. Battle of Civitella (Civitate, Northwestern Apulia). Normans, led by **Humphrey Guiscard** and his brother Robert, defeated the army of Pope **Leo IX,** which was attempting to relieve the region of Norman depredations. The Pope was captured.

1059. Rise of Robert Guiscard. In return for assistance against German Emperor **Henry IV,** Pope **Nicholas II** appointed Robert Duke of Calabria and Apulia, and authorized him to conquer Sicily.

1060–1091. Norman Conquest of Sicily.

1071. Capture of Bari. Robert eliminated the last Byzantine foothold in Italy.

1081–1085. Invasion of Byzantine Empire. Robert, aspiring to the imperial throne, led a Norman fleet and army across the Adriatic to capture Corfu (1081) and laid siege to Durazzo, defended by **George Palaeologus.** Despite defeat in the naval **Battle of Durazzo** (1081) by a combined Byzantine-Venetian fleet, Robert, who had less than 20,000 men, maintained his position around Durazzo during the winter, but was forced to lift the siege upon the arrival of a relieving army of some 50,000 led by Byzantine Emperor **Alexius Comnenus.**

1082. (Land) Battle of Durazzo* (or Dyrrhachium). Robert, assisted by his son, **Bohemund,** converted defeat to victory by a desperate surprise cavalry attack, smashing and routing Alexius' famed Varangian Guard. Robert's wife, **Sicelgaeta,** apparently participated actively and helped her husband rally his troops before the final decisive charge. In this battle the Norman cavalrymen were faced by Anglo-Saxon axmen, some of them veterans of Hastings, who had taken service with the Byzantines after the Norman conquest of Britain. (Comparison of the two battles is interesting; in both cases the tough, courageous Anglo-Saxons, in their monolithic phalanxes, advanced prematurely, without orders, and were eventually overwhelmed by the combined efforts of bowmen and a furious Norman heavy-cavalry charge.) Robert reinvested Durazzo, which soon surrendered. The Normans now advanced into Thessaly. Soon after this Robert returned to Italy in response to an appeal from

---

* Erroneously called the Battle of Pharsalus by some historians who have misread Gibbon's cryptic comments.

Pope Gregory VII for assistance against Emperor Henry IV, leaving Bohemund in command in Greece. Bohemund advanced to the Vardar River, but while besieging **Larissa** was defeated by Alexius, who drove him back into Epirus.

1083–1084. Relief of Rome. While Henry IV was besieging Gregory VII in Castel San Angelo, in Rome, Robert collected an army of Normans, Lombards, and Moslems, then marched north. Henry retired into north Italy, without risking a major battle (May 1084). The Norman soldiers then sacked Rome, causing the local population to rise against their deliverers and against the Pope who had called them. Robert escorted Gregory to Salerno, then rejoined Bohemund in Epirus.

1084. Battle of Corfu. The Norman fleet fought an indecisive battle with the Byzantines and Venetians.

1085. Death of Robert. Bohemund was forced to return to Italy.

1096–1104. Participation of Bohemund and Tancred in the Crusades. (See p. 338)

1104–1108. Bohemund's Operations in Epirus. (See p. 331)

1105–1150. Reign of Roger II of Sicily. A period of confused warfare against **Rainulf** of Apulia; Roger was eventually successful, establishing sovereignty over all southern Italy as well as Sicily (1137–1139). During this conflict he was at one time or another at war with the Pope, Emperor Lothair of Germany, Pisa, Genoa, Venice, and the Byzantines; he repulsed all invasions, and eventually (1139) obtained papal sanction for his earlier claim to the title of king (1130). Meanwhile, he had been engaged in frequent overseas wars against the Moslem rulers of North Africa. His admiral, **George of Antioch,** captured Malta, Tripoli, Mahadia, and other North African coastal cities to establish a Norman colony extending from Barca to Kairouan (1146–1152). During this time also George's fleet operated successfully along the European coast of the Byzantine Empire, capturing Corfu and sacking Athens, Thebes, and Corinth (1147–1149). George actually brought his fleet within bowshot of the imperial palace at Constantinople (1149), but sailed away when Emperor **Manuel** returned to his capital with a large army.

1154–1166. Reign of William I of Sicily. His fleets were defeated by the Byzantines under Palaeologus, who invaded Apulia (1155; see p. 331). William won an overwhelming land victory at the **Battle of Brindisi** (1156), driving the Byzantines from Italy. He could not prevent the steady conquest of his North African dominions by the Almohads (1147–1160; see p. 333).

1166–1189. Reign of William II of Sicily. Despite energetic efforts, William was generally unsuccessful militarily. He failed in several efforts to reconquer the lost North African colonies and was repulsed at Alexandria (1174). He invaded the Byzantine Empire and captured Durazzo and Thessalonika (1185). Advancing on Constantinople, he was decisively defeated by Emperor **Isaac II Angelus** at the **Battle of the Strymon** (September 7, 1185), and was forced to make peace and to abandon his conquests (1191; see p. 331).

1190–1194. Reign of Tancred (illegitimate grandson of Roger II). His rule was disputed by Emperor **Henry VI,** husband of Roger's legitimate daughter, **Constance.** Tancred repulsed one invasion by Henry (1191) and subdued baronial revolts (1192–1193).

1194. Hohenstaufen Usurpation. Upon Tancred's early death, Henry seized the throne and was succeeded by his son, **Frederick II,** under the regency of Constance (1198–1208).

SPAIN

The slow process of Christian reconquest continued, despite a number of setbacks at the hands of periodically resurgent Moslem leaders. While constant warfare occurred along the shifting Moslem-Christian frontiers, violent internal struggles took place on both sides of the border, resulting in the establishment of a number of virtually independent principalities, kingdoms, and emirates, only occasionally subject to central authority. The principal events were:

1000–1035. Reign of Sancho III, the Great, of Navarre. He had also inherited Aragon, and in a series of wars against his neighbors conquered Castile, León, and Barcelona. As a result he became ruler of all Christian Spain (1027). Upon his death he divided his kingdom among his sons.

1002–1086. Decline of Moslem Spain. After the death of Al-Mansour (see p. 281), his son, **Abdulmalik-al-Mozaffar** briefly maintained Moslem ascendancy in Spain, but his early death led to a succession of disputes and widespread civil war. The caliphate collapsed completely (1031), breaking into a number of minor principalities. The most important was the Abbasid Dynasty of Seville, acknowledged as suzerain by many of the others. During this period of decline, most of the Moslem dynasties, including the Abbasids, became tributary to the aggressive successors of Sancho.

1035–1065. Reign of Ferdinand I, the Great, of Castile (son of Sancho). After seizing León from one brother (1037), he reconquered former Castilian territory inherited by another brother, **Garcia** of Navarre. He won numerous victories over the Moslems, forcing the emirs of Saragossa, Toledo, and Seville to become tributaries.

1035–1065. Reign of Ramiro I of Aragon. Another son of Sancho, Ramiro was the first king of independent Aragon, and expanded his small country at the expense of Christian and Moslem neighbors.

1065–1072. Dynastic Civil War in Castile and León. Sancho, eldest son of Ferdinand I, was at first successful, thanks largely to the inspired generalship of **Rodrigo Díaz de Bivar.** When Sancho was killed in a siege, his younger brother, **Alfonso,** became unchallenged ruler of Castile.

1072–1109. Reign of Alfonso VI of Castile. He was successful in his early wars against the Moslems, his armies being led to victory by Rodrigo Díaz, who soon became known to the Moslems as "sidi" (lord), known in history as **El Cid Campeador.** Ordered into exile by the jealous king (1081), the Cid offered his services to the Moorish emir of Saragossa, winning consistent victories over the Christian rulers of Aragon and Barcelona.

c. 1074–1104. Reign of Pedro I of Aragon. Aragon continued to grow and prosper. He captured the fortified town of **Huesca** by assault (1096) and made it his capital.

1085. Castilian Conquest of Toledo. Mohammed al-Motamid, Emir of Seville, was so alarmed by the loss of Toledo that he called for assistance from the Almoravids of Morocco. **Yusuf ibn Tashfin** thereupon brought a Moorish army to Spain (see p. 333).

1088, October 23. Battle of Zallaka (near Badajoz). Yusuf decisively defeated Alfonso VI. The Castilian king pardoned the Cid, who returned to reorganize the shattered armies of Castile. After another dispute with the king, however, the Cid was banished again (1089).

1086–1091. Almoravid Conquest of Moslem Spain. On the verge of defeat by his erstwhile allies, al-Motamid endeavored to ally himself with Alfonso, but was soon overwhelmed and imprisoned by the Almoravids (1091), who had overrun all of Moselm Spain except Saragossa.

1089–1094. Conquest of Valencia by the Cid. Raising a private army of Christians and Moslems, Diaz made himself virtually an independent monarch. He repulsed invasions by the Almoravids at the **Battle of Cuarte** (1094) and again at the **Battle of Bairen** (1097). After his death (1099), the Almoravids reconquered Valencia (1102).

1104–1134. Reign of Alfonso, I, the Warrior, of Aragon. His brief marriage with **Urraca,** Queen of Castile, involved him in Castilian affairs. Quarreling with his wife, he defeated her adherents at the **Battle of Sepulveda** (1111). He expanded Aragon below the Ebro, capturing Saragossa (1118).

1126–1157. Reign of Alfonso VII of Castile and León (son of Urraca). He was generally successful in wars against Christian and Moslem neighbors, and forced Aragon and Navarre to accept his sovereignty. However, his cousin, **Alfonso (Affonso) Henriques,** asserted the independence of Portugal (1140; see below). To suppress Moslem resurgence under the Almohads (see below), Alfonso of Castile invaded southern Spain, but was repulsed, and died in the **Battle of Muradel** during the withdrawal (1157).

1145–1150. Almohad Conquest of Moslem Spain. The Moorish Almohads (see p. 333), under **'Abd al-Mu'min al-Kumi,** overwhelmed the Almoravids.

1158–1214. Reign of Alfonso VIII of Castile (grandson of Alfonso VII). An infant at accession, Alfonso as a young man had to fight refractory nobles to establish his authority. He then undertook several successful campaigns against the Moslems until he was overwhelmingly defeated by Almohad Caliph **Abu-Yusuf Ya'qub al-Mansour** at the **Battle of Alarcos** (near Ciudad Real, 1195). Following this disaster, Castile was invaded by León and Navarre, as well as by the Moslems, but Alfonso repulsed the invaders.

PORTUGAL, C. 1100–1200

The country of Portugal came into existence on the north bank of the Douro River in the 10th century as the Kingdom of León pressed the Moslems southward on the west coast of the Iberian Peninsula. Separatist tendencies appeared early in the 12th century during the stormy reign of Queen Urraca of Castile and León (see p. 324). The strong Count Alfonso (Affonso) Henriques (1112–1185) won numerous victories over the Moslems, culminating (1139) at the **Battle of Ourique** (location uncertain, probably not modern Ourique). Affonso then declared his independence from his cousin, Alfonso VII of Castile and León, who reluctantly recognized him as King of Portugal (1140). Affonso continued to press the Moslems southward, capturing Lisbon and establishing the Tagus as the southern boundary of Portugal (1147). In his operations against Lisbon he was assisted by Crusader contingents from England and the Low Countries, en route to the Holy Land for the Second Crusade (see p. 343). In the latter years of his reign, Affonso repelled repeated Almohad invasions. He also raided frequently into Moslem territory; in one of these he was captured at Badajoz (1169), but was later released. He captured **Santarem** (1171) and repulsed a Moorish effort to recapture the city (1184).

EASTERN EUROPE*

POLAND

During the first century and a half of the period, Poland gradually expanded under a number of able and martial rulers. During the latter part of the 13th century, however, the country suffered from internal discord, dynastic disputes, and weak leadership. The principal events were:

992–1025. Reign of Boleslav I, the Brave. (See p. 283.) This great king continued the process of centralization and expansion. He fought a series of wars against **Henry II of Germany (1002–1019)** in which he was generally successful, conquering the province of

---

* Bohemia, Hungary, Moravia, Croatia, and Bulgaria are omitted as separate sections during this period, save as they appear in the sections of other nations. The many wars of these small states had little lasting military or historical significance.

Lusatia and displacing the German emperor as the suzerain of Bohemia. He then turned eastward, defeated **Yaroslav,** Prince of Kiev, in the **Battle of the Bug** (1020), and temporarily occupied Kiev. At his death Poland extended from the Elbe to the Bug, and from the Baltic to the Danube.

1025–1034. Reign of Mieszko II (son of Boleslav). A period of decline. Slovakia and Moravia broke away (1027, 1031). Ruthenia was reconquered by Yaroslav of Kiev, and Pomerania was lost to Canute of Denmark (1031).

1034–1040. Dynastic Struggles. Silesia was lost to **Bratislav I** of Bohemia (1038), who also temporarily occupied Cracow (1039).

1038–1058. Reign of Casimir I, the Restorer (son of Mieszko). After reestablishing order in Poland, Casimir reconquered Silesia (1054).

1058–1079. Reign of Boleslav II, the Bold (son of Casimir). A strong and vigorous ruler, Boleslav failed in two attempts to reconquer Bohemia, though he did reconquer Upper Slovakia (1061–1063). He twice occupied Kiev (1069, 1078), and regained Ruthenia for Poland. He was overthrown by a rebellion of nobles.

1079–1102. Period of Decline.

1102–1138. Reign of Boleslav III. A strong ruler and good soldier, he repulsed an invasion of Silesia by German Emperor Henry V at the **Battle of Glogau** or **Hundsfeld,** near Breslau (1109). He reconquered Pomerania by a great victory over the Pomeranians at the **Battle of Naklo** (1109). He subdued a Pomeranian revolt under Prince **Warcislav** of Sczecin (1119–1124). He failed in invasions of Bohemia and Hungary (1132–1135).

1138–1200. Decline of Poland. The German nobles **Albert the Bear,** Margrave of Brandenburg, and **Henry the Lion,** Duke of Saxony, steadily drove the Poles east of the Vistula, while the Danes, under Valdemar and Absalon, seized the Baltic coastal region (see pp. 316 and 320).

RUSSIA

The history of Russia during this period was dominated by the presence or by the memory of two great princes of Kiev: **Vladimir the Saint** (see p. 283) and his son **Yaroslav.** They and their direct descendants ruled Russia for nearly three centuries. The principal events were:

978–1015. Reign of Vladimir the Saint.

1015–1019. Dynastic War between the Sons of Vladimir. Yaroslav was eventually successful, defeating **Sviatopolk,** and driving **Mstislav** east of the Dnieper, where he remained semi-independent until his death (1036).

1019–1054. Reign of Yaroslav the Wise. He defended his domains successfully against the Poles (despite some setbacks; see above), as well as against the Pechenegs.

1054–1136. Arrival of the Cumans. This Turkish tribe, even more ferocious than the Pechenegs, contributed to the decline of Kiev. At the same time the principalities of Novgorod and Suzdal, nominally subject to Kiev, prospered. The rulers of all three states were descendants of Yaroslav.

1113–1125. Reign of Vladimir II, Monomakh. The last great ruler of Kiev, he was engaged in almost ceaseless war against the Cumans.

1157–1174. Reign of Andrei Bogoliubski of Suzdal. He conquered Kiev from his cousin, **Mstislav II, the Daring.**

EURASIA AND THE MIDDLE EAST

BYZANTINE EMPIRE

During the first part of this period the Byzantine Empire continued to demonstrate the remarkable vigor which had returned it to preeminence in the Middle East in the preceding two centuries. This was brought to a sudden and catastrophic conclusion by the appearance of the Seljuk Turks under **Alp Arslan,** who inflicted the disastrous defeat of **Manzikert** on the Emperor **Romanus.** Almost overnight practically all of the Asiatic dominions of the empire were lost to the Turks, the Byzantines retaining only a handful of fortified seaports on the coast of Anatolia. With considerable assistance (as well as some opposition) from the First Crusaders (see p. 339), the able soldier-emperor Alexius Comnenus began a slow and painful process of resurgence, fighting more or less successfully against Normans, Bulgars, Turks, and occasionally Crusaders. Despite the amazing vitality and resilience demonstrated under Alexius and his successors, the empire never fully recovered from Manzikert. Yet at the end of this period it was still the wealthiest and most powerful single nation of Eastern Europe and Southwestern Asia. The principal events were:

Continued Byzantine Revival, 1000–1067

976–1025. Reign of Basil II. (See p. 286.)

996–1018. Conquest of Bulgaria. Basil defeated Bulgarian Tsar **Samuel** at the **Battle of the Spercheios** (996), and reconquered Greece and Macedonia from the Bulgarians. Soon after this Samuel invaded Macedonia again, capturing and sacking Adrianople (1003). Resuming the offensive, Basil finally ejected the Bulgarians from Thrace and Macedonia (1007); his advance into Bulgaria culminated in victory at the decisive **Battle of Balathista** (1014). Basil blinded his 15,000 Bulgarian captives, then sent them back to Samuel, who died of shock. After this, Bulgarian resistance crumbled; Basil soon conquered the entire country (1018).

1018. Battle of Cannae. Byzantine victory over Lombards and Normans attempting to invade south Italy.

1020. Annexation of Armenia. The Armenians asked protection against the Seljuk Turks (see p. 332).

1025–1028. Reign of Constantine VIII. This was marked by a Pecheneg invasion across the Danube. The barbarians were defeated and driven back across the river by **Constantine Diogenes** (1027).

1028–1050. Reign of the Empress Zoë. Her husband, **Romanus,** was defeated by the Moslems in Syria (1030). The precarious situation in Syria was soon restored, however, by a series of victories won by the great general, **George Maniakes** (1031).

1032–1035. Naval Wars against Moslem Pirates. In combination with the Ragusans, a Byzantine fleet defeated Moslem pirates in the Adriatic (1032), completing the work begun earlier by the Venetians (see p. 321). A Byzantine fleet, largely manned by Norse mercenaries under Harald Hardraade (see pp. 309, 316), swept Moslem pirates from much of the Mediterranean, and harassed North Africa.

1038–1040. Invasion of Sicily by George Maniakes. His army included Harald Hardraade's mercenaries and Normans from south Italy. Maniakes captured Messina (1038) and won important victories at the battles of **Rametta** (1038) and **Dragina** (1040). When he was called back to Constantinople to deal with a Bulgarian revolt, the invasion collapsed; the Moslems recovered the island (1040).

1040–1041. Unsuccessful Revolt in Bulgaria. Rebellion, led by **Peter Deljan,** was suppressed by Maniakes.

1041–1042. Revolt of Michael V. After temporarily usurping the throne from Zoë, he was overthrown by the Byzantine nobility.

1042. Battle of Monopoli. Maniakes decisively defeated Norman invaders of Byzantine southern Italy.

1042–1059. Decline. During the reigns of **Constantine IX** (1042–1054), second husband of Zoë), **Theodora** (1054–1056, Zoë's sister), **Michael VI** (1056–1057), and **Isaac I Comnenus** (1057–1059), the army and fleet were seriously neglected and allowed to deteriorate.

1043. Revolt of Maniakes. He crossed the Adriatic from Italy, but died in an accident while marching on Constantinople.

1048–1049. Appearance of the Seljuk Turks. They defeated the Armenians at **Kars** and occupied Ardzen, west of Lake Van. They fought a drawn battle with the Byzantines at **Kapitron,** but were defeated at **Stragna** and repulsed at **Manzikert** (Malazgirt).

1048–1054. Pecheneg Raids. Repeated forays across the Danube.

1059–1067. Reign of Constantine X. Continuing neglect of the military resulted in a series of defeats. In Italy, the Normans captured Rhegium, and completed the conquest of Calabria (1060). The Seljuks, under **Alp Arslan,** overran most of Armenia (1064) and began raiding deep into Anatolia (1065–1067). A Cuman raid across the Danube reached Thessalonika (1064–1065).

Romanus and Manzikert, 1067–1071

1067–1071. Reign of Romanus Diogenes. An able general, Romanus was married by the Empress **Eudocia,** widow of Constantine X, apparently under pressure from the Byzantine aristocracy, who wanted a strong military ruler to meet the mounting foreign threats to the Empire.

1068–1069. Early Campaigns of Romanus. After spending nearly a year in rebuilding the army, Romanus conducted a successful winter campaign against the Turks, who had boldly taken up winter quarters in Phrygia and Pontus. He defeated Alp Arslan at the **Battle of Sebastia** (Sivas), forcing him to withdraw to Armenia and Mesopotamia. Romanus then conducted a successful lightning campaign into Syria, where the Arabs had taken advantage of the Seljuk success to rise against the Byzantines. After subduing a mutiny among his mercenaries, Romanus marched back to Cappadocia, where the Turks had again begun raiding. Driving them off, he marched eastward to the Turkish stronghold of Akhlat (Ahlat) on Lake Van, which he besieged. He sent part of his army on a raid farther east, into Media, where it was defeated by Alp Arslan. Romanus abandoned the siege, and Alp raided into Anatolia toward Iconium. Romanus cleverly moved behind the Turks and defeated them at the **Battle of Heraclea** (Eregli). Alp withdrew to Aleppo. Save for a few fortresses in Armenia, Romanus had now driven the Turks completely out of the empire.

1068–1071. Operations in Italy. Taking advantage of Romanus' occupation with the Turks, the Normans renewed their attacks against Byzantine outposts in southern Italy, capturing Otranto (1068), and threatening Bari. Romanus hastened to Italy (1070), where he had mixed success against the Normans. But events in Anatolia soon forced him to return there. After his departure the Normans captured Bari, driving the Byzantines completely from Italy (1071).

1070. Renewed Turk Invasion. Two Turkish armies invaded the empire while Romanus was in Italy. One, under **Arisiaghi,** brother-in-law of Alp Arslan, defeated the main Byzantine army under **Manuel Comnenus** near **Sebastia.** Meanwhile, Alp captured the fortress of Manzikert, but was repulsed from Edessa (Urfa).

1071. Campaign of Manzikert. Early in the spring Romanus advanced east from Sebastia, via Theodosiopolis (Erzerum), with an army of 40,000–50,000 men. He sent part of his army ahead to the vicinity of Lake Van, under the general **Basilacius,** to ravage the country around Akhlat and to screen the main body. Romanus besieged and captured Manzikert. He then advanced to lay siege to Akhlat, while his covering force advanced toward Khoi, in Media, where Alp Arslan was gathering an army. Late in July or early August, Alp advanced with 50,000 or more men. He brushed aside Basilacius' covering force (probably 10,000–15,000 strong), which apparently withdrew to the southwest without informing Romanus. Since Basilacius' mission was to screen the main Byzantine army, he evidently was involved in a widespread treasonous conspiracy. (The conspirators' leader was the emperor's principal lieutenant, **Andronicus Ducas,** and Empress Eudocia was probably behind the plot.) In any event, Romanus was surprised, in mid-August, by the sudden appearance of Alp's army, which ambushed and annihilated a Byzantine force near Akhlat. Romanus hastily fell back toward Manzikert, and prepared to give battle between that fortress and Akhlat. During this withdrawal his mercenary Kipchak and Pecheneg light cavalry deserted, bringing his strength down to less than 35,000 men.

1071, August 19 (?). Battle of Manzikert. Romanus formed his army up in a typical Byzantine formation of two lines, in open, rolling country. He personally commanded the first line. Andronicus Ducas commanded the second. While the opposing light-cavalry units were skirmishing and the main Turkish army was forming, Alp sent a messenger offering peace. Considering this a ruse—which it probably was—Romanus refused to consider any terms save an immediate Turkish withdrawal from Byzantine territory. Soon after this the Turkish horse archers opened the battle at long range. Judiciously combining the fire of his own archers with the advance of his heavy cavalry, Romanus tried in vain to pin the Turks down. He did, however, drive his elusive foe back so far that in the late afternoon the Byzantines captured the Turkish camp. With dusk approaching and no decision reached, his army in an exposed position, several miles from his own lightly guarded camp, Romanus decided to withdraw. The Turks immediately turned to harass the retiring Byzantines, causing some disorder among the mercenaries. To prevent the confusion from spreading, Romanus ordered a halt and turned against the Turks with his first line. It is not clear whether he intended to bivouac under arms for the night or merely wanted to drive off the Turks before continuing the withdrawal. Despite the order to halt, Andronicus Ducas continued the retirement with the second line and the flank outpost units. The Turks immediately seized the opportunity and swarmed around Romanus, who had no time to reorganize his remaining troops for all-around defense. The two wings of the front line were quickly overwhelmed. As darkness fell the Turks closed in around the small body remaining with Romanus. Every man was killed or captured, the emperor himself being taken a prisoner to Alp Arslan.

1071, August–September. Outbreak of Civil War. In Constantinople, upon word of the defeat at Manzikert, **John Ducas** seized power and established his nephew, **Michael VII,** son of Constantine X, on the throne. Romanus, soon released by Alp Arslan upon promise to pay ransom, endeavored to regain his throne, but was defeated by his enemies, captured, and blinded by Andronicus Ducas; he died soon after. Pathetically, the last act of this tragic, noble figure was to gather all of the money he could raise to send to Alp Arslan in proof of his good faith in paying his ransom.

COMMENT. *Neither inferior combat capability nor poor generalship lost the Battle of Manzikert; the one real cause of defeat was treason. The consequences were stupendous. While rival claimants struggled for the Byzantine throne, the victorious Turks overran practically all of Anatolia, wiping out the heart of the empire. The Turks ravaged the country mercilessly, partly from barbarism, partly from policy. Anatolia became a virtual desert. A great proportion of the population perished; the survivors fled. When parts of Anatolia were later reconquered by the Byzantines, they were unable to raise any significant forces from the region. The most important single military result of Manzikert, then, was to eliminate the principal native recruiting ground of the empire, a region from which it had habitually raised armies of 120,000 men and more. Thenceforward the empire was forced to rely upon mercenaries for the bulk of its armed forces. Principal among them were West Europeans for heavy cavalry, Russians and Norsemen for infantry, Pechenegs and Cumans for light cavalry. The most important permanent component of the imperial armies became the Varangian Guard—the name revealing the Norse-Russian origin of most of its members. Yet, surprisingly, the empire survived for almost four more centuries, its continuing existence almost entirely due to the legacy of doctrine and professional skill of its amazing military system.*

Byzantium Struggles for Survival, 1071–1200

1071–1081. Internal Chaos. The Seljuk Turks overran all of Anatolia save for a few isolated coastal cities. The downfall of Michael came with the simultaneous revolts of Generals **Nicephorus Bryennius** in Epirus and **Nicephorus Botaniates** in the dwindling Anatolian provinces (1078). Botaniates, with Turk help, was successful, and became Nicephorus III. He was immediately faced with more insurrections. His general, Alexius Comnenus, subdued Nicephorus Briennus in the hard-fought cavalry **Battle of Calavryta,** in eastern Thessaly, in which a high order of leadership and professional skill was displayed on both sides (1079). Soon after this, however, the emperor's jealousy led Alexius to revolt and seize the throne (1081). Meanwhile, taking advantage of the inner turmoil of the empire, the Turks captured Nicaea (1080).

1081–1118. Reign of Alexius Comnenus. One of the most astute diplomats and resourceful rulers of history. With meager resources, with the empire falling apart from internal dissension, he was faced with external threats from all directions. The Turks were hammering at the Asiatic gates of Constantinople, the Pechenegs and Cumans were ranging south of the Danube, Bulgaria was in revolt, and the Normans under **Robert Guiscard** and **Bohe-** **mund** were driving toward Constantinople through Epirus and Thessaly. Yet surprisingly, despite many defeats and heartbreaking setbacks, he met these terrible dangers with a fortitude and skill worthy of a successor of Heraclius. He was particularly adroit at turning his enemies—internal and external—against each other. By the end of his reign the European boundaries of the empire had been restored nearly to their pre-Manzikert positions, while more than one-third of Anatolia—including the entire coast line—had been recovered, thanks in large part to the unintended and somewhat reluctant assistance of West European Crusaders (see p. 339).

1081–1085. War against the Normans. (See p. 322). In order to meet this threat Alexius was forced to make peace with the Seljuks, temporarily recognizing their conquests in Anatolia. He then employed a number of Turk mercenaries in his battles against the Normans.

1084. Fall of Antioch. The Seljuks captured the last Byzantine foothold in Syria.

1086–1091. Revolt of the Bogomils. This Christian heretical sect of Bulgaria formed an alliance with the Pechenegs and Cumans. The allies defeated Alexius at the **Battle of Dorostorum** (Silistra, 1086). Eventually Alexius suppressed the revolt, then defeated the Pechenegs at the **Battle of Leburnion** and drove them back across the Danube (1091).

1094. Revolt of Constantine Diogenes. He led a Cuman army across the Danube into Thrace, where he besieged Adrianople. Alexius defeated the pretender in the **Battle of Taurocomon.**

1098–1102. The First Crusade. (See p. 338.) This enabled Alexius to recover almost half of Anatolia from the Turks.

1098–1108. Renewed Norman War. The cause was Bohemund's refusal to return Antioch to the empire in accordance with the Crusaders' promise (see p. 339). The Byzantines, supported by **Raymond** of Toulouse and Tripoli, fought a desultory war against Bohemund and his nephew **Tancred** in northern Syria (1099–1104). Bohemund then returned to Italy, raised a polyglot army of West European volunteers (mostly Normans), and crossed the Adriatic. He besieged Durazzo, but was repulsed (1106). Alexius then defeated Bohemund in a number of minor engagements, finally forcing him to make peace and to give up his claim to Antioch (1108). This was meaningless, however, because Tancred refused to honor Bohemund's signature; sporadic fighting continued between Byzantines and Normans in northern Syria.

1110–1117. Renewed War with the Seljuks. The Turks were initially victorious, and again ravaged Byzantine Anatolia, raiding as far as the Bosporus. Alexius, however, waging war in typical, calculating manner, repulsed them, gaining a great victory at the **Battle of Philomelion** (Akshehr, 1116). The Turks sued for peace, and agreed to abandon the entire coast of Anatolia to the Byzantines.

1118–1143. Reign of John II Comnenus (son of Alexius). He defeated the Seljuks, recovering most of Anatolia (1120–1121). He crushed a Pecheneg invasion of Bulgaria so thoroughly that they were never again a threat to the Empire (1121–1122). He was unsuccessful in a prolonged naval war with Venice (1122–1126). He intervened in Hungary to settle a dynastic war, forcing the Hungarians to accept Byzantine suzerainty (1124–1126). He reconquered southeastern Cilicia (1134–1137), where the Armenians had maintained their independence against both Turks and Byzantines since Manzikert. He defeated Raymond of Antioch (who had assisted the Armenians) and was preparing to invade the Principality of Antioch when he died (1143).

1143–1180. Reign of Manuel Comnenus (son of John). Continuing his father's campaign in Syria, he defeated Raymond and forced him to swear allegiance to the empire (1144). He then drove the Turks from their mountain strongholds in Isauria (1145).

1147–1149. The Second Crusade. (See p. 343.)

1147–1158. War with Roger of Sicily. The Norman fleet captured Corfu (1147) and ravaged the coast of Greece. Manuel reestablished the old alliance with Venice, and the combined Byzantine-Venetian fleets defeated the Normans (1148) and reconquered Corfu (1149). Manuel then invaded Italy and seized Ancona (1151), which he held, despite Norman repulse of his further invasion attempts (1155–1156; see p. 323).

1150–1152. Serb Rebellion. Subdued by Manuel.

1151–1153 and 1155–1168. Byzantine Invasions of Hungary. Manuel's victory in the **Battle of Semlin** (Zemun, northeast Yugoslavia) forced Hungary to make peace and to cede Dalmatia and other territories to him (1168).

1158–1177. Renewed Wars with the Seljuks and Raynald of Antioch. Manuel forced **Kilij Arslan IV** to recognize Byzantine suzerainty (1161). Manuel was badly defeated in the **Battle of Myriocephalum** (1176), but recouped his losses with the assistance of **John Vatatzes** the following year in Bithynia and in the Meander Valley.

1169. Expedition to Egypt. A joint Byzantine-Crusader force under King **Amalric** of Jerusalem was repulsed from Egypt (see p. 335).

1170–1177. War with Venice and the Normans. The Venetians were driven completely from the Aegean (1170). However, with Norman assistance, the Venetians captured Chios and Ragusa (1171), but were repulsed at Ancona (1173). The war was a draw.

1180–1185. Regency and Reign of Andronicus Comnenus. This brilliant soldier and dissolute prince seized power from the child Emperor **Alexius II,** whom he soon put to death (1183). His strong rule antagonized the nobility, who invited **William of Sicily** to assist them (1184). William invaded Greece and captured Thessalonica. This prompted an uprising in Constantinople and Andronicus was overthrown and brutally killed. He was succeeded by **Isaac II Angelus.**

1180–1196. Serb Independence. A successful rebellion under **Stephen Nemanya.**

1185–1195. First Reign of Isaac Angelus. Isaac defeated William at the Battle of the Strymon and Byzantine Admiral-General **Alexius Branas** won a decisive victory off the

Greek coast at the **Battle of Demetritsa** (1185). The Normans were finally driven from their conquests (1191; see p. 323). Meanwhile, the Bulgarians, under **John** and **Peter Asen,** successfully revolted (1186–1187) and began to raid periodically into Thrace. They defeated Isaac at **Berrhoe** (1189) and retained their conquests despite a Byzantine success at **Arcadiopolis** (1194).

1186–1195. Internal Discord. Imperial power declined as Isaac was forced to defend his throne against numerous rival claimants, the first being Branas, who was defeated and killed by the emperor near Constantinople (1186). Isaac was overthrown, blinded, and imprisoned by his brother, **Alexius.**

1195–1203. Reign of Alexius III. The empire began to disintegrate; Turks raided at will through Byzantine Anatolia, while the Bulgars ravaged the European provinces.

MOSLEM DOMINIONS

The Middle East (Syria, Arabia, Mesopotamia, Anatolia)

There was utmost confusion and turmoil in the Moslem Middle East, with almost incessant warfare among minor Moslem potentates. Three major historical trends emerge from the bloody welter of minor wars. First was the overrunning of most of the Middle East by the Seljuk Turks during the latter half of the 11th century. Next was the appearance of the Crusaders, whose presence shaped events in the region during the entire 12th century. Finally, there was the temporary unification of Arab, Egyptian, and Turkish Moslems of the Middle East—a reaction to the Crusades—under the leadership of **Nur-ed-din** and his renowned successor, **Saladin.** The principal events were:

1000–1098. Chaos in Mesopotamia and Syria. Among the many contenders for supremacy were the Byzantines, who held the north and northwest, centered around Antioch (until 1087), the Buyids of Baghdad (until 1055), and the Fatimids around Damascus (until 1079).

1043–1055. Seljuk Conquest of Mesopotamia. (See p. 336.) Capture of Baghdad by Tughril Beg ended the Buyid Dynasty.

1055–1060. Revolt in Mesopotamia. This was led by **Al-Basasiri,** Turk general formerly under the Buyids, and supported by the Fatimids. While Tughril was putting down a mutiny in his own army, Al-Basasiri temporarily occupied Baghdad (1058), but Tughril soon crushed the revolt.

1063–1092. Seljuk Conquest of Syria and Anatolia. (See p. 336.)

1092–1098. Breakup of the Seljuk Empire. Taking advantage of a dynastic civil war in the heart of the Seljuk Empire (see p. 336), **Kilij Arslan,** son of Sulaiman ibn Kutalmish (see p. 336), reestablished his father's domain in Anatolia: the Sultanate of Roum or Rum, with its capital at Iconium. A rival regime was set up at Sivas under the Turk *atabeg* ("general" or "prince") **Danishmend.** In Syria and northern Mesopotamia, **Ridwan,** son of Tutush, was Emir of Aleppo.

1097–1102. The First Crusade. (See p. 338.)

1102–1144. Many-Sided Struggle for Control of Mesopotamia. The Crusaders held the northwestern portion around Edessa (until 1144). Ridwan and his successors were predominant, but their supremacy was disputed by the Seljuks of Rum and several other Moslem principalities. Kilij Arslan of Rum captured Mosul (1102), but was defeated and killed by Ridwan at the **Battle of the Khabur River** (1107).

1102–1127. Struggle for Syria. The Crusaders held the entire coastal region, and consolidated their control, but they never had the numerical strength to expand far beyond the coast. After Ridwan's death (1117), the Seljuks of Aleppo were overthrown by the Burid atabegs of Damascus, whose principal rivals were the Urtuqids of northeastern Syria. Throughout this period and the remainder of the century, the Assassins held a mountain stronghold northeast of Tripoli against all efforts, Christian and Moslem.

1127–1146. Reign of Imad-al-Din Zangi, Atabeg of Mosul. He extended his rule over all of northern Mesopotamia and northern Syria, capturing Edessa from the Crusaders (1144).

1146–1174. Reign of Nur-ed-din. (son of Zangi). He conquered all of Moslem Syria and defeated **Raymond** of Antioch (1149–1150). Later his general **Shirkuh** conquered Egypt (1163–1169; see p. 335).

1147–1149. The Second Crusade. (See p. 343)

1155–1194. Resurgence of the Abbasids. The Caliphs gained control of southern and central Mesopotamia, despite continued opposition from the declining Seljuk Dynasty.

1174–1186. Rise of Saladin. (See p. 335.) Expanding from Egypt into Syria and northern Mesopotamia, he conquered the entire empire formerly held by Nur-ed-din.

1187–1189. Saladin's Jihad (Holy War) against the Crusaders. This resulted in the conquest of Jerusalem and most of the Crusader kingdom.

1189–1192. The Third Crusade. (See p. 344.)

North Africa

MOROCCO AND WESTERN ALGERIA

Northwest Africa was fragmented under local Berber chieftains until veiled Tuareg tribesmen erupted from the Sahara in the mid-11th century to establish the vigorous Almoravid (or Murabit) Dynasty. They swept through Morocco and western Algeria, while simultaneously conquering the Negro kingdom of Ghana. They next conquered Moslem Spain and reconquered part of central Spain recently captured by the Christians. Slightly less than a century after their appearance, the Almoravids were swept from power, as abruptly as they had seized it, by a mountain Berber tribe which established the Almohad (Muwahhid) Dynasty in control of the same territory. The Almohads then expanded eastward. The principal events were:

1053–1056. Rise of the Touaregs. This Berber people inhabiting the Sahara gained control of most of the Western Sahara oases under **Yana ibn Omar,** a natural military genius, who established the Almoravid Dynasty and organized his people for war.

1054–1076. Almoravid Conquest of West Africa. The first penetration of Islam south of the Sahara.

1056–1080. Almoravid Conquest of Morocco, and Western and Central Algeria. This was under the leadership of **abu-Bakr ibn-'Umar** and his brilliant soldier cousin, **Yusuf ibn Tashfin.**

1086–1092. Almoravid Conquest of Moslem Spain. (See p. 324.)

c. 1120–1130. Rise of the Almohads. A militant Berber religious confederation was established in the Atlas Mountains by **Mohammed ibn-Tumart.**

1130–1147. Overthrow of the Almoravids. The Almohad leader was **'Abd Al-Mu'min ibn-'Ali (al-Kumi),** a follower of ibn-Tumart and a great soldier.

1145–1150. Almohad Conquest of Moslem Spain. (See p. 325.)

1147–1160. Almohad Conquest of Algeria, Tunisia, and Western Tripolitania. Following his victories, al-Mu'min took the title of caliph.

EASTERN ALGERIA, TUNISIA, AND TRIPOLITANIA

Central North Africa remained under the nominal rule of the Fatimid caliphs of Cairo until the middle of the 11th century, when the local Zirid Dynasty declared its independence. In retaliation the Fatimids sent a number of wild Arab nomadic tribes to devastate the region. The Pisans, Genoese, and Normans raided the coast frequently; the Normans held much of the littoral for several years. Zirids, Arabs, and Christians, however, were all overwhelmed by the violent Almohad expansion shortly after the middle of the 12th century. The principal events were:

1015–1016. Zirid Corsairs Driven from Sardinia. (See p. 321.)

1049. Zirid Independence. Emir Mu'izz ibn Badir declared independence from the Fatimids and abjured the Shi'ite faith, returning to Sunni teachings.

1058–1060. Fatimid Retaliation. Arab nomads were sent to ravage Zirid territory by Fatimid Caliph **al-Mustansir.** They overran most Zirid territory save the immediate vicinity of the capital, Mahadia (Mahdia).

1060–1091. Zirids Driven from Sicily. (See p. 322.)

1087. Temporary Conquest of Mahadia by Genoese and Pisans. (See p. 321.)

1135–1160. Norman Invasions of North Africa. (See p. 323.)

1147–1160. Almohad Conquest. (See p. 333.)

EGYPT AND CYRENAICA

Eastern North Africa was held by the Fatimid caliphs for most of this period, though during the 12th century actual power was usually exercised by the caliphs' viziers. In the middle of the 12th century a complicated struggle for control of Egypt took place between the Crusader King Amalric of Jerusalem and the Zangid Turkish armies of **Nur-ed-din** of Aleppo and Mosul. Eventually the Turkish general, **Saladin,** founder of the Ayyubid Dynasty, became the virtual ruler of Egypt, though still nominally subject to Nur-ed-din. Upon the death of Nur-ed-din, Saladin asserted his independence and conquered the former Zangid domains of Syria, Kurdistan, and Mesopotamia. He then captured Jerusalem and most of Palestine from the Crusaders, to precipitate the Third Crusade. He repulsed the Crusaders' efforts to retake Jerusalem, and negotiated a compromise peace with **Richard I** of England. The principal events were:

996–1020. Reign of al-Hakim. A brutal and oppressive tyrant, he founded the religious sect of Druses.

1020–1035. Reign of al-Zhir. This was the beginning of Fatimid decline. Egypt was raided by various Syrian insurgents, including **Salih ibn Mirdas** of Aleppo. Tenuous Fatimid control of Syria was reestablished by the victory of **Anush-takin al Dizbiri** over Mirdas at the **Battle of Ukhuwanah** (1029).

1035–1094. Reign of al-Mustansir. Continued Fatimid decline. Mirdas established his independent Mirdasid Dynasty at Aleppo (c. 1040). Mecca and Medina became independent (1047). The Zirids of Tunisia defected (1049; see above). Much of Syria and Palestine was overrun by the Seljuks (1060–1071). Egypt was wracked by uprisings and revolts of Turkish elements of the army (1060–1074). Order was gradually established in Egypt and parts of southern Syria by the Armenian general and vizier, **Badr al-Jamali** (1074–1094).

1094–1101. Reign of al-Musta'li (son of al-Mustansir). He was supported in a successful dynastic war against his brother by **al-Afdal Shahinshah,** son of Badr, who then recaptured Jerusalem and other cities of Palestine and southern Syria (1098). The Crusaders soon took Jerusalem, however, and decisively defeated al-Afdal at the **Battle of Ascalon** (1099; see p. 341).

1101–1121. Rule of al-Afdal. He exercised control in the name of puppet caliphs. His attempts to reconquer Palestine were repulsed by the Crusaders (see p. 342).

1121–1163. Civil War and Dissensions. Ascalon, last Egyptian foothold in Syria, was lost to the Crusaders (1153).

1163–1169. Crusader and Zangid Struggle for Egypt. A Turkish-Syrian army under Nur-ed-din's general **Asad ud-Din Shirkuh** arrived in Egypt to assist the Egyptian vizier, **Shawar ibn Mujir,** suppress an insurrection (1163). At the same time the Crusaders from Jerusalem were raiding the Nile Delta. Discovering that Shirkuh intended to seize control of Egypt for Nur-ed-din, Shawar asked for help from King Amalric I of Jerusalem. The Crusaders helped Shawar defeat Shirkuh near Cairo (April 11, 1167). Shirkuh's nephew, **Saladin,** distinguished himself in the battle. The Crusaders established themselves firmly in Cairo, which was repeatedly threatened by Shirkuh, but were unsuccessful in besieging Saladin at Alexandria. After negotiations, both antagonists agreed to evacuate Egypt, save for the Christian garrison of Cairo. Moslem disorders in Cairo caused Amalric to return to Egypt (1168). He had formed an alliance with Byzantine Emperor Manuel Comnenus, who sent a fleet and small army to Egypt by sea (early 1169). Shirkuh immediately returned also. Skilful political and military maneuvering by Shirkuh and Saladin, combined with bad luck and suspicion between Crusaders and Byzantines, inhibited cooperation, and both armies withdrew. Shirkuh, although still subject to Nur-ed-din, became vizier to the Fatimid caliph, but died (May, 1169). He was succeeded by Saladin, who became virtual ruler of Egypt.

1169–1193. Regency and Reign of Salah-al-din Yusuf ibn-Ayyub (Saladin). A Kurd of Turkish descent, he founded the Ayyubid Dynasty. He deposed the Fatimid Caliph and restored the Sunnite Moslem faith in Egypt (1171). He conquered Tripolitania from the Almohads (1172). Though nominally loyal to Nur-ed-din, he was virtually independent, and a coolness arose between them before Nur-ed-din's death (1174). Saladin then asserted his claim to all of the Zangid dominions, and gradually made this good in a series of campaigns against a number of other claimants (1174–1183). His successes were due primarily to his well-trained regular army of Turkish slaves (Mamelukes), primarily horse archers, but also including shock-action lancers. During this time he had had frequent minor conflicts with the Crusaders; as his power increased, and completely encircled the Latin states, these wars grew in intensity (1183–1187).

1187–1192. Saladin's Jihad (Holy War). Saladin conquered Jerusalem and most of Palestine, leading to the Third Crusade (see p. 344). Despite some defeats, Saladin retained Jerusalem and most of his Palestine conquests; but in a treaty with Richard I of England, granted the Christians special rights in Jerusalem (1192).

1193. Death of Saladin. The kingdom was divided between his sons: **al-Aziz** became ruler of Egypt; **al-Afdal** became ruler of Syria.

1196–1200. Dynastic War between Saladin's Successors. The brother's uncle, **Abul Bakr Malik al-Adil,** fought on both sides and eventually gained full control of Saladin's empire as a result of a victory over al-Afdal at the **Battle of Bilbeis** (Egypt, January 1200).

Persia and South-Central Asia

Under **Mahmud** of Ghazni, the Ghaznevids expanded their control from Khorasan southward, to include all of eastern Persia, and northward to gain undisputed control of Khwarezm (Khiva) and much of Turkestan between the Oxus and Syr Darya rivers. His most important campaigns, however, were in India (see p. 347). After his death the Ghaznevids declined and soon lost their western dominions to the Seljuks, under **Tughril Beg** and **Chaghrai Beg.** Tughril then went on to conquer western Persia and Mesopotamia from the Buyids, and to begin Turkish penetration of Anatolia (see pp. 328, 332). The Seljuk practice of assigning semiautonomous control of provinces to their generals, or atabegs, soon resulted in the disintegration of the Seljuk Empire (save for Khorasan) into a number of warring principalities. During this time the small Shi'ite sect of Assassins prospered at Dailam, on the heights of Mt. Alamut in the Elburz range. Shortly after the middle of the 12th century, migrations of Turkish tribes from Turkestan overthrew **Sanjar,** Sultan of Khorasan and last of the direct Seljuk line. Following this, the Turkish shahs of Khwarezm (Khiva) conquered Khorasan and Isfahan to establish a new and powerful Persian empire. The principal events were:

PERIOD OF GHAZNEVID ASCENDANCY, 1000–1040

1006–1007. Ilak Invasion of Khorasan. Ilak Khan **Nasr I** from Transoxiana was repulsed by Mahmud, who defeated the invaders near Balkh.

1011–1016. Uprisings in Ghor and Khwarezm. Mahmud suppressed these and strengthened his control.

1029. Expedition into Persia. Mahmud defeated the Buyids and annexed their eastern territories.

PERIOD OF SELJUK ASCENDANCY, 1040–1150

1034–1055. Rise of the Seljuk Turks. Tughril Beg and **Chaghrai Beg,** grandsons of Seljuk (see p. 291), rose against the Ghaznevids. Crossing the Oxus, they occupied most of Khorasan, decisively defeating **Mas'ud,** son of Mahmud, at **Nishapur** (1038) and near **Merv** (1040). Chaghrai and the main body of Turks remained in Khorasan, operating successfully against the Ghaznevids until his death (1063). Tughril formed a standing army of Mamelukes and began moving westward against the Buyids and Byzantines.

1043–1055. Seljuk Conquest of Mesopotamia. Tughril conquered the Buyid regions of eastern Persia and northern Mesopotamia, and began raiding into Armenia and Byzantine Anatolia (1048). He captured Baghdad, ending the Buyid Dynasty (1055).

1063–1072. Reign of Alp Arslan (son of Chaghrai Beg). He invaded Armenia, capturing Ani, the capital (1064). He began a determined invasion of the Byzantine Empire (see p. 328), while at the same time expanding into northern Syria (1068–1071). He defeated the Byzantine Emperor Romanus in the decisive **Battle of Manzikert** (1071; see p. 329). Jerusalem was captured from the Fatimids that same year.

1072. Khwarezmian Invasion. Alp defeated **Yakub** of Khwarezm at the **Battle of Berzem.** Yakub was captured, but he assassinated Alp in an interview after the battle.

1072–1092. Reign of Sultan Malik-shah (son of Alp Arslan). Malik continued the conquest of Anatolia (1072–1081; see p. 330). Despite numerous revolts, he held his empire together and even expanded it in Central Asia (see p. 354) with the assistance of his able vizier, **Nizam-al-Mulk,** who conquered Transoxiana and advanced as far as Kashgar. Malik's principal rival was **Sulaiman ibn Kutalmish,** who briefly held most of Anatolia (1080–1086), and who captured Antioch from the Byzantines (1084). Sulaiman was defeated and killed near **Aleppo** by **Tutush,** Malik's brother (1086). **Atsiz** the Khwarezmian led one of Malik's armies through Syria, capturing Damascus (1086), and then on to the Nile, where he was repulsed by the Fatimids.

1092–1117. Decline of the Seljuk Empire. While the sons of Malik (**Sanjar, Barkiyarok,**

and **Mohammed**) fought over the succession in the heart of the empire (Iraq and Persia), outlying provinces fell away.

1117–1157. Reign of Sanjar, Seljuk Sultan of Khorasan (son of Malik-shah). He occupied Ghazni (1117) and later returned to put down a revolt there (1134). He had earlier suppressed a revolt of the Turk Khan of Transoxiana (1130). Sanjar's principal rival was **Atsiz,** his viceroy in Khwarezm, who repeatedly revolted against Sanjar, with mixed success (1138–1152). During this period an invasion of Kara-Khitai Tartars was joined by local tribes; the combined forces decisively defeated Sanjar near **Samarkand** (1141), forcing him to abandon Transoxiana. He defeated and temporarily captured **Alaud-din Jihansuz** of Ghor (1150). The continuing influx of Turkoman into Khorasan led to widespread revolts, which caused the collapse of Sanjar's kingdom (1153–1157).

1148–1152. War between Ghor and Ghazni. The initial successes of **Saif ud-din Suir** and Ala-ud-din of Ghor over **Bahram Shah** of Ghazni were temporarily offset by the capture of Ghor by Sanjar (1150; see above). Bahram recaptured Ghazni from Saif ud-din, who was made prisoner and killed. Ala-ud-din then invaded Ghazni again, destroying the city completely and wrecking the remaining power of Ghazni (1152). This set the stage for a rapid expansion of Ghori power into India (see p. 348) and Khorasan.

PERIOD OF KHWAREZMIAN ASCENDANCY, 1150–1200

1172–1199. Reign of Takash, Shah of Khwarezm. A period of steady Khwarezmian expansion through Khorasan and Isfahan, culminating in the conquest of Mesopotamia (1194). Throughout this period, however, Khwarezm's control over southern and eastern Khorasan was constantly challenged by Ghor.

1173–1203. Reign of Ghiyas-ud-Din of Ghor. He had established himself as unquestioned master of all modern Afghanistan. He then devoted himself to constant inconclusive warfare against Khwarezm in Khorasan, appointing his brother **Muizz ad-Din Mohammed—** known to history as Muhammad of Ghor—as governor of Ghazni with the mission of extending Ghori rule into India (see p. 348).

THE CRUSADES

The Crusades were military expeditions undertaken by West Europeans for primarily religious purposes, but in which political considerations frequently played an important part. The immediate, direct, or ostensible object of the Crusades was the liberation or preservation of the Holy Land (and particularly the Holy Places in Jerusalem) from Moslem control. It should be noted, however, that during the Middle Ages the term "crusade" was frequently applied to other military expeditions against non-Christian foes who were pagans (as the German eastward expansion into Slavic territory; see p. 347) or Moslems (as the wars of reconquest in Spain; see p. 323) or heretics (as the Albigensian Crusade; see p. 397).

The forces which brought about the Crusades were really set in motion by the victory of the Moslem Seljuk Turks over the Byzantines at Manzikert (see p. 329), and Seljuk conquest of Jerusalem that same year from the more tolerant Fatimid caliphs of Cairo (1071). The subsequent Seljuk conquest of practically all of Anatolia from the Byzantines, combined with persecution of Christian pilgrims to Jerusalem, aroused Christendom. After several vain appeals for help from Byzantine emperors to popes and to West European monarchs, Pope **Urban II** eloquently called for action at the Synod of Clermont (1095). This evoked from his listeners spontaneous cries of "God wills it!"—which became the watchword of the Crusaders.

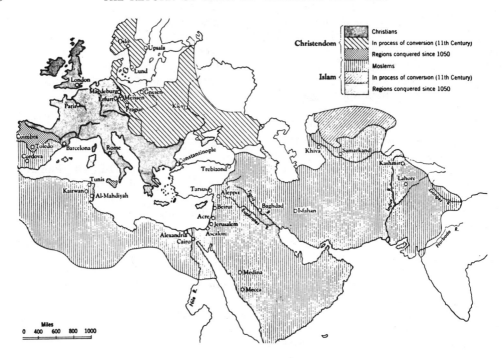

Islam and Christendom on the eve of the Crusades

The First Crusade, 1096–1099

1095–1096. The Leaders. The most prominent were French Bishop **Adhemar du Puy,** a courageous, wise soldier-priest who was appointed papal legate and was the mediator in the many disputes among his belligerent fellow leaders, Norman Duke **Bohemund** of Taranto (son of Robert Guiscard; see p. 322), his nephew **Tancred,** Count **Raymond** of Toulouse, Duke **Godfrey** de Bouillon of Lorraine, his brother **Baldwin,** Duke **Hugh** of Vermandois (brother of the King of France), Duke **Robert** of Normandy, Count **Stephen** of Blois, and Count **Robert** of Flanders.

1096, April–October. The People's Crusade. A crowd of unarmed pilgrims under **Peter the Hermit** marched overland toward the Holy Land. Many died of starvation; most of the remainder were massacred by the Turks in Anatolia.

1096–1097. The Assembly of Forces. The var-

ious contingents moved toward the agreed rallying point of Constantinople in four main groups. Godfrey and Baldwin, with their own and other German units, followed the Danube Valley through Hungary, Serbia, and Bulgaria, thence over the Balkan mountains, having several armed brushes with local forces en route. This was the first contingent to reach Constantinople; they camped outside the city through the winter. Bishop Adhemar, Count Raymond, and others from southern France proceeded through north Italy, continued down the barren Dalmatian coast of Durazzo in a grueling march, thence east to Constantinople. Hugh, the two Roberts, and Stephen, with contingents from England and northern France, marched across the Alps, and then down the Italian peninsula. While his companions wintered in south Italy, Hugh took ship to Constantinople, was shipwrecked, but was rescued by the Byzantines and taken to Constantinople, where he became a virtual hostage to Emperor Alexius

Comnenus. The following spring the two Roberts and Stephen crossed the Adriatic to Durazzo, and then marched east to Constantinople. This was also the route followed by the Sicilian Normans of Bohemund and Tancred.

1096–1097. Byzantine-Crusader Friction. Alexius had hoped his appeal for help would net him a few thousand mercenaries for his depleted armies. He neither expected nor desired an independent and unruly army of upward of 50,000 men to collect at his capital. Long-standing religious and political differences between Byzantines and West Europeans made him suspicious of Crusader motives and intentions, particularly since Bohemund had recently been an active and exceedingly dangerous enemy (see p. 323). Alexius, desirous only of regaining his Asiatic dominions from the Turks, had little interest in the Crusader objective of capturing Jerusalem. The Crusaders were equally suspicious of the Byzantines and their wily diplomacy. They had no desire to serve as pawns of Alexius in the reconquest of his empire from the Turks. These mutual suspicions seriously affected the outcome of this and most subsequent Crusades. The first manifestation was in intermittent skirmishing which went on between Crusaders and watchful Byzantine guards on the outskirts of Constantinople during the winter.

1097, Spring. Agreements between Alexius and the Crusaders. By a combination of firmness and diplomacy, Alexius avoided serious disturbances. In return for his assurance of assistance, he obtained oaths of fealty from the Crusader leaders, and their promise that they would help him to recapture Nicaea (Iznik) from the Turks and return to him any other former Byzantine possessions which they conquered. Alexius then transported them across the Bosporus—being careful not to allow any large gathering of Crusader contingents inside the walls of his capital. He also provided them with food and with an escort of Byzantine troops to guide them toward their objectives (and incidentally to prevent Crusader plundering).

1097, May 14–June 19. Siege of Nicaea. Accompanied by Alexius and his main army, the Crusaders invested Nicaea. By a combination of skillful fighting and skillful diplomacy, Alexius arranged for the surrender of the city to him, following a successful Crusader-Byzantine assault on the outer walls. The Crusaders were affronted by Alexius' refusal to grant them permission to sack the city. They then continued the advance southeastward, marching in two parallel columns. There was no single commander; decisions were reached by council of war, with Bishop Adhemar serving as mediator.

1097, July 1. Battle of Dorylaeum (Eskisehir). The left-hand column, which was led by Bohemund, was suddenly attacked by a Turkish cavalry army under the personal command of Kilij Arslan, Seljuk Sultan of Rum (see p. 332). Using their traditional mounted-archer tactics, the Turks (probably more than 50,000 strong) severely punished the outnumbered Crusaders, who were unable to come to grips with their elusive, mobile foes. Bohemund's contingent was close to disintegration when the heavy cavalry of the right-hand column, under Godfrey and Raymond, smashed into the Turkish left rear. Kilij Arslan had failed to secure the approach from the south. Caught in a vise, about 3,000 Turks were killed and the rest scattered in rout. Total Crusader losses were about 4,000.

1097, July–November. Advance to Syria. The Crusaders resumed their advance and captured Iconium (Konya), Kilij Arslan's capital. (Meanwhile, behind them, Alexius and his Byzantine army were reoccupying the western portions of Anatolia from the shaken Turks.) After one more battle, near **Heraclea** (Eregli), the Crusaders continued on through the Taurus Mountains toward Antioch. During this advance a detachment under Tancred and Baldwin had a hard fight at **Tarsus.** Then Baldwin left the main column to embark on a private career of conquest, crossing the Euphrates and seizing Edessa, which became the center of an independent Christian domain.

1097–1098, October 21–June 3. The Crusader Siege of Antioch. The city was held by

Emir **Yagi Siyan,** who conducted a skillful, energetic defense. Soon after the investment began, the Turks made a successful sortie, causing heavy casualties among the uncoordinated Christian contingents. Skirmishing outside the walls was frequent. Two Syrian relieving armies were driven off in the **Battles of Harenc** (December 31, 1097; February 9, 1098). For a while the Crusaders were close to starvation, since they had no logistical organization and had made no arrangements for supply. They were saved, however, by the fortuitous arrival of small English and Pisan flotillas, which seized the ports of Laodicea (Latakia) and St. Simeon (Samandag) and brought provisions. During the 7 months of the siege, the bickering among the leaders became more intense, particularly between Bohemund and Raymond. Finally, mainly from the initiative of Bohemund and the treachery of a Turkish officer, Antioch was captured (June 3). It was none too soon; a relieving army at least 75,000 strong, commanded by Emir **Kerboga** of Mosul, was only two days' march distant. Stephen of Blois, feeling the situation was hopeless, fled from Antioch.

1098, June 5–28. Kerboga's Siege of Antioch. The Crusaders were now besieged and cut off from their seaports. Yagi Siyan still held the citadel. The Crusaders were again on the verge of starvation; the misery of the population of Antioch was intense. Alexius, who had been leading his army through the Taurus Mountains to occupy Antioch in accordance with his agreement with the Crusaders, met Stephen of Blois, who assured him the Crusaders were doomed. The Byzantine army accordingly withdrew into Anatolia. Despair in the city was suddenly dissipated by the discovery of the supposed Holy Lance (the weapon which had pierced Jesus' side during the Crucifixion). Few historians or theologians believe that this supposedly miraculous discovery was valid (in fact there were doubts among many of the Crusaders), but it nonetheless had a truly miraculous effect. Confident of victory, the Crusaders sallied out.

1098, June 28. Battle of the Orontes. The starving Crusaders were able to muster only 15,000 men fit for combat, with less than 1,000 of these mounted as cavalry. Under the coordinating direction of Bohemund, they crossed the Orontes under the eyes of the surprised Moslems. Then, repulsing Kerboga's attacks, the Christians counterattacked. Hemmed in between the river and nearby mountains, the Turks were unable to maneuver, and they could not stand against the determined Crusader charges. The Moslem army fled, after suffering heavy casualties.

1098, July–August. Pestilence in Antioch. One of the victims was Bishop Adhemar. After his death the disputes among the Christian leaders became more acute, particularly between Bohemund (who was determined to keep control of Antioch) and Raymond (who insisted that the Crusaders should return the city to Byzantium in accord with their oath to Alexius).

1099, January–June. Advance to Jerusalem. After much bickering, all the Crusaders except Bohemund and his Normans agreed to continue on to Jerusalem. (Bohemund stayed in Antioch, where he established an independent principality.) With the logistical support of the Pisan fleet, the Crusaders—now about 12,000 strong—advanced slowly down the coast to Jaffa, then turned inland to Jerusalem.

1099, June 9–July 18. Siege of Jerusalem. The city was defended by a strong force of Fatimid Moslems considerably more numerous than the besiegers. By this time Godfrey was generally accepted as the principal leader, assisted by Raymond and Tancred. The Crusaders were not numerous enough to blockade the city completely, and so they could not hope to starve it into submission. Despite a severe water shortage, they grimly prepared to assault the powerful fortifications, building a large wooden siege tower and a battering ram. Under heavy missile fire from the walls, the Crusaders pushed the tower against the city wall and Godfrey led the assault over a wooden drawbridge, while other Crusaders scaled the wall with ladders. This was apparently the only fully coordinated operation of the entire 2-year campaign. Fighting their way into the city, the Crusaders brutally massacred the garrison and the Moslem and Jewish population. Godfrey was elected Guardian of Jerusalem. (He refused the title of king.)

1099, August 12. Battle of Ascalon. Learning of the approach of a relieving army of 50,000 men from Egypt under Emir al-Afdal, Godfrey led his remaining 10,000 Crusaders out to meet them. Unlike the Turks, who were mainly mounted archers, the Fatimids relied upon the combination of fanaticism and shock action, which had been so successful in Islam's earliest days. They were at a hopeless disadvantage against the heavily armored and armed Crusaders. Godfrey won an overwhelming victory, culminating in a crushing cavalry charge.

The Crusader States, 1099–1148

Most of the Crusaders returned home, but a handful remained in Jerusalem with Godfrey; a few others joined the banners of the three other leaders remaining in Syria: Baldwin in the county of

ASIA MINOR,
and the STATES
OF THE CRUSADERS
in SYRIA, about 1140.

Sites of important events thus (•).
: Ruins. Last of the Christian
possessions to be surrendered,
thus: St. Jean d'Acre.
The dates are those of conquest,
or period of retention, by the
Crusaders. C.-County;
L.-Lordship; P.-Principality.
Scale 1:10 000 000

Scale 1:5 000 000

Edessa, Bohemund in the principality of Antioch, Raymond in the county of Tripoli (so called, even though the Moslems still held Tripoli itself, which Raymond besieged). Godfrey died less than a year after his great victories (July 1100). He was succeeded by his brother Baldwin, who assumed the title of King of Jerusalem. The other three Crusader states, nominally subject to his feudal authority, actually were completely independent. All occupied themselves at once in consolidating their positions by besieging the remaining Moslem towns and fortifications in the coastal regions, and by constant fighting with the neighboring Moslem principalities. Successive tides of Crusaders set out from Europe to help in this activity. The first of these, an expedition almost as large as that of the First Crusade, met disaster in Anatolia. But many others came in smaller groups, by sea, and gave some help before returning to their homes. The Crusaders won most of the pitched battles, which were not frequent, but they lacked the manpower to exploit success, and were never able to cut Moslem communications between Syria and Egypt. Their ability to maintain themselves was simply due to the lack of unity among the bickering Moslem principalities. After three decades, however, **Imad-al-Din-Zangi** unified the Moslems of northeastern Syria and Northwestern Mesopotamia, and captured Edessa from the Crusaders (see p. 333). The principal events were:

1100. Capture of Bohemund. He was ambushed and captured by the Turks near Aleppo, remaining a prisoner at Sivas for 3 years.

1101–1102. Crusader Disasters in Anatolia. Three expeditions started across Anatolia under the leadership of **William** of Poitiers, Raymond of Toulouse, Stephen of Blois, and Hugh of Vermandois. The first, attempting to rescue Bohemund, was smashed at **Mersevan** by Turks under **Mohammed ibn Danishmend** (1101). Only a few survivors, including Raymond, escaped. The second group was overwhelmed at **Heraclea** (1101). The third was almost wiped out a few months later, also at **Heraclea** (1102).

1101. First Battle of Ramleh. Baldwin, with 1,100 men, defeated 32,000 Egyptians under **Saad el-Dawleh** at Ramleh, an important road junction west of Jerusalem.

1102. Second Battle of Ramleh. Overconfident, Baldwin with 200 men attacked an Egyptian army of 30,000 and was crushed. He escaped, collected another army of 8,000, and defeated the Egyptians at the **Battle of Jaffa,** then pursued them to Ascalon.

1102–1103. Danish Crusade. Eric I, the Good, of Denmark marched overland to Constantinople through Russia, then went by sea to Cyprus, where he died. His wife, **Bothilda,** then took command of the little expedition and continued to Jaffa.

1104. Battle of Carrhae. Bohemund, recently released from prison, was defeated in a battle reminiscent of the defeat of Crassus (see p. 129).

1107–1109. Norwegian Crusade. Sigurd I of Norway led an expedition by sea, harassing Moorish Spain en route. He was the first European king to reach the Holy Land. With the Venetians he helped Baldwin capture Sidon; they were repulsed at Tyre (1109).

1116–1117. Expedition to the Red Sea. Baldwin led an expedition to the Gulf of Aqaba, where he built the fortress of Ailath (Eilat).

1118. Invasion of Egypt. Baldwin, with less than 1,000 men, crossed the Sinai Desert, but he died during the campaign. His little army returned to Palestine, accomplishing nothing.

1119. Battle of Antioch. A force under Count **Roger** of Salerno, marching to meet an invading army under **Ilghazi** at Aleppo, was surprised and wiped out.

1124. Attack on Jerusalem. The Crusaders repulsed an Egyptian raid.

1144. Fall of Edessa. Zangi captured Edessa from Count **Joscelin II.** This Crusader disaster led to a papal call for another Crusade.

1146. Expedition into Arabia. Successful Crusader raid on Bosra, 100 miles east of the Jordan.

The Second Crusade, 1147–1149

The principal leaders were Emperor Conrad III of Germany and King Louis VII of France. Both set out from Constantinople by separate routes (1147). The Germans followed the same general route as the First Crusade, but ran out of food near **Dorylaeum.** There the starving Crusaders were overwhelmed by a Turk attack. Conrad and a handful of survivors got back to Nicaea, then went by sea to Palestine. The French had taken a longer route, closer to the coast so as to remain in Byzantine territory as long as possible. Checked by the Turks in an indecisive battle east of **Laodicea,** Louis embarked his cavalry and went by sea to Palestine. The infantry continued eastward on foot and were annihilated by the Turks. Upon arrival in Palestine, Conrad and Louis joined **Baldwin III** of Jerusalem in an expedition against Damascus (1148). After investing the city, the Crusader army soon fell apart due to dissension among the three leaders. This ended the ill-starred Second Crusade.

The Crusader States, 1149–1189

Following the death of Zangi, the Crusader states were subjected to ever-increasing pressure from his son, **Nur-ed-Din** (see p. 333). To prevent his armies from conquering Egypt, King Amalric I of Jerusalem campaigned repeatedly in the Nile Valley, but was eventually repulsed. Soon after this, the Moslems of Syria and Egypt were united under the strong rule of **Saladin,** who after some preliminary campaigns declared a holy war against the Christians. He won a great victory over King **Guy** (of Lusignan) of Jerusalem at the **Battle of Hattin,** then captured Jerusalem and most of the Christian towns and forts of Palestine. Tyre was the only important seaport remaining in Christian hands, thanks to the fortuitous arrival by sea of Marquis **Conrad** of Montferrat and a small contingent of new Crusaders. While the pope issued a call for a new Crusade, Conrad and Guy continued to fight against Saladin. The principal events were:

1146–1174. Consolidation of Moslem Syria by Nur-ed-Din. (See p. 332)

1153. Capture of Ascalon. Baldwin III gained control of the entire Palestine coast.

1156. Battle of Jacob's Ford (near Sea of Galilee). The Moslems defeated separated Crusader infantry and cavalry contingents while crossing the Jordan.

1163–1169. Struggle for Egypt. (See pp. 333, 334.)

1169–1193. Reign of Saladin. (See p. 335.)

1174–1187. Crusader Dissension. The leprosy of childless King **Baldwin IV** caused a struggle for succession. Saladin profited from this by completing his conquest of Syria, while constantly harassing the Crusaders, although he was defeated at **Ramleh** (1177). Count **Raymond** of Tripoli was the most able Crusader leader, but his enemy, Guy of Lusignan, be-

came king by marrying Baldwin's sister. Meanwhile, unscrupulous **Reginald** of Châtillon and Kerak twice broke Crusader truces with Saladin, provoking him into declaring a holy war.

1187, June. Saladin's Invasion of Palestine. He besieged Tiberias with an army of about 20,000. Guy gathered all available Crusader manpower, raising an army of almost equal size, and advanced against Saladin. Ignoring Raymond's advice, Guy led the army into a waterless area, where he was attacked and surrounded by the Moslems.

1187, July 4. Battle of Hattin. Saladin separated the Crusader infantry from the cavalry, then overwhelmed the detachments. Raymond and a small force of cavalry cut their way through the Moslems; the remainder of the Crusaders were killed or captured. Saladin later released Guy on parole not to fight again. Meanwhile, Raymond died of wounds.

1187, July–September. Saladin's Conquest of Palestine. He captured Tiberias, Acre, Ascalon, and other cities. The garrisons of these places had been captured at Hattin. Saladin was advancing on Tyre when **Conrad** of Montferrat and a body of Crusaders fortunately arrived by sea, providing a garrison. Saladin was repulsed.

1187, September 20–October 2. Saladin's Capture of Jerusalem. He then turned against the Crusaders around Tyre. There was cautious skirmishing for more than a year.

1189, August. Investment of Acre. Guy of Lusignan boldly led a small Crusader force to besiege Acre. He was joined by Conrad and other Crusaders recently arrived from Europe, driving off relieving forces sent by Saladin.

1189–1191. Battles around Acre. Nine major engagements and innumerable minor skirmishes were fought between the Crusaders besiegers and Saladin, whose army surrounded the Crusaders by land. The Genoese and Pisan fleets controlled the sea, assuring an uninterrupted flow of supplies and reinforcements to the Crusaders.

The Third Crusade, 1189–1192

Responding to the appeal of Pope Clement III for a new Crusade were the three most powerful rulers of Europe: Emperor **Frederick I,** Barbarossa, of Germany; King **Philip II,** Augustus, of France; and King **Richard I,** the Lionhearted, of England. All were able and experienced soldiers. The principal events were:

1189–1190. Frederick's March. Departing Germany in the spring, Federick marched overland to Constantinople. While spending the winter there (1189–1190), he became involved in a dispute with Byzantine Emperor **Isaac Angelus.** Their troubles were patched up, and Frederick continued across Anatolia. His force of 30,000 was better organized than his predecessors on this route, and he repulsed all Turk attacks. His crossbowmen were particularly effective against the horse archers. He captured Iconium by storm, after which the Sultan of Rum made peace and offered no further resistance to the German advance. Late in the summer, in Cilicia, the old emperor was drowned, either in a river crossing or while bathing in a river at the end of a hot march. His son, **Frederick** of Swabia, continued, but, lacking his father's ability, he lost most of his men through starvation, disease, and local Moslem harassment. With little more than 1,000 men he joined the Crusader army besieging Acre late in the year.

1190–1191. Voyages of Richard and Philip. A year after Frederick started, Philip and Richard went by sea to Sicily, where they spent the winter quarrelling continuously. Next spring Philip sailed directly to Acre. Richard stopped en route to conquer Cyprus, which he wisely wanted as a base. He then sailed to Acre.

1191, July 12. The Fall of Acre. Upon Richard's arrival (June 8), his leadership was tacitly accepted by all Crusaders. He drove off Saladin's relieving army, then pressed the siege with such vigor that the Moslem garrison surrendered, ending a defense of 2 years. Soon after this increasing dissension between Richard and Philip resulted in the return of the French king to France. Considerable bickering continued among the Crusaders, particularly between Richard and Conrad.

1191, August–September. The March to Ascalon. Richard was determined to press on to Jerusalem. He organized the polyglot Crusader force (probably less than 50,000 strong) into a unified army and started out on a march in which he demonstrated exceptional strategic and tactical ability, as well as the effect of his dominant personality in enforcing responsiveness from unruly knights and barons of many lands. His staff and logistics work was far superior to that normally found in West European medieval armies. He even had a laundry corps to keep clothes clean to prevent disease. In short daily marches to avoid fatigue, he marched slowly down the coast, accompanied by his fleet offshore. Saladin's army hung on the inland flank, harassing the Crusaders, and seeking opportunities to cut off stragglers or to break into the Christian formation as they had at Hattin. But Richard's carefully planned and organized march column offered no such opportunities. He forbade his knights to respond to the Turkish harassment; all of Saladin's efforts to entice them to break ranks proved unavailing. Richard kept the Turkish horse archers at a distance by disposing crossbowmen throughout the column.

1191, September 7. Battle of Arsouf. Saladin set an ambush near the coast at Arsouf, then launched a strong attack at the rear of Richard's column to try to force the knights of the rear guard to charge against their tormentors. Richard at first refused permission, and the column continued its dogged march. Then as the Turks grew bolder, and the pressure on the rear guard became more than his knights could bear, Richard gave the prearranged trumpet signal for attack. The astonished and unsuspecting Turks were taken completely by surprise by the overwhelming coordinated charge. In a few minutes the battle was over. Following Richard's orders, the Crusaders overcame the temptation to scatter in pursuit of the defeated foe. About 7,000 Turks were cut down, and the remainder were scattered in flight. The Crusaders lost 700 men. Saladin never again attempted a battle in the open against Richard.

1192. Advance on Jerusalem. While the Crusader army wintered at Ascalon, Conrad was murdered by one of the Assassins. Soon after this Richard advanced on Jerusalem. Saladin retired before him, carrying out a "scorched earth" policy, destroying all crops and grazing land and poisoning all wells. The lack of water, absence of fodder for the horses, and growing dissension within his multinational army forced Richard to the reluctant conclusion that he could not besiege Jerusalem without risking almost certain destruction of his army. Reluctantly he withdrew to the coast. There were numerous minor engagements during the remainder of the year, Richard distinguishing himself as a heroic knight as well as a tactical leader.

1192. Treaty with Saladin. Abandoning hope of capturing Jerusalem, Richard concluded a treaty with Saladin in which special rights and privileges were granted to Christian pilgrims to Jerusalem.

1193. Death of Saladin. The breakup of Saladin's empire (see p. 335) gave the Crusader states a brief respite.

MILITARY SIGNIFICANCE OF THE CRUSADES

For centuries Western Europe survived the collapse of Rome only because it was shielded from the assaults of Islam by the sturdy Byzantine Empire. During those centuries a vigorous new Germano-Latin society had risen and begun to flourish. The Crusades were a natural reaction of this society to protect Europe from a renewed western thrust of Islam when Byzantium was faltering.

At this time the relative crudity of western military methods was such that the First Crusade would almost certainly have failed had the Moslems of the Middle East not been hopelessly divided by internal squabbling. That initial success, however, and the consequent creation of the Kingdom of Jerusalem and the other Crusader Latin states of the East had far-reaching results. There, to an extent not matched either in Spain or Sicily, took place a meeting and mingling of three distinct civilizations. The sophisticated, cultured, cynical, and resilient Byzantine civilization had already had fruitful interactions with the equally cultured and intellectual civilization of the Moslem East. Both eastern societies looked with a mixture of awe, amusement, and disgust at the rough, brutal, crude European society whose military spearhead literally bludgeoned its way into their midst.

Though there was never real peace among these three societies during the two centuries of the Crusading era, nonetheless there was considerable social contact, facilitated by frequently shifting alliances in their wars against one another and in the inevitable meddling of neighbors in the incessant internal disorders of each.

From these contacts the Crusaders profited most—since they had the most to learn. In the process, they became fatally enervated by their contacts with the more subtle civilizations in which they had placed themselves. But, fortunately for the future of Western Europe, this enervation in Syria and Palestine did not affect the basic, vigorous home societies which became the beneficiaries of the lessons learned in the East, brought to Europe by returning Crusaders. The military lessons were as important to the West as were those in culture, science, and economics.

Among the tactical lessons learned by the Crusaders were the use of maneuver in the form of envelopment and ambuscade, the employment of light cavalry for reconnaissance and for screening, the use of mounted firepower in the form of horsearchers, and, above all, the importance of the coordinated employment of the combined arms of infantry and cavalry, and of missiles and shock action, when dealing with a resourceful, mobile foe.

The most obvious military effect of the Crusaders' eastern experience was seen in European fortifications. The Westerners were particularly impressed with the powerful Byzantine walled cities and fortresses, with double or triple concentric lines of massive turreted walls. There was nothing like this in the West at the time. The result was a complete revolution in castle construction and city defense in Western Europe in the 12th century. The most impressive single manifestation was Château Gaillard, built by Richard the Lionhearted in Normandy after his return from the Third Crusade. This was a tribute not only to his powers of observation but to his ingenuity and competence as a military engineer, since he improved on eastern fortresses.

The Crusaders learned little new about siegecraft, but improved the methods and machines which they already used. Nor did they learn much about weapons, save for increased emphasis on archery. The one aspect of military activity in which they probably taught more than they learned was in arms and armor. Yet even here they profited, learning better methods of manufacture and construction so as to obtain comparable protection, or equal striking power, with lighter equipment.

One of the important lessons was the importance of logistics, an art which had practically disappeared in the West after the fall of Rome. European armies lived off the countryside, or they evaporated. Because of the limited time of obligatory feudal service, campaigns were rarely long, save for sieges and for the operations of the relatively small mercenary standing forces of kings and nobles. But in protracted campaigns in the East, with long marches over barren country, the Crusaders had to learn logistical organization or perish. In the First and Second Crusades, in fact, more perished from starvation, or from lack of fodder for their horses, than from any other single cause—including Turkish swords and arrows. Richard, in particular, showed how well he had learned this lesson by the establishment of an intermediate supply base at Cyprus, by exploiting the logistical potentialities of sea power, by the excellent logistical arrangements of his march from Acre to Ascalon, and by his refusal to embark on a protracted siege of Jerusalem with inadequate logistical facilities.

There were several unrelated military implications in the role of the Church in the Crusades. In the first place, many churchmen considered the Crusades as a means of diverting the inherent bellicosity and brutality of the feudal gentry from private and domestic wars to the more laudable purposes of supporting religious objectives. As one historian has observed, in reference to the unkept "Truce of God" (see p. 307), it was "easier to consecrate the fighting instinct than to curb it."

One other religious-military development was the appearance of military orders of monks: the Knights Templars, the Knights of St. John (or Knights Hospitalers), and the Teutonic Knights. This last order, although established in the Holy Land (1190), soon moved to Prussia, where it did its crusading against heathen Slavs (see p. 402). The Templars and Hospitalers, however, maintained themselves in the Holy Land and provided the nucleus of the regular fighting forces of the Kings of Jerusalem; their fortresses long stemmed the resurgent tide of Islam under Zangi, Nur-ed-din, and Saladin. By the end of the 12th century these two orders probably had the most efficient military establishments in the world. Their members were highly valued as military advisers and staff officers to rulers and generals of West European armies.

The other important religious development with military implications was the increasing friction between the Roman Catholic Church and the Greek Orthodox Church of Constantinople. The suspicions and fighting between Crusaders and Byzantines inflamed the long-smoldering disputes between the two churches. The consequences would be momentous and tragic early in the following century (see pp. 412–414).

SOUTH ASIA

NORTH INDIA

During the early years of the 11th century, North India was dominated in repeated campaigns by one of the most able warriors of Asiatic history, Mahmud of Ghazni, son of Sabuktagin (see p. 292). But although he forced the Hindu princes of the Punjab to swear allegiance, Mahmud made no effort to annex the regions he overran there and elsewhere in India. He did, however, weaken the power of the Hindu states of the north, and completely wrecked the Pratihara Dynasty of Kanauj, which had so long held off the threat of Islam (see p. 291). For more than a century after the death of Mahmud, there were no more major incursions from the northwest, where the various Turk tribes and dynasties were struggling for control of Persia and Central Asia. Late in the 12th century, **Muizz ad-Din Mohammed** (Muhammad of Ghor) undertook a series of invasions which led to a complete Moslem conquest of north India by the end of the century. The principal events were:

977–1030. Reign of Mahmud of Ghazni. He made about 17 raids into India (1000–1030). Although he extended his father's foothold in the Punjab, Mahmud's principal interest was not conquest but rather plunder and—incidentally—gaining religious merit. Mahmud used the plunder from these raids to support his conquests in Central Asia, and at his death his suzerainty included all of modern Afghanistan, eastern Persia, and Transoxiana, as well as the Punjab. He defeated the larger Hindu armies with mobile forces, primarily horse archers, which he directed with great skill, courage, and vigor.

1001. Victory over Jaipal, Raja of Lahore. In subsequent years Mahmud ranged east and south over the Punjab.

1009. Battle of Peshawar. Mahmud was opposed by a coalition of north Indian Hindu princes under the leadership of **Anang-pal,** son of Jaipal. Mahmud's victory was reminiscent of Alexander's at the nearby Hydaspes, since he created panic among the Hindus' elephants. In this and the following years Mahmud marched far and wide over north India, fighting, killing, looting, and destroying Hindu temples. Particularly notable were the sack of Thaneswar (1014) and of Kanauj (1018).

1025. Raid to the Coast of Gujarat. Mahmud's major purpose of this expedition across the desert from Multan was to destroy the phallic idol of a famed Hindu temple at Somnath. He also killed perhaps 50,000 Hindus.

1030–1175. Internal Wars of North India. The numerous north Indian princes continued their interminable wars with each other, employing the same unwieldy armies that Mahmud had consistently cut to ribbons. They had learned nothing from their defeats.

1175. First Indian Expedition of Muhammad of Ghor. He overcame the Ghaznevid rulers of the Punjab. Soon after this he sought to emulate Mahmud by expeditions farther afield.

1178. Defeat in Gujarat. Following the route taken by Mahmud, Muhammad was decisively defeated by the Hindu Raja of Gujarat and forced to retire.

1179–1191. Muhammad's Consolidation of the Punjab. He prepared for a renewed invasion. He began more cautiously with the capture of Peshawar (1179) and Lahore (1186), and raids into the upper Ganges Valley (1187–1190).

1191. First Battle of Tarain (Tirawari). Muhammad, leading an expedition to Thaneswar, was badly defeated by **Prithviraja,** Raja of Delhi and Ajmer. Badly wounded, Muhammad returned to Ghazni, where he made great preparations for another expedition to Thaneswar.

1192. Second Battle of Tarain. On almost exactly the same battleground as that of the year before, Muhammad decisively defeated the Hindu allies, capturing Prithviraja, whom he had killed. In this battle Muhammad exploited to the utmost the superior mobility of his Turkish horse archers. North India was now at the mercy of the Moslems.

1193–1199. Ghurid Conquests in India. After capturing Delhi, Muhammad returned to Ghazni, leaving his forces in India to his able lieutenant and slave, **Qutb-ud-Din.** Qutb conquered Benares (1194), fought the Rajputs in Gujarat (1195–1198), and captured Badaun and Kannauj (1198–1199).

SOUTH INDIA

During the first half of the 11th century, Chola expanded its already substantial influence by land and sea to dominate the entire Deccan. In the middle of the century, the Chola were forced on the defense by a coalition of Hindu states, led by the Chalukya. Chola resurgence in the latter part of the century was perpetuated into the next by the emergence of a Chalukya-Chola dynasty as the result of a marriage alliance. At the same time, the Chalukya Dynasty of Kalyani returned to a position of predominance in the northwestern Deccan. Both of these dynasties declined rapidly, however. The principal events were:

985–1014. Reign of Rajaraja I of Chola. Continued expansion of Chola power in the Deccan and along the shores of the Bay of Bengal.

1014–1042. Reign of Rajendra I (son of Rajaraja). This was the height of Chola power. Rajendra suppressed a revolt in Ceylon. His armies dominated the Deccan. In a combined land and sea expedition he defeated the Pala King of Bengal on the banks of the lower Ganges (1022). His overseas expeditions may have brought tribute from the Mon kings of Pegu (Burma), but it is more likely that Chola merely established trading settlements on the Burmese coast. His fleets attacked the Indonesian kingdom of Srivijaya (1025), which had harrassed Chola trade with China, and thereby re-opened the trade routes.

1042–1052. Reign of Rajadahira I (son of Rajendra). He spent most of his reign fighting a coalition of Hindu princes eager to gain revenge for their earlier humiliations. The principal leader of this alliance was the Chalukya ruler, **Somesvara I** (1040–1068). Another was the Chalukya Princess **Akkadevi,** one of the few female generals of history, who participated in a number of battles and sieges. Somesvara defeated and killed Rajadahira at the **Battle of Koppan** (1052).

1062–1070. Reign of Virarajendra of Chola. He defeated Chalukya, and reestablished Tamil supremacy in south India. An overseas expedition assisted Kedah (in Malaya) against Srivijaya. Upon his death his two sons killed each other off in a struggle for supremacy.

1070–1122. Reign of Rajendra Chola Kullotunga. This eastern Chalukya prince had married the daughter of Virarajendra, and established the Chalukya-Chola Dynasty controlling the eastern and southern Deccan.

1076–1127. Reign of Vikramaditya VI. He began the rise of Chalukya of Kalyani in the northwestern Deccan.

1076–1147. Reign of Anantavarman Choda Ganga of Kalinga (or Orissa). He raised his country to a position of leadership on the east coast between the Godavari and Ganges rivers.

c. 1100–1150. Rise of the Hoysala Dynasty (modern Mysore). They began to challenge the hegemony of Chalukya and Chalukya-Chola.

c. 1130–1200. Resurgence of Pandya. They gained independence from Chola.

1156–1183. Religious Civil War in the Chalukya Dominions. This weakened the dynasty's power; during most of this time, the rebels held Kalyani.

1173–1220. Reign of Vira Bellala II of Hoysala. In alliance with other Hindu neighbors, he overthrew the Chalukya of Kalyani (1190). During this war the Hoysala king was assisted by his warrior-queen **Umadevi,** who apparently exercised command in at least 2 campaigns against the Chalukya and their allies.

CEYLON (SRI LANKA)

During the 11th and 12th centuries Ceylon was invaded frequently by the Tamil kingdoms of the southern tip of India. From about 1017 to about 1070 the Chola dominated the island, but were plagued by frequent uprisings. **Vijaya Bahu** (1065–1120) finally expelled the Chola; (1070); and brought prosperity to a united, independent kingdom. The zenith of Sinhalese glory was under the great King **Parakram Bahu** (1153–1186), who repelled a Tamil invasion (1168) and in turn invaded Madura on the mainland (c. 1170), taking advantage of the struggle between Pandya and Chola.

SOUTHEAST ASIA, 1000–1200

THE TAI

The Kingdom of Nanchao was relatively quiescent during this period. It remained independent of China, but evidently attempted few foreign adventures. Tai mercenary soldiers were common in the armies of the Khmer, Mon, and Burmese; most of these probably came from Nanchao.

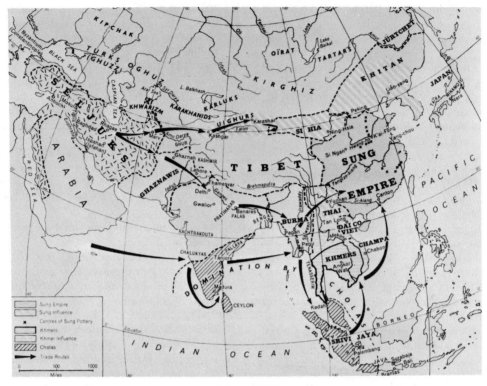

Asia, c. 1030

The gradual southern migration of the Tai people continued as they inched their way into the regions that their descendants now occupy in Thailand, eastern Burma, and Laos. A number of small independent Tai states began to appear in these areas late in the 11th century, the first being Raheng (1096) at the junction of the Mep'ing and Mewang rivers (northwestern Thailand).

VIETNAMESE REGION

During most of the 11th century, Champa and Annam were at war with each other, a continuation of the disputes arising from conflicting claims to three border provinces which had been seized by Champa in the preceding century (see p. 294). The Annamese generally had the best of these wars. Even more bitter was the warfare between Champa and the Khmer kings of Angkor, which raged during most of the following century and continued through the close of the period. The principal events were:

1000–1044. Intermittent War between Champa and Annam. The Annamese usually held the initiative, finally capturing Vijaya (Binh Dinh), the Cham capital. They killed the King of Champa and evidently reannexed the disputed provinces (1044).

1068–1069. Cham Counterinvasion of Annam. This was eventually repulsed.

1074–c. 1100. Reign of Harivarman IV of Champa. He repulsed Annamese and Khmer attacks (1070–1076), and later raided successfully into Cambodia.

1103. Cham Defeat. The Annamese repulsed an attempt to recover the disputed provinces.

c. 1130–1132. Khmer Invasion of Champa. Suryavarman II of Angkor forced the Cham (probably not unwilling) to assist him in his concurrent unsuccessful invasion of Annam.

1145–1149. War between Champa and Angkor. Suryavarman conquered Champa, but was then expelled by a Cham uprising.

1149–1160. Rebellion and Disorder in Champa.

1150. Khmer Invasion of Annam. The Khmer were disastrously defeated near Tonkin.

1167–1190. Constant Warfare between Champa and Angkor. A Cham invasion of Cambodia was initially successful, thanks largely to the effective use of mounted crossbowmen. Angkor was captured and sacked (1177). The Khmer rallied under **Jayavarman VII,** who repulsed the Cham, then conquered Champa, which he divided into two puppet vassal regimes (1190).

1191–1192. Civil War in Champa. Suryavarman, one of the puppet rulers, united the country, then expelled Khmer occupation forces.

1192–1203. Continued War between Cambodia and Champa.

CAMBODIA

This period of Khmer history was dominated by three great warriors kings of Angkor— **Suryavarman I, Suryavarman II,** and **Jayavarman VII**—who distinguished themselves as much by their magnificent construction programs at Angkor as by their extensive foreign conquests. Each of these glorious reigns, however, exhausted and weakened the country. The principal events were:

1002–1050. Reign of Suryavarman I. He consolidated his initially tenuous control over the country by victories in civil wars during the first 10 years of his reign. He them expanded his domain northward by conquering the Mekong Valley as far north as Luang Prabang.

1050–1066. Reign of Udayadityavarman II. Plagued by constant revolts, some of which were inspired by neighboring Champa. The revolts were suppressed by the brilliant Khmer general, **Sangrama.**

1066–1113. Internal Disorder.

1113–1150. Reign of Suryavarman II. Despite repeated victories over Champa, he was unsuccessful in attempts to conquer Annam, and was eventually repelled from Champa as well (see above). He was more successful in his campaigns to the west, apparently overrunning all of the small Tai states between Cambodia and the plains of Burma. His embassies to China were received with respect and honor.

1150–1177. Internal Turmoil. Champa exploited this by capturing and sacking Angkor (1177; see above).

1177–1218. Reign of Jayavarman VII. He unified his discouraged people, then defeated the Chams in a great naval river battle, probably on the Mekong River or Tonle Sap Lake (1178?). He restored internal order (1178–1181) and suppressed a revolt of the vassal Kingdom of Malyang (Battambang). He made good use of the military skill of refugee Cham Prince **Sri Vidyananda,** who also played an important part in the subsequent conquest of Champa (1190). Jayavarman expanded Khmer control of the Mekong Valley northward to Vientiane and to the south, down the Kra Isthmus. He built the powerful fortified city of Ankor Thom.

1192. Renewed War with Champa. (See above.)

BURMA

A Burmese nation, covering much of the area of modern Burma, first emerged in the middle of the 11th century under the great ruler and conqueror, **Anawrahta.** His successors were able to maintain the kingdom more or less intact, despite considerable internal disorder and the revolts of subject peoples. The principal events were:

1044–1077. Reign of Anawrahta. After annexing Arakan and Lower Burma, he conquered the Mon Kingdom of Thaton in the lower Irrawaddy and Salween Valleys (1057), checking Khmer probes. He established his capital at Pagan. He conducted a number of punitive expeditions in the Shan (Tai) states to the east, and may have raided eastward as far as the Chao Phraya Valley of modern Thailand. He built a number of fortifications along his eastern frontier to prevent Shan depredations.

1077–1084. Reign of Sawlu (son of Anawrahta). A prolonged Mon rebellion was initially successful, resulting in the temporary capture and sacking of Pagan and the overthrow of Sawlu.

1084–1112. Reign of Kyanzittha (another son of Anawrahta). He repulsed the Mons and eventually suppressed their rebellion.

1112–1173. Internal Disorder.

1173–1210. Reign of Narapatisittu. Order and peace were restored.

AFRICA

EAST AFRICA

Early in the 11th century, the Zagwe Dynasty from central Ethiopia absorbed the waning kingdom of Aksum. Their kingdom became a refuge for Christian monks and clerics fleeing Moslem Egypt, sparking the construction of many churches and monasteries. As Moslem expansion limited access to the Red Sea, Ethiopia became isolated, but Ethiopian princes still ruled a few coastal enclaves in Yemen as late as 1200. Meanwhile, Moslem traders continued to expand their influence along the East African coast, and Mogadishu became a major trading city.

c. 1075–1200. Somali expansion. Led by Arabic chieftains, Somali tribesmen migrated southward, extending their influence throughout the "Horn of Africa," destroying the fading Bantu kingdom of Zanj. Moslem Somali and pagan Galla warred against the Christians of highland Ethiopia.

1150–1200. Shirazi Migration from the Persian Gulf. These immigrants from Oman settled at the harbor of Kilwa, and also on Pemba, Zanzibar, and in the Comoro Islands.

WEST AFRICA

1054–1076. Almoravid Conquests in West Africa. The Almoravids, a militant Berber sect of Islam (see p. 333), overran much of West Africa. Although they sacked the Ghanian capital (1076), and apparently forced the conversion of its rulers, they did not subjugate the entire kingdom. Despite the Almoravids, Ghana's rulers continued to observe rites of worship for ancestors and agricultural spirits, closely connected with their authority. Almoravid influence ebbed by the early 1100s and Ghanian power began to wane afterward.

c. 1085. Conversion of Kanem. In the last decades of the 11th century, the *mai* (king) **Umme** (later known as **Ibn 'Abd al-Jalil**) converted to Islam, although some of his subjects remained pagans for centuries.

SOUTH AFRICA

During the 11th century, the structures at Zimbabwe (Bantu for "stone dwellings") became a ceremonial and religious center for the great inland empire of Karanga, which covered an irregular area about 300 kilometers across. It traded gold with Arab merchants on the coast for glass and porcelain from China.

EAST AND CENTRAL ASIA

CHINA

At the outset of the 11th century, the Sung Dynasty of China was involved in unsuccessful wars with neighbors to the north and northwest: the Liao Dynasty of the Khitan Mongols of Manchuria and the newly established Western Hsia Kingdom in Kansu. Fortunately for the Sung, these northern barbarian kingdoms were also engaged in more or less continuous warfare with each other. The latter part of the century was relatively peaceful, thanks to annual tribute paid to the northern barbarian kingdoms. Early in the 12th century, the Sung joined with the Juchen Mongols to overwhelm the Khitan. Almost immediately, however, the Juchen (later known as the Chin Dynasty) turned against the Sung and drove them from northern China. During the remainder of the century, the Sung and Chin were engaged in several wars; the Sung were unable to reconquer north China, but kept the Chin from penetrating south of the Yangtze. The principal events were:

1000–1004. Khitan Invasion. This culminated a long-drawn-out war between the Khitan and Sung dynasties. The Khitan drove to the Yellow River near the Sung capital of Pien Liang (modern Kaifeng). The Sung sued for peace, agreeing to pay a large annual tribute.

1001–1003. War with the Western Hsia Kingdom. This was inconclusive.

1038–1043. Renewed War between Sung and Hsia. With the assistance of the Uighurs of Sinkiang, the Sung achieved a face-saving peace whereby the Hsia agreed to continue nominal vassalage to China in return for annual tribute.

1070. Military Reorganization. This was part of a sweeping program of reform of the entire governmental and financial systems by **Wang An-shih,** Sung prime minister. Due to con-stant external threats and frontier incidents with Khitans and Hsia, the Sung standing army had grown to about 1,100,000 men. This, combined with the tribute to the northern neighbors, was an economic burden that was bankrupting the country. Wang created a conscript militia system designed to provide for more effective border defense and more effective internal and local security, while permitting a reduction of the standing army to about 500,000. Within 6 years the militia force numbered 7 million.

1115–1122. Sung Alliance with the Juchen Mongols. The allies destroyed the Khitan Liao Kingdom.

1123–1127. Juchen Mongol Invasion. They reached Pien Liang, but were repulsed (1126). The following year, however, they stormed the

Sung capital and captured the emperor, **Hui Tsung,** and his eldest son, **Ch'in Tsung.** A younger son, **Kao Tsung,** escaped and fled south to the Yangtze River to establish the Southern Sung Dynasty, with its capital at Nanking. The Juchen (now the Chin Dynasty) pursued, crossed the Yangtze to capture Nanking, and drove Kao Tsung to Hangchow (1127).

1128–1140. Yangtze Campaigns. Sung General **Yueh Fei,** with the cooperation of the Sung fleet in the Yangtze River, defeated the Chin (Juchen) army, and drove it north of the river. In a series of campaigns, Yueh Fei and other Chinese generals pushed the Chin north to the vicinity of Pien Liang. A palace intrigue then resulted in the recall of Yueh Fei, who was executed by Kao Tsung (1141).

1141. Uneasy Peace. The boundary between the Southern Sung and the Chin was established along the Hwai and upper Han rivers.

1161. Chin Invasion of Southern China. This was led by Chin Emperor **Liang.** He was halted by the Sung army and fleet at the Yangtze River. Sung General **Yu Yun-wen** employed explosives, apparently for the first time in warfare, in defeating the Chin at the **Battle of Ts'ai-shih** (near Nanking).

INNER ASIA

The nomadic barbarians of North and Central Asia inhabited a great arc of land extending from the sea of Japan, Korea, and the Yellow Sea in the east, across modern Manchuria, Mongolia, Sinkiang, and Russian Turkestan roughly to the line of the Syr Darya River. This had been an area of constant ferment, and the origin of numberless migrations and invasions to the southeast (into China), to the southwest (into Transoxiana, Persia, and India), and west (across Scythia toward Europe) since the dawn of history. In the earliest days the inhabitants of the western portion of this region had evidently been mainly Aryans; the inhabitants of the eastern regions had been primarily Mongoloid in their ethnologic characteristics. As the centuries passed, and as the tribes roamed and fought over the vast area, there was considerable mixing of the races. The principal Aryan group, the Yueh Chih, had earlier migrated toward Europe, Persia, and India (see p. 132).

Thus, by the beginning of the 11th century, this vast arc of deserts, mountains, and pastureland was inhabited by swarthy or yellow-skinned people resembling each other in racial, cultural, and linguistic characteristics, and ethnologically more Mongoloid than Aryan. This similarity among the Mongols, Turks, Tartars (Tatars), and Tanguts who inhabited the vast region makes for considerable ethnic and historical confusion. Generally speaking, the Mongols and Tartars inhabited the northern and eastern areas; the Turks (who had already begun to spread over western Asia and southeastern Europe) were in the southwest; the Tanguts, who were more closely related to the Tibetans than were the other nomads, were generally in eastern Sinkiang and Kansu.

The Khitan Mongols of Manchuria comprised a homogeneous nation, and were beginning to lose their nomadic characteristics. To their west and northwest were many other Mongol tribes, linked together in various amorphous alliances and groupings, but with little national cohesiveness. In Kansu and Eastern Sinkiang the Tanguts had recently formed a nation—the Western Hsia—still nominally under Chinese rule. In Sinkiang were the Uighur Turks, who were still loosely allied with the Chinese. Farther west, between the Tien Shan Mountains and the Aral Sea, were the so-called Guzz Turks, comprising many independent tribes and tribal groups, including those of the Seljuks and the Ilak khans. (Farther north and west were the Kipchak

Turks—or Cumans—of the steppes of Western Asia and Caucasia, and to their west were the Pechenegs of the South Russian steppes; these Turkish peoples, however, were by this time more involved in Eurasian history.)

During much of the 11th and 12th centuries the Khitan and Western Hsia were frequently at war with each other and with the Sung Dynasty of China. The Uighurs were often involved in these wars, aiding their Chinese allies against the Hsia. Early in the 12th century a new Mongol people—the Juchen—moved southeastward from Mongolia to clash with the Khitan, whom they soon overthrew with Chinese assistance. The Juchen did not pause, however, and immediately invaded the territory of their former allies, the Sung Chinese, to precipitate a series of wars that continued through the rest of the century. Toward the end of the century, in central Mongolia, another new Mongol kingdom was coalescing around the tribe of a brilliant young chieftain named **Temujin.** The principal events were:

1000–1038. Intermittent Wars of the Khitan and the Tangut of Kansu.

1038–1043. Revolt of Western Hsia. They obtained complete independence from China, under **Yuan Ho** as emperor. Despite Uighur assistance, the Chinese were unable to suppress the revolt. In return for Chinese payment of tribute, Yuan Ho agreed to consider himself a Sung vassal.

1044. Renewed War between Khitan and Western Hsia. This was precipitated by Tangut invasion of Manchuria, repulsed by the Khitan Liao Dynasty.

1080–1090. Seljuk Expansion. (See p. 336.)

1115–1123. Juchen Mongol Conquest of Khitan. Accomplished with Chinese assistance. The Liao ruler, **Ye-lu Ta-shih,** fled with the small remnant of his army to the Tarim Basin, where he allied himself to the Uighurs to establish the Kingdom of Kara Khitai.

1122. Proclamation of the Chin Dynasty. The Juchen leader **Tsu** took the imperial name **T'ien Fu.** He then began an invasion of China (1123; see p. 353).

1126–1141. Expansion of the Kara Khitai. The former Khitans, now also called the Western Liao, controlled both sides of the Pamirs.

Ye-lu Ta-shih defeated Sanjar, Seljuk Sultan of Persia, and drove him from Transoxiana (1141; see p. 337).

c. 1132. The First Gunpowder Weapon. The first gunpowder weapon was a very long-barrelled bamboo musket invented by General **Ch'en Gui,** commander of the garrison of Anlu, Hopei Province.

1135. Mongol Invasion of Chin. The Mongol Khan **Kabul** led this, the first of a series of Mongol raids into the new empire.

1161. Chin Invasion of Sung China. Chin Emperor Liang was defeated near Nanking (see p. 354). When he ordered his defeated troops to renew the attack, they mutinied and hanged him, then withdrew across the Yangtze.

1162. Birth of Temujin. He was the son of Mongol chieftain **Yesugai,** in the region southeast of Lake Baikal.

1180–1190. Rise of Temujin's New Mongol Kingdom. Created by a combination of alliances and conquests of neighboring tribes.

1194. Temujin Campaigns against the Tartars. In alliance with the Chin and with the Kerait Mongols, he conducted a successful campaign southwest of the Gobi.

KOREA

The Wang Dynasty (see p. 299) continued to rule Koryo throughout this period. They had several wars with the Khitan Liao Kingdom of Manchuria during the 11th century. An effort to prevent Khitan depredations by construction of a great wall across the narrow neck of the peninsula (1044) was only partially successful; the Wang paid tribute to the Khitan until Liao was overwhelmed by the Juchen (1123; see p. 353). The Wang immediately acknowledged Juchen suzeranity. The regime barely survived military revolt late in the 12th century (1170).

JAPAN

The Fujiwara clan continued to dominate Japan during the first half of the 11th century. Although this was a period of relative peace and prosperity for the country as a whole, the two warrior clans of Taira and Minamoto (see p. 299) clashed in many small conflicts between feudal lords. Both clans were also called upon frequently by the Fujiwara regent-dictators to deal with an unusual class of lawbreakers. These were Buddhist monks, who had become increasingly secular and arrogant, and who were wont to achieve their demands by force of arms if necessary. The increasing reliance of the Fujiwaras upon the warrior clans impaired their own prestige, permitting the imperial family to regain its power. From the middle of the 11th century until the middle of the 12th, the emperors ruled as well as reigned. This period was ended by two civil wars resulting from a dynastic dispute between rival emperors. The Taira clan emerged predominant from these wars. Their longstanding rivalry with the Minamoto soon became bitter and acute. The result was further civil war in which the Minamoto clan virtually annihilated the Taira and seized power. By the end of the century Japan was under a feudal military dictatorship. The principal events were:

1000–1068. Zenith and Decline of the Fujiwara Clan.

1051–1062. The Earlier Nine Years' War. This was an uprising of the Abe clan of northern Japan, suppressed by the Minamoto.

1068–1156. Period of Direct Imperial Rule. Actual power was usually exercised by an abdicated or retired emperor.

1083–1087. The Later Three Years' War. The Minamoto clan destroyed the dissident Kiyowara clan of northern Japan.

1156. The Hogen War. This was the result of a complicated dispute over imperial power between two brothers, **Goshirakawa II** and **Sokotu,** who had formerly reigned as emperor. Members of the Taira, Minamoto, and Fujiwara clans fought on both sides of this war. Mainly through Taira assistance, Goshirakawa was successful and Sokotu was exiled.

1159–1160. The Heiji War. The Minamoto clan temporarily seized power in a surprise coup, but were then decisively defeated by the Taira.

1160–1181. Predominance of Taira Kiyomori. He was practically a dictator.

1180–1184. War of the Taira and Minamoto. Under the leadership of **Minamoto Yoritomo** the Minamoto clan rose against the Tairas. A prolonged and bitter war ensued. At first the Taira were successful, but their arrogant rule alienated some of their feudal vassals, who joined the Minamoto. The Taira were driven from Kyoto (1183). The Minamoto won an important victory at the **Battle of Yashima** on Shikoku. The climax of the war came in the naval **Battle of Dannoura,** at the western exit of the Inland Sea; a land battle took place simultaneously on the adjacent shore. The Minamoto, under the command of Yoritomo's brother, **Yoshitsune,** were victorious, and the Taira virtually annihilated. Also distinguishing themselves in the war were Yoritomo's uncle, **Yoshiie,** and his cousin, **Yoshinaka.**

1185–1199. Military Dictatorship of Minamoto Yoritomo. Jealous and suspicious of the military successes of his relatives— Yoshitsune, Yoshiie, and Yoshinaka—Yorimoto had them all assassinated (1189). He then accused the Fujiwaras (whose power was increasing) of responsibility for the crimes, and took this opportunity to destroy the Fujiwara clan. He was the first Japanese dictator to hold the title of **Shogun** (generalissimo).

XI

THE AGE OF THE MONGOLS:

1200—1400

MILITARY TRENDS

During most of these two centuries the military and political history of mankind was dominated by one power. The Mongols—or their Tartar vassals—conquered or ravaged every major region of the known world save Western Europe. It was caused by the unique military and political genius of an illiterate Mongol chieftain, **Genghis** (or **Jenghiz**) **Khan,** one of the greatest military leaders the world has seen. No other leader of this period can be ranked with the great Khakhan of the Mongols, though comparable strategical and tactical skill was demonstrated by subordinates and successors such as **Subotai, Chepe, Mangu,** and **Kublai.** There were, of course, other able generals, of whom **Edward I** of England, **Du Guesclin** of France, **Tamerlane** the Tartar, **Baibars** the Mameluke, and **Murad** the Ottoman were outstanding. Although he was short on strategic ability, the tactical skill and success of English King **Edward III** probably warrant his inclusion with this group.

The continuing importance of the Turks in warfare during this period is demonstrated by the fact that three of the generals mentioned above were Turanians or Turks. By the same token, the inclusion of two English names is also significant.

Although cavalry domination of warfare reached its zenith in the Mongol conquests, this period also saw the return of the infantry soldier to predominance on the battlefield. The Battle of Crécy is generally accepted as the symbol of this return, just as the Battle of Adrianople marked the earlier ascendancy of cavalry over footmen. Edward III, building on a tactical system created by his grandfather Edward I, accomplished this by combining the staunch defensive capability of heavily armored pikemen with the maneuverability and firepower of light archers wielding the formidable English longbow. This, plus his use of cavalry for counterattack, demonstrated once more that military success is most likely to crown the efforts of those who employ the capabilities of combined arms to their maximum.

Although it did not have a significant effect on military operations during this period, the appearance of gunpowder weapons on the battlefield marked the beginning of one of the most momentous military trends of history.

This period also saw the conclusion of two great eras of military history. The Byzantine Empire, once a bastion of Christian Europe, had virtually disappeared by the close of the 14th century. Even earlier, the fall of the last Christian footholds in Palestine and Syria to the Mamelukes closed the age of the Crusades.

ARMOR

The steadily improving metallurgical skill of medieval European armorers permitted the introduction of plate armor in the 13th century. At first these iron plates—covering vital and vulnerable areas such as shoulders and thighs—were worn under chain mail. By the middle of the century they were being worn over the mail, or in its place, covering shoulders, elbows, kneecaps, shins, and thighs. Late in the 13th century plate cuirasses or breastplates began to replace chain shirts.

Early combinations of plate and mail sometimes resulted in inadequate protection at the junction of the two, and on the inside of elbow, shoulder, and knee joints. This led armorers in the 14th century to develop cleverly constructed complete suits of plate armor that began to replace mail.

Skillful European smiths introduced mail mittens early in the 13th century, soon followed by mail gloves.

As plate armor made the knight or man-at-arms increasingly impervious to the weapons of the day, the shield became reduced in size to a relatively small triangle on which were emblazoned armorial bearings. Heraldry was becoming important to medieval nobility and gentry, and armorial bearings were also often embroidered on surcoats, decorative as well as useful for identification.

Most knights adopted the pot helmet with its eyeholes and breathing holes, though a number were still satisfied with the lighter and more comfortable simple casque worn over a mail coif. During the 14th century movable visors began to appear on the pot helmets.

These innovations still further increased the weight of the knight's armor. If knocked down, or unhorsed, he could rarely rise without assistance; this put a premium upon disabling heavy mounts, which in turn led to increasing the armor protection of horses. By the end of the 14th century, then, the heavy cavalry horse was usually carrying a total weight of at least 150 pounds of armor and equipment—its own and its rider's—in addition to the man's basic weight. This meant that only ponderous, slow horses could be used for heavy-cavalry work, and even these could charge only at a trot or a slow canter.

One effect of these improvements and additions to armor was to reduce greatly the overall numbers of combat casualties, but it sometimes led to massacres of defeated armies. (One important mitigation of this latter effect, however, was in the handsome ransoms which could be obtained for titled prisoners.) Another effect was to sacrifice mobility for protection—and mobility was the essential characteristic of cavalry. This sacrifice was not serious so long as a relative superiority of mobility could be combined with relative invulnerability. But events of the 14th century foretold the knell of armored heavy cavalry. Crécy and the subsequent introduction of gunpowder weapons showed that neither superior mobility nor invulnerability could be maintained by increasingly ponderous armored horsemen.

WEAPONS

Pre-Gunpowder Weapons

One important new weapon appeared shortly before the introduction of gunpowder: the English longbow. Originally a Welsh weapon, introduced in England in the 12th century, its potentialities were first fully appreciated by Edward I in his campaigns in Wales (see p. 391). He adopted it as the basic arm of the English yeoman infantry—both militia and professional—and employed it successfully against the Scots. Methods for tactical employment of the longbow were perfected in the Scottish campaigns of Edward III, who then used it with such stunning effect at Crécy.

The longbow not only had twice the range of the crossbow—up to 400 yards maximum, and an effective range approaching 250 yards—but also a far more rapid rate of fire. In the hands of English professional soldiers it was more accurate than, and apparently had penetrating power comparable to, the crossbow. It was lighter, more easily handled, and adaptable for skirmishing or for volley fire. It was the most effective and most versatile individual weapon yet to appear on the battlefield.

English longbowman

There were no other significant developments or improvements in individual weapons or war machines.

The Advent of Gunpowder

The origin of gunpowder weapons is obscure. Apparently the Chinese had been using some kind of explosives as early as the 9th century (see p. 298), but mostly as a noisemaker and for demolition. The first actual gunpowder weapon appears to have been a Chinese bamboo musket, introduced in 1132; its effectiveness is unknown (see p. 355). There is also some evidence that the Chinese had developed rudimentary rockets in their long wars against the Mongols in the 13th century, and that these were employed at least experimentally by some of the successors of Genghis Khan. There seems to be little doubt, however, that the first serious and effective use of gunpowder as a missile propellant was in Europe in the mid-14th century. Both the Englishman Roger Bacon and German monk Berthold Schwarz have been credited with the discovery of gunpowder in Europe.

It is commonly accepted that the first use of gunpowder weapons in battle in Europe was by the English at the Battle of Crécy (1346), although guns may have been used at Metz (1324) or at Algeciras (1342). Edward is reported to have had three or five **roundelades** or *pots de fer* (iron jugs), so called because of their shape, like round iron bottles. Soon after this, crude cannon were being used in the Hundred Years' War by both English and French on the battlefield and in siege warfare. They appeared in Germany and Italy about the same time.

The first gunpowder weapons were small metal pots or tubes inaccurately propelling arrow-like bolts. Because of the problems of weight, size, and the difficulties of holding such weapons while firing with a match or fuse, the first crude handguns—really small cannon—were unsuccessful. There were rapid and dramatic improvements in larger artillery pieces, however. As a result, by the end of the century artillery had become a permanent element of all European armies, and had even appeared in battles between the Lithuanians and the "Golden Horde" in Russia.

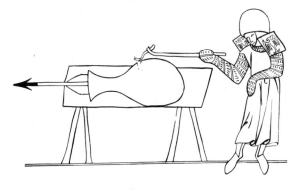

Early cannon

FORTIFICATION AND SIEGECRAFT

There were no basic improvements either in fortification or siegecraft in this period. Powerful castles appeared in increasing numbers throughout Europe patterned after the Byzantine fortifications of Anatolia and the magnificent castles which the Crusaders continued to build in Syria and Palestine.

The technique of siegecraft lagged far behind the art of fortification. Human ingenuity and mechanical skill had seemingly exhausted themselves. No important new machines or siege weapons could be devised until the human brain found some means for propulsion more powerful than the forces of torsion, tension, and counterpoise, or more destructive than fire.

The capture of Château Gaillard by Philip II proved that a powerful and wealthy monarch who was able and willing to devote sufficient resources to the task could in time overcome even the most powerful defenses. But such patience, wealth, determination, and skill were rare in the Middle Ages. Feudal armies could rarely be maintained long enough to undertake the prolonged sieges necessary to capture such powerful forts; feudal levies could be called to action for only a few weeks of the year, and mercenary armies were extremely expensive to maintain in the field indefinitely.

As a result, defensive strategy could usually be counted upon to offset offensive capabilities. Accordingly, this period probably had relatively fewer pitched battles in the open field than any era before or since. The weaker side could usually count on being able to retire behind fortifications to win—or at least not to lose. In general, only the foolhardy or inept, or unlucky, would risk a battle when not completely confident of superior combat power.

Twelfth-century castle

TACTICS OF LAND WARFARE

The return of infantry to predominance in warfare as the principal element of a balanced combined-arms infantry-cavalry team—a trend evident in previous centuries—reached its climax at the Battle of Crécy only because of a new and improved firepower weapon—the longbow—supporting disciplined armored pikemen.

The full significance of this was not appreciated even by the English. As for other 14th century leaders, they sought in vain for the elusive key to victory by following the English example of dismounting their heavy cavalry in battle. They failed to realize that the secret of English success was not merely their dismounted knights and their archers, but the judicious combination of these two with one another and with mounted cavalrymen to obtain a flexible combination of missile fire-power, defensive staying power, and mobile shock action.

The successes of the English, as well as those of Flemish and Swiss pikemen, resulted in giving to the defense a substantial tactical superiority over the offense in European warfare. This complemented and reinforced the strategic superiority of the defense resulting from the power of fortifications noted above.

It is interesting to speculate on what might have occurred if the 14th century combined-arms defensive tactics of Edward III had ever been opposed by the 13th century cavalry-offensive tactics of Genghis Khan.

English dismounted man-at-arms   Swiss halbardier

Western Soldiery in the Later Middle Ages*

Reliance upon the ill-trained, hastily raised feudal levies continued to decline while mercenary forces waxed in numbers, size, and influence. In England, however, and later in France, there was a partial reconciliation of these two inconsistent systems.

* A discussion of the Mongol military system appears elsewhere in this chapter (see p. 367).

The Plantagenet kings introduced the method of "indenture" as a means of combining the militia and mercenary systems. They encouraged their vassals to provide standing contingents of troops for wars and for garrisoning royal fortifications in return for set payments and allowances. Although nominally this had no connection with the vassal's feudal obligation, the English kings rarely called out the militia.

These standing contingents were, in effect, the genesis of a permanent royal English Army. The system avoided compulsion, since the barons were paid for the forces they raised. These were professional soldiers, in both the Roman and modern sense of the term; they bore allegiance to the king, even though commanded and raised by the nobles, and they thought of themselves as English soldiers. They were, however, also mercenaries, who were not easily controlled or utilized in times of peace, when they often turned their unruly natures and military skills to plundering and terrorizing the civil populace. Since English standing units were usually quartered in the French domains of the Plantagenets, this aspect of the indenture system had less impact on England than it did on France and those other European countries using the method.

It is perhaps a paradox that every soldier in many of the English armies, which dominated European warfare by virtue of their infantry superiority, was a horseman. This explains the ability of Edward III and his son, the Black Prince, to raid fast and far through France, more than matching the mobility of the French cavalry armies.

EAST AND CENTRAL ASIA

MONGOLIA AND THE MONGOL EMPIRE

The Early Conquests of Genghis Khan, 1190–1217

1190–1206. Unification of Mongolia. By conquest of neighboring tribes, and by adroit diplomacy, **Temujin** (see p. 355) created a large and homogeneous Mongol empire in East-Central Asia, around the Gobi Desert. Establishing his capital at Karakorum, he proclaimed himself **Genghis** (Jenghiz) **Khan** (meaning "Perfect War Emperor," or "Supreme Emperor").

1206–1209. First Wars against the Western Hsia Empire. Genghis began his conquest of China in a series of campaigns against the Western Hsia (1205, 1207, 1209). The Western Hsia emperor, with reduced dominion, acknowledged the suzerainty of Genghis Khan.

1208. Battle of the Irtysh. Genghis overcame the last resistance in Mongolia by defeating **Kushluk** (Guchluk), leader of the Naiman tribe, who fled to seek refuge with the Kara-Khitai Tartars (see p. 355).

1211–1215. First War against the Chin Empire. At first frustrated by the strong fortifications of the northern Chinese cities, and by his own lack of siegecraft, Genghis gradually built up a siege train and conquered Chin territory as far as the Great Wall (1213). He then advanced with three armies into the heart of Chin territory, between the Wall and the Yellow River. He completely defeated the Chin field forces, spread fire and sword through north China, captured numerous cities, and finally besieged, captured, and sacked Peking (1215). The Chin emperor was forced to recognize the suzerainty of the Mongol conqueror. **Yeh-lu Ch'u-ta'ai,** a Khitan official in Chin service, was taken into Mongol service. Apparently his expertise in administering sedentary populations was utilized by Genghis in later conquests.

1209–1216. Kushluk and the Kara-Khitai. The fugitive Kushluk, supported by **Mo-**

hammed Shah of Khwarezm, treacherously overthrew the Khan of Kara-Khitai and seized the throne. He prepared for war with Genghis, whose spies kept him informed of these events.

1217. Conquest of Kara-Khitai. The Mongol army was exhausted by 10 years of continuous campaigning. So Genghis sent only two **toumans** (20,000 men) under brilliant young general **Chepe** (Jebei) against Kushluk. A Tartar revolt was incited, then Chepe overran the country. Kushluk's forces were defeated west of Kashgar; he was captured and executed; Kara-Khitai was annexed by Genghis.

Campaigns against Khwarezmian Empire, 1218–1224

1218. Outbreak of War. Following mistreatment of Mongol ambassadors by **Alaud-Din Mohammed** (Mohammed Shah), Khwarezmian-Turkish ruler of Persia, Genghis gathered a force of more than 200,000 men (including auxiliaries), divided into 4 main armies

commanded respectively by himself and his sons **Juji, Jagatai,** and **Ogatai.** He planned to advance into Persia from the northeast, while Chepe also moved in from Kara-Khitai with a small force.

1219. Battle of Jand. Juji and Chepe, with not more than 30,000 men between them, fought a drawn battle in the Ferghana Valley against Mohammed's vastly superior forces (perhaps 200,000).

1220. The Mongol Advance. The Mongols moved on a broad front. Juji, reinforced to about 50,000, advanced down the Ferghana Valley to besiege Khojend. Jagatai and Ogatai, with about 50,000 men each, marched in parallel columns past Lake Balkash to invest Otrar on the Syr Darya (Jaxartes) River. Genghis, with an army of similar size, swinging wide to the north of Lake Balkash, approached the Syr Darya near its Aral Sea mouth. Chepe, with 20,000 men, moving south through the Pamirs to the headwaters of the Amu Darya (Oxus) River, followed it into the heart of Transoxiana.

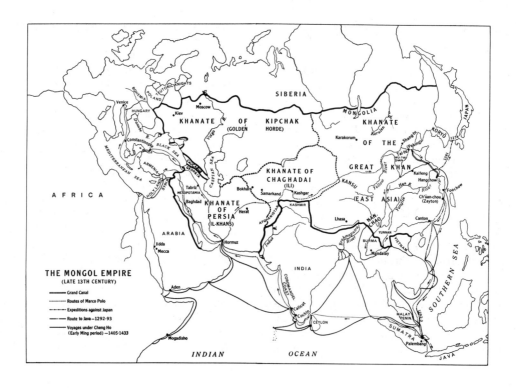

1220. Flight of Mohammed. The shah, near Samarkand, learned of the investments of Khojend and Otrar, then got alarming reports of Chepe's advance. As he shifted a large force (possibly 100,000 men) to the southeast to protect his lines of communication to Khorasan and Ghazni, the stunning news came that Genghis Khan himself was advancing against Bokhara from the west—a direction totally unexpected. The rapid Mongol movements and the utter devastation they spread for miles on each side of their lines of march caused the Khwarezmians greatly to exaggerate their strength. Mohammed wrongly believed the forces in the encircling armies outnumbered his own 500,000 men. In panic he threw most of the remainder of his great host into Bokhara, Samarkand, and other fortresses and fled south with his family and a small bodyguard.

1220. Conquest of Transoxiana. As Genghis was reaching Bokhara, his sons sacked and destroyed Khojend and Otrar. In the upper Amu Darya Valley, Chepe defeated the Turk force which Mohammed had sent against him and advanced toward Samarkand, pillaging, killing, and destroying. Bokhara surrendered to Genghis, who forced the population to raze the city's walls. Hastening onward, he joined his other forces outside Samarkand, garrisoned by 100,000 men. Despite desperate resistance, it was quickly stormed, cruelly sacked, and destroyed; the garrison and much of the population were massacred.

1220–1221. Pursuit. A force of about 30,000 men under generals Chepe and **Subotai** chased the Turkish emperor relentlessly for five months through his vast empire to the shore of the Caspian. On an islet off the coast, Mohammed died (it is said) of a broken heart (February 1221).

1220–1221. Consolidation. Genghis supervised the conquest of Khawarezm by Ogatai,

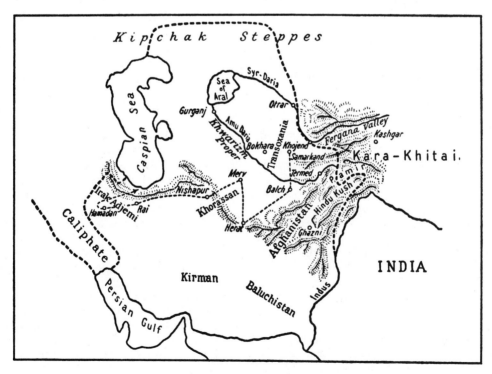

Khwarezmian Empire
Chepe's and Subotai's pursuit of Shah Mohammed

Juji, and Jagatai, and of Khorasan by his youngest son, **Tuli.** Moslem resistance was prolonged but ineffective at **Herat, Merv** and **Bamian.** Meanwhile, Genghis' scouts and spies discovered that **Jellaluddin,** son of Mohammed Shah, was raising a new army in Ghazni.

1221. Battle of Pirvan. While Genghis was gathering his sons' contingents together for another campaign, Jellaluddin with 120,000 men defeated an advance Mongol force of three toumans (30,000 men) in the Hindu Kush Mountains northwest of Ghazni. Genghis moved swiftly to avenge the defeat, and the Turkish prince, deserted by his Afghan allies, withdrew into the northern Punjab with about 30,000 men. Genghis followed with more than 50,000.

1221. Battle of the Indus. The Turks took up an excellent defensive position beside the Indus, their flanks protected by mountains and a bend of the river. A violent Turkish counterattack almost broke the center of the Mongol army, but Genghis rallied his men and sent a touman over apparently impassable mountains against Jellaluddin's flank. Struck from two directions, the Turkish defense collapsed. Jellaluddin himself rode his horse over the edge of a cliff to plunge into the river, escaping to the far bank under Genghis' admiring eyes. The Mongol sent a few toumans (see p. 368) in pursuit; these ravaged the Lahore, Peshawar, and Multan districts of the Punjab, but did not find the Turkish prince. Later, learning that Jellaluddin had been refused asylum by the Sultan of Delhi, Genghis wisely avoided overextending his resources by an invasion of India. He called back the pursuers and withdrew from the Punjab to consolidate his conquest of Ghazni (1222–1224).

The Expedition of Chepe and Subotai, 1221–1224

1221. Advance into the Caucasus. After the death of Mohammed Shah, Chepe and Subotai received permission from Genghis to make a reconnaissance in force through the Caucasus into Southern Europe. Reinforcements probably brought their total strength close to 40,000. Apparently Chepe was in command. After advancing through Azerbaijan, they probably spent the winter on the pleasant banks of the Araxes in eastern Armenia.

1222. Invasion of Russia. The Mongols decisively defeated a large Georgian army which had been gathered to join the Fifth Crusade, then continued northward into the steppes of Russia, smashing mountain tribes that endeavored to interfere with their passage. The Cumans briefly halted the invaders in a drawn battle in the foothills east of the Kuban River. Combining rapid movement with guile, the Mongols defeated them in detail, captured Astrakhan, then pursued across the Don. Penetrating into the Crimea, they stormed the Genoese fortress of Sudak on the southeast coast, then turned north into the Ukraine.

1222–1223. Winter by the Black Sea. The Mongol leaders now felt they had accomplished their mission. Before returning to Mongolia, however, they decided to rest their toumans and to gain more information about the lands to their north and west. Their spies were soon scattered all over Eastern and Central Europe.

1223. Battle of the Kalka River. Under the leadership of **Mstislav,** Prince of Kiev, a mixed Russian-Cuman army of 80,000 marched against the Mongols near the mouth of the Dnieper River. Chepe and Subotai sought peace, but when their envoys were brutally killed, they attacked and practically annihilated the Russians. After raiding a few hundred miles to the north, the Mongols turned eastward in compliance with a courier message from Genghis. As they were marching north of the Caspian Sea, Chepe sickened and died. Subotai brought the expedition back after a trek of 4,000 miles to a rendezvous with the main Mongol armies, now returning from their victories over the Khwarezmians (1224).

The Final Campaigns of Genghis Khan, 1225–1227

1224–1226. Resurgence of the Hsia and Chin Empires. The vassal Tangut emperor of the Western Hsia had refused to take part in the war against Mohammed Shah. The Hsia and Chin Empires (formerly bitter enemies) now formed an alliance against the Mongols.

1225–1226. Mongol Preparations. Feeling his age, Genghis selected Ogatai as successor to his throne, established the method of selection of subsequent Khakhans, studied intelligence reports from Hsia and Chin, and readied a force of 180,000 men for a new campaign.

1226. Invasion of Western Hsia. Late in the year, when the rivers were frozen, the Mongols struck southward with their customary speed and vigor. The Tanguts, well acquainted with Mongol methods, were ready, and the two armies met by the banks of the frozen Yellow River. The Hsia army numbered something more than 300,000 men.

1226. Battle of the Yellow River. Genghis enticed the Tangut cavalry to attack over the ice. As they crossed, his dismounted archers harassed them. The Tangut horsemen, soon thrown into confusion, were now charged by mounted and dismounted Mongols. At the same time several toumans swept past the struggle on the ice to attack the Hsia infantry on the far bank. The battle was soon over. The Mongols counted 300,000 dead; it is not known how many Tanguts escaped.

1227. Victory over Hsia and Chin. Pursuing with customary vigor, the Mongols killed the Hsia emperor in a mountain fortress. His son took refuge in the great walled city of Ninghsia, which the Mongols had vainly besieged in earlier wars. Leaving a third of his army to invest the Hsia capital, Genghis sent Ogatai eastward, across the great bend of the Yellow River, to drive the Chin from their last footholds north of the river. With the remainder, he marched southeast (evidently to eastern Szechwan), taking a position in the mountains where the Hsia, Chin, and Sung empires met, to prevent Sung reinforcements from reaching Ninghsia. Here he accepted the surrender of the new Hsia emperor, but rejected peace overtures from the Chin.

1227. The Death of Genghis Khan. A premonition of death caused the Khakhan (or Great Khan) to start back to Mongolia, but he died en route. On his deathbed he outlined to his son Tuli the plans which would later be used by his successors to complete the destruction of the Chin Empire.

THE MONGOL MILITARY SYSTEM, C. 1225

The Mongol "Hordes"

The word "horde," denoting a Mongol tribe or a field army, has become synonymous with vast numbers because the Mongols' Western foes refused to believe that they had been overwhelmed by small forces. Half to excuse their defeats, half because they never had the opportunity to understand the marvelous system that permitted the Mongols to strike with the speed and force of a hurricane, 13th-century Europeans sincerely but wrongly believed the Mongol armies to be tremendous, relatively undisciplined mobs that achieved their objectives solely by superior numbers.

Genghis Khan and his armies accomplished feats that would be hard, if not impossible, for modern armies to duplicate, principally because he had one of the best-organized, best-trained, and most thoroughly disciplined armies ever created. The Mongol army was usually much smaller than those of its principal opponents. The largest force Genghis Khan ever assembled was that with which he conquered Persia: less than 240,000 men. The Mongol armies which later conquered Russia and all of Eastern and Central Europe never exceeded 150,000 men.

Mongol Organization

Quality, not quantity, was the basis of Mongol success. The simplicity of their organization was its chief characteristic. Consisting entirely of cavalry—with the exception of some auxiliary elements—it was homogeneous. The organization was based on the decimal system. The largest independent unit was the *touman*, consisting of 10,000 men, corresponding roughly to a modern cavalry division. Three toumans normally constituted an army, or an army corps. The touman, in turn, was composed of 10 regiments of 1,000 men each. The regiments consisted of 10 squadrons, each comprising 10 troops of 10 men.

About 40 percent of a typical Mongol army consisted of heavy cavalry, for shock action. These men wore complete armor, usually of leather, or mail armor secured from defeated enemies. They wore a simple casque helmet such as was normally used by contemporary Chinese and Byzantines. The heavy cavalry horses also usually carried some leather armor protection. The main heavy-cavalry weapon was the lance.

Light-cavalry troopers, comprising about 60 percent of the army, wore no armor, save usually a helmet. Their chief battle weapons were the Asiatic bow, the javelin, and the lasso. The firepower of the Mongol reflex bows—only a little shorter than the English longbow—was particularly devastating, in the Scythian-Turkish tradition. Each light cavalryman carried two quivers of arrows and there were additional arrows in a supply train. The mission of the light cavalry was reconnaissance, screening, provision of firepower support to the heavy cavalry, mopping-up operations, and pursuit.

To assure and to enhance mobility, each Mongol trooper had one or more spare horses. These were herded along behind the columns and were available for quick change of mounts on the march, or even during battle. Changes were made in relays to maintain security and to assure minimum interference with accomplishment of assigned missions.

Both light and heavy cavalrymen carried either a scimitar or a battleax. Each man also had a shirt of strong raw silk to be donned just before action. Genghis found that an arrow would rarely penetrate such silk, but rather would drive it into the wound. Thus the conscript Chinese surgeons were able to extract arrow heads from wounded soldiers merely by pulling out the silk.

Training and Discipline

The individual Mongol troopers were the best-trained soldiers of the time. Hardened veterans of Spartan endurance and fortitude, highly skilled in the use of weapons, they had been riders from early youth, brought up in the harsh school of the Gobi Desert. Inured to hardships and extremes of weather, lacking luxuries and rich food, these men had strong bodies and needed little or no medical attention to keep fit for operations.

The commander of each unit was selected on the basis of individual ability and valor on the field of battle. He exercised absolute authority over his unit, subject to equally strict control and supervision from his superior. Instant obedience to orders was demanded and received. Discipline was of an order unknown elsewhere during the medieval period.

We know little of the details of the training system of Genghis Khan. We do know that each troop, squadron, and regiment was capable of precise performance of a kind of battle drill that formed the basis of Mongol small-unit tactics. Such precision required constant practice under close and demanding supervision. The battlefield coordination of units within the toumans, and between toumans and the larger hordes, is evidence of painstaking practice in precombat maneuvers by forces of all sizes.

Doctrine and Tactics

The mobility of Genghis Khan's troops has never been matched by other ground soldiery. He seems to have had an instinctive understanding that force is the product of mass and the square of velocity. No other commander in history has been more aware of the fundamental importance of seizing and maintaining the initiative—of always attacking, even when the strategic mission was defensive.

At the outset of a campaign, the Mongol toumans usually advanced rapidly on an extremely broad front, maintaining only courier contact between major elements. When an enemy force was found, it became the objective of all nearby Mongol units. Complete information regarding enemy location, strength, and direction of movement was immediately transmitted to central headquarters, and in turn disseminated to all field units. If the enemy force was small, it was dealt with promptly by the local commanders. If it was too large to be disposed of so readily, the main Mongol army would rapidly concentrate behind an active cavalry screen. Frequently a rapid advance would overwhelm separate sections of an enemy army before its concentration was complete.

Genghis and his able subordinates avoided stereotyped patterns of moving to combat. If the enemy's location was definitely determined, they might lead the bulk of their forces to strike him in the rear, or to turn his flank. Sometimes they would feign a retreat, only to return at the charge on fresh horses.

Most frequently, however, the Mongols would advance behind their screen of light horsemen in several roughly parallel columns, spread across a wide front. This permitted flexibility, particularly if the enemy was formidable, or if his exact location was not firmly determined. The column encountering the enemy's main force would then hold or retire, depending upon the situation. Meanwhile the others would continue to advance, occupying the country to the enemy's flank and rear. This would usually force him to fall back to protect his lines of communication. The Mongols would then close in to take advantage of any confusion or disorder in the enemy's retirement. This was usually followed by his eventual encirclement and destruction.

The cavalry squadrons, because of their precision, their concerted action, and their amazing mobility, were easily superior to all troops they encountered, even when these were more heavily or better armed, or more numerous. The rapidity of the Mongol movements invariably gave them superiority of force at the decisive point, the ultimate aim of all battle tactics. By seizing the initiative and exploiting their mobility to the utmost, the Mongol commanders, rather than their foes, almost always selected the point and time of decision.

The battle formation was composed of five lines, each of a single rank, with large intervals between lines. Heavy cavalry comprised the first two lines; the other three were light horsemen. Reconnaissance and screening were carried out in front of these lines by other light-cavalry units. As the opposing forces drew nearer to each other, the three rear ranks of light cavalry advanced through intervals in the two heavy lines to shower the enemy with a withering fire of well-aimed javelins and deadly arrows.

The intensive firepower preparation would shake even the staunchest of foes. Sometimes this harassment would scatter the enemy without need for shock action. When the touman commander felt that the enemy had been sufficiently confused by the preparation, the light horsemen would be ordered to retire, and synchronized signals would start the heavy cavalry on its charge.

In addition to combining fire and movement—and missile action with shock action—the Mongols also emphasized maneuver and diversions at all tactical levels and in all phases of combat. During the main engagement, a portion of the force usually held the enemy's attention by frontal attack. While the opposing commander was thus diverted, the main body would deliver a decisive blow on the flank or rear.

Siegecraft and Attack of Fortifications

In his early campaigns, Genghis Khan's cavalry armies were frequently frustrated by the strong walls of Chinese cities. After intensive analytical study—plus the adoption of Chinese weapons, equipment, and techniques, and considerable empirical experience—the Mongol leader in a few years developed a system for assaulting fortifications which was well-nigh irresistible. An important component of this system was a large, but mobile, siege train, with missile engines and other equipment carried in wagons and on pack animals. Genghis conscripted the best Chinese engineers to comprise the manpower of his siege train. Combining generous terms of service with the compulsion of force, he created an engineer corps at least as efficient as those of Alexander and Caesar.

In Genghis' later campaigns, and in those of his brilliant subordinate Subotai, no fortification could long stop the march of a Mongol army. Those of importance, and those which contained large garrisons, would usually be invested by a touman—supported by all or part of the engineer train—while the main force marched onward. Sometimes by stratagem, ruse, or audacious assault, the town would be quickly stormed. If this proved impossible, the besieging touman and the engineers began regular siege operations, while the main army sought out the principal field forces of the enemy. Once the inevitable victory had been achieved in the field, besieged towns and cities usually surrendered without further resistance. In such cases, the inhabitants were treated with only moderate severity.

But if the defenders of a city or fort were foolish enough to attempt to defy the besiegers, Genghis' amazingly efficient engineers would soon create a breach in the walls, or prepare other methods for a successful assault by the dismounted toumans. Then the conquered city, its garrison, and its inhabitants would be subjected to the pillaging and destruction which have made the name of Genghis Khan one of the most feared in history.

Sometimes even the strongest cities were overwhelmed and captured before they were fully aware that any large force of Mongols was in their vicinity. The leading Mongol light horsemen always attempted to pursue defeated enemies so closely and so vigorously as to ride through the gates before they could be closed. Even if the enemy was sufficiently alert to prevent this, he rarely anticipated the speed, efficiency, and vigor with which the Mongol engines of war—ballista and catapult—were put into action within a few minutes of the arrival of the leading cavalry units. The Mongols discovered that a prompt and vigorous assault, covered by a hail of fire from these machines, would often permit them to scale the walls and to seize a portion of the defenses before the surprised enemy was prepared to resist.

If the initial assaults were repulsed and regular siege operations were required, the ballistae and catapults provided fire cover for battering rams, siege towers, and all of the normal methods of siegecraft. When the Mongols were prepared to make their final major assault through a breach, or over the opposing fortifications, they frequently made use of a ruthless, heartless, but usually effective method of approach. Herding great numbers of helpless captives in front of them, the dismounted Mongol troopers would advance to the walls, forcing the defenders to kill their own countrymen in order to be able to bring fire against the attackers.

To add to the confusion and difficulties of the enemy, the Mongols often preceded their assaults by firing flaming arrows into besieged camps and cities to start fires. These arrows were fired by light cavalrymen dashing in front of the walls, as well as by catapults from farther away.

Staff and Command

We know little of the staff system of Genghis Khan, probably because the history of his operations was mostly written by his enemies, who rarely understood how he accomplished his victories. He seems to have been his own operations officer, although later he evidently made considerable use of the skill and genius of able subordinates like Subotai and Chepe. Strategy and tactics for every campaign were obviously prepared in painstaking detail in advance.

An essential element of Mongol planning was its intelligence system. Operations plans were always based on thorough study and evaluation of amazingly complete and accurate information. The Mongol intelligence network spread throughout the world; its thoroughness excelled all others of the Middle Ages. Spies generally operated under the guise of merchants or traders.

Once the intelligence had been evaluated, lines of operation were decided upon in advance for the entire campaign, and toumans were assigned to follow these lines, each with its own objective. Nevertheless, the widest latitude was given to each subordinate commander in accomplishing the specific objective assigned to him. Prior to a general engagement a touman commander was at liberty to maneuver and to meet the enemy at his discretion, and was required only to maintain general conformance with the overall plan. Orders and the exchange of combat intelligence information passed rapidly between the Khan's headquarters and his subordinate units by swift mounted couriers. Thus Genghis, to an extent rarely matched in history, was able to assure complete unity of command at all levels and yet retain close personal control over the most extensive operations.

The Mongols were particularly adept at psychological warfare. Tales of their ruthlessness, barbarity, and slaughter of recalcitrant foes were widely disseminated in a deliberate propaganda campaign to discourage resistance by the next intended victim. As a matter of fact, while there was considerable truth in this propaganda, it was equally true that the Mongols were most solicitous in their treatment of any foe who gave evidence of willingness to cooperate, particularly those who had skills which the great Khan thought might be operationally or administratively useful.

One evidence of the existence of a general staff is the way in which the Mongols responded to the lessons of their campaigns. They not only promptly analyzed each of their major actions;

they put the results of this analysis to work immediately in a systematic training program, which must have been directed by an alert, smoothly-functioning staff.

Administration

The gathering of supplies on campaign was apparently left to individual subordinate commanders. The Mongol toumans lived off the country through the most ruthless requisitioning practices. That this was done efficiently, however, and in a systematic and coordinated manner, is clearly evidenced by uniform success of the Mongols in maintaining their forces, even when operating in desert and mountain regions. The administrative staff of the great Khan apparently consisted largely of captured Chinese scholars, officials, and surgeons.

The food supply problem was facilitated by the Mongol practice of drinking mare's milk; most of their horses were mares.

A somewhat gruesome evidence of the thoroughness and efficiency of the Mongol administrative and staff operations is the manner in which they kept a record of enemy casualties. (Their own losses, of course, were easily determined by a roll call.) After a battle, specified units were responsible for cutting off the right ear of each enemy corpse on the battlefield. These were thrown into sacks so as to obtain at leisure an accurate count of enemy dead.

Communications

We have already noted the extensive use of couriers for long-range communications purposes. Tactical movements were controlled by black and white signal flags under the direction of squadron and regimental commanders. Thus there were no delays caused by poorly written orders or messages. (Probably few of the Mongol commanders could read or write.) The signal flags were particularly useful for coordinating the movements of units beyond the range of voice control.

When signal flags could not be seen, either because of darkness or intervening terrain features, the Mongols used flaming arrows.

Stratagems and Ruses

Unlike the sometimes foolishly chivalrous knights of Western Europe, the Mongols disdained no trickery which might give them an advantage or reduce their own losses or increase those of their foes. Some of their tricks were obviously the result of inspiration of the moment—encouraged by the latitude of initiative that Genghis Khan inculcated among his subordinates. Most, however, were apparently included in a kind of repertoire of stratagems, with which all commanders were acquainted. Two examples are worth noting.

The Mongols liked to operate in the winter, when their mobility was enhanced by frozen marshes and ice-covered rivers. A favorite way of finding out when the river ice would be thick enough to support the weight of their horses was to encourage the local population to test it for them. In late 1241, in Hungary, Mongols left untended cattle on the east bank of the Danube, in sight of the famished refugees they had driven across the river earlier in the year. When the

Hungarians were able to cross the river and bring the cattle back with them, the Mongols decided to start their next advance.

Another stratagem, which might more properly be called a tactical technique, was their use of smoke screens. It was common practice to send out small detachments to start great prairie fires, or fires in inhabited regions, both to deceive the enemy as to their intentions and to hide movements.

Military Government

Once armed resistance ceased in a conquered territory, the Mongols changed immediately from apparently wanton destruction to carefully calculated reconstruction. Genghis created what was probably the most carefully planned military government system to appear before the 20th century. The civil administration was usually left under a local leader satisfactory to the Mongols, who was supported, and closely supervised, by a Mongol occupation force. A census was taken, and efficient tax-collection machinery immediately established. Genghis Khan had no intention of allowing conquered territories to be a burden on his economy. On the contrary, the funds collected in this fashion not only maintained the local government, and its occupation troops, but also were used to pay tribute to Karakorum.

The Mongols absolutely forbade any continuation of local and internal squabbles in their conquered territory. Law and order were rigidly and ruthlessly maintained. As a consequence, regimented conquered regions were usually far more peaceful under the Mongol occupation than they had been before the invasion.

The Successors of Genghis Khan, 1227–1388

1227–1241. Reign of Ogatai Khan. The new ruler completed the conquest of the outlying territories of the Western Hsia. The expansion of the Mongol Empire continued in all directions. The first new victim was Korea (1231; see p. 381).

1231–1234. Conquest of the Chin Empire. Forming an alliance with the Sung, the Mongols undertook the final destruction of Chin. Tuli led a great army south, through Hsia territory, into the Sung province of Szechwan, then turned east through Hanchung (Nancheng) into Chin territory. In the middle of the campaign Tuli died, and Subotai took command (1232). He continued on to besiege the great city of Pien Liang (Kaifeng), the Chin capital. Despite skillful use of explosives by the defenders, the city finally fell to Mongol assault after a year's siege (1233). Subotai then completed the conquest of the Chin Empire. Ogatai refused to divide the conquered region with the Sung, who then attempted to seize the former Chin province of Honan. This was the signal for war.

1234–1279. War with the Sung Empire. This prolonged conflict was brought to a conclusion by Ogatai's nephews and successors, **Mangu** and **Kublai.** Operating under Mangu, Kublai conquered Yunnan from the Nanchao, vassals of the Sung (1252–1253). A subordinate then overran Tonkin, capturing Hanoi (1257). Sung resistance was based upon determined defense of their well-fortified, well-provisioned cities. The Chinese Empire crumbled, however, under the impact of a series of brilliant campaigns personally directed by Mangu (1257–1259). His sudden death from dysentery caused a lull in the war during a Mongol dynastic dispute (see below); the Sung revived, and the war dragged on. Finally, Kublai could turn his full attention to the war in China (1268). A series of campaigns, distinguished by the skill of the general **Bayan** (grandson of Subotai) were culminated by the capture of the capital city of Hangchow (1276). It took three more years to subdue the outlying provinces. The last action of the war was a naval battle in the Bay of Canton (1279). The Sung ships were all destroyed by a Mongol fleet. The Chinese admiral jumped into the waves, carrying the Sung boy emperor in his arms. Refugees on coastal islands were rounded up and slaughtered.

1235. Plan for Mongol Expansion. Following a *kuriltai* (conference) with the principal Mongol leaders, Ogatai authorized four simultaneous campaigns of conquest: (1) the ongoing war with the Sung Empire in China, (2) Korea, (3) Southeast Asia, and (4) Europe.

1237–1241. Mongol Conquests in East and Central Europe. (See p. 375.)

1239. Mongol Conquest of Tibet. This was done by **Godan,** son of Ogatai.

1241, December 11. Death of Ogotai. All royal princes were summoned to assemble at Krakorum for a kuriltai to elect a new khakhan. This temporarily ended all Mongol offensives. Due mainly to the rivalry of **Batu** and **Kuguk,** eldest sons of Ogatai, the Kuriltai was delayed for more than 4 years. Kuguk was elected khakan (January 1246). Batu was confirmed as khan of the western regions (Northwest Asia and Eastern Europe), now becoming known as the Khanate of the Golden Horde, due to the magnificence of Batu's camp.

1243. Battle of Kosedagh. The Mongols, after overrunning the remnants of the Khwarezmian Empire, defeated the Seljuk Turks and established suzerainty over Anatolia.

1246–1248. Reign of Kuguk. The new khakhan issued orders to renew the campaigns of conquest. He personally led a small army west to head a renewed invasion of Central Europe, but he died en route to meet Batu. Two kuriltais were held to elect a successor. At the first, held at Batu's camp, Batu was elected but declined the honor. He nominated Mangu, who was elected. To comply with Genghis Khan's will, the kuriltai reassembled at Karakorum. It was delayed by a conspiracy involving descendants of Ogatai and Jagatai, who planned to overthrown Mangu. The conspiracy was uncovered, the plotters executed, and Mangu's election was reaffirmed (1251).

c. 1250. Politico-Military Organization of the Empire. Ogatai and his successors adhered to the system of territorial vice-royalties established by Genghis before his death. Batu, son of Juji, was confirmed as ruler of the steppes of Northwestern Asia and Eastern Europe. First khan of the Golden Horde (1243–1256), Batu established his capital at Sarai on the Volga (see p. 379). Jagatai and his successors ruled the vast intermediate region roughly corresponding to modern Turkestan and Central Siberia. They were known as khans of the Jagatai (Chagatai) Mongols. Farther south, Transoxiana, Khorasan, and modern Afghanistan became the realm of Tuli. One branch of Tuli's family, through Mangu and Kublai, became the direct imperial line of khakhans and Yuan emperors of China. Another branch under Hulagu and his progeny completed the conquest of Southwest Asia and became the Ilkhans of Persia. The khans of the Golden Horde, the Jagatai, and the Ilkhans remained vassals and viceroys of the khakhans of Karakorum and Peking until the decline of the Yuan Dynasty in the mid-14th century (see p. 380).

1251–1259. Reign of Mangu (son of Tuli). He was an able warrior and administrator. Though he devoted most of his attention to prosecution of the war against China, he also sent his brother Hulagu to carry out the conquest of Southwest Asia.

1252–1287. Mongol Conquests of Nanchas and Annam. (See p. 432.)

1255–1260. Hulagu's Conquests in Southwest Asia. (See p. 423.)

1260–1294. Reign of Kublai Khan. Selection of Kublai as khakhan was violently opposed by his younger brother, **Arik-Buka,** precipitating a civil war, which Kublai won by vigorous action (1260–1261). Ogatai's grandson **Kaidu,** who had supported Arik-Buka, took refuge in the Altai Mountains and continued sporadic resistance until his death (1301). Upon the overthrow of the Sung Dynasty, Kublai declared himself emperor and established the Yuan Dynasty (see p. 380). He moved his capital to Peking (1264). One of the ablest of all the rulers of China, he reunited the country, which has ever since remained at least nominally unified (despite subsequent civil wars).

1261–1262. Civil War between the Ilkhans and the Golden Horde. (See p. 424.)

1274. First Mongol Invasion of Japan. (See p. 381.)

1281. Second Mongol Invasion of Japan. (See p. 381.)

1281. Battle of Homs. An Ilkhan Mongol army invading Syria was defeated and repulsed by Mameluke Sultan Kala'un (see p. 422).

1281–1287. Invasions of Annam and Champa. (See p. 431.)

1287–1300. Invasions of Burma. (See p. 433.)

1292–1293. Invasion of Java. (See p. 433.)

1294–1307. Reign of Timur. The grandson of Kublai, Timur (not to be confused with Timur the Lame, or Tamerlane, see p. 424) was the last recognized khakhan of the entire Mongol Empire.

1299–1300. Invasion of Syria. Ilkhan **Ghazan Mahmud** captured Damascus and overran most of Syria, then withdrew.

1303, April 20. Battle of Marj-as-Suffar. In their last important invasion of Syria, a Mongol Ilkhan army was defeated and repulsed by the Mamelukes.

1307–1388. Decline of the Yuan Dynasty. The Mongol Empire gradually fell apart. Ilkhans, Jagatais, and the Golden Horde became independent khanates; other Mongol princes became independent or were overthrown by the subject peoples.

1356–1388. Overthrow of the Mongols by the Chinese Ming Dynasty. (See p. 380.)

Mongol Campaigns in Europe, 1237–1242

INVASION OF EASTERN EUROPE, 1237–1240

In 1236–1237 Ogatai sent an army of 150,000 men to invade Eastern Europe. Nominally in command was Batu, son of Juji, but the real commander was Subotai who, after defeating the Bulgars, led his troops westward over the frozen Volga River in December 1237. As a result of his initial reconnaissance in 1223 (see p. 366), and subsequent accumulation of intelligence information about Russia, this invasion had been prepared with the same care that had characterized all the campaigns of his mentor, Genghis Khan. Sweeping over the frozen countryside like a whirlwind, the Mongols advanced through Moscow, Vladimir, Kaluga and completely destroyed the north Russian principalities in a few months. The campaign was concluded near a Volga tributary by the **Battle of the Sil River** (March 4) with the destruction of the Russian army under **Yuri II,** Grand Prince of Vladimir, who was killed.

Meanwhile, a contingent of the Mongol army, under Mangu (son of Tuli) crossed the lower Volga to overrun the eastern portions of the Cuman Empire in the northern Caucasus and north of the Black Sea. The reassembled army spent the summer and early fall in the Ukraine, reorganizing, re-equipping, and training fresh horses from Kazakhstan.

During the next two years Subotai consolidated control over eastern and southern Russia, at the same time accumulating detailed information about Europe to his west. Only the vaguest rumors of the Mongol conquest of southern and eastern Russia—largely carried by Cuman refugees—had reached the kingdoms of central and western Europe. Subotai, on the other hand, knew the political, economic, and social conditions in Europe in detail. Again moving in winter, to assure full mobility over the great rivers of the Ukraine, 150,000 Mongols crossed the frozen Dnieper at the end of November 1240. When the Prince of Kiev rejected the Mongol

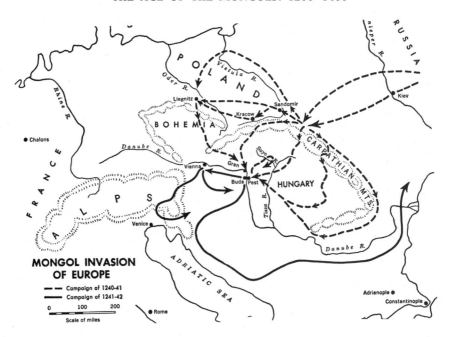

**MONGOL INVASION
OF EUROPE**

--- Campaign of 1240-41
—— Campaign of 1241-42

0   100   200

Scale of miles

demand for surrender, Subotai assaulted, captured the city, then destroyed it (December 6). Overrunning all of the regions southeast of the Carpathians and northwest of the Black Sea, Subotai was now ready to start his next campaign.

INVASION OF CENTRAL EUROPE, 1241

Subotai left 30,000 men to control the conquered regions and to protect his lines of communication, leaving about 120,000 men to invade Central Europe. He realized that Hungary, Poland, Bohemia, and Silesia could each raise forces larger than his own, and he was equally aware that an invasion of any one of these countries would bring him into immediate conflict with the other three, and with the Holy Roman Empire. His knowledge of European politics, however, made him confident that he could count on jealous bickering between the pope, the German emperor, and the kings of France and England to keep these more powerful nations from involvement until he had secured central Europe. He then intended to deal with them one at a time.

Subotai divided his army into four principal columns, or field armies, each of three toumans. One of these, under **Kaidu,** grandson of Ogatai, was to protect the northern flank. Another flank force, under **Kadan,** son of Ogatai, was to protect the southern flank while invading Hungary from the south, through Transylvania and the Danube Valley. The two remaining hordes, apparently advancing in parallel columns, under Batu and Subotai, were to force the passes over the central Carpathians. These two columns would meet on the Hungarian plains in front of the city of Pest, on the east bank of the Danube opposite the capital, Buda.

Kaidu, to attract the attention of the Poles, Bohemians, and Silesians from the main objective, moved out in early March, somewhat ahead of the other three columns. He swept through Poland and into Silesia, defeating three larger Polish armies as he went. A detached touman protected his right flank, swinging north through Lithuania, then west through East Prussia and along the Baltic coast into Pomerania. With the other two, Kaidu smashed the army of **Boleslav V** of Poland at **Cracow** (March 3).

As the toumans carried fire and sword through northeastern Europe, panic spread across the countryside. Terror-stricken refugees fled westward. As town after town was seized, destroyed, and burned, the swarm of refugees was augmented and the tale of horror was repeated and amplified. By the time Kaidu's two toumans of 20,000 men had reached Silesia in early April, the Europeans believed that his force was upward of 200,000 men.

The Battle of Liegnitz, April 9, 1241

Despite their belief in these wild exaggerations, the chivalry of North-Central Europe was prepared to fight desperately. Prince **Henry** (the Pious) of Silesia gathered a mixed army of some 40,000 Germans, Poles, and Teutonic Knights, and took up a defensive position at Liegnitz in the path of Kaidu's horde. King **Wenceslas** of Bohemia marched northward hastily with an army of 50,000 men to join Henry.

Kaidu struck while the Bohemians were still two days' march away. The Europeans fought bravely and stubbornly, but Henry's army was smashed; its broken remnants fled westward. Because the Mongols did not pursue, some European historians have incorrectly inferred that the **Battle of Liegnitz** (or Wahlstatt) was a drawn fight and that the Mongols had suffered so severely at the hands of their Western opponents that they decided not to press further into Germany.

Kaidu had no reason to pursue. He had accomplished his mission. All north-central Europe from the Baltic to the Carpathians had been devastated. All possible danger to the right flank of Subotai's army had been eliminated. The one remaining effective army of the region—the Bohemians of Wenceslaus—was withdrawing to the northwest to join the hastily gathered forces of other German nobles. Having so brilliantly carried out his part of Subotai's plan, Kaidu called in his detached touman from the Baltic coast and turned south to join the main body in Hungary, laying Moravia waste as he advanced.

Advance into Hungary

The southern column had been equally effective, although thawing snows and flooding rivers held up the advance. After three pitched battles, all resistance collapsed in Transylvania by April 11. Passing between the Danube and the Carpathians at the Iron Gates, Kadan drove northward into the Hungarian plain to meet Subotai.

Meanwhile, on March 12, the columns of the main body had broken through the Hungarian defenses at the Carpathian passes. King **Bela** of Hungary, receiving word of the Mongol advance through the passes, called a council of war in Buda, 200 miles away, to consider how to prevent the Mongols from continuing their invasion. While this council was in progress, on March 15, he received word that the Mongol advance guard had already arrived at the opposite bank of the river.

Bela did not panic. Within two weeks he had gathered nearly 100,000 men, while the Mongol advance was held up by the broad Danube River and the formidable fortifications of the city of Pest.

At the beginning of April, Bela marched eastward from Pest with his army, confident of repelling the invaders. As he advanced, the Mongols withdrew. After several days of cautious pursuit, he made contact with them late on April 10, near the Sajo River, almost 100 miles northeast of Budapest. Bela surprised Subotai by promptly and vigorously seizing a bridge over the Sajo from a weak Mongolian detachment. He established a strong bridgehead beyond the river, and encamped with the remainder of his army in a fortified camp of wagons chained together on the west bank. He had received accurate and complete information regarding the Mongol strength from loyal subjects, and knew that his army was considerably more numerous than the Mongols.

Battle of the Sajo River, April 11, 1241

Just before dawn the Hungarian bridgehead defenders found themselves under a hail of stones and arrows, "to the accompaniment of a thunderous noise and flashes of fire." That the Mongols were actually using the first cannon of European military history is doubtful. More likely, it was their normal employment of catapults and ballistae with terror-inspiring sound effects from early forms of Chinese firecrackers. In any event, this was a 13th-century version of a modern artillery preparation. It was followed closely, as in modern tactics, by fierce assault.

The defenders, stupefied by noise, death, and destruction, were quickly overwhelmed, and the Mongols streamed across the bridge. Bela's main army, aroused by the commotion, hastily sallied out of their fortified camp. A bitter battle ensued. Suddenly it became evident, however, that this was only a Mongol holding attack.

The main effort was made by three toumans, some 30,000 men, under the personal command of Subotai. In the predawn darkness he had led his troops through the cold waters of the Sajo River, south of the bridgehead, then turned northward to strike the Hungarians' right flank and rear. Unable to resist this devastating charge, the Europeans hastily fell back into their camp. By 7 A.M. it was completely invested by the Mongols. For several hours they bombarded it with stones, arrows, and burning naphtha.

It appeared to some desperate Hungarians that there was a gap to the west. A few men galloped out safely. As the intensity of the Mongol assault mounted elsewhere, more and more men slipped out. Soon a stream of men was pouring westward through the gap. As the defense collapsed, the survivors rushed to join those who had already escaped. Lacking all semblance of military formation, many of the fugitives threw away weapons and armor in order to flee better. Suddenly they discovered that they had fallen into a Mongol trap. Mounted on swift fresh horses, the Mongols appeared on all sides, cutting down the exhausted men, hunting them into marshes, storming the villages in which some of them attempted to take refuge. In a few hours of horrible butchery the Hungarian army was completely destroyed, losing between 40,000 and 70,000 dead. More serious, the defeat assured Mongol control of all Eastern Europe from the Dnieper to the Oder and from the Baltic Sea to the Danube. In four months they had overwhelmed Christian armies totaling five times their own strength.

WITHDRAWAL OF THE MONGOLS, 1242

During the summer of 1241, Subotai consolidated control of eastern Hungary, and made plans to invade Italy, Austria, and Germany the following winter. Frantic efforts of panic-stricken West Europeans were poorly coordinated, and little was accomplished to prepare an effective defense. Only chance—or Divine Providence—saved the remainder of Europe.

Just after Christmas of 1241, the Mongols started westward across the frozen Danube. Spearheads crossed the Julian Alps into north Italy, while scouts approached Vienna through the Danube Valley. Then came a message from Karakorum, 6,000 miles to the east. Ogatai, the son and successor of Genghis Khan, was dead.

The law of Genghis Khan explicitly provided that "after the death of the ruler all offspring of the house of Genghis Khan, wherever they might be, must return to Mongolia to take part in the election of the new Khakhan." Reluctantly Subotai reminded his three princes of their dynastic duty. From the outskirts of trembling Vienna and Venice, the toumans countermarched, never to return. Their ebb went through Dalmatia and Serbia, then eastward across northern Bulgaria, virtually destroying the kingdoms of Serbia and Bulgaria before they vanished across the lower Danube (see pp. 411, 412).

Khanate of the Golden Horde, 1243–1400

Under the Mongol system of administrative decentralization, **Batu,** grandson of Genghis Khan, reigned as the first khan of the Golden Horde, ruling most of Western Asia and practically all of Europe east of the Carpathians. By a combination of military superiority, wise diplomacy, and cunning intrigue, Batu and his Mongol or Tartar successors retained almost uninterrupted authority over Russia for a century and a half (see p. 410). They were unquestionably the most powerful rulers in Europe during this period. Dynastic troubles toward the end of the 14th century almost tore their empire apart. The White Horde, in the region extending northward from the Aral Sea, became practically independent, and the Russian principalities attempted to follow this example, at first successfully. After two decades of civil war and decay, however, Khan **Toktamish,** with the help of **Tamerlane** of the Jagatai Mongols (see p. 424), established himself first as the leader of the White Horde, then by conquest reunited the White and Golden Hordes. He ruthlessly suppressed the tentative independence of the Russian principalities. He made the mistake, however, of provoking a war with his benefactor Tamerlane. After conflicts lasting ten years, Toktamish was defeated and overthrown. This defeat contributed to the lingering decline of the Horde in subsequent centuries. The principal events were:

1242–1256. Reign of Batu. He established his capital at Sarai (Zarev) on the lower Volga. He accepted the vassalage of **Alexander Nevski,** Prince of Novgorod, who assisted in the establishment of Mongol suzerainty over all of Russia (see p. 410). In return, Alexander was treated with courtesy and respect by the Mongols.

1256–1263. Reign of Bereke (Batu's Brother). The first descendant of Genghis Khan to become a Moslem, he sent armies raiding deep into the Balkans, Hungary, and Poland. One of these raids, led by his generals **Tulubaga** and **Nogai,** reached as far as Silesia; the cities of Cracow, Sandomir, and Bythom were all captured and sacked (1259).

c. 1270–1296. Autonomy of Nogai. As governor of the south Ukrainian region, he became virtually independent. The rulers of Bulgaria and Serbia paid him tribute.

1290–1313. Reign of Toktu. For several years Toktu was engaged in a bitter war to establish central authority over Nogai. After one serious defeat near the Dnieper at the **Battle of Kukanlyk** (near modern Poltava), where Nogai was killed (1300).

1314–1340. Reign of Uzbeg. He was the last great khan of the Golden Horde. On his orders, a revolt of the Prince of Tver (Kalinin) was suppressed by Prince **Ivan** of Moscow (see p. 410) with a mixed Russian-Mongol army. Ivan's success established the princes of Moscow as the principal vassals and administrators for the Mongols in northern and central Russia. During much of his reign Uzbeg was at war with Il-khan **Abu-Said** (see p. 424).

1359–1379. Dynastic Civil War. There were at least 25 khans during this period of turmoil. During this period **Urus,** Khan of the White Horde, became virtually independent, and also took part in the continuing confused struggle for the khanate of the Golden Horde. One of the claimants, a chieftain named **Toktamish,** defeated by his rival **Mamai,** fled to seek safety and assistance from Tamerlane, by that time ruler of the Jagatai Tartars (see p. 424).

1378. Rise of Toktamish. With Tamerlane's assistance, Toktamish became the Khan of the White Horde.

1380. Battle of Kulikovo. Mamai, allied with the Lithuanians, attempted to reconquer the rebellious Russian principalities. He was defeated by **Dmitri** of Moscow, leader of the Russian princes, before the arrival of the Lithuanians, who then retreated.

1380. Battle of the Kalka River. Near the site of Subotai's great victory (see p. 366), Toktamish defeated Mamai to become Khan of the Golden Horde.

1381–1382. Russian Rebellion Suppressed. Toktamish campaigned across northern and central Russia, completely subduing the rebellion. He captured Moscow (August 23, 1382), massacring many inhabitants.

1385–1395. Wars between Toktamish and Tamerlane. (See p. 424.) Toktamish was defeated in three great battles: on the Syr Darya River (1389), at Kandurcha (1391), and at the Terek River (1395).

1396. War with Lithuania. In alliance with the fugitive Toktamish, **Vitov,** grand prince of Lithuania, supported by both Poles and the Teutonic Knights, invaded the Khanate of the Golden Horde. Vitov's army had a substantial number of artillery pieces. Despite the superiority in weapons of the invaders, they were decisively defeated by the Mongols at the **Battle of the Vorskla,** a tributary of the Dnieper.

CHINA

The history of China during these two centuries is essentially the record of the conquest of the country by Genghis Khan and his successors, of their establishment of the Mongol (Yuan) Dynasty, and of the eventual overthrow of that dynasty by the new Ming Dynasty of native Chinese. Important events were:

1205–1279. Mongol Conquest of China. (See pp. 363, 374.)

1279. Kublai Khan Establishes the Yuan Dynasty. (See p. 375.)

1307–1388. Decline of Yuan Dynasty. (See p. 375.)

1356–1368. Rise of the Ming. Chu Yuan-Chang, a Buddhist monk, led a popular uprising. He captured **Nanking** (1356), where he established a native Chinese government. He slowly expanded his control over most of China. For a while his efforts were hampered by rival Chinese rebels, whom he defeated at the **Battle of Lake Po-yang** (1360?). Having driven the Mongols completely from most of China, Chu later proclaimed himself to be the first emperor of the Ming Dynasty (1368).

1368–1388. Continuing Ming War against

the Mongols. Yunnan was the last province of China to be reconquered (1382). The Chinese then invaded Mongolia, driving the Mongols from their capital of Karakorum, winning a great victory in the **Battle of the Kerulen River** (1388).

1398–1402. Civil War. A struggle for succession after the death of Chu Yuan-Chang.

c. 1400. Ming Maritime Power. Chinese merchant vessels and warships dominated the South China Sea, and extended Chinese maritime influence to the Indian Ocean. An indication of the extent of this maritime power was payment of tribute by Ceylon.

KOREA

For more than half of this period Korea was a dependency of the Mongol Empire (1231–1356). The Wang Dynasty finally regained its independence after an eight-year war in which General **Li Taijo** finally ejected the Mongols (1364). This was followed by a period of internal violence which was ended only when Li deposed the last Wang emperor to establish his own dynasty (1392). He continued to recognize the suzerainty of the Ming, which the Wang had accepted (1369). During most of these two centuries Japanese pirates and adventurers conducted constant depredations against the Korean coast.

JAPAN

During the early years of the 13th century the Hojo family gradually replaced their kinsmen, the Minamotos, in control of the Kamakura Shogunate (1199–1219). The Hojos ruled over a generally peaceful and prosperous country. Toward the end of the century, however, Japan was constantly threatened by Mongol invasion. Two actual invasion attempts were made, but neither succeeded. The Kamakura Shogunate was overthrown in a two-year civil war, which left the nation and the imperial family hopelessly split. For more than half a century Japan was divided between two emperors, and the country was plagued by constant civil war. Peace was finally established by the strong Shogun **Yoshimitsu** of the Ashikaga family. The principal events were:

1221. Uprising against the Hojos. Several members of the imperial family, including retired emperor **Gotoba II,** were defeated.

1274. First Mongol Expedition to Japan. This was apparently only a reconnaissance in force. Using Korean ships, the Mongols captured the islands of Tsushima and Iki. They landed at Hakata Bay (Ajkozaki) in northern Kyushu. Japanese forces were hastily assembled, and attacked the invaders near the beaches. The Mongols easily defeated the numerically superior Japanese. Part of the invasion fleet was wrecked in a storm, and as more Japanese

gathered, the Mongols embarked to return to Korea.

1281. Second Mongol Expedition. The Japanese refusing to recognize Mongol suzerainty, and having mistreated and killed Mongol ambassadors, Kublai prepared a major expedition. In Mongol and Korean ships, the army sailed in two divisions from north China and Korea. Again the islands of Tsushima and Iki were captured to provide a base for operations. The invading army was probably less than 50,000. The Japanese were waiting, and a series of violent land and sea engagements took place in and off north Kyushu. The Japanese carried out a number of daring raids against the larger Mongol fleets. On land the Japanese were unable to drive the invaders into the sea, but by superior numbers and fanatic defense were able to limit their advance inland. A few days after the landing, a violent storm wrecked most of the invasion fleets. The Mongols, their supplies cut off, were soon defeated and only a few survived and escaped. Kublai planned another expedition, but never got around to it.

1331–1333. Civil War. Emperor **Go Daigo II** led a revolt against Hojo rule. He was assisted by the great samurais **Kitabatake Chikafusa** and **Kusoniki Masashige.** At first defeated and captured (1332), Go Daigo escaped and continued his revolt. He was finally successful when several Hojo generals, including **Ashi-**

kaga **Takauji,** deserted to join his cause, permitting him to capture **Kamakura,** the Hojo capital.

1335. Revolt of Takauji. Expecting to be shogun, the general revolted when the emperor insisted upon personal imperial rule.

1336–1392. Civil Wars of the Yoshino Period. Takauji expelled Go Daigo from the capital of Kyoto, set up another member of the imperial family on the throne, and established himself as military dictator and shogun (1338). Go Daigo, in the mountainous Yoshino region south of Nara, continued the war. It was prosecuted bitterly and vigorously on both sides for 56 years.

1392. Peace was established through the diplomacy of **Yoshimitsu.** The successor of Go Daigo agreed to abdicate.

WESTERN EUROPE

The Hundred Years' War—Phase One, 1337–1396

Actually this was a series of eight major periods of conflict between the royal houses of England and France, lasting nearly 120 years (1337–1453). Four of these periods of military activity took place in the 14th century, four in the 15th century (see p. 444).

Although this war, or series of wars, started out as a typical feudal dispute, it had from the outset nationalistic aspects unseen in Europe since the Fall of Rome, growing into an Anglo-French military-political rivalry which persisted for more than five and a half centuries. There were three principal causes of the war:

1. The feudal relationship of the kings of France and of England. As dukes of Aquitaine and barons of other French lands, the kings of England were vassals; as rulers of England, they were sovereign. The French kings rightly feared that the English rulers would endeavor to consolidate their French lands with their English dominions; the English kings disliked their partially subordinate position.
2. Growing English commercial dominance in the wealthy industrial County of Flanders. The French kings had eliminated German influence from Flanders in the 13th century only to discover that their hegemony was becoming increasingly threatened by growing English commercial relationships with the burghers and weavers of Flemish towns.
3. French influence in Scotland and the assistance and support the French rendered to the Scots in their almost continuous wars with England.

The Sluys Period, 1337–1343

Philip VI of France announced the forfeiture of English fiefs south of the Loire River, then halfheartedly invaded them (1337). In response, **Edward III** of England sent raiding parties into northern and northeastern France from England and Flanders (1337–1338). There was only one decisive action: a great English naval victory over the French fleet at the **Battle of Sluys.** After that battle, the war petered out and was suspended by a two years' truce. The principal events were:

1337–1338. English Bases Established in Flanders. These were provided by Flemish weavers under **Jacques van Artevelde,** who had revolted against Count **Louis** of Flanders.

1338. Edward Proclaims Himself King of France. His claim of direct succession, through the daughter of Philip IV, was recognized by Emperor **Louis IV** of Germany.

1339. Edward in Northern France. Raiding from Flanders, he withdrew upon the approach of Philip's much larger army.

1340, June 24. Battle of Sluys. Edward, with an English fleet of about 150 vessels, engaged a French fleet of some 190 ships at the entrance of Sluys harbor. English archers and war engines placed a devastating fire on the French fleet, which was virtually annihilated, 166 French vessels being captured or sunk. Edward landed, besieged Tournai, but later was forced to raise the siege and make a truce with Philip.

1341–1343. Dynastic War in Brittany. Philip supported **Charles of Blois,** Edward assisted **John de Montfort.** Inconclusive operations ended in another truce (1343).

The Crécy Period, 1345–1347

1345. Renewal of War in Brittany. Pro-English forces gained minor advantages. Edward began to raise a large expeditionary force in England.

1346. French Invasion of Gascony. A French army under Philip's son, **Duke John of Normandy,** invaded Gascony. The French were held up by the stubborn defense of the castle of Aiguillon, at the junction of the Lot and Garonne Rivers (April–August).

EDWARD'S EXPEDITION TO FRANCE, 1346

Learning of the siege of Aiguillon, Edward sailed from Portsmouth for northern France (July 11?) to relieve pressure in Gascony, and to assist his hard-pressed allies in Flanders and Brittany. Edward's army probably comprised about 3,000 knights and men-at-arms (heavy cavalry), 10,000 English archers, and 4,000 Welsh light infantry, of whom half may have been archers. The squires and retainers of the knights and men-at-arms probably numbered an additional 3,000 men, serving as light cavalry for reconnaissance, screening, and combat. This was no feudal levy. Raised under the indenture system (see p. 363), these were experienced soldiers, veterans of the Scottish wars. While owing allegiance to the king through his vassals, they were paid fighting men whose term of service was for the "duration." This was probably the best-organized, most-experienced, and best-disciplined force gathered in Western Europe since the days of the Roman legion.

Edward landed at La Hogue, near Cherbourg (probably July 12). For reasons not clear, his fleet sailed back to England shortly after the army was disembarked, leaving the small English army on its own in hostile country.

Edward marched inland via Carentan and St. Lô (July 18). After a sharp engagement, he captured Caen (July 27). The army continued slowly toward the Seine, pillaging as it went. Edward learned that John of Normandy had raised the siege of Aiguillon and was marching north from Gascony, while Philip was collecting a large army near Paris. Edward therefore decided to cross north of the Seine, so as to have a free line of retreat to Flanders if need be.

Reaching the Seine near Rouen, Edward discovered that the French had destroyed all bridges across the lower river save for one in strongly defended Rouen. Increasing his pace, he marched rapidly upriver toward Paris, seeking a crossing. Evidently he was becoming nervous, since if Philip came out from Paris he would be forced to fight south of the river. At Poissy, only a few miles from Paris, he found a repairable bridge. While Philip and his army unaccountably sat idle at St. Denis, the English crossed the Seine (August 16).

Only then, too late to strike the English astride the river, did Philip begin to move. Edward, still in danger of being cut off, began forced marches northward. About one day ahead of the pursuing French, he reached the Somme River (August 22) to find that its bridges (save in fortified towns) had also been destroyed. While the English sought a crossing, the French reached Amiens to the east. Finally English scouts discovered a ford near the mouth of the river, passable only at low tide. The English army forced its way across against light opposition just as the tide was rising and as the French army appeared on the south shore (August 24).

THE BATTLE OF CRÉCY, AUGUST 26, 1346

Having passed the last major obstacle to a further retreat into Flanders, Edward decided to fight. While the French army crossed the Somme at Abbeville, he discovered a suitable battleground a few miles farther north near the village of Crécy-en-Ponthieu (August 25). Here a gentle slope overlooked the route which the French army would have to take.

During the next day the English organized their position carefully. The right flank, near Crécy, was protected by a river. The left flank, in front of the village of Wadicourt, was covered by trees and by ditches dug by the English infantry. The army was formed in three main divisions, or "battles," each of about equal strength. English strength was probably about 20,000.

The division on the right was nominally commanded by 17-year-old **Edward,** Prince of Wales, later known as the **"Black Prince";** actual command was evidently exercised by the veteran Earl Marshal **Warwick.** About 300 yards to the northeast, and echeloned slightly to the rear, was the left division, commanded by the Earls of **Arundel** and **Northampton.** A few hundred yards to the rear and covering the gap between the two front rank divisions was the third "battle," under Edward's personal command. The king took a position in a windmill midway between his division and that of the Prince of Wales, from which he could observe all of the action and send orders to his subordinate commanders.

The core of each division was a solid phalanx of some 1,000 dismounted men-at-arms, probably 6 ranks deep and about 250 yards long. The archers were ranged on the outer flanks of each division and echeloned forward so as to obtain clear, converging fields of fire. In front of the center of the army, the flank archers of the two front divisions met in an inverted V pointed toward the enemy. It is not clear whether the Welsh light infantry were interspersed with the archers or were ranged in the central phalanx with the dismounted men-at-arms. Behind the center of each division was a small reserve of mounted heavy cavalry prepared to counterattack if any French assault should break through the front lines. During the day it appears that the English and Welsh infantrymen dug a number of small holes in the rolling fields to their immediate front as traps for the French cavalry horses. Crécy was probably the first European battle in which gunpowder weapons were employed (see p. 360), but they had no influence on the outcome of the battle.

The French army, estimated at nearly 60,000 fighting men, was composed of approximately 12,000 heavy cavalry—knights and men-at-arms—the flower of French chivalry, about 6,000 Genoese mercenary crossbowmen, some 17,000 additional light cavalry (the retainers of the chivalry), and more than 25,000 communal levies—an undisciplined rabble of footmen straggling along in the rear.

This force, strung in an interminably long march column without any reconnoitering screen, bumped unexpectedly into the English line of battle about 6 o'clock in the late afternoon. Philip endeavored to halt the mass and concentrate. He was apparently able to get his crossbowmen into the lead, but the impetuous knights, filled with pride and the valor of ignorance, could not be controlled; so the French vanguard began to pile forward in a confused mass behind the Genoese. A quick shower and thunderstorm momentarily swept the field, making the footing

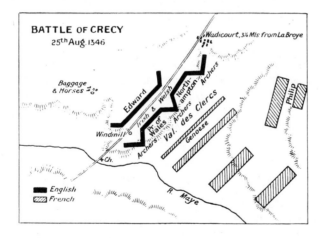

slippery. Then the sinking sun shone out again as the disciplined Genoese, deployed in firm alignment, crossed the valley and started up the slope. Halting about 150 yards from the English front, they fired their bolts, most of which fell short. Then they moved on again to meet the full blast of cloth-yard English arrows sheeting like a snowstorm on their line.

Shattered, the Genoese reeled away from this devastating fire. The French van, impatient, put spurs to their mounts and rolled through and over them, a ponderous, disorganized avalanche. In a moment the slippery slope was covered by a churning mass of mailed men and horses, pounding and stumbling through the unfortunate Genoese, while English arrows rained down on all. The impetus of the assault carried some of the French as far as the English line, where a sharp fight developed for a few moments. Then, repulsed by the stout English line, French survivors were driven back by Prince Edward's mounted detachment.

Without rhyme or reason, each successive element of the French column came rushing into this horrible welter, to be caught in turn by the devastating arrows. Apparently Edward had made excellent arrangements for a resupply of arrows, and evidently as each French charge ebbed away scuttling archers would rush across the field to collect arrows for reuse.

The slaughter continued into the night; some 15 or 16 separate French waves dashed themselves to fragments in that ghastly valley. Then the French gave up. The English stood in their formations until dawn.

In dreadful piles across the little valley lay 1,542 dead lords and knights (including King John of Bohemia; see p. 409); between 10,000 and 20,000 men-at-arms, crossbowmen, and infantrymen; and thousands of horses. The French king and many others of his lords and knights were wounded; there is no accurate total. The English loss was probably about 200 dead and wounded; the killed included 2 English knights, 40 men-at-arms and archers, and "a few dozen" Welsh infantrymen. Again we have no accurate figures.

COMMENT. *Europe was stunned by the English victory over a French force nearly three times its size. Continentals, knowing nothing of the fierce wars with the Welsh and the Scots, were unaware of English tactical, technical, and organizational developments. In a few earlier European battles, as at Legnano (see p. 320), Courtrai (see p. 399), and in the conflicts between the Austrians and the Swiss (see p. 408), infantry had gained some successes against feudal heavy cavalry. In each of these earlier battles, however, some special circumstances contributed to the outcome. Crécy was different. Here was a clear-cut victory in the open field of steady, disciplined infantrymen over the finest cavalry in Europe—even though atrociously led. Edward, who had scant strategical skill, proved himself the master tactician of his time. Understanding the value of disciplined infantry against cavalry, and aware of the devastating firepower of his longbowmen, he made the optimum use of the force at his disposal. Within a century the political significance of Crécy would be negated by other factors (see p. 452). In a purely military sense, however, this was one of the most decisive battles of world history. After almost a millennium in which cavalry had dominated warfare, the verdict of Adrianople had been reversed. Since the time of Crécy, infantry has remained the primary element of ground combat forces.*

AFTER CRÉCY, 1346–1354

1346–1347. Siege and Capture of Calais. Resuming a leisurely march northward, Edward reached Calais (September 4, 1346), which he invested and captured after nearly a year's siege (August 4, 1347). This was followed by a truce (September 28).

1347–1354. Truce. Both sides were prostrated by the ravages of the Black Death.

The Poitiers Period, 1355–1360

1355. Resumption of the War. Negotiations for a permanent peace having failed, Edward crossed the Channel and led an army in devastating raids across northern France. At the same time, the Black Prince raided deep into Languedoc from Bordeaux, causing great damage and collecting much booty. Edward's second son, **John of Gaunt** (and of Lancaster) raided into Normandy from Brittany (1356).

1356, August–September. Raid by the Black Prince. Starting from Bergerac (August 4), young Edward conducted a deliberate and devasting raid into central France. His army consisted of about 4,000 heavy cavalry, probably another 4,000 light cavalry, some 3,000 English archers, and another 1,000 light infantry. All were professional indentured troops, but (save for the archers) most were French Aquitanians, vassals of the English royal family. The prince avoided fortified places and concentrated on plundering as far as Tours (September 3).

1356, September 11–17. French Pursuit; English Retreat. Learning that King **John** (former Duke of Normandy) had crossed the Loire at Blois (September 8), the Prince started south as rapidly as his great train of booty could move. Unencumbered, the French moved more rapidly, converging to intercept the English at Poitiers. Apparently neither side conducted adequate reconnaissance, and neither had any idea of the other's location. Bypassing Poitiers, the English advance guard unexpectedly ran into the French rear guard less than three miles east of the city (late afternoon, September 17). Edward, realizing that his tired and burdened army could no longer elude the French, spent the evening and next morning locating a good defensive position.

1356, September 18. Preparation for Battle. While their armies rested and prepared for battle, the French king and English prince conducted fruitless negotiations through intermediaries. Edward had found an excellent position overlooking a gently sloping vineyard facing north. His left flank was protected by a marsh and creek; his front of 1,000 yards was covered by a hedge traversed only by a few narrow sunken lanes. His exposed right flank was protected by a wagon park. The hedge, the sunken lanes, and wagon park were lined with archers; other archers were extended forward as skirmishers in the vineyard and in the marsh to the left. As at Crécy, Edward's men-at-arms were dismounted behind the hedge, save for a small mounted reserve which he posted near his exposed right flank. The French army consisted of about 8,000 men-at-arms, 8,000 light cavalry, 4,500 professional mercenary infantry (including 2,000 crossbowmen), and possibly 15,000 unreliable militia infantry. This mass was organized into four divisions or "battles," each comprising nearly 10,000 men. Save only for about 3,000 heavy and light cavalrymen in the first battle, all of the French horsemen were dismounted, on the supposition that this was the only way to beat the dismounted Englishmen. John did not understand that the secret of the English success at Crécy had been the use of dismounted men-at-arms only as a solid defensive base for the devastating firepower of the English archers. By dismounting his knights he deprived them of their principal assets for offensive action: mobility and shock. Because of the narrow English front, he planned to attack in a column of divisions, one behind the other. Apparently he gave no consideration to any kind of maneuver, to envelop, or to turn the outnumbered English from their exposed position.

BATTLE OF POITIERS, SEPTEMBER 19, 1356

Early in the morning Prince Edward sent much of his booty train across the creek, accompanied by about two-thirds of his army. Possibly he hoped to withdraw without a fight. He left one "battle" to hold the previously selected position. The French advance, however, caused him to hasten back to the scene with his two other battles. The first French division attacked prematurely, with typical feudal lack of coordination. The crossbowmen, lined up behind the French mounted cavalry, had no chance to engage the English archers. The first mad mounted dash was broken by the English archers just as at Crécy. In renewed assaults a few of the horsemen and some of the infantry reached the English line, to be thrown back by the stout dismounted defenders and by enfilading fire from the English archers in the marsh.

Some time later, the second French battle, under the Dauphin, approached the English line. Their armor clanking loudly, the French knights trudged up the gentle vineyard slope, to be met

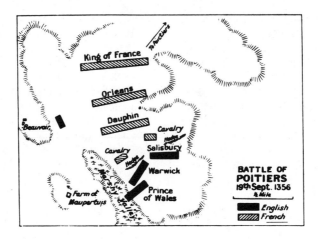

by a hail of English arrows. Despite heavy losses, the gallant French closed grimly against the English line. There was a prolonged, desperate struggle. The French came very near to breaking through, and the English line was reestablished only when Edward put his third division into the front line, holding out only a 400-man reserve. The French-English men-at-arms, aided by continuing harassment from the archers on the flank, finally repulsed the attackers. Both sides had suffered severely. If the remainder of the French had advanced promptly, or if they had enveloped the exposed English right, the Black Prince would have been decisively defeated. However, the sight of the disaster suffered by the Dauphin's division caused the third division, commanded by the **Duke of Orléans,** brother of the king, to lose heart and to flee before getting in range.

Next the fourth and largest French division, under John himself, approached. Although the French were exhausted by marching more than a mile in their heavy armor, the English were at least equally exhausted from the prolonged combat. Fearful that his outnumbered men could not sustain a determined French attack, the Prince seized the initiative. Putting his 400 fresh men-at-arms in front, he ordered the entire army to charge. He also sent about 200 mounted men, probably light cavalry, to attack the French left rear. The two advancing lines met with terrific impact in the vineyard. The English archers, having used up all of their arrows, joined in the vicious hand-to-hand combat. The issue was in doubt until the small English enveloping force suddenly struck the French rear. This was too much for the French, who began to flee. John, surrounded by the flower of French chivalry, fought on until he was overpowered and captured. The exhausted English did not pursue, but they captured thousands of prisoners who were too tired and encumbered to escape. In this bitterly contested battle, the French lost 2,500 dead and 2,600 prisoners, most of the casualties being knights and men-at-arms. English losses are unknown; they were probably over 1,000 killed and at least an equal number wounded.

1356, September—October. English Withdrawal. Making no effort to exploit his great victory, the Black Prince retired to Bordeaux with his booty and his prisoners.

1356–1360. English Raids. Recognizing their inability to meet the English in the open field, the French remained in their castles and fortified towns while the English ranged all over

France. King Edward himself made a final raid to the walls of Paris (1360).

1360, October 24. Peace of Bretigny. English holdings in southwestern France were augmented and France recognized English possession of Calais and Ponthieu. Edward gave up his claims to Normandy and to the French crown, while John was ransomed for 3 million gold crowns.

1360–1367. Nominal Peace. The war of succession in Brittany continued and one major battle was fought: an English army besieging **Auray** defeated a French relieving army and took the town (1364). **Charles the Bad** of Navarre attempted to take advantage of French troubles by inroads into southern France. Disbanded French and English mercenaries participated in both of these wars. Still others took part in a civil war in Castile (see p. 407), where the French and English continued the Hundred Years' War by proxy. Still others of the disbanded mercenary companies ranged over France, ravaging the countryside.

The Du Guesclin Period, 1368–1396

A revolt of Gascon nobles against the Black Prince (who was also Duke of Aquitaine) provided French King **Charles V** with an excuse to intervene with a new, improved French military force (see p. 399). Leading the French armies was Bertrand Du Guesclin, Constable of France and one of the great soldiers of the age, who was able to reconcile chivalry with common sense. Recognizing the superiority of English archery firepower, Du Guesclin avoided attacks against the English in prepared positions. He seized all possible opportunities to force the English to fight at a disadvantage. He excelled particularly in night attacks, despite English objections that these were "unknightly." Determined and skillful in siegecraft, he captured the English towns and castles in Poitou and Aquitaine one after another, reducing English holdings in France to small regions around Bordeaux, Bayonne, Brest, Calais, and Cherbourg. Desultory, truce-studded warfare continued for 16 years after his death. The French retained what he had gained; France, however, was in ruins. The important events were:

1368. Noble Revolt in Gascony. In reponse to Du Guesclin's assistance to the rebels, Edward III reclaimed the throne of France.

1370. Massacre of Limoges. The Black Prince captured and sacked Limoges, slaughtering many of the inhabitants. Soon afterward he returned to England (1371).

1372. Battle of La Rochelle. With Castilian assistance, the French fleet defeated an English fleet, regaining control of the western and northern coasts of France.

1375–1383. Truce. Spasmodic fighting continued between French and English adherents.

1376. Death of the Black Prince.

1377. Death of Edward III.

1380. Death of Du Guesclin and Charles V.

1386. French-Planned Invasion of England. Never too serious, the plan was abandoned after the combined French-Castilian fleet was defeated off **Margate** by the English (1387).

1389–1396. Period of Truces. These were interspersed with fighting.

1396. Peace of Paris. Richard II of England and **Charles VI** of France agreed to peace for 30 years; England retained only Calais and that part of Gascony between Bordeaux and Bayonne.

THE BRITISH ISLES

England

During these two centuries of violence, unrest, and endemic civil war, there were two relatively long periods of peace and prosperity under two of England's greatest monarchs—**Edward I** and **Edward III**—who were able to divert the energies of their vigorous brawling barons into highly successful foreign wars. Edward I—who proved his tactical and strategic brilliance by defeating the nobles who had overthrown his father—later conquered Wales and Scotland. To him must go the credit for starting the tactical system with which his grandson revolutionized warfare in the Hundred Years' War (see p. 386). **Edward II,** an inept ruler, was driven from Scotland by Robert Bruce (see p. 396), and was later defeated and deposed by rebellious nobles. His young son, Edward III, made common cause with the barons, then gained ascendancy over them. In the Hundred Years' War, Edward proved himself the leading tactician of his age and made his country the most powerful in Europe. In his declining years, French resurgency reduced England's power on the Continent. Predeceased by his eldest son, the famous Black Prince, Edward was succeeded by his grandson, **Richard II,** whose reign was another period of foreign defeats and internal unrest. At the close of the century, Richard was defeated and deposed by his successor, **Henry IV.** The principal events were:

1199–1216. Reign of John I (brother of Richard I). John was constantly, and generally unsuccessfully, at war with his barons at home and with Philip II of France abroad. His reluctant signature of the Magna Carta was the most important result of his wars with the barons.

1202–1204. War with Philip of France. John lost Anjou, Brittany, Maine, Normandy, and Touraine. With the assistance of his vassals in Aquitaine, he repulsed Philip's efforts to seize his possessions southwest of the Loire.

1213–1214. Renewed War with France. In an effort to regain his lands north of the Loire, John made an alliance with **Otto IV** of Germany and Count **Ferdinand** of Flanders. By invading west-central France from Aquitaine, John hoped to so distract Philip as to permit a successful invasion of northeastern France by his allies. He failed, and his allies were crushed at the **Battle of Bouvines** (1214; see p. 398).

1215–1217. Intensified Civil War in England. Louis, Dauphin of France, was invited by a faction of English barons to replace John. Outnumbered, most of his kingdom in the hands of his enemies, and his own forces close to dissolution, John revealed a vigor and ability worthy of comparison with his great brother, Richard I. His determination and skill rallied to him the support of most of the people in England in opposition to the baronial party. Avoiding major battles, John consolidated control of western and central England, repulsed an invasion by King **Alexander** of Scotland (see p. 395), and forced Louis and the barons to the defensive in southeastern England. John died (1216), but his followers completed his victory on land at **Lincoln** and at sea off **Sandwich** (1217).

1216–1272. Reign of Henry III. A period of inept rule both before and after Henry came of age (1227). England lost control of most of Aquitaine and Poitou; Henry was repulsed in efforts to reestablish his authority. Meanwhile bickering in England flared into revolt.

1263–1265. Civil War. The rebels were led by Earl Simon de Montfort.*

1264, May 14. Battle of Lewes. Simon and the other barons were besieging Rochester when Henry, who had just captured Northampton,

* Son of the leader of the Albigensian Crusade; see p. 397.

arrived with a relieving army. In a cavalry battle, Henry was decisively defeated, though his young son Prince Edward (later Edward I) drove a portion of the baronial army off the field. Henry became a prisoner and puppet of Simon, who became virtual ruler of England. Prince Edward escaped to raise an army in western England. Simon made an alliance with **Llewellyn,** Prince of Wales, and marched against the royal insurgents.

1265, July 8. Battle of Newport. Culminating several minor setbacks, Simon was defeated by Edward's larger army. Simon retreated into Wales; his efforts to evade the royal army and to return to central England were blocked by Edward's skillful defense of the Severn River.

1265, July–August. Campaign of Evesham. Simon de Montfort the younger, son of Earl Simon, marched west from London with about 30,000 men to the relief of his father. Edward, with 20,000, now found himself caught between the two converging Montfort armies, whose combined strengths were approximately twice his own. Edward marched eastward from Worcester (July 31) by forced marches, making a surprise dawn attack on young Simon and practically annihilating his larger army in the brief **Battle of Kenilworth** (August 2). He immediately returned to Worcester (August 3). Meanwhile, Earl Simon had taken advantage of Edward's move by promptly throwing his army across the Severn, marching toward Stratford to join his son by way of Evesham (August 3). Edward at once left Worcester, driving his exhausted men to march southwestward through the night (August 3–4), planning to interpose himself between Simon and his line of retreat to the east. By dawn two-thirds of Edward's army had reached the Stratford-Evesham Road, just northeast of Evesham. The remainder approached Evesham from the west.

1265, August 4. Battle of Evesham. Simon, with only 7,000 men against Edward's 20,000, was cooped in a bend of the Avon River, and in a hopeless position. Nevertheless, he attempted a desperate charge against the central column of Edward's converging army. Despite ex-

haustion, Edward and his men were full of fight. Simon's army was annihilated. Edward became the virtual ruler of England, his father king in name only. Edward stamped out the embers of revolt. His forces crushed the remaining rebel barons at **Chesterfield** (1266) and in their refuges in fens or marshes at **Axholme** (1265) and **Ely** (1267).

COMMENT. *The Evesham campaign was one of the most brilliant operations of the Middle Ages—truly Napoleonic in concept, vigorous execution, and personal exercise of leadership. It alone is sufficient to include Edward I among the very greatest of English generals.*

1270–1272. Edward in the Crusades. He left England in the hands of trusted subordinates (see p. 418).

1272–1307. Reign of Edward I. As a ruler and soldier, he fulfilled the promise revealed at Evesham.

1278. Reorganization of England's Militia System. Edward increased centralized control.

1282–1284. Conquest of Wales. Wales had become virtually independent of England under able Prince Llewellyn. After Welsh incursions into England, Edward conquered and pacified the country, Llewellyn having been killed at **Radnor** (1282). During these campaigns Edward became familiar with the Welsh longbow and adopted it as a basic English weapon.

1293–1303. War with France. In an inconclusive struggle, Philip IV of France annexed Gascony and repulsed Edward's expeditions to reconquer the province (1294, 1296, 1297). The war was settled by negotiation, however, and Gascony was returned to Edward.

1295–1307. Scottish Wars. Border disputes and internal difficulties in Scotland led to war (see p. 395). Edward invaded, stormed and sacked Berwick-upon-Tweed, defeated King **John Baliol** at the **Battle of Dunbar,** and declared himself King of Scotland (1296). The Scots rose, under Sir **William Wallace,** who was decisively defeated by Edward at the **Battle of Falkirk** (1298; see page 395). In subsequent operations Wallace was captured and executed (1305), and Scotland again annexed to England. Scottish resistance continued, however,

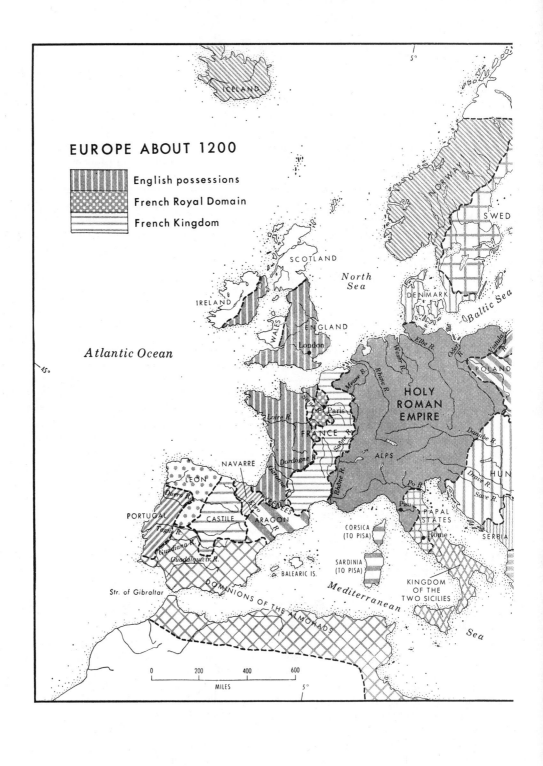

EUROPE ABOUT 1200

English possessions
French Royal Domain
French Kingdom

ICELAND

NORWAY
SWEDEN
DENMARK

North
Sea

Baltic Sea

SCOTLAND

IRELAND
WALES
ENGLAND
London

Atlantic Ocean

POLAND

HOLY
ROMAN
EMPIRE

Meuse R.
Seine R.
Paris
FRANCE
Loire R.
Dordogne R.
Garonne R.
Rhône R.
Rhine R.
Weser R.
Elbe R.
Oder R.
Vistula R.
Danube R.
Drave R.
Save R.
ALPS

HUN

NAVARRE
LEON
Duero R.
PORTUGAL
CASTILE
ARAGON
Tagus R.
Guadiana R.
Guadalquivir R.
PYRENEES

CORSICA
(TO PISA)

SARDINIA
(TO PISA)

BALEARIC IS.

Po R.
Pisa
PAPAL
STATES
Rome

SERBIA

KINGDOM
OF THE
TWO SICILIES

Mediterranean

Sea

Str. of Gibraltar

DOMINIONS OF THE ALMOHADS

| 0 | 200 | 400 | 600 |

MILES

45°

5°

5°

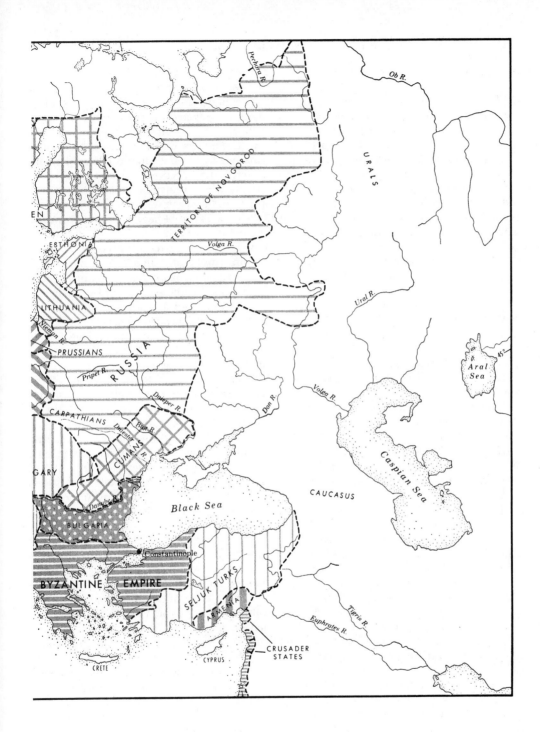

Pechora R.

Ob R.

U R A L S

TERRITORY OF NOVGOROD

Volga R.

EN

ESTHONIA

Ural R.

LITHUANIA

Aral Sea

45°

Niemen R.

PRUSSIANS

R U S S I A

Pripet R.

CARPATHIANS

Dnieper R.

Dniester R.

Bug R.

C U M A N S

Don R.

Volga R.

Caspian Sea

GARY

CAUCASUS

Danube R.

BULGARIA

Black Sea

Constantinople

BYZANTINE EMPIRE

SELJUK TURKS

Euphrates R.

Tigris R.

ARMENIA

CRUSADER STATES

CYPRUS

CRETE

under the leadership of **Robert Bruce** (see p. 395).

1307–1327. Reign of Edward II. A weak ruler. Scotland virtually reestablished its independence. Attempting to reconquer the country, Edward was decisively defeated by Robert Bruce in the **Battle of Bannockburn** (1314; see p. 396). England was forced to recognize Scottish independence. Continuing disorders in England broke into violent civil war in which Edward and his adherents defeated the **Duke of Lancaster** at the **Battle of Boroughbridge** (1322). Edward was a puppet in the hands of his favorites, **Hugh le Despenser,** Earl of Winchester, and his son, **Hugh;** they were defeated and overthrown by a revolutionary party led by Queen **Isabella** and her lover, **Roger Mortimer.** Edward was forced to abdicate, and was murdered (1327). Isabella and Mortimer became regents for her son, Edward.

1327–1377. Reign of Edward III. Fifteen years old when his father was deposed, Edward seized power by leading the barons in a revolt against Mortimer and Isabella (1330). His was one of the most momentous reigns of England, militarily and politically.

1332. Battle of Dupplin Muir. Under **Edward Baliol,** a mixed group of noble Scottish exiles ("the Disinherited") and English adventurers invaded Scotland, then technically at peace with England. The expedition consisted of about 1,000 knights and men-at-arms and 1,500 archers. They were opposed by about 2,000 Scottish heavy cavalry and 20,000 infantry under the Earl of Mar. The invaders made a daring and successful night attack across the Earn River. The Scots rallied and at dawn counterattacked in overwhelming force. The invading men-at-arms dismounted to form a small phalanx on a hilltop. The archers, probably echeloned forward in an irregular line, covered the flanks. The Scots were halted by the stout phalanx, then suffered so heavily from the archery fire on their flanks that they broke and fled. This battle provided Edward III with an excuse to invade Scotland, as well as a model for his tactics in subsequent battles.

1333. Battle of Halidon Hill. Edward, in a defensive position, formed his army into three battles or divisions; he dismounted his men-at-arms and covered the flanks of each division with archers. The attacking Scots were repulsed by the archers and the men-at-arms, then smashed by a counterattack. Edward had discovered the effective infantry tactics which would, a few years later, make England supreme in European warfare (see p. 386).

1337–1396. First Phase of the Hundred Years' War. (See p. 382).

1350–1367. English Involvement with Spain. This began with a Spanish naval expedition into the English Channel, during a lull in the Hundred Years' War. Edward led his fleet to victory in a hard-fought battle off **Winchelsea,** protecting his communications to Calais (1350). This undeclared war culminated in the great victory of the Black Prince at the **Battle of Navarrette** (see p. 407).

1377–1399. Reign of Richard II (age 10 at Accession). At the outset England was ruled by a regency in which his uncle **John of Gaunt,** Duke of Lancaster, was dominant (1377–1385). The government was nearly overthrown by a **Peasants' Revolt** (1381); London was seized and terrorized by a mob of 100,000 under **Jack Straw** and **Wat Tyler.** The youthful king negotiated with the rebels and granted most peasant demands, though harsh reprisals were later taken. Richard led a futile expedition to Scotland (1385). A brief revolt was led by **Thomas, Duke of Gloucester** who defeated the Royalists at **Radcot Bridge** and forced Richard to submit to the barons (1387). After a few years of relative calm, another dispute with his leading nobles led Richard to seize absolute power in defiance of Parliament (1397). This resulted in rebellion, in which he was at first supported, then opposed, by his cousin **Henry of Bolingbroke,** son of John of Gaunt. Richard's supporters deserted him; he was forced to abdicate. Henry of Bolingbroke ascended the throne as King **Henry IV** (1399).

Scotland

The 13th century was relatively peaceful. This was suddenly changed by a dynastic dispute, leading to English intervention and the outbreak of almost incessant war. This began with a prolonged struggle for independence, during which Scotland was twice conquered by Edward I, despite the gallant resistance of **William Wallace.** Later **Robert Bruce** led the Scots in a successful uprising that culminated in his great victory of Bannockburn and independence. Renewed outbreak of war after the death of Robert prompted French support, which in turn was one of the causes of the Hundred Years' War. For the remainder of the century there was continuous conflict with England. The English were generally on the defensive, since their major efforts were devoted to the struggle against France. The principal events were:

1214–1249. Reign of Alexander II. Hoping to take advantage of the barons' revolt against John of England, Alexander invaded England and attempted to seize Northumberland, but was repulsed near London by John (1216; see p. 390).

1249–1286. Reign of Alexander III. He drove the Norwegians from the islands of western Scotland by his victory in the **Battle of Largs** (1263).

1290. Dispute Succession. The death of Alexander (1286) and of his only direct descendant, **Margaret,** "Maid of Norway" (1290), resulted in claims to the throne by 12 different Scottish nobles. Edward I of England, asked to mediate, gained recognition of English overlordship, then chose John Baliol as king.

1295. Alliance with France. John entered this as a result of a dispute with Edward, who thereupon declared John's right forfeited and claimed the throne of Scotland for himself.

1296. English Invasion and Conquest. Edward defeated John at the **Battle of Dunbar** (April) and annexed the kingdom.

1297. Scottish Uprising. Sir William Wallace, the leader, defeated the **Earl of Warenne** in the **Battle of Cambuskenneth Bridge.** He then raided into Northumberland and Durham.

1298. English Invasion. Edward led an army of about 7,000 heavy cavalry, 3,000 light cavalry, and 15,000 infantry. Despite Wallace's efforts to avoid a major battle, Edward's aggressive advance forced the Scots to fight.

1298, July 22. Battle of Falkirk. Wallace had approximately 1,000 heavy cavalry, 2,500 light cavalry, and about 25,000 infantry. The steady Scottish pikemen, in four "schiltrouns" (circular phalanxes), repulsed English attacks. Driving the Scottish cavalry off the field, Edward brought up his archers to riddle the pikemen. After the Scots had suffered heavy losses, the English cavalry charged and broke the schiltrouns, annihilating Wallace's army. This was the first significant English use of the longbow. Edward reannexed Scotland, and in subsequent operations captured and executed Wallace (1305).

1306. Rebellion of Robert Bruce. He was a grandson of one of the earlier claimants to the throne, and had fought under Edward against Wallace at Falkirk. After some minor successes he was defeated by the **Earl of Pembroke** at **Methven** (1306, June 19), and **Dalry** (1306, Aug. 11). He won his first major victory over the Earl of Pembroke in the **Battle of Loudoun Hill** (1307, May).

1307–1314. Scottish Revival. Bruce, now King **Robert I** (1306–1329), gradually cleared the English from Scotland by energetic and resourceful operations, combining ambush with raids into northern England and against English strongholds in Scotland. As a result of these brilliant operations, only Stirling, Dunbar, and Berwick remained under English control (1314).

CAMPAIGN OF BANNOCKBURN

1314. Siege of Stirling. To rescue his garrison, Edward II finally took the field against Robert.

The English army consisted of at least 60,000 men, including nearly 5,000 heavy cavalry, 10,000 light cavalry, some 20,000 archers, and about 25,000 other infantry. Investing Stirling, Robert had an army of about 1,500 heavy cavalry and 40,000 infantry.

1314, June 23. Scottish Preparations for Battle. Robert carefully selected a position a few miles south of Stirling, placing his army on a slope behind Bannockburn. This sluggish, fordable stream passed through a marshy valley with only one old road affording a good passage. The battlefield was restricted by an impassable morass on the Scottish left, and by a forest to their right. The front was about a mile long, on a slope overlooking the stream. In front of his position Robert had had his men dig numerous small holes, covered with branches and grass, intended as traps for the English horses. Before the arrival of the English, he had his troops stake out their battle positions. The main army then withdrew behind the next hill to avoid harassment by English archers before the battle began. A line of skirmishers was left along the banks of the stream.

1314, June 24. Battle of Bannockburn. When the English began to cross the stream, the Scots resumed their positions. The English archers, crowded behind several contingents of cavalry, never had a chance to employ their weapons properly. A series of English cavalry charges were thrown back by the Scottish pikemen. As more and more English tried to crowd forward into the narrow battle area, they became hopelessly confused. The archers, attempting overhead fire, hit more of their own men in the back than they did Scots in the front. Efforts of the archers to deploy in the woods to their left flank were smashed by Robert's reserve. While the armies were locked in combat on the slope above the stream, the Scottish camp followers (probably not upon orders from Robert) decided to pretend an attack through the woods against the English left flank. Blowing horns, waving banners, and simulating a large combat force, they approached the English left, which began to crumble. Edward himself decided to leave the battlefield, and his craven example was soon followed by most of his army. The Scots pursued, slaughtering thousands of Englishmen trying to struggle back across the stream and the marsh. English losses were at least 15,000, those of the Scots probably about 4,000. "So ended," says Oman, "the most lamentable defeat which an English army ever suffered."

1314–1328. Scottish Victory. The Scots raided frequently and far into England, **James Douglas** winning a victory at Myton (1319). However, Robert was careful to avoid a major battle. When Edward attempted another invasion, Robert defeated him in a surprise dawn attack at the **Battle of Byland,** and the English king again fled the battlefield (1322). The war ended with the **Peace of Northampton** (1328). Scottish independence was recognized. The following year Robert died (1329).

1329–1371. Reign of David II (Bruce).

1332. Battle of Dupplin Muir. (See p. 394)

1333. English Invasion. Supporting Edward Baliol's claim to the throne, Edward III invaded Scotland and won the **Battle of Halidon Hill** (see p. 394). David fled to France. The outbreak of the Hundred Years' War in France forced the English to the defensive in Scotland, and permitted David to regain his throne.

1346. Battle of Neville's Cross. David invaded England, but was defeated and captured.

1350–1400. Desultory Warfare. A kind of sideshow to the Hundred Years' War. The most important episode was a Scottish-French invasion of northern England by Earl **James Douglas.** He defeated an English army under Sir **Henry Percy** at **Otterburn** (1388), where Douglas was killed.

Ireland

English colonial control of Ireland was shaken by a Scottish invasion under **Edward Bruce,** brother of Robert. Edward was crowned king (1316), but English rule was reestablished when Edward was defeated and killed at **Faughart** (1318). In following years English control weakened as the strength of the Irish chieftains increased. Richard II led an expedition to Ireland to pacify the country (1398–1399). This had no permanent results, and by the close of the century English control had become nominal.

FRANCE

During the 13th and early part of the 14 centuries, the power of the French kings steadily increased. This process was accomplished by rulers such as **Philip II** (Augustus), **Philip IV,** and to a lesser extent **Louis IX,** despite the frequently vigorous and violent opposition of the kings of England, who were the principal feudal landholders in France. One of the most important events of the early 13th century was the so-called Albigensian Crusade against heretical communities of southern France. The French kings exploited this "crusade" by greatly expanding their control and power in southern France at the expense of local barons. These royal gains were all undone in the early years of the Hundred Years' War during the latter part of the 14th century. The French monarchy was almost destroyed by the great victories of English King Edward III. The tide had turned by the end of the century as the result of the brilliant leadership of Bertrand Du Guesclin, but France had been ruined and impoverished. The principal events were:

1180–1223. Reign of Philip II, Augustus. (See p. 318)

1202–1204. War with John I of England. Philip conquered English fiefs north of the Loire (1202–1204). The outstanding event of this war was Philip's siege and capture of Château Gaillard (see p. 361) and Rouen in Normandy (1203–1204). During this and subsequent wars Philip established a semipermanent royal army, using the mercenary-indenture system (see p. 363).

THE ALBIGENSIAN CRUSADE, 1203–1226

1208–1213. First Phase. At the beginning of the 13th century the Catharist and Waldensian heresies flourished in southern France. The principal heretical centers were Toulouse and the Catharist stronghold of Albi. Efforts of the Catholic clergy to eliminate the heresies having failed, Pope **Innocent III** proclaimed a crusade against them. Philip II took no personal part in the crusade, but, with royal urging, the barons of northern France raised substantial forces. The principal leader of the crusading armies was half-English **Simon de Montfort.** Me-

thodically and ably Simon campaigned across western Languedoc, destroying most of the military forces of the southern French barons and capturing most of their strongholds. Albi and most of the other heretical centers were captured and their inhabitants ruthlessly slaughtered by the crusaders. Only the cities of Toulouse and Montauban still held out. The counts of Toulouse, Foix, and Comminges were the only southern nobles still daring to resist. Most of the northern nobles had returned to their homes, leaving Simon with relatively small forces to complete the task of conquest and religious conversion.

1213. Intervention of Pedro of Aragon. Fearful of the increase in French royal power in southern France, and anxious to protect his own feudal holdings north of the Pyrenees, Pedro, despite his own staunch Catholicism, led a Spanish army into Languedoc to join the heretics. Joining forces with Count **Raymond** of Toulouse, he captured several of Montfort's fortified posts. With a combined army of about 4,000 heavy cavalry and 30,000 infantry, Pedro and Raymond invested Muret, one of the most important of Simon's strongholds, garrisoned by 700 men. Just as Pedro and Raymond were arriving to besiege the town, Montfort and about 900 heavy cavalry joined the defenders.

1213, September 12. Battle of Muret. Simon discovered that Muret was low on provisions. He could not expect reinforcements to arrive from northern France before the garrison would be forced by starvation to surrender. He seized the initiative with a brilliant, daring plan. He enticed part of the besieging army to attack an apparently poorly defended gate on the southeastern side of the city. As the attackers rushed in, Montfort and his cavalry ambushed them just inside the gate and drove them back with heavy casualties. Then, while the attention of the enemy was attracted to this side of Muret, Montfort and his 900 cavalrymen dashed out of the city through the southwestern gate, causing the besiegers to think he was endeavoring to escape. In fact, riding around some low hills to the west of the city, Montfort and his cavalrymen turned north and crossed the Longe, which flowed north of Muret, dispersing a small force which was protecting the far bank. Montfort next surprised and smashed a far larger force under the Count of Foix. This brief action provided warning to Pedro and his Aragonese troops. They barely had time to form line when two-thirds of Montfort's force hit them in a violent frontal charge. Outnumbering the attackers by more than 30-to-1, the Aragonese quickly engulfed them. But while their attention was devoted to this attack, Montfort and his 300 remaining men completed a wide envelopment and came charging in behind the Aragonese army. The Spanish broke and fled, suffering heavy losses, including the death of King Pedro. After a very short pursuit, the Crusaders turned to strike the force of Count Raymond of Toulouse, the one remaining portion of the besieging army not yet engaged. Dismayed by what had happened, the Toulousans were quickly overwhelmed and almost annihilated.

1213–1226. Conclusion of the Crusade. Muret was the last important engagement of the crusade, but the heresies persisted for more than a decade while the crusaders systematically seized one fortified stronghold after another and ruthlessly put most of the heretics to the sword. Simon was killed during the siege of a stronghold near Toulouse (1222). But his successors carried on; the last embers of the heresy were stamped out (1226).

1213–1214. War against England, Germany, and Flanders. (See pp. 390, 402.) Philip was foiled in his efforts to conquer Flanders by an English naval victory at **Damme** (1213). However, he displayed his military ability by repulsing English King John I's diversion in western France and then turning to meet the main allied invasion of northeastern France under **Otto IV** of Germany.

1214, March–July. Campaign of Bouvines. Philip (with 11,000 heavy cavalry and 25,000 infantry) slightly outmaneuvered Otto (who had 11,000 heavy cavalry and 60,000 infantry). With German and Flemish reinforcements about to arrive, Otto endeavored to cut Philip's communications between Tournai and Paris. Pretending panic, Philip enticed a rash German attack on the good cavalry terrain east of Bouvines, where superior French cavalry would have an advantage. Otto, unexpectedly finding himself faced by Philip's army in order of battle, hastily formed his own.

1214, July 26. Battle of Bouvines. A French infantry attack was repulsed by Flemish and German pikemen. While part of his cavalry engaged the main imperial cavalry forces on the flanks, Philip led the remainder of his knights in determined converging attacks which finally smashed the center of Otto's army. At the same time his right flank cavalry, under **Garin** the Hospitaler, drove off superior imperial forces facing them. While the battle was still in doubt, Otto fled the field, permitting Philip to concentrate against the imperial right (where a small English contingent particularly distinguished itself), which was soon overwhelmed.

1223–1226. Reign of Louis VIII. Most of this short reign was devoted to a war with England over the lands southwest of the Loire. After some initial successes, Louis lost most of his conquests in Aquitaine, but retained Poitou, Limousin, and Périgord.

1226–1270. Reign of Louis IX (St. Louis). During Louis's minority (1226–1234), his mother, **Blanche of Castile,** was regent. She suppressed several feudal rebellions (1226–1231). Particularly renowned for his participation in two crusades (see pp. 416, 418), Louis was the epitome of chivalric valor and knightly prowess. He was an able ruler, as well as an essentially holy man who deserved subsequent canonization. He was partially successful in outlawing private warfare in France.

1242–1243. War with England. By a victory at **Saintes** (1242), Louis defeated an invasion by **Henry III,** and conquered most of Aquitaine and Toulouse.

1270–1285. Reign of Philip III. With papal support Philip invaded Aragon in an effort to dethrone excommunicated **Pedro III,** and to give the crown of Aragon to his son, **Charles** (1284). Disastrously repulsed, Philip died soon afterward.

1285–1314. Reign of Philip IV (the Fair). He greatly extended royal power and royal domains.

1294–1298. Inconclusive War with Edward I of England. This was over the province of Guienne in Aquitaine. Philip began the French alliance with Scotland (1295), with its far-reaching effects on France, Scotland, and England (see pp. 391, 395). The French repulsed Edward's invasion of northern France.

1300–1302. Renewed War against England and Flanders. The culmination of this otherwise indecisive war was a disastrous defeat of the French heavy cavalry by Flemish infantry at the **Battle of Courtrai** (July 11, 1302). The Flemish pikemen had taken up a position in a boggy area, cut up by canals and traversed by a few bridges. When an initial infantry and crossbow attack failed to shake the Flemish, the French knights charged recklessly over their own infantry, but were repulsed by a combination of the staunchness of the Flemish pikemen and the difficulty of horse movements over the marshy terrain. Caught in the morass, great numbers of the heavily armored French men-at-arms were massacred by the Flemish pikemen, who tore off the golden spurs from the victims' boots. (The battle is often called "The Battle of the Golden Spurs." For another "Battle of the Spurs," see p. 515.) This was the first

important victory of an essentially infantry army over cavalry in the Middle Ages, but its significance is diminished because of the special conditions. Two years later the French knights gained revenge at **Mons-en-Pévèle** (1304).

1328–1350. Reign of Philip VI. He was the first ruler of the House of Valois. His victory at **Cassel** (1328) brought Flanders again under French control.

1337–1453. Hundred Years' War. (See pp. 382, 444.)

1358. The Jacquerie. A bloody peasant uprising, centering in the Oise Valley north of Paris, was suppressed by nobles led by **Charles the Bad** of Navarre.

1364–1380. Reign of Charles V (the Wise). Thanks to his own sagacity, and with the aid of his great Constable of France Bertrand Du Guesclin, Charles saved his country from collapse after his predecessors Philip VI and John II (1350–1364) had suffered a series of stunning defeats. Drastic and far-reaching measures improved the military organization of France. Charles and Du Guesclin created new regular units of a permanent army, including the formation of artillery units. A new, permanent military staff was established; the French navy was reorganized. New walls were built around Paris.

1364–1367. Intervention in Civil War in Castile. This culminated in the **Battle of Navarrette** (1367; see p. 407).

1380–1422. Reign of Charles VI. The ineptitude of this intermittently insane king, as well as the continuing English pressure in the Hundred Years' War, made this a period of violence, unrest, and frequent revolts.

1382. Flemish Revolt. The rebels, led by **Philip van Artevelde,** were crushed by the French at the **Battle of Roosebeke.**

DENMARK

1202–1241. Reign of Valdemar II Sejr (the Victorious). Valdemar was an energetic general. He conquered most of the Pomeranian coast (c. 1203–1210) and launched a crusade against Estonia (1219). He built a great fortress at **Reval** (modern Tallinn), but was captured by Count **Heinrich** of Schwerin, a rebellious German vassal (1223), and bought his freedom by yielding all his conquests except Rügen and Estonia (1225). An offensive to regain the lost areas was decisively defeated at the **Battle of Bornhöved** (1227). This ended Danish domination of the Baltic. Valdemar concentrated on domestic matters.

1241–1252. Power Struggle among Valdemar's Sons. This ended when the youngest son was crowned as **Christopher I** (r. 1252–1259).

1259–1319. Church–Crown Struggle in Denmark. This began late in Christopher I's reign, and continued with varying intensity during the reigns of **Erik V Glipping** (1259–1286) and **Erik VI Menved** (1286–1319).

1232–1240. Interregnum. After the death of **Christopher II** (Erik VI's younger brother), the counts of Holstein (who had gained most of Denmark in pawn from Christopher II) maintained tenuous control.

1340–1375. Reign of Valdemar IV Atterdag. The son of Christopher II, **Valdemar** spent most of his reign reclaiming the kingdom, selling Estonia (1346) to buy back other lands. Valdemar quelled a revolt in Jutland (1350), regained Skane (Scania) from Sweden (1360), and conquered Gotland after winning the **Battle of Visby** (1361). This led to war with the Hanseatic League; although Copenhagen was sacked, Valdemar's great victory at the **Battle of Hälsingborg** (1362) forced the League to make peace (1363). By 1367, he faced an alliance of the League, Mecklenburg, Holstein, Sweden, and dissident Jutland nobles, and battlefield reverses forced him to flee Denmark (1368). The **Treaty of Stralsund** (1370) pawned Skåne to the Hanse for 15 years and gave them extensive trading rights in Denmark.

1397. Union of Kalmar. Valdemar IV died childless, and was succeeded by **Olaf III Haakonsson** (1375). Olaf became king of Norway as Olaf IV (1380) but died at age 17 (1387). His mother, **Margaret,** remained regent of Denmark and Norway, and was acclaimed Queen of Sweden after **Albert** of Mecklenburg was deposed (1389). Her greatnephew **Erik VII** of Pomerania became king of Sweden (1396), and was crowned king of all three Scandinavian kingdoms (June 1397).

NORWAY

1217–1263. Reign of Haakon IV Haakonsson the Old. The posthumous, illegitimate son of **Haakon III** (r. 1202–1204), he gained the throne at age 13, and ended an era of civil wars and internal unrest (see p. 317). Haakon undertook important reforms and defeated a major revolt by **Jarl Skuli** (1240). He later persuaded Iceland (1261) and Greenland (1262) to accept Norwegian sovereignty. A war with Scotland (1263) produced only a few skirmishes before a peace treaty gave Haakon closer control over the Shetland and Orkney islands while surrendering the Isle of Man and the Hebrides to the Scots. He died in the Orkneys soon after (December 1263).

1263–1319. Reigns of Magnus VI Lawmender (1263–1280) and Erik II (1280–1299). Norway was not involved in foreign wars, but internal development continued.

1299–1319. Reign of Haakon V Magnusson. Haakon built several fortresses, notably the Akershus at Oslo, which marked a shift of power from the west to the Oslo area. He lent aid to the Scots during their wars with the English (1297–1314), and was involved in several wars with Sweden and Denmark.

1319–1355. Reign of Magnus VII Eriksson. He inherited the throne through his mother **Ingeborg,** daughter of Haakon V. He spent most of his reign in Sweden, and was forced by Norwegian magnates to recognize his son **Haakon VI Magnusson** the Younger as king of Norway (1343), and then to abdicate in his favor (1355).

1355–1380. Reign of Haakon VI Magnusson the Younger.

1380–1387. Reign of Olaf IV Haakonsson. His mother Margaret (daughter of **Valdemar IV Atterdag)** became regent of Norway after he died at age 17.

1397. Union of Kalmar. Margaret's adopted

heir **Erik of Pomerania** (her great-nephew) was proclaimed heir apparent of Norway (1389), and was crowned King of Norway as well as of Denmark and Sweden (see below) at Kalmar, Sweden (June 1397).

SWEDEN

1210–1250. Reigns of Erik IV Knutsson and Erik V Eriksson. Peace returned as the civil wars ended.

1250–1275. Reign of Valdemar I Birgesson. The son of **Birger Jarl,** he was made the heir for childless King **Erik V Eriksson,** and became king at age 7 when Erik died (1250). He ruled with his father until Birger died (1266). He was deposed by his brother **Magnus** because Valdemar made major concessions to Pope Gregory X while on a pilgrimage to Rome (1275). He lived in exile in Norway until his death (December 1302).

1275–1290. Reign of Magnus I Ladulās (Barn-Lock). He overthrew his brother Valdemar and, by exempting landowners from taxes if they served in his army, created a feudal aristocracy. He also won a reputation for protecting commoners' property (hence his byname), and also gained tax relief for the Church.

1290–1318. Civil Disorder. This coincided with the reign of **Birger Magnusson,** who struggled unsuccessfully with the magnates and his brothers for control of the kingdom.

1318–1364. Reign of Magnus II Eriksson. At age 3 he was selected king by an assembly of magnates; he was also King of Norway as Magnus VII (see p. 400). Once he gained his majority at age 17 (1332), he mostly stayed in Sweden, and so lost the support of Norwegian magnates and the Norwegian throne (1355). Magnus struggled unsuccessfully with the Swedish magnates, and had to agree to the marriage of his son Haakon to **Margaret,** daughter of Valdemar IV Atterdag of Denmark (1359). The magnates imprisoned Magnus and offered the throne to **Albert** of Mecklenburg (1364). Magnus was released and went to Norway (1371) where he died (1374).

1364–1388. Reign of Albert of Mecklenburg. Even though Albert agreed to two charters sharply limiting his powers (1371, 1383), his efforts to reassert his powers provoked resistance, and the Swedish magnates called on **Margaret** for help (1388). The army she sent defeated and captured Albert (1389).

1389–1396. Regency of Margaret. A gifted diplomat and administrator, Margaret had to fend off attacks from Albert's allies. After repeated efforts her troops captured Stockholm from Albert's partisans (1398).

1397. Union of Kalmar. Margaret's great-nephew and adopted son Erik of Pomerania was simultaneously crowned king of Sweden, Denmark, and Norway at Kalmar, Sweden (June 1397). The Union lasted until 1523.

FINLAND

1191–1216. Danish–Swedish Rivalry in Finland. The Swedes gained papal support for their efforts (1216), and were authorized to create a bishopric in Finland (eventually sited at bo [Turku]).

1227–1229. Russian Intervention. Duke **Jaroslav** of Novgorod occupied areas of eastern Finland; the spread of Russian Orthodoxy was prevented only by a prompt Swedish response with papal support.

1240. Crusade against Novgorod. An army of crusaders under Birger Magnusson was defeated at the **Battle of the River Neva** (June? 1240) by **Aleksandr** of Novgorod, later bynamed **Nevsky** for his victory (see p. 410).

1249. Expedition to Tavastia. Birger led a force to Tavastia (now Häme) where he built a fortress.

1293–1323. Karelian War. Torgil Knutsson launched an expedition to conquer Karelia (1293), and built a fortress at Viipuri (Vyborg). This led to renewed war with Novgorod, which dragged on for three decades.

1374–1386. Grip Regime. Albert of Mecklenburg was unpopular in Finland, and by 1374 a Swedish nobleman, **Bo Jonsson Grip,** had gained control of most of the country. After

Grip's death (1386), Finland passed to Margaret's control (1388–1389), and eventually came under the Union of Kalmar (1397).

Haakon IV Haakonsson (see p. 400). Three years passed before all the chieftains had sworn allegiance to Haakon.

ICELAND

1261–1264. Iceland Accepts Norwegian Sovereignty. The *Althing* accepted the offer of

GERMANY

As previously, developments within Germany were still greatly affected by the obsession of the emperors with Italy, and their constant power struggle against the popes. The first half of the 13th century was dominated by brilliant, able, and cruel **Frederick II,** who combined the good and bad qualities of his Hohenstaufen and Italian Norman ancestors. Soon after his death Germany descended into the chaos of the "Great Interregnum," finally ended by the election of **Rudolph I,** first Hapsburg emperor. The 14th century was marked by almost incessant dynastic disputes between the houses of Hapsburg, Wittelsbach, and Luxembourg.

During these two turbulent centuries two strong, semiautonomous forces rose within Germany. Early in the 13th century, under the leadership of **Hermann von Salza,** the knights of the Teutonic Order began to expand Christianity and their own power through the heathen lands of Prussia and Lithuania. Farther west, the growing wealth and power of German cities—particularly those bordering the North and Baltic seas—were enhanced by banding together into the loosely organized Hanséatic League. Victories over German barons and against the Scandinavians brought the League to a position of preeminence in north Germany and the Baltic Sea by the middle of the 14th century. The principal events were:

1197–1214. Rival Emperors. Otto IV of Brunswick (Welf) and Philip II of Swabia (Waiblingen) were rivals until the assassination of Philip (1208). Frederick II then became Otto's rival. He allied himself with Philip II of France against Otto and John of England in the war which culminated in Otto's downfall at the **Battle of Bouvines** (see p. 398).

1210–1239. Rise of the Teutonic Knights.

1211–1250. Reign of Frederick II. He spent most of his reign in Italy opposing the popes. He also became King of Jerusalem (see p. 415). A high point in his many Italian wars was his victory over the Lombard League at the **Battle of Cortenuova** (1237; see p. 403). Frederick also twice fought the armies of Pope **Gregory IX** (1229–1230, and 1240–1241). His successful war against Pope **Innocent IV** (1244–1247) led the Pope to support antikings **Henry**

of Thuringia and **William of Holland** (1247–1256) in civil war against Frederick and his son **Conrad IV** in Germany (1246–1256).

1211–1224. The Teutonic Knights in Hungary. They operated against the Cuman Turks at the request of **Andrew** of Hungary. Alarmed by the growing strength of the Knights, Andrew later expelled them.

1226–1285. Crusade and Conquest of Prussia by the Teutonic Knights.

1241. Battle of Liegnitz (or Wahlstatt). Mongol victory over **Henry the Pious** of Silesia (see p. 377).

1250–1254. Reign of Conrad IV. This was marked by continuing civil war in Germany and struggles against the Pope and other independent states of Italy.

1254–1273. The Great Interregnum. Germany was in utter chaos.

1260. Battle of Durben. The Teutonic Knights were disastrously defeated by the Lithuanians.

1273–1291. Reign of Rudolph I (of Hapsburg). It began with wars against **Ottocar II** of Bohemia, who was defeated and killed at the **Battle of Marchfeld** (or Durnkrut, 1278; see pp. 409, 411).

1314–1347. Reign of Louis IV (of Wittelsbach). Civil war raged between Louis and **Frederick** of Hapsburg during the early years of his reign (1314–1325). Louis defeated and captured Frederick at **Mühldorf** (1322). He was deposed in a civil war against **Charles** of Bohemia and Luxembourg (1346–1347).

1347–1378. Reign of Charles IV (of Luxembourg). Internal unrest and anarchy continued in Germany.

1387–1389. The "Town War." This resulted from the jealousy of the German nobility over the rising wealth and power of the cities. The strength of the cities was impaired in this indecisive conflict.

ITALY

North and Central Italy

The interminable struggles among the petty states of northern and central Italy had two major unifying themes: the continuing struggle between Guelph and Ghibelline (papal party versus imperial party) and the commercial rivalry of the great maritime states of Venice and Genoa. The principal events were:

1204. Participation of Venice in the Fourth Crusade. (See pp. 412–414.)

1237. Battle of Cortenuova. Victory of Frederick II over Guelph army of Milan and the Lombard League.

1242. Mongol Raids. (See p. 379.)

1253–1299. Wars between Venice and Genoa. These were caused by a commercial dispute over trading concessions at Acre in Palestine. After early Venetian successes, Genoa recovered by assisting the Byzantines in recapturing Constantinople from the Latin emperors (1261; see p. 420). By a naval victory in the **Battle of Trepani** near Sicily (1264), Venice regained ascendancy and reestablished trading rights in Constantinople. A Genoese victory at **Alexandretta** (1294) was offset that same year when a Venetian fleet under Admiral **Morosini** forced the passage of the Dardanelles and sacked **Galata,** the Genoese trading post at Constantinople. Peace was finally negotiated after a Genoese victory in the Adriatic at the **Battle of Curzola** (1299).

1312–1400. Growth of Milan. Led by the Visconti family, Milan conquered Verona, Vicenza, and Padua (1386–1388), and Pisa and Siena (1399).

1320–1323. War between Florence and Lucca. Lucca was victorious.

1351. War between Florence and Milan. Neither succeeded in gaining control of Tuscany.

1375–1378. Alliance of Florence and Milan. They opposed papal efforts to annex Tuscany.

1397–1398. Renewed War between Florence and Milan. Again the issue was control of Tuscany.

1353–1355. Renewed War between Venice and Genoa. Genoa won the decisive **Battle of Sapienza** (1354), in which most of the Venetian fleet was destroyed.

1378–1381. The War of Chiogga (between Venice and Genoa). Genoese Admiral **Luciano Doria** defeated the Venetians under **Vittorio Pisani** at **Pola,** but was killed. His brother **Pietro Doria** seized Chiogga, and blockaded Venice. Pisani in turn blockaded and captured the Genoese fleet (1380). Genoa, forced to make peace, never again seriously threatened Venetian maritime supremacy.

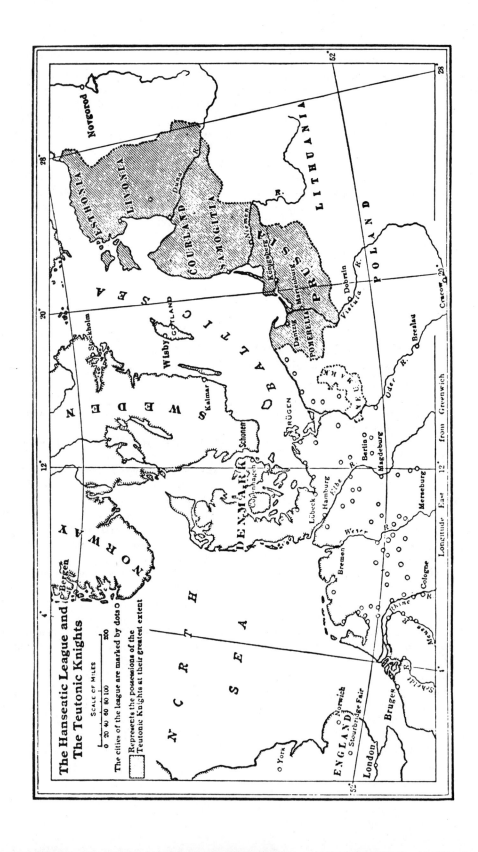

The Hanseatic League and
The Teutonic Knights

SCALE OF MILES

0 20 40 60 100   200

The cities of the league are marked by dots ○

☐ Represents the possessions of the
Teutonic Knights at their greatest extent

South Italy and Sicily, 1200–1400

During most of the 13th century, south Italy was the main Hohenstaufen base in the imperial-papal struggles. The papacy finally prevailed, through the instrumentality of **Charles** of Anjou. Like his Norman predecessors, Charles was then attracted toward the glittering prize of Constantinople, but during the last years of the century he and his Angevin successors were defeated and overthrown by the allied forces of decaying Byzantium and expanding Aragon. During the 14th century the Aragonese consolidated their control over southern Italy. The principal events were:

1221–1231. Frederick II Consolidates South Italy.

1228–1241. Wars of Gregory IX and Frederick II. Papal mercenary armies devastated Apulia (1228). Frederick, returning from his crusade, expelled the invaders, and forced the pope to make peace (1229). When war broke out again (1240), Frederick threatened Rome and annexed Tuscany (1241).

1244–1250. War between Frederick and Pope Innocent IV. Frederick successfully invaded Campagna (1244). The struggle spread throughout Italy. Although Frederick was repulsed after a long siege of Parma (1248), he was generally successful until his sudden death. His son, **Conrad IV,** continued the struggle (1250–1254).

1255–1265. Ascendancy of Manfred (Half-Brother of Conrad). He gained control of southern Italy (1255) and of Sicily (1256). In subsequent campaigns he established his control over most of Italy. Pope **Urban IV** offered the crown of Sicily to Prince **Charles** of Anjou, brother of Louis IX of France, and provided funds to assist Charles to raise an army against Manfred (1265).

ANGEVIN-HOHENSTAUFEN STRUGGLE, 1266–1268

1266, January. Angevin Invasion. Despite winter weather, Charles invaded Naples and crossed the Apennines to avoid Manfred's army holding the line of the Volturno River. Manfred pursued to Benevento. The two armies were closely matched, Manfred probably having a slight numerical superiority.

1266, February 26. Battle of Benevento. Manfred's failure to coordinate the operations of his archers and light and heavy cavalry permitted Charles's heavy cavalry to defeat the German and Neapolitan army in detail. Manfred was killed, and Charles established himself as King of Naples.

1268, July–August. Hohenstaufen Invasion. A Ghibelline army under **Conradin,** 15-year-old son of Conrad IV, invaded southern Italy after seizing Rome (July). The autocratic rule of Charles had already led to widespread revolt in southern Italy and Sicily. Conradin led his German army of 6,000 men across the Apennines into Apulia (August 18).

1268, August 25. Battle of Tagliacozzo. Charles, who had less than 5,000 men, took a defensive position near the Salto River, holding out approximately one-third of his army in reserve, hidden behind a hill. With more imagination than usual in medieval battles, Conradin enveloped the smaller Angevin army and dispersed all except the hidden reserve. After most of the German army had scattered in pursuit of the fleeing Angevins, Charles led his reserve against the victors, defeating each of its contingents in detail as they returned from their pursuit. Charles's victory was complete; Conradin was captured and executed, thus ending the Hohenstauffen line.

1266–1285. Reign of Charles I of the Two Sicilies (Naples and Sicily). Charles participated in his brother's crusade to Tunis (1270; see p. 418). In a subsequent expedition to Tunisia, Charles won the hard-fought **Battle of Carthage,** defeating the Moors in a brilliant cavalry maneuver after pretending flight

(1280). He made repeated expeditions to the west coast of the Balkans, establishing bases in anticipation of a major effort to overthrow the Byzantine Empire.

1282–1302. War of the Sicilian Vespers. A bloody uprising against Charles in Sicily was incited by the Byzantines, who also encouraged **Pedro** of Aragon to lead an expedition to the island. Charles had been assembling an army in Calabria, preparatory to an expedition against Constantinople. With this he rushed to Sicily, where he engaged in a war of maneuver with Pedro near Messina. At the same time Aragonese Admiral **Roger de Loria,** with a Catalan fleet, defeated the Angevin fleet in the naval **Battle of Messina** (1283). Loria defeated the Angevins again in the naval **Battle of Naples,** capturing **Charles,** son of the king (1284). The war dragged on for 20 years, Aragon finally conquering Sicily. French intervention on the side of Charles was abortive (1284–1285; see below).

The Iberian Peninsula

Spain

Christian expansion on the Iberian peninsula continued, the Moors being reduced to the small enclave of Granada in southeastern Spain. After a half-century of glory, Castile declined under mediocre rulers, while the power and influence of smaller Aragon grew steadily. Having no remaining frontier with the Moslems, vital Aragon began an eastward maritime expansion. Conquering the islands of the western Mediterranean plus Sicily and southern Italy, by the close of this period Aragon had become the most powerful state in the Mediterranean area. The principal events were:

1158–1214. Reign of Alfonso VIII of Castile. (See p. 325.) Continuing his earlier successes, Alfonso led a series of coordinated Christian offensives against the Almohads, culminating in his great victory at the **Battle of Las Navas de Tolosa** (1212), which gave Castile control of central Spain.

1196–1213. Reign of Pedro II of Aragon. His defeat and death at the **Battle of Muret** (1213; see p. 398) ended Spanish interests and influence north of the Pyrenees.

1213–1276. Reign of James I (the Conqueror) of Aragon. One of the great soldiers of Spanish history, he conquered the Balearic Islands (1229–1235) and Valencia (1233–1245) from the Moors. He took a leading part in the campaign which recovered Murcia from the Moslems for Castile (1266; see below).

1217–1252. Reign of Ferdinand III of Castile. He gained the throne after a prolonged dynastic war (1214–1217). He won a series of great victories over the Moors, capturing Cordova (1236), Seville (1248), Jaén (1246). By the close of his reign the Moors had been reduced to the small Emirate of Granada, acknowledging his suzerainty. Intermittent warfare with Granada continued for a century and a half.

1252–1284. Reign of Alfonso X of Castile. Generally unsuccessful in frequent wars with Granada, France, and Portugal, he regained his losses to the Moslems with the aid of James I of Aragon (see above).

1276–1285. Reign of Pedro III of Aragon. Forming an alliance with the Byzantine Empire, Pedro, pretending to sail on a crusade to Africa, actually landed at Sicily to begin the War of the Sicilian Vespers (1282; see above). He was excommunicated by Pope Martin IV, who offered the crown of Aragon to Charles of Valois, son of Philip III of France. Philip, allying himself with his uncle, Charles of Anjou, invaded Aragon, but was decisively repulsed by Pedro (1284).

1312–1350. Reign of Alfonso XI of Castile. He consolidated royal control of Castile. With the assistance of **Affonso IV** of Portugal he decisively defeated a combined army of Spanish and Moroccan Moslems at **Rio Salado** (October 30, 1340) ending the last serious Moslem threat to Spain. He besieged and captured **Algeciras** (1344) to gain control of the north shore of the Strait of Gibraltar.

1323–1324. Aragonese Conquest of Sardinia. Captured from Genoa and Pisa. The struggle for Sardinia continued for nearly 30 years.

1336–1387. Reign of Pedro IV of Aragon. Prolonged difficulties with his barons culminated in a violent civil war in which Pedro won a great victory at the **Battle of Eppila** (1348). He reconquered Sardinia from Genoa (1353).

1350–1369. Reign of Pedro the Cruel of Castile. He was constantly engaged in a dynastic civil war with his brother **Henry of Trastamara,** who held the throne on several occasions in this seesaw war. Pedro was assisted by the English, under Edward the Black Prince, while Henry was allied with the French, whose contingent was led by Du Guesclin. This, despite a truce in France, was really an episode of the Hundred Years' War (see p. 389).

Campaign and Battle of Navarrette, 1367

Edward the Black Prince, led an army across the Pyrenees (February 1367) to restore Pedro to his throne. Demonstrating greater strategical skill than in previous campaigns, the prince outmaneuvered the French and Castilian army, which attempted to block his advance through the Pass of Roncesvalles into the Ebro Valley. Rapidly marching around the Castilians, the Black Prince crossed the Ebro and forced Henry and Du Guesclin to retreat hastily south of the river. The Castilian army consisted of about 2,000 French heavy cavalry, 5,500 Castilian heavy cavalry, 4,000 light cavalry, 6,000 crossbowmen, and about 20,000 unreliable infantry. The Black Prince's army comprised probably about 10,000 heavy and light cavalry, plus 10,000 mercenary infantrymen, of whom perhaps half were English archers, the remainder being mixed crossbowmen and javelin men.

Facing each other near Nájera (Navarrette) south of the Ebro (April 3), each army formed in three lines and advanced, with no attempt to maneuver. Leading the Castilian vanguard were the dismounted French heavy cavalry under Du Guesclin. The English archers took a heavy toll of the Castilian cavalry, driving them off the field, but were unable to do serious damage to the heavily armored French contingent. Though the French fought staunchly and gallantly, their Spanish allies were soon dispersed. After a ferocious fight the French surrendered. The Castilian army lost 7,000 dead, including 400 French and 700 Spanish heavy cavalry, and about 6,000 of the fleeing infantry; they probably had an equal number of wounded. The English army probably lost no more than 100 dead and perhaps a few hundred wounded.

1369. Battle of Moutiel. A quarrel with Pedro caused Edward to leave Spain. Henry, again supported by Du Guesclin, rebelled. In a violent battle near Ciudad Real, Henry personally killed Pedro and gained the throne.

1369–1379. Reign of Henry II of Castile (and Trastamara). He fought several inconclusive wars with Portugal and Aragon. The Castilian fleet aided the French in their victory over the English at the **Battle of La Rochelle** (1372; see p. 389).

Portugal

During these centuries, the Portuguese continued incessant warfare with the Moors, sometimes alone, sometimes in coordination with Castile and Aragon. **Affonso II** supported the Castilians at Las Navas de Tolosa, and with Crusader assistance expanded southward as a result of a victory at **Alcácer do Sol** (1217). Wars against Castile were also frequent. Toward the end of the 14th century, when it seemed likely that Portugal would succumb to Castile, English support resulted in victory in the **Battle of Aljubarrota** (1385) by **John I** of Aviz. This was followed by the **Treaty of Windsor** (May 9, 1386), establishing an alliance between Portugal and England, which has continued to this day.

SWITZERLAND

During the latter part of the 13th century, the mountainous region that is now the heart of Switzerland became relatively autonomous, though nominally under the Hapsburg family of Swabia and Austria. Early in the 14th century, the Swiss opposed the Hapsburgs in the Welf-Waiblingen disputes. The Hapsburgs sent punitive expeditions into the mountains, and precipitated nearly a century of warfare, which eventually brought complete independence to the mountain cantons. The principal events were:

1315, November 15. Battle of Morgarten. An Austrian army of about 8,000 men, at least one-third heavy cavalry, under **Leopold** of Austria, invaded Switzerland. A Swiss force of about 1,500 pikemen and archers ambushed the Austrians in a defile between a lake and a mountain. The Austrians were thrown into confusion by boulders and trees hurled down the mountainside; the Swiss then practically annihilated the invaders with arrows and halberds.

1339. Battle of Laupen. Here the Swiss pikemen met a Burgundian invasion and first displayed their prowess and steadiness. Despite uneven terrain, the pikemen, supported by archers, drove the Burgundian infantry from the field, then repulsed a heavy-cavalry charge.

1386, July 9. Battle of Sempach. Some 1,600 Swiss pikemen were faced by approximately 6,000 Austrians under **Leopold III** of Swabia. Remembering the lessons of their previous defeats at the hands of the Swiss infantry, and influenced by English example in the Hundred Years' War, the Austrian heavy cavalry dismounted. Advancing across broken mountain fields as a line of heavily armored pikemen, they were at first successful in pushing back the less numerous and more lightly armored Swiss. The encumbered Austrians became exhausted, however, and some gaps appeared in their line.* The Swiss counterattacked, broke the Austrian line, and smashed Leopold's army. This was the decisive battle of the Swiss-Hapsburg war, although the Swiss won the subsequent **Battle of Näfels** (1388).

1394. Truce with the Hapsburgs. This in effect recognized the complete independence of the Swiss confederation within the German (or Holy Roman) Empire.

* Swiss legend says that the first gap was created by the self-sacrifice of Arnold Winkelried, who grasped a number of Austrian pikes, to plunge them into his own breast; his countrymen then exploited the gap thus made.

EASTERN EUROPE

BOHEMIA AND MORAVIA

Under the great rulers **Ottocar I** and **Ottocar II,** Bohemia in the 13th century was briefly the most powerful nation of Central Europe. Ottocar II was defeated, humbled, and finally killed in a series of wars against **Rudolph** of Hapsburg. Bohemia revived under **Wenceslas II,** who also became King of Poland. Under the rule of the House of Luxembourg, Bohemia again became a major European power. The principal events were:

1198–1230. Reign of Ottocar I. Made Duke of Bohemia by Emperor Henry VI (1192), he obtained the title of King and near-total autonomy for Bohemia from Emperor Philip of Swabia. This title was confirmed by Frederick II Hohenstaufen (1212), ensuring virtual Bohemian independence; this was cemented by Ottocar II.

1253–1278. Reign of Ottocar II (the Great). Ottocar joined the Teutonic Knights in a successful campaign against the Prussians (1255). He next conquered Styria from Hungary by a victory over Bela IV at **Kressenbrunn** (1260). He was unsuccessful in his efforts to establish domination over Poland and Lithuania. Following this, he took advantage of the Interregnum in Germany to annex Carinthia, Carniola, and Istria.

1274–1278. Wars against Rudolph of Hapsburg. After several defeats, Ottocar was forced to abandon his conquests between the Danube and the Adriatic, and to acknowledge the suzerainty of Rudolph over Bohemia and Moravia (1275). In a renewal of the war, Ottocar was defeated and killed in the **Battle of Marchfeld** (August 26, 1278; see above and p. 403).

1278–1305. Reign of Wenceslas II. He also became King of Poland (1290) and was briefly King of Hungary (1301–1304). After his death Poland reestablished its independence (see below).

1310–1346. Reign of John of Luxembourg (son of Henry VII of Germany). A typical valiant, foolhardy medieval warrior, John sought all opportunities to fight. Blinded in the process, he was generally successful, and greatly increased his own prestige as well as that of his nation. He frequently fought with the Teutonic Knights against the Lithuanians (1328–1345). He allied himself with Philip VI of France against Edward III, and was killed fighting with a contingent of French and Bohemian knights at the Battle of Crécy (see p. 386).

1347–1378. Reign of Charles I (also Charles IV of Germany, son of John). A time of great prosperity and expansion for Bohemia, which became the most powerful state within the Holy Roman Empire.

1378–1419. Reign of Wenceslas IV (son of Charles). As a result of internal conflicts, Bohemia began to decline.

POLAND

During much of the 13th century, Poland, weak, disunited, and frequently in a state of anarchy, was exploited by her neighbors, particularly the Teutonic Knights and Bohemia. Despite this, and despite internal difficulties due principally to an inadequate system of royal succession, Poland regained strength in the 14th century. By the end of the century, union with Lithuania established a basis for Poland's subsequent emergence as a great power. The principal events were:

1200–1279. Decline of Poland.

1228. Appearance of the Teutonic Knights in Prussia.

1241. The Mongol Invasion. (See p. 377)

1242–1253. Wars with the Teutonic Knights. Their continued expansion eventu-

ally cut off Poland from access to the Baltic Sea (1280).

1290–1305. Bohemian Domination. (See above.)

1305–1333. Reign of Ladislas IV. Poland regained its independence from Bohemia. During this reign the Teutonic Knights raided frequently into Poland.

1333–1370. Reign of Casimir (the Great). A competent soldier, who waged war successfully with most of his neighbors.

1370–1382. Reign of Louis of Anjou. His daughter **Jadwiga** (Hedwig) succeeded to the throne after a bitter civil war over the succession (1382–1384).

1384–1399. Reign of Jadwiga. She married **Jagiello,** Grand Duke of Lithuania, to unite the two countries.

LITHUANIA, 1300–1400

Largely as a result of pressure and influence by the Teutonic Knights, Lithuania emerged as a state early in the 14th century, and under Grand Duke **Gedymin** (1316–1341) expanded rapidly to the east and south at the expense of Russia. His successors warred successfully against the Russians and against the Cuman Tartars, extending their domains as far as the Black Sea between the Bug and the Dnieper rivers. Under Jagiello, who also became King of Poland upon the death of his wife Jadwiga, Lithuanian expansion and prosperity continued.

RUSSIA

These were dark centuries of decline for Russia. This was due in large part to the Mongol conquest. The principal events were:

1223. Battle of the Kalka River. First appearance of the Mongols (see p. 366).

1236–1263. Reign of Alexander Nevski of Novgorod. Nephew of **Yuri II** of Vladimir, he was a vassal of the Mongols (1240–1263). He defeated the Swedes under Birger Magnusson (see p. 401), at the **Battle of the Neva River,** ending Swedish attempts to conquer northern Novgorod (July 1240). He defeated the Teutonic Knights in the **Battle of Lake Peipus** (April 5, 1242).

1325–1353. Rise of Moscow. Princes **Ivan I** and **Simeon I** became the principal vassals of the Mongols (or Tartars as they became generally known). Ivan and Simeon were engaged in almost constant warfare against Lithuania, with little success.

1346. Start of the Black Death in Crimea. It spread through Russia (reaching Moscow in 1353) as well as into Egypt and Europe.

1359–1389. Reign of Dmitri Donskoi of Moscow. He was able to take advantage of a period of turmoil in the Golden Horde (see p. 380). His victory in the **Battle of Kulikovo** on the upper Don was the first important Russian victory over the Tartars (1380; see p. 401). He and his successors also fought a series of generally unsuccessful wars against the Lithuanians, who were sometimes allied with the Tartars.

HUNGARY

Hungary played a leading military and political role in Eastern Europe. Despite his disastrous defeat by the Mongols, King **Bela IV** later restored and enhanced his country's greatness. A prolonged dynastic war at the beginning of the 14th century was finally ended by the selection of **Charles I** (of Anjou) as king. He and his son **Louis** still further increased the power and prosperity of the country. The principal events were:

1206–1270. Reign of Bela IV. Decisively defeated by the Mongols at the **Battle of the Sajo River** (1241; see p. 378), Bela was ruthlessly pursued and finally found safety in the islands of Dalmatia. Following the withdrawal of the Mongols, Bela restored central authority and repulsed an invasion by **Frederick** of Austria (1246). He repulsed a second Mongol invasion (1261). Bela was less successful in his prolonged, indecisive wars with Ottocar II of Bohemia. His last years were plagued by disputes with his son, **Stephen.**

1272–1290. Reign of Ladislas IV. He allied himself with **Rudolph** of Hapsburg, helping to win a decisive victory over Ottocar at the **Battle of Marchfeld** or **Durnkrut** (1278; see p. 409).

1301–1308. Civil War over the Succession.

1308–1342. Reign of Charles I (of Anjou). He was forced to fight continuously during the first part of his reign to subdue unruly nobles (1308–1323).

1342–1382. Reign of Louis (the Great, son of Charles). He led a successful expedition to Italy to solidify Angevin control over Naples (1347). Allied with Genoa against Venice in prolonged wars, he forced Venice to cede Dalmatia and to pay tribute (1381). He temporarily stopped Turkish penetration into the Balkans in northern Bulgaria (1356). When selected King of Poland, during his final years Louis was the most powerful ruler in Eastern Europe (1370–1382).

1387–1437. Reign of Sigismund. Decline of the kingdom. During this reign occurred the disastrous Crusade of Nicopolis (1396, see p. 419).

SERBIA

During the 13th century, Serbia barely maintained itself against pressures from Hungary, Bulgaria, and the Byzantine Empire. During much of the 14th century, however, Serbia prospered under three great soldier kings. Then, by the end of the century, a series of crushing defeats at the hands of the Ottoman Turks, and the loss of the outlying regions to other neighbors, reduced Serbia to a vassal of the Ottoman Empire. The important events were:

1254. Defeat by Hungary. Loss of Bosnia and Herzegovina.

1281. Defeat by Byzantines. King **Dragutin** abdicated.

1281–1321. Reign of Milvutin (brother of Dragutin). He extended Serbian domains into Macedonia, at the expense of the declining Byzantines (see p. 420).

1321–1331. Reign of Stephen Dechanski. He gained control of most of the Vardar Valley through his decisive victory over a combined Byzantine-Bulgarian army at the **Battle of Kustendil,** or Velbuzhde (1330).

1331–1355. Reign of Stephen Dushan (Stephen Urosh IV). The son of Stephen Dechanski, whom he deposed and killed, Dushan was the greatest of Serbian medieval monarchs. Taking advantage of the continuing weakness of Byzantium, he subjected all of Macedonia, Albania, Thessaly, and Epirus and conquered Bulgaria. He invaded and annexed part of Bosnia (1349–1352). Victories over Louis of Hungary permitted Serbia to annex much of the western bank of the Danube River, including the site of modern Belgrade. Dushan, intending to seize the Byzantine throne, captured **Adrianople** and was advancing against Constantinople when he died (1355).

1371. Battle of Cernomen (or Maritza River). The Turks defeated Lazar I (1371–1389).

1376. Bosnian Aggression. Tvrtko I, ruler of Bosnia, proclaimed himself King of Serbia and Bosnia, annexing all of western Serbia including much of the Adriatic coast.

1389, June 20. Battle of Kossovo. (See p. 422.) Serbia became subject to the Turks.

1389–1427. Reign of Stephen Lazarevich

(son of Lazar). He fought loyally under the Turks at the **Battle of Nicopolis** (1396; see p. 419) and the **Battle of Angora** (1402; see p. 425).

BULGARIA

Bulgarian expansion during the early years of the 13th century was brought to a halt by crushing defeat at the hands of the Mongols (1241). For the remainder of this period Bulgaria played a minor role in the Balkans, a satellite in turn of the Byzantine Empire, Serbia, and the Turks. The principal events were:

1197–1207. Reign of Kaloyan. He expanded his nation as a result of victories over the Serbs, Hungarians, Byzantines, and the Latin Crusader States. At the **Battle of Adrianople** he captured Latin Emperor **Baldwin I** (1205; see p. 419).

1207–1217. Reign of Boril. He was defeated by the Crusaders, under **Henry I,** at the **Battle of Philippopolis** (Plovdiv; 1208). He was later besieged, captured and overthrown at **Trnovo** by **Ivan Asan** (1217–1218).

1218–1241. Reign of Ivan Asan II. He defeated **Theodore** of Epirus at the **Battle of Klokonitsa** on the Maritza River (1230; see p. 420). Later, in an alliance with the Empire of Nicaea against the Crusaders, Ivan unsuccessfully besieged Constantinople.

1241. Mongol Invasion. Bulgaria was crushed (see p. 377).

EURASIA AND THE MIDDLE EAST

THE LATER CRUSADES, 1200–1396

The Fourth Crusade, 1202–1204

BACKGROUND, 1195–1200

1195. Hohenstaufen Ambitions. Henry VI, German (Holy Roman) Emperor, and King of Sicily, inherited the Norman-Sicilian ambition to overthrow the Byzantine Empire and established a tenuous Hohenstaufen claim to the Byzantine succession. He also desired to emulate the exploits of his father, Frederick Barbarossa, as a Crusader (see p. 344).

1197–1198. Preliminary Moves. Henry sent an advanced party of German Crusaders to the Holy Land; they captured Beirut and other coastal towns (1198). Meanwhile Henry died (1197); plans for the Crusade were suspended.

1199. Renewed Papal Appeal. Pope **Innocent III** called upon Christendom to undertake a new effort to regain Jersualem from the Moslems.

PRELIMINARIES, 1200–1203

1200. The Leaders. Reponding to the papal appeal were **Theobald III,** Count of Champagne, Count **Louis** of Blois, Count **Baldwin** of Flanders and his brother **Henry,** Count **Boniface** of Montferrat, and the renowned French-English warrior **Simon de Montfort.** Theobald was the main leader. Philip of France and John of England ignored the appeal.

1201. Negotiations with Venice. Theobald, Louis, and Baldwin, agreeing with the pope that the Crusade should go by way of Egypt, negotiated with aged Venetian doge **Enrico Dandolo** to transport 25,000 Crusaders to Egypt, and to maintain them there for three years, in return for 85,000 marks and half of the Crusader conquests.

1201, May. Death of Theobald. Boniface of Montferrat was selected as the new Crusader leader. A cousin of Philip of Swabia, he was a strong Hohenstaufen supporter.

1201, December. Meeting at Hagenau. Philip persuaded Boniface to lead the Crusade to the Holy Land by way of Constantinople, in accordance with the plans of Henry VI. **Alexius,** son of deposed Isaac II of Constantinople and brother-in-law of Philip, was evidently at the conference, plus possibly a representative of Venice. (The idea of the expedition to Constantinople did not originate with the Venetians, even though they adopted it eagerly.)

1202. Assembly at Venice. During the summer and early fall, the Crusaders gathered. When they were unable to raise 85,000 marks, the Venetians agreed to transport them anyhow, in return for assistance in reconquering Zara, a former Venetian dependency now belonging to Hungary. The pope angrily denounced the deal, and threatened to excommunicate Crusaders who shed the blood of other Christians instead of going on to the Holy Land.

1202. Capture of Zara. Despite papal objections, the Crusader-Venetian expeditions captured and sacked Zara. Innocent excommunicated them.

EXPEDITION TO CONSTANTINOPLE, 1203–1204

1202–1203. The Final Decision. While at Zara the Crusader leaders negotiated with Dandolo about continuing on to Constantinople. Alexius joined the expedition and promised that in return for assistance in overthrowing Alexius III, he and his father would bring the Greek Church back into union with the Roman Church; that they would pay the Crusaders and Venetians 200,000 marks; and that they would provide active support for the expedition to the Holy Land. Ignoring papal demurrer, the Crusader and Venetian leaders agreed to go to Constantinople. Simon de Montfort and some other Crusaders refused to participate in the expedition and sailed directly to Palestine.

1203, June 23. Arrival at Constantinople. Sailing into the Bosporus, the Crusaders landed near Chalcedon, on the Asiatic shore. They then crossed to Galata, where, despite Byzantine efforts, they established a fortified camp. Soon after this the Venetian fleet forced its way into the Golden Horn.

1203, July 17. The First Assault. While the Crusader army attacked the land defenses of the city from the west, the Venetian fleet assaulted the sea wall. The Byzantines, led by **Theodore Lascaris,** repulsed the Crusaders, but the Venetians, inspired by the personal courageous leadership of their venerable Doge (95 years old), finally gained a foothold on the sea wall. Against vigorous opposition, by dark they captured several towers and held a portion of the city. During the night Alexius III fled the city. Next morning the Byzantine nobles released Isaac II from prison and replaced him on the throne. Young Alexius was elected coemperor as Alexius IV.

1203–1204. Delay at Constantinople. The Crusaders returned to their camp at Galata and the Venetians to their ships on the Golden Horn to await payment of the promised 200,000 marks. Isaac and Alexius had difficulty in raising the sum. Meanwhile, they asked the Crusaders to help them consolidate their shaky control of the Empire.

1204, January. Byzantine Revolt. Increasing resentment of the Byzantine nobles against the Crusaders and against their puppets, Isaac and Alexius, resulted in an insurrection led by **Alexius Ducas Mourtzouphlous,** son-in-law of Alexius III. Isaac was imprisoned, Alexius IV executed. Alexius Ducas seized the throne as Alexius V. The Crusaders now had an excuse for attacking Constantinople. Investing the city, the Crusaders and Venetians planned an assault (March–April).

1204, April 11–13. Conquest of Constantinople. For two days violent land and sea attacks were skillfully resisted by the Byzantine army. The Varangian Guard, composed primarily of English and Danish mercenaries, particularly distinguished itself. Catapults on the Venetian ships flung incendiaries into the town, starting a violent conflagration. Assisted

by this diversion, Crusaders and Venetians gained footholds on both the land and sea walls. An unaccountable panic seized part of the defending forces. The attackers redoubled their efforts and forced their way into the city, annihilating the Varangian Guard. For 3 days Constantinople was sacked in an orgy of violence, slaughter, loot, and rape.

COMMENT. *This, the first successful assault of Constantinople, was for all practical purposes the end of the Byzantine Empire. Remnants persisted, scattered along the coasts of Anatolia and Greece, and after 57 years even reestablished themselves in Constantinople. (See p. 419 for the Latin Empire of Constantinople and its struggles with the Byzantines.) Even though these last fragments would persist for almost 2 centuries, it would be only as pale shadows of the once-great Byzantine Empire. The Fourth Crusade smashed the bulwark of Christendom in the East.*

The Fifth Crusade, 1218–1221

Innocent III preached another Crusade in 1215, again urging that the expedition go by way of Egypt. To gain papal support in his struggle with **Otto IV** for the imperial throne in Germany (see p. 402), **Frederick II** promised to lead. Crusader contingents arrived in Acre from Hungary, Scandinavia, Austria, and Holland, where they joined **John of Brienne,** King of Jerusalem, and the Hospitalers, Templars, and feudal levies of the Crusader states (1217–1218). Landing near Damietta, they were vigorously opposed by a Moslem army under **Malik al-Kamil,** son of the Sultan, al-Adil. While the Genoese fleet defeated an Egyptian fleet in the mouth of the river, the Crusaders invested Damietta. The siege lasted for almost a year and a half before Damietta fell (November 1219).

Meanwhile, Cardinal **Pelagius,** papal legate, joined the crusading army and insisted upon assuming command. Somewhat reluctantly, King John acceded, but retained operational control. For more than a year, rejecting Egyptian peace overtures, the Crusaders waited in Damietta, momentarily expecting the arrival of Frederick II and reinforcements. Frederick never came; but early in 1221 **Hermann von Salza,** Grand Master of the Teutonic Order, and **Louis,** Duke of Bavaria, arrived with reinforcements. This brought the Crusader army to a total of about 46,000 men, of whom perhaps 10,000 were cavalry. Al-Kamil, who had become sultan upon the death of his father, had about 70,000. In June—despite intense heat and the flooded condition of the Delta—Pelagius ordered a march on Cairo. The advance was painfully slow, but by July the army had reached an Egyptian fortress overlooking the fork of the Damietta branch of the Nile, near the site of Mansura. Al-Kamil again offered peace, proposing to cede Jerusalem and other areas of the Holy Land to the Crusaders in return for Crusader evacuation of Damietta. John urged acceptance, but Pelagius demanded an additional indemnity and the cession of the fortresses of Kerak and Monreal. Negotiations broke down. The Crusaders attempted to cross the Ashmoun Canal and were decisively repulsed. At the same time the Moslem fleet seized control of the river between the Crusaders and Damietta, cutting their supply line. Facing starvation, Pelagius now agreed to give up Damietta in return for a free retreat and the face-saving cession of some holy relics by the Egyptians. An 8-year peace was agreed upon. The Fifth Crusade had been an abject failure.

The Sixth Crusade, 1228–1229

Shortly after this failure, Frederick II married **Yolande,** daughter of John of Brienne and heiress to the Kingdom of Jerusalem. Frederick, taking the title of King of Jerusalem, then began preparations for a Crusade. He was under considerable pressure from Rome, since the pope believed with some cause that the Fifth Crusade would have succeeded if he had participated as promised. Finally he sailed from Sicily in 1227. But when the fleet was at sea many of the Crusaders, including Frederick, were stricken with fever. He returned to Sicily to recuperate. Pope **Gregory IX,** assuming this was only more procrastination, excommunicated Frederick. Relations between Pope and Emperor were already strained as a result of Frederick's attempts to incorporate his Kingdom of South Italy into the Holy Roman Empire. Without bothering to ask the Pope's forgiveness, Frederick started again for the Holy Land in 1228. Considering that the Emperor was being insolent, Gregory renewed the excommunication, declared Frederick's lands in south Italy forfeited, and proclaimed a Crusade against the Crusader. Papal troops invaded Naples.

Frederick meanwhile arrived in the Holy Land, where his excommunicated status caused most other Crusaders to refuse cooperation. His orders were accepted only by his own troops and by Hermann von Salza's Teutonic Knights. Frederick opened negotiations with Al-Kamil. Through able and adroit diplomacy, Frederick obtained Jerusalem, Nazareth, and Bethlehem, with a land corridor connecting Jerusalem to the coast. Going to Jerusalem, he crowned himself king (February 18, 1229). Returning to Italy (May), he drove the papal armies from South Italy, then made peace with the Pope (August).

Because it culminated merely in a deal between two sharp traders, some historians refuse to recognize Frederick's expedition as a true Crusade. Nonetheless, he accomplished more in his few months in the Holy Land than any other Crusader leader since Godfrey de Bouillon.

The Crusader States, 1229–1247

Fifteen years after Frederick had regained Jerusalem, and while the local Crusaders were preoccupied with petty bickerings, Jerusalem fell unexpectedly to a new Moslem tide from the East. The Crusader states were also pressed from the south by resurgent Egyptian power. The important events were:

King Louis IX (later St. Louis)

1229–1243. Discord. Constant bickering between Frederick (who never returned to the Holy Land) and the local Crusader barons, who became increasingly independent. Frederick's position was further undermined when he was again excommunicated in another quarrel with the Pope (see p. 405).

1239–1241. Reinforcements. Count **Theobald IV** of Champagne (and King of Navarre) sailed to Acre, despite papal prohibition (1239). (The Pope felt that the arrival of new Crusader strength in the Holy Land would strengthen Frederick's position; he preferred to have Jerusalem fall to the Moslems rather than to have Frederick keep it.) **Richard** of Cornwall, brother of Henry III, also defied the prohibition and sailed to the Holy Land (1240). He joined Theobald in fortifying Ascalon and in conducting a few unimportant raids against Moslem territory. Nothing else was accomplished.

1244. Fall of Jerusalem. Khwarezmians, fleeing from the Mongol conquerors of Persia (see p. 374), captured and sacked Jerusalem, and allied themselves with Egypt. The local Crusaders, allied with the Emir of Damascus, were decisively defeated by the Egyptians and Khwarezmians at the **Battle of Gaza.** An important part in the victory was played by the young Tartar Mameluke leader, Baibars.

1247. Mameluke Capture of Ascalon.

The Seventh Crusade, 1248–1254

Christian reaction to the Moslem capture of Jerusalem led King Louis IX (later St. Louis) of France to lead still another Crusade, moving again by way of Egypt. After occupying Damietta, Louis was repulsed in an effort against Cairo; his army was routed and he was captured. Later ransomed, Louis went to Acre, where he tried in vain to reestablish the power of the Kingdom of Jerusalem. The principal events were:

1245. Papal Appeals for Crusades. Innocent IV preached a Crusade to recapture Jerusalem, and simultaneously preached another Crusade against excommunicated Frederick. Louis IX tried in vain to mediate between pope and emperor.

1248. Louis Sails for the East. His fleet of 1,800 ships carried 20,000 cavalry and 40,000 infantry, mostly French. He spent the winter in Cyprus.

1249. Occupation of Damietta. Louis landed against slight opposition. The garrison abandoned the city in panic (June 6). Possibly remembering the difficulties encountered by the Fifth Crusade in attempting to advance on Cairo in midsummer, Louis waited until fall. This delay, however, permitted Sultan **Malik-al-Salih,** despite severe illness, to restore confidence in his panic-stricken troops and to prepare opposition.

1249, November 20. Advance toward Cairo. The Crusaders traveled only 50 miles in 4 weeks. They were stopped in front of the Ashmoun Canal, at the same point where the Fifth Crusade had been halted. The Moslem army, probably about 70,000 men, under Emir **Fakr-ed-din,** held the line of the canal. The core of that army was a force of 10,000 Mamelukes (see p. 422). The Sultan died, but this was kept secret, and one of his wives, **Shajar-ud-Durr,** ruled in his name.

1249–1250, December–January. Operations along the Canal. When Louis endeavored to build a causeway, the Moslems harassed the French with war engines, at the same time widening the canal by digging away the opposite bank. A surprise Egyptian attack was repulsed on Christmas.

1250, February 8, Battle of Mansura.* During the night Louis secretly took his cavalry across a ford which had been discovered about 4 miles from the causeway. The Moslems were com-

* Actually the village where this action took place was named Mansura—Victory—after the battle.

pletely surprised; Fakr-ed-din was killed. Louis's brother, **Robert** of Artois, leading the advanced guard, was supposed to seize the canal bank, opposite the causeway, to await the arrival of the main body of the cavalry and to cover a river crossing by the waiting infantry. He disobeyed orders. Rashly attempting to win the battle by himself, he pursued the Moslems into Mansura, where the effectiveness of his heavy cavalry was limited in street fighting. The Moslems rallied, annihilated Robert's division, and almost overwhelmed the remainder of the French cavalry. Louis was saved only by the timely arrival of his infantry over a bridge hastily constructed from the end of the causeway.

1250, February 11. Egyptian Counterattack. The exhausted Crusaders, hemmed in their bridgehead, barely held their position, making effective use of Greek fire.

1250, March–April. Retreat. His line of communications cut, his army wracked by disease, Louis retreated toward Damietta, harassed by the Moslems.

1250, April 6. Battle of Fariskur. Attacked by their pursuers, the French infantry broke into panic. The Moslems closed in, annihilating most of the army, and capturing Louis.

1250–1254. Failure. Louis agreed to pay 800,000 gold livres in ransom and to abandon Damietta (May). Sending most of the survivors back to France, he sailed to Acre. His further efforts, including attempts to negotiate with the Mongols, accomplished nothing (1250–1254).

The Crusader States, 1254–1270

The Crusader states became pawns in a great struggle between the new Mameluke Empire of Egypt and the Mongol conquerors of Persia. When the Mongol tide receded from Syria, Baibars, now Sultan of Egypt (see p. 422), began a systematic conquest of the remaining Crusader footholds in the Holy Land. The principal events were:

1258–1260. Mongol Conquest of North Syria. Ilkhan **Hulagu** (see p. 423) was preparing to move south when the death of Khakhan **Mangu** caused him to return to Karakorum, in accordance with the law of Genghis Khan (see

pp. 379, 423). He left a detachment, under the general **Kitboga,** to hold the conquered area. Apparently he also directed Kitboga to assist the Crusaders against the Mamelukes. **Bohemund VI** of Antioch allied himself with

Kirboga. The other Crusaders held aloof, finding it difficult to choose between the Moslem Mamelukes and the pagan Mongols.

1260. Battle of Ain Jalut. Kitboga, supported by a few Crusaders, was decisively defeated by a much larger Mameluke army under **Kotuz** near Nazareth. This was the high-water mark of Mongol expansion.

1260–1268. Mameluke Conquests. Under Baibars the Mamelukes expanded steadily, aided by bickering among the Crusader barons and the wars between Genoa and Venice, and between Venice and the Latin Empire of Constantinople. After the fall of Antioch (1268) only Acre, Tripoli, and a few scattered castles remained in Christian hands.

1269. Aragonese Crusade. James the Conqueror, responding reluctantly to papal pressure, led a futile expedition to Asia Minor. Deterred by storms, he abandoned the effort.

The Eighth Crusade, 1270

The disasters in the Holy Land inspired Louis IX to take the cross again. Baibars' intrigues led the French king to believe that the Bey of Tunis could be converted to Christianity; he decided to go to the Holy Land via Tunis and Egypt. His brother, **Charles** of Anjou, newly established king of Sicily, reluctantly agreed to join. The Tunisians resisted, and Louis besieged Tunis. Soon after, this expedition was swept by an epidemic, in which Louis died. Charles took command. Entering negotiations with the Bey of Tunis, Charles obtained tribute for himself and for France. Prince **Edward** of England (later Edward I) arrived, to discover that the Crusade had ended without further hostilities.

The Crusader States, 1271–1291

The Moslem reconquest of the Crusader lands was completed. Although there was some later sporadic activity, the fall of Acre ended the real Crusader era after two momentous centuries. The principal events were:

1271. Capture of Krak. After a long siege, Baibars captured the Knights Hospitalers' castle near Tripoli, probably the most powerful fortification in the world.

1271–1272. Arrival of Edward of England. With 1,000 men, he made some raids against the Moslems—once reaching Nazareth—and attempted negotiations with the Mongols. He was almost a victim of the Assassins. Edward, virtually the last of the great Crusaders, accomplished little.

1281. Battle of Homs. A Mongol raiding army, 30,000–50,000 strong, swept into Syria, where it was joined by a substantial force of Crusaders. Sultan **Kala'un,** successor of Baibars, led an army probably 100,000 strong to meet the invaders in an indecisive battle. Soon afterward the Mongols withdrew.

1289. Fall of Tripoli. Captured by Kala'un.

1291. Fall of Acre. Sultan **Khalil,** son of Kala'un, captured Acre after a long siege and a gallant defense. This ended the Crusader dominions of the East.

Aftermath of the Crusades, 1291–1396

During the century after the fall of Acre, the crusading spirit died a lingering death. The most important events were:

1299–1301. Mongol Invasion of Syria. They apparently expected Crusader cooperation and intended to reinstall the Christians in Jerusalem (see p. 424). When Christian help was not forthcoming, they withdrew, never to return to Syria.

1310. Knights Hospitalers Capture Rhodes. The island would become for more than two centuries the strongest bulwark of Christendom against the Turks.

1311. Suppression of the Knights Templars. Its members were exterminated by Philip IV of France.

1344. Capture of Smyrna. A crusading expedition of Knights Hospitalers, Cypriots, and Venetians captured Smyrna from the Ottoman Turks. It remained in Christian hands until it was captured and sacked by the Tartars of Timur (1402; see p. 425).

1365–1369. Crusade of Peter I of Cyprus. He sacked Alexandria (1365) and continuously ravaged the coasts of Syria and Egypt. His project ended with his assassination (1369).

1396. Crusade of Nicopolis. Pope **Boniface IX** preached a crusade against the Ottoman expansion in the Balkans. Duke **John the Fearless** of Burgundy, **Jean Bouciquaut** (Marshal of France), and a number of other nobles (mostly French) advanced into the Danube Valley. They were decisively defeated by the Turks at the **Battle of Nicopolis** (Nikopol, Bulgaria; see p. 412).

THE BYZANTINE EMPIRE

Wrecked and fragmented by the Fourth Crusade (see p. 412–414), the Byzantine Empire was reunited in reduced and weakened form in the latter half of the 13th century. The declining empire was shaken by ever-weakening, dying convulsions during the remainder of the period, while its territory was slowly absorbed by the Ottoman Empire. Important events were:

1202–1204. Fall of Constantinople to the Fourth Crusade. (See p. 413.)

1204. Establishment of the Latin Empire. Count **Baldwin** of Flanders became Emperor Baldwin I of Constantinople (1204–1205). **Theodore Lascaris** established himself as Emperor of Nicaea, the largest remnant of the Byzantine Empire (1206–1222). **Alexius Comnenus** established the Empire of Trebizond; his brother **David Comnenus** established a short-lived empire on the northern shores of Bithynia and Paphlagonia.

1205. Battle of Adrianople. The Bulgarians under **Kaloyan** defeated Baldwin and Doge **Dandolo** (see p. 412). Baldwin, captured, died in captivity.

1207–1211. Byzantine Civil War. Theodore Lascaris defeated David and Alexius Comnenus. The Seljuk Turks of Rum were first allied with Theodore, and later with the Comneni.

1214–1230. Reign of Theodore Ducas Angelus of Epirus. In a series of wars with his Latin, Byzantine, and Slavic neighbors, Ducas extended his control over most of Macedonia and Thrace.

1222–1254. Reign of John III, Ducas Vatatzes of Nicaea. He was an able ruler and excellent soldier in the Byzantine tradition. In successful wars against all of his neighbors, he made Nicaea a powerful, prosperous country.

1224. Three-Way War. Theodore of Epirus, John of Nicaea, and **Robert** of Courtenay, Emperor of Constantinople, fought a confused struggle. John defeated the Crusaders at the **Battle of Poimanenon.** He then sent an army across the straits to seize Thrace from Theodore. Meanwhile, Theodore had defeated the Crusaders at the **Battle of Serres.** In Thrace he then repulsed the Nicaean effort to capture Adrianople.

1229–1237. Regency of John of Brienne. The former King of Jerusalem (see p. 414) became regent for the boy emperor **Baldwin II.** The ablest of all the Crusader leaders, John restored Latin control over the approaches to Constantinople, and became coemperor (1231).

1230. Battle of Klokonitsa. Ivan Asen of Bulgaria defeated and captured Theodore of Epirus (see p. 412).

1236. Alliance of John of Nicaea and Ivan of Bulgaria. Their plan to attack Constantinople was thwarted by the arrival of a Venetian fleet and a Latin army from Achaia.

1244. Seljuk Pressure Relaxed. The Nicaean Empire was relieved as a result of Mongol victories over the Seljuks (see p. 374).

1246. Nicaean Conquest of Macedonia and Thrace. John of Nicaea led an expedition across the straits.

1254–1258. Reign of Theodore II Lascaris of Nicaea. Michael Asen of Bulgaria, hoping to take advantage of the death of John III, occupied much of Thrace and Macedonia. Theodore defeated Michael at **Adrianople** (1255) and, despite ill health, in two brilliant campaigns reconquered the lost territories.

1259–1282. Reign of Michael VIII Paleologus of Nicaea. An able general, he seized the throne from the child emperor John Lascaris.

1259. Battle of Pelagonia. The Nicaeans defeated an alliance of the Byzantines of Epirus with Manfred of Sicily and the Latin prince of Achaia.

1261. Byzantine Seizure of Constantinople. A small Nicaean army under General **Alexius Stragopulos** captured Constantinople from the Latins by surprise, and almost without opposition, during the absence of the Venetian fleet. Michael reestablished the Byzantine Empire with its capital in Constantinople, and formed an alliance with Genoa against Venice (see p. 403).

1261–1265. Byzantine Successes. Campaigns against the Latins in Greece, the Bulgarians in Thrace and Macedonia, and Epírus.

1267–1281. Wars with Sicily. Charles of Anjou attempted to expand into the Balkans, with the intention of capturing Constantinople (see p. 406). Michael won a great victory over the Angevins at the **Battle of Berat,** Albania

(1281). Charles made an alliance with Venice, the Pope, the Serbs, and the Bulgars for the final overthrow of the Byzantine Empire. Michael formed an alliance with Pedro of Aragon, supporting him in the War of the Sicilian Vespers, thus relieving Angevin pressure against the Byzantine Empire (see p. 406).

1282–1328. Reign of Andronicus II. Renewed Angevin and Turkish threats against Byzantium caused Andronicus to hire a small army of Catalan mercenaries under **Roger de Flor** (1302). Although the Catalans repulsed Turkish attacks, they terrorized Constantinople and other Byzantine territories. After the murder of Roger de Flor (1305), his troops became the scourge of Thrace and Macedonia, later seizing Athens and setting up an independent state (1311).

1317–1326. Siege of Bursa. (See p. 421).

1321–1328. Civil War. This was between Andronicus and his grandson, also named **Andronicus.** Much of the empire was devastated. Young Andronicus established himself as coemperor (1325), and finally deposed his grandfather (1328).

1325–1341. Reign of Andronicus III. After an initial defeat by Serbia (1330; see p. 411) he was generally successful against European neighbors and in the Aegean. Andronicus lost most of his remaining Anatolian possessions to the Ottoman Turks.

1341–1347. Civil War. This was between the supporters of **John V,** son of Andronicus (under the regency of his mother **Anna** of Savoy), and **John Cantacuzene,** rival claimant to the throne. Thrace and Macedonia were ravaged. Both sides called in Serbs and Turks in complex shifting alliances. The result was the loss of most of the dwindling possessions of the empire and the devastation of that which remained under Byzantine control.

1347–1354. Reign of John VI (Cantacuzene). He captured Constantinople through treachery. He called for Turkish assistance against Stephen Dushan of Serbia. The Ottomans repulsed the Serbs and at the same time established a foothold on the European shore at Gallipoli.

1355–1391. Second Reign of John V. He recaptured Constantinople and forced Canta-cuzene to abdicate. A weak ruler, he lost Adrianople to the Turks (1365) and was captured by Czar **Shishman** of Bulgaria (1366).

1376–1392. Confused Dynastic Struggle. The participants were John V, his sons **Andronicus IV** (1376–1379) and **Manuel II,** and grandson, **John VII** (1390–1391, 1399–1402). The Turks took advantage of the chaos to overrun most remaining Byzantine territory.

1391–1425. Reign of Manuel II. An able ruler, he inherited an empire consisting only of Constantinople, Thessalonika, and the Morean Peninsula.

1391–1399. Ottoman Siege of Constantinople. French Marshal **Jean Bouciquaut,** leader of a small force of volunteer soldiers and sailors, gained a respite by repulsing repeated attacks of Ottoman Emperor **Bayazid I** on land and sea (1398–1399).

THE OTTOMAN EMPIRE, 1290–1402

Near the middle of the 13th century, the Seljuk Turks were overwhelmed by the Mongols, and Turkish Anatolia disintegrated into chaos. Soon afterward, a new Turkish tribe, under Osman I (Othman), emerged into prominence in Anatolia. After establishing their supremacy over neighboring Turkish tribes, the Ottomans steadily whittled away at the declining Byzantine power and territory, taking advantage of incessant internal discord and intrigue within Constantinople. Near the middle of the 14th century, the Ottomans established a foothold in Europe, then rapidly overran Thrace, Macedonia, Bulgaria, and much of Serbia. However, they continued to be frustrated by the powerful defenses of Constantinople, where the Byzantine Empire was eking out a few years of miserable existence. Just as Constantinople was on the verge of surrender, the Ottomans were smashed, apparently beyond hope of recovery, by the Tartar conqueror **Tamerlane** (1402; see p. 425). The principal events were:

1230. Battle of Erzinjan. Seljuk Sultan of Rum, **Ala ud-Din Kaikobad,** defeated Khwarezmian Shah Jellaluddin, who had retained the western remnant of his empire after his defeat at the Indus (see p. 366). This defeat facilitated the Mongol advance through western Persia and into Anatolia in the next 7 years (see p. 423).

1243. Battle of Kosedagh. The Mongols defeated the Seljuk Turks, and established suzerainty over Anatolia (see p. 374).

1290–1326. Reign of Osman I. He culminated his successes against the Byzantines by besieging Bursa (1317), which fell shortly after his death (1326).

1326–1359. Reign of Orkhan I. He made Bursa his capital. He was the real organizer of the Ottoman Empire. The first of his many victories over the Byzantines was over Andronicus III at the **Battle of Maltepe** (1329). In the following years the Byzantines were eliminated from

most of their remaining footholds in Anatolia by the capture of Nicaea (1331) and Nicomedia (1337). While assisting John Cantacuzene, the Ottomans conducted their first expeditions into Europe (1345) and established their first permanent European settlement at Gallipoli (1354). In subsequent years they expanded rapidly in Thrace. The Byzantine Empire became a virtual Ottoman fief.

1359–1389. Reign of Murad I. Capturing Adrianople (1365), Murad made this his capital.

1366. Crusader Interference. A Crusade led by **Amadeus** of Savoy and Louis of Hungary was initially successful. Amadeus captured Gallipoli, but was later driven out. Louis defeated the Turks near **Vidin,** but was later forced to withdraw.

1369–1372. Ottoman Conquest of Bulgaria. c. 1370. Establishment of the Janissary Corps. Murad created this elite body of soldiery, at first almost entirely infantry archers, and consisting of former Christians captured in childhood and brought up as fanatic Moslems. For more than 500 years the Janissaries would play a leading role in Ottoman history.*

1371. Battle of Cernomen. Turkish victory over the Serbs on the Maritza River resulted in the conquest of Macedonia (see p. 411).

1389, June 20. Battle of Kossovo. Victory of

Murad and his son Bayazid over a coalition of Serbs, Bulgars, Bosnians, Wallachians, and Albanians under the leadership of Lazar of Serbia, who was killed in the battle. Murad was later assassinated by a Serbian prisoner.

1389–1402. Reign of Bayazid I.

1391–1399. First Ottoman Siege of Constantinople. (See p. 421.)

1396. Crusade of Nicopolis. (See pp. 412, 419.)

1397. Turkish Invasion of Greece.

1402. Battle of Angora. (See p. 425.)

* Oman suggests that Orkhan, rather than Murad, established the Janissaries.

Egypt, Syria, and the Mamelukes

The Ayyubid successors of Saladin briefly expanded his empire, then quickly declined, to be overthrown and replaced by their Mameluke slave-soldiers. During the latter part of the 13th century, the Mamelukes were engaged in a desperate struggle against the Mongols. Through a combination of luck, able leadership, and adoption of Mongol military practices, the Mamelukes halted the invaders at the gates of Egypt, and retained possession of most of Syria. During lulls in this struggle, they evicted the Crusaders from Syria and Palestine. For most of the following century they prospered in Egypt and Syria, expanding their control southward up the Nile Valley into Nubia. At the very end of the century, however, Tamerlane inflicted a crushing defeat on the Mamelukes and drove them from Syria. The important events were:

1200–1238. Ayyubid Expansion. Military success, expansion, and prosperity under sultans Abul Bakr Malik al-Adil and his son Malik al-Kamil.

1218–1221. The Fifth Crusade. (See p. 414.)

1231–1232. Ayyubid Invasion of Anatolia. This was repulsed by the Seljuk Turks.

1238–1240. Dynastic Struggle.

1240–1249. Reign of Malik-al-Salih. He devoted himself to the reconquest of Syria.

1249–1250. The Seventh Crusade. (See p. 416.)

1250. Mameluke Revolt. Turanshah, son of al-Salih, was overthrown during the Seventh Crusade by a conspiracy of Mameluke leaders and Queen **Shajar ud-Durr,** al-Salih's widow. Shajar married the Mameluke leader **Aidik,** the first Mameluke sultan. She later murdered Aidik, but was in turn killed by his adherents (1257).

1257–1260. Reign of Kotuz. Assisted by his lieutenant, **Baibars,** Kotuz defeated the Mongol general, **Kitboga,** at the **Battle of Ain Jalut** (see p. 423). Kotuz was then murdered by Baibars.

1260–1277. Reign of Baibars. The greatest of the Mameluke sultans. Of Tartar origin, he had served in the Mongol armies in his youth, and recognized the deadly threat the Mongols posed to his empire. He formed an alliance with **Bereke,** Mongol Khan of the Golden Horde, who had become a Moslem (see p. 379). From Bereke he obtained Mongol instructors for his army. This permitted Baibars and his successors to fight on even terms with the Mongols in Syria. Baibars' efforts to restore the Abbasid Caliphate in Baghdad were repulsed by the Mongols. In a series of deliberate campaigns he recovered most of the Crusader territory in Palestine and Syria (see p. 418). All of the North African Moslem states were tributary to him.

1277–1290. Reign of Kala'un (father-in-law of Baibars' son). He repulsed a Mongol invasion of Syria by a narrow victory at the

Battle of Homs (1281). He continued the recovery of Crusader territory.

1290–1293. Reign of Khalil Malik al-Ashraf (son of Kala'un). He completed the eviction of the Crusaders from the Levant.

1294–1341. Reign of Malik al-Nasir. Twice deposed from the throne during his childhood and youth, Malik eventually established himself firmly (1310) and under his rule the Mameluke domains expanded. He made peace with the Mongol Ilkhans (1322). By frequent raids up the Nile Valley, he strengthened Mameluke influence in Nubia. There was a decline after his death.

1381–1399. Revival of the Mamelukes. This was under warrior-Sultan **Bartuk Malik-al-Zahir.**

1400–1402. Tamerlane's Conquest of Syria. This began with his overwhelming victory at the **Battle of Aleppo** (October 30, 1400). Syria never recovered from Tamerlane's devastation.

PERSIA AND TURKESTAN, 1200–1405

Khwarezmian and Mongol Periods, 1200–1381

The conquests of the Khwarezmian Turks, begun by Takash (see p. 337), were continued by his son, **Ala-ud-Din Mohammed,** who expanded from Khorasan and Khwarezm to create a vast empire by conquering all of Persia, Transoxiana, and Afghanistan. Mohammed's glory was smashed by the Mongols under Genghis Khan. Mohammed's son, **Jellaluddin,** was defeated by the Mongols in two vain efforts to reestablish the Khwarezmian Empire. The Mongols held Persia and Turkestan for a century and a quarter, the powerful, autonomous Ilkhan Dynasty being formally established by **Hulagu,** grandson of Genghis. The Ilkhans expanded westward to conquer Mesopotamia, Transcaucasia, most of Anatolia, and much of Syria, but failed in repeated efforts to destroy the Mameluke sultans of Egypt. They were occasionally involved in fierce but indecisive struggles with their cousins, the Khans of the Golden Horde and of the Jagatai Mongols. During the last four decades of the Ilkhan period, central authority declined, and their empire became a collection of virtually independent Turkish and Tartar principalities. The principal events were:

1199–1220. Reign of Ala ud-Din Mohammed, Shah of Khwarezm.

1205. Battle of Andkhui (south of the Oxus). Mohammed defeated **Muhammad of Ghor** to begin Khwarezmian expansion into modern Afghanistan (see p. 429) and thence to take all of Persia.

1220–1221. Mongol Conquest of Khwarezm. (See p. 364.)

1221. Battle of the Indus. Genghis defeated Jellaluddin (see p. 366). Jellal fled into India, then returned to southern Persia, where he was later defeated by Mongol troops of **Ogatai Khan.** Next defeated by the Seljuks (1230; see p. 421), he soon after was assassinated (1231).

1231–1236. Mongol Conquest of the Remainder of Persia. To include northern Mesopotamia, Azerbaijan, Georgia, and Armenia.

1243. Mongol Conquest of Anatolia. (See p. 374.)

1255–1260. Conquests of Hulagu. He was the first of the Ilkhans, subject to the loose suzerainty of his brother Mangu, in Karakorum (see p. 374). He conquered southern Mesopotamia, where he stormed **Baghdad** and extinguished the Abbasid Caliphate (1258). He had begun the conquest of Syria when he was recalled to Karakorum at the death of Mangu (1260).

1260. Battle of Ain Jalut. Hulagu left his general Kitboga (Ket-Buka) in command west of the Euphrates. Learning of Hulagu's departure, the Mameluke leader Kotuz hastily gathered an army of 120,000 men at Cairo and invaded Palestine. Kitboga moved with two or three toumans (20,000-30,000 men) to meet the Egyptians near Goliath Wells. The Mongols were close to victory, and pursuing the fleeing Egyptians, when they were ambushed by Baibars and the Mamelukes. Kitboga was killed and the Mongols routed. The Mongol army was small, but this Moslem victory had great psychological significance.

1260-1304. Mongol Raids into Syria. (See pp. 419, 422.)

1261-1262. Civil War between the Golden Horde and Ilkhans. Bereke, supporting the Mamelukes (see p. 422), marched south against Hulagu. Bereke withdrew, after inconclusive skirmishes in the Caucasus, when threatened by Kublai (see p. 375).

c. 1294. Conversion of the Ilkhans to Islam.

1316-1335. Reign of Ilkhan Abu-Said. During most of his reign he was at war with Uzbeg, Khan of the Golden Horde (see p. 380), at one time advancing through the Caucasus into Russia as far as the Terek River before being repulsed by Uzbeg (1325).

1335-1381. Decline of the Ilkhans.

The Period of Tamerlane, 1381-1405

Timur (Timur the Lame, or Tamerlane) was born at Kesh (Shahr-i-sabz) near Samarkand (1336). His claim of descent from Genghis Khan is doubtful; he was apparently a Tartar, not a Mongol. By fighting and intrigue, at the age of 28 Tamerlane became vizier, or prime minister, of the khan of the Jagatai Mongols (see p. 374). He then overthrew the khan (1369). During the next ten years he fought wars with Khwarezm and Jatah (Eastern Turkestan). The capture of Kashgar gave Tamerlane control of Jatah (1380), and was the beginning of a quarter of century of ruthless conquest. The principal events were:

1381-1387. Conquest of Persia. Beginning with the capture of Herat (1381), Tamerlane continued with the conquest of Khorasan (1382-1385). He later overran Fars, Iraq, Azerbaijan, and Armenia (1386-1387). He massacred ruthlessly.

1385-1386. First War with Toktamish. A former protégé of Tamerlane, Toktamish, now Khan of the Golden Horde (see p. 380), invaded Azerbaijan and defeated one of Tamerlane's armies in that area (1385). Tamerlane repulsed the northern Mongols, but did not pursue.

1388-1395. Second War with Toktamish. Taking advantage of Tamerlane's absence in Persia, Toktamish raided into Transoxiana, defeated one of Tamerlane's lieutenants, and advanced against Tamerlane's capital of Samarkand. By forced marches exceeding 50 miles a day, Tamerlane returned to Transoxania; Toktamish and his allies hastily retreated northward.

1389. Battle of the Syr-Darya. Toktamish returned during the winter with a larger army, but was outmaneuvered and decisively defeated by Tamerlane. Toktamish retreated with the remnants of his army. Tamerlane prepared to invade the Khanate of the Golden Horde (1389-1390).

1390-1391. Tamerlane's Invasion of Russia. With an army that probably exceeded 100,000 men, Tamerlane marched northward. He had no political or strategic objective such as always characterized the operations of Genghis Khan. His intention was merely to seek and to destroy Toktamish. He was relying upon his own tactical ability and the superb fighting qualities of his mercenary—but completely devoted—army. Toktamish apparently deliberately lured Tamerlane west of the Ural River to exhaust and discourage the invaders. His strength is not known, but it was undoubtedly substantially larger than that of Tamerlane.

1391. Battle of Kandurcha (or of the

Steppes). The two armies met somewhere east of the Volga and south of the Kama rivers. The battle lasted for 3 days, with Tamerlane often close to defeat. By the third day Toktamish had destroyed Tamerlane's left wing and personally led an enveloping force to attack from the rear. Tamerlane seems to have gained the victory eventually by a ruse, causing Toktamish's army to believe that their khan was dead when in fact he was on the verge of victory. Discouraged, the northern Mongols and Tartars scattered into the steppes. Casualties are unknown; based upon one chronicle, it may be estimated that Toktamish lost more than 70,000, Tamerlane more than 30,000. Probably realizing that he was dangerously overextended, Tamerlane made no effort to continue into Russia, but returned to Persia to deal with trouble there.

1392. Revolt in Persia. Tamerlane defeated and killed rebel Shah **Mansur** in the **Battle of Shiraz.**

1393–1394. Conquest of Mesopotamia and Georgia. He captured Baghdad (1393). While in Georgia, Tamerlane was attacked by Toktamish, but drove him back north of the Caucasus Mountains. Tamerlane pursued.

1395. Battle of the Terek. Again on the verge of defeat, Tamerlane rallied his troops and finally drove Toktamish from the field. Tamerlane pursued relentlessly, up to the Volga Valley, westward to the Ukraine, and across much of central Russia, south of Moscow. Defeating all Mongol detachments that he found, he massacred and ravaged ruthlessly. He destroyed the cities of Astrakhan and Sarai (capital of the Golden Horde). He established a vassal khan to replace the fleeing Toktamish; the Golden Horde never recovered.

1398–1399. Invasion of India. Tamerlane's advance guard and right wing, under his grandson **Pir-Mohammed,** debouched into the Punjab to seize Multan (spring 1398). The left wing, under another grandson, **Mohammed Sultan,** marched by way of Lahore. Tamerlane himself, with a small picked force, traversed the highest regions of the Hindu Kush before turning south to join his main body east of the Indus (September). Killing and plundering, he marched against Delhi. He destroyed the army of **Mahmud Tughluk** in the **Battle of Panipat** (December 17; see p. 430). After massacring 100,000 captive Indian soldiers, he stormed Delhi. After several horrible days of slaughter, rape, plunder, and destruction, Tamerlane marched north into the Himalayan foothills, storming supposedly impregnable **Meerut,** then westward back to the Punjab, destroying everything in his path, then disappeared from India as suddenly as he came (March 1399). Probably no more senseless, bloody, or devastating campaign has ever been fought.

1400. Invasion of Syria. Tamerlane destroyed the Mameluke army at the **Battle of Aleppo** (see p. 423). He captured both Aleppo and Damascus, slaughtering many of the inhabitants.

1401. Capture of Baghdad. Another horrible massacre, punishment for the city's uprising after its previous capture (see above).

1402. Invasion of Anatolia. He defeated the sultan Bayazid of the Ottoman Turks in the **Battle of Angora** (July 20; see p. 425). He then captured Smyrna from the Knights Hospitalers (see p. 419). After overrunning all of Anatolia, and having received tribute from the sultan of Egypt and the Byzantine emperor, he returned to Samarkand (1404).

1405. Death of Tamerlane. Hoping to conquer an empire even larger than that of Genghis Khan, Tamerlane started for China, but died en route at Otrar.

AFRICA

NORTH AFRICA

Morocco

Early in the 13th century, the Almohad Dynasty began to break up, the principal cause being the rise of the Berber tribe of Beni Marin, which became the Marinid Dynasty, with capital at Fez. After consolidating control of Morocco, the Marinids began to expand eastward, while also sending numerous expeditions to Spain, where they temporarily halted the advance of Christianity. After a prolonged struggle they conquered the Ziyanids of western Algeria and briefly overran the Hafsid dominions of western Algeria and Tunis. Overextended, they soon lost their conquests to the resurgent Hafsids and Ziyanids, barely retaining control of their own turbulent kingdom of Morocco. The principal events were:

1201. Almohad Conquest of the Balearic Islands.

1215–1258. Breakup of Almohad Empire.

1217–1258. Marinid Conquest of Morocco.

1229–1233. Aragonese Reconquest of the Balearic Islands.

1258–1286. Reign of Yakub II. Greatest of the Marinids, he campaigned against the Ziyanids of Algeria (1270–1286) and also in Spain (1261, 1275, 1277–1279, 1284). In the latter campaigns he was allied with Alfonso X of Castile and León against Alfonso's insurgent son **Sancho.** The death of Alfonso left Yakub temporarily in control of most of Spain, but he was soon driven out by Sancho.

1286–1307. Reign of Yusuf IV. He was constantly at war with the Ziyanids of Tlemcen. During a siege of Tlemcen (1300–1307) he was assassinated, and the siege was raised.

1335–1337. Siege of Tlemcen. The Ziyanid capital was captured by **Ali V,** son and successor of Yusuf.

1347. Invasion of Tunisia. Repulsed by Hafsids.

1351. Dynastic Civil War. Ali was defeated and overthrown by his son **Faris I.**

1351–1359. Conquest and Loss of Algeria and Tunisia. Faris conquered the remaining Ziyanid dominions (1351–1357), then overran much of the Hafsid lands, including Tunis itself (1357). The defeated dynasties made common cause, drove out the invaders, and recaptured Tlemcen (1359).

1359–1400. Decline of the Marinids. Morocco was wracked by internal violence.

1399. Sack of Tetuán. This Castilian raid was linked with operations against Granada (see p. 406).

Western Algeria

The Zenata tribe, under **Abu Yahia Yarmorasen,** drove out the Almohads. Abu established the Abd-el-Wahid or Ziyanid Dynasty, which retained control of the region for the remainder of the period, save for two decades of Marinid occupation. The principal events were:

1236–1248. Rise of the Ziyanids. Almohad resistance collapsed with fall of Tlemcen (1248).

1248–1282. Reign of Abu.

1270–1359. Wars with the Marinids. (See above.)

Tunisia and Eastern Algeria

The Prince of Tunis, **Abu Zakariya,** established the Hafsid Dynasty by ejecting the Almohads. The Hafsids retained control of the region (which included Tripolitania) despite frequent internal uprisings, and numerous land and sea invasions. The principal events were:

1236. Ejection of the Almohads.
1270. French Invasion. The short-lived Eighth Crusade of Louis IX (see p. 418).
1280. Angevin Invasion. (See p. 405.)
1347–1359. Wars with the Marinids. (See above.)

1390. Siege of Mahdia. During a truce in the Hundred Years' War, a joint Franco-English force, under Prince **Louis** of Bourbon, invaded by sea and besieged Mahdia unsuccessfully for 61 days.

WEST AFRICA

The once-flourishing Empire of Ghana never recovered from an Almoravid invasion which had sacked its capital of Kumbi (1076; see p. 352). The breakup of Ghana into several smaller states was soon followed by the coalescence of power in three of these—Mali, Soso, and Songhai—by the beginning of the 13th century. By this time Mali and Songhai had been converted to Islam. During the 13th century, after some setbacks, Mali gradually established virtually complete control over most of West Africa, and the culture of its capital of Timbuktu became known even in Europe and the Middle East. The principal events were:

c. 1200–c. 1230. Ascendancy of Soso. This was largely attributable to its warrior king, **Sumanguru,** who sacked the ancient Ghanaian capital of Kumbi (1203) and temporarily conquered Mali (1224).

1235. Revival of Mali. King **Sundiata** overthrew Soso control, and in following years steadily expanded his territory at the expense of Soso, Songhai, and other nearby kingdoms.

c. 1270–c. 1285. Origins of Benin. This state, in modern southwestern Nigeria, began its rise to prominence under *oba* (king) **Ewedo.**

1307–1332. Reign of Mansa (Emperor) Musa (Kankan Musa) of Mali. Under this ruler, the grandson or grandnephew of Sundiata, the Empire of Mali reached its height. Much of Musa's fame derives from his famous pilgrimage from his capital at Niani on the Niger to Mecca (1324–1325), reputedly accompanied by 60,000 retainers and 80 camels bearing 12 tons of gold. His lavish spending in Cairo depressed the price of gold there for over a decade. After his return, Musa built the magnificent Sankoré mosque in Timbuktu, where he also founded the University of Sankoré.

1325. Mali Conquers Songhai. During Musa's pilgrimage, his general **Sagmandia** (Sagaman-dir) conquered the Songhai Kingdom, capturing its capital at Gao and extending Mali's control below the Niger bend.

c. 1340–c. 1360. Rise of Hausa States. During mid-century several city-states in modern northern Nigeria, linked by a loose confederation, rose to regional prominence. The seven inner, or "true," states (*Hausa Bakwai*) included Biram, Daura, Kano, Katsina, Gobir, Rano, and Zaria; the seven outer satellites (*Banza Bakwai*) were Zamfara, Kebbi, Yauri, Gwari, Nupe, Kororofa, and Yoruba. They converted to Islam later in the century, but non-Islamic influences remained strong until the 1800s.

c. 1350. The Kongo Kingdom. This arose around modern Songololo, and was ruled by the *manikongo* (king), elected from adult males of suitable matrilineage.

c. 1380. Kanem-Bornu. Raids and attacks by the Bulala people forced the Sef Dynasty to abandon Kanem and move their kingdom to Bornu, west of Lake Chad. They made their capital at Birni Ngazargamu.

EAST AFRICA

During the 13th and 14th centuries Arab traders continued their activities along the East African coast, while Moslem armies from Egypt conquered more of the Sudan. Ethiopia remained independent, and fended off repeated Moslem attacks. Among the major events in the region were:

1270. End of the Zagwe Dynasty in Ethiopia. Amhara princes in the Lake Haik, Geshen, and Ambasel regions rebelled and deposed the last Zagwe king. The first monarch of a new dynasty, **Yekuno Amlak** (ruled 1270–1285), moved the capital south to Shewa (Shoa) province in central Ethiopia.

c. 1275–c. 1350. Arab Raids Ruin Maqurrah. These raids, combined with repeated Mameluke military expeditions, doomed the Christian Maqurrah kingdom. By the mid-13th century, Arab immigrants of the Juhaynah people were settling and intermarrying with the Nubians, ending old social patterns.

1314–1344. Reign of Amda Tseyon (Seyon). This redoubtable monarch made great efforts to expand Ethiopian dominions. He defeated four coalitions of Moslem rebels, conquered and made tributary the Moslem state of Ifat (eastern Shewa province to Zeila on the Red Sea) (1328), and intervened with the Mameluke Sultan Malik Al-Nasir (see p. 423) on behalf of Coptic Christians in Egypt.

SOUTH ASIA

NORTH INDIA

By the beginning of the 13th century, Muhammad of Ghor had conquered most of Hindustan. Soon after this, however, his empire was shaken by his defeat in northern Khorasan by Ala-ud-Din Mohammed, Shah of Khwarezm (see p. 423). While restoring order in Ghazni and India, he was assassinated and the Ghuri Empire collapsed. The northwestern regions were absorbed by Mohammed Shah, who installed his son Jellaluddin as Viceroy of Ghazni. In Hindustan, after a period of turmoil, Moslem control was perpetuated by the so-called Slave Dynasty of the Sultanate of Delhi. During the remainder of the century, periods of stability in north India were punctuated by anarchic dynastic changes, Mongol raids, Hindu uprisings, and endemic civil war among the Turkish nobility. At the turn of the century, under able Turkish Sultan **Ala-ud-Din,** they conquered the Deccan, to unite most of the subcontinent under one rule. During the subsequent reign of **Mohammed bin Tughluk,** this great empire began to fall apart. At the very end of the century a strong man appeared on the scene, but not to restore order. In six months **Tamerlane** inflicted on India more misery than had been caused by any other conquerer, and destroyed the Sultanate of Delhi. The principal events were:

1200–1206. Ikhtiya-ud-Din Muhammad Bakhtiyar Khalji. This military adventurer used his junior officer's salary from the Ghurid General Qutb-ud-Din (see p. 348) to gather a retinue. Based near Oudh and Banares, he raided aggressively into Magadha lands to the east. Raising an army with the loot, he gained Qutb's permission to continue operations on his own resources. He conquered Magadha and Bengal (1201–1203), but his campaign into Tibet was a disaster (1204–1205); he died in Bengal shortly after his return (1206).

1202. Conquest of Kalinja. Qutb-ud-Din defeated the forces of the Chandela (Candella) kingdom of Bundelkhand.

1203. Muhammad Becomes Emperor of Ghor. He succeeded his elder brother **Ghiyas ud-din.** His empire extended from the Caspian and Aral seas to the Arabian Sea and the Bay of Bengal.

1205. Battle of Andkhui. Muhammad was defeated south of the Oxus River by Mohammed Shah of Khwarezm. Later in the year Muhammad of Ghor marched to India, suppressing uprisings en route and defeating the rebellious Rajput Hindus (November).

1206. Assassination of Muhammad. This was followed by anarchy.

1207–1210. Establishment of the Sultanate of Delhi. This was done by Qutb-ud-Din, who took the name **Aibak** when he gained the throne. He was the first of the so-called Slave (or Mamluk) Dynasty. His death in a polo accident was followed by turmoil.

1210–1236. Reign of Iltutmish. After restoring order and reconquering rebellious provinces, this Turkish general and son-in-law of Qutb gained control of all North India as far south as the Narbada River.

1216. Battle of Taraori. In the Punjab, Iltutmish defeated the invading army of **Taj-ud-din Yildiz** of Ghazni (who had been driven from his kingdom by the Khwarezmians).

1221. Flight of Jellaluddin. Driven from Ghazni by Genghis Kahn (see p. 366), the Khwarezmian prince fled into the Punjab with the remnants of his army. Iltutmish refused him asylum and pursued Jellaluddin southward through the Indus Valley, then westward into Persia. After a few raids into the Punjab, Genghis Khan withdrew without further action.

1236–1246. Disorder in the Delhi Sultanate. In the decade after Iltutmish's death, 4 of his children (including 1 daughter, **Raziya**) gained the throne but were each in turn deposed. Real power rested with a group of Iltutmish's slaves called **the Forty,** who maintained state power while battling Rajput rebels and Mongol incursions.

1238–1264. Reign of Naruseinha I of Orissa. He established the only Hindu dynasty which successfully resisted later Turkish invasions of east and south India.

1241. Mongol Raid into the Punjab. Taking advantage of turmoil in North India, the raiders captured Lahore.

1246–1266. Reign of Nasir-ud-din. He restored order in north India, largely through the ability of his chief minister, **Ghiyas ud din Balban.** Balban repulsed several Mongol raids and suppressed a number of rebellions.

1266–1287. Reign of Balban. He revitalized the Sultanate, reducing the power of the nobles.

His reign was beset by a number of Mongol raids. Upon his death North India reverted to anarchy as the Turkish nobility struggled for power, and Hindu princes of Malwa and Gujarat, among others, threw off the hated Moslem yoke.

1290–1296. Reign of Jalal-ud-din Firuz Khalji. He was founder of the Khalji, or Afghan, Dynasty. He repelled another Mongol invasion (1292). His nephew and son-in-law, **Ala-ud-din Khalji,** led the first Moslem expedition south of the Narbada River, capturing and plundering Devagiri, the capital of Yadava (1294). Firuz heaped honors on Ala-ud-din, who demonstrated his gratitude by murdering the Sultan and seizing the throne.

1296–1316. Reign of Ala-ud-din. An able and ruthless monarch.

1297–1306. Mongol Invasions. The invaders ravaged Punjab (1297–1298), but were twice defeated before Delhi by Ala-ud-din (1299, 1303). The repulse of further invasions in 1305 and 1306 ended Mongol aggression in India.

1300–1305. Reconquest of Malwa and Gujerat. Noted for the hard-fought sieges of the Hindu fortresses Ramthanbor (1301) and Chitor (1303). Ala-ud-din finally captured Chitor after the hopeless defenders performed the rite of *jauhar,* casting their wives and children on a great funeral pyre, then sallying from the fortress to die in a suicidal attack on the besiegers.

1307–1311. Moslem Conquest of the Deccan. General **Malik Kafur** took Devagiri (1307), Warangal (1309–1310), Hoysala (1310), and Pandya (1311). Only Orissa remained independent of the Moslem Sultan of Delhi.

1316–1320. Disorder in North India. A succession struggle after Ala-ud-din's death ended with General **Malik Kufar** on the throne (1316), followed by Ala-ud-din's third son, **Qutb-ud-din** (1316–1320). Qutb was murdered by one of his generals, who was deposed by **Ghazi Malik** (made warden of the western marches by Jalal-ud-din Khalji), enthroned as **Ghiyas-ud-din Tughluk.**

1320–1325. Reign of Ghiyas-ud-din Tughluk. Accepted by the warring Turkish nobles, this strong general restored order, suppressing rebellions in Warangal and Bengal. He was killed in an "accident," reputedly engineered by his son, Mohammed.

1325–1351. Reign of Mohammed bin Tughluk. This ruthless intellectual was compelled to buy off an invading Mongol army under **Tarmashirin,** Khan of Transoxiana, which had advanced to the gates of Delhi (1329). Mohammed then planned to conquer Persia, but was unable to cope with the logistical problems of moving a great army across the intervening deserts and mountains. Later he sent another large army into the mountains between India and Tibet, forcing the hill tribes there to acknowledge his suzerainty, but losing nearly 100,000 men in the process (1337). Meanwhile, his dominions in the Deccan began to break away (1331–1347). Mohammed died while battling a revolt in Sind (1351).

1351–1378. Decline of the Sultanate of Delhi.

1398–1399. Tamerlane's Invasion. (See p. 425.) North India, never so thoroughly devastated before or since, was in utter chaos. Delhi did not rise from its ruins for more than a century.

SOUTH INDIA

The old multipartite conflicts continued between the various Hindu and Hindu-Dravidian dynasties of the Deccan. All of the competing Hindu dynasties disappeared, however, following the conquests of Ala-ud-din and his general Kafur at the beginning of the 14th century. The Moslem tide was reversed by the rise of the new Hindu Kingdom of Vijayanagar, continually at war during the latter part of the 14th century with the new Moslem Sultanate of Bahmani. These conflicts were typical of Moslem-Hindu warfare; the Bahmanis were essentially cavalrymen, while Vijayanagar, suffering from the historical difficulties of raising horses in the southern Deccan, relied primarily upon infantry. The principal events were:

1200–1294. Incessant Wars. Participants were the Yadava, Kakatiya, Hoysala, and Pandya; the Hoysala were the most successful.

1294–1306. Moslem Expeditions into the Northern Deccan. (See p. 429.)

1307–1311. Conquest of the Deccan. By Malik Kafur (see p. 429).

1311. Revolt of Madura. The Moslem governor established an independent sultanate in southern Deccan.

1336. Creation of the Vijayanagar Empire. Founded in modern Mysore by the brothers **Harihara** and **Bukka.**

1345–1346. Revolts in the Central Deccan. Moslem nobles revolted against the Dehli Sultanate, notably at Daulatabad (Deogir), Warangal, and Gulbarga.

1347. Independence of Bahmani Sultanate. Deccan rebels united under **Ala-ud-Din Bahman Shah,** who made his capital at Gulbarga.

1350–1400. Wars between Vijayanagar and the Bahmani. Though the Moslems won several notable victories (1367, 1377, 1398), the Hindu kingdom maintained its independence and prospered.

1370. Vijayanagar Conquest of Madura. Bakku (r. 1356–1377) thus gained control of all southern India.

CEYLON (SRI LANKA)

Native Sinhalese were restive under increasing south Indian influence and domination by rulers of foreign extraction in the capital of Polonnaravu (east-central Ceylon, about 60 kilometers southwest of Trincomalee). This led to a partial Sinhalese migration to relatively inaccessible southwest Ceylon, and the establishment of a new Sinhalese kingdom, with capital at Dambadeniya, about 110 kilometers southwest of Polonnaravu. Endemic hostilities ensued between the Sinhalese, in the south, and Indians (mainly Kalingas) in the north, with the Sinhalese generally successful. The principal events were:

1232–1236. Sinhalese in Southwest Ceylon. Under King **Vijiyabahu III,** the Sinhalese withdrew from Ceylon's dry zone in the north, and founded a new capital at Dambadeniya.

1236–1270. Reign of Parakamabahu II. The Dambadeniya Kingdom's power reached its height during his reign. He drove the Kalingas from northern Ceylon, and repelled invasions by the Malay ruler **Chandrabanu** (1242, 1258).

c. 1310. Tamil Kingdom at Jaffna. The Arya Chakavarti Dynasty founded a kingdom on the Jaffna peninsula in the early part of the century.

SOUTHEAST ASIA

THE TAI

The slow and gradual migration of the Tai into Southeast Asia was accelerated by the overthrow of the Nanchao (Tai) Kingdom in Yunnan by Kublai Khan (1252–1253; see p. 374). Pressing into the Irrawaddy and Menam valleys, the Tai (or Shan) pushed back Burmese, Mon, and Khmer. By the end of the 13th century, the many small Tai states in the Menam and Mekong valleys began to coalesce into larger states. The city and kingdom of Ayuthia were established in the middle of the 14th century, the genesis of modern Thailand. The latter part of the century was a period of continuous wars between Ayuthia and its neighbors, particularly against the Tai state of Chiengmai and the Khmer Kingdom of Angkor (see p. 351). The principal events were:

1283–1317. Reign of Rama Khamheng, of Sukhot'ai (in the Upper Menam Valley). This conqueror expanded to the southwest far down the Malay Peninsula (c. 1295) and northeast to the Mekong River near Luang Prabang. He was a loyal ally of Mongol China. His kingdom declined rapidly under successors.

1350–1369. Reign of Ramadhipati. He founded the city and kingdom of Ayuthia. He expanded rapidly in all directions, and conquered Sukhot'ai. He also warred frequently with the Tai state of Chiengmai and with the Khmer of Cambodia. He is considered the first King of Siam or Thailand.

1369–1388. Reign of Boromoraja I. He suppressed a rebellion of Sukhot'ai (1371–1378). His forces invading Chiengmai were defeated and repulsed at the **Battle of Sen Sanuk,** near Chiengmai.

VIETNAMESE REGION

The old and bitter conflict between Champa and Annam, quiescent for a century, was renewed and prosecuted bitterly on both sides. In a brief hiatus in the struggles late in the 13th century, the two small nations united to oppose successfully a series of Mongol invasions. The principal events were:

1203–1330. Khmer Occupation of Champa. This ended with the voluntary withdrawal of the Khmer, probably because they found themselves overextended in their struggles with the Tai (see above).

1220–1252. Reign of Jaya Parmesvaravarman II of Champa. He renewed the old war with Annam over the still-disputed frontier provinces (see p. 294, 350). The conflict was drawn out and indecisive. He was killed during an Annamese invasion by King **Tran Thaiton.** Both sides made peace.

1257. Mongol Invasion of Annam. Kublai sent his general **Sogatu** to conquer Champa. Apparently Sogatu was able to advance through Annam without serious opposition, but was unable to overcome the Cham, who retired into the mountains to carry out a prolonged guerrilla war.

1285. Mongol Disasters. Togan, son of Kublai,

led an army into Annam to assist Sogatu. He captured Hanoi, but then was defeated and repulsed by the Annamese. Meanwhile Sogatu, attempting to join Togan, was frustrated by the Cham and Annamese. Driven back to Champa, he was defeated and killed by the Cham.

1287. Final Mongol Invasion. The invaders captured Hanoi, but were unable to progress farther in the face of determined Annamese resistance under King **Tran Nhon-Ton** (1278–1293). Both sides agreed to a face-saving solution. The kings of Champa and Annam acknowledged the suzerainty of the emperor; Kublai was happy to stop the costly invasions.

1312–1326. Renewed War between Champa and Annam. Champa was defeated and annexed by Annam (1312). Combined Cham-Annamese forces then repulsed an invasion by the Tai of **Rama Khamheng** (1313). This was followed immediately by Cham rebellions against the Annamese. Finally **Che Anan** drove out the Annamese and became king. There was a quarter-century of peace between the two exhausted countries.

1353. Cham Invasion. The Annamese repulsed them from the disputed province of Hué.

1360–1390. Reign of Che Bong Nga of Champa. A great soldier, he was incessantly at war against Annam. He captured and sacked Hanoi (1371). The Annamese continued guerrilla resistance, and at one point King **Tran Due-Ton** attempted a counteroffensive into Champa, only to be defeated, killed, and his army annihilated by Che at the **Battle of Vijaya** (1377). Despite his many victories, Che never completely pacified Annam. He was killed in a sea battle off the coast against Annamese and Chinese pirates.

CAMBODIA

At the beginning of the century the great King of Angkor, **Jayavarman VII,** conquered Champa (see p. 351). A few years later, he was compelled to withdraw to meet the increasing pressure of the Tai from the west. During the remainder of these two centuries the kings of Angkor were constantly on the defensive against the Tai, losing their outlying provinces in the Mekong and Menam valleys, and hardpressed to maintain themselves in their home territories (see p. 431).

BURMA

The first half of the 13th century was relatively uneventful. Then the accelerated Tai (Shan) migration from the north (see p. 431) brought increasing friction between Burmese and Shan. Soon after this the King of Pagán refused to acknowledge the suzerainty of Kublai Khan. Almost simultaneously with an outbreak of internal war between Burmese and Shan, a Mongol invasion swept through Burma to capture Pagán and topple the Burmese dynasty. During the remainder of the period Burma was wracked by constant internal warfare. The principal events were:

1271. Burmese Refusal of Tribute to Kublai Khan. Two years later King **Narathihapate** executed a Mongol ambassador repeating Kubai's demands (1273). Kublai, occupied with other affairs in his vast empire, took no immediate punitive action.

1277. Burmese Raid into Kanngai. This was a border state owing allegiance to China. The Governor of Yunnan sent 12,000 Mongol troops, who decisively defeated 40,000 Burmese invaders at the **Battle of Ngasaunggyan.** Their horses frightened by Burmese elephants, the Mongol archers dismounted to fight on foot. After dispersing the elephants, they remounted to make a decisive charge. The Mongols raided into Burma as far as Bhamo.

1283. Battle of Kaungsin. Repulsing another Burmese border raid, the Mongols again invaded and defeated the Burmese near Bhamo, and established outposts along the upper Irrawaddy.

1287. Mongol Invasion. Prince **Ye-su Ti-mur,** grandson of Kublai, swept through Burma to capture Pagān and establish a puppet regime.

1299. Shan Revolt. The Burmese puppet of the Mongols was overthrown.

1300. Last Mongol Invasion. Planning to reestablish control, the small invading army was held up by the stubborn defense of the fortified Shan town of **Myinsaing.** Accepting a large Shan bribe, the Mongol commander withdrew. He was executed by the governor of Yunnan, but the Mongols did not return.

1365. Capital at Ava. The principal Shan chieftain, **Thadominbva,** established the new capital of a mixed Shan-Burmese kingdom on the middle Irrawaddy River. This became the true successor of the Kingdom of Pagān.

1368–1401. Reign of Mingyi Swasawke. Ava was almost constantly at war with Pegu, with its Shan neighbors, with a new Burmese kingdom at Toungoo, and with the Arakanese states on the western seacoast.

INDONESIA

During the 13th century the power of Srivijaya was reduced in the Malay Peninsula and in the islands partly by the Tai and partly due to revolts and encroachments of Indonesian neighbors. Of these, the Kingdom of Singosari of Java expanded over much of Indonesia in the late 13th century. Almost simultaneously with a Javanese revolt which overthrew this kingdom came a Mongol invasion. The Mongols, in alliance with the heir of the overthrown Singosari Kingdom, soon conquered Java, but then encountered such serious guerrilla opposition from their former allies that they finally withdrew to China. During the middle of the following century, under the great minister **Gaja Mada,** the Singosari-Majapahit Kingdom again established control over most of Indonesia. By the close of the century this empire was seriously threatened by the growing strength of Islam in the Malay Peninsula and western Indonesia. The principal events were:

1222. Rise of Singosari. It overthrew the Kediri Kingdom of western Java.

c. 1230–1270. Reign of Dharmaraja Chandrabam of Ligor (in Malaya). He evidently threw off domination by Srivijaya, and established a powerful maritime kingdom. He sent expeditions to Ceylon (1247 and 1270). His successors succumbed to the Tai conqueror Rama Khamheng of Sukhot'ai (c. 1290; see p. 431).

1275–1292. Reign of Kertanagara of Singosari. He evidently began a plan to conquer the island possessions of decaying Srivijaya.

1292. Kediri Revolt. Kertanagara was killed by Kediris under **Jayakatwang.**

1292–1293. Mongol Invasion of Java. Prince **Vijaya,** son-in-law of Kertanagara, formed an alliance with Mongol Admiral **Yi-k'o-mu-su** which overthrew Jayakatwang and conquered Java. Vijaya then turned against the Mongols, defeating several detachments in a guerrilla war. The Mongols finally withdrew after Vijaya recognized Mongol suzerainty. He established his capital at Majapahit.

1295–1328. Series of Rebellions. These were against Vijaya and his son **Jayanagara.** They were finally suppressed by the young officer **Gaja Mada** (1319).

1330–1364. Ascendancy of Gaja Mada. As prime minister, he was actual ruler of the kingdom. This was a period of conquest and expansion in neighboring islands of Indonesia and on the Malay Peninsula. By the time of his death, West Java, Madura, and Bali were firmly held, and a large number of coastal city states throughout Indonesia paid tribute.

XII

THE END OF
THE MIDDLE AGES:

1400–1500

MILITARY TRENDS

GENERAL

The 15th century was an epoch of change. The same scientific, cultural, economic, and social forces that inspired the Renaissance, the sudden initiation of overseas exploration and colonialism, and changing political patterns throughout the world—particularly in Europe—also definitely affected warfare. The result was an era of military uncertainty and of blundering experimentation.

No single nation or military leader dominated the world. There were, however, a number of competent soldiers, such as **Henry V** of England, **John Ziska** of Bohemia, **Janos Hunyadi** of Hungary, and the Ottoman Sultans **Murad II** and **Mohammed II. Joan of Arc** of France was not really a military person and—save for her ability to use the moral factor in war—lacked any real understanding of military affairs.

Historians generally agree that the Middle Ages ended with the close of this century. This was no abrupt change, however, and was symbolized by the following, primarily military, events: the fall of Constantinople, the end of the Hundred Years' War, the end of the Wars of the Roses, and Charles VIII's invasion of Italy.

Each of these climatic events was greatly influenced by gunpowder weapons—the Wars of the Roses less than the others. This growing effectiveness of gunpowder weapons was the main theme of the century's military history, although there were two others of importance: the reappearance of military professionalism in Western warfare after an absence of more than a millennium, and the growing importance of infantry. By the end of the period these three themes foretold the subsequent combined-arms concept of the cavalry-infantry-artillery team.

One example of the appearance of professionalism was the resumption of theoretical studies of warfare, almost unknown since the time of Vegetius. Representative of this new intellectual approach to military affairs were treatises on war and on chivalry by **Christine de Pisan.**

In the politico-military sphere, ancient Byzantium finally succumbed to revitalized Islam, opening Southern and Eastern Europe to the Moslems. But the breach was blocked for a century by the Hungarian descendants of the Magyars.

Armor

Personal armor reached the height of development at the very time when the three major military trends of the period were assuring its obsolescence. The still-increasing weight of the armor made the knight (or man-at-arms, or heavy cavalryman) ever more ponderous; yet after midcentury the tempered steel plates of even the finest German and Italian armor could no longer stop the new and increasingly powerful missile weapons of gunner and foot soldier. Furthermore, the frustrated, encumbered cavalryman needed more—not less—mobility and flexibility in dealing with dangerous hostile infantry formations, or working in cooperation with his own infantryman.

As a result of these imperatives, by the end of the century plate armor had become important more as a prestige symbol than for defense or protection. There can be no better example of the flux or confusion in military art than what was then happening to armor. It was at its very heaviest, in a vain effort to protect from small-arms fire, while it was being lightened or discarded by realistic cavalrymen seeking mobility. The polished plates of tempered steel were shaped to assure the greatest protection by providing a glancing surface, while (sometimes on the very same suit of armor) they were also being embossed and etched with magnificent designs, which necessarily reduced tensile strength and impaired the tangential characteristics.

Weapons

General

By the beginning of the century, the practice of the heavy cavalry of fighting on foot had caused some modification and diversity in the weapons used by men-at-arms when dismounted. The sword was generally left sheathed, and the scabbard was now often strapped to the saddle instead of on the knight's belt. Because of the greater protection of armor, the men-at-arms usually carried battering weapons combining great weight with a sharp cutting edge. Although the spiked mace was still used, even more popular was some kind of poleax, with the glaive—a sort of cleaver or broad-bladed ax—and the halberd among the most favored.

But fearsome though these implements of slaughter were, the new gunpowder weapons were rapidly proving to be even more effective in smashing through armor plate. By the end of the century even the most conservative knights had to admit that commoners serving gun-powder weapons had become the arbiters of the battlefield. The preeminence of the aristocracy, shaken by English yeomen at Crécy and by Swiss mountaineers at Sempach, was now ended forever. The power and potentialities of these new weapons were so obvious that they were generally adopted throughout Europe, despite the problems imposed by powder that had to be protected from the weather, by clumsy and cumbersome methods of combustion, and by the dangers of explosion posed to those who handled the weapons.

Strangely enough, the two military systems most responsible for restoring the ascendancy of the infantry were the slowest to adopt the new gunpowder weapons which reinforced that ascendancy. These were the English, satisfied with their still-efficient longbow; and the Swiss, who relied primarily upon their deadly pikes. The French, who developed the finest artillery and artillerymen of the era, were strangely slow in adopting gunpowder small arms; it would take a disaster at Pavia in the following century (see p. 517) to make them realize the dangers of their almost total dedication to cannon. The Spanish seem to have done more than other nationalities to exploit both cannon and small arms, while minimizing the limitations of these weapons by continued utilization of pike and sword, and by skillful employment of field fortifications. As a result, at the close of the century Spain had the finest and best-balanced military force in the world.

Small Arms

At the outset of the century there was no real distinction between artillery pieces and small arms, other than size and portability. Those iron tubes small enough to be carried onto the battlefield by an infantry soldier were called "fire sticks" or "hand cannon." They were usually about 2 or 3 feet in length, with a bore of a half inch, or greater, and frequently the tubes were strapped to a pike handle. Later, for convenience the butt ends of these hand cannon were often fitted with short wooden sleeves, or cases, which could be held under the armpit or, more usually, rested on the ground or on objects of convenient height, such as a rock, a window, a wagon, or a forked stick. The Hussites were apparently the first to attempt to aim such pieces from the shoulder.

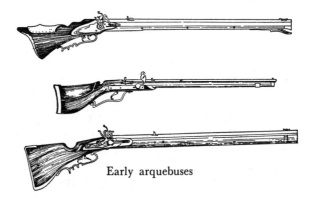

Early arquebuses

Powder, poured into the tube, was held in place by cloth or paper tamping, as was the projectile (an iron or leaden ball or slug), pushed in on top of the powder. At the beginning of the century, this powder charge was ignited, as in the case of larger artillery pieces, by means of a slow-burning "match": a cord or tightly twisted rag which had been soaked in saltpeter and then dried out. Like the "punk" of a modern Fourth of July, this would smolder (unless extinguished by rain) and would ignite priming powder sprinkled in the "pan" (a slight depression on top of the tube); the pan was connected to the powder charge inside by a touchhole, also filled with priming powder. The difficulties of loading, aiming, and firing such a weapon are self-evident. Accuracy in range and in direction was impossible. Thus these weapons had little direct effect on tactics or the conduct of war. They could supplement, but not replace, the more manageable longbow and crossbow.

Sometime before mid-15th century these crude noisemakers were transformed into weapons of significance by development of the matchlock. An S-shaped mechanism attached to the side of the gun, referred to as a "serpentine," held a burning match, usually of cord, which could be applied to the priming powder of the pan and touchhole (moved to the side of the gun for convenience and also for protection from rain) by means of a trigger. Thus the operator of the gun could hold the wooden butt piece against his shoulder, while aiming and firing the weapon in much the same manner as a modern shoulder gun. Sometime after midcentury this led to another improvement, in which the butt piece became a curved or bent stock, which could be held against the shoulder more conveniently and more comfortably. Because of this crooked stock, the modified weapon was called a "hackenbüsche" (in German), "hachbut" (in English), or "arquebus" (in French)—literally meaning "hookgun." Here, for the first time, was a truly usable small-arms gunpowder weapon, as manageable as crossbow or longbow, and with substantially greater destructive power than either—although its effective range of barely 100 yards was less than that of the bows. This remained the standard infantry weapon for the next century.

Artillery

By the beginning of the century practically every European army had some artillery weapons, and cannon had been introduced into east Asia by the Turks. These simple tubes—mostly cast iron or wrought iron, but sometimes built up from iron bars encased with hoops—were almost exclusively used for seige operations, or for the defense of fortifications. Moved from place to place on ox-drawn sledges, these bombards, as they were mostly called, had to be emplaced on mounds of earth or on locally constructed log platforms. Thus they were totally unsuited for mobile warfare and could not be considered as prototypes of field artillery. Projectiles were either iron or stone. The art of cutting stone cannon balls was well developed by midcentury.

As the techniques of casting and gunmaking improved and the methods of handling the crude and unstable powder mixture were perfected among "master gunners," bombards became increasingly effective in the attack and defense of fortified places. They grew in size so that the Turkish bombards at the siege of Constantinople included 70 heavy pieces, of which 12 were particularly large, including one 19-ton monster, firing stone balls up to 1,500 pounds in weight for a distance of more than a mile. Because of the effects of recoil, and the need to replace the cannon in position after each shot, these weapons could fire only about 7 rounds per day.

Most cannon were muzzle-loaders, but breechloading bombards were not uncommon. The breechloaders were prone to accident, however, due to lack of adequate sealing (or obturation, as modern artillerymen call it) to prevent explosive gases from escaping to the rear, and thus soon became unpopular. All weapons were smoothbore, although the first experiments with rifling may have begun in Germany at the end of the century.

The greatest advances in gunmaking and in gunnery throughout the century were made in

France. French artillery ascendancy was begun about 1440 by the Bureau brothers (see p. 461), and was a major factor in the sudden and overwhelming French victory in the Hundred Years' War. The Bureaus used their cannon to batter down the walls of English-held castles with amazing rapidity. The improvement in siege artillery during the early 15th century is demonstrated by the contrast of the length of the sieges in Henry V's campaigns in Normandy (exemplified by the 5- and 6-month investments of Rouen and Cherbourg in 1418 and 1419) with the brief sieges which marked the speedy French reconquest of the province in 1449–1450. By the end of the century, artillery had made medieval fortification obsolete—this, to a degree, was because castles and walled cities had found no way to emplace large cannon suitable for counterbattery.

Artillery also played a significant part in the two most important field battles of the concluding phase of the Hundred Years' War: Formigny and Castillon (see p. 453). It should be noted, however, that the culverins (relatively light long-range cannon) and bombards, which facilitated these French victories, had in each instance been previously emplaced for siege operations and were merely shifted in direction to repulse English relieving armies.

The one approach to the concept of field artillery in the first part of the century was in the wagon forts of John Ziska in the Hussite wars (see p. 470). He put bombards in the intervals between his wagons. The fire of these cannon, combined with that of infantrymen armed with handguns, invariably repulsed the attacks of German, Hungarian, and Polish cavalrymen and infantrymen, setting the stage for Ziska's typical climactic counterattack. But these guns were always emplaced in the *wagenburg* prior to battle, on ground selected in advance by Ziska; they were useless if the enemy did not attack.

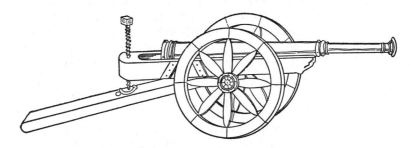

Fifteenth-century culverin

True field artillery made a sudden and dramatic appearance in the final decade of the century when the French—still preeminent artillerists—mounted new and relatively light cast-bronze cannon on two-wheeled carriages pulled by horses. The new mobile French field artillery could be quickly unharnessed and unlimbered on the battlefield and, by going promptly into action from a march column, played an important part of the significant French victory of Fornovo.

It was during this decade that the French also introduced the concept of the trunnion, which facilitated both the mounting of cannon on permanent wheeled carriages and relatively accurate aiming and ranging—in sharp distinction to the earlier awkward methods of raising and lowering the weapon's bore. Some, but certainly not all, of the French cannon at Fornova

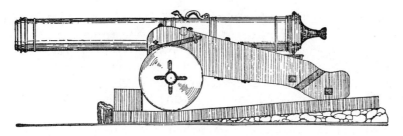

Medium bombard

probably had trunnions, and could be elevated or depressed for ranging without the need for digging holes under the trail or putting the wheels on blocks.

FORTIFICATION AND SIEGECRAFT

There were some limited developments in fortification during this century. The use of stone machicolations became widespread, replacing earlier wooden (and flammable) galleries, sometimes called hoardings. Stone machicolations permitted the defenders of a castle or town to fire on the besiegers, or drop rocks or boiling oil on scaling parties, from positions of relative safety.

By midcentury, the widespread use of firearms and cannon in sieges had a considerable effect on fortress design and on the conduct of siege operations, especially in France. Firearms were popular with both attackers and defenders, since the presence of walls and entrenchments allowed the gunners to reload in comparative safety, and these weapons' slow rate of fire was not as serious a drawback as on the battlefield.

Castles and fortified towns were built or remodelled with the employment of cannon a major consideration. Most common was the creation of cannon-ports in the lower floors of towers, and provision for the use of smaller guns on swivel mounts on upper levels. In an effort to reduce vulnerability to cannonfire, walls were made thicker at the base, and permanent outworks were constructed at a distance from the main walls, where cannon and lighter firearms could be emplaced by defenders during a siege. These modifications were not universal and generally were rare outside France and Italy. The French siege of Bordeaux, following the English defeat at Castillon, prosecuted by the full might of the French artillery train under Jean Bureau, lasted over ten weeks (late July–October 19, 1453), while the great Turkish siege of Constantinople lasted only 55 days, due both to the power of the Ottoman siege train and to poor Byzantine preparations for the defensive use of cannon and firearms.

TACTICS OF LAND WARFARE

General

The importance and value of well-trained, well-disciplined infantry continued to grow in this century, to the degree that infantry could undertake successful offensive action against cavalry, as well as against other infantry, either on its own or in combination with cavalry. In the case of the superbly disciplined and trained Swiss pikemen—and their German imitators, the

landsknechts—successful infantry offensive tactics were not in the slightest dependent upon missile weapons—bows, handguns or artillery. They were virtually a reincarnation of the Macedonian phalanx. By the close of the century, however, both Swiss and landsknechts were enhancing the morale and momentum effects of their terrifying, inexorable charges by the use of arquebusiers. The employers of these mercenary pikemen obtained the maximum value from them by fitting them into a combined-arms team with heavy and light cavalry. By the end of the century the French had made this into the prototype of the infantry-cavalry-artillery team, which dominated land warfare tactics for more than four centuries.

English and French Infantry Tactics

The actual and moral ascendancy of the English infantry defensive tactics over Continental chivalry was reaffirmed early in the century by the almost unbroken string of successes won by Henry V and his brother **John of Bedford.** Joan of Arc's subsequent inspirational success did not change the fundamental technical superiority of disciplined, heavily armored pikemen, flexibly supported by powerful missile weapons, in defensive formation against medieval heavy cavalry, on horse or on foot.

The final French successes of the Hundred Years' War, in fact, substantiated the superiority of good infantry on the defense, when well supported by missiles. The Bureau brothers merely added to the power of the missiles by substituting their new artillery weapons for longbows and crossbows.

The Hussite Tactical System

This same lesson was also being given to the chivalry of Germany and Eastern Europe by John Ziska and his Hussites. Ziska had apparently noted the effectiveness of Russian and Lithuanian wagon laagers in defense against Tartar, Polish, and Teutonic Knight cavalry. He also noted that the wagon defenders were lost if a combined cavalry and infantry attack could penetrate the laager. From this lesson he developed one of the simplest and most effective tactical systems in history.

The Hussites moved in columns of horse-drawn carts or wagons, most of which were armor-plated, the sides pierced with loopholes. Inside these protected wagons, or on other open four-wheeled carts, he carried a number of small bombards. His troops were mostly footmen, highly disciplined by the intensive training and control methods established by Ziska in his Mount Tabor stronghold; a number of these were armed with handguns; most were pikemen. In addition he had a small force of lightly armored cavalrymen, probably lancers on the Polish model, used for reconnaissance and for counterattack. Although good at siegecraft, Ziska always avoided an offensive battle in the open. His strategy was to penetrate as far as possible into enemy territory and then to select a good defensive position upon which to establish a wagon fort. Hussite raids from the wagon base inevitably forced the foe to disastrous attack.

The wagons were formed into a laager and linked together by chains. In front of this wall of wagons a ditch was dug by camp followers—who were also ammunition carriers. In the intervals between wagons he placed his bombards—possibly on their four-wheeled carts, but more likely on earthen mounds or heavy wooden platforms. Also in these intervals, and firing from the wagon loopholes, were his handgun operators and crossbowmen. Pikemen were available to protect the bombards and to prevent enemy infantry from cutting the chains holding the wagons together. They rarely had to perform these missions, however, since the attackers were more often than not repulsed by the firepower of the *wagenburg*. As soon as an attack was repulsed, Ziska's pikemen and cavalry charged to counterattack, sealing the victory.

Though Ziska did not introduce true field artillery, and did not use gunpowder weapons in a tactically offensive role, his was a most imaginative and offensive-minded use of field fortifications, and his battles were classics of defensive-offensive tactics. The *wagenburg*, however, could not be properly established if the enemy army was alert and aggressive. After the death of Ziska, his successors were unable to seize and maintain his strategic initiative. Furthermore, the wagon fort was extremely vulnerable to true field artillery and to efficient small arms; thus it was soon outmoded.

The Swiss Tactical System

It was the Swiss who really brought infantry back into offensive warfare for the first time since the decline of the Roman legion. A relatively poor mountain people, without horses, the lightly armed, lightly armored Swiss foot soldiers had been forced to fight their wars for independence against the Hapsburgs by making the best possible use of their local resources and of the difficulties of their mountain terrain. In the process they had discovered the benefits of mobility, which they gained through lack of encumbrances, and had also rediscovered the ancient Greek concept of the massed shock of a body of pikemen charging downhill. Like the Greeks and Macedonians, they had also learned that this same principle would work on level ground if the pikemen could maintain their massed formation without gaps and without faltering in the face of a cavalry charge. If they could press their own attack home with speed and vigor, the massed pikes could be more terrifying to horse and rider than the cavalryman was to the pikeman.

To exploit this lesson required excellent organization, rigorous training, and iron discipline of a sort unseen since Roman times. The determined Swiss met these challenges, and produced forces comparable to the Macedonian phalanxes in maneuverability, cohesiveness, and shock power. As a result, by midcentury they were universally recognized as incomparably the finest troops in the world.

As with the Macedonians, the principal Swiss weapon was a 21-foot pike, consisting of an 18-foot wooden shaft topped by a 3-foot iron shank which defied desperate slashes of attacking cavalry swords. They marched in cadence, often to music, the first to do so since the Romans. Again like the Macedonians, they had a variety of formations—such as line, wedge, square— which they could adopt in prompt, systematic, orderly fashion. For the most part, however, they marched and fought in heavy columns, with a frontage rarely exceeding 30 men, but often 50 to 100 men deep.

In addition to the basic pikemen, Swiss formations included a few crossbowmen—later handgunmen and arquebusiers—as skirmishers, and a number of halberdiers in the interior files of the column or phalanx. The principal task of the halberdiers was to cut down individual horsemen in a melee, when the pikemen had stopped or repulsed a cavalry charge.

Whenever possible, the Swiss marched directly and rapidly into combat from march column. There was no deploying, no delay of any sort in going through the formality of establishing a battle line. They usually fought in 3 columns, or phalanxes, echeloned to the left or right rear. If the countryside would permit, the individual columns would march on parallel roads, or across country, in the same relationship as they planned for action. If only one route was available, the first group would march directly ahead into the fight, the others peeling off to the left or right, as the case might be, then hurrying up on the flank of the first column.

The Swiss scorned cannon, and during the 15th century did not fully understand the vulnerability of their own massed columns to effective artillery fire—mainly because up to that time there had been no real field artillery. Charles the Bold of Burgundy recognized this Swiss vulnerability to cannon and small-arms fire, but his own rashness prevented him from exploiting his relatively advanced tactical ideas (see pp. 465–466).

No other infantry in Europe could stand up to the Swiss, and by the end of the century their moral as well as tactical ascendancy was complete. Although they hired themselves out as mercenaries indiscriminately, more often than not they provided the principal infantry component of the French armies of the late 15th century. Maximilian of Hapsburg, plagued by this as much as by his wars with Switzerland, tried to offset the Swiss infantry advantages by creating a German counterpart: the landsknechts. Organized, trained, and equipped just like their model, the landsknecht mercenaries were soon nearly as sought after as the Swiss themselves. Strangely, however, whenever Swiss and landsknechts were pitted against each other, although the struggle was always fierce and bloody, the Swiss were invariably victorious. They seem to have been even more ferocious and bloodthirsty against their German imitators than against any other foe. Both Swiss and landsknecht referred to conflicts between them as *"der böse Krieg"* (the bad war) because of mutual animosity, heavy casualties, and poor opportunities for booty from prisoners in such a clash.

Cavalry Tactics

The frustration experienced by West European heavy cavalry in action against English infantry in the 14th century was intensified by the appearance of the Swiss tactical system and the advent of firearms. No real solution to this problem had been found by the end of this century. Heavy cavalry lingered on as a major element of all of the West European armies, since in conjunction with other arms it was still effective in a climactic charge or counterattack against an enemy thrown into confusion by artillery fire, or by a prior infantry encounter. The disciplined French **gendarmes** were the best European heavy cavalry.

As a compromise, however, more lightly armed and armored horsemen began to appear in Western Europe. Some Westerners who had participated in the Turkish wars of Eastern Europe had noted the effectiveness of the relatively lightly armed and armored Hungarian, Turkish, and Albanian cavalry, who combined discipline and some shock power, on the one hand, with the mobility and flexibility of unarmored light cavalry, on the other. These were mixed horse arches and lancers, quite similar in organization, armament, and tactics to the old Byzantine cataphract, though less heavily armored. The Venetians seem to have introduced Albanian cavalry of this variety, called **stradiots,** into their wars in north Italy at the end of the century, and their example was quickly followed by the French. The term stradiot began to be used in Western Europe for any cavalry of this intermediate genre, whether or not they were actually Albanians. The similar Spanish **genitours** (see p. 305) also began to appear in Italy about the same time.

This was the first step in a series of major transformations in European cavalry, which did not really regain effectiveness until the next century.

MILITARY ORGANIZATION

At the beginning of the century only the English and the Ottoman Turks had armies which approximated regular standing forces. Both accomplished this by modifying, without eliminating, the traditional pattern of feudal levies.

English armies were raised by contract, along the lines established by Edward III (see p. 363). Each knight or noble who had the influence, wealth, or ability contracted with the king to raise a force of given size for foreign operations. In addition to the funds disbursed by the royal treasury for such contracts, the leaders and their men were also attracted by prospects of loot. The English kings and great nobles usually had a substantial number of English fighting men engaged in local conflicts in France, as well as along the Scottish border. This led to the growth of a reserve of battle-experienced men in England, who would flock to the king's colors for any major expedition to the north or across the Channel.

The Ottoman sultans avoided any hereditary feudal relationship insofar as possible. In particular, they made a point of bestowing newly conquered land in Europe or Asia to the most deserving of their warriors. Thus a noble who raised a large effective force for the sultan's wars could hope to share in the distribution of lands gained by that war. Similarly, young, impecunious soldiers could gain lands and riches by performing well in combat. The result was that the sultans had no difficulty in raising excellent enthusiastic forces, with war naturally feeding upon conquest.

The Turkish armies were composed mostly of cavalry, the feudal levies being lightly armored horse archers or lancers of the stradiot variety. Constant experience in warfare assured competence, and although these feudal levies could not stand up against the heavily armored Western men-at-arms, their lesser shock capability was somewhat offset by greater mobility and maneuverability. The nuclei of the Turkish armies were the highly trained and disciplined Janissary infantry units—never more than 10,000 strong in this century—and the sultan's horse guards. These cavalry guardsmen, called Spahis of the Porte, were lancers somewhat more heavily armored than the feudal cavalry. Turkish armies also included large numbers of irregular light horse and light infantry, useful for reconnaissance and pursuit but of little use in pitched battles.

In Western Europe—and particularly in war-torn France—nobles and kings relied extensively upon mercenary contingents to provide the continuity and professional competence and experience missing from their short-term feudal levies. In time of peace, however, these free companies, as they were called, were deadly menaces, pillaging, looting, and murdering indiscriminately. They were too strong to be dealt with by the feudal levies, and so rulers tried to entice them out of the country or to hire them for foreign expeditions which hopefully would provide enough loot to pay their wages.

The French solution to the problem of the free companies, apparently recommended to Charles VII by **Arthur de Bretagne,** Constable de Richemont, and other advisors, was to establish a standing army. This force consisted of the *compagnies d'ordonnance,* numbering at first some 1,500–1,800 lances, but later expanded under Louis XI to include over 3,000. These were grouped in companies of 30–100 lances, each company under the command of a reliable captain; there were 20–25 companies in the later Hundred Years' War. Each lance consisted of a knight or man-at-arms (a *gendarme* in French usage), a squire, 2 archers, and 2 pages or valets, 1 for the gendarme and squire, and 1 for the archers. The gendarme and the squire were both equipped as heavy cavalry of the period; the 2 archers were originally mounted infantrymen, but as time went on their battlefield role often approximated that of the gendarme and squire, though they rarely carried lances. The valets, who were often barely into their teens (or younger), were not counted as combatants, but sometimes served as foragers, scouts, and pickets. All the men of the *compagnies* were well paid and provided with liberal rations. There was thus no need to pillage. Their sole loyalty was to the king.

These companies were quartered in various regions of France (190 lances in companies of 30, 60, and 100 were quartered in the county of Poitou in 1445–1446), and absorbed a number of men of the free companies, both en masse and individually. Indeed, over half of the captains of the new *compagnies* were foreigners. Quickly they established law and order, the remaining mercenaries going elsewhere—mainly to *condottieri* companies in Italy. The pattern of establishing state control over mercenary forces had been going on for some time (albeit more gradually) in Italy, as the *condottieri* were transformed from military entrepreneurs to contract-soldiers with ties to a particular state.

Establishment of the *compagnies d'ordonnance* was the deathknell of feudalism as a military and social system, and the true beginning of standing armies and professionalism in warfare. Less successful was the simultaneous French effort (1448) to set up a centralized semipermanent militia force of some 8,000 (later doubled in size by Louis XI). These were called "free archers" (*franc-archers*) since they were initially armed with bows and were exempt from taxes. This represented another step away from feudalism (and was widely resented by the militia's aristocratic officers), but due to shortcomings in training, administration, and recruitment the "free archers" proved ineffective and were disbanded after their dismal showing at the Battle of Guinegate in 1479.

WESTERN EUROPE

THE HUNDRED YEARS' WAR—PHASE TWO, 1396–1457

The Uneasy Truce, 1396–1413

The 30-year Truce of Paris (1396) was intended by the rival monarchs—**Richard II** of England and **Charles VI** of France—to terminate the war. Internal instability in both countries, however, contributed to continuing small-scale violence, each ruler implicitly encouraging raids and revolts. In England central authority was soon reestablished under strong kings **Henry IV** and **Henry V** (see p. 454). In France, however, the intermittent insanity of Charles VI prompted a power struggle between the rival dukedoms of Orléans and Burgundy, which soon grew to full-scale civil war. The principal events were:

1402. French Support to Scotland. French troops (primarily Orléanist) assisted a Scottish invasion of England (see p. 454).

1403. French Raids on the English Coast. Plymouth and other English Channel ports were ravaged while Henry IV was preoccupied with scattered revolts (see p. 454).

1405. French Support to Wales. French troops

assisted Welsh rebel **Owen Glendower** (see p. 454).

1406. French Attacks on English Possessions in France. The Burgundians and Orléanists temporarily collaborated in operations in Vienne and against Calais.

1407, November 24. Burgundian Assassination of Duke Louis of Orléans. This led to full-scale war between the Burgundians and Orléanists. An important anti-Burgundian role was played by Count **Bernard of Armagnac;** thus the royal (or Valois, or Orléanist) faction was also called Armagnac.

1411. French Appeals to England. Both Orléanists and Burgundians sought aid from Henry IV.

The Period of Henry V, 1413–1428

Young King Henry V, one of the strongest and most vigorous rulers of English history, seized the opportunity created by anarchy in France to attempt ending the war on the basis envisaged by his great-grandfather (Edward III). After adroit diplomacy assured the neutrality of **John the Fearless,** Duke of Burgundy, Henry invaded France and won an overwhelming victory against a far larger French army at Agincourt, a battle in the Crécy and Poitiers pattern. Most Orléanist leaders were killed in the disaster, permitting the Burgundians to gain ascendancy in strife-torn France. During the 4½ years following Agincourt, Henry systematically conquered Normandy, then advanced on Paris. The French government, then dominated by John of Burgundy, submitted to Henry in the Treaty of Troyes, in which Charles VI disinherited his son, the Dauphin **Charles,** declared Henry his heir, and gave Henry his daughter **Catherine** in marriage. Henry spent the following 2 years consolidating northern France, but died suddenly as he was preparing to move into south-central France, where the Dauphin still had some following. Charles VI died soon after; infant **Henry VI** was declared King of France as well as of England, and John, Duke of Bedford, younger brother of Henry V, acting as regent in France, continued consolidation of English authority. The Dauphin set up a rival court in the city of Bourges, and delayed English conquest of France with Breton assistance. Bedford defeated the combined armies of the Dauphin and the Duke of Brittany at Avranches, thus finally eliminating all opposition to English control in northern and western France. He then renewed preparations for an invasion of the province of Berry, south of the Loire, in order to overthrow the last center of Valois resistance. The principal events were:

1413, May. Alliance of Henry V with John of Burgundy. John promised neutrality in return for increased territory as Henry's vassal. Henry began preparations for an invasion of France.

1415, April. Henry's Declaration of War on Charles VI of France. He sailed for Normandy with an army of 12,000 men (August 10).

1415, August 13–September 22. Siege of Harfleur. Henry finally captured the city after his army had been depleted by casualties and disease.

THE CAMPAIGN OF AGINCOURT, OCTOBER 10–24, 1415

For obscure reasons, Henry decided to march overland from Harfleur to Calais (October 10). He may have hoped to entice the French into a battle, although—considering the circumstances of his army (now about 6,000 men), his selected line of march, the fact that he took no artillery or heavy baggage, and the speed with which he moved—this is very doubtful. On the other hand, the apparent rashness of the move is hard to reconcile with Henry's normally deliberate nature.

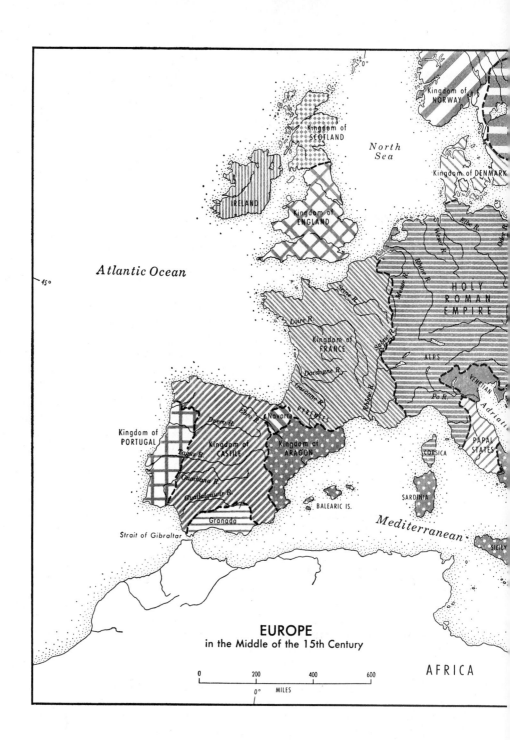

EUROPE
in the Middle of the 15th Century

Kingdom of NORWAY

Kingdom of SCOTLAND

North Sea

Kingdom of DENMARK

IRELAND

Kingdom of ENGLAND

Atlantic Ocean

45°

HOLY ROMAN EMPIRE

Loire R.

Seine R.

Elbe R.

Weser R.

Oder R.

Kingdom of FRANCE

Dordogne R.

Garonne R.

Rhone R.

ALPS

Po R.

VENETIAN

Adriatic

PYRENEES

Novarre

Kingdom of PORTUGAL

Duero R.

Ebro R.

CORSICA

PAPAL STATES

Kingdom of CASTILE

Kingdom of ARAGON

Tagus R.

Guadiana R.

SARDINIA

Guadalquivir R.

BALEARIC IS.

Mediterranean

Granada

Strait of Gibraltar

SICILY

| 0 | 200 | 400 | 600 |

0°   MILES

AFRICA

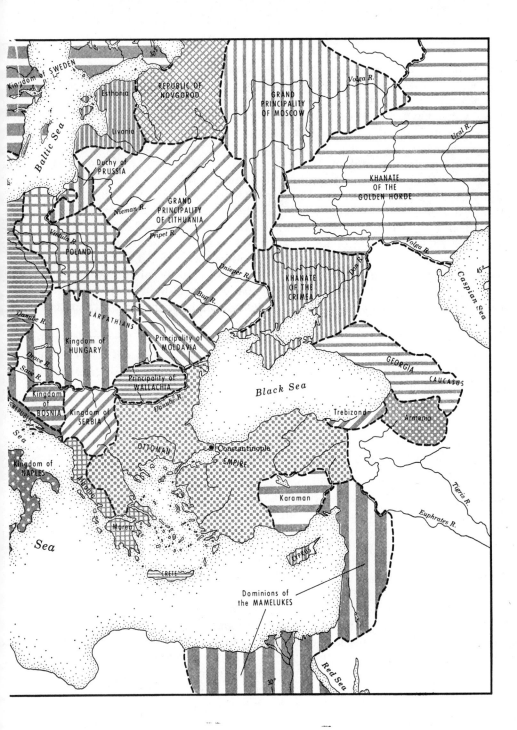

Moving more than 14 miles a day, despite torrential rains, he headed directly for the Ford of Blanchetaque, where Edward III had crossed in the Crécy campaign. Finding this blocked by local feudal levies, he headed eastward, up the Somme River, seeking an undefended crossing. All the bridges were down, or strongly defended, and the fords were either defended or impassable due to high waters. Finally at Athies, 10 miles southeast of Péronne, he found an undefended crossing (October 19). Meanwhile, a French army of more than 30,000 men, under the Constable of France, **Charles d'Albret,** had pursued from Rouen in two divisions. Barely failing to intercept the English, the French crossed the Somme at Abbéville and Amiens, joined at Corbie, and then moved northeast via Péronne and Bapaume. Apparently well informed of Henry's advance, d'Albret took a blocking position on the main road north toward Calais near the castle of Agincourt. Here the French were found by the English on the afternoon of October 24. Henry was apparently ready to negotiate. His army was isolated, its line of retreat blocked by the French. The English had been short of food, and had barely subsisted by plundering the countryside. In turn the local inhabitants had harassed the English column, killing many stragglers and pillagers. There was some inconclusive parleying, but the French, confident of victory, were in no mood to offer concessions, and Henry would not give up his claim to the French throne.

It has been suggested that d'Albret should have attacked as soon as the English army appeared. Save for a few hours' rest which this gave the English, it is doubtful if the delay contributed to the factors which determined the outcome the following day. D'Albret, a disciple of Du Guesclin, had planned a defensive battle. During the evening, however, the more impetuous French nobility pleaded with him to attack, confident of their more than 3 to 1 superiority and aware of the generally low physical condition of the English army.

BATTLE OF AGINCOURT, OCTOBER 25, 1415

At daybreak the English army was in position at the sourthern end of a defile—slightly over 1,000 yards in width—formed by heavy woods flanking the main road north to Calais. The open fields, between road and woods, were freshly plowed and sodden from recent heavy rains. The English—some 800 men-at-arms and 5,000 archers—were drawn up in three "battles" in the tactical formation proved so effective by Edward III at Crécy. The French army was also deployed in 3 "battles," one behind the other due to the narrowness of the defile. The first 2 were composed of dismounted men, with some crossbowmen in support of each, while the 3rd "battle" was mostly mounted men. There was a small body of mounted troops on each flank of the lead battle, perhaps 400 or 500 horsemen all told. For some three hours the two armies faced each other at opposite ends of the wooded defile, probably a little more than a mile apart. D'Albret, apparently relying on Henry's unfounded reputation for youthful recklessness, seems still to have hoped for an English attack. Henry, deciding late in the morning to entice a French assault, ordered a cautious advance of approximately half a mile. Coming to a halt, the English resumed their usual formation, archers echeloned forward in Vs in the intervals between the battles of dismounted men-at-arms and pikemen. Archers on the flanks evidently infiltrated forward for 100 yards or more in the edges of the woods. Those in the two Vs in the center of the line quickly pounded stakes into the ground to form a palisade to impede a mounted attack.

The English advance had the desired effect. Unable to control his eager and undisciplined

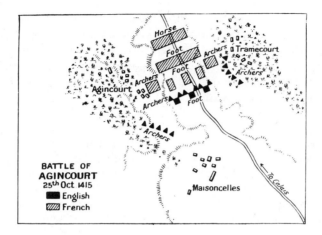

fellow nobles, who had "forgotten nothing and learned nothing," the French constable reluctantly ordered an advance. As the first battle of dismounted men-at-arms moved ponderously forward in their heavy armor, mounted contingents on each flank galloped past them toward the English lines. The experience of Crécy and Poitiers was repeated. A large proportion of these men, or their horses, was cut down by English arrows. The remainder, thrown into confusion by the hail of arrows, and moving slowly because of the soggy ground, were thrown back with heavy loss when they reached the thin but solid English line.

The cavalry attack had been completely repulsed before the first French battle, led by d'Albret himself, labored ponderously within range of the archers. The combination of the weight of their armor—even heavier than at the time of Poitiers—and the very difficult movement over the wet plowed fields had exhausted the French knights before they came within range of the English archers. French losses were heavy as they struggled through the mud for the last 100 yards, but nevertheless most of them reached the English line, and sheer weight of numbers bore the defenders back. But now the English archers, wielding swords and hatchets and axes, came out from behind their palisades and from the edges of the woods into gaps in the French line from the flank and rear, causing fearful destruction among the weary, terribly immobile French knights and men-at-arms. Within a few minutes every man in the first French battle had been cut down or captured.

Now, completely uncoordinated with their predecessors, the French second battle surged forward. This attack, however, was apparently not pushed with the same vigor or determination, and although losses were heavy the French knights fell back to re-form. Apparently joined by some of the mounted men of the third battle, they prepared for one final effort. At this moment Henry received a report that the French had attacked his camp, a mile to the rear. He ordered the slaughter of his prisoners, since he did not think his army was strong enough both to guard the prisoners and to meet attacks from front and rear. It turned out, however, that the feared envelopment had been only a rabble of French peasants seeking plunder. The third assault, furthermore, was halfhearted and was easily thrown back. The climax of this final phase of the battle was a charge by a few hundred English mounted men under Henry personally, completely dispersing the remnants of the French army.

French losses were at least 5,000 men of noble or gentle birth killed and another 1,000 captured. D'Albret was among the killed; the **Duke of Orléans** and the famed Marshal **Jean Bouciquaut** (see p. 419) were among the captured. Most of the important leaders of the Orléans-Armagnac faction were killed. Thus an important result of the Battle of Agincourt was to assure the ascendancy of the Burgundians in France. English losses were reportedly only 13 men-at-arms and about 100 footmen killed; it is likely that casualties were substantially greater than this.

1415, November. Return to England. Henry, taking no strategic advantage of his overwhelming tactical victory, marched to Calais and shortly thereafter returned to England. It is not clear whether this was due to a lack of strategic competence or to other factors. Evidently he was concerned by a threat to control the English Channel posed by a Genoese fleet, allies of the French. Also he seems to have developed considerable respect for the guerrilla fighting capabilities of the French peasants, who had caused him so much trouble on the march from Harfleur to Agincourt.

1416. A Year of Preparation. Henry built up a powerful fleet and drove the Genoese from the Channel. His recognition of the importance of control of the sea and his actions to secure it have led some historians to suggest him as the founder of the Royal Navy. He proved his skill and ability as a diplomat by obtaining the neutrality of Emperor **Sigismund,** formerly an ally of France.

1417–1419. Return to France. Henry systematically conquered Normandy in 3 well-planned campaigns, notably different from the uncoordinated raids of Edward III three-quarters of a century earlier. The principal feature in these otherwise uneventful campaigns was the 5 months' siege of Rouen (September 1418–January 1419). In the following 6 months Henry completed consolidation of his conquest of Normandy save for the coastal fortress of Mont-Saint-Michel. He was now ready to move inland.

1418, May 29. Burgundian Capture of Paris. John of Burgundy, after taking the capital, massacred most of the remaining Armagnac and Orléanist leaders, with the notable exception of the Dauphin, Charles, who escaped to the south.

1419, September 10. Assassination of John of Burgundy. This took place during a truce, while he was in conference with the Dauphin on a bridge at Montereau. As a consequence, the Burgundians—now led by **Philip the Good** and sustained by Queen **Isabella,** wife of insane Charles VI—were confirmed in their support of the English alliance against the Dauphin and the Orléans-Armagnac faction.

1420, May. Henry Marches on Paris. He took advantage of the renewed bitterness in the French civil war.

1420, May 21. Treaty of Troyes. Henry became heir of Charles VI and virtual ruler of France.

1420–1422. Henry's Consolidation of Northern France. He intended to eliminate systematically all remaining Valois opposition north of the Loire before moving directly against the Dauphin. This culminated in the successful siege of Meaux (October 1421–May 1422), where Henry became ill.

1421, March 21. Battle of Baugé. A French raiding force surprised and defeated a small body of English under **Thomas, Duke of Clarence** (Henry's brother) in southern Normandy. Clarence was killed.

1422, August 31. Death of Henry V. His 9-month-old son, **Henry VI,** became King of England and heir of France.

1422, October 21. Death of Charles VI. Henry VI was declared King of France; John, Duke of Bedford was regent.

1422, October 31. The Dauphin Proclaimed Charles VII at Bourges. He made no other immediate effort to regain his birthright.

1422–1428. Bedford's Consolidation of Northern France. While John pursued his brother's strategy, Burgundian enthusiasm in support of English objectives waned.

1423, July 21. Battle of Cravant. A small English and Burgundian army was victorious over a slightly larger French force, which included a Scottish contingent.

1424, August 17. Battle of Verneuil. Bedford defeated a larger French army in a battle fought in the Crécy-Agincourt pattern, although the French did attempt an envelopment, which was repulsed by archers of the English baggage guard. The principal French leaders were the **Duke of Alençon** and the Scots **John Stuart** and the **Earl Archibald of Douglas.** Alençon was captured and Douglas was killed.

1426, March 6. Battle of St. James (near Avranches). Bedford defeated **Arthur de Richemont,** Constable of France, forcing de Richemont's brother, the **Duke of Brittany,** to submit to the English.

1426–1428. Bedford Completes Consolidation of Northern France.

1428, September. English Advance to the South. The **Earl of Salisbury** moved from Paris with about 5,000 men to seize the Loire River crossing at Orléans as a preliminary to invasion of Berry, the principal stronghold of the Dauphin.

1428–1429. Siege of Orléans. Salisbury commenced operations (October 12) by eliminating the fortified French bridgehead south of the Loire. Mortally wounded by a cannon shot from across the river in the culmination of these assaults (October 24), he was succeeded by the **Earl of Suffolk.** In subsequent months Suffolk constructed a line of redoubts (**bastilles**) around the city on the north bank of the river. The English were so weak numerically that the blockade was incomplete, permitting a trickle of supplies and reinforcements to get into the beleaguered city, mainly via the river. The defenders, under **Jean, Count of Dunois** (the "Bastard of Orléans"), considerably outnumbered the besiegers, but French morale was so low, and the terror inspired by English victories so effective, that the French made no serious offensive effort.

1429, February 12. Battle of Rouvray (or "of the Herrings"). Sir **John Fastolf** (model for Shakespeare's Falstaff) successfully defended a supply convoy carrying Lenten rations to Suffolk's army from a French attack under the **Count of Clermont** and John Stuart. Fastolf formed his wagons, filled with salted herrings, into a laager and beat off the attackers.

French Revival under Joan of Arc, 1429–1444

A deeply religious peasant girl from Champagne, 17-year-old **Joan of Arc,** partially convinced the Dauphin that she had a divinely inspired mission to help him to expel the English from France and to have him crowned as rightful king. Despite his doubts and the intense jealousy of most of his court, Charles put Joan in command of an army to relieve Orléans. Her inspirational, if untutored, leadership drove the English from their siege lines in defeat, and ended the myth of English invincibility. Joan won numerous subsequent victories, and engineered the triumphant coronation of the Dauphin as Charles VII, although she was subsequently repulsed in a rash attack on Paris. The following year she was captured by the Burgundians near Compiègne. Turned over to the English by the Burgundians, she was tried by a religious court on the orders of Bedford, who thus hoped to discredit Charles' coronation. Charles, who could have bargained for her release, made no effort to help her, and Joan was convicted and burned at the stake as a heretic. Her spirit, however, continued to inspire French fighting men. English morale, in turn, was shaken by the fear that they had killed a saint. Within 5 years of Joan's death, the Burgundians and Orléanists had ended their civil war, and the French had recaptured Paris. In the following 8 years the French were generally successful in small actions and sieges in northern France, causing the English to agree to a 5-year truce. The principal events were:

1429, February–April. The Emergence of Joan of Arc. She inspired new hope and determination in the Dauphin.

1429, April 27–29. Joan Reaches Orléans. Accompanied by the **Duke of Alençon,** she led a relief army of 3,000–4,000 men, plus supplies, from Blois, slipping past the British siege lines by boat along the river.

1429, May 5. The First Sortie. Joan led a successful attack on an English bastille north of the river.

1429, May 7. Relief of Orléans. Joan led an attack against the British-held fortified bridgehead on the south bank of the Loire. Despite a serious arrow wound, she remained in the forefront of desperate assaults, and was successful by nightfall. The British blockade was broken. Suffolk withdrew the next morning. Mistakenly, he divided his army into garrisons to retain English-held towns in the Loire Valley. In subsequent weeks Joan retook most of these towns.

1429, June 19. Battle of Patay. In a surprise attack, Joan's army smashed the forces of Lord **John Talbot** and Sir John Fastolf, culminating efforts to drive the English from the Loire Valley. Talbot was captured.

1429, June–July. French Resurgence. Under Joan's inspiration French morale soared, while that of the English plummeted. The peasants rose in guerrilla war against the English, despite Charles' lethargy and inaction. This unprecedented popular resistance marked the beginning of French nationalism. Joan captured Troyes, Châlons, and Rheims (July).

1429, July 16. Coronation of Charles VII at Rheims. This followed Joan's bold invasion of territory formerly dominated by the English. Henceforward Charles and his jealous court did everything in their power to hamper Joan and to restrict her influence.

1429, August. Arrival of English Reinforcements at Paris.

1429, September 8. Attack on Paris. Despite lack of royal support, Joan with a small force made an unsuccessful attack, in which she was wounded.

1430, May 23. Capture of Joan. She led a small relief force to Compiègne, which the English and Burgundians were besieging as part of Bedford's efforts to reestablish English control of the central Seine Valley. Leading a sortie that same day, she was captured by the Burgundians.

1431, May 30. Execution of Joan of Arc.

1431–1444. Continued Effect of Joan's Inspiration. Nobles and peasants continued to wage a partisan war against the English despite the apparent indifference and lethargy of Charles VII. Bedford manfully and skillfully opposed the French resurgence until his death (1435).

1435, September 21. Peace of Arras. This ended the Burgundian-Armagnac civil war, disrupting the Burgundian-English alliance.

1436, April. Recapture of Paris. The English garrison took refuge in the Bastille, but was starved into surrender.

1444, April 16. Truce of Tours. A 5-year peace.

The Triumph of Professionalism, 1444–1453

The end of the Burgundian-Armagnac civil war in France foreshadowed French victory, since England's population of 2½ million could scarcely hope to overcome a united nation of 15 million. The slight continuing English tactical advantage conferred by the longbow was offset by the higher morale and greater numbers of the French. Only a condition of internal anarchy in France—the result of depredations of thousands of soldiers demobilized at the end of the civil war—gave the English any hope of salvaging a substantial remnant of their former vast holdings in the north and west. But this was speedily dashed when otherwise inept Charles VII followed wise military advice to establish the basis of a disciplined standing army, leading to the emergence of history's first efficient and scientific artillery organization. When the 5-year Truce of Tours expired, the new French military organization had not only enforced internal peace and stability but had achieved a substantial technical and tactical superiority over the English. It took this revitalized, professional French army only 4 years to sweep the English completely out of France, save for one tiny foothold at Calais. The principal events were:

1444–1449. Reorganization of the French Army. (See p. 460.)

1449–1450. French Reconquest of Normandy. The **Duke of Somerset,** incompetent English commander in Normandy, surrendered **Rouen** (1449) and soon after was besieged in Caen (March, 1450).

1450, April 15. Battle of Formigny (near Bayeux). Constable de Richemont, learning of the advance of an English relieving army of some 4,500 men toward Caen under Sir **Thomas Kyriel** and Sir **Matthew Gough,** sent the **Count of Clermont** with about 5,000 men to block the English advance. The English formed up in typical defensive formation, but the French, refusing to follow the Crécy-Agincourt pattern, moved up 2 culverins beyond each flank of the waiting English army. Out of longbow range, the cannon began a deliberate enfilade bombardment of the English line, causing considerable damage. Spontaneously the English archers on both flanks charged against the guns, which they briefly seized. At this point the French dismounted men-at-arms and infantry counterattacked to recapture the cannon, bringing on a general engagement. While the issue was still in doubt, French reinforcements struck the English flank. In the resulting rout, Gough and a few survivors fought their way to safety; most of the English, nearly 4,000 men, were killed.

1450, July 6. Surrender of Caen. De Richemont immediately advanced to invest Cherbourg.

1450, August 12. Surrender of Cherbourg. English rule in Normandy was ended.

1451. French Invasion of Guyenne. The Count of Dunois, with 6,000 men, made efficient use of an excellent artillery siege train, and soon captured Bordeaux (June 30) and Bayonne (August 20). Considerable overt and underground resistance continued, however, on the part of the nobles of Aquitaine, still loyal to the kings of England and feeling little identity with their French cousins.

1452, October. English Resurgence. In response to Aquitanian appeals, John Talbot, now **Earl of Shrewsbury,** landed near the mouth of the Garonne with 3,000 men. Much of the countryside rose to meet him and support him. Bordeaux opened its gates to the English.

1453, July 17. Battle of Castillon. Shrewsbury led an army to raise the siege of Castillon, then being attacked by Jean Bureau, French master of artillery. Shrewsbury, deceived by the departure of the French army's horses into believing that most of their troops had withdrawn from their entrenched camp, launched a hasty attack without waiting for his footsoldiers to arrive. The English assault was halted by French cannon and small arms fire, and Shrewsbury and many of his men were killed in the French counterattack. The remnants of the English army fled in disorder. This was Crécy, Poitiers, and Agincourt in reverse.

1453, October 19. Fall of Bordeaux. This virtually ended the Hundred Years' War, although there was some minor coastal activity, particularly French raids on the English coast, for the next 4 years. England retained only Calais on the coast of northern France.

THE BRITISH ISLES

England

THE LANCASTRIAN ERA, 1400–1455

Henry IV, who had seized the shaky throne of Richard II (see p. 394), dealt firmly and vigorously with continuing unrest. His brilliant son, **Henry V,** brought England to the pinnacle of power and prestige in the Middle Ages. His great victory at Agincourt (see p. 448), followed by a cautious and sound campaign of conquest, made him virtual ruler of northern and western France, and recognized heir to the French throne. His early death, however, caused a sudden reversal in the fortunes of his nation and his family. Infant **Henry VI** grew up a pious, weak, and mentally unbalanced ruler. At home the country was racked with anarchy, while abroad the Hundred Years' War was lost. Strong-willed Queen **Margaret** seized the reins of government during her husband's fits of insanity, but she and her favorites were unpopular with most English nobles, who turned to the leadership of **Richard, Duke of York,** the king's cousin. The principal events were:

1399–1413. Reign of Henry IV. He soon suppressed a revolt of adherents of Richard II (1400).

1402, September 14. Battle of Homildon Hill. The northern nobles, under the leadership of Lord **Henry** (Harry "Hotspur") **Percy,** overwhelmingly repulsed a Scottish raiding army under the Earl of Douglas in a typical English longbow defensive battle.

1402–1409. Revolt of Owen Glendower in Wales. A master of ambush and surprise, Glendower made use of his knowledge of the Welsh hills to fight a highly successful partisan war.

1402. Battle of Pilleth. One of Glendower's rare battles, in which he surprised and routed an English force under Sir **Edmund Mortimer** in a mountain defile.

1403–1408. Revolt of the Percys. Northern rebels, under "Hotspur" Percy, advanced deep into central England with some 4,000 men, planning to join Glendower. Henry interposed his army of 5,000 between the two rebellious allies, and at the **Battle of Shrewsbury** (July 21, 1403) defeated the Percys before Glendower could come up. Hotspur was killed. His father submitted, then rebelled again, and was defeated and killed at **Bramham Moor** (1408).

1405. French Landing in Wales. Little was accomplished by this expedition in support of Glendower, and the disappointed French returned home toward the close of the year.

1405. Rebellion of Archbishop Scrope of York. This was promptly subdued by Henry.

1409. Defeat of Glendower. The English captured his stronghold of Harlech. Glendower disappeared, and Henry's troops gradually pacified Wales.

1413–1422. Reign of Henry V. A brave, resourceful, and calculating soldier and king. At the outset of the reign he suppressed a Lollard (heretical) uprising (1413–1414).

1413–1457. Conclusion of the Hundred Years' War. (See p. 453.)

1422–1461. Reign of Henry VI. This began under the regency of Duke **Humphrey of Gloucester** in England and John, Duke of Bedford in France, the infant king's uncles. Gloucester's rashness and ineptitude contributed greatly to the unrest which plagued the reign.

1437–1450. A Period of Growing Disorder.

1450. Cade's Rebellion in Kent and Sussex. **Jack Cade,** an ex-soldier, briefly occupied and pillaged London after defeating royalists in an ambush at the **Battle of Seven Oaks** (June 18). The rebels were driven from London by troops from the Tower, assisted by aroused London militiamen. A royal general pardon brought about collapse of the rebellion; Cade was later captured and executed for treason. **Richard, Duke of York,** governor of Ireland, returned to protect himself against accusations that he had inspired the rebellion.

1454, March 27. Richard Elected Protector by Parliament. This was during Henry VI's first fit of insanity.

1454, December. Richard Dismissed. The king having regained his sanity, he was persuaded by Queen Margaret to dismiss Richard from the government.

The Wars of the Roses, 1455–1485

The dismissal of Richard of York from the King's Council, and the assumption of dictatorial powers by Queen Margaret and the **Duke of Somerset,** prompted a revolt by Richard and his adherents, who included wealthy, powerful **Richard Neville, Earl of Warwick.** Five years of alternate violent conflict and political maneuvering followed, marked by treachery, murder, and wildly fluctuating fortunes. Richard was killed in battle, but his son Edward proclaimed himself **Edward IV,** and then crushed the Lancastrians in the sanguinary Battle of Towton. After nearly 10 years of relative peace, Edward and Warwick fell out due to Warwick's efforts to assume dictatorial powers. Edward cleverly outmaneuvered Warwick politically and militarily; the earl thereupon joined forces with Margaret to lead an invasion army from France, and temporarily replaced Henry VI on the throne. Warwick ("the kingmaker") was killed in the decisive Battle of Barnet. Edward reigned for 12 more peaceful and prosperous years, succeeded upon his death by his 13-year-old son, **Edward V. Richard, Duke of Gloucester,** younger brother of Edward IV, then usurped the throne. The House of York, left securely in control of England by Edward IV, was now threatened by a series of uprisings, which culminated in an invasion led by **Henry Tudor, Earl of Richmond,** leader of the revived Lancastrians. Richard, an able warrior, was defeated and killed by Tudor at the Battle of Bosworth because of the defection of most of his army. Henry Tudor ascended the throne as Henry VII.

One factor contributing to the length and bitterness of the Wars of the Roses was the pernicious effects of "livery and maintenance," sometimes called "bastard feudalism." Livery and maintenance grew out of an expansion of the money economy, so that great nobles maintained their followers (who wore the noble's badge, or livery) with cash stipends (termed "maintenance") rather than estates. As a result, the retainers had no geographic ties, little interest in the status quo, and little incentive to remain in the same lord's service should a better opportunity arise. This development, begun in the mid-14th century, coupled with the perception that a noble's political influence depended on the number of his retainers, magnified the basic political conflicts between York and Lancaster.

Edward IV and his brother Richard were the best generals in this generally dull, sanguinary series of wars. Both sides adhered to the policy of "slay the nobles, spare the commons," which resulted in exceptionally high loss of life among the leaders. Both sides generally avoided pillage and massacre of civilians. The principal events were:

1455, May 22. First Battle of St. Albans. Richard of York and Warwick, marching on London with 3,000 men, defeated and killed Somerset, leading 2,000 Lancastrians. Richard seized Henry VI, had himself appointed Constable of England, and assumed almost dictatorial power.

1456–1459. Political Maneuvering. Mainly a power struggle between Queen Margaret and Richard. Henry, intermittently sane, was a pawn. Margaret and her adherents regained control of the king; York and his supporters were again dismissed from the government.

1459, September 3. Battle of Blore Heath. The **Earl of Salisbury** (father-in-law of Richard and father of Warwick) defeated and practically destroyed a Lancastrian force near Market Drayton.

1459, October. Flight of Richard and Warwick. A royal army under Henry and Margaret approached Richard's stronghold of Ludlow Castle; a number of his adherents treacherously abandoned him, forcing Richard to flee to Ireland (October 12); Warwick fled to Calais.

1460, June. Return of Warwick. Warwick and

Edward, Earl of March (second son of Richard and later Edward IV) landed at Sandridge with a small army, then marched triumphantly to London, where the townspeople favored the Yorkists.

1460, July 18. Battle of Northampton. Warwick and Edward, aided by treachery and by a rainstorm which drenched the Lancastrians' gunpowder, defeated a royal army marching on London, and recaptured King Henry. Richard returned from Ireland, somewhat hesitantly claiming the crown. In a compromise, Parliament proclaimed him heir to the throne, which Henry was to retain during his life. This disinherited **Edward,** Prince of Wales, the infant son of Margaret and Henry. Margaret, having fled to Wales, went to northern England and raised a new Lancastrian army. Richard and his eldest son **Edmund** led an army north to deal with the main Lancastrian threat, while his second son, Edward, took another force to subdue uprisings in the west.

1460, December. Battle of Wakefield. In the last days of the year, Richard, Edmund, and their army sallied from their base at Sandal Castle, and were attacked by the larger Lancastrian army under **Henry Beaufort,** Duke of Somerset and **Henry Percy,** 3rd Earl of Northumberland based at nearby Pontefract Castle. Richard was killed in the battle itself, while Edmund and many others were killed in the pursuit, and the Yorkist remnants scattered. Details are sketchy, but most sources indicate that the Lancastrians employed several ruses to gain their success, including getting some 400 of their own troops into Sandal disguised in the livery of a Yorkist adherent.

1461, February 2. Battle of Mortimer's Cross (near Leominster). Edward, now Duke of York as the result of the death of his father and brother, defeated a Lancastrian army under the **Earl of Pembroke** and pursued the fugitives into the mountains. A number of captured Lancastrian nobles were summarily executed, initiating the brutality which marked the remainder of this war.

1461, February 17. Second Battle of St. Albans. Warwick's Yorkist army intercepted Margaret's troops advancing toward London. Warwick was defeated, and King Henry recaptured by the Lancastrians. Warwick withdrew to the southwest, where he joined young Edward, returning hastily toward London (February 22). While the Lancastrians continued a slow, triumphant advance toward London, Edward and Warwick rushed to the capital, where Edward proclaimed himself King (March 4). In the face of this unexpected development, Margaret and the Lancastrian army withdrew to Yorkshire. Edward pursued (March 19), defeating a delaying force at **Ferrybridge** on the Aire River (March 28).

1461, March 29. Battle of Towton. Palm Sunday was cold and windy with snow on the ground when the two armies met just south of Towton. Details are sketchy, as no eyewitness accounts have survived, but fighting raged most of the day, favoring the Yorkists only after the Duke of Norfolk's troops arrived to attack the Lancastrian left about midafternoon. The Lancastrian line collapsed, and thousands were slaughtered in the rout. Contemporary sources reckon the armies at about 30,000 each (but they were probably barely half that size) and give Yorkist casualties totalling 8,000 and Lancastrian losses as some 20,000. This was probably the most sanguinary battle ever fought on English soil.

1461–1471. Sporadic Lancastrian Uprisings. All were suppressed by Edward and Warwick.

1464, April 25. Battle of Hedgeley Moore (near Wooler). Lord Montague (Warwick's brother) defeated Sir **Ralph Percy,** who was killed.

1464, May 15. Battle of Hexham. Montague defeated another Lancastrian army, accompanied by both Henry VI and Queen Margaret. Following this battle, widespread executions crushed the Lancastrian cause. Margaret and Prince Edward fled to France; Henry went into hiding in a monastery in northern England, where he was found a year later, and imprisoned in the Tower of London.

1469. Rebellion of Robin of Redesdale. This Lancastrian rising in the north was really inspired by Warwick, who, with Edward's younger brother, **George, Duke of Clarence,** was plotting to seize power. Edward left London, planning to join a hurriedly raised army in the north, assembling under the **Earl of Pembroke.** Before Edward arrived, however, the rebels had defeated Pembroke and the **Earl of Devon** at the **Battle of Edgecote,** near Banbury. Edward, joined by Warwick and Clarence, who had followed him north with a large army, now became a virtual prisoner of the plotters. During subsequent months, however, Edward made skillful political use of his influence with the Yorkist nobility to reestablish power. As a result Warwick began to negotiate secretly with the Lancastrians.

1470, March. Lancastrian Uprising in the North. This was defeated near Empington, Rutlandshire, by a royal army at the **Battle of "Lose-Coat" Field.** The rebel leader, Sir **Robert Welles,** was captured and confessed that the rising had been provoked by Warwick, with the intention of making Clarence the king. Warwick and Clarence fled to France. There, through the mediation of King **Louis XI,** Warwick was reconciled with Margaret and plans made for a new Lancastrian effort with French support.

1470, September–October. Uprising in Yorkshire. This was inspired by Warwick, to attract Edward away from London. With the king in the north, Warwick and Clarence landed at Dartmouth and marched rapidly to London. Warwick released Henry VI from prison in the Tower and restored him to the throne. Edward, finding himself caught between the northern rebels and Warwick's growing army in the south, fled to Flanders to seek support from his brother-in-law, **Charles the Bold** of Burgundy.

1471, March. Return of Edward. Sailing from Flushing with about 1,500 men—mostly German and Flemish mercenaries—he eluded a Lancastrian fleet and landed at Ravenspur (in the Humber estuary). He marched quickly south, evading an army under the **Earl of Northumberland** (one of Warwick's many brothers), to Nottingham (March 23), where he learned that Warwick was assembling an army

at Coventry. The **Earl of Oxford** was to his left at Newark with another Lancastrian army. Feinting toward Newark, then toward Coventry (March 29), Edward slipped south, gathering adherents as he went. Of these the most important was his brother, the Duke of Clarence, who again changed sides to rejoin him (April 3). Continuing southward, Edward entered London (April 11). Warwick, whose army was now combined with those of Northumberland and Oxford, followed, and Edward marched out to meet him (April 12).

1471, April 14. Battle of Barnet. Edward had approximately 9,000–10,000 troops, Warwick 12,000–15,000. The battle began in the early morning, in a heavy fog. As the two armies groped for each other blindly, the right wing of each overlapped the left flank of the other. The Lancastrian right, under Oxford, defeated the Yorkist left, under **Lord Hastings,** while the Yorkist right under Richard of Gloucester (Edward's younger brother) was having comparable success in enveloping the Lancastrian left. Oxford's men pursued Hastings' routed wing. With considerable difficulty Oxford reassembled some of his men and attempted to rejoin the battle. Because of the mutual envelopments of the opposing left flanks, both armies had swung about at almost a right angle to their original front. In the fog Oxford's men blundered into the rear of his own center, under the **Earl of Somerset,** causing most of the Lancastrians to assume that they had been betrayed by treachery within their own ranks. In the ensuing rout, Warwick was cut down and killed. Margaret and Prince Edward landed in southwestern England that same day.

1471, April–May. Tewkesbury Campaign. Learning of the Barnet disaster and joining forces with the remnants of Warwick's army, now under Somerset, Margaret and Prince Edward attempted to reach Lancastrian strongholds in Wales. Edward hastily remobilized his army and marched rapidly from London, hoping to prevent the Lancastrian army from reaching Wales. He caught up with them near Tewkesbury (May 3) after both armies had made a series of grueling forced marches.

1471, May 4. Battle of Tewkesbury. The outnumbered Lancastrians, with perhaps 3,000 men, took up a strong position on a slope amid a warren of lanes, hedges, and thickets, between two streams, and awaited attack. Edward, with perhaps 4,000–4,500 men, opened the battle by bombarding the enemy position with artillery. Somerset reacted vigorously, using a hidden lane to attack the Yorkist left center. The struggle was fierce for a time, but **Richard of Gloucester** arrived to help his brother, and with 200 men-at-arms whom Edward had deployed in a wood on the Yorkist left struck Somerset's troops. The Lancastrian center under **Lord Wenlock** and **Prince Edward** failed to advance, and Somerset's troops broke. While Richard pursued Somerset's men, Edward attacked the Lancastrian center, and this force too broke and fled after a short melee. Prince Edward of Wales was killed in the rout, as were Lord Wenlock and **John Courtenay,** Earl of Devon, who had led the Lancastrian left wing; in all some 2,000 Lancastrians were killed. Somerset was captured and then executed, and Queen Margaret was also taken. After Tewkesbury, the Lancastrian cause was virtually extinct.

1471, May. Fauconberg's Revolt in Kent. **Thomas Fauconberg,** a bastard son of **William Neville,** landed in Kent in early May with the Calais garrison and a contingent of sailors from his fleet of over 40 ships. Gathering support in Kent, he marched on London and arrived before Southward with about 8,000 men (May 12). London and its Yorkist garrison under **Anthony Woodville,** Earl Rivers, stood firm against him, and a determined and well-planned assault (May 14) failed. Fauconberg withdrew to Blackheath (May 16), and by the time King Edward arrived in London (May 21), the rebel army had dispersed. Fauconberg surrendered his fleet 6 days later. Soon after, old King Henry VI was murdered in the Tower on Edward's orders.

1475. Edward Declares War against France. Allied with Charles the Bold of Burgundy, Edward led an invading army across the Channel. Following negotiations with adroit Louis XI,

however, he abandoned the expedition in return for a substantial cash payment.

1483, April 9. Death of Edward. He was succeeded by his 13-year-old son, **Edward V.** Richard of Gloucester became regent.

1483, June 26. Richard III's Usurpation of the Throne. He was evidently supported by most English nobles, who did not wish to be ruled by a child king.

1483, October. Revolt by the Duke of Buckingham. This was apparently in conspiracy with **Henry Tudor,** Earl of Richmond (now head of the House of Lancaster), and was promptly and efficiently crushed by Richard.

1483–1485. Growing Unrest. Widespread unpopularity of Richard, suspected of having murdered Edward V and his young brother, **Richard,** Duke of York, in the Tower.

1485, August. Invasion by Henry Tudor. With approximately 3,000 French mercenaries, he landed at Milford Haven. Picking up considerable support in Wales and western England, Henry marched into central England, avoiding interception by royalist (Yorkist) forces under the brothers Lord **Thomas** and Sir **William Stanley.** By the time Henry reached the vicinity of Bosworth, his army had grown to about 5,000 men. Richard, with about 10,000, marched rapidly northwest from London to meet him. Unknown to Richard, the Stanleys were in communication with Henry, and were merely waiting a suitable opportunity to betray the king.

1485, August 22. Battle of Bosworth. With the army of Sir William Stanley standing strangely aloof on high ground to the northwest, while that of Lord Stanley hovered to the southeast, the armies of Richard and Henry Tudor deployed in the open area between Sutton Cheney and Stanton. When Richard ordered an advance, his entire left wing, under the **Earl of Northumberland,** refused to move. Simultaneously the Stanleys moved in to join the Lancastrian army. With only a portion of his army remaining loyal, Richard led a handful of adherents in a vigorous charge into the center of the Lancastrian army, apparently hoping either to reach Henry and to win the battle singlehanded or to die like a king. Despite his gallantry, Richard was soon overwhelmed by superior numbers and killed. His death ended the Wars of the Roses.

THE BEGINNING OF THE TUDOR ERA, 1485–1500

Henry Tudor, crowned as **Henry VII,** reestablished peace in England and laid the foundations for four centuries of English glory. For more than a decade there were periodic instances of Yorkist insurgency, but these were mainly led by imposters, who were unable to obtain any substantial support. The principal events were:

1487. Uprising of Lambert Simnel. Pretending to be **Edward, Earl of Warwick** (son of the Duke of Clarence), Simnel was supported by Margaret of Burgundy, sister of Edward IV and Richard III. Simnel landed in Ireland with about 2,000 German landsknecht mercenaries, then crossed to England, where he was defeated and captured by royal troops in the hard-fought **Battle of Stokes** (June 16).

1488–1499. Insurrections of Perkin Warbeck. He impersonated Richard, younger of the murdered sons of Edward IV. He too was supported by Queen Margaret. With considerable foreign recognition and support, Perkin made several unsuccessful efforts to land in England and Ireland. After participating in an unsuccessful revolt in Cornwall (1497), he was captured and later executed 1499).

Scotland

As in past centuries, border warfare against England was more or less continuous during this period, but was less intense than in previous centuries. The entire period was marked by frequent struggles between the kings and recalcitrant nobles, led mostly by the House of Douglas. The principal events were:

1390–1406. Reign of Robert III.

1406. Capture of Prince James. The heir to the throne was captured by the English at sea on his way to school in France, just before the death of his father.

1406–1424. Imprisonment of James I in England. During most of this time Scotland was under the regency of the **Duke of Albany,** the young king's uncle. Albany, taking advantage of English occupation in the renewed activity of the Hundred Years' War in France, succeeded in recovering important areas of southern Scotland from the English. He also sent substantial contingents of Scottish soldiers to assist the French in the war against the English.

1411. Battle of Harlaw (near Inverurie). Donald, Lord of the Isles, invading Aberdeenshire, in an alliance with northern English nobles, was defeated by the **Earl of Mar.**

1424–1437. Effective Reign of James I. Having been ransomed (1423), for several years James maintained peace with England. Finally renewing the war, during the regency of Humphrey of Gloucester, James besieged Roxburgh, but was repulsed (1436).

1437–1460. Reign of James II. Constant bickering along the English border. James raided Northumberland (1456), and later successfully besieged Roxburgh, which surrendered shortly after the king was killed by the explosion of a siege cannon (1460).

1460–1488. Reign of James III. During this time the Scots took advantage of the Wars of the Roses to recover all remaining areas of southern Scotland still in English hands.

1464–1482. Truce with England.

1482. Renewal of the War against England. Edward IV supported the king's younger brother, **Alexander, Duke of Albany.** Richard, Duke of Gloucester (Richard III), accompanied Albany with an army to invade southern Scotland, recapturing Berwick from the Scots. James's efforts to defend his country were frustrated by a noble revolt, permitting Albany and Gloucester to capture Edinburgh and seize control of the country. Albany was later forced to flee.

1484. English Invasion. Albany and the 9th **Earl of Douglas** invaded Scotland with a small English force, but were repulsed.

1488. Death of James III. He was killed at the **Battle of Sauchieburn,** near Stirling, in an effort to suppress a noble's revolt.

FRANCE

The Crisis of the French Monarchy, 1400–1444

The reign of intermittently insane Charles VI brought France to the verge of dissolution. While the nation was being torn to pieces by violent civil war between two major factions of the nobility, Henry V of England established himself as the legally recognized heir to the French crown, and was on the verge of bringing the entire nation under English control (see p. 450). The trend of history in all Western Europe would have been vastly different had it not been for two events: the early and unexpected death of Henry, closely followed by the meteoric appearance of Joan of Arc, who ignited the spark of French nationalism. The Dauphin, who became Charles VII, was a weak, unenterprising prince, totally incapable of rallying his people to stop the steady, systematic English conquests. Despite himself, however, Charles was carried along to victory and the consolidation of the French monarchy by Joan of Arc and the French enthusiasm which she inspired. The principal events were:

1380–1422. Reign of Charles VI. An era of chaos and economic distress.

1411–1435. Civil War between Burgundians and the Armagnacs. (See p. 444)

1413–1414. Cabochian Revolt in Paris. (So-called from **Simon Caboch,** leader of the Parisian butchers.) This was connected with the great civil war raging between the noble factions. Paris and the French court were terrorized.

1413–1453. Conclusion of the Hundred Year's War. (See p. 445.)

1420. Treaty of Troyes. Henry V became virtual ruler of France (see p. 450).

1422. Deaths of Henry V and Charles VI.

1422–1429. Disintegration of France. The English and their Burgundian allies continued a

slow and systematic conquest, without effective opposition from the uncrowned Dauphin.

1429. Spiritual and Nationalistic Revival of France. This was initiated by Joan of Arc (see p. 451).

1440. The Praguerie. A noble revolt which followed the pattern of a previous similar uprising in Prague, Bohemia. The rebels, who included the Dauphin, later Louis XI, were defeated in a campaign of maneuver in Poitou by Charles' able constable, Arthur de Richemont, but were then pardoned by the king.

1443. Revolt of Count of Armagnac. This was suppressed by Dauphin Louis.

1444. Expedition against Switzerland. (See p. 465.)

The Rise of Military Professionalism, 1445–1500

The anarchy created by the disbanded companies of mercenary soldiers threatened the recovery of France brought about by the victories of Joan of Arc. This led Charles VII and his advisers to undertake a series of military reform measures of the utmost significance. The *compagnies d'ordonnance* were created as a permanent military establishment to maintain order in France (see p. 443). These companies, which included the most reliable of the officers and men of the disbanded companies roaming the country, soon restored order. At the same time two of the king's ablest military advisers, brothers **Jean** and **Gaspard Bureau,** established a permanent artillery organization, which their energy, scientific interests, and military skill soon made into an organization that was technically and organizationally far superior to any artillery elsewhere in the world. The creation of a French standing army, combined with this development of a superb—for that period—artillery organization within that army, enabled the French to eject the English from their country with ease when the war was resumed (see p. 453). Upon the death of Charles VII, his son Louis XI, one of the ablest kings in French history, continued the military and political reforms initiated in the previous reign, to consolidate royal power and to return the nation to peace and prosperity.

Louis's measures to increase central control were opposed by most of the great nobles, particularly the Dukes of Burgundy, whose realm extended from the Swiss Alps to Flanders on the North Sea. Charles the Bold (or the Rash) of Burgundy seemed to be on the verge of establishing a new and independent West European kingdom, rivaling in wealth and power both France and the Empire. He was frustrated, however, by Louis's diplomatic skill and by the military prowess of the Swiss (see p. 465). The power of the French monarchy continued to increase during the subsequent reign of Charles VIII, culminating in a French expedition to Italy, which for the first time revealed to the rest of Europe the efficiency of the new French army. This, the first appearance of what was essentially a modern-type professional army, marked the end of the Middle Ages and the dawn of the modern era. The principal events were:

1445–1449. Military Reforms. The establishment of the first standing army in Europe since the days of the Romans.

1449–1453. Expulsion of the English from France. (See p. 453.)

1461–1483. Reign of Louis XI. Feudal anarchy ended in France.

1465, July 13. Battle of Montlhéry. A drawn battle between Louis and a coalition of great nobles called the League of the Public Weal, led by Charles the Bold of Burgundy and the Dukes of Alençon, Berry, Bourbon, and Lorraine. This was primarily a cavalry battle, which ended with both armies in great confusion. Louis was forced to sign the unfavorable **Treaty of Conflans,** which he then evaded, splitting the noble league by adroit diplomacy.

1467–1477. Reign of Charles the Bold, Duke of Burgundy. He married Margaret of York (1468) to establish a new Anglo-Burgundian alliance with Edward IV of England, then treacherously captured Louis XI at a conference at Péronne. To regain his freedom Louis was forced to submit temporarily to the demands of Charles.

1471–1472. Charles the Bold's Invasion of Normandy and Île de France. He was repulsed at **Beauvais.**

1472–1475. Intermittent Warfare with Aragon. Operations in the Pyrenees region ended with French conquest of Roussillon, but their repulse from Catalonia.

1474–1477. War of Charles the Bold with the Swiss. The Swiss received diplomatic support from Louis XI. This resulted in the defeat, humiliation, and death of Charles the Bold (see pp. 465–466).

1475. Invasion of France by Edward IV. This ended without conflict by the **Treaty of Picquigny** (August 29), in which Louis XI bought off the English king (see p. 458).

1479, August 7. Battle of Guinegate. Louis, hoping to take advantage of the death of Charles, sent an army to invade the Netherlands, but the French were defeated by **Maximilian** of Hapsburg, son-in-law of Charles. French cavalry defeated the imperial horse, but Flemish infantry held the field, Maximilian fighting on foot in their midst. The French "free archers" (a new militia force) fled.

1483–1498. Reign of Charles VIII.

1488–1491. The "Mad War." A revolt of the Dukes of Orléans and Brittany supported by Henry VII of England and German Archduke Maximilian. Royal forces defeated the rebels at **St.-Aubin-du-Cormier** (1488), which settled the war, though inconclusive skirmishing continued for more than 2 years.

1493. Treaty of Barcelona. Charles obtained Spanish neutrality in preparation for his planned invasion of Italy.

1494–1496. Expedition to Italy. Charles conquered Naples as the first step in a contemplated crusade to recapture Constantinople or Jerusalem from the Moslems. His army of more than 25,000 men included 8,000 Swiss mercenaries. Most of the remainder of Europe, alarmed by the French success, formed a "Holy League" including Emperor Maximilian, Pope Alexander VI, Spain, England, Venice, and Milan. With his line of communications to France threatened, Charles left half of his army to hold Naples while he marched to northern Italy to make a junction with a small army under **Louis, Duke of Orléans** in Piedmont. Italian condottiere **Giovanni Francesco Gonzaga** of Mantua, commanding a combined mercenary army of Milan and Venice, moved to intercept the French retreat in the Apennines (July, 1495).

1495, July 6. Battle of Fornovo. Gonzaga, who had approximately 4,000 men-at-arms, 2,500 light cavalry, and 15,000 infantry (mostly crossbowmen) and some artillery, blocked the northern side of the Pass of Pontremoli, where there was adequate room for his condottiere cavalry to maneuver. The French army consisted of approximately 950 French and 100 Italian lances, plus 200 archers of the Royal Guard (4,200 cavalry), with 4,300 French and 3,000 Swiss infantry, plus the highly efficient field artillery corps. The French, realizing that the Italians would attempt to interfere with their passage, marched through the pass prepared for battle.

When the Italians attempted a flank attack in the defile, they were hampered by the river Taro's steep banks and its high waters, swollen by a heavy thunderstorm the day before, which had also muddied the ground. The alert French unlimbered their artillery and counterattacked vigorously. The Italians, used to a highly formalized and relatively apathetic form of combat (see p. 466), were dismayed by the violence of the French onslaught and were quickly routed. Total French losses were 200 killed and possibly an equal number wounded. The Italians lost 3,350 men killed, most of their wounded being cut down by the vigorously attacking French. Charles marched on into northern Italy and back to France.

1495–1498. Franco-Spanish Conflict for Control of Naples. (See p. 512.)

1498–1515. Reign of Louis XII.

1499. Invasion of Naples by Louis XII. (See p. 512.)

SCANDINAVIA

The Union of Kalmar continued, with some interruptions, throughout the century. Despite this, there were several civil wars and some external conflicts, and by the century's close the Union was showing signs of strain.

1397–1412. Reign of Margaret. After **Erik VII** was crowned (June 1397), Margaret was officially only regent, but remained the effective ruler of all three kingdoms until she died at Flensburg (October 28, 1412).

1412–1439. Reign of Erik VII of Pomerania.

Crowned in 1397, he did not gain full power until Margaret died (1412). He waged war against the Counts of Holstein for control of Schleswig (1416–1422). The Second Holstein War led to war with the Hanseatic League (1426–1435). Erik defeated a Hanse fleet and then imposed the first tolls for ships plying the Danish–Swedish straits (1428). Schleswig fell to a Hanseatic siege (1432), and the Hanse naval blockade led to a Swedish miners' revolt led by **Engelbrecht Engelbrechtsson** (1434). Erik was forced to sue for peace. Swedish resentment of Erik's pro-Danish stance led to demands for constitutional reform, and when he refused these he was deposed in Denmark and Sweden (1439), then Norway (1442), temporarily breaking the Union of Kalmar. Erik's bid to regain his thrones failed (1449), and he retired to Pomerania, dying there (June? 1459).

1439–1448. Christopher III of Bavaria. He succeeded his maternal uncle, Erik VII, and was crowned in Denmark (1439), Sweden (1441), and Norway (1442), reestablishing the Union. He suppressed a peasant rebellion in Jutland (1441), but was compelled to grant further concessions to the Hanseatic League, and died without heirs, breaking the Union of Kalmar for the second time in a decade (1448).

1448–1481. Reign of Christian I Oldenburg. The son of Count **Dietrich the Happy** of Oldenburg, Christian was elected to succeed Christopher III in Denmark (1448) and Norway (1450). He fought a war with Sweden (1451–1457), eventually deposing **Charles VIII.** He struggled with the Swedes for over a decade to reestablish his rule there, but his hopes were dashed when **Sten Sture** the Elder defeated Christian at the **Battle of Brunkeberg** (October 10, 1471). Scandinavian rule in the Orkney and Shetland islands ended with the marriage of Christian's daughter **Margaret** to **James III** of Scotland (1469). Christian founded the University of Copenhagen (1479) before his death (May 1481).

1448–1457. First Reign of Charles VIII of Sweden. After Christopher III was deposed, the Swedes chose **Karl Knutsson Bonde** for their king. He was soon at war with Christian I (1451), and his domestic policies aroused opposition among the nobility. The Oxenstiernas and Vasas deposed him, and elected **Christian I** king (1457).

1464–1465. Second Reign of Charles VIII. The anti-Danish noble faction, led by the Axelsson of Tott and Sture families, brought Charles back from exile for a brief reign.

1467–1470. Third Reign of Charles VIII. The Tott Sture faction brought Charles back again, but he was merely a figurehead. He died in Stockholm (May 1470).

1468–1474. The First "Cod War." English trade with Iceland and English fishing in Icelandic waters provoked a series of naval clashes between the English and the Danes and their Hanseatic allies. Similar naval clashes, generally over fishing rights, have marked British-Icelandic relations ever since, and are colloquially known as "Cod Wars."

1470–1497. Regency of Sten Sture the Elder. After Charles VIII's death, Sture became ruler of Sweden. He remained in power until King **Hans** of Norway and Denmark was crowned king of Sweden as part of the peace settlement between Sweden and Denmark (1497), reestablishing the Union.

1481–1513. Reign of Hans (John) I. He succeeded his father Christian I as king of both Denmark and Norway (1483). Accepted as king of Sweden (1483), he was not crowned until the end of a war with Sweden (1493–1497).

GERMANY

During this century the power and importance of the Holy Roman Emperors declined markedly. There were several reasons, not least being generally inept leadership, permitting the usually unruly German nobility to exercise even more independence than usual. The wealth and power of the towns increased, at the expense of both emperors and nobles. The Hanseatic League remained influential, although its strength declined in the face of growing Scandinavian power in the Baltic. Germany was also overshadowed during much of the century by the resurgency of neighboring countries under strong leadership. Bohemia reasserted its independence, while the increasingly warlike Swiss smashed every effort to reestablish imperial control over their cantons. The principal events were:

1400–1410. Four-Sided Civil War. There were four rival emperors: **Wenceslas, Rupert III, Sigismund,** and **Jobst.** Sigismund of Luxembourg, King of Hungary, was eventually triumphant.

1410, July 10. Battle of Tannenberg. The Teutonic Knights were decisively defeated by a Polish-Lithuanian army, starting the decline of the order (see p. 471).

1410–1437. Reign of Sigismund. He was also King of Hungary and (nominally) of Bohemia.

1420–1431. Hussite Wars in Bohemia. (See p. 470).

1438–1439. Reign of Albert II (son-in-law of Sigismund). This began the Hapsburg line of emperors. An able soldier, he spent most of his brief reign defending Hungary against the Turks.

1440–1493. Reign of Frederick III.

1451. Revolts in Austria and Hungary. These were suppressed by Frederick's subordinates.

1463–1485. Bohemian and Hungarian Raids on Austria. George of Podebrad, King of Bohemia, and **Mathias Corvinus,** King of Hungary, ravaged Austria, culminating in the expulsion of Frederick from Vienna by the Hungarians (see pp. 471, 474). Archduke **Maximilian,** Frederick's son, elected King of

the Romans (1486), assumed most of the imperial authority from his inept father.

1479. Battle of Guinegate. (See p. 461.) This was the beginning of the protracted Valois-Hapsburg struggle (see p. 511).

1486–1489. Renewed War with France. Maximilian supported the revolt of **Francis II,** Duke of Brittany (see p. 462). Maximilian, preparing to operate against France from the Netherlands, was forced to spend most of his time suppressing opponents of this unpopular war in the Low Countries. Maximilian created a permanent Dutch admiralty office, thereby laying the administrative foundations of the Dutch navy (1488).

c. 1486. Creation of the Landsknechts in Germany. This followed the Swiss example (see p. 442). It was the beginning of an imperial standing army.

1491. Maximilian's Reconquest of Vienna. He drove the Hungarians from Austria.

1492. Battle of Villach. Maximilian repulsed an invading Turk army (see p. 477).

1493–1519. Reign of Maximilian I.

1494. Rebellion in the Netherlands. Suppressed by Maximilian.

1499. Maximilian's Invasion of Switzerland (See p. 466.)

SWITZERLAND

During the early part of this century the Swiss so perfected their infantry organization of pikemen, halberdiers, and crossbowmen that Switzerland became the leading military power of Europe. The ascendancy of infantry over cavalry, begun by the defensive tactics of Edward III at Crécy, were brought to culmination by the offensive capability of the highly disciplined, well-trained, maneuverable Swiss columns. That the Swiss did not utilize this tremendous military superiority for wars of foreign conquest was not due to lack of ambition, or any early tendency toward neutrality. It was, rather, a reflection of the imperfect Swiss federal system, which did not permit any unified national foreign policy, or even the development of a consistent and cohesive military strategy. Tactically, during the 15th century, the Swiss were practically invincible. Strategically, most Swiss operations, other than in defense of their homeland, were usually abortive. As a result this nation of warriors had little influence on the overall history of Europe during this century, save to the extent that more able politicians and strategists of other nations were able to use Swiss mercenaries to further their own ambitions. The Swiss reputation of invincibility, combined with their ruthless slaughter of foes, inspired such terror that by the end of the 15th century Swiss battles were usually won before the engagement began. The principal events were:

1403–1416. War with Savoy. The Swiss gained control of the southern Alpine passes into northern Italy.

1415. Conquest of the Aargau Region. The Swiss defeated Austria, with the support of Emperor Sigismund.

1422–1426. Intermittent Warfare with Milan. Duke **Filippo Maria Visconti** intrigued to prevent the Swiss cantons from acting in concert. The result was the temporary loss of Bellinzona to Milan after Swiss defeat at the hands of the Milanese army under **Carmagnola Arbedo** (June 30, 1422).

1436–1450. Civil War. Zurich fought neighboring cantons, with frequent intervention and participation by Germans and French.

1444, August 24. Battle of St. Jakob (near Basel). Frederick III, hoping to take advantage of the civil war to reestablish Hapsburg control over Switzerland, obtained assistance from Charles VII of France, who was desperately seeking employment for troops disbanded during a long truce in the Hundred Years' War (see p. 452). The Dauphin Louis later Louis XI) collected an army of 30,000 and invaded Switzerland. Near Basel the French veterans were met and attacked by approximately 1,500 Swiss. In the ensuing battle, the Swiss force was annihilated, but they killed approximately 3,000 French attackers. The determination and efficiency of the Swiss infantry so discouraged the French that they withdrew, and turned instead to harass Frederick's domains in Alsace. This defeat made the reputation of the Swiss as Europe's finest soldiers.

1460. Renewed War with Austria. The Swiss conquered Thurgau.

1468. Renewed Civil War. The Austrians again attempted to exploit civil war in Switzerland. Patching up internal peace, the Swiss again repulsed the Austrians.

1474–1477. War with Charles the Bold of Burgundy. The Swiss were allied with the south German cities and the Hapsburgs (who were fearful of Charles' ambitions of conquest in Alsace), and were supported by subsidies from Louis XI of France, bitter enemy of Charles the Bold. After repulsing Charles at **Héricourt** (November 13, 1474), the Swiss and their allies occupied Burgundian territories on the borders of Switzerland and Alsace (1474–1475). Charles recaptured Granson (February 1476) and hanged the entire Swiss garrison.

1476, March 2. Battle of Granson. A Swiss army of 18,000 men, without artillery, attacked the Burgundian army of 15,000. The Swiss, eager to avenge their slaughtered countrymen, advanced in 3 heavy columns, echeloned to the left rear, moving directly into combat without deploying, in typical Swiss fashion (see p. 441). The speed of the Swiss advance did not give the Burgundians time to make much use of their numerous artillery and missile units. Charles attempted a double envelopment of the leading Swiss column before the other 2 arrived, but as his troops were shifting to make this attack, they caught sight of the other Swiss columns and retreated in panic. The Burgundians lost about 1,000 men, the Swiss only 200.

1476, June 22. Battle of Morat. Charles, having rallied and reorganized his army, moved to drive the Swiss from the Burgundian territory they had occupied. With 20,000 men he besieged Morat, near Bern and Fribourg, establishing a heavily entrenched and palisaded defensive position to cover his besieging troops. Inexplicably, he established no outposts or patrols. The weather being rainy, only about one-fifth of the Burgundians were in the entrenchments; the remainder were in their nearby camp. Moving with typical rapidity, a Swiss army of 25,000 infantry, plus about 1,000 Austrian and German allied cavalry, advanced unexpectedly against the Burgundian entrenchments, again moving directly into battle without taking time to deploy. The few Burgundians holding the entrenchments were driven out in confusion, just as the remainder of the Burgundian army, belatedly alerted, was rushing to join in the defense. The Swiss simply overran the confused Burgundians. About one-third of Charles' army, with their backs to the lake, were cut off by the Swiss advance and were slaughtered to a man. Burgundian losses were between 7,000 and 10,000 men killed; Swiss losses were negligible.

1477, January 5. Battle of Nancy. The Swiss and their allies advanced into Burgundian territory, where Charles met them with a new and reorganized army. Again the vigorous Swiss infantry assault, combined with an envelopment of the Burgundian left flank, resulted in a complete victory. Charles was killed as he fought bravely to try to cover his army's retreat.

1478. War with Milan. This ended with a Swiss victory at the **Battle of Giornico** (December 28).

1499. War with Emperor Maximilian. This was the result of a boundary dispute on the Austrian border. The emperor was supported by the south German cities; the Swiss were allied with the French. For a few months fighting raged all along the northern and eastern frontier, the principal battles being fought at **Hard, Bruderholz, Schwaderloh, Frastenz,** and **Calven.** The decisive battle was fought at **Dornach** (July 22), when the Swiss defeated Maximilian himself, who was forced by the **Treaty of Basel** (September 22) to grant to Switzerland virtual independence from the Empire.

ITALY

Military events in Italy in this century were of little tactical interest, but witnessed administrative and strategic innovations which had great significance for future military developments in Western Europe. The many small states of Italy were often at war with each other and with neighboring countries, and in this respect Italy has been compared with Greece in the 3rd and 4th centuries B.C. Militarily, however, there were a number of differences. The wars in Greece were dominated by armies of hoplites (citizen heavy infantry), supported by small forces of cavalry and larger numbers of mercenary or allied light troops. The armies of 15th-century Italy placed much greater importance on heavily armored men-at-arms, almost entirely in mercenary units.

These mercenaries were called *condottieri,* after the Italian word for contract, *condotta.* The first *condottieri* of the 1320s and 1330s were foreigners, but after the late 14th century most of them were native Italians. The first few generations of *condottieri* were freelances, who changed employers as they chose and generally sold their loyalties to the highest bidder. Some, like **John Hawkwood,** were remarkably faithful to one city; in Hawkwood's case this was Florence. By the 1420s and 1430s, however, the fundamental relationship between *condottiere* and employer had changed from one of temporary expediency to a long-term tie of patron and client. By midcentury, *condottieri* generally spent their entire careers in the service of one government (or occasionally of its allies). Those few who retained more entrepreneurial autonomy, like **Frederigo da Montefeltro** of Urbino, had an independent power base, usually a small state, and even those usually served a limited circle of employers.

The traditional—and essentially accurate—view of warfare under the *condottieri,* exemplified by the acerbic comments of **Niccolo Macchiavelli,** is that it was desultory, nearly bloodless, and generally pointless. There were, however, several reasons. Mercenary commanders were generally reluctant to incur heavy casualties, since their soldiers represented their operating capital and their only method of generating income. Further, warfare in Renaissance Italy was highly politicized, and commanders were often directed not to prosecute operations too vigorously, and occasionally were even bribed to proceed slowly. This latter course was naturally discouraged by employers: **Francesco da Bussone,** called Carmagnola, was arrested and executed by his Venetian employers for lack of military zeal (1431). To be fair, Carmagnola was ill, and was concerned for the welfare of relatives in Milan, Venice's principal enemy. Finally, battles among the *condottieri* sometimes lacked the decisiveness, and casualty lists, of engagements elsewhere in Europe because Italian troops were especially well-equipped and well-armored; and second, the proliferation of castles and fortified towns made success in the field indecisive, since a beaten army could simply take refuge in a fortress.

All of these factors produced a military environment which encouraged strategems, subtlety, and small-scale actions, and avoided major field battles. The military environment also placed an emphasis on geographical objectives. It did produce some notable innovations, including the use of field entrenchments and light cavalry, as well as large-scale employment of infantry (and sometimes mounted troops) equipped with firearms.

The fundamental weakness of the *condottieri* system was political rather than military. Italy's persistent state of disunion, and the willingness of one state or another to call on outside powers for aid, ensured a nearly endless round of warfare among the Italian states. These largely bloodless conflicts continued into the 16th century, and ended only with the close of the Hapsburg-Valois wars after the **Treaty of Cateau-Cambésis** (1559; see p. 521). The *condottieri*'s house of cards collapsed when Charles VIII of France led a hard-fighting professional army into Italy toward the close of the century (see p. 512). The principal events were:

1402–1454. Series of Wars between Milan, Venice, and Florence.

1402–1405. Fragmentation of Milan. This followed the death of **Gian Galeazzo Visconti.** Venice seized the former Milanese possessions of Padua, Bassano, Vicenza, and Verona to control most of northeast Italy.

1405–1406. Conquest of Pisa by Florence.

1414–1435. Reign of Joanna II of Naples. This was a period of incessant conflict between Joanna and her favorites, on the one hand, and the barons, on the other, with frequent intervention by Aragon and the papacy.

1416–1453. Intermittent Turkish-Venetian Naval Warfare. The goal was domination of the Aegean area.

1420. Venetian Conquest of Fruili (from the Emperor Sigismund).

1423–1454. Wars of Milan against Florence and Venice. The result of this intermittent warfare was to the general advantage of Venice and Florence.

1435–1442. Conquest of Naples by Alfonso V of Aragon.

1447–1450. Civil War in Milan. This ended in the victory of former condottiere **Francisco Sforza,** who began his military career in Naples (his family was from the Romagna, and originally named "Attendolo").

1463–1479. War between Venice and Turkey. The Turks were victorious; beginning of the decline of Venice (see p. 477).

1482–1484. Ferrarese War. Milan, Florence (under **Lorenzo de' Medici**), and Naples combined to prevent Venice from seizing Ferrara; Pope **Sixtus IV** was allied with Venice.

1485–1486. Neapolitan Revolt. The Neapolitan barons, supported by Pope **Innocent VIII,** rose against King **Ferdinand I.** Ferdinand was supported successfully by Florence.

1494–1495. French Invasion of Naples. Charles VIII led the invaders, following the death of Ferdinand I (see p. 512).

1492–1503. Reign of Pope Alexander VI. This was marked by pacification of Romagna by his son, **Cesare Borgia,** breaking the power of the great Roman families (1492).

1495–1498. Spanish Intervention in Naples. Spanish forces were led by **Gonzalo de Cordova** (see p. 512), who was defeated by **Gilbert, Duke of Montpensier** at the **Battle of Seminara** (1495). Cordova's subsequent cautious defensive-offensive tactics were successful against Montpensier at **Atella** (July 1496).

1499. Renewed French Invasion of Naples. (See p. 512.)

IBERIAN PENINSULA

During the first three-quarters of the 15th century, Castile was wracked by anarchy, while Aragon, despite some internal disturbances, maintained its position as the principal maritime power of the western Mediterranean. During much of this time Aragon was involved in a series of struggles for the control of Naples (see above). The last quarter of the century witnessed the beginning of an amazing new era in a united and revitalized country, which in a few years would lead Spain to a preeminent global position of power and wealth. This was the result of the marriage of **Ferdinand II** of Aragon with Queen **Isabella** of Castile, followed by their joint conquest of Granada, completing the expulsion of the Moslem Moors from Spain. This, in turn, was followed immediately by a period of vigorous overseas exploration and expansion, initiated by the earth-shaking voyage of Columbus to discover the Western Hemisphere. The principal events were:

1406–1474. Decline of Castile. The country came close to disintegration in repeated civil wars, in which Aragon and Navarre frequently intervened; royal power was tenuously preserved only by the support of the towns against the rebellious nobility.

1409–1442. Aragon's Involvement in Naples. Intermittent warfare, at first unsuccessful, finally ended in conquest by **Alfonso V** (see p. 467).

1458–1479. Unrest and Intermittent Revolt in Catalonia. This was during the reign of **John II** of Aragon.

1474–1479. Civil War of Succession in Castile. After the death of Henry IV, rival claimants were his sister, **Isabella,** wife of Prince **Ferdinand of Aragon,** and **Joan,** Henry's doubtfully legitimate daughter, who was wife of **Affonso V** of Portugal. Portuguese intervention was ended by Ferdinand's victory at the **Battle of Toro** (1476). Ferdinand and Isabella then triumphed over Joan's adherents in Castile.

1479–1504. Joint Reigns of Isabella of Castile (1474–1504) and Ferdinand of Aragon (1479–1516). Pope Innocent VIII gave them the title of "Catholic Kings."

Final Christian-Moslem War in Spain, 1481–1492

This was begun by the raid of **Muley Abu'l Hassan,** bellicose King of Granada, to capture Zahara, near Ronda (December 26, 1481). Marquis **Rodrigo Ponce de León** of Cadiz raided back to seize Alhama, 25 miles from Granada (February 28, 1482). When Hassan invested Alhama, Ferdinand and Isabella came to the assistance of Ponce de León, forcing the Moslems to raise the siege (May). Incessant war continued for 10 years. Christian successes were facilitated by protracted internal civil war between Hassan and two rivals for the throne of Granada: his brother, **Abdullah el Zagal,** and his son, **Abu Abdulla Mohammed Boabdil.** The principal events were:

1482, July 1. Battle of Loja. Ferdinand was ambushed and defeated.

1483. Capture of Boabdil. Captured by the Christians at **Lucena,** he obtained his release by acknowledging complete suzerainty of Ferdinand and Isabella over Granada. This was not accepted, however, by Hassan or Zagal.

1483–1487. Spanish Military Reorganization. Centered around a permanent constabulary, loosely modeled on the French *compagnies d'ordonnance* (see p. 443). A contingent of Swiss mercenaries became a model for the new constabulary infantry. Constant campaigning against the Moors, and the strict discipline demanded by the sovereigns, resulted in the emergence of one of the finest armies in Europe. The Spanish fleet was also built up, preventing intervention from Morocco.

1485–1487. Campaigns against Málaga. Following the capture of Loja (1486) and systematic Spanish conquest of the surrounding region, Málaga was besieged and captured (May–August 1487).

1488–1490. Approach to Granada. The surrounding mountain regions were systematically conquered and Granada isolated by the capture of the fortresses of **Baza** and **Almería** (1489).

1491, April–1492, January 2. The Siege of Granada. The Christians, in overwhelming force, repulsed all desperate Moorish sorties. The final surrender of the city by Boabdil concluded 8 centuries of Moslem-Christian struggle for control of Spain.

1492–1502. The Four Voyages of Columbus. He began the colonization of America on the island of Hispaniola (1493). Following a revolt against the authority of Columbus there (1498), royal authority of the Indies was established under the first governor, **Francisco de Bobadilla** (1499).

Portugal

The conclusion of long-drawn-out wars with the Moslem Moors and with Castile late in the 14th century brought unexpected and unwanted inactivity to the warlike, vigorous Portuguese. An outlet for this energy was soon found in a series of overseas expeditions of conquest and exploration, which made Portugal the leading maritime nation of the world at the close of the 15th century. The principal events were:

1385–1433. Reign of John I. The first ruler of the House of Avis, which brought Portugal to glory.

1415. Conquest of Ceuta from the Moors. Prince **Henry,** third son of John, later renowned as "the Navigator," particularly distinguished himself.

1418–1460. Beginning of Portuguese Explorations. Africa and the South Atlantic were explored under the sponsorship of Henry the Navigator.

1433–1438. Reign of Edward I.

1437. Repulse at Tangier. An invasion expedition was disastrously repulsed by the Moors.

1438–1481. Reign of Affonso V (infant son of Edward). During his minority there was a three-way struggle for power and the regency between Affonso's mother, **Eleanora;** his uncle, **Peter;** and the powerful family of Braganza. This culminated in Affonso's seizing power himself and turning against the able Peter, who was defeated and killed at the **Battle of Alfarrobeira** (May 1449).

1463–1476. War with Morocco. The Portuguese captured Tangier and Casablanca.

1474–1476. Intervention in Castile. (See p. 468.)

1481–1495. Reign of John II. He energetically revived the program of exploration which had lagged following the death of Henry (1469).

1481–1483. Noble Revolt. Ferdinand of Braganza, the rebel leader, was decisively defeated by John.

EASTERN EUROPE

BOHEMIA

A period of turmoil and civil war in Bohemia, the fierce Hussite Wars foreshadowing the violent religious struggles of the Reformation. The early years of these wars were dominated by the military genius of John Ziska. The principal events were:

1378–1419. Reign of Wenceslas IV. This was marked by almost continuous conflicts with the barons. For much of his reign **Sigismund** (stepbrother of Wenceslas and later emperor) actually ruled the country.

1419–1437. Reign of Sigismund. He was not recognized by the Bohemians for 17 years (1436).

1419–1436. Hussite Wars. These resulted from the persecution of the religious followers of **John Huss,** executed by Sigismund (1415). The Hussites were considered heretics by the majority of the Catholic population of Bohemia. Under the leadership of **John Ziska,** an able soldier with much experience as a mercenary in Poland's wars against the Teutonic Knights, they established a fortified mountain community at Tabor, near Usti, which served as a base of operations. Ziska developed a new and effective defensive-offensive tactical system, built around a **wagenburg,** or wagon fortress, equipped with artillery (see p. 440). When not on campaign, Ziska drilled his soldiers endlessly.

1419. Battle of Sudoner. Ziska, leading the Hussites on their exodus from Prague to Tabor, repulsed pursuing Catholics. Soon after, Pope **Martin V** declared a Bohemian Crusade against the Hussites (March 17, 1420).

1419, June 30–July 20. Siege of Prague. Sigismund, asserting his claim to the throne of Bohemia, invested Prague. The citizens appealed to the Hussites for assistance.

1419, July 30. Battle of Prague. Ziska led an army to Prague, taking up a *wagenburg* position on Vitkov Hill, where he decisively repulsed an attack by Sigismund, who was forced to withdraw from Bohemia. The continuing religious struggle between Catholics and Hussites now became inextricably involved with the political struggle between the Bohemians and the unwanted ruler Sigismund.

1421. Return of Sigismund. He was defeated by Ziska and the wagon defenses at the **Battles of Lutitz** and **Kuttenberg.** During this campaign Ziska lost his remaining eye in the Siege of Rabi. (He had lost his other eye in the earlier civil wars against Wenceslas.) Despite total blindness, Ziska retained control of the Hussites.

1422, January. Sigismund Returns Again. Again he was severely defeated by Ziska in the **Battles of Nebovid** (January 6) and of **Nemecky Brod** (January 10).

1423. Civil War among the Hussites. The Utraquist sect and the citizens of Prague opposed the Taborites. Ziska reestablished Taborite supremacy by victories at the **Battles of Horid** (April 27) and **Strachov** (August 4).

1423, September–October. Ziska's Invasion of Hungary. This failed, despite initial success.

1424. Renewal of Hussite Civil War. Ziska reestablished peace by decisive victories over Utraquists and Praguers at the **Battles of Skalic** (January 6) and **Malesov** (June 7). Ziska then planned an invasion of Moravia, but died before it could be undertaken. Military command of the Hussites was assumed by **Prokop "the Great,"** a priest and former Utraquist.

1426–1427. Renewed German Invasions. The Germans were decisively defeated by Prokop at the **Battles of Aussig** (1426) and **Tachau** (1427).

1427–1432. Hussite Raids. Repeated expeditions went into Germany, Hungary, Misnia, and Silesia, and ranged as far as the Baltic.

1433–1434. Renewed Civil War. The Taborites opposed the combined forces of nobles and Utraquists. The Taborite leaders, Prokop the Great and unrelated **Prokop the Less,** were decisively defeated and killed at the **Battle of Lipan** (June 16).

1436. Peace in Bohemia. All factions accepted Sigismund as king.

1437–1439. Reign of Albert of Austria. He died during a war with Ladislas of Poland in a confused struggle for succession to the thrones of Bohemia, Poland, and Hungary (see p. 473).

1439–1457. Reign of Ladislas Posthumous (son of Albert, born after his father's death). Bohemia was racked by civil wars during a period of ineffectual guardianship by Emperor Frederick III.

1448. Emergence of George of Podebrad. He seized power and proclaimed himself regent; recognized as such by Frederick III. Podebrad captured Gabor, to assure the ascendancy of the Utraquist faction of the Hussite religion as the official religion of Bohemia (1452).

1459–1471. Reign of George of Podebrad. He took the crown after the death of Ladislas.

1462–1471. Renewed Civil War in Bohemia. Pope **Paul II** excommunicated George and all Hussites as heretics (1465). King **Mathias** of Hungary supported the Catholic rebels and had himself proclaimed King of Bohemia (1469).

1471–1516. Reign of Ladislas II (son of Casimir of Poland). This began with continuing war against Mathias of Hungary, was ended by the **Treaty of Olomouc;** Mathias recognized Ladislas as King of Bohemia, but received Moravia, Silesia, and Licesia in compensation (1478). During the remainder of this reign the power of the nobility grew greatly at the expense of the authority of the weak king.

POLAND

This was a century of transition for Poland, during which the crude, semibarbarous society matured and foreign threats from the northeast and south were repelled. The century ended with Poland emerging as a leading power in Europe. The principal events were:

1386–1434. Reign of Jagiello (Ladislas II).* The former Grand Duke of Lithuania, he had become King of Poland by marriage (see p. 410). During his long reign Jagiello had to contend with three formidable antagonistic forces: the turbulence of the unruly Polish nobles; the separatist tendencies of Lithuania under his cousin **Witowt,** who succeeded him as grand duke; and the ruthless aggression of the Order of Teutonic Knights. By a combination of diplomacy and generalship, Jagiello maintained ascendancy over each of these forces, although he could not eliminate any of them.

1410, July 15. Battle of Tannenberg. The combined Polish and Lithuanian forces of Jagiello and Witowt decisively defeated the Teutonic Knights. Included in the Polish army was a force of Bohemian mercenaries under John Ziska (see p. 470) and some Russian and Tartar contingents. Following the victory the Poles and Lithuanians devastated the holdings of the Teutonic Order in Prussia, but, lacking support from his unruly Polish nobles, Jagiello concluded the **First Peace of Thorn** with the Knights, permitting them to retain most of their lands. Nonetheless, the terrible losses suffered at Tannenberg, combined with the disastrous blow to prestige, marked the beginning of the order's decline.

1434–1444. Reign of Ladislas III (son of Ja-

---

* Some authorities give him the title of Ladislas V.

giello). The first years were under the regency of Cardinal **Olesnicki,** a great diplomat and administrator.

1435. Lithuanian Revolt. Despite support from the Teutonic Knights, the rebels were subdued by a Polish army under Olesnicki.

1444. Death of Ladislas. He was defeated and killed at the **Battle of Varna** against the Turks (see p. 474).

1447–1492. Reign of Casimir IV (younger brother of Ladislas).

1454–1466. War with the Teutonic Order. The Poles supported a Prussian revolt. After desultory campaigns, Casimir won a decisive victory at the **Battle of Puck** (September 17, 1462). The war was concluded by the **Second Peace of Thorn,** which gave Poland an outlet to the Baltic by cession of the province of Pomerania and the region around the mouth of the Vistula. The Teutonic Knights, retaining Prussia, acknowledged the suzerainty of Poland.

1471–1478. War with Hungary. This resulted from the election of **Ladislas,** son of Casimir, as King of Bohemia, in opposition to the claims of Mathias of Hungary (see pp. 471, 474).

1478–1493. Semiwar with Hungary. This followed Mathias' efforts to stir up border troubles, to encourage the Teutonic Order to rebel against Poland, and to incite the Tartars to ravage Lithuania.

1475–1476. Turkish Invasion of Moldavia. This was repelled by **Stephen,** Voivode of Moldavia, a vassal of Poland.

1484–1485. Expedition to Moldavia. A new Turkish invasion caused Casimir to lead an army to assist Stephen in expelling the Turks. Casimir was leading an army of 20,000 men to cross the Pruth River into Turkish territory when Ottoman Emperor **Bayazid II,** then engaged in war in Egypt, made a truce (see p. 477).

1487–1491. War with the Golden Horde. At the urging of the Ottoman Sultan, the now-tiny remnant of the Golden Horde attacked Poland-Lithuania. This required a complex operation through the territory of Moscow-Russia, and across Moscow's lines of communications in the ongoing wars with Viatka and Kazan. Ivan of Moscow, however, initially encouraged the Horde's offensive, since it was bound to weaken one of Moscow's principal rivals. The Crimean Tartars and the Turks simultaneously attacked Polish-Lithuanian frontier outposts in southern Ukraine, while Ivan also conducted raids against Lithuania's eastern frontier. Most of the operations took place in Poldavia and Galicia, with the Tartars at one time reaching Lublin. The war was concluded by a decisive Polish victory at **Zaslavl** in Poldavia (1491). The Golden Horde was near collapse.

1492–1501. Reign of John Albert (son of Casimir). He increased the authority of the monarchy by reducing the power of the great nobles in the cities.

1497–1498. Renewed Struggle with the Turks in Moldavia. Stephen played the Turks off against the Poles, who were repulsed from Moldavia. Turkish forces raided deep into Poland (see p. 477). The independence of Moldavia was recognized by both Poles and Turks (1499).

RUSSIA

During this century the Grand Dukes of Moscow struggled to obtain increasing independence from the suzerainty of the Tartars of the Golden Horde. At the same time they were engaged in a continuing series of frontier wars with the Khanate of Kazan and with Poland-Lithuania. Shortly before the end of the century, **Ivan III** (the Great) of Moscow threw off the Tartar yoke. The principal events were:

1389–1425. Reign of Basil I of Moscow. He annexed Nijni-Novgorod, and was engaged in practically continuous warfare with both Tartars and Lithuanians.

1425–1462. Reign of Basil II. This was marked by anarchy and civil war against rival princes **Yuri** and **Shemyaka.**

1447, 1451. Tartar Invasions. Both failed, though the second reached the walls of Moscow.

1456. Moscow-Novgorod War. Basil's victory at the **Battle of Rusa** hastened the decline of Novgorod and the rise of Moscow.

1462–1505. Reign of Ivan III. He was the first national sovereign of Russia. He defeated and then conquered Novgorod (1471–1478), later annexing Tver (1485). Hostilities against Lithuania, Viatka, and the Khanate of Kazan were endemic throughout his reign. By his marriage to **Zoë,** only niece of the last Byzantine Emperor, he established a claim for himself and subsequent Russian monarchs as the successors of the Byzantine emperors.

1478–1489. War with Lithuania. Frequent raids made by both sides, with occasional inconclusive major operations. Ivan was supported by the Crimean Tartars.

1480. Independence from the Tartars. Ivan refused to pay tribute; he then repulsed poorly coordinated Tartar efforts to reestablish their sovereignty. Ivan benefitted from hostility between the Great Horde (Golden Horde) and the Crimean Tartars.

1492–1503. Intermittent War with Lithuania. Ivan, generally successful, was supported by **Mengli Girai,** Tartar Khan of the Crimea.

1495. Alliance with Hans I of Denmark. Ivan made common cause with Hans against Sweden, and then invaded Swedish Finland.

1495, June–December. Russian Invasion of Finland. A Russian army invaded southeastern Finland (June), and then besieged the frontier fortress of Vyborg (Viipuri) (September). Sten Sture, regent of Sweden, led an army of 10,000 men to Vyborg's relief, but the Russians withdrew as he approached (December 1495).

1496, April–September. Further Campaigning in Finland. Another Russian army invaded Finland (April), and reached the Gulf of Bothnia before it withdrew. Meanwhile, Sten Sture attacked, captured, and sacked Ivangorod (September?).

1497. Peace. Sweden and Russia signed a 6-year peace treaty, paralleling the settlement between Denmark and Sweden (see p. 463).

HUNGARY

Hungary succeeded the Byzantine Empire as the bulwark of Christendom against the rising tide of the Moslem Turks. Although their kings were constantly distracted by feudal claims and dynastic adventures elsewhere in Central and Eastern Europe, the Hungarians fought the Turks to a standstill. This was due primarily to two great soldiers: **John Hunyadi,** the national hero of Hungary, and his son **Mathias Corvinus,** who reigned as King of Hungary for nearly a third of a century. The principal events were:

1387–1437. Reign of Sigismund of Luxembourg. He was also German Emperor and King of Bohemia. Despite the distraction of his responsibilities and struggles in Germany, and his protracted efforts to subdue the Hussites in Bohemia, Sigismund devoted himself energetically to defending Hungary, and all of Christendom, against the Turkish threat, which he perceived better than most of his contemporaries. He was unable, however, to prevent the Turkish conquest of most of Serbia. He also failed to suppress revolts in the former Hungarian provinces of Bosnia, Wallachia, and Moldavia. These setbacks, and the loss of Dalmatia to Venice, were the result of Sigismund's prolonged absences from his country.

1437–1439. Reign of Albert of Hapsburg. An able warrior, his short tenure was plagued by increasing troubles with the Hungarian nobles and disputes over succession to the thrones of Poland, Bohemia, and Hungary (see p. 471).

1437. Emergence of John Hunyadi. A Transylvanian nobleman, he had served under Sigismund in his wars in Central Europe and against the Turks. In the first of many victories, he evicted the Turks from **Semendria** (Smederevo).

1439–1440. Civil War in Hungary. Ladislas I (Ladislas III of Poland; see pp. 471–472) was successful, due largely to Hunyadi's support.

1440–1444. Reign of Ladislas. This was marked by Hunyadi's victorious campaigns against the Turks (1441–1443). He won the **Battles of Semendria** (1441), **Hermannstadt** (1442), and the **Iron Gates** (1442). In Turkish Europe he captured Nish and Sofia, finally—in combination with King Ladislas—winning a major victory against Sultan Murad II at the **Battle of Snaim** (or Kustinitza, 1443). Turkish power was temporarily smashed in the Balkans. Murad agreed to a 10-year truce (1443).

1444. Ladislas' Crusade. In alliance with Venice, Ladislas broke the truce, invading Bulgaria (July). The Venetian fleet failed in its mission to prevent Murad from crossing back into Europe from Asia with his army (see p. 476).

1444, November 10. Battle of Varna. The Hungarians were completely routed, Ladislas was killed, and Hunyadi escaped with the remnants of the Hungarian army (see p. 476). This ended the last real Crusade.

1444–1457. Reign of Ladislas V (son of Albert). This began under the regency of Hunyadi (till 1452). Hunyadi invaded Austria to force German Emperor Frederick III, protector of Ladislas as King of Bohemia, to permit the young king to come to Hungary (1446). The ungrateful and jealous monarch spent most of the next decade hampering Hunyadi's efforts to defend Hungary against the Turks.

1448, October 17. Second Battle of Kossovo.

Hunyadi, with 25,000 men, was defeated by Turkish Sultan Murad II, with 100,000 men, in a 2-day battle. This was due to the treachery of **Dan,** ruler of Wallachia, and Hunyadi's traditional enemy, Prince **George Brankovic** of Serbia. Hunyadi made excellent use of German and Bohemian mercenary infantry, armed with handguns, who were opposed by Janissary archers. Infantry on both sides used palisades to protect themselves in prolonged exchanges of missile fire at a range of about 100 yards. Casualties were enormous, the Hungarians losing half their army, the Turks one-third.

1449. Invasion of Serbia. Hunyadi led a successful punitive expedition against Brankovic.

1445–1456. Turkish Invasion. Hunyadi sent a force to hold Belgrade, while raising a relief army and river flotilla in Hungary.

1456, July 14. Naval Battle of Belgrade. Hunyadi defeated the Turkish fleet on the Danube.

1456, July 21–22. Battle of Belgrade. Hunyadi routed the investing Turkish army, forcing Sultan **Mohammed** to withdraw to Constantinople. He was planning to carry the war back into Turkey when he died (August 11). Ladislas died the following year.

1458–1490. Reign of Mathias Corvinus (son of Hunyadi). He spent more time warring in Central Europe than in taking advantage of his father's victories over the Turks. He did undertake 2 limited invasions of Turkey (1463 and 1475).

1468–1478. War in Bohemia. (See p. 471.)

1477–1485. Wars against the Emperor Frederick III. (See p. 464.) Following an unsuccessful siege of Vienna (1477), Mathias eventually captured the city (1485), at the same time annexing Austria, Styria, and Carinthia. Hungary now dominated central and southeastern Europe.

1490–1516. Reign of Ladislas VI (Ladislas II of Bohemia). A weak ruler; Hungary's power declined rapidly.

SERBIA

Although the rulers of Serbia were vassals of the Turkish sultans at the outset of this century, with Hungarian assistance Serbia briefly reestablished its independence in midcentury. This was followed by Turkish reconquest; Serbia again was completely submerged. The principal events were:

1389–1427. Reign of Stephen Lazarevich. As a vassal of the Turks, he fought against Tamerlane at the Battle of Angora (1402; see p. 425).

1427–1456. Reign of George Brankovic (Despot of Serbia). Reasserting independence, Brankovic was driven from his fortress capital of Semendria by the Turks (1439). Despite a personal feud with John Hunyadi of Hungary, Brankovic was restored to his dominions by the victories of Hunyadi and Ladislas over the Turks (1443; see p. 473).

1459. Turkish Conquest. End of Serbia's tenuous independence.

1463. Turkish Conquest of Bosnia.

1483. Turkish Conquest of Herzegovina.

1499. Turkish Conquest of Montenegro.

EURASIA, THE MIDDLE EAST, AND AFRICA

THE BYZANTINE EMPIRE, 1400–1461

Although the Byzantine Empire had declined to insignificance in size and power due to internal decay and the steady encroachments of the Ottoman Turks, even in its dying years it demonstrated some flashes of its ancient vitality. But the old, crumbling fortifications of Constantinople, which had defied assault for so many centuries, could not stand up against the bombardment of the massive siege artillery of **Mohammed II.** The last Emperor, **Constantine XI,** died a hero's death in a vain but gallant defense of the breach pounded in the walls by the Turkish guns. The principal events were:

1391–1425. Reign of Manuel II. An able ruler, but his domains were limited to the cities of Constantinople and Thessalonika, and part of the Morean Peninsula (the ancient Peloponnese) in southern Greece.

1422. Turkish Attack on Constantinople. Repulsed by Manuel.

1425–1448. Reign of John VIII (son of Manuel). His younger brother, **Constantine** (later Constantine XI), and **Thomas Palaeologus** conquered Frankish Morea (1428).

1446. Byzantine Invasions of Central Greece. Constantine Palaeologus was repulsed by Sultan Murad II.

1448–1453. Reign of Constantine XI.

1453, April–May. Siege and Capture of Constantinople. (See p. 476.)

1460. Turkish Conquest of Morea.

1461. Turkish Conquest of Trebizond. This ended the last vestiges of the Roman and Byzantine empires.

THE OTTOMAN EMPIRE

Despite the catastrophic defeat of Bayazid by Timur (see p. 421), the Ottoman Empire made a remarkably quick recovery during the early years of the 15th century. This was principally due to the skill and vigor of **Mohammed I,** son of Bayazid. His son **Murad II** resumed the Turkish career of conquest in southern Europe interrupted by the Tartar invasion. This was brought to an abrupt halt, however, by stubborn and vigorous Hungarian resistance. Their northern advance being thus checked, the Turks now tried to consolidate their control over areas of southern Europe and the Levant which they had originally bypassed. First and foremost was Constantinople. The remaining fragments of the once-great Byzantine Empire were absorbed, while the growing Turkish fleet gradually gained ascendancy over the Venetians and Genoese in the Aegean, plucking off their island and coastal colonies one by one. By the end of the century, steady Turkish expansion had conquered all the Balkans save for Hungary. The principal events were:

1400–1403. Tamerlane's Invasion of Anatolia. (See p. 425.)

1403–1413. Three-Way Civil War. The sons of Bayazid fought for succession to the Ottoman throne. Mohammed defeated and killed in turn his brothers **Suleiman** (1411) and **Musa** (1413). While this was going on, Ottoman holdings in Europe shrank to Thrace, around Adrianople.

1413–1421. Reign of Mohammed I ("the Restorer"). He reestablished central authority. He expanded his dominions in Asia Minor and in Europe, notably by the conquest of Wallachia (1415).

1416. Naval War with Venice. The Doge **Loredano** defeated and destroyed a Turkish fleet off **Gallipoli,** causing Mohammed to sue for peace.

1421–1451. Reign of Murad II. He briefly besieged Constantinople unsuccessfully (1422) in reprisal for aid the Byzantines had given to an unsuccessful rival.

1425–1430. War with Venice. Turkish fleets captured Thessalonika and Venetian possessions along the coasts of Albania and Epirus. Venice, also engaged in a war with Milan, made an unfavorable peace with the Turks.

1441–1442. Turkish Invasions of Serbia and Hungary. They were repulsed by Hunyadi (see p. 474).

1443–1444. The Last Crusade. This was led by Ladislas of Poland and Hungary and his brilliant general, Hunyadi. Most of the Crusaders came from Hungary, Poland, Bosnia, Wallachia, and Serbia. After Hunyadi captured Nish and Sofia, Murad made peace, abandoning his suzerainty over Serbia and Wallachia (1443). Murad then abdicated. When the Hungarians broke the truce and renewed their invasion by an advance to Varna, Murad resumed

the throne. A Venetian fleet was supposed to meet the Crusader army at Varna and to convoy it to Constantinople, meanwhile keeping the principal Ottoman armies from crossing the Straits from Anatolia to Europe. The Venetians failed to carry out their part of the campaign, and Murad with a great army marched to Varna, where he decisively defeated the Crusaders (November 10; see p. 474).

1443–1468. Albanian Wars of Independence. Their leader, **Skanderbeg (George Castricata),** had risen to prominence as a soldier (probably a Janissary) in the Turkish army in the wars against Serbia and Hungary. Skanderbeg established Albania's virtual independence and, with occasional and sporadic assistance from Venice and Naples, repulsed all Turkish invasions. Upon his death, however, the country was quickly reconquered by the Turks.

1448, October 17. Second Battle of Kossovo. (See p. 474.) Largely because of the experience of this hard-won victory over Hunyadi, the Janissaries began to adopt handguns to replace bow and crossbow.

1451–1481. Reign of Mohammed II ("the Conqueror").

1453, February–May. The Siege of Constantinople. Mohammed led an army of more than 80,000 men, with a siege train of 70 heavy cannon, commanded by **Urban,** a Hungarian renegade. To defend the city Constantine XI had less than 10,000 men, including some Genoese mercenaries under gallant **John Giustiniani.** A Venetian fleet provided some initial assistance, but was driven off by the Turkish navy, which completed the blockade of the city. The great siege batteries, including 12 superbombards, were then established and began to hammer against the more vulnerable parts of the ancient city wall from the west (April 2). Several breaches were made both by artillery and mining, and a number of unsuccessful assaults attempted, but the defenders, under the energetic and gallant leadership of Constantine, built palisades behind each breach, and vigorously counterattacked to drive off each Turkish assault. Most of the Turkish mines were detected and blocked before they reached the walls. A final and intensive bombardment leveled a great portion of the wall, which the defenders were unable to block off completely (May 29). Nevertheless, Constantine's gallant defense for several hours stopped a massive Turkish assault through the breach, until another Turkish force, gaining entrance through an unguarded section of the thinly manned wall, attacked the defenders in flank and rear. Disdaining to flee, Constantine fought on in the breach until he was overwhelmed and killed. The victorious Turks then pillaged the city for 3 days.

1456. Mohammed's Siege of Belgrade. He was driven off by Hunyadi (see p. 474).

1459–1483. Turkish Conquest of Serbia, Bosnia, and Herzegovina (see p. 475).

1461. Conquest of Greece and the Aegean. A Turkish fleet drove the Genoese from the Aegean, while at the same time Turkish land and sea forces conquered Morea.

1463–1479. War with Venice. The Turks raided Dalmatia and Croatia (1468), while a Turkish fleet and army invaded and conquered Negroponte (Euboea, 1470). Venetian diplomacy brought Persia into the war, but invading Persians were defeated at the **Battle of Erzin-** jan (see p. 479). Having reconquered Albania after the death of Skanderbeg (see p. 476), the Turks then captured most of the Venetian coastal posts in Albania, while Turk cavalry raiders crossed the Alps from Croatia into Venetia, terrorizing northeastern Italy. Thoroughly defeated, the Venetians made peace, recognizing the loss of all of the regions conquered by the Turks (1479). **Scutari,** whose Venetian garrison had repulsed repeated Turkish attacks (1478–1479), was ceded to the Ottomans.

1480. Turkish Expedition to Italy. They crossed the Adriatic to seize Otranto.

1480–1481. First Siege of Rhodes. Mohammed II besieged Rhodes, but was repulsed with heavy loss due to the gallant resistance and effective defense of the Knights of St. John.

1481–1512. Reign of Bayazid II. This started with a civil war of succession against his younger brother **Djem,** who also claimed the throne. Djem was defeated and took refuge on Rhodes. Essentially a peaceful man, Bayazid ordered the abandonment of the Turkish outpost in Otranto (1481), and generally relaxed pressure against Turkey's European neighbors.

1492–1494. Turkish Invasion of Carniola and Styria. This was repulsed by the Emperor Maximilian (see p. 464).

1495–1500. Inconclusive War against the Poles in Moldavia (see p. 472).

1499, July 28. First Battle of Lepanto. The Turkish navy won its greatest naval victory over the Venetians. It then captured several Venetian island and coastal possessions in the Aegean and Ionian seas (1499–1502).

EGYPT AND SYRIA

During this turbulent century the Mamelukes retained control over Egypt and Syria, but were generally unsuccessful in efforts to expand into Asia Minor and the Kurdistan highlands. The principal events were:

1400–1403. Tamerlane's Invasion of Syria. (See p. 425.)

1412–1421. Reign of Sultan Sheikh Mahmudi. This was a period of internal anarchy and turbulence, during which Sheikh Mahmudi established tenuous suzerainty over the Turkoman principalities in the mountainous regions of eastern Asia Minor and Armenia.

1422–1438. Reign of Sultan Barsbai. He continued Mameluke efforts to subdue the Turkoman states. These brought him into inconclusive conflict with **Shah Rukh** of Persia, son of Tamerlane.

1424–1426. Mameluke Invasions of Cyprus. These were repulsed after initial success and conquest.

1442–1444. Attacks on Rhodes. Repeated invasions by the Sultan **Malik al-Zahir** were repulsed by the Knights of St. John.

1468–1496. Reign of the Sultan Kaietbai. He reestablished order in the Mameluke domains after a long period of internal violence. His efforts to expand in eastern Asia Minor brought him into conflict with Ottoman Sultan Bayazid II.

1487–1491. War with the Ottomans. The Mamelukes were at first successful, then were repelled by the Turks from Adana and Tarsus.

1496–1501. Anarchy. A succession of rival sultans.

PERSIA AND TURKESTAN

After the death of Tamerlane (1405), much of his empire was split up among several sons and grandsons, while other portions seized the opportunity to reassert independence. Timurids (successors of Tamerlane) ruled at Herat, Fars, Tabriz, and Transoxiana, frequently in conflict with each other and with the Turkoman tribes of Armenia and Azerbaijan. **Shah Rukh,** a son of Tamerlane and ruler of Herat, established control over much of his father's empire near the middle of the century. Soon after this, however, the Turkomans began to spread across most of northern Persia and into Transoxiana. These Turkoman tribes were organized into two rival confederacies, the "Black Sheep" and "White Sheep" Turkomans. A three-way struggle ensued between the Timurids, the Black Sheep, and the White Sheep. The White Sheep, ultimately successful, quickly declined after a series of unsuccessful wars with the Ottoman Turks. The principal events were:

1390–1420. Rise of the Black Sheep Turkomans. They were led by **Kara Usuf,** who ruled most of Azerbaijan and Armenia.

1404–1447. Reign of Shah Rukh of Herat. He defeated his brothers and nephews in a series of wars to unify most of the southern dominions of Tamerlane. He also defeated Kara Usuf to reestablish Timurid control of Azerbaijan and Armenia.

1420–1467. Black Sheep Expansion. Kara Iskandar, son of Kara Usuf, repelled the Timurids from Armenia and extended his power into Azerbaijan and north Persia (1420–1435). His successor, **Jehan Shah** (1435–1467), conquered north Persia as far as Herat (1448).

1452–1469. Reign of Abu Said (nephew of Shah Rukh). He controlled eastern Persia and Transoxiana.

1453–1478. Rise of the White Sheep Confederacy. They were led by **Uzun Hasan.** He was

defeated by the Ottomans in his efforts to expand westward into Anatolia (1461).

1460–1488. Rise of the Turkoman Safawid Dynasty. The Safawids inhabited the mountainous region southwest of the Caspian Sea around Ardabil in northeastern Azerbaijan. Under their leader **Haidar,** they were intermittently in conflict with both Black and White Sheep confederacies. Haidar was defeated and killed by the White Sheep confederacy and the Georgians of Shirvan (1488).

1467–1469. Struggle for Persia. Jehan Shah was defeated and killed by Uzun Hasan (1467). Abu Said then invaded Azerbaijan, and was also defeated and killed (1469). Uzun Hasan and his White Sheep confederacy thus controlled Persia, Armenia, and Azerbaijan, the Timurids retaining only Transoxiana and Herat.

1473. War with Turkey. Uzun Hasan invaded

Anatolia, in an alliance with the Venetians against the Ottomans. He was defeated and repulsed by Mohammed II in the **Battle of Erzinjan.**

1478–1500. Decline of the White Sheep Confederacy. It began to break up into a number of petty states.

1499–1500. Revival of the Safawids. Ismail, son of Haidar, conquered Shirvan, captured Baku, and renewed the war with the White Sheep (1500).

c. 1500. Ascendancy of the Kazakhs in Turkestan. A pagan tribe of Tartar-Turkish origin, the Kazakhs, part of the Golden Horde, occupied the area south and east of the Urals. As the Golden Horde and the Timurids declined in the mid-15th century, the Kazakhs, without any overall central authority, became the virtually independent rulers of western Turkestan and of the Kirghiz steppes.

SOUTH ASIA

NORTH INDIA

Tamerlane's invasion and the consequent collapse of the Sultanate of Delhi completely and hopelessly fragmented north India for more than a half-century. A new Delhi sultanate, nominally subject to the Timurids of Persia, maintained tenuous control over the turbulent Moslem nobles of northwestern India. In the central Ganges Valley, between Delhi and the Sultanate of Bengal, was the Moslem Kingdom of Jaunpur, almost constantly at war with both of its neighbors. Hindu Orissa, which had successfully defied the Delhi sultans at the height of their power, and which had now become one of the leading states of India, expanded far southward along the east coast of the Deccan. In south-central and southwestern Hindustan were a number of small Moslem states, of which the most important were the kingdoms of Malwa and Gujarat. In the Rajputana Desert of west-central Hindustan, the Hindu princes who had been driven into this area by Muhammad of Ghor took advantage of Delhi's weakness to regain some of their old power and influence. The most important of these Rajput states was the Kingdom of Mewar. The principal events were:

1414. Establishment of the Sayyid Dynasty. This was by **Khizar Khan,** Tamerlane's governor of the Punjab, who seized Delhi, but remained nominally subject to Shah Rukh of Persia.

1414–1450. Intermittent Wars between Delhi and Jaunpur. A struggle for control of the central Ganges Valley.

1451–1489. Establishment of the Lodi Dynasty. Bulal Lodi overthrew the Sayyid Dynasty. Most of his reign was spent in a drawn-out struggle with Jaunpur, which the Lodis finally conquered (1487).

1458–1511. Reign of Mahmud Shah Begarha of Gujerat. He was an able soldier and magnificent builder, who brought his state to the pinnacle of its wealth and power.

1489–1517. Reign of Sikandar (son of Bulal). He added further to the revived power and prestige of the new Delhi Sultanate.

SOUTH INDIA AND CEYLON (SRI LANKA)

The central feature of the history of the Deccan in the 15th century was a continuation of the struggle between the Moslem sultans of Bahmani and the Hindu kings of Vijayanagar. During most of the century Bahmani was also engaged in frequent wars with its Moslem neighbors of Malwa and Gujerat to the north and northwest, and Hindu Orissa to the northeast. The Carnatic coast was also the scene of frequent contention between Vijayanagar and Orissa, with outcomes generally in favor of Vijayanagar. The principal events were:

1397–1422. Reign of Firuz Shah of Bahmani. He obtained regular tribute from Vijayanagar as a result of his victories in the early wars of his reign (1398–1406). Firuz was defeated, however, by a resurgent Vijayanagar at the **Battle of Pangul** (1420) and forced to abdicate (1422).

1411–1440. Chinese Conquest of Ceylon. An amphibious Chinese expedition conquered the island, which remained tributary to China for more than 30 years (see p. 481).

1422–1435. Reign of Ahmad Shah of Bahmani. He restored Bahmani superiority in the Deccan by capturing Vijayanagar and again imposing annual tribute (1423). He next defeated and annexed Warangal (1425). During the remainder of his reign he was engaged in wars against Malwa, in which he was successful, and against Gujerat, which repulsed the Bahmani armies.

1435–1457. Reign of Ala-ud-din of Bahmani. He again defeated Vijayanagar (1443).

1436–1469. Reign of Mahmud I of Malwa. This warrior king consistently defeated Bahmani and annexed its northern provinces (1457–1469). Had it not been for the interven-

tion of Mahmud Shah Begarha of Gujerat, who wished to preserve a balance of power, Mahmud of Malwa might have succeeded in completely overthrowing Bahmani during this period of weak rule.

1463–1482. Reign of Mohammed III of Bahmani. Bahmani power reached its height, due almost entirely to the military and administrative genius of Bahmani's chief minister, **Mahmud Gawan.** Gawan conquered and annexed to Bahmani the Hindu principalities of the Konkan coast, between Gujerat and Goa (1469). He next invaded and defeated Vijayanagar, annexing Goa (1475). Gawan then turned to the northeast, defeating Orissa and annexing much of the coastline northeast of the Godavari River (1478). He was next victorious in another war with Vijayanagar (1481), following which he was executed by his ungrateful, jealous master.

1481–1500. Decline and Collapse of Bahmani. This was caused by attacks from resurgent Vijayanagar and Orissa. As a result, by the end of the century, the frontiers of these two Hindu states met along the Krishna River.

EAST ASIA

CHINA

During the first half of this century the Ming Dynasty reached the height of its power. This was due primarily to the Emperor **Yung-lo** (or **Ch'eng Tsu**), who conducted an aggressive and uniformly successful foreign policy supported by excellent land and sea forces. A series of punitive expeditions kept the Mongols in check, while Chinese armies reestablished imperial authority in Upper Burma and in Annam. The great admiral **Cheng Ho** led a series of naval expeditions which gave China unchallenged control of Indonesian waters and of the Indian Ocean. This Chinese naval ascendancy was due in part to research and development of improved sea-going junks with watertight compartmented hulls. By the middle of the century, however, decline had set in. The outlying provinces in Southeast Asia were lost. By the end of the century the Mongols were again a serious menace in the north. Control of the sea was lost, and the coasts of China were at the mercy of ruthless Japanese pirates. The principal events were:

1398–1403. Dynastic Struggle. Chu Ti was finally successful, taking the royal name of **Ch'eng Tsu,** but is more commonly known by the name of his reign, **Yung-lo.** This civil war laid waste much of the land between the Yellow and Yangtze rivers.

1403–1424. Reign of Yung-lo.

1405–1407. Early Naval Expeditions of Cheng Ho. This Moslem eunuch seaman invaded Sumatra, conquered Palembang, and forced most of the Malay and Indonesian states to pay tribute to the emperor.

1408–1411. Invasion of Ceylon. Following an insult to a Chinese ambassador, Cheng Ho led a combined land and naval force that conquered Ceylon. The king and royal family were taken to Peking.

1410–1424. Punitive Expeditions into Outer Mongolia. These operations prevented incipient coalescence of Mongol power.

1412–1415. Indian Ocean Expedition. Cheng Ho led a naval expedition as far as Hormuz.

1416–1424. Further Expeditions of Cheng Ho. He extracted tribute from most of the important nations on the shores of the Indian Ocean.

1427. Revolt in Annam. This resulted in the loss of that province (1431; see p. 483).

1431–1433. Cheng Ho's Final Expedition. It included a cruise up the Red Sea, where he obtained tribute from Mecca.

1436–1449. Reign of Ying Tsung. This began the decline of the Ming Dynasty. The emperor was defeated and captured in a battle on the northern frontier by a Mongol Oirat army (1449).

1449–1457. Reign of Ching Ti (brother of Ying Tsung). He seized the throne during the captivity of his brother. When Ying Tsung was released by the Mongols (1450), a prolonged dynastic war followed.

1457–1464. Second Reign of Ying Tsung.

c. 1470–1543. Resurgence of Mongol Power under Dayan. (See p. 558.)

KOREA

This century was the golden age of Korea and probably the most peaceful epoch in the history of that strife-torn peninsula. A strong and enlightened central monarchy, nominally subject to the Ming emperors, maintained internal peace and protected the coasts against Japanese pirate depredations, which had plagued the country in previous centuries, and which would soon recur. An important factor in this success against the pirates was Korean seizure of the Tsushima Islands (1460).

JAPAN

The first half of this century in Japan was relatively quiet, though growing political unrest was a harbinger of violence to come. Internal wars for more than a century were precipitated by a struggle for succession to the Shogunate in the Ashikaga family. The principal events were:

1465. War of the Monks. Conflict between the monks of Enryakui and those of Honganji, the latter being defeated and their monastery destroyed.

1467–1477. Onin War. This raged primarily in the region immediately around the capital city of Kyoto. Nominally a war of succession in the Ashikaga family, in reality this was a struggle between two great warlords of western Japan: **Yamana Mochitoyo** and his son-in-law **Hosokawa Katsumoto.** Both died during the war (1473), but a senseless struggle continued between their adherents.

1477–1490. Continuing Unrest. Though relative peace returned to the Kyoto area after the return to power of former Shogun **Yoshimasa** (1449–1467, 1474–1490), violence spread through the provinces.

1493. Renewed Civil War. Hosokawa Masamoto led a revolt to drive Shogun **Yoshitame** from Kyoto. Masamoto then set up a puppet shogun; civil strife lasted intermittently through the end of the century.

SOUTHEAST ASIA

THE TAI

During this century there were two principal centers of power among the Tai peoples who inhabited what is now Thailand. In frequent wars the larger Kingdom of Ayuthia was unable to establish ascendancy over the smaller, vital Kingdom of Chiengmai in the jungled mountain ranges of northwestern Thailand. Even more intensive were fierce conflicts between Ayuthia and the Khmer of Angkor. Also during this century, the Tai consolidated control over most of the Malayan Peninsula, with the notable exception of the region around Malacca, where a new and vigorous kingdom fought the Tai to a standstill. The principal events were:

1408–1424. Reign of Int'araja. He intervened successfully in a succession dispute in Sukhot'ai (1410) to reassert the suzerainty of Ayuthia. He was also partially successful in an invasion of Chiengmai, capturing the town of Chiengrai, but being repulsed from the towns of P'ayao and Chiengmai itself (1411). Accounts of the battle near **P'ayao** have been interpreted (probably incorrectly) to imply that both sides used cannon.

1424–1448. Reign of Boromoraja II. He continued the wars with the Khmer. He captured Angkor after a long siege, but was then driven out (1430–1432; see p. 483). He twice tried and failed to conquer Chiengmai (1442, 1448).

1448–1488. Reign of Boromo Trailokanat. He strengthened and centralized control of the Kingdom of Ayuthia, establishing an organized military administration, which was by far the most advanced in Southeast Asia. He was involved in practically incessant war with Chiengmai. At one time the Ayuthians captured Chiengmai (1452), but were later forced to withdraw because of the intervention of Luang Prabang. A few years later, after several Chiengmai invasions, Ayuthian forces again advanced toward Chiengmai. This campaign came to a conclusion in a moonlight **Battle of Doi Ba,** near Chiengmai, where Chiengmai forces repulsed the Ayuthians (1463).

1455. Unsuccessful Siamese Attack on Malacca.

1494–1520. Renewed Intermittent War between Chiengmai and Ayuthia.

VIETNAMESE REGION

The continuing fierce struggle between Champa and Annam was interrupted early in the century by the temporary (20-year) conquest of Annam by the Chinese. A successful struggle for independence against China, however, by Annamese leader **Le Loi** was a signal for the resumption of the Cham-Annamese conflict, which continued without interruption until Annam finally conquered Champa late in the century. The principal events were:

1400–1407. Civil War in Annam. Despite this internal struggle, Annamese forces conquered the northern province of Champa.

1407. Chinese Conquest of Annam. The pretext was to restore order following internal unrest.

1418–1427. Guerrilla War against the Chinese. This was led by Le Loi.

1427–1428. Siege of the Chinese Garrison in Hanoi. Upon the surrender and withdrawal of the Chinese, Le Loi made himself king, then concluded peace with the Ming Dynasty, agreeing to nominal submission to the emperor (1431).

1441–1446. Civil War in Champa. Frequent Cham raids into Annam were repulsed.

1446–1471. Annamese Invasion of Champa. After initial success, and capture of the Cham capital of Vijaya, the Annamese were temporarily driven out, but returned to complete a systematic conquest of Champa as far south as Cape Varella. This ended the centuries-old war. An insignificant Cham kingdom persisted farther south as a buffer between the Annamese and the Khmer.

1460–1497. Reign of Le Thanh Ton. He was the conqueror of Champa.

CAMBODIA

This century was marked by almost continuous war between the Khmer and the Tai of Ayuthia. Though the Tai won the most spectacular success by capturing Angkor, they were subsequently driven out by the Khmer, and the war continued to rage throughout the century, with raids and invasions continuing on both sides, the Khmer frequently threatening Ayuthia. The principal events were:

1394–1401. Tai Invasions. They conquered much of western Cambodia, but were eventually repulsed.

1421–1426. Cham Invasion of the Mekong Delta. Despite the strain of the continuing war between Ayuthia and Angkor, the hard-pressed Khmer repulsed the Chams.

1430–1431. Siege of Angkor. Boromoraja II of

Ayuthia captured the city by treachery after 7 months.

1432. Khmer Counteroffensive. The Tai were expelled from Cambodia. The Khmer abandoned Angkor as a capital and reestablished their kingdom on a much reduced scale with Pnom Penh as capital.

BURMA

During this confused century of Burmese history, war was endemic between 6 major rival powers. Within Burma itself, the most important of these were the Burman Kingdom of Ava and the Mon Kingdom of Pegu, both of which were embroiled almost constantly with each other, with the smaller Burman Kingdom of Toungoo, and with the Shan (Tai) tribes of northeast Burma. Despite relative isolation by the coastal mountain ranges, the Arakanese were also frequently at war with both Ava and Pegu. During most of this century there were numerous Chinese interventions in the affairs of northern Burma, as the Ming attempted to pacify China's frontiers. For much of the time Ava and the Shan states were subject to the real or nominal authority of Ming governors. The principal events were:

1401–1422. Reign of Minhkaung of Ava. He had numerous desperate struggles with the Mons and Arakanese. He was repulsed in efforts to conquer Pegu.

1385–1423. Reign of Razadarit of Pegu. He was an excellent soldier and administrator, as well as an adroit diplomat. The early years of his reign were devoted to desperate defensive wars against Ava, Chiengmai, Ayuthia, and various smaller Tai and Shan principalities. The most dangerous threat to his embattled nation was that of the Burmans of Ava, whom he eventually repulsed with Arakanese assistance. After repulsing a final Burman invasion (1417), Razadarit initiated a long period of relative peace and prosperity for his country.

1404–1430. Conflicts of Ava and Arakan. This series of violent raids and counter-raids and counterraids was inspired by the intrigues of Razadarit.

1406. War of Ava and Mohnyin. Minhkaung sent a punitive expedition against the Shans of Mohnyin under his able general **Nawrahta.** Threat of Chinese intervention caused the Burmans to withdraw.

1413. Shan Invasion of Ava. The Sawba (lord) of Shenwi raided deep into Ava territory, reaching modern Maymyo, where he was defeated and repulsed by the Burmans.

1414–1415. Ava Invasion of Pegu. Under Prince **Minrekyawswa,** this brought the Mon Kingdom to the verge of disaster.

1415. Shan Raids into Ava. Inspired by the diplomacy of Razadarit, these raids threatened Ava, forcing Minhkaung to recall his son from his invasion of Pegu.

1416–1417. Renewed Ava Invasion of Pegu.

This ended with the death of Minrekyawswa in the Irrawaddy Delta. The Ava army again withdrew to meet renewed Shan threats, ending, for the time being, the struggle between Ava and Pegu.

1422–1426. Reign of Hsinbyushin Thihatu of Ava (son of Minhkaung). Leading a punitive expedition against the Shans, he was killed due to the treachery of his Shan wife.

1426–1440. Anarchy in Upper Burma. The Shans dominated the country.

1438–1465. Chinese Intervention. The Ming, wishing to stabilize their frontiers, reestablished order in Upper Burma. General **Wang Chi** conquered and subdued the Shan states.

1445. Chinese Invasion of Ava. They were defeated at **Tagaung** and Wang Chi was killed.

1446. Renewed Chinese Invasion. A large punitive expedition reached Ava, forcing King **Narapati** (1443–1469) to submit to Ming suzerainty. The Chinese then helped him restore order and suppress rebellions. Thirty-five years of relative peace followed.

1481–1500. Unrest and Violence in the North. There were frequent Shan raids into Upper Burma, as Chinese control in the Shan states weakened.

INDONESIA

The Majapahit Dynasty of Java declined and disappeared into obscurity about the end of the century. Early in the period some of the Indonesian states, including Majapahit, were forced to acknowledge the suzerainty of China, and to pay tribute, as a result of the naval expeditions of Chinese Admiral Cheng Ho (see p. 481).

MALAYA

During the first half of this century, practically all of the states of the Malayan Peninsula were under the domination of the Tai Kingdom of Ayuthia, which ejected remaining vestiges of Majapahit authority. By the middle of the century, however, the rapid expansion of Malacca, under its great leader **Tun Perak,** began to drive the Siamese from southwestern and southern Malaya, repulsing repeated Tai efforts to reestablish control. The principal events were:

c. 1402–1424. Reign of Paramesvara. A prince of Palembang, he was a fugitive of civil war in

Majapahit who established himself at Malacca. Creating essentially a pirate kingdom, he established control of the Straits of Malacca and extracted tolls from ships passing through, obtaining recognition by the Ming emperors as a ruler independent of Siam (1405).

1409. Chinese Visit to Malacca. A show of force by the war fleet of Cheng Ho.

c. 1414. Paramesvara's Conversion to Islam. He took the name of **Megat Iskandar Shah.** Malacca continued to expand and prosper until his death (1424).

c. 1450–c. 1498. Virtual Dictatorship of Tun Perak. This prime minister overshadowed 4 sultans, and by combined military and political genius greatly expanded Malacca's power at the expense of other neighbors in Malaya and Sumatra.

c. 1490. Conversion of Java to Islam. Begun by Malaccan traders.

AFRICA

NORTH AFRICA

Save for the beginnings of European colonialism along the Moroccan coast, this was a period of anarchy and of little military significance in North Africa. Petty squabbles between the minor dynasties continued, with the Marinids of Morocco being replaced by the Wattasids through a palace revolution shortly after the middle of the century. Farther east the Hafsids were slowly being expelled from Algeria by the rising Ziyanids, while their hold on Tunisia and Tripolitania was being constantly menaced by the Arab nomads of the desert to the south and east. Meanwhile, the Portuguese and Spanish took advantage of these internal squabbles to seize a number of coastal footholds. The principal events were:

1415. Capture of Ceuta by the Portuguese. (See p. 469.)

1437. Portuguese repulsed at Tangier. (See p. 469.)

1468. Sack of Casablanca by the Portuguese.

This temporarily destroyed the pirate base at Casablanca.

1470. Capture of Melilla by the Spanish.

1471. Capture of Tangier by the Portuguese.

EAST AFRICA

Warfare between Christian Ethiopia and surrounding Moslem areas continued throughout this period. By this time, only the Fung Kingdom of southern Sudan, and Ethiopia itself, remained Christian. Among the major events were:

1414–1429. Reign of Yeshaq I. He defeated Sultan S'adad-Din of Ifat (see p. 428), destroying that state and retaking Zeila on the Red Sea, near modern Djibouti (1415), and led several campaigns to bring the northeastern border area under control. He also campaigned against the rebellious Falashas (Ethiopian Jews). He sent diplomatic missions to King Affonso V of Aragon and Jean, duc de Berry.

c. 1450–c. 1500. Fall of 'Alwah. Under heavy pressure from the Moslem inhabitants of the northern Sudan, this kingdom was overrun by **Abd Allah Jamma** (c. 1500). Much of its territory was settled by the Funj, from the mountains along the modern Ethiopian border.

c. 1450–1500. Rise of Mombasa. Fueled by trade revenue, Mombasa became the leading East African coastal town under its Shirazi rulers. Other major settlements were Malindi, Kilwa, Pemba and Zanzibar islands, and Pate. Arab commercial activity supported the development of Swahili (a Bantu language with Arabic loan-words) for commerce with inland tribes.

c. 1450–1500. Rise of Bunyoro. Bito rulers of Luo origin created a state in the Bunyoro-Kitaro region, on the east shore of Lake Albert. This kingdom lasted well into the 19th century.

1468–1478. Reign of Baeda Maryam of Ethiopia. He campaigned against Moslem attacks, as did his successor **Eskender** (1478–1494).

WEST AFRICA

This region began to emerge from obscurity primarily because of the expansion of Portuguese maritime power southward down the Atlantic coast throughout this entire century. The Portuguese were unsuccessful, however, in eliminating a Spanish foothold off West Africa in the Canary Islands. During this century, the great Negro empire of the Songhoi, with capital at Timbuktu, achieved considerable power and splendor. The Songhoi, who had apparently originated in the Nile Valley, dominated the central Niger region from the 8th century on. They were converted to Islam in the 11th century. During the 15th century, under the great Kings **Sunni Ali** and **Askia Mohammed,** the Songhoi Empire dominated most of the western bulge of Africa. The principal events were:

c. 1400. Rebellion of Gao. This Songhai trading city rebelled against the weakening Empire of Mali, which had grown too large for its rulers to control effectively. Later, the Takrur and Wolof of Senegal threw off Mali rule (c. 1420). At about the same time, the Mossi tribes (in central Burkina Faso) began to raid the Niger valley with their light cavalry.

1402–1404. Spanish Conquest of the Canary Islands.

1425. Portuguese Expedition to the Canaries. The Portuguese, sent by Prince Henry, were repulsed by the Castilians.

1431. Tuareg Raid to Timbuktu. The city was captured and sacked by the nomads.

1434–1498. Portuguese Explorations. These went along the coast of Africa south of Cape Bojador.

c. 1440–1480. Reign of Ewaure the Great of Benin. This monarch, described in contemporary annals as a great warrior and magician, instituted a hereditary succession to the Benin throne. He also waged several wars which greatly expanded the kingdom's territories.

1450–1453. Renewed Portuguese Attacks on the Canaries. They were all repulsed.

1464–1492. Reign of Sonni Ali Ber (Sunni Ali) of Songhai.

1466–1473. Songhai War with Jenne. Sonni Ali defeated and conquered the rich trading state of Jenne (modern Djénné) on the Bali River.

1468. Conquest and Sack of Timbuktu. Sonni Ali had been invited by the city fathers to deliver the city from Tuareg domination.

1469–1488. Campaigns against the Mossi. The Mossi (see above) continued their raids into Songhai. Sonni Ali's 2 campaigns against this tribe (1469–1470, 1478–1479) made scant progress, but he repulsed a major Mossi raid (1480). A 3rd campaign culminated in a notable victory at the **Battle of Kobi** (1483). A 4th campaign was less successful (1488).

c. 1475. Kongo Kingdom. Late in the century, this kingdom (see p. 427) reached its greatest geographical extent, ruling the land between the Congo and Loge rivers inland nearly to the Kwango. Coastal districts to the south sent intermittent tribute.

c. 1481–c. 1504. Reign of Ozolua "the Conqueror" of Benin. Ozolua, Ewuare's son and a redoubtable warrior, also established close relations with the Portuguese, and sent envoys to Lisbon. The kingdom grew wealthy trading ivory, palm oil, and pepper to European merchants.

1483–1491. Portuguese Contact with Kongo. The Kongo were favorably impressed with the Portuguese, and some began to embrace Christianity.

1492. Songhai Campaign against the Fulani. Returning from this expedition, Sonni Ali drowned while crossing a stream.

1492–1493. Songhai Succession War. Sonni Ali's son **Sonni Baru** was challenged by **Muhammad ibn abi Bakr Ture,** one of Sonni Ali's ablest generals. Muhammad defeated Sonni Baru at the **Battle of Anfao** (April 12, 1493), and so secured the throne, later taking the name **Askia Muhammad** (Mohammed I Askia).

1493–1528. Reign of Askia Muhammad. The Songhai Empire reached its pinnacle of power and wealth. (For events after 1500, see p. 565.)

1498–1502. Songhai Conquest of the Mossi of Yatenga. After returning from a pilgrimage to Mecca (1495–1497), Askia Muhammad conquered the Mossi of Yatenga in a series of campaigns.

SOUTH AFRICA

Bantu-speaking peoples, who had been moving southward for centuries, continued their slow migration. They gradually displaced existing Khoisan inhabitants, who in turn had driven the San (Bushmen) into the desert or mountains. Bantu on the east coast, west of the Drakensberg Mountains, adopted a variety of clicks, derived from Khoisan and San, into their languages. These Nguni people were thus linguistically differentiated from the inland Sotho.

c. 1480. Foundation of the Matapa Empire. This state, between the Zambezi and Limpopo rivers, was founded by **Nyatsimba,** who took the title of *mwene matapa*, "ravager of the lands," which he passed to his successors. This gruesome honorific title provided the popular name of the Matapa Empire. Nyatsimba also moved the capital from the old center at Zimbabwe (see p. 353) north to Mount Fura on the Zambezi.

NORTH AMERICA

MEXICO

The Rise of the Aztecs

Early in the 14th century, the Mexica (Aztecs), a wandering warlike tribe from the north, arrived in Anahuac, the Valley of Mexico. They settled on two islands called Tenochtitlan and Tlatilulco in Lake Texcoco (c. 1325); this comprised a natural fortress, connected to the mainland by three causeways with removable bridges to prevent passage. Through successful wars and fortunate alliances, the aggressive Aztecs expanded their control in the Valley of Mexico.

1427–1440. Reign of Emperor Itzcoatl ("Black Serpent").

1428–1433. The Tepanaca War. The Tepanaca of the western shore of Lake Texcoco, fearing growing Aztec strength, blockaded the Aztec islands, cutting off much-needed supplies, including fresh water. After a desperate struggle, Emperor Itzcoatl defeated the Tepanaca and eventually beseiged them in their own capital. After overwhelming the Tepanaca (1428–1430), the Aztecs defeated the powerful allied city of Xochimilco (1430–1433).

1440–1468. Reign of Moctezuma (Montezuma I) Ilhuicamina ("Frowns like a Lord"). He campaigned mainly to the south of Mexico City. Under his rule the Aztecs expanded and prospered.

1468–1481. Reign of Axayactl. He conducted estensive campaigns to the east all the way to the Gulf of Mexico and to the southwest, to the Pacific Coast.

1473. Civil War ("War of Defilement"). The two island cities of Tenochtitlan and Tlatilulco were long-time rivals. Tlatilulco had allied itself with some other anti-Tenochtitlan cities out of fear and jealousy of its more powerful twin. War between the twin cities broke out and before any allies of Tlatilulco could send help, Tenochtitland crushed Tlatilulco in brief but intense fighting. Tenochtitlan took fierce revenge against Tlatilulco and its allies.

1478. Battle of Zamacuyahuac. The Three-City League of Tenochtitlan, Tlacopan, and Texcoco went to war against long-time enemies the Tarascans. When the League forces encountered the main Tarascan army, Aztec Emperor Axayacatl committed his forces in piecemeal fashion, to be defeated in detail by the Tarascans. After the failure of an ill-advised all-or-nothing charge on the second day of battle, League forces were disastrously defeated. The Tarascans relentlessly pursued League forces up to the border. Supposedly, only 4,000 of 24,000 League forces returned home. However, the Aztecs recovered quickly from this reverse.

1481–1486. Reign of Emperor Tizoc. He campaigned mainly southeast of Mexico City, with mixed results.

1486–1502. Reign of Emperor Ahuitzotol. He undertook extensive and successful campaigns, reaching both coasts. In a three-year campaign (1495–1498) he expanded Aztec control southward.

XIII

SPANISH SQUARE AND SHIP-OF-THE-LINE:

1500–1600

MILITARY TRENDS

This was a crucial century of world history, marking the beginning of the Reformation and of a period of bitter religious struggle which had lasting political, military, and cultural effects on the entire world. The tactical-technical flux caused by the introduction of gunpowder continued.

The most remarkable operational development of the century was the relatively abrupt change in the strategy and tactics of naval warfare, both of which had been relatively static for approximately 2,000 years. During those millennia, control of the seas had been exercised near the coastlines by war fleets of short-ranged row galleys. By the end of this century the row galley had been displaced in most parts of the world by the sailing warship, with its heavy broadsides of long-range cannon.

It is not surprising, therefore, that this century was more notable for its admirals than its generals—though standards of military competence on land were substantially higher than in previous centuries. Outstanding were two of the most farsighted innovators of naval warfare, who can be ranked among the handful of great admirals of history, though neither participated in the great naval tactical revolution in Western Europe: Portugal's Dom **Affonso de Albuquerque,** the father of modern naval strategy, and Korea's **Yi Sung Sin,** inventor and successful commander of the world's first armored warships. Hardly less capable were **Khair ed-Din** of Algiers, one of the last and greatest of galley admirals, and Britain's Sir **Francis Drake,** the man most responsible for introducing long-range naval gunnery tactics. Three Spanish admirals rank close behind: Italian-born **Andrea Doria,** the great rival of Khair ed-Din; **Don Juan** of Austria, who won the last great fleet engagement of war galleys; and the Marquis **Álvaro of Santa Cruz,** who distinguished himself in both galley and sailing warfare.

The honors in land generalship during this century were shared by Turks and Spaniards. The most outstanding Turks were the Ottoman Sultans **Selim** and his son **Suleiman,** and the Mogul conquerors of India, **Babur** and his grandson **Akbar.** Italian-born Duke **Alexander**

489

Farnese of Parma was the outstanding European soldier of the century, though his Spanish predecessors **Hernández Gonzalo de Córdoba** and conquistador **Hernando Cortez** were probably equally able. Parma's superiority among Europeans is based mainly upon his ability to outmaneuver and dominate two almost equally great opponents: French King **Henry IV** and Dutch Stadholder **Maurice** of Nassau. **Bayinnaung** of Burma and **Hideyoshi** of Japan deserve mention in the company of these other leading generals of the era.

During this century, gunpowder domination of the battlefield became complete in Europe, and was almost equally pronounced in Asia. Armor, of little use against either small arms or artillery fire, was fast disappearing. The great nobles continued to wear light armor, more for prestige reasons than anything else; many cavalrymen retained helmet and breastplate, useful in hand-to-hand combat; some infantry, particularly in the pike formations, also retained helmet and breastplate for the same reason.

SMALL-ARMS WEAPONS

Throughout this century the Spanish were consistently ahead of other nations in the development and employment of infantry small arms. This ascendancy began with the Italian campaigns of Gonzalo de Córdoba (El Gran Capitán) at the turn of the century. Recognizing the tactical importance of small-arms fire, the Spanish strove constantly to perfect and improve their weapons. By the close of the century the French, learning the lesson as a result of disastrous defeats at Spanish hands, were fast catching up.

European arquebusier

Shortly before midcentury, both Spanish and French began to standardize the calibers and mechanisms of their weapons in order to simplify ammunition supply and training procedures. The newly standardized arquebuses were often termed "calivers"—an English corruption of the word "caliber." This development was largely the result of efforts of the Florentine **Filippo Strozzi.**

During the latter part of the century the Spanish, endeavoring to enhance the solidarity of their infantry tactics, began to introduce a heavier small arm, called the musket, with a range up to 300 yards. (This was developed from the so-called *arquebus à croc*, really a light artillery weapon generally mounted on walls or ramparts for defensive fire. Count **Pedro Navarro** introduced the *arquebus à croc* into mobile warfare at the Battle of Ravenna by mounting some

30 of them on handcarts, which he fitted into Spanish field fortifications.) Appearance of the musket, which had to be fired from a fork rest, and which took longer than the arquebus to load and fire, added complexity to already complicated maneuver and loading drills. The sacrifice was accepted, however, because of greater range and striking power. By the end of the century it had largely replaced the old arquebus as the basic infantry weapon of Europe.

The English were the last important European nation to adopt officially gunpowder small arms. After prolonged debate, a Royal Ordinance of 1595 banned the longbow as the basic weapon of the militia train bands. Each soldier now had to supply himself with an arquebus, caliver, or musket. This did not end the debate in England, however, since archers frequently proved that they could shoot faster, farther, and more accurately than most musketeers (Benjamin Franklin would seriously urge adoption of the longbow as the basic American infantry weapon almost two centuries later).

Efforts to adapt gunpowder weapons to cavalry had resulted in the development of the small, light, horse arquebus early in the century. This prototype of the pistol was theoretically a one-hand weapon, but because of the complexities of handling the clumsy matchlock, two hands were really necessary. Thus horse arquebusiers usually had to choose between firepower and horsemanship; both were usually inadequate and chaotic.

Shortly before the middle of the century, however, the invention of the wheel lock brought about the development of the first true pistols. Possibly invented by the German **Johann Kiefuss,** this mechanism operated much like a modern cigarette lighter: a rough wheel rotated against a piece of iron pyrite to generate a spark which ignited the powder in the weapon's flash pan. The wheel rotated when a heavy spring was released by pulling a trigger. To assure a modicum of sustained fire-power, the horse pistoleer carried three weapons: two in holsters and one in the right boot. After all three pistols had been fired, the cavalryman either had to drop the last pistol and draw his sword or else retire to reload the pistols—an operation requiring both hands.

Because the wheel lock was more expensive and more delicate, it did not replace the rugged, reliable matchlock on infantry weapons during this century.

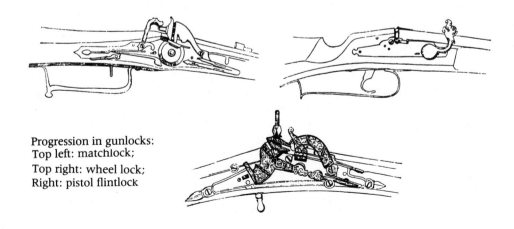

Progression in gunlocks:
Top left: matchlock;
Top right: wheel lock;
Right: pistol flintlock

ARTILLERY

Development of artillery weapons in this century failed to keep up with the progress of small arms mainly because artillerymen were unable to solve the problem of combining mobility with reliable long-range firepower. It had long been realized that long range, accuracy, and destructiveness were best achieved by guns that were 20 or more calibers in length (bore length 20 times the bore diameter), and with thick walls, which could withstand the pressures built up by detonation of a large powder charge. Pieces with thinner walls and lighter powder charges could fire equally heavy projectiles, but with significant reduction in accuracy and range. And even the lightest of these weapons was still clumsy, difficult to move, and took a long time to prepare for action.

Because of these limitations, the artillery supremacy achieved by the French at the end of the previous century was soon reversed by the dramatic Spanish improvements in infantry small arms and the tactics of their employment. As a result, artillery declined in importance during this century, save in the attack and defense of fortifications and in naval warfare. Few major battles were fought without artillery being employed, but in general, after the bloody Battle of Ravenna, small arms were more decisive.

At about the same time the French lost their superiority in artillery construction and techniques to more imaginative German gunmakers. These in turn were soon excelled by the Spanish, who enjoyed a clear-cut superiority in this, as in most other aspects of military science, for most of the century.

An interesting commentary on the status of artillery tactics and techniques during this century is contained in Manucy's *Artillery through the Ages:*

Although artillery had achieved some mobility, carriages were still cumbrous. To move a heavy English cannon, even over good ground, it took 23 horses; a culverin needed nine beasts.* Ammunition—mainly cast-iron round shot, the bomb (an iron shell filled with gunpowder),† canister (a can filled with small projectiles), and grape shot (cluster of iron balls)—was carried the primitive way, in wheelbarrows and carts or on a man's back. The gunner's pace was the measure of field artillery's speed: the gunner *walked* beside his gun! Furthermore, some of these experts were getting along in years. During Elizabeth's reign several of the gunners at the Tower of London were over 90 years old.

Lacking mobility, guns were captured and recaptured with every changing sweep of the battle; so for the artillerist generally, this was a difficult period. The actual commander of artillery was usually a soldier; but transport and drivers were still hired, and the drivers naturally had a layman's attitude toward battle. Even the gunners, those civilian artists who owed no special duty to the prince, were concerned mainly over the safety of their pieces—and their hides, since artillerists who stuck with their guns were apt to be picked off by an enemy musketeer. Fusilier companies were organized as artillery guards, but their job was as much to keep the gun crew from running away as to protect them from the enemy.‡

---

* More than 60 horses were required to move some basilisks.
† These first crude efforts to create explosive shell were not very successful.
‡ Manucy, Albert, *Artillery Through the Ages*, Washington, 1949.

Gunmakers experimented constantly with new designs and combinations of bore diameter, wall thickness, powder charges, and projectile weights. As a result there were almost as many types of artillery pieces as there were weapons. Ammunition supply became an impossible task, contributing to the decline of artillery's importance in field operations. To correct this situation, shortly before midcentury Emperor Charles V ordered standardization of all artillery weapons into 7 types. Soon afterward Henry II of France followed suit by establishing 6 standard models for French artillery. Experimentation continued, and many additional types were added to these basic standard models, but in a more orderly and systematic manner than previously. There were no fixed standards among the weapons of different nations, though in general Spanish leadership resulted in imitation by other nations.

By the end of the century the art of gunmaking had progressed to the point where the range, power, and major types of guns were to change little for nearly three centuries. Artillery modifications would be mainly limited to improved mobility, organization, tactics, and field-gunnery techniques.

The distinctions between the three major types of weapons as they existed at the end of the century were fundamentally the same as those that exist today. The first class comprised long-barreled (about 30 calibers), thick-walled pieces designed to fire accurately and at long range. This was the **culverin** type of weapon, roughly comparable to the modern **gun.**

The second class consisted of lighter, shorter pieces designed to fire relatively heavier projec-

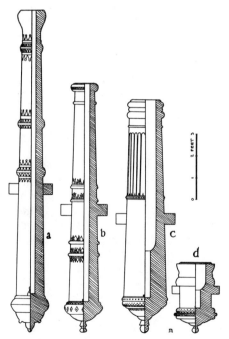

Sixteenth-century Spanish artillery:
(a) culverin; (b) cannon; (c) pedrero;
(d) **mortar**

tiles shorter distances, sacrificing range and accuracy in order to achieve more mobility without loss of smashing power. This was the so-called **cannon** type of weapon, about 20 calibers in length and roughly comparable to the modern **howitzer.**

The third class comprised short-barreled, thin-walled weapons, firing relatively heavy projectiles for shorter ranges. Within this class were two subcategories. First of these was the **pedrero,** so called because it fired a projectile of stone, which was much lighter than an iron projectile of the same diameter. Thus the pedrero (10 to 15 calibers in length) could have quite thin walls and still fire a rather large stone cannon ball almost as far as a cannon. The other subcategory was that of the **mortar,** identical in concept to the weapons class of the same name

16TH-CENTURY ARTILLERY PIECES
(Characteristics are indicative and approximate; records are incomplete, confusing, and contradictory)

Name	Piece Weight (lbs.)	Projectile Weight (lbs.)	Bore (in.)	Length (ft.)	Point-Blank or Effective Range (yds.)	Maximum Range (yds.)
CLASS I: CULVERIN TYPES (25–44 CALIBERS IN LENGTH)						
Esmeril (or rabinet)	200	.3	1.0	2.5	200	750
Serpentine	400	.5	1.5	3.0	250	1,000
Falconet	500	1.0	2.0	3.7	280	1,500
Falcon	800	3.0	2.5	6.0	400	2,500
Minion (or demisaker)	1,000	6.0	3.3	6.5	450	3,500
Pasavolante	3,000	6.0	3.3	10.0	1,000	4,500
Saker	1,600	9.0	4.0	6.9	500	4,000
Culverin bastard	3,000	12.0	4.6	8.5	600	4,000
Demiculverin	3,400	10.0	4.2	8.5	850	5,000
Culverin	4,800	18.0	5.2	11.0	1,700	6,700
Culverin royal	7,000	32.0	6.5	16.0	2,000	7,000
CLASS II: CANNON TYPES (15–28 CALIBERS IN LENGTH)						
Quarto-cannon	2,000	12.0	4.6	7.0	400	2,000
Demicannon	4,000	32.0	6.5	11.0	450	2,500
Bastard cannon	4,500	42.0	7.0	10.0	400	2,000
Cannon serpentine	6,000	42.0	7.0	12.0	500	3,000
Cannon	7,000	50.0	8.0	13.0	600	3,500
Cannon royal	8,000	60.0	8.5	12.0	750	4,000
Basilisk	12,000	90.0	10.0	10.0	750	4,000
CLASS III: PEDRERO AND MORTAR TYPES*						
Pedrero (medium)	3,000	30.0	10.0	9.0	500	2,500
Mortar (medium)	1,500	30.0	6.3	2.0	300	750
Mortar (heavy)	10,000	200.0	15.0	6.0	1,000	2,000

* Though variations were great, pedreros were usually 10–15 calibers in length, and fired projectiles up to 50 pounds in weight. Mortars were 3 to 5 calibers in length, and fired projectiles up to 200 pounds.

today. Mortars were and are short (10 calibers or less) weapons, firing relatively large projectiles for relatively short distances in a high, parabolic trajectory.

Bearing in mind the fact that even within nations standard types varied greatly, the table shows typical characteristics of the most common artillery weapons within each of the three above classes.

FORTIFICATION AND SIEGECRAFT

A revolution in fortification was in progress at the beginning of the century. The high masonry walls of even the most massive medieval fortifications had been rendered obsolete by the smashing power of heavy siege guns. Use of cannon by the defenders did little to rectify this situation; light guns mounted on the high walls could not reach long-range attacking guns. Heavier weapons—when they could be laboriously lifted to the tops of the ramparts—soon became counterproductive; the force of recoil shook the foundations, dangerously weakening the walls and making them easier to breach.

Consequently, walls became lower and thicker—as much to provide adequate emplacements for defending artillery as to make the breaching process more difficult for attacking siege guns. New fortifications were built with broad, low walls from which triangular bastions extended to permit defending artillery to sweep all approaches to the fort. Existing fortifications were modernized by the erection of new walls and bastions of this type; older walls were lowered and broadened where possible.

The remainder of the century saw an intensive struggle between the rapidly improving science of fortification and the equally rapid increase in power and range of siege artillery. The great basilisks and cannon royal could still breach the new walls by prolonged, concentrated fire. To make this more difficult, and to foil exploitation by besieging infantry, the ditches around fortifications were widened, and in turn protected by a counterscarp wall, where light artillery pieces could be emplaced beyond the ditch to keep the great siege guns at a distance. To provide clear fields of fire for the artillery and small-arms weapons of the counterscarp defenders, earth excavated from the ditch was spread in front of the counterscarp wall to create a gradually sloping terrace, or **glacis.** This open slope, descending from the counterscarp, added to the strength of that low wall, and at the same time complicated the task of the attackers in bringing effective fire to bear on counterscarp defenders.

Mining became difficult since tunnels now had to be too long to permit fresh air to reach the diggers. Another deterrent to mining was the costliness of gunpowder; when an opportunity for mining could be exploited, it was almost always by collapse rather than explosion, a reversion to the ancient technique of burning the tunnel support timbers.

The new scientific methods of fortification having outstripped gunnery development, sieges again became lengthy, difficult affairs. This, in turn, caused warfare once more to become a series of sieges, punctuated by battles only when some combination of maneuvering skill, confidence, or logistical pressures brought two armies face to face in the open.

This led to serious efforts to improve siegecraft by adapting it to the new problems posed by the new fortifications. The obvious solution to the problem posed by the counterscarp and the power of defensive artillery was to find a relatively safe method of getting attacking artillery and small arms close enough to the defenses to bring effective fire to bear. The old apparatus of

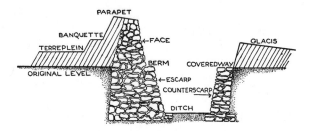

Elements of fortification

mantelets, siege towers, and the like were totally ineffective against defending gunpowder weapons. Attackers, accordingly, resorted to digging. Before the end of the century the concept of approach entrenchments was quite well developed—though crude in comparison with the refinements that would be introduced by Vauban in the following century.

Under the cover of long-range culverin-type guns, attacking engineers and infantry dug trenches toward a presumably vulnerable point in the defenses. When these trenches were within easy artillery battering range of the fortification's counterscarp—or other outworks—thick earthen walls were thrown up in front of wide, shallow trenches to create protected emplacements for siege guns. Under the cover of darkness the heavy weapons would be trundled into their emplacements, and would then begin the painstaking battering process. Under the cover of this fire, trenches would again be pushed forward until a combined artillery and infantry attack could overwhelm the counterscarp defenders. Again the big guns would be moved forward, this time to concentrate against the main fortifications.

It was a long, laborious, costly, and bloody process—almost prohibitive against active and alert defenders. A 16th-century fortress, if provided with adequate stocks of food and ammunition, was as impregnable as the 13th-century castle had been in its day.

The new fortifications, and the siege process to deal with them, greatly stimulated the long-lost art of field fortifications, largely dormant in Europe since Roman times. The principal stimulus, however, had already been provided by farsighted Spanish soldiers, led by Gonzalo de Córdoba, who was apparently the first to realize the potentialities of field fortifications in combination with the new firepower of gunpowder small arms. Following his example, Pedro Navarro and Alexander of Parma kept Spanish engineering and field fortification techniques preeminent, a major element of Spanish military supremacy.

TACTICS OF LAND WARFARE

General

Tactical experimentation continued as military men strove to adapt changing and improving gunpowder weapons to problems of combat. The result was a variety of organizational and tactical patterns, which began to crystallize by the close of the century. During most of this time Spanish military men were generally more imaginative and farsighted than those of other nations. By the close of the century, however, the French under Henry IV and the Dutch under

Maurice of Nassau—both following Spanish precedents—began to wrest tactical superiority from the over-extended Spanish.

As pointed out above, the increasing effectiveness of firearms, combined with the growing strength of fortifications, impelled generals to avoid battle unless success seemed to be assured. This practice, particularly evident after the hitherto aggressive French learned their lesson at the disastrous Battle of Pavia (see p. 517), was also due in part to the increased logistics problems posed by ammunition supply.

The effectiveness of small-arms fire, combined with the vulnerability of arquebusiers and musketeers when reloading their clumsy weapons, kept coordination of combined arms an ever-present challenge to 16th-century generals. Differing methods of employing cavalry shock action, cavalry small-arms attacks, artillery fire, pike assaults, and field fortifications were usually results of the experimental efforts to get the maximum advantage from infantry small-arms fire.

Infantry Tactics

In his campaigns in Naples at the turn of the century, Gonzalo de Córdoba led the way in recognizing and exploiting the potentialities of small-arms fire. He discovered that extensive frontages could be held by arquebusiers behind entrenchments, thus permitting him to meet, and to outmaneuver, much larger French forces. He also recognized the basic infantry problem that was to govern tactics for the remainder of the century: the need to protect arquebusiers in the open while they were reloading. His solution was to use pikemen to provide steadiness in the defense, and to exploit small-arms firepower by offensive shock action.

Córdoba's Spanish successors followed his example and improved on it. Arquebusiers were employed as skirmishers in front and on the flanks of pike and halberdier columns. The skirmishers—called *enfants perdus* by the French—were particularly important during the slow and painful process of forming up a conglomerate 16th century army for battle.

Spanish pikeman

As the century continued, the Spanish steadily increased the proportion of small arms men to pikemen, and began to employ them in solid formations of several ranks, intermixed with pike units. A steady volume of fire could be maintained by having front-rank men retire to reload, while those behind moved up to the firing line. The pikes continued to bear the brunt of both offensive and defensive shock action in this heavy kind of formation, sometimes called the "Spanish Square." The mutually supporting roles of highly disciplined Spanish arquebusiers and pikemen frustrated their enemies during much of the century.

Toward the close of the century the firepower of these heavy-infantry formations was still further increased by introduction of the musket. The lighter, more maneuverable arquebusiers were now almost completely relegated to skirmishing roles. This change seems to have been initiated almost simultaneously by Alexander of Parma and Maurice of Nassau. About the same time Henry of Navarre (at the Battle of Coutras) initiated the practice of having front-rank arquebusiers kneel in front of a standing second rank, so as to increase the steadiness of the formation and to double its firepower.

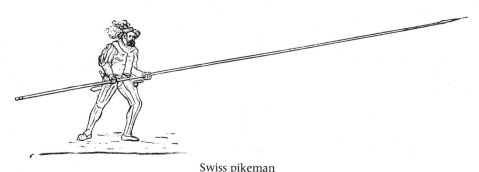

Swiss pikeman

Cavalry Tactics

During the first portion of the century, French heavy cavalry (mainly built around the *gendarmerie* of old *compagnies d'ordonnance;* see p. 442, 443) remained preeminent among European horsemen. The disaster of Pavia, however, impressed upon the French the obsolescence of these successors to the medieval knights. Like other European soldiers they began experimenting with greater proportions of light horsemen and attempting to find ways of adapting gunpowder firearms to cavalry.

During this period of experimentation the Germans developed a new species of heavy cavalry armed with the new wheel-lock pistols. These were organized into mercenary units, the men usually wearing black armor and accouterments, causing them to be called **reiters**— contraction of Schwartzreiter, or "black rider." The reiters were a kind of mounted counterpart of the famed Landsknechts. At first the reiter wore mail armor; later this became open helmet, breastplate, and heavy thigh-length leather boots. The reiters "charged" at a trot in a line of small, dense columns, each several ranks deep. As they approached a foe, the front-rank horsemen each emptied their three pistols, then swung away to flanks and rear in a tactic called the "caracole." While these men were reloading and joining the rear of their respective columns, the succeeding ranks continued the process of deliberate advance, pistol fire, and peeling off. Usually the caracole tactic was employed against pike elements of the opposing infantry for the purpose of knocking gaps in the line prior to a general advance. The caracole was a very difficult operation to carry out smoothly, and could easily be disrupted by a cavalry countercharge. The fact that it was used throughout the century, however, combined with the great demand for German mercenary reiters, is clear proof that the tactic was relatively effective.

During the latter part of the century, however, French cavalry regained its preeminence in Europe. Charging at the gallop in long lines two or three ranks deep, the French heavy cavalry fired their pistols as a prelude to shock action with the sword. The French discarded the lance completely. By the end of the century all European cavalry was armed with the pistol, or pistol and sword, save for a few Spanish and Polish lancers.

European cavalry tactics and organization were greatly influenced by the Turks, who in turn borrowed much from their Christian enemies. Save for the relatively small force of Janissaries, Turk armies were composed almost entirely of cavalry fairly equally divided between light, irregular skirmishers—armed mostly with the bow—and the heavier, lightly armored, well disciplined **timariot** (horsemen), feudal cavalry units in which lancers and horse archers were equally divided. At the beginning of the century this Turk feudal cavalry was capable of effective shock action against almost any cavalry opponent save for the French-type gendarmerie. The Turks, satisfied with the effectiveness of their archers, were slower than Western Europeans in adopting cavalry firearms.

The best Turkish cavalry were the **spahis** of the sultan's guard, more heavily armed and armored than the timariots, and capable of meeting the very best European infantry and cavalry. These, like the Janissaries, were permanent standing military formations.

The Turkish cavalry armies had two major weaknesses. In the first place, such an army had to move constantly in order to find enough forage for great numbers of horses. This made it most difficult for them to carry on sustained operations in any one region, and put them at a disadvantage when their conquests brought them up against the cavalry and infantry armies of central Europe, who could base themselves at powerful fortifications such as Vienna. Another weakness was the lack of steadiness of a fluid cavalry army in fighting a prolonged and hard-fought battle against a balanced professional infantry and cavalry force. The Turks got around this, to some extent, by steadily augmenting their small, well-disciplined corps of infantry Janissaries, and by using field entrenchments so that the Janissaries could provide a base of maneuver for the cavalry elements. The results were sufficiently effective as to pose a mortal threat to Central Europe for most of the century.

Envying, and baffled by, the Turkish light and intermediate cavalry, Germans and Hungarians put increasing emphasis on the development of **hussars**—effective light cavalry of the Turkish type—for screening, reconnaissance, and raiding. This trend—already started in western Europe by the Spanish **genitours** and French and Venetian **stradiots** (see p. 442)—resulted in the development of two clearly distinct types of European cavalry by the end of the century: helmeted, breastplated, pistol-carrying heavy cavalry of the reiter or French heavy-cavalry type, designed primarily for battlefield combat; and lighter, unarmored horsemen, armed with one or two pistols and a light sword, to undertake the other kinds of cavalry tasks.

MILITARY ORGANIZATION

Combat Formations

Though outmoded and inadequate, the medieval combat formation of three massive "battles"—dense blocks of mounted men and infantry—lingered on into the early years of the century. These unwieldy masses were particularly vulnerable to firearms and artillery.

The Spanish took the lead in efforts to solve the problem by thoughtful experimentation and improvisation. Based upon the experience of Córdoba (and possibly at his suggestion or instigation), in 1505 King Ferdinand created twenty units called **colunelas** (columns) each consisting of some 1,000–1,250 men: mixed pikemen, halberdiers, arquebusiers, and sword-and-buckler men, organized as five companies. This was the first clear-cut tactical formation based upon a coherent theory of weapons employment to be seen in Western Europe since the decline of the Roman cohort. The colunela was, for all practical purposes, the genesis of the modern battalion and regiment. It was commanded by a *cabo de colunela* (chief of column), or **colonel.**

Interestingly, this title soon became corrupted in the land of its origin. The colunelas, standing formations of the permanent Spanish royal army, or "crown" troops, were frequently called *coronelia.* Through inaccurate usage (similar to the corruption of the word "shrapnel" in our own time) colunelas were frequently called *coronelias,* and their commanders became commonly known as *coronels.*

The French soon copied the successful colunela concept, and also adopted the military title of rank which—uncorrupted—persists to this day in the French and English languages. For reasons quite unaccountable, the English language retains the uncorrupted Spanish word, as received through the French, but has adopted the corrupted Spanish pronunciation.

Over the next 30 years the Spanish gradually developed a larger organization called a **tercio,** which consisted of several colunelas—finally standardized at three—giving the tercio a total strength of slightly more than 3,000 men. It is not clear whether the term tercio came from this triangular formation or (more likely) originated because it comprised about one-third of the infantry component of the average Spanish army. By the time this formation became standardized the Spanish had eliminated the sword-and-buckler soldiers and halberdiers, leaving pikemen and arquebusiers as the components of a tercio, or "Spanish Square." This organization, like the ancient Roman legion and the modern division, became the basic combat unit of its army—a permanent formation, with a fixed chain of command, large and diverse enough to fight independent actions on its own.

The French, again following Spanish lead, soon organized permanent regional units at first called legions, and later regiments. These were somewhat smaller than the Spanish tercios, and not so well organized. This was the origin, however, of the famous French regional regiments of the 17th and 18th centuries.

Command Procedures and Military Rank

Following tradition tracing back to the Roman title of **imperator,** a European monarch was always the general of his country's army. His principal military assistant, in peace and war, was usually called a **constable,** a member of the nobility renowned for military prowess. Other outstanding noble warriors, particularly in France, frequently carried the honorific title of marshal. When the ruler was present in the field, he automatically exercised command as general. His second in command, who might or might not be the constable or one of the marshals, exercised his military functions as the **lieutenant general.** In the absence of the monarch, the lieutenant general commanded in the king's name.

Under the operational command of the monarch or his lieutenant general was a senior administrative officer known as **sergeant major general.** An experienced soldier, not necessarily a nobleman, the sergeant major general was in effect the chief of staff. He was responsible for supply, for organization, and for forming up the heterogeneous units of a 16th-century army for battle—a long, complicated process, with much shouting and confusion, considerably helped if the sergeant major general had a stentorian voice. In his administrative functions he was assisted in the subordinate units—national or mercenary—by administrative officers known as sergeants major and sergeants.

There was no permanent military hierarchy or chain of command below king and constable. Lieutenant generals and sergeant major generals were appointed for a campaign only.

The Composition of Armies

By the beginning of this century most monarchs had a number of permanent standing units to provide a nucleus for armies raised in war. The officers of these permanent professional units were almost always noblemen, though outstanding commoners could rise from the ranks, and some even became sergeant major generals.

The expense of maintaining a large, balanced standing army, however, was still prohibitive to the national economies of the time. Thus in wartime rulers relied primarily upon temporarily hired mercenaries to augment their relatively small permanent forces. The French permanent units, for instance, were mostly cavalry; they relied for infantry upon Swiss mercenaries or (on the few occasions when the Swiss were hostile) German landsknechts and reiters, and Italian infantry and cavalry condottiere companies.

Rulers relying so heavily on mercenaries had many headaches. Though usually well-disciplined and perfectly willing to fight (particularly the Swiss and Germans), the mercenaries had no patriotic feelings and were frequently untrustworthy. It was not uncommon for them to ask for bonuses on the eve of battle—threatening to abandon the army, or to join the enemy, if the demand was not met. This unreliability of mercenaries was one reason why generals were reluctant to risk a major battle. And the expense of paying the mercenaries was a primary factor in keeping armies small.

For foreign operations, an army would consist only of a relatively few national and mercenary professional units. The undisciplined, poorly organized feudal militia would have been a detriment rather than a help in any offensive operations, aside from the expense of paying them and of supporting them logistically. The feudal obligations of the nobility and the towns persisted, however, and the militia were levied, if necessary, particularly in times of foreign invasion; in a defensive role, they could provide some support to the professional units.

There were three important European exceptions to the operational pattern discussed above. One of these, the Turkish, has been discussed in a previous chapter (see p. 443).

Another exception was the appearance of a national regular army in Sweden in the latter part of the century, as the Vasa kings lifted their country to great power status. The Swedish rulers used no mercenaries. The nucleus of their armies was relatively numerous regular units, paid by the state, augmented in wartime as necessary by levies from a well-trained militia.

Maurice of Nassau raised a similar regular army in the Netherlands in the final years of the century. The ever-present Spanish threat enabled him to insist upon long-term enlistment of regular soldiers, and to impose the strictest discipline seen on the Continent since Roman times. Thanks to the sudden expansion of Dutch commerce, he was able to pay his soldiers well, and punctually. The result was a highly disciplined, homogeneous, responsive professional army, at least a match for the finest Spanish troops, and far superior to most of the Spanish mercenaries.

MILITARY THEORY

The increasing complexity of combat with gunpowder weapons, and the growing significance of economic and political considerations of waging war, attracted the attention of more and more men of intellectual bent. All aspects of military affairs were subjected to analysis in this revival of interest in the theory of warfare: strategy, tactics, organization, gunnery and ballistics, fortification.

Most important of the military theorists was **Niccolò Machiavelli** (1469–1527), Florentine statesman and political philosopher. Inactive most of the last 15 years of his life, because out of favor with the ruling Medici family of Florence, he devoted himself to study and writing. From a military point of view his masterpieces, *The Prince* and *The Art of War*, are the most important of his writings. His concepts were based upon his own experience, combined with intense study of classical military history. He was contemptuous of the *opéra bouffe* Italian wars of the 15th century and of the mercenaries who fought them. But he did counsel against taking unnecessary battle risks. He came to the conclusion that the old Roman legion should be the model for the military forces of his time—a conclusion which the Spanish and French were approaching more slowly and more pragmatically. He erred, however, in underrating the effectiveness of both firearms and cavalry. He devised a militia system for Florence, under procedures designed to assure the retention of civilian control, which was later largely adopted by the Medicis. The system's failure was due more to the inherent weakness of Florence than to Machiavelli's theoretical shortcomings.

More specialized theoretical contemporaries of Machiavelli included the famous German artist, **Albrecht Dürer** (1471–1528), who wrote on the theory of fortifications, and **Niccolò Tartaglia** (1500–1557), who published a number of works on the science of gunnery.

The French Huguenot **François de la Noue** (1531–1591) was an intellectual successor to Machiavelli. Like the Florentine, de la Noue combined a brilliant mind with extensive practical experience and an intense interest in military history. He was a principal Protestant leader in the early French Wars of Religion. His clear insight on strategy and tactics was reflected in his writings (mostly done while a prisoner of war), of which the most important was *Political and Military Discourses*.

DEVELOPMENTS IN NAVAL WARFARE

The Revolution in Tactics

This century witnessed an unprecedented revolution in naval warfare. The era of the galley, which had lasted for more than 2,100 years, ended shortly after it had reached its climax under three of the most able galley admirals of history: Khair ed-Din, Andrea Doria, and Don Juan of Austria. War galleys continued to operate in the Mediterranean for more than a century, but they were merely auxiliaries to the broadside battery sailing ship whose era began as the potentialities of naval gunfire were realized and exploited by such Spaniards as the Marquis of Santa Cruz. He died before he could meet Sir Francis Drake, who probably recognized the new developments in naval warfare even more clearly.

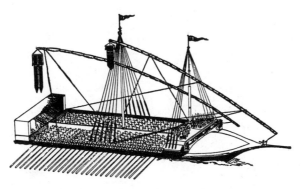

Mediterranean galley

Naval tactics had changed little between the Battles of Salamis and Lepanto (see pp. 30, 547). The objective of combat was either to ram or to board an opponent. The fragile galleys were not much different than those which the Romans had used in the Punic Wars. They were long, narrow, single-decked vessels, about 150 feet long and 20 feet in beam, propelled by about 54 oars, 27 to a side. In addition they had 2 or 3 lateen-rigged masts, useful to rest the oarsmen and to give added speed when the wind was favorable. There were 4 to 6 oarsmen—usually slaves—on each oar. In Christian vessels they were usually protected by mantelets; the Turks did not bother with such a consideration for galley slaves. The total crew consisted of some 400 men, including oarsmen, sailors, and a contingent of soldiers. Most Christian galleys at Lepanto had 5 small cannon mounted in the bow; the slightly smaller Turkish galleys had only 3 guns. Projecting forward from the bow, just above the water line, was a metal beak, some 10 to 20 feet long, for the purpose of ramming.

There were two important variants of these galleys at the time of Lepanto. The first was the Turkish **galiot,** a smaller, faster vessel, modeled after an earlier Byzantine type, with 18 to 24 oars and a crew of about 100. The other variant, in the other direction, was the **galleass,** introduced by the Venetians. This was a double-sized galley, slower but stronger, more seaworthy, and carrying more soldiers. It was a not very successful effort to reach a compromise between the fast Mediterranean war galley and the new multicannoned sailing vessels of

Northern Europe. The galleass carried 50 to 70 guns, but most of these were falcons or smaller, designed for man-killing rather than ship-smashing.

Up until the beginning of this century, the Northern European sailing warship, like the Mediterranean galley, had been considered primarily as a floating fort or platform, carrying soldiers who were to engage other soldiers on hostile ships. Naval battles were essentially fought like ground combat as soon as the vessels came within archery, or light cannon, range of each other, the conflict culminating in the boarding and capturing of one ship by the soldiers of the opponent. The vessels were still essentially transformed merchant "round" ships, barely twice as long as they were wide. The advent of gunpowder had merely added to the range of the fighting by the incorporation of small cannon on the fore and after castles, and along the railings of the upper deck. Heavy cannon could not be mounted on the castles or upper decks without risk of capsizing the vessel.

Spanish galleon

But at the beginning of the 16th century, someone—credit is sometimes given to one **Descharges,** a shipwright of Brest, France—invented the "port": an opening in the ship's side with a hinged cover facilitating the stowage of cargo in the hull without hoisting overside. English shipbuilders, spurred by Henry VIII's determination to mount heavy guns in his newly planned warships, seized on the idea as a way to permit a cannon to be fired from the lower decks of a ship. Thus the broadside battery came into existence—its weight safely distributed below the center of gravity. The Spanish soon followed this example. The resulting warship, barely 100 feet long and about 30 feet in beam, was called a galleon—probably because, like the galley, it was a vessel designed specifically for war, because of the trend toward a slimmer shape, and because it had a low beak, just above the water line, which facilitated ramming. This 3-masted, square-rigged sailing ship still carried castles fore and aft, with large numbers of small cannon mounted on the upper works. But its row of larger cannon in the main hull gave the galleon the ability to fight effectively at long range without necessarily closing for the traditional hand-to-hand climax of earlier naval engagements.

The principal shortcoming of the galleon, as compared with the galley, was that it was largely at the mercy of the wind. This was only partly offset by the fact that newer, narrower vessels were more maneuverable than the old round ships and, thanks to improved sails and rigging, were able to "beat" against the wind. Unlike the galley, the galleon had the seaworthiness to make long-range ocean voyages.

English sailors, as early as the time of Henry VIII—who greatly encouraged and stimulated the development of English warships—seem to have had a glimmering of the tactical change made possible by the introduction of broadside guns. Though boarding was still considered the main aim of battle, the English tended to put more and more emphasis on designing their ships for long-range gunnery. As a result, the fore and aft castles became lower and lower, and the beaks soon disappeared from English galleons. The proportion of big guns to small guns steadily increased. The Spanish, however, kept the galleon beak, and maintained a balance between man-killing and ship-smashing guns. They followed the English example of lowering the forecastle, but retained a towering aftercastle, on which was mounted a formidable array of small guns.

The Spanish still considered their ships primarily as floating fortresses, carrying garrisons of land soldiers. The English, on the other hand, emphasized seamanship, and their officers and sailors became more skillful than the Spanish. Rather than wasting space and man power by carrying a garrison of landlubber soldiers, the individual English sailor was trained to leave his gun, or to scramble down the rigging, to pick up pike or cutlass when the time came to board an enemy ship or to repel boarders.

These were the differences in naval tactical theory which led to the decisive Battle of the English Channel (see p. 508) in which the English repelled the Spanish Armada, and introduced the new era of the broadside, sailing warship in naval warfare—while at the same time staking out their claim to mastery of the seas.

The Emergence of Naval Strategy

Prior to this century naval strategy was largely an adjunct of land strategy. It is possible that the Chinese Admiral Cheng Ho had had some glimmering of the use of sea power as an instrument supporting national political and economic interests across wide expanses of ocean (see p. 481). There can be no question, however, that this concept of the employment of naval forces in support of national objectives was clearly in the mind of Portuguese **Affonso de Albuquerque** when he established a network of bases around the Indian Ocean, which gave Portugal virtual control of its sea routes and coast lines (see p. 555).

The Spanish probably did not understand sea power quite so clearly as the Portuguese. They employed it successfully, nonetheless, throughout most of this century, in consolidating their control of much of the Western Hemisphere, and in dominating the sea routes of the Atlantic and the eastern Pacific Oceans.

The significance of Spanish bases, and their control of major sea routes, was certainly evident to English seamen like Drake. Confident that the new tactics of broadside sailing ships gave him and his compatriots a clear-cut naval tactical advantage over the Spanish, Drake was probably more responsible than any other Englishman for deliberately initiating the chain of events that were to lead to the strategic supremacy of British sea power.

The First Ironclad Warships

Few men of relative obscurity have so directly influenced the course of history as did Admiral Yi Sung Sin of the semi-independent Korean state of Cholla. His two great victories over the Japanese fleets of Hideyoshi are all that saved Korea—and possibly China—from Japanese conquest at the close of the 16th century. The principal instruments of these victories were two or more ironclad warships which Yi himself had designed and built, and which he employed in battle with courage, skill, and determination.

Yi's "tortoise ships" were low-decked ironclad galleys, with a complete overhead covering of iron plates, ringed with spikes to prevent boarding. Since galleys were uncommon in the Far East, his use of oars, and oarsmen protected by armor plate, reveals the imaginative genius of a man who realized that his concept required a motive power other than the wind. The vessels were equipped with heavy iron rams, and were armed with two or more cannon that fired from gun ports in the armor plate. Other embrasures were used by archers, who fired incendiary arrows at the sails, rigging, and wooden hulls of enemy warships.

Yi died in the moment of his final victory. With his death disappeared the imaginative concept of using heavily gunned armored warships. Not until more than 2½ centuries later, after the Industrial Revolution had provided a new kind of motive power, did Yi's idea emerge again. It might be said that Yi was a visionary, like **Leonardo da Vinci,** whose imaginative ideas were unrealistically centuries ahead of his time. The difference is that Yi actually converted imaginative designs into successful instruments of war that changed the course of history.

WESTERN EUROPE

BRITISH ISLES

England

This was the century of the Tudors, dominated by lusty **Henry VIII** and his amazing daughter **Elizabeth I.** Between them they established the foundation for the great maritime British Empire, already in the making before Elizabeth's death. Endemic border warfare with Scotland persisted, though the potential Scottish threat did not seriously inhibit English overseas activities after their great victory at the Battle of Flodden early in the century. Sporadic warfare with France continued during the century, but with little important result, and largely incidental to France's greater external and internal conflicts (see pp. 511, 521). With the growth of English overseas commerce and the beginnings of the Royal Navy, insular Britain's obvious enemy was the nation then dominating the seas and monopolizing the most lucrative colonial trade: Spain. Efforts to avert this rivalry brought about the marriage of Queen **Mary I** with **Philip II** of Spain, but to no avail; in fact, it possibly hastened the crisis. Under Elizabeth there was a decade of undeclared war—on the high seas, in the Americas, and in the Netherlands—before the outbreak of formal hostilities led to the famed Spanish expedition of the Great Armada. Repulsed from England's shores in a series of tactically significant, but indecisive, combats, the

Armada was destroyed by the elements—not by the English Navy. But this heralded the decline of the Spanish Empire in Europe and in the Americas, and portended England's future mastery of the waves. The principal events were:

1485–1509. Reign of Henry VII. Central royal authority was strengthened and private feudal armies suppressed (see p. 459).

1509–1547. Reign of Henry VIII.

1511–1514. Wars with France and Scotland. (See pp. 513, 510.)

1520, June 7. Field of the Cloth of Gold (near Calais). Establishment of a short-lived alliance between Henry VIII and **Francis I** of France.

1522–1523. English Invasions of France. (See p. 515.) These were abortive.

1530–1534. Break with the Papacy. Henry established the independent Church of England, creating a basis for future internal and external conflicts. He was soon forced to suppress the first of several Catholic rebellions (1536).

1542–1550. Wars with France and Scotland. (See pp. 518, 510.) French landings on the English coast (1545–1546) led Henry to commence an intensive naval construction program, the true beginning of the modern Royal Navy. He also began a great coast-defense construction program.

1547–1553. Reign of Edward VI. He died before reaching his majority. The warrior **Edward Seymour, Duke of Somerset,** was Lord Protector at the beginning of the reign.

1549. Internal Unrest. A Catholic revolt in Devonshire was suppressed by Somerset. A peasant revolt in Norfolk, under **Robert Ket,** was also suppressed by royal forces led by **John Dudley, Earl of Warwick.**

1549–1550. Renewed War with France and Scotland. French successes outside Boulogne and Scottish recapture of Haddington, combined with internal religious and social unrest and noble dissatisfaction with Somerset's liberal ideas, permitted Warwick to force Somerset out of power (September 1549). Warwick (who became **Duke of Northumberland**) made peace with France, surrendering Boulogne in return for a cash payment.

1553, July 6–19. Insurrection of Northumberland (Warwick). Upon the death of Edward VI, he attempted to place his daughter-in-law, Lady **Jane Grey,** on the throne instead of the rightful successor, Edward's sister, **Mary.** Most of the nation rallied to Mary; Northumberland was captured, and Jane was deposed and executed after a reign of nine days.

1553–1558. Reign of Mary I. She reestablished Catholicism. Her marriage to Philip of Spain added to religious unrest, many English Catholics joining the Protestants in distrust of Spain and Spanish Catholicism.

1554. Insurrection in Kent. Led by Sir **Thomas Wyatt,** Sir **Thomas Carew,** and the **Duke of Suffolk,** this was to prevent Mary's marriage to Philip. Wyatt was defeated and overpowered while trying to take London; the rebellion collapsed and the leaders were executed.

1557–1559. War with France. Mary's marriage led to English involvement in Spain's endemic wars with France (see p. 519).

1558–1603. Reign of Elizabeth I (Sister of Edward VI and Mary). Return of England to Protestantism. She followed a general policy of avoiding involvement in major continental wars, though she intrigued constantly and sent several small expeditions to the Continent.

1559–1560. Intervention in Scotland. (See p. 511.)

1562. English Expeditions to France. These were to aid Huguenots (see p. 521).

1568–1572. Growing Hostility between England and Spain.

1573. Temporary Rapprochement with Spain. This was due to the ascendancy of the Guises in France (see p. 523).

1577. Alliance with the Netherlands Republic. The Netherlands was in rebellion against Spain (see p. 528), but Elizabeth refrained from outright war against Spain.

1577–1580. Sir Francis Drake Circumnavigates the Globe. He raided Spanish and Portuguese colonies and shipping en route (see p. 570).

1585. English Military Assistance to the Netherlands. (See p. 530.)

1585–1586. Drake's Expedition to the Caribbean. (See p. 566.)

THE SPANISH ARMADA, 1586–1588

1586, March. Spain Plans a Naval Expedition against England. This was recommended by Admiral Marquis de Santa Cruz. Philip II directed the Duke of Parma, Spanish commander in the Netherlands, to prepare to take his army to England under convoy by Santa Cruz's fleet.

1587, April–June. Drake's Expedition to Cádiz. He was aware of Spanish naval preparations, and sailed into Cádiz with a fleet of 23 ships, destroying 33 Spanish vessels of all sizes, "singeing the beard of the King of Spain" (April 19). On his return Drake harassed Spanish shipping off Cape St. Vincent, sacked Lisbon harbor, then captured a Spanish treasure galleon in the Azores (May–June).

1587–1588. Spanish Preparations. Santa Cruz diligently repaired the damage done by Drake, but died (January 30, 1588) before the expedition was ready. His death was probably the most important single factor in saving England from Spanish invasion. **Alonzo Pérez de Guzmán, Duke of Medina Sidonia,** replacing Santa Cruz, was a man of courage and ability, but without either naval or army experience. Admiral **Diego de Valdéz** was second in command. The Duke of Parma was to assume over-all command when the expedition reached the Netherlands.

1587–1588. English Preparations. Elizabeth selected Lord **Howard of Effingham** as the commander-in-chief of her fleet (December 21, 1578). Commoner Drake was appointed vice admiral. Howard had little naval experience, but relied upon good seamen serving under him, particularly Drake. Drake and Howard urged another expedition against Spanish ports to prevent the sailing of the Spanish Armada. Elizabeth overruled them (March). She did, however, raise a force of about 60,000 men to meet the threatened invasion (June).

1588, July 12. The Armada Leaves Corunna. The fleet, which had started from Lisbon (May 20), had stopped at Corunna to take refuge from a storm and to repair a number of unseaworthy vessels. It comprised 20 great galleons, 44 armed merchant ships, 23 transports, 35 smaller vessels, 4 galleasses, and 4 galleys. The fleet was manned by 8,500 seamen and galley slaves, and 19,000 troops; the warships mounted 2,431 guns. Of these 1,100 were heavy guns, including about 600 culverins; over half of the cannon mounted on Spanish ships were light antipersonnel weapons, based on age-old naval tactics of reaching decision by grappling and hand-to-hand fighting.

1588, July 19. Appearance of the Armada. It was sighted off Lizard Head by English scout vessels.*

* Dates are given here in Old Style, thus differ by 10 days from those in the best-known account: Garrett Mattingly, *The Armada,* New York, 1959.

1588, July 20. Howard Sails from Plymouth.
Naval forces available for the defense of England, under his over-all command, consisted of his own fleet of 34 ships, Drake's squadron of 34 (also based on Plymouth), a London squadron of 30 ships, and another squadron of 23 vessels under Lord **Henry Seymour** off the Downs in the eastern English Channel. There were some 50 additional vessels of varying types, carrying a few guns, mostly transports and supply ships that took little part in the subsequent action. The principal warships of the English fleet carried a total of approximately 1,800 heavy cannon, mostly long-range culverins.

1588, July 21. Engagement off Plymouth.
The English outsailed and outshot the Spanish, who lost one ship sunk, and suffered heavy losses and damage from English long-range fire.

1588, July 23. Action off the Devon Coast.
This all-day engagement followed a day of protracted calm and aimless maneuvering. There was no coordination between vessels on either side, much ammunition was consumed, no vital damage was done.

1588, July 25. Battle off Dorset. The English had replenished their ammunition; the Spanish had had no such opportunity. Medina Sidonia, abandoning his plan of landing on the Isle of Wight, headed for Calais, hoping to be able to replenish his empty ammunition magazines from Parma's supply depots.

1588, July 26–27. Cannonade off Calais. The Spanish fleet, now 124 vessels, anchored off Calais, unable to renew action without ammunition. Howard now had 136 ships of all types, organized into 4 squadrons under himself, Drake, Sir **John Hawkins,** and Sir **Martin Frobisher.** He contented himself with long-range fire, but anchored most of his fleet out of range. He knew that he could do no serious damage to the Spanish ships without coming so close as to be in danger of boarding by the Spanish soldiers. Medina Sidonia requested Parma to come to his assistance, but Parma was unable to leave Bruges, which was being closely blockaded by a Dutch fleet under **Justinus of Nassau.**

1588, July 28. Battle off the Flanders Coast.
Before dawn the English sent several fire ships into the Spanish fleet. To avoid the fire ships, Medina Sidonia ordered anchor cables cut. He planned to return to the anchorage after the danger was past, but his subordinates panicked in predawn darkness. Due to unfavorable winds, the Spanish, unable to concentrate, drifted northward in a straggling formation. The English pursued and closed in, to begin an all-day running fight at very close range, with Spanish small guns and arque-buses trying vainly to reply to the English culverins and other heavy cannon. The English kept to windward, with several ships concentrating against individual Spanish vessels, firing alternate broadsides. The Spaniards fought heroically but were unable to reply effectively, since they had no heavy ammunition. Despite severe damage and heavy loss of life on the Spanish ships, none was sunk, but the English were closing in toward the end of the day to cut off and capture 16 of the hardest hit. The Spanish vessels were saved by a sudden squall.

1588, July 29–30. Unfavorable Winds. The Spanish fleet was unable to break the Dutch blockade of the Flanders coast, where Medina Sidonia had planned to refit, obtain new ammunition, and to join Parma. After the fleet almost went aground on Zeeland, and with no change in wind, the Spanish lost all chance of reaching any Netherlands port. Medina Sidonia decided to return to Spain via the North Sea, completely encircling the British Isles.

1588, August 2. Past the Firth of Forth. The Spanish fleet was now fairly well concentrated. The English fleet, which had been following, ran short of provisions and returned to home ports.

1588, August–September. The Ordeal of the Armada. Terrible hardships and losses were suffered by the Spanish, partly due to storms, but even more to starvation and thirst. Thousands of men died. Out of 130 ships that started, 63 were known to be lost, the survivors straggling into Spanish ports in September. The English sank or captured about 15. Nineteen were wrecked on the Scottish or Irish coast. The fate of the remaining 33 vessels is unknown.

1589–1596. Small Expeditions to the Continent. Lord **Peregrine Bertie Willoughby** led an expedition of 4,000 men to Normandy to aid **Henry of Navarre** (1589). Small forces were landed at St.-Malo and Rouen (1591). Troops were landed during a raid on Cádiz (1596).

1594–1603. Tyrone Rebellion in Ireland. Endemic rebellion in Ireland erupted into full-scale war under the leadership of **Hugh O'Neil, Earl of Tyrone,** who joined already rebellious **"Red" Hugh O'Donnell.** O'Neil defeated the English at **Yellow Ford** on the Blackwater River (1598, August), and outmaneuvered Queen Elizabeth's favorite, the **Earl of Essex** (1599–1600). In response to Irish appeals for help, a Spanish army of 4,000 men under Don **Juan D'Aquila** arrived in Ireland and captured **Kinsale** (1601). A few weeks later, **Charles Blount, Lord Mountjoy** defeated the Irish-Spanish army at the **Battle of Kinsale.** The rebellion soon was suppressed. O'Neil surrendered and was pardoned by James I (1603).

Scotland

Fierce, incessant conflicts with England persisted through this century. The Scots continued to seek French assistance whenever possible, and to exploit all English preoccupation on the Continent. This was, however, the last century in which the armed forces of the two nations were ranged against each other in formal battle. The principal events were:

1488–1513. Reign of James IV. He greatly increased central authority.

1513. War with England. When Henry VIII led an invasion of Scotland's traditional ally, France, James promptly declared war (August 11) and invaded England.

1513, September 9. Battle of Flodden. **Thomas Howard, Earl of Surrey** hastily raised an army to defend England, aided by James's slow advance. Finding the Scottish army drawn up in an extremely strong defensive position on a hill known as Flodden Edge, Surrey boldly and rapidly moved his entire army around the Scottish position, and forced James to face to the rear. Surrey covered the deployment of his army and the arrival of his rearmost contingents by a long-range artillery bombardment and archery harassment of the Scottish position. Though James had planned to fight a defensive battle, this harassment caused the impatient Scots (mostly in masses of pikemen) to charge violently down the hill. The battle began in late afternoon and lasted till darkness. The English repulsed repeated Scottish charges and gradually gained ascendancy in a violent melee. Ten thousand Scots were killed, including James and most of his leading nobles. English losses were also heavy.

1513–1542. Reign of James V. During most of his reign there was indecisive border warfare with England (1513–1534).

1542–1550. Renewed War with England. This was due mainly to religious differences between James and Henry VIII, who renewed old claims of English suzerainty.

1542, November 25. Battle of Solway Moss. An English invading army inflicted a crushing defeat on weak and disorganized Scottish forces. The English failed to exploit.

1544–1547. Reign of Mary, "Queen of Scots" (one week old when her father died). English forces under **Edward Seymour, Earl of Hertford** inflicted much damage in southern Scotland (1544–1545). The Scots renewed their alliance with France, and French expeditionary forces soon arrived in Scotland. The Scots rejected the demand of Seymour (now Duke of Somerset and Lord Protector of England; see p. 507) to establish an English-Scottish alliance by the marriage of Edward VI (age 9) to Mary (age 5).

1547, September 10. Battle of Pinkie. An English army of 16,000 under Somerset invaded Scotland. A Scottish army of 23,000 under Earl **James Hamilton of Arran,** Regent of Scotland, met the English at the River Esk, on the Firth of Forth. Each side attempted to turn the other's left, but Somerset's superiority in cavalry, artillery, and arquebusiers gave him a decisive victory with the offshore assistance of an English fleet under Lord **Edward Clinton.** Scottish casualties exceeded 5,000, plus 1,500 captured; the English lost about 500. This was the last formal battle fought between the national armies of Scotland and England. The English occupied Edinburgh.

1547–1550. Guerrilla Warfare and Small-Unit Raids. The Scots with French assistance regained the territories captured earlier by the English, who abandoned Edinburgh (1550).

1550–1559. Growing French Influence. Many Scots became resentful of "foreign occupation."

1559–1560. Rising against the French. The Scots, assisted by English forces sent by Elizabeth, besieged the French contingents and forced them to surrender at Leith (February 1560).

1560, July 6. Treaty of Edinburgh (between England, France, and Scotland). The French agreed to end their intervention.

1567, June. Noble Revolt. Mary's adherents were defeated at the **Battle of Carberry Hill** (June). She was imprisoned and forced to abdicate; her half-brother, **James, Earl of Moray,** became regent for her infant son, **James VI.**

1567–1625. Reign of James VI (who became James I of England and Scotland). His minority was a period of turmoil, during which his mother's adherents, led by Arran initially (1567–1573), attempted to overthrow the regency.

1568, May. Battle of Langside. Mary escaped from prison; her adherents rose to support her, but were defeated by Moray. Mary fled to England, where she was imprisoned by Elizabeth in the Tower of London.

1582–1583. Edinburgh Raid. Adherents of Mary captured James VI, who was held prisoner for 10 months.

1585. James's Alliance with Elizabeth. This ended the national wars between England and Scotland.

1587, February 8. Execution of Mary. This was ordered by Elizabeth.

FRANCE, 1495–1600

During the first half of this century, France vainly attempted repeatedly to gain ascendancy in Italy and to become the principal power of Western Europe. It is not surprising that the Valois kings were unable to achieve their objectives against the united power of the Holy Roman Empire and Spain, with England and most of the Italian states also being generally allied against them; it is remarkable that they frequently came close to success, and that their determination never faltered, despite numerous defeats. There were several victories, as well. On balance, the French national spirit, created by Joan of Arc during the adversity of the Hundred Years' War, was strengthened and tempered in the constant struggle of the Valois-Hapsburg wars. French national territory actually grew slightly during this period at the expense of less united neighbors. The principal feature of the Valois-Hapsburg Wars was the intense personal feud of Francis I and Emperor Charles V.

These wars had barely ended, just after midcentury, when France was plunged into nearly four decades of bitter religious civil strife. Despite her neighbors' efforts to take advantage of this violent, divisive struggle, France emerged without serious or lasting effects. In part, this was due to the fact that the neighbors were also suffering from conflicts spawned by the Reformation; in part, it was because the French, so violently divided against each other religiously, and eager to accept foreign aid from those of like religious faith, nevertheless closed ranks periodically to prevent foreign political domination; finally, it was due to the genius of **Henry of Navarre,** who gained decisive military victories as a Protestant leader, but who brought political as well as military unity and peace to France by embracing the Catholic religion as King Henry IV. The principal events were:

Wars with Spain and the Empire, 1495–1559

ITALIAN WARS OF CHARLES VIII AND LOUIS XII, 1495–1515

FAILURES IN NAPLES AGAINST THE "GRAN CAPITÁN," 1495–1504

1495. Charles VIII's Invasion of Italy. (See p. 462.) A Spanish expedition of some 2,100 men was sent to support Aragonese King Ferrante of Naples. The Spanish commander, **Hernández Gonzalo de Córdoba,** landed in Calabria (May 26), one of the few parts of Naples still unsubdued by the French, whose overall commander was **Gilbert, Duke of Montpensier.**

1495, June 28. Battle of Seminara I. Córdoba and Ferrante, with a combined force of about 8,000 men, were defeated by a somewhat smaller French Army under **Bernard Stuart, seigneur d'Aubigny** (a member of the Scottish Stuart family). This was Córdoba's first and last defeat.

1495–1498. Victories of Córdoba. Adapting the Spanish-Moorish methods of operations to the French and Swiss military systems as he found them in Italy, and refusing to fight any large-scale engagements, Córdoba slowly reconquered most of Naples. He was assisted greatly by the length of the insecure French line of communications through northern Italy. His principal success was the siege and capture of Montpensier at Atella (July 1496). During this time Córdoba intensively trained a portion of his infantry in Swiss pike tactics.

1498. Departure of Córdoba. Upon the withdrawal of the French, he returned to Spain. Gonzalo's army was quartered in Messina, Sicily, for a brief period. Civil war broke out in Naples. Taking advantage of this, the French again invaded (1499).

1498–1515. Reign of Louis XII.

1500, September–1501, January. Córdoba's Expedition to Cephalonia. Taking advantage of the lull in the fighting in Naples, Córdoba moved his army to Cephalonia in the Ionian Sea, and reconquered the island from the Turks, materially assisting Spain's Venetian ally in its war with the Ottoman Empire. Córdoba then returned to Sicily.

1500, November 11. Treaty of Granada. The French and Spanish agreed to divide Naples between them.

1500–1502. Louis XII Invades Italy. The French easily conquered Milan, then advanced south to overrun Naples. At the same time a Spanish fleet seized Taranto (March 1502).

1502. Renewal of Franco-Spanish War in Naples. Córdoba with about 4,000 men was opposed by the **Duke of Nemours** with approximately 10,000. Córdoba was forced to withdraw into Apulia, to the seaport of Barletta, leaving most of Naples in the hands of the French.

1502, August-1503, April. Blockade of Barletta. The loosely maintained French investment was enlivened by frequent duels and tournaments. **Pierre du Terrail, seigneur de Bayard** (the *"chevalier sans peur et sans reproche"*) was the principal French champion.

1503, April 21. Battle of Cerignola. Reinforced by sea to 6,000 men, Córdoba marched out of Barletta and took a position on a hillside, behind a ditch and a staked palisade. The French attacked. At the outset, the Spanish artillery became useless due to the explosion of its powder supply. Supported by effective artillery fire, French heavy cavalry and Swiss pikemen attacked, but were disastrously repulsed by the effective fire of Spanish arquebusiers behind the ditch and palisade. With the attackers in confusion, the Spanish counterattacked to drive them off the field. Nemours was killed. This was probably the first battle in history won by gunpowder small arms.

1503, April 21. Battle of Seminara II. On the same day Córdoba won Cerignola, a Spanish army defeated Aubigny's French army at Seminara, scene of Aubigny's earlier victory over Córdoba.

1503, May 13. Fall of Naples. After taking the city, Córdoba advanced to Gaeta.

1503, June–October. Siege of Gaeta. Arrival of French and Italian allied reinforcements caused Córdoba to fall back to the line of the Garigliano River.

1503, August–December. French Probes in the Pyrenees. These were repulsed by the Spanish.

1503, October–December. Stalemate on the Garigliano. Because of illness of Marshal **Louis de La Trémoille** command of the French army passed to Italian condottiere allies: first was **Gonzago, Marquis of Mantua** (who had opposed the French at Fornovo; see p. 462), until he was in turn replaced due to illness by **Ludovico, Marquis of Saluzzo.** Neither of these Italians was trusted or respected by their French subordinates. Córdoba conducted frequent raids along the river, but avoided any general attacks.

1503, December 29. Battle of the Garigliano. When Córdoba's army was reinforced to about 15,000 men, he decided to attack the French-Italian army of some 23,000. Taking advantage of bad weather, his skillful engineers used secretly assembled bridging materials to make a surprise crossing of the swollen river immediately opposite the French winter cantonments. The Franco-Italian forces were driven in confusion into Gaeta with heavy losses in killed and captured.

1504, January 1. Fall of Gaeta. The terms permitted the French army to evacuate by sea.

1504–1507. Córdoba Viceroy of Naples.

1505. Treaty of Blois. Louis XII gave up his claims to Naples and made peace with Spain, which now controlled all of Naples.

1507. Córdoba's Recall to Spain. Due to the jealousy of Ferdinand, he was never given another opportunity to command. His death (December 1, 1515) preceded that of Ferdinand by 3 months.

LATER ITALIAN WARS OF LOUIS XII, 1508–1514

1508–1510. War of the League of Cambrai. Urged by Emperor Maximilian, France joined the Empire, the Papal States, and Spain in an alliance to break the power of Venice. Because of mutual jealousies the League did not prosecute the war intensively; there was no coordination of allied activities. The only important battle was a French victory at the **Battle of Agnadello** (May 14, 1509). Imperial, French, and Papal forces, acting separately, occupied most of Venetian territory in northern Italy. At the same time Spanish troops captured the Venetian Apulian towns, including Brindisi and Taranto. Taking advantage of allied disunity, the Venetians recaptured Padua and Vicenza (July). After a brief and unsuccessful siege of Padua by Maximilian, the war became a stalemate, due to allied mutual jealousies.

1510–1514. War of the Holy League. Pope **Julius II,** fearful of both French and Germans, joined Venice in a new alliance, later called the Holy League. Ferdinand of Spain, at first neutral, joined Venice before the end of the year, as did Henry VIII of England. Desultory military operations were almost entirely between French forces and those of the Holy League allies dominated by Spanish leaders.

CAMPAIGNS OF GASTON OF FOIX, 1511–1512

1511. Arrival of a New French Commander. This was Count **Gaston of Foix,** age 21, the new Duke of Nemours. The war was electrified by his energy.

1511–1512. Campaign of Bologna and Brescia. The French had captured Bologna (May 13, 1511), but a combined Papal and Spanish force under **Raymond of Cardona,** Viceroy of Naples, had laid siege to the city. Gaston marched to Bologna, where he drove the allied army off. He then turned north to defeat the Venetians near Brescia, which he then invested and captured by storm (February 1512).

1512, March–April. Campaign of Ravenna. With most of north Italy under his control, Gaston now marched boldly south to besiege Ravenna, hoping to entice the Spanish into battle. A combined Spanish-Papal army under Cardona moved cautiously toward Ravenna, to establish a strong defensive position nearby, threatening the French siege lines. Cardona had approximately 16,000 men; Foix had 23,000, including 8,500 German landsknechts. The garrison of Ravenna was probably about 5,000 men. Foix sent a formal invitation to battle to Cardona, who accepted with equal formality.

1512, April 11. The Battle of Ravenna. The Spanish position was selected with its back against the unfordable Ronco River, and its front protected by strong entrenchments and obstacles prepared by the excellent Spanish engineer, **Pedro Navarro,** who also commanded the Spanish infantry. Gaston had 54 artillery pieces; Cardona probably had about 30. The French army moved from the siege of Ravenna, leaving only 2,000 men to guard against a sortie by the garrison. They crossed the stream between Ravenna and the Spanish camp without interference, and drew up in a semicircle facing the Spanish trenches. The French artillery on the flanks began an intensive bombardment; the Spanish artillery returned the fire. This bombardment lasted for several hours, extremely heavy casualties being suffered by both sides. Navarro's infantry was protected by its entrenchments. The Spanish cavalry, less protected, finally could endure the bombardment no longer, and attacked without orders. These attacks, poorly coordinated, were repulsed by the French, and Gaston promptly counter-attacked. The struggle between the landsknechts and the Spanish infantry at the entrenchments was particularly bloody and indecisive, lasting for perhaps an hour. At this point two French cannon, which Gaston had sent back across the river behind the Spanish camp, opened fire from the rear, creating panic among the defenders. In the ensuing rout, the Spanish, endeavoring to cut their way out to the south, suffered great losses. Gaston, taking part in the pursuit, was killed in a minor skirmish with one stubbornly withdrawing unit of Spanish infantry. French losses were about 4,500 killed and at least an equal number wounded, the majority of these being among the landsknechts. Spanish losses were 9,000 killed and an unknown number wounded. Navarro was captured. (When the Spanish government later did not ransom him, in disgust he renounced his allegiance to Spain and entered the French service, performing with distinction.)

COMMENT. *The death of Gaston of Foix, as the result of his own impetuous participation in the pursuit, was a real disaster to the French. In the previous six months of campaigning he had demonstrated a potentiality of becoming one of the great captains of history.*

1512, May–July. The French Situation Becomes Desperate. The Emperor Maximilian and Switzerland both joined the Holy League. (This was the real beginning of the Valois-Hapsburg wars.) As a result, the German landsknechts withdrew from the French Army and joined imperial forces advancing across the Alps into north Italy. At the same time Swiss troops occupied Lombardy. **Jacques de Chabannes, Marshal La Palice,** raised the siege of Ravenna and withdrew behind the Alps. The Swiss established a virtual protectorate over the Duchy of Milan.

1512. English Invasion of Guyenne. An expedition under Marquis **Thomas of Dorset** ended in mutiny and failure.

1513, May–June. French Invasion of Italy. An army of 12,000 under Marshal **Louis de La Trémoille** crossed the Alps into Italy and captured Milan. He then besieged a Swiss garrison at Novara. Threatened by a Swiss relief army of some 5,000 men, La Trémoille withdrew from his investment lines and prepared for battle.

1513, June 6. Battle of Novara. The Swiss, sleeping only 3 hours, marched all during the night to join the garrison of Novara and to make a surprise dawn attack upon the unsuspecting French. Attacking in typical echelon form, in 3 heavy columns, the Swiss first struck the French left flank and smashed through to the center of the French camps, cutting the infantry to pieces, though most of the cavalry escaped. La Trémoille withdrew the remnants of his shattered army back behind the Alps into southern France.

1513, June. English Invasion of France. Henry VIII, with an army of about 28,000 English and continental mercenary troops, invaded France from Calais. A hastily assembled French army of approximately 15,000 men, under La Palice endeavored to harass the English besieging Thérouanne, but avoided battle.

1513, August 16. Battle of Guinegate (Battle of the Spurs).* A French effort to harass the English siege lines at Thérouanne was anticipated by the English, and the French, expecting to achieve surprise, were the ones who were surprised by fierce opposition. The French cavalry fled from the field in confusion. Casualties were slight; the battle received its nickname from the fact that many of the French knights lost their spurs during the flight, and these were later picked up by local townspeople. Among the prisoners captured by the English were La Palice and **Bayard,** one of the few Frenchmen who had fought bravely, and who had attempted to rally his panic-stricken comrades.

1513, August 22. Surrender of Thérouanne to the English.

1513, September. Swiss Invasion of France. At Dijon the Swiss accepted a French indemnity and made peace, thus letting down their English and German allies.

1513, October 7. Battle of La Motta. Cardona, Viceroy of Naples, defeated the Venetians,

---

* Not to be confused with "The Battle of the Golden Spurs," see p. 399.

largely due to the superior performance of the Spanish infantrymen—pikemen and arquebusiers—under **Fernando F. de Avalos, Marquis of Pescara.**

1513–1514. Allied Indecision. Despite victories at Novara and Guinegate, the allies could not agree upon a common strategy. Discouraged by the defection of the Swiss, one by one they made peace with France, first the Pope and Spain (December, 1513) and then the Empire and England (March, July 1514).

The Reign and Wars of Francis I, 1515–1547

1515, June. Invasion of Italy. Francis, inheriting his predecessor's obsession with Italy, formed an alliance with Henry VIII and with Venice against the Emperor, the Pope, Spain, Milan, Florence, and Switzerland. Leading an army of about 30,000 men, he suddenly invaded Italy by the high and difficult Argentière Pass; the difficult passage was facilitated by the skill of turncoat Spanish engineer Pedro Navarro (see p. 514).

1515, June–September. Advance on Milan. Francis at the same time attempted to bribe the Swiss to make peace. Nearly 10,000 of the Swiss in Lombardy accepted the bribe and marched off to Switzerland. This left approximately 15,000 Swiss in and around Milan, still faithful to their allies. Meanwhile a Venetian army of about 12,000 was facing a Spanish army of comparable size near Lodi.

1515, September 13–14. Battle of Marignano. The French were in an entrenched camp about 10 miles from Milan. Francis had called on **Bartolomeo de Alviano,** the Venetian commander, to join him as soon as possible. The Swiss, after a typical rapid advance, launched a surprise attack in midafternoon. The rapidity of the Swiss advance gave the French no opportunity to use their artillery. Repulsing a French cavalry attack, the Swiss smashed their way across the entrenchments, but were finally halted by a counterattack led by Francis himself, with Bayard at his side. A fierce fight raged at the edge of the French camp for some five hours, when both sides pulled back by mutual consent at about 10 P.M. By dawn both sides had reorganized; the battle was resumed by a violent Swiss attack. This time the French artillery was ready, but despite heavy casualties the Swiss pressed forward so rapidly that again an inconclusive, sanguinary hand-to-hand combat ensued. About 8 A.M. Alviano's Venetian army neared the Swiss rear after a forced march. The Swiss, sending a small force to delay the Venetians, undertook a deliberate and remarkable withdrawal from contact with the French. Total Swiss losses were about 6,000 killed as against 5,000 French dead. Francis im-

mediately occupied Milan, and Switzerland negotiated peace—which would endure until the French Revolution. The Pope soon sued for peace, and the anti-French alliance collapsed, leaving most of Lombardy in French hands.

1516, August 13. Treaty of Noyon. Charles I, newly crowned King of Spain (later Emperor Charles V), also made peace, recognizing French control of Milan in return for Francis' renunciation of his claim to Naples. Charles' grandfather, Emperor Maximilian, after an abortive attempt to invade Lombardy, also made peace at the **Treaty of Brussels** (December 4).

FIRST WAR BETWEEN CHARLES V AND FRANCIS, 1521–1526

The new German Emperor (also the King of Spain) claimed Milan and Burgundy, while Francis claimed Navarre and Naples. Italy was again the principal battleground, although considerable fighting took place in northeastern France and in the Pyrenees, particularly in Navarre. The principal events were:

1521. French Invasion of Luxembourg. This invasion, led by **Robert de la Marck, Duke of Bouillon,** precipitated undeclared border war between the Rhone and the North Sea.

1521, May–June. French Invasion of Navarre. André de Foix conquered part of Navarre, **Ignatius de Loyola** (later the founder of the Jesuits) being wounded in the unsuccessful defense of Pamplona. The invading army had barely installed the exiled king, Frenchman **Henri d'Albret,** on the throne when they were surprised and defeated by a Spanish force at the **Battle of Esquiroz,** near Pamplona (June 30), and driven out of the country.

1521, November 23. French Loss of Milan. Italian condottiere General **Prosper Colonna,** commanding the combined Spanish-German-Papal Army, outmaneuvered Marshal **Odet de Lautrec** and captured Milan in a surprise attack.

1522, April. Operations near Milan. Lautrec, with Swiss and French reinforcements bringing his strength to about 25,000 men, plus 10,000 allied Venetians, advanced on Milan. Colonna, with less than 20,000, took up a strong defensive position at Bicocca, near Milan. Lautrec's

8,000 Swiss troops, their pay in arrears, threatened departure unless paid immediately, then agreed to fight one battle before leaving. Lautrec, therefore, decided to attack.

1522, April 27. Battle of Bicocca. Lautrec carefully prepared a plan to make the maximum use of his artillery. The impatient Swiss, refusing to wait for the French artillery to get in position, attacked without orders. Held up at the entrenchments, they suffered heavy casualties from the Spanish arquebusiers and were repulsed with severe losses. In half an hour 3,000 Swiss were killed. French heavy cavalry attempted a diversion by striking at the rear of the allied position, but were driven off. Lautrec withdrew eastward into Venetian territory. Allied losses were merely a few hundred men. This battle struck a terrible blow at Swiss morale, and they never again attacked arquebusiers in their old confident fashion. The battle was a clear demonstration of the predominance of gunpowder small arms.

1522–1523. English Raids in Picardy. The English, based on Calais, were first under the **Earl of Surrey,** later under the **Duke of Suffolk.**

1523. Venice Makes Peace. The remnants of

the French Army were forced to withdraw to France. Francis planned to invade Italy again, but the treason of Prince **Charles of Bourbon,** Constable of France, interfered. Bourbon fled to Germany to join Charles V. Francis then sent the army to Italy under **William de Bonnivet,** Admiral of France, who had recently conquered Fontarabia, on the Spanish frontier. Bonnivet captured Novara, but was held there by Colonna's skillful maneuvering.

1524, March. Imperial Spring Offensive. A reinforced imperial army under **Charles de Lannoy,** Viceroy of Naples, surprised the dispersed French forces in their winter cantonments in northwestern Italy; Bonnivet withdrew in confusion.

1524, April 30. Battle of the Sesia. The French were routed by Lannoy, Bayard being killed in a vain effort to lead a counterattack. Bonnivet, wounded, was replaced by Count **François de St. Pol,** who retreated back across the Alps.

1524, July–October. Imperial Invasion of Southern France. An imperial army of 20,000 men invaded Provence through the Tenda Pass. While a Genoese fleet blockaded the sea approaches to Marseille, the imperials invested the city (August–September). They withdrew on the approach of a French relieving army of about 40,000 men under Francis moving down the Rhone Valley. The imperials hastily retreated eastward through the then-trackless Maritime Alps, suffering severely.

1524, October. French Invasion of Italy. Francis immediately marched into Italy, crossing by the Argentière Pass. Due to a plague in Milan, Lannoy withdrew to Pavia. The French, leaving a small force to observe Milan, followed (October 24).

CAMPAIGN OF PAVIA, 1524–1525

1524, October 28–1525, February 24. Siege of Pavia. Lannoy, who felt that his army was in no shape for battle, left a garrison in Pavia and retreated to the southeast to recruit. The French, unable to capture Pavia by storm, invested the city. Francis stupidly divided his

forces, sending an army of 15,000 men under Scottish **John Stuart, Duke of Albany** on an abortive expedition to conquer Naples. This left barely 25,000 French troops before Pavia.

1525, January–February. Approach of Imperial Army. Lannoy, with a reorganized and reinforced army of approximately 20,000 men, approached Pavia from the east (January 25). Though Lannoy was the nominal commander, the Marquis of Pescara appears to have exercised actual command. Francis shifted part of his army to face the imperial forces, building trenches to block the most likely route of advance, meanwhile continuing the investment of Pavia. The two armies, both entrenched, faced one another across a narrow, unfordable brook, harassing one another with intermittent artillery fire. During this period approximately 6,000 Swiss mercenaries departed from the French army to reopen lines of communication to Switzerland, now harassed by imperial detachments. Approximately 4,000 Swiss remained with Francis, whose army was now reduced to less than 20,000 men.

1525, February 23–24. Imperial Maneuver. On a stormy night, and under the cover of night bombardment, the imperial army marched to its right, crossed the brook 2 miles above the opposing lines of entrenchments, turning the French left. A few companies were left in the trenches to demonstrate, so as to fool the French. Breaking through an old park wall, on which Francis had counted to protect his left flank, by dawn the imperials were in order of battle about 1 mile north of the main French camp.

1525, February 24. Battle of Pavia. Francis, completely surprised, endeavored to reorganize his army to meet the threat. He personally led a cavalry charge against the imperial left center for the purpose of providing his army time to shift 90° to face to the north. The vigor and gallantry of the French charge surprised the attackers and temporarily halted them. If the French infantry, which straggled up slowly, had taken advantage of this momentary success, French victory might have been possible. As it was, the imperial troops rallied, drove the

French cavalry back, then overwhelmed the still-confused French infantry in a 2-hour battle. About a third of the French Army, under Duke **Charles of Alençon,** was never engaged. Francis fought gallantly, leading his dwindling heavy-cavalry detachment in charge after charge until his horse was killed under him, and he was badly wounded and captured. La Trémoille was killed. Alençon led the remnants of the French Army in a withdrawal to the west; he was thereafter treated with contempt by his countrymen. French losses in killed and wounded were about 8,000 men. Imperial losses were probably no more than 1,000. Artillery took little part in this fight, partly because Francis had charged so impetuously at the outset as to block the field of fire of the few French guns that were in position. Most French casualties came from Spanish arquebusiers. The French, who had very few arquebusiers or crossbowmen, were unable adequately to return the imperial small-arms fire.

1526, January 14. Treaty of Madrid. To obtain his freedom, Francis, a prisoner in Madrid, signed a treaty giving up all of his claims in Italy, and surrendering Burgundy, Artois, and Flanders to Charles V.

SECOND WAR OF FRANCIS AND
CHARLES V, 1526–1530

1526, May 22. Francis Repudiates the Treaty of Madrid. He claimed it had been made under duress. He formed the **League of Cognac** with the Pope, Milan, Venice, and Florence against Charles V.

1526, July 24. Fall of Milan. It was captured by Lannoy's Spanish army.

1527. Inconclusive Maneuvering by Lautrec and Lannoy. The French had the worst of a war of attrition. Spanish and German mercenaries sacked Rome, committing horrible atrocities (May 6).

1528. French Disasters. A revolt in Genoa, led by Andrea Doria, lost the most important French base in Italy. A French army under St.

Pol was decisively defeated at the **Battle of Landriano** (June 19).

1529. Treaty of Cambrai. Known as the "Ladies' Peace," reached through the instrumentality of Charles' aunt and Francis' mother. Charles V, alarmed by a Turkish threat (see p. 543), was anxious for peace, as was Francis. The *status quo* was restored with Francis agreeing to pay a nominal indemnity and again renouncing his claims in Italy, while Charles withdrew his claims to Burgundy.

1530. Defeat of Florence. Though abandoned by Francis' unexpected treaty, the Florentines continued an unequal contest with the Empire. They were led by an exceptionally able soldier, **Francisco Ferrucio;** the imperial forces by Prince **Philibert of Orange.** Both leaders were killed at the **Battle of Gavinana** (August 2). Florence soon surrendered (August 12).

1531. French Reorganization. Francis established infantry legions, standing units of 6,000 mixed pikemen and arquebusiers, of Picardy, Languedoc, Normandy, and Champagne.

THIRD AND FOURTH WARS OF FRANCIS AND
CHARLES V, 1536–1544

1536–1538. The Third War. A large French army unexpectedly invaded Italy, capturing Turin, but was unable to reach Milan. Charles V attempted two attacks against France, personally leading one through Provence and sending the other through Picardy. The northern invasion soon bogged down. Charles advanced as far as Aix, but when he discovered that Francis was prepared to fight at Avignon, the emperor declined the challenge and returned to Italy. A temporary peace was patched up by the **Truce of Nice,** intended to restore the *status quo* for 10 years. The French retained their foothold in northwest Italy.

1542–1544. The Fourth War. Taking advantage of a number of imperial setbacks, Francis concluded an alliance with Ottoman Sultan **Suleiman the Magnificent,** to the horror of many Christians in his own country and elsewhere in Europe. **Henry VIII** of England allied himself with Charles. For the first 2 years, fighting was inconclusive and centered mostly in north Italy and Roussillon.

1543. Sack of Nice. A combined Franco-Turkish fleet under Turkish Admiral **Khair ed-Din** (Barbarossa; see p. 545) bombarded, besieged, and sacked the imperial town of Nice.

1543, September–October. Charles Invades Picardy. He besieged Landrecies (September). Francis approached with a large army (October). After some inconclusive maneuvering, both sides retired to winter quarters.

1543–1544. English-Imperial Plan to Invade France. Henry and Charles agreed that the English would advance from Calais, while imperial forces invaded through Lorraine and Champagne.

1544, April 11. Battle of Ceresole. French forces under **Francis of Bourbon, Count of Enghien)** were maneuvering against imperial forces under Spanish **Marqués Del Vasto** south of Turin. The French army was approximately 15,000 strong, including 4,000 Swiss infantry, 7,000 French infantry, 2,000 Italian infantry, and 1,500 French and Italian heavy and light cavalry. The imperial army consisted of 7,000 landsknechts, 6,000 Italian infantry, 5,000 Spanish infantry, and approximately 1,000 cavalry. Each side had about 20 guns. After a prolonged arquebus skirmish and a long-range artillery duel, the principal infantry contingents became locked in a sanguinary conflict; the landsknechts were almost completely wiped out by the Swiss infantry and French cavalry. The French infantry, on Enghien's left, was simultaneously being overwhelmed by the Spanish and German veterans, but Enghien retrieved the situation there by an enveloping cavalry charge. Del Vasto was forced to retreat, having lost more than 6,000 dead and 3,200 prisoners. The French lost approximately 2,000 dead. The battle reaffirmed the lesson that infantry of the day—whether arquebusiers or Swiss and landsknechts pikemen—could repulse any cavalry attack; but that the employment of cavalry against the flank of infantry engaged against other infantry was likely to be decisive.

1544, May–August. Imperial Invasion. In accordance with the allied plan, Charles invaded eastern France. He was delayed by the gallant defense of St.-Dizier (June 19—August 18), giving Francis time to call back his army from Italy and to collect other forces to defend Paris. The imperial army halted after seizing Épernay, Château-Thierry, Soissons, and Meaux.

1544, July. English Invasion. This followed Henry's leisurely crossing to Calais, which gave Francis adequate warning. The English Army, about 40,000 men, included many foreign mercenaries.

1544, July 19–September 14. English Siege of Boulogne. Henry captured the city but made no attempt to coordinate actions with the Germans.

1544, September 18. Peace of Crépy. Charles, discouraged, was eager to accept Francis' offers of peace. The *status quo* was reestablished, with the French retaining northwest Italy. The English thus suddenly found themselves without allies. Henry returned to England, leaving a garrison in Boulogne.

1544, October 9. "Camisade of Boulogne." French troops under Dauphin Prince **Henry** (later Henry II) assaulted Boulogne, and were almost successful, but were driven out by rallying English.

1545–1546. Continuing War between France and England. The French held the initiative, but operations were limited to the Boulogne-Calais area and to coastal raids on both shores of the English Channel. After failure of a planned French invasion of the Isle of Wight, peace was concluded, the French recognizing English conquest of Boulogne (1546).

The Last Valois-Hapsburg War, 1547–1559

1547–1559. Reign of Henry II (son of Francis). He formed an alliance with Prince **Maurice of Saxony** to take advantage of Charles V's preoccupation with the Turks and the German Protestants (1551).

1552, April. French Invasion of Lorraine. They captured Metz, Toul, and Verdun.

1552, October–December. Seige of Metz. Charles, having patched up a peace with the Protestants, attempted to reconquer Lorraine, but was repulsed at Metz.

1552–1555. French Invasion of Tuscany. This was a failure. Marshal **Blaise de Monluc** was defeated by an imperial army under the **Marquis of Marignano** at the **Battle of Marciano** (August 2, 1553). He was cooped up in Siena, where he was forced to surrender after a long siege (1553–1554).

1553–1556. Inconclusive Maneuvering in Flanders. Charles personally led one invasion of Picardy, which was ended by French success in a small cavalry fight in the **Battle of Renty,** where **Francis, Duke of Guise** distinguished himself (August 12, 1554).

1556, February–November. Truce of Vaucelles. This was marked by the abdication of Charles V (October), broken in spirit by his failures. He was succeeded in Spain by his son **Philip II,** and in the Empire by his brother **Ferdinand** (see pp. 533, 526). The truce soon collapsed.

1556–1557. French Invasion of Southern Italy. Despite severe losses and hardships, the French, under Guise, established a foothold in northern Naples. They were forced to withdraw upon word of French disasters elsewhere (1557). The French retained northwestern Italy around Turin.

1557, June. England Joins the War. Mary of England, wife of Philip II, brought her country into the war against France.

1557, July–August. Philip's Invasion of Northern France. He had an army of 50,000, including an English contingent of 7,000 men. The senior commander was Duke **Emmanuel Philibert of Savoy.** Since the flower of the French Army was engaged in Italy, the Constable, Duke **Anne of Montmorency,** was able to raise a conglomerate force of only about 26,000 to oppose this invasion threat. Foolishly, the Spanish army paused to besiege St.-Quentin, where Admiral **Gaspard de Coligny** organized a small force for defense.

1557, August 10. Battle of St.-Quentin. Montmorency attempted to cross the Somme River to attack a detached portion of the Spanish army and to relieve St.-Quentin. While the French were crossing, Savoy ordered a counterattack by imperial cavalry. The French were disastrously routed; about half their army escaped; 6,000 were killed and about 6,000 captured, including Montmorency. Northern France was now completely defenseless. Savoy wished to march on Paris, but cautious Philip refused to bypass St.-Quentin.

1557, August 27. Fall of St.-Quentin. It was taken by storm in a hard fight.

1557, September. Philip Orders Withdrawal to the Netherlands. Savoy was disgusted. Meanwhile, Henry had called Guise back from Italy, and had begun desperate efforts to rebuild the French Army.

1557, November–December. Guise Seizes the Initiative. Taking advantage of allied vacillation and overcaution, Guise boldly sent raiding expeditions into Champagne and along the Netherlands frontier. Meanwhile, he assembled an army of 25,000 men at Abbeville.

1558, January 2–7. Fall of Calais. Guise undertook a bold and well-calculated assault against weakly held Calais, first storming the outworks, then capturing one fort after the other until the small garrison under Lord **Thomas Wentworth** capitulated. Calais, England's last foothold in France, was lost forever.

1558, May–June. The French Maintain the Initiative. Guise led a small army into northern Champagne to capture Thionville. At the same time Marshal **Paul des Thermes,** with about 10,500 men, raided from Calais into Spanish Flanders to capture Dunkirk (June 30). After gathering much loot, the French withdrew back toward Calais, closely followed by **Lamoral, Count of Egmont,** commanding Spanish forces in Flanders.

1558, July 13. Battle of Gravelines. Egmont, catching up with the withdrawing French on the sand dunes near Gravelines, promptly attacked. At the same time an English fleet just offshore opened fire on the French army. The combination routed the French and their landsknechts mercenaries; they lost about 5,000 dead and 5,000 prisoners. Only a handful escaped.

1558, July–August. Guise Rushes to Picardy. He retrieved the disaster by putting on such a bold front that the Spanish failed to exploit.

1559, April 3. Peace of Cateau Cambrésis. This ended the Hapsburg-Valois Wars and a projected attack on Spain by Admiral Coligny. France gave up all of its holdings and claims in Italy (save for the border region of Saluzzo) and the other frontier conquests made by Guise in the latter days of the war (save for the cities of Metz, Toul, Verdun, and Calais).

The Wars of Religion, 1560–1598

THE PERIOD OF CONDE, COLIGNY, AND CATHERINE DE' MEDICI, 1560–1574

1559–1560. Reign of Francis II. National policy was dominated by the towering figure of Francis of Guise, now the leading personality in France. Because of their suspicions of Guise, and his sponsorship of French meddling in Scotland, Philip of Spain and Elizabeth of England were temporarily united in opposition to him and to Mary of Scotland (see pp. 507, 512). Within France the political and religious policies of the ardently Catholic Guise family were firmly opposed by Protestant **Louis of Bourbon, Prince of Condé** and other Protestant nobles. Religious unrest in France became acute.

1560–1574. Reign of Charles IX (10-year-old younger brother of Francis II). This began under the guardianship of his mother, **Catherine de' Medici,** widow of Henry II and rival of Guise.

1560. Conspiracy of Amboise. This attempt of a group of the minor nobility (many of them Protestant) to overthrow Guise's government was brutally suppressed.

FIRST WAR OF RELIGION, 1562–1563

The principal Protestant (Huguenot) leaders were Condé and Count Gaspard of Coligny, Admiral of France. The principal Catholic leaders were Guise, the Constable Montmorency, and vacillating ex-Protestant Prince **Antoine of Bourbon, King of Navarre** (brother of Condé, father of Henry of Navarre). Because of the recent peace with Spain and the Empire, France was full of unemployed soldiers, ready and eager for fighting and loot.

1562, March 1. Massacre of Vassy. A number of Protestants were massacred by Guise troops.

1562, April–November. Huguenot Rising. Condé and Coligny seized Orléans and called for a national rising of Protestants. Irregular skirmishing and atrocities spread throughout France.

1562, September 20. English Intervention. A force under **John Dudley, of Warwick,** sent by Queen Elizabeth to assist the Huguenots, landed to capture Le Havre.

1562, October 26. Fall of Rouen. The Catholics stormed this Protestant center, Antoine of Bourbon being killed in the attack.

1562, December 19. Battle of Dreux. A Huguenot army of nearly 15,000 men, under Condé and Coligny, marching north to make contact with the English at Le Havre, unexpectedly ran into a Catholic army of about 19,000, nominally under Montmorency, actually under Guise. The Catholic army included about 5,000 Swiss mercenary pikemen; the Protestants had about an equal number of landsknechts. A hard-fought, confused, and indecisive battle ensued, in which gallant leadership by Guise saved the Catholics from disaster and held the field. A unique aspect of the battle was that the two opposing commanders, Condé and Montmorency, were each captured. Losses on each side were about 4,000.

1563, February 18. Assassination of Guise. He had been besieging Orléans.

1563, March. Peace of Amboise. This was negotiated between the two prisoners, Montmorency and Condé, at the instigation of Queen Catherine. The Catholics and Huguenots then immediately marched jointly against the English at Le Havre.

1563, July 28. Fall of Le Havre. Warwick's English army surrendered.

SECOND WAR OF RELIGION, 1567–1568

1567, September 29. Huguenot Uprising. Condé and Coligny and 500 Huguenot gentlemen failed in an attempt to seize the royal family at Meaux. Other Protestant bands, however, seized Orléans (captured by **François de La Noue** and 15 other Huguenot cavaliers), Auxerre, Vienne, Valence, Nîmes, Montpellier, and Montauban. Condé immediately gathered the scattered Protestant forces to threaten Paris (October–November).

1567, November 10. Battle of St.-Denis. Montmorency, then aged 74, took advantage of Huguenot dispersion. With 16,000 men he moved against Condé, who had only 3,500 at St.-Denis. Condé rashly stood to fight, and amazingly the Huguenots held the field for several hours before finally being driven off by sheer weight of numbers. Montmorency was killed. The moral effect of Condé's magnificent defense was demonstrated by the fact that the following day, when he scraped together almost 6,000 troops, he again offered battle, which the Catholics declined.

1568, March 23. Peace of Longjumeau. Substantial concessions were made to the Huguenots by Catherine.

THIRD WAR OF RELIGION, 1568–1570

1568, August 18. Royalist Treachery. An attempt to capture Condé and Coligny by treachery failed, but precipitated war. During the remainder of the year there was considerable indecisive maneuvering in the Loire Valley.

1569, March. Royalist Initiative. Marshal **Gaspard de Tavannes** outmanuevered Condé in the region between Angoulême and Cognac.

1569, March 13. Battle of Jarnac. Precipitated by Tavanne's surprise crossing of the Charente River near Châteauneuf. The Huguenot army,

unprepared, was badly defeated; Condé was captured and murdered. Coligny succeeded in withdrawing a substantial portion of the Protestant army in good order.

1569, June 10. Arrival of German Protestant Reinforcements. Some 13,000 reiters and landsknechts joined Coligny near Limoges.

1569, July–September. Protestant Siege of Poitiers. Prince **Henry of Navarre,** then 15, Protestant son of Antoine of Bourbon, was with Coligny at this siege.

1569, August 24. Battle of Orthez. Count **Gabriel of Montgomery,** detached by Coligny, repulsed a Royalist invasion of French Navarre, defeating Royalist General **Terride.**

1569, September. Relief of Poitiers. Tavanne, resuming the offensive, forced Coligny to raise the siege. A campaign of maneuver followed in the vicinity of Loudun.

1569, October 3. Battle of Moncontour. Tavanne's army consisted of 7,000 cavalry, 18,000 infantry, and 15 guns; Coligny had about 6,000 cavalry, 14,000 infantry, and 11 guns. The Royalist army included a substantial Swiss contingent. A combined attack by Catholic cavalry and the Swiss infantry put the Huguenot cavalry to flight. The Catholics then crushed the Huguenot infantry, the Swiss being particularly ruthless in slaughtering the landsknechts. Huguenot losses were nearly 8,000, Royalist losses probably no more than 1,000. Coligny was wounded; La Noue was captured.

1570, April–June. Protestant Initiative. Coligny, recovered from his wound, boldly marched through central France with 3,000 cavalry and 3,000 mounted infantry, threatening Paris in late June.

1570, August 8. Peace of St.-Germain. The Huguenots again were promised wide religious freedom and considerable military and political autonomy.

THE FOURTH WAR OF RELIGION, 1572–1573

1572, August 23–24. Massacre of St. Bartholomew's Eve. Coligny and thousands of other Protestants were murdered in Paris. Young Prince Henry of Navarre succeeded Coligny as the nominal Huguenot leader. Save for a prolonged siege of La Rochelle by a Royalist army under Prince **Henry,** younger brother of Charles IX, there was little military action. During this war the Protestants established political and military control over most of southwest France, which became virtually autonomous.

A new political party emerged, the "Politiques," moderate Catholics anxious to make concessions to the Protestants in order to restore national peace and unity. The royal princes, Henry (later Henry III) and **Francis of Alençon,** were leaders of the Politiques.

The Period of the Three Henrys, 1574–1589

1574–1589. Reign of Henry III. He was the third son of Henry II and Catherine de' Medici to become King of France. Since he was childless, as was his youngest brother, **Francis of Alençon,** the question of royal succession became mixed up in the Religious Wars, Henry of Navarre being next in line of succession to the throne of France. Henry of Navarre, not a very religious man, was largely motivated by natural political ambitions during this period. The principal Catholic leader was now **Henry, Duke of Guise,** son of Francis.

FIFTH WAR OF RELIGION, 1575–1576

1575. Outbreak of War. This followed widespread disorders. The Protestants had generally the best of scattered fighting throughout the country, though Guise was victorious in the one important action in the **Battle of Dormans** (October 10, 1575). Participation of Catholic Politiques on the Protestant side led King Henry, whose support of Guise was reluctant, to make peace and to renew pledges of religious freedom to the Protestants at the **Peace of Beaulieu** (May 5, 1576).

1576. The Holy League. Established by Guise, who refused to accept the Peace of Beaulieu, to take advantage of widespread Catholic resentment of favors granted the Protestants. He entered into conspiratorial negotiations with Philip of Spain, who agreed to intervene and place Guise on the throne.

SIXTH AND SEVENTH WARS OF RELIGION, 1576–1580

1576–1577. The Sixth War. This included only one brief, inconclusive campaign, and was

ended by the **Peace of Bergerac** (1577, reaffirming the Peace of Beaulieu).

1580. The Seventh War. This was called the "Lovers' War" because of indiscretions of Margaret, wife of Henry of Navarre. No important military actions took place.

1580–1585. Increasing Intrigue. Catholics, Protestants, and Politiques were all involved, due to the question of succession, with Henry of Navarre now the direct hereditary successor to the throne of France. Philip of Spain and the Guise family again reached an understanding (December 1584) to support Henry of Guise to be the successor of aged Cardinal **Charles of Bourbon,** who was to be proposed as the immediate successor of Henry III.

EIGHTH WAR OF RELIGION, 1585–1589

This was also known as the "War of the Three Henrys": Henry III, Henry of Navarre, and Henry of Guise.

1585. Royal Vacillation Precipitates War. Henry III, still partly under the influence of his scheming mother, nominally supported the Catholic League and withdrew religious freedom for Protestants. The Huguenots rose in revolt. In the subsequent confused and complicated struggle, the king endeavored to steer a middle course, undercutting the Catholic League whenever possible in order to reduce Guise influence.

1587, October 20. Battle of Coutras. A Catholic army of 10,000 men, under **Anne, Duke of Joyeuse,** endeavoring to link up with Catholics in Bordeaux, was intercepted by Henry of Navarre with 6,500 men. Henry, awaiting a charge, prepared a coordinated arquebus-cavalry counterattack. The overconfident Royalist cavalry was smashed, following which the combined Protestant infantry and cavalry cut the Royalist infantry to pieces. More than 3,000 Royalists including Joyeuse, were killed; Henry lost less than about 200 dead. Henry was in the forefront of the Huguenot countercharge, fighting boldly and well. He did not, however, exploit his victory.

1587, October–November. Catholic Successes. Guise, by victories at **Vimory** and **Auneau,** was able to turn back a force of German auxiliaries which had marched to the Loire Valley to make contact with the Huguenots.

1588, May 9–11. Guise Seizes Paris. King Henry fled, then later capitulated to Guise and the League (July).

1588, December 23. Murder of Guise. Inspired by King Henry. Two days later he also instigated the murder of Guise's brother, Cardinal **Louis of Guise.** This aroused the Catholic League against the king, who fled to refuge with Henry of Navarre.

Ascendancy of Henry of Navarre, 1589–1598

1589, August 2. Assassination of Henry III. The murderer was a Catholic monk, avenging the Guise brothers. The dying king designated Henry of Navarre as his successor. The formerly 3-way conflict now became simplified, though its intensity and ferocity in no way abated. It was now the Catholic League, probably supported by less than half the population of France, against Henry of Navarre, supported by Protestants and Catholic Politiques. Cardinal Charles of Bourbon, proclaimed king by the Catholic League, was actually a prisoner of Henry of Navarre and acknowledged him as king. Spanish forces moved into France to support the League against Henry. The new leader of the League, was **Charles, Duke of Mayenne,** younger brother of Henry of Guise.

1589, September 21. Battle of Arques. Near Dieppe, Henry, with 8,000 troops, lured Mayenne with 24,000 French Catholics and Spanish (from Flanders) into a defile of the Bethune River, which he had prepared with trenches and gun emplacements. The ambush, combined with superb tactics and excellent employment of his artillery, gave Henry a tremendous victory over great odds; Mayenne was driven off the field. He retreated to Amiens, to await Spanish reinforcements from the Duke of Parma in Flanders. Henry, with about 20,000 men, then dashed toward Paris (October 31).

1589, November 1. Repulse at Paris. Henry was repulsed by the Catholic garrison. Proclaiming himself Henry IV, he established a temporary capital at Tours. Civil war raged all over France.

1590, February–March. Siege of Dreux. Upon the approach of Mayenne's army, Henry raised the siege and prepared for battle near Evreux.

1590, March 14. Battle of Ivry. Henry had 11,000 men, Mayenne about 16,000. As at Coutras, Henry mixed his cavalry and arquebusiers to obtain the maximum advantage of the capabilities of each. In an extremely confused and hard-fought battle, the Huguenot arquebusiers and better artillery repelled Catholic cavalry charges, while Henry himself led a successful cavalry counterattack into the Catholic center. The entire Catholic army fled in confusion, save for the Swiss contingent, which stood in good order, threatening to fight to the death if quarter was not granted. Upon receiving honorable terms, they surrendered. Catholic losses were nearly 4,000 killed; the Huguenot-Royalist army lost only about 500 killed. Henry was slow in exploiting his victory, permitting the Catholics to rally and to prepare for the defense of Paris.

1590, May–August. Siege of Paris. The city was reduced to semistarvation. Henry was forced to raise the siege when the Duke of Parma joined Mayenne at Meaux. This gave Parma a combined Catholic army of some 26,000 men. Henry's army was about the same size.

1590, September–October. Inconclusive Maneuvering. After sending a relief force and substantial supplies into Paris, Parma withdrew to winter quarters (October).

1591–1592. War of Maneuver in Northern France. The opposing leaders were two of the outstanding soldiers of the western world: Henry of Navarre and Alexander, Duke of Parma. Parma, the abler strategist, was forced to divide his attention between the war in France, and a desperate struggle in the Netherlands (see p. 530). Henry succeeded in penning Parma's army up in a loop of the Seine River, only to have Parma's superb Spanish engineers bridge the river overnight, permitting 15,000 men to get across before dawn (May, 1592).

1593, July. Henry Formally Returns to the Catholic Faith. He thus reunited most of the nation against the threat of a full-scale Spanish invasion planned by Philip and the Catholic League.

1594, March 21. Henry's Triumphal Entry into Paris.

1595–1597. Inconclusive Warfare. Henry won the closely contested **Battle of Fontaine-Française** over the Spanish, but was almost killed due to his own rash gallantry (June 9, 1595). Next year the Spanish captured Calais by surprise (April 9, 1596). Taking advantage of Henry's internal preoccupations, the Spanish next captured Amiens (September 17). There was no more major campaigning.

1598, April 13. Edict of Nantes. Henry granted religious freedom to the Protestants, ending the Wars of Religion.

1598, May 2. Treaty of Vervins. End of the war with Spain, following secret negotiations in which Henry apparently double-crossed his English and Dutch allies.

GERMANY

During this century the Holy Roman Empire was involved not only with steadily increasing internal violence resulting from the religious Reformation, but also—later—with the personal enmity of Francis I of France and Charles V, which kept Western Europe constantly at war for more than a third of a century (see p. 515). At the same time, the Ottoman Turks were overrunning the Balkans and hammering at the gates of Central Europe, making Vienna a frontier town. The principal events were:

1493–1519. Reign of Maximilian I. Almost continuous war with France, opposing the ambitions of Charles VIII and Louis XII in Italy. At the same time, in the complicated Italian wars, Maximilian also unsuccessfully endeavored to break the power of Venice (see p. 515).

1517, October 31. Luther's 95 Theses. His action at Wittenberg is generally considered the beginning of the Reformation, resulting in the most violent and implacable hostilities in European history.

1519. Rivalry for the Imperial Throne. Upon the death of Maximilian, the German electors chose the Spanish-German Hapsburg Charles I of Spain in preference to Francis I of France, precipitating a lifelong enmity.

1519–1556. Reign of Charles V. He was ruler of the combined lands of Castile, Aragon, Naples, Burgundy, the Hapsburg lands in Germany, and the tremendous new colonial empire of Spain in the Americas.

1521–1544. Wars between Charles and Francis. (See pp. 516–519.)

1522. The Knights' War. Minor nobles, sympathetic with the Reformation, attempted to overthrow the power of the church principalities in Germany; they were crushed by imperial and church forces.

1524–1525. The Peasants' War. This was an uprising in Swabia and Franconia by adherents of Luther, who repudiated them. The rebels were defeated at **Frankenhausen**

(1525) and cruelly suppressed by the nobility.

1526. Collapse of the Teutonic Knights. The order, restricted to eastern Prussia as vassals of Poland since 1466, was evicted when its Grand Master, **Albert of Brandenburg-Anspach,** turned Protestant and secularized his territories. (Its remnants lingered in Germany proper until the French Revolution, when its Rhenish estates were absorbed; in 1809 the order was suppressed.)

1526–1534. War against the Turks. This was marked by the successful defense of Vienna against the Ottoman armies (see p. 543).

1535. Charles's Expedition to Tunis. (See p. 545.)

1541. Charles's Expedition against Algiers. (See p. 545.)

1546–1547. Schmalkaldic War. This was a revolt of Protestant nobles. Charles, having made a truce with the Turks, won a decisive victory over the Elector of Saxony and other Protestant nobles at the **Battle of Mühlberg** (April 24, 1547).

1547–1556. Charles's War with Henry II of France and Maurice of Saxony. (See p. 519.)

1553. Battle of Sievershausen. In a primarily cavalry battle, Maurice defeated Catholic forces under **Albert of Hohenzollern-Kulmbach.** Maurice, victorious, died of wounds.

1556, October. Abdication of Charles. (See p. 520.) Spain, Naples, Milan, Franche-Comté, and the Netherlands were inherited by his son, Philip II of Spain. The German Hapsburg lands were ceded to his brother, Ferdinand, who was elected emperor.

1556–1564. Reign of Ferdinand I. There was constant warfare over succession to the throne of Hungary, and against the Ottomans (see p. 546).

1564–1612. Reigns of Maximilian II and Rudolph II. There was continuing warfare against the Turks in southeastern Europe.

SCANDINAVIA

Denmark

The uneasy Union of Kalmar fell apart early in the century when the Swedes established their independence under **Gustavus Vasa.** A decade of civil war in Denmark followed, more or less mixed up with sporadic warfare against Lübeck and other Hanseatic towns. **Christian III,** after settling the civil war, further reduced dwindling Hanseatic power by a series of successes against Lübeck. Around the middle of the century he involved Denmark in the early religious wars in Germany, lending assistance to the Protestants. During the remainder of the century royal authority was strengthened at the expense of the nobility and of the towns. The principal events were:

1481–1513. Reign of Hans (John) I. (See p. 463.)

1506–1513. War with Sweden. This resulted from the efforts of King Hans to reestablish the Union of Kalmar by force.

1513–1532. Reign of Christian II.

1520–1523. Successful Revolt of Sweden.

1523–1537. Internal and External War. Intermittent warfare with Lübeck was complicated by civil wars in Denmark. **Frederick, Duke of Holstein** seized the throne (1523). Christian, endeavoring to recover it, was captured in Norway and spent the rest of his life in captivity (1532). Frederick, due largely to the ability of his general **Johan Rantzau,** defeated the Hanseatic League.

1533–1536. The "Counts' War." The nobles were opposed by burghers and peasants after the death of Frederick. **Christopher, Count of Oldenberg** led the popular forces with Hanseatic support, while **Christian of Holstein,** son of Frederick, was selected by the

Rigsraad to succeed his father. Johann Rantzau swept Jutland with Holstein and German troops. A Danish fleet under **Peder Skram** defeated Lübeck, and then conveyed the Holstein army to Zeeland (July 1535). Malmö and Copenhagen were besieged, and finally surrendered (April and July 1536). This ended the civil war and the politico-military power of the Hanseatic League.

1534–1558. Reign of Christian III.

1563–1570. War with Sweden. (See below.)

1588–1648. Reign of Christian IV. He expanded Danish power and possessions in the Baltic.

Sweden, 1520–1600

Gustavus Vasa reestablished the identity of the Swedish state, which began to expand on the eastern shores of the Baltic Sea, and by the end of the century was the leading state of Northern Europe. The principal events were:

1501–1503. Return of Sten Sture the Elder (see p. 463). The Swedes, unhappy with Hans' rule, recalled Sten Sture as regent, but he died soon after (1503). He was succeeded by **Svante Nilsson Sture** (a cousin of King Charles VIII), who held the regency until his death (January 1512).

1506–1512. Danish-Swedish War. The war wound down after the death of Svante Sture, when nobles favoring the union controlled the state council.

1513–1520. Regency of Sten Sture the Younger. This son of Svante Sture mounted a coup against the council, was acclaimed regent, and made peace with Denmark (1513). Quarrels between Sture's partisans and those of Archbishop **Trolle,** who supported the union, led to civil war (1516), and Trolle was defeated and imprisoned (1517). Sture also defeated a Danish army sent to aid Trolle (1518). A Danish mercenary army invaded, defeating and mortally wounding Sture at the **Battle of Lake Malaren** (January 1520), and Trolle became regent. Sture's widow, Kristina Gyllenstierna, continued resistance to the Danes until her surrender (September 1520).

1520–1523. Civil War within the Union of Kalmar. Gustavus Vasa, with the assistance of Lübeck (enemy of Denmark), threw off Swedish domination and was elected King of Sweden (June, 1523).

1523–1560. Reign of Gustavus I.

1531–1536. Alliance of Denmark and Sweden against Lübeck. With the assistance of its Hanseatic allies and Danish rebels (see above) Lübeck captured Copenhagen and Malmö. The Swedes assisted the Holstein army in ejecting Hanseatic forces and in reestablishing peace in Denmark.

1543. Revolt in Smaland. This was led by peasant **Nils Dacke.**

1560–1568. Reign of Erik XIV. His efforts to free Sweden's trade from Danish and German control were not wholly successful. Erik suppressed internal violence stirred up by Poland and Russia, after he imprisoned his half-brother **John** for making an unsanctioned marriage alliance with **Sigismund II Augustus** of Poland (1562). He suffered a supposed mental breakdown (1567), was deposed (1568), and died in prison (1577).

1561. Expansion in the Baltic. Upon the collapse of the Teutonic Knights (1526), Revel and neighboring regions in Estonia and Livonia joined Sweden, causing strained relations with Russia, Denmark and Poland, each of which also coveted this territory, and each of which had also gobbled up some of the Teutonic Knights' territory (see p. 526).

1563–1570. Seven Years' War of the North. This struggle was waged by Sweden against Denmark, Lübeck, and Poland. The Swedes won a major naval victory over the Danes off **Götland** (1564), and briefly occupied Trondheim that year, but were unable to capitalize on their successes. After **John III**

deposed Erik XIV, Sweden made peace with the allies (1570).

1568–1592. Reign of John III. He deposed Erik XIV (his half-brother) after Erik's mental collapse. After his death, the throne passed to his brother Charles.

1570–1595. War with Russia. Desultory operations resulted in Sweden's slow conquest of all Estonia and Livonia.

1592–1599. Reign of Sigismund. Following the death of **John III,** political and religious differences between **Sigismund** (son and heir of John) and **Charles** (John's brother) broke out. Sigismund, who had earlier been elected King of Poland, invaded Sweden with a Polish army, but was defeated at **Stängebro** (1598, September 25), driven back, and then dethroned by the Swedish Riksdag (1599).

1599–1611. Reign of Charles IX. (See p. 622.)

The Netherlands, 1566–1609

The Netherlands (including most of modern Netherlands, Belgium, Luxembourg, and the French provinces of Flanders and Artois), under domination of Spain since the days of the division of the empire of Charles V, boiled during this period. Arrogant Spanish centralized rule combined with religious differences caused open rebellion, led first by **William "the Silent,"** Prince of Orange and Count of Nassau, and (after his assassination) his son **Maurice of Nassau.** Opposing the rebels were two exceptional soldiers: **Fernando Álvarez de Toledo, Duke of Alva** and **Alexander Farnese, Duke of Parma.** Against these two and their disciplined Spanish forces, the determination of William and the later display of military genius by Maurice finally brought freedom to the stout Protestant burghers of the Northern Provinces (principally Holland). Spain retained the southern Netherlands. Spain's preoccupation elsewhere, particularly in England and France, played some part in the Dutch victory, but the most important single factor in the eventual outcome was the Dutch ability to create a small fleet at the outset of the war and maintain command of the coastal waters despite Spain's naval might. Like the English, the Dutch grasped the significance of the large ship-killing gun, which revolutionized naval warfare in the period. The Spanish did not. Of military interest, too, was the diversity of terrain. In the Northern Provinces the dikes, rivers, canals, marshes, and inlets lent themselves to partisan warfare. The more open country farther south could be more easily dominated by regular troops. The principal events were:

1566. Political and Religious Riots. These spread throughout the northern and southern Netherlands; William of Orange, stadtholder of the king in Holland and Zeeland, unsuccessfully tried to mediate between the government and the people. Philip sent the Duke of Alva and a Spanish army from Italy to the Netherlands to restore order.

1567. William of Orange Outlawed. He fled to Germany.

1568, January–April. William Raises an Army. He sent this into northern Netherlands under his brother **Louis of Nassau.**

1568, May 23. Battle of Heiligerlee. Louis, who had about 3,000 men, defeated a small German-Spanish force under **John, Duke of Arenberg.**

1568, July 21. Battle of Jemmingen. Alva moved into northern Netherlands, enticed Louis to attack, and overwhelmed him with superior firepower and a crushing counterattack from front and flanks. Forces were equal—about 15,000 each. The rebels' loss was between 6,000 and 7,000 killed; Royalist losses were 100 dead. The rebel cause was completely crushed in the northeastern Netherlands for several years.

1568, October 5. William Invades the South-

eastern Netherlands. With an army of 25,000 men, he forded the Meuse River at Stochen. In Brabant, Alva, with about 16,000 men, outmaneuvered him, destroying his rear guard in a skirmish near **Jodoigne** (October 20). Alva had so effectively terrorized the population that few Netherlanders rose to join William's dwindling army. He retreated southward into France, then back to Germany by way of Strasbourg (November).

1569–1572. The Netherlands Seethe under Alva's Stern Rule. The "Sea Beggars," Dutch privateers with commissions from William of Orange, began to prey on Spanish commerce, and raided the Netherlands' seacoast from bases in England and Friesland.

1572, April 1. The Sea Beggars Capture Bril. This precipitated renewed strife.

1572, April–July. Uprisings Sweep Low Countries. Save for Amsterdam and Middleburg, all of the provinces of Holland and Zeeland were swept clear of the Spanish. William and his brother Louis again invaded the southern provinces. Alva, besieging Louis in Mons, repulsed William's efforts to relieve the city.

1572–1573. Alva's Terrorism. Combining skillful military action with horrible atrocities, Alva slowly reestablished Spanish control in the southern and eastern provinces, besieging and capturing city after city, massacring captured garrisons and many civilians. William retreated to Delft, where he organized desperate resistance of the Maritime Provinces. Alva captured **Haarlem,** which had resisted heroically in a 7-month siege (December–July), and the Spanish slaughtered the garrison to a man. Alva was repulsed in his siege of **Alkmaar** (August–October) when he discovered that the Dutch were prepared to cut the dikes rather than submit. A Dutch fleet defeated a Spanish squadron on the **Zuyder Zee** (October 1573). Because of these setbacks, Alva resigned, and was replaced as viceroy by **Luis de Requesens** (November 1573). Early the following year the Sea Beggars smashed another Spanish fleet off **Walcheren** in the Schelde estuary (1574, January).

1573, October–1574, October. Spanish Siege of Leyden. Various relief efforts by William and Louis failed.

1573, April 14. Battle of Mookerheyde. Louis' army was routed, and he was killed by a Spanish veteran army of 6,000 men under **Sancho de Avila.**

1573, September–October. Relief of Leyden. With the city on the verge of starvation, William cut the dikes, flooding the Spanish from their trenches and camps. Admiral **Louis Boisot**'s Sea Beggar fleet sailed across the inundated countryside to relieve the starving city (October 31).

1575–1576. The Revolt Spreads. Ineffective de Requescens could not halt the spread from the north through most of the southern provinces. The only important Spanish success was the capture of **Zierikzee,** an island off Zeeland; Boisot was killed in a vain attempt to block a bold Spanish crossing from the mainland at low tide.

1576, October 3. "The Spanish Terror." Mutinous Spanish troops, whose pay was in arrears, captured and sacked **Antwerp** with terrible ferocity. This atrocity temporarily united all of the Netherlands, whose provincial representatives signed the **Pacification of Ghent** (November 8).

1577. Complicated Negotiations. These involved the States General (civil government of the rebellious Netherlands provinces), William of Orange, and Don **Juan of Austria,** the new Spanish viceroy. William stepped down from leadership as a gesture to the Catholics. The States General raised an army of 20,000 men under Sieur **Antoine de Goignies** when negotiations broke down and Don Juan prepared to reconquer the provinces.

1578, January 31. Battle of Gemblours. Juan, reinforced by additional Spanish troops under Alexander Farnese (later Duke of Parma), moved against the Dutch army at Gemblours (Gembloux) with some 20,000 men. As the Dutch fell back toward Brussels, Farnese enveloped their cavalry rear guard, then pursued the routed Dutch horsemen through the columns of retreating Dutch infantry. In the ensuing slaughter, the Dutch lost about 6,000 men dead, to about 20 Spanish killed.

1578, February–September. Juan Recovers Control of the South. Internal jealousies and bickering on both sides then brought operations to a halt.

1578, October 1. Death of Don Juan. He was succeeded by Farnese as the Spanish viceroy.

1579, January 29. The Union of Utrecht. The northern Netherlands provinces established a confederation, the beginning of the modern history of Holland, or of the Netherlands. William was appointed stadholder by the States General.

1579–1589. Farnese's Campaigns of Pacification. Ignoring the complicated political maneuvering which involved his own sovereign (Philip), King Henry of France, and **Francis, Duke of Anjou,** on the one side, and the quarreling Catholic and Protestant Netherlanders, on the other, Farnese systematically conquered city after city of the southern Netherlands, beginning with **Maastricht** (March, 1579) and culminating with the prolonged siege and capture of **Antwerp** (August 17, 1585). Soon after this he became the Duke of Parma (1586).

1584, July 10. Assassination of William of Orange. He was succeeded as stadholder by his 17-year-old son, **Maurice of Nassau.**

1585–1587. English Intervention. Robert Dudley, Earl of Leicester, brought 6,000 English troops to the Netherlands where the Dutch granted him almost dictatorial powers. However, he was inept, notably in his vain siege of Zutphen (1586). After the death of Sir **Philip Sidney,** Leicester's most able lieutenant (1586), the English withdrew (1587); Maurice was appointed by the States General as captain and admiral-general of the union (1588).

1587–1588. Parma Prepares to Invade England. The expedition of 25,000 men never got started, however, due to the disaster to the great Armada (see p. 509).

1589. Parma to France. Philip was confident that the reconquest of the Netherlands was close to completion (the Dutch held only the

islands of Zeeland, the area north of the Waal in Holland, plus Utrecht and a few isolated towns farther east). Parma was ordered to take most of his army to assist the French Catholics against Henry of Navarre (see p. 524).

1589–1590. Maurice Takes the Offensive. Aided by the absence of Parma, his principal exploit was the surprise capture of **Breda,** the assault force hidden in barges which moored at the town wharves. At the same time the Dutch fleet, aided greatly by the English victory over the Spanish Armada, now held unchallenged control of the Netherlands' seacoasts.

1591, June–July. Maurice Expands Control. By rapidity of movement and concentration of greatly superior artillery fire, he captured **Zutphen** in a siege of 7 days (June) and **Deventer** after 11 days (July).

1591, August–September. Maurice versus Parma. Parma, who had returned from France, threatened Utrecht. The two armies maneuvered cautiously on opposite banks of the Waal River near Arnhem. Parma was then ordered by Philip to return to France to relieve Rouen, being besieged by Henry of Navarre (see p. 525).

1591, September–October. Maurice Resumes the Offensive. Again taking advantage of Parma's absence, Maurice conducted another lightning campaign of movement by water and road, and brief sieges, marked by superb use of artillery. He captured **Hulst** (September 14) and **Nijmegen** (October 21).

1592–1594. Maurice Continues His Campaign of Conquest. Parma died of wounds in France (December 1592). The next Spanish viceroy, **Peter of Mansfeldt,** was unable to offer effective interference to Maurice.

1594–1596. Maurice Consolidates the Northern Netherlands. He conducted sporadic border raids and expeditions, and devoted all possible attention to training and improving his army for operations in the southern Netherlands.

1597, January 24. Battle of Turnhout. Maurice, with approximately 7,000 troops, marched suddenly, taking advantage of bad weather, to attack and rout an isolated Spanish force of 6,000 troops, under Count **Jean de Rie of Varas,** at Tournhout. Spanish losses were 2,000 (including Varas) killed and 500 made prisoner. The Dutch lost about 100 killed.

1597–1599. Military and Political Consolidation. Archduke **Albert of Austria,** Philip's son-in-law, became the Spanish viceroy (1599), but was unable to interfere with Maurice.

1600. Maurice Invades Flanders. The States General ordered Maurice, despite his objections, to complete the conquest of the coastal strip of Flanders as far as Nieuport and Dunkirk. Maurice crossed the Scheldt River (June 21–22) and drove off Spanish forces blockading Ostend. He then crossed the Yser and immediately prepared to besiege Nieuport (July 1).

1600, July 2. Battle of Nieuport (Battle of the Dunes). Maurice, learning of the approach of Albert's Spanish army, sent a detachment to delay it at the Leffingham Bridge over the Yser, since his main army could not cross the river estuary until low tide. The Spanish won the race for the bridge, destroying the Dutch detachment. Maurice forded the Yser estuary at low tide (8 A.M.), marching rapidly northeast along the exposed beach toward the Spanish. The armies met about noon, at the turn of the tide. At the same time Dutch warships opened long-range fire with little effect. Driven by the rising waters, both armies shifted into the sand dunes farther inland. A Dutch cavalry charge was repulsed and Albert threw in his reserve, forcing the Dutch back. Effective Dutch artillery fire and a charge by Maurice's reserve cavalry checked the Spanish advance. Then Maurice made a general counterattack. The Spaniards, exhausted by several days of hard marching, collapsed. Their losses were about 3,500 men, Dutch casualties some 2,000. Maurice, with his lines of communication still vulnerable to Spanish attack, made no attempt at pursuit. Raising the siege of Nieuport, he withdrew into Dutch Flanders.

1601–1604. Spanish Siege of Ostend. The Dutch lost about 30,000 men in this long siege, but were able to prevent starvation by sending in reinforcements and supplies by sea. Albert's investing Spanish army lost perhaps 60,000 men by casualty or disease. Meanwhile, both England and France had made peace with Spain, leaving the Dutch without allies. Though Maurice had numerous successes elsewhere, the drain of defending Ostend, and at the same time meeting generally increasing Spanish pressure, caused the States General to order its surrender after a siege of 3 years and 71 days (September 20, 1604).

1604–1607. Desultory Warfare in the Netherlands. The Dutch were able to hold their own only because of control of the sea. A series of Dutch naval successes, culminated by the victory of Admiral **Jacob van Heemskerck** in the **Battle of Gibraltar** (1607), caused the Spanish to agree to a temporary truce during peace negotiations.

SWITZERLAND

During this century Swiss infantry lost its preeminence in Europe, due to the vulnerability of the pike phalanxes to the growing strength of artillery and small arms. The Swiss belatedly modified their tactics, increasing the percentage of arquebusiers and musketeers. Despite disasters early in the century, Swiss mercenary pike columns were still much sought by other European armies, though their callous avarice led to the common saying: "When the money runs out, so do the Swiss." The Swiss had their own strong national discipline, but refused to be disciplined by their employers. Nor did they allow sentiment to affect business; Protestant Swiss fought against French Huguenots. Within Switzerland itself, civil strife resulting from the imperfect federal union continued sporadically, complicated and intensified by the interjection of the religion issue in midcentury, by which time most of the northern and western cantons were Protestant. Despite civil war at home, Swiss mercenaries refused to fight one another. The principal events were:

1510. Switzerland Joins the Holy League against France. The Swiss Confederation restored Count **Maximilian Sforza** to control in Milan, annexing Locarno, Lugano, and Ossola (1512).

1513, June 6. Battle of Novara. Swiss victory over the French (see p. 515).

1513–1515. Swiss Protectorate over the Duchy of Milan. They thus obtained the duchy's rich revenues.

1515, September 13–14. Battle of Marignano. French victory over the Swiss (see p. 515).

1531. War between Zurich and the Catholic Cantons. This was culminated by Zurich's defeat in the **Battle of Kappel** (October 11).

1536. Expansion of Bern. Vaud, Chablais, Lausanne, and other territories of Savoy were subdued and annexed.

1541–1555. Sporadic Religious Civil War.

1564–1602. Intermittent War between the Protestant Cantons and Savoy. Bern, Geneva, and Zurich formed an alliance against Savoy and the Catholic cantons (1584). Spain intervened on the side of the Catholics (1587). This drawn-out war was inconclusive.

ITALY

Italy—weak, fragmented, and wealthy—was the principal battleground of the great wars between France, Spain, and the Hapsburgs during the first half of this century. The almost continuous minor wars between the petty Italian states were merely incidents in these larger struggles, described elsewhere (see pp. 512–521). Venice, despite the decline of her maritime empire, barely maintained her position as the most important independent state in Italy.

THE IBERIAN PENINSULA

Spain

Within the century Spain reached, then passed, the apogee of world power. The highly efficient military force created for the expulsion of the Moslems (1492; see p. 468) was further improved by the organizational and tactical genius of Gonzalo de Córdoba, "El Gran Capitán," the first important European soldier to fully understand, and to utilize, small-arms firepower. For the remainder of the century the Spanish were the preeminent soldiers of Europe. In addition to Córdoba, Spain provided three more of the finest military leaders of the century: **Hernando Cortez,** conqueror of Mexico, Álvarez de Toledo, Duke of Alva, and Italian-born Alexander Farnese, Duke of Parma. The ambitions of Spanish monarchs in exploring, conquering, and settling most of the Western Hemisphere, endeavoring to establish command of the seas, fighting interminable wars against France and Turkey, and in supporting the cause of the imperial Hapsburgs elsewhere in Europe all constituted an intolerable strain on Spain's slender manpower resources. Yet this relatively small nation had an impact on the rest of the world, in a similarly short period of time, comparable to earlier Macedonia and Mongolia. The result was the same. A brilliant burst of glory, lasting less than a century, followed by a long, slow decline. The principal events were:

1474–1516. Reigns of Ferdinand and Isabella. After the death of Isabella (1504), Ferdinand retained control of united Spain despite the legal claim of his daughter, **Joanna,** to Castile.

1495–1559. Wars with France. These were mostly in Italy, but also along the Pyrenees and the Franco-Netherlands and Rhine frontiers (see p. 512).

1509–1511. North African Expeditions. Cardinal **Jiménez de Cisneros,** an able statesman and general as well as churchman, commanded these amphibious operations. He captured Oran, Bougie, and Tripoli; all Moslem rulers of Northwest Africa were forced to pay tribute to Spain.

1516–1556. Reign of Charles I. He was the son of Philip of Hapsburg and of Joanna, daughter of Ferdinand and Isabella, and thus sole heir of the houses of Hapsburg, Burgundy, Castile, and Aragon. Elected Holy Roman Emperor, as Charles V (1519), he became the most powerful ruler in Europe. His reign was notable militarily for almost continuous wars with Francis I of France and Suleiman of Turkey.

1521–1523. Uprising of the Comuneros. This was an urban bourgeois movement, covertly supported by France. This, combined with personal rivalry, led to Charles's first war against Francis (1521–1529; see p. 516). The rebels were finally crushed at the **Battle of Villalors** (April 1523).

1533. Exile of the Moriscos. Many Moslem Moors, refusing to become Christians, were exiled. The fleets of **Khair ed-Din** of Algiers ferried most of them to Morocco.

1535. Charles's Successful Expedition against Tunis. (See p. 545.)

1541. Charles's Disastrous Expedition against Algiers. (See p. 545.)

1556–1598. Reign of Philip II (son of Charles, after Charles's abdication). This meant the separation of Spain and the Holy Roman Empire. During this reign Spanish influence, prestige, and capability reached their apex, and began to decline.

1567–1609. Rebellion in the Netherlands. (See p. 528.)

1568–1571. Revolt of the Moriscos. These were converted Moslems who were still persecuted by Spanish Christians.

1571, October 7. Battle of Lepanto. (See p. 547.)

1580. Philip II Becomes King of Portugal. This followed the victory of the Duke of Alva at the **Battle of Alcántara** (August 25; see p. 534).

1582. Naval Battle of Terceira. (See below.)

1583. Naval Battle of San Miguel. (See p. 534.)

1587. Drake's Raid on Cádiz. (See p. 508.)

1588. The Great Armada. (See p. 508.)

1589–1598. War with France. Philip's intervention in French civil war (see p. 524).

Portugal, 1500–1589

Portugal's maritime glory, initiated by Henry the Navigator (see p. 469), reached its zenith in the early years of this century, when the Portuguese established the oldest and the most enduring of all European colonial systems on the shores of the Indian Ocean. Preeminent among Portuguese sailors was Affonso de Albuquerque, one of the great admirals of history, who employed sea power with an understanding, determination, and energy unmatched until the time of Nelson. Like more populous Spain, Portugal, too, was soon overextended by colonial and maritime endeavors; ambitions exceeded capabilities. The collapse of Portuguese maritime supremacy began with the defeat and death of King **Sebastian I** in an effort to conquer Morocco. Civil war broke out between 7 rival claimants to the throne. Philip II of Spain was successful, thanks to the military ability of the Duke of Alva. Spain and Portugal were combined, a union that lasted for more than half a century, with evil effects from which Portugal never fully recovered. The principal events were:

1495–1521. Reign of Manuel I, the Great. The height of Portuguese greatness and glory.

1500–1513. Almeida and Albuquerque in the Indies. (See p. 555.) These two viceroys established Portuguese ascendancy in the entire Indian Ocean.

1557–1578. Reign of Sebastian I.

1578. Battle of Al Kasr al Kebir. Sebastian, personally leading an expedition of conquest to Morocco, was disastrously defeated. Three kings died in the battle: Sebastian, the King of Fez, and the pretender to the throne of Fez (allied with Sebastian).

1578–1580. Civil War of Succession. Philip II of Spain was successful, due to the victory of his general, the Duke of Alva, over Don **Antonio de Crato of Beja** at the **Battle of Alcántara** (August 25, 1580). Antonio fled to England, then to France.

1580–1598. Reign of Philip.

1582. Naval Battle of Terceira (Azores). Spanish Marquis **Álvaro de Bazán of Santa Cruz** defeated a French fleet under **Filippo Strozzi,** which was attempting to seize the Azores for Antonio de Crato.

1583. Naval Battle of San Miguel (or Second Terceira). Crato again tried to seize the Azores with French support, intending to reconquer the throne. The Franco-Portuguese fleet, under **Aymard de Chaste,** was again defeated by Santa Cruz. Spanish superiority in the Atlantic was reaffirmed.

1589. English Intervention. Crato again attempted to reconquer Portugal, landing in his native land with English amphibious support, led respectively by Sir Francis Drake and Sir **John Norris.** There was disagreement among the leaders; the population failed to rise as expected, alienated by English plundering. Crato was disastrously defeated, and fled to France. The English returned home.

EASTERN EUROPE

POLAND

Poland became firmly united with Lithuania, and the combined monarchy, after a series of fierce struggles with the princes of Moscow, repulsed Russian and Tartar threats, pushing Poland's eastern frontier well beyond the Dnieper River in White Russia and the Ukraine, and establishing Poland as one of the great powers of Europe. The principal events were:

1501–1506. Reign of Alexander I. The union of Poland and Lithuania was broken, permitting **Ivan III** of Russia to conquer all the left bank of the Dnieper River (1503).

1506–1548. Reign of Sigismund I. This able and farsighted king reunited Lithuania with Poland. Sigismund labored intensively to build up the military strength and political unity of his weakened nation, but was defeated in a long war with **Basil III** of Russia, losing Smolensk (1514). Later, after creating an outstanding army, he fought a more successful, though inconclusive, war with Russia (1534–1536).

1548–1572. Reign of Sigismund II. Another able king. During his reign Lithuania and the Ukraine were permanently united with Poland.

1557–1571. Livonian War. Poland, Russia, Sweden, and Denmark had rival claims and aspirations in Livonia (modern Estonia, Latvia, and part of Lithuania). The war was precipitated by a Russian invasion after the collapse of the Teutonic Order (see p. 526). A confused 4-way war followed. The Russians were at first repulsed, with the Swedes occupying Estonia and Poland taking most of the remainder of Livonia, save for Kurland, which was seized by Denmark (1561). Poland's war with Russia continued fiercely for nearly 10 more years, with Poland losing part of its Livonian conquests to **Ivan the Terrible** (1563–1571).

1573–1574. Reign of Henry of Galois (later Henry III of France). A period of turmoil; weak Henry was unable to control the unruly Polish nobles. He left Poland to become King of France (see p. 523).

1575–1586. Reign of Stephen Bathory. A strong ruler and a capable warrior. Although plagued by the lack of discipline and unity amongst the great nobles, he won a series of substantial victories over Ivan the Terrible.

1577–1582. Renewed Livonian War. Stephen repulsed the Russians who were endeavoring to exploit internal unrest in Poland, recaptured Polotsk from Russia, and pressed the boundary of Livonia as far east as Pskov.

1582–1586. Reorganization of the Polish Military System. Stephen incorporated into it the fierce Ukrainian Cossacks. He made plans for a joint Polish-Russian crusade against the Turks to expel them from their footholds along the northern coast of the Black Sea, but died before anything came of the plans.

1587–1632. Reign of Sigismund III. He was the first Polish ruler from the Swedish Vasa family. He involved Poland in endless wars with Sweden, most of which took place in Livonia (see p. 528).

1593. Cossack Revolt in the Ukraine. This was suppressed by Poland.

HUNGARY AND BOHEMIA

Bohemia and Hungary were combined under a single ruler for most of this period, and both were jointly incorporated into the Hapsburg dominions. Weak leadership resulted in disaster early in the century, when **Louis II** was defeated and killed by the Turks under Suleiman in the decisive **Battle of Mohacs.** This led to the collapse of Hungary, which for more than a century had barred Turkish aspirations to penetrate north up the Danube Valley into Central Europe. Suleiman, taking advantage of dissension, overran and annexed most of the country, while **Ferdinand of Hapsburg,** brother of Charles V, established tenuous control of a strip of northern and western Hungary, for which he paid tribute to the Turkish sultan. During the last three-quarters of the century, warfare was endemic along this entire border between the Ottoman and Hapsburg portions of Hungary. The principal events were:

1471–1516. Reign of Ladislas II of Bohemia (Ladislas VI of Hungary after 1490). A weak ruler, both of his kingdoms declined while internal dissension arose. The Turks took advantage of this to encroach on Hungarian borders.

1514. Peasant Revolt in Hungary. This was led by **George Dozsa.** It was suppressed by the noble leader **John Zapolya** of Transylvania, after atrocities on both sides.

1516–1562. Reign of Louis II (son of Ladislas II). Another weakling.

1521. Turkish Capture of Belgrade. (See p. 540.)

1526, August 29. Battle of Mohacs. Louis was killed and the Hungarian Army smashed (see p. 542).

1526–1538. Civil War of Succession. This was between Ferdinand of Hapsburg and John Zapolya, who was supported by Suleiman (see p. 543).

1529. Siege of Vienna. (See p. 543.)

1538. Revolt in Moldavia. Peter Rarish, Prince of Moldavia and vassal of Turkey, allied himself with Ferdinand. Suleiman, aided by Krim (Crimean) Tartar vassals, suppressed the revolt brutally; Peter fled to exile.

1538. Treaty of Nagyvarad. Ferdinand recognized Zapolya as King of Hungary, while retaining the northern and western regions. The treaty recognized Ferdinand as the successor of Zapolya.

1540. Death of Zapolya. His adherents ignored the Treaty of Nagyvarad and elected Zapolya's son **John II** (1540–1571).

1540–1547. Renewed Civil War in Hungary. This caused Suleiman to invade and to annex all of Zapolya's former territories (see p. 546). Ferdinand agreed to pay tribute for the strip of northern and western Hungary which he retained.

1547–1568. Intermittent Ottoman-Hapsburg Warfare. Ferdinand invaded Transylvania, but was repulsed (1551; see p. 546). The war was officially ended by the **Truce of Adrianople** (see p. 547), but frontier warfare continued.

1593–1606. The "Long War" (see p. 550). Sigismund Bathory, Prince of Transylvania (1581–1602), in alliance with the Hapsburgs, conquered Wallachia by defeating Turkish General **Sinan Pasha** at the **Battle of Guirgevo** (October 28, 1595). Internal warfare broke out in Transylvania, however, and Bathory was defeated by Voivode **Michael of Moldavia** at the **Battle of Suceava** (1600). An imperial army under **George Basta** then occupied Transylvania and began intense persecution of Transylvanian Catholics. Under the leadership of **Stephen Bocskay,** the Transylvanians formed an alliance with the Turks (1602). The war ended in a stalemate (see p. 635), with the **Treaty of Zsitva-Torok** (November 11, 1606). The independence of Transylvania was recognized by both sides.

RUSSIA

In this century the Grand Duchy of Moscow was transformed into the Russian Empire. Under **Basil III** and **Ivan IV** (the Terrible), this great new Slavic empire was almost constantly at war with its neighbors to the west, east, and south. These wars were caused partly by the lack of natural geographical frontiers on the vast flatlands of East Europe, and partly by the early manifestations of the landlocked Russians' insatiable desire for an outlet on the sea. At first they had more than their share of successes against the Poles and Lithuanians to the west; but as the century ended, Russian westward expansion had been halted by both the Poles and the Swedes, who now began to push the frontiers eastward. Russian expansion to the east was dramatic as they established loose authority over much of Siberia at the expense of the disorganized Mongols. To the south, the Russians were also successful in their struggles against the Tartars, gaining control of the entire Volga and Don river valleys. The principal events were:

1501–1503. War with Lithuania. The Russians conquered parts of White Russia and Little Russia.

1502. End of the Golden Horde. The Muscovites and Crimean Tartars combined to defeat and kill **Ahmed Khan,** last ruler of the Golden Horde.

1505–1533. Reign of Basil III (son of Ivan III and Zoë; see p. 473.) He not only consolidated the vast conquests of his father but also annexed the only 2 remaining independent Russian states: Pskov (1510) and Riazan

(1517). He conquered Smolensk from the Lithuanians (1514).

1521. Tartar Invasion. Mohammed Girai, Khan of the Crimean Tartars, invaded as far as Moscow. Basil agreed to pay perpetual tribute.

1533–1584. Reign of Ivan IV, "the Terrible" (son of Basil). He ascended the throne at age 3. During his minority (1533–1547) the nobles (boyars) became dominant in Russia.

1547–1564. Subjugation of the Nobles. This culminated in Ivan's cruel suppression of a boyar revolt (1564).

1552–1557. Conquest of Kazan and Astrakhan. Russia gained control of the entire Volga Valley from the Tartars. **Devlet Girai,** Khan of the Crimea, instigated by Suleiman, invaded as far as Moscow. He captured the city, but was repelled from the Kremlin (1555).

1557–1582. Livonian Wars. (See p. 534.) Russian efforts to obtain an outlet on the Baltic were successful, but with Polish resurgence all these conquests were lost to Poland and Sweden."

1569. War with Turkey. The Russians repulsed Turkish efforts to take **Astrakhan.**

1570. Sack of Novgorod. Ivan suspected the people of pro-Polish sympathies.

1571–1572. War with the Crimean Tartars. A Tartar raid briefly captured and sacked Moscow. They were defeated by the Russians at the **Battle of Molodi,** 30 miles from Moscow.

1572. Revolt of the Volga Tartars. This was cruelly suppressed.

1578. Battle of Wenden. The Russians were defeated by the Swedes, resulting in the loss of Polotsk (1579).

c. 1580. Expansion East of the Urals. Russian traders, led by the merchant-noble family of **Strogonov,** began to expand eastward, establishing posts as far as the Amur.

1582. Revolt of the Volga Tartars. This was suppressed.

1584–1598. Reign of Theodore I (son of Ivan). He was a weak ruler. The boyars again became dominant, the principal noble leaders being **Nikita Romanov** and **Boris Godunov.**

1590–1593. War with Sweden. Boris Gudunov demanded that King Johan III of Sweden return the territories seized by Sweden in the Livonian War. Boris invaded Livonia (January 1590), but after initial success he was repulsed at **Narva.** However, by a truce Russia regained outlets to the Baltic Sea at Kaporye, Ivangorod, and Yani. The Swedes, expecting the Russians to be diverted by a Crimean Tartar invasion, renewed hostilities but met with no success (1591). The following year the Russians invaded Finland. Armistice negotiations soon followed, and Russia kept its gains in the peace treaty (1593).

1591–1593. Tartar Resurgence. The Crimean Tartars, taking advantage of the Russian war with Sweden, invaded and reached the gates of Moscow (July 1591). They were repulsed, but invaded again the following year, with temporary success.

1598–1605. Reign of Boris Godunov. He was elected czar by the nobles after the death of Theodore.

EURASIA AND THE MIDDLE EAST

THE OTTOMAN EMPIRE, 1500–1606

This was the era of Turkey's greatest glory and power. In a brief reign, Sultan **Selim I** nearly doubled its size as the result of decisive victories over the Persians and the Egyptian Mamelukes. Selim's better-known son, **Suleiman the Magnificent,** inherited much of his father's military and administrative genius, though without the same grim, flinty character.

Probably the most powerful ruler of the century, Suleiman's long reign roughly coincided with those of Francis I of France, Charles V of the Empire, and Henry VIII of England. Suleiman conquered most of Hungary, but his ambitions to extend Ottoman control into Central Europe were dashed by the staunch defense of Vienna and of other fortifications along the Austro-Hungarian border. With the assistance of his great admiral, **Khair ed-Din Barbarossa,** Suleiman extended Ottoman suzerainty along most of the coast of North Africa. Other naval expeditions established Turkish preeminence in the Red Sea, but the Portuguese repulsed all

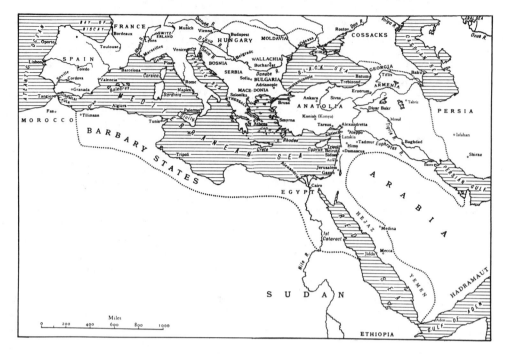

Ottoman Empire, c. 1550

efforts to gain control of the Arabian Sea (see p. 556). Victories over Persia resulted in Turkish conquest of Mesopotamia and much of Armenia. The Black Sea became a Turkish lake. The frequent diversions of Turkish strength to wars against Persia during the reign of Suleiman may have been the decisive factor in saving divided Europe from Ottoman domination. Charles V and Ferdinand wisely encouraged and fomented Persian opposition to Suleiman.

Six years after the death of Suleiman, the long, slow decline of the Ottoman Empire was initiated by a decisive naval defeat at the Battle of Lepanto by Don Juan of Austria. This decline is evident in retrospect; at the time it was less clear, since in the final years of the century the Turks decisively defeated the Persians, extending their conquests in the Caucasus, and also won slightly less conclusive victories over the Austrians in Transylvania and northern Hungary. The principal events were:

The Era of Selim, 1500–1520

1481–1512. Reign of Bayazid II. (See p. 477.)
1499–1503. War with Venice. The Turkish fleet under Admiral **Kemal Re'is** defeated the Venetian fleet in the **Second Battle of Lepanto,** or of Modon (1500). The Turks captured Modon (Methone), Lepanto, and Koron, though they lost Cephalonia. Turkish cavalry raided across the Julian Alps into Italy as far as Vicenza.

1501. Persian Raids. These were repulsed by the Bayazid's warlike son, Selim.
1502–1515. Undeclared Border War with the Mamelukes. Mameluke armies several times raided into Adana.
1509–1513. Civil War among the Sons of Bayazid. These were **Ahmed, Kortud,** and

Selim. Selim was successful; he forced his father to abdicate, and seized the throne (1512).

1513–1520. Reign of Selim I. He was the greatest soldier of the Ottoman sultans. His concentration on Asiatic campaigns, however, lost Turkey the chance to capitalize on the divisions of Europe, which were partly healed by accession of Charles V (see p. 525).

1514–1516. War with Persia. Selim provoked this war partly because the Persians had supported his brother Ahmed, partly because Shah **Ismail** (see p. 550) had raided into Ottoman territory in previous years, and partly because of his implacable hatred, as a devout Sunni Moslem, of the Shi'ite sect of Persia.

1515, June–August. Invasion of Persia. Selim marched from Sivas, then the eastern-most Ottoman city, with an army of more than 60,000 men, via Erzerum to the upper Euphrates. Despite the Persian "scorched earth" policy, Selim's logistical foresight permitted him to advance through the rugged mountains to Khoi, where the shah had assembled an army, probably less than 50,000 men, entirely cavalry—typical horse archers and lancers of southwest Asia, based generally on the old Mongol system. Turkish food supplies were consumed on the march.

1515, August 23. Battle of Chaldiran. Selim deployed his army on the Plains of Chaldiran in front of the Persians, protected by a screen of irregular cavalry and infantry. Behind this screen the infantry Janissaries (mixed archers and arquebusiers) formed behind a hastily dug trench. The flanks were protected by carts chained together, and in front of these carts, to the right and left front of the Janissaries, was the Turkish artillery, roped together wheel to wheel. On either side of this semifortified position were the royal cavalry guards (spahis), with the feudal Ottoman light cavalry (timariot) units extending the line farther to the flanks. The Persians attacked, drove off the screen of irregular cavalry, routed the European timariots on the Turkish right, but were repulsed in a hard fight by the Asiatic timariots of the left. The spahis and Janissaries stood fast, repulsing repeated attacks by the Persian left.

The Turkish left then swung around to engage the remainder of the Persian army in a violent struggle in which Ismail was wounded and his army routed. The Persians fled to Khoi, abandoning their camp and its provisions to the starving Turks. Casualties are unreported, but it is doubtful if the Persians lost more heavily than the Turks, since they were able to retreat rapidly and without effective Turkish pursuit.

1515, September 5. Capture of Tabriz. Selim took Ismail's capital without resistance. Soon after this, the Janissaries mutinied, refusing to advance farther into Persia. The timariots also wished to go home, so Selim reluctantly marched back by way of Erivan and Kars. The Persians reoccupied Tabriz. The Turks, however, kept their foothold on the upper Euphrates.

1515. Conquest of the Middle Euphrates. Selim's operations in this area, nominally subject to the Mamelukes, indicated contempt for both Persians and Mamelukes by thus engaging in war against both simultaneously. He then marched into Kurdistan, driving the Persians back into the Iranian highlands.

1516, July. Invasion of Syria. Selim, preparing for another invasion of Persia, learned that Mameluke Sultan **Kansu al-Gauri** had formed an alliance with Persia and was preparing to invade Turkey from Aleppo. Selim immediately turned south into Syria via the Euphrates Valley, bypassing the Taurus Mountains and surprising the Mamelukes (August). Kansu assembled his army, about 30,000 cavalrymen, at Merj-Dabik, 10 miles north of Aleppo. About half of these were Mameluke cavalry, mixed lancers, and bowmen, similar to the Persians, save that they wore turbans instead of helmets. The remainder were Arabic levies, unarmored, and less disciplined and less well organized than the Mamelukes. Kansu had neither infantry nor artillery. Selim had approximately 40,000 men, of whom 15,000 were timariot feudal levies, 8,000 Janissaries, 3,000 spahis, and perhaps 15,000 irregulars.

1516, August 24. Battle of Merj-Dabik. Selim drew up his army in the same formation as at Chaldiran. The Mamelukes attacked in a half-moon formation. Action on both flanks·was indecisive, but the central and heaviest Mameluke attack was thrown back in rout by the Turkish artillery in the center. Kansu was killed, and the Mameluke wings then fled. The battle was over so quickly that casualties were relatively light on both sides. The Mamelukes abandoned Syria and fell back to Egypt.

1516, August–October. Occupation of Syria. Selim prepared to invade Egypt. Meanwhile the Mamelukes elected **Touman Bey,** Kansu's nephew, as their sultan.

1516, October 28. Battle of Yaunis Khan. The Turkish advance guard under Grand Vizier **Sinan** Pasha defeated a Mameluke force near Gaza.

1517, January. Invasion of Egypt. Selim marched across the Gaza Desert, via Salihia (January 16), and advanced on Cairo. At El-Kankah, Touman blocked the road to Cairo, emplacing some artillery hastily created by buying guns from Venetian ships at Alexandria and by stripping seacoast defenses. Selim decided to envelop these fortifications. In a night march, the Turkish Army shifted to the left, forming before dawn in battle order at Ridanieh, on the edge of the desert.

1517, January 22. Battle of Ridanieh. Touman hastily shifted his army over 90° to its right, trying to drag his wheelless guns to new positions. Selim gave him no time to complete this movement, commencing an artillery bombardment for the purpose of provoking the Mamelukes to attack. As he intended, the Mamelukes charged the center of the Ottoman Army, while light Arab auxiliaries harassed the flanks. The desperate charge broke one wing of the Turkish army, but the reserve under Sinan Pasha plugged the hole, though Sinan was killed in the resulting struggle. The Mamelukes were driven off, having lost about 7,000 dead. The Turks lost about 6,000. Selim immediately marched into Cairo.

1517, January–February. Conquest of Egypt. Touman continued desperate resistance, including several days of bloody street fighting in Cairo (January 29–February 3) in a surprise effort to recapture the Egyptian capital. Selim proclaimed himself Sultan of Egypt, and then Caliph—this supreme Moslem title ostensibly conferred by the last of the Abbasids. He left the Mamelukes as nominal rulers of Egypt, but under a Turkish governor general.

1517. Occupation of Mecca and the West Arabian Coast.

1518–1519. Religious Outbreaks in Anatolia and Syria. These were suppressed by Selim.

1519. Alliance with Algiers. Khair ed-Din, Dey of Algiers, paid homage to Selim, and offered a fleet to the Ottoman Empire, in return for support against Spain.

1520, September 22. Death of Selim. He was preparing an expedition against Rhodes.

The Era of Suleiman, 1520–1566

1520–1566. Reign of Suleiman I, the Magnificent (son of Selim). During the early part of his reign, Suleiman was assisted greatly by his loyal vizier **Ibrahim** Pasha, who administered the empire, permitting Suleiman to devote himself almost exclusively to military conquest (1523–1536).

SULEIMAN'S EARLY WARS IN THE WEST, 1521–1544

1521, May–September. Invasion of Hungary. Suleiman advanced northward, ostensibly to redress pretended Hungarian slights. He captured **Shabotz** by assault (July 8), then besieged **Belgrade;** its small garrison surrendered after gallant defense (August 29). He sent raiding forces northward to terrorize Hungary and Central Europe. Finding that sea communications in his empire were hampered by the Christian fortresses of Rhodes, Crete, and Cyprus, he planned to seize them.

THE SIEGE OF RHODES, 1522

Suleiman continued his father's preparations against Rhodes, eastern Mediterranean stronghold of the Knights Hospitalers of St. John, and probably the most strongly fortified place in the world at that time. It was one of the earliest of the new bastion fortresses, with a broad ditch and a glacis, and (in some places) two interior walls. The port was protected by powerful tower forts at the entrance of the harbor, which was also blocked by massive bronze chains. Grand Master **Philip Villiers de L'Isle Adam** had 700 knights, plus 6,000 light local auxiliaries, including some cavalry, marines, pikemen, and arquebusiers. Artillery in substantial numbers was well emplaced. No reinforcements were available. Provisions and ammunition were plentiful, but no resupply could be expected. As Turkish preparations became obvious, Villiers had all available foodstuff collected from the remainder of the island within the fortified city, and gave refuge to the farm peasants.

1522, June 25. Turkish Landings. A beachhead was established without opposition. The Turkish army, approximately 100,000 men, was well trained, confident, and included an excellent engineer corps, as well as a powerful train of siege artillery.

1522, July 28. The Siege Begins. A contest of artillery fire and mining operations, conducted with great skill, determination, and energy on both sides. Defending artillery fire was devastatingly accurate, since the knights had exact ranges to all key points. Though the knights destroyed most Turkish mines, some reached the walls to blow great breaches (August).

1522, September–November. Repeated Turkish Assaults. All were repulsed with shocking slaughter. Nonetheless, the losses among the garrison were heavy under the Turkish bombardment and in the hand-to-hand fighting in the breaches and along the walls. The garrison dwindled to a strength inadequate to man the walls properly, while gunpowder was almost exhausted. Turkish casualties were replaced by reinforcements from the mainland.

1522, December 1–15. Negotiations. Suleiman realized the difficulties of the defenders, but was horrified by the terrible losses his own army had suffered. He offered honorable terms of surrender, guaranteeing that the population could leave or stay and that their religion would be respected. The negotiations broke down, however, and fighting renewed.

1522, December 17–18. Turkish Penetration into the City. The knights counterattacked vigorously and blocked the streets to contain the Turkish lodgement.

1522, December 20. Negotiations Reopened. Suleiman did not relish the thought of more costly street fighting. Hostages were exchanged; the Turks pulled out of their lodgement in the city and withdrew one mile from the walls. Suleiman's personal conduct in the negotiations, and in living up to the terms of the treaty, was exemplary.

1522, December 21–1523, January 1. Evacuation of Rhodes. Only 180 knights, plus 1,500 other troops, were left alive, and most of these were wounded. Turkish losses in the siege of Rhodes were at least 50,000 dead, and may have been as high as 100,000, out of a total force of almost 200,000 engaged. The tragedy of the siege, from the Christian standpoint, was the fact that Rhodes could probably have been held indefinitely had the knights received some reinforcements and supplies. Venice and Spain could have provided such assistance. Possibly it was contrition which caused Charles V later to give Tripoli, and then the island of Malta, to the Hospitalers, where they began to establish a new island fortress.

1522–1525. Intermittent Border Warfare

with the Hungarians. This was particularly intense in Wallachia.

1524. Neutrality Treaty with Poland. This freed Suleiman's hands for an invasion of Hungary.

THE CAMPAIGN OF MOHACS, 1526

1526, April 23. Advance from Constantinople. Suleiman's army was 70,000–80,000 strong.

1526, May–July. Hungarian Preparations. King Louis, aware of the danger, was unable to arouse interest among other European nations. He was slow in organizing his nation for defense, but was given additional time by the 2-week siege of **Peterwardein** (July 12–27). Slowly gathering troops, Louis moved south to Mohacs, where his army grew to about 12,000 cavalry and 13,000 infantry (August 15). At this point, rumors that Suleiman had 300,000 men caused Louis and some of his advisers to waver. He was persuaded to stand firm by the confident arguments of Archbishop **Tomori,** a formidable warrior, who correctly estimated Turkish strength and discounted their capabilities.

1526, August 28. Arrival of the Turks. Suleiman reached the southern edge of the Plain of Mohacs. His light reconnaissance cavalry discovered the Hungarian Army prepared for battle in the center of the plain southwest of Mohacs. It was an area ideal for cavalry combat, the principal arm on both sides.

1526, August 28–29. Hungarian Dispositions. The infantry, about half German and other foreign mercenary contingents, the remainder Hungarian, was formed in 3 large phalanxes in front of Mohacs, the left flank covered by marshes along the Danube River, the right flank in the air. The infantry included a substantial percentage of arquebusiers. Louis's artillery (about 20 cannon) was in front of the central square. Large cavalry detachments were placed in the intervals between the 3 squares; the remainder of the cavalry was drawn up into reserve lines to the rear.

1526, August 28–29. Turkish Dispositions. Suleiman formed his army in 3 lines, the first 2 consisting of feudal timariots. Behind them, providing a base of maneuver for the cavalry, Suleiman deployed Janissaries, with artillery in front and the spahis on their flanks. A detachment of 6,000 timariots was sent by a circuitous route to the west, taking advantage of undulations of the ground, to launch a surprise attack on the Hungarian right after the armies were fully engaged.

1526, August 29. Battle of Mohacs. The advancing Turks were met by a cannonade. The Hungarian first-line heavy cavalry then charged to drive the Turkish first line back. The remainder of the Hungarian army advanced, following this initial success, but the guns could not keep up. Just as the Hungarian cavalry charged Suleiman's second line, the Turkish enveloping force hit them on the right flank and threw them into considerable confusion. However, this flank attack was driven off by the second line of Hungarian cavalry. Both Hungarian cavalry lines joined in smashing the Turkish second line. The Hungarians then charged against the center of the third and final Turkish line, but suffered heavy casualties from the Turkish artillery, which they could not penetrate because the cannon were chained together. When the Janissaries and spahis counterattacked, the exhausted Hungarians broke and fled. Suleiman's army, which had suffered severely, attempted no organized pursuit; a few timariot units, rallied from their earlier defeat, harassed the fleeing Hungarians, whose losses were enormous: 10,000 infantry and 5,000 cavalry killed. The few prisoners captured by the Turks were beheaded. Louis, Tomori, and most of the other Hungarian leaders were killed. Without leadership, the fleeing remnants of the Hungarian Army simply scattered. Turkish losses are unknown, but they were probably at least as heavy as those of the Hungarians. Suleiman spent three days on the battlefield reorganizing his army.

1526, September 10. Occupation of Buda. This was unopposed. Deciding not to annex Hungary, Suleiman made it a tributary kingdom under **John Zapolya** of Transylvania (see p. 535), who had traitorously absented himself from Mohacs.

1526–1528. Civil War in Hungary. Zapolya, with Turkish assistance, established effective control over all of Hungary except a fringe to the north and west, which remained under the control of **Ferdinand of Hapsburg** (brother-in-law of Louis, brother of Charles V, and later emperor; see p. 526). Ferdinand, bringing a German contingent to form the core of a new Hungarian army, captured Raab, Gran, and Buda, pursuing John and defeating him at the **Battle of Tokay** (1527). John appealed to Suleiman for help.

1528. Suleiman Prepares Another Expedition into Hungary. His preparations included a secret alliance with France against the Hapsburgs (see p. 518).

THE VIENNA CAMPAIGN, 1529

1529, May 10–August 6. Turkish Invasion. With an army exceeding 80,000 men, Suleiman advanced to Essek, being joined en route by Zapolya with 6,000 Hungarians.

1529, September 3–8. Siege and Capture of Buda. Most of the garrison was massacred.

1529, September 10–23. Advance on Vienna. Accompanying Suleiman's army up the Danube was a Turkish flotilla. All of Lower Austria was ravaged by the Turkish light cavalry. Vienna was garrisoned by a force of 17,000 men commanded by **Philip, count palatine of Austria,** but the real leaders were aged **Nicholas, count of Salm** and Marshal **William von Roggendorf.** While the Turks were advancing, these men were energetically and effectively preparing the city for defense. Vienna had an ancient medieval wall, to which a few bastions for artillery had been added. The defenders hastily leveled all of the area around the walls in order to prepare fields of fire. Interior lines of entrenchments were dug within the walls to bolster the areas most vulnerable to artillery bombardment and mining. All wooden and thatched roofs in the city were ripped away.

1529, September 23. Land Investment of Vienna. There was skirmishing between cavalry patrols.

1529, September 27. Investment Completed on the Danube. The Turkish river flotilla sailed past the city to cut the line of communications to Bavaria and Bohemia.

1529, September 27–October 15. Siege of Vienna. Fought vigorously on both sides, incessant Turkish artillery bombardments and mining activities were punctuated by assault attempts (particularly October 9–15). The defenders, who discovered underground activity by observing agitation in bowls of water placed on the walls and outworks, energetically countermined. At the same time they conducted numerous sorties, severely damaging the Turkish entrenchments, mineheads, and artillery positions (September 29, October 2, October 6). With cold weather coming on, it became obvious to Suleiman that success was impossible. He withdrew, after killing all adult male prisoners (October 14–15).

1529, October–November. Austrian Pursuit. The Turks were vigorously harassed. Adding to the Turkish difficulties were premature snowstorms, making the roads so muddy that all wagons and carts had to be abandoned. The flotilla, carrying the Turkish siege artillery, suffered severe damage as it passed by the guns of Pressburg (Bratislava).

1530. Austro-Hungarian Initiative. They captured Gran, and raided as far as Buda-Pest.

1531. Austro-Hungarian Siege of Buda. This was unsuccessful. Widespread fighting along the frontier was generally inconclusive.

1532. Renewed Turkish Invasion. Suleiman led another large expedition north from Belgrade (June 25). Charles V moved to defend Vienna with a large army, but Suleiman avoided a direct conclusion, turning into southwest Austria by way of the Mur-Drava Valley. While Turkish light cavalry was demonstrating against Vienna and ravaging Lower Austria, Suleiman besieged **Guns** (Koszeg), and was repulsed. After a feint toward Vienna and further demonstrations in Styria (avoiding both Graz and Marburg), Suleiman withdrew down the Drava River, the expedition a dismal failure.

1533. Peace between Suleiman and Ferdinand. The Turks were anxious to devote full attention to war with Persia (see below). The "eternal" peace conceded to Ferdinand control of the northern and western strip, about 1/3 of Hungary. Ferdinand and John Zapolya both paid tribute to Suleiman. This treaty did not affect Charles V; the Empire continued at war with Turkey in the Mediterranean (see p. 545).

Suleiman's Wars in Asia, 1523–1559

1523–1525. Revolts in Egypt. These were suppressed by Ibrahim Pasha.

1525. Janissary Mutiny in Constantinople. Suppressed by Suleiman.

1526–1527. Insurrections in Anatolia. These were largely inspired by religious friction between the Sunni and Shi'ite sects. They were suppressed with difficulty, and forced Suleiman to hasten back to Turkey from Hungary.

1526–1555. War with Persia. Shah **Tahmasp** of Persia took advantage of Suleiman's European wars and of the Shi'ite uprisings (which he probably stimulated) to invade Turkish Armenia and to recapture Baghdad, Van, and other areas conquered by Selim (1526).

1533–1534. Turkish Invasion of Azerbaijan. Ibrahim Pasha captured Tabriz (July 13, 1534) against light opposition.

1534, December. Invasion of Mesopotamia. Suleiman joined Ibrahim at Tabriz and led the Turkish army into Mesopotamia via Hamadan to recapture Baghdad against light opposition. His army suffered heavy losses at the hands of Persian and Kurd guerrillas in its march.

1535, January 1. Persian Recapture of Tabriz. Tahmasp took advantage of Suleiman's absence in Mesopotamia.

1535, April–June. Turkish Recapture of Tabriz. Suleiman returned by forced marches from Baghdad. Tahmasp again retired without fighting.

1535, July–August. Invasion of Northern Persia. Suleiman pursued Tahmasp, but the shah avoided battle. Logistical difficulties caused Suleiman to return to Tabriz, which he destroyed (August), and then marched back to Constantinople by way of the Euphrates Valley and Aleppo.

1536, March 30. Assassination of Ibrahim Pasha. Suleiman ordered this, apparently upon the instigation of his Russian-born wife, **Roxelana,** who exercised great political influence over the Sultan.

1538. Suleiman Raids into Persia. Again he temporarily occupied Tabriz. In following years there was continuous border skirmishing.

1545–1549. Suleiman Renews Activity against Persia. Taking the field personally, he recaptured Van and Tabriz, the Persians again refusing to fight (1548). When the Turkish army went into winter quarters in Anatolia, the Persians once more reoccupied Tabriz, leaving the Turks with only minor gains (1548–1549).

1552–1555. Final Phase of the Persian War. This was initiated by a Persian offensive which captured Erzerum (1552). Suleiman took the field to recapture Erzerum. He then advanced into western Persia, which he ravaged (1553–1554). This time he established firm control over Erzerum, Erivan, Van, Tabriz, and Georgia, and established a tenuous Turkish foothold on the Caspian Sea. The war ended with the **Treaty of Amasia** (1555).

1553. Brief Janissary Revolt.

1559. Civil War between the Sons of Suleiman. Selim, with the support of his father, defeated his brother **Bayazid** at the **Battle of Konya.** Bayazid fled to Persia, but was executed by orders of the shah, who received a substantial payment from Suleiman.

Struggles in the Mediterranean, 1532–1565

1532. Andrea Doria Raids Morea. His imperial fleet captured Coron (Karoni), installing a Spanish garrison. This precipitated a prolonged Turkish-Christian struggle for control of the Mediterranean.

1533. Operations of Khair ed-Din. The Dey of Algiers, a vassal of the Ottoman Empire (see pp. 519, 563), was appointed by Suleiman as Admiral (Kapitan Pasha) of the Turkish fleet. He recaptured Coron and Patras.

1534. Khair ed-Din Seizes Tunis. The former sultan, **Mulai-Hassan,** fled to Europe, where he offered to become a vassal of Charles V in return for aid against the Turks and south Italy.

1535, June–July. Charles V's Amphibious Expedition to Tunisia. The emperor's army was protected by Andrea Doria's fleet. Doria decisively defeated Khair ed-Din in a battle off the coast. The imperial-Spanish Army landed and quickly captured Tunis, Bey Mulai-Hassan being reinstated. A Spanish protectorate controlled Tunis for nearly 40 years (1535–1574).

1536. Alliance of Francis I and Suleiman (March; see p. 518). The renewal of war in Western Europe forced Charles to abandon a plan to attack Algiers. Khair ed-Din, after rebuilding his fleet at Algiers, raided and devastated Minorca.

1537. Suleiman Declares War against Venice. This resulted from a presumed Venetian insult. He ordered Khair ed-Din to ravage Venetian territory. The admiral harried Apulia, Turkish forces raiding inland around Taranto (July), while Suleiman led an army to the Albanian coast opposite the Venetian island of Corfu.

1537, August 18–September 6. Siege of Corfu. Turkish land and naval forces received some French naval assistance. The approach of a combined imperial-Venetian fleet under Andrea Doria and Ferdinand's declaration of war against the Ottomans in Hungary, in compliance with orders from Charles, caused Suleiman to abandon the siege.

1538. Khair ed-Din Intensifies Naval Operations. He swept the Aegean and Adriatic, and raided the coasts of Sicily and south Italy. He rejected an effort by Charles V to bribe him from the service of Suleiman. Although the Venetian garrisons of Nauplia and Monenvasia in the Morea repulsed Turkish attacks, Khair ed-Din captured all other Venetian island and mainland Aegean outposts, and raided Crete.

He engaged in prolonged and inconclusive maneuvers against Doria and a combined imperial-Venetian fleet off the west coast of Greece and Albania (August–December).

1538, September 27. Battle of Preveza. Khair ed-Din had the best of an indecisive fleet action against Doria.

1539. Venice Makes Peace. She acknowledged her Aegean losses, giving up her two remaining footholds in the Morea and promising neutrality for 30 years. The only remaining important Venetian possessions outside of the Adriatic were Corfu, Zante, Crete, and Cyprus.

1541, September–October. Charles V's Expedition to Algiers. The Emperor embarked despite Doria's warnings of seasonal storms. The force of 21,000 men was landed 12 miles east of Algiers (October 20). Just as the investment of the city was beginning, a great storm wrecked most of the imperial fleet. The Moslem defenders of the city took advantage of the storm to conduct a sortie which caused heavy casualties (October 24–26). Charles and 14,000 survivors embarked on the remaining galleys and transports to return to Europe (October 27).

1542–1544. Khair ed-Din Terrorizes the Western Mediterranean. Part of the time he was accompanied by a French fleet under **Francis, Prince of Enghien.** Doria took refuge in Genoa. The French and Turks ravaged the coast of Catalonia, then besieged and sacked Nice (see p. 519). After wintering on the coast of Provence, Khair ed-Din continued his bold raids along the coast of Italy. When Francis suddenly concluded the Peace of Crépy, Khair ed-Din, losing his base in the western Mediterranean, returned to Constantinople, where he died 2 years later (1546).

1546. Truce between Suleiman and Charles V.

1551. Capture of Tripoli. Turkish land and naval forces ejected the Knights of St. John, after a gallant defense.

1551–1553. Turkish Fleet Sweeps the Western Mediterranean. Again, French forces cooperated. The Turks captured Bastia in Corsica.

1554–1556. Turkish Expansion in North

Africa. This was led by Admiral **Torghoud,** successor to Khair ed-Din. Most North African Moslem potentates acknowledged Ottoman suzerainty.

1555. Torghoud Captures Bugia (Bougie). The Spanish were ejected.

1558. Torghoud Captures Jerba Island. This brought the Turks close to Spanish-controlled Tunis. Turkish warships and corsairs terrorized the western Mediterranean; **Port Mahon,** Minorca, was seized and sacked.

1560. Doria Recaptures Jerba. He was then driven off by the Turkish fleet; Doria's garrison surrendered after a siege of 3 months (March–June).

1561. Turkish Naval Raids against Sicily.

1563. Assault on Oran. The Spanish repelled the Turks.

1565, May–September. Siege of Malta. The Turkish Army was under **Mustafa** Pasha. The defending Knights of St. John were commanded by Grand Master **Jean de la Valette.** The Turkish Army was initially 30,000; approximately an equal number of reinforcements arrived during the siege. De la Valette had 500 knights and about 8,500 other troops, including mercenaries and 4,000 Maltese levies. Reinforcements consisted of 80 knights and 600 other troops who slipped in during the siege. The struggle was fought with the same skill and intensity on both sides as the siege of Rhodes 43 years earlier. Turkish bombardment was almost continuous, assaults were frequent, but the defenses remained firm and the defenders fought boldly and well under a magnificent leader. The Turks abandoned the siege upon the arrival of a relieving Spanish fleet and army under **García of Toledo,** Viceroy of Naples. The Spanish did not interfere with the Turkish embarkation and withdrawal. Turkish losses were probably 24,000 killed. The defenders lost 240 knights and 5,000 other troops killed.

Suleiman's Later Wars in Europe, 1537–1566

1537, September–December. Austro-Hungarian Invasion of Turkish Hungary.

Ferdinand, ordered by Charles, joined the Empire, Venice, and Papal States against Turkey and France (see p. 518). He was repulsed at **Essek.** The Austrians, commanded by **John Katvianer,** withdrew into Austrian territory, harassed by the Turks and suffering severe losses from cold and snowy weather and from desertion as well. The Turks practically annihilated the remaining Austrians in a number of scattered engagements near **Valpo** (December 2). Approximately 20,000 Austrians were killed, the army entirely dispersed. Turkish forces raided the coast of Apulia.

1538, July–September. Operations in Moldavia. Suleiman, assisted by his Krim (Crimean) Tartar vassals, suppressed a revolt in Moldavia, which had allied itself with Ferdinand.

1540. Death of John Zapolya. Ferdinand invaded Turkish Hungary, but was repulsed at **Buda.** He sent an envoy to Persia urging the shah to attack Turkey (see p. 536).

1541. Suleiman Pacifies Hungary. He annexed the country to the Ottoman Empire.

1532. Ferdinand Invades Hungary. He was repulsed after a vain siege of **Pest.** Suleiman, preparing for another expedition up the Danube, renewed the Franco-Turkish alliance.

1543, April–September. Suleiman Invades Austria. He marched up the Danube and then up the Drava, capturing Gran (August) and Stuhlweissenberg (September). He made no effort to engage the principal Austrian army, which Ferdinand held to protect Vienna.

1544. Suleiman Makes Peace with Ferdinand. He gave up plans to invade Austria after being abandoned by his French ally (see p. 519). The situation in Hungary returned to the *status quo,* with Ferdinand agreeing to continue his tribute for his portion of Hungary (1545). This permitted Suleiman to renew his war with Persia (see p. 544).

1551–1553. Ferdinand Renews the Turkish War. He invaded Transylvania (see p. 536). The Austrians were repulsed, but stopped Turkish counterinvasion by successful defense of border fortresses. Intermittent border war continued (1553–1562).

1562. Peace of Prague. Ferdinand and Suleiman made peace, with no substantial changes in the *status quo;* Ferdinand continued to pay tribute. The Mediterranean war between the Ottoman and Holy Roman Empires continued (see p. 526).

1566, January. Renewed War in Hungary. This was precipitated by raids ordered by the new Emperor **Maximilian.**

1566, July–August. Invasion of Austria. Suleiman, despite his age (72 years) and suffering from gout, led an army of more than 100,000 men.

1566, August 5–September 8. Siege of Szigeth (Szigetvar). The fortress was gallantly defended by Count **Miklos Zrinyi.** When the walls had been battered to rubble and further resistance was useless, Zrinyi ignited time fuses to his powder magazine and led the survivors of his garrison in a desperate sortie, in which all were killed (September 8). Hundreds of Turks, rushing into the fortress, were killed when the magazine exploded. Suleiman had died 2 days earlier. The Turkish grand vizier, pretending Suleiman was still alive, marched the army back to Constantinople.

The Crest of the Ottoman Flood, 1566–1600

1566–1574. Reign of Selim II ("the Sot").

1568, February. Treaty of Adrianople. Peace with the Hapsburgs (see p. 536)

1569. Revolt in Turkish Arabia. This was suppressed by local Ottoman governors.

1569. Capture of Tunis. Ouloudj Ali, Bey of Algiers, drove out the Spanish.

1569. War with Russia. The Turks and Krim Tartars failed in efforts to conquer Astrakhan.

1570, January. War with Venice. The republic had refused to cede Cyprus to Turkey. Selim sent a fleet under Admiral **Piale** Pasha and an army of 50,000 men under General **Lala Mustafa** to seize Cyprus. The garrison of 10,000

was commanded by Governor **Nicolo Dandolo,** who divided his forces between the two cities of Nicosia and Famagusta.

1570, July 22–September 9. Siege of Nicosia. The garrison was not strong enough to man the modern defenses of the city, which were taken by storm. Most of the population and the entire garrison were massacred.

1570, September 18–1571, August 1. Siege of Famagusta. There was little activity other than blockade during the winter. Some reinforcements and supplies of munitions slipped past the Turkish blockading fleet (January). The governor, **Marcantonio Brigadino,** made good use of the time to strengthen the antiquated defenses, manned by a total force of 5,400 men. Fighting was fierce during the spring and summer; Brigadino's defense was masterly. After all his ammunition was consumed and his garrison reduced to 2,000, Brigadino accepted terms offered by Mustafa, who then treacherously massacred the defenders (August 4).

1570–1571. Pope Pius V Establishes the Holy League. Its purpose was to conduct a crusade against the Turks for the relief of Famagusta, but the forces were slow in gathering, due to mutual suspicions of Spain and Venice. Famagusta fell while Christian forces gathered at Messina.

1571, August–September. Turkish Naval Operations. Admiral Pasha **Ali Monizindade** ravaged Venetian possessions in the Aegean and Ionian seas, then sailed into the Adriatic to within sight of Venice. Learning of Christian naval preparations at Messina, Ali rushed back to the Ionian Sea.

1571, September 22. Formation of the Allied Christian Fleet. This consisted of some 300 ships of all types, gathered at Messina, under the command of Don Juan of Austria. Next day they sailed for the Gulf of Corinth to seek the Turkish fleet.

THE BATTLE OF LEPANTO, OCTOBER 7, 1571

When the Turkish fleet was discovered at Lepanto, early on October 7, Juan's fleet available for battle was 108 Venetian galleys, 81 Spanish galleys, and 32 other provided by the pope and other small states. He also had 6 giant Venetian galleasses.

The Turkish fleet immediately moved out from Lepanto. Ali had 270 galleys, on the average somewhat smaller than the Christians'. The crews were probably not so experienced as those of the Christian ships.

The two fleets formed up in a traditional battle formation which had varied little since the Battle of Actium: each in a long line of 3 divisions, with a reserve in the rear. Save that the Turkish left wing was larger than the right wing, thus indicating a planned envelopment, neither side had any real tactical plan other than a crude melee, to be won by ramming and boarding. For this latter purpose there were 20,000 soldiers (out of a total strength of 84,000) on board the Christian fleet, and about 16,000 Turkish soldiers (of a total of 88,000). The only real difference between these two fleets and those which had fought in the Punic Wars, nearly 2,000 years earlier, was that a few small cannon were mounted in the bows of the galleys, and in the broadsides of the galleasses. The Christian marines, who included a number of arquebusiers, wore light armor. Few of the Turks had armor, and most were armed with bows or crossbows.

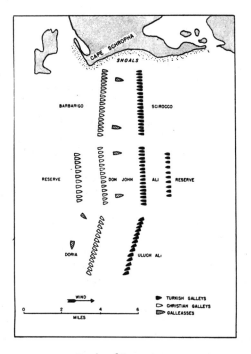

Battle of Lepanto

The fleets, each stretched out over a 5-mile front, met in a series of great clashes, beginning about 10:30 A.M. By noon the main bodies of both sides were completely engaged. The galleasses broke the Turkish line, but this was not decisive. Confused fighting raged for about 3 hours, during which the superior skill of the Christian sailors, and the superior armament of the Christian soldiers, gradually asserted themselves. An additional advantage, though not of great significance, was that the wind was favorable to the Christians.

The Turkish right flank, which had not been able to get very far from land, was driven back against the shore and exterminated. The fight in the center lasted somewhat longer, but here too the Christian superiority finally overwhelmed the defending line. The Turkish left, which was far more numerous and led by **Ouloudj Ali,** Dey of Algiers, the best of the Moslem commanders, did better. The fight to the southwest was quite even until Ouloudj Ali discovered what had happened to the remainder of the Turkish fleet. He disengaged and escaped with 47 of his 95 vessels, plus one captured Venetian galley. These were the only Turkish survivors of the battle. Sixty other galleys had gone aground, 53 had been sunk, and 117 had been captured. Fifteen thousand Christian rower slaves were freed from captured or sunken Turkish vessels, though at least 10,000 more must have gone down with their ships. At least 15,000–20,000 Turkish sailors and soldiers were killed or drowned. Only 300 prisoners were taken. The Christians lost 13 galleys, 7,566 dead, and nearly 8,000 wounded (among these was **Miguel Cervantes,** who lost his left hand).

Because of the lateness of the season and the likelihood of seasonal tempests, the Christian fleet returned to Italy to await good weather the following spring before trying to exploit the great victory. Lepanto was one of the world's decisive battles. This success of temporarily united Christendom, ended the growing Turkish domination of the central and western Mediterranean, and marked the high tide of Islam's second great threat against Christian Europe.

1571–1572. Turkish Shipbuilding Effort. This was inspired by Grand Vizier **Mohammed Sakalli,** who dominated the government of the inept Sultan Selim. Nearly 200 new Turkish vessels were built during the winter. Though many of these were poorly constructed and crews were inexperienced, nevertheless Ouloudj Ali, now the senior Turkish admiral, sailed from the Dardanelles with 160 galleys in the spring (June).

1572, June–August. Allied Indecision and Lack of Coordination. Philip of Spain, expecting war with France, refused to permit Juan to take his fleet into the Aegean. Finally Venetian Admiral **Jacopo Fascarini** sailed with about 150 ships, only 20 of which were Spanish (August). Off Cape Matapan he sighted the new Turkish fleet and, amazed at its size, fell back to Corfu to await reinforcements.

1572, August. Spanish Recapture of Tunis. Don Juan restored the Hafsid beys to the throne.

1572, September–October. Renewed Christian Naval Operations. Juan, finally gaining Philip's permission to operate against the Turks, joined Fascarini at Corfu with about 60 ships. The combined fleet then sailed to meet the Turks. Ouloudj, knowing that his new fleet had no hope of success against the veteran Christians, took refuge in the fortified harbor of Modar, where Juan refused to risk battle. He landed his troops under Alexander Farnese of Parma to attack the port from land, while the Christian fleet blockaded the harbor. Farnese's landing force was held off by the Turkish garrison, which was reinforced from central Greece. Juan re-embarked his troops and sailed for Italy.

1573, March. Venetians Make Peace. Disgusted by Philip's vacillation, the Venetians abandoned Cyprus. The treaty was not signed for almost a year (February 1574).

1574–1595. Reign of Murad III.

1574, August–September. Moslem Recapture of Tunis. Ouloudj Ali again, and finally, captured Tunis and Galeta. Juan sailed to do battle, but was forced to abandon the enterprise when his fleet was badly damaged by a storm en route (September).

1574–1581. Inconclusive Naval War. There were no major actions, and peace was established.

1577–1590. Renewed War with Persia. This was initiated by Murad. The Turks conquered

Shirvan, Tiflis, Daghestan, and Luristan. The principal Turkish leader was Lala Mustafa, the conqueror of Cyprus.

1590. Land War Renewed with the Hapsburgs. Emperor Rudolph refused tribute for the Austrian-controlled strip of Hungary. No major operations were conducted by either side, though border activity was incessant.

1593, June 20. Battle of Sissek. An Austrian army defeated and annihilated the army of **Hassan,** Ottoman governor of Bosnia. This infuriated **Sinan** Pasha, grand vizier under weak Sultan Murad III, who led an army to invade Austria and Hungary.

1593, October 13. Capture of Vesprism (Beszprem). Apparently Sinan planned to continue on toward Vienna, but the Janissaries refused further operations during the winter, and he was forced to return to Belgrade.

1593–1594. Austrian Raids into Turkish Hungary. They captured Neograd and other frontier places.

1594. Sinan Invades Northern Hungary. He forced the Austrians to raise their siege of Gran (June 1). He then captured Raab and some other border towns. He was repulsed in a long siege of Komorn (Komarno), an important Danube fortification.

1595. Revolts in Transylvania, Moldavia, and Wallachia. The Turks were temporarily driven out. **Charles of Mansfeldt** took advantage of this situation to lead an Austrian army in invasion of Hungary, defeating local Turkish forces in the **Battle of Gran** (August 4). The entire northern frontier of the Ottoman Empire was crumbling. Wallachian raiders crossed the Danube to sack Silistria and Varna.

1596. The Turks Regain Control of Hungary. New Sultan **Mohammed III** and his Vizier **Ibrahim** Pasha led a Turkish army north and repelled the Austrian invasions. They invaded Austrian Hungary to capture Erlau (September). An Austrian army of nearly 40,000 men under the Archduke **Maximilian** marched to recapture Erlau. Included in the army was **Sigismund Bathory** (nephew of the great Stephen) and a Transylvanian contingent (see p. 536). The Turk army, approximately 80,000 strong, met them 12 miles southeast of Erlau.

1596, October 24–26. Battle of Kerestes. The Austrians, entrenched behind a brook, repulsed strong Turkish attacks on the first day of the battle. After a day's lull, the Turkish attack was renewed at dawn. The Austrians repulsed the Turks and counterattacked to drive them back to their camp. At this moment an encircling detachment of Turkish cavalry under **Cicala** Pasha struck the Austrian rear. The Austrians broke in panic, abandoning their 97 cannon. The Turks rallied to pursue and inflicted great slaughter upon the retreating Austrians. Austrian casualties were about 23,000; Turk losses were probably nearly as great. The disorganized and badly punished Turkish army made no effort to pursue.

1597–1598. Renewed Austrian Invasions. They captured Raab and Vesprism. They were repulsed, however, by Turkish defenders of Buda (October 1598).

1598. Revolt in Iraq. Suppressed by the Ottomans.

PERSIA

At the outset of this century Shah **Ismail I,** the founder of the Safawid Dynasty, established himself as the ruler of Azerbaijan and western Persia. In following years he expanded his authority over all of Persia, but in the process came in conflict with the powerful Ottomans to the west and the Uzbek Tartars to the northeast. The remainder of the century was devoted to recurring wars against both foes, with the Persians generally defeated by the more powerful and more modern Turkish Army, but more successful in repelling Uzbek incursions into Khorasan. As the century ended, Persia, under the leadership of her greatest Moslem ruler, Shah **Abbas I,** was building up a modern army, balanced between cavalry, infantry, and artillery and ready to seek vengeance against the Turks. The principal events were:

1500–1524. Reign of Ismail I. At the outset he attacked Shirvan and captured Baku from the Tartars (1500).

1500–1507. Uzbek Expansion. The Turkomans conquered Transoxiana, Khurasan, and Herat under the leadership of **Shaibani Khan,** a descendant of Genghis.

1501. Battle of Shurer. Ismail defeated **Alwand** of the White Sheep Confederacy. He then captured Tabriz and established himself in control of Azerbaijan and northwestern Persia (1502).

1502–1510. Safawid Expansion. Ismail consolidated control over western and central Persia.

1510, December. War with the Uzbeks. Ismail was victorious in a battle near **Merv,** Shaibani being killed. The Uzbeks were driven from Herat and Khorasan, but retained control of Transoxiana.

1514–1555. War with Turkey. (See p. 539.)

1587–1629. Reign of Abbas I. He soon established order in Persia. He quickly made peace with Turkey in order to consolidate his position, and to deal with an Uzbek invasion of Khorasan.

1590–1598. War with the Uzbeks. Under the leadership of **Abdullah II** the Turkomans had captured Herat, Meshed, and much of Khorasan. Abbas slowly drove the Uzbeks from most of Khorasan. He was severely defeated, however, near **Balkh** (1598). Both sides were exhausted, and so peace was made, with the Uzbeks retaining a small foothold in Khorasan around Balkh.

1598–1600. Reorganization of the Persian Army. Abbas, who wanted to build up a military force comparable to that of the Turks, was assisted by the English adventurers Sir **Anthony Shirley,** his brother **Robert,** and 26 other Europeans, sent on an unofficial mission to induce Persia to combine with the Christian nations of Europe against Turkey. Evidently the most useful of these foreign advisers was Robert Shirley, an artillery expert. In a short time an excellent artillery organization was created, and also a strong contingent of musket-armed infantry who comprised a regular standing force. At the same time, to reduce the power of the tribal chiefs who had contributed so much to Persia's past weakness, Abbas created a new tribe for the sole purpose of establishing a semipermanent cavalry force directly responsible to him. The best cavalry warriors of the nation flocked to join the new tribe and to pledge allegiance to the Shah. As a result, by the end of the century, a new and formidable military power was growing in Western Asia, anxious to settle old scores with the Ottoman Turks.

SOUTH ASIA

NORTH INDIA

In the early years of the century the Rajput revival continued, threatening Moslem hegemony of north India. Leader of this revival was **Rana Sanga,** King of Mewar (or Chitor), who took advantage of squabbles amongst the Afghan and Turkish nobility of the Delhi Sultanate. The anarchy in the Moslem states of north India also attracted the attention of **Babur,** King of Kabul, a descendant of Genghis Khan and Tamerlane. After some exploratory raids, Babur invaded and conquered north India to establish the Mogul Empire (so named because of his Mongol ancestry). Babur died soon after. His son, **Humayun,** was overthrown by **Sher Shah,** an elderly military genius; but, after the death of Sher Shah, Humayun reconquered north India and reestablished the Mogul Dynasty. His son, Akbar, actually carried out the conquests planned by Babur and Sher Shah. Though perhaps not so brilliant a soldier as either of these predecessors, he was nevertheless one of the great warriors in Indian history. The principal events were:

1488–1517. Reign of Sikandar, Sultan of Delhi. His persecution of the Hindus led to frequent conflicts with the Rajput Hindu principalities, which, combined with internal dissension in the Sultanate, contributed to the steady decline of the Lodi Dynasty.

1509–1529. Reign of Rana Sanga, King of Mewar. He increased Rajput power by victories over the Lodi sultans of Delhi and the Moslem kings of Malwa and Gujerat.

1515–1523. Babur Raids into the Punjab. These were reconnaissances in force.

1516–1526. Reign of Ibrahim, Sultan of Delhi.

1524. Babur's Lahore Campaign. Taking advantage of an uprising in the Punjab, he seized Lahore, but was driven out by the Lodi governor of the Punjab.

1525–1526. Babur Invades North India. He conquered the Punjab (1525). Then he advanced on Delhi (March–April, 1526). He had about 10,000 men, allegedly including an Ottoman Turkish contingent of artillery* and musketeers. The remainder of his army consisted of the veteran Central Asiatic cavalrymen who had fought with him against the Uzbeks (see p. 557). Arriving at Panipat, about 30 miles north of Delhi, Babur learned that Ibrahim's army (probably 30,000–40,000 men) was just ahead. Babur immediately prepared a defensive position with his infantry and artillery sheltered behind a line of baggage carts tied together. In gaps purposely left in the line of carts, he placed his Turkish guns, tied together by chains in the typical Ottoman fashion. Other gaps were left for counterattacks by reserve cavalry. The Mogul army had been reinforced by a few thousand Hindu and Moslem Indians anxious to assist in overthrowing the Delhi Sultanate. Babur's total strength was probably about 15,000.

* Some authorities doubt whether Babur had any artillery.

1526, April 20. Battle of Panipat. After several days of indecision, Ibrahim attacked. The attackers were held up by the protected infantry and artillery. Babur then sent his own cavalry to drive in the flanks of the Delhi army. Fighting in 3 directions, the Delhi troops soon broke and fled, suffering severe casualties. Ibrahim and 15,000 of his men were killed.

1526, April 27. Babur Occupies Delhi. This began the Mogul Empire.

1526–1537. Reign of Bahadur Shah of Gujerat.

1527, March 16. Battle of Khanua (or Fatehpur Sikri). Rana Sanga led a confederated Rajput army of nearly 100,000 men against Babur. The Mogul, with less than 20,000 men, marched to meet the Rajputs 40 miles west of Agra. Again Babur used his artillery and musketeers as a base of maneuver for his mobile cavalrymen. Rana Sanga was seriously wounded. A violent Mogul counterattack broke the Rajput army, which fled in panic, suffering heavy casualties.

1528–1529. Conquest of Bihar and Bengal. Babur's operations against the Turko-Afghans was culminated by victory in the 3-day **Battle of the Gogra River** (1529) near Patna. The Mogul Empire now stretched from the Oxus to the Brahmaputra. Babur died before he could expand further (1530).

COMMENT. *The hardships, disappointments, and defeats of his early years had taught the Mogul chieftain many lessons, and had made him cautious without dampening his ardor or love of adventure. His three great victories in North India, each against excellent armies greatly outnumbering his own, established him as one of the greatest warriors of his times. He was also a man of great literary skill, as proven by his admirable memoirs and a number of outstanding poems.*

1530–1540. First Reign of Humayun (son of Babur).

1531–1536. Mogul War against Gujerat. Humayun defeated Bahadur Shah at Chitor, and then captured Mandu and Chanpanir. But Bahadur energetically collected another army and drove the Moguls off.

1537–1539. Revolt of Sher Khan. This 65-year-old Afghan-Turk, of relatively humble origin, had risen through luck and ability to the position of governor of Bihar. He annexed Bengal to his own province, and became a rallying point for the Afghan-Turk nobles who opposed the Moguls. Humayun invaded Bihar, but Sher Khan avoided battle, while constantly harassing the Mogul line of communications. As more and more of the north Indian Moslem dissidents were attracted to his banner, Sher Khan finally felt strong enough to meet Humayun in battle, defeating him at the **Battle of Chaunsha** (or Buxar). Sher Khan then pursued the defeated Moguls up the Ganges Valley, winning another great victory at the **Battle of Kanauj.** Humayun fled to asylum in Persia. Sher Khan seized the imperial throne as **Sher Shah.**

1539–1545. Reign of Sher Shah. He quickly and firmly established control over all of northern Hindustan, including the Punjab. Sher Shah extended the Delhi dominions southward, conquering Malwa and Marwar (1541–1545). He was killed by an accidental gunpowder explosion at the **Siege of Kaninjar** (1545). During his brief reign he created an efficient standing army based on sound military policy, and had time to institute far-reaching governmental reforms. His tremendous military and administrative accomplishments as emperor were achieved between the ages of 68 and 73.

1545–1555. North India in Anarchy. The Hindu General **Hemu** became virtual dictator in Delhi (1552–1555).

1555. Return of Humayun. He had become ruler of Afghanistan (see p. 558). Seeing an opportunity to regain his old empire, he marched through the Punjab and recaptured Delhi. He died soon after in an accident (1556).

1556–1557. Turmoil in the Mogul Empire. The oldest legitimate son of Humayun, 14-year-old Akbar, was in the Punjab with his able adviser **Bairam.** In the Ganges Valley the Afghan–Turks had united under the leadership of Hemu, who again seized power in Delhi. In Afghanistan, Akbar's older half-brother **Mizar Mohammed Hakim** was virtually independent. Akbar and Bairam collected an army in the Punjab, then marched toward Delhi (October 1556).

1556, November 5. Second Battle of Panipat. In a bitterly contested battle, Hemu's numerically superior army was close to victory, but was thrown into confusion when he was wounded by a chance arrow. Bairam and young Akbar counterattacked, to win the battle and to reinstate the Mogul Empire.

1556–1605. Reign of Akbar. This began with a 4-year regency of Bairam, during which Mogul control over northern Hindustan was firmly consolidated. Akbar then dismissed the regent and began to govern for himself (1560).

1561–1562. Conquest of Malwa.

1562–1567. Conquest of Rajputana. This culminated in Akbar's capture of Chitor (1567). He now conciliated the Rajput princes, confirming them on their thrones, repealing the discriminatory laws which had favored Moslems over Hindus under the old Delhi Sultanate. Save for the Mewar hero, **Pratap,** who continued resistance in the desert and mountain fastnesses of Rajputana, the Rajput princes soon became the most loyal supporters of the Moguls.

1566. Afghan Raid. Akbar's half-brother, Mohammed Hakim, raided into the Punjab, but withdrew in the face of Akbar's threats.

1573. Conquest of Gujerat. Akbar first came in contact with the Portuguese.

1574–1576. Conquest of Bihar and Bengal.

1576. Expedition into the Deccan. This was under Akbar's son **Murad.** The Moslem sultans of the northern Deccan banded together and drove the Mogul invaders out. In subsequent years Akbar campaigned almost constantly to gain control of all Hindustan.

1581. Hakim Again Raids into the Punjab. Akbar immediately marched to meet the threat, drove the Afghans out, invaded and conquered Afghanistan.

1586–1595. Conquest of Kashmir, Sind, Orissa, Baluchistan.

1596–1600. Operations in the Deccan. Akbar conquered Berar, Ahmadnagar, and Kandesh.

1601–1603. Revolt of Akbar's Son Salim. The old emperor returned to the Ganges Valley, defeated and captured Salim, and then pardoned him. When Akbar died 2 years later, it is likely that he was poisoned by this ingrate son (1605).

COMMENT. *The main characteristics of the reign of Akbar were conquest, justice, and tolerance. He built on the military and administrative foundations established by Sher Shah, and owed much to the accomplishments of his father's great rival. Akbar's standing army was organized on the same line as Sher Shah's with loyal, well-paid garrisons to hold the key hill forts. Like his grandfather, Babur, Akbar favored the use of artillery and a corps of 12,000 musketeers as the central components of his field forces. Most of the remainder of his army was highly mobile cavalry, the major portion of which, paradoxically, consisted of Rajput lancers. His most able general was an exceptional Hindu, **Raja Todar Malla,** who was also his chief minister and financial expert.*

SOUTH INDIA

During the first part of this century, while the Bahmani Sultanate collapsed and disappeared, the Hindu Kingdom of Vijayanagar achieved its greatest glory during the reign of **Krishna Deva,** who consistently defeated the neighboring Moslem kingdoms to the north, greatly expanding his dominions. The kingdom declined rapidly under his successor, but by midcentury King **Rama Raya** restored order to some extent, and revived Vijayanagar's waning power by taking advantage of the almost incessant wars between the Moslem Sultanates of Berar, Bijapur, Bidar, Golconda, and Ahmadnagar (the successors of the Bahmani Sultanate). The Moslems, however, finally composed their quarrels and temporarily united to overthrow Rama Raya and to smash the power of the Hindu kingdom. Vijayanagar, with diminished territory, prestige, and prosperity, continued to exist for nearly a century. The principal events were:

1509–1530. Reign of Krishna Deva of Vijayanagar. He defeated Bijapur, annexing much of the region between the Tungabedra and Krishna rivers (1512).

1512. Golconda Independent of Bahmani.

1513–1515. Expansion of Vijayanagar. Krishna Deva defeated Orissa, Golconda, and Bidar, extending his empire up the east coast of the Deccan as far as Vizagapatam.

1520. Vijayanagar Victory over Bijapur. Krishna Deva's boundary now reached the Portuguese colony of Goa. He was friendly with the Portuguese, who were also frequently at war with Bijapur. The Europeans imported horses for the Hindu cavalry, since Vijayanagar was poor horse-raising country (see p. 430).

1542–1565. Reign of Rama Raya of Vijayanagar. He restored order after a period of turmoil, and revived Vijayanagar's influence among its neighbors. He was involved in a bewildering series of alliances and wars with the northern sultanates, and became something of an arbiter in their disputes.

1565. Moslem Alliance against Rama Raya. His arrogance and power led the sultans of Bijapur, Bidar, Golconda, and Ahmadnagar to unite temporarily against him. Rama Raya was defeated and killed at the **Battle of Talikot.** The victors then sacked the capital city of Vijayanagar, which was never rebuilt.

1576. First Mogul Expedition into the Northern Deccan. (See p. 553.)

1596–1603. Akbar's Conquest of the Northern Deccan. (See above.)

PORTUGUESE EMPIRE OF THE INDIAN OCEAN

The Indian Ocean had long been an important avenue of commerce between India and lands farther east and west. To a lesser extent it had also on occasion been a route of conquest for the maritime-minded Cholas of southeast India. To the east the major sea routes led from the Coromandel coast and from Bengal to Malaya, the Straits of Malacca, and the South China Sea. To the west the routes extended from Gujerat and the Malabar coast to the Red Sea and the Persian Gulf, thence overland through Egypt and Persia to the Mediterranean and Europe.

The arrival on the Malabar coast in 1498 of a small squadron of Portuguese vessels, which **Vasco da Gama** had brought around the Cape of Good Hope, shattered the old patterns of trade and piracy. By the beginning of the century, regular trade had been established between western India and Portugal, and the Portuguese had begun to assert themselves as a force along the entire coast from the mouth of the Indus to Cape Comorin.

Rarely in history have so few men been able to affect so profoundly the fortunes of millions, and to change so completely the course of history. The principal Portuguese objective was trade, particularly in pepper and other spices, which were literally worth their weight in gold in 16th-century Europe. Next in importance was missionary zeal in spreading the Christian faith. Hardly less significant was the fierce pride of these Portuguese in carrying the power and prestige of their tiny kingdom to the furthest corners of the earth.

In a very few years a number of Portuguese trading posts—known as factories—had been established either with the permission of the local Hindu or Moslem rulers or by force.

Outstanding among the adventurous Portuguese who established these outposts were **Francisco de Almeida** and Affonso de Albuquerque, who soon became rivals. When Almeida finally accepted the orders of the Portuguese king to turn over his post as Viceroy of Portuguese India to Albuquerque, he had, with Albuquerque's help, established complete Portuguese dominance over the west coast of India.

Albuquerque realized the significance of sea power, and was the first man to apply it systematically. Though his appointment as Viceroy of Portuguese India lasted only 6 years, in that short time he established complete maritime dominance over the Indian Ocean. This supremacy on the ocean, and along the west coast of India, continued through the century, though Portuguese vigor declined rapidly in subsequent years. The principal events were:

1500. Appearance of a Portuguese Fleet at Calicut.

1500–1505. Establishment of Portuguese Trading Posts. These were along the west coast of India; the most important was at Cochin (1503).

1505. Almeida Appointed First Viceroy of Portuguese India. He established bases along the east coast of Africa, then brought a large fleet to the Malabar coast, where he established several small forts. One of his principal assistants was Albuquerque, who returned to Portugal (1506). Almeida's son, **Lorenzo,** established a settlement on Ceylon and negotiated a commercial treaty with Malacca in the name of his father (1507).

1507. Albuquerque Appointed Viceroy of India. Almeida knew nothing of this appointment. Albuquerque sailed from Lisbon with a small squadron, en route capturing the island of Socotra near the entrance to the Red Sea. He then seized the island of Ormuz, commanding the entrance to the Persian Gulf and then one of the main trade centers of the East (1508). Lacking sufficient force to hold this vital point, however, he then sailed on to Cochin.

1508. Gujerat-Egyptian Alliance. Sultan **Mahmud Begarha** of Gujerat formed an alliance with **Kansu al-Gauri** of Egypt (see p. 539) in an attempt to eliminate arrogant Portuguese interference with Moslem trade between India and the Red Sea.

1508. Naval Battle of Dabul. The combined Mameluke-Gujerat fleet attacked Lorenzo Almeida's small Portuguese squadron near Chaul; Lorenzo was killed in the inconclusive engagement.

1508, December. Arrival of Albuquerque. He reached Cochin while Almeida was in the midst of plans to avenge the death of his son. Almeida refused to recognize Albuquerque's orders, imprisoned his rival, and sailed north.

1509, January–February. Almeida's Vengeance. He captured and burned a number of Moslem ports along the coast, including Goa and Dabul.

1509, February. Battle of Diu. Almeida discovered the allied fleet near Diu. He attacked vigorously, completely destroying the Moslem fleet, then captured and sacked Diu. He returned to Cochin. Not until a new fleet arrived from Lisbon did he accept the Portuguese king's decree to turn his post over to Albuquerque.

1509. Expedition to Malaya. Albuquerque sent **Diego Lopez de Siquiera** to establish a factory at Malacca.

1510. Albuquerque Captures Goa. Seized from the Sultan of Bijapur in a bold, vigorous attack.

1511. Albuquerque Captures Malacca. He stayed there for a year to consolidate this vital outpost controlling the main eastern approach to the Indian Ocean. From Malacca, he later sent expeditions to the Moluccas (1512–1514).

1512, September. Revolt at Goa. Suppressed by Albuquerque upon his return from Malacca.

1513. Expedition to Aden. Albuquerque besieged the port, but was unable to capture it.

1515. Albuquerque Recaptures Ormuz. It remained in Portuguese hands for a century and a half.

1515. Recall of Albuquerque. Unjustly deposed from his post, he died on the way home.

COMMENT. *Albuquerque was apparently the first man to realize that sea power is founded upon bases and upon merchant shipping as much as it is upon the fighting capabilities of a fleet of warships. From the outset he was obsessed with the need for gaining control of all of the major entrances into the Indian Ocean in order to be able to dominate shipping in the broad waters of that ocean. Only at Aden was he foiled in this effort, but he was nevertheless able to control trade between India and the Red Sea from Socotra and his many other bases along the coast of the Arabian Sea. Goa was the key to maintenance of this chain of bases, which were scattered along the coast in such a way as to dominate all the major seaports of western India.*

1518. Portuguese Build Fort at Columbo, Ceylon.

1519. Portuguese in Burma. A trading station was opened at Martaban.

1528. Conquest of Diu. Captured by **Nunho da Cunha**, Portuguese viceroy.

1536–1537. Gujerat-Ottoman Alliance. Sultan Bahadur of Gujerat established an alliance with the Ottoman Turks to eliminate Portuguese control over trade between India and the West. Da Cunha, alarmed by the appearance of a large and powerful Ottoman fleet, entered into negotiations with Bahadur. The Sultan, relying upon the Portuguese reputation for integrity established by Albuquerque, visited da Cunha's flagship, where he and his entourage were seized and treacherously murdered.

1538. Siege of Diu. The Ottoman fleet and a Gujerat army blockaded and besieged Diu, but were repulsed by the Portuguese.

1546. Indian-Ottoman Alliance. Again the rulers of western India and of the Ottomans failed in attempts to capture Diu and other Portuguese posts in India.

1557–1600. Portuguese Conquer Coastal Ceylon. Through guile as well as force, the Portuguese gradually won control of the Ceylonese coast.

1559. Seizure of Daman. This port, on the east side of the Gulf of Cambay, was taken by **Constantine de Braganza.** Opposite Diu, it ensured Portuguese control of the Gulf.

1580. Expedition to Kandy. An overland Portuguese attack from Colombo on the Ceylonese capital was unsuccessful.

1594. Second Expedition to Kandy. Another Portuguese attack on Kandy was repulsed by King **Vimalla Dharma Surya.**

CENTRAL AND EAST ASIA

INNER ASIA

At the outset of this century, Inner Asia was still in a state of flux. In the extreme west the tiny remnant of the Khanate of the Golden Horde in the middle Volga Valley was about to succumb finally to the onslaughts of its former vassals: the Christian principality of Moscow and the Moslem-Tartar khanates of Crimea, Astrakhan, and Kazan. In the vast central and eastern region, between the Ural Mountains and Lake Baikal, the Kalmuk, or Oirat, Mongols were linked together in a loose confederation of four tribes, dominated by the Oirat chieftain **Dayan Khan.** To the south the Kirghiz maintained themselves in the mountains of the Tien Shan, surrounded by the Turkic descendants of the Chagatai Mongols who occupied the lowlands to the east and west in the three independent sultanates of Yarkand (modern Chinese Turkestan), tiny Ferghana, and Khwarezm (or Khiva, or Transoxiana) with capital at Samarkand.

The most vigorous of the nomadic peoples of Inner Asia at this time were the Uzbek Tartars, who had broken loose from the authority of the Golden Horde in the previous century, and who had been slowly gathering strength in the steppes southeast of the Volga and between the Aral and Caspian seas. At the beginning of the century, under their great leader **Mohammed Esh Shaibani Khan,** the Uzbeks swept south and east to overrun the three independent Chagatai khanates. They continued southwest across the Oxus into Persian Khorasan, but Shaibani was defeated and killed by Shah Ismail of Persia in a great battle near **Merv** (see p. 551). The Uzbeks were forced to withdraw from Khorasan, but firmly consolidated their control over the regions now known as Chinese and Russian Turkestan. In a series of bitterly contested wars they repulsed efforts of Babur, former Sultan of Ferghana, to recover his kingdom. Babur conquered a new domain in the Hindu Kush, around Kabul (eastern Afghanistan), from which he later continued southeast to conquer north India and to establish the Mogul Empire (see p. 552).

During the remainder of the century only the Uzbek khanate remained relatively stable in Central Asia. The eastern Kalmuck Mongols disintegrated into insignificant tribal groupings, though the western Kalmucks retained considerable vitality and raided frequently across the Urals into the Volga Valley, which by now had been completely conquered by the new Russian Empire (see p. 537). The principal events were:

1494–1509. Uzbek Conquest of Western and Eastern Turkestan. Principal opponent among the Chagatai Tartars of Turkestan was Babur, Sultan of Ferghana. His ambitions of expanding from Ferghana into Transoxiana precipitated a series of violent clashes for control of Samarkand.

1497. Babur Captures and Loses Samarkand. He was driven out by Shaibani.

1500–1501. Babur Again Captures Samarkand. Shaibani defeated Babur in the **Battle of Sar-i-pul,** then conquered Ferghana, driving Babur south into the Hindu Kush mountains.

c. 1500–1543. Unification of Mongolia under Dayan Khan. He was a descendant of Genghis.

1502. Collapse of the Golden Horde. The Horde was defeated by the allied forces of the Krim Tartars and Muscovites.

1509–1510. Uzbek Invasion of Khorasan. This culminated in the **Battle of Merv;** Ismail of Persia was victorious; Shaibani was killed (see p. 551). The Uzbeks were driven from all of Khorasan, save for the region immediately around Balkh, where Ismail was repulsed.

1511–1512. Babur's Invasion of Transoxiana. Now based in his new kingdom of Kabul, Babur entered an alliance with Ismail of Persia. He quickly recaptured Samarkand, but the Uzbeks, already rallying from their disaster in Persia, defeated him in the **Battle of Ghazdivan,** again driving him back to Kabul (1512).

1526. Babur's Invasion and Conquest of Northern India. (See p. 552.)

c. 1540. Humayun's Conquest of Afghanistan. He took the region from his brother, Komran, after his expulsion from India by Sher Shah (see p. 553).

c. 1550–1597. Reign of Uzbek Abdullah Khan. He revitalized his declining nation to create a strong centralized state in western Turkestan.

1555. Humayun's Reconquest of North India. (See p. 553.)

1597–1600. Breakup of the Uzbek State.

CHINA AND MANCHURA

The Ming Dynasty, declining politically and militarily, was constantly on the defensive against Mongols and Tartars to the north and west, Manchus to the northeast, and Japanese pirates along the east coast. The century closed just after partial Chinese victories repulsed Japanese invasions of Korea. This success, however, was due primarily to Korean prowess at sea and to Japanese overextension on land. During this century, Europeans first appeared in force in East Asia: the Portuguese obtained a settlement on the coast of China, while Russian adventurers penetrated across the Amur River before being driven back by the Manchus. The principal events were:

c. 1514. Portuguese Appearance at Canton. Apparently there had been previous coastal contacts (c. 1411).

1520–1522. First Portuguese Mission to Peking. The Portuguese were expelled from the Chinese capital due to depredations of some of their ships along the coast of China.

1522–1566. Reign of Shih Tsung. China was plagued by pirate raids along the seacoast. The greatest threat was posed by the Mongols of **Altan Khan,** Prince of Ordos, within the great bend of the Hwang Ho (Yellow River; particularly in 1542 and 1550).

1525. Russian Adventurers Cross the Amur. They were repulsed by the Manchus.

1555. Siege of Nanking by Japanese Pirates.

1557. The Portuguese Establish a Base at Macao.

1567–1620. Reign of Shen Tsung. This was a period of considerable cultural development in China, but of accelerated military decline. China's borders were steadily eroded to the west and north.

c. 1560–1626. Rise and Consolidation of the Manchus. They were the descendants of the Juchen or Chin People (see p. 355), whose home was in the valley of the Sungari, in present central Manchuria. By the end of the century their leader, **Nurhachi,** had begun expansion of his territory to the Yellow Sea and to the Amur.

1592–1598. Japanese Invasions of Korea. The Chinese assisted the Koreans in repelling the invaders (see p. 559).

KOREA

During most of this century Korea was plagued by persistent coastal raids of Japanese pirates. Internally there were a number of minor disturbances; central authority declined steadily. With little central direction or coordination of defensive efforts against the Japanese, the military and naval contingents of the various provinces became almost autonomous and developed considerable combat skill. This stood the nation in good stead when Japanese dictator Hideyoshi decided to invade and conquer China by way of Korea. With some Chinese assistance, but particularly due to their superiority at sea and the genius of Admiral Yi Sung Sin, the Koreans repelled two Japanese invasions. The principal events were:

1592. Japanese Conquest of South and Central Korea. The Koreans, faithful vassals of the Ming Dynasty, refused free Japanese passage to China; Hideyoshi sent an army across the Straits of Tsushima to assault and capture Pusan (May 1592). Advancing north, the Japanese captured Seoul and Pyongyang (June–July).

1592, July. Battle of the Yellow Sea. Yi Sung Sin, commanding the Cholla provincial navy, probably assisted by other provincial flotillas, met a Japanese fleet carrying reinforcements to the Japanese army at Pyongyang. Yi's fleet included at least two "tortoise ships," low-decked ironclad galleys of his own design. Apparently almost singlehanded these novel warships, the first ironclad vessels in history, won a smashing victory over the Japanese fleet, at least 59 Japanese ships being sunk or burned. The reinforcement convoy was scattered and destroyed.

1592–1593. Chinese Intervention. This was at the request of the Korean government (October 1592). As the Chinese advanced south, the Japanese, short of supplies, without reinforcements, and constantly harassed by Korean guerrillas, withdrew slowly. After a long, dull campaign, the Japanese entrenched themselves in a relatively small beachhead perimeter around Pusan (October 1593).

1594–1596. Inconclusive Peace Negotiations. The Chinese evacuated Korea, leaving the Koreans to observe the Japanese Pusan perimeter. There were a number of minor skirmishes. Breakdown of negotiations led Hideyoshi to send reinforcements to Korea.

1597–1598. Japanese Offensive. They broke out of their perimeter to advance northward. The Chinese sent another army to the assistance of the Koreans, but the allies, defeated in a number of battles, were unable to halt the steady Japanese advance through south Korea.

1598, November. Naval Battle of Chinhae Bay. Admiral Yi Sung Sin attacked the Japanese fleet and won another great victory, in which he himself was killed. Nearly half of the Japanese fleet of 400 ships was sunk; the survivors fled to Kyushu.

1598, December. Armistice. With their line of communications to Japan again cut, and Hideyoshi having died in Japan, the Japanese made peace and evacuated Korea.

JAPAN

The violence which had split Japan into a land of independent warring feudal nobles continued into the early part of this century. Reunification began, however, shortly after the middle of the century with the emergence of the strong warrior **Oda Nobunaga.** Systematically, he began the unification of Honshu by force. With the support of the emperor, he overthrew the shogun and made himself dictator. He was assisted in his unification efforts by the young **Tokugawa Ieyasu** and his principal subordinate, General **Toyotomi Hideyoshi,** a commoner.

Nobunaga died during the revolt of a vassal, and was succeeded by Hideyoshi, who—assisted by Ieyasu—completed the unification and pacification of Japan. Hideyoshi then initiated a grandiose plan for the conquest of Korea, China, and most of Asia. He was frustrated, however, by the unexpected vigor of Korean resistance, particularly at sea, combined with the support of a large Chinese army. The principal events were:

c. 1542. Arrival of the Portuguese in Kyushu. The Portuguese introduced the musket, which soon modified Japanese warfare.

1549–1587. Expansion of Christianity in Japan. This was stimulated by Portuguese missionaries, of whom **St. Francis Xavier** was the first.

1560–1568. Nobunaga Begins the Reunification of Japan. He established himself in control of most of Honshu. He installed **Ashikaga Yashiaki** as shogun.

1571–1580. Overthrow of the Buddhist Monasteries. These had become militarily powerful. Nobunaga systematically defeated and destroyed them.

1573. Overthrow of Yashiaki. With secret support from the emperor, Nobunaga deposed the shogun, who was plotting against him.

1576. Nobunaga Begins Era of Castle Building.

1577–1582. Rise of Hideyoshi. As Nobunaga's principal general, he conquered most of western Japan from the Mori family.

1582–1584. Civil War. This was provoked by the revolt of General **Akechi Mitsuhibi.** Nobunaga, surprised and surrounded in his capital, Kyoto, committed suicide (1582). Hideyoshi defeated and killed Mitsuhibi (1582). With the assistance of Tokugawa Ieyasu, he defeated the Oda family to gain control of all central Japan.

1585–1590. Hideyoshi Unifies Japan.

1592. Hideyoshi's Plan of Conquest. He planned to take over all of East Asia. China was to be invaded through Korea and conquered in two years, at which time the Japanese court would move to Peking.

1592–1598. Korean War. (See p. 559.)

1598, August. Death of Hideyoshi.

1598–1600. Power Struggle. The leading vassals of Hideyoshi contended for control. Ieyasu was successful (see p. 650).

SOUTHEAST ASIA

BURMA

During the first half of this century, the long warfare between the Shan and Burman states came to a climax in the conquest and sack of Ava by the ruler of the Shan state of Mohnyin (1527). While the Shans gained control of northern and central Burma, the Burman rulers of Toungoo made a bid for the overall leadership of Burma. The first attempt to reunite Burma was made by **Tabinshwehti** (1531–1550), who planned to deal with the Shans only after he had conquered the rich Mon Kingdom of Pegu. He was successful in this (1535–1541), but then dissipated his efforts in fruitless attempts to conquer Arakan and Siam (1547–1548). A Mon revolt brought his reign to an end (1550). Burma was only rescued from chaos by his foster brother, **Bayinnaung** (1551–1581), whom one historian has called "the greatest explosion of human energy ever seen in Burma." A special feature of Bayinnaung's wars was the participation of Portuguese mercenaries. In a period of 30 years Bayinnaung reunited Burma, destroyed the power of the Shans, expanded Burma to approximately its modern frontiers, and conquered the three Tai states of Chiengmai, Ayuthia, and Laos. His son **Nanda Bayin** had to face a strong movement of Siamese national independence led by **Pra Naret,** who threw off the Burmese yoke (1587) and became King **Naresuen** (1590–1605). Nanda Bayin repeatedly invaded Siam, but failed to reestablish control. Naresuen's counterattacks upon Burma caused Nanda Bayin's downfall and murder (1600). The Arakanese invaded Lower Burma, and the Upper Burma principalities threw off their allegiance. Burma relapsed into chaos for some years.

SIAM

During the first half of this century the two principal Tai kingdoms—Ayuthia (or Siam) and Chiengmai—were almost constantly at war with each other, save for one brief respite of about 5 years. In the second half of the century these internal Tai struggles were merged into, and overshadowed by, the Burmese wars of conquest. Both countries, as well as the related Tai peoples of Laos, were conquered by Burmese King Bayinnaung. After his death, however, first Laos, then Ayuthia, and finally Chiengmai broke loose from Burmese control. By the end of the century the Siamese, under strong and able King Naresuen, had conquered substantial portions of southeast Burma, and had gained unchallenged suzerainty over Chiengmai. The principal events were:

1491–1529. Reign of Rama T'ibodi II. He received the first Portuguese envoy to Siam, **Duarte Fernandez,** conqueror of Malacca. He signed treaties with Fernandez, giving the Portuguese certain commercial rights in Siamese ports.

c. 1500–1530. War with Chiengmai. Smaller Chiengmai maintained the initiative, making repeated raids and invasions of Ayuthian territory, particularly against Sukhot'ai. Ayuthia undertook repeated punitive expeditions against Chiengmai. Because of this constant warfare, Rama T'ibodi created a comprehensive military system based on compulsory military service and regional military areas. The most important single action of this prolonged war was his victory in the **Battle of the Mewang River,** near Nakhon Lamp'ang. T'ibodi then ejected Chiengmai troops from Sukhot'ai.

1548. Burmese Invasion of Siam Repulsed.

1563–1584. Burmese Invasions, Conquest, and Occupation of Siam. (See p. 560.)

1571. Return of Prince Pra Naret to Siam. A hostage in Burma, Bayinnaung sent him back to serve as a puppet in Burmese-occupied Siam.

1584–1587. Revolt of Pra Naret. The Burmese were expelled from Siam (see above).

1587. Invasion by King Satt'a of Cambodia. This diversion permitted the defeated Burmese army to escape pursuit at the hands of Pra Naret. He drove the Cambodians out, and pursued as far as their capital, Lovek.

1590. Pra Naret Ascends Throne as Naresuen.

1590–1600. Continuous War between Siam and Burma. Naresuen repelled invasions of his country by Nanda Bayin of Burma, then turned to invade Burma, capturing Tavoy and Tenasserim (1593), and then Moulmein and Martaban (1594).

1593–1594. Invasion of Cambodia. Naresuen captured Lovek after a desperate struggle, and established Tai overlordship.

1595. Laotian Invasion of Chiengmai. Tharrawaddy, son of Bayinnaung, who had installed him as ruler of Chiengmai, became a vassal of Naresuen in return for Siamese help in driving out Laotian invaders.

1595. Naresuen Invades Burma. He was repulsed from Pegu.

1599–1600. Naresuen Again Invades Burma. He took advantage of the collapse of Nanda

Bayin's kingdom. The Siamese occupied Pegu, which had earlier been destroyed during the civil war in Burma, but were repulsed from Toungoo when the Burmese rebels, having overthrown Nanda Bayin, closed ranks against the Siamese invaders (see p. 561).

LAOS

During the first half of the century Laos had on occasion become involved in the disputes between her sister Tai states of Ayuthia and Chiengmai. A few years later, when the able Laotian ruler **Sett'at'irat** (1547–1571) unsuccessfully attempted to eject the Burmese forces of Bayinnaung from Chiengmai (1558), he gained the implacable enmity of the great Burmese conqueror. Twice Burmese armies overran Laos (1564–1565 and 1572–1573), the first time capturing the capital of Vien Chang. Determined Laotian guerrilla resistance each time forced the Burmese to withdraw. But Sett'at'irat's frequent efforts to assist Ayuthia against the Burmese, and to try to drive the Burmese from the Tai and Shan states, were invariably crushed by Bayinnaung. Sett'at'irat died during the prolonged war (1571) and Bayinnaung finally conquered the country (1575). But after his death Laos threw off the Burmese yoke and established its independence (1592). Later Laotian efforts to recapture Chiengmai were repulsed by the Siamese (1594).

VIETNAMESE REGION

This was a period of general decline in Annam, the result of weak rulers and endemic civil wars. Before the middle of the century Tonkin and Annam had broken apart, though Tonkin was still nominally subject to the impotent rulers of Annam. Soon after the middle of the century the southern provinces—former Champa—had also become virtually independent. Late in the century the dictator **Trinh Tong** of central Annam conquered most of Tonkin, to reunite more than two-thirds of the country under the strongest and most stable rule Vietnam had during this century (1592).

MALAYA AND INDONESIA

The principal feature of the history of Malaysia during this century was the Portuguese capture of Malacca, and their subsequent establishment of fortified bases in the Spice Islands and intermediate islands of Indonesia, in the early part of the century. The remainder of the century was a constant and unsuccessful struggle on the part of the predominantly Moslem rulers of Malaya and Indonesia to eject the Portuguese. Subsidiary to this primary struggle was the continuing expansion of Islam in the archipelago. The principal events were:

1511. Conquest of Malacca by Albuquerque. (See p. 556). Sultan **Mahmud** of Malacca escaped and set up a new capital on the island of Bintang, from which he incessantly endeavored to regain Malacca.

1513–1521. Portuguese Bases in the Spice Islands. This was facilitated by intermittent warfare between the sultans of Ternate and Tidor.

1517–1520. Alliance of Acheh and Bintang. The Sumatran Sultanate of Acheh assisted Mahmud of Bintang in unsuccessful efforts to recapture Malacca.

c. 1520. Final Collapse of Majapahit. It fell

under attacks from a coalition of Moslem states: Madura, Tuban, Surabaya, and Demak. Demak became the most important of the many small states on Java.

1521. Magellan Claims the Philippines for Spain. He was killed in a battle with the natives on the island of Mactan.

1521–1530. Portuguese-Spanish Friction. This came from conflicting claims to the Spice Islands.

1525–1526. Rise of the Moslem Sultanate of Bantam. Bantam conquered Sunda Kalapa, where the Portuguese had obtained trading rights, and renamed the city Jacatra (Djakarta). The hostile Moslems refused the Portuguese permission to establish a factory.

1526. Portuguese Capture of Bintang. Acheh became the leading state among the Portuguese enemies around Malacca.

1529–1587. Repeated Achenese Attacks on Malacca.

1535–1600. Portuguese Wars in Northern Java. The small Moslem states were steadily weakened by consistent defeats at the hands of the Portuguese.

1550–1587. War with Ternate. The Portuguese murdered their archenemy Sultan **Hairun** of Ternate (1570), causing the war to intensify under his successor **Baabullah,** who sought revenge. Baabullah captured the Portuguese fort on Ternate and massacred the garrison after a long siege (1570–1574). The Portuguese on Malacca were unable to send assistance, since they were at this time subjected to intensive attacks by the Achenese and the Moslems of Java.

1558. Achenese Siege of Malacca. A force of 15,000 Achenese, supported by 400 Turk artillerymen, were repulsed by the Portuguese after a siege of a month.

1568–1595. Expansion of Bantam. The Sultanate controlled all of western Java.

1570. Spain Begins to Colonize the Philippines. Miguel López de Legaspi began the conquest. He established the Spanish headquarters at Manila (1571).

1570–1580. Intermittent Spanish-Portuguese Warfare. This was for control of the Philippines and neighboring Indonesian islands. It was concluded by Spanish annexation of Portugal (see p. 533), though the two colonial systems remained separate.

1574. Javanese Attack on Malacca. The Moslem attackers were driven off only by the arrival of Portuguese reinforcements from Goa.

1595. Arrival of a Dutch Fleet in the Spice Islands. This was the beginning of colonial rivalry with the Portuguese.

AFRICA

North Africa

During the early part of this century, Portuguese and Spanish influence became widespread over North Africa. Both nations soon found themselves directly challenged, however, by the expanding power of the Ottoman Empire, spearheaded by **Khair ed-Din** (Barbarossa), Dey of Algiers. The struggle between Suleiman and Charles V for control of the Mediterranean was largely centered on the coast of North Africa. The Ottomans were successful, and by the end of the century most of the Iberian footholds in North Africa had been eliminated. The principal events were:

1509. Spanish Capture of Oran and Mers El Kebir.

1510. Pedro Navarro Captures Tripoli for Spain.

1517. Ottoman Conquest of Egypt. (See p. 540.)

1518. Rise of Khair ed-Din. He and his brother **Harush,** Moslem Greeks, became leaders of a

North African pirate group; they defeated and expelled the Spanish from Algiers, Harush being killed in the battle for the city.

1518–1546. Reign of Khair ed-Din, Dey of Algiers. (See p. 545.)

1533. Khair ed-Din Ravages Sicilian and Italian Coasts.

1534. Ottoman Capture of Tunis. Khair ed-Din, acting as admiral of the Ottoman navy, captured Tunis from **Mulai-Hassan** of the decadent Hafsid Dynasty.

1535. Charles V Conquers Tunis. (See p. 545.)

1541. Charles V Repulsed from Algiers. (See p. 545.)

1574. Expulsion of the Spanish from Tunis. (See p. 549.)

1578. Battle of Al Kasr al Kebir. King **Sebastian** of Portugal was defeated and killed while intervening in a Moroccan dynastic dispute (see p. 534).

1580. Spanish Occupation of Ceuta.

1590. Janissary Revolt in Tunis. They seized power, while retaining nominal allegiance to Constantinople.

1591. Moroccan Expedition to West Africa. The force, composed predominantly of Spanish and Portuguese mercenaries, crossed the western Sahara to inflict a crushing defeat on the Songhai Empire. Gao was destroyed, Timbuktu temporarily occupied, and the country generally devastated. (See p. 566.)

EAST AFRICA

The history of East Africa during this century was primarily a struggle between the Coptic Christians of Ethiopia and the Moslem coastal tribes of Somalia, in which Portugal became involved. The principal events were:

1502. Portuguese Compel Tribute from Kilwa.

1505–1506. Portuguese Sack Kilwa and Mombasa. They also struck at Pate, Brava, and Lamu.

1520. Portuguese Embassy to Ethiopia. This began a Portuguese-Ethiopian alliance, and helped offset Ottoman influence in the Horn of Africa.

1523–1543. Revolt of Harer. Moslems in this area revolted against the Ethiopian King **Lebna Dengal,** and immediately gained massive Arabian and Ottoman support.

1528. Second Portuguese Sack of Mombasa.

c. 1540–1600. Expansion of Galla and Somalis. The Somalis displaced the old Arabic Moslem states in the Horn of Africa, ruined by the strains of the war effort directed against Ethiopia (1525–1543) by **Ahmad Gran** (Ahmad ibn Ibrahim al-Ghazi) the "Left-Handed," of Adal. The pagan Galla spread into most of Ethiopia and became especially prominent in the southern highlands, where they conquered or displaced the former inhabitants.

1541–1543. Portuguese Intervention. A small but well-armed Portuguese force under **Cristovao da Gama** (son of Vasco da Gama) landed at Massawa to aid the Ethiopians. Da Gama helped them raise the siege of Kassala, and decisively defeated the Moslem-Arab coalition led by Ahmad Gran at the **Battle of Waim Dega** (February 21, 1543), where Gran was killed.

c. 1550–1600. Rise of Buganda. This state, a rival to the Bunyoro Kingdom (see p. 486), grew to great power by the mid-18th century.

1559. Death of King Galawdewos (Claudius). This Ethiopian king, who had succeeded Lebna Dengal (1540), was killed in battle against a renewed assault by the Harer rebels.

1563–1597. Reign of Sartsa Dengal. This able warrior-king mounted several counteroffensives against Ottoman incursions in Eritrea, winning notable victories in 1578 and 1589.

1567. Sack of Harer. As part of their migrations, a Galla army laid waste this city.

1585–1589. Ottoman Intervention in East Africa. In 1585, an Ottoman fleet moved south down the Indian Ocean coast, promising aid if towns revolted against the Portuguese. Much of the coast north of Pemba rose, but a Portuguese fleet sent from Goa restored order (1587). The Ottomans mounted another effort but met with little success (1588). The Portuguese and their Simba allies sacked Mombasa a third time (1589). Following these events, the Portuguese built massive Fort Jesus at Mombasa (1593–1594). Despite Fort Jesus, the Portuguese presence in East Africa was small, never amounting to more than 1,000 men, and Portugal exercised little direct control.

1587. Destruction of Sofala. This city, enriched by the gold trade, had its populace massacred by the marauding Simba tribe, originally from the Zambezi Valley.

WEST AFRICA

Three principal Negro empires flourished in West Africa during this century. The Songhai Empire was still powerful, though declining, largely due to defeat by the Hausa Confederation, under **Kebbi.** The Hausas became dominant east of the Niger River. To the northeast the empire of Kanen, or Bornu, expanded around Lake Chad. During the latter part of the century it flourished under Emperor **Idris III** (1571–1603).

c. 1504–c. 1530. Reign of Esigie of Benin. The son of Ozolua (see p. 486) continued his father's policies and also enjoyed good relations with the Portuguese.

1505–1506. Conquest of the Tuareg of Aïr. Askia Muhammad (see p. 487) subjugated these people, settled around modern Agadez, Nigeria, in two campaigns.

1506–1543. Rule of Afonso I of Kongo. In addition to adopting a Christian name (like his father and predecessor, John I) and faith, this ruler encouraged the Christianization of his subjects.

1507–1514. Songhai War with the Fulani and the Borgu. This was directed against the Fulani in the Niger Valley, and the Borgu along the Niger-Nigeria border; the Songhais met with some success but failed to conquer their opponents.

1512. Songhai Conquest of Diara. This area in modern northwestern Nigeria was subjugated between Askia Muhammad's other campaigns.

1516–1517. Revolt of the Karta of Kabi (Kebbi). The Karta, one of Askia Muhammad's chief lieutenants, led a revolt, which was eventually suppressed by loyal troops.

1528. Songhai Succession Crisis. Askia Muhammad's eldest son **Askia Musa** deposed his father and banished him (now nearly blind) to an island in the Niger. **Ismail** (a younger son of Askia Muhammad) brought him out of internal exile, and he died soon after (1538).

1540–1591. Decline of the Songhai Empire. Continuing succession struggles and internal unrest gradually sapped the empire's strength.

1550. Hausa Conquest of Oyo. The Hausa states of Borgu and Nupe conquered the small state of Oyo to their south.

c. 1571–1603. Reign of Idris Alawma of Kanem-Bornu. Under this able monarch, Kanem-Bornu reached its greatest power, and had great influence over the easternmost Hausa states.

1570–1571. Attacks on Kongo. Raids by the Jaga warrior tribe of Angola compelled the Kongo monarch to appeal to Portugal for aid. The Portuguese sent a small army from Sao Tomé (1571), and after some hard campaigning drove out the Jaga.

c. 1590. Oyo Independence. The Oyo *alafin* (king) **Orompoto** threw off Hausa control. Following the example of his former suzerains, he created an effective cavalry army. The Oyo state swiftly became a major regional power, with its capital at Old Oyo (Katunga).

1591. Moroccan Invasion of Songhai. A Moroccan army 4,000 strong, composed largely of Spanish and Portuguese mercenaries equipped with firearms (hitherto unknown in inland West Africa) invaded the Songhai Empire and decisively defeated its armies. The invaders sacked Timbuktu and destroyed the Songhai capital of Gao, thereby ending the Songhai Empire. (See p. 564.)

THE AMERICAS, 1492–1600

This was the century of the *conquistadores,* following the discovery of America by Columbus. Towering over all of the others as a military man was **Hernando Cortez,** who conquered Mexico for Spain. Equally bold, perhaps, but an unprincipled adventurer who lacked the military and administrative genius of Cortez, was **Francisco Pizarro,** who conquered the Inca Empire of Peru. Struggles among the Spanish were almost as frequent as were their wars of conquest against the Indians. The principal events were:

CARIBBEAN REGION

1492, October 12. Discovery of America by Christopher Columbus. He explored numerous Caribbean islands, including Hispaniola (Santo Domingo), all of which he claimed for Spain.

1493. Columbus' Second Voyage. He established a permanent colony on Hispaniola, as the first Spanish Viceroy of the New Indies.

1493–1502. Explorations, Discoveries, and Colonization by Columbus. He discovered many more islands, the coast of South America, and Central America, all of which were claimed for Spain. Spanish adventurers rushed to the New World.

1494, June 7. Treaty of Tordesillas. This pact between Spain and Portugal established a line of demarcation 370 leagues west of the Cape Verde Islands; Spain was to have exclusive rights of exploration and colonization west of the line, Portugal exclusive rights east of the line. This modified Pope Alexander VI's original line, 100 leagues west of the islands (1493).

1508–1511. Conquest of Puerto Rico. A Spanish expedition from Hispaniola defeated the Carib Indians.

1511–1515. Conquest of Cuba. The leader was **Diego de Velázquez,** governor of Hispaniola.

1521–1535. Colonization of South America. The Spanish gained footholds along the coast of Venezuela and New Granada (Colombia).

1522–1523. Conquest of Nicaragua. The leaders were **Gíl González Dávila** and **Alonzo Niño.**

1523–1530. Spanish Squabbles. Conflict between Dávila and **Francisco Hernandez de Córdoba,** a subordinate of **Pedro Arias de Ávila,** the governor of Darien. Dávila defeated Córdoba, but in turn was defeated by another subordinate of Ávila's, **Cristobal de Olid,** thus permitting Ávila to gain control of Nicaragua.

1536–1538. Conquest of the Chibcha Indian Empire. Gonzalo Jiménez de Quesada took this region centered around modern Bogotá.

1585–1586. Drake's Raids. An English expedition under Sir Francis Drake sacked Santo Domingo, Cartagena, and St. Augustine (see p. 508), and generally terrorized the West Indies.

MEXICO

1502–1520. Reign of Moctezuma (Montezuma II) Xocoyotl ("the Younger"). He campaigned, mainly to the east and southwest of Mexico City. Due to distance and logistical

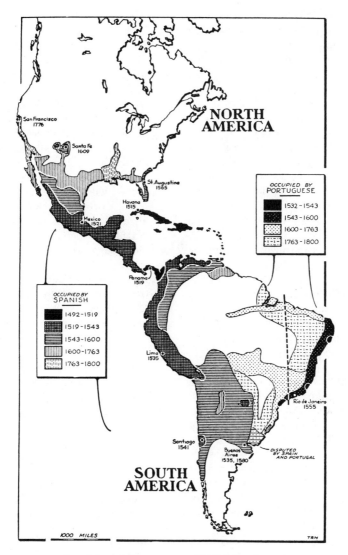

OCCUPIED BY
PORTUGUESE

■	1532 - 1543
▨	1543 - 1600
░	1600 - 1763
▤	1763 - 1800

OCCUPIED BY
SPANISH

■	1492 - 1519
▨	1519 - 1543
▦	1543 - 1600
▥	1600 - 1763
▨	1763 - 1800

1000 MILES

Spanish and Portuguese colonial empires

difficulties, he enjoyed mixed success. Aztec political miscalculations ensured that former allies were less willing to assist them.

1515–1519. Tlaxcala Campaigns. Several expeditions were sent against Tlaxcala, long-time rival in the east, but none were decisive. There were strains in the Three-City League due to the dominance of Tenochtitlan. Montezuma

was indecisive in dealing with these problems, relying for advice on sorcerers and magicians. He was further perturbed on receipt of word that gods had appeared in floating temples far to the east on the Gulf of Mexico. This was the Spanish fleet of Hernando Cortez.

1518–1519. Expedition of Hernando Cortez.
He was nominally under the command of

Diego de Velázquez. In defiance of Velázquez' orders he led an expedition of 600 men, 17 horses, and 10 cannon from Hispaniola.

1519, August. Cortez Burns His Ships. After burning his fleet to prevent desertion, Cortez established Vera Cruz, subjugated the independent Mexican Indian Kingdom of Tabasco, and opened negotiations with Montezuma, Emperor of the Aztecs.

1519, September. Invasion of Central Mexico. After establishing an alliance with the Totonac tribe, Cortez invaded the interior of Mexico, where he overcame the independent state of Tlaxcala, which then allied itself with him against the Aztecs. From Tlaxcala he marched on Tenochtitlán. Despite a few skirmishes en route, the Spanish were permitted by Montezuma to enter the city as friends (November 8, 1519).

1519, December. Cortez Seizes Montezuma. This gave him virtual control of Tenochtitlán.

1520. Expedition of Pánfilo de Narváez. He was sent by Velázquez with a force of 1,500 men to subdue and punish Cortez for disobedience of orders. Cortez, learning of the arrival of this expedition at Veracruz, left **Pedro de Alvarado** in command at Tenochtitlán, marched to the coast with part of his force, defeated Narváez, and enlisted the survivors of Narváez' command. Cortez then marched back to Tenochtitlán, where Alvarado's excessive harshness had caused an Aztec revolt against the Spanish.

1520, June 30. Evacuation of Tenochtitlán. After hard fighting Cortez withdrew after nightfall, suffering heavy losses in the process. During the confusion Montezuma was murdered by the Spanish. Cortez retreated west across Lake Texcoco over a narrow causeway to Chapultepec but was spotted by the Aztecs. A furious night battle developed, and the Spanish were attacked from all sides. Unable to bring their cavalry or firepower to bear on the Indians in canoes and on the narrow causeway, Cortez suffered many casualties, particularly among his Indian allies. The Aztecs failed to follow up on their victory, content to sacrifice captured prisoners (and horses) to the war god,

Huitzilopochtli. After this "night of sadness," Cortez reached the north end of the lake.

1520, July 7. Battle of Otumba. Attacked by the Aztecs, Cortez fought another desperate running battle, which he finally won decisively.

1520, August–1521, May. Preparations for Renewed Campaign. Cortez withdrew to Tlaxcala and made extensive preparations to retake Tenochtitlan. Realizing that naval power was the key to taking the island city, Cortez established a base at Texcoco and began to build a fleet of 13 vessels, brigantines each armed with a small cannon. Meanwhile, Emperor **Cuitlahuac** (Montezuma II's successor) died of smallpox, a disease brought by the Spanish, which was devastating the Indians. Former Aztec allies, sensing Aztec defeat, switched allegiance to the Spanish.

1521, May 31–August 14. Seige of Tenochtitlan. After making thorough preparations, Cortez invested the city. He first cut off all supplies. His European forces, approximately 1,000 soldiers, less than 100 horses, and 18 cannon, were formed into 4 contingents: 3 land units and a group to man the 13 brigantines on the lake. There were not enough firearms for all the European soldiers, so many were equipped with bows and arrows, pikes, and swords. There were also several thousand Indian allies. The land forces attacked respectively south, east, and west into the city, closely and effectively supported by the brigantines, which cleared the lake of Aztec war canoes and provided firepower at critical points during the assault. Against desperate resistance, the Spanish and their allies slowly fought their way into the heart of the city. The Aztec empire had lasted 152 years; it ended in the smoking ruin of the city. Emperor **Cuahtemoc** (who had succeeded Cuitlahuac) was captured (and hanged 3 years later in Guatemala). Perhaps 100,000 people died in combat, or from disease or starvation during this campaign. Cortez established a new Mexico City atop the ruins of the old one.

COMMENT. *The siege of Tenochtitlan was one of the most successful sieges of history. Cortez proved himself one of the great generals of his era, capturing an empire at small cost in a well-executed plan. The victory assured Spanish mastery of much of the New World. The wealth of the region was plundered and brought back to Spain, contributing to Spanish pre-eminence in Europe and the world for over a century.*

1522–1539. Conquest of South Mexico and of Northern Central America. Cortez advanced as far as Salvador. Mayan resistance was particularly violent in Honduras.

1524–1526. Expedition into Honduras. Cortez failed to achieve lasting results.

1524–1555. Conquest of North Mexico.

1529. Revolt of Tabasco. This was suppressed by **Francisco de Montejo,** successor of Cortez.

1539. Subjection of the Mayans. Montejo was successful after repeated campaigns.

1540–1542. Coronado's Expedition. Francisco Vázquez de Coronado, with 350 soldiers, horses, cannon, and Indian slaves and allies, as well as a large livestock herd for food, marched north from Mexico in search of the fabled Seven Cities of Gold. Near **Zuni Pueblo** (New Mexico) Coronado fought the first recorded military action in the American southwest, repulsing an Indian attack and then capturing the pueblo (July 1540). At first welcomed by Indians, the expedition wintered at Tigeux Pueblo near modern Albuquerque. As supplies ran low, the Spanish commandeered food, resulting in several clashes with the natives.

1546. Mayan Revolt. This was cruelly suppressed.

1550–1600. Extensive Exploration and Conquest. Expeditions ranged along the Gulf and Pacific coasts to Florida and to California.

1565. Founding of St. Augustine. It was established by **Pedro Menéndez de Avilés** as a base of operations against the recently established French colony on the St. Johns River (see p. 570). He then captured the French post, massacring the garrison. The French threat to Spanish Florida was eliminated.

1598, November 8. Battle of Acoma Pueblo. Indians in modern New Mexico, resentful of harsh treatment, attacked a party of Spanish soldiers. After reinforcements arrived, the Spanish took back the pueblo in a hard-fought battle.

PERU AND THE WEST COAST

1500–1530. Power and Prosperity of the Inca Empire.

1531. Expedition of Francisco Pizarro. He had earlier explored along the northwest coast of South America. He now sailed south from Darien with an expedition of 180 men, 27 horses, and 2 cannon. He landed at Tumbes (San Miguel). Here he waited for reinforcements. He then advanced inland with a force of 62 cavalry and 102 infantry.

1532, November 16. Seizure of Atahualpa. Pizarro seized the Inca emperor by treachery at Cajamarca.

1533, August 29. Murder of Atahualpa. Despite payment of a ransom for his release, Atahualpa was murdered by the Spanish, who then established control over the territory of the Inca Empire.

1535–1536. Inca Revolt. During the absence of Pizarro, the Inca leader **Manco** revolted. The Incas besieged the Spanish in Cuzco, but were repulsed.

1537–1548. Civil Wars among the Spanish. These began with a revolt of **Diego del Almagro** against Pizarro. After some initial success, Almagro was defeated and executed (1538). In subsequent unrest and violence Pizarro was also killed (June 26, 1541).

1540–1561. Conquest of Chile. This was begun by **Pedro de Valdivia,** and was marked by a succession of wars against the warlike Aracanian Indians. After Valdivia was killed in battle (1553), the conquest was continued by **García Hurtado de Mendoza.**

1546. Royal Intervention. Charles V appointed **Pedro de la Gasca** as his viceroy in Peru. Gasca, assisted by Valdivia, defeated **González Pizarro** (younger brother of Francisco) in the **Battle of Xaquixaguana** (1548), which resulted in partial restoration of royal authority in Peru. The younger Pizarro was executed.

1557–1569. Continued Violence among the Spanish in Peru.

1569–1581. Royal Authority Reestablished. Violence was suppressed by Viceroy **Francisco de Toledo.** (See p. 507.)

1579. Drake's Raids along the Peruvian Coast. (See p. 507.)

1587–1588. Cruise of Thomas Cavendish. His squadron of three English ships captured 20 Spanish galleons in the eastern Pacific.

PORTUGUESE AMERICA, 1494–1600

1494, June 7. Treaty of Tordesillas. This established Portuguese rights to the eastern tip of modern Brazil (see p. 566).

1500–1520. French and Portuguese Rivalry. Their expeditions explored the coast of northeast South America, with several armed clashes.

1521–1555. Portuguese Expansion. The coast of modern Brazil was systematically colonized.

1555. French Colony. This was established at Rio de Janeiro.

1556–1557. Portuguese Attack on Rio de Janeiro. Mem De Sa destroyed the French colony.

FRENCH AMERICA

1555–1557. Colony in Brazil. (See above.)

1564. Colony in Florida. Jean de Ribaut and **René de Laudonnière** established Fort Caroline at the mouth of the St. John's River.

1565. Spanish Destruction of Fort Caroline. The Spanish were led by Avilés (see p. 569).

1567. Raid on Fort Caroline. Chevalier de Gourgues temporarily seized the fort, massacring the Spanish garrison in revenge for Avilés' massacre 2 years earlier. Realizing their inability to cope with the Spanish in Florida, the French then abandoned the fort.

XIV

THE BEGINNING OF MODERN WARFARE:

1600–1700

MILITARY TRENDS

In the 17th century the transition from the Middle Ages to the Modern Era was completed insofar as military weaponry, tactics, and organization were concerned. As the period opened, the musket and the pike were complementary rivals in land warfare; basic battle formations differed little from the Greek phalanx of two millennia earlier; the armored horsemen of the gentry had still not accepted the fact that gunpowder had destroyed chivalry; and artillery was essentially immobile. At the close of the period, the pike had practically disappeared, infantry was fighting in the linear formations which would persist into the 20th century, and mobile artillery had become a major combat arm in coordination with infantry and cavalry. Furthermore, standing armies in the modern sense of the term had come into being, with organizations and hierarchical ranks comparable to today's tables of organization.

One individual—King **Gustavus II Adolphus** of Sweden—was mainly responsible for most of these changes. Their acceptance throughout Europe was facilitated by a series of wars—the Thirty Years' War, the War of the Grand Alliance, and the Dutch War—which transformed Europe into a vast, bloody proving ground.

On the high seas also, tactical concepts crystallized—to change relatively little in technical detail for two centuries to come. England's **Robert Blake** deserves most of the credit.

MILITARY LEADERSHIP

Gustavus Adolphus was the one great captain of this century. There was, however, a host of eminent leaders, of whom perhaps the best were Holland's **Maurice of Nassau,** France's exceptionally gifted **Henri de La Tour d'Auvergne, vicomte de Turenne,** and the brilliant Manchu Emperor **K'ang-hsi.** Although the strategical genius of France, Cardinal Duke

571

Armand Jean du Plessis, cardinal-duc de Richelieu was not a soldier, his understanding and use of military strategy were unexcelled.

Among other outstanding soldiers of the century were the Belgian **Johan Tserclaes, comte de Tilly,** the Czech **Albrecht von Wallenstein,** France's **Louis II, prince de Condé, Francois Henry de Montmorency, duc de Luxembourg,** the great military engineer **Sébastien Le Prestre, marquis de Vauban,** England's **Oliver Cromwell,** Italy's Prince **Raimondo Montecuccoli,** the Maratha **Sivaji,** the Manchu **Nurhachi,** and Japan's **Tokugawa Ieyasu.**

At sea England's Blake was outstanding, though rivaled by such men as **George Monck** (soldier turned sailor) and **William Penn.** But Holland's **Maarten H. Tromp,** his son **Cornelis,** and **Michael A. de Ruyter,** and France's **Jean Bart** and **Anne Hilarion Tourville** were worthy opponents of England's best.

SMALL ARMS

The transformation of the musket was significant. Gustavus Adolphus found it a clumsy weapon weighing from 15 to 25 pounds, and fired from a forked rest. He trimmed the weight to 11 pounds. He also adopted, if he did not invent, the cartridge—a fixed charge, with powder carefully measured (affording ballistic uniformity) and the ball attached. The result was a lighter, handier weapon, easier to load, with rate of fire doubled to one round per minute.

Early cartridge

The snaphance (or snaphaunce) lock, deriving a spark from flint on steel, had been introduced during the 16th century. The true **fusil,** or flintlock musket, invented by French gunsmith **Le Bourgeoys** in 1615, was perfected as a sporting weapon about 1630. However, adoption as a military weapon was slow, partly because of increased cost of manufacture, and partly because of traditional conservatism of military leaders satisfied with the matchlock. France armed one regiment entirely with the flintlock musket in 1670; by 1699, it was standard in European armies.

A plug bayonet, inserted into the muzzle of the musket, was widely in use by midcentury as partial replacement of the pike. However, since it rendered the weapon inoperable as a firearm, the pikeman was still a necessary adjunct to the infantry formation to ensure continued firepower. But about 1680, someone—possibly Vauban—invented the ring bayonet, which left the bore clear for firing. This was soon improved by a socket in the handle, which firmly locked the bayonet to a stud on the musket barrel. By the end of the century this had been adopted by all European armies. The musketeer became his own pikeman; the pikeman himself was fading from the scene.

ARTILLERY

Gustavus Adolphus was the father of modern field artillery and of the concept of massed, mobile artillery fire. Deploring the fact that artillery's heavy firepower potential could not support infantry and cavalry maneuver, Gustavus devoted much time and effort to solving the problem.

He reduced weight by decreasing the amount of metal in the barrel: shortening the tube and reducing its thickness. Then he introduced improved, standardized gunpowder, permitting greater accuracy despite the shortened barrel, and also restoring ballistic power lost by lightening of the tube.

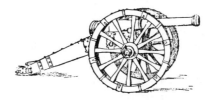

Swedish 3-pounder regimental gun

He allowed only three standard calibers: a 24-pounder, a 12-pounder, and the light, sturdy 3-pounder regimental gun. Then, to ensure command control over the weapons, he abandoned the old system of hiring civilian contract gunners and established military units of cannoneers, as responsive to discipline and to training as were his infantry and cavalry.

SIEGECRAFT AND FORTIFICATION

One man—Vauban—dominated the culminating developments of the two opposed functions of siegecraft and fortification. Through him these approached the ultimate capabilities for military forces limited to muzzle-loading weapons and black powder. This was extremely important in a century in which sieges were the most common combat activity.

The Vauban system of siegecraft* provided for systematic approach of the attackers and their artillery to a fortification by means of entrenchments. The ultimate objective was to permit the attacking siege artillery to blast a breach in the defensive wall and the covering obstacles through which an infantry column could make an assault. Sometimes a successful assault could be made under the cover of fire from the approach trenches without waiting for the siege artillery to breach the walls. The attackers would then have to be prepared to cut their way through the defending obstacles across the moat (if there was one), and climb the wall, under the cover of artillery and small-arms fire from the approach trenches. Fascines (bundles of twigs or brushwood) were usually used to fill up ditches and the moat before such an assault.

The method of approach was standardized. A **first parallel** was dug some 600–700 yards from the fortification. This was a trench parallel to the line of defenses at the selected point of

---

* The remaining paragraphs of this section are based upon Appendix IV, "Fortifications and Siegecraft," from *The Compact History of the Revolutionary War*, New York, 1963, with the permission of the publishers, Hawthorn Books, Inc.

assault. This precluded enfilade fire by the defenders down the length of the trench. The distance was fixed because it was close to the maximum effective range of defending and attacking artillery of the age. When the parallel was dug, additional earthworks were thrown up in front of it as protection for siege-artillery emplacements. To expedite construction, fascines were used as the basis of the earthworks. Under the cover of fire from these guns the attacking engineers began to dig "saps," or approach trenches, toward the fort (thus the origin of the word "sapper" for engineer). These were dug at an angle, and zig-zagged back and forth, again to reduce the defender's opportunity for enfilade fire. The sappers were protected from direct fire by shelters called gabions: wicker baskets filled with earth and frequently put on wheels so that they could be pushed in front of the sap.

When the approaches progressed to a point about 300 yards from the defenses, a **second parallel** was dug, and new artillery emplacements prepared. From these positions the siege guns could begin an intensified bombardment to drive the defenders from the ramparts, to silence their artillery, and to begin to batter a breach. The defenders would, if possible, sortie in limited counterattacks to prevent the completion of the second parallel and to try to destroy or to "spike" the attacking guns. (Guns were "spiked" by driving spikes, nails, or bayonets down their touch-holes, thus rendering them useless until the spike could be removed.) The attackers had to be ready for such sorties, and strong forces of infantry were maintained constantly in the parallels to protect the guns and the cannoneers.

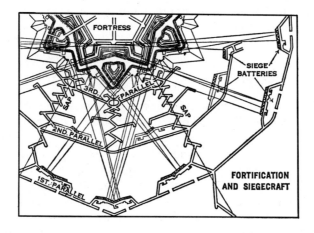

If the defenders persisted, and if the attackers did not believe they could assault successfully from the second parallel, approach saps were again pushed forward, this time in the face of small-arms fire from the defenders, to within a few yards of the ditch or moat at the base of the walls. Here a **third parallel** was constructed. While fire from attacking infantry prevented the defenders from manning the ramparts, the breaching batteries were emplaced to batter the walls at point-blank range. Sometimes improved mining techniques (based fundamentally on the old principles) were also employed, either to help knock down the wall or to permit small groups of attackers to debouch inside the fortification. The defenders, of course, would normally drive countermines.

A day or two of siege-gun pounding from the third parallel would usually knock a breach in the wall. The assault then followed, if the garrison had not already surrendered.

TACTICS OF LAND WARFARE

At the outset of the century the "Spanish square" dominated the battlefields of Europe. Actually, as we have seen (p. 500), these squares were broad columns of pikemen and musketeers, several usually arrayed abreast in a line of battle. The relatively immobile artillery was usually lined up in front of the columns, generally protected by cavalry. More cavalry protected the flanks of the ponderous "squares."

This columnar concept of infantry combat was changed completely and forever by Gustavus. Thanks to increased firepower and rate of fire, he arranged his musketeers and his reduced number of pikemen in relatively thin lines, no more than 6 men deep. A number of individual units, deployed abreast in such lines, with intervals between units, formed a longer line. Undoubtedly following the example of the Roman legion, a Swedish army usually comprised two such lines of infantry, with a third line in reserve. Gustavus usually had some cavalry and artillery deployed in the intervals in these infantry lines. For flank protection, there was usually cavalry on the flanks, but the small, mobile infantry units were also perfectly capable of protecting their own flanks when the need arose.

Swedish musketeer

This was the genesis of modern linear tactics, which remained basically unchanged—though constantly modified—until World War I. Even today, despite drastic tactical changes of two world wars and the advent of nuclear weapons, vestiges of Gustavus' linear concept remain in modern infantry doctrine.

Gustavus also modified cavalry tactics, combining the best features of the cumbersome German "caracole" and the traditional French saber charge, so that his cavalry had both firepower and shock capability. And he trained his infantry, artillery, and cavalry to work closely together. This concept of tactical teamwork among the three combat arms was Gustavus' greatest tactical contribution to warfare.

MILITARY ORGANIZATION AND THEORY

Combat Formations

The Thirty Years' War marked a major turning point in methods of warfare and military organization. This undoubtedly would have occurred without the impact of a Gustavus Adolphus, but he sparked not only the tactical changes; he also transformed organization. His was the first true professional standing army. The internal structure within that army established prototypes for small-sized military units today. (For the larger units—divisions and corps—see p. 805).

Gustavus' infantry squadron, consisting of approximately 500 men, was called a battalion by the French, and this designation has persisted to this day for the basic combat command. The battalion usually consisted of 4 companies. Three battalions were combined into brigades (equivalent of modern regiments or brigades) for combat. This was not an entirely new concept—what was new was Gustavus' decision to make the brigade a permanent unit with a permanent command hierarchy, the origin of the modern regimental officer system. The regiment or brigade was commanded by a colonel, the battalion by a lieutenant colonel, and the companies—whose origin was the free companies of earlier centuries—by captains. Subordinate officers were lieutenants.

The Regimental Proprietary System

The establishment of permanent units hastened the almost universal adoption of the proprietary system, which had already begun to replace the vestiges of feudalism and of free companies. The permanent colonel was the proprietor of his regiment, accepted by the king as a permanent officer and authorized (personally, and through him his captains) to raise men. Initially, with armies being raised only for a campaign and disbanded afterward, the troops raised by the proprietary system were volunteers, more or less carefully selected from the available and willing man power. But as the armies became permanent, the standing units were not disbanded and were kept up to strength by regular influx of recruits, usually provided by the crown. This, combined with the financial considerations in maintaining year-round units, gave the crown increasing rights of supervision over the administration and training of the regiments, and thus somewhat restricted the proprietary rights previously exercised by colonels and captains.

This proprietary system could be profitable. A commander was paid for the number of men he mustered, as well as for their weapons, equipment, and subsistence. In addition to the profit to be derived from economical exercise of his proprietorship (to say nothing of the possibilities offered by parsimony and fraud), an officer could sell his proprietary interest when he retired. Thus officers' commissions were valuable, and could be purchased. This custom of purchase of commissions continued in some armies—notably that of England—long after the proprietary system itself had disappeared.

The Hierarchy of Command

Later in the century the French—who adopted the Swedish system almost in its entirety—extended the concept of a permanent combat military hierarchy upward from the regiment to the army commander. Hitherto the king or a prince of the realm was usually the titular "general" of an army. The second in command—and often the actual field commander—was the lieutenant general, almost invariably a nobleman, who usually exercised direct command over the aristocratic cavalry. The infantry was usually commanded by a senior professional soldier, who was not necessarily a nobleman and was usually called the sergeant major general, or major general. He was charged with the responsibility for forming up the army for battle and with care of the various administrative duties of command. When the army was disbanded at the end of a campaign, as was usually the case until the very end of this century, the lieutenant general and major general lost both command and rank. Usually they reverted to their proprietary positions as colonels of permanent regiments.

By the end of the century, however, the old title of marshal became a permanent rank in the French army for a commander of a field army in the absence of the king. Since **Louis XIV** usually had several armies in the field at once, this resulted in the establishment of a permanent list of officers, each with sufficient experience and distinction to warrant him to serve as major general, lieutenant general, or marshal of an army. In time the relative position of officers on this list established precedence for command. Thus, for the first time since the fall of Rome, one of the most significant aspects of modern military professionalism appeared: the permanent classification of an officer by rank, and not by the temporary command which he happened to be exercising. In due course the national general army lists were extended to the grades of colonel and below, eventually undermining the proprietary system of independent regiments, which mostly disappeared in the 18th century.

Supporting Services

This century also saw the real beginning of the militarization of the supporting arms and services. This was to some extent a natural concomitant of the emergence of the standing army. To Gustavus Adolphus' militarization of artillery personnel was later added the militarization of the teamsters who hauled the army's supplies. This was carried further, throughout the military supply system, by France's **Francois Michel le Tellier, marquis of Louvois,** King Louis XIV's Minister of War during 1666–1691.

Vauban was responsible for carrying to fruition the systematization of the engineering tasks of an army begun by Gustavus. He consolidated the responsibility for these tasks under specially trained engineer officers and men, rather than leaving such things to individual line units and to civilian contractors. Similar militarization occurred in military medicine and military law.

Armies and Society*

This was an age of absolute monarchy—though absolutism was rejected in England—and war was indeed "the trade of kings," as Dryden, a product of the age, well put it. A military system based on strict discipline, centralized administration, and long-term, highly trained troops was particularly congenial to such form of government. Even the civilian economy was being centralized, since war's cost created the crown's need for money. The trend, which included

---

* This section is adapted, with permission, from an unpublished study, "Historical Trends Related to Weapon Lethality," made for the U.S. Army by the Historical Evaluation and Research Organization, 1964.

higher taxation, was begun in France by Louis's great ministers Louvois and **Jean Baptiste Colbert.**

The effect on society was profound, since the steady procession of wars created new demands for man power. War ceased to be of concern to the upper classes only. The cavalry, once the exclusive domain of aristocracy, opened its ranks to all who could ride a horse. Mercenary regiments drew heavily on the lower classes; the press gang and universal service were just around the corner.

Science and technology were increasingly being put to the service of war, particularly in the newly militarized artillery and the engineers. Maurice and Gustavus used portable telescopes; cartography for military purposes became essential. Soldiering, particularly for the officer, was becoming a profession; systematized instruction increased in importance. The first military academy of modern times was established in 1617 by **John of Nassau.**

There was a marked increase in the size of armies and the scope of warfare. Gustavus had about 30,000 men in his field army in 1631; his opponents had more in their pay, but rarely employed all of them in battle. Richelieu had more than 200,000 men under arms before his death, and Louis XIV maintained a military establishment of 400,000, with field armies sometimes approaching 100,000 men. It was estimated that a country could support an army of about 1 percent of its population, approximately the ratio in France at the end of the century.

THE MILITARY SYSTEM OF GUSTAVUS ADOLPHUS, C. 1631*

At the time Gustavus Adolphus assumed the Swedish crown in 1611, the Swedish Army was poorly equipped, poorly organized, under strength, and badly led. Gustavus' first task was to rebuild his army. He provided for continuous training from the moment a recruit entered the ranks. There were frequent maneuvers for small and large units. Discipline was strict; every regimental commander read the Articles of War to his troops once a month. Punishment for infractions was heavy, and Gustavus' soldiers had a reputation for good behavior unusual for troops of that day.

Infantry

Gustavus' basic tactical unit was the squadron, consisting of 408 men—216 pikemen and the remainder musketeers. The pikes were formed in a central block, 6 deep, and the musketeers in 2 wings of 96 men, also 6 deep, on each side of the pikemen. Attached to each squadron was an additional element of 96 musketeers. He grouped 3 or 4 squadrons to form a brigade. Since the attached musketeers were employed frequently for outpost, reconnaissance, etc., they were often not available to the squadron.

In combining firepower with the pike, and missile with shock, Gustavus put principal emphasis on firepower. He employed the "countermarch" concept introduced by the Spanish, in which front-rank musketeers moved to the rear to reload after firing; but since he had speeded the loading process, he was able not only to reduce the number of ranks to 6 but to have 2 ranks of musketeers fire simultaneously before countermarching. Further, the counter-

* *Ibid.*

march was so executed that the formation moved forward; the fire was, in effect, a small-arms rolling barrage. During this movement, the musketeers were protected by the pikes while they reloaded. Later, Gustavus introduced the salve, or salvo, to increase further the fire-power of his line. In the salvo, 3 ranks fired simultaneously.

Since salvo fire rendered the musketeers impotent while they reloaded, the role of the pike was enhanced in the Gustavian system. But the pike had a broader mission. It was to deliver the decisive blow, the salvo itself being but the prelude to the pike assault, as it was to the cavalry charge. And the best protection for the musket was the offensive action by the pike. Gustavus had his pikemen wear light armor, which was on the way out in other countries. He shortened the pike from 16 to 11 feet and sheathed its foremost part with iron so that it could not be severed by a sword blow. Thus the obsolescent pike became in Gustavus' hands an offensive weapon, combined with missile power.

Cavalry

Gustavus' cavalry, armed with pistol and saber, formed in 6 ranks (later in 3). The pistol was a gesture; the real effect came from the saber charge. The first rank fired when it was close to the enemy, the other 2 held fire, retaining the pistol for emergency use. Detached musketeers stationed between cavalry squadrons provided the firepower that shook the enemy line. While the cavalry charged, the musketeers would reload, to be ready to fire another volley for a second charge or to cover a retreat. To this was added the fire from the regimental cannon.

There was an obvious disadvantage to this system; by tying the cavalry to the infantry and artillery, Gustavus sacrificed the speed and momentum of the horse except for the final distance of the charge. But it was better than anything yet devised, and it was successful. As a result, it was imitated widely.

Artillery

Overall artillery direction was provided by 27-year-old **Lennart Torstensson,** the best artilleryman of his time. The artillery was organized into permanent regiments of 6 companies. Of the 6 companies, 4 consisted of gunners, 1 of sappers, and 1 of men with special exploding devices. Thus the artillery was organized as a distinct and regular branch of the army, manned almost entirely by Swedish troops. After some experimentation Gustavus adopted the 3-pounder "regimental gun," which revolutionized the role of artillery; each regiment of foot and horse in Gustavus' army had 1 (later 2) of these cannon. The enormous advantage which this provided in battle was soon imitated in other armies.

Summation

Most of Gustavus' innovations were adapted from others, and he was not the only one to improve the military system. But no one else so surely bridged the gap between conception and achievement; no other fitted his innovations into a completely integrated system with its own set of unifying principles. His accomplishments were many: he gave to infantry and cavalry the capacity for the offense, he increased firepower and made it the preliminary for shock, he made artillery mobile, he made linear formations more flexible and responsive to the commander's will, he solved the problem of combined arms, and he made the small-unit commander the key to action. In him, the military revolution that began in the middle of the 16th century was most completely realized, although it did not find fullest expression until the time of Louis XIV.

THE FRENCH SYSTEM OF LOUIS XIV

During the reigns of Louis XIII (1610–1643) and Louis XIV (1643–1715), which spanned a century, the French Army emerged as Europe's dominant land force, supplanting that of Spain. The two French kings, aided by able ministers, were able to consolidate political power in their persons to an unprecedented degree, systematically breaking down and subordinating all internal opposition, whether that of dissident great nobles or of the Protestant religious minority—the Huguenots.

In a dizzying series of conflicts, both internal and external, the principal instrument of the Crown's power—"the last argument of kings"—was the Army. Practically nonexistent at the accession of Louis XIII, it numbered 440,000 in its heyday, in 1693. Ably led by a succession of great captains, among them Turenne, Condé, Luxembourg, and Villars, it was a military instrument at once respected and feared.

The French Navy too enjoyed a renascence, its growth carefully nurtured by Richelieu and later by Colbert. For a variety of reasons, however, it was never as powerful or as professional as its principal antagonist, England's Royal Navy.

The French Army adopted from the Swedes the basic infantry formation—a battalion (or regiment) of 600–800 men. This unit was usually organized in 1 line, 6 deep, with the pikes in the center and the muskets on the flank, and occupied a front of about 100 yards. In battle, several lines were formed, with the battalions arrayed in checkerboard fashion not unlike the Roman *quincunx* (see p. 80). Two-thirds of the men were musketeers, and from this group a detachment supported the cavalry. The interval between battalions was supposed to be equal to their front, so that the second line, usually 300 to 400 paces behind, could pass through. The reserve was kept twice that distance behind the second line.

On his accession, Louis XIV possessed an army of 139 regiments, 20 of which were foreign; about 30 were cavalry. But they were far from disciplined, and administration was poor. The task of reorganizing and training the Army was the work of Louvois. He hampered field commanders with deadening restrictions, but his talents paid off in other ways: his improved administration and the fortifications he built. Administrative reforms included control of the proprietary system (see p. 576) and frequent reviews and inspections.

A chain of fortresses, fully equipped and stored with all the supplies needed by an army, was constructed. An army on the march could base at any one of these posts, certain of finding there everything it needed, including heavy artillery. At the same time, an enemy army would find the task of breaching these forts, one after the other, an overwhelming job. Fortress construction was largely the work of Vauban (see p. 573). Altogether, Vauban built 33 new fortresses and remodeled 3,000 others.

Louis's cavalry consisted of *gendarmerie* (heavy cavalry), carabineers, light cavalry, and dragoons. The carabineers, numbering at the turn of the century about 3,000 men, were armed with rifled carbines and swords; the dragoons used the musket with the newly developed bayonet and carried an entrenching tool on their saddles. They combined the advantages of infantry and cavalry, and, being very mobile, proved very useful. From one regiment in 1650, the number of dragoons increased until, by 1690, there were 43 such regiments in the French Army.

DEVELOPMENTS IN NAVAL WARFARE

Because of their revolutionary concepts of naval firepower, so strikingly demonstrated in the wars against Spain at the end of the previous century (see pp. 507–510), English sailors at the beginning of the century were tactically far ahead of all possible rivals. Surprisingly, none of these rivals, not even the Spanish victims or the aggressive and imaginative Dutch, seem to have grasped the secret of English success: broadside firepower. And, with no major naval wars in the first half of the century, there was no opportunity for the others to learn, or for the English to improve.

Early ship-of-the-line

The one important development of the early part of the century, apparently another English discovery, was how to harness a gun's recoil with ropes in such a way that it would be brought to rest far enough inboard from the gunport to permit easy reloading. Hitherto guns had been tightly bound to bulkheads to inhibit recoil. This had made loading very difficult—almost impossible in the heat of action. Thus evolved the English practice, followed in the Armada battle, of a small group of about 5 ships following each other in a circle, only one at a time firing its broadsides at the enemy. Harnessing the recoil, therefore, not only greatly increased the rate of fire of each ship; for all practical purposes it potentially multiplied this increased firepower by a factor of about 5.

From this development, and from experience in the First Dutch War, England's great Blake seems to have formalized the concept of the line-ahead formation—all ships in single column, at regular intervals, so as to achieve maximum firepower by broadside fire and, at the same time, maximum control in an orderly formation responsive to the admiral's will.

The difficulty of controlling a great number of ships, stretched across several miles of sea, created massive combined problems of naval tactics and seamanship. Rudimentary flag signals had been devised, but even when these reached the peak of sophistication more than a century later, they were totally inadequate for an admiral to communicate precise orders to his subordinates. Even if the flags could have transmitted exactly what he wanted, and transmitted it quickly (neither of which was possible), distance, fog, gunsmoke, and confusion of battle all made this a most uncertain means of communication at best. So the English Navy developed its system of "Fighting Instructions," which tried to establish a common, understood doctrinal procedure for dealing with every contingency. These instructions were augmented by further detailed orders given by an admiral to his subordinates before they put to sea, and (usually) before a battle was expected. But since no two battles could ever be exactly alike, and no two enemies would react in exactly the same way, contingencies constantly arose which the "Fighting Instructions" did not cover.

This led to the emergence in England, between the First and Second Dutch Wars, of two tactical schools of thought. Both agreed on entering battle with the line-ahead formation, endeavoring to be to the windward of the enemy so as to have the choice of closing or pulling away, as the circumstances of the battle might dictate. But once the battle was joined, the two schools differed on how it should be fought. Somewhat oversimplified, the differences were as follows:

The "formal" school believed in adhering to the line-ahead formation at practically any cost, until or unless complete victory was achieved. Each ship would engage with its guns the enemy closest to it, but would at the same time follow the course of the preceding vessel. Thus the admiral would always know where his ships were, could be assured that they were hammering away at their nearest opponent, and could pull them all out together, if necessary.

Those of the "melee" school, however, believed that if an opportunity arose the admiral should be able to release individual squadron and ship commanders to move out of the line in mass attacks against the enemy. They counted on the judgment and experience of subordinate commanders, and the traditional fighting spirit of the Royal Navy, to make the most of such opportunities, since it was impossible to give adequate orders on the spur of the moment.

By the end of the century both systems had been tested, each with mixed success and failure. For a variety of reasons, the formalists were in the ascendancy as the period ended, and they remained ascendant for more than a century.

The fighting Dutch admirals—even the two Tromps and de Ruyter—were always behind the British tactically. Although they adopted the British line-ahead firepower system in the Second and Third Anglo-Dutch Wars, they always preferred to board and fight hand to hand. In sheer seamanship these three Dutchmen were equal to the very best English admiral, and perhaps better. Their challenge to English sea power was vigorous and nearly successful. But the English refused to be outfought, and their margin of victory was the clear gunnery superiority they maintained.

By the 1680s the French were prepared to challenge the British at sea. The British Navy had been administratively reorganized by **James, Duke of York** with the very able assistance of efficient **Samuel Pepys** as Secretary of the Admiralty. Between them they established methods for the direction, administration, and organization of the Royal Navy, whose principles have, in large measure, persisted to this day.

But Louis XIV had in Jean Colbert a minister at least equally efficient. Colbert, furthermore, had been given a free hand by Louis in developing French sea power and finance, while James, before ascending the throne, had seen much of his work wasted due to the frivolous policies of his elder brother, **Charles II.** On top of this, Colbert had encouraged the development of scientific ship design and ship-building in French shipyards. The French adopted the best the English had done, and then improved on it. Ship for ship, the vessels of the French Navy were faster and better than those of the Royal Navy. When the War of the Grand Alliance began, the French Navy was the best in the world, and was numerically equal to the combined fleets of England and Holland. The senior French Admiral **de Tourville** was at least as courageous and probably more able than his mentor and counterpart, the English **Arthur Herbert, Earl of Torrington.**

By all logic, French sea power should have swept the British fleet from the Channel in the War of the Grand Alliance. That this did not occur was due primarily to the failure of Louis XIV to recognize the opportunity, and his refusal to let Tourville fight the war as he wished. France never again had a comparable opportunity.

By the end of the century France was able to threaten or annoy England at sea only by a *guerre de course* (destruction of commerce by attacks on merchant shipping). **Jean Bart,** the fighting French sailor who terrorized England's coasts in the War of the Grand Alliance, at least gave his successors a gallant example of an inherently defeatist method of naval war.

MAJOR WARS

THE THIRTY YEARS' WAR, 1618–1648

The Thirty Years' War started as a religious war, a product of the struggle between Roman Catholics and Protestants in Germany. Although the religious problem was always present, the war became increasingly a political struggle, with the Hapsburg Dynasty of the Holy Roman Empire attempting to control as much as possible of Europe while other powers—particularly the Bourbon Dynasty of France—were determined to contain the Hapsburgs. On the Catholic-Hapsburg side of the conflict were Austria, most of the German Catholic princes, and Spain. The Protestant princes of Germany and the Protestant kingdoms of Denmark and Sweden, and Catholic France were in opposition. The line-up changed from time to time; many German princes moved from one side to another, or attempted to stay out of the quarrel altogether. Several separate wars—the Empire with Transylvania, Spain with France, Spain with Holland, Sweden with Poland, and Sweden with Denmark—all became involved in the great struggle. The fighting ranged all over Germany and into France, Spain, Italy, Poland, and the Netherlands. Much of the area—particularly Germany—was plundered and replundered by undisciplined and motley armies, whose major source of supply was the land over which they fought. The physical devastation of this war, and the loss of life among civilians, were the most severe in Europe since the Mongol invasion.

Background

1608. Evangelical Union. The Protestant princes of Germany, led by **Frederick IV,** Elector Palatine, and **Christian** of Anhalt, combined in protest against the occupation by **Maximilian** of Bavaria of the free city of Donauwörth.

GERMANY DURING THE THIRTY YEARS' WAR.

Route of Gustavus Adolphus →

1609. Holy Catholic League. Created by Maximilian and the other Catholic princes to offset the Evangelical Union.

1609. Freedom of Religion in Bohemia. King **Rudolph** guaranteed religious freedom to his Protestant subjects; it was to be safeguarded by a body of men known as ''the Defenders.'' As a result, Rudolph was deposed by his brother **Matthias** (1611), who disapproved of religious freedom.

1612. Matthias Elected Emperor.

1617, June 17. Bohemian Succession Election. Because he was childless, Matthias' Catholic councilors elected his cousin Archduke **Ferdinand** of Styria as heir to the throne. The Protestants, led by Count **Matthias of Thurn,** refused to recognize Ferdinand.

1618, May 22. Defenestration of Prague. A Diet summoned by the Defenders met at Prague. Two of the king's most trusted councilors, **Martinitz** and **Slawata,** were thrown out of the windows of the Castle of Hradschin. The Bohemians then established a rebellious provisional government and proceeded to raise an army under Count Thurn.

The Bohemian Period, 1618–1625

1618, July. Hostilities Begin. Thurn attacked and captured Krummau.

1618, November 21. Capture of Pilsen. Taken by Protestant Count **Ernst von Mansfeld,** with 20,000 mercenaries, in a desperate struggle.

1618–1619. Catholic Reinforcements. One army came from Flanders, supported by Spanish money; a second came from Austria. In the following winter months, native Bohemian troops under Count Thurn and Count **Andreas Schlick** kept the Flemish army shut up in Budweis and laid waste the Austrian border.

1619, March 20. Death of Matthias. The imperial and Bohemian thrones were vacant.

1619, May–June. Invasion of Austria. Thurn marched on Vienna, where he was joined by a

Transylvanian army under **Bethlen Gabor** (see p. 635).

1619, June 10. Battle of Zablat. Mansfeld was badly defeated in Bohemia by Imperial forces under Count **Charles B. de L. Bucquoi.** Thurn was recalled from Austria; Gabor returned to Hungary.

1619, August 26. Bohemian Protestants Elect a New King. They chose **Frederick** Elector Palatine. They had already declared Ferdinand's election invalid. Bohemia was supported by Lusatia, Silesia, and Moravia.

1619, August 29. Ferdinand II Elected Emperor. This was by the Imperial Electoral College, at Frankfurt. In following months Ferdinand's efforts to eject Frederick from the throne in Bohemia were supported by Maximilian of Bavaria, **John George** of Saxony, and **Philip III** of Spain.

1620, April 30. Imperial Mandate Directs Frederick to Withdraw. His disregard of this mandate amounted to a declaration of war.

1620, July 3. Treaty of Ulm. Princes of the Evangelical Union, as jealous of fellow-Protestant Frederick as they were fearful of the Emperor, declared neutrality. Meanwhile, Bohemian troops had entered Austria, aided by an uprising of Austrian nobles.

1620, July 23. Intervention of Maximilian of Bavaria. With the army of the Catholic League, 25,000 men under the General **Johan Tserclaes, Count Tilly,** Maximilian crossed the Austrian frontier to support the Emperor against the Bohemians and his unruly nobles.

1620, July–August. Spanish Intervention. Pursuant to a plan formed in Brussels, Vienna, and Madrid during the winter-spring of 1619–1620, Marquis **Ambrogio di Spinola** set out from Flanders for the Lower Palatinate with 25,000 men (July). Spinola's secret objective was to keep the princes of the Protestant Union off balance so that they would keep their own forces in hand and not commit them to Frederick in Bohemia. His march created much anxiety among the Emperor's enemies. Crossing the Rhine at Coblenz, and feinting in the direction of Bohemia, Spinola then occupied the cities of the Palatinate. Frederick, in Bo-

hemia, was unable to protect his lands. **The** intervention assisted Ferdinand and at the same time removed Protestant "choke points" along the vulnerable "Spanish road" connecting Lombardy with Flanders.

1620, August 4. Maximilian Victorious in Austria. He forced submission of the Austrian Estates at Linz.

1620, November 8. Battle of the Weisser Berg (White Mountain). Maximilian and Tilly, coming from Linz, joined an Imperial army under Bucquoi and entered Bohemia. Near Prague they were opposed by a Bohemian army, 15,000 strong, under Prince **Christian of Anhalt-Bernberg,** drawn up on the Weisser Berg. The Catholics, about 20,000, attacked at dawn. The Bohemians were routed. Frederick fled to Breslau. The League and Imperial armies under Maximilian entered Prague in triumph and sacked the city. For rebel Bohemia a severe retribution followed—the price of its allegiance to the "winter king." This was the end of Bohemian independence for 300 years, until the emergence of Czechoslovakia after World War I (November 1918; see p. 1091).

1621, January 29. Frederick Banned. His lands were confiscated. This action was protested by the princes of the Evangelical Union. Ferdinand rejected the protest. Meanwhile, Mansfeld rallied an army of Bohemian refugees and mercenaries in the Palatinate, and declared allegiance to Frederick. Penniless, Mansfeld supported his army by pillaging the Rhine Valley. Thus began the depredations of the war.

1621, April 27. Frederick's Alliance with Holland. The Dutch agreed to support his efforts to reconquer his lands on the Rhine. Both Frederick and Ferdinand refused to accept Spanish and English mediation efforts.

1622. Protestant Reinforcements. Duke **Christian** of Brunswick and Margrave **George Frederick** of Baden-Durlach supported Frederick and Mansfeld.

1622, April 27. Battle of Mingolsheim. Mansfeld defeated Tilly and delayed his union with a Spanish force from the Netherlands under General **Gonzales de Córdoba.**

1622, May 6. Battle of Wimpfen. George Frederick, after initial success, was defeated by the combined forces of Tilly and Córdoba. Córdoba then pursued Mansfeld into Alsace; Tilly marched north to oppose Christian on the Main.

1622, June 20. Battle of Höchst. Christian, attempting to join Mansfeld, was intercepted by Tilly and Córdoba while crossing the Main River. Despite severe losses, Christian succeeded in joining Mansfeld with the remnants of his army.

1622, July–August. Protestant Withdrawal. Mansfeld, Christian, and Frederick retreated into Alsace, and then into Lorraine, living off the land, destroying towns and villages as they went. Frederick, after a dispute with his two army commanders, revoked their commissions, then retired to Sedan, leaving them with armies but without active employment for them. They decided to join the Dutch.

1622, August 29. Battle of Fleurus. The Spanish under Spinola had invaded Holland and laid siege to Bergen-op-Zoom. Mansfeld and Christian, marching to its rescue, were intercepted at Fleurus by Córdoba. After desperate fighting, in which Christian was badly wounded, the Mansfelders extricated themselves from the combat and pushed on toward Bergen-op-Zoom, pursued part way by Spanish cavalry and abandoning guns and booty. Both sides claimed victory. The unexpected (and unwanted) appearance of the Mansfelders before Bergen-op-Zoom caused Spinola to raise the siege. The Mansfelders then moved to unspoiled East Friesland, where foraging was good.

1622, September 19. Tilly Takes Heidelberg. This followed a siege of 11 weeks. He had already conquered most of the Palatinate.

1623, February 23. Frederick Deposed as Elector Palatine. This was by the Electoral College, at Ferdinand's behest. The title was given to Maximilian of Bavaria. The electors of Saxony and Brandenburg refused to recognize Maximilian.

1623, August 6. Battle of Stadtlohn. Christian, having left Friesland, was defeated by Tilly near the Dutch border, losing 6,000 dead and 4,000 prisoners. He fled into Holland with 2,000 survivors.

1623, August 27. Armistice. Frederick, after English urging, sought peace with Ferdinand. Mansfeld, however, as well as other Protestant princes, refused to make peace.

1624, January. France Enters the War. Already at war with Spain, Cardinal Richelieu brought his nation into the war against the Hapsburgs.

1624, June 10. Treaty of Compiègne. Alliance between France and Holland against the Hapsburgs. They were soon joined by England (June 15), Sweden and Denmark (July 9), and Savoy and Venice (July 11). France, Savoy, and Venice agreed on joint operations against the Valtelline, Alpine pass on the Spanish line of communications from Italy to the Netherlands, in order to prevent effective Spanish-Hapsburg cooperation.

1624, August 28–1625, June 5. Siege of Breda. This key fortress guarding the roads to Utrecht and Amsterdam was captured by Spinola.

The Danish Period, 1625–1629

1625, April. Wallenstein on the Scene. Ferdinand hired mercenary general Count **Albrecht von Wallenstein,** who agreed to raise and maintain 20,000 men to defend the Hapsburg lands. In June Wallenstein was charged with covering the whole Empire.

1625, Summer. Danish Invasion of Germany. Christian IV of Denmark advanced down the Weser and tried to gather support from the Protestant German princes, from the Dutch, and from England, with limited success. Wallenstein moved north to join Tilly to oppose Christian.

1626, March 26. The Peace of Monzon. France withdrew from the war; a Huguenot revolt required Richelieu to recall French troops from operations against Spain.

1626, April 25. Battle of Dessau Bridge. Mansfeld, heading for Magdeburg, attempted to cross the Elbe with 12,000 men. He was repulsed with heavy losses by Wallenstein, who pursued Mansfeld into Silesia. Mansfeld died soon afterward.

1626, August 27. Battle of Lutter. Christian of Denmark was badly defeated by Tilly. Christian, having lost about half his army, fled. Many Protestant princes made peace. The fortunes of the Protestant or anti-Hapsburg cause were at a low ebb.

1626–1627. Ordeal of Germany. The harvest had failed, famine, plague, and other diseases were widespread, violence was everywhere, as the various armies lived off the land where they were.

1627. Catholic King of Bohemia. The Archduke **Ferdinand,** Emperor Ferdinand's son, was crowned the first hereditary King of Bohemia.

1627. Peace between France and Spain. They formed an alliance against England.

1627, September and October. Repulse of the Danes. Tilly and Wallenstein marched down the Elbe and drove Christian over the frontiers of Holstein.

1628. Honors to Wallenstein. Ferdinand gave his successful general the duchies of Mecklenburg and Pomerania. This exercise of illegal imperial authority alarmed the German princes. The electors failed in attempts to get Wallenstein removed from command of the Imperial armies.

1628, February–July. Siege of Stralsund. Wallenstein attempted to gain control of the Baltic coast, but the defense was vigorous and, in face of Swedish threats, Wallenstein withdrew.

1628, September 2. Battle of Wolgast. Wallenstein badly defeated Christian of Denmark.

1629, March 6. Edict of Restitution. Ferdinand ordered the restoration of all church lands which had been seized or acquired by Protestants.

1629, June 7. Peace of Lübeck. Denmark withdrew from the war, giving up the north German bishoprics of Holstein, Stormarn, and Ditmarschen, and agreeing not to meddle in the territorial arrangements of Lower Saxony.

1629, October 5. Truce of Altmark. Richelieu, alarmed by Imperial victories, arranged a truce between Poland and Sweden to release Gustavus Adolphus of Sweden to enter the war in Germany.

The Swedish Period, 1630–1634

1630, July 4. Arrival of Gustavus Adolphus. He landed at Usedom with 13,000 men and advanced on Stettin and into Mecklenburg. His army grew to 40,000.

1630, August 24. Dismissal of Wallenstein. Under pressure from the Protestant electors, meeting at Regenburg, Ferdinand, alarmed by the Swedish invasion, agreed to dismiss his favorite general in order to achieve German unity.

1630, November–1631, May. Siege of Magdeburg. This had been planned by Wallenstein, and was carried out by Count **Gottfried H. zu Pappenheim** and Tilly. The garrison was strong and well supplied, repulsing repeated attacks. The besiegers, finding it difficult to get food from the denuded countryside, suffered more than the besieged.

1631, January 23. Treaty of Bärwalde. Alliance of Gustavus with France, which was to provide financial support. He guaranteed freedom of worship for Catholics in Germany and to make no separate peace for 5 years.

1631, March 28. Leipzig Manifesto. Encouraged by the Swedish entry into the war, the Protestant princes, led by John George of Saxony, demanded that Ferdinand take steps to remedy many evils: the Edict of Restitution, depredations of Imperial and Catholic armies, the decay of the rights of the German princes, disregard of the constitution, and the dreadful situation in the country. Ferdinand ignored the protests. The Protestant princes established an army under **Hans Georg von Arnim** (May 14).

1631, April 13. Capture of Frankfurt. Gustavus, after a brilliant surprise march, stormed Frankfurt on the Oder. He hoped this would cause Tilly to raise the siege of Magdeburg. When Tilly hung on grimly, Gustavus prepared to move directly to Magdeburg. Because of the alliance between Maximilian of Bavaria and France, Tilly renounced his allegiance to Maximilian and gave his loyalty to the Emperor (May).

1631, May 20. Sack of Magdeburg. Tilly and Pappenheim stormed the city, which was thoroughly sacked by the enraged and hungry troops. Only about 5,000 of 30,000 inhabitants survived the holocaust.

1631, July–August. Gustavus vs. Tilly. The Swedish Army, dangerously weakened by lack of supplies, entrenched at Werben. Tilly attacked twice and was repulsed with heavy losses. Tilly then marched into Saxony, laying it waste. In an effort to save Saxony from Tilly, John George placed his troops under Gustavus (September 11).

1631, September 15. Tilly Takes Leipzig. He had been reinforced to about 36,000 men. The Swedish and Saxon Armies, respectively 26,000 and 16,000 strong, joined at Düben, 25 miles to the north.

THE BATTLE OF BREITENFELD, SEPTEMBER 17, 1631

Tilly, influenced by Pappenheim, took a stand at Breitenfeld, 4 miles north of Leipzig. Drawing up his army with the infantry in the center and the cavalry on the wings, Tilly commanded the center and right, Pappenheim the left. Gustavus, under harassing fire from the imperialists, drew up his lines with the Saxons, under John George, on the left, Swedish infantry and other German infantry in the center, and Gustavus' cavalry on the right wing.

Pappenheim opened the battle, sweeping around behind the main body of the Swedish cavalry and attacking Gustavus' reserve. The maneuverable Swedish horse wheeled and pinned Pappenheim between them and the reserves, forcing him to flee in disorder. Tilly's right wing meanwhile attacked the Saxons, driving them from the field. Tilly now attempted to swing against the exposed Swedish left flank. The Swedes, shifting to the left with agility, countered the movement, repulsing the imperial attack. Gustavus then personally led his right-wing cavalry, followed closely by infantry, around Tilly's left flank, recapturing the Saxon guns as well as the unwieldy Imperial artillery, which had been left behind when Tilly attacked. Cut off from their communications to Leipzig, and under fire from their own and the quick-firing Swedish artillery, the Imperial army broke and fled when Gustavus pressed home his attack.

The Swedes pursued till nightfall, when the remnants halted under the cover of Pappenheim's rallied cavalry. Tilly's army lost 7,000 dead and 6,000 prisoners. He was badly wounded. The remnants of the imperial army retreated from Leipzig next day, and Gustavus

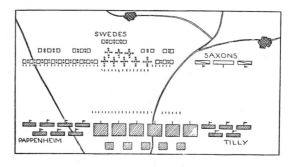

Battle of Breitenfeld

entered the city. His army had lost 2,100 killed and wounded. The Saxons lost about 4,000 men.

1631, September–December. Gustavus Adolphus Moves into the Rhineland. Because of the insecurity of the Protestant alliance, and thus of his base and his line of communications, Gustavus rejected suggestions that he advance on Vienna. Within 3 months he controlled all northwest Germany. Saxon troops had seized Prague. By Christmas, Gustavus and his allies had nearly 80,000 men within the shaking Empire. He captured **Mainz** (December 22) and spent the rest of the winter there.

1632, April. Ferdinand Recalls Wallenstein to Command. Wallenstein's terms made him a virtual viceroy. He quickly began to raise a new army.

1632, April 7. Gustavus Advances into South Germany. He crossed the Danube at Donauwörth and marched eastward into Bavaria, where the reorganized Imperial army waited, under Tilly and Maximilian of Bavaria.

1632, April 15–16. Battle of the Lech. Gustavus sent his army across the river on a bridge of boats and quickly attacked Tilly's entrenched camp. Tilly was mortally wounded, and Maximilian led the rest of the army in retreat, leaving most of the artillery and baggage on the field. Gustavus occupied Augsburg, Munich, and all southern Bavaria.

1632, July 11. Wallenstein Joins Maximilian at Schwabach. He entrenched his Imperial army of 60,000 near Fürth and the castle of Alte Veste (August). Gustavus, with 20,000, was entrenched near Nuremberg and sent for reinforcements. Their arrival brought his strength to 45,000.

1632, August 31–September 4. Battle of Alte Veste. Gustavus repeatedly attacked Wallenstein's entrenchments. The rough ground and heavy scrub of Wallenstein's well-chosen position made it impossible to bring either the redoubtable Swedish cavalry or artillery into action, and Gustavus withdrew with a heavy loss. The region around Nuremberg being denuded of food, both armies withdrew, Gustavus to the northwest, Wallenstein to the north.

1632, September–October. Wallenstein's Invasion of Saxony. Gustavus, responding to this threat to his communications, marched north at once. Wallenstein, who now had about 30,000 men, occupied the region around Leipzig. He vainly attempted to block Gustavus' crossing of the Saale with about 20,000 men (November 9).

1632, November 9–15. Gustavus Entrenches at Naumburg. He awaited reinforcements. Unaccountably, Wallenstein sent Pappenheim and a large detachment to Halle. Hearing of this, Gustavus marched rapidly to attack.

BATTLE OF LÜTZEN, NOVEMBER 16, 1632

As Gustavus approached (late November 15), Wallenstein sent an urgent summons to Pappenheim. He drew up his remaining forces, about 20,000 strong, in a defensive formation east of Lützen, facing south across a ditched road. The armies spent the night in battle formation.

Next morning Gustavus, held up by fog, was not able to attack until 11 A.M. He formed up his 18,000 troops south of the road, his right wing against a small treed area, with both wings composed, as at Breitenfeld, of interspersed cavalry and infantry units, **Bernard** of Saxe-Weimar commanding on the left wing, and Gustavus himself on the right, facing Count **Heinrich Holk.**

Gustavus charged Holk's cavalry, drove the musketeers from the ditch, and pushed the cavalry back upon the artillery. Wallenstein set fire to Lützen. The smoke blew into the faces of

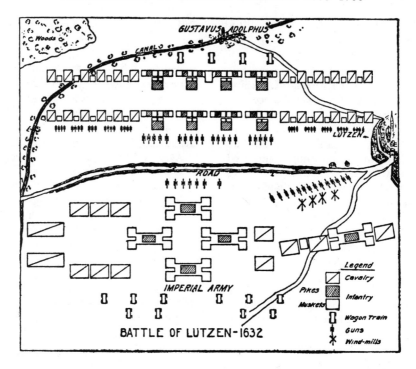

BATTLE OF LUTZEN – 1632

the Swedish center, which was then shaken by Wallenstein's surprise cavalry charge. It held, however, and Gustavus galloped to help rally the troops. In a confused cavalry engagement Gustavus was killed. Bernard took command.

At this juncture Pappenheim and his 8,000 men arrived, counterattacked, and again forced the Swedes back across the ditch. In the fray, Pappenheim was fatally wounded. Bernard drove Wallenstein's men back on Lützen, seizing the imperial artillery, clearing the ditch again, and drove Pappenheim's cavalry off. Wallenstein withdrew to Leipzig, leaving artillery and baggage behind. Wallenstein had about 12,000 casualties; the Swedes about 10,000.

1632, November 29. Death of Frederick. He died of the plague.

1633, March–April. The League of Heilbronn. This was established to defend the Protestant cause under the direction of Count **Axel Oxenstierna,** Chancellor of Sweden, who had succeeded Gustavus in directing Swedish policy in Germany. He also renewed the Franco-Swedish alliance.

1633, September–December. Wallenstein's Bid for Power. Wallenstein, in Bohemia after futile overtures for peace to Arnim and to Bernard, defeated Thurn and conquered all of Si-

lesia (October). While Bernard took Regensburg and occupied Bavaria, Wallenstein set up winter quarters in Bohemia, where he conspired to be King of Bohemia. Ferdinand, meanwhile, had again agreed to dismiss him.

1634, February 24. End of Wallenstein. His *coup d'état* failing, he fled with a few followers. Cornered at Eger, he was assassinated by his own officers, led by Scottish and Irish mercenaries, who were richly rewarded by the Emperor. King **Ferdinand** of Hungary (son of the Emperor) was named to the nominal chief command of the Imperial forces, with Count

Matthias Gallas, one of the chief conspirators against Wallenstein, as field general.

1634, July. Campaign in Bavaria. Bernard and Swedish General **Gustavus Horn,** with about 20,000 men between them, moved from Augsburg toward the Bavarian-Bohemian border, hoping to divert Ferdinand and Gallas, who were moving toward Regensburg. The Protestant commanders took Landshut, but lost Regensburg and Donauwörth to Ferdinand and Gallas, who laid siege to Nördlingen.

1634. Appearance of the Cardinal-Infante. Youthful Prince **Ferdinand** of Spain, brother of **Philip IV,** was leading an army of 20,000 from Italy to Spain. Having been appointed a cardinal, Ferdinand was called the *Cardinal-Infante* (Cardinal-prince). He was ordered by Madrid to join the imperial army in Bavaria.

1634, September 2. Junction of the Cardinal-Infante and Ferdinand at Nördlingen. The combined army of these young royal cousins—both named Ferdinand—was 35,000 strong. Horn and Bernard had arrived near Nördlingen with about 16,000 foot and 9,000 horse.

1634, September 6. Battle of Nördlingen. The Swedish-Protestant plan was poorly conceived and poorly executed. Horn, on the left, was to attack the Imperial right, under King Ferdinand, entrenched on a commanding hill. Bernard, in the plain, was to contain the imperial left flank, comprising the Spanish contingent. Horn was at first successful, his veteran Swedes storming the entrenchments and capturing the Imperial batteries. They fell into confusion, however, which was heightened when the captured powder magazine exploded. The unoccupied Cardinal-Infante turned most of his command to counterattack the hill. As the Swedes prepared to fire a volley, the alert Spanish infantry knelt and let the bullets pass over them, rising to fire themselves before the Swedes could reload. The Swedes were driven from the hill, Horn sending word to Bernard to cover his retirement across the valley. The Spanish and Imperialists now charged Bernard's troops, who broke and fled. The Swedes were overwhelmed and almost annihilated;

Horn was taken prisoner. The Protestant losses were about 12,000, while the Catholic armies lost 2,000. The battle was a catastrophe for Sweden. Politically, it forced Catholic Cardinal Richelieu to assume direction of the Protestant cause, bringing the struggle between Bourbon and Hapsburg into the open.

French Period, 1634–1648

1634, November 1. Treaty of Paris. France offered the German Protestants 12,000 men and half a million livres, in return for which Bernard and the Heilbronn League guaranteed the Catholic faith in Germany and ceded some lands in Alsace. France was not bound to enter the war openly, but no truce or peace was to be made without her.

1635, April 30. Treaty of Compiègne. Signed by Oxenstierna and Richelieu, Sweden recognized French ownership of the left bank of the Rhine from Breisach to Strasbourg. In return France accepted Sweden as an equally ally, and recognized Swedish control of Worms, Mainz, and Benfeld. France agreed to declare war on Spain and to make no separate peace.

1635, May 21. France Declares War on Spain. Meanwhile, much of Sweden's forces were engaged in a war with Poland.

1635, May 30. Peace of Prague. A partial reconciliation between the Protestants and the Catholic-Imperialists. From this point on, the war was primarily political rather than religious in nature.

1635, June–October. Richelieu's Strategy. His basic aim was to separate the Spanish Netherlands from Spanish Lombardy, with the related aim of eliminating all Spanish footholds in the intermediate areas of what is now eastern France and western Germany. French forces, totaling some 130,000 men, were divided into five main armies, having the following missions: (1) from Upper Alsace to invade and seize most of Spanish-controlled Franche-Comté; (2) to occupy Lorraine and repel Spanish invasion; (3) march across Switzerland to seize the key Valtelline Pass; (4) in alliance with Duke **Victor Amadeus of Savoy,** to invade Milan; and (5) in alliance with Dutch **Frederick Henry of Orange,** to invade the Spanish Netherlands. The first 3 missions were successfully accomplished by French armies; the 2 allied efforts were failures. At the end of the year Bernard and his Protestant "Weimar Army," after vague maneuvering in the Rhine and Main valleys, were incorporated into the French armed forces.

1635, August–November. Swedish Operations in East Germany. The Saxons and other signatories of the Peace of Prague turned against the small Swedish Army under **Johan Baner,** who completely outgeneraled his much more numerous opponents. Receiving reinforcements from Poland, he seized Domitz on the Elbe, then defeated the Saxons at the **Battle of Goldberg** (November 1).

1636. Invasions of France. Taking advantage of Richelieu's risky strategy of divergent lines of operation and of the retreat of Frederick Henry back to Holland, and with western and central Germany relatively peaceful for the first time, the Imperialists and Spanish decided to invade France. A Spanish and Bavarian army under the Cardinal-Infante drove through the scattered and outnumbered French forces in the northeast, while Gallas and the main Imperial army marched westward into Burgundy. The Spanish captured Corbie, crossed the Somme, and advanced on Compiègne. Despite initial panic in Paris, Richelieu and **Louis XIII** raised a new—largely militia—army of 50,000 and rushed with it to Compiègne. The Spanish and Bavarians withdrew to the Netherlands, where Frederick Henry had resumed the offensive. Meanwhile, Bernard's Weimar Army entrenched at Dijon and blocked the advance of Gallas, who, after attempting to go into winter quarters in Burgundy, was forced by desertions and French pressure to withdraw.

1636. Operations in Italy. A combined French-Savoyard army under Victor Amadeus and Marshal **Charles de Créqui** defeated the Spanish in the hard-fought **Battle of Tornavento** (June 22). But the duke refused to advance beyond the Ticino against Milan.

1636, October 4. Battle of Wittstock. Baner decisively defeated the combined Imperial-Saxon army, commanded by Count **Melchior von Hatzfeld** and Elector **John Georg.** Baner was assisted by Swedish Count **Lennart Torstensson** and Scottish soldiers-of-fortune **Sir Alexander Leslie, James King,** and **Sir Patrick Ruthven.**

1637. Operations on the French Frontiers. A Spanish invading army crossed the Pyrenees from Catalonia into Languedoc, but was halted at the rugged fortress of **Leucate,** and then badly defeated and driven back into Spain by a relieving French army under **Henri Charles de Schomberg, duke of Halluin.** On the Netherlands frontier the French captured several important fortresses, largely because the Spanish were giving primary attention to Frederick Henry of Orange and attempting to relieve their beleaguered garrison of Breda. In the east, Bernard defeated Duke **Charles of Lorraine** on the Saône River (June) and advanced into Alsace, still held by Imperial forces under Duke **Ottavio Piccolomini.** Piccolomini, however, was joined by a Bavarian army under **Johann von Werth,** who had cleared French detachments out of the Rhine Valley, and Bernard withdrew.

1637, February 15. Death of Ferdinand II. He was succeeded by his son and heir, Ferdinand III, King of Hungary.

1637, May. Swedish Advance in Germany. Armies under Baner and Torstensson reoccupied Brandenburg and threatened Leipzig and Thuringia. Their advance was halted by the arrival of Gallas' Imperial army in Saxony.

1637, October 10. Fall of Breda. Frederick Henry, after a year-long siege, regained the city, which had been held by the Spanish for 12 years.

1638. French Frontier Operations. These were generally unsuccessful, save in Alsace (see p. 593). The French were driven back in Italy and repulsed in the Spanish Netherlands; a large army under inept Prince **Henry II of Condé** advanced from Bayonne with Madrid as its objective, but was halted and repulsed at the small frontier fortress of **Fuenterrabia.**

1638. Swedish Operations in Eastern Germany. Imperial forces under Bavarian General **Gottfried von Geleen** forced Baner's outnumbered Swedish Army to withdraw from the Elbe to the Oder, where he expected reinforcements under Count **Karl Gustav Wrangel.** Instead, he was cut off by the main Imperial army under Gallas. Barely escaping the trap, he joined Wrangel in Pomerania, where he spent a cold, miserable winter, which seriously reduced the strength of his army.

1638, February 6–December 17. Siege of Breisach. Bernard, joined by a force under **Jean Baptiste Budes, Count of Guébriant,** invested the city and drove off Imperial attempts to relieve it. After eating all dogs, cats, and rats, the city surrendered (December 17), giving to the French the key to the Rhine.

1638, February 28–March 1. Operations in Alsace; Battle of Rheinfelden. Bernard with his Weimar Army moved to cross the Rhine at Rheinfelden, near Basel. He and his advanced guard were cut off from their main army, west of the river, by a surprise attack of Imperialists under Count **Savelli** and Werth. Bernard withdrew, crossed back at Laufenburg, and joined the rest of his army. Then he marched on Rheinfelden, breaking up and defeating Savelli's forces, which were attempting to defend it. Werth was captured.

1639. French Frontier Operations. On the Netherlands frontier the French captured the fortress of Hesdin, but a French army investing **Thionville** was smashed by Piccolomini (June 7). Condé was again unable to make any progress into Spain, but an invasion of Roussillon made some progress. French forces in north Italy, under the Duke of **Harcourt** (Cadet la Perle), were more successful. Although Savoy had now joined the Imperialists, Harcourt defeated an Imperial-Savoyard army at the **Battle of the Route de Quiers.** Turenne served under Harcourt in this campaign.

1639. Swedish Operations in Eastern Germany. Baner, reinforced, advanced into Saxony, Gallas retiring in front of him. The Swedes crossed the Elbe, defeated John George with a Saxon-Imperial army at the **Battle of Chem-**nitz (April 14), and overran western Saxony. Baner then invaded Bohemia, but was repulsed in an effort to capture Prague. He lived off the countryside, further devastating the region, retiring into the Erz Gebirge Mountains for the winter.

1639, February–July. Operations in Alsace. Bernard, having conquered Alsace, demanded possession with the title of Duke of Alsace. Richelieu refused. Bernard died of a fever as he was preparing to march to join Baner. His army was then commanded by Count **Jean de Guébriant.**

1639, October 21. Battle of the Downs. A strong Spanish fleet under **António de Oquendo** was destroyed by a Dutch squadron under **Maarten Tromp** (see p. 624).

1640. Minor Operations. Louis XIII captured Arras (August 8). In Italy, Harcourt with 10,000 men defeated 20,000 under Spanish General **Leganez** at the **Battle of Casale** (April 29). He then besieged another Spanish army in Turin, which was besieging the French garrison of the citadel. Leganez's reinforced army in turn tried to invest Harcourt, resulting in a situation unique in military history. Harcourt repulsed Leganez (July 11), then relieved the citadel and captured the city. In Germany, Baner was relatively inactive. Spain was in serious straits due to revolts in Catalonia and in Portugal (see p. 626).

1640, July 22. Naval Battle off Cadiz. A French fleet commanded by **Jean Armand de Maillé, marquis de Brézé** (nephew of Richelieu), defeated the Spanish Cadiz Fleet under **Gomez de Sandoval.**

1640, September 13–1641, October 10. Diet of Regensburg. Ferdinand, anxious for peace without concessions, tried to line up support from all the German princes. It was agreed that representatives should be chosen to negotiate on the basis of the Peace of Prague and a general amnesty.

1641. Revolts in Spain and France. These prevented major operations, though Harcourt continued to be successful in Italy. French troops in Catalonia assisted the rebels.

1641, May 20. Death of Baner. Lennart

Torstensson took command of the Swedish Army in Germany (November).

1642. French Operations. Unrest and revolt in France and Spain still handicapped the principal contestants. The Spanish had slightly the best of the continued frontier fortress sieges and countersieges on the Flemish border. The French had a similar advantage in northern Italy. Louis XIII led an army which conquered Roussillon on the Spanish Frontier. Richelieu crushed a Spanish-inspired conspiracy in Paris, while the French in Catalonia under Marshal **La Mothe Houdancourt** helped the insurgents hold the province in a successful campaign culminated by a victory over Lleganez at **Lérida** (October 7).

1642, June 30–July 2. Naval battle off Barcelona. A French fleet under Maillé-Brézé defeated a Spanish fleet under **Ciudad Real.**

1642. Swedish Operations. In the spring Torstensson crossed the Elbe and besieged Leipzig, at the same time overrunning much of Saxony. An Imperial army under Archduke **Leopold William** (brother of Ferdinand III) and Piccolomini approached, forcing Torstensson to raise the siege and retire northward toward Breitenfeld (November 1).

1642, November 2. Second Battle of Breitenfeld. While the Imperial army was forming up under the cover of a cannonade (featuring one of the first uses of chain shot), Torstensson led a cavalry charge against the Imperial left, which was smashed before it could form. The remainder of the Swedish army had meanwhile advanced, but was unable to gain any success until Torstensson and his cavalry hit the exposed left flank of the infantry of the Imperial center. The Imperial infantry was soon driven off the field, leaving isolated the cavalry of their right wing, initially successful against the Swedish left. Many of these were surrounded and captured; the remainder fled.

1642–1643. Deaths of Richelieu and Louis XIII. (See p. 613.) Before he died, Richelieu appointed the 21-year-old Duc **Louis d'Enghien,** son of the Prince of Condé, commander of the northeastern armies.

1643, May. Invasion of France. A Spanish-Imperial army from the Netherlands advanced through the Ardennes toward Paris. The invaders, some 27,000 strong, under **Francisco de Mello,** stopped to besiege Rocroi. Enghien, now barely 22, advanced with his army of 23,000 to defend the line of the Meuse. Finding the Spanish besieging Rocroi, and knowing that Spanish reinforcements of some 6,000 were en route to join them, Enghien decided to attack at once (May 18). The only approach to the town was through a defile between woods and marshes, which the Spanish had failed to block. Against the advice of elderly subordinates, Enghien marched through the defile and in the late afternoon took up position on a ridge overlooking the besieged city. The Spanish immediately formed up between the French and Rocroi. The armies bivouacked in their lines through the night.

1643, May 19. Battle of Rocroi. Shortly after dawn Enghien attacked, personally leading a successful cavalry charge against the Spanish left. Sending some squadrons in pursuit, Enghien led the remainder of the cavalry against the exposed left flank of the Spanish infantry. Meanwhile, the cavalry of the French left flank, against Enghien's orders, attacked the Spanish right, and were repulsed. Counterattacking Spanish cavalry scattered the French horse, but were halted by the French reserve. Enghien, learning of the disaster on his left, swung behind the Spanish lines, then cut his way directly through the rear center of the Spanish infantry to attack the rear of the Spanish cavalry facing his reserve. These horsemen also scattered, leaving the 18,000 Spanish infantry on the field, somewhat shaken from the unexpected developments, but still the most formidable foot troops in the world. Twice the French attacked the Spanish squares and twice were repulsed. Enghien now assembled all of his guns and the captured Spanish guns, and began to hammer the Spanish infantry mercilessly. The outcome inevitable, the Spanish asked for quarter; Enghien and his staff advanced between the lines to receive their surrender. But some of the Spanish infantry, thinking this was the beginning of another charge, opened fire at him. Enraged at this apparent treachery, the French army hurled itself at the squares, completely overwhelming the defenders and cutting down all whom they encountered. The Spanish army was virtually annihilated; 8,000 were killed, 7,000 captured. French casualties were about 4,000.

1643, June 18–23. Capture of Thionville. Enghien, anxious to secure Lorraine and the Rhine Valley, besieged and captured this important fortress. Ferdinand began to explore peace possibilities.

1643, August–November. Operations in Swabia. The Weimar army, under Marshal Guébriant, crossed the Rhine and advanced through the Black Forest into Württemburg. Guébriant was killed in the successful siege of **Rottweil** (November). He was succeeded by General **Josias von Rantzau,** who was defeated and captured by an Imperial army under Baron **Franz von Mercy** and General von Werth on the upper Danube at the **Battle of Tuttlingen** (November 24). Turenne succeeded to command of the army and retreated to Alsace.

1643. Swedish Operations. Torstensson ravaged Bohemia and Moravia without any effective interference from the Imperial army under Gallas. On outbreak of war between Denmark and Sweden, Torstensson hastily retreated to the Baltic, and spent the winter in Holstein.

1643, September 3. Naval Battle of Cartagena. A French fleet under Maillé-Brézé defeated a Spanish fleet near the seaport on the southeast coast of Spain.

1644. Early French Operations. Save for some fighting around Dunkirk, there was little activity on the northern frontier. Italy was also quiet. In Spain the French were driven from Lérida. Along the Rhine, Turenne, commanding the old Weimar army, crossed the Rhine (May) and advanced into the Black Forest, where he was opposed by a numerically superior Bavarian army under von Mercy. He withdrew to Breisach, leaving a garrison in Freiburg, which was besieged by Mercy (June). Turenne awaited reinforcements under Enghien.

1644, August 2. Junction of Enghien and Turenne. Enghien assumed command over the combined army of 17,000 men at Breisach. Mercy also had 17,000. Meanwhile Freiburg had surrendered.

1644, August 3–10. Battle of Freiburg. The combined French armies recaptured Freiburg in bitter fighting. There were three major engagements.

1644, August 3. French Attacks on Mercy's Lines of Circumvallation. The French initiated the first of a series of sanguine attacks against the Bavarian lines in difficult, mountainous terrain south of Freiburg. Enghien led a frontal attack against the Bavarian redoubts on the steep, wooded "Bohl," while Turenne, advancing through a narrow defile, attempted to turn Mercy's left. After severe fighting Mercy withdrew to fortifications on the Lorettoberg-Wonnhalde mountain, just south of the city.

1644, August 5. French Assaults on the Lorettoberg-Wonnhalde. After resting for a day before the Bavarian lines, the French on the 5th made a series of uncoordinated frontal assaults which were repulsed with great slaughter. The French and Weimar infantry were severely depleted.

1644, August 10. Rear-guard action at St. Peter. Enghein attempted a turning movement, sending Turenne and most of the army by mountain paths to get behind Mercy. Suspecting the French intention, Mercy withdrew from his entrenchments just before Turenne arrived. Covered by a sharp rear-guard action, at the St. Peter monastery in the Glotterthal valley northeast of Freiburg, the Bavarians retreated in good order, leaving the Rhine Valley to the French. Despite loss of half of his army, Enghien (with excellent help from Turenne) had gained another great victory. Leaving Turenne and the Weimar Army at Speyer, Enghien methodically captured the fortresses of the middle Rhine Valley.

1644. Swedish Operations. Torstensson completely outmaneuvered the Danish and Imperial armies under Gallas. The discouraged Gallas retreated into Bohemia, closely followed by Torstensson.

1644, October 13. Naval Battle of Fehmarn. In the Baltic Sea's Fehmarn Strait the Swedish fleet, commanded by **Karl Gustav Wrangel,** defeated the Danish fleet, commanded by **Pros Mund.** The Danes lost 1,200 men; the Swedes 100.

1644. Congress of Münster. Representatives of the warring powers met in an effort to achieve peace on the basis of preliminary diplomatic arrangements begun by Ferdinand.

1645. Swedish Operations. Torstensson, advancing toward Prague, was intercepted by a force of Imperialists and Bavarians under von Werth, which he decisively defeated in the **Battle of Jankau** (March 6). But the invaders (under Wrangel, who relieved Torstensson) were held up by a 5-month-long siege at Brünn and, unable to support themselves in impoverished Bohemia, they finally withdrew to Hesse, followed by Archduke Leopold.

1645, May 2. Battle of Mergentheim. Turenne, invading central Germany, was surprised and badly defeated by Mercy. He withdrew toward the Rhine, as Enghien and a Swedish contingent rushed to reinforce him.

1645, August 3. Battle of Nördlingen (or Allerheim). The combined armies of Turenne and Enghien, invading Bavaria, won a costly victory from Mercy, who was killed in the desperate struggle.

1646, January–July. Preliminary Maneuvering. In Hesse the rival armies of Wrangel and Leopold cautiously maneuvered around each other, most of their energies spent in search for food and forage for men and horses from the desolate, denuded countryside. Meanwhile, Turenne, not wishing to risk further losses, decided on a war of maneuver.

1646, August. Turenne and Wrangel Meet. Leopold's movement to the southwest in the Rhine-Main Valley, to seek food for the Imperial army, gave Turenne an opportunity. Secretly and rapidly he marched down the Rhine to Wesel, then southeast to join Wrangel near Giessen (August 10). Leopold, fearful that the larger allied army, 19,000 strong, would cut him off from central Germany, fell back to Fulda.

1646, September–November. Invasion of Bavaria. The Franco-Swedish Army by-passed Leopold, marching southeast into Bavaria and on to Munich, laying waste the countryside. Bavaria, which had recovered from the devastation suffered early in the war, was again ravaged mercilessly until Maximilian, disgusted by lack of Imperial support, sued for an armistice, resulting in the **Truce of Ulm** (signed March 1647).

1646. Other Operations. In Flanders, the French, under **Gaston of Orléans** and Enghien, took several frontier fortresses, including Dunkirk. In Italy, the French attempted to advance into Tuscany with little success. In Spain, Harcourt was outmaneuvered by Leganez in efforts to take Lérida and by the end of the year was replaced by Enghien, who had succeeded to the title of **Prince of Condé,** or "the Great Condé."

1646, June 14. Naval Battle of Orbetello. The French fleet of Maillé-Brézé, covering the siege of the Tuscan port, was attacked and defeated by a Spanish fleet under **Linharès.** Maillé-Brézé, at age 27 one of the greatest admirals of the time, was killed in the fighting.

1647. Operations in Germany. The increasing barrenness of the country made it more difficult for the armies to survive. Wrangel returned to Hesse, raided unsuccessfully into Bohemia, and retired to the Baltic coast. Turenne, ordered to Luxembourg to help repel a Spanish invasion from the Netherlands, was faced with a mutiny in the Weimar army, which he suppressed after a brief fight; he disbanded the army, then moved to Luxembourg with his French troops. Late in the year Maximilian again made peace with the Emperor and put his armies in the field against the French and Swedes, but too late for any action before winter.

1647. Other Operations. In the Flanders-Meuse area, save for one abortive Spanish attempt at invasion, the war of sieges continued. French forces in Italy, under Marshal **du Plessis-Praslin** (Duke César de Choiseul), won the **Battle of the Oglio** against the Spanish (July 4), but gained little from the victory. In Spain, Condé besieged Lérida, but was repulsed and had to withdraw toward the coast. Disgusted by lack of support from home and from the Catalonians, he resigned his command and returned to France.

1648, January 30. The Peace of Münster. End of the war between Spain and Holland (see p. 624), first result of the drawn-out negotiations begun in 1643. The other negotiations continued while the war went on in Germany.

1648, May. Invasion of Bavaria. Turenne and Wrangel again joined in Hesse and marched on into Bavaria, the Imperial-Bavarian army of 30,000 under Marshal **Peter Melander** retreating before them to the Danube. The allies destroyed and burned as they advanced.

1648, May 17. Battle of Zusmarshausen. The allies caught up with the Imperial rear guard and virtually destroyed it. 14 miles west of Augsburg. Melander was killed and the remainder of his army retreated in confusion. The allies continued on to the Inn, where they were halted by the reorganized Imperialists, now under able Piccolomini. A flood of the Inn halted operations for several weeks.

1648, June–October. Swedish Invasion of Bohemia. Hans Christoffer, Count von Koenigsmarck (ancestor of the future French military great, Marshal Saxe, and of George Sand, the French romantic novelist) led an army into Bohemia and besieged Prague. Piccolomini was withdrawn from Bavaria to protect Imperial lands from this more immediate threat. Königsmarck continued the siege, driv-

ing off Imperial relief efforts, and was preparing to assault the city in late October.

1648, July–September. Advance to Munich. The withdrawal of Piccolomini permitted Turenne and Wrangel to invest the Bavarian capital. But with peace negotiations close to culmination, Cardinal **Giulio Mazarin** of France (Richelieu's successor) ordered the allied army to withdraw into Swabia.

1648, August 10. Battle of Lens. Condé again commanded the northern army. He decisively defeated Archduke Leopold William in a final Hapsburg-Spanish attempt to invade France through Artois. Spanish military ascendancy was ended north of the Pyrenees.

1648, October 24. Peace of Westphalia. The Empire and France signed the **Treaty of Münster;** the Empire, Sweden, and the German Protestants signed the **Treaty of Osnabrück.** (The war between France and Spain continued; see p. 615.) Sweden received an indemnity and substantial Baltic coastal territories. France received Alsace, most of Lorraine, and the right to garrison Philippsburg on the east bank of the Rhine. In addition to minor territorial changes among the German principalities, there was a general amnesty, and the sovereign rights of the German princes were recognized by the Empire. Catholic and Protestant states were given complete equality in the Empire, with ownership of ecclesiastical states set as of 1624. Both treaties were guaranteed by France and Sweden.

WAR OF THE GRAND ALLIANCE, OR WAR OF THE LEAGUE OF AUGSBURG, 1688–1697

By 1680, France under Louis XIV threatened to secure hegemony over the continent. Her population, approaching 19 million, was 3 times that of England or of Spain and nearly 8 times that of the United Provinces. The organizational reforms of Louvois, Minister of War, had created the most powerful army in Europe. Under the administrative and financial genius of Colbert (see p. 583), the French Navy was one of the three major powers in the Atlantic and, after the withdrawal of the English from Tangier (1684), was supreme in the Mediterranean. Unlike the United Provinces (the Netherlands) and Spain, France's economy was not dependent upon overseas commerce. Thus, Anglo-Dutch efforts to apply economic pressures on France failed.

Louis's obviously expansionist objectives, particularly his efforts to gain influence over Spain, had aroused widespread fears. The Emperor **Leopold,** completing a successful war against Turkey, was strengthening the Imperial position in Central Europe to prevent French encroachment. The Protestant princes of Germany (notably **Frederick William** of Brandenburg) had begun aligning themselves with **William of Orange** of Holland in opposition to Louis. And in England, William's largely Protestant supporters were maneuvering to overthrow **James II,** a friend of Louis's.

Louis's revocation of the Edict of Nantes had made it easy for William to establish an anti-French coalition (League of Augsburg; July 9, 1686). Two years later, Louis provoked the League into preparation for war by claiming the Palatinate and meddling in the election of the Bishop of Cologne. Almost simultaneously, a revolution in England brought William of Orange to the English throne.

The war was fought mostly in the Netherlands, and was distinguished by the military incompetence of William of Orange and the naval myopia of Louis. Combat on land generally assumed the form of siege warfare. Naval engagements and maritime policy were essentially reflexes of the campaigns on land. After 9 years of inconclusive combat, economic and political pressures led the antagonists to conclude peace. Though French influence had increased on land, and English at sea, the basic struggles of Bourbon versus Hapsburg and England versus France were still undecided. The principal events were:

1688, September 25. Invasion of Germany. Louis, to encourage Turkey to continue its war with Leopold, sent a French army under the Dauphin into Germany before the German princes could be united in an effective anti-French coalition. Employing Vauban's techniques (see pp. 573–575), the French besieged and captured Philippsburg (October 29). This aggressive thrust, and subsequent devastation of the Rhine territory, consolidated German support of the Emperor.

1688, November–December. Revolution in England. (See p. 611.) The accession of William III immediately reversed England's pro-French foreign policy. An Anglo-Dutch military alliance was concluded, which resulted in Dutch naval support under English command, and English troops in the Netherlands.

1688–1689. Devastation of the Palatinate. To economize forces, Louis evacuated the Palatinate; to prevent the region being used as a base of operations against France, he ordered it devastated. Heidelberg, Mannheim, Speyer, and Worms were destroyed.

1689, April. Louis Declares War on Spain.

1689, May 1. Anglo-French Naval Skirmish at Bantry Bay. Thirty-nine French men-of-war, under Marquis **François de Châteaurenault,** clashed inconclusively with Admiral Earl **Arthur Herbert of Torrington's** 22 sail at Bantry Bay (Ireland).

1689, May 7. England and France Declare War.

1689, May 12. Treaty of Vienna. Alliance between the Netherlands and the Empire against France. England, Spain, Savoy, Brandenburg, Saxony, Hanover, and Bavaria soon subscribed to the provisions, to establish the "Grand Alliance."

1689, August 25. Battle of Walcourt. In Flanders, Prince **George Frederick of Waldeck,** commanding an army of 35,000 (including an English contingent of 8,000 led by **John Churchill, Earl of Marlborough**), defeated a French army under Duke **Louis de C. d'Humières** and Duke **Claude de Villars.** D'Humières was soon replaced by Marshal François de Luxembourg.

1690, July 1. Battle of Fleurus. In a bold double envelopment, using his cavalry superbly, Luxembourg with 45,000 men defeated Waldeck's English, Spanish, and German force of 37,000. The French lost 2,500 killed; the allies lost 6,000 dead and 8,000 prisoners. Luxembourg wished to exploit the victory by striking deep into the Netherlands and Germany, but Louis ordered him to keep abreast of other French armies maneuvering on the Meuse and Moselle.

1690, July 10. Naval Battle of Beachy Head. Admiral de Tourville with 75 ships decisively defeated Torrington's Anglo-Dutch fleet of 59 sail. Torrington, who had wished to avoid battle against a superior force in order to maintain a "fleet in being," had been ordered to fight by the Admiralty. The allies lost 12 warships. Subsequently, Torrington was tried; he was unanimously acquitted, but was dismissed from his command. Louis rejected recommendations that he should exploit the victory by sending a powerful army to Ireland and by pressing vigorously at sea for control of the Channel. This was the worst military decision he ever made. Britain and the Netherlands feverishly built up naval strength unhindered.

1690, August–September. French Advance in Savoy. At the **Battle of Staffarda** (Piedmont), Duke **Victor Amadeus II** of Savoy was defeated by a French army commanded by **Nicolas de Catinat** (August 18). The French advanced through Savoy with little resistance.

1690, October. Recapture of Belgrade by the Turks. (See p. 641.) This eliminated the possibility of an early peace on the eastern front; Leopold's main army remained tied down on the Danube.

1691, April 8. Capture of Mons. In Louis's presence, Luxembourg stormed Mons. William, operating ineptly from Brussels, failed to interfere. Luxembourg soon took Hal (June).

1691, July. Death of Louvois.

1691, September 20. Battle of Leuze. After a rapid night march, Luxembourg's cavalry routed Waldeck's army as it prepared to retire to winter quarters.

1691. Other Operations. In Piedmont, Catinat continued to progress. In Spain, the French Army of Catalonia, commanded by Duke **Jules de Noailles,** captured Ripoli and Urgel (September).

1692, May 25–June 5. Siege of Namur. It was taken by Vauban, under Louis's direct com-

mand, and supported by Luxembourg. William, attempting to relieve the besieged Dutch-Spanish-German garrison, was hampered by persistent rains and the flooding of the Mehaigne (a tributary of the Meuse).

1692, May 29–June 3. Naval Battle of La Hogue (Barfleur). De Tourville's fleet of 44 men-of-war and 38 fireboats, mounting 3,240 guns, engaged an allied force of 63 English and 36 Dutch warships and 38 fireboats, mounting 6,736 guns, under the command of Admirals **Edward Russell** and **George Rooke.** Unfavorable winds had prevented the French Toulon fleet from making a planned juncture with de Tourville. Despite inferiority in numbers, the French attacked and were decisively defeated, losing 15 men-of-war. This was the decisive battle of the war. Louis abandoned plans for an invasion of England by 30,000 troops under Marshal **Marquis B. C. de Bellefonds.** Henceforth, England dominated the Channel—although French privateers, like **Jean Bart,** inflicted severe losses on English shipping.

1692, July 24–August 3. Battle of Steenkerke. After unsuccessful attempts to engage Luxembourg throughout June and July, William attacked the main French force entrenched at Steenkerke. The allies, after initial success, were beaten back by French counterattacks. William ordered withdrawal under cover provided by the British Guards. Each army lost about 3,000 men.

1693, June 27–28. Naval Battle of Lagos. Off Lagos, de Tourville intercepted an Anglo-Dutch convoy proceeding toward Smyrna. He defeated Rooke's protective squadron, then destroyed nearly 100 of the 400 allied vessels.

1693, July 29. Battle of Neerwinden (Landen). William, entrenched near Landen with 70,000 troops, dispatched 20,000 to support Liége, which was threatened by de Noailles. Luxembourg with 80,000 men attacked the allied position. After 3 unsuccessful assaults, the French cavalry penetrated the defenses; William's army was routed. The French suffered 9,000 casualties, the allies 19,000. Again

Luxembourg failed to pursue, or William might have been forced to sue for peace.

1693, October 4. Battle of Marsaglia. Catinat's French army again decisively defeated the Duke of Savoy. He advanced into Piedmont.

1693, October 11. Capture of Charleroi. The French consolidated control over the Sambre southwest of Namur.

1693. Other Operations. In Catalonia the French took Rosas. The allied cause suffered another setback when an Austrian attack on Belgrade was repulsed (October; see p. 641).

1694, May–August. Operations in Catalonia. De Noailles captured Palomas, Gerona, Ostabich, and Castel-Follit. Supported by de Tourville's fleet, he besieged Barcelona.

1694, June–November. The English Fleet in the Mediterranean. In order to cut the French interior lines by land and sea, Russell, reappointed commander after an Admiralty purge, sailed for the Mediterranean (June). The arrival of the English fleet caused the French fleet to withdraw to Toulon harbor; de Noailles' army raised its siege of Barcelona.

1694, July. The English Bombard Dieppe and Le Havre.

1695, January. Death of Luxembourg. The undefeated French general was succeeded by the incompetent Duke **François de N. de Villeroi.**

1695, June–September. William's First Successful Campaign. He captured Dixmude and Huy (June). He then besieged Namur, manned by 14,000 troops under Duke **Louis de Boufflers.** After 2 months the French surrendered (September 1). The economics of the campaign were more significant than its uninspired military character; it was financed by the proceeds of the newly established Bank of England (1694).

1696, Spring. French Invasion Threat. The Mediterranean fleet, under Sir George Rooke (he succeeded Russell, September, 1695), was recalled to England early in the year when the appearance of Marquis **François L. R. de Châteaurenault's** fleet near Brest stimulated fears of invasion. This diversion enabled de Tourville to proceed from Toulon to Brest, increasing the apparent threat to England.

1696, June. Amphibious Failure at Camaret Bay. A landing near Brest, under English General **Thomas Talmash,** found the French prepared, and failed completely.

1696, June. Treaty of Turin. The Duke of Savoy reached an accord with Louis, who returned all conquests. This neutralized Italy and enabled Catinat to proceed to the northern front with his 30,000 soldiers.

1696, Summer-Fall. William Immobilized in Flanders. The French, reinforced by troops from Savoy, thwarted William's inept offensive efforts. Discouraged, he began secret peace negotiations with Louis (1696–1697).

1697, May–August. French Successes in the Mediterranean. With the English fleet gone, the French Navy raided Cartagena (May) and seized Barcelona (August). Leopold decided to seek peace.

1697, September 20–October 30. Treaty of Ryswick. France, Spain, England the Netherlands and the German principalities agreed to restore all territories acquired since 1679 (**Treaties of Nijmegen,** see pp. 620). Significantly—in anticipation of his Spanish ambitions—Louis surrendered fortifications at Mons, Luxembourg, and Courtrai to Spain and permitted the Netherlands to assume control of vital strongholds in the Spanish Netherlands (for example, Namur and Ypres). In the following month, Leopold and Louis arrived at a similar accord (October 30).

WESTERN EUROPE

ENGLAND, SCOTLAND, AND IRELAND

The Early Stuarts, 1603–1642

1603–1625. Reign of James I. James VI of Scotland, son of Mary Queen of Scots.

1604, October 24. Unification of Britain. The union of the crowns of England and Scotland eliminated internal frontiers and, consequently, a primary need for a standing army; this reduction of the crown's forces augmented Parliamentary power at the expense of royal authority.

1624–1625. English Participation in Thirty Years' War. James decided to assist Frederick in the Thirty Years' War (see p. 586). James also agreed to provide financial support to the operations of Christian IV of Denmark in Germany. Ernst Mansfeld, a German mercenary, led a small English force of 1,200 men to the Netherlands. There the expedition collapsed, suffering severe hardships as a result of inadequate training, supplies, and equipment. Spain considered dispatch of this expedition as an act of war.

1625–1649. Reign of Charles I.

1625, October 8. Failure at Cádiz. An ill-equipped force of 8,000, under Viscount **Edward Cecil of Wimbledon,** was repulsed.

1626–1630, April 24. Anglo-French War. The **Duke of Buckingham's** expedition to the **Isle of Ré,** near La Rochelle, to support Huguenot forces besieged by Richelieu was a fiasco (1627). Since Parliament refused to appropriate funds, Charles secured a forced loan to finance the French campaigns and to satisfy his debt to Christian IV. Buckingham was assassinated while preparing a second expedition against the French (1628).

1628, May. Petition of Right. A parliamentary list of demands and grievances, including protests against royal taxation without Parliamentary assent, the imposition of martial law during times of peace, and the mandatory billeting of troops in private homes.

1630, November 5. Peace with France and Spain.

1639. First Bishops' War. Scottish unrest, stemming from the restoration of church lands (1625) and efforts to introduce an Anglican liturgy into Scottish Presbyterian services, erupted in rioting. Insurgents seized Edinburgh Castle. Charles led an army to the vicinity of Berwick; however, continuing lack of funds, and Scottish military competence (many had fought under Gustavus Adolphus), compelled Charles to agree to a compromise settlement in **Pacification of Dunse** (June 18).

1640. Second Bishops' War. With hostilities renewed in Scotland, Charles convened the "Short Parliament" (April–May, 1640) to secure financial support. When the Commons refused, Charles dissolved Parliament. The Scots defeated Charles's troops at **Newburn** (August 28), then penetrated Northumberland and seized Durham. The **Treaty of Ripon** brought hostilities to a halt (November).

1640–1660. The Long Parliament. Charles, again in need of funds to pay his own army and the Scots, reassembled Parliament (November 3). The presence of Scottish forces in northern England, the crown's pressing monetary needs, and riots in London enhanced the power of Commons. Charles acquiesced in the execution of his Lord Lieutenant of Ireland, the **Earl of Strafford** (see below), and passage of the Triennial Act (May 1641) permitting Parliament to assemble every 3 years without royal initiative.

1641, October. Outbreak of the Irish War. Irish opposition to Strafford's policies and those of his successors culminated in confused armed rebellion, which continued intermittently for nearly a decade (see p. 606).

1641–1642. Prelude to the Great Rebellion. The crown, in desperate financial straits, required forces to cope with the Irish rising. Puritan dissidents in Parliament, like **John Pym** and **John Hampden,** feared the likely consequences of rejuvenated royal power in the form of a standing army. They secured passage of the **Grand Remonstrance** (December 1), a compilation of grievances spanning Charles's reign. Charles's inept attempt to arrest 5 leaders of Commons (January 3) prompted legislation providing for Parliamentary control of the armed forces and the ouster

of ecclesiastics from the Lords. Charles, who had hastily established a military base of operations at York, rejected the bills (March). Parliament reiterated its earlier demands and also demanded Parliamentary consent to the appointment of officers commanding fortifications (June 2). Parliament established a force of 20,000 infantrymen and 4,000 cavalrymen under Earl **Robert Devereux of Essex** (July). Charles began to raise an army at Nottingham (August 22).

The First Civil War, 1642–1646

Royalist and Parliamentary partisanship in the Great Rebellion cut across social, economic, religious, and family lines. The aristocracy, the peasants, and the Anglican establishment in general supported Charles; the emergent middle and commercial classes and the Navy supported Parliament. The north and west tended to be Royalists; the south, the Midlands, and London inclined toward Parliament. At the commencement of the war, Parliament's army was somewhat larger and better equipped and organized. Both forces suffered from the lack of experienced troops (however, both possessed experienced officers, many of whom had served in the Continental war); they were plagued by the reluctance of local militia to leave their home grounds; most critically, both suffered from a dearth of weapons and material. (The Navy's allegiance to Parliament prevented Queen **Henrietta Maria** from shipping supplies for the Royalists from the Continent.) On both sides, superfluous garrisons proliferated; supplies proved inadequate, pay tardy, and recruiting (from 1643, impressment) insufficient. The principal events were:

1642, October. Emergence of Oliver Cromwell. Parliament appointed him to command a military association of 6 counties.

1642, October 23. Battle of Edgehill. Charles with 11,000 men clashed with Essex, who had about 13,000. Prince **Rupert,** son of the Elector Palatine, beginning his dashing career as a leader of the superior Royalist cavalry, drove the Parliamentary cavalry off the field, then pursued. Essex, rallying his forces, counterattacked successfully, but was forced to withdraw when Rupert and the Royalist horse finally returned to the field at night-fall. Essex then fell back toward London. Charles moved his headquarters to Oxford.

1642, October 29. Charles Marches on London.

1642, November 13. Battle of Turnham Green. Following a skirmish at Brentford (November 12), Essex, reinforced by London militia to about 20,000, faced the king. After a brief skirmish Charles withdrew to Oxford.

1643, March–September. Widespread Inconsequential Strife. Oliver Cromwell had the best of a cavalry engagement at **Grantham,** east of Nottingham (March). The Cavaliers (Royalists) were successful in Yorkshire under Duke **William Cavendish of Newcastle** (April–July). Royalists under **Sir Ralph Hopton** were successful against **Sir William Waller** at **Stratton** in Cornwall (May), at **Lansdowne** near Bath (July 5), and **Roundway Down,** near Devizes (July 13). Rupert won a victory at **Chalgrove Field** near Oxford

(June 18), and captured Bristol (July 26). Charles was repulsed from Gloucester (August 5–September 10).

1643, August 10. Conscription. Parliament empowered local authorities to impress troops. The Royalists followed suit.

1643, September 20. First Battle of Newbury. Charles, having retreated from Gloucester to Newbury in the face of Essex's advance, turned to fight an inconclusive engagement. He then returned to Oxford.

1643, September 25. The Solemn League and Covenant. To enlist Scottish military support, Parliament agreed to protect Presbyterianism in Scotland, with virtual establishment of Presbyterianism throughout Britain.

1644, January. Participation of Scottish Rebels. A Scottish army (18,000 infantrymen and 3,000 cavalrymen) under Earl **Alexander Leslie of Leven** crossed the River Tweed.

1644, January. "The Cessation" in Ireland. Charles concluded a temporary peace with the Irish confederation, permitting him to withdraw his troops from Ireland and to recruit Irish Royalists as reinforcements for his cause in England.

1644, January 25. Battle of Nantwich. Irish Royalists and local troops laid siege to Nantwich, Parliament's only stronghold in Cheshire. Parliamentary reinforcements, led by Sir **Thomas Fairfax,** routed the Irish, almost half of whom joined the victorious force.

1644, February–April. Operations in the North. The Scottish army advanced into northern England, ending Cavalier supremacy in Yorkshire. Newcastle marched to stop them (February). Taking advantage of Newcastle's move to the north, Fairfax defeated his lieutenant, Lord **Bellasis,** at the **Battle of Selby,** thus threatening Newcastle's rear (April 11).

1644, March–June. Operations in the South. Waller finally defeated his rival Hopton at the **Battle of Cheriton** (March 29) near Dover. Charles outmaneuvered Essex and Waller, inflicting a sharp defeat on Waller at **Cropredy Bridge,** north of Oxford (June 6).

1644, April–June. Siege of York. Newcastle moved from Durham to York (April 18), where he was besieged by the Scots and Fairfax. Charles ordered Rupert to proceed from Shrewsbury with reinforcements (May 16). Rupert's advance caused the Parliamentarians to abandon the siege and to march to the vicinity of Long Marston in an attempt to head off the Royalists. There they were joined by a Roundhead (Parliamentarian) army under Earl **Edward Montagu of Manchester,** with Cromwell second in command. Newcastle joined Rupert and the joint force (under Rupert's command) proceeded to Marston Moor, 7 miles west of York.

1644, July 2. Battle of Marston Moor. Both armies had approximately 7,000 cavalry; however, Parliament had about 20,000 infantry to the Royalist 11,000. Manchester, Fairfax, and Leven jointly commanded the Parliamentary Scottish army, with Manchester evidently "first among equals." Cromwell, commanding his "Ironsides" and a Scottish force, defeated part of Rupert's cavalry on the Royalist right. On the left, however, Lord **George Goring's** horse turned back Fairfax and smashed the Scottish infantry, but became disordered in the process. Cromwell, swinging his disciplined Iron-sides to their right, routed Goring, then helped the remaining Parliamentary infantry crush the Royalist center. The Royalists lost nearly 3,000 dead, the Roundheads nearly 2,000. The defeat led to the capitulation of Royalist northern strongholds like York (July 16) and Newcastle (October 16). Rupert, with 6,000 survivors, escaped through Lancashire to rejoin Charles in the south.

1644, July–September. Charles's Success in the South. After Cropredy Bridge, Charles took the initiative from his opponents. He suddenly turned southwest, where Essex was attempting to consolidate Parliamentary control. Essex, trapped at **Lostwithiel** in Cornwall, escaped with his cavalry, leaving 8,000 infantry and his artillery to surrender (September 2). Southwest England was held for the Royalists. However, Manchester and Cromwell now joined Waller to threaten Charles's hold on south-central England (September–October).

1644, August 28–1645, September 13. Campaigns of Montrose in Scotland. James

Graham, Marquess of Montrose, raised the royal standard at Blair Atholl on August 28. His army numbered just 3,000 men and consisted of a crack brigade of Irish Macdonnels under **Alasdair MacColla,** Highland infantry, and a handful of cavalry. In a brilliant campaign, marked by victories at **Tippermuir** (September 1), **Aberdeen** (September 13), **Inverlochy** (February 2, 1645), **Auldearn** (May 9), **Alford** (July 2), and **Kilsyth** (August 15), Montrose defeated several armies of the Covenant, its militia, and allies. His final lopsided victory over General **William Baillie** at Kilsyth (6,000 casualties in a force of 6,800; the Royalists lost a few dozen men) threatened Edinburgh and alarmed Scottish commander the Earl of Leven in England. General David Leslie and a powerful force, mostly cavalry, were detached from Leven's army to crush Montrose. Montrose's force was surprised at **Philiphaugh,** near Selkirk, and surrendered on terms after a brief resistance (September 13, 1645). The Covenanters then massacred them, including the camp followers. Montrose escaped, but the Royalist cause in Scotland collapsed.

1644, October 27. Second Battle of Newbury. Charles, with about 10,000 men, met Manchester, Waller, and Cromwell, with about 22,000. The inability of the numerically superior Parliamentary forces to coordinate their attacks allowed Charles to retreat to Oxford after an inconclusive battle. Manchester refused to pursue.

1645, January–February. Treaty of Uxbridge. A brief truce, terminated when Charles rejected a set of Parliamentary proposals.

1645, January–March. The New Model Army. Meanwhile, Cromwell, who had long felt the need for major military reform, had urged Parliament to adopt a "frame or model of the whole militia," a proposal accepted by the Commons (January) and the Lords (February), which provided for a standing army of 22,000. The force was to be raised by impressment and supported by regularized taxation. The infantry, 12 regiments, totaled about 14,000 men; the cavalry's 11 regiments totaled 6,600 men, plus 1,000 dragoons (mounted infantrymen).

Artillery, a stepchild of the British weapons family, underwent some reorganization and expansion, largely on the basis of the continental experience. Modeled loosely after Cromwell's Ironsides, the new scheme introduced several innovations as well as vastly improved organizational principles. For the first time, red garb became the general uniform for a British army. Most significantly, the traditional aversion of the local militia to campaigns beyond their home territories was overcome in the creation of this mobile professional force.

1645, April 3. The Self-Denying Ordinance. It required members of Parliament to relinquish their civil and military commands. Fairfax succeeded Essex as captain general; Cromwell was granted a dispensation and became lieutenant general while still a member of Commons.

1645, May 7, Royalist Dissension. Charles resolved the growing rivalry between Rupert and Goring by splitting his 11,000-man army, sending Rupert to the north and Goring to the west. Parliament sent Fairfax to besiege Oxford.

1645, May 31. Battle of Leicester. Despite severe losses, the Royalists carried Leicester by storm in an effort to divert Fairfax from Oxford. Charles then proceeded toward Oxford. Parliament ordered Fairfax to raise the siege of Oxford and to "march to defend the association." Fairfax was reinforced by a cavalry contingent under Cromwell (June 13). Charles, surprised by the approach of the Roundheads, elected to fight.

1645, June 14. Battle of Naseby. Charles had about 7,500 men; the Parliamentary strength was 13,500. Rupert's Royalist cavalry defeated the Parliamentary left and charged off the battlefield in pursuit. Cromwell, whose mounted charge had defeated the cavalry on the Royalist left, exploited the prince's reckless pursuit and smashed the Royalist infantry, most of whom surrendered. Upon their return to the melee, Rupert's men refused to engage the Ironsides. Charles fled to Leicester. The victory of Naseby contributed decisively not only to Charles's eventual defeat but also to the ascendancy of Cromwell and the Independents within the Parliamentary party.

1645, June–1646, May 5. Royalist Collapse. In the southwest Fairfax and Cromwell overwhelmed Goring at **Langport** (July 10). Royalist strongholds (Leicester, Bristol, Bridgewater, Winchester) yielded one by one. After his final field force was crushed at **Stow-on-the-Wald,** west of Oxford, Charles surrendered to a Scottish, in preference to a Parliamentary, force (May 5, 1646).

Interlude between Civil Wars, 1646–1648

Charles's Scottish captors turned him over to Parliament for £400,000, about one-third of the amount owed the Scottish army (January 1647). Meanwhile, tension between Presbyterians in Parliament and Puritan Independents in the army became an overt schism. Parliament voted virtual disbandment of the army, which was owed several months' pay. The army refused to disband (March). Under Cromwell's virtual leadership, the army seized Charles and occupied London. Cromwell pressed the King to accept the army's proposals; Charles meanwhile conducted clandestine negotiations with the Scottish Presbyterians (August–October). The King escaped to the Isle of Wight (November). He concluded an **Engagement** with the Scots, promising to impose Presbyterianism for a 3-year period in return for military support in a campaign to regain his kingdom (December 26). Parliament renounced allegiance to the king (January 15, 1648).

The Second Civil War, 1648–1651

Overthrow of the Monarchy

Churchill has written: "The story of the Second Civil War is short and simple. King, Lords and Commons, landlords and merchants, the City and the countryside, bishops and presbyters, the Scottish army, the Welsh people, and the English Fleet, all now turned against the New Model Army. The Army beat the lot."* The principal events were:

1648, March—August. Uprisings in Wales and South England. Dissident Welsh Parliamentarians proclaimed allegiance to the King. Cromwell besieged and captured their stronghold at Pembroke Castle (July). A simultaneous revolt in Kent and Essex was suppressed by Fairfax; the insurgents took refuge at Colchester, but eventually capitulated (August).

1648, July—August. Scottish Invasion. James, Duke of Hamilton led 24,000 Scots into England. From south Wales, Cromwell marched with 8,500 men toward Wigan and Preston. He moved too fast for his artillery to keep up; the Scots had no artillery. Although numerically superior, the Royalists were poorly equipped (they were compelled to gather horses en route to serve as ammunition carriers); moreover, Hamilton permitted his units to straggle for nearly 50 miles.

1648, August 17–19. Battle of Preston. Cromwell attacked Hamilton's advance guard at Preston, defeating the Scottish-Royalist contingents in detail in 2 days of hand-to-hand combat, in which the main reliance of each side was on its pikemen. The Royalist survivors were split into disorganized bands; the army was supreme in England.

1648, August—December. Army against Parliament. Learning that Parliament reopened negotiations with Charles (July), the Army occupied London and seized the King in order to bring him to trial. When Parliament defied the military, Colonel **Thomas Pride**'s

* *History of the English Speaking People*, Vol. 2, p. 274.

troops prevented the entry of over 100 members of the Commons ("Pride's Purge," December 6–7). Those members permitted to sit (the "Rump Parliament") instituted a High Court, which tried the king for treason.

1649, January 30. Execution of Charles. After a summary trial by the Army-dominated Rump Parliament, the King was beheaded. The Crown was abolished and a Commonwealth established under military control.

THE IRISH CAMPAIGN, 1649–1650

From 1641 to 1649, against the background of Royalist-Roundhead conflict in the south, Ireland was subject to incessant strife between the English Protestant Royalists, who held Dublin under Earl **James Butler of Ormonde,** and Irish dissidents. In turn, the Irish rebels were split into factions—a Catholic confederacy, led by Papal Nuncio Archbishop **John Rinuccini,** and the established Anglo-Irish gentry—reflected in the schism between **Owen Roe O'Neill**'s Ulster army and Earl **Richard Preston of Desmond**'s Leinster forces. O'Neill defeated a Roundhead force under Sir **George Monroe** at the **Battle of Benburb** (1647), while Ormonde surrendered Dublin to the Parliamentarians commanded by Colonel **Michael Jones.** Subsequent events were:

1649, August. Royalist-Catholic Alliance. Ormonde, leading the coalition, established tenuous control of Ireland in behalf of the Crown. This was disputed by the Roundheads under Michael Jones.

1649, September. Cromwell to Ireland. Parliament ordered Cromwell to reestablish Commonwealth authority. After quelling a mutiny, Cromwell embarked.

1649, September. Battle of Rathmines. Jones's Roundhead army defeated the Royalists under Ormonde. Thus, when Cromwell arrived, the Catholic-Royalists had taken refuge in fortified towns.

1649, September–1650, May. Cromwell's Reign of Terror. Immediately initiating systematic siege warfare, Cromwell captured the principal Catholic-Royalist strongholds: Drogheda (September), Wexford (October), Clonmel (May 1650). He ruthlessly massacred the defenders of every captured fortress, the most horrible atrocities in British history.

1650, May. Cromwell's Return to England. He entrusted the continuation of the campaign to **Henry Ireton, Edmond Ludlow,** and **Charles Fleetwood.** Two years later the last Royalist stronghold at Galway surrendered (May 1652).

UPRISING IN SCOTLAND, 1650–1651

1650–1651. King and Covenant. Upon the execution of Charles I, Scottish Presbyterians proclaimed his son, **Charles II,** King of Great Britain. But before accepting Charles, they demanded that he accept the Solemn League and Covenant (see p. 603). Montrose had meanwhile brought a Royalist mercenary force from France. Suspected of being a Royalist without being a Covenanter, he was defeated by the Scots at **Carbiesdale** (April 27, 1650), captured, and executed. Charles, however, after entering into an agreement with the Presbyterians and agreeing to accept the Solemn League and Covenant, was invited to Scotland and crowned (January 1, 1651).

1650, July–September. Cromwell's Invasion of Scotland. Fairfax refused to lead a Scottish expedition. Cromwell was installed as commander in chief and crossed the border with a force of 16,000, including nearly 6,000 cavalry, and proceeded toward Edinburgh. David Leslie, commanding 18,000 Scottish infantrymen and 8,000 cavalrymen, maneuvered cautiously and defensively, his scorched-earch strategy forcing the Roundheads to rely on supply by naval transport. Cromwell vainly sought a decision. Disease and exposure reduced Cromwell's army to about 11,000; the Roundheads pulled back to Musselburgh and then to Dunbar, the Scottish army ringing the surrounding hills. The Scots then moved at night down from the hills in expectation of an attempted Roundhead evacuation by sea (September 2–3).

1650, September 3. Battle of Dunbar. Cromwell gathered the cavalry on his left and struck at dawn. The surprise charge, led by **John Lambert,** overwhelmed the Scottish right wing. (At midnight, in the mistaken belief that no action was imminent, Leslie had ordered 5 of every 6 Royalist musketeers to extinguish their matches.) The victory was total: the Scots suffered 3,000 killed, while the Roundheads had less than 30 dead and captured 9,000 prisoners. Leslie retreated to Stirling, leaving Cromwell in control of southern Scotland and the route to Edinburgh.

1650–1651. Interlude. The illness of Cromwell, combined with discord between Scottish factions, caused a hiatus.

1651, June–September. Stirling-Worcester Campaign. Upon Cromwell's recovery, and following unsuccessful overtures to the Scots for a pacific settlement, Cromwell led his forces to Perth, threatening Royalist supply lines to Stirling, but also leaving the road to London deliberately unobstructed. Charles, thinking he could regain his throne in a bold move, fell into Cromwell's trap. With Leslie's army he marched southward, proceeding along the west coast (July 31). Cromwell instructed Lambert and his horse to follow the Royalists, ordered another force to move from Newcastle to Warrington, and ordered the Midlands militia northward. Then, after capturing **Perth** (August 2), Cromwell led the main Roundhead force at the rate of 20 miles a day, down the east coast, collecting volunteer militiamen en route.

The four English contingents converged on the Scots at Worcester, Cromwell's engineers erecting bridges across the Teame and Severn.

1651, September 3. Battle of Worcester. Outnumbered, but not outfought, 16,000 Scots struggled hopelessly against 30,000 English, of whom 20,000 were the New Model Army. Few survivors reached Scotland, among them Charles II, who had fought with distinction. But he was soon forced to flee to France. Scattered Royalist strongholds held out until the capitulation of Dunettar Castle in May, 1652. The Civil Wars were ended.

The Commonwealth, 1649–1660

1649–1652. Naval Reform. Under Cromwell, zeal for military reform found expression in the Navy as well as the New Model militia. During the life of the Commonwealth, 207 ships were added to the Navy (40 between 1649 and 1651). The Royalist office of Lord High Admiral and the wartime Parliamentary Navy Board were replaced by the Committee of Admirals (1649). The committee improved food and instituted regularized wages and an incentive system of "prize money" for the capture or destruction of enemy vessels. It appointed three "Generals at Sea," former land commanders during the Civil War—**Edward Popham, Richard Deane,** and **Robert Blake**—the most important being Blake. The committee was transformed into the Commissioners of the Admiralty (1652); this body produced the original Articles of War.

1651, October 9. First Navigation Act. This anti-Dutch legislation forbade importation of goods transported by ships other than those of England or the vendor country.

FIRST ANGLO-DUTCH WAR, 1652–1654

The essentially naval conflict arose primarily from maritime competition, particularly in respect to the East Indies trade.

1652, May 19. Action off Dover (or Goodwin Sands). War began when Blake, patrolling the Channel with 20 ships (later reinforced to over 40) engaged a Dutch fleet under **Maarten Tromp** after he was refused permission to search a Dutch East Indies convoy. The Dutch withdrew after the loss of two vessels.

1652, July. Declarations of War.

1652, September 28. Battle of Kentish Knock. Blake with 60 English ships defeated an equal number of Dutch; personal antagonism of Dutch **Cornelius Witte de Witt**'s subordinates impeded his efforts.

1652, November 30. Battle of Dungeness. Maarten Tromp, restored to command, led a reinforced Dutch fleet of 80 vessels, twice the size of Blake's fleet, in a decisive victory off England's south coast.

1653, February 18. Action off Portland. Blake, with 70 English ships, stalked a Dutch merchant convoy escorted by Tromp's men-of-war. Blake's squadron, separated from the main fleet by a dense fog, was engaged by 80 Dutch sail off Portland, the Dutch retreating upon the appearance of the main English force (February 18).

1653, February 20. Battle of Beachy Head. A 2-day running fight culminated in a bitter battle. Tromp escaped with difficulty after losing 17 men-of-war and over 50 merchant vessels. The English lost about 10 warships. Blake was badly wounded.

1653, March. First Issuance of the Fighting Instructions. This officially provided for line-ahead formations to make optimal use of the broadside.

1653, June 2–3. Battle of the Gabbard Bank. An English fleet, commanded by Deane and **George Monck** (another general turned admiral), followed the Fighting Instructions in an engagement with Tromp. The arrival of Blake with 13 ships and their subsequent loss of 20 vessels caused the Dutch to withdraw. The English pursued to the Dutch coast.

1653, June–July. Blockade of the Netherlands. Finally Tromp managed to engage Monck in a diversionary action, slipped away, and joined Cornelius Witte de Witt at sea (July 25).

1653, July 31. Battle of Scheveningen (or Texel). The combined Dutch fleet of 100 ships attempted to shatter the blockade. After a preliminary inconclusive engagement (July 30), the English fleet under Monck, about equal in strength, won a bitter 12-hour contest. Tromp and 1,600 other Dutch sailors were killed and the Dutch lost 30 men-of-war. English losses were less than half as great. There were no other operations of significance.

1654, April 5. The Treaty of Westminster. Holland indemnified England and agreed to respect the Navigation Act.

COMMENT. *This conflict was of particular interest not only because of the official introduction of line-ahead, close-hauled formations with 100-yard intervals separating ships but also in light of the English government's use of propaganda to consolidate public support.*

STRUGGLE BETWEEN ARMY AND PARLIAMENT, 1653–1660

1653, April 20. Dissolution of the Rump. The growing schism between the Army and the Rump—culminating in Parliamentary proposals to drastically reduce the military establishment—led Cromwell to dissolve the assembly and the council of state, and to establish a new Council and the Little, or "Barebones," Parliament.

1654–1655. Operations against the Tunisian Corsairs. Blake, with 24 ships, was ordered to the Mediterranean (November). After securing an indemnity from the Grand Duke of Tuscany, Blake engaged the corsairs, winning a brilliant victory at **Porto Farina.**

1655, March–May. Penrudock's Revolt. Colonel **John Penrudock** led a short-lived Royalist revolt in Wiltshire.

1655. Military Reorganization. Cromwell organized England and Wales into 11 military districts, each supporting an armed force (financed by 10 percent taxes on Royalist property) and commanded by a major general.

1655. Capture of Jamaica. During underclared hostilities with Spain, an expedition of 2,500 soldiers, ill equipped, poorly disciplined, and afflicted with dysentery, commanded by Admiral William Penn and General **Robert Venables,** earlier dispatched to Barbados (December 1654), seized Jamaica (May). This led Spain to declare war.

1656–1659. Anglo-Spanish War. England joined France in an attempt to challenge Spanish hegemony over the Indies (see p. 615).

1656, September 9. Capture of Spanish Treasure Fleet. Seized by Captain **Richard Stayner** off Cádiz.

1656–1657. Blake's Last Victories. During the winter Blake successfully imposed one of the first extensive naval blockades in history on the Spanish coast. In the spring, in the most effective action of the war, he took his fleet into **Santa Cruz** harbor (Tenerife, Canary Islands) and destroyed 6 treasure transports, 10 escorts, and 6 forts (April 20, 1657). He died on the return voyage to England. (For land operations, see p. 615).

1658, September 3. Accession of Richard Cromwell. Upon the death of Oliver Cromwell, his son was proclaimed Lord Protector.

Relations between the Army and the Parliament (convened in January 1650) deteriorated. After a military show of force, Cromwell dissolved the Short Parliament (April 22).

1659, May 7. The Rump (Long) Parliament Reconvenes. Richard Cromwell, unable to control either Army or Parliament, resigned. The Army, suppressing Royalist uprisings, seized control and dismissed Parliament (October).

1659, November 7. Treaty of the Pyrenees. (See p. 616.)

1660, February 3. Monck Assumes Control. In response to popular dissatisfaction with the extramilitary activities of the Army, Monck seized power. He restored the Rump Parliament and the ascendancy of civil over military authority (February–March 1660).

The Later Stuarts, 1660–1700

1660, May. The Restoration of Charles II. Charles II returned to the throne, thanks to the support of Monck. Prince James, Duke of York, was appointed Lord High Admiral (later assisted by efficient Samuel Pepys) and Monck Captain General. The Army was reduced to 5,000 men (October). A Covenanters' revolt in Scotland was easily suppressed a few years later (1666).

SECOND ANGLO-DUTCH WAR, 1665–1667

Again commercial competition fostered Anglo-Dutch conflict. In 1663, the English had attacked the source of the Dutch slave trade, the West African ports (October), and the following year seized New Amsterdam (see p. 663).

1665, May. Declaration of War. This followed Dutch recapture of the African ports and **Michiel de Ruyter's** attack on Barbados.

1665, June 3. Battle of Lowestoft. Lord **Jacob van Wassenaar Opdam** led a Dutch fleet of over 100 ships in the seizure of English supply vessels returning from Hamburg. A powerful English fleet of 150 ships, commanded by Prince James (assisted by Prince Rupert, Sir William Penn, and Earl **Edward Montagu of Sandwich**), defeated Opdam in a sanguinary

battle, in which Opdam, Dutch Admiral **Kortenaer,** and English Admiral **John Lawson** were killed. Over 30 Dutch vessels were sunk. **Cornelis Tromp** (son of Maarten Tromp) skillfully covered the Dutch retreat. James's refusal to pursue the survivors led to his ouster from command; he was succeeded by Sandwich.

1665, August. Battle of Bergen. Sandwich, pursuing a valuable Dutch merchant convoy from the Indies into Bergen harbor, was repulsed by Danish shore batteries. England declared war on Denmark. de Ruyter arrived soon thereafter and escorted the convoy to Texel.

1666, January. France Enters the War against England. (See p. 616.)

1666, June 1–4. Four Days' Battle. Monck, commanding 80 sail, detached Prince Rupert's squadron of 25 ships to intercept a French force mistakenly believed to be arriving from the Mediterranean (late May). De Ruyter sailed against Monck with 80 ships. The Four Days' Battle commenced off the North Foreland with an English attack (June 1); the arrival of Dutch reinforcements caused Monck to begin withdrawal (June 2). Rupert's squadron after being delayed by adverse winds returned (June 3, evening). Following a fierce engagement, in which 20 English ships were lost, Monck and Rupert retreated to the Thames (June 4), which de Ruyter blockaded.

1666, July 25. Battle of the North Foreland.* Having repaired and refitted, Monck attacked, broke the blockade, and defeated de Ruyter, who lost about 20 ships. Monck proceeded to the coast of the Netherlands and attacked **Vlie Channel,** destroying 160 anchored merchantmen.

1666, August–1667, June. Peace Negotiations. The Dutch were now as ready for peace as were the British, still suffering from the Great Plague (1665–1666). While negotiations were in progress, Charles, responding to the effects of the Plague and the Great London Fire (September 2–9, 1666) and against Monck's adamant advice, moored the fleet and disbanded the crews. The Dutch made a token gesture at disarmament.

1667, June. De Ruyter's Raid in the Medway. The Dutch fleet, in a surprise raid into the Thames estuary, advanced up the Medway to within 20 miles of London, devastating shipping and several men-of-war en route. This highly effective strike and the ravages of the domestic catastrophes led England to seek peace in earnest.

1667, July 21. Treaty of Breda. The terms were a standoff, slightly in favor of the Dutch, who did, however, acknowledge English possession of New Amsterdam. England received some West Indies islands from France; France received Acadia from England (see p. 661).

1668, January 13. The Triple Alliance. (See p. 617.) Prompted by English, Swedish, and Dutch fears of French aspirations in the Spanish Netherlands.

1670, May. Treaty of Dover. Charles executed an about-face and concluded a secret treaty with Louis XIV. This provided for English naval support of proposed French operations against the Netherlands (and the establishment of Roman Catholicism in Britain) in return for 200,000 per annum. The ships of the Royal Navy were hurriedly put back in commission.

* Sometimes called St. James' Day Battle.

THE THIRD ANGLO-DUTCH WAR, 1672–1674

This was a deliberately provoked war of aggression by England and France. Holland had no desire to fight either nation, and certainly not both simultaneously.

1672, March 13. English Attack on Dutch Smyrna Convoy. Sir **Robert Holmes** attacked Dutch craft in the Channel, but the convoy escaped.

1672, March–May. Preparations for War. (For land operations, see p. 617.) While de Ruyter organized his fleet, British and French fleets, totaling 98 warships, concentrated in the Channel under the command of Prince James and completed their preparations off Sole Bay, Suffolk.

1672, May 28. Battle of Sole Bay. De Ruyter with 75 ships surprised the French and English. The rapid withdrawal of the 35 French vessels under Admiral Count **Jean d'Estrées** enabled de Ruyter to mass his ships against the slightly smaller English fleet. The arrival of English reinforcements forced the Dutch to retire, but only after they had badly mauled the English. The Duke of York was twice forced to abandon his flagship. Lord Sandwich was killed.

1672–1673. Lull in Naval Actions. Parliament opposed a war of aggression in alliance with France. During an internal political debate the Duke of York was forced to resign as High Admiral because of his conversion to Catholicism. Prince Rupert replaced him.

1673, May 28, Battle of Schooneveld Channel. In preparation for an English invasion of the Netherlands Rupert attacked de Ruyter's fleet in its coastal anchorage. The Dutch were ready. De Ruyter counterattacked, driving off the British with heavy loss.

1673, June–August. De Ruyter Takes the Offensive. In a series of minor engagements he forced the English fleet to retire to the Thames. His attempt to impose a blockade was frustrated by the outbreak of plague on his ships. An allied fleet then blockaded the Netherlands while the French prepared to invade by land.

1673, August 11. Battle of Texel. William of Orange ordered de Ruyter to sortie to protect an East Indies convoy. The French contingent, after a sharp fight, retired; following a more effective English defensive action, the French returned and the Dutch retreated. De Ruyter, however, had gained a clear-cut victory. He brought the convoy in, and he frustrated allied plans for a seaborne invasion.

1674, February 19. Treaty of Westminster. Due to growing public opposition to the war, the English government made peace. The war between Holland and France continued (see p. 618).

1679. Convenanter Rebellion. A Scottish rebel force defeated a Royalist force under **John Graham, Viscount Dundee** at **Drumclog,** south-central Scotland (June 1). **James Scott, Duke of Monmouth** (illegitimate son of Charles II) led a Royalist army that crushed the rebels at **Bothwell Bridge** (June 22). Monmouth then treated the rebels leniently.

JAMES II AND THE GLORIOUS REVOLUTION, 1685–1689

1685, February 6. Accession of James II. Upon his death (February 6), Charles achieved a final political victory—the accession of his brother, James II, a Catholic and former Lord High Admiral.

1685, July 6. Monmouth's Rebellion. The Duke of Monmouth, leading a Protestant revolt as pretender to the English throne, landed in Dorsetshire with 82 men. After recruiting considerable local support, he led a desperate attack against a Royal army commanded by **Louis Duras, Earl of Feversham** at **Sedgemoor** (July 6), was repulsed and the rebellion crushed. Monmouth was captured and beheaded.

1685–1688. Unrest in England. English Protestants resented James's efforts to restore Catholicism to equality.

1688, November–1689, February. Accession of William III. Seven Tory and Anglican leaders solicited the assistance of William of Orange, Stadholder of the Netherlands (see p. 624; married to **Mary,** daughter of James II) and his army in countering James's efforts to assert royal authority over the Church of England. William accepted the invitation, with the intention of leading England into the anti-French League of Augsburg (see p. 597). He landed in November and advanced on London. James fled to France (December 11). William and Mary were proclaimed joint sovereigns of England (February 13, 1689).

1689, April 3. Mutiny Act. Renewable semiannually, it extended the control of Parliament over the Army by the grant for 6 months of the right of military commanders to try soldiers by court-martial. Parliament also made the year-to-year existence of the Army dependent on Parliament by passing the Army Act, which provided for the Army's annual appropriation (1689).

The Irish War, 1689–1691

This war, the British phase of the continental War of the Grand Alliance (see p. 597), was the product of disparate motives united by common opposition to William III: James's desire to regain the English throne, Louis's desire to divert William from operations in Flanders in order to secure unopposed access to the Netherlands, and Irish support for a Catholic king against the hated Protestant English.

1689, March–April. James II in Ireland. James, accompanied by a small French force, landed at Kinsale. With the exception of several Protestant strongholds in the north, a Jacobite army of 40,000, under Earl **Richard Talbot of Tyrconnel,** held Ireland. Protestants in Enniskillen and Londonderry immediately affirmed their allegiance to William. James and Tyrconnel proceeded northward.

1689, April–August. Sieges of Enniskillen and Londonderry. Initial English efforts to raise the Irish-French investment failed. Later Captain **John Leake** led a naval convoy with reinforcements through the Londonderry blockade and terminated the siege (August 9–10). Soon after, local militia, commanded by Colonel **William Wolseley,** compelled the Jacobites to raise the siege of Enniskillen.

1689, June–July. Jacobite Rising in Scotland. While the joint Irish-French forces besieged Enniskillen and Londonderry, **John Graham of Claverhouse, Viscount Dundee** ("Bonnie Dundee"), James's leading Scottish partisan, attempted to foment a Jacobite rising in the Highlands. While defeating a Williamite army under Major General **Hugh Mackay** at the **Battle of Killiecrankie** (July 27) Dundee was killed. The next month the Jacobites' new commander, Colonel **Alexander Cannon,** led an advance on Perth, but was checked at **Dunkeld,** by the Earl of Angus's newly-recruited infantry regiment (later the **Cameronians**) in sharp street fighting (August 21). This repulse broke the Highland army, and Cannon's remnants were finally defeated at **Cromdale** (May 1, 1690). Within months, most of the Highland chiefs swore allegiance to William III. The ensuing English pacification of the Highlands was brutal

and marked by atrocity, the worst of which was the **Glencoe Massacre** (February 13, 1692). This treatment contributed to further Jacobite risings in Scotland in 1715, 1719, and 1745.

1689, September–November. English Offensives. English troops advanced into Ulster. Meanwhile, under orders from William, a small force led by Duke Friedrich **Hermann of Schomberg** landed near Belfast, but was repulsed at **Dundalk.**

1690, March–July. William in Ireland. William personally led an army of 35,000–40,000, including continental mercenaries, toward Dublin (June). Meanwhile, to the north, James's army—raised to 21,000 by the arrival of a relatively small contingent of French troops under **Antonin Nompar de Caumont, duc de Lauzun** (March)—encamped near the Boyne River.

1690, July 11. Battle of the Boyne. While the central infantry contingents of the 2 armies were engaged indecisively, the much more numerous English cavalry began an envelopment. James, competent and intrepid as an admiral, was unable to shift his Irish horse to block the English envelopment. He fled the field, and to France. His army fought well, and despite severe losses withdrew in good order to continue the war for another year.

1690–1691. William Consolidates. With the exception of western and southwestern Ireland, the island fell before William's advance. On the west coast, the Jacobite stronghold at Limerick, commanded by Earl **Patrick Sarsfield of Lucan,** resisted an English siege throughout the summer and autumn. However, Cork and Kinsale readily yielded to the Earl of Marlborough (September–October).

Louis denied James's request for support in a proposed invasion of England.

1691, July 12. Battle of Aughrim. Godert de Ginkel led a Williamite army against an Irish-French force of 25,000 under the Marquis de **Saint-Ruth,** who was killed. The allied effort collapsed. The English suffered some 700 casualties, while the allied army lost 7,000. This defeat ended James's Irish aspirations and the war, save for the continuing siege of Limerick.

1691, October 13. The Pacification of Limerick. The terms of the surrender of Limerick provided for the voluntary transport of Irish soldiers to France, freedom of religious (Catholic) belief in Ireland, and an amnesty. It was ratified by the English Parliament; however, the predominantly Protestant Irish assembly rejected the liberal terms and enacted a harshly anti-Catholic Penal Code.

COMMENT. *Had Louis XIV fully exploited his command of the sea prior to 1692, and sent an adequate force to Ireland, English and Irish history would have followed a far different course, and France might have achieved military hegemony on the Continent by the end of the century.*

FRANCE

The Age of Richelieu, 1610–1642

1610, May 14. Assassination of Henry IV. He had been preparing for war against the Hapsburgs on the Rhine to settle a succession dispute in Jülich-Cleves.

1610–1643. Reign of Louis XIII (9 years old). The queen mother, **Marie de Medici,** was regent during his minority.

1619–1624. Rise of Richelieu. The Cardinal Duke **Armand Jean du Plessis** became virtual dictator of France after political triumphs over the queen mother.

1625. Protestant Revolt. The Huguenots rose under the leadership of Duke **Henri of Rohan** and his brother **Benjamin of Rohan and Soubise.** Based upon La Rochelle, the revolt seemed destined to tear France apart, as had occurred in the previous century. Duke **Henry of Montmorency,** High Admiral of France,

defeated a Protestant fleet under Soubise off **La Rochelle,** to begin a partial blockade of that Protestant stronghold (September).

1626–1630. English Intervention. (See p. 601.)

1627–1628. Siege of La Rochelle. Richelieu personally supervised the 14-month siege. After 3 months' blockade, a land investment was begun (November). Duke **Henry II of Lorraine and Guise,** with the royal fleet, carried out a maritime blockade while stone dikes were built to close the port. By spring the dikes and sunken hulks blocked the channel. Two English relief fleets were repulsed (May, September). After heroic resistance the garrison capitulated (October 29), and the Protestant city lost all of its former privileges.

1628–1629. Defeat of the Huguenots. With the defeat of the Duke of Rohan by Montmorency in Languedoc, the Huguenots were crushed as a major and semi-independent political-military force in France.

1628–1630. War with Savoy. Richelieu indirectly attacked Spanish power by conquering Savoy and reestablishing control of the Valtelline Valley of the upper Adda River, above Lake Como, leading to the Stelvio Pass, linking the Hapsburg possessions of Milan and the Tyrol, thus blocking the main military and logistical route between Spanish Italy and the Spanish Netherlands. The climactic victory was won by Montmorency at **Avigliana** (1630).

1631–1648. French Participation in the Thirty Years' War. (See p. 591.)

1632. Noble Revolt. This was led by the King's brother, and enemy of Richelieu, Duke **Gaston of Orléans.** Montmorency led the rebel army, which was defeated by Marshal **Henry Schomburg** at **Castelnaudary** (September 1). Montmorency was captured and executed.

1642, December 4. Death of Richelieu.

1643, May 14. Death of Louis XIII.

The Early Reign of Louis XIV, 1643–1661

1643–1661. Mazarin's Ministry. Louis was 5 years old. His mother, **Anne of Austria,**

became regent, relying completely on Cardinal **Giulio Mazarin.** (Unlike Richelieu, Italian-born Mazarin was a political cardinal, not a priest.)

THE FRONDE, 1648–1653

Resentment of nobles and people against taxation and the accumulation of royal power under Richelieu, and even greater resentment of Mazarin, his replacement and continuer of the policies, led to full-scale revolt as soon as the Thirty Years' War was over, even though the war with Spain continued.

1648, July 12. Beginnings of the First Fronde. Demands of the Parliament of Paris, addressed to Louis and his mother, were rejected by Anne and Mazarin.

1648, July–December. Unrest and Rioting in Paris.

1649, January 8. Rebellion of Parliament. Mazarin was outlawed by Parliament, which also ordered seizure of royal lands. The royal family and Mazarin had fled to St.-Germain. Civil war broke out with many nobles, including **Prince Armand of Bourbon and Conti,** on the rebel side.

1649, January–February. Condé Supports the Queen. Partly because of a feud against his younger Bourbon brother, Armand of Conti, the Great Condé led his army against Paris, captured the fortress of Charenton, and laid siege to the city. When Conti and other rebel nobles appealed to Spain for help, Parliament sought peace.

1649, March 11. Peace of Rueil. Parliament dismissed its troops and made peace with the court. Mazarin was reinstated. The Queen declared a general amnesty.

1649, August–December. Friction between Condé and Mazarin. The general, thinking his services warranted a greater voice in affairs, quarreled with Mazarin, then began negotiating with the formerly rebellious nobles.

1650, January 18. Imprisonment of Condé. Mazarin also imprisoned his brother, Conti, and brother-in-law, Duke **Henry of Longueville.**

1650, February–April. The Second Fronde Begins. While nobles raised insurrections in Normandy, Burgundy, and Bordeaux, Turenne took over military leadership of the Fronde, and agreed to join a joint Spanish-Frondist army to invade France from the Spanish Netherlands.

1650, June–October. Inconclusive Maneuvering in Northeastern France. The invasion failed to take advantage of an opportunity to march on Paris largely through mutual suspicions of Turenne and the Spanish leaders.

1650, October 15. Battle of Champ Blanc. Turenne, vainly attempting to relieve besieged Frondeurs in Rethel, was defeated by the much larger and better-trained royal army under Marshal **César de Choiseul, Comte du Plessis-Praslin.** Turenne escaped with only a small fragment of his army.

1651, February–September. Amnesty and Lull. In a general amnesty the princes were released from prison (February 15) and Turenne returned to Paris. Mazarin fled to Germany.

1651, September. Renewal of Civil War. Condé, Conti, and other nobles again revolted, then captured Bordeaux and made it their base. Condé signed an alliance with Spain. Turenne remained loyal to the crown, but the revolt spread throughout France. Louis called Mazarin back to Paris (November).

1652, March–July. Operations between the Loire and Seine. A most complicated and fruitless series of maneuvers and engagements took place between Turenne and Condé.

1652, July 5. Battle of St.-Antoine. A new royal army, brought up from the frontier, joined Turenne to trap Condé outside the walls of Paris, now benevolently neutral toward the royal cause. Condé, after a bitter struggle, was on the verge of defeat when fickle Paris opened her gates to let him escape. He then fled to join Spanish forces and imperial troops of the Duke Charles IV of Lorraine invading France's denuded northeastern frontier. Turenne immediately marched north.

1652–1653. Operations North and East of Paris. Turenne outmaneuvered superior forces of the Spanish, Frondeurs, and Lorrainers, protecting the capital while the rebellion collapsed. Louis re-entered Paris (October 21, 1652). Mazarin returned (February 6, 1653). The Fronde was over, though the war with Spain went on.

CONTINUATION OF THE SPANISH WAR

1653–1657. Condé against Turenne. The pattern of the previous campaign governed. Despite superior strength, Condé, now a Spanish generalissimo, was unable to draw Turenne into combat under unfavorable conditions. In a war of maneuver and siege in northern France, Turenne won a victory at **Arras** (August 25, 1654), and was defeated at **Valenciennes** (July 16, 1656), but slightly outgeneraled his able opponent, who was hampered by Spanish suspicions and lack of cooperation.

1657. Franco-British Alliance. France and England being separately at war with Spain, Cromwell and Mazarin concluded a treaty of alliance whereby their forces were to attack jointly the coast towns of Gravelines, Dunkirk, and Mardyck. Dunkirk was to be ceded to England.

1657, Fall. Capture of Mardyck.

1658, May–June. Siege of Dunkirk. Turenne advanced rapidly from Mardyck on Dunkirk despite opening of the dikes by the Spanish. Investing the city—garrisoned by 3,000 men— he was joined by 3,000 English troops under Sir **William Lockhart.** The arrival of the English fleet boosted his force to 21,000 as siege operations began. Don **Juan of Austria,** viceroy of the Spanish Netherlands, gathered a relief army at Ypres under himself and Condé. This army consisted of 6,000 cavalry and 8,000 infantry plus 2,000 English Jacobites under **James, Duke of York** (later James II), furnished by pretender Charles II of England.

1658, June 7–13. Advance on Dunkirk. Don John and Condé moved into a camp site on the dunes, between beach and pasture lands northeast of Dunkirk.

BATTLE OF THE DUNES, JUNE 14, 1658

Turenne decided to seize the initiative by a surprise attack. During the afternoon of the 13th, he collected 9,000 cavalry and 6,000 infantry for battle, leaving 6,000 men guarding the siegeworks. There was some skirmishing before dark.

Early in the morning, just before low tide, Turenne's army—deployed in two lines and a reserve—advanced across the dunes; the English infantry was on the left; the cavalry was divided between his flanks; half on the beach and half in the meadows on the land side of the dunes. Turenne's advance was slow, giving the Spanish time to prepare. This slowness was deliberate because his plan was based upon the changing of the tide.

The Spanish infantry held a line across the dunes, right flank resting on the beach. A cavalry contingent covered the left flank; the bulk of the cavalry was in reserve; none was on the beach, for fear it would be destroyed by fire from the English fleet maneuvering offshore. Don Juan commanded the right, Condé the left. Against Condé's advice, Juan had advanced so rapidly that he had left his artillery behind.

Initial contact was made on the Spanish right, where the English attacked, supported by guns of their fleet. The reserve Spanish cavalry belatedly tried to support the infantry but was badly beaten by the French left-flank cavalry, led by the Marquis **Jacques de Castelnau.** In the center the French infantry slowly pushed the Spanish back. On the Spanish left, Condé's cavalry attack against the Marquis **François de B. de Créqui** was foiled by arrival of Turenne with cavalry from his left wing. Turenne's stratagem of using the change of the tide on the beach to enable him to carry out a cavalry envelopment around the Spanish land flank was a manifestation of his skilled adaptation of battle formations suited to the peculiarities of the battlefield. The battle, which lasted from 8 A.M. to noon, was a complete victory for Turenne, who only lost about 400 men. Don Juan's toll was 1,000 killed and 5,000 taken prisoner; the Spanish tercios were virtually annihilated. Turenne pursued vigorously until nightfall. Dunkirk subsequently surrendered and was ceded to Cromwell. It remained in English hands until Charles II sold it to Louis XIV in 1662.

1659, November 7. Peace of the Pyrenees.
Spain ceded much of Flanders and other frontier regions. The Spanish Empire was crippled, the French monarchy stabilized. Louis XIV married Maria Theresa, daughter of Philip IV of Spain. The following year Condé asked and received the forgiveness of Louis XIV.

The Personal Reign of Louis XIV, 1661–1700*

Upon the death of Mazarin, who had ruled in his name for almost 18 years, Louis XIV began his personal reign. His theory of government had emerged from the writings of Richelieu, the teaching of Mazarin, and his own experience. He was not inclined to share his authority with anyone—*''L'état, c'est moi.''* The nation in turn empathized with the person of the king. Louis's finance minister Jean-Baptiste Colbert sought to balance the need for a strong army with a stabilized and increased wealth. Emulating the Dutch, he promoted commerce and navigation, thereby initiating an extensive colonial policy. The colonial companies were represented to the king as armies participating in an economic war. A virtually new Navy was created for the defense of the colonies, and Vauban constructed great fortified naval bases at Brest, Dunkirk, Le Havre, Rochefort, and Toulon. Colbert had 20 warships in 1661; there were 196 vessels in 1671, and 270 in 1677. The Army was reorganized through the efforts of Louvois and Vauban (see p. 577).

The reorganization of the Army, the creation of the Navy, and the increased wealth were instigated by Louis's ambition for fame and his desire for an increase in territory. His secretary for war, Louvois, was an opportunist who played to Louis's weakness for military expeditions. A feud developed between the practical Colbert, who frowned on military expeditions, and Louvois, who pointed out the failures of commercial enterprises. The most important events were:

1664–1666. Operations against the Barbary Pirates. Because of damage to the commerce of Marseilles, Louis dispatched a naval squadron to bombard Djidjelli in Algeria and made subsequent attacks on Algiers until an agreement was reached with the Dey of Algiers.

1666. War with England. By an agreement of 1662 Louis was obligated to aid the United Provinces in their second war with England (see p. 610).

* The last 15 years of the reign (1700–1715) are covered in the next chapter.

THE WAR OF DEVOLUTION, 1667–1668

Upon the death of his father-in-law, Philip IV of Spain (1665), Louis demanded the inheritance right of devolution, or succession, and claimed the whole of the Spanish Netherlands. Spain refused; **Charles II** became king (see p. 626).

1667, May 24. Invasion of the Netherlands. Turenne quickly subdued a part of Flanders and Hainault.

1668, January. The Triple Alliance. The United Provinces, England, and Sweden concluded a treaty to check French expansion.

1668, February. Invasion of Franche-Comté. Condé, leading French troops, completed the occupation in 14 days.

1668, March–April. Triple Alliance Demands Peace. Louis entered negotiations.

1668, May 2. Treaty of Aix-la-Chapelle. Franche-Comté was returned to Spain, but France was left with a number of fortified towns on the Spanish Netherlands' frontier and some strongholds in Flanders. The unsettled succession question was deferred.

THE DUTCH WAR, 1672–1678

Louis set about isolating the Netherlands from her allies. The Triple Alliance was undermined by the secret **Treaty of Dover** with Charles II of England (May 22, 1670; see p. 610). A similar treaty was concluded with Sweden (1672), as well as treaties with Cologne and Münster.

1672, March. Declaration of War. The French army of 130,000 men, under Louis's personal command, marched down the Meuse, bypassing the Dutch fortress at Maastricht, to establish a base at Düsseldorf, on allied territory. Louis's army was well armed and equipped. The growing French Navy had the support of the English fleet. The Dutch Navy had 130 ships, but the land army in 1671 comprised only around 27,000 men, badly armed and commanded. When danger of war appeared imminent in February (1672), William III of Orange began to raise the effective force to 80,000 men.

1672, May–July. Invasion of the Netherlands. Three columns invaded the United Provinces: Turenne, with over 50,000 men, down the left bank of the Rhine; Condé, with an equal force, down the right bank; Luxembourg, with an army consisting primarily of Louis's German allies (Cologne and Münster), advanced from Westphalia toward Overijssel and Groningen. William and a weak field army attempted to hold the Ijssel-Rhine line, but the French rapidly forced passage at **Tolhuis** (June 12). Nijmegen, Gorinchem, and other towns fell. General Count **Henri L. d'A. Rochefort** led a cavalry raid into central the Netherlands, took Amersfoort and Naarden, was checked at Muyden, then captured Utrecht. The French advance threatened Amsterdam. However, the Dutch people saved Amsterdam by cutting the dikes and flooding the countryside. Meanwhile, Luxembourg had been stopped at Groningen.

1672, August. Revolution in the Netherlands. William of Orange was placed at the head of the government (see p. 625).

1672, August–September. Support for the Netherlands. Fearful of Louis's ambitions, several princes formed a coalition against France: Elector Frederick William of Brandenburg, the Emperor, and Charles II of Spain.

1672, August–September. Dispersion of the French Army. Against the professional advice of Turenne and Condé, Louis now parceled his army into small packets, trying to deal with all of his enemies at once. Turenne had urged that the Germans be smashed first by the concentrated French Army. Instead, he was sent to Westphalia, Condé to Alsace, other units to the Spanish Netherlands frontier, and the remainder of the Army waited in the Netherlands for the flood waters to freeze.

1672, September–December. Dutch Raid on Charleroi. William hoped to exploit the French dispersion, but was repulsed.

1672, September–1673, January. Turenne on the Middle Rhine. With forces never exceeding 20,000 men, Turenne out-maneuvered the much larger combined imperial and Brandenberg armies under Count **Raimondo Montecuccoli** and Frederick William. The latter was so discouraged that he sought peace (**Treaty of Vasem,** June 6, 1673).

1672–1673. Winter Operations in the Netherlands. Luxembourg, threatening Leyden and The Hague, was forced to withdraw to Utrecht by a sudden thaw. Condé had similar trouble attempting to take Amsterdam.

1673, June 29. Capture of Maastricht. Louis personally commanded an army of 40,000; Vauban's professional skill overcame the Dutch fortress after a brief siege. Louis then invaded Lorraine and conquered the electorate of Trier (Treves).

1673, July–November. Turenne and Montecuccoli. On advice of Louvois, Louis ordered Turenne, with 23,000 men, to prevent Montecuccoli (who had 25,000) from joining the Dutch. At the same time Turenne was to protect Alsace and do nothing to anger neutral states. In vain Turenne protested that these instructions were inconsistent. He did his best, but able Montecuccoli avoided battle and joined William at Bonn. The allies besieged and captured Bonn (November 12). This isolated Louis's allies (Cologne and Münster), who made peace.

1674, January. Growth of the Coalition. Denmark joined the coalition, many German princes joined the Emperor, the Great Elector (Frederick William of Brandenburg) re-entered the war, and England made peace with the allies by the **Treaty of Westminster** (February 19; see p. 611). Louis withdrew from the Netherlands so as to concentrate against the Spanish Netherlands and Franche-Comté, and to protect Alsace from the Germans.

1674. Other Operations. A French expedition to Sicily surprised the Spanish and gained success. In Roussillon a small French force under Marshal Schomburg repulsed Spanish attacks.

1674, May–June. Campaign in Franche-Comté. Louis, with Vauban providing the professional skill, recovered Franche-Comté in 6 weeks. Besançon fell after a siege of only 9 days.

1674, May–August. Campaign in the Spanish Netherlands. Condé held the line of the Meuse with about 45,000 men. William advanced against him with about 65,000, proposing to invade France.

1674, August 11. Battle of Seneffe. Seeing an opportunity to strike the allied army on the march, Condé attacked boldly with only half of his army and without artillery. The more-numerous allies rallied after an initial setback and recovered some lost ground. Both armies withdrew during the night. Condé, being joined by the remainder of his army, decided to renew the battle next day. But the allies had withdrawn completely; he could justly claim tactical as well as strategic victory. The allied plan of invasion was foiled. Allied losses were more than 10,000 killed, 5,000 captured (mostly wounded), and about 15,000 additional wounded. French casualties were nearly 10,000.

TURENNE'S RHINELAND CAMPAIGN, 1674–1675

Turenne was again given the mission of protecting Alsace. The Elector of Brandenburg and a number of other German princes had now entered the war against France. His principal opponents were Imperial Generals **Enea Sylvio Caprara,** Prince **Alexandre-Hippolyte of Bournonville,** and Duke Charles Leopold of Lorraine, assembling armies north and south of Heidelberg.

1674, June 14. Turenne Crosses the Rhine. Deciding to strike before Imperial forces could concentrate, he crossed at Philippsburg.

1674, June 16. Battle of Sinsheim. Caprara and Lorraine had moved with 9,000 men (7,000 being cavalry) to block Turenne's advance. Caprara thought his position behind the Elsenz River was inaccessible. Turenne thought otherwise. He crossed the river, drove the imperial outposts out of Sinsheim, then forced a way up rugged heights to a plateau where Caprara's main force was deployed. After a hard struggle, Turenne won a complete victory. The French lost 1,200 men killed, the imperialists 2,000 killed and 600 captured. Under orders from Paris, Turenne soon thereafter retired back across the Rhine.

1674, July–September. Summer Maneuvers. Reinforced to 16,000 men, Turenne again crossed at Philippsburg (July 3). The allies withdrew behind the Main after an engagement near **Heidelberg** (July 7). Under orders from Paris, Turenne devastated the region between the Main and the Neckar. Then he returned across the Rhine to block a threatened invasion along the Moselle. His army was reinforced to 20,000 men. The allied army, now under Bournonville, crossed and recrossed the Rhine near Speyer, then, while Turenne was occupied to the northwest, seized neutral Strasbourg (September 24). Turenne immediately moved to the vicinity, seeking an opportunity to attack his more numerous, but cautious, foes.

1674, October 4. Battle of Enzheim. Turenne, with 22,000 men and 30 guns, deliberately attacked Bournonville's 38,000, with 50 guns in defensive position covered by entrenchments. One English regiment was there under John Churchill, later Duke of Marlborough. Although his army had been marching for two days and two nights, Turenne held the initiative throughout most of the day, but was unable to drive the allies from their position. He withdrew from the inconclusive battle at the end of the day. The allies withdrew simultaneously. French casualties were about 3,500, those of the allies about 3,000.

1674, October–November. Watchful Waiting. Allied forces in Alsace were augmented to 57,000 by the arrival of the Great Elector. They made their winter quarters in all the towns from Belfort to Strasbourg; Turenne camped at Dettweiler. To mislead the enemy he displayed activity by placing the fortresses of middle Alsace in a state of defense, while he was quietly taking his whole field army (28,000 men) to Lorraine.

1674, November 29–December 29. Advance into Alsace. Turenne marched secretly southward behind the Vosges; in the last stages of the movement, he split his forces into many small bodies to further misguide enemy spies. After having marched over snow-covered mountains, the French reunited near Belfort and hastened into Alsace from the south. Bournonville tried unsuccessfully to stop them at **Mulhouse** (December 29). The scattered Imperial forces could only retreat toward Strasbourg. Turenne's advance continued to Colmar, where the Great Elector stood with forces slightly greater than his.

1675, January 5. Battle of Turckheim. In a battle reminiscent of Enzheim, Turenne attacked vigorously, despite exhaustion of his troops. After perfunctory resistance and light casualties the allies retreated, leaving the field to the victorious Turenne. Thus was completed one of the most brilliant campaigns in military annals.

1675. General Operations. The French recaptured some fortresses in the Meuse Valley and took Liége and Limburg. The expeditionary forces in Sicily continued to be successful, and Schomburg invaded Catalonia.

1675, June–July. Operations in Germany. The dejected Elector had withdrawn to Brandenburg and Montecuccoli resumed command of the Imperial armies. In a war of maneuver east of the Rhine between Philippsburg and Strasbourg, Turenne prevented Montecuccoli from retaking Strasbourg.

1675, July 27. Death of Turenne. Forcing Montecuccoli's more numerous army to withdraw, Turenne was typically preparing to attack the Imperial army in defensive position at Nieder-Sasbach, when he was killed by a cannonball (July 27). Thereupon the French were driven by Montecuccoli across the Rhine almost to the Vosges.

1675, August–October. Operations in Lorraine. French Marshal Créqui was defeated and captured by Charles of Lorraine at the **Battle of Conzer-Brucke** (August 11) on the Moselle. The allies then recaptured Trier (September 6) and prepared to invade France. Condé left Luxembourg in charge in Flanders and proceeded to take over the command of Turenne's and Créqui's armies. Condé forced Montecuccoli to retreat across the Rhine. These 2 men, the best generals of the era after Turenne, both retired from their commands at the end of the year.

1676, January–April. Operations in and around Sicily. After Admiral Marquis **Abraham Duquesne** repulsed De Ruyter's Hispano-Dutch naval attack on Messina (January), French Marshal **Louis Victor de R. Vivonne** defeated the Spanish Army in Sicily in the **Battle of Messina** (March 25) and was named viceroy of Sicily. De Ruyter was defeated by Duquesne and mortally wounded in the **Naval Battle of Messina** (April 22).

1676, April–August. Operations in the North. Again Louis and Vauban conducted the campaign. They captured Condé-sur-l'Escaut (April 26), then maneuvered against William III near Valenciennes. William was repulsed from Maastricht by Schomburg (August). Rochefort still held the Meuse Valley.

1676. Operations in the Rhineland. Créqui, who had been exchanged, held Lorraine, but lost Philippsburg to the Imperialists (September 17). The French now laid waste the land between the Meuse and Moselle to prevent it from supporting an invasion army. At the end of the year, Charles of Lorraine was engaged in a war of maneuver with Luxembourg on the upper Rhine.

1677. Operations in the North. Against the advice of most of his marshals, Louis approved Vauban's bold plan for an assault on besieged Valenciennes, which was completely successful (March 17). The French then besieged St.-Omer. William's relieving army was defeated by Duke **Philippe d'Orléans** at the **Battle of Mont Cassel** (April 11) and St.-Omer soon surrendered. During the summer William and Luxembourg carried on an indecisive campaign of maneuver, the French having the better of it.

1677. Operations in the Rhineland. Créqui drove Charles of Lorraine to the Rhine. Another Imperial army, attempting to cross the Rhine at Philippsburg, was isolated on a river island. Lorraine sent a large relieving force, which was smashed by Créqui in the **Battle of Kochersberg,** near Strasbourg (October 17). The invested army was then forced to surrender. Créqui followed up his successes by capturing Freiburg (November 14).

1678. French Successes in the North. Louis captured Ghent and Ypres (March). With his armies generally victorious, but threatened with the intervention of England on the side of the coalition, and informed by Colbert that the war was financially disastrous, Louis decided to make peace. Over William's objections the States General entered into negotations.

1678, July 6. Battle of Rheinfelden. Créqui defeated the Imperial armies.

1678, July 23. Battle of Gengenbach. Créqui was again victorious, and seized Kehl.

1678, August 10. Treaty of Nijmegen with the Netherlands. The Netherlands pledged its neutrality for the return of its territories.

1678, August 14. Battle of Saint Denis (near Mons). Presumably unaware that the treaty had been signed, William attacked Luxembourg and was defeated after a fierce struggle.

1678, September 17. Treaty of Nijmegen with Spain. France retained Franche-Comté, Valenciennes, Cambrai, Cambresis, Aire, St.-Omer, Ypres, Condé, Bouchain, Maubeuge, and other frontier fortresses. Spain received back Charleroi, Binche, Oudenarde, Ath, Courtrai, Limburg, Ghent, and others.

1679, February 6. Treaty of Nijmegen with the Emperor. France received Freiburg and gave up the right to garrison Philippsburg. Charles of Lorraine was permitted to regain his duchy, but refused to accept French conditions.

1679, June 29. Treaty of St. Germain with Brandenburg. The Elector surrendered nearly all of his conquests in Pomerania to Sweden (see p. 624).

Louis at the Height of his Power, 1679–1700

1680–1684. Chambers of Reunion. Louis, taking advantage of his superior position and the weakness of the Empire, set up courts of claims (chambers of reunion) to determine what dependencies had belonged to the towns and territories France had received in the peace treaties. Based on the decisions of his tribunals, he annexed many additional towns (such as Strasbourg). This policy of reunion involved an invasion of the Spanish Netherlands and the occupation of Luxembourg and Trier (1684). Lorraine was permanently annexed.

1684. Treaty of Regensburg. Louis, the Emperor, and the Empire concluded a 20-year truce; Louis retained everything he had claimed by reunion up to August 1, 1681.

1685, October 18. Repeal of the Edict of Nantes. This denial of Protestant freedom of worship led to the emigration of more than 50,000 families.

1688–1697. The War of the Grand Alliance (League of Augsburg; see p. 597).

Germany

1576–1612. Reign of Rudolph. Marked by continuing warfare against the Turks in Southeastern Europe and by Catholic-Protestant dissension.

1612–1619. Reign of Matthias. Growing tension between the Protestants and Catholics.

1618–1648. The Thirty Years' War. (See p. 583.)

1619–1637. Reign of Ferdinand II (of Styria). An ardent Catholic, his accession brought problems to his hereditary dominions and precipitated a serious Bohemian crisis.

1658–1705. Reign of Leopold I. He had been destined for the church and was very reluctant to become a temporal ruler. His prime interest was the promotion of the Catholic Church; the interests of the house of Hapsburg and Austria were secondary.

1658, August 16–1667. The Confederation of the Rhine. A defensive alliance of Catholic and Protestant princes to execute the Peace of Westphalia, to prevent foreign wars, and to defend the Rhineland.

1662–1664. War with the Turks. (See p. 636.)

1672–1679. The French Wars, or Dutch War. (See p. 617.)

1682–1699. War against the Turks for the Liberation of Hungary. (See p. 637.)

1688–1697. The War of the Grand Alliance. (See p. 597.)

Scandinavia

Denmark and Norway

1588–1648. Reign of Christian IV.

1611–1613. War of Kalmar. Caused by rivalry with Sweden in the Baltic, and by Swedish efforts to gain control of Finnmark. Christian sent forces to the mouth of the Gota and to Kalmar, which fell to the Danes after a long siege, despite relief attempts of **Charles IX of Sweden.** Charles died on his way home after the summer campaign, during which 16-year-old Prince **Gustavus Adolphus** of Sweden distinguished himself (October 1611). The following campaign was inconclusive; border provinces of both sides were harried. Mediation of James I of England resulted in the **Peace of Knarod** (January 1613). Finnmark was given to Denmark and Norway.

1618–1648. The Thirty Years' War. (See p. 583.)

1648–1670. Reign of Frederick III. A monarchical *coup d'état* (1660) supported by the clergy and burghers transformed the king into a hereditary and absolute ruler.

1656–1660. Danish Participation in the First Northern War. (See p. 622.)

1670–1699. Reign of Christian V.

1675–1679. The Dutch War. (See p. 617.) The Danes, allied to the Dutch, saw an opportunity to regain lost territory from Sweden, France's ally. Though the Danes were successful at sea, this was offset by Swedish land victories at **Lund** (1676) and elsewhere.

Sweden

THE RISE OF SWEDEN, 1600–1655

1599–1611. Reign of Charles IX. Charles led the intervention of Sweden in Russia during the Time of Troubles (see p. 631). His attempt to extend Swedish rule from Finnmark up to the Arctic led to war with Denmark.

1611–1613. War of Kalmar. (See 621.)

1611–1632. Reign of Gustavus II (Adolphus). Height of Swedish power.

1613–1617. War with Russia. (See p. 632.)

1617–1629. Wars in Poland. This resulted from a dynastic dispute with **Sigismund of Poland** (see pp. 528, 628).

1630–1648. Swedish Participation in the Thirty Years' War. (See p. 587.)

1644–1654. Reign of Christina. Decline of Swedish Power.

1654–1660. Reign of Charles X Gustavus.

THE FIRST NORTHERN WAR, 1655–1660

To extend Sweden's possessions in the southern Baltic, Charles X declared war against Poland. He was encouraged by Russia's current invasion of Poland and the internal dissension and weakness of Poland. His war plan provided for a double offensive: from Pomerania against western Poland, and from Livonia against Lithuania. His pretext was the failure of **John Casimir,** King of Poland, to recognize his own special status of "Protector of Poland."

1655, Summer. Invasion of Poland. Count **Arfwid von Wittenberg,** with a force of 17,000, occupied western Poland. Charles arrived soon after with reinforcements, increasing the Swedish Army to 32,000 men. He advanced into central Poland.

1655, September 8. Swedish Capture of Warsaw. Cracow fell soon afterward. John Casimir fled to Silesia. Meanwhile, Lithuania, simultaneously invaded by the Russians and the Swedes, had been surrendered by Hetman **Janusz Radziwill** to Sweden.

1655, October–January. Intervention of Brandenburg. Frederick William, taking advantage of Poland's collapse, seized west Prussia, a Polish fief. Charles immediately invaded Brandenburg, besieged Berlin, and forced the Great Elector to sign the **Treaty of Königsberg,** whereby east Prussia became a fief of Poland and the Elector became vassal of Charles as "Protector of Poland."

1655–1656. Revival of Polish National Resistance. Swedish atrocities, desecration of churches, and plunder of public and private property inflamed the Poles against the Swedes. The gentry formed military detachments all over the country; the partisans harassed the Swedes. John Casimir returned to Poland and assumed the leadership of the struggle.

1656, Spring. Charles X Returns to Poland. At the news of John Casimir's return to Poland, Charles rushed back from Prussia with 10,000 men. After some minor successes he was repulsed from the fortress of Zamosc, and near **Sandomierz** was surrounded by superior Polish forces in a triangle formed by the Vistula and the San rivers. In an extraordinary feat of bravery and skill, but with heavy loss, Charles broke out of encirclement and retreated to Prussia harassed by the cavalry of Hetman **Stefan Czarniecki** and ambushed in the forests by merciless peasant partisans. He reached Prussia with only 4,000 men. His remaining forces were dispersed as garrisons in Polish cities and fortresses.

1656, June. John Casimir Retakes Warsaw. He captured its Swedish garrison under Wittenberg. By the end of the same month, however, Charles X, reinforced by the army (18,000) of his new vassal and ally, Frederick William of Brandenburg, invaded Poland.

1656, July. Battle of Warsaw. Casimir and General Stefan Czarniecki, with 50,000 men, were defeated in a 3-day battle in which Brandenburg General **Georg von Derfflinger** distinguished himself. Charles reoccupied Warsaw, but the Elector refused to send his army deeper into seething Poland.

1657, February–July. Transylvanian Intervention. Another Swedish ally, Prince **George Rakoczy** of Transylvania, crossed the Carpathians and invaded Poland with 30,000 men. By summer, however, he was surrounded by Polish forces and capitulated. He left Poland with only 400 horsemen, his army being destroyed during the retreat by the Tartars.

1657, Spring. Russia Enters the War. The Russians invaded Livonia and besieged Riga.

1657, Spring. Support for Poland. The Emperor made an alliance with John Casimir, sending 12,000 troops to help the Poles recapture Cracow. Denmark declared war on Sweden (June 1) and was soon joined by the Elector, who deserted the Swedish cause. The overextended Swedes withdrew from Poland to protect threatened possessions around the Baltic.

1657, July. Swedish Invasion of Mainland Denmark. Charles and 6,000 troops crossed the frontier of Holstein and expelled the Danes from Bremen.

1657, October 24. Battle of Frederiksodde. Swedish Count Karl Gustav Wrangel and 4,000 men stormed the city. Danish Marshal **Bilde** was killed and more than 3,000 of his men surrendered. The Swedes now commanded the mainland of Denmark.

1658, January. Invasion of Insular Denmark. Charles led a daring advance over the ice from Jutland to Fyn and then to Zeeland. Denmark sued for peace.

1658, February 27. Treaty of Roeskilde. Denmark relinquished its possessions in the Swedish peninsula. Both countries agreed to make common cause to keep enemy fleets out of the Sound.

1658, June. Renewal of the Danish War. Danish delays in carrying out the peace terms and Dutch activity in Denmark made Charles fearful that more hostilities would be forthcoming, so he decided to crush Denmark completely.

1658, July. Repulse at Copenhagen. Charles ordered Wrangel to attack Copenhagen, Kronberg (Elsinore), and Christiania. A national patriotic movement enabled Copenhagen to withstand the Swedish assault, and Charles faced a bloody siege.

1658, September. Fall of Kronberg. International interest in the Baltic Straits now became important in frustrating Charles's design to control the Sound.

1658, October 29. Dutch-English Intervention. A fleet of 35 ships under Admiral **Opdam** joined the Danes and temporarily relieved Copenhagen.

1659, February 11. Dutch Intervention. Following renewal of the Swedish blockade, a Dutch fleet arrived to relieve Copenhagen, driving the Swedish fleet away.

1659, November. Battle of Nyborg. Michael de Ruyter transported 9,000 Danish troops from Jutland to Fyn. They defeated **Philip of Sulzbach** and 6,000 Swedish troops.

1660, February 13. Death of Charles X. Sweden, under a regency of nobles, immediately sought peace.

1660, May. Treaty of Oliva. Peace between Sweden and Poland. Both gave up respective monarchical-dynastic claims. Livonia was confirmed to Sweden. Brandenburg received full title to East Prussia.

1660, June. Treaty of Copenhagen. Denmark formally surrendered Skane, the southern portion of the Scandinavian peninsula. She and Sweden agreed to guarantee access of foreign vessels to the Sound.

1661. Treaty of Kardis. This confirmed the *status quo ante bellum* between Sweden, Poland, and Russia.

THE REIGN OF CHARLES XI, 1660–1697

1672–1679. The Dutch War. (See p. 617.) Sweden was allied with Louis XIV of France.

1675, June 28. Battle of Fehrbellin. An army

of 12,000 Swedes invading Brandenburg were defeated by Field Marshal Derfflinger. The Brandenburgers then took the offensive and invaded Swedish Pomerania, taking Stettin, Stralsund, and **Griefswald.**

1676, May 25. Battle of Jasmund. The Swedish Navy was defeated by the Danes under Admiral **Niels Juel.**

1676–1679. Campaigns of Charles XI. The young king took the field to win impressive victories against Danes and Brandenburgers, and retain Sweden's preeminence in northern Europe. He decisively defeated the Danes in the campaigns of 1677 and 1678. Danish Admiral Juel, however, again defeated the Swedes, under **Evert Horn,** near Copenhagen at **Kjöge Bight** June 30, 1677).

1679, June 29. Peace of St. Germain-en-Laye. The Elector returned to Sweden nearly all of the conquests he had made in Pomerania.

1697–1718. Reign of Charles XII. (See p. 704.)

THE NETHERLANDS

1609–1621. The Twelve Years' Truce. The desultory, indecisive fighting with Spain ended temporarily and the northern provinces consolidated their independence.

1621, August 9. Renewal of the War with Spain. As soon as the 12 years' truce expired, hostilities broke out. This became part of the Thirty Years' War (see p. 583). Maurice commanded the Dutch; Spinola again commanded the Spanish.

1624, August 28–1625, June 5. Spinola's Siege and Capture of Breda. Maurice was repulsed in attempts to relieve the siege.

1625, April 23. Death of Maurice. He was succeeded as stadholder by his brother **Frederick Henry.**

1626–1629. Dutch Successes. The Dutch captured Oldenzaal (1626), Grol (1627), 's Hertogenbosch, and Wesel (1629). These latter successes interrupted Spanish and Imperial communications along the Rhine.

1631. Naval Battle of the Slaak. A Dutch fleet defeated a Spanish invasion fleet.

1632. Dutch Capture of Maastricht. Peace negotiations were undertaken, but without results.

1635. Dutch-French Alliance.

1637. Dutch Recapture of Breda.

1639, October 21. Battle of the Downs. Maarten Tromp engaged and dealt a blow to a fleet of Spanish warships and transports commanded by **António de Oquendo.** The Dutch now held complete mastery of the seas.

1643. Battle of Rocroi. (See p. 594.) Peace negotiations were begun anew.

1644–1645. Dutch Successes. Capture of Sas van Ghent gave Frederick Henry a foothold on the southern bank of the Scheldt (1644). Next year the Dutch captured Hulst.

1648. Treaty of Münster (Part of the Peace of Westphalia). This ended the long "Eighty Years" war with Spain. The frontiers that had been established were recognized, as was the independence of the United Provinces.

1647–1650. William II as Stadholder. He disapproved of the Treaty of Münster because it left France and Spain at war. He entered into secret negotiations to reenter the war in alliance with France, but he died suddenly of smallpox (November 6, 1650).

1650–1672. Holland Without a Stadholder. William's posthumous son **William III** was now head of the House of Orange. **Jan de Witt** was the strong man who became the Grand Pensionary (1653) and maintained Dutch prestige for the next 20 years.

1652, May–1654, April 5. First Anglo-Dutch War. (See p. 607.)

1657–1660. The First Northern War. Holland intervened on the side of Denmark to prevent the takeover of the entrance of the Baltic by Sweden (see p. 623).

1657–1661. War with Portugal. This was over the expulsion of the Dutch from Brazil (1654; see p. 665). The Dutch were unsuccessful.

1664. Undeclared Hostilities with England. England seized Dutch posts on the West African coast and the New Netherlands Colony, whose capital, New Amsterdam, was renamed New York.

1665–1667. Second Anglo-Dutch War. (See p. 609.)

1668, January 13. The Triple Alliance. (See p. 617.)

1672–1678. War with France and England. (See p. 610.)

1672, August 27. Overthrow and Murder of de Witt. A popular uprising established the young son of William II as stadholder.

1672–1702. William II as Stadholder. Aided by the Emperor, the Elector of Brandenburg, and the withdrawal of the British (1674), William held off the French and emerged from the **Treaty of Nijmegen** (1678) without losses (see p. 620).

1678, March. Anglo-Dutch Defense Alliance.

1686. The League of Augsburg Formed by William. (See p. 611.)

1688, November 5. The Glorious Revolution. William, whose wife was Mary, daughter of James II, landed in Torbay, England, in response to an appeal from the opponents of James II (see p. 611). William and Mary became rulers of England.

1689–1697. War of the Grand Alliance (League of Augsburg; see p. 597.)

SWITZERLAND

The Swiss Confederation was a very loose kind of union with the cantons divided among themselves on religious questions. Fighting was the main occupation and Swiss mercenaries continued to be employed by foreign states, especially France. The principal events were:

1602. Savoyard Invasion. The invaders, supported by Spain, were repulsed from Geneva.

1618–1648. The Thirty Years' War. The Swiss Confederation was officially neutral. Acceptance of the principle of neutrality was significant. Swiss mercenaries were hired by both sides and much material was sold to the belligerents at a profit.

1620. Spanish Seizure of the Valtelline Passes. This was the most convenient link in communications to the Spanish Netherlands (and also to Hapsburg Austria) from the Spanish Hapsburg possessions around Milan, Italy. A Swiss effort to expel the Spanish was unsuccessful (1621).

1625. Swiss Seizure of the Valtelline. The Swiss were led by the French **Marquis de Coeuvres.**

1626. Treaty of Monzon. As a result of the mediation of Pope **Urban VIII,** France and Spain agreed on rights in the Valtelline. Spain retained control.

1637. Swiss Seizure of the Valtelline. A subsequent treaty with Spain (1639) left the passes open to the use of Spanish troops, but reinstated Grisons' sovereignty.

1648. The Treaty of Westphalia. European recognition of the Swiss Confederation's independence of the German Empire.

1656, January 24. First Villmergen War. The Catholic cantons under **Christopher Pfyffer** defeated the Protestants from Bern and Zurich.

ITALY

Italy continued to be divided into numerous political entities, many under foreign rule. There was a general decline in prosperity. The papacy tried to be neutral in the conflicts between the Bourbons and the Hapsburgs. In Savoy the French and Spanish factions met in sporadic conflicts, which brought about a decisive weakening of the ducal power. Naples continued under Spanish rule. Venice maintained independence and became an advocate of peace and neutrality despite her wars with the Turks. The *status quo* was restored in Italy by the **Peace of the Pyrenees** (1659) as Spain kept her traditional territories and France retained Pinerolo with its control over Savoy.

IBERIAN PENINSULA

Spain

1598–1621. Philip III. He devoted himself to the needs of the church rather than the country. Spain entered a period of decline and decadence.

1609, April 9. Twelve Years' Truce with the Netherlands. This followed 45 years of wasteful warfare (see p. 624).

1609–1614. Expulsion of the Moriscos. (See p. 533.)

1618–1648. The Thirty Years' War. (See p. 583.)

1621–1665. Reign of Philip IV. The military and economic decline of Spain accelerated.

1640. Independence of Portugal. (See below.)

1648. The Treaty of Münster. This ended the Thirty Years' War and the long war with the United Provinces; war with France continued (see p. 597).

1659, November 7. The Peace of the Pyrenees. (See p. 616.) This treaty marked the end of Spanish ascendancy in Europe and the beginning of French ascendancy. For the first time in 40 years, Spain was free from foreign war.

1665–1700. Reign of Charles II. (See p. 617.) A half-mad monarch, the last of the Spanish Hapsburgs.

1667–1668. War of Devolution. (See p. 617.)

1673–1678. The Dutch War, or War of the First Coalition against Louis XIV. (See p. 617.) Disastrous to Spain.

Portugal

1640, November–December. Portuguese Independence. Revolt against Spanish rule. Spain was unsuccessful in attempts to regain control (1640–1668).

1640–1656. Reign of John IV. First king of the house of Braganza. He was recognized by France and Holland, but Spain continued sporadic hostilities.

1641, January. Surrender of Malacca to the Dutch. (See p. 653.)

1644, May 26. Battle of Montijo. A Portuguese force commanded by General **Mathias d'Albuquerque** and backed by England and France successfully invaded Spain. After this victory near Badajoz, Portugal was left in peace for several years.

1657–1661. War with Holland. The Dutch were unable to regain the foothold in Brazil from which they had been driven (1654).

1661–1663. Spanish Invasion. Philip IV of Spain attempted to reconquer his lost kingdom. Don Juan and some 20,000 men crossed the border from Estremadura while a smaller force entered Portugal from the north. The Portuguese, aided by an English auxiliary force under Schomberg, were successful in halting the Spanish. Between 1662 and 1663 Don Juan was able to overrun Alentejo Province.

1662. English Assistance. Charles II of England married **Catherine,** the daughter of John IV, and in return for the dowry provided men and arms for the war with Spain.

1663, June 8. Battle of Ameixal. Sancho de Vita Flor defeated Don Juan.

1663, June 17. Battle of Montes Claras. Antonio de Marialva routed the Spanish.

1664–1665. Portuguese Successes. Desultory warfare.

1665, June. Battle of Villa Viciosa. Marialva and Schomberg defeated Count **Caracina** in a bitter 8-hour hand-to-hand battle. This ended Spain's efforts to reconquer Portugal.

1668, February 13. Peace with Spain. Through the mediation of Charles II of England, Spain finally recognized Portugal's independence.

EASTERN EUROPE

POLAND

This was a century of almost continuous wars, offensive in the first half of the century and defensive in the second. These exhausting conflicts contributed to Poland's downfall in the 18th century.

At a time of infantry predominance in western Europe, cavalry remained the principal arm in Poland, and the charges of Polish cavalry armed with lance and heavy sword broke many pike squares. Polish heavy cavalry wore armor (mail shirts), discarded in the West. At the same time, Polish horsemen were highly skilled in warfare against the elusive Tartars. Camps fortified by several rows of chained carts were widely used as field fortresses. The principal events were:

1587–1632. Reign of Sigismund (Zygmunt) III Vasa. (See p. 535.) His refusal to abandon his claim to the Swedish throne led to incessant Polish-Swedish wars.

First Polish-Swedish War for Livonia, 1600–1611

1600. Swedish Invasion of Estonia. Swedish forces under **Charles, Duke of Södermanland** (uncle of Sigismund; later Charles IX) invaded and occupied Estonia (except for Riga) and most of Livonia (today Latvia).

1601. Polish Counteroffensive. Successfully defending Riga, the Poles reconquered most of Livonia. In subsequent campaigns Hetman **Jan** Karo Chodkiewicz was victorious at **Dorpat, Revel,** and **Weisenstein.**

1604, September 27. Battle of Kircholm. Newly crowned Swedish King Charles IX landed in Estonia with a fresh army of 14,000 men and marched toward Riga. Chodkiewicz with 8,000 men, two-thirds horsemen, routed the Swedish army with an impetuous charge. Charles narrowly escaped captivity. The Swedes lost 9,000 killed.

1604–1611. Intermittent Operations. A truce was agreed to after the death of Charles IX (1611).

1605–1609. Polish Intervention in Russia. (See p. 631.)

1609–1618. Polish-Russian War. (See p. 631.)

Polish-Turkish War, 1614–1621

This was provoked by Polish Cossack raids upon Turkish ports along the western and southern Black Sea coasts, and by Polish support of revolts in Moldavia and Walachia (present Rumania). In retaliation, devastating Turkish and Tartar raids overran the Polish Ukraine, while Poland was primarily occupied in Russia (1614–1618).

1620, September 20. Battle of Jassy. Hetman **Stanislas Zolkiewski** with about 10,000 men decisively defeated a much larger Turkish and Tartar army. As a result the Ottoman Sultan, **Osman II,** led a large army north from Constantinople. Zolkiewski began to retreat.

1620, December. Battle of Cecora. Zolkiewski and his army of 9,000 men were annihilated in Moldavia.

1621, Fall. Battle of Chocim (Khotin, or Hotin). A Polish army of 46,000 under Hetmen Chodkiewicz and **Stanislas Lubomirski** had the better of an inconclusive struggle on the Dniester River against Sultan Osman's

Turkish Army of some 100,000. This was followed by a truce confirming the *status quo*.

Nevertheless, Cossack raids upon Turkey and Tartar raids upon Poland continued.

The Second Polish-Swedish War, 1617–1629

Gustavus Adolphus took advantage of Poland's simultaneous involvement in major wars with Turkey and Moscow by reconquering the eastern littoral of the Baltic Sea.

1617. Gustavus Victorious in Livonia. He captured several Baltic ports in Livonia, and compelled Hetman **Krzysztof Radziwill** to conclude an armistice until 1620.

1621. Gustavus Invades Estonia. His 12,000 Swedes were reinforced by 4,000 Estonians. He besieged and captured **Riga,** the political and commercial center of Livonia (September 15).

1622–1625. Armistice. Radziwill was again forced to conclude an armistice.

1625. Gustavus Occupies Livonia. After landing in Riga, he occupied all Livonia and Kurland against minor resistance.

1626. Swedish Invasion of Prussia. Gustavus landed at Pillau (vassal of Poland) with 15,000 men and quickly occupied all northern Prussia, threatening Poland's access to the Baltic. Only Gdansk (Danzig) rejected his demand of surrender. Gustavus left forces to blockade Danzig and returned to Sweden.

1626–1627. Polish Counteroffensive. Under Hetman **Alexander Koniecpolski** the Poles attempted to reopen the Vistula and relieve blockaded Danzig. Koniecpolski captured the fortified port of Puck (a fortress on the Vistula south of Tczew and west of Danzig), but could not cut the Swedish line of communication with Pillau. He intercepted and captured 4,000 Germans recruited in Germany for Gustavus.

1627, May. Gustavus Adolphus Returns from Sweden. He brought reinforcements, giving him 14,000 against Koniecpolski's 9,000. Koniecpolski was repulsed at the **Battle of Tczew** (Dirschau), where Gustavus was wounded. For the first time Swedish cavalry fought Polish cavalry on equal terms. Sigismund III organized a small fleet of privateers; these harassed the Swedish fleet, but Sweden retained control of the sea.

1628. Swedish Advance. With his army increased to 32,000 and recovered from his wounds, Gustavus forced Koniecpolski to retreat south. The Poles fought a war of harassment. By winter Gustavus withdrew to Prussia.

1629. Imperial Support from Germany. The Emperor sent Sigismund III a corps of 7,000 men to prevent Gustavus' proposed march from Prussia to Germany to enter the Thirty Years' War. Koniecpolski immediately started an offensive, surprising Gustavus and a detachment of cavalry on the march.

1629. Battle of Sztum. The Swedes had the worst of an inconclusive cavalry action. Gustavus, wounded in the back, narrowly escaped capture. Because of the 2 wounds he received during this war, Gustavus could never wear armor again. The Swedish hold on the Baltic coast was unshaken.

1629. Truce of Altmark. Poland needed peace and Gustavus wanted to be free to enter the war in Germany. Poland lost Livonia north of the Dvina River. Sweden was permitted to use all Prussian ports—except for Königsberg, Danzig, and Puck—for 6 years. The lower Vistula ports (Marienburg, Sztum, and Glova) were temporarily left in the hands of the Elector of Brandenburg.

1632–1634. Russo-Polish War. (See p. 632.)

1632–1648. Reign of Ladislas IV.

1634–1653. "Bloodless War" with Sweden. Taking advantage of Swedish involvement in the Thirty Years' War (see p. 587), Ladislas assembled an army of 24,000 men on the lower Vistula and, while threatening Swedish ports in Prussia, began negotiations with Sweden with regard to the eastern littoral of the Baltic. This led to the **Treaty of Sztumsdorff,** which provided for a 26-year truce and restored all Prussia to Poland; most of Livonia south of the Dvina River was returned to Poland.

1648–1668. Reign of John II Casimir.

Cossack Uprising in the Ukraine, 1648–1654

The rebellious Cossacks were led by **Bogdan Chmielnicki.** There was great ferocity and cruelty on both sides; the Ukraine was devastated and depopulated. Chmielnicki recruited not only Cossacks but also masses of rebellious Ukrainian peasants. He formed an alliance with the Crimean khan and was assisted by a Tartar army.

1648. Cossacks Victories. These were at the **Battles of Zolte Wody** and **Korsun.**

1649. Battle of Pilawce. A Polish force of 36,000 was defeated by the Cossacks.

1648–1649. Poland on the Defensive. The Poles held their strategic fortresses of Lwow (1648), Zamosc (1648), and Zbaraz (1649) against Cossacks attacks.

1649. Treaty of Zborow. A temporary Polish-Cossack peace.

1651. Resumption of Hostilities. At the **Battle of Beresteczko,** 34,000 Poles under John Casimir decisively defeated 200,000 Cossacks and Tartars under Chmielnicki (July 1).

1651. Treaty of Biala Cerkew. The Cossacks made peace. Chmielnicki, however, continued the war with a few followers.

1654. Treaty of Pereiaslavl. The Cossacks placed the Ukraine under the protection of Czar **Alexis,** provoking a war between Poland and Moscow.

1654–1656. Polish-Russian War. (See p. 632.)

1655–1660. First Northern War. War with Sweden (see p. 622).

1657. Transylvanian Invasion. (See p. 636.)

1658. Treaty of Hadziak. Regent **Wyhowski** of the Ukraine, alarmed by the Russo-Polish alliance in the Northern War, reached an agreement with Poland, which proclaimed equality of the Ukrainian, Polish, and Lithuanian peoples and creation of an autonomous Ukrainian duchy ruled by a Hetman confirmed by the Polish king, and sending senators and representatives to the Polish Diet. This provoked war with Moscow at a time when Poland was suffering the ordeal of repeated Swedish invasions.

1658–1666. Polish-Russian War. (See p. 633.)

1668. Abdication of John Casimir. A violent struggle for the throne ensued.

1669–1673. Reign of Michael Wisniowiecki. This was a period of turmoil and struggle between partisans and enemies of the incompetent monarch. A principal leader of the opponents was **John Sobieski.**

Polish-Turkish War, 1671–1677

The Ukraine had been divided along the Dnieper River between Russia and Poland after a long war (1667; see p. 634). The Cossack chieftain **Peter Doroshenko,** in Polish Poland, refused to accept Polish authority and acknowledged Turkish sovereignty (1668). As a result of Polish efforts to suppress the Cossack revolt, as they considered it, Turkey demanded cession of the Ukraine. Though unprepared for war, Poland refused.

1671, December 9. Declaration of War by Sultan Mohammed IV.

1672, Summer. Invasion of Poland. Grand Vizier **Ahmed** led a huge army of more than 200,000 into southeastern Poland. The Turks and their Tartar vassals and Cossack allies besieged and captured the strategic fortress of Kamieniec. Most of the garrison of 1,100 perished under the debris of the castle, which they blew up after a 12-day siege. The Turks then took Lublin against little resistance.

1672, October. Treaty of Buczacz. Michael, unable to organize his disintegrating kingdom, ceded to Turkey the province of Podolia.

Doroshenko was granted independence in the western Ukraine. Poland also agreed to pay an annual tribute. The treaty was never ratified by the Polish Diet.

1672–1673. Sobieski Rallies Poland. In the face of the deadly Turkish menace, Grand Hetman John Sobieski, renowned for earlier victories over Tartars, formed an army of 40,000. He marched against the Turks.

1673, November 11. Second Battle of Chocin (Chotyn). Sobieski annihilated a Turkish army of 30,000 men. He then took the fortress of Chocim. Other Turkish forces withdrew from Poland without fighting. King Michael had died one day before the battle. On learning this, Sobieski suspended further offensive operations, and disposed his forces for the protection of the frontiers. He then hastened to Warsaw, where he was elected king (May 21, 1674).

1674–1696. Reign of John III Sobieski. Internal turmoil and intrigue continued.

1675. Turkish Invasion. Some 60,000 Turks and 100,000 Tartars invaded Podolia and the Ukraine. They recaptured Chocim and threatened Lwow.

1675. Battle of Lwow. Sobieski defeated the Turks. Gradually he drove the invaders back, winning several small battles and liberating all of Poland except Kamieniec. Internal difficulties hampered his efforts to raise a large army to invade Turkey.

1676, September. Turkish Invasion. An army of 200,000 under **Ibrahim Pasha** again invaded southeastern Poland.

1676, September–October. Battle of Zorawno. Sobieski with 16,000 men repulsed the Turks from his strongly fortified camp on the Dniester in an inconclusive 2-week battle.

1676, October 16. Treaty of Zorawno. The Turks agreed to the return of the western Ukraine to Poland. They retained Podolia, including Kamientiec and Chocim.

1676–1681. Continuing War in the Ukraine. Meanwhile, Moscow, also engaged in an unrelated war against Turkey (see p. 634), had frequently violated the truce agreement of Andrussovo (see p. 634) and invaded the Polish Ukraine, devastating it and inciting the Cossacks to continue rebellion against Poland. The continuing Russo-Turkish war, largely waged in the eastern part of the Polish Ukraine, turned this rich land into a desert.

1683, March 31. Alliance with the Empire. Turkey, having waged inconclusive wars with Poland and Moscow, turned against Austria. John Sobieski believed that if Austria was defeated, the next Turkish blow would be directed against Cracow, the heart of Poland. He concluded a defensive-offensive alliance with the Empire and prepared for war.

1683, July–September. Siege of Vienna. (See p. 637.) The high point in Sobieski's career.

1683–1699. Desultory Warfare. The war against the Turks dragged on with varying results. Sobieski, unable to reach any agreement with the Russians and obtaining no assistance from jealous Emperor Leopold, bore the brunt of the war and was hard put to hold the Ukraine. Meanwhile, Poland continued to be wracked by internal intrigue and violence.

1696, June 17. Death of Sobieski.

RUSSIA

The military power of Moscow developed in relative isolation from the West, and reflected the general backwardness of the country. At the beginning of the 17th century, Moscow's army, comprised of relatively ill-armed and poorly trained levies, was often defeated by considerably smaller Swedish and Polish forces. However, in the defense of fortresses the Russians showed tenacity, resourcefulness, and courage. In 1612, Polish and Swedish invaders of Muscovy were forced to retreat by masses of popular militia and not by the army.

By the middle of the century a new system of recruitment was introduced, based on compulsory service of peasants and city dwellers, and permanent infantry and cavalry regiments replaced the old levy. In addition to these semiregular troops, there were nearly 150,000 Cossacks eager for war. By the end of the century, Moscow in emergency was able to mobilize an armed force of over 300,000 men, mostly poorly trained.

Early in the 17th century, Moscow suffered its "Time of Troubles." A combination of violent internal upheavals and Polish and Swedish interventions ruined the country and led to territorial losses. By the middle of the century, however, the czardom had recovered its political and economic strength, and, owing to Poland's mounting difficulties, in the second half of the century Moscow conquered much of the northeastern Ukraine from Poland.

The "Time of Troubles," 1604–1613

This began with the appearance of a false **Dmitri,** a pretender who claimed to be the murdered son of Ivan IV and who was supported by the Poles and Cossacks. After the death of Boris Godunov (1605; see p. 537), the struggle for power among various boyar factions threw the country into a state of extreme confusion and brought Polish and Swedish intervention. The principal events were:

1605. Brief Reign of Theodore (or Fëdor, son of Boris). He was murdered by hostile boyars. The false Dmitri briefly took the throne, but also was soon murdered by the boyars.

1605–1609. Polish Intervention. Several Polish magnates supported the claims of the false Dmitri. As chaos mounted in the Moscow czardom, Polish intervention was increasingly persistent.

1606–1610. Reign of Basil Shuisky. He was bitterly opposed internally and by the Poles.

1608. Battle of Balkhov. A Polish army, reinforced by 45,000 rebellious Cossacks, defeated Basil. Soon after, the Poles and Cossacks defeated Basil again at the **Battle of the Chadyuka River,** bringing a second "false Dmitri" to power. This was followed by an uprising of the Cossacks and peasants in the east and south in support of the pretender. Basil requested Swedish aid at the price of the cession of Karelia to Sweden.

War with Poland, 1609–1618

1609. Sigismund III of Poland Declares War. Claiming the disputed throne of Moscow for himself, he invaded Russia.

1609–1611. The Siege of Smolensk. Michael Shein, commander of Smolensk, stubbornly and skillfully defended the fortress against repeated assaults. The garrison of 12,000 was reinforced by 70,000 civilians.

1610, September. The Battle of Klushino. A Russian army of 30,000 men, including 8,000 Swedes, led by **Dmitri Shuisky** marching to relieve Smolensk, was surprised and annihilated by a small Polish corps of 4,000 (3,800 horsemen and 200 infantry with 2 small cannon) led by Hetman Stanislas Zolkiewski. The Russians left 15,000 killed on the battlefield. Shuisky fled to Moscow. Zolkiewski, reinforced to 23,000, moved on Moscow. Smolensk surrendered early the following year.

1610, October 8. Capture of Moscow. Polish troops entered Moscow and the Kremlin. The boyars deposed Czar Basil Shuisky.

1610–1611. Protracted Negotiations. The boyars were prepared to elect Ladislas (son of Sigismund) to the Moscow throne, but negotiations broke down when Sigismund insisted on having the throne for himself.

1611. Rising of Russia. Rebellion of the populace against the invaders broke out in Moscow. A Polish garrison was besieged in the Kremlin. The uprising spread in the northern and eastern part of the country under the leadership of

Prince **Pazharsky.** The second "false Dmitri" died.

1612. Polish Withdrawal. The Polish army, unable to relieve the Polish garrison of the Kremlin, withdrew from Moscow. Desultory frontier fighting continued. The Poles in the Kremlin surrendered and were massacred.

The Swedish and Polish Wars, 1613–1667

1613–1645. Reign of Michael Romanov. This weak ruler founded Russia's greatest dynasty.

Russo-Swedish War, 1613–1617

1613. Novgorod Expedition. Moscow sent troops against Novgorod, occupied by the Swedes. Swedish troops intercepted and severely defeated the Russian corps.

1614. Invasion of Moscow. Gustavus Adolphus transferred military operations into Moscow territory. Pskov, the strongest Russian frontier fortress, resisted his assault successfully and, following a 7-month siege, Gustavus withdrew to Swedish territory. Since Moscow expected a Polish invasion and Gustavus was preparing an attack on Polish Livonia, both Moscow and Sweden were anxious to conclude peace.

1617. Treaty of Stolbovo. This terminated the Russo-Swedish War. Sweden restored Novgorod and several other towns to Moscow, and the Russians ceded to Sweden their last footholds on the Gulf of Finland. Thus Moscow lost its outlet to the Baltic.

1617. Renewed Polish Invasion. Ladislas marched to Moscow, claiming the Moscow throne as czar-elect. After the rejection of his claim by Michael, Hetman Chodkiewicz tried to capture Moscow by direct attack, but failed. Peace negotiations followed.

1618. Truce of Devlin. Poland was confirmed in possession of all border towns conquered in 1609–1611 (including Smolensk).

1618–1632. Military Reorganization. Michael hired foreign mercenaries in England, Holland, Denmark, and Sweden and formed of them several foreign regiments. Several Russian regiments were also trained by foreigners and according to foreign regulations. Guns, arms, and powder were purchased.

Russo-Polish War, 1632–1634

1632, September. Russian Invasion of Poland. Boris Shein with 30,000 men advanced to besiege Smolensk, which had a garrison of only 3,000.

1632–1633. Siege of Smolensk. After a gallant defense of 11 months, the provisions of the garrison were nearly exhausted.

1633, September. Battle of Smolensk. Ladislas IV with an army of 40,000 men decisively defeated the Russians, then surrounded Shein's battered army, which was besieged for 6 months before it surrendered (February 1634). Shein was permitted to withdraw his troops to Moscow, where he was unjustly accused of treason and executed.

1634. Treaty of Polianovo. This confirmed the Treaty of Deulin. Ladislas renounced his claims to the czardom and recognized Michael as czar.

1637. Capture of Azov. The Cossacks seized the fortified port of Azov from the Crimean Tartars and offered it to the Czar, but Michael, afraid of a conflict with Turkey, refused the offer. Azov therefore was returned to the Tartars.

1645–1676. Reign of Czar Alexis.

Russo-Polish War, 1654–1656

Seeking revenge for past defeats at the hands of Poland, the Russians took advantage of Poland's deep involvement in the struggle with the Cossacks in the Ukraine (see p. 629). The Cossack leader, Bogdan Chmielnicki, offered to surrender the Ukraine to the Czar.

1654, July–October. Operations in Lithuania. Alexis personally led an army of over 100,000 into Lithuania. He captured Smolensk after a 3-month siege (July 2–September 26), then at the **Battle of Borisov** overwhelmed a small Polish corps. The Russians then occupied defenseless Lithuania as far as the Berezina River until winter stopped the operations.

1654. Operations in the Ukraine. Simultaneously, a smaller Russian force of 40,000 invaded the Ukraine and occupied Kiev. Between the two Russian armies, an allied Cossack force under Hetman **Zolotarenko** utterly devastated the area of the Pripet Marshes and massacred its population.

1655, January. Polish Counteroffensive in the Ukraine. In spite of bitter cold, Hetman **Stanislas Lanckoronski** defeated a much larger joint Russo-Cossack Army at the **Battle of Okhamatov.** The Russians promptly retreated.

1655, May–October. Resumed Russian Offensives. In Lithuania the Russian advance was facilitated by simultaneous Swedish attacks on Poland from Livonia (see pp. 622, 629). Alexis advanced to the Niemen River and proclaimed himself Grand Duke of Lithuania. His troops, however, were harassed by partisans. Simultaneously, Russo-Cossack forces reoccupied most of the Ukraine and even sent raids into central Poland. They were unable to take the fortified cities of Kamieniec and Lwow.

1655, November. Tartar Intervention. Poland's ally, Crimean Khan **Mahmet Girei,** invaded the Ukraine with 150,000 men. He defeated and captured Chmielnicki at **Zalozce,** and forced the Russians to withdraw precipitously toward the north.

1656, November 3. Treaty of Nimieza. After prolonged negotiations the war was terminated by the conclusion of a three-year truce and an anti-Swedish alliance between Russia and Poland.

Russo-Swedish War, 1656–1658

Alarmed by Sweden's expansion in the Baltic area, Alexis, rejecting Swedish offers of alliance against Poland, instead concluded alliances with Poland and Denmark against Sweden. This was part of the First Great Northern War (see p. 622).

1656. Russian Northern Offensive. Advancing toward the Gulf of Finland, the Russians captured the strong fortresses of Schlüsselburg and Nienshanz, near the Neva River's mouth.

1656, July–August. Battle of Riga. Alexis advanced from Polotsk down the Dvina River, captured Dinaburg (whose defenders were massacred), and then took the fortress of Kokenhuzen. Riga, whose small garrison was commanded by **Magnus De La Gardie,** was besieged. After the arrival of considerable reinforcements, De La Gardie made a sortie, defeating the much more numerous entrenched Russians, who lost 8,000 killed and 14,000 wounded and prisoners. Alexis retreated, pursued energetically.

1657. Swedish Counteroffensives. The Swedes devastated Karelia. At the same time Swedish troops advanced overland from Finland to Riga to strengthen its defense.

1658. Renewed Russian Offensive. A small Russian force captured Yamburg and besieged Narva, but soon was ejected from the Baltic and the Gulf of Finland by the Swedes, who recaptured all previously lost towns.

1658. Three-Year Truce. Alexis conceded defeat, and decided to make peace so he could renew the war with Poland. The Swedes kept all the towns along the Baltic coast and the Gulf of Finland.

Russo-Polish War, 1658–1666

The Russo-Polish Truce of Nimieza (see above) expired. Moscow resumed hostilities.

1658–1659. Invasion of Lithuania. Prince **George Dolgoruki** surprised and captured Hetman **Wincenty Gosiewski** at Werki and occupied Wilno, which he devastated with fire and sword. Prince **Chowansky** captured Grodno, Nowogrodek (1659), and besieged the fortress of Lachowicze.

1658–1659. Invasion of the Ukraine. A Russian force of 150,000, under command of Prince **Trubetskoi**, was temporarily successful, but was severely beaten by the Ukrainian Hetman **John Wyhowski** in the **Battle of Konotop,** with the loss of 30,000 killed (1659).

1660. Operations in Lithuania. The Russians took the initiative, but were defeated by increased Polish forces, freed by the peace with Sweden. Hetmen Stefan Czarniecki and **Paul Sapieha** defeated Dolgoruki and Chowansky, and forced them to withdraw to Polotsk and Smolensk. The Poles not only liberated Lithuania but invaded the area of Vitebsk, Polotsk, and Velikie Luki.

1660. Operations in the Ukraine. Russian General **Sheremetiev** led 60,000 well-trained and equipped troops against Lwow; he was defeated and captured at the **Battle of Lubar** (1660) by Hetmen Czarniecki and George Lubomirski, who had 20,000 Poles assisted by 20,000 Tartars. Another army of pro-Moscow Cossacks—numbering 40,000 and led by Chmielnicki—attempting to join Sheremetiev, was defeated by **George Lubomirski** at the **Battle of Slobodyszcze.** Chmielnicki pledged obedience to the Polish king.

1661–1667. Polish Initiative. There followed several years of sporadic frontier fighting.

1664. Cossack Raid into Persia (see p. 642).

1667. Treaty of Andrusovo. The Ukraine was partitioned between Poland and Moscow with the Dnieper River as the dividing line. Kiev, though located on the southern Polish bank, was ceded to Moscow for 2 years (it never returned to Poland). Smolensk was also ceded to Moscow.

Beginnings of the Turkish Wars, 1667–1700

1667–1678. Friction with Turkey. The acquisition of northeastern Ukraine brought Moscow into contact with Turkish possessions, and frontier friction soon began.

1676–1682. Reign of Theodore III (or Fëdor, son of Alexis).

1678–1681. First Russo-Turkish War. Cossack Hetman Doroshenko was dissatisfied with the results of the Treaty of Zorawno (see p. 630), and asked Russian assistance against Poland and Turkey. At the same time, Sultan Mohammed IV claimed the Ukraine. He sent a large army under Vizier **Kara Mustafa** to drive out the Poles and Russians. The area of operations was mostly in the eastern part of the Polish Ukraine, which had been utterly devastated in the Polish war against Turkey (see p. 630). Russo-Cossack forces under the command of Prince **Romodanowsky** withdrew without offering serious opposition. The Turks captured and destroyed a number of Ukrainian fortresses and cities.

1681, January 8. Treaty of Radzin. Turkey renounced her claims to the Ukraine, but under condition that the population of the area between the rivers Bug and Dniester should be evacuated and this area should remain unpopulated. Podolia was returned to Poland. The Cossacks were to have special rights throughout the Ukraine.

1682–1689. Joint Reigns of Peter I and Ivan V. Ivan was the brother of Theodore III; he and his younger half-brother were co-rulers until the death of Ivan (1689).

1689–1725. Reign of Peter I, the Great. He initiated Russia's struggle for access to the sea, beginning with the Black Sea.

1695–1700. Russo-Turkish War for Azov. The first Russian offensive, under the personal command of Peter, reached Azov via the Volga and Don rivers, but was repulsed with heavy losses. Having no navy, the Russians were unable to blockade the water approaches to the fortress.

1695–1696. Peter Builds a Navy. Peter built a fleet near Voronezh and sent it to Azov down the Don River in the spring.

1696, July 28. Capture of Azov. The Russian army, increased to 75,000, advanced partly by land, partly by river. This time the fortress was captured by combined operations of ground and naval forces which repulsed the Turkish warships and blockaded approach to Azov from the sea.

COMMENT. *The possession of Azov, which was located on the Sea of Azov and was separated from the Black Sea by the Strait of Kerch, solidly blocked by Turkey, could not solve the problem of Russian access to the Black Sea. Its selection as the strategic objective of the war can be explained only by lack of means for an attack on the Crimea. The cost of Azov in heavy Russian losses was disproportionate to its relative importance to Russia.*

1698. Mutiny in Moscow. Peter, who had been traveling incognito in Western Europe (1697–1698), returned to suppress an army mutiny and to reform the Russian government and nation. **1700. Truce with Turkey.** The war was terminated by a truce for 30 years.

HUNGARY

During most of the 17th century, Hungary was the main battleground in the continuing but sporadic struggle between the Hapsburg and Ottoman empires. The Hapsburgs' involvement in the Thirty Years' War and in other conflicts of Central and Western Europe long prevented them from taking advantage of the gradual decline of Turkish power. During much of this period, Transylvania, nominally a fief of the Ottoman Empire, was virtually independent and played the two empires and Poland off against each other. Toward the end of the century, brief Turkish resurgence enabled them to reconquer Transylvania and to overrun practically all of Hapsburg Hungary, culminating in a siege of Vienna. Repulsed from Vienna, thanks to the intervention of Poland's John Sobieski, the Turkish decline accelerated. By the end of the century, practically all of Hungary was controlled by the Hapsburgs. The principal events were:

The Rise and Fall of Transylvania, 1593–1662

1593–1606. The Long War. (See p. 536.)
1599–1601. Struggle for Transylvania. The country was briefly dominated by **Michael the Brave,** Voivode of Wallachia and Rumania's national hero. Driven out of Transylvania by a Magyar revolt (1601), Michael regained power briefly by a victory at the **Battle of Goroslau,** won with the assistance of Hapsburg General **George Basta.** Soon afterward, however, Basta procured Michael's murder, and began a reign of terror, which led to the rise of **Stephen Bocskay** and his successful alliance with the Turks (see p. 536).
1606, June 23. Treaty of Vienna. Stephen Bocskay was recognized as sovereign Prince of Transylvania by Emperor **Rudolf.** Rights of Transylvanian (and other) Protestants in Hapsburg Hungary were guaranteed.
1606, November 11. Treaty of Zsitva-Torok. End of the Long War. The territorial status of Hungary was unchanged, but the Emperor was no longer required to pay tribute to the Sultan. Transylvania was still nominally under Ottoman suzerainty.
1606, December 29. Death of Bocskay. He was probably poisoned; a scramble for power followed.
1608–1613. Reign of Gabriel Bathory of Transylvania (Son of Stephen; see p. 535.) Cruel and tyrannical, he was overthrown and murdered.
1613–1629. Reign of Bethlen Gabor of Transylvania. He was a wise prince and able soldier.
1618–1629. Transylvanian Participation in the Thirty Years' War. Bethlen briefly besieged Vienna (1619) and, though driven away from the Austrian capital (see p. 585), overran most of Hapsburg Hungary (1620). The Hapsburgs recognized him as Prince of Transylvania, and ceded much of Hapsburg Hungary to him in the **Treaty of Nikolsburg** (December 21, 1621). Subsequent campaigns were inconclusive.
1630–1648. Reign of George Rakoczy I of

Transylvania. He increased the power, size, and independence of Transylvania.

1645. Renewed War with the Hapsburgs. George defeated the Hapsburgs and (by the **Treaty of Linz,** December 16, 1645) forced renewed grant of religious freedom to Protestants as well as Catholics in Hapsburg Hungary.

1648–1660. Reign of George Rakoczy II of Transylvania. In an undeclared frontier war between Turks and Austrians Rakoczy supported his nominal suzerain, the Sultan of Turkey (1648–1652).

1657. Transylvanian Invasion of Poland. Hoping to take advantage of Poland's early defeats in the First Northern War (see p. 622), Rakoczy invaded southern Poland, hoping to seize the throne. This move, without his permission, enraged the Sultan, who still coveted the Ukraine. The Khan of the Crimean Tartars, a Turkish vassal, was ordered to eject Rakoczy from Poland. Rakoczy briefly occupied Warsaw, but, deserted by the Swedes, was forced to retreat. During the withdrawal his rear guard was overwhelmed by the Tartars at the **Battle of Trembowla,** near Tarnopol (July 31), and its commander, **Janos Kemeny,** was captured. Rakoczy reached Transylvania with only a fraction of his army.

1657–1662. War with Turkey. Mohammed Koprulu, Grand Vizier of Turkey, attempted to depose Rakoczy, but failed. Rakoczy at first defeated Koprulu's invading army at the **Battle of Lippa** (May 1658), but was driven out of Transylvania to his estates in Hapsburg Hungary by the convergence of the Ottoman army and that of their Crimean Tartar vassals. **Akos Barcsay** was proclaimed Prince of Transylvania by Koprulu (November, 1658). Rakoczy drove Barcsay out of the country (1659), but was in turn so hard-pressed by the Turks that he appealed for Hapsburg assistance in return for the cession of some border provinces.

1660. Battle of Fenes. Attempting to halt an invasion by **Ahmed Sidi,** Pasha of Buda, Rakoczy was defeated and mortally wounded. He died in Nagyvarad, where his forces were besieged by the Turks. After a 4-month siege, Nagyvarad finally fell to Turkish assault (August 27), while an Austrian army stood idly nearby.

1661. Janos Kemeny Elected Prince of Transylvania. He had recently returned from imprisonment in the Crimea. Despite some assistance from a small Austrian army under Raimondo Montecuccoli, Kemeny was unable to repel a Turkish invasion.

1662, January 22. Battle of Nagyszollos. Kemeny was defeated and killed by a Turkish army under **Mehmed Kucuk.** Turkish control was restored to Transylvania and **Mihaly Apafi** was proclaimed Prince by the Turks.

Renewal of the Hapsburg-Ottoman Struggle for Hungary, 1662–1683

1663–1664. Turkish Invasion of Hapsburg Hungary. The new Grand Vizier, **Fazil Ahmed Koprulu Pasha** (son of Koprulu), invaded Hapsburg Hungary with the intention of advancing on Vienna. The Turks were halted by the gallant defense of Neuhause, under **Adam Forgach.** When Neuhause finally surrendered (September 1663), Ahmed decided to postpone the advance on Vienna until the following spring. But when he renewed the advance, the Austrians were better prepared, and he was again slowed by resistance of fortresses. Peace negotiations began in late summer at Vasvar.

1664, August 1. Battle of the Raab River (or of St. Gotthard Abbey). Ahmed, attempting to force the Austrians to accept the peace terms, crossed the river to attack Montecuccoli's army. Despite greatly superior numbers, the Turks were repulsed, and retreated to Buda; the Austrians did not pursue.

1664, August 11. Peace of Vasvar. A 20-year peace was agreed. Autonomous Transylvania remained under Turkish suzerainty. The Turks were also allowed to retain the frontier fortresses they had captured.

1664–1673. Unrest in Hapsburg Hungary. Partly because of bitterness at the unfavorable terms of the Treaty of Vasvar, partly because of religious persecution, Magyar conspiracies against the Hapsburgs in Hungary were widespread, even to the extent of seeking Turkish assistance. The Austrians subdued the unrest and minor uprisings with ease, and repression increased.

1678–1682. Revolt against Austria. Rebels under the leadership of Count **Imre Thokoly** gained control of most of northern and western Hungary. But popular support faded as Leopold made concessions to the Magyars; Thokoly turned to the Turks for aid. A Turkish army, dispatched by the Grand Vizier, Kara Mustafa, helped Thokoly to overrun northeastern Hungary. Thokoly was declared King of Hungary, as a vassal of the Sultan. Unable to gain concessions from Austria, the Turks prepared for an invasion the following year. Leopold hastily made peace with Thokoly, leaving him virtual master of all Hungary, save for Transylvania.

1683, March 31. Austro-Polish Alliance. Austria and Poland, both at war with Turkey, and aware of Kara Mustafa's forces gathering at Adrianople and Belgrade, became allies in mutual defense against invasion.

Vienna Campaign, 1683

1683, March–May. Turkish Advance. Sultan **Mohammed IV** personally led the Turkish Army from Adrianople toward Belgrade (March 31). Continuing north from Belgrade, under the actual leadership of Kara Mustafa, the Turks were joined by a Transylvanian force under Apafi, bringing their total strength to nearly 200,000 men. Meanwhile, a Hungarian army under Thokoly advanced against the Austrians in Slovakia, but was repulsed at Pressburg.

1683, June. Invasion of Austria. Leaving a force to besiege Gyor, the Turkish army, now about 150,000 men, continued up the Danube into Austria. Leopold and his court fled to Passau. Duke Charles of Lorraine, with an Austrian army of about 30,000, retreated from Vienna to Linz. Count **Rudiger von Star-**hemberg,** governor of Vienna, was left with a garrison of about 15,000 men.

1683, July 14. Turks Reach Vienna. Immediate investment of the city began.

1683, July 17–September 12. Siege of Vienna. The Turks were handicapped by insufficient heavy artillery; only a few big guns had been brought up the river by barge. The defending guns were superior in both quality and quantity. The Turks were also hampered by the vigor of the defenders, who made frequent sorties. Meanwhile, Charles of Lorraine kept contact with the defenders via the river, and blocked Turkish raids up the Danube Valley. However, the energetic Turks opened several breaches in the walls and forced their way into portions of the city, where their advance was stopped by hastily erected fortifications (September 1). The defenders by this time had lost about half of their strength and were running short of ammunition and other supplies.

1683, August–September. Arrival of Sobieski. Faithful to his treaty, and also responding to an appeal from the Pope, John Sobieski marched to the relief of Vienna with 30,000 men. His main body made the march of 220 miles from Warsaw in 15 days, an unusual speed in those times. The arrival of the Poles and their juncture with the Austrians and Germans west of Vienna completely surprised Kara Mustafa. The allied army advanced on the Turkish camp (September 11).

1683, September 12. Battle of Vienna. Sobieski assumed supreme command of the allied army of 76,000 (his Poles and 46,000 Austrians and Germans). The Poles were on the right wing, the Austrians under Charles of Lorraine on the left, and various German corps in the center. The allies advanced slowly because of difficult terrain; Sobieski planned to attack the next day. Noticing, however, that Turkish resistance was weak, at 5 P.M. he ordered a general attack. At the same time the Vienna garrison attacked the siege lines. For an hour an inconclusive infantry struggle raged between the river and the Turkish camp. Sobieski and his Polish cavalry charged toward Kara Mustafa's headquarters. Kara Mustafa and his army fled in panic; Turkish losses were extremely heavy. Sobieski suspended the pursuit because of fear of ambush in the dark, and the Turkish army escaped destruction. Sobieski sent the captured Imperial Banner of the Prophet to the Pope.

1683, September–December. Sobieski's Pursuit. Leopold, jealous of Sobieski's victory, was cool to his savior, who nevertheless led a pursuit of the Turks and helped liberate Grau (October) and much of northwestern Hungary.

Hapsburg Conquest of Hungary, 1683–1688

1684, March 31. Treaty of Linz. At the instigation of Pope **Innocent XI,** a crusading **Holy League** against Turkey was created. Original members were the Empire, Poland, and Venice. Moscow later joined (April 1686).

1684–1685. Inconclusive Frontier Warfare. The Hapsburgs retained the upper hand.

1686, September 2. Capture of Buda. Charles of Lorraine took the Turkish capital of Hungary after a siege of nearly 2 months.

1687. Battle of Nagyharsany (Harkány). Charles decisively defeated the Turks near the site of the Battle of Mohacs (see p. 542), establishing Hapsburg control over all of southern Hungary and much of Transylvania. The Hungarians were forced to accept the Hapsburgs as hereditary monarchs.

1688, September 6. Capture of Belgrade. The stronghold was taken after a siege, immortalized by the anonymous alliterative poem which begins: "An Austrian army awfully arrayed/Boldly by battery besieged Belgrade." Most of Serbia was taken by the Austrians.

Turkish Efforts to Reconquer Hungary, 1688–1699

1690. Turkish Invasion of Transylvania. Sultan **Suleiman II,** having nominated Thokoly as Prince of Transylvania after the death of Apafi, sent him into Transylvania. He defeated a combined Hapsburg-Transylvanian army at the **Battle of Zernyest** (August). **Louis of Baden** then marched from Belgrade with the main Imperial army, overrunning much of Transylvania, Serbia, and parts of Macedonia.

1690, August–October. Turkish Reconquest of Serbia. The new Turkish Grand Vizier, **Mustafa Koprulu,** took advantage of Louis's expedition to conduct a lightning campaign, recapturing Nish, Smederevo, Vidin, and Belgrade. Thousands of Serbian refugees flocked into southern Hungary.

1691, August 19. Battle of Szalánkemen. Mustafa and Thokoly were decisively defeated by Louis of Baden. Mustafa and 20,000 Turks were killed. Transylvania was now firmly in Hapsburg hands.

1691–1696. Inconclusive Frontier Campaigns.

1697. Battle of Zenta. Prince **Eugene of Savoy** marched to oppose a major Turkish invasion of Hungary from Belgrade under the personal command of Sultan **Mustafa II.** The armies met near the Theiss River; Eugene attacked and practically annihilated the Turkish army. This was the last serious Turkish threat to Hungary.

1699, January 26. Treaty of Karlowitz. Austria received all of Hungary and Transylvania except the Banat. Venice obtained the Peloponnesus and much of Dalmatia. Poland recovered Podolia. Turkey retained Belgrade. Austria was now the predominant power of southeastern Europe, despite considerable internal Hungarian unrest and discontent with the Hapsburg monarchy which had been imposed upon her.

EURASIA AND THE MIDDLE EAST

The Ottoman Empire

The slow decline of the Ottoman Empire continued. Yet at midcentury the empire stretched over three continents; the frontier in Europe was only 80 miles from Vienna; it included all of North Africa except Morocco; the Black Sea and the Red Sea were Turkish lakes, and in the east it stretched to the shores of the Caspian and the Persian Gulf. The heart of the empire included the provinces of Rumelia, Anatolia, and the territories along the coasts of the Aegean and the Mediterranean. The Christian principalities of Moldavia, Wallachia, and Transylvania were autonomous; they were administered locally and were not garrisoned by Turkish forces, but their princes were appointed by the Porte and they had to provide corn and sheep for the Turkish Army and pay tribute. The Khanate of the Crimea, an autonomous Moslem state, had similar status; the Tartar khans paid no tribute but had to participate in Turkish campaigns with some 20,000 to 30,000 horsemen, and were responsible for defending Ottoman territories against Cossack raids. The Barbary States (Tripoli, Tunis, and Algiers) were more or less independent, having their own military and administrative organizations. They pursued their piracy without concern for Ottoman alliances or policies.

The military strength on which the Ottoman Empire depended was deteriorating. The regular army consisted of the Janissaries—originally slaves of the Sultan, but now a privileged social class, recruited from the Moslem population. Lax in their duties and unruly, their popular support precluded disbanding. The spahis and the feudal cavalry were declining both in numbers and in quality. As a result, the Porte was forced to place increasing reliance on Tartar horsemen from the Crimea, on untrained levies, and on undisciplined volunteers.

These factors all contributed to speeding the decline of Turkish power by the end of the century. The principal events were:

1595–1606. War with Austria. So-called Long War (see p. 536).

1602–1612. War with Persia. (See p. 641.)

1603–1606. Revolt of Kurds and Druses. Janbulad, Kurdish governor of Klis, and **Fakhr-ad-Din,** Prince of the Druses, joined in revolutionary alliance. Janbulad was defeated by Ottoman forces and fled to Fakhr-ad-Din in Lebanon. The Druse prince, maintaining himself in his mountainous territory, made a temporary truce with the Porte, and paid tribute.

1603–1617. Reign of Ahmed I (14 years old). A period of revolt, turmoil, and discord within the empire.

1606. Treaty of Zsitva-Torok. Peace with Austria, which the Ottomans were forced to recognize as an equal, nontributary power.

1610–1613. Renewed Druse Revolt. Fakhr-ad-Din had plotted with European Christian princes—including the Pope and the leaders of the Empire, Spain, and Tuscany—to help them recover the Holy Land. He seized Baalbek and threatened Damascus. Overthrown by a combined Turkish land and sea invasion, he fled to Italy.

1614–1621. War with Poland. (See p. 627.)

1616–1618. War with Persia. (See p. 642.) The Turks lost Azerbaijan and Georgia.

1617–1621. Reign of Osman II.

1621–1622. Janissary Revolt. Osman attempted to reform the Janissaries, but they revolted and overthrew him and put his imbecile brother **Mustafa I** on the throne. Mustafa soon abdicated in favor of Ahmed I's fifth son, **Murad IV.**

1623–1640. Reign of Murad IV. Eleven years

old at accession, the first 9 years of his reign were under the regency of his mother, **Kosem Sultan,** while the mutinous Janissaries virtually ruled the country.

1623–1638. War with Persia. (See p. 642.)

1631–1635. Renewed Druse Revolt. Fakhr-ad-Din returned from Italy and began fomenting revolt (1618). During the Persian war he gained control of much of Syria, and defeated a Turkish army attempting to go into winter quarters in Syria between campaigns against Persia (1631). Again the Turks mounted a land and sea invasion (1633), and Fakhr-ad-Din's son was defeated and killed in a decisive battle north of Damascus. Fakhr-ad-Din took refuge in the mountains again, but was captured and beheaded in Constantinople (April 13, 1633).

1638–1640. Military Reforms of Murad. Last of the warrior sultans, he was strong enough to initiate long-needed reforms in the Janissary organization. The reforms ended with his death (1640).

1640–1648. Reign of Ibrahim I. Another son of Ahmed, his reign was dominated by strong-willed Kosem Sultan. Turkey's decline continued.

1645–1670. War with Venice over Crete. This resulted from capture of Ibrahim's wives by Maltese corsairs, based on Venetian ports in Crete. A Turkish expedition of 50,000 men aided by the Greek inhabitants, who hated the Venetians, captured Canea (August 1645) and Retino (1646).

1648–1669. Siege of Candia. The Turks had great difficulty in protecting their supply lines to Crete against the more modern Venetian fleet, which was able to keep Candia supplied and reinforced. Venice also captured Clissa, in Dalmatia, and the islands of Lemnos and Tenedos, blockading the Dardanelles (1648).

1648. Revolt in Constantinople. This was partly stimulated by famine caused by the Venetian blockade. Ibrahim was overthrown and killed. His 7-year-old son **Mohammed IV** was placed on the throne, and Kosem Sultan continued to rule.

1649. First Naval Battle of the Dardanelles.

A Turkish fleet, attempting to open the blockade, was defeated by the Venetians.

1651. Death of Kosem Sultan. She was strangled at the instigation of the mother of Mohammed IV, **Turhan Sultan,** who became regent.

1656. Mohammed Koprulu Becomes Grand Vizier. An Albanian pasha, Koprulu was appointed by Turhan Sultan. He halted the decline of the empire by his energy, firmness, and wisdom.

1656. Second Naval Battle of the Dardanelles. Another Venetian victory maintained the blockade. A Venetian fleet briefly threatened Constantinople. This led Koprulu to devote a massive effort to rebuilding the Turkish Navy (1656–1657).

1656–1661. Reorganization of the Ottoman Government. Koprulu succeeded in a major reform of the civil and military sides of the government. He ruthlessly suppressed a revolt and executed all who opposed him. At least 50,000 people were killed, mostly rebels.

1657. Revolt in Transylvania. (See p. 636.)

1657. Third Naval Battle of the Dardanelles. The Turks were victorious, then reconquered Tenedos and Lemnos. They then proceeded to take several other Venetian islands in the Aegean (1657–1658). Regaining control of the Aegean, the Turks sent substantial reinforcements to their hard-pressed army on Crete.

1661. Death of Koprulu. He was succeeded as grand vizier by his son, **Fazil Ahmed** Koprulu.

1662–1664. War with Austria. (See p. 636.) Despite Fazil Ahmed's defeat in the Battle of the Raab, he obtained favorable peace terms.

1664–1669. French Operations against the Barbary Pirates. (See p. 616.) Because of this involvement against Turkish vassals, the Venetians attempted to persuade France to enter an alliance against the Ottoman Empire. Louis XIV did send a fleet and a force of 7,000 men to Candia (1669).

1666–1669. Turkish Efforts Intensified on Crete. Ahmed went to Crete and assumed command of the siege of Candia.

1669, September 6. Fall of Candia. Seeing defeat inevitable, the French contingent withdrew, and Venetian General **Francesco**

Morosini was forced to surrender. In the subsequent treaty, Venice lost all of Crete save for 3 small seaports. Venice also lost most of its Aegean islands and much of Dalmatia to Turkey.

1671–1677. War with Poland over the Ukraine. (See p. 629.)

1678–1681. Revolt of the Cossacks and War with Russia. (See p. 630.)

1678. Kara Mustafa Becomes Grand Vizier.

1682–1699. War with Austria and Poland and the "Holy League." (See pp. 630, 637, 638.) The Turkish defeat at Vienna precipitated a series of disasters which eventually caused Turkey to lose all of Hungary.

1687. Turmoil in Constantinople. Following news of the defeat at the **Battle of Nagyharsany** (see p. 638), there was panic in Constantinople; the Janissaries mutinied. **Mohammed IV** abdicated and was succeeded by his brother.

1687–1691. Reign of Suleiman II.

1685–1688. Venetian Offensives. Under **Fran-** **cesco Morosini** the Venetians were successful in Dalmatia and southern Greece. He overran the Peloponnesus (1686); advanced through the Isthmus of Corinth; besieged and captured Athens (September 1687). He later was forced to abandon Athens, but held most of the Peloponnesus.

1690. Mustafa Koprulu Becomes Grand Vizier. Temporarily he stopped the disasters and retook the lost fortress of Belgrade (see pp. 599, 638).

1691. Venetian Capture of Chios.

1693. Austrians Repulsed at Belgrade.

1695. Turkish Successes. Chios was recaptured from the Venetians. At the **Battle of Lippa,** Austrian Field Marshal **Friedrich von Veterani** was defeated and killed.

1697, September 11. Battle of Zenta. (See p. 638.) Climactic Austrian victory.

1699. Treaty of Karlowitz. (See p. 638.)

PERSIA

Under brilliant soldier, statesman, and administrator Shah Abbas, Safawid Persia briefly became the most powerful nation of southern and western Asia. His successors, however, soon lost what he had gained from the Turks and Uzbeks, and by the end of the century the Safawid Dynasty was on the verge of collapse. The principal events were:

1587–1629. Reign of Shah Abbas the Great. (See p. 551.)

1602–1612. War with Turkey. Abbas took advantage of Turkish internal troubles and wars in Europe. Marching rapidly from Isfahan (the new Persian capital), his efficient modern army easily defeated local Turkish forces and besieged Tabriz, which capitulated after a long siege (October 21, 1603).

1603–1604. Reconquest of South Caucasus Region. Abbas captured Erivan (after a 6-month siege), Shirvan, and Kars. In little over a year he had regained the territories surrendered to Turkey a decade earlier.

1605–1606. Consolidation and Preparation. While Abbas consolidated his control over the reconquered territories, the Turks, under young and vigorous Sultan Ahmed, prepared to get them back.

1606. Battle of Sis (or Urmia). Ahmed with an army of 100,000 met Abbas with 62,000 near Lake Urmia. The Turks followed their traditional battle plan against the Persians: a meeting engagement of cavalry, designed to draw the essentially cavalry Persian army within range of the guns of Turkish infantry and artillery. Abbas, however, sent a detachment to carry out a demonstration behind the Turks. This caused much of the Turkish army to face to the rear, thinking this the main Persian threat. Then, with his main force of disciplined cavalry, infantry, and artillery, Abbas made his main effort. The Turks were routed, losing more than 20,000 dead. Following this the Persians occupied all of Azerbaijan, Kurdistan, Baghdad, Mosul, and Diarbekh.

1612. Peace with Turkey. Turkey renounced all claim to the conquests of Sultans Murad and Mohammed III. In a face-saving gesture, Abbas agreed to give the Sultan 200 camel loads of silk annually.

1613–1615. Expedition against Georgia. Georgia acknowledged Abbas' suzerainty. This being an area considered in the Ottoman sphere of interest, and since Abbas had made no payment of silk, Turkey prepared for war.

1616–1618. War with Turkey. A powerful Turkish army laid siege to Erivan, but was repulsed. Forced by winter weather to withdraw, the Turks lost heavily from the cold and from Persian harassment (1616–1617). After a year's lull, the Turks again invaded, advancing rapidly on Tabriz. A portion of their army was ambushed by a Persian force and suffered severely, but the main Turkish army continued to advance, at a reduced rate. Peace negotiations opened; the terms of the treaty of 1612 were reaffirmed, save that the silk tribute from Abbas was reduced to 100 camel loads annually.

1622–1623. War with the Mogul Empire. Abbas led an army that captured Kandahar (see p. 643).

1623–1638. War with Turkey. Turkey again invaded the areas lost to Abbas, this time attempting to recapture Baghdad. After a 6-month siege, Abbas arrived with a relieving army (1625). Fierce fighting was inconclusive, until a mutiny forced the Turks to withdraw (1626). They suffered severe losses in their retreat. For several years there were no campaigns of importance, but incessant border warfare. Renewing active warfare, Sultan Murad IV sent an invasion army through Kurdistan (1630). Capturing Mosul en route, the Turks defeated a Persian army, then besieged and captured Hamadan. They continued on to Baghdad, but were repulsed and retreated to Mosul (1630). Internal unrest in Turkey and a series of mutinies in the Turkish Army provided a respite to the Persians (1631–1634). Murad led an invading army in person to capture Erivan and Tabriz (1635). Murad having returned to Constantinople, **Shah Safi** (see below) besieged Erivan and after a long winter siege captured the town (1636). Following another lull, Murad took the field again and, showing great personal energy and skill, assaulted and captured Baghdad (1638). Peace was now made; the Turks kept Baghdad and the Persians kept Erivan.

1629–1642. Reign of Shah Safi (grandson of Abbas). He was a weak and cruel ruler. He lost Kandahar to the Uzbeks (1630; see p. 646).

1638. War with the Moguls. Shah Jahan's army reconquered Kandahar (see p. 646).

1642–1667. Reign of Shah Abbas II. He halted the decline of the Safawid Dynasty.

1649–1653. War with the Moguls. Abbas personally led the army that recaptured Kandahar (1649). His troops then repulsed Aurangzeb's repeated attempts to reconquer the city (see p. 643).

1659. Revolt in Georgia. Abbas repressed an uprising led by **Tahmurath** Khan.

1664. Cossack Raid into Mazandaran. Instigated by Czar Alexis of Moscow, the Cossacks raided deep into Persia, at the base of the Caspian Sea, but were driven out after causing considerable loss and damage. This was the first Russian aggression against Persia.

1667–1694. Reign of Shah Suleiman. Persia declined rapidly during this drunken reign. The Dutch seized the port of Kishm and the Uzbeks made numerous inroads into Khorasan.

SOUTH ASIA

NORTH INDIA

1605. Death of Akbar. Probably the result of poisoning by his son Salim, who ascended the throne with the title **Jahangir** (1605–1627).

1606. Revolt in the Punjab. Jahangir's example of filial insubordination was followed by his own son **Khusru,** who seized Lahore. Jahangir promptly marched to the Punjab and at **Jal-lundur** defeated and captured his errant son, whom he caused to be partially blinded and imprisoned.

1607–1612. Rebellion in Bengal. Smoldering Afghan opposition to the Moguls broke into open warfare in Bengal, but was suppressed by Jahangir.

1610–1629. War with Ahmadnagar. During most of his reign Jahangir was involved in a desultory war in the Deccan with Ahmadnagar. Its prime minister and general, an Abyssinian named **Malik Ambar,** realizing that his small forces could not meet those of Jahangir in a pitched battle, resorted successfully to guerrilla tactics.

1622–1623. War with Persia. The Persian Emperor, Shah Abbas, captured Kandahar (1622; see p. 642). Rebellion of Jahangir's second son, **Khurram,** prevented Mogul retaliation.

1623. Khurram's Rebellion. He was defeated at the **Battle of Balochpur,** and fled (1623).

1625. Revolt. This unsuccessful revolt was led by Jahangir's chief general, **Mahabat Khan.**

1627–1658. Reign of Shah Jahan. He was creator of the Taj Mahal, erected as a tomb for his lovely wife, **Mumtaz Mahal.**

1629–1636. Conquest in the Deccan. Shah Jahan renewed his predecessors' efforts to dominate the Deccan. After the death of Malik Ambar, he conquered and annexed Ahmadnagar (1633–1636). Golconda and Bijapur were defeated and forced to acknowledge Mogul sovereignty (1635–1636).

1631–1632. War with the Portuguese. The Portuguese were meddling in Bengal and had also encouraged piracy against all merchant ships but their own. Shah Jahan sent an army which captured the Portuguese fort at Hooghly, in the Ganges-Brahmaputra Delta; the defenders were all killed or brought to Agra as captive slaves (1632).

1637–1646. Unrest in the Deccan. The conquered territories were pacified by Shah Jahan's young second son, Aurangzeb.

1638. Recapture of Kandahar from the Persians and Uzbeks. (See pp. 642, 646.)

1649–1653. War with Persia over Kandahar.

The Persians again seized Kandahar (1649). Three attempts by Aurangzeb to recapture the fortress failed (1650, 1652, 1653).

1653–1658. Revolt in the Deccan. Due to the Maratha Revolt (see below), Jahan sent Aurangzeb back to the Deccan. He reconquered Golconda and Bijapur (1658), but the Maratha leader **Sivaji** was still unconquered.

1657–1659. War of Succession. The illness of Shah Jahan was followed by a fratricidal struggle among his 4 sons. Aurangzeb, prevailing after the **Battles of Samugarh** (May 1658) and **Khajwa** (January 1659), imprisoned his father in Agra Fort (where he remained, within sight of the Taj Mahal, until his death in 1666).

1658–1707. Reign of Aurangzeb. He crowned himself Emperor at Delhi (July 1658). Soon afterward he conquered Assam (1661–1663) and annexed Chittagong (1666).

1664–1707. Persecution of Non-Moslems. Aurangzeb's revival of persecution and discrimination alienated the Hindu masses and encouraged revolts. The consequent unrest resulted in constant turmoil during most of his long reign.

1675–1681. Rajput Revolt. Aurangzeb's persecution of the Sikhs turned that peace-loving religious group into a community of fanatic warriors dedicated to the destruction of Moslem rule. His policies caused the same reaction among the formerly conciliated Rajputs. Under a popular Mewar leader named **Durgadas,** the Rajputs drove out most of the Mogul garrisons (1675–1679).

1681. Revolt of Akbar. Aurangzeb's son Akbar joined the Rajput rebels, but was defeated by his father and forced to flee to Persia.

1681–1707. Continuing Conflict in Rajputana. This resulted in the elimination from the imperial army of the Rajput cavalry, one of its most effective components. This loss was particularly serious at a time when the empire was becoming engaged in a major military effort in the south.

1681–1689. Renewed War in the Deccan. The continuous exploits and victories of Sivaji and the consequent rise of the Marathas (see below) now caused Aurangzeb to take the field personally with his Grand Army in the Deccan. He overran the Maratha country, captured and executed Sivaji's son **Sambhuji** (1689), and recaptured both Bijapur (1686) and Golconda (1687). The Marathas refused to acknowledge defeat and resorted to protracted guerrilla war, leaving the land unconquered and unconquerable (see below).

1689–1707. The Decline of Aurangzeb. His large empire bankrupt and exhausted, many of his soldiers mutinous, his policy of expansion and conquest unconsolidated, it was only through tremendous personal vitality and will power that Aurangzeb succeeded in holding his realm together until his death.

SOUTH INDIA AND CEYLON

1610–1629. Revival of Ahmadnagar. Though nominally the prime minister, the virtual ruler of Ahmadnagar for nearly 2 decades was the Abyssinian General Malik Ambar. He defeated the Moguls, then retained the independence of the nation in a protracted series of partisan wars (see p. 643).

1636. Mogul Reconquest of Ahmadnagar. After the death of Malik Ambar the Moguls reconquered the country.

1646–1680. Rise of the Marathas; Reign of Sivaji. The Marathas, a hardy, frugal Hindu people living generally in arid mountains of the west-central Deccan, east and southeast of modern Bombay, were essentially the descendants of the former Yavada Kingdom of Devagiri; their major employment had been as mercenary soldiers in other states. A growing feeling of religious and national consciousness was sparked into a flame by the rise of the young Yavada Prince **Sivaji.** At the age of 19 he rose against the Moslem principality of Bijapur and established an independent Maratha principality in the western Ghat Mountains.

1646–1664. Expansion of the Marathas. Slowly Sivaji increased his strength and power, initially at the expense of Bijapur, and then through incursions and conquests in the Mogul provinces to the north. He created a navy (1659), which enabled him to carry the war to Mogul lands bordering the Arabian Sea. It proved itself very effective as a bulwark against the Portuguese and English fleets and as a mercantile force trading between India and Arabia.

1664. Sack of Surat. Sivaji captured and sacked the principal Mogul port. Subsequently defeated by a Mogul army under the Rajput **Jai Singh,** he made peace, and was sent to Delhi as a nominal ambassador but virtual prisoner. After more than a year in Delhi, Sivaji made a dramatic escape and returned to the Deccan.

1670–1674. Renewed Maratha-Mogul War. Sivaji was successful and established himself as king of an independent Maratha kingdom.

1674–1680. Expansion of the Maratha Kingdom. A first-rate soldier and able adminstrator, Sivaji continued to gain power and prestige until his untimely death (1680). His military success was due to his use of inaccessible hill forts as bases of operations for guerrilla warfare, and to the exceptional efficiency of the regular army he created, compensating his lack of quantity by quality. Since this arid country could not support such an army, he found provender and the wealth he needed for his military budget by constantly raiding Moslem lands, or receiving "protection" tribute from those who wished to escape his depredations. The Marathas, good horsemen, rode light, devoid of luxuries, heavy baggage, equipment, and tents. In the actual conduct of operations Sivaji followed the traditions of Hindu chivalry in his treatment of defeated enemies and of noncombatants.

1683–1689. Renewed Maratha-Mogul War. (See p. 643.)

1689–1707. Resurgence of the Marathas. Slowly the Marathas, in desperate guerrilla warfare, wore down the strength of the Mogul Grand Army. They received an assist from nature when most of the Mogul army was wiped out in a flood of the Bhima River (1695). Aurangzeb returned personally to the field in his old age, but died in Ahmadnagar (1707). Following this, the Grand Army withdrew northward, leaving the Marathas again dominant in the Deccan.

THE EUROPEANS IN INDIA AND THE INDIAN OCEAN

1601–1603. Arrival of the Dutch. Appearance of Dutch ships on the Gujerat and Coromandel coasts ended Portuguese monopoly of Indian overseas trade.

1602. Portuguese Loss of Bahrein. Arab vassals of Persia recaptured the island.

1608. Arrival of the English. William Hawkins of the British East India Company arrived at Surat on the Gujerat coast and tried to gain a trading-concessions treaty with Jahangir.

1612–1630. English-Portuguese Conflict. Trading rights were obtained at Surat when **Thomas Best** defeated a much superior Portuguese fleet (1612). Hostilities persisted, but Portuguese warships were once again defeated by **Nicholas Dowton** (1614). In following years the Dutch and British East India Companies established trading posts scattered along the coast. The Indians welcomed the newcomers because they envisaged an opportunity to play off the 3 European nations against each other.

1616. England Begins Trade with Persia. The first merchant ship of the East India Company evaded Portuguese warships and arrived at the Persian port of Jask.

1618. English-Portuguese Conflict in the Persian Gulf. A small English squadron, operating from Surat for the purpose of keeping the Persian Gulf open to English trade, defeated a larger Portuguese fleet in two engagements off **Jask.** This gave the English superiority in the Arabian Sea.

1620. Kandyan-Dutch Alliance. (see p. 557.) After years of negotiation, Singhalese King Vimala's successor **Senerat** concluded an alliance of Kandy with the Dutch against the Portuguese.

1621. English Recognition by the Moguls. Jahangir granted the English additional trade rights, and established the East India Company as a naval auxiliary of the Mogul Empire.

1621–1622. English Capture of Ormuz. An English fleet, cooperating with a Persian land army, captured the major Portuguese base of Ormuz. After this disaster the Portuguese moved their Persian Gulf base to Muscat.

1624–1630. English-Dutch Cooperation against the Portuguese. A major Portuguese effort to reestablish supremacy in the Indian Ocean caused the English and Dutch to cooperate. After some indecisive actions in the Arabian Sea and Persian Gulf (1624–1625), Portuguese power continued its slow decline. A Portuguese effort to retake Ormuz failed completely (1630).

NOTE: *The ancient alliance between England and Portugal was ignored in their colonial relationships.*

1633. Dutch and Kandyan forces capture Trincomalee.

1640. First Fortified English Post. Fort Saint George (the site of modern Madras) became the company's headquarters on the Coromandel coast.

1641–1663. Dutch Encroachment against the Portuguese. The Dutch attempted to eliminate Portuguese land bases in the Indian Ocean area, capturing Malacca (1641), Colombo (1656), Negapatam (1658), and Cochin (1663). Goa held out against several determined Dutch attacks.

1650. Portuguese Loss of Muscat. The base was captured by the Imam of Oman.

1651. English in Bengal. A post was established at Hooghly. They later decided to fortify in the face of Mogul threats (1681).

1656–1658. Ceylonese Isolation. After the Dutch victory over Portugal, the kingdom of Kandy realized it had merely exchanged one foreign influence for another, and withdrew into the mountainous Ceylonese interior, yielding the coastal areas to the Dutch.

1666. Arrival of the French. The French entered the Indian trade later than their Dutch and English rivals, but they established a number of posts, the most important being Chandernagore (1670) on the Ganges Delta near Hooghly, and Pondichery (1673) on the Coromandel coast near Fort Saint George.

1667. Establishment of Bombay. The British

established a fort and trading post on the island of Bombay.

1686. War in Bengal. Desultory fighting erupted between the British at Hooghly and the Mogul governor of Bengal. This resulted in the building and fortification of Fort William lower down the river (the site of modern Calcutta).

EAST AND CENTRAL ASIA

INNER ASIA

At the outset of the 17th century, Inner Asia was inhabited by 3 major groups of largely nomadic peoples: the Turks, the Mongols and Tartars, and the Tibetans. The 2 most important of the Turkic peoples were the Uzbeks of southern and central Turkestan and the Kazakhs of northern Turkestan. The Uzbeks dominated the Turkoman nomads, who inhabited the desert region between the Caspian and Aral seas, and the Tajiks, who lived in the highlands north of the Hindu Kush Mountains. Most important of the many Mongol tribes were the Khalkas, who lived north of the Gobi Desert; the Kalmucks, who inhabited western Mongolia and the Altai region; and the Kirghiz, who lived in the valleys of the Tien Shan Mountains.

During this century began a process which would, in the following century, completely extinguish the traditional independence of these warlike peoples of Inner Asia. This was the steady expansion of Czarist Russia from the west and of Manchu (or Ch'ing) China from the east. That the Mongols and Turks, traditionally conquerors, should themselves now be conquered was not due to any decline in their warlike proclivities, but rather to the fact that the evolution of the art of war had progressed beyond the capacity of essentially nomadic peoples. Their economic resources would not permit the production or purchase of muskets and cannon, and their cavalry could not stand up to modern musketry and artillery.

For our purposes we shall deal with the principal events of Inner Asia in this century under three headings: the Uzbeks, the Mongols, and Russian expansion. Chinese expansion is dealt with under China.

The Uzbeks

1599. Independence of Bokhara. Bokhara, under the Ashtarkanids, threw off the rule of the decaying Shaibanid Dynasty. Bokhara now again became the principal seat of Uzbek power, but its emirs shared control of southern Turkestan with the khans of Khiva (successors of the Shaibanids).

1608–1646. Reign of Iman Kuli of Bokhara. Bokharan Uzbeks raided occasionally into Persian Khorasan and Afghanistan. Sporadic efforts to conquer the Kirghiz were invariably repulsed by the mountain Mongol tribes.

1630–1634. Conquest of Kandahar. Bokhara seized Kandahar from the Persians (see p. 642) but the Uzbeks were soon driven out by Mogul Shah Jahan (1638; see p. 643).

1688. Conquest of Khiva by the Persians. The Khan of Khiva retained autonomy as a vassal of the Persian Shah.

Mongolia

The innumerable wars between the great Mongol tribal groupings, among the tribes within the groupings, and with the neighboring Tibetans, Chinese, Tartars, Turks, and Russians were too many, and too confused, for simple and concise summarization. During most of this century the Mongols dominated Tibet. The principal event were:

1623–1633. Conquest of Inner Mongolia by the Manchus. (See p. 648.) The Ordos Mongols became virtually assimilated by the Chinese.

1636. Migration of the Western Kalmucks. They fought their way through Kirghiz and Kazakh territory to cross the Emba River. They subsequently settled in the Trans-Volga steppe, raiding Russian settlements on both sides of the river. Submitting to Russia (1646), they maintained autonomy under their own khan. They became an excellent source of light cavalry for the Russians, who later used them in their campaigns against the Krim Tartars and in Central Asia.

1672. Raid of the Torgut Mongols. Ayuka Khan of the Torgut Mongols raided through western Siberia, across the Urals and the Volga, as far as Kazan in Russia. He then made peace with the Russians and, supported by this, was able to maintain his lands in relative peace for the remainder of the century.

c. 1680–1688. Expansion of Galdan's Dzungar Empire. Under the leadership of **Galdan** (Kaldan) Khan (also known as Bushtu Khan), the Dzungars—a subdivision of the Kalmucks inhabiting the region around Lake Balkash—conquered most of Kashgar, Yarkand, and Khotan from the Kirghiz, and also expanded into Kazakh territory.

1688–1689. Galdan Attacks the Khalkas. The Khalka Mongols, hard-pressed by the Dzungars, appealed to the Manchus for aid. A Manchu army helped them.

1689. Congress of Dolonor. (See p. 650.)

1690–1695. Galdan Harasses the Khalkas and the Chinese.

1696. Battle of Chao-Modo (Urga). (See p. 650.)

1697. Death of Galdan. (See p. 650.)

Expansion of Russia into Central Asia

The initial movements of the Russians east of the Urals late in the 16th century had been made in the high latitudes, carefully avoiding the still-formidable Kalmucks, who inhabited the region that now comprises central Siberia.

1604. Founding of Tomsk. Russian expansion extended eastward as far as the middle Yenisei River.

1618. Russians Reach the Upper Yenisei. They established the fort of Yaniseisk.

1628. Russians Reach the Middle Lena River.

1630. Founding of Kirensk.

1632. Founding of Yakutsk.

1639. Russians Reach the Sea of Okhotsk. This was their first foothold on the Pacific.

1641–1652. Conquest of the Buryat Mongols. This gave the Russians control of the region around Lake Baikal.

1643–1646. Penetration into the Amur Valley. A Russian expedition from Yakutsk under **Vasily Poyarkov** made the first contact with Chinese civilization.

1644. Russians at the Mouth of the Kolyma River. This brought them to the Arctic Ocean.

1648. Founding of Okhotsk.

1651. Founding of Irkutsk and Khabarovsk.

1651–1653. Occupation of Amur Valley. Yerofey Khabarov established Russian forts in the Amur Valley and in the Daur land to the north.

1653–1685. Clashes with China. The Manchus, relatively uninterested in the Amur Valley, were aroused by Russian persecution of Chinese and Manchu settlers in the region. Occasional Manchu raids made the Russian position precarious.

1660. Manchus Eject the Russians from the Amur Valley. (See p. 649.)

1683–1685. War with the Manchus. The Russians were again ejected from the Amur Valley (see pp. 649–650).

1689. Treaty of Nerchinsk. (See p. 650.)

CHINA, 1600–1700

1600–1615. Emergence of a Manchu State. **Nurhachi** erected a 3- or 4-walled fortress in his capital, Liaoyang, and welded the former Ming garrison system (*wei*) into a primitive military and administrative bureaucracy of 4 (later 8) "banners" (1600–1615). Manchu nobles formed an officer corps, usurping the traditional command functions of tribal chieftains. By 1644, the Manchu army included 278 Manchu, 120 Mongol, and 165 Chinese companies, each company (*niru*) consisting of approximately 300 men.

1618. Outbreak of War between Manchus and Mings. Nurhachi's army seized a Ming stronghold at Fushun, then defeated a Ming retaliatory force. The Ming recruited 20,000 Koreans, traditional allies of the Chinese dynasty, to help put down the Manchu revolt.

1621. Capture of Mukden (Shenyang). The Manchus drove the Ming from the Liao Basin and attacked Mukden. The city capitulated after Ming troops, dispatched from the garrison to meet Nurhachi's advance, were trapped when traitors within Mukden destroyed the single bridge spanning the fortress moat.

1623. Manchu Advance Halted. Nurhachi was defeated near the **Great Wall** by a Ming provincial governor, **Yuang Ch'unghuan;** the Ming artillery, provided by European Jesuit missionaries, overpowering the Manchu longbows. The Manchus thereupon turned west and began to expand into Mongolia.

1624–1625. The Dutch Reach Taiwan. They established several posts on Taiwan.

1626. Death of Nurhachi. He was succeeded by his son, **Abahai** (or T'ai Tsung), who immediately augmented the banners.

1627, February. Manchu Invasion of Korea.

They crossed the frozen Yalu and subdued the Ming ally prior to invading China.

1628. Manchus Again Repulsed. A renewed effort to invade China was again halted by Yuang with his Ming artillery.

1629–1634. Raids into Northern China. Repeated Manchu raids into northern China through the Jehol Pass were repulsed by the Ming. Manchu forces also raided Shansi Province (1632 and 1634). During this period, Abahai began developing his own highly effective artillery to permit his troops to face the Ming army in open battle.

1633. Manchu Conquest of Inner Mongolia. This was facilitated by Mongol soldiers defecting to escape harsh conditions in the Mongolian army. Subsequently, there was large-scale assimilation of Mongol troops into the Manchu banners.

1635–1644. Decline of the Ming. Widespread rebellions throughout China.

1636–1637. Conquest of Korea. Upon the failure of the Koreans to pay tribute and their reluctance to participate in campaigns against the Ming, Abahai led 100,000 Manchus into Korea and compelled a formal renunciation of the Chinese dynasty.

1636–1644. Consolidation of the Amur Basin. In 4 expeditions, Abahai secured control over the entire Amur region.

1636. Establishment of Ch'ing Dynasty. At Mukden, the Manchus proclaimed an imperial dynasty; Abahai took the title of **Ch'ung Teh.**

1643. Death of Abahai. His 5-year-old son, **Shun Chih,** assumed the imperial title. Actual power passed to Abahai's brother, the regent, Prince **Dorgon.**

1643–1661. Reign of Shun Chih. Ming resistance in the south persisted throughout Shun Chih's reign.

1643–1646. Russian Explorations in the Amur Region.

1644, May 26. Collapse of the Ming Dynasty. A rebel brigand, **Li Tzu-ch'eng,** seized Peking. **Wu San-kuei,** a Ming general (and later viceroy), enlisted Manchu aid to overthrow the rebel regime. In a great battle just south of the

Great Wall the Manchus defeated Li and seized Peking.

1644–1645. Expansion to the Yangtze. Prince **Fu** of the Ming Dynasty set up a new government at Nanking, defying the invaders. Prince Dorgon defeated the Ming army in a 7-day battle in and around **Yangchow.** A bloody and indiscriminate massacre of the defeated army and the inhabitants followed. Nanking soon fell, with little struggle, and Prince Fu fled into oblivion. Several Ming cousins, however, claimed the imperial throne and continued scattered resistance to the invaders.

1645–1647. Conquest of Fukien. Taking advantage of dissension among the Ming claimants, the Manchus swept through Fukien Province (1645–1646). They next seized Canton (1647). Strong resistance continued, however, in Shensi, Shansi, and Szechwan.

1648–1651. Campaigns of Kuei Wang. Prince **Kuei Wang,** last of the Mings, now became the rallying point for Chinese opposition to the invaders. An able soldier, he briefly controlled most of south China. Dorgon quickly consolidated control of the Yangtze Valley, however, then systematically overran most of south China. Kuei Wang fled to the mountainous southwest, where he briefly retained control of Kweichow and Yunnan.

1651–1659. Conquest of the Southwest. Despite desperate resistance and difficulties of terrain, the Manchus established control over most of the southwest, although they were unable to capture Kuei Wang before his death (1662).

1652–1662. War against the Pirates. A family of pirates, whose most renowned member was **Cheng Ch'eng-kung** (known to Europeans as **Koxinga**), conducted incessant warfare on behalf of the Ming against both Manchus and Europeans in China. Koxinga captured Amoy (1653) and Ch'ung-ming Island (1656) and unsuccessfully assaulted Nanking (1657). He attacked Taiwan with 900 vessels and besieged Fort Zelanda (at Anping), compelling the Dutch to capitulate (1661–1662). The Ch'ing Dynasty evacuated 6 coastal provinces and installed their inhabitants behind a guarded barrier some 10 miles from the sea (1661). Pirate power faded, however, after Koxinga's death (1662).

1659–1683. Manchu Consolidation. The conquerors strengthened the garrison at Peking, surrounding it with about 25 smaller posts, and established stout garrisons in 9 provincial capitals and at other strategic sites. "Tartar generals" (*chiang-chun*) and Manchu brigade generals commanded these fortifications on a rotational basis. Local police operations in 14 provinces were conducted by the Chinese Army of the Green Standard under Chinese officers.

1660. Clash with the Russians. Manchu troops forced the Russians to evacuate their posts in the Amur and lower Sungari River regions.

1662–1722. Reign of K'ang-hsi. The height of modern Chinese culture and glory.

1663–1664. Dutch Operations against the Pirates. With Manchu support, a Dutch fleet forced **Cheng Chin,** Koxinga's son, to retire from his Fukien coastal stronghold to Taiwan.

1674–1681. Revolt of the Three Viceroys. K'ang-hsi, wary of the growing power of the governors (or viceroys) of Yunnan, Fukien, and Kiangsi, decreed their removal (1673). Wu San-kuei of Yunnan (the former Ming general) resisted the imperial decree and held Szechuan, Kweichow, Hunan, and Kwangsi (1674–1679). He was joined in revolt by the other 2 governors. Ultimately the uprising was quelled.

1675. Revolt in Chahar. Suppressed by the Manchus.

1683. Taiwan Annexed. Cheng K'e-shuang, the son and successor of Cheng Chin, surrendered Taiwan to the Manchus.

1683–1685. Clash with the Russians. The Russians had reestablished themselves securely in the Amur Valley (1670) while the Manchus were engaged in suppressing the revolt in southern China. A Manchu military commission studied the Russian penetration (1682) and concluded that about 2,000 troops would be required to counter the advance; 1,500 were dispatched to the Zeya River (1683). The following year the Russians were expelled from the lower Sungari area. A Manchu army then marched on the Russian stronghold at Albasin and forced the garrison to surrender (1685). When the Manchus withdrew, however, later that year, another Russian contingent reconstructed the fortifications. The Manchus began to prepare for a more extensive war.

1689, August. Treaty of Nerchinsk. Distracted by events in Russia and Europe (see pp. 634–635), Peter the Great was anxious to resolve the Albasin dispute by pacific means. Chinese war preparations and the skillful diplomacy of Jean François Gerbillon, a Jesuit member of the Manchu delegation, caused the Russians, led by **Fyodor Golovin,** to compromise their original rigid demands and to abandon Albasin and the area north of the Amur River. The terms of the Treaty of Nerchinsk, supplemented by the primarily commercial **Treaty of Kiakhta**

(1727), regulated Russian-Chinese relations for 175 years.

1688–1689. Dzungar Raids into Mongolia. (See p. 647.)

1689. Congress of Dolonor. The Mongols formally accepted Manchu suzerainty. Prolonged resistance of elements of the Khalka and Kalmuck tribes continued, but K'ang-hsi's operations in the Gobi finally prevailed (1721).

1696. War in the Ili Valley and Turkestan. After 5 years of desultory raiding by the Dzungars into central Mongolia, K'ang-hsi led 7 columns of 80,000 troops across Mongolia and, with the decisive use of the Manchu artillery, crushed Galdan at **Chao-Modo** (Urga). The Dzungar leader fled and committed suicide (1697). Mongol bands continued to wage sporadic war in the northwest.

JAPAN, 1600–1700

Within 2 years of the death of Hideyoshi (1598), the joint regency of his most powerful vassals had disintegrated into a struggle for supremacy between Ieyasu and the combined forces of Ishida, Uesugi, Mori, Ukita, Konishi, and Shimazu.

1600, October 21. Battle of Sekigahara (Barrier Field). Ieyasu's 74,000 troops decisively defeated a coalition force of 82,000, torn by internal dissension. About 40,000 confederates were reportedly killed. This victory, and Mori's subsequent surrender of Osaka Castle, enabled Ieyasu to consolidate control over Honshu.

1603. Ieyasu Proclaimed Shogun. The capital was established at Edo (Tokyo). His Tokugawa Shogunate was to rule Japan until the second half of the 19th century.

1606–1614. Tokugawa Exclusion Policy. Fear of the political objectives of missionary activities—and of possible intervention of Spain and Portugal in a violent Franciscan-Jesuit controversy—led to the wide-scale deportation of European priests. Persecution of Christians was intensified under **Iyemitsu** (1623–1651).

1609, June. Sinking of the Madre de Dios. Following a Japanese-Portuguese flare-up at

Macao, Ieyasu ordered seizure of a Portuguese merchant ship, the *Madre de Dios,* moored in Nagasaki Harbor. A Japanese fleet, carrying over 1,200 troops, surrounded the vessel. After 3 days of combat, the Portuguese captain, **Pessoa,** blew up his ship, its crew, many of his assailants, and over 1 million crowns' worth of cargo.

1614, December–1615, June. Siege of Osaka Castle. Hideyoshi's son, **Hideyori,** had retired to Osaka Castle after the Battle of Sekigahara. During the first decade of the Tokugawa Shogunate, the castle became the center of potential resistance as some 90,000 alienated warriors (*ronin*) gathered at Osaka. A siege of the castle, conducted by Edo forces between December and January, was terminated by a truce. This respite enabled the besieging force to fill the outer moats and, in the following June, to successfully storm the stronghold.

1616, June 1. Death of Ieyasu. He was

succeeded by his son, **Hidetada** (1616–1623).

1623–1651. Shogunate of Iyemitsu.

1624. Japanese Isolation Policy. Following the voluntary withdrawal of English traders (1623), the Spanish were forcibly driven from Japan. The shogun ordered the destruction of all large Japanese vessels in order to prevent the mass embarkation of any disaffected subjects.

1637–1638, April 12. Shimabara Revolt. Christianized peasants on the Shimabara peninsula and Amakusa Islands rose against the oppressive religious and agrarian practices of the Tokugawa regime. Between 20,000 and 37,000 insurgents gathered at the ancient fortress of Hara; for nearly 3 months they were besieged by 100,000 Edo troops supported by a Dutch man-of-war dispatched from Hirado. Exhaustion of supplies and ammunition compelled the insurgents to capitulate; most were massacred.

1637–1638. Exclusion Edict. Iyemitsu, suspecting that Portuguese traders had encouraged the insurgents, demanded their expulsion from Japan. Only 2 foreign settlements and trade outposts remained in Japan: the Dutch outpost at Hirado and the Chinese at Nagasaki.

1651–1680. Shogunate of Ietsune. Two attempted coups, the last until the 19th century, were readily suppressed at Edo (1651–1652).

1653–1700. Internal Peace in Japan.

KOREA

1606. Peace with Japan. Good relations were restored after the Tokugawa shoguns came to power (1600) and lasted without a break until the second half of the 19th century.

1624. Military Uprising. This was caused by renewed factional dispute in Korea. It swept the northern half of the peninsula before it was suppressed.

1627–1637. Wars with the Manchus. The Koreans, as vassals of the Ming, became involved in the wars with the rising Manchu power on the northern frontiers of Korea (see p. 648).

c. 1630. The Hermit Kingdom. A policy of national seclusion was introduced. Although the Yi Dynasty maintained its traditional tributary relationship with Manchu China, Korea become known as the Hermit Kingdom, stagnated.

1650–1700. Internal Factional Strife.

SOUTHEAST ASIA

BURMA

1600. Destruction of Pegu. As a result of the incursions of Arakan, Ayuthia, and Toungoo (see p. 561), Burma was broken up into several small states.

1601–1605. Revival of Burma. The **Nyaungyan** Prince, descendant of the old royal family, held Ava. He embarked on a series of campaigns to reestablish authority in Upper Burma and the Shan States.

1602. Arrival of the Portuguese. Philip de Brito established a base at Syriam.

1605–1628. Reign of Anaukpetlun. Son of the Nyaungyan Prince, Anaukpetlun reconquered much of south Burma. He took Prome (1607) and Toungoo (1610).

1613. Conquest of Syriam. Anaukpetlun captured the Portuguese base; de Brito was impaled.

1613–1614. Repulse from Tenasserim. Anaukpetlun was driven out by Siamese and Portuguese forces.

1628. Murder of Anaukpetlun. A brief civil war followed.

1629–1648. Reign of Thalun. He transferred the capital from Pegu to Ava.

1648–1661. Reign of Pindale.

1658–1661. War with China. Prince **Yung Li,** last of the Mings, fled from Yunnan to Burma after defeat by the Manchus (1658). Pindale gave political asylum to the prince in Sagaing, and for 3 years Yunnan was harassed by repeated raids of Ming supporters in Upper Burma.

1660–1662. War with Siam. (See below.)

1661–1672. Reign of Pye (Brother and Murderer of Pindale). At the outset Pye was confronted by a Manchu army of 20,000 men and was forced to surrender the Ming prince (1662).

SIAM

1600, May. Repulse from Toungoo. King **Naresuen** returned to Siam, leaving Burma in a state of chaos (see p. 561).

1603–1618. Expedition to Cambodia. Prince **Srisup'anma,** accompanied by an army of 6,000 men, became King of Cambodia as a vassal of Siam.

1604–1605. Naresuen's Last Campaign. Naresuen went north with a large army to meet the Nyaungyan Prince of Burma, who was attempting to reconquer the Shan States. Naresuen fell ill and died. His successors abandoned the effort, and the Shan States returned to Burmese domination.

1610–1628. Reign of Songt'am the Just. He suppressed an insurrection incited by Japanese traders (1610). He then turned to defeat a simultaneous invasion from Laos.

1612. Portuguese-Siamese Dispute. The Siamese governor of Pegu and Philip de Brito quarreled over their alliance against Toungoo and withdrew their support from each other. As a result de Brito and Syriam fell to Anaukpetlun (see p. 651) and Siam lost most of its Peguan possessions.

1614. Repulse of Burmese. Siamese forces, aided by Portuguese mercenaries, defended Tenasserim against Anaukpetlun's invasion (see p. 651).

1618. Truce between Burma and Siam.

1622. War with Cambodia. Srisup'anma died and his successor proclaimed Cambodia independent. Songt'am invaded Cambodia by land and by sea. The Siamese fleet returned to Siam without seeing any action, but the Siamese army suffered heavy losses in men, horses, and elephants, and withdrew.

1630–1656. Reign of Prasat T'ong. He was a usurper who soon established friendly relations with the Dutch, who were given special trading privileges.

1634–1636. Rebellion of the Patani. This state, in the lower Malay Peninsula, refused to recognize Prasat T'ong. Two Siamese expeditions to Patani were total failures, and Patani became virtually independent.

1657–1688. Reign of Narai.

1660–1662. Campaigns around Chiengmai. King Narai, taking advantage of Burma's preoccupation with the Ming raiders, marched upon Chiengmai. After one repulse (1660), another effort was made, capturing Chiengmai (March 1662).

1662. Repulse of Burmese. The Mons, who were rebelling against the Burmese and fleeing into Siam, called on King Narai to protect them from pursuing Burmese troops. Narai repulsed the pursuers, then undertook a series of punitive raids into Lower Burma (see above).

1687, August. War against the British East India Company. This followed a series of disputes with the British and the Dutch. The French East India Company supported Narai, sending a force of some 600 men to help man coastal forts. The French soon were at odds with the Siamese.

1688–1703. Reign of P'ra P'etraja. He reached understandings with both the English and French. The reign was plagued by serious internal rebellions in the provinces.

LAOS

1624–1637. War of Succession. Finally ended when **Souligna-Vongsa** defeated 4 rivals.

1637–1694. Reign of Souligna-Vongsa.

1651–1652. War with Tran Ninn. Souligna, denied the hand of the daughter of the king of Tran Ninn (a small state between Annam and Laos), sent a military force to bring back the princess. After an initial repulse Xieng-Khouang, capital of Tran Ninn, was captured and the princess seized.

1694–1700. Turmoil in Laos. Three-way struggle for the throne.

VIETNAM

While the Le Dynasty was recognized as the only legitimate ruling force in Vietnam, rivalry and partition between the Trinh family (north) and the Nguyen family (south) led to seven campaigns. The Trinh had an army and fleet of 100,000 men, 500 elephants, 500 large junks, and cannons. The Nguyen had numerically inferior forces, but as early as 1615 had been producing heavy guns under Portuguese auspices. The shipments of modern weapons from Portugal and the Portuguese military advisers enabled the Nguyen to successfully resist the Trinh offensive. The Nguyen reinforced their natural defenses by constructing 2 huge walls across the main avenues of approach, north of Hué. The Truong-duc wall was 6 miles long, contained a camp for troops, and was an obstacle for passage up the Nhat-Le River. The Dong-hoi wall was 11 miles long and fortified with heavy cannons. In over 50 years of fighting, the Trinh never managed to break through both of these walls. The Nguyen also began constructing arsenals, cannon foundries, rifle ranges, and training grounds for infantry, cavalry, and elephants (1631). The battles generally took place south of Ha-Tinh and north of Hué, or in the wall region of Dong-hoi. The principal encounters were:

1633. Naval Battle of Nhat-Le. The Nguyen defeated the Trinh fleet.

1642–1643. Expeditions of Trinh-Trang. He and his Dutch allies were defeated on land and sea by the Nguyen.

1648. Battle of Truong-duc. The Trinh were repulsed with heavy losses at the wall.

1658–1660. Nguyen Offensives. There followed 2 years of inconclusive attacks and counterattacks.

1673. Peace. After the sixth and seventh campaigns, the Linh River was recognized as the boundary line between the 2 territories.

CAMBODIA

1603–1618. Reign of Srisup'anma. (See p. 652.) He was a puppet of Naresuen of Siam.

1618. Accession of Jai Jett'a (son of Srisup'anma). He proclaimed Cambodia to be independent of Siam.

1622. Siamese Invasion. (See p. 652.) This was repulsed.

1679. War of Succession. Sri Jai Jett'a's accession to the throne was disputed by **Nak Norr.** With assistance of Narai of Siam, Sri Jai Jett'a defeated Nak Norr, who fled.

MALAYA AND INDONESIA

1600–1641. Three-Way War. Bickering between the Achenese, the Sultan of Johore, and the Portuguese.

1606–1607. Dutch Repulsed from Portuguese Malacca. Despite this failure, the Dutch did succeed in ousting the Portuguese from the Spice Islands of east Indonesia and strangling their trade in the Straits by cutting off Malacca from the main sea routes.

1607. Portuguese Repulsed from Johore.

1607–1629. Rise and Expansion of Acheh. A new ruler, **Iskandar** Shah (1607–1636) controlled most of Sumatra and several mainland territories.

1613–1615. Achenese Sackings of Johore.

1616. Acheh and Johore Repulsed from Malacca.

1629. Naval Battle of Malacca. The Achenese were defeated by combined fleets of Portugal, Johore, and Patani. Decline of Acheh begins.

1637. Alliance of Johore with the Dutch.

1640–1641. Dutch Capture of Malacca. The Malaccan fortress **A Famosa** had walls 32 feet high, 24 feet thick. It was garrisoned by 260 Portuguese and 2,000 or 3,000 Asiatics. The Dutch, aided by Johore's fleet, blockaded the port (June 1640). The siege began in August. A Famosa did not surrender until January 1641, after an estimated 7,000 had died in battle or by sickness and famine. The Dutch, establishing a garrison at Malacca, enforced their trade monopoly of the Straits.

1666–1689. War between Johore and Jambi. Johore's power was reduced.

1681. Arrival of the Bugis. The Bugis came from Celebes as a result of their growing maritime strength and the benefits of an alliance with the Dutch. A warlike people, using chain armor, they gained a reputation as sea fighters—invincible on sea and on land. They were primarily river pirates, mercenaries, and homeless adventurers whose presence was disturbing to the Malays.

AFRICA

NORTH AFRICA

All 4 of the North African nations supported themselves largely by piracy and the slave trade. The Barbary pirates became the terror of European mariners; their galleys and sailing vessels ranged over much of the Mediterranean and the eastern Atlantic Ocean. The Moroccans and Algerians, and to a lesser extent the Tunisians, also sent raiding parties across the Sahara to capture slaves from the Negro nations to the south. During this period Tripoli was more directly controlled by Turkey than were Algeria and Tunis.

Morocco

1608–1628. Reign of Sultan Zidan. The decline of the Sa'ad Dynasty began.

1618. Abandonment of Timbuktu. Expense in man power and other resources in retaining control of the sub-Saharan colony did not bring any real financial return to Morocco. Zidan wisely withdrew his forces; Moroccan colonists remained, however (see p. 655). Slave raiding continued.

c. 1645–1668. Civil War. The Hassani Berber tribes, living on the edge of the desert, rose against the Sa'adis.

1649. Hassani Capture of Fez. Mohammed XIV established the Hassani Dynasty, still ruling in Morocco. It was almost 2 decades before the conquest was completed by the Hassani capture of Marrakech (1668).

1662. Portugal Cedes Tangier to Britain.

1668–1672. Consolidation. This was done by **Rashid II.**

1672–1727. Reign of Sultan Mulay Ismail. Known as "the Bloodthirsty." The relatively stable history of Morocco during this and subsequent reigns was due to his creation of a special Negro bodyguard, the Bukharis, who were, in effect, a small and efficient standing army. Tied to the Sultan by religious oaths and his paternal care, they not only protected the royal person and palace but also were rotated on garrison duty at forts scattered throughout the empire.

1684. British Abandon Tangier to Morocco.

Algeria

Algeria, nominally ruled by the Ottomans, was in fact almost completely independent by the close of the century. The autonomous coastal towns and the central government at Algiers lived mainly on the proceeds of piratical plunder. During the latter part of the century, the ruling deys established a licensing system for all Algerian pirates whereby the dey was paid a percentage of all loot and ransom money. Toward the end of the century, more or less successful French punitive expeditions against the Barbary pirates forced the deys of Algiers to release their Christian slaves (1684–1688).

Tunis

During the early years of the century, the deys of Tunis were elected by Ottoman Janissaries, who at first were an army of occupation. The beys, originally administrative subordinates of the ruling deys, and responsible for administering the inland tribes, became increasingly powerful in Tunis toward midcentury.

1650. Ali Bey Ruler of Tunis. The deys were supplanted.

1655. Blake's Expedition against Porto Farina. English Admiral Blake bombarded the corsair base, destroying many of the pirate vessels and the coastal fortifications.

WEST AFRICA

In the interior savannah region, south of the Sahara and north of the tropical rain-forest belt, there was a more or less continuous struggle between Negro and Arab-Negro nomadic kingdoms and empires—some Moslem, some pagan. Along the narrow coastal ledge, the European colonial powers fought each other for bases and favored positions in the growing and profitable slave trade. In between were the jungle peoples, generally more fragmented and less land-hungry than their nomadic neighbors to the north and northeast, but less easily exploited by European slave traders than were the relatively docile tribes along the narrow coastal lowlands. The principal events were:

1591–1618. Moroccan Conquests. The Moroccan conquerors of the Songhai (see p. 566) gradually spread their control over the central Niger Valley. They also sacked the great university at Timbuktu, which threatened their politico-religious authority. However, the cost of colonization was too great for Moroccan Sultan Zidan, and he withdrew his forces (1618; see p. 654).

1618–c. 1700. Turmoil in the Central Niger Valley. The withdrawal of Moroccan forces left political chaos, which lasted over 80 years. Some Moroccan military adventurers stayed behind as petty princes, but their power waned under pressure from Tuareg nomads, Hausa, and Fulani.

c. 1620–1654. Dutch-Portuguese Wars. The growing maritime power of the Netherlands began to take advantage of the earlier incorporation of Portugal into Spain (see p. 533) to raid and capture Portuguese bases on both sides of the Atlantic. Although the Dutch abandoned the initiative after the independence of Portugal (1640; see p. 626), the Portuguese continued the conflict for several more years until they had recovered a number of their lost bases.

1621. Dutch Seizure of the Portuguese Bases of Arguin and Goree.

c. 1625. Rise of Luba Kingdom. This tribe, lying west-northwest of the Lunda, established a state under the leadership of the warrior-chieftain **Kongolo.** He was deposed by a rival's son, **Ilunga Kalala** (also an able warrior) (c. 1650). The Luba were later absorbed by the Lunda.

1626. Establishment of St. Louis. This was the first French base at the mouth of the Senegal River.

1631. Establishment of Kormantine. This was the first English base on the African coast.

1637–1645. Portuguese-Dutch Conflicts. The Dutch seized Elmina from Portugal. In retaliation the Portuguese reconquered São Thome and Loanda from the Dutch (1640–1648), while the Dutch seized St. Helena (1645).

1641. Soyo Independence. This Kongo province broke away from the kingdom, marking the start of Kongo's decline.

c. 1645–1685. Reign of Wegbaja of Abomey. This monarch, the grandson of founding monarch **Do-Aklin,** built Abomey into a powerful centralized state, and broke free of the Yoruba kingdom of Oyo to the east. Wegbaja began a regular census in 1680 to regulate conscription (of both sexes) for the army. He was succeeded by **Akaba** (1685–1708).

c. 1650. Foundation of Bambara Kingdom of Segu. Traditionally founded by two brothers, **Barama** and **Nia Ngoli,** who began as bandit chieftains but gained control of the market town of Segu.

c. 1652–1682. Reign of Kaladian Kulibali of Segu. This Bambara ruler expanded the kingdom to include Timbuktu, but it fell apart after his death.

1662. British Base at the Mouth of the Gambia River.

1664–1665. Anglo-Dutch War. The British took Cape Coast Castle (10 miles from Elmina) from the Dutch (1664). Next year they took St. Helena from the Dutch, who captured Kormantine from the British (1665).

1665. Battle of Ambuila. Portuguese intervention supporting Kongo against a seceding province was defeated, and Kongo's decline accelerated. The Kongo capital of San Salvador (modern M'Banza Congo, in Angola) was abandoned (between 1678 and 1703).

1670. Defeat of Mandingo and Timbuktu by the Bambaras. The remnants of the old Mali and Moroccan empires became tributary to Segu and Kaarta.

c. 1670–c. 1700. Rise of the Lunda Kingdom. Traditionally founded when Luba Prince **Cibinda Ilunga** married a Lunda chief's daughter (c. 1670). By century's end, the Lunda Kingdom stretched from the Kasai River in the southeast to the Kwango River in the west. The Lunda were politically very adaptable, and tribes brought under Lunda control by bands of warrior-adventurers found Lunda rule beneficial and conciliatory.

1677. French Capture Arguin and Goree from the Dutch. Goree became the main French base on the West African coast.

c. 1670–1700. Rise of the Ashanti. The Ashanti, a tribal confederation in modern southern Ghana, struggled for decades against the suzerainty of the Denkyera state, and only under the leadership of **Osei Tutu** (c. 1675) did they break away after a series of military campaigns. Osei Tutu became *Asantehene* (king), at the new capital of Kumasi. He was enthroned on the *sika 'dwa* (Golden Stool), thereafter the symbol of Ashanti royal authority.

1697. French Begin Expansion up the Senegal River.

EAST AFRICA

While Bantu tribes, migrating from the north and west, broke up the old Monomotopa Empire and simultaneously disrupted the neighboring Portuguese coastal possessions, the Portuguese found themselves facing a more lasting threat from the north. During most of this century the Portuguese and the Omanis from Arabia struggled bitterly for control of the entire western coast of the Indian Ocean, its raw-materials resources, and the even more valuable African slave trade with Asia. The principal events were:

1590–1620. Turmoil in the Monomotopa Empire. The Sotho and Ngoni Bantu tribes completely broke up the empire and left it in chaos as they continued their great migration southward to and across the Limpopo River (see below).

1606–1607. Funj Consolidate Control of Central Sudan. (See p. 485.)

1626. First French Settlements on Madagascar.

1633. Expulsion of Portuguese from Ethiopia. Facing civil war because of Catholic missionary efforts to absorb the Ethiopian church, the *negus* (emperor) **Fasilides** closed the country to Catholic missionaries, and established a new capital at Gondar (c. 1640).

1643. The French Settle Réunion Island.

1644–1680. Reign of Badi II Abu Dagn. This Funj monarch brought that kingdom to its greatest power, extending his authority into Kordofan and northward along the Nile to the Egyptian border.

1650–1700. Coastal War of Oman and Portugal. After expelling the Portuguese from Muscat (1650), the Omanis began to attack Portuguese commercial posts in East Africa.

They supported rebellions at Pate and Zanzibar (1652), attacked Faza and Mombasa (1660), and narrowly failed to capture Mozambique (1669). Portuguese countermoves (1678–1679) were largely ineffective, and although the Portuguese suppressed a rebellion in Pate (1686), they were unable to prevent the Omani siege and capture of Mombasa (December 1698), and later of Zanzibar (1701).

1686–1706. Reign of Iyasu I of Ethiopia. Iyasu (Jesus), based at Gondar, waged repeated campaigns in the southern Ethiopian highlands to establish control over the pagan Galla, who had invaded and settled there (see p. 564).

1686. French Annexation of Madagascar. Proclaimed by Louis XIV. France did not exercise any effective control over the vast island.

1699. French Embassy to Abyssinia. The French arrived at Gondar via Cairo and the Nile.

SOUTH AFRICA

Southern Africa, inhabited by primitive Khoisan (Hottentot) and San (Bushmen), was the scene of near-simultaneous migrations from both north and south. For centuries, Bantu pastoralists (who also worked iron) had been moving southward and by midcentury had reached what is now Orange Free State and Transkei. By sea, to the southern tip of the continent, came Dutch colonists. The migrating races, unaware of their approaching historic confrontation, did not come into any substantial contact during this century.

1629. Portuguese War with Mwene Matapa Empire. The *mwene matapa,* annoyed by interference from Portuguese traders, tried to expel them. They invaded, deposed the king, and compelled his successor to grant them extensive economic privileges. This marked the start of the empire's decline.

1652, April 6. Arrival of the Dutch. Colonists in 3 ships, under **Jan van Riebeeck,** arrived at the site of modern Cape Town, and established the first European settlement in South Africa.

1652–1700. Dutch Expansion Inland. By the end of the century, Dutch settlers had established themselves more than 250 miles inland.

1658. Arrival of the First Slaves at the Cape.

1684–1695. Reign of Changamire Dobo I of Rozwi. As the Mwene Matapa Empire declined, this *changamire* (ruler) raised his small state (around Fort Victoria and Shabani in modern Zimbabwe) to a position of regional preeminence. In the 1690s **Dobo I** drove the Portuguese from the Zambezi Valley. The nature of relations between the *changamires* and the Mwene Matapa Empire is unclear.

1686. Dutch Coastal Expansion. They reached Mossel Bay. Intermittent conflict began with the Bushmen.

1688. Arrival of Huguenot Refugees at the Cape.

THE AMERICAS

During the early 17th century, the increasingly competitive commercial and colonial activities of European states in North and South America, and their challenge to the native Indians, led to frequent conflicts which, in some localities, were incessant. The very process of colonization and combat between the colonial powers on the American continents often was a reflex or phase of European strife. Thus, **Jean de Ribaut**'s attempt to establish a colony in Florida (1562) had been an element in Admiral Coligny's projected attack on Spain (see p. 521). At times, European forces enlisted Indian support in their campaigns (for example, the French counterattack on Spanish-held St. Augustine, August 22, 1567–June 2, 1568; see p. 570). Expeditions conducted ostensibly for pacific purposes of exploration, colonization, and commerce often employed military means or sought military ends. These had been initiated by Drake's circumnavigation (1577–1580), which constituted a sustained assault upon Spanish shipping (see p. 507), and by Sir Walter Raleigh's expedition of 1584, which served both to explore the mainland and to reconnoiter Spain's Caribbean defenses.

INDIAN WARFARE

The character and duration of warfare between Indians and Europeans varied according to region. At the time of the first European settlements in the Americas, there were some 16 million native inhabitants scattered from Hudson Bay to Cape Horn, comprising a vast array of tribes, cultures, and military attitudes, practices, and values. By far the most civilized, and possessing the most advanced military techniques, had been the Mexican tribes (particularly the Aztecs) and the Incas of Peru. As we have seen, these were smashed by the Spanish in the 16th century (see pp. 566–570). In North America the Iroquois confederacy constituted a remarkably civilized form of politico-military organization. Similarly, Iroquois linear and assault tactics proved relatively sophisticated.

Lack of discipline and cohesiveness, far more than crudity of weapons, put the Indians at great disadvantage in opposing the Europeans, although a long series of successful raids and massacres testify to some degree of adroitness and, perhaps more significantly, colonial unpreparedness. Several tribes proved invaluable allies (for example, Mohawk intervention in the Dutch-Algonquin conflict of 1645) and some became tenacious enemies, as the French discovered after Champlain's folly in firing upon a band of Iroquois in 1609. However, from the first, Indian prospects were doomed; increasingly effective European weapons, the curious reluctance of the Indians to exploit tactical victories, and an inability to cope with the fundamental problem of supply proved fatal.

GENERAL COLONIAL MILITARY EXPERIENCE

The British home government, like the French, afforded its North American colonies scant military assistance, whether in money, supplies, or men. For example, New York was the only colony to house British troops regularly throughout the period of British suzerainty. Similarly, at no point did the French post large regular forces in Canada (estimates vary from between

2,000 and 5,000); a request for reinforcements during a Seneca uprising (1686) merely brought an admonition from Paris that peace was desirable. The Spanish military establishment, also operating with small numbers in the Americas, attempted to shield the sprawling empire; thus, with Spanish troops already stationed in Cuba, Mexico, and Peru, 2 companies totaling 280 men were dispatched to the stone fort at St. Augustine (1672) to hold Florida. However, as piracy and marauding intensified, the inadequacy of numbers in the face of such extensive geographic demands became apparent (see p. 570).

The initial reluctance of the colonial powers to provide aid is understandable when one considers (1) the potentially enormous economic drain (the colonies expended over £90,000 in King Philip's War, a considerable sum in the 17th century), (2) crucial commitments in Europe, (3) the distraction of domestic discord (for instance, the English Civil War and the French Fronde), and (4) the relative autonomy of the colonizing companies. This philosophy of colonial self-sufficiency was articulated explicitly in a later British declaration (1742), cautioning the colonies that they must bear the responsibility for their own defense.

Confronted with what Burke labeled "salutary neglect," the early settlers developed militia in accordance with familiar European models. In the British colonies, the tradition of the citizen-soldier was transplanted from England as early as 1638; in 1643, Massachusetts instituted a compulsory policy requiring 4 to 6 days of military training for all males (with the exception of doctors, ministers, educators, the Harvard student body, and other peripheral groups). The other British colonies, except Pennsylvania, followed suit, creating forces whose composition was usually one-third pikemen and two-thirds musketeers. As the pike proved itself inapplicable to combat in the wilderness, the pikemen became musketeers. The colonial militia was characterized by the absence of a staff hierarchy. Generally, the governor assumed military command (although in Massachusetts, the sergeant major general performed the function). Lack of intercolonial military coordination was one of the most critical problems facing the English colonies. The failure of effective cooperation was exposed in the Pequot War; an attempt to rectify it by the creation of a New England Confederation proved futile. Furthermore, there was the inherited English distrust of a standing and powerful military establishment as a threat to civil institutions. This fear was expressed as early as 1638 at the formation of the Ancient and Honorable Artillery Company.

The fascinating problem of the adaptability of European concepts of war to American conditions recurs throughout early colonial history. For example, although dense forest growth undermined the utility of cavalry, it remained the most prestigious branch of the New England militia. Between 1607 and 1689, the need to adapt and modify at times served as a catalyst to improvement of weapons. Thus the inadequacies of the matchlock were underscored in the North American setting and the flintlock was adopted extensively before it was in general use by European armies. By 1650, European armor, although effective against Indian projectiles, had been largely abandoned (first in New England, in the course of the Pequot War, and later in Virginia) and replaced by lighter and less cumbersome protective garb of cloth and leather.

By the time of the first major clashes between European forces in North America, much of the wilderness in the British New England, Middle Atlantic, and Southern colonies had been at least partially subdued, and physical conditions more closely approximated those of Europe. After the 1680s, and into the 18th century, only minor military improvements were made as a result of the colonial experience; the muzzle-loading, smoothbore musket and pistol reigned for nearly a century, colonial units clinging to European traditions.

BRITISH AMERICA

Virginia and Maryland

1607. The First Colony. A London Company expedition settled at Jamestown.

1622, March 22. Indian Massacre at Jamestown. This led to reprisals by the colonists.

1635–1644. Intermittent Warfare between Virginia and Maryland. William Claiborne, secretary of the Virginia colony, established a trade post on Kent Island, a parcel included in the Maryland tract granted to the first Lord Baltimore. His refusal to acknowledge the legitimacy of the second **Lord Baltimore's** proprietary claim led to intermittent hostilities, supported in the Council of Virginia, and strained his relations with **Leonard Calvert,** brother of Lord Baltimore and the first governor of the colony. Clashes occurred at sea (1635). Claiborne's unsuccessful appeal to the crown was followed by his seizure of Kent Island (1644).

1644–1646. Religious War in Maryland. A Protestant trader, **Richard Ingle,** captured the predominantly Catholic settlement of St. Mary's. Governor Calvert fled to Virginia; enlisted the aid of Sir **William Berkeley,** governor of Virginia; and recaptured the settlement.

1649–1689. The Civil Wars and Glorious Revolution. Both colonies suffered considerable unrest during this period of strife and turmoil in the mother country.

1675–1676. Indian Depredations. Susquehannock warriors crossed the Potomac into Virginia, conducted a series of raids, and repulsed a joint Maryland and Virginia retaliatory force at Piscataway Creek. The raids were intensified, resulting in the abandonment of frontier homes. Critics of Berkeley attacked his refusal to dispatch a punitive force, alleging a personal interest in the preservation of the fur trade.

1676, May 10–December. Bacon's Rebellion. Nathaniel Bacon, a civilian, disgusted with Berkeley's inaction, led a force to the Roanoke River and destroyed a band of Susquehannocks. He was declared a traitor, arrested, and eventually pardoned by Berkeley. Bacon then marched on Jamestown with a force of 500 men and secured a commission. Again he was declared a traitor. Failing to consolidate military support, Berkeley fled. On September 18, the insurgents ousted Berkeley's troops from Jamestown and burned the settlement. One month later Bacon died of malaria. His sudden death fragmented his following; during November and December, rebel groups were captured or chose to surrender in response to a promise of amnesty. The subsequent execution of 23 insurgents elicited royal censure of Berkeley.

New England

1620. Colony Established at Plymouth.

1633. English and Dutch Claims in the Connecticut Valley. Following exploration of the Connecticut Valley by **Edward Winslow** of Plymouth, the Dutch claimed the region and sent a ship up the Connecticut River. A Dutch fort and trading post were erected at Fort Good Hope (Hartford).

1636, July 30–1637, July 28. Pequot War. The murder of a trader by the Pequots called forth retaliation by the colonists and, in turn, reprisals by the Indians. A Connecticut band destroyed the major Pequot base near Stonington (May 26). On July 28, near New Haven, a force composed of contingents from Plymouth, Massachusetts, and Connecticut annihilated the fleeing survivors of the May attack.

1638. The First American Military Unit. The Ancient and Honorable Artillery Company was created in Boston. Fears were expressed in some quarters regarding the possible ascendancy of military over political leadership.

1643, May 19. The New England Confederation. The Pequot War had revealed the inadequacy of military coordination between colonies. This realization, coupled with the threat of Dutch expansion, led to a union of Massachusetts, Plymouth, Connecticut, and New Haven (the United Colonies of New England). Military expenses were to be shared by the colonies in proportion to their male populations. Save for a brief period during King Philip's War (see below), the confederation accomplished little.

1675, June 20–1678, April 12. King Philip's War. Colonial expansion brought conflict with the Wampanoags under **Philip.** Following raids in southern New England, the New England Confederation declared war. At first Indian attacks were frequent and generally successful. However, a severe food shortage later drained Wampanoag strength (spring 1676). Widespread surrender followed a successful campaign waged by a combined Connecticut Valley and Mohegan force (June 1676). Philip was killed (August). Although major operations soon ceased, Indian raids in the north persisted until peace was concluded (April 1678). During the 3-year conflict, over £90,000 was devoted to the war effort, 500 settlers were captured or killed, and between 10 and 20 towns were destroyed or abandoned and many others damaged.

1689. The "Glorious Revolution." Upon news of the flight of James II (see p. 611), the people of Boston revolted and imprisoned unpopular Governor Sir **Edmund Andros.**

The Carolinas

By 1653, settlers from Virginia had established the Albemarle Colony (North Carolina). English expansion in the south was stimulated by the English victory over Spanish forces on Jamaica in 1655 (see p. 664). Stirred by suspicions of a Spanish and Indian conspiracy, Charles Town colonists attacked and defeated the Kusso tribe (1671). The following years were marked by continued English expansion, occasional Indian strife, and intermittent conflict with the Spanish.

FRENCH AMERICA

1609–1627. French Penetration of the St. Lawrence. With the aid of Montaignais, Algonquins, and Hurons, the French drove the Iroquois from the valley. **Samuel de Champlain** led joint French-Algonquin and French-Huron expeditions in an attempt to push the Five Nations southward and to preserve contact between the allies (1609, 1616).

1613. English Raids. French settlements bordering the Bay of Fundy were destroyed.

1629, July 20. Capture of Quebec. The combined attacking force, under Sir **William Alexander** and Sir **David Kirke,** was unaware of a truce in the Anglo-French War (see p. 601).

1632, March 29. Peace Settlement. England returned Acadia and the St. Lawrence settlements to France.

1654–1670. New England Occupation of Acadia. A New England force under Major **Robert Sedgwick** departed from Boston for the purpose of attacking New Netherlands (the first Anglo-Dutch War; see p. 607). Instead, the colonists struck Acadia (July 1) in an attempt to destroy French competition in the fur trade. The English retained the province until 1670, when it was restored to France in accordance with the Treaty of Breda (see p. 610).

1642–1653. Iroquois War. The Dutch, seeking to attract the Algonquin and Huron fur trade, provided the Iroquois with arms to use against the French and their Indian allies. The Five Nations struck first against the Hurons, ultimately extending their raids as far as Montreal. After a brief truce (July 4, 1645), the Iroquois set fire to Fort Richelieu, drove the Hurons westward, and forced the Jesuits to abandon missions in Huron country. Attacks along the St. Lawrence continued until the Five Nations concluded a peace treaty with the French (November 5, 1653).

1661–1681. French Expansion in the Caribbean. They established themselves in the Lesser Antilles, a base for piracy against Spain.

1665–1666. Renewed Iroquois War. After allying themselves with the English, the Iroquois resumed hostilities against the French. The flare-up proved brief and the Five Nations were compelled to seek peace in 1666.

1668–1688. Conflict over Hudson Bay. An expedition, supported by an English syndicate and led by French explorer **Pierre-Esprit Radisson,** sailed to Hudson Bay (June 1668). Later Radisson rejected his sponsors and pledged his allegiance to France; he then changed his mind and reaffirmed loyalty to the company (1680). A French force, led by Sieur **Pierre Le Moyne d'Iberville,** captured the company posts at James Bay (1688). The English retained isolated posts at the mouths of the Hayes and Severn rivers.

1684, July. La Salle's Expedition to Texas. During the Franco-Spanish conflict of 1683–1684 (see p. 621), Sieur **Robert Cavelier de La Salle** sailed from France to the Gulf of Mexico to establish a colony at the mouth of the Mississippi, near the center of Spanish colonial power. He missed the Mississippi and landed at Matagorda Bay. The disobedience of **Beaujeu,** naval commander of the expedition, and a series of misfortunes led to insubordination and La Salle's assassination (1687). His colony was destroyed by the Indians (1689).

1684–1689. Renewed Iroquois War. The Indians were incited by **Thomas Dongan,** governor of New York. Raiding parties reached the Mississippi and swept across Lakes Erie and Ontario, disrupting the flow of the French lake trade. An unsuccessful attempt at retaliation provoked Indian reprisals in the St. Lawrence Valley; 200 French were killed and 90 captured.

1689, May 12–1697, September 20. King William's War. (War of the League of Augsburg; see p. 597.) In America the war encompassed conflict in Hudson Bay, in the St. Lawrence Valley and upper Hudson Valley, and in Acadia. The English were aided by the Iroquois, while the French were assisted by most of the other Indian tribes.

1690–1697. Frontenac's Indian Raids. Count **Louis de Buade of Pallau and Frontenac** led a series of combined French and Indian assaults along the entire northwestern frontier of the English colonies. The Abenaki, French allies, raided independently in New York and Massachusetts. Frontenac also conducted a western campaign against the Iroquois (1693–1696).

1690, May 1. Albany Conference. Representatives from Plymouth, Massachusetts, Connecticut, and New York elected to invade Canada with 2 land forces and to send a heterogeneous fleet of 34 vessels up the St. Lawrence (see below).

1690, May 11. Capture of Port Royal. New England (Massachusetts) troops under Sir **William Phips** seized the capital of French Acadia. This was the single effective Anglo-American operation of the war.

1690–1691. Abortive Expeditions against the St. Lawrence. The planned colonial expedition against Quebec and St. Lawrence by the New England and New York colonies failed. The French retook Port Royal, but lost it in the Treaty of Ryswick (see p. 600).

1690–1694. Hudson Bay Operations. Iberville drove the English from their posts at the mouths of the Severn (1690) and the Hayes (1694). However, the English regained the James Bay area (1693).

NEW NETHERLANDS AND NEW YORK

1614. Dutch Establish Fort Nassau Near Albany.

1624. Additional Dutch Settlements. They founded their first settlements on Manhattan and at Fort Nassau on the Delaware (Gloucester, N.J.).

1638, March. New Sweden. In defiance of

Dutch protest, a Swedish expedition reached the Delaware (abandoned by the Dutch in 1627) and established Fort Christina (Wilmington).

1641–1645. Algonquin War. The Algonquins, provoked by colonial expansion, raided Staten Island and Manhattan. Reprisals and counter-reprisals resulted in damage to Dutch and nearby English settlements and plantations. The Dutch, aided by the Mohawks, imposed peace (August 9, 1645).

1647–1655. Dutch-Swedish Hostilities. John Bjornsson Printz, governor of New Sweden, constructed a chain of blockhouses on Tinicum Island (1634) and Fort New Krisholm (1647). In response, the Dutch built forts overlooking the approaches to New Sweden by the Delaware and Schuylkill rivers. The Swedes burned Fort Beversrede on the Schuylkill (May and November 1648) and seized Fort Casimir (Newcastle, 1653). The Dutch, under Governor **Peter Stuyvesant,** held Beversrede and retook Casimir. They then captured the Swedish stronghold, Fort Christina (September 1655). This ended Swedish attempts to colonize North America.

1650. Treaty of Hartford. Following encroachments of English settlers on Dutch settlements, a number of incidents led to this boundary settlement between the Dutch and the New England Confederation.

1652–1654. First Anglo-Dutch War. (See p. 607.) There was no formal declaration of war, but the English seized some Dutch frontier posts. Connecticut forces entered Westchester and Long Island, compelling Stuyvesant to acknowledge English dominion over English settlements on Long Island (1653).

1655–1664. Dutch-Indian Wars. There were 3 major episodes, the first extending from Manhattan to Esopus and Long Island (1655). The second was centered near Esopus (1663). The wars ended with the Indian surrender of the Esopus Valley after the third episode (1664).

1664–1667. Second Anglo-Dutch War. (See p. 609.) Even before formal outbreak of war in Europe, James, Duke of York, directed Colonel **Richard Nicolls** to capture New Netherlands. The arrival of 4 English frigates in New Amsterdam Harbor in late August resulted in the surrender of New Amsterdam (September 7). The English took Fort George and replaced the Dutch in their alliance with the Iroquois.

1672, March–1674, October 31. Third Anglo-Dutch War. (See p. 610.) Dutch naval and land forces seized New York (August 1673). Subsequently, Albany, New Jersey, and western Long Island yielded to the Dutch. However, the **Treaty of Westminster** (see p. 611) restored the settlements to England.

1689–1691. Leisler's Rebellion. Reports of the arrest of Sir Edmund Andros in Boston (see p. 661) encouraged anti-Catholic agitation in New York City and on Long Island (May 1689). The militia captured Fort James and turned it over to **Jacob Leisler,** the rebel leader, who declared his allegiance to William and Mary. He refused to give up his authority and was seized by British troops (March 30, 1691). He was executed (May 26).

SPANISH AMERICA

North America

Spanish Jesuits extended their missionary activities on the Pacific Coast. Settlements around the missions in the Mayo and Yaqui river valleys led to Indian attacks, which were met by punitive expeditions under Captain **Diego Martínez de Hurdiade** (c. 1625).

English expansion and raiding in the south—Florida and the Carolinas—led to Spanish reprisals against English settlements.

Mexico–New Mexico

1680–1692. Pueblo Indian Revolt. Spanish mistreatment stimulated a Pueblo uprising in the northern Rio Grande Valley. The revolt, led by **Pope,** a shaman (or medicine man) who had been imprisoned by the Spanish for witchcraft (1675), caught the Spanish by surprise (August 10, 1680). At least 400 Spanish soldiers and settlers were killed, and most of the 2,000 survivors were beseiged in Santa Fé. After 10 days, they abandoned Santa Fé and retreated south to El Paso, Texas. With the Spanish gone, Pope and his followers restored traditional customs, but internal dissension and a series of poor harvests weakened Pope's popularity. Following Pope's death (1692), Spanish Governor Don Diego de Vargas (1692–1696) decided to reoccupy the colony, in part to protect Spanish holdings from growing French influence in the Mississippi Valley. The Pueblo, harrassed by Apache raids, put up no resistance and never again seriously challenged Spanish control in the area.

COMMENT. *Unlike the nomadic plains Indians to the east, the peaceful Pueblo Indians were no match for even weak Spanish military forces. They lacked military organization, and had no command system; they owed their initial success to surprise. Lacking metals, firearms, and horses, they were outclassed by the Spanish and the Apache.*

Central and South America

During this century Spain assumed a defensive posture to preserve the empire which she had acquired in the preceding century. Encroachments by France, England, and the Netherlands (particularly in the West Indies) both contributed to, and were symptoms of, the decline of Spanish power. Conflict in the Caribbean region and farther south, as in North America, reflected the shifting alliances and antagonisms of Europe. Three prime characteristics marked the period: (1) Indian unrest, particularly among the Araucanians of southern Chile, the Piajos of Peru, and the Chiriguanes, between Las Charcas and Paraguay; (2) the *Corsarios Lutheranos,* or Protestant pirates (often encouraged if not directly subsidized by the European powers, particularly England); and (3) relatedly, an extensive contraband traffic. By the end of the century, Spain had attempted to compensate for her naval inadequacy by heavy fortification of the coastal and Caribbean communities. Typical of this defensive policy was the atrophy of highly vulnerable Santo Domingo and the development of Cartagena with its round fortress, walls of a 40-foot diameter, and an easily blocked channel. The principal events were:

1624–1629. War with the Dutch. The Dutch captured Bahia (1624), but were dispossessed in the following year by a Spanish naval force. However, soon after, as a preliminary to an invasion of Pernambuco, the Dutch recaptured Bahia (1627). Dutch buccaneers, under **Piet Heyn,** seized a Spanish silver shipment off the Cuban coast (1628). Subsequently, Heyn was made an admiral in the Dutch Navy. A Dutch force of 60 vessels captured Pernambuco and Recife (1629).

1629–1660. Punitive Expeditions against Other Colonies. The Spanish drove English settlers from the islands of Nevis and St. Christopher (1629). They returned upon the departure of the Spaniards. English Puritans were forced from the Providence Islands, off Central America (1641). In reprisal, a band organized by the Providence Company and the Earl of Warwick assaulted the Venezuelan coast and briefly seized Jamaica (1643). Some years later the Spanish ousted English and French colonists in the Lesser Antilles (1660).

1654. English Seizure of Jamaica. Cromwell ordered Admiral **William Penn** to capture Santo Domingo; the operation failed and Penn proceeded to take Jamaica.

1668–1671. Morgan's Panama Raids. Led by

Henry Morgan, a force of 2,000 buccaneers plundered Porto Bello and Panama City. Morgan was knighted by Charles II.

1670. Treaty of Madrid. England promised to halt piracy in the Caribbean. Spain recognized English sovereignty over Jamaica.

1679. English Pirates Raid Panama.

PORTUGUESE AMERICA

1609–1642. Portuguese Slave Traders Repulsed from Paraguay. Spanish Jesuits armed mission Indians and successfully resisted the penetration of Paraguay by Portuguese slave traders.

1615. French Expelled from Brazil. The Portuguese seized the French colony on Maranhão Island and subsequently settled Belém, ruining French efforts to colonize South America.

c. 1620–1654. Dutch-Portuguese Wars. (See p. 626.) Insurrection of the Portuguese population in Brazil ended in the withdrawal of the Dutch (1654). Holland relinquished all claim to the region (1661).

1680. Conflict over the Rio de la Plata. Portugal established a fort, Colonia del Sacramento, across from Buenos Aires, asserting its claim to the northern and eastern bank of the Rio de la Plata. The Spanish seized the post. English intervention (in Europe) resulted in the return of the fort to Portugal (recognized formally in 1701).

XV

THE MILITARY SUPREMACY OF EUROPE:

1700–1750

MILITARY TRENDS

During this half-century Europe became preeminent in world affairs. Save for the meteorlike passage of Nadir Shah's Persian Empire, only three major non-European powers still existed—Manchu China, Mogul India, and Ottoman Turkey—all declining in strength and incapable of exerting great influence outside of their borders. Tiny Europe, on the other hand, contained five major powers with intercontinental interests, influence, or ambitions—Britain, France, Spain, Russia, and Austria—and two other powers with modest strength in Europe, but great influence elsewhere: Portugal and the Netherlands. And there was in Europe one new power—Prussia—with wholly continental interests and contacts, but whose military prowess was beginning to have worldwide repercussions.

In military and nonmilitary affairs, Europe set the pace; others followed if they could. Europe led the world in part by default and in part because of a growing, and apparently irresistible, superiority in technology.

All of the important military leaders of the period were European—again with the significant exception of **Nadir Shah,** last of the great Asian conquerors. There was one general of great-captain rank: **Frederick the Great** of Prussia. And there were four other generals of truly outstanding ability: Prince **Eugene of Savoy** and of Austria, **John Churchill, Duke of Marlborough** of England, **Charles XII** of Sweden, and **Claude, Duke of Villars** of France. Hardly less competent were **Maurice de Saxe,** the German who fought for France; **James, Duke of Berwick,** the English expatriate who fought for France; **Louis, Duke of Vendôme,** also of France; and **Peter the Great** of Russia (tactically weak but strategically brilliant). Otherwise, even though this was a period of almost incessant—though extremely limited—warfare, leadership was for the most part unimaginative and pedantic.

Naval affairs were dominated by Britain to an extent almost unmatched in prior or (after a century) subsequent history, although imaginative leadership was largely stultified in the Royal

Navy (however for reasons quite different from the pedantry on land). At least two British admirals deserve recognition for achieving imaginative victories despite this stultification: **George, Lord Anson** and **Edward, Baron Hawke.**

EIGHTEENTH-CENTURY WARFARE

The example of Gustavus Adolphus was the military basis of 18th-century European warfare. A handful of his imitators had sufficient ability to apply his example with imagination and energy comparable to his. Most, however, were satisfied to adopt the external aspects of Gustavus' military organization with little or no understanding of the fact that his success was due to flexibility in the employment of novel weapons and organization in accordance with old principles. Here is a perfect example of what Toynbee has called "a cycle of invention, triumph, lethargy and [eventually] disaster."*

There were, however, social, political, and economic factors which affected the conduct of 18th-century warfare at least as much as this military cycle.

The economic aspects in themselves were perhaps least important. They reflected relatively little change from previous eras, although they were soon to be completely overturned by the effects of the Industrial Revolution. The economic limitations on war in the 18th century, as earlier, were primarily those of agricultural and industrial production. The vast majority of people were employed in agriculture, producing sufficient food for themselves, with but little left over for the much smaller percentage of the population engaged in handicraft industries, in services, or in the military forces. In time of war it was simply impossible to raise very large armies. The soldiers came mostly from the farms, thus simultaneously reducing the number of producers while increasing the demand for food for nonproducers. In addition, the limited industrial capacities of nations prior to the introduction of mass-production techniques, could not provide sufficient weapons, equipment, and other munitions to equip large armies engaged in active and wasteful warfare.

The colonial expansion of Europe, with its increased trade and growing mercantilism, only served to accentuate these major economic limitations. The manpower demands of colonialism and mercantilism attracted farmers and handicrafters from their old tasks at a time when the expansion of the economy was placing ever-increasing demands upon agricultural and industrial production.

The social conditions existing in 18th-century Europe exerted an influence upon military systems very closely related to the economic considerations. The manpower demands of the economic expansion dictated the recruitment of armies from among the least productive elements of the populations. The officers were from the nobility, a generally unproductive stratum of the society, while the soldiers came from riffraff and the unemployed, who were also unproductive.

Social considerations, in turn, were closely related to the political factors which governed the conduct of wars. For all practical purposes governments were organizational structures used by monarchs in the administration of their domains. Wars were fought for personal or dynastic objectives. While the growing sense of nationalism and the widespread existence of

* Arnold J. Toynbee, *A Study of History*, Somervell Abridgment (Vols. 1–6) (Oxford: N.Y., 1947, p. 336.

patriotic sentiments cannot be dismissed, these were linked to monarchical loyalty, and otherwise were relatively unimportant in the rigidly stratified political and social organizations of society. Rigidity was indeed the dominant characteristic of the era, an economic, social, and cultural rigidity which reinforced and contributed to unimaginative military pedantry.

Take, for instance, discipline. Discipline and precision had been fundamental to the successes of Gustavus; therefore they were considered by his imitators to be goals in themselves. Furthermore, only through harsh, ironbound discipline and the most rigid physical and mental controls could the riffraff of the average European army be controlled and trained. The dregs of society could not be allowed to wander freely among the productive communities; either they would terrorize the civilians through the exercise of their newly learned military skills or, more likely, they would desert. Because of the relatively small segment of a nation's man power available for military service, because there was difficulty in recruiting new soldiers, and because it took time to train men in the precise military drills necessitated by existing weaponry, losses by desertion were as serious as battle losses.

When on campaign armies could not be allowed to forage off the country—not to spare the citizenry, but because the civilian economy could not stand ravaging and looting by hordes of soldiers, and the military economy could not stand the inevitable desertions. Therefore, a campaign could not be initiated until the commanders had assembled large stores of food and equipment in storage depots called magazines; armies could not be allowed to maneuver any great distance—usually four or five days' march at the most—from these magazines.

It was also important to avoid the risk of losing expensively trained soldiers in combat. Consequently, a commander was unwilling to fight a battle unless he calculated his chances of success to be near certainty. His opponent, who was capable of making a roughly comparable estimate of the situation, would naturally seek to avoid battle under such circumstances, and would do so by placing himself more than four or five days' distance from the enemy's nearest magazine. Under such circumstances battles were infrequent, and decisive actions resulted only from the energy of exceptionally good commanders or the foolhardiness of exceptionally stupid ones.

But once battles were joined, they were fought with ferocity and tenacity. Grim testimony to this are the casualty figures for such engagements as Blenheim, Malplaquet, and Fontenoy. But, given the nature of the warfare and the composition of the armies, such losses did not create serious strains on the national economies or societies of the opponents.

It must not be assumed, however, that the era (including the late 17th century) did not have its share of capable military commanders (as noted above, p. 666). But none of these men had the genius—and perhaps also lacked opportunity—to overcome the rigid military system within which they operated and to which, therefore, they adapted themselves and their strategies. Not even Frederick the Great—one of the undoubted geniuses of warfare—was able completely to break the rigid shackles. Frederick's greatest military weakness—his failure to pursue after a decisive victory—probably reflected a fear that his army would melt away in desertion were he to unleash it in pursuit.

It would take the cataclysm of the French Revolution, combined with the impact of the growing Industrial Revolution, harnessed by the ruthless genius of Napoleon, to shatter completely the rigid system of 18th-century warfare.

WEAPONS DEVELOPMENT

As might be expected in this era of conventionality and conservatism, significant developments or innovations in weapons were few. What technological progress there was lay primarily in the refinement of existing weapons.

The basic small-arms weapon was the flintlock musket—typified by Britain's "Brown Bess"—with its offset ring bayonet. The Prussians produced two refinements to improve its utility in combat. They introduced a double-ended iron ramrod (about 1718), much more reliable (and thus faster) than the relatively insubstantial wooden ramrod. The Prussians also developed the funnel-shaped touchhole, making it easier to prime the musket, particularly in the stress of battle. When these innovations were combined with the speedy and precise loading drill demanded by Frederick the Great, Prussian infantry soldiers could fire about twice as fast as those in other armies.

In artillery also there was little real change. The French improved on their earlier standardization efforts, and abolished all mobile pieces larger than 24-pounders. The major effort toward artillery improvement was made in Austria following the War of Austrian Succession (See p. 689). The artillery reforms of Prince **Joseph Wenzel von Lichtenstein** became a model for future artillery development. Lichtenstein not only standardized the Austrian artillery into 3-, 6-, and 12-pounder guns and 7-, and 10-pounder howitzers, he also standardized the design of carriages, wheels, battery forges, and other artillery equipments. Lichtenstein also advanced the science of ballistics by the establishment of an experimental test facility near Budweis in Bohemia.

TACTICS OF LAND WARFARE

With warfare primarily governed by convention, there was relatively little new in tactics. As we have seen, most generals preferred to avoid battle, relying solely on ponderous maneuvering. Tactical formations were generally the same as those that had emerged in the previous century, except that they tended to be somewhat more elaborate and somewhat slower. Tactical perfection was considered to consist in the ability to form perfect lines, and to maintain these lines even during the heat of battle. The lines themselves, due to increased rates of fire, were often reduced to four ranks instead of six.

The principal exception to these generalities was to be found in the tactical developments of Frederick the Great, which were beginning to make themselves felt by the close of the War of the Austrian Succession at midcentury. Frederick was just as concerned as other 18th-century generals with discipline and drill, and with precision in tactical infantry maneuver. But with him these were means to an end, rather than conformance or acceptance of convention. In fact his Prussian armies had even more ruthless discipline, had a greater emphasis on drill, and were required to meet higher standards of precision than any other army in Europe. But Frederick added two significant elements: speed and battlefield maneuver. The result was an instrument of war probably unsurpassed in technical military perfection for its time by any army save possibly that of Alexander of Macedon.

Frederick carried one step farther the artillery tactical developments of Gustavus Adolphus. During the Seven Years' War he created the concept of horse artillery (as opposed to conventional horse-*drawn* artillery) in which every cannoneer and ammunition handler was mounted, so that the light guns could keep up with the fast-moving, hard-riding Prussian cavalry. He also exploited the high trajectory of the howitzer by striking at enemy reserves concealed behind trees and hills, and in the process gave to artillerymen their first glimmering of the concept of indirect fire.

The only other significant tactical development of the period was improved organization and employment of light infantry. In Europe this was primarily an Austrian-Hungarian innovation and was to some extent a result of necessity in the War of the Austrian Succession, when Maria Theresa's generals adapted the Croatian *grenzer* (borderer), accustomed to informal skirmishing with the Turks, to operations against the Prussians and French. But there was also a trend toward comparable light-infantry development in the operations in the American colonies, a trend that would become increasingly important by the end of the century.

MILITARY ORGANIZATION

The armies of the early 18th century were generally small, were highly professional, to a large extent were still mercenary, and the troops had little enthusiastic passion for the dynastic cause for which they fought. In organization and structure there was little change from the new systems which had emerged at the close of the preceding century.

As we have seen, armies were largely composed of vagabonds and nobles. As a consequence there was probably a greater gulf between officers and men than at any other time in history. It is not much of an overstatement to suggest that soldiers fought in battle mainly because they feared their own officers more than they did the foe. This was because armies of this sort could be held together and kept from deserting only by iron, ruthless discipline.

NAVAL WARFARE

Tactics

Up to 1750, naval tactics were as blunt as a sledgehammer. Opposing fleets, each sailing in a single column (line ahead) on parallel courses, attempted to close with one another and engage ship against ship. Individual vessels fought broadside to broadside, sometimes actually side by side, pounding away until one or the other gave in. England dominated the seas, not through new concepts of conflict, but through the momentum gained by her great admirals of an earlier era, by numerical superiority in numbers of fighting ships, and by seamanship.

The French Navy was second, having clearly surpassed the Dutch. Ship for ship, French war vessels were better built than the British, but they were usually fewer in number—and in general English seamanship was the better. The basic distinction between English and French battle tactics was that the French, usually outnumbered, were forced to save their vessels. In fleet actions they followed the principle that "He who fights and runs away will live to fight another day." In consequence French admirals preferred to enter action on the leeward side of their adversaries, which would enable a quick breakoff of conflict if necessary, and their gunnery was directed at the spars and rigging of the enemy to slow him down. Contrariwise, the

English sought the "weather gauge" (position to windward of the adversary), facilitating closing in; their fire was directed at the enemy's hull, seeking to sink or pulverize him.

Broadside firepower dominated the action; seamanship and shiphandling were part and parcel of the art of naval gunnery, for firepower was merely a statistic unless the ship herself was so maneuvered as to bring her almost rigid guns to bear on the target. One might have thought that under such conditions individualism in battle would be a *sine qua non* among the admirals and captains of the Royal Navy. Far from it; the "Fighting Instructions" of the Royal Navy had become frozen into law (1691), and the line (a rigid single file) was gospel. Woe betide the captain who broke from its frozen mediocrity. Since no other navy had a system any better, or even as good, only an exceptional naval commander was able to achieve any major success against a force even approximately the strength of his own. There were few decisive naval actions. Attempts to experiment or innovate were discouraged; frequently they led to court-martial and disgrace. In fact, it would seem that almost every important engagement of the Royal Navy during this period culminated in a court-martial.

The Ships

This was the golden age of the ship of the line with her multitiered broadside batteries. As her name signified, she was large enough and carried sufficient armament to slug it out in the line of battle. She averaged about 200 feet in length, with a maximum tonnage of 2,500. The biggest of these vessels carried a crew of about 1,000 men. By the end of this period warships had become assorted into six "rates." The first three of these classes were ships of the line: a 1st-rate carried 100 or more guns on three decks; a 2nd-rate had about 90 guns on three decks; the 3rd-rate—the workhorse of the battle fleet—carried from 64 to 74 guns on two decks. The 4th-rater was the smallest line-of-battle ship, a 50-gun vessel (two gun decks) usually called a frigate, which could be a razee (a converted 3rd-rate with one gun deck removed) or a purpose-built vessel. Like all compromises in naval construction, she was not powerful enough to play her part in the line, and she was usually too clumsy to act as a cruiser. The real cruisers were the smaller frigates—5th- and 6th-raters—carrying from 24 to 40 guns, all on one gun deck. These vessels, relatively lighter and faster than ships of the line, were built for commerce destruction, scouting, and screening. All the foregoing vessels were, in the true nautical term, ships—square-rigged three-masters.

Below these rates came the sloops of war (the term "sloop" had nothing to do with the rig), carrying from 16 to 24 guns, which were sometimes square-rigged, three-masted ships, but also brigs (two masts, square-rigged) or brigantines (two masts, square-rigged on the fore only). And finally came cutters and other small craft, usually known by the name of their rig (sloop, schooner, or ketch).

Naval Ordnance

The main batteries of ships of the line and frigates had become fairly standardized by this time: 16-, 18-, and 24-pounder cannon. The multitiered vessels usually carried 16s on their upper decks, 24s on the lower. Lighter craft mounted 4-, 6-, and 9-pounders, and the big ships frequently carried some lighter guns in addition to their normal armament.

Naval Developments

Only two innovations appeared during the period, but they were important. The tiller, the great beam projecting inboard from the rudder by which the ship was steered, by 1700 had been rigged by cables to a steering wheel mounted on the quarter-deck, greatly facilitating the conning of the ship. And underwater sheathing of copper was being introduced, protecting the oaken bottoms to a great extent from the ravages of the dreaded teredo (marine worm attacking wood), and barnacles.

Morale

Living conditions on board ship in all navies of the period were abominable. It is hard to imagine, for instance, how 1,000 men could exist in a 2,500-ton vessel, carrying, in addition to ammunition, enough food and water for extended cruises of a year or more without entering port or reprovisioning. The nature of the food, and its condition after months at sea, can be imagined. The principal item of diet was weevil-infested, brick-hard biscuits, with a few sips of brackish water. The addition of some rum to the water—to make "grog"—helped. The physical effects of the diet were apparently severe. Sailors who survived battle usually aged quickly and died young. The English discovered that lime juice inhibited scurvy, and from their use of limes comes the word "Limey," meaning an Englishman.

To these cruel living conditions was added even more cruel discipline. Punishments were atrocious. The gulf between officers and men was at least as great as that in the ground forces, and commanders exercised almost unquestioned life-and-death authority over their men. This, of course, was probably the only way of assuring obedience from men who to a large extent had not volunteered, but who had been impressed into service against their wills.

As a social institution, the principal characteristics of the Royal Navy of this time have been aptly summed up by Winston Churchill: "Rum, buggery, and the lash!"

MAJOR WARS

THE GREAT NORTHERN WAR, 1700–1721

1698–1699. Alliance against Sweden. Peter I of Russia, **Augustus II** of Poland (also Elector of Saxony), and **Frederick IV** of Denmark entered into a secret alliance against Sweden, hoping to take advantage of the youth of **Charles XII** (age 16) to end Sweden's dominance in the Baltic.

1700, April. Danish Invasion of Schleswig. Conquest from the Duke of Holstein-Gottorp, ally of Sweden.

1700, June. Polish-Saxon Invasion of Livonia. Augustus' Saxon troops besieged Riga.

1700, August. Russian Invasion of Ingria. Peter's army of 40,000 besieged Narva (October 4).

1700, August 4. Charles Invades Zealand (Sjlland). Against the advice of his admiral, the 18-year-old Swedish monarch boldly took his fleet and army across supposedly unnavigable waters, and immediately advanced on Copenhagen. The Danes sued for peace.

1700, August 18. Treaty of Travendal. Denmark returned Schleswig to the Duke of

Holstein-Gottorp; agreed not to engage in further hostilities against Sweden.

1700, October 6. Charles Lands in Livonia. Arriving at Pernau with a few men, he had planned to relieve Riga. While awaiting reinforcements he decided instead to march to the relief of hard-pressed Narva.

1700, November 13–19. Advance on Narva. Charles with 8,000 men scattered Russian forces attempting to block the approaches (November 18). Revealing more perception than courage, Peter fled.

1700, November 20, Battle of Narva. In a 2-hour battle, during a snowstorm, Charles virtually annihilated the Russian army, nearly 5 times as numerous as his own. He lost 2,000 men. During the winter he prepared to march on Livonia.

1701, June 17. Battle of Riga. Charles defeated a Russian-Polish-Saxon army and relieved Riga, besieged for almost a year.

1701, July. Swedish Invasion of Poland. Considering Augustus and Poland to be more formidable foes than Peter and Russia (reasonable under the circumstances but wrong in retrospect), Charles decided to deal with Poland first. He crossed the Dvina, defeated a Saxon-Russian army at the **Battle of Dunamunde** (July 9), occupied and annexed Courland, then moved into Lithuania.

1702, January–May. Advance on Warsaw. Charles occupied the undefended Polish capital (May 14). He then boldly marched westward, seeking battle with Augustus.

1702, January–December. Russian Invasion of Ingria. Peter defeated Swedish General **Schlippenbach** at the **Battle of Errestfer** (January 7). He defeated the Swedes again at the **Battle of Hummselsdorf** (July 18), and seized the Neva Valley (December).

1702, July 2. Battle of Kliszow. Charles routed a much larger Polish-Saxon army. He then marched on through hostile country to seize Cracow.

1702–1703. Consolidation of Poland. Charles systematically eliminated Augustus' control.

1703, April 13. Battle of Pultusk. Charles defeated another much larger Augustan army, then besieged and captured Thorn (May–December).

1703, May 16. Peter Founds St. Petersburg. Russian troops occupied and fortified the mouth of the Neva River, regaining an outlet on the Baltic.

1704. Charles Places Stanislas Leszczynski on Polish Throne. Civil war raged through Poland between the adherents of Augustus and Stanislas.

1704, July 4. Peter Recaptures Dorpat.

1704, August 9. Peter Recaptures Narva. All Swedish inhabitants were massacred.

1705–1706. Charles Pacifies Poland. Defeating the Saxons at the **Battles of Punitz** and **Wszowa** (1705), Charles then hastened to Lithuania to meet a Russian army under Scottish General **Ogilvie** which had occupied Courland (1705). The Russians avoided major battle, but Charles finally chased them out of Lithuania and as far as Pinsk, where he halted (July 1706). Meanwhile, Augustus attempted to retake Poland, but was routed by a much smaller Swedish army under General **Karl G. Rehnskjold** at the **Battle of Franstadt** (February 3, 1706).

1706, August–September. Charles Invades Saxony. The Swedish Army seized Leipzig, practically without opposition. Augustus sued for peace.

1706, September 24. Treaty of Altranstadt. Augustus abdicated the Polish throne, recognized Stanislas as king, and broke his alliance with Russia. Peter sued for peace, but Charles, determined on revenge, refused to negotiate.

1706–1707. Quarrel with the Empire. Charles, poised menacingly in Saxony, demanded restitution for mistreatment of Silesian Protestants under the provisions of the Treaty of Osnabrück (see p. 597). Fearful that Sweden would join France in the War of the Spanish Succession (see p. 676), the allies persuaded the Emperor to accede to Swedish demands (August 1707). Charles at once marched from Saxony into Poland to prepare for an invasion of Russia. Peter concentrated an army at Grodno and Minsk.

1708, January 1. Invasion of Russia. Charles,

waiting until the rivers were frozen, crossed the Vistula with an army of 45,000 men, the largest he ever commanded. He took Grodno, abandoned by Peter (January 26), and continued toward Moscow. He waited near Minsk during the spring thaw (March–June).

1708, June 29. Charles Crosses the Berezina at Borisov.

1708, July 4. Battle of Holowczyn. Assaulting a Russian army defending the line of the Bibitch River, Charles penetrated and scattered the defenders. He then advanced to the Dnieper at Mogilev (July 8).

1708, July–October. Scorched Earth. The Russians retreated slowly, destroying food and crops as they withdrew. They refused battle, though Charles forced one brief engagement at **Dobry** (September 9). The Swedes, short of food and fodder, began to suffer severely. Charles decided not to retreat, but to go into the Ukraine to join **Ivan Mazeppa,** the Cossack leader, who secretly agreed to rise against the Russians with 30,000 men. Charles ordered a small relief army under General **Adam Loewenhaupt** to march in from Livonia, with a tremendous wagon train of supplies, to join him in the Ukraine. This was Charles' worst military decision. He should have concentrated his forces and made adequate logistical arrangements before marching deeper into Russia.

1708, September–October. Repulse at St. Petersburg. Swedish General **Lybecker** was forced to retreat into Finland.

1708, October. Ousting of Mazeppa. Learning of Mazeppa's planned revolt, Peter sent a Russian army into the Ukraine to seize and burn his capital, Baturin. Mazeppa and a small force fled to join Charles at Horki in Severia (November 6).

1708, October 9–10. Battle of Lyesna. Loewenhaupt's army of 11,000 men was attacked and defeated by much larger Russian forces in a 2-day battle just east of the Dnieper. Loewenhaupt, forced to burn all of his supply train, reached Charles with 6,000 exhausted and hungry survivors (October 21).

1708–1709. Winter Struggle for Survival.

The Swedes, harassed by the Russians, were forced to fight continuously. This, perhaps the coldest winter ever experienced in Europe, caused severe hardships and losses to the Swedes. Charles did wonders to hold his army together, but by spring he barely had about 20,000 men fit for action. He had only 34 artillery pieces and almost no powder for his guns.

1709, May–July. Siege of Poltava. Charles advanced on Voronezh, but stopped to capture the Russian fortification at Poltava (May 2). Peter immediately marched on Poltava with a relieving army of 80,000 men and more than 100 artillery pieces. The 2 armies maneuvered cautiously, preparing for battle. In preliminary skirmishing Charles was badly wounded in the foot and had to be carried in a litter (June 17).

1709, June 28. Battle of Poltava. Seriously short of ammunition, Charles needed a quick victory. His initial attacks went according to plan; the Swedish left and center were successful. Peter was on the verge of retreat, when he realized that there was little or no coordination among the 3 major elements of the overextended Swedish army. Rallying his army, the Czar awaited the climactic Swedish charge—7,000 strong—with 40,000 fresh troops. Overwhelming strength, more and better weapons, and adequate ammunition soon reversed the tide of battle. The Swedish infantry was practically annihilated, the cavalry scattered. Charles was lifted from his litter and placed on a horse. Accompanied by Mazeppa and about 1,500 horsemen—half Cossacks and half Swedes—he fled to Turkish Moldavia. The 12,000 Swedish cavalry survivors surrendered to the Russians at Perevolchna, on the Dnieper, 2 days after the battle. The Russians lost 1,345 killed and 3,290 wounded. The Swedes lost 9,234 killed and wounded and 18,794 prisoners.

1709, August–December. Russian Occupation of Poland. Peter marched westward to occupy Poland, placing back on the throne his ally Augustus II, who had renounced the Treaty of Altranstadt and returned to Poland when he heard the news of Poltava. By virtue of military occupation, however, Peter was the real ruler of Poland. Stanislas fled to Swedish Pomerania; his former adherents were ruthlessly persecuted.

1709–1710. The Allies Collect the Spoils. The Danes took Schleswig and the Swedish possessions of Bremen and Verden, which were in turn given to Hanover as an inducement to enter the war. A Danish army overran Skane, in southern Sweden. Another Danish force and a Polish-Saxon army invaded Swedish Pomerania, but were held at bay. Meanwhile, the Russians occupied Carelia, Livonia, Estonia, and the remainder of Ingria.

1710, February. Battle of Helsingborg. Swedish General **Magnus Stenbock** defeated the Danes in southern Sweden and forced them to withdraw across the Sound. Stenbock then proceeded to Germany to protect Swedish possessions there.

1710, October. Turkey Enters the War as Ally of Sweden. Responding to the appeals of Charles XII, still a refugee at Bender in Moldavia, the Sultan declared war on Russia and sent **Baltaji Mehmet,** the grand vizier, to the Russian frontier with an army nearly 200,000 strong.

1711, March–July. Moldavian Campaign. Overconfident after Poltava, Peter invaded Moldavia with about 60,000 men, to be outmaneuvered by the larger Turkish army and driven back against the Pruth River, where his starving army entrenched itself and was surrounded. Instead of investing the Russians, Baltaji opened negotiations. Had he persevered, Peter would have been forced to surrender, with historical consequences unimaginable.

1711, July 21. Treaty of the Pruth. Peter returned Azov to Turkey, dismantled Taganrog and other fortresses, and agreed to withdraw from Poland and to refrain from interfering in Polish internal affairs. (He never carried out the latter two provisions.) Charles, who had been granted free passage to Sweden, was bitterly disappointed at the easy escape of his rival. Refusing to leave Turkey for 3 years, he continued vain efforts to get Turkey to undertake major operations against Russia. He was finally

imprisoned by his Turkish allies after a fierce hand-to-hand struggle (February 1, 1713).

1712–1713. Operations in Northern Germany. Stenbock defeated a Danish army at the **Battle of Gadebusch** (1712), but was soon afterward set upon by a much larger allied army and forced to capitulate at **Toningen** (1713). Meanwhile, the Saxons and Russians were being repulsed from Stralsund (1713). Despite disasters, Sweden continued the war, and the Swedish Navy still dominated the Baltic.

1713. Peace of Adrianople. Peace between Russia and Turkey.

1713–1714. Russian Conquest of Finland.

1714. Naval Battle of Gangut (Hanko). Peter's newly created Russian fleet defeated the Swedes to gain control of the Gulf of Finland, facilitating the conquest of Finland. Swedish predominance in the Baltic was ended, although the Russians did not yet feel strong enough to offer a climactic challenge.

1714, November 11. Arrival of Charles at Stralsund. Finally despairing of Turkish action, Charles, fleeing from virtual house arrest in Turkey, made a daring journey across Europe with 1 servant.

1714–1715. Charles Revitalizes Sweden. Unwisely refusing to consider opportunities for peace negotiations, Charles would not compromise his determination to achieve complete victory. After repulsing a series of allied assaults on Stralsund, he returned to Sweden (December 1715), now threatened by another invasion from Denmark and Norway. He raised an army of 20,000 and moved into Skane, causing his enemies to abandon their invasion plans (1716).

1717–1718. Norwegian Campaigns. Charles, taking the initiative, invaded Danish Norway. He hoped that the conquest of Norway would give him a stronger base for further operations, as well as territory which he could trade in eventual peace negotiations.

1718, December 11. Death of Charles. As his army was preparing to assault besieged Fredriksten, Charles, peering over the front-trench parapet, was shot through the head by a Danish bullet.

1719–1720. Russian Raids against Sweden. The Russian fleet now controlled the Baltic. Repeated landings on the Swedish coast near Stockholm were repulsed by the exhausted Swedes with the greatest difficulty; the Swedish Riksdag sued for peace.

1719–1721. Treaties of Stockholm. The *status quo* was restored between Sweden, Saxony, and Poland. Prussia kept Stettin and part of Pomerania, and also paid an indemnity. Denmark gave up her conquests, save for Schleswig, but received an indemnity from Sweden.

1721, August 30. Treaty of Nystad. Russia kept Livonia, Estonia, Ingria, part of Carelia, and many Baltic islands. She returned Finland (except for Vyborg) and paid an indemnity to Sweden. By this treaty Russia supplanted Sweden as the dominant power in the Baltic, and became a major European power.

WAR OF THE SPANISH SUCCESSION, 1701–1714

Charles II, last of the Spanish Hapsburg line, was childless. There were two major contenders for the throne of Spain: (1) **Louis XIV** of France, son of the elder daughter of Philip III of Spain and husband of the elder daughter of Philip IV; (2) **Leopold I,** Hapsburg emperor, son of the younger daughter of Philip III and husband of the younger daughter of Philip IV. Neither of the maritime powers—England and the Netherlands—was willing to see Spain united with either France or Austria, and because of this Louis was claiming the Spanish throne for his second grandson, **Philip of Anjou,** while Leopold's claim was for his second son, the **Archduke Charles.**

1700, March 13. Charles's Will. Philip of Anjou was declared heir to the Spanish throne.

1700, November 1. Death of Charles. Philip of Anjou was proclaimed Philip V, and assumed the throne in Spain.

Operations in Western and Central Europe

1701, March. French Occupy Fortresses of Spanish Netherlands. England and Holland prepared for war. Since Leopold still claimed the Spanish Netherlands, Austria also prepared for war. Prince **Eugene** assembled an Austrian army in the Tyrol. A larger French army under Marshal **Nicolas de Catinat** occupied the defile of Rivoli (May).

1701, May 28. Eugene Arrives in Italy. Traveling little-used roads in a surprise and secret march, the Austrians bypassed Catinat and reached Vicenza. The French withdrew westward to avoid being cut off.

1701, May–August. Maneuvers in Lombardy. Despite the superior size of the French Army, Eugene maneuvered Catinat back to the Oglio. Paris replaced Catinat by aging Marshal Duke **François de Villeroi** (August).

1701, September 1. Battle of Chiari. The French attacked Eugene's fortified position and were repulsed with heavy loss. They withdrew to Cremona and went into winter quarters. Eugene went into winter quarters farther east, blockading the French-Spanish garrison of Mantua.

1701, September 7. Establishment of the Grand Alliance. The enemies of France were England, the Netherlands, Austria, Prussia, most of the other German states, and (later) Portugal. At this time French allies were Savoy, Mantua, Cologne, and (later) Bavaria. (Savoy soon changed sides.)

1702, February 1. Battle of Cremona. Eugene made a surprise raid, captured Villeroi, then withdrew after causing considerable damage. Paris sent Marshal **Louis Josef, Duke of Vendôme** to replace Villeroi.

1702, May 15. England Declares War. John Churchill, **Earl of Marlborough** was sent to

Holland as captain general of combined English and Dutch forces.

1702, June–July. Marlborough Invades the Spanish Netherlands. With an army of 50,000 (12,000 being British), Marlborough attempted to force French Marshal **Louis, Duke of Boufflers** into battle. To his intense disgust, the Dutch governmental deputies who had veto power over his use of Dutch troops refused on 4 different occasions to permit him to fight.

1702, July. Austrian Threat to Alsace. Prince **Louis, Margrave of Baden,** led an imperial army across the Rhine at Spires. Marshal Catinat was sent to protect Strasbourg.

1702, July 29–September 12. Siege of Landau. Louis of Baden prepared to move into Alsace after capturing the town.

1702, August 15. Battle of Luzzara. Drawn battle between Eugene and Vendôme. No other important operations in Italy, as Eugene skillfully maintained himself despite greatly superior French and Spanish strength.

1702, September–October. Marlborough on the Meuse. The allies opened the lower Rhine and Meuse Rivers. Since there was no risk of battle, and since their security and trade routes were both involved, the Dutch agreed to Marlborough's efforts against the Meuse fortresses; he captured Venloo (September 15), Ruremonde (October 7), and Liége (October 15). As a result he was made Duke of Marlborough.

1702, September. Bavarian Declaration of War. Bavarian troops seized Ulm. Louis of Baden recrossed the Rhine to protect his country. Catinat sent a small army after him under the command of General **Claude L. H. de Villars.**

1702, October 14. Battle of Friedlingen. Villars defeated Louis of Baden. He was promoted to the rank of marshal.

1703. Opposing Plans. Marlborough planned to reopen communications with Austria via the Rhine, and then to penetrate and shatter the French cordon of forts in the Spanish Netherlands and to seize Antwerp. The French planned to send Villars to join the Elector of Bavaria on the Danube, and then march to capture Vienna.

1703, April–May. Villars Marches through the Black Forest. He first seized Kehl, to have a base east of the Rhine. He then plunged into the Black Forest. Arriving at Ulm, he left a small force to contain Louis of Baden, joined the Elector (May 8), and urged an immediate advance on Vienna. But timid Maximilian insisted instead upon a junction with Vendôme in the Tyrol before attempting to take Vienna.

1703, May. Marlborough Takes Bonn. This opened the Rhine Valley, and Marlborough then turned to his second objective, Antwerp.

1703, June–August. Bavarian Tyrol Adventure. Maximilian occupied the Tyrol, while Villars remained in the Danube Valley to hold off Louis of Baden and an Austrian army under Count **Hermann Styrum.** Vendôme was slow in starting from Italy, and the Austrians in the Tyrol, supported by an aroused local population, drove the Bavarians out (August). Vendôme, who had been marching toward the Brenner, remained in Italy.

1703, June–October. Failure in the Netherlands. Marlborough was frustrated in the Netherlands by a combination of superior French forces under Villeroi (who had been exchanged) and by lack of Dutch cooperation.

1703, July–September. Villars in the Danube Valley. Louis of Baden and Styrum advanced separately into the Danube Valley, Villars skillfully concentrating against each in turn. He repulsed Louis's efforts against Augsburg at the **Battle of Munderkingen** (July 31), then, in collaboration with the Elector, defeated Styrum decisively at the **Battle of Hochstadt,** losing only 1,000 men, while the Austrians lost 11,000 out of 20,000. Louis had taken advantage of this brief campaign to seize Augsburg (September 6), but hastily withdrew and went into winter quarters as Villars approached. Villars again urged Maximilian to join him in a dash to Vienna, which was simultaneously being menaced by Hungarian rebels from the east (see p. 702). But the fainthearted Elector refused, and after a violent quarrel Villars gave up his command and returned to France. He was replaced by Marshal **Ferdinand of Marsin.**

1703, August–November. Operations along

the Middle Rhine. To assure better communications with the army in Bavaria, Marshal Count **Camille de Tallard** was given the mission of securing the middle Rhine. An army under the aging Vauban besieged and captured **Alt Breisach,** most important fortification on the right bank of the Rhine south of Kehl (September 6). Tallard cleared the allies from the Bavarian Palatinate by a victory at **Speyer,** and recaptured **Landau** (November 12).

1703, October 25. Savoy Changes Sides. Duke **Victor Amadeus,** distrustful of the French, allied himself with the Emperor.

1703, November–December. Recall of Eugene from Italy. Leopold, alarmed by the situation in southern Germany, and with his southern flank now guarded by Savoy, recalled Eugene from Italy to protect Vienna.

1704. Opposing Plans. The French planned to send Tallard to reinforce Marsin and the elector for a powerful thrust at Vienna. Villeroi, meanwhile, was expected to contain Marlborough along the Netherlands frontiers. Marlborough, as captain general of Anglo-Dutch forces, decided to send the Archduke Charles and an expeditionary force for an invasion of Spain to Lisbon. Admiral Sir **George Rooke** was then to take the fleet, and additional troops, to attack Toulon and to support Huguenot rebels (the Camisards) in southern France. These operations, however, were primarily diversionary. The allied main effort was to be a concentration of forces under Eugene and Marlborough in the Danube Valley to save Vienna, to drive the French out of Germany, and to knock Bavaria out of the war. It is not clear whether this bold concept was Marlborough's or Eugene's; more likely it was the Austrian's, though either general was capable of it.

Blenheim Campaign, 1704

1704, April. French Reinforcements to Bavaria. The strength of Maximilian and Marsin, concentrated around Ulm, was brought to about 55,000 men. Tallard with 30,000 was at Strasbourg, preparing to march east. The Austrians had 30,000 around Vienna, Eugene had about 10,000 men south of Ulm, and Louis of Baden had about 30,000 at Stollhofen, on the Rhine north of Kehl.

1704, May–June. Marlborough's March to the Danube. Leaving about 60,000 men to protect the Netherlands, Marlborough, without informing the Dutch of his intentions, marched toward the Rhine Valley (May 20) with 35,000 men, of whom less than one-third were British and the remainder German forces in British pay. At Mondelsheim he met Eugene and Louis of Baden for a planning conference (June 10–13). Marlborough and Baden continued toward the Danube, while Eugene returned to Stollhofen to prevent Tallard (now at Landau) and Villeroi (now at Strasbourg) from reinforcing Maximilian and Marsin. Distribution of forces was as follows (June 30): Allies: Marlborough and Baden advancing eastward toward Donauwörth, 70,000 men; Eugene at Stollhofen, 30,000. French and Bavarians: Maximilian and Marsin, near Ulm, 60,000; Villeroi at Strasbourg, 60,000; Tallard, advancing east across the Rhine from Landau, 30,000.

1704, July 2. Battle of the Schellenberg. An audacious surprise attack by Marlborough on the fortified Schellenberg Hill overlooking Donauwörth. Despite heavy losses, Marlborough took the hill, and then Donauwörth, without a siege.

1704, July–August. Maneuvers in Southern Germany. Marsin and Maximilian moved to Augsburg to block Marlborough's approach to Munich. Since Maximilian and Marsin refused battle until they received reinforcements, Marlborough and Baden began to devastate western Bavaria. Meanwhile, Tallard had marched to Ulm (July 29), and soon thereafter joined Marsin and Maximilian south of the Danube. Eugene, misleading Villeroi into thinking he was staying near the Rhine, marched hastily eastward with about 20,000 men to join Marlborough. Villeroi, not knowing where Eugene had gone, decided to stay near Strasbourg to protect Alsace. With battle approaching, Marlborough sent Baden (for whom he had low regard) to besiege Ingolstadt, and prepared to march west to join Eugene.

1704, August 10. French and Bavarians Cross the Danube. Knowing Eugene had arrived near Donauwörth, and that Marlborough was still in western Bavaria, Tallard (real commander of Maximilian's Franco-Bavarian army) moved to attack Eugene, or at least force him and Marlborough to retreat northward to protect their lines of communication. Eugene sent a message urging Marlborough to join him.

1704, August 12. Marlborough and Eugene Join Forces. The combined allied force totaled about 56,000 men. The Franco-Bavarians, encamped behind a little stream near Blenheim, just north of the Danube, totaled about 60,000, almost equally divided between the armies of Tallard and of Marsin-Maximilian.

1704, August 12–13. The Allied Advance. Marlborough and Eugene decided to attack. Marlborough's larger army, on the left, was to make the main effort against Tallard, while Eugene was to contain Marsin and Maximilian with an aggressive holding attack. Following carefully prepared plans, the allied armies, deployed in battle order, marched from their camps near Donauwörth at 2 A.M. in 9 columns. Marlborough's 5 columns, going over easy ground, arrived on the heights east of the Nebel stream at 7 A.M., and began skirmishing with French outposts and exchanging long-range fire with the artillery of the surprised and hastily deploying Franco-Bavarians. Eugene's troops, having a longer and more difficult route, did not reach their appointed positions until nearly noon.

1704, August 13. Battle of Blenheim. Marlborough and Eugene attacked simultaneously across the Nebel at 12:30. Marlborough initially attacked the villages of Blenheim (beside the river) and Oberglau (about 2 miles inland) to pin down French reserves. The British lost heavily, but the purpose was accomplished; Tallard poured in his reserves. Learning that Eugene was advancing very slowly in heavy fighting on the right, in accordance with plan, Marlborough at 4:30 launched his cavalry squadrons in an attack which after about an hour of tough fighting broke through Tallard's

center. Tallard's army was shattered and he was captured. Many of the fleeing troops drowned in the Danube. Before Marlborough could complete a right turn, in the dusk Marsin and Maximilian were able to extract most of their army from the jaws of Eugene's and Marlborough's double envelopment. Allied losses were 4,500 killed and 7,500 wounded. The French and Bavarians lost a total of 38,600 in killed, wounded, and prisoners.

COMMENT. *There is probably no finer example in history of allied coordination and cooperation than that of Eugene and Marlborough in this campaign and battle. The prestige of France and French armies was shattered; the Elector of Bavaria had to flee his country, which was annexed by Austria.*

1705. Stalemate in the Low Countries. Marlborough, back in the Netherlands, was opposed in a desultory campaign by Villeroi and the Elector of Bavaria. His offensive inclinations were again frustrated by the overcautious Dutch.

1705. Stalemate on the Rhine. Equally inconclusive were the maneuverings of Villars and Louis of Baden.

1705. Stalemate in Italy. After Vendôme had overrun much of Savoy, Eugene was sent to the aid of Victor Amadeus. No further results were achieved. The only major action was the drawn **Battle of Cassano** (August 16).

THE RAMILLIES CAMPAIGN, 1706

1706, April–May. Convergence on Namur. Villeroi, rightly believing that Marlborough intended to seize Namur, moved with about 60,000 men to protect the city. He was intercepted at Ramillies by the allied army in about equal strength.

1706, May 23. Battle of Ramillies. The French drew themselves up on high ground in a defensive position, partially entrenched. Marlborough, in a typical maneuver, feinted against the French left, causing Villeroi to shift his reserves and to draw some units out of his right wing. Marlborough then launched his main attack against the weakened French right. The

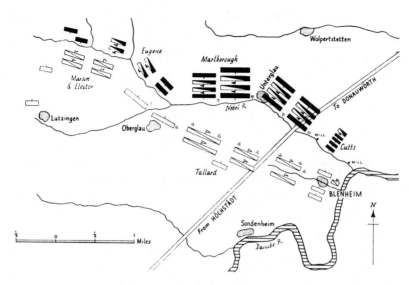

Battle of Blenheim

outnumbered French right defended itself gallantly, but was overwhelmed by the strength and power of the allied attack. The entire French line wavered, and Marlborough ordered a general attack, driving the French off the field in great disorder. He pursued vigorously, inflicting heavy losses. French casualties were about 8,000 killed and wounded and 7,000 prisoners. The allies lost 1,066 men killed and 3,633 wounded.

1706, June–October. Marlborough Consolidates the Spanish Netherlands. He captured the following important fortifications, among others: Antwerp (June 6), Dunkirk (July 6), Menin (August 22), Dendermonde (September 5), Ath (October 4). In these operations the allies took about 14,000 additional prisoners.

1706, August. Vendôme to Command in the North. Hurriedly called north from Italy (see below), he replaced the incompetent Villeroi. Louis XIV also made tentative peace overtures.

The Turin Campaign, 1706

1706, April 19. Vendôme Seizes the Initiative. With an army of over 100,000, opposed

to about 30,000 Savoyards and less than 40,000 Austrians, he boldly divided his field forces into 2 main groups of about 40,000 each. With one he drove the Austrians, under Count **Reventlau,** from central Lombardy back to the Adige River. The remaining French besieged Turin, capital of Savoy.

1706, April 22. Arrival of Eugene. Eugene, who had been in Vienna, took command and halted the Austrian retreat. He remained on the Adige, maneuvering against Vendôme, for more than a month.

1706, July–August. Eugene's Advance on Turin. Having received reinforcements, which brought his strength about equal to that of the French on the Adige, Eugene responded to appeals from Victor Amadeus by adroitly marching south with 25,000 men, leaving 18,000 on the Adige in front of Vendôme. Swinging around Vendôme's right, he threatened the French line of communications, causing Vendôme to fall back. Vendôme was now called north to retrieve the disaster of Ramillies (see above). He left the young **Duke of Orléans** in command, with Marshal Marsin as his second in command. Suddenly, to the amazement of the French, Eugene marched due west, abandoning his line of communications and living off the countryside, something almost unheard of in 18th-century warfare. He seized Parma (August 15), then turned northwestward. Orléans, confused by Eugene's maneuvers, rushed back to Turin, just as Eugene met Victor Amadeus on the upper Po (August 31). The combined allied army was then about 36,000, with another 6,000 Piedmontese east of Turin, and about 15,000 in the besieged city. Orléans had about 60,000 in the lines west of the Po at Turin, with another 20,000 facing the Piedmontese detachment. Another 20,000 or more French were scattered through northwest Italy in garrisons. The French lines of circumvallation around Turin were very strong. Eugene, in a favorable position on the French lines of communication, but with only half the strength of his strongly entrenched enemy, decided to attack.

1706, September 7. Battle of Turin. Selecting a relatively isolated portion of the French lines between the Dura and Stura rivers, tributaries of the Po, Eugene and the Piedmontese at-tacked vigorously, smashing through the defending lines before Orléans could send adequate reinforcements. As more French troops arrived, however, the allied attack was halted. Eugene reformed, then vigorously resumed his bold attack against the more numerous enemy just as the garrison made a sortie. The French could not cope with these converging attacks. Orléans was wounded, Marsin mortally wounded, and the French army collapsed and fled, abandoning all of its baggage, 100 pieces of artillery, and heavy equipment. French losses were 2,000 killed, 1,200 wounded, and 6,000 prisoners. The allies lost 950 killed and 2,300 wounded. The routed French fled to France, pursued by Austrian and Piedmontese detachments.

1706, September–December. Eugene Sweeps the French from Italy. By the end of the year Eugene had captured all of the remaining French garrisons in north Italy.

COMMENT. *This was probably the most brilliant campaign of the war.*

INCONCLUSIVE OPERATIONS, 1707

Vendôme held the Flanders frontier, while Marlborough and the Dutch continued their debates. In the most daring operation of the year, Villars, on the Rhine, drove the Austrians from their lines at **Stollhofen** (May 22), then raided into south-central Germany as far as Bavaria, living off the countryside (June–September). In Italy Eugene and Victor Amadeus invaded southern France, but without result (see p. 685).

OUDENARDE CAMPAIGN, 1708

Deployment of Forces. The French Flanders army, about 100,000 strong, under the nominal command of the young **Duke of Burgundy,** but with Vendôme exercising actual operational command, was located around Mons. The **Duke of Berwick*** and **Maximilian** of Bavaria had another French army of about 50,000 at Strasbourg, while Villars had about 20,000 in the Vosges near Switzerland. On the allied side Marlborough had nearly 70,000 at Brussels, Eugene had 35,000 at Coblenz, and the **Duke of Hanover** had about 50,000 opposite Strasbourg.

1708, April. Allied Planning Conference. Marlborough and Eugene, at The Hague, agreed that Eugene's army would join Marlborough in the Netherlands when another allied army, gathering in central Germany,

* Nephew of Marlborough, illegitimate son of James II and Arabella Churchill.

arrived on the Rhine to replace him. Eugene was under strict Imperial orders not to leave the Rhine until these reinforcements arrived.

1708, May–June. Maneuvers in Flanders. Vendôme aggressively seized the initiative. He gave Marlborough no opportunity to attack, even when he approached allied positions near Louvain (late June).

1708, July 1–9. Vendôme Threatens Oudenarde. The French army suddenly moved westward to capture Ghent (July 4) and Bruges (July 5) in a *coup de main*. Vendôme then turned south to threaten the allied garrison of Oudenarde. Knowing that Eugene's army had started westward from Coblenz, he was anxious for battle before the Austrians arrived. Marlborough, realizing he had been outmaneuvered, and with Dutch morale plummeting, also sought battle. He had been reinforced and his field strength was about 80,000, about equal to that of Burgundy and Vendôme, west of the Scheldt. Furthermore, although Eugene's army was still marching from the Rhine, Eugene himself arrived to join the allied army. He also urged battle.

1708, July 10. French Command Troubles. With battle obviously imminent, young Burgundy became nervous and insisted that Vendôme avoid battle. While the prince and the general argued, the French army stood aimlessly in scattered units north of Oudenarde, instead of moving to block the allied passage of the Scheldt.

1708, July 10–11. Marlborough's Advance. The allied army marched 28 miles in 22 hours. Marlborough turned the right wing of his army over to Eugene. They hastily prepared a plan roughly similar to Blenheim.

1708, July 11. Battle of Oudenarde. While the allies swarmed across the river in the morning and early afternoon, Burgundy ordered a retreat. However, Vendôme finally convinced the prince that he would be cut off from France if he did not stand and fight. Reluctantly Burgundy gave his permission for a defensive battle; Vendôme hastily deployed the army on the heights north of Oudenarde. Despite the royal orders, Vendôme ordered an attack just as Marlborough and Eugene were beginning their own attack (about 2 P.M.). There was little plan on either side, and the result was a long, confused, bloody struggle, in which the allies held a slight advantage due to the vigor, determination, and field generalship of their joint commanders; the French had perhaps been discouraged by the obvious vacillation in their command. By dusk Marlborough had achieved an envelopment of the French right and, as Eugene continued to press forward, the allies drove the French from the field in the gathering darkness. French losses were 4,000 killed, 2,000 wounded, 9,000 prisoners, and 3,000 missing (deserters). Allied losses were 2,000 killed and 5,000 wounded.

1708, July 12. Allied Repulse at Ghent. Vendôme, brilliantly rallying his defeated troops, repulsed the pursuing allies at Ghent. As a result, the French retained control of western Flanders, and regained a secure line of communications to France.

COMMENT. *It is interesting to speculate on the result if Vendôme had been initially free to fight this campaign as he wished.*

1708, July 15. Arrival of Eugene's Army. The allies now had a clear-cut superiority of force, totaling about 120,000 men. Vendôme, reinforced by Berwick, had 96,000. Marlborough's plan of invading France was vetoed by the Dutch. Surprising Vendôme, Marlborough moved suddenly south against Lille (August).

1708, August 14–December 11. Siege of Lille. The siege was conducted by Eugene with 40,000 men, while Marlborough covered with 70,000 in strong defensive positions. Boufflers gave up the city due to lack of food, but withdrew into the citadel with his garrison (October 22). At this point Vendôme marched northeastward to threaten Brussels, hoping this would force the allies to raise the siege, but Marlborough moved after him vigorously and Vendôme fell back (November 24). Boufflers was finally forced to surrender (December 11). Because of his skill and gallantry, Marlborough and Eugene let him dictate his own surrender terms.

1708, December–1709, January. French Withdrawal. Marlborough, refusing to go into winter quarters, decided to force the French out of western Flanders. He besieged and captured Ghent (January 2), and then took Bruges. The French withdrew to their own borders, and both sides went into winter quarters.

1708–1709. Fruitless Peace Negotiations. Louis again sought terms, but refused harsh allied conditions. The French people gave full support to Louis's decision to continue the war.

MALPLAQUET CAMPAIGN, 1709

1709, Spring. Deployments in Flanders. The French army, about 112,000 strong, under Villars, with Boufflers, his senior, voluntarily serving under his command, was entrenched generally north of the border. Villars was instructed from Paris to avoid general battle if possible. The allied army, about 110,000, under Marlborough and Eugene, occupied most of the Spanish Netherlands. Their objective was to try to get past, or through, the French fortified line, or to entice the French to leave that line for battle on conditions favorable to the allies.

1709, June–August. Inconclusive Maneuvering. The most important of many minor incidents was the siege and capture of Tournai by the allies (June 28–July 29). Gradually the allies were able to reach a position to threaten Mons, held by the French.

1709, September 4–October 26. Siege of Mons. Villars now received orders from Paris to fight to hold Mons. He immediately concentrated 90,000 men at Malplaquet, where he entrenched, knowing that this threat to the besieging forces would attract them to attack him. Marlborough and Eugene, leaving about 20,000 men to continue the siege, eagerly advanced with about 90,000 to accept Villars' challenge (September 9–10).

1709, September 11. Approach to Battle. The allies advanced with, as usual, Eugene on the right and Marlborough on the left. Eugene was to make a holding attack on the French left, while part of Marlborough's force was making a secondary holding attack on the French right; the remainder of Marlborough's command would make the main effort near the French center after French reserves had been committed. This simple tactical scheme had been effective at Blenheim and Ramillies, and something similar had won the day at Oudenarde. This, however, was the first time allied generals had been opposed by Villars, a commander of ability comparable to their own.

1709, September 11. Battle of Malplaquet. The vigor of the early allied flank attacks led Villars to weaken his center, mainly to oppose Eugene. In the bitter fighting on the French left and allied right, Eugene was twice wounded, but refused to leave the field; Villars was so badly wounded while leading a counterattack that he had to turn the command over to Boufflers. During the early afternoon Marlborough launched his typical hammer-blow cavalry-infantry attack against the French center, and broke through. Boufflers, following Villars' plan, promptly counterattacked with his last reserves, re-establishing the line. Marlborough and Eugene now committed their remaining reserves and renewed the vigor of their attacks, again penetrating the French center. Boufflers thereupon ordered a general withdrawal, which was carried out in good order. The allies, who had suffered fearful losses, were unable to pursue. French losses were about 4,500 killed and 8,000 wounded. The allies lost about 6,500 killed and more than 14,000 wounded. Both sides had fought superbly, and the leadership was excellent on both sides at all levels. The allied victory was primarily due to the absolute determination of Eugene and Marlborough, who persisted in attacks after most generals would have admitted defeat. The battle had no results other than to permit the allies to continue the siege of Mons, which they eventually captured (October 26). Both armies then went into winter quarters.

1709–1710. The "Ne Plus Ultra" Lines. During the winter the French built a strong line of fortifications along the frontier.

1710–1711. Inconclusive Maneuvering. Marlborough and Eugene captured a few more border fortresses, including Douai (June 10, 1710) and Bethune (August 30). Nothing further of real significance occurred during that year. The following year Eugene was ordered to Frankfurt to protect that city during the coronation of new Emperor **Charles VI**—the former Archduke Charles. Because none of the allies would accept Charles both as Emperor and as

King of Spain, the alliance began to fall apart. In Flanders, Villars and Marlborough treated each other with cautious respect.

1711, December 31. Recall of Marlborough. The new Tory cabinet relieved him of his command, ending his military career.

1712. Peace Negotiations. While negotiators met at Utrecht, Eugene assumed command in the Netherlands. He was no more successful than Marlborough in getting the Dutch to agree to aggressive action, particularly with negotiations under way. Eugene nevertheless crossed the Scheldt with 120,000 men (May) to try to force a battle on Villars, who was entrenched from Cambrai to Arras with 100,000 men. Unknown to Eugene, however, his English contingent, now under **James Butler, Duke of Ormonde,** had been ordered not to fight, and suddenly was withdrawn. The remainder of the allied army was forced to halt, while Eugene attempted to obtain reinforcements.

1712. Battle of Denain. Villars, by prompt and energetic action, overwhelmed a portion of Eugene's army in a bayonet attack. About 8,000 allied soldiers were killed or drowned in the Scheldt; the French lost 500. Eugene, who arrived too late to take part in the action, decided to withdraw across the river.

1712, August–October. Villars on the Offensive. The French now advanced, besieging and recapturing a number of the fortresses they had lost in previous years, including Douai, Quesnoy, and Bouchain. These successes materially aided the French negotiators at Utrecht.

1713, April 11. Treaty of Utrecht. Agreed to by all participants save the Emperor, who continued the war. France recognized the Protestant succession in England, and ceded Newfoundland and parts of Canada to England. Philip V was recognized as King of Spain, but France gave assurances that the crowns of France and Spain would remain permanently separated. Savoy received Sicily and some lands in northern Italy from Spain. Prussia was recognized as a monarchy. Portugal received a favorable correction of colonial boundaries in South America. The Empire was to receive the Spanish Netherlands (meanwhile administered by the Netherlands and all Spanish possessions in Italy, save those given to Savoy.

1713, May–November. Operations on the Rhine. Eugene, given only 60,000 men and poorly supported by Austria, was opposed by Villars with 130,000. Villars besieged and captured Speyer, Landau, and Freiburg.

1713–1714. Peace Negotiations. These were concluded by Eugene and Villars.

1714, March–September. Treaties of Rastatt and Baden. At Rastatt Austrian Emperor Charles VI made peace with France. As Holy Roman Emperor, he made peace at Baden. Generally the terms confirmed Utrecht. The Emperor still refused to recognize Bourbon rule in Spain.

Operations in Spain and the Mediterranean

1702, August–September. Allied Repulse at Cádiz. An Anglo-Dutch force of 50 ships and 15,000 men under Rooke (land force under the Duke of Ormonde) was repulsed in a fiasco of mismanagement.

1702, October 12. Naval Action at Vigo Bay. Rooke's allied force, returning from the failure at Cádiz, discovered in Vigo Bay the Spanish treasure fleet from the River Plate, protected by a Franco-Spanish squadron including 24 French ships of the line. In a combined action, the Duke of Ormonde's troops took the fortifications overlooking the bay while the fleet forced the boom and entered the harbor. Rooke destroyed or captured the entire treasure fleet and its convoy. The booty of 2 million was perhaps the largest ever taken in one action.

1703, May. Portugal Joins the Alliance. The Portuguese agreed to make bases available to the allies for operations against Spain by land and by sea (known as the **Methuen Treaty**).

1704, February. Charles at Lisbon. The Archduke Charles and 2,000 English and Dutch troops were landed at Lisbon by Rooke's English fleet.

1704, March–June. Rooke in the Mediterranean. Rooke passed through the Straits of

Gibraltar, hoping to join the Duke of Savoy in an attack on Toulon, but Victor Amadeus did not cooperate. Arrival of the French Brest fleet, which outsailed the English to reach Toulon unscathed, gave the French numerical superiority in the Mediterranean, and Rooke returned through the Straits to join another Anglo-Dutch fleet under Admiral **Clowdisley Shovell** near Cádiz. The combined allied fleet then returned to the Mediterranean.

1704, July 23–24. Capture of Gibraltar. An English fleet under Vice-Admiral Viscount **George Byng of Torrington** bombarded Gibraltar, which was seized next day by 1,800 English marines, commanded by Prince **George of Hesse-Darmstadt.**

1704, August 13. Naval Battle of Malaga. The allied (English and Dutch) fleet under Rooke defeated the combined French fleet under the **Count of Toulouse.** This secured the British capture of Gibraltar.

1704, August–1705, March. Siege of Gibraltar. A Franco-Spanish force under Marshal Count **R. de Fromlay Tessé** besieged Gibraltar, which Prince George held successfully with only 900 marines. He was reinforced and re-supplied (December), and after a supporting French naval force was destroyed (see below), Tessé raised the siege.

1705, March 10. Naval Battle of Marbella. An English naval squadron under Admiral Sir **John Leake** destroyed a French squadron under Admiral **de Pointis** near Gibraltar.

1705, June. Allied Landing in Catalonia. Admiral Shovell and **Charles Mordaunt,** Lord **Peterborough,** as joint commanders, landed in Catalonia, accompanied by Charles. They besieged and captured Barcelona (October 3). Prince George, who had joined at Gibraltar, was killed during the final assault. The main fleet returned to the Atlantic, leaving a squadron under Leake to support the land force.

1705, November–1706, April. Siege of Barcelona. Philip and Tessé marched from Madrid to besiege Barcelona, supported by the French Toulon fleet. Leake, out-maneuvering the French, forced them to return to Toulon, and the French army abandoned the siege (April

30), to return westward to face an invasion of Spain from Portugal under English Lord **Henry Galway** (see below).

1706, May–September. Leake's Coastal Operations. The English fleet captured Cartagena (June 1), Alicante (August 24), and the Balearic islands of Mallorca and Iviza (September).

1706, June 26. Allied Capture of Madrid. Galway's invading force from Portugal seized the capital and proclaimed Charles king.

1706, October. French Recapture Madrid. The Duke of Berwick took command of French and Spanish forces in Spain. Accompanied by Philip, he forced the allies to abandon Madrid. Galway and the Portuguese returned directly to Portugal, but Charles and other allied contingents retreated to Valencia.

1707, April. Allied Advance toward Madrid. Galway, who had returned to the east coast by sea, led an army of 33,000 from Valencia toward Madrid. Berwick, with an equal force, marched to meet him.

1707, April 25. Battle of Almansa. Berwick decisively defeated Galway. French losses were slight; the allies lost 5,000 killed and wounded and 10,000 prisoners. Galway retreated to Valencia, closely pursued by Berwick. With only 16,000 men left, Galway continued his retreat to Catalonia (October). The French now had effective control over practically all of Spain. Both armies went into winter quarters near Barcelona.

1707, July–August. Allied Invasion of France; Blockade of Toulon. The allied fleet under Shovell blockaded Toulon by sea, while Eugene and Victor Amadeus led an Imperial-Savoy army across the Maritime Alps. Marshal Tessé, however, who had just arrived from Spain with veterans of Almanza, repulsed the allied land force, thanks mainly to the Duke of Savoy's refusal to cooperate closely with Eugene (July 26). The French, expecting to lose Toulon, had sunk their 50 ships in the harbor, giving the allies uncontested control of the Mediterranean. The allied army returned to Italy, while the main allied fleet returned to the Atlantic.

1708, August. Allied Capture of Sardinia.
Admiral Leake, his marines, and a contingent of Savoy troops captured the island.

1708, September. Capture of Minorca. The English force was commanded by General **James Stanhope.**

1708-1709. Stalemate in Catalonia. Berwick, recalled to the Netherlands after the French disasters at Ramillies and Turin, left the Duke of Orléans in command. Despite French superiority, there was little action. The allies in Catalonia were now commanded by Stanhope.

1710, May-December. Madrid Campaign. Stanhope's force, about 26,000 men, was opposed by about 35,000 under Philip in Catalonia. A Portuguese army of 33,000 was at Elvas. Advancing rapidly, and expecting to meet the Portuguese at Madrid, Stanhope won victories at **Almenar, Lérida,** and **Saragossa,** then marched on Madrid, accompanied by Charles. The Portuguese, however, had withdrawn when opposed by a much smaller Franco-Spanish force. The Spanish people made clear their preference of Philip over Charles. Vendôme arrived in Spain, collected the scattered French forces, and advanced on Saragossa with 27,000 men, cutting Stanhope's line of communications (October). As the French approached Madrid, Stanhope, who had only 16,000, retreated toward Valencia, closely pursued by Vendôme. The French caught up with the allied rear guard, under Stanhope's personal command, defeating and capturing him at the **Battle of Brihuega** (December 10). The remainder of the allied force, under General **Guido von Starhemberg,** retreating to Barcelona, was again defeated en route at the **Battle of Villaviciosa.**

1714, September 11. Fall of Barcelona. Finally captured by the French under Berwick; last major action of the war.

1715, February. Treaty of Madrid. Peace with Portugal, officially ending the war.

War of the Quadruple Alliance, 1718-1720

After the death of Louis XIV of France (1715), his grandson, Philip of Spain, uncle of **Louis XV,** hoped to gain the crown of France himself. These ambitions seem to have been fanned by the designs and intrigues of his powerful and unscrupulous prime minister, Cardinal **Giulio Alberoni.** At the same time other, and equally aggressive, ambitions were being entertained by Philip's second wife, **Elizabeth Farnese** of Parma, who wished her children to inherit the family's possessions in Italy, and perhaps all Italy. The maritime powers—England and the Netherlands—were unalterably opposed to any union of France and Spain. The Emperor Charles VI, aside from his continuing enmity toward Philip as a result of the War of the Spanish Succession, had equally strong dynastic reasons to oppose any return of Spanish power to Italy. The Empire, however, was engaged in a major war against Turkey, hence Charles was unprepared to precipitate another war in Italy or Western Europe.

1717, January 4. The Triple Alliance. England, France, and the Netherlands agreed to oppose the ambitions of Philip and of Spain both in France and in Italy.

1717, November. Spanish Occupation of Sardinia. Taking advantage of Austria's war with Turkey, Philip sent an army to seize Sardinia, formerly a Spanish possession, but granted to Austria by the Treaties of Utrecht and Rastatt (see p. 684).

1718, July. Spanish Occupation of Sicily. This former Spanish colony, awarded to Savoy by the Treaty of Utrecht, provided a second base for further political and military action in Italy. It was seized by an army of 30,000 under the **Marquis of Lede.**

1718, August 2. Formation of the Quadruple Alliance. Austria joined the Triple Alliance. France, Austria, and England (the Netherlands, though in the alliance, was not at this time active against Spain) demanded Spanish withdrawal from Sicily. Spain refused. An English fleet immediately landed a force of 3,000 Austrian troops near Messina, where they blockaded the Spanish garrison (August).

1718, August 11. Naval Battle of Cape Passaro (off Syracuse). Byng's English fleet of 21 ships of the line completely overwhelmed a somewhat smaller, but much weaker, Spanish fleet under Admiral Castenada. The British sank or captured 16 Spanish ships of the line and 4 frigates; only 2 or 3 Spanish ships escaped.

1718, December. Declarations of War. The Netherlands later joined her allies in the war (August 1719).

1718–1719. Operations in Sicily. Austrian troops, supported by an English fleet, operated against Spanish forces based in Messina. The Austrians captured Messina (October 1719).

1719, March–June. Spanish Expeditions to England and Scotland. Alberoni sent 2 expeditions to aid a planned uprising of English and Scottish Jacobites. A force of 29 ships and 5,000 troops led by Jacobite exile James Butler, Duke of Ormonde, sailed from Cadiz but was wrecked by a storm (March). A smaller force, under Jacobite **George Keith, Earl of Marischal,** left Corunna for Scotland with about 300 Spanish troops and a few Jacobite exiles. Only 1,000 Highlanders under **Rob Roy Macgregor** and other chiefs joined Marischal (April). Marischal made a stand against government troops led by **General Joseph Wightman** at **Glenshiel** (June 10). After a brief bombardment the Highlanders bolted, and the isolated Spanish surrendered.

1719, April. French Invasion of Spain. A French army of 30,000, under Berwick, invaded the Basque provinces of Spain. Berwick ranged through northern Spain almost unopposed before disease and bad weather caused him to return to France (November).

1719, October. English Operations along the Galician Coast. Amphibious forces captured Vigo and Pontevedra.

1719, December. Dismissal of Alberoni. Philip finally recognized that the recent series of disasters to Spanish arms had largely been the result of the intrigues of his prime minister.

1720, February 17. Treaty of The Hague. Philip abandoned his Italian claims in return for Austrian agreement that his son Charles could succeed to the duchies of Parma, Piacenza, and Tuscany upon the impending extinction of the male Farnese line. Sicily was ceded to Austria; Savoy got Sardinia in return. Victor Amadeus of Savoy was recognized as King of Sardinia.

WAR OF THE POLISH SUCCESSION, 1733–1738*

During the Great Northern War (see p. 672), Stanislas Leszczynski was briefly installed as King of Poland by Charles XII, but the eventual defeat of Charles left Augustus II (protégé of Peter the Great of Russia) on the throne. As the death of Augustus II approached, there were 2 major claimants to the throne of Poland: Stanislas Leszczynski, now the father-in-law of Louis XV, and thus supported by France (and also by Spain and Sardinia); and **Augustus III of Saxony,** son of Augustus II, supported by Austria and Russia.

1733, February. Death of Augustus II.

1733, September. Arrival of Stanislas in Warsaw. The former king was popularly reelected to his former throne.

1733, October. Russian Invasion. An army of 30,000 Russians, later joined by 10,000 Saxons, marched on Warsaw to support the claims of Augustus III. There was no effective Polish Army, and Stanislas fled to Danzig.

1734, January–June. Siege of Danzig. Despite the arrival by sea of a French relief force of 2,200 men, the Russians and Saxons cap-

* Since the Polish monarchy was elective and not hereditary, Polish historians designate this more accurately as the "War for the Polish Throne."

tured the town. Stanislas escaped to Prussia just before the capitulation.

1734, April–September. Operations in the Rhineland. The only important event of this inconclusive campaign was the French capture of Philippsburg, after overrunning Lorraine. Of greater historical interest is the fact that it was the last campaign of 2 great soldiers: Prince Eugene of Savoy, the magnificent Frenchman who had fought all his life for Austria, and the Duke of Berwick, the gifted Englishman who had fought all of his life for France. Berwick was killed during the siege of Phillippsburg (June 12). Also significant was the first appearance of a Russian army in the Rhine Valley after the campaign had come to its languid close and peace negotiations had begun.

1734–1735. Operations in Italy. Following French and Spanish invasions of Lombardy, Naples, and Sicily, the Austrians won the **Battle of Parma** (June 29), the French the **Battle of Luzzara** (September 19), and the Austrians had slightly the better of the final engagements of the war at the **Battle of Bitonto** (May 25, 1735).

1734–1735. Civil War in Poland. Stanislas, more popular among the Poles than his rival Augustus, attempted to revive his fallen fortunes, but received little help from France, whose prime minister, Cardinal **André H. de Fleury,** was anxious to restore peace on terms favorable to France.

1738, November. Treaty of Vienna. Stanislas abdicated the throne of Poland; the coronation of Augustus (which had taken place at Cracow, September 1734) was recognized, but Leszczynski was created Duke of Lorraine (and permitted to retain the title of King of Poland), while possession of Lorraine was to go to the French crown upon his death. **Charles Bourbon** (son of Philip of Spain) was recognized as King of the Two Sicilies, in return for which his 3 duchies were ceded to Austria.

AUSTRO-RUSSIAN-TURKISH WAR, 1736–1739

This was an outgrowth of the War of the Polish Succession (see above). France urged Turkey to join her in the war against the Ottoman Empire's traditional enemies: Russia and Austria. Turkey postponed her participation in the European war until she had concluded an ongoing war against Persia. By this time the War of the Polish Succession was over, and so a new and different war resulted: Russia, learning of Turkish intentions, and anxious to gain revenge for the Treaty of the Pruth (see p. 675), declared war (late 1735).

1736. Operations in the Ukraine. Two Russian armies invaded the Turkish-Tartar regions north of the Black Sea. The army under Marshal Count **Peter Lacy** captured Azov after a fierce struggle, but lost so heavily in the campaign that Lacy abandoned the fortress and retreated into the Russian Ukraine. The army under Marshal Count **Burkhardt C. von Munnich** invaded the Crimea and captured the Tartar capital. Losses from disease, exhaustion, and starvation, however, resulted in a mutiny, and Munnich retreated, having lost 30,000 dead out of his army of 58,000.

1736–1737. Tartar Raids in the Ukraine. Retaliating for the invasion of their territory, an army of 100,000 Tartars devastated much of the Russian Ukraine.

1737, January. Austrian Declaration of War. Revealing a secret treaty with Russia, Austria sent invading armies under Marshal Count **Friederich H. von Seckendorf** into Bosnia, Wallachia, and southern Serbia, capturing Nish. A Turkish counter-offensive, however, threw the Austrians out of most of the regions they had conquered. The Austrians and Russians made no attempt to coordinate their operations.

1737. Renewed Russian Offensives. While one Russian army raided into the Ukraine and another reoccupied Azov without serious opposition, the main army under Munnich overran most of the Turkish Ukraine and captured Ochakov, at the mouth of the Bug River. Again the Russians were stricken with disease, and Munnich was fortunate in being able to withdraw without serious opposition from large Turkish forces waiting passively in Moldavia and Bessarabia.

1738. Operations in the Balkans. The Turks, retaining the initiative, by the end of the campaign had thrown an army across the Danube into the Banat, and were threatening Belgrade. There were no major battles, as Austrian Count **Lothar J. G. Konigsegg-Rothenfels** conducted a typical 18th-century campaign.

1738. Operations in the Ukraine. One Russian force conducted an inconclusive and meaningless raid into the Crimea, while Munnich threatened Moldavia. Repulsed at the **Battle of Bendery,** in an effort to cross the Dniester River, he was once more forced to retreat, leaving half of his army dead (from battle, disease, and starvation) while abandoning almost all of his artillery.

1739. Operations in the Balkans. An Austrian army commanded by General Count **Georg O. von Wallis** was forced to fight a greatly superior Turkish army advancing on Belgrade. Wallis, decisively defeated in the **Battle of Kroszka,** fell back into Belgrade, where he was besieged.

1739. Operations in Moldavia. Munnich's fourth offensive was finally successful. Advancing through Polish Podolia, he invaded Moldavia with an army of 68,000. He defeated 90,000 Turks in the **Battle of Khotin** (or Stavuchany, August 17); captured the fortress of Khotin; seized Jassy, capital of Moldavia; and prepared for an advance on Constantinople.

1739, September. Negotiations at Belgrade. The Austrians, dismayed by the successes of their Russian allies, immediately entered negotiations with the Turks at Belgrade.

1739, September 18. Treaty of Belgrade. Austria abandoned all of the gains of the Treaty of Passarowitz (see p. 703) save for the Banat, giving up northern Serbia, Belgrade, and por-

tions of Bosnia and Wallachia. The victorious Turkish army now moved to strike the right flank of Munnich's armies in Moldavia, eastern Wallachia, and Bessarabia.

1739, October 3. Treaty of Nissa. Abandoned by Austria and threatened by the convergence of two larger Turkish armies, the Russians made peace. They surrendered all of their important conquests save for Azov, where they agreed to demolish the fortifications. They also agreed not to build a navy or merchant marine in the Black Sea. As a result of the war, however, the effective Russian frontier on the Ukrainian steppes was pushed about 50 miles closer to the Black Sea.

WAR OF THE AUSTRIAN SUCCESSION, 1740–1748

Background

1711. Death of Emperor Joseph I. He was succeeded as Emperor and as heir to the Hapsburg dominions by his half-brother, **Charles VI,** who had been one of the two major claimants to the Spanish succession (see pp. 676, 702).

1713, April 19. The Pragmatic Sanction. Having no male children, Charles established the order of succession to be followed after his death, so as to avoid a disputed succession. The main provisions were (1) the lands of the Hapsburg Austrian Empire were to be held intact; (2) if he had no male heirs, the succession should go to his daughters—the eldest being **Maria Theresa**—and their heirs in accordance with the laws of primogeniture; (3) if this line should fail, the succession would go to the daughters of Joseph I and their heirs. This Pragmatic Sanction was accepted by Hungary (1723) and by the major European powers except Bavaria.

1725–1740. Claim of Charles Albert, Elector of Bavaria. As a descendant of the eldest daughter of Ferdinand I, he interpreted Ferdinand's will to assure him the succession in the event of a failure of Ferdinand's male descendants.

1740, October 20. Death of Charles VI. Maria Theresa inherited the Hapsburg Empire, in accordance with the Pragmatic Sanction. Her succession was disputed by the following claimants: Charles Albert of Bavaria, Philip V of Spain on grounds largely arising from his successful claim to the Spanish throne (see p. 684), and Augustus III of Saxony, husband of the eldest daughter of Joseph I.

1740, November. Involvement of Frederick II of Prussia. He recognized Maria Theresa's succession, offered his aid against the other claimants, but announced that in return he would occupy Silesia, pending settlement of an old Brandenburg claim for the province. Maria Theresa refused.

1740, December 16. Frederick Invades Silesia.

First Silesian War, 1740–1742

This war had nothing to do with the Austrian succession. It did, however, precipitate the wider conflict. Maria Theresa appealed to all guarantors of the Pragmatic Sanction to assist her against Frederick. There were no responses.

1741, January–February. Prussian and Austrian Actions. Frederick consolidated control over Silesia, save for a few towns held by Austrian garrisons. Investing Neisse and Glogau, Frederick put the remainder of his army in winter quarters. Meanwhile, Count **Adam A. Neipperg** was quietly collecting an Austrian army in Bohemia.

1741, March 9. Assault on Glogau. Prince Leopold of Anhalt-Dessau (son of **Prince Leopold I** [see p. 704] the "Old Dessauer") led the efficient and successful assault.

1741, March–April. Austrian Invasion of Silesia. Unprepared for Neipperg's move while the passes from Bohemia were still covered with snow, Frederick hastily collected his scattered forces. Meanwhile Neipperg overran much of the province. By relieving the beleaguered garrison of Neisse, he cut Frederick off from Prussia.

1741, April 10. Battle of Mollwitz. At the outset of the battle superior Austrian cavalry drove the Prussian right-wing cavalry off the field. Frederick, commanding on that wing, was reluctantly persuaded by Marshal Count **Kurt C. von Schwerin** to flee with his men. Unfortunately for the Austrians, however, the commander of their own left wing cavalry was killed in the charge, throwing the Austrian cavalry into confusion.

Schwerin, with the magnificent Prussian infantry, smashed the Austrian cavalry and then drove the Austrian infantry from the field. This was the first and last time Frederick left a still-disputed battlefield.

1741, May–August. The War Spreads. Charles Albert of Bavaria, gambling to gain the Imperial crown and the Hapsburg lands, sent an army to invade Bohemia. France, allied to Bavaria, sent an army under Marshal **François M. de Broglie** into southern Germany to support the Bavarians. This force was described by the French both as "volunteers" and as "auxiliaries"; most of the officers wore Bavarian insignia. Saxony and Savoy both joined Bavaria in the war against Austria. England and the Netherlands then immediately announced their support for Maria Theresa, and prepared for war. Sweden, influenced by France, supported Prussia and used this as an excuse to attack Russia (which supported the Pragmatic Sanction) in revenge for the Great Northern War (see p. 672).

1741, April–October. Maneuvering in Silesia. Frederick recovered much of the territory he had lost before Mollwitz, reopened his line of communications to Prussia, and cautiously maneuvered against Neipperg.

1741, July–September. Bavarian Invasion of Upper Austria. Charles Albert seized Passau (July) and was soon joined by a French army. They rejected Frederick's suggestion of a converging advance on Vienna.

1741, October 9. Truce of Klein Schnellendorf. This was a secret truce, following several weeks of secret Austrian-Prussian negotiations. In return for Frederick's agreement not to carry out further operations against the Austrians, he was to be allowed to capture Neisse after a mock siege, and was to be left in virtual control of Silesia. Neipperg immediately withdrew his army from Silesia to join the gathering Austrian forces preparing for campaigns in Bohemia and Bavaria.

1741, October. Franco-Bavarian Invasion of Bohemia. The allies marched on Prague, hoping to be joined by Frederick.

1741, November–December. Austrian Mobilization. An army under the Grand Duke **Francis of Lorraine** (Maria Theresa's consort) ineptly opposed the French-Bavarian invasion of Bohemia; additional forces gathered in Vienna. Maria Theresa, as Queen of Hungary, had made a successful appeal to the Hungarian nobles, who contributed substantial forces, mostly light troops, in return for substantial political concessions. Additional forces were gathered from Austria and other Hapsburg dominions. The principal army, under Marshal Count **Ludwig A. von Khevenhüller,** prepared to invade Bavaria. Additional forces, under Marshal Prince **Charles of Lorraine** (younger brother of Francis) prepared to go to Bohemia.

1741, November 26. Fall of Prague. The French and Bavarians quickly consolidated control of western and central Bohemia. Charles Albert was crowned King of Bohemia (December 19).

1741, December. Renewal of War in Silesia. With her forces mobilized, Maria Theresa thought she could recover Silesia from Frederick, as well as defeat the Bavarians and French. She divulged the terms of the secret Truce of Klein Schnellendorf to embarrass Frederick with his allies. Frederick immediately sent Marshal Schwerin to Bohemia, and soon followed himself, intending to operate in conjunction with the Bavarians and French.

1741, December 27. Khevenhüller Invades Bavaria. Pushing aside weak Bavarian forces (their main army was in Bohemia), Khevenhüller invested Linz and marched on to capture Munich (January 24).

1742, January 24. Election of Charles Albert as Emperor. The same day he became Charles VII, the new Emperor's capital (Munich) was being captured by the Austrians (see above).

1742, January–April. Operations in Bohemia. Planned cooperation of the Prussians, Bavarians, and French collapsed when the Bavarians had to rush back to protect what was left of their country from the Austrians. The French, under Broglie, were too weak to leave the vicinity of Prague. Leaving a force to observe the French, Charles of Lorraine, now commanding in Bohemia, moved against Frederick, who was overrunning Moravia and had even sent cavalry raiders to the outskirts of Vienna. But Charles's threats to his line of communications, combined with a Hungarian invasion of Silesia, forced Frederick to return to Silesia. Charles followed closely.

1742, May 17. Battle of Chotusitz. In a hard-fought battle Frederick was victorious, thanks largely to the results of his intensive efforts to improve the combat capability of his cavalry.

1742, May 27. Battle of Sahay. Taking advantage of the absence of Charles, Broglie attacked and defeated the Austrian covering force in an engagement near Budweis.

1742, June 11. Treaty of Breslau. Practical Maria Theresa decided to make peace with Prussia, ceding Silesia to Frederick. She hoped that a later opportunity would permit her to recover the province. This ended the First Silesian War and (temporarily) Frederick's participation in the War of the Austrian Succession.

Operations in Bohemia and South Germany, 1742–1743

1742, June–September. Blockade of Prague. Charles, returning from Silesia, virtually besieged the French in Prague.

1742, August–September. French Invasion of Franconia. Marshal **Jean B. F. D. Maillebois** advanced from the Rhine toward

Amberg. Charles raised the siege of Prague (September 14) and marched to meet the French, calling Khevenhüller to join him.

1742, October–December. Maneuvers in Bavaria and Bohemia. After an abortive demonstration toward Prague, Maillebois timidly fell back to the Danube west of Regensburg. As the Austrians again concentrated against Prague, Broglie was ordered to leave the forces there under the command of Marshal Duke **Charles L. A. F. Belle-Isle** and to take over Maillebois's army. Broglie did so, and quickly overran most of Bavaria. Charles and Khevenhüller fell back to the line between Linz and Passau, to cover the approaches to Vienna, leaving a force under Prince **Johann G. Lobkowitz** to cover Prague. Lobkowitz denuded the countryside, so that the French could obtain no provisions. Belle-Isle therefore withdrew from Bohemia (December 16–26), leaving a garrison in Prague under General **François de Chevert.** (Early the following year, after a gallant defense of the city, Chevert was allowed to evacuate with all honors of war and to rejoin the main French army.) Meanwhile, Charles VII and an Imperial-Bavarian army under General Seckendorf joined Broglie in the Danube Valley in eastern Bavaria.

1743, April–May. Austrian Convergent Invasions of Bavaria. As Prince Charles advanced up the Danube, Khevenhüller marched into southern Bavaria from Salzburg, and Lobkowitz moved from Bohemia down the Naab. Broglie and Seckendorf could not agree on a defensive plan or strategy. After Khevenhüller defeated Seckendorf at the **Battle of Braunau** (May 9), the French and Bavarians withdrew westward from Bavaria.

1743, May–June. Advance of the Pragmatic Army. King **George II** of England, who was also the Elector of Hanover, had collected a multinational army on the lower Rhine. The principal elements were Hanoverian, English, and Dutch, with contingents from other German allies of Austria and guarantors of the Pragmatic Sanction. This army, about 40,000 strong, advanced slowly up the Rhine and into the Main and Neckar valleys. A French army of 30,000, under Marshal Duke **Adrien M. Noailles,** advanced from the middle Rhine to block the Pragmatic advance, and to protect the withdrawal of Broglie's command. The armies of George and Noailles approached each other in the Main Valley, between Hanau and Aschaffenburg. Noailles, far more skillful than George II, soon had the Pragmatic army virtually blockaded in the Main River defiles.

1743, June 27. Battle of Dettingen. In a battle typical both of English sturdiness and Hanoverian royal stubbornness, George extricated himself from the dangerous situation into which he had ineptly brought his army. At the outset a French cavalry charge came close to overwhelming the allied left wing. When his horse bolted and tried to gallop off the field, George dismounted and, sword in hand, led his English and Hanoverian infantry in counterattack. When the day was over, the allies held the field and Noailles was forced to retreat. This was the last time that an English monarch personally commanded and led his troops on the battlefield. England was still not officially at war with France.

1743, July–October. Operations along the Rhine. Threats by both Prince Charles and King George to invade France were both frustrated by inept attempts to cross the Rhine. All armies went into winter quarters.

1743–1744. French Plans to Invade Britain. An army under Marshal **Maurice de Saxe** was assembled at Dunkirk to go to England with Prince **Charles Stuart** (pretender to the English throne). The movement was foiled by weather and the Royal Navy (see p. 697).

1744, April. French Declaration of War. France now officially entered the war. A French army of 90,000, under the personal supervision of Louis XV, prepared to invade the Spanish Netherlands. Saxe and Prince Charles Stuart were still at Dunkirk, but Saxe was soon shifted to take operational command of the main army under Louis. Another army, under Marshal **François Coigny,** collected on the middle Rhine opposite Prince Charles. Still another under Prince **Louis François de Bourbon-**

Conti prepared to join the Spanish against the Austrians in northern Italy (see p. 696).

1744, June–August. Operations along the Rhine. Charles seized the initiative, crossed the Rhine near Philippsburg (July 1), and advanced on Weissenburg, cutting Coigny off from his base in Alsace. Louis XV, who had barely begun his invasion of the Austrian Netherlands, now left Saxe in Flanders and moved southeast with half of his army into Lorraine, while Coigny in a series of confused running fights fought his way through the Austrian Army at Weissenberg and reestablished himself near Strasbourg. At this point, with Charles threatened by the convergence of the 2 French armies, Louis XV became ill and all French activity ceased.

Second Silesian War, 1744–1745

1744, August. Frederick Reenters the War. Concerned by the completeness of the Austrian and allied victories of 1743, and realizing that Maria Theresa still hoped to regain Silesia, Frederick made an alliance with Louis XV.

1744, August–September. Prussian Invasion of Bohemia. Believing Charles to be in danger in northern Alsace, Frederick suddenly advanced with 80,000 men in 3 columns through Saxony, Lusatia, and Silesia into Bohemia, converging on Prague, which he quickly besieged and captured (September 2–6). He at once moved due south on Budweis, from whence he could threaten the Danube Valley and Vienna.

1744, September–November. Austrian Reaction. Inspired by Maria Theresa, new units were raised in Austria and Hungary, and regular garrisons hastily assembled to cover the approaches to Vienna. Marshal **Otto F. von Traun,** in combination with Saxon forces, held Frederick in check, while Prince Charles marched hastily east across the Rhine toward Bohemia. Frederick, finding himself opposed by the entire might of Austria and her allies, while his French allies sat on their hands, began a reluctant retreat into Silesia. Meanwhile, in the west, Louis recovered, secured the Rhine Valley, then returned to Flanders, where his army went into winter quarters.

1744, December 27. Death of Charles VII. He was succeeded as Elector of Bavaria by his son, **Maximilian Joseph,** who decided not to contend for the Imperial election.

1745, January 7. Battle of Amberg. The Austrians, in a surprise invasion of Bavaria, caught the Bavarian army in winter quarters and in a few weeks overran most of the country (January–March).

1745, January 8. The Quadruple Alliance. At Warsaw, Austria, Saxony, England, and the Netherlands united themselves formally against France, Bavaria, and Prussia.

1745, April 22. Treaty of Füssen. Peace between Austria and Bavaria. Maximilian renounced all pretensions to the Austrian crown and promised his vote to Francis Stephen, husband of Maria Theresa, in the coming Imperial elections. In return, Austria restored all of his possessions to the young elector. Frederick was now completely isolated from his only allies, the French, who seemed interested only in operations in Flanders.

FONTENOY CAMPAIGN, 1745

1745, May. French Advance in Flanders. Marshal Saxe (illegitimate son of Augustus II of Saxony) led a French army of about 70,000 in an advance on Tournai, which the French invested (May 10). The army was accompanied by Louis XV and the Dauphin. Opposed to Saxe was the Pragmatic Army, about 50,000 strong, commanded by the young **William Augustus, Duke of Cumberland** (son of George II), who moved slowly to relieve Tournai.

1745, May 10. Battle of Fontenoy. Leaving part of his army to continue the siege, Saxe drew up about 52,000 in position at Fontenoy to block the allied advance. Dominating the French line of entrenchments were 3 hastily constructed redoubts. Saxe was so ill with the dropsy that he had himself carried to the field on a litter, but refused to relinquish command.

The allies mounted an unimaginative frontal attack shortly after dawn, but were soon halted by the French defensive line. After some inconclusive skirmishing, Cumberland decided to attempt to smash his way through the center of the French position with a force of about 15,000 infantry, drawn up in 3 lines, the first 2 English (mainly consisting of the Guards Brigade), the third Hanoverian. While skirmishing continued along the line, Cumberland led these troops toward the center of the French line, between one of the redoubts and Fontenoy. As the ponderous mass, an almost square column, approached the French lines, Cumberland halted them to dress ranks and reorganize prior to a final assault. Lt. Col. Lord **Charles Hay,** commanding the 1st Foot Guards (later the Grenadier Guards), walked out between the lines; the French infantry and artillery fire slackened and came to a virtual halt. Facing the French Guards, Hay pulled out a flask, drank a toast, shouted a polite taunt, then saluted, led his troops in 3 hearty cheers, and dashed back to his own lines. (He almost certainly did not invite the French Guards to fire first, as apocryphal legend tells us.) As the amazed French were returning the salute and cheers, a tremendous volley was fired from the English line, which then resumed the advance, smashing through the shattered first French line. Louis XV, urged to flee, refused and stood fast as panicking soldiers ran past him. Saxe roused himself from his litter, mounted his horse, and established a second line which, in an exchange of smashing volleys, brought the English attack to a halt. Prominent among the French units in this new line was the Irish Brigade, refugees and supporters of the Stuarts. Saxe now brought up artillery, and soon a combined artillery-infantry-cavalry assault smashed the great English-Hanoverian square. The survivors withdrew in small groups, stubbornly returning the French fire and repulsing pursuing French cavalry. Cumberland withdrew his army in good order. Allied losses were reported as 7,500 killed and wounded, but the assaulting column alone must have lost that number; French losses were 7,200 killed and wounded.

1745, May–September. French Conquest of Flanders. Following up his victory, Saxe took Tournai, Ghent, Bruges, Oudenarde, Ostend, and Brussels.

HOHENFRIEDBERG CAMPAIGN, 1745

1745, April–May. Skirmishing in Silesia. Austrian and Hungarian irregular troops harassed the Prussian army. Frederick concentrated 60,000 men at Frankenstein, between Glatz and Neisse. At the same time Charles marched across the passes from Bohemia with an army of about 80,000. He concentrated at Landshut, in the mountains of western Silesia, threatening Breslau. Frederick at once marched north to Striegau.

1745, June 3. Austrian Advance to Hohenfriedberg. Charles, not realizing that Frederick had left southern Silesia, marched toward Breslau and camped near Hohenfriedberg. After dark Frederick marched quickly and secretly to Hohenfriedberg, and before dawn drew his army up in order of battle.

1745, June 4. Battle of Hohenfriedberg. At dawn the Prussians struck, completely overwhelming the Austrians and Saxons. By 8 A.M. the battle was over. The allies lost more than 9,000 killed and wounded, about 7,000 prisoners, and 66 guns. Frederick's total losses were barely 1,000 men.

1745, June–September. Pursuit into Bohemia. As the Austrian and Saxon refugees fled into Bohemia, Frederick pursued aggressively with about half of his army, leaving the remainder to deal with the Austrian and Hungarian irregulars in the south. In Bohemia, Charles was reinforced, and soon reorganized his shaken troops. He was reluctant to risk another battle with Frederick, however, and the Prussian king did not have enough strength to force the Austrians to fight. After 3 months of inconclusive maneuvers along the upper Elbe in northeastern Bohemia, Frederick, whose army had gradually shrunk to 18,000 men, began to withdraw to Silesia, followed by Charles, who had about 39,000 men. Frederick halted at Sohr (Soor).

1745, September 30. Battle of Sohr. Expecting an Austrian attack, Frederick formed his army for battle, only to find that Charles had outmaneuvered him, and in a surprise night march had seized the heights to the right rear of his army. The Prussian line of retreat was cut. Frederick immediately swung his army in a great right wheel, under heavy Austrian fire. In the middle of this maneuver the Prussian pivot suddenly advanced against the Austrian left wing. The result was an oblique formation which suddenly threw itself on the Austrians, who had no thought that Frederick would be foolhardy enough to attack. Overwhelming Prussian strength smashed the Austrian left. The dazed Austrians retreated to the northwest, leaving the passes to Silesia open. Frederick's losses were 3,876, while the Austrians lost 7,444 killed and wounded, and abandoned 22 guns to the Prussians. Frederick and his army then returned deliberately to Silesia.

1745, October–November. Allied Invasion of Prussia. Collecting reinforcements, Charles marched north into the territory of his Saxon allies, while another Austrian army from the west joined the main Saxon army under Marshal **Rutowski** in western Saxony. The two allied forces then slowly advanced toward Berlin. Frederick immediately marched from Silesia into Saxony toward Dresden, forcing Charles to halt and face eastward.

1745, November 24–25. Battle of Katholisch Hennersdorf and Görlitz. Before the Austrian general could concentrate his scattered marching columns, they were struck 2 sharp blows by Frederick on successive days. Charles retreated back into Bohemia.

1745, November–December. Operations on the Elbe. Meanwhile, another Prussian army under elderly Leopold of Anhalt-Dessau (the "Old Dessauer") marched up the Elbe from Magdeburg to meet Rutowski and his allied army. Rutowski, though considerably superior in numbers, took up a defensive position between Meissen and Dresden.

1745, December 14. Battle of Kesselsdorf. Leopold, marching against the allied defensive position in battle order, surprised and over- whelmed Rutowski. The allies retreated in disorder, after suffering heavy losses.

1745, December 25. Treaty of Dresden. This unbroken series of Prussian victories, following so soon after Saxe's successes in Flanders and the withdrawal of the English army to deal with civil war at home (see p. 699), caused Maria Theresa to seek immediate peace. She again recognized Frederick's conquest of Silesia, while he recognized the election of her husband as Emperor. (This had taken place on September 13.)

Final Campaigns in the Netherlands, 1746–1748

1746. French Successes in the Austrian Netherlands. Saxe continued his deliberate operations against the major fortresses of the Austrian Netherlands, taking Antwerp, then clearing the region between Brussels and the Meuse. Charles of Lorraine, who had been sent to oppose the French, made only one major effort to interfere, and was defeated in the **Battle of Raucoux** (Rocourt), near Liége (October 11). The armies then went into winter quarters, on opposite sides of the Meuse.

1747. French Invasion of Holland. Saxe was now opposed by the Prince of Orange and by the Duke of Cumberland, who had returned from England after suppressing the insurrection of "the '45" (see p. 699). Saxe met and defeated the allied army at the **Battle of Lauffeld,** near Maastricht (July 2). He then sent a corps under General Count **Ulrich F. V. de Lowendahl** to besiege and capture Bergen-op-Zoom (September 18), while with his main body he virtually isolated Maastricht before the armies again went into winter quarters.

1748. Concluding Operations. While a large Russian army marched across Germany to join the allies in the Netherlands, Maastricht fell to Saxe's assault (May 7). There were no further operations of importance while the diplomats concluded peace negotiations.

1748, October 18. Treaty of Aix-la-Chapelle. With a few exceptions, all conquests were restored on both sides. In Italy, Parma, Piacenza, and Guastella were ceded to Spanish Prince **Don Philip.** Also reaffirmed were Prussian conquest of Silesia, the Pragmatic Sanction in Austria, and the retention of both the electorate of Hanover and the English throne by the House of Hanover.

Operations in Italy, 1741–1748

1741–1743. Desultory Warfare. Spanish and Neapolitan armies, hoping to conquer the Austrian Duchy of Milan, were generally outmaneuvered by Austrian Marshal Traun. The Austrians were joined by Sardinia (Savoy, 1742), while at the same time the Neapolitans were forced to return home to protect their country against British amphibious threats.

1744. Intensification of the War. The tempo of the war speeded up as the French army in Dauphiné under the Prince of Conti attempted to join the main Spanish army on the lower Po River under General Count **John B. D. de Gages.** King **Charles Emmanuel I** of Sardinia, with considerable Austrian assistance, succeeded in holding up the French advance. The French besieged **Cuneo,** where they were joined by a Spanish force. Charles Emmanuel, attempting to relieve Cuneo, suffered a series of defeats at **Villefranche** and **Montalban** (April), **Peyre-Longue** (July 18), and in the major **Battle of Madonna del Olmo** near Cuneo (September 30). Conti, however, was repulsed in his efforts to take Cuneo, and retired into Dauphiné for winter quarters. The Austrians, now under Lobkowitz, had driven the main Spanish army southward out of northern Italy, but were halted and defeated by a combined Spanish-Neapolitan army in the inconclusive **Battle of Velletri** (August 11). Lobkowitz withdrew northward, but the Spanish were too weak to follow him closely. The year ended with the Austrians and Sardinians firmly in control of the Po Valley.

1745, March. Genoa Enters the War. She joined France, Spain, and Naples. De Gages and a French army under Maillebois joined forces south of Piacenza. Outmaneuvering Lobkowitz, the allies decisively defeated the Sardinians at the **Battle of Bassignano** (September 27), then quickly captured Alessandria and other Po fortresses.

1746, April–May. The Pendulum Swings. Reinforcements from Austria, released by the peace with Frederick, again permitted the Austrians, now under Marshal Count **Maximilian U. von Browne,** to regain the initiative.

1746, June 16. Battle of Piacenza. After a series of extremely confused maneuvers, in which Maillebois distinguished himself, the combined French and Spanish army, under Infante Don Philip, engaged the Austrians and Sardinians in an inconclusive battle. Interference from Don Philip prevented Maillebois from winning; as it was, the French and Spanish armies were separated and forced to retreat.

1746, August 12. Battle of Rottofreddo. Maillebois repulsed the pursuing Austrians, and made good his escape to Genoa.

1746, September. Austrian Conquest of Genoa. The arrival of Browne's army caused Maillebois to retreat along the Riviera to France; the Austrians took the city.

1746, December 5–11. Genoese Uprising. The Austrians were driven out. Communications were briefly restored with France, but the city was besieged by an Austrian army early the next year.

1747–1748. French Initiative. The French, now under Marshal Belle-Isle, advanced across the Maritime Alps and relieved besieged Genoa (July 1747). The Austrians withdrew into Lombardy, followed by Belle-Isle. There were no further operations of importance. When the war ended, the Austrians still held most of the Duchy of Milan.

Operations at Sea

1739–1741. War of Jenkins's Ear. (See p. 723.) This colonial war in the Caribbean was soon engulfed in the larger war.

1739, November. Vernon Captures Porto Bello. (See p. 723.)

1740. Vernon Repulsed at Cartagena. (See p. 723.)

1740, September–1744, July. Anson's Cruise. A squadron under Commodore **George Anson** was dispatched to raid the Spanish Pacific coast possessions. Only his flagship, the **Centurion,** succeeded in getting around Cape Horn. In the 2 following years Anson ravaged the west coast of the Americas, then sailed west to intercept and capture the famed Manila Galleon. He then continued westward, circumnavigating the globe.

1742–1744. Blockade of Toulon. Although France and England were not yet officially at war, the British maintained a loose blockade of Toulon, where a combined French and Spanish fleet lay idle. The principal mission of the English blockaders was to prevent the Spanish from reinforcing their land forces in Italy by seaborne convoys.

1744, February 11. Battle of Toulon. The Franco-Spanish fleet sailed south from Toulon under Admirals **de la Bruyère de Court** and **Don José Navarro.** The British fleet, under Admiral **Thomas Matthews,** approximately equal in strength to the French and Spanish, sailed to intercept. Matthews believed that de Court was trying to lure him away from the coast, so that a Spanish troop convoy could take reinforcements to Italy. He ordered an immediate attack without waiting to form his fleet in line ahead, since he feared that otherwise de Court would get away. His orders were not clearly understood; his principal subordinate, Admiral **Richard Lestock** probably deliberately went out of his way not to understand. The result was an inconclusive fight, in which the French and Spanish inflicted somewhat more damage than they received. Matthews was courtmartialed and dismissed from the Royal Navy for having failed to adhere to its rigid and formal "Fighting Instructions." Lestock, who should have been convicted of gross disobedience of orders was acquitted through political influence.

1744, March–April. French Threats in the Channel. A convoy of French troops at Dunkirk was ready to sail for England (see p. 692) and a French fleet of 20 ships of the line on at least 2 occasions sailed from Brest, and then from Cherbourg, to convoy the troops to invade England. Admiral Sir **John Norris,** age 84, stayed at sea in the Channel with 25 British ships, despite wintry gales, and the French did not attempt to fight their way through.

1745. Expedition to Louisburg. (See p. 723.)

1746, July 25. Battle of Negapatam. French Admiral **Mahé de la Bourdonnais** outmaneuvered Commodore **Edward Peyton** in an inconclusive action in which the English suffered more damage than the French. Peyton sailed away, permitting the French to capture Madras. (See p. 714.)

1747, May 3. First Battle of Finisterre. Anson with 14 ships intercepted a French convoy escorted by a French squadron of 9 ships under Admiral **de la Jonquière.** In a bold, brilliant action, the English fleet captured every French warship and several of the convoy.

1747, October. Second Battle of Finisterre. Under similar circumstances, Admiral **Edward Hawke** with 15 ships encountered a French convoy heading for the West Indies, escorted by a squadron of 9 commanded by Admiral **de l'Étenduère.** The odds were less than they might seem, because 8 of the French ships were bigger, better, and faster than any of the British. Hawke, fighting a bold battle, largely ignoring the "Fighting Instructions," captured all but 2 of the French warships. The convoy escaped, heading for the West Indies. But Hawke sent a sloop to warn the British West Indies fleet to be on the lookout; most of the convoy was captured in the Leeward Islands.

1748, October. Battle of Havana. Admiral **Charles Knowles** defeated a Spanish fleet under Admiral **Reggio** in a close-fought but relatively inconclusive engagement.

WESTERN EUROPE

GREAT BRITAIN

In this half-century modern Great Britain emerged as a political entity, and as the predominant world power. Also during this period the last important English dynastic struggle came to a close, and with it the end of serious threats to the internal security of Great Britain. The principal events were:

1699, February. Disbanding Act. Parliament, to limit involvement of William III in European wars, reduced the standing army to a total of 7,000 men.

1701, June 12. Act of Settlement. Succession to the crown (after **Anne,** second daughter of James II) was established by Parliament to **Sophia,** Princess of Hanover (granddaughter of James I) and her issue. All future sovereigns were to be Protestant (thus barring James II and his male, Catholic, issue) and (in rebuke to William) were not to leave the country without Parliament's permission, were not to involve England in war for defense of their foreign possessions, and were not to grant office to foreigners. The act asserted the right of ministers to be responsible for the actions of the sovereign.

1701, September 16. Death of James II in France. He was succeeded as the Stuart Pretender to the English throne by **James Edward,** known to English history as the "Old Pretender." He was recognized as King of England by Louis XIV.

1701–1713. War of the Spanish Succession. (See p. 676.)

1702, March 8. Death of William III. Since Mary, his wife, was dead (December 28, 1694), he was succeeded by Mary's younger sister, Anne.

1702–1714. Reign of Anne. Last Stuart sovereign of England.

1707, May 1. Union of England with Scotland. This established the United Kingdom of Great Britain.

1707, October 21. Shovell's Disaster. Six ships of his squadron returning from Toulon (see p. 685) crashed on a Scilly Islands reef. Shovell went down with his flagship and 2 other ships of the line.

1708, March. Landing of James Edward in Scotland. He was disappointed both in the lack of a spontaneous rising in his favor and by the failure of a French expeditionary force to follow him, although part of the French fleet did reach the Firth of Forth before being scattered by a storm. He returned to France.

1710, August. Fall of the Whig Ministry. Tory enemies of Marlborough came to power.

1711, December. Dismissal of Marlborough. The new Tory Parliament acted for political reasons as well as on not completely disproved charges of financial peculation on the part of the Duke while captain general on the Continent. He was replaced by the Duke of Ormonde as commander of British land forces in Europe, but for all practical purposes the new Parliament withdrew the nation from the war (see p. 684).

1713, April 11. Treaty of Utrecht. (See p. 684.) Louis XIV renounced the Pretender (who left France) and recognized the Protestant succession in England. He also dismantled (temporarily) Dunkirk's fortifications, and ceded to England French American possessions of Hudson Bay, Acadia, Newfoundland, and St. Kitts. Spain ceded Gibraltar, and also gave Britain limited slave-trading rights in the Spanish-American colonies.

1714, August 1. Death of Anne. She was succeeded by **George,** Elector of Hanover, eldest son of Sophia.

1715, September–1716, February. "The Fifteen." Jacobite rebellion in Scotland, led by

John Erskine, Earl of Mar, who raised an army of 4,000 Scotsmen.

1715, November. Battles of Preston and Sheriffmuir. While Mar fought an inconclusive battle at Sheriffmuir (November 13) with loyal troops under **John Campbell, Duke of Argyll,** other regular troops under General **Wills** recaptured the barricaded town of Preston from a Jacobite force commanded by Tory M.P. **Thomas Forster,** who surrendered after a desultory three-day resistance (November 12–14).

1715, December. Arrival of James Edward. The Pretender landed at Peterhead and started south, but the advance of Argyll caused him to retreat to Montrose and then, as his dispirited Highlanders dispersed, he sailed again for France (February 5, 1716).

1718–1720. War of the Quadruple Alliance. (See p. 686.)

1719, March–June. Spanish Expeditions to England and Scotland. (See p. 686.)

1727–1760. Reign of George II.

1727–1729. War with Spain. (See p. 700.)

1739–1748. War with Spain (War of Jenkins's Ear). For all practical purposes this colonial and naval war became a part of the War of the Austrian Succession (see p. 689.)

1740–1748. War of the Austrian Succession.

"The Forty-five," 1745–1746

Jacobite rebellion in Scotland, inspired against the advice of his followers by young Prince **Charles Edward Stuart** ("Bonnie Prince Charlie" or "the Young Pretender"), son of James Edward.

1745, July 13. Charles Sails from Nantes. He arrived, practically alone, in the Hebrides (August 3).

1745, August–September. Rising of the Clans. With a Highland army of about 2,000, Charles marched on Edinburgh. Tactical command was exercised by able Lord **George Murray.**

1745, September 17. Capture of Edinburgh. Charles occupied the city, and Holyrood Palace, but the English garrison, under General **Joshua Guest** held Edinburgh Castle.

1745, September 20. Battle of Prestonpans. Charles and Murray decisively defeated a British army of about 3,000 under General Sir **John Cope.**

1745, November–December. Invasion of England. Charles and Murray, with about 5,000 men, hoped both for a rising of English Stuart sympathizers, and for direct aid from France. Charles was disappointed on both counts. He did, however, capture Carlisle and Manchester, and reached Derby (December 4). Many of his Highlanders had meanwhile deserted, and 2 strong English armies were advancing against him. Charles retreated north (December 6), closely followed by the Duke of Cumberland, who had returned from France (see p. 695).

1745, December–1746, February. Siege of Stirling. Cumberland having stopped at the border due to bad weather, Charles vainly tried to capture Stirling.

1745, December 18. Battle of Penrith. Jacobite rear-guard victory over an English detachment, while Charles' main army invested Stirling.

1746, January 17. Battle of Falkirk. English General **Henry Hawley,** who had recaptured Edinburgh, advanced to try to raise the siege of Stirling, but was defeated by Charles and Murray.

1746, February–March. Inconclusive Operations in Northern Scotland.

1746, April 8. Cumberland Advances from Aberdeen. With 9,000 regulars Cumberland advanced on Inverness, where Charles had about 5,000 men (practically all Highlanders).

1746, April 16. Battle of Culloden. Charles and Murray, attempting to surprise Cumberland, made a night march, but found the English ready for them at dawn. Under heavy artillery fire, the tired Highlanders nonetheless attacked, but were repulsed by the steady English infantry, then cut down and routed by cavalry charges. Charles fled. The Highlanders lost about 1,000 killed and about 1,000 were captured. Most of the prisoners were killed out of hand, or later summarily executed, earning for Cumberland the sobriquet of "Butcher."

1746, April–September. Charles a Fugitive. Although his cause had been smashed forever at Culloden, Charles stayed in Scotland, spurned by most of his formerly ardent supporters, who were now terrorized by Cumberland's ruthless executions of most of the leaders who had taken part in the rebellion. Charles finally returned to France (September 20).

FRANCE

1701–1714. War of the Spanish Succession. (See p. 676.)

1715, September 1. Death of Louis XIV.

1715–1774. Reign of Louis XV (5 years old; great-grandson of Louis XIV). The early years of the reign (1715–1723) were under the regency of Duke **Philippe of Orléans,** nephew of Louis XIV, who reversed his uncle's policies and entered alliances with England and Holland.

1718–1720. War of the Quadruple Alliance. (See p. 686.)

1727–1729. War with Spain. (See below.)

1733–1738. War of the Polish Succession. (See p. 687.) France gained Lorraine.

1740–1748. War of the Austrian Succession. (See p. 689.)

THE IBERIAN PENINSULA

Spain

1700, November 1. Death of Charles II. Without issue, the king's death precipitated the War of the Spanish Succession.

1700–1724. First Reign of Philip V.

1701–1715. War of the Spanish Succession. (See p. 676.)

1718–1720. War of the Quadruple Alliance. (See p. 686.)

1724. Abdication of Philip. The king abdicated in favor of his son, **Louis I,** apparently to pursue his efforts to gain the French crown.

1724–1746. Second Reign of Philip V. He returned to the Spanish throne when Louis died.

1727–1729. War with England and France. An almost bloodless war resulted from refusal of England and France to permit Charles (son of Philip) to go to Italy to take over the duchies to which he had received succession rights in the Treaty of The Hague (see p. 687). Spanish troops besieged Gibraltar; minor naval engagements took place in the West Indies. Due to the mediation of Cardinal Fleury of France, however, overt hostilities ended almost immediately (May 1727) and peace negotiations soon began (March 1728).

1729, November. Treaty of Seville. France and England again agreed to succession of Prince Charles to the Italian duchies, as in the Treaty of The Hague. Spain recognized British possession of Gibraltar.

1731. Charles Inherits the Farnese Duchies (Parma, Piacenza, and Tuscany).

1733–1738. War of the Polish Succession. (See p. 687.) Philip took advantage of Austrian occupation in Central Europe and the Balkans to expand his family's dominions, and his own interests, in Italy.

1733. Spanish Conquest of Sicily and Naples. Charles was crowned King of the Two Sicilies. By treaty with Austria, he relinquished his 3 duchies in order to obtain recognition of his new crown (1735). Philip and Charles agreed that the crowns of Spain and the Two Sicilies would never be united.

1738, November 13. Treaty of Vienna. (See p. 688.) The Austrian-Spanish agreements on Italy and Sicily were ratified.

1739–1741. The War of Jenkin's Ear. (See p. 723.)

1740–1748. War of the Austrian Succession. (See p. 689.)

1746–1759. Reign of Ferdinand VI (third son of Philip). He brought needed and welcome peace to a country which had been at war during 40 of the 46 years of his father's reign.

Portugal

Save for unsuccessful participation in the War of the Spanish Succession, this was a period of relative peace and great prosperity for Portugal. The prosperity came from a combination of expanding trade with Britain as a result of the Methuen Treaty (see p. 684) and the discovery of great natural wealth in Brazil.

THE NETHERLANDS

In the War of the Spanish Succession the Dutch Republic declined from great-power status. The war exhausted the country physically and financially. The principal events were:

1702, March 19. Death of William III. The States General refused to accede to William's will; for nearly 45 years there was no Stadholder (1702–1747).

1702–1713. War of the Spanish Succession. (See p. 676.)

1743–1748. War of the Austrian Succession. The Dutch played a minor role militarily in the early part of the conflict. (See p. 689.)

1747. French Invasion of the Netherlands. (See p. 695.) In desperation, in the face of Saxe's advance, the Dutch people rose in a bloodless revolution against the States General and installed **William IV** of Orange, a distant cousin of William III, as stadholder. Simultaneously the war was drawing to a close with the **Treaty of Aix-la-Chapelle,** which practically ignored the Netherlands (see p. 696).

ITALY

As in previous centuries, Italy was again a battleground of the neighboring great powers. The decline of Venice and Genoa continued, while Savoy—which became the Kingdom of Sardinia—began to emerge as the strongest and most aggressive of the many Italian principalities. The principal events were:

1675–1730. Reign of Victor Amadeus II of Savoy and Sardinia. The beginning of the steady rise of the House of Savoy that would lead, more than a century later, to the unification of Italy.

1701–1713. War of the Spanish Succession. (See p. 676.)

1707. Austrian Occupation of Naples. Spanish occupying garrisons were defeated and ejected.

1713. Treaty of Utrecht. (See p. 684.) Savoy received formerly Spanish Sicily; Victor Amadeus was recognized as King of Sicily. Austria received Naples and most other Spanish possessions in Italy.

1718. Treaty of Passarowitz. (See p. 703.) Venice lost the Morea to Turkey, but gained bases in Albania and Dalmatia.

1718–1720. War of the Quadruple Alliance. (See p. 686.) Savoy remained virtually neutral although, despite Spanish seizure of Sicily, Victor Amadeus concluded an alliance with Spain.

1720. Treaty of The Hague. (See p. 687.) Savoy surrendered Sicily to Austria in return for Sardinia; Victor Amadeus changed his title to King of Sardinia.

1730–1773. Reign of Charles Emmanuel of Savoy and Sardinia.

1730–1768. Endemic Revolt in Corsica. Local Corsican chieftains became virtually independent of Genoa, which finally requested and received French assistance in reconquering the island, which Genoa then sold to France (1768).

1733–1738. War of the Polish Succession. (See p. 687.) Austria ceded Naples and Sicily to the Spanish Bourbons (1735; see p. 700).

1742–1748. War of the Austrian Succession. (See p. 689.)

SWITZERLAND

Switzerland played little part in international events of this period. Swiss mercenaries were still an important source of foreign exchange for the country, and fought on both sides in the War of the Spanish Succession. But no longer did the Swiss troops have an influence on the outcome of battles disproportionate to their numbers. Internally, the country was still rent by Protestant-Catholic conflict, which gradually declined by the middle of the century.

GERMANY

Central Europe was still a mass of small, unimportant, and relatively ineffectual principalities. To a greater extent than previously, however, a coalescence of German interests and rudimentary nationalism had begun. This was to a considerable extent inspired by a common hatred of France and the French, arising largely from the repeated invasions of Germany by the armies of Louis XIV and Louis XV.

Austria (or the Empire)

The House of Hapsburg became increasingly identified with the still-expanding Austrian Empire, which it controlled directly from its capital at Vienna. At the same time the Hapsburgs' nominal responsibilities as emperors of the almost meaningless Holy Roman (or German) Empire became less important. Expansion continued in the Balkans at the expense of the Ottoman Empire. Although the military and political fortunes of the Hapsburgs were mixed elsewhere in Europe, Austrian power and influence reached their zenith during the reign of Maria Theresa. The principal events were:

1658–1705. Reign of Leopold I.

1701–1714. War of the Spanish Succession. (See p. 676.)

1703–1711. Revolt in Hungary. Hungarian nationalists, resentful of Austrian control over their country, revolted under the leadership of **Francis II Rakoczy.** The rebels gained control over most of the country, and on several occasions directly threatened Vienna. The Austrians slowly gained the upper hand, however, particularly after Rakoczy was defeated at the **Battle of Trencin** (1708). Guerrilla war continued.

1705–1711. Reign of Joseph I (son of Leopold).

1711–1740. Reign of Charles VI (brother of Joseph). He had been the Hapsburg claimant to the throne of Spain in the War of the Spanish Succession (see p. 676). His accession to the Imperial and Austrian crowns caused virtual withdrawal of England and the Netherlands from the alliance that had been supporting his claims to the Spanish throne (see pp. 683–684).

1711, May. Peace of Szatinar. The Hungarians accepted Charles VI's assurances that Austria would respect Hungary's national rights and liberties. Rakoczy, still coveting the Hungarian throne, fled to Turkey.

1716–1718. War with Turkey. Austria joined Venice, already at war with Turkey (since 1714; see p. 708). With Turkish support, Rakoczy returned to Hungary to lead an insurrection, but failed completely.

1716, August 5. Battle of Peterwardein. Prince Eugene with 60,000 men met and (after a fierce fight) routed a Turkish army of nearly 150,000. The Turks lost 6,000 killed, an unknown number of wounded, and all of their 164 guns. The Austrians lost 3,000 killed and 2,000 wounded.

1716, August—October. Siege and Capture of Temesvár (Timisoara). Eugene captured the last Turkish stronghold in Hungary after a 5-week siege.

1717, July—August. Siege of Belgrade.

Eugene, with about 20,000 men, besieged Belgrade, the strongest Turkish fortification in the Balkans, garrisoned by 30,000 troops. While he was preparing for an assault, a Turkish relieving army of 200,000 approached under Grand Vizier **Khahil Pasha.**

1717, August 16. Battle of Belgrade. While a detachment repulsed an attempted sortie by the garrison, Eugene typically and boldly attacked the main Turkish host. Although wounded (for the 13th and last time), Eugene remained on the field, and his army smashed the Turks, whose casualties were estimated at over 20,000. Austrian losses were 2,000. Belgrade soon surrendered (August 21).

1717–1718. Austrian Successes in the Balkans. Eugene and other Austrian commanders occupied much of Serbia, Wallachia, and the Banat. Eugene was actually contemplating an advance on Constantinople when Turkey sued for peace. Despite Venetian desire to continue the war (which had earlier gone badly for Venice in the Morea), Charles was also anxious for peace in order to respond to Spanish aggression and threats in Sardinia and Italy (See pp. 686–687).

1718, July 21. Treaty of Passarowitz. Austria gained Temesvar (thus completely liberating Hungary), part of Wallachia, and Belgrade. Venice gave up the Morea, but was given new coastal strongholds in Albania and Dalmatia.

1718–1720. War of the Quadruple Alliance. (See p. 686.)

1733–1738. War of the Polish Succession. (See p. 687.)

1737–1739. War with Turkey. (See p. 688.) Most of the gains of Passarowitz were lost.

1740–1780. Reign of Maria Theresa. Hapsburg Queen of Bohemia and of Hungary, and Archduchess of Austria. Her husband later became Emperor (see below), and thus for all practical purposes she was Empress Regnant as well.

1740–1748. War of the Austrian Succession. (See p. 689.)

1742–1745. Reign of Charles VII as Emperor. Charles Albert of Bavaria, principal rival of Maria Theresa, reigned briefly after his election. Had he not died during the war, he almost certainly would have been deposed.

1745–1765. Reign of Francis I as Emperor. Francis of Lorraine, husband of Maria Theresa, founded the House of Lorraine-Hapsburg, the line which followed Maria Theresa.

Prussia

By the end of the 17th century, the Electorate of Brandenburg had become the most powerful state of northern Germany, thanks to 2 strong Hohenzollern rulers: **Frederick William** ("the Great Elector") and his son, **Frederick I,** who crowned himself King of Prussia early in the 18th century. Frederick's son, **Frederick William,** the third strong ruler of the line, increased the strength and influence of Prussia. Frederick William contemptuously considered his eldest son a weakling, but as **Frederick II** ("the Great") that son was destined to make Prussia a great power, establishing the base upon which modern Germany is founded. By the middle of the century, Frederick had taught himself to be one of the greatest soldiers of history in 2 small, successful wars. His greatest achievements, and their consolidation, took place in the latter half of the century. The principal events up to 1750 were:

1688–1713. Reign of Frederick I. He started his reign as Elector of Brandenburg.

1701, January 18. Establishment of the Kingdom of Prussia. Frederick proclaimed himself king.

1701–1713. War of the Spanish Succession. (See p. 676.) Prussian contingents particularly distinguished themselves in Marlborough's great victories: Blenheim, Ramillies, Oudenarde, and Malplaquet. Prussia's status as a kingdom was recognized by the Treaty of Utrecht, signed shortly after Frederick's death (see p. 684).

1713–1740. Reign of Frederick William. He built up the strength of the Prussian Army to 80,000 men, largest per capita in Europe, thanks largely to his parsimony. His men were recruited (or shanghaied) from the dregs of society; his officers were rather reluctantly conscripted from the nobility. By iron discipline and brutal tyranny the monarch (assisted by his able training genius, Prince **Louis of Anhalt-Dessau**—the "Old Dessauer") transformed this unlikely material into the best army in Europe.

1713, January 24. Birth of Karl Frederick. The 4th of 14 children, destined to become Frederick the Great.

1730. Frederick Sentenced to Death. Frederick William decreed that his "weakling" son should be strangled for "desertion" when he ran away from home. He was dissuaded only by threats of mutiny among his courtiers, and by the representations of foreign ambassadors.

1740–1786. Reign of Frederick II. His personality, accomplishments, and genius defy ready or simple analysis. He was a sensitive, cultured intellectual, and at the same time a ruthless, coldhearted disciplinarian. A man of great personal honor, as a monarch he was a sly, treacherous, and untrustworthy foe and ally. He was in many respects a typical 18th-century monarch and a typical 18th-century soldier. He accepted the military system as he found it (unlike Gustavus and Napoleon). But he recognized its tactical weaknesses: slowness, ponderousness, lack of imagination, slow rates of fire. So he became a conservative innovator. He injected mobility, speed, and rapidity of fire, thus roughly doubling his infantry effectiveness. He used cavalry vigorously, like Marlborough and Eugene, but he acted much more quickly than either of them did, particularly in the approach to battle and in the early stages. He was always inferior in strength to his enemies; he always attacked first. He created horse artillery to give increased firepower to his fast-moving cavalry. He emphasized the howitzer for two reasons: its lightness made it more mobile than a gun, its higher trajectory enabled it to get at enemy reserves concealed behind hills. He learned that by speed and agility he could concentrate superior power at a critical point before his more ponderous foes could react effectively. He achieved his mobility and speed by reemphasizing the drill and disciplinary methods inherited from his father.

1740—1748. War of the Austrian Succession. (See p. 689.) Precipitated by Frederick, in the **First Silesian War.** After gaining Silesia, Frederick made peace, then re-entered the conflict in the **Second Silesian War** (see p. 693) to prevent Austria from recovering sufficient strength to oust him from Silesia. Both foes and allies had reason to respect both the military prowess and slippery political double-dealing of the young king.

SCANDINAVIA

Sweden

The undisciplined and rash military genius of Charles XII of Sweden was the direct cause of his nation's rapid decline from great-power status. Treacherously attacked by his neighbors in the Great Northern War, Charles defeated all of them decisively in a series of campaigns that are among the most brilliant in military annals. These great victories did not satisfy his thirst for vengeance, or his ambitions for military honor and conquest. By rejecting his foes' peace overtures, Charles assured his own eventual defeat at the hands of emerging Russia. After his death Sweden made peace—with honor but with great loss of territory and prestige. One subsequent attempt to gain vengeance against Russia led to further losses and decline of influence. The principal events were:

1697–1718. Reign of Charles XII.
1700–1721. The Great Northern War. (See p. 672.)
1718–1720. Reign of Ulrica Eleonora (sister of Charles).

1720–1751. Reign of Frederick I (husband of Ulrica).
1741–1743. War with Russia. (See p. 707.)

Norway and Denmark

Save for participation in the Great Northern War (1700–1721; see p. 672), this half-century was relatively uneventful militarily in Denmark and Norway.

EASTERN EUROPE

POLAND

This was a period of rapid political, military, economic, and moral decline in Poland. Impoverished and exhausted by the wars and uprisings of the 17th century, her population diminished by one-third, Poland found herself surrounded by 3 powerful and wealthy enemies: Austria, Russia, and Prussia. Her central government was weak and ineffective; there were deep distrust and suspicion between the foreign king (Augustus II) and the Poles; there was dissension among the leading magnates; the economy was backward; and the treasury was empty. The imposition by Russia of a second Saxon king (Augustus III) merely made Poland weaker and more divided. The principal events were:

1697–1733. Reign of Augustus II. One of the least attractive figures of world history.
1700–1721. Great Northern War. (See p. 672.) Poland was ruined economically and demoralized politically by this war, during which the nation was a battleground for Saxony, Sweden, and Russia.
1715–1717. Revolt against Augustus II. This rising of the gentry was provoked by huge and arbitrary Saxon requisitions of food and fodder, and by the murder of two Polish officials. Hostilities were terminated by a Russian-supported agreement which was approved without discussion by the "mute" Diet (1717).
1718–1719. Anti-Russian Unrest. Augustus, whose relations with Peter of Russia had deteriorated, concluded an anti-Russian alliance with Austria and England, the so-called **Treaty**

of **Vienna** (1719), providing for enforced evacuation of the Russian troops.
1720. Russian Withdrawal from Poland. Under this pressure the czar withdrew his troops from Poland, but simultaneously concluded in Potsdam the first secret Russo-Prussian alliance.
1721. Treaty of Nystad. (See p. 676.) Without Polish participation, the treaty provided for cession of the eastern coast of the Baltic, including Polish Livonia, to Russia.
1727. Russian Occupation of Polish Courland.
1733–1738. War of the Polish Succession. (See p. 687.)
1734–1763. Reign of Augustus III. Continued decline of Poland under an inept and uninterested foreign king.

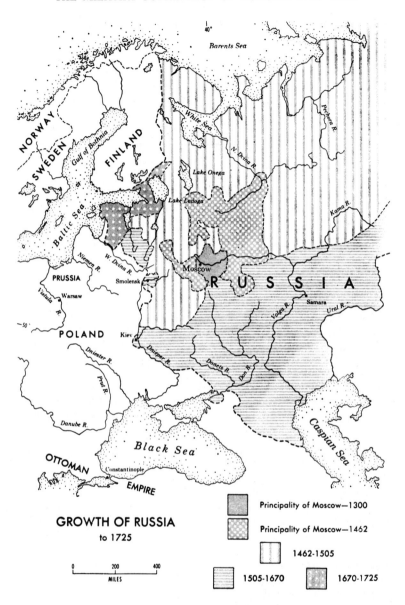

GROWTH OF RUSSIA
to 1725

0   200   400
MILES

Principality of Moscow—1300

Principality of Moscow—1462

1462-1505

1505-1670   1670-1725

RUSSIA

During the first quarter of the 18th century, Russia under Peter the Great rose to be one of the great military and political powers of Europe. Following this came a decline during the reigns of Peter's mediocre successors. The principal events were:

1689–1725. Reign of Peter I. This cruel, ruthless ruler reformed the archaic and Asiatic government, economy, and society of Russia, primarily to assure an adequate civilian base for military conquest. He reformed and modernized the army, and learned the art of war from his archenemy, Charles XII of Sweden. He conquered the eastern coast of the Baltic and relegated Sweden to the position of a second-rate power. He ruined Poland, his other main rival. He built a navy, but failed to gain unrestricted access to the Black Sea. When he died, he left Russia exhausted, her population diminished by 20 percent, but equipped with a regular army of 212,000 veterans hardened in continuous campaigns and reinforced by 110,000 Cossacks, and a strong navy. There was no immediate successor who could use this powerful instrument for further conquests and expansion of the Russian Empire until the reign of Catherine II in the second half of the 18th century.

1700–1721. Great Northern War. (See p. 672.)

1722–1723. War with Persia. (See p. 709.)

1725–1741. Successors of Peter I the Great. There was a series of short reigns of weak rulers, preventing Russia from conducting a consistent and effective foreign policy. They also neglected the Russian military forces. In only one policy was there Russian consistency: keeping Poland weak.

1725–1736. Guerrilla Rebellion in the Caucasus. Native partisans engaged in constant guerrilla warfare against Russia in the lands recently conquered from Persia. Soon one-quarter of the Russian Army was engaged in the Caucasus-Caspian region (1730).

1732–1736. Restoration of Land to Persia. Russian restored to Persia the Caspian littoral, with the towns of Derbent and Baku, because their defense against native partisans became too heavy a drain on the Russian Army. (See pp. 710, 711.)

1736–1739. War with Turkey. (See p. 629.)

1741–1762. Reign of Elizabeth I. One of Russia's strong rulers.

1741–1743. Russo-Swedish War. Under the influence of France, Sweden saw in the War of the Austrian Succession an opportunity to gain revenge on Russia, which had declared its support of Maria Theresa and the Pragmatic Sanction (see p. 689). The timing was bad for Sweden, whose army was only 15,000 men, because after the conclusion of peace with Turkey (1739) the entire Russian Army could be used against Sweden.

1741. The Battle of Wilmanstrand. The Swedes, 6,000 strong, were defeated with great loss by 10,000 Russians. Swedish losses were 3,300 killed and wounded and 1,300 prisoners; Russian losses were 2,400 killed and wounded.

1742. Russian Invasion of Finland. The Russians cut the road of retreat of the main Swedish Army at Helsinki. The Swedes (17,000) surrendered, virtually ending the war.

1743, August 7. Peace with Sweden. Russia obtained some new territory in Finland, where the frontier was fixed on the Kymmene River.

EURASIA AND THE MIDDLE EAST

THE OTTOMAN EMPIRE

The decline of the Ottoman Empire continued, but at a slower pace than in the previous century. In part this was due to the fact that England and France, separately and without coordination, were becoming aware of the possible threat to their Mediterranean interests, and to the European balance of power, if either Russia or Austria were to gain control of the Straits and of the eastern Mediterranean. Thus French and English diplomacy—while not pro-Turkish, and rarely deliberately coordinated—was directed toward maintaining the *status quo* in the Balkans and Anatolia. Another reason for the deceleration of Ottoman decline was the jealousy of Austria and Russia, each afraid that the other might displace Turkey in control of the Balkans and of the Straits. Perhaps the most important reason for Turkey's continuing, even though weakened, power was the inherent toughness and fighting qualities of the Turkish soldier, displayed on battlefield after battlefield against Austrians, Russians, Venetians, and Persians. The principal events were:

1703. Janissary Revolt. Sultan Mustafa II was forced to abdicate.

1703–1730. Reign of Ahmed II (brother of Mustafa II).

1710–1711. War with Russia. (See p. 679.) Ended by favorable **Treaty of the Pruth.**

War with Venice and Austria, 1714–1718

1714. Revolt in Montenegro. Quelled by the Turks, who believed it had been instigated by Venice.

1714. Turkey Declares War on Venice

1715. Turkish Offensives. Energetic and skillful Grand Vizier **Damad Ali** conquered the Peloponnesus (Morea) in a 100-day campaign of successful sieges without any pitched battles. All Venetian fortresses were taken, their 8,000 defenders captured. The Ottoman fleet, reinforced by ships from Egypt and the Barbary States, drove the Venetians out of the Aegean Islands. The remaining Venetian fortresses on Crete were taken.

1716, January–December. Venetian Gains. The Venetians repulsed a Turkish attack on Corfu and received naval reinforcements from Spain, Portugal, and some other Italian states.

1716, January. Austria Joins Venice. The grand vizier marched north from Belgrade toward Peterwardein (June–July).

1716, August 5. Battle of Peterwardein. (See p. 702.)

1717. Belgrade Campaign. (See pp. 702–703.)

1718, July 21. Treaty of Passarowitz. (See p. 703.) Unfavorable to Turkey.

1722–1727. Expansion in Persia. Both Turkey and Russia took advantage of Persian weakness and involvement with an Afghan invasion to seize and divide between them large portions of northwestern Persia (see p. 709). Division of the spoils was agreed on by the **Treaty of Constantinople** (1724).

1730–1736. War with Persia. (See p. 710.)

1730, September–October. Revolt in Constantinople. News of the disastrous Turkish defeat at Hamadan (see p. 710) led to a popular rising. The grand vizier was killed; the sultan abdicated. The city continued to be plagued by unrest and riots for 2 years.

1730–1754. Reign of Mahmud I.

1733, 1736. Peace of Baghdad. (See pp. 710, 711.) Ended the inconclusive war with Nadir Shah of Persia; Turkey lost some of the land seized earlier.

1736–1739. War with Austria and Russia. (See p. 688.) Ended by the favorable **Treaty of Belgrade.**

1743–1747. War with Persia. (See p. 711.) Turkey's last major war with Persia. Turkey was forced to give up all the lands seized 20 years earlier.

PERSIA

In this short half-century Persia virtually collapsed, conquered by Afghan invaders and her provinces divided at the conference table between Russia and Turkey. Then briefly and astonishingly Persia revived for one of her most glorious military episodes, before another disastrous collapse. The period was dominated by **Nadir Kuli Beg,** later **Nadir Shah,** one of the greatest soldiers in Persian history and the last great Asian conqueror. He drove out, then conquered, the Afghan invaders, repeatedly defeated the Turks, overawed the Russians, conquered Mogul

India, and conquered the Uzebek khanates of the Oxus region. His empire collapsed at his death. The principal events were:

1694–1722. Reign of Shah Husain. His strong emphasis on Shi'i doctrine aroused resentment among Sunni subjects and neighbors, reopening the wounds of ancient Moslem religious differences.

1709. Afghan Revolt at Kandahar. Under the leadership of **Mir Vais,** leader of the Ghilzai Afghans, a separate Afghan state established its independence. Repeated Persian efforts to reconquer Kandahar were repulsed (1709–1711). The struggle was embittered by the religious dispute.

1711. Siege of Kandahar. A Persian army under **Khusru Khan** defeated Mir Vais and besieged Kandahar. Mir Vais was ready to surrender, but Persian insistence upon unconditional surrender stimulated the defenders to greater efforts. When the besiegers began to suffer from food shortages, Mir Vais sortied and defeated them. Khusru was killed; only 1,000 of the 25,000 Persians escaped. Subsequent Persian reconquest efforts were repulsed.

1715. Death of Mir Vais. He was succeeded by his brother **Abdulla,** and then by his son **Mahmud Khan** (1717).

1717. Afghan Revolt at Herat. Under the leadership of **Asadullah Khan,** the Abdali Afghans revolted. They joined with the Uzbeks to plunder Khorasan.

1719. Battle of Herat. A Persian army of 30,000 under **Safi Kuli Khan** attempted to reconquer Herat. Asadullah met them with 15,000. In a hard-fought battle, Asadullah was saved by confusion in the Persian army, which resulted in the defeat and capture of Safi.

1720. Afghan Raid into Persia. Mahmud Khan (son of Mir Vais) invaded Persia and captured Kerman. The Afghans were routed when capable Persian general **Lutf Ali Khan** surprised their camp and pursued them back to Kandahar. Lutf was preparing for a full-scale invasion when he was dismissed due to the jealousy of the grand vizier, his brother-in-law.

1721. Frontier Raids. The Abdalis raided un-

checked in Khorasan, while the Lesgians of Daghestan sacked Shamaka, capital of Shirvan. Mahmud prepared for a major invasion of Persia.

1722, January. Mahmud's Invasion. Advancing with about 20,000 men, he again captured Kerman, but was repulsed from its citadel. Moving on to Yezd, he was again repulsed. He continued on to the walls of the capital, Isfahan. He rejected a handsome ransom to return to Afghanistan; with about 10,000 men he encamped at Gulnabad, 11 miles east of Isfahan, and waited for a higher offer.

1722, March. Battle of Gulnabad. A Persian army of about 30,000 now advanced against Mahmud. The Afghans attacked, defeated the Persians, and, after capturing some outlying fortifications, invested Isfahan.

1722, March—October. Siege of Isfahan. Mahmud repelled, or bribed off, a few half-hearted efforts to relieve the city. The Shah's son **Tahmasp** fought his way out with 600 men, but in the city garrison and people began to starve. The shah surrendered, abdicating in favor of Mahmud.

1722–1725. Reign of Mahmud. The Afghan conqueror controlled only part of the country. His accession was challenged by Tahmasp, who raised resistance in Mazandaran.

1722–1727. Russian and Turkish Aggression. Peter of Russia, nominally supporting Tahmasp, seized Derbent (1722), and then Resht and Baku (1723). In return for promised Russian support, Tahmasp agreed to cede Shirvan, Daghestan, Gilan, Mazandaran, and Astrabad to the Russians. Turkey, anxious to regain the provinces lost to Shah Abbas, and fearful of Russian penetration, also moved into northwest Persia, seizing Tiflis, capital of Georgia (1723). By the **Treaty of Constantinople** (1724), Russia and Turkey agreed to divide northern and western Persia. The Turks then occupied Tabriz, Hamadan, and Kermanshah (1724–1725).

1724–1725. Chaos in Isfahan. Mahmud

became insane and was eventually put to death by his followers.

1725–1730. Reign of Ashraf Shah (cousin of Mahmud). Order was reestablished in central and southern Persia.

1726–1727. Turkish Invasion. Ashraf repulsed the invaders at the **Battle of Isfahan** (1726). He then made peace, allowing the Turks to retain the extensive border provinces they had seized in return for Ottoman recognition of his accession (1727).

1726–1729. Rise of Tahmasp. Supported by **Nadir Kuli Beg,** an obscure Khorasan chieftain, Tahmasp conquered Meshed and Herat (1728). Marching on Isfahan, he and Nadir defeated Ashraf at the **Battle of the Mehmandost,** near Dourghan, and again at the **Battle of Murchakhar,** near Isfahan, and then seized the capital (1729). Ashraf fell back on Shiraz, pursued by Nadir, who decisively defeated the Afghans at the **Battle of Zarghan** (1730). In trying to return to Kandahar, Ashraf was murdered.

1730–1732. Reign of Tahmasp. The shah was a figurehead; the real ruler of Persia was brilliant soldier Nadir Kuli Beg, who was made viceroy of most of northern and eastern Persia by the grateful shah.

1730. Conquest of Meshed. Nadir marched on Meshed, where an independent Afghan chieftain, **Malik Mahmud,** had established himself (1722) and now refused to submit to the new shah. Nadir defeated Malik outside the city, which he then captured through treachery. Malik was captured and later killed.

1730–1736. War with Turkey. The Turks refusing to give up the Persian territories they had seized, Nadir decided to fight to recover the lost provinces. He marched on Hamadan.

1730. Battle of Hamadan. Nadir decisively defeated the Turks. He quickly occupied Iraq and Azerbaijan, and laid siege to Erivan.

1731–1732. Campaign against the Abdalis. With Nadir now occupied in the west, there were outbreaks in Khorasan, largely inspired and exploited by the Abdali Afghans of Herat. Nadir raised the siege of Erivan, marched 1,400 miles east, defeated the Abdalis, invested and captured Herat (1732).

1731. Second Battle of Hamadan. Tahmasp, hoping to complete the reconquest of the western provinces from Turkey during the absence of Nadir, took the field himself and invested Erivan. He withdrew when a Turkish relief army appeared. The Turks pursued and decisively defeated him. Tahmasp, now having lost all the territories conquered by Nadir the previous year, made peace to forestall a Turkish invasion, recognizing the Turkish conquests.

1732. Tahmasp Deposed. Nadir, returning from the east, deposed Tahmasp, installed his 8-month-old son **Abbas** on the throne, repudiated the treaty with the Turks, and renewed the war.

1732. Treaty of Resht. The Russians gave up their claims to Gilan, Mazandaran, and Astrabad (see p. 707).

1732–1736. Reign of Abbas III. Nadir, as regent, was the real ruler in this reign, the last of the Safawids.

1733. Nadir's Invasion of Mesopotamia. Nadir defeated the Turkish governor, **Ahmed Pasha,** near **Baghdad.** He invested the city with a portion of his army, then turned to meet a powerful Turkish relief army under **Topal Osman.** At the **Battle of Karkuk,** near Samarra, the Turks decisively defeated Nadir's smaller Persian army. The garrison of Baghdad then sortied to overwhelm the small Persian detachment Nadir had left in the trenches around the city. Despite these twin disasters, Nadir rallied his troops and held off Topal while awaiting reinforcements. When these arrived he again took the offensive, defeated Topal at the **Battle of Leilan,** near Karkuk, then marched again on Baghdad. News of a revolt in Fars led him to make a favorable peace with Ahmed Pasha (**Treaty of Baghdad**) and he then returned to suppress the revolt in Fars.

1734–1735. Operations in Transcaucasia. Turkish Sultan Mahmud refused to recognize the Treaty of Baghdad. A Turkish army of 80,000 under **Abdulla Koprulu** assembled near Kars. On the approach of Nadir the Turks went into an entrenched camp. Nadir threatened Tiflis, Erivan, and Ganja, but was unable to entice the Turks into battle.

1735. Treaty with Russia. Nadir, knowing

Russia was planning war against Turkey in Europe, sent troops against Baku and Derbent, threatening to join Turkey in war against Russia. The Russians quickly negotiated an alliance with Nadir, returning Baku and Derbent to Persia. (See p. 707.)

1735. Battle of Baghavand. Nadir invaded Turkey. When he sent off a number of detachments from his main army in the plain of Baghavand, near Kars, Koprulu decided to risk battle. By so doing, he fell into a trap set by Nadir, who counterattacked and won a complete victory. Nadir then quickly captured Tiflis, Erivan, and Ganja.

1736. Peace with Turkey. Sultan Mahmud, anxious to make peace so that he could deal with Russia and Austria, recognized the terms of the Treaty of Baghdad; Persia recovered part of the lost provinces. Nadir agreed not to join Russia in the new war.

1736. Death of Abbas. Nadir was elected shah by the Persian chieftains. A devout Sunni, he accepted only on condition that the Shi'i heresy be abandoned. The condition was nominally accepted, but had little practical effect.

1736–1747. Reign of Nadir Shah. He devoted himself to restoring the old frontiers of Persia.

1737–1738. Invasion of Afghanistan. Nadir Shah took Kandahar after a 9-month siege (1738). During the siege Nadir sent smaller forces to occupy and pacify the former Persian provinces of Balkh and Baluchistan. When Kandahar surrendered, he treated the Afghans leniently, and many joined his army.

1738–1739. Invasion of India. To punish the Mogul Emperor, **Mohammed Shah,** for aiding the Afghans, Nadir invaded India. He captured Ghazni and Kabul (September 1738), then turned toward the Punjab. With 50,000 men he bypassed a Mogul army holding the Khyber Pass, instead went over the nearby Tsatsobi Pass (where Alexander had crossed), and then turned to attack the Moguls in the rear at the successful **Battle of the Khyber Pass.** He then advanced rapidly into India, seizing Peshawar and Lahore, crossing the Indus to Attock.

1739, February. Battle of Karnal. Mohammed marched from Delhi to meet the Persians with an army of 80,000 men. Nadir, enticing the Moguls to battle from their entrenched camp, defeated them, then besieged the camp. Mohammed surrendered. Nadir, treating him well, occupied Delhi.

1739, March. Massacre of Delhi. Nadir ruthlessly suppressed a rising of the population. Then, leaving Mohammed on the throne, but taking an indemnity of more than $100 million in precious metals and jewels and annexing all of India west and north of the Indus, he returned to Persia.

1740. Conquest of Bokhara and Khiva. Defeating the Uzbeks at the **Battles of Charjui** and **Khiva,** Nadir annexed the region south of the Aral Sea.

1741. Failure in Daghestan. Nadir attempted to subdue a rising of the Lesgians, but was frustrated when the people took to the mountains in guerrilla war.

1742–1747. Growing Unrest. Nadir's cruelty, his poor administration, and his efforts to force unpopular religious beliefs on his people caused bitterness and hatred to spread through the country.

1743–1747. War with Turkey. Ostensibly for religious reasons, but mainly to exploit the growing unrest in Persia, Turkey again invaded Persia. Nadir, outmaneuvering the more numerous invaders, blocked their advance east of Kars.

1745. Battle of Kars. Nadir once more decisively defeated the Turks, and occupied most of Armenia.

1747. Peace with Turkey. The boundaries were reestablished as they had been at the time of Turkish Sultan Murad (1640).

1747. Death of Nadir. He was assassinated by his own bodyguard.

1747–1750. Anarchy in Persia. The empire fell apart in civil war. A new Afghan state was created by **Ahmad Khan Durani,** one of Nadir's leading generals. Georgia declared its independence.

SOUTH ASIA

AFGHANISTAN

During this century, modern Afghanistan emerged from the ruins of the Persian and Mogul empires. For most of the period the history of Afghanistan was inextricably bound up with that of Persia. But upon the death of Nadir Shah, Ahmad Khan Durani, a young Afghan of the Abdali tribe who had joined Nadir Shah after the Persian conquest of Kandahar (1738) and had

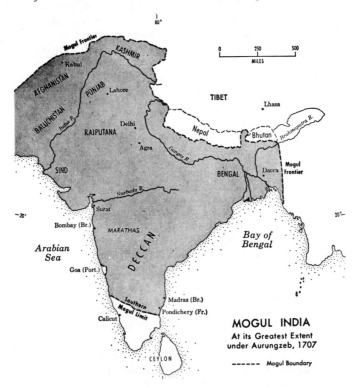

MOGUL INDIA

At its Greatest Extent
under Aurungzeb, 1707

------ Mogul Boundary

become a Persian general, seized control of most of the region now known as Afghanistan. Known as Ahmad Shah Durani (or Abdali), he quickly began to expand his small kingdom at the expense of both Persia and India. Having been with Nadir during the invasion of India, he was well aware of the weakness of the Mogul Empire. He had no sooner seized control of Afghanistan than he made the first of his 10 major invasions of India. This first invasion (1747) was unsuccessful, but in later years he became the strongest ruler in northern India, and annexed much of the remains of the Mogul Empire (see pp. 711, 715).

INDIA

During this half-century the Mogul Empire declined rapidly. The decline was hastened by Nadir Shah's successful invasion, and by the steady advance of Maratha power from the Deccan into northern India. As the period ended there were 4 major powers competing for a dominant place in India to supplant the nearly extinct Mogul Empire. These were the Maratha Confederacy, the Afghan empire of Ahmad Shah, the French, and the English. The principal events were:

c. 1700. Rise of the Sikhs. A militant Hindu sect, they became powerful in Rajputana and the Punjab.

1700–1707. Maratha Struggle for Existence. The Marathas, with difficulty, maintained their independence against the last powerful efforts of aged Mogul Emperor Aurangzeb (see pp. 643–644).

1700. Consolidation of British East India Company Bases in Bengal. Sir **Charles Eyre** became governor.

1701. Expansion of French East India Company. A new base was established at Calicut. French bases were already established at Surat, Pondichery, Masulipatam, Chandernagore, Balasore, and Kasimbazar.

1707. Death of Aurangzeb.

1707–1712. Reign of Bahadur Shah (son of Aurangzeb). He assumed the throne in the absence of his elder brother, **Muazim,** governor of Kabul. He made peace with the Marathas and Rajputs, and repressed the rising ambitions of the Sikhs.

1708. Battle of Agra. Bahadur defeated and killed his brother Muazim, who was returning to claim the throne.

1708–1748. Reign of Sahu, the Maratha. Son of Sivaji, he had been imprisoned by Aurangzeb. He was released by Bahadur and set on the Maratha throne. His allegiance to the Moguls soon became less than nominal.

1710. Campaign against the Sikhs. Bahadur, concerned by the rise of Sikh power, defeated them in an inconclusive campaign.

1712–1719. Mogul Interregnum. After the death of Bahadur the situation in the empire became anarchic; in a series of bloody struggles for the throne, some 5 puppets reigned in Delhi, but the real power was in the hands of the nobles and court officials.

1712. Appointment of the First Maratha Peshwa. Sahu appointed able **Balaji Visvanath Bhat** as his prime minister, with title of Peshwa. From this time until his death (1720) Balaji was the virtual ruler of what had become a confederacy of the Maratha kingdoms of Baroda, Gwalior, Indore, Nagpur, and the Peshwa's dominions. The position of Peshwa became hereditary, and his successors continued to be the real leaders of the Maratha Confederacy.

1712–1720. Spread of Maratha Power. The Marathas gained strength and spread northward into Hindustan, taking advantage of anarchy in the Mogul Empire.

1713. Nizam ul-Mulk Viceroy of the Deccan. During the struggle in Delhi, **Asaf Jah** (also known as Chin Kilich Khan), a Turkoman who had been one of Aurangzeb's best generals, quietly had himself appointed as subahdar, or viceroy, of the Mogul dominions in the Deccan, with the title of **Nizam ul-Mulk.** Not troubling the autonomous Marathas, he reestablished firm Mogul control over the central and eastern Deccan.

1719–1748. Reign of Mohammed Shah as Mogul Emperor. He was supported by Nizam ul-Mulk and the Marathas in gaining the throne. The Nizam became the wazir, or prime minister, but remained subahdar of the Deccan (1720).

1720–1740. Rule of Peshwa Baji Rao Bhat over Marathas. The Maratha Confederacy continued its expansion northward and eastward at the expense of the Mogul Empire.

1721. French Base Established at Mahé.

1724–1748. Reign of Nizam ul-Mulk over Hyderabad. With the Mogul Empire obviously approaching collapse, Nizam ul-Mulk left

Delhi (1722) and established the independent Kingdom of Hyderabad in his Deccan viceroyalty. He appointed **Dost Ali** as nawab, or governor, of the Carnatic.

1737. Battle of Delhi. Baji Rao and the Marathas defeated Mohammed Shah's imperial army outside of Delhi. The emperor made peace, ceding Malwa to the Marathas. In the following years they also seized Gujerat, Orissa, and Bundelkhand.

1739. Nadir Shah's Invasion. (See p. 711.) This accelerated the decline of the Mogul Empire.

1739. Baji Rao Takes Bassein from the Portuguese.

1740. Bengal Independent of the Mogul Empire.

1740–1761. Rule of Peshwa Balaji Baji Rao Bhat over Marathas.

1741. Marquis Joseph Dupleix Appointed French Governor General. He had been governor of Chandernagore; he moved to Pondichery.

1743. Death of Dost Ali. The Nizam appointed **Anwar-ud-din** the new Nawab of the Carnatic, refusing the claim of **Chanda Sahib,** Dost Ali's son-in-law.

First Carnatic War, 1744–1748

This was part of the War of the Austrian Succession (see p. 689). Word of hostilities in Europe having reached India, Dupleix sought neutrality between the French and British East India Companies.

1745. Arrival of a British Fleet. Commodore **Curtis Barnett** swept French shipping off the nearby seas. Dupleix sent an urgent plea for help to Count **Mahé de la Bourdonnais,** an able French admiral, commanding at Mauritius.

1746, June. Arrival of la Bourdonnais. A fleet of 8 ships of the line, carrying 1,200 French troops, arrived at Pondichery.

1746, July 25. Naval Battle of Negapatam. (See p. 697.) Commodore **Edward Peyton** (who had succeeded to command after Barnett's death) was outmaneuvered and driven away by la Bourdonnais. Peyton sailed to Hooghly.

1746, September 2–10. Seizure of Madras. Dupleix invested the main British base, while la Bourdonnais blockaded it by sea. Madras surrendered after brief resistance.

1746, September 21. Battle of Madras. Nawab Anwar-ud-din, who had allied himself with the British, arrived near Madras with a large army. The French sortied from the captured city and easily defeated the Nawab's army.

1746, November 3. Battle of St. Thomé. A French detachment of 230 Europeans and 730 sepoys (native troops) attacked and routed a force of 10,000 of the Nawab's troops near Madras.

1746, November–1748, April. Siege of Fort St. George. The French tried unsuccessfully for 18 months to take the British base near Madras. Dupleix finally had to raise the siege due to the arrival of a new British fleet, under Admiral **Edward Boscawen,** with reinforcements.

1747. Ahmad Shah's First Invasion of India. The Afghans were halted at the inconclusive **Battle of Sirkind** by a mixed Mogul-Rajput force, and forced to retreat to Afghanistan. (See p. 712.)

1748, August—October. Siege of Pondichery. Boscawen unsuccessfully tried to take Pondichery, ably defended by Dupleix. Onset of the monsoon forced the British fleet to withdraw.

1748. Treaty of Aix-la-Chapelle. (See p. 695.) News of the treaty reached India at the end of the year. Madras was returned to Britain, in return for Louisbourg in Nova Scotia.

1749. Ahmad Shah's Second Invasion of India. This was a combined raid and reconnaissance in force, but led him to believe that he could conquer the Punjab and Kashmir.

Second Carnatic War, 1749–1754

Despite formal peace between France and Britain, hostilities between the 2 East India companies continued, through their involvement in Indian conflicts. Following the death of Nizam ul-Mulk (1748), Chanda Sahib began to conspire for the nawabship of the Carnatic, and was supported by Dupleix. At the same time, **Muzaffar Jang,** grandson of Nizam ul-Mulk, had received the blessing of the new and ineffectual Mogul emperor to succeed as the Nizam of Hyderabad, instead of his father **Nasir Jang,** who had automatically succeeded to the throne as Nizam. Dupleix also supported Muzaffar Jang. The British East India Company, under the able leadership of **Thomas Saunders,** supported the incumbents: Nasir Jang, the new Nizam; and Anwar-ud-din, the Nawab of the Carnatic.

1749. Battle of Ambur. The combined forces of Chanda, Muzaffar, and the French under the Marquis **Charles de Bussy** defeated and killed Anwar-ud-din. Chanda was proclaimed Nawab of the Carnatic, with Dupleix the actual ruler. However, **Mohammed Ali,** claiming to be the successor to Anwar-ud-din, retained control of Trichinopoly, where he was recognized and supported by the British.

1749–1750. Nasir Jang in the Carnatic. With English support, Nasir Jang invaded the Carnatic in a partially successful effort to reestablish Hyderabad control.

1750, December. Assassination of Nasir Jang. Following the death of his father (for which he was probably responsible), Muzaffar Jang seized power as the new nizam. He was proclaimed Subahdar of the Deccan by Dupleix. In return he appointed Dupleix governor of all former Mogul lands in the eastern Deccan from the Krishna River to Cape Comorin. Dupleix put his small but efficient army, commanded by de Bussy, at the disposal of the new Nizam. (For operations after 1750, see p. 764.)

EAST AND CENTRAL ASIA

CHINA

The Manchu Dynasty of China reached its height of power, prestige, and influence during this period. During the latter years of his reign, the great Emperor K'ang-hsi consolidated the Manchu conquest of China and Mongolia, and continued the expansion of the empire into Tibet. His grandson, Ch'ien-lung, continued the expansion of China into neighboring areas, and by midcentury had become the most powerful monarch of China since Kublai Khan. The principal events were:

1662–1722. Reign of K'ang-hsi. (See also p. 649.)

1700. Chinese Troops to Tibetan Frontier. Chinese troops occupied parts of the border region between Tibet and China because of K'ang-hsi's concern about the growing influence in Tibet of the Eleuth (or Dzungar) Mongols (part of the Oirat Mongols).

1705. Chinese Intervention in Tibet. Chinese troops installed a new Dalai Lama, against the opposition of most Tibetans.

1716–1718. Dzungar Intervention in Tibet. A Dzungar force of 6,000 men, under Galdan's nephew, **Chewanlaputan** (Tsewang Araptan), invaded Tibet to join in the continuing dispute over the succession to the Dalai Lama. They seized Lhasa and imprisoned the incumbent Dalai Lama. A Manchu force, coming to his assistance, was ambushed and destroyed (1718).

1720. Manchu Conquest of Tibet. K'ang-hsi sent 2 armies into Tibet, 1 from Kansu and 1 from Szechwan. They smashed the Dzungars and drove out the survivors. Meanwhile another Chinese army advanced into Dzungaria to capture Urumchi and Turfan. This was the first war in which Mongol forces made extensive use of musketry; they were not very effective, however, against the larger, better-armed, better-equipped Chinese troops. A new and more popular Dalai Lama was installed by K'ang-hsi, and a Manchu garrison was left in Lhasa.

1721. Revolt in Formosa. Rebels, led by **Chu I-kuei,** were quickly overwhelmed by Chinese troops.

1723–1735. Reign of Yung-cheng (son of K'ang-hsi). China was peaceful, but there was considerable conflict along the western frontiers with Central Asian tribes.

1727. Treaty of Kiakhta. A treaty with Russia, delineating frontiers and regulating trade and other relations.

1727–1728. Civil War in Tibet. A Chinese army of 15,000 restored order. The Dalai Lama was exiled for 7 years. The Chinese garrison was increased.

1729–1735. Campaigns against the Dzungars. Continuing Dzungar opposition to Chinese policy in Tibet and frontier depredations in western China resulted in a series of Chinese punitive expeditions. At first, largely due to poor management, the Chinese were not very successful. After reorganization (1732), the Dzungars were soundly defeated.

1736–1795. Reign of Ch'ien Lung (or Kâo Tsung). Height of Manchu power and prestige.

1747–1749. Tibetan Frontier Campaigns. Renewed unrest along the Tibetan frontier was suppressed in hard, inconclusive partisan warfare.

INNER ASIA

The Russian and Chinese empires continued their expansions into Inner Asia. Apparently oblivious of what was happening to them, the Mongol and Turkic tribes continued their interminable bitter quarrels with each other, in the process weakening themselves and their neighbors so that they could not offer any truly sustained opposition to the colonial encroachment from east and west. The principal events were:

1714–1717. Russian Expedition against Khiva. Peter the Great sent the expedition, expecting that Khiva would accept a protectorate. Instead, the Khivans ambushed and annihilated the Russian expedition.

1720. Chinese Expedition in Dzungaria. (See above.)

1722. Peace of Astrakhan. A meeting between Peter of Russia and the aged Torgut Khan **Ayuka** confirmed the submission of the Torguts to Russia.

1727. Treaty of Kiakhta. Russia and China (see above).

1729–1735. Chinese Campaigns in Dzungaria. (See above.)

1730–1731. Kazakh Acceptance of Russian

Suzerainty. Russia thereby gained control of tribes inhabiting areas claimed by Khiva and Bokhara.

1738. Conquest of Balkh by Persia. (See p. 711.)

1740. Conquest of Khiva and Bokhara by Persia. (See p. 711.)

1741. Discovery of Bering Strait. This was by **Vitus Bering,** a Danish navigator in Russian service.

1747. Khiva Regains Independence. This was the result of the collapse of Nadir Shah's empire (see p. 711).

JAPAN

Under the Tokugawa Shogunate, Japan continued to exclude herself from practically all contact with the outside world, while internal peace reigned within the kingdom.

SOUTHEAST ASIA

BURMA

This was a period of weakness and decline in Burma. The country was ravaged by raids from weak and small neighbors (Manipur and Arakan), as well as by unrest and revolt at home. There was no leadership capable of dealing with these problems. The principal events were:

1714–1749. Manipuri Raids. Under Raja **Gharib Newaz** of Manipur, repeated cavalry raids into Upper Burma devastated the country. The Manipuris swept to the very gates of Ava (most notably, 1738), plundering and pillaging at will.

1740–1752. Mon Revolt. The Mons, or Talaings, of Lower Burma, seeing the weakness and ineptitude of the kings of Ava, revolted to set up an independent kingdom. They seized Toungoo, original home of the ruling family, and raided frequently into Upper Burma as far as the walls of Ava. Finally, under the leadership of King **Binnya Dala,** they began a systematic conquest of Upper Burma, ending in the capture of Ava and overthrow of the Toungoo Dynasty (1752; see p. 767).

SIAM

Stability marked this period in Siam. There was one major foreign war, there was one major civil war, and there were the usual sort of border bickerings which were practically continuous in Southeast Asia. In relative terms, however, it was a quiet period. The principal events were:

1717. Invasion of Cambodia. (See p. 718.) Although 1 of 2 Siamese armies was destroyed, the expedition was eventually successful.

1733–1758. Reign of King Boromokot. This began with a violent struggle between Boromokot, younger brother of previous King **T'ai Sra,** and the son of the dead king. Eventually successful, Boromokot ruthlessly revenged himself by killing all of his principal enemies, then ruled wisely and well.

LAOS

Laos was rocked by violence and civil war at the outset of the century. This soon led to the division of the kingdom into the 2 hostile kingdoms of Luang Prabang and Vientiane. Although a condition of war was endemic in the country, there were no outstanding events.

VIETNAM

This was a period of internal truce in Vietnam between the Trinh and Nguyen families. The Nguyens took advantage of this internal peace to extend the power and territory of Vietnam to the south and west at the expense of Cambodia and Laos. The major military activity centered around Vietnamese ambitions in Cambodia, resulting in 3 major wars: 1714–1717, 1739, and 1749 (see below).

CAMBODIA

During the first part of this period Cambodia was the victim of expansionist aims of Vietnam and of Siam. In the process Vietnam annexed a considerable part of Cambodia. Later, as a measure of strength and stability was restored, Cambodia attempted to regain her lost provinces, but was unsuccessful, and in the subsequent prolonged fighting was forced to cede more territory to Vietnam. The principal events were:

1714–1716. Civil War and Vietnamese Intervention. In a struggle for succession, King **Prea Srey Thomea** was driven from his throne by his uncle, **Keo Fa,** who had the support of a Vietnamese army and a small Laotian force. Prea Srey Thomea fled to Siam to request help from King T'ai Sra (1709–1733).

1717. Siamese Invasion. Two large Siamese expeditionary forces entered Cambodia. The southern force was disastrously defeated at the **Battle of Bantea Meas,** mostly because of a panic apparently caused by the destruction of the supporting Siamese naval force in a storm at sea. The northern expedition, however, defeated the Cambodians and their Vietnamese allies in a series of engagements and reached Udong, Keo Fa's capital. Keo Fa thereupon offered his allegiance to Siam. This was accepted, and Prea Srey Thomea's cause was abandoned. Meanwhile, however, the Vietnamese had annexed several small border provinces of Cambodia in the Mekong River region.

1739–1749. War with Vietnam. A Cambodian army attempted to regain the coastal region of Ha-tien, which had been taken by Vietnam during the previous war. The Cambodians were repulsed. In retaliation, Vietnamese forces again invaded Cambodia. Further parcels of Cambodian territory were annexed in the Mekong River area.

MALAYA

A period of confused warfare in the Strait of Malacca area, involving the principality of Siak (in Sumatra) and the Malay states of Johore, Kedah, Perak, and Selangor. The Dutch avoided involvement. Taking advantage of internal troubles in Johore, and of warfare between Siak and Johore, the martial Buginese pirates (see p. 654), who had established a pirate maritime state in the Riau Archipelago, soon established their ascendancy over Johore, Kedah, Perak, and Selangor. The principal Bugi leader was **Daing Parani,** who was ably supported by 4 equally daring and ruthless brothers. By the end of the period, however, the Buginese were facing the possibility of trouble with the Dutch, who found the growing Buginese power and maritime interests a threat to their own position in Indonesia and in the Malacca Strait.

INDONESIA

During this period the Dutch established virtually complete control over Java, and also expanded their holdings elsewhere in the islands of the archipelago. The principal events were:

1704–1705. First Javanese War of Succession. Finding their trade, and their position in Batavia, threatened by a dispute within the royal family of Mataram (which still held most of eastern and central Java), the Dutch intervened, deposed an unfriendly monarch, and placed a puppet on the throne. They also extended the area of their dominion in western Java and annexed the eastern half of Madura.

1706–1707. Dutch War with Surapati. This Balinese soldier of fortune had carved out a kingdom for himself in the northeastern portion of Java, near Surabaya, at the expense of Mataram. He had played an important role in the intrigues which had led to the First Javanese War of Succession. His overthrow and death gave virtual control of northeast Java to the Dutch.

1712–1719. Continued Turmoil in Northeast Java. Followers and successors of Surapati tried to eject the Dutch, but were finally wiped out.

1719–1723. Second Javanese War of Succession. Another struggle for the succession to the throne of Mataram caused chaos on Java and again led to Dutch intervention and subsequent territorial expansion.

1740–1743. Chinese War. This was a rebellion, started by Dutch persecution of Chinese merchant residents of Batavia and other Javanese cities, the initial incident being a panicky massacre of Chinese in Batavia. The Chinese survivors fled to the country and, linking up with Javanese dissidents, stirred up a major rebellion. The Dutch finally suppressed this, and annexed the entire north coast of Java and the remainder of Madura.

1743–1744. Rebellion in Madura. Madurans, who had fought beside the Dutch in the Chinese War, expected thereby to gain their independence. When this was not forthcoming, they revolted, but were suppressed after hard fighting.

1749–1757. Third Javanese War of Succession. A conflict was stirred up when Baron **Gustaaf van Imhoff,** Dutch governor general, became unnecessarily involved in an internal dynastic feud in the remnant of Mataram. This was the most bitter of all the Java wars, and was complicated for the Dutch by a simultaneous revolt in western Java. The Dutch were defeated by **Mangku Bumi,** the opposing Mataram ruler, at the **Battle of the Bogowonto River** (1751) and the Dutch commander, **De Clercq,** was killed. Determined fighting and internal discord among the Javanese finally brought Dutch victory.

AFRICA

NORTH AFRICA

During this period Tripolitania for all practical purposes became independent of Turkey. Tripolitania, Morocco, and Algeria (and Tunisia to a lesser extent) maintained themselves by piracy and by shipborne and overland slave trade. The heyday of Barbary piracy was beginning to pass, however, due to active and aggressive opposition of the European Mediterranean powers.

Morocco

Morocco was by far the most powerful and most extensive of the North African states at the beginning of the century. Under the strong and able rule of Sultan **Mulay Ismail** ("The Bloodthirsty"), Moroccan influence and territorial claims extended in all directions across the Sahara Desert, reaching as far as the Sudan (1672–1727). After Mulay Ismail's death, however, the country was thrown into turmoil by struggles among the oligarchical military caste which he had created. Moroccan power and influence practically disappeared beyond the Atlas Mountains.

Algeria

This was a period of Algerian decline. Central rule was weak, and European naval activity reduced the activity and profits of the corsairs. Oran was captured from Spain (1708), which was then busily engaged in wars north of the Mediterranean, but was recaptured by Spanish forces a few years later (1732). Algeria was still nominally a Turkish possession.

Tunisia

Tunisia was the first of the Barbary States to turn from piracy to normal maritime trade relations. This trend was begun by **Hussein ben Ali,** the leading soldier of Tunisia, who made himself bey, overthrew the last of the deys, and virtually ended Turkish rule (1705). He entered into trade treaties with the main European Mediterranean powers, but this led to opposition, and his nephew, **Ali Pasha,** rebelled, seized power, and beheaded Hussein. During the subsequent reign of Ali Pasha (1710–1756) relatively peaceful relations with Europe were retained, but with less cordiality than during the brief reign of Hussein.

Tripolitania

Tripolitania became virtually independent from Turkey (1714) under the leadership of **Ahmed Pasha Karamanli.** Nominal tribute to Constantinople continued, paid out of receipts from piracy.

EGYPT

Under the Mamelukes, Egypt, like the three North African lands farther west, became virtually independent of Turkey, though continuing nominal allegiance. It was a half-century of intrigue and conflict between the 2 major Mameluke factions, the Kasimites and the Fikarites. By the middle of the century one of the greatest of Egypt's Mameluke rulers, **Ali Bey,** was fighting his way to a position of preeminence in Egypt (1750).

WEST AFRICA

The slave trade, which had begun in the previous century, expanded rapidly to meet the requirements for agricultural labor in the Americas. Many coastal states and tribes raided inland for slaves, which they exchanged for European weapons and trade goods. The political situation along the Niger remained largely unchanged, but among major events elsewhere were:

c. 1700. Lunda Domination. The Lunda Kingdom (see p. 656) became the dominant power in southern Zaire and northeastern Angola, supported partly by access to European firearms.

c. 1712–c. 1720. Unrest in Ashanti. Factional strife followed the death of Osei Tutu, and was ended when **Opuku Ware** gained the throne.

c. 1712–1755. Reign of Mamari Kulibali of Segu. Known as "the Commander," this Bambara ruler resuscitated the Bambara Segu Kingdom (see p. 656). He created a professional army and river navy, using them to conquer Timbuktu and Djénné in the northeast, and Bamako in the southwest. He also fought off an attack by the King of Kong (c. 1730).

c. 1720–1750. Reign of Opuku Ware. This able monarch led Ashanti expansion into the northern interior, and defeated the Akim tribe.

1724–1727. Abomey Conquests. King Agaja of Abomey (see p. 656) conquered Allada (1724) and Whydah (1727) on the Guinea coast, to ensure access to European arms. The enlarged kingdom was called Dahomey.

c. 1726–1730. Oyo War with Dahomey. The Oyo Kingdom, grown wealthy from control of trade routes to the interior, used its riches to support a large army, and defeated Dahomey, whose expansion had worried Oyo.

1738–1748. Second Oyo-Dahomey War. The Oyo again defeated Dahomey, and compelled payment of tribute until 1840; peace lasted for several decades.

SOUTH AFRICA

The converging advances of the Boers from the southwest and the Bantus from the north and east continued. By midcentury these 2 dynamic and aggressive peoples were beginning to meet in an arc extending from Orange River Valley in the north and west to the coast more than 550 miles northeast of Capetown. Although the Boers had been engaged in continuing minor military pacifications against Hottentots and bushmen in the interior, there had been no major clash with the Bantus before midcentury.

EAST AFRICA

The Sultanate of Oman maintained its hegemony over coastal East Africa, fending off feeble Portuguese efforts to restore their supremacy. The Omanis grew wealthy from the slave trade, which catered to markets in India, Egypt, and the Persian Gulf.

c. 1700. Masai Migration. The Masai, a pastoral warrior tribe of Nilotic stock, began to move south and east into modern Kenya and Tanzania, conquering or displacing existing Bantu inhabitants (and some Galla). This migration continued into the late 19th century.

1727. Pate Alliance with Portugal. The trading city-state of Pate (near modern Lamu) allied with the Portuguese to throw off Omani domination.

1728–29. Portuguese Recapture Mombasa. With the help of Pate, Portuguese forces recaptured Mombasa (1728), but were driven out by the Mombasans themselves the next year.

1730–1755. Reign of Iyasu II. This Ethiopian monarch warred with the Funj of Sudan, but his armies suffered a costly defeat at **Sennar.**

1746. 'Ali ibn Uthman al-Mazrui. This Mombasan leader of Omani descent led an uprising that drove the Omanis from Mombasa. Soon after, he seized Pemba and would probably have captured Zanzibar but for his premature death. The Omanis regained control of Mombasa and Pemba soon after.

THE AMERICAS

NORTH AMERICA

Queen Anne's War, 1702–1713

This was the American version of the War of the Spanish Succession (see p. 676).

1702, December. Sack of St. Augustine. Hostilities were begun in a destructive raid by a combined force of Carolinians and Indians.

1703–1704. Operations along the Southern Frontier. Carolinian-Indian attacks destroyed Spanish missions in the Apalachee region; however, Choctaw resistance prevented raids on French communities in the Gulf area.

1703–1707. Operations in the Northeast. Abenaki and French bands conducted raids from Maine to the Connecticut Valley. Five hundred New Englanders were dispatched to Acadia in an attempt to disrupt Abenaki supply lines and to capture fisheries (1704). English-colonial forces twice vainly attempted to take the main French base at Port Royal, Nova Scotia (1704, 1707). Farther north, a French and Indian attack destroyed the Newfoundland settlement of Bonavista (August 18–29, 1704); later, the French consolidated control of the area by seizing St. John's (1708).

1710, October 16. Capture of Port Royal. A British force of 4,000, with naval support, seized Port Royal which was renamed Annapolis Royal.

1711. Failure in the St. Lawrence. In planned land and naval attacks, **Francis Nicholson** was to lead a mixed group of colonists and Iroquois against Montreal; simultaneously, an English force, including 7 crack Marlborough regiments, was to strike Quebec by sea. However, the loss of 10 ships on the St. Lawrence rendered the joint campaign stillborn.

1713. Treaty of Utrecht. An uneasy peace in North America was established. (See p. 684.)

1711–1712. Tuscarora War. A Tuscarora uprising in the Carolinas was suppressed after the slaughter of 200 settlers. Tuscarora survivors moved to New York State and became the sixth nation of the Iroquois Confederacy.

1715–1728. Yamassee War. The Yamassees and lower Creeks drove South Carolina settlers from the territory west of Savannah. A concerted Carolinian and Cherokee effort pushed the Yamassee confederation into Florida (1716). The colonists erected fortifications on the Altamaha, Savannah, and Santee rivers to protect the Carolina communities from French and Spanish aggression (1716–1721). The inefficiency of the proprietary regime in prosecuting the Yamassee War (and generally in providing defenses against Indian raids and piracy) led the Board of Trade to substitute royal for proprietary government (Act of Parliament, 1729).

1720–1738. Fortification of the Great Lakes. The French erected Fort Niagara (1720) as a headquarters for the Iroquois campaigns and in order to gain control of the lower Great Lakes. In response, the British built Fort Oswego on Lake Ontario (1725). New England settlers fortified the frontier in anticipation of Abenaki raids.

1733–1739. English Expansion in Georgia. James Oglethorpe, one of 20 trustees of the territory extending from the Savannah to the Altamaha rivers, supervised construction of a series of forts on St. Simon's, St. Andrew's, Cumberland, and Amelia islands. Oglethorpe was also instrumental in establishing peaceful relations with the Creeks and lesser tribes.

1739–1743. War of Jenkins's Ear. Prolonged Anglo-Spanish discord concerning the boundaries of Florida, the commercial provisions of the Treaty of Utrecht and the **Assiento** (a contract providing for an English monopoly of the slave trade with the Spanish colonies), and Spanish abuse of British seamen caused England to declare war (October 19). The immediate cause of war was the claim of a British merchant ship captain, named Jenkins, that his ear had been cut off by Spanish officials. British forces under Admiral **Edward Vernon** seized Porto Bello (1739) and Chagres (1740). An assault on Cartagena was repulsed (1740). Under Oglethorpe, Carolina, Georgia, and Virginia contingents—protected on the west by the Creeks, Cherokees, and Chickasaws—captured 2 forts on the San Juan River (January 1740) and unsuccessfully assaulted St. Augustine (May–July). The following year, an English attack on Cuba failed. In the **Battle of Bloody Swamp** (1742), a Spanish attack on St. Simon's Island was repulsed. Oglethorpe's troops again attempted to seize St. Augustine, but were compelled to withdraw from Florida (1743).

King George's War, 1740–1748

This was the American version of the War of the Austrian Succession (see p. 689).

1744. French Attack on Annapolis. The French were repulsed from Annapolis Royal (formerly Port Royal), Nova Scotia.

1745, April–June. Siege of Louisbourg. A New England expedition under **William Pepperell,** supported by British vessels, engaged in the most notable American operation of the war, a siege of the heavily fortified stronghold at Louisbourg. The fort surrendered (June 16). This marked one of the first uses of field artillery in North America.

1746. French Expedition against Nova Scotia and Cape Breton Island. The assault force was demolished in a storm off the coast of Nova Scotia.

1746–1748. Operations on the Northern Frontier. A New England campaign to penetrate into Canada provoked French and Indian raids on Maine settlements. **William Johnson,** New York commissioner for Indian affairs, and **Conrad Weiser,** who pressed Iroquois land claims at the expense of the Delawares, secured the support of the Six Nations.

1748. Treaty of Aix-la-Chapelle. An uneasy peace settled over northern North America. To the disgust of the colonists, Louisbourg was restored to the French in return for Madras. (See p. 695.)

LATIN AMERICA

Spanish America

1702–1713. War of the Spanish Succession. (See pp. 676, 722.) In its American phase, the conflict was restricted primarily to the North American continent and to naval operations in the Caribbean and among the Antilles.

1718–1720. French and Spanish War on the Gulf Coast. This was a part of the War of the Quadruple Alliance (see p. 686). Clashes occurred in Florida and Texas.

1720–1722. Spanish Occupation of Texas. This was intended to prevent French expansion from Louisiana.

1721–1725. Rebellion in Paraguay. José de Antiquera, governor of Asunción, disturbed by the increasing dominance of Buenos Aires traders, rebelled. He refused to accept a new governor sent from Lima, ousted the Jesuits, and defeated a force from Buenos Aires (1724). Although Antiquera was captured (1726), the insurrection persisted for 9 more years until an expedition under Governor **Zabala** of Buenos Aires crushed the rebels.

1723. Araucanian Revolt. The restless and warlike Araucanian Indians (see p. 569) attempted unsuccessfully to oust the Spaniards from southern Chile.

1726. Establishment of Montevideo. This was to prevent Portuguese expansion in the Banda Oriental (Uruguay), a region disputed between Spain and Portugal because of differing interpretations of the Treaty of Tordesillas (see p. 566).

1736–1737. Oruro Revolt. Harsh labor conditions in the mines of central Peru caused the Oruros to revolt. Led by **Juan Santos,** they seriously damaged the city of Oruro before the uprising was quelled.

1739–1748. War of Jenkins's Ear and War of the Austrian Succession. (See pp. 689, 723.) Although combat extended over a wide area, the emphasis remained on North America. Nevertheless, British forces operated in the Caribbean: Porto Bello and Chagres were captured while attacks on Cuba and Cartagena failed (see pp. 696, 723). In addition, there was British raiding along the coast of Venezuela.

1749. Insurrection in Venezuela. Uprisings among the Venezuelan Creoles were suppressed by force.

Portuguese America

1701–1713. War of the Spanish Succession. (See p. 676.) The French made repeated raids and attacks along the Brazilian coast. Colonia was captured and held temporarily (1702), and Rio de Janeiro was captured, sacked, and held for ransom by able French Admiral **René Duguay-Trouin** (1711).

1708–1709. War of the Emboabas. The Portuguese government suppressed and brought under control the adventurous, expansionist, and semi-independent **Paulistas** (Portuguese inhabitants of São Paulo).

1710–1711. War of the Mascates. A small-scale conflict between the inhabitants of Olinda and Recife, in Pernambuco, over the primacy of the 2 towns.

1735–1737. Spanish-Portuguese War. This was a part of the War of the Polish Succession (see p. 687). Spain seized Colonia (1735), but was persuaded by Britain to return the port to Portugal (1737). Colonia, the main Portuguese foothold on the River Plate (since 1680), continued to be a source of friction between Portugal and Spain in their dispute over the Banda Oriental (Uruguay).

XVI

THE DOMINANCE OF MANEUVER ON LAND AND SEA:

1750–1800

MILITARY TRENDS

Maneuver—the tactical manipulation of fire and movement on the battlefield—was the predominant military characteristic of this period on both land and sea. This was due in part to improvements in weaponry, but mainly evolved through the genius of 4 of history's greatest captains.

On land, the period opened with **Frederick II** of Prussia—**Frederick the Great**—in the full flower of his military capacity. **George Washington*** emerged in midperiod, and before the half-century closed **Napoleon Bonaparte's** star was shining over the horizon of war. On the sea, **Horatio Nelson,** exploring down the path previously opened by France's **Pierre André Suffren** and England's **John Jervis** and **George Brydges Rodney,** shattered the old order of naval tactics.

The trend toward maneuver progressed through the 3 major wars of the period: the Seven Years' War, 1756–1763; the American Revolution, 1775–1783; and the French Revolutionary Wars, 1791–1800. These embraced not only Europe and the North American continent but also much of the rest of the world. As a result of these wars and their subsidiary operations, the French colonial empires in both North America and India were destroyed, a worldwide British Empire was firmly established, and the United States of America came into being. Paradoxically, the close of the period saw France the predominant military land power in Europe.

* Military and other historians have tended to underrate Washington's generalship. Those who disagree with the authors' firm conviction that he merits comparison with Frederick, Napoleon, and Nelson are referred to their *Compact History of the Revolutionary War*, New York, 1963, Appendix V, "The Historians and the Generalship of George Washington."

Sea power's importance became firmly established, although it is doubtful whether many people realized it at the time. Certainly **William Pitt** the elder, George Washington, and Horatio Nelson did. Frederick the Great, whose operations were confined to the land, had no reason to understand sea power. Bonaparte never did adequately appreciate it, a blind spot in his genius which would be most costly to him.

WEAPONS AND TACTICS ON LAND

The continued employment of the flintlock musket and its increased rate of fire (see p. 669) caused a still further thinning of Gustavus' 6-man line into one which was at first 4 men deep, then 3, and finally, in the British Army, 2. This created a radical new problem in control. The already rigid discipline of early 18th-century warfare became even more stringent. It was maintained in all European armies. Frederick the Great went further. The individual Prussian soldier, through incessant, brutal, rigid drill discipline, including the cadenced step, became an automaton. Prussian units could change direction or front simultaneously or by small groups in succession; could shift into battle formation from marching column or vice versa, even over broken and irregular terrain. In addition, fire by platoon replaced the more formalized volley fire by larger formations.

The net result was a mobile infantry which could be so shifted and massed at will on the battlefield as to produce the maximum effect of fire and shock action at a chosen spot. This rendered possible Frederick's oblique attack, with Leuthen (see p. 732) as prime example.

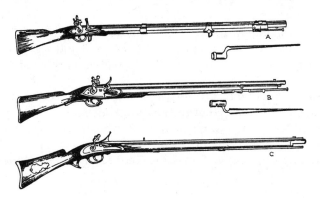

Eighteenth-century flintlocks: (a) French regulation musket; (b) British "Brown Bess" musket; (c) American frontier rifle

Another weapon affecting infantry tactics was the rifle, which came slowly into military use toward the end of the period. Originally a sporting weapon, this heavy, cumbersome hand arm, whose grooved barrel imparted a twist to its bullet, achieved amazing accuracy and range as compared to the smoothbore musket. It had crept oversea to North America from its original habitat, the Rhineland, where the huntsman—the *jäger* or "yager"—had used it for nearly 200 years. It skipped New England, but German craftsmen in Pennsylvania began turning out a somewhat lighter and longer-barreled rifle for the woodsman. The rifle had a slower rate of fire than the musket, since to load it each bullet (wrapped in a greased patch) had to be hammered down into the grooved barrel with a mallet. It was an individual arm, used by individualists along the western fringes of the thirteen colonies. It carried no bayonet. As a result of experience in the American Revolution, the rifle and the rifleman had become an element in European warfare as the period closed. The German and Austrian *jäger* battalions were armed with the rifle.

Out of the French and Indian War also came further mobility. Skirmishers—light troops covering the front of the field of battle—had always been present in one way or another. Previous to 1756, in Europe these units were usually "expendables," irregulars. But as a result of Braddock's defeat (see p. 771), experiments were made in the British Army leading to the establishment in each foot regiment (or battalion) of a "light" company, usually detached from its battalion for covering the advance, or for some other special mission. By the time of the American Revolution it was British practice to separate the light companies from their regiments for action, organizing them into provisional units. In addition, the grenadier companies—also one to each regiment—were separated and gathered into special units in combat. These grenadier companies were composed of the largest men in the regiment—a selection dating from the time of the 3-or-more-pound hand grenade (discarded by this time) best hurled by powerful men. The fault of this British system was that the grenadier (not to be confused with the 1st Foot Guards (later Grenadier Guards) Regiment) and light companies became an elite, thus inducing a feeling of inferiority in the remainder of each regiment—the "line" companies. However, the use of regular troops for skirmishing also led to the establishment of "light" and "rifle" regiments in the British service. At the beginning of the French Revolutionary Wars (1792), the untrained French conscripts hurriedly gathered wholesale by the *levée en masse* were literally herded into battle. Perforce, "line" tactics were abandoned and the French infantry, with bayonets fixed, advanced to the assault in deep, dense formations— somewhat erroneously called "columns"—covered by a cloud of skirmishers. This forced compromise would remain, even though the later well-disciplined armies of Bonaparte would include regiments of *chasseurs à pied* and *voltigeurs* (light infantry), while the elite were organized into grenadier regiments.

The role of the cavalry was clarified in Western Europe. During the previous century it had become common practice to intermingle masses of cavalry and infantry in the battle line. Save in Sweden and France, where the saber and shock action were often employed, the horsemen, armed with light muskets and pistols, employed fire action while advancing, and when and if they charged it was only at a slow trot. During the early part of the 18th century, shock action was more generally employed, but infantry and cavalry were still intermingled. This hybrid cavalry-infantry formation was last used in 1759 at Minden (see p. 735), where British infantry broke the French cavalry.

Frederick, however, long before Minden, had restored his cavalry to its original functions: shock action on the battlefield and reconnaissance off the battlefield. He prohibited the firearm as a cavalry weapon (excepting always dragoons, who were taught to fight on foot as well as on horseback). Frederick's cavalry, who went into action first in a 3-rank and later a 2-rank formation, charged boot-to-boot at full gallop. They were never interspersed among their own infantry; their principal weapon was the saber.

Field artillery, despite the efficient Frederick's attempts to improve its mobility by lightening tubes and by introducing horse artillery (see pp. 669–670), was still a cumbersome arm. (In the early stages of the Seven Years' War, Frederick experimented with extremely light guns, which utilized a chambered breech and a reduced powder charge. These proved to have insufficient range and hitting power compared to the well-designed Austrian guns, the fruits of Prince von Lichtenstein's developments at Budweis in the 1750s.) But in 1765 a Frenchman, **Jean Baptiste de Gribeauval,** following service with the Austrian Army during the war, built on Lichtenstein's system and began to revolutionize the arm, which a quarter-century later would really come into its own under Napoleon's direction. Gribeauval standardized the French field artillery into 4-, 8- and 12-pounder guns and 6-inch howitzers, smoothbores cast of iron or bronze. The pieces were lightened, carriages strengthened; horses were harnessed in pairs instead of in tandem; drivers—heretofore usually hired civilians (despite the example of Gustavus) unreliable on the battlefield—became soldiers; caissons and limbers were provided; tangent scales and elevating screws increased ease and accuracy of laying. For nearly a century to come, Gribeauval's 12-pounder gun—the lineal descendent of which, the M1857 gun-howitzer, achieved fame in the American Civil War as the "Napoleon," so known in honor of its designer, Napoleon III—would dominate battlefields the world over.

Toward the end of the period, two inventions—one British, the other American—radically affected future warfare. In 1784, Lieutenant **Henry Shrapnel,** R.A., invented the artillery shell which would bear his name—in essence a spherical shell filled with lead bullets surrounded by a bursting charge. Exploding in air, shrapnel spread its hail over an extended area, lethal to troops in the open. In 1798, **Eli Whitney,** New England inventor of the cotton gin, turned his hand to making small arms. By invention of a jig and the distribution of his labor force, he manufactured muskets whose various parts were interchangeable; here was the initial mass production of weapons. An earlier approach had been made in Britain in 1702–1704 by the Tower of London armories in producing the Brown Bess (or Tower) musket. Two other refinements were the semaphore telegraph of **Claude Chappe** and the military observation balloon, both developed by the French during the Wars of the French Revolution.

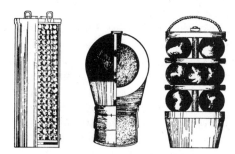

Types of artillery ammunition: (a) caseshot, or canister; (b) shell; (c) grapeshot

Logistically, it was Frederick who during the period evolved the basic principles of modern military supply. Breaking away from slavish dependence on depots, 3 days' rations were carried in the Prussian soldier's knapsack, 8 days' bread supply in the regimental trains and a month's supply in the army trains. A fairly well-organized transport system linked Frederick's armies to such depots as he did organize. His troops, too, were prepared to live off the country when necessary.

TACTICS AND WEAPONS AT SEA

Brilliant English naval iconoclasts smashed the "Permanent Fighting Instructions" of the Royal Navy, which up to this period had frozen initiative by limiting battle formations to a rigid line ahead (see p. 581). Until a new school of British naval officers dared to try it and succeeded, it was unthinkable that vessels should leave the line of battle and sail in, bow-on, toward enemy ships (which could rake them with entire broadsides), break through the opposing line, and, falling upon separated parts of the hostile formation, destroy them in succession by concentration of superior firepower. But by 1800 this method—the Nelsonian touch—had been added to naval warfare. And this was why, as the period ended, Britannia indeed "ruled the waves."

Naval gunnery of the period was still of two schools (see p. 670). The British so aimed their guns as to hull the enemy; smash in her oaken sides and sink or disable her. The French aimed for the enemy's top hamper, to immobilize her by shooting away masts and rigging. To increase firepower, the British introduced the carronade, a short, squatty piece hurling a 32-pound or larger ball; its smashing power at close range was far superior to that of the long 12-, 16-, and 24-pounder guns comprising the normal armament of ships of the line and frigates. The carronade was cheaper to manufacture than the long gun and, being lighter, easier to handle aboard ship. It had much to do with British victories during the period, but it had one major drawback which would not become critical until the War of 1812: the carronade-armed ship had to be much faster and handier than an opponent armed with long guns, or she would be demolished before her carronades came in range.

British technical ingenuity during the period brought forth several innovations in naval gunnery, contributing to superior firepower. These included a flintlock device, flashing a spark into the touchhole, instead of the loose powder priming and the linstock—the slow match previously used; also improvements in powder bags, and the wetting of the wads between powder and ball to prevent premature firings. Metal springs were added to the rope breechings which held the gun in recoil, and inclined planes of wood were placed under the carriage wheels further to ease recoil. Block-and-tackle purchases enabled the traversing of individual guns to right or left—a tremendous advance, this, in naval gunnery, since one no longer had to steer the entire vessel at a right angle to the target when firing. Another invention was the firing of red-hot cannon shot at wooden ships, a procedure first introduced by the British at Gibraltar in 1782. This incendiary weapon, fired with relative accuracy, was a vast improvement over the employment of drifting fire ships and fire rafts, necessarily chancy and haphazard in results. The crowning development was in tactical control: an improved flag signal code, whereby for the first time in naval history a commander could maintain control and issue orders right up until battle was joined. These improvements appeared gradually during the period; by its end all had proved their efficacy.

The period closed with a massive mutiny in England of the enlisted personnel of the Royal Navy (1797). This, as it turned out, was a blessing in disguise, for public attention was drawn to the injustices and horrors suffered by the sailors and caused immediate remedial action (see p. 672).

MAJOR EUROPEAN WARS*

Seven Years' War, 1756–1763

Land Operations in Europe

The Holy Roman Empire (Austria), France, Russia, Sweden, and Saxony, alarmed by Prussia's growing power and territorial expansion under Frederick, joined in a coalition to cripple or destroy Prussia (May–June 1756). England, already involved in colonial and maritime war with France in the French and Indian War in North America (see p. 770) and in India (see p. 764), supported Prussia. The 7 principal land campaigns of this war—also known as the Third Silesian War—were fought in Europe east of the Rhine, but naval operations spread over the Atlantic and Indian Oceans and the Mediterranean and Caribbean Seas.

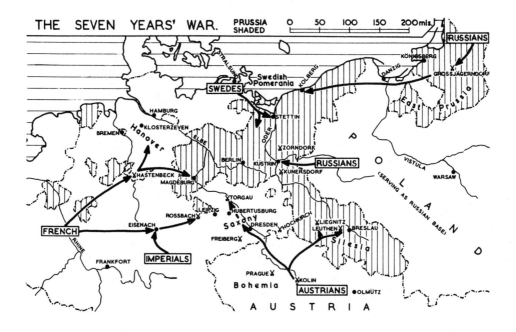

1756, August 29. Frederick Strikes First. Learning the intentions of his enemies, Frederick crossed the Saxon frontier with 70,000 men, the remaining 80,000 men of his army guarding the northern and eastern frontiers. He occupied Dresden (September 10), Saxon forces, only 14,000 strong, falling back to the fortified camp of Pirna, on the Elbe, where they were blockaded.

1756, October 1. Battle of Lobositz (Lovocize). Frederick, learning that an Austrian army of 50,000 men under Marshal **Maximilian U. von Browne** was approaching to relieve the Saxons, marched south with equal

---

* The French and Indian War (an extension of the Seven Years' War) and the American Revolution (which had European and worldwide repercussions) are covered below, pp. 770–791.

strength and defeated the Austrians in the rugged Erzgebirge. Casualties on each side were approximately 3,000. The Saxons at Pirna surrendered. Saxony fell into Frederick's possession, and the Saxon troops were incorporated into his army.

1757, April. Invasion of Bohemia. The coalition had 132,000 Austrian troops in northern Bohemia; none of the other allies were ready. Frederick now had nearly 175,000 men under arms. Almost half of these were scattered along the Bohemian frontier in three groups: the left under Marshal **Kurt C. von Schwerin** and **August Wilhelm, Duke of Brunswick-Bevern,** and the center and right under Frederick's personal command. An additional force of about 40,000 Prussians and 10,000 English under **William Augustus, Duke of Cumberland** stood in Hanover to guard against any French move; another 50,000 guarded the frontiers of Swedish Pomerania and Russia. Frederick sent his left wing south across the mountains east of the Elbe, while with the bulk of his army (about 65,000 men) he pressed from Pirna on Prague, where the Austrians had concentrated about 70,000 troops under Prince Charles of Lorraine.

1757, May 6. Battle of Prague (Praha). The Austrians were defeated in a short, savage battle by Frederick. His initial attack against the Austrian right having been repulsed, he sent his cavalry to envelop the enemy right flank. A gap appeared in the Austrian formation as they tried to meet this envelopment. Penetrating the gap (in the "Prague Maneuver"), Frederick broke the Austrian Army in two, and threw it back into the city, which he invested. Losses on both sides were heavy: 13,400 Austrians killed, wounded, and captured; 14,300 Prussian casualties included Marshal von Schwerin, who was killed while leading the first infantry assault.

1757, May–June. Approach of Austrian Relief Army. Marshal **Leopold J. von Daun** with 60,000 fresh troops moved leisurely to the relief of beleaguered Prague.

1757, June 18. Battle of Kolin. Frederick gathered all the forces he could spare from the siege—32,000 men—and attacked Daun's position. The Prussians were repulsed, losing 13,768 men, while the Austrians had 9,000 casualties. Frederick was forced to raise the siege of Prague and to evacuate Bohemia.

1757, July. Allied Invasions of Prussia and Hanover. A French army of 100,000 men, under Marshal **Louis d'Estrées,** invaded Hanover. Simultaneously another 24,000 French under **Charles, Duke of Soubise** and 60,000

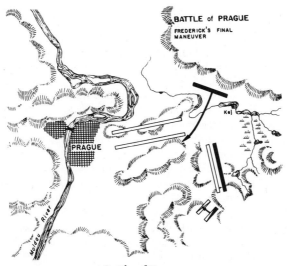

Battle of Prague

Austrians under Prince **Joseph of Saxe-Hildburghausen** were moving northeast from Franconia to join d'Estrées. The main Austrian army, 110,000 strong, under Prince Charles of Lorraine and Marshal Daun, was in Bohemia, advancing toward Prussia. Also a Russian army of 100,000 under Marshal **Stepan Apraksin** was invading East Prussia, and 16,000 Swedes had landed in Pomerania.

1757, July 26. Battle of Hastenbeck. The Duke of Cumberland with 54,000 men was defeated and driven from Hanover by d'Estrées.

1757, July 30. Battle of Gross-Jägersdorf. Prussian General **Hans von Lehwald,** with 30,000 men, was defeated by Apraksin's Russians. The road to Berlin was open. Poor Russian supply arrangements kept them from exploiting promptly.

ROSSBACH-LEUTHEN CAMPAIGN, 1757

1757, September–October. The Allies Converge on Berlin. Frederick, leaving a small force in Silesia under the Duke of Brunswick-Bevern, moved west by forced marches with only 23,000 men to meet the most pressing threat: the juncture, in Prussian Saxony, of the main French army under the **Duke of Richelieu** (who had replaced d'Estrées) and Hildburghausen's and Soubise's Austro-French army. But Richelieu's army did not move, and Soubise and Hildburghausen—who had taken Magdeburg—declined to give battle and retreated to Eisenach. Frederick then dashed back to Silesia to stop Charles' and Daun's attempted Austrian invasion from Bohemia, and to block the lethargic Russian advance on Berlin. He learned that Austrian raiders had entered and plundered Berlin (October 16).

1757, October 18–November 4. Maneuvers in Saxony. While a small force was detached to rescue his capital, Frederick again marched westward, having learned that Soubise and Hildburghausen were again advancing. By the end of the month Frederick was west of the Saale River, again endeavoring to entice his

enemies into battle, in the vicinity of the village of Rossbach.

1757, November 5. Battle of Rossbach. Hildburghausen and Soubise had 64,000 men on commanding ground, facing east. The allies decided to envelop Frederick's left flank. The king diagnosed his enemy's intentions. As 3 parallel allied columns marched southward, 41,000 strong, Frederick pretended to withdraw his 21,000 men to the east from Rossbach. His cavalry, 38 squadrons under General **Friederick Wilhelm von Seydlitz,** swung wide to the east, while the infantry, much more mobile than those of the allies, shortly changed direction to the south, screened by hills from the sight of the allies. As a result, when the allied army completed its circling around the original Prussian flank and headed north to deploy, it was met head on by heavy Prussian artillery fire, supported by 7 battalions of infantry. At the same time the Prussian cavalry charged into the allied right flank, throwing the columns into confusion. The Prussian main effort then attacked the enemy mass, in echelon (overlapping succession) from the left. In less than an hour and a half the allied army was completely routed, with loss of about 10,150 men. The fugitives were joined by the 20,000-odd allied troops that had not been engaged. Frederick's casualties totaled 548 including Seydlitz, who was badly wounded.

1757, November 6–December 5. Return to Silesia. Frederick immediately hurried back to Silesia, where the Austrians were advancing toward Breslau (Wroclaw). Enroute he learned that Charles of Lorraine and Marshal Daun had defeated and captured August Wilhelm of Brunswick-Bevern at **Breslau** (November 22), and had driven Brunswick-Bevern's troops west of the Oder. With 13,000 men Frederick covered 170 miles in 12 days, to unite with Brunswick-Bevern's survivors near Liegnitz (Legnica). The king, who now had about 33,000 men, at once took the offensive, marching east.

1757, December 6. Battle of Leuthen. Prince Charles' Austrian Army, nearly 65,000 strong, lay in a 5-mile-long line of battle, facing west in

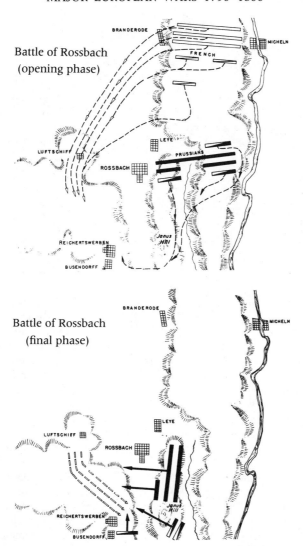

Battle of Rossbach (opening phase)

Battle of Rossbach (final phase)

undulating country a few miles from Breslau. Cavalry protected both flanks. The Austrian reserves lay behind the left wing, in anticipation that Frederick would attempt his favorite enveloping maneuver on favorable terrain. The right flank lay on a marsh. Frederick moved toward the Austrian right, in 4 columns, the inner 2 columns made up of infantry, the outer ones composed of cavalry. Under cover of a range of low hills, he changed direction obliquely to the right, leaving his left-hand col-

umn of cavalry in his rear, to begin a demonstration against the Austrian right. Still out of sight of the Austrians, the infantry was thus marching past the enemy front in 2 columns, the remainder of the cavalry having taken the lead. Charles, meanwhile, had moved his reserves to bolster his apparently threatened right wing. When Frederick's marching columns, still concealed behind the hills, began to overlap the Austrian left, the king faced his infantry to the left and attacked in 2 lines echeloned

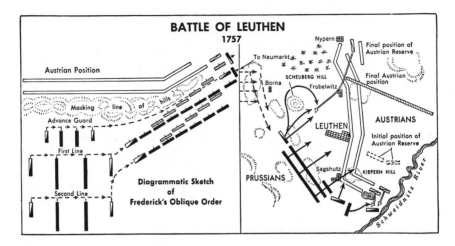

BATTLE OF LEUTHEN
1757

Diagrammatic Sketch of Frederick's Oblique Order

from the right, the pressure increasing as each successive battalion moved in. At the same time, massed Prussian artillery fire shot at the apex of the Austrian left flank, ranged in a vee. The Prussian cavalry on the right flank, under General **Hans J. von Ziethen,** charged into the broken Austrian left and threw it back on the center. Charles attempted to form a new line against this attack, at the same time throwing his right-wing cavalry against the Prussian left flank. But the Austrian horse was caught and scattered by the rear half of Frederick's cavalry, who then charged in on the Austrian right flank. Thus caught off balance, the Austrians never rallied, despite efforts by Charles and Daun. Nightfall alone enabled the escape of the vanquished across the Schwiednitz River to Breslau. Charles's army had been ruined; more than 12,000 were captured, 6,750 killed or wounded, additional thousands scattered to the four winds. Frederick captured 116 guns and 51 colors. Only about half of the original Austrian strength returned to Bohemia and winter quarters. Prussian losses in this desperate fight were 6,150 killed and wounded. Breslau surrendered 5 days later, 17,000 more Austrians being made prisoner, ending the campaign.

COMMENT. *"Masterpiece of maneuver and resolution,"* said Napoleon. *Leuthen is generally consid-*

ered to be the chef-d'oeuvre of the man who was perhaps the ablest tactician of military history.

1758, January. Renewed Russian Invasion. The Russians, now under the command of General Count **Wilhelm Fermor,** attempted to occupy east Prussia, but were soon immobilized by muddy roads.

1758, May–July. Operations in Moravia. Frederick, certain that he had nothing to fear from the Russians or the Swedes until midsummer, again opened operations by moving into Moravia against the Austrians. An attempt to invest Olmütz (Olomouc) was unsuccessful. Frederick now learned that the Russians, marching into Prussia from the east, were approaching the Oder River. Raising the siege (July 1), by skillful maneuver Frederick deceived the Austrians and marched against the Russians. Meanwhile in the west, Duke **Ferdinand of Brunswick** drove the French back over the Rhine, defeating them at **Crefeld** (June 23).

1758, August 20. Across the Oder River. Frederick arrived at the Oder, across the river from Kustrin (Kostrzyn), which Fermor with 45,000 men was besieging. Feinting a concentration and crossing at that point, Frederick instead moved northward in a rapid, brilliantly conceived night march, crossed the river, and put himself on Fermor's line of communications in a wide, right turning movement. Fermor, raising the siege, turned and moved to a defensive position at the hamlet of Zorndorf, on high ground. The Russians were arrayed in 3 irregular masses, beyond mutual supporting distances, facing north.

1758, August 25. Battle of Zorndorf. Frederick, who had 36,000 men, attacked again in his oblique formation, passing completely around the Russian front to fall on their right (east) flank. Fermor, forming a new front, repulsed the Prussian attack until Seydlitz with his cavalry crossed marshy ground on the Russian right to charge into their masses, smashing the square. Frederick, shifting his attack to the new Russian right (southwest) flank, was checked until Seydlitz, having reformed his cavalry, again rode into the Russian infantry. An inconclusive butchery followed, ending at nightfall. Almost half of the Russian Army (some 18,500 men) had fallen, while Frederick had lost some 12,797. Next day Fermor retreated to Königsberg (Kaliningrad); the Prussians were too exhausted to pursue. The Russians threat was ended for the time being.

1758, September–October. Meeting the Austrians in Saxony. Frederick, learning that Daun was threatening his brother, Prince **Henry of Prussia,** not far from Dresden, hurried back with part of his army by forced marches, making 22 miles a day. On his arrival the Austrians withdrew (September 12). After a short rest, Frederick with 31,000 men took the offensive again. Overconfident, he was surprised by Daun, who surrounded the Prussian Army with 80,000 men during the night and attacked at dawn.

1758, October 14. Battle of Hochkirch. Frederick's army fought hard against overwhelming force. An escape route was opened and held by General **Hans J. von Ziethen**'s cavalry. Through it the Prussian army retreated, leaving behind 101 guns and some 9,097 men killed, wounded, and prisoners. Daun's army, with losses of about 7,587 was so shaken that no pursuit was made.

1758, October–November. Frederick Regains the Initiative. Daun now laid siege to Dresden, but on learning that Frederick had been reinforced and was moving to attack him, withdrew to the fortified camp at Pirna. The close of the campaign found Frederick in firm possession of Saxony and Silesia, while both the Russians and the small Swedish invading force had evacuated Prussia.

COMMENT. *This had been a costly campaign. Frederick could still keep 150,000 men in the field, but the combat value of the Prussian Army was deteriorating. In 3 years of campaigning he had lost 100,000 of his superbly trained and disciplined troops.*

MINDEN CAMPAIGN, 1759

1759, April. Prussian Failure in West-Central Germany. Duke Ferdinand of Brunswick, reinforced by British troops, advanced with 30,000 men from the line Münster-Paderborn-Cassel, held during the winter, to drive out the French, based on Frankfurt and Wesel. Rebuffed at **Bergen** (near Frankfurt) by sheer weight of numbers, he was forced to retire again, and the French, seizing the bridges across the Weser at Minden, occupied a position too strong for direct attack.

1759, August 1. Battle of Minden. Marquis **Louis de Contades,** commanding both his own troops and those of Duke **Victor de Broglie,** had 60,000 men near Minden. Ferdinand, who now had 45,000, feinted against the French right with a fraction of his command. Contades moved out in force to overwhelm it. As he did, Ferdinand launched a counter-assault, in 8 columns. Bad weather delayed the movement, and the French, reforming, were overwhelming the Prussians when 2 British and 1 Hanoverian infantry brigades attacked the French cavalry, who were drawn up in the center of the line of battle and protected by artillery. The British were met by a cavalry charge; with an amazing discipline and gallantry (of 4,434 men, 1,330 became casualties) they not only drove off the French horse by fire but then advanced with the bayonet against French infantry and pierced the center of Contades' line of battle. Ferdinand, realizing that victory was in his hand, at once ordered the British cavalry, 5 regiments strong, to charge and complete the rout. Lieutenant General **George, Lord Sackville** thrice refused to move. As a result the French, though disorganized and beaten, and having lost 10,000 prisoners and 115 guns, retired from the field unhindered. Ferdinand, following them almost to the Rhine, checked his advance only when he had to send reinforcements to assist King Frederick (see below).

COMMENT. *Sackville, tried by courtmartial for his disobedience, was cashiered from the British Army. However, 12 years later he returned as a favorite of King George III, to become Lord* **Germain** *and contribute mightily to the ineptness of the British military effort in the American Revolution (see p. 779).*

1759, July–August. Russian and Austrian Invasion. The Russians (now under Count **Peter S. Soltikov**) defeated the Prussian covering forces along the Oder line at **Kay** (July 23) and moved on Frankfurt, linking with 35,000 men of Daun's Austrian army under Lieutenant General **Gideon E. von Loudon** Frederick, gathering all available forces, moved to meet the combination.

1759, August 12. Battle of Kunersdorf. The Austro-Russian armies 80,000 strong, lay entrenched in the sand hills 4 miles east of Frankfurt. Frederick, with 50,000 men, crossed the Oder and attempted a double envelopment. But his columns lost their way in the woods and their attacks were delivered piecemeal. Thrown back at all points, Frederick insisted upon continuing the vain attacks. He lost more than 19,100 men, 172 guns, and 28 colors in 6 hours—the greatest calamity ever to befall him. Completely discouraged, he contemplated abdication. Fortunately for the Prussians, their enemies (who had lost 15,700 men) were too sluggish to reap advantage from their victory.

1759, August–November. Frederick Retrieves the Initiative. Frederick, receiving reinforcements from Ferdinand, recovered his determination and reorganized his army in the field. The Russians, having exhausted all the food and forage in the vicinity, later retired to their own frontier, and Frederick turned his attention to Daun, who had occupied Dresden (September 4).

1759, November 21. Battle of Maxen. Frederick sent a detachment of 13,000 men under General **Frederick von Finck** to envelop Daun's army. But the Austrians concentrated 42,000 to overwhelm the detachment; the survivors surrendered. Both sides now spent the winter in relative quiet. This had been Frederick's worst year.

1760. Renewed Allied Convergence on Prussia. The allies planned to move concentrically upon Frederick; Austria's Daun with 100,000 men in Saxony, and Loudon with 50,000 more in Silesia, would work in unison with Russian's Soltikov (50,000 strong) in East Prussia. If Frederick turned on one, the others would converge on Berlin. Frederick with 40,000 faced Daun on the Elbe; Prince Henry with 34,000 was in Silesia, and 15,000 additional Prussians were covering additional Russian and Swedish forces ravaging Pomerania. In the west, Duke Ferdinand's Prussian-British army of 70,000 faced 125,000 French troops in Hanover.

1760, June–October. Ferdinand Maneuvers in the West. Ferdinand, following the instructions and example of his king, played a successful game of march, maneuver, and quick battle against the French armies of the Duke de Broglie and **Claude, Count of St. Germain.** In this campaign the British contingent, commanded by Lieutenant General Sir **John M. Granby** (successor to the incompetent Sackville), played a conspicuous part. Highlight of the British effort was the storming of Warburg (July 31) by Granby's cavalry, who took 1,500 prisoners and 10 guns. Ferdinand outfoxed the French, driving them back to the Rhine, but there he was repulsed at **Kloster-Kamp** (October 16) and withdrew to Lippstadt and Warburg for the winter.

1760, June–August. Frederick Maneuvers in the East. While Prince Henry engaged the Russians in operations sterile for both sides, and Loudon destroyed a Prussian detachment at **Landeshut** (June 23), Frederick with his main army began shuttling between Loudon—who was besieging Glatz (Klodzko)—and Daun. When Daun marched to help Loudon. Frederick turned back to besiege Dresden (July 12). Daun came hurrying to its relief and almost surrounded Frederick, who slipped away in the nick of time (July 29). That same day Glatz fell to Loudon. Frederick, summoning Prince Henry's army to join him, now plunged into Silesia with 30,000 troops in a remarkable series of forced marches.

1760, August 15. Battle of Liegnitz. Loudon and Daun, with 60,000 men and 30,000 Russians under Czernichev not far behind, closed in on Frederick. The Austrians planned a double-envelopment of the Prussian position. In a brilliant operation Frederick, using the night to conceal his movement, turned on Loudon's column (24,000 men). Loudon attempted to attack in the pre-dawn hours, but was enveloped and crushed by Frederick's counterattack before Daun could arrive. Frederick then cut his way to safety, leaving 8,500 Austrian casualties on the field and capturing 80 guns, the Prussian loss was 3,394. Tricking Czernichev by a false message, which led him to believe the Austrians had been totally defeated, Frederick resumed his maneuvering against Daun, while the Russians hurriedly retreated, no more to appear in this campaign.

1760, October 9. Capture of Berlin. Learning that Austrian and Russian raiders had again seized and partly burned his capital, Frederick turned to its relief. On his approach the allies hurriedly evacuated the city (October 12) and the main Austrian army, 53,000 strong, began concentrating near Torgau. Frederick, gathering 50,000 men, moved against the Austrian entrenched camp.

1760, November 3. Battle of Torgau. The Austrians, under Daun, lay in a formidable position on high ground west of the Elbe. Frederick, advancing from the south, planned to move half of his army entirely around the Austrian right, through dense woods, and attack their rear. Ziethen, with the other half, was to make a simultaneous frontal attack. Through a combination of errors and bad weather, Fred-

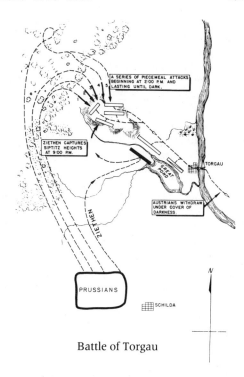

Battle of Torgau

erick's wing became disorganized. Surprise was lost and the Austrians regrouped their forces to meet the threat. Meanwhile, Ziethen's advance guard encountered a small force of Austrian light infantry and opened fire. Frederick, hearing the cannonade, assumed Ziethen was assaulting the main Austrian position and threw his own forces in piecemeal. For 2½ hours all his attacks were repulsed. By that time all Frederick's reserves had been committed. It seemed to be another Kunersdorf. But as dusk was falling Ziethen finally reached his appointed position and attacked. Frederick renewed his assaults in darkness. Austrian resistance collapsed; Daun withdrew from the position and across the river. Frederick had squandered 16,670 men, while Austrian losses were more than 8,697 killed and wounded and 7,000 more made prisoner. Both sides having fought themselves to exhaustion, the campaign ended for the year.

1761. Operations in the West. Duke Ferdinand had some success against the French as the campaign opened, but by October they had pushed him eastward to Brunswick.

1761. Frederick at Bay in the East. He could scrape together barely 100,000 troops, and was facing thrice that number of Austrians and Russians in Silesia. Despite brilliant maneuvering, he was finally cut off from Prussia by the junction of the armies of Loudon and Russian General **Alexander Buturlin,** near Liegnitz. Frederick retired (August 20) to the entrenched camp of Bunzelwitz, 20 miles east of Glatz, in the vicinity of Jauernick (Javornik). Situated in the Eulen Gebirge hills on what is now the northern frontier of Czechoslovakia, the Bunzelwitz camp was a natural fortress, improved by elaborate field fortifications extemporized in 10 days and nights of frantic effort—time granted by the vacillation of the allied commanders. Loudon wanted to attack, Buturlin refused. Then (September 9) the Russians withdrew permanently, and the Austrians soon fell back into winter quarters. Frederick, the pressure relieved, did the same.

1761, December. Frederick on the Brink of Defeat. Although Prince Henry in Saxony had held his own against Daun, Ferdinand's army in the west was dwindling; **George III** of England, succeeding George II in 1760, had begun withdrawing part of the British contingent, and now threatened to cut off all subsidies. Frederick could only muster 60,000 troops. He was at the end of his rope.

1762, January 5. Death of Elizabeth of Russia. Her successor, **Peter III,** an admirer of Frederick, immediately began peace negotiations.

1762, May 15. The Treaty of St. Petersburg. This ended the war between Russia and Prussia, and saved Frederick from almost certain defeat. Peter loaned a Russian army corps to Frederick.

1762, May 22. Treaty of Hamburg. This brought peace between Prussia and Sweden. Frederick could now concentrate against the Austrians while Ferdinand held off the French.

1762, June 24. Battle of Wilhelmstal. Duke Ferdinand's Prusso-British army defeated the French in Westphalia, Granby's British brigades delivering the vital assault.

1762, July 9. Collapse of the Russian Alliance. Peter was deposed. His wife, **Catherine II** of Russia, broke off the alliance with Prussia, but did not renew the war.

1762, July 21. Battle of Burkersdorf. In Silesia, Frederick defeated Daun. The exhausted armies maneuvered inconclusively, but Frederick steadily strengthened his hold on the province.

1762, October 29. Battle of Freiberg. In Saxony, Prince Henry, assisted by Seydlitz, defeated Austrian Marshal **Serbelloni.**

1762, October–November. Ferdinand Drives the French across the Rhine.

1762, November. Armistice. All concerned were exhausted by the years of protracted conflict. France was already making peace with England.

1763, February 16. Treaty of Hubertusburg. Austria agreed to the return of the *status quo* in Europe. Frederick retained Silesia.

Naval Operations, 1755–1763

PRE-WAR CONFLICT, 1755–1756

1755. Franco-British Rivalry in North America. This led to the French and Indian War (see p. 770); it also brought about naval warfare nearly a year before formal declaration of hostilities. Shortly after Britain sent General **Edward Braddock** to the American colonies, the French sent a naval squadron and transports with reinforcements and supplies for Canada. A British fleet under Admiral **Edward Boscawen** was sent to intercept the French squadron before it reached the St. Lawrence River.

1755, June 8. Action of the Strait of Belle Isle. Boscawen found the French squadron, but was able to capture only 2 transports; the remainder of the French ships escaped and reached Quebec. The English now began to seize all French ships wherever found. British Admiral Sir **Edward Hawke,** cruising off the coast of France, captured a large number of ships in a convoy from the French colonies in America. In retaliation, **Louis XV**'s government seemingly prepared for an invasion of the British Isles, but still without declaring war. Britain, alarmed by this threat, paid little attention to a French joint army-navy expeditionary force preparing at Toulon.

1756, April 9. British Squadron to the Mediterranean. Belatedly concerned about the Toulon preparations, the British Admiralty sent Admiral **John Byng** (son of Viscount George Byng of Torrington; see p. 685) with 10 ships of the line to the Mediterranean. He was to protect the British possessions of Gibraltar and Minorca.

1756, April 12. French Expedition to Minorca. Admiral Count **Augustin de la Gallissonnière,** with 12 ships of the line, convoying 150 transports carrying 15,000 troops under Marshal Duke **Louis de Richelieu,** sailed from Toulon for Minorca. One week later (while Byng was still out on the Atlantic Ocean) the French landed and invested Port Mahon, the island's capital, while the fleet blockaded the port. The British garrison of 3,000 men was in peril when Byng finally appeared off the island (May 19).

1756, May 20. Battle of Minorca. Byng, who now had 13 ships, tried to bring his column (in line ahead) into action against the French, ship for ship. But he came in on an angle. One of his center vessels, disabled by French fire, held up the rest. Meanwhile the 5 leading British warships were badly hammered by the entire French squadron. Instead of letting his ships sail on individually to join the van of his squadron, Byng now attempted to carry on in his original formation, which gave the French squadron time to sail away. Byng gathered up his damaged vessels, making no attempt to land the army battalion he had on board, which he felt was insignificant in comparison to the French army. He returned to Gibraltar. A victim of restrictive Admiralty instructions, and later of Whig-Tory politics, he was tried by court-martial, convicted of failure to do his utmost, and was shot. Minorca surrendered (May 28).

WARTIME OPERATIONS, 1756–1763

1756. May 17. War Formally Declared. The British reinforced squadrons in the Mediterranean and western Atlantic. Britain had 130 ships of the line; France had 63; Spain had 46. The British were superior in equipment and personnel.

1757. Aggressive British Naval Policy. The reins of government were in the hands of the elder **William Pitt** (later Earl of Chatham), who knew sea power's value.

1757–1758. Planning in North America. Admiral Boscawen and General **Jeffrey Amherst** prepared a joint expedition against France's fortress of Louisbourg on Cape Breton Island which would end successfully (see p. 772).

1758, April 29. Battle of the Bay of Bengal. Admiral Sir **George Pocock,** with 7 ships of the line, engaged French Commodore Count **Anne Antoine d'Aché**'s 8 ships in an inconclusive action off Fort St. David. The Frenchman was carrying reinforcements to Pondichery while the British fleet was vainly attempting to relieve the besieged port.

1758, August 3. Battle of Negapatam. In a renewed battle between Pocock and d'Aché near the site of the previous battle, the French squadron was so cut up that d'Aché withdrew, to the detriment of French land forces in India (see p. 765).

1758, September. British Repulsed at St. Cas Bay. This concluded a series of fruitless minor British expeditions against the French coast.

1759. Operations in North America. British army-navy cooperation resulted in the capture of Quebec and the conquest of Canada (see pp. 772–773).

1759–1760. Operations in the West Indies. A British joint task force, repulsed at **Martinique** (January), took **Guadeloupe** in a long campaign (February–April). Another force took **Dominica (1760).**

1759, August 18. Battle of Lagos. Meanwhile a French plan to invade England and Scotland was in the making, its initial success dependent upon the junction at Brest of 2 French squadrons. Commodore **de la Clue** left Toulon for Brest with 12 ships of the line (August 15) and reached the Straits of Gibraltar on the 17th, unaware that Boscawen, now commanding in the Mediterranean, had taken his squadron into Gibraltar to refit. A patrolling English frigate sighted the French and gave the alarm. Boscawen, with 14 ships of the line, gave chase. De la Clue headed for the open sea, but 5 of the ships broke away and fled to Cádiz. The remainder got away momentarily, thanks to the gallantry of the rear ship, whose captain fought 5 English ships for 5 hours. But de la Clue's case was hopeless, so he ran the remainder of his squadron ashore on the Portuguese coast, near Lagos. Boscawen, with the usual British disdain for neutrality, captured them.

1759, September 10. Naval Battle of Pondichery. D'Aché reappeared off the Coromandel coast with 11 ships to fight a third tactically inconclusive battle with Pocock. The French ships were so badly damaged that d'Aché departed, never to return. With him went the last hope for French forces in India (see p. 765).

1759, November. Operations off the Coast of Brittany. The French squadron in Brest, 21 ships of the line and several frigates, was under Admiral **Hubert de Conflans.** Off the port stood British Admiral Edward Hawke with 25 sail of the line, plus 4 50-gun ships and 9 frigates. A tremendous gale drove Hawke off station and into Torbay for shelter, whereupon Conflans put to sea. Off Quiberon Bay on the Breton coast, he fell in with some British ships on blockade duty, and at the same time Hawke, who had hurried out when the storm subsided, caught up with him. Conflans, outnumbered by at least 2 ships of the line and the 4 50-gun ships, fled for Quiberon Bay, hoping that his opponent would not dare to follow him through the treacherous banks and reefs of the entrance. But Hawke did just that.

1759, November 20. Battle of Quiberon Bay. Engaging his enemy in the crowded bay, on a lee shore, in heavy weather, Hawke destroyed or captured 7 French ships and the remainder were so scattered that they were unable to reassemble. Hawke lost 2 ships, wrecked on a reef; his losses in action were negligible. This battle ended France's naval power for the remainder of the war.

1760–1763. England Supreme at Sea. Although French privateers had taken heavy toll—10 percent—of British merchant vessels, French commerce, with the French Navy swept from the sea, was completely destroyed.

1761. Spanish Intervention. Charles III of Spain chose this unpropitious time to ally himself with France, and broke off diplomatic relations with England. France gave Minorca to Spain, which soon declared war (January 4).

1762, January 4. British West Indies Offensive. Admiral **George B. Rodney** proceeded against Martinique, the last French stronghold in the West Indies. Its surrender (February 12) was followed by seizure of Grenada, St. Lucia, and St. Vincent. While Rodney was securing the West Indies, Pocock, back from the East Indies, promptly sailed from England for Havana with 19 ships of the line and transports bearing 10,000 army troops (March).

1762, June 20–August 10. Operations against Havana. Pocock and Rodney, joining forces, sailed to Havana (June). Morro Castle was taken (July 30) after a 40-day siege by troops under **George Keppel, Earl of Albemarle.** The Spaniards then surrendered the port, with 12 ships of the line and more than $15 million in money and merchandise. It was a body blow to Spain's economy and prestige, since all Spanish-American commerce funneled through Havana.

1762, October 5. Capture of Manila. Another British joint expedition, mounted in India and commanded by Admiral Sir **Samuel Cornish** and Lieutenant General Sir **William Draper,** sailed into Manila Bay and captured the city. A ransom of $4 million was levied (in lieu of the normal practice of looting). The Acapulco Galleon with $3 million on board also fell into the expedition's hands. As an additional jolt, out in the Atlantic an English squadron bagged a treasure ship from Lima carrying $4 million in silver to Spain. King Charles at once joined

France in negotiations for peace; his brief excursion into war had cost Spain much.

1763, February 10. Treaty of Paris. France renounced to England all claims to Canada, Nova Scotia, and the St. Lawrence River islands, the Ohio Valley, and all territory east of the Mississippi except the city of New Orleans. Spain traded Florida (all Spanish possessions east of the Mississippi) to England for the return of Havana. She also returned Minorca to England. Spain had already been compensated for her alliance in the secret **Treaty of San Ildefonso*** (November 3, 1762) by receiving from France New Orleans and all Louisiana west of the Mississippi. In the West Indies, Guadeloupe and Martinique were returned to France. Of the "neutral" islands of the Lesser Antilles all but St. Lucia went to England. In India, the British East India Company had become the supreme power in Bengal. Pondichery was returned to France, but its fortifications were razed, and limits were set on French military strength on the Coromandel coast; French stations in Bengal were to be strictly commercial.

Commentary

On land, the Seven Years' War decided nothing politically. Militarily, it proved that loose-knit, bumbling coalitions, regardless of superiority of numbers, are at a disadvantage when waging war against a determined, disciplined, single-purposed nation under capable military leadership. Frederick the Great, through indomitable courage, tactical genius, and brilliant strategic use of interior lines, preserved the integrity of Prussia against amazing odds. He placed his country in the ranks of the great powers, laying the foundation for a united Germany. His campaigns in the Seven Years' War proved him to be a master tactician, one of the great captains of history. At sea, the Royal Navy firmly established the British Empire and destroyed the colonial pretensions of France and severely shook those of Spain—now Britain's only overseas rival. Sea power had come into its own.

WARS OF THE FRENCH REVOLUTION, 1792–1800

War of the First Coalition, 1792–1798

BACKGROUND

1791, August 2. Declaration of Pillnitz, Saxony. The growing strength and violence of the French Revolutionary movement, and the parlous situation of **Louis XVI,** led King **Frederick William II** of Prussia and the Emperor **Leopold II** of Austria to declare themselves ready to join with all the other European powers to restore the monarchy to power in France. Russia and Sweden agreed to raise contingents (subsidized by Spain). England declined to join. The French Legislative Assembly, alarmed by this threat, hastily organized armies to protect the eastern frontiers.

1792, February 7. Austro-Prussian Alliance. Partly in response to pleas of French monarchical émigrés, and partly to take military and political advantage of the chaos in France, Austria and Prussia formed an alliance and began to move troops toward the frontiers. The Kingdom of Sardinia (or Piedmont) joined the alliance soon afterward.

LAND OPERATIONS, 1792–1793

1792, April 20. The French Assembly Declares War against Austria. Fighting began at once along the Flanders frontier. The French Army, though disorganized by the republican movement, still retained some of its previous

* One of three Treaties of San Ildefonso in this century; see pp. 753, 793.

discipline, as well as a few of its professional officers. Patriotic enthusiasm brought flocks of volunteers to the colors. The result was an uneasy mixture of professionals and amateurs (National Guard and volunteers) easily brushed aside by the Austrians in the opening skirmishes near Lille, which the Austrians besieged. Disgusted with the conduct of his troops, Marshal Comte **Jean Baptiste de Rochambeau** resigned command of the Army of the North, and was succeeded by the Marquis **Marie Joseph du M. de Lafayette,** who had been in command of the Army of the Center.

1792, July 11. Assembly of an Allied Army. Karl Wilhelm, Duke of Brunswick had 80,000 men (42,000 Prussians, 30,000 Austrians, and small contingents of Hessians and French émigrés) gathering at Coblenz.

1792, August 10. Storming of the Tuileries. A Paris mob stormed the royal palace, massacring the Swiss Guards. Louis took refuge with the Assembly and was deprived of his few remaining powers. Lafayette, who had briefly thought of marching on Paris to restore order, was relieved of his command, and fled across the frontier to surrender to the Austrians (August 20).

1792, August 19. Allied Invasion of France. Capturing the fortresses of Longwy and Verdun, Brunswick's army moved very slowly through the Argonne Forest, heading toward Paris, weakly opposed by the small French Army of the Center, commanded by General **François C. Kellermann.**

1792, August–September. French Preparations for Defense. General **Charles C. Dumouriez** was appointed by the Assembly to replace Lafayette in command of the Army of the North. With part of this army he hastened to join Kellermann in an attempt to halt the advance of Brunswick's army.

1792, September 20. Battle of Valmy. Dumouriez and Kellermann were able to bring together 59,000 troops, divided almost equally between distrusted prewar regulars and enthusiastic, but poorly trained, volunteers to oppose 35,000 veteran regulars of Brunswick. (The remainder of the allies were spread out along the

line of communications.) Fortunately for the French, the main bodies never became engaged, the Prussian infantry being thrown back by accurate fire from 54 French guns—manned by regulars of the old army. Brunswick, who had never been enthusiastic about the invasion, was so discouraged by this cannonade that he withdrew to Germany. Losses had been fewer than 300 for each army.

1792. Other Operations. In the south, French forces invaded Piedmont, overrunning Savoy and capturing Nice. From Alsace, General **Adam Philippe Custine** led a French offensive which captured Mainz and got into Germany as far as Frankfurt. Dumouriez returned north and pushed into Flanders. Meanwhile (September 21), the newly elected National Convention installed itself as the government of France and abolished the monarchy.

1792, November 6. Battle of Jemappes. On Dumouriez's approach the Austrians raised the siege of Lille and went into winter quarters at Jemappes, just across the border in Belgium (the Austrian Netherlands). The French army of 45,000 men and 100 guns followed and defeated the Austrians, who numbered only 13,000. Austrian losses were 1,500 men and 8 guns, the French lost 2,000. This victory stimulated the French republicans; Brussels was captured (November 16) and a French squadron pushed up the Scheldt to besiege Antwerp, arousing fear in England.

1792, December. Allied Successes. Brunswick, with 43,000 men, drove Custine back to the Rhine. Frankfurt was retaken (December 2). Custine retired to Mainz and winter quarters.

1793, January 21. Execution of Louis XVI. The beheading of the French king sparked England to oust the French ambassador, whereupon France, already at war with Austria, Prussia, and Piedmont, declared war on England, the Netherlands, and Spain. National conscription was decreed, Belgium annexed to France, and Dumouriez ordered to invade Holland. But the allies were already on the offensive. **Friedrich Josias, Prince of Saxe-Coburg** with 43,000 Austrians crossed the

Meuse to recover Belgium; Brunswick's 60,000 Prussians besieged Custine in Mainz. Other allied troops took positions along the Rhine and in Luxembourg.

1793, March 18. Battle of Neerwinden. Dumouriez with 41,000 men attacked Saxe-Coburg, attempting to turn his left. The French, advancing in 8 columns, were repulsed in disorder. Austrian losses were 3,000, the French lost 4,000.

1793, April–July. French Defeats and Demoralization. The Austrians recaptured Brussels. Dumouriez, falsely accused of treason, turned against France and fled to the allies; with him went the **Duke of Chartres** (Prince **Louis Philippe**). General **Picot Dampierre** assumed command and tried to stem the advance of Coburg's allied army, but was killed in action near besieged **Condé** (May 8). Custine was now placed in command of the demoralized French Army of the North. A defeat near besieged **Valenciennes** (May 21–23) resulted in the beheading of Custine by order of the Committee of Public Safety, establishing a pattern which would thereafter menace all French republican commanders. Condé was captured by Coburg (July 10) and Valenciennes fell soon afterward (July 29). The Army of the North retreated to Arras under its new commander, **Jean Nicolas Houchard.**

1793, July–August. The Reign of Terror. France was now assailed from within and without, while the Reign of Terror was beginning in Paris under **Maximilien F. de Robespierre** and the Committee of Public Safety.

1793, August. France Approaches Collapse. The Allies recaptured Mainz. In the Vendée a monarchist counterrevolution was in full swing. Lyon and Marseilles declared for the monarchy. Prince **Frederick Augustus, Duke of York,** with a British-Hanoverian army, was investing Dunkirk, while the **Prince of Orange**'s Netherlands forces linked York to the Austrians farther east. A British-Spanish fleet closed in on Toulon, which declared for the monarchy (August 21).

1793, August 23. The Levée en Masse. The Committee of Public Safety decreed conscription of the entire male population of France. Fourteen armies were hastily put into the field in the next few weeks. Symbolic of the revitalization of France was the recovery of Marseille by the Republicans on the day of the decree (August 23).

1793, September 8. Battle of Hondschoote. With 24,000 men Houchard assailed the Duke of York's 16,000 disciplined English and Hanoverian troops just east of Dunkirk. The uncoordinated but desperate charges of the French recruits forced York back by sheer weight of numbers; he extricated himself, although losing his siege artillery. Losses were about 3,000 on each side.

1793, September 13. Battle of Menin. Houchard routed the Prince of Orange, but he failed in efforts to maneuver the Austrians out of eastern France. He was removed and guillotined, General **Jean Baptiste Jourdan** taking his place.

1793, September. Lazare Nicolas Carnot. The war minister of the Committee of Public Safety joined the army, and ordered Jourdan to relieve Maubeuge, besieged by 30,000 Austrians under Saxe-Coburg.

1793, October 15–16. Battle of Wattignies. The French, 45,000 strong, were at first repulsed by the better-disciplined allies. Next day, however, Jourdan enveloped and drove back Saxe-Coburg's left. The Austrians abandoned the siege of Maubeuge and retired eastward. French losses were 5,000, the allies lost about 3,000.

COMMENT. *These 3 battles (Hondschoote, Menin, Wattignies) are remarkable because, beginning with Hondschoote, not only did they demonstrate an amazing will to win on the part of the new French levies, but in addition, at Wattignies, the new army for the first time used maneuver on the battlefield. A peculiar mixture of patriotism and terror was transforming armed mobs into soldiers. What had happened was mainly due to the incessant efforts of one man: Lazare Carnot, a stern, uncompromising soldier of the old regime, who amalgamated the raw conscripts with the remnants of the old professional army (1793–1794), and brought discipline to enthusiastic hordes of Frenchmen filled with the valor of ignorance. Carnot built the prototype of the instrument that Bonaparte would use later on. In his 2-year tenure of office, Carnot's efforts in organizing the national defense caused him to become known as the "Organizer of Victory."*

1793, October–December. French Revival. Elsewhere the Vendée uprising was finally suppressed, and Lyon was retaken (October 20) by Kellermann. Toulon, besieged by the Republicans, fell (December 19; see below). Along the Rhine and in Alsace an indeterminate campaign ended when General **Louis Lazare Hoche,** after being checked by Brunswick at **Kaiserslautern** (November 28–30), defeated the Prussians at **Fröschwiller** (December 22), then turned on Austrian General **Dagobert Wurmser** in Alsace and defeated him at **Geisberg** (December 26). By the year's end the invaders were back across the Rhine; Alsace and the Palatinate had been cleared and Mainz recaptured. Fighting along the Spanish and Italian frontiers was sporadic and inconclusive.

War at Sea, 1792–1793

1792. Sad State of the French Navy. Its professional officer corps had been purged; its enlisted force, disrupted by a succession of mutinies, was a mob. Merchant marine officers, recruited for the emergency, had no knowledge of naval tactics. Administratively, ships, navy yards, and stores had been completely neglected. Less than half of the 76 ships of the line could be manned.

1792. State of the Allied Navies. Spain had only 56 serviceable capital ships; its personnel were of less than mediocre ability. Holland's 49 ships were good, but light. Portugal had 6 capital ships, Naples 4. The British Navy consisted of 115 ships of the line, in good shape, commanded by professionals, but many were undermanned. The 2 principal British operational forces were the Channel Fleet under Admiral Lord **Richard Howe** and the Mediterranean Fleet under Admiral Sir **Samuel Hood.**

1792. No Naval Operations of Significance.

1793, August—September. Inconclusive Maneuvering. The French fleet, under Admiral **Morad de Galles,** tentatively put to sea off the Atlantic coast of France. He avoided battle with Howe. De Galles was a capable sailor, but was unable to control his unruly crews. By the end of the summer he had given up his command; the fleet remained in port.

1793, August 27–December 19. Siege of Toulon. Hood, with 21 ships of the line, entered the port, which at once declared for the monarchy. The British seized the great naval arsenal together with about 70 vessels, 30 of them ships of the line, nearly half of the French Navy. A Spanish squadron under Admiral **Juan de Langara** accompanied the British. Republican land forces invested the fortress (September 7), but little was done until 2 inept commanders were replaced by General **Jacques E. Dugommier,** who adopted a plan prepared by young Colonel **Napoleon Bonaparte,** commanding an artillery brigade. The forts commanding the anchorage were soon carried by the French troops (December 16). The allied naval forces hastily left (December 19). The British destroyed or carried away part of the French fleet, but 15 ships of the line were left untouched, through Spanish neglect. The Spanish Navy took no further part in the war as an ally of Britain. Colonel Bonaparte was promoted to brigadier general.

Operations on Land, 1794–1795

1794. Opposing Plans in the North. The allied "plan of annihilation" was proposed by Baron **Karl Mack von Leiberich,** Austrian chief of staff. It was matched by Carnot's plan to clear the invaders from French territory.

1794, May 11. Courtrai. The French forces of General **Charles Pichegru**'s Army of the North defeated the Allied English and Austrian Army of Field Marshal **Count Charles von Clerfayt.**

1794, May 18. Battle of Tourcoing. Pichegru's army (temporarily commanded by General **Joseph Souham**) of 70,000 men defeated Saxe-Coburg's badly managed Austro-British-Hanoverian force of 74,000 men. French losses were 3,000 men and 7 guns, the allies lost 5,500 men and 60 guns.

1794, May 23. Battle of Tournai. A fiercely fought drawn battle, between forces almost equal in strength (about 50,000 allies and 45,000 French). Both sides retreated, losses were 3,000 for the allies, the French lost 6,000 men and 7 guns.

1794, June 17. Battle of Hooglede. A fierce French counterattack retrieved victory from defeat. The allies began to withdraw northward.

1794, June. Formation of the Army of the Sambre and Meuse. Jourdan's Army of the Moselle had moved north and invested Charleroi (June 12). He was now placed in command of the combined armies, totaling 75,000–80,000 men.

1794, June 26. Battle of Fleurus. Coburg, with 46,000 men, moved to relieve Charleroi. Unaware the town had fallen (June 25), 5 allied columns struck the 81,000 French grouped in a 20-mile circumference around the town. At first successful all along the line, the allied right was halted by French counterattacks, General **Jean B. Kléber** driving the Prince of Orange from the field. Jourdan led a counterblow in the center. Despite the success of his left against the overextended French right, Saxe-Coburg called off the attack. Next day he retreated. Jourdan pursued at once. Losses were about 5,000 each.

1794, July–August. French Victories in Belgium and on the Rhine. They entered Brussels (July 10), Antwerp was occupied (July 27), and the British contigent sailed home. Belgium was abandoned forever by the Austrians. Jourdan followed Coburg across the Roer and cleared the left bank of the Rhine in desultory operations against the Prussians.

1794, September–December. French Victories Elsewhere. General **Jean Victor Moreau,** who had distinguished himself under Jourdan's command in Belgium, was given command of a newly formed Army of the Rhine and Moselle; in late summer he advanced into the Rhineland and pressed forward despite a setback near **Kaiserslautern;** by October he had forced the allies to cross the upper Rhine and had laid siege to Mainz. An army under Pichegru invaded Holland. In the Alps and along the Mediterranean coast the French had also had minor successes, driving the allies completely out of Savoy, and pressing eastward as far as Savona on the Mediterranean coast. In the southwest French armies had penetrated both flanks of the Pyrenees.

1795, January–March. French Occupation of the Netherlands. Pichegru's advance was facilitated by the frozen rivers, lakes, and estuaries. The Dutch fleet, frozen in the Texel, was captured by French cavalry, an exploit unique in military history. Holland now became the Batavian Republic, a satellite and active military ally of republican France.

1795, April 5–June 22. Treaties of Basel. Prussia, financially exhausted, made peace with France. Saxony, Hanover, and Hesse-Cassel followed suit. So did Spain, ceding Santo Domingo to France.

1795, June 27. Landing on Quiberon. French émigrés and a small contingent of British troops, totaling 17,000 men, landed from British warships, attempted to invade Brittany. They were disastrously defeated by the waiting French Army of the West under Hoche with 13,000 men (July 16–20). The émigré forces lost 8,200 men, mostly captured, while French losses were less than 500. The British troops were safely reembarked (July 20–21).

1795, August 22. The Directory. Republican France rocked internally when a new government—a 5-man group—was established by the Constitution of 1795. After a monarchist outbreak in Paris was put down by General Bonaparte (October 5), the Convention was dissolved. Military operations in the field continued, Carnot remaining in control.

1795. Operations along the Rhine. Jourdan's Army of the Sambre and Meuse, about 100,000 strong, was west of Coblenz. Pichegru's Army of the Rhine and Moselle, about 90,000 men, was in Alsace and blockading Mainz in the Palatinate. Another force captured Luxembourg after a siege of 8 months (June 26). Opposite Jourdan were about 100,000 allies under Austrian Marshal Clerfayt; Wurmser had 85,000 east of the upper Rhine. Jourdan attempted an invasion of Germany (September 3–6), which

Wurmser ignored. Jourdan, advancing toward Frankfurt, was outmaneuvered near **Höchst** by Clerfayt (early October). Clerfayt then defeated Pichegru's forces blockading **Mainz** (October 29) and invaded the Palatinate. Then, unaccountably, he consented to a general armistice (December 21). Jourdan's defeat was attributable to the treachery of Pichegru, who, involved in the royalist plot, had defected to the Austrians, after betraying the invasion plans.

1795. Operations in Italy. The French Army of Italy, under the over all command of General **Barthélemy Schérer,** had progressed slightly eastward and northeastward along the Mediterranean coast. The principal action in this area was at **Loano** (November 23–25), where General **André Masséna** disloged allied forces from their positions in the mountains.

1795–1796. Operations in the Vendée. Revolt blazed briefly again. A British expedition with 2,500 troops approached the Ile d'Yeu off the Poitou coast. Before it could disembark, French troops crushed the uprising and the British withdrew (November 15). By next spring pacification of the Vendée had been brutally completed by Hoche.

War at Sea, 1794–1795

1794, May 29–June 1. Battle of the First of June. In an extraordinary effort to avert famine in France, 130 merchant ships laden with purchased foodstuffs sailed from the United States (April). To protect its safe arrival at Brest, Admiral **Louis T. Villaret de Joyeuse** put to sea with his fleet. Howe intercepted the French fleet in the Atlantic, 400-odd miles off Ushant. When the fight started the opponents had each 26 ships of the line, but before it ended 4 additional French vessels joined Villaret. The 4-day running fight ended when Howe flung his entire fleet into the midst of the French. In an epic melee, 6 French ships were taken and a seventh

vessel sunk. The French fleet broke off combat, returning to Brest, to find the food convoy had arrived safely. Tactically a British victory, the French admiral had nevertheless accomplished his mission by drawing the British from the convoy's path. No other actions of major importance occurred in the Atlantic for the rest of the year.

1794. Operations in the West Indies. Admiral Sir **John Jervis** completed the occupation of Martinique, St. Lucia, and Guadeloupe (April). These islands were later recovered by the French governor, **Victor Hugues.** In the flux of naval maneuvering, however, the British Navy retained the upper hand, destroying a great part of the French merchant marine.

1794, August 10. Capture of Corsica. Accomplished by a British joint operation.

1795. Operations in the Atlantic. Only 2 minor actions took place, both in connection with the ill-fated landings in Quiberon Bay (see p. 745). Admiral **William Cornwallis** with 5 blockading ships escaped Villaret's 12 in the Bay of Biscay (June 8–12). Villaret was prevented from attacking the convoy of troop transports only by luck and the boldness of Admiral **John Warren,** the convoy commander. Shortly after this (June 23), Admiral Lord **Alexander Hood of Bridport** (younger brother of Samuel Hood) with 12 ships engaged Villaret off Ile de Groix, capturing three French vessels. For most of the year, both sides engaged in commerce destruction, the British Channel Fleet being held in port, while the French fleet at Brest was in too poor condition to put to sea.

1795. Operations in the Mediterranean. Admiral Lord **William Hotham** twice clashed inconclusively with Admiral **Pierre Martin's** refurbished Toulon fleet. Tactically, the British were favored; strategically, the honors went to France. In November, efficient Sir John Jervis succeeded indolent Hotham.

Operations on Land, 1796–1798

War was waged in two major theaters—Germany and Italy—under the overall direction of the efficient Carnot, who was now in effect both minister of war and chief of staff. The character of the conflict changed from a defense of national territory to a war of conquest by France. The motivation was partly missionary zeal in spreading the gospel of "Liberty, Equality, Fraternity" throughout Europe, but there was also a definite economic-logistic impulse: France could no longer feed the vast citizen armies she had raised for defense. Carnot's plan was a pincers movement against Austria: the armies of the Sambre and Meuse and of the Rhine and Moselle, driving through Germany, would unite at Vienna with the Army of Italy.

GERMANY; THE NORTHERN FRONT

East of the Rhine a vast war of maneuver took place in the triangle Dusseldorf-Basel-Ratisbon, with the youthful **Archduke Charles** of Austria, utilizing interior lines, pitted against Jourdan and Jean Victor Moreau. Charles had complete liberty of action; the French generals, independent one of the other, were hampered by the orders of the Directory. Jourdan's mission was to attract Charles to the north so as to permit Moreau to invade Bavaria. Jourdan would then join Moreau in an advance into Austria.

1796, June–July. Jourdan's Invasion. With 72,000 men, he crossed the Rhine at Düsseldorf (June 10), penetrated to Wetzlar. Charles repulsed the French advance at **Wetzlar** (June 16). His mission accomplished, Jourdan fell back across the Rhine (July).

1796, June–August. Moreau's Invasion. With 78,000 troops, he crossed at Strasbourg (June 23–27). Charles, leaving General **Alexander H. Wartensleben** with 36,000 men to observe Jourdan, marched with 20,000 against Moreau. According to plan, Jourdan recrossed the Rhine and drove Wartensleben back. After the indecisive **Battle of Malsch** (July 9), Charles fought an inconclusive battle at **Neresheim** (August 11), and was pushed across the Danube by Moreau, between Ulm and Donauwörth (August 12). Jourdan again crossed the Rhine.

1796, August 24. Battles of Amberg and Friedberg. Charles, reinforced, recrossed the Danube, left 30,000 men under General **Latour** to watch Moreau, and hurried north with 27,000 to find Jourdan with 34,000 still pressing Wartensleben near Amberg. Charles struck the French right flank while Wartensleben's 19,000 attacked frontally. Jourdan, suffering losses of 2,000 to the Austrian 500, was decisively defeated and retired. That same day, Moreau attacked and defeated Latour at **Friedberg.** Jourdan, learning this, regrouped near Würzburg on the Main. Charles pressed after him.

1796, September 3. Battle of Würzburg. Charles, with an advantage of 44,000 to 30,000, enveloped both French flanks in a brilliant manuever, the Austrian cavalry distinguishing itself. Jourdan managed to disengage after losing 3,000 men and 7 guns and fell back to the Rhine in a series of running fights. The Austrian loss was 1,500. An armistice was concluded here, and Charles turned his attention to Moreau in Bavaria, who, learning of Jourdan's defeat, broke off his pursuit of Latour and retired. Reaching the Rhine at Hunningen, he crossed to safety (October 26).

1797, April 18. New French Offensive. Hoche, who had superseded Jourdan as commander of the Army of the Sambre and Meuse, moved up the right bank of the Rhine to assist Moreau's Army of the Rhine and Moselle in a planned crossing at Kehl. Charles had been sent to meet Bonaparte's threat in Italy (see p. 751). Latour commanded the Austrians opposite Moreau; under him General **Werneck** held the lower Rhine. Crossing the Rhine between Düsseldorf and Coblenz, Hoche struck Werneck near Neuwied, and defeated him in the **Battle of the Lahn** (April 18). Moreau crossed the river two days later. The Austrians were driven back to Rastatt. Operations ended with the **Peace of Leoben** (see p. 751).

ITALY; THE SOUTHERN FRONT

1796, March 27. Bonaparte Commands the Army of Italy. This army, some 45,000 ill-fed, poorly clothed men, lay along the Riviera from Nice almost to Genoa. British sea power blockaded the coast ports, while beyond the hills to the north on the edge of the Lombardy Plain lay 2 allied armies: Baron **Colli** (Piedmontese) with 25,000 men and **Jean Pierre Beaulieu** (Austrian) with 35,000. Their widely separated dispositions invited attack. Bonaparte concentrated for an offensive just as Beaulieu extended his own left to the sea at Voltri.

1796, April 12. Battle of Montenotte. Striking north between his opponents, Bonaparte with 10,000 men drove in Beaulieu's right flank, defeating an Austrian force of 4,500 at Montenotte. Bonaparte then continued to press on, widening the gap between the allied armies.

1796, April 14–15. Battle of Dego. Pressing his advantage, Bonaparte drove the Austrians out of the town, but a counterattack regained it. Bonaparte in person gathered reserves who threw the Austrians out again. Beaulieu retreated northeast to Acqui. Instead of pursuing, Bonaparte turned west, on Colli's army, which had begun retirement.

1796, April 21. Battle of Mondovi. Colli, with 13,000 men, attempting to stand against Bonaparte's 17,500, held back 1 French assault, but was finally driven out. Bonaparte's rapid pursuit ended when Colli sued for an armistice (April 23), which put Piedmont out of the war (April 28).

1796, April 23–May 6. Advance to the Po. Bonaparte shifted to the northeast. On the north bank of the river, Beaulieu had spread his army out in cordon defense, covering all possible crossing places along a 60-mile front.

1796, May 7–8. Crossing at Piacenza. Bonaparte, having demonstrated on a wide front along the river line, suddenly surprised Beaulieu by a swift crossing, jeopardizing the Austrian left and its line of communications with Mantua. Beaulieu retreated east in haste, abandoning Pavia and Milan.

1796, May 10. Battle of Lodi. The Austrian rear guard bitterly contested a bridge over the Adda. Putting himself at the head of a massed column of infantry, Bonaparte led a bayonet charge across the bridge, which decided the day and won from his soldiers his soubriquet of "Little Corporal." While Beaulieu continued his retreat toward the Tyrol, Bonaparte entered Milan in triumph (May 15). The Austrian garrison of Milan surrendered the citadel 6 weeks later (June 29). Piedmont made peace with France (May 21); King Victor Amadeus II surrendered Savoy and Nice to France, and granted the French Army the right to garrison fortresses in Piedmont.

COMMENT. *In 17 days Bonaparte had defeated 2 enemies in detail and had conquered Lombardy. After a few days of rest and reorganization, he proceeded after the Austrians.*

1796, May 30. Passage of the Mincio. Beaulieu had reorganized his own forces, holding with 19,000 men the line of the Mincio River from Lake Garda south to Mantua, where he had an additional 11,000. Bonaparte, who had arrived at the Mincio (May 29) with 28,000, concentrated his forces and thrust through this cordon defense at Borghetto. Beaulieu, barely saving himself from destruction, hurriedly retreated across the Adige and into the Tyrol. Except for Mantua, with a garrison of 13,000 men, all north Italy was in Bonaparte's hands. He invested Mantua (June 4); most of the French army was deployed along the Adige to cover the siege operations.

1796, July–August. Arrival of Wurmser. He hurried south from the Tyrol with a new Austrian army. Sending General **Quasdanovich** with 18,000 men down the west side of Lake Garda, toward Brescia, to cut the French communications, the elderly Wurmser himself with 24,000 came down the east side, through the Adige Valley, to relieve Mantua; another detachment of 5,000 came down the Brenta Valley. Abandoning the siege of Mantua and recalling his troops from the Adige, Bonaparte, who had 47,000 men in all, concentrated southwest of Lake Garda against Quasdanovich, leaving Wurmser to move south unopposed to join the 13,000-man garrison of

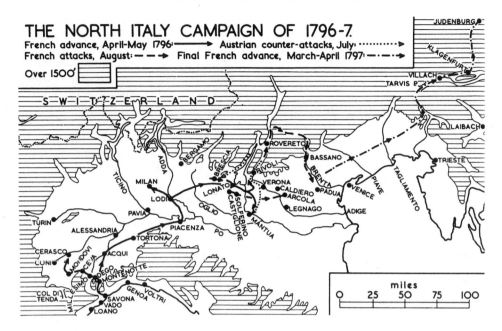

THE NORTH ITALY CAMPAIGN OF 1796-7.
French advance, April-May 1796: ⟶ Austrian counter-attacks, July: ·········▷
French attacks, August: — — ➔ Final French advance, March-April 1797: —·—·➔

Mantua. Quasdanovich had split his force in 3 columns; he cut the Milan-Mantua road (July 31). Advancing southeast, he expected to unite with Wurmser, who by this time had come around the south end of the lake, hoping to complete a double envelopment of the French.

1796, August 3. Battle of Lonato. While General **Pierre F. C. Augereau**'s division delayed Wurmser, Bonaparte threw the remainder of his army against Quasdanovich north of Lonato, defeating the 3 columns in detail. One of Quasdanovich's columns surrendered; the shattered remainder retreated north to the head of the lake.

1796, August 5. Battle of Castiglione. Bonaparte's entire army now turned upon Wurmser. Attacked on both flanks and struck in the rear by a turning movement around his left, the Austrian retreated with difficulty across the Mincio and began a general retirement to the Tyrol, having lost 16,000 casualties.

COMMENT. *Again Bonaparte, throwing himself between 2 enemy forces, had concentrated on each in turn and defeated them in detail.*

1796, August 24. Resumption of the Siege of Mantua. The Austrian garrison, reinforced by fugitives from Wurmser's army, was now 17,000 men.

1796, September. French Advance on Trent. Bonaparte moved north with 34,000 men, leaving 8,000 to invest Mantua. At the same time, Wurmser, who had regrouped, also decided to resume the offensive, but again he divided his forces. General **Paul Davidovich** with 20,000 would defend the Tyrol, while Wurmser himself with 26,000 would march down the Brenta Valley on Mantua. Bonaparte, unaware of the Austrian plan, defeated Davidovich at **Caliano** on the upper Adige and drove him back through Trent (September 2–5). Learning of Wurmser's movement down the Brenta, Bonaparte followed, overhauling the Austrians at Bassano after a series of forced marches.

1796, September 8. Battle of Bassano. Augereau and Masséna each attacked an Austrian flank; 1 entire Austrian division was cut off and surrendered, only a few units escaping to the east. Wurmser, however, reached Mantua with the remainder of his army and fought his way into the city, swelling its garrison to 28,000 (September 13). He attempted to expand the defended area by a sortie, but was driven back into the city by blockading French troops under Masséna and General **Charles Kilmaine** (September 15).

1796, November 1. Third Austrian Attempt to Relieve Mantua. Baron **Josef Alvintzy** sent Davidovich with 16,000 men down the Adige, while he moved on Vicenza with 27,000, east of the Brenta. Alvintzy expected to join Davidovich at Verona, then move south of Mantua. Bonaparte, with 30,000 men (not counting the 9,000 containing Wurmser's 28,000 in Mantua), interposed. Leaving General **Pierre A. Dubois** with 8,000 to delay Davidovich, he hurried east with 18,000 to meet Alvintzy; 4,000 were in reserve at Verona.

1796, November 12. Battle of Caldiero. Attacking Alvintzy's advance guard, the French were checked by a rapid Austrian build-up and finally withdrew to Verona, having lost 2,000 men. Alvintzy, however, hesitated to advance.

1796, November 15–17. Battle of Arcola. Bonaparte, swinging south around Alvintzy, crossed the Adige, then turned north, moving through a marshy area, to cut the Austrian communications. Despite Napoleon's personal attempt—flag in hand—to storm a bridge over the Alpone at Arcola, continuous French attacks across the causeway were repulsed. But on the third day, Augereau crossed by a trestle bridge below the village, Masséna made another assault on the main bridge, and at the same time a detachment of French cavalry rode to the rear of the Austrian position, blowing bugles. Fearing that they were encircled, Alvintzy's troops broke and retreated. That same day Davidovich drove Dubois back toward Verona, but Bonaparte, shifting from pursuit of Alvintzy, turned on Davidovich and chased him back into Trent (November 19).

1797, January. Fourth Attempt to Relieve Mantua. Once more Alvintzy divided his forces. With 28,000 men, he marched south down the Adige Valley, sending 15,000 more toward Verona and Mantua from the east. Bonaparte ordered the concentration of the bulk of his forces on the plateau of Rivoli, between the Adige and Lake Garda. He was there with 10,000 men under General **Barthélemy C. Joubert,** while Masséna with 6,000 more was hurrying up from Verona, and Generals **Antoine Rey** and **Claude P. Victor,** with 6,000 additional, were also approaching. Augereau with 9,000 at Verona was holding the line of the Adige against the threat from the east. General **Jean M.P. Serurier** with 8,000 continued the investment of Wurmser's 28,000 in Mantua.

1797, January 14. Battle of Rivoli. Alvintzy began a complicated attack in 6 columns; 3 of these were to attack Bonaparte's position on the Rivoli plateau frontally, while 1 each would envelop the French flanks; the sixth column would advance down the east bank of the river to cross over at Rivoli behind Bonaparte's main body. Joubert's division, despite a stout defense and the inspiration of Bonaparte's personal presence, was pressed back by the main Austrian assault, and the eastern Austrian flanking column had reached the plateau to threaten the French right flank when Masséna arrived. The penetration on the French right flank was smothered; the right of the Austrian center was assailed, and the western Austrian flanking column surrounded. By late afternoon, Alvintzy had been driven from the field. Bonaparte pursued immediately, sending Rey's division through the mountains to strike the flank and rear of the retreating Austrians. Spearhead of this pursuit was a regiment of infantry under Colonel **Joachim Murat** that had been ferried across Lake Garda, marching across Monte Baldo just in time to smash Alvintzy's efforts to reorganize. Austrian losses were over 12,000, including 8,000 prisoners-of-war and 8 guns, the French loss was about 3,200.

1797, January 14–February 2. Operations around Mantua. Meanwhile, 2 Austrian columns advanced toward Verona and Mantua from the east. One of these, 9,000 strong, under General **Provera,** slipped around Augereau and moved west across the Adige. Reaching the outskirts of Mantua (January 16), Provera at-

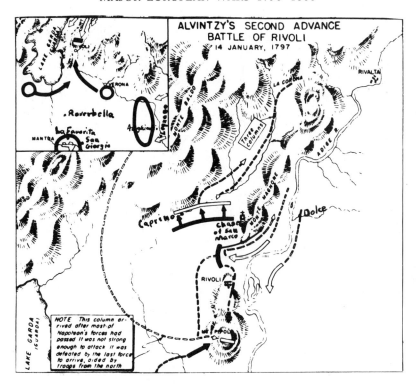

tacked the French besiegers, while Wurmser simultaneously attempted a sortie. While Serurier drove Wurmser back into Mantua, Bonaparte hurried up from Rivoli with Masséna and, joining Augereau, surrounded Provera's column and forced him to surrender. Wurmser's garrison (now 16,000) in Mantua capitulated (February 2). In this campaign the French captured 39,000 prisoners, nearly 1,600 guns, and 24 colors. During the siege 18,000 Austrians had died, mostly from disease.

1797, March. Arrival of Archduke Charles. He replaced Alvintzy. In the Tyrol he had 14,000 regulars and 10,000 Tyrolese volunteer riflemen, with 27,000 more under his immediate command along the Tagliamento.

1797, March 10. Invasion of Austria. Having been reinforced from France, again Bonaparte interposed himself between two enemy forces. With 41,000 men he moved against Charles on the Tagliamento, leaving Joubert with 12,000 to operate in the Tyrol. Charles retired behind

the Isonzo (March 18), but had to fall back again to avoid envelopment after Masséna smashed a detachment at **Malborghetto** (March 23). Bonaparte, in pursuit, crossed the snowy passes of the Julian and Carnic Alps in 3 columns. These united at Klagenfurt and advanced to Leoben, 95 miles from Vienna (April 6). At the same time Joubert, having driven in the Austrian forces opposing him in the Tyrol, had reached Lienz on the way to join Bonaparte in front of Vienna. The emperor sent emissaries to discuss peace with the Republican general.

1797, April 18. Preliminary Peace of Leoben. Bonaparte, knowing well that behind him lay a conquered country whose seething inhabitants were threatening his communications, boldly dictated truce terms without any reference to the Directory. They were accepted; Bonaparte then returned to Paris.

1797, October 17. Treaty of Campo Formio. The terms laid down by Bonaparte at Leoben were formally accepted and the face of Europe changed. Belgium became a part of France, and a northern Italian (Cisalpine) Republic was recognized by Austria, which in return for its losses was given the Republic of Venice. From Basel to Aldernach the left bank of the Rhine was French.

COMMENT. *The French victory in this War of the First Coalition was due to a number of complex factors. The most important of these were (1) the sweeping social and political changes brought about by the French Revolution, which were epitomized by the levée en masse of free citizens whose patriotic obligation brought them together in a people's army to defend the state; (2) revolutionary fervor which inspired the rank and file of the French Army to prodigies of valor and energy, which could not be matched by the better-trained, better-disciplined, stolid professionals of 18th-century Europe; (3) the first flowering of the Industrial Revolution, permitting the mobilized state to raise and support a larger proportion of its population than heretofore had been possible in purely agrarian economies; (4) the organizing genius of Carnot in taking advantage of these revolutionary changes; (5) the strategic and tactical genius of young Bonaparte, who applied classical principles of war with skill, ingenuity, and audacity that entirely baffled his 18th-century pedestrian opponents, and whose brilliant campaign culminated in the dictation of peace terms in the heart of Austria.*

1798, January. Continuing War with Britain. Bonaparte was placed in command of an Army of England assembling at Dunkirk for proposed invasion. Convinced that this was a very doubtful project in the light of British command of the sea, Bonaparte suggested to the Directory (February) an expedition to Egypt, which he recognized as being a crossroads of the world. With Egypt under French control, he visualized an outlet to the Orient and a base from which the British could be driven from India.

1798, February. French Occupation of Rome. The pope was captured; a Roman Republic was proclaimed.

1798, April. Occupation of Switzerland. A new Helvetian Republic was established.

EGYPTIAN CAMPAIGN, 1798–1800

1798, April 12. Establishment of the Army of the Orient. The French Directory appointed Bonaparte to command. The politicians were glad to have this dynamic, popular leader out of the country. Immediately he began to assemble his new army at Toulon; meanwhile, the concentration at Dunkirk continued as a feint to keep the British fleet out of the Mediterranean.

1798, May 19–July 1. Voyage to Egypt. Bonaparte sailed from Toulon with 40,000 men, escorted by Admiral **François P. Brueys** with 13 ships of the line. Capturing Malta en route (June 12), the expedition landed near Alexandria.

1798, July 2. Storming of Alexandria. The French advanced on Cairo.

1798, July 21. Battle of the Pyramids. Marching up the left bank of the Nile, Bonaparte was harassed by the Egyptian forces, whose main component consisted of the Mamelukes— courageous horsemen unaccustomed to organized modern warfare. The Egyptians, 60,000 strong, concentrated under **Murad** and **Ibrahim** near Embabeh, with their left flank resting on the Pyramids at Gizeh. The French Army of 23,000 attacked to turn the Egyptian left, its divisions formed in squares in checkerboard fashion. A wild charge of the Mameluke horsemen was repulsed, their entrenched camp stormed, and the Egyptians dispersed with great loss.

1798, July–August. Occupation of Egypt. Cairo was seized (July 22). French forces moved in pursuit of Ibrahim, who had retreated toward Syria.

1798, August–December. Allied Reaction. Nelson's annihilation of the French fleet (see p. 754) disrupted Bonaparte's entire plan. The French army in Egypt was completely isolated from the homeland and surrounded by an actively hostile population. A Turkish army under **Achmed Pasha** (the "Butcher") gathered in Syria; another assembled at Rhodes to invade Egypt under escort of an English squadron.

1799, January 31. Bonaparte Takes the Offensive. He marched into Syria with 8,000 men. El Arish (February 14–15) and Jaffa (March 3–7) were taken in turn, and Acre invested (March 17). British Navy Captain Sir **Sidney Smith** took command of the garrison.

1799, April 17. Battle of Mount Tabor. Achmed's army attempted to relieve Acre. Kléber's division, in hollow squares, was surrounded, but Bonaparte enveloped the Turkish force with the remainder of his troops. Under a combined attack the Turks were defeated and driven across the Jordan.

1799, May 20–June 14. Retreat from Acre. After Acre resisted several assaults, and plague broke out in his army, Napoleon retired. After a grueling retreat, continually harassed, the French reached Cairo (June 14). He had lost 2,200 dead, nearly half from disease.

1799, July 25. Battle of Aboukir. Under English escort, the Turkish force from Rhodes, 18,000 strong, landed at Aboukir and entrenched (July 15). Bonaparte at once concentrated all available forces—6,000 men—and moved to the Delta. The Turks lay in 2 concentric positions. The outer position was rapidly brushed away by French attack. An assault on the inner ring was repulsed, but a second assault, led by General **Jean Lannes,** was successful. Murat's cavalry at the same time broke through, and the Turks were annihilated. The citadel of Aboukir surrendered (August 2). Napoleon's losses were 900 killed and wounded.

1799, August 23–October 9. Return to France. Bonaparte, satisfied that he had stabilized the Egyptian situation, realizing that further conquest and glory were unlikely without possibility of reinforcements from France, and alarmed by developments in Europe, turned over the command to Kléber. He returned to France in a fast frigate, eluding British naval forces on the way.

War at Sea, 1796–1800

1796–1798. The War of Commerce Destruction. This was continuous. Frigates of both sides preyed on the commerce of the other, with honors rather evenly divided.

1796. Inconclusive Operations. The British rather ineffectively blockaded the main French naval bases of Brest and Toulon, and at the same time observed both Dutch and Spanish activities. French Admiral **Richery** broke out of Toulon (September 14) with a small squadron, cruised to Newfoundland where he harassed the Canadian fishing industry, captured a number of merchantmen, and returned safely. A more imposing joint force left to invade Ireland from Brest (December 15); 43 warships and transports, under Admiral de Galles, carried some 13,000 troops under Hoche's command. Ill-fitted and badly managed, the flotilla was scattered by bad weather. A portion arrived at Bantry Bay (December 21), but foul weather prevented landing and the project was abandoned. Five ships were lost in storms, 6 more captured by the British; the remainder finally got back. The only significant British naval activity in the Mediterranean was Captain **Horatio Nelson's** singlehanded harassment and commerce destruction along the Franco-Italian coast, which hampered the sea communications of the French Army of Italy.

1796, August 19. Treaty of San Ildefonso. Spain joined France in the war against Britain. The British position in the Mediterranean was gravely threatened, and it was decided to withdraw. Corsica and Elba were evacuated, and the Mediterranean fleet, now under Jervis, moved to Gibraltar (December 1) and transferred its activities to the Atlantic coasts of Spain and Portugal.

1797, February 14. Battle of Cape St. Vincent. Spanish Admiral **José de Córdova,** with 27 ships of the line, sailed from the Mediterranean past Gibraltar en route to Brest and a planned junction with the French fleet, and a proposed invasion of England. Jervis with the former British Mediterranean fleet, 15 ships of the line, cruising off the Portuguese coast to prevent such a move, met the Spanish off Cape St. Vincent. Cordova's ships—in 2 groups of 9 and 18, respectively—were good, but their officers and crews relatively untrained. Jervis sailed between them in line, then turned on the larger group, which might have escaped had not Nelson, commanding HMS *Captain*, broken away from the British line without orders and after a wide sweep placed his vessel in the path of the Spanish flotilla. He engaged the 7 leading ships until the rest of Jervis' squadron swarmed among them. In the resulting melee 4 Spanish ships were captured—2 by Nelson, who boarded them in quick succession. The shattered Spanish fleet escaped next day to Cádiz. Jervis' victory marked the end of the Spanish threat to England. It was also significant in definitely confirming the tactical superiority of the melee over the more formal line of battle, at least when initiated by a more aggressive force with superior seamanship.

1797, April–August. "The Breeze at Spithead." The Royal Navy in home ports was rocked by mutinies at Spithead (April 16), the Nore (May 12), and other home ports—revolts of the enlisted men against existing brutality, poor food, and poor pay. The mutineers asserted willingness to return to duty if the French fleet put to sea. Suppressed after several weeks of uproar, the mutinies nevertheless resulted in great improvement in the British seaman's status.

1797, October 11. Battle of Camperdown. British Admiral **Adam Duncan,** with 16 capital ships, attacked Admiral **Jan Willem de Winter's** Dutch fleet of 15 as it emerged from the Texel. Duncan charged the Dutch line in 2 columns, broke it into 3 groups, and in the resulting melee captured 9 vessels, including the flagship. The Dutch put up bitter opposition; the British suffered severe damage and the prizes were so badly battered as to be completely unusable.

1798, May–June. British Return to the Mediterranean. News of Bonaparte's troop movement to Egypt in June caused the Admiralty to send a small British fleet to the Mediterranean under young Admiral Nelson. The French expedition evaded his search; but shortly after its landing in Egypt (see p. 752), Nelson discovered the French fleet of 13 ships of the line anchored in the Bay of Aboukir.

1798, August 1. Battle of the Nile. French Admiral Brueys was entirely unprepared; part of his crews were ashore. Nelson attacked in the late afternoon, a portion of his 13 ships carefully sailing between the French line and the shallow coastal waters, the remainder closing from the outer side, bringing the French between 2 fires. In the dusk and through the night the British fleet cruised slowly down the French line, concentrating on ship after ship in succession. By dawn only 3 French vessels remained; one was run aground and burned by her crew and the other 2 escaped. Brueys was among the killed in this catastrophic defeat.

1798, August 22. Invasion of Ireland. General **Jean J. A. Humbert** landed with 1,200 troops at Killala Bay. The invaders were surrounded and forced to surrender by Lord **Charles Cornwallis** (September 8). A reinforcing squadron under Admiral **J. B. F. Bompard** was overtaken and destroyed by Sir **John Warren** (October 12).

1798–1800. Operations on Malta. A Maltese uprising broke out against the French occupation force (September 1798). Nelson sent ships and men to assist the Maltese guerrillas against French General **C. H. Vaubois,** who was driven into Valetta and besieged. Starvation finally forced the French to surrender (September 5, 1800).

1798–1800. British Domination of the Seas.

War of the Second Coalition, 1798–1800

1798, December 24. Russo-British Alliance. While Napoleon's Egyptian campaign was continuing, Emperor **Paul I** of Russia organized the Second Coalition, England being his principal partner. Austria, Portugal, Naples, the Vatican, and the Ottoman Empire joined (these latter 3 already being at war with France). Preliminary operations had been going on in Italy: French General Joubert overran Piedmont in November and early December, while a Neapolitan army, commanded by Austrian General Karl Mack von Leiberich, had attacked the Roman Republic and captured Rome (November

29), only to be driven out by French General **Jean Etienne Championnet** (December 15). The over-all allied plan contemplated that an Anglo-Russian army under the **Duke of York** would drive the French from the Netherlands, an Austrian army under the Archduke Charles was to oust them from Germany and Switzerland, and a Russo-Austrian army under 70-year-old Marshal **Alexander Suvarov** would expel them from Italy. Available allied forces totaled about 300,000 men, not counting Mack's 60,000 unreliable Neapolitans. French strength was about 200,000, in 5 separate armies: Jourdan with 46,000 was on the upper Rhine; Masséna had 30,000 in Switzerland; Scherer with 80,000 was in northern Italy; Championnet had about 30,000 mixed French and Italian troops invading Naples in southern Italy; approximately 24,000 (including the satellite Batavian army) were in Holland under General **Guillaume Brune.** Already France was incorporating vassal legions in her armies: Dutch, Piedmontese, Italian, and Swiss. (In Egypt, Bonaparte had recruited Mameluke, Maltese, and Coptic groups.) Despite the allied preponderance of force, Carnot ordered an early advance on all fronts.

ITALY—THE SOUTHERN FRONT

1799, January 11. Capitulation at Capua. Mack fled to the French lines to save his life from his mutinous Neapolitan troops. Championnet then stormed Naples (January 24). Establishment of the satellite Parthenopean Republic followed.

1799, March. Action along the Adige. Farther north, Scherer with 41,000 men moved against General **Paul Kray**'s 46,000 Austrians, hoping to win a victory before Suvarov's Russian army and Austrian reinforcements could join Kray. Late in March, French probes around Verona were repulsed. Both armies now moved to attack each other south of Verona.

1799, April 5. Battle of Magnano. Scherer's attack, at first successful, was checked. Kray threw his reserves at the French right flank and broke it in. Scherer retired in considerable dis-

order. Flamboyant, eccentric Suvarov arrived soon after. Assuming command of the allied army, he advanced westward with 90,000 men, driving the French before him. Leaving Kray with 20,000 men to besiege Mantua and Peschiera, he caught up with the French midway between Brescia and Milan.

1799, April 27. Battle of Cassano. Moreau, who had replaced Scherer, had about 30,000 men to oppose nearly 65,000 allies. After desperate resistance, he was driven off the field. Suvarov entered Milan in triumph (April 28), then seized Turin. Largely because of differences with the Austrian government, Suvarov gave up his pursuit of the French, and allied forces were scattered around northern Italy besieging French garrisons. Meanwhile, General **Jacques E. J. A. Macdonald,** who now commanded the French army in southern Italy, was hastening north with 33,000 men. Moreau, based on Genoa, attempted with some success to distract Suvarov's attention from this reinforcement. Suddenly, in mid-June, the Russian general, who had about 40,000 men scattered around Alessandria, found himself between the 2 French armies.

1799, June 17–19. Battle of the Trebbia. Concentrating 37,000 men unexpectedly against Macdonald, Suvarov defeated him in a hard-fought engagement. French losses were some 11,500; allied casualties 5,500. Macdonald, closely pursued through hostile country by energetic Suvarov, finally joined Moreau near Genoa, losing another 5,000 men en route. Suvarov now pressed into the Apennines to drive the Army of Italy back to the Riviera. Moreau was relieved of his command by the Directory and replaced by Joubert (August 5).

1799, August 15. Battle of Novi. Joubert with 35,000 men attacked Suvarov with 50,000. But Joubert was killed, and his army decisively defeated with the loss of 11,000 men. Suvarov, who had lost 9,000, promptly pursued and drove the French across the Apennines. The pursuit was halted, however, when he discovered that the French Army of the Alps, some 30,000 strong, under Championnet, had entered Italy via the Mt. Cenis Pass. Suvarov turned northward, but before he could deal with Championnet, he was ordered to march to Switzerland with 20,000 Russians, leaving operations in Italy to Marshal **Michael Melas,** with about 60,000 Austrians.

1799, November 4. Battle of Genoa. Melas defeated Championnet, driving the French back across the Alps.

COMMENT. *By year's end practically all of Bonaparte's gains of 1796–1797 had been erased; the credit goes almost entirely to doughty old Suvarov.*

GERMANY AND THE NETHERLANDS

1799, March. Jourdan's Invasion of Germany. The Army of Mayence, 40,000 strong, crossed the Rhine at Kehl and commenced operations against the Archduke Charles, who had about 80,000. Checked at **Ostrach** (March 21), Jourdan boldly determined to attack the archduke at Stockach, with his main effort against the Austrian right. Charles, unaware of Jourdan's intentions, at the same time was leading half of his army forward on a reconnaissance in force.

1799, March 25. Battle of Stockach. Jourdan's initial assault gained some ground, and at first he thought he had won a victory. But Charles held on grimly and sent for the remainder of his army, giving him 46,000 men on the field to Jourdan's 38,000. The Austrian counterattack smashed through the center of Jourdan's loosely linked divisions. Austrian losses were 6,000 against the French 3,600, but Jourdan's offensive had been wrecked. He withdrew his shaken army at nightfall and made good his retreat to the Rhine. There he resigned his post. His army was put under Masséna (see below). There was little further action in Germany for the remainder of the year, the attention of both sides being focused instead on operations south of the upper Rhine in Switzerland, and west of the lower Rhine in the Netherlands.

1799, August. Operations in the Netherlands. A British force, 27,000 strong, under the **Duke of York,** landed on the tip of Holland's peninsula, south of the Texel. The Dutch Republican fleet in the Texel surrendered to a British squadron without fighting (August 30).

Reinforced by 2 Russian divisions, the Duke of York took the offensive.

1799, September 16. First Battle of Bergen. The combined Russo-British army, 35,000 strong, was met by Brune with 22,000 Franco-Batavian troops. The Russians and English could not coordinate their operations, and were defeated. Allied losses were 4,000; the French had 3,000 casualties. Two weeks later the allied forces advanced again.

1799, October 2. Second Battle of Bergen. This time the allies drove the republican forces from the field, each side losing about 2,000 men. Pressing his advantage, York continued south.

1799, October 6. Battle of Castricum. The allies were checked in a spirited action along the sand dunes of the North Sea coast, losing 3,500 men to the republicans' 2,500. Lack of Russo-British liaison again contributed to York's defeat. Realizing that he did not have sufficient force to drive the French from Holland, and since his principal mission—destruction of the Dutch fleet—had been accomplished, he withdrew northward.

1799, October 18. The Convention of Alkmaar. This ended hostilities in the Netherlands. The allied army withdrew from Dutch territory; 8,000 French and Dutch prisoners held in England were returned, but the Dutch fleet remained in British hands.

SWITZERLAND AND THE ALPS

1799, March. Operations in Switzerland. Masséna had nearly 30,000 men in central Switzerland. In accordance with Carnot's overall plan, he was to advance through the mountainous Vorarlberg and Grisons regions, covering the right flank of Jourdan's army. Surprising the Austrians by his early advance through the snowy passes, Masséna quickly crossed the upper Rhine near Mayenfeld, and captured most of the 7,000-man force around Chur in the Grisons. The Austrian garrison of Feldkirch, on Masséna's left flank, however, repulsed 2 efforts to capture the town (March 7 and 23). Masséna now held up his advance to await developments north of the Rhine. Meanwhile, he sent a column of 10,000 men under General **Claude Lecourbe,** from the vicinity of the Splugen Pass, raiding down the valley of the upper Inn into the Tyrol. Here Lecourbe was joined by a small French column which had marched north from the Army of Italy, the combined force creating consternation in the western Tyrol.

1799, April. Increasing Austrian Pressure. The defeats of the French armies in Germany and Italy exposed Masséna's northern and southern flanks. Austrian Generals **Heinrich J. J. Bellegarde** and **Friedrich von Hotze** brought overwhelming numbers to bear on Lecourbe's force in the Inn Valley. This was driven back to the upper Rhine.

1799, May. Formation of the Army of the Danube. Masséna took over Jourdan's defeated army as well as his own. He was responsible for defense of the Rhine south of Mainz, as well as of Switzerland. With his main body of 45,000 he withdrew slowly toward Zurich, followed by the Archduke Charles and Hotze, with about 55,000 men. Masséna repulsed an Austrian attack on his entrenchments at **Zurich** (June 4), but the pressure of superior numbers and the doubtful loyalty of the Swiss caused him to withdraw farther westward (June 7). Things remained quiet in Switzerland for the next 2 months, since Charles did not feel his strength permitted further advance.

1799, August. Masséna Resumes the Initiative. He defeated Charles's left wing in the rugged mountains of the upper Rhine and Rhone valleys. He then advanced against **Zurich,** but was repulsed (August 14). Two days later he repulsed an Austrian attack against his left flank at **Dottingen.**

1799, August–September. New Allied Plans. Charles was to march north through western Germany to join the Russo-British force of the Duke of York in driving the French from the Netherlands. Suvarov with his Russians would march north from Italy to Switzerland to drive Masséna back into France. The departure of Charles left less than 40,000 allies in Switzerland under General **Aleksander M. Korsakov,** with Suvarov still advancing from Italy. While Lecourbe with 12,000 men held the St. Gotthard Pass, Masséna with 40,000 struck in central Switzerland.

1799, September 25. Third Battle of Zurich. Masséna with 33,500 swept Korsakov's 20,000 from the field, inflicting 8,000 casualties and loss of 100 guns, French losses were 4,000. Suvarov, arriving at the foot of the St. Gotthard, found himself unexpectedly blocked. He fought his way through the pass, at great cost, only to learn of Korsakov's disaster. He managed to struggle back across the central Alpine spine to Ilanz, on the upper Rhine. Flighty, unstable Czar Paul relieved him of command, a sad ending for the last and most glorious campaign of a great soldier. Meanwhile, the disaster at Zurich, combined with the Russo-British defeat in the Netherlands, forced Charles to abandon his proposed march toward the Low Countries.

COMMENT. *The campaign of 1799 had been disappointing to both sides. The French had been driven from Italy, though they managed to hold their own elsewhere, despite several costly defeats. The allies, on the other hand, had failed miserably to take advantage of their opportunities, which would not come again for many years. Russia in disgust withdrew from the coalition.*

EUROPE

GREAT BRITAIN

1756–1763. The Seven Years' War. (See p. 730.)

1760–1820. Reign of George III.

1775–1783. War of the American Revolution. (See p. 774.)

1792–1800. Wars of the French Revolution. (See p. 741)

1795–1797. United Irishmen Revolt. These nationalists, inspired by **Wolf Tone, Napper Tandy,** and others, rose in armed rebellion to secure separation from England. They expected French help, but what little materialized was abortive and weak (see pp. 753, 754). The uprising was savagely repressed (1796–1797) by British General **Gerard Lake.**

1797, April–August. The Naval Mutinies. (See p. 754.)

1798, June 12. Battle of Vinegar Hill. The Irish revolt flared again when an armed mob without competent military direction stormed and burned Enniscorthy, County Wexford, after which they encamped on Vinegar Hill, overlooking the town. Surrounded by Lake's troops, the rebels surrendered after a short struggle. Again violence and brutality marked the suppression of the revolt.

FRANCE

1715–1774. Reign of Louis XV.

1756–1763. The Seven Years' War. (See p. 730.)

1774–1792. Reign of Louis XVI.

1778–1783. French Intervention in the American Revolution. (See p. 781.)

1789, July 14. The Storming of the Bastille. Traditional date for the beginning of the French Revolution.

1792–1800. Wars of the French Revolution. (See p. 741.)

IBERIAN PENINSULA

Spain

1759–1788. Reign of Charles III.

1761–1763. Spanish Participation in Seven Years' War. (See p. 740.) Plans were made to invade Portugal, England's ally.

1761–1762. Spain Invades Portugal. Bragança and Almeida were seized. The Portuguese Army, reorganized under command of **William of Lippe** and reinforced by a British contingent under General **John Burgoyne,** repelled the invasion.

1779–1783. American Revolution. (See p. 782.)

1793–1795. War with France. (See p. 742.)

1796–1799. War with England. (See p. 753.)

Portugal

1761–1763. War with Spain. (See above.)

THE NETHERLANDS

1780–1784. Dutch-English War. This was an extension of the American Revolution (see p. 784). The Netherlands lost several possessions in both the West and East Indies.

1785–1787. Civil War. Emergence of the Patriot party (pro-French) brought internal conflict between Stadholder **William V** and the States General. William called in Prussian troops (1787) to restore his authority.

1793–1795. War with France. (See p. 742.) The Netherlands was conquered and became the Batavian Republic. William V fled to England.

SCANDINAVIA

Sweden

1751–1771. Reign of Adolphus Frederick.

1771–1792. Reign of Gustavus III. A brilliant, although controversial, ruler and soldier. He was also one of Sweden's greatest authors.

1772, August 19. Royal Coup d'État. Since the reign of Ulrica Eleonora (1718; see p. 705), Sweden had been a constitutional monarchy, with most power shared by the Riksdag and the Royal Council. Gustavus seized power in a series of well-planned military moves. The remainder of his reign was plagued by continuous internal unrest. The *coup d'état* also aroused fears in Denmark and Russia; they contemplated war, but were deterred by Gustavus' readiness for war.

1788–1790. War with Russia and Denmark. (See p. 761.) Denmark played little part.

1792, March 29. Assassination of Gustavus. This was the culmination of an aristocratic conspiracy.

1792–1809. Reign of Gustavus IV.

Denmark and Norway

This was a period of relatively minor military interest or activity in Denmark. Danish participation as Russia's ally in the war against Sweden (see pp. 758, 761) was nominal.

GERMANY

Austria and Hungary

The Hapsburgs continued to wear 2 imperial crowns: that of the Holy Roman Empire (or Germany) and that of their hereditary Danubian lands, now commonly known as the Austrian Empire. Austria continued to be involved as a primary participant in all of the major wars of Western and central Europe, and was also still engaged in frequent wars in the Balkans, mostly against Turkey. The principal events were:

1740–1780. Reign of Maria Theresa.

1756–1763. The Seven Years' War. (See p. 730.)

1765–1790. Reign of Joseph II (as Holy Roman Emperor). He was coregent with his mother over Hapsburg lands, but with little influence until her death.

1772. First Partition of Poland. (See p. 760.)

1785. Revolt in Transylvania. The Rumanian peasants rose and massacred many Magyar nobles. The rising was crushed ruthlessly.

WAR WITH TURKEY, 1787–1791

1787. Austrian Alliance with Russia.

1788. Austrian Failures. Joseph, personally leading his armies, was completely unsuccessful in efforts in Transylvania and Serbia.

1789. Capture of Belgrade. After repulsing a Turkish invasion of Bosnia, the Austrians captured Belgrade.

1790. Peace Negotiations. Threats of Prussian intervention, the weakening of Russian cooperation because of her war with Sweden, revolt in the Austrian Netherlands (see below), and widespread unrest throughout the empire led Joseph and his successor, **Leopold,** to seek peace and to negotiate an armistice with Turkey (July–September).

1789–1790. Brabant Revolt. Imperial infringement of established civil rights in the Austrian Netherlands resulted in armed insurrection, led by **Henry van der Noot.** The rebellion was suppressed, but its objectives were achieved; Joseph restored the former constitution and privileges.

1790–1792. Reign of Leopold II.

1791, August 4. Treaty of Sistova. Austria returned Belgrade to Turkey in return for a strip of northern Bosnia.

1792–1800. Wars of the French Revolution. (See p. 741.) These were disastrous to Austria.

1792–1835. Reign of Francis II.

1793. The Second Partition of Poland. (See p. 760.)

1795. The Third Partition of Poland. (See p. 760.)

Prussia

By virtue of his victories in the War of the Austrian Succession, Frederick the Great made his nation a major power and a rival to Austria in Germany and the Empire.

1756–1763. The Seven Years' War. (See p. 730.) Frederick and Prussia emerged with enhanced power and prestige.

1772. First Partition of Poland. (See below.)

1777–1779. War of the Bavarian Succession. Death of Elector **Maximilian III** caused a crisis when Austria and Prussia selected different candidates for succession. Frederick mobilized and marched into Bohemia (July 1778). An Austrian army marched to oppose him, and for more than a year the 2 armies faced each other without battle. The bloodless conflict (also known as the "Potato War" because of the soldiers' diet) was ended by the **Treaty of Teschen** (May 13, 1779).

1786–1797. Reign of Frederick William II.

1790. Preparations to Intervene in the Austro-Turkish War. Frederick was satisfied when Austria made peace with Turkey (see p. 759).

1792–1800. Wars of the French Revolution. (See p. 741.)

1793. Second Partition of Poland. (See below.)

1795. Third Partition of Poland. (See below.)

1797–1840. Reign of Frederick William III.

POLAND

1768–1776. Civil War and Russian Invasion. A group of Polish noblemen ("Confederation of the Bar") organized an armed movement to defend the country against Russian religious and political aggression. Despite the efforts of French General Dumouriez, sent by France to command the dissidents, the movement failed, largely because of Russian armed intervention in support of the pro-Russian government.

1772. First Partition of Poland. Engineered by Frederick the Great to prevent Austria, jealous of Russian successes against Turkey, from going to war. Guerrilla war continued throughout Poland.

1792. Russian Invasion of Poland. This was caused by constitutional changes that would have stabilized the weak Polish monarchy.

Prussia at once also invaded to prevent Russian conquest.

1793. Second Partition of Poland. Russia and Prussia made a bargain at Poland's expense.

1794. National Uprising. The Polish nation rose in revolt against Russian and Prussian domination. **Thaddeus Kosciusko,** whose democratic principles had brought him to America to take part in the American Revolution, led a popular uprising.

1794, April 3. Battle of Raclawice. Kosciusko, with 4,000 troops and some 2,000 peasants armed only with scythes and pikes, defeated a force of 5,000 Russian troops. The Russian garrison of Warsaw (Warszawa) was driven out (April 17) after several days of street fighting.

1794, April–August. Russian and Prussian Invasions. The allied forces converged on the unfortunate country and after several desperate encounters Kosciusko's forces were invested in Warsaw by Frederick William of Prussia with 25,000 men and 179 guns, and Russian General **Fersen** with 65,000 men and 74 guns. Another Russian force of 11,000 occupied the right bank of the Vistula (Wisla) across from the capital.

1794, August 26–September 6. Defense of Warsaw. The Poles, 35,000 strong, with 200 guns, resisted 2 successive assaults and the siege was lifted. There was, however, no cohesion between the Polish forces in the field.

1794, October 10. Battle of Maciejowice. Kosciusko, with only 7,000 men, was crushed by Fersen's 16,000 Russians when 1 of his divisions failed to support him. Kosciusko, seriously wounded, was made prisoner. The entire uprising now collapsed.

1795. Third Partition of Poland. The ancient Polish nation disappeared; Russia had a permanent hold in Central Europe.

RUSSIA

1741–1762. Reign of Elizabeth (youngest daughter of Peter the Great).

1756–1763. Seven Years' War. (See p. 730.)

1762. Revolution. Peter III was deposed by his nobles and generals after a 6-month reign.

1762–1796. Reign of Catherine II (the Great). She was the German-born wife of Peter III.

1766–1772. Russian Intervention in Poland. (See p. 760.)

1768–1774. First War of Catherine the Great against Turkey. (See below.)

1773–1774. Peasant and Cossack Revolt. Russian efforts against Turkey were hampered by **Emelyan I. Pugachev**'s revolt in southeast Russia. This was put down only with difficulty by Suvarov.

1783. Annexation of the Crimea. England and Austria persuaded Turkey not to go to war.

1787–1792. Second War of Catherine the Great against Turkey. (See p. 762.)

WAR WITH SWEDEN, 1788–1790

1788. Swedish Invasion. Gustavus III invaded Russian Finland to offset possible future moves against his country by either Russia or Prussia. Bold assaults on land and sea brought success to the Swedes at first, causing consternation in Catherine's court, already occupied by the Turkish war. Gustavus' efforts on land ended with a Swedish repulse before the fortress of **Svataipol,** and a sudden Danish attack on Sweden further hampered him.

1789, August 24. First Naval Battle of Svensksund. The Swedish fleet, under Count **Karl A. Ehrensvard,** was decisively defeated by the Russian Admiral **Krose,** with loss of 33 ships. However, within a year, the Swedish Navy had been completely restored by the intensive efforts of Gustavus.

1790, July 2–9. Second Battle of Svensksund. The Swedish fleet, under the **Duke of Sudermania,** attacking in single file in a narrow channel, successfully ran the gantlet of the Russian fleet until the explosion of a powder ship threw the line into confusion, and the Swedes withdrew to the protection of the guns of Sveaborg (then a Swedish fortress). On the 9th, the reorganized Swedish fleet again put to sea to meet the Russians, who advanced in a crescent formation. The excellent Swedish gunnery chopped gaps in the Russian line and the Swedes, who had in all 195 ships to the Russians' 151, defeated their enemy piecemeal in the greatest sea fight in Scandinavian history. Russian losses were 53 ships sunk or captured. The stupidity and poor leadership of **Charles H. N. O. Nassau-Siegen,** an international adventurer in the Russian service, contributed to the disaster. British **William Sidney Smith** served brilliantly with the Swedes.

1790, August 15. Treaty of Wereloe. Peace with Sweden followed, restoring the *status quo ante.*

1792–1800. Wars of the French Revolution. (See p. 741.)

1793. Second Partition of Poland. (See p. 760.)

1795. Third Partition of Poland. (See p. 760.)

EURASIA AND THE MIDDLE EAST

TURKEY

First War against Catherine of Russia, 1768–1774

1768, October. Turkey Declares War. Poland pleaded for assistance against Russian intervention (see p. 760), but Turkey merely protested until Russian troops pursued Poles into Turkish territory and destroyed the Turkish town of Balta. The Crimean Tartars at once invaded the Ukraine.

1769. Russian Invasion of the Caucasus. The Turks were defeated in the Kabardia and in Georgia.

1769. Russian Invasion of the Balkans. Russian Count **Peter Rumiantsev** defeated the main Turkish army on the banks of the **Dniester.** He then seized Jassy and overran Moldavia and Wallachia, including the occupation of Bucharest.

1769–1773. Revolt in Egypt. Ali Bey, governor of Egypt, revolted and declared Egypt independent. He received Russian support.

1770. Russians Incite and Support Rebellion in Greece. Admiral **Aleksei G. Orlov**'s Russian Baltic fleet sailed to the Mediterranean and captured Navarino (April) and several other small places in the Morea. The Turks, however, assembled a large army, mostly Albanian, ruthlessly suppressed the revolt, and drove the Russians away (June).

1770, July 6. Naval Battle of Chesme (Çeşme). Orlov defeated the Turkish fleet near the island of Chios, off the coast of Anatolia. Scottish Admiral **Samuel Greig,** serving under Orlov, was largely responsible for the victory.

1770, August. Battle of Karkal. A Turkish and Tartar army, attempting to drive the Russians from Moldavia, were badly defeated by Rumiantsev. The Turks were forced to retreat behind the Danube. The Russians then systematically captured the Turkish fortresses along the Danube and Pruth Rivers.

1771. Egyptian Invasion of Syria. The Egyptians, under **Abu'l Dhahab,** captured Damascus. Abu'l then negotiated secretly with the Turks, evacuated Syria, and marched back to attack Egypt for Turkey (see below).

1771. Invasion of the Crimea. A Russian army under Prince **Vasily Dolgoruky** stormed the Isthmus of Perekop and conquered the Crimea.

1772–1773. Truce. Peace negotiations were fruitless. The Turks took advantage of the lull to reorganize and revitalize their army.

1772–1773. Confused War in Egypt and Syria. As Abu'l returned to Egypt to reconquer it for Turkey, Ali marched to Syria to reconquer it. Abu'l consolidated Egypt, and Ali consolidated Palestine.

1773, April 19–21. Battle of Salihia. After initial success, Ali was defeated and captured. Egypt returned to Turkish control. Abu'l was governor.

1773. Operations on the Danube. Count Rumiantsev, commanding Russian forces on the Danube, advanced south toward the main Turkish army under the Grand Vizier **Muhsinzade.** The Turks fell back on Shumla (Shumen or Kolarovgrad). Other Russian forces besieged the Turkish fortresses of Varna and Silistria.

-1773. Peace Negotiations. The grand vizier began to negotiate for peace with Rumiantsev. The Russians, too, were eager for peace because of a vast Cossack and peasant revolt in southeast Russia (see p. 761).

1774, June. Battle of Kozludzha. Suvarov, subordinate to Rumiantsev, defeated a large portion of the Turkish Army near Shumla.

1774, July 16. Treaty of Kuchuk Kainarji. Russia returned Wallachia, Moldavia, and Bessarabia, but was given the right to protect Christians in the Ottoman Empire, and to intervene in Wallachia and Moldavia in case of Turkish misrule. The Crimea was declared independent, but the sultan remained the religious leader of the Tartars as the Moslem caliph. Russia gained Kabardia in the Caucasus, unlimited sovereignty over the port of Azov, part of the Kuban near Azov, the Kerch peninsula in the Crimea, and the land between the Bug and Dnieper rivers and at the mouth of the Dnieper. Russian merchant vessels were to be allowed passage of the Dardanelles. Russia thus gained 2 outlets to the Black Sea, which was no longer a Turkish lake. Russian troops were withdrawn north of the Danube. Suvarov was sent to deal with the Cossacks.

1783. Russian Annexation of the Crimea. (See p. 761.)

1786. Revolt in Egypt. Suppressed by an Ottoman army.

Second War with Catherine of Russia, 1787–1792

1787. Russia Precipitates War. This resulted from encroachments in Georgia and demands that Turkey recognize a Russian protectorate over Georgia. Turkey was also intriguing with the Crimean Tartars to rise against Russia. Austria joined Russia on the basis of a secret alliance. Prince **Peter** of Montenegro rebelled against Turkey (1788).

1788. Turkish Repulse at Kinburn. Suvarov defeated a Turkish attempt to reconquer the Crimea.

1788. Russian Invasion of Moldavia. Rumiantsev captured the cities of Chocim and Jassy, then advanced to the seacoast. After **John Paul Jones's** naval victories (see below), Prince **Grigori Potemkin** took Ochakov, at the mouth of the Danube, opposite Kinburn. He seems to have ordered the massacre of all Turkish inhabitants of the captured cities, men, women, and children.

1788, June 17 and 27. Naval Battles of the Liman (Lagoons near the Mouth of the Dnieper). John Paul Jones, commanding the Russian Black Sea fleet, defeated **Hasan el Ghasi**'s Turkish flotilla in 2 sharp actions. Nassau-Siegen, who was vying with Jones for command, botched the opening of the first battle, but Jones's leadership retrieved the situation, and the Turkish ships drew off. The second battle brought disaster to the Turks, who lost 15 ships, 3,000 men killed, and more than 1,600 taken prisoner. Russian losses were 1 frigate, 18 men killed, and 67 wounded.

1788. Operations in Serbia and Transylvania. The Turks repulsed the Austrian armies under Joseph II. (See p. 759.)

1789. Operations in Moldavia and Wallachia. The Russians invaded Moldavia from the north, the Austrians from the west. The Austrians and 2 of the Russian armies (under Prince Potemkin and Count Rumiantsev) were generally unsuccessful against a revitalized Turkish army. But a Russo-Austrian army under Suvarov and Saxe-Coburg smashed the Turks at **Focsani** (August 1) and **Rimnik** (September 22). Suvarov's victorious advance then forced the Turks to withdraw to the Danube.

1789. Operations in Serbia. The Austrians, now commanded by General Gideon E. von Laudon, repulsed a Turkish invasion of Bosnia, then besieged and captured Belgrade by assault.

1790, July 27. Truce with Austria. (See p. 759.)

1790. Revolt in Greece. This seriously interfered with Turkish operations for the remainder of the war.

1790, December. Storming of Ismail. After a prolonged siege, Suvarov captured this fortress at the mouth of the Danube.

1791. Treaty of Sistova. (See p. 759.) Peace with Austria.

1791. Peace Negotiations. Russia was anxious for peace because of Prussian activity in Poland.

1792, January 9. Treaty of Jassy. Russia returned Moldavia and Bessarabia to Turkey. She retained all conquered territory east of the Dniester River, including the port of Ochakov.

1798–1800. War with France; Bonaparte's Invasion of Egypt. (See p. 752.)

PERSIA, 1747–1800

1747–1750. Struggle for the Throne. After the death of **Nadir Shah,** a protracted 3-cornered fight for control of the empire ensued between **Mohammed Husain,** who held the Caspian provinces; **Karim Khan,** strong in the south; and **Azad** in Azerbaijan. Karim Khan was successful and established unquestioned control over Persia.

1750–1779. Reign of Karim Khan. A period of relative peace and prosperity.

1779–1794. Civil War. This was a revolt of **Agha Mohammed,** an able eunuch general, against the successors of Karim Khan.

1794–1797. Reign of Agha Mohammed. A cruel and brutal ruler.

1795–1796. Invasion of Georgia. This nation had fallen away from Persian rule after the death of Nadir Shah. Defeating the Georgian King **Heraclius,** Agha Mohammed brutally reimposed Persian sovereignty over Georgia.

1797. Assassination of Agha Mohammed.

1798. Invasion of Afghanistan. The new Shah, **Fath Ali,** was incited by the British. This began a long-drawn-out series of wars weakening both Asian kingdoms.

SOUTH ASIA

AFGHANISTAN

Ahmad Shah, assuming control of the Afghan provinces of Persia after the death of **Nadir Shah** of Persia (see p. 711), established the Durani Dynasty. His military operations extended over much of Central Asia, including 9 invasions of India. The high point of his Indian incursions was a great victory over a numerically superior Maratha army at the **Battle of Panipat** (see p. 765). On his death the Afghan Empire extended over eastern Persia and all the region south of the Oxus, including Baluchistan, Kashmir, the Punjab, and Sind. His son, **Timur Shah** (1773–1793), weak and ineffectual, permitted loss of Sind and other Indian territory and the slow disintegration of Afghan sovereignty over the entire empire. Timur's son, **Zaman Murza** (1793–1799) in 1799 ceded Lahore to Sikh **Ranjit Singh** (see p. 860). Meanwhile, a protracted war with Persia had begun (1798; see p. 763).

INDIA

When the period opened, 4 European nations were milking India for their own purposes. Waning Dutch exploitation was soon eliminated (1759), and Portugal's influence was frozen in her possessions of Goa, Diu, and Daman. Only England and France continued their battle for supremacy amid the wreckage of the Mogul Empire, while the Marathas were simultaneously rising to power. Both European nations took advantage of native rivalries in an unabashed and almost continual warfare with but one objective—the exploitation of India's riches for the benefit of European investors. Both masked their national rivalry through agencies—for France the **Compagnie des Indes,** for England the **British East India Company** ("John Company"). This latter remarkable organization had the delegated power of a sovereign state, including the making of both war and peace "with any non-Christian nation." It had its own army, composed of European adventurers and native troops, under English commanders. Further English military influence was furnished (1754) by an English regular regiment, the 39th Foot (later the Dorsetshire Regiment), at Madras. This became the backbone of English military operations in India. Many of its officers and men later transferred to the company's service. At midcentury France and her native allies controlled a great part of southern India; **Joseph François Dupleix** (see p. 714) was governor general of all French establishments in India; his subordinate, General **Charles J. P. Bussy-Castelnau,** was operating in Hyderabad in the Deccan.

1751, September–October. Capture and Siege of Arcot. Chanda Sahib, France's puppet Nawab of the Carnatic, was besieging a small English garrison at Trichinopoly (Tiruchirappalli). Young **Robert Clive,** former East India Company clerk turned soldier, with 500 men and 3 guns moved suddenly from Madras and captured Arcot, Chanda Sahib's capital, as a diversion to relieve the siege (September 12). The nawab's son, **Raja** Sahib, with 10,000 men, at once returned to invest Arcot. After a siege of 50 days, in which Clive's troops were reduced to 120 Europeans and 200 sepoys, the Indians assaulted, driving ahead a herd of elephants, their heads armed with iron plates, to batter down the gates. But the elephants stampeded when they came under musket fire. Another assault across the moat was repelled, and then the assailants withdrew after an hour's fight, having lost 400 casualties. Clive's losses were only 5 or 6 men. This exploit greatly enhanced England's prestige, to the disadvantage of the French, whose campaign now broke down. Dupleix, a most capable administrator, whose greatest trouble had been lack of understanding by and support from the home government, was summarily relieved in 1754.

COMMENT. *This was the beginning of Clive's career, which would lead to England's domination of India and his own fame as one of Britain's foremost captains and colonizers.*

1756, June 20. Capture of Calcutta. Suraja Dowla, Nawab of Bengal, after a 4-day siege, seized the town, founded by the British and now the focal point of their activities. Europeans unable to escape were forced into a small subterranean dungeon—the **Black Hole**—in which it has been alleged 123 out of 146 perished overnight. Historians today are still not certain of the numbers actually involved.

1757, January 2. Recapture of Calcutta. This was by a joint expedition from Madras under Clive and Admiral **Charles Watson.**

1757, March 23. Capture of Chandernagore. This French post was seized by Clive after a spirited fight. This permitted Clive to march inland against Suraja without fear for his line of communications with Calcutta.

1757, June 23. Battle of Plassey. Clive, pursuing Suraja, found him entrenched on the far side of the Bhagirathi River near the village of Plassey. The Indian Army was 50,000 strong, with 53 guns. Clive had 1,100 Europeans, 2,100 native troops, and 10 guns. He crossed the river and massed his force in a mango grove. The Nawab's army moved to surround it in a large semicircle, its guns—under French command—massed on the right flank. A sudden rainstorm wet the Nawab's powder. As the firing ceased, a cavalry charge swept up to the British position, but Clive's gunners had covered their ammunition from the rain and re-

pulsed the charge, inflicting great loss. Clive, counting rightly on the disaffection of part of the Indian army, advanced to cannonade the Nawab's entrenchments at short range. An Indian infantry sortie was repulsed and Clive now assaulted the position, in which only the French gunners, under M. **St. Frais,** continued fighting to the last. Plassey decided the fate of Bengal. Suraja was assassinated a few days later and was succeeded by Clive's ally, **Mir Jafar.**

1758. Arrival of French Reinforcements. An expeditionary force under Irish-born Baron **Thomas Lally** arrived at Pondicherry, capital of French interests. He immediately marched to besiege and capture British **Fort St. David,** south of Pondicherry (March–June 2).

1758, December–1759, February. Siege of Madras. Lally failed in efforts to take the city. He was hampered by lack of naval support, while the British garrison was supplied by sea (see p. 739).

1759, January 25. Battle of Masulipatam. A British relieving force, arriving by sea, under **Francis Forde** defeated Lally, forcing him to raise the siege of Madras.

1759. Operations against the Dutch. A Dutch naval expedition was repulsed by the British at the mouth of the Hooghly River, and Forde captured the Dutch port of **Chinsura.**

1760, January 22. Battle of Wandiwash. Lally was defeated by General Sir **Eyre Coote** and forced to retire to Pondicherry. There he was besieged.

1760, August-1761, January 15. Fall of Pondicherry. Lally capitulated to Clive's troops (Clive had returned ill to England) there being no hope for reinforcements from France (see p. 739).

1761, January 14. Battle of Panipat. The Afghan armies of **Ahmad Shah Durani** had been ravaging northern and central India (since 1747). Ahmad reached and captured Delhi (1757). The **Maratha** revolt which had demolished the Mogul Dynasty had swept north into the Punjab (1753), but Ahmad, again invading, slowly drove the Mahrathas back, routing them completely at Panipat. Ahmad's later withdrawal, caused by a Sikh insurrection, left India in complete chaos, facilitating British consolidation of their position in Bengal.

1763. Treaty of Paris. Pondicherry was restored to France (see p. 741), but the *Compagnie des Indes* was soon dissolved (1769).

1763–1765. Operations in Bengal. Sporadic conflicts between the British East India Company's troops and the native rulers continued. A mutiny in the Company's Bengal Army, near Patna (1763), inspired by the Nawab of Bengal, was summarily quelled at **Buxar** (October 23, 1764), and 24 of the ringleaders blown from gun muzzles by Major **Hector Munro** to avenge the massacre at Patna.

1766–1769. First Mysore War. Haidar Ali, ruler of Mysore, fought British troops to a standstill, then signed a treaty of defensive alliance with the East India Company.

1771. Mysore-Maratha War. Inconclusive.

1774. Rohilla War. Governor General **Warren Hastings** loaned a brigade of East India Company troops to **Shuju-ud-Dowla,** Wazir of Oudh and a British ally, to help conquer the territory of the Rohillas, a fierce Afghan tribe in the Ganges-Jumna river region west of Oudh (April). He offset the threat of growing Maratha strength, but his employment of Company troops in a native war was contrary to British and Company policy.

1779–1782. First Maratha War. Inconclusive despite British success in Gujarat and the capture of the fortress of Gwalior (1780).

Second Mysore War, 1780–1783

1780. Haidar Ali Declares War. He was smarting from failure of the British to support him in his own war with the Marathas. After British troops moved against French holdings in 1778 (see p. 781), he joined France. The British had taken Pondicherry and Mahé from the French. But Haidar Ali, in the Carnatic, attacked and cut to pieces a small British force at **Perambakam** (September 10), and swept up to the gates of Madras.

1781, June 1. Battle of Porto Novo. Coote, with 8,000 men sent by sea from Bengal, attacked and defeated Ali's 60,000, saving the Madras Presidency.

1781, August–September. British Victories. Haidar was defeated at **Pollilur** (August 27) and at **Sholingarh** a month later.

1782, August 30. Capture of Trincomalee. French Admiral **Pierre de Suffren** captured the port from the British and was thus able to send help to Haidar Ali. He retained tenuous control of the Indian Ocean until the signing of peace between England and France in 1783 (see p. 791).

1783. Withdrawal of French Aid Ends the War. Haidar Ali having died, his son **Tippoo Sahib** made peace.

1789–1792. Third Mysore War. Tippoo attacked Travancore. **Lord Cornwallis,** who had succeeded Hastings as Governor General of India, invaded Mysore; stormed the fortress of Bangalore, its capital; and drove Tippoo into Seringapatam, where he was besieged. Tippoo made peace (March 16, 1792) by ceding half his dominions to the British.

1795–1796. British Expedition to Ceylon. The Dutch were easily defeated. The independent King of Kandy soon recognized British sovereignty, and Ceylon became a crown colony (1798).

1799. Fourth Mysore War. Governor General **Richard Wellesley** (then Lord Mornington) as directed by Pitt, moved actively to suppress the last vestiges of French influence in India, lest Bonaparte invade it. French troops in the employ of the Maratha Confederacy were disbanded on Wellesley's demand. Tippoo, however, was carrying on secret correspondence with the French government; on his refusal to cooperate with the British, Wellesley sent 2 armies into Mysore. General **George Harris** drove Tippoo once again into Seringapatam and then took the place by storm. Tippoo was killed while fighting bravely in the breached wall. Wellesley's younger brother, General **Arthur Wellesley** (later Duke of Wellington) distinguished himself in the campaign and was appointed governor of Seringapatam.

SOUTHEAST ASIA

BURMA

1740–1752. Mon Revolt. The inhabitants of Lower Burma rose up against the Burman Toungoo Dynasty, drove them out, and overran much of Upper Burma, capturing Ava (1752). The Toungoo Dynasty was completely destroyed. The Mons (or Talaings) reestablished their former capital at Pegu (see p. 717).

1752–1757. Rise of Alaungpaya. This obscure lord of the Shwebo district united the demoralized Burmans. He began to repel the Mons from Upper Burma, and reconquered most of Lower Burma (1754–1755). The Mons were able to hold only the regions around Syriam and Pegu. In these operations Alaungpaya proved himself probably the ablest military leader of Burmese history. His successes, thanks also to strong lieutenants, established a strong dynasty.

1753. British Post at Negrais. This small fortified trading post was unrecognized by the Mons, but Alaungpaya recognized it.

1755, 1758. Alaungpaya's Operations in Manipur. This was a punitive response to Manipuri raids (1715–1749). When the local inhabitants rose up against the garrisons he left there, Alaungpaya returned to devastate Manipur (1758).

1755–1756. Siege of Syriam. It was captured after a long struggle, and despite French assistance to the Mons. Alaungpaya also captured 2 French warships in the narrow river channel near the town. He killed the officers but spared the crews, incorporating them into his own army.

1756–1757. Siege of Pegu. After driving the remaining Mon army into the city, Alaungpaya finally wore down the defenders and captured the town by assault. This, and the ensuing massacre, ended Mon national existence in Burma, though there was an abortive uprising (1758) while Alaungpaya was absent on his punitive expedition to Manipur (see below).

1759. Capture of Negrais. Alaungpaya massacred most of the small English and Mon garrison.

1760. Invasion of Siam. Alaungpaya advanced to besiege Ayuthia, the capital. While directing the fire of his siege batteries, he was seriously wounded by a gun explosion. Because of either this or an epidemic of malaria, the siege was raised and the army retreated to Burma, suffering some losses from Siamese harassment. The king died en route.

1764. Invasion of Manipur. Hsinbyushin (1763–1776), the second son of Alaungpaya, carried off much of the population into slavery.

1764–1767. Invasion of Siam. This culminated in the capture of Ayuthia. The operation was very well conducted, and the victory was due mainly to the able Burmese General **Maha Nawraha,** who died just before the surrender of Ayuthia.

1765–1769. Chinese Invasion. The Chinese, aggravated by Burmese frontier forays, invaded Burma in great force. They seized most of eastern Burma from Bhamo south to Lashio and west almost to the Irrawaddy at Singaung. The Burmese, however, refused to accept defeat. Holding on to a number of strategically located fortified stockades in the jungles, avoiding open battle against the larger Chinese armies, and aggressively and incessantly raiding the Chinese lines of communication, they soon had the Chinese armies cut up into several segments and isolated in north-central Burma. As a result of heavy casualties, of their inability to come to grips with the elusive Burmese in the jungles, and of losses from disease and starvation, the Chinese asked for terms. Burmese General **Maha Thihathura,** realizing that, if the war continued, China had inexhaustible supplies of man power, granted terms to the Chinese, who withdrew.

COMMENT. *This was one of the most glorious episodes of Burmese military history.*

1768–1776. Defeat in Siam. Exhausted by the struggle with the Chinese, and unable to follow an effective and consistent pollicy in Siam, the Burmese were being expelled from Siam by **P'ya Taksin** (see below).

1770. Invasion of Manipur. Maha Thihathura and other Burmese leaders, to divert the king's mind from foolhardy operations against the Siamese or Chinese, and fearful of execution if they tried to reason with him, diverted their armies into operations against a minor foe.

1775–1776. War with Siam. (See below.)

1782–1819. Reign of Bodawpaya (youngest son of Alaungpaya). He gained the throne after a violent struggle with several other rivals.

1784–1785. Conquest of Arakan.

1785–1792. Renewed War with Siam. A Burmese invasion (1785–1786) was generally unsuccessful, probably largely due to the utterly incompetent leadership of the king. Burma did, however, annex the regions of Tavoy and Tenasserim after sporadic border warfare.

Siam

1760. Alaungpaya's Invasion. (See p. 767.)

1764. Burmese Invasion of Southern Siam. General P'ya Taksin finally halted the invasion at **P'etchaburi,** but only after most of the Malay Peninsula provinces of Siam had been overrun.

1764–1767. Burmese Invasion of Central Siam. This resulted in the capture and destruction of Ayuthia (see p. 767). P'ya Taksin, however, had been able to fight his way through the Burmese besieging lines with a few troops, and escaped southward to raise a new army.

1767–1768. Burmese Ejected from Central Siam. P'ya Taksin assumed the throne. The country, however, was in turmoil, with two other claimants to royal power; at the same time the Burmese were planning to invade the country again, and trouble with Vietnam was brewing in Cambodia, an area which had long been under Siamese suzerainty.

1769. Repulse at Chiengmai. P'ya Taksin failed in an effort to win the region back from Burma. He was, however, able to defeat his 2 major rivals and to secure his rule over the nation.

1770–1773. Operations in Cambodia. The fortunes of war fluctuated widely. The Siamese defeated a Vietnamese-supported puppet ruler and regained full control of Cambodia.

1775–1776. Renewed War with Burma. P'ya Taksin reconquered Chiengmai (1775). He then repulsed an attempted Burmese invasion (1776).

1778. Invasion of Laos. The Siamese seized Vientiane, which recognized Siamese suzerainty.

1780–1782. Intervention in Vietnam. This ended with the death of P'ya Taksin (see below).

1782. Death of P'ya Taksin. He had become insane, and was killed in an uprising. General **Chakri** took the throne as King **Rama I** (1782–1809). He founded the modern city of Bangkok as his capital.

1785–1792. Sporadic Warfare with Burma. (See above.)

Vietnam

1755–1760. Expansion in Cambodia. The Khmer, their Siamese allies occupied with a struggle against Burma, were helpless against the Vietnamese aggressors.

1769–1773. War with Siam over Cambodia. After initial success, the Vietnamese were driven out. (See above.)

1773–1801. Civil War. Long and complicated, it resulted in the elimination of the Trinh family and a briefly emergent Tay Son family, with **Nguyen Anh** eventually victorious and assuming the imperial title of **Gia Long** at the capital of Hué (1802). During this struggle, Nguyen Anh received considerable assistance from Siam, with the result that Siamese control over Cambodia was strengthened and Siamese influence became a political issue in Vietnam.

CAMBODIA

Cambodia, as we have seen, was essentially simply a football to the conflicting ambitions of Siam and Vietnam.

LAOS

Laos, divided into the 2 principalities of Luang Prabang and Vientiane, was a pawn in the great struggle between Burma and Siam. Dominated at first by Burma as a result of the victories of Alaungpaya and his successors, toward the end of the century the 2 principalities of Laos were generally under Siamese control, though some southeastern regions were under strong Vietnamese influence.

MALAYA AND INDONESIA

There were no significant military activities in Malaya or Indonesia during this period. The Dutch-English trade rivalry continued, with the Dutch consolidating their hold on most of Indonesia, save for strong British influence in Borneo; at the same time, the English established a foothold in Malaya at Penang (1771) to initiate a challenge to the powerful Dutch position in the peninsula.

The conquest of the Netherlands by France in 1795, during the War of the First Coalition (see p. 745), gave Britain an opportunity to break the Dutch trade monopoly in this area. At the "request" of the Dutch government in exile, England took Cape Colony (the important Dutch base at the top of Africa, controlling the southwest approach to the Indian Ocean), Ceylon (the main intermediate base), all the Dutch posts in India and on the west coast of Sumatra, the Spice Islands, and Malacca. A solemn promise was made to return these to the rightful Dutch government when peace was made.

EAST AND CENTRAL ASIA

CHINA

1736–1796. Reign of Ch'ien Lung (or Kao Tsung). Under this able and energetic emperor, the Manchu Empire reached its greatest geographical extent and the zenith of its power. In particular, imperial Chinese control of Central Asia was tightened.

1751. Invasion of Tibet. Tibet, nominally under Chinese control since 1662, had revolted (1750). Ch'ien Lung sent an invasion force, which quickly captured Lhasa and forced the Dalai Lama to submit to even stricter Chinese control.

1755–1757. Mongol Revolt in the Ili Valley. This was promptly suppressed by General **Chao Hui,** who seized this opportunity to strengthen overall Chinese control over western Mongolia.

1758–1759. Conquest of Kashgaria. Chao Hui now turned southwestward to establish Chinese control once more over the Tarim Basin and the surrounding regions. Sinkiang was now organized as a Chinese province.

1774–1797. Minor Rebellions. These plagued the latter years of the reign of Emperor Ch'ien Lung. All were quickly suppressed, but they presaged a general weakening of power of the

MANCHU EMPIRE

At its Greatest Extent
under Ch'ien Lung, 1795

Ch'ing Dynasty. These revolts were in Shantung (1774), Kansu (1781 and 1784), Formosa
(1786–1787), and Hunan and Kweichow
(1795–1797).

KOREA AND JAPAN

This was a period of relative peace and prosperity in both of these isolated East Asiatic
kingdoms. No military operations of any significance took place.

NORTH AMERICA

FRENCH AND INDIAN WAR, 1754–1763

**1754, February–July. Fort Necessity Cam-
paign.** Alarmed by increasing French influx
from Canada into the Ohio Valley, Lieutenant

Governor **Robert Dinwiddie** of Virginia sent
Lieutenant Colonel **George Washington** with
a detachment of Virginia militia to construct a
fort at the confluence of the Allegheny and
Monongahela rivers (present Pittsburgh). Finding the French had already erected Fort

Duquesne on this site, Washington built Fort Necessity at Great Meadows (near modern Uniontown, Pa.). Here he was attacked by the French. After vigorous resistance, he was forced to capitulate to superior numbers (July 3). He was allowed to march out with the honors of war.

1755, April 14. Arrival of Major General Edward Braddock in Virginia. The new commander in chief in America brought 2 British regiments. A conference with several provincial governors resulted in a plan to attack the French with 4 separate expeditions. Braddock himself, with 1,400 regulars and 450 colonials, moved to attack Fort Duquesne and to establish British control of the Ohio basin.

1755, June. Arrival of French Reinforcements. A French fleet slipped into the St. Lawrence past the blockade of Admiral **Edward Boscawen.**

1755, June. Bay of Fundy Expedition. From Boston, a provincial force accompanied by a few English regulars, under Colonels **Robert Monckton** and **John Winslow,** sailed into the Bay of Fundy, capturing Forts St. John and Beausejour (June 19). All the Fundy area was soon in British hands (June 30). The French Acadians were cruelly exiled (October).

1755, July 9. Battle of the Monongahela. As his force approached Fort Duquesne, Braddock was surprised and routed by 900 French and Indians. He was killed with more than half of his force. Washington, who accompanied the expedition as a volunteer, helped to lead the remnants back to Virginia. The rigid linear formations of European warfare had succumbed to the elusive individualism of wilderness combat.

1755, August–September. Crown Point Expeditions. A force of 3,500 provincials and 300 Indians advanced from Albany toward Crown Point under **William Johnson,** a civilian with wide influence among the Indian tribes, commissioned a brigadier general. A mixed force of 2,000 French regulars, Canadians, and Indians under Baron **Ludwig A. Dieskau** simultaneously moved up the Richelieu River to defend Crown Point. The 2 forces met at the head of Lake George.

1755, September 8. Battle of Lake George. Dieskau was defeated and captured, but Johnson, plagued by militia discontent, contented himself with building Fort William Henry; he left a small garrison there and the remainder of his force dispersed to their homes. The French entrenched themselves at Ticonderoga.

1755, August–September. Expedition against Fort Niagara. Governor **William Shirley** of Massachusetts, who had replaced Braddock in command, marched with 1,500 men up the Mohawk Valley to Oswego. There Shirley decided the task was too great against French reinforcements, and returned to Albany.

1755–1756. Establishment of the Royal Americans. Spurred by Braddock's defeat, the British War Office took steps to meet the tactical situation imposed by wilderness warfare. A 4-battalion regiment of light infantry was authorized (1755). At Governors Island, N.Y., the Royal Americans (now the King's Royal Riffle Corps)—actually composed of equal parts of German-American settlers and Germans recruited abroad—began light-infantry training (1756). Many of the officers were soldiers of fortune.

1756. Seven Years' War Begins in Europe. (See p. 730.)

1756, May 11. Arrival of Marquis Louis Joseph de Montcalm. The new commander of all French forces in Canada brought reinforcements.

1756, July 23. Arrival of General John Campbell (Earl of Loudoun). He was appointed to the command in North America.

1756, August. Montcalm Takes the Offensive. He crossed Lake Ontario and captured Oswego, destroying the settlement. Returning to Montreal, he took station finally at Ticonderoga with a force of 5,000 regulars and militia. Loudoun, who had 10,000 men between Albany and Lake George, made no offensive move.

1756–1757. Winter Quarters. The regulars on both sides withdrew, leaving garrisons at Forts Ticonderoga and William Henry, respectively. The provincial militiamen went home.

1757, June–September. Expedition against

Louisbourg. Loudoun arrived at Halifax to meet a British squadron that would take part in the expedition (June 30) only to learn that Louisbourg had been strongly reinforced and a French fleet was in the harbor. The British blockading squadron, under Admiral **Francis Holborne,** was scattered by a storm and the expedition abandoned (September 24). Loudoun's 12,000 troops returned to New York.

1757, August 9. Massacre at Fort William Henry. Montcalm, taking advantage of the absence of the British troops, had meanwhile moved from Ticonderoga with 4,000 troops and 1,000 Indians to besiege and capture Fort William Henry. The garrison, under Colonel **Monro,** marching out with the honors of war, was treacherously attacked by Montcalm's Indians and many were killed. Content with destroying the fort, Montcalm withdrew, and both sides went into winter quarters.

1757–1758. British Plans. The British government, now headed by **William Pitt,** decided to make a major effort to eliminate France from North America. General **James Abercrombie** arrived to replace ineffective Loudoun (December 30). General **Jeffrey Amherst** soon followed (February 19) with troops to take Louisbourg. Abercrombie was directed also to attack Forts Ticonderoga and Duquesne.

1758, May 30–July 27. Louisbourg Expedition. Amherst with young Brigadier General **James Wolfe** took 9,000 British regulars and 500 colonials from Halifax to Louisbourg, escorted by Admiral Boscawen's squadron. Twelve French sail of the line were shut up in the harbor, while the troops invested the fortress (June 2). It surrendered (July 27) after considerable fighting. Wolfe particularly distinguished himself.

1758, July 8. Battle of Fort Ticonderoga. Abercrombie, with 12,000 men, 6,000 being British regulars, moved from Lake George (late June) to attack Fort Ticonderoga. Montcalm, with but 3,000, defended a ridge in front of the fort. Abercrombie's frontal attack was repulsed with the loss of some 1,600 men killed or wounded. He withdrew. He was later relieved and replaced by Amherst (September 18).

1758, July–November. Fort Duquesne Expedition. Brigadier General **John Forbes,** retracing Braddock's steps, cut a road through the Blue Ridge and after some hard fighting advanced upon Fort Duquesne, which the French blew up and abandoned (November 24). Backbone of Forber's force was a battalion of the Royal Americans, under Major **Henri Bouquet,** Swiss soldier of fortune in British service. The former French post now became Fort Pitt. Forbes's operations were methodical and efficient, Bouquet's new system of light-infantry tactics in extended order and rapid deployment proved its worth, and Colonel George Washington, who commanded a Virginia regiment, assisted in planning the campaign.

1758, August 27. Capture of Fort Frontenac. This fortress (modern Kingston) at the entrance to the St. Lawrence, on Lake Ontario, was captured by a provincial force under Colonel **John Bradstreet** which marched up the Mohawk.

1758–1759. French Situation. The year ended with the French hard-pressed on both flanks of their North American colony, but clinging firmly to Ticonderoga in the center.

1759. Pitt's Plan. He ordered a 3-pronged campaign to drive the French out of Canada: (1) capture of Fort Niagara to cut western Canada off from the St. Lawrence, (2) an offensive through the Champlain Valley to the St. Lawrence, and (3) an amphibious assault against Quebec.

1759, June–July. Fort Niagara Operations. Brigadier General **John Prideaux** with 2,000 British regulars marched up the Mohawk Valley, reoccupied Oswego and, moving by water on Lake Ontario, captured Fort Niagara (July 25). Prideaux was killed during the siege.

1759, June–July. Ticonderoga Campaign. Amherst, commanding 11,000 regulars and provincials, captured Ticonderoga (July 26) and Crown Point (July 31), where he spent the winter.

1759, June–September. Quebec Campaign. Wolfe, young military perfectionist, with some 9,000 men, escorted by Admiral **Charles Saunders'** squadron, departed from Louisbourg to disembark on Orléans Island, just below

Quebec (June 26). Montcalm, with some 14,000 troops and some Indians, defended an almost impregnable fortress, standing high above the St. Lawrence River. For 2 months all Wolfe's efforts to gain a foothold were foiled. Admiral Saunders, fearing that his ships might be caught in winter ice, threatened to leave. Then a footpath winding up the precipitous cliffs just north of the city was discovered. Wolfe sent his 1 battalion of provincial rangers (light infantry), under Colonel **William Howe,** followed by 4 regular battalions, up this path during the night of September 12. They gained the plateau above, and the main body followed. By dawn Wolfe's command—4,800 strong—was in line of battle in front of the city.

1759, September 13. Battle of the Plains of Abraham. Montcalm attacked at once. He had about 4,500 men, but lacked artillery—**Pierre François de Vaudreuil,** the French governor, would not release guns to him. Excellent British musketry threw the French back into the city. Both commanders were mortally wounded. Wolfe died during the battle, Montcalm that night. Quebec capitulated (September 18), and the backbone of French resistance in Canada had been broken. Leaving a garrison in the city, the expedition sailed away.

1760, April. French Attack on Quebec. General **François G. de Lévis** led an expedition from Montreal. He invested Quebec, but the siege was broken by the timely arrival of a British naval squadron.

1760, September. Advance on Montreal. Three columns converged: Amherst, moving down the St. Lawrence from Oswego, Colonel **William Haviland** from Crown Point down the Richelieu, and Brigadier General **James Murray,** with the Quebec command, up the St. Lawrence.

1760, September 8. Surrender of Canada. Governor Vaudreuil capitulated, ending French dominion of Canada.

1762, January–February. British Conquest of Martinique. The island was captured by Admiral George Rodney.

COMMENT. *The effects of the French and Indian War upon British infantry tactics and techniques were* *enormous. The colonists' successful combination of discipline and loose-knit Indian fighting impressed the English soldiers. The Royal American Regiment under its 2 Swiss-born battalion commanders, Henri Bouquet and* **Frederick Haldimand,** *showed its efficiency during all the campaigns after 1756; Colonel* **Robert Rogers** *of Connecticut, whose battalion of Rangers became a component of the British Army in America, and Howe at Quebec further proved the worth of disciplined light infantry.*

1763, February 10. Treaty of Paris. (See p. 741.)

PRE-REVOLUTION, 1763–1774

1763, May 23–November 28. Pontiac's Rebellion. Unhappy over the substitution of British control for French along the northwestern frontier, and fearful of a growing influx of settlers, the Indian tribes became restless, then went on the warpath. **Pontiac,** chief of the Ottawa, after a surprise attack upon Detroit failed, destroyed nearly every other British fort west of the Niagara except Fort Pitt.

1763, August 4–6. Battle of Bushy Run. Colonel Bouquet, moving to Fort Pitt from Carlisle, Pa., with a relief expedition composed of his Royal Americans, the Black Watch (42nd Foot), and other light companies, 500 men, was surprised as Braddock had been in 1755. He formed a circle about his train, where the troops lay all night. Next morning Bouquet enticed the Indians to attack, then cleverly fell on their flank and routed them. Fort Pitt was relieved (August 10). Pontiac, abandoning the siege of Detroit, later submitted to British authority.

1769–1807. Yankee-Pennamite Wars. Connecticut's claim to transcontinental dominion "from sea to sea" led to bitter strife between Connecticut and Pennsylvania settlers in the Wyoming Valley of the Susquehanna River. Drafting of Connecticut settlers for the Revolutionary War so depleted manpower that the settlement fell easy prey in 1778 to a bloody massacre by New York Tories (see p. 782). Sporadic bickering continued until 1807. (See p. 791.)

1770–1774. Growing Violence and Unrest.
Colonial resentment of royal rule led to repressive measures by British troops in the so-called **"Battle of Golden Hill"** in New York and the **"Boston Massacre"** (January, March 1770). The "Boston Tea Party" (December 16, 1773), caused Parliament to put Boston under military rule. To unify colonial reaction and protest against this and other "intolerable" acts, the first Continental Congress met at Philadelphia (September 5–October 26, 1774).

1771. Uprising in Western North Carolina. A group of frontier settlers, calling themselves "Regulators," defied royal rule. Governor **William Tryon** sent a force of militia, which suppressed the rebellion at **Alamance Creek** (May 16).

1774. Lord Dunmore's War. Governor **John Murray, Earl of Dunmore,** of Virginia, sent a force of militia under General **Andrew Lewis** to suppress a Shawnee uprising on the Virginia-Kentucky frontier. A contingent from the Watauga Association, from the area where Virginia, North Carolina, and Tennessee now meet, took part in this campaign. The Indians made peace after Lewis won a decisive victory at **Point Pleasant** (mouth of Great Kanahwa; October 10).

WAR OF THE AMERICAN REVOLUTION, 1775–1783

THE WAR ON LAND, 1775–1776

1775, April 19. Lexington and Concord.
Long-standing and ever-increasing differences between the thirteen colonies and the motherland flamed into open rebellion when General **Thomas Gage,** governor of Massachusetts, sent 700 men from the British garrison in Boston (April 18) to capture arms and ammunition gathered by the colonists at Concord. Forewarned by the "midnight ride" of **Paul Revere, William Dawes,** and Dr. **Samuel Prescott,** "minute companies" of militia began gathering. The British advance guard under Major **John Pitcairn** met Captain **John Parker**'s company of 70 men assembled on Lexington Common. The "shot heard 'round the world"* (no one knows who fired it) brought British volleys. The Americans scattered, leaving 8 men dead and 10 wounded. Proceeding to Concord, the British found most of the supplies removed, but they destroyed what remained. Harassed all the way back to Boston by swarms of militiamen, British losses were 73 killed, 174 wounded and 26 missing; in all, 93 Americans were killed, wounded, or missing.

1775–1776. Siege of Boston. The Massachusetts Provincial Congress authorized the raising of 13,600 militia, Major General **Artemas Ward** to command. Calls for assistance were answered by Rhode Island, Connecticut, and New Hampshire. Boston and its British garrison were besieged by some 15,000 colonists.

1775, May 10. Capture of Ticonderoga. Colonel **Benedict Arnold** was authorized to attack the fort on Lake Champlain. Colonel **Ethan Allen** of Vermont with his Green Mountain Boys, refusing to accept Arnold's command, captured the post by surprise (May 10). Arnold accompanied the expedition. Crown Point was soon captured (May 12).

1775, June 12. Seizure of the *Margaretta*. At Machias Bay, Me., a party of lumbermen led by **Jeremiah O'Brien** captured the British armed cutter *Margaretta,* the first naval action of the war.

1775, June 15. Washington Commands an Army. The second Continental Congress at Philadelphia (since May 10), accepting the colonial forces besieging Boston as a Continental Army, authorized the raising of 6 companies of riflemen and commissioned Colonel George Washington of Virginia as a general, to command the whole. He started north on June 23.

1775, June 17. Battle of Bunker Hill. Gage had been reinforced (May 25) by British troops from overseas; with them came Generals **John Burgoyne, William Howe,** and **Henry Clinton.** The garrison of 7,000 was sufficient to hold the city but too few to attack the besiegers. The Massachusetts Committee of Safety or-

---

* Poetically, Emerson attributes this to Concord.

dered Ward (June 15) to fortify Bunker (Bunker's) Hill on Charlestown Heights overlooking Boston Harbor. Next night 1,200 men under Colonel **William Prescott** occupied not Bunker Hill but Breed's Hill, a lower and more vulnerable eminence, and threw up an earthen fort. Gage responded at dawn by a naval bombardment and an amphibious attack. Howe, crossing the bay with a picked force of 2,200 men, was twice repelled. A third assault, with fixed bayonets, was successful when the defenders' ammunition gave out. The affair, badly mismanaged by the Americans, was distinguished by gallantry on both sides. The British lost practically half of their strength; of less than 1,500 colonials actually engaged, 140 were killed, 271 wounded, and 30 captured.

COMMENT. *Tactically a British victory, the result left the local situation unchanged. But psychologically the effect on the colonies was tremendous; Americans had met British regulars in battle and had been defeated only when their ammunition was exhausted. The significance of the earthworks was largely overlooked.*

1775, July 3. Birth of the Continental Army. Washington, on Cambridge Common, assumed command of all Continental forces and began the stupendous job of transforming an armed rabble into an army while maintaining the siege of Boston throughout the remainder of the year.

NORTHERN CAMPAIGN; EXPEDITIONS AGAINST CANADA

1775, August-November. Operations against Montreal. General **Philip Schuyler,** with 1,000 men, invaded Canada from Ticonderoga, laying siege to St. Johns on the Richelieu River (September 6); falling ill, he was replaced (September 13) by General **Richard Montgomery.** St. John's 600 defenders capitulated (November 2). Meanwhile, a harebrained attempt by Ethan Allen to capture Montreal was repulsed and Allen captured. Montgomery pushed on to Montreal, occupying it (November 13) and capturing a British

river flotilla. Sir **Guy Carleton,** British governor general of Canada, withdrew to Quebec.

1775, September–December. Arnold's Expedition to Quebec. With Washington's approval, Arnold and 1,100 volunteers left Cambridge (September 12). After an amazing autumn march through the freezing Maine wilderness, the expedition arrived on the St. Lawrence, opposite Quebec (November 8). Only 600 men completed the trek. Crossing the river (November 13), Arnold was joined by Montgomery from Montreal with 300 men (December 3). Montgomery took command.

1775, December 31. Assault on Quebec. Attacking under cover of a driving snowstorm, the Americans were disastrously repulsed by Carleton's 1,800-man garrison. Montgomery was killed, Arnold wounded; nearly 100 Americans were killed or wounded and 300 more captured.

1776, January–May. Retreat to Montreal. The survivors, still under Arnold's command, maintained a tenuous hold outside the city for the next 5 months, but retired to Montreal upon the arrival from England (May 6) of General Burgoyne with British and German reinforcements for Carleton. General **John Sullivan,** now commanding the Americans, received additional troops from General Washington and attempted a counterstroke to stabilize conditions on the upper St. Lawrence.

1776, June 8. Trois Rivières. American General **William Thompson,** with 2,000 men, discovered that instead of a supposed garrison of 600, Burgoyne's 8,000 were concentrating there. The Americans were completely dispersed.

1776, June–July. Retreat from Canada. Sullivan, abandoning Montreal, hastily retired first to Crown Point and then to Ticonderoga. Both sides now occupied themselves with extemporizing naval forces to dispute possession of Lake Champlain as Carleton prepared to march into New York.

1776, October 11. Battle of Valcour Island. The makeshift flotilla built and manned by American soldiers through Benedict Arnold's efforts was attacked by Carleton's flotilla—also

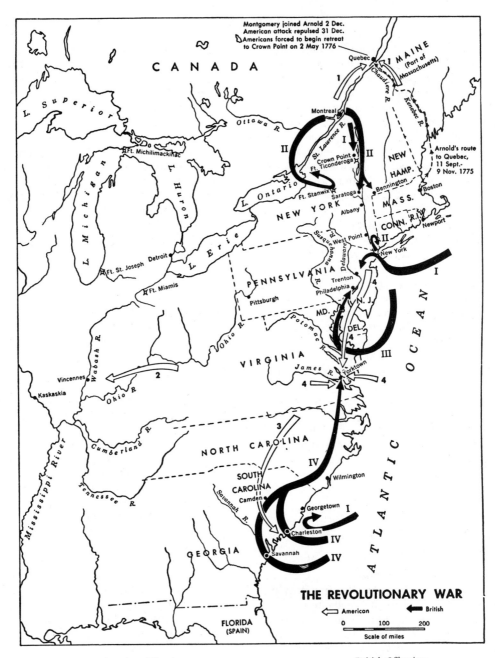

Montgomery joined Arnold 2 Dec.
American attack repulsed 31 Dec.
Americans forced to begin retreat
to Crown Point on 2 May 1776

THE REVOLUTIONARY WAR

⇦ American   ◄━ British

0   100   200

Scale of miles

American Offensives

(1) 1775—Invasion of Canada by Montgomery and Arnold
(2) 1779—Clark in the West
(3) 1781—Greene in the South
(4) 1781—Yorktown Campaign

British Offensives

(I) 1776—Three-pronged offensive: Lake Champlain, New York, Charleston
(II) 1777—Converging drives on Albany by Burgoyne, St. Leger, Clinton
(III) 1777—Howe's offensive to Philadelphia
(IV) 1780–1781—Invasion of South by Clinton and Cornwallis

makeshift, but manned by sailors drafted from the transports which had carried Burgoyne's troops, and more heavily armed. Most of the American vessels were destroyed or crippled. Arnold slipped away in the night with the remainder, but at Split Rock (October 13) Carleton's ships caught up and destroyed them. The American flotilla had, however, served its purpose, delaying Carleton's advance for so long that after occupying Crown Point briefly he postponed further advance and withdrew into Canada.

Central Campaign; New York and the Jerseys

1776, March 17. Evacuation of Boston. Threatened by the growing strength of Washington's army (20,000 Continentals, 6,000 militia, and heavy ordnance dragged from Ticonderoga), Howe, now commanding at Boston, evacuated the city, sailing to Halifax. Washington shortly began transferring his forces to New York as the most likely place for the next British move.

1776. French and Spanish Assistance. Louis XVI of France authorized the supplying of munitions to the Americans; **Charles III** of Spain followed suit.

1776, April. British Plans. The British government decided to take the offensive and crush the revolution. General Howe was to sail to New York from Halifax with considerable reinforcement from England under escort of a fleet commanded by his brother, Admiral **Richard Howe.**

1776, July 4. Declaration of Independence.

1776, July–August. Concentrations Near New York. After arriving off New York Harbor (July 2), Howe landed his 32,000 men on Staten Island. These included 9,000 Hessian mercenaries. Washington, instructed by the Congress to hold New York, attempted to guard against attack by land or water or both. Earthworks were thrown up across Brooklyn Heights and batteries manned on lower Manhattan and Governors Island. Half of Washington's army of 13,000, under General **Israel Putnam,** was deployed across the Flatbush area of Long Island; the remainder held on Manhattan.

1776, August 27. Battle of Long Island. Howe, landing 20,000 men on Long Island (August 22–25), skillfully turned Putnam's left flank. The Americans were thrown back into Brooklyn Heights with losses of 200 killed and nearly 1,000 captured. British casualties were 400. While Howe slowly approached the Brooklyn Heights line, Washington evacuated Long Island in a brilliant operation (August 29–30).

1776, September 6–7. The "American Turtle." Sergeant **Ezra Lee** in **David Bushnell**'s 1-man submarine attacked the British fleet off Staten Island. Though unsuccessful, he created much alarm. This was the first use of the submarine in war.

1776, September 12. Washington Decides to Abandon New York.

1776, September 15. Action at Kip's Bay. American troops fled as British troops assaulted across the East River from Brooklyn.

1776, September 16. Battle of Harlem Heights. Washington halted Howe's slow pursuit. But the British advance up the East River endangered Washington's communications and he fell back again.

1776, October 28. Battle of White Plains. After obstinate resistance the British regulars drove the Americans off the field.

1776, November 16–20. Forts Washington and Lee. Fort Washington on northern Manhattan, overlooking the Hudson, fell with the capture of 2,800 Americans. Fort Lee on the west bank was evacuated (November 20), with the loss of much badly needed materiel.

1776, November–December. Retreat through New Jersey. Washington with the bulk of his remaining troops retreated southward. General **Charles Lee,** left behind to cover the retreat, allowed himself and many of his 4,000 men to be captured near Morristown. Washington and the remaining 3,000 men of the Continental Army crossed the Delaware River into Pennsylvania. Congress, fleeing from

Philadelphia to Baltimore (December 12), in desperation turned over dictatorial powers to Washington. Confident of a speedy conclusion, Howe went into winter quarters, the bulk of his army at New York with garrisons scattered in several sections of southern New Jersey.

1776, December 26. Battle of Trenton. Washington, whose little army was disintegrating as enlistments expired, took a desperate chance. Crossing the Delaware 9 miles north of Trenton with 2,400 men on Christmas night, during a snowstorm, at dawn he fell on the Hessian garrison at Trenton. His victory was complete. Of 1,400 Hessians, nearly 1,000 were captured and a great amount of booty—small arms, cannon, and other munitions—taken. Thirty Hessians, including their commander, Colonel **Johann Rall,** were killed. American losses were 2 men frozen to death and 5 more wounded. In immediate reaction, 8,000 British troops under Lord Cornwallis moved to box Washington's army between the Delaware River and the Atlantic Ocean.

1777, January 2. Confrontation at Trenton. Cornwallis faced the American position with 5,000 men, while behind him at Princeton, 12 miles away, were 2,500 more under orders to join. Cornwallis ordered an assault for the next day. Washington now had 1,600 regulars and 3,600 unreliable militia. Leaving his campfires burning, he slipped away to the eastward in the night over a disused road, behind Cornwallis.

1777, January 3. Battle of Princeton. Washington met and defeated British reinforcements advancing from Princeton, captured a large quantity of military stores in that town, and hastened away to Morristown before fuming Cornwallis could return from Trenton.

COMMENT. *Washington's outstanding personal leadership had in 10 short days fanned the dying embers of the Revolution into lively flame. Frederick the Great characterized these operations as one of the most brilliant campaigns in military history. With Washington now flanking their communications, all British garrisons in central and western New Jersey were evacuated.*

SOUTHERN CAMPAIGN; THE CAROLINAS

1775, December. Uprising in Virginia. After a minor engagement at **Great Bridge,** near Norfolk, Governor Dunmore fled to take refuge on a British warship.

1776, February 27. Battle of Moores Creek Bridge. A force of 1,100 North Carolina patriots defeated about 1,800 Tories marching toward Wilmington, where they had hoped to establish a British coastal base.

1776, June 4. Charleston Expedition. General Sir **Henry Clinton** with troops from Boston and reinforcements from England arrived off Charleston, S.C., after abandoning his plan to land at Wilmington as a result of the Battle of Moores Creek Bridge (see above). General Charles Lee had earlier been sent by Congress to organize the defense of the most important American port south of Philadelphia.

1776, June 28. Battle of Sullivan's Island. The palmetto-log fortification on this island was the key to the harbor defenses. Admiral Sir **Peter Parker's** squadron, which had convoyed the British expedition, attacked the fort. But American artillery under Colonel **William Moultrie** did amazing damage to the British ships; the squadron withdrew and shortly afterward Clinton reembarked his troops and left to join Howe at New York. (Rechristened Fort Moultrie, the island post so well defended was destined to further fame in the Civil War.) There would be no further operations by the regular British forces in the Carolinas for 2 years to come.

War at Sea, 1776–1777

1776, February 17. Bahamas Cruise. Commodore **Esek Hopkins** with a squadron of 6 small extemporized warships sailed from Philadelphia for the Bahamas. He attacked New Providence, sending a landing party ashore and capturing a quantity of cannon and powder (March 3–4). Hopkins returned to Providence, R.I., where the squadron remained for the rest of the year.

1776, March 23. Congress Authorizes Privateering. This became a profitable business throughout the war.

1776–1777. The New Navy Is Blockaded. Thirteen frigates authorized by Congress were building. However, the overwhelming strength of the Royal Navy dominated the entire eastern seaboard.

War on Land, 1777–1779

NORTHERN CAMPAIGN, 1777

1777. British Plans. They intended to seize the Hudson River Valley, thus splitting the colonies. Burgoyne was to move south from Canada via Lake Champlain, while Howe was to move north from New York, joining Burgoyne at Albany. A third column under Colonel **Barry St. Leger,** moving up the St. Lawrence to Lake Ontario, would land at Oswego and, in conjunction with Iroquois Indians and Tories under Sir **John Johnson,** sweep down the Mohawk Valley to unite with the others at Albany. While perhaps an attractive plan on the map, success was entirely dependent upon close coordination and timing, which would be most difficult to achieve under the circumstances of the time in the trackless forests of North America. Furthermore, the entire affair—originally conceived by Burgoyne—was bungled by Lord **George Germain** (see p. 736), who as secretary of state for the colonies in Lord **North**'s cabinet was in control of operations in America. There was no coordination. Burgoyne had been ordered to meet Howe, but Howe's operations were left to his own discretion. The results were disastrous to England.

1777, April 25. Danbury Raid. A British force under General William Tryon, Governor of New York, destroyed an American supply depot at Danbury, but was severely harassed by local militia gathered and led by General Benedict Arnold.

1777, July 5. Capture of Ticonderoga. Burgoyne, with 7,200 regulars—including 4,000 British and 3,200 Hessian-Brunswicker mercenaries—as well as a small number of Tories and Indians, reached Fort Ticonderoga July 1, emplacing his artillery on a dominating height. American General **Arthur St. Clair** evacuated the post the night of July 5–6, the 2,500 poorly armed and equipped Americans retreating into Vermont over a bridge spanning Lake Champlain; the sick and the baggage went by boat to Skenesboro (now Whitehall). Burgoyne pursued immediately, his main body up the lake, his advance corps over the road.

1777, July 7. Battle of Hubbardton. The American rear guard was overtaken and defeated; the baggage and supplies at Skenesboro captured. However, **Seth Warner**'s delaying action saved St. Clair's army. He fled to Rutland, thence south to join General Schuyler at Fort Edward.

1777, July–August. Burgoyne's Pursuit. He sent his heavy equipage from Ticonderoga via Lake George, but the Americans had completely devastated the rough roads southward, so that it took him 3 weeks to reach Fort Edward. By that time Schuyler had retreated to Stillwater. The American forces had been strengthened to some 3,000 Continentals and 1,600 militia, and Washington had also sent Generals Benedict Arnold and **Benjamin Lincoln** to assist Schuyler as well as Colonel **Daniel Morgan**'s rifle regiment.

1777, August 3. Burgoyne Learns of Howe's Nonparticipation. Howe was bound south, hunting Washington. Burgoyne decided to strike on for Albany anyway, hoping to meet St. Leger there.

1777, July 25–August 25. St. Leger on the Mohawk. Arriving at Oswego, St. Leger, with 875 British, Tory, and Hessian troops and 1,000 Iroquois under **Joseph Brant,** invested Fort Stanwix (also called Fort Schuyler) on the present site of Rome, N.Y., garrisoned by 750 Continentals and militia. General **Nicholas Herkimer** of the county militia hurried to the rescue with 800 men.

1777, August 6. Battle of Oriskany. Herkimer was ambushed by Brant with Indians and Tories 6 miles from the fort. The Americans successfully fought off the attack, but Herkimer was mortally wounded; having lost nearly half its strength, the relief column retreated. While the action was on, a sortie from the fort ravaged St. Leger's camp. The siege was resumed. At Stillwater, Schuyler, with Burgoyne only 24 miles away, detached 1,000 volunteers under Arnold to relieve the fort.

1777, August 16. Battle of Bennington. Burgoyne, short of supplies, meanwhile had sent 700 Brunswickers under Colonel **Friedrich Baum** to capture military stores at Bennington, and to scour the countryside for horses. Colonel **John Stark,** with 2,000 New Hampshire, Vermont, and Massachusetts militiamen, surrounded Baum and destroyed his force. Reinforcements for Baum, 650 Brunswickers under Colonel **Heinrich von Breymann,** arriving later that day, were defeated by Stark and Seth Warner's 400 Green Mountain Boys, who had made a forced march from Manchester. Breymann escaped with less than two-thirds of his force. British casualties were 207 dead and 700 captured; the Americans lost 30 killed and 40 wounded. Much-needed small arms, cannon, and wagons were captured.

1777, August 23. Relief of Fort Stanwix. St. Leger's Indians were put in panic through a ruse, and he abandoned the siege before Arnold's arrival, leaving behind him tents, cannon, and stores as he retreated precipitately to Oswego.

1777, September 13. Burgoyne Crosses the Hudson Near Saratoga. Having reached the point of no return, he crossed to attack the American forces entrenched on Bemis Heights, with General **Horatio Gates** now in command. Burgoyne also sent an urgent call for help to General Clinton in New York, commanding in Howe's absence.

1777, September 19. Battle of Freeman's Farm. Burgoyne's attack on the American left was repulsed with 600 casualties; American losses were 300. Arnold, who distinguished himself, requested reinforcements to counterattack, but these were refused by Gates.

1777, October 3. Clinton's Gesture. Down in New York, Clinton, with 7,000 remaining British and Germans, took 4,000 men up the Hud-

son, capturing Forts Clinton and Montgomery in the highlands (October 6). Satisfied that he had created a diversion helpful to Burgoyne, Clinton then returned to New York after his ships had burned Esopus (Kingston, N.Y.).

1777, October 7. Battle of Bemis Heights. Burgoyne, who had entrenched in the hope that Clinton would come to his help, made one last desperate attempt to turn the American left. Arnold, who had been deprived of command by Gates, nevertheless led an American counterattack, which drove the British back against the river with loss of 600 men to the Americans' 150. Morgan's riflemen played a leading part in the victory. Withdrawing toward Saratoga, Burgoyne was soon surrounded by a force thrice his remaining strength of 5,700.

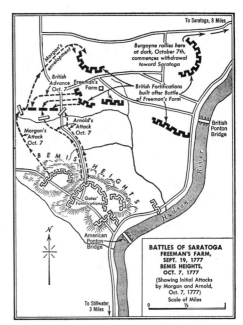

BATTLES OF SARATOGA
FREEMAN'S FARM,
SEPT. 19, 1777
BEMIS HEIGHTS,
OCT. 7, 1777
(Showing Initial Attacks by Morgan and Arnold, Oct. 7, 1777)
Scale of Miles

1777, October 17. Burgoyne's Surrender. By the terms of the "Convention of Saratoga," the disarmed British Army was to be marched to Boston and shipped to England on parole. These terms were later shamefully repudiated by the American Congress; the British and German soldiers were not permitted to return home.

COMMENT. *The Saratoga victory was the turning point in the Revolution. To the American people it more than balanced the reverses of the central campaign related below. British forces were redistributed. Ticonderoga and Crown Point were evacuated; Clinton abandoned the Hudson highlands. The British held only New York City, part of Rhode Island, and Philadelphia. Overseas, France recognized the independence of the United States, a forerunner of her active participation in the war.*

CENTRAL CAMPAIGN, 1777–1778

1777. Maneuvers in New Jersey. Spring and early summer were taken up by fruitless maneuvers as Howe attempted to coax Washington into battle.

1777, July 23. Howe Sails from New York. He took some 18,000 men in transports. Washington, puzzled and anxious about Burgoyne's expedition in the north, detached Arnold and Morgan with reinforcements to Schuyler, but held his army north of the Delaware until he received word that Howe's expedition had sailed up Chesapeake Bay and was landing (August 25) at the Head of Elk (Elkton). Washington, with 10,500 men, then barred the way to Philadelphia on the east bank of Brandywine Creek. Howe advanced, brushing aside light American resistance at **Cooch's Bridge** (south of Wilmington; September 2).

1777, September 11. Battle of the Brandywine. Howe skillfully turned the American right, forcing Washington back toward Philadelphia with losses of 1,000 men; the British lost 576.

1777, September 21. Action at Paoli. A surprise British night attack routed the brigade of General **Anthony Wayne.** Philadelphia was evacuated, the Congress fleeing to Lancaster and then to York. Again it left dictatorial powers to Washington.

1777, September 26. Howe Occupies Philadelphia. With the assistance of his brother's fleet, he cleared the Delaware River as a supply route despite the desperate and gallant American defense of **Forts Mifflin** and **Mercer** (October 22–November 20).

1777, September 25–December 23. Conway Cabal. Colonel **Thomas Conway,** Irish soldier of fortune who left the French service to join the Americans, attempted to stir congressional dissatisfaction with the direction of the war into the elevation of Gates over Washington. Unsuccessful, Conway resigned (December 23).

1777, October 4. Battle of Germantown. Washington, reinforced to a strength of 13,000, attempted a complicated movement against Howe's main encampment at Germantown, but through the blunders of his officers was defeated, incurring 700 casualties and losing 400 prisoners.

1777–1778. Ordeal at Valley Forge. The American army spent an agonizing winter, while the British enjoyed the comforts of Philadelphia.

1778, January–June. The Impact of Baron Augustus H. F. von Steuben. This German volunteer instilled discipline and tactical skill in the remnants of the starving little Continental Army at Valley Forge.

1778, February 6. Franco-American Alliance. Two treaties were signed: one of amity and commerce, the other an alliance effective if and when war broke out between France and England.

1778, June 17. Outbreak of War between France and England.

1778, June 18. British Evacuation of Philadelphia. Sir Henry Clinton, succeeding to command from Howe, removed his 13,000 troops over land toward New York, intending to institute a more vigorous campaign. Washington's army, recruited to equal strength and hardened by von Steuben's training, followed by a more northern road, seeking a chance to give battle under favorable circumstances.

1778, June 28. Battle of Monmouth. On Washington's orders General Charles Lee,* heading the American advance guard, moved toward Clinton's rear guard, but at a critical moment fell back, throwing the army into confusion. Clinton turned to give battle. Washing-

---

* He had been exchanged as a prisoner of war.

ton's personal leadership rallied his army, repulsing repeated British attacks. An all-day fight in intense heat finished with both armies bivouacking on the field, but during the night Clinton hurriedly withdrew. Reported losses were about 350 on each side; they were undoubtedly much greater. The Continental troops had proved themselves equal to the British regulars. Clinton reached New York safely, where he was blockaded by Washington, who took position at White Plains July 30. Lee was cashiered for his shameful conduct.

NORTHERN, CENTRAL, AND WESTERN CAMPAIGNS, 1778–1779

1778–1781. Operations in the North. There were no major operations in the northern or central theaters, though there was hard fighting, both in organized warfare and in guerrilla actions. Tory and Indian atrocities rocked northern Pennsylvania and New York: the **Wyoming Valley** (July 3, 1778) and **Cherry Valley** (November 11, 1778) **Massacres** were monuments to the bloodthirsty **Johnson** family, the **Butlers—John** and **Walter—**and Chief **Joseph (Thayendanegea) Brant**'s Indians. An expedition into northwestern New York, under Generals Sullivan and James Clinton, reduced the Tory-Indian threat (August–September 1779). Simultaneously Colonel. **Daniel Brodhead** led an expedition from Fort

Pitt to ravage Indian villages in the Allegheny Valley (August 11–September 14). A Franco-American expedition against **Newport** (August 1778) failed because of bad weather and faulty cooperation between General Sullivan's American land forces and Admiral **Jean Baptiste d'Estaing**'s French fleet. In the lower Hudson Valley the British capture of **Stony Point** (May 31, 1779) was quickly followed by Wayne's brilliant and successful bayonet assault (July 15–16), which recaptured the post and ended a threat to West Point, key American fortress on the Hudson. A similar American attack by Major **Henry "Lighthorse Harry" Lee** on **Paulus Hook** on the New Jersey shore of New York harbor, was also successful (August 19).

1778–1779. Operations of George Rogers Clarke in the West. With a handful of militia, he waged a campaign tactically minor but strategically most important. His final capture of Vincennes, Ind. (February 25, 1779), secured to the United States in the later peace treaty the entire region from the Alleghenies to the Mississippi, a territory larger than the thirteen original colonies.

1779, June 21. Spain Declares War against England. King **Charles III,** however, and his wily Prime Minister, Count **José de Floridablanca,** refused to recognize American independence.

SOUTHERN CAMPAIGNS, 1778–1779

For nearly 2 years, the southern colonies had writhed in a continuous guerrilla warfare between Tories and patriots, but no major operations took place until Clinton, after the evacuation of Philadelphia, was ordered to the Carolinas.

1778, December 29. Capture of Savannah. Lieutenant Colonel **Archibald Campbell,** with 3,500 men of Clinton's command, crushed 1,000 American militia under General **Robert Howe** and occupied the city.

1779, January–June. War Flames in the South. Tory hopes rose as British General **Augustine Prevost** moved up from Florida,

while Campbell seized Augusta, Ga. (January 29). In South Carolina, Moultrie at **Port Royal** repulsed an attack by Prevost (February 3), and **Andrew Picken's** militia defeated a Tory brigade at **Kettle Creek** (February 14). An American attempt to recapture Augusta failed at **Briar Creek** (March 3). General Lincoln, sent by Washington to command the Southern

Department, arrived in the area with a force of Continentals. Prevost failed in an effort to take Charleston. He withdrew and, despite Lincoln's unsuccessful attack at **Stono Ferry** (June 19), safely regained Savannah. In isolated actions, North Carolina and Virginia militia raided Chickamauga Indian villages in Tennessee, while British Admiral Sir **George Collier** sacked and burned Portsmouth and neighboring towns on the Virginia coast (May).

1779, September 3–October 28. Siege of Savannah. D'Estaing's fleet with 4,000 French troops, returning from the West Indies (see below and p. 784), arrived off the port, capturing two British warships and two supply ships. Disembarking his French troops (September 12), d'Estaing invested the place. Lincoln with 600 Continentals and 750 militia joined him (September 16–23). Prevost, commanding the British garrison, had some 3,500 men and his defenses were strong. Discouraged by slow progress of the siege, and anxious lest his fleet meet disaster in seasonal storms, the French admiral after a futile bombardment insisted on an assault (October 8). At dawn next day, five allied columns advanced. Warned by a deserter, the British were waiting for them and the assaults were thrown back after hand-to-hand fighting in the most severe conflict of the war since Bunker Hill. Allied losses were more than 800 killed and wounded, among them Polish count **Casimir Pulaski.** British losses amounted to 150. Refusing to continue the siege, d'Estaing embarked his troops and sailed away. Lincoln returned to Charleston.

COMMENT. *A serious blow had been struck against the cause of independence; Tory spirits soared, while the disheartened patriot's confidence in the French dropped sharply.*

1779, December 26. Clinton Sails South. After withdrawing British troops from Newport, R. I. (October 11), and placing German General **Wilhelm von Knyphausen** in command at New York, Clinton sailed with 8,000 men to attack Charleston. All available British pressure would now be put on the South.

Operations at Sea, 1778–1779

1778, April–May. Exploits of John Paul Jones. In USS *Ranger,* after terrorizing British shipping in the Irish Sea, he landed at Whitehaven in England, spiked the guns of the local fort, then, crossing Solway Firth to St. Mary's Island, seized the residence of the **Earl of Selkirk** (April 23). This was the first purely foreign invasion of England since the Norman Conquest. Jones then cruised over to northern Ireland and, after a 1-hour fight with HMS *Drake* off Carrickfergus, captured her (April 24) and took her into Brest. Jones's activities threw the British into a ferment.

1778, July 11–22. Arrival of the French Fleet. Admiral Count d'Estaing arrived from France off Sandy Hook, blockading New York for a short time. He then sailed for Narragansett Bay, lying offshore and making contact with American General Sullivan for planned operations against Newport (see p. 782).

1778, July 27. Battle of Ushant. British Admiral **Augustus Keppel,** with 30 ships of the line, met and clashed briefly and inconclusively off the coast of Brittany with French Admiral Count **d'Orvilliers'** squadron of equal strength.

1778, August. Maneuvers of Howe and d'Estaing. Howe's fleet, based at New York, tried to come to grips with d'Estaing's fleet off Newport, but a storm dispersed both squadrons before they could come into general contact with each other (August 11). D'Estaing then withdrew from Newport to repair storm damage at Boston; as a consequence, Sullivan, without naval support, had to abandon the Newport operation.

1778, November 4. Departure of d'Estaing. He left Boston, cruising to the West Indies, where local French forces had seized Dominica (September 8), and British Admiral **Samuel Barrington** took St. Lucia (November 13).

1779. Operations in the West Indies. D'Estaing captured St. Vincent (June 16) and Grenada (July 4) while British Admiral **John Byron** was hunting for him.

1779, June 21–1783, February 6. Siege of

Gibraltar. The successful defense of the fortress by General **George Augustus Eliott** with the assistance of the Royal Navy was one of the great combined arms exploits of the British armed forces.

1779, July 6. Battle of Grenada. Off Georgetown Byron's squadron was roughly handled by d'Estaing's superior French force and 4 British ships were dismasted. D'Estaing sailed to Savannah (see p. 783).

1779, July 25–August 14. Penobscot Expedition. An amphibious force under Continental Navy Commodore Dudley Saltonstall and Massachusetts militia Brigadier General Solomon Lovell tried ineptly to seize the British timber base on Penobscot Bay, Maine. A British squadron destroyed the flotilla; the land force dispersed in the Maine wilderness.

1779, September 23, USS *Bonhomme Richard* verses HMS *Serapis*. John Paul Jones in a converted 42-gun Indiaman, off Flamborough Head, defeated the British frigate *Serapis*, 44, in one of history's most remarkable single-ship actions. Jones's flaming ship was sinking, but he refused British Captain **Richard Pearson's** hail to surrender with, ''I have not yet begun to fight!'' Pearson, his mainmast shot away, his gun deck swept by a powder explosion touched off by an American grenade, at last surrendered. Jones moved his crew to the *Serapis* and brought his prize into Texel, Netherlands (October 3).

Spanish Operations in the South and West 1779–1781

1779, August–September. Expedition up the Lower Mississippi. Don **Bernardo de Galvez,** Spanish governor of Louisiana, recognized the threat to New Orleans of Britain's West Florida posts on the Mississippi above the city. After Spain declared war, with a few troops he quickly took **Manchac** (September 7), **Baton Rouge** (September 20), and **Natchez** (September 30).

1780, February–March. Mobile Campaign. Galvez landed troops from New Orleans and captured Mobile, capital of British West Florida (March 14). Major General **John Campbell,** coming from Pensacola with reinforcements, turned back.

1780–1781. Raid into Michigan. Galvez sent Captain **Eugenio Pourré** up the Mississippi from St. Louis. He surprised and captured the British garrison of **Fort St. Joseph,** on Lake Michigan (January 1781,) and returned to St. Louis.

1781, February–May. Pensacola Campaign. With troops from Havana and Mobile and his naval squadron from New Orleans, Galvez besieged **Fort St. George** near Pensacola (March 10). After a Spanish hit blew up the fort's powder magazine, General Campbell surrendered (May 9).

War on Land, 1779–1783

SOUTHERN CAMPAIGN, 1780–1781

American Disasters, 1780

1780, February 28. Russia's Armed Neutrality. Britain's blockade of French and Spanish supply to the United States suffered when **Catherine II** declared her navy would protect her trade against all belligerents. Denmark (July 9) and Sweden (August 1) accepted her invitation to join; within 2 years the Netherlands, Prussia, Portugal, Austria, and the Kingdom of the Two Sicilies (Naples) joined.

1780, July 10. French Army at Newport. Lieutenant General Count **Jean Baptiste de Rochambeau** with 5,000 men, convoyed by 7 ships of the line under Admiral **De Ternay,** landed at Newport. British ships blockaded the harbor, preventing proposed Franco-American operations against New York.

1780, September 23. Treason of Benedict Arnold. His plans to deliver his command, West Point, guardian of the Hudson Valley, to Clinton were captured with British Major **John André.** Arnold escaped and received a British brigadier general's commission and substantial cash. Thereafter he fought for England. André was hanged as a spy.

1780, December 20. England Declares War on the Netherlands. This was because of clandestine Dutch trade with the American states.

1780–1781. American Mutinies. Continental currency depreciation and supply deficiencies led to six quickly suppressed mutinies, at West Point (January 1, 1780), Morristown (May 25, 1780), Fort Stanwix (June 1780), Morristown (January 1, 1781), Pompton (January 20, 1781), and again Morristown (May 1781).

1780–1781. Blockade of New York. Washington maintained a cordon about the British in New York.

1780, February 11–May 12, Siege of Charleston. Clinton's expedition disembarked near Charleston, bringing the British forces in the area to 14,000. Despite Lincoln's energetic efforts, Clinton slowly completed the investment (April 11) and began siege operations. Admiral **Marriot Arbuthnot's** squadron in the harbor added to the heavy bombardment from the land. Lincoln finally surrendered with 5,400 men, plus cannon, small arms, and ammunition, in the worst American disaster of the war (May 12). Leaving Cornwallis with 8,000 men in South Carolina, Clinton returned to New York.

1780, May–August. Pacification and Reaction. Cornwallis's severe measures, particularly the brutality of Sir **Banastre Tarleton's** Tory cavalry, caused strong guerrilla opposition, led by **Francis Marion, Thomas Sumter,** and **Andrew Pickens.** Tarleton's massacre of a small American force at **Waxhaw Creek** (May 29) aroused the countryside. Washington sent Brigadier General **Johann de Kalb** and 900 Continentals to South Carolina. Congress put Horatio Gates in command of the Southern Department, independent of Washington.

1780, August 16. Battle of Camden. Gates with some 3,000 men, mostly militia, met Cornwallis with 2,400 British and Tory regulars at dawn. The militia broke; Tarleton's cavalry charged from the rear. The Continentals, struggling gallantly, were overwhelmed. Gates fled 160 miles to Hillsboro, N.C., followed by the remnants of his army. Nearly 900, including De Kalb, were killed and 1,000 captured. Two days later Tarleton smashed Sumter's guerrilla force at **Fishing Creek.**

The Turn of the Tide, 1780–1781

1780, October 6. Congress Returns Control to Washington. He at once appointed **Nathanael Greene,** his "right arm," to command in the south.

1780, October 7. Battle of King's Mountain. British Colonel **Patrick Ferguson**'s corps of Tory riflemen, 1,100 strong, was demolished by 1,400 sharpshooting Carolinian "mountain men" and Virginians—all militia—led by Colonels **Isaac Shelby** and **Richard Campbell.** Ferguson was killed.

COMMENT. *The fight was remarkable in that all participants on both sides were Americans, save Ferguson himself. This disaster, combined with widespread disorders in South Carolina, led Cornwallis to abandon a proposed invasion of North Carolina. He retreated to winter quarters at Winnsborough, N.C., and took stern measures against rebellious colonists.*

1780, December 2. Greene Assumes Command. His army at Charlotte had 1,482 men, 949 of them Continentals, all ill-clothed and badly equipped. By mid-December this force had grown to some 3,000–1,400 of whom were Continentals—thanks to reinforcements sent by Washington. Meanwhile, Cornwallis had also been reinforced to a strength of 4,000 well-equipped, well-trained regulars.

1780, December 20. Greene Takes the Offensive. He deliberately (and mistakenly) divided his forces, sending Brigadier General **Daniel Morgan** with 1,000 men on a wide western sweep. Greene moved with the remainder to a camp at Cheraw Hill, S.C., nearer to Charleston than Cornwallis. The American forces were now separated by 140 miles. Cornwallis, instead of concentrating against one of the American forces, unwisely split his own command. He sent Tarleton with 1,100 men against Morgan, while General **Alexander Leslie** with a comparable force was to contain the Americans at Cheraw Hill. Cornwallis with the main body followed Tarleton.

1781, January 17. Battle of Cowpens. Tarleton, catching up with Morgan, was decisively defeated in a brilliant double envelopment. The British lost 110 killed and 830 captured of their 1,100 strength; Morgan lost 12 killed and 61 wounded.

COMMENT. *This American "Cannae" was one of the most brilliant tactical operations ever fought on American soil.*

1781, January–February. Retreat to the Dan. Greene and Morgan, reuniting, hastily retreated into southern Virginia, closely pursued by Cornwallis. The British general, destroying his baggage in order to increase his mobility, pursued to the unfordable Dan River, then retired to Hillsboro again. Greene turned and followed him, his army having been rein-

forced to a total of about 4,400 men, of whom more than two-thirds were untrained militia or newly raised Continentals. Crippling arthritis prevented Morgan from remaining in the field.

1781, March 15. Battle of Guilford Courthouse. Cornwallis, with 1,900 effectives, attacked Greene, who had chosen his ground carefully. The outnumbered British with great gallantry gained the upper hand and drove the militia from the field; Greene broke off the action in order to avoid disaster and retreated in good order with the remainder of his army. Cornwallis' victory was costly: 93 killed and 439 wounded, against 78 Americans killed and 183 wounded. The British general marched eastward to Wilmington, N.C., then, deciding he could no longer hold Georgia and the Carolinas, proceeded with 1,500 men into Virginia (see below and p. 787).

Greene's Later Operations, 1781

Left behind in the Carolinas were several isolated British garrisons, against whom Greene now proceeded, his operations aided and supported by the guerrilla activities of Marion, Sumter, and Pickens.

1781, April 19. Battle of Hobkirk's Hill (near Camden). Greene was repulsed by a British force under Colonel **Francis Rawdon.** As Greene prepared to renew his advance, the British withdrew toward Charleston.

1781, May 22–June 19. Siege of Fort Ninety-six. Greene was again unsuccessful when a relieving force rescued the garrison and then withdrew to Charleston. Most of Carolina was now liberated from the British.

1781, September 8. Battle of Eutaw Springs. Approaching Charleston, Greene with 2,400 men attacked Lieutenant Colonel **Alexander Stewart** with some 2,000. The Americans were at first successful, but the British rallied and repulsed the Americans. Greene withdrew. Next day, Stewart, whose losses had been very heavy, fell back on Charleston, which, with Savannah, was 1 of only 2 remaining British footholds south of Virginia.

COMMENT. *Greene's strategical task had been competently accomplished, despite his failure to win a single tactical victory.*

YORKTOWN CAMPAIGN, 1781

Preliminaries in Virginia

1780, December 30-1781, March 26. Arnold's Operations in Virginia. The American traitor, with a force of 1,600 men, arrived at Hampton Roads to carry out Clinton's instructions to destroy military stores, to prevent reinforcements from reaching Greene, and to rally Tories. He seized and devastated Richmond (January 5). He then returned to Portsmouth, maneuvering inconclusively against von Steuben.

1781, March–May. Phipps's Operations in Virginia. British Major General **William Phipps** was sent with reinforcements to command in Virginia over Arnold. After some further destruction of supplies and goods, Phipps and Arnold marched southwest to meet Cornwallis, arriving from North Carolina (see p. 787). Phipps died at Petersburg (May 10).

1781, April 29. Lafayette Reaches Rich-

mond. He had been sent from New York by Washington with reinforcements to take command in Virginia. His total force, including new troops being trained by Steuben, was 3,550 men, of whom 1,200 were veteran Continentals. Washington sent Wayne with his brigade of 1,000 Continentals from Pennsylvania to join Lafayette (June 10). Von Steuben was also sent to Virginia, where he gave a poor performance as a field commander but again demonstrated his superb training talents.

1781, May 20. Arrival of Cornwallis at Petersburg. He took command of British forces in Virginia. Counting the garrison of Portsmouth, these totaled 8,000.

1781, May–July. Maneuvers around Virginia. Cornwallis attempted to bring the Americans to battle, but Lafayette, keeping out of reach, led the British in a chase around eastern Virginia. Cornwallis, ordered by Clinton to sent part of his force back to New York, reluctantly returned to Portsmouth, closely followed by Lafayette. In a well-planned ambush near **Jamestown Ford** (July 6) on the banks of the James River, Cornwallis caught Wayne's brigade by surprise, but the Americans, despite heavy losses, repulsed the British attack, counterattacked, then retreated in good order. Cornwallis continued on to Portsmouth.

1781, August 4. Cornwallis Moves to Yorktown. New orders from Clinton (July 20) permitted Cornwallis to keep all of his troops, and directed him to occupy the tip of the Virginia Peninsula. He moved by water to Yorktown with more than 7,000 troops, planning to keep a link by sea with Clinton in New York. Lafayette, who now had 4,500, cautiously moved to nearby West Point, Va. closely observing the British. He sent a report to Washington.

Washington's Plans and Movements, May–September, 1781

1781, May 21. Washington and Rochambeau Confer near New York. They reached the joint conclusion that British strength now lay in 2 points: New York and Chesapeake Bay. In the West Indies at this time was Admiral **François**

J. P. de Grasse, with a powerful French fleet (see p. 790). Were de Grasse to cut the sea communications between these 2 British centers, joint land and sea operations against either would be favored. A French frigate was sent to de Grasse with the word from Washington that sea power was essential to success. At the same time Rochambeau's army moved from Newport to join Washington's outside New York.

1781, August 13. De Grasse Sails North. His fleet headed for Chesapeake Bay, carrying an additional 3,000 French troops. He had earlier sent word to Washington that the fleet would be available until mid-October. Washington received the message (August 14) just after getting Lafayette's report of Cornwallis' move to Yorktown. Washington at once recognized that if de Grasse could keep command of the sea, Cornwallis' army could be destroyed. Rochambeau agreed.

1781, August 21. Washington Marches South. Leaving **William Heath** with 2,000 men to contain Clinton's 17,000 in New York, Washington led the allied armies south by forced marches. They reached Philadelphia before Clinton, deceived by a pretended attack, realized what was happening.

1781, August 30. De Grasse Arrives off Yorktown. He disembarked his troops, who reinforced Lafayette. A British fleet under Admiral **Thomas Graves** rushed from New York, appearing off the entrance of the bay.

1781, September 5–9. Battle of the Capes. (See p. 790.) More French ships arrived from Newport, strengthening victorious de Grasse and bringing siege artillery for the allies. Graves sailed away for New York, leaving command of the sea to the French and sealing Cornwallis' fate.

1781, September 14–26. Arrival of Washington's and Rochambeau's Troops. Most were transported by French ships from Head of Elk, Baltimore, and Annapolis, and put ashore at Williamsburg.

Operations against Yorktown

1781, September 28. Investment of Yorktown. Washington had some 9,500 Americans and 7,800 splendidly equipped French regulars (Rochambeau had put himself without reservation under Washington's orders) with a variety of artillery—including the Gribeauval field guns of the French, here to receive their baptism of fire. Cornwallis had 8,000 men in all.

1781, September–October. Siege of Yorktown. Hoping to receive aid soon from Clinton, Cornwallis withdrew to his inner fortifications (September 30), thus enabling the allies to bring their siege artillery within range of his entire position. Bombardment soon began (October 9). Two key redoubts were stormed by Franco-American detachments (October 14) and new batteries established. A British counterattack was sharply repulsed (October 16). Cornwallis' situation was now impossible. A desperate scheme to evacuate part of his troops across the York River was thwarted by storm.

1781, October 17. Cornwallis Opens Negotiations for Capitulation. Washington allowed 2 days for written proposals, but insisted on complete surrender.

1781, October 19. Surrender of Cornwallis. Yorktown's garrison marched out and laid down all their arms. To all intents and purposes the war was over.

1781, October 24. Clinton's Belated Arrival. Graves convoyed him and 7,000 reinforcements to Chesapeake Bay. But, deterred by the presence of de Grasse's fleet and realizing that he was too late, Clinton put back to New York. De Grasse, fearful of hurricanes, declined to participate in Washington's suggestion of an attack on Charleston or the British base at Wilmington, and sailed for the West Indies.

COMMENT. *Sea power, and Washington's brilliant strategic grasp of its importance, combined with superb allied tactics, had won the decisive battle of the war.*

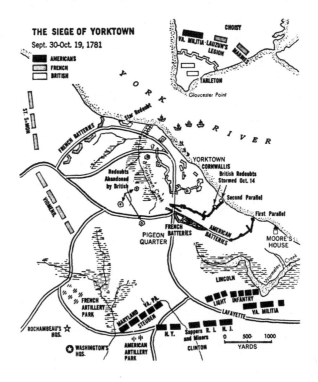

THE SIEGE OF YORKTOWN
Sept. 30-Oct. 19, 1781

AMERICANS
FRENCH
BRITISH

FINAL LAND OPERATIONS, 1781–1783

1781, November. Washington Marches Back to the Investment of New York. He established his headquarters at Newburgh.

1781–1782. French Activities. Rochambeau's army spent the winter in Virginia, returned to Rhode Island (fall, 1782), and embarked from Boston for France (December 24, 1782).

1781–1782. Greene Secures the South. He moved to invest British-held Charleston, while sending detachments to secure Georgia in a few minor engagements.

1782. Britain Sues for Peace. Lord North's ministry collapsed (March 20) and peace negotiations were opened (April 12). Sir Guy Carleton succeeded Clinton in the United States (May 9). All British troops were concentrated in New York. (Wilmington had been evacuated in January; Savannah was cleared July 11 and Charleston December 14.)

1782, August. Canadian and Indian Raid into Kentucky. An irregular force of about 240 men under American turncoat **Simon Girty** raided across the Ohio to attack **Bryan's Station** (August 15). Repulsed, they ambushed a pursing American force at **Blue Licks** (August 19). Colonel **Daniel Boone,** who had warned his companions of a possible ambush, distinguished himself. The victorious raiders withdrew.

1782, November 30. The Treaty of Paris. This treaty recognized the independence of the United States, and concluded the war between the United States and Great Britain; it was to become effective upon conclusion of Britain's war with France and Spain, now being pursued in the West Indies, off the Indian coast, and in the Mediterranean.

1783, April 15. Congress Ratifies the Treaty. Hostilities elsewhere having ended, the war was now officially concluded. Disbandment of the Continental Army (June) set a pattern for hasty demobilization, which would be followed in the United States thereafter.

1783, November 25. British Evacuation of New York. The last English troops embarked; Washington with Governor George Clinton entered the city. The British transports stood out to sea (December 4). That same day Washington took leave of his officers at Fraunces' Tavern.

War at Sea, 1780–1783

RODNEY'S OPERATIONS, 1780

1780, January 16. The "Moonlight Battle." British Admiral **George B. Rodney,** with 22 ships of the line and a large convoy of transports under orders to assist Gibraltar (under siege by the Spaniards on land since July 1779), fell in with Admiral **Langara**'s squadron of 11 Spanish ships off Cape St. Vincent. The Spaniards were defeated in a night action, 1 ship being sunk and 6 captured. After reinforcing Gibraltar, Rodney made for the West Indies, where he was to take command.

1780, March–August. Operations in the West Indies. French Admiral **Lucurbain de Guichen** met Rodney in equal strength in 3 indecisive actions (April). The Frenchman was then reinforced by a Spanish squadron, but through disease and indecision the allied effort aborted. De Guichen sailed for France (August).

1780, September. Rodney Sails North. Leaving half his fleet, he sailed with the remainder for New York, foiling Washington's plan for a joint Franco-American land and sea assault on the city. Returning to the West Indies (December), Rodney seized the Dutch islands of St. Eustatius and St. Martin (February), crippling the contraband trade on which the United States depended. He then sailed for England, leaving Admiral **Samuel Hood** in command.

THE PERIOD OF FRENCH PREDOMINANCE

1781, March 16. First Battle of the Virginia Capes. When Washington sent Lafayette to Virginia to catch Arnold, French Commodore **Sochet Destouches** at Newport sailed for Chesapeake Bay with 8 ships of the line and a detachment of French troops to join Lafayette.

British Admiral Arbuthnot, off New York, gave chase with 8 ships. Destouches had the better of the resultant engagement; 3 British ships were crippled. But Destouches returned to Newport, leaving command of the sea to the British. (Sometimes called Battle of the Chesapeake Capes.)

1781, March 22. De Grasse Leaves Brest. He had 26 ships of the line and a large convoy of troop transports. He headed for the West Indies and, possibly, the United States. Off the Azores, Admiral **Pierre André de Suffren,** with 5 ships, parted company, bound for the East Indies. De Grasse arrived off Martinique (April 28).

1781, April–August. Fencing in the West Indies. De Grasse had somewhat the better of inconclusive operations against Hood. He received Washington's message (see p. 787) off Haiti.

1781, August. The Fleets Move North. Promptly gathering all available French troops (3,500) on Haiti, de Grasse sailed by a roundabout route, arriving at Lynnhaven with 28 ships of the line (August 30). Hood, following de Grasse, had actually reached Chesapeake Bay 2 days before the main French fleet, but finding it empty went on to New York to join Admiral **Thomas Graves,** who as senior officer took command of the combined force, 19 ships of the line. Graves returned to the Chesapeake to find de Grasse already there.

1781, September 5–9. Second Battle of the Capes. De Grasse with 24 ships promptly put to sea to meet Graves. The 8 leading French ships rounded Cape Henry well ahead of the remainder, but Graves failed to seize the opportunity to crush them. Some British ships were roughly handled in an inconclusive action, and one was abandoned. Due to hidebound adherence to outmoded tactics, Graves never closed in. After 5 days of fruitless maneuvering, he returned to New York, leaving command of the sea to the French admiral. At this time the French Newport squadron—8 ships of the line—now under Commodore **de Barras,** arrived at the Capes, convoying transports carrying Rochambeau's siege artillery. (Sometimes called Battle of the Virginia Capes.)

1782, February 5. French-Spanish Conquest of Minorca. An allied force under General **Duke Louis de Crillon** had occupied all of Minorca except Port Mahon (July–August 1781). Port Mahon fell after a six-months' siege.

AMERICAN OPERATIONS, 1780–1781

1780, June 1. *USS Trumbull vs HMP Watt.* The *Trumbull,* 28, fought a bloody, drawn battle with the **34-gun** British privateer.

1781. American Privateering at Its Peak. During this year there was a total of 449 American privateers in action.

1781, May 29. Cruise of the *Alliance.* Sole American regular naval action this year was that of Captain **John Barry** in the *Alliance* frigate, which cruised to France and back, fighting and taking, on the return journey, the British sloops of war *Atlanta* and *Trepassy.* Barry also captured 2 British privateers.

SUFFREN'S OPERATIONS, 1781–1782

1781, April 16. Battle of Porto Praya. Suffren, who was bound for the Cape of Good Hope to preserve that Dutch colony from England, found British Commodore **George Johnstone** with 5 ships and 35 transports at anchor in Porto Praya in the Cape Verde Islands. Johnstone was also on his way to the Cape. Suffren, disregarding neutrality, attacked and so injured Johnstone's flotilla that the British expedition was called off. Suffren made for India (see p. 766).

1782, February–September. Operations off India. The first of 5 violent actions to be fought against British Admiral Sir **Edward Hughes** took place south of **Madras** (February 17). In this, as in the 4 subsequent actions (April 12, near **Trincomalee;** July 6, off **Cuddalore;** September 3, again near **Trincomalee;** and April 20, 1783 again near **Cuddalore**) Suffren had the best of the encounter, although no ships were lost or captured by either side. Suffren received less than full support from several of his subordinates; nonetheless he attacked with such vigor that the British were invariably forced on the defensive and in each instance broke off the action. As a result Suffren was able to give considerable support to French land forces, thus offsetting the great British superiority in India. The most amazing aspect of Suffren's operations in the Indian Ocean was that he was able to maintain his squadron without the help of a port in which he could refit, and that he at least temporarily checked the rise of British supremacy in India by actually seizing the strategically important anchorage of Trincomalee (August 30, 1782). In these operations, the French admiral proved himself to be the greatest naval man of his nation and one of the outstanding sailors of history.

BRITAIN REGAINS MASTERY OF THE SEAS, 1781–1782

1781, August 5. Battle of Dogger Bank. Sir Hyde Parker with 7 ships of the line, escorting a merchant convoy from the Baltic, encountered Admiral **Johann A. Zoutman**'s Dutch squadron, also 7 ships, convoying a merchant fleet to the Baltic. Parker had the best of the engagement; the Dutch ships returned to port.

1781, December 12. Second Battle of Ushant. Following months of maneuvering by Franco-Spanish fleets against English forces along the Atlantic coast of Europe, British Admiral **Richard Kempenfelt** defeated de Guichen's squadron escorting merchantmen and supply vessels to the West Indies, capturing 20 transports.

1782, April 12. Battle of the Saints. After 2 clashes with Rodney and Hood (January), de Grasse met their combined squadrons between Dominica and Guadeloupe. The British fleet, 34 ships of the line, and the French, 29 ships,

each maneuvered in line ahead until Rodney unorthodoxically burst through the middle of the French line. In a short time de Grasse's fleet was broken into 3 detachments and utterly defeated. Seven ships were captured, including the French flagship; 2 more fell to Hood a week later as the scattered French vessels were pursued. But this restoration of British sea command was too late to affect the outcome of the Revolution.

1782, September 13–14. Franco-Spanish Attack on Gibraltar. For 3 years, the fortress garrison of 7,000 men under Sir **George Augustus Eliott** had maintained a grim defense against Spanish blockade and bombardment by land and sea. Twice British squadrons had sent in reinforcements and supplies (see pp. 789–790). Following the capture of Port Mahon and Minorca (see p. 790), the Duke of Crillon launched an attack on Gibraltar. Ten floating batteries, with overhead and side armor of green wood some 6 feet thick, moved close to the walls of the fortress and opened heavy fire. British projectiles failed to pierce the armor until **Eliott's** gunners began, for the first time in history, to use red-hot shot. The balls, heated on grates and handled by tongs, were rammed home against wet wads protecting the powder charges, then fired at once. The device was most successful. In all, 8,300 rounds were fired. Before noon next day all 10 of the armored batteries had either been blown up or burned to the water's edge. The siege continued, but the garrison, close to starvation, was reinforced and supplied for the third time by a fleet under Lord Howe (October).

1783, January 20. Treaty of Versailles. Peace between England and the Franco-Spanish Alliance. England proclaimed cessation of hostilities (February 4).

UNITED STATES, 1783–1800

Disbanding of the Continental Army following the Revolution left the United States practically without Federal forces as Indian unrest again swept the frontier. An uneasy truce halted bloody Yankee-Pennamite fighting in the Wyoming Valley (see p. 773), finally leading to settlement of the dispute in Pennsylvania's favor (1782). Organization of a tiny Regular Army to supplement militia levies in protection of the frontier was begun (June 3, 1784). A navy did not come into being until 1794, when construction of warships to protect commerce—particularly against the pirates of the Barbary coast—slowly began. Meanwhile, internal troubles brought hastily organized levies of militia troops into the field during the abortive **Shays' Rebellion** in New England (August 1786–February 1787) and again in the **Whisky Rebellion** in Pennsylvania (July–November, 1794). The principal events were:

1790, October 18–22. Harmar's Defeat. Depredations of the Maumee Indians in the Ohio Valley brought a punitive expedition into the field under Brigadier General **Josiah Harmar.** Of his 1,133 men, only 320 were regulars; the remainder were untrained militia, poorly armed and officered. Leaving Fort Washington (site of Cincinnati, Ohio), the expedition engaged in 3 disastrous clashes in the wilderness, not far from the present site of Fort Wayne, Ind. In each, the militia ran away and the few regulars present were slaughtered. The expedition returned abjectly with losses of 200 men.

1791, November 4. St. Clair's Defeat. General **Arthur St. Clair,** with the entire Regular Army of 600 men—and some 1,500 militia—took the field in October, slowly cutting his way for 100 miles through dense forests north from Fort Washington to the present site of Fort Recovery, Ohio, on the banks of the upper Wabash. Here, after part of his militia had deserted, he was surprised and his force practically destroyed, with great slaughter. The remaining militia decamped, most of the regulars were killed or wounded; more than 900 men and women in all were butchered.

1792, June. New Army; New Leader. Aroused by these reverses, Congress authorized a larger army, which was called the Legion of the United States. President Washington called General Anthony Wayne out of retirement; he began to train this new army near Pittsburgh.

1793–1794. Wayne Prepares. He marched with his command to Greeneville, some 80 miles north of Fort Washington, to establish a new training camp and to continue his rigorous training activities through the following winter.

1794, August. Advance into the Wilderness. Moving north from Greeneville with what was possibly the best-trained command in the history of the United States Army, Wayne set out over St. Clair's ill-fated route.

1794, August 20. Battle of Fallen Timbers. Near present Toledo, Ohio, Wayne's 3,500 troops finally caught up with the main force of the Maumee Indians, gathered in a dense forest slashing. Wayne's infantry now attacked with the bayonet, while his cavalry moved in on both flanks of the Indians in a perfectly executed double envelopment. The result was an overwhelming victory. Wayne lost 33 killed and 100 wounded; the Indian stronghold and nearby villages were destroyed, bringing peace to the frontier for some years. The action took place within view of a British garrison still maintained illegally on American soil.

1798–1800. Quasi War with France. Operations in the West Indies during the Wars of the French Revolution brought strained relations between France and the United States. French interference with American shipping resulted in violence and the threat of war. Washington was called back to command the army, and a navy department was established (May 3). When an American schooner was captured by a French warship off Guadeloupe (November 20), U.S. naval vessels moved into West Indian waters. Captain **Thomas Truxton** in USS *Constellation*, 36 (guns), met and captured the French frigate *Insurgente*, 40, after an hour's engagement off Nevis (February 9, 1799). *Insurgente* served thereafter as a vessel of the U.S. Navy until 1800, when she was lost at sea. A year later (February 1, 1800), Truxton met the French *Vengeance*, 52, off Guadeloupe in a 5-hour night battle. The French ship, much more powerful than the American frigate, was partly dismasted and her guns silenced; she escaped in the night. In all, 85 French vessels, mostly privateers, were taken before peace was made. The last engagement was that of USS *Boston*, 28, with the French privateer *Berceau*.

LATIN AMERICA

SPANISH AMERICA*

1750. Treaty of Madrid. Portugal agreed to cede Colonia (on the River Plate) to Spain, in return for seven Spanish-Jesuit mission villages (called *reductions*) on the east bank of the Uruguay River. (The treaty also recognized Portuguese sovereignty over the Amazon and Paraná river basins, and Spanish sovereignty over the Philippines.) The Guarani Indians, incited by the Jesuits, refused to accept Portuguese control over the seven ceded *reductions* (see below). Portugal therefore held on to Colonia. Charles III of Spain then annulled the treaty (1761).

1754-1763. The Seven Years' War. (See p. 730.) The war came late to the Iberian Peninsula (see p. 758), and thus to Latin America (see below).

1762. Spanish Invade Portuguese Possessions in the Banda Oriental. Colonia and Portuguese holdings on the lower Uruguay River were captured, and part of southern Brazil was occupied.

1762. British Expeditions to Cuba and the Philippines. Havana was captured and Manila was also occupied by the British.

1763. Treaty of Paris. (See p. 789.) Havana was returned to Spain by Britain; Florida was ceded to Britain. To compensate for the loss of Florida, France ceded Louisiana to Spain. Manila was later returned to Spain by Britain.

1771. Falkland Islands Dispute. Conflicting Spanish and British claims almost led to war. Spain backed down when France refused to support her.

1776-1777. Spanish-Portuguese War. Span-

* Spain administered the Philippines as an element of her American empire.

ish troops captured Colonia and seized other Portuguese territories in the Banda Oriental and in southern Brazil. The **Treaty of San Ildefonso** (see p. 741) awarded Colonia and the Banda Oriental to Spain. Portugal retained the upper Uruguay River and Brazil.

1779-1783. Spanish Participation in the American Revolutionary War. Spanish forces captured Mobile, Pensacola, and the Bahamas. British efforts to gain control of the Mississippi River were blocked by Spanish forces in Louisiana. The **Treaty of Versailles** (see p. 791) gave Florida to Spain; returned the Bahamas to Britain.

1789. Pacific Coast Dispute with Britain. Spanish forces seized Nootka Island (claimed by both nations) and some British ships in Nootka Sound (between Nootka and Vancouver islands). Again Spain backed down from war when French support was not forthcoming.

1793-1795. Spanish Participation in the War of the 1st Coalition. (See p. 679.) By the **Treaty of Basel** (see p. 745), Spain ceded Santo Domingo to France. Soon afterward, Spain ceded Trinidad to Britain (1797).

PORTUGUESE AMERICA

1750. Treaty of Madrid. (See above.)

1752-1756. War of the Seven *Reductions*. Portugal finally subdued the Guarani Indians and consolidated control over the seven villages ceded by Spain in the Treaty of Madrid (see above).

1754-1763. The Seven Years' War. (See pp. 730, 758, and above.)

1776-1777. Spanish-Portuguese War. (See above.)

AFRICA

NORTH AFRICA

The Barbary States

The internal and external affairs of the Barbary States (Morocco, Algeria, Tunisia, and Tripoli) in this half-century retained the same general pattern of the previous period. Morocco was completely independent; the other 3 were nominally vassals of the Ottoman Empire, but were virtually free from Turk interference during this period. Aside from elementary husbandry, the principal occupations of all 4 states remained piracy and the slave trade, although Tunisia was somewhat less dependent upon these lucrative pursuits than the other 3. During most of the period central authority was fairly strong in Morocco under the benevolent rule of **Sidi Mohammed;** the other 3 countries were less strongly ruled. The principal events were:

1757–1790. Reign of Sidi Mohammed (Mohammed XVI) of Morocco. He did much to establish law and order over his unruly subjects, and was successful in introducing a modicum of Western culture.

1765. French Bombardment of Larache and Salé. These attacks somewhat reduced piratical attacks upon French vessels from these rover bases, and gained some respect for France from Sidi Mohammed.

1767. Treaty with France. Morocco recognized special status of French consuls, and undertook to discourage the pirates from attacking French vessels.

1777. Abolition of Christian Slavery in Morocco. French influence had much to do with this decree by Sidi Mohammed.

1791. Spain Abandons Oran. The Spanish still retained a few small coastal towns in Morocco and western Algeria, of which Ceuta and Melilla were the most important.

Egypt

The effectiveness of Ottoman control over Egypt fluctuated greatly. Even when Turkish sovereignty was most firmly enforced, the Mameluke governors had considerable autonomy. The principal events were:

1750–1769. Rise of Ali Bey. This low-born Mameluke gained ascendancy in Egypt by force of arms, and was recognized as governor by Constantinople. Save for a period when forced to flee the country (1766), he ruled with an iron hand.

1769. Ali Bey Declares Egyptian Independence. After Ali had been ordered to raise a force of 12,000 troops for the war against Russia (see p. 762), Constantinople began to entertain doubts about his loyalty, fearing he would use this force to seize Syria. Secret instructions were sent to Egypt to have Ali assassinated. Ali intercepted the message and, with Mameluke support, declared independence.

1769–1770. Conquest of Western Arabia. Ali conquered most of the Red Sea coast and much of the interior of Arabia from nominal Turk rule.

1771–1772. Expedition to Syria. Ali sent a small army under his General **Abu'l Dhahab** into Syria. After arriving at Damascus, Abu'l opened negotiations with the Porte, which commissioned him to overthrow Egypt and replace Ali.

1772, April 8. Flight of Ali. One day before

Abu'l's army reached Cairo, Ali fled, seeking support from the Pasha of Acre. At Acre he received money, supplies, and men (3,000 Albanian soldiers) from Russian naval vessels.

1772–1773. Ali's Return to Egypt. Enroute he recaptured Jaffa and Gaza. He then advanced across the Sinai Desert toward Egypt (February).

1773, April 19–21. Battle of Salihia (Sa- lihiyeh). In a climactic battle with Abu'l, Ali, at first successful, was finally defeated and captured; he died soon afterward.

1785–1786. Revolt in Egypt. This was suppressed by an Ottoman army sent overland and by ship.

1798–1801. Bonaparte's Expedition to Egypt. (See p. 752.)

WEST AFRICA

Increasing European attention to the West African coast—stimulated by the slave trade—had relatively little impact upon the continuing turbulence of the interior jungle and upland regions. The remnants of the old Mali Empire persisted around the headwaters of the Niger and Senegal rivers. Farther east in the upland regions were the Kaarta, Mossi, Oyo, Hausa, and Bornu states, of which the Oyo was the only one still flourishing by the end of the century. The others, suffering from internal disorders, disputes with neighbors, and particularly plagued by depredations of the Saharan Tuaregs, were in varying stages of decline, decay, or collapse. Of the many tribal groupings in the intermediate jungle region, the Ashanti continued to prosper and expand until, by the end of the century, they had expanded into the open uplands of the middle Volta region and farther west, and had also approached the Ivory and Gold Coasts. Meanwhile, along the coast, the European wars of France and England were reflected in raid and counterraid. The principal events were:

c. 1750–c. 1775. Tuareg Invasions. Possibly because of the "Great Drought" of 1735–56, Tuareg nomads from the north conquered much of the central Niger valley, including Timbuktu.

1752–1781. Reign of Osei Kojo of the Ashanti. This warrior leader greatly expanded the Ashanti dominions in all directions.

1753. Reestablishment of Portuguese Base at Bissan. This was Portuguese Guinea.

c. 1753. Founding of Kaarta. Bambaran exiles founded this city, near the ancient Ghanian capital of Kumbi. This became the leading city of a neighboring group of Bambara states.

1755–1766. Disorder in Segu. The death of King Mamari Kulibali (see p. 721) sparked a succession struggle, which ended only when **Ngolo Diara** seized the throne in 1766. He restored order to the empire, and ruled until the mid-1790s.

1757. British Conquest of French Senegal Posts. The Seven Years' War in Africa.

1763. Treaty of Paris. (See p. 789.) Britain returned Goree to France, retaining the remaining posts in the Senegal region.

1764–1777. Reign of Osei Kwadwo of Ashanti. Osei Kwadwo and his successor **Osei Kwame** (r. 1777–c. 1801) established a strong, centralized, and efficient state, with government officials selected by merit.

c. 1770–1789. Reign of Abiodun of Oyo. This *alafin* (king) settled the long-standing internal Oyo dispute, between parties supporting economic growth versus territorial expansion, in favor of the mercantilists, and allowed the army to wither.

c. 1775. Beginning of Decline of Benin. Wracked by succession disputes which often erupted into civil war, Benin's prosperity waned, and its political influence weakened.

1776. Tukulor Confederacy. Moslem Tukulor clerics created a theocratic confederacy in Fouta-Toro (north-central Senegal).

1778–1779. French Reconquest of the Senegal Posts. The African version of the War of the American Revolution (see pp. 774–791).

1783. The Treaty of Paris. (See p. 789.) The Senegal posts were divided between England and France; France retained St. Louis.

c. 1785. British Domination of the Slave Trade.

1787. British Acquisition of Sierre Leone.

1792–1800. Renewed Anglo-French Struggle over Senegal Coast. The Wars of the French Revolution in Africa. Under the command of **F. Blanchot de Verly,** the French recovered, then partially lost, the Senegal posts. Verly repulsed repeated British assaults on St. Louis.

c. 1790. Renewed Dahomey Pressure on Oyo. Dahomey had avoided conflict with Oyo for 40 years after the 1738–1748 war (see p. 721). During the reign of **Agonglo** (r. 1789–1797), Dahomey's well-organized armies, with European weapons, again threatened Oyo, although Agonglo continued tribute payments.

1793. Oyo Attack on Ife. *Alafin* **Awole** of Oyo raided the Yoruba client-state of Ife to collect slaves. Ife resistance was unexpectedly strong, and the ensuing defeat was a serious blow to the Oyo state.

SOUTH AFRICA

About midway in this half-century the 2 great South African migrations met and began a series of violent clashes. The Dutch settlers who had started from the area of Capetown almost 2 centuries earlier had now become a new colonial people known to themselves as **Boers,** or farmers. The spearhead of the movement of the Bantu people was the Xhosa tribe, which met the Boers about 400 miles to the east and northeast of Capetown. The typical self-reliant Boer tendency toward unruly independence was also beginning to manifest itself along the new frontier, directed first against the administration of the Dutch East India Company and later against the British, after their seizure of the colony during the French Revolutionary Wars. The principal events were:

1750. The Bantu Reach the Keiskama River. This was the Xhosa tribe.

1760. The Boers Cross the Orange River. They begin to penetrate the region known as the Great Namaqualand.

1770. The Boers Reach Graaff-Reinet Area. This was 400 miles from Capetown.

c. 1775–1795. Boer Conflict with Khoisa and San. Despite desperate native resistance, particularly in the mountains, the Boers had little trouble in overcoming them. Many were slaughtered; many more were enslaved.

1778. Boer and Bantu Meet. There had been long-range contact for several decades, but now the outpost settlements of both peoples came into close contact.

1779–1781. First Kaffir War. This struggle of white Boer against black Bantu (called Kaffirs by the Boers) was inconclusive. It led the frontier peoples in the Graaff-Reinet and Swellendam regions to appeal to the company for help.

1793–1795. Second Kaffir War. This was also inconclusive.

1795. Independence Declared by Graaff-Reinet and Swellendam. The frontier Boers were disgusted by lack of support from the company and by the Capetown Boers. Without effective government, these frontier regions were in chaotic condition.

1795, September. British Conquest. There was little opposition, since the Boers accepted the British as enemies of the French, who had conquered Holland.

1796. Order Restored on the Frontier. The British established control over the entire colony, including the two "independent" regions.

1799–1801. Renewed Anarchy on the Frontier. Unrest and chaos resulting from Boer refusal to accept centralized control from Cape-town were exploited by Xosa raids and by Hottentot uprisings.

EAST AFRICA

By the middle of the 18th century, the Portuguese had abandoned efforts to reestablish control over the East African coast. Moslem Arab-Swahili-Bantu coastal principalities, owing nominal allegiance to the Sultan of Oman, flourished north of Mozambique. The Portuguese held all the coast of modern Portuguese East Africa. Inland, the southern migration of the Bantu was slowly ending as the northern-most tribes closed up into a region extending roughly from Lake Nyasa to the Orange River. Farther north, the unending conflict of Christian Abyssinia (or Ethiopia) continued against the various surrounding Moslem tribes and states (to the north, east, and west) and pagan tribes (to the south). The principal events were:

1750–1800. Uganda Annexes the Budda. The well-organized state of Buganda (see p. 564), on the northwestern shore of Lake Victoria, annexed the Budda region to the southwest, its last major territorial acquisition.

1768–1773. James Bruce in Ethiopia. This Scottish traveler wrote a vivid record of contemporary Ethiopia (including the Battle of Sarbakussa, below, in which he took part), which many contemporaries refused to credit.

c. 1768–c. 1780. Ascendancy of Mikael Sehul in Ethiopia. The leading noble of Tigré province, Mikael was called upon by King Iyasu II, who had allied himself with the Galla nobility, to assist in thwarting Galla efforts to seize power and supplant Christianity with Islam. Mikael occupied Gondar, the capital, and declared himself regent. When Iyasu tried to have him assassinated, Mikael deposed and assassinated the king, then placed **Yohannes II,** and then **Tekle-Haimanot II,** on the throne (1769). Sources are contradictory as to subsequent events. According to Bruce, Mikael was defeated by Galla nobles at the **Battle of Sarbakussa** (May 1771), yet he seems to have retained power until again defeated by a coalition under Galla leadership (1784). He returned to Tigré, as governor, and died soon afterwards. The Galla-Tigré struggle for supremacy in Ethiopia continued after his death, and the power of provincial chiefs and nobles increased at the throne's expense. Ethiopians call this feudal age, lasting about a century, "the time of princes," because the country was dominated by Gallan or Tigréan princes.

1785. Capture of Kilwa. The Al Bu Sa'idi clan of Muscat captured this trading city.

XVII

THE ERA OF NAPOLEON:

1800–1850

MILITARY TRENDS

In this period military history crossed the second of its three great watersheds. Under the direction or stimulus of Napoleon Bonaparte, the weapons of the age of gunpowder were finally assimilated into consistent patterns of military theory and practice. For the first time since gunpowder had appeared on the battlefield, there was a substantial congruence among weapons, tactics, and doctrine. The bayoneted flintlock musket and the smoothbore cannon had each been perfected to a point closely approaching its maximum potential. After centuries of experimentation, the tactical means of employing these weapons in combination with each other and with cavalry had been refined to the point where a skillful commander could exploit the full potential of his weapons and his arms to achieve decisive results with minimum cost. The last time that commanders had been able to exercise comparable discriminating control over the means available to them had been in the 13th century, in the Mongol and English tactical systems.

Yet just as those two tactical systems had approached perfection in the employment of men and weapons at a time when the systems were doomed to early obsolescence because of the emergence of gunpowder weapons, so the principal tactical systems of the early 19th century (French and English on land, and English at sea) would be equally short-lived under the impact of the Industrial Revolution. (Interestingly, the phenomenon would be repeated again in mid-20th century as weapons and tactics of the Industrial Revolution era approached congruence just as the third great military watershed was reached with the dawn of the Nuclear Age.)

The congruence of weapons, tactics, and doctrine was bound to come during this half-century as a logical result of earlier developments. But the achievement was probably hastened, and certainly made more significant, through the genius of one man: **Napoleon Bonaparte.** No man has more indelibly stamped his personality on an era than did Napoleon. In his own time and for more than a century to come, military theory and practice were measured against his standards and related to his concepts of warmaking.

To a lesser degree, the same kinds of generalizations can be made about the influence of

Horatio Nelson upon naval thinking and the exercise of the military art at sea. Yet if Nelson's impact upon his time and upon warfare was unquestionably less significant than that of Napoleon, his is the more unique and unchallengeable position with respect to other practitioners of the naval art. There has only been one Nelson; Napoleon's historical greatness as a soldier must always be compared with predecessors like Alexander, Hannibal, and Genghis Khan.

There were other excellent generals during this period, of whom the most outstanding were the Englishmen, **Arthur Wellesley, Duke of Wellington,** and Sir **John Moore,** the Prussian **Gebhard L. von Blücher, Prince of Wahlstatt,** the Austrian **Archduke Charles,** and the French Marshal **Louis Nicolas Davout.** Among Napoleon's marshals, in addition to Davout, at least two others proved themselves capable generals in their own rights: **Nicolas Jean de D. Soult,** and **André Masséna.** Three Americans warrant consideration with this company: **Andrew Jackson, Winfield Scott,** and **Zachary Taylor.**

For reasons noted below, the Industrial Revolution had little direct effect upon the battlefield and upon generalship in this era; that would come in following decades. But it had already had an impact upon manufacturing and upon agriculture which, in combination with the democratization of war that had occurred during the French Revolution, permitted nations to raise, equip, and supply larger armies than had hitherto been possible. And the Industrial Revolution, combined with the stimulation provided by Napoleon, contributed to the development of military theory and military professionalism in this period (see pp. 808–811).

WEAPONS

Great strides were taken in ordance development. In 1810 **Friedrich Krupp** started the small Prussian forge which would blossom into a great steel and ordnance empire. In the United States, **Robert P. Parrott, John A. Dahlgren,** and **Thomas J. Rodman;** in England, **William G. A. Strong** and **Joseph Whitworth;** in France, **Henri Joseph Paixhans;** and in Sardinia, **Giovanni Cavalli,** in arsenal and foundry were evolving innovations and improvements in cannon manufacture which would begin a revolution in the science of gunnery.

In small arms, the percussion cap would supplant the flintlock, and make possible the invention in 1835 of the revolver. The Minié bullet, with expanding base, an invention of 1849, would improve the accuracy and range of the muzzle-loading rifle, which would then quickly displace the smoothbore musket as the primary infantry weapon.

In practice, however, few of these innovations reached the fighting man on sea or land in this half-century. There was little change in service arms and ammunition. There was one notable exception: the rocket, which emerged as a lethal weapon from its previous long existence as a pyrotechnic oddity, thanks to the efforts of Sir **William Congreve,** English ordance expert. It found almost immediate favor, both in the United States and in Europe, as an intermediate-range weapon bridging the gap between the still-standard flintlock musket and the 12-pounder field gun. But although its relative economy vis-à-vis conventional ordnance was attractive, its notorious inaccuracy and limited maximum range of about 1,500 yards soon forced its disappearance from the battlefield.

TACTICS

Napoleonic Tactical Concepts

By the end of the 18th century, European warfare was characterized by large-scale formal battle waged primarily on the principles of linear tactics, and sieges conducted according to set patterns. There were no revolutionary new weapons and no major changes in the basic linear organization, although there had been steady development in both. Small arms had improved, and infantry was clearly the dominant arm. Rates of fire had been increased as a result of drill and improvement in firing mechanisms. Rifled firearms had been developed, but were used only by special troops. The flintlock with the bayonet was the standard arm of the infantry. The cavalry primarily used the saber, but some were armed with carbines. Artillery provided support to the extent of its ability. The guns were not yet capable of the rate of fire, accuracy, or trajectory required for close support of troops in the attack. Napoleon did not change these tactical methods so much as he revitalized them, while at the same time integrating his tactics into his broader and more revolutionary strategic concepts.

In war Napoleon always sought a general battle as a means of destroying the enemy's armed force after having gained a strategic advantage by maneuver. Tactically, he usually directed his main blow against the enemy's flank while simultaneously attacking his front; or launched his main thrust against the center of the enemy's battle front with the aim of breaking through, at the same time carrying on an enveloping maneuver against a hostile flank. The divisions attacking important objectives were often supported by massed fire from Napoleon's artillery reserve. Divisions with exposed flanks were protected by corps cavalry or even by the cavalry reserve.

Only after tactical destruction of the main armed force of the enemy did Napoleon bother about occupying the principal strategic and political centers of the enemy's country.

Infantry Tactics

LIGHT INFANTRY

As we have seen, light infantry had been reintroduced into European warfare during the 18th century. It was not a new development; light infantry had been used by all ancient armies and in various forms it had accompanied armies throughout the ages. Almost without exception, however, these had been irregular troops: archers, slingers, javelin men, and various others, who usually opened battles and then moved aside during the main action. With the introduction of gunpowder weapons, similar groups were armed with firearms, but the troops were usually undisciplined and did not form part of a regular army.

The rigid linear tactics of the 18th century had prescribed a fixed and inflexible role for the regular infantry. During the considerable time it took infantry battalions to take up their battle stations, they were vulnerable and needed to be screened from enemy action. In addition, the supply depots and convoys that were needed to support the armies were highly vulnerable to enemy attacks. The regular infantrymen of the enlarged armies, recruited as they were from the rejects of society and subjected to a rigorous discipline, could not be entrusted with detached operations. Consequently, to furnish the necessary support, to carry out operations against the enemy's lines of communications, to raid, and to take prisoners, while at the same time providing for the security of their own depots and convoys, and to screen the main army against surprises, light troops, mainly infantry, were reintroduced into the European armies about midcentury. Within a short time additional functions, above all, individual or group fire missions in advance or on the flanks of the main line, were added to their tasks.

The first large-scale appearance of light troops in Europe had occurred during the War of the Austrian Succession (1740–1748). In 1740, Maria Theresa found herself attacked by the superior strength of Frederick of Prussia and his French and Bavarian allies. She had to muster all the forces at her disposal and did not hesitate to call upon the Borderers, the "wild Croats and Pandours," who had been part of the Austrian frontier defenses against the Turks, to defend her realm. The effectiveness of these light troops impelled the other powers to introduce or augment similar forces. Prussia hastily increased her light cavalry and raised some irregular "free" battalions to counter the Croats, and in France several light regiments as well as a number of combined infantry-cavalry units—usually called "legions"—were raised.

The British Army had no light troops until the line battalions serving in America during the 1750s raised some light companies on an *ad hoc* basis. These units differed significantly from the irregulars, Borderers, free battalions, and free corps by being trained and disciplined troops, usable in the line as well as on detached operations, such as advance guards, assault parties, and also, occasionally, as raiders. They differed in function, but not in equipment and discipline, from the rest of the army, and more often than not were used as line infantry. The inspiration for the formation of these troops derived in part from the painful experiences of the war in America and in part from the continental European developments. After 1770, a light company as well as a grenadier company became part of the permanent establishment in each line battalion. Both companies rapidly assumed elite status and as "flank companies" were often used for special missions during the American Revolution.

However, light infantry never became a dominant element. In Prussia, Frederick II retained his reliance on the massed volleys of the line and spent most of his resources to speed up the fire. Prussia formed a number of fusilier units, but these were trained and equipped as line infantry. The same development took place in Austria, where the regiments of Borderers were drilled in linear tactics. In the British Army, too, there came a sharp reaction against light infantry and a determined attempt to return to the linear system at the turn of the century.

The only country which did not follow this backward evolution was France. Here, even before the Revolution, there was wide agreement that shock action should be delivered by formations having greater depth than the firing line. The main controversy concerned the extent of fire preceding and supporting the assault and whether this fire should be delivered by line, line and skirmishers, or skirmisher swarms. Circumstances and combat leaders together ultimately fashioned that combination of close-order columns and loose-order skirmishers which constituted the new tactics of the Revolutionary and Napoleonic infantry. Skirmishers would so occupy the enemy that the deeper assault formations (see below) could move up without being unduly exposed to the fire of the enemy line.

In the War of the First Coalition (1792–1798), the habit of skirmishing spread throughout the French infantry, and by 1793 all battalions were acting as light infantry, dissolving into skirmisher swarms as soon as action was joined. These fighting methods, sometimes called "horde tactics," were in turn superseded after 1795 by a tendency to return to properly controlled, deeper assault formations, preceded by skirmishers to scout the ground and disturb the enemy by individual aimed fire.

The important point about the French skirmishing action during this period was that it was performed not by special light troops but by integral parts of the regular bodies. Infantry became more flexible and to some observers it appeared as if specialized light troops would soon be eliminated by one all-purpose infantry. But special light troops, not only brought up to the standard of the line but in some cases excelling it in performance and capable of winning a decision in battle, remained in French service for another 50 years.

In their political implications, the new light-infantry tactics were revolutionary. Both the French Revolutionary armies and the British system abandoned the brutal and degrading discipline of the 18th-century armies. The light infantryman (or the all-purpose infantryman), fighting often as a relatively autonomous individual in small formations in open order, was much less under the direct supervision of his officers. Brutal treatment and close control gave way to appeals to regimental pride, revolutionary *élan,* and the spirit of nationalism.

The character of light infantry and of warfare were greatly changed by the introduction of the rifle. For fighting individually in dispersed order, an accurate missile weapon was particularly valuable to light infantry. Rifles, however, were expensive and, more serious, slow to load. Therefore only select units and select individuals in line companies were thus equipped until well into the 19th century.

THE FRENCH INFANTRY "COLUMN"

The introduction of the so-called attack column as a standard combat formation in the Wars of the French Revolution was in part a result of tactical experimentation begun by Marshal Saxe in the middle of the previous century, and in part a natural formation adopted to make the maximum use of the poorly trained hordes produced by the *levée en masse,* troops that lacked the training and discipline to stand and fight effectively in the then linear system of Frederick.

Nevertheless, this French "column" as developed under Carnot, and perfected by Napoleon, was in no sense a reversion to the phalanx or the Swiss pike column. In its most sophisticated form, the *ordre mixte,* the French column was actually a versatile combination of battalion columns and lines, protected by a strong screen of skirmishers. This deployment allowed for flexible combinations of fire and shock tactics. As the battalions forming the columns in the formation normally deployed no more than 9 ranks deep, the depth was more psychological than physical. The individual units could still operate in a linear formation if desired. It was not until much later in the wars, after attrition had diluted the quality of the French infantry, that Napoleon utilized massed battalion columns as battering rams.

The great tactical value of the column lay in its flexibility and versatility. It permitted the commander to move large numbers of men over the battlefield with better control and far more rapidly than had been possible before. The column could operate in hilly terrain. It could easily change into different formations. The deployment from marching column to attack column, in particular, took far less time than had the development of linear formations from the marching column. Skirmishers could be detached without necessitating major readjustments in the formation. Two- or three-rank firing lines and squares could be formed rapidly. The earlier need to maintain tight flank connections between units in line became less important; the tactical situation opened up and became more dynamic.

The attack column had two main functions. First, it could be used to bring men in close order rapidly to the enemy. The success of such an action was largely dependent on adequate preparation by artillery and skirmishers, and it was they who inflicted most of the casualties rather than the column itself, which possessed little or no firepower once it started to move. Bayonet charges actually driven home against a steady enemy were rare.

The far more common employment of the attack column was as a sustaining force. The column sent out skirmishers to start the fire fight and served as a replacement pool for the skirmishers and as their immediate tactical reserve. If it encountered firm resistance, the column might deploy into lines to carry on the fight with volleys. Once the enemy wavered, these lines could resume the advance, or they might again reduce their front and move forward in column.

THE ENGLISH LINE

The most effective answer to the French system, as well as the most effective form of light infantry, was provided by the British. Their system was largely based on the effects of controlled, aimed musketry, delivered by troops combining as far as possible the mobility of skirmishers with the steadiness of the line. Under Sir John Moore and Sir Arthur Wellesley (later Duke of Wellington), the British began to take advantage of cover, usually crouched behind the crest of a ridge, and then, formed only two deep, arose to deliver a devastating fire against the French columns.

The English light infantry—partially armed with rifles which could deliver rapid fire by the use of subcaliber bullets, or individual aimed fire when using regular-sized bullets—able to operate individually or in close order, represented in essence the all-purpose infantry of the future. The musket of the light infantry was of a special type, a lightweight piece, constructed for this particular purpose. It was somewhat more accurate than the "Brown Bess," with better sights, but shorter in length. The line battalions used the Brown Bess, which was considered superior to muskets used on the Continent. The bayonet was long and triangular, and when fixed made accurate firing difficult. Sergeants did not carry muskets; each had a sword and a pike or a halberd, which served as a signaling instrument and a rallying point.

Despite the early successes of the French system, the British retained the two-deep line—in which every man would employ his weapons—to produce a relatively greater volume of fire than could the column. Wellington's success was due undoubtedly in part to this, but it was due also to his tactics. He decided he could overcome Napoleon's tactics by three means: not to expose his line to artillery until the action opened, to protect it against the skirmishers, and to secure his flanks. The first he achieved by placing his infantry whenever possible on reverse slopes; the second by building up his light troops; the third was accomplished by natural obstacles or by his relatively weak cavalry.

In part because of his chronic shortage of cavalry, Wellington paid considerable attention to defense against French cavalry. The steady line and accurate fire of the British infantrymen were usually able to repulse a cavalry charge. On one occasion in the Peninsula, an infantry line advanced against cavalry and drove it from the field. In a square formation, British infantry was practically unbreakable; there is recorded the instance of the Light Division, formed into 5 squares, retreating for 2 miles with only 35 casualties, under attack by 4 brigades of cavalry.

Cavalry Tactics

Cavalry remained the shock arm, with lance and saber the principal hand weapons. However, the division between "heavy" cavalry—partly armored men on big horses—and "light"— more agile troopers on smaller mounts, who could harass as well as shock—again became marked during the period.

Napoleon's cavalry, provided with horse artillery and used in great but articulate masses and in surprise operations against the enemy's cavalry and infantry, was very effective. It was usually thrown against the enemy infantry already shaken or shattered by massive artillery fire, or by infantry attacks. It was particularly effective against retreating infantry. It was less successful against fresh infantry which had time to form squares. Under outstanding leaders and by its impetuous charges, French cavalry usually proved superior to the best cavalry of other European nations. By its lightning action in pursuit, French cavalry exploited victory with minimum losses to its own army. Napoleon also used his cavalry very effectively for reconnaissance and for screening.

During the early Napoleonic wars, French cavalry was unexcelled. Later, as casualties and the passage of years took their toll, Napoleon found it difficult to maintain the same high standards of cavalry performance. At the same time, his enemies steadily improved their cavalry, in part by devoting more attention to its organization and training, and in part by copying the French organization, tactics, and methods. In the Peninsula, for instance, cavalry played a minor role in Wellington's campaigns. At Waterloo, however, it was the English cavalry which smashed the initial French infantry assaults and later assisted in repulsing the final attack of Napoleon's Old Guard.

Artillery Tactics and Techniques

The French Revolutionary Army inherited Gribeauval's excellent field-artillery system from the monarchy. The main feature of this artillery was mobility, obtained by reducing the length and weight of the barrel and the weight of the gun carriage; the latter was also provided with iron axletrees and wheels of large diameter. Range and precision were preserved by more precise manufacture of the projectile (balls of true sphericity and correct diameter, which also made possible a reduction in powder charge). Prefabricated cartridges, which replaced the old loose powder and shot, increased the rate of firing. Draft horses were disposed in double files instead of single. Six or 8 horses sufficed to draw the 12-pounder in good conditions, while 4 or 6 were used for smaller guns: 8- and 4-pounders and the 6-inch howitzer. However, Napoleon attempted to increase the mobility of the artillery still further. Like Frederick the Great, he further lightened the piece by shortening the barrel and by employing a chambered breech and smaller powder charge. Like the earlier Prussian system, the French system (known as the System of the Year XI, or 1803) was a failure, lacking range and hitting power. As a stop-gap measure to replace these pieces, large numbers of the excellent Austrian and Prussian 6-pounders, captured in the course of the wars, were used by the French.

Napoleon took full advantage of the maneuverability of the French artillery and made out of it the most important tool of his warfare. One of his favorite techniques, particularly employed in later years as the quality of his troops declined, was employment of the *grande batterie*, physically massing a preponderance of artillery fire in support of his main effort on the battlefield, literally blasting the enemy line to shreds to permit his infantry to advance.

British artillery equipment also improved. The standard double-block trail carriage construction was replaced by a lighter and more manueverable single-block trail carriage. This simple design change made the heavy British 9-pounder gun as manueverable as the French 4- and 6-pounders, the ideal that Napoleon had sought. The British "Shrapnel" (more properly "spherical case") shell was usable by all the British guns and was detested by the French who had nothing to compare to it. Wellington employed this artillery selectively, in small numbers and individual batteries at carefully chosen sites, to be used at critical moments. They were placed all along the front as support for the infantry and played a minor but important role in his defensive-offensive tactics.

MILITARY ORGANIZATION

Origin of the Division and the Corps

The infantry division as a large permanent tactical and administrative formation appeared in France in the 18th century. In 1759, the Duc de Broglie introduced in the French Army a divisional organization, permanent mixed bodies of infantry and artillery.

In 1794, Carnot, the Revolutionary minister of war, developed the division embracing all 3 arms, infantry, cavalry, and artillery, and capable of carrying out independent operations. By 1796, the divisional system became universal in the French Army. It was Napoleon Bonaparte, however, who developed all the potentialities of the divisional system and used it in mobile warfare and tactics of fast maneuver. The men were trained in fast marching and the supply system was improved to support them wherever they went. The mobility of the division was also enhanced by artillery which could follow infantry and maneuver on the battlefield.

When army sizes approached 200,000, it became necessary to group divisions into army corps for administration and control. The first such organization was made in 1800, when Moreau grouped the 11 divisions of the Army of the Rhine into 4 corps. It was, however, not until 1804 that Napoleon introduced permanent army corps in the French Army, employing them as he had previously used divisions. However, the division remained the major tactical unit, composed of infantry and artillery, and entrusted with a definite mission. The corps included cavalry as well, which conducted reconnaissance for the whole corps. In addition, Napoleon formed cavalry divisions and cavalry corps.

Napoleon's infantry division consisted of 2 or 3 infantry brigades (each comprising 2 to 5 battalions in 1 or more regiments) and 1 artillery brigade, consisting of 1 or more batteries, each with 4 to 6 field guns and usually 2 howitzers.

British Organization

The British did not adopt the division until 1807, and Wellington's army in the Peninsula in 1809 was still composed of independent brigades.

The British Army was a volunteer force and necessarily smaller than the French. But it had

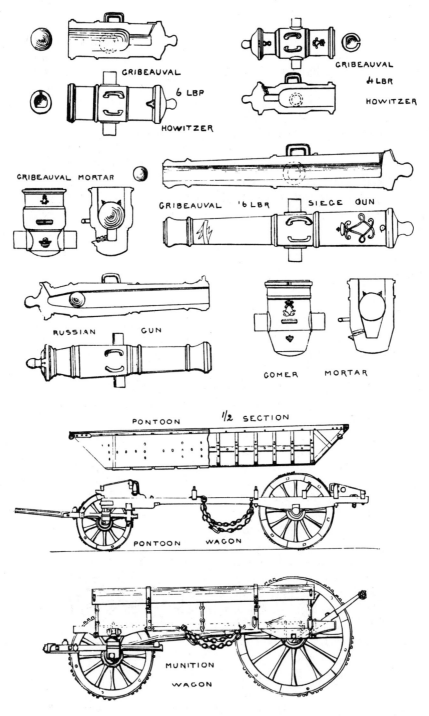

CRIBEAUVAL

CRIBEAUVAL 4LBR HOWITZER

6 LBR HOWITZER

CRIBEAUVAL MORTAR

CRIBEAUVAL '6LBR SIEGE GUN

RUSSIAN GUN

GOMER MORTAR

PONTOON ¹/2 SECTION

PONTOON WAGON

MUNITION WAGON

Artillery of the Napoleonic era

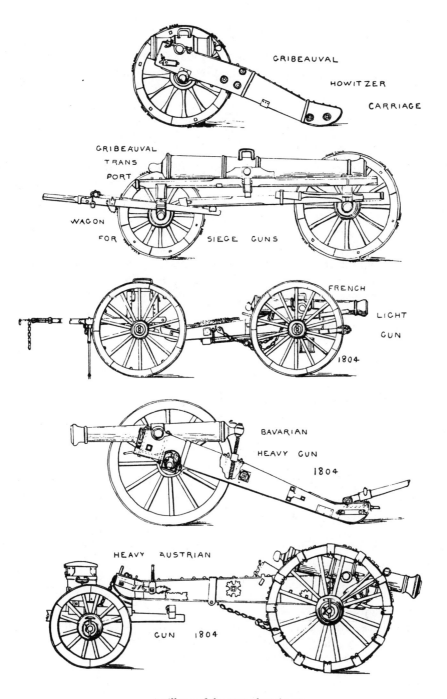

Artillery of the Napoleonic era

the advantage of more training and drill. The infantry was also the superior of any other in the excellence of its musketry, an advantage enhanced by its 2-rank line.

During the Peninsular campaigns (1809–1814), Wellington's army at first was organized into 8 brigades of 2 or 3 battalions each. Reorganized as its size increased, it consisted finally of 7 divisions, a light division, and the cavalry under separate command. Although the elements of the divisions varied, they were composed ordinarily of 2 British brigades (usually with 3 battalions each) and 1 Portuguese brigade (usually with 5 battalions), about 6,000 each. The 1st Division differed in that 2 of its brigades were composed of battalions of the Foot Guards, while the 3rd brigade was composed of battalions of the King's German Legion (KGL), made up of expatriate Germans in British service. The cavalry was organized as a division with a variable number of brigades and regiments. The light division served as a protective screen for the entire army, operating far to the front, and excelled at advance- and rear-guard operations.

One of the more interesting and important aspects of Wellington's organization grew out of his efforts to secure a strong screen of skirmishers to meet the French *tirailleurs*. Wellington added to every brigade in his army an extra company of light riflemen to reinforce the light company, now standard in each of the British battalions of the brigade. In addition, usually 1 of the Portuguese battalions in the brigade were *cacadores* or light infantry. Further, the light division was composed entirely of British and Portuguese light infantry battalions, as well as 1 or 2 battalions of riflemen.

LOGISTICS

Napoleon was a master of planned and improvised supply. Logistical planning, which provided, in advance, depots and distribution points for the supply of armies, was brought to a higher and more efficient plane than heretofore envisaged.

Insofar as possible, Napoleon's armies lived off the countryside through which they were marching or fighting. Troops were often billeted in towns and villages, where the local population was required to provide food. The soldiers and supply columns following the troops each carried 4 days' provisions, to be consumed only in emergency. In addition, provisions were stored at the main base and intermediate depots, the latter moved forward with the advance of the troops.

Through this forethought, French armies moved with amazing rapidity. The most notable instance was the 500-mile march in 1805 from the northern coast of France across Western Europe to Ulm, Vienna, and Austerlitz. A force of nearly 200,000 men kept up an average of 12 to 15 miles per day for 5 weeks, the fastest sustained march of comparable length by an army since the days of Genghis Khan.

This system of logistics proved very satisfactory until the Russian campaign of 1812, when it completely broke down because of bad roads in Russia, the poverty and devastation of the country, and activities of the Russian partisans.

MILITARY THEORY AND STRATEGY

Napoleon

The first coherent new concept of warmaking to manifest itself since Genghis Khan had been demonstrated in the early campaigns of young Napoleon Bonaparte in Italy and Egypt. In his hands it continued to dominate warfare directly for the first decade and a half of this century. Although his enemies copied the Napoleonic system to the best of their abilities, they never fully understood the concept which underlay Napoleon's tremendous revolution in warmaking. Even at the time they were overwhelming him, in 1814, respectful and awed enemies like Blücher or the Duke of Wellington could say (the statement, attributed to both, was probably Blücher's) that Napoleon's mere presence on the battlefield was worth 40,000 men.

Napoleon never committed his concepts to paper in systematic form. From his writings and remarks have been collected a number (varying, but the most authoritative collection has 115) of maxims which, one way or the other, encompass most of his fundamental ideas on strategy and tactics. These maxims are, however, very uneven; sometimes they are merely aphorisms expressing something which is relatively obvious or unimportant; often they deal with tactical and technical details of significance only to the times and places where Napoleon fought.

But Napoleon's methods of warfare and the concepts underlying those methods were deducible from the record of his accomplishments. Friends and enemies began to tackle the task of analysis even before his downfall, and the relatively superficial early insights which his enemies thus derived unquestionably played some part in that downfall. In subsequent decades three great theorists devoted themselves to more systematic and objective analysis, as noted below.

Napoleon avoided stereotypes and attempted to develop his plans for every campaign and every battle in such a way that his enemies could never know what to expect from him. Nevertheless, as he himself remarked, his theories were based upon simple principles, and certain broad patterns of strategic concepts and methods can be discerned from his many campaigns and battles. (One historian has asserted that he fought more battles than Alexander, Hannibal, and Caesar combined; while this depends upon the definition of a "battle," nonetheless it is indicative of the richness of the Napoleonic record.)

Insofar as possible, Napoleon tried to win a campaign before the first battle was fought. Whenever there was an opportunity, he would combine rapid marching and skillful deception to pass around the enemy's flanks to reach the hostile line of communications, and then turn to make the enemy fight at a disadvantage. Outstanding among the campaigns which were thus won before the fighting began were Marengo, Ulm, and Jena.

To deceive and confuse his enemies, whose combined military strength almost always exceeded his, as well as to permit rapidity of movement and efficient foraging, Napoleon kept his forces spread out until the last possible moment. Then, concentrating rapidly, he would bring superior forces to bear at some critical point. Rivoli, Friedland, and Dresden are typical examples. In a favorite variant, Napoleon would endeavor to place his concentrated army between two hostile armies, defeating them in turn. His first and last campaigns are good examples: Montenotte and Waterloo; his failure in the latter was due to the failure of performance (his and his subordinates') to match his superb strategic concept.

Napoleon's Enemies

After the first defeats inflicted on them by Napoleon, other European military leaders tried to imitate him. They gradually introduced divisions and army corps into their armies, replaced linear tactics by deep combat formations (except for the English), applied concentration of forces on the battlefield in general and in its decisive areas in particular, and formed reserves. But although they learned much and greatly improved their military instruments over the years, his opponents could never match the great master and never really grasped the secrets of his genius. They finally overwhelmed him through numerical superiority and the attrition of war on France, both traceable to Napoleon's diplomatic failures.

Jomini

Antoine Henri Jomini, a Swiss by birth (1779), served as a junior officer under Napoleon, being a protégé of courageous (if not brilliant) Marshal **Michel Ney.** An injustice at the hands of Napoleon's chief of staff, Marshal **Louis Alexander Berthier,** led Jomini to resign and to transfer his allegiance to Russia in the 1813 campaign; he refused, however, to take part directly or indirectly in any operations against Napoleon. For the next 56 years he served with distinction as a general officer in the Russian Army. During those years he devoted himself to study and to writing, mostly based upon his analysis of Napoleonic operations.

Jomini's writings were voluminous. His first book, *Treatise on Great Military Operations,* written while in the French Army, led Napoleon to remark, in alarm, "It teaches my whole system of war to my enemies!" It presented for the first time in writing the fundamental principles of warfare which today are taken for granted by all military men. His *Summary of the Art of War* is his most complete and comprehensive study, and the work most often quoted.

Clausewitz

Karl von Clausewitz was born in Magdeburg, Prussia (1780). He fought with the Prussian Army in all of the campaigns against Napoleon, from disastrous Jena to victorious Waterloo. From 1818 to 1830, he was administrative director of the Kriegsakademie in Berlin, and devoted as much time as possible to writing. Most of his works were studies of military campaigns. His best-known work, however, is *On War,* which embodies his theories and doctrines. It is a book which has cast an indelible stamp on all subsequent military thought. And, as with Jomini, his principal source of inspiration was the genius of his former enemy, Napoleon.

Although Clausewitz believed that he had discovered the fundamental laws of war, he insisted that in practice these are always subject to an almost infinite number of modifications from external influences, of which psychological and moral influences are the most important. He was firmly opposed to any effort to codify doctrine on the basis of the laws or principles of war. He was afraid that abstract rules would be applied dogmatically on the battlefield, and with disastrous results.

Clausewitz was the greatest philosopher of war. His inspiration has been strong in every army during the century and more after his death. Probably no other military writer is so widely quoted as he. Even his critics—discarding his basic theories mostly because these theories have been so distorted by German military men in the 20th century—still repeat many of his observations.

Dennis Hart Mahan

The fame of **Alfred Thayer Mahan** as a military and naval theorist (see p. 899) has tended to obscure the reputation of his father: **Dennis Hart Mahan.** The elder Mahan graduated at the top of the class of 1824 at West Point, and was immediately appointed an assistant professor. After 2 years of teaching and 4 years of further study in France, he was appointed Professor of Engineering at West Point (1830). There being no texts available for the courses he taught (there were then no other engineering schools in the United States), he wrote his own, which became the standard American engineering texts for many years.

Nevertheless, it was the art of war which particularly fascinated Mahan. As a scientist, he spent all of his spare time in analyzing military operations of the past, and particularly those of Napoleon. These studies became the basis of a course of lectures on the art of war—the only formal instruction which American officers received in military theory (see below). He wrote several books on the subject, the most important of which was the text for his course at West Point, which was carried and studied by most of the top leaders on both sides in the American Civil War. This book, horrendously entitled *Advanced Guard, Outpost and Detachment Service of Troops, with the Essential Principles of Strategy and Grand Tactics,* was dubbed *"Outpost"* by his students.

Unlike Prussia, the United States had no war college for its officers; their formal education stopped with graduation from West Point. Mahan and his writings became the war college of the American Army.

MILITARY PROFESSIONALISM

The emergence of military professionalism came hand in hand with the appearance of coherent and scientific military theory. This theory became the basis for systematic military education, at graduate and undergraduate level, an essential element for professionalism in the modern sense. The products of this educational system then became members of a highly specialized group of professionals, the practitioners of the theory.

As with the other modern professions, this development was in large part the result of the successive impacts of the Age of Enlightenment and of the Industrial Revolution upon human affairs. It was first manifested by the appearance of military schools for youths preparing themselves for a military life. The British Royal Military College at Sandhurst (1802), the French St. Cyr (1808), and the American West Point (1802) were early examples of this development. Prussia, which had several such cadet schools even earlier, went farther, however, and established a Kriegsakademie, or War Academy (1810). This was to provide the intellectual stimulation for the Prussian General Staff and thus, more than any single factor, to spark Prussian (later German) preeminence in land warfare for more than a century.

WAR AT SEA

Britain continued to dominate the oceans of the world. The trend away from the "formalist" school to the "melee" school was completed by Nelson's dramatically successful adaptation of his means to the circumstances, without regard to custom, tradition, or "Fighting Instructions." His ability to do this was greatly facilitated by a greatly improved signal-flag system developed toward the close of the previous century by Admiral Sir **Home Popham** and officially introduced at the beginning of this period (1800).

Nelson's tactics, combined with Popham's means of control, brought to perfection methods of warfare at sea with sailing vessels. No significant changes or improvements were possible without radical new developments in science and technology. Yet, in fact, these developments were already at hand. The first steamship was launched in France 18 years before the Battle of Trafalgar.

Ship of the line and frigate at the time of Trafalgar

But professional naval men were most reluctant to accept the new technology. Maritime nations, freed from dependence upon the wind for motive power, found themselves hampered by the limitations in cruising range imposed by steam. The necessity for coaling stations stimulated a scramble for colonies. Dubious eyebrows were raised in the matter of iron hulls versus wood. In a naval era still wedded tactically to the exchange of broadsides, the vulnerability of paddlewheels to gunfire loomed large. **John Ericsson**'s screw propeller, adopted in midperiod, would remove this particular objection, but, except for a few keen-minded and progressive enthusiasts, an overriding fear remained, particularly in England, lest steam propulsion cancel out the advantages of good seamanship as a factor for victory.

The consensus of naval though was best expressed in the words of the British Admiralty in 1828: "Their Lordships find it their bounden duty to discourage to the best of their ability the employment of steam vessels, as they consider the introduction of steam is calculated to strike a fatal blow at the naval supremacy of the Empire."

THE NAPOLEONIC* WARS, 1800–1815

War of the Second Coalition (continued)

On November 9, 1799, **Napoleon Bonaparte** became dictator of France. By *coup d'état* he placed himself in complete control as First Consul. He offered peace to the allies, but his gesture was rebuffed. Though Russia had withdrawn, the war continued. He prepared to wage it aggressively.

* Napoleon Bonaparte's accession to power as First Consul is taken as the dividing point between the Wars of the French Revolution and the Napoleonic Wars.

Operations on Land, 1800–1802

SECOND ITALIAN CAMPAIGN, 1800

The Austrians planned to drive the French from their remaining footholds in Italy. General **Paul Kray von Krajowa,** with 120,000 men, was to hold Germany against any offensive movements by **Moreau,** who had about 120,000 along the upper Rhine in Switzerland and Alsace. Baron **Michael Melas,** with another Austrian army, 100,000 strong, was to overwhelm **Masséna,** who with 40,000 men held the Riviera coast of Italy.

1800, March 8. Bonaparate Raises a New Army. The Army of Reserve assembled at Dijon while he weighed 2 possible courses of action: to combine with Moreau in an offensive through Switzerland into Germany to cut off Kray from his communications with Vienna, or to invade Italy through Switzerland and crush Melas between his army and Masséna's. He decided on the latter plan.

1800, April 6–20. Austrian Victories in Italy. Masséna's army was scattered; he with 12,000 men was driven into Genoa and besieged by General **Karl Ott** with 24,000 men. Melas pursued the remainder of the French, under **Louis Gabriel Suchet,** beyond Nice into the valley of the Var. Bonaparte hastened the concentration of the Army of Reserve at Geneva.

1800, May 14–24. Bonaparte Crosses the Great St. Bernard Pass. He had only 37,000 men. He ordered Moreau to send 15,000 to join him in Lombardy via the Simplon and St. Gotthard passes. To divert Melas' attention, an additional 5,000 men moved through the Mt. Cenis Pass. Bonaparte's advance guard brushed aside Austrian resistance at Fort Bard and Ivrea, and debouched into the Lombardy Plain (May 24). He seized Milan and Pavia and advanced toward Brescia, Cremona, and Piacenza, hoping to relieve Masséna. Melas, hearing of Bonaparte's arrival in Italy, hurried back from Nice.

1800, June 4. Capitulation of Genoa. Masséna, after a protracted and terrible siege, capitulated to Ott, the 7,000 survivors of the French garrison marching out with the honors of war.

1800, June 7. Melas Cut Off from Austria. The Austrian general, at Turin, discovered that Bonaparte was on his line of communication,

advancing west against him. Melas moved east, concentrating 31,000 men at Alessandria (June 13).

June 9. Battle of Montebello. General **Jean Lannes,** whose corps numbered 8,000 men, unexpectedly met Ott, who with 18,000 had marched north from Genoa. Lannes attacked furiously; joined by General **Claude Victor**'s corps of 6,000, he drove the Austrians in confusion toward Alessandria.

1800, June 14. Battle of Marengo. Bonaparte, thinking Melas still at Turin, approached carelessly, with his troops widely separated. Unexpectedly he found himself engaged by superior numbers at Marengo, a mile east of Alessandria. He had only 18,000 men—the corps of **Claude Victor** and **Jean Lannes,** and **Joachim Murat**'s cavalry reserve. **Louis C. A. Desaix**'s corps and other units were scattered farther east and south, gathering supplies. Melas enveloped the French right. By 1 o'clock the Austrians had driven the French back 2 miles. Thinking he had won the battle, Melas ordered a march formation, proceeding east leisurely. But Bonaparte, undismayed by apparent defeat, had rallied his troops and had sent for reinforcements—about 10,000, mainly Desaix's corps, only a few miles away. At 5 P.M. he counterattacked; the Austrian advance guard was struck by Desaix from the front, while the French cavalry (under **François E. Kellermann,** son of the hero of Valmy) swept down on the north flank of the Austrian main body, followed by Lannes's corps and the Guard. Melas' forces crumbled; half of his army was scattered, cut down, or made prisoner. The remainder fled back into Allessandria. Austrian

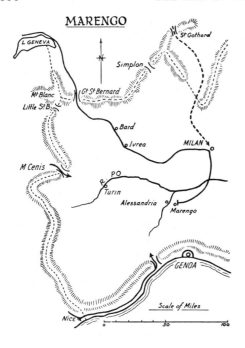

MARENGO

L GENEVA

St Gothard

Simplon

Mt Blanc

Gt St Bernard

Little St B.

Bard

Ivrea

MILAN

M Cenis

PO

Turin

Alessandria

Marengo

GENOA

Scale of Miles

Nice   0   50   100

losses were about 11,000 French 8,000. Desaix was among the dead.

1800, June 15. Melas Capitulates. The Second Italian Campaign had virtually ended. Bonaparte returned to Paris, ordering General **Guillaume M. A. Brune** to advance on Mantua to consolidate the victory.

Germany, 1800–1801

1800, May–June. Moreau Drives Kray into Bavaria. He won victories at **Stockach** (May 3), **Möskirch** (May 5), **Ulm** (May 16), and **Hochstadt** (June 19). Kray retired behind the Inn. Moreau slowly advanced toward Munich.

1800, July 15–November 13. Armistice. Kray was replaced by the youthful **Archduke John** when operations renewed.

1800, December 3. Battle of Hohenlinden. The 2 armies, each seriously overextended, clashed in mud, snow, and rain. Moreau had about 90,000 men available for battle; John had about 83,000, but both committed their

forces piecemeal, about 55,000 French and 57,000 Austrians and Bavarians were actually engaged. The superior speed and energy of the French, combined with considerable good luck, enabled them to surround large portions of the Austrian Army, which was completely crushed, losing 14,000 to a French loss of 2,500. Moreau marched eastward toward Vienna; another French army under General **Jacques Macdonald** invaded the Tyrol from Switzerland; Brune's army in Italy advanced toward the passes over the Julian Alps.

1800, December 25. Austrians Sue for Peace.

1801, February 9. Treaty of Lunéville. This reaffirmed the terms of Leoben and Campo Formio. In addition Spain ceded Louisiana to France, which in turn soon sold the vast territory to the United States (1803).

Operations in Egypt, 1800–1801

1800, January 21. Convention of El Arish. **Kléber,** attacked by both the English and the Turks, agreed to evacuate Egypt on guarantee of free passage of his troops to France. England later disavowing the agreement, Kléber attacked and defeated the Turks at **Heliopolis** (March 20) and recovered Cairo. He was assassinated (June 14), the day Bonaparte was winning at Marengo; **Jacques F. de Menou** assumed command.

1801, March 8. Amphibious Landing at Aboukir. A British-Turkish army (18,000 strong) under Sir **Ralph Abercromby** landed in a brilliant operation.

1801, March 20. Battle of Aboukir. The allies defeated Menou; Abercromby was killed.

1801, April–August. Reconquest of Egypt. The allies took Cairo (July), Alexandria (August). British sea power preventing all French attempts to reinforce the Egyptian expedition, Menou capitulated (August 31). His 26,000 troops were given immediate free passage to France (September). British sea power had wrecked Bonaparte's dream of Oriental conquest.

War at Sea, 1800–1802

1800–1801. Routine Operations. Aside from blockade and convoy operations, the major effort of the Royal Navy was in support of operations against the French in Egypt. The principal exploit was the support of Abercromby's landing at Aboukir by the fleet of Sir **William Keith** (see p. 814).

1801, February. Neutral League. Following the Treaty of Lunéville, Russia, Prussia, Denmark, and Sweden joined to protect their shipping from British belligerent claims.

1801, March. British Reaction. A squadron of 53 sail—18 of them ships of the line—entered the Baltic under command of Admiral Sir **Hyde Parker,** with Sir **Horatio Nelson** as second in command.

1801, April 2. Battle of Copenhagen. Nelson, with 12 ships of the line, boldly sailed into the harbor of Copenhagen. Ignoring orders from Parker, he engaged Admiral **Fischer's** Danish flotilla consisting of warships, armed hulks and floating batteries at anchor. In a 5-hour fire fight, fierce Danish resistance was smashed. Nelson hoped to induce Parker to continue to Revel (Tallin) to destroy the Russian fleet, but an armistice suspended further operations. Meanwhile, Czar **Paul** had been assassinated (March 24), and his successor, **Alexander I,** signed a convention terminating all necessity for further hostilities (June 17).

1801, July 6 and 12. Naval Battles of Algeciras. Sir **James Saumarez** suffered a setback in an inconclusive action with a small French squadron raiding near Gibraltar (July 6), but retrieved his reputation by a victory against 2-to-1 odds after the French had been reinforced by a much larger Spanish fleet.

Peace of Amiens, 1802–1803

1802, March 27. Treaty of Amiens. Peace between France and England brought general peace to Europe for the first time in a decade.

1802, August 2. Bonaparte Proclaimed Consul for Life.

FRANCO-BRITISH WAR, 1803–1805

1803, May 16. Resumption of Hostilities. Britain imposed a naval blockade of the Continent. Bonaparte began preparations for an invasion of England.

1803–1805. French Threaten Invasion of England. If Bonaparte really intended to invade, all his efforts were thwarted by British sea power. His invasion flotilla gathered at the Rhine mouths was seriously mauled by **Sidney Smith**'s fire and explosion ships (October 2, 1804). British squadrons also balked French naval domination of the Mediterranean, the East Indies, and the West Indies. As **Alfred Mahan** puts it: "Those far distant, storm-beaten ships, upon which the Grand Army never looked, stood between it and the domination of the world."*

1804, December 2. Bonaparte's Coronation as Napoleon, Emperor of the French. He continued concentrating his Grand Army on the northern coast. Numerous landing craft were constructed; there was much apprehension in England. But British diplomatic maneuverings were bringing results.

WAR OF THE THIRD COALITION, 1805–1807

1805. Britain Gains Allies. Austria, Russia, and Sweden prepared to revenge themselves. The bulk of the French Army was assembled near Boulogne; the only other important French concentration was in north Italy: Masséna, with 50,000 men. The allies planned first to destroy Masséna in Italy, then to move westward with overwhelming forces on the northern side of the Alps toward the Rhine and France.

* Alfred T. Mahan, *Influence of Sea Power upon the French Revolution and Empire* (Boston: Little, Brown, 1893) II, 118.

War on Land, 1805

ULM CAMPAIGN

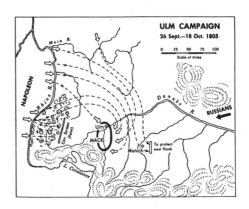

1805, August 31. The Grand Army Marches Eastward. Napoleon, learning of his enemies' intentions, took the initiative, discarding plans for invading England. The Grand Army, totaling more than 200,000 men, secretly began to march eastward from the Boulogne area.

1805, September 2. Austrian Invasion of Bavaria. General **Mack von Leiberich,** knowing nothing of the French movement, marched toward Ulm with 50,000 troops. The **Archduke Charles** with 100,000 men prepared to attack Masséna in Italy. A Russian army of 120,000 began to march westward. An additional contingent was promised by Sweden. When they joined Mack, the second stage of the allied plan would begin.

1805, September 26. French Cross the Rhine. The advance continued on a wide front. Mack, still unaware that Napoleon had left Boulogne, was between Ulm and Munich; the **Archduke John** with 33,000 Austrians was concentrated at Innsbruck; the Archduke Charles was in the Adige Valley, between Trent and Venice. Far behind them, the Russian generals **Mikhail I. Kutuzov** with 55,000 men and **Friedrich Wilhelm Buxhöwden** with 40,000 more were approaching the Carpathians.

1805, October 6. French Reach the Danube. While Murat's cavalry, thrusting through the Black Forest, demonstrated in front of Mack at Ulm, the Grand Army, in 6 great columns, swept north and east of the Austrian general in a wide concentric arc. Too late Mack realized the French were behind him.

1805, October 17. Capitulation of Ulm. After one futile attempt to break through the encirclement at **Elchingen** (October 14), Mack surrendered with nearly 30,000 men, 40 standards and 65 guns.

COMMENT. *This campaign opened the most brilliant year of Napoleon's career. His army had been trained to perfection; his plans were faultless. Sweeping through Western Europe on a wide front, he con-*

centrated the magnificent machine on Mack's line of communications in one of the finest historical examples of a turning movement. Ulm was not a battle; it was a strategic victory so complete and so overwhelming that the issue was never seriously contested in tactical combat.

THE ADVANCE ON VIENNA

1805, October. Napoleon Advances East. He peeled off several army corps to the south to prevent interference by Austrian movements through the Alps from Italy.

1805, October 30. Battle of Caldiero. Masséna in Italy attacked the Archduke Charles, who retired eastward, hotly pursued by Masséna. Joined by Archduke John, retreating from the Tyrol, Charles withdrew across the Julian Alps.

1805, November 1–14. Invasion of Austria. Napoleon, leaving some 50,000 men to guard his line of communications, drove the Russian Kutuzov before him and occupied the Austrian capital. The Russian fought effective delaying actions at **Dürrenstein** (November 11) and **Hollabrünn** (November 15–16).

AUSTERLITZ CAMPAIGN

1805, November 15. Napoleon Advances North. Leaving 20,000 men in Vienna, he began concentrating the major portion of his remaining army near Brünn, some 70 miles from Vienna.

1805, November 20–28. The Strategic Situation. Napoleon with 65,000 men, in the midst of his enemies, waited for them to move. **Archduke Ferdinand** with 18,000 men lay at Prague, to the northwest; Emperors Alexander of Russia and **Francis II** of Austria at Olmütz, to the northeast (Kutuzov actually commanded) with 90,000 Russians and Austrians. The Archdukes Charles and John, with 80,000 more, were blocked from crossing the Alps (by about 20,000 men of the corps of **Michel Ney** and **Auguste F. L. Marmont,** who guarded the passes) and harassed by Masséna's 35,000 as they withdrew to Austria through Hungary south of the Alps. Napoleon's problem was to prevent the junction of these numerically

TO ILLUSTRATE THE GERMAN AND AUSTRIAN CAMPAIGNS OF NAPOLEON: 1805-7, 1809, & 1813.

Austrian & Prussian Frontiers as in 1805.

superior allied forces while preserving his line of communications through Vienna to France. His apparently exposed situation was enticing to his enemies. The emperors and Kutuzov planned to move south from Olmütz, circle his right flank, and cut his communications. This was exactly what he anticipated.

1805, November 28. The Allied Army Begins to Move. The French army lay facing east, about 2 miles west of the village of Austerlitz. Deliberately, Napoleon placed his army on low ground and over-extended his right wing—1 division—for 2 miles in plain view of allied scouts, while the bulk of his troops were collected east of Brünn, near the road to Olmütz.

1805, December 1, P.M. The Allies Reach Austerlitz. At once they noted the weakened right flank of the French Army. They planned to crush it and get between Napoleon and Vienna.

1805, December 2. Battle of Austerlitz. At dawn the weight of the allied main effort fell upon the French right. Although reinforced by the arrival of Louis N. Davout with 8,000 additional men, the French right was forced back. By 9 A.M. fully one-third of the allied army was pressing against this wing, with more troops moving laterally across the French front to join in the assault. Napoleon then sprung his trap. **Nicolas J. Soult**'s corps, in the French center, stormed the heights of Pratzen and split the allied front. Soult then encircled the allied left, rolled it up, and, assisted by Davout, drove it in confused retreat. French artillery fire broke ice on frozen ponds, and many Russians were drowned. Meanwhile, the corps of **Jean Baptiste J. Bernadotte** assaulted directly east, through the gap made by Soult, while on the French left Lannes's corps drove full tilt against the allied right, on the Brünn-Olmütz road. The allied right, under Russian Prince **Peter I. Bagration,** resisted fiercely until enveloped by Bernadotte from the south. By nightfall the allied army had ceased to exist. French losses were nearly 10,000 men; the Austro-Russian losses were 36,000 men, 45 standards, and 185 guns; the remnants of the allied army were hopelessly scattered.

COMMENT. *Austerlitz stands as a tactical masterpiece ranking with Arbela, Cannae, and Leuthen.*

1805, December 4. Austrian Capitulation. Emperor Francis agreed to an unconditional surrender. The shattered forces of Czar Alexander retreated to Russia.

1805, December 26. Treaty of Pressburg. Austria withdrew from the war, surrendering

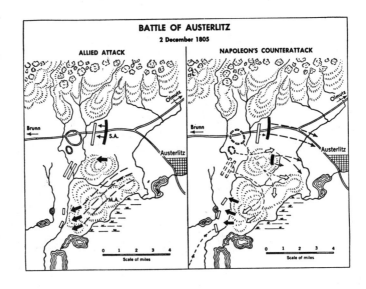

BATTLE OF AUSTERLITZ
2 December 1805
ALLIED ATTACK   NAPOLEON'S COUNTERATTACK

territory in Germany and Italy; France gained virtual domination over western and southern Germany. Napoleon had changed the political face of Europe.

War at Sea, 1805

1805, April–July. Nelson's Pursuit of Villeneuve. Admiral **Pierre Villeneuve,** escaping Nelson's blockade of Toulon, sailed into the Atlantic, was joined by a Spanish force, and sailed for the West Indies with about 20 ships. Nelson, who had 10 ships, pursued. After some maneuvering in the West Indies, Villeneuve fled back across the Atlantic, again pursued by Nelson.

1805, July 22. Action off Cape Finisterre. British Sir **Robert Calder**'s squadron of 18 ships clashed with Villeneuve's combined Franco-Spanish fleet, capturing 2 Spanish ships. Villeneuve sailed to Cádiz, where he was reinforced.

1805, August. Napoleon's Naval Plan. Villeneuve was to take his fleet from Cádiz to the Mediterranean. There he would unite with other French ships at Cartagena and move on southern Italy to support Masséna's campaign. Villeneuve knew that Britain's Nelson, now with 29 ships of the line (actually only 27 were present), lay in the offing, a fact unknown to Napoleon; but Villeneuve, smarting under the threat of removal from command for cowardice, complied with his orders (September 27).

1805, October 21. Battle of Trafalgar. Nelson, off Cape Spartel, learned of the allies' move and sailed to meet them. Off Cape Trafalgar the fleets clashed. The Franco-Spanish force of 33 ships turned back toward Cádiz in a 5-mile-long irregular line. In accordance with a pre-arranged plan, Nelson, in 2 divisions, each in single column, on a course at right angles to his adversary's, drove directly into the center of the long allied column, cutting it in two. In a 5-hour battle, 18 allied ships were taken; the remainder fled, only 11 reaching Cádiz. No English ship was lost. Nelson was mortally wounded as his flagship *Victory* closed in furious combat with the French *Redoutable*.

COMMENT. *Nelson had destroyed French naval power and established Britain as the mistress of the seas in the most decisive major naval victory—tactically and strategically—of history.*

Jena Campaign, 1806

1806. Napoleon Controls Central and Western Germany. He established the Confederation of the Rhine. By August, the last vestige of German unity, the Holy Roman Empire, expired; its emperor, **Francis II,** becoming **Francis I,** Emperor of Austria. Prussia, which until the French victory of Austerlitz had contemplated joining the coalition against Napoleon, was now so alarmed by the state of events that, with England's encouragement, she secretly prepared for war. Saxony joined Prussia.

1806, September. Opposing Plans. Napoleon's Grand Army of 200,000 was still mostly in southern Germany. He prepared to invade Prussia. Secretly he concentrated far to the east, in northeastern Bavaria, close to the Austrian border, but without violating Austrian neutrality. The Prussian-Saxon field army comprised about 130,000 men under the overall command of **Karl Wilhelm Ferdinand, Duke of Brunswick.**

1806, October 8. Napoleon Starts North. The rapidity of his strategic concentration and of his advance was remarkable. Preceded by a cavalry screen, the Grand Army moved in 3 parallel columns on a front of about 30 miles at a rate of

BATTLE OF
TRAFALGAR
21st October 1805
British   Allies

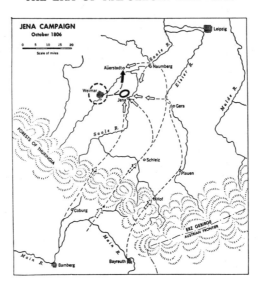

15 miles per day. Roughly forming a great square, the army was prepared for tactical concentration in any direction.

1806, October 12. Strategic Turning Movement. The French advance bypassed the Prussian left flank, completely surprising the Prussians. Their first intimation of the situation came when the corps of Marshal Lannes, on the left vertix of the French square, overwhelmed a smaller Prussian force at **Saalfeld** (October 10). Napoleon was now closer to Berlin than they were. Knowing that the bulk of the Prussian Army was to his left, he ordered Davout and Bernadotte to move west from Naumburg to cut the Prussian line of communications. The remainder of his army advanced toward Jena. Davout responded correctly; Bernadotte, misinterpreting his orders, moved southwest instead of west.

1806, October 14. Battle of Jena. Meanwhile the Prussians, alarmed by news of the French advance, changed their own plans. The **Duke of Brunswick,** with 50,000 men, moved northeast toward Auerstadt, 15 miles north of Jena. Prince **Friedrich Ludwig Hohenlohe,** and 64,000 more were scattered on a 15-mile front between Weimar and Jena to protect Brunswick's rear. Napoleon, shortly after dawn, struck Hohenlohe, concentrating 100,000 men at Jena. By noon, he completely swept the Prussians from the field. Prussian losses were 27,000 and 112 guns, the French loss was 6,000.

1806, October 14. Battle of Auerstadt. Davout with 27,000 men engaged Brunswick's 50,000. In an epic defensive battle, Davout, on the Prussian line of communications, withstood repeated assaults for more than 6 hours. Brunswick was mortally wounded and King **Frederick William III** assumed command. Prussian morale was weakened by rumors of the French success at Jena. As the Prussian effort slackened, Davout counterattacked. Bernadotte, near Dornburg—far from both battles—now realized something was wrong and marched to the sound of the guns, but did not arrive on the field in time to enter the battle. Crumbling under Davout's counterattack, Frederick's wavering command gave way, and the entire Prussian Army now disintegrated. They lost 25,000 killed and wounded, and nearly 25,000 more made prisoner. French casualties were approximately 7,000 on both fields.

COMMENT. *Once again Napoleon, by rapidity of movement and skillful combinations, had won a strategic victory before tactical operations ever began.*

1806, October–November. Pursuit. The Grand Army immediately swept northward after the remnants of the Prussian Army. Murat's cavalry seized Erfurt (October 15), 10,000 men, remnants of Frederick's command, capitulated. Berlin was seized (October 24), and at Prenzlau, Prince Hohenlohe surrendered with another 10,000 (October 28). A further 32,000 surrendered (mostly to small French cavalry detachments) in various fortresses in November. The last element of Prussian resistance, **Gebhard L. von Blücher**'s command, surrendered near Lübeck, nearly 150 miles northwest of Berlin (November 24). Frederick William fled to Russia.

1806, November 30. Advance into Poland. Napoleon moved east with 80,000 men to the line of the Vistula (Wisla) River, occupying Warsaw (Warszawa) to thwart any Russian attempt to sustain Prussia. Count **Lévin A. Bennigsen,** with about 100,000 Russians and some Prussian fragments, lay at Pultusk.

1806, December 30. Winter Quarters. After desperate rear-guard fighting at **Pultusk** (December 26), Bennigsen evaded French attempts to corner him. Napoleon, faced by bitter cold and the exhaustion of his own troops, went into winter quarters. His corps were spread across northern Poland and east Prussia from the Bug River to Ebling (Elblag) on the Baltic.

Eylau Campaign, 1807

1807, January. Russian Offensive. Bennigsen attacked Ney's cantonments south of Königsberg (Kaliningrad) and forced him to withdraw. Pursuing into East Prussia, Bennigsen's communications were menaced by Napoleon's rapid concentration and advance. The Russian withdrew hurriedly. Napoleon caught up with him February 7 at Preussisch-Eylau (Bagrationowski).

1807, February 8. Battle of Eylau. Napoleon hastily attacked, with only part of his army in hand. Russian strength was 74,500 (with a Prussian corps of 8,500 more moving rapidly toward them); Napoleon had less than 50,000 (Ney's and Davout's corps, with an additional 25,000, were expected by noon). In a driving snowstorm, a French assault was checked. Davout arrived and turned the Russian left. But he, in turn, was checked by the arrival of **C. Anton Wilhelm Lestocq**'s Prussian corps. Ney now arrived, but neither side could gain a decisive advantage. That night Bennigsen withdrew. Losses on both sides were enormous.

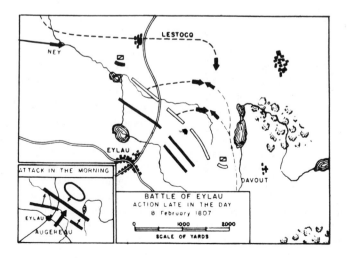

BATTLE OF EYLAU
ACTION LATE IN THE DAY
8 February 1807
SCALE OF YARDS

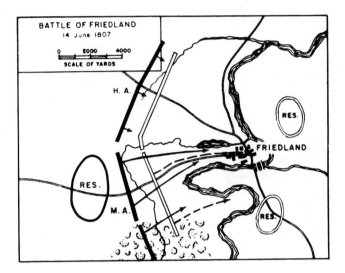

Russian casualties were 23,000 men killed and wounded, with some 3,000 men and 24 guns captured. The French lost nearly 22,000 men, with 1,000 more captured. Both armies returned to winter quarters; both brought up reinforcements and refitted.

Friedland Campaign

1807, March 15–April 27. Siege of Danzig. The French captured the city, beating off futile Russian and Prussian relief efforts. Napoleon planned a spring offensive to begin June 10.

1807, June 5. Bennigsen Resumes the Offensive. Again he hoped to overwhelm Ney in a surprise assault. Ney withdrew as Napoleon concentrated north of Allenstein.

1807, June 10. Battle of Heilsburg. Napoleon repulsed the Russians, who retreated north. Moving by parallel roads, Napoleon placed the bulk of his army between Bennigsen at Friedland and Lestocq at Künigsberg (June 13).

1807, June 14. Battle of Friedland. Napoleon sent Lannes, with 17,000 men, to pin Bennigsen down while the remainder of the army concentrated to the west. The Russian, with 61,000 men in hand and 20,000 more nearby, crossed the Alle River and attacked Lannes with 46,000, leaving the remainder in reserve east of the river. Lannes's delaying action halted the Russians after a 3-mile advance. Napoleon, taking personal command of the battle as his concentrations progressed, launched his main attack at 5 P.M. with 80,000 men. Within 2 hours the Russians' left flank had disintegrated and they were driven back into Friedland. Their resistance stiffened, but by 8 P.M. Napoleon had driven them across the river in great disorder, leaving 20,000 dead and wounded on the field and 80 guns. French casualties were about 12,000 men.

1807, June 15. Evacuation of Königsberg. Lestocq, who was being pressed by Murat, retreated to Tilsit with 25,000 men when he learned of Bennigsen's defeat.

1807, June 19. Napoleon Occupies Tilsit (Sovetsk). The Russians asked for a truce, which he granted.

1807, July 7–9. Treaties of Tilsit. Napoleon met his enemies—Alexander of Russia and Frederick William III of Prussia—on a raft in the middle of the Niemen (Neman). Prussia gave up to the Grand Duchy of Poland all land taken in the partitions of Poland (see pp. 760). She gave to Napoleon and the Confederation of the Rhine all her territory between the Elbe and Rhine rivers. Her army was reduced to 42,000 men. She was to pay an indemnity of

140 million francs; until paid, French troops would occupy the country. Russia recognized the Grand Duchy of Warsaw and agreed to an alliance with France against Britain.

COMMENT. *Napoleon was the virtual ruler of Western and Central Europe.*

THE PENINSULAR WAR— CAMPAIGNS OF 1807–1808

1807, July–October. England Alone in Opposition to Napoleon. Her coastline was guarded by her navy; her offensive operations were concentrated in a stringent naval blockade of the entire European coastline. By economic pressures England hoped that internal stresses on the Continent would eventually lead to a new coalition to challenge French control. Napoleon had imposed a counterblockade on Britain—his **Continental System**—aimed to throttle British trade (November 21, 1806). After Tilsit, neutral Portugal was the only access route (save by smuggling) for British trade with the Continent. Napoleon now turned his attention to the Iberian Peninsula.

1807, November–December. French Invasion of Portugal. With Spain's permission, **Andoche Junot** led an army into Portugal, capturing Lisbon (December 1). The Portuguese royal family fled to Brazil.

1807, December 17. Milan Decree. This reaffirmed the Continental System. British trade was forbidden from all Europe. Actually, smuggling became rampant, particularly through Spanish ports.

1808, March. French Invasion of Spain. Under pretext of guarding the Spanish coast, Murat led a French army of 100,000 into Spain. **Charles IV** and his son **Ferdinand** were forced to renounce the throne; Napoleon's brother **Joseph** (who had been King of Naples) was crowned King of Spain. (Murat succeeded him as King of Naples.)

1808, May. Insurrection Flares in Spain. French garrisons became islands in a sea of guerrilla warfare, accompanied by intense cru-

elty. Murat temporarily fell back to the Ebro. British arms, equipment, and money poured in to help the Portuguese and Spanish peoples. The British government decided to send an expeditionary force to Portugal, commanded by Sir Arthur Wellesley.

1808, June 15–August 15. First Siege of Saragossa. The Spanish garrison, assisted by the aroused population, resisted French attempts to reopen the main line of communications into central Spain and Portugal. After Murat's withdrawal to the Ebro, the French raised the siege.

1808, July 19. Battle of Baylen. General **Pierre Dupont**'s 20,000 men, surrounded by some 32,000 Spanish regulars and levies, attempted to break out, but failed. Dupont's army was forced to capitulate (July 22). Promise of safe conduct to France was immediately violated; the unarmed French not butchered on the spot were thrown into prison hulks, where most of them perished. The disaster—the first surrender of a Napoleonic army—stimulated Spanish resistance and shocked French morale. Junot's army in Portugal was isolated.

1808, July 20. French Reoccupy Madrid. This partially offset Baylen.

1808, August 1. The Wellesley Expedition. Wellesley landed north of Lisbon.

1808, August 21. Battle of Vimeiro. Wellesley, with 17,000 men, repulsed Junot's attack with 14,000.

1808, August 30. Convention of Cintra. Junot's position had been rendered untenable by Wellesley's victory and by a popular uprising in Lisbon. Following negotiations Junot capitulated to Wellesley's superiors, Generals Hew Dalrymple and Harry Burrard, the terms providing that the French be evacuated to France by sea on British ships. News of these terms sparked a political uproar in Britain; Wellesley, who was not responsible for the terms, was temporarily recalled to London. Sir John Moore, with reinforcements, landed and took command of British forces in Portugal.

1808, September. British Invasion of Spain. Moore's British army of 35,000 advanced to seize Madrid, while 125,000 Spanish regulars, levies, and irregulars drove the French back to

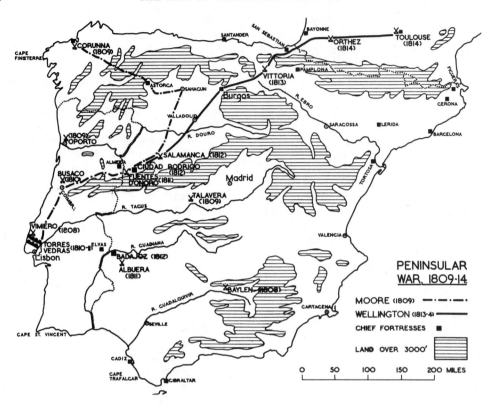

PENINSULAR WAR, 1809-14

MOORE (1809) — · — · —
WELLINGTON (1813-4) ——————
CHIEF FORTRESSES ■

LAND OVER 3000'

0   50   100   150   200 MILES

the Ebro River. Napoleon, in Paris, commented that his army in Spain seemed to be commanded by "post-office inspectors." He sent reinforcements to Spain, and prepared to take the field there himself.

1808, November 5. Napoleon Joins His Army in Spain. He advanced immediately with 194,000 men.

1808, December 4. Capture of Madrid. Napoleon then turned northwest to cut off Moore's British forces from Portugal. Moore, resting at Salamanca, received word of Napoleon's intentions from the Spanish, and retreated northwest to a new base established at Corunna on the Biscay coast. The British retreated skillfully in a series of rear-guard actions, but were hampered by a lack of supplies and by a breakdown of discipline brought on by the logistical problems and exacerbated by the harsh Spanish winter.

1808, December 20–1809, February 20. Second Siege of Saragossa. French assaults under Lannes penetrated the defenses (January 27), but desperate resistance continued for 3 weeks. Finally, those defenders who had survived disease, starvation, and bombardment capitulated.

1809, January 1. Napoleon Returns to Paris. Spain appeared to be pacified and under control. Serious developments in Central Europe demanded his immediate attention (see p. 825). He took much of his army with him, leaving the pursuit of Sir John Moore to Soult.

1809, January 16. Battle of Corunna. Moore reached the coast successfully with 15,000 men. He repulsed Soult's 20,000 in a stiff battle, each side suffering about 1,000 casualties. Moore, mortally wounded, died on the field and was buried on the ramparts of Corunna. The British expedition was safely evacuated by sea.

WAR AGAINST AUSTRIA, 1809

1809, January. Austria Prepares for War. Encouraged by events in Spain, Austria decided to gain vengeance and to liberate Germany from the French yoke. Word of Austrian preparations caused Napoleon to leave Spain.

1809, April 9. Austrian Invasion of Bavaria. The Archduke Charles marched on Ratisbon (Regensburg). Another force moved southwestward from Bohemia.

1809, April. Operations in Italy. An Austrian army of 50,000 men under Archduke John crossed the Julian Alps to invade Italy. Prince **Eugène de Beauharnais,** Napoleon's stepson, attacked with 37,000 men at **Sacile,** east of the Tagliamento River, but was repulsed (April 16). Eugène retreated behind the Piave.

1809, April–1810, February. Revolt and Guerrilla War in the Tyrol. A popular uprising against the Bavarian garrison (French allies) freed most of the region. Archduke John sent troops to assist the guerrillas. After the Wagram Campaign (see below), French troops under Eugène slowly reconquered the region.

1809, April 16. Napoleon Arrives at Stuttgart. He had rushed from Spain via Paris. He found his armies in Germany—176,000 men ineptly commanded by his chief of staff, **Louis Alexandre Berthier**—had been split by the Austrian invasion forces. Napoleon took command and seized the initiative, while at the same time correcting Berthier's faulty dispositions.

1809, April 19–20. Battle of Abensberg. Having collected more than half of his army west of Ratisbon, Napoleon crossed the Danube River to penetrate the center of the extended Austrian Army—about 186,000 men. The Austrian right flank was forced back toward Ratisbon, while the left was driven back toward Landeshut.

1809, April 21. Battle of Landeshut. The Austrian left wing, under Baron **Johann Hiller,** pursued by the bulk of Napoleon's army, was in danger of being cut off completely by the arrival of Masséna's corps from the west. Masséna's indecision, however, permitted the Austrians to escape eastward across the Isar River after a brisk fight. Napoleon, despite annoyance at less than complete success, immediately turned northward to join Davout, who with 36,000 men had been maintaining pressure against the remainder of the Austrian Army, about 80,000 under Archduke Charles, south of Ratisbon.

1809, April 22. Battle of Eckmühl Charles attacked the left of Davout's isolated force, planning to cut Napoleon's line of communications. The Austrian attack was lethargic; Davout held firm. Napoleon and the van of his main body arrived on Davout's right early in the afternoon. By midafternoon the Austrian left had been crushed and Charles was retreating, after a loss of 11,000 men. French losses were 3,000. Napoleon was unable to pursue; his troops were utterly exhausted by marching, fighting, and heat.

1809, April 23. Battle of Ratisbon. Charles now devoted every effort to escape northward across the Danube. To cover this retreat, he defended the walled city of Ratisbon. Due to the exhaustion of his troops after a week of hard fighting and marching, Napoleon's pursuit was not energetic. By the time the French fought their way into the town of Ratisbon, most of the Austrian army had escaped. Napoleon was slightly wounded.

COMMENT. *In 7 days of brilliant marching, fighting, and maneuvering, Napoleon had taken a French army from the verge of defeat to brilliant victory. Davout again distinguished himself. Estimated Austrian losses were 30,000 men, including over 11,000 prisoners; the French lost some 15,000.*

1809, May 13. Capture of Vienna. There was no opposition. Archduke John withdrew from north Italy, followed closely by Eugene.

1809, May 21–22. Battle of Aspern-Essling. The Austrians having concentrated on the north bank of the Danube, Napoleon attempted to smash his way across at the island of Lobau against fierce resistance. Part of the French Army (about 66,000) was able to cross the river to face 99,000 Austrians. Plagued by the complete breakdown of the inadequate bridging arrangements, Napoleon, unable to reinforce his troops north of the river, withdrew. This was his first defeat. Austrian losses were 23,000;

French were about 25,000, including Marshal Lannes, who was killed.

1809, June. Napoleon Makes Detailed Plans. He gathered nearly 200,000 men in the vicinity of Vienna and Lobau, and assembled adequate bridging material. Meanwhile, Eugène pursued Archduke John into Hungary against fierce Austrian rearguard resistance. Eugène defeated John at **Raab** (June 14) and marched north to join Napoleon at Vienna. John continued his retreat to Pressburg (Bratislava).

1809, July 4–5. Night Danube Crossing. The French move surprised Charles, who had 130,000 men. Once more Napoleon had placed himself between 2 adversaries, for the Archduke John was approaching from the east with nearly 15,000 men. Napoleon had approximately 160,000 men north of the river; the remainder held his line of communications. Napoleon decided to attack at once before John could arrive with reinforcements.

1809, July 5–6. Battle of Wagram. To prevent Charles from moving eastward to join John, Napoleon made his main attack against the Austrian eastern or left wing. Charles tried to turn Napoleon's left so as to cut him off from his Danube bridgehead. Results were indecisive the first day. On the second day, Napoleon planned to continue the envelopment of the Austrian left. Charles attempted to turn the French left so as to cut them off from the Danube bridgehead. Starting first, the Austrian attack was initially successful, driving back Bernadotte's Saxon allied corps in considerable confusion. The Austrian attack was finally halted by the indecision of Charles' subordinates and by flanking fire from French batteries on Lobau Island. Meanwhile, Davout's attack on the Austrian left continued to make progress, despite fierce Austrian counterattacks. Finally, Napoleon massed his guns against the Austrian center in the greatest concentration of artillery ever made to that time. He launched a heavy infantry assault in the center, while Davout redoubled his efforts to turn the Austrian left. Charles's center was penetrated, his flank thrown back. He withdrew. Although the retreat was made in moderately good order, the defeat was decisive. The Archduke had lost 26,000 men, 19,000 killed and wounded and 7,000 captured. A further 19,000 were lost to desertion and straggling. French losses were about 37,000.

1809, July 10. Austrians Ask for Armistice. Austrian troops had been driven from the Grand Duchy of Warsaw by Polish troops under Prince **Poniatowski;** Russia had not joined the coalition, as hoped; British landings in Holland and Belgium had been contained and repulsed (see p. 834). It was evident to Emperor Francis that further conflict would be disastrous to Austria.

1809, October 14. Treaty of Schönbrunn. Napoleon's preeminence in Europe was reaffirmed. Austria ceded 32,000 square miles of territory, with 3,500,000 inhabitants, to France and her satellites. She agreed to join the Conti-

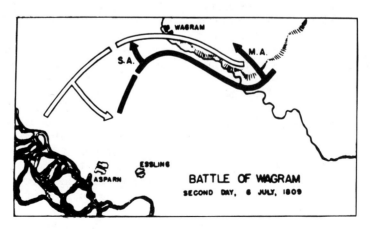

BATTLE OF WAGRAM
SECOND DAY, 6 JULY, 1809

nental System and to break off all connections with England. Except for the fighting still continuing on the Iberian Peninsula (see p. 835)

and at sea (see p. 837), and for a few minor rebellions (particularly in the Tyrol), an uneasy peace settled on Europe.

WAR WITH RUSSIA, 1812

Franco-Russian relations had frayed steadily. Czar Alexander, Napoleon's only major rival on the Continent, resented the revival of Poland through the establishment of the Grand Duchy of Warsaw. Napoleon's refusal to join him in trying to drive the Turks out of Southern Europe (see p. 850) made matters worse (1811). England, aware of this situation, made overtures to Russia. She made peace with both Sweden and Russia (June 1812), and got both nations to renounce the Continental System, a serious economic blow to France.

1812, May–June. Napoleon Prepares for War. He assembled an army group of 450,000

men in Poland. He planned to invade and crush Russia from a base of communications reaching

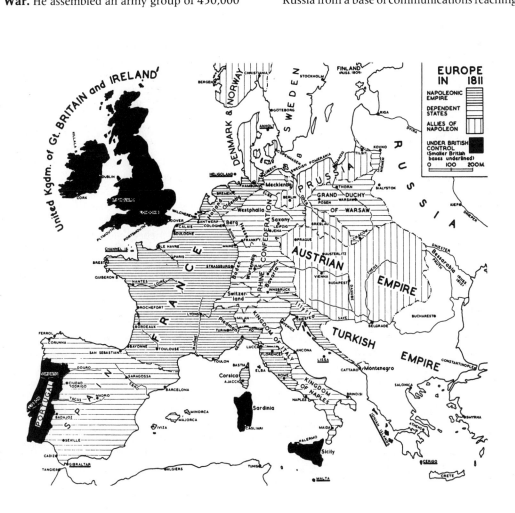

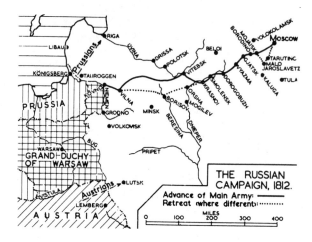

from the Pripet Marshes on the south to the Baltic Sea on the north. His right flank was protected by the Austrian army of Prince **Karl Philipp von Schwarzenberg,** nearly 40,000 strong. The left flank was covered by Marshal **Jacques E. J. A. Macdonald**'s army, of equal strength, with its principal element a Prussian corps. In the center stood 3 French armies. Napoleon's left wing of 220,000 men was just west of the Niemen (Neman) River, between Kovno (Kaunas) and Grodno. Echeloned to his right rear, to secure the line of communications, were the armies of **Eugène de Beauharnais** and of Napoleon's incompetent brother **Jérôme,** each about 80,000 strong. This Grand Army of 1812 contained less than 200,000 Frenchmen; the remainder were German, Austrian, Polish, and Italian contingents. Napoleon was now depending upon conscripts and upon the soldiery of uneasy allies and seething nations whom he had ground under his heel.

1812, June. Russian Preparations. In a cordon defense along the Russian northwestern frontier were 2 armies: **Barclay de Tolly** (Prince **Michael Andreas Bogdanovich**) with 127,000 men was north of the Niemen; Bagration had 48,000 between the Niemen and the Pripet Marshes. South of the marshes was a third Russian army, 43,000 strong, under General **A. P. Tormassov,** guarding the south-

western frontier. More than 200,000 additional Russian troops were scattered throughout western and central Russia.

1812, June 24. Napoleon Crosses the Niemen. Penetrating between the 2 main Russian armies, he planned to crush them in succession. But Jérôme failed to carry out instructions, unusual heat sapped French strength, and much of the cavalry was incapacitated by a colic epidemic among the horses. Meanwhile, Napoleon relieved Jérôme and put Davout in his place. Davout blocked Bagration's effort to join Barclay by a victory at **Mogilev** (July 23), but Bagration retreated eastward across the Dnieper (Dneper) to join Barclay near Smolensk (August 3). Barclay assumed command.

1812, August 7–19. Maneuvers near Smolensk. Barclay planned to fight, but inefficient Russian staff work slowed the Russian movement. Ironically, the errors foiled a turning movement by Napoleon which would have crushed both Russian armies, since he had anticipated their plan. They escaped eastward, with Napoleon—who now had about 230,000 men—again trying to turn their flank by crossing the Dnieper (Dneper) River south of Smolensk. After hard-fought but unsuccessful defensive battles at **Smolensk** (August 17) and nearby **Valutino** (August 19), the Russians escaped Napoleon's trap, largely because of errors by French Marshals Junot and Murat. French casualties were over 10,000; Russian were at least 15,000.

1812, August 29. Kutuzov Commands Russian Armies. He continued to retreat. Reluctantly Napoleon followed. He had planned to spend the winter at Smolensk, but now realized that his supply arrangements were not working as planned. A complete victory in 1812 was essential.

1812, September 7. Battle of Borodino (or of the Moskva). With 122,000 men, Kutuzov made a stand 60 miles west of Moscow. Napoleon concentrated slightly more than 124,000 men and started enveloping the Russian left flank. In the midst of the fighting, he unaccountably gave up personal control of the battle. This was the first of recurrent seizures, never fully explained medically, which hampered a number of subsequent operations. The bitterly contested struggle continued; by nightfall the Russians had been forced back with nearly 50,000 casualties including the death of Bagration. The French lost some 28,000 men.

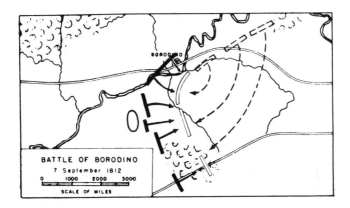

BATTLE OF BORODINO
7 September 1812
0   1000   2000   3000
SCALE OF MILES

Kutuzov withdrew to Moscow. There was no energetic pursuit, due to Napoleon's illness.

1812, September 14. Napoleon Enters Moscow. The city, which had been evacuated of its inhabitants, began to burn, set fire by the Russians. Most of the wooden houses were soon in ashes, and the French army had to bivouac in the suburbs. Napoleon had some 95,000 men near Moscow, most of them utterly exhausted. The remainder of his great army was scattered in an elongated triangle with its base on the Niemen River, and extending from Riga through Kovno to Brest Litovsk (Brest). French morale was deteriorating; the allied troops were becoming unreliable. South of Moscow near Kaluga, Kutuzov's army of 110,000 was still intact and full of fight. The entire Russian nation, aroused by needless French cruelty, united in the defense of its homeland. North of Polotsk the French line of communication was threatened by an army under Count **Ludwig Adolf Wittgenstein.** The Russians south of the Pripet Marshes, now under General (or Admiral) **Tshitshagov,** moved west, threatening the French flank at Brest Litovsk. Desperately short of food, and Czar Alexander having ignored peace overtures, Napoleon decided to withdraw to Smolensk (October 19).

1812, October 24. Battle of Maloyaroslavets. Moving southwest from Moscow, Napoleon planned to destroy Kutuzov's army. Following a sharply fought, but indecisive, meeting engagement, Napoleon decided to turn back to the northwest and west, retracing the old route of advance from Smolensk and Borodino. This decision was fatal in that it forced the Grand Army to traverse a countryside ravaged by the French and Russian depradations of the summer. With no forage available, the overstrained logistical system collapsed, leaving the army starving. Czar Alexander rejected a truce. As winter closed down over central Russia, Napoleon continued to withdraw.

1812, October–December. Retreat from Moscow. Snow, followed by bitter cold, began to impede the march (November 4). Surrounded by swarms of regular and irregular Russian forces, the freezing, starving Grand Army, all of its commissariat broken down, marched through snow to disintegration. Separate French corps fought off repeated attacks. When Smolensk was reached, discipline broke down as the starving troops ransacked the depots, destroying most of the supplies in their frenzy. Napoleon decided to continue the retreat from Smolensk (November 12). Most of the army became a disorganized mob.

1812, November 16–17. Battle of Krasnoi (Krasnoye). West of Smolensk, Kutuzov's advance guard, which had circled west of the French, barred the road. Napoleon collected his few effective elements and, in a brilliant display of leadership, drove the Russians off. Ney's corps, reduced to 9,000 men, sacrificed itself in a desperate all-day rear-guard action to save the rest of the army, losing all but 800 men.

1812, November 26–28. Crossing of the Berezina. Near Borisov on the Berezina, Napoleon was cornered between Tshitshagov and Kutuzov, who had 144,000 men between them, of whom 60,000–64,000 became engaged. French effectives were about 33,000 men. Despite Russian attacks on both sides of the river, pontoon bridges were built across the ice-laden Berezina. By the night of the 27th only Victor's corps remained on the east bank, together with a horde of disorganized stragglers. Oudinot and Ney, on the west bank, were battling desperately against Tshitshagov to keep the bridgehead open. Victor, with 10,000 men, repulsed several attacks of 40,000 Russians with support from French artillery on the west bank. The French disengaged through the night, but thousands of wounded men and stragglers still remained on the far bank when the bridges were blown up at dawn; most of these were massacred by the Cossacks. French losses were over 30,000; Russian at least 10,000.

1812, December 8. Napoleon Leaves for Paris. For all practical purposes the Grand Army was no more; only 10,000 effectives remained. Napoleon rushed home to raise a new army. Murat, left in command, brought the pitiful remnants—as well as Macdonald's troops from Riga—back to Posen (Posnan) in western Poland. The exhausted Russians stopped their pursuit at the Niemen River. French (and allied) casualties in the campaign exceeded 300,000; Russian losses were at least 250,000.

1813, January 1. Revolt of Prussia. The Prussian contingent of **Hans D. H. Yorck von Wartenburg,** which had been with Macdonald at Riga, broke away to join the Russians. The remainder of Prussia rose, the army was hastily expanded with recalled reservists and *Landwehr* (militia), an expansion that was possible mainly due to the efforts of General **Gerhard von Scharnhorst** and his new Prussian General Staff. Leaving French garrisons at Danzig (Gdansk) and Thorn (Torun) on the Vistula, and at Stettin (Szczecin), Kustrin (Kostrzyn), and Frankfurt on the Oder, Eugène, who had replaced Murat in command, withdrew to Magdeburg on the Elbe. Late in January reinforcements sent by Napoleon brought his strength to 68,000 men. Meanwhile Schwarzenberg's Austrians had retreated to Warsaw and then turned into Bohemia, one more defection from Napoleon's army.

LEIPZIG CAMPAIGN, 1813

1813, February–March. New Coalition. Russia, Prussia, Sweden, and Britain united to end the Napoleonic grip on Europe. Almost 100,000 veteran allied troops were spread in the Elbe Valley, between Magdeburg and Dresden.

1813, April. Napoleon Returns to Germany. He had a new army 200,000 strong—gathered with amazing speed, but woefully inexperienced. He hastened from the Rhine to join remnants of the old Grand Army.

1813, April 30. Napoleon Crosses the Saale. He moved on Leipzig in 3 columns, preceded by a strong advance guard. Typically, he planned to penetrate the allied cordon and to defeat his enemies in detail. Faulty reconnaissance by his inexperienced cavalry left him unaware that Wittgenstein, with 93,000 allied troops, was concentrating on his south flank.

1813, May 2. Battle of Lützen. The French advance guard drove a small allied force into the outskirts of Leipzig. To the southwest, south of Lutzen, Wittgenstein attacked, surprising Ney's corps (the advance guard of the southern column) on the road. Napoleon, hearing the sound of artillery as he stood on the historic battlefield of Lützen (see p. 589), galloped to the scene. He was the Napoleon of old, brilliant and at his best in tactical improvisation on the field. He concentrated his army with a great mass of artillery opposite Wittgenstein's center. Leading an overwhelming counterattack in person, Napoleon split the allied lines. Had his green troops not been exhausted, there might have been a repetition of Austerlitz. Wittgenstein withdrew in fairly good order. Casualties were heavy. French losses were about 22,000, mostly from Ney's corps; the allies were about 12,000. Scharnhorst, serving as Wittgenstein's Chief of Staff, was among the wounded; he died of infection.

1813, May 20–21. Battle of Bautzen. Napoleon captured **Dresden** (May 7–8), then followed the retreating allies east of the Elbe. Sending Ney, with nearly half of his army, on a wide turning movement 50 miles north of Dresden, Napoleon pursued Wittgenstein with the remainder. The Russian stood in a formidable position on the east bank of the Spree. Napoleon attacked across the river with 115,000 men, driving the 97,000 allied troops from their defenses. Ney, coming down from the north with 52,000 men after dark, was in position to fall on the allied flank and rear next morning. The Napoleonic strategic concept was brilliant. Ney stupidly failed to grasp the situation, attacked late, and made no move toward the enemy rear and his communications. Napoleon, waiting to launch his reserve until Ney had sprung the trap, realized too late that Wittgenstein, in rapid retreat, had gotten safely away eastward into Silesia. There were about 25,000 French casualties; the allies lost about 11,000.

1813, May 22–June 1. Increasing Allied Strength. Napoleon, pursuing east of the Elbe, found his enemies growing stronger. Bernadotte, former marshal of France and now Crown Prince of Sweden, was nearing Berlin with a Prusso-Swedish army 120,000 strong. Austria was preparing to enter the war on the allied side. Schwarzenberg, with 240,000 Austrians, was mobilizing in northern Bohemia ominously close to the French line of communications. Blücher, who had taken Wittgenstein's place, was reorganizing that army, checking further French advance.

1813, June 4–August 16. Napoleon Obtains an Armistice. He used this respite to improve training, particularly his cavalry. The allies also used this time to prepare; Austria was able to complete its mobilization; the Prussian Army was strengthened by further levies, and equipment shortages were made good from British stocks; and the Russian Army was reinforced.

1813, August 12. Austria Declares War. The allies, with British subsidies, fielded 3 armies: the Bohemian, 230,000 men under Schwarzenberg; the Silesian, 195,000 under Blücher; and the Northern, 110,000 under Bernadotte. A token British contingent, including a rocket troop, accompanied Bernadotte.

1813, August 16. Opposing Plans. Napoleon had placed Davout's corps at Hamburg, transformed into a fortress, as a permanent threat to allied movements westward through Prussia. Dresden was occupied by **St. Cyr**'s corps as a potential pivot of maneuver. Between these two strong points, the emperor had withdrawn the greater part of his field forces (except **Jean Rapp**'s corps, besieged in Danzig) to positions between the Elbe and the Oder, preparing to operate on interior lines against his enemies. He had now in hand some 300,000 men; the allies totaled more than 450,000. Their strategy was to avoid battle with Napoleon, but to attack his lieutenants wherever possible. Bernadotte drove Oudinot (who had 60,000 troops) south of Berlin in the **Battle of**

Grossbeeren (August 23), and Blücher inflicted a severe defeat on MacDonald at the **Katzbach,** 30,000 of 60,000 French were lost, while Prussian losses were about 4,000 (August 26).

1813, August 26–27. Battle of Dresden. Schwarzenberg's Austrian army (accompanied by the Emperors of Russia and Austria, and the King of Prussia) attacked St. Cyr. Napoleon, arriving rapidly and unexpectedly with reinforcements, repulsed the assault. Next day, although outnumbered 2 to 1, Napoleon attacked, turned the allied left flank, and won a brilliant tactical victory. Then he went into one of his torpors, and left the field. By this time the allies had lost some 38,000 men killed, wounded, and captured and 40 guns. French casualties were about 10,000. Schwarzenberg disengaged hurriedly, narrowly escaping encirclement. Without Napoleon to direct the pursuit, only one of his subordinates, **Dominique René Vandamme,** realized the opportunity. Rapidly crossing the mountains into Bohemia on Schwarzenberg's east flank, he flung his corps across the Austrian line of communications.

1813, August 29–30. Battle of Kulm. No other French corps commander had followed Vandamme. With 37,000 men, he found himself unsupported and pinned between Austrian,

Russian, and Prussian forces totaling more than 100,000 men. Vandamme was captured, his corps lost 17,000 men, while the rest were dispersed.

1813, September. Blücher Avoids Battle. Having recovered, Napoleon hastened east again, vainly trying to entice Blücher into battle in Silesia. The Prussian, however, withdrew rapidly (September 4).

1813, September–October. French Disasters. Ney, who had succeeded Oudinot, attempted to take Berlin, but was defeated by Bernadotte at **Dennewitz** (September 6) when his Saxon divisions fled. Bavaria withdrew from the Rhine Confederation to join the alliance (**Treaty of Ried,** October 8).

1813, October 1–15. The Allies Close In. The French army, tired and discouraged, pressed in between Dresden and Leipzig, had been reduced to 200,000 men. Blücher, abandoning his line of communications, crossed the Elbe north of Leipzig to threaten Napoleon's rear. Schwarzenberg, in coordination, marched north to link with Blücher. Bernadotte, several miles north of Blücher, stood indecisive while French units opposite him joined Napoleon's main army at Leipzig. Napoleon, leaving 2 army corps at Dresden, turned to attack Blücher as Schwarzenberg advanced from the south (October 15).

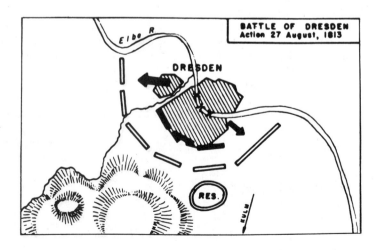

1813, October 16–19. "Battle of the Nations" (Battle of Leipzig). For 3 days the French fought Prussians on the northwest and both Russians and Austrians on the south. Then Bernadotte moved in on the east, nearly surrounding the French Army (October 18). The allies now began massive frontal attacks. Napoleon was driven into Leipzig, though his lines remained intact. The Saxon corps now deserted the French Army, ending all possibility of victory. Having been able to keep his line of communications open, despite Blücher's repeated efforts, Napoleon withdrew after a frenzied conflict outside the city (October 19). The bridge over the Elster River was blown prematurely. Prince **Joseph Anthony Poniatowski** (who had received his marshal's baton just the previous day for his actions on the first 2 days of the battle) and Marshal Macdonald rode their horses into the river to escape; Poniatowski was drowned; Macdonald got safely across. The allies had won a tremendous victory, but should never have allowed Napoleon to escape the net. As it was, the survivors of the Grand Army, having suffered overall losses of 60,000 men, 325 guns, and 500 wagons, withdrew toward the Rhine. Allied losses had been more than 80,000 men.

1813, October 30–31. Battle of Hanau. An Austro-Bavarian army, under Prince **Karl Philipp von Wrede,** 40,000 strong, marched to cut off the French retreat. With the allied armies slowly pursuing from the east, Napoleon was apparently trapped. In a flash of his old spirit, Napoleon, brilliantly maneuvering his artillery in support of the attack, routed the Bavarians, whose losses were 10,000 men and several guns; the French lost less than 6,000, although a further 4,000 stragglers were captured by the desultory allied pursuit. The Grand Army continued on to cross the Rhine (November 1–5).

DEFENSE OF FRANCE, 1814

1813, November 8. Allies Offer Peace. French boundaries would be restricted behind the Alps and the Rhine; Napoleon foolishly rejected this. The allies thereupon decided to invade France (December 1). By this time the Netherlands had revolted and the Rhine Confederation had dissolved (November). The French garrisons of Dresden and Danzig surrendered (November 11, December 30).

1813, December 21. Allies Cross the Rhine at Mannheim and Coblenz.

1814, January 1. French Dispositions and Plans. Some 50,000 men were in German garrisons, most under Davout, at Hamburg. Another 100,000 were in Spain fighting the English and Spaniards. In northeastern Italy, Eugène, with 50,000 men, was facing Austria's

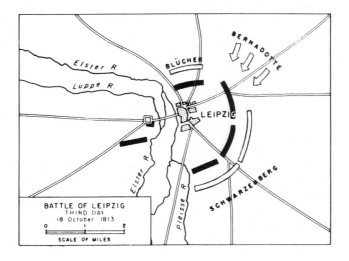

BATTLE OF LEIPZIG
THIRD DAY
18 October 1813
0   1   2
SCALE OF MILES

Count **Heinrich J. J. Bellegarde** with equal strength. None of these outlying forces, except Oudinot's corps, which would be withdrawn from Spain, would take part in the coming campaign in France. There Napoleon mustered nearly 118,000 men west of the Rhine, from Antwerp to Lyon. Napoleon, operating on interior lines, proved that his strategic brains were unimpaired, but the tactical quality of his troops was low. Except for a small proportion of veterans, the ranks had been filled by boy conscripts and by untrained national guards.

1814, January 1. Allied Dispositions and Plans. There were 3 allied invading armies: Bernadotte, with 60,000, was moving west through the Low Countries; Blücher, with 75,000, was advancing up the Moselle Valley into Lorraine; Schwarzenberg, with 210,000, was crossing Switzerland and moving through the Belfort Gap. Their combined objective was Paris.

1814, January 29–February 1. Operations around La Rothière. Moving with 36,000 men to crush Blücher before he could join Schwarzenberg, Napoleon won engagements at **Brienne** (January 29) and **La Rothière** (January 30) but barely escaped a vast allied envelopment as Blücher returned to attack La Rothière (February 1). He lost more than 5,000 men, the allies some 8,000. The allies pushed on toward Paris—Blücher down the Marne Valley, Schwarzenberg down the Seine.

1814, February 10–14. The "Five Days." Napoleon turned on Blücher, defeating him at **Champaubert** (February 10), **Montmirail** (February 11), **Château-Thierry** (February 12), and **Vauchamps** (February 14), a series of brilliant maneuvers directed by the emperor in person. The allies lost in all some 9,000 men in these operations, the French about 2,000. The Prussians retreated north of the Marne.

1814, February 18. Battle of Montereau. Marching rapidly south, Napoleon then attacked Schwarzenberg and drove him back 40 miles despite Austrian superiority of more than 2 to 1. French losses were 2,500, those of the allies 6,000. Blücher meanwhile, regrouping, had pushed on to La Forté (February 27), only

25 miles from Paris. Napoleon turned north to meet this threat, leaving Macdonald on the Aube to contain Schwarzenberg.

1814, March 7. Battle of Craonne. Blücher was driven north, and his rear guard—a Russian corps—was decisively defeated, although each side suffered over 5,000 casualties. Falling back to Laon, Blücher was reinforced by 2 of Bernadotte's army corps. He now had 100,000 to oppose Napoleon's 30,000. Meanwhile, Schwarzenberg was forcing Macdonald back toward Paris after a victory at **Barsur-Aube** (February 27).

1814, March 9–10. Battle of Laon. Despite the inferiority of his force, Napoleon fell on Blücher in a 2-day fight. Sturdily resisting in strong positions, Blücher finally made a night counterattack which drove 1 French corps from the field in panic. The French lost 6,000 men, the allies about 3,000. Napoleon's reckless assaults had failed; he withdrew to Soissons.

1814, March 13. Battle of Rheims. After a brief rest, Napoleon boldly marched across the front of Blücher's army to smash an isolated Prussian corps at Rheims. He recaptured the city, suffering 700 casualties; the Prussians lost 6,000 men. Napoleon now marched south to strike at Schwarzenberg's line of communications, hoping the Austrian would turn back.

1814, March 20–21. Battle of Arcis-surAube. Schwarzenberg repulsed the French strike and continued westward, as did Blücher. Napoleon ordered the corps of Marmont and Mortier to join him as he prepared for another attack.

1814, March 25. Battle of La Fére-Champenoise. Schwarzenberg's army defeated the greatly outnumbered corps of Marmont and Mortier, and drove them toward Paris. The allied advance continued; Blücher and Schwarzenberg united in front of Paris (March 28), leaving Napoleon's main army far to the east.

1814, March 30. Battle of Paris. Marmont and Mortier, with about 22,000 men, vigorously resisted the advance of the allied army—about 110,000 men. Despite commendable efforts, the French were forced back to Montmartre, bringing allied artillery within range of the city.

Marmont agreed to an armistice, surrendering Paris to the allies the next day (March 31). French losses had been 4,000 men, those of the allies 8,000. Napoleon, who had been hurrying to defend his capital, now halted at Fontainebleau.

1814, April 6–11. Abdication of Napoleon. At the urging of his marshals, he abdicated unconditionally (April 11). He was granted the principality of the little island of Elba, where he retired (May 4). The allies installed **Louis XVIII** (brother of Louis XVI) as King of France.

COMMENT. *Though he was crushingly defeated, Napoleon's military luster had never shone more brilliantly than in the 1814 campaign. Despite recurring bouts of illness and the poor quality of his troops, he postponed the inevitable in a series of maneuvers and battles which aroused the grudging admiration of his opponents.*

1814, May 30. First Treaty of Paris. France was reduced to the frontiers of 1792, and recognized the independence of other areas of Napoleon's empire. England restored most of the colonies seized from France and Holland.

THE PENINSULAR WAR, 1809–1814

The "Spanish Ulcer," as Napoleon termed the war in Spain, was one of the principal factors leading to his downfall. It diverted French military resources badly needed for operations elsewhere. Characterized by appallingly cruel and brutal excesses on both sides, it was a prime example of guerrilla warfare—the almost spontaneous response of a population to invasion and the military problems immediately encountered both by the invaders and the regular foreign forces assisting resistance. The war also emphasized the importance of sea power; for had not England commanded the seas, her troops could neither have landed on the Iberian peninsula nor been maintained to achieve the final victory.

1809, January–February. Guerrilla Warfare in Spain. After the French capture of Saragossa (see p. 824), their position in Spain was reestablished. Guerrilla warfare continued throughout the country, however.

1809, March–May. French Invasion of Portugal. Wellesley, in the Lisbon area, had some 26,000 British and Hanoverian troops, plus 16,000 Portuguese under General **William Beresford.** Soult took and sacked **Oporto** (March 29), but was driven out by Wellesley in the **Battle of Oporto** (May 12).

1809, June. Wellesley Invades Spain. In theory he was supported by some 100,000 Spanish irregulars. Actually, jealousies and political jugglings among the junta controlling patriot Spain prevented unified command; Wellesley found the Spaniards poor allies.

1809, July 28. Battle of Talavera. Attacked by the combined 47,000 troops of Victor and King Joseph Bonaparte, Wellesley with 54,000 allied troops fought a drawn battle. Strategically

it was a British victory, since the French later retired on Madrid. Allied losses were 6,500—5,300 of them British—the French lost 7,400. Irritable and jealous, Spanish General **Gregorio García de la Cuesta** now withdrew his contingent and Wellesley, his rear menaced by Soult and other French forces, fell back into Portugal. Soon after this he received the title of Viscount Wellington.

1809, November 19. Battle of Ocana. Spanish General **Areizago,** who replaced Cuesta, led an army of 53,000 Spanish troops against 30,000 French under Joseph and Soult. The Spanish were completely defeated, losing 5,000 killed and wounded and 20,000 prisoners; French losses were only 1,700.

1810, February–1812, August. Siege of Cádiz. The Spanish garrison was reinforced by 8,000 men sent by sea by Wellington. The city became the capital of free Spain.

1810. Wellington on the Defensive. Since England was again abandoned by all her allies, during the winter he began the construction of the Torres Vedras defensive lines, north of Lisbon. These constituted a massive 3-line system of field fortifications, mounting 600 guns and stretching for some 30 miles from the Tagus River to the sea. In Spain, Masséna had been placed in command of the Army of Portugal, 65,000 strong. Soult commanded the Army of Andalusia, of equal strength. As Masséna advanced toward the Portuguese frontier, Wellington took a force of 32,000 men (18,000 English, the remainder Portuguese) to confront him. Masséna besieged and captured Ciudad Rodrigo (July 10).

1810, July. French Invasion of Portugal. Wellington withdrew slowly. Meantime, in Spain, Soult solidified French control of Andalusia.

1810, September 27. Battle of Bussaco. To secure his retirement into the Torres Vedras lines, Wellington offered battle from a strong position on high ground. Masséna's attacks were repulsed, and Wellington continued his retreat. French losses were 4,500, those of the English 1,300.

1810, October 10. Wellington Occupies the Torres Vedras Lines. Masséna probed the defenses, but found them impregnable. Running out of food, he retired into Spain with his army in poor condition (November).

1810–1811. Frontier Warfare. During the bitter winter, both sides strove for possession of the main passes on the Spanish Portuguese frontier—through Ciudad Rodrigo and Badajoz. There were also several flurries of activity outside of besieged Cádiz as a result of British amphibious raids or guerrilla attacks.

1811. French Offensives. Masséna advanced to relieve Almeida, blockaded by Wellington. At the same time, Soult moved toward Badajoz, besieged by a British force under General Beresford.

1811, May 5. Battle of Fuentes de Oñoro. Masséna fought a drawn battle with Wellington, marked by the resistance of British infantry squares to French cavalry, and by the exploit of a British horse artillery battery. Captain **Norman Ramsey,** cut off in the melee,

limbered up and charged with his guns through the French horsemen. Allied losses were 1,800 out of 35,000 men engaged; the French lost 2,700 out of 45,000 on the field.

1811, May 16. Battle of Albuera. Soult, with 18,000 men, probing at Badajoz, was defeated by Beresford, with 32,000 men, including 7,000 British. The Spanish contingent on the allied right was driven in, but the British infantry retrieved the day. Allied losses were 7,000, French 8,000.

1811, May–December. Inconclusive Operations. Both French commanders fell back. Masséna was relieved in disgrace; Marmont replaced him. A series of indecisive but hard-fought actions followed, the net result being that at the end of the year the French still held the frontier gateways. Meanwhile, in southern Spain, Marshal **Louis Suchet** took **Tarragona** (July 28) and later **Valencia** (January 9, 1812) in a successful antiguerrilla campaign.

1812, January–July. Wellington Takes the Offensive. He stormed **Ciudad Rodrigo** (January 19) and **Badajoz** (April 19). He captured siege and pontoon trains of both Marmont and Soult in his rapid advance.

1812, July 15–21. Marching and Manuevering. Wellington, with about 47,000 men, met and was outmanuevered by Marmont, who had 42,000. A full week was spent in hard marching as the 2 each attempted to get an advantage. The faster-marching French finally forced Wellington to consider retreat. In preparation he ordered his baggage to the rear. Marmont, learning of this, believed that Wellington was already in full retreat and ordered a headlong pursuit.

1812, July 22. Battle of Salamanca. Wellington rapidly assessed the situation, realized that the French were overextended, turned and attacked the French advance guard in front and flank and rolled it back onto the French main body. Marmont was wounded, and his army driven off the field with losses of nearly 13,000 men. The allies lost about 5,200. Joseph evacuated Madrid.

1812, August 12. Wellington Takes Madrid.

He captured 1,700 men, 180 guns, and a quantity of stores.

1812, August–November. Allied Setbacks. After being repulsed at **Burgos** (November), Wellington was forced to fall back as Soult and Marmont concentrated against him. In a bitter retreat to cantonments near Ciudad Rodrigo, the allies lost 7,000 men. But by November, the French, unable to live off the country, were forced to disperse.

1813. Wellington Again Takes the Offensive. He was now in supreme command of allied forces in Spain, with 172,000 men to oppose 200,000 French. In a series of brilliant maneuvers Wellington put the French on the defensive on all fronts. At the same time, in eastern Spain, a British force under Sir **John Murray** forced Suchet back into Tarragona. Joseph abandoned Madrid again (May 17) and fell back to the line of the Ebro. North of that river and south of Vittorio, he occupied an extended position with 60,000 men and 150 guns.

1813, June 21. Battle of Vittorio. Wellington, with 90,000 men and 90 guns, attacked in 4 columns, all within mutual supporting distance. Joseph's center was pierced, both flanks turned. Wellington's left column neared the Bayonne road—the French communication link—and Joseph's army, after a determined defensive stand, broke and fled toward Pampeluna. This was the decisive battle of the Peninsular War. Joseph lost 7,000 men and 143 guns, while his treasury of $5 million and great masses of stores fell into Wellington's hands. Joseph retired across the Pyrenees into France, harried by allied pursuit. Soult took over the French command. Suchet abandoned Tar-

ragona and also slowly fell back into France (March, 1814).

1813, July 26–August 1. Battle of Sorauren. Soult, with 30,000 men, attempting to return to Spain, encountered Wellington with 16,000. After repulsing the initial French attack, Wellington resumed the offensive. Both armies were reinforced, but after 6 days of fighting the French had been driven back across the Pyrenees, having suffered losses of 13,000 men. Wellington had lost 7,000. Soon after this the British captured San Sebastian (August 31). Both sides rested and reorganized.

1813, October–November. Wellington's Invasion of France. Wellington crossed the Bidassoa River (October 7–9) and advanced into France. Soult, his army reduced to less than 50,000 by drafts for Napoleon's main army, could only fight delaying actions. At Nivelle (November 10), Soult attempted to overwhelm a detachment of Wellington's army. Hard fighting and Wellington's quick reaction saved the detachment. Losses were heavy: 5,300 of 45,000 British and allies in action, 4,500 of 18,000 French.

1814, February 27. Battle of Orthez. Wellington, in an amphibious operation, invested Bayonne and attacked Soult, who fell back to Toulouse to avoid encirclement, losing 4,000 men. Wellington's loss was 2,000. The allies advanced to seize Bordeaux (March 17). Wellington then turned toward Toulouse.

1814, April 10. Battle of Toulouse. Wellington assaulted the city, losing 6,700 casualties. Soult, who had lost 4,000 men, was driven out. Two days later, news of Napoleon's abdication was received, and a convention was agreed to between Wellington and Soult.

WAR AT SEA, 1806–1814

Despite Britain's overwhelming naval supremacy, there were a number of French challenges. At the same time the Royal Navy was busy in supporting the anti-Napoleonic effort on the Continent.

1806, January 8. British Capture of Cape-town. Dutch and French defenders were de-

feated in an amphibious assault by General Sir **David Baird.**

1806, February 6. Battle of Santo Domingo. Sir **John Duckworth**'s British squadron destroyed Admiral **Laissaque**'s French West Indian squadron. Soon after this, halfway around the world in the Indian Ocean, Admiral **C. A. L. Durand Linois** surrendered his 2 ships to a British squadron (March 14).

1806–1807. Operations at Buenos Aires and Montevideo. (See p. 889.)

1806, June–July. British Amphibious Raid on Calabria. A force of 5,000 troops under General Sir **John Stuart** was put ashore by Admiral **Sidney Smith** to help guerrillas opposing King Joseph Bonaparte. The British defeated a French force of 6,000 under General **E. Reynier** at **Maida** (July 4). The British soon withdrew to Sicily.

1806, October 8. Attack on Boulogne. British Admiral Sidney Smith attacked with rockets.

1807, February–March. British Repulsed at Constantinople. Duckworth's squadron forced a passage through the Dardanelles after a spirited action against the Turkish fleet, Sidney Smith, leading the rear division, taking a prominent part. Duckworth delivered a 24-hour ultimatum to the Porte to make peace with Russia and to dismiss the French ambassador (see p. 850). Sultan **Selim III** decided to resist. Within a day the entire population of Constantinople collected 1,000 guns along the seawall and opened fire on the British fleet. Having suffered considerable damage, Duckworth withdrew to the Mediterranean, receiving further mauling in repassing the Dardanelles.

1807, September 2–7. Second Battle of Copenhagen. Fearful that Denmark and her fleet would join the Franco-Russian alliance after the Treaty of Tilsit, Britain sent a powerful combined force to Zealand under Admiral **James Gambier** and General Lord **William S. Cathcart.** A landing force under Sir Arthur Wellesley invested Copenhagen. When the Danish government refused to negotiate, the British land and naval forces severely bombarded the Danish capital. Congreve rockets were used in great quantities. The Danish fleet surrendered.

1809, June 30. Capture of Martinique and Santo Domingo. French Admiral **Villaret de Joyeuse** surrendered the islands to a Hispano-British expedition after a stout defense. Spain controlled Martinique until 1814, Santo Domingo until 1821.

1809, July–October. Walcheren Expedition. In an effort to divert Napoleon's attention from Central Europe, a British expedition of 35 ships of the line, escorting 200 transports carrying 40,000 men, attempted to take Antwerp. The expedition, commanded by the **Earl of Chatham** (the younger **Pitt**) was a dismal failure, frittered away in the capture of Flushing (August 16), by which time King Louis Bonaparte and Marshal Bernadotte had heavily reinforced Antwerp. Chatham departed, leaving a garrison of 15,000 on Walcheren Island, 7,000 of whom died during an epidemic of malaria. The remnants were later returned to England.

1810–1814, Operations in the West Indies. The British captured **Guadeloupe** and held it until the end of the wars (1816) despite French efforts to recapture (1814).

1810, December. Capture of Mauritius and Réunion. A British naval squadron seized the two French islands, bases for French commerce destroyers operating against British merchant vessels in the Indian Ocean.

WATERLOO CAMPAIGN— "THE HUNDRED DAYS," MARCH– JUNE, 1815

1815, March 1. Napoleon's Return from Elba. He landed at Cannes and marched to Paris, where he resumed power (March 20). At the Congress of Vienna, the allies declared him to be an outlaw (March 13). They renewed their mutual pledges and began gathering forces to invade France.

1815, March–June. Napoleon Prepares for War. Displaying once more his amazing efficiency (and at the same time bleeding France white), he extemporized a field force of regulars some 188,000 strong, with another 100,000 in forts and depots. More than 300,000 additional hastily raised levies were in training around the country. Napoleon's own Army of the North, 124,000 men, concentrated around Paris. Other contingents guarded the rest of the frontier. In Italy, to Napoleon's intense annoyance, Murat, who was still King of Naples, declared his support of Napoleon and marched into central Italy. He was defeated by an Austrian army at **Tolentino** (May 2) and fled to France, where Napoleon refused to see him.

1815, June 1. Allied Concentrations on the French Border. In Belgium were Wellington's Anglo-Dutch army of 95,000 men and Blücher's Prussians, 124,000 strong (not including a corps of 26,000 in Luxembourg under General **Friedrich Emil Kleist**). Schwarzenberg, with 210,000 men, lay along the Rhine from Mannheim to Basle. In northeast Italy, Austrian **Johann M. P. von Frimont** had 75,000. In central Germany, a Russian army of 167,000 under Barclay moved slowly westward.

1815, June 1. Napoleon's Plans. Knowing that his enemies could not exercise their full power before mid-July, he decided to take the offensive and to defeat them in detail. He planned to crush the nearest menace first: the allied armies in Belgium. The plan was vigorous and bold, in full Napoleonic brilliance. His ablest subordinate, Davout, Napoleon mistakenly left in command of the defenses of Paris. Murat, in disgrace, was rebuffed when he requested a field command. Soult, a capable field commander, was miscast as chief of staff. So too were brave but flighty Ney and vacillating Marshal **Emmanuel de Grouchy,** selected to command the army's wings.

1815, June 11. Napoleon Leaves Paris. His army, secretly concentrated near Charleroi (June 14), poised to strike Wellington and Blücher before either dreamed that the French had left their cantonments near Paris.

1815, June 15. Seizure of Charleroi. Blücher reacted promptly; by evening 3 of his corps were assembled near Sombreffe, 10 miles northeast of Charleroi. Wellington, more cautious, and fearing for his line of communications, began concentrating 15 miles to the west. The key point was the little crossroads village of Quatre-Bras, linking the two allied armies.

1815, June 16. Battles of Ligny and Quatre-Bras. Napoleon had ordered Ney to seize Quatre-Bras with the left wing of the French Army. With the center and right wing (about 71,000 men) Napoleon fell on Blücher's 84,000 at Ligny. By afternoon of the 16th he had rocked the Prussians back and ordered Ney—who he thought had already seized Quatre-Bras—to strike Blücher's right flank and complete the victory.

At **Quatre-Bras,** Ney, procrastinating, had permitted his 24,000 men to be held off by a gallant brigade of Nassauers (mercenaries in Dutch service) until early afternoon, when British reinforcements began to arrive. By evening Wellington had concentrated 32,000 men; he counterattacked and threw Ney back. Meanwhile Marshal **Jean Baptiste d'Erlon**'s corps of 20,000 men, moving from reserve toward Ney, through a confusion of orders (for which Ney was responsible) spent the afternoon marching to and fro between the 2 French armies, and aiding neither. Ney lost about 4,400; Wellington about 5,400.

At **Ligny,** Napoleon penetrated Blücher's center, and the Prussians began to retreat. Had aid arrived from Ney, Blücher would have been destroyed, and Napoleon could then have fallen on Wellington's left flank and rolled him up. As it was, defeated Blücher (who had been wounded) retreated to Wavre, with the important assistance of his chief of staff, General **August Neidhart von Gneisenau.** In these two battles, allied losses were 21,400; the French lost 16,400.

1815, June 17–18. Delayed French Pursuit. In the morning Napoleon tardily sent Grouchy with the right wing of 33,000 men after the Prussian Army. With the remainder of his forces he turned toward Wellington, who then retired from Quatre-Bras toward Brussels. A torrential rainstorm and an energetic British rearguard hampered the French pursuit. By late evening of the 17th, Wellington had taken up a defensive position south of Waterloo, following assurances by Blücher that the Prussian Army

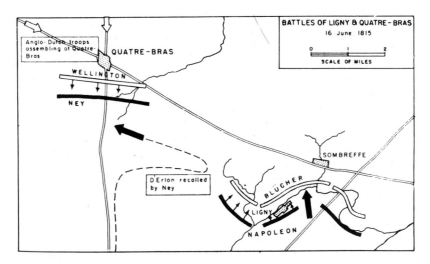

BATTLES OF LIGNY & QUATRE-BRAS
16 June 1815
SCALE OF MILES

would reinforce him. A French reconnaissance revealed the British positions. Napoleon prepared to attack on the morning of the 18th. In the morning Napoleon delayed for several hours to permit the ground to dry. The delay was fatal, for Grouchy had as yet failed to make contact with the Prussians at Wavre. Blücher, unknown to either French commander, had ably rallied his command from the defeat at Ligny. Overriding the objections of Gneisenau, who feared that the British would not stand, Blücher moved west toward Wellington, less than 9 miles away. A single Prussian corps was left behind to delay Grouchy.

1815, June 18. Battle of Waterloo. Napoleon with 72,000 men attacked Wellington's 67,000 at noon. By 4 P.M., despite obstinate resistance, the Anglo-Dutch army had been pushed back all along the line. The French cavalry, led by Ney in person, charged into the center of the British positions. Had the Imperial Guard infantry supported this cavalry attack, Wellington's center would have been pierced. But Napoleon, worried by the ominous appearance of Prussian elements on his right flank, held the Guards back, and the unsupported horsemen could not break the squares of the magnificent British infantry. Although the Prussian advance guard had appeared as early as 1 P.M., Gneisenau, still worried that the British would

disengage, had deliberately delayed the Prussian march. However, by about 4 P.M., Blücher's strength began to be felt. General Georges Lobau's corps, hurriedly shifted to oppose the Prussians, was driven in. Napoleon made one last effort to break Wellington's line. The Imperial Guard—without cavalry support—struck the British center, made some progress, but was then checked by the "thin red line." Blücher's full strength (53,000) was now engaged, overlapping the French right. As the Guard fell back, Wellington counterattacked. The French army collapsed into a mob of refugees, harassed through the night in their flight by the Prussians. Losses were great on both sides: Wellington, 15,000 men; Blücher, 7,000. The French lost 32,000 of whom 7,000 were captured.

1815, June 18. Battle of Wavre. Grouchy, meanwhile, had caught up with Blücher's rear guard (about 20,000). He was content to fight an unimaginative and partially successful battle while the French Empire was tumbling into ruins 8 miles to his west. However, Grouchy's withdrawal from Belgium was skillful. His force of 33,000 men remainded intact. Napoleon, at first refusing to accept reality, planned to use this force as the core of a new army with which to prosecute the war.

1815, June 19–21. Return to Paris. Seeking to

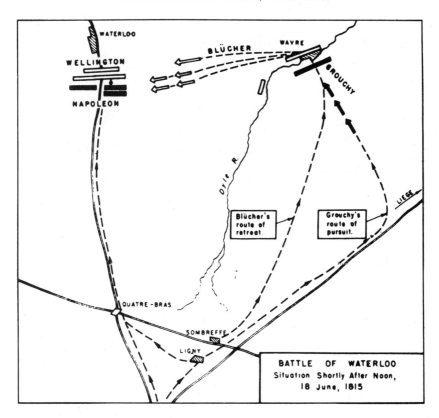

BATTLE OF WATERLOO
Situation Shortly After Noon,
18 June, 1815

raise a new army as either emperor or simply general, Napoleon was met by an ultimatum from the Chamber of Deputies: abdicate or be deposed. Meanwhile, Davout, with an army composed of Grouchy's force and contingents of raw levies, maneuvered skillfully against the slowly advancing allies, attempting to gain time for the Emperor.

1815, June 22. Napoleon Abdicates. The next day the Deputies elected a Council of Regency for Napoleon's son, the infant King of Italy. The regency lasted until the return of Louis XVIII to Paris (July 8). Davout, after savaging the allied advance guard, received news of the abdication and resigned in disgust. As the allies ap-

proached Paris, Napoleon escaped to Rochefort (July 2). He then surrendered himself to the British. HMS *Bellerophon* received him on board (July 15) and he started for his exile on the island of St. Helena, where he would die (May 5, 1821).

1815, November 20. Second Peace of Paris. France lost some more territory, her boundary generally that of 1790. She was forced to pay an indemnity.

1815, November 20. Quadruple Alliance. The victors—England, Austria, Prussia, and Russia—agreed to assure the execution of the Peace of Paris. In effect, they established themselves as the arbiters of Europe.

WESTERN EUROPE

BRITISH ISLES

Except for the Napoleonic Wars and the War of 1812, whose respective fringes touched the British Isles and their territorial waters, all of the military and naval operations of the British Empire took place away from the British Isles. These are discussed under their appropriate territorial subheads.

FRANCE

1800–1815. Napoleonic Wars. (See p. 812.)

1814–1824. Reign of Louis XVIII.

1823. Invasion of Spain. This was in behalf of **Ferdinand VII** (see below).

1824–1830. Reign of Charles X.

1830, July 28. Revolution in Paris. The monarchy of the Restoration toppled, and **Louis Philippe,** cousin of the deposed king, headed a new constitutional monarchy. The new regime used troops to suppress insurrections in Lyon (1831), and again in both Paris and Lyon (1834).

1830–1848. Conquest of North Africa. (See p. 855.)

1848, February 22–24. Revolution in Paris. A series of uprisings culminated in the abdication of Louis Philippe and establishment of the Second Republic, which tottered in a ferment of social and political unrest.

1848, June 23–26. Insurrection of June. A bloody uprising of well-organized Paris workers was suppressed by General **Louis Eugène Cavaignac,** temporary dictator of a provisional government. He executed rebel leaders without mercy, then resigned his dictatorship.

1848, December 20. Prince Louis Napoleon Bonaparte Elected President. He was the nephew of Napoleon I.

1849, April 24. Intervention in Italy. (See p. 846.)

IBERIAN PENINSULA

Spain

1807–1814. Peninsular War. (See pp. 823, 835.)

1820, January. Mutiny at Cádiz. The revolt, led by Colonel **Rafael del Riego y Núñez,** spread; King Ferdinand VII was taken prisoner.

1822, October. Congress of Verona. The Quadruple Alliance authorized France to intervene to restore peace and the monarchy in Spain.

1823, April 17. French Invasion. An army under Duke **Louis d'Angoulême** crossed the Bidassoa River on the western flank of the Pyrenees, seized Madrid, then marched to Cádiz, where the rebel government had fled.

1823, August 31. Battle of the Trocadero. Riego's rebel forces, defending 2 forts 8 miles from Cádiz, were routed by the French. The revolution was ended; the king, freed, took drastic measures of reprisal.

1834–1839. The Carlist War. Ferdinand, before his death (September 29, 1833), designated his infant daughter **Isabella** as his successor under the regency of his widow, **Maria Christina,** depriving his brother Don **Carlos** of his Salic law right to the throne. Carlos led a revolution. France, England, and Portugal entered an alliance with the Spanish government to assist in suppressing the revolt (April 22). From England went the so-called Spanish Legion, made up of 9,600 mercenaries recruited under special authority of Parliament and led by Sir **George de Lacy Evans.** France rented her entire Foreign Legion to Spain. Five years of ferocious guerrilla warfare followed; the French Foreign Legion was the backbone of the otherwise ill-handled Spanish loyalist forces. The Legion particularly distinguished itself in the decisive **Battles of Terapegui** (near Pampeluna; April 26, 1836) and **Huesca** (March 24, 1837). When the Legion left Spain after suppression of the revolt (January 17, 1839), it had lost 50 percent of its initial strength. The war was concluded by the **Convention of Vergara** (August 31, 1839). Don Carlos took refuge in France.

1840–1843. Civil War. General **Baldomero Espartero** led a revolt, seizing dictatorial power and driving Maria Christina and Isabella from the country. He suppressed uprisings (October 1841; November–December 1842). A counter-uprising under General **Ramón Narváez** was successful (July–August 1843). Espartero fled, and Isabella was restored to the throne under the guidance of Narváez (1843–1851).

Portugal

1820, August 29. Oporto Revolution. This was sparked by the Spanish revolution (see above). The insurgents expelled the regency established under England's aid during the absence of King **John VI** in Brazil (see p. 894).

1821, July 4. Return of King John. He accepted the insurgents' invitation to return as constitutional monarch.

1823–1824. Civil War. John's second son, **Miguel,** failed in an attempt to restore an absolute monarch.

1826, March 10. Death of John. His son Dom **Pedro** (now Emperor of Brazil; see p. 894) nominally succeeded him as **Peter IV,** but refused to leave Brazil, handing the Portuguese throne to his infant daughter, **Maria de Gloria,** with Miguel as regent. Civil war flared between the followers of Miguel, who seized Lisbon, and the constitutional government, supported by General **João Carlos de Saldanha** at Oporto.

1827, January. British Intervention. An expeditionary force of 5,000 men under Sir **William Clinton** landed in support of Saldanha's constitutional army. Miguel bowed to force, and the British withdrew (April 28, 1828).

1828, July 7. Miguel Seizes the Throne. Queen Maria was taken to Brazil.

1828–1834. Miguelite Wars. Adherents of the queen organized revolt in the Azores, reinforced by volunteers from Brazil, England, and France. The Miguelite fleet was defeated in **Praia Bay** (August 12, 1828) in an attempt to take the Azores. Maria's supporters, obtaining a British loan, purchased a small squadron in England. Dom Pedro, abdicating in Brazil, joined his daughter's forces (April 1831). France, following persecution of French subjects in Portugal, seized the Miguelite fleet in the Tagus (July 1831).

1832, July. Seizure of Oporto. Dom Pedro's "Liberation" expedition from England, 6,500 strong, occupied Oporto, but was besieged by Miguel with 80,000 troops. Britain's Admiral Sir **Charles Napier,** commanding the "Liberation" squadron, defeated a Miguelite flotilla off **Cape St. Vincent** (July 5, 1833), then captured Lisbon (July 24) and held it against the Miguelites. The Quadruple Alliance (see p. 841) solidified anti-Miguel resistance.

1834, May 16. Battle of Santarém. Saldanha's and Dom Pedro's allied liberation forces defeated Dom Miguel, who surrendered 10 days later. His banishment concluded the wars.

THE LOW COUNTRIES

1813, November 30. House of Orange Returns. Prince William of Orange, son of the late stadtholder William V, returned to the Netherlands. The Dutch crowned him as their sovereign, and French troops withdrew to the south.

1815, August 24. Unified Kingdom of the Netherlands. Established by the Quadruple Alliance with **William I** (Prince of Orange) as king. The Netherlands and Belgium were joined.

1830, August 25. Belgian Revolution. A revolt against William I started in Brussels with fighting between civilians and Dutch troops.

1830, October 4. Belgian Declaration of Independence.

1830, October 27. Bombardment of Antwerp. The city was held by the Belgian populace, but the citadel was occupied by a Dutch force under General **David Hendryk Chassé.** His bombardment of the city strengthened anti-Dutch feeling among the Belgians.

1830, November 4. Armistice. Ordered by the great-power **Conference of London.**

1830, December 20. Independence of Belgium. The great powers at London declared the dissolution of the Kingdom of the Netherlands.

1831, June 4. Leopold I (of Saxe-Coburg) Elected King of the Belgians.

1831, August 2. Dutch Invasion of Belgium. William led an army of 50,000 troops and was at first successful against extemporized Belgian forces. A French army of 60,000, under Marshal **Étienne Maurice Gérard,** forced the Dutch to retire. Chassé, still holding the citadel of Antwerp, was besieged by the French.

1832, November–December. The French army, assisted by a Franco-British naval squadron, forced Chassé to surrender. An armistice followed (May 21, 1833).

1839, April 19. Dutch Recognition of Belgian Independence.

SCANDINAVIA

1800–1815. Napoleonic Wars. (See p. 812.) These resulted in several changes of national status in Scandinavia.

1808. War in Finland. After their alliance at the Treaty of Tilsit, Napoleon and Alexander of Russia called on Sweden to forsake the coalition with England. When Sweden refused to declare war on England, a Russian army under **Count F. W. Buxhöwden** invaded Finland (February). After inconclusive warfare, the Swedes evacuated Finland (December). By the **Treaty of Frederikshavn** (September, 1809) Sweden ceded Finland and the Aland Islands to Russia.

1815. Act of Union. Norway was given the status of an independent kingdom, united with Sweden under one ruler.

1848, March–August. Schleswig-Holstein Revolt. This was inspired by Prussia. Prussian troops under General **Friedrich Heinrich von Wrangel** marched in to occupy the provinces. Sweden sent troops to assist Denmark to reestablish its control: England threatened

naval action against Prussia. Austria and Russia also gave diplomatic support to Denmark. The **Convention of Malmo** (August 26, 1848) brought about quasi peace. Desultory operations continued until the **Treaty of Berlin** (1850) restored full Danish rights in the disputed provinces.

AUSTRIA-HUNGARY

1800–1815. Napoleonic Wars. (See p. 812.)

1806, August 6. Dissolution of the Holy Roman Empire. The Austrian Empire took its place (see p. 000). The new empire, a mélange of German, Slav, Magyar, Bohemian, and Latin elements, was dominated by its foreign minister (virtually prime minister) Prince **Klemens W. N. L. von Metternich.** Internal discord boiled in the Italian, Bohemian, and Hungarian provinces, and to some extent in Austria itself.

1810–1850. Austrian Involvement in Italy and the Balkans. (See pp. 846, 848.)

1848, March 13. Revolt in Vienna. Metternich resigned. Emperor **Ferdinand I** promised constitutional reforms and relaxation of suppressive efforts throughout the empire.

1848, March 18–22. Revolt in Milan. (See p. 846.)

1848–1849. Austro-Sardinian War. (See p. 846.)

1848, April 10. Hungarian Independence. Under the leadership of **Lajos Kossuth,** Hungary became virtually independent.

1848, June 17. Bombardment of Prague. A Czech uprising was quelled by an Austrian force under Marshal **Alfred zu Windischgrätz.** This was followed by martial control over all Bohemia.

1848, September 17. Austrian Invasion of Hungary. Croatian forces under Count **Josef Jellachich** invaded Hungary to suppress the independence movement.

1848, October 3. Hungarian Invasion of Austria. After repulsing Jellachich, the Hungarians advanced toward Vienna.

1848, October 6. Revolt in Vienna. In sympathy with the Hungarians, the populace of the

Austrian capital again rose in revolt, but this was promptly suppressed by Windischgrätz (October 31), who then repelled the Hungarians and pursued them. Soon after this Ferdinand abdicated and was succeeded by his nephew, **Franz Josef** (December 2).

1849, January 5. Occupation of Budapest. Windischgrätz captured the Hungarian capital, and drove Hungarian General **Arthur von Görgei** into the mountains north of Budapest. Görgei having broken with Kossuth, Polish General **Henry Dembiński,** was placed in command of Hungarian forces.

1849, February 26–27. Battle of Kápolna. Dembiński was defeated by Windischgrätz.

1849, April 13. Hungarian Republic Established. Kossuth was elected president.

1849, April–May. Hungarian Victories. Görgei, recalled to command, defeated Windischgrätz in a series of smart actions, forcing the Austrians to evacuate Hungary.

1849, June 17. Russian Intervention. Russian troops, led by General **Ivan Paskievich,** assisting Austria, invaded Hungary from the north. At the same time Austrian General **Julius von Haynau,** replacing Windischgrätz, invaded from the west.

1849, June–July. Hungarian Defeats. Görgei was driven southeast, toward Transylvanìa, where General **Jozef Bem,** a Pole in Hungarian patriot service, had been operating successfully.

1849, July 31. Battle of Segesvar. Bem was badly defeated by Haynau's and Paskievich's converging armies. With the remnants of his troops, Bem joined Görgei for a last stand.

1849, August 9. Battle of Temesvár (Timisoara). Haynau overwhelmed the Hungarians. Görgei managed to withdraw in good order, but recognizing inevitable defeat, surrendered to the Russians (August 11) with 22,000 men. Haynau then ferociously stamped out the last embers of Hungarian rebellion.

GERMANY

Following the dissolution of the Holy Roman Empire, the German states began, with much internal friction, a coalescence around Prussia as a nucleus. The sole external conflict was Prussia's quasi war with Denmark over Schleswig and Holstein (see p. 844).

1805–1807, 1812–1815. Napoleonic Wars. (See p. 812.)

1807–1814. Reorganization of the Prussian Army. Driving spirit in extensive reforms was General Gerhard von Scharnhorst. Principal among his reformer subordinates were August Neidhart von Gneisenau, **Hermann von Boyen, Karl von Grolman,** and Karl von Clausewitz.

1809, March 1. Establishment of Prussian General Staff. Scharnhorst, first Chief of the General Staff, was virtually Minister of Defense in a new War Department.

1821, January 11. General Staff Reorganized. Under General **Karl von Müffling** it took the general form it retained for 120 years.

1847–1848 Unrest in Germany. It was part of the revolutionary ferment in Europe.

1848, March 15–31. March Days in Berlin. A popular uprising, with bloodshed, forced concessions from King **Frederick William IV,** but was suppressed when General von Wrangel occupied Berlin (November 10).

1848, April–August. Intervention in Schleswig-Holstein. (See p. 844.)

1849, March 27. Frankfurt Constitution. A federal German state was to be established under a hereditary "Emperor of the Germans." However, Frederick William IV of Prussia rejected the crown.

1850, November 29. Treaty of Olmütz. Prussia abandoned a plan for a new German Confederation excluding Austria; Austrian primacy in Germany was reaffirmed.

ITALY

Revolutionary Ferment, 1814–1847

Following the collapse of the Napoleonic structure, the Italian states—dominated by overt Austrian influence—were reconstituted under their former near-medieval autocratic status. This recrudescence of autocracy was opposed by the popular urge for independence and constitutional government fermenting throughout Europe, and particularly stimulated in the Italian peninsula by Napoleon's enlightened reorganization and government.

1820, July 2. Neapolitan Revolution. General **Guglielmo Pepe** led an army revolt, which was crushed at **Rieti** by an Austrian army coming to the aid of King **Ferdinand IV** (March 7, 1821).

1821, March 10. Sardinian Revolution. This was crushed at **Novara** (April 8) by a combined Austro-Sardinian royalist force.

1831, February–March. Revolts in Modena, Parma, and the Papal States. These were inspired largely by **Giuseppe Mazzini,** and were put down with Austrian military assistance.

1832–1834. Renewed Revolts. Austrian troops occupied Romagna (January 1832). French troops took Ancona (March). In another Mazzini uprising in Piedmont and Savoy, a young Sardinian sailor, **Giuseppe Garibaldi,** fled when the uprising failed (1834). Matters in Italy went from bad to worse, as abortive revolts flared for several years.

1847, July 17. Austrian Occupation of Ferrara. This united the Italians in growing hatred of Austrian overlordship.

Italian War of Independence, 1848–1849

1848, March 18–22. Milan Revolt. The "Five Days" revolt ended when Austrian Marshal **Josef Radetzky,** 82 years old, withdrew from the city. He concentrated his army of occupation into the fortress "Quadrilateral": Mantua, Verona, Peschiera, and Legnago. He went on the defensive as a coalition of Italian forces gathered in north Italy.

1848, March 22. Sardinian Declaration of War on Austria. King **Charles Albert** of Sardinia took command of the allied Italian forces. Radetzky, with some 70,000 men, fought a brilliant offensive-defensive campaign on interior lines against enemies double his strength, but lacking his skill.

1848–1849. Spread of Italian Independence Movement. Inspired by the Milan Revolt and the Sardinian declaration of war, Italian patriots in Venice, under **Daniele Manin,** declared an independent republic (March 26, 1848). The unrest spread to the Papal States, where Giuseppe Mazzini eventually ousted the Pope and declared the independence of a Roman Republic (February 9, 1849).

1848, July 24–25. Battle of Custozza. Massing superior forces against a portion of the Sardinian army, Radetzky overwhelmed it and drove Charles Albert out of Lombardy. He reoccupied Milan and besieged Venice. Garibaldi, who had formed a volunteer army fighting in the Alps, fled to Switzerland. Following a brief armistice, hostilities were resumed.

1849, March 23. Battle of Novara. Duplicating Napoleon's maneuver at Marengo, Radetzky with 70,000 men completely defeated Charles Albert and his Polish chief of staff, General **Albert Chrzanowski,** who could put only part of their 100,000 men on the field. Charles Albert abdicated in favor of his son, **Victor Emmanuel II.**

1849, April 24. French Intervention. An expeditionary force, 8,000 strong, under General **Nicolas Oudinot** (son of Napoleon's marshal), landing at Civitavecchia, moved on Rome.

1849, April–June. Siege of Rome. The 20,000

garrison included 5,000 of Garibaldi's red-shirted "Legion." After an initial repulse (April 29), Oudinot persisted, and forced a capitulation (June 29). Garibaldi and his legion retreated toward Venice, hoping to join its defenders, but hot pursuit by French, Austrian, and Italian loyalist forces scattered it. Garibaldi himself fled to America.

1849, May–August. Siege of Venice. Radet-zky's army had seized all of Venice's mainland territory after Novara. After terrible suffering from bombardment, starvation, and disease, Manin capitulated (August 24).

1849, August 9. Sardinia Makes Peace. The revolution was quelled; Italy writhed under the cruelty of its conquerors—particularly the Austrian Baron **Julius von Haynau.**

EASTERN EUROPE

RUSSIA

1800–1815. Napoleonic Wars. (See p. 812.)
1804–1813. War with Persia. (See p. 852.)
1806–1812. War with Turkey. (See p. 850.)
1808. Conquest of Finland. (See p. 844.)
1825–1828. War with Persia. (See p. 853.)

War with Turkey, 1828–1829

1828, April 26. Declaration of War. Russia came to the support of the Greeks in their war for independence (see p. 849). This also was an opportunity for further Russian expansion at Turkey's expense.

1828, May–December. Operations in the Balkans. A 3-pronged offensive under the general direction of Czar **Nicholas I** crossed the Danube, to be checked temporarily by the Turkish fortresses. Silistra was invested. Varna (Stalin) was captured after a siege of 3 months (October 12).

1828, May–December. Operations in the Caucasus. An army under General **Ivan Fedorovich Paskievich** advanced from Tiflis, captured Kars, then moved north to defeat the Turks at **Akhalzic** (Akhaltzikke, August 27), while the Black Sea fleet captured Poti. Fierce Kurdish resistance along the upper Euphrates then checked further Russian advance.

1829. Advance on Adrianople. General **Hans K. F. A. von Diebitsch-Zabalkansky** took over command of the Balkan offensive. He moved south against Turkish Grand Vizier **Mustafa Reshid Pasha,** leaving a force to continue the siege of Silistra (which fell June 30), and defeated Reshid at **Kulevcha** (June 11). Diebitsch then boldly penetrated the passes of the Balkan (Haemus) Mountains, completely outmaneuvering the Turks by the rapidity of his advance. He took Adrianople (August 20). With Constantinople threatened, Turkey sued for peace.

1829, September 16. Treaty of Adrianople. Russian gained the mouth of the Danube River and the eastern coast of the Black Sea.

1830–1832. Polish Insurrection. (See p. 848.)
1833. Russian Intervention in Turkey. (See p. 851.)
1839–1847. Khivan Conquest. To extend Russian boundaries southward into Turkestan and to open up trade, an army under General **Basil A. Perovsky** moved from Orenburg into the Khanate of Khiva, in what are now the Kazakh, Uzbek, and Turkoman S.S.R.s. The expedition ended in disaster (1839). The Russians returned to more gradual efforts. A Russian fort was established on the northeastern edge of the Aral Sea, at the Jaxartes (Syr Darya) River mouth (1847). Other Russian spearheads slowly pushed southeastward toward Tashkent.

1849. Intervention in Hungary. (See p. 845.)

POLAND

1815, June 9. New Partition of Poland. The Congress of Vienna divided the Grand Duchy of

Warsaw between Prussia, Russia, and Austria, once more dissolving Poland. A tiny fragment—the Cracow (Krakow) area—for a time remained a republic. Russian Poland became the so-called Congress Kingdom, with the czar as its king. There was recurrent friction between the Poles and Russians.

1830, November 29. Polish Insurrection. This began in Warsaw. Polish patriots, among them numbers of veterans of the Napoleonic armies, gathered forces totaling 80,000 men and 158 guns. Russian generals Diebitsch and Paskievich, whose troops totaled 114,000, brought fire and sword throughout the region, while Polish factions fought with one another for supremacy.

1831, February 20. Battle of Grochow. Diebitsch was halted on the outskirts of Warsaw in a sanguinary battle by a Polish army under Prince **Michael Radziwill.** Diebitsch withdrew eastward.

1831, May 26. Battle of Ostrolenka (Ostroleka). Polish General **Jan Zigmunt Skrzneki** fought a long, bloody, but indecisive battle with Diebitsch on the Narew River. The Polish artillery was well handled by General Bem, but the Poles were forced to retreat. Skrzneki gradually fell back on Warsaw, where he was relieved by General Dembiński.

1831, September 6–8. Battle of Warsaw. The capture of the gallantly defended Polish capital by Paskievich's troops was accompanied by great bloodshed. The insurrection was stamped out violently.

1846, February. Cracow Insurrection. A local armed movement against Austria—intended as another general Polish uprising—was speedily suppressed by Austrian troops; the state of Cracow was absorbed by Austria.

THE BALKANS

During this half-century the Balkan Peninsula was in a constant turmoil of war and internal disorder. The Ottoman Empire's decay gave hope of freedom to the oppressed peoples of the peninsula, who as a consequence were ceaselessly engaged in plotting or waging revolutions. And even while fighting or intriguing against the Turks, they were fighting and intriguing among themselves.

Encouraging this internal unrest, and attempting to profit from it both politically and territorially, were the neighboring empires of Russia and Austria, both hoping to inherit all of the Balkan Peninsula and the strategic straits of the Dardanelles and the Bosporus when, as they expected, the Turkish Empire reached an early collapse. The aspirations of both the Hapsburg and Romanoff empires were viewed with alarm and suspicion by both Britain and France. During the Napoleonic Wars, Russia and Austria were somewhat reluctant allies in the Balkans, while Napoleon directly or indirectly supported or encouraged the Porte. For the same reason, Britain—also reluctantly—was frequently allied with Russia and Austria against Turkey, but nonetheless endeavored to direct the main efforts of the alliance against Napoleon. After Napoleon's defeat, Britain continued this policy of maintaining the balance of power in the Balkans, even though, on a few occasions, other policy considerations (such as supporting the independence of Greece) required her to participate in limited military operations against Turkey. The principal events were:

1799. Independence of Montenegro. Achieved by King **Peter I** through his alliance with Russia against Turkey (see p. 763) and support of both Russians and British against French efforts in Dalmatia.

1804–1813. Serbian Insurrection. The Serbs

were led by **George Petrovich** (known as **Kara George,** and founder of the Karageorgevich family). Under his leadership Turkish troops were driven from Belgrade and Serb independence proclaimed (December, 1806). Apparently successful, the independent state collapsed when Russia made peace with Turkey (1812) and Turkish armies reoccupied the country; Kara George fled to the mountains.

1812. Bessarabia to Russia. This was a provision of the **Treaty of Bucharest** (see p. 851). The remainder of Moldavia remained under Turkish suzerainty.

1815–1817. Serbian Insurrection. The principal leader was **Milosh Obrenovich,** an opponent of Kara George. The murder of Kara George by Obrenovich adherents resulted in a blood feud between the families, which lasted more than a century. The Turks suppressed the revolt, but recognized Obrenovich as hereditary prince of Serbia when he agreed to Turkish suzerainty.

Greek War of Independence, 1821–1832

1821, October 5. Massacre of Tripolitsa. The beginning of revolt in Morea resulted in the massacre of the 10,000-man Turkish garrison. Immediate, savage Turkish reprisals followed, and all Greece flamed.

1822, January 13. Independence Proclaimed at Epidauros.

1822, April–June. Occupation of Chios (Scio). A Turkish squadron under **Kara Ali** took the island, massacring or enslaving its entire population.

1822, June 18–19. Battle of Chios. Greek sailor-patriot **Constantine Kanaris,** with 2 fire ships, sailed into the midst of the Turkish squadron and blew up the Turkish flagship with all on board.

1822, July–1823, January. First Siege of Missolonghi. A Turkish army, invading the peninsula north of the Gulf of Corinth, was stopped before the Greek fort. **Mustai Pasha** hurried with reinforcements for the besiegers.

1822, August 21. Battle of Karpenizi. With 300 men, **Marco Bozzaris** surprised Mustai's advance guard in a night attack. Bozzaris was killed, but the 4,000 Turks were routed, losing most of their leaders.

1823, January. Turkish Withdrawal. Mustai gave up the siege of Missolonghi and retired.

1823–1825. Greek Dissension. Instead of pressing their advantage, the Greeks began quarreling among themselves.

1825, February 24. Egyptian Involvement. Answering Sultan **Mahmud II**'s call for help, **Mohammed Ali** (see p. 852) sent a fleet and an army. **Ibrahim,** son of Mohammed, landed in the Morea, with some 5,000 well-disciplined troops, and quickly overran the peninsula. Meanwhile, a new Turkish army, under Reshid Pasha, penetrated from the north and again invested Missolonghi.

1825, May–1826, April 23. Second Siege of Missolonghi. The starving garrison was destroyed in a final desperate sortie.

1825, May–June. Siege of the Acropolis. Reshid moved on Athens and besieged the Acropolis. European interest by now focused on assisting Greek independence; Lord **Byron** had already arrived (1822). He died soon after (1824) but was followed by a number of volunteer adventurers and idealists, the most prominent of whom were Admiral Lord **Cochrane** and General Sir **George Church.** Cochrane was placed in command of the Greek Navy, Church of the Army, but both were paralyzed by Greek internecine intrigue.

1825, June 5. Capitulation of the Acropolis. Greek inefficiency negated the plans of both Cochrane and Church, who fled to the safety of warships in the bay. Reshid accorded the honors of war to the garrison. Continental Greece was again under Turkish control.

1827, July 6. Treaty of London. Forced by pro-Hellenic public opinion, the governments of England, France, and Russia demanded that the Egyptians withdraw and that the Turks agree to an armistice. Both refused. A large Egyptian squadron landed reinforcements at Navarino (September 8), where a Turkish squadron also lay. The 3 allied governments sent naval forces to Greece; they rendezvoused off the harbor of Navarino.

1827, October 20. Battle of Navarino. Learning that Egyptian Ibrahim was carrying out depredations ashore, British Admiral Sir **Edward Codrington,** the senior allied commander, sailed into the harbor where the Egyptian-Turkish fleet was at anchor. Following Codrington's 3 ships of the line and 4 frigates were French Admiral **Henri G. de Rigny**'s 4 ships of the line and 1 frigate. In a second line came Russian Admiral Count **Heiden**'s 4 ships of the line and 4 frigates. **Tahir Pasha**'s 3 ships of the line, 15 frigates, and some 50-odd smaller craft lay in a long horseshoe formation; the flanks were protected by shore batteries. The allied fleet dropped anchor in their midst. A Turkish ship fired on a British dispatch boat, killing an officer and several men. Codrington at once opened fire and his allies joined in. The Turko-Egyptian fleet, heavily outgunned, was destroyed, three-fourths of its vessels either sunk or fired by their own crews. Allied losses were 696 killed and wounded. The Turko-Egyptian loss, never officially reported, must have been great.

COMMENT. *Navarino, which really decided Greek independence, was a gun duel, pure and simple, between floating batteries.*

1828, April 26. Russia Declares War. (See p. 847.)

1828, August 9. Egyptian Evacuation of Greece. This was supervised by a French expeditionary force. The evacuation was completed early next year (1829). This virtually ended the war.

1832, May 7. Treaty of London. This created the independent Kingdom of Greece.

1829. Autonomy of Serbia, Wallachia, and Moldavia. This was a result of the Treaty of Adrianople (see p. 847).

1829–1834. Russian Occupation of Wallachia and Moldavia.

1843, September 14. Revolt in Greece. King **Otto I** was forced to grant constitutional government.

1848. Revolt in Wallachia. This was related to continent-wide political unrest. Russia, with Turkish approval, invaded the principality and subdued the insurrection.

EURASIA AND THE MIDDLE EAST

THE OTTOMAN EMPIRE

The Ottoman Empire, football of France, Russia, and England during the Napoleonic Wars, began to disintegrate.

War with Russia, 1806–1812

A smoldering dispute with Russia over the status of the principalities of Wallachia, Moldavia, and Bessarabia was fanned into flame by the French ambassador to the Porte.

1806, November 6. Turkey Declares War. Russian troops invaded Wallachia and Moldavia.

1807, February–March. British Naval Attack on Constantinople Repulsed. (See p. 838.)

1807, June 30. Battle of Lemnos. A Russian fleet under Admiral **Dmitri Seniavin** defeated a slightly larger Turkish fleet.

1807, August. Armistice. This was largely due to mediation by Napoleon at Tilsit. Russian forces withdrew from Wallachia and

Moldavia; the Turkish Army retired to Adrianople.

1809–1812. Desultory Warfare. Russia was generally more successful than Turkey.

1812, May 28. Treaty of Bucharest. This resulted from British mediation. Bessarabia went to Russia, but Moldavia and Wallachia remained under Turkish control.

1808–1809. Internal Unrest and Mutinies. Sultan **Selim** was dethroned (1807) and killed (1808) when adherents attempted to reinstate him. His successor **Mustafa IV** was quickly dethroned in turn (1808) and succeeded by **Mahmud II.**

1821–1832. Greek War of Independence. (See p. 849.)

1821–1823. War with Persia. (See p. 853.)

1826, June 15–16. Massacre of the Janissaries. Sultan Mahmud, enraged by the inefficiency of his troops, attempted to displace the Janissaries, who rose in revolt. Loyal troops bombarded their barracks in Constantinople, and Turkish mobs ended the job by a wholesale massacre; more than 6,000 were killed.

First Turko-Egyptian War, 1832–1833

Mohammed Ali demanded control of Syria as reward for his assistance against Greece. This was rejected by the sultan.

1832. Egyptian Invasion of Syria. Ibrahim captured Acre (May 27), Damascus (June 15), and Aleppo (July 16).

1832, December 21. Battle of Koniah (Konya). Invading Anatolia, Ibrahim met and defeated Reshid's main Turkish army, capturing Reshid. Turkey in desperation called on Russia for aid.

1833, February. Defensive Alliance with Russia. A Russian squadron from the Black Sea arrived at Constantinople (February 20, 1833).

1833, May 14. Convention of Kutahya. England and France, alarmed by Russian influence in Turkey, persuaded the sultan to give Syria and Adana to Egypt.

1833–1839. Internal and External Confusion. The great powers then began to squabble among themselves, France siding with Egypt. Turkey, meanwhile, rebuilt its army with Prussian assistance. Captain **Helmuth von Moltke** was one of the advisors (1835–1839).

Second Turko-Egyptian War, 1839–1841

1839. Turkish Invasion of Syria. Hafiz Pasha, accompanied by von Moltke, was in command.

1839, June 24. Battle of Nezib. The Turks were defeated by Ibrahim, partly because Hafiz ignored Moltke's advice. The Turkish fleet went to Alexandria to surrender to Egypt (July 1). With the Ottoman Empire apparently about to disintegrate, the powers—except France—decided to shore it up.

1840, September–November. British and Austrian Intervention. A combined naval force cut Ibrahim's sea communications with Egypt. This was followed by British bombardment and occupation of **Beirut** (October 10) and **Acre** (November 3).

1840, November 27. Convention of Alexandria. British Admiral **Charles Napier** reached an agreement with the Egyptians whereby they abandoned claims to Syria and returned the Turkish fleet.

1841, February. Ibrahim Evacuates Syria. He returned to Egypt.

1841, July 13. Straits Convention. The great powers agreed that the Bosporus and Dardanelles should be closed to all foreign warships in time of peace.

1848, September. Revolt in Wallachia. (See p. 850.)

EGYPT AND THE SUDAN

1805. Rise of Mohammed Ali. The governor of Egypt for the Ottoman Empire, by intrigue and violence he established himself as an independent potentate. Violence continuing, the British decided to intervene in favor of one of his opponents.

1807, March 17. British Expedition. A force of some 5,000 men under General **A. Mackenzie Fraser** occupied Alexandria without resistance, but found their protégé had died. In attempting to take Rosetta and Rahmanieh to secure supplies, a British detachment was badly cut up, retreating to Aboukir and Alexandria with loss of 185 killed and 218 wounded. Another attempt on Rosetta (April 20) failed, with losses of some 900 men out of 2,500 engaged. Many British soldiers were massacred. Alexandria was then evacuated (September 14).

1811, March 1. Massacre of the Mamelukes. Mohammed Ali by treachery wiped out most of the Mameluke chieftains, and then their troops. He consolidated his position; made peace with the Porte, whose suzerainty he recognized; and reorganized the country, building up a powerful army (composed largely of Albanian mercenaries) and a fleet with the assistance of European military advisors. At Turkish command he now instituted a religious war in Arabia.

1811–1818. War with the Wahhabis. This Moslem sect had occupied Mecca, Medina, and Jidda, and threatened Syria. During 7 years of serious combat the Moslem holy places were recovered and the eastern Red Sea coast placed under Egyptian rule.

1815. Mutiny in Cairo. Mainly by Albanian units, which resented Mohammed Ali's modernization program.

1820–1839. Conquest of the Sudan. Egyptian forces under **Hussein** (son of Mohammed) moved into the eastern Sudan and extended Mohammed's control over a great part of the western Red Sea coast. Egyptian troops pushed up the Nile as far as Gondokoro (modern Equatoria, Sudan).

1823–1827. Operations in Greece. Egyptian warships and troops assisted in checking the Greek struggle for liberty from Turkish rule (see p. 849). As a result, Egypt gained control of Crete.

1832–1833. First Turko-Egyptian War. (See p. 851.) This resulted in Egyptian control of Syria and Adana, and extension of Egyptian influence eastward to the Persian Gulf.

1839–1841. Second Turko-Egyptian War. (See p. 851.) Although losing both Crete and Syria, Mohammed Ali secured hereditary rule over Egypt.

1848. Death of Ibrahim.

1849. Death of Mohammed Ali.

PERSIA (IRAN)

Russia's policy of expansion to the east and southeast, and the clash of Franco-British interests, as well as internal rebellions, kept Persia in turmoil throughout the period—a striking exemplification of the impact of power politics on underdeveloped regions at the height of European colonial expansion. The unfortunate country was tugged hither and yon by French, British, and Russian interests. The influence of Britain (and the East India Company) was generally predominant. A number of British officers served from time to time in the Persian forces. The principal events were:

War with Russia, 1804–1813

This resulted from Russian annexation of Georgia (1800), which had been nominally under the suzerainty of Persia. Persian assistance to Georgian factions resisting the annexation aroused Russian ire.

1804. Russian Invasion of Persia. The Russian Army was under General **Sisianoff.**

1804. Battle of Echmiadzin. An inconclusive 3-day battle. Persian forces were under able Persian prince **Abbas Mirza.**

1804. Siege of Erivan. Shah **Fath Ali** brought up Persian reinforcements, and Abbas Mirza later during the same year relieved the city, forcing the Russians to retire.

1805–1813. Sporadic Warfare. This dragged on throughout the Caucasus and along the Caspian coast.

1812, October 31. Battle of Aslanduz. The Russians surprised and routed Abbas Mirza on the Aras River.

1813, October 12. Treaty of Gulistan. Persia agreed to the cession of Georgia and other trans-Caucasian provinces to Russia.

1816. Persian Invasion of Afghanistan. (See p. 864.)

1821–1823. War with Turkey. The conflict was prompted by Russian intrigue and by Turkish protection of rebellious tribes fleeing from Azerbaijan. Abbas Mirza moved west into Turkey (1821) to the Lake Van region. In retaliation, Turkish forces under the Pasha of Baghdad struck east into Persia, but were repelled. The Persian invasion in the north culminated in the **Battle of Erzerum,** where Abbas Mirza with 30,000 men defeated a Turkish army estimated at 52,000 (1821). Peace was finally established by the **Treaty of Erzerum;** both sides agreed to maintain the *status quo* (1823).

1825–1828. War with Russia. Following a boundary dispute, Persian forces were successful along the Caspian littoral, reaching the gates of Tiflis (Tbilisi). Russian General Ivan Fedorovich Paskievich then initiated a brilliant counteroffensive campaign.

1826, September 26. Battle of Ganja (or Kirovabad). Abbas Mirza with 30,000 men met Paskievich with 15,000. After initial Persian success, they were routed by the Russians.

1827. Russian Victories. Several indecisive actions were followed by the storming of Erivan by Paskievich and the surrender of Tabriz.

1828, February 22. Treaty of Turkomanchi. This ended the war. It also ended Persia's role as a major power.

1836–1838. War with Afghanistan. Again through Russian intrigue, prompted by British penetration into Afghanistan, Mohammed Shah of Persia undertook to invade that region. Mobilizing his army at Shahrud, he moved slowly through Khorasan (Khurasan) via Meshed into Herat Province, wasting nearly a year in bickerings with the Turkoman tribesmen on his northern flank. Captain **Eldred Potter** of the East India Company's Bombay Army turned up in Herat City, in disguise, as Mohammed's troops began investment of the city (November 23). He offered his services to the Afghan commander, and so organized the defenses that a Persian assault—delivered simultaneously in 5 places—was driven off. After a siege of nearly 10 months, Mohammed withdrew (September 28, 1838).

1849–1854. Russian Conquest of the Syr Darya Valley. This ended nominal Persian suzerainty over this region of Central Asia, and brought Russian power to the northern border of Persian Khorasan.

AFRICA

NORTH AFRICA

The 4 Barbary States—Algiers, Tripoli and Tunis (semi-independent satellites of the Ottoman Empire), and the independent Kingdom of Morocco—continued to afflict the Mediterranean and its Atlantic approaches by their piratical activities. War-torn Europe had found it practical to pay tribute to them for the protection of its commerce. The new United States had followed suit, but undertook punitive action after extreme provocation.

1801–1805. Tripolitan-American War. The Pasha of Tripoli raised his protection ante. He declared war against the United States (May 14) because his demands were not satisfied. President **Thomas Jefferson** sent a punitive expedition to the Mediterranean (July). For the next 2 years a small American naval force blockaded Tripoli, but only lackadaisically and ineffectually (1801–1803).

1803, August. Arrival of Commodore Edward Preble. A more vigorous policy was followed; American warships hunted down Tripolitan corsairs.

1803, October 31. Capture of USS *Philadelphia*. This frigate, commanded by Captain **William Bainbridge,** pursuing a Tripolitan frigate, ran aground off the harbor. At once surrounded by a score of Tripolitan vessels, Bainbridge was forced to surrender his helpless ship. Later the *Philadelphia* was floated and brought into Tripoli in triumph. .

1804, February 16. *Philadelphia* Destroyed. Lieutenant **Stephen Decatur** with 73 officers and men, all volunteers, boldly sailed a small captured Tripolitan vessel into the harbor, boarded, and burned the *Philadelphia* as she lay anchored under the guns of the castle, then escaped. Preble's squadron then systematically bombarded the port, but could not enter through the tortuous, rock-strewn channel.

1804–1805. Eaton's Expedition. U.S. naval agent William Eaton started from Alexandria with 8 Marines, a navy midshipman, and about 100 mercenaries to restore **Hamet Karamanli,** the rightful but dispossessed ruler of Tripoli, to his throne (November 14). After a 600-mile march across the Libyan desert, he reached and, aided by a naval bombardment, stormed the city of Derna (April 26, 1805). For another 6 weeks Eaton held the city against Tripolitan counterattacks. But Commodore **Samuel Barron,** who had succeeded Preble in command of the Mediterranean squadron, had meanwhile made peace with the reigning Pasha of Tripoli—the usurper **Yusuf Karamanli.** In return for $60,000 cash bounty, the *Philadelphia*'s crew were released from prison and all further ransom payments ceased.

Eaton's expedition—America's first overseas land operation—was withdrawn.

1805, August 1. American Peace with Tunis. Commodore **John Rodgers,** in the USS *Constitution,* sailed into Tunis and forced its bey to make peace. Algiers alone still received American tribute. Trouble, then war, with England delayed American action.

1815, March 3–June 30. American War with Algiers. All U.S. warships having withdrawn from the Mediterranean during the War of 1812, the Dey of Algiers preyed on U.S. commerce. He threw out the U.S. consul, enslaved U.S. nationals, and finally declared war, on the grounds that he was not being paid sufficient tribute. Commodore Stephen Decatur, with a squadron of 10 vessels, entered the Mediterranean, captured 2 Algerian warships, and sailed into Algiers harbor. At cannon mouth he demanded and received cancellation of all tribute and release of all U.S. prisoners without ransom. Cessation of piracy was guaranteed by the dey. Decatur then forced similar guarantees from Tunis and Tripoli, with compensation for U.S. vessels seized by Britain in those waters during the War of 1812. Decatur's forceful action ended U.S. participation in the Barbary Wars.

1816, August 26. Allied Bombardment of Algiers. Following a succession of outrages against British shipping and subjects, a squadron under Admiral **Edward Pellew, Lord Exmouth** destroyed the port fortifications and the Algerian fleet in a 9-hour engagement. A Dutch squadron under Admiral **Theodore F. van der Capellen** assisted. The Algerians surrendered some 3,000 European prisoners.

1819. British Demonstration on the Barbary Coast. This was to discourage recurrent piracy.

1824. Bombardment of Algiers. This was by a British squadron under Admiral Sir **Harry Neal.** Algerian piracy was still not completely suppressed, however.

1826. Independence of Constantine. The city and surrounding region successfully defied the Dey of Algiers.

1827–1829. French Blockade of Algiers. Increasing friction between France and the Algerian pirates culminated when the Dey of Algiers slapped the French consul's face (April 30, 1827). During the subsequent blockade, a French ship bearing negotiators and flying a flag of truce was fired upon by the Algerians (August 3, 1829).

French Conquest of Algeria, 1830–1847

1830, June–July. French Invasion. France sent an army of 37,000 men and 83 guns to Algeria under Marshal **Louis A. V. de Bourmont.**

1830, July 5. Capture of Algiers. The dey was expelled; important ports and key inland towns were occupied.

1832–1837. Wars against Abd el Kader, Emir of Mascara. He emerged as leader of Islam against the Christian invaders. A succession of French generals failed to curb the skillful Algerian warrior.

1837, June 1. Treaty of Tafna. Peace was restored, with France controlling only a few of the major coastal ports. Limited French military operations continued, however.

1837, October. Storming of Constantine. Marshal **C. M. D. Damremont,** leading a final assault on the walled city, which for a year had resisted all French attacks, was killed as his troops won victory. Abd el Kader, claiming a violation of the peace treaty, took the field (December) with a well-trained force of 8,000 infantry and 2,000 cavalry, supplemented by some 50,000 gouma (irregular horse). Indecisive but bitter fighting continued.

1840, December. Arrival of Marshal Thomas R. Bugeaud. He brought large reinforcements. The French Army in Africa now amounted to 59,000 men, later to be augmented to 160,000. Bugeaud was the first major French African colonial organizer. His forceful leadership combined great military and administrative qualities. He reorganized the African army, adding indigenous elements, and established a tactical square combat assault formation, which he called the "boar's head." The colorful names and uniforms of Bugeaud's troops—zouaves, voltigeurs, chasseurs d'Afrique, spahis, and the famous Foreign Legion—intrigued and influenced American amateur military organizations and stimulated romantic writers the world over. The French strategic pacification policy initiated by Bugeaud and carried on successfully in North Africa for almost 100 years more consisted of a few fixed bases from which flying columns issued. These mobile elements were as fluid and fast-moving as the Berbers who opposed them.

1843, May 10. Battle of Smala. Duke **Henri d'Aumale,** commanding a flying column of less than 2,000 men, surprised and attacked, headlong, Abd el Kader's army of 40,000, dispersing it.

1843–1844. French Successes. Systematically and progressively Kader was driven west into Morocco. A French squadron dampened Moorish sympathy by bombardment of Tangier and Mogadir, while Bugeaud, with 6,000 infantry, 1,500 cavalry, and some artillery, crossed the border in pursuit.

1844, August 14. Battle of Isly. Adb el Kader, with 45,000 men, was camped on the Isly River. Bugeaud crossed the river and attacked. His "boar's head" square repulsed Kader's counterattacks. Finally the French cavalry, the "boar's tusks," charged in 2 columns, overrunning Kader's camp. After a stubborn struggle Kader's forces gave way. Isly was the decisive battle of the war, although some sharp encounters occurred later.

1844, September 10. Treaty of Tangier. The French withdrew from Morocco.

1847, December 23. Surrender of Abd el Kader.

COMMENT. *Abd el Kader deserves a place in history not only as a chivalrous soldier of great ability but also as one of the great champions of men of letters of the 19th-century Islamic world. Imprisoned by the French for several years, he was later released and returned to the Near East.*

1835–1836. Ottoman Control Reestablished in Tripolitania. An Ottoman fleet and army took control.

WEST AFRICA

The related Moslem Fulani and Tukulor peoples became predominant in the western Sudan highlands between the upper Senegal River and Lake Chad. To the south the warlike Ashanti and Dahomey tribes expanded their areas of control in the jungled regions west and east of the Volta River. Meanwhile, along the coast, Britain and France extended their seaport colonial control at the expense of the Portuguese and Danes, as well as the coastal tribes. The principal events were:

1804–1810. The Fulani Jihad. Usman dan Fodio, a Moslem divine who had quarreled with the King of Gobir, preached a *jihad* (Holy war) to renew Islam. Supported by many Hausa as well as Fulani, his armies overran northern Nigeria, northern Cameroon, southeastern Niger, and northeastern Benin, conquering the Hausa city-states (1804–1808). He divided the realm between his son **Muhammad Bello** (at Sokoto) with the northwest, and his brother **Abdullahi dan Fodio** (at nearby Gwandu) with the southern and western emirates (1809).

c. 1805–c. 1835. End of Oyo. Oyo was pressured by Dahomey to the east, and weakened by the effects of Abiodun's anti-army policies (see p. 795). The northern half of the kingdom was overrun by a Fulani army (c. 1817), which also created the emirate of Ilorin. A rump Oyo kingdom survived in the south, around Oyo, which became the new capital (c. 1835); some Oyo refugees settled near Ife.

1806–1807. Ashanti Conquest of the Gold Coast. Under their King **Osei Bonsu** (r. c. 1801–1824), the Ashanti extended their dominions to the coast.

1808–1811. Fulani Attack on Bornu. Forces of **Usman dan Fodio**'s Fulani *jihad* invaded Kanem-Bornu and defeated its army (1808). *Mai* **Ahmad** was driven from his capital and appealed to a local scholar, cleric, and warrior, **Muhammad al-Kanami.** Al-Kanami led a spirited resistance, and by 1811 had defeated the Fulani.

1809. British Capture St. Louis (Senegal) from the French.

1810. Revolution in Macina. Shehu **Ahmadu Lobbo,** a Fulani cleric, overthrew the ruling Fulani Dynasty of Macina, and created a new state with its capital at Hamdallahi.

1811. Britain Abolishes the Slave Trade. Similar action was taken by Holland (1814), France (1815), and most other European nations.

1816. Britain Returns the Senegal Posts to France.

1817. The Sokoto Caliphate. After the death of **Usman dan Fodio,** his son **Muhammad Bello** succeeded him as *sarkin musulmi* ("commander of the faithful") and caliph, making his seat at Sokoto, and becoming the spiritual and political leader of the Fulani.

1818. Macina Jihad. At Macina, **Ahmadu ibn Hammadi,** a disciple of **Usman dan Fodio,** began preaching a *jihad.* One of the first conquests was the kingdom of Segu, ruled by **Da Kaba,** grandson of Ngolo Diara (see p. 795). The forces of this *jihad* also conquered Djénné and Timbuktu, but Hammadi's puritanical brand of Islam alienated many of the local tribes.

1818–1858. Reign of Gezo of Dahomey. After deposing King Adandozan (r. 1797–1818), this ruler led Dahomey to its greatest power. The Dahomeyan army, like its government, contained male officials and regiments in the countryside, matched by female counterparts at court. The female royal guard (called "Amazons" by Europeans) were the elite shock troops of the army. Nearly all the female troops, and many male units, were equipped with modern firearms. The end of the slave trade after 1840 hurt Dahomey, which had grown rich selling slaves taken in its campaigns to Europeans; a slave-plantation system to produce palm oil was less lucrative.

1822–1847. Establishment of Liberia. Freed slaves from America established a colony, despite constant conflict with neighboring tribes.

1824–1831. First Ashanti War. Sir Charles M'Carthy, governor of British posts on the Gold Coast, attempted to protect the coastal tribes from the Ashanti, but was defeated and killed. The British sent reinforcements and finally defeated the Ashanti (1827). Peace was established, with the Ashanti surrendering their suzerainty over the coastal region (1831).

1826. Muhammad al-Kanami in Bornu. By this time, al-Kanami had cemented his control of the kingdom and made it into a vigorous Islamic state, although he did not displace the ruling Sef Dynasty.

1842–1843. French Occupation of Ivory Coast Ports.

1846. Civil Unrest in Bornu. The puppet *mai* rebelled against **Muhammad al-Kanami's** son and successor **'Umar,** but he was defeated and killed, leaving the kingdom in 'Umar's capable hands.

1847, July 26. Independence of Liberia. The capital, Monrovia, had been founded by freed slaves from the United States (1822).

1848–1850. Rise of al-Hajj Umar. Umar had made the pilgrimage to Mecca (hence the honorific "al-Hajj"), joined the moderate Tijaniyah *tariqa* (religious brotherhood), and lived for a time in Sokoto. He established a refuge and religious community at Dinguiraye in northern Guinea for his followers among the Tukulor people (close cousins to the Fulani). There he raised an army equipped with European weapons, and laid the basis of a theocratic state (see p. 936).

SOUTH AFRICA

1800–1814. British Annexation of Cape Colony. The Dutch settlement, occupied by the British (1796), was returned to Dutch control (1803), but renewal of the Napoleonic Wars led to its reoccupation (January, 1805). The **Treaty of Paris** (1814) granted the colony to Britain.

c. 1807–1818. The Mtetwa Empire. The Mtetwa were a large clan, dwelling near the coast of modern Zululand and northern Natal. Their chieftain **Dingiswayo,** who took the throne about 1807, undertook a program of expansion. He brought neighboring tribes under his control (more often through diplomacy than force), but allowed them to retain some autonomy. The Mtetwa army embodied the *amabutho,* an age-graded regimental system also characteristic of the later Zulu army (see below and pp. 931–932). Drafts from subject clans were incorporated into the army. Dingiswayo was killed in a war with **Zwide,** chieftain of the powerful Ndwandwe clan (1818). A half-brother succeeded him, but failed to maintain Mtetwa power.

c. 1811. Zulus Brought under Mtetwa Control. The Zulus were a small upcountry clan, and their chieftain **Senzangakona** submitted to Dingiswayo after the more powerful nearby Butelezi clan was defeated.

1811–1819. Frontier Warfare. The British and Boers drove the Xosa Bantu people beyond the Fish and Keiskama rivers.

1814–1835. Growing Boer-British Friction in the Cape Colony. This was exacerbated by abolition of slavery in the British Empire (1833); the Boers felt they were not adequately compensated for the slaves they freed. Also they felt that British territorial policy was preventing them from occupying frontier lands.

1816–1819. Rise of Shaka of the Zulus. He was Senzangakona's illegitimate son; after an unsettled childhood **Shaka** lived among the Mtetwa, and joined the iziCwe regiment of the Mtetwa army (c. 1810). After distinguished service, Dingiswayo appointed him chief of the Zulus after Senzangakona died (1816). Shaka reformed the Zulu army (which numbered less than 400 when he arrived), forging it into a disciplined and efficient fighting force. Within 2 years he had forcibly absorbed several neighboring clans and led an army of 2,000 warriors.

1818–1819. Zulu-Ndwandwe War. About a year after Dingiswayo's death, Zwide of the Ndwandwe attacked the smaller Zulu clan, but Shaka defeated him at the **Battle of Gqokli**

Hill (late 1818). Zwide tried again but, hampered by lack of supplies, the exhausted Ndwandwe army was destroyed in a climactic 2-day battle as it tried to withdraw (May 1819). Shaka sent a force to destroy Zwide's kraal, but Zwide escaped and fled north, where he died (c. 1821).

1819–1828. Shaka Creates the Zulu Empire. Shaka used his army, which had grown to over 30,000 by the late 1820s, to conquer neighboring clans. The vanquished were forcibly absorbed by the victors and became Zulus. His rule was autocratic and often cruel; after the death of his mother Nandi (October 1827), he became capricious and irrational. After some months of plotting, Shaka's half-brothers **Dingane** and **Mhlangana** assassinated him (September 22, 1828). After a quarrel, Mhlangana fled and Dingane took the throne.

c. 1820–1835. The Mfecane. During his war with Shaka, Zwide attacked the neighboring emaNgwaneni clan and drove them from their land. They eventually fled over the Drakensberg to the interior plateau. With both coastal and interior Bantu clans so closely packed there was no room to spare. These events unleashed a chain reaction of attack, marauding, and dispossession, which virtually depopulated the interior plateau. Necessarily crude estimates put the death toll between 1 and 2 million. The effect on the coastal clans was significant, but nowhere near as devastating. These events were christened *Mfecane*, or "the Crushing," an apt description of its effect on Bantu society. Some effects of the *Mfecane* reached as far north as Tanzania.

1834. Boer Frontier War with Bantu. The Bantu tribes (mainly Xosa), enraged by Boer encroachment, invaded the frontier regions and were repulsed with difficulty.

1835–1837. Great Trek of the Boers. Some 12,000 Boers, chafing under British control, decided to make a mass migration northward to establish their own independent states.

Some, under **A. H. Potgieter,** headed north toward the Vaal River; the others, under **Piet Retief,** went northeastward toward Natal and Zululand. Potgieter and his people, after a fierce struggle, defeated the Matabele Bantu tribe in the Vaal region, and settled down (1837).

1838, February. Durban Massacre. Retief and other Boer leaders were massacred by Dingane, who then attacked and killed many of the settlers.

1838, December 16. Battle of Blood River. A Boer force under **Andreas Pretorius** decisively defeated Dingane. The Zulus, beset by internal discord, did not seriously interfere with Boer colonization.

1840, January. Battle of Magango. Boers, allied with **Mpande,** brother of Dingane, defeated the Zulu ruler, who fled to Swaziland; Mpande took his place and made peace with the Boers, granting them all of southern Natal.

1839. Pretorius Proclaims Boer Republic of Natal.

1842–1843. British-Boer Conflict in Natal. The British barely repulsed a Boer attack on their post at Durban. They then occupied Natal and annexed it (1843). Friction continued.

1846–1847. War of the Ax. Conflict broke out between the British and Kaffirs in the region between the Keiskama and Great Kei rivers. The natives were protesting against both British and Boer colonist encroachment. The British were successful in a 21-month campaign.

1846–1850. British-Boer Friction. There was considerable unrest and some fighting along the frontier near the Great Kei River and in the region between the Orange and Vaal rivers.

1848, August 29. Battle of Boomplaats. Strained relations between Boers in the Orange River colony and the British resulted in an armed clash. A force of British under Sir **Harry Smith** defeated the Boers under Pretorius. The disgruntled Boers retreated across the Vaal River.

EAST AFRICA

The great native migrations continued within the interior of central East Africa. Along the coast the Arabs retained and strengthened their control, while in the highlands of Ethiopia internal chaos continued. The principal events were:

c. 1810. Battle of Shela. Pate and Mombasa were defeated by Lamu; as a result, Pate's power in the area was destroyed and Mombasa's influence in the area ended.

1822. Sayyid Sa'id Captures Pemba. This British ally and ruler of Muscat began an expansion program in East Africa. That same year, under British pressure, he agreed to end slave exports to "Christian" countries, although slaves continued to be taken for Zanzibar and the Persian Gulf markets. He placed a garrison in Pate (1824). He established a major base at Zoazibar (1828).

1824–1826. British in Mombasa. The Mombasans, desperate to avoid Sa'id's rule, invited a British protectorate. This was withdrawn to prevent British conflict with their ally.

1829–1833. Mombasa Defies Muscat. Sayyid Sa'id twice assaulted Mombasa but was repulsed both times (1829, 1833).

1837. Sayyid Sa'id Gains Mombasa. A succession crisis (1835–1836) enabled Sa'id to gain control of the city and place a garrison there. Sa'id moved his capital to Zanzibar in 1840.

1839. British in Aden. As part of a policy to secure ports along the route to India, the British occupied the port of Aden in Yemen.

1841–1850. Rise of Lij Kassa of Gojjam of Ethiopia. An able and ambitious minor chieftain, he parlayed his well-organized brigand band into an army and so gained political power in the region west of Lake Tana. By 1847 (at the age of 29) he had married the daughter of Ras Ali, lord of Gojjam and Amhara provinces.

MADAGASCAR

The island became a shuttlecock in the Franco-English conflict of the Napoleonic era and later, a situation complicated by the hostility of the natives to both nations. A French force reestablished French control at Tamatave (1803), but in 1810 the British seized all French stations on the island. A French effort to regain power (1818–1819) was finally repulsed by the Hovas, the principal tribe (1825); a French squadron bombarded Tamatave (1829) without much effect. In 1846 Franco-British squadrons—now in conjunction—futilely bombarded Tamatave in retaliation for native hostility to all foreigners.

SOUTH ASIA

INDIA

The British conquest of the country progressed under joint crown and East India Company auspices as the British flag followed British trade, and England's leaders parried Bonaparte's thrusts to Egypt and India. Differences with the native rulers were heightened by French intrigue, particularly with the powerful Marathas, whose leaders—**Doulut Rao Sindhia** and **Jaswant Rao Holkar**—had large armies of well-equipped and French-trained soldiers.

1799–1802. Rise of Ranjit Singh. In 3 years of warfare, at the age of 23 he united the Sikhs to control most of the Punjab.

1802. Civil War among the Marathas. Baji Rao II, the Peshwa, hereditary ruler of the Marathas, was defeated and overthrown by Holkar of Indore at the **Battle of Poona.** The British demanded that their ally, the Peshwa, be restored to his throne. Holkar refused.

1802–1803. British Attack on Kandy. Although the British had conquered coastal Ceylon from the Dutch in 1796 (see p. 766), the interior remained unsubdued. A British expedition captured Kandy in late 1802, but was driven out the following year.

Second Maratha War, 1803–1805

Simultaneous British offensives into the Deccan and Hindustan were led respectively by Sir Arthur Wellesley (later Duke of Wellington, and not to be confused with his brother, Marquis **Richard C. Wellesley,** then governor general in India) and General **Gerard Lake.**

Deccan Campaign, 1803

1803, March 20. Capture of Poona. Wellesley restored the Peshwa to power without opposition. Meanwhile, Sindhia's main army had moved south. When he refused to withdraw, Wellesley's army, some 9,000 regulars and 5,000 additional native contingents, penetrated farther into the Maratha homeland (August 6).

1803, August 11. Capture of Ahmadnagar. Following this, Wellesley continued to the junction of the Jua and Kelna rivers, where he unexpectedly found the army of Sindhia and the Raja of Berar—30,000 horse, 10,000 French-trained infantry, and 200 guns (September 12). Wellesley had with him 4,500 regulars, 2,200 of them cavalry; the remainder of his expedition was beyond supporting distance.

1803, September 23. Battle of Assaye. Wellesley found an unprotected ford over the swollen Kelna. He promptly crossed and attacked Sindhia's army, which faced to its flank to meet him. Wellesley's disciplined infantry—English and Indian—pierced the center of Sindhia's line, only to be attacked on both flanks by the horde of Maratha cavalry. These were repulsed in turn by Wellesley's cavalry, and despite a determined counterattack Sindhia's masses finally broke and ran, leaving 1,200 dead and 98 guns on the field. Wellesley's losses were extremely heavy, more than 2,000 killed and wounded. Wellesley resumed his offensive.

1803, November 28. Battle of Argaon. Sindhia made another stand, but his disheartened troops were easily defeated. The storming of Gawilarh (December 15) ended the campaign. COMMENT. *This campaign was the first step in Wellington's march to military fame.*

Hindustan Campaign, 1803

Farther north, Lake had commenced operations with the so-called Grand Army, 10,500 strong, mostly East India Company native troops with a small English contingent. He was opposed by a Maratha army of 43,000 men, with 464 guns, under the command of **Pierre Cuillier Perron,** French adventurer, successor to Count **Benoît de Boigne,** who had established European military training. A number of other French adventurers with previous military experience were also in the Maratha forces.

1803, September 4. Capture of Aligarh. The walled city was stormed by the British at a cost of 260 killed and wounded. Maratha losses, while heavy, are unknown. While Lake was proceeding toward Delhi, Perron and some of his French officers rode into Lake's lines and surrendered. **Louis Bourquien** now became the chief foreign officer with the Marathas.

1803, September 16. Battle of Delhi. Lake attacked the Marathas and drove them off the field, capturing 63 guns and much treasure. British losses were 477; again Maratha losses were unknown but heavy. After occupying Delhi, Lake moved south to capture Agra. Anxious to settle the campaign, he pushed rapidly with his cavalry after the retreating Marathas.

1803, November 1. Battle of Laswari. The Maratha army was drawn up behind a line of cannon, linked together by chains. Lake's cavalry charged over the line, but the Maratha infantry rallied. Lake's infantry, which had covered 65 miles in 48 hours, arrived in time to effect the complete destruction of the Maratha army. British losses were 834; Maratha losses again were unknown. This victory, coupled with Wellesley's success in the Deccan, brought Sindhia to submission (December 20).

Holkar's Offensive, 1804

1804, August 24–29. Intervention by Holkar. His troops ambushed a small British detachment under Colonel **William Monson** and wiped it out (August 24–29). Holkar then threatened Delhi (September).

1804, October 1. Relief of Delhi. Lake, who had pulled his army out of summer-monsoon quarters, advanced toward Delhi. Holkar retreated.

1804, November 17. Battle of Farrukhabad. Lake, after a 350-mile cavalry pursuit, dispersed the Maratha cavalry. Holkar fled to the Punjab. Lake continued into Indore to capture Dig (December 25), then turned on Bhurtpore, whose rajah had sided with Holkar.

1805, January–April. Siege of Bhurtpore. Investing the city (January 1), Lake made 4 successive assaults, all of which were repulsed

with heavy losses: 103 officers and 3,100 men. The rajah, however, later made peace (April 17). Lake was superseded as commander in chief by Lord **Cornwallis** (July), but returned to command after Cornwallis' death (October).

1805, October–December. Pursuit of Holkar. Lake pursued into the Punjab, and Holkar surrendered at Amritsar (December). An uneasy peace settled over India.

1806. Sepoy Mutiny at Vellore. Religious pressures led to the outbreak—foreshadowing the Great Mutiny (see p. 939). It was quickly suppressed.

1809, April 15. Treaty of Amritsar. Growing friction between the British and Ranjit Singh was settled by agreement on the Sutlej River as the boundary between the Sikh territories and territory the British had seized from the Marathas.

1810–1820. Sikh Conquest of the Punjab. Ranjit Singh conquered all of the Punjab from the Afghans and local princes, consolidating his lands and, with help of French and Italian officers, developing the most powerful and efficient native army in India. He overran Kashmir (1819; see p. 864).

1814–1816. Gurkha War. Expeditionary forces from the British Indian Army were sent into Nepal (November 14) to stop Gurkha raids into northern India. The ferocity of the Gurkhas repelled initial attempts, but General **David Ochterlony** campaigned systematically through the southern portion of the rugged hill country, taking mountain forts one by one (1815). He penetrated the Katmandu Valley and forced peace on the warlike Gurkhas (1816). Since that time the Gurkhas have furnished the British Army with one of its most famous corps.

1815–1818. British Consolidation on Ceylon. British forces captured Kandy in 1815, and finally subdued the island after suppressing a Sinhalese revolt (1818).

Third Maratha (and Pindari) War, 1817–1818

A vast horde of Pindaris (outlaws and freebooters of all castes and tribes, mostly former Maratha soldiers) had begun widespread organized depredations in central and southern India. The

Maratha chieftains, though giving lip service to Britain, supported the Pindari raids. Two British forces—the Army of the Deccan, under Sir **Thomas Hyslop,** and the Grand Army, under the British commander in chief, Lord General **Francis Rawdon-Hastings of Hastings and Moira,** nearly 20,000 in all—prepared to take the field against the Pindaris, while keeping a watchful eye on the Marathas, who could muster nearly 200,000 men and 500 guns. Meanwhile, several outbreaks against British garrisons occurred, the most notable at Nagpur and Kirkee (a suburb of Poona). Though badly outnumbered, the local garrisons held out, even after Holkar and the Peshwa took the field (November, 1817).

1817, December 21. Battle of Mahidput. Hyslop with 5,500 men crushed Holkar's army of 30,000 horse, 5,000 infantry, and 100 guns.

1818, January–June. Hastings' Campaign. His Grand Army hunted down the Pindaris and remaining Marathas. The campaign ended with the surrender of the Peshwa of the Marathas (June 2, 1818). It marked the end of Maratha political power and the real beginning of the East India Company's paramountcy.

1825–1826. British Intervention in Bhurt-pore. Lord **Combermere,** with 1 cavalry and 2 infantry divisions, plus a large artillery train, invested Bhurtpore to settle a disputed succession. Heretofore considered impregnable, the city's strong defenses were successfully assaulted by the British after a desperate conflict (January 18, 1826). The British lost nearly 1,000 killed and wounded; Indian losses were an estimated 8,000.

1839–1842. First Afghan War. (See p. 864.)

Conquest of Sind, 1843

There had been increasing friction between the Baluch rulers of Sind and the British during the First Afghan War.

1843, February 15. Attack on the Residency at Hyderabad. Resentful of humiliating terms demanded by British Governor Generals **Auckland** and **Ellenborough,** and outraged by British military threats, 8,000 Baluchis attacked the British residency. The residency was defended by a handful of men under a young British officer, **James Outram.** General Sir **Charles James Napier** immediately initiated a whirlwind campaign to come to the relief of Outram and his gallant garrison.

1843, February 17. Battle of Miani. Napier, with 2,800 men, attacked and defeated 30,000 Baluchis in savage hand-to-hand combat; the 61-year-old general, musket in hand, personally led his troops.

1843, February–March. Advance to Hyderabad. Leaving his transport behind, Napier made a forced march through the desert country in intense heat, with his British infantry mounted 2 men each on camels.

1843, March. Battle of Hyderabad. Napier dispersed the Baluchis at the gates of the city and relieved the British residency.

1843, March–May. Conquest of Sind. Napier, who never had more than 5,000 men, marched 600 miles and fought numerous minor actions and 2 battles to defeat a total of 60,000 enemies. At the conclusion of the campaign he sent his famous message to the governor general: "Peccavi" ("I have sinned").

1843, June–August. Consolidation. Napier marched another 280 miles and dispersed a force of 12,000. Total Baluchi casualties during the campaign are estimated at 12,000. This campaign stabilized India's western frontier, securing the Indus waterway. Napier also constructed the port of Karachi, which replaced Hyderabad as the capital of Sind.

First Sikh War, 1845–1846

Since the death of Ranjit Singh (1839), friction between the British and the Sikhs of the Punjab increased, as did internal disorder in the Punjab.

1845, December 11. Sikh Invasion. An army of 20,000 crossed the Sutlej into British Indian territory. Sir **Hugh Gough** marched to meet them with 10,000 men.

1845, December 18. Battle of Mudki. The Sikhs attacked the British in late afternoon, but were repulsed with heavy losses.

1845, December 21–22. Battle of Ferozeshah. This battle was unique because the British governor general, Sir **Henry Hardinge,** volunteered to serve as second in command under Gough. It was not clear who was giving orders to whom. This was another late-afternoon battle, delayed due to disagreement between the generals. The British finally attacked the well-entrenched army of **Lal Singh,** 50,000 strong, and, after several repulses, finally took the position in one of the most bitterly contested battles ever fought in India. Early next morning Sikh reinforcements under **Tej Singh** attacked halfheartedly and were repulsed. The Sikhs then withdrew across the Sutlej. British losses were 2,400; Sikh casualties were at least 10,000.

1846, January 28. Battle of Aliwal. A raiding Sikh army under **Runjoor Singh** had a brief brush with a British force under Sir **Harry Smith** at **Ludhiana** (January 21). Smith, reinforced, attacked and decisively defeated the Sikhs in another vicious battle.

1846, February 10. Battle of Sobraon. Gough, having crossed the Sutlej with about 20,000 men, attacked the Sikhs, about 50,000 strong, entrenched in a bend of the river. In another sanguinary engagement the British smashed the Sikhs, who lost 10,000 men and 67 guns. Gough had 2,300 casualties. This ended the war. Gough marched to Lahore.

1846, March 11. Treaty of Lahore. The Punjab became a British protectorate.

Second Sikh War, 1848–1849

Knowing they had come closer to victory over the British than had any other Indian forces, the Sikhs wanted revenge. The opportunity came in an incident at Multan (April 20, 1848), in which 2 British officers were killed. The Punjab rose in revolt. At first the Sikh government in Lahore attempted to suppress the revolt, with British assistance, then turned against the British (August).

1848, November–1849, January. Siege of Multan. This was carried out by part of the British Army under General **William S. Whish.**

1848, November 9. Invasion of the Punjab. Gough's main army, about 15,000 men, crossed the Sutlej. The Sikhs under **Shere Singh** moved to meet the British with about 30,000.

1848, November 22. Battle of Ramnagar. Gough's cavalry was repulsed in an effort to cross the Chenab River. After this inconclusive skirmish, Gough decided to wait for Whish to take Multan and join him. When the siege dragged on, he renewed his advance.

1849, January 13. Battle of Chilianwala. Both sides attacked simultaneously. The British had slightly the best of an all-day fight, but lost 2,800 men. The Sikhs lost about 8,000. When the British retired, the Sikhs retook their captured guns. When news of the battle and its losses reached England, Gough was relieved and was to be replaced by Napier.

1849, February 21. Battle of Gujrat. Before his

orders arrived, Gough, reinforced by Whish, now had 24,000 men and artillery superiority. He crushed the Sikhs, about 50,000 strong, who were assisted by **Dost Muhammad** of Afghanistan (see below and p. 865). Preceded by a 2½-hour artillery bombardment, a general British advance overwhelmed the town and camp, capturing all the Sikh guns. Cavalry charges on both flanks completed the victory. British losses were only 96 killed and 700 wounded. Sikh losses were estimated at more than 2,000. Their leaders soon surrendered, and the Punjab was annexed (March 12).

AFGHANISTAN

Harrassed by its western neighbor, Persia, and by Sikh invasions from the Punjab on the east, Afghanistan was also torn by internal revolts and violent usurpations of power. The principal antagonists during this period were **Mahmud Shah,** his brother **Shah Shuja,** and Dost Muhammad. In addition, Russian and British intriguers were also working to control the country.

1799–1803. Reign of Mahmud Shah (brother of Zaman Shah). He overthrew and blinded his brother.

1803–1810. Reign of Shah Shuja (younger brother of Mahmud Shah). He overthrew and imprisoned his brother.

1810–1829. Second Reign of Mahmud Shah. Escaping from prison, he defeated Shuja, who fled. Mahmud was greatly assisted by his faithful vizier and general, **Fath Ali.**

1816. Persian Invasion. They captured Herat, but were soon thereafter driven off by local Afghans (see p. 853).

1818. Revolt and Chaos. This was caused by Mahmud's dismissal of Fath Ali.

1819. Sikh Conquest of Kashmir. This province of Afghanistan was conquered by Ranjit Singh, Sikh ruler of the Punjab (see p. 861).

1826. Rise of Dost Muhammad (brother of Fath Ali). He captured Kabul, then gradually extended his rule over the rest of the country.

1835–1839. Reign of Dost Muhammad. He became friendly with the Russians, arousing British fears.

1836–1838. Persian Invasion. Their objective was to annex the Herat region (see p. 853).

1838, July 29. Tripartite Treaty. The British East India Company, anxious to block Persian and Russian encroachments, reached an agreement with Ranjit Singh and with Shah Shuja.

Shah Shuja was to be restored to the throne of Afghanistan.

First Afghan War, 1839–1842

1839. Opening Campaign. Sir **John Keane** led the Army of the Indus, 21,000 strong, into Afghanistan. He occupied Kandahar (April), stormed Ghazni (July 21), and captured Kabul (August 7). Dost Muhammad was captured and sent to India; Shah Shuja replaced him. Keane returned to India, leaving a garrison at Kabul, to support 2 British civilian envoys: Sir **William Macnaghten** and Sir **Alexander Burnes.**

1841, November. Rising against the British in Kabul. This was led by **Akbar Khan,** son of Dost Muhammad, and resulted in the murder of both British agents. The British garrison, under elderly Major General **William G. K. Elphinstone,** was surrounded.

1842, January 6. British Capitulation at Kabul. Elphinstone's force of 4,500 men and 12,000 refugees was permitted to return to India under safe conduct.

1842, January 13. Gandamak Massacre. Elphinstone's demoralized command was surrounded by Akbar's troops in a defile on the Khyber Pass road. After ineffectual resistance, most of the British and Indian soldiers and noncombatants were massacred; a few were taken prisoner, but only a handful survived in captivity. The small British garrison at Ghazni had also been forced to surrender. Other British detachments at Kandahar and Jalabad held out.

1842, April. Shuja Assassinated.

1842, April–September. British Invasion. Sir **George Pollock,** with a punitive force from India, stormed the Khyber Pass and relieved Jalalabad (April 16). After reorganizing, he pushed on to Kabul, where he rescued 95 prisoners (September 15). Stern reprisal for the massacre followed. The citadel and central bazaar of Kabul were destroyed. The East India Company decided, however, that continued occupation of Afghanistan would be both unprofitable and dangerous, so the country was shortly evacuated (December), and Dost Muhammad was permitted to resume the throne.

1842–1863. Second Reign of Dost Muhammad.

1848–1849. Second Sikh War. Dost Muhammad allied himself with the Sikhs (see p. 863). His cavalry contingent was routed at the **Battle of Gujrat** (February 1949). The British annexed the Peshawar territories, but peace was not officially signed until 6 years later (1855).

SOUTHEAST ASIA

BURMA

Militant Burma continued its expansion north, west, and south, encroaching frequently on India and continuing intermittent warfare with Siam.

1811–1815. Arakan Uprising. Arakanese rebels, based in British territory near Chittagong, tried to drive out the Burmese, who protested British failure to stop use of their territory as a rebel base.

1819. Burmese Conquest of Assam. Assamese refugees fled into Manipur, under British protectorate, and again British territory was used as a base for attacks on territory occupied by the Burmese (1820–1822). Burmese complaints got no satisfaction from the British East India Company.

1822. Burmese Invasion of Manipur and Cachar. This punitive expedition was engaged inconclusively by British forces sent to help the native rulers.

First War with Britain, 1823–1826

1823. Burmese Invasion of India. Infuriated by British action, **Maha Bandula,** Burma's great general, and governor of Assam, planned a 2-pronged attack on Bengal from Assam and Arakan. Burmese troops were soon threatening Chittagong.

1824, March 5. British Declare War on Burma. There were operations in Arakan, Assam, Manipur, and Cachar. The decisive operations were in Burma itself.

1824, April. British Preparations. An expeditionary force of 5,000 British and Indian regular troops, under Major General Sir **Archibald Campbell,** was mobilized in the Andaman Islands.

1824, May 10. Occupation of Rangoon. The city was seized without opposition. The British, soon ravaged by disease, then found themselves ringed by a determined enemy.

1824, May–December. Investment of Rangoon. Savage fighting continued (June–July). Maha Bandula's army arrived from Arakan after a forced march across monsoon-flooded country (August). British reinforcements arrived, including a rocket battery (October–November).

1824, December 1. Burmese Assault. Bandula was decisively repulsed.

1824, December 15. British Assault. They broke Bandula's cordon investment. The Burmese, no match for the British discipline, retreated.

1825, February 13. Advance up the Irrawaddy. Campbell's column of 2,500 men was supported by a flotilla of 60 boats, manned by British sailors, and carrying 1,500 additional troops.

1825, April 2. Battle of Danubyu. Checked by Bandula, Campbell's rockets broke up a Burmese attack; a British counterattack swept the Burmese from the field. Bandula was killed.

1825, April 25. Occupation of Prome. Campbell went into quarters for the monsoon season, behind entrenchments, while the Burmese army—now under **Maha Nemyo**—again surrounded his position with field fortifications.

1825, November 30–December 2. Battle of Prome. Following repulse of a Burmese attack (November 10), Campbell launched an attack in 2 columns, supported by the flotilla. The Burmese investing lines were ruptured and rolled up; Nemyo was killed and the Burmese army disintegrated after 3 days of intense fighting. The British advance continued upriver to the vicinity of Yandabo (some 70 miles from Ava), the capital, where Burmese envoys, under a flag of truce, sought peace (January).

1826, February 24. Treaty of Yandabo. Burma surrendered Assam, Arakan, and the Tenasserim coast, and paid a large indemnity. Campbell's expedition then withdrew.

COMMENT. *This war ended Burmese military predominance in Southeast Asia.*

1838. Mon Rebellion. This was crushed by King **Tharawaddy.**

1844–1845. Civil War. Misrule by insane Tharawaddy led to revolt, which was suppressed by Tharawaddy's son, **Pagan Min,** who became king.

1845–1850. Deterioration in Anglo-Burmese Relations.

SIAM

Siam continued to be nominally at war with Burma (1785–1826), but this was confined mostly to frontier raids in the Tenasserim area and occasional clashes of expeditionary forces in northern Malaya and on the Kra Isthmus, since Burma was more seriously occupied to the west. Keeping a wary eye on her most dangerous rival, Siam turned her attention mainly to the south, where she coveted Malaya; to the east, where she periodically clashed with Vietnam over control of Cambodia; and to the northeast, where she strengthened her control over most of Laos. The principal events were:

1812. Intervention in Cambodia. (See p. 867.)

1821. Invasion of Kedah. Siamese conquest of this Malayan sultanate alarmed the British.

1826, June 20. Treaty with Britain. This established trade relations, and Siam agreed to halt its effort to penetrate farther into Malaya.

1826–1829. War with Vientiane. This ended with Siam virtually sovereign over most of Laos (see below).

1831–1834. Invasion of Cambodia. A Siamese army under able general **P'ya Bodin** was initially successful. The Khmer were defeated at the **Battle of Kompong Chhang.** Cambodian King **Ang Chan** fled to Vietnam. Guerrilla resistance arose in eastern Laos. Intervention of a Vietnamese army of 15,000, combined with a general uprising throughout the country, forced the Siamese to withdraw, leaving Vietnam in virtual control of Cambodia.

1841–1845. War with Vietnam over Cambodia. The Cambodians, revolting against Vietnamese rule, asked Siam for help. Aging P'ya Bodin led another army into Cambodia. The Siamese had somewhat the best of 4 years of fierce warfare. A compromise peace was agreed, with both Siam and Vietnam nominally sharing a protectorate over Cambodia, but with Siam predominant.

LAOS

Able King **Chao Anou** of Vientiane was determined to free himself from Siamese vassalage. He built up a strong army, but maintained friendly relations until Siam's troubles with the British seemed to provide an opportunity to revolt. The principal events were:

1819. Revolt. Chao Anou suppressed this, increasing his own power.

1826. Invasion of Siam. Chao Anou led his army in 3 columns into the heart of Siam, approaching within 30 miles of Bangkok before the surprised Siamese, under P'ya Bodin, halted and repulsed the invaders.

1826–1827. Siamese Invasion. P'ya Bodin led the Siamese army.

1827. Battle of Nong-Bona-Lamp'on. In a fierce 7-day struggle the Siamese forced their way across the Mekong River. The Laotian army was destroyed; Chao Anou fled to Vietnam. The Siamese destroyed the capital and devastated the country, deporting entire populations to regions of Siam which had been depopulated in the wars with Burma. They annexed Vientiane (1828).

1829. Return of Chao Anou. The former king and his Vietnamese troops were defeated and scattered by the Siamese. Chao Anou was captured.

1832. Vietnam Conquest of Xieng-Khouang. The Vietnamese annexed this east Laotian kingdom.

CAMBODIA

This weak nation continued to be a football for Siam and Vietnam. The principal events were:

1802–1812. Joint Siamese-Vietnamese Protectorate. King Ang Chan sought to bring peace to his country by paying tribute to both neighbors.

1812. Rebellion and Invasion. Ang Snguon, brother of Ang Chan, revolted, seeking Siamese help. **Rama II** of Siam sent in an army which quickly overran the country. Ang Chan fled to Vietnam. Emperor **Gia Long** of Vietnam immediately sent a large army to restore Ang Chan. The Siamese withdrew without battle. Vietnam became predominant in Cambodia.

1831–1834. Siamese Invasion and Siamese-Vietnamese War. (See p. 866.)

1841–1845. Siamese-Vietnamese War over Cambodia. (See p. 866.)

VIETNAM

Prolonged civil war finally came to an end early in the century, with successful **Nguyen Anh** ascending the throne in Hué as Emperor **Gia-Long** (1802). During the remainder of the period the Vietnamese were occupied with affairs in Cambodia to the west, and with increasing French intervention along the seacoast to the east. The principal events were:

1812. Intervention in Cambodia. (See p. 866.)

1824–1847. French Interventions. Persecution of Christians by Emperor **Minh-Mang** prompted several interventions by French naval commanders, with some armed clashes on land and sea. These culminated with an incident at **Tourane,** where 2 French warships destroyed a swarm of attacking Vietnamese vessels (1847).

1831–1845. Involvement in Cambodia. (See p. 866.)

MALAYA

1802–1824. Malacca Changes Hands. It was returned by England to the Dutch by the **Treaty of Amiens** (1802), and was repossessed (1811) and used by England as a base for an expedition against Java (see below). Malacca, Java, and Sumatra were restored to the Dutch (1814–1818) by the **Treaty of Vienna.** Malacca again came under English rule—traded by the Dutch (March 17, 1824) in exchange for Benkgulen, on Sumatra.

1819. Founding of Singapore by Sir Stamford Raffles. It soon outstripped the port of Malacca, to become the strategic and commercial center of the area, and Britain's bastion of naval power in the East.

1821–1826. British-Siamese-Burmese Rivalry in Malaya. (See pp. 865–866.)

1826, June 20. British Treaty with Siam. (See p. 866.)

1826. Establishment of Straits Settlement. Centralization of British colonial holdings in Malaya.

1837–1860. Operations against Pirates. (See below.)

INDONESIA

Despite British interference during the Napoleonic Wars, the Dutch were able to consolidate their position in Java and the Spice Islands, and to expand their control over most of the archipelago. The principal events were:

1808. Pacification of Bantam. This brought all of western Java under Dutch sovereignty.

1810–1811. Colonial War with Britain. A large British expedition under Lord **Gilbert Elliot of Minto,** governor general of India, sailed against Java. Batavia was seized (August 1811). A month later (September 17) the Dutch signed the **Capitulation of Semarang,** in which they ceded Java, Palembang, Timor, and Macassar to the British.

1812. Capture of Palembang. The British seized the town after the local sultan refused to accept the terms of the Capitulation of Semarang.

1816. Britain Returns Indonesian Possessions to the Dutch. This was in accordance with the treaties settling the Napoleonic Wars.

1825–1830. Great Java War. The last native prince, **Dipo Negara,** revolted against Dutch rule. The insurrection was suppressed after 5 years of continuous guerrilla warfare, in which 15,000 Dutch soldiers and an unknown number of Javanese were killed.

1837–1860. Operations against the Pirates. Piratical depredations throughout Indonesia and neighboring waters spurred the Dutch, the British, and the Spanish to action and occasional cooperation.

1841. The White Rajah of Sarawak. Sir **James Brooke,** leading British efforts against the pirates, was granted territory along the west coast of Borneo—and the title of rajah—by the grateful Sultan of Brunei for suppressing a revolt.

1849. Revolt in Java. Suppressed by the Dutch, who increased their control.

1839–1849. Dutch Pacification of Bali.

THE PHILIPPINES

The Philippines had been economically linked with the Spanish colonies in the Americas. Loss of these colonies resulted in stagnation and decadence in the Philippines. There was little of military significance save for operations against the pirates (see above). A French naval expedition, anxious to establish a base for operations against Vietnam, attempted to seize Basilan, in the Sulu Islands, but was ignominiously repulsed (1845). France did not repeat the attempt due to Spanish protests.

EAST AND CENTRAL ASIA

CHINA

Sporadic rebellions and army mutiny, and increasing piratical depredations along the south coast of China, indicated a weakening of central control and a decline of the Manchu Dynasty. The principal events were:

1796–1804. Revolt in Western China. Uprising of the White Lotus Society in Hupei, Szechuan, and Shensi was suppressed with difficulty.

1825–1831. Moslem Invasion of Kashgaria. Inroads from western Turkestan into Kashgaria in western Sinkiang threatened for a time the Chinese hold on that area (captured in 1758). The invaders were finally ejected.

First Opium War, 1839–1842

Disagreements between Chinese officials and British merchants trading at Canton—particularly concerning the importation of opium into China—led to Chinese action against the European community (November 1839).

1840. British Expedition to China. A force of some 4,000 men—partly English regulars, partly East India Company native troops—under Sir **Hugh Gough** arrived in Chinese waters escorted by a British naval squadron. They occupied the island of Chusan, at the entrance of Hangchow Bay. Moving south, the squadron blockaded Hong Kong and Canton.

1841, February 26. Capture of the Bogue Forts. A British amphibious operation captured the Pearl River fortifications, key to Canton. The flotilla then moved up the Pearl River.

1841, May 24. Capture of Canton. An amphibious assault was followed by the storming of the forts surrounding the city. Temporary peace followed.

1841, August–October. British Coastal Operations. The expeditionary force moved up the China coast. Amoy was bombarded and captured (August 26); Ningpo (Ninghsien) fell (October 13).

1841–1842. British Administrative Mismanagement. Operations halted for the winter, during which the British troops suffered heavily from improper administration and supply. Although British naval and military combat operations were efficient, other arrange-

ments were shockingly poor. At times as many as 50 percent of the troops were disease-ridden. Many transports were unseaworthy; several foundered in typhoons, with much loss of life. Food and medical attention were unsatisfactory. But China was far away; the stockholders of the British East India Company were profiting; the British War Office and Admiralty paid little heed.

1842. Yangtze River Operations. The capture of Shanghai (June 19) was followed by a move upriver. The fall of Ching-kiang (July 21) menaced Nanking, and China sued for peace.

1842, August 29. Treaty of Nanking. China was required (1) to cede Hong Kong; (2) to open treaty ports—Canton, Amoy, Foochow (Minhow), Ningpo, and Shang-hai—to British trade; and (3) to pay an indemnity of $20 million.

1847. Renewed Moslem Invasion of Kashgaria. Incursions from western Turkestan were combined with uprisings among the local Mohammedan population. Severe Chinese reprisals were followed by a mass exodus of Moslems.

JAPAN AND KOREA

Self-isolated Japan continued her efforts to bar foreign contacts, despite several efforts to break down the barrier. Russia's **Nicolai Petrovich de Rezanov** of the Russian-American Company (analogous to Britain's East India Company) reached Nagasaki (1804), but after 6 months of effort failed to obtain a treaty. Raids on Sakhalin and the Kuriles by expeditions of the Russian company were repulsed (1806–1807). Japanese gunfire greeted an American ship attempting to land a party of missionaries and merchants at **Naha** in the Ryukyu Islands (1837). An American naval mission was rebuffed in 1846.

In Korea, no events of military importance transpired during the period.

NORTH AMERICA

THE UNITED STATES, 1800–1812

1801–1805. Tripolitan War. (See p. 854.)
1803. Louisiana Purchase.
1810, October 27. Annexation of West Florida. This was proclaimed by President Madison after American frontiersmen seized the Spanish fort at Baton Rouge and neighboring regions and proclaimed the independent State of West Florida. The dispute over whether West Florida was included in the Louisiana Purchase was settled when Spain renounced sovereignty (1819).
1811. Indian Trouble in the Northwest. Tecumseh, chief of the Shawnees, attempted to organize a confederation of tribes to oust the white settlers encroaching on their land. He was supported by British-Canadian fur interests. As friction increased, Brigadier General **William Henry Harrison,** governor of Indiana Territory, moved with 1,000 troops against the Indian capital, "Prophetstown," 150 miles north of Vincennes.
1811, November 7. Battle of Tippecanoe. Harrison's force, bivouacked on Tippecanoe Creek, was attacked at dawn by the Indians, some 700 strong. The assault was repulsed. Harrison's well-trained troops counterattacked and routed the Indians, then destroyed Prophetstown, shattering Tecumseh's power temporarily. American losses were 37 killed and 150-odd wounded. The Indians left 36 dead on the field, with an unknown number of dead and wounded carried off.

1811–1812. Friction with Britain. The Tippecanoe campaign brought to boiling point an ever-increasing anti-British sentiment in the United States, engendered in part by Canadian support of the Indians, in part by American expansionists determined to annex Canada. The tension had been increased by a succession of encroachments on American neutrality by England and France in the continuing Napoleonic Wars. Both nations preyed on American commerce in their mutual blockading efforts. British warships continuously impressed seamen from American merchantmen. (More than 6,200 American citizens were thus handled.) The British frigate HMS *Leopard* opened fire on USS *Chesapeake* (June 22, 1807) off Norfolk Roads, killed and wounded several men, and then took off 4 alleged deserters. The incident aroused national indignation. Four years later the American frigate USS *President*, searching for a British frigate which had impressed more American sailors, overhauled the British sloop of war HMS *Little Belt*, disabling her with gunfire and killing or wounding 32 men (May 16, 1811). This retaliation for the *Leopard-Chesapeake* affair was widely acclaimed. Expansionism was in the air; the "War Hawks" urged conquest of Canada while England was busy at war with France.

THE WAR OF 1812 (1812–1815)

1812, June 19. Declaration of War. This was ostensibly to defend the doctrine of "freedom of the seas."

Land Operations, 1812

1812, June. American Plans. A poorly planned 3-pronged invasion of Canada speedily aborted as other events caused Americans to go on the defensive.

1812, July 17. British Capture of Fort Mackinac.

1812, August 15. British Capture of Fort Dearborn (present site of Chicago). The garrison, a company of infantry, was massacred by the Indians after surrendering.

1812, August 16. Surrender of Detroit. Brigadier General **William Hull** had moved from Dayton, Ohio, to Detroit with some 2,500 men. There he abjectly surrendered, without firing a shot, to British General Sir **Isaac Brock,** who had but 730 Canadians and 600 Indians (under Tecumseh).

1812, October 13. Battle of Queenston. A force of some 2,270 militia and 900 regulars, under New York militia Major General **Stephen Van Rensselaer,** attempted to invade across the Niagara River at Queenston. Part got across, mostly regulars. But Brock, who had hurried from Detroit to command the defense, brought up reinforcements and pinned them down. The militia contingent still on the American side, refusing to leave the territorial limits of the U.S., stood idly by while Brock's 600

British regulars and 400 Canadian militia overwhelmed their comrades. American losses were 250 killed and wounded and 700 made prisoner. The British lost 14 killed (including Brock) and 96 wounded.

1812, November. Lake Champlain Expedition. The American main effort, 5,000 men under Major General **Henry Dearborn,** moved down the lake from Plattsburg to Rouses Point. The militia contingent now stood on their constitutional rights and refused to cross into Canada (November 19). Dearborn returned to winter quarters.

Naval Actions, 1812

In bright contrast to the dismal army performance, the small but efficient U.S. Navy (14 seaworthy vessels) took to the high seas against overwhelming British sea power (1,048 war vessels) and engaged in a number of brilliant single-ship actions.

1812, August 19. USS *Constitution* vs. HMS. *Guerrière*. Captain **Isaac Hull** in the frigate *Constitution*, 44, after eluding the blockade and escaping from the clutch of a British squadron in a remarkable stern chase, demolished the British frigate *Guerrière*, 38, in a half-hour action off the Nova Scotia coast. Hull arrived at Boston with news of the victory the same day that word was received of his uncle's ignominious surrender at Detroit (see p. 871).

1812, October 18. USS *Wasp* vs. HMS *Frolic*. These 18-gun sloops of war clashed off the Virginia coast in a 43-minute broadside-to-broadside hammering. When it was over the *Frolic* was a helpless wreck. But the *Wasp*'s rigging was so shot up that when HMS *Poictiers*, 74, came over the horizon, the American vessel, unable to make sail, was forced to strike her colors.

1812, October 25. USS *United States* vs. HMS *Macedonian*. Captain **Stephen Decatur**'s frigate *United States*, 44, off Madeira, systematically smashed the British frigate by long-range gunfire in 90 minutes. Partly dismasted, the *Macedonian*, 38, was brought back to Newport, refitted, commissioned in U.S. service, and took further part in the war against her former flag.

1812, December 29. USS *Constitution* vs. HMS *Java*. Now under the command of Captain **William Bainbridge,** "*Old Ironsides*" encountered the British frigate off Bahia, Brazil. After a 2-hour contest of maneuver and gunfire, the *Java*, 38, was a total wreck and surrendered. COMMENT. *Ship for ship and crew for crew the superiority of American design, seamanship, and gunnery was proven. But the strangulating pressure of British sea power was closing in on the Atlantic coastline of the United States. In all, nearly 100 vessels were now engaged in this blockade duty: 11 ships of the line, 34 frigates, and the remainder sloops of war or smaller.*

Land Operations, 1813

ACTION IN THE WEST

1813, January–August. Preliminaries. British General **Henry A. Proctor** with a force of British regulars, Canadian militia, and Indians under Tecumseh (commissioned a British brigadier general) held the initiative. Major General Harrison had received a directive to retake Detroit, but his means were scanty. While he was gathering and training troops, his outposts were attacked by Proctor, who took **Frenchtown** on the Raisin River (January 22). American losses were 197 killed and wounded; another 737 were captured. Many of the wounded were murdered by the Indians. The British also besieged Fort Meigs on the Maumee River, and Fort Stephenson on the Sandusky without success.

1813, September. Advance on Detroit. Harrison was now ready with 7,000 well-trained men. But American control of Lake Erie was essential to his plans against Detroit. Soon (September 10) he received a message from Commodore **Oliver Hazard Perry** reporting American victory in the Battle of Lake Erie (see p. 875). Harrison at once started north, sending his cavalry 100 miles overland around the lake, and moving his infantry and artillery in Perry's ships from Fort Stephenson across to Amherstburg. Both forces converged on Detroit, accompanied by Perry's squadron. Proctor withdrew into Canada.

1813, September 29. Recapture of Detroit. Harrison pushed up the Thames River in pursuit of the British, accompanied by 3 of Perry's lightest ships. When the water shoaled, the commodore came ashore, accompanying Harrison as a volunteer aide.

1813, October 5. Battle of the Thames. Proctor with 800 regulars and 1,000 Indians was brought to bay at Chatham. Harrison, with 3,500 men, attacked methodically. While his infantry assaulted frontally, his cavalry charged the British right. Proctor's command collapsed, losing 12 killed, 22 wounded, and 477 taken prisoner. Tecumseh's Indians held firm until their chief was killed, then they fled, leaving 35 dead. American losses were 29 killed and wounded.

COMMENT. *Highlights of this campaign were the close military-naval cooperation and the prowess of the American cavalry—a regiment of Kentucky mounted riflemen. The victory brought about the collapse of the Indian confederacy, England's allies. It would later bring "Old Tippecanoe" to the White House, but at the moment its fruits were dissipated. The War Department ordered Harrison's militia disbanded and sent home; his regulars were to join the American concentration in the Niagara area. Harrison, enraged, resigned his commission and returned to civil life.*

OPERATIONS ON THE NORTHERN FRONT

On the Niagara front, General **John Armstrong,** Secretary of War, planned an ambitious 2-pronged invasion of Upper Canada. As a preliminary, an amphibious operation on Lake Ontario was mounted by Major General **Henry Dearborn** and Commodore **Isaac Chauncey.**

1813, April–May. Burning of York. A force of 1,600 troops under Brigadier General **Zebulon M. Pike,** embarking in Chauncey's flotilla at Sackets Harbor, captured York (now Toronto), capital of Upper Canada (April 27). The explosion of a powder magazine resulted in killing or wounding 320 of the invaders, Pike being among those killed. Despite Dearborn's orders, the public buildings of York were burned. The expedition returned (May 8), having accomplished nothing except to harden Canadian resentment.

1813, May 27. Capture of Fort George. During the illness of the ineffectual Dearborn, Colonel **Winfield Scott,** his adjutant general, in cooperation with Perry (then on Lake Ontario under Chauncey), mounted an amphibious operation against Fort George at the mouth of the Niagara River. Some 4,000 men were landed in the rear of the British post, taking it by assault. As a result of the move, the British garrison of Fort Erie (opposite Buffalo) withdrew, liberating American vessels bottled up in the Black Rock navy yard. Scott's operation was nullified when British General **John Vincent,** retreating from Fort George with 700 men, turned on an American pursuing force 3 times his size at **Stony Creek** (June 6) and threw it back. Generals **William H. Winder** and **John Chandler,** political appointees, were bagged in the disaster. The British reoccupied Fort Erie.

1813, May 28–29. Battle of Sackets Harbor. Amphibious assault by Sir **John Prevost,** governor general of Upper Canada, was repulsed by the small garrison of Brigadier General **Jacob J. Brown.**

1813, September–November. Planned Operation against Montreal. Brigadier General

James Wilkinson, replacing Dearborn, was to move down the St. Lawrence from Sackets Harbor. Brigadier General **Wade Hampton** would push north from Lake Champlain. Uniting, they were to take Montreal, defended by some 15,000 British troops.

1813, October 25. Battle of the Chateaugay. Hampton, entering Canada with 4,000 men, moved west to the Chateaugay River and established himself about 15 miles from its mouth (October 22). Attacking a much smaller British force, he entangled himself in a swamp. The British commander caused several bugles to be blown, simulating an envelopment, whereupon Hampton, abandoning any attempt to join Wilkinson—for whom he had a heartily reciprocated antipathy—retreated to Plattsburg and went into winter quarters.

1813, November 11. Battle of Chrysler's Farm. Wilkinson, with 8,000 men, moved down the St. Lawrence by bateaux, with flank guards marching on both banks. A British force of some 800 regulars and Indians under Colonel **J. W. Morrison** menaced his rear about 90 miles from Montreal. Attempting a bungling, piecemeal attack, Wilkinson was routed, losing 102 killed, 237 wounded, and more than 100 others captured. Wilkinson now withdrew to winter quarters at French Mills, on the Salmon River, where his troops suffered great privation.

1813, December 18. British Capture of Fort Niagara. The Indian allies ravaged the countryside. The fort would remain in British hands for the rest of the war.

1813, December 29–30. Burning of Buffalo. A column of 1,500 men under General **Gordon Drummond** moved south down the Niagara River, burning the city of Buffalo and the Black Rock navy yard, inflicting much damage.

Naval Actions, 1813

At Sea

The inferiority of American naval strength reduced the U.S. Navy to a policy of commerce destruction, individual ships eluding the British blockade to cruise the oceans. A great number of privateers assisted in the operations.

1813, February 24. USS *Hornet* vs. HMS *Peacock*. The sloops of war clashed off the Brazilian coast, the British ship being sunk after 11 minutes of combat.

1813, June 1. HMS *Shannon* vs. USS *Chesapeake*. American Captain **James Lawrence,** unwisely accepting the challenge of Captain Sir **Philip Broke,** took his untried frigate, 38, and newly raised crew out from Boston to meet a vessel reputed to be the most efficient in the Royal Navy. In a few minutes the British frigate, 38, had so raked the *Chesapeake*'s decks that nearly one-third of her crew were killed or wounded. Lawrence, mortally wounded, cried, "Don't give up the ship!" but the Americans surrendered after Broke brought his ship alongside and boarded. In all, 146 Americans, including most of the officers, were casualties; 83 Britishers fell. The *Chesapeake,* brought into Halifax as a prize, would remain on the Royal Navy list for many years to come.

1813, August 14. HMS *Pelican* vs. USS *Argus*. The sloop of war *Argus*, 18, after a successful cruise in the English Channel, in which she captured and destroyed 20 prizes, was sunk by the British sloop of war, 20.

1813, September 3. USS *Enterprise* vs. HMS *Boxer*. The American sloop met and captured the *Boxer,* 14, off the New England coast.

OPERATIONS ON THE LAKES

When the war broke out, England possessed several warships on lakes Erie and Ontario, while the U.S. had none. American Commodore Chauncey had begun the construction of vessels at

Sacketts Harbor on Lake Ontario, and at Erie and Black Rock on Lake Erie. In a remarkably short time, experienced ship carpenters constructed an inland navy, which was manned by seamen sent up from the Atlantic Coast. Chauncey retained command of operations on Ontario, but delegated command on Lake Erie to Captain Oliver H. Perry. By early August, Perry had in commission 2 brigs, 6 schooners, and a sloop ready for action in Put in Bay, near Erie. The British flotilla on Lake Erie, under Captain **Robert Barclay,** consisted of 2 ships, 2 brigs, 1 schooner, and a sloop.

1813, September 10. Battle of Lake Erie. The British squadron approached the American anchorage about noon, and Perry went out to meet it. His flagship, the brig *Lawrence,* 20, met the British fire before the rest of his squadron could close, and was put out of action. Perry, in a rowboat, transferred under fire to the brig *Niagara.* In Nelsonian fashion the *Niagara* broke the British line; the remainder of the American squadron closed in. The *Niagara* engaged and put out of action 3 of the British craft, including Barclay's flagship. A detachment of General Harrison's sharpshooting soldiers acted as marines. By 3 P.M. Barclay's entire squadron had struck. British losses were 41 killed and 91 wounded. The Americans lost 27 killed and 93 wounded, most of these (22 killed and 61 wounded) on board the battered *Lawrence.*

COMMENT. *Perry's victory was the turning point in the war in the northwest. His message (beginning "We have met the enemy and they are ours . . .") triggered Harrison's operations leading to the recapture of Detroit (see p. 873).*

Land Operations, 1814–1815

NIAGARA FRONT

1814, February–March. Wilkinson Renews the Offensive. He advanced from Plattsburg and Sacketts Harbor with 4,000 men.

1814, March 30. Repulse at La Colle Mill. Wilkinson's assault on a small border fort was repulsed by its garrison of 600 men. Wilkinson, who before starting had announced his determination "to return victorious or not at all," at once fell back on Plattsburg. He was summarily relieved of his command (April 12).

1814, April–July. New American Command Team. Jacob Brown, the new commander, at once started reorganizing and training his troops. The task was efficiently performed by Scott, now a brigadier general. Brown's troops were well equipped and newly uniformed—Scott's own brigade was dressed in gray since the contractors could not supply blue cloth.

1814, July 2–3. Invasion of Canada. Brown, with 3,500 troops, crossed the Niagara River, seizing Fort Erie. He then started north.

1814, July 5. Battle of Chippewa. General **Phineas Riall** with 1,700 British regulars and a small number of Canadian militia and Indians, held a defensive position on the north bank of the unfordable Chippewa River, 16 miles north of Fort Erie. Brown bivouacked on the south bank of Street's Creek, a mile south of the river (July 4). A flat plain lay between the two forces. Riall crossed the Chippewa next day, driving in a militia and Indian force of Brown's left, only to meet headlong Scott's brigade, 1,300 strong. Riall, noting the gray uniforms of the Americans, believed them to be militia. But when, under fire, they formed line with parade-ground precision and moved to meet him with fixed bayonets, he is said to have exclaimed: "These are regulars, by God!" Scott led a charge that drove the British back in complete defeat across the river and into their entrenchments. British losses were 236 killed, 322 wounded, and 46 captured. Scott's losses were 61 killed, 255 wounded, and 19 captured.

COMMENT. *This was the first time that regular forces of both sides met in close combat on even terms. The gray full-dress uniform of the U.S. Military Academy at West Point commemorates this victory.*

1814, July 25. Battle of Lundy's Lane. Brown moved north to Queenston, Riall falling back before him. But Commodore Chauncey failed to cooperate on the lake with Brown's movement. British reinforcements from Europe having reached Canada, Sir Gordon Drummond took over Riall's command, bringing with him additional troops. Near Niagara Falls the opponents clashed: Brown with 2,600, Drummond with 3,000. A 5-hour sanguinary contest ended in a draw. American losses were 171 killed, 572 wounded, and 110 missing. The British lost 84 killed, 559 wounded, and 235 prisoners and missing. Brown, Scott, and Drummond were wounded; Riall was taken prisoner. Brown fell back on Fort Erie, which Drummond besieged.

1814, September 17. Battle of Fort Erie. An American sortie from Fort Erie broke the deadlock. British losses were 609 in all; American, 511. Fort Erie was later abandoned and with it all further effort to invade Canada (November 5). No further actions of importance occurred on the Niagara front.

LAKE CHAMPLAIN FRONT

1814, August 31. British Invasion. Prevost, with 14,000 of Wellington's veterans just arrived from Europe, advanced on Lake Champlain from Montreal. The only ground force available to oppose him consisted of 1,500 green regulars and about 3,000 raw militia, under General **Alexander Macomb,** manning field fortifications at Plattsburg. Once they had been dispersed, the road would be open all the way to New York, to complete successfully the invasion of the Hudson Valley bungled by Burgoyne 37 years earlier. Prevost planned to assault the fortifications at Plattsburg frontally, while a naval force took care of an American flotilla protecting the lake flank of the American position.

1814, September 11. Battle of Plattsburg. Prevost made his combined ground and naval assault on Plattsburg. The British naval squadron, however, was completely defeated by the American flotilla and was forced to surrender (see p. 879). Prevost thereupon hastily broke off the land action, even though one of his assault columns had made good progress. Be-

lieving, with good reason, that his proposed invasion would fail without command of the lake, he retired in some disorder, abandoning a quantity of stores. There was no further action on this front during the war.

CHESAPEAKE BAY AREA

1814, July–August. British Threat to Washington. On the east coast of the United States, Admiral Sir **John Cockburn**'s British squadron, which had been harassing the seacoast (see pp. 878–879), was joined by a force of 5,400 British veteran troops released by the conclusion of the Peninsular War. Under Major General **Robert Ross,** they were landed on the Patuxent and advanced on the American capital, 40 miles away (August 19).

1814, August 24. Battle of Bladensburg. The British advance was opposed by 6,500 untrained militia levies and a minuscule force of 400 sailors and marines, all under incompetent political Major General **William H. Winder.** Ross's advance guard, 1,500 strong, routed them in disgraceful panic. Only the naval contingent, of gunners under Commodore **Joshua Barney** and a handful of regular army troops contested the attack. American losses were 100 killed and wounded and 100 captured. The British had 294 casualties.

1814, August 24–25. British Occupation of Washington. The American government had fled. The British burned the Capitol, the White House, and several other public and private buildings—retaliation for the burning at York. The invaders then marched back to their ships. National indignation resulted in the ousting of War Secretary, General **John Armstrong,** who was replaced by **James Monroe.**

1814, September 12–14. Attack on Baltimore. The British expedition then sailed north, and penetrated the Patapsco River. Ross landed his troops 16 miles from Baltimore, while the naval force attacked Fort McHenry. Local militia forces, from behind entrenchments, repulsed the land attack. Ross was mortally wounded. Fort McHenry successfully

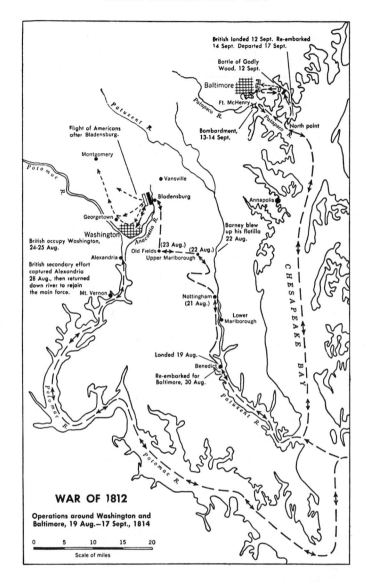

British landed 12 Sept. Re-embarked
14 Sept. Departed 17 Sept.

Battle of Godly
Wood, 12 Sept.

Baltimore

Ft. McHenry

North point

Bombardment,
13-14 Sept.

Patuxent R.

Patapsco R.

Patapsco R.

Flight of Americans
after Bladensburg.

Montgomery

Potomac R.

Vansville

Annapolis

Bladensburg

Georgetown

Barney blew
up his flotilla
22 Aug.

Anacostia R.

Washington

British occupy Washington,
24-25 Aug.

Old Fields ⟵—⟶ (22 Aug.)
(23 Aug.)
Upper Marlborough

Alexandria

British secondary effort
captured Alexandria
28 Aug., then returned
down river to rejoin
the main force.

Mt. Vernon

Nottingham
(21 Aug.)

Lower
Marlborough

C H E S A P E A K E   B A Y

Landed 19 Aug.

Benedict

Re-embarked for
Baltimore, 30 Aug.

Potomac R.

Patuxent R.

Potomac R.

WAR OF 1812

Operations around Washington and
Baltimore, 19 Aug.–17 Sept., 1814

0   5   10   15   20

Scale of miles

withstood the bombardment, inspiring **Francis Scott Key** to write "The Star-Spangled Banner." The expedition departed October 14.

SOUTHERN FRONT

1813–1814. The Creek War. The Creek Indians in Alabama allied themselves with England (July, 1813). The militia garrison of Fort Mims, 35 miles above Mobile, was surprised and more than half of the 550 soldiers and refugees massacred (August 30, 1813). **Andrew Jackson,** then a major general of militia, organized volunteer forces and at **Tallasahatchee** (November 3), and again at **Talladega** (November 9), inflicted severe punishment on the Indians. His force was then disbanded, however, and the Indians again became a serious menace.

1814, February. Jackson Again Takes the Offensive. He had a reorganized force of volunteers and a small increment of regulars.

1814, March 27. Battle of Horseshoe Bend. The main Creek force of 900 warriors was overwhelmed on the Tallapoosa River by Jackson's 2,000. Some 700 Indians were killed. Jackson lost 201. The Creek War was ended by the **Treaty of Fort Jackson** (August 9). Jackson, commissioned a major general in the Regular Army (May 22), was put in command of the Gulf Coast area.

1814, November 7. Seizure of Pensacola. Pensacola, in Spanish East Florida, had become a British base, supplying the Creeks. The U.S. Administration, fearing that neutral Spain might be brought into the war against the United States, forbade any movement against it. However, when a British-Indian attack mounted from Pensacola struck at Fort Bowyer, Jackson took matters into his own hands. Invading Spanish Florida, he seized and occupied Pensacola.

New Orleans Campaign, 1814–1815

1814, November 22. Jackson Goes to New Orleans. Rumors of a British concentration in the Caribbean—Gulf of Mexico area caused him to go to New Orleans to investigate its defenses.

1814, November 26. British Expeditions against New Orleans. A force of 7,500 Peninsular War veterans, under Major General Sir **Edward Pakenham,** sailed from Jamaica, its objective the seizure of New Orleans and control of the Mississippi River Valley.

1814, December 1. Jackson Arrives at New Orleans. Confirmation of the British preparations at Jamaica led him to prepare a defensive position at Baton Rouge in case he was unable to hold New Orleans.

1814, December 13. British Arrival. They landed in the Lake Borgne area. Jackson concentrated his energies on the defense of New Orleans, where he declared martial law to control the largely anti-American Creole population. Pakenham's advance guard penetrated to

within 7 miles of the city. Jackson swiftly organized a defensive line along the Rodriguez Canal, an abandoned waterway south of the city. His right flank rested on the Mississippi, his left on a cypress swamp. His forces consisted of a small core of regulars, his own Tennessee and Kentucky veteran volunteers, and some New Orleans volunteer militia, 3,100 in all. Some 2,000 additional untrained, ill-armed militia were also gathered, but he kept these in the background.

1814, December 23–24. American Night Raid. This reconnaissance in force rocked the British concentration, but was soon repulsed.

1814, December 25–1815, January 7. Opposing Preparations. Jackson continued to strengthen his defenses, piling earth, timbers, and cotton bales along the canal. A British probe of the defenses was hurled back by American artillery and small-arms fire (December 28). Effectiveness and staunchness of American response to a tremendous artillery bombardment (January 1) led Pakenham to wait further action until all of his troops were ready. Attempts by a British naval force to sail up the Mississippi and flank the position were thwarted by the successful defense of **Fort St. Philip,** 65 miles downstream. Finally, all his troops available, overconfident Pakenham determined to assault Jackson's position, unaware, of course, that the **Peace of Ghent** had been signed (December 24).

1815, January 8. Battle of New Orleans. The British infantry, in serried ranks, twice dared the aimed fire of Jackson's riflemen behind their entrenchments, only to be completely repulsed. Some 2,100 of the attackers were killed or wounded; an additional 500 were captured. Pakenham and his 2 senior subordinates were killed leading their men. The survivors withdrew, Jackson wisely making no attempt at counterattack with his motley command. He lost 7 killed and 6 more wounded. One week later the British force retreated to its boats.

Naval Actions, 1814–1815

1814. British Blockade. The effective restrictions on trade and manufacture brought the

U.S. Treasury close to bankruptcy. Depredations of British landing parties brought havoc along the coast. British general orders (July 18) were "to destroy and lay waste such towns and districts upon the coast as you may find assailable."

1814. American Guerre de Course. Although constrained by the blockade, a few U.S. warships and many privateers dared it successfully to prey on British commerce over the world. By midsummer more than 800 British merchantmen had been captured and most traffic about the English and Irish coasts moved only under naval convoy.

1814, April 29. USS *Peacock* vs. HMS *Epervier*. The American sloop, 18, named to commemorate the British vessel sunk by the *Hornet* (see p. 874) fell in with the *Epervier*, 18, off the coast of Florida. The British vessel struck after a 45-minute gun battle.

1814, March 21. Capture of the USS *Essex*. Captain **David Porter,** in the frigate USS *Essex*, 38, had left the Delaware River October 28, 1812, rounding Cape Horn on a fantastic 17-month cruise that terrorized British commerce up and down the Pacific, capturing or destroying more than 40 merchantmen and whalers. Shortly after entering Valparaiso, Chile (February 3), in company with *Essex Junior*, a prize converted into a man of war, Porter was blockaded by a British frigate, HMS *Phoebe*, 36, and *Cherub*, 18, a sloop of war. Porter, leaving his consort, put to sea in a heavy wind, hoping to draw off the slower British vessels. His main topmast snapped in a gust, and Porter returned to the sanctuary of the neutral port. The British ships, standing off his anchorage out of range of *Essex*'s carronades, with their long guns battered the helpless American frigate to a bloody, burning hulk. Porter struck after more than 3 hours, having lost 58 killed, 31 drowned, and some 70 wounded of the ship's 255-man complement. British losses were 5 killed and 10 wounded.

1814, June 28. USS *Wasp* vs. HMS *Reindeer*. The *Wasp*, 18, cruising in the English Channel, met and virtually destroyed the *Reindeer*, 18, in a half-hour fight. Continuing her cruise, the *Wasp* (second vessel of the name) captured 13 merchantmen.

1814, September 1. USS *Wasp* vs. HMS *Avon*. The British sloop, 18, was sunk by the American vessel in a night battle.

BATTLE OF LAKE CHAMPLAIN, SEPTEMBER 11, 1814

Paralleling the advance of Prevost's army on Plattsburg (see p. 876) was Captain **George Downie**'s hastily built British squadron of 4 ships and 12 armed galleys. Protecting the lake front of the American position at Plattsburg was the flotilla of Lieutenant **Thomas Macdonough,** who also had 4 ships and 12 galleys, as hastily constructed as their British opponents. The American vessels were anchored close to shore. In weight of metal the 2 squadrons were practically equal, but the Americans had a preponderance in long guns. Prevost's assault on the land side started as the British squadron rounded Cumberland Head and stood in to engage the Americans, ship to ship, taking a galling fire from Macdonough's long guns as they did. A closely contested 2-hour battle when Macdonough, whose frigate *Saratoga*, 26, was being badly battered by Downie's flagship, the frigate *Confiance*, 37, cleverly swung his ship by means of a stern anchor and presented a fresh broadside to his enemy. Downie was killed, some 180 of his crew dead or wounded, when the splintered *Confiance* struck her flag. The remainder of the British squadron then surrendered. American losses are estimated at 200, British casualties 300. A British veteran who had been at Trafalgar said it was "child's play" compared to Lake Champlain.

COMMENT. *This battle, ending all danger of British invasion from the north, was undoubtedly the decisive action of the war, saving America from possible conquest or dismemberment. Its influence was strongly felt at the Ghent Peace Conference when the news arrived there (October 21).*

1814, September 26–27. Stand of the *General Armstrong*. This 9-gun privateer brig for 2 days fought off successive boarding parties from a British squadron which attacked her in neutral Fayal, Azores. Captain **Samuel C. Reid** then put his crew ashore and blew up his brig. Portuguese authorities prevented a land assault by the British, who then sailed away.

1814, October. End of the USS *Wasp*. The *Wasp* was last seen by a Swedish vessel (October 9), then disappeared forever, one of the mysteries of the sea.

1815, January 15. Capture of USS *President*. Like the following actions, this took place after the peace had been signed. The American frigate, 44, commanded by Decatur, attempted to run the British blockade of New York. Several British vessels chased her. For some hours HMS *Endymion*, 50, and the *President* fought side by side until the Britisher was disabled. By that time 2 more British ships had caught up, and 2 more were in sight. Decatur, his ship crippled and 75 men killed or wounded, surrendered. The *President* became a unit of the **Royal** Navy.

1815, February 20. USS *Constitution* vs. HMS *Cyane* and *Levant*. Off Madeira, Captain **James Stewart** in "Old Ironsides," 44, met the British corvette, 34, and sloop of war, 21. By skillful maneuver, he fought them separately, capturing both.

1815, March 23. USS *Hornet* vs. HMS *Penguin*. The American sloop, 18, captured the British sloop, 18, off Tristan da Cunha Island in the war's last naval action.

End of the War

1814, December 24. Treaty of Ghent. Ignoring the war's basic issues, it provided for release of prisoners, restoration of territory, and arbitration of boundary disputes. The news reached New York February 11. The treaty was ratified February 15.

COMMENT. *The war decided nothing, and, until the final battle at New Orleans, the British had had somewhat the best of it. They never repeated their* *highhanded conduct at sea; the Americans never again tried to conquer Canada.*

THE UNITED STATES, 1815–1846

1815. Decatur's Mediterranean Expedition. This concluded U.S. troubles with the Barbary States (see p. 854).

1818. First Seminole War. The Seminoles had sided with the British during the War of 1812. After its end, they continued their depredations. Their so-called Negro Fort (largely manned by escaped slaves) on the Apalachicola River was destroyed by U.S. regular troops (November–December 1817). The Indians in Spanish Florida raided, massacred, and pillaged in U.S. territory. General Jackson moved overland from Nashville with 1,800 regulars and 6,000 volunteers and relieved the besieged garrison of Fort Scott on the Georgia-Florida border (March 9).

1818, April–May. Invasion of Florida. Jackson, with some 1,200 men, occupied the Spanish post of St. Marks (April 7) and the Spanish capital, Pensacola (May 24). The governor fled to Fort Barrancas, where Jackson promptly shelled him out and captured the post. Meanwhile, other American columns had been destroying Indian villages, breaking up Seminole power. Two British subjects, traders, found assisting the Indians at St. Marks, were summarily court-martialed, convicted, and executed. The net result was American occupation of Spanish Florida—without authority—and a near war with England. But Seminole depredations had been ended. Jackson's highhanded action was upheld by the government. The posts were returned to Spain, but negotiations were begun to obtain all Florida from Spain.

1819, February 22. Adams-Onís Treaty. Spain surrendered all claims to West Florida, and ceded east Florida to the United States. All Spanish claims to the Pacific Northwest were also canceled. In the southwest, the treaty defined the boundary line of Spanish Mexico from the mouth of the Sabine River in the Gulf of Mexico, northwest to the Pacific Ocean.

1823, December 2. Monroe Doctrine. The declaration of U.S. intention to protect America, for the Americans, and against foreign aggression, was announced.

1825–1832. American Expansion Westward. This steadily encroached on land the Indians considered their own. The Army, in its role of exploration and protection of the settlements, began expansion of its string of frontier posts. Previous treaties with various tribes were nullified, both legally and illegally, and the long process of moving the red man from areas coveted by the white began. This led to a continuous succession of small wars, which would last for three-quarters of a century more.

1832, April–August. Black Hawk War. The Sac and Fox tribes, along the Mississippi in Illinois and Wisconsin, led by Chief **Black Hawk,** tried to regain their lands. Colonel **Zachary Taylor,** operating under Brigadier General **Henry Atkinson,** led 400 regulars and 900 militia through swampland to defeat the Indians at the **Battle of the Bad Ax** (August 2).

Second Seminole War, 1835–1843

The Seminoles and Creeks of Georgia, Alabama, and Florida resisted removal to the west. Under **Osceola** they ravaged frontier settlements in Florida. Army garrisons were too small to quell the outbreak.

1835, December 28. Dade Massacre. A detachment of 150 regulars under Major **Francis L. Dade** was ambushed in the Wahoo Swamp of the Withlacoochie River. Only 3 men escaped.

1836, February–April. Scott's Campaign against the Seminoles. Major General Winfield Scott, commanding the Eastern Department, sent to suppress the uprising, rescued his rival, Major General **Edmund P. Gaines,** commanding the Western Department, from a Seminole ambush on the **Withlacoochie River,** where he had come without authority (February 27–March 6). Scott then swept twice through Seminole country with regulars and volunteers until malaria halted operations. He reported that victory would require a large regular force.

1836, June–July. Scott's Campaign against the Creeks. The War Department (in May) ordered Scott to arrange for suppression of the uprising of the Creek Indians of western Georgia and Alabama. Helped by Brigadier General **Thomas S. Jesup,** Scott suppressed it quickly with little loss of life.

1836–1837. Court of Inquiry on Scott and Gaines. Because of a misunderstanding regarding Scott's conduct of these operations and charges brought by Gaines, President Andrew Jackson ordered a court of inquiry, headed by Commanding General **Alexander Macomb.** It commended Scott's performance highly but strongly criticized Gaines (March 14, 1837).

1837, December 25. Battle of Lake Okeechobee. Meanwhile, Brigadier General Zachary Taylor was placed in command in Florida. Finding the Seminole stronghold in the Everglades, he attacked with about 1,000 men—half regulars, half volunteers—through swampland in 2 lines, the volunteers leading. When the volunteers broke as expected, the regulars held and won the day. Taylor lost 26 killed and 112 wounded. Organized Seminole resistance collapsed, but costly, bloody guerrilla actions continued for 4 years.

1837, December 29. *Caroline* Incident. Canadian militia invaded U.S. territory and seized the steamer *Caroline* on the Niagara River. Troops were concentrated on the border. General Scott's tactful diplomacy averted hostilities with Britain (see Mackenzie's Rebellion, pp. 888–889).

1838, February 12. Aroostock "War." Bickering between Maine and New Brunswick over lumbering on U.S. land again roused fears of war. Again Scott's diplomacy brought a settlement. Border disputes were resolved by the **Webster-Ashburton Treaty** (1842).

1844, February 28. The *Princeton* Disaster. On one of the first voyages of USS *Princeton*, first warship driven by a screw propeller, a gun designed by Commodore **Robert F. Stockton,** commander and co-designer of the ship, exploded, killing the Secretary of State, the Secretary of the Navy, and several congressmen, injuring a number of others. President **John Tyler,** aboard, was uninjured.

1846. June 15. Oregon Treaty. The longstanding Anglo–U.S. argument over Oregon Territory boundaries, which had led to the expansionist slogan, "Fifty-four forty, or fight!" was settled.

War of Texan Independence, 1835–1836

1835, June 30. Armed Insurrection. American settlers in Texas revolted against the Mexican government.

1836, February 23–March 6. Siege of the Alamo. General **Antonio López de Santa Anna,** President of Mexico, with a force of 3,000 Mexican regulars, laid siege to 188 Texans in the Alamo, San Antonio. After several repulses, the Mexican troops finally assaulted the place, massacring all within. The Mexicans lost more than 1,500 men. Santa Anna marched eastward, where an insurgent force was gathering.

1836, March 2. Texan Declaration of Independence. Texas was proclaimed a republic; an army was raised under **Sam Houston.**

1836. Massacre at Goliad. Some 300 Texans defending the place were killed.

1836, April 21. Battle of the San Jacinto. Sam Houston with 740 men surprised and routed Santa Anna's army of 1,600. Santa Anna was captured and recognized the Texan Republic. (This was repudiated by the Mexican government.) Houston was elected president of Texas. U.S. recognition of Texas (July 4, 1836) aroused Mexican resentment.

U.S.–Mexican War, 1846–1848

After long discussion and negotiations, Texas—which had asked for it—was formally annexed to the United States (March 1, 1845), despite Mexico's threat that this would mean war. Expansionist sentiment in the United States crystallized under the slogan of "Manifest Destiny," referring to the American "right ... to spread over this whole continent." Mexico contended that her territory extended north to the Nueces River; the United States claimed down to the Rio Grande.

Northern Campaign

1846, March 24. American Advance to the Rio Grande. General Zachary Taylor, with 3,500 men (two-thirds of the Regular Army), advanced and established Camp Texas opposite Matamoros, where some 5,700 Mexican troops were concentrated.

1846, April 25. Cavalry Encounter. Mexican cavalry, 1,600 strong, sweeping north of the river near Matamoros, overwhelmed an American reconnaissance force of 63 dragoons. Taylor announced next day that hostilities had

commenced and called for volunteers from the governors of Texas and Louisiana. He marched to protect his supply base at the mouth of the Rio Grande from the Mexican cavalry (May 1), leaving his camp opposite Matamoros under command of Major **Jacob Brown.**

1846, May 1. Mexican Crossing of the Rio Grande. Mexican General **Mariano Arista** led 6,000 troops and advanced on Camp Texas.

1846, May 3–8. Siege of Camp Texas. Major Brown was killed in the successful defense— the site became Brownsville.

1846, May 8. Battle of Palo Alto. Taylor, hav-

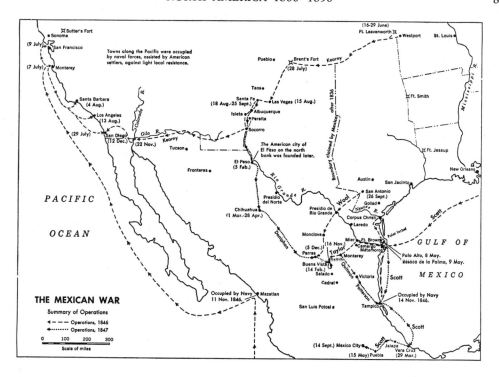

THE MEXICAN WAR

Summary of Operations

◄— — — Operations, 1846
◄········· Operations, 1847

0   100   200   300
Scale of miles

ing fortified his base at Point Isabel, rushed back to relieve Camp Texas. He had 2,200 men. Arista, with about 4,500, moved to meet him on flat, open ground. As the 2 lines faced each other, American artillery caused severe casualties among the Mexicans. A Mexican cavalry charge was repulsed and the shaken Mexicans fled as the American infantry line advanced.

1846, May 9. Battle of Resaca de la Palma. Taylor boldly attacked a strong Mexican defensive position, broke through after a brief but intense fight, and routed Arista's army completely. The Mexicans fled back across the Rio Grande. Mexican losses in the 2 days' battles were about 1,100; American casualties were 170 killed and wounded.

1846, May 13. Declaration of War. The United States declared war on Mexico.

1846, May 18. Taylor Crosses into Mexico.

1846, May–July. Delay in Matamoros. Taylor had to wait 3 months for transportation, which had been promised him by the United States government. Meanwhile, he trained volunteer reinforcements hurriedly raised by Congress.

1846, August. Taylor Moves South. He took 6,000 men—half regulars, half volunteers. The remainder of his force—about 6,000 volunteers—he left behind for additional training.

1846, September 20–24. Battle of Monterrey. Mexican General **Pedro de Ampudia,** with 7,000 regulars and 3,000 militia, stood on the defensive, holding the fortified city. In a 3-day battle the Americans, storming position after position, penned Ampudia in the central area of the city. Ampudia capitulated, marching out with the honors of war, but Taylor acceded to a Mexican request for an 8-week armistice. Mexican casualties were 367; American losses were greater: 120 killed, 368 wounded. The armistice was later disapproved by President **James K. Polk.** Taylor thereupon informed the Mexicans, and again advanced south.

1846, November 16. Occupation of Saltillo. There Taylor was joined by Brigadier General **John E. Wool**'s expedition, 3,000 strong, which had marched 600 miles overland from San Antonio (December 21).

1846–1847. American Strategic Debate. General Winfield Scott, commanding the U.S. Army, had planned an invasion of central Mexico via Veracruz. President Polk, a Democrat, was not anxious that Scott, a Whig, should become the war hero. This and other political bickering in Washington delayed matters until Taylor, at Saltillo, declined as militarily unsound the administration's suggestion that he march 300 miles across the desert to San Luis Potosí and then move on Mexico City. Taylor instead recommended a movement via Veracruz along the lines already planned by Scott. Finally Polk ordered Scott to undertake the campaign. Scott left Washington (November 24, 1846). Stopping to gather some of Taylor's troops at Point Isabel, he sailed to Tampico, where he established his headquarters (February 18).

1847, January–March. Scott's Plans and Preparations. The amphibious expedition, mounted in the Tampico area, was 10,000 strong. It included a large proportion of Taylor's veteran troops, leaving the latter with but 5,000 men, of whom only a handful were regulars. Taylor's role was to be defensive, while Scott's mass of maneuver executed a bold turning movement into central Mexico.

1847, January–February. Santa Anna's Plans. The Mexican president, learning of Scott's plans through a captured message, decided to crush Taylor before Scott could land. He moved with 20,000 men across the desert from San Luis Potosí—the very march which Taylor had rightly hesitated to take. Santa Anna's army arrived near Saltillo (February 19) after having lost some 4,000 men in the grueling midwinter desert march.

1847, February 22–23. Battle of Buena Vista. Taylor, though surprised by the Mexican approach, elected to defend, taking position in a narrow mountain gap 8 miles south of Saltillo. Santa Anna, sending a cavalry brigade on a wide northeastern detour to cut Taylor's line of communications, drove in Taylor's outposts on the first day. Next day he attacked the American position with his main body. Initially successful, in a most complicated and disjointed battle, the Mexican attack captured 2 guns, but was held up by magnificent efforts of the American regular artillery after volunteer infantry regiments had recoiled. Here Taylor is supposed to have uttered his historic: "A little more grape, Captain Bragg!"* The gallant Mexican infantry almost reached the guns before they broke under the American fire. Colonel **Jefferson Davis'** 1st Mississippi Volunteers, supported by regular artillery, meanwhile had checked an attack driving in Taylor's left flank. A vigorous counterattack then drove the Mexicans from the field. Meanwhile, the Mexican cavalry envelopment of Taylor's rear collapsed after a short conflict with the American rear guard at Saltillo. Santa Anna withdrew to the south, ending the northern campaign of the war. American losses were 267 killed, 456 wounded, 23 missing, together with the 2 guns captured Estimated Mexican losses were 500 dead and 1,000 wounded.

---

* Almost certainly apocryphal; he probably said: "Double shot your guns and give 'em hell!"

Operations in the West

KEARNY'S EXPEDITION

California had been disturbed since U.S. Commodore **Thomas ap C. Jones** had landed a naval force at Monterey (1842). This act had been repudiated by the American government. Upon the outbreak of war, revolts of American settlers were fomented (June) by army Captain **John C. Frémont,** heading a surveying expedition in the west.

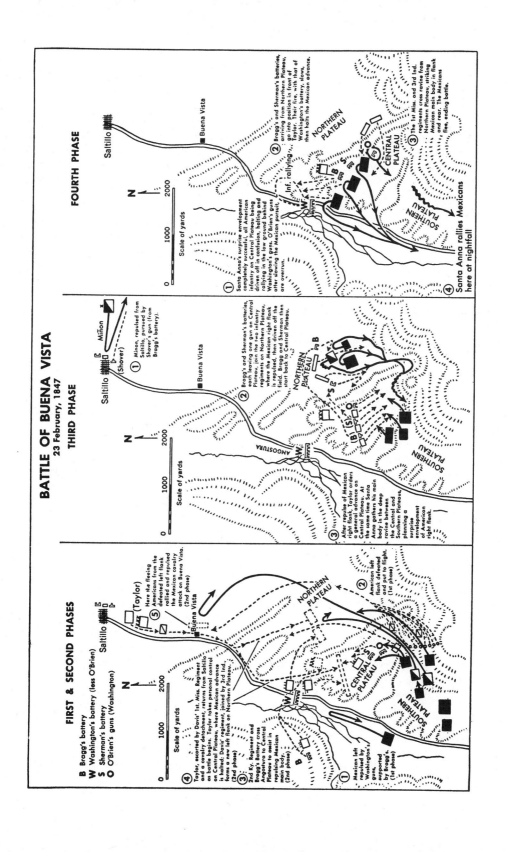

BATTLE OF BUENA VISTA

23 February, 1847

FIRST & SECOND PHASES

B Bragg's battery
W Washington's battery (less O'Brien)
S Sherman's battery
O O'Brien's guns (Washington)

Scale of yards
0   1000   2000

Saltillo

(Taylor)

Buena Vista

Here the fleeing Americans from the defeated left flank rallied and repulsed the Mexican cavalry attack on Buena Vista. (2nd phase)

NORTHERN PLATEAU

CENTRAL PLATEAU

SOUTHERN PLATEAU

④ Taylor, escorted by Davis' 1st Miss. Regiment and a cavalry detachment, returns from Saltillo as battle begins. Taylor takes personal control on Central Plateau, where Mexican advance is halted. Davis' regiment, joined by 3rd Ind., forms a new left flank on Northern Plateau. (2nd phase)

③ 2nd Ky. Regiment and Bragg's battery cross Angostura to Central Plateau to assist in repulsing Mexican main body. (2nd phase)

② American left flank defeated and put to flight. (1st phase)

① Mexican left repulsed by Washington's guns, supported by Bragg's. (1st phase)

THIRD PHASE

Scale of yards
0   1000   2000

Saltillo   Miñon

(Shover)

Buena Vista

ANGOSTURA

NORTHERN PLATEAU

SOUTHERN PLATEAU

① Miñon, repulsed from Saltillo, pursued by Shover's guns (from Bragg's battery).

② Bragg's and Sherman's batteries, each leaving one gun on Central Plateau, join the two infantry regiments on Northern Plateau, where the Mexican right flank is repulsed, then driven off the field. Bragg and Sherman then start back to Central Plateau.

③ After repulse of Mexican right flank, Taylor orders a general advance on Central Plateau. At the same time Santa Anna gathers his main body in the deep ravine between the Central and Southern Plateaus, planning a surprise envelopment of American right flank.

FOURTH PHASE

Scale of yards
0   1000   2000

Saltillo

Buena Vista

NORTHERN PLATEAU

CENTRAL PLATEAU

SOUTHERN PLATEAU

① Santa Anna's surprise envelopment completely successful, all American infantry on Central Plateau being driven off in confusion, halting and rallying in the low ground behind Washington's battery. O'Brien's guns, after slowing the Mexican pursuit, are overrun.

② Bragg's and Sherman's batteries, arriving from Northern Plateau, go into position in front of Taylor. Their fire, with that of Washington's battery, slows, then halts the Mexican advance.

③ The 1st Miss. and 3rd Ind. regiments cross ravine from Northern Plateau, striking Mexican main body in flank and rear. The Mexicans flee, ending battle.

④ Santa Anna rallies Mexicans here at nightfall

1846, June–December. Kearny's March. Brigadier General **Stephen Watts Kearny** was ordered to occupy New Mexico and California. Marching overland from Fort Leavenworth with 1,700 men, he established control over New Mexico. He received a message that California had been pacified, and so decided to leave most of his troops to hold New Mexico. Leaving Santa Fe (September 25) with 120 dragoons, Kearny reached southeastern California to learn of the revolts and to find the road to San Diego blocked by about 500 Mexican cavalry.

1846, July 7. Occupation of Monterey. Commodore **John D. Sloat** sent a naval force ashore. His successor, Commodore Robert F. Stockton, expanded the occupation and Frémont was named governor of California. Soon after this the Mexican-California populace rose against the Americans and controlled most of the province.

1846, December 6. Battle of San Pascual. This was a violent, inconclusive action in which Kearny held the field, but found his path to San Diego still blocked. However, he made contact with Stockton's sailors and marines, who marched out to meet him (December 10). Kearny's wagon train (the first to cross the Rockies) and a battalion of infantry, under Major **Philip St. George Cooke,** taking another trail, joined him later (January 29, 1847).

1847, January 8. Battle of San Gabriel. Kearny's combined army-navy command defeated the main Mexican force outside of Los Angeles.

1847, January 9. Battle of the Mesa. Kearny won a climactic victory, breaking the back of Mexican-Californian resistance. He at once occupied Los Angeles.

Doniphan's Expedition, 1846–1847

An offshoot of Kearny's operations was Colonel **Alexander W. Doniphan's** march with 850 Missouri mounted riflemen from Santa Fe (December 12) to Chihuahua.

1847, February 28. Battle of the Sacramento River. Near Chihuahua, Doniphan defeated a much larger Mexican force. Mexican losses were 600; American, 7 killed and wounded. Doniphan then continued south.

1847, May 21. Doniphan Arrives at Saltillo. This overland march, like that of Kearny, had been an amazing feat. Each of them marched more than 2,000 miles over desert territory, never previously traversed by organized military forces.

Central Mexican Campaign

1847, March 9. Scott Arrives Near Veracruz. The transports, carrying 10,000 troops, escorted by Commodore **David Connor's** squadron, began an unopposed landing which completely invested the fortress city.

1847, March 27. Capture of Veracruz. A 5-day bombardment by land and sea brought the surrender of the 5,000-man garrison. Losses on both sides were light: 82 Americans killed and wounded; 180 Mexicans, of whom 100 were civilians. Scott at once moved westward with 8,500 men toward Jalapa to get away from the low coastal area, where yellow fever was feared.

1847, April 18. Battle of Cerro Gordo. Santa Anna, with more than 12,000 men, held the fortified defile, whose flanks were seemingly impregnable. Scott's engineers (who included Captains **Robert E. Lee, George B. McClellan,** and **Joseph E. Johnston,** and Lieutenant **Pierre G. T. Beauregard**) discovered a flanking mountain trail over which Scott hurried his main body, enveloping most of the Mexican force. After a sharp fight the Mexicans were completely routed, losing 1,000 killed and wounded and 3,000 prisoners. Forty-odd guns and 4,000 small arms were captured. American losses were 64 killed and 353 wounded.

1847, May 15. Occupation of Puebla. Here

Scott halted, 75 miles from Mexico City. By this time he had but 5,820 men, having had to send home 4,000 others whose enlistments had expired.

1847, May–August. Wait at Puebla. Scott's strength was rebuilt by new troops, hurriedly raised in the United States, to almost 11,000 effectives and an additional 3,000 sick.

1847, August 7. Advance on Mexico City. Leaving his invalids and a small garrison at Puebla, Scott deliberately cut all communications with the outside world. Wellington's comment, when he learned of this, exemplified foreign military opinion: "Scott is lost. He has been carried away by successes. He cannot capture the city and he cannot fall back on his base." Santa Anna defended the capital with some 30,000 men. Scott, finding the direct route through Ayutla to be impracticable (August 12), circled to the south, where he discovered strong Mexican concentrations near Contreras and Churubusco, commanding the roads into the city. Scott, with only 4 days' rations left, decided to attack both positions (August 19), his decision guided by the effective reconnaissances of his engineers, who once again had found unmapped trails leading into the Mexican positions.

1847, August 20. Battle of Contreras. The hill position was stormed in a surprise dawn attack and its defenders routed. The 2 guns lost at Buena Vista were recaptured here.

1847, August 20. Battle of Churubusco. Almost simultaneously another American column burst through, storming a fortified convent. Losses were heavy in these 2 battles: 137 Americans killed, 877 wounded and 38 missing; while the Mexicans lost an estimated 4,297 killed and wounded, 2,637 prisoners, and some 3,000 missing. Scott had lost one-seventh of his effectives, Santa Anna one-third of his. The Mexicans retired within the city walls, except for strong detachments in and near the fortress of Chapultepec.

1847, August 25–September 6. Armistice. Discussion of peace proposals ended in disagreement; Scott immediately resumed his offensive.

1847, September 8. Battle of Molino del Rey. Some 12,000 Mexicans garrisoning a gun foundry and an old fort near Chapultepec were finally defeated after a day-long attack in which the Americans lost 116 killed and 665 wounded. Mexican losses were about 2,000 killed and wounded and 700 captured. Scott now had less than 7,500 effectives.

1847, September 13. Battle of Chapultepec. The fortified hill, last major obstacle outside Mexico City, was bombarded (September 12) and taken by storm next day, the rocky slopes and walls being scaled by ladders. Heroic features of the defense was the resistance of some 51 Mexican Military College cadets on the crest. At the same time other American columns drove along the causeways below the hill to the San Cosmé and Belén gates of the city, under heavy fire. American casualties were 130 killed and 703 wounded. Mexican losses are estimated at 1,800 in and around Chapultepec alone; their total losses are undetermined. Among many Americans distinguishing themselves, Lieutenants **U. S. Grant** and **T. J. Jackson** were outstanding.

1847, September 14. Capture of Mexico City. Scott's troops worked through the night, preparing for a final assault. Santa Anna's main force, however, had withdrawn after dark. At dawn, as bombardment was about to begin, the remaining garrison surrendered. Scott immediately occupied the city.

1847, September 14-October 12. Siege of Puebla. A Mexican force besieged the American garrison, but was driven off by troops from Veracruz.

1847, September-1848, February. Peace Negotiations. While these dragged on at Guadalupe Hidalgo, political chicanery in Washington intervened, Scott being summarily removed from command and ordered home to stand trumped-up charges, which were later withdrawn. He was received as a hero by the nation, and Congress bestowed a gold medal on him for his successful invasion. Ironically, Polk had to present the medal.

1848, February 2. Treaty of Guadalupe Hidalgo. This confirmed the southern bound-

THE MEXICAN WAR

SCOTT'S CAMPAIGN
12 Aug.—14 Sept. 1847

Operations at Mexico City

◀- - - - Scott's Route

0   1   2   3   4   5

Scale of miles

ary of Texas and transferred to the United States the regions now comprising California, Nevada, Utah, most of Arizona and New Mexico, and parts of Colorado and Wyoming. The U.S. paid Mexico $15 million. U.S. forces evacuated Mexico City (June 12) and Veracruz (August 2). Meanwhile the treaty was ratified (July 4).

CANADA

1812–1815. War of 1812. (See p. 871.)

1837, November. Papineau's Rebellion. French-Canadian leader **Louis Joseph Papineau,** believing that the grievances of French Canadians were being ignored by the British crown, led a brief uprising in Lower Canada (modern Quebec Province), which was sup-

pressed by British forces in a clash at **St. Denis** (November 22). Papineau fled to the United States, and then to France, returning to Canada after a general amnesty (1847).

1837, December. Mackenzie's Rebellion. Advocate of a republican government for Upper Canada (modern Ontario Province) or all of Canada, **William Mackenzie** had been in correspondence with Papineau, although there was evidently no collusion between their entirely separate rebellions. His call for an uprising and establishment of a provisional government was foiled at **Toronto** by Sir **John Colborne** (December 6), and he fled to the United States. In Buffalo he collected a group of Canadian malcontents, and seized Navy Island, in the Niagara River. His followers made a few raids into Canada, but all fled when Canadian forces approached. He returned to Canada after the Amnesty Act of 1849.

1838. Aroostook "War." (See p. 881.)

1840. Act of Union. Upper and Lower Canada were united as a single, partially self-governing colony.

LATIN AMERICA

When the period opened, the greater portion of the Americas roughly south of the Missouri River and 42° latitude belonged to Spain and Portugal. The young United States had as western boundary the line of the Mississippi down to the Floridas. England, the Netherlands, France, and Denmark each claimed some of the West Indies islands. England, too, had a tiny toehold—Belize—on the Caribbean coast just south of Yucatan, and was casting envious eyes on Guiana, where France and Holland held coastal slices. Portugal's share of this rich Latin immensity was Brazil, whose boundaries stand today generally as they were then. The rest was Spain's.

When the period closed, all that remained of these 2 Latin colonial empires were Cuba and Puerto Rico, both Spanish. A group of new nations, self-carved by armed revolution against European overlords and by their own internecine wars, were protected by the Monroe Doctrine (see p. 881) against further overseas encroachments.

The independence movements, ignited by the American Revolution, were further stimulated by the French Revolution, by Napoleon's incursions into Spain and Portugal, and by later revolution in Spain. Turmoil and dissidence, beginning in 1790, smoldered uncertainly until British operations against Spanish South America, in the Napoleonic Wars, kindled the flame.

Wars of Independence, 1800–1825

Spanish South America

1806, June 17. British Occupation of Buenos Aires. Admiral Sir **Home Popham**'s squadron landed some 500-odd troops under General William Beresford. **Jacques de Liniers** rallied the colonial militia, who forced the capitulation of the invaders (August 12). Popham, who had acted without specific orders, was later courtmartialed.

1807, July. British Occupation of Montevideo. An expedition of 8,000 men under General **John Whitelocke** occupied the city, then seized Buenos Aires. Again Liniers led a popular uprising, forcing capitulation of the invaders, after serious street fighting.

1807–1810. Insurrections throughout South America. These were triggered by the 2 successes against foreign soldiery in Buenos Aires.

Spanish troops suppressed all except in the River Plate viceroyalty (Argentina), which became practically independent (1810).

1810–1814. Unsuccessful Rebellion in Chile. After a brief struggle Chile won independence under the leadership of **José Miguel Carrera** (1811). Because of incompetence he was replaced by his rival, **Bernardo O'Higgins.** Spanish forces from Peru defeated O'Higgins at **Rancagua** (October 7, 1814) due largely to Carrera's noncooperation. O'Higgins fled to Argentina, and the Spanish reestablished control of Chile.

1811, August 14. Independence of Paraguay. This little inland subdivision of the River Plate viceroyalty declared its independence, following a popular uprising, breaking its ties both with Spain and with the revolutionary movement in Buenos Aires. Paraguay was soon under the iron rule of **José Rodríguez de Francia,** who assumed perpetual dictatorship with a policy of armed isolation (1816).

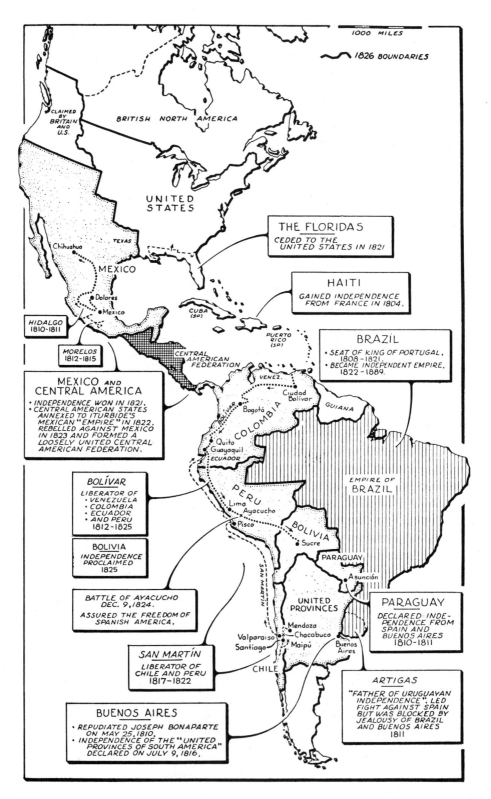

Independence movements in Latin America

1811–1821. Banda Oriental (Uruguay). The area between the River Plate and the Brazilian border—claimed by Portugal but actually under Spanish control—was liberated by Argentine and local revolutionaries, led by **José Artigas** (1811). He was ousted, in turn, by Brazil (1816–1820), which annexed the Banda.

1811–1825. Bolívar and San Martín. Simón Bolívar in Venezuela and **José de San Martín** in Argentina were 2 strong men who initiated separate and amazing campaigns of liberation from Spanish rule. Bolívar's area of operations in the north was mainly New Granada (Venezuela, Colombia, and Ecuador). San Martín's operations in the south included the River Plate (Argentina), Chile, and Peru. When these leaders met at the equator at Guayaquil (1822), Spain's domination of South America was doomed. Both men were able to enlist foreign soldiery—English, Irish, German, and North American—in their forces. The foreign adventurers included a remarkable British sailor, **Thomas Cochrane,** 10th Earl of Dundonald, perhaps the stormiest petrel the Royal Navy ever produced.

BOLIVAR'S CAMPAIGNS

Simón Bolívar (1783–1830), born in Caracas, spent most of his early life in Europe. He returned to Venezuela (1809) and became involved in the revolutionary movement (1811).

1811, July 5. Venezuelan Independence Declared. Bolívar was put in command of Puerto Cabello, but was driven from it by Spanish General **Juan Domingo Monteverde,** who momentarily crushed the entire revolt, capturing **Francisco Miranda,** its leader. Bolívar fled to Curaçao.

1813, May. Bolívar's Return. He took the field again (May 1813), defeating Monteverde at **Lastaguanes,** and then capturing Caracas (August 6, 1813). He won another victory over the royalists at **Araure** (December 5, 1813). At **La Victoria** (February 1814) and **San Mateo** (March 1814), Bolívar defeated General **José Tomás Boves.** He won another victory at **Carabobo** (May 1814), but was defeated by Boves at **La Puerta** (July 1814). With Venezuela again under Spanish control, Bolívar fled to New Granada (Colombia), where he was given another revolutionary command.

1815. Bolívar Defeated. After some preliminary successes, he was defeated by General **Pablo Morillo** at **Santa Mara.** Spanish troops, released by the conclusion of the Napoleonic Wars, now reconquered most of the rebellious provinces. Bolívar fled to Jamaica and Haiti. He led two unsuccessful raids against the coast of Venezuela.

1816, December. Bolívar Returns to Venezuela. He led a new movement against Morillo, winning an important victory near **Barcelona** (February 16, 1817). Now recognized as commander in chief of the patriot forces, Bolívar attempted to raise the country against the Spanish.

1818, March 15. Battle of La Puerta. On the site of an earlier defeat, Bolívar's army was routed by Morillo. Without losing heart, Bolívar continued the insurrection. He decided to try to drive the Spanish from New Granada.

1819, June 11. Crossing of the Andes. Leaving Angostura (Ciudad Bolívar) with 2,500 men, well armed but practically in rags, Bolívar marched across Venezuela, over 7 rivers, and then through the frigid passes of the Andes, debouching in the valley of Sagamose (July 6).

1819, August 7. Battle of Boyaca. Spanish Colonel **Barreiro,** with 2,000 infantry and 400 horse, defended the approaches to Bogotá. Bolívar outmaneuvered him, placing his troops between Barreiro and the capital. While his right drove against the Spanish left flank, Bolívar's British Legion, all veterans, moved in frontal assault, repulsing the Spanish cavalry. Barreiro's troops were completely routed, losing 100 killed, 1,600 prisoners, and all their heavy equipment. This was the decisive battle of the revolution in northern South America. Bolívar entered Bogotá in triumph (August 10), establishing the Republic of Colombia, with himself as its president.

1820–1821. Operations in Northwest Venezuela. Bolívar and other patriot leaders campaigned indecisively against royalist forces. An armistice was effected (November 25, 1820–April 28, 1821) at Trujillo. Upon the conclusion of this truce, Bolívar with 6,000 men once more moved through the Cordillera, this time southeastward.

1821, June 25. Battle of Carabobo. Spanish General **Miguel de La Torre,** with 5,000 men, held the valley of the Carabobo, at the southern foot of the pass. Bolívar, brushing aside Spanish outposts in the pass, found another route through the mountains and sent his cavalry and the British Legion through the detour to fall on the Spanish right flank. La Torre assaulted this force as it debouched with a detachment of his own, but the British Legion held firm and the cavalry broke up the attack. La Torre's error in dividing his force led to defeat in detail. The Spaniards were driven from the field in a 20-mile-long rout. Bolívar entered Caracas. Spanish strength unraveled. **Cartagena** capitulated (October 1) after a 21-month siege, and other garrisons began to fall.

1822. Operations in Quito Province. Bolívar moved south to join his lieutenant, General **Antonio de Sucre,** who, with 2,000 men, had cornered **Melchior Aymerich,** Spanish governor general of Quito, in the mountains near the city. At **Bomboná,** a royalist force of some 2,500 so bitterly contested Bolívar's advance through the mountains that he was forced to halt (April 7).

1822, May 24. Battle of Pichincha. Aymerich turned on Sucre, attempting to storm the slopes of the volcano of Pichincha, where Sucre's forces lay. Although at first successful, the Spanish force was defeated by a flank attack and driven down the mountain, retreating into Quito, where Aymerich surrendered next day. His losses had been 400 killed, 190 wounded, and 14 guns; Sucre's losses were 200 killed and 140 wounded.

San Martín's Campaigns

San Martín (1778–1850), of Spanish descent, had served in Spain's army during the Peninsular War. He returned to Buenos Aires (1810).

1814–1816. San Martín Commands the Revolutionary Army. For 3 years he organized and trained the Army of the Andes—partly Argentine, partly Chilean—at Mendoza, in preparation for an invasion of the west coast. Meanwhile, Argentina declared its independence (July 9, 1816).

1817, January 24. March over the Andes. San Martín led an army of 3,000 infantry, 700 cavalry, and 21 guns across the snowy passes of the Gran Cordillera—a feat never before attempted—debouching into the plain between the coastal range and main Andean chain (February 8). O'Higgins commanded a Chilean contingent.

1817, February 12–13. Battle of Chacabuco. Spanish Colonel **Maroto,** with 1,500 infantry, 500 cavalry, and 7 guns, disputed the advance. San Martín attacked in 2 columns. O'Higgins, moving in the moonlight of early morning, contained the Spanish force while San Martín turned its left, completely routing it. Spanish losses were 500 killed, 600 prisoners, and all their artillery. San Martín lost 12 killed and 120 wounded. Santiago was then occupied (February 15), and Chilean independence became fact (proclaimed a year later, February 12, 1818). Indecisive fighting continued through the remainder of the year, while Spanish General **Mariano Osorio** with 9,000 men moved down from Peru.

1818, March 16. Battle of Cancha-Rayada.

San Martín with 6,000 men was defeated by Osorio, losing 120 killed and 22 of his guns. Spanish losses were 200 killed and wounded, and they were much shaken by the encounter.

1818, April 5. Battle of the Maipo. San Martín, regrouping after his defeat, attacked the Spaniards on the bank of the Maipo River. Hurling his main effort against the Spanish left, San Martín turned it and the royalist army collapsed, losing 1,000 killed, 12 guns, and 2,300 prisoners. San Martín's losses were about 100 killed and wounded. Osorio, giving up further contest, retreated into Peru, and San Martín prepared to follow. But further revolutionary invasion was impossible unless Spanish command of the sea were broken. **Manuel Blanco Encalada,** a Chilean and a former Spanish naval officer, had gathered a squadron of sorts, including a Spanish frigate, cleverly captured.

1818, November 28. Arrival of Cochrane at Valparaiso. After he had resigned from the British Navy under a cloud, the Chilean revolutionary government invited him to command its navy.

1819, January–February. Cochrane Takes to Sea. With 4 ships, he attempted to draw the Spanish squadron based on Callao and Valdivia into action. Unsuccessful in this, he cruised the coast.

1820, June 18. Capture of Valdivia. Cochrane sailed his flagship *O'Higgins,* 50, former Spanish heavy frigate, into the harbor. He bombarded the forts and landed a party to capture the defenses. This broke the last tenuous Spanish hold on the Chilean coast. San Martín now felt able to carry out his proposed invasion of Peru.

1820, September 8. Invasion of Peru. Cochrane's little squadron escorted north 16 transports carrying San Martín's forces. They were landed at Pisco; then Cochrane blockaded Callao, where the Spanish squadron lay.

1820, November 5. Capture of the *Esmeralda*. In a daring cutting-out expedition Cochrane in person took 250 sailors and marines in small boats into the harbor and captured the frigate *Esmeralda*. Had his plans been carried out, all the other Spanish ships would

then have been taken in turn, but a subordinate cut the frigate's cable and Cochrane had to be content with his one prize.

1821, July. San Martín Takes Lima. After this, Callao capitulated (September 21) and Peruvian independence was established. San Martín turned his eyes northward again, toward Quito. He apparently decided then to coordinate with Bolívar, who had simultaneously been advancing from the north.

CONCLUDING CAMPAIGNS AGAINST SPAIN

1822, July 26–27. Meeting at Guayaquil. Bolívar and San Martín officially linked their forces. The nature of the conference has always been a mystery, but San Martín turned over control to Bolívar and retired from all further revolutionary activities. The 2 men were completely incompatible: Bolívar with dictatorial ambitions, San Martín a soldier-patriot of high military genius. Bolívar, organizing the Republic of Peru, continued the drive to expel Spanish domination, despite mutinies and vigorous Spanish reaction.

1824, August 6. Battle of Junin. Bolívar and Sucre, with about 9,000 men, met a Spanish army of equal size under General **José Canterac** about 100 miles northeast of Lima. In a purely cavalry battle, about 2,000 men on each side being engaged, and not a shot fired, the revolutionaries were victorious. The Spanish army retreated into the highlands southeast of Lima. Sucre pursued, while Bolívar organized a government at Lima.

1824, December 9. Battle of Ayacucho. Spanish General **José de La Serna** with 10,000 men and 10 guns attacked Sucre with 7,000 and 2 guns. Sucre's left and center contained La Serna's attack, while his right, mostly cavalry, turned the Spanish left. Sucre now threw in a reserve division and the Spanish army collapsed. La Serna, who had received 6 wounds, was captured. Next day General Canterac, La Serna's successor, capitulated. Spanish losses were 1,400 killed and 700 wounded; the republicans lost 300 killed and 600 wounded.

Known as the "Battle of the Generals" (14 Spanish generals were captured), Ayacucho ended Spain's grip on South America. All Spanish forces remaining in Peru were withdrawn by the terms of the capitulation. Sucre then moved eastward into the Presidency of Charcas, bordering on Brazil, and established the Republic of Bolivia (August 6, 1825).

Brazil

1807–1821. Brazil Becomes the Seat of Portuguese Government. The Portuguese royal family fled there from the Napoleonic invasion (1807–1808; see p. 843). The example of revolt in neighboring Spanish America caused an urge for liberty to permeate Brazil. After much dissension, John VI of Portugal returned to his country (1821), leaving his son Dom Pedro as prince regent.

1808. Brazil Occupies French Guiana. This was returned by the Treaty of Paris (1814).

1816–1821. Occupation of Uruguay. (See p. 891.)

1822, September 7. Independence of Brazil. This culminated growing tension between Brazil and Portugal and between John and Dom Pedro, who led the independence movement. He was soon crowned emperor (December 1).

1823, March 21. Cochrane Assumes Command of the Brazilian Navy. This was a small squadron of warships obtained by purchase and capture. Portuguese naval strength lay at Bahia, which Cochrane blockaded.

1823, July 2. Portuguese Evacuation of Bahia. Portuguese forces left the port in a convoy of some 60-odd transports escorted by 13 warships—a force far too strong for Cochrane to attack with only 2 frigates. The convoy headed north along the coast for Maranhão (São Luiz).

1823, July. Cochrane Wins the War. In his flagship, the heavy frigate *Pedro Primiero*, 50, the admiral—by superior seamanship—avoided conflict with poorly handled Portuguese war vessels while cutting out a number of transports. Then, surmising its destination,

he sailed right to Maranhão ahead of the convoy and captured the port, permitting the garrison to embark for Portugal. The main convoy now had no place to go but Portugal. The combined Portuguese convoys were harassed all the way across the Atlantic by several of Cochrane's ships.

1825. Portugal Recognizes Brazilian Independence.

Mexico

1810. Peasant Revolt. The first real revolution in New Spain began with the revolt of **Miguel Hidalgo y Costilla,** a priest interested in social reform. With a horde of 80,000 *paisanos,* he swept through the southern provinces and threatened Mexico City, but his untrained masses were repulsed (November 6) by Spanish General **Félix Calleja.**

1811, January 17. Battle of the Bridge of Calderón. Near Guadalajara, Calleja, with 6,000 men, crushed Hidalgo, who was captured and executed. Hidalgo's lieutenant, **José María Morelos,** also a priest, revived the insurrection, but after 3 years of increasingly hopeless fighting was captured and executed by General **Agustín de Iturbide,** a Mexican-born Spaniard.

1821, January–February. Renewed Rebellion. This time the leader was General Iturbide, who rebelled after the revolution of 1820 in Spain.

1821, February 24. Declaration of Independence. Iturbide occupied Mexico City and was crowned **Agustín I,** Emperor of Mexico (July 21, 1822).

1823, March 19. Republican Revolution. Iturbide was forced from the throne. One of the republican leaders was **Antonio López de Santa Anna.** A federal republic was established (October 4, 1824).

Central America

1821–1823. Mexican Domination. The Spanish captain-generalcy of Guatemala crumbled in a series of revolutions gradually growing in intensity (1811–1821). The uprisings were generally stimulated by the independence movements to the north and south. Following Mexico's independence, the area fell under Mexican domination.

1823, July 1. The United Provinces. After the overthrow of Agustín, a constitutional assembly called by Mexican General **Vicente Filosola** in Guatemala City declared the confederation of Guatemala, El Salvador, Nicaragua, Honduras, and Costa Rica as the United Provinces of Central America.

Haiti and Santo Domingo

1791–1801. Chaos in Haiti. The island of Haiti (Santo Domingo or Hispaniola) was the center of brutal war, revolution, and rapine following a slave rebellion. **Pierre Dominique Toussaint L'Ouverture,** a Negro of military genius, finally restored order (1800), but his efforts aroused the suspicion of Bonaparte, who resolved to restore complete French control over the island and to reinstitute slavery.

1801–1803. French Domination Overthrown. French General **Victor Emmanuel Leclerc,** with 25,000 men, debarked and despite Toussaint's resistance occupied the seaports. He could not, however, compete with the savage resistance of the Negro troops in inland guerrilla warfare. His army wasted by disease, Leclerc offered amnesty, and Toussaint laid down his arms in order to negotiate (June 1802). He was treacherously seized and taken to France, where he died in prison (1803). Under **Jean Jacques Dessalines,** the Negroes resumed fighting. Leclerc died of yellow fever, and the demoralized French Army finally evacuated the island (November 1803).

1804–1806. Supremacy of Dessalines. He crowned himself Emperor Alexandre I (October 8) and instituted a general massacre of all remaining whites.

1806–1820. Struggle for Power. Following the assassination of Dessalines, the island was divided between **Henri Christophe** in the north and **Alexandre Sabes Pétion** in the south, who warred with one another until Pétion's death (1818). Christophe, a man of undoubted genius, was assassinated (1820). **Jean Pierre Boyer** became President of Haiti.

1808–1809. Revolt of Santo Domingo. With British help, the Spanish eastern portion of the island threw off the control of the French-speaking blacks of the west.

1814. Spanish Control Restored in Santo Domingo.

1822. Haitian Reconquest of Santo Domingo. Boyer drove out the Spaniards to reunite the island under one government.

AFTERMATH OF INDEPENDENCE, 1825–1850

Wars and revolutions raged throughout Latin America as the new republics struggled within and among themselves. Dictators rose and fell.

1825–1828. Brazil-Argentine War. Argentina, assisting a revolt in the Banda Oriental (Uruguay) against Brazil, went to war to acquire the province. At the **Battle of Ituzaingo** (February 20, 1827), Brazilian forces were defeated by an Argentine-Uruguayan combination. Blockade of Buenos Aires by a Brazilian naval squadron brought friction with Europe. British mediation brought peace and the establishment of independent Uruguay.

1827–1829. Peruvian Aggressions. José **Lamar,** President of Peru, launched his nation on a policy of expansionism. A Peruvian invasion of Bolivia (1827) forced the withdrawal of Sucre, who had become president after the departure of Bolívar. Lamar then invaded Ecuador. A Peruvian naval squadron captured Guayaquil (January 1829). But Sucre, who had taken refuge in Ecuador, and Ecuadorian General **Juan José Flores** led an army which defeated the invading Peruvians at the **Battle of Tarqui** (February 27, 1829). Guayaquil was recaptured the next day. Lamar's plans for expansion to the north now crumbled.

1829. Spanish Invasion of Mexico. An expedition from Cuba seized Tampico (August 18). Santa Anna promptly led a force against the invaders, who were soon forced to surrender (September 11).

1830. Independence of Ecuador. Flores led this withdrawal from Grañ Colombia, and became first president of Ecuador.

1836–1839. War of the Peruvian-Bolivian Confederation. Bolivian dictator **Andreas Santa Cruz,** with the agreement of Peruvian President **Luis Orbegosa,** established a combination of the two countries (1835). Chile and Argentina opposed the confederation, and Chile declared war (November 11, 1836). A Chilean army, commanded by General **Manuel Bulnes,** decisively defeated Santa Cruz at the **Battle of Yungay** (January 20, 1839), breaking up the confederation.

1838–1839. French Expedition to Mexico. To obtain satisfaction for claims of its citizens, France sent an expedition to Mexico. The French fleet blockaded the Atlantic coast of Mexico and seized Veracruz (April 16, 1838). When the claims were satisfied, the French withdrew (March 9, 1839).

1839–1840. Disintegration of Central American Confederation. Continual internal strife finally caused the dissolution of the confederation into its components. Later attempts at confederation failed.

1841. Peruvian Invasion of Bolivia. President **Agustín Gamarra** of Peru, attempting to annex part of Bolivia, was defeated and killed by the Bolivians under President **José Ballivián** at the **Battle of Ingavi** (November 20).

1844. Independence of Santo Domingo. The predominantly Spanish-speaking inhabitants of the eastern portion of the island of Hispaniola, or Haiti, revolted successfully against the French-speaking rulers of the Republic of Haiti.

1843–1852. Argentine Intervention in Uruguay. A confusing struggle broke out involving internal strife in Argentina and Uruguay, and the intervention of Brazil, France, and England. Argentine dictator **Juan Manuel de Rosas** attempted to take advantage of a civil war in Uruguay to annex that country. He supported **Manuel Oribe** of Uruguay, who for 8 years maintained an ineffective siege of **Fructuoso Rivera** in Montevideo (1843–1851). Giuseppe Garibaldi was among the defenders of besieged Montevideo. To check Rosas, France and England sent forces to occupy parts of Uruguay and to blockade the River Plate (1845–1849). Meanwhile, Argentine **Justo José de Urquiza** revolted against Rosas. Brazil, fearful of Argentine annexation of Uruguay, supported Urquiza, who was also encouraged by France and England. At the **Battle of Monte Caseros** (February 3, 1852), Urquiza with a force of Argentines, Brazilians, and Uruguayans defeated Rosas, who fled to England.

1846–1848. U.S.-Mexican War. This war, discussed elsewhere (see p. 882), sharpened the line of political demarcation and distrust between Anglo-Saxon North America and the Latin nations in both the northern and southern divisions of the hemisphere.

AUSTRALASIA

AUSTRALIA

1804. Irish Convict Rebellion. New South Wales had been established (1788) as a penal colony ruled by a military governor—usually an officer of the Royal Navy—and policed by the New South Wales Corps, a military body recruited in England. This "Praetorian Guard" waxed fat on land grants, monopoly of cargoes arriving in the colony, and the rum trade. A considerable proportion of the convicts were Irishmen, political prisoners deported after the suppression of the revolution of 1798. This group rose against the local government, but the mutiny was suppressed with great severity.

1806. Renewed Mutiny. After this was suppressed, Captain **William Bligh,** R.N., of *Bounty* mutiny fame, was appointed governor, the home government expecting that his stern discipline would bring both the N.S.W. Corps and the convicts into line. His drastic methods soon resulted in another mutiny.

1808, January 26. The Rum Rebellion. A group of N.S.W. Corps officers, led by Major **George Johnston,** a regular, arrested Bligh as unfit for office and held him prisoner pending the arrival of a new governor. The home government, while condemning the action, accepted it. Colonel **Lachlin Macquarie,** an able administrator, arrived (1809) and Bligh was sent home (to become a rear admiral). Macquarie broke up the N.S.W. Corps.

New Zealand

1843–1848. First Maori War. Native resentment of land expropriation broke into conflicts between British settlers and the native Maoris in a guerrilla war that raged for 5 years.

XVIII

THE EMERGENCE OF THE PROFESSIONAL:

1850–1900

MILITARY TRENDS

A multiplicity of technological developments forced sometimes reluctant military professionals to broaden their horizons and to create new standards. Epitome of this new professionalism was Prussian **Helmuth von Moltke.** He was a competent, sometimes brilliant, soldier who guided rather than led, and his reputation has paled somewhat in the face of growing modern opinion that during the latter half of the 19th century two men alone displayed military genius ranking them among the great captains of the world: **Ulysses S. Grant** and **Robert E. Lee.**

Though neither was completely successful in the effort, these two surpassed their contemporaries in coping in their very different ways with the unprecedented changes in warfare brought about by the Industrial Revolution. The new technological means for waging and supporting warfare meant that for the first time the French Revolutionary concept of the "nation in arms" had been eclipsed by a new concept: the "nation at war." With the national economies on both sides fully integrated into their respective war efforts, the American Civil War was truly the first modern war, and the first "total" war in the modern sense.

It was a period of colonial expansion and conflict of interest among the great powers. War raged practically all over the world, except in the British Isles and in the Scandinavian Peninsula. But the expert juggling of British diplomacy, self-interested in the maintenance of the balance of power, went far to prevent the numerous minor conflicts from spreading into international conflagrations, such as those of the periods immediately preceding and following. Queen Victoria's reign would go down in history as that of the "Pax Britannica."

MILITARY THEORY

The example of Napoleon Bonaparte still dominated military theory on both sides of the Atlantic Ocean. In the United States, **Dennis Hart Mahan** was still the apostle of Napoleonic concepts; more than any other single factor, Mahan's teaching influenced the strategical and tactical thinking of the leading generals on both sides of the American Civil War.

In Europe the seed of the famous—and much misunderstood—Prussian Army General Staff had been planted by Frederick the Great, and its early growth had been nurtured by both Scharnhorst and Clausewitz. But the quiet organizing genius of Moltke brought that staff of its full flower in the second half of the 19th century. The combination in the Prussian General Staff of scholarly research, meticulously detailed planning, and thorough indoctrination in a logical concept of war was the major factor in stunning victories over Austria and France, causing Prussia to be acknowledged as the leading land military power of the world. That Prussia and Germany still held this position at the end of the century was due mainly to Moltke's great successor, **Alfred von Schlieffen.** Military historian Schlieffen's studies of ancient and modern campaigns corroborated the theories of Napoleon and of Clausewitz. He developed a simple, flexible strategic concept for a two-front war against France and Russia. In the hands of a less able successor, this so-called Schlieffen Plan ironically was to become the very epitome of rigidity.

Among the significant military writings of this period, the works of the French officer **Charles J. J. J. Ardant du Picq** were a major contribution to the development of modern combat theory. Killed in battle at the age of 49 in 1870, Colonel Ardant du Picq was chiefly concerned with the behavorial element in warfare. His masterpiece, *Battle Studies* (published posthumously), had immense influence on the generation of officers who led France in World War I, particularly Marshal **Ferdinand Foch.**

By the end of the century an American sailor was beginning to be acknowledged throughout the world as the leading military theorist of the age. **Alfred Thayer Mahan** had become the Jomini and Clausewitz of naval theory through lucid, logical analyses of the fundamentals of sea power. His theories, which related naval strength to national policy and strategy, were as respected by soldiers, politicians, and emperors as they were by naval colleagues of all nations. Interestingly, his writings were appreciated and respected in Europe before they were thoroughly understood by his own countrymen.

THE IMPACT OF TECHNOLOGY

Steam power became a critical consideration in national strategic planning. This impact of steam power on nautical affairs stimulated Mahan's writings and caused them to be influential in maritime nations.

For such countries the location of a coaling station not only became a limiting factor but also could directly determine the direction and extent of colonial expansion. Two examples were provided in the Spanish-American War. Lacking coaling stations, Admiral **George Dewey** could move his squadron from Hong Kong to Manila only by the outright purchase of British colliers and their cargoes. The long cruise of the U.S.S. *Oregon* from San Francisco around Cape Horn to join the fleet in Cuban waters (nearly 13,000 nautical miles) sparked the later American acquisition of the Panama Canal Zone and the construction of the canal itself.

The railroad became a logistical weapon for both sides during the American Civil War, an example used later to good advantage by the Prussian General Staff in its preparations for the Franco-Prussian War.

Another far-reaching development was the electric telegraph and cable, affecting both strategy and tactics. Beginning with the Crimean War, this was a two-edged sword so far as the overall direction of war was concerned. Cable and telegraph ensured rapid communication between field commander and government; they also permitted political armchair strategists to hinder field operations. Almost instantaneous transmission of war correspondents' dispatches, immediately appearing in the daily press, brought a new threat to security by revelation of operational plans and moves. This was most pernicious in the American Civil War. But news dispatches also promoted the common soldier from a digit to the status of an individual by emblazoning his hardships for all to see. The result was improvement of the soldier's welfare and of his morale.

Tactically, the field telegraph augmented other improved means of signal communication: the semaphore "wigwag" flags and the heliograph, which already ensured contact between field headquarters and lower commands, and spanned hostile-held terrain to link with isolated elements and reconnaissance detachments. Long before the end of the Civil War the telegraph was in common use in the Union armies and less extensively in Confederate forces. Telegraph wires linked aerial observers in balloons with the ground. The wires also created a communications web, including—as Grant wrote—"each division, each corps, each army and . . . my headquarters."

In addition to the direct effect of new and improved instruments of warfare, industrialization had other significant consequences in the conduct of war. The change from an essentially agricultural economy released a greater percentage of national man power to both the armed forces and to the developing war industries. Larger armies could be raised and supported than had been true in the past, while the development of steam transportation and the electric telegraph facilitated the movement and control of these larger forces.

But it was in the refinement and proliferation of weapons that the new technology had its greatest effect.

WEAPONS AND TACTICS ON LAND

The transition from muzzle-loading to breechloading artillery pieces was complete by the end of the period. So, too, was the refinement of armor-piercing and lighter-walled antipersonnel projectiles. Solid shot became a thing of the past; elongated, streamlined shells, with explosive charges, detonated by improved fuses—both time and impact—were the rule. The smoothbore cannon—which was the predominant artillery type both ashore and afloat as the period opened—had been replaced by the rifled gun, a change made more reluctantly, perhaps, in the world's navies than in their armies. The problem of cannon recoil had been solved, first by ingeniously controlled springs, and later by more sophisticated hydropneumatic apparatus.

For coastal defense the disappearing gun had come into being, its principal proponent the United States. It rose by means of counterweights to fire from above its parapet and then sank by force of recoil, controlled by pneumatic brakes.

In small arms, the single-shot muzzle-loader was replaced by the repeating magazine rifle, and the Minié ball gave way to the elongated conical bullet. The range, accuracy, and volume of fire of individual weapons were greatly increased. Though its potentialities were still scarcely realized, the machine gun had become an important weapon in the arsenals of all nations.

Smokeless powder had become the generally accepted propellant for small arms as well as artillery by the end of the century, facilitating concealment in broken country.

Field mines and booby traps were used in the American Civil War by the Confederates as early as 1862, at the opening of the Peninsula Campaign. Though the ideas were not entirely new, modern prototypes of trench mortars and hand grenades were used by both sides of that conflict in the extended trench warfare of 1863–1865.

Percussion cap pistol

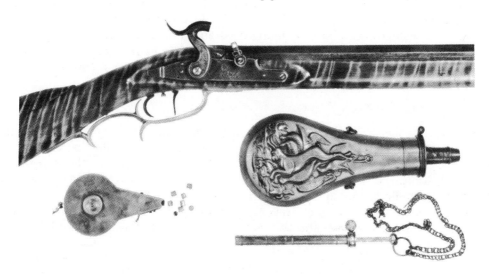

Firing assembly of a percussion cap rifle, with a percussion capper and eight percussion caps (lower left), and a powder flask above a syringe-like powder measure on a chain

The overall effect of the new weapons was to introduce an entirely different order of firepower on the battlefield, where improved weapons combined with more numerous armies to sweep the front lines with unprecedented hails of iron, steel, and lead missiles. This in turn produced a far-reaching tactical revolution, which was most dramatically and most clearly manifested in the American Civil War.

Neither infantry nor cavalry could attack frontally in the face of the combined firepower of artillery and small arms. There were four immediate effects. First was dispersal; battlefield formations became progressively more spread out and more flexible. Next, maneuver became the order of the day as men sought for decision without suicidal frontal assaults.

Trench warfare was perhaps the most obvious manifestation of this tactical revolution. Field fortifications—heretofore reserved almost exclusively for siege operations—became an integral aspect of infantry tactics. Troops under fire immediately dug in. Foxholes and rifle pits were soon expanded into trench systems, which became the bases for maneuver by both sides. It was in the utilization of improvised field fortifications that Lee surpassed all of his contemporaries; most of his victories were the result of his ability to use hasty entrenchments as a base for the aggressive employment of fire and movement.

Another manifestation was in cavalry combat. By the end of the period cavalry shock action had practically disappeared from the battlefield. With frontal charges against infantry unthinkable, mounted cavalry operations were limited almost entirely to screening and reconnaissance missions. The horse, however, provided mobility to mounted riflemen, whose normal mode of combat was as dismounted skirmishers.

The lessons of 1861–1865 were not learned readily elsewhere. It took their own bitter experiences for Europeans to realize the significance of the American example. A decade passed before the old order fell in a welter of broken men and horses during the Franco-Prussian War. The last gasps of the old order came in British colonial wars in Africa. By participating in the climactic cavalry charge at Omdurman, young Winston Churchill experienced the glory of the premodern era. A few years later, as the century was closing, he recognized that the era had ended, when massed British infantry were slaughtered by mobile Boer sharpshooters.

Weapons and Tactics Afloat

At Sinope the Russian Navy first showed the devastating potentialities of improved naval ordnance by smashing a Turkish fleet with shell projectiles. Armored floating batteries were also employed in the Crimean War. A logical evolution led to the creation of the French armor-plated *Gloire*—causing a panic in Britain, which forced the reluctant Admiralty to rush construction of its own first armored warship, HMS *Warrior*.

But it was the armor-plated Confederate *Merrimack* (CSS *Virginia*) which tolled the bell for wooden warships; one day later the USS *Monitor* bowed in the age of turret ships. Naval

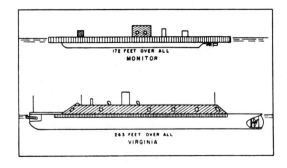

USS *Monitor* and CSS *Virginia*

architects here and abroad soon solved the problem of wedding Ericson's revolving battery—introduced on the *Monitor*—to a seaworthy armored hull. By the end of the period the battleship was queen of the seas.

Confederate ingenuity also evolved the submarine, though not until the turn of the century would the gasoline engine and electrical storage battery solve the problem of propulsion. The ram, Civil War rebirth of an old device, enjoyed a fleeting popularity, but soon proved of dubious tactical value, though its vestigial influence lingered long in naval design.

Submarine mines (originally called "torpedoes") came into common use in the Civil War, and were used effectively by the Confederates to protect their ports and coastal defenses from Union warships. The South and North both used such mines offensively by attaching them to the ends of spars or booms attached to submarines or other small craft. By the end of the century, these spar torpedoes had been transformed by compressed-air self-propulsion into a much more lethal threat as prototypes of the modern torpedo. With this development there came into being in the world's navies the fast but fragile torpedo boat, and close on its heels the larger, faster torpedo-boat destroyer.

In combat the line ahead (ships in single file) was still the normal tactical formation and broadside fire was optimum. But the tacticians were toying with "crossing the T," a new version of raking and of Nelsonian massing of forces, in which (by superior speed) a fleet in single file could concentrate its fire upon the lead ships of the slower line. And an important function of the speedy torpedo boat and the destroyer was the disruption of the hostile line by sinking individual ships.

Naval ordnance, striding from the muzzle-loader to the recoil-controlled breech-loader, dallied for a brief moment with the dynamite gun—hurling its charge by compressed air. By 1898, one vessel of the U.S. Navy—the *Vesuvius*—was armed with this novel piece. However, advances in explosives—TNT and similar substances which could be propelled by conventional gunpowder—put the relatively shortranged dynamite gun in the discard (although on land in the Spanish-American War one such piece was briefly used). All naval ordnance specialists were simultaneously developing the armor-piercing projectile as the answer to armored warships.

MAJOR EUROPEAN WARS

CRIMEAN WAR, 1853–1856

1853. Prelude. A monkish squabble over jurisdiction within the Holy Places of Turkish-ruled Jerusalem brought France (protector of the Catholics) and Russia (protector of the Orthodox clergy) into diplomatic controversy, with Turkey squeezed between. Czar **Nicholas I** saw an opportunity to dominate Turkey and secure entrance into the Mediterranean through the Turkish Straits. A Russian army began occupation of Turkey's Rumanian principalities (July). France had no intention of letting her rival wax more powerful in the Near East, while England opposed any change in the balance of power. Accordingly, British and French fleets arrived at Constantinople to encourage the Turks.

1853, October 4. Turkey Declares War. A Turkish army under **Omar** Pasha (Croatian-born **Michael Lattas,** a good soldier) crossed the Danube.

1853, November 4. Battle of Oltenitza. Omar defeated the Russians in southern Rumania, near the Danube.

1853, November 30. Destruction of Turkish Flotilla at Sinope (Sinop). Russian Admiral **Paul S. Nakhimov,** with 6 ships of the line, 3 frigates, and several smaller craft, attacked Turkish Admiral **Hussein** Pasha's 7 frigates, 3 corvettes, and 2 small steamers lying in the harbor. The Turkish flotilla was destroyed in a 6-hour engagement, the outnumbered and outgunned Turks fighting to the end. Significant aspect of the encounter was the enormous damage done by the Russian shell guns—a new type of naval ordnance making its first bow in warfare.

1854, January 3. Franco-British Fleet in the Black Sea. Both nations, forgetting momentarily their mutual jealousies and suspicions, then allied themselves with Turkey to protect the Turkish coast and shipping (March 12).

1854, March 20. Russians Cross the Danube. A strong Russian army under Marshal **Ivan Paskievich** invaded Bulgaria.

1854, March 28. British and French Declaration of War. They then concluded a mutual alliance (April 10). A Franco-British expeditionary force moved to Varna (Stalin) to assist in repelling the Russian invasion, which had now reached and was besieging Silistria (Silistra). The British frigate *Furious* having been fired on while trying to enter Odessa under a flag of truce (April 16), a Franco-British squadron bombarded the shore batteries, inflicting serious damage.

1854, April 20. Threatened Austrian Intervention. Austria massed an army of 50,000 men in Galicia and Transylvania, after entering into a defensive alliance with Prussia against Russia. With Turkish permission, Austria moved into Turkey's Danube principalities. In face of this threat, Russia abandoned the siege of Silistria (June 9) and later withdrew her forces from the area (August 2), but rejected the joint peace conditions set by England, France, Prussia, and Austria (Vienna Four Points, August 8) that Russia must keep hands off the Ottoman Empire.

Crimea and the Siege of Sevastopol

1854, September. Planned Invasion of the Crimea. The Russian evacuation of the Balkans achieved the principal objective of the allied expeditionary force at Varna, which was now rotting with cholera. But London and Paris decided that Russian power in the Black Sea must be broken by crippling the great naval base at Sevastopol. The expedition was decided upon without any consideration of the magnitude of the task, and without any adequate prior reconnaissance. British Major General **Fitzroy James Henry Somerset, Lord Raglan,** 66, and French Marshal **Armand J. L. de Saint-Arnaud,** 53 (who was already seriously stricken by cholera), commanded jointly.

1854, September 7. Departure from Varna. The allied force was transported to the Crimean Peninsula in a great convoy of 150 warships and transports. Typical of the haphazard conduct of campaign was the fact that not until the convoy lay offshore was any decision made as to the point of debarkation. Equally typical was the fact that no attempt was made by Prince **Alexander Sergeievich Menshikov,** Russian commander in the Crimea, to oppose the landing.

1854, September 13–18. Landing at Old Fort. This was on an open beach with no harbor, some 30 miles north of Sevastopol. Bad weather and the weakened condition of the troops combined to delay the debarkation.

1854, September 19. Advance toward Sevastopol. The expeditionary force—51,000 British, French, and Turkish infantry, 1,000 British cavalry, and 128 guns—moved south, the British on the landward flank. The fleet kept pace along the seacoast.

1854, September 20. Battle of the Alma. Menshikov, with 36,400 men, elected to defend on the bank of the Alma River, his left flank refused out of range of the allied fleet, his right anchored on a hill ridge. The allied attack crossed the river without much difficulty, but the British then found themselves faced by a steep slope, which was carried only after a hard fight. Menshikov then withdrew without molestation. Allied losses (mostly British) were about 3,000 men; Russian, 5,709.

1854, September 25–26. Movements around Sevastopol. The allies were in sight of Sevastopol, whose harbor channel had been blocked by sunken ships, rendering naval cooperation in an assault from the north impossible. Without a base, siege was impractical. The only solution was a flank march around the fortress to the south side, and establishment of bases at the ports of Kamiesch and Balaklava. As the allies, abandoning their line of retreat, made the 15-mile circuit with their flank unmolested, Menshikov left a garrison within the works and wisely marched the remainder of his army north to Bakhchisarai to join Russian reinforcements now moving in. Actually, the opponents marched across one another's front without knowing it. The allied army safely made the circuit and regained contact with the fleet, the British based on Balaklava and the French on Kamiesch.

1854, October 8–16. Siege of Sevastopol Opens. The southern defenses of Sevastopol had not yet been fully completed. An immediate assault might have been successful. Instead, while the British contingent and one French army corps covered the operation from attack by the Russian field army, a siege corps was extemporized and investment began. St. Arnaud died of cholera (September 29) and General **François Certain Canrobert** succeeded to command. By October 17, when the first bombardment opened, Colonel **Frants E. I. Todleben,** the Russian chief engineer, had done an amazing job of fortification. The bombardment and counterbattery fire caused serious losses on both sides, but no permanent damage to the works. The port was blockaded by the fleet of Admiral Sir **Edmund Lyons.**

1854, October 25. Battle of Balaklava. Menshikov's field army attempted to drive between the besieging lines and the British base at Balaklava. A penetration was made and some Turkish guns taken. Russian cavalry attempting to exploit the breakthrough were repelled by the British Heavy Cavalry Brigade and the stand of the 93rd Highlanders in "the thin red line." The Light Cavalry Brigade, through circumstances never satisfactorily explained, now charged the Russian field batteries to their front, riding up a narrow, mile-long valley, exposed at the same time to fire from the captured Turkish guns on their right flank and other Russian guns on their left. They reached the guns, rode through them, clashed with Russian cavalry beyond, and then the survivors rode back through the cross fire of the "Valley of Death" made famous by Tennyson's poem. The charge will stand forever as a monument to gallant soldiers doomed to death by the arrant stupidity of Brigadier General **James Thomas Brudenell, Lord Cardigan,** commanding the brigade, and Major General **G. C. Bingham, Lord Lucan,** commanding the Cavalry Division. The return of the survivors was assisted by the equally gallant charge of the 4th French **Chasseurs d'Afrique,** who rode down part of the flanking Russian artillery. Of 673 mounted officers and men entering the 20-minute-long Light Brigade charge, 247 men and 497 horses were lost. Fitting epitaph was the remark of French General **Pierre F. J. Bosquet,** witnessing the charge: "It is magnificent, but it is not war." The Russians retained possession of the Vorontosov ridge, commanding the Balaklava-Sevastopol road. The allies retained Balaklava.

1854, November 5. Battle of Inkerman. Menshikov again tried to break through between the besieging troops and their field support. The brunt of the action fell on the British, in an all-day struggle during which all coordinate control was lost on both sides. The arrival of Bosquet's French division finally tipped the balance, and Menshikov withdrew, with loss of 12,000 men. The allied loss—mostly British—was 3,300 men.

1854, November–1855, March. Dismal Winter. Theoretically, the allies, with sea communications unhindered, should have had little difficulty in their siege operations, whereas the Russians, although their northern line of communications was open, had a long, tenuous overland supply problem. Actually, the allies were totally unprepared for a winter campaign, while a heavy storm (November 14), which wrecked some 30 transports lying at Balaklava, destroyed most of the existing stores of rations, forage, and clothing. To make matters worse, the Russian field army still sat astride the only paved road from Balaklava to the siege lines. Wagon haulage over the muddy plain was almost impossible. The British troops, without shelter or adequate winter clothing, were also actually semistarved. Cholera raged; men died like flies in shockingly inadequate medical facilities. By February, British effectives were down to 12,000. Canrobert, whose administration was much better handled, had 78,000 men in hand, and took over part of the British sector. But the cable and telegraph were pouring out the grim story as seen by **William Howard Russell,** war correspondent for the London *Times.* An outraged British public forced the fall of **George H. Gordon, Lord Aberdeen**'s government, with immediate remedial results, one of the most important being the establishment of proper medical and hospital facilities under **Florence Nightingale.** Meanwhile, despite serious casualties to the working parties from the allied bombardments, Todleben's incessant activity and engineering genius countered the damage done by the allied guns, and the defensive works grew in strength.

1855, January 26. Arrival of a Sardinian Contingent. These 10,000 troops were commanded by General **Alfonso Ferrero di La Marmora.**

1855, January–February. Improved British Logistical Support. A new road and a railroad over the mud plain linked the Balaklava base to the siege corps.

1855, February 17. Battle of Eupatoria (Yevpatoriya). The Russian field army, now commanded by Prince **Michael Gorchakov** (replacing Menshikov), made a halfhearted attempt to interfere. This was repulsed by the Turks. The siege lines drew closer.

1855, April 8–18. Easter Bombardment. A major part of the Russian defenses was destroyed. Russian troops drawn up to meet an expected assault lost more than 6,000 men. But the attack never came. Allied field commanders and home governments were wrangling via telegraph over conduct of the operations.

Canrobert, enraged by the interference, resigned his command. His successor, General **Aimable Jean Jacques Pélissier,** a veteran of the Algerian wars like his predecessor, brought new vigor to the allied operations.

1855, May 24. Capture of Kerch. A well-handled joint expedition cleared the Sea of Azov (Azovskoe More) and severed Russian communications with the interior.

1855, June 7. Allied Assault. Part of the Russian outer defenses were seized; there were 8,500 Russian casualties and about 6,900 allied.

1855, June 17–18. Renewed Allied Assault. The objectives were 2 principal Russian strong points, the **Malakoff** and the **Redan.** Lack of coordination brought utter failure. The French attack on the Malakoff dwindled into an indecisive fire fight, while the British assault on the Redan was caught in the crossfire of 100 heavy guns and thrown back with great loss. The allies lost 4,000 men in this effort, the Russians 5,400. Raglan, heartbroken, died 10 days later, General Sir **James Simpson** succeeding him.

1855, July–August. Attrition. Russian losses through bombardment were draining Sevastopol's strength—some 350 a day during July. The Russian field army decided to make one final effort to break through the allied curtain between Balaklava and the fortress.

1855, August 16. Battle of the Traktir Ridge. Two corps of Gorchakov's army were thrown against some 37,000 French and Sardinian troops on the height above the Chernaya River. A 5-hour combat ended in Russian defeat, ending the last hope for relieving Sevastopol, despite the dogged determination of the Russian infantry. Russian losses were 3,229 dead and some 5,000 wounded. The allies lost about 1,700 killed and wounded.

1855, September 8. Storming of the Malakoff. The one perfectly planned and executed operation of the war. After an intense bombardment (September 5–8) softened the defenses, the French launched a long-prepared mass assault by Bosquet's entire corps. Meticulous attention to detail included a last-minute check by staff officers to ensure that each of the 3 assaulting columns would have easy egress from the trenches, now only 30 yards from the Russian strong point. To preserve secrecy, no signal was given for the assault. Synchronization of watches—for the first time, perhaps, in military history—governed the move. On the stroke of noon the corps surged forward, each column led by its commanding general, while Bosquet established his command post on the outermost French trench. The assault gained the outer wall and swept into the inner defenses, where the Russians disputed every casemate and traverse in hand-to-hand combat. By nightfall the Malakoff was safely in French hands. A simultaneous British assault on the Redan was thrown back, but from the Malakoff the French now turned their fire on the Rus- sians in the Redan and drove them out with heavy loss. That night Gorchakov evacuated Sevastopol, after blowing up the remainder of the fortifications. Next day the allies occupied the city. In all, the allies lost more than 10,000 casualties in this final assault, the Russians 13,000.

1855, October 16. First Appearance of Iron-clad Warships. Bombardment of Kinburn, at the mouth of the Bug River, was anticlimactic, but 3 French steam floating batteries made history, demolishing heavy masonry works while Russian round shot and shell spent themselves harmlessly on their iron plates at ranges of 1,000 yards or less.

The Caucasus Front, 1854–1855

Severe but indecisive fighting between Turks and Russians took place in the Caucasus and Transcaucasus. The principal operation of note was the **Siege of Kars** by Russian General **Michael Muraviev.** The Turkish garrison, commanded by Sir **William Fenwick Williams** (Williams Pasha), British commissioner and a lieutenant general in the Turkish Army, repelled a savage Russian assault (September 29, 1855). Omar Pasha, after the fall of Sevastopol, took a force of 15,000 men to the relief of Kars, but the fortress succumbed to starvation and disease before his arrival, Williams surrendering (November 26, 1855).

Naval Operations in the Baltic

1854, August 7–16. Landing on the Alands. A French squadron landed 10,000 men under General **Achille Baraguay d'Hilliers** at Bomarsund, Aland Islands. After an 8-day siege, during which naval gunfire from Sir Charles Napier's allied fleet took joint part, the 2,400-man garrison surrendered and the fortress was destroyed.

1855, August 7–11. Bombardment of Sveaborg. A Franco-British fleet, after demonstrating before the fortress of Kronstadt (Kronshtadt) bombarded Sveaborg (fortress in Helsinki harbor) without success.

1856, February 1. Preliminary Peace Conditions. Agreed to at Vienna. Final ratification took place at the Congress of Paris (February 28–March 30).

COMMENT. *The outstanding aspect of the war was its abysmal mismanagement on both sides, featur-* *ing indifference in governments and senility and incompetence on the part of field commanders. Russia, fighting England, France, Turkey, and Sardinia, lost—from all causes—some 256,000 men. The allies lost about 252,600. Actual Russian battle deaths were an estimated 128,700; those of the allies, 70,000. Disease—mainly cholera—accounted for the rest. The combatants themselves displayed raw courage under great handicaps. Revolutionary was the awakening of national interest in the welfare of the troops, particularly so in Great Britain, where war correspondents' accounts spread the news of shocking conditions on the fighting front.*

WAR OF AUSTRIA WITH FRANCE AND PIEDMONT, 1859

1859, March 9. Piedmont (Kingdom of Sardinia) Mobilizes. King **Victor Emmanuel II** and his prime minister, Count **Camillo Benso di Cavour,** saw an opportunity to renew the struggle for Italian independence (see p. 846). They had been assured, by a secret treaty, of French support in a war to expel Austria from north Italy. Austria also mobilized (April 9).

1859, April 23. Austrian Ultimatum. Immediate Piedmontese demobilization was demanded. This provided France an excuse to intervene, and Piedmont rejected the ultimatum.

1859, April 29. Austrian Invasion of Piedmont. The half-hearted movements of General Count **Franz Gyulai** gave time for French troops to arrive.

1859, May 30. Battle of Palestro. This was a Piedmontese victory. The allied forces, under personal command of **Napoleon III,** then invaded Lombardy.

1859, June 4. Battle of Magenta. In a series of blundering meeting engagements, the inept commanders on both sides managed to engage portions of their respective commands: 54,000 French against 58,000 Austrians. The *élan* of the French troops, despite the stupidity of command, brought them victory. French losses were 4,000 killed and wounded and 600 missing. The Austrians lost 5,700 killed and wounded and 4,500 missing. Gyulai retired toward the protection of the famous Quadrilateral: the fortified cities of Mantua, Peschiera, Verona, and Legnago. Napoleon and King **Victor Emmanuel** entered Milan in triumph (June 8).

1859, June 24. Battle of Solferino. Austrian Emperor **Franz Josef,** dismissing Gyulai, assumed personal command and moved to meet the Franco-Piedmontese armies. Both sides were approximately equal in number—about 160,000 strong. Again both blundered into a series of meeting engagements along the Mincio River in which the respective high commands lost all control of their forces. Again the spirit of the French soldiery and the vigor of individual corps commanders—French generals **Marie E. P. M. de MacMahon,** François C. Canrobert, **Adolphe Niel,** and Achille Baraguay d'Hilliers—decided the issue after a sanguinary all-day battle. The Austrian Army was saved from total rout by the dogged rear-guard

action of General **Ludwig A. von Benedek.** Allied losses totaled 17,191 (5,521) Piedmontese); Austrian, 22,000.

1859, July 11. Conference of Villafranca. Napoleon and Franz Josef agreed that most of Lombardy (save for the fortress cities of Mantua and Peschiera) should go to Piedmont. Austria was to keep Venetia, protected by the Quadrilateral. This was ratified by the **Treaty of Zurich** (November 10). Most Italians, who wanted Venetia under Italian control, were enraged. The entire Italian peninsula continued in revolution, which gradually coalesced about Piedmont (see p. 920).

AUSTRO-PRUSSIAN (SEVEN WEEKS') WAR, 1866

1866, June 14. Austrian Denunciation of Prussian Power Politics. At the Frankfurt Diet, Austria condemned Prussia's occupation of Holstein (see p. 919) and Chancellor **Otto von Bismarck**'s secret treaty with France. Most German states—including Bavaria, Saxony, and Hanover—concurred. Bismarck, who had concluded (April 8) an offensive-defensive treaty with Italy, dissolved the Germanic Confederation and mobilized for immediate war against Austria and her south German supporting states. Italy declared war on Austria (see p. 921 for operations).

1866, June 16. Von Moltke Strikes. Utilizing the railway net to fullest capacity, Moltke caught his opponents off balance. General **Vogel von Falkenstein** with some 50,000 men entered Hanover on the west, while Prussian Crown Prince **Friedrich Wilhelm**'s Second Army (near Landeshut), Prince **Friedrich Karl**'s First Army (near Görlitz) and General **Karl E. Herwarth von Bittenfeld**'s Army of the Elbe (near Torgau) moved south through Silesia and Saxony. Moltke's scheme was to advance on the widest of fronts and concentrate on the battlefield, subordinates being given the greatest latitude in initiative, provided it be offensive. The strategy effected surprise, but had the drawback of a succession of

THE WAR OF 1866.
Prussian & Italian attacks on Austria:
Prussian campaign against Confederacy:

Italo-Prussian Alliance

0   50   100 M.

meeting engagements where reinforcements built up piecemeal, moving to the sound of the guns.

1866, June 27–29. Battle of Langensalza. In western Germany, General **Alexander von Arentschildt**'s Hanoverian army, repelling an attack by one of Falkenstein's widespread corps, found itself surrounded by Prussian battlefield concentration and was forced to surrender. Meanwhile, to the southeast, the vital contest of the war was about to take place.

1866, June–July. Converging in Bohemia. Moltke's intelligence reports indicated Austrian intention to concentrate northwest of Olmötz (Olomouc). His eastern armies, nominally un-

der the over all command of King **Wilhelm I,** converged on Gitschin (Jičin). The Army of the Elbe occupied Dresden (June 19), then advanced to unite with the First Army in the Bohemian mountain passes. The 2 armies then brushed back advance Austrian elements and retreating Saxons first at **Münchengrätz** (June 27) then at **Gitschin** (June 29). The Second Army engaged in 2 sharp encounters at **Trautenau** (Trutnov) and **Nachod** (both June 27) and halted just east of Gitschin (June 30). General von Benedek, commanding the Austrians, and endeavoring to concentrate along the upper Elbe, north of Königgrätz (Hradec Králové), was more concerned with subsistence than with combat. Moltke, who was in telegraphic touch with all his armies, sought for a Cannae (see p. 72). The Army of the Elbe was ordered to circle to the south and then attack north. The First Army would attack due east, while the Second Army, driving south down the Elbe Valley, would seal the rest of the circle. It did not work out quite that way. The converging Prussian armies totaled about 220,000 men; Benedek's army included 190,000 Austrians and 25,000 Saxons.

Battle of Königgrätz (or Sadowa), July 3, 1866

In a driving rain, the Elbe and First Armies attacked at dawn, but through a telegraphic failure the Second Army had not received Moltke's first attack order and did not move. Overeager, the Elbe Army did not extend its front sufficiently, and its advance crossed the path of the First Army. The confused Prussians were met by a savage Austrian counterattack and massed artillery fire. By 11 A.M. the Prussian advance had been checked and their reserves drawn into what had now become a densely packed frontal attack. Had the Austrians pressed home a cavalry charge at this time, the Prussians might have been driven from the field. But Benedek held his cavalry immobile.

Meanwhile, after a courier had galloped 20 miles to deliver the king's (really Moltke's) peremptory order to the crown prince, the Second Army moved up. At 2:30 P.M. its attack hit the Austrian northern sector, and at the same time the Prussian Guard reserve artillery, pushed forward brilliantly by Prince **Kraft zu Hohenlohe-Ingelfingen,** opened a devastating fire into the Austrian center. The superiority of the Prussian breech-loading needle rifle over the Austrian muzzle-loading rifle musket gave Moltke's infantry a decisive fire superiority. The entire situation changed. Benedek began a dogged withdrawal, covered by his artillery. The Austrians were decisively defeated, but not routed. Prussian losses were nearly 10,000 men; the Austrians lost 45,000, including 20,000 prisoners. Despite the Prussian victory, Königgrätz stands as an example of the dangers to be expected in a battlefield concentration, where success is dependent upon delicate coordination and timing.

1866, June–July. Operations in Italy. (See p. 921).

1866, July 5. French Mediation. Napoleon III, who had expected both contestants to wear themselves out in a long-drawn-out campaign, now offered mediation, which Bismarck accepted—on his own terms.

1866, August 23. Treaty of Prague. Austria was excluded from other German affairs; the German states north of the Main River would form a North German Confederation under Prussian leadership, the South German states, remaining independent, could form a separate confederation.

FRANCO-PRUSSIAN WAR, 1870–1871

Bismarck's diplomatic rally of the north German states into the anti-French North German Confederation was unexpected by Napoleon III. A Prussian effort to place a Hohenzollern prince on the Spanish throne in mid-1870 threatened France with the possibility of a 2-front war. Napoleon, thinking the French Army invincible, decided to precipitate a war he believed inevitable.

1870, July 15. French Declare War. Immediate mobilization followed in both countries. German mobilization and troop concentrations followed a well-directed plan, using the railway net to the full. French mobilization was haphazard and incomplete.

1870, July 16–17. South German States Join Coalition against France. Bavaria, Baden, and Württemberg began mobilization.

1870, July 31. Prussian Concentration and Plan. Three well-equipped German armies—380,000 men—were concentrated on the frontier west of the Rhine: the First, 60,000, under General **Karl F. von Steinmetz,** between Trier and Saarbrücken; the Second, 175,000, under Prince Friedrich Karl, between Bingen and Mannheim; the Third, 145,000, under Crown Prince Friedrich Wilhelm, between Landau and Germersheim. Nominally commanded by King Wilhelm I, the actual direction was that of General Moltke and his efficient general staff. Prussian intelligence had determined the complete French order of battle. The objective was the destruction of the French armies in the field, to be followed by capture of Paris. An additional 95,000 troops were held back until it was certain that Austria would not intervene.

1870, July 31. French Concentration and Plan. By contrast, the French, some 224,000 men in 8 separate army corps, lay behind the frontier from Thionville to Strasbourg and echeloned back to the fortress line Metz-Nancy-Belfort. Transport was improvised, munitions scanty, units below war strength. Napoleon III, at Metz with his incompetent War Minister Marshal **Edmond Leboeuf,** was in command. The only plan was the cry of the French populace: "On to Berlin!" French intelligence was nonexistent. Napoleon ordered a general advance.

1870, August 2. Battle of Saarbrücken. This skirmish between units of the German First Army and French II Corps alerted the French only to the fact that the enemy was near. Napoleon belatedly directed the grouping of his troops in 2 armies: the Army of Alsace (consisting of the 3 southernmost corps under Marshal MacMahon) and the Army of Lorraine (the remaining 5 corps under Marshal **Achille F. Bazaine**). No army staffs existed; the 2 commanders had to function with their own individual corps staffs.

1870, August 4. Battle of Weissenburg. The crown prince's army, advancing in 4 columns, surprised the leading division of MacMahon's corps in early morning on the Lauter River. The other 2 corps had not yet joined him, although 1 division came up during the day. After a sharp action in which the badly outnumbered French lost 1,600 men killed and wounded and 700 prisoners, against German casualties of 1,550. MacMahon pulled back and concentrated defensively on a wooded plateau fronting on the Lauter.

1870, August 6. Battle of Fröschwiller (Wörth). A German reconnaissance in force was repulsed by MacMahon's right. The crown prince built up his strength, overlapping both French flanks, with the main effort against the right, supported by the fire of 150 guns. MacMahon sacrificed his cavalry in gallant, suicidal charges, but was unable to halt the envelopment. He fell back on Fröschwiller, covered by his reserve artillery. There he clung until nightfall, then retreated without interference to Châlons-sur-Marne (August 7–14). Of 125,000 men and 312 guns engaged, German losses amounted to 8,200 killed and wounded and 1,373 missing. The French, of 46,500 men and 119 guns engaged, lost 10,760 killed and wounded and 6,200 prisoners. The Vosges barrier had been pierced, the road to Paris opened. The crown prince's army marched methodically toward the Meuse. The pattern of tactical operation had been set; the French chassepot rifle was superior to the needle gun in accuracy and volume of fire, but the French artillery, thanks to a mistaken reliance on machine guns (mitrailleuses) in place of cannon (about one-fifth of French artillery pieces were mitrailleuses), was far inferior to the German.

1870, August 6. Battle of Spichern. The German First and Second Armies moved into Lorraine, where Bazaine's army was spread in 3 areas, out of mutual supporting distance.

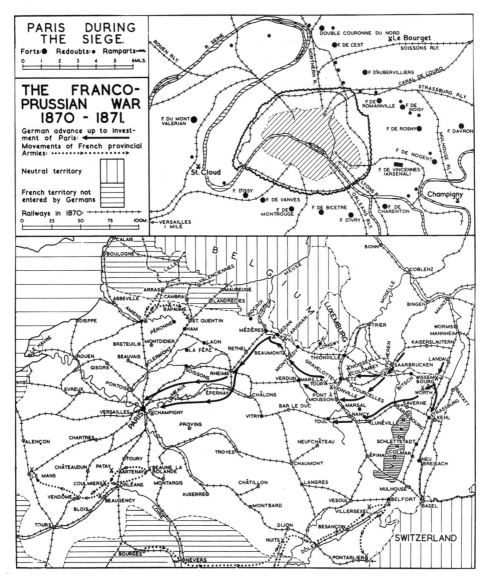

Attacked by Steinmetz and a corps of Friedrich Karl's army, General **Charles Auguste Frossard**'s II Corps held the heights of Spichern, southeast of Saarbrücken, for an entire day until threatened by envelopment on both flanks as the German piecemeal attack gradually built up. Bazaine made no attempt to reinforce him. French losses, of 29,980 men engaged, were 1,982 killed and wounded and 1,096 missing. The Germans, who had put 45,000 men into action, lost 4,491 killed and wounded and 372 missing. There was no pursuit by the exhausted Germans.

1870, August 6–15. German Pursuit. Moltke, sending the Third Army after MacMahon, followed hard on Bazaine's trail with the First and

Mitrailleuse

Second Armies, on the widest of fronts. The rapidity of the German advance detachments and the boldness of their operations gave the French no respite. The Prussian sweep was a strategic penetration between the 2 French armies, threatening Bazaine's line of communications.

1870, August 12. Napoleon Relinquishes Command. Shaken by these defeats, he surrendered all initiative and betook himself to Verdun. Leboeuf was relieved, General **Charles G. M. Cousin-Montauban,** Count of Palikao, replacing him. Bazaine, put in full command of a reorganized Army of the Rhine, fell back on the fortress of Metz, while Mac-Mahon regrouped at Châlons.

1870, August 15. Battle of Borny. The Prussian First Army forced Bazaine to retire across the Moselle. He hoped to gain Verdun and a juncture with MacMahon. But the German Second Army, crossing at Pont-à-Mousson, cut him off. Still hoping to break out, Bazaine concentrated between the Orne and the Moselle, facing south with his left resting on Metz.

1870, August 16. Battles of Mars-la-Tour, Vionville, and Rezonville. Friedrich Karl, moving north at dawn across the Verdun-Metz highway, collided with the French. His leading corps at once attacked, while the remainder of the army hurried to the sound of the guns. A French cavalry charge was repulsed with much loss. The German offensive built up in the usual German pattern of concentration on the battle-field, and a piecemeal engagement developed into a full-blown battle. Successive cavalry charges on both sides ended in a cavalry duel in the afternoon; great masses of horsemen mingled in an almost aimless melee for nearly an hour until both sides broke off in exhaustion. Friedrich Karl finally assaulted along his entire front, pushing into Rezonville. Actually, the combat—or series of combats—was a drawn battle, for both sides bivouacked on the field after the hardest-fought engagement of the entire war. German losses were some 17,000; the French lost more than 16,000. Next day Bazaine, giving up hope for a breakout, retired unmolested on Metz, pivoting about his left flank, and his army of 115,000 men took up a new position, some 6 miles long, facing west on a ridge between the Moselle and the Orne. The bulk of the German armies—some 200,000 men, now between Bazaine and Paris—began movement into this area, only a reinforced corps remaining in observation east of Metz.

1870, August 18. Battle of Gravelotte-St. Privat. Moltke, who had taken personal charge of operations, attacked Bazaine, making his main effort on his own left with the Second Army. The walled village of **St. Privat la Montaigne** became the key point of combat. Friedrich Karl squandered the Prussian Guard in a series of charges against the hamlet, which was defended by Marshal Canrobert's VI Corps. From early morning until dusk, Canrobert's 23,000 men held out against 100,000 assailants, while Bazaine ignored his requests for reinforcement. Then a Saxon corps reached Roncourt, to the north, outflanking the French and threatening their rear. After a house-to-house combat through the village, Canrobert pulled the remnants of his troops back into Metz. Meanwhile, on the German right, an almost independent combat was fought. Two German corps battered their way east of Gravelotte, then became entangled in a ravine beyond. Attempts at disengagement turned into panic, and hordes of refugees poured west through Gravelotte. A brilliant French counterattack was checked only by Hohenlohe-Ingelfingen's artillery and the personal efforts of Moltke, who led reinforcements and averted disaster. Not until midnight, when news of the success at St. Privat reached Moltke, were the Germans sure of victory. Had Bazaine made a general counterattack with the forces still at his disposal, he might have broken free. Instead, he remained passive, relinquishing all control to his corps commanders. Moltke, after waiting for a counterattack which never came, proceeded to seal him within his perimeter.

1870, August 21–28. Advance of Mac-Mahon. MacMahon, meanwhile, responding to the frantic appeal of the government, moved out of Châlons with 120,000 men and 393 guns to the relief of Bazaine, his strength and movements widely advertised in the press. Napoleon III accompanied him. MacMahon's stupid choice of a northerly route invited a turning movement. Moltke accepted. While the German First Army and part of the Second—all under Friedrich Karl—invested Metz, the remainder of the Second Army—called the Army of the Meuse, under Crown Prince **Albert** of Saxony—struck west to cooperate with Friedrich Wilhelm, whose Third Army was driving rapidly through the Argonne Forest to bar Mac-Mahon's advance.

1870, August 29–31. Engagements on the Meuse. MacMahon threw part of his army across the Meuse at Douzy. The Prussian Army of the Meuse, advancing on both sides of the river, forced him northward toward Sedan after sharp clashes at **Nouart** (August 29) and

Beaumont (August 30). Another clash at **Bazeilles** (August 31), where MacMahon was wounded, forced the French into a bend of the river at Sedan itself. Again the Prussians lay between a French army and Paris. The crown prince, arriving from the southeast through Wadlincourt and Doncherry on the left bank of the Meuse, crossed the river on ponton bridges and moved into the plain north of Sedan, completing the envelopment of the French army. Meanwhile, Bazaine's halfhearted attempt to break out of Metz was repulsed by Friedrich Karl (August 31).

1870, September 1. Battle of Sedan. General **Auguste Ducrot,** replacing MacMahon in command, found himself with his back to the Belgian frontier, while nearly 200,000 German troops under Moltke pressed in on the south, west, and north. In a desperate attempt to break out, the French cavalry was shattered by German infantry fire, while 426 German guns, ranged in a semicircle on the heights above Sedan, raked the French positions in a day-long bombardment. German cavalry charges were repelled in turn by the French machine guns (mitrailleuses). Thwarted in his northwesterly drive, Ducrot in the afternoon attempted a southerly assault. This was repulsed. By 5 P.M. the day was lost; the French army crowded into the fortress and town, which was pummeled by devastating artillery fire. General **Emmanuel F. de Wimpffen,** who had succeeded to command, urged the emperor to place himself at the head of his troops and make one last charge. Napoleon, refusing to have any more of his men sacrificed, now drove out under a white flag and surrendered as an individual to the King of Prussia. Wimpffen then surrendered the army: 83,000 men and 449 guns. French losses were 17,000; German, 9,000.

1870, September. German Advance on Paris. The war, it seemed, was over. Half of France's regular organized field forces had been captured, the other half immured in Metz. All that remained were the other fortresses studded along the eastern frontier—Strasbourg, Verdun, and Belfort being the most important. While German reinforcements methodically went about reducing these, and the First and Second Armies tightened their iron ring about Bazaine in Metz, the Third Army and the Army of the Meuse rolled on toward Paris. But as they marched, all France flamed in an amazing demonstration of patriotic resiliency.

1870, September 4. The Third Republic. In Paris the empire toppled in popular uprising. A provisional government rose, with **Léon Gambetta** its torchbearer and General **Louis Jules Trochu** its president. Trochu, as military governor of the city, manned its forts with 120,000 hastily recruited soldiers (including many veterans, reservists, and 20,000 regular marine infantry), 80,000 *gardes mobiles* (untrained recruits under 30) and 300,000 highly volatile and anarchistic-minded *gardes nationales* (recruits between 30 and 50).

1870, September 19. Siege of Paris Begins. Moltke had no intention of squandering his troops in assaults upon the deep-sited and massive system of fortifications ringing the city in 2 belts. Elaborate siege works sealed Paris, King Wilhelm established his headquarters at Versailles, and Moltke waited for starvation to bring the great metropolis into his hands. Much to his astonishment, he found his line of communications harassed by *francs-tireurs* (guerrillas) and a new French army gathering in the Loire Valley. Gambetta, escaping from Paris by balloon—the sole link with the outside world—organized nationwide resistance from Tours (October 11), where the provisional government functioned. Moltke now found himself concurrently engaged in 2 major sieges, a field campaign, and a constant guerrilla warfare along a long line of communications—severely straining the efficient German war machine.

1870, October 27. Fall of Metz. Bazaine's army of 173,000 men surrendered after a 54-day siege; it lost more through the vacillation of its commander and by starvation than by force of arms. After the war Bazaine was court-martialed, convicted of treason, and imprisoned.

1870, October–December. French Initiative. Moltke at once utilized the besieging vet-

erans in what had now become a large-scale operation in the Loire and Sarthe valleys as the inexperienced French Army of the Loire made several gallant but unsuccessful attempts to move to the relief of Paris. Fighting continued through the winter, marked also by harsh treatment of the guerrillas on the German line of communications.

1870, October–December. Operations around Paris. Despite famine, Trochu's forces harassed the besiegers. The task was complicated by the mutinous behavior of the *gardes nationales,* whose revolt (October 31) seriously compromised the defense. Two major sorties (November 29–30, December 21) were repulsed after some initial success.

1870, November 9. Battle of Coulmiers. A French victory over a Bavarian corps caused German withdrawal from Orléans, but the French advance was then checked by Prussian reinforcements.

1870, December 2–4. Battle of Orléans. Two days of heavy fighting between General **Louis J. B. d'Aurelle de Paladines'** French army of the Loire and the army of Friedrich Karl ended with the unwise division of the French forces and reoccupation of Orléans by the Germans. While General **Charles D. S. Bourbaki** hurried to the east to assist the garrison of Belfort, which was under German investment, General **Antoine E. A. Chanzy** with the remainder of the Army of the Loire kept up a constant struggle against much superior force.

1871, January 5. Commencement of Bombardment of Paris. The German shelling caused more resentment than damage, while the war in the provinces continued unabated.

1871, January. Campaign in the North. General **Louis L. C. Faidherbe,** with mixed success, had opposed German efforts to pacify northern France in the **Battle of the Hallue** (December 23). He fought another drawn battle with General **August Karl von Goeben** at **Bapaume** (January 2–3). At the **Battle of St.-Quentin,** however, he was severely defeated by Goeben (January 19). Faidherbe withdrew in good order and repulsed German pursuit. He immediately prepared to renew his offensive, adding to the alarm of the Germans, now seriously overextended by the unexpectedly effective French resistance in all of the outlying provinces.

Krupp gun

1871, January 10–12. Battle of Le Mans. In the Loire Valley, the Germans repulsed a desperate offensive effort by Chanzy. The untrustworthiness of his troops forced Chanzy to retreat to the west, but he still threatened the German hold on the Loire.

1871, January 15–17. Battle of Belfort. In the east, Belfort was the only important French frontier fortress still resisting. Bourbaki with a woefully inexperienced army of 150,000 men threw General **Karl Wilhelm F. A. L. Werder,** with 60,000 German besiegers, onto the defensive. Bourbaki attacked Werder's positions on the Lisaine River, within cannon shot of the fortress. Thanks to his own ineptness and that of **Giuseppe Garibaldi** (a volunteer fighting for France), Bourbaki was defeated after 3 days of furious fighting. German losses were nearly 1,900, French more than 6,000. Bourbaki failed in attempted suicide and was succeeded by General **Justin Clinchant.** With the arrival of a German reinforcing army under General **Edwin von Manteuffel,** Clinchant found himself pinned between the 2 German armies, with the Swiss frontier at his back. With 83,000 men he moved into Switzerland and hospitable interment at Pontarlier (February 1).

1871, January 26. Armistice at Paris. A third and final sortie of the Paris garrison had been decisively thrown back when the *gardes nationales* treacherously fired at their comrades (January 19). With all hope lost and the population on the verge of starvation, Trochu obtained an armistice.

1871, January 28. Convention of Versailles; Capitulation of Paris. All French regular troops of the garrison, and the *gardes mobiles,* became prisoners of war; the forts around the city were occupied by the Germans. At French request—unwisely, as it turned out—the terms did not include disarmament of the *gardes nationales,* who were supposed to become a police force to control the restive population of Paris. The victors marched triumphantly into the city (March 1).

1871, January–February. Belfort the Invincible. Colonel **Pierre M. P. A. Denfert-Rochereau,** commanding the fortress, had resisted siege since November 3, 1870. An engineer officer who had been in command of the ancient fortress for 6 years, he utilized the existing works, extended an outer line of resistance, and with a garrison of some 17,600 men, mainly *gardes mobiles* and *gardes nationales,* carried on an elastic defense. Not until late January were the Germans established within cannon range of the inner fortress. Even then, their progress was slow. Only upon an imperative order from the French General Assembly at Bordeaux (February 15) did Denfert-Rochereau capitulate. The garrison marched out with the honors of war—under arms, colors flying—with all their baggage and mobile equipment. In a siege of 105 days, French losses were some 4,800, while 336 of the townsfolk had been killed by bombardment. German losses were about 2,000. The defense of Belfort was an epic of the French Army.

1871, May 10. Treaty of Frankfurt. France agreed to cede Alsace and northwestern Lorraine to Germany, and to pay an indemnity of 5 billion francs ($1 billion); a German army of occupation was to remain in France until the indemnity was paid.

WESTERN AND CENTRAL EUROPE

BRITISH ISLES

Great Britain's wars during the period took place overseas; they are noted in the regions directly concerned. Except for the Crimean War (see p. 903) and the Boer War (see p. 933), both of which demanded British man power and logistical support for their successful prosecution, the economy of the British Isles was hardly disturbed. However, British diplomacy and British objectives to maintain an equilibrium in the balance of power were directly concerned in almost every conflict which occurred.

THE LOW COUNTRIES

Neither Belgium nor the Netherlands was directly involved in warfare. Both nations successfully maintained their neutrality.

FRANCE

1851, December 2–4. Coup d'État. President (Prince) Napoleon Bonaparte seized control of Paris with army support. A brief republican uprising was suppressed in the "massacre of the boulevards." Napoleon seized dictatorial powers.

1852, December 2. Accession of Napoleon III. Establishment of the Second Empire brought a resurgence of Napoleonic militaristic spirit.

1854. Crimean War. (See p. 903.)

1859. War of France and Piedmont against Austria. (See p. 907.)

1861–1867. Mexican Expedition. (See p. 996.)

1870–1871. Franco-Prussian War. (See p. 910.)

The Paris Commune, 1871

1871, March 18–May 28. Reign of Terror in Paris. The *gardes nationales* overthrew municipal government and proclaimed the city a free town. Pillage and murder were widespread. The National Assembly fled to Versailles with the disarmed regular troops, and the metropolis was given up to the ravages of the mob. Not until the armies of Sedan and Metz—captured in the Franco-Prussian War (see pp. 914, 915)—returned to Versailles and had been rearmed was remedial action possible. Under Marshal MacMahon, the Versailles troops went into action (April 2) from Fort Valerian on the left bank of the Seine. The Communards were swept progressively from the forts they held. The German army of occupation observed without interference.

1871, May 21–28. The Bloody Week. Entering Paris itself, the government troops fought the rebels from street to street, from barricade to barricade. Ruthlessly, as the troops advanced, the Communards murdered hostages they had held, among them the Archbishop of Paris. They burned the Hôtel de Ville and other prominent buildings and attempted to blow up the cathedral of Notre Dame and the Panthéon. The government troops were equally ruthless in suppressing the revolt. An estimated 20,000 Parisians were killed in the grim but not unexpected reprisals.

1873, September 16. German Evacuation. Payment of the war indemnity ended German occupation.

1873–1895. Indochina Expansion. For extension of the French colonial empire in the Far East, see p. 941.

1881. Occupation of Tunis. This was the most significant step of continuing French expansion in Africa.

1886–1889. The Boulanger Crisis. The threatened overthrow of the government by General **Georges Boulanger** never materialized, but widened the existing social fissures in France and in her army.

1894–1906. The Dreyfus Affair. Captain **Alfred Dreyfus,** a Jew, was accused and convicted of treason for espionage actually done by a fellow member of the general staff, Major Count **Charles Walsin-Esterhazy.** When the true facts were discovered, they were suppressed by the aristocratic faction in the army. Finally, thanks largely to the eloquence of **Émile Zola,** the true facts resulted in Dreyfus' exoneration and release from Devil's Island. The French Army was almost torn apart in the violent and bitter controversy.

SPAIN

Internal Disorders

1854, July. Revolution. Generals **Leopoldo O'Donnell** and **Baldomero Espartero** overthrew the government of Queen-Mother **Christina,** who was forced to leave Spain. There followed 14 insurrection-wracked years.

1868, September 18. Revolution. The scandalous and despotic reign of Queen **Isabella II** prompted another uprising, led by Admiral **Juan B. Topete.** The rebel army, commanded by General **Francisco Serrano,** defeated loyal troops at **Alcolea** (near Córdova). The queen, fleeing, was formally deposed. The nation seethed in semianarchy, while a military junta headed by Generals Serrano and **Juan Prim** attempted to keep order.

1869, February. Meeting of a Constituent Cortes. The delegates voted for continuation of a monarchy. The throne was offered to a number of European royalty, who declined. Germany's efforts to place Prince **Leopold of Hohenzollern-Sigmaringen** on the Spanish throne helped to spark the Franco-Prussian War (see p. 910).

1871–1873. Amedeo I. The Duke of Aosta accepted the crown, but abdicated when the Spanish people refused to acknowledge his rule.

1873–1876. Renewed Carlist War. (See p. 842.) The nation was again torn by civil war, fought with extreme brutality.

1873–1874. Republican Proclamation. The **First Spanish Republic** was proclaimed by the radicals, only to collapse. Serrano returned to power. The Carlist disorders continued.

1874, November 24. Accession of Alfonso XII. The son of Isabella was placed on the throne through a military coup, and a constitutional monarchy was established.

1876, February. Flight of Don Carlos. Spain slowly returned to peace.

External Troubles

1859–1860. War with Morocco. Incidents along the frontiers of Spain's Ceuta enclave caused Spain to declare war on Morocco (October 1859). A Spanish army occupied Tetuan (February 1860). Under British pressure peace was restored (April), with Spain expanding its Ceuta enclave, receiving an indemnity, and gaining vague rights on Morocco's Atlantic coast which eventually became the Ifni enclave.

1861–1862. Intervention in Mexico. (See p. 996.)

1861–1864. Unsuccessful Annexation of Santo Domingo. (See p. 1002.)

1864–1865. War with Peru. (See p. 999.)

1865–1866. War with Chile. (See p. 999.)

1895–1898. Cuban Revolution. This culmination of decades of unrest led directly to the **Spanish-American War** (see p. 994) and Spain's loss of Cuba, Puerto Rico, and the Philippine Islands.

PRUSSIA AND GERMANY

1850–1871. Unification of Germany. Powerful economic, political, and cultural forces within fractionated Germany had long been working toward unification. Realization of this dream of most Germans had been greatly hampered by the polarization of the many petty, jealous German states around the rival powers of Austria-Hungary, Prussia, and neighboring France. The accomplishment of unification, and the resultant emergence of the German Empire, were finally brought about primarily by the joint efforts of 2 Prussian aristocrats:

politician Count Otto von Bismarck, chancellor of Prussia, and General Helmuth C. B. von Moltke, chief of the Prussian general staff. Bismarck's careful but bold machinations brought about 3 wars, which were efficiently implemented through the military genius of Moltke. Together, with Bismarck calling the tune, they exemplified Clausewitz' dictum that war is the extension of diplomacy.

Schleswig-Holstein War, 1864

Prussia, with Austria as a somewhat reluctant ally, settled by force the question of Schleswig-Holstein, which had long vexed the great powers (see p. 844).

1864, February 1. Invasion of Denmark. Prussian troops under Prince Friedrich Karl, with some Austrian assistance, swept northward through Schleswig-Holstein. The only major Austrian participation in the war was by the small naval squadron of Admiral **Wilhelm von Tegetthoff,** which broke a Danish blockade of the Elbe and Weser rivers.

1864, March 15–April 17. Siege of Dybböl. Friedrich Karl invested the Danish fortress, finally capturing it by assault. Danish losses were 1,800 killed and wounded and 3,400 prisoners. German losses were negligible. An amphibious assault then captured the island of Als.

1864, April 25–June 25. Truce. British efforts to bring about peace were foiled by Bismarck.

1864, June 26. Renewed Invasion of Denmark.

1864, August 1. Denmark Sues for Peace. She renounced her right to the disputed provinces to Prussia and Austria.

1864, October 30. Treaty of Vienna. This ratified the August agreement.

1865–1866. Friction between Austria and Prussia. Bismarck next moved to gain complete control of the captured provinces and to cement Prussian ascendancy in north Germany. He created a *casus belli* by pressures to oust his former Austrian partner from joint occupation of Schleswig-Holstein. The instrument lay in hand: the Prussian Army, revamped through the efforts of Moltke and

Generals **Albrecht T. M. von Roon** (Minister of War) and Manteuffel. It was equipped with breech-loading artillery and small arms (the needle gun), and the brief Danish War had provided its baptism of fire.

1866. Austro-Prussian War. The second of Bismarck's steps toward his dream of a German empire.

1886–1870. Friction between France and Prussia. France was the major obstacle remaining between Bismarck and his dream. The wily chancellor secretly gained the support of the smaller German states, and jockeyed Napoleon III into premature action.

1870–1871. Franco-Prussian War. (See p. 910.)

1871, January 18. Foundation of the German Empire. The efforts of Bismarck achieved their goal when King Wilhelm of Prussia was proclaimed emperor of a united Germany (not including Austria) in the Palace of Versailles while the Siege of Paris was still in progress.

1879, October 7. Austro-German Alliance. Bismarck brought about a reconciliation of the 2 major Germanic powers in order to establish a powerful military bloc in Central Europe.

1881, June 18. Alliance of the Three Emperors. This loose alliance of the German, Russian, and Austro-Hungarian empires was anti-British and ostensibly pro-Turkish. The conclusion of the agreements was delayed by the assassination of Czar **Alexander II** (March 13, 1881); Russian association in the alliance was halfhearted.

1882, May 20. Triple Alliance. This treaty, between Germany, Austria, and Italy, was primarily anti-French; it was, however, also anti-Russian and (save for the Italians) anti-British.

1890, March 16. Dismissal of Bismarck. Bismarck's alliance systems were the capstone of his career as empire builder. He was dropped from the Germanic helm by Kaiser **Wilhelm II.** The militarist young kaiser then extended German colonial hold in China, Africa, and—through military and economic assistance to Turkey—the Near East.

1895, June. Completion of the Kiel Canal. This gave Germany a strategic bypass between the North and Baltic seas and enabled her to avoid the Scandinavian straits.

AUSTRO-HUNGARIAN EMPIRE

1850–1900. Austrian Involvement in the Balkans. Austria expanded her influence and power in the Balkans, generally at the expense of Turkey and, to a lesser extent, of Russia. Austrian influence was great in all of the Balkan nations that became independent during this period (see pp. 922–924).

1859. War with France and Piedmont. (See p. 907.)

1866. Seven Weeks' War. (See pp. 908, 921.)

ITALY

1850–1870. Unification of Italy. The pressures for unification of Italy were similar to those which were working simultaneously in Germany. A complication—and also a stimulus to cohesion among volatile and suspicious Italian patriots—was the continued occupation of most of north Italy by the Austro-Hungarian Empire, and the consequent political, economic, and military predominance of Austria over the entire Italian peninsula. Piedmont (Kingdom of Sardinia), most stable of the Italian states, became the nucleus of efforts to end Austrian predominance. The premier of Piedmont, Count **Camillo Benso di Cavour,** whose interests initially were solely for increasing the power and prestige of his small country, became the architect of a new state. In this effort he was soon overshadowed in the public mind by his principal instrument, the Italian patriot and soldier of fortune, Giuseppe Garibaldi.

1858, July 20. Alliance of France and Piedmont. Having gained British and French good will by joining with them in the Crimean War (see p. 906), Cavour brought about a secret formal treaty with Napoleon III, whereby France would join Piedmont in driving Austria from Italy if this could be done without opening France to the charge of aggression. Following expulsion of Austria, the Italian states were to be amalgamated into a federation of 4 under the pope's presidency. France, as a *quid pro quo,* would gain Savoy and Nice.

1859, April–July. War of Austria with France and Piedmont. (See p. 907.)

1860, April 4. Uprisings in Naples. Revolts in the Kingdom of Naples were severely suppressed, arousing widespread revulsion in Italy and abroad.

1860, May 11. Garibaldi's Invasion of Sicily. With covert support from Cavour and King Victor Emmanuel II of Piedmont, Garibaldi and his "Thousand Redshirts" landed at Marsala in Sicily, having sailed from Genoa (May 5). He marched inland, rallying the inhabitants to revolt against the Kingdom of Naples. He defeated Neapolitan forces at **Calatafimi** (May 15) and took Palermo (May 27). Marching eastward, he defeated the Neapolitans again at **Milazzo** (near Messina; July 20).

1860, August 22. Passage of the Strait of Messina. With British assistance, Garibaldi crossed to the Italian mainland and marched on Naples.

1860, September 7. Capture of Naples. Seizure of Naples was accomplished against negligible opposition. Garibaldi prepared to march on Rome and Venetia.

1860, September 10. Piedmontese Invasion of the Papal States. Unrest in the Papal States provided an excuse for Cavour to send troops across the border. Papal forces were decisively defeated at **Castelfidardo** (September 18); the Piedmontese marched south to link up with Garibaldi in Neapolitan territory. Meanwhile, French troops occupied Rome, and a French fleet took station along the Neapolitan coast, to preclude any Italian attack against the immediate papal domains.

1860, October 26. Battle of the Volturno. After one repulse (October 1), Garibaldi gained another victory over the demoralized Neapolitan forces and advanced on Gaeta.

1860, November 3–1861, February 13. Siege of Gaeta. King **Francis II of Naples,** with 12,000 men, made his last stand against the united Italian forces. The withdrawal of the French fleet (January 19) made it possible for Piedmontese warships to bombard Gaeta from the sea, forcing Neapolitan surrender.

1861, March 17. The Kingdom of Italy. An all-Italian parliament (not including representation from the papal territories around Rome) proclaimed a united Kingdom of Italy, with Victor Emmanuel as its first constitutional monarch. Less than 3 months after the realization of his dream, Cavour died. The hopes of Garibaldi to include all of the Papal States in the kingdom were foiled by continued French occupation of Rome.

1862, March–October. Garibaldi's Operations against Rome. Sporadic and ineffectual attempts to seize Rome were made by Italian patriots—with the covert support of Victor Emmanuel. Garibaldi, rallying volunteers to his standard, marched from the Strait of Messina toward Rome (August 24). The Italian government, unable to countenance an overt act of war against French forces in Rome, ordered its forces to intercept Garibaldi in the toe of Italy.

1862, August 29. Battle of Aspromonte. Regular Italian troops defeated Garibaldi and his volunteers. Garibaldi was wounded and captured. He and his men were soon released.

1866, May 12. Italian-Prussian Alliance. This was concluded with the approval of Napoleon III.

War with Austria, 1866

1866, June 20. Italy Declares War. This followed the outbreak of the Austro-Prussian War (see p. 908).

1866, June 24. Second Battle of Custozza. An Austrian army, 80,000 strong, under Archduke **Albert,** met the Italian army, under Victor Emmanuel, 120,000 strong. The Italians, poorly led, were defeated piecemeal. The outstanding feature of the battle was the spirited shock action of 2 improvised Austrian cavalry brigades that broke up repeated Italian assaults. The Italians retreated across the Mincio in disorder, having lost 3,800 killed and wounded and 4,300 missing. Austrian losses were 4,600 killed and wounded and 1,000 missing. The Austrian army was shortly withdrawn for the defense of Vienna against the Prussians.

1866, July. Garibaldi's Alpine Campaign. Leading a small volunteer army, Garibaldi won several minor successes at **Lodrone** (July 3), **Monte Asello** (July 10), **Condino** (July 16), **Ampola** (July 19), and **Bezzecca** (July 21). He was ordered to withdraw when Bismarck made clear to Italy that he would not consent to Italian occupation of the Trentine Tyrol.

1866, July 20. Naval Battle of Lissa. An Italian squadron of 10 ironclads and 22 wooden vessels, under Admiral **Carlo T. di Persano,** was attacked off the island of Lissa (Vis) in the Gulf of Venice by the Austrian squadron of Admiral Count **Wilhelm von Tegetthoff,** 7 smaller ironclads and 14 wooden vessels. The Italians were in a long line-ahead formation. The Austrian commander struck the line in a wedge-shaped formation, ignoring Italian fire until within point-blank range. Breaking through the Italian line, Tegetthoff sank 3 ironclads, with a loss of 1,000 men. The remainder of the Italian squadron broke off action and steamed away.

1866, October 12. Treaty of Vienna. The conclusion of the war resulted in ratification of Italian annexation of Venetia, brought about earlier by the good offices of Napoleon III (July 3).

1866, December. French Troops Withdraw

from Rome. Garibaldi seized the opportunity to lead a futile volunteer invasion of the Papal States (January–September 1867). This, as well as attempts—both overt and covert—by the Italian government to overthrow the papal dominion, led Napoleon III to send French forces back less than a year later.

1867, October 26. French Expedition to Rome. Some 2,000 men under Major General **Charles A. de Failly,** having landed at Civitavecchia, arrived in Rome. Meanwhile Garibaldi, renewing his own invasion, had reached **Monte Rotondo,** where he had defeated a small papal force (October 24).

1867, November 3. Battle of Mentana. Garibaldi's bands, some 4,000 strong, met a combined force of 3,000 papal troops and de Failly's 2,000 French regulars, who were armed with the new **chassepot** rifle. The redshirts were mowed down with heavy losses, and streamed back over the Italian border, leaving 800 prisoners. They were all arrested by the Italian authorities.

1870, September 20. Occupation of Rome. French forces having been withdrawn because of the Franco-Prussian War (see p. 910), an Italian army of 60,000 men under General **Raffaele Cadorna** invested it. After a short bombardment, a breach was effected. The invaders began to stream in and Pope **Pius IX** ordered his troops to cease fire.

1870, October 2. Formal Annexation of Rome by Italy. Following a plebiscite, the city was declared the capital of the nation.

1887–1896. Operations against Ethiopia. (See p. 929.)

EAST EUROPE

RUSSIA

1853–1856. Crimean War. (See p. 903.)

1861, February 2. Warsaw Massacre. Unrest in Russian Poland culminated in a mass demonstration in the Polish capital. Russian troops fired on the mob, killing many. Sporadic uprisings flared through the area.

1863–1864. Second Polish Revolution. Polish guerrilla bands took the field, and the revolt quickly spread into Lithuania and White Russia (January 1863). Despite protracted resistance, harsh repressive measures finally quelled the uprising (May 1864).

1864–1876. Conquest of Kokand, Bokhara, and Khiva. Russian colonial expansion followed steady military advances in Central Asia. Subjugation of Kirghiz and Turkoman tribes eventually resulted in the annexation of the entire Transcaspian region (1881).

1877–1878. Russo-Turkish War. (See p. 924.)

1884–1885. Conquest of Merv. The last independent Moslem principality of Central Asia was finally absorbed by Russia, establishing a common frontier with Afghanistan. The result was an increase in British-Russian colonial rivalry and tension.

1885. Frontier Incidents with Afghanistan. Fighting between Russian and Afghan troops came close to full-scale war, and was settled peacefully only after Britain made clear her determination to intervene on the side of Afghanistan (see p. 938).

1891–1894. Franco-Russian Alliance. This was a foil to the Triple Alliance of Germany, Austria, and Italy (see p. 920).

THE BALKANS

Greece

1850, January–March. British Blockade. The purpose was to force Greece to pay interest on an international loan and to pay compensation due some British citizens. A compromise was reached and Britain lifted the blockade.

1854, January–February. Greek Invasion of Thessaly and Epirus. Greece took advantage of Turkish participation in the Crimean War.

1854, April–1857, February. British and

French Occupation of the Piraeus. The allies prevented Greece from joining Russia in the Crimean War.

1862, February 13–October 23. Revolution. Otto was deposed. He was replaced, after a long delay, by **George I,** a Danish prince selected by the great powers.

1878, February 2. Greece Declares War on Turkey. Again she was prevented from taking advantage of a Russo-Turkish War (see p. 924) by intervention of the other powers.

1886, May–June. British and Allied Blockade. Again the powers prevented Greece from taking advantage of Turkish troubles.

1896–1897. War with Turkey. (See p. 926.)

Serbia

All internal and external affairs of Serbia were affected by the continuing bitter feud between the Obrenovich and Karageorgevich families (see p. 849). The principal events were:

1862. Turk Bombardment of Belgrade. This was an outgrowth of troubles between Serbians and the Turkish garrisons.

1867, April. Withdrawal of the Turkish Garrisons. This was forced by the great powers.

1876, July. War with Turkey. (See p. 924.) Serbia was quickly overwhelmed. This was one cause of the Russo-Turkish War (see p. 924).

1877, December 14. Renewed War with Turkey. Serbia took advantage of Turk setbacks in the war with Russia.

1878, March 3. Treaty of San Stefano. Turkey recognized Serbian independence and ceded substantial territory (see p. 926).

1878, July 13. Treaty of Berlin. (See p. 924.) Serbian independence was reaffirmed, but territorial cuts caused much bitterness in Serbia, particularly against Austria, which virtually annexed Bosnia and Herzegovina, largely populated by Serbs.

1883, November. Revolt. King **Milan** savagely suppressed it. Unrest continued.

1885–1886. Serbo-Bulgarian War.

1885, November 13. Serbia Declares War. King Milan demanded territory to compensate for Bulgaria's growth after union with Eastern Rumelia. Bulgaria refused, so Milan invaded, hoping war would unify troubled Serbia.

1885, November 17–19. Battle of Slivnitza. Milan's army was defeated by Bulgarian Prince **Alexander's** hurriedly concentrated forces. The Bulgars then invaded Serbia.

1885, November 26–27. Battle of Pirot. Prince Alexander's outstanding leadership brought another victory after 48 hours of serious conflict between armies each about 40,000 strong. Austrian intervention saved the Serbs (January 1886).

1886, March 3. Treaty of Bucharest. The *status quo* was restored.

Montenegro

1852–1853. War with Turkey. A Turkish army under Omar Pasha invaded Montenegro. Defeated by Prince **Danilo II,** near **Ostrag,** the Turks withdrew in the face of Austrian threats. Border friction erupted in open warfare (1853), and invading Turks were defeated at **Grahovo** by **Mirko Petrovitch,** brother of Danilo; again the Turks withdrew. Montenegro's independence was not recognized abroad.

1861–1862. Renewed War with Turkey. Montenegro supported a revolt in Herzegovina (1860–1861); again Omar Pasha invaded. Despite Mirko's heroic defense of Ostrag, Montenegro was quickly overrun and forced to acknowledge Turkish suzerainty at the **Convention of Scutari** (August 31).

1876–1878. War with Turkey. Montenegro and Serbia shared defeat, triumph, and then disappointment, but gained recognition of independence by the Treaty of Berlin (see above). However, hostilities on the Albanian frontier continued until 1880.

Bulgaria, 1875–1900

1875, September. Bulgarian Revolt. Suppressed by Turkey.

1876, April–August. Bulgarian Revolt. Again suppressed by Turkey. This, however, helped precipitate the Russo-Turkish War, in which Bulgaria was a major battleground (see below).

1878, March 3. Treaty of San Stefano. Bulgarian independence was recognized.

1878, July 13. Treaty of Berlin. Because the other powers feared that Bulgaria would dominate the Balkans and be under Russian influence, the principality was reduced to less than half of the territory recognized in the Treaty of San Stefano. Parts of Macedonia were given to Serbia (in return for lands taken by Austria); most of Macedonia remained Turkish, and much of Thrace was formed into the autonomous Turkish province of Eastern Rumelia. The Bulgarians were bitter against the powers. The treaty also led to friction with Serbia.

1885, November 13. Annexation of Eastern Rumelia. A popular uprising was followed by absorption of the neighboring province. Serbia demanded compensation to offset this increase in Bulgarian power; Bulgaria refused.

1885–1886. Serbo-Bulgarian War. (See p. 923.)

Rumania

1853, July. Russian Occupation. This precipitated the Crimean War (see p. 903).

1854, August–1857, March. Austrian Occupation. Russia withdrew under Austrian pressure; the Austrian occupation continued during the Crimean War (see p. 904).

1877–1878. Russo-Turkish War. (See below.) This was precipitated by Russian invasion of Rumania (April 24, 1877).

1877, May 21. Rumanian Declaration of Independence. Rumania immediately entered the war against Turkey, and gave the Russians some assistance at the siege of Plevna (see p. 925).

1878, July 13. Treaty of Berlin. This reaffirmed the terms of the Treaty of San Stefano, so far as Rumania was concerned. She reluctantly surrendered southern Bessarabia to Russia, but received northern Dobruja, to assure an outlet to the sea.

EURASIA AND THE MIDDLE EAST

THE OTTOMAN EMPIRE

1852–1853. War with Montenegro.

1853–1856. The Crimean War. (See p. 903.)

1861–1862. Second Montenegrin War. (See p. 923.)

1866–1868. Cretan Insurrection. This uprising, instigated by Greeks, was quelled after furious fighting.

1876. War with Serbia and Montenegro. Cruel Turkish suppression of Christian insurrections in Herzegovina and Bosnia (1875) led to Serbian and Montenegran declarations of war. Serb forces (including Russian volunteers) under Russian leadership were defeated by **Suleiman** Pasha at **Alexinatz** (September 1) and **Djunis** (October 29). Russia, champion of the pan-Slav movement in the Balkans, began to mobilize along her southern frontier.

Russo-Turkish War, 1877–1878

1877, April 24. Russia Declares War. She immediately invaded Rumania, while mobilizing 275,000 men of all arms, plus 850 fieldpieces and 400 siege guns. Another 70,000 invaded Turkey's Caucasian provinces. The Russians expected easy victory over the ill-assorted and scattered Turkish forces of some 135,000 men and 450 guns in Europe and about the same number in Asia. The Grand Duke **Nicholas**

(brother of Czar Alexander II) commanded the Russian armies in Europe, **Abdul Kerim** the Turks. The Russian plan was to advance across the Danube, cross the Balkans, and move on Adrianople (Edirne). A flank corps entering the Dobruja would protect the Russian left, while detachments were to mask the Turkish fortress quadrilateral of Ruschuk (Ruse), Silistria (Silistra), Shumla (Kolarovgrad), and Varna (Stalin). Due to Rumania's declaration of independence from Turkey (see p. 924), the Russian advance was made from friendly territory (though Russia delayed recognition of Rumania's independence). From the outset, the operations of the high command on both sides were remarkable for indecision and ineptitude.

1877, June. Clearing the Danube. In a series of combined army-navy operations, Russia quickly seized control of the Danube; bridges were built for the crossing.

1877, June–July. Invasion of Bulgaria. The Russian advance guard of General **Ossip V. Gourko,** with 31 squadrons of cavalry, 10 battalions of infantry, and 32 guns, plunged across the Danube (June 23), followed by the main army. Gourko, learning that the Turks had garrisoned the key Shipka Pass in the Balkans, led his force through the smaller and undefended Khainkoi Pass to the east and forced the immediate retreat (July 18–19) of the Shipka defenders. Meanwhile, his patrols raided to within 90 miles of Adrianople, destroying railway and telegraph lines. He had, in fact, penetrated the Turkish cordon defense of the Balkans; a Russian advance in force would probably have resulted in overwhelming victory. Meanwhile, the fortresses of **Svistov** and **Nikopol** were seized by the main army.

1877, July 19–December 10. Siege of Plevna. **Osman** Pasha, commanding the fortress of Vidin, well west of the Russian advance, hurried to Plevna (Pleven), whence he threatened the Russian right flank and indirectly threatened the Danube bridges. Instead of merely containing him, the Russians halted to take Plevna, which Osman and his engineer officer, **Tewfik** Pasha, were massively fortifying.

Thrice the Russians assaulted (July 20, July 30, and September 11) and were repulsed with total losses of 30,000 men. Young Russian General **Mikhail D. Skobelev** distinguished himself in the last of these attacks, but was not properly supported by his colleagues, while Osman took advantage of the poor Russian coordination. The Russian investment was then put under the command of Todleben (hero of Sevastopol; see p. 905) and systematic siege operations began. Short of provisions, Osman attempted a final sortie (December 9), but was wounded and the effort collapsed. Plevna capitulated (December 10). The Russian advance had been held up for 5 months, completely negating the potentialities of Gourko's brilliant seizure of the Shipka Pass.

1877, July–December. Operations in Central Bulgaria. Meanwhile, Abdul Kerim had been replaced as Turkish commander in chief by **Mohammed Ali;** the army of Suleiman Pasha had been recalled from Montenegro by sea and was opposing Gourko south of the Shipka. The Russians, fearful of overextension while Plevna was still resisting, called Gourko back north of the mountains, but the Shipka Pass was fortified. The Turks established a fortified camp 2 miles to the south, at Senova, under **Vessil** Pasha. The efforts of Mohammed Ali to drive the Russians back to the Danube, and to relieve Plevna, were futile and poorly coordinated; Russian countermoves were no more effective. Gourko, however, assured Russian retention of the initiative by capturing Sofia at the end of the year.

1877, August–1878, January. Operations in the Caucasus. The opposing armies in the east, each about 70,000 men, under Grand Duke **Michael** and **Mukhtar** Pasha, undertook no major operations until the Russians received sufficient reinforcements to permit an advance. The Russians gained a victory at **Aladja Dagh** (October 15), forcing the Turkish army to fall back on its 2 main fortresses of Kars and Erzerum. The Russians immediately invested **Kars,** which they took by assault a month later in a gallant, well-planned operation (November 18). Despite the bitter cold of winter in the mountains, the Russians continued their advance, and by the end of the year were heavily engaged against the Turkish defenses of **Erzerum.** They were seriously threatening the fortress when operations ceased as a result of an armistice (January 31).

1878, January. Final Campaign in Europe. Deciding upon a winter campaign in the Balkans, the Russians turned on Suleiman Pasha, who now commanded the main Turkish Army south of the Shipka. Advancing on a wide front, Russian generals **Fëdor Radetsky,** Skobelev, Gourko, and Prince **Imeretinsky** outmaneuvered the badly extended Turks.

1878, January 8–9. Battle of Senova. Concentrating his columns against Vessil Pasha, south of the Shipka Pass, Skobelev defeated and encircled the entire Turkish force, capturing 36,000 men. The Russians pressed on to Plovdiv and Adrianople (January 17 and 19), cutting off Suleiman, who, with 50,000 men, was still farther west. After some minor engagements, the Turkish general retreated across the mountains to the Aegean at Enos (January 28). His forces went by sea to Constantinople. The Russians had advanced to the Chatalja lines, just outside the Turkish capital (January 30). An armistice immediately followed (January 31).

1878, March 3. Treaty of San Stefano. Turkey paid a large indemnity and conceded independence of Montenegro, Serbia, and Rumania. Bulgaria became an autonomous state under internal Russian control. Russia gained Ardahan, Kars, Batum, Bayazid.

1882. Germany Military Mission Arrives. After the death of the original mission chief, Colonel **Kolmar von der Goltz** headed the mission (1883–1895); his influence on Turkey was significant for the next 33 years.

1896, February. Insurrection in Crete. Local Greeks rose against Turkish rule. Greece intervened, precipitating war.

1897, April 17–September 18. Greco-Turkish War. Greek forces, led by Crown Prince **Constantine,** were consistently defeated by **Edhem** Pasha in parallel campaigns in Thessaly and Epirus. The ineptness of both high commands was remarkable. Through the mediation of the Russian Czar, an armistice was arranged (May 19) and a peace treaty effected (September 18).

PERSIA

1850–1854. Russian Conquests in the Syr Darya Valley. Muscovite influence reached Persia's Central Asian frontier.

1855. Invasion of Afghanistan. After complicated negotiations between Shah **Nasr ed-Din** and local Afghan provincial rulers, and despite British warning, Persian troops occupied Herat, Afghanistan (1856).

1856, November–1857, April. War with Britain. Britain then declared war on Persia (November 1). Britain seized the port of Bushire on the Persian Gulf (January). Sir **James Outram** with 2 Indian Army divisions invaded Persia, which soon sued for peace and evacuated Afghanistan (Treaty of Paris).

1878. Russian Military Influence. A Russian training mission helped establish a Cossack brigade, tangible indication of growing Russian influence in Persia during the conquest of Central Asia (see p. 922).

AFRICA

The powers of Europe were engaged in a bargain-counter rush, each trying to carve out a colonial parcel of the Dark Continent for its own selfish purposes. France and Great Britain were in keen competition not only with one another but also with the newcomers, Germany and Italy. Spain and Portugal, with toeholds long established, strove to maintain those positions. The slowly decaying Ottoman Empire was also clinging—ineffectually—to its possessions in North Africa. Military actions were necessary for these colonial powers to maintain and increase their holdings. A host of little wars flickered around the African periphery throughout the period.

EGYPT AND THE SUDAN

1863–1879. Reign of Khedive Ismail. Nominally a vassal of the Sultan of Turkey, Ismail attempted to modernize Egypt, but in so doing saddled his country with an enormous debt. As a result, Britain and France exerted a dual control, particularly after the British government purchased from Ismail all his interest in the Suez Canal (completed 1869). Ismail, with the aid of British and American army officers in his service, attempted—with mixed success—the conquest of the Red Sea coast (1865–1875) and of the Sudan (1871–1875).

1874–1879. The Upper Nile. Egyptian military control was established as far as Unyoro on Lake Albert Nyanza by a succession of minor expeditions against Arab slave traders and the indigenous tribal kingdoms of the region.

1875–1879. War with Abyssinia. Egyptian expansion eastward, beginning with the occupation of **Suakim** and **Massawa** (1865) by direction of the Ottoman Empire, threatened to cut off Abyssinia from the sea. When the Egyptians militarily occupied **Harar** and adjacent seacoast ports (1872–1875), King **John** of Abyssinia declared war. His troops moved against Ismail's forces, who were almost annihilated at **Gundet** (November 13, 1875). A second Egyptian expedition met defeat at **Gura** (March 25, 1876).

1879, June 25. Deposition of Ismail. The sultan appointed Tewfik, Ismail's son, as his successor.

1881, February 1. Rebellion of Ahmet Arabi. This revolt against Turkish and foreign control spread into an anti-Christian conflict.

1882, May–June. Disorder in Alexandria. French and British naval squadrons converged on Alexandria (May) and were present when some 50-odd Europeans were massacred by a native mob (June 11).

1882, July 11. British Bombardment of Al- exandria. The French refused to cooperate with Admiral Sir **Frederick B. P. Seymour**'s reprisal. Britain now landed 25,000 troops at Ismailia under Sir **Garnet Wolseley.**

1882, September 13. Battle of Tell el-Kebir. Arabi's forces—38,000 men and 60 guns—were intrenched along the railway and sweetwater canal. After some preliminary skirmishing, the British launched a surprise night attack, driving the Egyptians off in disorder with losses of 2,000 killed and 500 wounded. British casualties were 58 killed, 379 wounded, and 22 missing. An immediate pursuit resulted in complete collapse and surrender of Arabi's forces. Britain dominated the Egyptian government. All but 10,000 of the British force were now sent home.

1883. Mahdist Uprising. Mahdi **Mohammed Ahmed of Dongola,** announcing himself to be a prophet, led a dervish uprising to set the Sudan in flames. About 10,000 Egyptian troops under General **William Hicks,** a former Indian Army officer, were completely wiped out at **El Obeid** (November 3), while **Osman Digna,** the Mahdi's lieutenant, moved successfully against Egypt's Red Sea ports, wiping out another Anglo-Egyptian force at **El Teb** (near Suakim; February 4, 1884). Britain ordered Egypt to evacuate the Sudan.

1884, January 18. Gordon to Khartoum. General **Charles ("Chinese") Gordon** (see p. 946) was sent by the British government to Khartoum, at the confluence of the White and Blue Nile rivers, to supervise the evacuation of the Sudan by Egyptian forces. Mahdist forces invested the city (February).

1884–1885. Siege of Khartoum. For nearly a year, Gordon and a small garrison were besieged, while public opinion in England mounted against the British government's indecision in rescuing them. Finally a relief expedition under General Wolseley struck south from Wadi Halfa (October 1884), too late.

1885, January 26. Fall of Khartoum. The Mahdi's forces swept over the last defenses of Khartoum, massacring the entire garrison. Wolseley's force, after several engagements, reached the city 2 days later, but was ordered to withdraw.

1885, June 21. Death of the Mahdi. His wild dervish followers under his successor, the Kalifa **Abdullah,** soon completed control of the entire Sudan.

1896–1898. Reconquest of the Sudan. Concerned by increasing French and Italian colonial interest in the Nile Valley, Britain decided to reoccupy the Sudan. Major General Sir **Horatio Kitchener,** sirdar of the Egyptian Army, commenced a methodical reconquest with a mixed force of British and Egyptian troops. Moving up the Nile, escorted by a river gunboat flotilla, he constructed a railway as he went along to ensure his logistical support. Highlights of the step-by-step advance, against fanatical opposition, were the capture of **Dongola** (September 21) and **Abu Hamed** (August 7, 1897) and defeat of Mahdist forces at the **Battle of the Atbara River** (April 8, 1898). Abdullah and Osman Digna concentrated their strength about the strong fortress position of Omdurman on the Nile, just north of Khartoum.

1898, September 2. Battle of Omdurman. Kitchener's army, some 26,000 men—about half British regulars and half well-trained Egyptian troops—were attacked in their fortified encampment at Egeiga, 4 miles from Omdurman, by Abdullah's 40,000 men. The savage bravery of the tribesmen was unavailing against machine guns and modern small arms, and they were repulsed with great slaughter. Kitchener then counterattacked toward Omdurman. The dervishes, rallying, fell on the British right and rear, Sir **Hector A. Mac-Donald**'s Sudanese brigade repelling 2 successive charges. At almost the same time, a dervish force of some 2,000 men leaped from concealment on Kitchener's right flank. In an old-fashioned cavalry charge, the 21st Lancers swept them away, though losing some 25 percent of its strength. This ended the battle, the

Kalifa's forces streaming from the field. While British and Egyptian cavalry took up pursuit, Kitchener's main body marched into Omdurman. Mahdist losses were more than 10,000 killed, an estimated equal number of wounded, and 5,000 prisoners. Kitchener's casualties amounted to approximately 500 in all.

COMMENT. *Omdurman's importance was twofold: it reestablished Anglo-Egyptian influence and control over the vast Nile watershed region, and it was the first real demonstration of the firepower latent in the machine gun (Kitchener's artillery included 20 of them). Kitchener became a British national hero.*

1898, September–November. Fashoda Incident. A small French military and exploratory expedition under Major **Jean B. Marchand** had reached the Nile River at Fashoda in the southern Sudan (July 1898), apparently assuring France of a foothold on the left bank of the Nile. Kitchener moved up the Nile to Fashoda, while Britain threatened war with France if the expedition did not withdraw. Under orders from Paris, Marchand evacuated Fashoda (November 3).

ETHIOPIA (ABYSSINIA) AND THE RED SEA

1854–1855. Lij Kassa Becomes Emperor Theodore. Lij Kassa (see p. 000) defeated the allied forces of Ras **Ali** and Ras **Ubie** at **Gorgora** on Lake Tana (1854) and so gained effective control over all northern and central Ethiopia. The next year he deposed Emperor **John III** and conquered Shewa (Shoa) province, and had himself crowned Emperor as Tewodoros (Theodore) II.

1855–1868. Reign of Theodore II. He made energetic efforts to modernize the country and reform the government, moving the capital from Gondar to Magdala (1856). The death of his wife and of 2 trusted British advisers contributed to a conditional of mental instability, and his later reign was marked by excessive and erratic actions.

1867–1868. British Abyssinian Expedition. As a result of the imprisonment and murder of consular officials, some 32,000 well-equipped troops under Sir **Robert Napier** landed at Mulkutto (Zula) on the Red Sea and moved inland. Theodore's troops were defeated at the **Battle of Arogee;** Theodore committed suicide (April 10, 1868). **Magdala,** his capital, was stormed and destroyed (April 13). The expedition, having accomplished its purpose, left the country (May).

1869–1872. Civil War. After prolonged conflict, **Kassai** of Tigré defeated several rivals and reunified the country as **John IV.**

1875–1879. War with Egypt. (See p. 927.)

1878. Conquest of Shoa.

1882. War between John and Menelek of Shoa. John, victorious, designated Menelek as his successor.

1882. Italian Colony at Assab, Eritrea. They expanded to Massawa (1885).

1884. War with the Mahdi. John's troops helped the British withdraw from Gallabat and Kassal.

1884. British and French Protectorates on Somali Coast.

1887. War with Italy. Italian penetration from the Red Sea littoral led to war. An Italian force of 500 men was surrounded and most of them killed by John's forces at **Dogali** (January 26). The matter was temporarily composed after an Italian expeditionary force arrived next year. Italy now permanently garrisoned her holdings in Eritrea.

1885–1889. Mahdist Incursions. This resulted in sporadic frontier warfare.

1889, March 12. Battle of Metemma (Gallabat). In response to a Mahdist invasion, Emperor John with a huge force stormed and took Gallabat, defended by 60,000 dervishes under Amir **Zeki Kumal.** The emperor having been killed in the fighting, the Abyssinian army then withdrew.

1889–1909. Reign of Menelek. During the first 10 years of this reign there was sporadic civil war with Ras **Mangasha,** illegitimate son of John, who claimed the throne. At the outset of his reign Menelek negotiated the **Treaty of Uccialli** with Italy; the Italians considered that this gave them a protectorate over Abyssinia.

Meanwhile, Menelek steadily expanded the empire, and defeated Ras **Mangasha** (1899).

1895–1896. War with Italy. This resulted from disagreement over the Treaty of Uccialli, followed by renewed Italian encroachment.

1896, March 1. Battle of Aduwa (Adowa). An Italian force of 20,000 men—Italians and natives—under General **Oreste Baratieri** rashly attacked Emperor Menelek's 90,000 warriors in mountainous country. The leading Italian brigade pushed too far ahead of the main body, was surrounded and destroyed, and the other three brigades then were overwhelmed in detail. Despite the odds and poor leadership, Baratieri's troops fought well. Some 6,500 (4,500 Italian) were killed and wounded; 2,500 prisoners (1,600 Italian) were captured. Despite propaganda of Abyssinian atrocities, Menelek apparently treated his captives humanely.

1896, October 26. Treaty of Addis Ababa. Italy recognized Abyssinian independence.

1899–1905. The Mad Mullah. Mohammed ben Abdullah, claiming supernatural powers, began harassing British and Italian Somaliland, and part of Abyssinia, from his base at Burao, 50 miles south and east of Berbera. Despite all efforts of organized soldiery, mostly British (with considerable Abyssinian assistance), to quell his depredations, Mohammed and his followers kept up a successful guerrilla warfare until he was granted a semi-independent territorial area in Italian Somaliland.

NORTH AFRICA

1859–1860. Spanish-Moroccan War. (See p. 918.)

1865–1880. Franco-Italian Rivalry in Tunisia. France, continuing the pacification of the Algerian hinterland, found Italy her rival in exploitation of Tunisia. Matters came to a head when Tunisian tribesmen began raiding across the Algerian frontier.

1881, April. Seizure of Bizerte. A French naval force occupied Bizerte while French land forces moved across the Tunisian border. The Bey of

Tunis was forced to accept a French protectorate (**Treaty of Bardo,** Kasr es-Said, May 19).

1881, June–July. Insurrections in Southern Tunisia and Algeria. These were finally composed by French arms.

1893. Spanish Operations in the Riff. For most of the year incursions of Berber tribesmen from the Riff mountains menaced Spanish Morocco's coastal areas. The Riffians gathered in force around Melilla, then Spain's most important North African port, besieging it on the land side. An expeditionary force of 25,000 Spanish troops finally drove the tribesmen back into the hills.

SOUTH AFRICA

British vs. Boers vs. Blacks, 1850–1880

THREE-WAY STRUGGLE

1850–1878. Kaffir Wars. Britain's development of the Cape Colony was accompanied by a continuation of the tribal outbreaks which had been going on for three-quarters of a century. Most of the operations were those of volunteer units recruited from among the Boer and English settlers. The most serious of these was the **8th Kaffir War** (1850–1853), provoked when the Cape governor, Sir **Harry Smith,** reduced the power of the native chiefs in the eastern regions of the Cape Colony. Shortly after this (1856–1857), the desperate Kaffirs (mainly the Xosa tribes), believing that their actions would call back their ancestral heroes to help drive out the white men, slaughtered their cattle and destroyed their crops. The result was virtual self-destruction of the Kaffirs, about two-thirds of them dying of starvation. Despite this, when their strength had been built up again, after a generation, the Kaffirs rose in the last and **9th Kaffir War** (1877–1878), which was suppressed, and Britain annexed all of Kaffraria.

1852–1856. Emergence of the Boer Republics. Britain renounced sovereignty over the Transvaal region (1852), which later became the South African Republic, with **Marthinus Pretorius** as the first president (December 16, 1856). The Orange Free State was established, and Britain recognized its independence (February 17, 1854).

1854–1877. Zulu-Boer Border Disputes. The Zulus resisted Boer efforts to expand eastward to the sea. There was no overt warfare, but frequent frontier clashes.

1856. Civil War in Zululand. This was the result of a dispute over the succession between **Cetewayo** and **Mbulazi,** sons of **Mpande.** Cetewayo defeated and killed his brother in the decisive **Battle of the Tugela River** (December).

1858–1868. Basuto Wars of the Orange Free State. Under their king, **Mosheshu,** the Basutos resisted and repelled Boer encroachment (1858). A few years later the war was renewed; this time the Boers were successful, and annexed large parts of Basutoland (1864–1866). Friction continued, however, and the Orange Free State annexed more territory after another war (1867–1868). Britain finally annexed Basutoland to prevent further Boer expansion (1868–1871).

1862–1864. Civil War in the Transvaal. The obstreperous Boers, it seemed, could not only not get along with their neighbors, they could not get along with each other. Pretorius and his principal aide, **S. J. Paul Kruger,** finally suppressed the insurgency.

1867. Discovery of Diamonds along the Orange River. This led to British reannexation of the Hopetown region (1871), despite protests of the Orange Free State.

1872. Accession of Cetewayo as Zulu Ruler. The new king immediately began to rebuild Zulu military strength, which had declined somewhat since the death of his uncle, Shaka (see p. 858).

1877, April 12. British Annex Transvaal. This was done in an effort to bring about federation of all of South Africa. The Boers protested vigorously, and began to plan rebellion.

THE ZULU WAR, 1879

After annexation of the South African Republic, the British inherited the Boer border disputes with the Zulus. Cetewayo, rightly convinced that his position was legally and morally sound, was as adamant with the British as he had been with the Boers. At the same time, he continued to develop his nation's fighting power.

1878, December 11. British Ultimatum. Britain demanded a virtual protectorate over Zulu land. Cetewayo ignored the British demands.

1879, January 11. British Invasion of Zululand. To force Zulu compliance, General **F. A. Thesiger, Viscount Chelmsford,** led a force of about 5,000 British and 8,200 native troops into Zululand in 3 widely dispersed columns. Cetewayo had a force of 40,000 trained, fanatically brave professional warriors, organized in impis (regiments) and highly disciplined. While some had firearms, the principal Zulu weapon was the assagai, or spear, of which there were 2 types—throwing and stabbing. The normal Zulu battle formation was a crescent; while the center engaged the enemy, the horns swung in for a double envelopment. All movements were made on foot, at a lope that enabled the warrior to cover ground as fast as a mounted man. Reminiscent of the tactics of the Roman legion, the Zulu attack was carried home with the stabbing assagai, under a shower of throwing assagais.

1879, January 22. Battle of Isandhlwana. Chelmsford's center column—1,800 Europeans and 1,000 natives—established a camp at Isandhlwana. While Chelmsford and about half the Europeans were away, trying to intercept a Zulu force, the camp was hit early in the morning in a surprise attack by 10,000 Zulus. All but 55 Europeans and about 300 natives were killed. Returning next day, Chelmsford discovered the ruined camp and bodies. He fell back through Rorke's Drift to defensive positions. Cetewayo's forces pressed hard against them.

1879, January 22–23. Defense of Rorke's Drift. About half of the Zulu victors of Isandhlwana struck this British base camp, which had a garrison of about 85 able-bodied soldiers, late in the afternoon. In one of the British Army's great epics, the defenders drove off 6 full-scale Zulu attacks that lasted through the night. Early next day the Zulus withdrew, leaving about 400 dead. The British lost 17 killed and 10 seriously wounded.

1879, January 28–April 4. Siege of Eshowe. Chelmsford's right flank (coastal) column, under Colonel **C. K. Pearson,** was besieged at Eshowe. After a hard fight at **Gingindhlovu** (April 3), Chelmsford's relief column reached Eshowe and relieved it (April 4).

1879, March 28. Battle of Hlobane. A mounted force from Colonel **Sir Henry Evelyn Wood's** northern column, led by Colonel (later General) **Redvers H. Buller,** was attacked near Hlobane mountain by a large Zulu force; the column of 400 Europeans and 300-odd native auxiliaries suffered over 100 casualties in a confused running fight.

1879, March 29. Battle of Kambula. Some 20,000 warriors attacked Sir Evelyn Wood's column in a desperate 4-hour attack, but were finally beaten off. Wood lost 84 out of 1,800 Europeans and 900 natives engaged; Zulu losses were an estimated 1,000.

1879, April–May. Arrival of Reinforcements. These were hurried out from England. By the end of May, Chelmsford prepared to take the offensive again.

1879, June 1. Death of the Prince Imperial. Young **Louis J. J. Bonaparte,** son of Napoleon III, serving with the British as a volunteer, was killed during a Zulu ambush while out on reconnaissance. His death focused international attention on what up to that time had been just one more of England's innumerable little wars.

1879, July 4. Battle of Ulundi. Chelmsford, with 4,200 Europeans and 1,000 natives, reached the vicinity of Cetewayo's capital. Attacked by more than 10,000 Zulus, the force formed a hollow square, with its 600 cavalry (although only 290-odd were regular British cavalry) inside. The Zulu charges were broken up by British musketry and bayonets, and the small English cavalry detachment then charged into the disorganized mass. Victory was complete. Cetewayo fled. British casualties were about 100, while the tribesmen lost some 1,500 killed. This action to all intents and purposes ended Zulu power. Cetewayo became a fugitive, but was captured (August 28).

British vs. Boers, 1880–1899

Transvaal Revolt (or First Boer War), 1880–1881

Diamonds, gold, and greed were the principal factors affecting the fortunes of Boer settlements lying between the Orange and Limpopo rivers in South Africa.

1880, December 30. Boer Republic Proclaimed. The Boers were led by Kruger, **Petrus Jacobus Joubert,** and Pretorius. Several small bodies of British troops were cut up.

1881, January 28. Battle of Laing's Nek. Some 2,000 Boers under Joubert invaded Natal. General Sir **George Colley,** who completely underestimated his opponents, met them at Laing's Nek in the Drakensberg mountains with 1,400 British regulars and was defeated.

1881, February 27. Battle of Majuba Hill. The hill overlooked the principal pass through the mountains. Occupying the hill with part of his force—550 men—Colley was overwhelmed by the Boer riflemen. He and 91 others were killed; 134 more were wounded and 59 captured. Boer casualties were negligible.

1881, April 5. Treaty of Pretoria. This granted independence to the South African Republic, under British suzerainty. Paul Kruger, grim protagonist of Boer independence, became its president (April 16, 1883).

Rising Stakes of Colonial Economics, 1880–1899

1880–1881. The Gun War. The Basutos revolted when ordered to give up their arms. The revolt was suppressed.

1883–1884. Zulu Civil War. Bickering between rival rulers flared into civil war when Cetewayo was restored (January 29, 1883). After almost a year of fighting, he was overthrown by **Zibelu** (December 1883), who soon afterward was overthrown by **Dinuzulu,** son of Cetewayo.

1884. Establishment of German Colony of Southwest Africa.

1886. Discovery of Gold in the Witwatersrand. This was followed by British diplomatic maneuvers, resulting in annexation of Zululand. The Boers were in fact being cut off from all communication with the sea. While **Afrikanders** and **Uitlanders** (foreigners settled in the area) battled with one another for control of the Transvaal government, **Cecil Rhodes,** then prime minister of the Cape Colony and architect of a plan for British domination of what had now become one of the richest areas in the world, schemed to throttle the Afrikanders. They, on the other hand, were stubbornly resolved to check any foreign aggrandizement.

1887. Zulu Rebellion. After the British annexed Zululand, Dinizulu revolted (June–August), but was defeated.

1893, July–November. Matabele-Mashona War. Lobengula, king of the Matabeles, tried to conquer the Mashonas. The British intervened and defeated the Matabeles near their capital at **Bulawayo** (October 23).

1895–1896. Jameson Raid. Dr. **L. Starr Jameson,** close friend of Rhodes, and 500 adventurers undertook a fantastic dash into the

Transvaal to spark an Uitlander uprising in Johannesburg. With 500 men, Jameson rode into Boer territory from Mafeking, 140 miles away (December 29); an *opéra-bouffe* performance, completely lacking in military acumen. Three days later, at Krugersdorp, hard-riding, hard-shooting Boer commandos rounded them up like so many cattle, while the proposed revolt at Johannesburg failed to materialize. Jameson and his men were handed over to the British government, tried and convicted under the Foreign Enlistment Act, and given a slap on the wrist. Conditions worsened in the Transvaal; Boer-British relations grew more strained.

1896, March–October. Matabele Uprising. Suppressed by the British.

1899, October 9. Kruger Ultimatum. Realizing that a British expeditionary force was in the making in Natal, Kruger gave the British government 48 hours to disband all military preparations. The ultimatum was refused. The Orange Free State announced its alliance with the South African Republic.

Boer War, 1899–1902

Boer military organization was extremely sketchy, a localized militia system grouped into so-called commandos varying in strength with the population from which they were recruited. But every individual was a marksman, armed with a modern repeating rifle, and every man was mounted. The riders of the veldt were hunters, trained from childhood to take advantage of cover and terrain. The result was an irregular firepower capability which could pulverize—from concealed positions—the ranks of any close-order formation. These irregulars were also capable of disappearing from the field when seriously threatened. The Boers also had a small quantity of modern German and French field artillery, on the whole well served. On the other hand, these Boer individuals lacked disciplinary control and most of their leaders had no real concept of tactics and strategy.

1899, October. The Boer Offensive. Fast-moving Boer columns advanced, both east and west. Transvaal General **Piet A. Cronjé** invested **Mafeking** (October 13), valiantly defended by Colonel **Robert S. S. Baden-Powell.** Free State forces besieged **Kimberley** (October 15). The Boer main effort, 15,000 strong, under Transvaal General Joubert, pushed through the Natal Defense Force, equal in number, under General Sir **George White,** at **Laing's Nek** (October 12), and after brushes at **Talana Hill** (October 20), **Elandslaagte** (October 21), and **Nicholson's Nek** (October 30), bottled up White's troops in **Ladysmith** (November 2). British relieving forces were unwisely divided up by General Sir **Redvers Buller,** who tried to check the Boers everywhere at once.

1899, November 28. Battle of the Modder River. General Lord **Paul Methuen**'s column, nearly 10,000 men with 16 guns, moved to the relief of Kimberley. Transvaal and Free State Boer commandos, about 7,000 strong, under Transvaal Generals Cronjé and **Jacobus H. De La Rey,** contested the advance in a series of delaying actions. Methuen finally won through to the Modder River, after losing 72 men killed and 396 wounded, but his troops were so exhausted that he paused to await reinforcements. Boer casualties were negligible.

1899, December 10, Battle of Stormberg. A British force under General Sir **William Gatacre** got lost in a night move against a Boer spearhead 70 miles from Queenstown and was ambushed, losing heavily.

1899, December 10–11. Battle of Magersfontein. The Boers under Cronjé near the Modder River, now about 8,000 strong, occupied an entrenched hill. Methuen attacked frontally, in mass formation, at dawn in the rain. He was

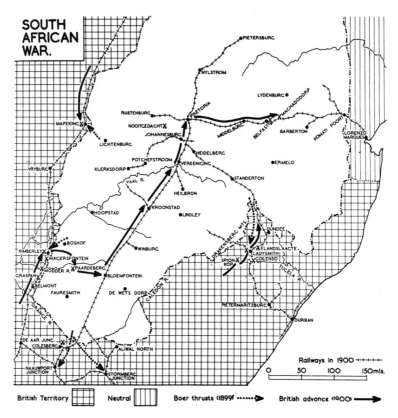

SOUTH AFRICAN WAR.

Railways in 1900 +++++

British Territory ▯▯▯   Neutral ▯▯▯   Boer thrusts (1899) ●●●●●▷   British advance (1900) ➔

defeated with loss of 210 men killed (including one general officer), 675 wounded, and 63 missing. Again, Boer casualties were negligible.

1899, December 15. Battle of Colenso. Buller himself led 21,000 men of all arms to relieve Ladysmith. Crossing the Tugela River, he attempted to turn the left flank of Free State General **Louis Botha,** entrenched with 6,000 men. The British flank attack, entangled in difficult terrain, was decimated by small-arms fire. British batteries unlimbering to support the frontal attack found themselves at the mercy of a concealed force of Boers. The British were driven back with losses of 143 killed, 756 wounded, and 220 men and 11 guns captured. Boer losses are estimated at not more than 50 men. At the end of Britain's "Black Week," Buller was so badly beaten that he advocated the surrender of Ladysmith. He was at once

relieved of the supreme command and replaced by Field Marshal **Frederick Sleigh, Viscount Roberts,** with General Kitchener as his chief of staff.

1900, January 10. Robert's Reorganization. Realizing at once that mobility was the keynote for success against the Boers, Roberts and Kitchener began revamping British field forces. To meet the Boer fluidity of fire and movement, a progressive build-up of mounted infantry began around the existing yeomanry militia units, a long and arduous task against conservative British military opinion. Brigadier General **John D. P. French,** with 2 small brigades of cavalry, kept up a spirited campaign against De La Rey and Free State General **Christiaan R. De Wet,** who were proving themselves to be natural leaders of light cavalry.

1900, January–February. Buller Repulsed

at the Tugela. He failed in 2 successive attempts, **Spion Kop** (January 23) and **Vaal Kranz** (February 5). British losses inflicted by small Boer forces of marksmen were 408 killed, 1,390 wounded, and 311 missing. The Boers lost some 40 killed and 50 wounded.

1900, February 15. Relief of Kimberley. French reached Kimberley, bringing the Boer siege to an end.

1900, February 15. Roberts Bypasses Cronjé. Roberts had set out toward Kimberley (late January) with 30,000 men. While French was driving directly on Kimberley, the main British force marched past Cronjé's left flank at Magersfontein, threatening his communications. The Boer leader began a slow withdrawal (February 16).

1900, February 18. Battle of Paardeberg Drift. Cronjé's retreat across the Modder River was blocked by French, rushing back from Kimberley. As the main British army approached, Roberts, temporarily sick, turned his command over to Kitchener, who made a tempestuous frontal piecemeal attack on the Boers' fortified laager (wagon train). The British were repulsed with losses of 320 killed and 942 men wounded.

1900, February 19–27. Siege of Paardeberg. Roberts, recovering, took command again and began a systematic encirclement and bombardment of the Boer laager. Cronjé, who might have broken out with his 4,000 mounted men, stubbornly refused to abandon his wounded and his train. He was starved into surrender (February 27).

1900, February 28. Relief of Ladysmith. Buller, on the Tugela, made a third attack (February 17–18) and succeeded. As he advanced toward Ladysmith, the besiegers withdrew and the relieving force made contact with the garrison. The tide had turned.

1900, March–September. Roberts' Cleanup in the Orange Free State. The British, now heavily reinforced, advanced on all fronts. Roberts took **Bloemfontein,** capital of the Orange Free State (March 13) and reached **Kroonstad** (May 12). Buller in Natal swept Boer resistance away at **Glencoe** and **Dundee** (May 15). The Orange Free State was annexed by Britain (May 24).

1900, May 17–18. Relief of Mafeking. A flying column of cavalry and mounted infantry under Major General **Bryan T. Mahon** relieved the garrison after a siege of more than 7 months.

1900, May–June. Invasion of the Transvaal. Johannesburg fell (May 31), then Pretoria (June 5). Roberts and Buller joined forces at Vlakfontein (July 4), ending all formal resistance. Kruger fled to Portuguese and Dutch protection. Annexation of the Transvaal was announced (September 3). Roberts went home (December), Kitchener being left in command. But the war was far from being over.

1900, November–1902, May. Guerrilla Warfare. De Wet, De La Rey, Botha, and some minor leaders, rallying to their respective commands the disbanded burgher forces, for 18 months played hob with British communications and defied all attempts to corner them. Erection of a line of blockhouses to protect the rail and other communication lines was the first British remedial action. But the raiders seemed to plunge at will through this cordon defense. Kitchener then copied Spanish procedure in Cuba. The country was swept by flying columns of mounted infantry; the farms on which the Boer raiders depended for sustenance were burned and some 120,000 Boer women and children were herded into concentration camps, in which an estimated 20,000 of them died of disease and neglect. Under these harsh measures, all resistance collapsed. The guerrilla leaders capitulated.

1902, May 31. Treaty of Vereeniging. The Boers accepted British sovereignty. As part of the very lenient terms, Britain granted them 3 million compensation for the destroyed farms. Total British casualties were 5,774 killed and 22,829 wounded. The Boers lost an estimated 4,000 killed; there is no accurate toll of the wounded. About 40,000 Boer soldiers had been captured.

COMMENT. *It took the British Empire 2 years and 8 months to subdue a foe whose man-power potential was 83,000 males of fighting age, and which never had in the field at one time more than approximately*

40,000 men. British forces engaged in the beginning totaled not more than 25,000, but before it was ended some 500,000 men were in South Africa—drawn from empire resources around the world. What happened was that the British Army, for the first time since the War of 1812, met hostile mounted riflemen and aimed small-arms fire. The experience of some 85 years of formal and little wars in Europe and around the world went into the discard, and an entire new system of tactics and techniques had to be evolved on the battlefield.

EAST AFRICA

1862. Increasing British Influence in Zanzibar.

1869. Buganda Intervenes in Bunyoro. This intervention in a Bunyoro succession crisis had no long-term results.

1870–1890. Bunyoro Recovery. Bunyoro began a recovery, largely based on its new and more efficient *abarasura* military formations. During the same era, **Buganda** built a fleet of war canoes on Lake Victoria.

1885. German Protectorate Established over Coast of Tanganyika. Protests of the Sultan of Zanzibar were overcome by German naval demonstrations.

1885–1898. British Wars with Arab Slave Traders in Nyasaland. This was the first, and longest-lasting, of intermittent wars of the various colonial powers with the Arab slave traders in East, West, and Central Africa.

1885–1889. Religious Wars in Uganda. Initially these were mainly between Moslems and the Catholic Christians under King **Mwanga**. The eventual victory of Mwanga over the Moslems (1889) was followed by continuing unrest resulting from Protestant-Catholic disputes.

1887. British Establish Control of Kenya Coast. A 50-year lease was negotiated with the Sultan of Zanzibar, largely to prevent further German expansion along the coast.

1888–1890. Uprising of Coastal Arabs in German East Africa. This was suppressed by the Germans with some naval assistance from Britain.

1890. British Protectorate Established over Zanzibar. This was recognized by Germany and France in return for Britain's cession of Helgoland to Germany and recognition of French claims in Madagascar.

1891–1893. Wahehe War in German East Africa. The rebellion was suppressed by the Germans.

1893. British Protectorate over Uganda.

1896. Uprising in Zanzibar. Suppressed by a British naval bombardment.

1897–1901. Mutiny in Uganda. Continued sporadic unrest in the British protectorate, fomented by rival French interests and religious conflicts, was complicated by the mutiny of British Sudanese troops. King Mwanga was defeated and captured (1899). Not until Indian troops were brought into the area was the outbreak controlled (1900–1901).

WEST AFRICA

c. 1850–c. 1880. Lunda Power Wanes. Pressured by Arab slavers from East Africa, and by invading tribes from the southeast, the Lunda paramountcy weakened (see p. 721).

1850–1864. Al-Hajj Umar's Jihad. Umar led his forces from their base in Guinea (see p. 857) in the conquest of nearby Bambara chiefdoms (1852). Umar overran the gold fields of the upper Sénégal, and conquered the Bambara Kingdom of Kaarta (1854). Umar's efforts to conquer his homeland around Fouta Toro led to war with the French (1856–1859), and his forces were repulsed by **Colonel Louis L. C. Faídherbe** in Senegal. Turning east, Umar's armies overran Segu, Macina (1861), base for the earlier *jihad* of Ahmadu ibn Hammadi (see p. 856), and then Timbuktu (1863). Al-Hajj-Umar was killed suppressing a revolt in Macina (1864).

1860. German Posts Established on the Cameroon Coast.

1864–1890. End of the Tukulor Empire. Hampered by relatives and former officers competing for control, the son of al-Hajj Umar, **Ahmadu Seku** was unable to establish firm control of the widespread empire. He began to disband the army in the 1880s to reduce disorder, so the French were able to infiltrate and construct forts without serious opposition (1885–1890). They overran most of the Tukulor Empire in 1890–1892; Seku was captured in 1893 and died in 1898.

1873–1874. Second Ashanti War. After years of dispute over the coastal districts (1863–1872), war broke out when Ashanti troops invaded the coastal areas. The invaders withdrew after a British expedition, under **General Sir Garnet Wolseley,** landed and marched inland to the Ashanti capital of Kumasi, which was razed (February 1874). Although not seriously hurt, the Ashanti state was shocked by its defeat, and some of its subject tribes broke away.

1877–1885. Belgian Explorations in the Congo. Initially these explorations were supported internationally, but the Belgians dominated the region, which was declared an independent state (1885), over which King **Leopold of Belgium** assumed personal sovereignty. The French stepped up their colonial activity to the west and north of the Congo to limit the Belgian-controlled region.

1883–1890. French Expansion. This activity was centered mainly along the frontiers of Dahomey and in the upper Niger, where the Tukulors and other tribes were conquered.

1883–1893. Rise of Rabih az-Zubayr. This half-Arab, half-Negro adventurer, who had fought for and against the British and Egyptians in the Sudan, conquered the ancient Sultanate of Bornu. He then began to expand into nearby regions, and clashed with the French.

1884. German Protectorates Established. They took control of the Togoland and Cameroon coasts.

1885. Spanish and British Protectorates. The Spanish took control of Rio de Oro and Spanish Guinea. The British proclaimed control of the lower Niger River region, to forestall further French expansion.

1885–1886. First Mandingo-French War. The French defeated **Samori,** warlike ruler of the Ivory Coast tribes. They established a protectorate (1889).

1889–1892. French Conquest of Dahomey. After **Behanzin** succeeded Glele as king (1889), a brief war broke out (1889–1890). The French later launched a major expedition (1892), when Colonel **Alfred-Amédée Dodds**'s troops defeated the Dahomeyan army and set up a protectorate. Behanzin was exiled to the West Indies and died there (1894). Scattered unrest and insurrection in Dahomey plagued the French (1893–1899).

1892–1893. Rising of Arab Slave Traders on the Upper Congo. The sporadic conflicts taking place farther east spread as the Arabs became disturbed by European interference with their trade. Belgian troops finally quelled the disturbances.

1893. French Operations against the Tuaregs on the Niger. The French occupied Jenne and Timbuktu.

1893–1894. Third Ashanti War. Depredations of Ashantis under their new ruler, **Prempeh,** were repulsed, and the Ashantis forced to accept a British protectorate.

1894–1895. Second Mandingo-French War. Samori threw off French control and defeated French efforts to reestablish the protectorate.

1895–1896. Fourth Ashanti War. Under heavy pressure, King **Prempeh** submitted to a British ultimatum, which limited Ashanti autonomy (1896), yet the British did not fully occupy the area until after suppression of a major revolt (see p. 1100).

1897–1899. British Conquest of Northern Nigeria. Granted the authority to raise troops, company forces under **Sir Frederick Lugard** overran Bida and the Fulani emirate of Ilorin.

1897–1898. Anglo-French Dispute over Nigeria. The 2 countries came close to war, but finally resolved their differences, with the British retaining the area of modern Nigeria (see also **Fashoda Incident,** p. 928).

1897–1900. Batetela War on the Upper Congo. This warlike tribe was finally pacified by Belgian troops.

1898. Third Mandingo-French War. The French defeated and captured Samori, and completely broke Mandingo power (September 29).

1899–1900. French Conquest of Chad Region. Forces from the Congo and from Algeria converged to defeat Rabah Zobeir (April 22, 1900) after bitter warfare.

MADAGASCAR

1859. French Protectorate over Coastal Areas.

1863. Revolt in Madagascar. The pro-European king, **Radama II,** was overthrown and killed.

1882. French Claim a Protectorate. This was rejected by the Hova government (see p. 859).

1883–1885. War with France. Majunga and Tamatave were bombarded by a French squadron and troops landed (June 13, 1885). After desultory fighting, the French protectorate was accepted. Unrest persisted, however, and Hova troops were trained by and to some extent officered by Englishmen.

1894, June 16. Extended French Claims Rejected. In consequence, Tamatave was again bombarded (December 12). A French expeditionary force of 15,000 men under General **Jacques C. R. A. Duchesne** embarked for Madagascar.

1895, February. French Conquest. Duchesne began landing at Majunga. The French force moved inland slowly, constructing a viable line of communications. **Tananarive,** the capital, was bombarded and at once surrendered (September 30). A violent insurrection then broke out in the interior.

1896, August 6. Madagascar Proclaimed a French Colony. General **Joseph S. Galiéni** set up a military government, deposed Queen **Ranavalona II,** and, actively campaigning, stamped out all revolt. Complete pacification was accomplished by 1905.

SOUTH ASIA

AFGHANISTAN

1850–1855. Recovery of Dissident Regions. **Dost Muhammad** consolidated control of outlying areas, principal among them Balkh (1850) and Kandahar (1855).

1855. Treaty of Peshawar. This officially concluded Afghan participation in the 2nd Sikh War (see p. 864), confirmed British annexation of the Peshawar territories, and assured British support of Afghanistan against Persian designs on Herat (see p. 926). Persian withdrawal from Herat (1857) having left the region under independent local rule, Dost Muhammad immediately moved to gain control. During the Indian Mutiny (see p. 939), he remained faithful to his alliance with Britain.

1863, May 26. Capture of Herat. After a long-drawn-out local conflict and siege, Herat was brought under central control. Dost Muhammad died shortly afterward (June 9), being succeeded by his son **Sher Ali** Khan.

1863–1879. Internal Disorders. Serious dynastic struggles shook the country as Sher Ali was repeatedly defeated by his warrior cousin **Abdur Rahman** Khan. Sher Ali, courting Russian favor, became increasingly hostile to the British.

Second Afghan War, 1878–1880

1878, November. British Invasion. When Sher Ali ignored British warnings against his pro-Russian stand, British troops, in 3 columns, under Sir **Frederick** (later Lord) **Roberts,** advanced from India and seized the frontier passes, defeating Sher Ali's troops at **Peiwar Kotal** (December 2). Sher Ali fled; his son **Yakub** Khan, replacing him, signed a treaty with Britain (May 1879). Sir **Louis Cavagnari** was installed as British resident at Kabul.

1879, September 5. Assassination of Cavagnari. In reprisal, a field force under Roberts marched on Kabul from Kurram and Kanda-har.

1879, October 6. Battle of Charasia. Roberts' force, 7,500 men and 22 guns, defeated an Afghan army of 8,000, then occupied Kabul (October 12). Yakub Khan, abdicating, fled to British protection.

1879, December 23. Battle of Sherpur. Afghan levies amounting to 100,000 men rallied to the call for a holy war against the infidel British. Roberts' cantonments were surrounded. The British broke out and then Roberts, falling on the Afghan flank, completely dispersed them. Pacification continued, without major opposition, under the over all command of Sir **Donald Stewart.**

1880, June. Ayub Khan's Offensive. Ayub Khan, brother of Yakub, had seized control of Herat early in the war (January 1879). Now claiming the throne, he marched on Kandahar with 25,000 men. Lieutenant General **James Primrose,** commanding at Kandahar, sent an Anglo-Indian brigade, 2,500 strong, under Brigadier General **G. R. S. Burroughs,** to **Maiwand,** about 50 miles northwest, to oppose the Afghan advance.

1880, July 27. Battle of Maiwand. Burroughs attacked the Afghan position, but the British artillery expended all its ammunition and a flanking movement by Ayub then shattered the Indian troops, who fled. The one British infantry battalion present was surrounded and prac-

tically annihilated; about half the remainder of the command escaped. Ayub now advanced to besiege Kandahar.

1880, August 9. Roberts' March. Roberts moved out of Kabul to intercept Ayub. With him were 10,000 men and a transport corps specifically organized on his own plan. He reached Kandahar in 22 days, covering 313 miles over mountainous country.

1880, September 1. Battle of Kandahar. Roberts at once attacked, completely routing the Afghan Army and capturing all its guns and equipage. Ayub fled back to Herat. This action ended the campaign. A pro-British government was established under Abdur Rahman (see p. 938). The British evacuated the country in 1881.

1881. Ayub Khan's Second Bid for Power. With British troops out of the way, Ayub led his followers again on Kandahar, defeating an army sent by Abdur Rahman, and capturing the city (July 27). Abdur Rahman, taking the field in person, crushed the rebels (September 22). Ayub fled to Persia. Abdur Rahman now established a firm, centralized government.

1885, March 30. The Penjdeh Incident. Following the occupation of Merv (1884), Russian troops crossed the disputed Afghan frontier and clashed with Abdur's forces, inflicting heavy losses. The incident brought Britain and Russia close to war (see p. 922).

INDIA

The Great Mutiny, 1857–1858

British military power in India at this time consisted of two elements: the native armies of the East India Company and a comparatively few regular British Army units. This armed force was controlled by the governor general, an official of the company, appointed with crown approval. In 1857 the 3 company armies—Bengal, Bombay, and Madras—consisted of 233,000 Asians (sepoys) and 36,000 Britishers, commanded by a British officer corps, commissioned by the company. Due largely to poor administration and command, considerable unrest existed among the native contingents. The introduction of the Minié rifle cartridge about this time provided the spark that changed unrest to violence. The paper cartridge, which had to be bitten for loading, was greased. Disaffected elements in the armies claimed—with some truth, as later investigation showed—that the grease used included the fat of cows (sacred to Hindus) and of pigs (unclean to Moslems).

1857, May 10. Outbreak at Meerut. The climax came in the Bengal army garrison at Meerut, some 25 miles from Delhi, after 85 cavalry sepoys who had refused the new cartridge had been disgraced and imprisoned. While the British units were at church, the Indian regiment released the imprisoned soldiers, then killed as many of their European officers and other European men, women, and children as they could find. Before the British troops could take action, the mutineers had fled to Delhi. There was no effective pursuit.

1857, May 11. Massacre at Delhi. Upon arrival at Delhi, the Meerut mutineers were joined by the native garrison, who, with the city rabble, began to butcher all Europeans they could find. All who could escape joined the British garrison, holding cantonments outside of town. In the city a few British officers and men held the arsenal as long as possible, then blew up themselves and the ammunition magazine, causing great damage to the attacking native soldiers and mob. The rebels declared **Bahadur** Shah, last of the decadent moguls, as their ruler. The surviving Europeans fled to Meerut and Umballa, a large military post in the Punjab, 120 miles northwest of Delhi. Word of the uprising spread throughout India, causing the entire Bengal army, from Delhi to Calcutta, to revolt against the British.

1857, May–July. British Reaction. Delayed by the scattering of British units to hill stations for the summer, General **George Anson,** British commander in chief, was unable for several days to get a force together at Umballa to march on Delhi. Shortly after starting the march (May 17), Anson died of cholera (May 27). The British advance ceased briefly. Sir **John Lawrence,** efficient chief commissioner in the Punjab, now assumed control of operations in the north. He immediately rushed a force of 3,000 British troops to besiege Delhi, and promptly gathered more British and loyal Punjabi troops to follow them. At Lucknow, in the center of the dissident area, his elder brother, Sir **Henry Lawrence,** organized the defense of the residency against an ever-increasing rebel force. Sir Henry had 1,720 fighting men,

including 712 faithful native troops, plus 1,280 additional noncombatants. At Calcutta, the governor general, Sir **Charles John Canning,** after some temporizing, undertook action to gather British reinforcements.

1857, June 27. Cawnpore Massacre. Dandu Panth (Nana Sahib), Rajah of Bitpur, led the rebellious native troops of the garrison in a 3-week siege of the British contingent (June 6–26). He persuaded Sir **Hugh Wheeler,** the British commander, to surrender his handful of British troops and more than 200 noncombatants, mostly English women and children, under promise of safe conduct to Allahabad. As the disarmed British were embarking on river boats, Nana's men murdered the men and imprisoned the women and children.

1857, July 7–16. Havelock Marches toward Lucknow. Sir **Henry Havelock,** with 2,500 men, mostly British, moved from Allahabad to the relief of Lucknow. Marching 126 miles in 9 days, during the hottest season of the year, Havelock met Nana Sahib's forces, defeating them at **Fatehpur** (July 12), **Aong** (July 15), and **Cawnpore** itself (July 16), where Nana's troops were completely shattered. On entering Cawnpore, the British discovered the bodies of the English women and children prisoners, hacked to death and thrown into a well by Nana's orders the day previous. Nana succeeded in escaping the berserk Britishers, although most of his followers were not so fortunate. (Nana's fate is obscure; he fled to Nepal, where he is reported to have died in 1859.) Havelock was now compelled to await reinforcements and supplies before continuing his advance on Lucknow (September 20).

1857, September 14–20. Storming of Delhi. The initial besieging force had taken the Badli-ki-Serai ridge dominating the city (June 8), but not until early August had its strength been sufficient for offensive action. After a 3-day artillery preparation, the assault force, 4,000 strong, moved against the walls in 4 columns, gaining a foothold inside after taking severe losses, including its commander, Brigadier General **John Nicholson,** who was killed. Six days of bitter street fighting followed before the city was finally subdued. Bahadur Shah was captured, and his sons shot out of hand after they had surrendered, in revenge for the murdered women and children of Cawnpore. The British took casualties totaling 1,574 men killed and wounded during the action. The capture of Delhi ended for all time the dream of a revival of the Mogul Empire.

1857, September 25. First Relief of Lucknow. Sir Henry Lawrence had been killed by artillery fire (July 4) and had been succeeded by Brigadier General **John Inglis.** Havelock's relief force, after an arduous campaign up from Cawnpore and bitter street fighting in Lucknow itself, broke through the 60,000 rebels besieging the residency, losing 535 officers and men out of a 2,500 strength. The rebels closed in again. Sir **James Outram,** who had joined the relief force as a volunteer (August 15), took command of the besieged garrison, since he was senior to Havelock. For 6 weeks more the obstinate defense of the residency continued.

1857, November 16. Second Relief of Lucknow. Another relief column, 4,500 strong, under Sir **Colin Campbell,** advanced from Cawnpore (November 12). On the 14th, the defenders heard Scottish bagpipes playing "The Campbells Are Coming." Two days later, the relief column crashed through. The residency was safely evacuated to Cawnpore (November 22), but Havelock was killed during the operation. Campbell decisively defeated a large rebel force outside of **Cawnpore** (December 6).

1858, January–March. Commencement of Central Indian Campaign. Sir **Hugh Rose,** with 2 brigades (about 3,000 men) from the loyal Bombay Army, in a whirlwind campaign relieved **Saugor** (February 3), forced the pass of **Madanpur** (March 3), and invested **Jhansi,** the rebel stronghold (March 21).

1858, March 16. Recapture of Lucknow. Having been reinforced, Campbell again marched on rebel-held Lucknow. He was joined by **Bang Bahadur,** prime minister of Nepal, with 10,000 Gurkhas. After a week of bitter fighting, the city was finally cleared of the rebels.

1858, April–June. Rose in Central India.

While Rose was investing Jhansi, 20,000 rebels under **Tantia Topi,** a lieutenant of Nana Sahib and the most capable military leader of the rebels, arrived on the scene. Rose boldly diverted half of his command to defeat Tantia Topi (April 1), then carried Jhansi itself by assault (April 3). Following this, he defeated rebel forces at **Kunch** (May 1) and **Kalpi** (May 22). In 5 months he had defeated the rebels in 13 actions. Tantia Topi, escaping, rallied another force and with the amazon **Rani of Jhansi** subverted (June 9) the troops of **Sindhia,** who had faithfully kept his treaty with England.

1858, June 19. Battle of Gwalior. Rose overwhelmed Tantia Topi and the Rani, who was killed. Topi escaped to the jungle, where he was later captured and executed (April 18, 1859). COMMENT. *The victory at Gwalior ended the Mutiny, although mopping-up operations continued. British repressive measures were harsh.*

1858, September 1. End of the Company. The government of India was transferred to the British Crown, ending the century-long rule of the East India Company.

1865. Bhutan War. Long-standing frontier disorders forced the British to send a small force to Bhutan (January). The Bhutanese surprised and drove out the British garrison of **Dewangiri.** Sir **Henry Tombs** pressed a punitive action, resulting in Bhutanese suit for peace (November 11).

1878–1881. Second Afghan War. (See p. 938.)

1885. Third Burmese War. (See p. 942.)

1895, March 4–April 20. Siege and Relief of Chitral. A small force of about 350 Indian Army troops, under Captain (later General) Charles V. F. Townshend were besieged in a tiny mud fort at Chitral on the North-West frontier by local tribesmen under **Sher Afzal.** A relief expedition under Major General, Sir **Robert Low** assembled at Peshawar and then marched for Chitral (early April), fighting several battles along the way. The beleaguered fort was relieved after a 6-week siege (April 20).

1895, August–1898, March. Pathan Uprising. As part of a general uprising on the North-West frontier, Pathan tribesmen captured the

Khyber Pass (August 27). The British retook it (March 1898). In the course of restoring order in the area, the British mounted several expeditions, including the Malakand Field Force and the Tirah Expedition (see below). During this period, British forces suffered some 1,300 casualties.

1896, October–December. Tirah Expedition. Lieutenant General Sir **William S. A. Lockhart** led a force of 44,000 Indian and British troops to restore order in Tirah, a region south and west of the Khyber (October). Following 2 sharp battles with the Pathans for the **Dargai Heights** (October 18, 20) and several smaller actions, the Pathans accepted British terms (December 13).

1896–1897. Malakand Campaign. A British expedition to the Swat Valley suppressed violence resulting from civil war in Chitral.

SOUTHEAST ASIA

BURMA

1852–1853. Second Burmese War. Following continued friction between British interests and the Burmese government, an amphibious expedition of 8,100 men under General Sir **H. T. Godwin** seized Rangoon (April 12). The Burmese army retired northward. Godwin seized Martaban at the mouth of the Salween River and Bassein in the Irrawaddy delta (May). After a halt until the monsoon ceased (October), Godwin occupied Pegu after a sharp engagement at the **Shwe-maw-daw Pagoda,** then seized Prome (October 9).

1852, December 20. Annexation of South Burma. Britain (the East India Company, actually) announced the annexation of Pegu Province. Meanwhile, a revolution at Amarapura (Burmese capital, near modern Mandalay) had overthrown King **Pagan Min,** who was replaced by his half-brother, **Mindon Min,** who tacitly acquiesced in the cession of Pegu, though no formal peace treaty was signed.

1878. Accession of Thibaw. One of Mindon's many sons, **Thibaw Min,** ascended the throne on his father's death. Internal order in Burma began to break down under Thibaw's inept rule, while a border dispute with the British flared on the Manipur frontier. Thibaw then tried to offset British influence in Burma by entering into trade and diplomatic negotiations with the French. Burmese interference with the British teak trade became an immediate cause of conflict (1885).

1885. Third Burmese War. A British amphibious force of 9,000 troops and 2,800 native followers, under General **H. N. D. Prendergast,** moved up the Irrawaddy from Thayetmo in 55 river steamers and barges manned by the Royal Navy (November 14). As the expedition rapidly approached his capital, Ava, Thibaw surrendered (November 27). Britain annexed Burma (January 1, 1886).

1885–1895. Guerrilla Warfare. Burma was eventually pacified, although banditry was endemic in the outlying regions.

SIAM

1850–1863. Frontier Incidents in Malaya. Northern Malayan princes attempted to play off Siam and Britain (see p. 944). Firm British military and diplomatic action in the frontier regions blocked Siam's efforts to expand southward.

1863–1867. Siam Ejected from Cambodia. French military intervention in that country forced Siam to choose between war and peace; King **Rama V** reluctantly withdrew his forces from Cambodia.

1871–1872. Expedition to Eastern Laos. Siamese military intervention limited Chinese Black Flag bandit depredations in Luang Prabang, but failed to eject the Chinese from the eastern section (see p. 944).

1883–1887. Renewed Expeditions to Eastern Laos. Concerned by French expansion in Vietnam, Siam sent a second expedition to pacify eastern Laos, so as to eliminate any excuse for French intervention. This force was defeated by the Chinese (1883). A larger army was sent (1885), which eventually pacified much of eastern Laos (1887).

1893. The French Threat. Continued French expansion (see p. 944) led to strained relations and serious incidents in disputed territories in Laos between French and Siamese forces (1892–1893). French warships entered the Menam River, fought their way past the Paknam fort, and anchored off Bangkok (July). This created a serious crisis between Britain and France. Britain, however, unwilling to risk war with France, advised Siam to accept French terms, which included recognition of a French protectorate over Laos and Siamese abandonment of her eastern provinces, which had formerly belonged to Cambodia. Siam agreed (October).

INDOCHINA

Vietnam

1851–1857. Clashes with the French. Under King **Tu Duc** a series of repressive measures against Christians, and sporadic murders of French missionaries, inspired French protests and occasional bombardment of Vietnamese ports by French warships.

1858–1862. French Invasion of Cochin China. Murder of the Spanish Bishop of Tongking caused a Franco-Spanish naval task force to bombard and to occupy Tourane (August–September 1858). Shortages of food caused the allied force to move to Saigon (February 1859). Sporadic operations continued, the French being primarily engaged in a war with China (see p. 944). A Franco-Spanish garrison of 1,000 held Saigon against a large besieging force for almost a year (March 1860–February 1861). A French relief force, under Admiral **Leonard V.J. Charner,** defeated the besiegers at the **Battle of Chi-hoa** (February 25, 1861). The French now steadily expanded their control of Cochin China. Tu Duc sued for peace, and ceded the three eastern provinces of Cochin China to France.

1862–1873. Unrest and Guerrilla War. Tu Duc was plagued by a series of rebellions. French intervention and pacification led to annexation of the 3 southwestern provinces of Cochin China. Internal disorders continued, but French control was strengthened.

1873–1874. Hanoi Incident. Without authority the French governor of Cochin sent a small force to intervene in civil war in Tongking. The French seized Hanoi, but evacuated it after forcing Tu Duc to grant numerous trade and diplomatic concessions. The relative interests of France, Vietnam, and Cochin China in Tongking resulted in continuing tension.

1882–1883. Renewed Vietnamese War with France. French expeditions seized Hanoi and the forts at Hué. Vietnam sued for peace, recognizing a French protectorate, and ceded additional areas to France. China's objections to this soon led to hostilities.

1884. Undeclared War between France and China. (See p. 946)

1885–1895. Widespread Revolts. The Vietnamese population supported Prince **Si Vattha** in an uprising against the French. Order was soon restored in most areas of Vietnam (1886), but the last embers of revolt were not extinguished until Si Vattha surrendered (1892). At the same time, guerrilla warfare engaged the French for several years in Cochin China and Tongking. Relative peace was restored, however, as the result of strenuous French efforts (1895).

1887. Administrative Consolidation of French Indochina. This affirmed the establishment of a permanent French colony.

Cambodia

1861–1862. Dynastic Disorders. King **Norodom** was forced to flee to Siam by a revolt of his brother, **Si Votha.** With Siamese assistance Norodom regained his throne.

1863. Limited French Protectorate. This was established despite the objections of Siam, supported by Britain (see p. 942). France agreed to Siam's retention of some nominal claims to suzerainty over Cambodia (1864).

1866–1867. Revolt of Po Kombo. After some initial success, this pretender was defeated, then chased down by French and Cambodian forces.

1887. Cambodia Part of Indochina. (See p. 943.)

1893. French Annexation of Western Cambodia from Siam. (See p. 943.)

Laos

1871–1872. Invasion by Chinese Refugees. Bandit refugees from China's internal wars (see p. 946) overran eastern Laos (under Siamese suzerainty) and threatened Luang Prabang.

Depredations were limited by Siamese forces, but the Chinese retained their hold on eastern Laos and western Vietnam (roughly the region between Xieng-Khouang and the Black River Valley of Tongking).

1883–1887. Siamese Expedition to Eastern Laos. (See p. 943.)

1887–1889. French Influence in Eastern Laos. The French took advantage of disorders in Laos to intervene, and to annex the region between Xieng-Khouang and the Black River Valley, despite Siamese protests (see p. 943).

1893. French Annexation of Laos. After a number of frontier incidents in Laos, probably engineered by local French officials, hostilities broke out between France and Siam (see p. 943). As part of the terms of settlement of the crisis, Siam recognized a French protectorate over Laos, which was then incorporated into Indochina.

MALAYA AND INDONESIA

Both British (in Malaya and parts of Borneo) and Dutch (in Indonesia) were engaged in numerous military activities in their respective colonies, dealing with piracy, native dynastic disputes, and occasional armed revolts. Of these the most serious was a bitter continuing war of the Dutch against the Sultan of Atjeh (1873–1908).

EAST ASIA

CHINA

1850–1860. Taiping Rebellion (First Phase). Accession to the throne of the **HsienFeng** emperor was followed by tyranny. Anti-Manchu revolts organized by **Hung Hsiu-ch'uan** broke out in Kwangsi, Hupei, and Hunan provinces. Hung's military leader, **Yang Hsiu-ch'ing,** led rebel armies that swept the Yangtze Valley, capturing **Wuch'ang** and **Nanking.** At the latter city the revolutionary capital was established and a new dynasty—**Taiping T'ienkuo**—announced with Hung as its ruler. Agrarian unrest aggravated revolt; Shanghai was captured by rebel supporters (September 7, 1853). China was virtually split in two.

1853–1881. Regional Insurrections. Collapse of central authority encouraged widespread insurrections. A bandit regime called **Nien Fei** terrorized Anhwei, northern Kiangsu, and Shantung (1853–1868) before its suppression by a rejuvenated central government. A Moslem (Panthay) regime established (1855) at Tali in Yunnan was finally suppressed (1873). A revolt of the Miao tribe in Kweichow (1855) lasted even longer (1881).

Second Opium War, 1856–1860

1856–1858. Anglo-Chinese Hostilities. Chinese seizure of the British ship **Arrow** at Canton (October 1856) brought armed reprisal: bombardment of Chinese ports. Later, an Anglo-French force under Admiral Sir **Mi-**

chael **Seymour** occupied Canton (December 1857), then cruised north to capture briefly the Taku forts near Tientsin (May 1858).

1858, June 26–29. Treaties of Tientsin. Negotiations between China and England, France, the United States, and Russia theoretically brought peace, China agreeing to open more treaty ports, to receive legations at Peking and missionaries in the interior, to legalize opium importation, and to establish a foreign-inspected maritime customs service. The **Treaty of Aigun (May 1858)** had already ceded the left bank of the Amur River to Russia. But China soon abrogated the Franco-British treaties.

1859, June 25. Attack on the Taku Forts. Refusal to admit foreign diplomats to Peking was followed by British Admiral Sir **James Hope's** bombardment of the forts guarding the mouth of the Peiho (Hai) River, below Tientsin. The squadron was severely mauled and British landing parties repulsed by a surprisingly efficient Chinese garrison. Commodore **Josiah Tattnall,** commanding the U.S. Asiatic Squadron, declaring "blood is thicker than water," assisted the British to withdraw. England and France agreed on joint action against China.

1860. Renewed Hostilities. Anglo-French forces gathered at Hong Kong (May 13). A joint amphibious expedition moved north to the Gulf of Chihli (Po Hai), 11,000 British under Lieutenant General Sir **James Hope Grant** and 7,000 French under Lieutenant General **Cousin-Montauban.** Landings, uncontested, were made at Pei-Tang (August 1) and the Taku forts taken by assault (August 21) with the assistance of the naval flotilla. Moving upriver to Tientsin, the expedition started for Peking—the British on the right bank of the Peiho, the French on the left. Chinese pleas for an armistice and parley resulted in the sending of a delegation headed by Sir **Harry Smith Parkes** into their lines. The party was seized (September 18), imprisoned, and—as learned later—horribly tortured. About half of its number died. Meanwhile, the expedition pushed ahead, brushing some 30,000 Chinese aside in 2 sharp actions and arriving before the walls of Peking (September 26). Preparations for assault commenced, and the old Summer Palace (Yuan Ming Yuan) was occupied and looted (October 6).

1860, October 18. Treaty of Peking. Another Chinese bid for peace was accepted; the price was return of the survivors of the Parkes party, surrender of Kowloon on the mainland opposite Hong Kong, and an indemnity of 8 million taels. General Grant then burned the Yuan Ming Yuan Palace (October 24) in reprisal for the mistreatment of the Parkes party. (Completely destroyed, the palace was never rebuilt; the new Summer Palace was later built nearby by Dowager Empress **Tzu Hsi;** see p. 1103.) Leaving a garrison at Tientsin, the expedition withdrew from China.

1860–1861. Russian Expansion. Russia now took advantage of China's prostration to extort from her the Maritime Provinces, where the port of Vladivostok was founded (1860–1861).

Taiping Rebellion (Final Phase), 1860–1864

Following the close of the war with the foreign powers, patriotic Chinese—including particularly **Tseng Kuo-fan** (who became viceroy) and **Li Hung-chang**—led a revitalized government and the Chinese people to organize against the dictatorial Taiping rebel regime in the south.

1860. Appearance of Ward. In the Shanghai area, Chinese merchants financed an army of liberation under **Frederick Townsend Ward,** American merchant marine officer and soldier of fortune from Salem, Mass. Ward won a series of victories, which earned for his force the title of "Ever Victorious Army" and for him a commission as brigadier general in the Imperial Chinese forces (1861).

1862, August 20. Battle of Tzeki. Having

cleared a 30-mile zone about Shanghai, winning 11 victories in a 4-month period (with the assistance of British and French forces returned from the Peking expedition), Ward attacked the walled city of Tzeki. Here he was mortally wounded while directing the assault.

1863. "Chinese" Gordon. Captain Charles G. Gordon, R.E., was detailed by the British government—at Chinese request—to succeed Ward in command of the Ever Victorious Army. Gordon moved north, along the line of the Grand Canal, taking Soochow (Wuhsien, December 4) and bottling up the Taiping forces in Nanking. Gordon, disgusted by the treacherous murder by Chinese government forces of rebel prisoners who had surrendered to him, threw up the command of his army, which was soon disbanded.

1864, July 10. Fall of Nanking. Chinese forces under Viceroy Tseng ended the Taiping Rebellion, although its last embers were not stamped out until the following year.

COMMENT. *This was perhaps the most destructive war of the entire 19th century. It has been estimated that 20 million people died directly or indirectly due to the war between 1850 and 1864.*

1865–1881. Suppression of Regional Revolts. Tseng supervised the operations of Chinese forces to suppress the revolts of Nien Fei and the Miaos (see p. 944).

1883–1885. Undeclared War with France. This was the result of French expansion in Tongking. Chinese troops, sent to Tongking from Yunnan, were driven out of **Pallen** and **Son-Tay** (1883), following which the French established themselves securely in the lower Black River Valley. Peace negotiations at Peking failed (May 1884). Chinese troops defeated a French force at **Bac-le** (June 23, 1884). A French naval squadron then entered the harbor of Foochow, destroying a Chinese naval squadron there, as well as the land fortifications and naval arsenal (August 23, 1884). The French then steamed to Formosa, bombarding **Chi-lung** (October 23, 1884). The forts there held out against repeated French attacks, but surrendered (March 1885), following which the French occupied the port briefly. They demonstrated at Tamsui, captured the Pescadores, and maintained a semiblockade of the island for 7 months. Meanwhile, in Tongking, a major French offensive, after hard fighting, culminated in the capture of **Langson** (February 13, 1885). Six weeks later, however, the French suffered a severe defeat outside Langson and precipitously evacuated the region, abandoning most of their equipment. When news of this reached Paris, the government fell, and its successor instituted peace negotiations. The **Treaty of Tientsin** (June 9, 1885) brought peace, restoring the *status quo;* France restored to China the areas occupied in Formosa and the Pescadores.

Sino-Japanese War, 1894–1895

Chinese and Japanese rivalry over predominance in Korea (semi-independent vassal of the Chinese Empire) had grown intense (see p. 949).

1894, June. Japanese-Fomented Riots in Seoul. China sent troops by sea to Asan (about 40 miles southwest of Seoul) to restore order at the request of the Korean government. Japan responded by rushing troops directly to Seoul through Chemulpo (Inchon). Meanwhile, the Korean government suppressed the disorders, but neither China nor Japan would withdraw troops until the other did.

1894, July 20. Japan Seizes Control of the Korean Government.

1894, July 25–29. Preliminary Clashes. Near **Phung-Tao,** on the west coast of Korea, Japanese Admiral **Tsuboi** attacked a Chinese troop convoy bringing reinforcements to Asan, sinking one transport and severely damaging Chinese naval escorts (July 25). At the same time, Major General **Oshima,** commanding Japanese troops in Seoul, advanced on Asan, defeating the Chinese at **Sŏnghwan** (July 29).

1894, August 1. Declaration of War. Both sides declared war and rushed reinforcements to Korea by sea, the Chinese to the Yalu River and northern ports, the Japanese through Pusan and Chemulpo. Neither navy attempted to interfere with the other's convoys.

1894, September 15. Battle of Pyongyang. General **Michitsura Nozu,** with an army in excess of 20,000 men (partly landed by sea, partly overland from Seoul), attacked and defeated a Chinese force of 14,000 defending Pyongyang. The remnants of Chinese forces withdrew to the Yalu River, where the main Chinese fleet, under Admiral **Ting Ju-ch'ang,** had just escorted a troop convoy with reinforcements.

1894, September 17. Naval Battle of the Yalu (Haiyang). Admiral **Yuko Ito,** with the main Japanese fleet, having convoyed troops to take part in the Pyongyang battle, then headed north to seek Ting's squadron. The two fleets met between the mouth of the Yalu and Haiyang Island. The Chinese squadron consisted of 2 ironclad battleships, 4 light cruisers, and 6 torpedo boats. The Japanese force comprised 4 heavy cruisers, 4 fast light cruisers, and 2 other old armored cruisers. The Japanese had a great preponderance of heavy (over 5″ caliber) quick-firing guns. Japanese gunnery and ship handling were far superior to the Chinese. Five Chinese ships were sunk; the remainder limped into Port Arthur, all severely damaged. The Japanese sustained considerable damage, but lost no ships. Ito, respectful of the 2 powerful Chinese battleships, had failed to close decisively with his defeated enemy, or to pursue aggressively; otherwise the Chinese squadron—out of ammunition at the close of the battle—would have been annihilated.

1894, October 24–25. Passage of the Yalu. After Pyongyang, Nozu's force—advancing to the Yalu—was augmented and reorganized as the First Army, under Marshal **Aritomo Yamagata.** With over 20,000 men, the Japanese crossed the Yalu without opposition.

1894, October 24–November 19. Operations against Port Arthur. The Japanese Second Army (26,000 troops and 13,000 auxiliaries), under Marshal **Iwao Oyama,** was convoyed to the Liaotung Peninsula, landing north of Port Arthur at Pitzuwu (October 24). Opposition to the advance down the peninsula to Port Arthur was negligible. Unaccountably, the Japanese Navy permitted Admiral Ting and his squadron to escape across the Strait of Pohai to Weihaiwei. Oyama's successful dawn attack on the fortifications of Port Arthur (November 19) was led by General **Maresuke Nogi;** the 10,000 defenders made only feeble resistance.

1894, October–1895, January. Operations in Manchuria. After winning some minor actions (of which **Tsao-ho-ku,** November 30, was particularly hard-fought), Yamagata's advance from the Yalu into Manchuria was stopped by a combination of supply problems and cold weather. The Japanese were forced to fight hard to defend their advanced base at **Haicheng** against Chinese General **Sung Ch'ing** (December–January).

1895, January. Weihaiwei Campaign. The Japanese decided to eliminate any possibility of further Chinese naval interference. The Third Army—comprising a number of units from the Second Army, and commanded by Oyama—was convoyed to Jungcheng, on the eastern tip of the Shantung Peninsula, about 20 miles east of the Chinese naval base at Weihaiwei (January 19). After an unopposed landing, Oyama promptly marched west and, while the Japanese fleet engaged the fortifications from the sea, captured Weihaiwei and its fortifications in 2 days of bitterly cold winter fighting (January 30–31).

1895, February 2–12. Naval Battle of Weihaiwei. The Chinese squadron (2 battleships, 4 cruisers, 6 gunboats, and 15 small torpedo boats) was now caught between Ito's blockading Japanese fleet and the Japanese Third Army, manning the Weihaiwei forts. Ito commenced a series of night torpedo-boat attacks against the Chinese fleet, while the captured land batteries hammered them by day. Admiral Ting and his men put up the stoutest

Chinese resistance of the war, but the situation was hopeless. Torpedo-boat attacks destroyed one of the battleships (February 5). Ting sent his torpedo boats on a dash for freedom, but only 2 escaped (February 7). With most of the fleet destroyed, Ting committed suicide; his remaining men and battered vessels surrendered (February 12).

1895, February–March. Advance into Manchuria. Meanwhile, the Japanese First Army (now under Nozu's command) had made contact with the Second Army in the Liaotung Peninsula, while continuing to beat off repeated Chinese attacks on Hai-cheng. Despite bitter cold, Nozu now advanced, defeating Sung at **Tapingshan** (February 21–23). Chinese forces now began to crumble, and the last effective resistance in Manchuria was eliminated by the Japanese near **Yingkow** (March 9). The Japanese paused for good weather before marching on Peking. Chinese Viceroy Li Hung-chang was sent to Japan to negotiate peace.

1895, April 17. Treaty of Shimonoseki. China recognized Korean independence, agreed to pay a 300-million-tael indemnity, and ceded Formosa, the Pescadores, and the Liaotung Peninsula to Japan. Japan had proven herself a major military power.

1895–1900. Grab for Concessions in China. Russia, France, Germany, and Britain now began a scramble for concessions from supine China and to keep Japan in check. The 3 first-named edged Japan from Port Arthur and the Liaotung Peninsula, which Russia then gobbled up, together with rights for the Chinese Eastern Railway, which would link Port Arthur with the trans-Siberian line. Germany occupied Kiaochow (November 14), while Britain secured a 99-year lease of the Kowloon territory on the mainland opposite Hong Kong, and a lease on Weihaiwei—to run so long as Russia occupied Port Arthur.

1899, September 6. The "Open Door" Policy. The United States alone stood firm on its policy of equal commercial opportunity in, and the territorial and administrative integrity of, China, as stated in a note to the other powers from American Secretary of State **John Hay.**

COMMENT. *European imperialist diplomacy sowed the seeds of future war.*

JAPAN

1853–1854. Perry Opens Japan. A U.S. squadron of 4 vessels arrived in Tokyo Bay to negotiate a treaty to end Japan's voluntary isolation from the rest of the world. The Japanese respected Commodore **Matthew C. Perry's** firmness and conciliation. In the **Treaty of Kanagawa** (March 31, 1854) they agreed to open trade relations with the U.S. and to protect shipwrecked American sailors. Similar treaties were later negotiated by England, Russia, and Holland.

1863, June. Shimonoseki Incidents. Forts of the Choshu clan fired on American, and later on French and Dutch, vessels, bringing immediate American and French reprisal bombardments.

1863, August 15–16. British Bombardment of Kagoshima. Unable to obtain satisfaction from the Satsuma clan for the murder of a British subject (September 1862), a British squadron bombarded the Satsuma capital.

1863–1868. Internal Strife. The western clans (Choshu and Satsuma) tried to influence the Tokugawa shogun to expel foreigners, and to obtain the emperor's support against the Tokugawas. Repeated demonstrations of Occidental military might caused the western clans to cease their antiforeign efforts (1864), but not their attempts to break the power of the Tokugawa shogunate. The struggle combined intrigue, murder, and civil war, which included clashes at **Kyoto** (September 1863; August 1864) and an unsuccessful Tokugawa expedition into Choshu territory, a conflict referred to by the Japanese as the "Four-Corners War" (July–October, 1866). The death of **Iemochi,** the old shogun (September 1866), was followed soon after by that of the emperor (February 1867). The new shogun, **Keiki,** resigned (November 1867), but the continuing turmoil erupted into the *Boshin* (or Restoration) War (January 1868). This conflict pitted supporters of the emperor, especially the samurai of the Choshu and Satsuma fiefs, against the conservative backers of the shogunal regime. The war lasted for 7 months, ending with the conservatives' defeat at the **Battle of Veno** near Tokyo (July 4). The fall of Tokyo ended the power of the Tokugawa, leaving the western clans predominant.

1864, September 5–8. Bombardment of Shimonoseki. An allied naval expedition—Dutch, French, British, and American—silenced the forts in reprisal for continued hostility by the Choshu clan. This demonstration of military superiority ended the antiforeign movement in western Japan.

1869, January 3. Restoration of Imperial Power. Emperor **Mutshuhito** (Meiji emperor) assumed control of the government, supported primarily by the western clans. This was the start of Japan's emergence from feudalism.

1874, April–October. Expedition to Formosa. Japan, claiming sovereignty over the Ryukyu Islands, sent an expedition to Formosa to punish natives there for the murder of Ryukyu sailors.

1875–1885. Japanese Interventions in Korea. (See below.)

1877, July–September. Satsuma Rebellion. The samurai (feudal warrior class), protesting against modern innovations and particularly against the raising of a national conscript army, rose against the government (but not against the emperor). The most serious threat was the march of 40,000 Satsuma warriors on Tokyo. They were stopped, then defeated, by the new national army at **Kumamoto.**

1879. Annexation of the Ryukyus. Despite Chinese protests, Japan began its overseas expansion.

1894–1895. Sino-Japanese War. (See p. 946.)

KOREA

1866. French Expedition. To punish Korea for the murder of French missionaries, a French naval force occupied **Kanghwa** at the mouth of the Han River. The Korean government refused to negotiate and the French, after inconclusive skirmishes, withdrew.

1871, May. American Expedition. To investigate the imprisonment and murder of some American sailors in earlier years, an American naval force captured the **Kanghwa** forts in a small-scale amphibious assault. After vain efforts to negotiate, the Americans withdrew.

1875–1876. Japanese Intervention. A Japanese vessel having been fired on in Korean waters, a Japanese naval expedition forced Korea to sign a treaty (February 25, 1876) establishing trade relations and opening several ports to Japanese vessels. Significantly, this treaty recognized Korean sovereignty, ignoring nominal Chinese suzerainty. Similar treaties were later concluded (but through Chinese mediation) by the U.S. and other western powers.

1882–1885. Intervention of China and Japan. Several internal disorders culminated in an attack on the Japanese legation. A Japanese military force entered Seoul, obtained reparations, and remained as a powerful legation guard. China then sent troops to reestablish her suzerain position in Korea; **Yuan Shih-k'ai** became the Chinese resident (1883). Renewed disorders led to the dispatch of additional Chinese troops to restore order. At the same time, Japanese troops arrived to protect Japan's interests. War between China and Japan was avoided only by the **Convention of Tientsin** (April 18, 1885) between Chinese Viceroy Li Hung-chang and Japanese Count Ito. Both nations agreed to withdraw troops and to notify each other if further intervention should become necessary.

1894. Tong-hak Rising. An antiforeign society, the Tong-haks, rebelled (March) and for 3 months successfully defied Korean government efforts at suppression. At the request of Korea, Chinese troops entered the country to help restore order. Japan, claiming a violation of the Tientsin Convention, sent a strong force to seize Chemulpo (Inchon) and Seoul. War inevitably resulted.

1894–1895. Sino-Japanese War. (See p. 946.)

1895–1900. Unrest in Korea. Continuing

disorders provided Japan and Russia with opportunities to improve their respective positions in Korea. The Russians to some extent supplanted Japanese influence in Seoul; tension between the 2 nations was increasing at the close of the century.

THE PHILIPPINES

1896, August. Insurrection against Spain. The rebellion, led by **Emilio Aguinaldo,** was settled peacefully when Spain promised substantial government and social reforms. Aguinaldo was exiled.

1898, May 1. Dewey's Victory at Manila Bay. (See p. 994.) Aguinaldo, brought to Luzon by Dewey, began to raise a Filipino patriot army to assist the U.S. against Spain.

1898, December 10. Treaty of Paris. The Philippines were sold by Spain to the U.S.

1899, January 20. Aguinaldo Proclaims Filipino Independence. Believing he had been betrayed by U.S. failure to grant independence to the Philippines, Aguinaldo refused to recognize the Treaty of Paris. He called on his compatriots to establish an independent government, under his leadership. His army undertook a virtual siege of Manila, held by U.S. troops at the close of the Spanish-American War.

1899, February 4. Commencement of Philippine Insurrection. Hostilities against the U.S. started when Filipino insurgents fired on an American patrol in the outskirts of Manila. U.S. troops immediately drove the Filipino forces away from the city. The campaign spread through Luzon. Under the field command of Major General **Arthur MacArthur,** punitive columns swept both Luzon and the Visayas as the U.S. poured men and supplies into the islands. The rebellion was still active as the century ended (see p. 1106).

NORTH AMERICA

THE UNITED STATES, 1850–1861

1850–1865. Indian Wars. During this period there were at least 30 separate "wars" or major disturbances involving conflict with the Indians. During the decade before the Civil War nine-tenths of the U.S. Army strength of 16,000–17,000 troops were spread out over an area of more than 1 million square miles west of the Mississippi, engaged in almost constant patrolling or combat operations against the Indians. Small detachments were also active in Florida, where there were recurrent troubles with the Seminoles. In these operations the Army was engaged at one time or another against Apache, Sioux, Cheyenne, Navajo, Mojave, Arapaho, Kiowa, Comanche, the various tribes of the Pacific Northwest, and others. In the year 1857 alone official records report 37 expeditions or operations involving combat, and many more without fighting. Although the tempo of these small wars subsided somewhat during the Civil War, because the nation's attention was diverted from the West, frontier hostilities continued. There were atrocities on both sides, although the disciplined troops of the Regular Army were rarely guilty of dishonorable treatment of the Indians. The most serious incident involving unquestionable white treachery was the so-called **Sand Creek Massacre,** in Colorado, where militiamen attacked a village of Cheyenne and Arapaho Indians while their leaders were negotiating a treaty, killing 300 Indians, mostly women and children (November 29, 1864). This disgraceful affair was probably the principal cause of the increasing ferocity of the Plains Indians in the 2 following decades.

1856, May 21–September 15. Civil War in Kansas. Sacking of **Lawrence** by a proslavery mob (May 21) was followed by abolitionist fanatic **John Brown**'s murderous raid on **Pottawatomie** (May 24–25). **Franklin** was seized (August 13) by Free Staters. Brown and his followers were driven out of **Osawatomie** and the settlement sacked by proslavery guerrillas (August 30). In all, some 20 persons were killed and an estimated $2 million worth of property destroyed before the arrival of Federal troops restored temporary calm in Kansas (September 16).

1857–1858. Mormon Expedition. The Mormons had established a practically independent state in Utah, ignoring government authority, and inciting the Indians against other settlers. A force of less than 1,000 men under Brevet Brigadier General **Albert Sidney Johnston** moved into Utah (October 1857). Despite provocations by the hostile Mormons, the disciplined troops reestablished U.S. government authority in Salt Lake City and the surrounding countryside without bloodshed (June 1858).

1859, October 16–18. John Brown's Raid. With 18 followers, Brown seized the Federal arsenal and armory at Harpers Ferry, and captured local citizens as hostages. The slave insurrection he hoped to spark failed to materialize. A detachment of U.S. Marines, commanded by Army Colonel **R. E. Lee,** captured Brown and his men. He and 6 of his men were convicted of treason and criminal conspiracy and hanged (December 2–March 16).

CIVIL WAR, 1861–1865

Election of Republican **Abraham Lincoln** to the presidency (November 6, 1860) precipitated Southern secession as the only way in which to preserve the southern agricultural economy, based on slave labor. South Carolina seceded (December 20).

1861, January–April. Confederate States of America. In quick succession Mississippi, Florida, Alabama, Georgia, Louisiana, and Texas followed South Carolina's lead. A convention held at Montgomery, Ala., framed a constitution and set up a provisional government (February 8). **Jefferson Davis** was elected Confederate President (February 9). (Outgoing U.S. President **James Buchanan** made no attempt to forestall the subsequent general seizure of U.S. military installations throughout the South by seceding state militias, but 4 U.S. Regular Army officers, each acting promptly on his own volition, saved 4 key points on the southern seacoasts—Forts Sumter, Charleston, S.C.; Pickens, Pensacola, Fla.; Taylor, Key West, Fla.; and Jefferson, Dry Tortugas, Fla.) Two days after Lincoln's inauguration, Davis, with Confederate congressional authorization, called for 100,000 volunteers for 1 year's military service (March 6). By mid-April the Confederacy had 35,000 men under arms—a force twice as large as the then-existing U.S. Army.

1861, April 12–14. Bombardment of Fort Sumter. South Carolinian forces under Major General **Pierre G. T. Beauregard** opened fire on Fort Sumter in Charleston Harbor, commanded by Major **Robert Anderson.** After 34 hours of bloodless bombardment, Anderson, his supplies exhausted, capitulated. He and his garrison, 7 officers and 76 enlisted men, marched out with the honors of war, and were shipped up North.

1861, April 15. Lincoln Calls for Volunteers. He called for 75,000 volunteers for 3 months' service to suppress insurrection.

1861, April 17. Virginia Secedes. She claimed the President's call for volunteers was an act of war against the seceded states. Virginia's secession was followed by Arkansas, Tennessee, and North Carolina. Eleven states were not in armed, open rebellion (May 20). Kentucky declared "neutrality" (May 24).

1861, April–May. Scott's "Anaconda Plan." As a nation *de facto,* the Confederacy had now only to maintain its sovereignty within its own borders. To restore the *status quo ante bellum,* the remainder of the U.S. must invade and destroy all armed resistance. Lieutenant General **Winfield Scott,** General in Chief of the U.S. Army (and a native of Virginia) had foreseen the difficulties and had earlier (October 1860) advocated adequate reinforcement of all

Federal garrisons in the South. Infirm Scott, 74 years old, but still possessing a sound and brilliant mind, now urged immediate blockade of Southern ports and the raising of an army of 25,000 Regulars and 60,000 3-year volunteers. Once this force was equipped and trained, Scott argued, it should invade the South down the Mississippi Valley to the Gulf of Mexico, thus isolating the Confederacy. This long-range concept of squeezing the South into submission was unpalatable to politicians who wanted quick, cheap victory, and derisively dubbed it the "Anaconda Plan." Although initially rejected, it was this plan, with a manifold increase in manpower, that eventually because the general strategy employed by the Union.

1861, April 19. Lincoln Proclaims a Blockade. This was the only part of Scott's plan that was adopted.

The Line-up

The 22 Northern states (including 3 border states with divided loyalties) had a population of some 22 million; the 11 seceding states had 5.5 million whites and 3.5 million Negro slaves. Actually, the available man power—since the Negro slaves performed labor which in the North had to be done by white men—favored the North by a 5:2 ratio. The North was an industrial-agricultural entity, potentially capable of supporting a protracted war. The South was agricultural, its crops of tobacco, cotton, and sugar mainly selling abroad; its manufacture was small. To succeed, it must market its produce overseas, and import war materiel.

The U.S. Army strength in 1861 was 16,367 officers and men, scattered along the western frontier and in seacoast forts. Concentration in force was impossible; recourse was had to militia and volunteers, poorly trained and originally of short enlistment terms. The 313 U.S. Regular Army officers who resigned and went South were wisely used as leaven for its officer corps. The 785 Regular officers who remained loyal were for the most part kept in their units during the early months of the war, depriving the mobilizing Union armies of the men best qualified to lead, command, and train them. Exacerbating this was the fact that, in the early days of the war, an officer holding a Regular commission was required to resign in order to be able to accept a higher commision in the Volunteers. Most Regulars were loathe to do this, as there was considerable doubt that they would be able to regain Regular status after the war. Thus, political appointees and elected officers, usually without military experience, filled most of the officer ranks.

The Confederacy had no naval vessels, but it had 322 of the 1,300 U.S. Navy officer corps and a vigorous Secretary of the Navy—**Stephen R. Mallory** of Florida, who left no stone unturned to build a fighting naval force. The U.S. Navy had some 90 wooden ships—sail and steam—42 of them in commission, and only 4 in Northern waters when the war broke out. **Gideon Welles** of Connecticut, competent U.S. Secretary of the Navy, set about extemporizing an enormous force, backed by the vast maritime resources of the North—resources which did not exist in the South. His assistant, **Gustavus Vasa Fox,** a former U.S. Navy officer, was of tremendous assistance in the task.

Confederate President Jefferson Davis was a West Pointer, a Mexican War hero, and had been Secretary of War; he had also been chairman of the Senate's Military Affairs Committee. He was well qualified for his position as Commander in Chief. U.S. President Abraham Lincoln had had no real military experience. Under political pressure he had rejected the recommendations of his senior military adviser. On the president fell the onus of producing and maintaining in the North the will to win, without which Southern victory would be inevitable. Yet he had no concerted plan of campaign. Lincoln's call for volunteers and his arbitrary immediate strengthening of the Regular Army were countered by Davis' authorization of Confederate privateering. Operations began on both sides of the Appalachians, at first piecemeal, later in force.

War in the East, 1861

1861, April 20. Loss of Norfolk Navy Yard.
Virginia militia, seizing this important station, secured for the Confederacy the fine steam frigate *Merrimack*, and a great quantity of naval stores and armament, including 52 modern 9-inch guns.

1861, May 24. Washington Menaced. Virginia militia occupied the Alexandria area and hastily installed Confederate batteries controlled the Potomac approaches.

1861, June. Operations in West Virginia and Virginia. Major General **George B. McClellan** with Ohio volunteers cleared western Virginia of local Confederate forces at **Philippi** (June 3) and **Rich Mountain-Carrick's Ford** (July 11–14). In eastern Virginia, there was a skirmish at **Big Bethel,** near Union-held Fortress Monroe (June 10).

1861, June–July. Focus on Washington and Richmond. The main bodies of both Union and Confederacy gathered between the opposing capitals. Brevet Major General **Irvin McDowell,** with 38,000 Union troops—less than 2,000 professional soldiers among them—moved from Washington (July 19) to attack a Confederate force of 20,000, under Beauregard, near Centreville, Va.

1861, July 21. First Battle of Bull Run.
McDowell found his enemy lying along Bull Run. Unaware that General **Joseph E. Johnston** had already arrived with most of his 12,000 men from the lower Shenandoah Valley to reinforce Beauregard, McDowell, who had 28,500 in the field, attempted to turn the Confederate left by a movement too complicated for his untrained troops and staff. After some progress, his attack was checked by the stand of Brigadier General **Thomas J. Jackson**'s Virginia brigade, Jackson winning his sobriquet of

"Stonewall." Johnston's skillful shift of troops from the original battle front culminated in success with the arrival of his last brigade by rail from Winchester. The Union enveloping force was itself enveloped and most of McDowell's army fled the field in panic, covered only by the dogged rear-guard action of Major **George Sykes**'s infantry battalion and Major **Innis N. Palmer**'s cavalry squadron—both regular units. Union losses were 2,706; Confederate, 1,981. Confederate President Jefferson Davis' refusal to authorize pursuit has been much criticized.

1861, July 22. New Northern Field Commander. McClellan was hurriedly called to Washington to command in place of McDowell.

1861, July–December. McClellan Trains an Army. He began organization and training of what would become the **Army of the Potomac,** while in Virginia across the river the Confederate forces under Johnston remained in the Manassas area.

1861, September 12–13. Battle of Cheat Mountain. Confederate General Robert E. Lee, attempting to recover western Virginia, was repulsed by Union troops.

1861, October 21. Battle of Ball's Bluff. A Union reconnaissance across the Potomac near Leesburg, Va., resulted in disaster when Colonel **Edward D. Baker,** U.S. senator transformed into soldier, foolishly engaged his troops piecemeal against superior Confederate forces led by Colonel **Nathan G. (Shanks) Evans.** Baker was killed. His troops, pinned on a precipitous river bank, were crushed, losing 237 killed and wounded and 714 captured or missing. Confederate losses were 149. Though the battle was minor in tactical importance, political repercussions in Washington were violent. A Joint Congressional Committee on the

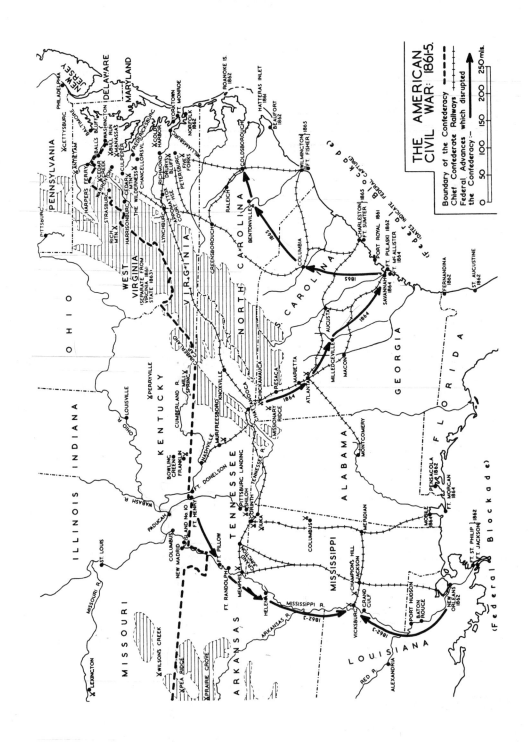

THE AMERICAN CIVIL WAR: 1861-5

Boundary of the Confederacy
Chief Confederate Railways
Federal Advances which disrupted
the Confederacy:—

0   50   100   150   200   250 mls.

Conduct of War, headed by abolitionist Senator **Benjamin F. Wade** of Massachusetts, caused Brigadier General **Charles P. Stone,** originator of the reconnaissance, to be imprisoned for 8 months without charges or trial. For the rest of the war, the committee became an incubus on the Union military.

1861, November 1. New Northern General in Chief. McClellan, promoted on the retirement of old General Scott, made no further movement, and the North began to grumble at delay.

War in the West, 1861

1861, June 1. Battle of Booneville. Brigadier General **Nathaniel F. Lyon,** after quelling a Southern threat at St. Louis (May 10), on his own initiative moved on Jefferson City and drove Governor **Claiborne Jackson**'s proSouthern militia down toward Arkansas. Missouri militia under Brigadier General **Sterling Price** was reinforced by Arkansas militia under Brigadier General **Ben McCullough,** and then attempted to retake Missouri for the Confederates.

1861, August 10. Battle of Wilson's Creek. Lyon, with 4,500 men, attacked McCullough and Price, who now had 11,600. Lyon almost succeeded with an ambitious but overcomplex plan to attack the Confederates in front and rear, but was forced on the defensive when the column under Colonel Franz Sigel, attacking the Confederate rear, was routed. In a desperate attempt to rally his forces, Lyon was killed and the Northerners retreated in confusion. The disorganized Confederates were unable to pursue. Confederate losses were 2,084; Union, 1,235.

COMMENT. *Although this was a tactical victory for the South, Lyon's aggressive operations had saved Missouri for the Union. "Neutral" Kentucky now became the target for both sides.*

1861, August–December. Grant and Johnston in Kentucky. Union Brigadier General **Ulysses S. Grant,** learning that General Albert Sidney Johnston, commanding Confederate

forces in the West, was sending Major General **Leonidas Polk** to seize key points in Kentucky, forestalled him by occupying Paducah (August 6), thus safeguarding Cairo, the most strategically important spot in the U.S. at that moment. Johnston then established a line across Kentucky to block river movements southward down the Mississippi and up the Tennessee and Cumberland rivers. Western anchor of the line was on the Mississippi at Columbus, Ky., with an outpost at Belmont, Mo., on the west bank of the river.

1861, November 7. Battle of Belmont. Grant, with some 3,000 men, made a successful hit-and-run attack on Belmont to relieve growing Confederate pressure in western Missouri. Federal losses were 498; the Confederates, who had more than 5,000 men on the field, lost some 900.

1861, November 18. Halleck Commands in the West. Major General **Henry W. Halleck** came West to supplant vacillating Major General **John C. Frémont,** but Union command was still divided, with Major General **Don Carlos Buell** commanding in central and eastern Kentucky. As the year ended, Johnston's Confederate cordon from the Mississippi to the Cumberland Gap held firm.

Naval Operations, 1861

1861, May. Appearance of Confederate Privateers. A few hastily converted merchant ships, mounting a few guns, began to prowl the Atlantic coast line.

1861, May. Blockade Begins. An improvised and growing U.S. Navy—converted merchant vessels for the most part—took up the task of blockade from the Potomac River to the Gulf of Mexico.

1861, August 27. Action at Hatteras Inlet. A Union flotilla, under Flag Officer **Silas H. Stringham,** reduced Southern fortifications protecting this back door to the Confederacy, landed 800 troops from Fortress Monroe under Major General **Benjamin F. Butler,** and established a blockade base.

1861, November 7. Action at Port Royal. Flag Officer **Samuel F. du Pont** led a joint expedition to Port Royal Sound, important inland waterway connecting Savannah and Charleston. His 9 warships overwhelmed the defensive forts, and 17,000 troops under Brigadier General **Thomas W. Sherman** were landed. Port Royal became an important naval base for the Union blockading squadrons.

1861, November 8. The *Trent* **Affair.** Captain **Charles Wilkes** in USS *San Jacinto* arbitrarily overhauled the British packet *Trent* in the Bahamas, forcibly removing Confederate commissioners **James M. Mason** and **John Slidell,** on their way to England to represent President Davis. The high-handed action almost precipitated war with England, but was composed through diplomacy; Wilkes's action was disavowed by the U.S., and the commissioners were turned over to a British warship to continue their voyage.

War in the East, 1862

1862, January–March. McClellan's Plans. Disturbed by the new General in Chief's lack of aggressiveness, Lincoln finally goaded him to action by an imperative order (March 8). McClellan had about 180,000 men in and near Washington. Misled by reports that J. E. Johnston had over 100,000 men in the Manassas area (actually he had only about 50,000), McClellan decided to turn Johnston's positions by shipping his army down the Chesapeake Bay to Urbana, 45 miles from Richmond. Johnston, informed of the plan, at once countered by retiring behind the Rappahannock. McClellan then decided to move by water to Fortress Monroe, thence overland up the Peninsula—between the York and James rivers—toward Richmond. Lincoln, who doubted the estimates of Johnston's strength, reluctantly approved. Presence of the Confederate ironclad ram *Virginia* (erstwhile steam frigate USS *Merrimack,* seized at Norfolk Navy Yard and hurriedly armorplated) threatened this proposed water movement.

1862, February–April. North Carolina Coastal Operations. Union troops under Major General **Ambrose E. Burnside,** convoyed by Flag Officer **Louis Goldsborough,** landed on **Roanoke Island** (February 7). After defeating the Confederate garrison (February 8), Burnside captured **New Bern** (March 14) and **Beaufort** (April 26).

1862, March 8. The Battle of Hampton Roads (First Day). The *Virginia,* debouching from Norfolk, attacked the Union blockading squadron—all wooden vessels—awaiting her. The heavily armed, cumbrous ironclad, impervious to Union cannonballs, rammed and sank the USS *Cumberland* (24), then attacked and sank USS *Congress* (50). USS *Minnesota* (50), trying to join the battle, ran aground. The *Virginia* returned to Norfolk before dark to repair superficial damage, as her crew prepared to return next day to finish off 3 remaining Union vessels. That evening USS *Monitor,* a small, revolutionary armored warship invented by **John Ericsson,** arrived from New York and took station near the helpless *Minnesota.*

1862, March 9. Monitor vs. Virginia (Merrimack). The Confederate ironclad reappeared and the *Monitor* moved to meet her. The *Virginia's* hull was covered by a heavy iron and oak carapace; she carried 10 large guns, 7- and 9-inch. The *Monitor,* a low-lying, metal "cheesebox on a raft," mounted two 11-inch guns in a heavily armored revolving turret. For 4 hours the ironclads hammered each other unsuccessfully, then the *Virginia* returned to Norfolk. The *Monitor* retained control of Hampton Roads.

COMMENT. *This inconclusive battle revolutionized naval construction by rendering wooden ships obsolete.*

1862, March 10. McClellan Prepares for Action. Satisfied that the *Virginia* was no longer a menace to his transports, McClellan began immediate preparations to move his army to fortress Monroe.

1862, March 11. Lincoln Assumes Direction of the War. To ensure McClellan's undivided attention to the Richmond theater, Lincoln temporarily relieved him as General in Chief and personally assumed overall military command. Lincoln's reasoning was sound; the results were disastrous, since neither he nor his Secretary of War, **Edwin M. Stanton,** had the professional experience required.

1862, March 22. McClellan Moves to the Peninsula. McClellan assured Lincoln that after the departure of the Army of the Potomac—130,000 strong—the security of Washington would be maintained by 75,000 men, including 23,000 under General **N. P. Banks,** who was responsible for stopping any threat to Washington through the Shenandoah Valley.

1862, March 23. Battle of Kernstown. Opposed to Banks in the lower Shenandoah Valley was Major General Stonewall Jackson, who had retired south of Strasburg with his 4,300 men. Banks, satisfied that Jackson posed no threat to Washington, left 9,000 men around Winchester under Brigadier General **James Shields** and moved east toward Manassas to cooperate with McClellan's operations in the Peninsula. Jackson, unaware of Shields's strength, attacked him at Kernstown, and was repulsed. So vigorous was the Confederate attack, however, that Shields reported that Jackson must have been reinforced.

1862, March 24. Lincoln Orders Banks Back to the Valley. He also discovered that McClellan had left only 50,000 men to cover Washington, not 75,000. To McClellan's disgust, therefore, Lincoln detained McDowell's corps of 30,000 at Washington as it was about to embark for Fortress Monroe. Lincoln ordered McDowell to join 10,000 Union troops already near Fredericksburg; he assured McClellan that McDowell would be permitted to march overland to join him near Richmond as soon as any possible threat to Washington disappeared. Few minor engagements have had as much significance as the Battle of Kernstown.

1862, April. Lee's Strategic Plan. General Robert E. Lee, military adviser to President Davis, realized that Johnston had only 60,000 men available to protect Richmond against McClellan's 100,000 assembling at Fortress Monroe and McDowell's 40,000 at Fredericksburg. He also knew of Lincoln's sensitivity about

Washington. Lee recommended that Jackson—reinforced to nearly 18,000 men—make a demonstration in the Valley so as to divert the largest possible Union forces from McClellan. Johnston agreed, and Davis approved.

THE VALLEY CAMPAIGN

1862, May 1. Jackson's Situation. He was at Swift Run Gap, menaced by 2 threats: Banks coming up the Valley and Frémont with about 15,000 approaching from West Virginia. Leaving Brigadier General **Richard S. Ewell** and 8,000 men to contain Banks, Jackson moved westward into the Alleghenies.

1862, May 8. Battle of McDowell. Jackson surprised and threw back the leading elements of Frémont's army. Frémont halted all further advance. Jackson returned to the valley.

1862, May 9–30. Jackson's Advance. He hurried north, down the Valley. Utilizing the Massanutten Mountain—a 50-mile-long spine bisecting the Shenandoah Valley—as a screen, the turned east through its one pass, then swung northward again.

1862, May 23. Battle of Front Royal. Jackson threw a Federal garrison out of Front Royal and threatened Banks's rear.

1862, May 25. First Battle of Winchester. Jackson pursued Banks to Winchester, defeated him there again, and chased him back across the Potomac. At Harpers Ferry, Banks lost some 3,000 prisoners, 10,000 stands of small arms, some cannon, and a vast quantity of stores. Union casualties were about 1,500; Confederate, 400.

1862, May 26. Lincoln's Reaction. The President diverted McDowell at Fredericksburg from his planned junction with McClellan, and hurried him toward the Valley. Frémont, too, was spurred to resume his offensive. Jackson, now on the Potomac, thus found himself in a trap. Frémont's 15,000 men were only 25 miles southwest of Winchester, advancing east; Shields, with 10,000 men of McDowell's command, had captured Front Royal and was

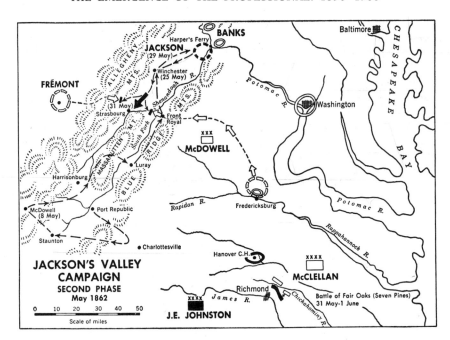

JACKSON'S VALLEY CAMPAIGN
SECOND PHASE
May 1862

0   10   20   30   40   50
Scale of miles

advancing west with 10,000 more Federals behind him; Banks, across the Potomac, was regrouping and preparing to move south again.

1862, May 30–June 7. Jackson's Withdrawal. With 15,000 men, 2,000 prisoners, and a double column of wagons containing booty, he withdrew rapidly, sending his cavalry ahead to hold Frémont. Clearing Winchester (May 31), he reached Strasburg before the jaws of the trap could close. While Frémont pursued west of the Massanutten and Shields up the east side, Jackson reached Harrisonburg (June 5). Leaving Ewell at Cross Keys to hold Frémont, Jackson turned on Shields with the rest of his command.

1862, June 8. Battle of Cross Keys. Ewell repulsed Frémont's halfhearted attack and burned the 1 bridge over which the Union general could advance.

1862, June 9. Battle of Port Republic. Jackson fell on Shields at Port Republic. A 5-hour battle ended when Ewell arrived to reinforce Jackson. Shields's force was pushed back some 20 miles. By midnight, Jackson sat safely at Brown's Gap in the Blue Ridge while his wagon train and its booty were rolling on to Richmond. Frémont had retired to Harrisonburg, Shields was rallying at Luray, and McDowell, definitively divorced from McClellan, was occupying Front Royal with orders to remain in the Valley.

COMMENT. *With less than 18,000 men, Jackson— between April 30 and June 9—had stymied 70,000 Federal troops and had changed the complexion of the entire Federal plan of campaign.*

THE PENINSULA CAMPAIGN

1862, April 4–May 4. Siege of Yorktown. McClellan, moving toward Richmond with 90,000 men, was faced by a 10-mile entrenched line, extending across the Peninsula southwest from Yorktown, manned by 15,000 Confederates under Brigadier General **John B. Magruder.** McClellan, deceived by many dummy guns and feigned activity, began a formal investment and ordered siege artillery from Washington. Johnston, arriving from the Rappahannock, brought Confederate strength to 60,000, but retired toward Richmond 2 days before McClellan proposed to open a massive bombardment. McClellan pursued slowly.

1862, May 5. Battle of Williamsburg. The Army of the Potomac was held up by Johnston's rear guard, under Major General **James Longstreet,** in a delaying action. The slow Union advance continued.

1862, May 25. McClellan in Sight of Richmond. McClellan deployed 3 corps north of the Chickahominy River to facilitate the expected junction with McDowell, keeping 3 south of the swollen river, facing Confederate entrenchments in front of Richmond.

1862, May 31–June 1. Battle of Seven Pines (Fair Oaks). Johnston, taking advantage of the division of the Union army, fell on McClellan's isolated left. Poor Confederate staff work turned a planned double envelopment into piecemeal frontal attacks, which were re-

pulsed after Federal reinforcements crossed the river. Johnston was seriously wounded, Major General **Gustavus W. Smith** taking his place. Next day the attacks were again repulsed. Union losses were about 5,000; Confederate, 6,100. McClellan, although he had now learned that McDowell would never join him, only rectified his line in part. Major General **Fitzjohn Porter**'s corps was left on the north side of the Chickahominy.

1862, June 1. Birth of the Army of Northern Virginia. President Davis appointed Lee to replace Johnston.

1862, June 12–15. Stuart's Raid. Brigadier General **J. E. B. Stuart**'s cavalry division circled the entire Federal Army, destroying large quantities of stores near McClellan's headquarters at White House, and brought Lee invaluable information on Union dispositions.

THE SEVEN DAYS' BATTLES, JUNE 25–JULY 1, 1862

Lee ordered Jackson to join him and prepared to attack Porter's 30,000 men; the remainder of the Union Army would be contained in front of Richmond by Magruder. Both armies totaled about 90,000.

1862, June 25. Confederate Concentration. Three-fourths of Lee's army concentrated west of Mechanicsville. Jackson reached Ashland Station.

1862, June 26. Battle of Mechanicsville. Inexperience and faulty staff work marred the Confederate assault. Jackson, probably suffering from exhaustion stemming from the arduous Valley Campaign, moved lethargically and never arrived on the field. Porter threw back piecemeal assaults. Then, learning of Jackson's approach, he retired that night to Gaines's Mill.

1862, June 27. Battle of Gaines's Mill. Lee with 65,000 men again attacked, but Jackson was late and failed to get behind Porter's right flank. Porter, utilizing the terrain and hasty earthworks to improve his right, unexpectedly found that his left (believed to be the strongest part of his line) was unexpectedly penetrated by a Confederate assault. With Jackson's tardy

approach on the right threatening a double-envelopment, Porter was forced to withdraw. Reinforcements sent on the initiative of the other Union corps commanders enabled Porter to fall back in good order. That night, on McClellan's orders he withdrew to the south bank of the Chickahominy. McClellan, convinced the Southern main effort was coming on his left and center, ordered a general withdrawal to the James River (and protection of the Union fleet) despite the objections of Brigadier Generals **Phil Kearny** and **Joseph E. Hooker,** who assured him the thin crust of Magruder's demonstration could be pierced, and that Richmond was his for the taking.

1862, June 29–30. Battles of Peach Orchard, Savage Station, White Oak Swamp, Glendale-Frayser's Farm. Lee, discovering McClellan's withdrawal, pursued. His sharp attacks were rebuffed by the individual Union corps commanders in a series of delaying actions while the Union trains successfully crossed White Oak Swamp. An envelopment of the Union right by Jackson was again unsuccessful, as Jackson continued to display an uncharacteristic lethargy. McClellan personally went on ahead to inspect his new base on the James.

1862, July 1. Battle of Malvern Hill. In McClellan's absence, Porter carried out a brilliant defensive action from excellent positions with good fields of fire. Lee's assaults broke down into uncoordinated attacks, which were beaten back with heavy loss—more than 5,000 men in about 2 hours. Malvern Hill was a clear-cut Union success, but McClellan ordered immediate retreat to Harrison's Landing and to the protection of Union gunboats.

COMMENT. *Union losses in the Seven Days were 1,734 killed, 8,062 wounded, and 6,053 missing (mostly prisoners). Southern losses were 3,478 killed, 16,261 wounded, and 875 missing. All Confederate attacks (except at Gaines's Mill) were repulsed. Yet when it was over, the Union Army had been expelled from Richmond's outskirts and huddled in defeat on the banks of the James.*

SECOND BULL RUN (MANASSAS) CAMPAIGN

1862, June 26. The Army of Virginia. President Lincoln called Major General **John Pope** from the West to command the new Army of Virginia, comprising the commands of McDowell, Frémont, and Banks.

1862, July 11–August 3. Union Reorganization and Regrouping. Realizing the need for professional generalship, Lincoln brought Halleck from the West to become General in Chief

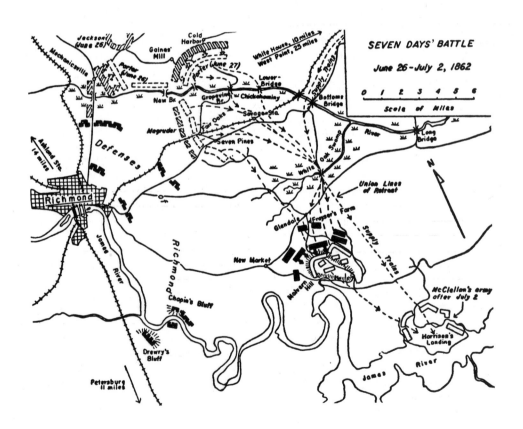

(July 11). The Army of the Potomac was ordered back to Washington (August 3).

1862, July–August. Confederate Reaction.
Lee, concerned by these shifts, sent Jackson with 12,000 men to observe Pope in central Virginia. Lee followed slowly, leaving 20,000 to garrison Richmond.

1862, August 9. Battle of Cedar Mountain.
Jackson, attempting to halt Pope's movement across the Rappahannock, was impetuously attacked by Banks, commanding the leading Union corps, with some 10,000 men. Jackson's personal efforts saved his troops from defeat; by nightfall he had driven Banks back on Culpeper. Jackson then withdrew across the Rapidan (August 11).

1862, August 24. Lee's Plan. Lee learned that McClellan's army was coming back from the Peninsula to join Pope. The combined Union armies would total 150,000 men against his 55,000. In a daring plan, he determined to crush Pope before such a junction could occur. Jackson was ordered to march north, then east, to get behind Pope's army; Lee, with Longstreet's corps, would follow, to join Jackson and to strike Pope somewhere near Manassas.

1862, August 25–26. Jackson Moves North.
He marched 54 miles in 2 midsummer days, falling on Pope's supply depot at Manassas, which was looted and burned (August 27). Jackson then took up a hidden defensive position just west of the old Bull Run battlefield.

1862, August 29–30. Second Battle of Bull Run. Pope, enraged at the raid on his depot, hastened north to hunt the invaders. Finally discovering Jackson's position (when Jackson deliberately attacked a Union division near **Groveton**—August 28—to attract attention from Longstreet's advance), Pope concentrated to destroy him (August 29). Jackson's line, protected by the ready-made entrenchment provided by the embankment of an unfinished railroad, held against Pope's poorly coordinated attacks. Meanwhile Lee, with Longstreet's wing, was advancing via Thoroughfare Gap, arriving near Jackson's right, between Gainesville and Groveton by early afternoon. Pope, refusing to believe reports of Longstreet's arrival, resumed the attack, concentrating on Jackson's right (August 30). The final Union attack foundered under the flanking fire of

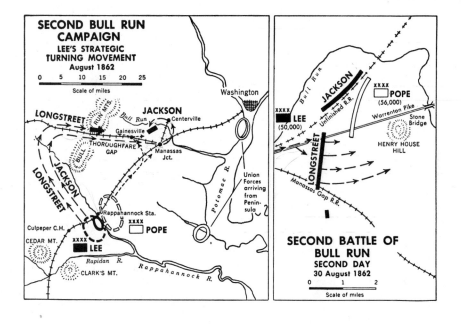

Longstreet's massed artillery, whereupon Longstreet's wing was unleashed in a devastating flank attack, rolling up the Union Army and throwing it back against Bull Run. The Union withdrawal across Bull Run, however, was in good order, covered by Sykes's Regular Army division of Porter's corps.

1862, September 1. Battle of Chantilly. Jackson, following retreating Union forces toward Washington by a parallel road, collided with Pope's rear guard near Chantilly. In a sharp twilight battle, fought partially in a violent thunderstorm and over rainswept fields and woods, Jackson's progress was checked. Union generals Phil Kearny and Isaac I. Stevens were killed in the confusion of the fight. Lee made no further attempt to press closer to heavily defended Washington. Confederate losses in the campaign were 9,197. Pope's losses were more than 16,000.

Antietam (Sharpsburg) Campaign

1862, September 4–9. Lee's First Invasion of the North. Reinforced to 55,000 men, Lee crossed the Potomac near Leesburg. His objective, approved by President Davis, was to carry the war to the North, and by a successful invasion to encourage Britain and France to recognize—and possibly assist—the Confederacy. Screened by Stuart's cavalry, he concentrated at Frederick, Md. (September 7). He decided to move northward into Pennsylvania behind the protection of the Catoctin Mountains. He issued orders (September 9) for Longstreet's corps to advance toward Hagerstown while Jackson returned across the Potomac to capture Harpers Ferry, and to secure a new line of communications through the Shenandoah Valley.

1862, September 7. McClellan Follows Lee. Meanwhile, Lincoln had dissolved the Army of Virginia, whose units were absorbed in McClellan's Army of the Potomac, now back in Washington. (Pope was sent to an administrative command in the northwest.) With 97,000 men, McClellan moved out from Washington on Lee's trail.

1862, September 12. The Lost Order. McClellan's advance guard reached Frederick, where a copy of Lee's order of June 9, detailing Confederate dispositions and movements, fell into McClellan's hands.

COMMENT. *Lee's army, spread out over 25 miles and split by an unfordable river, was at McClellan's mercy. Yet it took McClellan 2 days to advance 10 miles from Frederick to the passes over South Mountain.*

1862, September 14. Battle of South Mountain. Learning that his order was in Union hands, Lee hurriedly prepared to concentrate the army at Sharpsburg. Covering forces at Turner's, Fox's, and Crampton's Gaps delayed McClellan's ponderous advance in a series of small but intense engagements.

1862, September 14–15. Battle of Harpers Ferry. After a half-hearted fight, the Union garrison withdrew into the town's defenses. Confederate control of the high ground surrounding the town placed the garrison in an untenable position. Colonel **Dixon S. Miles,** the Union garrison commander, refused to attempt a breakout (that breakout was possible was evidenced by the escape of 1,500 Union cavalry under Colonel **Benjamin F. Davis** on the night of September 14) and opened negotiations for surrender (September 15). Miles was mortally wounded when he went forward to conduct the negotiations. Leaving the division of General **A. P. Hill** to secure the 15,000 prisoners and captured stores, Jackson hurried to join Lee at Sharpsburg. McClellan, who could have struck Lee's 20,000 with more than 50,000 on the 15th, incredibly still did not attack on the 16th, permitting the Southerners—now about 45,000 strong—to prepare well-organized defensive positions along Antietam Creek, with the Potomac to their backs.

1862, September 17. Battle of Antietam Creek (Sharpsburg). McClellan planned to roll up Lee's left with 3 corps, while another drove across the creek to pin the Confederate right. Two more corps were in reserve. The Union 3-corps hammer blow degenerated into separate, disjointed attacks against Jackson, with the most sanguinary action along the "Bloody Lane," a sunken road. By midday all

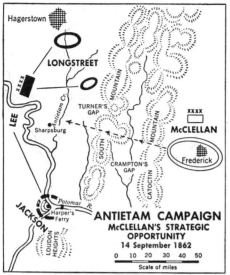

ANTIETAM CAMPAIGN
McCLELLAN'S STRATEGIC
OPPORTUNITY
14 September 1862

0   10 20 30 40 50
Scale of miles

BATTLE OF ANTIETAM
17 September 1862

0   1   2
Scale of miles

Lee's reserves were expended, but the Federal attacks had ground to a halt. The action now shifted to Longstreet's sector. Major General **Ambrose E. Burnside**'s Union corps made three desperate attacks to carry a bridge over the fordable creek. It then took him 2 hours more to reorganize his troops and move to the assault of Sharpsburg. All the while, McClellan had held more than 20,000 men unemployed in reserve. Burnside's final effort had reached the crest of the defense when A. P. Hill's division of Jackson's corps arrived from Harpers Ferry and charged into Burnside's left flank, driving the Federals back across the creek and ending the bloodiest 1-day battle of the entire war.

COMMENT. *With less than 50,000 effectives, Lee had stopped McClellan's 90,000—of whom only 70,000 took part in the battle. Federal losses were 12,400; the Confederates lost 13,700. Lee calmly faced his opponent through that night and through the next day. Then, unmolested, he withdrew across the Potomac (September 19). Tactically, Antietam was a Confederate victory; strategically, since Lee's invasion was stopped, it was a Union victory, which Lincoln used as a springboard for his Preliminary Emancipation Proclamation (September 23).*

FREDERICKSBURG CAMPAIGN

1862, November 7. Relief of McClellan. The Union general did not follow Lee across the Potomac until peremptorily ordered to do so by Lincoln (October 5). He then advanced cautiously to Warrenton, where he was between Longstreet at Culpeper and Jackson in the Valley. When, despite urging, McClellan failed to move further, Lincoln relieved him, placing Burnside in command of the Army of the Potomac.

1862, November–December. Burnside's Plan. He planned another drive upon Richmond. Advancing toward Fredericksburg on the fordable Rappahannock, he waited on the northeast bank for bridging equipment. Lee calmly concentrated on the other side, entrenching his army along the arc of hills overlooking Fredericksburg (from which civilians had been evacuated), Longstreet above the town, Jackson, on the right, farther south.

1862, December 13. Battle of Fredericksburg. Burnside launched a 2-pronged offensive, crossing the river opposite Fredericksburg and south of the town. The assaulting columns of the Union left initially enjoyed some success, the division of Brigadier General **George G. Meade** drove back the brigade of Brigadier General **Maxcy Gregg** in some confusion (Gregg was mortally wounded). However, a galling flank fire from Stuart's artillery and a violent counterattack by Jackson's reserves threw the Union attack back to its start line. By evening Jackson's position was intact, the Union troops holding defensive positions on the west bank of the river. On the north, the Federal assault debouched from the ruins of the town to meet devastating artillery and small-arms fire from Longstreet's corps on Marye's Heights beyond. Fourteen successive Union charges melted away under the Confederate fire. Along the entire line, more than 12,500 Union soldiers lay in windrows, 6,000 of them in front of Longstreet's position. Confederate losses were 5,309. Burnside, dissuaded by his corps commanders from making another suicidal attack next day, withdrew across the river (December 15). Burnside blamed his subordinates for the debacle.

1863, January 26. Relief of Burnside. He was removed from command after having made another abortive attempt to cross the Rappahannock above Fredericksburg (January 20), despite Lincoln's order to halt. This "Mud March" bogged down in torrential rains.

War in the West, 1862

WEST OF THE MISSISSIPPI

1862, March 7–8. Battle of Pea Ridge. Confederate Major General **Earl Van Dorn** with 16,000 men tried to encircle the right flank of Union Major General **Samuel R. Curtis,** whose 12,000 men held a defensive position 30 miles northeast of Fayetteville, Ark. Curtis, discovering the movement, spoiled the offensive, and Van Dorn withdrew after 2 days of stubborn fighting. Federal losses were about 1,300; Confederate, upward of 800.

1862, April 15. Battle of Peralta. Confederate Brigadier General **Henry Hopkins Sibley,** moving up the Rio Grande from Fort Bliss, Tex., with some 4,000 Texan volunteers to in-

vade California, had defeated a Union force of equal strength under Colonel **Edward R. S. Canby** at **Valverde,** N.M. (February 21), and occupied in turn Albuquerque and Santa Fe. But lack of food forced Sibley's retirement and Canby, reinforced, fell on him at Peralta, 20 miles south of Albuquerque, and in a series of running fights forced Sibley and his disintegrating force back to Fort Bliss (April–May).

1862, December 7. Battle of Prairie Grove. A Confederate force of 10,000 under Major General **T. C. Hindman** was repulsed by a Union force of equal size under Brigadier Generals **James G. Blunt** and **Francis J. Herron,** leaving northern Arkansas in Union hands.

EAST OF THE MISSISSIPPI

1862, January 19–20. Battle of Mill Springs. In eastern Kentucky a Confederate thrust of 4,000 men from Cumberland Gap, under Generals **Felix Zollicoffer** and **George Crittenden,** was repulsed decisively by an equal number of Union troops under Brigadier General **George H. Thomas,** part of the Army of the Ohio, commanded by Major General **Don Carlos Buell.** The Confederates lost all their artillery, most of their equipment, and many men in retreating across the Cumberland River. Zollicoffer was killed in the battle.

1862, February. Henry-Donelson Campaign. Grant led an amphibious operation from Cairo, Ill., to split A. S. Johnston's cordon defense. Moving up the Tennessee River, Flag Officer **Andrew Foote**'s escorting squadron of ironclad river gunboats' fire overwhelmed low-lying Fort Henry (February 6). Grant then moved overland to Fort Donelson on the Cumberland River while the gunboats retraced their route and advanced up the Cumberland to support him. Donelson was held by 15,000 men commanded by Major General **John B. Floyd,** assisted by Major Generals **Gideon Pillow** and **Simon Bolivar Buckner.** A Union gunboat attack was repulsed (February 14). Next day the Confederates attacked, almost breaking through the Union encirclement, but Grant's counterattack swept

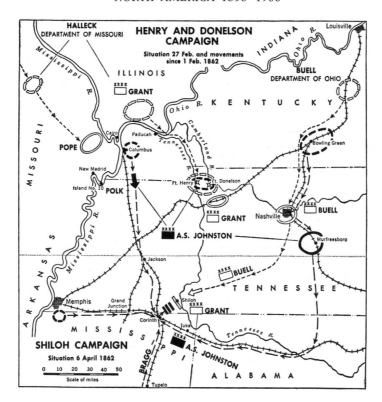

HALLECK
DEPARTMENT OF MISSOURI

HENRY AND DONELSON CAMPAIGN

Situation 27 Feb. and movements since 1 Feb. 1862

BUELL
DEPARTMENT OF OHIO

SHILOH CAMPAIGN

Situation 6 April 1862

0 10 20 30 40 50

Scale of miles

---

them back into the fort. Floyd and Pillow escaped by water; Buckner's request for an armistice was answered by Grant's famous ultimatum: "No terms except an unconditional and immediate surrender can be accepted." Buckner surrendered (February 16). Union losses were 2,832; Confederate—including prisoners—16,623.

1862, February–March. Collapse of the Confederate Cordon. Attempting to form a new line in Tennessee, Johnston gave up the project when Grant and Buell seized Nashville (February 25). Johnston then concentrated in the vicinity of Corinth, Miss., with 40,000 men. Grant's planned move on Corinth was halted on the Tennessee River at Savannah, Tenn., by Halleck's order, requiring him to wait until Buell should arrive from Nashville.

1862, April 6–7. Battle of Shiloh. Johnston made a surprise attack on Grant's inadequately outposted bivouacs, in the vicinity of Pittsburg Landing–Shiloh Church, driving in the badly shaken Union forces. Grant's vigorous personal efforts to retrieve the day were finally successful at dusk. The Confederates, all their reserves expended, were first halted by Grant's artillery, then driven back as 2 Union gunboats added their flank fire to the defense. Johnston was killed while leading a last-ditch charge; Beauregard succeeded to command. That night, Buell's army arrived on the east bank of the river and part of it was ferried across. Shortly after dawn next day, Grant counterattacked. By noon Beauregard gave up the struggle and retreated to Corinth. Confederate losses, of some 40,000 men in action, were 10,600. Grant, out of 62,700 engaged, lost 13,000. Criticism of Grant in the North was loud and to a great extent baseless. Lincoln's answer to demands for Grant's relief was: "I can't spare this man. He fights."

1862, May—June. Halleck's Advance. Taking personal command of the combined armies, Halleck moved very slowly to Corinth. Beauregard, falling back in front of him, evacuated it (May 29), retreating to Tupelo, where he reorganized his army. Confederate evacuation of Fort Pillow followed (June 5). Ponderous Halleck now divided his forces. Sending Buell eastward into Tennessee, he restricted Grant to defensive operations in the Mississippi Valley. Meanwhile (June 17), Beauregard was replaced by General **Braxton Bragg** in command of the Confederacy's western theater. To stop Buell's threat to Chattanooga, Bragg shifted most of his forces rapidly to southeastern Tennessee.

1862, July 11. Halleck to Supreme Command. Halleck, appointed General in Chief, left for Washington (July 17), turning over his western command to Grant and to Buell, in separate commands (Army of the Tennessee and Army of the Ohio).

GRANT'S OPERATIONS, JULY–NOVEMBER, 1862

1862, July–September. Grant on the Defensive. His slender forces (45,000 men) widespread on a long line of communications to Cairo, Ill., Grant, at Corinth with about 25,000 men, was menaced by Price (17,000 men) and Van Dorn (15,000 men), who attempted to unite against him.

1862, September 19–20. Battle of Iuka. Grant planned a double envelopment of Price's advancing forces, but was foiled by a number of circumstances. Not least was the error of a subordinate, Brigadier **General William S. Rosecrans,** who marched by the wrong route and became embroiled in a confused meeting engagement, failing to cut the Confederate line of retreat. Curiously, Grant, with the remainder of the army a few miles away, heard nothing of Rosecrans' fight due to a quirk of acoustics caused by the weather. Price and Van Dorn,

uniting, moved against Corinth under the latter's command.

1862, October 3–4. Battle of Corinth. Rosecrans, now commanding the field forces of Grant's army, occupied an entrenched position at Corinth. Van Dorn, after pushing in the Union outposts, attempted to envelop the position. The Confederate attacks degenerated into a series of piecemeal assaults, all of which were repulsed. Federal losses were 2,520 (of 23,000 engaged); Confederate, 4,233 (of 22,000 engaged). The net result was firm federal retention of Memphis and Corinth, both strategic outposts. Grant now moved toward Grand Junction, planning an advance against Vicksburg.

BUELL AND ROSECRANS VS. BRAGG,
JULY—DECEMBER, 1862

1862, July–October. Bragg's Invasion of Kentucky. Bragg moved north from Chattanooga while his subordinate, General **Edmund Kirby Smith,** advanced from Knoxville, overwhelming a small Union force at **Richmond** (August 30). The combined Confederate force totaled about 50,000 men. Buell—with about the same number—finding his communications with Nashville menaced, recoiled to the banks of the Ohio River. Bragg now occupied most of Kentucky. Buell, substantially reinforced, and threatened by Washington with replacement, finally advanced from Louisville.

1862, October 8. Battle of Perryville. Buell's advance resulted in a confused encounter near Perryville. Part of Bragg's army (some 16,000 men) under Generals Leonidas Polk and **William J. Hardee** met one of Buell's corps when elements of both armies attempted to secure a source of water. Although both sides claimed success, the result tactically was a draw. Strategically, it was a Union victory as Bragg, having concentrated his, army, now inexplicably turned and marched back through Cumberland Gap into Tennessee. Buell, equally negligent, failed to pursue him.

1862, October 23. Relief of Buell. Disgusted with Buell's lack of aggressiveness, Lincoln replaced him with Rosecrans, who occupied Nashville (November 6).

1862, November–December. Stalemate. For several weeks Bragg and Rosecrans confronted each other without action between Nashville and Murfreesboro, while both Washington and Richmond berated their procrastinating generals. Meanwhile, Confederate cavalry leaders **Nathan Bedford Forrest** and **John H. Morgan** bedeviled both Rosecrans and Grant by raids. Finally, Rosecrans began his advance, forcing Bragg to retreat to a position near Murfreesboro (December 26–30).

1862, December 31–1863, January 2. Battle of Stones River (Murfreesboro). Spurred to take action, both commanders attacked, each planning to envelop the other's right flank. Bragg, with 38,000 men, struck first, with outstanding initial success, but dilatory and blundering use of his reserve spoiled his exploitation. Rosecrans, who had some 44,000 troops, was able to hold a new defensive line. Both sides bivouacked on the field that night, and held their respective positions next day. Bragg made one more attack, but was repulsed (January 2). He withdrew the next day. The tactically drawn battle was a strategic victory for the North. Federal losses were 12,906; Confederate, 11,740.

UNION ADVANCE DOWN THE MISSISSIPPI, FEBRUARY–JUNE, 1862

1862, February–March. New Madrid and Island No. 10. The northern attack started as an amphibious operation; Major General John Pope, with some 23,000 men, moved on New Madrid and Island No. 10, supported by Flag Officer Foote's gunboats. Pope drove the Confederate garrison out of New Madrid (March 13). Two gunboats made daring runs past the heavy batteries on Island No. 10 (April 4–5, 6–7) and escorted Pope's troops across the river, cutting land communication with the island, whose garrison surrendered (April 8). Sixty

miles to the south downriver lay Fort Pillow, north of Memphis. Pope's army, ferried down, invested it, while gunboats and mortar schooners opened bombardment. Halleck withdrew Pope's troops for the Corinth campaign (see p. 966), but the naval bombardment continued. (His success here led to Pope's later appointment to command in the East; see p. 960).

1862, May 9. Battle of Plum Point. A flotilla of Confederate gunboats lying below the fort made a dawn attack on the Federal flotilla, now commanded by Captain **Charles H. Davis,** U.S.N. (Foote, wounded during the Donelson operation, had been evacuated.) After seriously damaging 2 Union gunboats, the Confederate flotilla withdrew. Because of Halleck's advance, Fort Pillow's garrison was withdrawn (June 5).

1862, June 6. Battle of Memphis. The Federal flotilla moved down to attack the southern gunboats off Memphis, destroying all but 1 of the Confederate vessels. Memphis surrendered and Davis' flotilla steamed down to the mouth of the Yazoo River, above Vicksburg.

UNION ADVANCE FROM THE GULF, MARCH–JULY, 1862

1862, March–April. Farragut at the River Mouth. Meanwhile, Commodore **David Glasgow Farragut** had moved into the Mississippi River estuary from the Gulf with 8 steam sloops and corvettes, a 20-ship mortar flotilla, and 9 gunboats. General Butler with 10,000 men accompanied the expedition. The river was defended by **Forts St. Philip** and **Jackson,** mounting between them 115 heavy guns. The Confederate ironclad *Louisiana,* her engines uncompleted, was moored as a floating battery just above St. Philip. A barrier of hulks and logs below the forts barred the river to traffic.

1862, April 24. Naval Battle of New Orleans. While his mortar boats bombarded the forts, Farragut's squadron crashed through the barrier and successfully ran by the forts. Above the Forts, 11 Confederate gunboats attacked. Nine of them were sunk. Next day Farragut steamed up and took New Orleans, which had been evacuated by all Confederate troops. Forts St. Philip and Jackson—bombarded by mortar boats and surrounded by Butler's troops—surrendered (April 27). Butler's troops then occupied New Orleans. Farragut moved upriver. Baton Rouge and Natchez surrendered, but Vicksburg, heavily fortified, repelled his squadron (May 23). Returning to New Orleans, Farragut received peremptory orders from Washington to open up the river, and the squadron retraced its steps to Vicksburg, accompanied by mortar boats.

1862, June 28. Run-by at Vicksburg. Farragut's squadron ran a 3-mile gantlet, with some casualties and considerable damage to the vessels, linking with Davis' flotilla off the mouth of the Yazoo River (July 1).

1862, July 15. The *Arkansas* vs. the Union Fleet. The Confederate ironclad ram *Arkansas* debouched from the Yazoo (a strategic error, for so long as she remained in place she menaced Union control of the river) and battered her way to Vicksburg through the entire Union squadron. Farragut ran by the forts again in an unsuccessful attempt to destroy the *Arkansas* (July 21–22) and then returned to the Gulf, to resume command of the Gulf blockading squadron. The misused *Arkansas* was later destroyed by her crew near Baton Rouge, after her engines broke down.

LAND OPERATIONS AGAINST VICKSBURG, NOVEMBER–DECEMBER, 1862

1862, November–December. Grant's Advance. Grant having concentrated the Army of the Tennessee at Grand Junction, Tenn., started south toward Vicksburg (November 13), pushing back General **John C. Pemberton,** now commanding all Confederate forces in the region. Grant also sent Major General **William T. Sherman** with 40,000 men down toward Vicksburg in an amphibious operation supported by Rear Admiral **David D. Porter,** now commanding naval units in the Memphis area

(December 20). But Grant's own advance was harassed by Forrest's raids in northwest Tennessee, and Van Dorn's capture of the Federal supply depot at **Holly Springs,** Miss. (December 20).

1862, December 14. Butler Replaced by Banks. Meanwhile, at New Orleans, General Nathaniel P. Banks had replaced Butler, and in coordination with Farragut was moving slowly up the river to threaten Port Hudson.

1862, December 25–29. Chickasaw Bluffs Operation. The Sherman-Porter expedition entered the Yazoo River and Sherman attempted unsuccessfully to assault the Chickasaw Bluffs north of Vicksburg, manned by 13,800 of Pemberton's veterans. After a 3-day action (December 27–29), Sherman admitted defeat, retiring to the mouth of the Yazoo. Federal losses were 1,776; Confederate, only 207. Thus the year ended with the Mississippi from Vicksburg to Baton Rouge still under Confederate control.

War in the East, 1863

1863, January 26. Hooker to Command. He was appointed to command the Army of the Potomac in place of Burnside.

CHANCELLORSVILLE CAMPAIGN, APRIL–MAY

1863, April 27–30. Preliminaries to Chancellorsville. Hooker moved to attack Lee from the Federal positions east of the Rappahannock and opposite Fredericksburg. Major General **John Sedgwick** with 40,000 men was to demonstrate by forcing a crossing of the river at Fredericksburg itself. Meanwhile, Hooker with 73,000 more would cross the river above, into the Wilderness, and circle Lee's left. The Federal cavalry under Major General **George Stoneman** would ride wide westward, then sweep south to cut Lee's communications with Richmond. Lee's Army of Northern Virginia was reduced to 60,000—high in morale, but short of equipment, rations, and clothing.

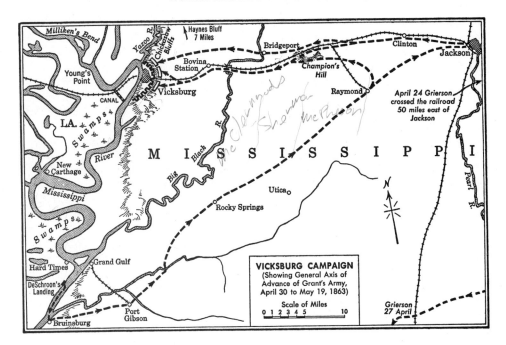

VICKSBURG CAMPAIGN
(Showing General Axis of
Advance of Grant's Army,
April 30 to May 19, 1863)

Scale of Miles
0 1 2 3 4 5   10

Longstreet's corps and much of Stuart's cavalry was away, gathering provisions and forage in the lower James River area. Stuart's remaining cavalry harrassed and contained Stoneman's men and quickly identified Hooker's intentions. Lee, although surprised by the skill and rapidity of this movement, reacted swiftly.

1863, May 1–6. Battle of Chancellorsville. Leaving Major General **Jubal Early** with 10,000 men on Marye's Heights to contain Sedgwick's Fredericksburg attack, Lee moved rapidly west to meet Hooker's main offensive. At first contact (May 1), Hooker, despite his tremendous superiority in numbers, halted and went on the defensive. Early next day, Lee sent Jackson with 26,000 men in a wide envelopment while Lee with only 17,000 held the front. Just before dusk Jackson assaulted—his

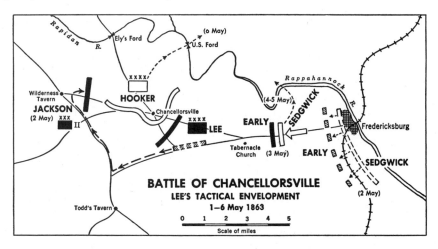

BATTLE OF CHANCELLORSVILLE
LEE'S TACTICAL ENVELOPMENT
1–6 May 1863

0   1   2   3   4   5
Scale of miles

line perpendicular to the right flank of the Federal entrenchments. He rolled up the Union right flank, completely demoralizing the entire right wing of Hooker's army. Attempting to complete the operation and cut Hooker from his Rappahannock ford communications, Jackson rode into the night in person to reconnoiter and was mortally wounded by the fire of some of his own troops. Next day (May 3) the Confederate attacks resumed, but were slowed by increasing Federal resistance, while at Fredericksburg Sedgwick, reduced to 28,000 men by detachment to Hooker's main body, was driving Early's small force from Marye's Heights. Hooker, with only part of his troops engaged, could still have retrieved the situation, but he had no fight left in him. Leaving Stuart to hold Hooker near Chancellorsville, Lee hastened east to defeat Sedgwick at **Salem Church** (May 4). The Army of the Potomac withdrew across the Rappahannock (May 5–6). Union losses were 16,792 killed, wounded, or captured. The South lost a total of 12,754. Lee had won a tremendous victory, but had lost the one man he could not spare; Jackson died of his wounds (May 10).

Gettysburg Campaign, June–July

1863, June 3. Lee Prepares to Invade the North. Lee reorganized his army, replacing the former 2 corps (wings) with a 3-corps structure, elevating Generals Richard Ewell and A. P. Hill to corps command. Some analysts believe that the inexperience of Ewell and Hill as corps commanders had a strong influence on the outcome of the campaign. (This is doubtful.) Determined to maintain the initiative, Lee moved over to the Shenandoah Valley, leaving Stuart to screen his movements.

1863, June 9. Battle of Brandy Station. Hooker discovered Lee's move when his cavalry, under Major General **Alfred Pleasonton** crossed the Rappahannock and surprised Stuart's cantonment. Rallying, the Confederates fought the attackers to a standstill in the largest cavalry battle of the war. Pleasonton of 12,000 men engaged, incurred more than 900

casualties. Stuart, out of 10,000, lost almost 500 men.

1863, June 13–15. Second Battle of Winchester. Ewell's II Corps (formerly Jackson's) smashed General **Robert H. Milroy**'s Union forces in the lower Shenandoah Valley.

1863, June 15–24. Lee Crosses the Potomac. The Southern army—80,000 strong—pushed northward into Pennsylvania through Chambersburg to reach Carlisle and York (June 28).

1863, June 13–27. Hooker Follows Lee. When finally convinced that Lee was making a major movement, Hooker and his 113,000 men began to move north (June 13). The movement of the Army of the Potomac was rapidly and efficiently carried out, Pleasonton's cavalry skillfully screened this move in a succession of small actions with Stuart's cavalry along the line of the Blue Ridge Mountains. Although Hooker missed repeated opportunities to strike at the stretched-out Confederate army, his army was well concentrated and deployed to protect Washington. More importantly, Hooker's whereabouts were almost unknown to Lee. Hooker's army reached Frederick at the same time that he discovered that Halleck (who remembered the discrepancy between Hooker's plan and Hooker's actions at Chancellorsville) had overruled his plan to trap Lee with a 2-pronged strategic envelopment (June 27). Hooker offered his resignation, which—to his surprise—was accepted before dawn next day.

1863, June 26–July 2. Stuart's Raid. Screening Lee's right flank during the crossing of the Potomac, Stuart took advantage of discretionary orders to take his 3 best brigades and swing wide on a raid around the rear of the Union army in central Maryland. Two brigades remained behind to guard the Blue Ridge passes, leaving 2 overworked brigades to operate with Lee's main body. Stuart became entangled in the marching columns of the Army of the Potomac, whose movement north had interposed itself between Stuart and Lee. Determined to complete his circuit of the Union army, Stuart pushed on, slowed down by 125 captured wagons which he insisted on keeping.

1863, June 28. Major General George C.

Meade Takes Command. He was the Army of the Potomac's fourth commander in 10 months. Lee, deprived of reconnaissance intelligence due to Stuart's absence, learned the general whereabouts of the Army of the Potomac—and its change of command—late the same day. Immediately he began concentration near Cashtown.

1863, June 30. Meeting Engagement. Meade, trying to coax the Confederates to give battle south of the Susquehanna, probed cautiously toward Emmitsburg and Hanover. A Confederate infantry brigade, moving southeast toward Gettysburg, met a Union cavalry brigade reconnoitering to the northwest. From that unexpected clash built up, piecemeal, the greatest battle ever to be fought on American soil.

1863, July 1. Battle of Gettysburg (First Day). A. P. Hill's Confederate III Corps opposed **John F. Reynolds'** Union I Corps and **Oliver O. Howard**'s XI Corps, north and northwest of Gettysburg itself. Ewell's II Corps, moving south from Heidlersburg, outflanked the Union right. The Union troops rallied south of the town on Cemetery Hill and Culp's Hill, while the Confederates held Gettysburg and Seminary Ridge to the southwest. Through the night the remainder of the Army of the Potomac marched up and moved into position facing west along Cemetery Ridge. The famous Union "fishhook" was taking shape (see map). Lee decided that he would envelop the Union left next day, using Longstreet's I Corps, which had not yet arrived on the field.

1863, July 2. Battle of Gettysburg (Second Day). Longstreet's attack was delayed until afternoon, but drove back **Daniel E. Sickles'** III Corps from an exposed position in a peach orchard and almost turned the Union left. Brigadier General **Gouverneur K. Warren,** Meade's engineer officer, discovering that Little Round Top was unoccupied, deftly diverted a nearby brigade and artillery battery to this key point and saved the situation. Desperate fighting to the south ended with the Union troops anchored firmly along the ridge from Round Top to Culp's Hill—which latter point Ewell had failed to wrest from Union hands.

1863, July 3. Battle of Gettysburg (Third

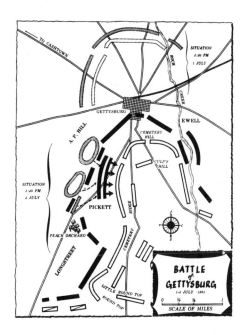

Day). Lee decided on a penetration of the Union center. Covered by a tremendous artillery bombardment, 10 brigades were launched in the assault known as "Pickett's Charge," although Major General **George E. Pickett** was but 1 of 3 division commanders involved. Some 15,000 men pressed for a half-mile through Union cannonading that blew great gaps in their line. Reaching the Union line, the surviving Confederates briefly broke in only to be thrown back by Meade's reserves. Survivors of the charge (barely half of those who started) reeled back to Seminary Ridge, where Lee, accepting responsibility for the disaster, waited for a counterattack which never came. Union casualties for the 3 days were 3,155 killed, 14,529 wounded, and 5,365 missing. The Confederate army lost 3,903 killed, 18,735 wounded, and 5,425 missing.

1863, July 4–14. Lee's Retreat. This was accomplished without serious hindrance through mud and rain, and despite the swollen Potomac. Meade followed cautiously. Lincoln was gravely disappointed by Meade's failure to exploit his victory.

1863, August—September. Stalemate on the Rapidan-Rappahannock.

1863, October 9–14. Lee Advances to Manassas. He jabbed at Meade's left in a long sweep, but was checked at **Bristoe Station** (October 14), and the armies returned to the Rapidan-Rappahannock triangle (November).

1863, November 27–December 1. Mine Run Campaign. Meade made several inconclusive thrusts across the Rappahannock at Lee's position on the west bank of Mine Run in the Wilderness. Union losses were 1,653; Southern casualties, 745.

War in the West, 1863

VICKSBURG CAMPAIGN

1863, January—March. Probing at Vicksburg. The Holly Springs experience (see p. 968) had convinced Grant that, with the forces available to him, overland operations against Vicksburg were impracticable at the end of a long line of communications from Cairo. He distrusted the capabilities of Major General **John A. McClernand,** who, as senior in rank, had superseded Sherman as commander of the river expedition against Vicksburg (see p. 968). McClernand had then gone on a useless diversion up the Arkansas River to capture **Fort Hindman** at **Arkansas Post** (January 11). Ordering McClernand to return to the vicinity of Vicksburg, Grant went down the river to assume personal command at Milliken's Bend, La. (January 30). Through the winter and early spring Grant probed at the increasingly formidable defenses of Vicksburg, keeping his army busy, but accomplishing little else. Actually he was waiting for the cessation of the rains and falling of the flooded rivers.

1863, February–March. Gunboats on the River. Admiral Porter, working closely with Grant in the probes at Vicksburg, also sent the gunboats *Queen of the West* and *Indianola* in daring runs past the Vicksburg batteries to interfere with Confederate river traffic between Vicksburg and Port Hudson. Honors were about even; Confederate river activities were seriously disrupted, but the *Queen* was eventually captured and the *Indianola* destroyed.

1863, March 14. Port Hudson. Carrying out his part of a jointly planned operation with General Banks, Farragut steamed upriver past the powerful defenses of Port Hudson. Only his flagship, USS *Hartford,* and 1 other vessel were able to fight their way past, all other Union ships being repulsed. Banks failed to cooperate. But Farragut with his 2 ships quickly swept southern shipping off the central Mississippi.

1863, April. Grant Completes His Preparations. Pemberton's forces defending Vicksburg numbered about 50,000, scattered on the east bank of the Mississippi from Haines's Bluff, 10 miles above Vicksburg, to Grand Gulf, 40 miles below. (Farther south, the 15,000-man garrison at Port Hudson was immobilized by the threats of Farragut and Banks.) Grant had about 50,000 men on the west bank. Leaving Sherman's corps above Vicksburg, Grant moved the rest of his army overland to Hard Times, below Vicksburg, opposite Grand Gulf. At about the same time (April 17), in a series of spectacular night actions, Porter began to run gunboats and transports past the Vicksburg batteries to Grant's new base at Hard Times.

1863, April 18–May 3. Grierson's Raid. To further confuse the Confederates, Grant dispatched Colonel **Benjamin H. Grierson,** with 3 regiments of cavalry, from La Grange, Tenn., on a skillful, destructive ride through the entire length of Mississippi to the Union base at Baton Rouge.

1863, April 30. Crossing the Mississippi. While Sherman's corps and some gunboats demonstrated up the Yazoo, near the Chickasaw Bluffs above Vicksburg, the remainder of Grant's army was ferried unopposed by Porter's vessels to Bruinsburg, 10 miles below Grand Gulf (which Porter's gunboats had bombarded the previous day).

1863, May 1. Battle of Port Gibson. The Confederate Grand Gulf garrison, attempting to resist the Union advance inland, was driven off.

Grant then waited while Sherman hurried down from his demonstration to join the remainder of the army east of the Mississippi (May 3).

1863, May 7–19. Big Black River Campaign. In addition to Pemberton's 35,000 in the vicinity of Vicksburg, Joseph E. Johnston (recovered from his wounds and now overall Confederate commander in the West) had gathered 9,000 men at Jackson, Miss., 45 miles east. Grant, with 41,000 men, now abandoned his communications and marched eastward between these two Confederate forces. While Pemberton probed vaguely for the nonexistent Union line of communications, Grant drove Johnston out of **Jackson** (May 14). Leaving Sherman there to complete destruction of supplies and railroads, and to block Johnston, Grant then turned west again on Pemberton.

1863, May 16. Battle of Champion's Hill. Pemberton, with 22,000 men, held a strong position east of the Big Black. Grant, to rectify errors of McClernand, his leading corps commander, took personal command and drove the Confederates from the field. Pemberton for a few hours held a bridgehead over the Big Black River, but next day McPherson's corps stormed it. At the same time Sherman crossed the river farther north. Pemberton retreated into the Vicksburg defenses (May 19).

COMMENT. *In 19 days, Grant had marched 200 miles, living off the country, and defeated detachments of a numerically superior enemy in 5 distinct engagements and several skirmishes. He had inflicted losses of about 8,000 men at a loss to himself of less than 4,400 casualties. He had now locked up his principal opponent and some 30,000 men in a fortress. The Big Black River campaign was the most brilliant ever fought on American soil.*

1863, May 19–July 4. Siege of Vicksburg. After 2 unsuccessful assaults (May 19 and 22), Grant buckled down to siege operations, leaving Sherman's corps as a mobile force shielding any possible attempt by Johnston to relieve the city. The Federal investment closed in gradually, with continual bombardment from siege guns on land and from Porter's ironclads on the river. Vicksburg's defenders and the civilian population, burrowing in caves and threatened by starvation, kept up a gallant but hopeless fight, which excited as much admiration in the North as in the South.

1863, July 4. Surrender of Vicksburg. Learning that Grant was preparing a general assault, Pemberton surrendered. Grant immediately rushed rations to the famished civilians and soldiers in the city. Repulse of a Confederate attack on **Helena,** Ark., that same day was an anticlimax, as was the surrender of Port Hudson to Banks (July 9).

COMMENT. *As Lincoln put it, the Mississippi now "flowed unvexed" to the sea; the Confederacy had been split in two.*

TULLAHOMA, CHICKAMAUGA, AND CHATTANOOGA CAMPAIGN.

1863, June 23–July 2. Tullahoma Campaign. After 6 months of inaction at Murfreesboro, while Bragg sat at Tullahoma—and Grant campaigned near Vicksburg—Rosecrans was finally prodded to action by Halleck's threats to replace him. In a well-planned, well-conducted, bloodless maneuver, Rosecrans forced Bragg to withdraw to Chattanooga.

1863, August 16. Offensives of Rosecrans and Burnside. After another long delay, Rosecrans advanced from Tullahoma. That same day Burnside, with the Army of the Ohio, moved from Lexington, Ky., toward Knoxville. Swinging wide to the southwest, Rosecrans crossed the Tennessee River near Bridgeport, Ala., threatening Bragg's line of communications to Atlanta. The Confederates abandoned Chattanooga (September 7); Rosecrans, with 60,000 men, pursued into northwestern Georgia. President Davis now rushed Longstreet's corps from Virginia by rail; these and other reinforcements brought Bragg's army up to 70,000.

1863, September 19–20. Battle of Chickamauga. Knowing that Rosecrans was widely strung out in pursuit, Bragg attacked across Chickamauga Creek, attempting to turn the Union army's left and cut its line of communications to Chattanooga. Both armies were hampered by the densely wooded terrain; Rosecrans held firm on the 19th. Confederate assaults continued next day. While Rosecrans was shifting divisions, a garbled order left a gap in his center through which Longstreet plunged, cutting the Union army in two and driving its center and right from the field in disorder. Only the resolute resistance of the left, under Major General **George H. Thomas** (who held the field until nightfall to win the soubriquet of "Rock of Chickamauga"), and the timely arrival of Brigadier General G. Gordon Granger's reserve corps, which marched without orders to Thomas's assistance, prevented total disaster. Bragg, irresolute, made no effort to pursue, and Rosecrans' Army of the Cumberland clung to Chattanooga. Federal casualties were 1,657 killed, 9,756 wounded, and 4,757 missing. Bragg's Army of Tennessee lost 2,312 killed, 14,674 wounded, and 1,486 missing.

1863, September–October. Siege of Chattanooga. Bragg invested Chattanooga, cutting Rosecrans off from all communication either with Burnside at Knoxville, Tenn., or with Hooker's 2 corps of the Army of the Potomac, rushed by rail 1,192 miles to reinforce him. Hooker's troops remained at Bridgeport, 30 miles from Chattanooga. The Union-held mountain trails north of the Tennessee River were too rugged to support a supply line. Only a telegraph wire linked Rosecrans' army with the North. Rosecrans' troops and the inhabitants of the city neared starvation.

1863, September–October. Burnside Stalled at Knoxville. His army could not obtain supplies over the atrocious roads back through Cumberland Gap and was unable to advance on Chattanooga.

1863, October 17. Grant to Overall Command in the West. Lincoln put him over all Union forces between the Mississippi and the Alleghenies.

1863, October 17–27. Relief of Chattanooga. Grant hastened to Chattanooga, en route relieving Rosecrans by telegram, replacing him with Thomas. Arriving by mule trail in Chattanooga (October 23), 4 days later Grant

slipped a portion of Thomas' troops on pontoon boats past the Confederates on Lookout Mountain and opened a gap in the blockading line. Hooker's troops advanced through this gap, bringing with them an adequate supply of food. At the same time Grant ordered Sherman up from Memphis with the Army of the Tennessee.

1863, November–December. Siege of Knoxville. Bragg, meanwhile, was so confident of the impregnability of his fortifications on Lookout Mountain and Missionary Ridge, overlooking Chattanooga, that he sent Longstreet and 20,000 men to besiege Burnside, whose forces were soon brought to the verge of starvation. Halleck and Lincoln wired frantic appeals to Grant to send relief to Burnside.

1863, November 24. Battle of Lookout Mountain (First Day, Battle of Chattanooga). With the arrival of Sherman—after a month's march—Grant had 61,000 men to oppose Bragg's well-entrenched 40,000. He immediately sent Hooker's 2 corps, on his right, to attack Lookout Mountain, while Sherman—without a pause to rest his weary troops—moved against the northern end of Missionary Ridge. Sherman was repulsed, but Hooker stormed Lookout Mountain in "the battle above the clouds," against light resistance.

1863, November 25. Battle of Missionary Ridge (Second Day, Battle of Chattanooga). Sherman renewed his assaults against the Confederate right, while Hooker somewhat dilatorily advanced against Bragg's left. In the afternoon Thomas' Army of the Cumberland began a limited attack against the front of the 3-tiered Confederate works. Taking the first line, Thomas' men, without orders, swept to the top of the ridge in a spontaneous assault. Bragg's troops, seized with panic, fled. Federal losses in the two days' combat were 753 killed, 4,722 wounded, and 349 missing. The Confederates lost 361 killed, 2,160 wounded, and 4,146 missing. Next day Grant sent Sherman's army to relieve Burnside at Knoxville.

1863, December 6. Relief of Knoxville. Sherman arrived to find that Longstreet had abandoned the siege 2 days previously, after the failure of a desperate assault on the Federal

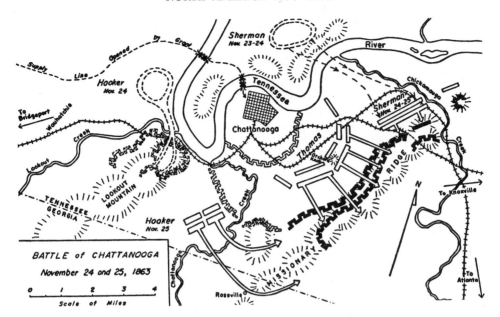

BATTLE of CHATTANOOGA
November 24 and 25, 1863

0 1 2 3 4
Scale of Miles

fortifications (this assault was noteworthy in that it was the first occasion in which wire entanglements were used to impede an attacker's movements; telegraph wire had been strung at knee-level in front of the Union fortifications). With Tennessee now cleared of southern troops, the way was open for the next move—marching into Georgia.

Naval Operations, 1863

COASTAL

By this time the Union naval blockade was slowly strangling the Southern economy, despite the handicaps plaguing the blockaders. Monotonous cruising in all weathers was relieved only by hurried trips to coaling stations. Deep-water ships and sailors had no love for shoal waters and treacherous banks over which shallow-draft blockade runners could slip from hot pursuit. Up to June 1863, Lincoln's net had bagged 885 runners, but the mesh was wide and blockade running had become big business. Specially built, speedy ships were operating out of Nassau in the Bahamas, Bermuda, Halifax, N.S., and Havana, Cuba. While the gulf ports took some of this traffic, Charleston and Wilmington, N.C., had become the principal ports of entry. It was evident that to dry up the flood the ports must be occupied. Union interest now centered on Charleston in particular, whose retention had become a symbol in Southern eyes. Its defense was ably planned and vigorously conducted by General Beauregard.

1863, April 7. Repulse at Charleston. Rear Admiral Samuel F. du Pont attacked the fortified harbor with a squadron of 9 Union ironclads, but was decisively repulsed by the forts and shore batteries, bristling with guns.

1863, July–September. Union Bombardment. Rear Admiral **John A. Dahlgren,** succeeding du Pont (July 10), kept up bombardment of the Charleston defenses, in cooperation with land attacks by the troops of Major

General **Quincy A. Gillmore.** Fort Wagner, principal target, repulsed several assaults with sanguinary losses (July–August).

1863, September 6. Occupation of Fort Wagner. it was evacuated by the Confederates. Its loss diminished Charleston's value as a blockade-runner's haven.

1863, September–December. Continued Bombardment. The other fortifications held out. Southern ingenuity now turned to the submarine boat.

1863, October 5. Submarine Attack on *New Ironsides.* CSS *David,* a semisubmersible cigar-shaped craft, steam-driven, manned by a crew of 4 men, approached USS *New Ironsides* at night and exploded a 60-pound copper-cased torpedo, manipulated on a long spar. Extensive damage was done to the Federal ship. The *David* returned safely to port, though 2 of her crew, washed off by the spouting water, were captured.

THE HIGH SEAS

Confederate commerce destruction during the period centered principally on the operations of 2 vessels constructed especially for the service in British shipyards and then turned over to Southern crews.

CSS **Florida** had ravaged the Atlantic coast since August, 1862. One of *Florida*'s prizes, converted to a warship, actually sailed into the harbor of Portland, Me., and blew up a Federal revenue cutter before being captured. Finally, *Florida,* a steam corvette, was cornered in Bahia Harbor (October 7) and, in disregard of Brazilian neutrality, was boarded and captured by USS *Wachusett.*

CSS **Alabama,** single-screw sloop of war, whose career also started in August, 1862, was still ranging the high seas on a history-making cruise under daring Captain **Raphael Semmes** as the year drew to a close.

Total War, 1864

1864, March 9. Grant General in Chief. Already partly choked by the constantly increasing pressure of naval blockade, the South now faced dismemberment by a gigantic coordinated turning movement planned by Grant. In the East, Meade's Army of the Potomac (under Grant's immediate supervision) would engage the Army of Northern Virginia while the western armies, under Sherman, swept into the deep South. Grant immediately prepared to bring the war to the speediest possible conclusion. His order (April 17) ending prisoner exchange was a severe blow to Southern manpower capability. Lincoln, realizing that at last he had found a general who could translate political objectives into vigorous, effective military strategy, wrote Grant: "The particulars of your plan I neither know nor seek to know . . .

I wish not to obtrude any constraints or restraints upon you."

WAR IN THE EAST, 1864

1864, February 20. Battle of Olustee. In Florida, Confederate General **Joseph Finegan,** with 5,200 men, defeated an invading Union force of equal strength under Brigadier General **Truman Seymour.**

1864, February 28–March 2. Kilpatrick-Dahlgren Raid. An ill-conceived dash on Richmond by Major General **Judson Kilpatrick** with 4,500 cavalrymen ended in disaster. Publication of papers allegedly found on the body of Colonel **Ulric Dahlgren,** indicating intention to burn Richmond and assassinate Jefferson Davis and his cabinet, aroused keen Southern resentment. Meade, queried by

Lee under a flag of truce, disclaimed such intentions. The papers were probably forgeries.

WILDERNESS—SPOTSYLVANIA—COLD HARBOR CAMPAIGN

1864, April–May. Grant's Plan. He intended to lead Meade's Army of the Potomac—almost 105,000 strong—into the dense thickets of the Wilderness, seeking Lee's Army of Northern Virginia—about 61,000 men deployed watchfully along the Rapidan-Rappahannock line. Leaving administrative duties to Halleck, his chief of staff in Washington, Grant planned to operate directly against Lee's army "wherever it may be found," seeking its right flank and dislocating it from Richmond. His own line of communications, based on tidewater Virginian ports, would be secured through the assistance of the Union Navy. Far on the Union flanks—to prevent concentration of outlying Confederate forces against the main effort—Butler (from Fortress Monroe) would move up the James toward Richmond, while Major General **Franz Sigel** was to advance southward up the Shenandoah Valley.

1864, May 4. Crossing the Rapidan. Grant began to implement his plan.

1864, May 5–6. Battle of the Wilderness. As soon as Grant moved, Lee began to concentrate. The armies met while Grant was still entangled in second-growth thickets of the Wilderness, 14 miles long by 10 miles wide. The first day's fighting consisted of 2 separate indecisive engagements between bewildered units. Grant forced a general offensive next morning, to be met by determined counterattacks. Longstreet was seriously wounded near the spot where Jackson had fallen a year previously. By nightfall the confused, furious combat was still undecided; the contenders spent the next day quelling forest fires and attempting to rescue wounded trapped in the blazing brush. The Union Army had lost 2,246 killed, 12,037 wounded, and 3,383 missing. Confederate casualties, unrecorded, are estimated at over 12,000.

1864, May 7. Grant Sticks to His Plan. He slipped around Lee's right and moved south. Lee, correctly estimating the situation, had rushed a covering force to Spotsylvania Court House before the Union advance elements arrived at dawn next day.

1864, May 8–18. Battle of Spotsylvania. Now on more maneuverable terrain, both armies struggled for 5 days (May 8–12) in a series of piecemeal engagements, while Lee's position built up into an immense V, with its apex—the "Mule Shoe"—pointing north. A promising envelopment of the Confederate left, by Hancock's corps, was unaccountably aborted in favor of an assault on the Confederate center by Warren's corps. This was decisively repulsed. At dusk, a well-planned attack by Colonel Emory Upton penetrated the "Mule Shoe," but was forced to withdraw when supporting troops failed to advance (May 10). After a day's lull, a grand assault was made by Hancock's entire corps. Initially successful, Hancock was stopped by a Confederate counterattack. Stalemated, the 2 armies battled for hours across the width of the earthworks. Casualties were so great that the "Mule Shoe" became known as the "Bloody Angle." Grant, searching for the hostile flanks, was for 6 more days thwarted by Lee's splendid shifting of his reserves. Union losses were 14,267; Confederate, over 10,000.

1864, May 11. Battle of Yellow Tavern. During the Spotsylvania struggle, Major General **Philip H. Sheridan**'s cavalry corps, 10,000 strong, raided south toward Richmond. At Yellow Tavern, on the outskirts of the city's fortifications, he met Stuart, with 4,500 sabers, in pitched battle. The Confederate horsemen were driven from the field and Stuart himself was mortally wounded, a body blow to Lee. Sheridan lost some 400 men; Confederate losses—untabulated—were about 1,000.

1864, May 20. Grant Again Moves South. As before, he went around Lee's right, toward Hanover Junction. Lee, recognizing Grant's stubborn intention to cut him off from Richmond, cleverly shifted from Spotsylvania to a strong position on the North Anna River (May 22) one day before Grant arrived.

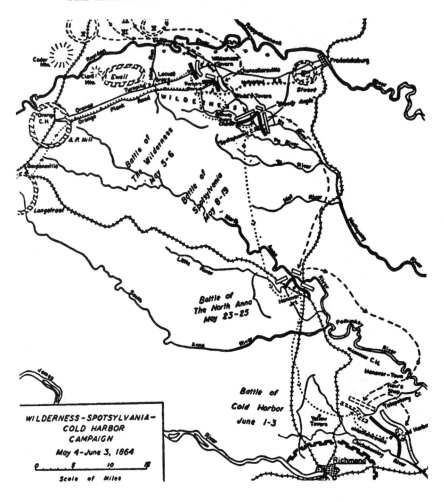

WILDERNESS–SPOTSYLVANIA–
COLD HARBOR
CAMPAIGN
May 4–June 3, 1864

Scale of Miles

1864, May 23–31. North Anna and Haw's Shop. Grant probed the position, but found it too strong for a major assault (May 23). Once more he sidestepped eastward around Lee's right flank. A 2-day delay crossing the Pamunkey River (May 27–29) gave Lee time to interpose once again between Grant and Richmond. Sheridan's cavalry engaged with entrenched Southern horsemen near Haw's Shop (May 28), while Lee was taking new positions south of Totopotomy Creek, near Mechanicsville. Again Grant probed and again he shifted to find Lee's right flank—this time to the

vicinity of Cold Harbor, close to the Seven Days' battlefields of 1862.

1864, June 3–12. Battle of Cold Harbor. Grant, reinforced to a strength of 100,000 men, determined to split Lee's army, which he believed to be overextended on a 6-mile line (June 2). Delays in troop movements postponed the attack for 24 hours. Lee, feverishly improving his field fortifications, was himself reinforced by 14,000 men, drawn from the successful opponents of Sigel in the Valley and Butler on the James (see p. 979). Grant's assaults were snubbed by confident, veteran defenders in a well-fortified position, strong in artillery. After losing nearly 7,000 men in less than an hour, Grant called off the attack. The defenders suffered fewer than 1,500 casualties. Other Union probes were also beaten back. Over all losses in the 10-day battle: Union, 13,078; Confederate, approximately 3,000.

FAILURES OF BUTLER AND SIGEL

1864, May 5–16. Bermuda Hundred. Butler, with the Army of the James, 30,000 strong, moved (May 4) from Fortress Monroe upriver by boat. Charged with the occupation of City Point as preliminary to his final objective— Richmond (in coordination with Grant's advance)—Butler instead landed at Bermuda Hundred. His timid and vacillating movements now did little more than stimulate the close-in defense of Richmond.

1864, May 15. Battle of Drewry's Bluff. Cleverly defeated by Beauregard, Butler was thrown back to his Bermuda Hundred position, where he remained corked up until Grant advanced south of the James.

1864, May 15. Battle of New Market. In the Shenandoah Valley, Sigel, with 10,000 Union troops, had started south in coordination with Grant's drive (May 4). Advancing from Cedar Creek, he was attacked by Confederate Major General **John Breckenridge,** with 5,000 men. Sigel, who made use of only half his original strength, was driven back and failed to maintain further pressure in the Valley. Noteworthy was the presence in the Confederate line of the cadet corps of Virginia Military Institute.

FROM COLD HARBOR TO PETERSBURG

1864, June 11–12. Battle of Trevilian Station. Grant, determination unshaken, prepared to cross the James River, again circling Lee's right. As a diversion, Sheridan's cavalry raided westward toward Charlottesville with the mission of uniting with old General **David Hunter,** who had replaced Sigel and had pushed up the Valley to Staunton after a successful engagement at nearby **Piedmont** (June 5). The combined force would then drive southeast to meet Grant. Lee matched the threats by sending **Wade Hampton** with 2 cavalry divisions after Sheridan and transferring Early's corps to the Valley to check Hunter. The opposing cavalry forces met (June 11) at **Trevilian Station.** After a confused 2-day conflict, Sheridan withdrew. Hunter, menaced by the arrival of Early, began withdrawal westward across the Alleghenies after some skirmishing near Lynchburg (June 17–18).

1864, June 13–18. Crossing of the James. While Warren's V Corps and Wilson's cavalry demonstrated as if to attack Richmond, the Army of the Potomac faded south from its Cold Harbor positions to the James River. One corps was ferried across by naval transport, another shipped to Bermuda Hundred; but the bulk of the troops crossed on a pontoon bridge erected (June 14) by the Union engineers in 8 hours, a feat unsurpassed in military engineering. Lee was still deployed in expectation of a full-scale Union attack over the old Seven Days' battlefields when the last of the Army of the Potomac had passed south of the James in one of the great feats of the war.

1864, June 15–18. Battle of Petersburg. While the James crossing was in full swing, Grant ordered Butler, with his Army of the James at Bermuda Hundred to capture fortified Petersburg, 8 miles away and—at the moment—scantily garrisoned. Butler and his subordinate **W. F. Smith** botched the job in 2 piecemeal attempts, while Beauregard, Confederate commander south of the Richmond defenses, rushed all available reinforcements. Grant, arriving on the scene, ordered a major attack as the remainder of the Army of the Potomac began coming up. Meanwhile Lee, finally realizing that Grant had crossed the James, hurried down. Grant, attacking with 65,000 men (June 18), was stopped by 40,000 veterans manning fortified lines. The siege of Petersburg began. This 3-day conflict cost the Union 8,150 casualties. Confederate losses totaled 4,752.

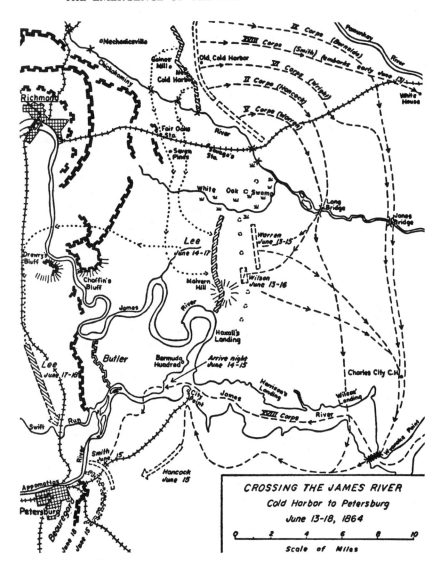

CROSSING THE JAMES RIVER
Cold Harbor to Petersburg
June 13-18, 1864

Scale of Miles

EARLY'S VALLEY CAMPAIGN

1864, July 2. Early Crosses the Potomac. Driving remaining Union forces before him from Winchester to the Potomac, Early's II Corps—Jackson's old command, some 13,000 strong—invaded Maryland; he levied from Hagerstown $20,000, from Frederick $200,000.

1864, July 9. Battle of the Monocacy. Early then struck toward terrified Washington, sweeping out of his way Major General **Lew Wallace**'s extemporized force of 6,000 on the Monocacy River. While Grant hurried the VI Corps up from the Richmond-Petersburg area to protect Washington, Early reached Fort Stevens, inside the District line (July 11).

1864, July 11–12. Action in Washington's

Outskirts (Battle of Fort Stevens). The next day elements of the VI Corps arrived, reinforcing the ill-assorted units of the XIX Corps, the normal garrison of the city. Early, after a demonstration, hurriedly retreated, leaving 400 wounded behind him, but carrying his cash booty safely back up the Valley. Procrastinating and contradictory orders from Washington prevented any coordinated pursuit.

1864, July 23–30. Early Invades Again. This time he smashed Major General **George Crook**'s Army of West Virginia aside at **Kernstown** and **Winchester** (July 23–24), then he lunged as far as Chambersburg, which he burned (after the town refused to pay $100,000 in gold) as reparation for the burning of Virginia Military Institute by Hunter.

1864, August 7. Sheridan's Army of the Shenandoah. Grant, realizing that so long as a Confederate army stood in the Valley the capital was unsafe, secretly ordered the VI and XIX Corps (48,000) to leave the defenses of Washington and to concentrate near Harpers Ferry, safe from politicians' meddling. He placed Sheridan in command. Sheridan's mission was to destroy Early's command and to "eat out Virginia clean and clear . . . so that crows flying over it for the balance of the season will have to carry their own provender."

1864, September 19. Battle of Opequon Creek (Third Battle of Winchester). Sheridan's 41,000 men began with cautious probings. Early started north with some 19,000 men, meeting Sheridan headlong. After initial Confederate success, the Union cavalry turned Early's left, while Sheridan's personal leadership sparked the main Federal effort. Early, attempting to stand just outside Winchester, was thrown back to Fisher's Hill, 12 miles south. Union losses were 5,000; Confederate, 4,600.

1864, September 22. Battle of Fisher's Hill. Sheridan assaulted the Southern position in frontal attack, while his cavalry smashed in Early's left. The Confederates were driven back to Waynesboro; they lost about 1,250 men; Union casualties were negligible.

1864, September–October. Devastation in the Valley. Sheridan moved back down the Valley, systematically turning it into a vale of desolation. Confederate guerillas on his flanks were extremely annoying, but neither they nor Early's troops, hanging on the Union rear, had any effectual result. By mid-October the 31,000-man Army of the Shenandoah, its work completed, bivouacked along Cedar Creek, 20 miles south of Winchester, Sheridan left his command for a conference in Washington about the future disposition of his troops.

1864, October 19. Battle of Cedar Creek. Early, reinforced to 18,400-man strength, sneaked his main effort through the Massanutten Mountain nose to surprise the Federal left and rear at dawn. His cavalry at the same time fell on both Union flanks, and his massed artillery sprayed the front. The daring move was almost successful. The Union VIII Corps stampeded (and would not be rallied for 48 hours). Major General **H. G. Wright,** commanding in Sheridan's absence, fought the XIX and VI Corps in 3 successive delaying positions. By noon, the Union army was defeated, it seemed, but Sheridan, returning from Washington, galloped up the pike from Winchester, "twenty miles away." He rallied his men in an extraordinary display of personal leadership. Regrouping on the battlefield, Sheridan's counterattack drove Early all the way back to New Market. Union casualties were some 5,600. Confederate losses were only 2,900, but the back of Southern resistance in the Shenandoah Valley had been broken. Early's effort, however, had prolonged the war in the East for some 6 months. Most of his troops now returned to Lee at Petersburg.

1865, March 2. Battle of Waynesboro. Early, with but a shadow of his command—some 1,000-odd men—was overrun by Major General **George A. Custer**'s cavalry.

SIEGE OF PETERSBURG—PHASE ONE, 1864

1864, June 19–December 31. Operations South of Petersburg. Grant immediately began to move against the railroads south of Petersburg to isolate Lee from his sources of supply. Operations dragged on through the rest of the year, combining trench warfare with spectacular sorties and maneuvers as Grant slowly extended his partial encirclement of the Richmond-Petersburg complex.

1864, June 22–23. Weldon R.R., Burke's Station, Ream's Station. A Union infantry attack under Generals **D. B. Birney** and H. G. Wright from the Jerusalem Plank Road toward Globe Tavern on the Weldon R.R. was combined with a longer-range cavalry raid of Generals **J. H. Wilson** and **A. V. Kautz** to Burke's Station, 15 miles farther west. The infantry were thrown back by A. P. Hill's counterattack, while the Union cavalrymen, initially successful, were severely handled and barely escaped capture as they returned by way of Ream's Station.

1864, July 30. The Petersburg Mine (Battle of the Crater). A cleverly devised Union mining operation ended with the explosion of 4 tons of gunpowder under the Petersburg defense systems. The blast—a complete surprise—ripped a great crater, through which Burnside's IX Corps was to assault. Shockingly mismanaged, the assaulting troops gathered in the crater without competent leadership. The rallying defenders poured cannon and rifle fire into the milling mass and sealed the break. This operation was one of the great tragic fiascos of the war, comparable to Burnside's previous fiasco at Fredericksburg. (In fairness to Burnside, last minute changes to the plan, dictated by Grant, had created some of the problems.) Union casualties were 3,793; Confederate (including the victims of the blast), 1,182.

1864, August 14–20. Deep Bottom. To keep the Confederates off balance, Grant tested the Richmond defenses north of the James. Lee shifted troops from Petersburg to meet the threat. Six days of hammering along Bailey's Creek cost the Union 2,900 men. Confederate losses are not recorded.

1864, August 18–21. Globe Tavern. Grant now struck southwest of Petersburg again, overrunning the Weldon R.R., despite A. P. Hill's bitter counterattack. Union losses were 4,455, against Confederate casualties of 1,619, but a vital supply line had been cut.

1864, August 25. Ream's Station. An extemporized Southern railhead farther south was cut off by a Union infantry-artillery force under Hancock. Lee's immediate reaction was a counterattack by Major General Wade Hampton, surprising the Union force as it was wrecking the line. Hancock, rallying, threw him back and completed destruction of a long stretch of railroad.

1864, September 29–30. Chaffin's Bluff. Covered by a demonstration on the southern end of the investment, 2 Union corps attacked the Richmond sector. Lee withdrew forces from Petersburg to counterattack, but Fort Harrison, at the Bluff, fell into Union hands.

1864, September 30. Peeble's Farm. A simultaneous attack southwest of Petersburg inched the Union investment a bit farther east, but could not penetrate the main defensive position.

1864, October 27–28. Boydton Plank Road. Grant's full-scale blow to cut the Southside R.R., last Confederate rail link, was repulsed by Hill. Butler's diversionary demonstration east of Richmond was rebuffed by convalescing Longstreet. Further mobile operations ceased as winter closed down, and siege became an artillery duel while the semistarved military and civilian population of the Richmond-Petersburg area tightened their belts. Behind the entrenchments of the well-fed, well-housed besiegers, a military railroad belt line, 21 miles long, linked every inch of the front to Grant's huge base at City Point.

War in the West, 1864

WEST OF THE MISSISSIPPI

1864, March–May. Red River Expedition. Before Grant became General in Chief, Halleck (with Lincoln's approval) ordered General Banks and Admiral Porter to undertake an overland invasion of Texas via the Red River. One objective was to discourage the French (then controlling Mexico; see p. 000) from any intervention through Texas. Porter managed to get 12 of his Mississippi gunboats above the

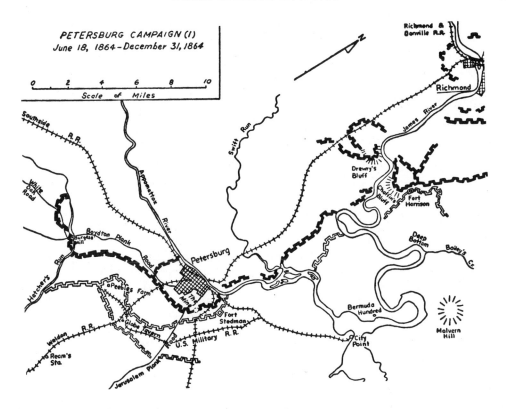

PETERSBURG CAMPAIGN (I)
June 18, 1864–December 31, 1864

rapids at Alexandria, La., and headed up the winding stream toward Shreveport, while Banks with 30,000 men marched along the river bank. At Grand Ecore, Banks struck across country toward Shreveport. Ambushed at **Sabine Cross Roads** (April 8) by Generals **Richard Taylor** and Edmund Kirby Smith, Banks retreated hastily to Alexandria, while part of his command delayed Confederate pursuit at **Pleasant Hill** (April 9). Banks lost 3,500 out of 25,000 men engaged in both actions; Southern losses were 2,000 out of 14,300. Porter, abandoned by Banks, steamed back on a falling river, sniped at by Confederate cavalry and artillery. At Alexandria he discovered the water over the rapids was now too low to float his vessels. Banks, bluntly ordered by Grant to remain at Alexandria until Porter got out, assisted in constructing a flume and dam through which the Union gunboats shot to safety. Alexandria was then evacuated (May 14).

1864, September–October. Price in Missouri. During 1863 and 1864, Confederate Major General **Sterling Price** had been operating in Arkansas against Union Major General **Frederick Steele** with mixed success. He now decided to invade Missouri. Crossing the Arkansas River between Little Rock and Fort Smith (September 1), Price with 13,000 veterans and 20 guns headed for St. Louis. Finding that city unexpectedly reinforced, he then struck west toward Kansas City. He defeated Major General **James G. Blunt** at **Lexington** (October 19). Continuing west, Price again drove Blunt back at **Independence** (October 22). But Blunt's stubborn resistance permitted Union troops, pursuing from St. Louis, to reach the scene.

1864, October 23. Battle of Westport. Still engaged with Blunt along the Missouri-Kansas line, Price was struck in the rear by Pleasanton's cavalry and defeated. He was driven back into Arkansas, ending major operations west of the Mississippi.

OPERATIONS OF N. B. FORREST,
FEBRUARY–OCTOBER, 1864

1864, February 22. Battle of Okolona. Forrest had a handful of Confederate troops in northern Mississippi. Union General **W. Sooy Smith,** with 7,000 cavalry, two to three times Forrest's strength, moved from Memphis (February 11) into northern Mississippi, planning to meet General Sherman at Meridian. Forrest, entrenched amid wooded swamps, stopped Smith, who withdrew (February 20). Forrest pursued, catching the Federals near Okolona, drubbing them in an all-day running fight.

1864, March–April. Raid into Kentucky. Sweeping as far as Paducah, Forrest assaulted and captured **Fort Pillow** on his return (April 12). That black soldiers of the U.S. Colored Troops there were killed after surrendering remains a matter of hot dispute. It is possible that the "massacre" was the result of the frenzy typical of the aftermath of assaults on fortified positions throughout history. However, it is also likely, and inexcusable, that at least a part of this killing frenzy was a result of Confederate hatred stemming from the Union use of black troops.

1864, June 10. Battle of Brice's Cross Roads. Union Major General Samuel D. Sturgis, under orders from Sherman to find and defeat Forrest, advanced from Memphis with 3,400 cavalry, 2,000 infantry, 12 guns, and a train of 250 wagons. Forrest, who had about 3,000 men, held the Union troops with a thin skirmish line, undertook a double envelopment, and sent a detachment on a wide envelopment to strike the Union rear. The Northerners were routed, abandoning all their wagons and all but 2 guns; they suffered 617 casualties, in addition to 1,623 captured by Forrest, who lost about 500.

1864, July 14–15. Battle of Tupelo. General **A. J. Smith,** with 14,000 troops, undertook another expedition from Memphis against Forrest, who had been reinforced to 10,000. Forrest again seized the tactical offensive in a bitterly contested 2-day drawn battle, but was repulsed. Smith withdrew, having lost 674 men; Forrest's casualties were over 1,300; he was wounded.

1864, August 21. Memphis Raid. Forrest penetrated the Union defenses at dawn, reaching the Union headquarters. The commander, Major General **C. C. Washburn,** escaped in his nightclothes, but Forrest rode off with his uniform.

1864, August–October. Forrest Moves East and North. This was partly due to increased concentration of Federal troops in western Mississippi, and partly in response to the crisis resulting from Sherman's capture of Atlanta (see p. 986). Forrest, in company with young General **Joseph Wheeler,** raided vigorously against Sherman's line of communications from Atlanta back to Nashville, and thence to the Ohio River.

1864, October 29–November 4. Naval Operations on the Tennessee. Raiding into northwestern Tennessee, Forrest ambushed Federal naval forces on the river near Johnsonville, capturing a gunboat and 5 transports. His troops manned these river craft for several days, but abandoned them after the Union Navy concentrated overwhelming force on the lower Tennessee. Forrest was now ordered to go east again to join Hood in the Franklin-Nashville Campaign, in which he again distinguished himself (see p. 987).

ATLANTA CAMPAIGN, MAY–AUGUST, 1864

1864, May 5. Advance into Georgia. Sherman's army group left Chattanooga: Thomas' Army of the Cumberland, 61,000 strong; **James B. McPherson's** Army of the Tennessee, 24,500 men; and **John M. Schofield's** Army of the Ohio, mustering 13,500. In front of Sherman was Johnston's Army of Tennessee, 60,000 strong. In his rear, from the Mississippi to the Appalachians, Confederate cavalrymen **John H. Morgan** and Nathan B. Forrest roamed, hampering communications and be-

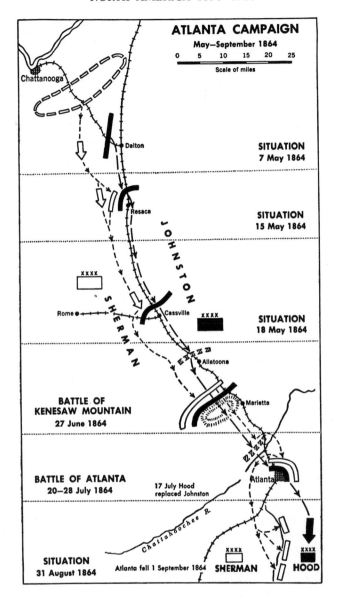

ATLANTA CAMPAIGN

May–September 1864

0   5   10   15   20   25

Scale of miles

Chattanooga

Dalton

**SITUATION
7 May 1864**

Resaca

**SITUATION
15 May 1864**

JOHNSTON

XXXX

SHERMAN

Rome

Cassville

XXXX

**SITUATION
18 May 1864**

Allatoona

**BATTLE OF
KENESAW MOUNTAIN
27 June 1864**

Marietta

**BATTLE OF ATLANTA
20–28 July 1864**

17 July Hood
replaced Johnston

Atlanta

Chattahoochee R.

**SITUATION
31 August 1864**

Atlanta fell 1 September 1864

XXXX

SHERMAN

XXXX

HOOD

deviling Union garrisons. Johnston, outnumbered, skillfully opposed Sherman in a series of delaying positions. Sherman, equally skillful, outmaneuvered him by turning movements at **Dalton** (May 9), **Resaca** (May 15), and **Cassville** (May 19). Each time the maneuver was the same: a holding force in front and a wide

Union turning movement around the Confederate left. Then Sherman drove due south, bypassing Johnston's position at **Allatoona** (May 24). Johnston, retiring to Marietta, placed himself directly in Sherman's path.
1864, June 27. Battle of Kenesaw Mountain.
After a series of indecisive combats near **Dallas**

and **New Hope Church** (May 25–28), Sherman made a frontal assault on Kenesaw Mountain, the key to Johnston's positions. The attacks were repulsed at all points, Sherman losing some 3,000 men, while Johnston's losses were only 800. Once again Sherman (July 2) turned his opponent's left, and Johnston (July 4) took up a powerful entrenched line north of the Chattahoochie River.

1864, July 9. Crossing of the Chattahoochie. Again Sherman turned the Confederate position. Johnston fell back on Peachtree Creek, just north of Atlanta, and prepared for a counterattack. He was summarily relieved from command (July 17) as an ungrateful administration's reward for a really remarkable delaying campaign against far superior forces. For 2½ months he had, with a minimum of losses, held Sherman to an average advance of 1 mile per day. Impulsive **John B. Hood** succeeded him.

1864, July 20. Battle of Peachtree Creek. Johnston had already foreseen that Sherman's advance on Atlanta, on a 10-mile front, offered possibility for a successful counterstroke. Hood seized the opportunity, falling on Thomas' army. Although surprised, the Union forces were alert and the attack was repulsed. Some 20,000 men on each side were involved. Southern casualties were about 2,500; Union losses were 1,600. The Union advance continued (July 21), forcing Hood to withdraw into the Atlanta defenses. Sherman, hoping to follow his foe into the city, sent his left-flank cavalry division eastward to cut the railway. Hood, however, had recoiled to strike again.

1864, July 22. Battle of Atlanta. William J. Hardee's corps—the elite of Hood's army—together with **Joseph Wheeler**'s cavalry division hit the open left flank of McPherson's army. Surprise was complete, but the veteran Federal troops reformed, despite the death of McPherson in the melee. The assault was repulsed with Confederate losses of some 8,000 men. Federal casualties were 3,722. Sherman—his strength insufficient for a siege—determined to swing entirely around to the westerly side of Atlanta and operate against the railroads. Sending most of his cavalry raiding south (July 27), he started the move next day.

1864, July 28. Battle of Ezra Church. An assault by Hood was repelled—mainly by the Army of the Tennessee—with 4,300 southern casualties against 632 Union losses.

1864, July 28–August 22. Cavalry Raids. Part of Sherman's cavalry—6,000 strong, moving around both sides of Atlanta—failed in its dual mission: cutting of the railroad and liberation of Union prisoners at Andersonville. Major General **George Stoneman** and some 2,000 men were surrounded and captured (August 4). Meanwhile, Sherman's strength built up on Atlanta's western face. Another Union cavalry raid ended (August 22) in failure to cut rail communication with Atlanta.

1864, August 27–31. Fall of Atlanta. Leaving 1 army corps to guard his own communications, Sherman swung his 3 armies forward in a great wheel toward the railroad lines south of the city, driving Wheeler's cavalry before them. Hood sent Major General **W. J. Hardee,** with half of his army, to hold the railroads, but Hardee was thrown back at **Jonesboro** (August 31). Hood's communications line was cut. Destroying ammunition and supply stores, Hood evacuated Atlanta that night, moving east and south. Next morning Sherman's troops marched in.

THE MARCH TO THE SEA

1864, September–October. Maneuvering around Atlanta. Sherman, making Atlanta a military base, found further movement hampered by the need to protect his 400-mile line of communications to Nashville. In addition to the daring and successful depredations of Forrest and Wheeler, Hood moved west and north with his entire army (October 1) to attack these communications, hoping to force Sherman's withdrawal. After a vain chase of Hood through **Allatoona** (October 5) as far as Baylesville, Ala. (October 22), Sherman came to the conclusion that further efforts to get to grips with the elusive Confederates would nullify Grant's giant pincers concept.

1864, November 15–December 8. March from Atlanta. Sherman solved the problem—

with the somewhat reluctant approval of Grant—by sending Thomas' Army of the Cumberland back to Nashville and Chattanooga, while he deliberately abandoned his line of communications and marched eastward from Atlanta toward Savannah with 68,000 veterans. With him were 2,500 wagons and 600 ambulances carrying supplies (mostly ammunition); otherwise, his men lived off the country. With practically no opposition, he cut a 50-mile-wide swath of "scorched earth" to the sea, 300 miles away. He was deliberately making "Georgia howl" as he devastated crops and the war-supporting economy of central Georgia; noncombatants were scrupulously respected, although depradations by the so-called "bummers" (stragglers, deserters, and other outlaws) following in his wake were less scrupulously conducted. He ignored Hood's efforts to distract him by a full-scale invasion of Tennessee (see below). To his front, Beauregard, assisted by Hardee, endeavored to organize effective resistance to protect Savannah and Charleston.

1864, December 9–21. Operations against Savannah. Arriving in eastern Georgia, Sherman discovered that Hardee held fortified Savannah with 15,000 men. Sherman stormed **Fort McAllister** at the mouth of the Ogeechee River, 15 miles from Savannah (December 13). Then, establishing communications with Union naval forces, he began an investment of the city. With his lines of communication about to be cut, Hardee evacuated, and Sherman moved in at once (December 21), presenting the city (in a ship-borne and telegraph message) as "a Christmas gift" to Lincoln.

FRANKLIN AND NASHVILLE CAMPAIGN

1864, November 14. Hood Invades Tennessee. Reinforced by Forrest's cavalry, Hood crossed the Tennessee River and moved rapidly north toward Nashville with 39,000 veteran troops. Thomas, building an extemporized army at Nashville about his own hard core of veterans, did not wish to withdraw garrisons from key points in Tennessee; he sparred for time. Major General John M. Schofield, with

2 corps and Wilson's cavalry division—about 34,000 men—was directed to delay the Southern advance. Schofield, slipping away from Hood's attempts to box him in at **Columbia** (November 26–27), and fighting his way through Confederate enveloping forces in a night battle at **Spring Hill** (November 29), stood in previously prepared defenses at Franklin, 15 miles south of Nashville.

1864, November 30. Battle of Franklin. Hood, impetuous, attacked piecemeal, with but two-thirds of his army. He was thrown back with shocking casualties—6,300 out of some 27,000 men actually engaged. Included in the losses were 12 general officers, including Major General Patrick Cleburne, the Irish-born "Stonewall of the West." This loss of general officers was unequalled in any other battle of the war. Schofield out of 32,000 men, lost 2,300. Having successfully performed his mission, Schofield retired that night to Nashville.

1864, December 15–16. Battle of Nashville. Hood stood before the Nashville defenses from December 2, while Washington fumed at the seeming procrastination of Thomas, whose numerical strength was superior. But methodical Thomas, busy training his largely recruit army—particularly Wilson's new cavalry corps—would not be budged. Finally he struck, shattering Hood's left flank, exposed because Hood had sent Forrest away raiding toward Murfreesboro. Attempting to continue the fight next day, Hood found Thomas enveloping both his flanks. Wilson's cavalry, striking behind the Confederate left, delivered the final blow. Federal losses, out of 49,773 men engaged, were 3,061; Confederate, 5,350 out of 31,000 on the field, but Hood's army had become a fleeing rabble.

COMMENT. *This was the most decisive tactical victory gained by either side in a major engagement in the war.*

Coastal and High Seas Operations, 1864–1865

1864, February 17. Sinking of USS *Housatonic.* Submarine warfare claimed its first victim when CSS *H. L. Hunley,* a real submarine boat, torpedoed and sank the steam sloop *Housatonic* of the Federal squadron blockading Charleston Harbor. The *Hunley* sank with her prey.

1864, June 19. *Kearsarge* vs. *Alabama.* Semmes's Confederate sea raider was sunk by U. S. S. *Kearsarge,* off Cherbourg Harbor, after a 1-hour battle. The *Alabama* had just completed a 75,000-mile cruise, practically around the world, in 23 months. She had sunk 1 Federal warship and captured 62 prizes. While the ships were theoretically of nearly equal strength, the *Alabama* was a tired ship and her gun crews lacked practice. The *Kearsarge,* kept in peak condition by Captain **John A. Winslow,** riddled her opponent. Three men only of the *Kearsarge*'s complement were wounded. Forty-odd of sinking *Alabama*'s crew were casualties; 70 were rescued by the *Kearsarge.* The remaining 42, including Captain Semmes, were rescued by a British yacht and avoided capture.

1864, August 5. Battle of Mobile Bay. Lashed to the rigging of his flagship *Hartford,* Admiral Farragut, with 4 ironclad monitors and 14 wooden warships, ran by the cross fire of Mobile Harbor's heavily gunned defenses. When his leading monitor, *Tecumseh,* was blown up by a mine, stopping the fleet under the guns of **Fort Morgan,** Farragut turned his own ship into the minefield (his "Damn the torpedoes!" has gone down in naval history) to clear the way. The other mines failed to go off and the Union ships passed safely into the bay. There Confederate Admiral **Franklin Buchanan** in the ironclad ram *Tennessee* attempted to take on the entire Union squadron, but was finally pounded into submission. Farragut's victory practically ended blockade running in the Gulf of Mexico.

1864, October 27. Destruction of CSS *Albemarle.* This powerful ironclad ram, constructed up the Roanoke River, had dominated the North Carolina sounds for several months. She took a major part in a successful joint Confederate Army-Navy assault on the Union blockade base at **Plymouth,** N.C., sinking 1 Union gunboat, driving off another, and supporting the land attack (April 18). Later the *Albemarle* dispersed a squadron of 7 Union gunboats (May 5). She was a menace to all Union coastwise operations, since her shallow draft enabled her to evade large warships and she outgunned Union small craft. Lieutenant **William B. Cushing,** U.S.N., in a specially designed launch carrying a powerful torpedo, daringly attacked and sank the *Albemarle* at Plymouth.

1864, December. First Expedition against Fort Fisher. Wilmington, N.C., last remaining Atlantic blockade-runners' haven, had become the overseas supply link of the Army of Northern Virginia. With the *Albemarle* disposed of, Admiral Porter moved against Wilmington's elaborate defenses. Two divisions of Butler's army, under Butler's personal command, took part. An attempt to destroy the principal fortification—Fort Fisher—by exploding a powder ship under its walls failed (December 23).

1864, December 23–25. First Attack on Fort Fisher. Porter's squadron opened a bombardment supposed to cover a landing by Butler's command of 6,500 men, 2,000 of whom went ashore Christmas Day. After a brief demonstration, Butler hurriedly reembarked and sailed back to Fort Monroe. Porter was furious. Grant sacked Butler and provided Porter with able and cooperative Major General **Alfred H. Terry.**

1865, January 13–15. Assault of Fort Fisher. Terry's force, now augmented to 8,000, landed north of the fort (January 13) and prepared to attack in coordination with an intensive naval bombardment. While a navy-marine landing force assaulting the sea face was repulsed, Terry's troops swept over the outer works, covered by naval gunfire. Colonel **William Lamb** and his remaining 2,000 men surrendered after 7 hours of bitter fighting. Union losses were 1,341; Confederate, 500-odd. With Fort Fisher's fall, the last sea gate of the Confederacy closed. Richmond and the Army of Northern Virginia were doomed.

Closing Campaigns, 1865

The year opened with Grant still constricting Lee in the Petersburg-Richmond sector, while Sherman started north through the Carolinas from Savannah.

1865, January 31. Abortive Peace Conference. President Lincoln met Confederate peace commissioners, led by Vice President **Alexander H. Stephens,** on a Union warship in Hampton Roads. Lincoln offered complete amnesty if the South rejoined the Union and abolished slavery. The Confederates insisted on independence. The conference failed.

1865, February 3. Lee General in Chief. With the Union pincer jaws closing around the Confederacy, Jefferson Davis (far too late) appointed Lee to supreme command of the Confederate Army. Lee at once restored Johnston to command scattered southern elements in the Carolinas.

Campaign in the Carolinas

1865, February 17. Burning of Columbia, S.C. The city was almost entirely destroyed the night that Sherman's troops marched in, probably ignited by cotton stores fired by the departing defenders, but the South blamed Sherman. That same day Hardee abandoned Charleston. (His troops, like those of Bragg and Beauregard, were now placed under Johnston's new command.) Sherman, at Fayetteville, learning of Johnston's return, cut loose from the seacoast (March 15) and marched on Goldsboro to break up the Confederate concentration.

1865, February 22. Fall of Wilmington. Grant, maintaining unremitting pressure against the collapsing Confederacy and exploiting Union command of the sea, sent Schofield, with troops from Thomas' army, to land at Fort Fisher to operate against Bragg near Wilmington. Driving the Confederates from the city, Schofield began to advance into central North Carolina to join Sherman.

1865, March 19–20. Battle of Bentonville. Sherman, moving north with his 60,000 men on a broad front, was surprised by Johnston, who had only 27,000, but who hoped to defeat the Union left wing before Sherman could concentrate. Sherman's left was driven in, as planned, but Johnston was unable to smash the Union veterans or to fully exploit the success. Sherman concentrated early the next day, and Johnston withdrew. Union casualties were 1,646, Confederate losses probably twice as great. Sherman, continuing the advance, reached Goldsboro (March 23), where he was joined by Schofield. Sherman, whose troops had marched and fought for 425 miles since leaving Savannah, decided to rest for 3 weeks, planning to join Grant near Petersburg in mid-April after muddy roads had dried somewhat.

Selma Campaign

1865, March 18. Advance into Alabama. Wilson's cavalry corps—13,500 strong—was sent by Thomas into Alabama to destroy the important Confederate supply depot at Selma. Crossing the Tennessee River at Eastport, Ala., Wilson's swift-moving expedition brushed Forrest's 3,000 men aside from 3 delaying positions. Forrest fell back on Selma, a fortified city ringed by bastioned earthworks. Here he stood, reinforced to a strength of 7,000 by local militia.

1865, April 2. Battle of Selma. Wilson's cavalry assaulted the works. Fighting on foot, they clambered over the earthworks, gained a toehold, and won the town. Forrest, with a few of his men, escaped. Union casualties were some 400; Confederate, including prisoners, were about 4,000. Wilson, after destroying foundries, ammunition dumps, and stores, crossed the Alabama River (April 10). When hostilities ceased, he had already swept through Montgomery into Georgia as far as Macon. His troopers later captured Jefferson Davis (May 10).

PETERSBURG CAMPAIGN (CONCLUSION)

1865, January–March. Grant Increases the Pressure. While supervising the hammer blows his subordinates were dealing around the periphery of the dwindling Confederacy, Grant maintained about 90,000 men in the Armies of the Potomac and of the James around the defenses of Petersburg and Richmond. Lee, who was able to scrape together about 60,000 poorly fed, poorly equipped soldiers, held a front of 37 miles of entrenchments. Their continuing high quality, however, was demonstrated when A. P. Hill threw back still another Union probe on the Southern extreme right flank at **Hatcher's Run** (February 5–7). But it was obvious to Lee that he could not stretch his overextended forces further.

1865, March 25. Battle of Fort Stedman. Lee, in a desperate gamble to break Grant's grip, threw nearly half his mobile force in a surprise assault on the Union right opposite Petersburg. He hoped that this would force Grant to so weaken his left that the Army of Northern Virginia could slip out, west of Petersburg, to join Johnston in North Carolina. Major General **John B. Gordon** conducted the assault. The fort was taken, but Union veterans promptly rallied and regained the position after a desperate 4-hour battle, in which they also captured several Confederate advanced posts. Lee's situation was now worse than ever. Confederate losses were about 4,000, Union losses half as great.

1865, March 26. Arrival of Sheridan. Next day Sheridan—his work in the valley completed—joined Grant with his cavalry corps and an infantry corps. Grant now had about 122,000 fighting men, Lee less than 60,000. Grant decided to "end the matter."

1865, March 29–31. Battle of Dinwiddie Courthouse and White Oak Road. While 2 Union infantry corps struck the Confederate right, Sheridan's cavalry corps rode wide to the west to encircle the flank. Lee's brilliant reaction was to encircle the encirclers. Pickett, with 2 divisions and **Fitzhugh Lee**'s cavalry, marched west outside of Sheridan's orbit to attack his left flank at Dinwiddie Courthouse. At the same time A. P. Hill hit the left flank of the Union infantry on White Oak Road. The Union advance was checked, but Sheridan increased his infantry and cavalry pressure against Pickett, who retired to Five Forks and entrenched.

1865, April 1. Battle of Five Forks. Sheridan, with infantry reinforcements, struck Pickett in front and flank with far superior force; the Southern position collapsed in panic, exposing the entire right of the Confederate line. Grant at once ordered a general assault.

1865, April 2. Assault on Petersburg. While Sheridan swept northwest over and behind the vital Southside R.R., 3 Union infantry spearheads broke-.through Lee's thin lines west of Petersburg. Gallant A. P. Hill was killed trying to rally his men. Only Gordon, at Petersburg itself, was able to hold his positions. Lee rushed Longstreet down from Richmond to bolster Gordon until nightfall. He then commenced an orderly evacuation of the Richmond-Petersburg defenses.

APPOMATTOX CAMPAIGN

1865, April 3. Lee's Withdrawal. The Confederates converged on Amelia Courthouse in abominable weather. Lee's desperate plan was to join Johnston south of Danville, where Jefferson Davis had set up a temporary capital. Grant, blocking the move, marched west—south of and parallel to Lee's movement. Both armies were exhausted, but both responded magnificently to the inspiration of their leaders. Sheridan reached Jeetersville just in time to block Lee's planned movement to the southwest (April 5). Lee, in a night march, continued westward. Sheridan's cavalry harassed the retreat incessantly.

1865, April 6–7. Battle of Sayler's (Saylor's) Creek. Anderson's and Ewell's corps took up a delaying position, hoping to check the pursuit long enough to let the Confederate artillery and trains get safely across the Appomattox River. The halt was fatal. Sheridan's cavalry and horse artillery dashed behind the Confederate position along the creek, while the Union VI Corps pressed frontally. Ewell surrendered his entire corps—barely 4,000 strong—and a third of Anderson's 6,000 was also captured. Lee's strength was now reduced to less than 30,000. Thanks to his foresight, rations were waiting at Farmville for his half-starved men. He resumed the march, pointing for Lynchburg, via Appomattox. Sheridan, realizing Lee's intentions, reached Appomattox as Lee's advanced guard approached Appomattox Courthouse, 2 miles northeast (April 8).

1865, April 9. Battle of Appomattox. Fitzhugh Lee and Gordon, at Lee's orders, attacked the Union cavalry, but the attack was broken off when masses of Union infantry, arriving after an all-night march, deployed for battle. Lee sent a request to Grant for a cease-fire and a conference on terms.

1865, April 9. Lee's Surrender. The leaders met at Appomattox Courthouse, in the home of Wilmer McLean (who had fled from Manassas in 1861 to get out of the path of war). The Army of Northern Virginia, 28,356 strong, was surrendered at 3:45 P.M. To all intents and purposes the war was over.

1865, April 14. Assassination of President Lincoln. Shot by John Wilkes Booth, Lincoln died next day. The outrage curdled Northern sensibilities and interfered with acceptance of Johnston's surrender to Sherman (April 18) until April 26. All other Confederate forces had laid down their arms by May 26.

1865, May 29. Proclamation of Amnesty. President Andrew Johnston's proclamation officially ended the Civil War.

Epilogue

1864–1865. Cruise of the *Shenandoah*. British-built raider CSS. *Shenandoah,* turned over to the Confederate Navy at Funchal, Azores (October 1864), under Commander **James I. Waddell,** ranged the high seas around the Cape of Good Hope to Australia, thence north to the Bering Sea. There Waddell destroyed the U.S. whaling industry, capturing 32 whalers and merchantmen, burning 27, and sending the remainder under cartel, with prisoners, to San Francisco. His last 8 victims were destroyed 7 weeks after Appomattox (June 28). Waddell, learning from a British ship off the California coast (August 2) that the war was over, dismounted his guns and sailed for England via Cape Horn. He arrived in Liverpool and surrendered to British authorities 6 months after the war's end (November 6). The *Shenandoah* was turned over to the U.S.; the crew was offered asylum in England.

1872, August 25. Settlement of the *Alabama* Claims. An international arbitration tribunal (representatives of Italy, Switzerland, Brazil, the U.S., and Great Britain) awarded the U.S. $14.5 million in damages from Great Britain in partial reparation for the depredations of British-built raiders (*Florida, Alabama,* and *Shenandoah*) upon the U.S. merchant marine.

THE UNITED STATES, 1865–1900

Wars with the Indians, 1865–1898

National expansion westward brought about incessant clashes between Indian and white man. Between 1865 and 1898, the Regular Army fought 943 actions in 12 separate campaigns and a number of disconnected bickerings. The official campaign listings are:

1865–1868—Southern Oregon, Idaho, northern California, and Nevada.

1867–1875—Against Comanches, Sioux, and confederated tribes in Kansas, Wyoming, Colorado, Texas, New Mexico, and Indian Territory. (Noteworthy were a series of bitter battles near Fort Phil Kearny in northern Wyoming; Captain **William Fetterman** and 82 other soldiers were overrun and massacred near here by 2,000 Sioux under **Crazy Horse** and **Red Cloud** [December 21, 1866]; Red Cloud and 1,500 warriors were repulsed by Captain **James Powell** and 31 men in the **"Wagon Box Fight"** [August 2, 1867]).

1870–1871—Against the Apaches in Arizona and New Mexico.

1872–1873—The Modoc War. (Major General E. R. S. Canby was treacherously murdered during preliminary negotiations.)

1873—Against the Apaches in Arizona. (Major General **George Crook** gained special distinction and won the admiration and respect of the Indians.)

1876–1877—Against the Northern Cheyennes and Sioux (see below).

1877—Nez Percé War (see p. 993).

1878—Bannock War.

1878–1879—Against the Northern Cheyennes.

1879—Against the Sheep-eaters, Piutes, and Bannocks.

1885–1886—Against the Apaches in Arizona and New Mexico (principal leaders were Apache chieftain **Geronimo** and U.S. Generals Crook and **Nelson A. Miles**).

1890–1891. Against the Sioux in South Dakota (see p. 993).

All this was guerrilla-type, fast-moving, light-marching warfare against a savage, brave, and crafty enemy. The Plains Indians—Sioux, Cheyennes, and Comanches—were as good light cavalry as the world has seen. The Apache preferred to fight on foot, using his horse as transportation. All could cover ground at an amazing rate of march. Three campaigns are worthy of specific note.

Sioux and Northern Cheyenne War, 1876–1877

1876, February–June. Powder River–Big Horn–Yellowstone Campaign. The refusal of the northern Sioux tribes to go onto their assigned reservations led to the most serious Indian uprising since that of Tecumseh (see p. 870). The uprising was led by Chief Crazy Horse of the Oglala Sioux; a leading instigator was the Prairie Sioux chief and medicine man, **Sitting Bull.** The Sioux were joined by the related Cheyennes. Taking the field with a mounted force of 800 men in bitterly cold weather, Brigadier General George Crook made a long, rapid march through the Big Horn Mountains to mount a surprise attack on Crazy Horse's winter hideout at **Slim Buttes** on the Powder River (March 17). The Indian camp was destroyed and the Sioux scattered, when a panicky subordinate ordered a withdrawal. Crazy Horse rallied his braves and renewed the fight. Seriously outnumbered and short of supplies, Crook was forced to retire. Overall command of the campaign was now entrusted to Major General Alfred H. Terry, who ordered 2 other columns to converge with Crook on the Yellowstone River.

1876, June 17. Battle of the Rosebud. Crook again caught up with Crazy Horse, who had now assembled a force of 4,000–6,000 warriors. Against odds of at least 5-1, Crook fought a bitterly contested drawn battle. The Indians, aware of the convergence of Terry's forces, withdrew; Crook also fell back to refit and reorganize his badly mauled command. Terry, out of communication with Crook and unaware of the Rosebud battle, crossed Crazy Horse's trail. He sent Lieutenant Colonel George A. Custer's 7th Cavalry to pursue and to get south of the Indians, to assure that they would be boxed in between the converging columns.

1876, June 25. Battle of the Little Big Horn. Custer impetuously followed the trail and caught up with Crazy Horse, confidently waiting battle. Dividing his small command (about 600 men) into 3 columns, Custer led 1 of these into the center of the Indians. He and the entire column were wiped out (212 officers and men); the remainder of the regiment took refuge in nearby hills and despite heavy losses were able to hold out for 2 days until Terry and the main body caught up. Crazy Horse and his jubilant warriors escaped; Terry's entire campaign had been ruined. For several months, despite intensive efforts, the U.S. forces were unable to regain contact with the elusive Indians.

1876, November 25–26. Battle of Crazy Woman Fork. Crook, discovering a large Indian encampment, sent Colonel **Ranald S. Mackenzie,** with 10 troops of cavalry, to destroy it. Mackenzie, in a surprise night attack in subzero weather, accomplished his mission.

1877, January 8. Battle of Wolf Mountains. Colonel Nelson A. Miles, with some 500 infantrymen and 2 light guns mounted in covered wagons, located Crazy Horse's village on a commanding bluff. The surprise shelling of the height by Miles's guns stampeded the Indians and ended the campaign, Crazy Horse surrendering shortly afterward.

Nez Perce War, 1877

1877, June. Defiance of the Nez Percés. Under their young leader, **Chief Joseph** (Thunder-Rolling-Over-the-Mountain), the highly civilized Nez Percés defied government orders to evacuate their rich homeland in the Wallowa Valley, Ore., coveted by greedy whites. Continuing incidents led to the death of 3 whites; a small Army detachment attempted to herd the Nez Percés into a reservation in Idaho. Joseph defeated the white soldiers in **White Bird Canyon** (June 17), then with his entire tribe of 700 people—only 300 warriors—he marched eastward into Idaho and Montana, looking for a new home.

1877, July–August. Battles of the Clearwater and of the Big Hole Basin. After repulsing a pursuing force of soldiers under Brigadier

General **Oliver O. Howard** on the Clearwater (July 11), Joseph and his people crossed the Bitterroot Mountains into Montana. In this and other minor engagements, the white soldiers discovered to their surprise that the Nez Percés and their leader fought with the skill and discipline of regular soldiers, while retaining the traditional elusiveness, guile, and stoic courage of Indians. In the Big Hole Basin, Joseph and his people were surprised by the attack of a new white force under Colonel **John Gibbon,** coming unexpectedly from the east. Rallying, in an unparalleled example of Indian discipline, Joseph and his men surrounded and besieged Gibbon's force, withdrawing only when Howard and his men arrived (August 8–11).

1877, August–October. March through Yellowstone and Montana. Joseph now tried to escape to Canada. Eluding or fighting his way past and around more Army forces trying to close in on him, Joseph reached the Bear Paw Mountains, in northern Montana, only 30 miles from the Canadian border. Here, on **Eagle Creek,** after an anabasis of nearly 2,000 miles, he was brought to bay by Generals Nelson A. Miles and Howard. After fighting against 10-1 odds for 4 days, Joseph surrendered (October 5).

COMMENT. *Chief Joseph must be ranked among great American military leaders.*

Sioux War in South Dakota, 1890–1891

1890, December 15. Death of Sitting Bull. The aging warrior again stirred up unrest among his people against the white man. Trying to escape capture, he was killed in a skirmish on the Grand River.

1890, December 29. Battle of Wounded Knee. Leadership of the Sioux now devolved on Chief **Big Foot.** His braves resisted efforts of Colonel **James W. Forsyth** on the 7th Cavalry to return them peacefully to the reservation, and were defeated in the ensuing fight, following a botched attempt to disarm them. Frequently mis-termed a "massacre," a more accurate term would be "tragedy." Wounded Knee was a murderous, close combat, with an estimated 146 Indians (62 noncombatants) killed and 51 wounded. The 7th Cavalry suffered 25 killed and 39 wounded. Forsyth was relieved by General Nelson A. Miles for bungling the affair. However, Forsyth was later reinstated by the War Department.

END OF THE INDIAN WARS

1898, October 4–7. Engagement at Leech Lake, Minnesota. This final Indian uprising—a minor disturbance—was suppressed by the U.S. 3rd Infantry.

Spanish-American War, 1898

1898, February 15. Blowing Up of the *Maine*. The American people, already exercised over Spanish cruelty in suppressing a Cuban revolt (see p. 1001), were enraged when the battleship USS *Maine*, moored in Havana Harbor, was blown up and sunk by a mysterious explosion, with the loss of 260 men. A naval court of inquiry asserted the explosion came from outside the battleship's hull (March 21). Spain was blamed.

1898, April 25. American Declaration of War. The U.S. War Department, completely unprepared, increased the Regular Army from 28,000 to 60,000 men, and tried to organize and train these recruits as well as 200,000 volunteers flocking to the colors. The U.S. Navy, ready, immediately went into action.

PACIFIC OPERATIONS

1898, April 27. Dewey Sails for the Philippines. The U.S. Asiatic Squadron—5 cruisers and 2 gunboats—under Commodore **George Dewey,** left China waters and steamed into Manila Bay at night (April 30), ignoring the danger of submarine mines.

1898, May 1. Battle of Manila Bay. Dewey's squadron attacked the Spanish squadron of Admiral **Patricio Montojo**—4 cruisers, 3 gun-

boats, and 3 other decrepit vessels. The Spanish squadron, inferior in armament, was anchored off the fortified naval yard of Cavite, supported by land batteries. Dewey's squadron completely destroyed the Spanish force, with only 8 men wounded; Spanish losses were 381 sailors killed or wounded. After shelling the shore batteries into submission, Dewey took possession of Cavite. Blockading the City of Manila, he then awaited the arrival of sufficient troops to capture it.

1898, June 30. Arrival of the Army. General **Wesley Merritt,** with 10,000 men—part regulars, part volunteers—arrived in Manila Bay from San Francisco and began debarkation at Cavite.

1898, August 13. Manila Capitulates. Merritt's force, supported by Filipino guerrillas under Emilio Aguinaldo, having invested the city, attacked, supported by the fire of Dewey's squadron. After a short, nominal defense, to satisfy Spanish honor, the city surrendered.

ATLANTIC OPERATIONS

1898, April 29. Cervera Sails for Cuba. Admiral **Pascual Cervera,** with the cream of the Spanish fleet—4 modern cruisers and 3 destroyers—left the Cape Verde islands bound for the Caribbean. Rear Admiral **William T. Sampson,** commanding the U.S. Atlantic Fleet, who had instituted a blockade of Havana, was unsuccessful in intercepting Cervera, who gained the safety of fortified **Santiago de Cuba** (May 19). Sampson had 5 battleships, 2 armored cruisers, and several smaller vessels, including the "Flying Squadron" of Commodore **Winfield Scott Schley.**

1898, May–July. Blockade of Santiago. Sampson at once blockaded the harbor. Lieutenant **Richmond P. Hobson** and a crew of 7 volunteers entered the harbor mouth with the collier *Merrimac,* which they sank there in a gallant but unsuccessful attempt to stopper the channel (June 3).

1898, June 14. Expedition to Cuba. The V Corps, 16,888 strong, consisting of 3 extemporized divisions commanded by Major General **William R. Shafter,** left Tampa under naval escort. The hurriedly mounted expedition included most of the available Regular Army troops (15 regiments), plus 3 regiments of volunteers (including the 1st Volunteer Cavalry, "Roosevelt's Rough Riders"). There were few horses for the 6 cavalry regiments—they would have to fight on foot—and there were serious shortages of equipment.

1898, June 22–25. Debarkation near Santiago. This took place at Daiquirí, an open roadstead, utilizing the small boats of the transports and naval craft. There was no opposition. Shafter then moved on Santiago, after a slight skirmish at **Las Guásimas** (June 23). Shafter and Sampson were unable to agree on coordinated action against Santiago, so Shafter decided to attack without naval assistance.

1898, July 1. Battle of San Juan and El Caney. Although there were some 200,000 Spanish troops on the island, less than 35,000 were in the Santiago area, and the actual garrison of the fortified city was but 13,000. General **Arsenio Linares,** passively defensive, made no attempt to concentrate his forces against the American threat. Barring the road to Santiago was the San Juan Ridge, with Kettle Hill in front of it, an organized position manned by 1,200 defenders. Northeast of the city, on the American right, was the isolated position of El Caney, held by 500 men. Shafter's plan was to assault both positions simultaneously. The troops were slow in maneuvering and tardy in assault, but finally took both works. The colorful charge of the "Rough Riders" up Kettle Hill, led by Lieutenant Colonel **Theodore Roosevelt,** would in particular become an American epic, although the assault was delivered with equal gallantry by all units engaged. By evening the Americans held the ridge and the fate of Santiago was sealed. The total cost of both battles was 1,572 Americans killed and wounded. Estimated Spanish losses were 850. Cuban patriot forces in the area took no part in the action. The entire operation was characterized by poor reconnaissance and lack of control, compensated by the raw gallantry of the troops and

American unit commanders' reckless and inspiring leadership.

1898, July 3. Battle of Santiago Bay. Cervera led his squadron toward the open sea. Overwhelmed by the firepower of the superior American force, the 4 Spanish cruisers and 2 destroyers were forced ashore. The American ships turned to the work of rescue. Spanish losses: 474 killed and wounded and 1,750 prisoners. U.S. losses: 1 killed, 1 wounded. In the subsequent bitter "Sampson-Schley Controversy," Commodore Schley claimed dubious credit for the victory.

1898, July 17. Santiago Capitulates. General **José Toral,** who had succeeded Linares (wounded in the San Juan battle), surrendered, unaware that the American forces surrounding the city were rapidly disintegrating through yellow fever, malarial fever, and dysentery.

1898, July 25. Landing on Puerto Rico. Major General Nelson A. Miles, with some 5,000 men, landed and, in a well-planned, well-executed operation, had almost eliminated Spanish forces when hostilities ended (August 13).

1898, December 10. Treaty of Paris. Spain relinquished sovereignty over Cuba, ceded Puerto Rico and Guam to the U.S., and sold the Philippine Islands to the U.S. for $20 million.

1899–1902. Philippine Insurrection. (See pp. 950, 1106.)

CANADA

1866, June 18. Battle of Fort Erie. Fenian raiders into Canada from Buffalo, N.Y., were arrested by U.S. troops but released due to political pressure.

1867, July 1. Federal Union. Canada became fully self-governing, the prototype of later British dominions.

1870, May 25–27. Fenian Raids Fail. U.S. and Canadian troops halted attempts to invade Canada near Frankfort, Vt., and Malone, N.H.

1869–1870. First Riel Rebellion. Louis Riel, French Canadian with Indian blood, led a group, mostly half-breeds, in armed revolt

(October) when the Hudson's Bay Company turned over what is now Manitoba to Canada. They seized **Fort Garry** (Winnipeg; November), and set up a provisional government with Riel as President (December 29). There were occasional clashes with settlers of English descent in the region. An expedition under Colonel Garnet Wolseley captured Fort Garry without opposition (August 24, 1870); Riel had fled. No effort was made to arrest him, and he later became a member of Parliament.

1885, March–May. Second Riel Rebellion.

Upon appeal of the frontier half-breeds (who had now moved to the region of Saskatchewan), Riel came to their aid in western Canada. Again he denounced the authority of the dominion, appealed to the Indians to join his revolt, and set up a provisional government. After several armed clashes and considerable loss of life, the rebellion was suppressed by government troops under General **Frederick Middleton** in a severe action near **Batoche,** Saskatchewan (May 12); Riel was captured (May 15). He was tried and hanged for treason.

LATIN AMERICA

Instability and unrest marked the growth of Latin America during the period, from the Rio Grande down to the tip of Cape Horn. Of the interminable succession of wars, insurrections, and clashes with foreign powers which racked the area, a few are worthy of note.

MEXICO

1857–1860. Civil War in Mexico. Struggle between the Conservative party, under **Félix Zuloaga,** and the Liberal party, under **Benito Juárez.** The Conservatives held the seat of government at Mexico City; the Liberals established a government at Veracruz. Juárez was recognized by the U.S. (April 6, 1859). The Conservative party was defeated at the **Battle of Calpulalpam** and Juárez assumed full control of the country (December 22, 1860). Mexico's finances had been disorganized by the war, and payments of foreign debts were suspended.

1861, December 17. Allied Occupation of Veracruz. To protect their interests and to obtain satisfaction for their debts, Spain, France, and Britain sent a joint expeditionary force to Mexico. The foreign troops moved to Orizaba. Napoleon III, desirous of establishing a puppet Mexican empire, seized the opportunity presented by U.S. involvement in the Civil War to meddle in Mexican politics.

1862, April 8. Spanish and British Withdrawal. This resulted from Napoleon's intrigues and was followed by further French reinforcements.

1862, May 5. Battle of Puebla (or Cinco de Mayo). The French advance toward Mexico City was checked by Mexican Generals **Ignacio Zaragoza** and **Porfirio Díaz.**

1862, September. Reinforcement of the French Expedition. An additional 30,000 French troops, under General **Elie F. Forey,** arrived in Mexico.

1863, February 17. French Offensive. Advancing from Orizaba, the French again moved on Puebla, defended by Zaragoza. Investing Puebla, the French found themselves involved in incessant guerrilla operations on their line of communications from Veracruz.

1863, April 30. Defense of Camerone. Heroic defense of a homestead by a company of the Foreign Legion. For 10 hours, 3 officers and 62 men fought off a Mexican force of some 2,000 men. In contrast to the Alamo massacre (see p. 882), the Mexicans hailed the last 3 surviving legionnaires as heroes and returned them to French control.

1863, June 12. Enthronement of Maximilian. Having captured Puebla (May 17),

Napoleon's troops entered Mexico City (June 7). He placed Archduke **Maximilian,** brother of Austrian Emperor Francis Joseph, on the Mexican throne. The puppet emperor, an honest man, entered on his duties in all sincerity, but the majority of the Mexican people, led by Juárez, wanted none of him.

1863–1865. Continuing Guerrilla War. This was conducted with savagery on both sides. The U.S. refused to recognize Maximilian, and continued to recognize Juárez, who had taken refuge in the United States, and then returned to Mexico to lead the war with his principal military officer, General **Mariano Escobedo.**

1865–1866. U.S. Demands French Withdrawal. The U.S. threatened to intervene, backing its demand by mobilizing a 50,000-man veteran army under General Sheridan on the Rio Grande. After prolonged negotiations, Napoleon, realizing U.S. determination and impatience, backed down.

1867, February 5. French Withdrawal Begins. Marshal Bazaine, who had superseded Forey, evacuated all French troops, and Maximilian's puppet regime began to crumble.

1867, March–June. Defeat of Maximilian. He refused to abdicate and desert his Mexican followers. Besieged at **Querétaro** by Escobedo he was betrayed by one of his own men, captured (May 14), courtmartialed, and executed (June 19). Juárez restored order and reassumed his post as President until his death (1872).

1871–1877. Turmoil in Mexico. A revolt of Díaz against Juárez was almost suppressed when Juárez died. Disorder continued until Díaz, having defeated his principal opponent, **Sebastian Lerdo de Tejada,** at **Tecovac** (November 16, 1876), was elected President (May 12, 1877).

1877–1911. Dictatorship of Díaz. Mexico's "strong man" ruled a prosperous Mexico, maintaining law and order (see p. 1107).

CENTRAL AMERICA

1855–1860. William Walker in Nicaragua and Honduras. Walker, a U.S. soldier of for- tune from Nashville, Tenn., with 56 followers, went to Nicaragua on invitation of one of several local factions (May 4). Seizing a U.S. steamer on Lake Nicaragua, he took control of Granada in a surprise move, proclaimed himself president, and opened Nicaragua to slavery. His regime was recognized by President Franklin Pierce (May 1856). **Cornelius Vanderbilt,** who wanted to monopolize travel to California and its gold across Nicaragua, incited opposition. Walker finally surrendered to U.S. naval control (May 1, 1857). Taken to San Francisco, he soon returned (November 25, 1857) and was again deported. Entering Honduras (August 1860) in another bid for power, he was arrested by British naval authorities and turned over to Honduran control. He was tried and executed by a firing squad (September 12, 1860).

COMMENT. *The Walker episode was one of the principal instances of meddling by U.S. citizens in Latin American affairs, prime cause for continued resentment against ''Yankee'' aggression.*

1871. Revolution in Guatemala. Miguel Garcia Granados and **Justo Rufino Barrios** overthrew the government of General **Vicente Cerna.** Granados soon retired, leaving Barrios as President.

1885. Barrios Attempts to Unify Central America. After several peaceful efforts to bring about a Central American union, Barrios tried force. Conflict rose between Guatemala and Honduras on the one hand and Costa Rica, Nicaragua, and El Salvador on the other. Barrios was killed at **Chalchuapa** (April 2) during an invasion of Salvador, and his forces defeated.

SOUTH AMERICA

1843–1852. Argentine Intervention in Uruguay. (See p. 896.)

1860–1861. Civil War in Colombia (New Grenada). Victorious rebel leader **Tomás Cipriano de Mosquera** proclaimed himself president (July 1861).

1863. Ecuador-Colombia War. Long-standing

border disputes between Colombia then New Grenada) and Ecuador culminated in war after President Cipriano de Mosquera supported Ecuadorans rebelling against conservative dictator **Gabriel Garcia Moreno.** Elderly General Juan José Flores invaded New Grenada but was repulsed (December 6). Fighting ended; a treaty composed differences.

War of the Triple Alliance (Lopez War), 1864–1870

Francisco Solano Lopez, having become perpetual dictator of Paraguay (1862), turned his country into an isolated despotism. He then plunged into war against Argentina, the Banda Oriental (Uruguay), and Brazil. The war offers no military lesson, but does demonstrate the remarkable bravery of the Paraguayan people. In the end, they were conquered by sheer pressure of numbers and by economic strangulation. Lopez' megalomania reduced the Paraguayan population from about 1,400,000 to some 221,000 (29,000 adult males, 106,000 women, and 86,000 children). His allied opponents lost an estimated 1,000,000 men. The combat area was the Paraguay River Valley from Corrientes north to Coimbra; the principal actions occurred between the confluence of the Paraguay and Paraná Rivers and Asunción.

1864, December 26. Invasion of Brazil. Lopez, who had raised an army 80,000 strong, seized the upper Paraguay River and invaded Brazil at Corumbá, in an area disputed by both nations. This move isolated the Matto Grosso region from communication with eastern Brazil, overland access being impossible across the intervening mountains and marshy jungle land.

1865, March 18. War on Argentina Declared. When Argentina refused to permit Paraguayan troops to cross its territory to invade southern Brazil, Lopez declared war. A Paraguayan river expedition, moving in captured Argentinian vessels, took **Corrientes** (April 13), and Colonel **Antonio L. Estigarribia** then moved overland through Encarnación to invade southern Brazil.

1865, May 1. Establishment of the Triple Alliance. Brazil, Uruguay, and Argentina agreed to destroy the Lopez government and to open the Paraguay and Paraná rivers to free navigation. The 3 nations began organizing new armies for joint operations under Argentina's **Bartolomé Mitre,** Uruguay's **Venancio Flores,** and Brazil's **Louis Osorio.** Naval operations were confided to Brazilian Vice Admiral **Viscount Tamandare.**

1865, July–October. Operations in Southern Brazil. Estigarribia's force reached the Uruguay River and moved south to **Uruguayana,** where it was surrounded by an allied force of 30,000 and a gunboat flotilla; part of his force was destroyed. He surrendered (September 18). The allied army moved north, threatening Paraguayan communications at Corrientes, which was evacuated at once. Lopez concentrated 25,000 men below the Paraguayan fortress of Humaita, in the peninsula between the Paraguay and Paraná rivers.

1866, January–May. Allied Invasion. The allied army, now 45,000 strong, crossed the Paraná and advanced slowly. Lopez attacked (May 2), but was thrown back.

1866, May 24. Battle of Paso de Patria. Surrounded on both flanks by superior force, Lopez' army was shattered after an all-day battle, losing 13,000 casualties. Allied losses were 8,000. No attempt was made to exploit this allied success, and Lopez was allowed to reorganize.

1866, September 22. Battle of Curupayty. After long delay and establishment of a base, the allies attempted to drive Lopez out of his entrenchments at Curupayty, on the Paraguay River, but were repulsed with great loss; 9,000 were killed and wounded. Paraguayan losses were but 54 men. Lopez failed to exploit his victory. For 14 months, operations dragged in stalemate; cholera ravaged both sides.

1867, May. Brazilian Invasion in the North. A Brazilian attempt to invade northwestern Paraguay, through the Matto Grosso, was a complete failure.

1867, July–August. Operations around Humaita. Despite Lopez' efforts, Brazilian monitors moved up the Paraguay, running Humaita's batteries. Asunción was abandoned. Lopez instituted a "scorched earth" policy as he withdrew his mobile forces north from Humaita to Angostura, with his right flank on the river, his left on Lake Ypoa. Humaita, now garrisoned by only 3,000 men, surrendered after an attempt to evacuate proved unsuccessful (August 2). The allied army moved slowly up both banks of the Paraguay, the Brazilians on the east, the Argentines on the west, the gunboat and monitor flotilla accompanying. The advance was stopped by the Angostura defenses, south of Asunción.

1867, August–December. Operations around Angostura. North of Angostura, at Ypacarai, Lopez concentrated his last 10,000 troops. Allied pressure slowly increased. Balloon observation pinpointed the Paraguayan batteries. The allied flotilla turned the Angostura position, while a Brazilian column, passed to the west side by boat, marched up through the Chaco and was then retransported to the east, threatening Lopez' rear at Ypacarai. A succession of assaults on the Paraguayan position was repelled at great cost to both sides.

1867, December 25. Battle at Ypacarai. A general assault breached Lopez' position and, except for the batteries in Angostura itself, the defense collapsed. Lopez fled north with a handful of cavalry. Angostura surrendered (December 30), and Asunción was occupied and sacked by the Brazilians next day.

1868–1869. Partisan War. Lopez continued a largely partisan campaign, maintaining himself in northern and eastern Paraguay. He rallied several thousand followers in the Cordillera area of eastern Paraguay, briefly threatening Asunción (1869). By this time a provisional government had been organized, and Argentina had withdrawn her forces. The Brazilian army of occupation, under **Gaston de Bourbon, Comte d'Eu,** instituted a mopping-up campaign, which slowly drove Lopez north to the Aquidaban River.

1870, March 1. Death of Lopez. The dictator was surrounded by a detachment of Brazilian lancers and killed. Peace came finally through U.S. mediation (June 20). Paraguay ceded some 55,000 square miles of territory to Brazil and Argentina.

Peruvian Hostilities with Spain, 1864–1866

Long-standing differences with Peru, whose independence Spain had not recognized, came to climax when a Spanish squadron under Admiral **Pinzon** seized the Chincha Islands, 12 miles off Pisco on the Peruvian coast, in retaliation for maltreatment of Basque immigrants (April 14, 1864). Spain later (January 27, 1865) concluded a treaty with President **Juan Antonio Perez,** but its provisions aroused Peruvian resentment.

1866, January 14. Peruvian Declaration of War. Perez was driven from office and his successor, General **Mariano Ignacio Prado,** declared war. Defensive alliances were concluded with Chile, Bolivia, and Ecuador. Wholesale deportations of Spanish subjects followed.

1866, March 31. Bombardment of Valparaiso by a Spanish squadron under Admiral **Nuñez** caused considerable damage. The Spaniards lost 1 gunboat to a Chilean cutting-out party.

1866, May 2. Bombardment of Callao. Nuñez attacked the port, but its shore defenses, under Prado's direct command, repelled the attackers. Spain now ceased hostilities (May 9). U.S. mediation brought peace (1871). The net result was a hurried naval program instituted by Chile to modernize and improve her sea power.

War of the Pacific, 1879–1884

This was a struggle for control of the guano—and nitrate-producing provinces of Tacna, Arica, and Tarapacá (Peru) and of Atacama (Bolivia). Atacama was Bolivia's only outlet on the coast. Chilean companies engaging in exploitation of nitrate were so heavily taxed by Peru and Bolivia that Chile went to war.

1879, February 14. Chile Opens Hostilities. A Chilean naval expedition seized Antofagasta; the nearby country was quickly occupied. Chilean blockade of coastal ports followed. Captain **Miguel Grau** in the Peruvian turret ironclad *Huascar* having successfully harassed Chilean warships off Iquique, the entire Chilean Navy hunted him down.

1879, October 8. End of the *Huascar*. Five Chilean warships—2 of them ironclads—surrounded Grau off Antofagasta. In a running fight lasting for 1½ hours, the *Huascar* was battered into submission, with her turret disabled, steering gear shot away, and 64 of her 193-man crew casualties. Among the killed were Grau and the 4 other officers who rapidly succeeded to command in the action. Command of the sea now fell to Chile.

1879, November 1. Chilean Invasion of Bolivia and Southern Peru. Landing an expeditionary force at Pisagua, the Chileans pushed through heavy Bolivian-Peruvian opposition at **Sán Francisco** (November 16). At **Tarapacá** (November 27) they suffered a repulse, but the allies made no attempt to exploit their success. With the Bolivian seacoast and all of Peru's Tarapacá Province in their hands, the Chileans turned their attention northward.

1880, January–June. Blockade of Arica and Callao. Peruvian defensive measures included submarine mines and ineffective torpedos. A Chilean force of 12,000 men landed at Pacocha near Arica and moved inland. At **Questa de Los Angeles** (March 22), Bolivian General **Narciso Campero** with 9,000 allied troops defended a defile until outflanked by Chilean General **Manuel Baquedano**. Allied casualties were 530; Chilean, 687.

1880, May 26. Battle of Tacna. Initial Chilean piecemeal assaults were at first repulsed, but Baquedano finally concentrated his effort against the Peruvian left, and swept the field. Casualties were severe: nearly 3,000 allied and 2,000 Chilean. With allied opposition in this area completely dissolved, the Chilean army moved on **Arica,** taking it in a general assault (June 7).

1880, November–December. Lima Campaign. Disembarking at Pisco (November 18) with more than 22,000 men, Baquedano 6 weeks later began moving up the coast on Lima, where some 22,000 defenders were installed behind 2 successive lines whose flanks could not be turned. In a frontal attack, the Chileans overran **Chorrillos,** the first position, taking losses of 3,000 men. Two days later, while negotiations for peace were in progress, an inadvertent Chilean movement resulted in a daring Peruvian counterattack. The action, hotly contested for 4 hours, ended with complete Chilean victory. Total casualties in the 2 battles: Chilean, 5,443; Peruvian, 9,000. Lima was occupied (December 17). Peruvian resistance having now completely broken down, the Chilean army was withdrawn, leaving only a small occupation force for mopping up.

1883, October 20. Treaty of Ancón. Peru formally ceded Tarapacá Province to Chile and agreed to Chilean occupation of Tacna and Arica for a 10-year period, to be followed by a plebiscite. (Tacna, as it turned out, would be returned to Peru in 1929; see p. 1149.) The **Treaty of Valparaiso** (April 4, 1884) recognized Chile's permanent possession of the Bolivian littoral.

1889, November 15. Brazilian Empire Ended. Led by General **Manuel Deodoro da Fonséca,** a military junta quietly removed Emperor Dom **Pedro II** from his palace and shipped him and his family off to Portugal. A provisional government was quickly followed by establishment of the United States of Brazil. Unrest continued, however, culminating in an insurrection in the province of Rio Grande do Sul (September 1891).

1893–1895, September 6. Naval Revolt at Rio. Admiral **Custodio de Mello,** commanding Brazilian naval vessels in the harbor of Rio de Janeiro, demanded the resignation of President **Floriano Peixoto.** On refusal, Mello shelled the city. Mediation by foreign ministers prevented further destruction. Mello joined forces with the insurgents in the south, led by **Gumercindo Saraiva.** The government purchased warships abroad and mobilized the national guard. Admiral **Saldanha de Gama,** now commanding the rebellious squadron in Rio Harbor, abandoned his ships to take refuge with his men on board two Portuguese warships, which transported them to Montevideo. Mello and Saraiva fell out. The rebel army was dispersed. Mello, after an unsuccessful attack on the town of Rio Grande do Sul, steamed to Buenos Aires and surrendered to Argentine authorities. Government troops put down the last traces of revolt in the south (July 1895), de Gama—who had returned from Montevideo— being killed.

1895–1896. Venezuelan-British Boundary Dispute. Disagreements about the boundary between British Guiana and Venezuela led to a crisis between the U.S. and Great Britain (December 1895). This was settled by negotiations; an international board of arbitration agreed generally with the British claims.

WEST INDIES

Cuba

1849–1851. Lopez Expeditions. Narciso I. **Lopez,** a Spanish general and leader of Cuban refugees in the U.S., mounted 3 abortive revolutionary expeditions from the U.S. The first collapsed (August 11, 1849) when U.S. federal authorities intervened. The second, which included a number of Southern volunteers, made a landing at Cárdenas (May 19, 1850), but was driven off. Lopez' third attempt, mounted from New Orleans—again with a number of American volunteers—effected a landing near Havana (August 11–21, 1850), but when no sympathetic Cuban uprising materialized, the filibusters were all captured. Lopez and 51 Americans were tried by court-martial and executed at Havana.

1868–1878. Ten Years' War. Cuban revolts attracted American sympathy and support. The former Confederate blockade runner *Virginius,* sold to a group of Cuban supporters, was captured by the Spanish warship *Tornado* off Morant Bay, Jamaica (October 1, 1873).

1895–1898. Cuban Insurrection. Discontent with the superficial reformation of Spanish colonial policy came to a head when constitutional guarantees were suspended (February 23). Open revolt brought stringent but unsuccessful retaliation; compartmentation of the island by *trochas*—lines of barbed wire, entrenchments, and blockhouses—isolating the insurgents. Finally Spanish Captain General **Valeriano Weyler** established concentration camps in which noncombatants, swept from their homes, were confined under abominable conditions. In the U.S., where sympathy with the rebels ran high, popular indignation was followed by a formal request for Weyler's removal. Spain acquiesced (October 1897), but rising tension between the U.S. and Spain was climaxed by the still-mysterious blowing up of the battleship USS *Maine* in Havana Harbor (February 15, 1898; see p. 994).

1898, April 25. Spanish-American War. (See p. 994.)

1898, December 10. Treaty of Paris. Spain relinquished Cuba to the U.S. in trust for its inhabitants. Military government under General **Leonard Wood** began grooming Cuba for permanent independence.

Haiti and the Dominican Republic

Revolutionary ferment continued throughout the entire island. Haiti's successive revolutions—in 1867, 1870, 1874, 1876, and 1888–1889—ended in partial restoration of law and order. In the Dominican Republic, estranged from its western neighbor, fear of Haiti resulted in another Spanish annexation (1861), requested by the people. However, revolution soon broke out (1864) as the result of harsh rule, and Spain then relinquished all claim (1865). A Dominican effort to obtain annexation to the U.S. was rejected by the U.S. Senate (1868–1870). For the remainder of the period, the country was relatively tranquil.

AUSTRALASIA

NEW ZEALAND

1860–1870. Second Maori War. The Maori tribes—sturdy, belligerent, and freedom-loving—rose again against continued encroachments by British settlers on their preserves. The result was a 10-year guerrilla war, reminiscent in principle of the U.S. struggle with the Plains Indians. Both regular British troops and local militia were involved in this wilderness war. In the end it was quelled more through diplomacy than by force of arms. As a result, the Maoris became a respected element of an integrated nation.

XIX

WORLD WAR I AND THE ERA OF TOTAL WAR:

1900–1925

MILITARY TRENDS

GENERAL

The period was one of amazing transitions and sharp contrasts. The Russo-Japanese War (see p. 1008) was typical limited war—fought with armies of unprecedented size. World War I, with even larger armies, was total war. The appearance of the gasoline internal-combusiton engine, combined with the sharp upward curve of destructiveness of improved weaponry, brought about a quickening of tempo and enlargement of scope that strained man's physical and intellectual capability to wage war.

LEADERSHIP

The quality of leadership developed during the period will always be moot. On the civilian side, France's premier, **Georges Clemenceau**—"the Tiger"—was the embodiment of ferocity and indomitable patriotism, capable of rallying an entire nation in its fight for existence. Outstanding among the military were France's **Joffre** and **Foch;** America's **Pershing;** Britain's **Haig** and **Allenby;** Germany's **Hindenburg, Lundendorff,** and **Falkenhayn;** Russia's **Nicholas** and **Brusilov;** Turkey's **Kemal;** Japan's **Oyama** and **Togo;** and Poland's **Pilsudski.** The administrative and military capability of Soviet Russia's **Leon Trotsky** slowly forged a capable Red Army from a mass of ignorant peasants and disgruntled soldiery. In general terms, the professional standards of military leadership of the German Army were unquestionably the highest in the world in this period. French, British, and American standards, however, were not far behind.

By the time World War I began, all of the major powers had adopted a general-staff system more or less along lines of the concept used so successfully by Prussia and Germany under von

Moltke (see p. 898). Both Britain and America had long resisted the introduction of a general staff system, since to many people in the Anglo-Saxon democracies this appeared to be a step toward militarism. But embarrassing evidences of military inefficiency appeared in both countries in relatively minor wars at the turn of the century. In Britain after the Boer War, the Esher Committee and Lord Haldane were able to convince most of their countrymen that a general staff was a military necessity that would not jeopardize the fundamental liberties of a free people (see p. 1087). In the United States, at about the same time, Secretary of War Elihu Root achieved a similar result, permitting action to redress the inadequacies so clearly demonstrated in the Spanish-American War (see p. 994).

STRATEGY

The Napoleonic concept of the nation in arms was replaced by that of the nation at war. The ability to the nation to produce and supply its fighting forces with weapons and food became more important than mere man power in uniform. This was demonstrated in World War I by Russia, whose millions of cannon fodder were relatively ineffectual when they could not be armed and maintained on the battlefield. Britain's surface blockade of the Central Powers—and Germany's submarine campaign against the maritime pipelines of Allied supply—equaled, if they did not in fact exceed, the importance of purely military might, since they were intended to strangle the respective populations on whom the armies and navies depended for food and munitions.

This meant that political and economic considerations inevitably dictated military decisions; no longer could war be reserved to the military. Success in total war depended as much on the farmer and the factory worker as on the warrior. That lesson, clear in the American Civil War, was relearned.

As exemplified by the stalemate on the Western Front, strategy in the field was greatly inhibited by weapon power. It would not be correct to conclude from this, however, that the deadlock of trench warfare meant either a decline in competence of military leadership or a decay in the broad strategic concept of maneuver, using both mobility and firepower to effect a decision.

The bloody stalemate on the Western Front was caused as much by territorial limitations as by the tactical effect of improved firepower. The war started in the West with maneuver (the Schlieffen Plan), and its first decisive clash was won by maneuver (Joffre at the Marne). Only when the maneuvering opponents came to the end of the available land mass (after the "Race to the Sea") did the struggle turn into a stalemate, with frontal attack and penetration the only possible solutions.

Through costly trial and error the tactical penetration evolved into a strategic maneuver, employed by both sides, yet never decisively. To be successful against defense in depth required not only the initial power to break through the defense but also the supplemental power, through the use of reserves and continuing firepower support, to erode the flanks of the gap— much as flood waters enlarge the original break in dam or levee—then to sweep with undiminished ferocity over the countryside beyond. Yet despite several instances of momentary breakthrough, the theoretical analogy to the flood was never fully realized on the Western Front. With existing means of transportation, neither side was able to solve the problems of combat movement, mobile firepower support, or logistical support in order to carry tactical exploitation to its logical, decisive conclusion before the defender could shift reserves to close the gap. On other major fronts there were comparable stalemates; but because of vast distances, tactical breakthrough and strategic maneuver were more feasible on the Eastern Front.

WEAPONS, DOCTRINE, AND TACTICS ON LAND

Military men had all noted the increasing deadlines of weapons during the latter half of the 19th century. With few exceptions, however, the leading military thinkers of Europe paid inadequate attention to the lessons of the American Civil War (and failed to note that these lessons were corroborated by the Russo-Japanese War). Their attention was focused on European experience, which they generally misinterpreted, due in part to the brevity of European wars during that half-century.

At the outset of the 20th century, the tactical doctrines of the two leading military nations of Europe—Germany and France—stressed the importance of seizing and maintaining the initiative in battle. Their example was followed by most other armies. The *élan,* or offensive spirit, of troops was to be nurtured. The greatest exponent of this doctrine of the offensive was General **Ferdinand Foch,** Commandant of the French École Supérieure de la Guerre, who was convinced that an indomitable will to win was the major ingredient for victory, and that enhanced fire-power was as advantageous to attacker as to defender. Foch also insisted upon the importance of high professional standards of training in employment of weapons, use of cover, security, and tactical maneuver. His disciples, however—led by Colonel **Louis Loizeau de Grandmaison**—perverted this concept of the importance of the offensive into a blind doctrine of attack at all costs, at all times, and under any circumstances—*l'offensive à l'outrance.* German doctrine, though also stressing the importance of offensive action, never reached such extremes as existed in France in the years just before World War I.

A few soldiers had some reservations about a doctrine which envisaged mass frontal attacks against modern firepower. American observers of the Russo-Japanese War noted, for instance, that neither the fanatically aggressive Japanese nor the stubbornly stolid Russian could be persuaded to repeat suicidal frontal attacks against field fortifications. As in the American Civil War, a few costly lessons forced commanders to resort to tactical maneuver in seeking a decision.

But most military men ignored the doubts expressed by a civilian—Warsaw banker and economist **Ivan S. Bloch**—in a 7-volume book, *The Future of War in Its Economic and Political Relations: Is War Now Impossible?* This was published in St. Petersburg in 1898, and began to circulate in the West in the early years of the 20th century. After study of the best works on military affairs, the author came to the conclusion that the increased power of modern firearms had made war impossible, "except at the price of suicide." Bloch overestimated the power of the weapons of his day. But he was probably closer to the mark than military men of the Grandmaison ilk.

The deadlines of modern weapons caused vital changes in tactics during World War I—though, as usual, the lessons were not fully digested until after the war was ended. The machine gun and the modern artillery piece, in combination with field fortifications and barbed wire, inhibited frontal attack and ended forever the shock value of horse cavalry. Organization in depth became the *sine qua non* of the defense. Fire and movement—as in the American Civil War—were the fundamental basis of the Hutier tactics (see p. 1064), which finally broke the trench stalemate. The infantry-artillery team became the basic tactical element of land warfare in the first half of the 20th century.

The tank—born during the war—emerged as the most important new development in land warfare. Poison gas was another innovation, although prompt countermeasures reduced gas to a weapon of harassment.

Motor transport became of great importance, although horse transport still predominated. In World War I, psychological warfare was for the first time used systematically, although, by present standards, amateurishly. Propaganda campaigns stimulated efforts on the respective home fronts, and subversive leaflets, dropped over enemy lines, were intended to weaken the soldiers' morale. The German long-range gun which bombarded Paris (see p. 1071) was a psychological weapon.

War became three-dimensional as the airplane and the dirigible developed from their initial role of reconnaissance into lethal weapons. By the end of the period the air arm had become a major factor in both land and sea combat.

French light tank, 1918

WEAPONS AND TACTICS AT SEA

Britain's *Dreadnought* (see p. 1087), (prototype of a class of heavy warships ever afterwards called "dreadnoughts"), combining heavy armor, immense weight of metal hurled by a homogeneous group of large caliber guns, 12-inch or more, and all-around fire capabilities, rendered obsolete all previously constructed battleships. In consequence, the great powers—including England herself—were forced into an armament race. Naval designers, searching to combine firepower and mobility, then brought out the battle cruiser, gunned to near equality with the dreadnought but lacking—a sacrifice to achieve speed—its protective armor belt. But the battle cruiser, like most naval hybrids, proved its inefficiency at Jutland, and by the end of the period was disappearing from the world's navies.

H.M.S. *Queen Elizabeth,* superdreadnought battleship

The most significant development in naval warfare, however, was the introduction of the submarine as a weapon of blockade and counterblockade. Although British supremacy on the surface of the waves was only challenged once—and was never seriously jeopardized—the German U-boat offensive against merchant shipping came close to bringing Britain to her knees in 1917. Sinkings by submarines at one point actually threatened to starve England into

German submarine

surrender. The introduction of the convoy system, however, permitted the British to ride out the crisis.

THE CONCEPT OF DISARMAMENT

The increasing cost, pervasiveness, and frightfulness of war caused many to consider seriously and hopefully man's age-old dreams of creating an eternally peaceful world. It soon became obvious, however, that the urge for peace was less compelling on nations and statesmen than the prior claims of assuring national security. The **First Hague Peace Conference** (May–July 1899) had been called by the Czar of Russia, whose interest in disarmament, it soon became clear, was mainly to save economically backward Russia from matching the military expenditures of Germany and Austria. International jealousies and suspicions prevented this, or its successor, the **Second Hague Peace Conference** (June–October 1907), from doing little more than defining and codifying some of the laws of war and establishing an international court of arbitration. Additional rules, pertaining to naval warfare, were elaborated at the **London Naval Conference** (1908–1909), but what the convention agreed to at the conference was never ratified by the participating governments.

The fourth of Wilson's Fourteen Points (see p. 1069) gave hope to a war-weary world (January 1918) that some kind of regulation of armaments would be achieved to establish perpetual peace after World War I. Embodied in the Covenant of the League of Nations, it stimulated the League to a series of disarmament studies in the postwar period. Some regulations on the size and numbers of warships of the leading naval powers were agreed to at the **Washington Conference** (1921–1922).

Realistic statesmen were dubious if true control of armaments would ever be possible. But, heeding popular demands for peace, they tried—just so long as their respective nation's security could be guaranteed absolutely. But, somehow, such guarantees could never be reconciled with arms-control measures satisfactory to potential foes.

MAJOR WARS PRIOR TO 1914

RUSSO-JAPANESE WAR, 1904–1905

Background

Between 1900 and 1903, Japan prepared for a limited war in Korea and Manchuria to crush growing Russian power there, to gain revenge for Russian interference after the Sino-Japanese War (see p. 946), and to ensure her own hegemony over Korea. By 1904 Japan was ready to act. Japanese deployment on the mainland was dependent upon command of the sea, hence it was deemed an essential first step to destroy the Russian Far East Fleet and capture its base, Port Arthur, on the tip of Manchuria's Liaotung Peninsula. Since Port Arthur was Russia's only year-round ice-free port on the Pacific coast, its capture would also deprive the Russians of any winter naval base should they send their Baltic Fleet into the Pacific. The second step in the Japanese plan was to destroy Russian land forces in Manchuria, thus inducing Russia to abandon the war. The Japanese well knew the sole Russian supply line was the Trans-Siberian

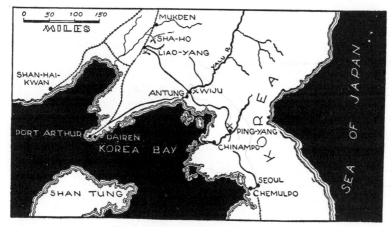

Russo-Japanese War, 1904–1905

Railway—a 5,500-mile single-track line between Moscow and Port Arthur. A 100-mile gap in the line at Lake Baikal complicated Russian logistical difficulties. Despite Russia's tremendous man power (the overall strength of her army was 4,500,000 men), east of Lake Baikal the Russians could dispose immediately only 83,000 field troops with 196 guns, plus some 50,000 garrison troops and railway guards. Given command of the sea, Japan could quickly place on the mainland against this force her entire standing army of 283,000 men and 870 guns, and soon reinforce this with 400,000 trained reserves. Russian naval strength in the Far East consisted of 7 elderly battleships, 9 armored cruisers, 25 destroyers, and some 30-odd smaller craft. The main fleet was based on Port Arthur; 2 cruisers lay at Chemulpo (Inchon), Korea, and 4 more cruisers at Vladivostok. Japanese naval strength consisted of 6 up-to-date battleships, homogeneous in type and carrying 12-inch guns. There were also 1 older battleship, 8 fine armored cruisers, 25 lighter cruisers, 19 destroyers, and 85 torpedo boats, plus 16 smaller craft. The Japanese Army and Navy both were superior to the Russians in doctrine, training, and leadership.

Opening Moves

1904, February 8. The Attack on Port Arthur. Without previous declaration of war, Japanese torpedo boats launched a surprise attack on the Russian fleet at anchor in the harbor of Port Arthur, causing severe damage. At the same time, the main Japanese battle fleet of Vice-Admiral **Heihachiro Togo** appeared off the port to engage the Russian shore batteries and fleet at long range. Togo then instituted a close blockade of the port.

1904, February 9. Naval Action at Chemulpo. The Japanese armored cruiser squadron of Vice-Admiral **Hikonojo Kamimura,** es-

corting the transports of an expeditionary force, entered the harbor of Chemulpo (Inchon), Korea, and attacked 2 Russian cruisers there. One of these was sunk, the other so severely damaged that it was scuttled by its crew.

1904, February 10. Declaration of War.

1904, February 17. Chemulpo Landing. General **Tamesada Kuroki's** Japanese First Army began debarkation, followed by a northward advance through Korea to the Yalu River, to cover operations at Port Arthur.

1904, March 8–April 13. Naval Operations off Port Arthur. Energetic and capable Russian Admiral **Stepan Makarov** arrived from Russia to take command of the fleet (March 8).

At once he began a series of sorties to harass the blockading Japanese cruisers, while avoiding Togo's battle fleet. Returning from one of these sorties, Makarov's flagship, *Petropavlovsk*, struck a Japanese mine and sank with all on board (April 13). Thereafter the Russian ships remained passively in port. The loss of Makarov was a catastrophe for the Russians.

1904, April–May. Russian Dispositions and Plans. General **Alexei Kuropatkin**, who had been Russian minister of war, and who assumed command of all field forces in the Far East, recognized Russian unreadiness for war. He anticipated Japanese efforts to obtain an early victory. He began concentrating all available forces in 3 groups south of Mukden. General **Stakelberg**, with 35,000 men, lay in the area Hai-cheng-Kaiping, directly north of Port Arthur. General Count **Keller**, who had 30,000, guarded the passes west of the Yalu River, with an advance guard of 7,000 under General **Zasulich** covering the Yalu crossings. Kuropatkin himself, with 40,000 men, lay in reserve at Liaoyang. An additional force of nearly 40,000 under General **Anatoli M. Stësel** comprised the garrison of the powerful fortress of Port Arthur. Recognizing the initial Japanese numerical advantage, Kuropatkin planned to permit the Japanese temporarily to besiege Port Arthur—which he felt sure could hold out for several months—while he fell back slowly toward Harbin, delaying the Japanese advance into Manchuria until reinforcements arrived from Russia. He anticipated that the Trans-Siberian Railway could bring him about 40,000 men per month. By the end of the summer, therefore, he felt he would be strong enough to return southward to relieve Port Arthur, and to drive the Japanese from Manchuria. (However, incompetent Admiral **Evgeni Alekseev**, Viceroy of the Far East, appointed as Russian generalissimo by the Czar, insisted upon an immediate offensive, ordering Kuropatkin to abandon his sound defensive-offensive plan).

1904, April 30–May 1. Battle of the Yalu. Kuroki's First Japanese Army, arriving at the Yalu near Wiju (Uiju or Gishu), was confronted by Zasulich, who stupidly gave battle against overwhelming odds. Zasulich was routed, losing 2,500 casualties; Japanese losses were 1,100 out of the 40,000 engaged. Kuroki advanced into Manchuria.

1904, May 5–19. Japanese Landings on the Liaotung Peninsula. The Second Army, under General **Yasukata Oku**, began landing at Pitzuwu, only 40 miles northeast of Port Arthur. Moving south, he was halted by a powerful Russian defensive position based on Nanshan Hill, at the narrowest part of the peninsula. While this was taking place, the Japanese Fourth Army, under General **Michitsura Nozu**, began disembarking at Takushan, west of the Yalu River. As the Japanese net tightened around Port Arthur, Admiral Alekseev fled north to Kuropatkin's headquarters at Liaoyang.

The Siege of Port Arthur, 1904–1905

1904, May 25. Battle of Nanshan. Nanshan Hill, outpost of the Port Arthur defenses, was garrisoned by some 3,000 men. Oku's troops, assaulting frontally, were repulsed. Then the Japanese right wing, wading through the surf, turned the Russian left; the defenders were forced to withdraw hastily. In ferocity the fight was a prototype of future Japanese assaults. Oku's losses were 4,500 men out of 30,000 engaged. Russian losses were 1,500. The loss of Nanshan Hill uncovered the port of Dalny (Dairen), which became a Japanese base. Port Arthur was now ringed both by land and by sea. The Japanese Third Army, under General **Maresuke Nogi** (captor of Port Arthur from the Chinese in 1894), began concentrating at Dalny. Nogi was entrusted with the investment of Port Arthur, while Oku's Second Army turned northward to confront Stakelberg's offensive, reluctantly initiated by Kuropatkin on Alekseev's order (see p. 1012).

1904, June 1–22. The Opposing Forces at Port Arthur. While Nogi's strength built up, Stësel—a most incompetent man—awaited attack in a dither. The fortress complex consisted of 3 main lines: an entrenchment surrounding the old town itself; the so-called Chinese Wall, some 4,000 yards beyond, composed of a ring of permanent concrete forts linked by a network of strong points and entrenchments; and beyond that, outer works consisting of a series of fortified hills—some fully organized, others still incomplete. The garrison, not counting the fleet personnel, amounted to about 40,000 men and 506 guns. The food supply was insufficient for a long siege, but danger of starvation was far from immediate. Outside the fortress, Nogi's strength was gradually increasing. By the end of July, the Third Army had waxed to more than 80,000 men with 474 field and siege guns, an imposing force, but—given a competent opponent—insufficient for a successful assault against such a formidable fortified position.

1904, June 15. Japanese Naval Losses. Accidents and the loss of 2 battleships to Russian mines left Togo with only 4 battleships and a reduced cruiser force.

1904, June 23. Russian Naval Sortie. Admiral **Vilgelm Vitgeft,** Makarov's successor, whose damaged ships had been repaired, made a sortie, causing Togo some uneasiness due to his weakened strength. He prepared to meet the Russians, but Vitgeft evaded action and returned to port.

1904, June 26. Russian Land Sortie. Stësel attempted a sortie, which was quickly checked.

1904, July 3–4, 27–28. Japanese Probings. These led to heavy but inconclusive fighting along the outer ring of forts.

1904, August 7–8. First Japanese Assault. Spurred by the evidence that the Russian fleet was still capable of action, Nogi attacked the eastern hill masses of the outer defense, which were captured after furious fighting.

1904, August 10. Naval Battle of the Yellow Sea. The Czar now ordered Vitgeft to break out and join the Vladivostok squadron, which was still at large, despite the searches by Kamimura's armored-cruiser squadron. Vitgeft steamed out with 6 battleships, 5 cruisers, and 8 destroyers. Togo closed with him by afternoon. Japanese gunnery was far superior to the

Russian, and his 4 modern battleships threw more metal than their Russian counterparts. Both fleets suffered severe damage. After an hour and a half of action, a 12-inch shell struck Vitgeft's flagship *Czarevich,* killing the admiral. A confusion of orders followed and the Russian ships fled in disorder. One cruiser was sunk; several others ran for neutral ports and were interned. Most of the Russian ships got back into Port Arthur.

1904, August 14. Naval Battle of Ulsan. Kamimura's 4 armored cruisers fell on the 3 remaining ships of Admiral **Jessen**'s Vladivostok squadron in the Korea Strait, sinking the cruiser *Rurik.* The other 2 ships got away. Japan now had complete command of the sea.

1904, August 19–24. Second Assault. In close-packed frontal attack, the Japanese struck both the Chinese Wall fortifications on the northeast and 174 Meter Hill on the northwest. Russian machine-gun fire mowed the attackers down again and again. Much of the fighting was at night, but Russian searchlights and rockets sought out the Japanese. Both sides fought with reckless bravery. Nogi called off the attacks after losing more than 15,000 men. He had captured 174 Meter Hill and one of the outlying batteries of the eastern defenses. Otherwise the Russian position was unimpaired. Russian losses were but 3,000 men. Nogi, calling for heavy siege artillery, set himself to systematic sapping and mining of the fortifications.

1904, September 15–30. Third Assault. Nogi, having pushed siegeworks close to outlying hill positions on the north and northwestern faces of the fortress, made another close-massed frontal assault. The northern objectives were carried (September 19) and next day one of the northwestern positions. But 203 Meter Hill, the key point of the entire defense system, resisted all attacks. The assault columns were swept away again and again until the hill slopes were covered with Japanese corpses.

1904, October 1. Arrival of Japanese Siege Artillery. This included 19 28-cm. howitzers throwing 500-lb. projectiles 10,000 yards. Continuous bombardment hammered the

Russian defenses, while sapping and mining operations lapped about 203 Meter Hill. On the eastern face, Nogi prepared for a mass frontal attack.

1904, October 30–November 1. Renewed Assault. Beginning at 9 A.M., the Japanese struck the northern and eastern works simultaneously. Once more the Japanese infantry, in close columns, attempted to claw through a rain of machine-gun, artillery, and hand-grenade fire. Once more they were repelled with tremendous losses. The same slaughter operation was repeated next day. Within the fortress food was running short, while the sick list mounted, but news that the Russian Baltic Fleet had steamed from Libau (October 15) brought some cheer. The same news spurred the besiegers; at all costs the ships in Port Arthur must be destroyed lest the fleets unite and defeat Togo.

1904, November 26. Fifth General Assault. This was repulsed by the Russians at all points, costing 12,000 more Japanese casualties. Nogi now concentrated on the 203 Meter Hill position, which consisted of a huge redoubt, surrounded by barbed wire and flanked on both sides by smaller hills, also fortified. Some 2,200 men, under Colonel **Tretyakov,** manned the complex, which looked down on the harbor, only 4,000 yards away. In Japanese hands it would seal the fate of the Russian fleet.

1904, November 27–December 5. Capture of 203 Meter Hill. Following an all-day bombardment, the assault troops advanced in the dusk, reaching the barbed-wire entanglements, where they managed to hold all the next day, despite Russian fire. Meanwhile, bombardment of the crest continued. Until December 4, in assault after assault, the living attackers came stumbling over the dead in successive waves. Twice Russian counterattacks brushed back detachments that had gained footholds in the fort. The last handful of defenders was overrun after about 11,000 Japanese had died. Next day Japanese artillery fire from the hill began demolishing the Russian fleet in the harbor below. Togo's fleet steamed home to refit, preparatory to meeting the Baltic Fleet.

1905, January 2. Surrender of Port Arthur. Japanese assaults continued against the northern defenses of the fortress, despite freezing weather and heavy snows. The last fort fell on New Year's Day. Stësel surrendered about 10,000 able-bodied but starving survivors of his garrison next day. The Japanese captured a vast quantity of guns, small arms, and foodstuff (a shocking indictment of Stësel's mismanagement). Japanese losses in the siege totaled some 59,000 killed, wounded, and missing, with approximately 34,000 more sick. The Russians had lost 31,000. Nogi prepared to join the other Japanese armies in the north.

Operations in Central Manchuria

1904, June 14–15. Battle of Telissu. Stakelberg's advance toward Port Arthur was halted, and he entrenched, when confronted by Oku's Second Army. Oku attacked. Following a sharp encounter Stakelberg retreated to avoid envelopment. Russian losses were 3,600 out of 25,000 men engaged; Japanese casualties were 1,000 out of 35,000. Oku, advancing, attacked a Russian covering force at **Tashichia** under General **Zarubayev,** who successfully beat off the Japanese in a delaying action (June 24). Oku lost 1,200 men in the fight.

1904, July 17–31. Battle of the Moteinlung River. Keller's defensive group southeast of Liaoyang attacked Kuroki, who was advancing from the Yalu, but was repulsed. Kuroki, attacking in turn (July 31), met sharp opposition.

1904, August 1–25. Russian Retirement. Kuropatkin began pulling back all his advance detachments on Liaoyang, toward which point Kuroki, Nodzu (who had moved up from the coast), and Oku converged.

1904, August 25–September 3. Battle of Liaoyang. Field Marshal **Iwao Oyama,** now commanding all Japanese field forces, gathered his 3 converging armies against Kuropatkin's well-organized positions about Liaoyang. Oyama's strength totaled 125,000 against the 158,000 Russians, who had been reinforced by an army corps from Europe. Kuropatkin took the offensive, but was checked within his own outpost zone. The Russian launched another assault, massing his main effort against Kuroki's First Army, on the Japanese left. Through mismanagement, the stroke was repulsed. As usual, aggressive Japanese tactics overcame preponderance in numbers. The results were indecisive; the Japanese had lost 23,000 men against the Russian 19,000. However, Kuropatkin believed himself defeated and began a systematic, well-managed withdrawal north toward Mukden. Oyama followed, his efforts at pursuit repulsed by effective Russian rear guards.

1904, October 5–17. Battle of the Sha-Ho. Kuropatkin, reinforced to 200,000 men, turned on Oyama's 170,000, concentrating his main effort against Kuroki's First Army, now on the Japanese right. While Kuroki dug in to hold the Russian attack, Oyama threw the weight of his strength violently against the weakened Russian center. The Japanese assault was so severe that Kuropatkin checked his own assault to reestablish his center (October 13). Both sides soon renewed their efforts (October 16–17) without decisive results. Russian losses totaled 40,000 men; the Japanese lost 20,000. Exhausted, both armies dug in.

1905, January 26–27. Battle of Sandepu (Heikoutai). Reinforced to 300,000 men, and with his troops organized in 3 armies—**Linievich, Grippenberg,** and **Kaulbars**—Kuropatkin took the offensive in an effort to crush Oyama's 3 armies, 220,000 strong, before Nogi's Third Army arrived from Port Arthur. Attacking in a heavy snowstorm, the Russians came close to victory. Had Kuropatkin pressed his initial advantage vigorously, the outcome of the war might have been completely different. As it was, after 2 days of most bitter action, Oyama's counterattacks brought a temporary stalemate.

1905, February 21–March 10. Battle of Mukden. The opponents, each entrenched and each about 310,000 strong, faced one another on a 40-mile front. Oyama, seeking an envelopment, attempted to turn the Russian right with Nogi's Third Army. By the end of the

first day's fighting, the Russian right—Kaulbars' army—was forced back until it faced west, instead of south. Attack and counterattack followed in quick succession, Kuropatkin shifting his reserves to backstop his crumbling right. Oyama's envelopment was unsuccessful, though Japanese troops actually entered Mukden during two weeks of violent battle. Bringing up his reserves, Oyama reinforced Nogi, permitting him to again try to envelop Kaulbars. After 3 days of fighting (March 6–8), the Russian right flank had been pushed back so far that Kuropatkin feared for his line of communications. He disengaged in workmanlike manner and fell back on Tieling (Teihling) and Harbin, defeated but not routed. Some 100,000 Russians had fallen; much matériel was abandoned. Japanese losses were 70,000 or more. There was no further concerted action on land.

The Naval Campaign of Tsushima

1904, October 15–1905, May 26. Voyage of the Baltic Fleet. Commanded by incapable Admiral **Zinovy P. Rozhdestvenski,** the Baltic Fleet left its home ports of Revel (Tallin) and Libau (Liepaja). It met its first mishap a few days later in the North Sea, where a false alarm of Japanese torpedo attack brought down a hail of Russian gunfire on a British fishing fleet near Dogger Bank. Several British trawlers were damaged, at least 7 fishermen killed. This incident almost brought war with Britain, and British cruisers trailed the Russian armada until it had passed the Bay of Biscay. Two battleships and 3 cruisers were detached to pass through the Suez Canal, while the main body went around the Cape of Good Hope. Problems of coaling and repairs in neutral ports continually vexed the Russians' creaking progress. Reuniting at Madagascar, after a prolonged delay, the fleet finally started across the Indian Ocean (March 16). One last stop was made at Van Fong Bay in French Indochina, where the Russians prepared for battle. Then heading for Vladivostok, and accompanied by supply ships and colliers, the fleet sailed north (May 14). As it approached Tsushima Strait, Rozhdestvenski's fleet comprised 8 battleships, 8 cruisers, 9 destroyers, and several smaller craft. Although imposing in paper strength, the force was a conglomeration of obsolescent or obsolete vessels, whose personnel was inferior in gunnery, discipline, and leadership to Togo's waiting fleet, which consisted of 4 battleships, 8 cruisers, 21 destroyers, and 60 torpedo boats.

1905, May 27. Battle of Tsushima. Rozhdestvenski entered the strait in line-ahead formation. To the northwest, Togo was steaming in similar formation. Both admirals led their respective main bodies—Rozhdestvenski in *Suvorov*, Togo in *Mikasa*. The Japanese turned to head northeast, hoping by superior speed to cross the Russian "T." Rozhdestvenski altered course to the northeast and then east, to avoid being raked. The action opened in early afternoon at 6,400 yards' range. Togo, at 15-knot speed, overhauled the 9-knot Russians and in less than 2 hours put 2 battleships and a cruiser out of action. The toll mounted as Togo brilliantly maneuvered his faster force around the hapless Russians. By nightfall, Rozhdestvenski had been wounded, 3 battleships (including his flagship) were sunk, and the surviving Russians, now under Admiral **Nebogatov,** were fleeing in confusion. Togo turned loose Kamimura's armored cruisers, the destroyers, and the torpedo boats to harry the exhausted Russians through the night. Next day, the destruction was completed. One cruiser and 2 destroyers escaped, to reach Vladivostok; 3 destroyers got to Manila and internment. The remainder of the Russian fleet was sunk or captured. The Japanese lost 3 torpedo boats. Russian casualties mounted to 10,000 killed and wounded; the Japanese lost less than 1,000 men in all.

Conclusions

1905, September 6. Treaty of Portsmouth, N.H. Both sides were ready to make peace. Japan's limited war objectives had been won,

while Russia, seething with internal discontent, had no stomach for continuing. Through the efforts of President Theodore Roosevelt, peace negotiations led to a treaty. Russia surrendered Port Arthur and one-half of Sakhalin, and evacuated Manchuria. Korea was recognized as being within Japan's sphere of influence.

COMMENT. *Tactically, the war on land made plain the enormous defensive value of the machine gun and the offensive value of indirect artillery fire. Strangely, western observers failed to grasp fully the lesson of the machine gun. The Russian soldier once more proved his stoic courage in adversity, regardless of the incapacity of most of his officers. The Japanese displayed considerable professional skill and fanatical devotion to duty. The Battle of Tsushima—first and last great fleet action of the ironclad predreadnought era—was also the greatest naval battle of annihilation since Trafalgar. It emphasized that both seamanship and gunnery were still essential to victory at sea. Psychologically and politically, Japan's victory in the war marked a turning point in world history. Asia woke to the fact that the European was not always invincible; "white supremacy," as such, became a shibboleth.*

Italo-Turkish War, 1911–1912

1911, September 29. Italy Declares War. Italy's objective was to conquer Libya and gain a foothold in North Africa, so as to counterbalance the French colonial empire in Algeria, Tunisia, and Morocco.

1911, September 29–30. Italian Bombardment of Preveza. During the attack on the Epirus coast town, the Italian fleet sank several Turkish torpedo boats.

1911, October 3–5. Italian Bombardment of Tripoli. During the bombardment, Turkish forces evacuated the Libyan capital; an Italian naval landing force then seized the town (October 5). At the same time another naval force was occupying Tobruk (October 4).

1911, October 11. Italian Expeditionary Force Reaches Tripoli. Italian troops took over from the navy. At the same time other landings were made at Homs, Derna, and Benghazi. Turkish resistance was spotty.

1911–1912. Stalemate in Libya. Turkish propaganda so inflamed the Moslem population that cautious Italian General **Carlo Caneva** confined his activities to the coastal zone.

1912, April 16–19. Italian Naval Demonstration off the Dardanelles. The Turks closed the straits and hastily prepared for an invasion; the Italians withdrew without further action.

1912, May. Seizure of the Dodecanese. Italian naval forces occupied Rhodes and other islands.

1912, July–October. Italian Offensive in Libya. Finally pushing out of their coastal enclaves, the Italians began a systematic expansion of their control in Libya, which culminated in clear-cut victories over outnumbered Turkish forces at **Derna** and **Sidi Bilal** (near Zanzur).

1912, October 15. Treaty of Ouchy. The threatened outbreak of the Balkan War (see below) caused Turkey to seek peace, which was concluded after 2 months of negotiations. Libya and (after further negotiations) Rhodes and the Dodecanese Islands were ceded to Italy. The conduct of the war against negligible opposition had not enhanced Italian military prestige.

THE BALKAN WARS, 1912–1913

The First Balkan War, 1912–1913

OUTBREAK OF THE WAR

1912. Formation of the Balkan League. Bulgaria, Serbia, and Greece, seeking to eliminate Turkish power in the Balkans and to increase their own territorial areas, entered into military alliance in the hope of taking advantage of Turkey's war with Italy. Tiny Montenegro was informally associated. The pretext for war was Turkish misrule in Macedonia. Turkey had about 140,000 troops in Macedonia, Albania, and Epirus; another 100,000 were in Thrace. The allies were rightly confident that Greek command of the Aegean would prevent rapid and direct transfer of other Turkish forces to the Balkans. Bulgar active strength was approximately 180,000, Serb 80,000, and Greek

50,000, and each had about an equal number of trained, readily mobilizable reserves. Montenegrin militia strength, capable only of guerrilla operations, was about 30,000. Courage and stamina of the opposing forces were equal, but Turkish tactical leadership was inferior to that of the allies, despite recent German assistance in reorganization of the Ottoman Army.

1912, October 17–20. The Allied Invasions. Almost simultaneously the allies moved into Turkey's European provinces. Three Bulgar armies under General **Radko Dimitriev** invaded Thrace, moving generally on Adrianople. In Macedonia, General **Radomir Putnik**'s 3 Serbian armies from the north and Crown Prince **Constantine**'s Greek army from the south converged on the Vardar Valley with the intention of compressing hastily grouping Turkish elements between them.

OPERATIONS IN MACEDONIA

1912, October 20–November 5. Greek Advance. While a small force invaded Epirus in the west, Constantine's main army pressed on to the lower Vardar Valley. He defeated the Turks at **Elasson** (October 23). Most of the Turkish force withdrew toward Monastir, but Constantine did not pursue since, contrary to prior agreements, a Bulgarian division (ostensibly aiding the Serbian invasion) was advancing toward Salonika, which was coveted by both Bulgars and Greeks. Constantine headed eastward to try to forestall the Bulgarians. Turkish resistance in unexpected strength at **Venije Vardar** at first held up the Greek advance (November 2–3). At the same time other Turkish units defeated Constantine's flank detachments at **Kastoria** and **Banitsa**. Despite these setbacks, Constantine finally overwhelmed the Turks at **Venije** (November 5) and pressed on to Salonika. The isolated Turks to his northwest withdrew to Yannina (Ioannina).

1912, October 20–November 4. Serbian Advance. The Serbs met and defeated a Turkish covering force at **Kumanovo** (October 24). Turkish resistance stiffened in the **Babuna**

Pass, near Prilep, and checked the Serbs until a threatened double envelopment in the hills forced the Turks to evacuate Skoplje and to retreat on Monastir. There, reinforced to a strength of 40,000, they again gave battle.

1912, November 5. Battle of Monastir. A Serb division impetuously stormed commanding ground to threaten an envelopment of the Turkish left. An Ottoman counterattack, with reinforcements drawn from the center of their line, retook the height, almost annihilating the Serb division. But the Turk center was so weakened that a Serbian frontal attack broke through. Faced with a threatened Greek advance from the south, Turkish resistance collapsed. Nearly 20,000 Turks were killed or captured. The remainder, scattering to the west and south, finally reached the fortress of Yannina, where they were besieged by the Greeks.

1912, November 9. Capture of Salonika. In the face of Greek preparations for an all-out assault, the Turkish garrison of 20,000 surrendered. Constantine occupied the city 1 day before the frustrated Bulgarian division arrived. This incident, and the subsequent dispute over possession of Salonika, worsened relations between the Bulgarians and the Greeks.

1912–1913. Sieges of Yannina and Scutari. By the end of the year, the only Turkish forces still holding out west of the Vardar were the garrisons of Yannina (besieged by the Greeks) and Scutari (Shkodër; besieged by the Montenegrins).

Operations in Thrace

1912, October 22–December 3. Bulgarian Advance. The First, Second, and Third Bulgarian armies advanced on a broad front. Simultaneous meeting engagements with Turkish forces under **Abdalla Pasha** took place at **Seliolu** and **Kirk Kilissa** (October 22–25). The fighting at Kirk Kilissa was particularly severe, but the Turks were defeated in both actions. Falling black, Abdalla regrouped, facing west along the 35-mile-long line from Lüle' Burgas (Lüleburgaz) to Bunar Hisar (Pinahisar), with his right anchored in the mountains. The Bulgarian Second Army, on the right, invested Adrianople, while the other 2 armies wheeled eastward against the Turkish position.

1912, October 28–30. Battle of Lüle' Burgas. A Bulgarian piecemeal attack on the north was repulsed, but as the battle spread along the overextended Turkish front, Abdalla was forced to fall back. The Turks reorganized behind the permanent fortifications of the Chatalja (Çatalca) Line, between the Black Sea and the Sea of Marmora, protecting Constantinople.

1912, November–December. Siege of Constantinople. The Bulgars, launching a premature assault, were driven back with heavy loss (November 17–18). A stalemate continued along the Chatalja Line until an armistice temporarily ended hostilities (December 3). Adrianople remained in Turkish possession. Greece and Montenegro ignored the armistice.

Concluding Operations

1912, December 17–1913, January 13. London Peace Conference. Representatives of the combatants and of the European Great Powers vainly endeavored to settle their conflicting aims with respect to the Balkans and the crumbling Turkish Empire; the conference collapsed.

1913, January 23. Coup d'État at Constantinople. The Turkish government was overthrown by the Young Turk nationalistic group, led by **Enver Bey.** The Young Turks at once denounced the armistice, and hostilities resumed (February 3).

1913, March 3. Fall of Yannina. The Turkish garrison of 30,000 surrendered to Crown Prince Constantine.

1913, March 26. Fall of Adrianople. A combined Bulgar-Serb siege operation ended with a 2-day assault against the eastern face of the fortress, breaching the Turk lines, despite an allied loss of 9,500 men. **Shukri Pasha** surrendered his garrison of 60,000.

1913, April 22. Fall of Scutari. A Serb force had

come to the assistance of the irregular Montenegrin besiegers, but left after a month of continuing disagreements (April 16). The Turks then surrendered to the Montenegrins.

1913, May 30. Treaty of London. The Great Powers finally imposed an uneasy peace on the combatants. Turkey lost all of her European possessions save the tiny Chatalja and Gallipoli peninsulas. Bitter squabbles broke out between Bulgaria, on the one hand, and the Greeks and Serbs, on the other, over the division of conquered Macedonia. Montenegro was forced to abandon Scutari to the newly established state of Albania.

Second Balkan War, 1913

Bulgaria's 5 armies were arranged as follows: the First faced the Serbs between Vidin and Borkovitsa, with the Fifth on its left. The Third lay above Kustendil, the Fourth about Koccani and Radaviste (Radovic). The Second faced the Greeks between Strumitsa (Strumica) and Serres (Serrai). The Serb Second Army was on the old Serb-Bulgar frontier; the First, in the center, at Kumanovo and Kriva Palanka. The Third, on the right of the First, was concentrated along the Bregalnica. The Greeks were assembled between the lower Vardar and the mouth of the Struma.

1913, May 30–June 30. Bulgarian Attacks. The Fourth and Third armies, moving to the Vardar, attacked the Serbs without a declaration of war. The Second Army drove in Greek advance elements. Both Serbs and Greeks were disposed in depth, however, and the Bulgar attack lost its momentum.

1913, July 2. Allied Counterattack. Putnik, responsible for the Serbian defensive success, now seized the initiative. While the Third Serbian Army checked the Bulgars on the upper Bregalnica, the First Army broke through, driving on Kyustendil and pushing the Bulgars back in a northeasterly direction. The Greeks (July 3–4) forced the Second Bulgarian Army back, outflanking the Bulgar left (July 7) and driving them north up the Struma Valley. A counteroffensive by the Third and Fourth Bulgarian armies against the Serbian Third, toward the upper Bregalnica, was soon checked (July 10).

1913, July–August. Rumanian and Turkish Intervention. Rumania declared war against Bulgaria (July 15) and moved her troops, practically unopposed, toward Sofia. At the same time the Turks issued from the Chatalja Line and from Bulair (Bolayr) to reoccupy Adrianople. A Bulgar attempt to regroup and attack the Greeks in the Struma Valley was unsuccessful. Bulgaria sued for peace (July 13) and hostilities ended (**Treaty of Bucharest**, August 10).

COMMENT. *Bulgaria's success in the First Balkan War caused her to underestimate her former allies' military capacity. The end result of the Second Balkan War was to deprive the Bulgars of all gains made in the previous conflict.*

WORLD WAR I

THE BACKGROUND

1914, June 28. Assassination of Archduke Franz Ferdinand. The heir to the Austro-Hungarian throne and his wife were murdered by a Serb terrorist in Bosnian Sarajevo. This toppled the power balance between Europe's 2 armed camps—the Triple Alliance of Germany, Austria-Hungary, and Italy vs. the Triple Entente of France, Russia, and Great Britain. Austria-Hungary, eager to expand in the Balkans and relying on German support, delivered an ultimatum to Serbia (July 23), which was accepted only in part. Serbia mobilized (July 25).

1914, July 28. Austro-Hungarian Declaration of War.

1914, July 28–August 6. The Initial Line-up. Russia ordered mobilization against Austria, whereupon Germany declared war against her (August 1), but (in accordance with her Schlieffen Plan; see p. 1020) began invasion to the west through neutral Luxembourg and Belgium, and declared war against France (August 3). Great Britain declared war on Germany because of invasion of neutral Belgium (August 4), Austria-Hungary against Russia (August 6). Italy temporarily remained neutral, (cynically pitting her desire to be on the winning side against her obligations to the Triple Alliance, which she claimed were void due to Austrian initiation of the war.) Overseas, the United States, stirred by conflicting ancestral urges, became a much-confused spectator.

The Opposing Forces

Except for Britain, the troops of each of the combatants were conscripts, welded by a hard core of career officers and noncoms. Fifty percent of German youth reaching military age were conscripted for 2 years' service; France, with a smaller population, called 80 percent of her available youth for 3 years' service. These men remained in reserve cadres after leaving active service; the German system was more highly organized than was the French.

The normal continental infantry division was approximately 16,000 strong, organized in 2 infantry brigades (4 3-battalion regiments), an artillery brigade of 12 6-gun batteries of light and medium guns, and supporting engineer, signal, medical, and supply units. Usually, 1 cavalry regiment was attached. An army corps consisted of 2 or 3 divisions. Small independent cavalry divisions, normally organized in corps of 2 or more divisions each, were standard in continental armies.

The best-equipped and trained army of 1914 was the German; its preponderance of available medium and heavy artillery was a great advantage. Behind the army was a well-knit industrial organization. Despite an offensive doctrine, units were also trained in defensive tactics.

The French Army, with one vital exception, ranked second to the German in general efficiency. French doctrine of the period was based on complete dedication to the offensive; defensive tactics were almost completely disregarded. The French were unskilled in defensive organization of the ground, field fortification, and defensive application of machine-gun fire.

The French did possess the finest field gun of its time—the famous French 75-mm.—prototype of 20th-century pre-atomic artillery. But, secure in the knowledge that this weapon surpassed anything possessed by their opponents, the French had neglected medium and heavy artillery. Actually, in the beginning there were but some 300 French pieces of larger caliber than 75-mm. to oppose 3,500 German heavier weapons. French battery organization—4 guns against 6—was another handicap.

The Austro-Hungarian Army was patterned on the German, but a poor general staff and the language barrier—75 percent of the officers were of Germanic origin, while only some 25 percent of the enlisted men could understand the language—were handicaps. Greatest bar to efficiency was the poor morale of its many discontented Slavic racial groups, with little or no loyalty to the Hapsburg Crown, many of them sympathizing with Russia.

The Russian Army, strong in docile, hardy, and fatalistically brave man power, suffered from severe shortages in matériel and munitions. Its high command, except for Grand Duke

French 75 and American artillerymen

Nicholas, who was suddenly made commander in chief on August 3, and a very few others, was careless and incompetent, as was its general staff.

In sharp contrast to the continental forces, the British Regular Army was composed of volunteers enlisted for a 7-year period. Under well-qualified career officers, its morale, discipline, and steadiness were high; individual marksmanship and fire discipline were excellent. The British Expeditionary Force (BEF) of 150,000 men—6 infantry and 1 cavalry divisions,

German machine gun

with supporting troops—would play a role in the opening campaign far beyond that indicated by its strength. Behind the regulars was the Territorial Army, a volunteer militia, inadequately trained and equipped.

All these armies entered the war with tactical indoctrination differing little from that of the Franco-Prussian War, but with new and improved weapons. Despite the example of the Russo-Japanese War, the latent power of the machine gun had not been appreciated—2 to each infantry battalion being the general rule. But the thorough Germans had conducted exercises in defense and in the attack and reduction of fortifications. And they had an almost perfect strategical plan, evolved in 1905.

Plans and Preparations

GERMAN PREPARATIONS; THE SCHLIEFFEN PLAN

Germany put into effect a modified version of the plan which had been evolved by General Count **Alfred von Schlieffen,** former chief of the German General Staff (1891–1906), for use in the event of a 2-front war against France and Russia. It was to hold the slowly mobilizing Russians on the east in check with a minimum of force, while the full weight of the German Army would crush France, the more dangerous enemy, in the west.

Schlieffen, who retired in 1906 shortly after his plan was adopted, correctly deduced that upon a declaration of war France would attack immediately; she would not violate the neutral territories of the Low Countries or Switzerland; and French concentration would take place generally between Belfort and Sedan, with the objective of seizing Alsace and German Lorraine.

Schlieffen planned a German feint and then withdrawal in Alsace-Lorraine to further entice the French into a major offensive there. Meanwhile, the bulk of the German field forces, 35 1/2 corps in 5 armies, pivoting on the fortified region of Thionville-Metz, would envelop the entire French Army by a wide sickle movement, which would drive through Belgium and the Netherlands and whose outer tip would pass well to the west of Paris. The French, thus attacked from the rear behind their left flank, would be rolled up to destruction against the German fortified positions in Alsace-Lorraine, or driven into Switzerland and internment.

Meanwhile, in East Prussia, small forces would withdraw slowly in the face of a Russian advance. By the time these had been forced back to the Vistula, Schlieffen expected that France would be defeated. The main German armies would then be transported eastward by Germany's excellent railroad system to crush the Russians in another lightning campaign.

If carried out as conceived, this plan might have ended the war in a few weeks. However, Schlieffen's successor, General **Helmuth von Moltke** (nephew of the hero of the Franco-Prussian War), began tinkering with it. His first modification was to refrain from violating Dutch neutrality. Thus the 2 northernmost (and strongest) of the 5 enveloping German armies would be crowded and slowed up through the Liége bottleneck in Belgium. Then Moltke, reluctant to surrender any German territory, limited the proposed strategic withdrawal of the left-wing armies before the French advance in Alsace-Lorraine. For similar reasons, he planned diversion of additional forces to East Prussia, which was to be defended near the frontiers, instead of a slow withdrawal to the Vistula in the delaying campaign envisaged by Schlieffen. As a result, the revised plan on which the Germans operated put only 60 percent of German

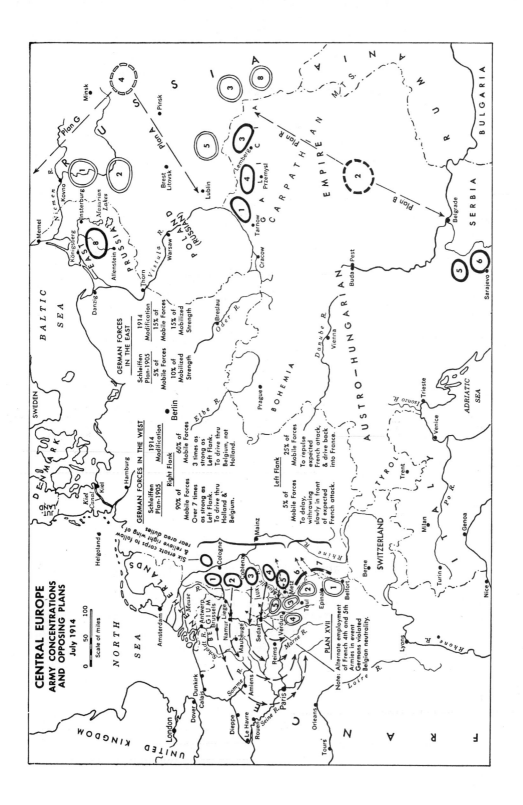

CENTRAL EUROPE
ARMY CONCENTRATIONS
AND OPPOSING PLANS

July 1914

Scale of miles
0   50   100

GERMAN FORCES IN THE WEST

Schlieffen Plan–1905	1914 Modification
Right Flank	
90% of Mobile Forces	60% of Mobile Forces
Over 7 times as strong as Left Flank.	3 times as strong as Left Flank.
To drive thru Holland & Belgium.	To drive thru Belgium, not Holland.
Left Flank	
5% of Mobile Forces	25% of Mobile Forces
To delay, withrowing slowly in front of expected French attack.	To repulse expected French attack, & drive back into France.

GERMAN FORCES IN THE EAST

Schlieffen Plan–1905	1914 Modification
5% of Mobile Forces	15% of Mobile Forces
10% of Mobilized Strength	15% of Mobilized Strength

Six ersatz corps to follow & relieve right wing of rear area duties.

PLAN XVII

Note: Alternate employment of French 4th and 5th Armies in event Germans violated Belgian neutrality.

mobile field forces in the right-wing blow against France instead of Schlieffen's proposed 90 percent. The right wing totaled about 1,500,000 men; the left-wing armies in Alsace and Lorraine were 500,000 strong; there were about 200,000 men in the army defending East Prussia, and another 200,000 along the remainder of Germany's eastern frontiers.

FRENCH PREPARATIONS; PLAN XVII

France planned on a concentration along her eastern frontier and an immediate attack through Alsace-Lorraine, exactly as Schlieffen had anticipated. This concept was the essence of French "Plan XVI" in 1911. Then General **Victor Michel,** the new commander in chief of the French Army, believing that any German invasion would come through Belgium rather than across the difficult terrain of Alsace-Lorraine, proposed a radical reorganization. He projected a French invasion through Belgium, regardless of her neutrality, to forestall the Germans. Had the Michel Plan been accepted, the opposing major offensives would have met head on. But this concept was rejected and Michel was relieved of command, General **Joseph J. C. Joffre** replacing him. A watered-down straddle, "Plan XVII," was adopted. The French main effort would be an immediate 2-pronged offensive into Alsace-Lorraine, passing both north and south of the Thionville-Metz area. At the same time, 2 French armies were to be prepared to shift westward if the Germans did in fact violate Belgian neutrality.

Joffre placed too much reliance on the ability of the Russian Army to engage Germany in the east. He relied also on close coordination with the BEF, which was expected to reinforce the French left flank. He also gambled—successfully—that Italy would leave the Triple Alliance. But above all, he and his general staff grossly underestimated the immediately available German strength. The French believed the German active Army to be too weak to drive west of the Meuse, and did not believe that they would employ their reserves without some refresher training. (Actually, the German reserve elements were in such advanced training that they were immediately available for first-line use, thus giving them a substantial and unexpected 3-2 numerical edge over the 1,300,000 French in the opening battles.)

THE RUSSIAN PLAN

This called for immediate, simultaneous offensives against both Germany and Austria, disregarding the fact that slow Russian mobilization could not be completed for 3 months. After the outbreak of war, Grand Duke **Nicholas,** the Czar's newly appointed commander, was coaxed by pressure from the French high command into speeding up these precipitate offensives. The result was to be disastrous defeat in East Prussia, but the premature Russian offensives into Galicia would be partially successful because of greater Austrian ineptitude.

AUSTRIAN PLANS

General **Franz Conrad von Hötzendorf,** dynamic Austrian Chief of Staff, had evolved 2 plans: Plan B for war against Serbia alone, Plan R for war against both Serbia and Russia. He chose the former, then changed his mind after concentration began. As a result, 1 of his 6 armies would not be available for action on either front when Conrad ordered advances into Serbia and into Galicia against the Russians. Even if it had been, it is doubtful that Austria would have had sufficient superiority of force to justify 2 simultaneous offensives.

The Naval Situation

The British Navy commanded the seas—28 dreadnoughts and battle cruisers to Germany's 18. Above all, the British Navy was ready to fight, thanks to the fore sight of First Lord of the Admiralty **Winston Churchill** and the First Sea Lord, Prince **Louis of Battenberg.** The German Navy had comparable standards of efficiency, but was hampered by deficiencies in bases as well as numbers. The Russian, French, and Austrian navies were limited in strength and were to play only minor roles.

COMPARATIVE NAVAL STRENGTHS—AUGUST, 1914

	BRITISH			GERMAN		
	Total	Home Waters	Grand Fleet[a]	Total	Home Waters	High Seas Fleet
Dreadnoughts[b] (modern battleships)	20	(20)	(20)	13	(13)	(13)
Battle Cruisers[c]	8	(4)	(4)	5	(4)	(4)
Old Battleships[b]	40	(38)	(10)	22	(22)	(10)
Cruisers[d]	102	(48)	(21)	41	(32)	(17)
Destroyers	301	(270)	(50)	144	(144)	(80)
Submarines	78	(65)	(9)	30	(30)	(24)[e]

[a] Numbers of cruisers and destroyers in the Grand Fleet are approximate, and varied considerably during early days of war.

[b] Dreadnoughts were modern, big-gun ships, completely outclassing all old battleships (see discussion in text). Britain had 2 more dreadnoughts completed, but not yet ready for action, and 15 under construction (including 3 that had been intended for other nations). Germany had 3 more completed, but not yet ready for action, and 4 under construction.

[c] Vessels carrying heavy guns similar to dreadnoughts, but with thinner armor and greater speed. Britain and Germany each had one additional battle cruiser completed but not yet ready for action. Britain had 1 building; Germany had 2 building.

[d] Included old armored cruisers and smaller, but faster and better armed, light cruisers.

[e] Records are conflicting on exact number of German submarines in commission. This figure is approximately correct.

Air Power in 1914

Negligible at the outset, aviation gradually would develop from a reconnaissance into a combat role. Lighter-than-air craft, particularly the German Zeppelin rigid dirigibles, were used from the beginning for both reconnaissance and bombing roles. Captive balloons for artillery observation were also used from the outset.

German Zeppelin dirigible

OPERATIONS IN 1914

Western Front

THE OPENING BATTLES

1914, August 3–20. Belgium Overrun. A specially trained German Second Army task force of about 30,000 men under General **Otto von Emmich** crossed the Belgian frontier between the Ardennes and the Dutch border, a narrow corridor guarded by Liége, one of the strongest fortresses of Europe. A night attack (August 5–6) penetrated the ring of 12 outlying forts. Heavy fighting followed, in which German Major General **Erich F. Ludendorff** distinguished himself, as did the Belgian commander, General **Gérard M. Leman.** German bombardment by 42-cm. howitzers (heaviest used to this time) systematically reduced the concrete and steel cupolaed defenses. Liége surrendered (August 16). The German First Army (General **Alexander von Kluck**) and the Second (General **Karl von Bülow**) poured through the Liége corridor and across the Meuse. Hastily mobilized Belgian field forces were brushed aside to the north of **Tirlemont** (August 18–19) and Brussels occupied (August 20). After some skirmishing along the Meuse (August 12–16), the Belgians, personally commanded by King **Albert,** fell back on the fortress of Antwerp.

1914, August 14–25. Battles of the Frontiers. The Germans and the Anglo-French armies met each other head on in 4 almost simultaneous actions.

1914, August 14–22. Battle of Lorraine. An early advance to Mulhouse in Alsace (August 8) by the French right-wing Army of Alsace (General **Paul Pau**) was followed by a full-scale offensive southeast of Metz by the French First (General **Auguste Dubail**) and Second (General **Noël de Castelnau**) armies (August 14–18). After planned withdrawals, the German Sixth (Prince **Rupprecht** of Bavaria) and Seventh (General **Josias von Heeringen**) armies turned in violent converging counterattacks. The French were thrown back to the fortified heights of **Nancy,** where they barely managed to stop the German drive. The French XX Corps, under General Ferdinand Foch, played a decisive role in holding Nancy.

1914, August 20–25. Battle of the Ardennes. The advancing French Third (General **Pierre Ruffey**) and Fourth (General **Fernand de Langle de Cary**) armies met headlong the German Fourth (Duke **Albrecht of Württemberg**) and Fifth (Crown Prince **Wilhelm**) armies, comprising the pivot of the Schlieffen Plan maneuver. After 4 days of furious fighting, the outnumbered French were repulsed with shocking losses, falling back to reorganize west of the Meuse, with their right flank on the fortress of Verdun.

1914, August 22–23. Battle of the Sambre. To the north, the German First, Second, and Third (General **Max von Hausen**) armies

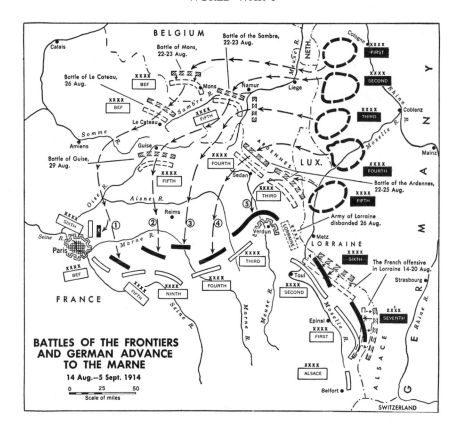

BATTLES OF THE FRONTIERS
AND GERMAN ADVANCE
TO THE MARNE
14 Aug.–5 Sept. 1914
0   25   50
Scale of miles

were beginning to sweep west and southwest. In accordance with the contingency provisions of Plan XVII, Joffre ordered the French Fifth Army (General **Charles Lanrezac**) into the Sambre-Meuse angle to meet this unexpected move. The German Second and Third armies struck Lanrezac southwest of Namur, defeating him and forcing him to retreat. The Belgian defenders of **Namur** were hammered into submission by some of Bülow's troops and siege guns after a brief siege (August 20–25).

1914, August 23. Battle of Mons. The British Expeditionary Force (Field Marshal Sir **John French**), 4 divisions and over 100,000 strong, had promptly and efficiently crossed the Channel and concentrated in the vicinity of Le Cateau, left of the French Fifth Army. Upon Joffre's request, the BEF moved into Belgium in

cooperation with Lanrezac's advance toward Namur (August 21). Near Mons the British were struck by the full weight of aggressive von Kluck's First German Army. Outnumbered, the British fought back stoutly, their fire discipline taking heavy toll of the close German formations. Sir John French was prepared to continue the fight next day, but the retreat of Lanrezac's Fifth Army from the Sambre left him without support; the BEF therefore withdrew during the night. French was bitter about Lanrezac's unannounced withdrawal, which he believed had jeopardized the existence of his own BEF.

COMMENT. *The French offensive had failed completely—at a cost of some 300,000 casualties. But Moltke overestimated the extent of the German victory. His communications with his armies were poor, his information faulty. Believing that the success in*

Lorraine was a decisive victory, he ordered his left to continue its offensive against the fortified Nancy heights, hoping thus to obtain a double envelopment of the entire French field forces. The Ardennes and Sambre battles he also considered decisive, and so he renewed the orders for his right-wing armies to continue their sicklelike sweep, with the First Army still to swing west of Paris. He decided to send to the Sixth and Seventh armies reinforcements originally intended for the right-wing armies, to provide more weight to his new offensive in Lorraine. Confident that the French armies were on the verge of destruction, he also detached 2 corps from the right to hasten by railroad to the Eastern Front, where the Russians had shown unexpected initiative. (Ironically, these 2 corps, whose absence would vitally affect the outcome of the Battle of the Marne, were still en route at the time the Battle of Tannenberg made their presence unnecessary in the east.) As a result of these and other detachments to contain the Belgian Army at Antwerp and to besiege the French fortress of Maubeuge, the 3 German right-wing armies had been bled from a total strength of 16 corps to 11. The already watered-down Schlieffen Plan—dependent upon a right-wing hammer blow—was thus still further modified from the concept of its creator.

*Joffre, on the other hand, had kept close touch with his subordinate commanders and was well aware of the actual situation. He knew that, despite tactical defeats, morale of his troops was still high. He was now also aware of the German plan. Seemingly oblivious of the disastrous results of his own Plan XVII, he calmly prepared for a counterattack. This would be a Schlieffen Plan in reverse, pivoting about Verdun and the Nancy heights, where his First and Second armies were ordered to hold on at all costs. While the Third, Fourth, and Fifth armies and the BEF were to continue their southwesterly withdrawals, Joffre drew units from his embattled right flank and from reserves in the interior of France to create 2 new armies. The Sixth, under General **Michel J. Maunoury,** was to assemble—first near Amiens, later in and around Paris—west of the German right wing, prepared to attack east. The Ninth, under General Foch, would be gathered in close support behind and between the Fourth and Fifth armies to provide weight for a counterattack against the German main effort. This attack was to be launched when the 4 Allied left-flank armies had fallen back to the general line of the Somme River-Verdun.*

1914, August 25–27. Battle of Le Cateau. Marshal French's BEF, hard-pressed by the German First Army, fought daily rear-guard actions. Attempting a stand (August 27) to relieve his exhausted II Corps troops, General **Horace Smith-Dorrien** became engaged in the biggest battle the British Army had fought since Waterloo. This corps fought off a double envelopment by the full strength of Kluck's army; the survivors successfully disengaged when night fell. The price was high: 7,800 casualties out of 40,000 men engaged.

1914, August 29. Battle of Guise. Joffre, to relieve German pressure on the BEF, ordered the Fifth French Army, itself pressed hard by the German Second Army, to make a 90° shift westward to attack the left flank of the German First Army. The initial attack got nowhere, but General **Louis Franchet d'Esperey,** commanding Lanrezac's I Corps, smartly moved from reserve to hit and halt the pursuing German Second Army, thus achieving the first French tactical success in the campaign. Bülow called on Kluck (August 30) for help.

1914, August 30–September 2. Kluck's First Dilemma. The German First Army had driven the BEF from its front; for the time being—as Kluck saw it—the British were out of the picture. On the right, some slight clashes had occurred with French troops (actually part of Maunoury's assembling Sixth Army, but in Kluck's opinion unimportant scattered elements). Bülow on the left had called for help. Aggressive Kluck, thinking the French Fifth Army now to be the left-flank unit of the opposing field forces, and unable to communicate with Moltke, threw the remnants of the Schlieffen Plan into the discard. He shifted his direction of march to the southeast to roll up the Fifth Army (August 31). This change would cause him to pass east of Paris; he knew nothing of the French concentration in the fortified area of the capital. By September 2, Kluck's left flank was on the Marne at Château-Thierry, his right on the Oise, near Chantilly.

1914, September 1–2. Joffre's Reaction. Aware of the German change in direction through air reconnaissance, Joffre ordered the Sixth Army to complete its concentration in the Paris area. He ordered the general retirement to continue until the Fifth Army was out of immediate danger of envelopment. Thus he was forced to abandon his originally planned counterstroke from the Somme-Verdun line. Foch's newly forming Ninth Army continued its concentration between the Fourth and Fifth armies. Joffre was concerned by British lack of responsiveness to his orders, but a visit to Field Marshal French by the British War Minister, Field Marshal **Lord Kitchener,** soon changed Sir John's attitude and he began to cooperate.

1914, September 3–4. Kluck's Second Dilemma. Belatedly Moltke sent a message to Kluck, agreeing to the move east of Paris, but complicating matters by ordering Kluck to guard the right flank of the Second Army, which would thus become the spearhead of the modified German wheel. But Moltke, whose intelligence had informed him of the French concentration near Paris, did not realize that his First Army had been moving at amazing speed under Kluck's driving leadership, and that its advance units were much farther south than those of the slower-moving Second Army. And Moltke failed to explain the reason for his order. For Kluck to have obeyed the order would have meant halting his army for 2 days, which he believed would permit the French either to escape or to rally. Again being unable to communicate directly with his commander, unaware of the situation in Paris, and trying to act in accordance with the apparent intention of Moltke's order, Kluck reasoned that its purpose was to assure that the French were driven southeast of Paris. His own First Army was ideally situated for this task. Accordingly, pugnacious Kluck continued southward, across the Marne, his right flank wide open, just east of Paris.

BATTLE OF THE MARNE, SEPTEMBER 5–10

Joffre's counterattack order (September 4) directed the Sixth Army to attack eastward toward Château-Thierry; the BEF was to move on Montmirail, with the Fifth Army, supported by the Ninth, prepared to conform. The Fourth Army would hold, prepared to advance, and the Third would strike westward from Verdun. On the success of this proposed double envelopment of the German right wing, as Joffre well knew, rested the fate of France. September 6 was to be D day.

Meanwhile, Maunoury's Sixth Army, temporarily under the regional command of General **Joseph S. Galliéni,** energetic military governor of Paris, had begun to carry out Joffre's warning orders by an advance from Paris toward the Ourcq River, where Kluck's right flank lay invitingly open. Only the aggressive initiative of the German right-flank corps commander, General **Hans von Gronau,** saved Kluck's army from surprise envelopment (September 5). As it was, Kluck believed that the French activity on his right was only a spoiling attack and merely detached one additional corps to help Gronau to repel it, while pressing southward with the rest of his army in pursuit of the BEF and the French Fifth Army. Not until this **Battle of the Ourcq** had raged for 2 days did Kluck realize the French intentions (September 7). By this time most of his army was south of the Marne. Pulling back north of the river, Kluck rapidly changed his front and turned his entire army westward in savage counterattacks that halted the French and forced Maunoury to fall back on the defensive (September 7–9). Only the arrival of reinforcements rushed from Paris by Galliéni—some in commandeered taxicabs—permitted Maunoury to stem the impetuous German advance.

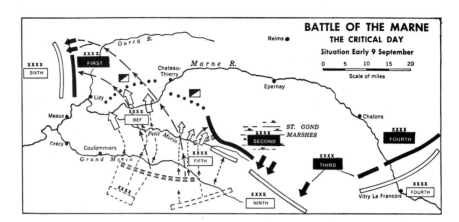

By this time the action had become general along the entire front west of Verdun. Kluck's westward shift, undertaken on the assumption that the BEF was no longer a threat, widened the already existing gap between his army and that of von Bülow, which was still moving south. Into this gap now moved the BEF, slowly, since Marshal French underrated the recuperative powers of his troops. Franchet d'Esperey's Fifth Army (Lanrezac had been relieved) battered at part of the German Second Army along the **Petit Morin.**

Farther southeast, Foch's Ninth Army, attacking north at **St.-Gond,** found itself confronting the rest of the Second Army while Hausen's Third Army struck its right. A surprise night bayonet attack by 4 divisions of Hausen's army threw part of Foch's army into confusion (September 8). Foch's response was to order an immediate renewal of his own assault; the German advance was halted, but Foch's position was precarious.

At **Vitry-le-François,** Langle de Cary's Fourth Army battled desperately but indecisively with the Duke of Württemberg's Fourth Army and part of the Third. At Revigny in the Argonne Forest, General **Maurice Sarrail's** Third Army (Ruffey had been relieved) stopped the Crown Prince's Fifth Army, while at **Nancy** and along the Alsace frontier the French First and Second armies—even though attenuated by drafts for Joffre's new formations to the west—clung successfully to the heights, despite a succession of attacks by the reinforced German Sixth and Seventh armies. (Schlieffen had warned against any such attacks.)

Moltke, worried by rumor and pessimistic fragmentary reports from his subordinates, sent a general staff officer, Lieutenant Colonel **Richard Hentsch,** to inspect the front (September 8). Hentsch's orders were oral; they still remain somewhat of a mystery. He arrived at the Second Army's headquarters just as news was received that its right flank was being turned by a vigorous night attack by Franchet d'Esperey's Fifth Army. This was probably the turning point of the battle. Bülow—personally defeated—was about to retreat. Kluck's First Army was making headway in the northwest against Maunoury's left, but the BEF's advance through the gap threatened Kluck's own left and rear.

Hentsch tacitly approved Bülow's planned retreat and, later the same day, in Moltke's name ordered Kluck also to withdraw (September 9). Moltke, now realizing that his offensive had failed, ordered a general retirement to the line Noyon-Verdun. Within 5 days the Germans, having disengaged without serious interference from the exhausted Allies, were organizing their new positions. The Battle of the Marne thus ended as a strategic Allied victory and Joffre emerged as savior of France. That same day Moltke was relieved, General **Erich von Falkenhayn** replacing him (September 14).

COMMENT. *France's initial offensive plan had failed because it was entirely unrealistic in concept and in execution. The German plan—sound and workable—failed because of the inefficiency of Moltke, who first emasculated the plan, then lost all personal touch with his army commanders and with their progress. Joffre, on the other hand, emerged as a strong and capable leader, who kept in close touch with his subordinates. His reconstruction of a counterattack upon the wreckage of his initial plan was masterful, its execution assisted by the marvelous resiliency of the French Army. The BEF's part was that of a sound professional soldiery. The clash of personalities and mutual distrust existing between Sir John French and Lanrezac prevented better use of the BEF, as did French's excessive caution in the counterattack. Casualties on both sides were enormous: the Allies lost about 250,000 men; German losses were somewhat greater. In 3 weeks of war, each side had lost more than half a million men in killed, wounded, and captured. The Battle of the Marne, tactically indecisive, was a clear-cut strategic victory for the Allies. Had it ended differently, the history of the 20th century would have been altered fundamentally. It was the world's most decisive battle since Waterloo.*

THE "RACE TO THE SEA," SEPTEMBER 15–NOVEMBER 24

1914, September 15–18. First Battle of the Aisne. Slow in their pursuit, the Allied armies, seeking to envelop the German right, were rebuffed from the hastily prepared German field fortifications. Both sides now extended their operations northward, attempting each to outflank the other. Both failed, in bitter fighting in **Picardy** (September 22–26) and **Artois** (September 27–October 10). Meanwhile, behind the German lines, beleaguered **Maubeuge** had fallen (September 8) and the fortress of **Antwerp,** systematically bombarded (October 1–9), surrendered. The Belgian Army fell back to the west along the coast. An extemporized British naval division, rushed to reinforce the Antwerp garrison, also escaped, but with loss of 1 of its 3 brigades.

1914, September 22–25. Verdun and St.-Mihiel. Farther south, repeated German attacks against Verdun were repulsed (September 22–25), but the Germans did seize the strategic **St.-Mihiel** salient (September 24), to which they would cling until 1918.

1914, October 18–November 24. Battles in Flanders. The final actions of the "Race to the Sea" were the **Battle of the Yser** (October 18–November 30) and the bloody **First Battle of Ypres** (October 30–November 24), in which the BEF was nearly wiped out in a successful, gallant defense against a heavily reinforced German drive, ordered by Falkenhayn, who expected to capture the Channel ports. The British were aided by French troops, under Foch, rushed north by Joffre.

1914, December 14–24. General Allied Attack. From Nieuport to Verdun an allied offensive beat unsuccessfully for 10 red days against the rapidly growing German system of field fortifications. The era of stabilized trench warfare had begun: the spade, the machine gun, and barbed wire ringing down the curtain on maneuver, from the North Sea to the Swiss border. A costly French attempt at breaking through in Champagne—the **First Battle of Champagne** (December 20)—was still in progress as the year ended. By this time, operations on the Western Front had cost the Allies nearly 1 million casualties. German losses were almost as great.

Eastern Fronts

OPERATIONS IN EAST PRUSSIA

THE RUSSIAN OFFENSIVE

1914, August 17–19. Invasions of East Prussia. The Russian Northwest Army Group under

French poilus march single file through a dense field of barbed wire

General **Yakov Grigorievich Jilinsky,** consisting of General **Pavel K. Rennenkampf's** First and General **Alexander Samsonov's** Second armies, advanced into East Prussia. Opposing them was German General **Max von Prittwitz'** Eighth Army, widely disposed from the Baltic south to Frankenau, and based on the fortress of Königsberg (Kaliningrad). Its mission was one of elastic defense and delay in accordance with the modified Shclieffen Plan.

1914, August 17. Battle of Stallupönen. The center of Rennenkampf's widely strung advance met General **Hermann K. von François's** I German Corps, was badly mauled by the alert François, and was thrown back to the frontier with loss of 3,000 men. François then retired on Gumbinnen.

1914, August 20. Battle of Gumbinnen. Slowly the Russians advanced again. Prittwitz, aware also of the Russian Second Army's advance far to his southern flank, feared envelopment. Aggressive François persuaded him to attack. François's own corps smashed in the Russian right flank, driving it back for 5 miles.

Other German attacks were not successful, and a drawn battle resulted.

THE TANNENBERG CAMPAIGN

1914, August 20. German Change in Command. Prittwitz, in near panic after his unsuccessful attack against Rennenkampf, and with Samsonov's army posing a potential threat to his line of communications, telephoned Moltke, at Coblenz, to report his decision to withdraw to the Vistula and to request reinforcements to be able to hold that river line. Moltke at once relieved Prittwitz of command, appointing in his place elderly General **Paul von Hindenburg,** called from retirement, with brilliant General Erich Ludendorff, hero of Liége (see p. 1024), as his chief of staff. Thus was created a team destined for world renown.

1914, August 22. Ludendorff's Plan. After studying reports from the east, Ludendorff telegraphed orders to the individual corps commanders, directing a concentration against Samsonov's Second Army, while delaying

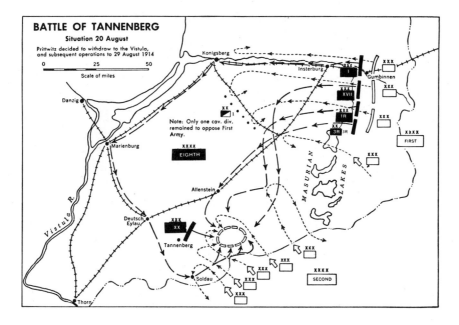

BATTLE OF TANNENBERG
Situation 20 August
Prittwitz decided to withdraw to the Vistula,
and subsequent operations to 29 August 1914

Rennenkampf's First Army farther east. Joining Hindenburg later that day for the rail trip east, Ludendorff reported his actions; Hindenburg approved. When they arrived at Marienburg, Eighth Army Headquarters, next day, they discovered that Lieutenant Colonel **Max Hoffmann,** Prittwitz' capable chief of operations, had already prepared for practically the same movements and dispositions that Ludendorff had ordered (August 20). (The coincidence is especially interesting as evidence of the uniform thought process of the German Army General Staff in dealing with an unexpected situation.) While one lone cavalry division was delaying fumbling Rennenkampf, the bulk of the German army was shifting south, by rail and road, against the equally incompetent Samsonov.

1914, August 24. Battle of Orlau-Frankenau.
Advancing without reconnaissance or cavalry screen, Samsonov's central corps suddenly ran into entrenched units of the German XX Corps. Severe fighting raged all day, but the Russian center was unable to advance. Next day, while

the Russian army rested, the XX Corps withdrew from Frankenau to Tannenberg, while other units of the Eighth Army hastened up to its right and left. The Germans, who had been listening to Samsonov's uncoded radio messages, now knew the locations of all Russian units, and were aware of their projected moves for the next day.

1914, August 26–31. Battle of Tannenberg.
Samsonov's right flank was pushed in from the north by the German XVII and I Reserve Corps; his left was enveloped and turned by François's hard-driving I Corps; his center was struck by the XX Corps. By nightfall of August 29, the encirclement was complete as François stretched his corps across the entire Russian rear. The rest was butchery of disorganized streams of rabble trying to escape the net. Not until the 27th had Jilinsky realized that his Second Army was in real danger; his orders to Rennenkampf to move to its assistance were obeyed only in shadow. Samsonov disappeared the night of the 29th; evidently he committed suicide. Russian losses totaled 125,000 men

and 500 guns; the Germans lost between 10,000 and 15,000 men. Aside from its strategic significance, the German victory was a tremendous psychological coup; Allied confidence in Russia was shattered, while the German nation was roused to such a pitch of enthusiasm that the true significance of the Battle of the Marne, which ended 2 weeks later, was overlooked.

1914, September 9–14. First Battle of the Masurian Lakes. Turning northeast, the German Eighth Army promptly moved against the Russian First Army. Again the vigorous François and his I Corps provided the *coup de grâce,* driving in the Russian left. Rennenkampf finally disengaged under cover of a stout 2-division counterattack, spoiling the German effort to gain another double envelopment. The Russians retreated, having lost 125,000 men, 150 guns, and half their transport. German losses were about 40,000.

COMMENT. *Incompetent leadership, faulty reconnaissance, lack of secrecy, and poor communications, added to an astounding state of unpreparedness and shortage of matériel, all contributed to the Russian defeats. Russia never completely recovered from these disasters.*

Austrian Invasions of Serbia

1914, July 29. Bombardment of Belgrade. The first military action of the war, this Austrian bombardment of the Serbian capital had little effect other than to enrage the Serbs.

1914, August 12–21. Battle of the Jadar. Austrian forces totaling more than 200,000 men, commanded by General **Oskar Potiorek,** crossed the Save and Drina rivers to invade Serbia from the west and northwest. They were opposed by slightly smaller numbers of tough, hardy Serb troops, inadequately equipped but battlewise from their Balkan Wars experience (see p. 1015), commanded by able Marshal **Radomir Putnik.** A Serb counterattack (August 16) punished the Austrians so severely that Potiorek withdrew across the Drina.

1914, September 7–8. Renewed Austrian Invasion. Ignoring a bold, but limited, Serbian invasion of Austrian Bosnia (September 6), Potiorek made a night attack across the Drina. Putnik withdrew his troops from Bosnia and strongly counterattacked the Austrian bridgeheads.

1914, September 8–17. Battle of the Drina. Unable to eliminate the Austrian bridgeheads in 10 days of vicious, bitter fighting, and running short of ammunition, Putnik withdrew to more defensible positions southwest of Belgrade.

1914, November 5–30. Austrian Offensive. In the face of an offensive by the reinforced Austrian armies and short of ammunition, Putnik withdrew slowly and deliberately, planning to counterattack after the Austrians became overextended in the rough mountain country. He evacuated Belgrade, which the Austrians occupied (December 2). At the end of the month, ammunition, sent from France, arrived by rail from Salonika.

1914, December 3–9. Battle of Kolubra. With the Kolubra River, behind the Austrian front, in flood, Putnik launched a vigorous counterattack. The Austrian forces collapsed in the face of determined Serb assaults and were driven from Serbian terrain, their retreat covered by Austrian monitors on the Danube and Save rivers. Belgrade was recaptured (December 15). Potiorek was relieved of his command and replaced by Archduke **Eugene.** Austrian casualties in this savagely fought campaign were some 227,000 out of 450,000 engaged. Serbian losses were approximately 170,000 out of 400,000.

Operations in Poland

THE GALICIAN BATTLES

1914, August 23–September 2. Austrian Offensive. The advance into Russian Poland from Galicia was coordinated directly by General Conrad, the Austrian chief of staff. The

First, Fourth, and Third armies (from left to right) moved north and east from the vicinity of Lemberg (Lvov) on a 200-mile front, to clash headlong with General **Nikolai Ivanov**'s Southwestern Russian Army Group (the Fourth, Fifth, Third, and Eighth armies) southwest of the Pripet Marshes. At the **Battle of Krasnik** (August 23–24) on the northern flank, the Russian Fourth Army was driven back by the Austrian First. The conflict spread as the Austrian Fourth Army struck and drove back the Russian Fifth in the **Battle of Zamosc-Komarów** (August 26–September 1). But on the southern flank, the Austrian Third Army (with some elements of the Second, belatedly arriving from the Serbian front) was thrown back on Lemberg by the Russian Third and Eighth in the Battle of **Gnila Lipa** (August 26–30). The Austrians, held on the defensive, were defeated again—this time decisively—at **Rava Ruska** (September 3–11), when the Russian Fifth Army penetrated between the Austrian First and Fourth. Abandoning Lemberg, the Austrians fell back 100 miles to the Carpathian Mountains, leaving a garrison in the key fortress of Przemysl. With all the remainder of Galicia now in their hands, the Russians prepared for further advances into the Carpathians. Austrian losses in this campaign amounted to more than 250,000 killed and wounded and 100,000 prisoners. There is no record of Russian losses, which must have been comparable.

OPERATIONS IN WESTERN POLAND

1914, September 17–28. German Movement to Assist Austrians. Falkenhayn ordered Hindenburg to assist the defeated Austrians in Galicia and to prevent a Russian invasion of Silesia. With extraordinary efficiency, 4 corps were transferred by rail in 750 trains from the Eighth Army to the vicinity of the Austrian north flank, near Cracow. There they became the Ninth German Army, commanded directly by Hindenburg. A general Austro-German advance followed (September 28). Meanwhile, as the Germans expected, the Grand Duke Nicholas, reorganizing his armies, was preparing for a general offensive through Poland into Silesia, the heart of Germany's mineral resources. His proposed movement consisted of the Fifth, Fourth, and Ninth armies.

1914, September 28–October 31. Hindenburg's Southwest Poland Offensive. The German Ninth Army hit the Russians west of the Vistula (September 30), attaining the river line south of Warsaw (October 9). Here, with but 18 divisions to oppose the Russian 60, the German offensive was checked (October 12). Hindenburg withdrew skillfully (October 17), leaving behind him a countryside systematically ravaged, vastly impeding fumbling Russian pursuit. Meanwhile, the Austrians to the south had made some advance, but were checked at the River **San.** By the end of October, the Austro-German armies had fallen back to their original line, but had seriously delayed the projected Russian advance.

1914, November 1. Hindenburg Appointed Commander in Chief of the Austro-German Eastern Front. He was told he could expect no reinforcements, despite the fact that the Russians had renewed their forward movement. Skillfully, in accordance with Ludendorff's plan, the German Ninth Army—the only mass of maneuver available—was shifted northwest again to the Posen-Thorn area, leaving another wide gap of "scorched earth" in front of the overwhelming Russian concentration southwest of Warsaw.

1914, November 11–25. Battle of Lódź. The German Ninth Army, now commanded by General **August von Mackensen,** struck southeast between the First and Second Russian armies, which were protecting the northern flank of the Grand Duke's planned offensive. The Russian First Army (still under Rennenkampf) was crushed and the Second, near Lódž, was embraced by an attempted double envelopment. The key element of the German stroke was the XXV Reserve Corps, commanded by General **Reinhard von Scheffer-Boyadel.** It rolled through the gap between the Russian armies and turned south and west. The movement was foiled by the Grand Duke's prompt counterattack. The Russian Fifth Army from the south and an improvised group from the northern forces checked the German advance. Scheffer's corps was completely surrounded. In an amazing display of leadership, Scheffer not only broke through to safety but also brought back with him 16,000 prisoners and 64 captured guns. This corps marched and fought continuously for 9 days in subzero weather with a net loss of 1,500 killed and 2,800 wounded, who were also brought safely back. While Lódź was tactically a Russian victory, for the Germans it was a strategic success, since the Russian offensive was now called off, Lódź was evacuated, and the Russians fell back in a general retirement, never again to menace the German homeland. German losses in the Lódź campaign were about 35,000 killed and wounded. Russian losses are not known; a conservative estimate would be 90,000 in all. The year ended in stalemate on the Eastern Front.

The War at Sea, 1914

The British Grand Fleet, poised in its bases at Scapa Flow and Rosyth, kept the German High Seas Fleet bottled up behind the highly fortified Heligoland-Jade littoral in the North Sea. Neutral Denmark locked the Baltic gateway to both contestants by mining the Skagerrak.

Germany had 10 major warships at large around the world, based on far-flung colonial ports. Six small British overseas expeditions—4 from England and 2 from Australia and New Zealand—moved (August) to dry up the German naval bases. **Togoland,** the **Cameroons, Southwest Africa, Samoa,** and some of the German Pacific islands were taken in late 1914 or early 1915. In **German East Africa,** however, Colonel (soon General) **Paul von Lettow-Vorbeck** repulsed a British landing effort at **Tanga** (November 3–4) and would carry on an amazing offensive-defensive campaign for 4 years. Except as an example of indomitable leadership, Lettow-Vorbeck's operations were of no significance in the war. Japan, entering the war on the Allied side (August 23), besieged **Tsingtao,** the only German base on the China coast, taking it November 7. Japan also occupied Germany's Marshall, Mariana, Palau, and Caroline island groups.

The 10 scattered German cruisers, deprived of their bases, waged a gallant and aggressive hit-and-run war until overwhelmed or blockaded. The High Seas Fleet at home was limited to sporadic cruiser raids, mine sowing, and submarine warfare.

1914, August 4. The *Goeben-Breslau* **Incident.** In the Mediterranean, the German battle cruiser *Goeben* and light cruiser *Breslau,* under the command of Vice Admiral **Wilhelm von Souchon,** shelled the French Algerian ports of Bône and Philippeville. Then, in anticipation of a British declaration of war, the German vessels headed east toward Turkey, Germany's secret ally. They met the British battle cruisers *Indomitable* and *Indefatigable* steaming west. The British commander, following instructions from Vice-Admiral Sir **A. Berkeley Milne,** made no effort to stop the Germans, knowing that, although France and Germany were already at war, Britain's ultimatum to Berlin would not expire until midnight. Admiral von Souchon, knowing that the British warships had superior firepower, also refrained from action. The opposing squadrons passed one another at close range without saluting, like two strange dogs; each side was cleared for action, each had shotted guns trained on the other. The British then attempted to follow Souchon, but he skillfully eluded them after a brush with the cruiser squadron of Rear Admiral **E. C. Troubridge** southwest of Greece (August 6–7) and reached the Dardanelles safely (August 10). Both German ships and their personnel then passed into the Turkish Navy, giving that nation naval supremacy in the Black Sea and contributing to her later entry into the war on the side of the Central Powers (October 29; see p. 1036).

1914, August 28. Battle of Heligoland Bight. British light cruisers raided into German waters, coaxing a fight. German cruisers emerged, whereupon Vice Admiral Sir **David Beatty**'s battle-cruiser squadron, lurking in support of his light cruisers, drove them back with loss of 4 ships and some 1,000 men.

1914, August–October. Cruise of the *Königsberg*. The German light cruiser, stationed on the East African coast, engaged and sank the British light cruiser *Pegasus* off Mombasa (August 6). She was later cornered by other British naval units and forced to seek refuge up the Rufiji River, German East Africa (October 30), where she remained blockaded.

1914, August–November. Cruise of the *Emden*. This fast light cruiser was detached from the China Squadron of German Admiral **Maximilian von Spee** (August 22). Under daring Captain **Karl von Müller,** she sailed into the Indian Ocean, where she harassed British shipping, taking 21 prizes and destroying ships and cargo valued at over $10 million. She bombarded Madras (September 22). The end of this gallant lone-wolf cruiser came when she was sunk in a hard-fought action with the Australian cruiser *Sydney* at the Cocos Islands (November 9).

1914, September–October. German Submarine Operations. *U-9,* off the Dutch coast, sank in quick succession the British cruisers *Aboukir, Hogue,* and *Cressy,* with a loss of 1,400 lives (September 22). A U-boat raid on Scapa Flow (October 18), while unsuccessful, resulted in the temporary transfer of the British Grand Fleet to Rosyth on the Scottish coast, while anti-submarine nets were installed at Scapa. The cruiser H.M.S. *Hawk* was torpedoed and sunk (October 15). The battleship *Audacious* struck a German mine, laid by a submarine off the Irish coast, and sank (October 27).

1914, November 1. Battle of Coronel. After the outbreak of war, Admiral von Spee's China Squadron—2 heavy and 3 light cruisers—crossed the Pacific to the Chilean coast, refueling from colliers. Off the west coast of South

German cruiser *Emden*

America, British Vice Admiral Sir **Christopher Cradock,** with 2 elderly heavy cruisers and 1 light cruiser, plus a converted merchant-ship auxiliary cruiser, searched for the German squadron. Despite Admiralty orders, Cradock had discarded the old battleship *Canopus* as being too slow to hunt the speedy Germans. The 2 squadrons met off Coronel. On paper the fire-power of the 2 forces was about equal, but Cradock had only 2 9.2-inch guns (on his flag-ship, the *Good Hope*), while Spee had 16 8.2-inchers on the *Scharnhorst* and *Gneisenau.* The more numerous British light guns never had a chance to get into the fight because Spee—foiling all efforts of the British ships to close—systematically battered them at long range with his 2 heavy cruisers. He sank the 2 British heavy cruisers with all on board. The light cruiser *Glasgow,* and the auxiliary cruiser *Otranto* obeyed Cradock's orders to run for it and escaped. News of the disaster shocked Brit-ain. The Admiralty feared that Spee would take his squadron around Cape Horn into the South Atlantic. The battlecruisers *Invincible* and *Inflex-ible,* under Vice Admiral Sir **F. D. Sturdee,** were rushed from home waters to seek Spee.

1914, November 3, December 16. German Coastal Raids against Britain. German cruisers raided the British east coast off Gorleston, but were driven off (November 3). In another raid, German heavy cruisers bom-barded Scarborough and Hartlepool (Decem-ber 16). These "hit-and-run" affairs killed and wounded many civilians.

1914, December 8. Battle of the Falkland Islands. Spee, planning to run into Port Stan-ley, Falkland Islands, to raid the British wireless and coaling station, discovered Sturdee's squadron there, refueling. The surprised Ger-mans took to their heels. Sturdee, pursuing, destroyed the German ships at long range. The light cruiser *Dresden* escaped and remained at large for 3 more months. Most of the crews of the sunken German ships perished—some 1,800 men—although a few were rescued by the British.

1914, December 25. Loss of the *Jean Bart.* The French battleship was torpedoed by an Austrian submarine in the Straits of Ot-ranto.

SUMMARY. *By year's end, except for the High Seas Fleet in the Jade, and the Baltic command based on Kiel, the German flag had been practically swept from the seas. Allied maritime traffic was uninterrupted, while Germany was already feeling the pinch of naval blockade. The German naval high command now fo-cused its attention on the one major weapon left to it on the high seas: the submarine.*

The Turkish Fronts, 1914

1914, October 29. Turkish Declaration of War against the Allies. This was proclaimed by the guns of the Turkish fleet (including the erstwhile German *Goeben* and *Breslau*), now commanded by German Admiral von Souchon, in a bombardment without warning of Odessa, Sevastopol, and Theodosia on the Russian Black Sea coast. This Turkish align-ment with the Central Powers closed the Dar-danelles to the Allies, thus physically separat-ing Russia from them.

CAUCASUS FRONT

1914, November–December. Turkish Of-fensive. Against the sage advice of General **Otto Liman von Sanders,** chief of the Ger-man military mission to Turkey, **Enver** Pasha, Turkish war minister, began an invasion of the Russian Caucasus.

1914, December 29. Battle of Sarikamish. The Turkish advance toward Kars was halted and rebuffed with severe losses by Russian General **Vorontsov** in winter snows. The struggle here continued as the year ended.

MEDITERRANEAN REGION

1914, November–December. British Reac-tion. Britain announced the annexation of Cyprus (November 5). Declaring a protectorate over Egypt (December 18), the British began moving troops there for the defense of the Suez Canal. Meanwhile, British cruisers shelled the Dardanelles forts without effect (November 30).

MESOPOTAMIAN FRONT

1914, October 23. British Landings. British Indian Army troops, who had already been rushed to Bahrain to protect oil refineries there, began an invasion of southern Mesopotamia. Local Turkish garrisons were driven back; **Basra** was captured by the British (November 23).

OPERATIONS IN 1915

The Global Situation

Turkey's entrance had changed the war's complexion. Russia, already shaken by the reverses of 1914, was now almost completely cut off from Franco-British war supplies, upon which she was dependent for a long-continued war. The western Allies, at the same time, were anxious to regain access to the Ukrainian grain fields. These considerations prompted a strategic debate in Britain between "Easterners" and "Westerners." A strident segment of British officialdom, led by capable and energetic Winston Churchill, First Lord of the Admiralty, urged immediate action to seize the Dardanelles and to restore the vital Mediterranean—Black Sea supply route to Russia through the Turkish Straits. British War Minister Field Marshal Horatio Herbert, Lord Kitchener, was equally insistent that a decision be obtained on the Western Front, and deplored any diminution of strength there for a peripheral operation in the east. He was strongly supported in this position by French military and political opinion. Nevertheless, in early January, after lengthy and heated debate in the British War Council, an amphibious operation against the Dardanelles was grudgingly approved.

In the Central Powers' camp also, strategical opinion was divided. The Hindenburg-Ludendorff team urged an all-out effort against faltering Russia. Falkenhayn, though reconciled to the fact that the war had become one of attrition, believed that it would have to be won in the west; he predicted that tactical victories in the east would be meaningless because of the space of Russia and her vast man-power resources. The Kaiser sided with Hindenburg. Accordingly, the Germans adopted a defensive posture in the west, while seeking a decision against Russia.

Western Front

1915, January 1–March 30. Allied Offensive in Artois and Champagne. This, a continuation of the **First Battle of Champagne** (see p. 1029), was a major effort by Joffre to liberate the extensive and valuable areas of France held by the Germans. A series of attacks against the western face of the Noyon salient and in the area between Reims and Verdun were unsuccessful. Limited German counterattacks along the La Bassé Canal and near Soissons stabilized the situation (January 8–February 5). Re-newed Allied assaults (March) made little headway. The British made an initial breakthrough in a well-planned attack at **Neuve Chapelle** (March 10), but poor management prevented an adequate follow-up; the Germans quickly reestablished the line (March 13). French casualties approached 400,000 during this period; British and German losses were also heavy.

1915, January 19–20. First German Air Raids on England. Bombing attacks by Zeppelin dirigible airships (under German Navy control) caused relatively minor casualties and

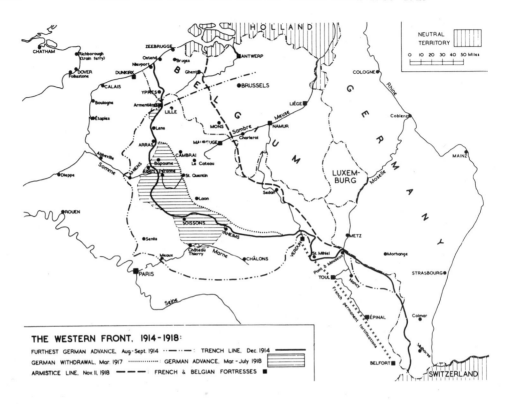

THE WESTERN FRONT, 1914–1918:

FURTHEST GERMAN ADVANCE, Aug.-Sept. 1914 ····—·—   TRENCH LINE, Dec. 1914 ————
GERMAN WITHDRAWAL, Mar. 1917 ················   GERMAN ADVANCE, Mar.-July 1918 ▤▤▤
ARMISTICE LINE, Nov. II, 1918 ——————   FRENCH & BELGIAN FORTRESSES ■

more anger than panic. Eighteen more such raids occurred during the year. The largest of these was a mass attack on London (October 13).

1915, April 6–15. Battle of the Woëvre. Repeated French assaults against the north face of the St.-Mihiel salient were repulsed with heavy losses.

1915, April 22–May 25. Second Battle of Ypres. Allied preparations for another coordinated offensive were spoiled by a surprise German attack preceded by a cloud of chlorine gas emitted from some 5,000 cylinders. This was the first use of poison gas in the west. Two German corps drove through 2 terrorized French divisions and bit deeply into British lines, creating a wide gap. The Germans, however, had made no preparations to exploit such a breakthrough and had few reserves available because of their build-up in the east. Local

counterattacks by the British Second Army finally stemmed the German advance after bitter fighting. German losses were some 35,000 men; the British lost 60,000, the French about 10,000.

1915, May–June. Battles of Festubert and Souchez (Second Battle of Artois). After limited gains, the British were stopped near Festubert (May 9–26). The French did only slightly better in their efforts to seize the commanding height of **Vimy Ridge** near Souchez (May 16–June 30). The Allies, exhausted by their costly and unsuccessful assaults during the first half of the year, spent the rest of the summer in resting, reorganizing, and reinforcing. The Germans, who had also suffered severely, were happy to take advantage of the lull, and by the end of the summer had also reinforced the west with troops from their successful operations in the east. Both sides had come perilously close to expending their ammunition reserves and were now waiting for munitions production to catch up with consumption.

1915, September 25–November 6. Renewed Allied Offensives in Artois and Champagne. This was another major coordinated effort planned by Joffre, and was again unsuccessful. In the **Second Battle of Champagne** the French lost more than 100,000 men and the Germans some 75,000. At the same time, in the **Third Battle of Artois,** the French continued their attacks against **Vimy Ridge** (September 25–October 30) while the British, a few miles north, smashed at **Loos** (September 25–October 14). The minor gains made were out of proportion to the casualties suffered: more than 100,000 French, 60,000 British, 65,000 German.

1915, December 17. Change in British Command. Blamed for the failure at Loos, Field Marshal French was relieved and General Sir **Douglas Haig** was placed in command of the BEF, which now comprised 3 armies.

COMMENT. *Increase of lethal firepower, both machine gun and field artillery, had revolutionized combat tactics and had given the advantage to the defense, which was able to bring up reserves to limit a penetration before the attackers could move forward sufficient reserves and artillery to exploit a breakthrough. This was particularly critical on the Western Front, where a continuous battle line prevented classical offensive maneuvers. The Germans, recognizing the change long before the Allies, had adopted an elastic defense, in 2 or more widely separated lines, highly organized with entrenchments and barbed wire, heavy in machine guns, and supported by artillery echeloned in depth. Assaulting troops broke through the first line only to be decimated by the fire from the succeeding lines and pounded by artillery beyond the range of their own guns.*

Appalling losses had been suffered during 1915 on both sides: 612,000 German, 1,292,000 French, and 279,000 British. The year ended with no appreciable shift in the hostile battle lines scarring the land from the North Sea to the Swiss Alps.

The Italian Front

1915, May 23. Italy Declares War on Austria. Adroit Allied diplomacy, offering substantial territorial gains, caused Italy to abrogate the

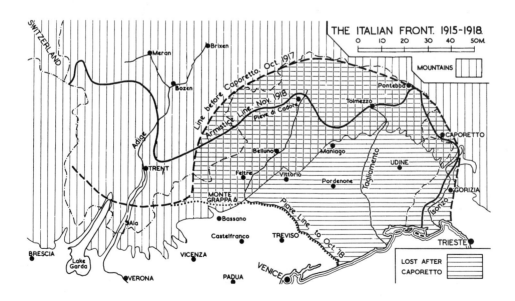

THE ITALIAN FRONT. 1915–1918.

0   10   20   30   40   50 M.

MOUNTAINS

LOST AFTER CAPORETTO

Triple Alliance and to enter the war. The total strength of the Italian Army, commanded by General **Luigi Cadorna,** was about 875,000, but it was deficient in artillery, transport, and ammunition reserves. The Italian plan was to hold the Trentino salient into Italy by offensive-defensive action, while operating eastward offensively in the Isonzo salient projecting into Austrian territory. The immediate objective was Gorizia, but Italian military men dreamed of advancing through Trieste to Vienna.

Austrian Dispositions. Despite the Triple Alliance, Austria had heavily fortified the entire mountain frontier with Italy. Austrian Archduke Eugene was in overall command of the Italian front. General **Svetozan Borojevic von Bojna,** with some 100,000 men, held the critical Isonzo sector.

1915, June 23–July 7. First Battle of the Isonzo. The Italian Second Army (General **Pietro Frugoni)** and Third Army **(Emanuele Filiberto, Duke of Aosta**), totaling approximately 200,000 men and 200 guns, battered in vain against the Austrian defenses.

1915, July 18–August 3. Second Isonzo. Cadorna, bringing up more artillery, tried again. The Austrians, reinforced by 2 additional divisions, held firm. The Italians broke off the struggle when their artillery ammunition gave out. Italian losses in these two battles amounted to about 60,000 men; the Austrian casualties totaled nearly 45,000.

1915, October 18–November 4. Third Isonzo. The Italians, reorganized and strengthened, and supported now by 1,200 guns, struck once more at Gorizia and were again repulsed.

1915, November 10–December 2. Fourth Isonzo. This was really a continuation of the third battle. When the offensive broke off, no material gain had been made to show for the Italian loss of 117,000 men in the 2 battles. The Austrians had lost almost 72,000 men.

COMMENT. *As in France, the invulnerability of highly organized positions to frontal assault had been proven. The Austrian defense was skillful; the Italian offensive tactics were inept, despite the gallantry shown by their infantry on many occasions. The Italian strategic objective—capture of Trieste, thence on the road to Vienna through the Ljubljana Gap and the Danubian Plain—was sound. It was, in fact, the only offensive open to Italy, since successful northern movement of major forces through the Alps not only led nowhere but was patently impossible. Germany took no part in this campaign, since Italy was technically not at war with her; this did not increase Austro-German cordiality.*

The Eastern Front

HINDENBURG'S WINTER OFFENSIVE, January–March

The Central Powers, reinforcing their armies, launched a great double offensive under Hindenburg. The Austro-German South Army (German General **Alexander von Linsingen**) struck northwest through the Carpathians on Lemberg. On its left the Austrian Third Army (General von Borojevic) was to rescue Przemysl from siege, while on the right the Austrian Seventh Army (General **Karl von Pflanzer-Baltin**) supported the main effort. From East Prussia the German Eighth and Tenth armies under Hindenburg's direct control struck east from the Masurian Lakes.

1915, January 31. Battle of Bolimov. This was a feint by the German Ninth Army, aimed at Warsaw, designed to distract Russian attention. Poison gas was used for the first time. The inno-

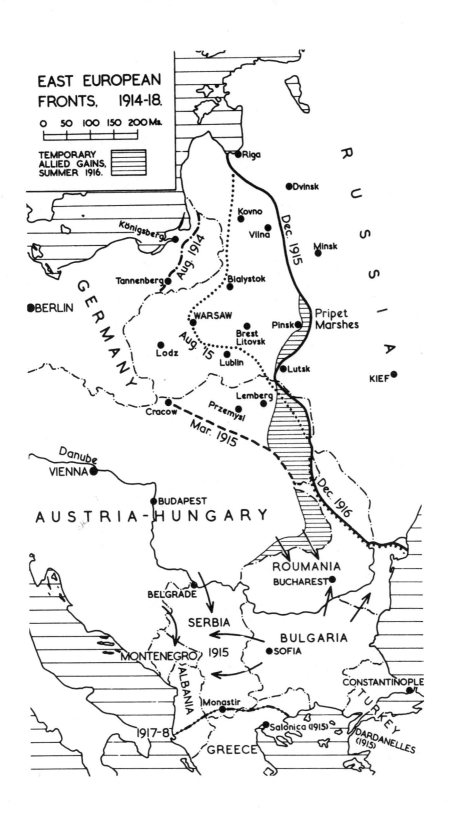

EAST EUROPEAN
FRONTS, 1914-18.

0   50   100 150 200 Ms.

TEMPORARY
ALLIED GAINS,
SUMMER 1916.

R U S S I A

Riga
Dvinsk
Kovno
Vilna
Minsk
Königsberg
Dec. 1915
Aug. 1914
Tannenberg
Bialystok
BERLIN
WARSAW
Pinsk
Pripet
Marshes
Aug. 15
Brest
Litovsk
Lodz
Lublin
Lutsk
KIEF
Lemberg
Cracow
Przemysl
Mar. 1915
Danube
VIENNA
Dec. 1916
BUDAPEST
A U S T R I A - H U N G A R Y
ROUMANIA
BUCHAREST
BELGRADE
SERBIA
1915
BULGARIA
MONTENEGRO
SOFIA
ALBANIA
CONSTANTINOPLE
Monastir
1917-8
Salonica (1915)
DARDANELLES
(1915)
GREECE
TURKEY
GERMANY

vation was not impressive; the gas was not very effective in freezing temperatures and the Russians did not report the gas attack to the other Allies (see p. 1038).

1915, February 7–21. The Winter Battle (or Masuria, or Second Masurian Lakes). Farther north, the Eighth Army, in a blinding snowstorm, began Hindenburg's offensive by hitting the left flank of Baron **Siever's** Russian Tenth Army (February 7). Next day the new German Tenth Army (General **Hermann von Eichhorn**), to the north, rolled up the Russian right. Despite desperate defense, the Russians were rapidly driven back into the **Augustow Forest.** There the Russian XX Corps, after resisting heroically, was surrounded and surrendered (February 21); its defense had enabled the other 3 corps of its Tenth Army to escape encirclement. Some 90,000 prisoners were taken. In all, Russian casualties were about 200,000 men in this campaign, an impressive tactical victory for the Germans but without great strategic value. A newly formed Russian army—the Twelfth (General **Wenzel von Plehve**)—counterattacking Hindenburg's right (February 22), halted further German progress after a 70-mile advance.

1915, February–March. Austrian Failure in Galicia. Linsingen's Carpathian advance broke down in freezing snow on a poor road net. Pflanzer-Baltin was initially more successful, capturing **Czernowitz** with 60,000 prisoners (February 17), but his further progress was halted by a Russian counterattack. Borojevic's efforts to relieve Przemysl failed.

1915, March 22. Surrender of Przemysl. After a siege of 194 days, the fortress and its garrison of 110,000 men surrendered to the Russians.

1915, March–April. Russian Counteroffensive. Following these defensive successes, the Russians resumed their advance through the Carpathians, but were checked by the Austro-German South Army (April 2–25).

The German Spring–Summer Offensive, May–August

1915, April. German Build-up in the East. Under instructions from the Kaiser, Fal-

kenhayn now gave full priority to the Eastern Front. Sending reinforcements, he came east to assume direct overall command. While Hindenburg's army group kept the Russians busy north of Warsaw, the new Eleventh German Army, under General August von Mackensen, supported by Austrian units, was to make the main effort farther south between Tarnow and Gorlice.

1915, May 2–June 27. The Gorlice-Tarnow Breakthrough. Concentrating superior force for the main effort, the Austro-German armies crashed through the Russian Third Army on a 28-mile front, following a 4-hour bombardment. The southern face of the great Russian Polish-Galician salient began to crumble. Przemysl was retaken (June 3), Lemberg occupied (June 22), and the Dniester crossed (June 23–27).

1915, June–September. Russian Retreat. Thrusting into northern Poland, General **Max von Gallwitz'** new German Twelfth Army advanced toward Warsaw, which was abandoned by the Russians (August 4–7). The entire Russian front was in complete collapse; another Cannae seemed to be inevitable as the salient melted back to the Bug (August 18). Brest Litovsk fell (August 25) and Grodno (September 2). The occupation of Vilna (September 19) marked the high-water mark of the colossal 300-mile advance. Skillfully, however, Grand Duke Nicholas preserved his armies intact, and they withdrew in fairly good order, evading German attempts at envelopment. Autumn rains finally turned roads into quagmires, and the reeling Russians were able to halt the German advance. By year's end the Eastern Front was a line running north and south from Riga on the Baltic to the eastern end of the Carpathians.

COMMENT. *In sharp distinction from the Western Front, this had been a war of movement on a grand scale, part of it in mountainous terrain, all of it hampered by primitive road conditions. The winter campaign was particularly arduous. German operations had been both methodical and brilliant. Excellent staff coordination enabled surprise, facilitated by employing short, intensive artillery preparations and by moving assault troops into position only at the last possible moment—usually at night. Austrian operations were spotty, due partly to lower professional standards and partly to friction between the German and Austrian high commands. The Austrians resented German arrogance. On the Russian side, poor troop leadership and lack of weapons, munitions, and supplies were jointly responsible for defeat. Only the marvelous capacity of the Russian soldier to fight while enduring incredible hardships, and the ability of the Grand Duke Nicholas to piece and patch together his baffled,*

defeated armies, prevented loss of the war in mid-1915. His reward was to be relieved of command (August 21) and banished to the Caucasus front. Czar **Nicholas II** *assumed personal command of the Eastern Front, with General* **Mikhail Alekseyev** *his chief of staff. Russian casualties on this front in 1915 were more than 2 million men, of whom about half had been captured. Combined German and Austrian casualties were in excess of 1 million.*

The Balkan Front, 1915

Direct communication between Turkey and her allies was essential to the Central Powers if the Turkish Straits were to be held and Russia kept isolated from the western Allies. The direct railway line passing through Serbia had been closed since the beginning of the war; munitions from Germany passed through neutral Rumania until June, when Rumania closed the channel. Both sides pressured Bulgaria to join the war; Germany won when the Gallipoli invasion was repulsed (August 15; see p. 1046). The Central Powers at once prepared to renew the attack on Serbia. Her call for assistance was answered by Greek mobilization against threatening Bulgaria; Greece requested 150,000 Allied troops to aid her in assisting Serbia. Small French and British expeditionary forces soon disembarked at Salonika (October 9). On the same day a political upheaval in Greece completely changed the situation; pro-German King **Constantine** dismissed his pro-Allied prime minister, **Eleutherios Venizelos,** and announced he would keep Greece neutral.

1915, October 6. Austro-German-Bulgarian Invasion of Serbia. Meanwhile 2 armies—1 Austrian (General **Hermann Kövess von Kövessháza**) and 1 German (General von Gallwitz)—drove south across the Serbian Save-Danube border. Two Bulgarian armies struck west (October 11); 1 on Nish, the other on Skoplje. Newly promoted Field Marshal von Mackensen was in overall command of the joint campaign; the plan was Falkenhayn's. Putnik's Serbian Army, struck in front and flank by forces nearly double its strength— 330,000 men—escaped envelopment but was rolled up and driven southwest in a series of bitter encounters. General **Maurice P. E. Sarrail,** commanding the French element of the allied force at Salonika, made a tentative stroke

up the Vardar Valley to link with the Serbs, but was turned back by superior Bulgarian forces. General Sir **Bryan T. Mahon,** commanding the British element, got no farther than the Bulgarian border.

1915, November 24. Serbian Retreat. While the Allied Salonika forces entrenched against Bulgarian pressure, the Serbian Army withdrew into Montenegro and Albania. After a dismal retreat through the snow-covered mountains, the remnants of the Serbian Army, accompanied by a horde of civilian refugees, reached the Adriatic, pursued by Austrian General Kövess. Serbian losses were more than 100,000 killed or wounded, 160,000 taken prisoner, and 900 guns. The survivors were transferred by Italian and French ships to the island of Corfu (January), where they were re-fitted by the Allies. Montenegro was also occupied by the Austrians.

COMMENT. *Serbia was doomed when Bulgaria entered the war. Allied intervention attempts were too little, too late, and further complicated by disunity of command. Sarrail and Mahon at Salonika were acting under separate instructions from their respective governments. To make matters worse for them, they were in the unenviable situation of occupying neutral territory, with the Greek Army potentially threatening their rear.*

The Turkish Fronts

THE DARDANELLES AND GALLIPOLI

THE PRELIMINARIES

1915, February—March 18. Allied Naval Assault. A Franco-British fleet under British Vice Admiral **Sackville Carden** attempted a systematic reduction of the formidable fortifications lining both sides of the narrow straits. More than 100 medium- and heavy-caliber Turkish guns swept the surface in cross fire, with 11 mine belts and an antisubmarine net below the surface. Long-range Allied bombardment of the outer forts (February 19) was followed by a second (February 25), which silenced them. After preliminary shelling and mine sweeping, the principal fortifications at the Narrows were attacked (March 18). Carden having broken down physically, Rear Admiral **John de Robeck** took command. He had, besides smaller craft, 16 battleships—including the *Queen Elizabeth,* most powerful superdreadnought then afloat, carrying 8 15-inch guns. By midafternoon the Turkish Narrows batteries were also silenced. Then 3 old battleships were suddenly sunk in an undetected minefield and 3 others were disabled, 1 of them by gunfire. Actually, at this time the Turks were at the end of their resources; their ammunition was nearly expended, some batteries had been demolished, and all fire-control communications put out of action. But de Robeck did not know that.

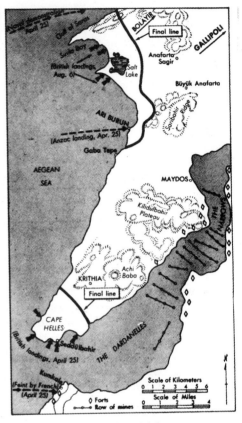

Gallipoli campaign

He called off the attack and hurried his 10 remaining capital ships out of the Strait.

1915, March–April. Assembling of the Army Expedition. Meanwhile, a hastily gathered British expeditionary force (including 1 French division) under General **Ian Hamilton** was en route from England and Egypt to the Gallipoli Peninsula. Hamilton himself was at this moment an observer on board a British warship. The total strength of his force was 78,000 men. As the first elements gathered at Mudros Bay on the island of Lemnos (mid-March), it was discovered that the contingent from England had been loaded haphazardly, with guns and ammunition on separate ships. The transports had to move (March 25) to Alexandria, Egypt, where they were combat-loaded (by unit) with men, their guns, ammunition, and equipment all on the same ship. This caused a month's delay, while the Turks, fully alerted to the impending landings, improved their dispositions. German General Liman von Sanders, in command, had some 60,000 men on the peninsula, disposed in an elastic defense.

THE FIRST LANDINGS, APRIL 25

The plan provided for 2 daylight assaults, with naval gunfire support. One of these landings was to be at Cape Helles on the tip of the peninsula; the other—beyond mutual-support distance—was to be about 15 miles farther north at Ari Burnu, on the western side of the peninsula. At the same time, the French division was to make a diversionary landing on the Asiatic side of the Strait, while a naval demonstration at Bulair—the neck of the peninsula, 50 miles northeast of Helles—distracted Sanders' attention.

At Ari Burnu the Anzacs (Australian and New Zealand Army Corps), landing in force, moved up the slopes toward **Chunuk Bair,** a height dominating the entire peninsula and the Narrows beyond. But a young Turkish reserve division commander with an eye for terrain moved first. Personally leading a battalion, **Mustafa Kemal** rushed for the height, the remainder of his division quickly following. Even though his troops were outnumbered, Kemal's vicious counterattack from the ridge drove the Anzacs back to the beach with a loss of 5,000 men. Here, ordered by Hamilton to hold on, the Anzacs dug in to retain a narrow beachhead.

At Helles, the 29th British Division landed on 5 beaches in a welter of mismanagement, incurring murderous losses. But a part of the division, overwhelming the Turkish beach defense, made its way gallantly almost to its objective, **Achi Baba,** a dominant hill mass. The division commander, safe on his command ship offshore, was not on the spot to handle the situation. Lacking further orders, the British troops stopped to brew tea below the still-unoccupied height. When they renewed their advance, it was too late. The Turks had occupied the hill in force.

Chunuk Bair and Achi Baba would never be taken by the British. Without either of these 2 critical heights the landings were doomed to failure. As the Turks ringed the tiny beachheads with entrenchments, the British found themselves involved in the same kind of trench warfare they had known on the Western Front—but with even less room for maneuver.

THE SECOND LANDING, AUGUST 6–8

Following 3 months of the bitterest kind of fighting on the rocky slopes of the peninsula, Hamilton, reinforced by 3 more British divisions, attempted a coordinated assault. Meanwhile, although British submarines had actually penetrated the Straits and sunk some Turkish craft, German submarines and Turkish destroyers accounted for 3 more British battleships off the Gallipoli coast. The Allied capital ships sought protected harbors; the *Queen Elizabeth* herself was ordered home. So Hamilton's second assault was made without adequate naval gunfire support.

The Anzacs at Ari Burnu were to make the main effort with a night attack, driving for Chunuk Bair and other heights on the Sari Bair ridge. The new divisions, landing at Suvla Bay to the north, were to make a secondary attack. From the Helles position a holding attack was to pin down Turkish reserves.

The holding attack fulfilled its mission, but the Anzac attack bogged down in the darkness. Only the Suvla Bay landing, made without serious opposition, promised success. But General Sir **Frederick Stopford,** the corps commander, lacked vigor and drive. The advance lagged until Turkish reinforcements had time to come up, and again it was too late.

The entire operation had failed. Russia was permanently cut off from her allies. Hamilton was relieved (October 15), General Sir **Charles Monro** taking his place. Monro's recommendation for an evacuation was approved (November 23). Excellent planning, staff coordination, and prompt execution did the rest. By December 10, all supplies and many of the troops had been moved out. The remaining crust of 35,000 men slipped away under the eyes of the unsuspecting Turks, completing (January 8–9, 1916) a masterpiece of deception without loss of a single man.

Allied casualties for the entire Dardanelles campaign amounted to 252,000. The Turks lost almost as many: 251,000, but of these 21,000 died of disease.

COMMENT. *With possible exception of the Crimean War, the Gallipoli expedition was the most poorly mounted and ineptly controlled operation in modern British military history. Surprise had been lost even before the inception of the plan, for a premature bombardment of the outer Dardanelles defenses by a British naval force early in the war (November 1914) had awakened both Turks and Germans to the danger. Under Liman von Sanders' competent direction the fortifications had been vastly strengthened. The prize was a rich one, for success would mean keeping Russia in the war and probably knocking Turkey out; its attainment should have been confided to a single commander, provided with the best of means. Instead, with improvised organizations, both naval and army commanders worked independently.*

De Robeck's decision to break off the naval action (March 18) when success was practically in his hands was particularly questionable. He had suddenly lost 6 of his 16 capital ships. But he knew a supporting expeditionary force was on the way. Under those conditions, if only one battleship had survived to run the gantlet and place Constantinople under its guns, his mission would have been fulfilled. But Nelsons and Farraguts are few and far between.

Back in the War Office, Kitchener failed to assure compliance with logistical, strategic, and tactical fundamentals in mounting the expeditionary force. And he refused Hamilton's requests for adequate forces and staff officers.

Hamilton possibly erred in the landing operations by trying to direct them by remote control from a warship. It has been suggested that he should have been ashore to see for himself and to whip his erring subordinates into decisive action. He should not have put up with the incompetence of some of these. However, it can be argued that he did as competent a job as was possible under the circumstances, and with the limited means and subordinates provided him by Kitchener.

Sharply contrasting with all the previous fumbling was the planning of the evacuation by General Monro and its execution by General **William Birdwood.**

On the Turkish side, German Liman von Sanders conducted a brilliant active defense. Mustafa Kemal, his chief subordinate, shone as an aggressive field commander. The fighting men on both sides performed wonders under most arduous conditions.

THE CAUCASUS FRONT

1915, January 1–3. Battle of Sarikamish (continued). Russian General Vorontsov, with about 100,000 men, lay in the vicinity of Kars to oppose Enver's advance. Colossal mismanagement of a winter campaign had frittered away at least 15,000 of Enver's 95,000 strength through frostbite and desertion before the battle had begun. The Turk's dream of a wide envelopment of the Russians was spoiled by a Russian counterattack, which smashed his army (January 3). The Turks lost 30,000 dead, while thousands more froze to death in retreat. Only about 18,000 effectives reached Erzerum. Enver gave up the field command and returned to Constantinople. Vorontsov failed to seize advantage by a pursuit; he was replaced by General **Nikolai Yudenich,** who had more initiative.

1915, April–May. Armenian Revolt. Turkish massacres of Armenians, on suspicion they

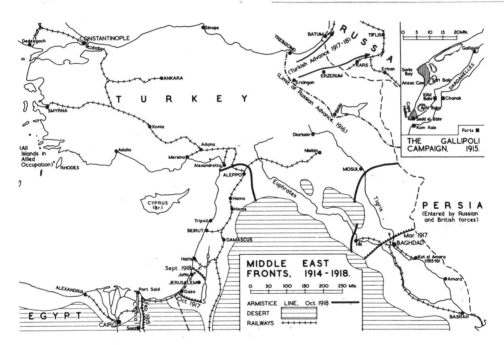

THE GALLIPOLI CAMPAIGN, 1915.

MIDDLE EAST FRONTS, 1914–1918.

were aiding the Russians, precipitated an Armenian revolt. The rebels seized the fortress of Van (April 20), holding it until the arrival of the Russians (May 19).

1915, July. Battles of Malazgirt and Kara Killisse (Karakose). The new Turkish commander, **Abdul Kerim,** struck and defeated a Russian corps north of Lake Van (July 16), near the site of the Battle of Manzikert (see p. 329). The Turks cautiously advanced eastward, but Yudenich sent a force of 22,000, mainly Cossacks, under General **N. N. Baratov,** to strike the Turk left, causing Kerim to withdraw after sustaining heavy casualties.

1915, August 5. Turkish Recapture of Van. In the ebb and flow of fighting in the Armenian mountains, the Russians were forced to evacuate Van, which the Turks promptly reoccupied.

1915, September 24. Arrival of Grand Duke Nicholas. Appointed Viceroy of Caucasia, Nicholas retained Yudenich in field command, though taking active part in planning. Preparation for a large-scale offensive began.

EGYPT AND PALESTINE

By the opening of 1915, British defense of the vital Suez Canal was partly organized. However, General Sir **John Maxwell**'s plans utilized the canal more as an obstacle to possible invasion than as a base for future operations to the northeast.

1915, January 14–February 3. Thrust against the Canal. Djemal Pasha, Turkish Minister of Marine, personally led a force of 22,000 men secretly across the Sinai Peninsula from Beersheba. German General Baron **Friedrich Kress von Kressenstein,** Djemal's chief of staff, had set up an efficient organization. Advance elements of the force struck across the canal in German-type ponton boats (February 2), but the assault was broken up by the defenders. Djemal retired to Beersheba with a loss of 2,000 men. No further Turkish attack was made against the canal, but the threat held much-needed British reinforcements back from Gallipoli.

MESOPOTAMIAN FRONT, 1915

1915, January–June. Tigris River Expedition. British forces in the Tigris Valley of lower Mesopotamia were reinforced to a strength of 2 infantry divisions, a cavalry brigade, and some heavy artillery, all commanded by General Sir **John E. Nixon,** at Basra. A Turkish attack on **Qurna,** a fortified post west of Basra, was repulsed (April 12–14). Another Turkish attack on **Ahwaz,** an outpost some 50 miles north of Basra, was driven off (April 24). Nixon, under orders from India to explore the possibility of an advance on Baghdad, sent Major General **Charles V. F. Townshend** with a reinforced division and a small naval flotilla up the Tigris. After overwhelming a Turkish outpost near Qurna in an amphibious assault (May 31), Townshend continued upriver to Amara, occupying it (June 3).

1915, July 24. Battle of Nasiriya. Major General **George F. Gorringe** with a small expeditionary force moved up the Euphrates to protect Townshend's flank and communications by wiping out a Turkish defensive position at **Nasiriya.** Turkish resistance was strong and only after a month of campaigning was Gorringe successful. British casualties were 533; the Turkish unknown. But they left 1,000 prisoners and 17 guns.

1915, August–September. Advance to Kut. Townshend was now reinforced and ordered to take Kut-el-Amara at the confluence of the Tigris and Shatt-el-Hai rivers. By September 16 his attack force was concentrated on the south bank of the Tigris near Sannaiyet, just below Kut. His communications were strained by insufficient river transport over the almost 300-mile winding Tigris, and one of his 2 infantry divisions was scattered downriver to protect the route. **Nur-ud-Din** Pasha, Turkish commander, with 10,000 men and 38 guns, held an entrenched position astride the Tigris below Kut. Townshend had 11,000 men and 28 guns.

1915, September 27–28. Battle of Kut. After receiving supplies, Townshend made a demonstration on the south bank (September 26), then crossed the river, driving the Turks from their positions with the loss of 5,300 men (including 1,300 prisoners) and all their guns. They retreated unmolested to a prepared position at Ctesiphon, midway between Kut and Baghdad. Townshend's troops, lacking land transport, could not pursue, and the naval flotilla was delayed by the shallows of the dry season. British losses were 1,230. After lengthy debate, the British high command finally ordered Nixon and Townshend to continue on to Baghdad, despite Townshend's plaint that his force was inadequate. There were no reinforcements available, however.

1915, November 11–22. Advance to Ctesiphon. Townshend advanced (November 11) with improvised camel and donkey transport supplementing the river boats. Arriving before Ctesiphon he discovered that the Turks had fortified extensively and that Nur-ud-Din had been reinforced to a strength of 18,000 regulars and additional Arabs, with 45 guns. Townshend, having added one of the communications line brigades to his forces, mustered some 10,000 infantry, 1,000 cavalry, and 30 guns. He also had, for the first time in that theater, a squadron of 7 airplanes.

1915, November 22–26. Battle of Ctesiphon. Townshend attacked savagely to turn the Turkish left. But he pressed his luck too far, for he held out no reserve. The first Turkish line was taken and held against repeated counterattacks, but Townshend's troops could do no more. After 4 days of battle, during which more Turkish reinforcements arrived, Townshend decided to withdraw after sending back his wounded. He had lost 4,600 men; the Turks,

6,200. Turkish pursuit was not vigorous, and after a rear-guard action at **Umm-at-Tubal** (December 1) the force arrived at Kut (December 3). Feeling now that his infantrymen were too exhausted to retreat farther, Townshend sent his cavalry away, remaining at Kut to await reinforcements. He had more than 2 months' food supply. Kut was soon invested by the Turks (December 7).

COMMENT. *Townshend's campaign is an example of military ''absentee landlordism'' at its worst. Operations in Mesopotamia were originally under the direction of the Indian Army. In the last stages the British War Office stepped in. Neither agency was in touch with the situation; both relied on the recommendations of Nixon at Basra, who, to say the least, held an exaggerated view of the capabilities of men in protracted combat at the end of an insufficient supply line.*

Persian Front

At the outset of the war, Russian troops occupied most of northern Persia, despite Persian declaration of neutrality. When Turkey entered the war, the British occupied the northwestern Persian Gulf coast to protect their oil interests and to obtain a base for operations in Mesopotamia. At the same time the Turks seized most of Persian Kurdistan. During the Battle of Sarikamish (see p. 1047), Turkish forces seized **Tabriz** from the Russians (January 7), but were soon forced to withdraw (January 30). Sporadic minor fighting continued through the year in western Persia on the flanks of the operations taking place in the Caucasus and Mesopotamia.

The War at Sea

1915, January 24. Dogger Bank Action. The German battle-cruiser squadron under Vice Admiral **Franz von Hipper** moved out (January 23) to raid the English coast and harass the British fishing fleet. Warned by radio intercepts, the Grand Fleet steamed to meet an expected full-dress attack. British Admiral **David Beatty**'s battle-cruiser squadron fell in with Hipper off the Dogger Bank, midway between England and Germany. Beatty had 5 ships, Hipper 3. Both were accompanied by lighter cruisers and destroyers. Hipper wisely fled. Beatty, with superior speed, overhauled him, disabling a heavy cruiser and damaging Hipper's flagship, *Seydlitz*. Then HMS *Lion*, Beatty's

flagship, was damaged and fell out of the line. Through misunderstanding of signals, the remainder of the British squadron contented itself with sinking the disabled heavy cruiser, allowing Hipper's remaining ships to get away. The Kaiser ordered his fleet to avoid further risks of losing major warships.

1915, February 4. Initiation of German Submarine Campaign. This was directed against merchantmen in waters surrounding the British Isles. Neutrals were also attacked; a Norwegian ship was sunk (February 19) and—despite American warning to Germany—the American tanker *Gulflight* torpedoed, causing the death of 2 of her crew (May 1).

1915, May 7. Sinking of the *Lusitania*. The British luxury liner was torpedoed without warning by *U-20* off the Irish coast. Among the 1,198 lost were 124 Americans. Feeling in the U.S. ran high, despite the facts that the liner carried a war cargo, including gold and ammunition; was under orders not to halt if hailed; and, prior to her departure from New York, the German Embassy in Washington had publicly warned Americans not to travel in the ship. A vehement U.S. protest was filed.

1915, August 19. Sinking of the *Arabic*. When this British liner was sunk with loss of 4 more Americans, reaction became so hostile that Germany announced (September 1) cessation of unlimited submarine war. However, by year's end German U-boats, operating all over, had accounted for almost one million tons of Allied shipping.

Operations in 1916

The Global Situation

The year opened with the Central Powers and the Allies at approximately equal strength, although Germany was perhaps better organized. The man-power drain in France was serious. Britain was on the verge of instituting compulsory service to fill her expanding armies. Unrest in Ireland was approaching rebellion. Russia, with plenty of man power, hoped for time to reorganize and supply it.

In a reversal of the previous year's strategy, Germany now sought decision on the Western Front because, as Falkenhayn told the Kaiser, even if the goal were not reached, France would be "bled white" to prevent it. Germany, too, expected to act aggressively at sea, including submarine warfare.

Joffre, feeling that the Allies' worst failure had been lack of coordination, had held a conference at Chantilly (December 1915) and succeeded in getting agreement from Britain, Russia, Italy, and Rumania that coordinated offensives would be launched on the Western, Eastern, and Italian fronts, probably about June, when Russia would be ready. Mutual-support measures were also framed.

In Britain a totally new and revolutionary weapon was being developed: the armored, track-laying tank, conceived partly by Colonel **Ernest Swinton** and partly by imaginative First Lord of the Admiralty Winston Churchill, who supported the new project when the War Office refused to sponsor development.

The Western Front

1916, January–December. Zeppelin Raids on England. The German aerial attacks continued, causing increasing numbers of casualties as their size and effectiveness increased. It was not until late in the year that the first Zeppelin was shot down over England by a British plane (September 3). Two more were soon afterward shot down by antiaircraft fire (September 26). As more were lost to British planes and antiaircraft artillery, German air raids over England were drastically reduced in numbers.

The Battle of Verdun, February 21–December 18

Both Joffre and Falkenhayn planned great offensives to break the deadlock in the West. But the Germans struck first.

Following an enormous bombardment, the Crown Prince's Fifth German Army attacked the fortified, but lightly garrisoned, region of Verdun, lying in the middle of a salient jutting into the German zone, whose southern face was framed by the countersalient of St.-Mihiel. Through the city ran the Meuse gorge, with the rocky escarpment of the Meuse Heights to the east. Beyond lay the Woëvre, a flat clay plain. The front line itself ran some 3 miles beyond the outer forts on the Meuse Heights.

The German assault, on an 8-mile front, bit through the outlying mobile defense zone on the east. Failure of the retiring French to secure a key position, **Fort Douaumont,** was almost fatal. But Joffre, prohibiting further retreat, determined to hold Verdun at all costs as a symbol of French determination as much as to retain an anchor for his battle lines. He sent General **Henri Philippe Pétain** to assume command and retrieve the situation (February 26). By that time the German assault had reached its first limited objective and halted. Pétain, reorganizing his command, brought up large reinforcements and more artillery.

The next German attack (March 6), launched against the western face of the salient, was at first successful, but was checked when Pétain ordered counterattacks to regain "every piece of ground lost." For the rest of the month a series of attacks and counterattacks heaped the ground with corpses. Rapid rotation of units restored French combat troops exhausted in battle, and Pétain got 2 capable lieutenants: Generals **Robert Nivelle** and **Charles Mangin.** Pétain's watchword for the defense (also attributed to Nivelle) became France's motto for the remainder of the war: *"Ils ne passeront pas!"* ("They shall not pass!")

With all other communications—road and rail—cut off, Verdun's only link with the French rear areas was a 40-mile-long secondary road to Bar-le-Duc. Over it Pétain methodically organized a supply system. Despite persistent German artillery harassment, an endless chain of military trucks moved over this road—called *"La Voie Sacrée"* (Sacred Way)—at 14-second intervals in and out of the salient. Permanent road gangs filled shell craters as fast as they were made.

On April 9 the third German offensive struck both sides of the salient, only to be checked again. Successive attacks and counterattacks followed, until the western German effort petered out (May 29). Pétain, meanwhile, had been promoted to army-group command; he was succeeded by Nivelle at Verdun (April). Savage German assaults continued on the east against **Fort Vaux** and **Thiaumont Farm.** Vaux finally capitulated (June 9) after its water had given out and its interior was churned to rubble. The German Crown Prince personally congratulated its commander, Major **Raynal,** on his heroic defense.

Renewed German assaults on the western salient face in late June and early July almost broke the French line. Phosgene gas was used here for the first time. Pétain recommended abandonment of the western Meuse line, but Joffre refused to permit it. The Somme Offensive was about to begin (see below), and he could not afford to lose the Meuse. The French clung to their positions, and the Germans hesitated. Pressing demands for replacements to meet the Brusilov Offensive on the Eastern Front (see p. 1053) then drained 15 German divisions from Verdun. Falkenhayn was relieved of command a little later (August 29) and the Hindenburg-Ludendorff team, replacing him, decided to cut their losses and to go on the defensive in the west.

In the fall the French—now under Mangin—went over to the offensive, retaking Forts Douaumont (October 24) and Vaux (November 2). After a lull of several weeks, Mangin again attacked, pushing the French front forward almost to the lines held in February, capturing over 11,000 prisoners and 115 guns (December 15–18). This brought the campaign to a close. The casualties in this bitterly fought battle were approximately 542,000 French and 434,000 German.

THE FIRST BATTLE OF THE SOMME, JUNE 24–NOVEMBER 13 1916/ 4 months 20 days

Joffre's long-planned offensive—delayed for several months by the crisis at Verdun—was finally launched by a stupendous 7-day artillery preparation, but—again due to Verdun—with the British playing the leading role, instead of the joint operation initially planned by Joffre. The main effort was to be made by the British Fourth Army (General **Henry S. Rawlinson**), north of the Somme, with General **Edmund Allenby**'s Third Army farther north also attacking. South of the river, armies of Foch's Army Group of the North would make a holding attack.

On July 1 the British infantry, following a rolling artillery barrage, dashed themselves against the highly organized defensive positions of the German Second Army. Small gains were made, but by nightfall the British had lost about 60,000 men, 19,000 of them dead—the greatest one-day loss in the history of the British Army.* The French, surprisingly, made greater advances, since the Germans had not expected them to take part in the initial assault and so were surprised by the attacks south of the Somme.

Despite the appalling losses of the first day, the British continued to forge ahead in a series of small, limited attacks. Falkenhayn, determined to check the threat, began shifting reinforcements from the Verdun front. To this extent, therefore, Haig had accomplished one objective of the offensive.

A British night attack on July 13 cracked the German second line. British cavalry rode into the gap—the last time horse cavalry was used on a large scale in Western Europe. But other reserves were slow to arrive, and the horsemen were soon mowed down by machine guns, then engulfed by a German counterattack, which again sealed the line. The Allied offensive deteriorated into a succession of minor but costly small actions.

Haig launched another major offensive on September 15, southwest of Bapaume. The British tanks had been secretly shipped to the front, and spearheaded the attack. Despite the surprise their appearance caused to the Germans, the tanks were underpowered, unreliable, too slow, and too few in number to gain a decisive victory. (Out of 47 brought up, only 11 got into the battle.) The British made substantial gains, but a breakthrough eluded them. Nevertheless, British and French continued attacks gained small bits of ground through mid-November.

British losses in this campaign were 420,000; French, 195,000. German casualties, which included a great proportion of prewar officers and noncommissioned officers, came to a shocking total of some 650,000.

COMMENT. *The British armies—brave, well-equipped, and gallantly led—could not stand up to machine-gun fire interlacing a defensive zone echeloned in depth for miles. In 4 1/2 months of almost continuous attacks, they were able to advance only a little more than 8 miles. Aerial operations were the most extensive yet seen in war. The British use of tanks was premature, like the German use of gas at Ypres the year before. In both cases the user did not anticipate the effect of a new and fearsome weapon, and so was unprepared to exploit. The German defensive role was magnificent, but repeated German*

---

*A comparison with the Normandy landing, largest assault operation of World War II, reveals that the combined Anglo-American armies fought for 20 days in Normandy before sustaining 60,000 casualties.

counterattacks proved even more costly than Allied assaults. <u>The Somme bled Germany white in experienced small-unit leaders. That army would never be the same again.</u>

The Italian Front

1916, March 11–29. Fifth Battle of the Isonzo. Like its predecessors, this was a succession of blunt-nosed inconclusive conflicts. The Italian offensive was broken off when the Austrians attacked elsewhere.

1916, May 15–June 17. Austrian Trentino (Asiago) Offensive. Long planned, the attack caught the Italians unprepared. Archduke Eugene's Eleventh and Third armies overran General **Roberto Brusati**'s First Army. Terrain difficulties and Italian reinforcements finally checked the drive (June 10). An Italian counteroffensive and the need to rush troops eastward to stem the Brusilov Offensive (see below) caused Eugene to withdraw from some of the conquered terrain to defensive positions. Italian losses were more than 147,000 (including 40,000 prisoners), 300 guns, and great stores of supplies. Austrian losses were 81,000 (including 26,000 prisoners).

1916, August 6–17. Sixth Battle of the Isonzo. Cadorna, rapidly shifting forces on interior lines, struck the Austrian Isonzo front, depleted for the Trentino Offensive. **Gorizia** was taken, but no breakthrough effected. Psychologically, the operation boosted Italian morale, lowered by the previous heavy losses in the Trentino. Italian losses were approximately 51,000 against the Austrians' 40,000.

1916, September 14–November 14. Continued Italian Offensives. Three more Italian assaults, dubbed the **Seventh** (September 14–26), **Eighth** (October 10–12), and **Ninth** (November 1–14) **Battles of the Isonzo,** did little more than to exhaust Austrian powers of resistance. The Italians lost in these operations 75,000, the Austrians 63,000, including more than 20,000 prisoners.

The Eastern Front

1916, March 18. Battle of Lake Naroch. Responding to French appeals, the Russians launched a 2-pronged drive in the Vilna-Naroch area to counter the German Verdun assault in the west. Despite a 2-day preliminary bombardment—the heaviest yet seen on the Eastern Front—the Russian assault broke down in the mud of the spring thaw. Its cost—between 70,000 and 100,000 casualties and 10,000 prisoners—did not improve Russian morale. German losses were about 20,000 men.

1916, June 4–September 20. Brusilov Offensive. The Austrian spring offensive against Italy (see above) brought another appeal to Czar Nicholas for help. In response, capable and courageous General **Alexei A. Brusilov,** commanding the Russian Southwestern Army Group, attacked on a 300-mile front, forgoing any prior massing of troops, or preliminary artillery preparation, in order to gain surprise. Well planned, rehearsed, and executed, his assaults bit through the Austro-German line in 2 places. The Austrian Fourth Army was routed, the Seventh unraveled, 70,000 prisoners were taken, the vital rail junction of Kowel endangered. However, Brusilov received little or no help or cooperation from the 2 other Russian army groups on the front, and a counteroffensive by German General Alexander von Linsingen's army group (June 16) checked his northern thrust. Under orders from Alekseyev, Brusilov renewed the offensive (July 28), making further gains, until slowed down by ammunition shortages. His third assault (August 7–September 20) brought him into the Carpathian foothills. The offensive ended in sheer exhaustion, as German reinforcements hurried from Verdun (see p. 000) bolstered the shattered Austrians. Had it not been for these reinforcements, Austria probably would have been knocked out of the war in 1916.

COMMENT. *The Brusilov Offensive was the most competent Russian operation of World War I. Its strategic consequences may be summed up as weakening the Central Powers' offensives in Italy and at Verdun, contributing to the downfall of Falkenhayn, and elim-*

inating Austria forever as a major military power. But the Russians had lost one million casualties, more even than that populous nation could afford. The Brusilov Offensive did not cause the Russian Revolution, but it probably made revolution inevitable. Austrian losses were even greater, and the defeat contributed more than any other single factor to the disintegration of the Hapsburg Empire.

1916, August–December. Rumanian Participation in the War. After long haggling with the Allies for a promise of rich territorial gain, Rumania was so impressed by the early success of the Brusilov Offensive that she declared war on Germany and Austria (August 27). Rumanian armies at once invaded coveted Transylvania. Falkenhayn, demoted from Chief of the Imperial German General Staff to army commander, met them with the German Ninth Army and threw them back, while Mackensen, up from the Salonika front with the German-reinforced Bulgarian Danube Army, drove north through the Dobruja and crossed the Danube (November 23). Penned in a salient, Rumanian General **Alexandru Averescu** attempted to envelop Mackensen's left flank with a portion of his forces, while the remainder checked Falkenhayn's advance. Russian cooperation was essential to success, but was not forthcoming. The German armies linked; the Rumanians were disastrously defeated in the **Battle of the Arges River** (December 1–4). Bucharest was occupied (December 6), and by year's end the remnants of the Rumanian armies had been driven north into Russia, holding one tiny foothold in their own country. The bulk of Rumanian grain- and oil-producing areas were in German hands. Rumanian losses are estimated at from 300,000 to 400,000 (more than half non-battle casualties), and the campaign cost the Central Powers some 60,000 combat casualties and almost the same number of sick.

The Balkan Front, 1916

The Allied forces now held a fortified position—the "Bird Cage"—around Salonika. Sarrail was technically in command, but the British took orders from their home government. The Central Powers' strategy was one of containment. The Greek political situation complicated matters. Sickness decimated the Allied forces. The reconstituted Serbian Army, 118,000 strong, joined (July) and with additional reinforcements the Allied strength rose to more than 250,000. Sarrail decided on an offensive up the Vardar Valley. But the Bulgarians struck first.

1916, August 17–27. Battle of Florina. Bulgar-German attacks drove in the Allied forces to the **Struma River** line.

1916, September 10–November 19. Allied Counteroffensive. Sarrail slowly pushed northward, taking **Monastir** (Bitolje, November 19). Operations dwindled to a stop as Sarrail bickered with his subordinates. The year's campaign had cost the allies some 50,000 men. Bulgar-German losses were approximately 60,000.

1916, July–November. Operations in Albania. During the year an Austrian corps opposed an Italian corps along the Vayusa River in indecisive semi-independent operations. The Italians finally (November 10) pushed the Austrians north and linked with Sarrail's main body at **Lake Ochrida.**

The Turkish Fronts

EGYPT-PALESTINE-ARABIA

1916, January–July. Sinai Bridgehead. British operations in Egypt under General Sir **Archibald Murray** focused upon eastward extension of Suez Canal defenses into the Sinai Desert, a tremendous plan involving water supply, communications, and fortifications. Meanwhile, insurrection of the Senussi tribes in western Egypt necessitated a diversion of force until it was suppressed (March 14). Several minor actions occurred in the Sinai as British covering troops met Turkish resistance.

1916, June 5. Arab Revolt in the Hejaz. This was the result of Franco-British political negotiations with Arab chieftains, aimed at undermining Turkish strength. The Turkish garrison at **Medina** was attacked. **Hussein,** Grand Sherif of Mecca, proclaimed Arab independence. **Mecca**'s garrison surrendered (June 10) and the Turkish forces opposing Murray now found themselves hampered by Arab dissidence threatening their entire line of land communications north through Syria to the Taurus Mountains.

1916, August 3. Battle of Rumani. German General Kress von Kressenstein with 15,000 Turkish troops and German machine gunners struck the British Sinai railhead in a surprise attack. He was repelled, losing more than 5,000 men; British casualties were 1,100. As the year ended, the British had progressed to El Arish.

MESOPOTAMIA

1916, April 29. Fall of Kut-el-Amara. Townshend's besieged force (see p. 1049) vainly waited for rescue while British General **Fenton J. Aylmer** with 2 newly arrived Indian divisions battered futilely against the heavy Turkish contravallation (January). General **George F. Gorringe,** succeeding Aylmer, attempted a surprise attack (March 7) on the south bank of the Tigris, but was repulsed by elderly (73) but capable German General **Kolmar von der Goltz,** commanding the Turkish Sixth Army. Continued efforts to break through were unsuccessful. The food supply in Kut failed. With starvation near, Townshend capitulated, surrendering 2,070 British and some 6,000 Indian troops as a half-hearted Russian attempt to move from Persia on Baghdad bogged down. The unsuccessful British relief force suffered more than 21,000 casualties. Von der Goltz died of cholera just prior to the surrender; **Halil** Pasha succeeded him.

1916, December 13. British Advance. British General Sir **Frederick S. Maude,** appointed to the Mesopotamian command (August), had found himself reduced to a defensive role while the War Office and the Indian Army command debated pros and cons of possible British withdrawal from the theater. After chafing on the defensive for more than 2 months, Maude received permission to resume the offensive. He now began movement up both banks of the Tigris with a combat strength of 166,000 men, two-thirds of them Indians.

THE CAUCASUS

1916, January–April. Russian Winter Offensive. General Yudenich (see p. 1047), one of the few really capable Russian commanders, advanced from Kars on Erzurum on a broad front (January 11). The Turkish Third Army (Abdul Kerim) was rapidly rolled up at **Köprukoy** (January 18), narrowly escaping envelopment. Kerim retreated to Erzurum, with loss of some 25,000 men—a great number by frostbite in the subzero mountain climate. Yudenich stormed **Erzerum,** rapidly breaking through its ring of forts in a 3-day battle (February 13–16). Meanwhile, he launched a subsidiary offensive along the Black Sea coast, supported by Russian naval craft. **Trebizond** (Trabzon) was captured (April 18), facilitating Russian logistical support.

1916, June–August. Turkish Counteroffensive. Enver Pasha contemplated a dual drive: the Third Army (now under **Vehip** Pasha) along the Black Sea littoral, and a new Second Army (**Ahmet Izzim** Pasha) to advance on Bitlis and turn Yudenich's left. Yudenich, moving with characteristic rapidity and judgment (July 2), split the Third Army at **Erzinjan,** routing it completely (July 25). The Turks lost 34,000 casualties. Yudenich then turned on the Turkish Second Army. Kemal, hero of Gallipoli and now a corps commander, scored the only Turkish successes, capturing **Mus** and **Bitlis** (August 15), but Yudenich retook them (August 24). Both sides went into winter quarters early.

The Persian Front

Russian forces in this area were comparatively small; General N. N. Baratov, with some 20,000 men and 38 guns, moved on Kermanshah to divert Turkish forces from Mesopotamia. Reaching Karind (March 12), he announced his intention to move on Baghdad. The day he started, Kut fell (April 29; see p. 1055). **Halil** Pasha thereupon shifted most of his forces from the Kut area to protect Baghdad. Baratov attacked at **Khanikin** (June 1), was repulsed, and withdrew to Karind. Halil took the offensive and by August had retaken Kermanshah. Then this front quieted down again.

The War at Sea, 1916

THE SUBMARINE CAMPAIGNS AND CRUISER RAIDS, JANUARY–MAY

1916, February 21. Extended German Submarine Campaign. Germany proclaimed that after March 1 armed merchantmen would be treated as warships. Sinking of the British passenger liner *Sussex* (March 24), with loss of several American lives, brought vigorous protest from the U.S.

1916, May 10. Sussex Pledge. Germany announced abandonment of the extended campaign, and stated that passenger ships would not be sunk without warning.

1916, April 24–25. German Cruiser Raids. Yarmouth and Lowestoft were bombarded in hit-and-run raids.

THE BATTLE OF JUTLAND, MAY 31–JUNE I

The German High Seas Fleet under Vice Admiral **Reinhard Scheer** put to sea, cruising north toward the Skagerrak (May 30). Von Hipper's scouting fleet, 40 fast vessels built around a nucleus of 5 battle cruisers, led the way. Well behind was the main fleet of 59 ships, 16 of them dreadnoughts and 6 others older battleships. Warned of the sortie by imprudent German radio chatter, the Grand Fleet under Admiral Sir **John R. Jellicoe** at once put to sea. Leading was Beatty's scouting force of 52 ships, including his 6 battle cruisers and Admiral **Hugh Evan-Thomas'** squadron of 4 new superdreadnoughts. Jellicoe's main fleet, following, was composed of 99 vessels, 24 of them dreadnoughts. Over all, the British had 37 capital ships at sea; 28 dreadnoughts, and 9 battle cruisers; the Germans had 27 capital ships: 16 dreadnoughts, 6 older battleships, and 5 battle cruisers.

Beatty's 2 divisions steaming east, line ahead, with his battle cruisers on the right and Evan-Thomas' dreadnoughts to the left, next afternoon sighted Hipper's force, also in line ahead, steaming south (he had already sighted Beatty and was returning toward the German main fleet). It was now 3:31 P.M. As Hipper hoped, Beatty turned on a parallel course to the German squadron, signaling Evan-Thomas, whom Hipper had not yet sighted, to follow. Both battle-cruiser forces opened fire at 16,500-yard range, with the German gunnery more accurate. Beatty's flagship, *Lion,* leading, received several hits. Then in succession came mortal blows to the thin-skinned British battle cruisers. A salvo from *Von der Tann* tore into *Indefatigable*'s vitals; she blew up and capsized. *Derfflinger*'s accurate salvos sent *Queen Mary* to the bottom 20 minutes later.

THE FLEETS—BATTLE OF JUTLAND

Type	NUMBER AT JUTLAND	
	British	German
Battleships[a]	28	16
Battle Cruisers[b]	9	5
Battleships, Second-Line[c]	0	6
Armored Cruisers	8	0
Light Cruisers	26	11
Destroyer Leaders (Big Destroyers)	5	0
Destroyers	73	61
Minelayers	1	0
Seaplane Carriers	1	0
Submarines	[d]	[d]
Total	151	99

[a] These were the modern, fast, heavily armored all big-gun ships known as dreadnoughts, after the first one commissioned by the British in 1906: HMS *Dreadnought*. They carried 8 or 10 guns of 12- to 15-inch caliber.

[b] These were very fast versions of the dreadnoughts, carrying fewer, slightly smaller guns—usually eight 11.5- to 13-inch caliber—and with lighter armor protection to give them more speed.

[c] These were older vessels, still big and powerful, but with fewer big guns, and slower in speed.

[d] The British and Germans each had about 45 submarines available, but none were used.

Beatty, with but 4 ships left to oppose the German 5, and Evan-Thomas still out of range, tersely signaled: "Engage the enemy closer." He remarked to his flag captain: "There seems to be something wrong with our bloody ships today. Turn two points to port." and *Lion,* badly damaged but still fighting, swung closer to the Germans.

Soon afterward, at 4:42, Beatty sighted the German main fleet approaching; he at once turned north to join Jellicoe, in his turn hoping to lead the German fleet behind him. Hipper had already turned and was firing accurately at Beatty's ships and those of Evan-Thomas, who was slow in turning and was now being pounded by Scheer's main battle line. For over an hour the chase to the north continued, much damage being done on both sides. Shortly after 6 P.M., Beatty sighted Jellicoe's 6 divisions approaching from the northwest in parallel columns, preceded by Rear Admiral Sir **Horace Hood**'s battle squadron—3 battle cruisers and 2 light cruisers. The British main fleet was still over the horizon from the Germans, but Beatty, still heavily engaged, turned generally eastward, in front of the Germans, to get himself into line in front of Jellicoe, whose 6 columns now also turned behind Beatty. Both British admirals were hoping to swing entirely around Scheer and block him from his base. Shortly before 6:30, Scheer sighted Hood's squadron to his right front, just as British dreadnought shells began to fall around the German battle line. Within minutes practically every major ship in both fleets was within range and a furious general engagement was taking place. The German battle cruisers caught the worst of the storm; Hipper's flagship *Lutzow* was hammered out of action. On the British side, at 6:34 Hood's flagship was sunk with all on board by *Derfflinger's* accurate gunnery, and the cruisers *Defence* and *Warrior* also went down.

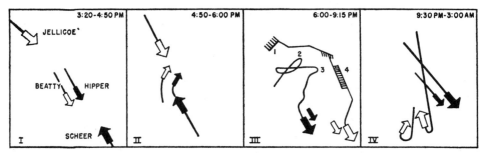

THE PHASES OF THE BATTLE OF JUTLAND

Phase   I: The Battle Cruiser Action—the Run to the South.

Phase   II: The Run to the North.

Phase III: The Main Fleet Action:

    1. Jellicoe's battle line deploys into column.

    2. Jellicoe caps Scheer's "T."

    3. The Crisis. Jellicoe again caps Scheer's "T."

    4. Jellicore turns away.

Phase IV: The Night Action.

The High Seas Fleet was now inside the converging arc of the Grand Fleet and taking heavy punishment. At 6:35, Scheer, under cover of a smoke screen and destroyer attacks, suddenly reversed course by a difficult and perfectly executed simultaneous 180-degree turn, headed west, and in a few minutes his ships were out of range of most of the surprised British. Jellicoe, instead of pursuing, continued southward, since he knew his fleet was now between the Germans and their bases. Then, at 6:55, Scheer made another 180-degree fleet turn back toward the British, apparently thinking Jellicoe had divided his fleet. Suddenly the entire German fleet was again under the guns of the entire Grand Fleet. This time it seemed that the Germans could not escape destruction in the hail of great projectiles.

A second time Scheer made a simultaneous turn away, while the 4 remaining German battle cruisers, under Captain **Hartog** of *Derfflinger,* most gallantly charged toward the British line to cover the withdrawal. (Hipper had transferred to a destroyer and was trying to catch up with his battle cruisers.) *Von der Tann,* her guns already out of action, remained in line only to spread the British fire. Both *Seydlitz* and *Derfflinger* broke into flames but remained in action as the German battle cruisers swung past the British battle line at short range. Then German destroyers sped in toward Jellicoe's battleships to make a torpedo attack and spread a smoke screen. Jellicoe, wary of torpedoes, saved Scheer by himself turning away. By the time he had resumed his battle line, the German High Seas Fleet had disappeared westward into the dusk as Scheer made another 180° turn. Amazingly, none of the German battle cruisers had been sunk in their courageous "death ride."

But the battle was not over. Scheer knew that the British fleet was now between his fleet and its home ports, and that Jellicoe was steaming to cover the entrances to those ports. Scheer also knew his fleet could not survive a renewed general battle. After dark he boldly turned to the southeast, deliberately crashing into the formation of light cruisers at the tail of Jellicoe's southbound fleet. He finally battered his way through in a chaotic midnight battle of collisions, sinkings, and gunfire. The British cruiser *Black Prince,* suddenly engulfed in the midst of the Germans, was sunk in 4 minutes. The German predreadnought battleship *Pommern* was cut in two. By dawn, Scheer was shepherding his cripples toward the Jade anchorage, and Jellicoe realized that his quarry had escaped.

The British now turned back to their bases. They had lost 3 battle cruisers, 3 cruisers, and 8 destroyers; they had 6,784 casualties. The Germans lost 1 old battleship, 1 battle cruiser, 4 light cruisers, and 5 destroyers; casualties were 3,039.

COMMENT. *Jutland marked the end of an epoch in naval warfare. It was the last great fleet action in which the opponents slugged it out within eyesight of one another. A drawn battle tactically, it made no change in the strategic situation, other than to make the Germans realize that they had no chance of defeating the Grand Fleet. Of the commanders engaged, Beatty, Hipper, and Hartog stand out, gifted with that "Nelsonian touch," which neither Jellicoe nor Scheer (both able professionals) appeared to have. In general, both sides behaved with the utmost gallantry.*

SUBSEQUENT NAVAL OPERATIONS, JUNE–DECEMBER

The remainder of the year saw one timid sortie of the High Seas Fleet (August 18), which ended as a fiasco, both opponents running home without making contact—Scheer deceived by a false airship report, Jellicoe because he feared a submarine ambush. Two German light-cruiser raids were made on the British coast (August 19 and October 26–27), and several auxiliary cruisers slipped through the British blockade to ravage Atlantic commerce. But in the main, German naval effort was now concentrated on submarine activities. Tremendous toll was taken of Allied shipping: 300,000 tons per month by December.

OPERATIONS IN 1917

Global Situation

Allied strength had grown during 1916. Toward the end of the year, at another Allied conference called by Joffre at Chantilly, there had been general agreement to continue a policy of joint Anglo-French large-scale operations on the Western Front in conjunction with simultaneous Russian and Italian offensives. These would have priority over all operations elsewhere, although new British Prime Minister **David Lloyd George** decided to undertake a major campaign in Palestine as well.

The western Allies at this time did not realize the extent of Russia's instability. The retirement of Joffre (December 31, 1916), who was succeeded by Nivelle, the hero of Verdun, immediately complicated the coordination of the Allied operations. Unity of command was nonexistent. Nivelle, planning a giant joint Anglo-French offensive, to be carried out with "violence, brutality, and rapidity," clashed with Haig on their command relationship. The French government supported Nivelle and the British were divided. British Prime Minister Lloyd George, who distrusted Haig and admired charming, English-speaking Nivelle, placed the BEF under Nivelle's command, to the horror of Haig and of Sir **William Robertson,** the new Chief of the Imperial General Staff. Through this bickering, and Nivelle's own imprudent announcements, secrecy was lost.

Ludendorff, aware of the Allied preparations and particularly fearing for over-extended German lines in the west, deliberately chose a defensive attitude on both major fronts, while forcing Austria (with German assistance) to take decisive action against Italy, which he believed could be defeated in 1917. The Kaiser approved this strategic concept, and also concurred in the inauguration of unrestricted submarine warfare, regardless of American opinion. He virtually granted unlimited authority to the military high command.

United States Entry

1917, January 31. Germany Proclaims Unrestricted Submarine Warfare. To offset growing hostility in the U.S., covert negotiations were already in process by German diplomats for a German-Mexican-Japanese alliance.

1917, February 3. The U.S. Severs Relations with Germany. This was a protest against unrestricted submarine warfare. Brazil, Bolivia, Peru, and other Latin American nations followed suit, as did China (March 14).

1917, March 1. Zimmermann Note. Publication of a proposed German defensive alliance with Mexico in case of war between Germany and the U.S., with the proviso "that Mexico is to reconquer the lost territory in New Mexico, Texas, and Arizona" caused a wave of American fury. **Alfred Zimmermann,** German Foreign Secretary, had sent the coded proposition, which contained the further suggestion that Mexico urge Japan to join the Central Powers, to **von Eckhardt,** German Minister to Mexico (January 19). British naval intelligence, intercepting and decoding it, gave a copy (February 24) to **Walter Hines Page,** U.S. ambassador to Brittain. He immediately turned it over to the State Department, which released it to the press (March 1). U.S. intelligence sources later verified the authenticity of the note.

1917, March 13. U.S. Merchantmen Armed. President Wilson's decision to arm for self-defense all vessels passing through war zones was announced by the State and Navy departments.

1917, April 6. The U.S. Declares War against Germany. This followed the sinking of several American ships and President Wilson's war message to Congress (April 2). War against

Austria-Hungary was not declared until 8 months later (December 7).

1917, April–June. U.S. Preparations. The Army would have to be built. Major General **John J. Pershing** was selected to command the American Expeditionary Force (AEF) and the 1st Division (an amalgamation of existing Regular Army units) was shipped to France (June). Pershing's plan called for a 1-million-man army overseas by May 1918, with long-range provision for 3 million men in Europe later. A draft law—the Selective Service Act— was passed (May 19) and the nation went into high gear. The Navy was ready (see p. 1067).

The Western Front

1917, February 23–April 5. German Withdrawal. Ludendorff had prepared a much shorter, highly organized defensive zone—the Hindenburg Line, or Siegfried Zone—some 20 miles behind the winding, overextended line from Arras to Soissons. Hindenburg approved, and decided to withdraw to the new line, which could be held with fewer divisions, thus providing a larger and more flexible reserve. Behind a lightly held outpost line heavily sown with machine guns lay 2 successive defensive positions, heavily fortified. Behind these again lay the German reserves concentrated and prepared for counterattack. Each successive defensive line was so spaced in depth that, should one be taken, the attackers' artillery would have to displace forward before progressing against the next. Between the original line and the new zone, the countryside had been devastated; towns and villages were razed, forests leveled, water sources contaminated, and roads destroyed. The actual withdrawal, conducted in great secrecy, began February 23 and was completed by April 5.

1917, April 9–15. Battle of Arras. This was the British preliminary to the Nivelle Offensive. The British First (General **H. S. Horne**) and Third (General Sir Edmund Allenby) armies, following a heavy bombardment and gas attack, crashed into the positions of the German Sixth Army (General **L. von Falkenhausen**). British air supremacy was rapidly gained. Canadian troops stormed and took **Vimy Ridge** the first day. The British Fifth Army (**Hubert Gough**), assisting on the south, made little progress. The British advance was finally slowed down in succeeding days of battle. Although this was a British tactical victory, there was no break-through. British casualties were 84,000; German, about 75,000.

1917, April 16–20. Nivelle Offensive (Second Battle of the Aisne, Third Battle of Champagne). The French Reserve Army Group (**Alfred Micheler**), heavily reinforced, assaulted on a 40-mile front between Soissons and Reims to take the **Chemin des Dames,** a series of wooded, rocky ridges paralleling the front. The Sixth (Mangin) and Fifth (**Olivier Mazel**) armies were closely supported by the Tenth (**Denis Duchêne**), and backed by the First (**M. E. Fayolle**). French strength in the attacking armies totaled 1,200,000 men and 7,000 guns. The German Seventh (**Max von Boehn**) and First (**Fritz von Below**) armies held the sector, fully cognizant of French plans as a result of Nivelle's confident public boasts of victory. Just before the attack, German flyers swept the sky of French aerial observation and German artillery fire destroyed French tanks still in march column. The French rolling artillery barrage moved too fast for the infantry, who met preplanned artillery and machine-gun fire, and sectional counterattacks. With exceptional gallantry, however, the French managed to reach and take the first German line, but were then stopped. Repeated attacks gained little ground. The whole affair was a colossal failure, costing the French nearly 120,000 men in 5 days. German losses, despite 21,000 captured, were much less.

Compared with similar attacks in previous years, such losses might not have seemed excessive, had Nivelle not promised a breakthrough and victory.

1917, April 29–May 20. Outbreak of Mutiny in French Armies. Widespread mutiny followed the Nivelle Offensive disaster. Political repercussions simultaneously shook the nation. Nivelle was replaced by Pétain (May 15). After a 2-week period in which the entire Western Front was nearly denuded of French combat troops, Pétain quelled the mutiny and restored the situation with a combination of tact, firmness, and justice. By amazingly efficient censorship control, French counterintelligence agencies completely blotted out all news of the mutiny. When it finally trickled to Ludendorff, it was too late; renewed British attacks to distract his attention had already drawn German reserves to the northern front. The full extent of the mutiny was not known to the outside world for more than a decade.

1917, June–July. British Offensive in Flanders. Haig, after an abortive renewal of the fighting around Arras to relieve German pressure on the French, had determined to break through between the North Sea and the Lys River. The Ypres salient was selected, but success could only be gained after first taking the dominating Messines Ridge. Plans for an assault had been begun many months earlier by competent, methodical General Sir **Herbert Plumer,** Second Army commander.

1917, June 7. Battle of Messines. After a 17-day general bombardment, British mines packed with 1 million pounds of high explosive tore a wide gap in the German lines on the Ridge. Under cover of this surprise and of British aerial superiority, in a carefully planned and organized attack, Plumer's Second Army successfully gained the position at cost of 17,000 casualties. German losses were 25,000, including 7,500 prisoners. Elbowroom had been gained for the main offensive, and the clear-cut victory bolstered British morale.

1917, July 31–November 10. Third Battle of Ypres (Passchendaele). Following an inten-

German pursuit planes attacking
Allied observation planes

sive bombardment, the British Fifth Army (Gough) assaulted northeast against the German Fourth Army (**Friedrich Sixt von Armin**). The French First Army (**François Anthoine**), on the left, was the pivot of maneuver; on the right, Plumer's Second Army covered the main effort. The low ground, sodden with rain, had been churned to a quagmire by a 3-day bombardment. Overhead the Allies had won temporary air superiority. All surprise had been lost, however, by the long preparation, and the German defense in depth was well organized. After some early gains, the attack literally bogged down. Haig now placed Plumer in command of the operation. After typical careful planning, a series of limited attacks on narrow fronts began (September 20); the British inched forward against determined counterattacks. Mustard gas was used here by the Germans for the first time, while German planes flew low to strafe British infantry with machine guns. The taking of Passchendaele Ridge and Passchendaele village (November 6) concluded the offensive. The British-held Ypres salient had been deepened for about 5 miles, at great cost—some 300,000 British and 8,528 French casualties. German losses are estimated at 260,000. But Haig, still determined to keep pressure on the Germans to permit the French armies to recover from the mutiny, had another card to play.

1917, November 20–December 3. Battle of

Cambrai. General **J. H. G. Byng**'s British Third Army struck General **Georg von der Marwitz'** German Second Army positions in front of Cambrai in complete surprise and under most favorable terrain conditions. At dawn, some 200 tanks followed a sudden burst of artillery fire into the German wire. Behind them moved wave after wave of infantry. The German defense collapsed temporarily and the assault bit through the Hindenburg Line for 5 miles on a 6-mile front, except at **Flesquières,** where German artillery knocked out tanks and the British infantry was unable to close in support. Although 2 cavalry divisions were poised to exploit the breakthrough, infantry reserves were weak, and too many tanks had been put in the first waves. Crown Prince Rupprecht of Bavaria, commanding the defending army group, rushed reinforcements to Marwitz. A large proportion of the British leading tanks became casualties—more from mechanical breakdown than by artillery fire—and the advance slowed down. German counterattacks fell on the salient (November 30) and Haig ordered a partial withdrawal (December 3). Casualties on both sides were approximately equal: about 45,000. The British took 11,000 prisoners; the Germans, 9,000. Cambrai marked a turning point in Western Front tactics on 2 counts: successful assault without preliminary bombardment and the first mass use of tanks.

COMMENT. *The most important lesson emerging from the entire western campaign of 1917 was the necessity for unity of command. Haig and Nivelle between them in two disjointed offensives had squandered more than one-half million men and exhausted the resources of 2 splendid war machines without appreciable effect. In Haig's defense, however, it should be noted that his persistent costly attacks in Flanders and Artois were largely intended to attract German attention from the weakness of the French armies farther south; in this he was successful, and to him must go at least part of the credit for France's survival through 1917.*

As the year ended, acquisitive eyes in both Britain and France turned to the as yet untouched human resources of the United States.

The Italian Front

1917, April. Allied Planning. Cadorna feared that the Germans would send troops to aid the Austrians in an offensive on the Italian front. Because of this, Nivelle sent Foch to meet Cadorna to work out plans for French and British assistance in such an event. Franco-British-Italian staff officers worked out a program for reinforcements to be rushed into Italy in emergency.

1917, May 12–June 8. Tenth Battle of the Isonzo. Cadorna, despite promises to aid the Allied offensive, did not get started until after the battles of Arras and the Aisne were over. Once again the Italians attempted to batter their way through, over mountainous terrain. After a 17-day battle, gains were small but losses huge: some 157,000 Italian casualties against about 75,000 Austrians. Following some minor give and take on both Isonzo and Trentino fronts, Cadorna decided to make a supreme effort with 52 divisions and 5,000 guns.

1917, August 18–September 15. Eleventh Battle of the Isonzo. The Italian Second Army (General **Luigi Capello**), heavily reinforced, assaulted north of Gorizia, while the Third (Duke of Aosta), to its south, drove into the rocky hills between Gorizia and Trieste. The southern assault was speedily stopped by the left wing of Austrian General Borojevic's Fifth Army, but Capello's Second Army on the north made a clear-cut advance, capturing the strategically important Bainsizza Plateau. Outrunning their artillery and supply, the Italians were then forced to stop. The net result was an incipient collapse of Austrian arms. The Austrians asked for German help.

1917, October 24–November 12. Battle of Caporetto (Twelfth Battle of the Isonzo). A new Fourteenth Austrian Army (7 of its divisions and much of its artillery were German), under German General **Otto von Below,** was concentrated behind the Tolmino-Caporetto-Plezza zone. Using novel "Hutier tactics" (see p. 1064), it suddenly crashed against the Italian Second Army. Surprise bombardment, with clouds of gas and smoke shells, disrupted Italian signal communications. Then the German assault elements loomed through mist and rain on the demoralized defenders. Cadorna, having learned of the projected assault, had ordered defense in depth, but Capello—a capable officer—was ill and the acting commander of the Second Army ignored the instructions. Bypassing strong points which would be mopped up later by reserves, the German assault elements streamed through the zone, uprooting the Second Army. The Austrian Tenth Army on the right and the Fifth Army on the left supported the main effort. The Italian Third Army withdrew in good order along the coast, but part of the so-called Carnic Force on the northern Alpine fringe was trapped. Farther west the Italian Fourth Army hurriedly fell back to conform with the situation as the battered Second Army was driven in succession from defensive lines along the Tagliamento and Livenza rivers. By November 12, Cadorna managed to stabilize his defense from Mt. Pasubia, south of Trent, to the Piave and along that river to the Gulf of Venice. There the Austro-German offensive slowly ground to a halt, having outdistanced its supply. The catastrophe cost the Italians 40,000 killed and wounded plus 275,000 prisoners, 2,500 guns, and huge stores of goods and munitions. Austro-German losses were about 20,000. By this time French and British reinforcements, in accord with the plan prepared earlier in the year, were moving in, 11 divisions in all, under British General Plumer. Cadorna was now removed from command, being replaced by General **Armando Diaz.**

COMMENT. *Caporetto is a prime example of the military principles (or virtues) of surprise, objective, mass, and economy of force. Below had but 35 divisions in all against the Italian 41, but was far superior in strength at the point of impact. Had he possessed cavalry and armored cars to exploit his success, the battle might have been decisive. As it was, the Italians were badly shaken, but still capable of carrying on the war. A direct result of this disaster to Allied arms was the Rapallo Conference (November 5), which set up a* **Supreme War Council,** *the first attempt to attain overall Allied unity of command.*

The Eastern Front

1917, March 12. Russian Revolution. Mutiny of the Petrograd garrison (March 10) was followed by establishment of a Provisional Government. Nicholas II abdicated (see p. 1098). The new regime, bickering with the Bolshevik-dominated **Petrograd Soviet** (Council of Workers and Soldiers' Deputies), pledged itself to continue war against the Central Powers until an Allied "victorious end" was attained. The Soviet (March 14), fearing counterrevolutionary measures by the officer corps, issued on its own authority the notorious "Order No. 1," depriving officers of disciplinary authority. Broadcast throughout the armed forces, and despite the counterorders of the Provisional Government, it produced the result desired by the Bolsheviks—breakdown of all military discipline. The Russian Army and Navy collapsed like an ice jam in a spring thaw. Mutinous soldiers and sailors murdered many officers; others were simply deposed by soldiers' councils. By mid-April an estimated 50 percent of the officer corps had been eliminated, among them most of the best men. **Nikolai Lenin** and other Bolshevik agitators were smuggled into Russia by Germany to undermine the Provisional Government. **Leon Trotsky** (Bronstein) joined them. Germany, halting all offensive movements on the Eastern Front lest the Russians reunite in defense of the homeland, took advantage of the lull to send troops to the Western and Italian fronts. Despite all the turmoil, **Alexander Kerensky,** appointed minister of war (May 16), and pressured by the alarmed Allies, attempted to mount an offensive on the Galician front. Brusilov, now chief of staff, commanded.

1917, July 1. Kerensky (or Second Brusilov) Offensive. Brusilov attacked toward Lemberg with the few troops still capable of combat operations. The Eleventh and Seventh Russian armies penetrated German Count **Felix von Bothmer**'s composite South Army (4 German, 3 Austrian, and 1 Turkish division) for 30 miles on a 100-mile front. On the Russian northern flank, the Austrian Second Army was roughly handled. On the south, General **Lavr Kornilov**'s Russian Eighth Army rocked the Austrian Third Army and threatened the oil fields of **Drohobycz** (July 5). But Russian enthusiasm and discipline faded quickly as German resistance stiffened and their own supply system broke down.

1917, July 19. German Counterattack. General **Max Hoffmann,** commanding on the Eastern Front, had quickly obtained reinforcements from the west. Preceded by a most intensive bombardment, the German assault, beginning on the northern flank, rolled up the now demoralized Russian armies in quick succession. They disintegrated; south of the Pripet Marshes no Russian Army now existed. The Germans halted their advance on the Galician border simply because they lacked sufficient reserves and logistical resources to occupy more territory.

1917, September 1. Hutier's Riga Offensive. German General **Oscar von Hutier**'s Eighth Army attacked the northern anchor of the Russian front. While a holding attack on the west bank of the Dvina River threatened Riga, 3 divisions crossed the river to the north on pontoon bridges, encircling the fortress, while exploiting elements poured eastward. Actually, this highly successful attack was but a dress rehearsal of new German assault techniques to be used again, 6 weeks later, at Caporetto. Long preliminary bombardment was eliminated. Instead, a short, sharp concentration of fire was followed immediately by infantry assaults, both guns and troops being brought into position at the last possible moment to ensure surprise. Heavy concentrations of gas and smoke shells masked known enemy strong points, while infiltrating elements—infantry and light guns—bypassed them. This was the first application of what would become known as "Hutier tactics." The Russian Twelfth Army streamed eastward in complete panic. Only 9,000 prisoners were taken; casualties on both sides were minimal. At this same time a small German amphibious force occupied Osel and Dago islands in the Gulf of Riga, and effected a landing on the mainland.

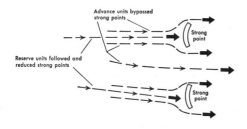

1917, September–October. Chaos in Russia.
The Kerensky government (Kerensky had become head of the Provisional Government July 20) fled Petrograd for Moscow, and the Bolsheviks began to take over. A brief flurry of counterrevolution under General Kornilov petered out (September 9–14).

1917, November 7 (October 25 O.S.) Bolshevik Revolution. Lenin and Trotsky seized power and began dickering for peace with Germany.

1917, December 15. Armistice of Brest Litovsk (Brest). After 12 days of bickering between Bolshevik and German negotiators, armistice terms were agreed, ending hostilities on the Eastern Front. (For further operations, including the Allied invasions of north Russia and Siberia, see p. 1092.) Russia had been permanently erased from the Allied ranks.

The Balkan Front, 1917

1917, January–June. Stalemate. Sarrail, with some 600,000 men on paper, could only muster about 100,000 for combat duty, as malaria and other disease kept the hospitals full. Behind his lines Greece seethed as King Constantine's government continued to conciliate the Central Powers. Several inconclusive Allied attacks failed, at **Monastir** and the **Battle of Lake Prespa (Djoran)** (March 11–17) and the **Battle of the Vardar** (May 5–19). French and British commanders were at cross purposes, while the reconstituted Serbian Army distrusted both. German aerial supremacy added to the Allies' woes.

1917, June 12. Abdication of Constantine.
This was the result of Allied pressure. New King **Alexander** appointed pro-Allied **Venizelos** as premier, clarifying the situation (June 26). At the same time, Allied troops moved into Thessaly. A French force occupied the Isthmus of Corinth.

1917, June 27. Greece Enters the War. Greek troops began to swell the Allied strength. However, no real offensive developed.

1917, December 10. Relief of Sarrail.
Georges Clemenceau, new French premier, appointed General **M. L. A. Guillaumat,** competent and battle-tried commander, who at once began reorganizing the Allied forces.

The Turkish Fronts

THE CAUCASUS

1917, March. End of Russian Pressure. With the first Russian Revolution in March, Turkish troops were freed to support other fronts.

PALESTINE

1917, January 8–9. Battle of Magruntein.
The Sinai Peninsula was cleared of all organized Turkish forces. British losses were 487; they captured 1,600 prisoners and a few guns. Sir Archibald Murray was now authorized to begin a limited offensive into Palestine, where the Turks were established in defensive positions along the ridges between Gaza and Beersheba, the 2 natural gateways to the region.

1917, March 26. First Battle of Gaza. General Sir **Charles M. Dobell** attacked, but, through defective staff work and a communications breakdown between his mounted force and infantry, the attack failed. The British withdrew. Out of forces nearly equal in strength (about 16,000), British losses were 3,967; Turkish, 2,447. Murray's report, unfortunately, presented the action as a British victory. Accordingly, he was ordered to advance without delay and take Jerusalem.

1917, April 17–19. Second Battle of Gaza.
Dobell tried again, this time in frontal assault

against the now well-prepared defensive Turkish position, and was thrown back, losing 6,444 men against the Turkish loss of about 2,000. Murray relieved Dobell. An exasperated War Office in turn relieved Murray. In his place came General Allenby, fighting cavalryman with the gift of leadership and tactical ability. His instructions were to take "Jerusalem before Christmas." His first step was to move British headquarters from Shepheard's Hotel in Cairo to the fighting front. Insisting on reinforcements, Allenby built up an efficient combat force of 7 infantry divisions and a cavalry element—the Desert Mounted Corps (horse cavalry and camel). Total British strength was 88,000.

Opposing the British was the Turkish Eighth (Kress von Kressenstein) and the incomplete Seventh Army. With some additional German machine-gun, artillery, and technical units, this force totaled 35,000. Basic Turkish weakness was the long and tenuous supply line from Constantinople. Basic British weakness was water supply for the offense; the efficient transport and pipeline extended only to their rear echelon.

1917, October 31. Third Battle of Gaza (Battle of Beersheba). Reversing his predecessor's plans, Allenby left 3 divisions demonstrating in front of Gaza and secretly moved against Beersheba. Surprise was complete, but success was predicated on the ability of the attackers to capture the city's wells; failure would mean collapse of the mounted elements and probably the entire offense. While the infantry assailed Turkish defenses in frontal attack, the Desert Mounted Corps swung wide to the east, then turned on the city. An all-day battle culminated at dusk in the mounted charge of an Australian cavalry brigade through and over the Turkish wire and trenches into Beersheba itself, capturing the coveted water supply. Hastily evacuating, the Turkish Seventh Army now lay with its left flank open. Allenby struck north (November 6), splitting the 2 Turkish armies, and launched the Desert Mounted Corps across country toward the sea. The Turks evacuated Gaza in time to avoid the trap, the Eighth Army retreating up the coast, the Seventh falling back on Jerusalem.

1917, November 13–14. Battle of Junction Station. Pursuing closely, despite logistical difficulties and shortage of water, Allenby struck a hastily established line of the Eighth Army, driving the Turks back north along the railroad. Turning now toward Jerusalem, Allenby was held up by the appearance of Turkish reserves from the Aleppo area (the so-called Yilderim Force) and the arrival of General von Falkenhayn to assume command. Falkenhayn reestablished a front from the sea to Jerusalem. A bitter slugging match in the Judean hills followed.

1917, December 9. Fall of Jerusalem. In the face of Allenby's determined attack against Jerusalem (December 8), the Turks evacuated the city, which was occupied by the British next day. A Turkish counterattack was repulsed (December 26). During this campaign Allenby had 18,000 casualties; Turkish losses were about 25,000, including 12,000 captured and 100 guns lost.

MESOPOTAMIA

1917, February 22–23. Second Battle of Kut. Maude skillfully assaulted after preliminary feints against the Turkish left. He crossed the Tigris on the Turkish right and then pressed the attack on both flanks. **Kara Bekr** Bey, Turkish sector commander, fell back to the vicinity of Baghdad, interposing skillful rear-guard resistance to Maude's pursuit. Then the Turkish Sixth Army (Halil Pasha) attempted a stand.

1917, March 11. Fall of Baghdad. After several days of fighting along the **Diyala River,** Maude entered the city, the Turkish forces retreating in some disorder. Maude now launched 3 exploiting columns up the Tigris, Euphrates, and Diyala rivers, securing his hold on Baghdad. A Turkish effort to retake Mesopotamia collapsed when the troops earmarked for the expedition—Yilderim Force—were diverted to bolster the Palestine front (see above).

1917, September 27–28. Battle of Ramadi. As the summer heat subsided, Maude struck sharply northwestward up the Euphrates River. He pursued the Turkish survivors into central Mesopotamia. Brilliant Maude, his objective the oil fields of Mosul, prepared to continue his

advance, but died of cholera (November 18), General Sir **William R. Marshall** succeeding to command.

PERSIA

Russian forces in Persia were of concern to the Turks in the beginning of the year, having moved to the vicinity of Saqqiz-Kermanshah, where they were a potential threat to the Mosul oil fields. Maude, in liaison with the Russians, hoped for joint operations against the Turkish Sixth Army, but this hope faded following the Russian Revolution.

War at Sea

THE SUBMARINE CRISIS

1917, February–April. German Unrestricted Submarine Campaign. After careful calculations, the German naval command had come to the conclusion that unrestricted submarine warfare would force Britain to sue for peace in 5 months. The danger of American intervention was recognized (see p. 1060), but the potential influence of the U.S. on the war at sea or on land was evaluated as negligible for 2 or more years. By then the submarine campaign, combined with operations on land, was expected to have brought victory to the Central Powers. It almost worked. British shipping losses soared to 875,000 tons per month (April). British and neutral merchant sailors began to refuse to sail. The diversion of most of his light warships from the Grand Fleet for the purpose of seeking and sinking submarines caused great concern to Admiral Beatty (now commanding that fleet) in event of a new sortie of the German fleet, but no such sortie was made. Recommendations for instituting convoys were rejected by the Admiralty as an unsound waste of available cruisers and destroyers. The efforts to sink submarines were disappointing, however. Admiral Jellicoe (now First Sea Lord) calculated that Britain would run out of food and other needed raw materials by July.

1917, May 10. Institution of Convoy System. Insistence of Prime Minister Lloyd George, combined with the strong recommendations of

American Admiral **William S. Sims** and of Beatty, finally forced adoption of the convoy system. The results were spectacular. British escort vessels, joined by American destroyers (May), provided adequate protection to merchant ships and at the same time were able to sink more submarines, since these were forced to attack the convoys. Unquestionably the convoys saved Britain. Although shipping losses by the end of the year exceeded 8 million tons, Allied shipbuilding programs more than offset the losses.

OTHER NAVAL ACTIONS

1917, February–April. German Destroyer Raids in the Channel. The first of these 3 raids was uneventful (February 25). In the second the Germans sank 2 British destroyers and a small coastal merchant vessel, with no loss to themselves (March 17). The third was foiled by the gallantry of Commander **E. R. G. R. Evans,** whose destroyer HMS *Broke* sank 2 German destroyers singlehanded in a thrilling sea fight.

1917, May–June. British Raids on Ostend and Zeebrugge. Naval bombardment of the German destroyer and submarine bases caused some damage, but failed to interfere seriously with German U-boat operations.

1917, May 15. Action off Valona. In the Adriatic, Austrian Captain **Miklós Horthy** (later dictator of Hungary) led an Austrian squadron on a raid against Italian transports off the Albanian coast. The Austrians were able to sink 14 of the small merchantmen, and then to escape successfully after a spirited action with British, French, and Italian warships.

1917, July 15. British Raid on German Coastal Shipping. British destroyers sank 2 German merchant ships and captured 4 more off the Dutch coast, stopping coastal trade with Rotterdam.

1917, November 17. Action off Heligoland. Poor British staff work aborted a battle-cruiser raid against German minesweeping operations in Heligoland Bight; the German squadron protecting the minesweepers was able to withdraw without damage.

1917, December. German Raids on British Scandinavian Convoys. These inflicted serious losses on British merchant shipping, forcing Beatty to use a squadron of battleships as a covering force for future convoys.

OPERATIONS IN 1918

Global Situation

The Allies entered the year in a state of frustration. The rosy promises of early 1917 had been unfulfilled. Except in the Near East, where Allenby's dynamic leadership had culminated in the capture of Jerusalem—with its tremendous psychological uplift to Christendom—Allied offensives had bogged down in a welter of cross-purpose and disunity of command. Russia had collapsed. The German U-boat campaign still threatened the maritime pipeline of supply from America. Finally, many months would still pass before American armed forces could bolster up lost Allied man power. Both Britain and France were therefore on the defensive. The Supreme War Council did no serious planning. Haig (who had been refused reinforcement by Lloyd George) and Pétain agreed among themselves on mutual support should a German offensive be launched. Some attempt at organization of defense in depth was made.

Nor had the Central Powers been successful. They all felt the strangulation of Allied naval blockade. Austria was at the end of her resources, Turkey and Bulgaria were wobbling, and the burden of the war fell heavier and heavier on Germany. Hindenburg and Ludendorff had established a virtual military dictatorship over Germany, and exercised almost as complete authority over the subservient governments of Austria, Bulgaria, and Turkey.

The American Build-up

Having entered the war without previous preparation, the U.S. was faced with organizing, equipping, training, transporting, and supplying an expeditionary force in Europe. The little Regular Army provided the leaven for 2 successive waves of man power: the National Guard and the draftees produced by the Selective Service Act (May 19, 1917). From a strength of 200,000 men and 9,000 officers (including 65,000 National Guardsmen then serving on the Mexican border), the Army swelled to over 4 million men, including 200,000 officers. Some 2 million in all served overseas. Based on Pershing's recommendations, a divisional organization of approximately 28,000-man strength was adopted. It consisted of 2 infantry brigades of 2 regiments each, an artillery brigade, an engineer regiment, 3 machine-gun battalions, and trains and supporting services. Forty-two of these divisions, which were nearly double the strength of their European counterparts, reached France. Though Pershing understood the need and importance of entrenchments, he eschewed what he considered to be a defeatist concept of trench warfare. Training was predicated on the spirit of the offensive—mobile combat—with stress on individual marksmanship.

Overseas, the Service of Supply became an empire in itself, manning 9 base sections. Pershing chose the Lorraine area east of Verdun as the American combat zone. The pipeline of supply from the United States went to ports in southwestern France, and movement overland conflicted little with the Allied efforts farther north. Except for small arms, ordnance needs were filled by America's allies. So too with airplanes; American production was limited to the Liberty engine.

Overseas transportation, the province of the U.S. Navy, was in part provided by the German merchant fleet seized in American ports, plus an improvised fleet of the American merchant marine—much of it built with remarkable celerity, some British ships, and neutral shipping sequestrated or leased. The combined fleet carried more than a million American soldiers to France without loss of a single vessel—on eastbound voyages. (The remaining million shipped overseas went on Allied ships, mostly British.)

The Navy, whose personnel waxed to 800,000, was primarily concerned in antisubmarine and convoy activities, though a division of 5 battleships joined the British Grand Fleet and 3 other battleships operated in Irish waters against surface raiders. In all, some 79 American destroyers took part in convoy work, and 135 subchasers also operated in European waters. An important part of U.S. Navy participation was in the laying of 56,000 of the 70,000 mines comprising the North Sea mine belt—from Scotland to Norway. Naval air squadrons took part in bombings of German submarine bases along the Belgian coast. A Marine brigade became part of the AEF.

American combat participation in World War I was based on "co-operation," as Pershing's directive put it. The U.S. was not technically an ally. Its expeditionary force was to be "a separate and distinct component of the combined forces, the identity of which must be preserved." Pershing's directive ran counter to the Allies' desires. They distrusted the inexperienced Americans' military ability, and they were short of man power. From the beginning Pershing was cajoled, coaxed, and finally threatened, in fruitless efforts to have him turn the AEF over *in toto* as a replacement reservoir for the French and British armies. War Secretary Newton D. Baker and President Wilson upheld Pershing when Clemenceau and Lloyd George went over his head to Washington with their demands.

The Fourteen Points

In an address to Congress on January 8, 1918, President Wilson laid down his "only possible program" for peace. The policy included (1) open covenants, openly arrived at; (2) freedom of the seas in war and peace; (3) removal of trade barriers; (4) national armament reductions; (5) impartial adjustment of colonial claims; (6) evacuation of Russian territory and independent solution by Russia of her political development and national policy; (7) evacuation and restoration of Belgium; (8) evacuation and restoration of all occupied French territory and return of Alsace-Lorraine; (9) readjustment of Italian frontiers on lines of nationality; (10) autonomy for the peoples of Austria-Hungary; (11) evacuation of Rumania, Serbia, and Montenegro, restoration of occupied territories, and Serbian access to the sea; (12) Turkish portions of the Ottoman Empire to be assured secure sovereignty, but other nationalities under

Turkish domination to be freed; (13) independence of Poland, to include territories with predominantly Polish population, with free Polish access to the sea; (14) formation of an association of nations ensuring liberty and territorial integrity of great and small alike.

Operations on the Western Front

LUDENDORFF'S OFFENSIVES

During the winter of 1917–1918, Ludendorff realized that Germany's only hope of winning the war lay in a decisive victory in the west in 1918, before the weight of American man power could have a significant effect. With Russia knocked out of the war, he believed that this could be done. Shifting most German forces from the east, he instituted an intensive training program in preparation for an all-out offensive to be launched as early as possible in the spring. The best units were developed into "shock troops," to be spearheads of the planned assaults. His intention was to smash the Allied armies in a series of hammer blows. Recognizing the divergent interests of the French (concerned with protection of Paris) and the British (interested in maintaining their lines of communications with the Channel ports), he intended to drive a wedge between the two Allied armies and then destroy the British in subsequent assaults. Preparations were made with remarkable efficiency.

THE SOMME OFFENSIVE

1918, March 21. The First Offensive. The Germans began their drive at dawn in heavy fog. Three German armies—Seventeenth (Otto von Below), Second (Marwitz), and Eighteenth (Hutier), from north to south—struck the right flank of the British sector—the Third (Byng) and Fifth (Gough) armies—on a 60-mile front between Arras and La Fère. The objective was to break through, dislocate, and roll up the British, wheeling to the north and splitting them from the French on their right. Following a surprise 5-hour bombardment by more than 6,000 cannon, the specially trained German shock elements rolled through the fog, using

"Hutier tactics"—infiltration behind a rolling barrage and passing of strong points which would be later mopped up by reserves, accompanied by artillery neutralization of battery positions and observation posts (see p. 1064). No limits were set to the advance; each division pressed as far and as fast as possible, with close-support elements passing through and taking up the advance whenever a local assault should bog down. Gough's Fifth Army, spread thin on a 42-mile front lately taken over from the French, collapsed, exposing the Third Army's right and forcing its withdrawal, but Byng, better organized in depth, held the German Seventeenth and Second armies to limited gains. Hutier, continuing on Gough's heels, reached and passed the Somme. All British reserves were committed to plug the gap and some French units also reinforced. But Pétain was more concerned with protecting Paris than he was with assisting Haig. The British commander hastily appealed to the new British Chief of Staff, General Sir **Henry Wilson,** and the War Minister, Lord **Milner,** for the appointment of "Foch or some other French general who will fight" to take supreme command.

1918, March 23–August 7. Artillery Bombardment of Paris. A remarkable long-range German cannon began a sporadic bombardment of Paris from a position 65 miles away. This amazing achievement of German ordnance technology seriously hurt morale of Parisians and inflicted 876 casualties, but did not significantly affect the war. Actually there were 7 "Paris Guns," with a caliber of about 9 inches, the barrels 117 feet long, with a maximum range of 80 miles.

1918, March 26. Foch Appointed Allied Coordinator. In an emergency meeting of the Supreme War Council at Doullens, Foch was appointed coordinator for the Western Front.

1918, April 3. Foch to Supreme Command. At Beauvais, the War Council appointed Foch commander in chief of the Allied forces in France. Pershing, who had already (March 27) generously offered his 8 available divisions in France to Foch in the emergency, agreed in principle to the appointment.

1918, April 5. End of the Offensive. Meanwhile, the German drive, after gaining a 40-mile-deep salient, lost momentum. Paris had been bombarded by long-range artillery (75 miles; March 21-April 6). Foch's shifting of reserves checked the German assault after it reached Montdidier, and Ludendorff brought it to a halt. Allied losses mounted to about 240,000 casualties (163,000 British, 77,000 French), including 70,000 prisoners and 1,100 guns. German casualties were almost as high, most of them in the specially trained shock divisions. Over Haig's protests, Gough was relieved by the British government; his shattered Fifth Army was taken over by General Sir **Henry Rawlinson**'s Fourth Army headquarters.

COMMENT. *The most serious consequence of the offensive, from the German point of view, had been the institution of an Allied unified command. Thus, despite its initial brilliant tactical success, the offensive was a strategic failure. There were 3 main reasons for this: (1) Lack of logistical mobility. Once a breakthrough had been made, the Germans found themselves advancing across land devastated by 4 years of war, particularly by their own "scorched earth" measures at the time of the withdrawal to the Hindenburg Line (see p. 1060). They did not have the means of keeping up a flow of ammunition, food, and other supplies to their troops advancing through a veritable quagmire. (2) Lack of strategic mobility. The same problem prevented them from fully exploiting the gap with fast-moving mobile forces, or even from providing adequate reinforcements and replacements to the breakthrough troops. (3) Lack of mobile tactical fire support. Once the breakthrough was made, the front-line infantry quickly outran its artillery, which was unable to advance in any significant numbers through the roadless morass. Thus, when the British were finally able to move reserves into the gap, the Germans lacked sufficient firepower to maintain the momentum of their drive or to deal adequately with the British fighter planes strafing them.*

THE LYS OFFENSIVE

1918, April 9. Ludendorff's Second Offensive. Again the Germans struck the British sector, this time in Flanders on a narrower front, threatening the Channel ports. The German Fourth Army (Sixt von Armin) struck Plumer's Second Army in a Hutier-type attack. (Plumer had returned from Italy at Haig's request.) **Ferdinand von Quast**'s German Sixth Army on its left clawed through the positions of Horne's First Army, demolishing a Portuguese division.

1918, April 12. "Backs to the Wall." Haig's order forbidding retirement galvanized British resistance. The German drive was halted (April 17) after a 10-mile advance which included recapture of Messines Ridge. Foch, gathering a reserve force behind the British, placed only part of it in the line (April 21), much to Haig's dissatisfaction. After a series of further attacks and counterattacks, Ludendorff finally called the operation off. Again, and for the same reasons as before, he had achieved tactical success but strategical failure. No breakthrough had been effected, and the Channel ports were safe. The cost had been great—another 100,000 British casualties—but again German losses had been almost as great. Ludendorff's carefully trained and prepared shock troops were sadly depleted, the morale of the survivors badly shaken.

THE AISNE OFFENSIVE

1918, May 27. Third German Offensive. This time Ludendorff struck along the Chemin des Dames, a diversion against the French preparatory to a planned final and decisive blow to be struck against the British in Flanders. The German First (**Bruno von Mudra**) and Seventh (Boehn) armies attacked the French Sixth Army (Duchêne) with 17 divisions in the assault, preceded by tanks. Duchêne's 12 divisions (3 of them British) were surprised in shallow defenses along a lightly held 25-mile front and collapsed. By noon the Germans were crossing the Aisne; by evening they were crossing the Vesle, west of Fismes, and reached the Marne (May 30).

1918, May 28. Battle of Cantigny. Meanwhile, as Pershing was rushing the 2nd (Major General **Omar Bundy**) and 3rd (Major General

J. T. Dickman) divisions to reinforce the French, the first American offensive of the war took place at **Cantigny,** 50 miles northwest. The 1st U.S. Division (Major General **Robert Lee Bullard**) attacked the village, a strongly fortified German observation point, taking all its objectives, and then repulsed a series of violent German counterattacks (May 28 and 29). While only a local operation, its success, against veteran troops of Hutier's Eighteenth Army, boosted Allied morale.

1918, May 30–June 17. Battles of Château-Thierry and Belleau Wood. The U.S. 2nd and 3rd divisions were flung against the nose of the German offensive along the Marne, moving into position through the retiring troops of the French Sixth Army. The 3rd Division held the bridges at Château-Thierry against German assaults, then counterattacked, and, with assistance from rallying French troops, drove the Germans back across the Marne at Jaulgonne. The 2nd Division, taking over the sector of the French XXI Corps between Vaux and Belleau, west of Château-Thierry, checked German attacks. Ludendorff called off his offensive (June 4). The 2nd Division then counterattacked, spearheaded by its Marine brigade. In 6 successive assaults the Germans were uprooted from positions at Vaux, Bouresches, and Belleau Wood, losing some 9,500 men and more than 1,600 prisoners.

COMMENT. *The net result of the third German drive had been to make a serious dent in the Allied front, a salient some 30 miles wide and more than 20 miles deep. Ludendorff determined to exploit this success by another diversionary drive, prior to his proposed Flanders stroke. It would be a 2-pronged affair converging on Compiègne, the Eighteenth Army attacking southwesterly, the Seventh Army westerly.*

FOURTH AND FIFTH GERMAN OFFENSIVES

1918, June 9–13. Noyon-Montdidier (Fourth) Offensive. Forewarned by German deserters, Foch and Pétain were ready. French defenses were organized in depth. A counter-preparation artillery bombardment disrupted the Eighteenth Army's assault. Some gains were made, but a Franco-American counterattack halted the advance (June 11). The Seventh Army's attack was quickly snubbed (June 12). By this time, 25 American divisions were in France, 7 of them at the front. French and British leaders were making strenuous efforts to incorporate American troops into their respective armies permanently; Pershing was resisting this.

1918, July 15–19. Champagne-Marne (Fifth) Offensive. Ludendorff, clinging to his plan for an all-out drive against the British in Flanders, attempted one more preliminary offensive in Champagne to pinch out the strongly fortified Reims area. Boehn's Seventh Army would advance up the Marne through Epernay to meet Mudra's First Army and **Karl von Einem's** Third attacking south toward Châlons. Foch, already planning a major counteroffensive, was again warned of the blow by deserters, aerial reconnaissance, and prisoners. German shock troops were tripped by an Allied artillery counterpreparation (night of July 14–15). East of Reims the attack was halted in a few hours by **Henri Gouraud's** French Fourth Army.

1918, July 15–17. Second Battle of the Marne. West of Reims, where the defenses were neither so strong nor so deep, the German Seventh Army penetration carried to the Marne, some 14 divisions crossing the river. The stout defense of the U.S. 3rd Division again snubbed the attack there. Then Allied aircraft and artillery destroyed the German bridges, disrupting supply and forcing the attack to halt. Ludendorff, admitting defeat, now prepared for a general withdrawal from the Soissons-Château-Thierry-Reims salient to reduce the front held by his depleted forces. In 5 months he had lost half a million casualties. Allied losses had been somewhat greater, but American troops were now arriving at a rate of 300,000 a month.

THE ALLIED COUNTEROFFENSIVE

THE AISNE-MARNE OFFENSIVE

1918, July 18–August 5. Allied Aisne-Marne Offensive. The French Tenth (Mangin), Sixth (**Jean M. J. Degoutte**), and Fifth (**Henri M.**

Berthelot) armies, from left to right, assaulted the Marne salient. The Ninth Army (**M. A. H. de Mitry**) was in reserve. In a series of smashing attacks, the Germans were rolled back all along the line, despite desperate resistance and skillful handling. The U.S. 1st and 2nd divisions spearheaded the Tenth Army's attack—the main effort. The 1st Division captured 3,800 prisoners and 70 guns from the 7 German divisions it encountered. Its casualties were 1,000 killed and 6,000 wounded. The 2nd Division, capturing 3,000 prisoners and 75 guns, suffered 5,000 casualties in all. Six other American divisions also took part—the 4th, 26th, and 42nd in Major General **Hunter Liggett's** I Corps with the French Sixth Army, and the 3rd, 28th, and 32nd in Major General Bullard's III Corps with the Ninth Army (which moved into line between the Sixth and Fifth armies). Ludendorff called off his proposed Flanders drive (July 20), concentrating his efforts to stabilize the situation along the Vesle. The Marne salient no longer existed. In reward for the victory, Clemenceau promoted Foch to Marshal of France (August 6).

COMMENT. *The entire July operation, German offensive and Allied counteroffensive, is sometimes called the* **Second Battle of the Marne.** *Strategically, it was the turn of the tide; the initiative had been wrested from the Germans. Ludendorff's gamble to conclude the war successfully had failed. The front had been shortened by 28 miles, the important Paris-Châlons railway line reestablished, and all menace to Paris ended. On the Allied side, troops of 4 nations— France, Great Britain, the United States, and Italy— had successfully participated in a unified operation. Allied morale soared as German dropped. Ludendorff had lost 30,000 more prisoners, more than 600 guns, 200 mine throwers, and 3,000 machine guns.*

THE AMIENS OFFENSIVE, AUGUST 8–SEPTEMBER 4

1918, August 8–11. First Phase. Haig, in conjunction with the French Aisne-Marne offensive, threw Rawlinson's British Fourth Army and the French First Army (**M. Eugène Debeny,** attached by Foch to Rawlinson's

The Paris gun

command) against the German Eighteenth (Hutier) and Second (Marwitz) armies. Expecting an Allied attack farther north in Flanders, the Germans were caught off guard by a well-mounted assault secretly prepared. The Canadian and Anzac corps jumped off without preliminary bombardment, preceded by tanks, and bit deep through a dense fog. More than 15,000 prisoners and 400 guns were captured. On their right, the French bombarded first, then advanced. Despite near panic among their front-line troops, the Germans managed to reestablish a position 10 miles behind the former nose of the salient. The French Third Army (**Georges Humbert**), on the right of the First, entered the action (August 10), forcing the evacuation of Montdidier. Haig cautiously paused (August 11) to regroup, despite Foch's wishes to maintain unremitting pressure on the Germans. Both Allied and German air forces took part in the initial fighting after the fog cleared.

1918, August 21–September 4. Second Phase. Progressively, the British Third Army on the left and the French armies on the right took up the assault. The British Fourth Army in the center joined in (August 22), followed by the British First Army (Horne) on the far left. Ludendorff ordered a general withdrawal from both the Lys salient in Flanders and the Amiens area. His plans were disrupted when the Anzacs penetrated across the Somme (August 30–31), taking **Péronne** and threatening **St.-Quentin.** The Canadian corps, shifted to the north flank, broke through near **Quéant** (September 2). The entire German situation deteriorated, necessitating retirement to the final position—the Hindenburg Line. By this time Haig had expended his reserves and could not further exploit his victory. German casualties were more than 100,000, including some 30,000 prisoners. Allied losses were 22,000 British and 20,000 French. Tactically and strategically, the Allies had gained another major victory, cracking German morale.

COMMENT. *Ludendorff's bitter statement that August 8 had been the "Black Day" of the German Army tells the story. He said flatly: "The war must be ended!"*

ST.-MIHIEL OFFENSIVE, SEPTEMBER 12–16

Pershing's insistence on a separate and distinct United States Army operating on its own assigned front was reluctantly accepted by Foch (July 24). Reduction of the St.-Mihiel salient was the first mission. The U.S. First Army, with the French II Colonial Corps attached, took over the sector (August 30). Foch, planning an all-out Allied offensive, then attempted to change Pershing's plan and divide part of the American forces between the French Second and Fourth armies. After sharp disagreement, Foch accepted Pershing's position, but the American agreed to shift his army and attack with the French in the Argonne Forest immediately upon conclusion of the St.-Mihiel operation.

Ludendorff, well aware of the threat, started evacuation of the salient (September 8).

Supported by a conglomerate Allied air force of some 600 planes—American, French, Italian, and Portuguese—under American Colonel **William Mitchell,** the First Army attacked both faces of the salient (September 12). The French corps held the nose. The assault—both ground and air—was completely successful; the converging attacks met at Hattonchatel by nightfall on the first day, and the salient was entirely cleared (September 16); more than 15,000 prisoners and some 250 guns were taken. American casualties numbered 7,000. The strategic importance of the victory was great; since 1914 the St.-Mihiel salient in German hands had constituted a standing threat to any Allied movements in Champagne. In addition, the First Army proved itself to both friend and foe to be a competent entity. This was the largest American operation since the Civil War. Pershing at once turned to the tremendous job of shifting his entire army some 60 miles, and entering another major offensive without any rest.

Foch's Final Offensives

THE CONCEPT

Foch planned a double penetration, in 2 major assaults. One of these was to be a Franco-American drive from the Verdun area toward Mézières, a vital German supply center and railroad junction. The other was to be a British offensive between Péronne and Lens, with the railroad junction of Aulnoye as its objective. Seizure of these 2 vital railroad junctions would jeopardize the entire German logistical situation on the Western Front. Supplemental assaults would be made in Flanders by a combined British-Belgian-French army group, and between La Fère and Péronne by another Franco-British force.

THE MEUSE-ARGONNE OFFENSIVE,
SEPTEMBER 26–NOVEMBER 11

1918, September 26–October 3. First Phase.
Having efficiently shifted by night more than a million men with tanks and guns over an inadequate road and rail net, Pershing launched the First Army—3 corps abreast—in attack at 5:25 A.M. On its left the French Fourth Army (**H. J. E. Gouraud**) attacked also. The American zone lay astride the Meuse Valley, including the Argonne Forest on its left, the Aire Valley, and the heights on both sides of the Meuse. The German defenses (Gallwitz's army group to the east, the Crown Prince's to the west) consisted of 3 heavily fortified lines taking clever advantage of the rugged and heavily wooded terrain. Initial rapid advance was finally slowed in the Argonne Forest and in front of **Montfaucon** as the Germans rushed in reinforcements. The

Lieutenant Frank Luke, American ace, with his Spad

American drive lost momentum on the line **Apremont-Brieulles** (October 3), having penetrated the first 2 German positions.

1918, October 4–31. Second Phase. Replacing a number of his assault divisions by veteran troops from the St.-Mihiel operation, Pershing renewed the offensive. There was no room for maneuver; the First Army battered its way slowly forward in a series of costly frontal attacks, and the actual combat zone was widened to include the east bank of the Meuse, where the Germans had excellent observation from the Heights of the Meuse. The Argonne Forest was cleared, facilitating the advance of the French Fourth Army, on the left, to the Aisne River. Pershing regrouped his forces into a group of 2 armies (October 12). The newly constituted Second Army, commanded by Bullard, prepared for an offensive northeast, between the Meuse and the Moselle, while the First Army, now under Liggett, continued its slow northward battering-ram progression. Clemenceau, exasperated by the Americans' slow progress, tried unsuccessfully to have Pershing relieved. Foch, aware of the nature of the opposition, well knowing that the American offensive—threatening the part of the front most vital to the Germans—was drawing all available German reserves from elsewhere for its defense, declined to support Clemenceau. As October ended, the First Army had punched through most of the third and final German line.

1918, November 1–11. Final Phase. With rested divisions replacing tired ones, the First Army jumped off again, smashing through the last German positions northeast and west of Buzancy, thus enabling the French Fourth Army to cross the Aisne. In the open now, American spearheads raced up the Meuse Valley, brushing aside last-ditch German defensive stands, reaching the Meuse before Sedan (November 6) and placing destructive artillery fire on the Mézières-Montmédy rail line, vital artery of supply for the entire German front. A spectacular drive on Sedan by the U.S. 1st Division was abruptly checked by orders from higher authority, to permit the French the honor of taking the city and erasing the stain of the 1870 disaster (see p. 914). Bullard's Second Army launched its final attack (November 10), driving for Montmédy. Next day the armistice ended all hostilities.

FINAL BRITISH, FRENCH, AND
BELGIAN OFFENSIVES

1918, September 27–October 17. Storming the Hindenburg Line. One day after the be-

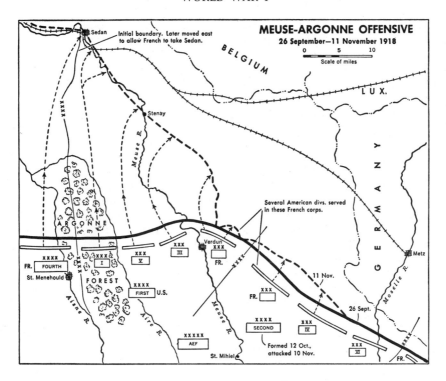

MEUSE-ARGONNE OFFENSIVE
26 September–11 November 1918

ginning of the American offensive, Haig's army group flung itself against the Hindenburg Line. Trading space for time on this front, Boehn's army group managed to withdraw after a succession of costly and gallant British attacks drove through the last of the Hindenburg Line positions (October 5). To Haig's surprise, he had been unable to achieve a complete breakthrough, and the momentum of his drive slowed down in the face of skillful German defense.

1918, September 28–October 14. Offensive in Flanders. British-Belgian troops of King Albert's army group swept over the Ypres Ridge, but then slowed down as swampy country choked all supply, and Rupprecht's army group fought back grimly.

1918, October 17–November 11. Advance to the Sambre and the Scheldt. Because of American progress in the Meuse-Argonne, a German retreat all along the line became necessary. Ludendorff hoped that he could reestablish a new line west of the German border

and by a determined defense through the winter force the Allies to grant generous terms. But his hopes were foiled by the pressure being maintained all along the Allied lines. In a renewed British assault, Rawlinson's Fourth Army broke through German defenses on the Selle River (October 17). Byng's Third Army forced a crossing lower down (October 20). The drive threw back Boehn's army group with the loss of 20,000 prisoners. At the same time the Belgians and British began to move again in Flanders. The German Army began to crack.

THE GERMAN COLLAPSE

1918, October 6. Request for an Armistice. As the front lines began to crumble, the new German chancellor, Prince **Max of Baden,** sent a message to President Wilson, requesting an armistice on the basis of Wilson's Fourteen Points (see p. 1069). An exchange of messages concluded (October 23) with Wilson's insistence that the U.S. (and the Allies) would not negotiate an armistice with the existing military dictatorship.

1918, October 27. Resignation of Ludendorff. Just before formal dismissal, Ludendorff resigned to permit the desperate German government to comply with Wilson's demand. Hindenburg, however, retained his post as German commander in chief, with General **Wilhelm Groener** replacing Ludendorff as Quartermaster General (Chief of Staff).

1918, October 29–November 10. Revolution in Germany. Inspired by the Communists and sparked by a mutiny of the High Seas Fleet, disorders, revolts, and mutinies flared inside Germany. A new Socialist government took power and proclaimed a republic (November 9). The Kaiser fled to Holland (November 10).

1918, November 7–11. Armistice Negotiations. A German delegation, headed by a civilian, **Matthias Erzberger,** negotiated an armistice with Foch in his railway coach headquarters on a siding at Compiègne. Agreement was finally reached at 5 A.M., November 11, 1918. The terms, which were in effect a German surrender, provided that the German Army must immediately evacuate all occupied territory and Alsace-Lorraine; immediately surrender great quantities of war matériel (including 5,000 guns and 25,000 machine guns); evacuate German territory west of the Rhine, and three bridgeheads over the Rhine, to be occupied by the Allies; surrender all submarines; intern all other surface warships as directed by the Allies.

1918, November 11. The Armistice. Hostilities ceased at 11 A.M.; the terms of the armistice immediately became effective.

COMMENT. *Comparisons are invidious. The American Expeditionary Force was the vital factor in the final Allied victory; the Meuse-Argonne offensive was decisive; 6 other American divisions played important spearhead roles elsewhere on the front during the final Allied advances. But the question whether Allied victory could have been achieved without the Americans should not be debated. The American role was to add a final increment of numbers and fresh initiative, permitting the much larger, and more experienced, Allied*

armies to achieve equally spectacular successes in the final weeks of the war.

The Italian Front

1918, June 15–22. Austrian Offensive. Germany during the spring transferred her troops in Italy to the Western Front, insisting that the Austrians crush Italy singlehanded. The argument had weight, since Russia was out of the war. Both Conrad (now commanding on the Trentino front) and Borojevic, on the Piave, demanded command of the decisive effort. A compromise decision by Archduke **Joseph** permitted them to attack simultaneously. Since the mountainous terrain and lack of lateral communications would prevent mutual support, the available reserves were split between them. The result was that neither commander had sufficient strength to exploit any initial success.

1918, June 15. Battle of the Piave. Following a diversionary attack in the west at the **Tonale Pass,** which was repulsed (June 13), the twin Austrian drives were launched; Conrad's objective was Verona; Borojevic's, Padua. Forewarned by deserters, Diaz was well prepared. Conrad's Eleventh Army, striking the Italian Sixth and Fourth armies, made some slight gains, but was then checked and thrown back by counterattacks. His troops took little further part in the offensive. Borojevic, attacking along the lower reaches of the Piave, forced a wide crossing and penetrated positions of the Italian Third Army for about 3 miles. The mirage of success failed when unexpected high water and Italian aerial bombing attacks disrupted the Austrian supply line. Diaz, who had kept one entire army in reserve—the Ninth—promptly shifted reinforcements over lateral lines and snubbed the attack. Borojevic, unable to obtain reinforcements from Conrad, withdrew during the night (June 22–23). Diaz, much to Foch's disgust, made no counterattack.

1918, July–October. Italian Counteroffensive Preparations. Diaz, marking time until sure of eventual Allied success on other fronts, finally prepared a double offensive. The Italian

Fourth Army was to penetrate the center of the Austrian front. The Eighth Army, supported by the new Tenth and Twelfth armies (mostly containing British and French divisions), was to attack across the Piave on Vittorio Veneto. This diversion of force was based on the demoralized condition of the Austrian forces, whose home government, falling to pieces, was requesting an armistice.

1918, October 23. Battle of Monte Grappo. Showing unexpected determination, the Austrian Belluno Group, defending the key point on the center front, threw back the Italian Fourth Army with heavy loss.

1918, October 24–November 4. The Battle of Vittorio Veneto. The Austrian Sixth Army halted the Italian Eighth on the Piave River line. However, French troops of the Twelfth Army (commanded by French General **Jean Graziani**) clawed a footing on the left, while on the right British troops of the Tenth Army (commanded by British General **Frederic Lambert, Earl of Cavan**) gained a large bridgehead (October 28), throwing back part of the Austrian Fifth Army and splitting the front. One American regiment, the 332nd Infantry, took part in this action. The penetration reached **Sacile** (October 30). Next day, as Italian reinforcements exploited the ever-widening gap, Austrian resistance collapsed. Belluno was reached (November 1) and the Tagliamento (November 2), while in the western zone British and French troops of the Sixth Army drove through to Trent (November 3). Some 300,000 Austrians became prisoners.

1918, November 3. Capture of Trieste. The city was seized by an Allied naval expedition in the Gulf of Venice.

1918, November 3. Armistice Signed. Hostilities were to be concluded the next day (November 4).

The Balkan Front

1918, September 15–29. Battle of the Vardar. Brilliant French General Franchet d'Esperey, who succeeded the capable organizer Guillaumat (July), was given grudging assent by the Supreme War Council to mount a major offensive. He nominally had nearly 600,000 men—Serb, Czech, Italian, French, and British—of whom some 200,000 were available for duty. Opposing him were about 400,000 Bulgars (practically all German troops had been withdrawn except for command and staff). Covered by heavy artillery support, the First and Second Serb armies attacked the center of the front, debouching between elements of the French Orient Army. The penetration was successful, the Serbs pushing north while the French exploited the gap on both flanks. A British diversionary attack (September 18) on the right gained some ground. Gaining momentum, the penetrating assault reached the Vardar (September 25), splitting the Bulgarian front. The British drive reached Strumitsa (September 26), and French cavalry, passing through the main effort, took **Skoplje** (September 29). Allied air forces brought panic to the fleeing Bulgars.

1918, September 29. Bulgarian Armistice. Energetic Franchet d'Esperey kept his Serbian and French troops moving north.

1918, November 10–11. Crossing of the Danube. Franchet d'Esperey had freed the Bulkans and was prepared to march on Budapest and Dresden when Germany's armistice halted hostilities.

The Eastern Front

Although hostilities had ended, Russian-German wrangling dragged on for 2 months over the conference tables at Brest-Litovsk. Germany insisted upon autonomy for the former Russian territories of Poland, Finland, Estonia, Latvia, Lithuania, and the Ukraine, all of which were in revolt against the Bolshevik government of Russia (see p. 1092). Trotsky, the chief Russian negotiator, refused to accept these terms and to sign a peace treaty. Finally the Germans called the Soviet bluff and marched eastward (February 18). Unable to offer military opposition, the Bolsheviks hastily signed a peace treaty (March 3) whose terms were even stiffer than the Germans had originally offered. German troops continued on to occupy the Ukraine, which provided grain to save the German people from starvation in 1918. As Cyril Falls has written: "Russia lay at the mercy of the Central Powers for the rest of the war. It was Foch and Haig who saved Bolshevik Russia."*

The Turkish Fronts

THE CAUCASUS

1918, January 27. British Advance on Baku. Under Major General **L. C. Dunsterville,** British troops moved northeast from Baghdad, reaching **Enzeli** (Pahlevi) on the Caspian (February 17). An advance battalion crossed the Russian border and entered **Baku** (August 4).

1918, February 24. Turkish Advance into Armenia. The Turks raced the British for possession of the oil wells of western Persia after the Russian collapse. Reoccupying **Trebizond,** the Turks in quick succession reached **Erzerum** (March 12), **Van** (April 5), **Batum** (April 15), and **Kars** (April 27). A German force, landing at **Poti** on the Black Sea (June 12), pressed east to take **Tiflis** (Tbilisi) (June 12), while the Turks reached **Tabriz. Baku** was abandoned by the British after vigorous Turkish attacks (August 26–September 14).

PALESTINE AND SYRIA

1918, January–September. Lawrence of Arabia. While Allenby at Jerusalem was restricted to minor operations because of drafts on his force to the Western Front, Arabia to the south and east was in flames. Colonel **T. E. Lawrence,** with a small group of other British officers, reaped a harvest of Arab rebellion against Turkish rule. Lawrence's guerrillas played hob with the Hejaz Railway, running for some 600 miles from Amman, Palestine, to Medina (Al Madinah) in Arabia, the southernmost Turkish garrison. In all, Lawrence's activities kept more than 25,000 Turkish troops pinned

* *The Great War, 1914–1918, New York, 1959, p. 287.*

down to blockhouses and posts along this line. By September, Lawrence, with Emir **Faisal,** son of Sherif **Hussein,** self-styled "King of the Hejaz," had isolated Medina by destroying the railway line and was moving north to operate on Allenby's right flank. The Arab strength totaled about 6,000 men. The Arab irregulars had been reinforced with small British armored-car detachments and light guns, and British gold coaxed the fickle Arabs to remain in ranks.

1918, September 18–October 30. Allenby's Offensive. Reinforced during the late summer, Allenby prepared meticulously for what was to be the decisive blow. The Turkish defensive line, skillfully fortified, lay from the Mediterranean, north of Jaffa, to the Jordan Valley. Liman von Sanders, who had replaced Falkenhayn in command, had the Turkish Eighth, Seventh, and Fourth armies in line from right to left, a total of some 36,000 men and 350 guns. Allenby's plan was to mass his main effort on the seashore, burst open a gap, and then let his cavalry corps through while the entire British line swung north and east like a gate, pivoting on the Jordan Valley. Utmost secrecy was kept. British planes cleared the sky of all enemy observation. Dummy camps and horse lines indicated to the Turks that the British cavalry was concentrated near Jerusalem. Elaborate plans for a race meet on the day of the offensive were widely publicized. Allenby's strength was 57,000 infantry, 12,000 cavalry, and 540 guns. Of these, 35,000 infantry and 400 guns were concentrated in the main effort against 8,000 Turkish infantry and 130 guns.

1918, September 19–21. Battle of Megiddo. At 4:30 A.M. an artillery concentration fell, followed immediately by the advance of the entire British line. The XXI Corps on the left tore a wide gap along the seacoast, through which the Desert Mounted Corps poured. At the same time the RAF bombed rail junctions and all Turkish army headquarters, completely paralyzing communications. The cavalry raced northward through the disintegrating Turkish Eighth Army, reaching the plain of Esdraelon early next morning. The great eastward wheel of the infantry was completed by dawn (September 20). By this time the Turkish Eighth Army (**Jerad** Pasha) had ceased to exist, and the Seventh (Kemal) was falling back eastward in disorder toward the Jordan. The British cavalry then swept through Nazareth—Sanders himself narrowly escaping capture—and turned east to reach the Jordan just south of the Sea of Galilee (September 21). Alone of the Sanders army group, the German Asia Corps (2 regiments) retained some semblance of order in the rout that followed.

1918, September 22–October 30. The Pursuit. Hordes of fugitives were bombed by the RAF as the Turkish Fourth Army in the Jordan Valley joined the torrent. On the desert flank to the east, Lawrence and Faisal cut the railway line at Deraa (September 27), while Allenby pressed to take Damascus (October 1) and Beirut (October 2). The Desert Mounted Corps continued to spearhead the advance, reaching Homs (October 16) and Aleppo (October 25).

1918, October 30. Turkey Leaves the War. Turkey signed an armistice (October 30) at Mudros.

COMMENT. *Allenby's victory at Megiddo was a set piece of fire and movement, perfectly planned and executed. It was one of the most brilliant operations in the history of the British Army. In 38 days, Allenby's troops had advanced 360 miles, fighting continuously. Three Turkish armies were destroyed, 76,000 prisoners (4,000 of them Germans and Austrians), 360 guns, and a vast amount of other material captured. British casualties—mostly incurred during the initial breakthrough—were 853 killed, 4,482 wounded, and 385 missing.*

MESOPOTAMIA

1918, January–September. Dunsterville Operation against Baku. (See p. 1080.)

1918, October 23. Tigris Offensive. A British force of all arms was hurriedly pushed north from Baghdad to secure the Mosul oil fields as a *fait accompli* prior to the expected Turkish collapse. Lieutenant General **A. S. Cobbe** dislodged the Turkish Tigris Group (General **Ismael Hakki**) from its outpost position at Fat-ha. Hakki fell back to the vicinity of **Sharqat,** where Cobbe's forces converged on him after some sharp fighting. A British assault (October 29) was only partially successful, but next day Hakki surrendered, with 11,322 prisoners and 51 guns. British casualties were 1,886. Cobbe now hurried his cavalry to the outskirts of Mosul (November 1). Despite the provisions of the October 30 armistice, Cobbe was ordered to take the place. After some squabbling, the Turkish garrison of Halil Pasha agreed to march out and the British remained (November 14).

COMMENT. *The entire checkered Mesopotamian campaign had hinged upon the oil fields and their protection. The war's end found Britain in possession, at a total cost of 80,007 casualties. Of these 15,814 had been killed in action; an additional 12,807 died from disease.*

1918, November. Capture of Baku. A British flotilla, extemporized on the Caspian Sea, drove the Turks from Baku.

1918, November 12. Envoi. The Allied fleet steamed through the Dardanelles, to arrive off **Constantinople** (Istanbul) next day.

War at Sea

1918, April 23. Zeebrugge and Ostend Raids. The German submarine warfare, contained though it was by this time by the Allied convoy system, was nevertheless still a menace. U-boats operated from bases at Zeebrugge and Ostend, and from the shelter of the canal port of Bruges. British Rear Admiral **R. J. B. Keyes,** commanding the Dover Patrol, organized a raid against the bases, some 75 ships taking part. With the utmost gallantry the light cruiser *Vindictive* (Captain **A. F. B. Carpenter**) dashed into Zeebrugge, with destroyer and submarine escort. The ship laid alongside the mole and demolition parties debarked. At the same time a British submarine loaded with high explosive was blown up against the lock gates and two blockships were also sunk. The *Vindictive* and her landing parties got away after inflicting some damage, but the base was not entirely blocked. A raid against Ostend at the same moment failed. Another raid (May 9) into Zeebrugge was made by the *Vindictive;* the cruiser was deliberately sunk against the lock gates. While not a complete success, these exploits went far to lower German morale. Several German light cruiser and destroyer raids against the British coast and against North Sea convoys were but minor irritations to the British.

1918, January 20. Sortie from the Dardan- **elles.** The *Goeben* and *Breslau* sailed into the Aegean Sea, but the voyage ended in disaster; the *Goeben* was badly damaged by British mines, and the *Breslau* was sunk. However, the *Goeben,* despite British aerial bombing, was salvaged.

1918, June 9. Austrian Sortie from Pola. This effort against Allied blockading craft was repelled when 2 Italian midget submarines attacked the Austrian squadron, sinking the battleship *Szent-Istvan,* the only dreadnought sunk in action during the war.

1918, October 29. Mutiny of the High Seas Fleet. Hipper (who had succeeded Scheer in August) planned a desperate sortie to bring on a final battle with the British Grand Fleet. The crews mutinied, refused to sail, and seized control of the warships, ending the war at sea.

Operations in East Africa

Despite the most intensive efforts, the British had been unable to overcome elusive and brilliant von Lettow-Vorbeck in 4 years of continuous search and pursuit. They drove him into Portuguese East Africa (1917), where he continued an active and aggressive guerrilla campaign, capturing Portuguese military posts and maintaining his small command by captured supplies. He then reentered German East Africa and, though he had only 4,000 men and was opposed by forces totaling 130,000, succeeded in capturing several small posts before marching into British Northern Rhodesia. Finally, after the British were able to inform him of the Armistice, he stopped hostilities (November 14) and surrendered his command (November 23).

POSTARMISTICE

1918, November 17. German Evacuation of Allied Territory. Allied troops began to reoccupy those portions of France and Belgium which had been held by the Germans since 1914.

1918, November 21. The Internment of the High Seas Fleet. The German fleet sailed into the Firth of Forth, between the lines of the British Grand Fleet. It later was shifted to Scapa Flow.

1918, December 1. Allied Movement into the Rhineland. In accordance with the Armistice terms, Allied and American troops fol-

lowed the withdrawing Germans into Germany.

1918, December 9. Bridgehead Occupation. The British were at Cologne, the Americans at Coblenz, the French at Mainz. The bridgeheads were 18 miles deep in radius beyond the Rhine.

1919, May 7–June 28. Treaty of Versailles. The terms as presented to and signed by the German representatives (no discussion was permitted them): (1) admission of German war guilt; (2) Germany was stripped of her colonies, Alsace-Lorraine, the Saar Basin (subject to a plebiscite in 1935), Posen, and parts of Schleswig and Silesia; (3) reparations were required, later fixed at $56 billion (4) Germany was dis-

THE COST OF THE WAR[a]

	Total Force Mobilized	Military Battle Deaths[b]	Military Wounded	Civilian Dead[c]	Economic and Financial Cost[d] ($ million)
ALLIES					
France	8,410,000	1,357,800	4,266,000	40,000	49,877
British Empire	8,904,467	908,371	2,090,212	30,633[e]	51,975
Russia	12,000,000	1,700,000	4,950,000	2,000,000[f]	25,600
Italy	5,615,000	462,391	953,886	g	18,143
United States	4,355,000	50,585	205,690	g	32,320
Belgium	267,000	13,715	44,686	30,000	10,195
Serbia	707,343	45,000[h]	133,148	650,000	2,400
Montenegro	50,000	3,000	10,000	g	g
Rumania	750,000	335,706	120,000	275,000	2,601
Greece	230,000	5,000	21,000	132,000	556
Portugal	100,000	7,222	13,751	g	g
Japan	800,000	300	907	g	g
Total	42,188,810	4,888,891	12,809,280	3,157,633	193,899
CENTRAL POWERS					
Germany	11,000,000	1,808,546	4,247,143	760,000[i]	58,072
Austria-Hungary	7,800,000	922,500	3,620,000	300,000[j]	23,706
Turkey	2,850,000	325,000	400,000	2,150,000[k]	3,445
Bulgaria	1,200,000	75,844[l]	152,390	275,000	1,015
Total	22,850,000	3,131,889	8,419,533	3,485,000	86,238
COST TO NEUTRAL NATIONS					
					1,750
Grand total	65,038,810	8,020,780	21,228,813	6,642,633	281,887

[a]Many of these figures (compiled from various sources) are approximations or estimates, since official figures are often misleading, missing, or contradictory.

[b]Includes only killed in action or died of wounds.

[c]Figures vary greatly; deaths from epidemic disease and malnutrition, probably not completely attributable to the war, are included in some instances and not in others. See specific notes below.

[d]Includes war expenditures, property losses, and merchant-shipping losses.

[e]About two-thirds of these were lost to U-boats, the remainder to naval and aerial bombardment.

[f]Includes approximately 500,000 Poles and Lithuanians.

[g]No reliable figures available; there was relatively small loss.

[h]Approximately 80,000 additional were nonbattle deaths: typhus, influenza, malnutrition, frostbite.

[i]Asserted by German sources to be due to the Allied blockade through 1919; a handful of deaths was caused by Allied air raids.

[j]At least two-thirds of these were Polish; many of the remainder have been attributed to Allied blockade.

[k]More than half of these were Armenian; most of the remainder were Syrian or Iraqi.

[l]At least 25,500 additional were nonbattle deaths.

armed. The Covenant of the proposed League of Nations was attached to the treaty. The U.S. Senate rejected the League Covenant despite the frantic efforts of President Wilson, which resulted in his physical breakdown. Finally (May 20), the Congress by joint resolution declared the end of war with Germany and Austria-Hungary, reserving for the U.S. any rights secured by the Armistice, the Versailles Treaty, or as a result of the war. Separate U.S. treaties concluding peace with Germany, Austria, and Hungary were later ratified (October 18, 1920).

1919, June 21. Scuttling of the High Seas Fleet. Most of the ships of the German fleet were scuttled by their own crews at Scapa Flow, in defiance of the terms of the Versailles Treaty.

1923, January. End of the American Occupation in Germany. The U.S. Occupation Force evacuated Coblenz to return home. The British and French still remained in occupation of the Rhineland as the period ended.

MAJOR WARS AFTER 1918

Russo-Polish War, 1920

1920. Prelude. During the Russian Civil War, Poland (see p. 1092) had occupied areas of mixed Polish-Russian population in undefined frontier areas bordering White Russia and the Ukraine. The Soviet government now disputed this Polish action, and began to build up its forces in the west. The Poles decided to attack before the Red Army seized the initiative. North of the Pripet Marshes was Russian Marshal **Mikhail N. Tukhachevski**'s Bolshevik Army of the West, opposed by a smaller Polish army under General **Wladyslaw Sikorski.** South of the Marshes, Russian General **Yegorov**'s Army of the Southwest was opposed by the numerically superior army of General **Józef Pilsudski,** overall Polish commander. Total forces on each side approached 200,000.

1920, April 25–May 7. Pilsudski's Offensive. The Polish general drove for Kiev, with the support of a mixed force of anti-Bolshevik Ukrainians under Hetman **Simon Petlyura** on his right flank. Capturing **Kiev** (May 7), Pilsudski prepared to swing north, behind the Pripet Marshes, to hit Tukhachevski's left rear. His plan, however, was too ambitious for the force and logistical backup available to him.

1920, May 15. Tukhachevski's Offensive. Striking southwestward, the Russian Army of the North pinned back Pilsudski's left wing. At the same time, General **Semën M. Budënny** of Yegorov's Army of the Southwest drove northwest against Pilsudski's right flank, with a Cossack cavalry corps some 16,000 strong with 48 guns. Budënny reached **Zitomir,** southwest of Kiev, almost bagging Pilsudski's right wing. By June 13, the Polish left was also in full retreat as the Cossack horde swept on to the outskirts of **Lemberg** (Lvov). North of the Pripet area, Tukhachevski advanced (July 4) westerly toward Warsaw. Polish troops fell back on both fronts. Tukhachevski reached Vilna (July 14) and Grodno (July 19), while Budënny's Cossacks kept up pressure on the southern front. By July 25 the Polish forces lay in 2 groups—1 near Warsaw, whose fall appeared imminent (Tukhachevski expected to take it August 14), the other around Lvov. In all, some 180,000 Poles faced approximately 200,000 Russians.

1920, July–August. Polish Counterattack Plan. France's General **Maxime Weygand** arrived at Warsaw to advise Pilsudski, who needed munitions more than advice. Weygand favored a counterattack north of Warsaw against the Russian right, launched from behind the defensive lines of the Vistula. Pilsudski had a better plan. The Bolshevik advance had outrun all supply; the Russians were living off the country, and could not stop. A halt would be disastrous. So, while the seemingly irresistible Russian right lapped westward, passing to the north of Warsaw, Pilsudski prepared a daring counterattack against Tukhachevski's center.

1920, August 16–25. Battle of Warsaw. Its weight concentrated at Deblin, 50 miles south of Warsaw, Pilsudski's assault crashed through the weakly held Russian center, along the axis of the Warsaw-Brest Litovsk road. Pilsudski accompanied it in person. The left of the Russian Sixteenth Army was shattered. Ignoring the Russian elements south of his penetration, Pilsudski turned north, threatening in turn the Third, Fifteenth, Sixteenth, and Fourth armies. At the same time the Polish forces north of Warsaw under Sikorski also advanced, despite desperate Russian resistance. Caught between the Polish pincers, Tukhachevski's command disintegrated. Some 30,000 or more fled north over the East Prussian border and were there disarmed. The remainder reeled eastward, closely pursued. Tukhachevski managed to rally (August 25) along the line Grodno-Brest Litovsk-Wlodawa, where the pursuit ended. The Poles captured 66,000 prisoners and more than 230 guns, 1,000 machine guns, and 10,000 vehicles. Total Russian casualties were approximately 150,000; the Poles lost 50,000 men.

COMMENT. *The Battle of Warsaw ranks among the decisive battles of the 20th century. It was as much a check to the Communism's first overt westward thrust as was Charles Martel's victory at Tours a check to the Moslem surge into Europe (see p. 222).*

1920, September 12—October 10. Continued Polish Offensive. The Poles continued their advance on broad fronts on both sides of the Pripet Marshes. Sikorski, now commanding in the south, reached Tarnopol (September 18). Pilsudski again defeated Tukhachevski in the **Battle of the Niemen** (September 26), driving the Russians from the river-line defensive positions and destroying their Third Army. The Poles next entered Grodno (September 26). Next day, in the **Battle of the Shchara** (Szczara), Pilsudski drove Tukhachevski's beaten troops back to Minsk. In these last 2 battles, the Russians lost some 50,000 prisoners and 160 guns.

1920, October 12. Armistice. Hostilities were brought to a close.

1921, March 18. Treaty of Riga. The Russians conceded all of Poland's territorial claims.

GRAECO-TURKISH WAR, 1920–1922

1918–1919. Allied Disagreements over Turkish Peace Settlement. While the Allies squabbled, Turkey was collapsing in anarchy. A Greek army was landed by the Allies at Smyrna (May 15, 1919) to act as an agent for Allied interests. An Italian force also landed in southwest Anatolia. Long-standing national antipathies precipitated incidents, leading to atrocities against the Turkish civilian population by the Greeks.

1919, June–September. Turkish Nationalist Movement. Aroused by Greek action, a group of patriotic Turks, led by General Mustafa Kemal, banded together to form a new national government. Despite opposition of the Allies (who still occupied Constantinople), Kemal set up his new government in Ankara (April 1920).

1919, June 22. Greek Offensives in Anatolia and Thrace. Greek Prime Minister Venizelos' offer to act as the Allies' agent in suppressing the Nationalist movement was warmly accepted by British Prime Minister Lloyd George. In Thrace, Greek troops occupied Adrianople (July 25). In Anatolia, a major eastward movement placed the Greeks at Ushaq (Usak), 125 miles from Smyrna (June 29). A flanking operation captured Panderma (Bandirma) and Brusa (Bursa), south of the Sea of Marmora (July 9). There was no effective Turkish opposition.

1919, August 20. Treaty of Sèvres. The Allies imposed terms on the Sultan's government; Turkey again lost all Asiatic possessions outside of Anatolia, and all of European Turkey except the Chatalja Peninsula and Constantinople. Armenia was established as an independent republic. The treaty was not recognized by the Nationalists.

1919, October. Nationalist Offensive against Armenia. Repeated Armenian raids into Anatolia caused the Nationalists to retaliate. A Turkish drive quickly resulted in the capture of **Kars** (October 21) and the massacre of a number of Armenians in counterrevenge for earlier vengeance atrocities against Turks. A truce was soon negotiated with the Armenian and Soviet governments, ending hostilities in northeast Anatolia.

1919, October 25. Political Changes in Greece. The unexpected death of King Alexander of Greece brought his warrior-father **Constantine** (hero of the Balkan Wars, villain of World War I) back to the Greek throne, changing the entire political complexion of the war. Pro-German Constantine was not trusted by the Allies, who withdrew their former support of Greek activities in Turkey. He nevertheless determined to continue operations to gain permanent Greek control of western Anatolia and Thrace.

1921, January. First Battle of Inönü. Greek General **Papoulas,** advancing on Eskisehir, was repulsed at Inönü, 20 miles to the west, by Turk forces under **Ismet** Pasha. The Greek offensive was temporarily halted.

1921, January–March. Kemal's Diplomacy. Negotiations by Kemal resulted in the withdrawal of Italian forces from Anatolia (March 13) and an understanding with the Soviet government (March 16). The Soviets received Batum, and Kemal recognized the independence of the Soviet Armenian Republic in exchange for the return of Kars and Ardahan to Turkey.

1921, March 23. Renewed Greek Offensive. Papoulas' strength had now been built up to 150,000, of whom two-thirds were disposed along his line of communications to Smyrna. He now launched a full-scale attack.

1921, March 28–30. Second Battle of Inönü. Papoulas drove the Turks (**Refet** Pasha) out of **Afyon Karahisar (Afyon).** However, Refet's right, under Ismet Pasha, at Inönü and Eskisehir, held its ground, threatening the Greek left. Three successive Greek attacks were thrown back and Papoulas broke off the offensive (April 2). Kemal replaced Refet by Ismet (who later adopted the name of the victory as his own, and is known to Turkish history as Ismet Inönü).

1921, July 16–17. Constantine's Offensive. The Greek king now assumed personal field command, and moved skillfully against the Turkish positions from Eskisehir to Afyon. A feint at the Turkish right distracted Ismet's attention. The Greek main effort then fell on his left flank at Afyon. The Greeks then shifted direction northward, rolling up the Turkish left and center in a combination of frontal assault and envelopment. Despite Ismet's counterattack, **Eskisehir** fell (July 17). Ismet prepared to fight to the finish, but Kemal sagely ordered a retreat. Ismet's army disengaged at great cost, falling back northward about 30 miles to the Sakkaria (Sakarya) River. As Kemal had expected, the Greeks were too exhausted to pursue. Turkish losses were over 11,000, Greek about 8,000. Constantine regrouped and moved against the Sakkaria line (August 10), which had now been reinforced through Kemal's strenuous efforts.

1921, August 24–September 16. Battle of the Sakkaria. When the attempted Greek turning movement against the Turkish left was stopped, the battle turned into a slugging match. The Greek main effort next shifted to the center, advancing 10 miles in as many days; then the offensive slowed. Kemal, assuming personal command, made a daring enveloping attack with a small force against the Greek left (September 10). Although the success was slight, the psychological effect of the move was immense. With his 350-mile line of communications threatened, his troops facing a deadlock, and the Anatolian winter coming on, Constantine disengaged. The Greeks fell back along the railway line toward Eskishehir and Afyon. Kemal did not pursue closely, undoubtedly fearing this might stimulate Allied intervention in favor of the Greeks.

1921–1922, October–July. Kemal Consolidates Politically and Militarily. While active operations lagged, Kemal built up military strength and mended diplomatic fences; France withdrew her troops from Cilicia (October 20). While the overextended Greek army in the field was weakening. Turkey's internal and international position rapidly improved.

1922, August 18–September 11. Turkish Counteroffensive. Kemal struck, taking **Afyon** (August 30). His main effort then drove westward along the railway toward Smyrna, while a large detachment attacked north, capturing **Brusa** (September 5). The shattered Greek army fell back in confusion, attempting to avenge their defeat in an orgy of burning, pillaging, and murdering. About 1 million Turkish civilians were made homeless, further embittering a conflict already notable for atrocities. Kemal pursued closely, and the Greek army soon disintegrated into a horde of terrified refugees. No arrangements had been made for evacuation by sea; thousands of Greek soldiers were captured—and many massacred—by the Turks. **Smyrna** was captured by assault (September 9–11); the Greek sections of the town were sacked and burned, and many Greek civilians were massacred by the Turks.

1922, September–October. Advance on Constantinople. Kemal, now confident of his own strength, advanced on Constantinople, which was still occupied by a small Allied garrison. The Allies were as anxious to avoid a clash as he was. Diplomatic negotiations at the **Convention of Mudania** (October 3–11) promised the restoration of Thrace and Adrianople to Turkey, and provided for neutralization of the Straits. Kemal proclaimed abolition of the Sultanate; the Sultan fled in a British warship (November 1).

1923, July 24. Treaty of Lausanne. This superseded the Treaty of Sèvres (see p. 1085), formally restoring to Turkey her Thracian territory to the Maritza River. The Allies then evacuated Constantinople (August 23). The Turkish Republic was officially established, with Mustafa Kemal, surnamed Atatürk ("Chief of the Turks"), as its first President (October 29).

COMMENT. *By his charismatic inspiration, firm determination, political skill, and brilliant military genius, Kemal had revived and unified his dismembered, prostrate nation, while at the same time carrying on a victorious war against Greece. Casualty figures for the war are not available. Turkish losses were probably less than 100,000 men. Greek casualties were upward of 200,000, of whom at least half were lost during the flight to Smyrna in August–September, 1922.*

EUROPE, 1900–1925

WESTERN EUROPE

The United Kingdom of Great Britain and Ireland

1904. Committee on Imperial Defense (Esher Committee). An evaluation of the British military system, in the light of Boer War experiences, resulted in recommendations for major reorganization.

1905–1910. Haldane Reforms. Richard Burdon, Viscount Haldane, civilian Secretary for War, achieved a military reorganization generally along the lines recommended by the Esher Committee. As a result, Britain developed the finest army for its size in the world. The vital role of the British Expeditionary Force in the initial Marne Campaign (see pp. 1025–1028) was due in large measure to the Haldane reforms.

1906, February 10. Launching of HMS. *Dreadnought.* This new warship, for which First Sea Lord Admiral Lord **John Fisher** was primarily responsible, was the first all-big-gun battleship, carrying 10 12-inch guns. She revolutionized the navies of the world.

1907, August 31. Anglo-Russian Entente. This fused earlier Anglo-French and Franco-Russian military agreements in the Triple Entente, offsetting the Triple Alliance of Germany, Austria, and Italy (see p. 920).

1914, March 20. Curragh "Mutiny." Brigadier General **Hubert Gough** and other officers at the Curragh military base in Ireland submitted their resignations rather than obey orders to force the loyal population of Ulster to accept Home Rule under the separatists of southern Ireland. This incident seriously shook the morale of the British Army on the eve of World War I. All officers were later reinstated, however, and Ulster was permitted to remain directly under the crown.

1914, July 26. Semimobilization of the British Fleet. After the conclusion of normal extended maneuvers in home waters, First Lord of the Admiralty **Winston Churchill** ordered the fleet not to disperse to peace-time stations. As the war clouds rolled over Europe, Churchill and the First Sea Lord, Admiral Prince **Louis of Battenberg,** sent the fleet to the wartime anchorage of Scapa Flow in the Orkneys.

1914, August 4. Britain Declares War on Germany. (See p. 1018.)

1914–1918. World War I. (See p. 1024.)

1916–1921. Anglo-Irish Civil War. After years of violence and unrest in Ireland, armed rebellion broke out in **Dublin** on Easter Monday (April 24, 1916), largely inspired by Sir **Roger Casement,** Irish nationalist leader, who had been landed on the Irish coast by a German submarine. British action soon suppressed this uprising (May 1); Casement and other leaders were tried and executed (August 3). Sporadic fighting continued throughout the country as the Sinn Fein, Irish patriotic society, fanned the flames of revolt, and later declared Irish independence (January 21, 1919). Open rebellion again broke out later in the year (November 26). As violence spread, the "Black and Tans" (a special British constabulary force) were moved to Ireland, along with British troops (1920). The insurrection continued with increased ferocity, marked by sabotage, arson, and murder, with the illicit Irish Republican Army active in guerrilla warfare against the British forces of law and order. Atrocities were committed by both sides. In all the British employed some 100,000 men—regulars and constabulary—in the conflict. The struggle between the Protestant northern provinces (Ulster) and the Catholic population of southern Ireland continued after peace was formally established (December 6, 1921) upon British promise of dominion status to southern Ireland. The Irish Free State was officially proclaimed a year later.

1919, January 11. Trenchard Appointed Chief of Air Staff. Major General Sir **Hugh Trenchard,** who had been first Chief of the Air Staff when the Royal Air Force was established (April 1918), and who had later commanded the Independent Air Force for the strategic bombardment of Germany (June–November 1918), returned to the head of the RAF. Trenchard's concepts for the exploitation of air power through strategic bombardment became the basis of RAF doctrine. He was later made the first Marshal of the RAF.

France

1894–1906. Dreyfus Affair. Captain **Alfred Dreyfus,** an officer of the French General Staff, unjustly accused of treason as a German spy, was tried and sentenced to imprisonment on Devil's Island. Important information that would have exonerated him was deliberately withheld from evidence in this and later trials by senior officers of the General Staff who were convinced of Dreyfus' guilt and refused to believe the evidence. The case became a *cause célèbre* within and without the army, involving overtones of anti-Semitism. Eventually it became a major political issue, particularly among liberals eager to sieze an opportunity to smash extreme conservatism and the arrogance of many aristocratic French army officers. With the support of politician **Georges Clemenceau** and the writer **Émile Zola,** the case was finally reopened; Dreyfus was exonerated and restored to duty as a major. The morale and cohesiveness of the French officer corps were severely shaken.

1900–1925. Colonial Wars. These are discussed under various geographic divisions in which the wars took place.

1904, April 8. The Anglo-French Entente. This agreement, which took almost 10 months to negotiate, resolved outstanding differences and colonial rivalries between the 2 traditional enemies, leading later to the Triple Entente (see p. 1087).

1906, January 10. Anglo-French Military Conversations. These established the basis for later military collaboration against Germany in World War I.

1911, July. Agadir Incident. The appearance of the German cruiser *Panther* at Agadir, Morocco, almost precipitated war between France and Germany over rival colonial interests in Morocco. Firm British support of France caused Germany to back down.

1914, August 3. Germany Declares War on France. (See p. 1018.)

1914–1918. World War I. (See p. 1024.)

Italy

1911–1912. Tripolitan War against Turkey. (See p. 1014.)

1914, August 3. Italian Declaration of Neutrality. Despite its membership in the Triple Alliance, Italy abjured its understanding with Austria and Germany.

1915, May 23. Italian Declaration of War against Austria. (See p. 1039.)

1915–1918. World War I. (See pp. 1018, 1039.)

1919, September 12. Fiume Coup. Gabriele d'Annunzio, Italian World War I hero, writer, and nationalist, with a band of volunteers seized **Fiume.** Italy disavowed his action. Diplomatic wrestling between Italy and Yugoslavia for the disputed area was ended temporarily by the **Treaty of Rapallo** (November 12), granting Fiume independence. D'Annunzio thereupon declared war on Italy. The *opéra bouffe*

ended when Italian troops bombarded Fiume and ejected him (December 27).

1921. Retirement of Douhet. Colonel **Giulio Douhet**'s criticisms of the Italian Army in World War I had led to his court martial and imprisonment (1916). The Caporetto disaster (see p. 1063) vindicated his criticisms; Douhet was released from prison and made chief of the Italian Army's aviation arm. He was promoted to general shortly before his retirement (1921). After retirement he published his book, *The Command of the Air,* elucidating his doctrine of winning wars by the use of airpower to break the enemy's will through violent and destructive bombardment of enemy cities. In order for airpower to perform its war-winning task, Douhet insisted that it must be completely independent of, and coequal with, surface forces. Through this and subsequent writings Douhet became the leading theoretician of air warfare, and the apostle of airpower.

1922. Rise of Fascism. Benito Mussolini's "Black Shirt" movement culminated with the **"March on Rome"** (October 28) and grant to him by the king and parliament—under armed pressure—of dictatorial powers (November 25). A virtual Fascist Reign of Terror followed as Mussolini consolidated his power (1923–1925).

1923, August–September. The Corfu Incident. Seizing upon the assassination of Italian members of an international border commission on the Greek-Albanian frontier, Italy sent warships to bombard and to occupy the Greek island of Corfu. After Greek appeal to the League of Nations, Italy withdrew from Corfu under pressure from Britain and other powers.

Spain

Not directly affected by World War I, Spain was involved militarily only in minor colonial operations during the early part of this period. Long-standing unrest in Spanish Morocco, however, broke out into full-scale revolt, in which Spanish arms suffered one disaster and several additional defeats (1921–1925; see pp. 1098–1099). One result of these setbacks was to revive internal frictions in Spain itself. A military revolt in Barcelona led to the assumption of dictatorial power by General **Miguel Primo de Rivera,** who was able to restore order temporarily (1923).

Portugal

Portugal was wracked by a series of minor revolts during most of this period. She joined the Allies in World War I (March 9, 1916).

CENTRAL AND NORTHERN EUROPE

Germany

1891–1906. Count Alfred von Schlieffen, Chief of General Staff. (See p. 1020.)

1900–1913. Colonial Activities. Military operations prior to World War I were confined to colonial areas, and are discussed under the respective geographical sections of this chapter. During this period there were repeated colonial crises in Morocco, leading to the brink of war with France (see p. 1098). Also Germany exerted an intensive, and generally successful, effort to achieve a position of predominating military and civilian influence in Turkey.

1900–1914. Prewar Civil-Military Relations. Militaristic Emperor **Wilhelm II** relied more and more on military advisers, ignoring competent German civilian statesmen. Since the General Staff was not subordinate to or responsible to the chancellor, or the defense minister, or any other element of the parliamentary government, the German Army became a "state within a state." Though most professional German soldiers eschewed politics, their single-minded devotion to the nation's military security had the gravest political implications, which they either did not realize or ignored.

1914–1918. World War I. (See p. 1024.)

1914–1918. Wartime Civil-Military Relations. The wartime requirements of military necessity still further increased the ascendancy of the military in Germany. Hindenburg and Ludendorff used the special constitutional status of the General Staff, and their own enormous prestige, to create a virtual military dictatorship (1916–1918; see p. 1060). Civilian governments were overruled or dismissed by the emperor at their demand. Their threats to resign always caused Wilhelm to comply abjectly with their wishes, thus making himself completely their puppet. Though the motives of these 2 able military men were patriotic and impersonal, their insistence upon military objectives ahead of the overall political interests of the state contributed greatly to the complete post-war collapse of Germany.

1919–1925. Civil-Military Relations under the Weimar Republic. The shaky government was undermined by a number of revolts, both Communistic and nationalistic. At the same time Allied demands under the Treaty of Versailles strangled the nation financially. During a particularly violent period of revolt and military mutiny, General **Hans von Seeckt** was appointed commander of the army. In effect he thus was Chief of the General Staff, even though this organization had officially been abolished under the terms of the Treaty of Versailles. A typically apolitical, aristocratic Prussian professional, Seeckt was distrustful of the republican regime. He was, however, loyal to Germany, and thus faithfully supported and defended the government. He was able to prevent the civilian chancellor and defense minister from exercising any authority over the army, and virtually recreated the prewar situation of an independent military. He was thus the most powerful figure in Germany, but made no effort to seize personal power. Although he hated Communism, he led Germany toward a *rapprochement* with Soviet Russia as a way eventually to regain the Polish provinces lost under the Treaty of Versailles, and thus to eliminate the hated Polish Corridor (see p. 1082). In personal understandings with the Russian commander, Marshal **M. N. Tukhachevski,** he undertook extensive secret military collaboration with the Red Army.

1923, January 11. Occupation of the Ruhr. Because of Germany's default in reparations payments, French and Belgian troops invaded and occupied the heartland of Germany's heavy industry. Britain abstained. Germany, financed by the **Dawes Plan,** met most of her financial obligations and the Ruhr was evacuated (August 1925). The **Treaty of Locarno** (October 5–16) seemingly guaranteed peace in Europe, but France started feverishly constructing the **Maginot Line** of fortifications along her eastern border for protection against future German aggression.

1923, November 8–11. The "Beer-Hall Putsch." Adolph Hitler, Austrian ne'er-do-well, leading a new German nationalist movement, together with General Ludendorff, representing the reactionary military element, tried to overthrow the Bavarian government in Munich. The rising was crushed and Hitler sent to jail, where he spent nearly a year, writing *Mein Kampf.* On release, Hitler resumed his activities.

1925, April 26. Hindenburg Elected President. With its war hero as its leader, the course of German military rehabilitation was accelerated.

Austria, 1918–1925

This tiny remnant of the once-mighty Hapsburg Austro-Hungarian Empire (dismembered by the **Treaty of St. Germain,** part of the overall Versailles settlement; September 10, 1919) became a republic after a series of Communist disorders (1919). There was strong sentiment in the country for union with Germany (forbidden by the Treaty of St. Germain), resulting in a quasi alliance with her northern neighbor.

Czechoslovakia, 1918–1925

1918, November 14. Republic Established at Prague. Tomás Masaryk became the first President.

1919, March. Conflict with Hungary. (See below.)

1919, May-1920, May. Teschen Dispute with Poland. Clashes between regular and irregular forces of the 2 republics continued sporadically until they agreed to divide the territory.

1920–1921. Creation of the Little Entente. This was an alliance concluded with Yugoslavia (August 14, 1920) and Rumania (April 23, 1921), directed primarily against Hungary and the possibility of a Hapsburg revival. Czechoslovakia built up respectable military strength as the strongest minor European state.

Hungary, 1918–1925

1918, November 16. Proclamation of a Republic. This followed a brief revolution under Count **Michael Karolyi,** who became first President.

1919, March 21. Establishment of Communist Regime. After the resignation of Karolyi, pro-Bolshevik **Béla Kun,** sent from Russia by Lenin, seized power and established a dictatorship.

1919, March 28. Declaration of War with Czechoslovakia. Hungarian troops invaded Slovakia.

1919, April 10. Rumanian Invasion of Hungary. This was begun to forestall threatened Hungarian efforts to reconquer Transylvania, which Rumania had occupied after World War I. With Hungary torn by an anti-Communist counterrevolution, Rumanian troops advanced rapidly.

1919, August 1. Flight of Béla Kun. The Communist leader fled to Vienna just before the Rumanians occupied Budapest (August 4).

1920, June 4. Treaty of Trianon. This confirmed Hungary's acceptance of the Versailles—St. Germain settlements.

1921, March and October. King Charles Attempts to Restore Monarchy. The last Hapsburg emperor failed in 2 abortive military attempts to regain his Hungarian throne. He was permanently exiled.

The Scandinavian States

Untouched by direct impact of World War I, neutral Denmark, Sweden, and Norway confined postwar military efforts to the development of mutual security.

EASTERN EUROPE

Russia

PREREVOLUTIONARY PERIOD, 1900–1917

1904–1905. Russo-Japanese War. (See p. 1009.)

1905, December 22–1906, January 1. Moscow Insurrection. Endemic unrest seethed into open, armed violence, largely Communist-inspired, as the Moscow workers revolted in protest against the government inefficiency revealed in the conduct of the Russo-Japanese War. The uprising was quelled by regular troops in a succession of street fights. Later revolutionary movements in the provinces were also put down by vigorous and ruthless army action.

1907–1914. Continuing Unrest. The growing revolutionary movement led to sporadic outbreaks throughout the country, suppressed by the army.

1914–1918. World War I. (See p. 1024.)

THE RUSSIAN REVOLUTION, 1917

1917, March 8–15. Overthrow of the Empire. Following the outbreak of strikes and riots in Petrograd (Leningrad, formerly St. Petersburg), the military garrison of the capital mutinied (March 10), joining the revolting workers. A provisional government was established (March 12), and Czar **Nicholas II** abdicated.

1917, March–September. Continuation of the War. The new regime, dominated by the Socialist Minister of War Alexander Kerensky, pledged itself to continue the war against Germany, but suffered severe reverses (see p. 1064).

1917, September 9–14. Kornilov Revolt. Alarmed by growing Bolshevik domination of the government, under **Vladimir Lenin** and **Leon Trotsky,** the army commander, General **Lavr Kornilov,** marched against Petrograd. He was defeated by a combination of his own defecting troops and by armed workers.

1917, November 6–7. The Bolshevik Revolution. Under the leadership of Lenin, troops and workers in Petrograd overthrew the Kerensky regime and established a new Soviet government. (See p. 1065.)

THE GREAT RUSSIAN CIVIL WAR, 1917–1922

THE GENERAL SITUATION

War seethed both inside Russia and around the periphery of her immense land mass, as the Bolshevik regime strove to establish itself. In the struggle were involved both the residual backwash of World War I (see p. 1093) and a number of serious counterrevolutionary movements, all of them uncoordinated. The principal areas of unrest were Poland (see pp. 1084, 1096), Finland, southern Russia and the Ukraine, White Russia and the Baltic States, and Siberia.

SOUTH RUSSIA AND THE UKRAINE

Here, within a 12,000-mile-long tortuous loop from the Black to the Caspian seas, were fought the major campaigns of the Civil War. The area was embraced within a winding arc from Odessa on the west, through Kiev, Orel, Voronezh, and Tsaritsin (Stalingrad) to Astrakhan and the Volga estuary.

1917, December 9. Revolt of the Don Cossacks. Expropriation of their lands by Bolshevist decree caused the first insurrection. Gathering in force to the standards of Generals **A. M. Kaledin** and Kornilov, the Cossacks moved north through the Kuban and the Don Basin, colliding with Red militia in a number of indecisive fights. Between them Kaledin and Kornilov organized their impromptu armies into a fighting force of sorts, while at the same time Trotsky's efforts were slowly welding the Bolshevik militia into what would become the Red Army.

1918, April–May. The Caucasus. Georgia, Armenia, and Azerbaijan declared their independence (April 22 and May 26). Bolshevik efforts to retain the oil lands of the area brought resentment, then sporadic revolt.

1918, November–December. The Ukraine. German-supported General **Pavel Skoropadski** had been overthrown by Ukrainian socialists under General **Simon Petlyura** after the withdrawal of German troops (November 15). France garrisoned Odessa as a base of supply to feed the growing counterrevolution (December 18).

1919, February 3. Communist Capture of Kiev. Bolshevik forces moving into the vacuum left by the German withdrawal (see p. 1083) reached and took Kiev. They then pushed down the valley of the Bug and, lapping westward, drove the French out of Odessa (April 8).

1919, January. The Caucasus. Bolshevik troop movements into the oil fields turned insurrection into full-fledged war. General **Anton Denikin** rallied the White counterrevolutionaries and drove the Bolsheviks out. The conflict had now spread throughout the wide expanse of southern Russia. When, in the central sector, Kaledin committed suicide (Feb-

ruary 13) and Kornilov was killed in action (April 13), Denikin came out of the Caucasus to assume—in name at least—command over the entire White fighting front. General **Peter N. Krasnov,** hetman of the Don Cossacks, assisted him.

1919, May. The Line-up. From left to right the Whites were now assembled in 4 groups—the Kiev, Volunteer, Don, and Causasus armies. Against them the Bolshevik array became the Twelfth, Fourteenth, Thirteenth, Ninth, and Tenth Red armies. Trotsky was now faced with a multifront major dilemma, for Kolchak's White offensive (see p. 1094) along the axis of the Trans-Siberian Railway had now penetrated the Urals and reached the line Ufa-Perm. Another White army was assembling in the Baltic provinces under General Nikolai Yudenich (see p. 1096). At the same time Allied forces had landed in some strength in north Russia and eastern Siberia (see pp. 1094–1095) with apparent intent to help the White Russians. Considering Kolchak to be the most pressing danger, Trotsky took the offensive in the east, with General Tukhachevski in command, and went on the defensive in south Russia and in the north.

1919, May–October. Denikin Offensive. The 4 White armies pushed northward in an ever-diverging course. **Kiev** was recaptured (September 2). General **Peter Wrangel**'s Caucasus Army drove for Tsaritsin, hoping to make contact with Kolchak. However, Wrangel was delayed by lack of supply, and by the time Tsaritsin fell (June 17), Tukhachevski's Red offensive had rolled Kolchak back through the Urals. The Reds now turned in force against Wrangel, who was in turn forced back. Krasnov's Don Army reached **Voronezh** (October 6), only to be hit in flank and rolled back

(October 24) by Tukhachevski, who then drove westward against the Volunteer Army, which had reached Orel (October 13). The White offensive crumbled. The Kiev Army was then pried out of Kiev (December 17) and the disorganized Whites driven back to the Black Sea.

1920, March 27. Evacuation of the White Armies. This was done by sea from Novorossisk, mostly in British ships hurried there for the purpose. Only Wrangel, with a small force, held out in the Crimea.

1920, April. Red Penetration of the Caucasus. Soviet armies reached Baku (April 28). Bolshevik efforts to gain control of the Caspian Sea were temporarily foiled by a British flotilla based on Persian Caspian seaports (see p. 1102).

1920, June–November. Wrangel Offensive and Defeat. Taking advantage of Red preoccupation in the Russo-Polish War (see p. 1084), indomitable Wrangel suddenly pressed north from the Sea of Azov, but it was too late. The war with the Poles had ended, and the Reds, concentrating again, drove Wrangel back into the Crimea (November 1). The remnants of his White forces were evacuated—again by the British—by sea to Constantinople (November 14). Soviet Russia had come of age.

COMMENT. *Disunity of command on the White side, and the administrative and strategic ability of Trotsky on the Red, decided the issue. Both sides started as hordes of peasant guerrillas and amateur militiamen. Trotsky welded his masses into a professional fighting force through a process of trial and error. The Whites, lacking competent overall leadership, and with a command shot through by corruption and stupidity, failed. Wrangel was probably the best of the Whites, and Tukhachevski stood out amid the mediocrity of the Red leadership.*

SIBERIA AND EAST RUSSIA

1918, June. The Czech Legion. Approximately 100,000 Czech, or Bohemian, prisoners of war from the Austro-Hungarian Army seized control of the Trans-Siberian Railway when their proposed repatriation via Vladivostok was in-

terfered with by the Soviets. Capturing arms from local Bolshevik units, the Czechs organized themselves into an efficient army and marched westward along the railroad into eastern Russia, where they captured **Ekaterinburg** (Sverdlovsk, July 26) just after the massacre of the Czar and his family there. They then opened negotiations with the Soviets and with the liberal anti-Bolshevik government which had been established at Omsk.

1918, November 18. Kolchak Seizes Power in Siberia. Admiral **Alexander Kolchak** of the Imperial Russian Navy seized control of the Omsk government and proclaimed himself "Supreme Ruler of Russia." He allied himself with the Czechs and advanced from Siberia with an army into eastern Russia, capturing Perm and Ufa (December).

1919–1920. Bolshevik Counteroffensive. Under Trotsky's dynamic leadership, the Bolsheviks counterattacked Kolchak's troops and recaptured Ekaterinburg (January 27). Despite the active support of the Czech Legion and the moral support of Britain, France, and Japan, Kolchak's troops were slowly driven back into Siberia. Omsk was captured (November 14); Kolchak lost control of the situation, and was later captured and executed by the Soviets (February 7, 1920). Upon the collapse of the Kolchak regime, the Czechs fought their way eastward along the Trans-Siberian Railway, brushing aside both Red and White Russian units in their way, to reach that portion of eastern Siberia controlled by an American expeditionary force (see below). They were then transported to Vladivostok and evacuated by ship. Though fighting continued, the Bolsheviks gradually consolidated their control over all those parts of Siberia not controlled by American and Japanese forces.

ALLIED INTERVENTION, 1917–1922

1917, December 30. Japanese Occupation of Vladivostok. A large Japanese force under General **Otani** landed at Vladivostok. The apparent Japanese intention of annexing the Russian Maritime Provinces caused alarm in Washington, London, and Paris.

1918–1919. Allied Invasion of North Russia. Small British-French-American expeditionary forces, under British command, seized Murmansk (June 23, 1918) and occupied Archangel (August 1–2). Their nominal objective was to retrieve Allied supplies and munitions which had been contributed to the Czarist government. Actually the ridiculous underlying concept was that these small forces would invade Russia, striking south and east to link with the Czech Legion at the Urals. This, some Allied visionaries argued, would contribute to a united White Russian counterrevolution and topple the Bolshevik regime. The American contingent was one reinforced infantry regiment. For more than a year of minor but hard-fought undeclared war, the Allied force bickered with Bolshevik troops along the Vologda River. The Americans were then evacuated (August 1919); the remaining Allied troops left shortly afterward (September–October).

1918, August. American Siberian Expedition. Partly to succor the Czechs, and also to prevent Japan from gobbling the Maritime Provinces, 2 American regiments, under Major General **William S. Graves,** landed at Vladivostok. Graves's instructions were explicit: to refrain from interference with Russian internal affairs, and to rescue the Czechs. He was soon at odds not only with the Japanese but also with British and French military missions, with the Bolsheviks, and with Kolchak's White Russian forces. Continuing tension between Americans and Japanese frequently came close to violence, but conflict was averted by Graves's combined firmness and diplomacy. Britain, France, and Japan all expected the Americans to join in support of Kolchak. Refusing, Graves hewed to his directive. American troops guarded the Trans-Siberian Railway from Lake Baikal to Vladivostok. Several clashes with partisan White and Red Russian forces occurred, the Americans successfully maintaining their control of the railroad in each instance. After the collapse of the Kolchak regime, the Americans held their positions until

the arrival of the Czechs, who were evacuated by sea from Vladivostok. The American expedition then left Siberia (April 1920). This left eastern Siberia, save for the vicinity of Vladivostok, under Soviet control. In the face of the growing strength of the Soviet Far Eastern Republic, the Japanese later evacuated (October 25, 1922).

Finland and the Baltic States

FINLAND

1917, December 6. Independence of Finland.

1918, January 18. Mannerheim to Command. The new Finnish government appointed Baron **Carl Gustaf Mannerheim,** former Russian cavalry general, to organize and command its army. Bolshevik forces had already seized Helsingfors (Helsinki) and a native Red Guard was in process of formation. Vasa (Vaasa), where Mannerheim had his headquarters, contained a Russian garrison.

1918, January 28. Outbreak of Hostilities. A Red revolution broke out throughout Finland, despite the opposition of most of the Finnish people. Mannerheim the same day seized Vasa from the ineffectual Russian garrison and armed his rising levies from its large stock of weapons and munitions. He then moved south with his extemporized army, taking **Tammerfors** (Tampera), but further progress was checked by a large Red Guard force (March 16).

1918, April 3. German Intervention. A force of 10,000 men, under General **von der Goltz,** landed at Hanko (Hangö) in a surprise move. The Germans drove through and seized Helsingfors (April 18), cutting the Red-held area in two, while Mannerheim's rapidly growing forces moved east and cut off the Karelian Isthmus from Russia (April 19). An attempt of the remaining Russian forces to break out near **Vyborg** ended with the surrender (April 29) of 12,000 Reds and much booty. Desultory fighting with Bolshevik forces along the frontier continued.

1920, October 14. Treaty of Dorpat (Tartu). This finally secured Finnish independence.

ESTONIA

1917–1918. Bolshevik and German Occupations. Following its declaration of independence (November 28, 1917), Estonia was trampled first by Bolshevik and then by German troops (see p. 1065). The latter were in occupation when the Treaty of Brest-Litovsk was effected (see p. 1065).

1918, November 22. Renewed Bolshevik Invasion. Following the German evacuation (November 11), the Soviets returned. Sharp resistance by Estonian forces, assisted by a British naval squadron in the Baltic, caused Russian withdrawal (January 1919).

1919, October. Yudenich Coup. Capable and energetic White Russian General **Nicolai Yudenich,** who had been gathering a counter-revolutionary army in northeast Estonia, crossed the Russian border (October 6) near Narva in a bold bid to capture Petrograd (Leningrad). Although he had no more than 20,000 men, he reached the outskirts of the city (October 19), causing near panic. Trotsky's frenzied rallying of all available forces, including workers, turned Yudenich back, and he retired into Estonia.

1920, February 2. Treaty of Dorpat. This brought freedom to Estonia through Russian recognition of independence.

LATVIA

1919, January. Russian Invasion. Immediately following Latvian declaration of independence (November 18, 1918), Bolshevik troops moved across the border. Riga was taken (January 4) and a Soviet government set up. With Allied approval, German-Latvian forces drove the Bolshevist invaders back (March). A German attempt to take over the Riga government resulted in confused fighting and occupation of Riga, ending in armistice (April 16–May 22). Fighting broke out again (October 20) between the Latvians and both German and Russian elements. Allied insistence on adherence to the provisions of the Treaty of Versailles (see p. 1082) forced German withdrawal (November 20).

1920, January. Russian Withdrawal. The last of the Bolshevik troops were expelled. An armistice with Russia followed (February 1).

1920, August 11. Treaty of Riga. Russia recognized Latvian independence.

LITHUANIA

1918. Bolshevik Invasions. Immediate Soviet invasion followed Lithuanian declaration of independence (February 16), but the Russians were soon driven out by the Germans (see p. 1065). However, the evacuation of German troops (November 11) was followed by another Bolshevik invasion.

1919, January 5. Russian Capture of Vilna. This led to Polish intervention and the outbreak of the Russo-Polish War (see p. 1084).

1920, July 12. Treaty of Moscow. This ended the Russo-Lithuanian hostilities; Russia recognized Lithuanian independence.

1920, October 9. Polish Seizure of Vilna. (See p. 1097.) League of Nations efforts to mediate and to hold a plebiscite satisfactory to both sides proved unsuccessful. Peace was reestablished 7 years later (December 1927).

1923, January 11. Insurrection in Memel (Klaipeda). A predominantly German city, Memel had been under inter-Allied control since 1918. Upon the outbreak of an uprising sponsored by Lithuania, Lithuanian troops occupied the city, forcing a French garrison to withdraw. The Allies later recognized this high-handed seizure.

Poland, 1914–1925

1914, August 16. Pilsudski's Legions. One of the principal prewar Polish revolutionaries against Russian rule of central Poland, **Józef Pilsudski,** organized Polish legions in Galicia (Austrian Poland) to fight beside the Central Powers against Russia.

1916, November 5. Central Powers Proclaim Polish Independence. This only applied to former Russian Poland. The country remained occupied by German and Austrian troops, however.

1917, March 30. Russia Recognizes Polish Independence. The Provisional Russian Government stated that its recognition applied to all lands with Polish majority populations, thus including German and Austrian Poland.

1918, November 11. De Facto Polish Independence. German troops were disarmed and expelled. Pilsudski, Commander in Chief of all Polish forces, became virtual dictator of the new country. With frontier wars already breaking out in all directions, he devoted all of his energies to building up an effective army; in this he was favored by the many Polish veterans of World War I, who had fought on all fronts and on both sides.

1918, November–1919, May. War with Ukraine. Ukrainian troops entered Galicia to establish a West Ukrainian Republic at Lvov. After 6 months of fighting, the Ukrainians were expelled from Galicia.

1918, December–1919, February. Fighting in Poznan. Threatened outbreak of war between Germany and Poland was halted by Allied pressure.

1919, January–February. Dispute over Teschen. Precipitated by a Czech attack on Polish forces in Teschen, the fighting was quickly ended by Allied pressure.

1919, January–November. Undeclared Conflict with Russia. Poland, insistent upon returning to pre-1772 boundaries, reacted violently when Soviet troops (following the withdrawing Germans) occupied Vilna (January 5) and continued on to the line of the Bug River (February). An undeclared war broke out, and the Poles drove the Russians back, reoccupying **Vilna** (April 1919). Desultory fighting continued along the front, which brought Polish troops to the Berezina River and into the northern Ukraine. A lull of several months followed.

1919, December 8. The "Curzon Line." The Allied Supreme Council established a provisional eastern frontier for Poland, generally following the Bug River. This line, later called the "Curzon Line," was totally unsatisfactory to

Poland. As Soviet forces began to build up in western Russia, the Poles determined to fight for the formerly Polish regions east of the Bug.

1920, April–October. Russo-Polish War. (See p. 1084.)

1920, October 9. Polish Reoccupation of Vilna. During the Russo-Polish War, the Soviets had ceded Vilna to Lithuania (July 12). Upon Soviet evacuation, after the Battle of Warsaw, the Lithuanians occupied the city (August 26). This precipitated a dispute between Poland and Lithuania, which led to fighting when Polish General **Lucian Zeligowski** occupied the city. A state of war, without active fighting, persisted between Poland and Lithuania for another 6 years (see p. 1096).

1921. Fighting in Silesia. A dispute between Germany and Poland over the results of the Silesian plebiscite broke out into open fighting (May 3). The Inter-Allied Commission forced a cessation of hostilities (June 24).

Turkey

1900–1909. Disintegration of the Ottoman Empire. The Young Turk nationalistic movement urged constitutional reform and fomented insurrection while the Great Powers and the Balkan States nibbled at the outer fringes of the empire of despotic Sultan **Abdul Hamid II.**

1909, April 13. Military Revolt at Constantinople. The I Army Corps (mostly Albanian) seized control of the capital. The uprising was put down by troops from Macedonia after a 5-hour fight within the city. Other disorders in Anatolia, involving Armenians, were ruthlessly suppressed by Turkish troops, who massacred a number of Armenian demonstrators.

1909, April 26. Deposition of Abdul Hamid. The Young Turks appointed his inept brother, **Mohammed V,** as sultan.

1910, April–June. Albanian Insurrection. The uprising was brutally stamped out by Turkish troops.

1911–1912. War with Italy. (See p. 1014.)

1912–1913. The Balkan Wars. (See p. 1015.)

1913, January 23. Young Turk Coup d'État. After brief turmoil, "strong men" Enver, **Taalat,** and Djemal came to power. They secretly wooed Germany.

1913, November–December. Liman von Sanders Crisis. The appointment of this German general to reorganize the Turkish Army aroused Russian and French resentment and suspicion.

1914, August 2. Secret Alliance with Germany. This foreshadowed Turkey's entry into the war on the side of the Central Powers.

1914–1918. World War I. (See p. 1024.)

1919. Internal Turmoil and Allied Intervention. Defeat in World War I still further weakened the power of the Sultan (now **Mohammed VI**). Increasing nationalist activity resulted from the Sultan's cooperation with the Allies. To curb the Nationalists, Italian troops were landed at Adalia (April 29) and Greek troops at Smyrna (May 14). Atrocities committed by the Greeks aroused Turkish resentment.

1920, March 16. Allied Occupation of Constantinople. An Allied force under British General **George F. Milne** landed to support the Turkish government against the Nationalists, to keep open the straits as a supply line for White Russian counterrevolutionaries, and to protect the Armenians.

1920, April 23. Provisional Nationalist Government at Ankara. Mustafa Kemal, war hero and Nationalist leader, was selected to be President. A military agreement was then reached with Soviet Russia to ensure supplies. (See p. 1086.)

1920, June 20. Treaty of Sèvres. This stripped Turkey of all European territory except Constantinople, and gave complete Armenian independence. The terms of the treaty were rejected by the Nationalists, increasing their popular support.

1921–1923. Greco-Turkish War. (See p. 1085.)

AFRICA

MOROCCO

1900–1906. Colonial Friction. Rival colonial interests in North Africa of France, Germany, England, Spain, and Italy were composed to some degree by the **Algeciras Conference** (January 16, 1906–April 7, 1907), recognizing special rights of both France and Spain in Morocco.

1907, July 20. Bombardment of Casablanca. Serious unrest was violently quelled when a French squadron opened fire on the city. French occupation of the Moroccan Atlantic coastal area followed.

1909, July–October. Riff Attacks against the Spanish. Riff Berber tribesmen clashed with the Spanish at **Melilla** on the Mediterranean.

1911, April 26. French Occupation of Fez. This followed a Berber attack on the city.

1911, July 1–November 4. The Agadir Incident. Germany sent the gunboat *Panther* to Agadir, allegedly to protect German nationals and their interests. War was averted by diplomatic horse-trading. In return for a free hand in Morocco, France ceded part of her Congo holdings to Germany.

1912, March 30. Treaty of Fez. This established a French protectorate over Morocco. General **Louis H. G. Lyautey,** forceful and tactful, appointed Resident General (May 24), began a successful, progressive program of political and economic betterment cementing French and Moroccan interests.

1912–1921. Troubles in Spanish Mediterranean Zone. In the west, bandit chieftain **Raisuli** (who gained international notoriety by kidnaping an American citizen in 1904) by 1920 was harassing the area south to Tetuán, while a Riff chieftain, **Abd el Krim,** was whipping up hostility south and east of Melilla. Spanish General **Dámaso Berenguer** attained success in pacifying the western area, but in the east General **Fernandes Silvestre** met disaster.

1921, July 21. Battle of Anual. Silvestre, with about 20,000 men, was moving southwest into the Riff mountains. Apparently reconnaissance was inadequate and security dispositions faulty. Abd el Krim's tribesmen had attacked and captured a frontier post at Abaran. The garrison of the next post fled to Anual, to meet Silvestre's column. At the same time the Riffians opened fire on both Silvestre's flanks. The resulting confusion turned to panic and then to slaughter. Silvestre and some 12,000 of his men were killed, several thousand more made prisoner. Dissidence flared throughout the Spanish zone, all frontier posts were either captured or abandoned, and the Spaniards were forced back to fortified coastal zones directly around Melilla and Tetuán. The Anual disaster rocked Spain, leading to the downfall of the government and rise to power (September 13, 1923) of "strong man" General **Primo de Rivera,** who became—with the king's blessing—virtual dictator of Span. Abd el Krim, establishing a "Republic of the Riff," prepared to drive out the French and control all Morocco. Between the tribesmen and a number of foreign adventurers he organized a force of 20,000 men, well armed and provided with artillery and machine guns captured from the Spanish.

1925, April 12. Riffian Advance Southward. Abd el Krim's forces, debouching secretly from the mountain border, swept south to overwhelm the chain of French border posts from Taza to Fez. Forty-three of the 66 blockhouses dotting the 50-mile span were captured and most of their garrisons killed after desperate and gallant resistance. Lyautey had expected attack, but his resources were limited. Skillful handling of his existing reserves checked the Riff advance within sight of Fez. France and Spain, forgetting their former rivalry in Morocco, agreed (July 26) to a joint counteroffensive. Reinforcements raised the French Army strength in Morocco to more than 150,000 men, while Spain prepared an expeditionary force of more than 50,000, commanded by General **José Sanjurjo.**

1925, September 8–9. Joint Counteroffensive. The Spanish expedition began landing in the **Bay of Alhucemas** under cover of French and Spanish warships. Ajdir was captured (October 2). The stage was set for an immense converging movement, with Targuist—Abd el Krim's headquarters—as target. While small Spanish forces struck from the Tetuán and Melilla zones respectively, Sanjurjo's main force advanced almost directly south. At the same time Marshal Pétain, commanding the French field forces, moved 6 converging columns, within mutual supporting distance of one another, into the Riff-held territory between Tafrant and Taza.

1925, September 24. Resignation of Lyautey. Exhausted in health, Lyautey's resignation ended a most brilliant career of pacification and organization. Pétain succeeded him, and General **Boichut** took field command. Despite desperate resistance, Abd el Krim's troops were progressively forced north by the French hammer against the slower-moving Spanish anvil. Delayed by seasonal winter rains, the advance was renewed in the spring.

1926, May 26. Surrender of Abd el Krim. A French spearhead under Colonel **André G. Corap** came thrusting into Targuist, and Abd el Krim gave up the fight.

COMMENT. *The original success of Abd el Krim testifies to the fallacy of cordon defense. Neither the French nor the Spanish frontier could withstand the initial blows, and the loss of one blockhouse led inevitably to the loss of the others. The European counteroffensive, particularly the French operations, consisted of a rapidly moving advance on a wide front, utilizing the terrain and taking advantage of crest lines to avoid ambush. Abd el Krim's own campaigns were an example of development of guerrilla warfare into that of a conventional force. He was overwhelmed by a combination of superior force and sage leadership.*

AFRICA

During this period there were almost continuous military operations on the part of the forces of Britain, France, Germany, Italy, and Portugal to stabilize and pacify their colonial empires in Africa. These were generally successful, and by the end of the period most of Africa was relatively peaceful. Listed below, in summary form, are some of these operations.

1899–1920. Operations against the "Mad Mullah" (Somaliland). Somali chieftain **Mohammed ben Abdullah** waged almost constant small war against the Italians, British, and Ethiopians. He and his fierce desert tribesmen proved themselves to be extremely able warriors. His raids ended only with his death.

1899–1902. The Boer War. (See p. 933.)

1900, March–November. Ashanti Uprising (Gold Coast, or Ghana). The Ashantis briefly besieged Kumasi before being suppressed by British troops. The British established a protectorate on January 1, 1902.

1900, April 22. Battle of Lakhta (Kusseri). French troops defeated **Rabah Zobeir,** raider and slave trader, in the eastern French Sudan, or Chad, region. (See p. 938.)

1900, May. French Conquest of Northern Sahara. Culminating protracted desert warfare, the French established themselves firmly in the main oases of the northern Sahara.

1900–1903. British Conquest of Northern Nigeria. The British revoked the Royal Niger Company's charter, and some 2,500 troops under Lagos Protectorate Governor-general Sir Frederick Lugard (see p. 937) invaded the Sokoto Caliphate, capturing Kano (February 3, 1903) and Sokoto itself (March 15) before a protectorate was proclaimed that autumn.

1902. Uprising in Angola. Suppressed by the Portuguese.

1903. Hottentot Uprising in German Southwest Africa.

1904. Insurrection in Southern Nigeria.

1904–1905. Insurrection in Cameroons. Suppressed by the Germans.

1904–1908. Uprising in German Southwest Africa. Led by the Herero tribe, this revolt was joined by many of the Hottentots. It was suppressed by the Germans only with difficulty. Its influence spread to Angola.

1905. Uprising in the French Congo.

1905. Insurrection in German East Africa.

1906. Religious Insurrection in Sokoto (Northwest Nigeria).

1907. Uprising in Angola. Largely inspired by the Herero uprising in German Southwest Africa.

1908–1909. French Conquest of Mauretania.

1909–1911. French Conquest of Wadai (mountainous, desert region of eastern Chad or central Sudan).

1914, October-1915, February. Boer Uprising. Boer extremists, led by former Boer General Christiaan De Wet and others, rose in protest against the Union of South Africa's declaration of war against Germany. The rising was suppressed by Prime Minister—and former Boer General—**Louis Botha,** assisted by former Boer commando leader **Jan Smuts.**

1914–1917. Unrest and Civil War in Ethiopia. Menelek nominated as his successor **Lij Yassu,** who came to the throne on Menelek's death in 1906. However, he proved an unpopular ruler, supporting Islam against the Coptic Church, as well as backing the "Mad Mullah" of Somalia (see p. 929 and above). Because of his support for Islam, Yassu was excommunicated by the *abuna* (primate of the Ethiopian Coptic Church) in September 1916, and the provincial chiefs and nobles (*ras*) formally deposed him soon after. Yassu's father Ras **Mikael** raised a Galla army to aid his son, but this was defeated at the **Battle of Sagalle** near Addis Ababa (October 27, 1916). Yassu fled (as did Ras Mikael), first to Danakil and then to Tigré province; Yassu was captured in 1921 and imprisoned until his death in 1935. The victorious rases selected Menelek's daughter **Zauditu** as empress (*nequs nagasti,* or "king of kings"), and made **Ras Tafari** (son of Menelek's nephew) regent.

SOUTH AND SOUTHWEST ASIA

THE ARAB STATES

Arabia, 1900–1925

1900–1919. Rise of Ibn Saud. A chief of the Wahhabi sect, or tribe, of southern Nejd (central Arabia), **Ibn Saud** established himself as undisputed ruler of the Nejd in a prolonged civil war (1900–1906). He then consolidated his control and annexed outlying regions to establish his authority over all non-Turkish portions of central and northern Arabia. He entered into an alliance with Britain during World War I and fought briefly against the Turks (1915).

1916, June. Arab Revolt against the Turks. Inspired by the British, the Arab tribes of Hejaz (western Arabia) revolted against the Turks under Hashemite Sherif **Hussein.** Under the leadership of his son, Emir **Faisal,** and British Captain (later Colonel) T. E. Lawrence, Arab forces fought successfully against the Turks for the remainder of the war (see p. 1055).

1919–1925. Struggle for Control of Arabia. A prolonged conflict broke out between Ibn Saud and the Hashemite Dynasty. After a series of Wahhabi victories over the Hashemites and independent Arab chieftains, Hussein abdicated (October 1924) in favor of his son **Ali.** Ibn Saud captured **Mecca** (October 13, 1924) and continued to press the war. He captured **Medina** (December 5, 1925), Sherif Ali abdicated (December 19), and Ibn Saud then captured **Jidda** to complete his conquest of Hejaz (December 23).

Syria, 1918–1925

1918, October 5. French Occupation of Beirut. A French naval squadron seized the port from the Turks as Allenby's troops were conquering the remainder of Syria (see p. 1081). By secret Allied understandings, Syria was to be France's share of the dismembered Ottoman Empire. French claims, however, were soon disputed by Emir **Faisal,** commander in chief of the Arab armies with T. E. Lawrence (see p. 1080 and above).

1919–1920. Dispute between France and Faisal. After the British formally turned the administration of Syria over to France (September 1919), open fighting broke out between Arabs and French (December). Though Faisal was proclaimed King of Syria (March 1920), he was unable to obtain support and was forced to flee. French troops occupied Damascus (July 1920). Faisal, for consolation, was made King of Iraq by the British (see below).

1925–1927. Insurrection of the Druses. An Arab tribe of southeastern Syria, the Druses became restive under French administration. They revolted (July 1925) and soon seized control of most of southern Syria. Aided by a revolt in Damascus, they forced the French to evacuate the city (October 1925). As the period closed the French were near defeat in Syria.

Iraq, 1919–1925

1919–1920. Unrest in Mesopotamia. Many Arabs objected to British occupation following the Mesopotamian campaign of World War I (see p. 1081).

1920, July–December. Arab Insurrection. Arab nationalists, hoping for the independence of Mesopotamia, rose against the British, who suppressed the uprising, but sought ways to satisfy Arab demands and aspirations.

1921, August 23. British Protectorate over Iraq. The British proclaimed Faisal (former King of Syria) as King of Iraq. The country was given a measure of self-government, but Britain retained control.

1922, June–1924, July. Insurrection in Kurdistan. The Kurds, traditionally semi-independent mountain people of the border region between Iraq, Turkey, and Iran, rose to throw off the control of the British and Arabs. The British, concerned about the rich Mosul oil fields, eventually suppressed the rebellion and granted considerable autonomy to the Kurds.

PERSIA (IRAN)

1905–1909. Persian Revolution. This was a combination of political activity and scattered insurrections against the corrupt and despotic rule of Shah **Mohammed Ali.** Open warfare against the central government was stimulated by a revolt in Tabriz (1908), which was then besieged by the Shah's army (1908–1909). To protect Russian interests, and nominally to assist the Shah, a Russian army intervened and captured Tabriz, brutally suppressing the revolt (March 1909). Meanwhile, other rebel forces took the field in northern and central Persia. A rebel army under **Ali Kuli Khan** captured Teheran (July 12, 1909), forcing the abdication of the Shah in favor of his 12-year-old son, Sultan **Ahmad** (July 16).

1911, June 17–September 5. Abortive Return of Mohammed Ali. With Russian connivance, Mohammed Ali landed on the Caspian coast of northeast Persia at Gumish Tepe, and was soon joined by dissident forces. He was quickly defeated by loyal troops and forced to flee again.

1911, November. Russian Occupation of Northern Persia. On the pretext of restoring order in Persia, and to protect their financial interests, Russian forces occupied the northern part of the country and established a virtual protectorate over it. To protect their oil interests, the British occupied much of southwest Persia, but despite strained relations did not break with their Russian allies because of the threat of European war.

1914–1918. World War I. (See p. 1024.) The war ended with British troops in occupation of much of western Persia.

1919, May 21. Battle of Alexandrovsk. A British flotilla on the Caspian (see p. 1094) defeated a Bolshevik naval force. The British flotilla was later turned over to White Russians, who were then defeated by the Bolsheviks (1920).

1920, May 18. Russian Naval Invasion. A Russian flotilla captured **Enzeli** (Pahlevi) and then seized Resht. British troops withdrew from the Caspian coastal region to avoid a clash. Persian troops briefly recaptured **Resht** (August 24), but were repulsed from Enzeli, then driven south of the mountains.

1921, January. British Withdrawal. Upon negotiation of a treaty between the Soviet government and Persia, British troops began to withdraw from northern Iran.

1921, February 21. Coup d'État of Reza Khan. Widespread disgust with the incompetence and corruption of the oligarchy in Teheran led Persian general **Reza Khan,** commanding the Cossack Brigade, to seize power in Teheran; he soon established himself as virtual dictator. He quickly ratified a treaty with Russia (February 26), whereupon Russian troops began withdrawing from the northern and coastal regions.

AFGHANISTAN

1914–1918. World War I. Neutral Afghanistan was in turmoil as Turkish and German interests provoked anti-Allied religious agitation. England, by subsidies, held the government in line, and after the war Russian interest was resumed.

1919, February. Accession of Amir Amanullah. Placed on the throne by the army and the Young Afghan radical party after the assassination of his father, **Habibullah** (February 19), Amanullah declared the country independent of all foreign control.

1919, May. War with Britain. Amanullah proclaimed a jihad (religious war) against Britain. Afghan levies broke across the Indian border near Landi Khana and occupied **Bagh** (May 3). Immediate mobilization of British Indian troops followed. A punitive expedition moved through the Khyber Pass to **Landi Kotal** and drove the invaders out of Bagh (May 11). The expedition then advanced into Afghanistan, and reached Dakka. British planes bombarded **Jalalabad** and **Kabul.** Amanullah sued for an armistice (May 31), and nominal peace resulted from the **Treaty of Rawalpindi** (August 8). Renewed British recognition of Afghanistan's independence (November 22, 1921), was accompanied by cessation of British subsidies. Sporadic guerrilla warfare continued along the Afghan-Indian border.

1926, August 31. Treaty with Soviet Union. The Soviet Union and Afghanistan signed a neutrality and nonaggression treaty in which they pledged benevolent neutrality and noninterference in each other's domestic politics.

INDIA, 1900–1925

1903–1904. Expedition to Tibet. Under the leadership of Colonel **Francis Younghusband,** a small British expedition entered Tibet to force the **Dalai Lama** to negotiate a treaty to stabilize the northern frontier of India. When the Tibetans refused to negotiate, Younghusband marched to Lhasa, which he reached after several severe engagements (August 3, 1904). A treaty was signed (September 7).

1914–1918, World War I. Indian troops—some 1,400,000 men in all—volunteered for service, taking active part on the Western Front, in the Near East and Africa.

1919, April 13. Amritsar Massacre. Religious conflicts between Moslems and Hindus, plus **Mahatma Gandhi**'s passive non-cooperation with the government, brought open rebellion in the Punjab. Several Europeans were killed during a riot at Amritsar (April 12). Brigadier General **Reginald Dyer,** commanding the garrison, paraded his troops to enforce order. When an unarmed mob failed to disperse, Dyer's troops fired on them, killing 379 and wounding 1,208 others. The incident aroused British public opinion. Dyer was denounced in the Commons, but the House of Lords upheld him on the ground that he had saved British rule in the Punjab. The Army Council, taking the middle ground, decided his act was "an error of judgment."

1919, May—August. Afghan War. (See p. 1102.)

1919, November. Revolt in Waziristan. Excesses committed by Masud tribesmen incited by the Afghans brought a punitive expedition of 30,000 men under General **S. H. Climo.** Concentrating on the **Tank Zam** (December 13), Climo's troops were attacked by the Masuds (December 17), who were repelled. The uprising was quelled (February 1, 1920).

1920–1925. Operations on the Northwest Frontier. Frequent small operations were undertaken to pacify Pathan tribesmen from Afghanistan, who periodically raided across the frontier.

EAST ASIA

CHINA

The Boxer Rebellion, 1900–1901

1899–1900. Rising of the Boxers. Antiforeign elements in the Chinese government, angered by the Great Powers' nibbling of her territory, incited a fanatical secret organization called the Society of the Righteous Harmonious Fists (thus Boxers) to wide-scale depredations against foreign missionaries and their Chinese converts to Christianity. The government of Dowager Empress **Tzu Hsi,** professing inability to control the Boxers, was actually inciting and supporting them. Protests of foreign diplomats brought further violence. Warships of foreign nations began gathering off Tientsin (June 1900) and military detachments of several nations, totaling 485 men, were sent to Peking to guard the legations.

1900, June 10–26. First Relief Expedition. As the situation worsened, a small allied force of some 2,000 marines and bluejackets, including 112 Americans, was landed under control of British Admiral **E. H. Seymour,** senior officer present. Its movement to Peking was repulsed by much superior Chinese strength at **Tang Ts'u.** The force returned to their ships (June 26), having suffered 300 casualties.

1900, June 17. Capture of the Taku Forts. After receiving an ultimatum, the forts guarding the river gate to Tientsin opened fire on the foreign ships. The fire was immediately returned, and landing parties captured the forts.

1900, June 20–August 14. Siege of the Peking Legations. Mob reaction culminated in the murder of Baron **Klemens von Kettler,** German Minister, and the siege of the legations. Russian, British, French, Japanese, and U.S. detachments were hurried to Taku, where an Allied Expeditionary Force was formed to go to the relief of Peking. (The first American contingent came from the Philippines, with reinforcements sent from San Francisco.)

1900, July 23. Taking of Tientsin. The expeditionary force, now numbering some 5,000 men, stormed the city walls and captured the fortress. The U.S. 9th Infantry Regiment in particular took heavy loss, including its commander, Colonel **E. H. Liscum.** By August 4, the expeditionary force included 4,800 Russians, 3,000 British, 2,500 U.S. (2 infantry and 1 cavalry regiments, a battalion of Marines, and a field battery, under Major General **Adna R. Chaffee**), and 800 French troops. Other detachments brought the total strength to 18,700. No overall commander was appointed; actions were taken on a cooperative basis.

1900, August 4. Advance on Peking. The Second Relief Expedition began its overland march along the rail and river line. A Chinese force estimated at 10,000 men was driven back at **Yang T'sun** (August 5–6). Here the French contingent remained to protect the line of communications. The other troops pushed on, driving off desultory resistance and arriving before Peking's outer walls (August 13). An immediate and precipitate attack by the Russians, who were in the lead, was thrown back from the Tung Pien Gate.

1900, August 14. Taking of Peking. A Japanese attack on the Ch'i Hua Gate was repulsed. The American force joined the Russians in front of the Tung Pien, and took it. At the same time 2 companies of the U.S. 14th Infantry assaulted the northeast corner of the outer wall. Bugler **Calvin P. Titus,** first to scale, hoisted the American flag. The defenders were driven off. The British troops waded under the wall through the Water Gate and the combined attack pushed on to relieve the diplomatic group, penned in the British Legation compound. This group of civilian men and women and the legation guards had resisted constant fire and assault for 8 weeks. Casualties among the civilian volunteers were 12 killed and 23 wounded, while the guard detachments lost 4 officers and 49 men killed and 9 officers and 136 men wounded. Rescued also were the defenders of the P'ei Tang, compound of the Catholic cathedral. There 40 French and Italian marines, a handful of priests and nuns, and some 3,000 of their Chinese Christian converts had also held out against incessant Boxer attacks. Seven marines, 4 priests, and an estimated 400 of the Christian Chinese had been killed in the fighting.

1900, August 15. Attack on the Imperial City. American field artillery blasted open the Ch'i Hua Gate, First Lieutenant **Charles P. Summerall** walking under Chinese fire to chalk an aiming spot on its timbers for his gunners. As a diplomatic sop to the Chinese government, the Imperial City was not immediately occupied, but the troops moved in later (August 28).

1900, September 4–October 10. Russian Occupation of Manchuria. This was a further blow to Chinese prestige. The Dowager Empress, from her refuge in Sian, accepted all the Allied demands (December 26).

1900, September–1901, May. Allied Punitive Missions near Peking. Of these, 35 were by German troops under Field Marshal Count **Alfred von Waldersee,** a late arrival (September 12).

1901, September 12. Boxer Protocol. Signed by 12 powers, this laid a crushing penalty on China, the most important item being an indemnity of 450 million taels (approximately $739 million). The United States set aside its share for education of Chinese students in the United States.

COMMENT. *Military operations in China, nominally under a joint command of mutual agreement between the various national commanders, was actually a hit-or-miss affair. Neither the Japanese nor Russian com-*

manders gave more than lip service to joint agree-
ments. Had the Chinese been well trained, well led,
and well armed, the expedition would have been
doomed to failure. It is interesting to note that the
British contingent included a Chinese regiment under
British officers; it performed well.

Turmoil in China, 1901–1925

1904–1905. Russo-Japanese War. This took place mainly on Chinese soil (see p. 1008).

1905–1910. Rise of Chinese Nationalism. Pressures of the European powers and of Japan following the Russo-Japanese War resulted in new territorial demands on China. National dissatisfaction with the government flared; revolutionary sentiment was widespread.

1911, October 10. Outbreak of the Chinese Revolution. Mutiny broke out among troops in Wuch'ang, leading to widespread revolt. Marshal **Yuan Shih-k'ai,** overall military commander, after some perfunctory moves to suppress the risings, joined the movement (December). This led to the abdication of the child emperor and the establishment of the Republic of China under the presidency of Yuan (February 12, 1912). During this time Tibet rose against Chinese rule and drove the Chinese garrison out of the country. Britain prevented any subsequent Chinese efforts to reestablish authority in Tibet by force.

1913, July–September. The "Summer Revolution." An uprising in the Yangtze Valley was easily suppressed by Yuan.

1915, December–1916, March. Rebellion against Restoration of Empire. In protest against plans of Yuan to re-establish the empire, with himself on the throne, rebellion broke out in Yunnan and spread through China. Yuan's announcement of a change in plans did not completely restore order, but the

issue was resolved by his death 3 months later (June 6, 1916). He was succeeded as President by **Li Yuan-hung.**

1917, May–August. Revolt of the Northern Military Governors. Dissatisfied with the parliamentary government operating under President Li, the northern military governors (*tuchuns*) revolted and established a rival government at Tientsin. One of their number, **Chang Hsün,** on the pretense of mediating with Li, overthrew the Peking government and briefly reestablished the Manchu Dynasty (July 1–12). He was repudiated by his military colleagues, who occupied Peking, overthrew the empire, forced Li to resign, and installed **Feng Kuo-chang** as the new President.

1917, August 4. China Declares War on Germany and Austria-Hungary. China took no active combat part in the war, but did send labor battalions to France, Mesopotamia, and Africa. In return China secured the termination of all German and Austro-Hungarian concessions and rights in China.

1920–1926. Rise of the War Lords. The military governors and other military leaders became almost completely independent of nominal central authority. They carried out a series of complex internecine struggles while the central government became weaker and weaker.

1924, January 21. Congress of the Kuomintang (National People's Party). Under the leadership of **Sun Yat-sen,** the Nationalists met at Canton to prepare for the liberation and unification of the country. Soviet political and military advisers were prominent. At about this time a new Whampoa Military Academy, with Russian and German advisers, was established by the Kuomintang, under young General **Chiang Kai-shek.**

MONGOLIA

The overthrow of the Manchu Empire in China led Mongolia, under the nominal leadership of the "Living Buddha" of Urga (Ulan Bator) to declare its independence. The following decades were chaotic in Mongolia, which felt the impact of both the Chinese and Russian Revolutions, as well as the tides of Japanese expansionism in East Asia. Russian Communism finally prevailed over the other conflicting forces attempting to take advantage of Mongolian weakness, and the Mongolian People's Republic became the first, and possibly most loyal, of Moscow's satellites. The principal events were:

1911, November 18. Outer Mongolia Proclaims Independence. A theocratic Lamaistic government was established under the *Jebtsun Damba Khutuktu* ("Living Buddha"). Russian influence increased, with Mongolia becoming a virtual Russian protectorate (November 3, 1912).

1919, October. Return of Chinese Control. The warlord government of Peking took advantage of confusion caused by the Russian Civil War to reassert control by sending a small army to occupy Urga.

1920, October–1921, July. White Russian Invasion and Occupation. A force under Baron **Roman von Ungern-Sternberg** invaded Outer Mongolia, and drove the Chinese out of Urga (February 3, 1921). At first welcomed as a liberator from the hated Chinese, sadistic Ungern-Sternberg instituted a reign of terror and aroused popular opposition.

1921, March 13. Establishment of Revolutionary Provisional Government of Mongolia. This was set up at Kiakhta, just inside Siberia, across the border from Mongolia, with Russian Communist support, by Nationalist patriots **Sükhe Bator, Dunzan,** and **Khorloghiyin Choibalsan.** A joint Russian-Mongolian military force was assembled; Sükhe Bator led the small Mongolian contingent.

1921, June–July. Russian-Mongolian Invasion of Mongolia. Ungern-Sternberg was defeated, **Urga** was captured (July 6). Later Ungern-Sternberg was captured and executed. Although nominally part of a joint force, the Soviet troops virtually occupied the country. Nevertheless, to avoid undue antagonism to China, the U.S.S.R. professed to recognize Chinese suzerainty over Outer Mongolia (May 31, 1924).

1924, November 26. Proclamation of Mongolian People's Republic. The death of the Living Buddha (May 20) provided the Mongolian Communists and their Soviet masters with an opportunity to complete the transition of Mongolia to a satellite and consolidate control (1924–1928). After the deaths of Sükhe Bator and Dunzan, Choibalsan became the sole leader of Communist Mongolia.

Japan

Japan participated in 3 military ventures during the period: Boxer Rebellion (1900–1901; see p. 1103); Russo-Japanese War (1904–1905; see p. 1008); and World War I (1914–1918; see p. 1034). Her participation in each was coldly calculated for her own aggrandizement. The Russo-Japanese War is particularly important from the standpoint of military history, since it marked the admittance of Japan into the coterie of the Great Powers. Japanese war potential was expanded as rapidly as possible under chauvinistic, expansionist, militaristic leadership.

The Philippines

1901–1902. Continuation of Philippine Insurrection. (See p. 950.) The capture of **Aguinaldo** through a ruse, by Brigadier General Frederick Funston and a detachment of Filipino troops in U.S. service (March 23, 1909), ended the formalized rebellion, though guerrilla warfare continued for more than a year. However, thanks to General **Arthur MacArthur's** wise and just military rule, the Christian Filipinos were gradually won over (1902).

1902–1905. Moro Campaigns. Operations in the southern islands, **Mindanao** and **Jolo** in particular, continued for 3 years. The fanatic Mohammedan tribesmen were not subdued until after serious, though small, campaigns successfully led by Colonel **John W. Duncan** and Captains **John J. Pershing** and **Frank R. McCoy,** among others.

COMMENT. *Suppression of the Philippine Insurrection necessitated the employment of 100,000 American troops, whose casualties amounted to 4,243 killed and* *2,818 wounded in action. Filipino losses were some 16,000 killed and approximately 100,000 more died of famine.*

NETHERLANDS EAST INDIES

1900–1908. Continuation of the Achenese War. (See p. 944.) The pacification of Sumatra was finally completed when the Dutch subdued rebellious forces in Acheh (December 1907).

THE AMERICAS

THE UNITED STATES

1899–1905. Philippine Insurrection. (See pp. 950 and 1106.)

1900, June 17. Boxer Rebellion in China. (See p. 1103.)

1900–1903. Army Reforms. Efforts of **Elihu Root,** secretary of war, brought about establishment of the Army War College (1900), the Command and General Staff School (1901), and the Army General Staff (1903).

1903, November 3. Panamanian Revolt against Colombia. (For U.S. involvement, see p. 1108.)

1906, October 2. Army of Cuban Pacification. (See p. 1109.)

1907–1909. U.S. Fleet Sails around the World.

1912. Intervention in Honduras and Nicaragua. (See p. 1108.)

1914, April 9–21. Tampico and Veracruz Incidents. U.S. military action against Mexico (see below).

1914, August 4. U.S. Neutrality Declared at Outbreak of World War I.

1915, July 29. Intervention in Haiti. (See p. 1109.)

1916–1917. Mexican Border Operations. (See p. 1108.)

1916, November 29. Intervention in Dominican Republic. (See p. 1109.)

1917, April 6. Declaration of War on Germany. (See p. 1160.)

1921, November 12–1922, February 6. Washington Naval Conference. (See p. 1008.)

MEXICO

1911–1914. Revolutionary Era. Overthrow of President **Díaz** (May 25, 1911) was followed by internecine war between numerous rival leaders. **Francisco Madero,** who succeeded Díaz, was defeated and killed by **Victoriano Huerta** (February 22, 1913). Huerta's regime (which was not recognized by the U.S.) was challenged by one of his many rivals, **Venustiano Carranza.**

1914, April 21. American Occupation of Veracruz. Arrest of unarmed U.S. sailors at Tampico (April 9) was followed by the shelling of Veracruz by a U.S. naval force and the landing of a small expeditionary force, which occupied the city. Huerta's government severed relations with the U.S. South American states attempted mediation. U.S. troops were withdrawn later in the year (November 25).

1914, August 15. Carranzists Capture Mexico City. Carranza assumed leadership of Mexico.

1914–1915. Revolt of Zapata and Villa. **Emiliano Zapata** and **Francisco (Pancho) Villa,** bandit-revolutionist leader who dominated northern Mexico, revolted. Villa briefly captured Mexico City, but was driven out by Carranza's General **Alvaro Obregón,** who won a climactic victory at **Celaya** (April

13–15). Carranza was recognized by the U.S. as the President of Mexico (October 15, 1915). Following a falling-out of Zapata and Villa, each continued separate, small-scale insurgencies until the death of Zapata in a government ambush (April 10, 1919) and the surrender of Villa (July 27, 1920; see below).

1916, March 9. Villa Raids across the U.S. Border. Villa's band of 500 men made a night attack on **Columbus,** N.M. Surprising the town and its garrison of U.S. cavalry, Villa killed 14 American soldiers and 10 civilians before he was driven off with a loss of 100 men. To prevent repetitions of the outrage, President Wilson sent Regular and National Guard troops to protect the border. (Eventually this force reached a strength of 158,000 men—most of the active military strength of the U.S. at the time.)

1916, March 15. Punitive Expedition. Under President Wilson's directive, Brigadier General **John J. Pershing** with 10,000 troops (mostly cavalry) struck into Mexican territory to pursue Villa. Carranza's reluctant consent soon turned to open antagonism. In addition to several skirmishes with Villistas, the Americans were also engaged on several occasions with regular Mexican troops, notably at **Carrizal** (June 21). Despite all his efforts, Pershing was not able to catch the elusive Villa. The expedition was withdrawn (February 5, 1917).

1916, July 24. The Zimmermann Note. (See p. 1060.) Though German efforts to exploit Mexican anti-American sentiment failed, coolness between the U.S. and Mexico persisted throughout World War I.

1920, April–July. Renewed Civil War. Carranza was overthrown and killed (May 21) in a revolt led by Generals Obregón, **Adolfo de la Huera,** and **Plutarco Elías Calles.** Villa later surrendered to the victors. Obregón was elected President (September 5) and was later recognized by the U.S. (August 31, 1923). Stabilization of the country began.

CENTRAL AMERICA

1903, November 3. Panama Revolution. U.S. efforts to purchase from Colombia the territory necessary for the proposed and congressionally authorized Panama Canal had been rebuffed (October 31). The Colombian province of Panama rose in revolt (November 3), while U.S. warships stood offshore to discourage effective Colombian reaction. The U.S. recognized Panamanian independence (November 6) and 10 days later received Panama's new Minister to the U.S., **Philippe Bunau-Varilla** (formerly associated with the old Panama Canal Co.). A treaty granting the present Canal Zone to the U.S. was signed November 18. The coup was obviously engineered by U.S. President **Theodore Roosevelt** in the paramount interests of national defense. The canal was intended as a bypass for strategical shifts of the U.S. Navy between the Atlantic and Pacific oceans.

1907, February–December. War between Nicaragua and Honduras. Honduras was defeated and **Tegucigalpa,** Honduran capital, occupied by the victorious Nicaraguans.

1909–1911. Civil War in Honduras. Former President **Manuel Bonilla** led a revolt against President **Miguel Danila.** A stalemate resulted in an armistice (February 8, 1911). Bonilla was elected president (October 29), but disorders continued.

1912, January. U.S. Marines Land in Honduras. Their mission was to protect U.S. property.

1912, July. Civil War in Nicaragua. U.S. Marines were landed to stop hostilities and assure free elections. This small contingent was not withdrawn until 1925.

1917–1918. World War I. All Central American nations declared war upon Germany, following U.S. entrance into the war.

1921, February–March. Panama—Costa Rica Dispute. Armed clashes took place in disputed territory, but U.S. political pressure averted war.

SOUTH AMERICA

During World War I, Argentina, Chile, Paraguay, Colombia, and Venezuela remained neutral. Brazil declared war against Germany (October 26, 1917). Her warships actively cooperated in antisubmarine activities, and much foodstuff was furnished the allies. Uruguay, Bolivia, Peru, and Ecuador severed relations with Germany, but did not declare war. The period was generally quiet in South America, save for civil wars or revolutions in Colombia (1900–1903), Peru (1914), Ecuador (1924–1925), and Brazil (1924).

WEST INDIES

Cuba

1906–1909. American Army of Cuban Pacification. Political unrest which created chaotic conditions caused President **Theodore Roosevelt** to send a force of 5 regiments of infantry, 2 of cavalry, and several field batteries to Cuba. Normal conditions were restored without friction or incident of any sort. The expedition evacuated the island.

1917, February–March. Revolt in Cuba. American forces landed at Santiago to restore order.

Dominican Republic

1916, May. U.S. Intervention. Following internal disorder threatening national bankruptcy, U.S. Marines were landed and U.S. officials took control of fiscal matters. The situation worsening, full military occupation began (November 29). The Marines were withdrawn in 1924.

Haiti

1915, July 3. U.S. Forces Intervention. This followed widespread disorders and economic claims against Haiti by European countries. A protectorate was proclaimed (September 16). Under Marine supervision, a constabulary was organized and conditions quieted.

1918, July 12. Haiti Declares War on Germany.

1918–1919. Revolt against the U.S. Occupation. Disorders were suppressed by the Marines.

XX

WORLD WAR II AND THE DAWN OF THE NUCLEAR AGE:

1925–1945

MILITARY TRENDS

GENERAL

A new era in warfare and a new era in history dawned in the closing days of this period: the nuclear age, ushered in by the first atomic bomb drop, on Hiroshima, August 6, 1945. Its mushroom-shaped cloud became a question mark. Was nuclear power to become arbiter of future world strategy and diplomacy, a possibility indicated by its unprecedented potentiality for cataclysmic destruction, or was this power merely another, more powerful, addendum to military arsenals?

Meanwhile, the dramatic surge of military technology, noted in the two previous periods, continued to accelerate. To mention only a few of the most obvious projects of this trend, one must note the perfections in internal-combustion engines which led to amazingly improved tanks and warplanes; the reappearance of rocket weapons, from the simple, hand-carried "bazooka" to the highly complex long-range German V-2; and, perhaps even more significant, the tremendous burgeoning of electronics, particularly in the form of radar and of improved radio communications.

Thanks to such developments and also to the vertical-assault potentialities of air warfare, the heretofore rigid compartmentalization of land and sea operations disappeared. Three-dimensional warfare multiplied possible strategic and tactical combinations. Success in this new, coordinated "triphibious" warfare depended upon ever-closer affiliation of civilian science and industry on the one hand, and military competence and genius on the other. Victory went to the nation which could best combine the disparate elements of such a team in exploitation of all available resources.

Improvements and refinements in weaponry, transport, and communications brought vast changes in tactics and techniques of warfare. However, man and his reasoning power—the ultimate weapons, and also the basic limitations, in warfare—were unchanged, as were the fundamental principles governing the application of force to human conflict.

A most remarkable development was the joint, unified Allied command, epitomized in the Anglo-American alliance of World War II. This encompassed a totally unprecedented pooling, by the British Empire and the United States, of leaders, staffs, troops, and resources. Never before had integrated armies, navies, and air forces entered combat together, willingly subordinating individual national characteristics, doctrine, and training to a unified command directed against common objectives.

Logistics developed into a science in itself, on land and sea. The U.S. Navy's Logistic Support Groups solved one of the most annoying problems of naval warfare: the necessity that vessels return to some land base for fuel, supply, and repair. The necessary withdrawal from action of naval fighting units became a matter of but a few days or even hours, instead of weeks and months, when sea trains of fast cargo ships and floating repair shops became components of each task force. On land, such things as the artificial harbors off the Normandy coast (see p. 1210), flexible fuel pipelines laid under water or along the ground, and the use of cargo planes extended the mobility of combat forces. Organized logistical elements, especially trained—like the Navy's Seabees, the Army's port troops and railroad and airborne Engineers—reduced to routine supply problems which, a short quarter-century before, would have been insuperable obstacles to mobility.

LEADERSHIP

Significant was the propensity of heads of state to exercise more and more the role of strategical military leadership. Britain's **Churchill** and America's **Roosevelt** dominated both their respective national military planners and the Combined Chiefs of Staff, not always for the best. **Stalin** in the U.S.S.R., **Chiang Kai-shek** in China, and Germany's **Hitler** exercised direct overall command of their respective armed forces. Hitler's frenetic vagaries, as we shall see, assisted greatly in bringing defeat to Germany.

In military leadership we believe that one man—**Douglas MacArthur**—may have risen to join the thin ranks of the great captains of history. Such evaluation is complicated by political and personal controversies which swirled about MacArthur in this and the subsequent period, and we are too close to the events to be certain that we can assess them with sufficient historical objectivity.

Two other excellent Allied soldiers—**Eisenhower** and **Montgomery**—achieved such notable and deserved victories as to present to their people ideal images of the quintessence of charismatic military leadership. We do not believe, however, that they can objectively be ranked above such other superb soldiers as America's **Bradley,** Britain's **Wavell,** and Germany's **Manstein, Model, Rundstedt,** and **Kesselring,** who were outstanding among the numerous leaders of army groups, nor indeed above America's **Patton** and Germany's **Rommel,** who as tacticians shone above all other army commanders.

Given the paucity of revealed evidence, it is impossible to evaluate the leadership of individual Russian commanders; but certainly someone—or some small group of men—possessed sufficient leadership qualities to bring about the remarkable achievements of the armies of the U.S.S.R. There seems to be reason to single out **Zhukov** as the man most responsible.

On the naval side, America's **Nimitz** as an overall commander, and **Spruance** as a fleet commander, were both outstanding exemplars of the highly competent officer corps of the U.S. Navy, as were superb Admirals **Cunningham** and **Ramsay** for the Royal Navy.

In air warfare, America's **Arnold** was the first individual who had an opportunity actually to apply the theories of long-range strategic air warfare which had been voiced by air prophets Trenchard, Douhet, and Mitchell a few decades earlier. The results were stupendous, though still inconclusive until the atomic bombs presaged weapons of hitherto-unimagined orders of magnitude. Germany's airmen, whose tactical and technical competence was perhaps unsurpassed, failed to visualize the opportunities for long-range strategic warfare, which were understood and exploited by British and American airmen like **Harris** and **Spaatz.** Other airmen particularly worthy of mention are Britain's **Dowding** (whose victory in the Battle of Britain was the most decisive of World War II) and **Tedder,** and America's **Kenney,** these latter two demonstrating exceptional competence not only in independent air operations but also in providing support to surface forces.

STRATEGY

Obviously, economic and political considerations dictated major strategy not only in World War II but also in the relatively minor wars which preceded it. The principle of the "nation at war," so clearly demonstrated in World War I (see p. 1004), was dominant during 1939–1945. So too were its limitations, though not immediately apparent. Japan, seeking the wherewithal to provide strategic material for prosecution of her aim, overreached herself totally in the advance into the Southern Resources Area. Hitler's preoccupation with the Russian oil fields was equally fatal (see p. 1191). On the Allied side curious contrasts are found. Churchill's determination to dispute control of the Mediterranean even when Britain was seemingly beaten to her knees (see pp. 1167–1168) was correct; it proved to be the salvation of the Allied cause. His fascination for attacking the "soft underbelly" of the Axis, which became a bone of contention between American and British planners throughout the war, was politically sound in its aim to curb Russian domination of Eastern Europe. However, the logistical problems would have been stupendous. The question is moot; it will be long debated, as will the opposing, perhaps shortsighted, American strategy, concerned with winning the immediate war rather than with long-range political objectives.

Roosevelt's dictum of "unconditional surrender" in all probability extended the war with Germany far beyond its normal end. Eisenhower's remark, at a press conference (February 28, 1945) that the policy faced the German high command with the choice of being hanged or jumping into a clump of bayonets, was a correct estimate.

Wars of the period were all wars of maneuver; World War II was one of maneuver vast in space, personnel, matériel, and logistics. Air power emerged as a combat force coequal with land power and sea power. Earlier, the Spanish Civil War (see p. 1127) became a testing ground for German, Italian, and Russian theories. The Douhet theory that decision could be attained by air power alone was tested during the siege of Madrid and found wanting, although its efficacy against a primitive nation (Ethiopia; see p. 1138) was high. The citizens of Spain's capital city surmounted their terror, and the decision was reached by ground forces in a war of movement along the Ebro River (see p. 1130).

While the value of fortifications—both fixed and temporary—as springboards for maneuver was again attested throughout the period, sieges *per se* had no appreciable effect other than psychological.

Japan's brilliant and successful opening moves in the Pacific in 1941–1942 (see p. 1232) furnished a lesson of utmost importance to military planners. Consensus of American military thought had been that the Philippines would be the initial target for any Japanese aggression. Defense plans were therefore predicated upon a stout defense of Luzon pending the irresistible counterattack of the U.S. Pacific Fleet. The elimination of that fleet at Pearl Harbor wrecked the entire strategic plan, necessitating an immediate reassessment and readjustment in the midst of war's confusion. British military thought presupposed an initial naval attack on Singapore, essential base for the logistically "short-legged" ships of the Royal Navy and the focal point for successful naval operations in defense of the eastern Indian Ocean and western Pacific areas. The Japanese assault upon the unprotected land side of the fortress (see p. 1236) pricked that balloon.

Both cases were tragic, and almost fatal, instances of the unfortunate tendency of military planners to rely on crystal-ball assumptions of *enemy probable intentions* rather than examining *the worst thing the enemy can do* and providing for such contingency. Proved again, too, was the dictum of Frederick the Great that there is no dishonor in hard-fought defeat, but that there can be no excuse for being surprised.

WEAPONS, DOCTRINE, AND TACTICS ON LAND

We have noted that amazing inventions and refinements in technique emerged from the crucible of World War II. Among these were such things as the proximity fuse, "shaped" charges, bazookas, recoilless rifles, rockets (returning from a century of oblivion), and concomitant refinements in artillery fire direction and control. Vastly improved mobile ordnance, fast tanks and tank destroyers, and other cross-country vehicles combined to produce a doctrine of mobile warfare at speeds heretofore impossible. With this came wide dispersal of units and elaboration of the "Hutier tactics" of World War I—the sweeping advance which bypassed strong points for later reduction by slower-moving elements; this, in turn, necessitated defensive zones of great depth, so-called "hedgehog" formations, which could themselves become points of support for fast counterattack.

Between the World Wars the Germans intensively studied ways and means of overcoming the deficiencies which had stopped their 1918 offensives just short of victory. They came to the conclusion that the tank provided the answer to maintaining the momentum of a breakthrough, and that self-propelled artillery and air support would provide the firepower required when the tanks and other attacking units moved too fast for conventional artillery to keep up. As to resupply and reinforcement, other track-laying vehicles and cross-country trucks were designed to assure resupply and reinforcement to the armored spearheads, while the speed of the advance was expected to avoid the shell-cratered morasses of World War I battlefields. As a result of these studies, when the war began the Germans were far ahead of any other country in tank doctrine. They also enjoyed a slight technical superiority in tank design and construction,

German 88

which they were able to maintain throughout the war, despite intensive efforts by the British, Americans, and Russians to catch up. The Germans also had a substantial superiority in antitank guns. Almost accidentally they discovered that their high-velocity 88-mm antiaircraft gun was the best antitank gun in the world, and was also useful in the normal artillery role in level or rolling areas where its flat trajectory was not a handicap. Probably the most effective use made of the 88-mm gun was by General Erwin Rommel in the desert fighting in North Africa, where he aggressively sent batteries of 88s forward with his tanks, to form deadly firepower bases around which his tanks maneuvered rapidly, to the dismay of his British opponents.

German tank used by the Afrika Korps

While the Germans were rectifying the shortcomings of their 1918 offensives, the major maritime powers were somewhat less intensively studying the Gallipoli campaign as an example of how not to combine land and sea forces in amphibious operations. The British, Americans, and Japanese, quite independently, began to devise weapons, equipment, and techniques to improve assault landings. All three developed shallow-draft, ramp-unloading landing craft that would get troops to the edge of the beach, and enable them to unload rapidly. (It is not known whether Byzantine examples had any influence on these developments.) During the war the British and Americans intensified this effort, even producing ocean-going vessels such as the Landing Ship Tank (LST) and Landing Ship Infantry (LSI) that such could carry assault troops across an ocean and land them in combat-ready formation on hostile beaches.

Although inhibited by lack of funds and shortages of personnel, the United States Army was making the most significant development in the enhancement of artillery firepower since the time of Gustavus Adolphus. Elaborating concepts and techniques improvised on the battlefield by Major General **Charles P. Summerall** in the closing days of World War I, the U.S. Field Artillery School developed the technique of massed fires, whereby a single Fire Direction Center could rapidly and accurately shift the fire of many batteries, and sometimes many battalions, across a wide front, multiplying artillery effectiveness many times. This capability, and particularly the massing of the multi-volleyed fires of large numbers of guns on one target in a demoralizing crescendo of destructiveness, gave the Americans a firepower superiority on the battlefield that was not matched by any other nation in the war.

Naval Weapons, Doctrine, and Tactics

The effectiveness of bomber and torpedo aircraft against surface warships was so pronounced from the outset of the war—in European waters as well as in the Pacific—that it soon became evident that air superiority also automatically included surface superiority, almost regardless of the relative strength in surface warships of opposing forces. This was demonstrated beyond all doubt in the dramatic American naval victory of Midway (see p. 1254). Thus it was apparent early in the war that carrier-based aircraft were not mere supporters of surface naval forces, but were in fact the primary striking element. The carrier, providing the weapon to destroy enemy surface forces—an extension of firepower—and thus quickly displaced the battleship as the capital ship of the fleet, at the very moment that the all-big-gun super-dreadnought reached its apogee of firepower and of invulnerability to gunfire in such formidable warships as the German *Bismarck* and the Japanese super-battleships *Yamato* and *Musashi*. The change started with the Japanese carrier blow at Pearl Harbor (see p. 1233). Before the period ended, great fleet actions were contested and won in the air by bombs and torpedoes delivered by airplanes launched from surface vessels which never sighted one another, nor fired other than antiaircraft artillery.

Yet, while the loss of HMS *Renown* and *Prince of Wales* (see p. 1235) early in the Pacific war demonstrated the inability of surface vessels to resist properly delivered air strikes—either land- or carrier-based—the period also saw a number of other naval actions in which surface vessels slugged it out with gunfire. Notable were the Battles of the River Plate (see p. 1153) and the Komandorski Islands (see p. 1272), while the Battle of Surigao Strait, in itself a contest of surface maneuver-*cum*-gunfire, was a component of the large-scale Battle of Leyte Gulf in which carrier air power played the major role (see p. 1288).

The submarine loomed large as a component of sea power, its primary mission being commerce destruction. Germany's U-boat Atlantic campaigns almost, but not quite, weighted

American LST (Landing Ship Tank) in action

the scales in favor of Nazi victory. In the Pacific, where Japan never quite understood the strategic employment of the submarine and ignored antisubmarine procedure, the U.S. submarine campaign strangulated the Nipponese merchant marine. What emerged from the conflict was the sound premise that submarine warfare, in both offense and defense, was a highly specialized affair and—like all military operations of the period, for that matter—demanded professional competence and vision of high order. As a result, despite the fact that at the outset of World War II Japan possessed in the "Long Lance" torpedo, technically the best weapon of its category, her underwater strategy was erroneous.

The use of submarines to evade surface blockade, first attempted by the U.S. Navy during the opening Luzon campaign, became highly developed by the Japanese in the Southwest Pacific. Underwater vessels carried troops and matériel, supplementing the use of fast destroyer-transports for the same purpose.

A striking development in surface operations was the use of naval firepower in support of landing operations. Refinements in fire control and direction enabled U.S. naval craft to put down most effective gunfire support to ground troops during the initial sensitive period prior to debarkation of the assaulters' artillery.

Technological inventions—radar and sonar in particular—played a great part in naval

Aircraft carrier USS *Lexington*

operations. Together with other electronic communications innovations, they constituted a vast and delicate refinement in command control. An important result of this development was the emergence of the command ship in U.S. Navy procedure. In fleet actions, no longer were the admiral and his staff an irritating excrescence on some unfortunate capital ship. In amphibious operations the floating command post enabled close personal cooperation between the naval commander, responsible for putting the ground forces ashore, and the ground commander, who assumed control once the troops gained a toehold on the beach.

Japanese battleship *Yamato*

Top: Messerschmitt ME-Bf109 E-4. A formidable opponent for the RAF's Spitfires and Hurricanes—but over-extended by the long-range flights over Britain, forced to cover ME-110 fighters as well as the bombers, and finally grossly misused as a fighter-bomber. *Speed:* 357 mph. *Max range:* 412 miles. *Armament:* two 20-mm cannons and two 7.92-mm machine-guns.

Middle: Junkers JU-87 A-1 Stuka. In close support of the German Army during the Blitzkrieg Campaigns in Poland and France, the Stuka had made history as "flying artillery." It could pinpoint targets with deadly accuracy, but only in the absence of fighter opposition—and it therefore suffered drastic losses in the Battle of Britain. *Crew:* two. *Speed:* 199 mph. *Max range:* 620 miles. *Bomb load:* 1,100 lbs (pilot only). *Armament:* two 7.92-mm machine-guns.

Bottom: Heinkel HE-111 H-3. The Battle of Britain was the first campaign which rammed home Germany's failure to develop a long-range heavy bomber to do the work done by planes such as the HE-111 medium bomber. Like the JU-88, it was also used as an anti-shipping strike aircraft. *Crew:* five. *Speed:* 254 mph. *Max range:* 1,100 miles. *Bomb load:* 4,000 lbs. *Armament:* five 7.92-mm machine-guns, one 20-mm cannon.

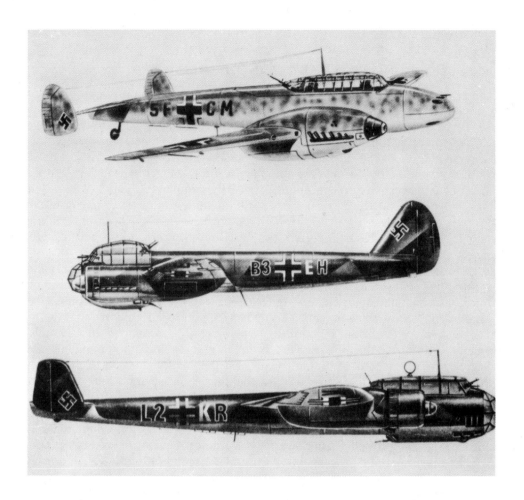

Top: Messerschmitt ME-Bf110 C-1. Goering hoped that "destroyer" formations of ME-110s would carve through all fighter opposition, cleaving a path for the bombers—but the 110 was far too heavy and sluggish to "mix it" with the Spitfire and Hurricane, and suffered accordingly. *Crew:* two. *Speed:* 349 mph. *Max range:* 565 miles. *Armament:* five 7.92-mm machine-guns and two 20-mm cannons.

Middle: Junkers JU-88 A-2. Maid-of-all-work for the Luftwaffe, serving as dive-bomber, level bomber, night-fighter, and reconnaissance. It also served with distinction in the torpedo-bombing role against Allied convoys. It suffered—like all German bombers—from a chronic weakness in defensive armament. *Crew:* four. *Speed:* 286 mph. *Max range:* 1,553 miles. *Bomb load:* 3,963 lbs. *Armament:* four 7.92-mm machine-guns.

Bottom: Dornier DO-17 Z-2. The DO-17 was a lighter, slimmer plane than the HE-111. Like the Heinkel, it had been blooded in the Spanish Civil War; like all Luftwaffe's bombers, the Battle of Britain forced it to carry out operations which proved the inadequacy of its design. *Crew:* five. *Speed:* 265 mph. *Max range:* 745 miles. *Bomb load:* 2,200 lbs. *Armament:* six 7.92-mm machine-guns.

WEAPONS, DOCTRINE, AND TACTICS IN THE AIR

The violence and effectiveness of air bombardment, and of close support of ground forces by fighter planes, were clear after the Italian conquest of Ethiopia (see p. 1138) and the Spanish Civil War (see p. 1127). But the full potentialities of air weapons against ground targets were first clearly demonstrated in the German blitzkrieg campaigns in Poland, Norway, the Low Countries, and France in 1939–1940 (see p. 1149 ff.). Yet for all of the superb tactical efficiency demonstrated by the Luftwaffe in these campaigns, the Germans had not grasped the full implications of air power as a new concept of warfare; they employed their air forces essentially as adjuncts of land forces. The British, though still fumbling in the development of their air doctrine, were nevertheless ahead of the Germans in their air-power concepts. It was this doctrinal advantage—as well as important technical and tactical factors—which more than anything else brought the RAF victory in the Battle of Britain.

As developed by the British, with some later American refinements, by the end of World War II air doctrine had resolved itself to encompass three closely related but nonetheless distinct major functions: command of the air, long-range (so-called "strategic") bombardment directed against the enemy's warmaking potential, and direct support to surface forces.

Command of the air, or "air superiority," was not only essential to effective offensive employment of air units in the other two functions but was also important in two negative, or defensive, aspects. Command of the air, or at least the capacity to effectively dispute the enemy's control of the air, was important for the defense of a nation's economic strength against long-range bombardment by the enemy, as well as for protecting surface forces against enemy air strikes. And because of the terrorizing effect of air attacks on both civilian and military personnel, command of the air was an important morale factor. Command of the air was achieved in several ways: defensive air combat, attrition of enemy fighter strength through repeated long-range strikes, attacks against air installations and (more slowly, and really as a part of the second function) against aircraft industry.

Strategic bombardment was the function which had been visualized as the decisive role of air power both by Douhet (see p. 1089) and by Mitchell (see p. 1147). Long-range bombardment aircraft permitted for the first time military operations against a nation's warmaking capacity by more direct and more rapid means than attrition and the traditional blockade. Despite their substantial numerical inferiority in aircraft, and despite the necessity for concentrating on the development of their defensive fighter strength in the early days of the war, the British never lost sight of their objective of offensive air warfare of this sort, and were actually conducting long-range strikes against German industrial and commercial targets even during the Battle of Britain. Quickly improved German fighter defenses, however, forced the British to conduct these attacks at night, when visual conditions were poor, and they had to be content with relatively inaccurate area bombardment of large industrial areas.

American bombers, protected by armor plate and carrying numerous guns—as epitomized in the so-called Flying Fortresses—were better able to bomb in daylight and, thanks also to a more effective bomb sight, were able to strike targets with considerably greater precision. An incidental function of such raids was to entice the German defensive fighters into the air and thus to help reduce German command of the air over Germany by attrition. American losses to the Germans in these daylight raids grew to alarming proportions, however, and could be contin-

British Spitfire fighters

ued in 1944 and 1945 only because of the development of new longer-range fighter aircraft that could accompany the bombers on their raids into the heart of Germany. Till the end of the war in Europe, therefore, Britain's RAF Bomber Command struck area targets at night, while American long-range units attacked during the day. The Combined Bomber Offensive was a major factor in hastening the collapse of Germany.

Though the details were different (see p. 1292), the development of American strategic bombardment tactics against Japan followed a similar course, with the final effectiveness of the air offensive being in large part due to the ability of long-range fighters to accompany and protect the bomber planes.

Flying Fortress—B-17F

Liberators—B-24

The tactical doctrine for close air support of ground forces, as it emerged from World War II, was largely the development of cooperative efforts of Air Marshal Tedder's Desert Air Force and General Montgomery's Eighth Army in 1942. The essential feature of this doctrine was in its command arrangements, whereby the air commander at all times retained full control over all his subordinate units, which were never attached to ground-force commanders. This permitted a degree of flexibility in meeting the ground-force requirements, and in coping with unexpected threats, which was impossible in the German system, for instance, where air units were generally assigned to ground-force command. At the same time, however, it inhibited full integration of air-support units into the land-warfare team.

It is important to note that an essential aspect of this doctrine of support was the dual capability of the fighter-bomber aircraft type. These light, speedy, maneuverable planes were not only the best implements for relatively precise, low-level bombing and strafing attacks against ground targets; they were also the only weapons which could be used with real effectiveness against comparable enemy types. This was why the first mission of these fighter aircraft—assisted by light and medium bombers attacking enemy air installations—had to be the achievement of local air superiority. Otherwise, neither they nor any other aircraft types (reconnaissance, cargo, troop-carrier, etc.) could carry on other missions in support of the ground troops without suffering prohibitive losses from enemy fighters.

At sea the aircraft mission of support of surface forces was similar in concept, though it differed in some important details. This is discussed under naval tactics (see p. 1115).

As the period ended, professional opinion upon the capabilities and limitations of air power was widely divided, despite the consensus that it had become an indispensable member of the combat team.

Meanwhile, although various types of aircraft proliferated, the basic categories of combat planes resolved into heavy long-range bombers for offensive purposes, fighters for defense of friendly bombers and use against hostile intruders, fighter-bombers for both air combat and close support, and a variety of reconnaissance, spotting, and other combat auxiliary types.

Airborne operations, initiated in and by the German vertical assault on Norway and Western Europe in 1940 (see pp. 1156, 1159) and on Crete during 1941 (see p. 1177), became a common practice later of U.S. and British offensives. The troops were transported in planes (to be parachuted to the ground) and towed gliders (for "crash" landings), and were escorted by combat aircraft.

The development of heavy cargo planes permitted logistical air support to ground forces. Allied operations in the Burma and China theaters completely depended upon it.

The weaponry of air combat resolved itself into machine guns, light cannon, bombs, rockets, and torpedoes. Development of a bomb sight of superior accuracy (American) and of radar (mainly British) were twin inventions of enormous importance.

As the war was coming to a close, both Britain and Germany were developing a radically new jet aircraft engine for interceptor aircraft, which would be far faster and more powerful than the conventional propeller-driven aircraft. The Germans actually succeeded in producing a number of such interceptors, which proved their effectiveness in combat against Allied bombers over Germany. Their appearance was too late, however, to prevent Germany's overall military collapse.

DISARMAMENT

During this period was initiated an intensive effort to achieve disarmament (or at least arms control) as a means to assuring world peace. The League of Nations, which had started the effort earlier (see p. 1008), continued to take leadership in this field. Although the United States was not a member of the League of Nations, it participated informally in most of the League activities during this period. The results of disarmament efforts were disappointing (see below).

INTERWAR INTERNATIONAL EFFORTS TO MAINTAIN PEACE, 1925–1939

Closely related to the new international attention being focused upon arms control and disarmament in general were the efforts of the major powers—mainly the victors of World War I—to ensure continuation of the postwar peace within the framework established by the Treaty of Versailles (see p. 1082). These efforts are summarized as follows:

1925, June 17. Arms Traffic Convention. This was convened by the League of Nations at Geneva in an effort to control international trade in arms and munitions. One result was the **Geneva Protocol,** prohibiting use of poison gas in warfare. The U.S. did not sign.

1925, October 5–16. Locarno Conference. The resulting treaties (signed December 1) included (1) a treaty of mutual guaranty of the Franco-German and Belgo-German frontiers signed by Germany, France, Belgium, and Great Britain, (2) arbitration treaties between Germany and Poland and Germany and Czechoslovakia, (3) arbitration treaties between Germany and Belgium and Germany and France, (4) a Franco-Polish and a Franco-

Czechoslovak treaty for mutual assistance in case of attack by Germany. The treaties helped create a sense of security among the European powers.

1925, December. Establishment of Disarmament Preparatory Commission. This was created by the League to prepare for a World Disarmament Conference. The commission wrestled for 7 years with issues that plagued disarmament efforts throughout the interwar period, including France's demand for security guarantees as a prerequisite for arms reduction, Germany's demand for equality in armaments, Great Britain's insistence on arms of the kind and number needed to defend her scattered possessions, and American opposition to the concept of collective security (entangling alliances) and inability to recognize the complexities of disarmament problems faced by other nations.

1927, January 31. Dissolution of Inter-Allied Military Control Commission. Germany had consistently flouted the injunctions of the commission and obstructed its procedures. None of the Allies was willing to use force to obtain compliance. The problem of German armament was henceforth placed under League jurisdiction.

1927, June 20–August 4. Three-Power Naval Conference. Great Britain, the United States, and Japan met at Geneva in an effort to agree on the question of establishing a ratio for strengths in cruisers, destroyers, and submarines as a step beyond the Washington Naval Treaties (see p. 1008). The conference failed to reach agreement.

1928, August 27. Kellogg-Briand Pact (Pact of Paris). This was signed at Paris by the United States, France, Great Britain, Germany, Italy, Japan, and a number of other states. This pact renounced aggressive war, but made no provisions for sanctions.

1930, January 21–April 22. London Naval Conference. This led to a treaty signed by Great Britain, the United States, France, Italy, and Japan, regularizing submarine warfare and limiting the tonnage and gun caliber of submarines. The limitation on aircraft carriers provided for by the Washington Treaty (see p. 1008) was extended. Great Britain, the United States, and Japan also agreed to scrap certain warships by 1933, and allocated tonnage to other categories. An "escalation clause," permitting an increase over specified tonnages if the national needs of any one signatory demanded it, was included. These agreements were to run until 1936. The treaty gave Britain and the United States equality in overall cruiser combat effectiveness, which was to be determined by formula. Japan's relative naval position was strengthened. The treaty was to expire at the end of 1936.

1932–1934. World Disarmament Conference at Geneva. The Preparatory Commission's draft treaty included provision for inspection and control, but no relative force ratios had been agreed on. The optimistic atmosphere created by the Kellogg-Briand Pact of 1928, by which all nations had renounced war, and by the London Naval Treaty, had been dissipated by Japan's 1931 invasion of Manchuria (see p. 1145), the international economic depression, and the rising strength of aggressive nationalism in Germany. France's "Tardieu plan," providing for an international police force, stronger security guarantees, and the placing of all powerful weapons "in escrow," to be used only at the direction of the League or to repel an invasion, was too extreme to win acceptance by the other powers. Germany made it clear that she would not accept this or any proposal that did not provide for German equality. Britain introduced a counterproposal providing for proportionate reduction and also "qualitative" reduction—prohibition of aggressive weapons. This proposal failed of adoption, as did a U.S. proposal for the abolition of all "offensive" weapons and another U.S. plan to reduce all armaments by one-third. Failure was certain after Japanese withdrawal from the League of Nations (March 1933; see p. 1142) and German withdrawal from the conference and the League (October 1933; see p. 1132). The conference soon adjourned without agreement (June 1934).

1934, December 19. Japan Denounces

Washington and London Naval Treaties. She gave the required 2 years' notice that she was withdrawing from the Washington Naval Treaty of 1922 and assurance that, when the London Treaty expired at the end of 1936, there would be no naval limitations agreement unless a new treaty could be achieved. This followed refusal of the U.S. and U.K. to agree to Japanese demands for equality.

1935, March 16. Germany Denounces Disarmament Clauses of Versailles Treaty. Germany claimed this was due to the failure of other nations to disarm. She announced a massive rearmament program.

1935, December. Five-Power Naval Confer- **ence.** Representatives of the United States, Great Britain, Japan, France, and Italy met in London with little hope of producing a treaty. Japan soon left the conference. America, Britain, and France did agree to "qualitative" limitation, with restrictions on certain kinds of ships and caliber of guns. Advance notification of construction programs was also to be given, and should one nation build in excess of treaty restrictions, others were freed of restrictions. After 2 years the treaty was discarded, as accelerated Japanese naval construction and the aggressive acts of the revisionist powers forced Great Britain and the United States to rearm as fast as possible.

WESTERN EUROPE BETWEEN WORLD WARS

GREAT BRITAIN

1935, June 18. Anglo-German Naval Agreement. German tonnage (including submarines) would not exceed 35 percent of British naval tonnage. This separate agreement estranged France from Britain.

1935, September. Britain and the Ethiopian Crisis. (See p. 1138.) Great Britain assumed guidance of the League of Nations in the imposition of sanctions, but shrank from adopting extreme measures (such as oil sanctions) and the League action failed.

1937, January 2. Anglo-Italian Mediterranean Agreement. This affirmed the independence and integrity of Spain, and international freedom of passage through the Mediterranean, but in the long run did not dispel British suspicion aroused by Italian activity in the Mediterranean and Near East.

1938, September 15–28. Attempts to Mediate German-Czech Crisis. (See p. 1136.) Prime Minister **Neville Chamberlain** twice flew to Germany to confer with Hitler.

1938, September 29. Munich Agreement. (See p. 1136.) Chamberlain returned with a peace pact with Germany, which he rated highly, though it was not universally viewed with such acclaim.

1939, March 31. British Guarantee to Poland. After German dictator **Adolf Hitler** seized Czechoslovakia (see p. 1136), the British government pledged aid in case of any aggressive action endangering Poland's independence. Following the Italian conquest of Albania (see p. 1138), similar guarantees were also given to Greece and Rumania, a mutual assistance pact was concluded with Turkey, and the British government began to try to bring Russia into the "peace front."

1939, September 3. Outbreak of War with Germany. (See p. 1149.)

IRELAND

1933–1938. Internal Disorder and Conflict.

1935, July. Anti-Catholic Riots in Belfast. They led to expulsion of Catholic families, to reprisals by the Free State government, and to increased friction with Britain.

1938, April 25. Agreement with Great Britain. This resolved a number of outstanding issues for at least 3 years, including turning over to the Free State (Eire) the coast defenses of Cobh, Bere Haven, and Lough Swilly. It improved relations between England and Eire.

FRANCE

1925, April. Beginning of Insurrection in Morocco. (See p. 1099.)

1925, July. Rising of the Druses in Lebanon. (See p. 1140.)

1935, January 7. Franco-Italian Agreement on Africa. France made several concessions to Italy in the hope of establishing a strong front against increasing German strength.

1935, May 2. Alliance with Russia. The French government, unable to bring Germany and Poland into an eastern pact (which would include Russia) to maintain the *status quo,* hastened to ally itself with Russia, after announcement of German rearmament. There was much opposition from conservative elements in France.

1936, June. France Begins Rearmament. This followed the German reoccupation of the Rhineland, the Italian victory in Ethiopia, the collapse of the League system (on which France had depended heavily), plus the outbreak of civil war in Spain. (See pp. 1127, 1132, 1139.)

1938, September. The German-Czechoslovak Crisis. (See p. 1136.) The French Army and Navy were partially mobilized and ready for war. However, the country was strongly in favor of a peaceful settlement and Premier **Edouard Daladier** received a warm reception on his return from Munich. French preponderance on the Continent had been definitely replaced by that of Germany.

1938, December 6. Franco-German Pact. This guaranteed existing frontiers.

1938, December. Franco-Italian Crisis. This arose from Italian demands for French colonies; France took an uncompromising attitude toward any cession of territory.

1939, March. Guarantees to Poland, Rumania, and Greece. France joined Britain in these guarantees, and used all her influence to draw Russia into the nonaggression system.

1939, August 20–September 1. The Danzig-Polish Crisis. (See p. 1136.)

1939, September 3. France Declares War on Germany. (See p. 1149.)

THE LOW COUNTRIES

Belgium

1936, October 14. Belgium Denounces Military Alliance with France. She resumed liberty of action, following German reoccupation of the Rhineland, in order not to become embroiled against Germany through connection with the Franco-Russian alliance.

1937, October 13. German Guarantee to Belgium. The inviolability and integrity of Belgium were guaranteed so long as the latter abstained from military action against Germany.

1939, August 23. King Leopold Appeals for Peace. This was on behalf of Belgium, Holland, and the Scandinavian states. The appeal was in vain. Belgium mobilized, but proclaimed neutrality in the European war that broke out on September 3.

The Netherlands

1926, November–1927, July. Serious Revolt in Java, Dutch East Indies.

1937–1938. Dutch Naval Forces Increased in Far East. Japan's advance in China exposed the Netherlands East Indies to possible Japanese aggression.

THE IBERIAN PENINSULA

Spain

PRELUDE TO REVOLUTION

1930, January 28. Primo de Rivera Resigns from Quasi Dictatorship. His replacement, General **Dámaso Berenguer,** irritated the Spanish people, who held him responsible for the disaster in Morocco at Anual (see p. 1099).

1930, December 12–13. Mutiny at Jaca. The garrison demanded a republic. The ringleaders were executed, but the virus of revolt spread among the military.

1931, April 14. Overthrow of King Alfonso XIII. After a near-bloodless revolt, **Alcalá**

Zamora, Republican leader, set up a provisional government and was elected President (December 10).

1932, August 10. Reactionary Revolt in Seville. This was led by General **José Sanjurjo;** it was speedily suppressed. Regional interests began a tussle for autonomy, especially in Aragon and the Basque country. Extreme parties of the right and left gained more influence and resorted increasingly to violence.

1933, January 8. Radical Uprising in Barcelona. This quickly spread to other large cities and was repressed only with difficulty by government troops.

October 6, 1934. Miners' Revolt in Asturias, Rising in Catalonia. An Anarchist-sponsored miner's revolt in Asturias was ruthlessly suppressed by the Conservative Madrid government, under news-blackout conditions. Claims of atrocities, uncorroborated by accurate news reports, inflamed emotions and hastened the political polarization of Spain. The Catalonian uprising was also repressed, but less harshly.

1936, February 16. Leftist Coalition Takes Power. Elections gave a majority to a Communist-influenced popular front. The new Socialist Cortes voted to remove Zamora for exceeding his powers, and replaced him with Left Republican **Manuel Azaña** (April 10).

THE SPANISH CIVIL WAR, 1936–1939

1936, July 18. Military Revolt. Simultaneously, the military garrisons in 12 cities of the mainland and 5 in Spanish Morocco rebelled, following a clash between a unit of the Spanish Foreign Legion and a Communist-led mob in Melilla, Spanish Morocco. General **Francisco Franco,** erstwhile chief of staff, but exiled by the Socialist government to command of the Canary Islands, flew to Melilla to take command. While Spanish troops in Morocco were air-lifted to Algeciras and La Línea, the garrisons of Burgos, Saragossa, and Huesca in northern Spain concentrated at Burgos, under General **Emilio Mola.** An uprising at Barcelona failed; its leader, General **Manuel Goded,** was captured by the Loyalists and shot. Madrid, the east-coast seaports, and part of the Basque regions remained loyal to the government, but Seville and Cádiz proclaimed for the rebels, as did the cadets of the military academy at **Toledo,** who rallied around their commander, Colonel **José Moscardo,** in the Alcazar fortress, which was lackadaisically besieged by Loyalist militiamen (July 20–September 28). About half the ships in the Spanish Navy joined the revolt.

1936, July–August. Advance on Madrid. Franco, with some 30,000 regular troops (Spaniards and Moroccans), struck north toward Badajoz; Mola with 15,000 more regulars began moving south. The rebel plan was to converge on Madrid.

1936, August 15. Rebels Capture Badajoz. They then began to advance eastward up the Tagus Valley and through Talavera and Toledo, which was relieved by troops under General **José Varela** (September 28) after a 10 weeks' siege.

1936, September. International Involvements. Soviet Russia supplied and reinforced the Loyalists; Germany and Italy lent material aid to Franco. The first tangible aid to arrive was a contingent of Luftwaffe Junker bombers and pursuit planes. An Italian expeditionary corps with light tanks and aircraft followed. Small anti-Communist volunteer contingents arrived from other countries as well. Meanwhile, Soviet aircraft, pilots, advisers, artillery, and tanks soon reached the east-coast ports to aid the Loyalists. The Soviets and other Communists also organized 6 "International Brigades," recruited from anti-fascist volunteers outside Spain. Some 2,900 Americans eventually served in the "Abraham Lincoln Battalion," part of the 35,000 who served in the Brigades at one time or another, including French, Italians, Germans, Poles, Britons, Czechs, Yugoslavs, and others.

1936, September 4. Popular Front Government Formed in Madrid. The new Loyalist leader was **Francisco Largo Caballero;** Catalan and Basque Nationalists were represented.

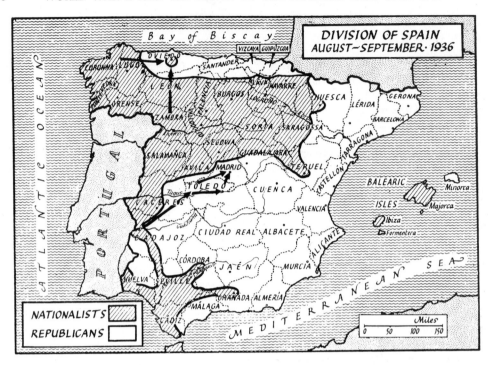

Anarchist-Syndicalists later were included in the government (November).

1936, October. Franco Appointed Chief of the Spanish State by the Insurgents. Four rebel columns under Mola continued to converge on Madrid; he said he had a "fifth column" in the city. This was the origin of this as a term for subversive activity.

1936, November 6. Insurgents Lay Siege to Madrid. The Loyalist government moved to Valencia. Despite heavy fighting in the suburbs of the city and intensive air bombardments, Loyalist troops under General **José Miaja** held the capital. After nearly four months of incessant combat, there was a brief lull in the struggle for the city (February–March 1937).

1936, November 18. Franco Government Recognized by Germany and Italy. France, Great Britain, and the United States continued their policy of nonintervention and nonassistance toward both factions. A nonintervention committee was set up in London; 29 nations, including Italy and Germany, made an attempt to patrol the Spanish coast and limit the war.

1937, February 8. Rebels Capture Málaga. Troops under General **Gonzalo Queipo de Llano** were assisted by Italian troops. Meanwhile, the siege of Madrid continued. Soviet equipment and the newly-arrived International Brigades went far to stiffen the defenses, and the Abraham Lincoln Battalion first saw action at **Jarama** (February 6–15).

1937, March 8–16. Italian Disaster of Guadalajara. Two Italian divisions, accompanied by 50 light tanks, made a surprise penetration of Loyalist lines, in an effort to isolate Madrid. Rain had turned the countryside into a sea of mud, so the Italians advanced in a long road column, with little opposition. Russian ground-attack fighters took them by surprise and after repeated attacks the Italians became a disorganized mob, scattered across the countryside. Most, however, escaped, and eventually returned to their own lines.

1937, March 18. Loyalist Victory over Italians at Brihuega. This was largely a result of

Guadalajara. Large stores were captured. The insurgents, frustrated in their efforts to cut off Madrid, turned north.

1937, April 1. Rebels Invest Bilbao. Surrounding Basque Loyalist resistance was soon crushed.

1937, April 25. Guernica Massacre. World opinion was shocked by the ruthless bombing of this northern village by Franco's Luftwaffe pilots with loss of many noncombatant lives. The towns of **Durango** and **Guernica** were occupied (April 28).

1937, April 30. Rebel Battleship *España* Sunk by Loyalist Planes.

1937, May 3–10. Anarchist Uprising in Barcelona. This was put down with considerable bloodshed by Loyalist troops, and caused a crisis in the government.

1937, May 17. The Largo Caballero Regime Falls. A new Loyalist government was formed under **Juan Negrín.**

1937, May 31. German Planes Bomb Almería. This was in retaliation for a Loyalist strike which damaged the German pocket-battleship *Deutschland* at Ibiza in the Balearic Isles.

1937, June 3. Death of General Mola. He died in an air crash.

1937, June 15 and 18. Submarine Attacks on German Cruiser *Leipzig*. These attacks, allegedly by Loyalist submarines, off Oran on the Algerian coast, resulted in Germany and Italy quitting the international patrol.

1937, June 18. Bilbao Captured by Rebels. This followed an 80-day siege. Rebel troops under General **Fidel Dávila** then moved to conquer **Santander** (August 25). By the end of the year all of northwestern Spain was under rebel control.

1937, July 6–25. Loyalist Offensive from Madrid. After some success, this was repulsed by Varela, commanding the investing forces. There was no further serious fighting on this front for nearly 2 years.

1937, August–September. Loyalist Offensive in Aragon. This was repulsed by the rebels at **Saragossa, Teruel,** and **Huesca.**

1937, September. Nyon Conference. The intense submarine attacks on ships destined for the Loyalists drew protests of piracy from France and England as these ships were supposed to be carrying non-contraband supplies. A 9-power conference was called at Nyon, Switzerland (Germany and Italy abstained). A new system of maritime zone patrols was established. An Anglo-French naval patrol was authorized to attack any submarine, surface vessel, or aircraft illegally attacking a non-Spanish vessel.

1937, October 28. Loyalist Government Moves to Barcelona. It had earlier taken over control of the Catalan government (August 12), following the Anarchist uprising in May.

1937, November 28. Franco Announces a Naval Blockade. This covered the entire Spanish coast, and was operated from the island of Mallorca.

1937–1938, December 5–February 20. Battle of Teruel. A Loyalist counteroffensive captured Teruel after a bitter struggle. This diverted the rebels from operations to the northeastward, but the government forces were unable to sustain the offensive. A rebel counteroffensive retook the city (February 15–20).

1938, February–June. Rebel Offensive. Franco's forces drove eastward in great strength, reaching the seacoast at **Vinaroz** (April 15), and isolating Catalonia from the remainder of Loyalist-held territory. An advance on Barcelona was checked by desperate Loyalist resistance along the Ebro River.

1938, July 24–November 16. Loyalist Ebro Counteroffensive. This effort by forces in Catalonia to restore communications with the rest of Republican Spain was hampered by insufficient resources. The attack was halted within a week, but the battle for the Loyalist bridgehead raged for 113 days, costing 70,000 Loyalist casualties before their tattered forces withdrew.

1938, August–November. Stalemate. The rebels, who now held most of Spain, prepared for a major offensive to end the war. Meanwhile, in accordance with the Anglo-Italian agreement, Mussolini withdrew some troops from Spain, but kept at least 40,000 troops there.

September 21, 1938. Withdrawal of International Brigades Announced. Bowing to British and French pressure, the Loyalist government announced the withdrawal of the International Brigades, badly battered in combat. Their farewell parade in Madrid took place on November 15.

1938, December 23. Beginning of Rebel Offensive in Catalonia. This cracked the Loyalist defense. Government troops retreated in growing disorder on Barcelona. Both sides were hampered by atrocious weather conditions.

1939, January 26. Rebels Capture Barcelona. They were assisted by Italians. Some 20,000 Loyalist troops, completely routed, fled across the French border, where they were disarmed and interned. Resistance continued at Madrid, Valencia, and in scattered areas of eastern Spain.

1939, February 27. France and Great Britain Recognize Franco Government.

1939, March 5–13. Disorder in Madrid. Following the loss of Catalonia, most Republican military officers, Socialists, and Republicans favored an end to the war, but Negrín and the Communists wanted to keep fighting. An anti-Communist coup installed the Republican-Socialist *junta* of Colonel **Sigismundo Casado** (March 5), and Communist resistance ended within a week.

1939, March 22–April 1. End of the War. Negotiations with the Burgos government failed (March 22–25). Nationalist troops entered Madrid (March 28), and all Republican forces surrendered (March 26–April 1), thereby ending the war. The struggle had cost some 120,000 combat deaths, 10,000 from air raids, 50,000 from disease and malnutrition, and perhaps 100,000 from wartime political reprisals on both sides. Franco at once set up special tribunals and courts which convicted and executed thousands of Loyalists, despite international efforts to ensure moderation. The final death toll for the war and its aftermath probably lies around 600,000.

COMMENT. *The war, prosecuted with the ferocity common to Spanish civil wars, presented at least one object lesson to military observers: the transformation in 3 short months of the Loyalist mob of militia into reasonably efficient soldiers (at least in defensive combat) thanks to Soviet instruction. The military of Germany, Italy, and the U.S.S.R. learned much from their respective participation—actually dress rehearsals for airing their doctrines and their matériel for the war to come.*

However, the lesson of Madrid, battered, bombed, and wrecked, with heavy loss of noncombatant life during a 28-month siege—all to no appreciable result—was not taken seriously; the Douhet theory of terror bombing as the ultimate weapon of war had been seemingly disproved, but its advocates the world over remained undaunted. The extent of foreign participation in the war was considerable. Foreign contingents in Spain numbered 40,000–60,000 Italian troops, some 20,000 German, and perhaps some 40,000 Soviet and Communist-inspired volunteers from various nations.

1939, April 7. Spain Joins the Anti-Comintern Pact. (See p. 1132.)

1939, September 3. Spanish Neutrality Proclaimed.

Portugal

1925–1932. Internal Disorders. This was a period of repeated revolts, military coups, and dictatorships, finally brought under control when **Oliveira Salazar** became premier and dictator (July 5, 1932).

1936, July. Portugal Aids Spanish Rebels. With the outbreak of the Spanish Civil War (see p. 1127), Salazar's government immediately sided with the rebels against the republican government. Portugal became one of the main routes by which supplies reached Franco from Germany and elsewhere. Subsequent British pressure forced Portugal to close her borders (April 1937), but by that time Franco was able to get his supplies through the north Spanish coast towns.

1939, March 18. Nonaggression Pact with Fascist Spain. Portugal soon after reaffirmed her traditional alliance with Britain (May 22).

1941–1945. Portuguese Neutrality in World War II. While ostensibly neutral during World War II, Portuguese grant of bases in the Azores

(October 13, 1943) to Great Britain and the U.S. was of much assistance to the Allies in combating the Atlantic submarine menace (see p. 1194); she finally severed diplomatic relations with Nazi Germany (May 6, 1945).

ITALY

1926–1930. Treaties of Friendship with Spain, Hungary, Greece, Ethiopia, and Austria. This was part of the Fascist policy of dictator **Benito Mussolini** of rallying the "revisionist" states against the Little Entente and its supporter, France.

1930, April 30. Italy Begins a Great Naval Program. This was the result of failure to secure recognition of Italian parity from France. During the following years, Italian naval and air forces were built up to imposing dimensions.

1934, March 17. Rome Protocols between Italy, Hungary, and Austria. These provided for closer trade relations and established a Danubian bloc under Fascist auspices to counterbalance French influence with the Little Entente.

1934, July. Abortive Nazi Coup in Vienna. (See p. 1133.)

1935–1936. Ethiopian War. (See p. 1138.)

1936, July. Outbreak of the Spanish Civil War. (See p. 1127.) Italian action in Spain aroused the apprehensions of Great Britain and France and served to increase the tension in the Mediterranean. Under these circumstances, Mussolini felt compelled to draw closer to Germany.

1936, October 27. Italian-German Agreement regarding Austria. This served as a foundation for Italo-German cooperation and may be regarded as the beginning of the Rome-Berlin Axis. Germany recognized the conquest of Ethiopia by Italy (October 25).

1937, January 2. Anglo-Italian Agreement. (See p. 1125.)

1937, March 25. Italian-Yugoslav Treaty. This guaranteed existing frontiers and the maintenance of the *status quo* in the Adriatic, ending a long period of friction between the 2 powers. It was a serious blow to the Little Entente structure and to French influence.

1937, November 6. Italy Adheres to the Anti-Comintern Pact. (See p. 1132.) This completed the triangle of states (Rome-Berlin-Tokyo Axis) engaged in upsetting the peace treaties and the *status quo*. In line with this policy, the Italian government announced the withdrawal of Italy from the League of Nations (December 11).

1938, March. Italy Accepts German Annexation of Austria. Italian commitments in Ethiopia, Libya, and Spain made any alternative course impossible. Mussolini was growing increasingly dependent on Germany.

1938, September. German-Czechoslovak Crisis. Mussolini remained in the background until the situation reached critical proportions, when he made a series of threatening speeches; in the final analysis, he appears to have done his utmost to bring about the Munich meeting and the accord, which entailed the sacrifice of a large part of Czechoslovakia to Germany.

1938, November 30. Franco-Italian Relations Strained. This resulted from agitated demands in the Italian Chamber for the cession by France of Corsica and Tunisia.

1939, April 7. Italian Invasion and Conquest of Albania. (See p. 1138.)

1939, May 22. Political and Military Alliance with Germany. Much exchange of military men and technical experts followed.

1939, September 1–3. Italy Declares Neutrality. This surprised most of the world, but one motive was evidently the thought of serving Germany as a channel for supplies.

GERMANY

1927, September 18. Hindenburg Repudiates German Responsibility for World War I. President Hindenburg in effect renounced a major provision of the Versailles Treaty (see p. 1082).

1928, January 29. Treaty with Lithuania. This confirmed frontiers, including the status of Memel as part of Lithuania, and provided for arbitration.

1929, February 6. Germany Accepts the Kellogg-Briand Pact. (See p. 1124.)

1929, June 7. Young Plan Agreement. Germany accepted the plan for stabilizing her finances, subject to promise of complete Allied evacuation of the Rhineland by June 1930.

1929, September–1930, June. Allied Evacuation of the Rhineland.

1930, September 14. Nazis Gain in Reichstag Elections. This marked the emergence of Hitler's National Socialists as a major party. Hitler's program was opposed to all provisions of the Versailles Treaty, particularly reparations. A period of disorder followed, with numerous clashes between the National Socialists (Nazis) and the Communists.

1931, December 12. Allied Evacuation of the Saar.

1932, April 13. Chancellor Brüning Bans Nazi Storm Troops.

1932, June 16. Ban on Nazi Storm Troops Lifted. The National Socialist movement gained new momentum, thanks to this decision by the government of Chancellor **Franz von Papen.**

1932, July 20. Coup d'État in Prussia. Von Papen removed the Socialist prime minister and other officials. The cities of Berlin and Brandenburg were put under martial law, the activities of the Nazi Storm Troops having made it almost impossible for the civil authorities to maintain order.

1932, November 17. Resignation of von Papen. This was the result of growing disorders and political unrest.

1932, November 24. Hitler Refuses Chancellorship. Hindenburg offered him the post, but would not grant Hitler's demand for full power.

1933, January 30. Hitler Becomes Chancellor. He accepted the post after 2 months of political crisis.

1933, February 27. Reichstag Fire. Hitler denounced the fire as a Communist plot to disrupt new elections (scheduled for March 5). As a result, President von Hindenburg issued emergency decrees outlawing the Communist party and suspending the constitutional liberties of free speech and free press as well as other liberties. This enabled Nazi Storm Troops to intimidate all opponents without fear of legal opposition. The result was a National Socialist victory in the elections. Later evidence showed that the Nazis were responsible for the fire.

1933–1938. Nazi Dictatorship Firmly Established. All opposing political parties were liquidated under government pressure. The German state governments were shorn of all effective power, and Germany became a national rather than a federal state. The legal system was radically changed, and dangerously wide powers were given to the People's Court, which was set up in May 1934. Concentration camps were set up for detainment of political opponents. Racial pressures increased upon the Jewish population and, to a lesser extent, on the Christian churches protesting this racial persecution.

1933, October 14. Germany Withdraws from the League of Nations. (See p. 1124.)

1934, January 26. Treaty with Poland. This provided for nonaggression and respect for existing territorial rights for 10 years.

1934, June 30. The Great Blood Purge. Some 77 persons, many of them leaders high in the Nazi party, were executed because of an alleged plot against Hitler and the regime.

1934, July 25. Nazi Putsch in Vienna. (See p. 1133.)

1935, March 16. Hitler Denounces Disarmament Terms of Versailles Treaty. (See p. 1125.)

1935, June 18. Anglo-German Naval Agreement. (See p. 1125.)

1936, March 7. Reoccupation of the Rhineland. Simultaneously, Germany denounced the Lacarno Pacts of 1925. France, without support from Britain, decided not to intervene. Later evidence shows that German generals would have deposed Hitler and withdrawn from the Rhineland if France had responded with force.

1936, October 27. Establishment of the Berlin-Rome Axis. (See p. 1131.)

1936, November 25. Conclusion of the German-Japanese Anti-Comintern Pact. This was an extension of the Berlin-Rome alignment to counterweight the Franco-Russian Alliance.

1937, November 5. Hitler's Goals. At a top-secret meeting with heads of armed forces and his foreign minister, he announced his intentions of gaining for Germany *Lebensraum* (territorial expansion) in Europe by force of arms. Austria and Czechoslovakia were to be seized, then Poland; finally, the Soviet Union. Target date—1938–1943.

1938, March 12–13. The German Invasion and Annexation of Austria. (See p. 1134.) This added over 6 million Germans to the Reich and paved the way for future expansion of influence in the Danube River Valley.

1938, September 7–29. The German-Czech Crisis. (See p. 1136.) This culminated with the Munich Agreement, annexing over 3 million Sudeten Germans to Germany. Hitler's success made Germany the dominant power on the Continent, politically shattered the Little Entente, and broke down the French alliance system in Eastern Europe.

1939, March 10–16. Annexation of Bohemia and Moravia. This blatant violation of the Munich Agreement obliterated the Czechoslovak state and left Slovakia nominally independent.

1939, March 23. Annexation of Memel. Following this, rigid demands were made on Poland regarding Danzig and Pomorze ("Polish Corridor"). Polish firmness, backed by a pledge of Anglo-French aid (March 31; see p. 1125) deterred immediate German action.

1939, May 23. Hitler Plans Attack on Poland. He informed his military commanders at a secret conference. He discounted danger of a conflict with England and France. A small special staff was set up in the General Staff (OKW) to implement the planning. Secret negotiations were begun for a nonaggression pact with the U.S.S.R. Military and naval preparations included calling of a quarter-million reservists to the colors for "training," and preparation to send the pocket battleships *Graf Spee* and *Deutschland* to sea, together with 21 submarines.

1939, June–August. Tense Relations with Poland. Frequent border incidents caused repeated warnings from France and England.

1939, August 22. Hitler Orders Invasion of Poland. While Foreign Minister **Joachim von Ribbentrop** rushed to Moscow to conclude a nonaggression pact, Hitler gathered his military commanders in a secret meeting at Berchtesgaden to order preparation for war. Troop concentration began behind the Polish border. Warships put to sea.

1939, August 23. German-Russian Pact Signed at Moscow.

1939, August 31. Gleiwitz Incident. A group of SS men, disguised in Polish uniforms, seized the radio station at Gleiwitz in Upper Silesia, just inside Germany, broadcast threats of invasion in the Polish language, then disappeared. One dying man (actually an inmate of a German concentration camp) was left behind as "evidence" of Polish aggression. The incident, long planned, was Hitler's fig leaf of respectability before world opinion.

1939, September 1. Germany Invades Poland. World War II began (see p. 1149).

1940, September 27. Rome-Berlin-Tokyo Axis. A 10-year mutual assistance alliance.

AUSTRIA

1927. Private Political Armies. The Christian Socialists organized their private army, the Heimwehr, while the Social Democrats organized the Schutzbund.

1930, February 6. Friendship Treaty with Italy.

1933. Nazi Unrest. This was stimulated by the National Socialists' triumphs in Germany. Nazi agitators staged occasional demonstrations until the Nazi party was dissolved (June 19). Campaigns of agitation and terrorism continued.

1934, July 25. Assassination of Prime Minister Engelbert Dollfuss. Growing Austro-German enmity was culminated by a Nazi *Putsch* in Vienna, resulting in the murder of Dollfuss. German intervention was stymied by the mobilization of Italian and Yugoslavian forces on the border. Hitler disavowed connection with the murder and repudiated his Austrian followers.

1936, July 11. German-Austrian Agreement. Mussolini instigated this agreement to assure himself of German support.

1936, October 10. Prime Minister Kurt Schuschnigg Assumes Dictatorial Pow- **ers.** He dissolved the Heimwehr and established a **Fatherland Front** militia. Renewal of Nazi activity led Schuschnigg to seek pledges of support from France and members of the Little Entente.

1938, March 11–12. German Invasion. Hitler took formal possession in Vienna, annexing Austria to Germany. A reign of terror followed.

THE SCANDINAVIAN AND BALTIC STATES

Denmark, Norway, and Sweden

With the rebirth of Germany as a powerful military state, Scandinavian efforts to achieve mutual cooperation and solidarity were intensified and collective security became a common goal.

Finland and the Baltic States

FINLAND

Finland followed primarily the same policies as the other Scandinavian states. Recognizing the altered situation in the Baltic region after Hitler's ascendancy, she attempted to maintain a balance between Germany and the U.S.S.R.

THE BALTIC STATES

The successes of Nazi Germany prompted Lithuania, Latvia, and Estonia to improve their relations with the Soviet Union for the purpose of establishing a strong front against possible German intervention in the name of the German minorities. All 3 Baltic states soon adopted some form of dictatorship in order to offer stronger regimes for resistance to Germany.

SOVIET RUSSIA

1924–1926. Power Struggle. Conflicts among Communist party leadership followed the death of Lenin (January 21, 1924).

1926, July–October. Triumph of Stalin. Trotsky, Zinoviev, Radek, and others were expelled from the political bureau of the party. Trotsky was later banished (January 1929).

1926, August 31. Treaty with Afghanistan. (See p. 1103.)

1929, February 9. Litvinov Protocol. This was an analogue of the Kellogg-Briand Pact, signed by the U.S.S.R., Poland, Rumania, Estonia, and Latvia at Moscow.

1929, December 22. Agreement with China. This settled disputed claims to the Chinese Eastern Railway (see p. 1144).

1932, July 25. Nonaggression Pacts with Poland, Estonia, Latvia, and Finland. A similar agreement with France followed (November 29). Uneasiness about the deterioration of relations with Japan prompted Stalin's government to amend relations with its European neighbors. Entrance into the world's political arena was further signaled by the U.S.S.R.'s active participation in the Disarmament Conference (see p. 1124) and general international cooperation.

1933, November 17. The United States Recognizes the U.S.S.R. Trade relations were opened.

1934, April 4. Extension of Nonaggression Pacts with Poland and the Baltic States. These became 10-year agreements.

1934, September 18. The U.S.S.R. Joins the League of Nations. She became an active exponent of collective security by supporting a French plan for an Eastern European pact.

1935, May 2. Franco-Russian Alliance. This was rightly regarded in Germany as a pact against Hitler's regime.

1935, May 16. Russian-Czechoslovakian Alliance. This further angered the Germans, who feared that the Soviets would make Czechoslovakia an air base for operations against Germany.

1935, July 25–August 20. Meeting of the Third International. The U.S.S.R. decided to side with the democracies against the Fascist states. Communist opposition to military appropriations in other countries was to cease and the governments were to be supported.

1936, July. Soviet Support to Loyalist Spain. (See p. 1127.)

1938–1939. Undeclared War with Japan. (See p. 1142.)

1938, September. German-Czech Crisis. The U.S.S.R. proffered assistance to the Czechs in resisting German demands (see below). Britain and France favored appeasement rather than Communist support and reached a compromise solution at Munich. As a result, the Franco-Soviet Alliance was all but abrogated and the U.S.S.R. was almost isolated in Europe.

1939, March–June. Britain Negotiates with Russia. The German annexation of Czechoslovakia and Memel caused this reversal in British policy. The U.S.S.R. was urged to join a common peace front against attack upon Poland and Rumania. The Russians negotiated for a complete offensive alliance as well as for guarantees for the Baltic States.

1939, May 3. Soviet Policy Changes. Foreign Minister **Maxim Litvinov,** generally considered to be pro-West, was suddenly replaced by **Vyacheslav Molotov,** whose public criticisms of Britain and France implied that the U.S.S.R. was considering a change of policy. Negotiations with England and France continued, but the U.S.S.R. voiced her distrust of the western powers by rejecting draft after draft at the conference table (June–August).

1939, August 27. Nonaggression Pact with Germany. This confirmed western suspicions of an unbelievable diplomatic union between 2 incompatible enemies.

1939, September 1. German Attack on Poland. The Soviet Union was a benevolent neutral.

1939, September 17. Soviet Invasion of Poland. That country was partitioned with Germany as the result of a secret treaty clause. (See p. 1151.)

1939, November 30. The U.S.S.R. Attacks Finland. (See p. 1154.)

1939, December 14. The U.S.S.R. Expelled from League of Nations.

1940, June 15–16. Soviet Occupation of Lithuania, Latvia, and Estonia.

1941, April 13. Nonaggression Treaty with Japan. Stalin's desire to assure peace in the East, in light of growing danger of German invasion, coincided with Japanese desires to obtain similar freedom to initiate adventures in Southeast Asia (see p. 1231).

CZECHOSLOVAKIA

1933, February. Reorganization of Little Entente. This was because of the potential danger of the new National Socialist Germany.

1933–1938. Subversion of Sudetenland. The 3 million Sudeten Germans in Czechoslovakia became a hotbed for Nazi agitation.

1935, May 16. Mutual Assistance Pact with the U.S.S.R. This guaranteed Soviet support against attack if it were preceded by French support (see above).

1936, September 10. Beginning of German Propaganda Campaign. Czechoslovakia was accused of basing Soviet planes for operations against Germany. The Czechs intensified their frontier defense program.

1938, March–September. Growing German-Czech Crisis. Recognizing the strategic danger created by the German annexation of Austria, Czechoslovakia undertook partial mobilization and a series of negotiations with the Sudeten German leaders.

1938, September 15. Hitler Demands Cession of Sudetenland. He threatened war otherwise. Chamberlain flew to Berchtesgaden in an attempt to mediate a peaceful settlement.

1938, September 24–29. Czechoslovakian Mobilization.

1938, September 29. Conference and Agreement at Munich. Hitler was accorded practically all his demands as Czechoslovakia was dismembered by Hitler, Ribbentrop, Mussolini, Ciano, Chamberlain, and Daladier. Czechoslovakia was deserted by the Little Entente and the larger powers. Only the U.S.S.R. (not represented at Munich) appeared ready to aid against German aggression. France and Britain, cognizant of their military inadequacies and unpreparedness (especially in air power), acquiesced in a move establishing German hegemony in Central Europe and subsequently the Danubian area.

1939, March 10–16. Hitler Annexes Remainder of Czechoslovakia. Britain and France were shocked by this breach of the Munich Agreement. Yet no resistance was offered; Hitler emerged undisputed victor of a bloodless battle.

HUNGARY

1927, April 5. Friendship Treaty with Italy.

1938, March. German Annexation of Austria. This brought the Reich to the frontier and increased the restlessness of the German minority in Hungary.

1938, September. German-Czech Crisis. Hungary supported the Germans, and was awarded 5,000 square miles of territory in southern Slovakia for her cooperation with Germany (November 2). Occasional conflict in the annexed region continued throughout the winter of 1938–1939.

1939, March 15. Annexation of Carpatho-Ukraine (Ruthenia). This result of cooperation with Germany in the destruction of Czechoslovakia gave Hungary a long-sought-after common frontier with Poland.

1939, April 11. Hungary Withdraws from the League of Nations.

POLAND

1926, May 12–14. Military Coup. Marshal Jósef Pilsudski established a dictatorship.

1929, February 9. Litvinov Protocol. (See p. 1134.)

1934, January 26. German-Polish Nonaggression Treaty. (See p. 1132.)

1938, October 2. Occupation of Teschen. Poland settled her old dispute with Czechoslovakia during the German-Czech crisis.

1939, March–April. German Demands for Polish Corridor. The German annexation of Memel (March 23) was accompanied by stiff demands on Poland for annexation of Danzig and the construction of an extraterritorial motor road through Pomorze in return for a German guarantee of Polish independence. These demands provoked the Anglo-French pledge of aid to Poland, which later became a pact of mutual assistance (April 6).

1939, April 28. Hitler Denounces the German-Polish Agreement of 1934 and the Anglo-German Naval Agreement of 1935.

1939, June–August. A Period of Rising Tension. (See p. 1133.)

1939, September 1. German Invasion of Poland. (See p. 1149.)

RUMANIA

1926, March 26. Alliance with Poland.

1926, June 10. Treaty of Alliance and Nonaggression with France.

1926, September 16. Friendship Treaty with Italy.

1929, February 9. Litvinov Protocol. (See p. 1134.)

1933, July. Nonaggression Pact with the U.S.S.R. This implied Soviet recognition of the Rumanian possession of Bessarabia.

1934, February 8–9. Balkan Pact. This treaty, concluded between Rumania, Yugoslavia, Greece, and Turkey, bound the signatories to consult with each other if their security was threatened.

1934, June 9. Mutual Guarantees with Czechoslovakia and the U.S.S.R. They agreed to guarantee each other's frontiers.

1939, April 13. Britain and France Guarantee Rumanian Independence. This followed the German annexation of Czechoslovakia. Soon afterward, Rumania straddled the fence and concluded a commercial agreement with Germany.

1939, September 4. Rumania Declares Its Neutrality.

BULGARIA

1925. Greek-Bulgarian Border Clashes. The League of Nations' mediation temporarily settled the crisis (October 21). Further incidents in 1931 (January, February) necessitated intervention by the powers.

1929, March 6. Friendship Treaty with Turkey.

1934, February 9. Bulgaria Rejects the Balkan Pact. She thus indicated that she did not endorse the *status quo* in the Balkans. (See above.)

1934, May 19. Military Coup d'État. The resulting military dictatorship was soon overthrown by King **Boris,** who established a royal dictatorship.

1937, January 24. Friendship Treaty with Yugoslavia.

1938, July 31. Rearmament Program. Bulgaria's right to rearm was recognized by the Balkan Entente through agreement with Greece. An Anglo-French loan of $10 million financed the program.

YUGOSLAVIA

1926, September 18. Friendship Treaty with Poland.

1927, May. Border Clashes with Albania.

1927, November 11. Treaty of Friendship with France.

1929, January 5. King Alexander Proclaims Dictatorship. This was the result of internal disorders, particularly Croatian nationalistic agitation.

1934, February 9. Balkan Pact. (See above.)

1934, October 9. Murder of King Alexander. The murderer, a Macedonian revolutionary, was believed to be in Hungarian pay, and the slaying brought Yugoslavia and Hungary to the verge of war. The League of Nations arranged a settlement.

1937, January 24. Treaty of Friendship with Bulgaria.

1937, March 25. Nonaggression and Arbitration Pact with Italy. This indicated closer ties with the Axis powers.

GREECE

1925, December 4. Greek-Bulgarian Border Clashes. (See above.)

1928, September 23. Friendship Treaty with Italy.

1930, October 30. Treaty of Ankara. This pact with Turkey advocated naval parity in the eastern Mediterranean and the acceptance of territorial *status quo.*

1933, September 15. Ten-Year Nonaggression Pact with Turkey.

1934, February 9. Balkan Pact. (See above.)

1935. Military Coup. King **George II** was restored to the throne.

1939, April 13. British and French Pledge of Support against Aggression. This followed the Italian conquest of Albania.

ALBANIA

1925, January 21. Albania Proclaimed a Republic.

1926, November 20–26. Rebellion in the North. Despite Yugoslavian aid to the rebels, this was suppressed.

1926, November 27. Treaty of Tirana. Italy and Albania pledged themselves to the maintenance of territorial *status quo.*

1927, November 22. Second Treaty of Tirana. This put Albania under Italian protection by establishing a 20-year defense alliance and providing for military cooperation. Italy obtained important rights, especially in oil, road construction, military supervision, and education.

1928, September 1. President Ahmed Bey Zogu Proclaimed King.

1934. Growing Friction with Italy.

1937, May 15–19. Rebellion in the South. This was in protest against the dictatorial measures of King **Zog.** It was suppressed.

1939, April 7. Italian Invasion. Albania was soon occupied and annexed by Italy.

AFRICA

EGYPT AND THE SUDAN

1927. Draft Treaty with Great Britain Rejected. This proposed British military occupation for 10 years, but was rejected by the Egyptian Parliament as inconsistent with Egyptian independence.

1931, April 22. Friendship Treaty between Egypt and Iraq.

1936, August 26. Treaty between Egypt and Great Britain. This granted full independence to Egypt, while retaining the minimum requirements for the strategic security of the British Empire. A 20-year defensive alliance provided for the withdrawal of British troops, except for 10,000 men of land and air forces who would be assigned to the Suez Canal Zone. England was permitted to maintain a naval base at Alexandria for a maximum of 8 years, and Egyptian troops were to return to the Sudan.

1937, May 26. Egypt Admitted to the League of Nations.

ETHIOPIA

1928. Coup. Ras Tafari (see p. 1100), backed by Harer provincial troops, mounted a successful coup and coerced Zauditu into yielding to him most executive power. Although Zauditu remained titular *neggus* until her death (1930), Ras Tafari had himself crowned *neggus* as **Haile Selassie I.**

1928, August 2. Twenty-Year Friendship Treaty with Italy. This gave Ethiopia a free zone in the Italian-controlled port of Assab in return for the concession of constructing certain roads.

1929. Reorganization of the Ethiopian Army. This was to be done by a Belgian military commission.

1934, December 5. Clash at Ualual. Ethiopian and Italian forces clashed in a disputed zone on the Italian Somaliland border. Approximately 100 Ethiopians and 30 Italian colonial troops were killed in the incident. The Ethiopian government requested an investigation of the incident and the Italians demanded reparation.

1935, September 3. Arbitration by the League of Nations. The arbitral board was unable to establish the responsibility for the clash at Ualual.

Ethiopian-Italian War, 1935–1936

1935, October 3. Outbreak of Hostilities. The Italians invaded Ethiopia without a declaration of war. Well supported by artillery and air forces, they captured Aduwa (October 6).

1935, October 7. Italy Declared to Be an Aggressor. The League debated the imposition of sanctions.

1935, November 8. Italians Capture Fortress of Makalle.

1935, November 18. Sanctions Imposed on Italy. Fifty-one nations joined in embargoes on arms, credit, and raw materials and the restriction of imports. The failure of the League to apply oil sanctions, to deny the movement of Italian troops and matériel through the Suez Canal, or to prevent the German reoccupation of the Rhineland (see p. 1132) gave Mussolini a free hand in Ethiopia with a united Italy backing him.

1935, December–1936, April. Lull in Operations. Field Marshal **Pietro Badoglio** reorganized the expeditionary forces.

1936, April–May. Italy Renews Offensive. Italian air forces, which were unimpeded, systematically spread terror and destruction on a brave but completely outclassed enemy, who fought in a medieval style against bombings and poison gas. Badoglio's ground forces advanced.

1936, May 5. Capture of Addis Ababa. Emperor **Haile Selassie I** fled; Ethiopian resistance collapsed.

1936, May 9. Italy Annexes Ethiopia. It was added to Eritrea and Italian Somaliland to make up Italian East Africa. The Italian king assumed the title of "Emperor of Ethiopia." Germany, Austria, and Hungary recognized the conquest at once, England and France a year later.

COMMENT. *The most important result of this war was to give both Mussolini and the Italian nation a much exaggerated opinion of their military prowess.*

NORTH AFRICA (MOROCCO, ALGERIA, TUNISIA, AND LIBYA)

The North African colonies suffered acutely from the economic depression of the 30s. The severity of the situation fostered the growth of unrest and nationalism.

WEST AFRICA

This was a period of consolidation of European colonial administrations.

LIBERIA

1942, March 31. Agreement with the U.S. Only native African state to retain independence during the colonial era, Liberia granted base rights to U.S. forces, both for air transit and for operations against German submarines.

SOUTH AFRICA AND SOUTHWEST AFRICA

In South Africa, the Union was established as a sovereign independent state by the Westminister Statue (1934).

MIDDLE EAST AND SOUTHWEST ASIA

TURKEY

1925, February–April. Insurrection in Kurdistan. Finally suppressed by **Mustafa Kemal.**

1925, December 17. Alliance with the U.S.S.R.. Close political and economic relations were established.

1928, May 30. Five-Year Nonaggression Pact with Italy.

1928, June 15. Treaty with Persia.

1929, March 6. Treaty with Bulgaria.

1929, December 17. Soviet Alliance Renewed. But Mustafa Kemal continued firm repression of communism.

1930, October 30. Treaty of Ankara. After an exchange of populations, Turkey and Greece agreed to territorial *status quo* and naval parity in the eastern Mediterranean.

1930, December 23. Dervish Revolt in Western Anatolia. This protest against Mustafa Kemal's westernization was soon suppressed.

1931, March 6. Russian-Turkish Naval Agreement. This prohibited any changes in the respective Black Sea fleets without 6 months' notice.

1931, October 30. Five-Year Extension of the Turco-Soviet Alliance.

1931, May 25. Five-Year Pact with Italy.

1931, July 18. Turkey Joins the League of Nations.

1932, January 23. Agreement with Persia. Outstanding border disputes were settled.

1933, September 15. Ten-Year Turco-Greek Nonaggression Pact.

1934, February 9. Balkan Treaty. A mutual security agreement of Balkan frontiers between Greece, Rumania, and Turkey.

1934, May. Turkish Rearmament. Turkey distrusted Italy's east Mediterranean policy.

1936, July 20. Treaty of Montreux. An international conference approved Turkey's request to refortify the straits.

1937, July 9. Southwest Asian Treaty. A nonaggression pact (**Saadabad** Pact) with Iran, Iraq, and Afghanistan in an Asiatic analogue of the Balkan Pact.

1937, December-1938, July. Alexandretta Crisis. Turk claims to Alexandretta brought tension and threat of war. The crisis ended in agreement to hold elections, which gave Turkey virtual control.

1938, November 10. Death of Kemal Atatürk.

1939, May 12. British-Turkish Mutual Assistance Agreement. This identified Turkey with the British bloc.

1939, June 23. Agreement with France. This resulted in the incorporation of Hatay (Alexandretta) into Turkey.

THE ARAB STATES

Arabia

1925–1930, May 20. Ibn Saud Consolidates Power. He now controlled most of Arabia.

1930, February 22. Peace in Northern Arabia. King Ibn Saud of Hejaz and Nejed made peace with King Faisal of Iraq (former Emir of Hejaz; see pp. 1080, 1101) on board HMS *Lupin* in the Persian Gulf, ending part of the Saudi-Hashemite dynastic feud.

1932, May–July. Unrest in Northwest Arabia. This came from opposition to the pro-western policies of Ibn Saud.

1932, September 22. Adoption of the Name Saudi Arabia.

1933, July 27. Treaty with Transjordan. This terminated years of dynastic animosity between the Saud and Hashemite families.

1934, February 14. Treaty of Sanaa. A 40-year treaty was signed with Great Britain.

1936, April 2. Nonaggression Treaty with Iraq. This became the basis of efforts to achieve Arab brotherhood and unity.

1936, May 7. Treaty with Egypt. This advanced Pan-Arabism and brought closer political cooperation.

Syria and Lebanon

1925, July 18. Beginning of Druse Rebellion. Under **Sultan el-Atrash** the Druses threatened French control.

1925, October. Bloodshed at Damascus. The French withdrew from the city (October 14). A 2-day bombardment (October 18–19) of the city, with air and tank attacks, took several hundred civilian lives.

1926, July 18. Druse Rebellion Again Sweeps Damascus. French forces, in fortified encampments outside the city, again unleashed a 48-hour period of artillery and aerial bombing, inflicting great damage and loss of life.

1927, June. Collapse of Druse Rebellion. Despite an amazing display of bravery, which included charges of horsemen against French tanks, the rebellious tribesmen were finally subdued, and Druse leaders fled to Transjordan.

1927, May 23. French Declare Lebanon a Republic.

1936, September 9. Franco-Syrian Treaty of Friendship and Alliance. This and a Franco-Lebanon treaty (November 13, 1936) brought comparative peace to the area. Syria and Lebanon were to be independent states after a 3-year period; France was to retain a privileged position for 25 years.

1937, September 8. Pan-Arab Conference in Syria. This was held to organize the defense of Arab interest in Palestine. Syria became a center of Palestine rebel activity.

Palestine and Transjordan

Jewish-Arab enmity kept the area in a turmoil with continuous fighting during this period. In Palestine, Great Britain was trying without much success to satisfy the opposite aspirations of Arabs and Zionists by an implementation of the Balfour Declaration (November 2, 1917) and the establishment of separate Jewish and Arab states.

1923, May 26. Autonomy for Transjordan. This Hashemite kingdom was relatively stable during the years preceding World War II because of the wisdom of King **Abdullah,** and because it was exempted from the clauses in the British mandate dealing with the establishment in Palestine of a national home for the Jewish people.

1936–1939. Arab Revolt. This resulted from opposition to a partition of Palestine. The Arabs controlled the country in most parts outside the large cities and the Jewish settlements. The revolt was suppressed by the British only after the loss of several thousand lives.

1932, April–June. Renewed Kurd Rising. This was suppressed by the Iraqi forces backed by the British air patrol.

1932, October 3. Iraq Admitted to the League of Nations.

1936, April. Nonaggression Treaty with Saudi Arabia.

1936, October 29. Military Revolt. General **Bakir Sidqi** established a military dictatorship. His assassination (1937) ended the direct intervention of the army in politics.

1937, July 8–9. Treaty of Saadabad. (See p. 1140.) Pact with Iran, Turkey, and Afghanistan.

Iraq

1930, November 16. Independence of Iraq. Great Britain agreed to support her admission to the League of Nations in 1932. Britain was to obtain a lease on new air bases, the use of Iraqi transportation and communication facilities, and British officers were to train the Iraqi Army.

1930, September 11–1931, April. Kurd Rebellion. Sheikh **Mahmud** led the uprising, which was eventually suppressed with British assistance.

Iran (Persia)

1926, April 22. Treaty with Turkey and Afghanistan.

1930, June–July. Kurd Uprising. This stimulated Persian-Turkish efforts to establish an agreed boundary.

1932, January 23. Treaty with Turkey. The border was revised around Mt. Ararat and relations between Turkey and Persia improved.

1937, July 8–9. Treaty of Saadabad. (See p. 1140.) Pact with Iraq, Turkey, and Afghanistan.

SOUTH ASIA

AFGHANISTAN

1928, November. Tribal Insurrection. This forced the abdication of the king.

1929, January–October. Civil War. A bandit leader, **Habibullah Ghazi,** captured Kabul (January), but was defeated and executed by General **Mohammed Nadir Khan,** who took the name **Nadir Shah.** Aided by the British, he reformed his army and restored stability.

1933, November 8. Assassination of Nadir Shah. He was succeeded by his son **Mohammed Zahir Shah.**

1934. Afghanistan Joins the League of Nations.

1937, July 8–9. Treaty of Saadabad. (See p. 1140.) Pact with Iran, Iraq, and Turkey.

INDIA

The years 1923–1932 witnessed political assassinations and renewed internationalist terrorist activity. In 1930, Gandhi began a second Civil Disobedience Campaign. This precipitated rioting, violence, and numerous arrests, which eventually led to the Government of India Act of 1935: Burma and Aden were separated from India and became crown colonies; Indian local government was reorganized.

EAST ASIA

JAPAN

1925, January 20. Treaty with the U.S.S.R. Diplomatic relations were established with the U.S.S.R. Japan evacuated North Sakhalin.

1927–1929. Japanese Interventions in Shantung. (See pp. 1143, 1144.)

1930, April 22. London Naval Treaty. (See p. 1124.) This was finally ratified by Japan, despite strong political and naval opposition (October).

1931, September 19. Mukden Incident. (See p. 1145.) Beginning of Japanese conquest of Manchuria.

1932, January–March. First Battle of Shanghai. (See p. 1145.)

1932, February 18. Nominal Independence of Manchukuo. This made Manchuria a virtual colony of Japan.

1933, May 27. Japan Withdraws from the League of Nations. (See p. 1124.)

1933, May 31. Japanese Invasion of Jehol. (See p. 1145.)

1936, February 26. Mutiny. A group of young army officers, impatient at apparent hesitation of politicians to press ahead with the conquest of China, attempted to set up a military dictatorship. Finance Minister **Makoto Saito** and several other high officials were assassinated. The rebellion was promptly suppressed.

1936, November 25. Anti-Comintern Pact with Germany. (See p. 1132.)

1937, July 7. Outbreak of War in China. (See p. 1229.)

1938, July 11–August 10. Undeclared Hostilities with the U.S.S.R. Severe fighting broke out because of a dispute over the poorly defined frontier where Manchuria, Korea, and Siberia meet. **Changkufeng Hill,** near the mouth of the Tumen River, had been occupied and fortified by Soviet troops. Japanese efforts to dislodge the Soviets failed; a truce ended the episode, with the Soviets retaining the hill.

1939, May–September. Nomonhan (Khalkin Gol) Incident. Renewed Japanese-Soviet hostilities resulted from a frontier dispute near the Khalkin Gol (River) claimed by the Japanese as the boundary between Manchuria and easternmost Outer Mongolia. After Soviet troops occupied the disputed territory between the river and Nomonhan, 20 kilometers to the east, the Japanese attacked with a reinforced division and were initially successful. In mid-August a Soviet counteroffensive by 3 divisions, 5 armored brigades, and some Mongolian units—commanded by General **Georgi K. Zhukov**—drove the outnumbered Japanese (28,000 to 65,000) back to Nomonhan. Japanese Kwantung Army commander, General **Kenkichi Ueda,** concentrated 3 fresh divisions for a counteroffensive, but a ceasefire agreement, forced cancellation (September 15). The Soviets admitted 9,824 casualties, but their losses were probably comparable to the Japanese admitted 17,405 casualties. This dispute and the Changkufeng dispute (see above) were settled by treaty (June, 1940).

1939, August 23. Japan Renounces the Anti-Comintern Pact with Germany. This result of outraged reaction to the Nazi-Soviet Treaty (see p. 1133) undoubtedly contributed to the Japanese decision to end hostilities with the U.S.S.R. in Outer Mongolia.

1941, April 13. Treaty with the U.S.S.R. (See p. 1231.)

CHINA

1925. "May Thirtieth Incident." The British used gunfire to break up student demonstrations in Shanghai (May 30) and Canton (June 23). As a result, a strike and boycott of British goods were in effect for over a year. A wave of antiforeignism swept the country.

1926, July. Nationalists Begin Northern Offensive. The Canton-based National Government sent its army, under General **Chiang Kai-shek,** with the military advice of Russian General **Vasily K. Blücher** (known as **B. K. Galin** to the Chinese), to the north to unify the country. Chiang advanced into the territory of warlord **Wu Pei-fu. Hankow** was captured (September 6); **Wuch'ang** was besieged (later captured, October 10). The National Government moved from Canton to Hankow. Chiang turned east into territory of warlord **Sun Ch'uan-fang,** in the lower Yangtze Valley. **Nanking** was captured by the Nationalists (March 24), amidst Communist-formented riots. Six foreigners were killed; the Chinese portion of Shanghai was seized by local Communists and trade unionists in the name of the National Government. (The International Settlement was protected by an international force of 40,000, mostly Japanese.)

1927, April 12. Chiang's Seizure of Shanghai. Suspicious of subversive collusion between Communist leaders in Shanghai and the radical elements (Communists and Kuomintang leftists) in the National Government at Hankow, Chiang seized the Chinese portion of Shanghai. All known Communists and trade union leaders (about 5,000) were killed.

1927, April 18. Split in Kuomintang. Denounced for the Shanghai incident by the leftist and Communist-influenced Hankow government, Chiang established a separate National Government at Nanking.

1927, April–May. Nationalist Campaigns against Northern Warlords. Independent campaigns north of Yangtze were fought by Hankow and Nanking regimes, against the northern warlords, now joined by Marshal **Chang Tso-lin,** warlord of Manchuria and northeast China. Hankow troops fought inconclusively against Wu Pei-fu and **Chang Hsueh-liang** (son of Chang Tso-lin) in northern Hupeh and southern Honan. Chiang advanced steadily through Anhwei toward Hsuchow against Sun Ch'uan-fang.

1927, June. Intervention of Feng Yu-hsiang. The so-called "Christian General" advanced southeastward from Shensi through the T'ungkuan Pass. Wu and Chang hastily retreated to northern Honan.

1927, June 21. Alliance of Feng with Chiang. Feng, visiting Chiang at recently captured Hsuchow, agreed to support Chiang's Nationalist regime. Temporary Japanese occupation of Shantung (May–June) blocked Chiang's northward advance.

1927, July. Hankow Nationalist Break with the U.S.S.R. and Communists. Aware of Communist plans to seize control, Kuomintang leaders of the Hankow regime purged Chinese Communists from the government and expelled Soviet political and military advisers.

1927, August 1. Nanchang Insurrection. Following the purge at Hankow, Communist elements of Hankow forces mutinied at Nanchang, hoping to spark a nationwide Communist revolution. Leaders were Generals **Yeh T'ing, Ho Lung,** and **Chu Teh.** They were driven out by loyal troops, and pursued by cooperating Hankow and Nanking Nationalist troops to southeast China coast, where the rebels dispersed. Chu Teh escaped to mountains of western Kiangsi with one small organized remnant. This was the real beginning of the 22-year Chinese Civil War.

1927, August–September. "Autumn Harvest Uprising." Attempts of Communist leader **Mao Tse-tung** to organize a peasant revolt in Hunan failed. He fled to the mountains of western Kiangsi, where he was later joined by Chu Teh.

1927, August 8. Chiang Resigns. Despite pleas of his adherents, Chiang refused to seek accommodation with Kuomintang radicals at Hankow. To permit party peace, and a unified front against resurgent northern warlords, he resigned and went to Japan. Soon afterward the Hankow regime moved to Nanking to join the regime there.

1927, September–October. Northern Counteroffensive. Taking advantage of the confusion in the south caused by Communist uprisings and the Kuomintang split, the northern warlords began a southward drive into Yangtze Valley. Sun Ch'uan-fang with 70,000 troops crossed the Yangtze west of Nanking, but was defeated by Nationalist General **Li Tsung-jen,** aided by Nationalist river gunboats, in the 5-day **Battle of Lungtan.** Sun, after losing 20,000 killed and 30,000 prisoners, retreated to Hsuchow. The warlords' offensive was called off.

1927, December 11–15. Canton Commune. A Communist uprising in Canton was ruthlessly suppressed by Nationalist troops.

1928, January 6. Chiang Returns. Reappointed Commander in Chief of the Nationalist Army, and Chairman of the Kuomintang Central Executive Committee, Chiang quickly restored stability, and prepared to renew the northern offensive to unify the nation.

1928, April 7–June 4. Nationalist Northward Drive. Under field commanders Li Tsung-jen, Feng Yu-hsiang, **Pai Ch'eng-hsi,** and **Ho Ying-ch'in,** and the newly allied warlord of Shansi, **Yen Hsi-shan,** the Nationalist armies, some 700,000 strong, defeated the three opposing northern warlords, who had about 500,000 troops, and advanced across the Yellow River. Despite Japanese interference (see below) the Nationalists continued on to capture **Peking** (June 4). The name of the city (which had meant "Northern Capital") was changed to Peiping (meaning "Northern Peace"). Retreating to Manchuria, Chang Tso-lin was assassinated when the Japanese blew up his private train near Mukden (June 4). His son, "Young Marshal" Chang Hsueh-liang, became warlord of Manchuria, and acknowledged Nationalist authority.

1928, May 3–11. Sino-Japanese Clash at Tsinan. The Japanese, again claiming special interests in Shantung, drove out the Nationalists and seized most of the province. Most Japanese troops were withdrawn a year later (May 20, 1929), after an agreement with the Chinese.

1928–1930. Consolidation. Chiang attempted to strengthen China for expected war with Japan. Simultaneously, the Communists were recovering strength in the mountain regions of Kiangsi and Fukien.

1929–1930, October–January. Dispute with Russia. As a result of conflicting claims to ownership and control of the Chinese Eastern Railway, Soviet troops invaded Manchuria, forcing China (and in particular the warlord governor, Chang Hsuehliang) to acknowledge that the U.S.S.R. retained Imperial Russia's share in control of the line. Russian troops withdrew.

1930, July–August. Li Li-san's Communist Revolt. Communist Party Chairman Li Li-san, believing that the time was ripe for a revolt of the Chinese urban proletariat, ordered the Communist guerrilla forces in Kiangsi and Fukien to seize the principal cities of central China. Mao Tse-tung, believing that Communist success depended upon a peasant uprising rather than an urban revolt, protested, but was overruled. Troops under Communist General **P'eng Teh-huai** briefly seized **Ch'ang-sha** (July 28), then withdrew. Li ordered a renewed and reinforced offensive against Ch'angsha, but forces under P'eng, Mao Tse-tung, Chu Teh, and Ho Lung were bloodily repulsed by reinforced Nationalist defenders. The Communists retreated to their mountain strongholds. Li was soon recalled to Moscow. Mao's influence increased.

1930–1934. Nationalist Anti-Communist "Extermination Campaigns." (Also called "bandit suppression" campaigns.) The first two efforts (December 1930–January 1931; and April–May 1931) were repulsed by aggressive Communist guerrilla tactics. Chiang personally led the third (July–September 1931), and Nationalists were converging on the Communist capital of Juichin when word of the Mukden Incident (see below) caused Chiang to halt the offensive. After agreement was reached with Japan, a peripheral offensive against Oyuwan Soviet (Anhwei-Honan-Hupeh border area) was successful (summer, 1932). The fourth main campaign (April–June 1933), was also disrupted by need to respond to renewed Japanese activity in the north, and by skillful Communist exploitation of this distraction. The fifth campaign (December 1933–September 1934) began after careful preparation, assisted by advice from a German military mission, led by General **Hans von Seeckt,** recently retired Chief of the German General Staff. This well-coordinated, converging offensive, using entrenchments and a chain of blockhouses, was successful. Communist losses were heavy. The Communist Central Committee approved Mao Tse-tung's recommendation to evacuate the area.

1931, September 19. The Mukden Incident. Alleging that the Chinese had plotted to blow up the railroad from Port Arthur to Mukden, Japan's Kwantung Army in "night maneuvers" seized the arsenal at Mukden and adjacent towns. Chinese troops were forced to withdraw. The Japanese continued their aggressive movements and in a few months all Manchuria was under their domination (February 1932). The only effective weapon the Chinese had was the boycott, which they invoked at great cost to Japanese trade.

1932, January 28–March 4. First Battle of Shanghai. In a move to stop the Chinese boycott, a Japanese army, 70,000 strong, landed at Shanghai. In a surprisingly effective and valiant resistance, the Chinese 19th Route Army held up the Japanese near the waterfront for about a month, but were finally driven out of their positions in the vicinity of the International Settlement. China agreed to end the boycott.

1932, February 18. Manchukuo (Manchuria) Declared Independent. The Japanese placed **Pu Yi** (former Emperor of China) on the throne of a puppet state, and announced a protectorate over it.

1933–1937. Growing Tension between China and Japan. Chiang Kai-shek attempted to unify and modernize his backward nation (while simultaneously fighting Communist and dissident warlords) in the face of increasingly aggressive Japanese actions.

1933, January–March. Japanese Invasion of Jehol. The pretext was that this Inner Mongolian province was really part of Manchuria. When Peiping (Peking) was threatened, the Chinese signed an armistice at Tangku (May 31), which required their evacuation of the Tientsin area and establishment of a demilitarized zone in eastern Hopei.

1934–1935. The Long March. The Communists, finally driven from their position by Chiang's Nationalist forces, organized a long retreat (October 1934), marching and fighting (against sporadic opposition) across southern and western China to northern Shensi. The longest distance, 6,000 miles, was covered by Chu Teh's First Front Army—accompanied by Mao Tse-tung as political commissar—in 13 months. This was the longest and fastest sustained march ever made under combat conditions by any army of foot troops, and has been exceeded in rate of march over a long distance only by a few Mongol expeditions of the 13th century. The Second Front Army, commanded by Ho Lung, and the Fourth Front Army, commanded by **Hsu Hsiang-ch'ien,** marched somewhat shorter distances. Total Communist strength at the beginning of the march was about 200,000 troops. They suffered over 100,000 casualties and noncombat losses during the march. About 40,000 men were left behind along the route of march as underground cadres. About 50,000 recruits joined on the march. The total Communist force in Shensi at the end of 1935 was nearly 100,000 men.

1934–1937. Japanese Expansion in Northern China. Japanese troops continued to press westward in the Inner Mongolian province of Chahar into northern Hopei.

1938–1939.   Japanese-Soviet   Frontier Clashes. (See p. 1142.)

1936, Summer. Renewed "Bandit Suppression Campaign." Following a Communist foray into Shansi, Chiang Kai-shek decided to renew efforts to eliminate the Communists. An army of 150,000 was assembled, under Marshal Chang Hsueh-liang, with headquarters at Sian.

1936, December 12–25. Sian Mutiny. Disappointed by inactivity of Chang's "bandit suppression" effort, Chiang flew to Sian (December 7). Chang urged Chiang to call off the campaign; he and his troops wanted to fight the Japanese rather than other Chinese. When Chiang refused, he was taken into "protective custody." The results of the subsequent negotiations, in which Communist **Chou En-lai** participated, have never been revealed. Chiang was released, taking Chang with him as a prisoner. He then called off the anti-Communist campaign.

1937, July 7. Outbreak of Hostilities between China and Japan. (See p. 1229.)

MONGOLIA

The Mongolian People's Republic continued to be a satellite of the U.S.S.R., although the Republic of China still stubbornly refused to recognize Mongolian independence from China. The principal events were:

1936, March 12. Soviet-Mongolian Mutual Assistance Pact. This 10-year treaty was a response to the growing Japanese threat in Manchuria and Inner Mongolia. With Soviet assistance, the Mongolian army was maintained at 90,000 men, 10 percent of the population.

1939, May–September. Border Hostilities with Japan. (See p. 1143.) Mongolian forces made up a major portion of the Soviet-Mongolian army, which repulsed the Japanese.

1941–1945. Mongolian Neutrality in World War II. Actually Mongolia gave full support to the Soviet Union.

1945, August 10. Declaration of War against Japan. Mongolian troops participated in the successful Soviet invasion of Manchuria and Inner Mongolia (see p. 1298).

1946, January 5. China Recognizes Mongolian Independence. China had promised to do this at the time of signing its treaty of friendship with the U.S.S.R. at the end of World War II (see p. 1423), but had insisted on a plebiscite first. The Mongols had voted virtually unanimously for independence (October 20, 1945).

THE AMERICAS

THE UNITED STATES

1925–1939. Military Retrenchment. The post–World War I reaction against American involvement overseas continued. In a period of pacifism and isolationism, the military services were largely ignored. Dwindling appropriations declined still more rapidly during the period of the Great Depression (1929–1938). Even the Navy, traditional "First Line of Defense," was neglected. In February 1929, a bill was passed authorizing the construction of 15 cruisers and 1 aircraft carrier in 3 years at a cost of $27 million. By March 1933, only 8 of the cruisers were completed. Although President **Franklin D. Roosevelt** obtained additional appropriations, construction still lagged far behind America's treaty rights. Once World War II broke in Europe, the U.S. Navy initiated a large-scale building program of warships, merchant ships, landing craft, and planes (1939–1941).

1925, December. Mitchell Trial. An avid crusader for the development of military aviation, Brigadier General **William Mitchell,** chief of the U.S. Army's Air Corps, provoked worldwide repercussions by challenging the traditional American military and naval hierarchy, and by his advocacy of the theories of Douhet (see p. 1089). His outspoken criticism of his superiors resulted in his demotion to colonel (April 1925). Later, Mitchell charged the War and Navy Departments with "incompetency, criminal negligence and almost treasonable administration of the national defense" (September). He was court-martialed by order of President Coolidge, found guilty, and ordered suspended from rank and duty for 5 years. He resigned from the Army (January 1926) and continued a writing and lecturing campaign in his effort to obtain public support for his concepts of air power.

1929, January 15. Pact of Paris Ratified. (See p. 1124.) The Senate did not consider the Kellogg-Briand Treaty to impair the right of self-defense.

1930, January 21–April 22. London Naval Conference. (See p. 1124.)

1933, November 17. U.S. Diplomatic Recognition of Russia.

1934–1936. Nye Committee Munitions Investigation. This gave the American public the impression that the entry of the U.S. into World War I had been to save the bankers and to protect the arms trade. These hearings had an appreciable adverse effect on national defense and helped to incite resurgent isolationism.

1935, April. Neutrality Act. This forbade Americans to furnish munitions or loans to foreign belligerents and refused protection to Americans sailing on belligerent ships. Japanese aggression in China, Italian aggression in Ethiopia, and the Spanish Civil War brought the program of neutrality into sharp focus, resulting in a compromise: the War Policy Act (May 1, 1937). The act retained the mandatory provisions of the Neutrality Act, but gave the president wide discretion, especially in determining if the law was to be put into effect by proclamation that a war actually existed.

1939, January 12. End of Military Retrenchment. President Roosevelt asked Congress for $552 million for defense spending.

1939, September 5. U.S. Neutrality Declared. This followed Roosevelt's fireside chat (September 3), stating, "This nation will remain a neutral nation, but I cannot ask that every American remain neutral in thought as well." He proclaimed a limited national emergency (September 8).

1939, November 4. Arms Embargo Lifted. Congress, removing the arms embargo at Roosevelt's urging, authorized "cash and carry" exports of munitions and arms to belligerent nations. The intent was to assist the Allies.

1940. Defense Measures. Roosevelt called for appropriations (January 3 and May 31) totaling approximately $3.4 billion for national defense, and in response to Churchill's call for help released to Britain more than $43 million worth of surplus stocks of arms, planes, and munitions (June 3). A National Defense Research Committee was established, headed by Dr. **Vannevar Bush** (June 15). **Henry L. Stimson** was named Secretary of War and **Frank Knox** Secretary of the Navy (June 20). Congress adopted defense tax measures and raised the national debt limit to $49 billion (June 22). In defense matters Roosevelt particularly relied upon the advice of U.S. Army Chief of Staff, General **George C. Marshall.**

1940, September. The U.S.–British Destroyer Base Agreement. Great Britain received 50 badly needed destroyers (see p. 1167) in return for leases for American naval bases at 8 points on British territory along the Atlantic coast from Newfoundland to British Guiana.

1940, September 16. Initiation of Conscription. The Selective Service Act was passed by Congress.

1940, December 20–29. Roosevelt Speeds Defense. The Office of Production Management was established to extend all aid short of war to Great Britain and all other anti-Nazi belligerents.

1941, March 11. Lend Lease. The act opened mutual aid between all anti-Nazi nations and the U.S.

1941, July 7. Defense of Iceland. American forces landed in Iceland at the invitation of the Danish and Icelandic governments, relieving British troops defending the island.

1941, September. Renewal of Selective Service. The act, violently opposed by isolationists, passed the House of Representatives by one vote.

1941, December 7. Attack on Pearl Harbor. (See p. 1235.)

MEXICO

1927, October. Insurrection. A rebellion spread through the provinces, but was crushed within 2 months.

1929, March–April. Renewed Rebellion. Insurrection broke out again in 1929 because of widespread political and religious discontent.

1934. Beginning of Stability. President **Lázaro Cárdenas,** supported by the army, established himself as the undisputed ruler of Mexico and undertook an ambitious program of reforms.

CENTRAL AMERICA (PANAMA, COSTA RICA, NICARAGUA, HONDURAS, SALVADOR, GUATEMALA)

Internal political conditions were unstable and numerous insurrections and depositions occurred. In Nicaragua (1925) a bloody civil war erupted; U.S. intervention resulted in a supervised election and a relatively clam administration until the renewal of rebel activities by General **César Augusto Sandino** (1931). U.S. Marines organized and trained the Guardia Nacional under General **Anastasio Somoza.** After the Marines withdrew (January 2, 1933), Somoza became president (1937) and was virtual dictator until his death (1956).

WEST INDIES (CUBA, HAITI, DOMINICAN REPUBLIC, PUERTO RICO, VIRGIN ISLANDS)

1934, May 29. The Platt Amendment Abrogated. The U.S. lifted limits on Cuba's sovereignty.

1934, August 6. Withdrawal of U.S. Marines from Haiti. The culmination of arrangements between the United States and the Haitian Assembly (see p. 1109).

1937, October. Border Dispute between Haiti and the Dominican Republic. This resulted in the death of many immigrant Haitians. An American-inspired conciliation treaty led to settlement.

1939. Expansion of Military Facilities in Puerto Rico. The island became the keystone of a U.S. Caribbean defense system.

SOUTH AMERICA (ARGENTINA, CHILE, PARAGUAY, URUGUAY, BOLIVIA, PERU, ECUADOR, COLOMBIA, VENEZUELA, BRAZIL)

The Latin American scene was punctuated with jealousy and war between neighbors, widespread internal turbulence, and 2 major international clashes.

1921–1929. Tacna-Arica Dispute. A legacy of the War of the Pacific (see p. 1000), Chile's dispute with Peru over this rich nitrate region resulted in several armed clashes. Bolivia, meanwhile, maintained that neither Chile nor Peru was entitled to the provinces. Through United States arbitration, the dispute was solved by compromise: Arica was awarded to Chile; Peru received Tacna. Bolivia's complaints were assuaged somewhat by allocation to her of a railway outlet to the Pacific between La Paz and Arica (1929).

1932, June–August. Disorders in Brazil. The federal government reestablished order in São Paulo State.

1932–1935. Chaco War. Armed clashes between Paraguay and Bolivia over the possession of the Chaco region began in 1928 and gradually developed into open war (1932). Efforts of other states and the League to mediate were unsuccessful and the struggle continued despite the economic strain on both sides. The Paraguayans soon gained *de facto* control over much of the region. Bolivia employed a German general, **Hans von Kundt,** to train and command its armies and to establish military posts in territory claimed by Paraguay. When Bolivia captured **Fort Lopez** (Pitiantuta) in central Chaco (June 15, 1932), she seemed to have accomplished her objective of gaining the Paraguay River as an outlet to the ocean. At this point the Paraguayans initiated a remarkable

national military effort. While frontier troops recaptured **Pitiantuta** (mid-July), the Paraguayan Army began an expansion from 3,000 to 60,000 men. A supply line was established via Puerto Casado in preparation for a major offensive. This was initiated by Colonel **José Felix Estigarribia,** and the Paraguayans steadily forged their way through the jungle region. In the next year and a half they conquered most of the disputed region, captured the Bolivian headquarters, and took more than 30,000 prisoners. A truce was then signed and active hostilities ceased (June 12, 1935). By the **Treaty of Buenos Aires** (July 21, 1938) an arbitral decision awarded Paraguay three-quarters of the disputed area; Bolivia, however, was provided an outlet to the Atlantic Ocean via the Paraguay River.

1936, December 1–23. Buenos Aires Conference. Meeting of the Pan-American Conference for the maintenance of peace accepted the principle of consultation in case the peace of the continent was threatened. A common policy of neutrality was drawn up in case of conflict between the American states.

1938, December 24. Declaration of Lima. This reaffirmed the absolute sovereignty of the various American states and expressed their determination to oppose "all foreign intervention or activity." This was manifest of a determination to defend Latin-American territory from both external aggression and internal subversion.

WORLD WAR II IN THE WEST

OPERATIONS IN 1939

Polish Campaign, September 1– October 5, 1939

1939, September 1–5. Invasion. Covered by predawn air bombardment, without declara-

tion of war, General **Walther von Brauchitsch**'s 1,250,000 men in 60 divisions—9 of them armored—struck from north, west, and south. The German plan was a double envelopment, gripping the 6 Polish armies, spread in cordon defense, between pincers closing down

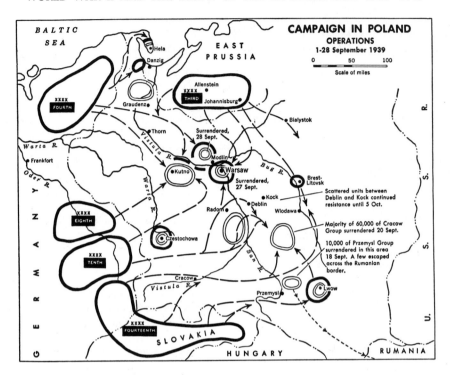

first on the Vistula River line, and later on the Bug, farther east. General **Fedor von Bock**'s army group—the Third and Fourth armies—converged from both sides of the Polish Corridor; General **Gerd von Rundstedt**'s group of 3 armies swept eastward and northeastward across Upper Silesia and Galicia. Some 1,600 Luftwaffe planes undertook terror bombings of all principal cities, destruction of air fields and railway centers, systematic sweeping of main highways to dislocate traffic, and close support to the ground troops. With 2 million German sympathizers in the Polish population, fifth-column and intelligence activities were well organized. Flat Poland, with no major terrain obstacles, in the pleasant, dry autumn weather offered ideal terrain for the use of tanks. German armored spearheads slashed through the thin screen of 6 Polish armies, some 800,000 men, commanded by Marshal **Edward Smigly-Rydz,** like knives through butter. By the third day the Polish air force ceased to exist. All communication between general headquar-

ters and the field armies was ended, further mobilization rendered impossible. Informed by spies, the Luftwaffe learned the location of Polish headquarters and bombed it continuously, even though the Poles moved frequently. In the Gulf of Danzig, a German naval force disposed of the little Polish Navy—4 destroyers, a mine sweeper, and some submarines. Poland was paralyzed, its civilian population terrified.

1939, September 5–17. Battles of Warsaw and Kutno-Lódź. On the north, Bock's left-flank army pressed down on Brest-Litovsk; on the south, Rundstedt's right-flank army streamed northeasterly past Cracow. In the center, Rundstedt's Tenth Army (General **Walther von Reichenau**), with the majority of the armored divisions, approached the Vistula below Warsaw. The interior ring of the double double envelopment was closing on the Vistula River line, the outer ring on the Bug. The Polish armies, fighting with futile gallantry (in some cases horse cavalry charged German tanks), were rolled up in several clusters, each completely surrounded and without any overall direction. Reichenau's Tenth Army tanks attempted to enter Warsaw (September 8), but local Polish counterattacks drove them off. Principal Polish resistance now centered around Warsaw-Modlin and farther west around Kutno and Lódź. The Polish Kutno-Lódź forces made a gallant but unsuccessful attempt to break out, then, after incessant air and ground attacks—their supplies and ammunition exhausted—surrendered (September 17). Meanwhile, the outer encirclement ring had closed, as German Third and Fourteenth armies met, south of Brest-Litovsk.

1939, September 17. Soviet Invasion of Poland. In conformity with the Nazi-Soviet Pact (see p. 1135), Soviet troops swept over Poland's eastern border, north and south of the Pripet Marsh area, ending any Polish hopes of a last-ditch stand in the southeast. The government and the high command took refuge in Rumania.

1939, September 17–October 5. Collapse of Poland. One by one the Islands of resistance succumbed. Warsaw, facing starvation and typhoid, fell (September 27); Modlin, just across the Vistula, next day. Up on the Baltic, the naval base of Hel capitulated (October 1). The last organized Polish resistance ended at Kock, where 17,000 Poles surrendered (October 5). Polish losses were about 66,000 killed and at least 200,000 wounded. The greater part of their armed forces was captured—some 694,000 men. A few escaped through Rumania. German losses were 10,570 killed, 30,322 wounded, and 3,400 missing. Germany and Russia divided the country between them, while the German armies slowly recoiled to prepare for a later move against Western Europe.

COMMENT. *The whirlwind campaign was a spectacular demonstration of fire and movement, utilizing the latest developments in weapons, aerial and ground. German adherence to the principles of the offensive, surprise, mass, and maneuver was irresistible. Weak spots or gaps in the Polish defensive ring were exploited by speedy armored columns with unlimited objectives and disregard for flank protection. Fast-moving infantry formations encircled Polish units isolated by the breakthroughs; the demoralized Poles never had time to recover. The pattern of blitzkrieg was set. Contributing factors to German success were the Polish tactical dispositions in cordon defense, and the inertia of France and Britain in the west, permitting undivided German attention to the eastern campaign. Thoroughly documented by scores of cameramen, the film story of the Polish debacle was distributed throughout Europe and the United States—a major weapon in Germany's propaganda war of terror. Hitler—convinced that he, rather than his professional soldiers, had achieved success—began to believe himself to be a military genius. But most western military men, deeply impressed by the German victory, nevertheless wrongly assumed that its magnitude was due mainly to Polish ineptitude.*

War in the West, 1939

Amazing inactivity marked the situation along Germany's western border. France and Great Britain began lethargic mobilization behind the partial shelter of the great concrete, steel-turreted Maginot Line, stretching from the Swiss border northward to Montmédy. From that point to the North Sea, along the Belgian border, extended an outmoded system of unconnected fortresses—relics of pre-World War I defenses. Belgium and Holland both maintained strict neutrality—a fatal concession to political expediency, for it prevented any cooperation with French and British military planners. The British Expeditionary Force, nearly 400,000 strong, was moved across the Channel and concentrated as part of the Allied line in the general area Arras-Lille. Across the German border the Westwall lay quiescent, with supporting mobile troops stripped to a minimum to the benefit of the Polish front. A strong Allied punch might well have broken through and ended Hitler's grandiose scheme of world conquest. Instead, the only overt move was a tentative French probe toward Saarbrücken. The Allies were relying on a policy of blockade, economic strangulation, and defensive fortification to exhaust German strength. From that groundwork, it was believed would eventually come an Allied offensive—as yet only a dream. "Sitzkrieg" and "phony war" were the derisive terms applied to this negation of Germany's blitzkrieg.

The Naval War, 1939

Great Britain proclaimed a naval blockade of Germany immediately after her declaration of war (September 3). The German Navy, under the direction of Grand Admiral **Erich Raeder,** was prepared to prosecute only one type of strategy: commerce destruction. There could be no thought of fleet action against Britain by Germany's small fleet (see table, below). But Germany had 98 submarines against Britain's 70, and the United Kingdom was dependent upon sea transport for both food and war matériel. Already 2 pocket battleships, the *Deutschland* and the *Graf Spee,* were at sea, together with a score of U-boats, with supply ships scattered strategically around the globe. The first German submarine blow shocked the world, and led immediately to Britain's adoption of the convoy system for her merchant marine.

COMPARATIVE NAVAL STRENGTHS[a]—SEPTEMBER, 1939

	Britain	France	Germany[b]	Italy
Battleships and Battle Cruisers[c]	18	11	4	6
"Pocket" Battleships[d]	—	—	3	—
Aircraft Carriers	10	1	1	—
Heavy Cruisers (8-inch guns or more)	15	18	4	7
Light Cruisers (6-inch guns or less)	62	32	6	15
Destroyers	205	34	25	59
Destroyer Escorts, Torpedo Boats, etc.	73	30	42	69
Motor Torpedo Boats	39	9	17	69
Submarines	70	72	98	115

[a] Includes ships built or nearing completion.

[b] All German vessels were newly built, and with few exceptions were more modern, faster, bigger, and generally more powerful than comparable types of other nations.

[c] Battle cruisers were as big as battleships and carried the same kind of heavy guns (usually 14- to 16-inch in caliber). But they carried less armor protection, and usually fewer heavy guns, so that, being lighter in weight, they could go faster than battleships. Thus they could hit as hard as battleships, but could not take as much punishment; they sacrificed protection for speed.

[d] Under the provisions of the Versailles Treaty after World War I, Germany was forbidden to build ships larger than 10,000 tons. The pocket battleships were really small battle cruisers; they carried 11-inch guns, were very fast, but did not have much armor, not to exceed the weight limit. (In fact, they displaced more than 11,000 tons.) They could beat any cruiser in the world, but could not stand up to a real battleship.

1939, September 3. Sinking of the *Athenia*. This British passenger liner, bound from Liverpool to Montreal with 1,400 passengers, was sunk without warning by *U-30,* 200 miles west of the Hebrides; 112 died.

1939, September 17. Loss of HMS *Coura-*

geous. Cruising with 4 escorting destroyers off the southwestern coast of England on antisubmarine duty, the carrier was torpedoed at dusk by *U-29,* sinking immediately with loss of 515 lives.

1939, September–December. Cruise of the *Graf Spee.* This German pocket battleship, under Captain **Hans Langsdorff,** cruised the South Atlantic between Pernambuco, Brazil, and Cape Town, South Africa, even venturing into the Indian Ocean, capturing and sinking 9 British merchantmen between September 17 and late November. Making secret rendezvous with the supply ship *Altmark,* she then took on fuel and supplies and transferred to her some 300 prisoners. Resuming his cruise, Langsdorff captured 2 more British ships. Meanwhile, a British aircraft carrier, a battle cruiser, and 6 cruisers, together with 2 French cruisers and some 10 Allied destroyers, were scouring the seas for the *Graf Spee.*

1939, October 14. Sinking of HMS *Royal Oak.* German Lieutenant Commander **Guenther Prien,** commanding *U-47,* at night daringly threaded elaborate antisubmarine defenses and treacherous tide rips to enter Scapa Flow, where the British Home Fleet lay concentrated. Firing 2 spreads of 4 torpedoes each against the nearest large vessel, he scored several hits on the battleship *Royal Oak.* She went down in 2 minutes, taking with her 786 officers and men. Prien then brought his submarine safely out of the harbor, successfully concluding a most gallant exploit.

1939, December 13. Battle of the River Plate. A British squadron—heavy cruiser *Exeter,* 6 8-in. guns, and the light cruisers *Ajax* and *Achilles,* 8 6-in. guns each—under Commodore **Henry Harwood** caught up with the *Graf Spee* off the mouth of the River Plate (early December 13). Langsdorff, confident that his 6 11-in. and 8 5.9-in. guns could blow the lighter British ships out of the water, closed with them, instead of standing off and destroying them with his heavy battery, which outranged their armament by nearly 10,000 yards. Harwood's squadron carried the fight to the German in preplanned maneuvers that prevented the *Graf*

Spee from concentrating fire on any one ship. At the end of 80 minutes, the *Exeter* was completely silenced and on fire, the *Ajax* had half of her guns out of action, and the *Achilles* was severely damaged. But the *Graf Spee,* too, had been badly hit, so Langsdorff broke off the fight and made for neutral Montevideo, where he hoped to land his wounded and make emergency repairs. Harwood, following with his 2 serviceable ships (the *Exeter* limped for the Falkland Islands), stood off the port and waited, radioing for reinforcements. Langsdorff was refused more than 72 hours sanctuary by the neutral Uruguayan government. Meanwhile, the British heavy cruiser *Cumberland,* 8 8-in. guns, arrived to join Harwood; other British vessels were converging. Langsdorff, leaving most of his crew on board a German merchantman in the harbor, steamed out with a skeleton crew and blew up his vessel. He and his party then returned in their lifeboats to internment. Heartbroken, Langsdorff committed suicide 3 days later.

1939, mid-December. Other German Raiders. In the North Atlantic the pocket battleship *Deutschland,* after some success in commerce destruction (among her prizes was the neutral American freighter *City of Flint,* whose seizure created an international incident), was forced by engine trouble to make for home. The German Naval Staff sent out the battle cruisers *Scharnhorst* and *Gneisenau* to cover the *Deutschland's* withdrawal and continue pressure against Atlantic shipping.

1939, December 23. Sacrifice of HMS *Rawalpindi.* The *Scharnhorst* and *Gneisenau* successfully evaded British patrols east of Scotland, but were sighted December 23 between the Faeroe Islands and Iceland by HMS *Rawalpindi,* a merchant liner converted into a cruiser and mounting 4 6-in. guns. Radioing an alarm, the *Rawalpindi* gallantly engaged the giant warships with her popguns, hoping to delay them sufficiently to ensure their interception. Literally blown to bits in a few moments, the *Rawalpindi* and her valiant crew nevertheless had fulfilled their mission, for the German warships returned immediately to port.

FINNISH-SOVIET WAR, NOVEMBER 30, 1939–MARCH 1, 1940

The U.S.S.R., following the division of Poland, and obviously fearful of Germany, began consolidating a Baltic sphere of interest. Mutual defense pacts with Latvia, Estonia, and Lithuania were followed by immediate inrush of Russian troops for their "defense" (October–November). From Finland, the U.S.S.R. demanded a similar agreement, including occupation of the southern portion of the Karelian Isthmus and other island and mainland base areas. Finland, rejecting these demands, mobilized forces along her frontier, under the command of aging, still brilliant Marshal Baron **Carl Mannerheim,** veteran of 3 wars. The U.S.S.R. demanded immediate withdrawal of Finnish troops.

1939, November 30. Bombardment of Helsinki and Viipuri. A Soviet air attack, without declaration of war, initiated hostilities.

1939, November 30–December 15. The Russian Invasion. Armies totaling nearly a million men smashed at Finland from east and southeast, and in amphibious invasions across the Gulf of Finland. They were opposed by Finnish forces totaling 300,000, of which about 80 percent were mobilized reservists. All the amphibious attacks on the southern coast were repulsed. In the far north a Soviet column seized Petsamo, pressed a short distance south, and then was stopped. The main Soviet attacking force, driving into the Karelian Isthmus, was hurled back with heavy loss at the Mannerheim Line—a World War I system of field fortifications, all cleverly knitted into the rugged terrain and heavily wooded areas. Other Soviet columns pressed into the vast lake and forest region of eastern and central Finland, opposed by mobile defense units of battalion or smaller size habituated to independent action in the forests. The heavy snows and sub-zero temperature of the Finnish winter had little effect on Finnish troops or their mobility. All were skiers; all were warmly clad. The Soviets soon bogged down. Meanwhile, Soviet air raids had been ineffective, and Communist fifth-column activities quickly collapsed.

1939, December—1940, January. Battle of Suomussalmi. In eastern Finland, the Soviet 163rd Division moved in 2 columns over narrow woods roads, which converged at the village of Suomussalmi. Deep snow impeded its advance and temperatures dropping to −40° bit deep into men fresh from the Ukraine and lacking arctic clothing. On the Soviet flanks, Finnish civil guard units in white smocks swooped silently on skis, sniping at supply vehicles and field kitchens. The Soviet columns joined at the village, but then paused to attempt reconnaissance (December 7–11). Arriving infantry elements of the Finnish 9th Division attacked without waiting for their artillery (December 11). The Russians clung to the village, but both their supply routes were blocked and ambushed, cutting off all supply. The Soviet 44th Division, motorized, attempted to cut through to the 163rd's assistance, but was itself immobilized by harassing civil-guard attacks, and dug in on the road some 5 miles east of Suomussalmi. On Christmas Eve both Soviet divisions vainly attempted to cut their way out. The Finnish 9th Division, its artillery now up, then made a coordinated assault upon the 163rd, and annihilated it (December 27–30). Turning then on the 44th Division, the Finns systematically cut it into smaller groups, each of which was in turn mopped up (January 1–8). Soviet losses were about 27,500 killed or frozen to death. Captured were 1,300 others, together with 50 tanks and the entire artillery and equipment of both divisions. Finnish casualties were 900 killed and 1,770 wounded.

1940, January 1–February 1. Soviet Regrouping. This followed humiliating repulses. (Soviet spearheads had been checked and repulsed in 6 areas along the eastern frontier, with astounding losses—a division and a tank brigade captured; 3 other divisions almost destroyed.) The Soviets now prepared to assault the Mannerheim Line, the only sector against which overwhelming strength could be concentrated. Meanwhile, France and England were preparing to mount a strong expeditionary force to Finland's assistance, but neutral Norway and Sweden refused to permit its passage.

1940, February 1–13. Assault on the Mannerheim Line. The Soviet Seventh and Thirteenth armies, totaling 54 divisions, began incessant attacks—4 or 5 per day—against the Mannerheim Line, covered by tremendous artillery bombardments and air support. Wave after wave of assault troops dared the devastating defense fire. The Finns were sickened by the slaughter they inflicted. Finally a breakthrough was effected near Summa (February 13). The Finnish right wing and center were rolled back to Viipuri (March 1).

1940, March 12. Finland Capitulates. Her armed forces were exhausted and all hope for

foreign assistance was gone. The Soviet terms of peace were practically identical with the orginal demands. Stalin had evidently no intention of tempting foreign intervention by further aggression, nor had he any stomach for a future guerrilla war in Finland. Finnish losses in the war were about 25,000 killed and 43,000 wounded. Soviet losses have never been published, but they probably were about 200,000 killed and 400,000 more wounded.

COMMENT. *The Finnish defense was conducted by well-led, disciplined soldiers, familiar with the terrain and weather conditions and utilizing tactics geared to those conditions. The Soviet offensive was amateurishly planned, without regard to terrain, weather, or the logistical problems involved. It was finally successful only because the Soviet government was willing to utilize cannon fodder in overwhelming masses without regard to casualties. As a result of the Soviet blundering, Hitler and his generals assumed that the Soviet Union, inferior in leadership, tactics, and weapons, would be a pushover for the German war machine. This opinion was unfortunately shared by many foreign observers, including the United States War Department.*

Operations in 1940

Conquest of Denmark and Norway

Sweden's iron ore, vital to German heavy industry, flowed in 2 channels: via the Baltic Sea and down the coast from Narvik in northern Norway. The latter route was threatened by the British naval blockade. German possession of Norway would permit use of land-based air power against the blockade and provide a springboard for aerial attacks against the British Isles. Hitler determined to seize both Norway and Denmark. The Germans began to make plans for a campaign against Scandinavia (January 1940), and Hitler gave the final order to go ahead on April 2.

Meanwhile, the British had made plans to mine Norwegian waters in order to force ore carriers from Narvik into the open sea where they could legitimately be attacked. The British and French had been frustrated by Norway's refusal to permit Allied aid to reach Finland during the Winter War (March 1940), on the legitimate grounds that it would compromise Norwegian neutrality.

1940, February 16. *Altmark* Incident. British blockaders discovered the *Graf Spee's* auxiliary heading home along the Norwegian coast with 299 British merchant seamen captured by the

Graf Spee. The Admiralty ordered their rescue at all costs. Captain **Philip Vian,** in HM destroyer *Cossack,* pursued the *Altmark* into Jossing Fjord near Stavanger and despite Norwegian protests boarded and captured her, releasing the prisoners. Norwegian protests of this violation of territorial waters died away in face of British proof that Norway had permitted an armed vessel to take refuge in neutral Norwegian waters.

1940, April 8–10. Naval Actions along the Norwegian Coast. In early April Churchill ordered mines laid in Norwegian territorial waters to deny their use by Germany. British destroyers approaching the Norwegian coast encountered German naval vessels convoying troop transports and warships toward Kristiansand, Stavanger, Bergen, Trondheim, and Narvik. Several violent engagements took place. The German warships successfully protected their convoys, but suffered losses. HM destroyer *Glowworm* deliberately rammed the heavy cruiser *Hipper,* damaging her severely, but was herself blown up. In a duel of battle cruisers the German *Gneisenau* was badly damaged by the *Renown.* Two other German cruisers were sunk, and the pocket battleship *Lützow* was damaged near Oslo by British ships and submarines. The German convoys all reached their objectives.

1940, April 9. Conquest of Denmark. German troops dashed across the border into Jutland. A battalion of infantry, hidden on a merchant ship in Copenhagen harbor, landed to seize king and government. Thus fell Denmark, almost without bloodshed.

1940, April 9. Invasion of Norway. The Germans planned simultaneous landings along the coast and parachute assaults at Oslo and Stavanger; they expected to paralyze the nation and to seize king and government by surprise as in Denmark. But the Norwegian armed forces, about 12,000 strong, were more substantial and better prepared than the Danes and had the brief warning provided by distance and size. The Army was a well-trained militia with a small cadre of permanent officers and men, in 6 infantry brigades and supporting artillery, plus local defense units. The Army needed 4 days to mobilize fully, and since the government was distracted by other events, not least of them the British minelaying operations (begun on

April 8), the order for mobilization was not issued until the small hours of April 9. The Navy, including coast defense artillery, was small and efficient, but ships and equipment were old. Support elements included some fighter and torpedo planes. There were trained reserves of 120,000.

1940, April 9–10. Capture of Oslo. Shore batteries in Oslo Fjord gallantly repulsed the German invasion flotilla; the German cruiser *Blücher* and a smaller vessel were sunk and 2 small craft were damaged. Airborne troops seized Oslo airport, over-whelming the small AA detachment. Norwegian fighter planes shot down 5 Germans before they were driven from the sky. The air-transported troops quickly seized the city, while troops landed from ships overran the coastal defenses.

1940, April 9–10. West Coast Operations. At Kristiansand and Bergen the Germans landed only after suffering heavy casualties at the hands of determined, outnumbered defenders. At Bergen, coast defense guns damaged the cruiser *Königsberg,* which British bombers then sank. Near Stavanger an infantry battalion vainly tried to stop history's first combat airborne assault. Only at Trondheim and Narvik did German efforts at surprise succeed.

1940, April 10–13. Actions at Narvik. Five British destroyers in surprise attack entered the fjord and sank 2 of 10 German destroyers which had carried the assault troops, then withdrew, having also lost 2. H.M. battleship *Warspite* with 9 destroyers went in and wiped out the remaining German destroyers (April 13). Seven U-boats remained.

1940, April 10–30. German Exploitation. With all initial objectives taken, German troops under General **Nikolaus von Falkenhorst** fanned out over the countryside. An intensive airlift brought more troops from Germany. Luftwaffe units from Norwegian airfields attacked British warships off the coast. Overcoming initial surprise and confusion, Norwegian troops under newly appointed Major General **Carl Otto Ruge** fought stubbornly. About 50,000 reservists were mobilized. For three weeks the Norwegians tried to hold south-central Norway. But the Germans were too strong and overwhelmed them or drove them into northern Norway. A small "fifth column" led by traitor **Vidkun Quisling** gave the Germans some help.

1940, April 14–19. Allied Landings. Some 10,000 French and British troops that had been assembled in British ports for a possible attempt to aid Finland (see p. 1154) were hastily embarked and landed at Namsos and Andalsnes to try to take Trondheim and retain a foothold in Norway. A smaller force landed near Narvik (April 14–15).

1940, April 20–May 2. German Reaction. Germans at Trondheim held off the Allies while reinforcements hurried from Oslo. The Luftwaffe struck Allied troops, landing areas, and support ships in continuous violent attacks. Unable to fight effectively, the Allies evacuated Namsos and Andalsnes (May 1–2). King **Haakon VII** and the government meanwhile had moved to northern Norway (May 2), as Norwegian forces withdrew from the central and southern parts of the country.

1940, April 24–May 26. Battle for Narvik. While the Allied force hesitated, the Germans prepared hastily for defense (April 14–23). On Churchill's urging, the attack began. Despite powerful naval support, the Allied force, now 25,000 men, was unable to dislodge the stubborn Germans. Norwegian troops, trained for mountain fighting, spearheaded the attack, but coordination between the Anglo-French force and the Norwegians was poor. Under Allied pressure the exhausted Germans withdrew and the Allies quickly occupied Narvik (May 28).

1940, June 7–9. Allied Evacuation of Narvik. Because of catastrophe in Western Europe (see below), the Allies evacuated Narvik, which was quickly reoccupied by the Germans, thus completing the conquest of Norway. The Allied force evacuating the area by sea was escorted by H.M. carrier *Glorious* and 2 destroyers. The *Glorious,* surprised by the German battle cruisers *Scharnhorst* and *Gneisenau,* with all her planes on deck, was promptly sunk. So, too, were the destroyers, but one of them succeeded in badly damaging the *Scharnhorst* by a torpedo. A few days later, a British submarine damaged the *Gneisenau.* Both German ships would be out of commission for 6 months. King Haakon and his government accompanied the withdrawing Allies and set up a government in exile in London.

COMMENT. *The German invasion was efficiently and daringly carried out in the face of possible destruction by the Royal Navy. Then speedy operation of the Luftwaffe from the captured bases gave the world its first realization of the value of air power in naval warfare. The net result of the audacious invasion was to loosen the British naval blockade of Germany, ensure the Reich's iron ore supply, and gain tremendous prestige for the German war machine. On the other hand, the German Navy was seriously crippled for several months to come. The performance of the Norwegians was commendable, particularly in their successful offensive efforts in and near Narvik.*

Campaign in the West, 1940

BACKGROUND

By early May some 2¹/₂ million Germans (104 infantry divisions, 9 motorized divisions, 10 armored divisions) had assembled along Germany's western borders. These were organized in 3 army groups: the northernmost—Army Group B, General Fedor von Bock—comprised 2 armies from the North Sea to Aachen. Army Group A, General Gerd von Rundstedt, consisted of 4 armies and a powerful armored (Panzer) group or army, in a relatively narrow zone between Aachen and Sarrebourg. Most of Germany's 2,574 tanks were concentrated in Army Group A. Army Group C, General **Wilhelm J. F. von Leeb,** consisted of 2 armies facing the French defenses in eastern Lorraine and along the Rhine River. Supporting the ground troops were 2 air fleets, consisting of 3,500 combat planes. Overall command was exercised by Hitler, as commander in chief, with General **Wilhelm Keitel** as his chief of staff. Directly commanding the army was Brauchitsch, who had also been in command of the invasion of Poland.

The opposing Allied forces (over 2 million men, mainly French) were assembled in 3 army groups behind the French borders. The First Army Group behind the French borders. The First Army Group (General **Gaston Billotte**), from the English Channel to Montmédy, consisted of 5 armies, including the British Expeditionary Force (General **John Vereker,** Lord **Gort**). The Second Army Group, General **André Gaston Prételat,** of 3 armies was behind the Maginot Line from Montmédy to Epinal. The Third Army Group, of 1 army (General **Besson**), occupied the Maginot Line defenses. Overall field commander was General **A. J. Georges,** who in turn was under the Allied commander in chief, General **Maurice G. Gamelin.** Although the Allies had 3,609 tanks (substantially more than the Germans, and many more powerful than the German tanks), these were scattered among 3 armored divisions and a number of separate tank battalions attached to other units. There were exactly 100 other divisions: 9 were British, 1 Polish, and 13 fortress troops incapable of operating outside of their Maginot Line defenses. In support was the French Air Force of some 1,400 combat planes, and about 290 British aircraft. Allied ground and air equipment was generally less modern than that of the Germans.

In the neutral Low Countries, Belgium nominally had some 600,000 men in 22 divisions, under the command of King **Leopold III.** The Dutch Army theoretically comprised some 400,000 men, under the command of General **Henri G. Winkelman.** Neither was assembled in full strength. Both countries had elaborate defensive systems, made more formidable by canal networks, with further arrangements for flooding great stretches of country by opening dikes. The troops, however, were not very well trained, and their equipment was less modern and less complete than that of the French and British. The most serious deficiency, however, was lack of adequate defensive plans. In hopes that they could remain neutral, neither Belgium nor the Netherlands had dared to discuss joint defensive plans with the French and British, and had rebuffed every Allied effort to carry on even informal discussions.

German Plans. Following overwhelming bombardment, Army Group B would overrun The Netherlands. Moving more slowly into Belgium to encourage the Allied left-flank armies to rush to the assistance of the Low Countries, Army Group A would then hurl an armored drive through the Ardennes Forest and via the Stenay Gap into France. Thus splitting the Allied armies (cutting off those which had advanced into Belgium), Army Group A would continue westward to Calais and roll the northern portion of the Anglo-British forces against the anvil of Army Group B in the Low Countries. Subsequent, prompt southward exploitation of the gap would then roll the southern French armies back upon the Maginot Line, where Army Group C would be waiting.

Allied Plans. The French were still thinking in terms of the Schlieffen Plan of 1914, a southwesterly sickle movement through Belgium. The Allied plan proposed, therefore—just as the Germans expected—to meet the expected invasion on the Dyle Line of Belgium, pivoting the First Army Group about the northern tip of the Maginot Line.

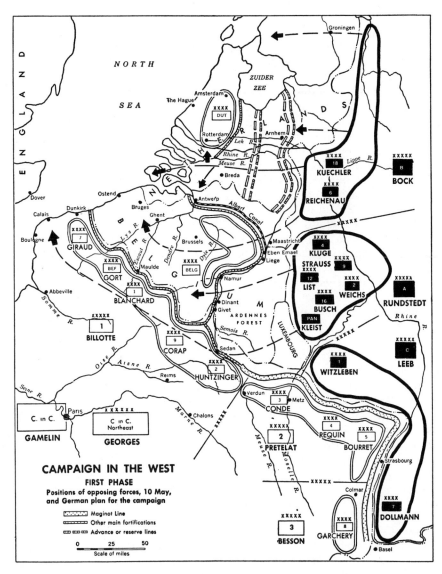

CAMPAIGN IN THE WEST
FIRST PHASE
Positions of opposing forces, 10 May,
and German plan for the campaign

- Maginot Line
- Other main fortifications
- Advance or reserve lines

0   25   50
Scale of miles

BATTLE OF FLANDERS, MAY 10–JUNE 4

INVASION OF THE LOW COUNTRIES

1940, May 10. The German Assault. Following predawn bombardments of all major Dutch and Belgian airfields, Army Groups A and B crossed the Belgian and Dutch frontiers. Initially the main effort appeared to be on the right, by Army Group B, in the Netherlands. Paratroop drops in the vicinity of Rotterdam, The Hague, Moerdijk, and Dortrecht quickly paralyzed the interior of the Netherlands. Early in the day, glider and parachute units landed on the top of powerful Fort **Eban Emael,** northern anchor of the main Belgian defense line, neutralizing it, while other German troops crossed the Albert Canal, which should have been defended by Eban Emael's guns. The violence and success of the initial German attacks, combined with bombings of the interior regions of both countries, threw their populaces into confusion and panic.

1940, May 10. Churchill Becomes Britain's Prime Minister. News of the early German successes aroused great alarm in Paris and London. Prime Minister Chamberlain, whose government had been tottering because of failures in Norway and general lack of popular support, resigned to permit lionhearted **Winston S. Churchill** to lead a coalition British government in the face of the German avalanche.

1940, May 11–14. Fall of Holland. Pressing its initial advantage, German Army Group B pressed steadily forward, despite frantic Dutch flooding of much of the countryside. By the 13th, German main elements had begun to force their way into the so-called Fortress of Holland, joining up with most of the paratroops, who had seized and held the key bridges over the Rhine estuary. At the same time, German spear-heads met advance elements of the French Seventh Army (**Henri Giraud**) near Breda, and drove them back toward Antwerp. The Queen of the Netherlands and her government escaped by ship to England from The Hague. Germany demanded complete surrender. The Luftwaffe brutally destroyed the entire business section of **Rotterdam** while negotiations were in process (May 14). Winkelman surrendered.''

1940, May 11–15. Fall of Belgium. Following a similar pattern of bombings, the German Sixth Army (Reichenau) drove southwest. Fort Eban Emael fell to its audacious attackers. As the Germans poured across the Albert Canal, the Belgian Army retired to the Dyle Line, to be reinforced (May 12) by elements of the BEF and the French First Army (**Georges Blanchard**). By the 15th, some 35 Allied divisions—including most of the BEF—were in the area Namur-Antwerp, with the German Sixth Army probing the Dyle Line in their front and the Eighteenth (**Georg von Kuechler**), now turning southward from Holland, threatening their left flank. At about the same time,

these Allied units realized that to their right rear the French center was being torn apart.

NORTHERN FRANCE

1940, May 10–12. Advance through the Ardennes. The German hammer blow— Rundstedt's Army Group A—moved through the difficult Ardennes simultaneously with the assaults on The Netherlands and Belgium, but by nature of the terrain and road net did not reach the Meuse until May 12. This calculated delay was sufficient to coax the Allied forces north of the Sambre in motion into Belgium. Leading the 3 German invading columns was General **Paul L. E. von Kleist**'s Panzer Group (5 armored and 3 mechanized divisions), to its north General **Hermann Hoth**'s Panzer Corps (2 armored divisions). Never dreaming that the Germans would make their main effort through the hilly, forested Ardennes, General **André-Georges Corap**'s French Ninth Army and General **Charles Huntziger**'s Second had their weakest elements in the Stenay Gap area, while the Ardennes Forest itself was screened only by small French cavalry and Belgian chasseur units, which were quickly brushed aside. With first word of the German advance, both French generals hurried their cavalry forward to cross the Meuse and delay until both armies could establish themselves on the river.

1940, May 13–15. Across the Meuse. But Corap was slow, and Huntziger's cavalry was outflanked. Supported by devastating dive-bombing attacks against accurate French artillery, one of Hoth's armored divisions forced a river crossing at Haux; General **Georg-Hans Reinhardt**'s corps of Kleist's Panzer Group was similarly getting over at Monthermé and General **Heinz Guderian**'s corps at Sedan (May 13). Despite the now frantic efforts of the French, the bridgeheads were quickly expanded. The French Ninth Army was completely shattered, and the Second Army's left pulverized (May 15). The German armor, spearheaded by Stuka dive bombers, roared west on a 50-mile front, while behind them

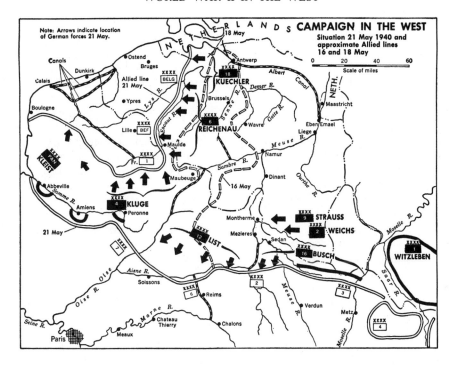

CAMPAIGN IN THE WEST
Situation 21 May 1940 and
approximate Allied lines
16 and 18 May

fast-moving German infantry poured through the gap.

1940, May 16–21. The Drive to the Channel.
All too late, Gamelin ordered up divisions from the French general reserve, and from the armies south of the German drive, into a new Sixth Army (General **Touchon**) to plug the gap. General Henri Giraud, succeeding the inefficient Corap, attempted to regroup the Ninth Army in the face of the tidal wave, but it was completely routed (May 17), and Giraud was captured. Brigadier General **Charles A. J. M. de Gaulle**'s 4th Armored Division made 3 successive punches into the German south flank from **Laon** (May 17–19), but after limited success (the only successful French attacks of the campaign) his gallant troops were turned back by dive bombers and counterattacks. Gamelin was relieved, General **Maxime Weygand** taking supreme Allied command (May 19). German armor reached the seacoast west of Abbeville, completely splitting the Allied forces and severing communications with the BEF's base port, Cherbourg (May 31). While the French to the south attempted to hold the line of the Somme and Aisne rivers, the severed northern grouping found itself being pinned against the sea.

1940, May 21–25. Exploitation in the North.
The German armor wheeled northward in 3 prongs, from the seacoast to Arras, while the Fourth (**Günther von Kluge**), Sixth, and Eighteenth Armies pressed in from the east on the French First Army, the BEF, and the Belgian Army. Lord Gort, on the First Army left, sent a task force south behind the French to bolster the right flank and to counterattack the German armor at Arras, but this effort was repulsed by General **Erwin Rommel**'s 7th Panzer Division (May 21). Guderian's armored corps captured Boulogne and isolated the British garrison of Calais (May 22–23). Dunkirk was chosen as substitute British base. The unbearable pressure of the German attack forced the Allies off the Escaut River line into an ever-shrinking perimeter, with the full force of the German armor knocking against the BEF detachment on the Allied right (May 25). Com-

plete and speedy annihilation of the penned-in Allies appeared certain.

1940, May 25–27. The Belgian Surrender. Meanwhile, on the Allied left, the Belgian Army was being pulverized by German attacks. King Leopold, deciding that further resistance was hopeless, surrendered to save further bloodshed, thus exposing the left flank of the Franco-British army to further assault. There could now be no hope of holding any part of Flanders. Churchill ordered the Royal Navy to help evacuate the British troops from Dunkirk.

1940, May 26–28. Hitler's Stop Order. By the Führer's command, the armored attack from the south was halted peremptorily. This incredible order permitted the hasty organization of perimeter defenses around Dunkirk and the equally hasty concentration of evacuation craft from the British Channel ports. The Luftwaffe was given the mission of pulverizing the Dunkirk perimeter. But the Germans in the air met an intensive, continuous attack by the RAF Fighter Command which, from bases in southern Britain, nullified German operations in a series of spectacular air battles.

1940, May 28–June 4. Evacuation from Dunkirk. Hitler rescinded his stop order and the German armor resumed assaults on the Allied right, to be checked by 3 British divisions aligned in deep zonal defense. A conglomeration of some 850 British vessels of every shape, size, and propulsion—most of them manned by civilian volunteers—converged on Dunkirk to begin the most amazing exodus in history. In 8 days, more than 338,000 men—among them 112,000 French and Belgian soldiers—were lifted. The troops streamed in orderly lines over wharves and beaches and through the surf, while overhead Spitfires of the Royal Air Force beat off most of the Luftwaffe's attempts at strafing, and their comrades along the ever-shrinking defensive perimeter held back German assaults. On the final night (June 4), General **Harold Alexander,** commanding the rear guard, personally toured the beaches and the harbor to verify the fact that the last living British soldier had been embarked, then himself got into a boat. Next morning the Germans overwhelmed the fragments of the French First Army gallantly screening the evacuation. The Battle of Flanders had ended.

COMMENT. *Aside from the duplicity and treachery of the Nazi attacks on The Netherlands and Belgium, the actual military operations of the German Army were, with one exception, clear-cut in ruthless efficiency. Hitler's strange stop order, arresting the armored assault on the boxed-in Allied armies in Flanders, cannot be charged against the German commanders. It appears to have been motivated by Goering's plea that the Luftwaffe be permitted to give the* coup de grâce *and thus have full share in the glory of victory. Added, perhaps, was Hitler's fear that miraculously the French might mount a counterattack from the south and wreck his plans of conquest.*

On the other side of the ledger, the Allied operations, having no strong, centralized control, either prior to or during the action, were disjointed and ineffective. The initial French troop distribution, with the weight of forces behind the Maginot Line defenses, was ridiculous. Friction and distrust between British and French commanders complicated the situation. Indecision was the most marked characteristic of the French high command. The over all handicap was the Allied reliance on fortifications per se, which throttled the spirit of the offensive. Much has been made of the decay of patriotic fiber in France, sapping the warrior spirit; but the troops of the French First Army, battling without hope in front of Dunkirk while their British comrades were being evacuated, certainly behaved most gallantly.

THE BATTLE OF FRANCE, JUNE 5–25, 1940

With amazing precision the German armies regrouped for the conquest of France, in accordance—except for minor changes—with previously prepared plans. Bock's Army Group B was poised on the line of the Somme extended east to Bourg. Rundstedt's Army Group A continued east to the Moselle in front of the Maginot Line, and Leeb's Army Group C stretched from there to the Swiss border. Facing it, behind the Somme, the Aisne, and the Maginot Line, the bewildered French forces were regrouping, with Army Group 3 (Besson was now on the left, Billotte having been killed in an auto accident) extending from the sea east to Rheims, Army Group 4 (Huntziger) continuing on to the Meuse and thence to Montmédy, and Prételat's Army Group 2 behind the Maginot Line. The best that Weygand could produce—with half of France's available strength already dissipated and the remainder shaken—was a defense in depth behind the Somme and Aisne. His concentration was hampered by incessant Luftwaffe bombings, dislocating rail centers and blocking troop movements on the roads. He had available only 65 divisions, 3 of them armored units already badly mauled, and 17 others fortress troops or second-line reserve units. All elements were under strength, all lacked equipment, and the general morale was very low.

1940, June 5–13. Renewed German Assault. Army Group B, spearheaded by Kleist's Panzers, struck from the Somme. Smashing through the French Tenth Army (**Félix Altmeyer**), the Germans reached the Seine west of Paris (June 9) and the armor turned westward to pin the French IX Corps and the British 51st Highland Division, one of the few remaining BEF elements still in France, against the sea at St.-Valery-en-Caux. This force surrendered (June 12). The French Seventh Army to the east put up a stiffer fight. But to restore his flank Weygand ordered Army Group 3 to withdraw to the Seine (June 8). Rundstedt's Army Group A launched its main-effort assault next day against the left of the French Group 4, east of Paris. His Panzer spearheads, under Guderian, were reinforced by Kleist's Panzers of Group B, rapidly shifted eastward. Despite valiant resistance in depth by the French Fourth Army (**Edouard Réquin**), and a series of counterattacks, Guderian's tanks crunched through at Châlons and roared southward. Kleist's armor crossed the Marne at Château-Thierry at the same time. The breakthrough was complete. The French government abandoned Paris for Bordeaux, toward which refugees in countless thousands were already pouring (June 10). Paris was declared an open city (June 13) and next day German troops marched in.

1940, June 10. Italy Enters the War. Mussolini, deciding now that France could not win, declared war and ordered an invasion of southern France.

1940, June 13–25. The Pursuit. The French armies disintegrated, while German columns spread west, south, and east. German armor swept the coastal ports from St.-Nazaire north to Cherbourg. Other Germans crossed the Loire (June 17) and easterly jabs reached the foothills of the French Alps southeast of Besançon, cutting off the French remnants behind the Maginot Line, now shattered by Army Group C's power drive (June 14–15). These isolated French troops soon surrendered (June 22).

1940, June 17. French Governmental Shuffle. After spurning Churchill's offer of an "indissoluble union" of France and Britain—to fight on forever if necessary—the French cabinet voted to sue for terms. Prime Minister **Paul Reynaud** resigned; Marshal **Henri Philippe Pétain** took his place and at once asked for an armistice.

1940, June 21. French Capitulation. General Huntziger, meeting Hitler in the railway car in Compiègne Forest where the 1917 armistice had been signed, capitulated. Hostilities officially ceased June 25. France surrendered three-fifths of her territory to German control; all French troops were disarmed. The one bright spot on France's battered military reputation shone in the Alpes Maritimes, where 6 French divisions threw back a 3-pronged invasion by 32 Italian divisions (June 21).

COMMENT. *Weygand's situation, beginning June 5, was hopeless insofar as the defense of metropolitan France was concerned. Whether he and Pétain should have accepted Churchill's offer of a Franco-English*

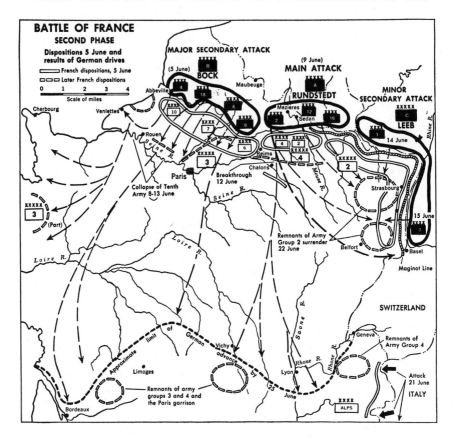

union, and carried on the war from Africa, is a moot question. Certainly the French fleet was intact, as was the French Army of Africa. The fall of France left Hitler the undisputed master of continental Western Europe. He now turned his eyes both northwest and east. Should he carry out his original plan and now conquer Soviet Russia, or should he tarry to invade and overwhelm seemingly helpless England?

Britain Embattled, June–December 1940

BRITAIN, FRANCE, AND THE FREE FRENCH

1940, June. Britain Alone. France's fall left the British Empire to face the might of the combined Rome-Berlin Axis powers. Her vital Atlantic sea lanes were menaced by German

submarines, her Mediterranean lifeline by Italy's fleet and army, with Soviet Russia an Axis silent partner on the sidelines. Furthermore, there was danger that Hitler would seize the French fleet, which would provide the Axis with a clear preponderance of sea power. The Royal Navy and Royal Air Force were intact though battered, but there were insufficient destroyers in the former to carry out antisubmarine, convoy, and patrolling activities. The British Army at home was in sad state. Although the majority of the BEF personnel had been brought back from Dunkirk, they had left all their arms and other matériel behind them in France, and the remaining army troops in Great Britain (29 divisions, with little armor or artillery) were not yet combat-ready. To most neutrals it appeared that Britain must bow to the inevitable and make peace. Hitler thought so, too, opening the door to a negotiated peace, which Churchill contemptuously slammed shut as Britain girded herself.

1940, June 23. De Gaulle and the Free French. A new element emerged from the debris of fallen France. Brigadier General de Gaulle, exponent of armored warfare, had been called from his division command (June 5) to become Undersecretary of State for National Defense in the short-lived cabinet of Paul Reynaud and had urged continued resistance. When Pétain capitulated, de Gaulle fled to England, calling on all true Frenchmen to rally and fight for freedom. The call was answered by French officers and soldiers from the world over—by driblets at first, then in mounting numbers. De Gaulle at once utilized the best of these men to organize outlying French territories. Captain **Jacques de Hautecloche**—who protected his family by changing his name to **Jean Leclerc**—was flown to French Equatorial Africa. General **Georges Catroux** went to Cairo as Free French commander in the Middle East. Colonel **Edgard de Larminat** organized the French Congo. In England, Admiral **Émile Muselier** began organization of a Free French Navy. Although supported to great extent by the British government, this Free French movement was as yet regarded with some misgivings by both Britain and the United States, whose leaders underestimated the dynamic driving force of de Gaulle.

1940, June 24. Arms from the U.S. In response to Churchill's urgent purchase request (June 3), a large shipment of small arms, machine guns, light artillery, and ammunition arrived in England.

1940, July 3–4. Seizure of French Warships. A British squadron appeared off **Oran,** Algeria, and demanded that the French squadron there choose (a) to join England and fight Germany, (b) to turn in at an English port for internment, or (c) to sink itself in Oran Harbor. On French refusal, Vice Admiral Sir **James F. Somerville** opened fire. In a short action, 3 French battleships were sunk, a fourth escaped, and 5 destroyers fled to Toulon. British damage was light. That same day the French squadron at Alexandria—1 battleship, 4 cruisers, and 3 destroyers—disarmed itself on orders of Admiral Sir **Andrew B. C. Cunningham,** commanding the British Eastern Mediterranean Fleet. In other English ports were 2 more submarines which, on British summons, joined "Free French" forces and served with the Royal Navy throughout the rest of the war. These coups accounted for a large proportion of the French Navy. Some ships remained at the naval base of Toulon, at Algiers, Casablanca, and Dakar. The conflict at Oran, however, embittered many Frenchmen. Pétain's French Vichy government severed diplomatic relations with Britain (July 5).

1940, September 27. Rome-Berlin-Tokyo "Axis." A 3-power pact was concluded at Berlin, each partner pledging the others total aid for 10 years. The treaty did not, however, require Japan to go to war against Britain or her allies.

The Battle of Britain

Operation "Sea Lion." Hitler, against the recommendations of his army and navy chiefs, but with the Luftwaffe's enthusiastic support, had decided to invade England (June 5). Control of the sea was essential. Having no adequate surface force to oppose British naval strength, the Luftwaffe's task was first to defeat the RAF and then to neutralize the Royal Navy. French, Low Country, and Norwegian airfields were developed to maximum capability. As prelude, harassing air raids were made daily against British coastal towns and shipping during July. Meanwhile, German armies were regrouping for embarkation, and the Navy was scouring Germany and the occupied countries for landing craft.

1940, August 8–18. First Phase. Goering mustered 2,800 planes, with capability of putting up 900 fighters and 1,300 bombers in 3 fleets—Marshal **Albert Kesselring**'s Air Fleet Two, flying from northern France; Marshal **Hugo Sperrle**'s Three, from Belgian and Dutch bases; and General **Hans-Jürgen Stumpff**'s Five (mainly bombers), based in Norway. Against this force, British Air Marshal Sir Hugh Dowding's Fighter Command mustered but 650 operational fighters in 52 squadrons. German strategy was to coax the British into combat, by strafing seaports and fighter bases, and then shoot them out of the sky. However, aided by Britain's newly developed radar, Dowding was able to concentrate superior force at vital spots and the Luftwaffe's massive daily day and night attacks—1,485 sorties the first day (August 8), rising to 1,786 sorties (August 15)—were roughly handled in combats ranging in a 500-mile arc from southwest to northeast England. In continuous fighting during the rest of the period, Fighter Command still dominated the air over Britain.

1940, August 24–September 5. Second Phase. German attacks were shifted to concentrate against main inland RAF bases. Large groups of bombers, each protected by 100 fighters, crashed through by sheer weight of numbers, inflicting great damage on airfields and communication and control centers. The German high command came close to cracking Fighter Command. More than 450 British fighters were destroyed; 103 pilots were killed and 128 wounded.

1940, August 24–29. Berlin Bombed. In retaliation for a bombing of London, the RAF Bomber Command staged a night raid on Berlin. For 3 hours, 81 British planes hovered unharmed over the fog-shrouded city. Damage was small, but the psychological effect immense. The raid was repeated (August 28 and 29), despite Berlin's 2 rings of antiaircraft batteries; numerous Berliners were killed or injured. Düsseldorf, Essen, and other German cities were also attacked. Hitler and Goering, in blind rage, after undergoing a week of British reprisals, again shifted their strategy. At the moment that British air defense had reached its lowest ebb, and with victory almost in German reach, the Luftwaffe was ordered to drop its assault against British airfields and control centers.

1940, September 7–30. Third Phase. London became the target for tremendous and incessant aerial bombardment. Fighter Command, its task simplified by the German singleness of objective, was thus able to concentrate its dwindling force. The bombing of London reached its crescendo (September 15) when more than 1,000 bombers and some 700 fighters swept all day over the city in wave after wave. By nightfall, 56 assaulting planes had been downed at the expense of only 26 British aircraft. British civilian casualties were heavy during this phase—from 300 to 600 lives lost and from 1,000 to 3,000 persons injured per day in the unrelenting assaults, while a considerable portion of the city was wrecked. But English spirit refused to falter, and the Luftwaffe's losses were so great that daylight bombing had to be dropped. The aerial tide had turned.

1940, September 14–15. British Counterblow. Bomber Command, in conjunction with light naval craft, destroyed nearly 200 barges in French and Low Country ports—one-tenth of the total gathered for the proposed invasion. Hitler suspended Operation Sea Lion—scheduled for September 27. Slowly the German aerial assaults tapered off. The last daylight raid occurred September 30.

1940, October 1–30. Final Phase. Sporadic German hit-and-run raids continued, doing relatively little damage. London was lashed by another intensive air raid (October 10). Hitler now canceled Operation Sea Lion (October 12). The Battle of Britain had been won by the RAF, of whom Churchill said: "Never, in the field of human conflict, was so much owed by so many to so few." Overall losses were 1,733 German planes shot down, to 915 British.

COMMENT. *Four factors decided British victory in the air: first was an indomitable will to win; second was radar, which pinpointed enemy presence, routes, and strength; third was a well-organized, efficient ground-control system, which enabled concentration of superior force at the right time and place; and fourth was the German's own strategical blunder: dispersion of effort. By mid-September, Luftwaffe losses were so great that decisive victory in the air became impossible, and without air victory there could be no invasion of England.*

1940, November. The Blitz Begins. While the Luftwaffe had given up its effort to gain permanent air control, sporadic night raids continued through the rest of the year. Coventry was struck (November 14–15) by about 500 German bombers and practically demolished. London was again swept by a devastating raid, causing many explosions and fires (December 29). Before the "blitz" ended (May 1941), more than 43,000 civilians—men, women, and children—had been killed and 51,000 others seriously injured.

Battle of the Atlantic

Despite the convoy system, and the addition to the British merchant marine of a number of Scandinavian vessels (most of these ships had escaped when their countries were overrun), the toll of British shipping sunk by the roving submarines kept mounting. By August 15, 2.5 million tons of shipping had been destroyed. Britain just did not have sufficient light warships to provide adequate protection for her merchant ships from the sea wolves, nor could the shipyards produce sufficient replacements.

1940, September 3. Trading Bases for Destroyers. Churchill expected his urgent shipbuilding program would in a few months produce destroyers in quantity, but until February 1941, the shortage would be tragic and perhaps fatal to Britain. U.S. President Franklin D. Roosevelt, alive to the worldwide threat imposed by Nazism, agreed to a momentous immediate exchange. For 50 old U.S. destroyers, Britain leased naval and air bases to the U.S. in its Western Hemisphere possessions—Newfoundland, Bermuda, the Bahamas, Jamaica, Antigua, St. Lucia, Trinidad, and British Guiana.

1940, September 22–25. Attack on Dakar. A British-Free French amphibious expedition attempted to take Dakar, French West Africa, to prevent its possible use by Germany as a submarine base for South Atlantic operations. General de Gaulle commanded the Free French force. Due to bad weather and Allied mistakes, the invaders were repelled by the Vichy French defenders. Churchill, hoping to avoid further bitterness between the Vichy French and Britain, ordered the attack canceled.

1940, October–December. The Surface Raiders. In late October the German pocket battleship *Admiral Scheer,* Captain **Theodor Krancke,** slipped through the blockade to pursue commerce destruction in the North Atlantic. She encountered (November 5) a 37-ship British convoy, escorted by the auxiliary cruiser *Jervis Bay,* with 4 6-in. guns, Captain **E. F. S. Fegen.** Fegen, radioing the alarm, ordered his convoy to scatter and deliberately attacked the *Scheer.* For more than an hour the unequal contest kept up until the *Jervis Bay* sank. However, 32 ships of the convoy got away. The German heavy cruiser *Hipper* also got out (November), but engine trouble later forced her into Brest for repairs. The presence of these 2 powerful vessels in the shipping lanes slowed down convoying and necessitated strengthening the escorts. The *Scheer* continued a destructive raid into the South Atlantic and Indian Oceans, returning safely to Germany 4 months later.

Operations in the Mediterranean Area, 1940

1940, June. The Situation. Control of the Mediterranean was vital to Britain. Through it ran the empire's "life line": the short sea route to

India and Australia–New Zealand. Egypt, where both the British Mediterranean Fleet and Middle East Command were based to protect the Suez Canal, was Britain's principal base securing this life line. Mussolini planned to seize the Suez Canal by a pincer movement: from Libya on the west and from Ethiopia and Italian Somaliland on the southeast. At the same time he prepared for invasion of Greece through Albania to secure the northern shore of the Mediterranean opposite Egypt. Recognizing the danger, even though Hitler was threatening to invade Britain, Churchill boldly and wisely rushed Britain's sole remaining armored division to Egypt.

Available British forces were General Sir **Archibald Wavell**'s Middle East Command, with 36,000 troops in Egypt (mostly administrative, plus the understrength armored division), 9,000 in the Sudan, 5,500 in Kenya, 1,475 in British Somaliland, 27,500 in Palestine, 2,500 in Aden, and 800 in Cyprus. His air-force contingent was very small. Admiral Sir Andrew B. C. Cunningham's Mediterranean Fleet consisted of 1 carrier, 3 battleships, 3 heavy and 5 light cruisers, and a number of destroyers. Against these forces were pitted the full strength of the Italian Navy (see table, p. 1152), the land-based Air Force, and much of the Italian Army. Italy itself lay geographically threatening Wavell's westward line of communications, while light Italian naval forces based on Ethiopian and Italian Somaliland coast ports threatened his eastern communications through the Red Sea and Indian Ocean. Mussolini's ground forces in East Africa—Ethiopia, Eritrea, and Italian Somaliland—numbered about 110,000, commanded by **Amadeo Umberto, Duke of Aosta,** while in Libya Marshal **Italo Balbo** had 200,000 men and a sizable air force.

1940, June 11. First Attack on Malta. Immediately upon declaring war, Mussolini launched 2 waves of bombers against the island of Malta—the first of many thousands of such raids.

1940, June–September. Italian Preparations. There were minor border clashes and Italian air raids as the Italians prepared for major operations against Egypt, the Sudan, Kenya, and British Somaliland.

First Western Desert Campaign, September–December 1940

1940, September 13. Graziani's Invasion of Egypt. Marshal **Rodolfo Graziani** (succeeding Balbo, killed in an air crash in June) entered Egypt with 5 divisions, moving on a narrow front along the coast. British covering forces fell back before him. Reaching Sidi Barrani (September 16), the Italians settled down in a series of fortified camps extending over a 50-mile area, while Wavell's forces—now 2 divisions strong—remained at Mersa Matruh, 75 miles east. Both sides received reinforcements. General Sir **Henry M. ("Jumbo") Wilson,** tactical commander in Egypt, made plans for attack, but operations were delayed when Wavell was ordered to occupy Crete and send part of his air force to Greece to assist in countering an Italian invasion there (see p. 1169).

1940, December 9. Wavell's Offensive. After a night approach march, Wavell's Western Desert Force, 31,000 men, 120 guns, and 275 tanks, commanded by Major General **Richard N. O'Connor,** ripped through a gap in the Italian chain of defenses. O'Connor's relatively small force—1 armored and 1 infantry division, 2 additional infantry brigades, and a battalion of the new British "I" tanks—hemstitched its way westward between the desert and the coast, gobbling in turn each Italian fortified area. Air and naval elements assisted. By mid-December, the Italians had been thrown completely out of Egypt, leaving 38,000 prisoners and great quantities of matériel in British hands. As the year ended, the Desert Force, after a pause of 2 weeks, was assaulting the perimeter of Bardia where Graziani's disorganized forces lay.

COMMENT. *This daring assault against a force 4 times its size was well planned and superbly executed. O'Connor's ground forces were ably supported by the RAF under Wing Commander* **R. Collishaw** *and by the long-range naval gunfire of Cunningham's warships along the coast. In principle the operation resembled Allenby's breakthrough in Palestine in 1917 (see* *p. 1066), in which Wavell had played a part. It must be noted, also, that at the time Wavell was launching his westward counterattack in Libya he was in process of mounting an equally daring assault into East Africa: a 2-pronged offensive to be launched simultaneously from the Sudan and from Kenya against Italian-held Abyssinia and Italian Somaliland (see p. 1173).*

GREECE AND THE BALKANS

1940, October 28. Italian Invasion of Greece. General **Sebastiano Visconti-Prasca,** who had an army of 10 divisions (about 162,000 men) in Albania, advanced into Greece with 8 divisions. The Italians immediately encountered unexpectedly fierce Greek resistance. General **Alexander Papagos,** whose army numbered about 150,000 men, had disposed them in a highly organized defensive zone through the difficult mountainous border area. The Italians were thrown back by determined counterattacks. The ineffective Visconti-Prasca was replaced by General **Ubaldo Soddu** (November 9).

1940, November 22–December 23. Greek Counteroffensive. Beginning with a successful assault on Italian-held Koritza, the Greeks rolled back all opposition. Assisted by Royal Air Force detachments sent by Wavell from Egypt, the Greeks advanced into Albania. As the year ended, the Italians were clinging desperately to the line Valona-Tepelino-Lake Ochrida, immense quantities of Italian war matériel had been captured, Italian military prestige had been degraded, and British naval units were bombarding the Valona base. Marshal Pietro Badoglio, Italian Chief of Staff, was forced to resign.

1940, June–October. Other Balkan Developments. During this year, Soviet Russia seized Bessarabia and northern Bukovina (June 26–28); the Rumanian oil-field area was occupied by "protecting" German troops (October 8), and British troops from Wavell's scanty forces occupied the island of Crete (October 30). The Mediterranean area was shaping up as a major theater of war.

NAVAL OPERATIONS IN THE MEDITERRANEAN, 1940

1940, July 9. Action off Calabria. Aggressive Admiral Cunningham carried the war into enemy waters. Cruising off the Straits of Messina with 3 battleships, an aircraft carrier, 5 light cruisers, and a number of destroyers, he fell in with Admiral **Angelo Campioni**'s squadron of 2 battleships, 6 heavy cruisers, 12 light cruisers, and destroyers, returning from troop convoy to Libya. Cunningham's immediate attack drove the Italians in flight for the Calabrian coast, with serious damage to a battleship and a cruiser. British injuries were negligible, despite the intervention of land-based Italian aviation. Ten days later, HM Australian light cruiser *Sydney,* with 4 destroyers, engaged 2 Italian light cruisers off the northwestern coast of Crete, sinking 1. Meanwhile, Malta—Britain's steppingstone for air between Egypt and Gibraltar—was being strengthened, and another battleship, a carrier, and 2 cruisers were added to Cunningham's command.

1940, November 11. Naval Air Assault on Taranto. Admiral Cunningham delivered a crushing attack from carrier HMS *Illustrious* on the Italian naval base of Taranto. Three Italian battleships were left in a sinking condition, 2 cruisers badly damaged, and 2 fleet auxiliaries sunk. The British lost but 2 planes out of 21. The assault reestablished British naval supremacy in the Mediterranean. However, Italian aircraft based on Sicily, on Sardinia, and on the African coast continued interfering with British water transport. Italian submarines played a very inefficient part in these operations.

OPERATIONS IN 1941

Battle of the Atlantic

THE SUBMARINE RAIDERS

The Wolf Packs. Under the direction of Admiral **Karl Doenitz,** German operations developed into the "wolf pack" pattern: groups of as many as 15 to 20 U-boats spread over the sea lanes approaching Britain. Any individual merchantmen were attacked at once. Convoys, however, were tracked by the discovering submarine until the pack could be assembled for several simultaneous assaults. Convoy losses mounted; by June 1941, some 5.7 million tons of British shipping had been sunk, whereas British shipyards could only build 800,000 tons of replacements. When the British intensified their long-range aerial searches, and began to use land-based air escorts in the eastern Atlantic, the wolf packs simply moved out to mid-Atlantic, beyond operational range of aircraft based in Northern Ireland or Great Britain. British bombing reprisals against U-boat bases in Germany, Norway, and France were frustrated by construction of so-called "pens," great roofs of reinforced concrete protecting submarine berths and shore installations.

The German Long-Range Bombers. Working in close coordination with the wolf packs were German long-range bombers based in Norway and France. These were able to scour the sea lanes closer to Britain, while still staying beyond the range of RAF fighter planes. Convoys that had been harassed by the wolf packs in mid-ocean suffered further serious losses from these planes as they approached Britain.

1940, September. Introduction of the Escort Carrier. Britain's answer to the combined menace of U-boats and bombers was to develop constant fighter-plane protection for convoys by means of escort carriers. The first of these, HMS *Audacity,* a merchant ship on which a flight deck was constructed, could carry 6 fighter planes. Working with destroyers and other surface warships, these planes somewhat reduced the wolf-pack menace in mid-ocean, and were able to drive off the bombers when the convoys approached the French coast. The *Audacity* was sunk in December, and U-boat sinkings began to mount again. But more escort carriers were being constructed for use in 1942.

THE SURFACE RAIDERS

1941, January–March. *Gneisenau* and *Scharnhorst.* These two battle cruisers under command of Vice-Admiral **Günther Lütjens** entered the North Atlantic. Five unescorted merchantmen were sunk, but the German efforts to find a convoy unescorted by British battleships were unsuccessful until March 18. Falling on a convoy just scattering from a submarine attack, the 2 German surface craft destroyed 16 vessels in a 2-day running chase, then fled for Brest as British battleships converged on the area. An RAF bombing attack found them there in harbor and so severely damaged them that they were out of action for several months.

1941, February–April. *Hipper.* Operating from Brest, the *Hipper* undertook a successful raid into the North Atlantic (February). Later she returned to Germany, slipping past British patrols between Iceland and Scotland.

CRUISE OF THE BISMARCK, MAY 18–28

1941, May 18. The *Bismarck* Leaves Gdynia. This giant new German battleship, largest and most powerful warship in the world at the time, in company with the new heavy cruiser *Prinz Eugen,* sailed to Bergenfjord.

1941, May 21. British Concentration. The *Bismarck* and *Prinz Eugen* were sighted in Bergenfjord by British reconnaissance planes. All available units of the Royal Navy—from Scapa Flow to Gibraltar—concentrated for their destruction.

1941, May 21. The *Bismarck* Puts to Sea. Admiral Lütjens took advantage of foggy weather; his 2-ship squadron sailed out, undetected by the British, heading for Denmark Strait.

1941, May 24. Battle of Denmark Strait. British cruisers sighted the German ships entering Denmark Strait (late May 23). Vice-Admiral **Launcelot Holland** in the battle cruiser HMS *Hood,* with the new battleship *Prince of Wales,* intercepted them early next morning. Fire opened at 25,000-yard range. A 15-inch shell from the *Bismarck* penetrated the *Hood's* vitals and the great ship blew up, taking all but 3 of her 1,500-man crew to the bottom with her. The *Bismarck's* excellent gunnery then severely damaged the *Prince of Wales,* forcing her to turn out of the fight. Despite damage to his own ship, Admiral Lütjens continued westward to the open sea. The damaged *Prince of Wales* and the light cruisers *Suffolk* and *Norfolk* trailed the German vessels, while other British warships gathered.

1941, May 24–26. The Chase. Lütjens turned east for Brest, ordering the *Prinz Eugen* to slip away. The British lost contact for more than a day, but a land-based patrol plane sighted the *Bismarck* again during the morning of May 26, 700 miles west of Brest.

1941, May 27–28. Death of the *Bismarck.* H.M.S. carrier *Ark Royal* successfully attacked to slow the German ship down and enable the battleships *Rodney* and *King George V* to catch up (May 27). That night a destroyer flotilla closed, inflicting further torpedo damage. Next morning the *Rodney* and *King George V* engaged the *Bismarck* in furious gun battle, finally silencing her. The great helpless, but still defiant, hulk resisted the British 14- and 16-inch shells. Finally, the cruiser *Dorsetshire*—just arrived—sent 2 torpedoes into the blazing *Bismarck* and sank her. Nearly 2,300 German sailors went down, including Admiral Lütjens. Only 110 survivors were picked up by the *Dorsetshire* and a destroyer before the arrival of a German U-boat prevented further rescue efforts.

"ULTRA"—READING GERMAN RADIO TRAFFIC

Radios were in use by armies and navies very early in the 20th century. Since wireless, or radio, transmissions can be received by friend and foe alike, the practice of encrypting messages was well established before World War I, to make them unreadable to all except intended recipients. After that war cryptography became dependent upon extremely complex electric coding machines, in many ways the forerunners of modern computers.

The most sophisticated of these was the German "Enigma" machine. The Germans remained confident to the end of the war that this machine could never be duplicated nor its codes broken. They were wrong! Furthermore, it was duplicated just before the war by scientists in Poland, a country scorned by the Germans. In a noteworthy act of allied loyalty and generosity, while Poland was being overwhelmed by Germany in 1939, Polish scientists delivered copies of their version of the Enigma machine to French and British colleagues. The Polish machine was not an exact duplicate of Enigma, but it was workable, and could read many German messages.

France was defeated and overrun by Germany in 1940 before French cryptographers received significant benefit from the Polish machines. The British, however, had improved the Polish Enigma, and were reading a substantial proportion of intercepted German radio messages. They had established a system—now given the code name "Ultra"—for getting important intelligence information from these messages rapidly into the hands of senior military commanders who could use it. Ultra information had already been very useful to the Royal Navy in combatting the German submarine campaign.

AMERICAN INVOLVEMENT, 1941

Waning of Isolationism in America. Despite reluctance to become involved in the war, most Americans began to realize the danger to the Western Hemisphere and to the world if Hitler should be able to starve Britain into surrender. President Roosevelt's cautious but deliberate steps to provide all possible assistance to Britain, short of declaration of war, were approved by most of his people.

1941, March 11. Lend Lease Act. The terms of this act empowered the President to provide goods and services to nations whose defense he considered to be vital to the defense of the United States.

1941, April 10. Protection of Greenland. Fearing the possibility that Germany might try to establish a Western Hemisphere base in this Danish possession, President Roosevelt announced that Greenland would be under U.S. protection.

1941, July 7. American Troops to Iceland. This step permitted the garrison which Britain had placed there to be withdrawn for active operations against Germany. It also paved the way for the next planned step.

1941, August 9–12. Atlantic Conference. Roosevelt and Churchill met on American and British warships in Placentia Bay, Newfoundland. In the Atlantic Charter, they pledged their countries to preserve world freedom and to improve world conditions after the war (published August 14). At the same time Roosevelt announced that American warships would escort all North Atlantic convoys west of Iceland.

1941, September 16. Beginning of American Convoys. The U.S. thus became a quasi belligerent in the war, and British warships which had been patrolling and escorting convoys west of Iceland were now free to intensify the anti-submarine war in the eastern North Atlantic. Germany promptly responded to this American quasi belligerency by attacking American warships on convoy duty. Two destroyers (USS *Kearney* and *Reuben James*) were torpedoed; the

Kearney limped to safety, the *Reuben James* was sunk (October 17, October 31).

1941, December 24. St. Pierre and Miquelon. Free French Admiral Muselier, with a cockleboat squadron, seized the islands off the Canadian coast, depriving the U-boats of a potential base, but causing some embarrassment to the U.S. government, which was keeping a tenuous link with Vichy.

Operations in the Mediterranean Area, 1941

WAVELL'S OFFENSIVES IN NORTH AND EAST AFRICA

1941, January 1–February 7. Campaign in Cyrenaica. The Western Desert Force resumed the offensive with the opening of the new year. **Bardia** was carried by assault (January 5) after an intensive naval bombardment, while the 7th Armored Division ("The Desert Rats") isolated Tobruk. Land assault against the southern face of **Tobruk**'s perimeter, accompanied by intensive air and naval bombardment, brought capitulation (January 22). The 7th Armored Division raced west across the desert bulge of Cyrenaica to cut off the remaining Italian troops, while the main British force pressed on along the coast road. The British armor closed the trap at **Beda Fomm,** on the Gulf of Sirte (February 5). Two days later, after an abortive effort to break through, General **Bergonzoli**'s demoralized Italians surrendered unconditionally.

COMMENT. *In 2 months' time, O'Connor's Western Desert Force had advanced 500 miles, destroyed 9 Italian divisions, and taken 130,000 prisoners, 400 tanks, and 1,290 guns. At no time did O'Connor employ more than 2 infantry divisions at once, though the "Desert Rats" were engaged throughout the operation. British casualties amounted to 500 killed and 1,373 wounded. It was a remarkable demonstration of fire and movement, aided by effective naval and air support, against an inept passive defense greatly superior in numbers.*

1941, January. British Plans against Italian East Africa. Simultaneously, Wavell launched a pincers offensive against the duke of Aosta's 110,000 troops in Ethiopia and Italian Somaliland. Lieutenant General **William Platt** with 2 South African divisions and indigenous troops was to strike east from Khartoum in the Sudan; Lieutenant General Sir **Alan A. Cunningham** in Kenya, with 1 South African and 2 African divisions and a conglomeration of Commonwealth and Free French forces, would move northeast. The combined forces totaled some 70,000 men.

1941, January 19–April 20. Invasion of Ethiopia and Eritrea. Platt crossed the frontier (January 19). He caught up with and defeated Italian General **Frusci**'s force at **Agordat** (January 31). He pursued to the defended defile of **Keren,** and broke through after severe fighting (March 3). Asmara and its seaport Massawa were occupied (April 1). Turning south, the British column now faced Amba Alagi, a 1,000-foot conical mountain, where Frusci's command planned a stand.

1941, January 24–March 30. Invasion of Somaliland and Ethiopia. Cunningham moved into Italian Somaliland (January 24), meeting little resistance. Occupying Mogadiscio on the Indian Ocean (February 25), he captured a huge supply depot of gasoline. The Italians began evacuating Somaliland. Reaching Jijiga (March 17), Cunningham pushed on to Harar (March 25).

1941, April 4. Capture of Addis Ababa. Turning southwest, Cunningham occupied the Ethiopian capital while the Duke of Aosta hurriedly retired northward. In an amazing advance of more than 1,000 miles, Cunningham had averaged 35 miles per day, captured 50,000 prisoners, and gained 360,000 square miles of enemy-held terrain at a cost of 135 men killed, 310 wounded, 52 missing, and 4 captured. He now moved north, occupying Dessie, 250 miles from Addis Ababa (April 20), after a brush at the **Combolcia Pass** with the Italian rear guard. Aosta's demoralized forces joined Frusci's at Amba Alagi.

1941, May 5. Return of Emperor Haile Selassie to Ethiopia.

1941, May 18. Italian Capitulation. Aosta surrendered as Platt and Cunningham pressed in.

COMMENT. *Another daring, fast-moving British*

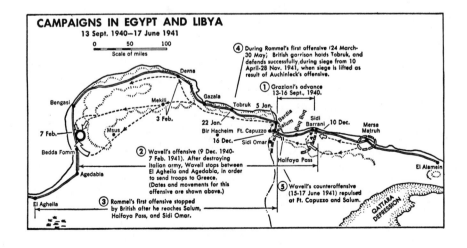

offensive had resulted in clear-cut victory. As in Libya, the Italian high command had proven to be inept. Wavell's daring strategy securely bolted his back door, eliminating any future threat to the Suez Canal from East Africa, and secured the Red Sea as an Allied supply channel.

GERMANY REDRESSES THE MEDITERRANEAN BALANCE

1941, January–March. Appearance of the Luftwaffe. Hitler, to bolster Mussolini, sent the Luftwaffe's Air Corps X (500 planes) from Norway to Sicily (January). By the end of February, the long-range German bombers barred use of the port of Benghazi as a base for Wavell's Desert Force. The necessity to send most of his combat troops to Greece and Crete (see p. 1175) had by this time reduced this force to 1 armored division, part of an infantry division, and a motorized brigade.

1941, March. Arrival of Rommel. Hitler sent General Rommel with his Panzer Afrika Korps to Tripolitania, charged with command of Axis operations against Egypt and the Suez Canal. (Titular commander, Italian General **Ettori Bastico,** was a mere figurehead.)

1941, March 24–May 30. Rommel's First Offensive. With the 21st Panzer Division and 2 Italian divisions—1 armored and 1 motorized—Rommel drove back the British covering force at **El Agheila** (March 24), then branched out in reverse of the original British offensive—the Italian column following the coast through Benghazi on Derna, the 21st Panzer cutting across the desert toward Tobruk. The British 2nd Armored Division, attempting to delay, was split; 1 brigade, forced into Derna because of gas shortage, was captured (April 6). Next day, most of the remainder of the division was surrounded and captured. The 9th Australian Division reached Tobruk. Brilliant General O'Connor, reconnoitering on the Barce-Derna road, was captured by a German patrol (April 17).

1941, April–December. Defense of Tobruk. Wavell determined to hold Tobruk at all costs—to deprive Rommel of a base port to support further advance into Egypt, and to threaten the Italo-German flank. By sea he threw in the 7th Australian Division and some tanks to reinforce the Tobruk garrison. Rommel (April 10) mounted an ill-prepared assault and was thrown back after a 3-day attack. Advancing part of his force to the Sollum escarpment east of Tobruk, Rommel invested the fortress, while overhauling his tanks and bringing up reinforcements. His supply problem was now serious, thanks to Wavell's decision to hold Tobruk and to the aggressive operations of the Royal Navy, which took continuous toll of Axis surface transport shuttling between Italy and Tripolitania.

1941, June 15–17. Wavell's Counteroffensive. Forced by political pressure from home to attempt the relief of Tobruk, Wavell launched a shoestring offensive against the Sollum-Halfaya passes held by Rommel—1 infantry division and an armored division participated. Split into 6 semi-independent task forces and committed piece-meal, the British attack was repulsed by Rommel after some minor successes.

1941, July 1. Wavell Superseded. General Sir **Claude Auchinleck** assumed the Middle East Command. General Wavell was transferred to India.

COMMENT. *Fifty-eight-year-old Wavell, a master of offensive-defensive strategy, was in effect made a scapegoat for the disastrous British intervention in Greece (see p. 1175), which had been added to his already monumental task of holding the Middle East for Britain.*

1941, July–October. Opposing Preparations. The Western Desert Force, renamed the Eighth Army, with General Alan Cunningham in command, was reinforced to 7 divisions and 700 tanks and prepared for the offensive. The Desert Air Force was built up to 1,000 planes. Meanwhile, Rommel was regrouping the Afrika Korps, which now comprised the 15th and 21st Panzer Divisions and the understrength 90th and 164th Light Infantry Divisions, and 6 weak Italian divisions. Bardia, Sollum, Halfaya, and Sidi Omar were organized into first-line defenses as springboards for invading Egypt. He had 260 German and 154 Italian tanks, and was supported by 120 German and 200 Italian planes.

1941, November 18–December 20. Auchinleck's Offensive. Cunningham's Eighth Army in surprise attack struck northwest from its base at Mersa Matruh. Its new American "Stuart" light tanks sought Rommel's army south of Tobruk, while the main force swung north to isolate the Italo-German Sollum-Bardia defensive zone. In a series of uncoordinated desert tank battles around **Sidi-Rezegh** (November 19–22), the British were checked, while a sortie from Tobruk was repulsed; seemingly, the British offensive was stalled. Rommel, with reinforced air support, gathered all available tanks and struck over the frontier into the British rear areas. Near panic resulted, and Cunningham contemplated withdrawal. But Auchinleck insisted on a stand. Rommel's momentum was checked, and junction between the Tobruk garrison and a New Zealand division then split the Afrika Korps. The surrounded Germans succeeded in breaking out, and Rommel then withdrew to the west and south of Tobruk (December 7–8). Auchinleck, having retrieved the situation, replaced Cunningham by General **N. M. Ritchie.** Under continued British pressure, Rommel then withdrew his mobile forces all the way back to his original position at El Agheila (December 28–30). The Italo-German garrisons at Bardia and Halfaya were captured (January 2–17, 1942). Italo-German losses were 24,500 killed and wounded and 36,500 prisoners. British losses were about 18,000 in all. But Rommel had also lost 386 tanks and 850 aircraft.

COMMENT. *General Auchinleck's determination had for the moment removed a most serious threat to the Suez Canal and the British Empire itself. Rommel was back where he had started.*

IRAQ AND SYRIA

1941, May. Operations in Iraq. Nazi-stimulated outbreaks in both countries added to General Wavell's troubles in the Middle East.

Rashid Ali, Prime Minister of Iraq, attacked British garrisons at **Basra** and **Habbaniya** (May 2), threatening the Iraqi oil preserves essential to the Allied war effort. At the same time Hitler sent in munitions, and finally an Italo-German air base was established at Mosul. British reinforcements from India and from Wavell's scanty Palestine garrison—already occupied in Syria (see below)—hurriedly assembled, took the offensive. RAF planes struck the Mosul air base, while ground troops moved on Baghdad. Rashid Ali fled the country, Baghdad was occupied (May 30), and stability was reestablished.

1941, June–July. Operations in Syria. Meanwhile, the 35,000 pro-German Vichy French forces in Syria, directed by German military and political advisers, threatened from the north. At Churchill's order, Wavell scraped together a 20,000-man force and invaded Syria from Palestine and Iraq (June 8). Free French troops under Catroux's command played an important part in the expedition. Damascus was soon captured (June 21). This caused the surrender of General **Henri Dentz,** French commander in Syria (July 12). Syria and the Lebanon were transferred to Free French control, most of the Vichy French troops volunteering to join Catroux's command.

Operations in the Balkans, 1941

1941, February–March. German Pressure on Yugoslavia. Hitler, planning an invasion of Russia in the summer of 1941, determined to secure his southern flank by first gaining control of Yugoslavia. (Hungary and Rumania were already German satellites.) Under intense diplomatic pressure and threats of force, Yugoslavia's Prince Regent **Paul** reluctantly agreed to join the Axis alliance (March 25).

1941, March 7–27. Arrival of British Troops in Greece. In anticipation of German intervention in the Balkans, and possibly against Greece, the British government had ordered Wavell to send his best Middle East Command troops to Greece. This 4-division force of

57,000 men was commanded by General Maitland Wilson.

YUGOSLAVIA

1941, March 27. Coup d'État. Patriotic anti-German elements in the Yugoslav Army, encouraged by the presence of British forces in Greece, overthrew Prince Paul, established a new government, and rejected the Axis alliance.

1941, March 27–April 6. German Preparations. Hitler, infuriated by events in Yugoslavia, which he felt resulted from the failure of Italy's invasion of Greece, ordered his military staff to mount an immediate attack against Yugoslavia and Greece. The German planners did so in 10 days—an amazing demonstration of military efficiency. Marshal **Wilhelm List**'s Twelfth Army and Kleist's First Panzer Group (both in Hungary and Rumania preparatory to the Russian invasion) were shifted to southwestern Rumania and Bulgaria, opposite the Yugoslav and Greek borders. The hurriedly organized Second Army was assembled in Austria and Hungary opposite northern Yugoslavia. The Italian Second Army, from Trieste, and the Hungarian Third Army were to assist.

1941, April 6–17. Invasion of Yugoslavia. After bombing that concentrated on Belgrade, paralyzing the Yugoslav high command, German armored and infantry forces slashed into the country from north, east, and southeast. Efforts to mobilize the million-man Yugoslav Army were never completed. Zagreb fell (April 10), Belgrade (April 12), and Sarajevo (April 15). Resistance ended in unconditional surrender (April 17). German losses totaled 558 men. Yugoslav losses are unknown; more than 300,000 were captured, perhaps 100,000 killed and wounded.

1941, May–December. Yugoslav Resistance Movements. Many Yugoslav officers and soldiers, refusing to surrender, fled to mountain hideouts from which they harassed the Germans. The principal leader of these refugees,

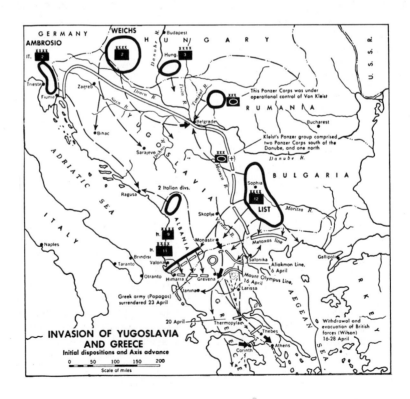

INVASION OF YUGOSLAVIA
AND GREECE
Initial dispositions and Axis advance

called **Cetnici** (Chetnicks), was Serbian Colonel **Draza Mihailovic.** One of his principal Montenegrin subordinates was Major **Aleksa J. Dujovic.** About the same time, the leader of the Yugoslav Communist Party, **Josip Broz**—known to his followers as **Tito**—established the National Liberation (or Partisan) Movement, operating then in areas other than Serbia and Montenegro. Partisans and Chetnicks hated each other as much as they did the Germans. By the end of the year a bitter, bloody 3-way war was raging in Yugoslavia.

GREECE

1941, April 6–9. Invasion of Greece. Simultaneously with the Yugoslav invasion, other elements of List's army smashed against Greece's fortified Metaxas Line, east of the Struma River, where Papagos mistakenly ordered his Second Army to hold. Penetrating quickly, the German spearheads reached the sea at Salonika (April 9). The Second Army, cut off, surrendered. The German Twelfth Army pressed westward toward the British, just arriving and preparing defensive positions between Mt. Olympus and Salonika. Kleist's armored group, driving south from Skoplje, struck through the Monastir Gap, held by a Greek division protecting the British left. The Greeks crumbled; after severe fighting, Wilson withdrew to new positions north of Mt. Olympus.

1941, April 12–20. Battle for Central Greece. German Spearheads forced their way through the rugged mountains of northern Greece between the British and the Greek First Army, now belatedly retreating from Albania. Gallant Papagos, convinced that success was impossible, recommended to Wilson that the British evacuate while he delayed the Germans as long as possible. Meanwhile, Wilson was forced to abandon his Mt. Olympus position, withdrawing to Thermopylae.

1941, April 21–27. Conquest of Greece. Assailed from north, east, and south, the Greek Army, after desperate resistance, was forced to surrender (April 23). Wilson abandoned Thermopylae (April 24), pulling back into the Peloponnesus, as Royal Navy warships, braving violent Luftwaffe attacks, began to move in to pick up British troops by night at eastern Greek ports. A German airborne drop at Corinth (April 26) almost closed the withdrawal route, but Wilson fought his way through with most of his remaining troops. British evacuation was completed on April 27, 43,000 troops being rescued, but all heavy equipment was abandoned. German losses in Greece were slightly over 4,500 men; the British had 11,840 casualties. Greek killed and wounded were more than 70,000; 270,000 were captured.

COMMENT. *The German operations in Yugoslavia and Greece were models of precision and efficiency, marked by bold use of tanks over terrain supposedly impassable to armor.*

CRETE

1941, April 27–May 10. Preparations. About 15,500 of the British troops evacuated from Greece were landed on Crete, expected to be the next German objective. There, joined by 12,000 reinforcements from Egypt and assisted by the 14,000-man Greek garrison, they feverishly prepared the island for defense, despite severe shortages in artillery and other equipment. In command was New Zealand Major General **Bernard C. Freyberg.** The small RAF fighter force on the island was soon forced to withdraw by increasingly severe German air raids, which daily mounted in intensity. In southern Greece, the German XI Airborne Corps, commanded by Lieutenant General **Kurt Student,** prepared for an airborne assault on the island.

1941, May 20–28. The Assault. Following heavy aerial bombardment, units of the German 7th Parachute Division began landing near the principal airports of Maleme, Rethymnon, and Herakleion. The alert defenders inflicted heavy casualties, and by nightfall still controlled the fields, though a few German units had established themselves near Maleme (May 20). Regardless of casualties, the remainder of the 7th Division dropped the next day and gained control of part of Maleme airfield. Student immediately began to send in elements of the 5th Mountain Division by plane. Most of the planes were destroyed, but enough crash-landed safely to permit the determined Germans to hold the field. The most intensive fighting continued in the following days, as German reinforcements continued to come in by plane. Two efforts to send reinforcements by sea were repulsed, with heavy casualties, by the Royal Navy (May 21–22). The airborne elements, however, with very effective air support, were able to force the British and Greeks back. Freyberg was finally forced to withdraw to the south coast (May 28).

1941, May 28–31. Conquest of Crete. Under heavy German pressure, most of the defenders withdrew to Sfakia, whence they were evacuated by the Royal Navy. Remaining elements, cut off by the Germans, were forced to surrender (May 31). British casualties were 17,325 (including 2,011 naval losses and 11,835 prisoners). The Germans admitted 5,670 casualties, mostly in the 7th Division; this does not include losses in the seaborne convoys.

COMMENT. *This was the first major airborne assault in history; despite its success, Hitler was so shocked by the losses that he never again ordered a comparable airborne operation.*

NAVAL OPERATIONS IN THE MEDITERRANEAN, 1941

1941, February 9. Surface Raid on North Italy. Daringly, British Admiral Somerville (February 9) with 2 battleships, a carrier, and a cruiser from the Gibraltar naval base swept the northeastern coast of Italy, bombarding Genoa and other ports.

1941, February–March. German Air Intervention. German planes sent to Sicily by Hitler (see p. 1174) established aerial superiority over the central Mediterranean, disrupting Allied surface and air activities. Malta was bombarded incessantly. Shipping from England had to be rerouted via the long Cape of Good Hope route into the Indian Ocean and the Red Sea. The carrier H.M.S. *Illustrious* was badly damaged and a cruiser was sunk.

1941, March 26–27. Sortie of the Italian Fleet. British troop movements from Egypt to Greece led Italian Admiral **Angelo Iachino** to attempt to intercept. The Italian fleet—3 battleships, 8 cruisers, and a number of destroyers and supporting craft—covered by Italian and German air, was divided into 3 detachments. Admiral Cunningham, learning (March 27) that the Italians were threatening several large convoys, ordered the transports back to Egypt and hurriedly steamed to meet the threat. He had 3 battleships, 1 carrier, 4 cruisers, and more than a dozen destroyers.

1941, March 28. Battle of Cape Matapan. Harassed by Cunningham's cruiser-carrier force, the Italians turned away. The battleship *Vittorio Veneto* and heavy cruiser *Pola* were damaged, however, and slowed down. After dark, Admiral Iachino sent 2 cruisers and 4 destroyers back to their assistance. At 10 P.M., Cunningham's pursuing battleship force fell in with the *Pola* and the rescuing Italian vessels, and in surprise attack sank them all. The injured *Vittorio Veneto* and the remainder of the Italian fleet escaped in the darkness. One British cruiser was slightly damaged and 1 airplane lost. From this time to the end of the war, Italian surface naval strength ceased to be an important factor; it remained mostly in port at La Spezia.

1941, April 26–June 1. Evacuation of Greece and Crete. Hitler's conquest of Greece and occupation of Crete (see above) placed an almost insuperable task on Admiral Cunningham's Mediterranean command: evacuating British troops by sea first from Greece and then from Crete. British surface craft, without air support, were subject to full-scale, determined, and incessant attack by Luftwaffe planes from Italy and Thessaly.

1941, April 26–30. Evacuation from Greece. Using all available shipping and operating under cover of darkness, the Royal Navy lifted some 43,000 of General Wilson's ill-fated Grecian expedition from the Piraeus and from Peloponnesian beaches. Some of the evacuated troops were landed on Crete. Two destroyers and 24 other vessels were lost to German air attacks, and severe damage was inflicted on other British ships.

1941, May 21–June 1. Operations off Crete.
A German amphibious operation (May 21–23) to reinforce the airborne assault on Crete met complete disaster. Cunningham's war vessels fell on the small craft attempting to shuttle troops from Greece, sinking most of them. Some 5,000 German troops were lost. Continuous air attacks harassed the British ships in the waters around Crete, sinking 2 cruisers and a destroyer, and damaging 2 battleships and 2 more cruisers. Despite these losses, Cunningham's ships dared the impossible to evacuate more than 15,000 of General Freyberg's Cretan command (May 28 and 31). During this period, 4 cruisers and 6 destroyers were sunk, and 1 carrier and 3 battleships so severely damaged they would be out of action for some time. British naval casualties included 2,000 men killed.

1941, June–December. Other Operations.
Arrival of 25 German U-boats in the Mediterranean further complicated the British naval situation. Cunningham's fleet, already crippled by its Greco-Cretan operations, and charged with assuring surface supply to Malta, was put on the defensive. Italian light craft attempted a raid on the Valetta harbor, but were repulsed by British coastal artillery (July). Also, German air attacks on Malta were intensified, for Malta—the "unsinkable aircraft carrier"—was severely punishing German trans-Mediterranean supply to Rommel. The German reinforcements soon made themselves felt. The *Ark Royal*, Cunningham's 1 aircraft carrier in service, was sunk while on escort duty—a disastrous loss (November).

1941, December 19. British Naval Disasters.
Three Italian "human torpedo" teams—midget submarines guided by 2 men each—boldly entered Alexandria Harbor and further damaged the battleships *Queen Elizabeth* and *Valiant,* putting them completely out of action for many months. The gallant Italians escaped. That same day, a task force of 3 British cruisers and a destroyer met disaster in a mine field. One cruiser and the destroyer were sunk and the other vessels crippled for months. On Christmas Day, the battleship H.M.S. *Barham,* Cunningham's 1 remaining capital ship, was torpedoed by German submarine *U-331* and sunk with heavy loss of life. Shortly afterward, another British cruiser was torpedoed.

COMMENT. *At year's end Cunningham's fleet, reduced to 3 cruisers and a handful of destroyers, was seemingly impotent. Supply to Rommel was unimpeded. One bold move by the Italian Navy could have swept British sea power from the Mediterranean. But the move never came. Cunningham kept his handful of warships constantly and aggressively at sea and British counterintelligence blacked-out news of the damage to the battleships at Alexandria.*

Invasion of the U.S.S.R., 1941

German Plans. Hitler considered the dismemberment of the Soviet Union as the final step essential to achieving his *Lebensraum* dreams (see p. 1133). Planning for an attack on the U.S.S.R., therefore, began as soon as France was conquered. Time was of the essence, since it was evident that Stalin planned westward expansion, both in the north—Finland—and in the south—Turkey and the Dardanelles. Formal planning began in December 1940, and troops were shifted to the eastern frontier. Originally set for May 15, 1941, the invasion timetable was seriously disarranged by the necessity for the Balkan campaigns to clear the German southern flank.

In Operation "Barbarossa," Rundstedt's Army Group South—4 armies (1 Rumanian) and Kleist's First Panzer Group—was to drive on Kiev and the Dnieper Valley to envelop and destroy all Russian forces between the Pripet Marshes and the Black Sea. Bock's Army Group Center—2 armies and 2 Panzer groups (Guderian's Second and Hoth's Third)—was to follow the traditional invasion path: Warsaw-Smolensk-Moscow. Its armored pincers were to meet on the upper Dnieper, then capture Moscow. Leeb's Army Group North—2 armies and Hoeppner's Fourth Panzer Group—was to swing northeast toward Leningrad, pinning the Soviets in its zone against the Baltic Sea. Finland allied itself with Germany, and Mannerheim's Finnish army group was to occupy the Karelian Isthmus-Lake Onega front, threatening Leningrad from the north, while still farther north, German General Falkenhorst's Norway Army was to cut the Soviets' Murmansk-Leningrad supply line. In all, 162 divisions of ground troops—approximately 3 million men—were involved, 200,000 of them from satellite nations. Hitler envisioned a typical blitzkrieg campaign of not more than 4 months' duration.

Russian Plans. Soviet forces were heavily concentrated along the western frontier in the annexed regions of Poland, Bessarabia, and the Baltic States. South of the Pripet Marshes lay Marshal **Semën M. Budënny**'s Southwest Front (Army Group), north of the Marshes and up to the Lithuanian border was Marshal **Semën K. Timoshenko**'s Western Front, and Marshal **Kliment Voroshilov**'s Northwest Front was in the Baltic countries. Soviet mobilized strength approximated 3 million on the western front, with another million scattered elsewhere. Reserves were available in great quantity. Russian matériel—tanks, cannon, and planes—was plentiful, but inferior in quality to German equipment.

1941, June 22–July 10. The Invasion. Simultaneously, along a 2,000-mile front, the onslaught started at 3 A.M. with the customary preliminary air bombardment. Though the Soviets expected an invasion, tactical surprise was complete. Center Army Group advanced spectacularly. Its armored pincers closed on **Minsk** by mid-July, enfolding 290,000 prisoners, 2,500 tanks, and 1,400 guns, then opened in another scoop as the Panzers crossed the Dnieper.

1941, July 10–19. Smolensk. Again the trap snapped, bagging an additional 100,000 prisoners, 2,000 tanks, and 1,900 guns. The remnants of Timoshenko's Western Front recoiled in disorder. One of Bock's spearheads reached Beloj, less than 200 miles west of Moscow. But Rundstedt's Army Group South had made slower progress, while difficult terrain and some errors slowed down Leeb's group on the north. The German supply system was strained to the breaking point by the immense distances, slowing the advance. Tanks and other vehicles were showing the strain of terrific pace and long marches. In all probability, however, these things would not have saved Moscow from the still fast-moving German Army Group Center

had Hitler himself not intervened to check its momentum.

1941, July 19–August 21. Hitler's Changes. To bolster the slower-moving flank armies; Hitler detached both Panzer groups and 1 army away from the Army Group Center, despite the vigorous protests of his field commanders. Guderian's Second Panzer Group and the Second Army (**Maximilian von Weichs**) were to support Rundstedt's Army Group South, which by this time had almost reached Kiev. Hoth's Third Panzer Group was to join Army Group North. The immediate results of this unexpected change were favorable.

1941, August 21–September 26. The Kiev "Pocket." Guderian's Panzers from the north, slicing around the eastern edges of the Pripet Marshes, met Kleist's Panzers from the south, to press 5 Russian armies inside the great bend of the Dnieper, near Kiev. German infantry sealed the trap. A period of confused fighting followed as some Soviet units battled to break out; others of Budënny's army group sought to rescue them. Kiev itself fell (September 19), and finally some 665,000 Soviet troops surrendered. Budënny's Southwest Group seemed to have been completely shattered. Meanwhile,

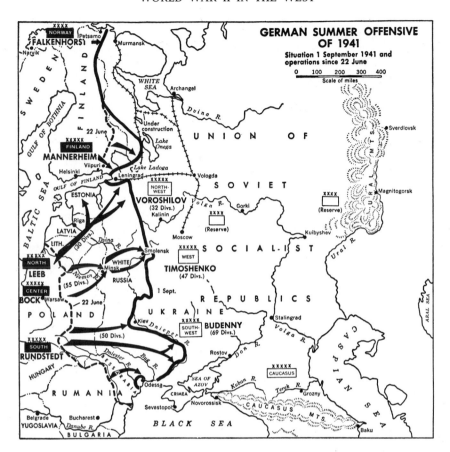

GERMAN SUMMER OFFENSIVE OF 1941
Situation 1 September 1941 and operations since 22 June

on the far southern flank, the Germans reached the Crimean Peninsula, where General **Erich von Manstein**'s newly formed Eleventh Army began a drive for Sevastopol (October). By this time the Luftwaffe had destroyed 4,500 Soviet planes, while losing less than 2,000.

1941, October–November. Time, Space, and "General Mud." Rundstedt reached the Don (October 15), threatening Rostov and Kharkov. Leeb slowly progressed toward Leningrad, which was invested (early October). But the Finnish Army, after reaching the original national boundaries, refused to press farther toward Leningrad, and in the far north Falkenhorst, after initial gains, bogged down in the tundra. The Murmansk-Leningrad supply line, though partly disrupted, was never closed entirely. Hitler, meanwhile, had changed plans

again and ordered an extreme effort against Moscow. Bock's Army Group Center, reinforced again by the Panzers and Luftwaffe elements taken from it to bolster the flanks, renewed its penetration, winning another tremendous victory at **Vyazma** (September 30–October 7), capturing more than 650,000 prisoners. Although rain now slowed the advance, the drive continued and reached Mozhaisk—40 miles from the Soviet capital (October 20). But the German armies were reaching the limit of their endurance. Armored units were reduced to 50 percent of their original combat efficiency; infantry divisions were scarcely in better shape. Autumn rains turned the roads into quagmires, and then the fierce Russian winter set in.

1941, November–December. Defense of

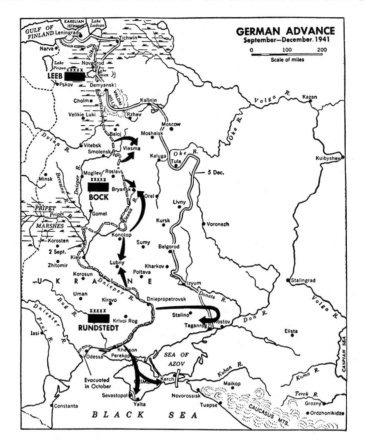

Moscow. Soviet resistance had stiffened. Budënny had been relieved, Timoshenko taking his command, and a newcomer—Marshal **Georgi K. Zhukov**—headed the central Russian forces barring the way to Moscow. These were mostly regular troops assembled from interior and eastern Russia. The losses elsewhere had been made good by newly mobilized but trained reserves, far inferior in quality to the Germans, but fresh, untired, and grimly determined. In the south a Soviet counterattack drove Rundstedt's spearheads out of **Rostov** (November 15), and Rundstedt resigned (just as Hitler relieved him), Reichenau taking his place (December 1). Kluge, now commanding the German Army Group Center, reached to within 25 miles of Moscow, and German patrols reached the suburbs in sight of the Krem-

lin. The Soviet government had moved to Kiubyshev. But temperatures had fallen to −40° F. The offensive ground to a halt all along the line. Hitler, infuriated, relieved Brauchitsch, Leeb, and other generals, and took personal command, by radio from Berlin.

COMMENT. *Hitler's decision virtually ruined the famed German General Staff (the Army Supreme Command, or Oberkommando des Heeres, or OKH). He kept General Franz Halder as Chief of Staff of OKH, but supervised it personally, limiting its responsibility to the Eastern Front. His own personal staff (Armed Forces Supreme Command, or Oberkommando der Wehrmacht, or OKW) was augmented and, under Field Marshal* **Wilhelm Keitel** *as Chief of Staff, directed operations in Western Europe and Africa. Thus Hitler had 2 separate general staffs, united only by his personal direction.*

1941, December 6. Russian Counteroffensive. Massively reinforced by 100 fresh divisions, by the year's end the Soviet armies were driving into the German armies at Kalinin, north of Moscow; at Tula, south of it; and at Izyum in the Ukraine. Soviet troops retook **Kalinin** (December 15) and **Kaluga** (December 26–30) despite desperate and skillful German resistance. As the year ended the stunned and frozen Germans were still yielding ground.

COMMENT. *Between July and November, Hitler's armies had accomplished one of the greatest sustained offensives in military history. About 3 million casualties had been inflicted on the Soviets (half of these prisoners), but German losses had also been tremendous—some 800,000. Moreover, the Soviet armies had not been destroyed; despite tremendous losses in men and matériel, space had been successfully traded for time. Time was what the Germans lacked, and space consumed their resources. Hitler's offensive was predicated upon quick annihilation of his enemy before the winter set in. No preparations had been made for a winter campaign. Soldiers were freezing to death in summer uniforms; tanks and trucks were immobilized as crankcases froze. German ignorance of Soviet capabilities was colossal. Hitler's own actions—the 3 weeks' delay to accomplish the Balkan sideshow, and his subsequent shifts of objectives when Moscow was almost within his grasp—combined to further handicap his worn generals and armies. Meanwhile, with most of the U.S.S.R. industrial area overrun or embattled, it was obvious to Britain and the United States that massive assistance in weapons, ammunition, and all kinds of military equipment would be essential to keep the U.S.S.R. in the war against Germany. British convoys to Murmansk started in August. U.S. Lend Lease to the U.S.S.R. was approved (November 6) a month before America became a full belligerent.*

United States Entry into the War, 1941–1942

1941, December 7. Pearl Harbor. Japan's attack on the U.S. base in Hawaii (see p. 1233) was followed by U.S. declaration of war (December 8).

1941, December 11. Germany and Italy Declare War on the U.S.

1941, December. U.S. Mobilization. Congress passed a $10-billion appropriation for defense and for Lend Lease aid (December 15), and 4 days later extended the legal age bracket for compulsory military service to include men from 20 to 44.

1941, December 22–1942, January 14. Arcadia Conference. Prime Minister Churchill and President Roosevelt met in Washington with their senior military advisers, who were officially designated as the Combined Chiefs of Staff (CCS). The new CCS began to plan and direct the Allied war effort under the two civilian leaders on February 6. Basic strategy for conducting global war was agreed: priority of effort was to be given to defeating the Western Axis; Japan was to be checked, but her defeat was to be sought only after Germany had been disposed of. Immediate and full cooperation between Britain and the U.S. in prosecuting the war followed, predicated upon the necessity for unity of command in each theater of war.

1942, July 24. Admiral W. D. Leahy, former Chief of Naval Operations, was appointed Chairman, Joint Chiefs of Staff, which assured a balance of Army and Naval Officers on the Joint Chiefs of Staff.

OPERATIONS IN 1942

Battle of the Atlantic

1942, January. German Operations against Allied Supply Routes. Immediately upon the entry of the U.S. into the war, German naval efforts concentrated on barring the vital sea lanes to Allied supply. While 64 U-boats went into the Atlantic to carry the war to the American coast, German surface warships gathered along the Norwegian coast to assist the submarine-Luftwaffe blockade of the only remaining sea route to the U.S.S.R.—via Murmansk—from England and the U.S., an essential logistical pipe line if the U.S.S.R. were to be kept in the war. The great new battleship *Tirpitz* (sister ship to the *Bismarck*) went to

Trondheim Fjord (January), to be followed later by several other warships. The German vessels at Brest were ordered home, and U-boat construction was intensified.

1942, January–April. The American Coast. The U-boats, supplied by "milch cows"—large cargo- and fuel-carrying submarines—which enabled them to operate for indefinite periods, at first found rich pickings. The U.S. Navy was not yet geared for convoy operations. Hapless merchantmen crowding the coastal sea lanes, silhouetted at night by the glare of lights from coastal ports and resorts, became easy prey. Some 80 merchant ships were destroyed in the first 4 months of 1942.

1942, February 11–13. Run-by in the Channel. The battle cruisers *Scharnhorst* and *Gneisenau* and the heavy cruiser *Prinz Eugen*, under Vice Admiral **Otto Ciliax,** with destroyer escort and Luftwaffe air cover, daringly swept through heavy fog from Brest through the English Channel into the North Sea despite belated British air and surface efforts to stop them.

1942, February-July. Mounting Losses on the Murmansk Run. Allied convoys to Murmansk (and in summer to Archangel) assembled off Iceland and thence moved eastward past North Cape. Despite strenuous efforts to protect them, German U-boats, surface craft, and land-based aviation from Norway took a grievous toll. Out of 39 convoys—533 ships—participating in 1941–1942, 69 vessels were lost, most during the spring of 1942. The most disastrous trip was that of convoy PQ-17 (June–July); 23 of 37 merchant ships in the convoy were sunk by German submarine, air, and surface attackers, which included the battleship *Tirpitz*, pocket battleship *Scheer*, and heavy cruiser *Hipper*.

1942, May–August. Surface-Submarine Contest off America's Coasts. As air-sea protection—Army, Navy, and Coast Guard—improved, U-boat losses mounted. So the U-boats transferred their operations to the Caribbean, intercepting oil tankers and bauxite carriers from South and Central American ports, and also preying on Panama Canal traffic. The submarines clustered across the Halifax-Iceland-Londonderry run; their operations extended across the South Atlantic to the Cape of Good Hope.

1942, August–December. The Balance Begins to Shift. The German submarine attacks were intensified, but so were Allied countermeasures. In all, Allied merchant losses in the Atlantic and the Arctic Oceans in 1942 amounted to 1,027,000 tons. But by year's end, some 85 U-boats had been sunk. The menace had not been reduced, but the Battle of the Atlantic was no longer a clear-cut German victory. Also the graph of new American ship construction—both naval and merchant—was coming near to balance with the losses. Meanwhile, the American war effort flowed overseas in increasing volume.

Around the Periphery of Fortress Europe

Strategy, Build-up, and Controversy. Anglo-American planning recognized the necessity of keeping Russia in the war by opening a second front. The British Isles would be the springboard, but beyond that the strategic concepts of British and American planners differed widely. The U.S. Chiefs of Staff desired the earliest possible invasion of Western Europe; the British Chiefs of Staff opposed this, fearing failure by premature blows against the efficient German war machine. Churchill believed in attacking the Axis through the "soft underbelly" of the Mediterranean basin. While the planners argued, American strength in the British Isles built up as the convoy system went into high gear. Air and ground troops began moving over in large numbers by mid-1942, and the first American strategic bombing operation began in a small way with the bombing of marshaling yards at Rouen (August 17). Meanwhile, embattled Britain was carrying the war to the enemy both by air and sea. Bomber Command's night blows against Germany grew heavier, and a succession of amphibious raids harassed the German-held coast of France. Three of these operations—1 by air and 2 by sea—are notable.

1942, March 28. St.-Nazaire. The destroyer *Campbeltown* (ex-American), loaded with explosives and accompanied by a small group of motor launches carrying commando troops, dashed into the harbor of St.-Nazaire under enemy fire. Despite very severe losses, the attackers wrecked the only dry dock in Europe outside of Germany capable of taking the giant *Tirpitz*.

1942, May 30–31. First 1,000-Plane Raid. A night assault by Bomber Command against the important railroad marshaling yards at Cologne caused heavy damage.

1942, August 19. Dieppe. A major amphibious raid by Anglo-Canadian troops, with some tanks, ended in complete failure after a gallant effort. Of 5,000 men put ashore, the attackers lost 3,350 killed and wounded, 28 tanks, and a number of landing craft. The lessons learned, however, were of immense value in the later Allied large-scale amphibious assaults. The disaster confirmed British views regarding German defensive capabilities and the risk that would result from a major invasion of Western Europe in 1942–1943.

Operations in the Mediterranean

EGYPT AND CYRENAICA

1942, January. Situation. Auchinleck's Eighth Army and Rommel's Italo-German army faced one another at El Aghelia in Cyrenaica, 800 miles from Cairo and 450 miles from the Axis base of Tripoli. The weakened British naval forces in the Mediterranean permitted German reinforcements; Rommel's build-up outdid Auchinleck's.

1942, January 21. Rommel's Second Offensive. Attacking on a narrow front, the German drive crashed through the British advanced screen, forcing immediate withdrawal. The Eighth Army, dispersed and still under strength, was rolled back beyond Benghazi and large quantities of stores fell into Rommel's hands.

1942, February 4–June 13. Stalemate at Gazala. Auchinleck made a stand, and Rommel's advance had outrun his supply line. For 4 months more both sides rested, each building up strength. The Eighth Army's fortified and heavily mined defensive positions stretched from Gazala on the coast to Bir Hacheim, nearly 40 miles south in the desert, where a Free French light division under General **J. P. Koenig** (up from French Equatorial Africa) held the British left flank. Behind Bir Hacheim, the British armor was concentrated to protect the open-desert left flank. British strength by this time was 125,000 men, with some 740 tanks and 700 aircraft. Rommel had 113,000 troops, with 570 tanks and 500 aircraft.

1942, May 28–June 13. Battle of the Gazala–Bir Hacheim Line. Rommel attacked, hoping to envelop the British desert flank and roll up the position. Italian troops were repulsed from Bir Hacheim by the French. But Rommel's Panzer divisions, circling to the south, turned north inside the British positions, despite vicious attacks by the RAF. The British line held, while behind it German and British armor battled repeatedly around a desert crossroad called Knightsbridge. Just as Rommel's tanks were running out of gas, Italian troops broke through the mine-fields between Bir Hacheim and the British Gazala positions (May 31), permitting the supply of Rommel's Panzers. Rommel created a fortress in an area called "the Cauldron" inside the British lines. The French had to evacuate Bir Hacheim (June 10–11). Then the German armor debouched from the Cauldron, threatening the rear of the entire Eighth Army. General Ritchie ordered withdrawal (June 13).

1942, June 14–30. Retreat into Egypt. The British fell back to Halfaya on the Egyptian

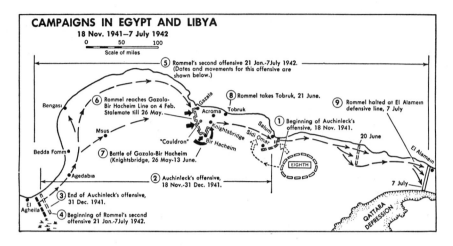

border, leaving the Tobruk fortress menacing the German advance. But **Tobruk** suddenly fell (June 21) to a skillfully coordinated ground and air attack by the Afrika Korps. The Eighth Army retreated in disorder as Rommel pursued. Auchinleck, assuming personal command again, rallied his troops and, after a bold delaying action at **Mersa Matruh** (June 28), fell back on the Alam Halfa ridge, a fortified line between El Alamein on the sea and the Qattara Depression some 40 miles inland. Rommel, following, was at the end of his endurance. There, after the entrenched British repulsed some tentative German probes (July), the 2 exhausted adversaries lay again in stalemate, only 60 miles west of Alexandria. Both Auchinleck and Ritchie were relieved by Churchill (August 13), General Harold R. L. G. Alexander taking the Middle East Command and Lieutenant General **Bernard L. Montgomery** the Eighth Army. COMMENT. *Rommel's advance was a brilliant display of armored force leadership and skillful use of his 88-mm. guns in coordination with his infantry and tanks. The British at Gazala frittered away their own armor in piecemeal counterattacks, losing all but 65 of their tanks. The disaster also cost the British 75,000 casualties (including the 33,000 garrison of Tobruk), the loss of great quantities of stores, and the elimination of Tobruk as a thorn in the side of Rommel's advances. The Germans and Italians lost about 40,000 men. Despite his success, Rommel's logistical situation*

was worsening as Allied naval and air strength in the Mediterranean again increased (see p. 1190), and Hitler—obsessed with the Russian invasion—urged him to advance to the capture of the Suez Canal, yet failed to provide adequate reinforcement or supply. Meanwhile, British strength was rebuilding.

1942, August 31–September 7. Battle of Alam Halfa. Rommel assaulted to try to gain another victory before British strength became too great. His plan, as at Gazala, was envelopment of the British left by the Afrika Korps. Montgomery was prepared. The Panzers, initially successful despite skillful delay by the British 7th Armored Division, reached the left rear of the British position and then thrust north, to be repulsed by a tank brigade dug in one the Alam Halfa ridge. Short of fuel and harassed by punishing British aerial attacks, Rommel began withdrawing his tanks (September 2). Montgomery refused to risk counterattack and the line stabilized again. But Rommel's failure had been a disaster—he had no further hope of offense; all he could do was defend.

1942, September–October. Preparations. Montgomery prepared methodically for attack. The Mediterranean situation had improved (see p. 1190). By October, the Eighth Army had been built up to an impressive strength— 150,000 men organized in 3 army corps—7 infantry and 3 armored divisions, with 7 additional armored brigades (1,114 tanks in all), plus corps and army troops; fuel and ammunition were plentiful. Rommel at this time had 96,000 men (half of them Italians) in 8 infantry and 4 armored divisions (nearly 600 tanks); shortages in fuel, ammunition, and other supplies were severe. (This shortage was in large part due to RAF sinking of German supply ships in the Mediterranean, based on information gained from Ultra.) Rommel, ill, had temporarily flown back to Germany for medical treatment, leaving General **Hans Stumme** in command. The opponents, each with unturnable flanks, were separated by a broad zone of mine fields. On the north was the Mediterranean, and 40 miles to the south was the Qattara Depression, impassable for either wheeled or tracked vehicles. Montgomery's plan was to effect a penetration, hold off Rommel's armor while eliminating the German infantry, and only then engage in a tank battle. The RAF's Desert Air Force had gained complete air superiority, and subjected Axis forces to intensifying punishment.

1942, October 23–November 4. Battle of El Alamein. At 9:40 P.M., 1,000 British guns opened along a 6-mile front near the sea. Twenty minutes later—under a full moon— the XXX Corps struck the Axis left, while to the south the XIII Corps began a diversionary effort near the Qattara Depression. Four hours later, the X Armored Corps advanced through 2 corridors in the mine fields opened by the XXX Corps infantry. Despite initial surprise, the Italian infantry put up obstinate resistance; an almost immediate counterattack by the 15th Panzer Division nearly stopped British progress. Nor did the diversionary attack of XIII Corps make much gain. Stumme died of a heart attack; Rommel, flying back at the first word of the battle, resumed command (October 25). Next day Montgomery, halting further effort on the south, threw his weight against the coastal area, where the 9th Australian Division threatened to pin the German 164th Division against the sea. For a week a ferocious tank battle raged in the mine fields south of the coastal road and railroad, as both sides brought their armored

units up from the south. The Axis armor, necessarily thrown in piecemeal, and under continuous aerial bombardment, shrank rapidly. Lack of replacements for damaged vehicles, together with shortages of fuel and ammunition, combined to take cruel toll. Meanwhile, the Australians had almost surrounded the German 164th Division along the coast. Rommel, whose existence depended on holding the coast road, committed his last reserve, extricated his infantry from encirclement, and dug in again 3 miles to the west (November 1). Montgomery quickly regrouped, then renewed his attempted breakthrough south of fortified Kidney Hill. The 2nd New Zealand Division, behind a rolling barrage, cleared a corridor through the mine fields for the British tanks (November 2). A desperate Panzer counterattack momentarily snubbed the breakthrough, but by the day's end only 35 German tanks remained in action, while British artillery and aerial bombardment neutralized the deadly German 88-mm. antitank guns. Rommel, his fuel and ammunition having reached the vanishing point, decided to withdraw, but was halted for 48 hours by Hitler's categorical and senseless command to hold at all costs. Montgomery hurled another assault against the Kidney Hill area, scoring a clean breakthrough. Rommel, disregarding Hitler's order, now disengaged the Afrika Korps, leaving the Italians behind. The entire Axis front crumbled. Cautious Montgomery delayed pursuit for 24 hours. Rommel's losses were enormous: some 59,000 men killed, wounded, and captured (34,000 of them German); 500 tanks, 400 guns, and a great quantity of other vehicles lost. The Eighth Army lost 13,000 killed, wounded, and missing, and 432 tanks had been put out of action.

COMMENT. *Strategically and psychologically, El Alamein ranks as a decisive battle of World War II. It initiated the Axis decline. The victory saved the Suez Canal, was a curtain raiser for the Anglo-American invasion of North Africa 4 days later, and was a prelude to the debacle of Stalingrad. Allied morale soared, particularly in the British Empire, proud to have at long last a victorious army and general; Axis morale correspondingly dipped. Hitler's order that*

Rommel should stand fast (rescinded 48 hours later, after the "Desert Fox" had already started to withdraw) contributed to the ruin of Rommel's army. Ultra played a significant role in this Allied victory.

1942, November 5–December 31. Pursuit. Montgomery's slightly delayed, methodical pursuit, planned to keep unremitting pressure on Rommel's forces, fell somewhat short of its goal. Although forced to stand several times, the Afrika Korps made good its escape. The RAF in particular has been criticized for lackadaisical operation. At **Mersa Matruh** (November 7), the 21st Panzer Division, lacking fuel, made a hedgehog stand, then abandoned its last tanks. At El Agheila (November 23–December 13), Montgomery stopped to open the port of Benghazi and establish a supply line, Rommel withdrawing when again threatened by encirclement. The year ended with another delaying action at the **Wadi Zem Zem,** near Buerat. Again, Rommel skillfully evaded entrapment while Montgomery wrestled with the logistical problems of his long communications line.

Invasion of North Africa, 1942

1942, August–November. Preparations for Operation "Torch." This invasion was planned for the purpose of seizing Morocco, Algeria, and Tunis as bases for further operations. It was a compromise between widely differing British and American military strategical concepts (see p. 1184). Roosevelt finally resolved the matter in favor of North Africa. It was the largest amphibious operation attempted to that time. Supreme commander was Lieutenant General **Dwight D. Eisenhower,** then commanding American troops in England. British Admiral Cunningham was Allied naval commander, and Eisenhower had an integrated American-British staff. The operation was divided into 3 main elements. Direct from the U.S. came Major General **George S. Patton**'s Western Task Force—35,000 men in 39 vessels, escorted by a powerful squadron under U.S. Rear Admiral **Henry K. Hewitt;** its main target was Casablanca on the Moroccan Atlantic coast. U.S. Major General **Lloyd R. Fredendall**'s Central Task Force, 39,000 men in 47 ships, came from England, its strong naval escort commanded by British Commodore **Thomas H. Troubridge;** the goal was Oran, on the Mediterranean. The Eastern Task Force, under U.S. Major General **Charles W. Ryder,** 33,000 men in 34 ships, also from England, was escorted by a naval force under British Vice-Admiral Sir **Harold M. Burrough;** its objective was Algiers. Except for a British contingent in the Eastern Task Force, all the troops were American, since it was believed that the French would be less favorably disposed toward the British. Prepared and lifted in the greatest secrecy, all 3 assault

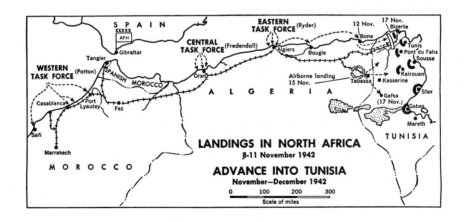

LANDINGS IN NORTH AFRICA
8-11 November 1942

ADVANCE INTO TUNISIA
November–December 1942

forces arrived off their respective landing zones after nightfall November 7. Despite some prior clandestine contacts, it was not known whether the French Army of Africa would resist or welcome the invasion.

1942, November 8. The Landings. *Western Task Force.* At 5:15 A.M., landings began at **Safi,** 125 miles southwest of Casablanca; at **Fedala,** 15 miles northeast of it; and at **Mehdia** and **Port Lyautey,** 70 miles to the northeast. Despite surprise, French troops resisted bravely, while at **Casablanca** itself French naval forces made a gallant sortie, promptly crushed by predominant U.S. strength. Despite French opposition, the landings were successful, and the 3 assault forces concentrated for an attack on Casablanca (November 11).

Central Task Force. Landings east and west of **Oran** met determined resistance. An attempt by 2 U.S. Coast Guard cutters at a run-by into the port itself failed dismally. An American airborne assault battalion, flying direct from England, was only partly successful in capturing the nearby airfields. However, the landings progressed; a coordinated attack penetrated Oran and the French capitulated (November 10).

Eastern Task Force. Forces landing on both sides of **Algiers** harbor converged on the city; a frontal naval assault met disaster, as at Oran. Algiers was soon ringed on the land side. The place capitulated (November 10).

1942, November 9–11. Franco-German Reactions. The Pétain regime broke diplomatic relations with the U.S. (November 9) and ordered French African forces to continue resistance. At the same time Hitler ordered immediate military occupation of "unoccupied" France, and began sending German troops by air into Tunisia.

1942, November 11. "Darlan Incident." Admiral Darlan, unexpectedly found in Algiers and taken into protective custody, broke with Vichy, ordered immediate cease-fire by all French troops, and agreed to cooperate with the Allies in driving the Germans out of Tu-

nisia. General **Henri H. Giraud** was put in command of French forces.

COMMENT. *This cooperation with a leading figure in the Vichy regime aroused furore in the U.S. and England, and affronted de Gaulle. From a military viewpoint, the bargain was a major factor in facilitating further Allied operations in North Africa, despite Darlan's later assassination (December 24).*

1942, November 17–December 31. The Race for Tunisia. With 1,000 German troops arriving in northern Tunisia each day, the Allies, unprepared as yet for a major overland operation, attempted a piecemeal eastward rush on Bizerte, while retaining sufficient force to oppose any possible Axis move from Spanish Morocco. British Lieutenant General **Kenneth A. N. Anderson,** with advance elements of what would become the British First Army, moved into the mountainous region southwest of Bizerte, while a screen of U.S. paratroops spread southeast. Aggressive German troops under General **Walther Nehring** checked the British advance, while mud and rain delayed Allied reinforcements from Algiers, 500 miles to the west. Allied spearheads reached to within 20 miles of Tunis (November 28), but were thrown back by German counterattacks. By December, Eisenhower conceded defeat in the race. The year ended with Anderson's First Army and the Fifth Panzer Army, now under General **Jürgen von Arnim,** facing one another in stalemate in north-central Tunisia.

COMMENT. *The invasion of North Africa succeeded because of strategic surprise, effective joint military-naval planning, and Darlan's check of French resistance. However, the confused French political situation and the prompt German reaction in Tunisia combined to impede the next move. Eisenhower, involved in political repercussions of the "Darlan incident," exercised little command supervision of the eastward thrust, and Anderson's initial moves were pedestrian. There was nothing now to prevent the eventual linking of Rommel, retiring skillfully westward from the Wadi Zem Zem, with von Arnim.*

MALTA AND MEDITERRANEAN AIR AND SEA OPERATIONS, 1942

Malta, Britain's "unsinkable aircraft carrier," was the major stumbling block for Axis supply to North Africa. Incessant aerial attacks and Axis surface and submarine operations almost throttled all efforts to supply it. Of 6 convoys in 1942, 2 were repulsed and only parts of the rest reached Malta. The Mediterranean Fleet, from Alexandria, and H Force, from Gibraltar, furnished what escort they could, but until September the situation was precarious. Submarines and 2 speedy mine-layers brought driblets of supply. Malta's defenses consisted of fighter planes and antiaircraft artillery. Some 350 fighters were flown in from aircraft carriers—notably the U.S.S. *Wasp*—to replace losses incurred on the open airstrips.

Malta's fighter squadrons played hob with Axis supply throughout the campaign. One-third of Axis supply ships from Italy were sunk in September, and in November British fighters sank three-quarters of the Axis surface craft. This toll of German supply ships was a direct result of the Allies' ability to read German secret radio messages, through the highly classified British code-breaking project known as "Ultra." The Allies profited from these radio intercepts in many operations, beginning with the Battle of Britain, but nowhere was the Ultra information more effectively used than in the sinking of German ships endeavoring to support German operations in North Africa.

Without possession of Malta the Allies could not have made such effective use of the Ultra information. Thus British denial of Malta to the Axis was a major factor in Rommel's defeat. The cost was high: an aircraft carrier, several cruisers and destroyers, hundreds of planes and thousands of men. But the tide had turned. Malta's role of cutting supply by surface ship forced the Germans to send reinforcements by air. By that time much Luftwaffe strength had gone to the Russian front and the rest in this area was suppuring Rommel and von Arnim in Tunisia. Furthermore, logistical support for the British Middle East Command through the Indian Ocean-Red Sea-Suez Canal route was now plentiful.

The Eastern Front, 1942

1942, January–February. Continuation of the Russian Counteroffensive. (See p. 1183.) Soviet attacks along the entire front, from Finland to the Crimea, forced the German armies into a stubborn defensive role. Only in Finland were the Russians repulsed, and the German besiegers of Leningrad still held stubbornly to their lines. Elsewhere, deep penetrations were made at many points. Hitler ordered all of his armies to stand fast. "Hedgehogs" (all-round defensive areas) checked the tide as the Soviet onslaught lost momentum. The danger of possible precipitate retreat, ending in a disaster such as Napoleon's retreat in 1812, was averted.

1942, March–May. Stalemate. Spring thaws and mud stymied movement. The Soviets had outrun their supply lines and the exhausted Germans were not yet prepared to resume the offensive.

1942, May 8–June 27. Preliminary German Offensive. Reinforced by 51 satellite divisions (Italian, Rumanian, Hungarian, and Slovak) and 1 Spanish volunteer division, the Germans resumed the offensive, to eliminate salients created by the Soviet winter drive. Despite the poor quality of replacements, and shortages of personnel and matériel in the German army units, substantial advances were made. In straightening out the lines, severe casualties were inflicted on the Soviets. General Manstein's Eleventh Army swept the Kerch Peninsula, inflicting 150,000 Russian casualties, and renewed the attacks on Sevastopol. Partisan groups hamstringing German movements in the rear areas were partly checked.

1942, June 7–July 2. Capture of Sevastopol. An amphibious assault climaxed the siege; the Soviets lost another 100,000 men.

1942, June 28–July 7. Opening of the Summer Offensive. Following plans personally developed by Hitler, Bock's Army Group South smashed eastward from the vicinity of Kursk to capture **Voronezh** (July 6). Hitler now reorganized Army Group South; the northern armies, designated Army Group B, were under General Weichs, the southern armies, designated Army Group A, under General List. It had been originally planned for these 2 groups to cooperate in a powerful and deliberate advance, first to clear the Don and Donets valleys, capturing Rostov and Stalingrad, then to move

southward to seize the Caucasus and its rich oil resources.

1942, July 13. Hitler Changes Plans. Hitler now decided to drive for Stalingrad and the Caucasus in simultaneous offensives by Army Groups B and A. This reduced the planned strength for both operations, dangerously overextending both combat and logistical organizations. Furthermore, with diverging objectives, the new plan inevitably would create a gap between the 2 army groups.

1942, July 13–August 23. Drive on Stalingrad. As the powerful attacks progressed, the growing gap between the army groups permitted most of the Soviet troops caught in the bend of the Don to escape. Although the German Sixth Army (**Friedrich Paulus**), was approaching Stalingrad, Army Group B's advance

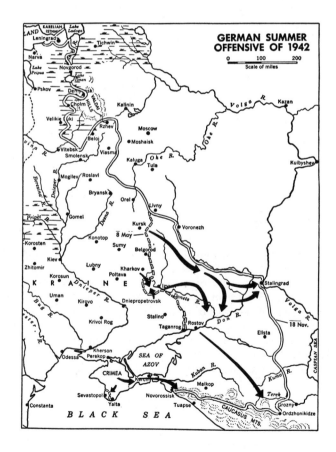

was slowed down by Hitler's diversion of Hoth's Fourth Panzer Army to join Army Group A's advance to the south. Nevertheless, with Luftwaffe aid, Group B cleared the bend of the Don, reached the Volga north of Stalingrad (August 23), and threatened the city.

July 13–August 23. Drive on the Caucasus. Army Group A captured **Rostov** (July 23), crossed the Don on a wide front, penetrated deep into the Caucasus Mountains, and had elements within 70 miles of the Caspian Sea near Astrakhan. But Hitler blundered again. Enraged at the relatively slow progress before Stalingrad, he yanked the Fourth Panzer Army back to join the Stalingrad struggle (August 1), leaving Army Group A, badly extended, trying to maintain an offensive in the Caucasus on a 500-mile front. Worse yet, Hitler also ordered Manstein's Eleventh Army—his sole reserve in this area—up north to reinforce the siege of Leningrad. General Franz Halder, OKH Chief of Staff, daring to protest these moves, was relieved. So, too, was List; Hitler took personal command of Army Group A (exercising command from his East Prussian command post more than 1,200 miles distant) in addition to his other responsibilities. The Caucasus campaign inched forward, some German patrols actually reaching the Caspian, but this success was inconsequential.

1942, August 24–December 31. Battle of Stalingrad. Hitler now concentrated on capture of the city's 30-mile perimeter astride the Volga, whose defenses were being strengthened every moment. In a tremendous battle of attrition, the Soviet defenders resisted stubbornly from house to house as the Germans gradually closed in and the Soviet winter freeze began. Most of Army Group B was now involved in the Stalingrad battle, while Army Group A to the south (now under Kleist) bogged down in stalemate. Between them, 1 single German motorized division held a 240-mile gap, and north of Stalingrad satellite forces screened the Don. German supply was meagerly maintained by a single rail line, completely inadequate for the situation. Hitler bypassed Weichs to issue personal commands to General Paulus, whose Sixth Army was in and around Stalingrad.

1942, November 19–23. Soviet Counterattack. Cleverly planned and timed to coincide with both the frost which enabled cross-country tank movements and the opening of the Allied landings in North Africa, 4 Soviet fronts or army groups, under the overall command of Zhukov, launched a double envelopment against the Germans crowded about Stalingrad. The Voronezh, Southwest, and Don Fronts slashed in north of the city, the Stalingrad Front on the south. Behind the armored spearheads swept great bodies of cavalry. The overextended German lines gave way, the trap closed. Inside was Paulus' Sixth Army and elements from units on its flanks that had been wrenched aside in the Soviet pincers move. The remainder of Weichs' Army Group B reeled back. Despite Weichs' urging that he fight his way out, Paulus waited for Hitler's command. The Fuehrer demanded he hold in place, assuring him of supply by air, and ordered Manstein, hurriedly rushed from the Leningrad front to Army Group Don (improvised from part of Group A), to recapture immediately the lost ground. The task was impossible. Paulus' situation worsened, the Luftwaffe being unable to deliver more than 70 tons of supply each day. Manstein finally managed to reorganize 3 understrength Panzer divisions and attempted to break through in relief, actually reaching to within 35 miles of Stalingrad (December 19). Manstein radioed Paulus to make a final effort to break out, but Paulus awaited Hitler's authorization. Then the German front collapsed. The year ended with Paulus and his men battling against starvation in an ever-tightening Soviet ring, while Weichs, Manstein, and Kleist fought desperately to maintain themselves as they slowly withdrew westward under constant Soviet pressure.

COMMENT. *Hitler's fantastic changes of objective, and his insistence first on an over-extended offensive, and later on senseless retention of terrain, had wrecked his eastern armies. Almost 65,000 prisoners were lost (aside from more than 100,000 surrounded in Stalingrad) and some 1,000 tanks destroyed or captured.*

His satellite forces—Italian and Rumanian—had proved to be unreliable, and his German elements, despite amazing resiliency under appalling conditions, were completely inadequate to oppose the growing Soviet armies. On the Soviet side, Marshal Zhukov and his 4 principal subordinates, Marshal **Konstantin K. Rokossovski** *and Generals* **Nikolai Vatutin, Vasily Chuikov,** *and* **Andrei I. Yeremenko,** *conducted their counterattack efficiently, with several simultaneous penetrations and almost immediate exploitation, which never gave the Germans opportunity to reestablish themselves. The defenders of Stalingrad suffered more casualties than the Germans, but proved once again the Russian soldier's ability to endure and to hold ground. Yet absolutely essential to this great Soviet victory, one of the decisive battles of history, was the availability of weapons and equipment. Despite the great industrial effort the U.S.S.R. was making, she could not have provided the materials needed for successful defense followed by successful offense without the great quantities of equipment supplied by her Anglo-Saxon allies, particularly the U.S., by the dangerous Murmansk route, and the longer and safer routes across the Pacific and through Iran.*

OPERATIONS IN 1943

Strategy and Policy

1943, January 14–23. Casablanca Conference. Roosevelt, Churchill, and the Combined Chiefs of Staff plotted their future operations. The net result was again a compromise. Britain agreed to U.S. exploitation of the Pacific initiative (see p. 1261). The U.S. agreed to postpone any cross-Channel invasion until 1944, and to the undertaking of another Mediterranean operation, with Sicily the initial objective. With the U-boat menace in the Atlantic again assuming formidable proportions, highest priority was ordered to construction of antisubmarine craft—escorts and carriers. During the conference Roosevelt proclaimed, and the British agreed to accept, nothing but the "unconditional surrender" of their enemies, a policy of dubious expediency.

1943, May 15–25. "Trident" Conference, Washington. Roosevelt, Churchill, and the Combined Chiefs of Staff reaffirmed the basic decision to concentrate primary effort against Germany in agreements (1) to greatly intensify the planned strategic bomber offensive against Germany and German-occupied Europe, and (2) to plan for a cross-Channel invasion of Europe, through France, for May 1, 1944. Other decisions were subsidiary, save for agreement for some further increase in effort against Japan.

1943, August 14–24. "Quadrant" (First Quebec) Conference. The principal decision of this conference was to wage the war against Japan with increasing force, but without relaxation of the war in Europe. Specific decisions relating to Europe were (1) to hasten the invasion of Italy because of the collapse of Mussolini (see p. 1199), (2) to draw the U.S.S.R. into full concert with the western Allies, and (3) to recognize de Gaulle's French Committee of National Liberation as representative of all Free French fighting the Axis.

1943, October 19–30. Moscow Conference. This meeting of the foreign ministers of the U.S.S.R., the U.K., and the U.S. was called primarily to lay the political groundwork for the forthcoming conference of heads of state at Teheran. They also agreed that China should be the fourth major member of the alliance, and agreed upon a postwar international organization (to become the U.N.).

1943, November–December. "Sextant-Eureka," Cairo-Teheran Conferences. Since the U.S.S.R. was not at war with Japan and was bound by a nonaggression treaty with her (April 1941), 2 separate heads-of-state conferences were necessary. At Cairo, Roosevelt and Churchill met with Chiang Kai-shek (November 22–26 and December 3–7). At Teheran, they met with Stalin (November 28–30). These were primarily political conferences. Stalin was informed that the cross-Channel invasion would take place in late May or early June 1944; he agreed to launch a general Soviet offensive at the same time.

Battle of the Atlantic, 1943

THE SUBMARINES

1943, January–March. Increasing U-Boat Toll. Admiral Raeder was relieved (early January) by an impatient Hitler. Submarine specialist Admiral Karl Doenitz took his place and stepped up an already-accelerated wolf-pack campaign. The German U-boat operational daily average rose to 116 craft. The U-boats crisscrossed the sea lanes, taking toll of every merchant convoy in a continuous battle. The worst winter storms in many years lashed the Atlantic, increasing the difficulties and dangers for surface craft. By the end of March, with 108 Allied merchantmen sunk, to a loss of only 15 U-boats, the campaign reached its climax. The United Kingdom's larder was reduced to a hand-to-mouth 3-month backlog of supply. The U-boat war was waged with success only against the slow-moving merchant convoys. The 2 other types—fast converted luxury liners operating individually without escort, and the troop convoys—were not touched. The first outran the U-boats; the second, heavily escorted, were too dangerous to attack.

1943, April 28–May 6. Gretton's Convoy. Commander **Peter W. Gretton,** R.N., commanding Convoy ONS-2—42 merchantmen—fought a running battle across the North Atlantic with 51 U-boats. Though he lost 13 of his convoy, 5 U-boats were sunk by Gretton's warships and 2 more were sunk by aircraft. This is considered to be the turning point of the Battle of the Atlantic.

1943, May. Focus on the Bay of Biscay. Air Vice-Marshal Sir **John Slessor** of Britain's Coastal Command put into full operation a scheme of air-sea cooperation against U-boats operating out of French bases. Bombers based on the south of England, in conjunction with RN corvettes, and guided by microwave radar, proved to be deadly. In this month 38 U-boats—12 more than Germany could build—were destroyed, and only 41 allied merchantmen lost.

1943, May 26. Doenitz Shifts U-Boat Operational Area. The wolf packs congregated in 3 areas: (1) west of the Azores, athwart the central Atlantic convoy route—life line for the Allied Mediterranean campaigns, (2) in the South Atlantic, and (3) in the Indian Ocean.

1943, June–December. Allied Hunter-Killer Campaign. The U.S. Navy's Tenth Fleet organized "killer groups," each comprising an escort carrier equipped with 24 fighter-bombers using bombs, depth charges, or torpedoes, and accompanied by several old destroyers or new destroyer escorts. Each group commander was given the widest latitude to hunt U-boats whenever a "fix" had been obtained. The result proved catastrophic to the U-boats and their "milch cows" not only along the central transatlantic lane but also in the South Atlantic. The German craft fought back savagely in a number of single-ship and airplane-ship actions. Remarkable was the running gun battle between USS *Borie* and *U-405,* halfway between Cape Race, Newfoundland, and Cape Clear, Ireland, in which the destroyer rammed the sub, tearing out her own plates. The German finally blew up and sank, the victor following the next day. Remarkable, too, was the raid of *U-516* in the Caribbean (November 5–December 25). The German, despite the efforts of all available American antisubmarine craft in the area, sank 6 vessels, including 2 tankers, then got safely away.

COMMENT. *By year's end the knell was sounding for the German U-boat; Allied "kills" exceeded German replacement ability. Between May and September, 3,546 merchantmen in 62 convoys had crossed the North Atlantic without loss of a single vessel. Allied merchant construction exceeded all enemy destruction by 6 million tons. The food crisis in the British Isles had ended. Acquisition of air bases in Portugal's Azores completed the aerial attack ringing the wolf packs (October 13).*

NAVAL SURFACE WARFARE

1943, January–March. Convoys to the U.S.S.R. Murmansk convoys, strongly escorted, ran the gantlet during the first 3 months of the year with comparatively little loss. When the Mediterranean later was opened, they were suspended; most supplies for the U.S.S.R. were sent via Iran to avoid exposing Arctic convoys to German air attacks in almost continuous daylight.

1943, March. The *Scharnhorst* Sails North. The battle cruiser, again in commission, slipped through to Norway without detection.

1943, September 6–9. Spitsbergen Raid. The *Tirpitz*, the *Scharnhorst*, and 10 destroyers successfully raided Spitsbergen, bombarding and damaging Allied coal mining and loading installations.

1943, September 21–22. Attack on the *Tirpitz*. The great ship was attacked in Alta Fjord by British midget submarines and again badly damaged.

1943, October–December. Arctic Operations. Harassing Allied carrier-borne aviation sank 9 German coastal vessels near Narvik. As the nights grew longer, Allied convoys to Murmansk resumed.

1943, December 24–26. Last Cruise of the *Scharnhorst*. Accompanied by 5 destroyers, the *Scharnhorst* steamed out of Alta Fjord to attack an Allied convoy, discovered 400 miles northwest of Trondheim. The German flotilla blundered into 2 British task forces west of the North Cape. In a thrilling chase, the *Scharnhorst*, hit at extreme range by the battleship *Duke of York*, was slowed down, battered by heavy gunfire, then finished off by a torpedo attack. Only 36 of her 1,900-man complement could be rescued. This disaster removed the last surface threat to Allied convoys on the Murmansk run.

The Combined Bomber Offensive, 1943

The U.S. Eighth Air Force and Britain's Bomber Command instituted a round-the-clock aerial offensive, codenamed operation "Pointblank," against Germany from British airfields. Bomber Command attacked the economic system and civilian morale by large-scale nighttime saturation bombing; the Eight Air Force concentrated on daylight precision bombing against aircraft industrial targets and the Luftwaffe itself. This was to be followed by later concentration of both forces against vital industries to destroy Germany's ability to make war. Both air forces used new and improved techniques—the "pathfinder" system, which located targets by radar and marked them, and "window" (myriad strips of metalized paper released in the air to jam and confuse Germany's electronic defenses).

British Air Assault

Bomber Command's night raids rocked German civilian morale by wide devastation of the Ruhr industrial area and of major cities. Highlights included:

1943, June 20–24. First Shuttle Bombing. RAF bombers from England struck Wilhelmshafen outbound, then landed in North Africa. They bombed the Italian naval base of Spezia on the return trip.

1943, July 26–29. Hamburg Incinerated. Two successive mass bombing assaults produced "a catastrophe the extent of which simply staggers the imagination" (Goebbels' diary). Much of the damage came from tornado-like fire storms. Most of the city was reduced to rubble; civilian casualties were enormous and 800,000 survivors made homeless.

1943, August 17–18. Peenemunde Bombed. British intelligence sources had revealed (May) German development of a pilotless jet-propelled aircraft (V-1) and a rocket (V-2). The experimental installation at Peenemunde was bombed, delaying the project.

1943, October–December. Attacks on Mysterious Launching Sites. Aerial reconnaissance revealed the construction of launching sites along the Channel coast, and these, too, were bombed and the majority destroyed (December to spring, 1944), causing a 5-month setback in German plans. Operations against these projects were labeled "Crossbow."

1943, November–December. Battle of Berlin. Bomber Command concentrated its efforts against the German capital in 16 massive raids which continued through the winter and until March 1944.

U.S. AIR ASSAULT

1943, June 11. Wilhelmshaven Attacked. Eighth Air Force assault on submarine construction works was only partly successful. German interceptors prevented accurate bombing; Allied fighters did not have enough range to accompany the big B-17s and B-24s, a problem plaguing Eighth Air Force operations throughout the year.

1943, July. Attacks on German Aircraft Industrial Targets. This produced a serious, though temporary, setback in German plane production.

1943, August 1. Bombing of Ploesti. Taking off from North African bases, 178 B-24 bombers of the Eighth and Ninth Air Forces struck deep behind the German lines in the east against the Rumanian oil fields of Ploesti, whence the Axis drew most of its fuel for both Luftwaffe and ground forces. The planes were assembled in secrecy near Benghazi for the 1,000-mile flight, the longest bombing raid yet attempted. Their transit was detected by German radar in Greece, Bulgaria, and Rumania. The bombers, in low-level attack, were met by determined and accurate antiaircraft fire, causing loss of 54 planes, plus 7 damaged planes which landed in Turkey. Extensive but, as it turned out, relatively superficial damage was done to the vast complex of refineries and storage tanks.

1943, October 14. Schweinfurt Raid. Eighth Air Force bombers struck the heart of the German ball-bearing industry a cruel blow. But of the 288 planes (accompanied by short-range fighter escort as far as Aachen), 62 were lost and 138 damaged by antiaircraft and Luftwaffe fighters. Casualties were 599 men killed and 40 wounded. Despite tactical success, the raid was so costly that a temporary halt was called on further daylight bombing until longer-range fighter escorts were available.

1943, December 5. First P-51 Escorts. With these new long-range fighters of the Ninth Air Force to accompany them, the big bombers renewed daylight raids.

COMMENT. *The year ended with the German home-front morale rudely rocked, but industrial production was still potent and the Luftwaffe had improved its defensive tactics.*

The Mediterranean Area, 1943

NORTH AFRICA

1943, January 1–February 13. Stalemate in Tunisia. Allied expeditionary forces, under Anderson's tactical command, were spread thin along the Tunisian Dorsal and Eastern Dorsal ranges from Cape Serrat on the Mediterranean to Gafsa in the south. The British First Army, with a provisional French corps and some U.S. elements, held the line from the north to Fondouk in central Tunisia; the bulk of Fredendall's U.S. II Corps stretched to the south. Arnim's Fifth Panzer Army held northeastern Tunisia; Rommel's Afrika Korps, completing a masterly withdrawal from Egypt, held an old French fortified zone at Mareth, with its left flank on the Gulf of Gabes and its right resting on the almost impassable salt marshes of the Chott Djerid; there they were reequipped. Montgomery slowly began concentrating in front of Mareth. There was no ground communication between Anderson and Montgomery. Arnim, in a series of aggressive blows, kept Anderson off balance, while the Luftwaffe dominated the air.

1943, February 14–22. Battle of Kasserine Pass. Rommel, fearful lest he be caught by an Allied drive in his rear while Montgomery assaulted his Mareth position from the south, launched a 2-pronged surprise attack against the U.S. II Corps sector. His armor thrust through Faid, Sidi-bou Zid, and Sbeitla, driving for Kasserine Pass, while another armored thrust rolled through Gafsa on the south to Feriana (February 17). Elements of the U.S. 1st Armored Division and attached units, scattered through the sector in small packets, were rolled up; a piecemeal counterattack was ambushed. Staged in the best blitzkrieg style, with the Luftwaffe closely supporting by dive bombings and strafings, the drive rolled on. Eisenhower, his vital supply base at Tebessa threatened, rushed reinforcements, while both U.S. and British troops on the front gained their second wind. Rommel's Panzers crashed through the Kasserine Pass (February 18) and fanned northward. Against increasing resistance, his spearheads reached into the Western Dorsal range at Djebel El Hanra, Thala, and Sbiba (February 20–22) before momentum was lost. An expected supporting attack by Arnim in the north failed to materialize. Rommel, unmolested, then retired to his Mareth position as hastily as he had come.

COMMENT. *The attack had accomplished its primary purpose: to wreck Allied plans to divide the German forces by a thrust eastward to the Gulf of Gabes. Had Arnim cooperated, serious damage would have been done, but Hitler had insisted on a divided command in Tunisia; not until February 23 did he place Rommel in overall command. The Allied reverse was what might have been expected from green troops faced by skillful and daring veterans with air superiority. It was compounded by a poorly organized and coordinated American air-support system, and by Anderson's faulty terrain appreciation and his patchwork dispositions— a cordon of units divided both in nationality and in tactical formation, with consequent improvisation of command chain. As it was, the American troops put up a remarkably tenacious fight. Eisenhower must share the blame; permitting himself to become too much involved in the Franco-Algerian political turmoil, he had neglected command supervision. His first visit to the front came but 24 hours before the German assault occurred. A shake-up of command followed. All Allied ground forces in Tunisia were combined (February 20) in the 18th Army Group, Alexander commanding, and the organizations were regrouped by nationality. Patton shortly succeeded Fredendall in command of U.S. II Corps (March 6). Upon advice from the RAF's Air Marshal Tedder, and from US-AAF officers, the Americans also adopted the British system of air support, which permitted more flexible and more effective employment of available air strength.*

1943, February 26. Arnim's Attack. Belatedly, the Fifth Panzer Army assaulted British First Army's positions in northern Tunisia and gained some ground. However, stubborn resistance cost Arnim much armor, and a British counterattack (March 28) restored the line.

1943, March 6. Rommel Repulsed. Testing Montgomery's positions in front of Mareth, Rommel was decisively repulsed at **Médenine** with much loss in matériel. Still ill, he left Africa. Arnim assumed over-all command in Tunisia; Italian General **Giovanni Messe** took over the Mareth front and General **Gustav von Vaerst** the Fifth Panzer Army in the north.

1943, March 20–26. Battle of Mareth. Montgomery, with greatly superior forces, assaulted the Italo-German position, the XXX Corps making the main effort on his right. Despite heavy bombardment, the 15th Panzer Division counterattacked in the moonlight, checking the British. Montgomery, who had sent the New Zealand Corps on a wide turning movement against the Axis right, now reinforced it with the X Corps. Messe, avoiding encirclement, withdrew to El Hamma. The Allies now held the initiative.

1943, April 1–22. Axis Withdrawal. Patton's II Corps threatened Messe's right; Montgomery lunged north (April 6). When they linked, Messe had escaped and was moving swiftly north. The British IX Corps (First Army) moved east through Fondouk, but Messe was going too fast. This second pincers movement failed also when it closed with Montgomery's pursuing Eighth Army at Kairouan. With Messe's arrival Arnim held a defensive line running generally south and then east from Cape Serrat on the Mediterranean to Enfidaville.

1943, April 22–May 3. Beginning the Final Offensive. Alexander planned a power thrust by First Army with Eighth on the right, and U.S. II Corps on the left, shifted north for this offensive. (Major General **Omar N. Bradley** had relieved Patton to command U.S. forces in the coming Sicilian operation.) Major General **Louis-Marie Koeltz's** French XIX Corps linked the First and Eighth armies. Vaerst's Fifth Panzer Army held the right, then came the Afrika Korps, then Messe's Italian First Army on the left of the Axis front. The Germans had been reinforced, but the Allies had air superiority. German supplies were desperately short, due largely to "ultra" radio eavesdropping that kept the Allies informed of the sailing of German supply ships, most of which were then sunk. An emergency German air supply effort did not help much, and most of the transports were shot down by Allied fighters. The First and Eighth armies inched forward against determined resistance; the II Corps, probing against the most difficult terrain on the front, moved more rapidly, driving the Germans from Jefna (May 1) and taking Mateur (May 3.)

1943, May 3–13. Battle of Tunisia. Preceded by heavy artillery and air preparation, the Allies penetrated the Axis perimeter, the II Corps north and south of Lake Bizerte, the First Army east from Medjez El Bab. With all reserves committed, and lacking air support (the Luftwaffe was withdrawing to Sicily), Arnim was unable to stem the tide. The U.S. 34th Division entered **Bizerte** (May 7); the 1st Armored Division rolled through Ferryville, cutting the German communications and linking (May 9) with the British 7th Armored Division, which had already entered **Tunis** (May 7), and turned north. To the south, the French corps and the Eighth Army surrounded the Italian First Army, and British armor cut off the Cape Bon Peninsula (May 11). Thus dislocated and canalized by Allied thrusts, Axis troops began surrendering in droves, Bradley garnering some 40,000 prisoners in his zone. In all, some 275,000 prisoners were taken, including the top commanders. The Italian Navy made no attempt at evacuation. The Axis hold on North Africa was ended.

COMMENT. *Alexander's power punch was well planned and executed. The U.S. II Corps, still stinging from its reverse at Kasserine, aggressively modified its planned secondary role to a major one. The Axis forces put up stout resistance until the last in a situation which they knew could have but one end. The North African victory was prototype for future Allied joint operations. Strategically, it provided a springboard for further Mediterranean operations, including invaluable sites for air bases. Axis losses in casualties and prisoners in the North African campaigns amounted to an estimated 620,000 soldiers, one-third of them Germans. British losses in North Africa—from the beginning of the war—amounted to 220,000 in all. From November 1942, to May 1943, French losses were about 20,000, and the Americans suffered some 18,500 casualties.*

SICILY

1943, June 11. Capture of Pantelleria. The Italian garrison of this rocky island surrendered after a week of incessant bombardment by the Northwest African Air Force, as an Allied landing force approached.

1943, July 9–10. Assault on Sicily. Following a month-long bombardment of Axis air bases in Sicily, Sardinia, and Italy, an amphibious assault by Montgomery's British Eighth Army and Patton's U.S. Seventh Army (elements of Alexander's 15th Army Group) gained a toehold on Sicily's southeast coast. Effective support was given by naval gunfire. The Italians, not expecting an attack in stormy weather, were caught by surprise.

1943, July 11–12. Axis Counterattacks. General **Alfred Guzzoni's** Italian Sixth Army struck back. German divisions, attacking American troops near Gela and Licata, were particularly effective, but were repulsed after nearly driving to the sea. Allied airborne land-

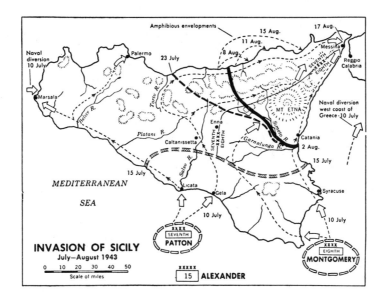

Amphibious envelopments · 15 Aug. · 17 Aug.

Naval diversion 10 July · Palermo · 23 July · 8 Aug. · 11 Aug. · Messina · Reggio Calabria

Marsala · Torto R. · SEVENTH XXXX EIGHTH · Naval diversion west coast of Greece-10 July

Belice R. · MT ETNA · Simeto R.

Platani R. · Enna · SEVENTH XXXX EIGHTH · Catania 2 Aug.

Caltanissetta · Garnalunga R. · 15 July

Salso R. · 15 July

15 July · MEDITERRANEAN · Licata · Gela · Syracuse

SEA · 10 July · 10 July

INVASION OF SICILY
July–August 1943

0 10 20 30 40 50
Scale of miles

XXXX SEVENTH **PATTON**

XXXX EIGHTH **MONTGOMERY**

XXXXX 15 **ALEXANDER**

ings were disrupted by bad weather and mistaken Allied antiaircraft fire.

1943, July 15–23. Clearing of Western Sicily. Though determined German resistance stopped Montgomery south of Catania, Patton's army swept through western Sicily, then turned east to assist the British.

1943, July 23–August 17. Advance to Messina. Despite spirited German resistance (now actually directed by one-armed German Colonel General **Hans Hube,** veteran of the Russian front), the Allies pressed steadily forward, aided by several small amphibious operations on the north coast east of San Stefano. The Italians began a mass exodus (August 3–16), followed by a more orderly German withdrawal (August 11–17), which succeeded in evacuating to the mainland over 100,000 men, 9,800 vehicles, and 50 tanks before the Allies occupied Messina (August 17).

COMMENT. *American casualties were 7,319; the British lost 9,353; Axis losses were over 164,000, including some 32,000 Germans. Great quantities of weapons and equipment were captured. Allied operations had been skillfully conducted. German resistance had been professional and tenacious, but the Italians showed little desire to fight. The Allies landed some*

160,000 men, who had been transported in 3,000 vessels. Total Axis strength had been about 350,000, of which perhaps one-third were German. The Allies were aided by a great preponderance of air power, 3,700 planes as opposed to 1,600. With the capture of Sicily, the Mediterranean was again opened as an Allied sea route.

ITALY

1943, July 24. Overthrow of Mussolini. The war-weary Italian nation toppled Mussolini from power during the Sicilian operation. His successor, Marshal Badoglio, though officially declaring intention to continue hostilities, began secret negotiations with the Allies through agents in Lisbon. Hitler, unwitting but suspicious, prepared to control and disarm the Italians, and began movement of German reinforcements into north Italy to protect the Alpine communications. Kesselring, German commander in southern Italy, correctly estimated that the Allies planned an early invasion, and that Salerno was a likely landing site.

1943, September 3. Armistice with Italy. This was signed secretly, to be effective September 8.

1943, September 3. Eighth Army Assault on Calabria. British troops landed on the toe of the Italian boot. Kesselring initiated delaying action in the south while still watching the Salerno area. He was unaware of the armistice.

1943, September 8. Publication of the Armistice. This was timed to coincide with Allied landings at Salerno. At the same time, the Italian fleet fled from Spezia to Allied protection at Malta, severely harassed on its way by the Luftwaffe. The alert Germans began to disarm and imprison all Italian forces.

1943, September 9. Assault at Salerno.* The U.S. Fifth Army, commanded by Lieutenant General **Mark Clark** and comprising the British X and American VI Corps, made assault landings in the Gulf of Salerno before dawn, to be met by alert German defenders, deployed in mobile defense. The British landed north of the Sele River (bisecting the crescent-shaped gulf shoreline), Americans to the south. Due to effective defense, and despite excellent naval gunfire support, by nightfall the Allies only held 4 narrow, unconnected beachheads. Kesselring began moving all available reserves to the Salerno area. To the south, 130 miles away, Montgomery's Eighth Army was slowly consolidating its hold of the Calabrian peninsula; a British division simultaneously landed to seize Taranto.

1943, September 10–14. Struggle for the Beachhead. Though both British and Americans were able to claw their way forward somewhat in their sectors, they still found it impossible to join their beachheads solidly at the Sele. German artillery, with perfect observation from the hills ringing the beaches, was particularly effective. Kesselring, having gathered most of 6 divisions, launched a violent counterattack against the Allied center (September 12). A last-ditch defense by American artillery units, supported by naval gunfire, prevented a breakthrough to the beaches (September 13). The Allied situation was desperate; sufficient reinforcements could not be put ashore to stop the Germans; 3 battalions of the

* Operation "Avalanche."

82nd Airborne Division were parachuted to bolster the crumbling beachhead. All the strength of the Northwest African Air Force and the carrier-based air and all the guns of the naval expedition combined to give close support. Alexander ordered Montgomery (still 50 miles away) to accelerate his advance, but skillful delaying action by a lone German battle group impeded his cautious advance through mountainous country.

1943, September 15–18. Securing the Beachhead. The hasty landing of more Allied reinforcements, combined with superb naval gunfire and air support, checked the German offensive. Montgomery's patrols made contact with the southeastern portion of the beachhead (September 16). Kesselring thereupon began a deliberate disengagement, retiring northward.

COMMENT. *Allied losses exceeded 15,000 men; German casualties were about 8,000. The outstanding feature of the battle had been the foresight, skill, and initiative of Kesselring, and the efficiency of his troops.*

1943, September 18–October 8. Consolidation of Southern Italy. Both Fifth and Eighth armies pressed on. West of the Apennines, the Fifth seized Naples (October 2) and pushed on, to be checked (October 8) by the swollen Volturno River, with all bridges destroyed. To the east, the Eighth Army took Foggia and its great air base (September 27), advancing to Termoli on the Adriatic, captured after a stiff fight (October 3). French troops, assisted by local Resistance elements, seized Corsica (September 11–October 4).

1943, October 12–November 14. Volturno River Campaign. Fifth Army, assaulting across the Volturno with both corps, made foot-by-foot progress over abominable trails in mountainous country drenched by the Italian autumn rains. The Eighth Army, having regrouped, also started north (October 22), forcing the Trigno River. Kesselring had selected the line Garigliano River–Sangro River to bar further Allied advance; his skillful, stubborn delaying actions granted him time to consolidate the position. The Allied armies, exhausted, found themselves facing ever-stiffening opposi-

INVASION OF ITALY
Situation 8 October 1943 and
Operations since 3 September

0   50   100
Scale of miles

tion. On November 15, Alexander halted to rest and regroup.

1943, November 20–December 31. "Winter Line" Campaign. Alexander's 15th Army Group resumed its advance on Rome. Kesselring by this time had established a formidable defensive zone (called the "Winter Line," or Gustav Line) 10 miles deep, running from the mouth of the swift-running Garigliano River on the Gulf of Gaeta, along its narrow tributary, the Rapido, then over the spine to the Adriatic north of the Sangro River. This zone was held by the Tenth Army (**Heinrich von Vietinghoff**). On the Mediterranean side the hill masses crowded down on the Liri Valley entrance, backstopped by Monte Cassino with the great Benedictine monastery frowning above the roaring Rapido. Into this zone the Fifth

Army attacked. Some progress was made, despite determined resistance, atrociously difficult terrain, and constant rain or snow, but the year ended in blizzard-bound stalemate, with the attackers still 5 miles southeast of the Rapido. The German defense was assisted by the fact that Fifth and Eighth armies alternated in their assaults, thus enabling the defenders to shift forces in opposition.

1943, December. Allied Reorganization. Both Eisenhower and Montgomery left for the coming offensive in Western Europe. British General Sir Henry M. ("Jumbo") Wilson assumed supreme command in the Mediterranean Theater, and General Sir **Oliver Leese** took command of the Eighth Army. Some veteran divisions were drawn off for the European invasion. The VI Corps, too, was out, regroup-

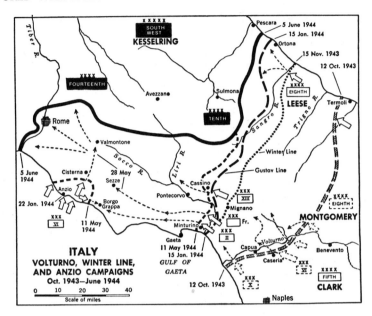

ITALY
VOLTURNO, WINTER LINE,
AND ANZIO CAMPAIGNS
Oct. 1943–June 1944

ing for another amphibious operation, and new troops—including Indian, French, Italian, and New Zealand forces—were moving into the now polyglot 15th Army Group.

AEGEAN

1943, September 12–November 16. Operations in the Dodecanese. British landings on **Kos, Samos,** and **Leros** (September 12) provoked a prompt German reaction. An airborne assault retook Kos (October 3–4); Samos was then evacuated. Leros was taken by a combined airborne-amphibious assault (November 12–16). British resources were not sufficient to continue operations in the Aegean without interfering with operations elsewhere approved by the Combined Chiefs of Staff.

The Eastern Front, 1943

1943, January–February. Soviet Pressure. All German forces in southern Russia were in jeopardy. East of Kharkov, the Hungarian and Italian satellite armies were disintegrating; the

Sixth Army at Stalingrad was in death agony; in the Don Valley in front of Rostov, Soviet assaults were threatening to cut off Army Group A's thin-spread First Panzer Army, now in retirement from the drive into the Caucasus oil fields.

1943, February 2. The End at Stalingrad. Paulus' Sixth Army, food and ammunition exhausted, surrendered to the relentless pressure of Rokossovski's Don Front (army group) after a last-ditch fight. Hitler's obstinacy had cost him in all 300,000 men. Paulus and 93,000 survivors surrendered. The First Panzer Army reached the Don at Rostov (February 1) to join Manstein's Army Group Don. The remainder of Kleist's Army Group A—Seventeenth Army—established a defensive bridgehead between the Sea of Azov and the Black Sea at Novorossisk.

1943, February 2–20. Soviet Drive across the Donets. Russian armor flooded over Kharkov, approaching the Dnieper bend, while partisans harassed the German rear area.

1943, February 18–March 20. Manstein's Counteroffensive. Manstein, skillfully employing reserves and shifting units, slowly checked, then threw back the Soviets, recaptur-

THE RUSSO-GERMAN FRONT, 1942-1943

BATTLE OF STALINGRAD

ing **Kharkov** (March 14) and restoring the line. Farther north, the line swayed back and forth in a succession of attacks and counterattacks. On the southern flank, Kleist's Army Group A maintained its bridgehead. The spring thaws then prohibited further mass movements by either side.

COMMENT. *The great Soviet winter offensive was a body blow to Axis power. The Soviets regained most of the territory lost in 1942. German losses are estimated at more than one million. The Soviets claimed capture or destruction of 5,000 aircraft, 9,000 tanks, 20,000 guns, and thousands of motor vehicles, railway cars, and locomotives. Russian losses were at least as great. Hitler's faulty strategy and almost complete incomprehension of logistical factors were his undoing. Large-scale guerrilla activity, engendered by SS atrocities in occupied terrain, greatly assisted Soviet leaders, who themselves had improved in military acumen. The most outstanding military performance had been Manstein's near miracle in halting the Soviet drive, despite odds of 7–1, then, amazingly, counterattacking successfully. This was one of the great military achievements of the war.*

1943, March–June. Recuperation and Reinforcement. German strength could no longer support another major offensive, as even Hitler realized. Soviet strength was now at least 4 times greater. By this time logistical aid furnished by the U.S. to the U.S.S.R. had mounted to 3,000-odd planes, 2,400 tanks, and 80,000 trucks.

1943, July 5–16. Battle of Kursk. Manstein's plan to strike a limited assault against the salient west of Kursk in combination with Kluge's Army Group Center was so long delayed by Hitler that when he finally launched it a Soviet counteroffensive checked it on the ground, while Soviet air power smothered Luftwaffe support. German losses were 70,000 killed or wounded, 3,000 tanks, 1,000 guns, 5,000 motor vehicles, and 1,400 planes. Russian losses in this largest of all tank battles were probably slightly less. This battle marked the end of German mass efforts in the east. Alarmed now by the Anglo-American Sicilian invasion (see p. 1198), Hitler broke off the move against Kursk and began transferring a number of Panzer divisions to the west, further weakening his force in the east.

1943, July 12–November 26. Soviet Summer Offensive. From Smolensk to the Black Sea, the Russians delivered a series of battering blows, featuring great masses of armor.

1943, August 2. Hitler's Orders to Hold in the East. Partly in reaction to the Soviet offensive, and partly because of fears for the safety of Ploesti created by the American air raid (see p. 1196), Hitler now ordered Manstein, who was conducting a masterly mobile defense, to freeze the defense of the Kharkov sector in place. A Soviet breakthrough (August 3) threatened disaster, and Manstein ignored Hitler's order, abandoning Kharkov (August 23), but kept his lines intact by skillful counterattacks as he fell back to the Dnieper.

1943, September–November. Continuation of the Soviet Offensive. Disregarding losses, the Soviets pressed ahead all along the front. When they finally paused, they had pressed Kluge's Army Group Center back to the edge of the Pripet Marshes, had recaptured Kiev (November 6), and Smolensk (September 25), had driven a bridgehead across the Dnieper in Manstein's sector, and had cut off the German Seventeenth Army in the Crimea, since Hitler had refused to follow Manstein's recommendations to evacuate the peninsula.

1943, December. Initiation of Soviet Winter Offensive of 1943–1944. Taking advantage of frozen ground, the Soviets launched a new winter offensive in the Pripet Marsh area and along the Dnieper. Although Manstein's Germans inflicted heavy casualties, they were forced to give up more ground.

THE WAR IN THE WEST, 1944

Strategy and Policy

The year opened with the Axis on the defensive. Hitler's Soviet gamble was lost, the eastern front crumbling. Italy was eliminated, North Africa cleared, and the Mediterranean sea lanes opened. In the United Kingdom, American strength was accumulating in astounding abundance. On the other side of the ledger, however, skillful German defense had halted the Allied advance midway up the Italian boot. The U-boats still menaced the Atlantic, Arctic, and Mediterranean sea lanes. Behind *Festung Europa's* barrier, the embattled German economy was producing war matériel at ever-increasing rates, despite the ravages of Allied bombers. The war was still far from a decision.

The strategic situation in Western and Northern Europe was plain to both Allied and Axis leaders, although never quite understood by their respective peoples. The Allies, from their bridgehead in the British Isles, held the priceless advantage of strategic interior lines. They were free, in principle, to attack Germany at any point in the long arc from Norway's North Cape to Brittany's Cape Finisterre. That they would invade was a foregone conclusion; there was no way to disguise the immense concentration of men and matériel. But where and when and how? This was the situation faced by Hitler, and this was the situation which—thanks to a deceptive cover plan whose details have never been fully revealed—the Allies maintained successfully until they were so firmly established in France that counterattack was futile.

1944, June–December. The V-bomb Threat to England. Soon after the invasion of Normandy, the first V-1 flying bomb struck England (June 13). German scientists had perfected this, the first of Hitler's "secret weapons," despite RAF attacks on the development and test site of Peenemunde (see p. 1195). Terror attacks by these weapons on

London severely shook English morale, despite the fact that about half the V-1s were shot down by antiaircraft or fighter defenses. The terror became more serious when the Germans introduced the V-2, the first effective long-range rocket missile (September 8). Because of their speed, these rockets could not be intercepted. The intensity of the attacks, however, was soon drastically reduced by the rapidity of the Allied ground advance along the French and Belgian coasts (see p. 1212). The Germans continued to fire V-2s against London from launching sites in Holland, though intensive Allied air bombardment prevented them from ever mounting a massive assault on England.

1944, July 20. Attempted Assassination of Hitler. A German Army plan to overthrow the Nazi regime failed when Hitler survived the explosion of a bomb in his headquarters. A ruthless purge of army officers ensued, and Hitler took absolute personal control over all military affairs. Rommel, a key figure in the plot, was forced to commit suicide (October 14).

1944, September 12–16. Octagon (Second Quebec) Conference. With Germany finally nearing collapse, principal attention was given to the war against Japan at this meeting of Roosevelt, Churchill, and the military Combined Chiefs of Staff. Despite Churchill's protests, America's increasing preponderance of military strength forced him to accede to American strategic concepts of concentrating all available strength directly against Germany, without consideration of possible "sideshows" in southern Europe, which Churchill felt would put the western Allies in a more favorable bargaining position with the Soviet Union at the time of a peace settlement. Because of such disagreements as to relative importance of political and basic military objectives, questions of postwar territorial settlements were postponed.

Battle of the Atlantic

1944, January–June. Foiling the U-Boats. Protection of the vast quantities of American men and matériel pouring across the seas, both to the United Kingdom and to the Mediterranean basin, became of paramount importance as the time for the cross-Channel invasion neared. German attempts to provide floating bases for the U-boats were scotched: a U.S. Navy hunter-killer group extirpated (February–March) "milch cow" submarines spotted in the Cape Verde Islands. Other groups hunted the elusive craft in both North and South Atlantic areas. The "schnorkel"—a tube providing air for diesel engines while submerged—assisted the latest German submarines to remain under water for long periods, but improved methods of detection prevailed. By the end of February, more than 1 million American troops had been convoyed to the United Kingdom, and by May the logistical lift had reached the colossal amount of 1.9 million tons per month.

1944, April–November. Target: *Tirpitz*. The British were determined to destroy the *Tirpitz*, hidden in Alta Fjord, Norway, before she could be repaired (see p. 1195). A surprise raid by carrier-based planes scored numerous hits with 1,000-lb. bombs, killing 300 of the *Tirpitz's* crew, damaging guns and radar equipment, but not harming her armored sides or deck (April 3). Subsequent attacks were less successful, and the Germans pressed on with repairs. A long-range bomber finally (September 15) hit the *Tirpitz* with a 6,000-lb. bomb, causing damage that forced the Germans to move the giant ship to better repair facilities at Tromsö. After several more long-range attacks with 6-ton "blockbusters" against the well-protected ship, she was finally hit by several of these great bombs and sunk, carrying 1,200 men to the bottom (November 12).

1944, November–December. Doenitz Tries Again. As the weather turned wintry and the days shorter, Doenitz sent large numbers of his schnorkel-equipped submarines from his Baltic and Norwegian bases (the French bases having all been captured by the Allies) in intensified attacks against Allied shipping, this time trying the new and rather effective technique of operating near Britain in shallow coastal waters where the Allied detection equipment was unreliable. Shipping losses again mounted, though never approaching the serious situation of 1941 and 1942.

The Combined Bomber Offensive

1944, January–May. Unremitting Pressure on Germany. The U.S. Strategic Air Forces (commanded by Lieutenant General **Carl Spaatz** and comprising the Eighth AF, based in England, and the Fifteenth AF, based in Italy) intensified operations, in continuing close cooperation with Air Marshal Harris' RAF Bomber Command. Thousand-plane raids were frequent, with the British bombers continuing to operate at night and the Americans by day, often striking the same targets in pulverizing one-two punches.

1944, February 20–26. The "Big Week." The Americans by day, the British by night, ranged over Germany in 5 days of coordinated strikes against Germany's airplane and antiaircraft factories and assembly plants, crisscrossing Leipzig, Regensburg, Augsburg, Fürth, and Stuttgart. The Luftwaffe rose to meet them in a last bid for supremacy. American losses alone in these operations were 244 heavy bombers and 33 fighter planes, but the Germans lost 692 planes in the air and many more on the ground.

1944, March–May. Attriting the Luftwaffe. The Germans lost 2,442 fighters in action and another 1,500 through accident or other causes. Although the German aircraft industry recovered from the "Big Week" blows, the loss in trained pilots during the period was irreparable. Allied air supremacy was henceforth unchallenged.

1944, May–June. Preparations for "Overlord." (See p. 1210.) The strategic bombers joined the tactical air forces in operations designed to "isolate the battlefield" over a great arc in France, the Low Countries, and western Germany—too widespread in direction and targets to betray the selected landing area. Bridges, tunnels, marshaling yards, and roads were plastered and the railway system largely paralyzed. All air bases in France in a 130-mile

radius from the assault beaches were neutralized. By June 6, Normandy and Brittany were in effect isolated from the remainder of France. In all, some 4 million tons of bombs were dropped during the period, nearly 60 percent by American aircraft.

1944, June–August. Shuttle Bombing from Russia. To permit American bombers to strike farther east into Germany with greater bomb loads, and also to confuse German air-defense efforts, arrangements were made for them to fly east to bases behind the rapidly advancing Russian front. Despite operational success, the coordination with the Russians was never very effective. Also the Russian air defenses were unable to cope effectively with German counterraids on the American bases in Russia, resulting in severe American losses. The effort was therefore soon abandoned.

1944, July–December. Climax of the Strategic Air Offensive. The long-range bombers returned to their primary mission of attacking Germany's warmaking capacity, save for a few diversionary attacks on tactical targets to assist in the breakout from the Normandy beachhead (July; see p. 1211). The pattern of operations was as before, save that the growing number of planes permitted even more intensive operations. Principal targets were Germany's oil-production facilities and transportation system. All large industrial areas were struck, however, including steel plants, electric-power facilities, and weapon factories. The attacks on oil production drastically reduced the available fuel for airplanes, tanks, trucks, and submarines, thus affecting Germany's ability to fight on land, in the air, and at sea. Lack of fuel reduced the new fighter-pilot training, creating a vicious cycle; inadequately trained German pilots were quickly shot down by the attacking bombers and their fighter escorts. The attack on German transportation also had a wide effect, choking the flow of raw materials to factories, and of finished products to the German fighting forces and population. Finally the curve of German production began to drop sharply; the Combined Bomber Offensive was wrecking Germany's capability to continue the war.

Operations in Italy, 1944

ANZIO-RAPIDO CAMPAIGN

1944, January 5–15. Drive to the Rapido. Stubborn assaults through the mountains, from the confluence of the Liri and Garigliano rivers north to the Apennines, advanced the Fifth Army nearly 7 miles to the final German Gustav Line along the Rapido, with Monte Cassino the key terrain obstacle in the bulge's center. Alexander planned a frontal attack, assisted by an amphibious landing at Anzio (Operation "Shingle"), some 60 miles from the Rapido front. The Anzio force would then advance inland to cut the German communications line. Although the 2 operations were beyond mutual-support capabilities, it was believed the dual operation would force evacuation of the Gustav Line. The Eighth Army, meanwhile, would continue its advance on Pescara, on the Adriatic coast.

1944, January 17–21. Rapido-Cassino Assaults. The X Corps attacked across the Garigliano, attaining a bridgehead. On its right the II Corps U.S. 36th Division attempted to force the Rapido, but was repulsed with heavy loss (January 17–19). The French corps nibbled north of Cassino to make slight but costly gains. As expected, German reserves were drawn to the Rapido front, and the amphibious operation was launched from Naples (January 21).

ANZIO OPERATIONS, JANUARY 22–FEBRUARY 29

1944, January 22. The Landings. Major General **John P. Lucas'** VI Corps—some 50,000 Anglo-American troops, with 5,200 vehicles—began landing without opposition. Forty-eight hours later, most of the troops were ashore, the initial objectives attained, and a beachhead established, 7 miles deep. Lucas, however, made no attempt to drive inland toward the Alban Hills—the vital terrain. Instead, he consolidated his position, awaiting the landing of heavy weapons, tanks, and additional supplies. General Clark, who was present, concurred. But Kesselring's quick reaction brought German reinforcements from the north as well as from quiet sectors of the Gustav Line.

1944, January 23–February 16. German Build-up. Under General **Hans Georg Mackensen** the quickly extemporized Fourteenth Army pinned Lucas to his beachhead.

1944, February 16–29. German Counterattacks. A series of brutal blows drove back the outlying Allied units. Lucas was relieved (February 23) by Clark, Major General **Lucius K. Truscott, Jr.,** U.S. 3rd Division commander, replacing him.

1944, March–May. Stalemate. The amphibious assault became a siege for 3 more months with all the elements of World War I trench warfare. All portions of the narrow beachhead were under continuous observation and fire, while the Luftwaffe swept the harbor area, disrupting supply and reinforcement efforts.

1944, February–May. Operations on the Rapido. Fifth Army battered at the Gustav Line. The U.S. 34th Division assault on Monte Cassino, the so-called **First Battle of Cassino,** was repulsed (February 12). The New Zealand Corps then tried, supported by aerial bombardment (General **Bernard C. Freyburg** mistakenly thought the Germans were using the monastery for observation), in the **Second Battle of Cassino** (February 15–18), and also failed. The Germans quickly occupied the ruined monastery and repulsed the New Zealanders. The most massive close air support attack attempted to date brought no different result in the **Third Battle of Cassino** (March 15–23).

COMMENT. *Since the objective of the Anzio-Rapido operation was to pry the Germans out of the Gustav Line by utilizing Allied sea power to cut their line of communications, Anzio should have been the main effort, the Rapido merely a holding attack. But insufficient sea transportation (because of the demands of the 2 coming amphibious invasions of France) was available to ensure a sledge-hammer blow at Anzio. So the joint operation—and the responsibility must rest on Alexander—became a weak planning compromise: 2 main efforts, entirely incapable of mutual support, neither of them powerful enough to do the job alone. It can be argued that had Lucas immediately and boldly pushed ahead to his final objective—the Alban Hills—the Gustav Line must have collapsed, with Rome quickly occupied. But Lucas' commander, General Clark, was ashore on D day and concurred in the decision to consolidate before driving inland. Some 23,860 American and 9,203 British casualties were evacuated during the 4-month hell on the beachhead. In the end the Gustav Line collapsed only as a result of the very type of frontal attack the amphibious operation was designed to avoid.*

1944, March 15–May 11. Operation "Strangle." U.S. Major General **Ira C. Eaker's** Anglo-American Mediterranean Allied Air Forces undertook a systematic air interdiction campaign to cut off supplies to German troops south of Rome. Despite severe punishment, the Germans did not withdraw as Allied air planners had hoped. However, the effect would soon be evident when intensive ground pressure was combined with the air interdiction campaign.

ROME CAMPAIGN

1944, May 11–25. Breakthrough. Regrouped to bring the weight of the 15th Army Group into his main effort, Alexander launched a full-scale surprise assault in the 20-mile zone between Cassino and the sea. The interdiction pressures of "Strangle" were intensified. In the combined air-ground offensive, "Diadem," French, Polish, British, Canadian, and U.S. units smashed through the German lines. The Poles took Cassino (May 17–18). At Anzio the reinforced VI Corps attacked (May 23) toward

the Alban Hills; contact was made between the two Allied forces two days later.

1944, May 26–June 4. Advance on Rome. General Clark's shift of the Fifth Army toward Rome now saved the German Tenth Army from possible envelopment. Skillfully handled rear guards checked American advances at Valmontone and Velletri (May 28–June 2), while the remainder of the Tenth Army fell back. Rome was entered (June 4), hot on the heels of a general German retirement.

1944, June–August. Advance to the Arno. The Allies pushed rapidly up the peninsula. But withdrawals of troops—both ground and air—to mount the invasion of southern France (see p. 1212) reduced Alexander's strength, while German reinforcements bolstered Kesselring. In a series of masterly delaying actions the Germans—despite Allied air superiority—retired to the Gothic Line, extending across the peninsula south of Bologna, its outposts running generally from Pisa, through Florence, to Ancona.

1944, August–December. Advance to the Gothic Line. The Fifth Army crossed the Arno (August 26). Leese's Eighth Army took Rimini (September 21), and Clark, committing all his reserves, made an unsuccessful bid for Bologna (October 1–20). Another Italian winter settled on an exhausted Allied army group. Alexander, promoted to Supreme Allied Commander in place of Wilson (transferred to head the British military mission in Washington), was replaced by Clark. Truscott took over the Fifth Army and Lieutenant General **Richard L. McCreery** took over the Eighth from Leese.

The Allied Invasion of Western Europe

THE PRELIMINARIES, MAY 1943–MAY 1944

ALLIED SITUATION AND PLANS

At the Trident Conference (see p. 1193), President Roosevelt and Prime Minister Churchill agreed on a major cross-Channel invasion of Europe in 1944. Planning was under Lieutenant General Sir **Frederick Morgan;** the target date was set as the first week of June 1944. A gigantic amphibious operation from southern England to France, with nearly 3 million men, was planned. After all possible landing sites were considered, the area east of the Cotentin

Peninsula of Normandy was selected, because of (a) its proximity to Allied fighter bases in England, (b) the short water distance for carrying supplies and reinforcements in limited numbers of landing craft, (c) the nature of the beaches, (d) the nature of the inland area, and (e) the German defenses. In February, General Eisenhower, designated to command the invasion, established Supreme Headquarters Allied Expeditionary Forces (SHAEF), approved the planning done by General Morgan's staff on Operation "Overload," as the invasion was designated, and continued plans and preparations. By May, 1944, he reported to the CCS that his force was ready.

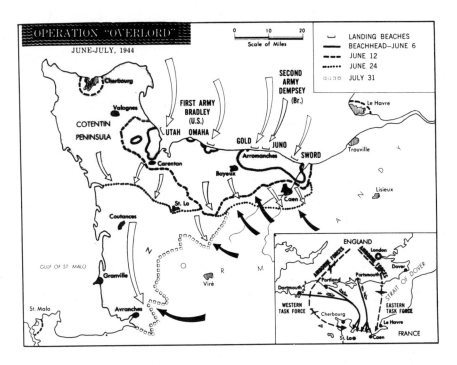

In essence, the Allied plan envisaged assault landings on 5 main beaches. Lieutenant General Omar Bradley's First U.S. Army was to land on Utah Beach, north-west of the impassable marshy Carentan estuary, and on Omaha Beach just east of the estuary. The decision to risk defeat in detail on these beaches was made because of the importance of Utah Beach as a base for a quick drive to seize the port of Cherbourg, essential to future logistical support. To reduce the danger of this decision, 2 airborne divisions were to be dropped well inland of the marsh country to facilitate early link-up of the 2 beachheads. Farther east General Sir **Miles Dempsey**'s British Second Army (including a Canadian corps) would land on 3 beaches—Juno, Sword, and Gold—west of the Orne River. Another airborne division was to land just east of the Orne to protect the British left flank. Overall commander of Allied ground forces was British General Sir **Bernard L. Montgomery.** The total combat strength of the invasion forces gathered in England was 45 divisions, totaling with supporting units about 1 million men (two-thirds American). Almost another million comprised the tremendous logistical and administrative support forces which would sustain the combat units.

The supporting naval and air elements totaled almost another million men. Commanding the invasion armada was British Admiral Sir **Bertram Ramsay.** British and American tactical air forces were commanded by RAF Air Marshal Sir **Trafford Leigh-Mallory.** For the period just before and after the attack, General Spaatz's long-range American bombers were also placed under Eisenhower's command. Eisenhower's deputy commander was RAF Air Chief Marshal Sir **Arthur Tedder.**

GERMAN SITUATION AND PLANS

Hitler's "Atlantic Wall" stretched from the North Sea coast to Brittany's Atlantic nose and thence southward to the Spanish border—a network of permanent fortifications laced by strong points and field positions, protected by mine fields and underwater obstacles. Inland areas favoring air drops were mined and strewn with obstacles. Manning this complex were 10 Panzer, 15 infantry, and 33 training or coast-defense divisions (of inferior quality) of the Western High Command (von Rundstedt) spread from Norway to the Mediterranean. German troops immediately available in the invasion area were portions of Army Group B (Seventh and Fifteenth Armies), commanded by Rommel—4 coast-defense divisions manning fortifications, 2 infantry divisions, the garrison of Cherbourg, and 3 Panzer divisions in reserve. The remainder of Rommel's troops were frozen, partly because of the disruptive Allied bombing offensive, partly because of Hitler's orders. Obsessed by the idea that the Allied thrust would come over the Pas de Calais—shortest distance from England—the Führer insisted that the Fifteenth Army (east of the designated landing area) rest in place, and that no Panzer divisions be released without his specific order. The brunt of the invasion would thus be borne by the Seventh Army, Colonel General **Friedrich Dollmann.** With the Luftwaffe already swept from the sky by Allied air, and no naval defense except light torpedo craft and submarines, the German situation was precarious, but far from hopeless.

OPERATION "OVERLORD"

1944, June 6. D Day. Preceded by the airborne drops—2 U.S. divisions on the west and 1 British division on the east—the greatest amphibious assault yet known to history began landing on the Normandy coast in complete tactical surprise. Some 4,000 ships and landing craft carried 176,000 troops and their matériel. Escorting the armada were 600 warships. The Allied air forces had earlier drenched the terrain with bombs; 10,000 tons of explosive were dropped by some 2,500 heavy bombers, while 7,000 fighters and fighter-bombers combed the area. Supporting the assault were the guns of the warships, large and small. By nightfall, 5 divisions were ashore and a comfortable toehold had been obtained at all beaches except Omaha, where German resistance had been heaviest and defending artillery fire best directed. There the initial assaults bogged down for a time. Meanwhile, offshore, an amazing conglomeration of concrete floating caissons—"Mulberries," a British invention spurred by Churchill's insistence—was being jockeyed into position to make 2 artificial ports, protected by lines of old vessels scuttled to provide breakwaters. (An unexpected storm of great ferocity was to destroy the American artificial port, June 19, and thereafter supply for both Allied forces moved through the British artificial harbor.)

1944, June 7–18. Expansion. Hitler's fixation that another Allied attack would come in the Pas de Calais area hampered Rundstedt and Rommel in their defensive strategy; reinforcements came in bits and pieces, and part of the Panzer strength was frittered away in piecemeal counterattacks. But the *bocage* (checkerboard of small fields boxed by deep hedgerows) reduced the Allied advance to a crawl. German resistance centered about Carentan and Caen. Montgomery's efforts to take Caen were rebuffed (June 13 and 18), while on the right the U.S. VII Corps, Major General **J. Lawton Collins,** was slowly battering its way across the Cotentin peninsula base. It turned north (June 18) and drove for Cherbourg, while the remainder of Bradley's army took up an aggressive defense.

1944, June 27. Fall of Cherbourg. The garrison surrendered after 5 days of desperate defense, and after demolishing harbor installations. The harbor would not be cleared until August 7, but beach unloading there began immediately. During this time Rommel had attempted to mass armor for a counterattack on the British, but was forced to commit his divisions piecemeal.

1944, July 1–24. Expansion of the Beachhead. Allied strength ashore built up to 1 million men, 150,000 vehicles, and half a million tons of supply. However, progress southward through the *bocage* was disappointngly slow and costly, and the Germans had caught their second wind. Rundstedt was relieved by Hitler, Marshal **Günther von Kluge** taking his place. SS General **Paul Hausser** (Dollmann had been killed) opposed the U.S. First Army's advance with approximately 7 veteran divisions of the Seventh Army, while Panzer Group West, General **Heinrich Eberbach,** barred the British advance with 7 armored and 2 infantry divisions. Montgomery made some progress (July 8) following a heavy bombing attack and renewed his effort, finally taking Caen (July 13). Then, after a costly repulse southeast of Caen (July 20), the attack halted. It had, however, drawn German armored strength to the British front. Bradley prepared for a breakthrough after having slowly crunched his way into St.-Lô (July 18), losing 11,000 casualties. Meanwhile, in the Channel, Allied sea power held open the crossing lanes against 56 German submarines and German light craft operating out of Le Havre. Rommel having been wounded (July 17) when an Allied plane strafed his car, Kluge took over his command in addition to his other duties. Thus far, Allied losses had been some 122,000 men, and the Germans had lost about 114,000, including 41,000 prisoners. The first phase of the invasion was complete.

OPERATION "COBRA"—BREAKOUT FROM THE BEACHHEAD

1944, July 25–31. Breakthrough. Bradley's U.S. First Army assaulted to penetrate the German line west of St.-Lô, with Collins' VII Corps making the main effort. A tremendous bomb carpet by Spaatz's long-range bombers—4,200 tons of explosive—opened a gap, although through miscalculation more than 500 Americans were killed or wounded by "shorts." (General **Lesley J. McNair,** commander of U.S. Army Ground Forces, present as an observer, was killed.) The assault advanced through heavy resistance to Coutances (July 28) and Collins' armor reached Avranches (July 31). Hitler released some Fifteenth Army divisions to Kluge (July 27).

1944, August 1. Allied Reorganization. Unveiled now was the U.S. Third Army, Patton commanding, taking the right of the Allied line. The U.S. 12th Army Group came into existence, Bradley commanding, with Lieutenant General **Courtney Hodges** leading the First Army. In the British zone Montgomery's 21st Army Group also expanded—comprising the Canadian First Army, Lieutenant General **Henry D. G. Crerar,** on the left of Dempsey's Second British Army. Montgomery remained over all ground-force commander.

1944, August 1–13. Breakout and Exploitation. Patton's Third Army whirled through the Avranches gap. His armor scoured Brittany, then turned south to the Loire and pointed

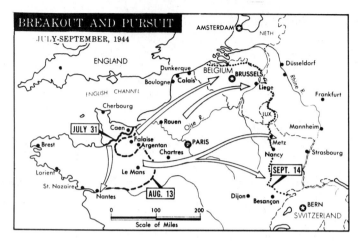

eastward. His infantry curved left toward Le Mans. On the inner ring, the First Army began pivoting to its left.

1944, August 6–10. Counterattack at Avranches. The First Army was momentarily halted by a vicious German counterattack ordered personally by Hitler. Kluge, gathering all available Panzer strength, hurled it westward at Mortain, toward Avranches, hoping to isolate the Third Army and—ultimately—to turn north and crush the Normandy beachhead. Bradley shifted reinforcements and—with the aid of air power, mostly British—Kluge's attack was halted (August 8), although, against his protests, Hitler ordered the effort continued for 2 more days. Meanwhile, oblivious to the threat, the Americans from the southwest and the British from the north bore down on the Germans in a pincers movement.

1944, August 13–19. Falaise-Argentan Pocket. Allied misunderstandings and some timidity contributed to provide a gap through which the now much-disorganized German Seventh and Fifth Panzer (former Panzer Group West) armies fled eastward. As it was, although a goodly part of the German armor got away, some 50,000 Germans were captured and 10,000 more lay dead. These operations cost the U.S. First Army 19,000 casualties; the Third Army suffered about 10,000.

1944, August 20–30. Pursuit. Kluge, in full retreat, made for the safety of the Seine bridges, with all 4 Allied armies in full cry behind him.

The Allies advanced to the Seine from Troyes to the Channel (August 25).

1944, August 25. Liberation of Paris. The local population had risen (August 23) and was liberated by the U.S. Fifth Corps, with Leclerc's French 2nd Armored Division spearheading the drive. Kluge's reward for a remarkable salvaging of the German remnants from the Falaise-Argentan gap was to be dismissed by Hitler; General **Walther Model** assumed command of Army Group B. Kluge committed suicide a short time later.

1944, August–September. Germans Isolated in Western France. At Hitler's orders, some German garrisons held French westcoast ports: Brest (captured September 18 after a long siege), Lorient, St.-Nazaire, La Rochelle, and the Gironde estuary (all of which held out for several more months). A wandering German army corps south of the Loire was hunting for Allied troops to surrender to, as the countryside boiled in the French Resistance movement—nominally at least under SHAEF direction.

OPERATION "ANVIL-DRAGOON"— SOUTHERN FRANCE

1944, August 15. The Landings. The U.S. Seventh Army (Lieutenant General **Alexander Patch**), consisting of the U.S. VI and French II corps, with a provisional airborne division, made an amphibious landing and air drop on the Côte d'Azur between Hyères and Cannes. The objective was to free Marseilles (for supply) and protect Eisenhower's southern flank. Mounted in the Mediterranean Theater, the operation was supported by tactical and strategic air forces, while ships of the Western Task Force escorted. Assault landings were made by General Truscott's VI Corps at small cost. By nightfall, some 94,000 men and more than 11,000 vehicles were ashore, the German coastal defense—2 second-line infantry divisions—broken, and over 2,000 prisoners taken. The assault casualties amounted to 183 men killed and wounded, with 49 nonbattle casualties. The French (General **Jean de Lattre de Tassigny**) began landing immediately behind the assault units.

1944, August 16–28. Pursuit. While de Lattre's French troops turned west toward Toulon and Marseille, Truscott's VI Corps boldly thrust

north in 2 columns, the main body up the Rhone Valley, the other through the foothills of the Alpes Maritimes. German General **Friedrich Wiese**'s Nineteenth Army was retreating in considerable disorder up the Rhone Valley. German strength consisted of 7 second-rate infantry divisions and the well-handled 11th Panzer Division. Operating on a logistical shoestring, Truscott's fast-moving right column (Task Force Butler and the 36th Division) raced and passed the Germans in the Rhone Valley, then turned west to close the trap at Montélimar (August 22).

1944, August 23–28. Battle of Montélimar. Indecision on the part of the American 36th Division and a vigorous Panzer counterattack kept the block open long enough to let most of the German Army fight its way past Task Force Butler, whose artillery, supported by tactical air, took a terrible toll. The road from Montélimar north to Loriol became a shambles.

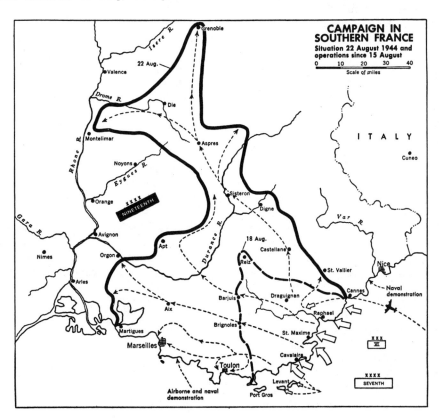

More than 15,000 prisoners were taken, while some 4,000 vehicles—tanks, guns, and trucks—were destroyed. The number of German dead is unknown.

COMMENT. *Truscott's Rhône Valley campaign is a shining example of boldness and initiative. In 14 days of incessant drive he had practically destroyed a German army with 1,316 of its 1,481 guns, captured 32,211 prisoners, and advanced some 175 miles, at a cost of 1,395 killed and 5,879 wounded and missing. The French component of the Seventh Army captured Toulon and Marseille (August 28), taking 47,717 prisoners, losing some 1,300 killed and 5,000 wounded and missing. German forces in the south of France had been eliminated.*

1944, August 29–September 15. Advance to the Vosges. Patch's Seventh Army moved north, the U.S. VI Corps leading, with the French I and II corps following on the right and left flanks respectively. Contact with Patton's Third Army was made west of Dijon (September 11). The southern invasion force then became the U.S. 6th Army Group (September 15), under Lieutenant General **Jacob M. Devers,** its components the U.S. Seventh Army and the newly established French First Army, General de Lattre de Tassigny.

ADVANCE TO THE WESTWALL, AUGUST 27–SEPTEMBER 14

1944, August 27–September 4. The British Pursuit. The AEF had crossed the Seine in hot pursuit of the ebbing German tide. But Eisenhower needed possession of Antwerp to provide supply—the 300-mile-long logistical line from the Normandy beaches was stretched almost to breaking point. He thus favored Montgomery's advance into the Low Countries. The British Second Army advanced rapidly. Brussels was entered (September 3) and Antwerp, with port facilities intact, was captured next day.

1944, August 27–September 4. The American Pursuit Falters. Gasoline for Bradley's army group was reduced to a trickle as supplies went to the British. Patton's Third Army crossed the Meuse (August 30), only to be halted there with empty fuel tanks. Hodges'

First Army, after bagging 25,000 prisoners at **Mons** (September 3), was also reduced to inchworm progress.

1944, September 3. Eisenhower Assumes Direct Command of Ground Operations. This permitted Montgomery to devote full attention to the 21st Army Group.

1944, September 4–14. The Germans Block Antwerp. Hitler ordered General Kurt Student's First Parachute Army to block further advance across the Albert Canal, and General **Gustav von Zangen**'s Fifteenth Army to hold the Scheldt estuary. Until that had been cleared, Antwerp was of no use as an Allied port. Clearing Antwerp became the main mission of the Second Army, while the Canadian First Army, after investing Le Havre, moved up the coast to clear the Channel ports and to overrun the V-1 missile sites in the Pas de Calais area.

1944, September 5. Recall of Rundstedt. Hitler again assigned Rundstedt to supreme command in the west. He prepared a stand along the Westwall fortified zone (Siegfried Line).

1944, September 14. Allies Close on the German Border. The line ran northward from Switzerland, through Lorraine to Aachen, then northwesterly through Maastricht, along the Albert Canal, and west to the Channel at Ostend. Bradley's 12th Army Group had renewed a creeping advance, still hampered by gasoline shortage.

COMMENT. *Since June 6, the Allies had put more than 2.1 million men on French soil, shattered Hitler's dream of a* Festung Europa, *and flung the German forces back to their own border. The stupendous drive through France had cost some 40,000 Allied killed, 165,000 wounded, and 20,000 missing. German losses had been a catastrophic half-million men in the field forces and an additional 200,000 in the coastal fortresses. Chewed, disrupted, and battered, the remnants now stood behind the dilapidated Siegfried Line, seemingly vulnerable to an Allied coup de grâce. However, their homogeneous staff structure was still intact, their discipline remained, and replacements were being rushed from Germany and the east. Rundstedt and his principal subordinates provided a high order of leadership. The war was still far from its end.*

INVASION OF GERMANY, 1944

Both Montgomery and Bradley were convinced that a single thrust into Germany, furnished with unlimited support, would end the war, but neither desired to play second fiddle. Eisenhower, straddling, approved Montgomery's plan to turn the northern German flank by seizing a northern bridgehead over the Maas (Meuse) before clearing the water approaches to Antwerp. The First Allied Airborne Army (3 divisions) under Major General **Lewis H. Brereton,** in reserve in England, was attached to Montgomery's command for the operations.

OPERATION "MARKET GARDEN"

Montgomery's bold plan was to drop 3 airborne divisions as stepping-stones behind the German lines along a narrow 60-mile causeway over marshy ground, seizing the bridges of three rivers—the Maas (Meuse), Waal (Rhine), and Lek (lower Rhine). The British Second Army—led by XXX Corps—would drive across this "airborne carpet" and turn the northern flank of the Siegfried Line.

1944, September 17. Air Drops at Arnhem, Nijmegen, and Eindhoven. The daylight drops were successful, but German resistance on the ground was stronger than expected. The U.S. 101st Division secured the Wilhelmina Canal crossing near Eindhoven. The U.S. 82nd captured the Maas bridge at Grave, but was unable to gain the Waal bridge at Nijmegen. The British 1st Airborne Division, with a Polish brigade attached, reached the Arnhem area north of the Lek (Neder Rijn), but at once became involved with unexpectedly strong German forces. The British ground advance, spearheaded by tanks, reached Eindhoven to join the 101st (September 18). The Nijmegen bridge was secured after another 24 hours of bitter fighting by a joint Anglo-American assault. Bad weather and lack of maneuver ground in the flooded countryside hampered further advance, though detachments reached the south bank of the Lek.

1944, September 17–26. Battle of Arnhem. The British 1st Airborne was pocketed by German reinforcements and forced into a small perimeter, though clinging for a while to the Lek bridgehead. Behind them the Germans now completely barred the way to further Allied advance and bad weather prevented dropping of reinforcements or supply. Ringed by close-in artillery and mortar fire, with food and ammunition exhausted, and forced away from the bridge, the defense collapsed. Some 2,200 survivors were evacuated across the Lek in assault boats during the night (September 25–26), leaving 7,000 men behind them killed, wounded, or captured.

COMMENT. *The stand of the 1st Airborne Division at Arnhem ranks high in the annals of the British Army. German reaction was prompt, skillful, and efficient. The Rhine barrier still faced the Allies and the German defense was still intact.*

THE SCHELDT ESTUARY CAMPAIGN, OCTOBER–NOVEMBER

Montgomery now turned to the task essential to any further Allied advance—liberation of the water gate to Antwerp. The keys to the Scheldt were the South Beveland peninsula and Walcheren Island at its western tip, both highly fortified.

1944, October 1–November 8. Battle for South Beveland and Walcheren Island. Furious British assaults (some American units participated), plus a small amphibious operation, resulted in capture of the peninsula (October 31) after both sides suffered heavy losses. Walcheren Island fell (November 8) to an amphibious assault, following on the flooding of the defenses by Allied air bombings of the sea dikes. The Scheldt was now free, for the south shore had been cleared already (October 22).

1944, November 4–26. Minesweeping the Scheldt. Even before the last gun was fired at Walcheren, one of the most difficult mine sweeps of the war had begun. The 70-mile channel to Antwerp was combed 16 times by 100 Allied vessels before the first Allied convoy entered the port (November 27).

BREAKING THE SIEGFRIED LINE,
OCTOBER–DECEMBER

1944, October 1–23. American Advance to the Westwall. Bradley's 12th Army Group pressed against the Siegfried Line from above Maastrich to Lunéville—3 armies now, for the new U.S. Ninth Army, Lieutenant General **William H. Simpson,** had joined (October 3). From Lunéville to the Swiss border, Devers' 6th Army Group penetrated Alsace and the Vosges Mountains. Winter was closing in. German opposition west of the Rhine was strong. Rundstedt's command, OB West, consisted from north to south of Student's Army Group H, from the North Sea to below Roermond; Model's Army Group B, reaching south to the line of the Moselle; and **Hermann Balck**'s Army Group G to Karlsruhe. An SS group, under SS Chief **Heinrich Himmler,** held the remainder of the line to Switzerland. Patton's Third Army reached and closed in on Metz (October 3). Hodges' First Army captured **Aachen** after furious fighting (October 21), the first breach of the Siegfried Line.

1944, October 28. Eisenhower Orders November Offensive. This was to destroy all German forces west of the Rhine, to establish bridgeheads across it, and to advance into Germany.

1944, November 2–7. Repulse at Schmidt. The U.S. 28th Division of the V Corps was repulsed in efforts to take Schmidt, just north of the Roer River dams.

1944, November 16–December 15. Roer River-Hürtgen Forest Operations. Bradley's Ninth and First Armies attacked against heavy opposition, over difficult terrain, enlarging the Aachen breakthrough. Major obstacle was the Hürtgen Forest. The attack, on a narrow front, reached the Roer River, but crossing could not be attempted until the dams near Schmidt had been seized to prevent the Germans from flooding the valley. A major offensive for this purpose was begun (December 13).

1944, November 16–December 15. Lorraine Operations. Patton's Third Army captured **Metz** (December 13) and battled its way across the Seille River.

1944, November 16–December 15. Alsace Operations. Devers' group made deep gains; Seventh Army's French 2nd Armored Division* thrust through the Saverne Gap of the Vosges Mountains to liberate Strasbourg (November 23), rousing French national morale to a peak. The French First Army overran Mulhouse. Devers was now on the Rhine from Karlsruhe to below Strasbourg, and again from Mulhouse to the Swiss border; but the deep Colmar pocket in between was still firmly held by Wiese's German Nineteenth Army.

---

*Bitter perosnal enmity existed between Leclerc and and de Lattre de Tassigny; hence the French 2nd Armored Division was never under the latter's command.

GERMAN ARDENNES OFFENSIVE (BATTLE OF THE BULGE), DECEMBER 1944–JANUARY 1945

The German Plan. Hitler had prepared a striking force to split the Allies. His armor would rip through to Antwerp, crippling their supply. He hoped to destroy all Allied forces north of the line Antwerp-Brussels-Bastogne, just as in 1940. Success depended on three elements: (1) a breakthrough, (2) seizure of Allied fuel supplies and the key focal points of communication in the area St.-Vith and Bastogne, and (3) widening of the initial gap to increase the flow of invasion. Hitler's commanders, though dubious of success, obeyed orders.

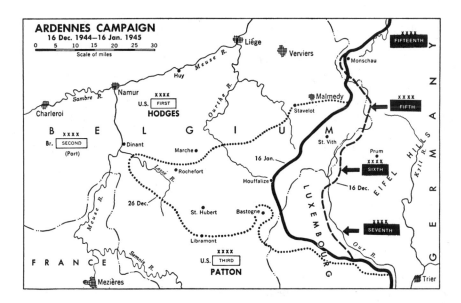

1944, December 16–19. The German Blow. The operation was launched after a period of fog, rain, and snow blanketed Allied aerial observation and hobbled combat capabilities. The striking force, from north to south, consisted of the Sixth SS (General **Sepp Dietrich**) and Fifth (General **Hasso von Manteuffel**) Panzer armies—24 divisions, 10 of them armored. The Seventh Army (General **Ernst Brandenberger**) was to cover the southern flank. The initial wave—5 Panzer, 12 infantry-type divisions—disrupted the U.S. VIII Corps. Tactical and strategic surprise was complete. (SHAEF intelligence estimates had dismissed all probability of any immediate major German offensive capability.) The 106th Division, just arrived on the front, and the 28th Division, recuperating from severe fighting at Schmidt, were shattered. A paratroop drop in the area Eupen-Monschau, and a spearhead force of English-speaking German soldiers in American uniforms, added to panic and confusion behind the assault zone. But on the north flank, the U.S. V Corps, halting its own offensive toward the Roer dams, held firm, as did the U.S. 4th Division on the south. Canalized between these shoulders, the attack roared on toward the Meuse. Two U.S. armored divisions were rushed in by Bradley as immediate reinforcement. Eisenhower then committed the SHAEF reserve—the 82nd and 101st Airborne divisions (recuperating near Reims from their Maas operation). Truckborne, they arrived (December 19)—the 101st (under Brigadier General **Anthony C. McAuliffe**) at Bastogne, a check to Fifth Panzer Army's progress, and the 82nd (Major General **Matthew B. Ridgway**) to bolster the northern flank. Montgomery began shifting 1 British corps to backstop the operation along the Meuse. At Bradley's order, Patton (December 18) halted his Third Army's advance in the Saar to begin an amazing 90° shift in direction to the north, to hit the German southern flank.

1944, December 20–26. Allied Recovery. Eisenhower transferred command of all U.S. troops north of the bulge to Montgomery, leaving only Patton's army under Bradley. Despite a desperate defense of **St.-Vith** by the U.S. 7th Armored Division (Brigadier General **R. W. Hasbrouck**), the Sixth Panzer Army forged slowly ahead (December 19–22), but the delay had been fatal to the German plan. The V Corps was still presenting an impenetrable front, while the U.S. VII Corps was hurrying southwest to seal the remainder of the northern flank. At **Bastogne,** the 101st Airborne, with some other units—some 18,000 men in all—resisted all efforts of the Fifth Panzer Army to overrun their perimeter. However, the invading tide, lapping around Bastogne, progressed northwest toward the Meuse. Model, commanding Army Group B, quite properly desired now to shift the weight of the German assault to Manteuffel's Fifth Panzer Army, but Hitler, obstinate and ignorant, insisted the decisive blow be struck by his SS pet, Dietrich. By December 22, Patton was attacking north toward beleaguered Bastogne on a 2-corps front, while Devers' 6th Army Group extended its left to cover his advance. Dietrich's penetration in the Manhay-Stavelot area, and Manteuffel's spearheads—Panzer Lehr and 2nd Panzer divisions—were grinding to a halt with empty fuel tanks at **Celles,** almost in sight of the Meuse, to be struck by American and British counterattacks (December 25–26). Hitler's gamble had failed. Patton's Third Army punched a hole through Manteuffel's troops to reach Bastogne (December 26), and, with the first clear weather, Allied air began pounding German supply trains west of St.-Vith.

1944, December 26–1945, January 2. The Battle for Bastogne. Hitler insisted on the capture of Bastogne, and a furious battle raged for a week while the German tide ebbed elsewhere in the Bulge under Allied pressure. Attempting to disrupt Allied air support, the Luftwaffe made its last offensive strike (January 1), some 800 planes attacking airfields in France, Belgium, and Holland, and destroying 156 Allied planes. The attack was repulsed with heavy losses to the Germans, and the Allied air offensive over the Ardennes area and German rear elements continued.

1945, January 3–16. Allied Counteroffensive. On the northern flank of the German penetration, Montgomery unleashed Hodges' U.S. First Army. German offensive efforts near Bastogne were repulsed, and Patton's increasing efforts, supported by XIX Tactical Air Force, shrank the southern face of the German penetration. Hitler permitted withdrawal of the Sixth Panzer Army (January 8; see p. 1228). The Bulge was eliminated (January 16). Hodges' First Army returned to Bradley's control (January 18), but Simpson's Ninth Army remained in Montgomery's 21st Army Group.

COMMENT. *Hitler's Ardennes offensive was a gamble, pure and simple. The blow was checked first by the resistance of the U.S. elements on both shoulders, next by Hasbrouck's stand at St.-Vith and McAuliffe's epic defense of Bastogne. Hitler's refusal to shift the weight of the attack to the flank making the best progress was stupid. When the German armor was unable to overrun Allied fuel depots to replenish its tanks, the end was inevitable. The net result was a delay of about 6 weeks to Allied operations in the west, while Hitler had expended the slim reserves with which he otherwise might have checked the coming Russian spring offensive. German losses were some 120,000 men killed, wounded, or missing, 600 tanks and assault guns, 1,600 planes, and 6,000 vehicles. Allied losses (mostly American) were approximately 7,000 killed, 33,400 wounded, 21,000 captured or missing, and 730-odd tanks and tank destroyers. Among the Americans were 86 prisoners captured by the 1st SS Panzer Division at Malmédy on December 17th, then lined up and ruthlessly machine-gunned to death.*

The Eastern Front

RUSSIAN WINTER OFFENSIVE

Following a series of probing attacks, the Soviet armies launched a concerted drive as winter hardened roads and froze the waterways.

1944, January 15–19. Liberation of Leningrad. Two Russian army groups fell on the German Eighteenth Army, investing Leningrad. General **L. A. Govorov**'s Leningrad Front, crossing the frozen Gulf of Finland, pierced the German left, while General **Kirill A. Meretzkov**'s Volkhov Front swept over frozen lakes and swamps to penetrate the German right. Novgorod was taken (January 19). German forces under General **Georg Linde-** mann escaped annihilation only by rapid withdrawal. A third Russian group—General **M. M. Popov**'s Second Baltic Front— threatened further envelopment and caused the retirement of General von Kuechler's entire German Army Group North. Model, who replaced Kuechler (January 31), checked the Soviet drive along the line Narva-Pskov-Polotsk (March 1), when the spring thaws impeded further Russian progress.

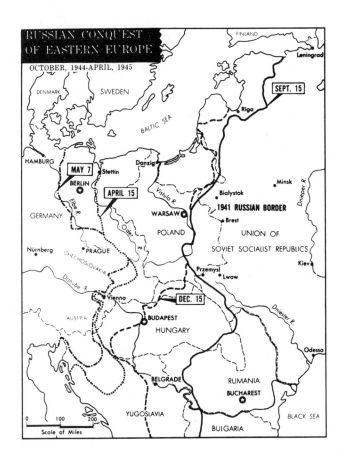

RUSSIAN CONQUEST OF EASTERN EUROPE

OCTOBER, 1944-APRIL, 1945

1944, January 29–February 17. Battle of Korsun. Farther south, the main Russian offensive hit Manstein's Army Group South along the Dnieper River. The First Ukrainian Front (now under Zhukov; Vatutin had been killed) from the north, and Konev's Second Ukrainian Front from the south, encircled the German salient at Korsun, trapping 2 army corps. Manstein's immediate counterattacks bogged down in blizzards and thaws, and attempts of the trapped troops to cut their way out were only partially successful. German casualties were about 100,000. The Russian advance continued southwest across the Bug and Dniester rivers, despite desperate counterattacks by Manstein.

1944, March 10–April 10. Isolation of the First Panzer Army. General Hube, commanding the First Panzer, and operating under instructions from Manstein, kept up a magnificent defensive-offensive behind the Russian lines, playing hob with communications. A hastily-improvised Luftwaffe airlift brought in supply, while the Fourth Panzer Army drove southeast (April 5) below Tarnapol to make contact. Hube attacked westward, fighting defensively to the flanks and rear, and brought out his command practically intact.

1944, March–April. Ukraine and Rumania. Kleist's Army Group A, under pressure both by Konev and by General **Rodion Y. Malinovskiy's** Third Ukrainian Front, was forced back after furious fighting, in which the German Sixth and Eighth armies were badly cut up, and Odessa evacuated (April 10).

1944, March 30. Relief of Manstein and Kleist. Infuriated by these reverses, Hitler relieved both his southern group commanders, Model replacing brilliant Manstein and General **Ferdinand Schoerner** succeeding Kleist. By mid-April, when the thaws and Model's counteroffensive from Lwow had slowed down the Russian advance, the entire western Ukraine had been cleared and Konev's spearheads were threatening the Carpathian passes.

COMMENT. *Several elements in this vast panorama of war of movement stand out: the methodical progression of the Russian masses in a war of attrition closely directed by the Soviet high command; the amazing tenacity and ability of the German commanders, fighting a campaign they all knew was being lost through their Führer's madness; and the simple devotion of the soldiers on both sides. The mobility and efficiency of the Soviet armies were due primarily to Allied logistical assistance provided through Lend Lease.*

THE SOVIET SUMMER OFFENSIVE

Hitler's obsession to hold at all costs all occupied terrain prevented any steps to rectify, consolidate, and fortify the Eastern Front during the respite granted by the spring thaws. In consequence, his thin-spread 1,400-mile-long cordon lacked reserves and depth when Stalin launched his next offensive, in step with the Anglo-American invasion of France.

1944, June 22–July 10. Battle for White Russia. Stalin (Marshal Zhukov, now deputy supreme commander, controlled the operation) struck first north of the Pripet Marsh area, after an outburst of guerrilla activity paralyzed General **Ernst Busch**'s communications in rear of Army Group Center (June 22). Supported by an immense mass of artillery (estimated at 400 guns per mile of front), the Russian Third, Second, and First White Russian Fronts assaulted on a 350-mile front along the axis Smolensk-Minsk-Warsaw. Air superiority was complete; the Luftwaffe had been drained for the Western Front. Soviet armor tore open a 250-mile-wide gap, encircling German strong points, which were then smothered by great masses of infantry. In turn, **Vitebsk** (June 25), **Bobruisk** (June 27), and **Minsk** (July 3) were captured. Army Group Center was completely shattered, some 25 of its 33 divisions trapped. Soviet estimates claimed 158,000 Germans captured, 381,000 killed, and the destruction or capture of more than 2,000 tanks, 10,000 guns, and 57,000 motor vehicles. Model was rushed north by Hitler to replace Busch.

1944, July 10–September 19. Defeat of Finland. A Soviet drive under Govorov had broken through the Mannerheim Line, capturing **Viipuri** (June 20). Now the eastern Finnish defenses were overrun in the Lake Onega region. Hostilities between Finland and the U.S.S.R. ceased by truce (September 4). The German Twentieth Army still clung in the north around the naval and air bases of Petsamo and Kirkenes.

1944, April 8–May 9. The Crimea. The Germans, after halting the first Soviet attack at the Perekop Isthmus, were forced back into Sevastopol's fortifications by General **F. I. Tolbukhin**'s Fourth Ukrainian Front, following an amphibious assault across the Kerch Strait. A 2-day assault, with heavy artillery support, cleared **Sevastopol** (May 9), but most of the German garrison was successfully evacuated by water.

1944, July 10–August 7. Drive into Poland. Methodically advancing on a widening front, Zhukov's masses enlarged both flanks of the gap, at the same time pressing on the main axis toward Warsaw. On the northern flank, a deep penetration toward Riga rolled up German Army Group North, threatening to pin it back against the Baltic Sea. Model scraped up sufficient reserves to counterattack and check the advance of Rokossovski's First White Russian Front just east of Warsaw, but farther south Konev's First Ukrainian Front surged through Lwow (July 27) to reach the upper Vistula at Baranov (August 7). The 450-mile advance by this time had overstrained Soviet supply capabilities, and the offensive on this front came to a temporary halt.

1944, August 1–September 30. Warsaw Revolt. Led by General **Tadeusz Bor-Komorowski,** Polish underground forces (anti-Communist) attempted to wrest the city from German control, hoping the Russians, just across the Vistula, would help them. But the Soviets lay there idle while a German SS force squelched the revolt in a bloody 2-month house-to-house battle.

1944, August 20–September 14. Conquest of Rumania. The Second and Third Ukrainian Fronts attacked across the Prut River, falling on General **Johannes Friessner**'s Army Group South Ukraine. Two of its 4 armies—Rumanian troops—allied themselves with the Soviets when Rumania capitulated (August 23); most of the German Sixth and Eighth armies were trapped. Soviet troops reached the Danube at Bucharest (September 1), having captured about 100,000 prisoners and much matériel. The entire German right flank collapsed, to reform in the Transylvanian mountains.

1944, September 8. Defection of Bulgaria. Bulgaria changed sides in the war when Soviet troops crossed the Danube.

1944, October 10–December 15. Drive to the Baltic. Despite efforts of Guderian (now German chief of staff) to pull Schoerner's Army Group North from its dangerous position in Latvia, Hitler delayed the movement until too late. The Soviet First Baltic Front drove through to the sea, investing Memel and barring Schoerner's retreat. However, a Russian assault into East Prussia was halted by counterattacks.

1944, October 20–December 31. The Balkans. Russian efforts to block movement of General von Weichs' Army Group F, moving from Greece into Yugoslavia to bolster the German right, were nearly successful. Tolbukhin's Third Ukrainian Front, with a Bulgarian army assisting on its left, took **Belgrade** (October 20), with **Tito**'s (**Josip Broz**'s) partisans fighting beside them. This forced Weichs to move through Sarajevo, to the west. He then linked with Friessner's Army Group South along the Drava, momentarily saving the German flank. The Russians, continuing northwest, reached the Danube and won a bridgehead (November 24). **Budapest,** encircled (December 24), was still holding out as the year ended.

COMMENT. *Again and again, Hitler, despite his generals' entreaties, prevented withdrawals and regroupings which would have improved his defensive capabilities. He resisted evacuating Greece until too late for Army Group F to take any but a minor defensive role in Yugoslavia. He bled white his Panzer strength by transferring its pick to the Western Front for the abortive Ardennes offensive (see p. 1217), and kept the 20 veteran divisions of Army Group North, a splendid reserve mass of maneuver, penned in Latvia. Not the least of his erratic mistakes was to shift generals continuously between commands, as exemplified by transferring Model—his efficient troubleshooter—back to command in the West (August 25).*

WAR IN THE WEST, 1945

Battle of the Atlantic

1945, January–April. U-Boat Resurgence.
Doenitz sent his new schnorkel-equipped submarines out into the Atlantic. Germany was producing approximately 27 per month, despite Allied air attacks on the submarine industry. U-boats accounted for the loss of 253,000 tons of shipping, much of it in British waters.

1945, April–May. Allied Countermeasures.
A double screen of U.S. destroyer escorts and carriers patrolling the Atlantic north of the Azores stamped out the final German effort. The last action in the Atlantic was the sinking of *U-853* by U.S.S. *Atherton* and *Moberly* off Block Island (May 6).

1945, May 28. Abolition of Convoys. A joint Admiralty—U.S. Chief of Naval Operations announcement so declared, adding that all merchant ships "at night will burn navigation lights at full brilliancy and need not darken ship." The order applied to the North and South Atlantic, the Arctic and Indian oceans. In the Pacific Ocean areas, of course, the war against Japan was still on.

COMMENT. *In all, 781 U-boats were destroyed by Allied naval and air efforts during the war, with loss of 32,000 German sailors. The submarines had taken toll of 2,575 Allied and neutral vessels, causing more than 50,000 Allied casualties, about three-fourths British. At the end, 398 U-boats were still in commission. Of these, 217 were destroyed by their own crews, and the others surrendered.*

The Combined Bomber Offensive

The British and American bombing attacks (alternating night and day as in previous years) mounted in intensity and effectiveness while German air defense efficiency declined as a result of the terrible attrition. Nevertheless, the Luftwaffe continued to fight vigorously, and the appearance of the Messerschmitt 262 twin-jet fighters early in the year partially redressed the balance because of their superior performance. But the German switch to jet fighters came too late to affect the outcome of the air war. Superior numbers of conventional planes, and the superior training and operational efficiency of the Allied pilots—who had no fuel shortage to worry about—prevented the new Messerschmitts from seriously threatening Allied air superiority.

1945, February 13–14. The Dresden Raids.
An RAF night attack, followed next day by a massive U.S. Eighth Air Force raid, created uncontrollable fire storms; at least 100,000 people perished in the most destructive bombardment of history.

1945, March 21–24. Crushing the Luftwaffe.
Partly in preparation for the Montgomery Rhine crossing at Wesel (see p. 1226), and partly as a culminating blow of the strategic air offensive, Allied strategic and tactical air forces flew a total of 42,000 sorties over Germany from bases in England, France, Italy, and Belgium. More than 1,200 heavy Eighth Air Force bombers smashed German jet bases within range of Wesel, while medium bombers and fighter bombers struck at all other Luftwaffe installations. This, combined with the collapse of the German ground front, ended further effective German air activity.

1945, April–May. Support of the Ground Offensive. The Strategic Air Forces had practically run out of targets. The few remaining operational German factories were within easy range of tactical air forces. Air Chief Marshal Harris and General Spaatz shifted their heavy bombers to direct support of the onrushing British and American armies.

Operations in Italy

1945, January–April. Continued Stalemate. German General **Heinrich S. von Vietinghoff** (Kesselring had been transferred in March to command the defense of western Germany) labored to strengthen the Gothic Line positions of what was now German Army Group Southwest. He wanted to retire north of the Po, but Hitler harshly refused permission. Despite Allied air superiority, reduced strength, and short supplies, German morale remained high in their deep fortified zone across the peninsula. To the west, from Monaco north to the Swiss border, Graziani's Fascist Italian Ligurian Army defended the Alpes Maritimes, a force of doubtful value, except for 2 German divisions.

The polyglot Allied 15th Army Group, reorganized and reequipped, prepared for a vigorous offensive. Alexander (theater commander) and Clark (army group commander) proposed a double penetration of the Gothic Line, British Eighth Army (Richard McCreery) on the right driving on Ferrara and U.S. Fifth Army (Truscott) on Bologna. It was in effect a one-two assault, the British jumping off first, the Americans following 5 days later, with air support massed in that same order. Several probing attacks, including an amphibious feint at the mouth of the Po, preceded.

1945, April 9. Eighth Army Assault. Accompanied by massive air and artillery blows, the British infantry attack struck the German Tenth Army, flame-throwing tanks in the lead. A small amphibious assault across Lake Comacchio on the extreme right enveloped the German left, and the assault moved into the Argentia Gap, southeast of Bologna.

1945, April 14–20. Fifth Army Offensive. With the weight of air support now shifted to the west, the Americans sliced into the German position. They soon broke into the open Po Valley (April 20). Vietinghoff, committing his last reserves, failed to stem the tide. German resistance collapsed as American armor pressed forward. While most of the German troops managed to cross the Po, practically all heavy equipment was abandoned.

1945, April 20–May 2. Pursuit. In the Fifth Army zone, Bologna was occupied (April 21). American armor, debouching from the main effort, now reached north through Milan, up to Lake Como, and encircled Lake Garda. Behind the German lines partisans captured Mussolini, attempting to flee to Germany, and killed him (April 28). On the far left the U.S. 92nd Division moved along the coast, occupying Genoa (April 28). It then reached west to meet, near Imperia, French troops advancing from Monaco, and north to take Alessandria (April 28), Turin, and Pavia. The Eighth Army's pursuit reached Verona, Padua, Venice, and crossed the Piave (April 29). Italian Fascist resistance faded, prisoners coming in in droves. Remnants of the German Fourteenth and Tenth Armies streamed north up the Adige and Piave valleys. General Vietinghoff at Caserta agreed (April 29) to unconditional surrender, effective May 2, and American spearheads moved into the Brenner Pass, linking (May 4) at Vipiteno with U.S. Seventh Army elements driving down from the north. New Zealand units of the Eighth Army had already rounded the northern tip of the Adriatic to take Trieste (May 2).

COMMENT. *In marked contrast to all previous operations of the 15th Army Group, the breaching of the Gothic Line and the subsequent pursuit of the broken German armies was tactically superb. Truscott's employment of the Fifth Army was particularly noteworthy.*

The Western Front

ADVANCE TO THE RHINE

1945, January 1–21. German Offensive in Lorraine and Alsace. To take advantage of

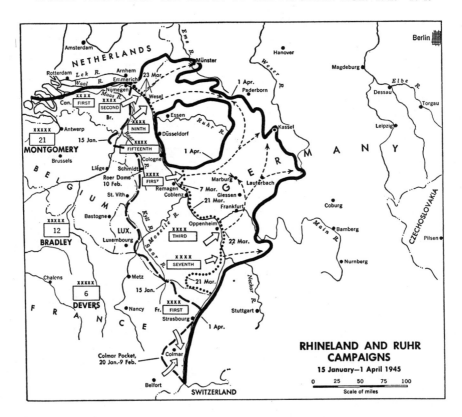

RHINELAND AND RUHR
CAMPAIGNS
15 January–1 April 1945

0   25   50   75   100
Scale of miles

shifting of Allied troops to meet the German Ardennes offensive, General **Johannes von Blaskowitz'** Army Group G launched a vicious drive against the extended U.S. Seventh Army in Devers' 6th Army Group. Before it was snubbed (January 21), Patch's troops had been forced back from their Rhine Valley salient near Karlsruhe to previously prepared positions stretching from Sarreguemines to Gambsheim, north of Strasbourg. De Gaulle, fearful of the effect on French morale should Strasbourg be evacuated, insisted the city be defended. Accordingly Eisenhower, who was prepared to give ground there, ordered Devers to hold it at all costs. He did. The German advance bogged down.

1945, January 17–February 7. Allied Advances in the North. Eisenhower prepared to resume his offensive to clear the left (west) bank of the Rhine. As preliminary, Montgomery's 21st Army Group pressed into the

Roermond area (January 15–26) and Bradley's 12th Army Group approached the upper Roer.

1945, January 20–February 9. End of the Colmar "Pocket." Devers threw the French First Army—reinforced by American divisions—against the Colmar "pocket" in the Vosges. The German Nineteenth Army was pinched out with heavy losses. The Allied southern flank was now solidly on the left bank of the Rhine from the Swiss border to Gambsheim, north of Strasbourg.

1945, February 8–March 10. Clearing the Rhineland—21 Army Group Sector. Montgomery launched a pincers move. Crerar's Canadian First Army attacked (February 8) southeasterly between the Maas (Meuse)—held all the way south to Roermond by Dempsey's Second British Army—and the Rhine. But the Canadians could make only slow progress over water-soaked terrain and in foul weather, and Simpson's Ninth Army attack across the Roer was held up by floods when the Germans suddenly emptied the Roer dams down the valley. His delayed assault (February 23) was most successful. Five days later his troops broke out into the open. The pincers met at Geldern (March 3), and German resistance began collapsing. The last German bridgehead, opposite Wesel, was wiped out (March 10) by the Canadians. Montogomery's armies stood on the Rhine.

1945, February 9–March 10. Clearing the Rhineland—12th Army Group Sector. Hodges' U.S. First Army, pushing through the Hürtgen Forest, protected Simpson's right flank and drove for the Rhine, rolling up Manteuffel's Fifth Panzer Army and remnants of Zangen's Fifteenth Army. Cologne was cleared (March 6–7) and task forces began probing south to contact Patton's Third Army. Patton, construing his mission of "active defense" as one of incessant attacks, had been battering his way through the rugged Eifel region, clearing the Westwall defenses from the Losheim Gap south to the Moselle. With Trier captured (March 5), his armored spearheads struck northeastward through disorganized German defense, the U.S. 4th Armored Division reaching the Rhine opposite Neuwied (March 7).

1945, March 7. The Remagen Bridge. A 2-battalion task force of the U.S. 9th Armored Division, probing east as part of the First Army's advance, unexpectedly found the Ludendorff Railroad Bridge over the Rhine, at Remagen, still standing. Daringly seizing it before it could be blown up, these Americans changed the entire course of Eisenhower's planned campaign. By nightfall the First Army held a rapidly swelling bridgehead on the east bank. Hitler, making another inept shuffling, displaced Rundstedt from the western command, putting Kesselring in his place (March 10).

1945, March 11–21. Clearing the Palatinate. While the First Army enlarged the Remagen bridgehead against a series of frantic piecemeal attacks, accompanied by heavy air, artillery, and V-2 bombing strikes against the original span and new bridges being thrown over the Rhine, a dazzling multipronged armored assault crushed the last remaining German forces west of the river. Patton's Third Army, driving southeast across the Moselle, and Patch's Seventh Army of Devers' 6th Army Group, attacking northeast from the Saar, each covered by its tactical air command, destroyed the dazed elements of General Brandenberger's German Seventh Army in an amazingly coordinated crisscross operation. Except for a quickly melting bridgehead held by General **Hermann Foertsch's** German First Army opposite Karlsruhe, the AEF lay solidly along the Rhine from Holland to Switzerland, with its Remagen bridgehead on the east bank now some 20 miles long, 8 miles deep, and linked by 6 bridges. The Rhineland campaign cost the Germans some 60,000 men killed or wounded and 250,000 taken prisoner, together with great quantities of matériel. Allied losses were less than 20,000.

THE RHINE CROSSINGS

Eisenhower, realizing that offensive operations would be much easier in the open North German Plain than in the mountainous region south of the Ruhr, had planned to have Montgomery's 21 Army Group make the main assault into the heart of Germany. Bradley's and Devers' army groups were to play secondary roles. The unexpected Remagen bridgehead had only partly changed this plan. To oppose 85 well-equipped, well-supplied Allied divisions, Kesselring now had east of the Rhine less than 60 half-strength divisions, short of equipment, weapons, fuel, and ammunition. However, they were defending their homeland.

1945, March 22. Oppenheim. Unexpectedly, Patton (who had long planned the move and brought bridging equipment and a navy detachment with landing craft in his army train) threw the U.S. 5th Division across the Rhine against negligible opposition. Neither air nor artillery support was used in this brilliant surprise assault. His losses were but 34 men killed or wounded. Forty-eight hours later, his bridges were completed, 4 divisions were across, armor was moving over, and another bridge at Boppard, 40 miles to the north, was also in being. Four days later, his spearheads reached Lauterbach, 100 miles east of the Rhine.

1945, March 23. Wesel. One day behind Patton, Montgomery launched his long-planned full-dress Rhine crossing north of the Ruhr. The British Second Army led off, behind the support of some 3,000 guns on a 20-mile front, and a heavy air attack (see p. 1222). The U.S. Ninth Army began its crossing at Dinslaken (March 24), while a daylight drop of 2 airborne divisions—U.S. 17th and British 6th—landed north of Wesel. German resistance was fierce but relatively ineffective. Montgomery's army group soon was pouring over 12 bridges (March 26) and 2 days later had broken through a final German stand at Haltern on the Lippe River.

1945, March 25. Remagen. Hodges' First Army broke out of its bridgehead and its armored spearheads reached Marburg, some 70 miles to the east (March 28).

1945, March 26. Worms-Mannheim. Devers' 6th Army Group began to move across the river. The Seventh Army crossed in the Worms-Mannheim area. De Lattre de Tassigny's First French Army soon followed at Gersheim (March 31).

THE GERMAN COLLAPSE

1945, March 28. Eisenhower's Change of Plan. Because of the dramatic and unexpected success of Patton's and Hodges' armies, Eisenhower decided to have Bradley's 12th Army Group make the main effort, driving east through central Germany on Leipzig, forgetting Berlin, toward which the Russians were advancing from the east (see p. 1228). Bradley's left and Montgomery's right (U.S. Ninth Army) would encircle the Ruhr, while the remainder of 21 Army Group covered the main advance by moving northeast on Hamburg and the Elbe. The advance would halt on the Elbe. SHAEF was concerned by the threat (later proven a fable) that a Nazi last-ditch guerrilla resistance—the "Werewolves"— would operate from a nebulous "National Redoubt" in the German-Austrian Alps. So while Montgomery and Bradley were to close and await the Russians on the line Elbe-Mulde-Erzgebirge, Devers' 6th Army Group would move down the Danube Valley to squelch all resistance. Eisenhower's decision to turn from what he considered to be a political objective— Berlin—to the apparently more immediate military objective of crushing the last dregs of Nazi armed force was received with bitter criticism by the British. The military and political issues and considerations were complex and confused. In retrospect, Eisenhower's decision appears to have been faulty, but it was evidently consistent with guidance from U.S. Army Chief of Staff General Marshall.

1945, March 28–April 18. Ruhr Encirclement. Armored spearheads of the Ninth and First armies met at Paderborn (April 1). The Ninth Army returned to Bradley's control. Rushing eastward, the Ninth Army reached the Elbe near **Magdeburg** (April 11), and encountered fierce resistance as it closed up to the river to north and south. The First Army's main thrust reached Warburg (April 4). That day Patton's troops were sweeping through the Thuringian Forest, nearly 150 miles east of the Rhine. In Devers' group, the U.S. Seventh Army was on the Main, and the French First Army was reaching through the Black Forest. Behind all, the newly organized U.S. Fifteenth Army, under Lieutenant General **Leonard T. Gerow,** was sealing the western face of the Ruhr area along the Rhine. Inside the 4,000-square-mile area of the Ruhr pocket were crowded more than 300,000 men—mostly the disorganized bulk of Model's Army Group B. To the north Blaskowitz' broken Army Group H was reeling back into Holland and northwest Germany, while to the east and south Army Group G (now under SS General **Paul Haus-**

ser) was putting up disorganized but often bitter resistance. Kesselring watched helplessly, having neither means nor communications to regroup or establish any covering force, while Hitler in Berlin kept thundering his usual senseless "hold in place" directives. Model, inside the pocket, conducted his typical skillful delaying resistance to the closing ring until simultaneous drives from north and south split his forces (April 14). In the face of a preposterous demand from Hitler that he now cut his way out, Model disbanded his troops and disappeared—apparently having committed suicide. Four days later the last resistance dissolved; some 317,000 German soldiers had entered Allied prison camps. The once mighty Wehrmacht was collapsing, although disjointed fanatical opposition was still stiff in places.

1945, April 18–May 7. Final Operations—21 Army Group Sector. In Holland, the Canadian First Army suspended hostilities in a partial truce, Blaskowitz agreeing to cease flooding the country and to permit food to reach the starving population; however, he still refused to surrender. The British Second Army reached the west bank of the Elbe (April 26); Lubeck and Wismar to the north were occupied (May 2); contact with advancing Russians was made the same day. Hamburg surrendered (May 3).

1945, April 18–May 7. Final Operations—12th Army Group Sector. The First Army crushed a determined pocket of resistance in the Harz Mountains area (April 14–21). Closing on the Mulde and Elbe rivers, Hodges's troops made the first contact with Soviet troops from the east at Torgau (April 25). The Allied front now froze on the Mulde-Elbe-Erzgebirge line, while the Third Army swept down the Danube Valley to occupy Linz (May 5). Its left, in Czechoslovakia, reached Pilsen (May 6) and moved on Prague until halted by SHAEF order next day.

1945, April 18–May 7. Final Operations—6th Army Group Sector. The U.S. Seventh Army, after a sharp struggle at Nuremberg, advanced southeast to force a crossing of the Danube, taking Berchtesgaden (Hitler's mountain retreat) and Innsbruck, and meeting in the Brenner Pass elements of the U.S. Fifth Army moving north from Italy. The Seventh Army's movements had been hampered by an annoying clash between Americans and French over Stuttgart. The squabble ended only when U.S. military authorities threatened to cut off further supply to French troops. Meanwhile, de Lattre's spearheads mopped up the Black Forest to the Swiss border.

1945, May 5–7. German Dissolution. Blaskowitz formally surrendered all German forces in Holland, Denmark, Schleswig-Holstein, and northwest Germany, together with the coastal islands, to Montgomery. On the southern flank, General Schulz surrendered Army Group G to Devers. Droves of individual German soldiers fled west over the Elbe from the Russians to surrender to the nearest Anglo-American units. Doenitz, successor to Hitler (see p. 1228), sent emissaries to Eisenhower at Rheims to negotiate final surrender of the Third Reich.

1945, May 7–8. Unconditional Surrender. Admiral **Hans von Friedeburg** and General **Alfred Jodl,** representing Doenitz, surrendered to Eisenhower—represented by his chief of staff, Lieutenant General **Walter Bedell Smith.** Next day, in Berlin, Marshal Keitel, Admiral Friedeburg, and Air General Hans-Jürgen Stumpff ratified the surrender to Soviet Zhukov and Tedder, Eisenhower's Deputy Supreme Commander. World War II in the West officially ended at midnight May 8–9, 1945.

The Eastern Front

SOVIET WINTER OFFENSIVE

1945, January 12–February 16. Drive into Germany. From the Baltic to the Carpathians, the front exploded, Konev's First Ukrainian Front on the left leading, the army groups to his right unfolding in turn (Zhukov's First White Russian; Rokossovski's Second White Russian; **Ivan Chernyakovski's** Third White Russian; **Ivan K. Bagramyan's** First Baltic; and Andrei Yeremenko's Second Baltic). Vastly outnumbered and lacking any secondary defensive preparations, the Germans fell back, with the isolated fortresses of Torun (Thorn), Poznan (Posen), and Breslau still battling in their rear. Zhukov's First White Russian Front reached the Oder River near Kustrin (January 31) after an advance of nearly 300 miles; Konev, to his left, gained the Oder-Neisse line (February 15). There the advance halted, its long communications lines strained beyond capacity. The invading tide turned north, where the Germans were pinned against the Baltic in 2 isolated areas—Army Group Center in East Prussia and Army Group Kurland (formerly Army Group North) in Latvia.

1945, January 12–April 16. The Danube Valley. South of the Carpathians the advance of the three Soviet fronts or army groups (**Ivan Petrov**'s Fourth Ukrainian; Malinovski's Second Ukrainian; and Tolbukhin's Third Ukrainian) was blocked at Budapest for more than a month. Completely surrounded, the city finally fell (February 13). Hitler, in a desperate effort to protect the Lake Balaton oil fields, ordered the rehabilitated Sixth Panzer Army into the area from the Western Front. Its counterattack (in March) checked the valley advance, but died when fuel gave out. Malinovskiy's Second Ukrainian Front then drove into Vienna (April 15).

1945, February 16–April 15. The Baltic Coast. Zhukov's army group and Rokossovski's Second White Russian Front moved north from the Oder near Stettin up to Königsberg, while Bagramyan, now commanding the Second Baltic Front, blocked the Kurland peninsula. More than 500,000 German troops were cut off, but not isolated, thanks to the German Navy's continuing control of the Baltic Sea. An estimated million and a half fugitives and 4 army divisions were evacuated from Kurland, including 157,000 wounded. In turn, Danzig and Gdynia (March), Königsberg and Pillau (April 25), and Kolberg (mid-April) were successfully evacuated by the German Navy. In these operations Allied air power slowly nibbled away German naval strength. By April

only the cruisers *Prinz Eugen* and *Nürnberg* were still afloat. But the task had been accomplished.

COMMENT. *The last stand of the German Navy in the Baltic presents an example of the value of sea power; a small, well-led, and well-trained squadron, in confined waters and without adequate air support, long denied victory to a more powerful enemy unfamiliar with the sea.*

1945, April 16–May 7. The End in Europe. For the final offensive against Berlin, General **Vassili Sokolovski** took over command of the First White Russian Front from Zhukov, who commanded a super army group of Konev's and Sokolovski's two fronts. The resumed Soviet northwest advance, despite desperate but scattered resistance, reached Berlin (April 22) and surrounded it (April 25). That same day, elements of Bradley's 12th Army Group and Konev's First Ukrainian Front made contact at Torgau on the Elbe. Farther north, Rokossovski, who had captured Stettin (April 26), made contact with American units of Montgomery's 21st Army Group at Wismar (May 3). In Berlin, meanwhile, Hitler—after appointing Doenitz as his successor—had committed suicide (April 30) before the Soviets, after desperate street fighting, finally stamped out all resistance (May 2). Farther south, Malinovskiy's Second Ukrainian Front was nearing Patton's advance down the Danube near Linz, and at Trieste the British Eighth Army had met Tito's partisans (May 1).

Russian estimates (probably correct) of German losses in the east during the last 3 months of the war were 1 million killed; 800,000 men, 6,000 aircraft, 12,000 tanks, and 23,000 guns were captured.

COMMENT. *Simpson's Ninth U.S. Army was on the Elbe, 60 miles west of Berlin, 5 days before the Russian offensive began, and 2 weeks before they surrounded the city. Since the Ninth Army had moved more than 120 miles in the previous 10 days, many critics assume that it could easily have reached and captured Berlin before the Russians.*

WORLD WAR II IN ASIA AND THE PACIFIC

THE EARLY YEARS, 1937–1941

1931–1937. Background. Japan had been pursuing a policy of aggressive expansion and domination in China (see pp. 1142–1146).

1937, July 7. Beginning of the "China Incident." Japanese troops in North China, ostensibly on night maneuvers, clashed with Chinese troops near the Marco Polo Bridge at **Lukouchiao,** near Peiping. This affair initiated a full-scale invasion of China, which the Japanese termed the "China Incident." It may be considered as the start of World War II.

The Opponents. Generalissimo Chiang Kaishek's National Government Army consisted of approximately 2 million poorly trained, poorly equipped troops. In addition, the Chinese Communist Army, in northwest China, then comprising about 150,000 guerrilla troops, nominally supported Chiang against the Japanese. The National Government supported 45,000 of these in a newly created Eighth Route Army, under Chu Teh. China had no navy, only a handful of outmoded aircraft with relatively inexperienced Chinese and foreign mercenary pilots, and no trained reserves existed. China's industry was incapable of supporting a major war effort.

The regular Japanese Army consisted of about 300,000 soldiers, equipped with the most modern military weapons. There were also about 150,000 moderately well-trained and equipped Manchurian and Mongolian troops under Japanese officers. There were more than 2 million trained reserves in Japan. The Japanese Navy was the third largest in the world and in many respects the most modern. The air forces of both army and navy were equipped with modern aircraft and manned by competent airmen. Japan was a modern industrial nation, capable of turning out great quantities of excellent war matériel.

1937, July–December. Operations North of the Yellow River. Japanese troops quickly captured Peiping (July 28) and Tientsin (July 29). In subsequent months the Japanese advanced west and south against relatively ineffective Chinese opposition to conquer Chahar and part of Suiyuan, reaching the upper bend of the Yellow River at Paotow. Their principal efforts, however, were southward down the railroad lines toward Nanking, Hankow, and Sian. Increasing effectiveness of the Chinese Army, growing unrest and resistance among the Chinese in the heavily populated conquered areas, and the long-distance logistical problems slowed the advance during the fall. The main Japanese drive, however, culminated in the capture of Tsinan, capital of Shantung (December 27). This gave them control of most of the area north of the Yellow River.

1937, August 8–November 8. The Second Battle of Shanghai. (See p. 1145.) Chinese resistance against an amphibious Japanese assault was as tenacious as it had been more than 5 years earlier. Japanese reinforcements were rushed to Shanghai to avoid defeat. Despite savage air bombardment and effective naval gunfire support, for several weeks the Japanese were pinned to their beachheads at the outskirts of the city. Finally, after 2 months, additional Japanese reinforcements permitted amphibious landings north and south of Shanghai, and they finally drove out the shattered defenders.

1937, September 25. Battle of P'inghsinkuan. The Japanese 5th Division, under General **Seishiro Itagaki,** was ambushed and defeated in the Wutai Mountains of northern Shansi by the Chinese 115th Division (Communist, of the Eighth Route Army), under General **Nieh Jung-chen.** This victory had great propaganda significance throughout China. It was the first and last division-sized engagement fought by the Chinese Communists during the war.

1937–1940. Communist Consolidation. They spent the next three years expanding and consolidating their control of northwest China, and establishing guerrilla bases behind the

Japanese lines. The Eighth Route Army made only a few minor raids against the Japanese front-line forces.

1937, November–December. Japanese Pursuit toward Nanking. The Japanese troops that captured Shanghai advanced up the Yangtze River toward the Chinese capital, meeting ineffectual resistance.

1937, December 12. The *Panay* Incident. Japanese aircraft made an unexpected and unprovoked attack on British and American gunboats moored in the river, near Nanking. The USS *Panay* was sunk by repeated dive bombings; the British vessel was badly damaged. The American public was outraged, but Japan later apologized and paid an indemnity.

1937, December 13. Fall of Nanking. The city was ravaged by Japanese troops for several days in a senseless orgy of slaughter, rape, and destruction. Chiang Kai-shek had meanwhile moved his capital westward to Hankow. To the surprise of the Japanese and of the rest of the world, the National Government did not collapse; and despite heavy military losses, both the Chinese Army and the Chinese people developed an amazing will to resist, and a moral unity.

1938, January–April. Renewed Offensive in the North. Completing the conquest of Shantung (January), the Japanese resumed their drive down the railroads toward Nanking and Hankow. Their advance was slow but steady; effective action by mobile Chinese regular forces and local guerrillas restricted their control to the vicinity of the railroad.

1938, April. Battle of Taierchwang. Chinese regular forces and guerrillas, under General Li Tsung-jen, probably exceeding 200,000 in numbers, cut off and surrounded a Japanese force of 60,000. Initial Japanese efforts to fight their way out were repulsed, and losses were heavy on both sides. The Japanese finally fought their way out to the north, but left 20,000 dead and large quantities of equipment behind them. This victory greatly bolstered Chinese morale, but had no lasting effects.

1938, May–June. Japanese Recovery. The Japanese quickly regrouped, reorganized, then renewed their attacks from the north, while another column moved up the railroad from Nanking. Hsuchow was captured (May 20); Kaifeng fell soon afterward (June 6). By the end of the month the Japanese had complete control of the Peking-Nanking railroad.

1938, May. Establishment of the New Fourth Army. Chiang Kai-shek agreed to support another Communist army in east-central China, mostly behind the Japanese lines south of the middle Yangtze. The commander was **Yeh T'ing.**

1938, June–July. Japanese Failure at Chengchow. The Japanese now advanced westward from Kaifeng to seize the important railroad junction of Chengchow, preparatory to an advance down the railroad to Hankow. The Chinese broke the Yellow River dikes, flooding the countryside and shifting the entire river course to a former bed, emptying into the Yellow Sea hundreds of miles from its recent mouth in the Gulf of Chihli. The Japanese advance was completely halted by this manmade catastrophe. Many troops were drowned, supplies destroyed, tanks, trucks, and guns covered with water or bogged down in the mud. The offensive was canceled.

1938, July–October. Renewed Advance on Hankow. Shifting their axis of advance farther south, the Japanese again began to threaten Chiang's capital, this time from the east, up the Yangtze River. Determined Chinese resistance resulted in the bloodiest fighting of the war in China. The Japanese, supported by their unopposed air force, captured the city (October 25). Chiang again shifted his capital westward up the Yangtze, this time to Chungking in mountainous Szechwan.

1938, October. Capture of Canton. Japanese amphibious forces landed near Hong Kong (October 12), then advanced inland to seize Canton (October 21). They now controlled China's two principal seaports.

1939. Revised Japanese Strategy. Japan, frustrated by inconclusive war and the problems of controlling a hostile population in occupied China, decided to shift to a strategy of attrition in China, which continued for several years. In

1939 they captured Hainan Island and most of China's remaining seaports, hoping to cut off all foreign supplies from China and thus force the collapse and surrender of Chiang's government. The Chinese, however, were able to keep open 2 supply routes, by means of which they could still obtain a trickle of military supplies. One was along the narrow-gauge railway from Haiphong, in French Indochina, to Kunming. The other was through British Burma, then over the narrow, twisting Burma Road to Kunming.

1940, March 30. Establishment of Puppet Chinese Government. The Japanese installed respected politician **Wang Chingwei** at Nanking as puppet ruler of occupied China. Their hopes that this government would obtain support of some of Chiang's followers were not fulfilled.

1940, June 25. Preliminary Moves in Indochina. Taking advantage of France's defeat in Europe, Japan demanded and received from the Vichy government the right to land forces. The arrival of Japanese warships in French Indochinese ports soon closed the Haiphong-Kunming supply route.

1940, July 18. Closing of the Burma Road. Hard-pressed by Germany, Prime Minister Churchill's British government acceded to Japanese demands for closing of the Burma Road. China was now virtually isolated, but Chiang and his people remained steadfast.

1940, August 20–November 30. Communist "Hundred Regiments Offensive." An intensive series of small-scale guerrilla raids ordered by Mao Tse-tung was carried out in Shansi, Chahar, Hopeh, and Honan. These attacks against Japanese outposts, roads, and railroads were highly successful, and disrupted the Japanese rear areas.

1940, September 4. American Warning to Japan. Secretary of State **Cordell Hull** warned Japan of the unfavorable reaction which would be aroused in the U.S. by aggressive moves against French Indochina.

1940, September 22. Japan Begins Occupation of Indochina. Japanese troops began to occupy northern Indochina. Air bases were established and ground forces began an offensive into China.

1940, September 26. American Embargo on Steel for Japan. President Roosevelt ordered an embargo on shipment of scrap iron and steel from the United States to Japan. Japan declared this to be an "unfriendly act" (October 8).

1940, September 27. Establishment of the Rome-Berlin-Tokyo Axis. (See p. 1165.)

1940, October 18. Reopening of the Burma Road. With encouragement from the U.S., which wished to be able to ship Lend Lease materials to China, and with the immediate German threat to Britain removed, Churchill reopened the Burma Road.

1941–1943. Japanese Reprisals against the Communists. In reprisal against the "Hundred Regiments Offensive" (see above) the Japanese began a series of savage punitive raids against the Chinese Communists. During 3 years, the Communist Eighth Route Army was constantly on the defense and lost about 100,000 casualties.

1941, January 1–7. Anhwei Incident. Chiang Kai-shek had ordered (December, 1940) the Communist New Fourth Army, operating south of the Yangtze in Anhwei, to operate against Japanese troops north of the river. The army commander Yeh T'ing ignored the order, which was apparently not consistent with his instructions from Mao Tse-tung in Yenan. Chiang repeated the order, and threateningly moved Nationalist troops to the vicinity. The Communists slowly began to cross (late December). When only 10,000 of New Fourth Army, including headquarters, remained south of the river, this element was attacked by the Nationalists; all Communists were killed or captured. Yeh T'ing was wounded and captured. Communist-Nationalist relations were greatly embittered by this incident. Mao appointed **Chen Yi** as new commander of New Fourth Army.

1941, April 13. Japanese-Russian Neutrality Treaty. This assured Japan that she would not be involved in a war in Siberia should German threats against Russia materialize, and at the same time assured Stalin that in such an event he would not have to fight a 2-front war.

1941, July 26. "Freezing" of Japanese Assets in the U.S. This was in retaliation for the continuing and spreading Japanese occupation of Indochina. Britain soon followed suit.

1941, August 17. Renewed American Warning. President Roosevelt warned Japan that further attempts to dominate Asia would force the U.S. to take appropriate steps to safeguard American rights and interests. Meanwhile, in Washington, negotiations were going on between the U.S. government and the Japanese ambassador to find ways to reduce the growing tension between the 2 nations.

1941, September–December. The "Flying Tigers." With the tacit approval of the U.S. government, retired U.S. Army Air Force Captain **Claire L. Chennault,** now a colonel in the Chinese Air Force, established a mercenary organization known as the American Volunteer Group for service in China against Japan. The AVG, later dubbed "Flying Tigers," comprised about 100 trained U.S. airmen (formerly Army, Navy, and Marine Corps officers), plus American maintenance personnel, and were equipped with P-40 airplanes provided China under Lend Lease. Chennault trained his group in his own unique, revolutionary air tactical concepts and procedures at an abandoned RAF base at Toungoo in Burma.

1941, October 17. Tojo Becomes Premier. A new militaristic government came into power under Lieutenant General **Hideki Tojo,** supported by Japan's principal military men: Marshal **Hajime Sugiyama,** Chief of the Army General Staff, and Admiral **Osami Nagano,** Chief of the Naval General Staff.

1941, November 5. Promulgation of Secret War Plans. The Japanese Imperial General Headquarters issued a secret plan for simultaneous offensives against the U.S. Pacific Fleet at Pearl Harbor, British Malaya, the American Philippines, and the Netherlands East Indies, to secure the entire Southern Resources Area. The plan was to be implemented only if the continuing negotiations in Washington failed to reach an agreement satisfactory to Japan.

1941, November 15. Arrival of Special Japanese Ambassador Kurusu in Washington. Ambassador (Admiral) **Kichisaburo Nomura,** who had been conducting Japan's negotiations since February, was joined by special envoy **Saburo Kurusu.** They endeavored to obtain American agreement to reopen trade negotiations.

1941, November 26. Secretary Hull's Proposals. Mr. Hull stated that the basis of agreement would have to include Japanese withdrawal from French Indochina and China, and recognition of the National Government of Chiang Kai-shek. This being totally unacceptable to Japan, Tojo's government decided to initiate the war plan as quickly as possible, meanwhile pretending to continue negotiations in Washington to facilitate military surprise.

1941, December 7. Pearl Harbor; War with the U.S. (See p. 1233.)

THE WAR AGAINST JAPAN, DECEMBER, 1941

Pearl Harbor and the Central Pacific

1941, November. The Japanese Strategic Plan. Japan realized that she could not match the industrial strength and resources of the Allies, or even of the United States alone. The Japanese believed that they could successfully employ, however, the same basic offensive-defensive strategic concept that had brought them victory over Russia in the Russo-Japanese War (see p. 1008). Their plan had 3 phases: Phase 1, neutralize the U.S. Pacific Fleet—the only major hostile force in the Pacific-East Asia region—by a surprise attack, while simultaneously seizing the Southern Resources Area and also strategic areas permitting establishment of a defensive perimeter around it; Phase 2, consolidate and strengthen the perimeter, so as to make any Allied attacks prohibitively costly; Phase 3, defeat and destroy any Allied efforts to penetrate the perimeter. The Japanese believed that the strength of their defenses, combined with the extremely long and vulnerable Allied lines of communications, would ensure success.

COMPARATIVE NAVAL STRENGTHS IN THE PACIFIC

December, 1941	Japan[a]	United States[b]	British Empire[c]	Netherlands	Total Allied
Battleships and Battle Cruisers	11[d]	9	2	—	11
Aircraft Carriers	11	3	—	—	3
Heavy Cruisers (8-inch guns or more)	18	13	1	—	14
Light Cruisers (6-inch guns or less)	23	11	7	3	21
Destroyers	129	80	13	7	100
Submarines	67	56	—	13	69

[a] Most Japanese ships were newer, faster, and more heavily armed than those of the Allies.

[b] These include the forces of the American Pacific and Asiatic Fleets. In addition, because of the threat of war against Germany's U-boat fleet, in the Atlantic the United States had 8 battleships, 4 carriers, 5 heavy cruisers, 8 light cruisers, 93 destroyers, and 56 submarines.

[c] Includes Australian and New Zealand vessels.

[d] Includes the *Yamato,* largest battleship ever built (with nine 18.1-inch guns, the most powerful in the world), completed in December 1941. Her sister ship, *Masashi,* was completed in July 1942.

1941, November 26. Departure of the Japanese First Air Fleet. Including 6 aircraft carriers, supported by battleships, heavy cruisers, and submarines, the fleet of Vice Admiral **Chuichi Nagumo** left the Kurile Islands under conditions of absolute secrecy (November 25, Washington time). He was informed by a code message at sea of the Japanese decision for war, and continued steaming east in accordance with the secret plan to strike the U.S. Fleet at its base at Pearl Harbor.

1941, November 26–December 7. American Readiness and Plans. Japan's preparations for war were well known to U.S. military and civilian authorities, who expected that the blow would fall on Malaya or the Philippines. U.S. intelligence, which had broken the Japanese secret radio code, was aware of the movements and location of most major Japanese army and navy units—with the notable exception of the First Air Fleet, which was moving under strict radio silence, while other radio stations in Japanese home waters simulated the call signals of the vessels of this fleet. (The Japanese did not guess that their secret messages were being decoded, but they knew that the intensity and origins of radio traffic were probably being monitored.)

1941, December 7. Pearl Harbor. In complete surprise, both strategic and tactical, 360 planes of Nagumo's fleet struck Oahu on Sunday morning, their targets the U.S. Pacific Fleet moored in Pearl Harbor and the military airfields on the island. Of the 8 battleships present, 3 were sunk, another capsized, and the remainder seriously damaged. Three light cruisers, 3 destroyers, and other vessels were also sunk or seriously damaged. On land, of 231 Army planes arrayed on airfields, only 166 remained intact or reparable; of Navy and Marine Corps planes, only 54 out of some 250 remained. Despite the surprise, both Navy and Army personnel fought back savagely. More than 3,000 Navy and Marine officers and men were killed and 876 wounded. Army losses were 226 killed and 396 wounded.

COMMENT. *The U.S. Pacific Fleet had been neutralized for at least a year to come. The fortuitous absence from Pearl Harbor of all 3 U.S. carriers of the fleet*—Enterprise, Lexington, *and* Saratoga—*alone prevented a major disaster from assuming the proportions of a national calamity. The issue of responsibility for the disaster has been controversial. However, both the U.S. Army Chief of Staff, General Marshall, and the U.S. Chief of Naval Operations, Admiral* **Harold R. Stark,** *had sufficient information, as a result of U.S. breaking of Japanese codes, to realize that war was imminent, and to recognize that an attack on*

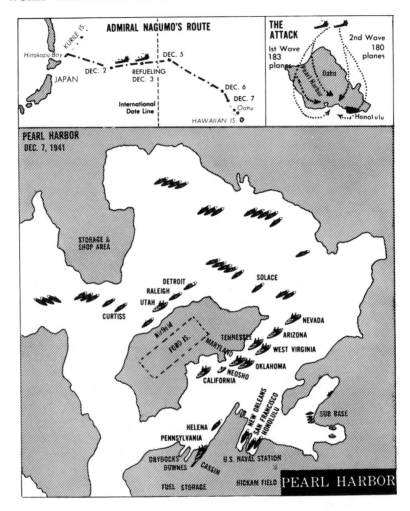

Pearl Harbor was a distinct possibility. They failed to satisfy themselves that their subordinates in Hawaii were sufficiently alert. On the other hand, those subordinates, Lieutenant General **Walter C. Short** *and Admiral* **Husband E. Kimmel** *had been alerted; their failure to be ready was inexcusable. Both were relieved of their commands.*

The above comment on "U.S. breaking of Japanese codes" requires elaboration. In the late 1930s cryptographers of the U.S. Army Signal Corps were able to break the secret codes of the Japanese Government (which the Americans called "Purple,") by building a machine that was virtually identical to the machines

used by the Japanese for encryption of their classified messages. Although totally unrelated, the process was similar to that used first by the Poles, then by the British, to break the German codes which used the "Enigma" machine. (See p. 1171; there was no relationship, other than incidental, in the Japanese development of the "Purple" machine and the German development of "Enigma"; nor was there any contact or collaboration between the American cryptographers and either the Poles or the British.) By 1939 the Americans were reading many messages of the Japanese Foreign Office, Army, and Navy. Like the Germans, the Japanese were so certain that their machine-

generated codes were unbreakable that, despite considerable evidence, during the war they never realized that their codes had been broken. This highly secret code-breaking process was known to the few people in the U.S. government who were aware of it, by the code-name "Magic," and continued to be of great value to the U.S. Armed Forces throughout the war. After the U.S. entered the war, American "Magic" became very much intermingled with British "Ultra."

If the U.S. Government was reading the Japanese codes, one might ask, why were the Americans surprised by the Japanese attack on Pearl Harbor? In fact the "Magic" process put into the hands of a very few people in the U.S. Government, the secret instructions of the Japanese Government to its 2 ambassadors in Washington in late November and early December. President Roosevelt and Secretary Stimson probably read before the ambassadors did the instructions they received to break off negotiations on December 7. But those instructions did not mention either war or Pearl Harbor. And the Japanese fleet kept total radio silence. U.S. Government leaders recognized that war was about to break out, they knew that there were Japanese invasion fleets in the South China Sea, but did not suspect (although they should have) the threat to Pearl Harbor.

1941, December 8–23. Wake Island. A gallant defense by U.S. Marines and Navy personnel under Major **James Devereux** and Commander **Winfield Cunningham** repulsed one Japanese assault (December 11). A second assault overwhelmed the defenders.

1941, December 10. Guam. A Japanese landing quickly overwhelmed a handful of Marines and sailors.

Operations in Asia

Hong Kong

1941, December 8–10. * Invasion of Kowloon. The Japanese 38th Division smashed its way through the mainland defenses of the Hong

* Note that events west of the International Date Line took place on December 8; that same day, east of the Date Line, was December 7.

Kong Colony. The British withdrew to Hong Kong Island.

1941, December 13. Japanese Demand for Surrender. The British commander, Major General **C. M. Maltby,** rejected the Japanese surrender demand. Japanese artillery, air units, and naval forces began an intensive bombardment of Hong Kong.

1941, December 18–25. The Assault. Amphibious landings on Hong Kong Island were resisted desperately but vainly by the British.

1941, December 25. Surrender. Out of water, and split into small pockets of resistance, the remnants of the 12,000-man British garrison surrendered. Japanese losses totaled fewer than 3,000.

Malaya

1941, December 8. Invasion of Northern Malaya. Following intensive predawn air attacks against RAF bases in Malaya and Singapore, the Japanese Twenty-fifth Army, commanded by Lieutenant General **Tomoyuki Yamashita,** made amphibious landings at Kota Bharu, near the northeast tip of Malaya, and at nearby Simgora and Patani, at the very southern extremity of Thailand's Kra Isthmus. The invaders, some 100,000 in number, were in 3 divisions, with substantial reinforcements of tanks and artillery. They quickly moved southward, on both sides of the peninsula, sweeping aside relatively light British covering units in the north. The bulk of British forces— slightly more than 100,000 men, also in 3 divisions—had been deployed to meet an anticipated invasion attempt farther south, directed against the great fortress of Singapore. The British commander, Lieutenant General **A. E. Percival,** regrouped to meet the Japanese drives.

1941, December 10. Loss of the *Prince of Wales* and *Repulse*. Admiral **Sir Tom Phillips** had raced north (December 8) from Singapore with HM battleship *Prince of Wales*, the battle cruiser *Repulse*, and a few destroyers, seeking to strike and destroy the amphibious armada

supporting the Japanese invasion. The RAF, hard hit by Japanese bombings, was unable to provide air support or reconnaissance. Unable to find the Japanese vessels, the British ships were returning south along the Malayan coast, when struck by Japanese aircraft based in southern Indochina. In the subsequent surface-air battle, which raged for about an hour, both great British ships, hit frequently by bombs and torpedoes, sank with heavy loss of life. Survivors were rescued by destroyers. Save for 3 American aircraft carriers, this left the Allies with no remaining serviceable capital ships in the Pacific.

1941, December 10–31. Japanese Drive to Southern Malaya. British troops, unaccustomed to jungle fighting, plagued by serious equipment shortages, and without adequate air support, were rolled up, infiltrated, and pushed back by the well-equipped Japanese—veterans of combat in China and trained for jungle fighting on Hainan Island. British efforts to stop the advance were further discomfited by a number of small amphibious landings, made on both coasts of the peninsula, by Japanese troops using captured fishing vessels. As the year ended, the demoralized British were being driven back relentlessly upon Singapore itself, now a naval base without a naval force.

The Philippines

American Situation and Plan. American and Filipino ground forces in the islands, under General Douglas MacArthur, consisted of some 130,000 men–22,400 U.S. Regulars (including 12,000 Philippine Scouts), 3,000 Philippine Constabulary, and the Philippine Army, 107,000 strong, but only partly organized, trained, and armed. The major portion of this force was on the island of Luzon; other Filipino elements were on the islands of Cebu, the Visayas, and Mindanao. MacArthur's U.S. Far East Air Force, under Major General **Lewis H. Brereton,** comprised about 125 combat planes, including 35 new B-17 "Flying Fortresses." Most of Admiral **Thomas C. Hart**'s U.S. Asiatic Fleet (1 heavy and 2 light cruisers, 13 destroyers, 28 submarines, and other craft) was being withdrawn to Java by Navy Department order; 4 destroyers, the submarines, a squadron of flying boats, and a flotilla of motor torpedo boats remained, together with a regiment of Marines. The major portion of the ground-force strength—Major General **Jonathan M. Wainwright**'s North Luzon Force—was disposed north of Manila to dispute an expected invasion via Lingayen Gulf. Brigadier General **George M. Parker**'s smaller South Luzon Force was south of Manila. MacArthur proposed to meet invasion by counterattack on the ground together with a B-17 strike at Formosa. Last-ditch defense contemplated withdrawal to the mountain jungle of the Bataan Peninsula, northern arm of Manila Bay, supported by the coast-defense complex at the bay mouth, with Corregidor its citadel. This eventuality had long been planned; the successful defense of the Philippines was predicated on the premise that the forces there would hold out until the U.S. Pacific Fleet opened the way for reinforcement.

Japanese Plans. These called for initial paralyzing air attack, mounted from Formosa and followed by amphibious assault by the Fourteenth Army under General **Masaharu Homma** (approximately 50,000 veteran troops). Discounting both the fidelity and the fighting capabilities of the Filipinos, a quick, easy victory was expected. The Japanese high command had allotted Homma 50 days to accomplish the task.

1941, November 27. Alert. With imminence of war with Japan evident to all, the U.S. and Philippine armed forces went on full war alert.

1941, December 8. Disaster at Clark Field. Preceded by minor blows at northern Luzon airfields, the main Japanese aerial attack—108 twin-engined bombers and 34 fighter planes from Formosan bases—struck the Clark Field-Iba airdrome complex in the Manila area at 12:15 P.M., to find the major portion of the American air force grounded, their crews at lunch or servicing their planes. Japanese success was complete and devastating. Eighteen of the 35 B-17s, 56 fighters, and a number of other aircraft were destroyed, together with part of the installations; the 1 U.S. fighter squadron in the air, though it took some toll of the invaders, was almost wiped out. Only 7 Japanese fighters were shot down.

COMMENT. *Responsibility for this tactical surprise has never been definitely fixed. News of the Pearl Harbor disaster had reached Manila at 2:30 A.M. (8 A.M. December 7, Hawaiian time) and was confirmed 3 hours later. At 9:30 the northern Luzon airfields had been bombed. Brereton had received a phone call from General H. H. Arnold in Washington, warning him of a likely Japanese attack. Yet, despite knowledge that Japan had struck the first blow, MacArthur's intended B-17 counterblow on Formosa had not materialized and Brereton's planes lay like sitting ducks on their fields at 12:15 P.M.*

1941, December 10–20. First Landings on Luzon. While Japanese bombers wrecked the naval base at Cavite, destroying ships, installations, and munitions, small amphibious landings were made at Aparri and Vigan, northern Luzon (December 10), and at Legaspi in the south (December 12). Air bases were established in the north, and planes transferred from Formosa. The remnants of the American air force were transferred to Mindanao; all remaining naval craft except the submarines and a flotilla of motor torpedo boats left for Java.

1941, December 20–31. Mindanao and Jolo. Japanese units gained a toehold on the southernmost island, Mindanao, Philippine Army troops present withdrawing to the hills. The invaders later seized the island of Jolo (December 25). By these moves, air and naval bases were gained for further operations against the Dutch East Indies (see p. 1244). Homma's amphibious force entered Lingayen Gulf and began landing, covered by fighter planes operating from northern Luzon bases.

1941, December 22. Main Invasion of Luzon.

1941, December 23–31. Withdrawal. Driving in the partly trained Philippine Army units opposing them, the Japanese on Luzon advanced rapidly south. Only the gallantry and steadiness of the few American and Philippine Scout units saved Wainwright's troops from disaster as he withdrew in successive positions. In the south, an additional Japanese landing in force at Limon Bay (December 24) threatened the South Luzon Force. MacArthur, between the closing pincer jaws, decided on withdrawal into Bataan. Manila was declared an open city (December 26). A counterattack by rallied Philippine Army units, supported by Regulars, threw back Homma's spearheads long enough to permit the South Luzon Force to withdraw westward across the unfordable Pampanga River. Thus the year ended with MacArthur's troops safe on the Bataan side, thanks to skillful training and excellent troop leading.

COMMENT. *Had MacArthur elected to defend Manila, as the Japanese expected him to do, his entire army would have been irrevocably lost. The Japanese could then have moved their troops southward to aid in seizing the entire Southern Resources Area. As it now stood, they were still confronted by a force in being, which would have to be conquered before the Manila Bay fortifications could be reduced and Manila converted to an advance base. Tactically, Homma had gained a victory; strategically, his campaign had just commenced.*

THE WAR AGAINST JAPAN, 1942

Strategy and Policy

1942, January–March. Struggle for the Malay Barrier. The year opened with Japan moving to victory on all fronts of her "Greater East-Asia Co-Prosperity Sphere." The Dutch East Indies alone remained untouched. There the remnants of Allied forces in the Far East rallied briefly under a short-lived, extemporized command—ABDA (American-British-Dutch-Australian)—authorized at the Arcadia Conference in Washington (see p. 1183) and inaugurated under command of British General Sir **Archibald Wavell** (January 15). His mission was to hold the so-called "Malay Barrier" (the Southeast Asia peninsula, and the projecting island chain of the Netherlands East Indies and New Guinea to northern Australia), while American naval, air, and land forces held open lines of communication across the Pacific to Australia and New Zealand. Hopefully, if the Malay Barrier could be held, Japan would be denied vital raw materials of the Southern Resources Area essential to her war effort.

1942, March. Allied Reorganization. The fall of Malaya, Singapore, and the Dutch East Indies and Japanese successes in the Philippines and Burma caused the collapse of ABDACOM and forced the Allied Combined Chiefs of Staff to reorganize for a last-ditch attempt to hold their most "vital interests" in the Pacific and Asia, while still concentrating primarily against Germany. Britain was to be strategically responsible for the operations in and near India and the Indian Ocean, the United States for the Pacific Ocean; China, under the supreme command of Chiang Kaishek, was to be within the area of American strategic and logistical responsibility, with American Lieutenant General **Joseph W. Stilwell** to be Chiang's Chief of Staff. (Later, in July, for administrative reasons, the U.S. created the China-Burma-India Command, or Theater, under Stilwell's command, responsible for logistical and combat support to China, for control—under the British—of American and Chinese combat troops in India and Burma; Stilwell was thus simultaneously responsible to the U.S. Joint Chiefs of Staff, to Chiang, and to the British commander in India.) The Pacific Theater was divided into two major commands. General MacArthur's Southwest Pacific Area included Australia, New Guinea, the Netherlands East Indies (save for Sumatra, within the British sphere), and the Philippines. Admiral **Chester**

W. Nimitz' Pacific Ocean Areas included the remainder of the ocean, and was in turn divided into three major combat zones: North, Central, and South Pacific Areas.

1942, April–May. Revised Japanese Plan. The second phase of the original Japanese war plan (see p. 1232) had contemplated the consolidation of a defensive perimeter anchored on the mountainous Burma-India border in the west, and in northern New Guinea, the Bismarcks and Gilberts, Wake, and their own Pacific islands on the southeast and east. The ease and rapidity of their first phase conquests, including the seizure of all of the planned island outposts, led them to plan an extension of the perimeter in the central and south Pacific: to seize Midway Island as a base from which to harass Hawaii, and southern New Guinea and the southern Solomon Islands as bases from which to harass Australia and to interfere with the Allied trans-Pacific supply routes from the U.S. and the Panama Canal. Furthermore, they felt that this extension of their perimeter would still further increase the magnitude of any Allied efforts to recover the conquered regions.

1942, May–August. Stemming the Japanese Tide. The failure of the Japanese to gain a victory in the drawn naval Battle of the Coral Sea (see p. 1252), their decisive defeat at the Battle of Midway (see p. 1253), and the repulse of their offensive in southern New Guinea (see p. 1247) prevented the Japanese from achieving any of the new objectives of their revised war plan, save for unopposed occupation of bases in the southern Solomon Islands.

1942, August–December. Limited Allied Offensives. Exploiting their first important successes of the Pacific War, the Allies began local offensives in New Guinea and the Solomon Islands. Since most Allied resources were earmarked for defeat of Germany (see p. 1183), these offensives were limited in scope and objectives, intended primarily to reduce Japanese threats to Allied lines of communication, to seize bases suitable for subsequent offensives, and to subject the Japanese to a war of attrition which they could not afford.

Land and Air Operations in Asia

MALAYA AND SINGAPORE

1942, January 1–31. British Defeat on the Peninsula. Completely outfought and outmaneuvered by Japanese infiltrating and flanking attacks, supported by overwhelming air power, British troops were rapidly forced back to the so-called Johore Line, some 25 miles north of Singapore. This was breached (January 15) and the defenders forced back (January 31) into the island fortress, separated from the mainland by the narrow Strait of Johore—less than a mile wide and bridged by a causeway, which the British partly demolished behind them.

1942, February 8–15. Conquest of Singapore. Japanese assaults followed a protracted air bombardment; crossings in armored barges were covered by intense artillery and machine-gun fire. British counterattacks were broken up by dive bombers. Tanks began moving across the causeway, which Japanese engineers had repaired, and despite foot-by-foot resistance the city's reservoirs were captured. The 70,000-man garrison thereupon surrendered unconditionally (February 15). Total casualties were 138,700 British (mostly prisoners) and 9,824 Japanese.

COMMENT. *Two unpardonable deficiencies brought about the fall of this so-called impregnable fortress-naval base: its fixed defenses and artillery were sited to repel naval attack only, and its field forces were shockingly ignorant of the terrain and of jungle fighting. The first was due to poor judgment, the second to inferior preparation and leadership. The high quality of Japanese planning and leadership, evident throughout, was clearly demonstrated by the availability of armored amphibious barges for the final assault across the Strait of Johore.*

BURMA

LAND OPERATIONS

Simultaneously with the various offensives begun on December 8, 1941, the Fifteenth Japanese Army quietly occupied Thailand. While the main body prepared for an overland invasion of Burma, a detachment seized an air base at Victoria Point, at Burma's southernmost tip, to cut the British air line of communications from India to Malaya (December 23).

1942, January 12–29. Invasion. Lieutenant General **Shojiro Iida**'s Fifteenth Army (initially 2 reinforced divisions, with heavy air support) began a westward advance on Moulmein and Tavoy from Thailand. The Japanese were accompanied by a small group of Burmese revolutionaries, who had been promised Burma's independence from Britain. In subsequent months, **Aung San** and his "Thirty Comrades" had some success in inciting minor uprisings against the British and in sabotage behind the British lines.

1942, January 30–31. Battle of Moulmein. Lieutenant General **Thomas Hutton**'s British forces in Burma (the equivalent of 2 small, ill-equipped divisions of British, Burmese, and Indian troops) were surprised, then driven out of Moulmein with heavy losses and forced to withdraw across the Salween.

1942, February 18–23. Battle of the Sittang. Japanese units crossing the unbridged Salween enveloped the British left, forcing them to withdraw to the Sittang River in a continuous running fight. Approximately half of Hutton's army had gotten across the Sittang's lone bridge when another surprise river-crossing envelopment threatened the bridge from the rear. The British blew up the bridge at once. Most of the cut-off men made their way back across the river on rafts, but all of their heavy equipment was lost.

1942, March 5. Alexander in Command. Lieutenant General Sir Harold R. L. Alexander arrived in Rangoon to replace Hutton. Although the arrival of a number of small reinforcement units from India had brought British ground strength back up to a strength of 2 small divisions, Alexander realized that his scattered and largely demoralized forces could not now hold the onrushing Japanese. He abandoned Rangoon, personally barely escaping capture by converging Japanese columns moving in to seize the capital (March 7). Burma, whose land frontiers were all mountain ranges traversed only by narrow trails, save for the Burma Road, was thus cut off from all the outside world except China.

1942, March 12. Arrival of Stilwell and the Chinese. With their forces in Burma crumbling, the British in February accepted an offer of assistance from Chiang Kai-shek, who at once sent his Fifth and Sixth armies marching down the Burma Road. (Chinese and Japanese armies were normally the equivalent of western corps; these 2 were undermanned and very short of artillery, trucks, and other equipment; the Fifth Army, comprising veteran troops, well led, had a combat efficiency comparable to a western or Japanese division; the Sixth was less reliable.) Chiang also sent his new American chief of staff, Lieutenant General Stilwell, to command this expeditionary force in Burma. Stilwell reported to Alexander at Maymyo, Burma's temporary capital.

1942, March 13–20. Reorganizing on the Prome-Toungoo Line. Alexander established a defensive line running generally south of Prome and Toungoo, and thence east to Loikaw, overlooking the Salween River. The British Burma Corps held the right, in the Irrawaddy Valley, covering Prome. The Chinese Fifth Army held the center, traversed by the Rangoon-Mandalay road and railroad; the Chinese Sixth Army held the mountainous, jungled region to the east. Competent British Major General **William Slim** arrived (March 19) to take command of the Burma Corps.

1942, March 21–30. Renewed Japanese Of-

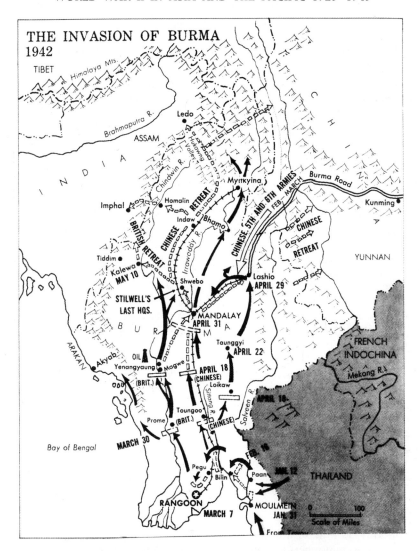

THE INVASION OF BURMA 1942

fensive. Iida's troops, reinforced and rested, made their main effort against the Chinese Fifth Army at Toungoo. They cut off the Chinese 200th Division, but determined counterattacks under Stilwell's direction, supported by Slim's British troops, permitted the trapped division to fight its way out. Japanese pressure, however, drove the Chinese slowly back up the railroad, while the British were forced to evacuate Prome.

1942, April 1–9. Reinforcements and Offensive Preparations. The coincidental arrival of reinforcements to both sides now caused a temporary lull. The arrival of part of the Chinese Sixty-sixth Army permitted Stilwell to bolster his troops holding the railroad. With the Japanese apparently stopped, he and Slim began to prepare for a counteroffensive. Iida, however, had just received 2 veteran divisions from the victorious Malaya-Singapore campaign. He prepared for a daring double envelopment; he was ready first.

1942, April 10–19. Battle of Yenangyaung. The Japanese plan called for the first blow to be made against the British Burma Corps while a holding attack kept the Chinese on the railroad occupied; the Sixth Army at Loikaw was temporarily left in peace. Despite a gallant and desperate British effort to hold Magwe and the Yenangyaung oil fields, the Japanese could not be stopped. The 1st Burma Division was cut off, but British counterattacks, combined with an attack by the Chinese 38th Division against the east flank of the Japanese encircling force, permitted Slim to rescue the division.

1942, April 18–23. Collapse of the Chinese Sixth Army. With Allied attention riveted on the desperate battle for the oil fields, the heavily reinforced Japanese 56th Division suddenly struck the Chinese Sixth Army in the Loikaw-Taunggyi area. The surprised Chinese were overwhelmed. Stilwell, personally leading the tough 200th Division, struck at Taunggyi from the west, only to discover that the Japanese had disappeared. Deliberately abandoning its own line of communications, the 56th Division had struck northward along the Salween toward Lashio, terminus of the railroad and starting point of the Burma Road. Meanwhile, the remainder of the Fifteenth Army maintained its pressure against the British and Chinese south of Mandalay, who now began to fall back because of the threat to their left rear.

1942, April 29. Fall of Lashio. Seizing Lashio, the 56th Division then turned southwestward, toward Maymyo and Mandalay. Alexander, who had already decided to abandon Mandalay, now hastened the withdrawal across the Irrawaddy.

1942, April 30–May 1. Fall of Mandalay. Holding off the converging Japanese drives during the night, the remnants of the Allies completed their withdrawal over the Ava railroad bridge, which was then blown up.

1942, May. Retreat from Burma. Pressed closely by the Japanese, Slim and his Burma Corps fell back quickly toward Tiddim and the Indian frontier. After a last desperate defensive battle at **Kalewa** (May 11), Slim got his remaining troops across the Chindwin River and into the border hill region; the retreat continued to Imphal. The Japanese pursuit stopped at

the river. To the north and east the Chinese were also scattering. The survivors of the Sixth Army had already fled to Yunnan. The scattered units of the Fifth and Sixty-sixth armies withdrew as best they could, some across the mountain borders to Yunnan, some to the Himalayan foothills of northern Burma, where they spent a miserable rainy season; a handful moved to the northwest, with a starving stream of civilian refugees, across the mountains to India. Stilwell, gathering his headquarters group, led an aggregation of 100-odd men and women—including American Dr. **Gordon Seagrave** and his Burmese field hospital personnel—on an amazing trek of 400-odd miles, by jeep and foot, across the mountain jungles to Imphal.

COMMENT. *The British had 30,000 casualties among the 42,000 involved in the First Burma Campaign. About half of these were "missing": Burmese who threw down their arms to return to home and family, or refugees who later turned up in India. Chinese losses defy estimation. Of the 95,000 Chinese troops engaged, only 1 division—the 38th, commanded by Major General* **Sun Li-jen** *(VMI and Purdue graduate)—withdrew as a fighting unit, but with heavy losses, cutting its way westward across the path of the Japanese advance and acting as a rear guard for the more demoralized British units retreating from Tiddim to Imphal. Japanese ground losses amounted to 7,000.*

1942, May–December. Japanese Consolidation in Burma. With the monsoon season beginning, the Japanese ended their pursuit and began to consolidate control of Burma, four-fifths of which was in their hands. They had accomplished their major objectives: to cut off China completely from surface communication with her allies, and to seize an area which could be developed into a powerful bastion to protect the western approaches to the Southern Resources Area.

1942, June–December. Allied Reorganization and Planning. Fearful of a possible Japanese invasion of India in the coming dry season, the British feverishly prepared for defense. Wavell, now commanding in India, knew that he could not consider an offensive effort to reconquer Burma for at least another year, until his newly raised British and Indian troops could be adequately equipped and trained. Meanwhile, for training and morale purposes, he planned for a limited invasion of the Arakan region, Burma's northwest seacoast region, isolated from the rest of Burma by mountains. Stilwell, more impatient to reconquer Burma, so as to reopen surface communication with China, was meanwhile building up a Chinese Army in India around the nucleus of the Chinese troops that had escaped from Burma across the Chin Hills. Reinforcements came to the training center at Ramgarh in empty "Hump" planes returning from China (see p. 1244), while equipment was arriving by ship from the U.S. In anticipation of his projected advance through north Burma, Stilwell began construction of a road eastward across the Chin Hills from Ledo, in Assam, generally following the principal northern refugee route from Burma.

AIR OPERATIONS

1941, December. Deployment of the "Flying Tigers." Upon the outbreak of war in Southeast Asia, Colonel Chennault sent 1 of the 3 squadrons of his American Volunteer Group (AVG, or "Flying Tigers"; see p. 1232) to Kunming to protect the Chinese terminus of the Burma Road from air attack. Another (upon British request) was sent to reinforce the weak RAF force (equipped with obsolescent planes) at Mingaladon Airport near Rangoon. The third squadron, kept in reserve, was rotated regularly with the others after they became involved in active operations. In its first action, the Kunming squadron intercepted Japanese bombers from Indochina on a raid against Kunming, shooting down 6 with no losses to themselves (December 20). Three days later the Japanese mounted their first air raid against Rangoon. Lacking the air warning system Chennault had established in China, the Flying Tigers and RAF fighters at Mingaladon were forced to take off during the raid, but

nevertheless shot down several of the attackers with slight loss of themselves (December 23).

1942, January–February. The Air Battle for Rangoon. The effectiveness of the Flying Tigers caused the Japanese to increase the number of fighter escorts for their bombers, but despite heavy losses (compensated by reinforcements from Malaya) they continued to bomb Rangoon regularly. Besides combating air raids on Rangoon, the American fighters frequently flew support missions for hard-pressed British ground forces, easing slightly the merciless pounding these were receiving at the hands of Japanese surface and air units. Meanwhile, RAF bombers based at Magwe made a few raids against the principal Japanese air bases in Thailand, with relatively little effect. By the end of February, the RAF fighters at Mingaladon had all been put out of effective action; but thanks to Chennault's superior aerial tactics, only 15 of his planes had been lost (8 of the pilots walked back to base to fly again), and they had shot down more than 100 Japanese planes. By rotating his squadrons into Mingaladon, he had been able to keep planes and men in good operational condition.

1942, March 1–20. Operations from Magwe. The approach of Japanese ground troops to Rangoon forced Chennault to pull back from Mingaladon to the RAF base at Magwe, whence he continued support of British and Chinese ground troops.

1942, March 21. Japanese Raid on Magwe. A surprise massive Japanese air raid caught most of the American and British planes at Magwe on the ground and destroyed the effectiveness of all air units based there. Chennault, able to salvage only 3 planes from the wreckage, withdrew north to Loiwing just inside the Chinese border. The RAF withdrew to fields in India.

1942, April–May. Long-Range Air Support for Burma. Flying Tigers and RAF fighters continued to give long-range support to the withdrawing Allied troops in Burma, but it was relatively ineffective due to inadequate communications with the ground and to the short time available for action over the battlefield. The arrival of some Spitfires from England, however, permitted the RAF fighter pilots to fight the Japanese on more even terms. With the loss of Burma and the closing of the Burma Road, Chennault, concentrating at Kunming, reduced his activities sharply to conserve his dwindling supplies of fuel and ammunition.

1942, June–December. Flying the "Hump." With ground communications to China severed, the Americans, under Stilwell's direction as commanding general of the China-Burma-India Theater, began a long-range supply airlift from bases in northeastern India to Kunming. Because of Japanese air bases at Myitkyina in northern Burma, the planes were forced to fly over the eastern Himalayas in southeastern Tibet, climbing to 21,000 feet and more to get over the mountains. This route over the world's highest mountains was dubbed the "Hump" by American pilots. Stilwell could get only a few additional transport planes, and so this initial airlift was a feeble and totally inadequate drop in the bucket. The newly activated U.S. Tenth Air Force, Major General **Howard C. Davidson,** was installed on Indian air bases to protect the Hump bases.

China

There was little formal combat activity in China during 1942. The Japanese, engaged elsewhere on an amazing series of peripheral offensives, were content to hold what they had. Guerrilla activity in areas behind the Japanese lines did not seriously threaten their firm occupation hold. The Chinese National Government armies were in no position to take advantage of Japanese inactivity, since the closing of the Burma Road had cut off their already meager supply of matériel and munitions. Stilwell's hitherto good relations with Chiang Kai-shek worsened when the Joint Chiefs of Staff turned down his requests for additional transport planes to increase supply over the Hump. (Other theaters had priority.) At the same time, friction rose between Stilwell and Chennault, now a brigadier general commanding the U.S. China Air Task Force (successors to the Flying Tigers, July 4). Chennault demanded that his fuel and ammunition requirements get priority over all other air-lifted supplies, while Stilwell insisted on balance between ground and air forces.

Land and Air Operations in the Pacific Areas

The Netherlands East Indies

1942, January–February. Japanese Invasion. A 3-pronged Japanese amphibious offensive moved south on the Dutch East Indies, with initial landings at Tarakan, oil-rich island off the east coast of Borneo, and at Menado on northern Celebes (January 11). The Japanese Eastern Force proceeded via Celebes, the Moluccas, and Timor toward Bali and Java. The Central Force advanced down the Macassar Strait along the east coast of Borneo toward Java. The Western Force moved through the South China Sea to north Borneo and Sumatra. Each force comprised a group of heavy cruisers and destroyers escorting transports carrying units of the Sixteenth Army and supported by powerful land-based air units. Admiral Nagumo's First Air Fleet provided additional far-flung air support. To oppose these forces, Wavell's ABDA Command consisted of vastly out-numbered American, British, and Dutch naval units, some 85,000 Dutch troops scattered on the main islands—most of them indigenous and of dubious loyalty—and a small Dutch air force equipped with obsolete planes. Covered by overwhelming and scarcely opposed air support, the Japanese forces moved from one key point to another, establishing new air bases to assure continued air superiority, then pressing generally southward again.

1942, February 25. Dissolution of ABDA-COM. The fall of Singapore (see p. 1239) and the rapid decline of Allied naval forces having sealed the doom of the Netherlands East Indies (see p. 1250), Wavell's command was dissolved and he returned to India to supervise the deteriorating situation in Burma. The Dutch retained responsibility for the continuing defense of Java, and remaining American, British, and Australian forces in the area were to continue to assist.

1942, February 28–March 9. Battle for Java. Following intensive air and naval bombardments, the Japanese Western Force landed near Batavia. In subsequent days, units of the Central and Eastern Forces also landed. Determined Dutch resistance was totally incapable of coping with Japanese land superiority and total control of the air. Remaining Dutch forces were finally forced to surrender, virtually completing the conquest of the Netherlands East Indies (March 9).

THE PHILIPPINES

1942, January 7–26. Defense of Bataan. MacArthur's troops lay astride the mountainous jungle at the base of the peninsula. Wainwright's I Corps held the left, Parker's II Corps the right, with the almost impassable height of Mt. Natib in the middle defended only by patrols. The supply situation was deplorable; although dumps had been established on the southern tip of the peninsula and Corregidor's ration reserves had been calculated for a 6-month stand, the mouths to be filled had been increased by more than 20,000 refugees from Manila crowding the area. Japanese command of sea and air precluded relief. The American command went on half-rations at once. Japanese attacks on both flanks were repulsed, but infiltrating elements, crossing Mt. Natib, threatened to split the American position, and after bitter fighting MacArthur withdrew to a more easily defended reserve position about 15 miles from the tip of the peninsula.

1942, January 26–February 8. Second Japanese Attack. While vicious assaults on the front were met and thrown back, amphibious landings on the western coast, well behind the battle line, became serious (January 23–26). They were finally repulsed by counterattack, by U.S. motor torpedo boat harassment, and by fire from Corregidor's artillery (January 29–February 13). Homma, withdrawing from the main front, waited for reinforcements from Japan.

1942, March 11. Departure of MacArthur. MacArthur, though unwilling, obeyed a peremptory order from President Roosevelt to leave for Australia and assume command of all Allied forces in the South Pacific. By PT boat and B-17 he arrived safely (March 17).

1942, March 11–April 2. Attrition. Wainwright now commanded, with Major General **Edward P. King** in command on Bataan. Despite good morale resulting from their success in repulsing all Japanese assaults, the efficiency of the troops was now sadly reduced. Rations had been cut to one-quarter, seriously affecting stamina; tropical diseases made great inroads. By the end of March, some 24,000 men were hospitalized or in convalescent areas.

1942, April 3–9. Japanese Breakthrough. Homma, reinforced and refitted, attacked under cover of incessant air and artillery bombardment. Bursting through the left flank of II Corps, the Japanese forced it back 10 miles in 48 hours. On the left, I Corps, bent back toward the sea, attempted counterattacks, but these were easily repulsed. II Corps disintegrated. King surrendered unconditionally (April 9). The Japanese, although permitting many of the Philippine Army personnel to return to their homes, hustled the Regulars, American and Filipino—in callous brutality—on a 90-mile "Death March" to Camp O'Donnell. American and Filipino losses in the Bataan campaign amounted to some 20,000-odd men, discounting deserters, detachments ferried to Corregidor, and those few brave souls who dared the jungle to become the nucleus for a guerrilla warfare which would rage until the end of the war.

1942, April 10–May 3. The Other Islands. Before his departure, MacArthur had left Briga-

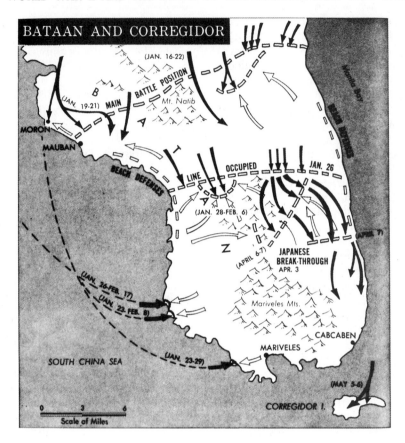

BATAAN AND CORREGIDOR

dier General **Bradford G. Chynoweth** in control of forces in Cebu, Panay, Negros, Leyte, and Samar with instructions to prepare for guerrilla operations. Brigadier General **William F. Sharp,** with a larger command, was to oppose any further landings on Mindanao. Homma moved 2 brigade-size task forces south. One seized Cebu (April 10), the other Panay (April 16), the defending troops retiring to mountain fastnesses. Additional Japanese landings on Mindanao (April 29–May 3) drove Sharp's troops also into the hill country.

1942, April 10–May 6. The Ordeal of Corregidor. The citadel and its satellite island forts at the mouth of Manila Bay had supported the defense of Bataan by the fire of large-caliber fixed armament. Now the full force of Japanese air and artillery from Bataan and from Cavite

swept the complex incessantly. Except for the 14-in. turret guns on Fort Drum—the "Concrete Battleship"—all these emplacements were open to air and plunging artillery fire. The only shelter on Corregidor was a tunnel carved into the rock of Malinta Hill, where the command post and hospital were installed. The bombardments destroyed all other installations.

1942, May 5–6. The Fall of Corregidor. The dazed defenders met Japanese amphibious assaults at the beach, their 155-mm. guns and lighter ordnance inflicting severe damage on the attackers, but a 2-battalion toehold was gained on the southern end of the island. At the end of his resources, with less than 3 days' supply of water remaining, Wainwright surrender unconditionally. At Homma's imperious demand, the surrender included all U.S. forces in the Philippines. American losses were some 2,000 killed and wounded and 11,500 made prisoner. Japanese losses in the final assault were more than 4,000. In compliance with Wainwright's orders, Sharp surrendered on Mindanao (May 10), and Chynoweth on Panay (May 18). However, individuals and groups of Americans and Filipinos continued guerrilla resistance for the next three years.

COMMENT. *The First Philippine Campaign, its outcome a foregone conclusion from the beginning, was a tribute to the skill, gallantry, and stoic determination of the officers and men involved in the defense. It delayed the Japanese timetable for 5 long months during which the U.S. was rallying its resources. The one glaring error in the otherwise efficient Japanese plan and prosecution of the campaign was a complete miscalculation of the resistance to be expected from the Filipino people. On all other major fronts of the Japanese assaults, the indigenous populations were either indifferent or even disloyal to the defense; the same condition was expected in the Philippines. Homma looked forward to opposition from the small American component of MacArthur's forces only. Instead, the Filipino Regulars, like the Americans, lived up to the finest traditions of the service. The poorly trained and equipped Philippine Army units, after their first panic, quickly settled down, becoming veterans almost overnight. Most of the Filipino people gave loyal support both during the campaign and, as they later showed, in the guerrilla operations which continued to plague the invaders to the end of the war.*

PAPUA, THE BISMARCKS, AND THE SOLOMONS

1942, January–March. Arrival of the Japanese. Amphibious units of the Japanese Fourth Fleet, overcoming brief but stout Australian resistance, seized Kavieng and Rabaul, the latter becoming the principal Japanese naval and air base in the South-west Pacific. Control of New Britain was consolidated during the following month. Amphibious landings were then made to seize Salamaua and Lae in Papua (March 6), and footholds on Bougainville (March 13).

1942, May–July. Expansion of the Japanese Perimeter. The Japanese expanded their control in the central and southern Solomons, culminating their efforts by establishing a base on Guadalcanal, where they began to build an airfield (July 6). Despite failure to gain naval control of the Coral Sea, Japanese plans to seize Port Moresby, principal city of southern Papua, were initiated by occupation of Gona (July 11) and Buna soon afterward. All land operations in Papua, the Bismarcks, and the Solomons were now under the Eighth Army, commanded by Lieutenant General **Hitoshi Imamura,** with headquarters in Rabaul. In response to repeated Japanese air attacks on Port Moresby, growing Allied air strength harassed Japanese bases.

1942, June–July. Strategic Controversy. Following the great American naval victory at Midway (see p. 1253), both General MacArthur and Admiral **Ernest J. King** (the new Navy chief in Washington) urged an offensive in the Bismarck-New Guinea area as a definite check to the Japanese menace to the U.S.-Australian supply line. King favored Navy-controlled island-hopping up the Solomons to Rabaul, MacArthur a direct thrust—Army-controlled—on Rabaul itself. This clash of personalities and of service jealousies was settled by compromise; the JCS ordered (July 2) a 3-phase operation: seizure of the southern Solomons by Vice Admiral **Robert L. Ghormley** (commanding the South Pacific Area under Nimitz), and seizure of the remainder of the island mass and of the northwest coast of New Guinea by MacArthur, whose final objective was the capture of Rabaul itself. MacArthur's initial move, however, was delayed by a continuation of the Japanese offensive.

OPERATIONS IN PAPUA, JULY 1942–
JANUARY 1943

1942, July 21–September 13. Advance on Port Moresby. Elements of the Japanese Eighteenth Army, under Major General **Tomitoro Horii,** pressed inland from Gona, drove in local Allied troops, and moved up the rugged Kokoda Trail to seize the key pass over the Owen Stanley Mountains (August 12). Pushing ahead, the Japanese reached to within 30 miles of Port Moresby before stiffening Australian and American resistance, under Australian Major General **Edmond F. Hering,** supported by tactical air forces with local air superiority, halted the advance.

1942, August 25–September 5. Operations at Milne Bay. A Japanese regimental-strength amphibious landing was contained and repulsed by Australians.

1942, September–November. The Return across the Owen Stanley Range. Partly under Allied pressure, and partly in compliance with defensive instructions from Imamura, the Japanese fell back stubbornly over the mountains, followed by the Australian 7th and U.S. 32nd divisions. The Allies were finally halted by a massive jungle fortress, cleverly constructed in the swampy region around the villages of Buna and Gona on the coast of the Solomon Sea (November 19).

1942, November 20–1943, January 22. Battle for Buna-Gona. The Allied offensive bogged down. The Americans and Australians were racked by disease; they were short of artillery and rations, and untrained in jungle warfare. U.S. Lieutenant General **Robert L. Eichelberger,** placed in command by MacArthur (December 1), quickly restored sagging Allied morale and rectified the serious logistical deficiencies. The Australians, on the Allied left, stormed Gona (December 9). More heavily fortified Buna, however, long resisted the American attacks, even after yard-by-yard advances had brought the attackers within sight of the shore. A final converging assault of Americans and Australians overran the defenders, some of whom were able to escape in surface craft, but most of the Japanese died. Total Japanese casualties in the Buna-Gona operation were more than 7,000 dead, an unknown number of wounded evacuated, and 350 wounded prisoners. The Allies lost 5,700 Australians and 2,783 Americans killed and wounded. Both sides lost heavily from disease, the Americans

alone having about 60 percent of 13,646 men incapacitated.

COMMENT. *One of the most important aspects of this relatively small operation was the proof that Allied troops could defeat the hitherto invincible Japanese in the jungle.*

OPERATIONS ON GUADALCANAL,
AUGUST 1942–FEBRUARY 1943

1942, July–August. American Preparations. Word of construction by the Japanese of an airfield on Guadalcanal hurried the planned American amphibious move into the southern Solomons. Admiral Ghormley, at Nouméa, sent an amphibious task force under Rear Admiral **Richmond K. Turner,** carrying Major General **Alexander A. Vandegrift**'s reinforced 1st Marine Division (19,000 men) and an air support force built around Admiral **Frank J. Fletcher**'s 3-carrier task force. Fletcher, as senior officer, commanded the expedition.

1942, August 7. Landings on Tulagi and Guadalcanal. The Marines began landing on Guadalcanal and Tulagi in complete surprise. The small Japanese garrisons—2,200 on Guadalcanal and 1,500 on Tulagi—were quickly scattered. While Vandegrift began a perimeter defense of the still-incomplete Guadalcanal airfield, unloading operations commenced from the transports huddled in the sound between Guadalcanal and Florida islands. But prompt and violent Japanese air attacks soon interfered with the unloading.

1942, August 9–19. Departure of the Navy. The combination of the intensive Japanese air attacks and a stunning Japanese night naval victory off Savo Island (see p. 1256) caused Admirals Fletcher and Turner to withdraw their naval forces from the sound and the nearby waters, leaving the Marines, short of supplies, completely isolated on Guadalcanal and Tulagi. However, long-range land-based planes from the New Hebrides flew frequent covering missions overhead, and fast destroyer transports brought additional supplies. Mean-

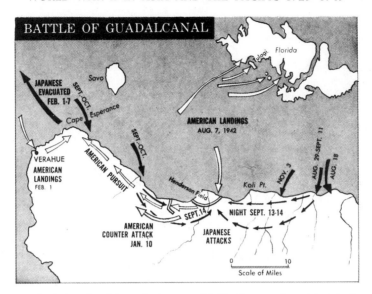

while, engineers rushed the Guadalcanal air-field, now renamed Henderson Field, to completion. Japanese reinforcements, brought down at night by destroyer, began to assemble near Taivu Point and Kokumbona, east and west of Henderson Field.

1942, August 20. Arrival of Air Support. Upon the completion of Henderson Field, a squadron of 31 Marine planes arrived to give direct support and air cover to the 1st Marine Division. This strength was gradually built up to about 100 aircraft.

1942, August 21–September 12. Skirmishing Around the Perimeter. Clashes between American and Japanese patrols (which had been sporadic since the landings) became more frequent and more intensive. Nightly rushes of Japanese destroyers down "the Slot" (the channel between the northeastern and southwestern chains of the Solomon Islands) brought Japanese reinforcements. In the daytime the waters in the vicinity of Guadalcanal and Tulagi belonged to the American Navy, covered by planes from Henderson Field and from the Hebrides; at night the "Tokyo Express"—Japanese destroyers and light cruisers—dashed into the sound to drop men and supplies on Guadalcanal, and to bombard the Marine positions and airfield.

1942, September 12–14. Battle of Bloody Ridge. Japanese forces, the equivalent of a regiment, vigorously attempted to seize positions on Lunga Ridge, overlooking the airfield from the south. Combination of infiltration and frontal assaults came close to breaking the Marine lines, but the attacks were repulsed. The Japanese left 600 dead on the field (their total casualties are unknown); American casualties were 143 killed and wounded.

1942, September 15–October 22. Continuing Build-ups on Guadalcanal. While violent air and naval battles took place over and around Guadalcanal, both sides built up their ground strength. By mid-October, Vandegrift's command on Guadalcanal alone had reached a strength exceeding 23,000. The Japanese Seventeenth Army, 2 divisions under Major General **Haruyoshi Hyakutake,** comprised at least 20,000. Patrol skirmishing and probes around the perimeter increased in intensity as the Japanese prepared for another assault.

1942, October 23–25. Land Battle of Guadalcanal. For 3 days the Japanese launched intensive but piecemeal and uncoordinated attacks at different points of the Marine perimeter. Though fighting was severe, none of the assaults came close to success, and all were repulsed with appalling losses. Japanese dead totaled over 2,000, with numbers of wounded unknown; American casualties were less than 300.

1942, October 26–December 8. Expansion of the Perimeter. Vandegrift immediately extended the American perimeter far enough to prevent Japanese artillery fire from reaching Henderson Field. During the following 6 weeks, the area under American control was gradually extended farther, as units of the 2nd Marine Division landed to reinforce and relieve tired units of the 1st Marine Division. The air and naval battles continued almost without interruption during this period, with Japanese air losses mounting rapidly.

1942, December 9. Relief of the 1st Marine Division. General **Alexander Patch,** commanding the Army American Division, relieved General Vandegrift in command of Allied operations on Guadalcanal. The 1st Marine Division was withdrawn.

1942, December 10–1943, January 9. American Build-up. The Japanese were now on the defensive, having established a heavily fortified jungle position about 6 miles west of Henderson Field, extending inland more than 4 miles from Pt. Cruz. This line was held by less than 20,000 Japanese, while American strength built up to 58,000 men in the XIV Corps, commanded by Patch, which included the Americal, 25th, and 2nd Marine divisions. The Japanese were short of supplies and riddled with disease; the Americans were well supplied, in good condition, and high morale, though less experienced in jungle fighting than the Japanese.

1943, January 10–February 7. The American Offensive. A vicious 2-week battle drove the Japanese from their jungle fortifications, which they held with tenacious bravery (January 10–23). Falling back under the cover of a desperate and well-conducted rear-guard action, the Japanese again attempted to stand at **Tassafaronga Point,** but were quickly driven northward toward Cape Esperance (January 31). A small American force was landed behind the Japanese lines west of Cape Esperance in an attempt to encircle the defenders and to prevent

evacuation by sea (February 1). The combination of dogged, skillful defense by the starving, defeated Japanese soldiers and brilliant support from the Japanese Navy frustrated the American attempt, however, and approximately 13,000 Japanese were evacuated in night operations (February 1–7).

COMMENT. *Guadalcanal was the first large-scale Allied victory against the Japanese. Both sides made a number of mistakes, but the Japanese, who had a preponderance of land, sea, and air forces at the outset, made more than the Americans. Their greatest failure was in persisting in piecemeal operations, strategically and tactically, instead of attempting to build up local superiority against the Americans, which might have been possible at any time during the first 2 months of the battle.*

Naval Operations in the Pacific

THE NETHERLANDS EAST INDIES

1942, January 11–22. Preliminary Japanese Penetrations. Powerful amphibious forces, under land-based air cover, moved into the northern areas of the Netherlands East Indies (see p. 1245). Supporting the numerous modern screening forces of cruisers and destroyers was Admiral Nagumo's First Air Fleet, veterans of Pearl Harbor. In opposition was a handful of American, British, and Dutch warships, mostly old World War I types. The Japanese air units exercised effective air superiority. The ABDA naval commander, under General Wavell, was American Admiral **Thomas Hart.**

1942, January 23. Battle of Macassar Strait. As the Japanese Central Force approached Balikpapan, it was surprised shortly after dark by 4 American destroyers. In an hour's battle the Americans sank 1 small Japanese warship and 4 heavily loaded troop transports, and damaged several other vessels. Then, having suffered little damage themselves, the Americans raced southward.

1942, January 24–February 3. Continuing Japanese Advance. The setback at Balikpapan did not delay the Japanese advance.

Rushing more warships and planes to the Macassar Strait area, all Japanese invasion forces continued to leapfrog southward. The main Dutch naval base at Amboina was captured (January 30).

1942, February 4. Battle of Madoera Strait. Japanese aircraft struck a combined American and Dutch squadron. The Japanese had few losses, but seriously damaged several Allied ships, including the U.S. cruisers *Houston* and *Marblehead.* The latter was forced to limp back to America for repairs, leaving the *Houston* as the only major U.S. ship in the area.

1942, February 13–14. Battle off Palembang. A U.S.-Dutch-British squadron, commanded by Dutch Rear Admiral **Karel Doorman,** attempted to prevent a Japanese landing at Palembang, Sumatra. Intercepted by Japanese planes, they were prevented from reaching the Japanese convoys.

1942, February 14. Change in Naval Command. Admiral Hart was succeeded by Dutch Vice Admiral **Conrad Helfrich.**

1942, February 19–20. Battle of Lombok (Bandoeng) Strait. Admiral Doorman's squadron, attempting to strike spearheads of the Japanese Eastern Force, approaching Java, was engaged by a smaller but skillfully handled Japanese destroyer squadron in a hard-fought night engagement. Damage was heavy on both sides, and 1 Dutch destroyer was sunk.

1942, February 19. Carrier Raid on Darwin. Admiral Nagumo's First Air Fleet struck a crippling blow against Darwin, main port in northern Australia, causing severe damage to shipping and shore installations.

1942, February 27. Battle of the Java Sea. Admiral Doorman's squadron of 5 cruisers and 10 destroyers made a bold but unsuccessful bid to attack the Japanese Eastern Force, which was escorted by Rear Admiral **Takeo Takagi**'s 4 cruisers and 13 destroyers. In a 7-hour running fight of crazy-quilt pattern, with the Dutch admiral gallantly forcing the fight to the end, Doorman's force was crushed; he went down with his flagship, RNNS *De Ruyter.* Only USS *Houston,* HMAS *Perth,* and HMS *Exeter,* with 5 destroyers (4 of them American), survived the

battle. Japanese control of air and sea was now undisputed.

1942, February 28–29. Action off Banten Bay. The *Houston, Perth,* and several destroyers, attempting to escape to Australia, encountered a Japanese landing force approaching Java. The Allied ships, despite a hopeless situation, attacked, but were themselves ringed by destroyers and cruisers in point-blank fire and sunk. HMS *Exeter* and a British and an American destroyer, attempting to thread their way to safety through minefields off Surabaya, were all sunk (March 1) by naval gunfire and carrier-based bombers. Four U.S. destroyers escaped to Australia.

1942, February 29–March 9. Conquest of the Indies. The entire Dutch East Indies was formally surrendered (March 9). Japan had gained control of the Southern Resources Area. The Malay Barrier now pierced, the Japanese were free to move west toward India and south toward Australia.

AMERICAN RAIDS IN THE CENTRAL PACIFIC

1942, January. Aftermath of Pearl Harbor. The loss of the U.S. Pacific Fleet battleships at Pearl Harbor meant that there could be no real counteroffensive in the Central Pacific for many months. However, the 3 aircraft carriers, USS *Lexington, Saratoga,* and *Enterprise,* had escaped the disaster. Admiral Ernest J. King, new Chief of Naval Operations in Washington, ordered Admiral Chester W. Nimitz, new commander of the Pacific Fleet, to use his carriers to harass the Japanese while American naval strength in the Pacific was being rebuilt. The task was complicated by severe damage to the *Saratoga,* struck by a Japanese submarine's torpedo (January 11). However, the USS *Yorktown* arrived from the Atlantic soon after this to join the Pacific Fleet.

1942, February 1. Attack on Gilbert and Marshall Islands. The carriers *Enterprise* and *Yorktown,* under Vice Admiral **William F. Halsey,** escorted by cruisers and destroyers, sent their planes to attack Japanese bases on the Gilbert and Marshall Islands. Some damage was inflicted. The task force returned safely to Pearl Harbor.

1942, February 20. Aborted Raid on Rabaul. A task force built around the carrier *Lexington,* under Vice Admiral **Wilson Brown,** approached Rabaul through the Solomon Sea in an attempt to strike a surprise blow at Japanese forces concentrating there. Discovered and attacked by Japanese planes, the task force withdrew, but inflicted heavy losses on the Japanese air units.

1942, February 24–March 4. Raids on Wake and Marcus Islands. Halsey's *Enterprise* task force struck Wake (February 24) and Marcus (March 4) islands, inflicting some damage, and returning safely later to Pearl Harbor.

1942, March 10. Strikes at Lae and Salamaua. Brown's *Lexington* task force, joined by the *Yorktown,* sent planes across Papua from the Coral Sea, sinking and damaging several small warships and transports.

1942, April 18. Carrying the War to Japan. Sixteen Army B-25s under Lieutenant Colonel **James H. Doolittle,** carried on the US carrier *Hornet,* escorted by Halsey's *Enterprise* task force, flew 800 miles to bomb Tokyo and other Japanese cities. Little physical damage was done, but the Japanese were greatly alarmed and Americans greatly heartened by the feat. All of the attacking planes were lost or forced to crash land in China, save for one which landed at Vladivostok, where plane and crew were interned. Most of the fliers survived, though 2, falling into Japanese hands, were beheaded. The most important effect of the raid was to influence the Japanese high command to attempt to expand its perimeter in the central and southern Pacific, with disastrous results.

BATTLE OF THE CORAL SEA

1942, May 1. Initiation of the New Japanese Plan. As the first step in the new Japanese plan, Admiral **Shigeyoshi Inouye** (commanding at Rabaul) had been ordered by the commander of the Japanese Combined Fleet,

Admiral **Isoroku Yamamoto,** to seize bases in the southern Solomons and to capture Port Moresby. The small carrier *Shoho,* 4 cruisers, and a destroyer escorted an assault force of several transports south through the Solomons toward Tulagi. Meanwhile, a larger force was being assembled at Rabaul to assault Port Moresby by sea. To provide additional support, a striking force built around the large carriers *Shokaku* and *Zuikaku,* under Rear Admiral Takagi, moved south from the Central Pacific to enter the Coral Sea from the east. Unknown to the Japanese, American intelligence was aware of most of these plans because of the breaking of the Japanese code (see p. 1234). Admiral Nimitz sent Admiral Fletcher with a task force built around the carriers *Lexington* and *Yorktown* to block the Japanese plan. Fletcher was joined by a small squadron of American and Australian cruisers and destroyers under Rear Admiral **J. G. Crace,** R.N.

1942, May 3. Occupation of Tulagi. Japaaese forces occupied the island without opposition. The *Shoho* task force sailed north to join the Port Moresby assault force, which was leaving Rabaul.

1942, May 4. *Yorktown* Attack on Tulagi. *Yorktown* planes, bombing Tulagi, had minor success.

1942, May 5–6. The Fleets Approach. the *Yorktown* rejoined Fletcher's fleet in the central Coral Sea, just as Takagi's squadron was steaming into the sea from the northeast and as the *Shoho* force and the Port Moresby assault force were approaching the Coral Sea from the Solomon Sea to the north. Both main carrier forces searched for each other without success.

1942, May 7–8. Battle of the Coral Sea. In a confused series of actions marked by serious errors on both sides, the 2 carrier forces found and struck at each other. The *Shoho* was sunk by American planes (May 7). An attempted Japanese night attack (May 7–8) resulted in severe plane losses; American losses were slight. An exchange of strikes the next morning, however, resulted in the sinking of the *Lexington* and some damage to the *Yorktown;* the *Shokaku* was severely damaged, but the

Zuikaku was unscathed. An American destroyer and oiler were also sunk. Meanwhile, the loss of the *Shoho,* and the threat of Admiral Crace's squadron of 3 cruisers and several destroyers (which survived without damage several Japanese air strikes and a mistaken attack by American land-based planes) moving to intercept the Port Moresby assault force, caused Admiral Inouye to call off the invasion plan. The assault force returned to Port Moresby. Because of the losses they had suffered, Admirals Fletcher and Takagi almost simultaneously decided to withdraw from further action.

COMMENT. *This was the first great carrier battle; no surface ship on either side sighted the enemy. Tactically a draw (the Japanese lost more planes, the Americans more ships), it constituted a major Allied strategic success by halting the proposed assault on Port Moresby.*

THE BATTLE OF MIDWAY

1942, May. Japanese Preparations. While Admiral Yamamoto, commander of the Combined Fleet, was preparing to carry out the revised Japanese strategic plan (see p. 1239), he learned that the American carriers *Enterprise* and *Hornet* were in the South Pacific (where Nimitz had rushed them in an effort to join in the Coral Sea battle). Believing that both the *Yorktown* and the *Lexington* had been sunk in that engagement, he was now certain that his planned blow at Midway would not be opposed by any American carriers. He nevertheless employed 165 warships, all of the available naval might of Japan, in the effort, the greatest armada yet assembled in the Pacific Ocean. Vice Admiral **Boshiro Hosogaya**'s Northern Area Force—2 light carriers, 7 cruisers, and 12 destroyers—was to make diversionary strikes against Alaskan bases just before the main forces struck Midway; after that, Hosogaya was to establish bases in the western Aleutians. The remainder of Yamamoto's fleet headed toward Midway in 3 separate forces: Nagumo's First Air Fleet (sadly missing the 2 carriers put out of action at the Coral Sea by damage or plane

loss); the Midway Occupation Force, under Vice Admiral **Nobutake Kondo,** consisting of the powerful Second Fleet of 2 battleships, 1 light carrier, 2 seaplane carriers, 7 cruisers, and 29 destroyers, escorting 12 transports carrying 51,000 troops; and Yamamoto's main body, comprising 7 battleships, 1 light carrier, 4 cruisers, and 12 destroyers. In addition, 18 submarines had been sent to the waters between Midway and Pearl Harbor to report American movements and to attack opportune targets.

1942, May. American Preparation. Nimitz, warned through Navy Intelligence's knowledge of Japanese codes, ordered Halsey's *Enterprise-Hornet* task force to rush back north from the South Pacific; also Fletcher's *Yorktown,* requiring repairs after Coral Sea estimated to take 3 months, was made combat ready in an amazing 48-hour period at Pearl Harbor. These 3 fleet carriers could carry approximately 250 planes, about the same number as on Nagumo's 4 fleet carriers. Nimitz, detaching about one-third of his 76 warships to protect Alaska, planned to fight the main battle within air range of Midway and thus to be able to employ 109 additional Army, Navy, and Marine land-based planes based on that small island. In late May, the Pacific Fleet of some 50 vessels and submarines assembled north of Midway, unsighted by Yamamoto's submarine force, which did not establish its screen until June 1. Halsey was sick in hospital; his *Enterprise-Hornet* force was commanded by Rear Admiral **Raymond A. Spruance.** Fletcher, on the *Yorktown,* was in command at sea under Nimitz' supervision from the Pearl Harbor base.

1942, June 3–7. Operations in the Aleutians. The Japanese Northern Area Force, outmaneuvering Rear Admiral **Robert A. Theobald**'s protective force, bombarded Dutch Harbor twice (June 3 and 4), avoided attacks by American land-based planes, then landed small forces on the islands of Kiska (June 6) and Attu (June 7). Had it not been for the American realization that this was a diversion, these operations would have constituted a great Japanese success. As it was, very little was accomplished except to leave some American military men with red faces, and to arouse further U.S. anger by this seizure of two bits of useless territory in the Western Hemisphere.

1942, June 4, 0300–0700. Battle of Midway—First Phase. Nagumo, confident in Yamamoto's conviction that there were no U.S. carriers in the Central Pacific and that most remaining American warships were rushing north to repel the Aleutian attack, launched half his attack force (108 planes) against Midway in early morning. From Midway, simultaneously, U.S. land-based planes rose both to strike the Japanese carriers and to attack the approaching bombers. Both these American efforts were repulsed with great loss, the American planes being inferior in speed to the Japanese aircraft. While half of the planes attacking the carriers were being shot down by Japanese defensive fighters, with no harm to

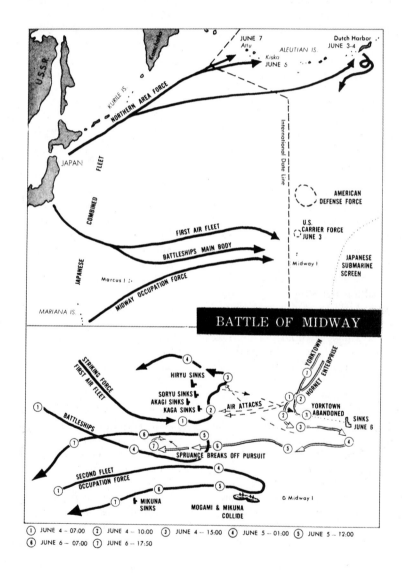

BATTLE OF MIDWAY

① JUNE 4 – 07:00   ② JUNE 4 – 10:00   ③ JUNE 4 – 15:00   ④ JUNE 5 – 01:00   ⑤ JUNE 5 – 12:00
⑥ JUNE 6 – 07:00   ⑦ JUNE 6 – 17:50

the carriers, Nagumo's bombers broke through the American screen at 6:30 A.M. and inflicted much damage on the island, sparing the runways, however, for their own expected later use.

1942, June 4, 0700–1700. Battle of Midway—Second Phase. Nagumo, receiving radio word from his Midway strike force that a second blow would be needed against the island, now began stripping his reserve planes of armor-piercing bombs and torpedoes, rearming with incendiary and fragmentation bombs—at least an hour's task. At that moment a search plane sighted and reported a formation of large American warships to the northeast. Startled Nagumo began rearming his reserve planes for naval action, while at the same time his Midway force was being retrieved on the carrier decks. He was steaming with his 4 carriers in box formation, each surrounded by protecting battleships, cruisers and destroyers. He shifted course 90° to the east to meet the new threat. Fletcher, meanwhile, had ordered Spruance's *Enterprise* and *Hornet* planes to be launched against the Japanese carriers at 7 A.M. The *Yorktown*, several miles to the east, launched half its planes at 7:30. Spruance's dive bombers, overshooting the new Japanese course, missed Nagumo's carriers in the overcast. But the American torpedo bombers found them and came in, without fighter cover, to be almost completely destroyed by antiaircraft fire and fighter planes at about 9:30. Not a single American hit was scored. Nagumo, assuming victory to be in his grasp, was still feverishly rearming and refueling planes on his carriers' cluttered decks when the American dive-bomber squadrons found him—*Enterprise, Hornet,* and *Yorktown* in turn. By 10:25, 3 Japanese carriers—*Akagi, Kaga,* and *Soryu*—were flaming wrecks. The fourth, *Hiryu,* undamaged, went steaming northeastward, sending her reserve planes to seek the American carriers. They found the *Yorktown* just after noon and hit her with 3 bombs. The *Hiryu,* meanwhile, had rearmed the planes which had struck Midway, and these also attacked the *Yorktown,* scoring 2 fatal torpedo hits. Flet-

cher abandoned his helpless and listing ship at 3 P.M. But searching American planes had located the Japanese carrier, and from the *Enterprise* a 24-dive-bomber strike—planes hurriedly retrieved from the earlier fight—mortally damaged the *Hiryu,* which burst into flames. It was now 5 P.M., the main battle over. Japan had lost her entire carrier force in being; the U.S. still had 2 in commission.

1942, June 4–5. Battle of Midway—Third Phase. Yamamoto, thunderstruck by a calamity in part due to his own separation of his fleet into several unsupporting forces, nevertheless hoped to retrieve the disaster. Pushing eastward during the night with all of his big ships, still infinitely superior in surface strength to the Americans, he hoped to lure them into a surface battle. Spruance (to whom Fletcher had relinquished command after the *Yorktown* was abandoned) was aware of this danger, and instead of attempting to pursue had turned eastward during the night. After midnight Yamamoto realized that he could not trap the Americans and, lacking air protection, he ordered a general retirement westward.

1942, June 5–6. Battle of Midway—Closing Actions. For 2 days the Americans pursued the retreating Japanese by day, but avoided possible traps at night. American planes caused additional damage, but no major actions took place. Finally, with fuel running low, Spruance ended the pursuit and turned back toward Pearl Harbor. Meanwhile, the crippled *Yorktown* had remained afloat, and was being slowly towed back to Pearl Harbor when she was sighted and, with an accompanying destroyer, sunk by the Japanese submarine *I-168.* These were the only American vessels lost during the battle. Also lost were 132 land- and carrier-based planes, with a total of 307 Americans killed in the battle. Japanese losses were 4 carriers, 1 heavy cruiser, 275 planes, and 3,500 men killed.

COMMENT. *Midway was one of the decisive battles of history. The loss of her fleet carrier force deprived Japan of the initiative; henceforward she was on the defensive—attempting to hold the great spread of the Southern Resources Area and contiguous regions she had so handily won. The psychological effect on the heretofore ever-victorious Japanese Navy was significant. Tactically, the American operations were almost faultless; Fletcher, the senior officer present, wisely gave Spruance full control over his 2-carrier task force after his own flagship was put out of action. And Spruance made good. Nagumo, on the other hand, was the victim of circumstance and faulty intelligence. His decision to rearm his reserve planes for a second strike at Midway—which resulted in his carriers being caught with their decks crowded with planes—was fatal but perhaps inevitable in view of his lack of vital information. Up to that time he too had conducted his operation faultlessly, and the final strikes of Hiryu planes against the Yorktown were planned and conducted with skill, gallantry, and determination. Two basic factors led to the result: first and foremost, American knowledge of the Japanese secret codes, which presented Nimitz with an accurate picture of Japanese intentions and dispositions; second, Yamamoto's original dispersion of his tremendous armada to fit his own estimate of probable American intentions and reactions. A minor but interesting sidelight to the battle was the successful transfer of command by both admirals from a sinking flagship to a cruiser: Nagumo from Akagi and Fletcher from Yorktown.*

Operations in the Solomons

1942, August 7. Amphibious Landings at Guadalcanal and Tulagi. (For background and land operations, see pp. 1247, 1248.) Over all commander of the Allied amphibious invasion of the Solomons was Vice Admiral Ghormley, with headquarters ashore at Nouméa. Commander afloat was Rear Admiral Fletcher, with a task force built round the carriers *Saratoga, Enterprise,* and *Wasp.* The amphibious force itself, comprising 4 U.S. cruisers, 3 Australian cruisers, 19 U.S. destroyers, and 19 troop transports, was commanded by Rear Admiral **R. K. Turner.** Surprise was complete; the Marines were put ashore on Tulagi and Guadalcanal, and within 24 hours had secured their objectives.

1942, August 7–8. Japanese Reaction. Admiral Inouye at Rabaul reacted promptly to reports of the Allied landings. Three separate air raids were made against the invasion force on

the first afternoon, and though these caused little damage, they seriously interfered with the landing of supplies and reinforcements to the troops ashore. Meanwhile, a naval task force of 7 cruisers and 1 destroyer, under Vice Admiral **Gunichi Mikawa,** rushed southward through the "Slot," or central sound between the central Solomon chains. The following day, Japanese planes continued their intensive attacks, and a series of violent air battles raged over the sound between Guadalcanal and Tulagi, and the seas nearby, as Japanese planes also attempted to hit the American carriers. The carrier planes took a heavy toll of the attackers, but suffered substantial losses themselves. Fletcher decided, therefore, that he would withdraw his carriers from the area, since he feared his losses would make the carriers more vulnerable and would prevent him from providing adequate support to Turner's force and the Marines ashore. As a result of this faulty decision, Turner decided (August 8, evening) that he could not keep his vessels near Guadalcanal without air cover, but decided to continue unloading supplies for 1 more day.

1942, August 9. Battle of Savo Island. Mikawa's force, passing south of Savo Island into the sound between Guadalcanal and Tulagi (later known as Iron-bottom Sound), surprised Turner's patrolling cruisers shortly after midnight. In 32 minutes of battle characterized by excellent Japanese gunnery and shiphandling, 4 of the 5 Allied heavy cruisers engaged—1 Australian and 3 U.S.—were sunk and the fifth damaged. Mikawa, fearing American air attack with the dawn (he did not know that Fletcher had withdrawn his carrier force), broke off the engagement and headed for Rabaul, his ships practically unscathed and with only 37 men killed and 57 wounded. Allied casualties were 4 cruisers and 1 destroyer sunk, 1,270 officers and men killed, and 709 wounded. On the way back, an American submarine, *S-44,* took the one substantial toll of Mikawa's command, the heavy cruiser *Kako,* small price for the most humiliating defeat ever suffered in fair fight by the U.S. Navy. Later that day, Admiral Turner departed with the remnant of his warships and all the transports, still only partly unloaded.

COMMENT. *Turner's disposition of his warships in 3 packets to guard all 3 entrances to Ironbottom Sound was faulty, setting up conditions for defeat in detail accomplished later by Mikawa's bold attack. Turner's controversial abandonment of the Marine division, taking with him the greater part of its supply and all heavy ordnance and construction equipment, was justified by Fletcher's prior departure with the carrier force (approved by Ghormley), leaving the amphibious flotilla bare to air attack.*

1942, August 10–23. Force Build-up on Guadalcanal. The Japanese sent in scattered reinforcements and supplies to their small force on Guadalcanal, mostly by fast destroyers which traveled through the Slot only by night. At the same time, the American Marines were receiving supplies by destroyer transport; the captured Japanese airfield (renamed Henderson Field) was completed, and planes brought in (August 20). During this time, long-range air battles were taking place over the island, Japanese planes coming from Rabaul and the northern Solomons, American planes from the New Hebrides. The Japanese failure to take advantage of the withdrawal of the American Navy and undertake an immediate, large-scale land build-up on Guadalcanal, and to make an overwhelming land and naval assault on the Marine positions, was a serious mistake.

1942, August 22–25. Battle of the Eastern Solomons. The next major action was precipitated by a Japanese decision to send a small convoy of destroyers and transports, under Rear Admiral **Raizo Tanaka,** carrying 1,500 reinforcements to Guadalcanal. To provide support and air cover, a Japanese squadron, built around the fleet carriers *Shokaku* and *Zuikaku* and light carrier *Ryujo,* under Admiral **Nobutake Kondo,** steamed toward the Solomons from Truk. The Japanese, because of Midway, had changed their code (although they were not certain that it had been compromised), but increased radio activity warned the Americans of a major operation impending. Ghormley ordered Fletcher's carrier force to intercept. Fletcher, misled by faulty intelligence, permitted the *Wasp* group to leave for a fueling rendezvous, reducing his strength by one-third. On making contact, Fletcher's planes sank the *Ryujo,* which was attacking Henderson Field. In so doing he left himself open to attack by the 2 large Japanese carriers (August 24). The American carriers were operating in independent groups. The Japanese air attack fell first on *Enterprise,* getting through the fighter air cover and doing much damage. No attackers reached *Saratoga,* and the new big battleship *North Carolina*'s antiaircraft fire drove off bombing thrusts. A small flight of *Saratoga*'s planes badly damaged a Japanese seaplane carrier, but the Americans could not find the Japanese fleet carriers. Fletcher now broke off contact, his loss being 17 planes and the damage to the *Enterprise.* Kondo gave up after a short pursuit and called his battleships and cruisers in. Meanwhile, Tanaka's destroyers, having put the reinforcements ashore, bombarded Henderson Field. Next day the Marine air group from Henderson Field damaged Tanaka's flagship, and an army B-17 sank a destroyer. Tanaka withdrew. Technically, the battle was a draw, since both naval carrier forces remained intact. The Japanese had accomplished their mission, but the loss of a light carrier and 90 carrier planes was a severe blow.

1942, August 31. Damage to the *Saratoga*. Torpedoed by the submarine *I-26,* the carrier was put out of action for 3 months, leaving only the *Wasp* fit for combat in the South Pacific.

1942, September 15. End of the *Wasp*. Admiral Ghormley had decided to reinforce the 1st Marine Division. His depleted carrier force escorted the 7th Marine Regiment, in 6 transports, leaving Espiritu Santo, New Hebrides (September 14). Japanese submarines *I-15* and *I-19* intercepted the force. The *I-15* damaged the battleship *North Carolina* and sank a destroyer, while *I-19* put 2 torpedoes into the *Wasp,* sending her to the bottom. Admiral Turner boldly continued to Guadalcanal, landing the reinforcements safely (September 18). Both sides now strove to reinforce their troops ashore. Their next efforts met head on.

1942, October 11–13. Battle of Cape Esperance. Admiral Turner with 2 big transports and 8 destroyer transports carried a reinforced regiment of the Army's American Division to Guadalcanal. Escorting was Rear Admiral **Norman Scott** with 5 cruisers and 5 destroyers. At the same time, a Japanese transport force—2 seaplane carriers and 6 destroyers carrying men, ammunition, and matériel—was coming down the Slot escorted by 3 heavy cruisers and 2 destroyers. Scott, forewarned, moved to intercept. He surprised the Japanese armada just before midnight, sinking 1 cruiser and a destroyer and crippling 2 more cruisers in gun and torpedo fight illuminated by American search planes. Rear Admiral **T. Joshima** safely landed his troops, losing two destroyers to land-based bombers from Henderson Field as he withdrew. Turner, arriving under cover of the naval battle, also safely landed his reinforcements.

1942, October 13–15. Surface Bombardment of Henderson Field. The battleships *Kongo* and *Haruna,* supported by several cruisers and destroyers, successfully bombarded the Marine perimeter on Guadalcanal, while more Japanese reinforcements were being landed on both sides of the position. A similar bombardment the following night by cruisers under Admiral Tanaka covered the arrival of more Japanese troops. The Japanese had reestablished effective control over the seas around Guadalcanal.

1942, October 18. Change in American Command. Vice Admiral Halsey replaced Ghormley as commander of the South Pacific Area. About the same time, Rear Admiral **Thomas C. Kinkaid** relieved Fletcher in command of the carrier force, which now comprised the *Hornet* and the hastily repaired *Enterprise* (this amazingly durable vessel becoming famous as the "Big E").

1942, October 23–25. Yamamoto's Plans. Yamamoto had placed 2 heavy and 2 light carriers under Kondo's command in the South Pacific; with these he expected to defeat the Americans in a climactic naval air battle for control of the southern Solomons. He was reluctant to risk the precious carriers in action,

however, without a good land airfield nearby, from which land-based planes could provide support and where valuable carrier pilots could take refuge if their carrier should be sunk or damaged. Because of their troop build-up on Guadalcanal, the Japanese were confident that Henderson Field could be wrested from the Marines. Once this was done, the carriers could close with the American fleet, while land-based planes rushed down from Rabaul. As the land forces engaged in a bitter battle for control of Henderson Field (see p. 1249), Kondo's fleet moved cautiously south from Truk. Kinkaid's force, with equal caution, circled north of the Santa Cruz Islands.

1942, October 26–27. Battle of the Santa Cruz Islands. While the Japanese land attack on Henderson Field was being repulsed (see p. 1249), Kinkaid obeyed Halsey's radioed orders to carry the fight to the Japanese fleet. In fact, both carrier forces launched simultaneous strikes against each other, the strike forces passing in midair; several American planes were shot down by Japanese escort fighters. The Americans damaged the light carrier *Zuiho* and severely damaged the *Shokaku* (put out of action for 9 months). The *Hornet* was hit even more seriously and had to be towed out of action by a cruiser. The *Enterprise* was hit and badly damaged by a second strike from the 2 remaining Japanese carriers. The Americans were now forced to withdraw. As the pursuing Japanese came closer, the *Hornet* had to be abandoned, and she was sunk by Japanese destroyers. Kondo then withdrew. The Japanese had won a clear-cut victory, but a Pyrrhic one, since they lost 100 planes, about half again as many as the American loss, and this still further depleted their thinning ranks of good carrier pilots. Furthermore, by failing to pursue and possibly to destroy the damaged "Big E," Kondo had made a serious mistake.

1942, November 12–13. Naval Battle of Guadalcanal—First Phase. A reinforcement contingent of 13,000 men in 11 transports, protected by 11 destroyers under Admiral Tanaka, was approaching Guadalcanal. To cover the approach of this reinforcement group, Admiral

Kondo sent a powerful squadron of 2 battle-ships, 2 cruisers, and 14 destroyers under Vice Admiral **Hiroaki Abe** to shell Henderson Field. The carrier force, steaming north of the Solomons, provided air cover. At the same time, American reinforcements arrived at Guadalcanal, convoyed by Admiral Turner, and escorted by 5 cruisers and 8 destroyers under Rear Admiral **Daniel J. Callaghan** (November 12). Learning of the approach of Tanaka's and Abe's forces, Callaghan, with Admiral Turner's approval, moved to intercept, counting on surprise to overcome his serious numerical disadvantage. Kinkaid's carrier force (with only one carrier, the damaged *Enterprise,* with a tender tied alongside actually making repairs as the force steamed toward Guadalcanal to cover Turner's retreating transports) was too far away to take part in the action. Though he had radar (which the Japanese did not), Callaghan misused it and was as surprised as the Japanese when Abe's force entered Ironbottom Sound shortly after midnight and suddenly discovered the Americans. One of the most confused and furious naval actions of history followed. The intermingled forces blazed away at each other in the darkness, sometimes at point-blank range. After 36 minutes, survivors of both sides pulled apart. The battleship *Hiei* was left helpless (to be sunk after dawn by *Enterprise* planes en route to Henderson Field), 2 Japanese cruisers were sunk, and all the other Japanese vessels were damaged. Two American cruisers and 4 destroyers were sunk, another cruiser and a destroyer were close to sinking, and all other American ships save 1 were damaged. Admiral Callaghan and Admiral Scott were both killed. Tactically a draw, this first phase of the battle was strategically an American success, since the planned bombardment of Henderson Field had been repulsed, and Tanaka's convoy had to turn back to the Shortland Islands.

1942, November 13–14. Naval Battle of Guadalcanal—Second Phase. During the following day, air activity was intensive on both sides. Before dark Tanaka again started south with his reinforcement convoy. After

dark, 2 Japanese heavy cruisers, part of a cruiser force under Admiral Mikawa, entered Ironbottom Sound, unopposed, to shell Henderson Field. The following morning the air activity stepped up further, with the Americans concentrating on Tanaka's convoy and Mikawa's cruisers. Seven of Tanaka's transports and 2 of Mikawa's cruisers were sunk. Tanaka persevered, however, continuing toward Guadalcanal with his remaining transports, while Admiral Kondo, with the battleship *Kirishima,* 4 cruisers, and 9 destroyers, raced south to cover Tanaka's approach, intending to bombard Henderson Field.

1942, November 14–15. Naval Battle of Guadalcanal—Third Phase. Kinkaid sent Rear Admiral **Willis A. Lee,** with the battleships *Washington* and *South Dakota* and 4 destroyers, to intercept the Japanese. The opponents met at close range in Ironbottom Sound, just south of Savo Island. Two of the American destroyers were quickly sunk and the other 2 put out of action. The 2 battleships, however, closed boldly with the Japanese, who concentrated their fire on the *South Dakota,* causing an electrical power failure and putting her out of action for the rest of the engagement. Briefly it was a battle between the *Washington* and 14 Japanese warships. The odds were quickly lessened, however, as Lee calmly concentrated the *Washington*'s radar-directed guns against the *Kirishima,* leaving her and a destroyer in sinking condition. Kondo now withdrew. The naval Battle of Guadalcanal was ended. Though Tanaka had succeeded in putting 4,000 troops ashore, and rescued 5,000 more on his return to Rabaul, the main result of this 3-day battle was to return to the Americans effective control of the waters around Guadalcanal, thus assuring final victory in this crucial land and naval campaign. The Japanese went on the defensive in the southern Solomons.

1942, November 30. Battle of Tassafaronga. Tenacious Tanaka continued to prove, however, that the victory would not come cheaply. An American force of 5 cruisers and 7 destroyers under Rear Admiral **Carleton H.**

Wright moved to meet Tanaka's "Tokyo Express," this time 8 destroyers carrying supplies for the Japanese garrison of the island. Warned by radar of the Japanese approach into Ironbottom Sound, the Americans opened fire and sank the leading destroyer. But Tanaka promptly and skillfully turned his entire squadron toward the Americans, fired several torpedoes each, then scampered north without further loss. They left 1 American cruiser sunk and 3 more badly damaged, with a total of 400 dead sailors.

1943, January 29–30. Battle of Rennell's Island. An American squadron of 6 cruisers, 8 destroyers, and 2 escort carriers, under Rear Admiral **Robert C. Giffin,** was cruising north of Guadalcanal in hopes of enticing gathering Japanese naval forces in the northern Solomons (actually preparing to cover the evacuation of Guadalcanal) into battle. The Japanese did not take the bait; instead, Yamamoto ordered 2 very successful land-based air strikes against the Americans. The first, at dusk, January 20, disabled the *Chicago.* Late the next afternoon, while the Americans were withdrawing southward, the Japanese again hit the *Chicago,* while in tow, and sank her.

1943, February 1–7. Japanese Evacuation of Guadalcanal. Rear Admiral **Koyanagi** (who had replaced Tanaka) brilliantly and skillfully evacuated about 13,000 Japanese troops from Guadalcanal without being detected by the Americans.

COMMENT. *Not since the Anglo-Dutch wars had 2 powerful navies engaged in such a prolonged, intensive, and destructive naval campaign as that which took place for 6 long months around Guadalcanal. Serious mistakes were made on both sides, and the honors were approximately even in skill and gallantry displayed. The margin of the American success was their eventual superiority of numbers and of equipment, both on the surface and in the air.*

Operations in the Indian Ocean

1942, March 23. Seizure of the Andamans. This provided the Japanese with a secure convoy route from Singapore to Rangoon.

1942, March 25. Appearance of the First Air Fleet. Admiral Nagumo with 5 carriers (veterans of Pearl Harbor and Darwin) accompanied by 4 fast battleships and several cruisers and destroyers sailed westward into the Indian Ocean. His mission was to strike British naval bases in Ceylon, and to destroy all British naval and merchant shipping he could find in the ocean and the Bay of Bengal.

1942, March 27–April 2. British Reaction. Admiral Sir James Somerville's British Far Eastern Force consisted of 5 battleships—4 of them old and slow—3 small carriers, 8 cruisers, and 15 destroyers. Upon learning of Nagumo's approach, Somerville sailed southeast from Ceylon to meet him, but the fleets passed each other without contact. Realizing that he had missed his enemy, Somerville turned back to Addu Atoll, in the Maldives, to refuel. Here he came to the conclusion that it would be suicidal to attempt a formal battle against the superior and faster Japanese fleet; accordingly, he sent his 4 old battleships back to the coast of Africa, attempting to harass the Japanese with his remaining vessels while avoiding a major battle.

1942, April 2–8. Action off the Coasts of Ceylon and India. Japanese carrier aircraft struck the bases of Trincomalee and Colombo, causing damage to the installations, but being heavily engaged by RAF fighter planes. During a week of confused fighting, the Japanese carrier aircraft sank 1 British carrier, 2 cruisers, a destroyer, and several merchant ships. In their several engagements with the RAF and with British carrier aircraft, the Japanese inflicted substantial losses, but suffered heavily themselves as well. Finally, satisfied that he had carried out his mission, and feeling the necessity to rebuild his depleted air strength, the Japanese admiral sailed back toward the Pacific (April 8).

COMMENT. *In 4 months of war, the First Air Fleet had operated across one-third of the globe and had sunk 5 battleships, 1 carrier, 2 cruisers, 7 destroyers, and damaged many more, to say nothing of other destruction, without one of its own vessels having been damaged.*

1942, April–May. British Concern about Africa. Nagumo's demonstration that Japan could control the Indian Ocean at will caused great worry to Churchill and British military leaders regarding the security of the African coast and of the sea routes to India. Fearing that the Japanese might try to gain control of French Madagascar as they had of Indochina, the British decided they would forestall them.

1942, May 5–7. Assault on Diégo-Suarez. A British amphibious force under Rear Admiral **Edward N. Syfret** assaulted and captured the Vichy French port at the northern tip of Madagascar. They hoped no further action on Madagascar would be necessary.

1942, May 29–30. Japanese Activity off Madagascar. A Japanese seaplane (submarine launched, though this was not known at the time) was sighted over Diégo-Suarez (May 29).

The following night a midget submarine (also launched from a large submarine) damaged HM battleship *Ramillies* and sank a tanker in the harbor. The British feared that the Japanese were using secret bases elsewhere on Madagascar, and so decided to seize the entire island.

1942, September–November. British Conquest of Madagascar. Following amphibious landings at **Majunga** (September 10) and **Tamatave** (September 18), the British, under Lieutenant General Sir William Platt, moved against the capital, Tananarive. The French were forced to surrender (November 5). The British later turned the island over to the Free French (January 8, 1943).

THE WAR AGAINST JAPAN, 1943

Strategy and Policy

By the end of 1942 the Axis tide was clearly receding in Europe and North Africa, and was ebbing slightly in the Pacific. But the Japanese realized (as did the Allies) that this was merely a shifting from the offensive to the defensive, in accordance with a plan that had been set before they went to war. The setbacks in Papua and Guadalcanal had been unexpectedly costly, but this was attributed primarily to hasty, last-minute modifications in the original strategic plan, and the Japanese intended to make no such error again.

The defeat at Midway had been another matter, however, and the Japanese naval high command was apprehensive about the consequences of the carrier losses in that disastrous battle. These losses, furthermore, had been compounded by extremely heavy casualties among the remaining trained carrier pilots in the air battles around Guadalcanal.

Japanese military leaders hoped, however, that they could make good these losses by initiating an intensive new training program, while the Americans and their allies dashed themselves in costly assaults against the increasingly powerful jungle fortresses rimming the widespread Japanese defensive perimeter. They expected that the U.S. would become discouraged and would make peace, leaving Japan with most of her conquests.

There was no thought of a negotiated peace in America or Britain, however. While concentrating most of their efforts to achieve the defeat of Germany, the Allies were sufficiently encouraged by their limited successes in late 1942 to increase slightly the scale of effort against Japan. Among the Allied leaders, Admiral King was particularly determined to increase the momentum of operations in the Pacific, even though he did not directly question or subvert the primary effort against Germany. But America's naval strength was growing more rapidly than had been expected, both in relative and absolute terms. There were limits to the extent to which naval force could be used against Germany, and so most of this growing naval strength was allocated to the Pacific areas. Because it was obviously uneconomic for this tremendously burgeoning power to remain idle, it was essential to match naval strength, at least to a limited degree, with additional land and air strength—particularly since these categories of military force were also increasing rapidly in America.

Thus, there was at least a partial deviation from the original strategy of remaining on the defensive against Japan, while concentrating full Allied strength against Germany. The increasing American effort was also recognized by a few Japanese as a demonstration of the United States' phenomenal industrial and operational war-making capability.

Land and Air Operations in Asia

BURMA

General Wavell, commanding Allied forces in India, realized that he would not be ready for a major invasion of Burma until the dry season of late 1943 or early 1944. But he was worried about the psychological effect of the defeats suffered in Malaya and Singapore in 1942, and hoped during the dry season of 1942–1943 to revive the spirit of British troops by success in 2 small, carefully prepared offensives.

THE FIRST ARAKAN CAMPAIGN

1942, December–1943, January. Advance into Arakan. The 14th Indian Division* advanced into Arakan, northwest coastal province of Burma, separated from the remainder of Burma by jungled mountains. Moving cautiously, the British gave the outnumbered Japanese time to construct strong defenses north of Akyab.

1943, January–March. Stalemate at Akyab. With the assistance of reinforcements rushed by General Iida, the Japanese repulsed repeated British attacks with heavy losses.

March–May. Japanese Counteroffensive. The Japanese 55th Division counterattacked from Akyab, while other Japanese units worked their way over supposedly impassable mountains to strike the British left and rear (March 13–17). The 14th Division retreated in considerable confusion. Though reinforcements were rushed to the front, the British could not stop the Japanese infiltrating tactics. The campaign ended where it had started, with the Japanese still occupying Arakan (May 12).

COMMENT. *Instead of rebuilding confidence, the First Arakan Campaign had made British troops still more fearful of Japanese jungle-fighting ability.*

*Elements of the Indian Army included British and Indian units, but were operationally controlled as integral British Army units.

FIRST CHINDIT RAID

Brigadier **Orde C. Wingate** obtained the approval of General Wavell to demonstrate his concept of "long-range penetration" with his 77th Indian Bridgade. It was his belief that small British ground forces, supplied by air, could operate for extended periods deep behind enemy lines, cutting communications, destroying supplies, and creating general confusion. This seemed to him, and to Wavell, a way of beating the Japanese at their own game of infiltration and encirclement.

1943, February 18. Crossing the Chindwin River. Entering Japanese-controlled territory, the 77th Brigade (known later as "Chindits")* split up into several small columns and set out to cut the Mandalay-Myitkyina and Mandalay-Lashio railroads.

1943, March 18. Crossing the Irrawaddy River. Having temporarily interrupted the Mandalay-Myitkyina railroad, Chindit columns pushed on to cross the Irrawaddy. They had now aroused the Japanese, however, and encountered increasing opposition.

1943, April. Withdrawal of the Chindits. As losses mounted, Wingate gave up his plan to cut the Mandalay-Lashio railroad. The Chindits retreated with considerable difficulty back to India in small groups. Losses totaled more than 1,000 men, over one-third of the total force engaged.

COMMENT. *The Chindit raid was a military failure, the losses far too heavy to justify the slight damage inflicted, which was quickly repaired by the Japanese. However, the fact that British troops had successfully raided behind the Japanese lines, then fought their way out again, was hailed by the newspapers as a great Allied victory. Thus, psychologically, the raid went far to offset the discouraging results of the Arakan operation. The total effect of the 2 operations, however, was to convince Wavell and other British leaders that no major British invasion of Burma was possible prior to the dry season of 1944–1945.*

OPERATIONS IN NORTH BURMA

1943, February. Chinese Troops to the India-Burma Border. As Japanese spearheads pressed closer to the India-Burma border in the Hukawng Valley area of north Burma,

* The nickname had two origins: their emblem was a *chinthe,* a mythical Burmese beast resembling a lion; they operated beyond the Chindwin River.

Stilwell feared that they might attempt to raid into upper Assam (northeast India), where American, Chinese, and Indian engineers were beginning to push a road toward the border from Ledo. He therefore moved most of the rebuilt, retrained Chinese 38th Division from Ramgarh to protect the road-building operations. There were minor clashes of patrols in the mountains just inside Burma. The Japanese withdrew back into the Hukawng Valley.

1943, October—November. Return to Burma. Bitterly disappointed by British and Chinese failure to undertake a major invasion of Burma in 1943–1944 to reopen the land route to China (see above and p. 1264), Stilwell decided that he would initiate the effort with the resources at his disposal. He obtained the reluctant approval of Wavell and Chiang Kai-shek. The 38th Division pushed southeastward through the mountain spine of the India-Burma border into the Hukawng Valley. All supplies were dropped by air. The 22nd Division was moved up from Ramgarh to Ledo and the border area. Work on the road from Ledo intensified as more American engineers arrived, and dry weather permitted more rapid progress.

1943, November 23–December 23. Stalemate. As elements of the Chinese 38th Division advanced in the Hukawng Valley, they were struck by a Japanese counterattack. Three battalions were surrounded and the advance came to a complete halt, though the Japanese were unable to overrun the 3 isolated units, which were maintained by American air supply.

1943, December 24–31. Stilwell Resumes the Advance. The arrival of General Stilwell, and also of some light pack artillery, inspired the Chinese to counterattack, relieving the beleaguered battalions and clearing the valley west of the Tarung River.

CHINA

The growing weakness of isolated China was repeatedly demonstrated by easy Japanese successes in minor local offensives. These operations were planned by the Japanese to give experience to newly raised units, and also to seize rice crops in unoccupied China. By these "rice offensives" they were able to get food easily for themselves, while denying it to Chiang's starving people and soldiers. Toward the end of the year, however, with increased American air support (see below) the Chinese repulsed one of these Japanese rice offensives in Hunan by a victory at the **Battle of Changteh** (November 23–December 9).

Although the arrival of additional American transport planes added greatly to the quantity of supplies flowing to China over the Hump during the year, these were totally inadequate to meet the many conflicting needs of China and of the war effort. This led directly to 2 serious disputes involving American leaders in China.

Chennault's China Air Task Force (the only American unit in China; see p. 1244) was having success comparable to that which he had earlier gained with his Flying Tigers, inflicting losses of more than 10 aircraft on the Japanese for every 1 they lost. But because of the restrictions on supplies over the Hump, Chennault felt that he had not had an opportunity to prove the potential of his force, which he believed could win the war in China unaided. He insisted that he should receive all, or most, of the Hump supplies. Stilwell, however, ordered that a substantial proportion of the Hump tonnage should go to the Chinese Army, since it would need supplies to attack north Burma from the east while his troops in India attacked from the west. Chiang, already displeased with Stilwell, supported Chennault, who finally won the argument when he also obtained the support of President Roosevelt. He was promoted to major general, and his command was enlarged and redesignated the Fourteenth Air Force (March 11).

With his increased share of the increasing Hump tonnage, Chennault's growing Fourteenth Air Force was able during the remainder of the year to gain air superiority over most of China. His bombers ranged as far as Formosa, inflicting severe losses on the Japanese.

Because of the diversion of supplies from the Chinese Army to the Fourteenth Air Force, Chiang would not permit Chinese troops in China to participate in Stilwell's planned 2-prong offensive into north Burma. Only with reluctance would he give his assent to the advance of Chinese troops from India, since this would not interfere with Hump supplies to China.

Deteriorating relations between the National Government and the Communist regime at Yenan had, by this time, led Chiang to establish a *de facto* blockade of Communist-held regions of China. This had required the diversion of a substantial number of divisions from the war effort against Japan. Stilwell complained about this diversion, which further increased the tensions between the 2 men.

Operations in the South and Southwest Pacific

Japanese Defensive Plans. Realizing that their defensive perimeter was threatened by Allied successes on Guadalcanal and Papua, the Japanese began to strengthen Rabaul, the southeastern anchor of their defensive perimeter, with outposts on the Huon Gulf in northeast New Guinea and in the Solomons as far south as New Georgia Island. New Guinea, the Bismarcks, and the Solomons were under navy control, with Vice Admiral **Jinichi Kosaka** commanding. Under him was the 8th Area Army (Lieutenant General **Hitoshi Imamura**), with the Eighteenth Army (Lieutenant General **Hotaze Adachi**) in New Guinea and the Seventeenth (Lieutenant General **Iwao Matsuda**) in the Solomons. Because Allied forces in this region were the most immediate threat to the Japanese perimeter, Admiral Yamamoto, commander of the Combined Fleet, with headquarters at Truk, gave close personal supervision to the defensive measures (see p. 1258).

Allied Offensive Plans. With Rabaul the obvious primary Allied objective, Admiral Halsey's forces of the South Pacific Area, now designated the U.S. Third Fleet, were shifted from the command of Admiral Nimitz to that of General MacArthur for a coordinated Allied two-pronged offensive against Rabaul. Halsey's forces were to drive northwestward through the Solomons, while General **Walter Krueger**'s U.S. Sixth Army, under the direct supervision of General MacArthur, was to advance northward through northeast New Guinea and the island of New Britain toward the Japanese base. Complicating the command arrangements was the fact that Halsey, while under the strategic direction of MacArthur, was still dependent upon his own naval commander in chief, Admiral Nimitz, for ships, troops, aircraft, and the logistic means necessary to operate tactically. That the 2 operations worked smoothly was due to the spirit of cooperation and tolerance displayed by all concerned, with examples set by MacArthur and Halsey.

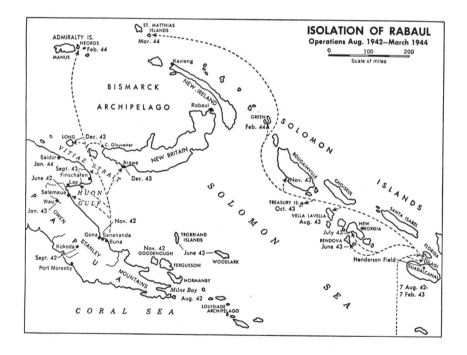

NEW GUINEA AND NEW BRITAIN, 1943

1943, January 9. Wau. A brigade of Australian General Hering's New Guinea Force, airlifted to a mountain airstrip at Wau, 30 miles west of Salamaua, established a forward base, threat-

ening that Japanese coastal holding while MacArthur built up strength for an offensive.

1943, June 30. Feint at Salamaua. An American amphibious landing at Nassau Bay, south of Salamaua, and demonstrations from Wau indicated a full-scale assault to come. At the same time, MacArthur dispatched elements of Krueger's Sixth Army to take and occupy the Trobriand-Woodlark island group in the Solomon Sea, north of Papua's eastern tip, securing new airfields.

1943, September 4–16. Lae and Salamaua. Australian General Sir **Thomas Blamey** took over command of the New Guinea Force. American and Australian troops made an amphibious landing east of Lae. Two days later, a U.S. paratroop regiment dropped at Nadzab in the Markham River Valley, to the northwest. The Australian 7th Division was immediately airlifted in to reinforce, and a converging approach began on Lae. Meanwhile, the Allied forces in front of Salamaua increased their activity. The Japanese garrisons began abandon-

ing both places. Salamaua was occupied (September 12) and Lae fell 4 days later.

1943, September 22–October 2. Finschhafen. A quick ground and amphibious advance around the tip of the Huon Peninsula encircled Finschhafen (September 22), which fell October 2. The way was paved for amphibious operations against New Britain.

COMMENT. *This campaign was a splendid example, on a small scale, of combined operations—land, sea and air, featuring deception and surprise. Lieutenant General* **George C. Kenney**'s *U.S. Fifth Air Force isolated the enemy from his strongholds at Madang, Wewak, and Rabaul, while also providing direct support to the ground forces. Of some 10,000 Japanese troops engaged, more than half had been killed, the rest dispersed in the jungle.*

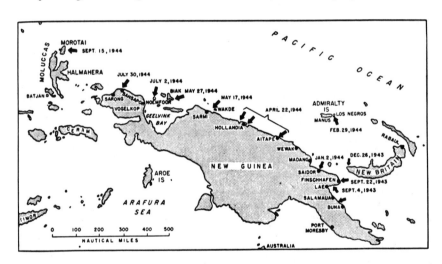

Operations on New Guinea

1943, October–December. New Britain Toehold. While Blamey's New Guinea Force blocked off the Huon Peninsula, protecting reorganization, Krueger's Sixth Army took over the next task. Elements of the U.S. 1st Cavalry Division* landed (December 15) at Arawe on southern New Britain, where a base was estab-

lished. The 1st Marine Division, put ashore (December 26) by the Seventh Amphibious Force, secured a successful beachhead and 2 airfields at Cape Gloucester on the northern side after 4 days of stiff fighting in which more than 1,000 Japanese were killed.

CENTRAL AND NORTHERN SOLOMONS

1943, February–June. Preparations. Following the conquest of Guadalcanal, there was a

---

* This division was actually an infantry formation, though retaining its prewar designations and organization.

lull in surface operations while Halsey reorganized his command and prepared to set in motion the eastern arm of MacArthur's pincers. Elements of the U.S. 43rd Division seized Russell Island (February 11).

1943, June 30. Capture of Rendova Island. Artillery was quickly landed in range of nearby New Georgia.

1943, July 2–August 25. Assault on New Georgia. Supported by guns from Rendova and by a naval task force, troops of the U.S. 37th and 43rd Divisions, reinforced by Marine battalions, landed near Munda. They were commanded by Major General **John H. Hester** (later by Major General **Oscar Griswold**). Since Munda, commanded by General **Noboru Sasaki,** was now the principal Japanese air base in the Solomons, the attacks were fiercely contested in the bitterest sort of jungle fighting. The U.S. 25th Division was also committed (July 25). Griswold, regrouping his forces, launched a coordinated assault on the airfield, which was captured (August 5). All resistance on New Georgia soon ended (August 25).

1943, August 15–October 7. Vella Lavella. A regimental combat team landed on Vella Lavella, leapfrogging Kolombangara, next Japanese airfield of importance, and an American advanced airbase was prepared. Elements of the New Zealand 3rd Division, replacing U.S. troops on Vella Lavella (mid-September), swept away the last remnants of the Japanese garrison. The Japanese began withdrawal from Kolombangara, and the Central Solomons operation came to an end (October 6–7). Allied losses (mostly American) in this campaign were 1,136 killed and 4,140 wounded. More than 2,500 Japanese dead were counted of the 8,000 engaged.

1943, October–December. Bougainville. South of Rabaul, the Japanese grip on the Solomons now included only Choiseul and Bougainville. A feint assault (October 27) on Choiseul by a small Marine expedition (quickly withdrawn) momentarily obscured intentions. That same day the Treasury Islands were seized as staging area for a move on Bougainville, and

from it (November 1) the 3rd Marine Division sprang ashore at Empress Augusta Bay against light resistance. By the year's end, Empress Augusta Bay had become an Allied naval base. Three airfields were in operation. All this was embedded in a defensive perimeter some 10 miles wide and 5 miles deep, now under General Griswold's Army control.

NAVAL OPERATIONS

OFF THE COAST OF NEW GUINEA

1943, January–June. The Seventh Fleet. The small American and Australian naval forces in MacArthur's Southwest Pacific Area were reorganized as the Seventh Fleet under the command of Vice Admiral **A. S. Carpender.** The principal missions of this fleet were to gain and keep control of the coastal waters of New Guinea, and to support MacArthur's ground troops in forthcoming amphibious operations. The actual conduct of the amphibious operations was a mission of the subordinate Seventh Amphibious Force (Rear Admiral **Daniel E. Barbey**), a collection of gunfire-support warships, troop transports, amphibious vessels, and landing craft, and including highly trained Army Engineers and Navy Seabees, whose task was to clear beach obstacles and to organize beachhead administration and supply.

1943, June–December. Amphibious Operations. The Seventh Fleet (commanded after August by Vice Admiral Thomas C. Kinkaid) and its Seventh Amphibious Force played key roles in the numerous amphibious operations along the coasts of New Guinea and New Britain (see pp. 1265–1266).

IN THE WATERS OF THE SOLOMONS

1943, February–June. Reorganization and Preparation. While waiting for reinforcements to permit him to continue operations northward from Guadalcanal, Admiral Halsey reorganized his South Pacific Forces into the Third Fleet. Initially the Third Amphibious

Force was commanded by Admiral Turner. Its organization and mission were similar to those of the Seventh Amphibious Force.

1943, June–December. Amphibious Operations. The Third Fleet and Third Amphibious Force (commanded by Rear Admiral **Theodore S. Wilkinson** after July) made leapfrog advances through the Solomons (see p. 1266). The Third Fleet was still opposed by substantial Japanese naval forces, leading to numerous surface actions, most of which (at Japanese initiative) took place at night.

1943, July 5–6. Battle of Kula Gulf. An American light cruiser squadron under Rear Admiral **W. L. Ainsworth** and a Japanese force of 10 destroyers under Rear Admiral **T. Akiyama** met. One U.S. cruiser was sunk; one Japanese destroyer was sunk and another ran aground (and was later destroyed). Despite U.S. superiority in gun power and armor, and the advantage of radar, superior Japanese night-fighting tactics and superior torpedoes gave them the best of this encounter.

1943, July 12–13. Battle of Kolombangara. In another night engagement, in almost the same waters, Japanese Rear Admiral **S. Izaki**'s squadron of 1 cruiser, 5 destroyers, and 4 destroyer transports met Ainsworth's force of 3 light cruisers and 10 destroyers. Again the Japanese inflicted more serious losses than they received: 3 Allied cruisers damaged and a destroyer sunk, while the Japanese lost their cruiser.

1943, August 6–7. Battle of Vella Gulf. An American squadron of 6 destroyers sank 3 out of 4 destroyers in a Japanese squadron without any loss to themselves.

1943, October 6–7. Battle of Vella Lavella. Rear Admiral **M. Ijuin,** with 9 destroyers, covering smaller craft evacuating Japanese troops from Vella Lavella, was encountered by 6 American destroyers under Captain **F. R. Walker.** One Japanese destroyer was sunk, as were 2 American vessels. With American reinforcements arriving, and with the troops evacuated from Vella Lavella, Ijuin withdrew. This was the last clear-cut Japanese naval success against the Americans.

1943, November 2. Battle of Empress Augusta Bay. This night engagement was the result of an effort by a Japanese squadron of 4 cruisers and 6 destroyers, under Rear Admiral **Sentaro Omori,** to interfere with the amphibious landings on Bougainville (see p. 1267). Rear Admiral **Stanton A. Merrill** with 4 cruisers and 8 destroyers was waiting. Making excellent use of their radar superiority, the Americans attacked Omori's force from front and flanks. The Japanese were assisted by flares dropped by supporting aircraft, but lost 1 cruiser and 1 destroyer, with most of the other vessels damaged. Only 1 American destroyer was badly damaged.

1943, November 5 and 11. Carrier Strikes on Rabaul. Planes of Rear Admiral **Frederick Sherman**'s task force—fleet carrier U.S.S. *Saratoga* and light carrier *Princeton*—severely mauled a powerful cruiser and destroyer force under Vice Admiral **Takeo Kurita,** staging from Truk through Rabaul for refueling in preparation for another blow at the American Bougainville beachhead (November 5). Kurita had 6 cruisers and destroyers badly damaged, and called off his planned attack. A few days later (November 11) another air strike was made by planes from Rear Admiral **A. E. Montgomery**'s task group of the carriers *Essex, Bunker Hill,* and *Independence.* Japanese planes intercepted the attackers, but were overwhelmed, and further severe damage was done to the port and to Kurita's ships, which soon thereafter limped back to Truk.

1943, November 25. Battle of Cape St. George. Captain **Arleigh Burke** with 5 destroyers southeast of New Ireland in early morning darkness, sinking 3 Japanese vessels without damage to his own squadron.

COMMENT. *Having learned from earlier costly experience, the Americans had now established a clear-cut tactical superiority over the Japanese in night naval operations. This, combined with their superiority in all types of equipment save torpedoes, meant that the Americans would never again encounter serious or dangerous opposition in small cruiser and destroyer actions, day or night.*

AIR OPERATIONS

NEW GUINEA AND THE BISMARCKS

1943, January–May. Struggle for Air Superiority. Lieutenant General **George C. Kenney**'s American-Australian Fifth Air Force was engaged during this period in a bitter struggle for air superiority with Japanese naval air units at Rabaul and with heavily reinforced air units of the Eighteenth Army. This resulted in a number of violent air battles and exchanges of strikes between the opposing airfields. By midspring the Allies had gained a clear-cut superiority.

1943, March 2–4. Battle of the Bismarck Sea. This was primarily an air-surface battle between Kenney's medium bombers and a Japanese squadron of 8 destroyers, escorting 8 troop transports with 7,000 soldiers bound from Rabaul to reinforce Lae. Despite persistent fog, which the Japanese had hoped would prevent effective air attack, the Allied planes came down close to the level of the sea to use a newly developed skip-bombing technique. The Allied planes sank 7 transports and 4 destroyers; the other transport was sunk by American PT boats. The 4 destroyers that escaped to Rabaul were badly damaged. The Allies lost 2 bombers and 3 fighters, the Japanese some 25 planes. More than 3,000 Japanese troops were drowned; about an equal number were rescued by the surviving destroyers and by Japanese submarines. The Japanese ceased further efforts to send merchant-ship convoys to New Guinea; all subsequent supply and reinforcement were done by fast destroyer transport.

1943, June–December. Air Support to Land Operations. During this period, Kenney's Fifth Air Force elaborated a pattern of support operations which it continued to follow in general throughout the remainder of the war. *First,* fighter planes and light bombers gained air superiority over the region where the next operation was planned to take place by attacking airfields and by endeavoring always to force Japanese planes to fight at a disadvantage. Simultaneously, longer-range bombers neutralized more distant Japanese air bases, sometimes in coordination with land-based or carrier aircraft from the South or Central Pacific areas. *Next,* Allied fighters and light bombers isolated the area to be attacked, making it impossible for Japanese troop transports or warships to land reinforcements. *Finally,* as MacArthur's troops advanced on land, or prepared for an airborne or amphibious operation, air attacks against ground targets in the area were intensified, these attacks continuing throughout the land battle. *In airborne operations,* troop-carrier aircraft flew paratroops to the objective, then shuttled back and forth to airlift supplies and reinforcements to the airborne troops. Meanwhile, engineers with the ground troops were engaged in improving existing airfields, or building new ones if necessary, so that fighter planes could move in to give more effective local air support. These fields also became the advanced bases for supporting the initial long-range air reconnaissance and air strikes for the next advance, the same general cycle of operations being repeated.

THE SOLOMONS

1943, January–April. Attrition of Japanese Air Strength. While ground operations in the Solomons temporarily came to a virtual halt after the American conquest of Guadalcanal, air activity continued intensively. The series of naval air losses that had begun at Midway, combined with growing American Army and Navy air strength in the Pacific, put the Japanese at a serious disadvantage. This was further accentuated by damaging Allied air blows from Henderson Field, from New Guinea, and from carrier raids against the complex of air bases around Rabaul.

1943, April 7–12. Japanese Air Counteroffensive. In an endeavor to regain air superiority and the initiative in the Solomons-Bismarcks area, Admiral Yamamoto sent most of the depleted carrier air squadrons to Rabaul from the fleet at Truk to take part in a proposedly devastating aerial offensive against Allied bases in Papua and the Solomons. The attacks, however, inflicted relatively minor damaged, while Japanese losses were heavy.

1943, April 18. End of Yamamoto. The Japanese naval commander in chief, visiting the area to inspect conditions personally, flew from Rabaul to Bougainville. Halsey, forewarned by the usual intercept of Japanese communications (the code had again been broken), sent 16 U.S. Army Air Force fighters from Henderson Field to meet him. The 2 bombers carrying the Japanese admiral and his staff were shot down in flames. The death of Yamamoto, Japan's foremost strategist, was tantamount to a major Japanese defeat. Yamamoto was replaced as Commander in Chief of the Combined Fleet by Admiral **Mineichi Koga.**

1943, May–December. Decline of the Japanese Naval Air Force. In a few months the carrier air groups which Yamamoto had sent to Rabaul were practically wiped out. Their replacements were not so well trained and, as pilot experience and ability declined, casualties mounted. Because of this vicious cycle, Japanese air losses became staggering. By the end of 1943, Japan had lost nearly 3,000 planes and pilots in the Solomons air struggle alone.

The Central Pacific

1943, January–October. American Naval Build-up. Men, ships, and planes in amazing numbers began to assemble in Hawaii, the Fijis, and the New Hebrides as Nimitz prepared for his coming westward offensive across the wide Pacific. By October he had the Fifth Fleet (Vice Admiral **Raymond A. Spruance**), consisting of 7 battleships, 7 heavy and 3 light cruisers, 8 carriers, and 34 destroyers—the largest fleet yet to be put in action by the U.S. He had the Fifth Amphibious Force (Turner), carrying more than 100,000 troops, and the V Amphibious Corps, commanded by Major General **Holland M. ("Howling Mad") Smith,** U.S.M.C. In addition to naval air, Nimitz had the Seventh Army Air Force (Major General **Willis A.**

Hale) operating from the Ellice Islands and (when its chore in the Solomons was completed; see p. 1269) the naval and Marine elements which had comprised Halsey's Third Fleet. Nimitz' initial target was the Gilbert Island group, whose seizure would provide advance air bases for the next step—capture of the Marshall Islands. In the Gilberts, Japanese defenses centered on Makin and Tarawa atolls—island clusters each ringed by a coral-reef barrier.

1943, November 13–20. Preassault Bombardment of the Gilberts. Tarawa and Makin were thoroughly strafed by Air Force bombers (November 13–17) and then subjected to heavy naval gunfire before the landing forces were committed.

1943, November 20–23. Makin. The 165th Infantry of the 27th Division, heavily reinforced, found little difficulty in crushing the island's 250 combat troops and several hundred workers, but the advance was too slow to suit General "Howling Mad" Smith. The delay necessitated subjecting the naval forces to counterattack by submarines rushed from Truk. The escort carrier *Liscome Bay* went down, with loss of 640 officers and men, though casualties suffered in the assault were but 66 killed and 152 wounded. Japanese casualties totaled some 500 killed; about 100 prisoners, nearly all of them Korean laborers, were taken.

1943, November 20–24. Tarawa. Betio Island, citadel of the atoll, in area only 300 acres, flat and sandy, had been honeycombed with underground shelters and some 400 concrete pillboxes, bunkers, and strong points. Defending artillery included 8-in. guns brought from Singapore. The island was ringed by a coral reef, its openings studded with submarine mines, and the beaches laced with barbed wire. Insufficient reconnaissance and faulty maps had not disclosed that the inner coral reef of Betio itself was too shallow for landing craft. Despite the heavy preliminary bombing and naval gunfire, 4,700 veteran combat troops under Rear Admiral **Keiji Shibasaki** emerged from underground shelters to man their defenses as the initial assault waves of the 2nd

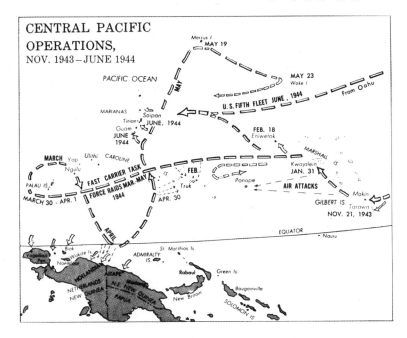

Marine Division approached. While some amphibious tractors reached the beach, most of the assault craft grounded on the inner reef and the assailants were forced to wade several hundreds of yards through crisscross fire, suffering shocking losses. Supporting naval fire ceased shortly after the initial wave hit the beach, lest it hit the Marines, who were thus left clinging to a few yards of terrain swept by artillery, machine-gun, and small-arms fire. Of 5,000 men put ashore by nightfall, some 1,500 were dead or wounded. That night tanks, guns, and ammunition got ashore. Next day the divisional reserve landed, losing 344 officers and men in the withering cross fire. Inch-by-inch frontal assaults—there could be no maneuver in that restricted area—supported now by air and by naval gunfire directed by observers ashore, gradually forced the defenders back to the eastern end of the island. Most of the surviving defenders lost their lives in a suicidal counterattack (night of November 22–23) and the last pocket of resistance was cleaned out next day.

Approximately 100 prisoners, only 17 of them combat soldiers, were taken. The remainder of the Japanese garrison were dead. Marine losses were 985 killed and 2,193 wounded. During these operations, Admiral Koga held his crippled fleet at Truk (see p. 1269), fearing the U.S. carrier force, but his land-based planes from Kwajalein twice unsuccessfully attacked warships of the Amphibious Force on Thanksgiving night (November 25).

COMMENT. *Tarawa, in ratio between casualties and troops engaged, ranks among the costliest battles in American military history. Its lessons brought about many improvements in amphibious assault techniques, correcting the manifest mistakes committed there: insufficient reconnaissance and insufficient artillery and air preparation fire, particularly the lack of the high-angle plunging fire necessary to obliterate well-constructed fortifications. Strategically, this epic of Marine Corps gallantry put Nimitz in position to combine land-based air and sea power in assaulting the Marshalls and eliminating or neutralizing Japan's great naval base at Truk.*

The North Pacific Area, 1943

The year opened with Japanese occupation forces on the 2 fortified western-most Aleutian Islands—Attu and Kiska—supported by land-based aviation from Paramushiro, off the southern tip of Kamchatka. U.S. efforts to dislodge the invaders stemmed from the naval and military base at Dutch Harbor on Umnak in the central Aleutians. Actually, operations in this fog-bound, stormy area had no effect on the war with Japan, other than irritating the U.S. But the North Pacific Theater was the scene of several sporadic clashes while the war was being decided in the momentous struggles of the Southwest and Central Pacific.

1943, March 26. Battle of the Komandorski Islands. Cruising in the mouth of the Bering Sea, Rear Admiral **Charles H. McMorris** with 2 elderly cruisers—1 light and 1 heavy—and 4 destroyers, met Japanese Vice Admiral Hosogaya's Northern Area Force of 2 heavy and 2 light cruisers and 4 destroyers, escorting reinforcements for Attu. In a long-range gun battle, U.S.S. *Salt Lake City* was crippled and the Japanese cruiser *Nachi* severely damaged. But as Hosogaya closed to smother the Americans, 3 of McMorris' destroyers charged in with a desperate torpedo attack. Hosogaya sheered away and then, fearing attack by American air from Dutch Harbor, turned and made for Paramushiro. (His timidity caused his relief from command.) Henceforth the Japanese relied upon submarines alone to resupply their Aleutian bases.

1943, May 11–29. Attu. Admiral Kinkaid's amphibious task force under Rear Admiral **Francis W. Rockwell** put the U.S. 7th Infantry Division ashore in heavy fog. For 18 days the Americans battled the garrison out of the usual Japanese prairie-dog complex of bunkers and underground shelters. Only 29 of the 2,500-man garrison were captured; the rest died fighting or killed themselves. U.S. losses were 561 killed and 1,136 wounded of the 12,000 men engaged, a costly price for the taking of an island written off by the Japanese high command.

1943, August 15. Kiska. A Canadian-American amphibious landing on Kiska (August 15) found it deserted; its 4,500-man garrison had been skillfully evacuated (July 29) in fog and darkness.

THE WAR AGAINST JAPAN, 1944

Strategy and Policy. In the Pacific, as across the Atlantic, the Axis sun was beginning to set as 1944 opened. U.S. resources—in men and machines—were increasing in strength and infinite technological variety. Japan was beginning to feel the pinch of U.S. submarine operations against her merchant marine, but had no antidote. Nearly 300 of her merchantmen were sunk in 1943 for a total of more than 1,300,000 tons, some 700,000 tons more than Nippon's replacement capability during the same period. U.S. submarine activities would be stepped up in the coming year to drain Japanese tanker strength, still greater than American. But, unwilling to admit overextension, Japan was still a formidable enemy. Her forces continued to strike with much tactical skill and tremendous—though sometimes misdirected—energy on all fronts.

Land and Air Operations in Asia. Both sides girded for offensive moves. Japanese objectives were to conquer all southeast China, eradicating the galling probes of Chennault's Fourteenth Air Force, and to invade India from Burma. The Allied objective was to remove the threat to India by driving the invaders out of Burma and, by opening the land routes of supply to China, to keep that nation in the war. General Stilwell, whose differences with Chiang Kai-shek were rapidly increasing in bitterness, hoped to be able to maintain a successful defense in China while mounting a vigorous offensive in Burma in joint operation with British General Slim's Fourteenth Army.

Burma

Allied Situation and Plans. In October 1943, the Southeast Asia Command, with Vice Admiral Lord **Louis Mountbatten** as Supreme Allied Commander, became operational, its headquarters at New Delhi (later Kandy, in Ceylon). Mountbatten assumed responsibility for all Allied operations in Burma. General Stilwell was appointed as deputy to Mountbatten, but since he still retained his other duties, spent little time at Kandy. All Allied ground forces operating against Burma from India were grouped under General Slim, including Stilwell's Northern Combat Area Command (2 Chinese divisions), actively engaged in combat in the Hukawng Valley of northern Burma. (Thus Stilwell, who was over Slim as deputy SACSEA, was under him as CG of NCAC, and at the same time was in a purely American chain of command as CG of the China-Burma-India Theater, and also was Chief of Staff to Chiang in the China Theater.)

Mountbatten and Slim felt that British forces in India would not be ready for a major invasion of Burma until the dry season of 1944–1945. They agreed, however, to assist Stilwell's NCAC by minor activities along the entire India-Burma border, as well as by 2 limited thrusts into Burma itself; one of these to be in Arakan, the other an augmented Chindit long-range-penetration effort in north-central Burma under General Wingate. (This raid had been specifically approved by Churchill and the CCS at Quebec.) Stilwell, with 3 (to be increased to 5) Chinese divisions, plus a small American combat unit, hoped to take the major north Burma city of Myitkyina, with its airfields, before the beginning of the 1944 monsoon.

Japanese Situation and Plans. The Japanese also had reorganized their forces in Burma. The new commander of the Burma Area Army was Lieutenant General **Shozo Kawabe,** who was under the overall command of Field Marshal **Count Terauchi,** with headquarters in Saigon, responsible for military operations in the entire Southern Resources Area. Kawabe had 6 divisions, 2 in southwest Burma, under his direct supervision; the other 4, to the north, were under Lieutenant General **Renya Mutaguchi**'s Fifteenth Army.

Kawabe had directed Mutaguchi to prepare for an offensive across the mountains into eastern India with 3 of his divisions, which, with attached units, totaled nearly 100,000 veteran combat troops. The objective was twofold: first, to seize the Imphal-Kohima plain of Manipur, the logical assembly area and base for any Allied invasion of central Burma from India; second, to cut the railroad line into Assam, which passed through Manipur, and which carried almost all of the supplies ferried to China over the Hump, as well as the supplies for Stilwell's NCAC divisions in north Burma.

THE SECOND ARAKAN CAMPAIGN

1943, December–1944, January. Allied Advance into Arakan. Three divisions of the British XV Corps, advancing toward Akyab, were soon halted by the Japanese 55th Division, which had fortified a mountain spur extending westward to the sea near Maungdaw, blocking the only possible overland route to Akyab. For nearly 2 months the British vainly hammered at this defensive position. Kawabe sent the 54th Division into Arakan as reinforcement.

1944, February 4–12. Japanese Counterattack. Using the same tactics which had defeated the British in the same area a year earlier, the Japanese 55th Division counterattacked, while elements of the 54th circled through the jungles to the east, crossing the mountains behind the British flank and cutting the lines of communication of both British front-line divisions, isolating them from each other and from many of their smaller formations. General Slim, refusing to permit any withdrawal, rushed reinforcements and initiated an emergency air supply to the beleaguered units.

1944, February 13–25. British Counterattack. The Japanese encircling forces now found themselves encircled by determined British and Indian units. The 2 front-line British divisions reestablished contact (February 24) and increased pressure on the trapped Japanese, most of whom were wiped out.

1944, March–April. Cracking the Maungdaw Position. In extremely bitter fighting, the British XV Corps fought its way gradually through the Maungdaw position. Having broken through, the corps was about to continue its advance on Akyab, when it was forced to halt and to send reinforcements to the Imphal front (see p. 1277).

1944, May–December. Monsoon Stalemate. No further activities of importance took place on this front until late in the year at the beginning of the 1944–1945 dry season (see p. 1294).

North Burma

HUKAWNG VALLEY OPERATIONS,
JANUARY–APRIL

1944, January–February. Stalemate in the Hukawng Valley. Effective resistance by the Japanese 18th Division, commanded by Major General **Shinichi Tanaka,** brought the advances of the Chinese 38th and 22nd divisions to a virtual standstill. Stilwell's return (late February) brought some increased Chinese activity

and a renewal of the advance against determined resistance. At this time there appeared a new American provisional infantry regiment, known to history as "Merrill's Marauder," from the name of its commander, Brigadier General **Frank D. Merrill.**

1944, March 3–7. Battle of Maingkwan-Walawbum. A Chinese frontal attack, combined with an envelopment by Merrill's Marauders, smashed the center of the Japanese 18th Division. Despite heavy losses, Tanaka skillfully extricated his division from near encirclement. The Allies, pressing close behind, were stopped at a new Japanese defensive position established along a jungle ridge separating the Hukawng and Mogaung valleys.

1944, March 28–April 1. Battle of Shaduzup. One battalion of Merrill's regiment and a Chinese regimental task force circled deep behind the Japanese lines to take up a blocking position behind the 18th Division at Shaduzup. Two of Tanaka's regiments were trapped, but fought their way out through obscure jungle trails after suffering severe losses and abandoning much equipment and ammunition. Tanaka, nevertheless, counterattacked and briefly isolated another of Merrill's battalions in the mountains southwest of Shaduzup (March 29–April 8).

THE SECOND CHINDIT EXPEDITION

1944, January–March. Preparations. For his renewed long-range penetration effort, Wingate was given 6 infantry brigades (only 5 actually participated) organized as the 3rd Indian Division, but more commonly called "Special Forces," or "Chindits" (see p. 1262). A total of 20 British, Indian, and West African infantry battalions, each divided into 2 "columns," were to be flown deep into north-central Burma, while the sixth was to march across the mountains, past Stilwell's left flank. Operating separately, but in accordance with a coordinated plan, the columns and brigades were to cut the Mandalay-Myitkyina railroad, and generally to disrupt the rear areas of Japanese forces facing Stilwell's NCAC and Slim's Fourteenth Army. The Chindits were to receive combat air support and air logistical support from an American air group commanded by Colonel **Philip C. Cochrane.**

1944, March 5–11. The Chindits Return to Burma. Transported initially in gliders, then in transport planes as a jungle air strip was hastily constructed, 3 brigades of Chindits were flown to preselected, isolated jungle spots dubbed "Broadway" and "Chowringhee." While a few columns spread out over Japanese rear areas to create confusion and to destroy supplies, the main body moved to Mawlu, where a strong defensive position was established, blocking the railroad line (March 16).

1944, March 17–May 6. White City. The blocking position at Mawlu was named "White City" (after a London amusement park) because it was soon strewn with supply parachutes dropped from Cochrane's supporting transport planes. The Japanese did not interrupt their major offensive into India because of this threat to their rear, but a number of determined local attacks were made; all were repulsed. Meanwhile, other Chindit columns continued to harass Japanese rear areas.

1944, March 25. Death of Wingate. The Chindit commander was killed in an airplane crash against a jungle mountain. He was replaced by Major General **W. D. A. Lentaigne.**

1944, May 6–25. Blackpool. As newly arrived Japanese reinforcements began to concentrate against White City, Lentaigne decided to abandon the base. Since he had been instructed to join Stilwell's command, a new base, "Blackpool," was established near Hopin, about midway on the rail line between Mawlu and Mogaung. Blackpool was soon attacked by strong Japanese forces, and a violent battle raged. The hard-pressed Chindits, close to exhaustion and having suffered heavy casualties, withdrew again, this time to the relative safety of the mountains farther west.

1944, June–July. Operations with NCAC. The exhausted and battered Chindits gathered in the vicinity of Indawgyi Lake, where British flying boats landed and began to ferry sick and wounded back to India. Meanwhile, the 2 bri-

gades that were least hard hit continued to operate with Stilwell's forces. One of these helped capture Mogaung (see p. 1276), and the other for a while protected the NCAC flank (see p. 1276). Relations between Lentaigne and Stilwell were not good; Stilwell thought the Chindits could have fought more vigorously; Lentaigne felt Stilwell was asking too much in the light of the ordeals they had suffered; both were probably right.

COMMENT. *This was a gallant but largely wasted effort. Although it had somewhat more success than Wingate's first expedition, the same results could have been obtained by smaller commando or guerrilla forces, assisted by air bombardment, without ruining the fighting effectiveness of 5 fine British brigades.*

MYITKYINA-MOGAUNG OPERATIONS,
APRIL–AUGUST

1944, April 28–May 17. Advance on Myitkyina. After a brief pause because of the Japanese threat to his lines of communications through Assam (see p. 1277), Stilwell decided that the British would repulse the Japanese offensive, and continued his efforts to reach Myitkyina before the beginning of the monsoon rains. Merrill's Marauders (now commanded by Colonel **C. N. Hunter,** due to Merrill's protracted illness), reduced by casualties and disease to 1,400 men, were sent with 2 Chinese regiments in a secret march over the high, extremely rugged ridge between the upper Mogaung and Irrawaddy valleys to take Myitkyina.

1944, May 17–18. Assault on Myitkyina. The American-Chinese attack quickly gained control of Myitkyina's 1 operational airfield, and supplies and reinforcements were rushed there by air while it was still under direct Japanese small-arms fire. Efforts of the exhausted and disease-ridden troops to seize the city were repulsed, however, and the Japanese, with reinforcements arriving from the east bank of the Irrawaddy, also repulsed the fresh air-transported Chinese reinforcements.

1944, May–June. Advance down the Mogaung Valley. Meanwhile, Stilwell's veteran

Chinese 22nd and 38th Divisions resumed a slow advance down the Mogaung Valley against Tanaka's continued skillful resistance. The 22nd Division (with reinforcements from China) captured **Kamaing** in an old-fashioned bayonet assault (June 16); the 38th, with assistance from a Chindit column (see p. 1275), soon afterward seized **Mogaung** (June 26). Both divisions, which (thanks to Stilwell's inspiration) had fought well and vigorously despite heavy casualties, were now given a rest as monsoon rains made further operations difficult.

1944, May 18–August 3. Siege of Myitkyina. Despite the monsoon, the Allies continued to press against the city, which was never completely invested, the Japanese retaining a line of communications across the Irrawaddy. Coordination between Chinese and Americans was poor; the efforts of commanders, staffs, and troops were frequently inept. The Japanese, forced to yield ground inch by inch, exacted a heavy price in Allied casualties. Finally, satisfied that they had imposed sufficient delay on the frustrated Allies, 700 semistarving Japanese survivors crossed the river and made their way south through the jungles to rejoin their main force, just as a final Allied assault, without opposition, swept over the city. Japanese losses were nearly 3,000 dead; the Allies had more than 5,000 casualties.

COMMENT. *The military performance of the Japanese, facing great odds, under constant air and artillery bombardment, was magnificent. But, despite relatively poor performance by his subordinates and troops, Stilwell had successfully completed a campaign which the British, and many Americans, had deemed impossible.*

1944, July–August. The "Railroad Corridor" Campaign. The 3rd West African Brigade of Chindits, now under Stilwell's command to protect his right flank against possible counterattack up the Mandalay-Mogaung railway line, was ordered to advance down the railroad from Mogaung to seize Pinbaw. The Africans were immediately stopped by Japanese defensive positions on the railroad south of Mogaung. The British 36th Division, under Major General **Francis W. Festing,** was trans-

ferred (July 7) to Stilwell's command to replace the now exhausted Chindits. Airlifted to Myitkyina, thence by foot to Mogaung, Festing's troops fought down the long, sheltered corridor against determined Japanese resistance, during the height of the monsoon season, to take **Pinbaw,** their objective (August 28). Aggressive Festing, obtaining Stilwell's permission to patrol southward, then struck 50 miles farther down the corridor, driving the Japanese out of a series of hastily prepared delaying positions.

COMMENT. *This 100-mile-long campaign was the first instance in modern military history of a large-scale offensive in Southeast Asia during the rain and mud of the monsoon season.*

STILWELL'S FINAL OFFENSIVE, SEPTEMBER–DECEMBER

1944, September–October. Pause and Preparation. Stilwell now had 5 Chinese divisions (3 good, 2 mediocre), the excellent British 36th Division, and the newly organized U.S. "Mars" Brigade (2 U.S. and 1 Chinese regiments and 2 light artillery battalions). He was opposed by the Japanese Thirty-third Army (Lieutenant General **Masaki Honda**) of 3 depleted divisions. As soon as the rains let up, and before the ground had time to fully dry, Stilwell planned a surprise offensive, featured by a sweeping envelopment in which 3 of his Chinese divisions and the American brigade would encircle the entire Japanese Thirty-third Army, which would then be trapped between his army in Burma and the Y-Force in Yunnan (see p. 1281).

1944, October 15. The Offensive Begins. With the 38th and the 30th divisions advancing from Myitkyina toward Bhamo against the main elements of the Thirty-third Army, and the 36th Division continuing down the Railroad Corridor to protect the NCAC right flank and rear, Stilwell's mass advanced southward from the Mogaung area to cross the Irrawaddy River near Shwegu, thence southeastward through the jungle with the objective of reaching the Burma Road near Lashio.

1944, October 15–December 31. Railroad Corridor (continued). The British 36th Division continued its advance, moving slowly but steadily down the railroad, despite intensifying Japanese resistance. By the end of the year, the division had seized the towns of Indaw and Katha, and halted temporarily because of developments farther east.

1944, October 18. Relief of General Stilwell. (See p. 1281.) The commander of the newly established India-Burma Theater was Lieutenant General **Dan I. Sultan.**

1944, November 6. Crossing the Irrawaddy. The Chinese 22nd Division began the crossing of the Irrawaddy as spearhead of the enveloping hammer blow.

1944, November 14–December 15. Siege of Bhamo. The Chinese 38th Division was held up for a month by the stubborn and determined defense of Bhamo. When further resistance was impossible, 800 surviving Japanese fought their way out at night to rejoin the 56th Division in the mountains between Bhamo and Namkham. The 30th and 38th Divisions pushed slowly after them.

1944, November 30–December 31. Change of Plans in North Burma. Due to disasters caused by a Japanese offensive in China (see p. 1281), Chiang Kai-shek decided to withdraw 2 divisions from Burma to help stop the dangerous Japanese drives in south China. The 22nd and 14th Divisions were flown out of Burma to Kunming in American planes from hastily constructed front-line airfields in the jungle region southeast of the Irrawaddy bend (December 5–10). Sultan was thereupon forced to modify Stilwell's plan of encircling the Japanese Thirty-third Army. The reduced mission was simply to open and secure the road that stretched east from Bhamo through Namkham toward the Burma Road.

Central Burma

THE JAPANESE INVASION OF INDIA, MARCH–JULY

1944, March 6. Crossing the Chindwin. Mutaguchi's Fifteenth Army began its projected offensive toward India by crossing the Chind-

win River on a broad front. One division headed for Kohima, 2 for Imphal.

1944, March 7–April 5. The Advance to Imphal and Kohima. Although the British had been expecting the offensive, they had underestimated the size of the Fifteenth Army; they were amazed by the speed and power of its advance. British outposts holding the Chin Hills around Tiddim and Fort White were cut off by the Japanese 33rd Division, but succeeded in breaking their way through Japanese road blocks to reach Imphal, just before the arrival outside that city of the Japanese 15th Division, which unexpectedly was approaching over rugged mountain trails from the east (April 5). The Japanese 31st Division had begun to invest Kohima the previous day. The British IV Corps, of 3 divisions, was now almost completely isolated, the bulk of the corps in and around Imphal, with a small garrison holding Kohima.

1944, April 5–20. The Sieges of Imphal and Kohima. While hastily assembled transport planes began an airlift to maintain some 50,000 men in the IV Corps and the 40,000 civilian inhabitants of Imphal and Kohima, General Slim assembled his XXXIII Corps on the railroad at Dimapur. Pushing back Japanese patrols, this corps began a drive to relieve the dangerously pressed garrison of Kohima. At the same time, Slim also began to fly in reinforcements to the IV Corps at Imphal. Bitter fighting flared continuously around both perimeters, and several times Kohima was close to collapse, the margin being air support from American and British fighter planes and medium bombers, which harassed the Japanese mercilessly. The situation became less precarious, however, when the XXXIII Corps broke through to relieve Kohima (April 20).

1944, April 20–June 22. The Siege of Imphal (continued). Further progress by the XXXIII Corps toward Imphal was painfully slow as the Japanese dug in and held with typical tenacity. Slim flew additional units in to Imphal from Arakan until the IV Corps strength rose to more than 100,000 men. Amazingly, the Japanese held back violent assaults against their lines by both British corps. They had failed, however, to capture the supplies on which they had been counting. The beginning of the monsoon, too, made their situation still more difficult. Because of hunger and disease, their fighting strength finally began to crumble, and the IV and XXXIII Corps were able to hack their way through the last remaining roadblocks (June 22) after a siege of 88 days.

1944, July–September. Collapse of the Fifteenth Army. The Japanese had not foreseen the possibility that the British would refuse to retreat and abandon their supplies. Now that

their logistical shoestring was broken, and with the monsoons making large-scale supply operations impossible in the mountain jungles, they had no choice but to retreat. Slowly and stubbornly, they fell back to the Chindwin Valley, harassed from the air and by pursuing British troops; amazingly, they never lost cohesion or combat effectiveness. The army had been virtually ruined, however, by a combination of battle casualties, malaria, and starvation. They lost 65,000 dead, less than half of whom were actual battle casualties.

BRITISH ADVANCE INTO CENTRAL BURMA, SEPTEMBER–DECEMBER

Allied Plans and Preparations. Pursuit of the Japanese Fifteenth Army had brought Slim's Fourteenth Army again to the edge of the Chindwin Valley. He now prepared for a broad-front crossing of the Chindwin, to take place in November. Over Slim was Lieutenant General Sir Oliver Leese, commanding the newly established headquarters of Allied Land Forces, Southeast Asia (ALFSEA). Under Leese, in addition to Slim, were Stilwell (later Sultan) and his NCAC, and the British XV Corps in Arakan.

Japanese Plans and Preparations. Because of the disaster to the Fifteenth Army, General **Hoyotaro Kimura** had replaced Kawabe in command of the Burma Area Army. He received reinforcements and he devoted the summer to reorganization, and in particular to the rehabilitation of the Fifteenth Army (now under Lieutenant General **Shihachi Katamura**) and the battered 18th Division. By fall his forces of 250,000 men were reorganized into 3 armies: the Thirty-third (3 divisions), holding northeastern Burma; the Twenty-eighth (3 divisions, under Lieutenant General **Seizo Sakurai**), responsible for the coast and Arakan; the Fifteenth (4 divisions), holding the west along the Chindwin. Kimura's strategy was to permit the Allies to reach central Burma, where their logistical difficulties would become increasingly acute, while those of the Japanese would be simplified in the proximity of their bases. Kimura was confident that his 10 divisions could smash the Allies in such circumstances. Accordingly, his orders were to harass and delay the Allied advance, but to avoid a finish fight until the British had been lured across the Irrawaddy near Mandalay.

1944, November 19–December 3. The British Cross the Chindwin. The IV Corps began the crossing at Sittaung; the XXXIII Corps followed soon afterward at Kalewa and Mawlaik. The Japanese fought delaying actions, but in accordance with plan did not attempt a firm defense.

1944, December 14. Linking of NCAC and Fourteenth Army. Patrols of the 19th Indian Division met those of Festing's 36th Division near Indaw. By the end of the year, the Four-

teenth Army was approaching the Irrawaddy on a broad front; the stage was set for a climactic struggle long foreseen by both Slim and Kimura.

AIR OPERATIONS OVER BURMA, 1943–1944

AIR COMBAT

1943, January–March. Japanese Raids into India. With unchallenged air superiority over

Burma, the Japanese were able to continue long-range bombardment attacks against Calcutta and to undertake a number of raids against the Hump air bases in Assam, in northeastern India. These raids were not very effective and did not do much damage, but they harassed the Allies considerably.

1943, March–June. Arrival of Allied Air Reinforcements. As British and American combat air strength built up in India, the Japanese were forced to abandon their raids across the mountains. The Allies then began to carry the war into Burma, and soon gained air superiority over much of the country.

1943, June–October. Monsoon Lull. Air operations were greatly curtailed by the rainy weather. The Japanese took advantage of the lull to repair damage caused by the Allied raids, to improve their antiaircraft defenses, and to build up their overall air strength in Burma.

1943, November–1944, May. Struggle for Air Superiority. With the advent of good weather, the Japanese were prepared to challenge Allied air superiority. They began both night and day raids against Calcutta and against the Hump air bases, while their fighters vigorously struck back against Allied air intrusions into Burma. Severe losses were suffered on the ground and in the air by both sides. Slowly, however, the greater numbers and greater skill of the Allied air forces began to assert themselves. By mid-1944, General **George E. Stratemeyer**'s Eastern Air Command completely dominated the skies over Burma; this superiority was never to be relinquished.

LOGISTICAL AIR SUPPORT

General Kimura's strategy for defending Burma in 1944 failed to appreciate Allied air capabilities. The Japanese invasion of India had failed largely because they could not keep up long supply lines through the jungle under the pressure of constant air attack. But the Allies did not depend upon such supply lines to support their troops, or to keep up the advance in the jungle.

Hundreds of Allied transport planes brought food, ammunition, and all manner of supplies directly to the front-line troops. If there were no nearby airfields where they could land, the airmen dropped these supplies into rice-paddy fields or jungle clearings. Anything that might break was dropped by parachute; everything else was free-dropped.

Thus the Allies' only supply line came through the air, which they controlled completely. And, having driven Japanese combat planes from the skies, the Allies had no worries about air strikes against their bases in India.

COMBAT AIR SUPPORT

The difficulties of surface transportation in jungle areas meant that ground troops had less artillery support than normal, while the potentialities for defense in the jungle increased their need for it. The Chindits, in particular, needed such support, for they had no artillery. This deficiency was made up, at least in part, by the extensive direct support which British and American fighter-bombers were able to provide, since they had no need to engage the nonexistent Japanese air force.

Most of the supporting missions were flown by fighter-bombers that dive-bombed strong Japanese positions, then strafed them just before ground attacks. In some instances, where Japanese defenses were particularly strong, light and medium bombers were used to support ground attacks.

China

1944–1945. Japanese-Chinese Communist De Facto Truce. The Japanese ended their punitive raids against the Communists. For the last 2 years of the war there was practically no activity on Chinese Communist fronts. Whether this truce was tacit or negotiated is not known. Both sides benefited: The Communists consolidated their control of northwest China. The Japanese diverted large forces for operations in the south.

1944, January–May. Japanese Plans and Preparations. The punishment they were receiving from Chennault's Fourteenth Air Force led the Japanese to decide to make an all-out effort to capture the American airfields by ground offensives. During the early months of the year, General **Yasuji Okamura**'s China Expeditionary Army, 820,000 strong, undertook preliminary operations to improve its railroad supply lines from northeast China.

1944, January–May. Allied Plans and Preparations. Chiang Kai-shek reconsidered his earlier veto on an advance against northeast Burma from Yunnan in the light of the successes being gained in northwest Burma by Stilwell's Chinese-American forces. He approved an offensive down the Burma Road by a small army group of 2 armies (called the "Y-Force"), consisting of 72,000 men commanded by Marshal **Wei Li-huang.** Because of shortages of equipment and weapons, this force was probably not much more powerful than the Japanese 56th Division, about 15,000 men, which held the portion of Yunnan west of the

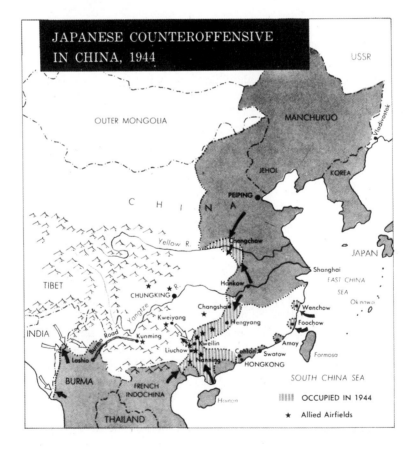

JAPANESE COUNTEROFFENSIVE IN CHINA, 1944

OUTER MONGOLIA

USSR

MANCHUKUO

Vladivostok

JEHOL

KOREA

C H I N A

PEIPING

Yellow R.

Chengchow

JAPAN

Shanghai

EAST CHINA SEA

TIBET

Hankow

CHUNGKING

Changsha

Okinawa

Yangtze

Kweiyang

Hengyang

Wenchow

Foochow

INDIA

Kunming

Kweilin

Amoy

Burma Road

Liuchow

Nanning

Canton

Swatow

Formosa

Lashio

HONGKONG

BURMA

FRENCH INDOCHINA

SOUTH CHINA SEA

Hainan

||||| OCCUPIED IN 1944

THAILAND

★ Allied Airfields

Salween River, its major elements holding the walled cities of Tenchung and Lun-ling (the latter on the Burma Road itself).

1944, May 7–November 30. Japanese East China Offensive. Under Okamura's direct supervision, the Japanese Eleventh Army, 250,000 strong, initiated a south-westward drive from Hankow on Changsha. The Japanese Twenty-third Army, 50,000 strong, that same day thrust west from the Canton area. Chinese resistance was spotty. Ch'ang-sha, abandoned, was occupied (June 19). First stiff Chinese opposition was at **Hengyang,** which fell only after an 11-day siege (July 28–August 9). Chinese resistance began to collapse. Methodically, despite the fierce aerial opposition of Chennault's flyers, 7 of the U.S. Fourteenth Air Force's 12 airfields were captured, and the Japanese movement then turned westward (November 15) to threaten Kunming and Chungking.

1944, May 11–September 30. Operations in Yunnan. The Chinese Y-Force advanced across the Salween in 2 major columns, driving back Japanese outposts. The southern column surrounded **Lun-ling** less than a month after crossing the Salween. A Japanese counterattack drove this column back (June 16). Rallying, the Chinese finally halted the smaller pursuing force and slowly reestablished a partial blockade of the city. The northern column was more aggressive. Closely investing **Tenchung** (early July), it penetrated the city walls through a breach made by supporting fighter-bombers of the American Fourteenth Air Force. After a bitter house-to-house battle, the Chinese annihilated the defenders of Ten-chung (September 15). The Japanese thereupon mounted another counterattack at Lun-ling, driving the Y-Force's southern column almost back to the Salween before events in Burma forced them to abandon their pursuit.

1944, October 18. Relief of Stilwell. As Chinese resistance crumbled in the face of the Japanese East China drives, Stilwell vainly recommended to Chiang Kai-shek various measures to reconstitute an effective defense. The American government became alarmed lest China collapse entirely, and President Roosevelt suggested that Chiang grant full command authority over all Chinese forces to Stilwell (October 17). Chiang flatly refused, and in turn demanded the recall of Stilwell. Reluctantly, Roosevelt ordered Stilwell back to the U.S. The China-Burma-India Theater was dissolved. Major General **Albert C. Wedemeyer** replaced Stilwell as commander of a new China Theater and as chief of staff to Chiang.

1944, December. Halting the Japanese Drives. Wedemeyer's tact succeeded where Stilwell's bluntness had failed; he persuaded Chiang to acquiesce in the transfer of only 2 veteran Chinese divisions from the Burma front (see p. 1277); Chiang had demanded the return of all 5. Airlifted, these 2 divisions became the backbone of a revitalized defense. The Japanese advance was blunted by Wedemeyer's reorganized and reinforced Chinese troops, supported by Chennault's air force. Their counterattack east of Kweiyang (December 10) stabilized the situation.

COMMENT. *The Japanese ground offensive disproved the fallacy—vainly disputed by Stilwell when Chennault obtained approval for a Hump-supported air build-up at the cost of Chinese ground troops (see p. 1264)—that air power, unsupported by ground power, could stop Japan in China. The error was costly, since to correct it necessitated abandonment of a full-scale offensive in Burma (see p. 1277) and—by elimination of Chennault's bases—for a time deprived both Nimitz and MacArthur of expected assistance in their respective offensives in the Pacific.*

Land, Naval, and Air Operations in the South and Southwest Pacific Areas, January–August

At the outset of the year, the objective of the combined forces of MacArthur's Southwest Pacific Area and Halsey's Third Fleet was still to isolate, and eventually to assault, Rabaul.

1944, January–May. Consolidation in the Solomons. Halsey, still under MacArthur's operational command, was still under Nimitz administratively. In addition to isolating and neutralizing enemy bases in the overall campaign against Rabaul (see below), Halsey's forces were engaged in hard fighting to contain the Japanese Seventeenth Army in the Solomons, and particularly on Bougainville, where the bulk of that army attempted repeated, uncoordinated assaults against the small Allied perimeter at Empress Augusta Bay. By May, however, when starvation and despair had worked sufficiently on the isolated Japanese forces, the threat declined.

1944, February 15. Green Island. New Zealand troops seized Green Island as a site for an air base to intensify the pounding of Rabaul.

1944, March 20. Emirau. The seizure of this island in the St. Mathias group completed the isolation of Rabaul.

1944, June. Return of the Third Fleet to Nimitz' Command. The isolation of Rabaul and the consolidation of the Solomons virtually ended further combat missions for the South Pacific Area. Halsey reverted to Nimitz' command.

SOUTHWEST PACIFIC AREA

1944, January 2. Saidor. A regimental combat team of the U.S. 32nd Division landed north of Saidor, enveloping the Japanese garrison. The Australians, moving up from Finschhafen, made contact (March 23); the garrison evaporated into the interior. This provided Kenney's Fifth Air Force with a base to support operations on Cape Gloucester.

1944, February 29. The Admiralties. A reconnaissance in force by the U.S. 1st Cavalry Division, accompanied by MacArthur in person, speedily turned into a division-strength invasion of Los Negros Island (cleared March 23) and establishment of advance air bases. In this fighting U.S. casualties were 290 killed and 1,976 wounded. Some 3,000 Japanese were dead, 89 made prisoner.

1944, January–March. New Britain. Across Vitiaz Straits the beachheads of Cape Gloucester and Arawe expanded by mid-March as the 1st Marine and 40th Army divisions moved east along the coasts of the banana-shaped island. U.S. losses were 493 killed, 1,402 wounded; the Japanese left 4,600 dead and 329 prisoners. Later, Blamey's Australians would take over the containment of Rabaul on the land side.

THE HOLLANDIA CAMPAIGN

The Japanese high command decided to make an all-out effort to hold western New Guinea. The Second Area Army (Lieutenant General **Jo Imura**) had established a major supply and maintenance base at Hollandia, 500 miles west of Saidor and beyond reach of Kenney's fighter aircraft, whose range was about 350 miles. They began to construct airfields for future defensive and offensive air operations. By April, 3 of these were ready, located several miles inland, behind coastal mountains. Knowing that MacArthur had never made an attack beyond the range of fighter aircraft, they had few security troops at Hollandia. Most of Adachi's Eighteenth Army—65,000 strong—was concentrated between Madang and Wewak.

MacArthur, however, had decided to by-pass the Eighteenth Army and to strike at Hollandia itself. With the approval of the JCS, Nimitz agreed to send Admiral **Marc Mitscher**'s Fast Carrier Task Force to provide air support and air cover until land-based air could take over. Because of the danger of operating so near to Japanese air bases in New Guinea and the Carolines, however, Nimitz insisted that the carriers could stay in the coastal waters of New

Guinea for only 4 days. Thus it was essential that MacArthur obtain a secure base for land-based fighters, in range of Hollandia, before the carriers withdrew. Yet any earlier attempt to seize such a base would alert the Japanese to the danger to their key base. Accordingly, MacArthur decided to seize Aitape—125 miles east of Hollandia—by amphibious assault at the same time that the main 2-division landing was made at Hollandia. Aitape was west of the main concentration of the Eighteenth Army, yet it was within extreme range of American airfields at Saidor and thus could be covered by land-based aircraft while the carriers were assisting the main landings. Then, before the carriers withdrew, he planned to shift land-based air units to Aitape to keep adequate air support at Hollandia.

1944, March 30–April 19. Preliminary Air Bombardment. Kenney's long-range bombers struck the Hollandia fields repeatedly as part of an intensified air effort against Japanese coastal installations. Some 120 Japanese planes were shot down and 400 destroyed on the ground; Allied losses were insignificant.

1944, April 1–20. Australian Ground Pressure against Madang. Intensified ground activity against the Eighteenth Army was so successful that reinforcements were drawn from Hollandia, still further weakening the garrison of the air and naval base region.

1944, April 22. Landings at Hollandia and Aitape. The American 24th and 41st divisions landed on beaches 25 miles apart, east and west of Hollandia. At the same time, 2 reinforced American regiments were placed ashore at Aitape, under long-range land-based air support. The Seventh Fleet and Seventh Amphibious Force were responsible for both operations.

1944, April 22–24. Battle for Aitape. A bitter 2-day struggle took place at Aitape, while American engineers hastily prepared an airfield for operational use. American losses were 450 killed and 2,500 wounded; the Japanese lost about 9,000 dead. The area was secure and the air base operational within 2 days (April 24).

1944, April 22–27. Battle for Hollandia. With excellent carrier support, the American divisions converged inland against the Japanese airfields. Japanese resistance was light, due to the complete surprise achieved. When the carriers withdrew (April 26), land-based fighters from Aitape were available to provide support and cover. American losses were about 100 dead and 1,000 wounded; the Japanese lost more than 5,000 dead; about an equivalent number of survivors fled to the jungle.

COMMENT. *The Hollandia operation was one of the most brilliant of World War II. An entire Japanese army was encircled and its effectiveness completely destroyed. Yet because of excellent planning and skillful coordination of land, sea, and air forces, the actual combat actions were relatively small, and only insignificant Japanese forces were engaged and defeated.*

1944, May 17. Wakde. An amphibious assault quickly secured this island and its airfields. As at Hollandia, however, the soil proved to be unsuitable to support extensive operations by Kenney's long-range heavy bombers.

1944, May 27–June 29. Biak. This coral and limestone island, already used by the Japanese for an airfield, seemed the most likely spot for a heavy-bomber base in the northern New Guinea area. The amphibious landing of the 41st Division encountered unexpectedly determined resistance. In one of the most bitterly contested battles of the Pacific war, the Americans suffered more than 2,700 casualties, the Japanese nearly 10,000.

1944, June 28–August 5. Operations around Wewak and Aitape. In a series of hard-fought actions, General Blamey's Australians repulsed a Japanese Eighteenth Army offensive against Allied coastal bases of north-central New Guinea. The Japanese army, cut off from supply by land or sea, disintegrated into the jungles of the interior, where its disorganized elements rotted for the remainder of the war.

1944, July 2–7. Noemfoor. Capture of this island, west of Biak, provided additional Allied air strips.

1944, July 30. Sansapor. The unopposed occupation of this cape, on the northwestern tip of New Guinea's Vogelkop Peninsula, brought the campaign to a close. Japan's power in New Guinea had been completely destroyed.

The Central Pacific, January–August

1944, January 29–February 7. Kwajalein. While MacArthur was starting his New Guinea drive, the next target of Nimitz' trans-Pacific offensive was the Marshall group, with Kwajalein atoll its citadel. Admiral Spruance's Fifth Fleet had operational responsibility. As preliminary, Mitscher's Fast Carrier Task Force (Task Force 58: 6 heavy and 6 light carriers, escorted by fast battleships, cruisers, and destroyers) swept ahead in 4 groups to neutralize Japanese air in the Marshalls. Turner's Fifth Amphibious Force, carrying General H. M. Smith's V Amphibious Corps of 1 Army and 2 Marine divisions, carried out the operation. A 3-day preliminary naval and aerial bombardment saturated the atoll. Then (February 1) the 7th Infantry Division was landed on Kwajalein Island, southeastern flank of the lagoon, and the 4th Marine Division was put ashore on Roi and Namur islands, some 50 miles away on the northern rim. As usual, the Japanese garrisons (commanded by Rear Admiral **M. Akiyama**) put up a savage fight, but, thanks to the lessons learned at Tarawa, the defenders were overrun with relatively light losses (February 7). Of 41,000 U.S. troops put ashore, casualties were 372 dead and some 1,000 wounded. Both Japanese garrisons fought to the death, 7,870 of them being killed out of 8,000. Admiral Koga, down at Truk, with no carriers available, dared not risk his Combined Fleet to oppose the assault, but sent up some submarines, which were driven off with loss of 4 of their number. However, a surprise Japanese bomber raid from Saipan later strafed the Roi-Namur complex (February 12), inflicting some damage.

1944, February 17–18. Bombing of Truk. Mitscher's carrier force hit the Japanese naval base. Koga managed to save most of the Combined Fleet by a hurried dispersal, but the attack found 50 merchant-men in harbor and 365 aircraft huddled on airstrips. One Japanese hit was scored on U.S. carrier *Intrepid.* By noon, Mitscher's planes had sunk 200,000 tons of shipping and destroyed 275 Japanese planes. Spruance, combing the vicinity with a battleship, 4 heavy cruisers, and 4 destroyers, sank 1 light cruiser and a destroyer. While the damage to Japanese naval strength was not fatal, the attack proved the vulnerability of Truk. Its usefulness ended, no further attempt was made to occupy the once-formidable naval base, and it was left to wither on the vine. Koga's fleet withdrew to the Philippine Sea behind the Caroline-Marianas islands screen.

1944, February 17. Eniwetok. Turner's amphibious force put the 22nd Marines and army troops ashore on Engebi Island on the northern rim of this atoll without interference by Japanese planes. But Eniwetok Island itself, and Parry Island on the southern rim, were sturdily defended by Major General **Nishida**'s 1st Amphibious Brigade, a veteran combat unit 2,200 strong. Three U.S. battalions fought for and reduced the islands in succession, losing 339 men dead. The Japanese garrisons were annihilated (February 21) in close-in fighting in which Japanese land mines and U.S. flame throwers took heavy toll.

1944, April–May. Operation A-Go. Admiral Koga was killed in an airplane accident (April 1). His successor, Admiral **Soemu Toyoda,** a stronger character, bent all efforts to renew the original Japanese plan to wipe out the Pacific Fleet. Accordingly, he ordered (May 3) Vice Admiral **Jisaburo Ozawa,** commanding both the now somewhat rejuvenated carrier striking force and the First Mobile Fleet, to lure the American naval strength into the area Palaus-Yap-Woleai, where Japanese land-based aviation could assist in its destruction. Ozawa's fleet rendezvoused (May 16) off the southern tip of the Sulu Archipelago, where it was harassed by U.S. submarines, which also later warned Spruance of Ozawa's movement into the Philippine Sea (see p. 1285).

1944, June–August. The Marianas. Turner's V Amphibious Force—530-odd warships and auxiliaries with over 127,000 troops of Smith's V Amphibious Corps aboard—arrived off Saipan (June 15), 1,000 miles from its rendezvous, Eniwetok. Ahead of it, Mitscher's Task Force 58 had softened up the island defenses by fighter and bombing sweeps that cost the Japanese an approximate 200 combat planes and a dozen or more cargo ships (June 11–12). Then Mitscher's battleships and destroyers pounded ground installations at long range. Commanding the Marianas defensive forces was Admiral Chuichi Nagumo (of Pearl Harbor fame), whose Central Pacific Fleet had no major ships; under him was the Thirty-first Army (**Hideyoshi Obata**).

1944, June 15–July 13. Saipan. Landings of 2nd and 4th Marine divisions on 8 beaches, abreast, met instant resistance from Lieutenant General **Yoshitsugo Saito**'s garrison (part of the Thirty-first Army) and Admiral Nagumo's 6,000 sailors ashore. By nightfall, a beachhead had been established, but continued heavy resistance necessitated commitment of the U.S. 27th Division to reinforce the assault. Departure of Mitscher's carriers (June 17) and—later—of all warships (for the Battle of the Philippine Sea; see below) deprived the attackers of much-needed naval air and gunfire support. The ground troops made small but continued progress at heavy cost over accidental terrain skillfully organized and defended by first-class troops. The island's airfield had fallen into U.S. hands by June 18, but not until July 9, following a last-ditch fanatical counter-attack by the 3,000 Japanese still surviving, did organized resistance end. American casualties on Saipan were 3,126 killed, 13,160 wounded, and 326 missing. Japanese losses were some 27,000 killed—including hundreds of Japanese civilians who committed suicide by jumping off the cliffs. Only about 2,000 were made prisoner. Both General Saito and Admiral Nagumo committed suicide.

COMMENT. *Slow progress of the 27th Division (committed June 19) particularly aroused "Howling Mad" Smith's ire. He relieved Major General Ralph C. Smith from command (June 24), precipitating an Army-Marine controversy to this day undecided, since it concerned radical difference between Marine and Army doctrine in assault tactics.*

Battle of the Philippine Sea, June 19–21, 1944

Ozawa's fleet had put to sea immediately upon receiving word of the assault on Saipan. Spruance, warned of the movement, gathered the Fifth Fleet under Admiral Mitscher's tactical command (June 18) 160 miles west of Tinian. Ozawa had 5 heavy and 4 light carriers, 5 battleships, 11 heavy and 2 light cruisers, and 28 destroyers. Spruance's fleet, augmented by Turner's Amphibious Force, numbered 7 heavy and 8 light carriers, 7 battleships, 8 heavy and 13 light cruisers, and 69 destroyers. U.S. planes numbered in all 956 to the Japanese 473; however, Ozawa would be giving battle within range of some 100 additional Japanese land-based planes based on Guam, Rota, and Yap. Also, Japanese planes had longer range than the Americans.

Ozawa's search planes picked up the Fifth Fleet by daybreak (June 19), 300 miles from his advance element of 4 light carriers and 500 miles from his main body. The opposing fleets were on southerly courses, Mitscher's carriers some 90 miles northwest of Guam and 110 miles southwest of Saipan in 4 carrier groups backed up by Spruance in the battle fleet. At about the same time, land-based planes from Marianas and Truk bases were repulsed with heavy loss. Following this, Ozawa attacked in 4 successive raids. Mitscher, discovering the first attacking group, loosed his interceptors and then sent the bombers aloft to keep his flight decks clear.

Misfortune struck Ozawa almost immediately. U.S. submarines following his movements picked off *Taiho,* biggest and newest of Japan's carriers, and *Shokaku,* one of the 2 surviving vessels of Pearl Harbor, sinking them both. Mitscher's interceptors took terrible toll of all the raiders, and most of those who did get through were then brought down by antiaircraft fire. The "Great Marianas Turkey Shoot" cost Ozawa 346 planes and 2 big carriers; U.S. losses were but 30 planes and some slight damage to a battleship from the single Japanese bomb which found its target.

Meanwhile, Mitscher's bombers strafed Guam and Rota, neutralizing the airfields. The battle was over by dark (6:45 P.M.). Ozawa had hauled off. Mitscher pursued, but contact was not made again until late afternoon next day. Mitscher launched 216 planes, which sank another carrier and 2 oil tankers and seriously damaged several other vessels. Twenty U.S. planes were lost, and Ozawa lost 65 more of his surviving aircraft. Homeward bound in the dark, the U.S. planes fell into major difficulties despite the fact that Mitscher boldly lighted up his ships to guide his birds home. Some 80 planes out of gas either ditched or crash-landed. Mitscher moved west, his destroyers combing the area to pick up 50-odd flyers floating in the water, reducing U.S. personnel losses that day to 16 pilots and 33 crewmen. Ozawa's crippled fleet made its getaway. The Fifth Fleet returned to Saipan.

COMMENT. *Ozawa's desperate attack in the face of formidable odds cost Japanese naval air power mortal injury. Aside from the material damage suffered in ship and plane losses, the death of more than 460 trained combat pilots was irreplaceable.*

1944, July 21–August 10. Guam. Guam fell to the 3rd Marine and 77th Army divisions and 1st Marine Brigade (August 10) after a tough struggle costing more than 1,400 killed and 5,600 more U.S. personnel wounded. Of the Japanese defenders, some 10,000 were killed during the operation, and several hundred more died rather than surrender during the long mopping-up that followed.

1944, July 25–August 2. Tinian. The island was invaded by the 2nd and 4th Marine divisions, and taken after the usual heavy resistance. U.S. losses were 389 killed and 1,816 wounded.

COMMENT. *Capture of the Marianas and the Battle of the Philippine Sea doomed Japan to defeat. The Tojo cabinet resigned (July 18).*

Return to the Philippines, July–December, 1944

STRATEGIC AND OPERATIONAL BACKGROUND

Strategy of the next move of the war was settled in principle at a conference at Pearl Harbor in July. President Roosevelt weighed 2 different concepts of approach to the final objective: an assault on Japan itself. Admiral Nimitz favored intermediate moves on Formosa or China; General MacArthur insisted on assault first to free the Philippines (for both military and political reasons), then on to Japan. Roosevelt accepted MacArthur's plan. Army and navy staffs, in close cooperation, plotted progressive moves: MacArthur on Mindanao, Nimitz on Yap, and a joint assault on Leyte. While MacArthur would then invade Luzon, Nimitz would take Iwo Jima and Okinawa.

As soon as the Marianas operation ended, Nimitz set Spruance and his Fifth Fleet staff to planning the future Iwo Jima and Okinawa operations, while Halsey and Third Fleet staff were allotted the immediate task against Yap. Actually, this was a shift in commanders and staffs only; the ships and men were unchanged. What had been the Fifth Fleet was now the Third;

Mitscher's Fast Carrier Task Force, TF 58, became TF 38; V Amphibious Force became III Amphibious Force; etc.

PRELIMINARIES

1944, September 15. Morotai and Peleliu. MacArthur landed troops on Morotai to establish a forward air base; opposition was negligible. Nimitz, in a coordinated move, struck Peleliu in the Palau Islands. Marines of Major General Roy S. Geiger's III Amphibious Corps (part of Vice Admiral **Theodore S. Wilkinson**'s III Amphibious Force) there met tough resistance from tenacious defenders under General **Sadal Inone** (September 15–October 13). Simultaneously Wilkinson, using army troops, assaulted and captured **Angaur** (September 17–20) and occupied, unopposed, Ulithi Atoll, about 100 miles west of Yap (September 23). Ulithi's magnificent harbor became a fleet base for the Halsey-Mitscher Third Fleet team. Mopping up of scattered resistance continued in the Palaus for several weeks (November 25).

1944, September 15. Change of Plan. In support of the Morotai-Peleliu operations, Halsey's Third Fleet carriers had struck Yap, Ulithi, and the Palaus to neutralize nearby Japanese bases and soften the defenses (September 6). They had then swept the Philippine coast (September 9–13). Encountering slight resistance, Halsey sent a message to Nimitz, recommending cancellation of the proposed intermediate landings on Mindanao and Yap, which seemed unnecessary, and urging that the Leyte assault be mounted as soon as possible. Nimitz, agreeing, queried the Joint Chiefs of Staff, then attending the Quebec Conference. Nimitz offered to "loan" MacArthur his III Amphibious Force and his Army XXIV Corps. MacArthur, queried by the JCS if he could step up his Leyte landings to October 20 instead of December 20, as originally planned, immediately agreed (September 15). Accordingly, a directive was issued that day.

COMMENT. *Never had American military men better displayed flexibility and initiative than did Halsey, Nimitz, MacArthur, and the JCS. MacArthur's decision to advance by 2 months the timetable of a fullscale amphibious assault was audacious in light of the monumental logistical problems involved.*

1944, October 7–16. Prelude to Invasion. The Third Fleet moved to paralyze the remnants of Japanese air power. While army land-based planes—the Fifth AF from New Guinea and the Seventh from the Marianas—attacked all enemy bases within range, long-range B-29s of XX Bomber Command bombed Formosa from bases in China (October 10). Meanwhile, Halsey hit shipping and shore installations in and around Okinawa, then turned south toward Formosa and Luzon (October 11).

1944, October 13–16. The Battle off Formosa. The Japanese air units on Formosa struck back fiercely; 2 cruisers were severely damaged, other vessels hit. Misled by fires on these ships and the flames of many shot-down Japanese planes, Japanese radio reports claimed a great naval victory. Halsey, hoping to decoy the Combined Fleet to action, left 1 carrier group to escort his cripples and took the rest of the Third Fleet east into the Philippines Sea (October 14–15). Toyoda, taking the bait, sent 600 of his carrier planes from Japan to Formosa airfields to complete the "destruction" of the American fleet. Halsey, returning toward Formosa, in 2 days destroyed about half these planes (October 15–16). The total score in these operations was more than 650 Japanese planes destroyed, many others crippled, and numerous shore installations smashed. Most serious was the wrecking of the rebuilt Japanese carrier air squadrons. U.S. losses were 2 cruisers badly damaged, several others hit, and 75 planes lost.

VISAYAS CAMPAIGN

The Philippines were defended by approximately 350,000 Japanese under the command of General Yamashita, conqueror of Malaya and Singapore. General Sosaku Suzuki's Thirty-fifth Army defended the southern and central Philippines.

1944, October 14–19. Approach to Leyte. A vast amphibious armada was closing on the Leyte coast, some 700 vessels, carrying 200,000 men of General Krueger's U.S. Sixth Army, in Admiral Kinkaid's Seventh Fleet. Wilkinson's III and Barbey's VII Amphibious Forces were supported by Rear Admiral **J. B. Oldendorf**'s gunfire-support group—6 battleships (reconditioned survivors of Pearl Harbor), 5 heavy cruisers (including **HMAS** *Shropshire*) and 4 light cruisers, and 66 destroyers (including convoy escorts). The air-support group (Rear Admiral **Thomas L. Sprague**) consisted of 16 escort carriers, 9 destroyers, and 11 destroyer escorts. Far in the offing (now neutralizing Luzon air bases and guarding San Bernardino and Surigao straits) was Halsey's Third Fleet, with Mitscher's fast carrier TF 38 its nucleus: 8 heavy and 8 light carriers, 6 fast new battleships, 6 heavy and 9 light cruisers, and 58 destroyers. Mitscher's planes numbered more than 1,000. Four American submarines were engaged.

COMMENT. *One operational blemish marred this magnificent amphibious array: command was joint, instead of being unified. MacArthur commanded the ground forces and Kinkaid's Seventh Fleet, while Halsey, still under Nimitz' direction, had a dual mission: first, to destroy the Japanese fleet should it put to sea; and second, to give all possible assistance to Kinkaid and the landings. The Americans had no reason to believe that these 2 missions could be contradictory, but the Japanese had a plan which the Americans had not foreseen.*

The "Sho" Plan. *Japan, feeling the effect of continuous air and submarine attacks against her shipping, had long realized that American seizure of the Philippines, Formosa, or the Ryukyus would split the empire in two; separate the Southern Resources Area, reservoir for fuel oil and gasoline, from the homeland. Accordingly, the "Sho" (Victory) Plan had been de-*

vised as a last-ditch gamble by the Combined Fleet. In principle it embodied use of a portion of the now scattered elements of the fleet to decoy the American carrier force, while the remainder, converging on any new landing area, would overwhelm the amphibious support and isolate the troops ashore. The plan, good in principle, had one major flaw. Without adequate air power it would not work, and the Battle off Formosa had ruined most of Japan's carefully reconstituted naval air strength. Toyoda hoped, however, that landbased air units in the Philippines might make up for this serious deficiency.

1944, October 20–22. Leyte Landings. Following intensive reconnaissance and a heavy naval bombardment, Sixth Army's X Corps (Major General **Franklin C. Sibert**) and Lieutenant General **John R. Hodge**'s XXIV Corps began landing. The Japanese 16th Division (General **Tomochika**), 16,000 strong, garrisoning the island, gave but little initial opposition. By midnight, 132,400 men and nearly 200,000 tons of supply and equipment were landed, and Kinkaid's combat ships were hurrying south to Surigao Strait to meet a Japanese naval threat. General MacArthur, accompanied by Philippine President **Sergio Osmeña,** came ashore, broadcasting his return (October 22).

THE BATTLE FOR LEYTE GULF

1944, October 17–23. The Preliminaries. Warned of American intentions, disclosed by the Leyte reconnaissance (October 17), Japan put the **Sho Plan** into immediate effect. The Combined Fleet moved from widely scattered bases. Ozawa's Northern Force (or Third Fleet)—the 4 remaining carriers, 2 battleships, 3 cruisers, and 8 destroyers—steamed from Japan toward Luzon to lure the U.S. Third Fleet away from the landing area; the carriers were phantoms of naval air power, carrying only 116 planes with half-trained pilots. Kurita's Center Force (or First Attack Force)—2 superbattleships, 3 other battleships, 12 cruisers, and 15 destroyers—moving northeast from Malaya, Borneo, and the China Sea, was to traverse the San Bernardino Strait. The Southern Force (or C Force) of Vice Admiral **Shoji Nishimura**'s 2 battleships, 1 heavy cruiser, and 4 destroyers (from Malaya and Borneo), backed up by Vice Admiral **Kiyohide Shima**'s Second Attack Force of 2 heavy and 1 light cruisers and 4 destroyers (from the Ryukyus), moved southeast and east to pass through the Surigao Strait between Mindanao and Leyte. These Center-Southern Force pincers were to destroy all U.S. amphibious forces in Leyte Gulf by a concerted attack, marooning the troops ashore.

1944, October 23–24. Battle of the Sibuyan Sea. U.S. submarines *Darter* and *Dace* discovered Kurita's Center Force as it entered Palawan Passage from the South China Sea. Flashing word to Halsey, the subs sank 2 heavy cruisers and damaged a third. Kurita, continuing into the Sibuyan Sea, was attacked by Mitscher's TF 38. The superbattleship *Musashi* was sunk after 2 days of incessant bombardment and several other vessels damaged. Kurita turned around and headed westward again (late October 24). Halsey assumed he was retreating. Meanwhile, Japanese land-based planes harassed a division of TF 38. Most were shot down, but USS *Princeton* (light carrier) was sunk and USS *Birmingham* (cruiser) severely damaged. Unknown to Halsey, after dark Kurita changed course again and headed doggedly for San Bernardino Strait.

1944, October 24–25. Battle of Surigao Strait. Warned of the approach of the Japanese Southern Force, Kinkaid placed Oldendorf's gun-support ships to intercept it. Reconnoitering PT boats picked up Nishimura's force that night, moving in single line ahead. Two converging destroyer attacks torpedoed the battleship *Fuso* and 4 destroyers. Nishimura's crippled line then ran into Oldendorf's main force, which crossed his "T." American gunfire and torpedoes sank all except

1 destroyer; Nishimura went down with his flagship, *Yamashiro*. Shima's force, bringing up the rear, now ran the gantlet of PT-boat attack, which crippled 1 light cruiser. Attempting now to retire, Shima's flagship collided with one of Nishimura's mortally wounded vessels. Oldendorf pursued through the strait. Between his fire and planes from Admiral Thomas L. Sprague's escort carriers and land-based army planes, another Japanese cruiser was sunk. The rest of Shima's force escaped as Oldendorf—his mission accomplished, and knowing he might have to fight another battle immediately—turned back.

1944, October 25, Dawn. San Bernardino Strait. Kurita, hoping still to join Nishimura in Leyte Gulf, debouched unopposed from San Bernardino Strait and turned south. Halsey, with his entire force, was rushing after Ozawa's decoy carrier force, which had been sighted to the far north. Word received of Kurita's retirement in the Sibuyan Sea the previous afternoon, combined with overenthusiastic reports from his flyers, led Halsey to believe Kurita was permanently out of the fight. So—his dual mission in mind—Halsey was rushing northward with all available elements of the Third Fleet to destroy the Japanese carrier force. He failed to inform Kinkaid.

1944, October 25, Early A.M. Battle off Samar. Kurita's southward advance completely surprised Rear Admiral **Clifton A. F. Sprague,** commanding an escort carrier group of Kinkaid's amphibious force, supporting land operations. Sprague, with 6 escort carriers, 3 destroyers, and 4 destroyer escorts, armed only with 5-in. guns, found himself giving battle to 4 battleships, 6 heavy cruisers, and about 10 destroyers. In an amazing running fight, during which his aircraft, armed with relatively harmless fragmentation bombs (for support of land operations), harassed the heavily gunned Japanese ships, while his destroyers nipped boldly at them, Sprague and his escort flattops fought off disaster. One carrier—USS *Gambier Bay*—was sunk, 2 destroyers and 1 destroyer escort went down, and Kurita's ships were closing in. But planes from other escort-carrier groups—

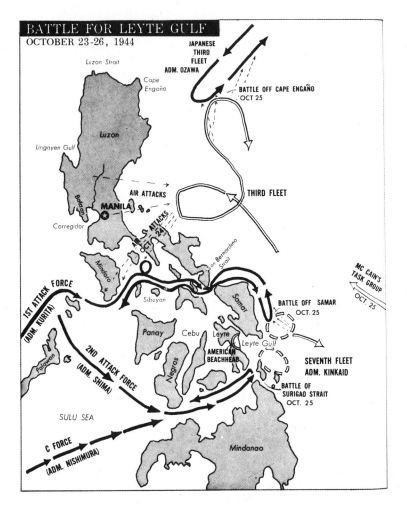

all fleeing south to rejoin Kinkaid's Seventh Fleet—were now attacking Center Force. Kurita, with victory in his hands, hauled off, thinking he was being attacked by TF 38, his decision bolstered by reports that Nishimura's Southern Force had been demolished. At first Kurita thought to reinforce Ozawa, then gave that up and retreated through San Bernardino Strait. He was now harried by American reinforcements—planes from Rear Admiral **John S. McCain**'s task group of Halsey's fleet—rushing back in response to Seventh Fleet calls for help. Meanwhile, Admiral Thomas L. Sprague's escort carriers and Olden-

dorf's force, returning from the Surigao Strait fight, were assailed by land-based planes—including the first Japanese kamikaze attacks, which sank USS *St. Lô* and damaged several other ships.

1944, October 25. Battle off Cape Engaño. Halsey's headlong northward rush to catch the Japanese carrier force bore fruit at 2:20 A.M., when Mitscher's search planes picked it up. With dawn came the first of 3 successive plane strikes. Ozawa, who had put nearly all his small force of planes ashore to operate from land bases, had only antiaircraft fire to oppose the U.S. planes and it was not enough. By nightfall,

all 4 Japanese carriers were sunk as well as 5 other of the remaining ships. Two battleships, 2 light cruisers, and 6 destroyers escaped. Halsey, infuriated by a query from Nimitz at Pearl Harbor, demanding explanation of his "failure" to help Kinkaid, went steaming southward at "flank speed" (full speed) with most of his fleet.

COMMENT. *The 4-phased naval Battle of Leyte Gulf ended the Japanese fleet as an organized fighting force. Four carriers, 3 battleships, 6 heavy and 4 light cruisers, 11 destroyers, and a submarine were sunk; nearly every other ship engaged was damaged. About 500 planes were lost; some 10,500 Japanese sailors and airmen were dead. American losses were 1 light carrier, 2 escort carriers, 2 destroyers, 1 destroyer escort, and more than 200 aircraft. About 2,800 Americans had been killed, some 1,000 additional wounded. In this, the greatest naval battle ever joined, 282 vessels were involved: 216 USN, 2 RAN, 64 Japanese. Estimated man-power strengths were 143,668 USN and RAN, 42,800 Japanese.*

Both Halsey and Kurita may be criticized, the one for abandoning the Leyte Gulf area to chase Ozawa, the other for his timidity in retreating at the moment of victory. In mitigation, one notes that Halsey was on the horns of his dual-mission dilemma; he had no way of knowing that Ozawa's carrier force was only a decoy. As for Kurita, he was bewildered by conflicting reports and already shaken by 2 previous severe joltings. For all he knew, he might be throwing the only Japanese naval strength that remained afloat into the jaws of the overwhelming force of the entire U.S. Navy. So he limped away, victim of an overly complicated plan and of his own fears.

STRUGGLE FOR LEYTE

1944, October 21–December 31. General Yamashita, Japanese commander in the Philippines, recovering from the strategic surprise of the landings, decided to fight for Leyte. While the veteran 16th Division carried on a skillful delaying action, reinforcements were hurried from Luzon and the Visayas to build up the Thirty-fifth Army. Between October 23 and December 11, some 45,000 men and 10,000 tons of supply arrived in small increments, carried mostly by fast-running destroyer transports. U.S. Navy and long-range Army planes then dried up the supply routes, despite fierce Japanese land-based air strikes, which only expended a great part of the remaining Japanese air power. Toward the end, only a trickle of sailing craft traversed the Leyte waters. Meanwhile, U.S. Sixth Army made slow progress against desperately defended delaying positions in the central mountain ridge. Progress was further impeded by tropical storms, which turned the few dirt roads into quagmires. On the right, X Corps, by overland and amphibious movements, was approaching Limon, the northern anchor of Suzuki's defense line (November 7). On the left, XXIV Corps hammered at the precipitous heights protecting Ormoc, the southern Japanese key point and principal port. An amphibious lift (December 7) threatened Suzuki's southern flank, while at the same time a desperate but ineffective Japanese air drop farther east was repelled. Krueger, reinforced, launched a double envelopment (December 7–8), taking both Limon and Ormoc (December 10). The 77th Division from the south and 1st Cavalry and 24th divisions from the north then met (December 20) at Libungao, surrounding Suzuki's combat elements and cutting him off from his last remaining port, Palompom. All organized resistance collapsed Christmas Day. The Japanese had suffered over 70,000 casualties in the campaign; American losses were 15,584. Meanwhile, a foothold had been gained on adjoining Samar (late October).

1944, December 15. Seizure of Mindoro. A bold amphibious operation put a task force of approximately brigade strength ashore on Mindoro in the northern Visayas, just south of Luzon, to establish an advance air base for MacArthur's coming assault on Luzon.

COMMENT. *Japan's election to force a decision in the Leyte area had cost her a fatally crippled fleet and air force, and a serious reduction in ground-force strength in the Philippines. As the year ended, Japan had lost the war. Her communications with the Southern Resources Area were severed, her merchant fleet reduced from a tonnage of 6 million in 1941 to 2.5 (60 percent of this loss caused by U.S. submarines). Approx-*

imately 135,000 of her soldiers were marooned, help-
less, behind the American advance. But as yet Japan
would not admit defeat.

The Strategic Air Offensive

OPERATIONS FROM INDIA AND CHINA

1944, April–May. B-29s to India. With the
agreement of Britain and China, new American
B-29 "Superfortress" bombers began to arrive
at bases near Calcutta. These planes of the XX
Bomber Command were to undertake long-
range strategic bombardment from India, and
were also to strike industrial targets in Manchuria and southern Japan itself from bases in
China. These powerful, heavily armed planes
could fly over 350 mph, and could carry 20,000
pounds of bombs against targets over 1,500
miles from their bases. They were each armed
with 12 .50-cal. machine guns and a 20-mm.
cannon.

1944, June 5. First B-29 Combat Mission.
This was an attack on railway targets at Bangkok, Thailand.

1944, June 15. First Strike against Japan.
Five airfields with extra-long runways had
been painstakingly built by coolie labor near

Ch'engtu, in western China. Sixty-eight planes,
staged to the Ch'engtu bases from Calcutta, hit
a steel plant on the Japanese island of Kyushu.

**1944, June–December. Continuing Strate-
gic Missions.** Because of the limitations on the
total amounts of supplies and equipment that
could be carried over the Hump, the B-29s
were not based permanently in China. Barracks, repair shops, and heavy support equipment were all near the Calcutta bases. The
B-29s carried their own fuel and bombs from
India to Ch'engtu in preparation for their periodic raids on Kyushu, southern Manchuria,
and Formosa. In between these raids, the great
planes continued to attack Japanese bases in
Southeast Asia from their fields near Calcutta.
Under Major General **Curtis E. LeMay,** the
XX Bomber Command became increasingly efficient.

OPERATIONS FROM THE MARIANAS

1944, October. B-29s to Saipan. As soon as the
Marianas islands of Saipan, Tinian, and Guam
had been captured (see pp. 1285–1286), army
B-17 and B-24 bombers moved into the captured Japanese airfields to strike at the Bonins
and Iwo Jima (August—September). At the

B-29 Superfortresses

same time, army engineers began to lengthen existing runways and to build new extra-long airfields on all available areas of the islands. As soon as the first of these was ready, B-29s of the XXI Bomber Command began to arrive.

1944, October 28. First B-29 Mission from the Marianas. This was the first of several operational-training missions against Truk.

1944, November 24. First Marianas Mission against Japan. More than 100 B-29s initiated the final strategic air offensive against Japan by a raid against an aircraft plant near Tokyo.

1944, November–December. Continuing the Strategic Offensive. About once every 5 days, 100 to 120 Superfortresses struck industrial targets in Japan. The raids usually began with a dawn take-off from the Marianas airfields, the planes climbing steadily until they reached an altitude of about 30,000 feet as they crossed the Japanese coast. Resistance was fierce. Fighter planes based on Iwo Jima struck at the bombers as they flew over in the mornings and as they returned in the early afternoons. Over Japan, 200 or more fighters, with the best pilots available, rose to meet them. The Superfortresses shot down many of the attackers, but suffered severely themselves. By the end of the year, XXI Bomber Command was losing about 6 percent of the planes that started on every raid, exceeding the theoretically acceptable maximum loss rate of 5 percent.

THE WAR AGAINST JAPAN, 1945

Strategy and Policy

The collapse of Japan's defensive perimeter in the east and south, resulting from the crushing defeats suffered in 1944 by Japan's land, sea, and air forces in the Southwest Pacific and Pacific Ocean Areas, made the outcome of the war against Japan a foregone conclusion to Allied military leadership. Surprisingly, at the outset of the year, it seemed even possible that Japan's defeat might come as early as Germany's. This was the result not so much of the Allies' failing to adhere to their basic strategy of "Germany first"—though the Americans *had* deviated from that strategy to some extent—but was rather the natural consequence of the amazing military might which America had been able to mobilize in 3 short years. No nation in history had ever before been able to exercise so much power over so wide an expanse of the globe, and with such efficiency, as that wielded by the United States at the beginning of 1945. So immense was this power, in fact, and so inevitable the outcome of the war, that the first steps toward U.S. demobilization were actually begun before either Germany or Japan completely collapsed.

Inevitable though the outcome now was, it was also evident that the final subjugation of Japan could be both time-consuming and costly in lives and money. Despite some setbacks, the Japanese hold on Burma had not been seriously impaired, and they still ruled all of the vast Southern Resources Area. And though their sea lines of communication to that rich area had been cut, they had opened a new overland route as a result of their successful campaigns in China in 1944. The American naval blockade of Japan, combined with a relatively small strategic air offensive, was putting a serious strain on Japan's home economy, but the vigorous Japanese air defense effort was already threatening the morale of B-29 airmen. Japanese tenacity, despite their many defeats, caused Allied military leaders to fear that Japan itself, defended by more than 2,000,000 hardy fighting men, could not be conquered without a costly and sanguinary struggle on the 4 main islands themselves. Possibly the most serious danger was that conquest of the home islands might not end the war at all. On the mainland of Asia, Japan controlled Korea, Manchuria, and the richest and most populous areas of China; this vast Asiatic region was garrisoned by a military force well in excess of 1,000,000 undefeated veteran troops. If the Japanese government and a substantial portion of the military strength remaining in Japan should be shifted to the continent, the war might be protracted interminably.

Accordingly, Allied military leaders—and particularly the American JCS—were extremely anxious to draw Russia into the war against Japan at the earliest possible date, solely to hasten the end of the war and avoid probable excessively high casualties. But the agreements reached at Yalta and later at Potsdam were inept and naïve. The Soviets' week-long participation in the war did not affect the outcome. By that time Japan had not only been brought to the verge of economic and social collapse by physical punishment; she had suffered a psychological shock so severe as to make it easy for a warrior people to acknowledge defeat for the first time in recorded Japanese history. The death of President Roosevelt (April 12) brought Vice President **Harry S Truman** to U.S. leadership, but there were no changes in policy or strategy.

Land and Air Operations in Asia

BURMA

As the year opened, 4 major Allied land forces were converging on Burma. Along the coast of the Bay of Bengal, the British XV Corps was once more pushing through Arakan toward Akyab. Slim's Fourteenth Army was advancing on a broad front through the jungled hills between the Chindwin and Irrawaddy rivers. In north Burma, forces of General Sultan's Northern Combat Area Command were approaching the Burma Road from the west, while Marshal Wei Li-huang's Y-Force was slowly moving down that road from Yunnan toward the Burma-China frontier. Though these Allied forces were encountering minor opposition, it was obvious that the 3 Japanese armies holding Burma were merely conducting delaying actions. Though few of the Allied leaders realized it—with the possible and notable exception of General Slim—General Kimura, overall commander of Japan's Burma Area Army, was deliberately luring his foes into central Burma for a climactic battle in an area where he expected to have a logistical superiority that would ensure his victory.

THIRD ARAKAN CAMPAIGN

The renewed advance of the XV Corps had begun on December 12, 1944. In accordance with Kimura's plan, opposition was negligible, and Akyab fell quickly (January 4). The XV Corps prepared for an amphibious invasion of south Burma. Meanwhile, its hold on Arakan was ensured by a number of minor operations to seize and secure the major islands along the Arakan coast.

NORTH BURMA AND YUNNAN

1945, January 1–27. Converging Advances of NCAC and Y-Forces. The Japanese 56th Division withdrew deliberately and skillfully in the face of these 2 converging Chinese drives, making the Chinese fight for every foot of ground. Although units of the 38th Division reached Chinese soil at Loiwing early in the month (January 6), and patrol contacts be-

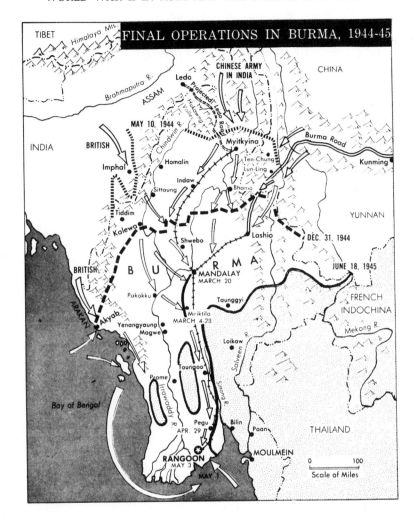

FINAL OPERATIONS IN BURMA, 1944-45

tween the 2 converging forces were continuous from that time, the Japanese stubbornly prevented a link along the road.

1945, January 27. Reopening the Burma Road. The Y-Force pushed across the Shweli River to Burmese soil at **Wanting,** just as troops of the Chinese New First Army (38th and 30th divisions) fought their way to the Burma Road at **Mong Yu;** so skillful was the Japanese withdrawal that the 2 Chinese forces became engaged in a fire fight with each other near Mong Yu before they realized that their elusive enemy had slipped away to new posi-

tions blocking further southward advance along the Burma Road toward Lashio.

1945, January 28–February 4. First Land Convoy to China. With great fanfare a truck convoy (manned largely by junketing American newspapermen) left Namkham in Burma for Kunming, arriving amidst rejoicing and celebration. In a magnanimous gesture, Chiang Kai-shek proposed that the combined Burma and Ledo roads be renamed the Stilwell Road, in honor of the man most responsible for breaking the land blockade of China.

1945, January–March. Climactic Opera-

tions in North Burma. Although the Chinese and American troops had accomplished their main mission, the war in north Burma was not yet over. Bitter fighting continued along the Burma Road as Allied troops fought their way toward Lashio against typically dogged Japanese resistance. A particularly fierce battle took place when the American Mars Brigade in the vicinity of **Namhpakka** vainly attempted to block the retreat of the Japanese 56th Division to Lashio (January 18–February 3). Equally fierce was the battle fought by the British 36th Division in forcing a crossing of the Shweli River at **Myitson** (January 31–February 21). A few days later, in the center of the NCAC zone, the Chinese 50th Division captured the silver and tin mines of **Namtu** after a sharp engagement (February 27). The Chinese New First Army captured **Lashio** after a desultory 2-day battle (March 6–7). The 50th Division captured **Hsipaw** (March 15), then repulsed Japanese counterattacks until joined by the 30th Division (March 24). The British 36th Division reached the road near Maymyo (March 30), returning to Fourteenth Army control and ending an 8-month period of uninterrupted combat.

CENTRAL BURMA

1945, January. Slim Sets a Trap. Because of the paucity of resistance to his advance toward the Irrawaddy, Slim divined Kimura's plan to counterattack the Fourteenth Army as it attempted to cross the river north of Mandalay. Accordingly, he left a dummy IV Corps headquarters near Shwebo, on his left flank, maintaining active radio contact with the 19th Indian Division, which continued to advance directly toward the river. Meanwhile, he secretly and rapidly moved the IV Corps and its 2 remaining divisions south, behind the XXXIII Corps, to make a surprise crossing of the Irrawaddy near Pakokku, 100 miles below Mandalay. While the 19th Division and XXXIII Corps were occupying Japanese attention by crossing the river farther north, Slim planned a rapid crossing by the IV Corps, which would then dash eastward, cut Kimura's line of communications to Rangoon, and encircle the Japanese Fifteenth and Thirty-third armies.

1945, January 14. Irrawaddy Crossing of the 19th Division. This crossing at **Kyauk-myaung** was immediately violently counterattacked, confirming Slim's estimate of Japanese intentions. With difficulty the division held its bridgehead, thanks to powerful artillery and air support.

1945, February 12. Crossing of the XXXIII Corps. Having deliberately attracted Japanese attention to their preparations, the divisions of this corps crossed the river against determined resistance just south of Mandalay; losses were heavy, but the bridgehead was established and maintained.

1945, February 13. Crossing of the IV Corps. Slim's planned encirclement got off to a brilliant start with a surprise crossing against negligible opposition. The 7th and 17th Divisions (the latter almost completely armored) established a bridgehead at Pagan and began a quick build-up.

1945, February 21–28. Dash for Meiktila. While the 7th Division held the Irrawaddy bridgehead at Pagan, the tanks of the 17th Division raced to seize the strategic road and railroad junctions at Meiktila and Thazi.

1945, February 28–March 4. Capture of Meiktila. Having seized the town against desperate resistance by surprised, outnumbered Japanese rear-area troops, the British prepared for the inevitable Japanese reaction. Reinforcements were rushed to the 17th Division by air, and an air-supply operation from India was initiated. Kimura, realizing what had happened, began to withdraw his reserves from the Mandalay area and prepared for an all-out assault on Meiktila. Violent, confused fighting raged around that town and Thazi (March 5–15).

1945, March 9–21. Battle of Mandalay. The 19th Division north of Mandalay and the XXXIII Corps to the south broke out of their bridgeheads (February 26) and converged on the old royal capital of Burma. While the XXXIII Corps kept the Japanese busy south of the city, the 19th Division captured it in a dogged house-to-house battle.

1945, March 15–31. The Battle of Meiktila. Elements of 3 Japanese divisions, commanded by capable Major General Tanaka transferred from north Burma, began an all-out assault on the 17th Indian Division in order to reopen communications with Rangoon. The Japanese closely invested the British and Indian troops, and for a while captured their airfield, forcing them to rely upon parachute drop for supply. Equally determined British counterattacks recovered the airfield, however. With the XXXIII Corps and remainder of the IV Corps moving rapidly toward Meiktila, Kimura realized that he could not fight his way into the town in time to avoid disaster (March 28). He therefore reopened a new line of retreat through Thazi. The IV Corps broke through to relieve the 17th Division, thus ending the climactic battle of the war in Burma.

COMMENT. *Slim's brilliant but simple plan went exactly as he had envisaged it. The plan, and its execution, warrant Slim's inclusion in the list of Britain's greatest generals.*

1945, March 26. Revolt of Aung San. Burmese General Aung San, who had been ordered by the Japanese to take his small Burma National Army from Rangoon to the assistance of the Japanese troops at Meiktila, left the city, ostensibly in compliance with orders, then ordered execution of long-laid plans for a Burmese revolt against the Japanese. Aung San had long been in secret contact with British agents, having become disillusioned with the Japanese. Though the effort was not decisive, nonetheless guerrilla harassment by these Burmese added significantly to the discomfiture of the Japanese.

1945, April 1–May 1. Race for Rangoon. Anticipating early arrival of the monsoon, which would immobilize his tanks, now able to maneuver so easily on the rock-hard, sun-baked rice paddies, Slim turned southward. Several sharp engagements were fought, but the British, spearheaded by the 17th Indian Division, had little difficulty in slashing through confused and uncoordinated Japanese units. The main problem was to get supplies of fuel and food to the troops by airlift; for the last 10 days of the month the 17th Division was on half-rations. Pegu, only 45 miles from Rangoon, was captured the same day the rains began (May 1).

1945, May 1. Amphibious Landing of the XV Corps. Anxious to beat the Fourteenth Army for the honor of capturing Rangoon, elements of the XV Corps landed at the mouth of the Rangoon River. The Japanese secretly evacuated the city that day.

1945, May 2. Capture of Rangoon. While the 17th Division was trying to push through mud north of the city and the XV Corps was organizing for a short advance upriver to capture the town from the south, an RAF pilot, flying over the city, could see no Japanese. He landed at Mingaladon, walked into the town, and, with the assistance of the local civilian population, released Allied prisoners from the city jail. He had himself rowed down the river to inform the British Army that Rangoon had been captured by the RAF.

1945, May–August. Pursuit and Consolidation. Despite the great difficulties of movement over the flooded countryside, the British pushed after the Japanese elements withdrawing in disorder toward Thailand. There were no further major engagements, though there were frequent patrol clashes between pursuers and pursued. Meanwhile, plans were being made by Admiral Mountbatten's SEAC headquarters for a major amphibious operation against Singapore later in the year.

CHINA

1945, January–February. Renewed Japanese Offensive in Southeast China. Japanese forces made wide gains in the coastal regions between Hankow and the French Indochina border. Three more U.S. Fourteenth Air Force bases fell.

1945, March 9, Japanese Consolidation in Indochina. French officials, heretofore permitted to retain some internal control despite Japanese occupation, were imprisoned.

1945, March–May. Japanese Offensive in Central China. A drive through the fertile area between the Yellow and Yangtze rivers netted the ripening crops, and the important Fourteenth Air Force base at **Laohokow** was overrun (March 26–April 8). The drive was checked (April 10) by the Chinese, as was a later attempt against **Changteh** and **Chihkiang** (May 8).

1945, April 5. U.S.S.R. Abrogates Treaty with Japan. As war in Europe neared its end, the Kremlin renounced the 1941 neutrality treaty (see p. 1231), which had served Stalin's purpose of avoiding a two-front war.

1945, April–August. Soviet Buildup in East Asia. Commanded by Marshal **Aleksandr M. Vasilevskiy,** the Soviet Far East Command consisted of 11 combined arms armies, 1 tank army, 3 air armies, and 3 air defense armies with a total strength of over 1,500,000 troops, 26,000 guns and mortars, 5,500 tanks, and 3,800 aircraft in three army groups. In Manchuria, Japanese General **Otoza Yamada**'s Kwantung Army had a paper strength of 925,000 men, of which 300,000 were unreliable Manchukuo puppet units with about 6,000 guns and mortars, 1,200 tanks, and 1,900 aircraft. Most veteran troops of the Kwantung Army had been sent to oppose the American offensives in the Pacific.

1945, May–August. Japanese Withdrawals. Realizing he was overextended, General Okamura began movements north to reinforce the Kwantung Army in Manchuria, now menaced by the Soviet threat to enter the war. Chinese counteroffensives cut the corridor to Indochina (May 30). By July 1, more than 100,000 Japanese troops were marooned in the Canton area, while some 100,000 additional had moved back into North China, harassed by the U.S. Fourteenth and Tenth Air Forces. The former American airfield at Kweilin was recaptured (July 27).

1945, August 8. U.S.S.R. Declares War on Japan. The U.S.S.R. declared war following the atomic bombing of Hiroshima to ensure participation in the war before Japan surrendered or collapsed.

1945, August 9–17. Soviet Offensive. Bad weather contributed to Soviet surprise. Marshal **Rodion Y. Malinovsky**'s 3rd Far Eastern Army Group, making the main effort, swept eastward from Outer Mongolia on a broad front. Marshal **Kirill A. Meretzkov**'s 1st FEAG moved west and south from the line Khabarovsk-Vladivostok. Between these pincers General **Maxim Purkhayev**'s 2nd FEAG advanced south from the Amur River line. Japanese resistance was bitter, except that the bold and unexpected crossing of the almost trackless Khingan Mountains by the Soviet Sixth Guards Tank Army was virtually unopposed. When this army, closely followed by the remainder of Malinovskiy's troops, reached the Central Manchurian Plain (August 13), a strategic envelopment of the main forces of the Kwantung Army had been achieved, and Japanese resistance began to collapse. Russian spearheads drove Japanese troops back across the Yalu into Korea. The advance into Korea continued after the official cessation of hostilities (August 15). Scattered fighting throughout Manchuria and northern Korea ended soon after.

1945, August 14. Japanese Surrender. (See p. 1309.)

Operations in the Southwest Pacific Area

THE PHILIPPINES

Yamashita's Defensive Strategy. General Yamashita had 250,000 ground troops on Luzon, deficient to some extent in equipment, but still formidable. His air strength had been dissipated, with little support expected from other sources. He delayed and finally retired to the mountainous central and western regions of Luzon, where formidable defensive works had been prepared. He did not contemplate defense of Manila.

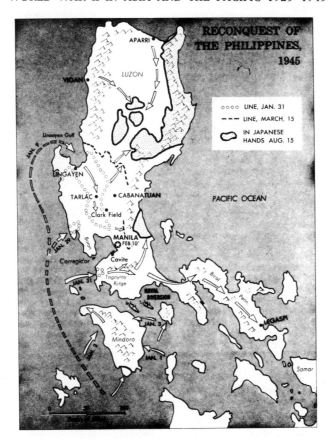

LUZON

1945, January 2–8. The Approach to Luzon.
Lifted by the Seventh Fleet, the Sixth Army
moved from Leyte through the Surigao Strait
and up the western coast of Luzon to Lingayen
Gulf. Planes from Halsey's fast carriers were
strafing Formosa, as was the China-based XX
Bomber Command. Kenney's planes from
Leyte and from the new-won base on Mindoro
(see p. 1291) attempted to neutralize Japanese
aircraft on Luzon, but kamikaze pilots took
heavy toll of Admiral Oldendorf's gunfire-
support group—both on the way and after it
started bombarding the Lingayen shore (January 6). One escort carrier was sunk, another
damaged. The battleship *New Mexico*, 1 heavy
and 4 light cruisers, and several other vessels
were severely damaged. The attacks petered
out as the Japanese eliminated themselves by
their suicide attacks, while Halsey's and Kenney's planes hammered their bases.

1945, January 9. Lingayen Landing.
Krueger's army landed, almost without opposition, on a 2-corps front, 4 divisions abreast. By
nightfall 68,000 men were ashore, from Lingayen to Damortis.

**1945, January 10–February 2. From Lin-
gayen to Manila Bay.** Major General **Oscar
W. Griswold**'s XIV Corps, on the right, pushed
aside Japanese delaying forces to reach the
Camp O'Donnell-Fort Stotsenberg-Clark Field
area (January 23), where it turned westward
for a week of serious fighting, while 1 regiment
pushed down to Calumpit (January 31) and
the marshy delta of the Pampanga River. Major

General **Innis P. Swift**'s I Corps, on the left, had more difficult going, fighting both to the east and south. Elements of his 1st Cavalry Division met the 37th Division at Plaridel (February 2).

1945, January 30–February 4. Arrival of the Eighth Army. Task forces of Eichelberger's Eighth Army made 2 amphibious landings north and south of Manila, Major General **Charles P. Hall**'s XI Corps in the Subic Bay area (January 30) seizing **Olongapo** to assist in sealing off the Bataan Peninsula, which was quickly occupied to prevent the Japanese from emulating the American defense of 1942. Two regiments of the 11th Airborne Division, under Major General **Joseph M. Swing,** landed at **Nasugbu** (January 31), southwest of Manila. Eichelberger then air-dropped the remaining regiment of the division on Tagaytay Ridge, 30 miles south of Manila (February 3). Next day, advance elements of the division reached Paranaque, on the southern outskirts of the city, where a stiff defense halted them.

1945, February 3–March 4. Recapture of Manila. While the 37th Division battered its way south into the city, the 1st Cavalry Division swung wide to east and then south, linking with the 11th Airborne Division. By February 22, the garrison had been driven into the old walled city (Intramuros) and Manila was isolated from Yamashita's main forces. American military and civilian prisoners held at Santo Tomás University and in Bilibid Prison were released. Inside Intramuros were the survivors of the fanatical Japanese garrison, some 18,000 strong—mainly naval personnel—commanded by Rear Admiral **Mitsuji Iwafuchi.** Ignoring Yamashita's instructions not to defend the city, Iwafuchi embarked on a suicidal defense, ruthlessly demolishing and burning many buildings. Street by street and house by house, the 37th Division now pried the defenders out. When the last resistance ended, historic old Intramuros was a mass of rubble, in which at least 16,665 Japanese lay dead.

1945, February 16–April 17. Clearing Manila Bay. Corregidor, "Gibraltar of the East," was assaulted by a skillful paratroop drop on its tiny

Topside golf course after intensive air and artillery bombardment (February 16). This was followed by an amphibious landing from Bataan and, after another suicidal defense, the fortress was cleared (February 27). Many of its garrison lay buried in blown-up Malinta Hill tunnel. Counted were 4,417 Japanese dead; 19 prisoners were taken. U.S. casualties were 209 killed and 725 wounded. Cargo began moving into the bay (March 15). **Fort Drum,** the "concrete battleship" on El Fraile rock, fell (April 13) when gasoline and fuel oil were poured down its ventilators and ignited, cooking the garrison. That same day a landing was made on **Caballo Island,** but its garrison was not subdued for 2 weeks. Carabao Island was taken without opposition (April 17).

1945, March 15–August 15. The Mountain Campaign. The weight of the Sixth Army's mass turned east, north, and south in an interminable series of assaults on Yamashita's mountain defenses. In the north, the North Luzon Guerrilla Force and newly reorganized Philippine Army assisted in pocketing the bulk of the Japanese forces in the Cordillera Central and the Sierra Madres. East of Manila, a smaller group held the ridges on the Pacific coast. When Yamashita surrendered at the end of the war (August 15), he still had an organized force of some 50,000 troops.

COMMENT. *The Luzon campaign cost the U.S. 7,933 killed and 32,732 wounded. Japanese losses were more than 192,000 killed and about 9,700 captured in combat. Yamashita, hampered by poor communications and incomplete information from his own high command, was still further impeded by the lack of command unity. It would appear that he did the best that could be expected under the circumstances. On the U.S. side, the ratio of battle deaths—24 Japanese to 1 American, a tremendous exception to the rule that the attacker must expect higher losses than the attacked—is tribute to the genius of MacArthur and the skill of Krueger.*

THE VISAYAS AND SOUTHERN ISLANDS

1945, February–August. Eichelberger's Campaign. Concurrently with the Luzon campaign, the Eighth Army was charged with the freeing of the remainder of the Philippines, garrisoned by elements of the Japanese Thirty-fifth Army—some 102,000 strong—under General Suzuki. He was directed by Yamashita to delay and defend as long as possible. He decided to concentrate his main defense on Mindanao. Beginning with Palawan, Eighth Army task forces made 50 amphibious landings. Barbey's VII Amphibious Force furnished part of the lifts, supplemented by Army Engineer amphibious special brigades. Air support was provided by Kenney's land-based planes. Each operation followed the same pattern: an assault landing, Japanese withdrawal to the interior, securing of the island, and departure of the troops, leaving mop-up to the extensive native guerrilla forces present.

1945, April 17–July 15. Mindanao. Here, with 2 divisions and some naval troops, Suzuki put up a stiff defense. Concentric landings from the north, east, and south finally pushed the Japanese into 2 groups in the center of the island, where they remained until conclusion of hostilities. Eighth Army casualties during the entire Philippine campaign totaled 2,556 killed and 9,412 wounded. Approximately 50,000 Japanese were killed in the Visayan-Southern Islands Campaign.

NEW GUINEA AND NETHERLANDS EAST INDIES

1945, January–August. The Bypassed Areas. Blamey's Australian troops were charged with containing and mopping up Japanese pockets of resistance in New Guinea and the Bismarck and Solomon archipelagos. Contenting himself with containment on New Britain, Blamey undertook aggressive operations on Bougainville in the Solomons and on New Guinea.

1945, May–August. Borneo. Tarakan Island on the northeast coast, Brunei Bay on the northwest coast, and Balikpapan on the southeast were all assaulted in amphibious moves lifted by Seventh Fleet task forces containing also units of the Royal Australian Navy. Air support was furnished by Kenney's air forces (also containing Australian elements).

1945, May 1–June 22. Tarakan. A brigade of the Australian 9th Division, with a detachment of Netherlands troops, was landed. Japanese resistance was determined, but finally overwhelmed.

1945, June 10. Brunei Bay. The Australian 9th Division quickly seized a wide perimeter and established airfields.

1945, July 1–10. Balikpapan. Following heavy naval and air bombardments, the Australian 7th Division landed and seized all major objectives.

COMMENT. *None of the operations in this area had any appreciable effect on the war. The Borneo operations were originally conceived as part of MacArthur's plan to recapture Java and restore the lawful government. As revised, its major objective was seizure of the principal oil-producing areas.*

Operations in the Pacific Ocean Areas

IWO JIMA

American Plans and Preparations. Possession of this pear-shaped rocky island in the Bonin group, only 8 square miles in area, was essential to U.S. advance toward Japan. In Japanese hands it menaced U.S. bombers from Saipan harassing the Japanese mainland at extreme range. In U.S. hands it would become a splendid forward air base. Nimitz's armada, now become the Fifth Fleet, with the Spruance-Turner command team replacing Halsey and his staff, moved against it, lifting Major General **Harry Schmidt**'s V Amphibious Corps—the 3rd, 4th, and 5th Marine divisions.

Japanese Plans and Preparations. Major General **Tadamichi Kuribayashi** commanded more than 22,000 Japanese army and naval troops garrisoning the island. Honeycombed with concealed gun emplacements, concrete pillboxes, and mine fields expertly interlaced, it was one of the most strongly fortified positions to be assaulted during the war. Kuribayashi elected a static defense, his troops protected in an elaborate underground cave system against the incessant bombings long preceding the assault and the 3-day naval bombardment immediately before it.

1945, February 19–March 24. Invasion. Schmidt came ashore (February 19) on the southeast coast, the 4th and 5th Marine divisions abreast and the 3rd in reserve, to be met by intensive fire—front, flank, and enfilade. Taking 2,420 casualties the first day, the Marines cut the island in 2, the 5th Marine Division turning left to assault Mt. Suribachi, the other divisions driving to the right. Naval gunfire gave continuous accurate support as the attackers inched forward through the volcanic dust. Mt. Suribachi was stormed (February 23) and the American flag raised on its crest. By March 11 the remnants of the defenders were

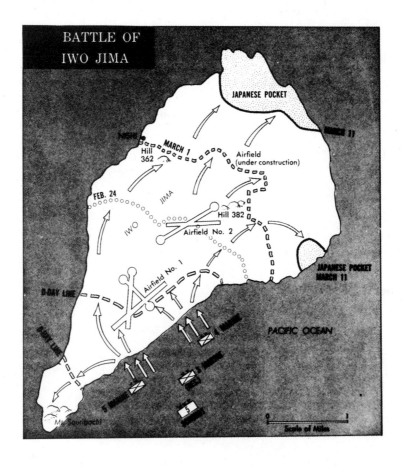

pinned on the northern tip of the island. Organized resistance ended (March 16). Next day, 16 B-29s returning from Japan made safe emergency landings, thanks to the U.S. Marine Corps.

COMMENT. *Slight Japanese aerial opposition was encountered. A kamikaze crashed on the carrier* Saratoga *(February 21), destroying 42 planes and killing 123 men; 192 more were wounded. Several other vessels were also damaged that day by kamikazes. Total Marine casualties were 6,891 killed and 18,070 wounded. Only 212 of the Japanese garrison surrendered. Counted dead were more than 21,000; many others were sealed in their underground shelters. Before the war ended, Iwo Jima had proved its usefulness; 2,251 B-29s had made emergency landings, sparing the lives of 24,761 U.S. airmen.*

OKINAWA

American Plans and Preparations. Final preliminary to invasion of the Japanese mainland was possession of the Ryukyu group, midway between Formosa and Kyushu, southernmost island of Japan. Operation "Iceberg" was organized for the task, with Okinawa, largest island in the group, the principal objective. Under Nimitz' command, Spruance's Fifth Fleet would lift U.S. Army Lieutenant General **Simon Bolivar Buckner**'s Tenth Army (some 180,000 in the assault force and additional reserves lying in New Caledonia): the XXIV Corps (Major General **John R. Hodge**) and III Marine Amphibious Corps (Major General **R. S. Geiger**). This would be the largest and most complicated amphibious expedition undertaken in the Pacific. Turner commanded the actual amphibious operation, while Mitscher's fast carrier force sought control of the air. Mitscher's TF 58 was supplemented by the Royal Navy's newly joined carrier force: 4 carriers and 1 battleship under Vice-Admiral **H. B. Rawlings,** R.N. The carrier forces' task was complicated by the fact that while Okinawa was beyond range of American land-based fighter planes, it was within reach of Japanese fighter planes from Formosa and Kyushu. Thus the carriers would have to supply cover from the expected kamikaze attacks, to which Japan was now evidently committed, as well as tactical support for the ground troops.

Japanese Plans and Preparations. On Okinawa was the Japanese Thirty-second Army, under Lieutenant General **Mitsuru Ushijima,** with some 130,000 men. As usual, an extensive defense system had been organized, particularly on the southern part of the island, in the midst of a native civilian population of more than 450,000.

1945, March 14–31. Preliminary Air Operations. To isolate the battlefield, Mitscher's carriers attacked Japanese air bases on Kyushu. Rawlings' British carrier force began neutralization of the Sakishima island group, midway between Formosa and Okinawa, strafing airfields and intercepting strikes destined for the battle area. Army long-range bombers attacked Formosa and industrial targets on Honshu. The big carriers received rough treatment from kamikazes; U.S.S. *Franklin* was so badly crippled that she had to be towed away by the cruiser *Pittsburgh,* and both the *Yorktown* and the *Wasp* were damaged. Special fire-fighting equipment devised by the New York City Fire Department saved all 3 vessels, but 825 officers and men were killed and 534 wounded. Temporary paralysis of the Japanese air bases was accomplished, and 169 of the 193 kamikazes committed were wiped out. (The steel-decked British carriers suffered less from kamikaze strikes than did the American teak-decked ships.)

1945, March 23–31. Preassault Operations. Admiral **W. H. P. Blandy**'s gunfire and escort-carrier ships of the amphibious force closed on

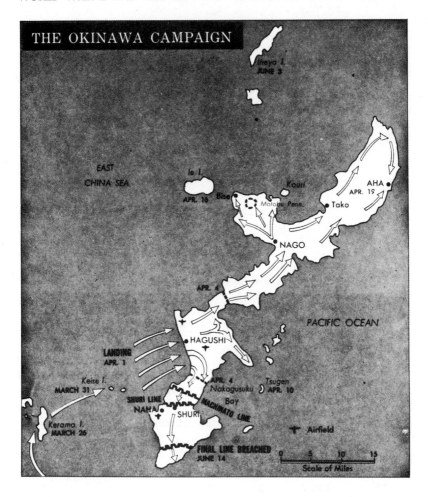

Okinawa and began (March 23) continuous air and artillery strafing. At the same time the 77th Division was lifted to the outlying Kerama Retto and Keise Shima island groups to secure a fleet anchorage and seaplane base in the Kerama roadstead and artillery positions on Keise. Capture on Kerama of some 350 small "suicide" boats—each carrying 2 depth bombs and some hand grenades, and manned by a 2- or 3-man crew—removed a potential hazard to the landing operations. The splendid roadstead at once became a valuable repair and supply base for the Navy.

1945, April 1–4. The Landings. While a simulated landing force demonstrated off the south-

east coast to distract Japanese attention, an 8-mile-wide column of landing craft disgorged from 1,300 naval vessels assembled off the Hagushi beaches on the western coast. It struck the beaches in 8 successive waves on a bright Easter Sunday morning—under cover of a tremendous air and artillery bombardment—and some 60,000 troops established their beachhead without any ground opposition. The Marine III Amphibious Corps, on the left, turned north; it would receive little opposition in clearing the entire northern area of the island (April 13) and nearby **Ie Shima** Island (April 16–20). The XXIV Corps, turning south, found itself (April 4) suddenly brought up short by the heavily organized **Machinato Line** of the Japanese Shuri Zone: an interlocking system of mountain defenses organized in great depth.

1945, April 6–7. Operation TEN-GO. The Japanese high command decided on a coordinated air-naval suicidal attack to check further American advance against the mainland. A kamikaze assault would cripple the amphibious force off Okinawa. The monster battleship *Yamato,* with 1 light cruiser and 8 destroyers, would crush what was left. The vessels, without any air cover, carrying only sufficient fuel for a one-way trip, but crammed with ammunition, hurried out of the Inland Sea (April 6) on the Japanese Navy's last desperate throw of the dice. Spruance, warned by U.S. submarines of the sortie, prepared to meet it with Mitscher's carrier force, backed up by his battle fleet.

1945, April 7. First Kamikaze Assault. Some 355 kamikaze and 340 additional orthodox dive bombers and torpedo planes swept in on the amphibious force. Although 383 of the attackers were lost, when the furious strike ended, 2 U.S. destroyers, 2 ammunition ships, a mine sweeper, and a landing ship had been sunk and 24 other vessels damaged.

1945, April 7. End of the *Yamato*. Vice Admiral **Sejichi Ito,** in the *Yamato,* leading his forlorn hope, was met by Mitscher's carrier planes shortly after noon. Wave after wave of American planes struck home on the monster's topsides with bombs and torpedoes. After nearly four hours of constant pummeling, the *Yamato* lay helpless on her beam ends. At 4:23 she slid under, with 2,488 officers and men. The cruiser, too, went down, as did 4 destroyers. The other 4, though damaged, managed to scuttle away. In all, 3,655 Japanese sailors met their deaths in this last fling of the Imperial Navy. U.S. losses were some 15 planes and 84 seamen and pilots.

COMMENT. *Had the Japanese high command concentrated its remaining air power to provide cover for* Yamato *until she reached Okinawa, she might have done severe damage to the amphibious force before her end came. The 42,000-yard range of her 9 18.1-in. guns far exceeded that of American battleships.*

1945, April 6–30. Advance and Stalemate.

The XXIV Corps battered its way south against increasing opposition. The Machinato Line was pierced (April 24), but finally the advance was brought to a standstill before the concealed positions of the **Shuri Zone**'s main line of resistance (April 28). General Buckner, with the 77th Division and the 1st and 6th Marine divisions of the III Amphibious Corps still in hand, was faced with a dilemma: to attempt a frontal attack or—as the Marines suggested—effect an envelopment by amphibious landings on the southern tip of the island behind the Japanese. He decided on a double envelopment—frontal attacks piercing both flanks—and reorganized on a 2-corps front, with the Marines on the right.

COMMENT. *The situation was analogous to that in Italy in 1943. The proposed Marine amphibious landing, like the Anzio operation, would have been beyond range of mutual support from the main front. On the other hand, Buckner had relatively more amphibious lift. The Marines and some army critics believe this was an error of overcaution.*

1945, April 12–13. Renewal of Kamikaze Assaults. Meanwhile, the Japanese suicide air strikes resumed. Nine more separate attacks were launched during the remainder of the campaign, until Japanese air strength had been expended. In all, more than 3,000 suicidal sorties hit the amphibious force and Mitscher's carriers, netting a total of 21 U.S. ships sunk, 43 additional permanently put out of action, and 23 more put out of action for a month or more.

1945, May 3–4. Japanese Counterattack. Ushijima hurled his 24th Division, supported by tanks and artillery, against the U.S. 7th Division on the eastern flank. A minor attack at the same time hit the 1st Marine Division on the western flank. A kamikaze attack from Kyushu supported the thrusts. Caught by American artillery and tactical aircraft, supplemented by naval gunfire, the Japanese offensive was crushed with loss of 5,000 men killed. U.S. losses were 1,066. In addition, the assault had disclosed the positions of heretofore hidden Japanese artillery.

1945, May 11–31. Buckner's Offensive. Under appalling weather conditions and despite bitter resistance, both flanks of the Japanese zone were finally pierced. Disengaging, Ushijima retreated slowly to the hill masses of the southern tip of the island.

1945, June 1–22. Final Operations. A climactic assault brought the 2 American prongs—Marine and Army—around and over the last organized position (June 22) and all resistance ceased. Neither Buckner nor Ushijima lived to see the end. Buckner was killed by an artillery shell as he peered from a Marine observation post on the front line (June 18) and Ushijima committed hara-kiri (ritual suicide), just before his headquarters was overrun, 4 days later.

Japanese losses on Okinawa totaled 107,500 known dead and probably 20,000 additional sealed in their caves during the fighting; 7,400 prisoners were taken. Tenth Army casualties were 7,374 killed and 32,056 wounded. Navy losses were some 5,000 killed and 4,600 wounded. About 4,000 Japanese planes had been lost in combat and an equal number destroyed from other causes. U.S. naval aircraft lost totaled 763. The Fifth Fleet lost 36 vessels sunk and 368 damaged.

COMMENT. *The campaign had cost Japan the remainder of her effective navy and much of her air force. Ushijima's tactical decision not to oppose the landings—probably dictated by the high command—was a grievous error. The tactical handling of the Fifth Fleet was brilliant. The logistical support of a great fleet at sea over a protracted period—unprecedented—was a most remarkable demonstration of efficiency.*

The Strategic Air Offensive

1945, January—February. Reappraisal. The B-29 attacks on Japan had not been as successful as the American JCS had hoped they would be. Japanese fighter defense had been more effective than anticipated, and the loss rate of 6 percent per mission was lowering morale and efficiency of the bomber crews. The weather over Japan at 30,000 feet was worse than had been expected; winds of several hundred miles per hour had thrown bombers off course and decreased effectiveness of their attacks; ice forming on the wings and on the windshields

affected plane performance, obscured pilots' vision, and reduced accuracy of instruments and bombsights. Fog often made visual bombing impossible, and radar was not sufficiently accurate for precise concentration of bombs on targets from a small number of planes. Nevertheless, the raids were having some effect on Japan. The strain of frequent alerts reduced worker efficiency and lowered the morale of the whole population. Furthermore, the concentration of attacks on aircraft plants had forced the dispersal of these plants, causing a decline in Japanese aircraft production. Since the operations from the Marianas were more effective and less costly than those through Ch'engtu in China, it was decided to shift the XX Bomber Command from India to the Marianas so as to bring the whole Twentieth Air Force (General **Nathan F. Twining**) together. Meanwhile, General LeMay, because of his experience and success with the XX Bomber Command, was ordered from India to the Marianas to command the XXI Bomber Command.

1945, February–March. Developing New Tactics. The dispersal of Japanese aircraft plants, plus continuing high losses among bomber crews, caused General LeMay and his staff to decide to change the tactics to try a new bombing method that LeMay and Chennault had worked out in joint operations in China: low-level incendiary attacks instead of high-level, high-explosive bombing attacks.

1945, March 9–10. The Tokyo Raid. A night attack by 334 B-29s, flying at about 7,000 feet, initiated the new tactics. A total of 1,667 tons of incendiary bombs was dropped in the most destructive single air raid in history. Widespread fires created a fire storm similar to those which had devastated Hamburg, Dresden, and Darmstadt in Germany. Fifteen square miles of the city were destroyed; more than 83,000 people were killed, nearly 100,000 injured.

1945, March 9–19. Inaugurating the New Phase of Strategic Bombardment. A new pattern of operations was created as the success of the Tokyo raid was repeated 4 times in 10 nights. The low-level attacks permitted the bombers to carry nearly three times heavier bomb loads than those they could lift to 30,000 feet. But because of the danger of Japanese antiaircraft fire and Japanese fighter defense, the raids were made at night. A total of 9,365 tons of incendiary bombs was dropped, 3 times the weight of bombs dropped on Japan during the previous 3½-month period. Thirty-two square miles of the most important industrial areas of Japan were destroyed. Perhaps most important, only 22 B-29s were lost on these raids—1.4 percent of 1,595 sorties. This dramatic reduction in losses was due to ineffectiveness of Japanese fighters in night operations, plus the use of Iwo Jima as a navigational aid and an emergency landing field. Morale of the Superfortress crews rose, and so did the efficiency of their operations.

1945, April 7. Appearance of American Escort Fighters. The VII Fighter Command, moved to Iwo Jima shortly after its capture, began to send its long-range P-47 Thunderbolt fighters, and its even more effective P-51 Mustangs, as escorts with the B-29s, which now began to mix medium-level daylight bombing raids with the low-level night raids. Losses of Japanese fighter planes mounted rapidly, while those of the B-29s continued to decline. American planes soon had clear air superiority over the heart of Japan (May–August).

1945, May–August. Climax of the Strategic Campaign. The arrival of the planes of the XX Bomber Command from India and China (April) increased the intensity of the raids. This continued to mount steadily as more air strips were built in the Marianas and as more planes arrived from the United States. The industrial areas of Tokyo, Nagoya, Kobe, Osaka, and Yokohama—Japan's 5 largest cities—were completely destroyed. The morale of the Japanese people plummeted. Millions had become homeless, and the government could not provide them with places to sleep. Food became scarce as the effects of the bombing on the transportation system, combined with Allied naval blockade, began to have its effect. And as the B-29s began to drop mines in Japanese ports, the blockade became even more effec-tive. By midsummer the strategic bombing offensive had brought Japan to the verge of economic and moral collapse.

The Collapse of Japan

RELENTLESS PRESSURE

1945, January–August. The Submarines and the Blockade. American submarines had been the first to bring the effects of the war home to the Japanese people through a stringent blockade, which began to have some effect in 1943. In these final months of the war, they roamed the waters of the China Sea and the coasts of Japan, making the blockade ever more effective. During the war they sank more than 1,100 Japanese merchant ships, totaling 4,800,000 tons of shipping—over 56 percent of the total lost by Japan. They also sank 201 warships, totaling 540,000 tons, including 1 battleship and 8 aircraft carriers. Some 52 American submarines were lost, most without a trace, unknown and unsung heroes of the war.

1945, June–August. Preparation for Invasion. The staffs of General MacArthur and Admiral Nimitz completed plans for 2 massive invasions of the main Japanese islands. One, Operation "Olympic" (November), was to strike southern Kyushu. Operation "Coronet" was to follow in March 1946, the mightiest amphibious force ever assembled, against the Tokyo plain of Honshu. Meanwhile, 3 main forces were setting the stage for these assaults by unrelenting onslaughts against the Japanese Empire: (1) the submarines of the Pacific Fleet (see above), (2) the B-29s of the Twentieth Air Force (see p. 1307), and (3) the Fast Carrier Task Force (see below).

1945, July 10–August 15. Last Raids of the Carriers. After its long ordeal off Okinawa (see p. 1306), Mitscher's Task Force 38 (once more under Halsey) rested briefly at its Ulithi anchorage, repairing damage and taking on supplies and replacements. Mitscher then sped back to the coast of Japan, where his planes added to the terrible punishment being delivered by the

B-29s. The Fast Carrier Task Force included more than 1,000 American and 244 British planes. During these 5 final weeks of operations, the Americans lost 290 planes, the British 72, mostly to Japanese antiaircraft fire. During that time, they destroyed more than 3,000 Japanese planes in the air and on the ground.

THE ATOMIC BOMBS

1945, July 16. Explosion at Alamogordo. The first experimental atomic bomb was exploded at Alamogordo, N.M. Meanwhile, the 509th Composite Group of the U.S. Army Air Forces was established, and carried out secret training, in preparation for dropping atomic bombs, in the desert areas of Utah.

1945, July. Debate on Employment of Atomic Weapons. Among the few U.S. officials who knew about the atomic-bomb possibility, a high-level debate took place on whether these terrible weapons should be used and, if so, how. Some felt it would be immoral to use such destructive weapons. Others felt that it would be unnecessary, since they expected Japan to collapse soon anyway because of the blockade and the strategic-bombing offensive. After careful study, however, the JCS and a group of senior officials headed by Secretary of War Henry L. Stimson recommended that the bombs be employed. They believed that the shock of the terrible explosions might convince Japanese military leaders that they could not hope to continue the struggle, and that the very destructiveness of the weapons would give Japanese military men an excuse to surrender. Otherwise, the Americans feared, Japan's 4 million undefeated soldiers would continue fighting in Japan, China, and Manchuria, killing many thousands of Allies as they went down to defeat. Thus, even though these bombs would cause terrible losses, American leaders believed that hastening the end of the war would eventually save more lives than the bombs would take.

1945, July 27. Potsdam Proclamation. At the final war meeting of the heads of state of the U.K., U.S., and U.S.S.R., Truman and Prime Minister **Clement Attlee** (who had just replaced Churchill) issued a virtual ultimatum to Japan. The circumstances indicated that Stalin approved, even though Russia and Japan were not at war. Japan was warned that if she did not surrender at once, it would mean "the inevitable and complete destruction of the Japanese armed forces and . . . the utter devastation of the Japanese homeland." Japan attempted to reply through the U.S.S.R., but the Kremlin did not relay the message. Since Japan seemingly did not reply, President Truman therefore approved the plans for dropping atomic bombs on Japan.

1945, August 6. Hiroshima. This city of more than 300,000 people was an important military headquarters and supply depot. It had not yet suffered severely in the bombing offensive. Early in the morning, an air-raid alert was sounded and most people took cover. Realizing, however, that there were only 2 or 3 planes overhead, many people came out, and so most of the population were unprotected when the bomb exploded over the center of the city. Two-thirds of the city was destroyed; 78,150 people were killed (most outright, in explosion or in fires, though some died later from radiation effects); nearly 70,000 others were injured and most of the remaining population suffered long-term radiation damage.

1945, August 9. Nagasaki. The second bomb was exploded over this seaport and industrial city of 230,000 people. Because hills protected portions of the town, less than half the city was destroyed; nearly 40,000 people were killed, about 25,000 injured.

1945, August 9. Soviet Invasion of Manchuria. (See p. 1298.)

THE SURRENDER OF JAPAN

1945, August 10. Japan Offers to Surrender. The effect of the atomic bombs was what Stimson's group and the JCS had expected. No one knows how long a fanatically militaristic Japan could have continued the war if the bombs had not been dropped. It is clear, however, that these weapons, combined with Soviet entry into the war, convinced the Japanese emperor and government that further resistance was hopeless. After brief negotiations by radio, Japan accepted Allied terms for "unconditional surrender," which actually included conditions the Japanese were eager to accept. They were permitted to retain their imperial form of government, and were assured of the integrity of the 4 main islands of the Japanese Empire.

1945, August 15. Cease-Fire. Japanese forces throughout Asia and the islands of the Western Pacific laid down their arms. General MacArthur and American troops began to fly to Japan (August 28).

1945, September 2. The Official Surrender. Representatives of the Japanese government surrendered on board the battleship USS *Missouri* in the heart of the Pacific Fleet, anchored in Tokyo Bay, while American planes filled the air overhead. General MacArthur, appointed by the Allied governments as Supreme Commander for the Allied Powers to initiate the occupation of Japan, received the official surrender of Foreign Minister **Mamoru Shigemetsu.** Among the observers of the ceremony were released prisoners American General Wainwright and British General Percival, whose respective commands in the Philippines and Malaya had been overrun at the outset of the war.

THE COST OF WORLD WAR II[a]

Nations	Total Forces Mobilized (million)	Military Dead	Military Wounded	Civilian Dead	Economic and Financial Costs ($ billion)
United States	14.9	292,100	571,822	Negligible	350
United Kingdom	6.2	397,762	475,000	65,000	150
France	6	210,671	400,000	108,000	100
Soviet Union	25	7,500,000	14,012,000	10–15,000,000	200
China	6–10	500,000	1,700,000	1,000,000	No estimate
Germany	12.5	2,850,000	7,250,000	500,000	300
Italy	4.5	77,500	120,000	40–100,000	50
Japan	7.4	1,506,000	500,000	300,000	100
All other participants	20	1,500,000	No estimate	14–17,000,000[b]	350
Total[c]	105	15,000,000	No estimate	26–34,000,000	1,600

[a] Many of these figures are approximations or estimates, since official figures are misleading, missing, or contradictory in many instances.

[b] This includes approximately 6,000,000 Jews of Germany and all occupied European nations, and approximately 4,500,000 Poles.

[c] In economic and financial costs, World War II was about 5 times as expensive as World War I: in military deaths alone, it was almost twice as costly; it was about 3 times as destructive in total deaths.

XXI

SUPERPOWERS IN THE NUCLEAR AGE:

1945–1975

MILITARY TRENDS

GENERAL HISTORICAL PATTERNS

The period opened at one of the most significant crossroads of world history: inauguration of the Nuclear Age. Man, it seemed, had discovered the power of mass self-destruction (see p. 1308). When the U.S.S.R. in 1949 detonated a nuclear explosion, canceling U.S. monopoly of the secret, the menace chilled the world, threatening as it did mutual annihilation should a general war arise between 2 major powers.

War as the extension of diplomacy, it seemed to many persons, had reached the end of its usefulness. Yet wars, and the threat of war, continued while military men struggled in a fog of experimentation and frustration, seeking on the one hand to harness this new element in weaponry, and on the other to maintain and improve postures of the so-called "conventional" means of warmaking without resort to the use of nuclear fission or fusion.

The confused, tangled, and momentous events of the decades following World War II do not readily yield to systematic historical analysis, partly because they are still too close to be seen in true historical perspective, partly because it was a time of rapid transition. In terms of policy and strategy, and their influence on military affairs, however, 8 major historical patterns seem to have emerged.

CONFRONTATION OF THE SUPERPOWERS

Political and military strategy of all nations was decisively influenced by the confrontation between the United States and the Soviet Union. The 2 superpowers towered over the other nations of the world, and the most important strategic fact of life for each was the existence and power of the other. The friction, or conflict, which resulted from this confrontation was called the "Cold War." Although, by the end of the period, Cold War grimness had been somewhat ameliorated by determined American and Soviet efforts to achieve a form of "peaceful coexistence," which both sides preferred to call "detente," the confrontation was still the major factor of world affairs, and relations were frequently chilly.

The Technological Revolution

The technological developments of the period may have been even more important historically than the confrontation between the U.S. and the U.S.S.R. In any event, the most significant development strategically was the emergence of nuclear power, although this was hardly more momentous than the beginning of the human conquest of space. The rocket power propelling a multiplicity of proliferating weapons systems was mainly responsible for thrusting vehicles into outer space. Manmade satellites girdled the globe; the U.S.S.R. and U.S. both put manned space ships into orbit and retrieved them safely; and American astronauts trod the moon in 1969 and subsequent years.

Other major technological developments included electronic communications equipment, enabling telegraph, telephone, and radio to be used in ways earlier undreamed of, reducing the time necessary to transmit ideas and the time available to react to events. The related developments of automation and computerization opened new areas of research and accomplishment; new and improved means of transportation reduced travel time, making the world seem ever smaller.

The Political and Economic Revolutions

The rapid break-up of colonial empires in a very brief span of time created many new states where often there was no group adequately prepared as leaders. Related to the end of colonial rule, and also manifested in countries other than the former colonial dependencies, was the phenomenon of "rising expectations" among the populations of the underdeveloped nations. But the ambitions of those poorer nations were frustrated in many ways. The result was increased unrest and instability in these nations and in vast regions of the globe.

The Conspiracy and Challenge of International Communism

The confrontation of 2 superpowers would undoubtedly have involved dangers and crises even if the idea of communism had never existed. Nevertheless, the scope of the Cold War, the violence and the brutality which marked the world political revolution, and a large measure of the bitterness and hatred in the postwar world stemmed directly from the conspiracy of international communism. There were, of course, a number of basic factors which generated communism and which made many people susceptible to its doctrines and promises: poverty, political and social injustice, and authoritarian attempts to maintain an intolerable *status quo*. The Soviet Union and, to an increasing extent, Communist China took advantage of the susceptibility to construct a net of Communist parties throughout the world, and to exploit disorders and unrest in the underdeveloped regions.

The Growing Importance of International and Regional Organizations

This was an era of internationalism. Despite its failures and weaknesses, the United Nations enjoyed a number of successes in restoring or maintaining peace in troubled regions of the world. The concept of regional political organization, more or less within the overall concept of the United Nations but involving only interested nations in problems that are primarily regional and local, was developed in several areas of the world.

The Trend toward Polycentrism

During the latter part of the period, centrifugal forces developed in both of the great sets of alliances which the superpowers built up in the years following World War II. Although both the United States and the Soviet Union were embarrassed and frustrated by this trend, the U.S.S.R. at first suffered more than did the U.S. This was due primarily to the fact that China, under aggressive and militant leadership, had become a leading Asian power and was explicitly challenging the leadership of the U.S.S.R. in the Communist world. The score was at least evened, however, by American loss of prestige in Southeast Asia, where the U.S. was unable to impose its will on a small and relatively underdeveloped nation, North Vietnam, in hostilities short of a major national effort—which the United States was unwilling to undertake.

Efforts toward Arms Control and Disarmament

Over the previous century there had been a number of attempts to establish controls and limitations on the employment of armed force among nations. There had been nothing to compare, however, with the worldwide intensity, sincerity, and sophistication of the search for arms control of the post–World War II era. This was in large part the result of growing realization by all mankind of the potentialities of nuclear weapons, and a feeling of desperate need to control the use of such weapons before they are allowed to destroy civilization.

New Elements in Military Strategy

If only because of the development of nuclear weapons, military men were faced with the greatest problems of transition in military concepts, doctrines, and strategies that have been known in the history of military affairs.

Weapons—or at least evaluations of weapons—for the first time became determinants of strategy instead of merely implements of strategists, possibly because strategic thinking had not yet reached a comparable development and sophistication. Certainly the very threat or possibility of the use of nuclear weapons was a principal factor in shaping the nature of 2 of the major clashes of the period—the Korean and Vietnamese wars—and influenced the adjudication of a host of minor wars and crises. In any event, from the availability of these weapons grew a new concept of limiting conflicts through possession of almost unlimited power. The concept of deterrence, known to man since force was first used against another human, achieved new importance, subtlety, and refinement.

Practically all of the other historical patterns noted above affected modern strategic thinking in one way or another. For instance, from the Communist conspiracy emerged a new concept of "wars of liberation," in which an ideology harnessed and modified old and well-known methods of making war. The technological revolution brought a weapon system—the long-range strategic bomber—to perfection and had already made it obsolescent. From the impingement of the technological revolution and the economic revolution upon military affairs emerged an essentially civilian concept of cost effectiveness to affect all modern strategic thinking.

As a result of these changes, the organizational arrangements for coordinating the military and political efforts of the nation, and for managing and directing the armed forces, were dramatically changed—particularly in the United States—during these years in an evolutionary process that was not always clearly logical and progressive, but which endeavored to adapt the armed forces and their civilian leadership to the changed environment of conflict.

STRATEGY AND CONFLICT MANAGEMENT

These new elements in military strategy did not seem to alter the fundamental nature of strategy,* but they immeasurably complicated and confused the processes of formulating strategy, and of exercising it in situations of conflict or likely conflict. Because of these complexities, the term "conflict management" began to be used to describe the control or direction of the resources at the disposal of the strategist or "manager of conflict."

The availability of nuclear weapons, and the almost certain catastrophe which would result from their use in any but the most controlled and limited fashion, divided warfare, and the concepts relating to warfare, into 2 distinct levels: the upper level of conflict, involving a general or "strategic" war of nuclear exchange between the superpowers; and a lower, or "conventional," level in which nuclear weapons would probably not be employed—but their possible use could never be discounted. This lower level, in turn, was considered to have various sublevels, or special forms, of violence. These differentiations were not at all new, but the distinctions took on a new importance not only because of the ever-present possibility of nuclear war but also because of the manner in which Communist theory and practice both exploited and fostered different kinds of violence in support of the nationalistic and international aims of the Communists.

There was not full agreement among military theorists and strategists as to the exact nature of these differentiations and distinctions, but both the significance and the complexities of the concepts can be understood by looking at a "spectrum of conflict" which was produced in one significant theoretical study at the end of the period.

The most significant aspect of strategic theory regarding the upper level of conflict was the concept of **deterrence:** the effort to make the cost of resort to nuclear weapons too high to be

---

* Based upon an intensive analysis of conflict experience and theoretical development during the period, in 1966 the Historical Evaluation and Research Organization defined 3 aspects of strategy as follows: Strategy—the art of employing all available resources for the purpose of achieving a sucessful outcome in a conflict of human wills. National Strategy—the art of employing all resources available to the highest national authority for the purpose of achieving a successful outcome in a conflict between nations. Military Strategy—the art of employing all resources available to a military commander in an actual or anticipated conflict against hostile, or potentially hostile, armed forces in order to carry out requirements established by national strategy.

profitable. In the United States during this period there was an intensive intellectual effort devoted to the development of theories of deterrence. The result was a series of deterrent strategies. At one extreme, and in vogue in the mid-1950s, was the brutal, direct concept of "massive retaliation": the threat to employ full nuclear force against the instigator of any form of aggression, direct or indirect, conventional or nuclear. The more subtle doctrine of "controlled and flexible response" prevailed in the United States at the close of the period.

The failure of "massive retaliation" demonstrated one of the serious strategic problems posed by the mere availability of nuclear weapons. It is not reasonable to employ truly cataclysmic means to combat threats or dangers which themselves are less than catastrophic. Neither friends nor enemies believed that the U.S. would attack Moscow in response to Kremlin-supported subversion in some remote region of the underdeveloped world. America's friends in Europe were not even certain that if the chips were down they could rely on America's assurances of immediate nuclear response in the event of Soviet aggression in Western Europe. These doubts were affecting both the vigor and stability of the western alliance at the close of the period.

The North Atlantic Treaty Organization (NATO) had emerged early in the period as counterfoil to Soviet aggression in Europe. NATO was unique in that its integrated armed forces, under a single command and backed by U.S. nuclear power controlled by the president of the U.S., constituted the first significant multinational force to arise in peacetime. Effective from the first as a deterrent, its usefulness was marred toward the end of the period by France's intransigent "independent" policy, which included a separate French striking force (*force de frappe*) with equally independent nuclear-power potential.

As the period closed, U.S. strategists had suffered a major defeat in Vietnam in an effort to cope with Communist aggression in an environment in which the existence of nuclear weapons automatically encouraged low-level violence under the "umbrella of mutual nuclear deterrence" caused by the nuclear stalemate between the superpowers.

Nevertheless, the United States had developed a strategic policy of military assistance to nations which wished to oppose Communist aggression. Despite the frustrations of Vietnam— and many other frustrations—this policy had been generally successful as a means of projecting U.S. power and influence in the exercise of national strategy in the nuclear age.

On the other hand, the opposing Communist strategy had also enjoyed considerable success. Both the U.S.S.R. and Communist China, exploiting the situation created by the umbrella of mutual nuclear deterrence, kept up incessant thrustings to produce turmoil, expand their respective territorial holdings, and repair damages. In brief, this consisted in exploitation of conflict in 3 classes: civil wars, wars for "national liberation," and wars in defense of Communist regimes. The possible effect of the Sino-Soviet "split" on this strategy was still unclear at the close of the period.

LEADERSHIP

One of most unique aspects of this period was the manner in which top-level political leadership became more and more involved in the detailed and day-to-day military operations of military forces in the field and at sea. Outstanding examples were President **Truman**'s relief of General MacArthur from command in Korea (see p. 1361), President Eisenhower's assumption of responsibility in the U-2 incident (see p. 1453), French President de Gaulle's intervention in

Unlimited Nuclear Exchange

Nuclear Threshold
The Firebreak

Threshold of Overt Hostilities

Threshold of Violence

Observable Conflict of Opposing Objectives

Nuclear War

Strategic War
- Countervalue Strategic Operations
- Counterforce Strategic Operations
- Limited Strategic War

Controlled General War

General War

Tactical Nuclear War

Subnuclear War

Major Power Confrontation Crisis (Direct or Indirect)

"Conventional" Limited War

"Unconventional" or "Sublimited" Conflict

Revolution   Insurgency

Civil War

Internal Conflict

Revolution

Guerrilla Warfare - - - →

Insurgency

Subversion

Terrorism

Guerrilla Warfare - - - →

Limited War

Nonviolent Conflict

Bloodless Coup d'Etat

Passive Resistance

Revolution   Insurgency

Low-Intensity Conflict
The Lower Level of Conflict

(Economic Warfare)
(Propaganda)
(Extension of Influence)
(Military Assistance)
(Show of Force)
(Stabilization Operations)

Grada-
tions

and

Special

Forms

of

Conflict

Major
Catego-
ries of
Conflict

Algeria (see p. 1433), the apparently unfortunate control exercised by President Kennedy and his civilian staff in the Cuban debacle of the Bay of Pigs (see p. 1460), Kennedy's control of the so-called "quarantine" which successfully eliminated the Soviet missile threat in Cuba (see p. 1384), and President Johnson's direction of the employment of air power against North Vietnam (see p. 1414).

These actions in no way suggested a reversion to the old tradition of the king-general. They resulted from political recognition of the fact that any conflict situation had within it the potentiality of escalating to uncontrolled nuclear war. Thus a military man, acting on his own initiative in response to a local military situation in traditional (or unconventional) military fashion, could "escalate" a conflict out of control. Civilian leadership's increasing control over operations in or near conflict situations was merely a reflection of the fact that responsible authority must retain to itself the capability to determine what risks are to be taken, and how.

On the purely military side, in this period, **Douglas MacArthur**'s stature as one of the world's great captains was confirmed by his handling of the early operations of the Korean War, with the Inchon landings the capstone of his career. He was later wrong in disputing strategy with his commander in chief and he must also bear considerable personal responsibility for the subsequent temporary defeat of his U.N. forces by unexpected intervention of Chinese Communist armies in late 1950. This reverse no more detracts from MacArthur's military genius than does Waterloo negate Napoleon's overall military stature or Gettysburg that of Robert E. Lee. **Matthew B. Ridgway**'s welding of the multinational U.S. Eighth Army in Korea into a competent war machine, and **James A. Van Fleet**'s later control of that army, were outstanding examples of successful troop leadership. The rapid defeat of the Chinese Nationalists by the Chinese Communists in the Chinese Civil War was evidence of **Mao Tse-tung**'s exceptional political-military genius.

Thus, before the period closed, it became quite evident that leadership in this new era of instantaneous decisions and cataclysmic potentialities was being exercised in a totally new way, a leadership linking field commanders to their political overlords in an increasingly complicated mesh of "command and control" systems far removed from Tennyson's "Theirs not to reason why, / Theirs but to do and die." Unless the field commander knew and adhered to the political objectives and directives of the civilian control, a tactical victory, it seemed, might be ephemeral. In other words, imaginative, creative military initiative must wear the harness of political limitation.

How to prevent subsequent galling and chafing from this tight-fitting harness, and still maintain military equilibrium on future battlefields, was at the period's close a basic problem confronting all commanders—from squad leader to general-issimo. One soldier who performed well under these constraints, however, was America's **Creighton W. Abrams** in Vietnam. Also performing well under similar constraints, although for far different reasons, was North Vietnam's brilliant **Vo Nguyen Giap,** the architect of Viet Minh victory over France and of North Vietnam's ultimate triumph over the United States and South Vietnam.

WEAPONS, DOCTRINE AND TACTICS

General

On land, sea, and in the air, rocketry, which had emerged during World War II after a century-long discard, became a major element of artillery. Its projectiles, capable of carrying either nuclear or high-explosive warheads, ranged from the giant intercontinental missile (ICBM), with practically worldwide range, to the one-man "bazooka" type of the foot soldier. Rocket missiles fell into 4 general classes: surface-to-surface, surface-to-air, air-to-surface, and air-to-air.

During midperiod, the fear of long-range nuclear bombing by long-range bombers or by intercontinental missiles resulted in abortive attempts to preserve the civil populations from holocaust. By the end of the period, the concept of civil defense had seemingly become less important for its potential saving of life—or reduction of catastrophic casualties—than as an element of the psychological considerations relating to nuclear-deterrent strategy.

Another new aspect of the terror of general war was furnished by the threat of contaminating an enemy country, poisoning atmosphere and vegetation by toxic measures: chemical, biological, or radiological. Communist propaganda played loudly on this theme as part of the strategy of terror. Factually, military scientists in all major nations were studying ways and means of prosecuting chemical and biological warfare as the period drew to a close.

On the Ground

Ground combat during the period produced a series of apparent abnormalities and contradictions, none of them vitiating fundamental principles of war, yet necessitating alterations in doctrine and tactics. The Korean War and later operations in Southeast Asia proved the serious limitations of mechanized-warfare operations in jungle and mountain terrain. Mobility of tactical maneuver—the concentration of fire by movement—depended in such regions on the ability of fleet-footed infantrymen to scramble across country.

The threat of nuclear warfare, with its consequent wide devastation, necessitated dispersion, yet ability for rapid concentration remained equally essential, conditions mutually contradictory. Furthermore, the multiplicity of communications essential to command control under such conditions became an enormous problem.

The helicopter, useful for reconnaissance, supply, transport of small packets of troops, and evacuation of wounded, supplanted the jeep as ubiquitous cross-country vehicle. In U.S. combat doctrine, so-called "sky cavalry" units utilized the helicopter to enhance the cross-country capability of the infantryman, and to provide him with flexible combat and logistical support. Through a confusing welter of statistical claims and counterclaims, the mobility of the helicopter, as well as its vulnerability to ground fire, were clearly demonstrated in the guerrilla-warfare operations in Vietnam.

Still another contradiction was to be found in an experimental trend in hand-weapon usage to the theory of the cone of fire rather than the individual aimed fire of the trained marksman. U.S. ordnance theorists seemingly turned the clock back a century and a half by equipping the American rifleman with a lighter weapon capable of firing 3 different loads: a .22-cal. bullet, lacking the shock action of the heavier calibers; a species of pocket shrapnel (a sheaf of fine steel arrows—*flechettes*); and a cartridge containing 2 bullets—a throwback to the "buck and ball" loads of smoothbore days. This trend seemed to ignore the fact that a host of electronic and optical improvements were increasing the lethal potential of the individual marksman.

The answer was that increased rates of fire were permitting the application to infantry firepower—consciously or not—of concepts of massed fire already pioneered by American field artillery. Reasonable accuracy, combined with such massing, assured effective neutralization of an enemy, and probably as much or more lethal effect as would have been provided by a smaller volume of more accurately aimed fire. The advantages were undeniable, so long as there was a logistical capability to "feed the battle." There remained, nonetheless, a need for precision, aimed fire among infantrymen as well as in the artillery. There was also evidence that the bayonet, as proven in the Korean War, was still a potential arbiter in hand-to-hand combat.

The days of the elite versus the mass, foreshadowed long ago by the grenadier and the light infantryman, and later by the *Sturmtruppen,* paratroopers, and Special Forces of World War II, seemed to be returning. The green beret of U.S. Special Forces in Vietnam became a status symbol, denoting a superior type of warrior. Whether or not this selectivity would survive cost-effectiveness scrutiny, or whether it might in the long run produce a deleterious inferiority complex in the run-of-the-mill foot soldier, were still moot questions.

The tank, which had emerged from World War II as the major ground battlefield weapon, retained its primacy, actual and potential, in those parts of the world where geography permitted its effective employment. This was demonstrated in the Indian-Pakistani battlefields in the arid Punjab-Kashmir region, as well in the lush farmlands of the Ganges Valley. Tanks dominated the operations of Arabs and Israelis in the Second and Third Arab-Israeli wars. This was only partially modified by the results of the Fourth Arab-Israeli War, in which the dominance was reduced, but not eliminated, by determined infantrymen armed with relatively inexpensive, effective antitank rockets and guided missiles. Despite a partial restoration of balance between infantry and armor in that war, it still seemed that—at least in the mid-1970's—the most effective antitank weapon was still the high-velocity gun of a hostile tank.

At Sea

The essential importance of sea power and its objectives remained unchanged during the period, as attested by the U.S. Navy's unchallenged control of the high seas. Significant trends were the projection of sea power inland and the logistical utilization of surface traffic, demonstrated during the Korean War and later operations in Southeast Asia and Middle East waters.

Nuclear power, combined with advances in rocketry, brought radical changes in both weaponry and propulsion, necessitating equally radical changes in ship construction. The nuclear-powered vessel, surface and submarine, could remain at sea indefinitely without replenishment of fuel if given periodic replenishment of victuals and ammunition from floating depots—an extension of the logistical use of the sea lanes developed by the U.S. Navy during World War II.

The carrier task force remained the queen of maritime weapons systems, although some observers doubted its survivability in general nuclear war. The nuclear-powered submarine, armed with Polaris-type missiles, launched subsurface, extended sea power inland as never before. To many sailors this was the prototype of the future capital ship.

Freedom of the seas from the always increasing threat of submarine warfare and the *guerre de course,* so menacing in 2 world wars, necessitated further development of the hunter-killer doctrine—fast, mobile weapons systems of surface and submarine craft, aided by naval air power, utilizing elaborate electronic instrumentation to detect and destroy enemy underwater craft.

During midperiod, rocket launchers appeared to be displacing tube artillery on naval craft.

Nuclear-powered attack submarine

However, as demonstrated both in Korean waters and very definitely later in the South China Sea, the precision fire of tube artillery was essential both in landing operations and in coastal actions involving small craft; as on land, the theorists were faced with the necessity for establishing proportional quotas of both massed and precision firepower, of both rocket and tube weapons.

Conventional landing operations faced almost certain destruction by nuclear weapons should general war break out. However, for smaller amphibious operations with so-called conventional weapons, the helicopter appeared to be the answer, with vertical assaults largely replacing the surface assaults. But the apparent vulnerability of this item of equipment placed an even greater requirement for control of the air than had been the case in World War II and Korea.

In the Air

Air power's principal problem centered about one crucial question—the respective capabilities of the manned bomber and the missile in delivering nuclear warheads. On the right choice rested, the efficacy of the free world's doctrine of deterrence. As the period ended, opinion was still divided and compromise governed.

Meanwhile, in both missile and manned-aircraft fields, important developments brought more speed and range in all areas of the air-power spectrum. Rate and range of climb increased. Concurrently, technological advances in optics went hand in hand with increased altitude for reconnaissance. The necessity that air and land power become a close-knit tactical team became more apparent. As a result, at least in the U.S., the Tactical Air Force drew closer to ground forces and its separation from the Strategic Air Force became wider.

The problem was far from being solved, however. The tactical use of fixed-wing aircraft in close support was complicated by the fact that the increased speed of jet-propelled combat aircraft restricted its firepower efficacy to a minimal fleeting moment over the target, thus greatly reducing the pilot's capability for identification and delivery of fire. This same condition also impeded the value of fast-flying aircraft in reconnaissance, while vast refinements in ground-force antiaircraft artillery range, fire control, and direction increased the capabilities of defense against air power, regardless of the speed of the aerial vehicles. The complexities of the problem were increased by controversial attitudes on the type and method of close air support and the necessity for an instantaneous and almost infallible system of command control if air superiority and isolation of the battlefield—an enormous area were nuclear warfare contemplated—could be attained.

Nevertheless, by the end of the period, largely as a result of experimentation in Korea and Vietnam, the United States had developed a variety of techniques of close air support for ground troops involving what was probably the most sophisticated system of interservice teamwork in the history of warfare. Most of this support was provided by the Air Force's Tactical Air Command; some was given by the Army's organic helicopter gunships; and some very effective support came from Navy carrier aircraft. The ability of American aircraft to provide such support, however, had been clearly inhibited in Vietnam by Soviet-made surface-to-air missiles (SAMs) and multiple-barrel, light, automatic AAA guns. The potential of these antiaircraft weapons was further demonstrated by their effective use by Egyptians and Syrians against the Israeli Air Force in the Fourth Arab-Israeli War.

There did not appear to be any ground weapon capable of offsetting the tremendous advantage accruing to an interservice team which possessed air superiority.

MAJOR WARS

The period was marked by 5 major, long-lasting confrontations, of which the most important was the "Cold War" of the superpowers, in which overt hostilities never took place, despite a number of crises and mobilizations. Of these near-conflict situations, the most important were the Berlin crises of 1948 and 1961 (see pp. 1378, 1379), the Cuban Missile Crisis (see p. 1453), and the 1973 October War alert (see p. 1353).

All 4 of the more traditional major "hot" shooting wars took place in Asia. Of these, probably the most important was the Chinese Civil War (see p. 1423). The other 3 major shooting wars—the Indo-China War, the Arab-Israeli Conflict, and the Korean War—are discussed in some detail below.

The 1967 Arab-Israeli War (and to a lesser extent the 1956 war) provides a classic example of **Preemptive War** or **Preventive War** in which one side (in this case Israel), having decided a potential enemy is planning to go to war, and that conflict is virtually inevitable, decides to seize the initiative by starting the war at the time and place of its own choosing.

THE INDO-CHINA WAR, 1945–1975

BACKGROUND

Future historians may refer to the conflict which raged throughout Indo-China from 1945 to 1975 as the **Second Thirty Years' War.*** Although the principal operations took place in Vietnam, both Laos and Cambodia were also extensively involved in conflicts distinct from, yet closely connected with, the fighting in Vietnam. The principal conflict had dragged on, with fluctuating intensity, as a bitter, sanguinary, but basically localized colonial war and civil war when the United States was drawn in as an active participant in 1965.

While the pattern was slightly different in Laos and Cambodia, the confused conflict can be considered as having 4 main periods: **the French Period** (1945–1954), **the Geneva Period** (1954–1965), **the United States Period** (1965–1973), and **the Cease-Fire Period** (1973–1975). Of these, the first 2 and the last were essentially local, internal conflicts, and are dealt with later in the Southeast Asia region under Vietnam, Laos, and Cambodia (see p. 1412). The third period, a major international war, directly involving 1 of the 2 superpowers, is discussed below.

THE UNITED STATES WAR IN VIETNAM, 1965–1973

January 1965 was a period of intense Communist activity throughout South Vietnam. Viet Cong and North Vietnamese forces launched assaults on cities, towns, and military bases along the length of the country, causing massive casualties among the South Vietnamese troops and temporarily cutting South Vietnam in two by means of a push eastward to the sea from the central highlands. In early February Viet Cong forces attacked a U.S. installation and helicopter base in the central highlands, and on February 8 President Johnson ordered South Vietnamese and U.S. Air Force and U.S. Navy carrier planes to begin a systematic bombardment of carefully selected military targets north of the demilitarized zone (DMZ), which divided North from South Vietnam. On March 8, the U.S. 9th Marine Brigade landed at Danang, second-largest city in South Vietnam, situated 70 miles south of the DMZ on the Tonkin Gulf, and went into immediate action to repulse Communist forces attacking U.S. military assistance installations.

---

* As a possible minor historical footnote, the first known public reference to "thirty years' war" was that of Eric Sevareid, on a CBS news broadcast on April 22, 1975.

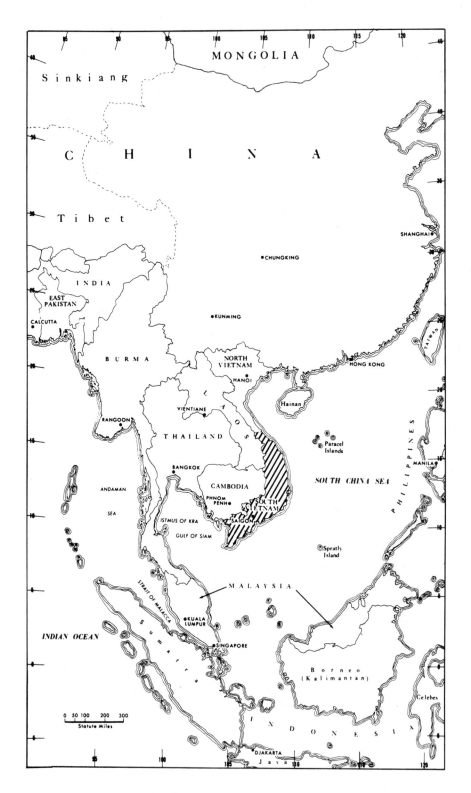

South Vietnam in Southeast Asia

Thus began the pyramiding of American forces pitted against the North Vietnamese and Viet Cong, both supplied by Communist China and the U.S.S.R. (independently and without coordination). The U.S. had entered into the longest, oddest, and by far the most unpopular war in its history. It was a war without a fixed front; the enemy was here, there, and everywhere. The hostile military effort was led by General Vo Nguyen Giap, leader of the Viet units in the earlier successful rebellion against the French; he had made the operations of guerrilla warfare a science. American ground forces held only the soil on which they stood in a war of thousands of savage engagements without a single major battle in the conventional modern sense. The Vietnam war, a phantasmagoria of brutal combat, political and social entanglements, and unceasing frustration, could be viewed on television in American homes; military personnel flew to the area in commercial aircraft; and the military effort was heavily influenced by political considerations in Washington and the frequently overoptimistic views of the Department of Defense, as well as by the State Department, USIA, AID, CIA, and other civilian agencies present in the embattled South Vietnam.

Despite the rapid commitment of some 7 additional U.S. ground battalions and an air squadron, and the commencement of raids on Communist bases in South Vietnam by U.S. B-52 bombers, by mid-1965 the military situation showed no improvement, and President Johnson decided that only a massive infusion of American combat troops could prevent the defeat of the Army of the Republic of Vietnam (ARVN).

The U.S. troops operated from fortified bases, carrying out search-and-destroy missions designed to eliminate Communist forces and base areas rather than capture and hold blocks of territory. Combat, except in sieges, consisted of clashes at platoon to no more than battalion strength, even in the occasional large-scale operations. The ubiquitous helicopter—gunship, personnel and cargo carrier, vehicle of rescue and evacuation of wounded—furnished amazing flexibility to the U.S. and allied troops.

In mid-1966 the U.S. forces launched their first prolonged offensive with the goals of locating and destroying major Communist units and bases. The offensive differed from earlier search-and-destroy missions in that it comprised lengthy and continuous sweeps rather than short, swift raids. Operations were carried out in the central highlands and in the provinces around Saigon. But when they were hard pressed, the Communists, relying on their knowledge of the terrain, were generally able to break contact and disappear into the jungle. When the allied sweeps receded, the Communist forces generally moved back into the territory they had held before.

Despite the execution of 2 massive operations in 1967, "Cedar Falls" and "Junction City," and the completion of the first U.S. base in the Mekong Delta, Communist forces were able to regroup and continue their operations against U.S. bases along the Laotian and Cambodian borders and south of the DMZ. These operations quickly gathered strength and momentum and climaxed in the Tet Offensive of early 1968.

This Communist offensive followed closely on the heels of well-publicized optimistic reports from field commanders; its size, scope, and fury shocked the American public and resulted in demands by many Americans for a U.S. withdrawal from Vietnam. Thus, although the Tet Offensive was a tactical military defeat for the Communists (since they suffered severe casualties and gained neither any substantial new territorial footholds nor increased support among the South Vietnamese), it was a major strategic victory for the North Vietnamese and Viet Cong.

Following the 1968 Communist offensive, General **William Westmoreland** stepped down as Commander, United States Military Assistance Command, Vietnam (USMACV), and returned to the United States to become U.S. Army Chief of Staff. General Creighton W. Abrams succeeded him in Vietnam. To exploit heavy Communist losses sustained in the recent offensive and in U.S. operations in the A Shau Valley, and to enhance improved ARVN combat capability, General Abrams undertook a "Vietnamization" program involving the supply of all South Vietnamese units with modern weapons and equipment, which would allow American forces to shift the bulk of the combat burden in the defense of their country to the South Vietnamese.

By 1969 the U.S. military presence in Vietnam had grown to the colossal strength of 543,482, supported by the Navy's Seventh Fleet in the Tonkin Gulf, and strategic aircraft flying from bases in Guam and Thailand. Harbors, roads, airfields, cantonments, and warehouses sprang up to supply the needs of more than 1 million men. Other SEATO nations had joined the U.S.-South Vietnam alliance, and when American ground forces in South Vietnam were approaching their peak, troops furnished by Australia, New Zealand, South Korea, Thailand, and the Philipines totalled some 62,000 men, while an Australian cruiser had joined the Seventh Fleet's Tonkin Gulf patrol. (These allies would sustain 8,500 casualties including 2,500 men killed in action.) Noncombatant elements came too from West Germany, Taiwan, and Malaysia. The U.S. furnished logistical support for all allied forces.

Concurrently, an enormous black market developed; South Vietnamese in collusion with corrupt members of the U.S. Armed Forces robbed commissaries, munitions dumps, and post exchanges. Drug traffic boomed, red-light ghettos thrived, venereal rates by 1972 soared to nearly 700 per thousand men. Disillusion with the war grew in the U.S., yet the war dragged on.

In a major effort to shorten the war and destroy large Communist supply bases in Cambodia, President Nixon ordered U.S. forces to move into the "Parrot's Beak" and "Fishhook" areas of Cambodia on April 30, 1970, in conjunction with an operation under execution by South Vietnamese troops. Although vast amounts of equipment were seized in the 2-month incursion, supplies continued to reach Communist forces in South Vietnam via the Ho Chi Minh Trail, which ran through Laos.

Antiwar sentiment in the U.S. skyrocketed as a result of this "invasion" of Cambodia. The U.S. government was never successful in justifying this action as part of a widespread war, in which Cambodia had already become involved. Tens of thousands of young American men joined many others who earlier had dodged the draft, resisted induction, or deserted the ranks of the uniformed services. These protesters fled into exile in Sweden and Canada, went "underground" inside the U.S., or went to jail. The size and scope of antiwar marches, rallies, and organizations grew, drawing into their ranks increasingly diverse segments of American society, ranging from sincere patriots to cynical supporters of international Communist propaganda. Meanwhile, American POWs remained in limbo, the Communists refusing all details of their number, identity, or condition. Peace talks in Paris brought no agreement.

Although by 1972 General Abrams's U.S. and ARVN forces had virtually won the land war in South Vietnam, drastically curtailing Viet Cong and North Vietnamese operations and inflicting unacceptable casualties on their troops, the American public had almost uniformly come to see the U.S. involvement in Vietnam as a tragic mistake. Through secret negotiations conducted since 1969 between presidential advisor **Henry Kissinger** and North Vietnamese envoy **Le**

Duc Tho, a peace agreement was signed on January 27, 1973, allowing for the release of U.S. POWs and the withdrawal of all U.S. forces from Vietnam. On March 29, 1973, the last American soldier left South Vietnam.

Included below are a few of the most important actions of the innumerable offensives, counteroffensives, and engagements of this war. See also the Statistical Summary (p. 1333).

OPERATIONS, 1965

1965, February 7. Viet Cong Attack U.S. Support Installations near Pleiku Air Base. President Johnson considered this a deliberately hostile act. North Vietnam having ignored U.S. warnings to cease its direct assistance to, and control of, the Viet Cong in their insurgency against the legal government of South Vietnam, and having instituted or sponsored a campaign of terrorism and murder against American advisors in South Vietnam, he ordered retaliation against North Vietnam. He also ordered that U.S. dependents be evacuated from South Vietnam.

1965, February 8. U.S. Air War Begins against North Vietnam. U.S. air units, in cooperation with South Vietnamese air forces, carried out joint retaliatory air attacks on bases at Dong Hoi on the coast just north of the Demilitarized Zone (DMZ). This began a systematic but limited offensive against carefully selected military targets in North Vietnam. In response, the Viet Cong intensified its terrorism and sabotage against U.S. installations and personnel.

1965, March 8. First U.S. Ground Combat Force in Vietnam. The 9th Marine Expeditionary Brigade landed at Danang, Quang Nam Province, in the northern coastal area.

1965, May 14. USMACV Authorizes Naval Gunfire Support.

1965, June 28. First Major U.S. Operation. The 173rd Airborne Brigade (arrived South Vietnam May 5) lifted 2 ARVN battalions and 2 battalions of the 503rd Infantry Brigade into battle in Bien Hoa Province, 20 miles northeast of Saigon.

1965, August 18–21. Operation "Star-light." This was the first important victory for U.S. forces. Over 5,000 Marines moved against the Viet Cong 1st Regiment south of Chu Lai, Quang Ngai Province, in the northern coastal region, trapping a large force of Viet Cong and destroying a major stronghold near **Van Tuong,** just south of Chu Lai.

1965, October 23–November 20. Battle of the Ia Drang Valley. The 1st Cavalry Division (Airmobile) defeated North Vietnamese forces which had gathered in western Pleiku Province in the central highlands with the goal of cutting South Vietnam in half. Both sides suffered heavy casualties.

1965, December 15. First U.S. Air Raid on a Major North Vietnamese Industrial Target. This was a thermal power plant at Uongbi, 14 miles north of Haiphong.

1965, December 31. U.S. Military Strength in South Vietnam: 154,000; U.S. combat deaths in Southeast Asia since January 1, 1961: 1,636.

OPERATIONS, 1966

1966, January 1–8. First American Unit in Mekong Delta. The 173rd Airborne Brigade, in Operation **"Marauder,"** successfully destroyed one Viet Cong battalion and the headquarters of a second in the **Plain of Reeds** area.

1966, January 19–February 21. Operation "Van Buren." The 1st Brigade of the 101st Airborne Division, the 2nd Republic of Korea (ROK) Marine Brigade, and the 4th ARVN Regiment carried out a search-and-destroy action to secure Phu Yen Province in the central coastal region.

1966, January 24–March 6. Operation **"Masher-White Wing."** 20,000 U.S., ARVN, and ROK forces conducted a massive sweep of Binh Dinh Province in the central coastal area. Search-and-destroy missions were carried out

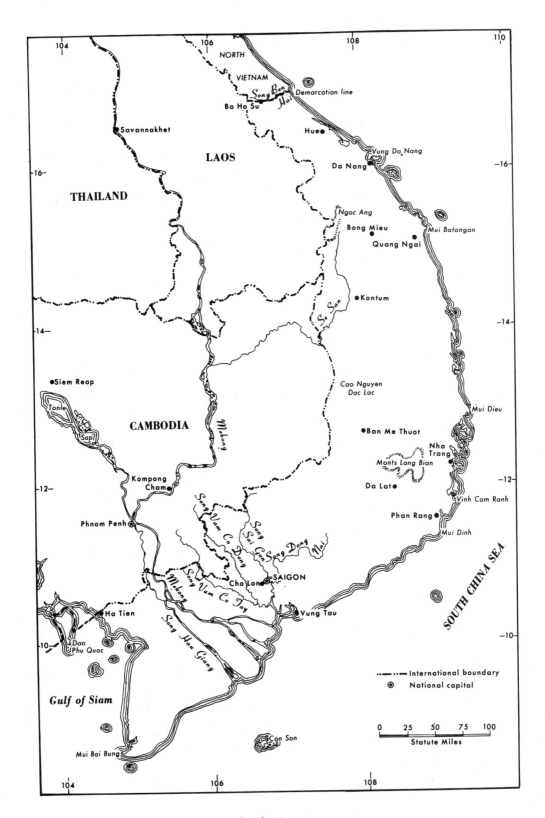

South Vietnam

against 2 Viet Cong and 2 North Vietnamese regiments before allied units linked up with U.S. Marine forces moving north from Phu Yen Province.

1966, March–October. U.S. Marines Constantly Engaged in Northern Provinces.

1966, April 12. First Involvement of B-52s Based on Guam. Strikes were carried out against North Vietnamese infiltration routes at the Mugia Pass near the North Vietnam-Laos border.

1966, May 1. First U.S. Combat Operation against Cambodian Territory. U.S. 1st Infantry Division shelled targets on the Cambodian bank of the Caibac River after American troops on the South Vietnam bank came under Communist fire from the Cambodian shore.

1966, May. Beginning of Operation "Game Warden." Naval force of air, sea, and ground units began coastal patrols and inshore surveillance to prevent Viet Cong infiltration into the Mekong Delta.

1966. May–August Operations in Central Highlands. To prevent infiltration from North Vietnam and Cambodia, U.S. 25th Infantry Division, 1st Cavalry Division (Airmobile) and ARVN forces carried out operations in Pleiku Province, while the 101st Airborne Division was heavily engaged in Kontum Province.

1966, June–July. Operations of U.S. 1st Division in Binh Long Province. The American troops were operating with the 5th ARVN Division about 70 miles north of Saigon.

1966, June 29. First U.S. Air Attacks near Hanoi and Haiphong. These were against oil installations.

1966, July. U.S. Aircraft Strike North Vietnamese Positions inside DMZ. The International Control Commission (ICC) said it would act to keep the zone free.

1966, August 3–1967. January 31. Operation "Prairie." The 3rd Marine Division was continuously engaged in the Con Thien-Gio Linh areas near the DMZ.

1966, August 26–1968. February 20. U.S. 1st Cavalry Division in the Central Coastal Region. It conducted a continuous series of one-battalion leap-frogging search-and-

destroy operations in Binh Thuan and Binh Dinh provinces.

1966. September 14–November 24. Operation "Attleboro." About 22,000 U.S. troops engaged Viet Cong and North Vietnamese forces in a series of small engagements in Tav Ninh Province next to the Cambodian border northwest of Saigon, with the goal of breaking up a planned Communist offensive. This was the largest U.S. operation to date.

1966, October 18–December 30. Operations in Central Highlands. The 4th Infantry Division and elements of the 25th Infantry and 1st Cavalry divisions continued operations in Pleiku Province to prevent infiltration.

1966, November 30–1967, December 14. Operations in and around Saigon. One battalion each from the 1st, 4th, and 25th Infantry divisions were engaged incessantly with Viet Cong forces November–December 1966). These units were replaced by the 199th Light Infantry Brigade and ARVN troops, who secured the area in less intensive combat (1967).

1966, December 31. U.S. Military Strength in South Vietnam: 389,000; U.S. combat deaths in Southeast Asia in 1966; 4,771.

OPERATIONS, 1967

1967, January 6. U.S. Troops Committed to Mekong Delta. Marines landed in Dinh Tuong Province and with the 9th Infantry Division established a base and carried out operations in Long An Province.

1967, January 8–26. Operation "Cedar Falls." The 1st and 25th Infantry divisions, 173rd Airborne Brigade, 11th Armored Cavalry Regiment, and ARVN forces operated against the Viet Cong regional military headquarters and major Communist base areas in the "Iron Triangle," some 25 miles northwest of Saigon.

1967, February–September. Marine Operations in the North. While the 3rd Marine Division continued operations along the DMZ, the 1st Marine Division was operating in Quang Nam and Quang Tin provinces.

1967, February 22–May 14. Operation "Junction City." In the largest operation of the war to date, 22 U.S. battalions and 4 ARVN battalions moved against Viet Cong and North Vietnamese forces in Tay Ninh and neighboring provinces north of Saigon along the Cambodian border.

1967, February 24. First U.S. Artillery Fire on North Vietnam. Marine gunners shelled antiaircraft positions in the DMZ and North Vietnam.

1967, February 27. U.S. Aircraft Drop First Mines in North Vietnam Rivers.

1967, March 22. Thailand Agrees to Use of U-Tapao Airfield as B-52 Base. The first B-52s arrived in Thailand April 10.

1967, May 3. 3rd Marine Division Seizes High Ground near Khe Sanh. This provided a dominating position over infiltration routes from Laos into the northernmost province of Quang Tri.

1967, May 14–December 7. 25th Infantry Division West of Saigon. Search-and-destroy operations were carried out in Hua Nghia Province.

1967, August–September. Seige of Con Thieu. Marines repelled Viet Cong efforts against a major base in Quang Tri Province near the DMZ.

1967, November 1–1968, March 31. Operations around Khe Sanh. The U.S. 3rd Marine Division probed enemy concentrations along the DMZ in western Quang Tri Province.

1967, November 11–1968, June 10. American Division Reinforces Marines in North. This involved search-and-destroy missions in South Vietnam's 5 northern provinces.

1967, December 8–18. U.S. 101st Airborne Division Undertakes Longest and Largest Airlift to Date. 10,024 men and more than 5,300 tons of equipment were carried in 369 C-141 Starfighters and 22 C-133 Cargomasters from Fort Campbell, Kentucky, directly into combat in Vietnam.

1967, December 31. U.S. Military Strength in South Vietnam: 480,000; U.S. combat deaths in Southeast Asia in 1967; 9,699.

OPERATIONS, 1968

1968, January 21–April 8. Seige of Khe Sanh. About 5,000 U.S. Marines were isolated in **Khe Sanh,** Quang Tri Province (northern sector) by some 20,000 Communist troops (see Operation "Pegasus," below).

1968, January 30–February 29. Communist Tet Offensive. Breaking the Tet holiday truce, an estimated 50,000 Viet Cong and North Vietnamese soldiers launched well-planned and simultaneous attacks on allied bases and major South Vietnamese cities and towns. **Saigon** and **Hue** were the principal targets of the Communist assault. Fighting went on block by block inside the city of Hue until February 25, and in Saigon combat reached inside the American Embassy grounds. All of the assaults were repelled, and fighting subsided in late February with the exception of combat around the U.S. Marine base at Khe Sanh, which remained under siege. Although the offensive was militarily unsuccessful, it was a psychological victory since the American and South Vietnamese forces were taken completely by surprise.

1968, February 24. Air Attack on Hanoi. Aircraft from the carrier USS *Enterprise* struck at the port area.

1968, March 16. My Lai (Song My) Massacre. During a search-and-destroy operation by a task force of the U.S. 23rd (American) Infantry Division, C Company, 20th U.S. Infantry, some 200 unarmed civilians—men, women, and children—were murdered in the village of My Lai 4 in Quang Ngai Province (northern sector). This outrage remained unpublicized for more than a year, and when it was disclosed it shocked the Free World and exacerbated antiwar sentiments in the United States.

1968, March 31. Cessation of Bombing of North Vietnam. President Johnson's announcement was a signal that the United States was seeking a negotiated solution.

1968, April 1–15. Operation "Pegasus." A force of 30,000 U.S. and ARVN soldiers, mainly from the 1st Cavalry Division (Airmobile), attacked to free the Marines besieged at Khe

Sanh. The North Vietnamese had already begun to withdraw into Laos.

1968, April 8. Operation "Complete Victory." Over 100,000 men from 42 U.S. and 37 ARVN battalions undertook an offensive against Communist forces in 11 provinces around Saigon.

1968, April 19–May 17. Operation "Delaware." An offensive was launched into Communist base areas in A Shau Valley, Quang Tri and Thua Thien provinces (northern sector), by 1st Cavalry Division (Airmobile), 101st Airborne Division, elements of the 196th Light Infantry Brigade, 1st ARVN Division, and ARVN Task Force Bravo (Airborne), designed to prevent an expected attack against Hue.

1968, April. Mobile Riverine Force Begins Operations. Activities took place along inland waterways of Mekong Delta.

1968, May 5–9. Communist Spring Offensive. Attacks were launched on 122 military installations, airfields, and towns throughout South Vietnam, including Saigon. The attacks were unsuccessful.

1968, May 10. Paris Peace Talks Begin. U.S. and North Vietnamese officials began discussions.

1968, May 19–June 21. Nightly Rocket Attacks against Saigon. More than 100,000 civilians were left homeless within the city.

1968, July 3. General Creighton W. Abrams Assumes Command. He replaced General William Westmoreland (who became Army Chief of Staff) as Commander of USMACV.

1968, July 14–18. Intensified B-52 Operations. They struck supply bases and troop concentrations 15 miles north of the DMZ (July 14) and North Vietnamese SAM sites for the first time (July 18).

1968, August 17. Third Communist Offensive. This was mounted throughout South Vietnam except the Delta; rocket attacks on Saigon resumed August 21.

1968, September 30. U.S.S. *New Jersey* Begins Combat Operations near DMZ. This was the first combat use of a U.S. battleship since July 1953.

1968, October 18. Operation "Sea Lords." This was launched by three U.S. naval task forces to interdict Viet Cong infiltration routes from Cambodia into the Mekong Delta and coordinate naval operations in the Delta.

1968, October 31. U.S. Ceases Attacks on North Vietnam. President Johnson ordered complete cessation of air, naval, and ground bombardment north of the DMZ, effective at 0800 EST November 1, in an effort to encourage peace negotiations.

1968, December 31. U.S. Military Strength in South Vietnam: 536,040; U.S. combat deaths in Southeast Asia in 1968: 14,437.

OPERATIONS, 1969

1969, January 25. First Substantive Peace Talks in Paris.

1969, February 6. American-Vietnamese Staff Organized to Facilitate "Vietnamization." The combined U.S.-Republic of Vietnam Armed Forces (RVNAF) Joint General Staff was to facilitate increasing Vietnamese responsibility for operations and to improve and modernize the RVNAF.

1969, February 23–March 29. Communist Offensive. This began with a series of rocket and mortar attacks against over 100 cities and bases throughout South Vietnam, including Saigon; attacks peaked on February 26 and March 6 and 16.

1969, April 24. Intensive B-52 Raids. Some 100 B-52s dropped bombs on targets northwest of Saigon near the Cambodian border.

1969, April 26. First "Vietnamization" Transfer. The 6th Battalion 77th Field Artillery completed turnover of equipment to the 213th ARVN artillery Battalion in ceremonies at Can Tho, Phong Dinh Province, in the Mekong Delta.

1969, April 30. Peak U.S. Troops Strength in South Vietnam: 543,482. (See Statistical Summary for additional forces in Thailand and at sea.)

1969, May 8–20. Battle of "Hamburger Hill." As part of an operation against North Vietnamese infiltration routes, U.S. troops took Hill

937 (Ap Bia Mountain or "Hamburger Hill") in the northern A Shau-Valley, Quang Tri Province, after fierce fighting and 10 attempts.

1969, May 11–14. Communist Summer Offensive. This began with coordinated ground attacks throughout South Vietnam.

1969, May 14. President Nixon Announces Planned Withdrawal from Vietnam.

1969, June 5. U.S. Aircraft Resume Bombardment of North Vietnam. Strikes on North Vietnamese targets were the first since the November 1968 bombing halt.

1969, June 17. Communists Retake "Hamburger Hill."

1969, July 8. U.S. Withdrawal Begins. The 3rd Battalion, 60th Infantry Brigade, 9th Infantry Division, left Tan Son Nhut Air Base near Saigon for Fort Lewis, Washington.

1969, July 25. U.S. Bombing Authorized in Laos. Prime Minister **Souvanna Phouma** announced that he had authorized U.S. bombing along the Ho Chi Minh Trail.

1969, September 3. Ho Chi Minh Dies in Hanoi.

1969, October 1. Vietnamese Forces Assume Responsibility for Saigon Area Defense.

1969, December 28. Peers in Vietnam. Lieutenant General **William R. Peers** arrived to lead the investigation into the My Lai massacre (see p. 1328).

1969, December 31. U.S. Military Strength in South Vietnam: 484,326; U.S. combat deaths in Southeast Asia in 1969: 6,727.

Operations, 1970

1970, January 20. United States Protests Execution of Two U.S. POWs. The U.S. government requested the International Red Cross to investigate the incident, which had occurred on September 30, 1966. The grave of the two soldiers was discovered by Marines (January 17) in Thua Thien Province (northern sector).

1970, January 26. POW Record. Lieutenant **Everett Alvarez, Jr., U.S.N.,** spent his 2,000th day in captivity, marking the longest time any American had spent as a prisoner of war.

1970, February 11. Partial Withdrawal from Thailand. Some 4,200 troops withdrew; 43,800 remained.

1970, February–March. U.S.-ARVN Offensive against Communist Supply Depots near Cambodian Border. Major caches were discovered. During this operation, U.S. troops entered Laos to establish an artillery base and mine the Ho Chi Minh Trail.

1970, March 14. U.S. Ammunition Ship Hijacked. USS *Columbia Eagle,* en route to Thailand, was hijacked by 2 multineers and taken to Cambodia; the ship was later released by the Cambodian government.

1970, April 1. Communist Spring Offensive. Viet Cong and North Vietnamese forces attacked more than 130 military bases and towns throughout South Vietnam. After 3 days the force of the assault diminished as the Communist troops encountered supply difficulties.

1970, April 19, U.S. Operations in Laos Acknowledged. A Senate subcommittee released official acknowledgment that U.S. forces had been indirectly involved in conflict in Laos, with 100 American civilians and military personnel killed there since 1962; since 1966 U.S. forces had served as air spotters in Laotian bombers.

1970, April 30–June 30. U.S. Troops Enter Cambodia. President Nixon announced that U.S. troops had begun a ground offensive against Communist bases in Cambodia; the U.S. 1st Cavalry Division and the ARVN Airborne Division, totaling more than 40,000 troops, launched the operation in the "Fish Hook" and "Parrot's Beak" areas immediately across the Vietnamese border in Cambodia. President Nixon later (May 5) announced that U.S. troops would penetrate no more than 21.7 miles into Cambodia and that all U.S. forces would be withdrawn from Cambodia by June 30.

1970, May 1–2. Intensified U.S. Air Raids. These were the largest north of the DMZ since the November 1968 bombing halt.

1970, June 1–7. Fierce Fighting Breaks Out along DMZ. Communist forces began a continuous and concentrated shelling of the U.S. base at **Danang,** Quang Nam Province (northern coastal sector).

1970, June 30. U.S. Troops Withdrawn from Cambodia. The U.S. Command reported capture of 155 tons of weapons, 1,786 tons of ammunition, and 6,877 tons of rice. U.S. forces lost 388 killed and 1,525 wounded; enemy dead were estimated at 4,766.

1970, August 1. U.S. Close Air Support for Cambodians Reported. U.S. air attacks reportedly enabled the Cambodians to free **Kampong Thom** in central Cambodia and the village of Skoun, a suburb of Phnom Penh.

1970, September. Daily B-52 Raids along Laotian Border. This was to prevent further Communist troop build-ups and to disperse the estimated 40,000 troops already gathered and poised for an offensive.

1970, November 21. Son Tay Raid. In an attempt to rescue U.S. POWs, a specially trained volunteer commando force carried out a helicopter raid on the Son Tay POW camp in North Vietnam, 23 miles from Hanoi; however, the camp was found abandoned.

1970, December 31. U.S. Military Strength in South Vietnam: 335,794; U.S. combat deaths in Southeast Asia in 1970: 7,171.

OPERATIONS, 1971

1971, January 12. Vietnamese Navy Convoys Supply Vessels for Cambodia River. This was the beginning of escorting merchant ships carrying fuel and ammunition to Phnom Penh.

1971, January 30. Operation "Dewey Canyon II." This operation by U.S. forces, just south of the DMZ, was to secure that area and establish lines of communication to support a planned South Vietnamese thrust into Laos.

1971, February 3. Renewed South Vietnamese Offensive in Cambodia. USMACV announced that it was providing full air support for the offensive into the Fish Hook and Parrot's Beak areas of Cambodia.

1971, February 8–April 9. South Vietnamese Operation "Lam Son 719" in Laos. This had the objective of disrupting North Vietnamese logistics along the Ho Chi Minh Trail. The U.S. Command in Vietnam announced that no U.S. ground forces or advisors would enter Laos. The South Vietnamese 1st Infantry Division and 1st Armored Brigade seized **Tchepone,** Laos (directly west of Quang Tri City), the main objective of their operation and the primary supply center for forces coming down the Ho Chi Minh Trail (March 6). Preliminary reports at the conclusion of the action on April 9 listed 13,462 Communists killed and 56 captured; 5,066 individual and 1,935 crew-served weapons, 106 tanks, 422 trucks, and 1,250 tons of rice were captured. Friendly losses were listed as 1,707 (176 U.S.) killed, 6,466 (1,042 U.S.) wounded, and 693 (42 U.S.) missing.

1971, July 9. Northern Province Defense Responsibility to Vietnamese Forces. The turnover of "Fire Base Charlie 2," 4 miles south of the DMZ, marked the completion of the transfer of the defense responsibility for that area by U.S. forces.

1971, August 11. All Ground Combat Responsibility Turned Over to South Vietnamese. Defense Secretary **Melvin Laird** announced the completion of the first phase of the Vietnamization program.

1971, December 26–31. U.S. Fighter-Bombers Attack North Vietnamese Targets. Airfields, missile sites, antiaircraft batteries, and supply depots were struck in retaliation for Communist attacks on Saigon, DMZ violations, and attacks on unarmed U.S. reconnaissance planes.

1971, December 31. U.S. Military Strength in South Vietnam: 158,119; U.S. combat deaths in Southeast Asia in 1971: 942.

OPERATIONS, 1972

1972, March 30. Communist Easter Offensive. In the biggest offensive since the 1968 Tet campaign, some 20,000 North Vietnamese forces launched a 4-pronged attack into South Vietnam across the DMZ with the goal of taking **Quang Tri City,** capital of South Vietnam's northernmost province, and driving the South Vietnamese 3rd Division from 15 border outposts. In retaliation U.S. aircraft and naval forces began bombing military supply facilities near Hanoi and Haiphong. Some 50,000 Communist troops, poised along the Cambodian and Laotian borders, drove into Binh Long Province north of Saigon (April 5), taking Loc Ninh (April 7) and securing half of **An Loc,** the provincial capital (April 13). In the central coastal region Communist forces attacked Binh Dinh Province in an effort to cut the country in two (April 18). Four North Vietnamese divisions attacked in the central highlands in Kontum Province (April 22), taking **Dak To** (April 24) and encircling the provincial capital of **Kontum** (April 29).

1972, April 7. Relief of U.S. Air Commander in Vietnam Announced. General **John Lavelle** was removed as Commander of the U.S. Seventh Air Force in Vietnam in March, retired, and demoted to lieutenant general when it was revealed that he had ordered some 20 unauthorized air strikes against North Vietnamese targets between November 1971 and March 1972.

1972, April 26–May 1. Battle of Quang Tri City. The city fell to North Vietnamese forces as the ARVN 3rd Division retreated to Hue. Heavy fighting continued elsewhere, particularly at An Loc and Kontum.

1972, May 8. President Nixon Orders Mining of North Vietnamese Harbors. Haiphong harbor and the harbors of 6 other North Vietnamese ports were to be mined and all land and sea routes interdicted.

1972, June 28. Appointment of General Frederick C. Weyand to Commander U.S. Forces in Vietnam. He replaced General Creighton W. Abrams.

1972, June 28–September 15. Second Battle of Quang Tri City. South Vietnamese attacked with 20,000 troops and the heaviest concentration of U.S. air support of the war; by July 26 U.S. and South Vietnamese Air Forces had flown more than 10,000 missions against North Vietnamese forces in support of the ARVN ground action; after Quang Tri City was retaken (September 15) fighting continued elsewhere in the province.

1972, July 19. South Vietnamese Binh Dinh Province Counteroffensive Begins. This was an effort to regain territory lost in the central coastal region during the Communist spring offensive.

1972, July 28. United States Admits Damage to North Vietnamese Dike System. A U.S. government report admitted U.S. bombing damage at 12 locations but maintained that the attacks were on military targets and dike damage had been unintentional.

1972, August 12. End of U.S. Ground Combat Role. This resulted from the withdrawal of the 3rd Battalion, 21st Infantry, from Danang.

1972, October 26. Kissinger Announces That "Peace Is at Hand." In a public report of his secret negotiations with the North Vietnamese, Kissinger said that final agreement could be worked out in one more conference.

1972, October 27. Temporary Halt in U.S. Bombing of North Vietnam above 20th Parallel. In recognition of North Vietnamese concessions in the secret negotiations, a bombing halt took effect at 1700 on October 23. (Bombing was continuing in North Vietnam below the 20th parallel, however.)

1972, October 30. U.S. Halts Naval Bombardment North of 20th Parallel. Simultaneously the U.S. stepped up shipment of war materiel to South Vietnam.

1972, December 18. Massive U.S. Air Attack Begins. Since no agreement had been reached, the U.S. resumed bombing strikes against Hanoi and Haiphong; North Vietnamese harbors were also mined. The Pentagon insisted that hits on hospitals and other civilian targets were unintended.

1972, December 30. Renewed Bombing Halt North of the 20th Parallel. This was as a result of North Vietnamese willingness for renewed peace talks.

1972, December 31. U.S. Military Strength in South Vietnam: 24,200; U.S. combat deaths in Southeast Asia in 1972: 531.

OPERATIONS, 1973

1973, January 8. Accidental U.S. Air Attack on Allied Air Base. USMACV acknowledged that U.S. fighter-bombers accidentally hit the base at Danang, Quang Nam Province (northern sector), causing several casualties.

1973, January 15. President Nixon Orders Halt to All Offensive Military Action. This included air strikes, shelling, and mining operations, and reflected progress in recent peace negotiations.

1973, January 23. Cease-fire Agreement Announced. Henry Kissinger and Le Duc Tho initialed a cease-fire agreement to go into effect at 0800 (Saigon time) on January 28. The final agreement was signed at ceremonies in Paris (January 27). Agreement provided for the release of all American POWs, the withdrawal of all remaining U.S. troops, and the establishment of an international force to supervise the truce.

1973, January 27. Defense Secretary Melvin Laird Announces End of Military Draft. He cited the Vietnam cease-fire agreement and the lack of further need for inductions as the reasons behind the decision.

1973, February 12. North Vietnam Releases First U.S. Prisoners. Starting Operation "Homecoming," 142 U.S. POWs were flown to Clark Air Force Base in the Philippines, then to Travis Air Force Base in California.

1973, February 21. Cease-Fire in Laos. The Laotian government and Communist-led Pathet Lao announced a cease-fire agreement to take effect at 1200 (Vientiane time) February 22, thereby ending 20 years of war. The U.S. immediately called on the Pathet Lao to release U.S. prisoners. On February 23 U.S. B-52s raided Communist positions in Laos at the request of the Royal Laotian Army.

1973, March 27. War Continues in Cambodia. The White House announced that the U.S. would continue to bomb in Cambodia until Communist forces suspended military operations and agree to a cease-fire.

1973, March 29. Last American Troops Leave Vietnam. Simultaneously North Vietnam released the last 67 U.S. prisoners, bringing the total number of prisoners released to 587. Only 8,500 U.S. civilian technicians remain in Vietnam.

1973, April 16–17. U.S. Air Strikes in Laos. U.S. B-52s and F-111s hit Communist positions at Tha Vieng at the southern end of the Plain of Jars in Laos.

1973, June 13. Supplementary Cease-Fire Agreement.

1973, August 14. U.S. Ceases Bombing in Cambodia. This was in compliance with a June Congressional decision. It officially ended more than 9 years of U.S. air combat activity in Indochina.

STATISTICAL SUMMARY: UNITED STATES WAR IN VIETNAM, 1965–1973

	Total U.S. Forces Worldwide	Maximum Deployed Strength		Total Combat Casualties	Killed and Died of Wounds	Wounded	Prisoners or Missing	Non-Battle Deaths
Total	3,300,000	625,866 (3/27/69)		205,023	46,226	153,311	5,486	10,326
Army	1,600,000	440,691	"	130,359	30,644	76,811	2,904	7,173
Navy	600,000	37,011	"	6,443	1,477	4,178	788	880
Marines	400,000	86,727	"	64,486	12,953	51,389	144	1,631
Air Force	400,000	61,137	"	3,735	1,152	933	1,650	592
RVN Forces		c. 1,000,000		c. 800,000	196,863	502,383	N/A*	N/A
Other Free World**(1969)		72,000		17,213	5,225	11,988	N/A	N/A
North Vietnam and Viet Cong (est.)		c. 1,000,000		c. 2,500,000	c. 900,000	c. 1,500,000	N/A	N/A

* Not available.

** Australia, South Korea, New Zealand, Philippines, Thailand.

THE ARAB-ISRAELI WARS

BACKGROUND

The roots of the conflict which has engulfed the Middle East for most of the period since World War II include the Diaspora of the Jews after the Roman subjugation of rebellious Palestine in the 1st century A.D., the Crusaders' conquest of Jerusalem in 1099, the Zionist movement beginning late in the 19th century, the Balfour Declaration of 1917, the Allies' denial of Arab expectations in the Versailles Treaty, and the Nazi efforts to exterminate the Jews of Europe during World War II. It is beyond the scope of this text to evaluate conflicting interpretations of these emotion-charged events, or to assess responsibility for the conflict. Both the Jewish Israelis and their Arab opponents are convinced of the religious, moral, and legal righteousness of their thus far irreconcilable causes. The military historian can only seek to illuminate what happened, and how it happened, rather than why it happened.

The years since 1945 have been marked by 8 distinct periods of hostility: (1) guerrilla warfare sparked by Jewish terrorism, 1945–1948; (2) the First Arab-Israeli War, or Israeli War of Independence, 1948–1949; (3) the Second Arab-Israeli or Sinai-Suez War, 1956; (4) the Third Arab-Israeli War or Six-Day War, 1967; (5) the War of Attrition, 1968–1970; (6) guerrilla warfare sparked largely by Arab terrorism, 1970–1973; (7) the Fourth Arab-Israeli War, or October War, 1973; and (8) strife in Lebanon (1975–1984) marked by 2 Israeli interventions in 1978 and 1982. The principal events of the 4 periods of overt or formal international hostilities (2), (3), (4), and (7), above, are covered in this section; the major events of the other 4 periods will be found in the sections on Palestine, Israel, and neighboring Arab states. For background through 1947, see p. 1392.

FIRST ARAB-ISRAELI WAR, OR ISRAEL'S WAR FOR INDEPENDENCE MAY 14, 1948–JANUARY 7, 1949

Despite previous debate, argument, and violence, it was not until the U.N. partition decision that the Arabs of Palestine and neighboring countries faced the reality of a Jewish state about to be established in their midst. Despite this, the Arabs did not systematically plan or organize to prevent this unacceptable occurrence. The highly organized Jewish Agency, under the leadership of **David Ben-Gurion,** prepared itself for the inevitable struggle. The Jews seized and maintained the initiative in the months of mounting violence that followed the U.N. partition decision. Early in 1948 the British government decided to withdraw from Palestine on May 14, rather than wait until October 1, the date in the U.N. plan. The Jewish population in Palestine was approximately 650,000, the Arab population approximately 1,200,000.

PRELIMINARY HOSTILITIES, 1948, JANUARY–MAY

1948, January 10. Appearance of the Arab Liberation Army (ALA). Volunteers from all the Arab world, including Palestine, under the loose and vaguely defined leadership of Iraqi General **Ismail Safwat** and Syrian leader **Fawzi Kaukji,** assembled in Syria. A raid from Syria on **Kfar Szold** by a force under Kaukji was repelled by British troops (January 10). Another raid on **Kfar Etzion** (south of Jerusalem) by Arab Palestinians (the Army of Salvation) under **Abdel Kader El-Husseini** was more successful, but was finally halted by troops of Palmach, a small elite striking force of the Haganah, the Jewish Agency's underground army.

1948, January–April. Increasing Violence. A 3-way war spread over much of Palestine. British troops attempted to retain order, but were primarily concerned with keeping open their own evacuation routes while incurring minimum casualties. The Jews systematically, and the Arabs in considerable confusion, were jockeying for position to seize areas not garrisoned by the British, and to be ready to move into British-controlled areas after the end of the Mandate. In the north, fighting was almost continuous in and around **Haifa,** despite the effort of British Major General **Hugh Stockwell** to maintain order; the Arab population finally evacuated the city after a climactic battle (April 21–23). The Jews were also successful at **Tiberias** (April 18) and **Safad** (May 6–12). In central Palestine there was continuous fighting along the Tel Aviv-Jerusalem Road, which ran through territory controlled by Arabs, as the Jewish Agency attempted to send weapons, food, and other supplies to the 100,000 Jews in the New City of Jerusalem. In fierce fighting on the road near **Kastel,** Husseini—perhaps the most effective Palestinian Arab leader—was killed (April 9). There was also hard fighting in and around the Arab city of **Jaffa** (April 25—May 13), finally ending in a mass Arab evacuation. South of Jerusalem the Arabs maintained siege lines around 4 Jewish settlements in the Kfar Etzion area, which was finally overrun by the Arab besiegers (May 14).

1948, April 9. Massacre of Deir Yassin. During the fighting for control of the Tel Aviv-Jerusalem Road, units of 2 Jewish terrorist organizations, the Irgun Zvai Leumi and the Stern Gang, slaughtered some 254 noncombatant Arab men, women, and children. Although most Palestinian Jews were as shocked by this incident as were the Arabs, it added greatly to Arab bitterness against the Jews and Zionism.

1948, April 25. Neighboring Arab States Agree to Invade Palestine. Lebanon, Syria, Iraq, Transjordan, and Egypt began to prepare for invasion of the Jewish regions of Palestine, under nominal but ineffective over all command of King Abdullah of Transjordan.

1948, May 14. Independence of Israel. David Ben Gurion became Prime Minister of the new state.

PHASE I: MAY 14–JUNE 11

OPPOSING FORCES

Jewish. The Haganah, the army of the new Jewish state, commanded by **Israel Galili,** mobilized approximately 30,000 men, including about 2,500 in the Palmach under Colonel **Yigael Allon.** Weapons were available for approximately 30,000 more trained men in the Haganah reserve. In addition each settlement had a well-trained armed guard. Both the Irgun with about 3,500 men and the Stern Gang with 500 were under the nominal direction of the Haganah high command.

Arab. The Arab forces were various. The ALA had approximately 4,000 poorly trained troops in 4 major groups. Lebanon provided 4 infantry battalions and 2 artillery batteries, about 2,500 men; Syria 2 infantry brigades, 1 tank battalion, and an air squadron, about 5,000 men; Iraq 4 infantry brigades, an armored battalion, and 2 air squadrons, about 10,000 troops. The Transjordanian Arab Legion, commanded by British General John Glubb, comprised 3 brigade groups and 4 armored car battalions, slightly less than 10,000 men, by far the best trained and most effective of all of the Arab contingents. Egypt provided initially 1 brigade group, 2 or 3 independent battalions, 2 air squadrons, and supporting units, about 7,500 troops. In addition to about 4,000 in the Army of Salvation, there was also an unknown number of armed Palestinian Arabs (at least 50,000) in local units which—despite poor organization and lack of

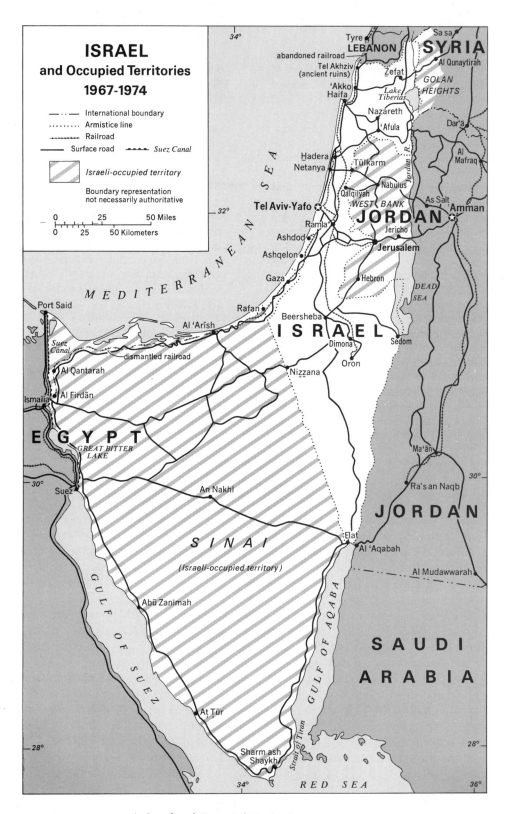

Israel and Occupied Territories, 1967–1974

training—were a constant threat to the Jewish community and tied down a substantial proportion of the mobilized Jewish strength.

OPERATIONS IN THE NORTH

1948, May 14–19. Syrian and Lebanese Invasions. Syrian invasion attempts were halted close to the frontier at **Degania** (where Major **Moshe Dayan** distinguished himself), **Zemach,** and **Mishmar Hayarden.** The Lebanese were stopped at **Malkya.**

1948, May 15–June 4. Iraqi Invasion. Advancing from Mafraq, through Irbid, the Iraqis crossed the Jordan south of the Sea of Galilee, but failed in repeated efforts to take **Geshir** (May 16–22). Meanwhile, other units moved into Jenin and occupied much of Samaria with armored spearheads approaching Natania (May 30). An Israeli effort to take **Jenin** was repulsed (June 1–4), but this forced the Iraqis to consolidate in the Nablus—Jenin area and withdraw from western Samaria.

1948, May 20. U.N. Mediator Appointed. The U.N. Security Council appointed Count **Folke Bernadotte** of Sweden as mediator between the Jews and Arabs.

1948, June 6–10. Renewed Syrian and Lebanese Offensive. The Syrians failed in their efforts to take **Ein Gev,** but did capture **Mishmar Hayarden.** The Lebanese, with some support from ALA and Syrian units, took **Matka** (June 6), **Ramat Naftali** (June 7), and **Kadesh** (June 7). Exploiting these successes, the ALA was able to overrun much of north-central Galilee.

OPERATIONS IN THE CENTRAL SECTOR

1948, May 15–25. Battle of Jerusalem. In hard fighting, General Glubb's Arab Legion seized and held the eastern and southern portions of New Jerusalem and occupied most of the Old City without opposition.

1948, May 15–28. Siege of the Jewish Quarter of Old Jerusalem. With some assistance from local Arabs, the Arab Legion systematically fought its way through the Jewish Quarter, repelling 2 Jewish relief attempts. The besieged garrison was finally forced to surrender.

1948, May 18–June 10. Struggle for the Tel Aviv–Jerusalem Road. Jewish forces under American volunteer Colonel **David "Micky" Marcus,** failed to break through the Arab-held road from Tel Aviv to Jerusalem, but succeeded in protecting the builders of a new road through the mountains farther south; this was opened 1 day before a U.N.-sponsored cease-fire (June 10). A few hours before the cease-fire went into effect, Marcus (who spoke no Hebrew) was killed by a Jewish sentry when he ignored a challenge.

1948, May 25–30. First Battle of Latrun (Bab el Wed). A Legion battalion under Lieutenant Colonel **Habis el Majali** occupied Latrun to secure the Tel Aviv-Jerusalem Road and pose a threat to Tel Aviv. Repeated Jewish efforts to drive them out were repulsed.

1948, June 9–10. Second Battle of Latrun. Again the Arab Legion repulsed Jewish attacks.

OPERATIONS IN THE SOUTH

1948, May 15. Egyptian Invasion. Two Egyptian brigade groups advanced into Palestine, under Major General **Ahmed Ali el-Mawawi.** The main force, based at El Arish, advanced up the coast to secure Gaza and threaten Tel Aviv. The smaller force, mostly ALA volunteers, advanced from Abu Ageila via Beersheba toward Jerusalem.

1948, May 16–June 7. Egyptian Coastal Offensive. The Egyptians advanced through Gaza, held by ALA units and local contingents (May 16). After a bitter battle this column captured **Yad Mordechai** (May 19–24), then advanced through Majdal and Ashkelon to seize **Ashdod** (May 29), only 25 miles from Tel Aviv. A Jewish attempt to recapture Ashdod was repulsed (June 2–3). The bypassed settlement of **Nitzanin** was then attacked and taken after a bitter fight (June 7), securing Egyptian communications back to El Arish.

1948, May 16–June 10. Egyptian Inland Offensive. The other column, advancing through El Auja, occupied Beersheba (May 20) and Hebron (May 21), and made contact with the Arab Legion at Bethlehem (May 22).

1948, June 11–July 9. The First Truce. Under U.N. auspices, both sides eagerly accepted the opportunity to rest, reorganize, and regroup. By now the Israelis had mobilized an army of 49,000. The first international police force, 49 uniformed U.N. guards, arrived from New York (June 20).

PHASE II: JULY 9–18, "THE TEN-DAY OFFENSIVE"

OPERATIONS IN THE NORTH

1948, July 9–14. Third Battle of Mishmar Hayarden. A serious Israeli effort failed to drive the Syrians out of their bridge-head west of the upper Jordan River.

1948, July 12–16. Israeli Nazareth Offensive. Israeli units expanded control of the coast north of Haifa, then turned their attention eastward toward Nazareth. An abortive offensive by Kaukji was thrown back, and the Israelis pressed on to seize his Nazareth base.

OPERATIONS IN THE CENTRAL SECTOR

1948, July 9–12. Israeli Lod (Lydda), Ramle Offensive. As part of a major effort to clear the Arab Legion from the coastal plain, and to secure the Tel Aviv-Jerusalem Corridor, the towns of Lydda (Lod) and Ramle and the nearby airport were occupied by the Israelis in tough fighting.

1948, July 9–18. Second Battle for Jerusalem. Israeli offensive efforts were repulsed by the Arab Legion.

1948, July 14–18. Third Battle of Latrun. Repeated Jewish efforts to take **Latrun** were unsuccessful.

OPERATIONS IN THE SOUTH

Israeli probes along the entire Egyptian front across central Palestine met with little success.

1948, July 18–October 15. The Second Truce. Once again both sides were eager to take advantage of the truce. The Israelis had taken the initiative in most areas of Palestine with limited success, and they felt the need for more time to prepare for a renewed offensive. The Arabs, shocked by the organization and equipment of the Jews and their ability to undertake major offensives in most regions of Palestine, were anxious to end the fighting as quickly as possible. The Israelis' total mobilized forces at the end of the truce were over 90,000, giving them a substantial numerical superiority over the Arabs.

1948, September 17. Assassination of U.N. Mediator. Confident of success in the war, many Jews were resentful of Count Bernadotte's peace efforts, which would have forced them to relinquish territory which they had conquered or which they hoped to occupy. He was killed by 3 unidentified men, apparently members of the Stern Gang. Dr. **Ralph Bunche** took over as U.N. mediator.

1948, September 29–October 15. Operations on the Egyptian Front. By mid-September neither side was paying much attention to the cease-fire along the Egyptian front. The Israeli pressure was particularly intense in the Faluja area, as they attempted to cut communications between the Egyptian coastal and inland corridors, and to open a secure line of communications into the Negev.

PHASE III: OCTOBER 6-NOVEMBER 5

OPERATIONS IN THE SOUTH

1948, October 15–19, Coastal Offensive. With Arab forces quiescent in the central and northern sectors, the Israelis concentrated their best forces against the Egyptians (now about 15,000 troops) in the south. An intensive effort to drive back the Egyptian forces in the Ashdod-Gaza area had some local successes but failed either to take any major positions or to cut the line of communications through Rafah.

1948, October 19–21. Israeli Beersheba Offensive. A drive to open firm communications with the Negev was successful. After severe fighting **Huleiqat** was taken (October 19) cutting communications between the 2 Egyptian corridors. Prompt exploitation of this success resulted in the capture of **Beersheba** (October 21). The road to the Negev was open. The Egyptian forces in the Hebron area and an Egyptian brigade group in Faluja were isolated.

1948, October 27–November 5. Egyptian Withdrawal. With their coastal line of communications now seriously threatened, the Egyptians withdrew from Ashdod (October 27) and Majdal (November 5), concentrating their remaining strength (other than forces in the Faluja and Hebron pockets) in the Asluj-Gaza-El Arish area.

CENTRAL SECTOR

1948, October–November. Limited Israeli Offensives. The Israelis succeeded in their efforts to widen the Jerusalem-Tel Aviv axis and to expand their control of regions on the central plateau, north and south of Jerusalem. They were defeated by the Arab Legion, however, in a probe near **Beit Gubrin** (between Faluja and Hebron).

1948, November 30. Israeli-Transjordan Cease-Fire.

1948, December 1. Union of Transjordan and Arab Palestine West of the Jordan. King Abdullah was proclaimed King of Arab Palestine, arousing the ire of most other Arab states. An official union of Arab Palestine and Transjordan was proclaimed in Amman, with the name Hashemite Kingdom of Jordan.

OPERATIONS IN THE NORTH

1948, October 22–31. North Galilee Offensive. Kaukji's ALA probed successfully at **Manara** (October 22), just south of the Lebanese border. In response, the Israelis mounted a major offensive with the dual aim of destroying the ALA and securing all of northern Galilee. After an initial Israeli setback at **Tarshia** (October 28), the ALA was driven back into Lebanon as a result of Israeli successes at

Gish (October 28) and **Sasa** (October 29). After occupying Tarshia (October 30), the Israelis cleared Arab resistance from the Hula Valley, retook Manara, and seized a strip of southern Lebanon. They then ceased operations to consolidate their gains.

1948, November 30. Cease-Fire in the North. Shaky local cease-fires were arranged by the Israelis with the opposing Syrian and Lebanese commanders.

PHASE IV: NOVEMBER 21– JANUARY 7

1948, November 19–December 7. Egyptian Offensive. Although the Egyptians failed in their effort to relieve the Faluja Pocket, they substantially expanded their holdings east of Gaza and in the Asluj area.

1948, December 20–1949, January 7. Israeli Sinai Offensive. With cease-fires in effect elsewhere around the periphery of Israel, the Israelis mounted a major offensive to force Egypt to withdraw from the war. In a series of encircling maneuvers, **Rafah** was isolated (December 22), then **Asluj** was taken (December 25), and next **Auja** (December 27). While one Israeli column turned west toward Rafah, the other, under Colonel Allon, struck south into the Sinai from Auja and took Abu Ageila after a short fight (December 28). Turning north toward El Arish, the Israelis seized the airfield and outlying villages, but determined Egyptian defense halted the drive east of the road and railroad (December 29). The Israelis turned northeast toward Rafah.

1949, January 7. Israeli-Egyptian Cease-Fire. As the Israelis prepared to attack Rafah, the Egyptians asked the UN Security Council to arrange an armistice, which was immediately granted. (See p. 1368.)

1949, January–July. Armistice Negotiations. These negotiations were held on the island of Rhodes, under the chairmanship of UN mediator **Dr. Ralph Bunche.** Israel signed armistices with Egypt (February 24), Lebanon (March 23), Jordan (April 3), and Syria (July 20). There were no peace agreements.

THE 1956 WAR (SUEZ OR SINAI WAR) OCTOBER 29–NOVEMBER 6, 1956

BACKGROUND

1956, July 4–26. Tension along Israeli Frontiers. Mounting tension, including a number of armed clashes along the Israeli-Jordan and Israeli-Gaza borders, caused U.N. Secretary General **Dag Hammarsk-jöld,** to visit the Middle East to attempt to restore the cease-fire.

1956, July 18. United States Withdraws Promised Aid to Egypt for Aswan Dam Project. This announcement by Secretary of State **John Foster Dulles,** shortly after Egyptian President **Gamal Abdel Nasser** had announced that Egypt would receive American and British help for the project, infuriated the Egyptian president. American aid had been conditional upon economic viability, and the rejection was based essentially on economic considerations; however, political considerations also undoubtedly contributed: American unhappiness over increasingly friendly relations of Egypt with the Soviet Union and other states of the Communist Bloc, and pro-Israeli sentiment in the United States.

1956, July 26. Nasser Announces Nationalization of the Suez Canal. In retaliation for the American reneging on Aswan, Nasser announced Canal revenues would be used for the dam's construction. Britain and France raised the issue in the U.N. Security Council and started secret plans for military action (August 3).

1956, August 16–30. New Violence along Israel's Frontiers with Egypt and Jordan. Increasing seriousness of incidents aroused international fears of outbreak of a new Middle East war.

1956, October 5. Security Council Debate Opens on Suez Canal. An Anglo-French proposal for a return to some measures of international control was approved by a majority but vetoed by the U.S.S.R.

1956, October 15. Jordan Accuses Israel of Aggression in U.N. Security Council. This was the result of more shooting incidents and increasing tension on the Israel-Jordan frontier.

1956, October 24. Secret Anglo-French-Israeli Agreement. At Sèvres final agreement was reached on covert coordination of military operations against Egypt.

OPPOSING PLANS

Israeli Plans

Ten brigades (6 infantry, 1 airborne, and 3 armored) were assigned to the Southern Command of Colonel **Asaf Simhoni;** 6 others were held in reserve in the central and northern sectors. Southern Command forces were divided into 4 task groups, each assigned to 1 of 4 corresponding land routes. The stated objectives of the Sinai Campaign, code-named Operation "Kadesh," were: to create a military threat to the Suez Canal by seizing the high ground to its east; to capture the Strait of Tiran; to create confusion in the ranks of the Egyptian Army, and to bring about its collapse. These objectives were directly related to an expected Anglo-French operation against the Suez Canal from the north.

Egyptian Plans

Evidence of the massing Anglo-French forces on Malta and Cyprus prompted Nasser to withdraw approximately half the Sinai garrison to the Delta region and the Canal Zone. Remaining in the Sinai were 2 infantry divisions and additional miscellaneous forces totaling some 30,000 men under the Eastern Command of Major General **Ali Amer,** most deployed in static defense positions in the northeast triangle formed by Rafah, Abu Ageila, and El Arish.

Anglo-French Plans

After Egyptian nationalization of the Suez Canal, British General Sir Hugh Stockwell was directed to formulate a plan for seizure of the Suez Canal by a joint Anglo-French expeditionary force. To avoid the appearance of imperialist aggression, the Anglo-French allies entered into secret negotiations with Israel to explore the possibility of relating their planned invasion to a renewed Arab-Israeli war. The modified plans called for Israel to create a threat to the Suez Canal, providing an excuse for Anglo-French intervention. The 2 governments would call on Israel and Egypt to withdraw 10 miles from the Canal, and when (as expected) the Egyptians refused, the allies would seize the Canal. Code-named "Musketeer," the operation was under the command of British General **Sir Charles Keightley,** with French Vice Admiral **Pierre Barjot** as deputy, General Stockwell commanding the land forces, French General **André Beaufre** as his deputy, Air Marshal **Denis Barnett** in command of air forces, and Admiral **Denis Dunford-Slayter** of naval forces. The necessity for coordination with Israel resulted in a delay of the planned operation from early September to early November.

OPERATIONS

1956, October 29. First Day. As the 1st Battalion of Colonel **Ariel Sharon's** 202nd Paratroop Brigade dropped near the eastern entrance to the **Mitla Pass,** the remainder of the brigade crossed the frontier and took **Kuntilla.** The brigade continued toward Thamad. Near Eilat, the border position of **Ras el-Nagb** was taken by the 9th Infantry Brigade, under Colonel **Avraham Yoffe,** in preparation for an overland march to Sharm el-Sheikh. Egyptian Commander in Chief, General **Abdel Hakim Amer,** ordered an infantry brigade and most of an armored division to cross the Canal to reinforce the Sinai garrison.

1956, October 30, Second Day. The Israeli paratroop battalion near Mitla Pass repulsed Egyptian ground and air attacks. Sharon's main body captured **Thamad** and linked up with the paratroop battalion before midnight. Farther north the Israeli Central Task Group (approximately a division), under Colonel **Yehuda Wallach,** captured **Sabha** and **Kusseima,** then began an attack on the formidable Egyptian defensive position of **Abu Ageila.** The Egyptian 4th Armored Division, having crossed the Canal, advanced to the vicinity of Bir Gifgafa and Bir Rud Salim.

1956, October 30. Anglo-French Ultimatum. At 1800 the British and French governments called on both sides for cessation of hostilities, withdrawal of all forces from the Canal, and approval of temporary Anglo-French occupation of Port Said, Ismailia, and Suez to guarantee freedom of transit in the Canal. Anticipating Egyptian rejection of the ultimatum, the Anglo-French invasion flotilla left Malta.

1956, October 31. Third Day. A large combat patrol of Sharon's paratroopers was ambushed in the **Mitla Pass.** Intense daylong fighting was climaxed by hand-to-hand combat for control of positions in caves in the walls of the Pass. Many of them were taken by the Israelis after dark, but the Egyptians still retained a substantial portion of the defile. At Abu Ageila, part of the Egyptian defense was overrun, but one Israeli brigade was repulsed with heavy losses. Israeli armored spearheads from the Central Task Group took **Bir Hassnah, Jebel Libni,** and **Bir Hama.**

1956, October 31. Israeli Naval Victory. Following the bombardment of naval and oil installations in Haifa, the Egyptian destroyer

Ibrahim al-Awal was captured by the Israeli Navy after a running battle at sea and was towed into Haifa harbor.

1956, October 31. Anglo-French Aerial Bombardment. All major Egyptian air bases were hit. President Nasser ordered a phased withdrawal of all forces in the Sinai to prevent them from being cut off by the expected allied invasion.

1956, November 1. Fourth Day. After another repulse at Abu Ageila, Israeli GHQ suspended further assaults on that stronghold. The Northern Task Group (approximately division strength) commanded by Colonel **Haim Laskov** captured **Rafah,** and Israeli armor advanced along the coastal road toward El Arish. By midnight all the Egyptian forces in the Sinai (except for the garrison at Sharm el-Sheikh, which lacked transportation, and the forces in Gaza, which were completely cut off by Israeli units) had begun to withdraw.

1956, November 2. Fifth Day. The 9th Infantry Brigade advanced down the coast of the Gulf of Aqaba toward Sharm el-Sheikh. Elements of the 202nd Paratroop Brigade and the 12th Infantry Brigade advanced to the Gulf of Suez from Mitla Pass and turned south also toward Sharm el-Sheikh. The Egyptians completed their evacuation of the Abu Ageila defenses and were pursued by the Israelis through Bir Gifgafa to a point 10 miles from the Canal. The 27th Armored Brigade under Colonel **Haim Bar Lev** captured **El Arish** and advanced as far as Romani. The 11th Infantry Brigade of Colonel **Aharon Doron** captured the northern half of the Gaza Strip, including the city of **Gaza.** All Egyptian forces not cut off by the Israelis completed the withdrawal from the Sinai.

1956, November 2. Egypt Accepts U.N. Call for a Cease-Fire with Israel. The Anglo-French bombardment of Egyptian air-fields and military installations continued unabated.

1956, November 3. Sixth Day. The Israeli 9th Brigade continued its tortuous progress down the Sinai coast. The paratroopers advancing down the Gulf of Suez coast captured the oil fields at Ras Sudar, Abu Zneima, and El Tur. The 11th Brigade captured Khan Yunis, completing its occupation of the Gaza Strip. The commander of the Sharm el-Sheikh garrison, **Colonel Raif Mahfouz Zaki,** withdrew his forces from Ras Nasrani and consolidated for defense.

1956, November 3–4. Israel Accepts and Rejects the U.N. Call for Cease-Fire. Israel at first assumed that by the time the cease-fire could take effect, Sharm el-Sheikh would have been captured. Cessation of hostilities, however, would reduce Anglo-French justification for intervention at the Canal. Under Anglo-French pressure, therefore, Israeli Prime Minister Ben Gurion withdrew Israel's acceptance by attaching impossible conditions to it.

1956, November 4. Seventh Day. The 9th Brigade continued slowly down the coast over very difficult terrain and passed through Ras Nasrani. Three miles north of Sharm el-Sheikh the brigade came under heavy fire.

1956, November 5. Eighth Day. The first assault on **Sharm el-Sheikh,** just at midnight, was stopped by a minefield and covering fire. At 0530 the attack was renewed with air and mortar support, and at 0930 Sharm el-Sheikh surrendered. The paratroops on the Gulf of Suez road arrived in time to take part in the final assault.

1956, November 5. Anglo-French Air Drop. At 0820, 500 allied paratroopers landed at Gamil Airfield near Port Said, soon followed by over 600 more near Port Fuad. They seized the waterworks and isolated the cities. The Chief of Staff of the Egyptian Eastern Command, Brigadier General **Salah ed-Din Moguy,** arranged a brief truce with Brigadier **M.A. H. Butler,** the paratroop commander, but fighting was renewed at 2230.

1956, November 6. Allied Amphibious Assault. At 0650 the amphibious forces began landing at Port Said. Moguy was captured at 1000, but refused to order a surrender. At 1930 Stockwell received an order to cease fire at midnight. On his order Butler advanced to El-Kap before the cease-fire was effected.

LOSSES

	Killed	Wounded	Captured	Aircraft
Israel	189	899	4	15
Egypt (in combat with Israel)	(30 officers) 1,000[a]	4,000[b]	6,000	8
Egypt (in combat with Anglo-French)	650[c]	900[c]	185	207[d]
Britain	16	96	0	4[e]
France	10	33	0	1

[a] The most widely accepted figure.

[b] Estimate.

[c] The official estimate. Others run as high as 3,000.

[d] Of these, 200 destroyed on the ground.

[e] One of these was shot down over Syria by Syrian AA fire.

THE SIX-DAY WAR
JUNE 5–10, 1967

Background

President Nasser of Egypt, who had concluded an alliance with Syria (November 1966), accused Israel of threatening aggression against Syria and promised to come to Syria's aid (May 16). He moved several divisions close to the Israeli-Egyptian border, in the eastern Sinai. On May 18 he demanded the withdrawal of the United Nations Emergency Force (UNEF), which had been patrolling the 1948–1956 cease-fire line. U.N. Secretary General **U Thant** ordered an immediate UNEF withdrawal. On May 22, Nasser, having placed a garrison at Sharm el-Sheikh, announced the blockade of the Strait of Tiran, effectively closing the Israeli port of Eilat. While the United States, within and without the U.N., endeavored to find a formula for peace, both sides seemed bent on war. On May 30 Egypt and Jordan signed a Mutual Security Treaty, and Egypt at once sent General **Abdul Moneim Riadh** to take command of allied Arab forces on the Jordan front. A chronology of prewar events follows:

May 16   Egypt declared a state of emergency.

May 17   Egypt and Syria announced "combat readiness," and Jordan announced mobilization.

May 18   Syria and Egypt placed troops on maximum alert; Iraq and Kuwait announced mobilization.

May 19   UNEF withdrawal.

May 20   Israel completed partial mobilization.

May 23   Saudi Arabian forces prepared to participate.

May 24   Jordanian mobilization completed.

May 28   Sudan mobilized.

May 29   Algerian units moved to Egypt.

May 30   Egypt and Jordan sign mutual security treaty.

May 31   Iraqi troops began moving to Jordan.

Israel had earlier announced that it would go to war under any of the following conditions: closing of the Strait of Tiran; sending of Iraqi troops to Jordan; signing of an Egyptian-Jordanian defense pact; withdrawal of UNEF forces. All of these conditions now existed. War thus was inevitable, although Nasser surprisingly did not think his actions would provoke an Israeli attack.

The Opposing Forces

The following table shows the numerical strengths of the principal participating forces. In most respects the Israeli Army was better trained and much more flexible than any of its opponents, with the possible exception of Jordan. Israeli first-line aircraft were more heavily armed, and the quality and training of Israeli pilots and support units were so much superior to those of the Arab pilots (who were in short supply) that the Israeli Air Force, under Major General **Mordechai Hod,** was at least two or three times more effective than the combined air forces of its Arab opponents.

ESTIMATED MOBILIZATION STRENGTHS
JUNE 5, 1967

	Mobilized Manpower	Division or Equivalent	Tanks	Artillery Pieces	Combat Aircraft	Naval Vessels[a]
Israel	230,000	8	1,100	200	260	22
Egypt	200,000	10	1,200	600	431	60
Syria	63,000	4	750	315	90	15
Jordan	56,000	3⅓	287	72	18	—
Iraq	90,000	5	200	500	110	15[b]

[a] Including MTBs.

[b] Not available for use against Israel.

OPERATIONS, SINAI FRONT

1967, June 5. Israeli Preemptive Air Strike. With the approval of Prime Minister **Levi Eshkol** and recently appointed Minister of Defense General Moshe Dayan, the Israeli General Staff, under Chief of Staff Lieutenant General **Itzhak Rabin,** decided that since war was inevitable, Israel should obtain the advantage of surprise by launching preemptive attack by air and ground. Early in the morning the Israeli Air Force, flying west over the Mediterranean, then south into Egypt, struck practically every Egyptian airfield and virtually wiped out the Egyptian Air Force. Later in the day, the IAF also destroyed the air forces of Jordan and Syria and—in retaliation for an Iraqi air strike against Israel—inflicted considerable damage on Iraqi air units based in the Mosul area.

1967, June 5. First Day. From north to south the Israeli forces comprised a reinforced mechanized brigade under Colonel **Yehuda Resheff,** a mechanized division commanded by Major General **Israel Tal,** an armored division under Major General **Avraham Yoffe,** a mechanized division under Major General Ariel Sharon; other smaller units were deployed along the frontier down to Eilat. Tal's division initiated the offensive by a drive into the Khan Yunis-Rafah-El Arish area. Resheff's brigade and attached elements drove into the Gaza Strip, and Sharon's division struck southward against the critical fortifications in the Abu Ageila-Kusseima area. Later in the day Yoffe struck southward between the divisions of Tal and Sharon, to penetrate into the heart of the Sinai, to cut off the Egyptian retreat.

1967, June 4–6. Accusations of US Participation. In a radio conversation monitored by the Israelis, King Hussein agreed with President Nasser to accuse the United States of collaborating with Israel, but quickly stopped the accusations after release of the taped radio conversation by Israel. Most Arab nations, except for Jordan, followed Nasser's lead and broke diplomatic relations with the U.S.

1967, June 6. Second Day. Gaza surrendered to Resheff in the afternoon. Tal, having secured Rafah and El Arish, sent a task force down the El Arish-Romani road toward the Suez Canal, and turned inland with the rest of his division and joined Yoffe. Sharon, having quickly captured Abu Ageila, sent part of his force to assist in the mopping up of Rafah and El Arish; with the remainder he struck southward toward Nakhl and Mitla Pass. Yoffe, after a brief engagement east of Bir **Lahfan,** successfully attacked the main Egyptian concentration in the central Sinai at **Jebel Libni** in a night assault. Unknown to the Israelis, Field Marshal Abdel Hakim Amer, the Egyptian commander in chief, had sent orders to all of the Sinai units to withdraw behind the Suez Canal; by this order he completed the demor-

alization begun by the Israeli attack the day before.

1967, June 7. Third Day. Tal's main body approached Bir Gifgafa; his northern task force moved past Romani. Yoffe's leading brigade reached the eastern end of the Mitla Pass, out of fuel and short of ammunition, and was quickly surrounded by withdrawing Egyptian units. Yoffe's other brigade was en route to relieve this isolated and hard-pressed unit. Sharon approached Nakhl. Other units cleared the northeastern Sinai, and airborne and amphibious forces seized **Sharm el-Sheikh.**

1967, June 8. Fourth Day. Egyptian armored units from Ismailia attempted to cover the general Egyptian withdrawal but were easily repulsed by Tal, who then pressed on to the Suez Canal between Kantara and Ismailia. Yoffe's division, having relieved its isolated brigade, pressed through the Mitla Pass and reached the Canal opposite Port Suez. After a grueling march through the desert Sharon's division took **Nakhl,** then followed Yoffe through the Mitla Pass. While there were a number of isolated Egyptian units still intact in the Sinai, for all practical purposes it was completely in the hands of the Israeli Army.

1967, June 9. Ceasefire. The UN Security Council reached a ceasefire which provided for Israeli control of the Sinai east of the Suez Canal. Israel accepted immediately, Egypt late the next day. (See also p. 1369.)

OPERATIONS, JORDANIAN FRONT

UNEASY NEUTRALISTS

Israeli strategy called for avoiding operations against Jordan and Syria until a decision had been reached on the Sinai front. At the same time Israel was strongly tempted to eliminate the deep Jordanian West Bank salient into the heart of Israel, and to wrest from the Jordanians control of the holiest places of Judaism in Jerusalem. There were good reasons for King Hussein to remain neutral; but Arab pressures to join in the war, combined with his recent agreement with Nasser, made neutrality difficult. Apparently he hoped that long-range artillery fire by 155mm "Long Tom" guns against Tel Aviv and other places would satisfy his Arab allies without provoking Israel to full-scale hostilities. However, Jordanian long-range artillery fire threatened to close down the runways at Ramot David, Israel's principal northern air base. Israel's leader would not allow this, and in mid-morning, June 5, decided on war against Jordan.

THE BATTLE FOR JERUSALEM

1967, June 5. First Day. Having decided that war against Jordan was unavoidable, Israel decided to take the initiative in Jerusalem, where sporadic firing, mostly by Jordanians, had already begun. Reinforcements were sent to Brigadier. General **Uzi Narkiss,** commanding Israel's Central Command, permitting him to take the offensive with 3 brigades, with the main effort by a parachute brigade under Colonel **Mordechai Gur.** The Israelis closed in on the old walled city of Jerusalem, whose garrison of 1 reinforced brigade was commanded by Jordanian Brigadier **Ata Ali.**

1967, June 6. Second Day. The Israeli advance against the Old City of Jerusalem slowed down in the face of stubborn opposition. Other Israeli units, however, tightened a wider ring around the city. The ridge to the east was seized, and a series of Jordanian relief efforts was smashed by combined ground and air forces. Elements of a tank brigade seized **Ramallah** to the north, while another brigade captured Latrun to the west. For the first time since 1947, the old Tel Aviv–Jerusalem Road was open to Jewish traffic.

1967, June 7. Third Day. Colonel **Gur** stormed into the Old City of **Jerusalem,** as the Jordanian garrison withdrew. **Bethlehem** was taken early in the afternoon, **Hebron** and **Etzion** soon afterward.

1967, June 7. Ceasefire. Both sides accepted the UN Security Council call for a ceasefire.

THE BATTLE OF JENIN-NABLUS

1967, June 5. First Day. The Israeli Northern Command, under Major General **David Elazar,** was roughly equivalent to 2½ divisions. Upon receipt of orders to seize Jenin and Nablus, and to push on to the Jordan River, Elazar committed one division and a reinforced armored brigade. By midnight armored and infantry units were approaching Jenin.

1967, June 6. Second Day. Converging Israeli columns took **Jenin** in hard fighting.

1967, June 7. Third Day. Despite repeated Jordanian counterattacks, the Israelis pressed on to **Nablus.** After another hard battle they secured Nablus just before dark. Although badly hurt and seriously depleted, the Jordanian forces withdrew across the Jordan River and were still intact when the Israeli and Jordanian governments agreed to a U.N. call for a ceasefire at 2000 hours.

OPERATIONS, SYRIAN FRONT

1967, June 5–8. First to Fourth Days. The Golan Heights were held by 6 Syrian brigades, with 6 more in reserve east of Kuneitra. For 4 days Elazar was permitted only long-range artillery duels with the Syrians, who obviously had no intention of seizing the initiative.

1967, June 9. Fifth Day. The U.N.-initiated cease-fire agreed to on June 8 was promptly violated by intensive artillery fire from both sides during the night of June 8/9. Elazar was ordered to begin a major offensive early on June 9. He concentrated his available forces for an initial advance through the Dan-Banyas area onto the northern Golan plateau, along the foothills of Mount Hermon. By nightfall these units had fought their way through the first line of Syrian defenses guarding the approaches to the northern Golan, and 3 brigades were poised to debouche onto the plateau early the following morning. Meanwhile, other units were forcing their way up the escarpment north of the Sea of Galilee, and Elazar had sent orders to the units recently engaged against the Jordanians in the Jenin-Nablus area to move north to strike into the Golan south of the Sea of Galilee.

1967, June 10. Sixth Day. The Israelis pushed through the crumbling Syrian defenses on the northern Golan early in the morning, then pressed ahead across the plateau to converge on Kuneitra from the north, west, and southwest. Meanwhile, the troops redeployed from the Jordan front had driven northeastward up the Yarmuk Valley to occupy the southern Golan and to threaten Kuneitra from the south. By dark Kuneitra was surrounded, and one armored unit was occupying the city.

1967, June 10. Cease-fire. The cease-fire again

became effective at 1830 hours, and this time was observed by both sides.

COMMENT. *This brief 3-front campaign clearly demonstrated the combat effectiveness superiority of the Israelis over their more numerous Arab foes. The scope, decisiveness, and speed of the victory was undoubtedly enhanced by the orders of the Egyptian com-* *mander, Field Marshal Amer, for a general Egyptian withdrawal on June 6, which turned an inevitable defeat into a disastrous rout. This contributed to an unwarranted Israeli contempt for the Egyptians, and an underestimation of their military potential which would have important consequences a few years later.*

The War at Sea

Naval operations were almost entirely between Egypt and Israel, since there was no naval activity by Jordan and practically none by Syria.

ESTIMATED LOSSES
June 5–10, 1967

	Killed	Wounded	Prisoners and Missing	Tanks	Combat Aircraft
Israel	800[a]	2,440	18	100	40[b]
Egypt	11,500[c]	15,000	5,500	700	264
Syria	700	3,500	500	105	58
Jordan	2,000	5,000	4,500[d]	125	22
Iraq	100	300	—	20	24

[a] Israel reported 679 dead and 2,563 wounded on June 11; it is assumed that about half of 225 seriously wounded later died.

[b] Only 2 in air-to-air combat; an air combat loss ratio of exactly 1–25.

[c] More than half were lost in the desert.

[d] Many probably were deserters with West Bank origins; Israel captured approximately 500 Jordanian prisoners of war.

1967, June 3–4. Israeli Deception. Four landing craft, on huge trucks, were ostentatiously sent by road from the Mediterranean to Eilat, then sent back at night to repeat the ostentatious daylight movement. Egyptian intelligence assumed that at least 8 (and probably more, due to multiple reports) of Israel's 18 LCTs were available for operations in the Gulf of Aqaba. During the night of June 4/5 several Egyptian vessels were sent from the Mediterranean through the Suez Canal to the Red Sea, to counter the anticipated Israeli threat. In this way Israel substantially reduced the imbalance of forces in the Mediterranean.

1967, June 5. Engagement off Port Said. An Israeli destroyer and several MTBs approached Port Said after dark. They were met outside the breakwater by 2 Egyptian *Osa*-class missile boats. After an inconclusive exchange of fire, with little damage to either side, the Egyptian vessels withdrew into the harbor. Israeli frogmen also entered the harbors of Port Said and Alexandria; some damage was done to Egyptian vessels in Alexandria, but all of the frogmen there were captured.

1967, June 6. Egyptian Withdrawal from Port Said. The intensity of Israeli air attacks, and the threat of General Tal's advance along the northern Sinai coast, caused the Egyptian Navy to withdraw to Alexandria all of its vessels based on Port Said.

1967, June 6–7. Egyptian Coastal Bombard-

ment. Three Egyptian submarines briefly shelled the Israeli coast, near Ashdod, and north and south of Haifa. They submerged and withdrew when attacked by Israeli air and naval forces.

1967, June 7. Israeli Seizure of Sharm el-Sheikh. A task force of 3 MTBs seized the Egyptian fortifications at Sharm el-Sheikh. After Israeli paratroops arrived, the naval vessels proceeded through the Strait of Tiran to the Red Sea without interference.

1967, June 8. The *Liberty* Incident. During the afternoon the USS *Liberty*, electronics surveillance vessel, 14 nautical miles north of El Arish, was attacked and seriously damaged by Israeli fighter-bombers and MTBs. The Israeli government's subsequent apology was accepted by the United States.

Economic Warfare

On June 5, at an Arab League meeting in Baghdad, representatives of Iraq, Saudi Arabia, Qatar, Bahrain, Kuwait, Libya, Algeria, Abu Dhabi, Egypt, Syria, and Lebanon agreed to stop the flow of oil to all nations they believed had attacked any Arab states. This included the United States, whom the Egyptians also accused of participating in the first air attack, Great Britain, and West Germany, as well as Israel. Only Kuwait, Iraq, and Algeria took serious measures to carry out this embargo.

THE "OCTOBER WAR" (YOM KIPPUR WAR OR THE WAR OF RAMADAN) OCTOBER 6–24, 1973

Background

President **Anwar Sadat** of Egypt apparently decided in November 1972 to go to war on the basis of readiness estimates supplied to him by the Egyptian Minister of War, General (later Field Marshal) **Ahmed Ismail Ali.** Both knew that Egypt had not reached tactical-technical military parity with Israel and that there might be another Israeli victory. Sadat, however, believed that Israel was satisfied with the *status quo,* and its *de facto* annexation of the territories conquered in 1967, and thus would make no moves toward reasonable negotiations without pressure from one or both of the great powers. The only possibility of moving toward a Middle East settlement seemed to be to precipitate action that would force the major powers and the U.N. to pay attention to the "no peace no war" situation in the Middle East.

1973, September 12. Arabs Select D-Day. Sadat, General Ismail, and President **Hafez Assad** of Syria met secretly during an Arab summit meeting in Cairo. Extraordinary and successful measures were taken to preserve secrecy of plans.

1973, September 26. Arab Concentrations; Israeli Alert. Egypt and Syria announced concentrations of troops for routine maneuvers. Although Israeli and U.S. intelligence believed there would be no war, a partial but perfunctory Israeli alert was ordered, including deployment of a second armored brigade to the Golan area (September 29).

1973, October 4–5. Partial Evacuation of Soviet Advisors. The hasty departure of some Soviet advisers and all dependents was noted by Israeli and American intelligence agencies, which again informed their governments that there would be no war (October 5).

1973, October 6, 0400 Hours. War Inevitable. Israeli Director of Intelligence, **General Elihau Zeira,** informed Lieutenant General David Elazar that the Arabs would attack at

1800 hours. Israeli mobilization was ordered at 0930.

SINAI FRONT

1973, October 6, 1405 Hours. Outbreak of War. A massive Egyptian air strike against Israeli artillery and command positions and a simultaneous intensive artillery bombardment of Bar Lev Line fortifications along the Canal achieved complete tactical surprise; Israeli frontline units had been only partially alerted.

1973, October 6–7. Egyptian Assault Crossing of Suez Canal. Egyptian commandos crossed the Canal at 1435, followed by infantry, engineers, and a few amphibious and ferried tanks. Engineers, opening approaches in the Bar Lev Line's sand embankment by demolitions and water jets, had bridges operational in the Second Army area before midnight October 7. In the Third Army area the bridge construction was not completed until the night of October 7/8. About 500 Egyptian tanks crossed the Canal. Two quickly mobilized reserve Israeli armored divisions under Generals Ariel (Arik) Sharon and **Abraham (Bren) Adan** approached the front, Adan near Romani, Sharon near Tasa.

1973, October 8. Israeli Counterattack Repulsed. Counterattacks against the Egyptian Second Army, by Adan's division and Sharon's division (only parts were engaged), were repulsed with heavy losses. The Israelis dug in and the Egyptians consolidated, linking up all their bridgeheads. Israeli close support aircraft suffered heavy losses from Egyptian antiaircraft defense using Soviet missiles and guns.

1973, October 11. Egyptians Plan Offensive. For several days General Ismail rejected subordinates' recommendations to attempt to drive deeper into Sinai. However, following appeals for help from the hardpressed Syrians, he reluctantly ordered an offensive to draw Israeli strength, particularly air power, to the Sinai front.

1973, October 14. Egyptian Offensive Repulsed. The Egyptians were thrown back with heavy losses, particularly in tanks.

1973, October 15–16. Israeli Thrust across Suez Canal. Sharon, hitting the boundary between the Egyptian Second and Third Armies, was able to establish a bridgehead with a brigade of paratroopers near Deversoir about midnight.

1973, October 16–18. Battle of the "Chinese Farm."* The Egyptian Second Army closed the corridor behind Sharon, isolating his division in small bridgeheads east and west of the Canal. In intensive fighting Adan's division broke through, bringing a bridge to the crossing point. The Second Army with some assistance from the Third Army, tried unsuccessfully but repeatedly to close the corridor leading to the Israeli crossing site. Egyptian tank losses again were heavy. Adan's division then crossed (night 17–18 October).

1973, October 18–19. Expansion of Israeli Bridgehead. Despite the Chinese Farm battle and continuing bridging problems, Adan's division pushed westward from Sharon's bridgehead, overrunning Egyptian rear areas, including AA missile sites, and Israeli planes attacked ground targets with less opposition. Sharon's attempt to seize **Ismailia** was repulsed (October 19).

1973, October 20–22. Israeli Breakout from Bridgehead. Repeated thrusts by Sharon to the northwest against Ismailia were contained by paratroop and armored reserves of the Second and Third Army, reinforced from Cairo. Adan's drive south met weaker resistance cutting the main Suez-Cairo road northeast of Suez (October 22).

1973, October 22. First Cease-Fire, 1852 Hours. Both sides promptly claimed violations of the U.N. cease-fire. Israel sent strong reinforcements across the Canal.

1973, October 23–24. Battle of Suez-Adabiya. Despite the cease-fire Adan was ordered to continue his southward drive to the Gulf of Suez, near Suez, isolating the Third Army. At the same time another Israeli division

* A former Japanese experimental agricultural station; occupying Israeli troops in 1967 had assumed the calligraphy was Chinese.

under General **Kalman Magen** followed Adan and continued on to reach **Adabiya** on the Gulf of Suez. The Israelis also endeavored to take Suez, but were repulsed (October 23–24).

1973, October 24. Second Cease-Fire, 0700 Hours. Activity and artillery fire continued, but cease-fire finally came into effect.

COMMENT. *The Egyptian plan for crossing the Suez Canal and its implementation were superb. The Egyptian failure to plan adequately for security of the boundary between the 2 armies was a costly blunder. The Egyptians have claimed that they had adequate reserves and resources to reopen communications with the Third Army if the fighting had continued. This is doubtful, and had the war lasted a few days longer, the Third Army would probably have been forced to surrender. General Ismail has been severely criticized for failure to exploit his initial successes, particularly after the victory of October 8. Such criticisms fail to recognize that the Israelis retained a tremendous superiority in air power and in the ability to conduct mobile warfare; his decision not to exploit was as sound as that of Andrew Jackson after New Orleans (see p. 878) and Montgomery after Alam Halfa (see p. 1186). Despite some technical and tactical shortcomings, the Israeli crossing of the Canal was also brilliant as a demonstration of flexible improvisation, based upon sound doctrine.*

Perhaps most significant was the return of Egyptian confidence as their infantry stood firm, even though not always successfully, in the face of Israeli tanks. Contributing to this confidence were the Egyptian's simple, reliable Soviet-built antitank missiles—the "Sagger" and the RPG7. It seems, however, that at least as many Israeli tanks were knocked out by tank and antitank guns as by these missiles.

GOLAN FRONT

1973, October 6, 1405 Hours. Outbreak of War. A massive Syrian air strike and artillery bombardment against Israeli positions and installations on the Golan achieved complete tactical surprise.

1973, October 6. Fall of Mount Hermon. Syrian commandos, in a ground-helicopter attack, captured the fortified Israeli observation post on **Mount Hermon,** overlooking the Golan Plateau and the Damascus Plain.

1973, October 6–7. Repulse at Amadiye. North of Kuneitra the Syrian 7th Infantry Division was repulsed by the Israeli 7th Armored Brigade; most of the Syrian tanks were destroyed. The 3rd Syrian Tank Division, committed to pass through the 7th Infantry Division, suffered a costly defeat in a renewed major tank battle west of **Amadiye** (October 7).

1973, October 6–7. Breakthrough at Rafid. Taking advantage of weaker opposition and more favorable terrain, the Syrian 5th Mechanized Division broke through the defenses of the Israeli 188th Armored Brigade. In 2 days of fighting the Israeli brigade was virtually destroyed; the Israeli Golan command post at Khushniye was surrounded. Spearheads of the 5th Mechanized Division, reinforced by the 1st Tank Division, halted near the western escarpment of the Golan, as much by the need for logistical replenishment as by the pressure of recently mobilized Israeli units fed piecemeal into the battle.

1973, October 8–9. Israeli Counterattack. Assisted by units of the 7th Armored Brigade, displaced from the north, newly-arrived Israeli units drove back the Syrian 5th and 1st Divisions, in several places to the original front line. Most of the Syrian tanks were lost, many because they had run out of fuel and ammunition. With difficulty the 7th Brigade halted a renewed Syrian drive north of Kuneitra (October 9).

1973, October 10–12. Israeli Counteroffensive. In a drive generally north of the Kuneitra-Damascus road, 3 Israeli divisions smashed through the first Syrian defensive zone east of the cease-fire line, and into the second zone, near Saassaa, in front of Damascus. The Israelis voluntarily halted their offensive and began to shift units to the Sinai front (October 12). The Iraqi 3rd Armored Division, on the south side of the Israeli salient (October 11), counterattacked but was ambushed and repulsed (October 12).

1973, October 15–19. Arab Counterattacks Repulsed. Another counterattack by the Iraqi

3rd Armored Division was repulsed (October 15). The Jordanian 40th Armored Brigade counterattacked beside the Iraqis but was also repulsed (October 16). Another general Arab counterattack, spearheaded by the Jordanians, was repulsed (October 19). The lines stabilized on the Damascus Plain.

1973, October 22. Israelis Retake Mount Hermon. Two Israeli efforts to retake their **Mount Hermon** position had been repulsed (October 8 and 21). In a final effort, just before the cease-fire became effective, helicopter-borne paratroops seized the original Syrian observation post, higher up than that of the Israelis, and the "Golani" Infantry Brigade finally retook the lost Israeli position.

1973, October 22, 1852 Hours. Save for some fighting continuing briefly on Mount Hermon, an uneasy lull came over the front.

COMMENT. *The Syrian attack was neither as well planned nor as well implemented as that of the Egyptians. Nevertheless Syrian fighting qualities, especially those of the 5th Mechanized Division, impressed the Israelis. Poor logistics, failure to recognize that a complete victory was in their grasp, and the self-sacrificing attacks of the Israeli Air Force were all that kept the Syrians from retaking the southern Golan on October 7. The Israelis were also impressed by Syrian tenacity on defense; there was no collapse and rout, as in 1967.*

The Air War

1973, October 6–8. Preeminence of the SAM. The first Israeli aircraft appeared over the Sinai and Golan fronts about 40 minutes after the Arab H-Hour. They immediately encountered Soviet-made Arab missiles in unexpected quantity and effectiveness. Before dark the Israelis lost more than 30 aircraft. In the following days Egyptian mobile SAM-6s claimed many Israeli planes, and the light, hand-carried Strela (SA-7) made many hits but damaged more planes than it destroyed. Israeli close air support was therefore negligible for several days.

1973, October 8–16. Disputed Skies over the Battlefields. Employing hastily devised tactics and utilizing chaff and electronic counter-measures (ECM), Israeli aircraft began to make a greater contribution to the ground battles. They claimed hits on Egyptian bridges over the Suez Canal and favorable results in strikes against Arab airfields. The Arabs, however, denied these claims.

1973, October 9–21. The Israeli Strategic Air Offensive against Syria. Claiming retaliation for Syrian "Frog" (long-range surface-to-surface missile) attacks on the Hula Valley, the IAF initiated an intensive and extremely effective strategic air bombardment campaign against targets (mostly industrial) deep within Syria, with a strike against the Syrian Defense Ministry in Damascus. Attacks on Syrian seaports, industrial plants, and fuel storage depots continued until the first ceasefire. The Syrian economy was severely affected.

1973, October 17–24. Israel Regains Air Preeminence over the Suez Canal. As General Adan's advancing tanks captured a number of Egyptian antiaircraft missile batteries and caused the more mobile SAM-6s to be moved hurriedly, a gap was created in the hitherto effective Egyptian anti-air network. Israeli aircraft quickly exploited this, and played a major role in the Israeli success in the Chinese Farm battle and in the Israeli breakout south from the west bank bridgehead.

COMMENT. *The SAMs—particularly SAM-6 and Strela—demonstrated that air superiority no longer assured as decisive an effect on the ground battle as had been the case since World War II. Desperate Israeli efforts to deal with the missile threat, particularly in the employment of ECM measures, regained only a slight (but nonetheless significant) measure of air superiority for the Israelis over the battlefield.*

The War at Sea

1973, October 6–25. Egyptian Blockade. Egypt declared waters of Israel's coasts were a "War Zone" severely curtailing Israeli commerce in the Mediterranean. A blockade by destroyers and submarines at the Strait of Bab el Mandeb stopped all traffic to Eilat.

1973, October 6. Action off Latakia. Israeli "Saar" missile boats striking at night at the Syrian seaport of **Latakia** were engaged by a Syrian squadron. The Israelis sank 4 Syrian vessels without loss to themselves. The surviving Syrians withdrew into the port.

1973, October 7–8. Second Action off Latakia. The results were inconclusive. The Syrian vessels again withdrew.

1973, October 7–8. Scattered Egyptian-Israeli Clashes. These occurred in the Mediterranean and Red seas. The results were inconclusive, with the Egyptians withdrawing in all instances.

1973, October 8–9. Action off Damietta. Egyptian vessels, coming out to meet Israeli raiding missile boats, suffered severe losses, and the survivors withdrew.

1973, October 9–10. Israeli Raids on Syrian Ports. Israeli missile vessels bombarded Latakia, Tartus, and Banias; there was no naval challenge by Syrian vessels.

1973, October 9–10. Action off Port Said. In an encounter between Israeli and Egyptian missile boats, 3 Egyptian vessels were sunk; the others withdrew to Damietta and Alexandria.

1973, October 12–13. Israeli Raids on Syrian Coast. Tartus and Latakia were again bombarded. There were inconclusive clashes with Syrian missile boats, which unsuccessfully attempted hit-and-run tactics.

1973, October 15–16. Nile Delta Raid. Israeli missile vessels sank a number of Egyptian landing craft.

1973, October 21–22. Israeli Attacks on Aboukir Bay and Alexandria. Two Egyptian patrol boats were sunk.

COMMENT. *Israeli "Saar" missile vessels, armed with the Israeli-made "Gabriel" missile, completely dominated the coastal waters off Syria and Egypt. Not a single hostile shell or missile was fired from the sea against the Israeli coast. On the other hand, the Egyptian blockade of the Strait of Bab el-Mandeb (southern entrance to Red Sea) cut off all commerce to and from Eilat, and the declared Egyptian blockade of Israel's Mediterranean coast also severely reduced ship traffic. However, the blockade had no noticeable effect on Israel during the relatively brief war.*

Involvement of the Superpowers

The association of the Soviet Union with Egypt and Syria had been only partially impaired by the Egyptian demand for withdrawal of most Soviet advisors and technicians in mid-1972 (see p. 1395). Both Egypt and Syria were almost completely dependent upon the U.S.S.R. for replacement and repair, equipment and parts. Despite close associations with the Soviet Union, however, the decision to go to war was certainly not dictated by the Soviets, and it is likely that the U.S.S.R. did not learn of the Arab plans to attack Israel until 2 or 3 days before the selected D-Day, when the warlike preparations of Egypt and Syria could no longer be concealed.

Israeli technological development was such that there was no need of assistance or advice in using, or maintaining, the wide variety of modern weapons and equipment which Israel had obtained from the United States. There was no collaboration in planning between the 2 countries. However, they maintained close contact in intelligence affairs, and both were very much aware of the Arab build-up in late September. Both sadly misestimated Arab intentions.

1973, October 6. Soviet Air Flights to Middle East Accelerate. This was primarily to return Soviets to the U.S.S.R. Both Syria and Egypt were so well supplied that resupply presumably seemed unnecessary.

1973, October 8. Israel Begins to Fly Supplies from the United States. The first of a number of flights from the United States to Israel by El Al aircraft took off from Oceana Naval Air Station, Virginia.

1973, October 9. Major Soviet Airlift to Egypt and Syria Begins. Flights went via and/or over Hungary and Yugoslavia. About ²/₃ of the flights were to Syria.

1973, October 13. American Airlift Begins. In response to urgent Israeli requests, the United States began to use American planes to supplement the El Al lift. The first 7 American C5A transport aircraft arrived in Israel, flying via the Azores (October 14).

1973, October 14–21. Massive Soviet and U.S. Airlifts Continue. By the time of the cease-fire the Soviets had airlifted about 15,000 tons, the U.S. more than 20,000.

1973, October 24. Alert of Soviet Airborne Troops. A force of about 7 divisions was alerted, presumably for airlift to Egypt if the Israelis did not loosen their stranglehold on the Egyptian Third Army. Apparently because of this alert, there was a decline in the Soviet resupply airlift.

1973, October 25. United States Military Forces Placed on "Precautionary Alert." Secretary of State Henry Kissinger announced that this was because "ambiguous" signs suggested the possibility of unilateral armed Soviet intervention in the Middle East. Unmistakably implied was the United States' determination to act militarily if necessary to prevent or respond to such intervention.

1973, October 27. United Nations Agreement Ends U.S.-Soviet Crisis. The U.N. Security Council, with both the U.S. and the U.S.S.R. voting affirmatively, agreed to establish for 6 months a 7,000-man international peace-keeping force to enforce the cease-fire in the Sinai and on the Golan. No permanent member of the Security Council would participate in this emergency force (UNEF); thus the U.S. and the U.S.S.R. agreed not to commit forces in the Middle East. Simultaneously, conversations were taking place on the west bank cease-fire line at Kilometer 101 (101 kilometers from Cairo) between Egyptian and Israeli military representatives, with the commander of the U.N. Truce Supervision Organization (UN-TSO, in existence since 1948). These talks resulted in agreement for Egypt to send noncombat supplies to its Third Army east of the Suez Canal.

1973, November 11. Egyptian-Israeli Prisoner of War Exchange Agreement. Having reached agreement on the location of the cease-fire line, the Egyptian and Israeli military representatives at Kilometer 101 agreed upon a prisoner of war exchange. Involved were 241 Israelis and 8,031 Egyptians. Following the exchange (completed November 22) the talks broke down because of inability of the 2 sides to agree on a disengagement formula (November 29).

1973, December 21–22. First Meeting of Geneva Peace Conference. Representatives of Egypt, Israel, the U.S. and the U.S.S.R. agreed that Egyptian-Israeli discussion of separation of forces should continue in Geneva (December 26).

1974, January 18. Israeli-Egyptian Disengagement Agreement. This followed a week of intensive "shuttle diplomacy" by Secretary of State Kissinger, flying back and forth between Egypt and Israel (January 11–17). Israeli troops were to withdraw within 40 days from their bridgehead to a line 15 to 20 kilometers east of the Canal. The Egyptians would remain on the East Bank, in a zone 8 to 10 kilometers deep. A buffer zone between the 2 forces, 5 to 8 kilometers wide, would be patrolled by UNEF. This left the Israelis in control of the Gidi and Mitla passes. The Israeli withdrawal began on January 24 and was completed March 4.

1974, February 28. The U.S. and Egypt Resume Diplomatic Relations. These had been broken since June 6, 1967 (see p. 1343).

1974, February–May. Syrian-Israeli "War of Attrition." Apparently as a tactic to place pressure on Israel to make territorial concessions on the Golan Heights, Syrian forces initiated a protracted artillery and small-arms duel along the entire cease-fire line between Kuneitra and Damascus.

1974, March 18. Arabs End Oil Embargo of the U.S. Libya and Algeria refused to vote with the majority of 7 Arab oil-producing nations (see p. 1401).

1974, April-May. Intensified Palestine Guerilla Attacks. Palestine guerrillas, attempting to place pressure on Israel, undertook a number of suicide raids across the border from Lebanon into Israel. Most notable were attacks on Qiryat Shemona (April 11) and Maalot (May 15). Israeli aircraft bombed Palestine guerrilla bases and camps in Lebanon in retaliation. Losses of life among the civilian population on both sides further inflamed Arab-Israeli hostilities.

1974, April 18. Egypt Abandons Reliance on Soviet Military Equipment. Announcing that the U.S.S.R. had failed for 6 months to honor Egyptian resupply requests, and that Soviet terms for renewed supply of arms to Egypt were "unacceptable" and an "instrument of policy leverage," President Sadat ended 18 years of Egyptian reliance upon Soviet arms deliveries.

1974, May 31. Israeli-Syrian Disengagement Agreement. After 32 days of shuttle diplomacy by Kissinger between Israel and Syria, Israel gave up all of the territory captured from Syria in the October 1973 war, plus 2 small strips taken in 1967, including the town of Kuneitra. Military forces were to be limited in zones on either side of the new cease-fire line, and a narrow buffer zone was to be patrolled by units of the UNEF.

THE KOREAN WAR, 1950–1953

BACKGROUND

Korea, annexed by Japan following the Russo-Japanese War (see p. 1008), was promised its freedom by the Allies at the Cairo Conference (December 1, 1943). The decision was reaffirmed in the Potsdam Proclamation (July 26, 1945). When Japan surrendered in World War II (see p. 1309), a hurried Allied agreement (August 15, 1945) established the 38th degree of latitude as an arbitrary dividing line, north of which the U.S.S.R. would accept surrender of Japanese forces in Korea; those Japanese south of the line would surrender to U.S. troops. Following the surrender, which took place with little friction, the U.S.S.R. held the 38th parallel to be a political boundary; along it the Iron Curtain dropped.

Two years of unsuccessful attempts to reach agreement were followed by U.S. referral of the problem to the U.N., which undertook the establishment of an independent Korean government following free nationwide elections. The U.S.S.R. refused to cooperate. In the southern zone, the Republic of Korea was established (August 15, 1947), with Seoul its capital. Declaring the action illegal, the U.S.S.R. set up a puppet government—the Democratic People's Republic of Korea, its capital at Pyongyang—and organized a North Korean Army (NKA). Allegedly, Soviet troops evacuated the north (December 1948). U.S. troops completed evacuation of the south (June 1949); a small American military advisory group remained to organize a Republic of Korea (ROK) Army. More than a year of continuous bickering—Communist propaganda, raids, sabotage, terrorism, and guerrilla action—harassed the south without breaking down the ROK government.

THE OPPOSING FORCES

Communist North Korea now had a well-trained and Russian-equipped army: 130,000 men in 10 divisions with a brigade of Russian T-34 medium tanks and supporting troops. Its hard core was composed of some 25,000 veterans of the Chinese Communist campaign in Manchuria (see p. 1424). The air force consisted of some 180 Russian Yak planes of World War II type. There were more than 100,000 trained reserves.

The ROK Army—little more than a national police force—consisted of about 100,000 men in 8 divisions with little supporting artillery. It lacked medium and heavy artillery, tanks, combat aircraft, and reserves.

Naval strength on both sides was negligible.

OPERATIONS, 1950

1950, June 25. Invasion. North Korean forces—7 infantry divisions, the tank brigade, and supporting troops, under Marshal **Choe Yong Gun**—crossed the border in 4 columns, driving on Seoul. Surprise was complete. The power punch, accompanied by radio broadcasts asserting it to be "national defense" against an alleged ROK "invasion," broke through the scattered resistance of elements of the 4 ROK divisions in the area. Its objective was to seize the capital and the entire South Korean peninsula, thus presenting the free world with a *fait accompli.*

1950, June 25–30. United Nations and United States Reactions. The Security Council, in emergency session (the U.S.S.R., boycotting the Council, had no representative present to veto the action), called for immediate end to hostilities and withdrawal of the NKA, asking member nations to assist. President **Harry S Truman** (June 27) ordered General MacArthur, commanding U.S. forces in the Far East, to support and cover ROK defense with air and sea forces. MacArthur effected naval blockade of the North Korean coast and furnished air support. Reconnoitering the front in person (June 28), as Seoul fell, he reported the ROK Army to be incapable of stopping the

invasion even with U.S. air support. Truman authorized use of U.S. ground troops (June 30).

MacArthur's Resources. Aside from the vessels of the U.S. Seventh Fleet and the Far East Air Force (8½ combat groups), U.S. ground forces—mostly in Japan—consisted of 4 understrength divisions organized in 2 skeleton army corps. Infantry and artillery units were each at ⅔ strength in personnel and cannon, and short of antitank weapons. Corps troops, such as medium tanks, artillery, and other supporting arms, did not exist.

1950, June 30. U.S. Forces Begin Move to Korea. The 24th Division (Major General **William F. Dean**) began movement piecemeal by sea and air into Korea; 2 more divisions were to follow.

1950, July 5. Task Force Smith. One understrength battalion (2 infantry companies) with 1 battery of artillery, under Lieutenant Colonel **Charles B. Smith,** joined the ROK Army near **Osan** (July 4). Next morning an NKA division, with 30 tanks, attacked. The ROK troops fled. Task Force Smith, completely surrounded, held out for 7 hours. Then, ammunition exhausted, the survivors cut their way out, abandoning all matériel.

1950, July 6–21. Dean's Delay. Throwing in the remainder of his division as fast as the units came up, General Dean partly snubbed the NKA advance down the peninsula, trading terrain for time, while the 1st Cavalry and 25th divisions were being rushed from Japan. A 5-day action at **Taejon** (July 16–20) ended when the NKA assaulted the 24th Division from 3 directions. Dean, personally commanding his rear guard while the remainder of the division withdrew, was captured. His battered troops were relieved by the 1st Cavalry Division (July 22), while the 25th Division on its right, together with reorganized ROK divisions, slowed the NKA advance in the center and on the north.

1950, July 7. MacArthur Named Commander in Chief United Nations Command. President Truman made the appointment in response to a Security Council request that a unified command be established under a U.S. officer.

1950, August 5–September 15. The Pusan Perimeter. Lieutenant General **Walton H. Walker,** commanding what had now become U.S. Eighth Army, stablized his defense on a thinly held line extending along the Naktong River some 90 miles north from Tsushima Strait, thence east for 60 more miles to the Sea of Japan. The area embraced the southeast edge of the Korean peninsula, including Pusan, the one available port. On the north, 5 ROK divisions, re-equipped but still shaken, attempted to contain the invaders, while the western flank, where the weight of incessant NKA attacks fell, was held by U.S. troops, now including 2 additional infantry regiments and a Marine brigade. The Seventh Fleet protected both sea flanks and harassed NKA movements along the coast, while the Far East Air Force (augmented by an Australian group), together with carrier-based naval air, hammered at NKA lines of communication and furnished much-needed close support. Thanks to the advantage of interior lines, Walker was able to shift a mobile reserve from point to point within the perimeter as the NKA attacks nibbled at his front. Several penetrations of the Naktong River line and a 20-mile NKA advance in the north (August 26) were checked. Choe's forces, now estimated at 14 infantry divisions supported by several tank regiments, continued a series of uncoordinated assaults all around the perimeter. A 3-division attack on the north (September 3) necessitated committing the entire U.N. reserve (the 24th Division) north of Kyongju. Arrival of the U.K. 27th Infantry Brigade (September 14) compensated for the withdrawal of the Marine brigade for duty elsewhere (see below).

1950, September 15–25. The Inchon Landing. At dawn, the U.S. X Corps, Major General **Edward M. Almond** commanding, began landing over the difficult and treacherous beaches at Inchon, on the west coast, more than 150 miles north of the battlefront, and west of Seoul. Strategic surprise was complete, although a 2-day preliminary bombardment

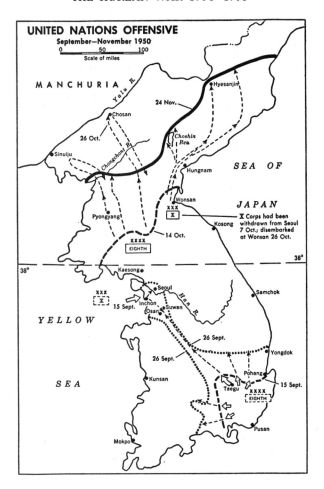

UNITED NATIONS OFFENSIVE
September–November 1950
0   50   100
Scale of miles

MANCHURIA

Yalu R.

Chosan

24 Nov.

Hyesanjin

26 Oct.

Chongchon R.

Sinuiju

Choshin
Res.

Hungnam

SEA OF

JAPAN

Wonsan

XXX
X

Pyongyang

Kosong

X Corps had been
withdrawn from Seoul
7 Oct.; disembarked
at Wonsan 26 Oct.

14 Oct.

XXXX
EIGHTH

38°

38°

Kaesong

38°

XXX
X

Seoul

Han R.

Samchok

15 Sept.

Inchon
Osan Suwan

YELLOW

26 Sept.

26 Sept.

Yongdok

Pohang

15 Sept.

SEA

Kunsan

Taegu

XXXX
EIGHTH

Pusan

Mokpo

had warned the few NKA detachments in and about Seoul. The 1st Marine Division swept through slight opposition, securing Kimpo airport (September 17). The 7th Infantry Division, following the Marines ashore, turned south, cutting the railroad and highway supplying the NKA in the south, and Seoul was surrounded.

1950, September 15–25. Breakout from the Perimeter. Simultaneously the Eighth Army broke out, the 1st Cavalry Division leading. Choe's NKA, its supplies cut off, and menaced from front and rear, disintegrated. The 1st Cavalry and 7th Infantry divisions met just as Seoul itself was liberated.

1950, September 26. Liberation of Seoul.

More than 125,000 prisoners were taken, together with most of its matériel, as the NKA scattered into the roadless, rugged countryside.

COMMENT. *The Inchon landing was one of the great strategic strokes of history, in conception, execution, and results. MacArthur's genius had transformed into a stunning victory a desperate defense seemingly doomed to disaster. The Communist grab at South Korea, which if successful might have meant the eventual absorption of all the Asian mainland, had been thwarted, the North Korean Army crushed. MacArthur's decision was taken while his Eighth Army was still clawing to maintain a toehold at Pusan. On August 12 his staff was ordered to prepare the operation, 1 month to accomplish what ordinarily would take several. His dynamic insistence overcame the doubts of the U.S. JCS (technically the Inchon area was most disadvantageous for an amphibious operation) and he was provided with the bulk of the 1 division of Marines he demanded. (One of its brigades, already there, came from his own Pusan perimeter.) The 7th Infantry Division was the last of his occupation troops in Japan, its ranks filled by more than 5,000 ROK soldiers hurried to Japan to train with it. Success depended on (a) the ability of the U.S. Navy to provide sufficient water transport, (b) the ability of Walker's hard-pressed Eighth Army to hold the Pusan perimeter until the stroke fell, and (c) perfect timing of the assault (a 30-foot tide variance permitted use of the beaches for only 6 of each 24 hours.)*

1950, October 1–November 24. Advance to the Yalu. As directed by both the U.N. and President Truman, MacArthur pushed north across the 38th parallel. ROK troops crossed the line (October 1); Eighth Army followed (October 9), leaving 2 divisions in the southern area to secure communication lines to Pusan and mop up roving remnants of the NKA. A serious military handicap was the injunction that under no circumstances were U.N. aircraft to fly north of the Yalu River. **Pyongyang,** North Korean capital, was overrun (October 20) by a combined airborne (187th Regimental Combat Team) landing and overland advance. The ROK 6th Division reached the Yalu at Chosan, and other ROK units fanned out behind it. By this time, other U.N. token forces had joined Eighth Army and were integrated in existing U.S. divisional elements: a Turkish brigade, and Canadian, Australian, Philippine, Netherland and Thai battalions.

1950, October 15. Wake Island Conference. President Truman and General MacArthur conferred at Wake Island on the course of the war. This later became an issue in the Senate investigation that followed the relief of General MacArthur (see p. 1361).

1950, October 16–26. Shift of U.S. X Corps to the East Coast. Embarked at Inchon, the corps was moved around to the east coast to Wonsan (October 19), which had already been captured by the ROK I Corps. A 7-day delay in landing was necessary to sweep the harbor clear of latest-type Soviet mines, sown under the direction of Soviet experts with the NKA.

1950, October–November. Threats from Communist China. Peking had threatened intervention should the 38th parallel be crossed by U.N. troops, and heavy concentrations of Chinese Communist troops were reported north of the Yalu, in Manchuria. Since aerial reconnaissance beyond the Yalu was prohibited, MacArthur knew neither the full strength nor the dispositions of these troops, nor—until his forward ROK divisions were ambushed by them and a U.S. regiment at Unsan was severely mauled (November 1)—was he aware that Chinese troops in considerable numbers were already south of the Yalu. Walker, confronted by the presence of this new element in the situation, recalled his leading Eighth Army units and consolidated temporarily along the Chongchon River.

MacArthur's Plans and Problems. His intention was to advance up the entire front of the peninsula, X Corps on the east coast, Eighth Army on the west, and make a sweeping envelopment. X Corps would turn west on reaching the Yalu and drive all enemy forces south of the border into the arms of Eighth Army. Since the rugged, desolate central massif precluded mutual support, these forces acted independently, their control and coordination directed by MacArthur in Tokyo. Almond's X Corps now consisted of the U.S. 1st Marine and 3rd and 7th divisions and the ROK I Corps (3rd and Capital divisions). The Eighth Army, 9 divisions strong, was grouped in 3 army corps: the U.S. I and IX and the ROK II, from left to right. The total combat strength of the entire command was about 200,000 men, with perhaps, 150,000 more in support functions in the rear. In addition to the ROK corps, some 21,000 more Korean troops were attached to or integrated in U.S. units.

MacArthur had believed that Communist China was bluffing; that she would not enter the conflict unless Manchuria itself were invaded. He had expressed this opinion to Truman at Wake Island (see above). The U.S. Central Intelligence Agency was of the same opinion. Yet now Red Chinese troops were in Korea. MacArthur considered that to suspend his advance would be a violation of his directive: "to destroy the North Korean armed forces." Aerial reconnaissance north of the Yalu being still prohibited, he decided the only remaining course was to clarify the situation by a bold advance. The decision was specifically approved by the JCS. Meanwhile, the X Corps had thrust north widely distributed over an immense front; the ROK Capital Division had reached Chongjin on the coast; the U.S. 7th Division was on the Yalu at Hyesanjin.

1950, November 24. Eighth Army Advance. MacArthur's "reconnaissance in force" began.

1950, November 25–26. Chinese Communist Counteroffensive. After advancing for 24 hours against practically no opposition, the Eighth Army was suddenly struck a massive blow, the main effort being directed against the U.N. force's right flank. Some 180,000 Chinese troops, in 18 divisions, shattered and ripped through the ROK II Corps, hit the U.S. 2nd Division on the right flank of IX Corps, and threatened envelopment of the entire Eighth Army. The 2nd Division, attempting to refuse its right flank, fell into an ambush at **Kunu-ri** as the Chinese envelopment trapped its columns while passing through a defile in march order. Some 4,000 men and most of the divisional artillery were lost while trying to fight their way out. Walker threw in his reserves, the U.S. 1st Cavalry Division and the Turkish and 27th Commonwealth brigades. They staved off the envelopment, the Turks in particular taking heavy losses, and the Eighth Army managed to disengage in comparatively good order. By December 5, the Eighth Army, its right flank refused, had completely extricated itself, and the Communist drive was beginning to lose momentum; but the central and east-coast area being wide open (see below), a stronger defensive position was essential. Walker accordingly withdrew to the general line of the 38th parallel, slightly north of Seoul and some 130 miles below the November 24 situation. There, as the year ended, the Eighth Army awaited a new Communist offensive.

1950, November 27-December 9. X Corps Withdrawal. In the eastern zone, an additional 120,000 Chinese troops, advancing on both sides of the Chosin reservoir, isolated the 1st Marine Division and drove in elements of the 3rd and 7th divisions. The ROK troops on the coastal flank were hurriedly withdrawn on Almond's order without much molestation. MacArthur ordered evacuation of the entire force, since the Communist drive, directed on the ports of Hungnam and Wonsan, threatened its piecemeal destruction. Navy transports were rushed to both ports. Defensive perimeters were established, manned by elements of the 3rd and 7th divisions, while the 1st Marine Division, under Major General **Oliver Smith,** consolidated south of the Chosin reservoir in subzero weather. Surrounded by 8 Communist divisions, General Smith, announcing to his troops that they were not retreating, but "attacking in another direction," moved southeast on Hungnam, supplied by the Far East Air Force. When a Communist blow destroyed the 1 bridge across a gorge otherwise impassable for the division's trucks and tanks, bridging material was flown in by air and the Marine southward "advance" continued. Thirteen days of running fight ended (December 9) when a relief column of 3rd Division troops met the Marine vanguard outside the Hungnam perimeter.

1950, December 5–15. X Corps Evacuation. Despite continued Communist attacks on the perimeters of both ports, evacuation by air and by sea went smoothly. Air Force and Navy carrier-plane support, together with naval gunfire, facilitated the final embarkation. In all, 105,000 ROK and U.S. troops were lifted by the Navy, together with 98,000 civilian refugees. Some 350,000 tons of cargo and 17,500 vehicles were also carried. The Far East Air Force evacuated 3,600 troops, 200 vehicles, and 1,300 tons of cargo. On arrival at Pusan, the X Corps came under Eighth Army control as a strategical reserve.

1950, December 23–26. Death of Walker; Arrival of Ridgway. Walker, killed in an automobile accident, was replaced by Lieutenant General **Matthew B. Ridgway.** MacArthur gave Ridgway command of all ground operations in Korea, retaining over all ground, air, and sea command.

COMMENT. *The U.N. forces in Korea had suffered a serious defeat, though disaster was averted through skillful troop leading and stubborn—in some cases phenomenal—resistance. MacArthur's critics were quick to blame his simultaneous advance in 2 independent zones on 1 front, and the inadequate security measures which had permitted the stunning surprise. His supporters, in rebuttal, pointed to the nature of the terrain, which rendered close ground liaison and mutual support between Eighth Army and X Corps impossible. They also blamed the artificial ground rules set up by the U.N. and the JCS prior to the Communist assault prohibiting aerial reconnaissance north of the Yalu. To add to MacArthur's problems, he had been denied his immediate request that he be now permitted to bomb the Yalu bridges and also the important North Korean entry port of Rachin, only 35 miles from Vladivostok, through which Soviet war matériel had long flowed freely to the NKA. Bombing of the southern ends of the Yalu bridges was finally permitted; this was an extremely hazardous and practically futile operation, since to approach them U.N. aircraft had to fly parallel to the river, exposed to antiaircraft fire from Manchuria and to the assaults of Communist attack planes. Communist traffic over the bridges was little interrupted.*

It would appear that MacArthur's original decision to advance—a decision concurred in by the JCS—was proper. But the method may be criticized. The JCS must take much blame for permitting the advance while prohibiting the prior air reconnaissance that was essential. There is no doubt that MacArthur expected that, should the Chinese Communist threat materialize, he would immediately be given full authority to extend his air operations over the border and choke the assault at its source, but the JCS had no intention of giving such authority. Finally, the actual local security measures taken just before and during the Eighth Army's move were inadequate under the circumstances. For this both MacArthur and Walker must share blame.

The most important lesson from this defeat was the necessity that U.N. troops relearn the rudiments of fire and movement on foot. Roadbound, they had found themselves too dependent upon supporting tanks, artillery, or aircraft. The Chinese troops, on the other hand, lightly equipped, utilized fluidity, surprise, and concealment in the rugged regions to compensate for their inferiority in firepower. They moved and attacked by night; lay camouflaged in daylight. Their attacks all followed the same pattern: infiltration, encirclement, and ambush. Frontal assaults were in effect holding attacks in small force, but the penetrations were deep. Each engagement was initially one of small units. It was a platoon commander's war. At no time was the U.N. able to employ its firepower superiority in full force.

OPERATIONS, 1951

1951, January 1–15. Second Communist Invasion. Long prepared and expected, the Communist assault crossed the 38th parallel at daybreak, its main effort in the western zone. Some 400,000 Chinese troops, with an additional 100,000 of the reconstituted NKA, pushed the 200,000-man Eighth Army back almost to Seoul. Then (January 3) a heavy penetration farther east, in the Chungpyong reservoir area, overran the ROK divisions on both flanks of the U.S. 2nd Division, which extricated itself only after serious fighting and the commitment of Ridgway's reserve—the 3rd and 7th divisions.

1951, January 4. Evacuation of Seoul. This was the third time the capital had changed hands. Stubborn resistance of ground troops, plus the Far East Air Force's close support and interdiction of the now exposed Communist lines of communications, slowly checked the momentum of the drive. The U.N. position stabilized some 50 miles south of the 38th parallel, from Pyongtaek on the west coast to Samchok on the east (January 15).

1951, January 25–February 10. U.N. Counteroffensive. Ridgway launched a series of limited-objective attacks, slowly driving north. A Communist counterattack near **Chipyong** and **Wonju** (February 11–18) checked the advance in the center, but on the west U.N. troops reached the outskirts of Seoul.

1951, March 7–31. Operation "Ripper." This was designed primarily to inflict casualties on the enemy, and secondarily to relieve Seoul and eliminate a large Communist supply base now built up at Chunchon. The main effort, in the center, forced the Communists back. The Han River was crossed east of Seoul.

1951, March 14. Reoccupation of Seoul. Patrols of the I Corps found it abandoned. Chinese resistance then stiffened, but an airborne drop by a reinforced regiment (187th Regimental Combat Team) at **Munsan,** 25 miles north of Seoul (March 23), forced a general Communist retirement. The Eighth Army was back roughly along the old 38th-parallel front (March 31). MacArthur and Ridgway decided on further advance, toward the "Iron Triangle"— Chorwan-Kumhwa-Pyonggang, the major as-

sembly and supply area, as well as communications center, for the Chinese.

1951, April 11. MacArthur Relieved. President Truman summarily ousted General MacArthur from his dual command of U.N. forces and of U.S. forces in the Far East. Ridgway was appointed in his place, and Lieutenant General James A. Van Fleet was hurried from the U.S. to command the Eighth Army.

COMMENT. *The President was exercising his legal prerogative as Commander in Chief. MacArthur was not in sympathy with the policy of limiting the war to the Korean peninsula and had not attempted to conceal his dissatisfaction with the restrictions placed on his operations. MacArthur had stated that "in war there is no substitute for victory" in a letter (March 20) to Representative* **Joseph W. Martin, Jr.** *(R., Mass.), which Martin promptly made public. Mac-*

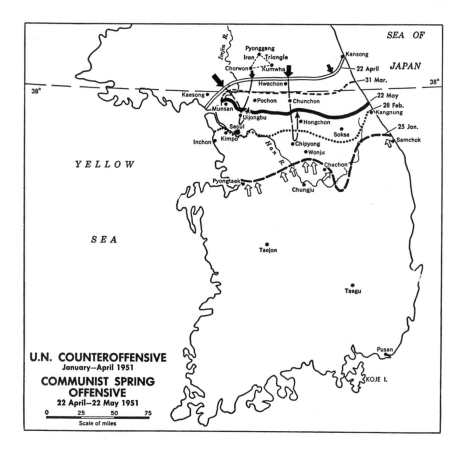

U.N. COUNTEROFFENSIVE
January–April 1951

COMMUNIST SPRING OFFENSIVE
22 April–22 May 1951

0   25   50   75
Scale of miles

Arthur was advocating neither the use of the atom bomb nor a land invasion of China. He did want to destroy, by conventional air attack, bases in Manchuria which were being used as springboards for invasion of Korea. He did urge the use of Chinese Nationalist troops in Korea, and also the "unleashing" of Chiang Kai-shek on the Chinese mainland. He believed that the Soviet Union could not afford to risk war by coming to the aid of Red China, but that if it did make such a mistake there could be no better time for the U.S. to face a showdown with the Kremlin. Without consulting Washington, he had called on the Chinese commander in Korea to surrender (March 25) and hinted that air and naval attacks against Communist China would be the probable consequence if the conflict continued.

To Truman such opinion and actions were anathema. MacArthur was defying presidential authority and debating national policy. The United States—listening to an anxious Free World opinion frightened by Soviet possession of the atomic bomb—had deliberately given up the idea of liberating all of Korea and was seeking merely to restore the status quo *in South Korea. So MacArthur had to go. The brusqueness of the ousting—the general learned it first through a news broadcast—offended many people. MacArthur returned to the U.S. to receive a hero's welcome and an invitation to address the Congress in joint session, which he did. A Senate investigation later (May–June 1951) aired all the policy issues; the results of the investigation were inconclusive; in general, however, U.S. policy became tougher subsequently.*

1951, April 12–21. Continued U.N. Advance. Aware of Communist preparations for a counteroffensive to blunt the threat to their "Iron Triangle," General Van Fleet continued forward movement, prepared to fall back, if necessary, to previously prepared defensive positions. There he would contain the enemy by his heavy firepower and then counterattack.

1951, April 22–May 1. Communist Spring Offensive—First Phase. The attack came on a moonlit night; the first assault broke through the ROK 6th Division, west of the Chungpyong reservoir. The U.S. 24th Division on the left and the 1st Marine Division on the right promptly refused their respective flanks, but the penetration compromised Van Fleet's general position

and he began withdrawal of his left—the I and IX Corps. The Chinese main effort developed against the I Corps, north of Seoul. Hasty withdrawal by the ROK 1st Division exposed the flank of the U.K. 29th Brigade on its right and a battalion of the Gloucestershire Regiment was cut off. After a heroic defense of their hill position, the survivors attacked *north,* to the momentary confusion of their assailants. Some 40 men escaped; the remainder were killed or captured. As usual, the Communist assault finally lost momentum and came to a pause (April 30). They broke contact, retiring in general beyond U.N. artillery range. Communist losses in this phase were at least 70,000 men, while Eighth Army casualties were about 7,000.

1951, May 14–20. Second Phase of Communist Offensive. Shifting the weight of their attack to the east, more than 20 divisions, with NKA divisions on their right and left, struck the right elements of X Corps—the ROK 5th and 7th Divisions. The U.S. 2nd Division, next on the left, stood firm, but the ROK III Corps, farther east, went to pieces under heavy assault. The ROK I Corps, on the extreme right, refused its flank against the Communist surge through this wide corridor. The U.S. 2nd Division (with French and Netherlands battalions attached) and the 1st Marine Division, on the west side, promptly counterattacked. Van Fleet had expected the blow in this area and had already shifted his reserves—the U.S. 3rd Division and 187th Regimental Combat Team. Their combined efforts snubbed the Communist offensive (May 20). Attacks on the west flank, north of Seoul, and in the center, down the Pukhon River, had been repulsed.

COMMENT. *As usual, initial Communist contact was by small units, attempting to infiltrate and terrorize. There were few tanks. The attacks, vigorous at night, ceased with daylight when U.N. artillery and tactical aircraft firepower could intervene. Communist losses in this 7-day phase were estimated at 90,000. They were now overextended, supplies expended and communications under continuous aerial attack.*

1951, May 22–31. U.N. Offensive. Preceded by limited attacks on the far left, anchoring the U.N. position on the Imjim River, north of

Munsan, the entire U.N. front moved north. The ROK Capital and 2nd divisions, on the extreme right, flashed up the east coast with little opposition, reaching Kansong. Advance was slower in the center, but was accelerating by month's end. Van Fleet was now ordered to halt. Despite his plea for approval of "hot pursuit" against an enemy on the verge of collapse, the JCS refused either increased means or permission for another drive northward. The U.S. government was concerned by Soviet threats and by consequent alarm elsewhere in the Free World. It was decided to do nothing to risk World War III.

1951, June 1–15. Consolidation of U.N. Position. With the rainy season on, Van Fleet decided to establish a defensive belt across Korea, from which springboard he could keep the enemy off balance by a succession of JCS-approved limited-objective moves. Some gains were made at the base of the "Iron Triangle"—thus denying its use to the enemy—and on the southern rim of the "Punchbowl," a fortified hill-circle northwest of Sohwa. Meanwhile, the

Communists were themselves organizing in depth to the north.

1951, June 23. Soviet Cease-Fire Proposal. This was made in the U.N. by Soviet Ambassador Malik. It confirmed that the Chinese had been badly hurt in the previous 6 months' fighting. Estimated enemy losses totaled 200,000 men, together with much matériel. Also, U.N. air attacks had foiled every attempt to install Communist air bases south of the Yalu. Delegations from both sides met at Kaesong.

1951, July–August. Negotiations. The Communists, taking advantage of the location of Kaesong just inside their lines, seized every opportunity to insult U.N. negotiators and to delay progress while playing for time to recuperate from their mauling. They used the negotiations as a sounding board for propaganda against the U.N. allies and the U.S. in particular. Meanwhile, clashes between patrols and outposts continued all along the firing line as both sides improved their positions. The Eighth Army improved its hold both on the

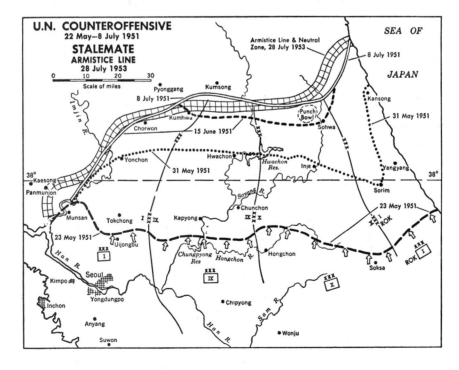

Iron Triangle and the Punchbowl. The negotiations broke down completely (late August).

1951, August–November. Resumption of U.N. Limited Attacks. Van Fleet's troops cleared the Iron Triangle and the Punchbowl, driving the Chinese back from the Hwachon reservoir and the Chorwan-Seoul railway line. These successes brought prompt Communist requests for resumption of armistice discussions.

1951, November 12. Discussions Begin at Panmunjon. This was a village between the lines in No Man's Land. General Ridgway (November 12) ordered offensive operations stopped, and Eighth Army went on a highly active defense.

OPERATIONS, 1952

While negotiations dragged out interminably at Panmunjon, minor actions flared continually all along the front. General Ridgway, ordered to NATO command (see p. 1372), was replaced (May) by General **Mark W. Clark.** The Communists continued building up their strength. By the year's end an estimated 800,000 Communist ground troops—$^3/_4$ of them Chinese—were in Korea, while heavy shipments of Soviet artillery were brought in, including excellent anti-aircraft guns, radar-controlled. However, U.N. command of the air was never seriously threatened, and the Communists were forced to continue their practice of taking shelter during daylight hours in concealed, deep-dug bunkers and other underground installations.

Prisoners of War. Communist wranglings at the Panmunjon conferences centered on the disposition of prisoners of war. About 92,000 U.N. troops had fallen into Communist hands: some 10,000 Americans, 80,000 Koreans, and 2,500 from other U.N. forces. Statistical computation is impossible; no one will ever know how many prisoners died of mistreatment or starvation. Communist boats in 1951 put their POW bag at 65,000, but at Panmunjon they admitted holding only 11,500. Consensus of reports by returned POWs indicated that about $^2/_3$ of U.S. prisoners died or were killed in the prison camps. No neutral or Red Cross inspections were ever permitted.

Some 171,000 Communist prisoners fell into U.N. hands, more than 20,000 of them Chinese. About 80,000 of them were assembled on the island of Koje, just off Pusan, where they were held under rather haphazard control, due to U.S. anxiety about our own captives in Communist hands, as well as humanitarian, if misguided, efforts to improve their conditions.

At Panmunjon, Communist negotiators insisted on total repatriation of all POWs, but at least 50,000 prisoners in U.N. hands were violently opposed to returning home. The U.N. command had no intention of forcing these unfortunates back into Communist hands. A deliberate build-up of hard-core Communists at Koje—organizers planted to become POWs—produced an organized revolt (May 7) when U.S. Brigadier General **Francis T. Dodd,** naïve camp commander, was captured through a ruse and held as hostage inside his own prison compound. His successor, U.S. Brigadier General **Charles F. Colson,** unwitting that he was abetting a Communist propaganda coup, bartered for Dodd's release by promising in effect that alleged abuses in treatment of Communist POWs (abuses which did not exist in fact) would be "corrected." General Clark ordered U.S. Brigadier General **Haydon Boatner** to clear up the situation. Boatner, combining military firmness with knowledge of the Chinese Communist mentality, swept the Communist hard-core recalcitrants into a separate compound and restored order.

All this while, a dreary succession of attacks and counterattacks—in reality tests of willpower—cost both sides great losses in flesh and blood. In October, the negotiations at Panmunjon again broke off, while the war became a political football in the U.S. presidential election. The American people, tired of the struggle, elected **Dwight D. Eisenhower,** who had promised to bring about an honorable conclusion.

OPERATIONS, 1953

1953, March 28. Communist Move for Peace. Unexpectedly, but apparently in tune with internal unrest in the Communist world following Stalin's death (March 5), Premier **Kim Il Sung** of North Korea and General **P'eng Teh-huai,** heading the Chinese "volunteers," informed General Clark of their agreement to his previously ignored proposal for mutual exchange of sick and wounded POWs. They also urged resumption of the Panmunjon conferences. Unquestionably, indications of extensive U.S. plans to renew offensive operations, and possibly to extend the war, were major factors in the Communist gambit.

1953, April. Operation "Little Switch." This was an exchange of 5,800 Communists for 471 ROKs, 149 Americans, and 64 other U.N. personnel.

1953, May–June. South Korean Intransigence. South Korea's President **Syngman Rhee** flatly refused to become a party to any agreement which left Korea divided. After Communist attacks against his troops (see below), he demanded resumption of the military offensive (June 18). At the same time he re-leased from his own prison camps 27,000 North Korean POWs unwilling to be repatriated.

1953, June 10–31. Chinese Communist Offensive. Massive attacks, mostly against ROK troops, were begun, obviously to bring about U.S. pressure on Syngman Rhee. When he released the prisoners, the Communists, accusing the U.N. of bad faith, again broke off negotiations and launched still another offensive (June 25), against the ROK sector. Some slight gains were made, but quick shifts of U.S. reinforcements and the usual Chinese inability to exploit their penetrations brought the attack to a halt, with loss of some 70,000 troops.

1953, July 10. Resumption of Negotiations. Following U.N. assurance to the Communists that no further ROK intransigence would occur, the negotiators hammered out a final armistice.

1953, July 27. Armistice Signed. The *de facto* boundary was the existing battle line. Exchange of prisoners who desired repatriation followed: 77,000 Communists, against 12,700 U.N. men—of whom 3,597 were Americans and 945 Britons.

SUMMARY

The Korean War cost the U.N. 118,515 men killed and 264,591 wounded; 92,987 were captured. (A great majority of these died of mistreatment or starvation.) The Communist armies suffered at least 1,600,000 battle casualties, 60 percent of them Chinese. An additional estimated 400,000 Communists were nonbattle casualties. US casualties were 33,629 killed and 103,284 wounded (see Statistical Summary). Of 10,218 Americans who fell into Communist hands, only 3,746 returned; the remainder (except 21 men who refused repatriation) either were murdered or died. In all, 357 U.N. soldiers refused repatriation. South Korea's toll—which can only be estimated—came to 70,000 killed, 150,000 wounded, and 80,000 captured. Approximately 3 million South Korean civilians died from causes directly attributable to the war.

This war was significant on several counts. It was the first major struggle of the nuclear age. While no nuclear weapons were employed, the threat of the atom bomb hung heavy over all concerned and throttled exploitation of success.

It was a war between 2 differing ideologies, a war of stratagem and deceit in which road-bound superior firepower was canceled out by lighter-armed fluidity over desolate, trackless wastes. All ethical standards of western civilization were scorned by the Communists.

STATISTICAL SUMMARY: U.S. FORCES IN THE KOREAN WAR, 1950–1953

	Total U.S. Forces Worldwide	Maximum Deployed Strength	Total Combat Casualties	Killed and Died of Wounds	Wounded	Prisoner or Missing	Nonbattle Deaths
Total	5,764,143	c. 440,000 (Apr–Jul '53)	147,131	33,629	103,284	10,218	20,617
Army	2,834,000	276,581 (July '53)	113,610	27,704	77,596	8,310	9,429
Navy	1,177,000	84,124 (Feb '53)	2,243	458	1,576	209	4,043
Marines	424,000	36,966 (Apr '53)	28,627	4,267	23,744	616	1,261
Air Force	1,285,000	46,388 (Dec '52)	2,651	1,200	368	1,083	5,884

U.N. Participation. Fourteen U.N. member nations besides the U.S. took part in this, the first war in which the U.N. had engaged. Britain and Turkey each contributed a brigade (each about ¹/₃ of a division). In addition, the U.K. furnished 1 aircraft carrier, 2 cruisers, and 8 destroyers, with marine and supporting units. Canada sent 1 brigade of infantry, 1 artillery group, 1 armored battalion; Australia, 2 infantry battalions, 1 each air fighter and transport squadrons, 1 aircraft carrier, 2 destroyers, 1 frigate; Thailand, 1 regimental combat team; France, 1 infantry battalion, 1 gunboat; Greece, 1 infantry battalion, 1 air transport squadron; New Zealand, 1 artillery group, 2 frigates; Netherlands, 1 infantry battalion, 1 destroyer; Colombia, 1 infantry battalion, 1 frigate; Belgium and Ethiopia, 1 infantry battalion each; Luxembourg, 1 infantry company; Union of South Africa, 1 fighter squadron. (Ground units from the British Commonwealth were combined into a Commonwealth Division.) All these elements, though merely token forces, bore themselves well and sustained heavy casualties. In addition, from Denmark, India, Italy, Norway, and Sweden came hospital or field-ambulance noncombat units.

Air Warfare. The conflict reaffirmed the critical importance of air power as an essential ingredient of successful combat; it also was a reminder that air power alone can neither assure adequate ground reconnaissance nor bring about final decision in land warfare. The immediate superiority achieved by the U.N. in the air necessitated bringing in Soviet Mig-15s—then the latest U.S.S.R. jet fighters, quite superior to America's F-84 and surpassing in some respects the F-86. Migs, first seen in Korea in late 1950, increased in number during 1951, but the training and competence of U.N. pilots—mostly American—compensated for any inferiority in maté-riel. While U.N. pilots were never permitted to hound the Migs across the Yalu in "hot pursuit," they were able to neutralize all Communist efforts to establish bases south of the river. In air-to-air combat, 1,108 Communist planes were destroyed, including 838 Mig-15s; probably destroyed were another 177 and severely damaged were an additional 1,027 planes, against a total U.N. loss of 114 aircraft. As the war drew to a close, U.S. F-86 jets were downing Mig-15s at the rate of 13 confirmed Communist losses to each F-86 shot down. U.N. plane losses to

Communist antiaircraft fire—while giving magnificent close support to ground troops—were 1,213.

The Helicopter. The potential of this new means of mobile transportation was clearly demonstrated. It was excellent for reconnaissance, evacuation, and rescue work.

The Navy's Role. American command of the sea was one of the principal handicaps to Communist success. Without this, the U.N. campaign in aid of South Korea would have been impossible. The U.S. Seventh Fleet gave valuable gunfire support along the coast and carried out amphibious operations, while naval and Marine air units participated in Air Force interdiction and close support to ground units. The Navy's blockade of the peninsula prevented any attempts to supply the Communist forces by water. Had it been possible to interdict ground-supply channels from Manchuria and Siberia in similar fashion, the war would have been over in short order.

Brainwashing. Not until the U.N. POWs returned home was the full extent of the Communist ideological warfare realized. Through brutality—physical and psychological torture (some 60 percent of American and British POW's died from torture or neglect)—many men unraveled. About 15 percent of Americans in captivity actively collaborated with the enemy; only about 5 percent of the total resisted categorically all Communist indoctrination and all efforts to use them for propaganda purposes. Consensus in the U.S. and Britain was that their soldiers had been mentally unprepared for such treatment.

INTERNATIONAL PEACEKEEPING

THE UNITED NATIONS

The U.N. Charter, formally ratified by 29 nations, came into force (October 24, 1945; the U.N. was later located in New York), composed of a General Assembly and an elected Security Council of 12, whose 5 permanent members were China, France, Great Britain, the U.S.S.R., and the U.S. Disarmament activities of the U.N. are treated separately below. Important highlights of its other activities during the period were:

1946, April 18. End of League of Nations. The League voted itself out of existence, transferring assets and responsibilities to the U.N.

1946, December 19. Greek Frontiers. A commission was established to investigate violations of the Greek frontiers with Albania, Yugoslavia, and Bulgaria (see p. 1388). A special commission was later established to observe the northern frontier of Greece (October 21, 1947).

1947, May 15. Palestine. A special commission was established to investigate the problem of Palestine (see p. 1392). This led to a plan for establishment of separate independent Jewish and Arab states, approved by the General Assembly (November 29, 1947).

1947, August 26. Indonesia. A Good Offices Commission was appointed to seek peaceful settlement of the war in Indonesia (see p. 1421).

1948, January 20. Jammu and Kashmir. Attempt was initiated to settle the India-Pakistan dispute by a U.N. mediation commission (see p. 1404).

1948, April 23. Palestine Truce Commission. This was established, after outbreak of Arab-Israeli war (see p. 1334), under the leadership of Count **Folke Bernadotte** of Sweden. An international Palestine U.N. Mediation and Observer Group was later created under Bernadotte's command (May 14). Soon after, Bernadotte was assassinated by Jewish terrorists (see p. 1338).

1948, October 25. Soviet Veto on Berlin. This blocked settlement by the Security Council of the Berlin Blockade (see p. 1378).

1948, December 11. Conciliation Commission for Palestine Established.

1948, December 12. Korea. The General Assembly appointed a commission to aid unification of Korea (see p. 1354).

1949, January 7. Israeli-Arab Armistice. (See p. 1339.) Supervision arrangements were made by the Security Council.

1949, January 28. Indonesia. The Security Council renewed efforts to halt hostilities and settle the Indonesian question; the Good Offices Commission (see p. 1421) was reconstituted as the U.N. Commission for Indonesia.

1950, January 3–August 15. Soviet Boycott. The U.S.S.R. boycotted all U.N. bodies on which Nationalist China was represented.

1950, June 27. Korean War. The Security Council voted 7-1 (Russia absent) to assist South Korea in repelling North Korean aggression (see p. 1355).

1950, September 14. U.N. Denounces North Korean Aggression. This was in a report by the Commission on Korea to the General Assembly.

1951, February 1. Communist China Named Aggressor in Korea. This was by General Assembly vote. Later the Assembly voted an embargo on arms to China (May 8).

1953, April 23. Investigation of Germ-Warfare Charges. The General Assembly adopted by a vote of 51-5 a resolution for impartial investigation of Communist charges that the United States had used germ warfare in Korea.

1953, November 11. U.S. Charges Communist Atrocities in Korea. A U.S. report to the General Assembly charged that at least 6,113 U.S. servicemen and a total of 11,622 U.N. servicemen had been murdered, tortured, or otherwise mistreated.

1954. Egyptian Blockade of Israel. U.S.S.R. vetoed a resolution in the U.N. Security Council to call on Egypt to end restrictions on passage through the Suez Canal of ships bound for Israel.

1955, September 29. Algeria. By 1 vote the General Assembly voted to investigate conditions in Algeria. France withdrew (September 30), but later returned to the Assembly (November 25).

1955, October 12. General Assembly Condemns South Africa's Racial Segregation. South Africa then withdrew from the General Assembly (November 9).

1956, April, July. Hammarskjöld to Middle East. On trips to the tense Middle East at request of Security Council, he attempted to calm down tempers and obtain cease-fire agreements from Israel and Arabs.

1956, October 28. Hungary Crisis. (See p. 1384.)

1956, October 29. Suez Crisis. (See p. 1340.)

1956, November 4–7. U.N. Emergency Force. At Egypt's request, following Israel's Sinai offensive and the abortive Anglo-French invasion of the Suez Canal area (see p. 1342), the Security Council sent a force of 6,000 men, selected from 10 nations, to supervise cessation of hostilities, protect Egyptian borders, and place a cordon sanitaire across the Negev, separating Israeli and Egyptian forces. The first units arrived in a few days (November 15).

1957, September 14. Soviet Intervention in Hungary Condemned. This was a General Assembly vote.

1958, June 11. Observer Teams for Lebanon and Jordan. (See pp. 1397, 1399.) These peace-preservation teams facilitated the withdrawal of U.S. and British troops (August 21).

1959, June 20. Korea. The U.N. command in Korea charged North Korea with violating the armistice by building military fortifications in the demilitarized zone.

1959, December 1. Antarctic Treaty. A 12-nation treaty established the continent of Antarctica as being available only for peaceful purposes.

1960, July 14. Establishment of Force for

Congo. The Security Council authorized the Secretary General to organize an *ad hoc* military force to help the Congolese government preserve order. These were mainly drafted from African nations into the Congo, under Swedish Major General **Karl Von Horn,** to preserve peace. Some 18,000 strong at its peak, this force actually battled dissident elements (see p. 1441) in hope of bringing about a peaceful settlement of the Congo question. The last U.N. troops left the Congo June 30, 1964, upon which rebellion flared again (see p. 1442).

1961, September 17. Death of Dag Hammarskjöld. The U.N. Secretary General was killed in a plane crash in nearby Northern Rhodesia as he was coming in to land at the airport at Elizabethville, Congo.

1962, September 21. West Irian. A security force was established to supervise transfer of western New Guinea from the Netherlands to Indonesia (see p. 1422).

1963, June 1. U.N. Observer Force for Yemen. (See p. 1402.) This force of 200 observers was established by the Security Council, initially for 2 months, under Swedish Major General **Karl von Horn.** Soon after the force's life was extended (August 1), von Horn resigned, citing lack of U.S. support and poor administration. He was replaced by Indian Lieutenant General **Prem Singh Gyani** (August 27).

1964, March 4. Cyprus. Establishment of a U.N. force (see p. 1390). A cease-fire was later obtained through U.N. efforts (August 11).

1964, July 3. U.N. Observation Mission in Yemen Extended. Secretary General U Thant announced that the extension to September 4 would facilitate negotiations for mutual withdrawal by the UAR and Saudi Arabia.

1967, May 18. U Thant Agrees to Withdrawal of UNEF from Sinai. (See p. 1343.)

1967, July 10. U.N. Observers Agreed for Suez Canal Cease-Fire Supervision. (See p. 1345.)

1973, October 27. UNEF Reestablished in Sinai. (See p. 1353.)

1974, May 31. UN Force Established on Golan Heights. (See p. 1354.) This was des-ignated UN Disengagement Observer Force (UNDOF).

DISARMAMENT ACTIVITIES

1945, November 15. U.S., Britain, and Canada Offer Atomic Information to U.N. President Harry S. Truman, Prime Minister **Clement Attlee,** and Prime Minister **Mackenzie King** offered to share information on atomic energy with other members of the U.N.

1946, June 14. Baruch Plan. At the first meeting of the U.N. Atomic Energy Commission, the U.S. delegate, **Bernard Baruch,** offered a plan whereby the United States would give up its store of atomic bombs and reveal its secrets of controlling atomic energy to an international Atomic Energy Development Authority. There could be no veto power in this international authority.

1946, June 19. Soviet Alternative Plan. The Soviet representative submitted an alternative plan to outlaw all atomic bombs, and insisted on the retention of the veto by the Big-Five Powers in all atomic matters. All atomic weapons would be destroyed within 3 months of ratification of an international agreement. Shortly after this (July 24), the Soviet Union rejected the Baruch Plan. The Russian representative, **Andrei Gromyko,** announced that Russia would not permit any form of inspection of atomic-energy projects within her borders.

1946, December 14. Resolution for Worldwide Disarmament. Adopted by acclamation by the General Assembly. This led to the establishment of a U.N. Disarmament Commission, which met sporadically and ineffectively, in subsequent years (1947–1960).

1946, December 30. Baruch Plan Approved. The basic points were adopted by the Atomic Energy Commission and it was referred to the Security Council, where it was killed by Soviet veto.

1958, March 31. Soviet Unilateral Test Ban. The Soviet Union announced that it was halting all tests of atomic and hydrogen bombs. Testing was renewed later, however (September 30).

1958, October 31. U.S. Unilateral Cessation of Nuclear Tests. President Eisenhower announced that this voluntary ban would last for 1 year. After completing a series of tests, the U.S.S.R. tacitly adopted the test ban (December).

1959, August 26. U.S. Extends Unilateral Nuclear Test Ban. A 2-month extension of its 1-year unilateral ban (due to expire October 31) was announced, in view of a 6-week recess in U.S.-British-Soviet nuclear-test talks at Geneva. When no agreement was reached, the U.S. announced that it had dropped its obligation to stop further nuclear-weapons tests (December 31). No tests were made, and it was implied that no atmospheric tests would be conducted so long as the U.S.S.R. refrained from such tests.

1960, March 15. First Meeting of Ten-Nation Disarmament Committee. Results of this effort at Geneva were as fruitless as those of the U.N. commission. It broke up by Soviet action (June 27).

1960, May 7. U.S. Plans to Resume Underground Nuclear Tests. This was for the purpose of improving methods of detecting underground nuclear tests.

1961, September 1. U.S.S.R. Begins Atmospheric Nuclear Tests. This unilateral action, without prior announcement, ended a $2\frac{1}{2}$-year implied atmospheric-test moratorium.

1961, September 5. U.S. Resumes atmospheric Tests. The U.S. began nuclear tests in response to the Soviet Union's unilateral resumption of testing.

1963, April 5. U.S.-Soviet "Hot Line" Agreement. This direct communications link between Washington and Moscow had been proposed at the 17-nation U.N. Disarmament Committee by the United States (March 15) to reduce the danger of accidental war. This action was inspired by the Cuban missile crisis (see p. 1454).

1963, July 25. Nuclear Test Ban Treaty Signed. This concluded more than 3 years of sporadic discussion. It prohibited all nuclear testing in the atmosphere, but permitted underground nuclear tests to continue. Subse-

quently, most of the nations of the world also signed the treaty, which went into effect in the fall (October 10).

1963, October 17. Renunciation of Weapons in Space. The U.N. General Assembly confirmed earlier unilateral declarations by the U.S. and the U.S.S.R. It was made binding by a formal **Treaty on the Demilitarization of Outer Space** (January 27, 1967; in force October 10, 1967).

1964, July 21. Denuclearization of Africa. Announced by a declaration of the heads of state and government of the Organization of African Unity at a summit conference at Cairo.

1967, February 14. Denuclearization of Latin America. This treaty, signed in Mexico City by representatives of most Latin American states, was approved by the U.N. General Assembly (December 5, 1967). United States endorsement went into effect May 12, 1971.

1967, March 2. U.S.-U.S.S.R. Agree to Nuclear Arms Limitation Negotiations. (See SALT entries below.)

1968, July 1. Treaty on Nonproliferation of Nuclear Weapons. Most nations of the world agreed to stop the spread of nuclear weapons to nations not yet possessing them. Although many non-nuclear nations protested that this committed them to perpetual second-class sovereignty, most signed the treaty (in force March 5, 1970). Among nations not signing were nuclear powers France and Communist China, and such potential nuclear powers as Argentina, Brazil, India, Israel, Pakistan, South Africa, and Spain.

1969, November 17. Beginning of U.S.-Soviet Strategic Arms Limitation Talks (SALT). Representatives of the 2 nations met at Helsinki, Finland, for a month to initiate a major effort to establish firm controls over nuclear weapons. Subsequent secret meetings alternated between Helsinki and Vienna.

1969, November 25. United States Renounces Biological Warfare and First Use of Chemical Warfare. President Richard M. Nixon renounced for the U.S. first use of chemical weapons, and all methods of biological warfare. Toxins—weapons on the borderline between chemical and biological agents—were later banned (February 14, 1970). He also requested the U.S. Senate to ratify the Geneva Protocol of 1925 (outlawing chemical warfare), never ratified by the U.S. This was eventually achieved (December 1974).

1971, February 11. Treaty to Denuclearize the Seabed. This treaty prohibited the emplacement of nuclear weapons and other weapons of mass destruction "on the seabed and the ocean floor and in the subsoil thereof."

1971, September 30. U.S.-Soviet Nuclear Accidents Agreement. This was the first published product of the SALT negotiations. To reduce the possible risk of accidental war, both nations pledged themselves to notify each other in the event of (a) an accident that might cause detonation of a nuclear weapon; (b) detection of suspicious activity by either nation's security warning system; or (c) planned missile launches in the direction of the other. The agreement, signed at Washington, entered into force immediately.

1972, May 26. U.S.-Soviet SALT Agreements. At a summit meeting in Moscow President Nixon and Soviet Communist Party Secretary **Leonid Brezhnev** signed the first major results of 3 years of SALT negotiations (see above). The first of these, **Treaty on the Limitation of Anti-Ballistic Missiles,** prohibited nationwide deployment of ABM systems but allowed each nation to establish limited ABM defenses for its national capital and 1 ICBM site. The problem of verification is not mentioned; both sides tacitly agreed that their observation satellites could assure compliance. The second, **Interim Agreement on Limitation of Strategic Offensive Weapons,** provided for a 5-year moratorium on deployment of strategic offensive rocket launchers; its objective was to freeze ICBM deployments at then-existing levels during the prolonged negotiations anticipated to be necessary to work out a comprehensive agreement limiting strategic nuclear weapons. At that time the U.S. had 1,056 launchers and the Soviets

had 1,618 launchers. The interim agreement did not prohibit qualitative improvements (such as accuracy) or the development of multiple independently targeted reentry vehicles (MIRVs).

1973, June 21. Second Round U.S.-Soviet SALT Agreements. At a summit meeting in Washington, President Nixon and Secretary Brezhnev signed 2 additional SALT agreements. The first of these, **Basic Principles of Negotiations on the Further Limitation of Strategic Offensive Weapons,** indicated some progress toward the comprehensive treaty needed to replace the Interim Agreement (see above). The second, **Agreement on the Prevention of Nuclear War,** provided that both nations would avoid "situations capable of causing a dangerous exacerbation of their relations, as to avoid military confrontations, and as to exclude the outbreak of nuclear war between them and between either of [them] and other countries."

1973, October 30. NATO-Warsaw Pact Negotiations on Mutual Force Reductions. These discussions began in Vienna.

1974, July 3. ABM Reduction Protocol. The U.S. and the Soviet Union agreed to reduce the number of ABM sites to 1 each.

1974, November 24. Vladivostok Agreement. A definite limit on all strategic launchers was set at 2,400 and of 1,320 on MIRVed missile launchers. Strategic bombers were to be counted as launchers.

COMMENT. *The SALT I treaty was criticized for a number of reasons, but in particular for limiting ABM development and deployment. Critics believed that the U.S. gave up a substantial advantage in the development of counterforce damage-limiting strategic systems. The treaty was also criticized as an example of a controversial nuclear doctrine of the 1960s–1970s deterrence through mutual assured destruction (MAD). These critics urged the development of a doctrine of deterrence through damage-limiting counterforce capabilities. The criticism seemed to be confirmed when, soon afterward, the U.S. deactivated its sole ABM site protecting missiles at South Forks, South Dakota.*

WESTERN EUROPE

DEFENSE ARRANGEMENTS; NATO 1948–1973

Immediately after World War II, Soviet truculence posed a threat to the postwar recovery of Western Europe. Thanks to American economic assistance through the Marshall Plan, Western Europe avoided the economic chaos which the Communists had hoped to exploit through subversion and revolution. But full economic recovery in these nations was hampered by fears that Soviet Russia, whose armed strength had increased rather than decreased after the war, would take by invasion what Communist agents were unable to subvert from within. Although the U.S. still possessed a nuclear monopoly, Western Europeans feared that a Soviet overland attack could overrun Western Europe in less than a week. They recognized their own military impotence and doubted that America would react in time to prevent a *fait accompli*.

On April 4, 1949, in Washington, a treaty was signed to establish (effective August 24) a North Atlantic Treaty Organization (NATO) by Belgium, Canada, Denmark, France, Iceland, Italy, Luxembourg, the Netherlands, Norway, Portugal, United Kingdom, and the U.S. Greece, Turkey, and West Germany joined later. They agreed to settle disputes by peaceful means, and to develop their individual and collective capacity to resist armed attack, to regard an attack on one as an attack on all, and to take necessary action to repel attack under Article 51 of the U.N. Charter. The principal NATO commands were the European Command (with headquarters at Supreme Headquarters Allied Powers, Europe—SHAPE), Atlantic Command, and a Channel Command. The principal relevant events were:

1948, March 17. Brussels Treaty. A 50-year military and economic assistance treaty between Britain, France, the Netherlands, Belgium, and Luxembourg. Field Marshal Viscount **Bernard L. Montgomery** was appointed to head a permanent "Western Union" defense organization (September 30). The U.S., Canada, and the Brussels Treaty nations began negotiations for a larger North Atlantic Security Treaty (July).

1950, December 19. Eisenhower Appointed Supreme Allied Commander, Europe. As SACEUR, General Eisenhower toured all NATO capitals from his headquarters in Paris to investigate possibilities of creating an effective peacetime force. Later he assumed command of all forces placed at his disposal in Europe by NATO members (April 2, 1951). Montgomery was Deputy SACEUR.

1951, April 27. Defense of Greenland. The U.S. and Denmark agreed on the joint defense

of Greenland for the duration of the North Atlantic Treaty.

1951, September 6. U.S. Rights in Azores. In Lisbon, the U.S. and Portugal agreed on continued U.S. rights in the Azores, within the NATO region.

1951, October 22. Greece and Turkey Admitted to NATO.

1952, April 11. Ridgway to Replace Eisenhower. General **Matthew B. Ridgway** was appointed SACEUR, effective May 30.

1952, May 27. Establishment of European Defense Community. A treaty establishing the EDC between France, West Germany, Italy, Belgium, Luxembourg, and the Netherlands. West Germany would be restored to complete independence, would be a participant in the defense community, and would supply troops to NATO.

1953, April. NATO Plans for Nuclear Weapons. SHAPE Chief of Staff, General **Al-**

fred M. Gruenther, testified before the Senate Foreign Relations Committee that the NATO defense plans in Europe called for limited use of ground troops and intensive use of atomic weapons.

1953, July 11. Gruenther Replaces Ridgway as SACEUR.

1954, April 16. U.S. Assurances to EDC. President Eisenhower assured the 6 EDC premiers that the U.S. would maintain forces in Europe so long as a threat of Soviet aggression continued to exist.

1954, August 30. France Rejects EDC. This vote of the French National Assembly ruined plans for a new European military pact and the rearmament of Germany.

1954, October 3. New Plan for German Rearmament. The U.S., Britain, France, West Germany, Canada, Italy, Belgium, the Netherlands, and Luxembourg agreed to integrate a rearmed West Germany militarily and politically into the Western European Union and NATO. This was approved by the French National Assembly (October 12) and by the North Atlantic Council (October 22).

1955, April 13. U.S. to Share Atomic Secrets with NATO. President Eisenhower approved an agreement to share information on atomic weapons.

1956, November 20. Norstad Appointed SACEUR. General **Lauris Norstad,** USAF, replaced Gruenther.

1957, January 24. German to Command NATO Ground Forces. General **Hans Speidel** was named commander of NATO forces in Central Europe.

1959, March 13. France Withdraws Fleet from NATO. She notified the North Atlantic Council that ⅓ of the French Mediterranean Fleet, which had been ear-marked for NATO command in wartime, would be retained under French control.

1959, July 8. U.S. Withdraws Planes from France. The decision to move 200 jet fighter bombers to Britain and West Germany was caused by French refusal to permit stockpiling of U.S. nuclear weapons in France unless under French control.

1960, December 16. U.S. Offers Nuclear Submarines to NATO. This offer of 5 nuclear submarines equipped with 80 Polaris missiles was conditional upon agreement by the NATO allies on a multilateral system of control of weapons and on the purchase of 100 additional Polaris missiles by the European NATO states.

1962, May 6. U.S. Commits Nuclear Submarines to NATO. These 5 submarines were to remain under U.S. control since no agreement had been reached on the previous U.S. offer for a multilateral force.

1963, January 2. Lemnitzer Appointed SACEUR. General Lyman L. Lemnitzer, USA, former chairman of the U.S. JCS, succeeded Norstad.

1963, January 14. De Gaulle Rejects U.S. Proposal for a Multilateral Nuclear Force.

1966, March 9. France Announces Military Withdrawal from NATO. The French Government announced withdrawal of its forces and staff officers from the integrated military commands of NATO, and demanded removal of all NATO bases and headquarters installations from French territory within 1 year. De Gaulle's announced reason was that conditions in 1966 were "fundamentally different from those of 1949." France made clear, however, that it was not withdrawing from the alliance, or from nonmilitary alliance activities. In a joint statement 9 days later, the other 14 members of NATO criticized the French action and reaffirmed their firm commitment to NATO's military structures. Within the next few months the NATO Secretary General's office was moved from Paris to Brussels; SHAPE headquarters was moved from Versailles to Casteau, Belgium; Allied Forces Central Europe headquarters was moved from Fontainebleau to Maastricht; and the NATO Staff College was shifted to Rome. U.S. forces in France, totaling nearly 30,000 troops and 60,000 dependents, began to move to neighboring NATO countries on June 30, 1966.

1969, July 1. Goodpaster Appointed SACEUR. General **Andrew J. Goodpaster,** U.S.A. former Deputy Commander in Vietnam, replaced General Lemnitzer as Supreme Commander Allied Powers Europe.

1974, December 15. Haig Appointed SACEUR. General **Alexander M. Haig,** **U.S.A.,** former Chief of Staff for President Nixon, replaced Goodpaster as Supreme Commander Allied Powers Europe.

UNITED KINGDOM

Britain, exhausted and on the verge of national bankruptcy in 1945, began drastic reduction of her international commitments. This was evidenced by dramatic British withdrawal from Greece and from other occupation roles (see p. 1388), as well as the reduction of its occupation force in Germany, and by initiating the liquidation of its vast colonial empire, beginning with India (see p. 1403) and Burma (see p. 1410).

Despite this curtailment, involving a frank though unpalatable acceptance of decline from truly great-power status, Britain still found itself almost continuously involved in military actions of one sort or another around the world throughout the period. The U.K. nevertheless remained the third most powerful nation in the world, despite economic constraint. The major events of military significance:

1945–1947. Pacification in Greece. (See p. 1387.)

1945–1965. Troubles in Arabia. Intermittent frontier conflicts in Aden and Arabian protectorates (see pp. 1400, 1401).

1947, March 4. Treaty of Dunkirk. A 50-year Anglo-French treaty of alliance in face of the Soviet threat to Western Europe.

1948, March 17. Brussels Treaty. Expansion of the Anglo-French alliance to include the Benelux nations (see p. 1372).

1948–1960. Communist Revolt in Malaya. (See p. 1420.)

1948–1949. Berlin Blockade. RAF units participated in the air shuttle (see p. 1378).

1949, April 18. Independence of Eire. Ireland broke off all ties with Great Britain and became a completely independent republic.

1950, March 29. Churchill Urges Rearmament of Germany. His suggestion that Germans share in defense of Western Europe was called "frightful" by Foreign Minister **Ernest Bevin.**

1950–1953. Korean War. British Commonwealth units (predominantly British) participated (see p. 1355).

1951, May 1. Armed Services Unification. Land, sea, and air forces were placed under the operational control of the 3-man Chiefs of Staff Committee with the ground-force commander, General Sir **Miles Dempsey,** as chairman.

1952–1956. Mau Mau Uprising in Kenya. (See p. 1436.)

1952, September 8. Establishment of Southeast Asia Treaty Organization (SEATO). (See p. 1410.)

1952, October 3. Britain Explodes Atomic Bomb. Through this explosion in northwest Australia, she became the third nation to possess nuclear weapons.

1952–1959. Civil War and Terrorism in Cyprus. (See p. 1390.)

1956, October 31–November 6. Franco-British Attack on the Suez Canal. (See p. 1342.)

1957, April 4. Drastic Change in Defense Policy. Recognizing that protection against nuclear weapons was impossible, Britain concentrated on deterrence by threat of nuclear retaliation. Overseas commitments were drastically reduced and armed forces cut by 40 percent over 5 years.

1957, May 15. Britain Explodes Its First Hydrogen Bomb.

1957, July–August. Assistance to Muscat and Oman. (See p. 1401.)

1960, February 16. New Defense Policy. The government announced a shift from reliance on ground-based nuclear weapons to ballistic nuclear missiles launched from aircraft and nuclear submarines.

1960, April 13. Britain Abandons Ballistic-Missile Development. A government decision was announced to abandon the military development of a fixed-site medium-range "Bluestreak" ballistic missile, and to rely on the RAF strategic V-bombers and the U.S.-designed "Skybolt" missile. This was formally agreed by the U.S. (June 6).

1960, November 1. U.S. Nuclear Submarine Base in Scotland. Prime Minister **Harold Macmillan** announced to the House of Commons an agreement for basing U.S. nuclear-powered Polaris missile submarines at Holy Loch, on the Firth of Clyde, in Scotland.

1962. Commonwealth Troops to Thailand. This was to fulfill a SEATO commitment, resulting from civil war in Laos (see p. 1417).

1962, December 21. Collapse of "Skybolt" Program. After discussions between President Kennedy and Prime Minister Macmillan at Nassau, Bahamas, it was announced that the U.S. had decided to cancel the "Skybolt" program (upon which British nuclear defense and deterrence plans were based), but had agreed on the development of a multilateral NATO nuclear force (see p. 1373).

1963–1965. Malaysia-Indonesia Crisis. Commonwealth forces (primarily British) supported the Federation of Malaysia against Indonesian encroachments.

1964, January. British Intervention in East Africa. Troops were dispatched to Tanzania, Kenya, and Uganda (at request of local governments) to suppress antigovernment uprisings (see pp. 1473, 1438).

1964, September 21. Independence to Malta. The tiny republic remained in the British Commonwealth.

1965, February 9. Retrenchment in Development of Weapons Systems. Aviation Minister **Roy Jenkins** told Parliament that Britain could not afford to develop its own expensive new weapons systems and in future must cooperate, particularly in aircraft projects, with other nations, primarily the United States.

1966, December 3. Reinforcement of Zambia. Britain sent a squadron of RAF jets, plus a ground force protective contingent from the RAF Regiment, in response to an appeal from President **Kenneth Kaunda** of Zambia, who feared an attack by Rhodesia on the Kariba Dam on the Zambia River.

1967, July–August. Violence in Hong Kong. Communist-inspired violence, bordering on insurrection, was suppressed by British troops after much bloodshed.

1967, November 2. Withdrawal from Arabia. Foreign Secretary **George Brown** announced that Britain would withdraw all forces (some 5,000 troops) from Aden and south Arabia, calling this "the end of the imperial era." Two months later (January 16 1968) it was announced that all British bases east of Suez would be closed by 1971; simultaneously Britain announced canceling an order of 50 F-111 jets from the United States.

1968, May 10. Britain Increases Forces "Earmarked" for NATO. Because of the withdrawal from east of Suez, Britain's share in land, sea, and air forces for NATO was increased by 40 percent.

Civil War in Northern Ireland, 1969–1974

1969–1974. Continuous Violence in Ulster. Catholic-Protestant terrorism kept the province in a bloody turmoil barely comtrolled by the equivalent of a British division plus local police. The Irish Republican Army (IRA) terrorist group, outlawed in Eire, was nonetheless based in Eire and attempted to force the British to agree to cession of Ulster to Eire. Some violence, in the form of bombings, spread to Britain. Ostensibly religious, the real issues were economic, social, and political.

1969, April 21. British Troops Protect Key Installations in Ulster. As a result of increased violence, more than 1,000 troops were deployed to protect reservoirs, telephone exchanges, and power stations.

1969, August 19. British Army Assumes Responsibility for Ulster Security. This was a response to increased violence in Ulster and to a suggestion from Eire that a British-Irish force, or a U.N. force, assume responsibility for security in Northern Ireland. The British government rejected the Irish proposal.

1971, November 1. ANZUK Force Replaces British Garrison in Singapore. (See p. 1421.)

1972–1973. Dispute with Iceland. The unilateral extension of Iceland's coastal fishing rights area from 12 to 50 miles (July 14) led to British protests, supported by the International Court of Justice (August 17). Iceland refused to recognize the ruling, and Icelandic police gunboats drove away British trawlers, leading to the bloodless collision of an Icelandic gunboat and a British frigate (September 10, 1973). Temporary British acceptance of the new limit did not settle the dispute.

1972, January 30. "Bloody Sunday" in Londonderry. Thirteen unarmed civilians were killed by British troops attacked by Catholic demonstrators. In response, an Irish mob in Dublin stormed and burned the British Embassy.

1972, July 28. British Army Occupies Catholic Areas of Belfast and Londonderry. Barricaded enclaves previously considered too dangerous to enter were swept in a 3-day operation. British forces in Northern Ireland were at a 17,000-man peak.

IRELAND

1956, December–1957, November. IRA Terrorist Campaign. Efforts by the outlawed Irish Republican Army to obtain the support of the Irish Government in efforts to unite Ireland and Ulster were partially successful and led to fruitless discussions between President **Eamon de Valera** and the British government (March 1958).

THE LOW COUNTRIES

Belgium and the Netherlands both assumed Western Union and NATO (see p. 1372) commitments. Other Belgian military action has been limited to occasional commitment in the Congo (see p. 1441), and the Netherlands has been involved in military action in Indonesia (1945–1949) and in defending New Guinea against Indonesian encroachments (1949–1963; see p. 1422). All 3 contributed combat units to the U.N. Command in Korea (see p. 1366).

FRANCE

The nation was plunged into political turmoil following World War II. General de Gaulle, elected president of the provisional government of the Fourth Republic (November 13, 1945), resigned in disgust as the feuding factions refused to compose their differences (January 20, 1946). He returned to power 12 years later, averting likely civil war. The following are the major events of military significance:

1946–1954. Indochina War. (See p. 1412.)

1947, March 4. Anglo-French Treaty of Alliance. (See p. 1374.)

1948, March 17. Brussels Treaty. (See p. 1372.)

1948, October 21–28. Revolution in Mahe (French India). A pro-Indian revolt was suppressed by French troops.

1949, April 4. NATO Treaty. (See p. 1372.)

1950–1953. Korean War. (See p. 1355.)

1954. Geneva Conference. (See p. 1414.) End of French rule in Indochina.

1954–1962. Algerian Revolt. (See p. 1433.)

1956, October 31–November 6. Franco-British Attack on the Suez Canal. (See p. 1342.)

1957–1958. To the Brink of Civil War. France slipped to the verge of civil war as increasingly impotent political leadership proved unable to cope either with the problems of the revolt in Algeria or the pressing political and economic issues at home.

1958, June 1. De Gaulle Returns to Power. With the French army in Algeria virtually in a state of mutiny against the government (see p. 1433), de Gaulle was recalled to power by popular clamor and the Fifth Republic was established with him as its first president (December 21).

1960, January–November. France Gives Up Colonial Empire. (See p. 1433.) De Gaulle offered the newly independent countries economically attractive terms to remain members of the French community. Most accepted.

1960, February 13. France Explodes Nuclear Weapon in the Sahara. She became the world's fourth nuclear power.

1960, December 6. France to Be Militarily Independent of U.S. De Gaulle formally announced a plan for establishing an independent French nuclear striking force. By development of its own independent military power, exemplified by the small nuclear-capable air bombardment striking force, or *force de frappe*, de Gaulle emphasized French military and political independence of the U.S. and NATO, while retaining a consistently western-oriented strategy (1960–1965).

1961, April 22–26. Army Mutiny in Algeria. This was forcefully suppressed by de Gaulle (see p. 1433).

1961, July 19–21. Clash with Tunis. Tunisian violence, caused by Algerian border dispute, erupted against French bases in Tunisia; Bizerte was besieged by Tunisian forces. Massive French counterattacks, by land, sea, and air, overwhelmed Tunisian besiegers; both sides accepted a U.N. cease-fire (see p. 1435).

1963, January 22. Franco-German Treaty of Friendship. This pledged cooperation between the 2 countries in foreign policy, defense, and cultural affairs.

1964, February 18. French Intervention in Gabon. (See p. 1444.)

1963, August 15. French Troops to Congo (Brazzaville). The French unit in the Congo was reinforced by troops airlifted to Chad during disorders related to a coup d'état (see p. 1444). President de Gaulle refused, however, to have French troops support ousted President **Fulbert Youlou.**

1965, July 14. Mirage IV Jet Bombers Revealed. Twelve flew over the Bastille Day Parade, demonstrating a French long-range nuclear weapon delivery capability.

1966, March 9. French Military Withdrawal from NATO. (See p. 1373.)

1966, July–October. French Nuclear Weapons Tests in the Pacific. Between July 2 and October 4, France exploded 5 weapons, up to 400 kilotons in size, at Mururoa Atoll, 750 miles southeast of Tahiti. France ignored protests from Australia, New Zealand, Japan, and 5 South American nations.

1967, November 16. Intervention in Central African Republic. (See p. 1440.)

1968, August 24. France Explodes Thermonuclear Weapon in the Pacific. The 2-megation device, suspended from a balloon above Mururoa Atoll, made France the world's fifth thermonuclear power. A second H-bomb was detonated on September 8.

1968, August 28. French Troops to Chad. The French contingent of 1,000 troops was ordered to assist the Chad government in putting down a rebellion in the northwest corner of the country. Approximately one battalion of French paratroopers was airlifted in to participate.

1970, January 9. Sale of Aircraft to Libya. (See p. 1434.)

1973, July. Nuclear Tests in Pacific.

WEST GERMANY

Threat of war between the U.S.S.R. and her satellites on the one hand and the nations of the Free World on the other hung over divided Germany throughout the period. The Federal Republic of Germany, with its capital at Bonn, was proclaimed May 23, 1949, the puppet Communist German Democratic Republic on October 7. Meanwhile, friction intensified in Berlin, lying within the Soviet zone of occupation but divided in a quadripartite Kommandatura, and access—except by air—was never established by formal agreement.

1945, November 20. Nuremberg War Crimes Trials Begin. Twenty-two Nazi leaders were convicted of war crimes, 11 being condemned to death by the International Military Tribunal (October 1, 1946): Hermann Goering, Joachim von Ribbentrop, Ernst Kaltenbrunner, Marshal Wilhelm Keitel, Alfred Rosenberg, Hans Frank, Wilhelm Frick, Julius Streicher, Fritz Sauckel, General Alfred Jodl, and Arthur Seyss-Inquart. Rudolph Hess, Walther Funk, and Admiral Erich Raeder were sentenced to life imprisonment. Baldur von Schirach and Albert Speer were sentenced to 20 years imprisonment, Konstantin von Neurath to 15 years, Admiral Karl Doenitz to 10. Hjalmar Schacht, Franz von Papen, and Hans Fritsche were acquitted. Ten of the convicted war criminals were hanged (October 15, 1946); Goering committed suicide by swallowing poison 2 hours before scheduled to be hanged.

1947, January 14. Beginning of Talks for German and Austrian Peace Treaties. In London, deputies of the Big-Four foreign ministers opened preliminary talks.

Berlin Blockade, 1948–1949

1948, March–June. Soviet Harassment of Western Powers in Berlin. This began when the Soviet delegation walked out of the Allied Control Council (March 20). Soon thereafter, the Soviets began interference and harassment of American and British access to Berlin from West Germany (April 1). The Soviet representative walked out (June 16) of the Kommandatura (4-power military commission in Berlin), virtually cutting off the Soviet military command in Berlin from the 3 western powers.

1948, June 22. Beginning of the Blockade. Soviet occupation authorities halted all railroad traffic between Berlin and the west. There was less than 1 month's food supply for the 2 million inhabitants of the western sectors of Berlin. U.S. General **Lucius D. Clay,** commanding U.S. occupation forces in Germany, urged that the Western Allied garrisons stay put, and that Berlin be supplied by air. His recommendations were upheld.

1948–1949, June 26–September 30. Operation "Vittles." Immediate mobilization of all Western Allied military aircraft available began, while Clay rallied Berlin civilian help to expand the 2 available air-fields. (A third was soon built.) Air Lift Task Force (Provisional), commanded by U.S. Major General **William H. Tunner** and composed mainly of U.S. planes and pilots, with smaller increments of British and French air forces, accomplished the most extraordinary military peacetime effort in history. Running on split-second schedule, through all sorts of weather, and harassed from time to time by "buzzing" of Soviet fighter planes, 277,264 flights were made, lifting a total of 2,343,315 tons of food and coal. The record day's lift was on Easter Sunday, April 16, 1949, when 1,398 flights brought 12,940 tons into Berlin. The operation cost the lives of 75 American and British airmen, including a collision when a Soviet pilot, bedeviling a passenger-loaded British plane, misjudged his distance and brought both aircraft down in the crash.

1948, July 26. Western Powers Halt All Trade with East Germany. This was retaliation for the blockade.

1949, May 12. Soviets End Blockade. Soviet

authorities, conceding defeat, officially lifted the blockade, but the air supply operation continued until September 30.

1949, April 14. End of the Nuremberg War Crimes Trials.

1949, September 7. The Federal Republic of Germany Established. Its capital was at Bonn. Dr. **Konrad Adenauer** was elected chancellor (September 15). The U.S., Britain, and France guaranteed the defense of West Germany (September 19) and ended their military government (September 21).

1949–1965. Intermittent Berlin Incidents. The U.S.S.R. and the East German Communist regime frequently tested Western Allied will and determination, and attempted to erode the occupation and access rights of the Western Allies to Berlin.

1952, March 1. Britain Returns Heligoland to West Germany.

1953, March. Allied-Soviet Air Incidents. An American plane was shot down over the U.S. zone (March 10), and a British bomber was shot down over the British zone (March 12). The U.S. ordered 25 of its latest Sabrejets to Germany, to counter the threat, while secret conciliation talks began between Britain and Russia (March 31), which later were attended by France and the U.S.

1954, October 23. Rearmament of Germany within NATO. The NATO Council admitted Germany to NATO. Next day, France specifically recognized the sovereignty of West Germany.

1955, May 5. Federal Republic of Germany Becomes a Sovereign State.

1961, July. Renewed Berlin Crisis. Prime Minister **Nikita Khrushchev,** renewing demands for Allied withdrawal from Berlin, announced suspension of planned troop reduction and an increase in the Soviet military budget (July 8). Britain, France, and the U.S. rejected Khrushchev's terms for the settlement of the Berlin and German questions (July 17). President Kennedy directed a build-up of U.S. military strength and mobilized 4 National Guard divisions (see p. 1454).

1961, August 12–13. The Wall. The East German government closed the borders between East and West Berlin to East Germans. A wall, built overnight, split the city. The Soviet Union rejected western protests against the sealing of the border (September 11). Soviet and U.S. tanks confronted one another at "Checkpoint Charlie," but tension died. The division continued at the end of the period.

1965, February 12. Germany Suspends Arms Shipments to Israel. This was in response to UAR warning that it would sever diplomatic relations if shipments were not halted.

1966, December 6. Luftwaffe's F-104G Star-fighters Grounded. This action was taken by General **Johannes Steinhoff** as a result of the Luftwaffe's 65th Starfighter crash since 1962. American airmen said that the German crashes were due to the German practice of putting more equipment into these aircraft than they were designed to carry.

1970, August 12. German-Soviet Non-aggression Treaty. This was signed in Moscow by German chancellor **Willy Brandt** and Soviet Premier **Alexei Kosygin.**

1970, December 7. Treaty with Poland. This recognized the post–World War II boundary imposed by the Allies, along the Oder-Neisse rivers. It confirmed cession of 40,000 square miles, $\frac{1}{4}$ of Germany, to Poland.

1971, August 23. New Agreement on Berlin. The U.S., Britain, France, and the U.S.S.R. signed a new agreement on access to Berlin and passage of West Berliners into East Germany. The agreement was endorsed by both East and West Germany. East Germans were not authorized to visit West Germany or West Berlin. This was followed (December 17) by an agreement between the 2 Germanies, authorizing West Germany free access to West Berlin.

1972, December 21. Treaty between East and West Germany. Although the possibility of future unity was left open, this meant full recognition by each of the sovereignty of the other.

SCANDINAVIAN STATES

All Scandinavian states were preoccupied during the period with the question of national security against Communist infiltration or outright aggression. While Denmark, Norway, and Iceland became members of NATO, Sweden maintained a strict—though western-oriented—neutrality. Finland—which clearly demonstrated an equally pro-western orientation—also maintained a neutral status, but was forced by geography and power realities to maintain closer relations with the U.S.S.R. (For Iceland's dispute with Britain, see p. 1376.)

AUSTRIA

After 10 years of Allied occupation marked by inability of the Big Four to agree on a peace treaty, the nation regained its independence by a treaty signed at Vienna by the U.S., U.K., France, and U.S.S.R. (May 15, 1955), restoring Austria's frontiers existing January 1, 1938. Under the treaty a small army and air force (about 60,000 strong) were permitted. The treaty requirement for neutrality was confirmed by Austria's official proclamation of permanent neutrality (act of October 26, 1955).

ITALY

1946, May 9. Abdication of King Victor Emmanuel III. He abdicated in favor of his son, **Humbert II.**

1946, June 2. Italian People Vote to End the Monarchy. The Italian government declared Italy a republic (June 2). After some outbreaks of violence between royalists and republicans, King Humbert left the country (June 13).

1946, July 29. Peace Conference Opens in Paris. Treaties were negotiated between the World War II victors and Italy, Hungary, Bulgaria, Rumania, and Finland. These allied peace treaties were eventually signed in Paris (February 10, 1947).

1947, September 16. Crisis at Trieste. A Yugoslav military force, menacing Trieste, was deterred by the deployment of an American battalion for combat. Tension remained high; incidents, including occasional small-arms fire, were frequent.

1947, December 14. End of Allied Occupation of Italy. British and U.S. units remained in Trieste, however, to prevent Yugoslav seizure.

1948, July 14–16. Communist Riots. These were sparked by an assassination attempt against Communist leader **Palmiro Togliatti.** Police and troops restored order after considerable bloodshed.

1952, March 20. Civil Disorder in Trieste. Pro-Italians attacked American and British installations in protest against continued Allied occupation of Trieste.

1953, August 29–31. Border Clash with Yugoslavia. Italy's dispute with Yugoslavia escalated as both sides held maneuvers near Trieste, and Yugoslav antiaircraft fired at Italian planes over the border. Shots also were fired between Yugoslav troops and the Anglo-American force occupying Trieste.

1953, October 8–December 5. De Facto Settlement at Trieste. Britain and the U.S. announced that they would withdraw their occupation forces (4,000 U.S. and 3,000 British) and return Zone A of Trieste to Italy. After an increase of tension between Italy and Yugoslavia, the 2 nations agreed to withdraw their troops from the border (December 5). An agreement was later signed between them (October 5, 1954).

1963, January. Nuclear Weapons Agreement with U.S. (See p. 1389.)

SPAIN

1949, August 4. Spain's Application for Marshall Plan Aid Rejected. The U.S. Senate refused to include such aid in the Marshall Plan appropriations.

1953, September 26. Ten-Year Defense Agreement with U.S. This gave the U.S. rights to Spanish naval and air bases in return for economic and military aid. The bases were to be: air bases near Madrid, Albacete, Valencia, and Seville; naval facilities at Ferrol, Cadiz, Cartagena, Valencia, and Port Mahon on Minorca.

1965, September 26. Renewal of Base Agreement with the United States. This provided for the continued lease of air and naval bases to the U.S., in exchange for economic and military aid.

1970, August 8. Renewal of Base Agreement with United States. In return the U.S. was to provide $153 million in economic and military aid, including the lease of 12 naval vessels.

PORTUGAL

1946, October 11. Unsuccessful Revolt Attempt. Most military rebels against the regime of dictator **Antonio de Oliveira Salazar** were captured in Estarreja or Lisbon.

1961, January 22–February 2. *Santa Maria* Incident. Portuguese rebels, planning a revolt against the Salazar government, seized the Portuguese passenger ship S.S. *Santa Maria.* The rebel leader was former navy Captain **Henrique Malta Galvao** and former General **Humberto da Silva Delgado** (ashore in Brazil). After warnings from the U.S. and British navies, Galvao took the *Santa Maria* to Recife and accepted asylum in Brazil.

1961–1974. Rebellion in Angola. (See p. 1443.)

1962, January 1. Unsuccessful Revolt Attempt. Loyal troops quickly suppressed an uprising by a small military unit in Beja.

1962–1974. Rebellion in Portuguese Guinea. (See p. 1445.)

1962–1974. Rebellion in Mozambique. (See p. 1448.)

1967, July–August. Violence in Macao. Communist-inspired violence, bordering on insurrection, was suppressed by local police only after the major center of violence, on Hong Kong, was eliminated by British troops (see p. 1375).

1970, November 22–24. Portuguese-supported Invasion of Guinea. (See p. 1443.)

(1974, March 16. Military Revolt Fails. Young officers, leading a small motor march on Lisbon, surrendered when other military units failed to join them. Reportedly they were supporters of General **Antonio de Spinola,** who was dismissed as Deputy Chief of Staff (March 14) as a result of a recent book critical of Portugal's military, diplomatic, economic, and colonial policies.

1974, April 25. Coup d'État. The government of Premier **Marcello Caetano** (successor of Salazar) was overthrown by a military uprising. General Spinola was the leading member of a six-officer junta, which pledged liberal freedom to Portugal after 40 years of repressive dictatorship. Spinola became provisional president (May 15).

1974, September 30. Spinola Resigns. This was because of his inability to control leftist tendencies in his new government represented by Premier (former general) **Vasco Gonçalves.** Spinola was replaced by General **Francisco da Costa Gomes,** who reappointed Gonçalves as premier.

EASTERN EUROPE

Soviet Bloc and Warsaw Pact Nations

1947, July. Communist Nations Reject the Marshall Plan. Although Poland and Czechoslovakia initially indicated an interest in joining in the Marshall Plan, they later rejected the offer, obviously under Soviet pressure, as did the East European nations of Hungary, Rumania, Albania, Bulgaria, Yugoslavia, along with Finland.

1947, October 5. Establishment of the Cominform. At a meeting in Moscow of the Communist parties of 9 European nations, a new Communist International, the Com-inform, was established.

1955, May 14. The Warsaw Pact. The Soviet Union and its satellites—Poland, Czechoslovakia, Hungary, Rumania, Bulgaria, East Germany, and Albania—signed a treaty of mutual friendship and defense at Warsaw. Yugoslavia refused to join. This nominally mutual-defense treaty was the Communist bloc's answer to NATO and the remilitarization of West Germany. This actually caused no changes in the relationship between Soviet and satellite forces.

Soviet Union

Despite near-catastrophic losses of man power, materials, and production facilities, at the close of World War II the U.S.S.R. was without question the second great power in a bipolar world. Soviet armies occupied all of Eastern Europe, much of Central Europe, northern Iran, Manchuria, and northern Korea. While the Western Allies demobilized their armies as quickly as possible after the war, Soviet military forces were maintained close to their wartime strength. At the same time, the Soviets were increasingly truculent in and out of the U.N., refused to carry out postwar agreements for the liberation of Eastern Europe, and were obviously determined to extend Soviet power and influence in any direction at any opportunity. This threatening and aggressive attitude was combined with the Soviet-directed efforts of international communism to take advantage of postwar chaos and dislocation to gain control of the governments of many nations in Europe and Asia.

It soon became evident that the combination of Soviet threats of external aggression and of internal subversion by indigenous Communists was creating pressures that few of the war-weakened nations in the world could withstand by their own individual efforts. Subservient Communist satellite governments were established throughout East Europe. Britain's sudden and unexpected decline caused the U.S. to undertake economic and military measures to assist nations threatened by Soviet Communist aggression (Truman Doctrine, Marshall Plan). This U.S. response to Soviet moves, with the aim of blocking Soviet expansion, triggered the so-called Cold War.

Thwarted by American counteraction—initially supported by the American monopoly of nuclear weapons and long-range bombardment capability—the U.S.S.R. devoted itself to unceasing efforts to improve and modernize its military capability in an effort to offset and, if possible, surpass that of the U.S. The result was unexpectedly early Soviet detonation of atomic and hydrogen bombs. Again the potentially more powerful U.S. slowly reacted to meet the armament challenge and Soviet-inspired aggression of North Korea against South Korea (see p. 1355). As a result, despite rapid modernization and sophistication of its weapons and armed forces, by the end of the period the Soviet Union had failed to achieve its aim of military parity or superiority over the U.S. (save possibly in the fields of rocket power and space exploration). Nevertheless, its substantial capability in nuclear weapons, and possession of long-range missiles to deliver such weapons, had brought the U.S.S.R. to a position of nuclear stalemate with the U.S., and permitted continuation of its policy of encouraging and supporting Communist subversion in underdeveloped nations throughout the world.

In the development of its nuclear capability, the U.S.S.R. had obviously come to understand the unprecedented destructive power of nuclear weapons, and apparently realized that an all-out nuclear exchange would result in the virtual destruction of the U.S.S.R. This, combined with serious ideological and nationalistic differences with increasingly powerful Communist China, had led to some diminution in the intensity of the Cold War (particularly after the Cuban missile crisis), though there was no evidence of any change in basic Soviet Communist objectives. The undeclared conflict continued at lower levels, and in different forms, as Communist agents fomented and carried on low-intensity "wars of national liberation" wherever the opportunity presented itself around the world. The principal military events of the period were:

1945, August 14. Treaty with China. (See p. 1423.)

1945–1946. Intervention in Azerbaijan. (See p. 1402.) This was the real beginning of the Cold War.

1945–1947. Soviet Bloc Established. Despite World War II agreements at Yalta and Potsdam, and subsequent peace treaties with Nazi East European satellite states, the U.S.S.R. established its own satellite governments in areas occupied by its forces: Albania, Bulgaria, Czechoslovakia, Hungary, Poland, Rumania, eastern Austria, and East Germany (including the eastern sector of Berlin). Yugoslavia, under Tito's Communist government, was initially included in this East and Central European Communist satellite bloc.

1947, March 4. Norway Rejects Soviet Demands on Spitsbergen. The U.S.S.R. had requested rights to establish a military base.

1947, July 7. U.S.S.R. Rejects the Marshall Plan.

1948, April 6. Treaty with Finland. A 10-year alliance was agreed.

1948–1949. Berlin Blockade. (See p. 1378.)

1948, June. Dispute with Yugoslavia. Amidst violent denunciation and counterdenunciation, Yugoslavia withdrew from the Soviet bloc (see p. 1387).

1949, September 23. U.S.S.R. Explodes Its

First Atomic Bomb. This ended the U.S. nuclear monopoly.

1950–1953. Korean War. The U.S.S.R. supported North Korea and (later) Communist China in the Korean War (see p. 1355).

1953, March 5. Death of Stalin. He was succeeded by a triumvirate of **Georgi Malenkov, Lavrenti Beria,** and **Nikita Khrushchev.** Beria, who apparently contemplated seizing power, was later overthrown (July) by his colleagues (with army support) and executed (December 23).

1953, June 16–17. Suppression of Uprising in East Berlin. (See p. 1386.)

1953, August 12. U.S.S.R. Detonates Its First Hydrogen Bomb.

1955, May 14. Establishment of the Warsaw Pact. (See p. 1382.)

1955, December 28. Soviet Defense Budget Reduction Announced. Later (May 14, 1956) it was announced that armed forces would be cut by 1,200,000 men over the following year.

1956, February 14–25. De-Stalinization. At the 20th Party Congress in Moscow, Khrushchev, by now unquestioned leader of the U.S.S.R., attacked the memory of Stalin, and to some extent liberalized life and governmental policy in the U.S.S.R. and in the satellites. This revived hopes of personal liberty, which created a wave of unrest in the U.S.S.R. and in the satellites.

1956, June 6. Bulganin Demands Reduction of Western Forces in Germany. President Eisenhower rejected the Soviet premier's demand (August 7).

1956, June 28. Unrest in Poland. (See p. 1385.)

1956, July 10. Charges of U.S. Military Aircraft Violation in Korea. A complaint was made to the U.N. Security Council. The U.S. denied the charges.

1956, October 23–November 4. Hungarian Revolt Suppressed by Soviet Forces. (See p. 1385.)

1957, October 4. Soviet Space Triumph. Soviet artificial satellite, Sputnik I, made the first successful penetration of space. It established Soviet preeminence in space exploration, which was being challenged, and perhaps overtaken, at the end of the period by belated American efforts to catch up. It also evidenced a Soviet military preeminence in long-range rocketry, and particularly in the power of its rocket boosters.

1958, March 27. Khrushchev Seizes Control. Ousting Bulganin as premier, he became virtual dictator.

1960, January 20. Soviet Ballistic Missile Achievement. The U.S.S.R. claimed it fired a missile 7,752 nautical miles to within 1.24 miles of its target in the Central Pacific.

1960, May 1. U-2 Incident. Soviet air-defense missiles in central Russia shot down an American U-2 reconnaissance plane piloted by **F. G. Powers,** employee of the U.S. CIA. (See p. 1453.)

1960, May 17. Khrushchev Wrecks Summit Conference. (See p. 1453.)

1961, April 12. First Manned Space Flight. Soviet Major **Yuri Gagarin** became the first human to travel in space. He landed safely in the Soviet Union after 1 circuit of the globe.

1961, August 12–13. The Berlin Wall. (See p. 1379.)

1961, December 9. Soviet Superbombs. At the close of the Soviet Union's post-moratorium test series (see p. 1369), Khrushchev publicly boasted that the Soviet Union had nuclear bombs more powerful than 100 megatons. American analysis of fallout seemed to substantiate the claim.

1962, October 17. Khrushchev Reveals Soviet-Sino Rift. During the 22nd Soviet Communist Party Congress in Moscow, Khrushchev revealed the existence of a Soviet-Sino ideological rift, mainly by attacking Albania, ideological ally of Communist China.

1962, October–November. Cuban Missile Crisis. (See p. 1454.)

1963, July 25. Limited Nuclear Test Ban Treaty. (See p. 1370.)

1964, October 12. First Multimanned Space Flight. The Soviet Union orbited a spacecraft, carrying 3 men, for 16 orbits.

1964, October 14–15. Khrushchev Deposed as Soviet Leader. He was replaced by **Leonid I. Brezhnev** and **Aleksei N. Kosygin.**

1964–1974. Border Clashes With Communist China. (See p. 1429.)

1967, November 3. Soviet Fractional Orbital Bombardment System. The development of this system was revealed in a statement by U.S. Defense Secretary Robert S. McNamara, who said that it would permit firing a nuclear weapon from an orbiting position at a target on the earth. McNamara said that the U.S. had rejected such a system, and did not believe that it would give the U.S.S.R. any advantage.

HUNGARY

Despite overwhelming anti-Communist popular sentiment, a Communist satellite government was installed with the assistance and protection of Soviet occupation forces (1945–1947). Communist control was unchallenged until a wave of unrest swept over Eastern Europe in 1956, following Khrushchev's de-Stalinization speech (see p. 1383).

1947, May 30. Communist Coup. The government of Premier **Ferenc Nagy** was overthrown.

1956, July 18–22. Shake-up in Communist Government. This was evidence of serious internal unrest in the party and the nation.

1956, October 23. Outbreak of Popular Revolt against Communism. This followed security police attempts to suppress a popular demonstration in Budapest. Soviet occupation forces fired on demonstrators (October 24–25). Revolutionary Councils sprang up throughout the country. The Communist government was toppled in a surprising constitutional parliamentary upheaval (October 25). Erstwhile moderate Communist **Imre Nagy** established a pro-western government. **Erno Gero,** ousted party secretary, called for Soviet troops. Fighting involving Hungarian Communist forces and Soviet troops spread across the country.

1956, October 28. Nagy Announces Soviet Agreement to Withdraw Troops. A lull in the fighting followed.

1956, November 1–4. Soviet Suppression of Revolt. By a combination of treachery and surprise, Soviet forces—with some 200,000 troops and 2,500 tanks and armored cars—surrounded Budapest. Nagy appealed for U.N. aid. The Soviets attacked, captured Nagy and his government (November 4), and swept through Budapest despite the valiant resistance of Hungarian troops and civilians. Approximately 25,000 Hungarians and 7,000 Russians were killed. One unfortunate aspect of the debacle was the stimulus of American broadcasting programs that led the patriots to believe that the U.S. would come to their aid.

1956, November 5–30. Unrest and Flight. Resistance and strikes persisted despite ruthless Soviet suppression. By the end of the month over 100,000 refugees had fled to the West.

POLAND

A Communist people's republic since 1947, when Stalin's repudiation of the free election promised at the Potsdam Conference brought a Communist puppet government into being, under the protection of Soviet occupation forces, Poland was an uneasy satellite of the U.S.S.R. during the period. Persecution of the Catholic Church and imprisonment—until 1956—of **Stefan Cardinal Wyszynski** added to the tension between the Polish people and their government.

1949. Russian Commands Army. Soviet Marshal **Konstantin Rokossovski** was appointed Minister of Defense and Commander in Chief of the Polish army.

1950, June 7. Frontier Agreement with East Germany. The Oder-Neisse line was accepted as the official boundary.

1956, June 28–29. Workers' Revolt in Poznan. This was suppressed by Russian troops and brought death to over 50, injury to hundreds more, and imprisonment of more than 1,000 persons. Unrest in Poland continued and tension mounted between people and Soviet occupation army.

1956, October. Wladyslaw Gomulka Becomes Premier. The moderate Polish Communist leader was released from jail and restored to party leadership. This reduced tension and unrest, brought amelioration of conditions, and slackening of Soviet restrictions. Gomulka defied Khrushchev's warnings that democratization was too rapid. Cardinal Wyszynski was released; cultural and financial relations with the West were initiated. Soviet Marshal Rokossovski was dismissed as Polish defense minister.

1956, November 18. Agreement with U.S.S.R. Poland was given greater independence.

1957, October 3. Riots in Warsaw. Suppressed by police and troops.

1970, December 15–20. Civil Disturbance.

Widespread disorders throughout much of Poland, particularly in the north, caused the death of more than 300 people in clashes between mobs and troops. Communist party Secretary **Wladyslaw Gomulka** resigned and was replaced by **Edward Gierek,** who ended the disturbances by promising revised economic policies.

EAST GERMANY

1949, October 7. Establishment of German Democratic Republic. Otto Grotewohl was established as Chancellor of this new Soviet satellite.

1953, June 16–17. Anti-Communist Riots in East Berlin and East Germany. These were suppressed by the Soviet Army.

1953, August 23. Soviet Moves to Strengthen Ties with East Germany. This included release of war prisoners, lowering of occupation costs, and an intensive propaganda campaign. The U.S.S.R. returned to East Germany 30 factories which had been seized as reparations after World War II (December 31). It was announced that all reparations were ended, and that further occupation costs would be limited to 5 percent of East Germany's national income.

1954, March 26. U.S.S.R. Announces East German Sovereignty. Soviet troops were to remain only for security functions and for the fulfillment of Soviet obligations under the Potsdam Agreement.

1955, January 10. Defections to West. The West German Refugee Ministry reported that 184,198 persons had left East Germany for the West in 1954.

1956, January 18. East Germany Rearms. Parliament approved creation of a defense ministry and a people's army.

1964, January–March. Air Incidents. Soviet aircraft shot down an unarmed U.S. reconnaissance plane (March 10) and an Air Force jet training plane which by mistake flew over East Germany (January 28). Both the U.S. and the U.S.S.R. protested.

1972, December 21. Treaty with West Germany. (See p. 1379.)

CZECHOSLOVAKIA

1948, February 24. Communist Coup d'État. President **Eduard Beneš** was forced to accept an ultimatum of Premier **Klement Gottwald,** putting Communists in charge of all branches of the government except the Foreign Ministry, where **Jan Masaryk** remained Foreign Minister, but all his aides were Communists.

1948, March 10. Death of Jan Masaryk. The Communist government announced this was suicide. There is little doubt that he was murdered by the Communists.

1968, July 15. Czechoslovakia Demands Revision of Warsaw Pact. Increasing coolness between the Kremlin and the "democratic" Communist government of Prime Minister **Alexander Dubcek** led the Czechs to request assurances that the treaty would not be used for political purposes, but only for the defense of Eastern Europe against foreign military aggression. A subsequent meeting of Czech and Soviet political leaders in the border town of Cernia did not reduce the tension.

1968, August 20. Warsaw Pact Invasion. Soviet, East German, Polish, Bulgarian, and Hungarian forces in overwhelming strength (at least 500,000 men) met little resistance as they occupied the country and deposed the government of Prime Minister Dubcek. After a safely pro-Soviet Communist government was firmly installed in power, most of the invading troops were withdrawn (beginning in October), but a garrison of several Soviet divisions (totaling more than 65,000 troops) remained.

YUGOSLAVIA

1946, March 24. Capture of General Draja Mikhailovich. He was tried and executed by a firing squad in Belgrade (July 17).

1946–1948. Intervention in Greece. Yugoslavia supported the rebels in the Greek Civil War (see p. 1387).

1947–1953. Crises over Trieste. (See p. 1380)

1948, June 28. Yugoslav-Soviet Rift. The Cominform denounced Marshal **Tito** and the Yugoslav Communist party for putting national interests above party interests. Tito and the Yugoslav Communist party, insisting that they were still Communists, also insisted that they were Yugoslav Communists and not Russian satellites. This marked the end of Yugoslavia's role as a Soviet satellite and the beginning of completely independent existence, although under President Tito the nation remained definitely, although individualistically, Communist.

1949, August–September. Yugoslav-Soviet Crisis. Each nation accused the other of preparing for war, and Russia denounced the 1945 treaty of friendship and mutual assistance with Yugoslavia (September 29).

1953, June 1. Political Commissars Abolished in the Armed Forces. This was announced by Marshal Tito personally.

1956, October 15. U.S. Assistance. Eisenhower authorized continuance of U.S. economic aid to Yugoslavia, but withheld heavy military equipment pending further study.

1956, November 11. Renewed Difficulties with U.S.S.R. This resulted from a Tito speech at a party meeting.

1961, October 13. U.S. Military Assistance. The U.S. confirmed that it was granting military aid to Yugoslavia, to include 130 jet fighter planes, and the training of Yugoslav fighter pilots in the United States.

ALBANIA

1946, October 22. Corfu Channel Incident. Two British destroyers were damaged by Albanian mines in the Corfu Channel, with 40 men killed or missing. (Earlier in the year, Albanian shore batteries had fired at British and Greek warships in the channel.) Britain cleared the mines from the channel, under naval protection, despite Albanian protests to the U.N. (November 12–13). The U.S.S.R. vetoed a Security Council resolution blaming Albania for the damage (March 25, 1947). In subsequent litigation before the Court of International Justice, Albania was found at fault (April 9, 1949).

1961, December 19. Diplomatic Relations with U.S.S.R. Broken. This was a protest against de-Stalinization. Albania became the first European satellite of Communist China.

RUMANIA

1966, May 10. Soviet-Rumanian Tension over Warsaw Pact. Cautious Rumanian moves toward a policy independent of the U.S.S.R., and suggestions that the terms of the Warsaw Pact be changed, led to a visit to Bucharest by Soviet Communist party First Secretary Leonid Brezhnev for discussion with independence-minded Rumanian First Secretary **Nicholas Ceausescu.** The result was a reduction in the stridency of Rumanian assertions of independence.

1968, August 20. Rumanian Nonparticipation in Invasion of Czechoslovakia. Apparently the Rumanian government was not informed of the Warsaw Pact plan.

MEDITERRANEAN AND SOUTHEASTERN EUROPE

GREECE

1944, December 3–January 11, 1945. Guerrilla Warfare. Communist resistance groups attempted to overthrow the reestablished legal government of Greece, then under the protection of British occupation forces. British troops suppressed the uprisings and established an uneasy truce between the rival factions.

1945, September 1. Monarchy Restored. The Greek people voted to return King **George II** to the throne.

1946, May–1949, October. Greek Civil War. Communist rebels under General **"Markos"**

Vafiades, with support from Albania, Yugoslavia, and Bulgaria, seized control of major northern border regions, while fighting flared throughout the nation. The Greek government received some support from Britain at the outset, but was barely able to maintain control of major cities and some portions of the countryside. Fighting was particularly intensive in the Vardar Valley.

1946, December 10. U.N. Begins Investigation. The Security Council began an investigation into Greek charges that Yugoslavia, Bulgaria, and Albania were supporting guerrilla forces on Greece's northern frontier. The Balkans Investigating Committee reported to the Security Council (May 23, 1947) that Yugoslavia, Bulgaria, and Albania had violated the U.N. Charter by aiding guerrilla uprisings in Greece.

1947, March 12. Truman Doctrine. Britain, close to economic collapse, was forced to suspend its assistance to Greece. President Truman announced American determination to assist Greece and Turkey against internal and external Communist threats. This resulted in extensive economic aid and provision of military equipment to the strife-torn nation. An American military advisory group trained the Greek Army in employment of U.S. military equipment, and also rendered combat advice. The Greek Army slowly regained the initiative, and suppressed the revolt throughout all of Greece (1947) save in the northern border regions, where rebels obtained direct assistance from the neighboring Soviet satellites.

1948, January 1. Relief of Konitsa. Greek government troops relieved Konitsa, long under rebel siege, driving the defeated guerrilla forces into Albania. A subsequent rebel effort to capture Konitsa was repulsed (January 25).

1948, June 19. Greek Army Offensive Begins. Greek efforts to capture the rebel stronghold of Vafiades were partially successful. Intensive fighting continued for several months in the **Mt. Grammos** region.

1948, November 27. U.N. Condemns Greece's Neighbors. The General Assembly condemned Albania, Bulgaria, and Yugoslavia

for continuing to give assistance to the Greek guerrillas. In fact, however, Yugoslav assistance had declined rapidly after her expulsion from the Cominform (see p. 1367). This greatly facilitated the task of government troops.

1949. General Markos Replaced. Markos Vafiades was replaced as commander of the Greek guerrillas by **John Ioannides.**

1949, June 25. U.N. Charges against Bulgaria. The U.N. Special Committee on the Balkans accused Bulgaria of permitting Greek guerrillas to establish fortifications within Bulgaria from which to fire on Greek troops in Greek territory.

1949, August 28. Greek Troops Clear Mt. Grammos. The principal rebel resistance in Greece was broken by a government assault which captured the northern ridge of Mt. Grammos.

1949, October 16. End of the Civil War.

1951, September 20. Greece Joins NATO.

1954, August 9. Treaty with Turkey and Yugoslavia. This was a 20-year treaty for military assistance and political cooperation, and marked a remarkable *rapprochement* among old enemies.

1955–1965. Strained Relations with Turkey. This was due to the Cyprus issue (see p. 1390).

1955–1959. Strained Relations with Britain. This was due to Cyprus (see p. 1390).

1967, April 21. Military Coup d'État. A junta of right-wing army officers, led by Colonel **George Papadopoulos,** seized control of the government to prevent elections which might have brought a leftist government to Greece under **George Papandreou.** The military leaders set up a puppet civilian government and claimed the support of King **Constantine,** who appeared unenthusiastic about the coup.

1967, December 13. Failure of King Constantine's Attempted Countercoup. A broadcast appeal failed to arouse popular and military support to overthrow the junta; Constantine fled by air to exile in Rome. Papadopoulos, with junta approval, appointed himself prime minister of a completely military government.

1973, June 1. Greece Becomes a Republic.

Papadopoulos proclaimed himself president. A popular referendum, held under strict military control, approved (July 29).

1973, November 25. Military Coup d'État Ousts Papadopoulos. A period of popular unrest, in which President Papadopoulos made some concessions to demonstrators, led militant right-wing military leaders, under secret police chief, Brigadier General **Demetrios Ioannides,** to seize control. Papadopoulos was placed in house arrest and replaced as president by Lieutenant General **Phaidon Gizikis.**

1974, July. Tension over Cyprus. (See p. 1391.)

1974, July 24. End of Military Dictatorship. The military junta turned control over to former Premier **Constantine Karamanlis,** who returned to Greece from self-imposed exile.

MALTA

1964, September 21. Independent Malta Guarantees British Bases. Becoming independent, Malta joined the British Commonwealth and guaranteed bases for British forces for 10 years.

1972, March 26. Treaty Renews British Base Rights. After demanding British and NATO withdrawal from Malta, and threatening to allow Soviet military forces to use facilities on Malta, Prime Minister **Dom Mintoff** agreed to a treaty extending limited British base rights for 7 years in return for $260 million.

TURKEY

1945–1947. Tension between Turkey and U.S.S.R. The Soviets unsuccessfully used diplomatic pressure and threats of force to gain concessions from Turkey in the Straits area and in Turkish Caucasus regions.

1946, August 12. Soviet Demands Dardanelles Rights. The U.S.S.R. demanded joint control of military bases along the Dardanelles, and proposed to Turkey that only Black Sea countries share in the administration of the Turkish Straits. Turkey rejected the Soviet demands (October 18).

1947, March 12. Truman Doctrine. American economic assistance and military equipment greatly strengthened Turkish resistance to Soviet pressures.

1950, September 20. Korean War. Turkey sent a major contingent (initially 4,500 men, later increased to about 8,000) to join the U.N. forces in Korea.

1951, September 20. Turkey Joins NATO.

1955–1965. Strained Relations with Greece over Cyprus. (See p. 1390.)

1955, September 6. Anti-Greek Riots. Riots against the Greeks broke out in Istanbul and Izmir as a result of troubles between the Greek and Turkish inhabitants in Cyprus and anti-Turkish demonstrations in Greece.

1957, October 8. Border Clash with Syria. This, like several previous incidents, arose from Syrian smuggling. This incident heightened tension which already existed, and was followed by other clashes (November 4, 9).

1960, April 28–May 21. Civil Disorder. Widespread rioting by students was put down by police and army with some bloodshed. Martial law was declared, but protests continued. Students from the Military Academy joined in protests, defying martial law (May 21).

1962, February 27. Attempted Coup. A Military Academy mutiny was suppressed without bloodshed.

1963, January 21. Nuclear Weapons Agreement with U.S. Turkey accepted U.S. offers to station Polaris missiles submarines in the Mediterranean Sea to replace Jupiter missiles stationed on Turkish soil. (Italy simultaneously agreed to the withdrawal of Jupiter missiles from Italian soil.)

1963, May 20. Attempted Coup. A Military Academy mutiny was suppressed with some bloodshed.

1964, August 7–9. Air Force Intervention in Cyprus. (See p. 1390.)

1966–1967. Intermittent Tension with Greece over Cyprus.

1971, March 12. Army Ultimatum Forces Government Change. Military leaders threatened to take over the government unless a strong coalition government was established to replace that of Prime Minister **Suleyman Demirel,** who had been unable to control sporadic unrest and violence. At the same time, several high-ranking officers who had favored an immediate coup were forced to retire. Prime Minister **Nihat Erim** established a coalition government two weeks later.

1974, July 20. Turkish Invasion of Cyprus. (See p. 1391.)

CYPRUS

For more than 70 years Cyprus was a British dependency or colony. The population of this east-Mediterranean island is about 80 percent Greek and nearly 20 percent Turkish. Long before World War II, there was a strong sentiment among the Greek population for *enosis,* or union with Greece. This was bitterly opposed by the Turkish minority, who believed that if under Greek control they would be deprived of the rights they enjoyed under British rule.

1952–1959. Guerrilla Warfare. Greek agitation for *enosis* was translated into terrorism directed against the Turkish minority, and guerrilla warfare combined with terrorism waged against the British occupation forces. Principal guerrilla leader was a Greek war hero, Colonel **Grivas.** Complete support to the *enosis* movement was given by Greek Orthodox Archbishop **Makarios,** who was exiled by the British.

1959, March 13. Cease-Fire in Cyprus. This followed an agreement between the British government and the Greek and Turkish communities on Cyprus, with the approval of Greece and Turkey (February 19). An independent republic of Cyprus was to be established, in which the rights of the Turkish minority would be clearly and constitutionally protected. Britain was to retain military bases on the island.

1959, December 14. Makarios Elected First President of Cyprus.

1960–1963. Agitation for *Enosis* Continues. Tension increased between the Greek and Turkish communities.

1963, December 21. Outbreak of Conflict. As a result of Makarios' efforts to reform the constitution and thus reduce the rights of the Turkish minority, armed clashes between Greeks and Turks spread throughout the island. Britain sent reinforcements to attempt to restore order, but widespread fighting continued.

1964, January–February. Unsuccessful Mediation. British and American efforts to establish an international peace-keeping force, under NATO or the U.N., were all rejected by Makarios, who meanwhile was building up his military forces.

1964, March 4. U.N. Intervention. After an impasse had been reached in U.N. discussion, the Security Council authorized Secretary General **U Thant** to establish a peace force and to appoint a mediator.

1964, March 27. A U.N. Peace Force Becomes Operational.

1964, August 7–9. Turkish Air Attacks. Following Greek Cypriote attacks on Turkish Cypriote villages, Turkish planes attacked Greek Cypriote positions. With Greece and Turkey close to war, the U.N. was able to get agreement of the Turkish and Cypriote governments to a cease-fire.

1966, August. Renewed Threats of Turkish Intervention. Unhappy over the condition of Turkish Cypriots in the uneasy stalemate on Cyprus, Turkey again threatened military intervention. Upon appeal from U.N. Secretary General U Thant, President Archbishop Makarios lifted restrictions which had been imposed upon Turkish Cypriots.

1967, November. Greek-Turkish Tension over Cyprus. Danger of war was averted by American mediation and strong pressures on Greece and Turkey. Greece began to withdraw its troops from Cyprus following the resultant agreement.

1974, July 15. Military Coup d'État. The Greek Cypriot National Guard, under leadership of Greek officers from Greece, and obviously abetted by the Greek military government, seized power. Archbishop Makarios (who had been reelected president on February 8, 1973) fled the country.

1974, July 20. Turkish Invasion. To forestall expected plans of the new Greek Cypriot government for "Enosis" (union) with Greece, Turkey seized the north coast of the island. Greece mobilized, but the military government of that country recognized its military weakness, combined with lack of popular support at home, and returned the government of Greece to civilian control (see p. 1389).

1974, July–August. Turkish Occupation Zone Expanded. Paying little attention to a series of cease-fire arrangements negotiated in the United Nations, Turkish forces expanded their control over the northeastern third of the island, establishing a *de facto* division of Cyprus into Greek and Turkish zones.

THE MIDDLE EAST

Unrest permeated the entire area, due to a unique combination of powerful and emotional forces: Cold War pressures, manifested in all parts of the world, particularly intense in this region due to its strategic geographic location and the untold wealth of vast oil reserves; aspirations of new nations breaking away from western colonialism; and the particularly virulent and irreconcilable strife between Israel and the Arab world. Least affected by these swirling tides within the region were the 2 non-Arab Moslem states of the region, Iran and Turkey.

THE ARAB LEAGUE

The Arab League was established shortly before the end of World War II (March 22, 1945) by the governments of Egypt, Iraq, Jordan, Lebanon, Saudi Arabia, Syria, and Yemen. The principal purpose was to prevent the British mandate in Palestine from becoming a separate and independent Jewish state. Later the League was joined by Algeria, Kuwait, Libya, Morocco, the Sudan, and Tunisia. Cairo has been the headquarters for the Secretary General of the League.

1947, January 5. Palestinian Talks. The Arab League accepted a British proposal to participate in a conference on the Palestinian problem to begin on January 26.

1948–1949. War with Israel. (See p. 1334.)

1950, June 17. Collective Security Pact. Egypt, Saudi Arabia, Syria, Lebanon, and Yemen signed a collective security pact in Alexandria. Iraq and Jordan did not join.

1954, June 11. Egypt and Saudi Arabia Reject Baghdad Pact. Under the terms of the Arab League Collective Security Pact, Egypt and Saudi Arabia agreed to pool defenses and military resources, rejecting western plans for a Middle East Defense Treaty against Communism (see below).

1956, March 3–11. Egypt, Syria, and Saudi Arabia Plan United Defense. Plans for combined action against Israel were agreed on by the heads of state at Cairo. King Hussein of Jordan refused to give up British subsidy.

1964, January 13–17. Arab League Conference. Leaders of the 13 Arab League nations met in Cairo and agreed to set up a joint military command for possible action against Israel.

1965, May 13. Severance of Diplomatic Relations with Germany. This was in protest over West Germany's recognition of Israel. Ten states (out of 13) participated: UAR, Jordan, Iraq, Syria, Lebanon, Yemen, Saudi Arabia, Algeria, Kuwait, and Sudan. Tunisia, Libya, and Morocco abstained.

1966, March 14. Arab League Conference in Cairo. Twelve of 13 members were represented; Tunisia was boycotting Arab League activities because of failure to consider compromise accommodations with Israel. The meeting agreed: (1) to continue severance of diplomatic relations with West Germany because of its recognition of Israel; (2) to denounce U.S. Arms shipments to Israel as endangering U.S.-Arab relationships; and (3) to raise money for a project to divert the Jordan River from Israel.

1966, December 10. Arab League Defense Council Agreements. Saudi Arabian and Iraqi troops, with Jordanian approval, were to be deployed in Jordan to bolster defenses against Israel. Jordan simultaneously announced that this agreement was conditioned upon UAR troops replacing U.N. forces in the Sinai and Gaza Strip, which in turn would require UAR withdrawal from Yemen. Jordan refused to permit Syrian troops or Palestine Liberation Organization (PLO) guerrillas in Jordan.

CENTRAL TREATY ORGANIZATION OF THE MIDDLE EAST

The Central Treaty Organization of the Middle East was a successor organization to the Middle East Treaty Organization, or Baghdad Pact, established (1955) as a result of initiatives begun by the United States (1954).

1954, April 2. Turkey-Pakistan Treaty. Signing a 5-year mutual-defense pact, the 2 nations invited neighboring nations, particularly Iraq and Iran, to join.

1955, February 24. Turkey-Iraq Treaty. A 5-year mutual-defense pact with a 5-year renewal clause, joined by Britain (April 4), to establish so-called Baghdad Pact.

1955, November 21. First Meeting of the Baghdad Pact. Representatives of Iran, Iraq, Pakistan, Turkey, and Great Britain met in Baghdad to establish the Middle East Treaty Organization (METO). The U.S. did not join, but sent official observers to the meeting.

1956, April 19. U.S. Partial Participation. At a meeting in Teheran, the U.S. agreed to establish a permanent liaison office, and to help support the permanent METO secretariat.

1959, March 5. U.S. Treaties with the "Northern Tier." The U.S. signed separate defense treaties with Iran, Turkey, and Pakistan, in Ankara, to assure the nations of the Baghdad Pact that the U.S. would come to their support in need of any Communist aggression.

1959, October 7. Reorganization of the Alli- ance. In view of the withdrawal of Iraq (see p. 1398), the remaining members—Britain, Turkey, Pakistan, and Iran—changed its name to Central Treaty Organization (CENTO).

PALESTINE AND ISRAEL

1945–1948. Guerrilla Warfare in Palestine. This bitter and bloody struggle was waged mainly by Jewish Zionists against the Arab population and against British occupation forces, in their efforts to achieve an independent Jewish nation. U.N. efforts to solve rival aspirations of Jews and Arabs proved futile.

1947, November 29. U.N. Decision to Partition Palestine. The General Assembly approved a plan presented to it by a special committee to partition Palestine into separate Jewish and Arab states effective October 1, 1948. The Arab states refused to accept the decision, and announced their determination to fight if necessary.

1948–1949. Arab-Israeli War. (See p. 1334.)

1948, May 14. Independence of Israel. Britain surrendered her mandate over Palestine and withdrew her armed forces. The Israeli nation, immediately recognized by the U.S., was at once attacked and invaded by troops of Egypt, Iraq, Lebanon, Syria, and Transjordan (or Jordan).

1950, May 19. Egypt Closes Suez Canal to Israel. Israeli ships and Israeli commerce were banned.

1951, April 5–16. Border Hostilities with Syria. A cease-fire was later reestablished (May 8).

1953–1956. Intermittent Frontier Clashes. Both sides appear to have been responsible for these various small and large outbreaks along the entire length of Israel's frontiers. However, the initiative appears in most cases to have been with the Arabs. Israeli punitive reprisals were largely stimulated by Arab breaches of the truce or by clandestine raids of Arab terrorists into Israel.

1955, December–1956, October. Increasing Tension along Israel's Borders. Raids of Arab terrorists and guerrillas were met by stern Israel punitive strikes at Egyptian, Syrian, and Jordanian border positions. Arab states, supported by the U.S.S.R., complained in the U.N., as tempers and emotions rose (see p. 1340). Both Israel and Arabs complained of U.S. arms shipments to the other side.

1956, October 29. Israel Strikes First. Hastily mobilized Israeli forces, commanded by General **Moshe Dayan,** plunged into the Sinai. Efficient mechanized columns scattered 4 Egyptian divisions in dismal, headlong rout. At cost of 180 men killed, the Israeli invasion pushed toward the Suez Canal, halting only 30 miles away when France and Britain gave both Egypt and Israel a 12-hour ultimatum to end hostilities (October 30), and then intervened (see p. 1342). U.N. denunciation of Israel, Britain, and France followed. Egyptian losses reported by Israel were 7,000 prisoners and an estimated 3,000 dead. Practically all the Egyptian equipment and matériel (worth $50 million) in the Sinai fell into Israeli hands.

1956, November 15. U.N. Intervention. (See p. 1368.) A curtain of U.N. troops moved into the Sinai between the adversaries. The Israeli invaders slowly retired to the Negev (March 1957), only after U.S. assurances of rights to use the Gulf of Aqaba as an international waterway. A U.N. unit occupied Sharm el Sheikh, overlooking the gulf entrance, which had been captured by the Israelis.

1956–1965. Continuing Tension. Frontier skirmishing continued along Israel's frontiers, with the economic boycott of the Arab League continuing, and growing threat of war over water rights in the Jordan River system. The Arab states, particularly Egypt, were already building up strength in hopes of revenge, obtaining plentiful Soviet equipment.

1962, September 26. U.S. Assistance to Israel. The State Department announced that the United States had agreed to sell defensive missiles to Israel in order to restore the threatened balance of power in the Middle East.

1965, January 30. German Military Equipment to Israel. As part of an agreement between West Germany and Israel, following German recognition of Israel, shipment began of some $80 million worth of military equipment, including helicopters, submarines, antiaircraft guns, and U.S.-made M48 "Patton" tanks. There were strong complaints by the Arab states, which threatened to recognize Communist East Germany if the shipments continued. Germany stopped the shipments, but not before most of the Arab states had broken diplomatic relations (see p. 1391).

1966, February 5. United States Assumes Responsibility for Arms to Israel. The U.S. took over the former West German obligation, to maintain arms "stabilization" in the Middle East, in the light of continuing Soviet arms shipments to Egypt and Syria. Included in the American shipments would be 200 M48 tanks. Soon afterward (May 20) the U.S. revealed that, in accordance with its stabilization policy, it was also providing A-4 Skyhawk tactical aircraft to Israel.

1969, December 25. Israeli Seizure of Embargoed Gunboats from France. Five fast missile gunboats, being constructed in France under contract with Israel, were held by the French government following President de

Gaulle's embargo on all arms shipments to Israel as a result of the 1967 Arab-Israeli War. In a daring feat, Israeli crews secretly seized the vessels and took them to Haifa, arriving New Year's Eve.

1970, January 9. Israel Protests French Aircraft Sale to Libya. (See p. 1434.)

1972, January 13. U.S.-Israeli Arms Assistance Agreement Revealed. By this agreement (November 1971) Israel was authorized to manufacture various kinds of American weapons and equipment, as part of the U.S. stabilization policy.

EGYPT—UNITED ARAB REPUBLIC

1945–1952. Internal and Anti-British Unrest. Egypt called for an end to British military occupation and abrogated its treaties with Britain (October 27, 1951). The U.K. acquiesced and began withdrawal, but rioting at Port Said and Ismailia, endangering Suez Canal operations, necessitated British military action and continued occupation.

1948–1949. Arab-Israeli War. (See p. 1334.)

1949–1956. Continuous Friction and Incidents along Israeli Border. (See above.)

1952, July 22. Military Coup d'État. King **Farouk** was dethroned by a military uprising under the leadership of General **Mohammed Naguib,** his right-hand man being Colonel **Gamal Abdal Nasser.**

1953, June 18. Egypt Proclaimed a Republic. Naguib became first president and premier.

1954, February 25–November 14. Naguib-Nasser Struggle for Control. After considerable internal maneuvering, Naguib was finally ousted and replaced by Nasser (November 14).

1955, September 27. Agreement with Czechoslovakia. Nasser announced a commercial agreement to exchange Egyptian cotton for armaments. He stated that western nations had refused Egyptian requests for arms and that Israel was buying French warplanes.

1955, October 20. U.S. Offers to Finance Aswan Dam. Egypt accepted financing by the International Bank for Reconstruction and Development and the U.S. (December 1955–February 1956), rejecting Soviet offers to supply more than 1/3 of the $.6- to $1.2-billion cost.

1956, June 13. Britain Completes Withdrawal from Egypt. This ended 74 years of British military occupation in Egypt.

1956, June 18. Soviet Renews Offer to Finance Aswan Dam. This time the U.S.S.R. agreed to furnish about $1.0 billion at 2 percent interest. President Nasser began to reopen bargaining with the U.S., Britain, and IRDB to get better terms.

1956, July 19. U.S. Withdraws Offer to Finance Aswan Dam. Disgusted by Egyptian anti-U.S. propaganda and negotiations with the U.S.S.R., Secretary **Dulles** withdrew the offer; Britain supported the U.S. action.

1956, July 26. Nationalization of Suez Canal. Egypt seized control of the canal from the private (primarily British) Suez Canal Corporation, announcing its nationalization. Hot debate in and out of the U.N. followed; France and Britain particularly considered the action as a threat to world peace. The U.S. was gravely concerned and began negotiations to achieve international control. The Communist bloc supported the Egyptian seizure. Nasser turned down western proposals and began to operate the canal (September 14).

1956, October 29. Israeli Invasion of Sinai. (See p. 1340.)

1956, October 31. Franco-British Intervention. Following Israel's assault in the Sinai, France and Britain issued an ultimatum calling on Israel and Egypt to cease fire in 12 hours. Israel accepted subject to Egyptian acceptance (October 30). Egypt rejected the ultimatum. Franco-British air forces began bombardment of Egyptian air bases.

1956, November 5–7. Franco-British Invasion of Canal Zone. British paratroops, flown by helicopter from Cyprus, made vertical combat landings at Port Said while Franco-British warships bombarded the port and landed troops in an amphibious assault. One day of street fighting brought the port into the attackers' hands, and the advance continued along the canal. Egyptian resistance concentrated on the sinking of stone-laden barges and other vessels in the canal itself, completely blocking it. Immediate U.N. reaction followed, with the U.S. and U.S.S.R. for once in agreement. A cease-fire demand was reluctantly obeyed by France and Britain (November 7). By this time the northern half of the canal, from Port Said to Ismailia, had fallen into the invaders' hands. Allied losses were 33 dead and 129 wounded; Egyptian losses are unknown. Following U.N. pressure and unilateral U.S. efforts, the Franco-British forces evacuated Egyptian territory (November 19–December 22).

COMMENT. *While U.S. denunciation of this bilateral assault, and the threats of the U.S.S.R. to use missiles against Britain and France and to furnish a "volunteer" army for the relief of Egypt, both played dominant roles in causing France and Britain to withdraw, a very practical military reason also underlay the decision. What might have become a* fait accompli, *had the operation been properly prepared, became a disastrous fumble when, due to inadequate warning and preparation, 5 inactive days elapsed between the initial air bombings and the amphibious assault. World opinion had time to react. The net result was a strengthening of Nasser's domination in Egypt and his affiliation with the U.S.S.R., and a weakening of accord amongst NATO members.*

1957, March 7. Suez Canal Reopens. U.N. salvage crews took only 69 days to clear hulks from the channel, sunk there by Egyptians during the crisis.

1958, February 1. Union with Syria. The United Arab Republic was established with Yemen as another partner. This was Nasser's next move to dominate the Middle East by a powerful Arab state. Anti-Egyptian opposition in Syria broke up the short-lived partnership (September 30, 1961), but Nasser's Egypt continued as the United Arab Republic.

1961–1967. Involvement in Yemen. Breaking relations with the monarchical government of Yemen (December 26, 1961), Egypt began subversive support of a republican movement, followed by active military support of the republican government against the Imam (see below). Egyptian losses were heavy; the experience was frustrating and costly to Egypt. Egyptian troops were withdrawn after the Sinai

defeat at the hands of Israel, to bolster Egypt's defense of the Suez Canal.

1964, May 26. Planned Unification with Iraq. President Nasser and Iraq's President **Abdel Salim Arif** signed an agreement in Cairo to establish a joint Egyptian-Iraqi military command in time of war, and to appoint immediately a joint presidential council to study ways of unifying the two governments.

1967, June. War with Israel. (See p. 1343.)

1970, September 28. Death of Nasser. This was the result of a heart attack.

1970, October 7. Sadat Appointed President. Vice President Anwar Sadat, a loyal assistant of Nasser, was appointed president by Egypt's National Assembly. This was confirmed by a national plebiscite (October 14).

1971, May 13. Sadat Purges Opponents. A coalition of militant anti-Israel and pro-Russian politicians, disappointed by Sadat's political moderation and skillful consolidation of government control, had been plotting to replace him when the president quickly dismissed and jailed dissident government leaders.

1971, May 28. Treaty of Friendship and Cooperation with the U.S.S.R. Signed in Cairo by President Sadat and President **Nikolai Podgorny** of the U.S.S.R., this assured continuing Soviet military assistance, but without the danger of a Communist takeover after Sadat had consolidated his political power by purging pro-Russian elements from his government (see above).

1971, August 20. Agreement for Union with Syria and Libya. This agreement theoretically went into effect when referenda in the 3 countries enthusiastically confirmed the agreement (September 1). Since each nation retained its national sovereignty, the agreement was virtually meaningless.

1972, July 18. Sadat Orders Withdrawal of Soviet Advisors. Sadat said that the decision to "terminate the mission of the Soviet military advisors and experts" was due to Soviet failure to provide arms for renewed war with Israel. Moscow announced that the withdrawal was by mutual agreement. Not all Soviet advisors were withdrawn, however.

1972–1973. Agreements for Union with Libya. (See p. 1434.)

1973, October 6–22. War with Israel (see p. 1348).

YEMEN

1948, February 19. Revolt. King **Hamid** was overthrown and replaced by Imam **Ahmed.**

1950, June 17. Yemen Joins Arab League Collective Security Pact. (See p. 1391.)

1954, May. Border Clashes with Aden.

1955, April 2–5. Military Revolt. Imam **Seif el-Islam Ahmad** and Crown Prince **al Badr Mohammed** and loyal tribesmen defeated an army revolt led by Emir **Seif al-Islam Abdullah.**

1956, April 3. Claim to Britain's Aden Protectorate. The nebulous boundary line became theater of a sporadic war continuing to the end of the period.

1957–1962. Communist Arms Provided to Yemen. This was through negotiations with the U.S.S.R. and later with Red China. The U.S.S.R. began extensive port construction at Hodeiya (Port Ahmed).

1961, December 26. U.A.R. Breaks Relations with Yemen. (See p. 1395.)

1962, September 19. Death of Imam Ahmed. Crown Prince **Saif al-Islam Mohammed al Badr** assumed the crown. A republican uprising followed.

1962, September 27. Proclamation of "Free Yemen Republic." The rebel government, supported by the U.A.R., was also recognized by Communist-bloc nations (September 29). Heading the rebels was Colonel **Abdullah al-Sallal** (proclaimed president, October 31).

1962–1965. Civil War. The rebel government was assisted by the U.A.R. Egyptian troops entered the country to assist the rebels. With Egyptian forces waxing to some 60,000 strength by 1965, Yemen became a battleground for pro- and anti-Communist Arabs. Saudi Arabia provided active military support to the monarchical faction, while British forces resisted a republican Yemenite invasion of Aden.

1965, August 24. Saudi-Egyptian Agreement and Cease-Fire. For the third time in two years President Nasser and King **Faisal** agreed to end their support of the republicans and royalists, who also agreed to a cease-fire and to a plebiscite to determine the future government.

1966, February 22. Egypt Supports Republicans in Renewed Fighting. A republican attack on a royalist tribe was supported by U.A.R. aircraft. That same day, in a response to Faisal's appeal to withdraw, President Nasser announced that his troops would stay in Yemen until a plebiscite was held. In the outburst of renewed fighting Saudi Arabia accused U.A.R. planes of bombing Saudi towns near the Saudi-Yemen frontier.

1967, November 5. Coup d'État in Republican Government. President **Abdullah el-Salal** was overthrown by dissident republicans under **Rahman al-Iryani,** who assumed title of Chairman of Presidential Council. Despite Iryani's apparent intention of seeking an understanding with the royalists, the war dragged on. U.A.R. troops had been withdrawn after Egypt's disastrous defeat in the Sinai (see p. 1343).

1968, February. Southern Yemen Intervenes. Participation of troops from Southern Yemen in attacks on royalist troops was announced by both the Southern Yemen and the Yemen republican government (February 19).

1968, February 28. Saudi Arabia Renews Aid to Royalists. The Saudi government announced that this was needed to offset assistance to the republicans from the U.S.S.R., Syria, and Southern Yemen.

1970, April 14. Agreement with Saudi Arabia Ends Civil War. King Faisal's government agreed to recognize the republican government of Yemen, in return for the inclusion of a number of royalists in the government.

1972–1973. Border Clashes with South Yemen.

1974, June 13. Military Coup d'État. Colonel **Ibrahim al-Hamidi** was the leader of an army group seizing control.

LEBANON

1945–1956. Precarious Existence. The republic strove to maintain independent status despite the coaxing and threats of its neighbors to force its participation in the Pan-Arab movement. There was constant friction with Syria.

1949, July 8. Brief Rebellion Suppressed.

1956, January 13. Military Defense Treaty with Syria.

1958, April 14–July 14. Insurrection. This was inspired by the U.A.R. Fighting spread to the streets of Beirut (June 14). U.N. observers, with **Dag Hammarskjöld,** secretary general, arrived.

1958, July 14. Lebanese Appeal for Help. President **Camille Chamoun** appealed to the U.S., Britain, and France, urging troop aid to seal the Syrian border as the revolution in Iraq (see p. 1398) set the Middle East aflame and stimulated renewed unrest in Lebanon.

1958, July 15. U.S. Intervention. Responding to Chamoun's appeal, a U.S. force of Marines and Army troops arrived by sea and by air from Europe and the U.S. to protect Lebanon from U.A.R. or Communist invasion.

1958, August 21. U.S. Begins Withdrawal. Stabilization of the situation having been effected, withdrawal of 14,300 U.S. troops began (completed October 25).

1961, December 31. Attempted Coup. Rightist elements, including some troops, were crushed by prompt reaction of the Army.

1969, October. Lebanese Army Clashes with Palestinian Commandos. A 6-month period of tension and occasional out-breaks of violence between Palestinian guerrillas and the Lebanese Army flared into virtual civil war. During three weeks of sporadic fighting the Palestinians seized control of 14 of 15 U.N.-operated refugee camps in Lebanon.

1969, November 3. Cease-Fire Agreement Signed in Cairo. The Lebanese government virtually accepted extraterritorial sovereignty of the Palestinian commandos. Following one more clash (November 20) an uneasy peace returned to most of Lebanon, with the Palestinian guerrillas free to continue their terrorist operations against Israel.

1973, May 2–8. Renewed Army Conflict with Palestinian Commandos. Confused fighting intensified when the Lebanese Army intercepted 1,000 guerrillas trying to cross into Lebanon from Syria. An uneasy cease-fire was finally arranged by representatives of Egypt, Iraq, and Algeria. Syria closed the border with Lebanon, denouncing the Lebanese government for conspiring with "foreign" powers in an anti-Palestinian conspiracy.

SYRIA

1946, April 15. Independence of Syria. A former League of Nations mandated territory under French control, Syria was granted independence in accordance with the United Nations Charter.

1948–1949. Arab-Israeli War. (See p. 1334.)

1949, March 30. Coup d'État.

1949, August 18. Coup d'État.

1949, December 17. Coup d'État.

1951, November 28. Coup d'État.

1953–1954, December–February. Jebel Druze Uprising. Suppressed by the Syrian Army after extensive fighting.

1954, February 25. Coup d'État.

1958, February 1. Union with Egypt. (See p. 1395.)

1961, September 28. Coup d'État. Elements opposed to the union with Egypt seized power.

1961, September 30. Dissolution of Union with Egypt. This was the result of the recent coup d'état.

1962, March 28. Coup d'État.

1962, April 1. Unsuccessful Military Revolt at Aleppo.

1963, March 8. Coup d'État.

1964, February 8–25. Civil Disorders. Bloody riots in Banias and Homs marked general unrest.

1964, April 13–15. Unsuccessful Insurrection in Hama. There was severe loss of life in the uprising, which apparently was aimed at overthrow of the government.

1966, February 23. Coup d'État. Premier **Salal al Bitar,** moderate leader of the Ba'ath Party, was overthrown by leftwing military dissidents of the party led by Major General **Salal Jedid.**

1967, June. War with Israel. (See p. 1343.)

1969, February 28. Coup d'État.

1970, September 19–23. Invasion of Jordan Repulsed. (See p. 1400.)

1970, November 13. Coup d'État. Right-wing army officer members of the Ba'ath party, led by Defense Minister Lieutenant General **Hafez el-Assad,** seized power.

1971, March 12. Assad Elected President. This was for a term of 7 years. Assad gave the country the most stable government it had had since independence.

1971, September 1. Union with Egypt and Libya. (See p. 1395.)

1973, October 6–22. War with Israel. (See p. 1348.)

IRAQ

Internal affairs in Iraq were chaotic throughout the period. The central government (in its various manifestations) was more or less constantly at war with the Kurdish minority in the northwestern mountain region. Principal military events of the period were:

1946, April. Uprising in Kurdistan. The Kurd rebellion had not been suppressed by the close of the period.

1948–1949. Arab-Israeli War. (See p. 1334.)

1955, November 21. Baghdad Pact. (See p. 1392.)

1958, July 14. Army Revolt. An army officer revolt, led by Brigadier General **Abdul Karim el Kassim,** overthrew the monarchy. King **Faisal II** and Premier **Nuri es Said** were brutally murdered. Initially closely tied to the U.A.R. and pan-Arab unity, Kassim's Republic of Iraq soon diverged from policies followed by Nasser's U.A.R.

1958, December. Iraq-Soviet Agreement Verified. U.S. sources determined that Premier Kassim had accepted Soviet arms and had entered into a working agreement with the U.S.S.R.

1959, March 8–9. Revolt Suppressed. This was an attempt by a group of officers in **Mosul.**

1959, March 24. Iraq Withdraws from the Baghdad Pact. Soon after this, Iraq abrogated her agreements with the U.S. and refused further U.S. military aid (June).

1959, July 14–20. Unsuccessful Rebellion. A confused uprising of Kurds, Communists, Moslem factions, and army troops in the Kirkuk region was put down with heavy loss of life by loyal troops.

1961, December 27. Britain Supports Kuwait. Britain dispatched naval reinforcements to the Persian Gulf to deter Kassim's threatened annexation of Kuwait (December 24).

1961–1970. Kurd Rebellion. Refusal of the Iraqi government to provide autonomy to their mountainous home land, in the region where Iraq, Iran, and Turkey meet, led to Kurdish defiance of the central government, and the outbreak of virtual civil war (September 17 1961). The Kurds were led by Mullah **Mustafa Barzani.**

1963, February 8. Revolt. Kassim was deposed and executed; his estranged partner in the 1958 revolt, **Abdul Salam Arif,** became the new president.

1963, November 18. Military Coup. The government was overthrown; President Arif, who had been a figurehead, seized control and pledged support to Nasser of Egypt.

1965, September 17. Partial Coup d'État. General **Abdel Rahman Arif** suppressed a revolt against his brother, President **Abdel Salim Arif,** then seized control of the government himself.

1968, July 17. Coup d'État. Major General **Ahmed Hassan al Bakr** led a group of army officers of the right wing of the Ba'ath party overthrowing President Abdel Rahman Arif.

1969–1975. Border Clashes with Iran. These began with a dispute over Iraqi control of navigation of the Shatt-al-Arab (see p. 1403), and with Iranian support of the Kurds.

1970, January 20. Attempted Coup d'État. The government said the crushed rebel forces were supported by Iran.

1970, March 11. Peace with the Kurds. Iraq agreed to grant autonomy and to make constitutional reforms to grant the Kurds a major role in the central government.

1972, April 9. Treaty with the U.S.S.R. This 15-year friendship pact carried with it a promise of more Soviet arms for Iraq.

1973, March 20–21. Border Clashes with Kuwait. Iraq again backed down in the longstanding dispute after the arrival of 20,000 Saudi Arabian troops to help Kuwait.

1973, June 30. Unsuccessful Coup Attempt.

1974–1975. Renewed Kurd Rebellion. The Kurd rebellion collapsed when Iran withdrew support as a result of a resolution of its border dispute with Iraq (March 1975).

JORDAN

1946, March 22. Independence of Transjordan. This former League of Nations mandate was granted independence by Britain. Emir **Abdullah** became the first king.

1948–1949. Arab-Israeli War. Transjordan, due mainly to its British-trained Arab Legion, was the only Arab state to perform creditably against the Jewish Army of Israel (see p. 1334).

1949, June 2. Hashemite Kingdom of Jordan Established. This change of name was a step in the process of annexation of Arab Palestine (April 24, 1950), occupied by the Arab Legion during the Arab-Israeli War. This annexation was strongly protested by most other Arab League states.

1951, July 20. Assassination of Abdullah. Murder of this moderate ruler by an Arab extremist was applauded in many other Arab countries. They feared that he was moving to an accommodation with Israel. He was briefly succeeded by his eccentric son, Emir **Tallal** (September 5).

1952, May 2. Accession of King Hussein. This young king (only 17 when he came to the throne) proved himself a wise, tough, durable ruler, who maintained control over his volatile nation (with the help and support of his Arab Legion), despite continued and overt efforts of Nasser and most other Arab League leaders to have him overthrown or assassinated (1955–1962). By the end of the period, Hussein and Nasser, however, had temporarily resolved their differences and were at least nominally in accord on their strong anti-Israel policies.

1954, May 2. Jordan Rejects British Suggestion of Peace Talks with Israel.

1956, March 2. Dismissal of Glubb. British Lieutenant General **John Bagot Glubb,** commander of the British-subsidized Arab Legion, was dismissed by Hussein for failure to prepare for an Israeli attack. Upon his return to Britain, Glubb was knighted (March 9).

1958, July 17-October 29. British Support to Hussein. Following the revolt in Iraq, and as combined U.A.R. and Communist pressure and threats against Hussein and Jordan mounted, British paratroops landed in Jordan, at Hussein's request, shortly after the landings of U.S. forces in Lebanon (see p. 1397). This strong Anglo-American cooperation restored comparative stability to the Middle East.

1962. Military Coordination of Jordan and Saudi Arabia. Merger of the armed forces of the 2 countries was announced as a show of strength and unity against the U.A.R. and other Arab League members. Though never officially abrogated, this merger was apparently never fully activated and was quickly ignored.

1967, June 5–10. War with Israel. (See p. 1345.)

1967, November 6. Hussein Suggests Peace Negotiations with Israel. The King said he was ready to recognize Israel if the boundary questions could be negotiated satisfactorily. He also suggested that Egypt was ready to permit Israeli ships in the Suez Canal, once a general agreement was reached. Egypt denied this latter statement, amid a storm of protests from most Arab nations.

1968, March 21. Battle of Kerama. An Israeli

raid against Palestine guerrilla bases in the Jordan Valley was met by Jordan Army units, who claimed success in driving the Israelis off after a sharp fight.

1970, February 10–12. Fighting between Jordanian Army and Palestinian Commandos. This resulted from Palestinians' refusal to obey a Jordanian directive forbidding them to carry arms openly in public, and ordering the surrender of arms and ammunition. Both sides backed down under pressure from other Arab states.

1970, June 7–11. Renewed Fighting with Commandos. A virtual 5-day civil war was again brought to a conclusion by pressure from other Arab states and appeals to the guerrillas from King Hussein and **Yasir Arafat,** leader of Al Fatah, the largest commando group. The Jordanian Army had the better of the fighting, but losses were heavy on both sides; each accused the other of numerous atrocities.

1970, September 17–26. Open Warfare with Palestinian Commandos. Apparently believing his monarchy could not survive if the commandos continued to defy his authority and Jordanian sovereignty, Hussein ordered an all-out attack on the commando bases. Despite interference from Syria (see below), the Jordanian Army was successful and commando losses severe. A cease-fire was arranged in Cairo (September 27), with the guerrilla survivors accepting practically all of the King's demands.

1970, September 19–23. Undeclared War with Syria. Following a day of artillery fire across the border, Syrian troops, mostly armored units, invaded northern Jordan and, with Palestinian commando assistance, captured Irbid, Jordan's second-largest city. A major tank battle took place near Irbid between Jordanians and Syrians (September 22). The Jordanians completely defeated the Syrians, who retreated across the border (September 23). Jordanian close support aircraft contributed largely to the victory. Syria broke diplomatic relations with Jordan. The Syrians were deterred from further action by a partial Israeli mobilization and clear indications that Israel would attack if the Syrians renewed the war.

1971, June 13–19. Final Suppression of Palestinian Commandos. Not satisfied with the commandos' adherence to the cease-fire terms, the Jordanian Army captured some 2,000 commandos and drove the remainder out of the country into Israel and Syria. Syria again fired artillery across the border, but made no further aggressive move.

1971, November 28. Palestinians Assassinate Jordanian Prime Minister. While attending an Arab League Defense Council meeting in Cairo, Prime Minister **Wafsi Tal** was killed by 3 Palestinian gunmen, who later said they were members of the Black September Organization, a Palestinian terrorist group whose name commemorated the Jordanian 1970 attack against the commandos.

SAUDI ARABIA

Involved like all the other nations in the Middle East in pan-Arab, anti-Jewish movements, Saudi Arabia, one of the great oil centers of the area, could be considered as at least partly amenable in policy to the **Arabian American Oil Company** (ARAMCO), which operated the fields and was the sole source of financial revenue. The death (November 9, 1953) of King **Ibn Saud,** a firm supporter of the West, momentarily loosed its ties with the Free World. His sons, **Saud Ibn Aziz** (1953–1965) and Faisal (who deposed his brother in 1964), continued, however, in attempts to maintain at least neutrality in the Cold War.

1952. Saudi Forces Invade the Buraimi Oasis. This territory was also claimed by the Sultan of Muscat and Oman. A clash with British interests in that area ended with the ousting of the invaders (October 26, 1955) by British-led troops of Muscat and Oman.

1956. Military Alliance with Egypt and Yemen. (See p. 1391.)

1957–1962. Friction with U.A.R. This brought about closer ties between Saudi Arabia and Jordan (see p. 1399).

1962–1970. Saudi Arabian Support of Royalists in Yemen. (See p. 1396.) This intensified friction with the U.A.R.

1964, March 28. King Saud II Stripped of Royal Power. This was by decree of a royal council, led by his younger brother, Prince Faisal, who became virtual regent, with Saud a figurehead.

1964, November 2. Saud Deposed; Faisal Named King.

1969, November 26-December 5. Border Clashes with Southern Yemen. (See p. 1402.)

1973, March 21. Troops Sent to Support Kuwait against Iraq. (See p. 1399.)

1973, October. Saudi Leadership in Oil Embargo. Supporting Egypt and Syria in the October War (see p. 1348), Faisal led the Arab oil-producing states in a highly successful embargo against all nations directly or indirectly supporting Israel.

ARAB GULF STATES

Britain retained a shaky hegemony over the southern and southeastern regions of Arabia, from Aden to the Persian (or Arabian) Gulf. British troops were involved in frontier fighting between Yemen and the Aden Protectorate, in suppressing internal disorders in Aden, and on the southeastern fringes of the Arabian desert in desultory warfare between the Sultanate of Muscat and Oman and Saudi Arabia, centering around the disputed Buraimi Oasis. British troops also helped the Sultan of Muscat and Oman suppress a serious revolt (July–August 1957).

1952–1955. Muscat, Abu Dhabi, Saudi Arabia Dispute over Buraimi Oasis. (See p. 1400.)

1967, November 2. Britain Announces Planned Withdrawal from Arabia. This was to take place in 1971 (see p. 1375).

1971, July 18. Establishment of United Arab Emirates. At a meeting in Dubai, 6 Persian Gulf emirs agreed to form a federation before the planned British withdrawal at the end of the year. Participants: Abu Dhabi, Sharjah, Ajman, Umm al Qaiwain, Fujairah, and Dubai.

1971, August 14. Independence of Bahrain.

1971, September 21. Independence of Qatar.

1971, November 30. Iranian Troops Seize Persian Gulf Islands. Ras al Khaimah, which claimed the Greater and Lesser Tunb Islands, protested the seizure. Abu Musa, also seized, was claimed by Sharjah, but the Iranian government agreed to respect Sharjah's sovereignty in return for base rights on the island.

1972, January 24. Unsuccessful Coup in Sharjah. Sheikh **Khalid bin Mohommed al Qasmi** was killed and was succeeded as emir by his brother, Sheikh **Saqr bin Mohommed al Qasmi.**

1972, February 11. Ras al Khaimah Joins United Arab Emirates. This brought membership to 7 states.

1972, February 22. Coup d'État in Qatar. Sheikh **Ahmed bin Ali al-Thani** was deposed and replaced by his cousin, Sheikh **Khalifa bin Hammad al-Thani.**

OMAN (MUSCAT AND OMAN)

1952–1955. Buraimi Oasis Dispute. (See p. 1400.)

1955, December 15. Sultan Reestablishes Authority over Oman. Sultan **Said bin Taimur** suppressed a conspiracy by the Imam of Oman, who was exiled.

1957, July–August. Imam Revolt against Sultan. The exiled Imam of Oman led a revolt against the Sultan's British-led army. After serious fighting near **Nizwa** (July 15), the Sultan requested British assistance, which was provided, mostly in air support, resulting in suppression of the rebellion (mid-August). An Arab League effort to censure Britain in the U.N. Security Council failed (August 20).

1968–1974. Revolt in Dhofar. Separatist rebels, Marxist and supported by the People's Republic of Yemen (Southern Yemen), waged constant guerrilla war against the government. Oman employed British officers to command counterinsurgency units. This buildup and Iranian assistance (beginning 1973) suppressed the insurgency.

1970, July 23. Palace Coup. Sultan Said bin Taimur was overthrown and replaced by his son, **Qabus bin Said,** on the basis of the former's "inability" to use the new-found oil "wealth of the country for the needs of the people." The new Sultan changed the name of the country to "Sultanate of Oman."

SOUTHERN YEMEN (PEOPLE'S DEMOCRATIC REPUBLIC OF YEMEN)

1945–1967. British Aden Protectorate. Violence and unrest in the protectorate. Despite a UN Observer Force (established 1 June, 1963), violence continued, resulting in virtual civil war (December, 1963), with the British and local sheikhs fighting leftist revolutionaries. Britain decided to withdraw.

1964, July 4. Britain Pledges Independence to the Federation of South Arabia. At a constitutional conference in London this was promised to take place "not later than 1968." Britain planned to retain its base in Aden, one of the 13 federated states. The leftist revolutionaries continued guerrilla warfare. Britain decided to withdraw completely. (See also p. 1369.)

1967, November 30. Independence of the People's Democratic Republic of Yemen. Violence and unrest continued, as rival factions vied for power.

1968–1974. Guerrilla Warfare against Oman in Dhofar. (See above.)

1968, March 20. Attempted Army Coup.

1968, May 13–16. Unsuccessful Revolt. This was led by dissident leaders of the ruling National Liberation Front.

1968, July 26–August 9. Unsuccessful Rebellion. Tribesmen south of Yemen rose under the leadership of members of the Front for Liberation of South Yemen. The rebellion was suppressed after fierce fighting. The government said the rebels were aided by Saudi Arabia.

1969, June 22. Coup d'État. A militant leftist government seized power.

1969, November 26–December 5. Border Clashes with Saudi Arabia. Saudi Arabia claimed victory in fierce fighting for the disputed oasis of **Al Wadeiah.**

1972–1973. Border Clashes with Yemen. (See p. 1396.)

1974, August 28. U.S. Alleges Soviet Naval Base in South Yemen. President Gerald Ford stated that the Soviet Union had a naval base in South Yemen. The Soviets denied this accusation.

IRAN

At the close of World War II, the U.S., British, and Soviet governments agreed to withdraw their forces from Iranian territory. The withdrawals were to be completed by March 2, 1946.

1945, November 18. Rebellion in Azerbaijan. A Communist-inspired revolt by the Tudeh party broke out. Efforts of the government to repress it were hampered by Soviet troops still in Iran. Prime Minister **Qavam** protested to the U.N. Security Council (January 19, 1946). The firm stand of the U.S. government in support of Iran brought withdrawal of Soviet troops (May 6), and the rebellion was later put down (December 6–11).

1946, September 15–October 7. Rebellion in Southern Iran. This was settled by agreement between government and rebellious tribesmen.

1951, April 29. Nationalization of Oil Industry. Mohammed Mossadegh, new Iranian premier, ordered nationalization of the oil industry. Violent repercussions in Britain and the U.S. followed, and oil production virtually ceased.

1953, August 19. Mossadegh Overthrown. The pro-Communist dictatorial prime minister was ousted and imprisoned by a coup supported by Shah **Mohammed Reza Pahlavi,** who became virtual prime minister himself.

1954, August 5. Iranian Oil Production Renewed. European and American oil interests agreed to operate the former Anglo-Iranian oil plant on a new royalty basis.

1955, November 21. Baghdad Pact. (See p. 1392.)

1969, April 19. Iran Abrogates Treaty with Iraq on Navigation of Shatt-al-Arab. Claiming that Iraq had violated provisions of a 1937 treaty giving most of the river to Iraq (and claiming the treaty was based on a series of unequal treaties imposed by Great Britain since 1847), Iran announced that it would enforce its claim to sovereignty up to the middle of the river (Arand-Rud in Iranian). Iraq massed troops on the border, as did Iran; Iran sent ships under armed escort to the ports of Abadan and Khorramshahr. There were no immediate hostilities, but this was the basis of later border clashes (see p. 1399 and below).

1969, September. Border Clashes with Iraq. (See below.)

1970–1974. Modernization of Iranian Army. A major program of military modernization, improvement, and build-up was undertaken by Shah Mohammed Reza Pahlavi.

1970, January 20. Border Clash with Iraq.

1971, November 30. Iran Seizes Persian Gulf Islands. (See p. 1401.)

1973–1974. Iranian Troops Assist Oman in Dhofar. (See p. 1402.)

1974, February 6–15. Border Clashes with Iraq. This ended a lull caused by the Arab-Israeli War.

1974, March 5–6. Border Clash with Iraq. Some 180 incidents occurred in 1973–1974.

1975, January–March. Border Clashes with Iraq. Iranian aid to the Kurds ended with signature of a boundary agreement (March 5).

SOUTH ASIA

BRITISH INDIA

The post–World War II events in South Asia were shaped almost entirely by the division of the Indian subcontinent on religious lines at the time Britain relinquished her colonial Empire of India in 1947: the essentially Hindu Dominion of India and the essentially Moslem Dominion of Pakistan, the latter being divided into 2 separate portions, about 1,000 miles apart, separated by Hindu India. Strife and bloodshed occurred across the country during the year before partition, and intensified immediately upon independence and partition. The principal events were:

1946–1947. Violence and Unrest Sweep India. With Britain clearly preparing to give independence, violence flared intermittently across all of India between Hindus and Moslems. The fighting was particularly intense and bloody in the Punjab, almost equally divided between Mohammedan and Hindu inhabitants. During one 4-month period (July–October 1946), the British government announced that 5,018 persons had been killed and 18,320 injured in the Hindu-Moslem rioting. The toll was even heavier in the early months of 1947.

1947, August 14. Independence of India and Pakistan. India and Pakistan both received their independence simultaneously, and both became dominions within the British Commonwealth of Nations. Rioting, violence, and death increased throughout both nations, and particularly in the Punjab. Mobs of religious majorities in both nations began to terrorize, rob, and murder the minority groups, most of whom took refuge beyond the partition frontier. After about 6 weeks of slaughter, relative peace returned, mostly because the persecuted minorities had been wiped out or chased away. No reliable statistics exist, but it is probable that close to a million people were massacred, while 10–15 million people were forced to flee from their homes.

KASHMIR DISPUTE, 1947–1965

1947, October. Moslem Uprising in Kashmir. The decision of the Hindu Raja of Kashmir to have his state join India (October 26) precipitated an uprising of the predominantly Moslem population, who wished to join Pakistan. Afridi and Mahsud tribesmen who had crossed into Kashmir from Pakistan joined in a march on Srinagar. Indian troops were flown into Kashmir to quell the uprising (October 27). Intensive fighting broke out between air and ground forces of the Indian government and the rebellious Moslems and their supporters from Pakistan. (October 28–30.)

1947, November–1949, December. Undeclared War in Kashmir. Pakistani troops crossed the border into Kashmir to assist the Moslem rebels, precipitating an undeclared war between India and Pakistan.

1948, February 8. Pathan Uprising in Kashmir. This was suppressed by Indian troops.

1949, January 1. Cease-Fire. U.N. mediation brought about an uneasy truce along the fighting front in Kashmir, ending 14 months of warfare.

1949–1954. Intermittent Negotiations between India and Pakistan. No firm agreements were reached.

1953, August 20. Plebiscite Agreement. Sheikh **Mohammad Abdullah,** prime minister of Kashmir, and **Jawaharlal Nehru,** prime minister of India, agreed to a plebiscite to settle the dispute over the state of Jammu in Kashmir. India later withdrew its agreement and imprisoned Mohammed Abdullah.

1954, February 6. Kashmir Ratifies Accession to India. Pakistan protested.

1954, October 4. Pakistan White Paper. This declared that negotiations with India had failed and asked the Security Council to settle the problem.

1957, January 26. India Annexes Kashmir. Pakistan protested; the U.N. disapproved. Henceforward Kashmir was merely one (although one of the most important) of a number of issues between India and Pakistan.

INDIA, 1947–1961

1947, August 14. Independence.

1948, January 30. Assassination of Gandhi in New Delhi. Rioting broke out across India.

1948, September 15–17. Indian Occupation of Hyderabad. After the Nizam of Hyderabad had refused to join the Dominion of India, Indian troops invaded and forced unconditional surrender.

1949, June–July. Disorders and Violence in Kerala. The Indian government intervened to dismiss the Communist government of Kerala State and to take direct control after widespread, bloody, anti-government riots.

1950, January 25. India Becomes a Republic. She retained membership in the British Commonwealth.

1950. Medical Unit to Korea. This was in support of the U.N. war effort (see p. 1366).

1953. Revolt in Nepal. Indian troops assisted in the suppression of a Communist-inspired revolt in Nepal.

1954–1974. Naga Revolts. India consistently refused demands for autonomy by these primitive tribesmen of the Northeast Frontier.

1954, April 29. Nonaggression Treaty with Communist China. This was *de facto* recognition of the Chinese seizure of Tibet (see p. 1429).

1954, July–August. India-Pakistan Dispute over Indus Valley Water. The dispute was settled by mediation by the International Reconstruction and Development Bank (August 5).

1954, July 22–1955, August 15. Border Clashes around Goa. Indian nationalists attempted to seize parts of the Portuguese possession, but were ejected by Portuguese troops (July–August 1954). After further violence around Goa (August 1955), India broke off diplomatic relations with Portugal.

1959, February 25. U.S. Arms Aid Rejected. Since Pakistan accepted preferred American assistance, Nehru demanded the withdrawal of U.S. members of the UN Ceasefire Commission in Kashmir.

1959, April 3. Arrival of the Dalai Lama in India. He was seeking refuge from Chinese persecution of Tibetans (see p. 1429).

1959, August 28. Border Dispute with China. Prime Minister Jawaharlal Nehru reported Chinese violations of India's frontiers with Tibet and China in the Longju and Ladakh areas.

1960, June 10. Himalayan Border Clash. India claimed Chinese troops were occupying Indian territory.

1961, March 14. Troops to the Congo. India sent troops to join the U.N. effort in the Congo (see p. 1441). The Indians were airlifted by U.S. cargo planes.

1961, December 18. Seizure of Goa. India seized the Portuguese enclaves of Goa, Damao, and Diu, which had been Portuguese possessions for 4½ centuries. There was little opposition.

Hostilities with China, October–November 1962

1962, October 20. Chinese Invasion. Chinese troops in massive surprise attacks defeated Indian frontier forces in Jammu and in the northeastern frontier region, on fronts 1,000 miles apart. The eastern drive, in particular, was spectacularly successful, and all Indian resistance north of the Brahmaputra Valley was overrun.

1962, November 21. Chinese Unilateral Cease-Fire. Having gained all of the border regions they had claimed, the Chinese suddenly declared a unilateral cease-fire and withdrew to lines which would assure their retention of these regions. Nehru rejected the Chinese terms for settling the dispute, but since the defeated Indians had no desire to renew the war, informal truce prevailed along the Himalayan frontiers at the end of the period.

1962–1965. Military Reform. India, receiving considerable military assistance from the U.S. and from other Commonwealth nations, attempted to revitalize her armed forces and remedy the many defects disclosed in the disastrous war with China.

Hostilities with Pakistan, May–September 1965

1965, April–May. Undeclared War in the Rann of Kutch. A frontier dispute with Pakistan, in a desolate region where the frontier had not been clearly defined, broke into full-scale hostilities for approximately 2 weeks. The Pakistanis seem to have had slightly the better of the struggle before monsoon rains ended operations.

1965, August 5–23. Border Clashes in Kashmir and Punjab. Following Pakistan's initiative in Kashmir, military and irregular infiltrators on both sides crossed the Kashmir cease-fire line and the nearby Punjab border in raids and counterraids.

1965, August 24. Indian Raid. Indian troops crossed the cease-fire line in considerable force; fighting raged along the northern frontier. U.N. truce observers in Kashmir brought about a temporary cease-fire.

1965, September 1–25. Major Hostilities. In retaliation for the Indian raid, Pakistan initiated a major invasion across the cease-fire line in Kashmir (September 1). Both sides undertook minor air raids against nearby Punjab cities, as well as against Karachi and New Delhi. Indian troops launched a major attack against Lahore (September 6). The attacks on both sides soon bogged down. In large-scale armored battles, Indian units achieved marginal success over Pakistani tanks. On balance, however, a stalemate resulted.

1965, September 7–8. U.S., U.K., and Australia Halt Arms Shipments to India and Pakistan. The British and Australian announcements followed that of the U.S. by one day.

1965, September 8. Communist China Threatens India. In the face of quiet but determined American and British diplomacy, China failed to carry out threatened actions against Indian border positions in the Himalayas.

1965, September 27. U.N. Cease-Fire Demands Honored. After accepting, then ignoring, an earlier U.N. demand (September 22), both sides agreed to abide by a Security Council cease-fire and began withdrawal to lines held on August 5. A new U.N. India-Pakistan Observation Mission (independent of the U.N. Observer Group in Kashmir) was established (September 25), and 75 observers from 8 nations came to the Punjab to supervise the cease-fire. Cease-fire violations by both sides were reported in the following weeks, but there were no further major hostilities.

1966, January 10. Declaration of Tashkent. India and Pakistan agreed to withdraw their troops from frontier confrontation positions. This pullback was completed on February 25.

1967, September 11–14. Clashes with Chinese on Sikkim-Tibet Frontier. Both India and China accused the other of border violations. However, the clashes were limited to exchanges of rifle and artillery fire, and no substantial movement was made into either Tibet or the Indian protectorate of Sikkim.

1971, August 9. Treaty of Friendship with the U.S.S.R. This 20-year pact greatly strengthened India's hand in the increasingly tense relations with Pakistan resulting from unrest in East Pakistan.

1971, December 3–17. War with Pakistan. (See p. 1407.)

1972, March 12. Last Indian Troops Withdraw from Bangladesh.

1972, March 19. Treaty of Friendship with Bangladesh. The provisions were very similar to those in the 1971 India-U.S.S.R. friendship treaty.

1972, July 3. Peace Treaty with Pakistan. (See p. 1408.)

1973, April 8. India Assumes Administrative Control of Sikkim. This was in response to a plea from Chogyal (Prince) **Palden Thondup Namgyal** of the protectorate, following two weeks of anti-government violence in Sikkim. This arrangement was ratified in an agreement between India and Sikkim one month later.

PAKISTAN

The military history of Pakistan during this period has been essentially that of her continuing friction with India. There have also been sporadic border disputes with Afghanistan. Pakistan is a member of the Central Treaty Organization (see p. 1392) and Southeast Asia Treaty Organization (see p. 1410).

1947, August 14. Independence. Viscount Mountbatten turned over the government of Pakistan to **Mohammed Ali Jinnah** in a ceremony in Karachi, as rioting, violence, and death spread through the Punjab and elsewhere along the border regions of India and Pakistan (see p. 1404).

1948, January 8. Unrest and Rioting in Karachi. A result of popular dissatisfaction with the government, the economic and political unrest generally increased during the following 10 years.

1954, May 19. U.S. Military Aid. Agreement between the U.S. and Pakistan for America to provide military supplies and technical assistance. Pakistan agreed to use the aid only for defense and participation in U.N. collective-security arrangements; India, however, proclaimed bitterly that this was giving Pakistan assistance for possible war with India.

1955, September 19. Pakistan Joins Baghdad Pact. (See p. 0000.)

1956, March 23. Pakistan Becomes a Republic. It remained in the British Commonwealth.

1958, October 7–27. Bloodless Coup d'État. Acting through President **Iskander Mirza,** General **Ayub Khan** dismissed the government, annulled the constitution, and established a "benign martial law." He immediately began sweeping economic and political reforms, reestablishing stability. Mirza soon resigned (October 27). Ayub Khan was elected president under a new constitution (February 17, 1960).

1959, February 25. U.S. Arms Aid Accepted. President Eisenhower reported that Pakistan would receive arms aid from the U.S. to strengthen the defensive capabilities of the Middle East. India rejected a comparable offer (see p. 1405).

1965, May–September. Hostilities with India. (See p. 1405.)

1969, March 25. Ayub Khan Resigns. As a result of increasing violence and unrest throughout the country, he turned over the government to General Agha **Mohommed Yahya Khan,** commander in chief of the Army.

1970, December 7. First General Election. This was to elect a National Assembly, which early in 1971 was to formulate a new constitution for the country. An absolute majority of Assembly seats was won by the Awami League of East Pakistan, headed by Sheikh **Mujibur Rahman.** Sheikh Mujibur and his League had long struggled to gain autonomy for East Pakistan. This was totally unacceptable to the dominant, but less numerous, people and administrators of West Pakistan.

1971, March 24. President Yahya Khan Proclaims Martial Law. In response to violence in East Pakistan after the President postponed the first scheduled session of the National Assembly. Sheikh Mujibur undertook a nonviolent civil disobedience campaign, giving him virtual control of the eastern province. President Yahya Khan flew to Dacca to negotiate with Sheikh Mujibur, but left abruptly after 11 days (March 24). He called the Sheikh a traitor, and outlawed the Awami League.

1971, March 25. Outbreak of Civil War. Initiation of suppressive action by Pakistani forces (almost all from West Pakistan) led Sheikh Mujibur to proclaim the independence of East Pakistan as Bangladesh (March 26). Mujibur was seized early in the struggle, and the well-equipped Pakistani Army brutally suppressed the revolt in about 5 weeks (by May 5). The number of East Pakistani casualties is unknown, but probably exceeded 100,000. During the fighting about 3 million refugees fled from East Pakistan into India, complicating the already serious food problem of that country. During the next month another 3 million Bengali refugees from East Pakistan fled into India. A few of the refugees attempted to harass the East Pakistan frontiers, with little success, until they began to receive covert Indian support (fall, 1971). The result was increasing frontier incidents and intensified Pakistani suppression of unrest in East Pakistan.

1971, June–November. Growing Tension with India. There were frequent instances of artillery fire across the border, and small raids by both sides. India gave considerable assistance to East Pakistan rebels based in eastern India.

1971, November 8. Cancellation of U.S. Arms Shipments to Pakistan. Although described as an action of "mutual consent," this was really the result of increasing American unhappiness about the use of American weapons to suppress the Bengalis of East Pakistan.

HOSTILITIES WITH INDIA, DECEMBER 3–16, 1971

1971, December 3. Outbreak of War. The Indian tactic of increasing aid to the Bangladesh rebels, while avoiding an overt Indian-Pakistani confrontation, finally accomplished the result India had obviously been seeking: goading Pakistan into taking the first hostile action. This was a massive air strike by the

Pakistani Air Force against most major Indian air bases, in hopes of achieving the kind of result accomplished by Israel at the outbreak of the 1967 Middle East War (see p. 1343). The Indians, aware that their policy of goading Pakistan to war demanded complete alertness, were ready, and the Pakistani air strikes were generally unsuccessful.

1971, December 3. Indian Invasion of East Pakistan. Fully prepared for the outbreak of hostilities, Indian forces at least triple the strength of the 90,000-man Pakistani garrison of East Pakistan began a major 2-pronged invasion of the province, from the north and the west, with 3 subsidiary attacks from north, west, and east.

1971, December 4. Pakistani Invasion of Indian Kashmir. The Pakistanis made minor gains, some as much as 10 miles, but their advance was soon halted by the alert Indians, who had expected this move.

1971, December 5–6. Soviets Veto U.N. Security Council Cease-Fire Resolutions. In support of their Indian allies, who were advancing rapidly into East Pakistan, the Soviets refused to agree to a cease-fire, which would have halted the advance. Pakistan Foreign Minister **Zulfikar Ali Bhutto** at once took the issue to the slow-moving General Assembly.

1971, December 6. India Recognizes Bangladesh.

1971, December 14–16. Battle for Dacca. This began when Indian ground troops, only 7 miles from the capital, advanced under cover of artillery fire and air attacks against the defenders in the city. The East Pakistani government resigned and took refuge in a neutral zone in the city, established by the Red Cross.

1971, December 15. U.N. General Assembly Demands Cease-Fire in East Pakistan. By this time most Pakistani resistance had collapsed in East Pakistan. Having piously called for a cease-fire, and having no enforcement authority, the General Assembly prepared to turn to other business. Foreign Minister Bhutto, denouncing the United Nations' "miserable, shameful" failure to take positive action, walked out of the building and flew home.

1971, December 15. Indian Attacks in the West. In intensified fighting the Indians recovered some captured territory on the Indian-Pakistan border in Kashmir and the Punjab and advanced into Pakistan at some points in Hyderabad and the Punjab.

1971, December 16. Pakistani Surrender in Dacca. Lieutenant General **A. A. K. Niazi** surrendered to General **S. H. F. J. Manekshaw,** Indian Army Chief of Staff. This virtually ended the war.

1971, December 17. Cease-Fire Accepted by Both Sides. Indian losses were approximately 2,400 killed, 6,200 wounded, and 2,100 captured; India lost 73 tanks and 45 aircraft. Pakistani losses were more than 4,000 dead and 10,000 wounded; most of the wounded were included among 93,000 prisoners of war.

1971, December 20. Yahya Khan Resigns; Bhutto Becomes President. As his first official act Bhutto dismissed the senior military leaders and put them and the former president under arrest.

1972, July 3. Interim Peace Settlement. President Bhutto and Prime Minister **Indira Gandhi** of India, after a 5-day meeting at Simla, India, agreed to a troop withdrawal along most of their joint frontier, but postponed action on the more difficult problems of the Kashmir frontier and the return of 93,000 Pakistani prisoners of war held by India.

1972, December 7. Agreement on Kashmir Truce Line. Both sides began to withdraw their troops.

1973, August 8. India Agrees to Release Pakistani Prisoners of War. Return was completed in April 1974.

AFGHANISTAN

1949–1965. Strained Relations with Pakistan. Pakistani refusal to honor Afghan claims for frontier revision resulted in tension and occasional border clashes.

1956, August–October. Soviet Military Aid. The U.S.S.R. provided guns, ammunition, and airplanes.

1973, July 17. Coup d'État. During the absence of King **Mohammed Zahir Shah,** his brother-in-law and Army commander, Lieutenant General **Mohammed Daoud Khan,** seized power and declared the country a republic with himself as president and premier.

(CEYLON) SRI LANKA

1948, February 4. Independence of Ceylon. It became an independent dominion within the British Commonwealth.

1953, August 12–19. Communist Terrorism. The government suppressed the disorders after receiving emergency powers from Parliament.

1956–1961, Sporadic "Language Riots." These stemmed from the desire of the minority Tamils to have their language accepted as an alternative to Sinhalese. These bloody riots intensified in 1958 (February–July) but were ended when limited official status was granted Tamil (August 5 1961).

1962, January 29. Attempted Coup.

1971, April 5–June 9. Rebellion of the People's Liberation Front. The PLF, impatient with the slow social progress of the leftist government of Prime Minister Mrs. **Sirimavo Bandaranaike,** attempted to seize control of Colombo and other cities. A plot to assassinate the Prime Minister failed. The revolt was soon suppressed in the cities (April 13), but fierce fighting continued in rural and jungle areas. The U.S.S.R. provided Ceylon with fighter aircraft and pilot training personnel, to help the government reestablish control. Both India and Pakistan provided helicopters and crews. Britain shipped weapons and ammunition. The rebellion was officially declared to be suppressed, and the nation's schools were reopened, 2 months after the outbreak of civil war (June 9).

1972, May 22. Ceylon Becomes Republic of Sri Lanka.

BANGLADESH

1971, March 26. Independence of Bangladesh Proclaimed. Following disputes with West Pakistan, and resentful of West Pakistan's domination, Sheikh Mujibur Rahman, political leader of East Pakistan, declared the independence of the province as Bangladesh. This resulted in prompt action by the government of Pakistan to reestablish firm control (see p. 1407).

1971, March–December. War for Independence. This was largely unsuccessful until intervention by India (see p. 1407).

1971, December 17. Independence Effective. This was the result of India's victory over Pakistan (see p. 1408).

NEPAL

1950, November 11–20. Insurrection. King **Tribhubarra Bir Bikram,** deposed by the government of Premier **Mohan Shumshere** (November 7), was supported by an uprising of the reform-minded Congress party. Loyal Gurkha troops defeated the rebels. The king was later invited back to resume his reign.

1952, January 24. Unsuccessful Revolt. The Communist party was outlawed.

1952, August 13. King Tribhubarra Takes Control. This was in an effort to end the unrest which had been plaguing the country. He restored parliamentary rule in 1953 (April 13).

1960, June 28. Border Incident with China. China apologized for "carelessness" in attacking Nepalese troops, but said they were on Tibetan territory.

1960, December 15. Royal Coup. King **Mahendra Bir Bikram** seized power, with army support, ousting the regime of Premier **B. P. Kirala.**

1961, March–December. Civil War. The revolt was suppressed.

SOUTHEAST ASIA

The area became the eastern battleground for the warring ideologies—communism vs. the Free World. The Communists, checked at least temporarily by the armistice in Korea, shifted their efforts to support existing, indigenous struggles already under way in Indochina, Malaya, and Indonesia. After the collapse of French colonial rule in Indochina (1954; see below), the U.S. took the lead in sponsoring an anti-Communist regional organization to prevent further Communist gains in the area.

SOUTHEAST ASIA TREATY ORGANIZATION (SEATO)

This treaty was established as part of the American effort to create a group of mutual-security pacts around the world after the 1954 Geneva Conference (see p. 1414). The 8 members were Australia, France, New Zealand, Pakistan, the Philippines, Thailand, the United Kingdom, and the United States. The treaty was set up for the purpose of providing for collective defense and economic cooperation in Southeast Asia, and to protect the weak nations of the region against aggression. Theoretically patterned after NATO, SEATO has been relatively helpless and ineffective, due to 3 major factors: lack of widespread support among Southeast Asian nations fearful of angering Communist China; skillful Communist subversion, diplomacy, and "agit-prop"; and French foot-dragging.

1954, September 8. Manila Treaty. The defense treaty for Southeast Asia was signed in Manila by representatives of the participating governments. This followed diplomatic initiatives by the ANZUS nations (Australia, New Zealand, and the U.S., beginning June 30).

1955, February 19. Southeast Asia Defense Treaty into Effect.

1964, April 15. SEATO Supports South Vietnam. The Ministerial Council of SEATO, meeting in Manila, issued a declaration of support of South Vietnam military efforts against the Viet Cong guerrillas. France abstained.

BURMA

Independence sentiment among the people, combined with results of Burmese independence activities during the war, directed against both the British and the Japanese (see pp. 1240, 1297), led to British agreement to grant independence to Burma.

1945–1946. Guerrilla Warfare. British troops were forced to wage a sporadic guerrilla warfare against armed dissidents, most of whom were bandits, throughout Burma.

1947, January 28. Britain Announces Plans for Burma's Independence.

1947, July 19. Assassination of General Aung San. The premier of Burma, the nation's war hero, and 5 members of his cabinet were assassinated by intruders during a cabinet meeting in Rangoon. The assassins were apprehended, tried, and executed (December 30).

1948, January 4. Independence of the Union of Burma. Burma, under Prime Minister U Nu, refused to join the British Commonwealth.

1948, March. Outbreak of Communist Revolt. This began in south-central Burma, mainly in the Irrawaddy Delta.

1948, August. Outbreak of Karen Revolt. The objective was to achieve an autonomous Karen state. At first successful, the Karens, in somewhat reluctant cooperation with the Communists, gained control of much of south-central Burma. They proclaimed their independence (June 14, 1949), with capital at Toungoo.

1949, January–February. Karen Rebels at Outskirts of Rangoon. They cut the Rangoon-Mandalay railroad and were within artillery range of parts of the area within Rangoon city limits.

1949–1950. Government Counteroffensive.

1950, March 19. Government Forces Capture Toungoo. The Karen revolt began to collapse. The Burmese government reestablished control over most of central Burma.

1950, May 19. Government Forces Capture Prome. This was the main Communist center of south-central Burma.

1950–1974. Continuous Guerrilla Warfare in Burma. After barely surviving collaboration between nationalistic and communistic rebels (1948–1949), loyal Burmese forces under General **Ne Win,** in methodical guerrilla warfare, reestablished law and order in most parts of the country (1954). Endemic rebellion and guerrilla warfare continued throughout many of the outlying provinces, however.

1953, April 23. U.N. Calls for Withdrawal of Chinese Nationalists. These were Chinese Nationalist refugee troops who were defying Burmese government authority in the northeastern portions of Burma, where they had withdrawn after the defeat of the National Government in China. Burma had complained to the U.N. (1950). The refugees refused to withdraw and the National Government of China refused to recognize them. However, after considerable U.S. pressure, some 2,000 of these Chinese Nationalist guerrillas were evacuated from Burma to Formosa (November). Early next year, 6,400 guerrillas and dependents were evacuated to Formosa (May). It was estimated, however, that at least 6,000 remained in the jungle region.

1956, July 31. Border Dispute with Communist China. Chinese troops seized 1,000 square miles of territory in northeast Burma.

1958, September 26. Military Coup. Deterioration of government control and a threatened Communist coup led General Ne Win to seize control of the government. He restored civil rule after national elections (February 6, 1960), after signing a nonaggression treaty with Communist China (January 28).

1962, March 2. Second Military Coup. Deterioration of civilian government again led Ne Win to establish a military dictatorship.

THAILAND

Despite considerable political turbulence, leading to several coups d'état and the murder of a king, Thailand as a nation has remained relatively stable, more so than any other in the region. It has been steadfastly anti-Communist, and is a member of SEATO (see p. 1410). As the period ended, Thailand was giving substantial assistance, including base rights, to the U.S. effort in Vietnam.

1946, May 26–30. Franco-Thai Frontier Dispute. After clashes along the Mekong River, Thailand appealed to the U.N. Security Council to halt French aggression, but France insisted that the so-called military activity was simply pursuit by Chinese troops of bandits from the Siamese side of the river that had been raiding east of the river.

1946, June 9. King Ananda Dies Under Suspicious Circumstances. He was succeeded by his brother, **Phumiphon.**

1947, November 9. Military Coup d'État.

Field Marshal **Luang Pibul Songgram** seized control of the government in Bangkok. Prime Minister **Pridi Phanomyong** fled the country.

1949, February 26–27. Insurrection. During a state of emergency caused by Communist activity along the Malayan border, units of the Thai Army and Navy accused each other of plotting a coup against the government of Premier **Pibul Songgram,** and extensive fighting took place in and around Bangkok.

1951, June 29–July 1. Naval Revolt Suppressed.

1951, November 29. Military Uprising Suppressed. In the confusion, however, a political coup forced Pibul to amend the constitution.

1957, September 17. Bloodless Coup d'État. Field Marshal **Sarit Thanarat** seized control; Field Marshal Pibul fled to Cambodia. Sarit later retired, establishing a caretaker government (April 1958).

1958, October 20. Sarit Seizes Control Again. He retained power and became prime minister (January 1959).

1958, November–December. Border Clashes with Cambodia.

1964–1970. Repeated Border Incidents with Cambodia. Thailand wanted to recover eastern frontier territory long disputed with France (see above, 1946, May 26–30).

1964–1974. Sporadic Communist Terrorism. This violence was confined to the extreme northern and southern areas of the country.

1967, March 22. U.S. B-52 Bomber Bases Permitted in Thailand. This enabled the U.S. to shift some heavy bombers from Guam to Thai bases much closer to targets in Vietnam.

1967, September. Thai Troops Committed to Combat in Vietnam. Their participation was financed by the U.S., under the terms of an executive agreement between Thailand and the U.S. (1965).

1967, December 1. Martial Law Declared in Five Provinces. This was the result of increased Communist guerrilla activity, and brought to 12 the number of provinces under martial law in the extreme north and south of the country.

1969, July 7. Secret Military Agreement with U.S. Revealed. U.S. Senator **J. W. Fullbright** revealed the agreement, which permitted stationing 47,000 U.S. military men in Thailand.

1970, February 2. Withdrawal of 4,200 American Military Personnel Announced. This left 43,800 American troops in Thailand.

1971, November 17. Government Coup. Prime Minister, General **Thanom Kittikachorn,** ended constitutional rule, seized full power, and declared martial law.

1972, February 4. Thai Troops Withdrawn from Vietnam.

1973, October 14. Resignation of Prime Minister Kittikachorn. This was the culmination of increasing unrest and popular dissatisfaction with the military dictatorship, which had erupted a few days earlier in bloody violence. King **Phumiphol Adulet** appointed **Sanya Dhamasakti,** Dean of Thammasat University (center of student protests against Kittikachorn) as Prime Minister.

INDOCHINA, 1945–1954

By the close of World War II, the guerrilla forces of Vietnamese nationalists and Communists, combined in an organization known as the Viet Minh, under the overall political leadership of Communist **Ho Chi Minh** and the military leadership of initially nationalist guerrilla leader Vo Nguyen Giap, had gained control of much of the jungle region of north Vietnam. This success had been achieved with the largely unwitting assistance of the Chinese National Government and of the U.S., both happy to receive assistance from the Vietnamese against Japan, and both willing to see France eliminated from Indochina, but neither fully realizing the international Communist ties of Ho. The Viet Minh declared their independence when Japan collapsed at the close of the war but the French, through their own efforts, and with some British assistance, moved immediately to reestablish their colonial rule over the area. The resultant conflict touched off the most prolonged warfare of the entire period since World War II.

1945, September 2. Vietnam Republic Proclaimed by Ho Chi Minh.

1946, March 6. France Recognizes Independence of Vietnam Republic. This was only as a free state within the Indochinese Federation and the French Union. Meanwhile, French military strength built up rapidly. French-imposed limitations on independence proving unacceptable to the Viet Minh, guerrilla warfare broke out, mostly in northern Vietnam, later in the year (December).

1947, January–February. Siege of Hué. After a siege of several weeks, French troops relieved the besieged garrison of Hué, driving off the Viet Minh guerrillas surrounding the ancient capital of the country.

1950, January. Viet Minh Recognized by Communist China and the U.S.S.R. Increasing military assistance was given to the Viet Minh guerrillas by China. Viet Minh troops received intensive training in southern China. American military aid to French Vietnam increased with the intensity of guerrilla warfare.

1950, October. French Setbacks. Well-trained, well-equipped Viet Minh troops, operating partly from China and partly from the jungled highlands of northern Vietnam, mounted a major assault against the French cordon of defenses in northern Tonkin, covering the Chinese border. French troops were badly defeated at **Fort Caobang,** near Langson (October 9). This, combined with increased activity by the Communist Pathet Lao insurgents in Laos (see p. 1417), forced the French to abandon most of northern Vietnam (October 21) and to establish a fortified perimeter around the Red River Delta in the north (December). The situation in southern and central Vietnam was not much better, with much of the Mekong Delta in Communist hands.

1950, December. De Lattre de Tassigny to Command. France sent her leading soldier to try to restore the situation. He soon reestab-lished French morale, regained the initiative, and reoccupied most of the areas lost in late 1950.

1950, December 23. Vietnam Sovereign within French Union. A treaty was signed at Saigon.

1951–1953. Continued Guerrilla Warfare. This was combined with anti-French terrorism in the major cities. Despite De Lattre's military successes, French control could be asserted only where major French forces were stationed.

1952, September. De Lattre Relieved. Seriously ill, he returned to France via the U.S., where he pleaded for more aid. He died a few months later.

1953, January–February. Intensified French Operations. In the biggest naval operation of the war, French troops (now under General **Raoul Salan**) seized **Quinhon,** a rebel base, and destroyed several Viet Minh war factories concealed in the jungles of south Vietnam.

1953, March–September. Increased U.S. Aid to France for the Indochina War.

1953, May 8. Navarre Relieves Salan. Pedestrian General Salan was relieved by pedestrian General **Henri-Eugène Navarre.**

1953, July 6. Increased Independence for Vietnam and Laos. They accepted a French offer to negotiate for greater self-government in the Associated States of Indochina. Cambodia refused.

1953, August–October. Negotiations between France and Cambodia. France gave the government of King **Norodom Sihanouk** almost complete military, political, and economic sovereignty, although France retained operational control of some military forces in eastern Cambodia for purposes of prosecuting the war against the Viet Minh.

1953, October–1954, April. Intensified Viet Minh Operations. French premier **Joseph Laniel** said his government would accept "any honorable" solution to the war in Indochina, and was not trying to force the Viet Minh to unconditional surrender (November 12).

1953, November 20–1954, May 7. Siege of Dienbienphu. General Navarre, hoping to decoy the Communists into 1 large pocket and then crush them, permitted Brigadier General **Christian de la Croix de Castries,** with some 15,000 men—French regulars, Foreign Legion, and indigenous troops—to fortify and hold the village and an airstrip, situated 220 miles west of Hanoi and near the Laotian border. General Giap, with 4 divisions of Chinese-trained Viet Minh troops, surrounded Dienbienphu with 2 divisions while the remainder of his force sealed it off and swept into Laos. Against the French artillery—24 105-mm. And 4 155-mm. howitzers—Giap assembled the overwhelming firepower of over 200 guns, including antiaircraft artillery and rocket launchers. A trickle of supply by air from Hanoi, little enough when the defenders still held the air field, ceased with its capture (March 27). Attempts at air drop failed; the Viet Minh antiaircraft artillery was too good. Of 420 French aircraft available for this purpose, 62 were shot down and 107 others damaged. One by one the outlying strong points of the Dienbienphu defense complex fell to a combination of mining, well-directed artillery fire, and direct assault. A final assault overran the starving defenders as their last ammunition was expended (May 7). Only 73 of the 15,094-man garrison escaped. Some 10,000—half of them wounded—were captured; the remainder were dead. Viet Minh losses were estimated at 25,000.

COMMENT. *The fall of Dienbienphu virtually ended French control over Indochina. At the same time, it proved the fallacy of cordon defense in jungle warfare, particularly when the opponents are well trained, armed, and supplied. French military thought in this instance was still clinging to methods used against guerrillas in North Africa and, in 1882–1885, in this very area. (See p. 943, the French garrison of Tuyen-Quang successfully resisted besiegement by ''Black Flag'' indigenous guerrillas from November 23, 1884, to February 28, 1885. But the ''Black Flags,'' while fanatically brave, had neither discipline nor resources,*

and were unable to hold up the French relief column advancing from Hanoi, only 50 miles away.)

1954, April 26–July 21. Geneva Conference. The Conference on Far Eastern Affairs of 19 nations (including Communist China) resulted in an agreement for a cease-fire and divided Tonkin and Annam into North (Communist) and South (anti-Communist) Vietnam as independent nations divided at the 17th parallel of North Latitude. Cambodia, which had proclaimed its independence of France (November 9, 1953), and Laos, its independence proclaimed (July 19, 1949), were both recognized as neutral independent states. The United States accepted the agreements, but refused to sign them, and reserved the right to take whatever action was necessary in the event that the agreements were breached. France withdrew her troops from Indochina, but continued military direction and instruction in South Vietnam, Laos, and Cambodia, while the U.S. assumed the chore of providing military equipment and instruction as well as economic aid.

1954, December 29. Independence of Indochina. Vietnam, Laos, and Cambodia signed agreements with France, giving them economic independence and virtually ending foreign control. The states granted each other freedom of navigation on the Mekong River.

VIETNAM

1954, July 7. Ngo Dinh Diem Appointed Premier.

1954, October 11. Communist Viet Minh Takes Control of North Vietnam.

1955, January 20. U.S. Military Aid. The U.S., France, and South Vietnam agreed to reorganize the Vietnamese Army with 100,000 active troops and 150,000 reserves. The U.S. was to send a training mission to operate under the direction of General **Paul Ely,** new French commander in Indochina.

1955, October 26. Republic Proclaimed. Diem was inaugurated president.

1956–1964. Continuous Insurrection. This was sponsored by the Communist bloc, despite efforts of the government to control rebellious factions (Viet Cong) supported by troops and equipment from North Vietnam in turn aided and supported by Communist China. U.S. efforts to strengthen the South Vietnamese military force consistently increased, without retrieving the situation.

1956, April 28. U.S. Military Assistance Advisory Group (MAAG) Assumes Responsibility for Training.

1960, November 11. Military Revolt against Diem. This was suppressed.

1961, October 11. U.S. Assistance Pledged. The U.S. agreed to support the government of South Vietnam against attacks by Communist Viet Cong guerrillas. General **Maxwell D. Taylor** was sent to Vietnam by President **John F. Kennedy** to determine the most effective means of help. President Kennedy sent a personal message to President Diem with a pledge to continue assistance (October 26).

1961, December 11. First U.S. Support Units Arrive. Two U.S. Army helicopter companies, the first direct military support for South Vietnam, arrived in Saigon aboard a U.S. aircraft carrier.

1962, February 8. U.S. Military Assistance Command Established. The purpose was to demonstrate U.S. determination to prevent a Communist takeover.

1962, March 22. Operation "Sunrise" Begins. This was designed to eliminate the Viet Cong. Operations began in Binh Duong Province.

1963, November 1–2. Military Coup d'État. The government of President Diem was overthrown; he and his brother were killed. A provisional government was established under former Vice-President **Nguyen Ngoc Tho,** and was recognized by the U.S. Actual control was under a military junta led by Major General **Duong Van Minh.**

1964, January 30. Military Coup d'État. The government was overthrown by Major General **Nguyen Khanh.**

1964, February 4–6. Viet Cong Launches Offensive in Tay Ninh Province and Mekong Delta.

1964, August 2–4. Action in the Gulf of Tonkin. Three North Vietnamese PT boats attacked a U.S. destroyer. The PT boats were repelled and damaged or sunk by the destroyer and U.S. planes. A similar incident occurred 2 days later (August 4).

1964, August 5. U.S. Air Strikes against North Vietnam. American carrier-based strikes against naval bases were ordered by President Johnson in retaliation for the PT-boat attacks.

1964, August 7. Gulf of Tonkin Resolution. Congress approved a resolution giving President **Lyndon B. Johnson** authority to take "all necessary measures to repel any armed attack" against U.S. armed forces. It also authorized him to take "all necessary steps, including the use of armed forces," to help any nation requesting aid "in defense of its freedom" under the Southeast Asia Collective Defense Treaty.

1964, August–September. Political Turmoil. The government of General Khanh survived riots and demonstrations, but only by promising to give early control to a civilian government and suppressing an attempted military revolt (September 13).

1964, November 1. Communist Guerrilla Attack on U.S. Support Base at Bien Hoa. Four Americans and 2 Vietnamese were killed; 12 Americans and 5 Vietnamese wounded. Several American and Vietnamese aircraft and helicopters were destroyed or damaged.

1964, November 4. Civilian Regime Installed. General Khanh resigned as **Tran Van Huong** became premier.

1964, December 19. Military Uprising. A Military Council retained Premier Huong in nominal control.

1965, January 27. Khanh Returns to Power. The Armed Forces Council deposed Premier Huong and returned General Nguyen Khanh to the head of the government.

1965–1973. The United States War in Vietnam. (See p. 1321.)

1965, February 21. Khanh Deposed. After a complicated series of moves, in which civilian **Phan Huy Quat** was installed as premier, with Khanh retaining behind-the-scenes control, the Armed Forces Council voted to oust Khanh as council chairman and armed forces commander. Quat remained premier.

1965, June 12–19. Bloodless Government Upheaval. Premier Quat resigned as a result of religious turmoil involving Buddhists and Catholics. The military took over and elected Air Vice-Marshal **Nguyen Cao Ky** (age 36) as premier. This was the eighth government since the overthrow of Diem (November 1963).

1967, September 3. Nguyen Van Theiu Elected President. He was inaugurated the first president of South Vietnam's Second Republic (October 31).

1973, January 23. Cease-Fire Agreement. (See p. 1333.)

1973, March 29. Departure of Last American Troops from Vietnam. (See p. 1333.)

1973, April 7. International Peace-Keeping Helicopter Shot Down by Communists. Attempting to investigate one of the increasingly numerous and serious cease-fire violations throughout South Vietnam, an international peacekeeping force team was shot down by a Communist missile in northern Quang Tri Province. All 9 aboard were killed.

1973, June 13. New Cease-Fire Agreement. Representatives of the U.S., North Vietnam, South Vietnam, and the Viet Cong signed a 14-point agreement calling for an end to all cease-fire violations. Among its provisions: U.S. reconnaissance flights over North Vietnam would end; U.S. minesweeping operations would be resumed in North Vietnamese waters; commanders of opposing troops in contact would meet to prevent further outbreaks of hostilities and to assure adequate medical supplies and care. For a while there was some reduction in the intensity of fighting, but the local ground commanders failed to meet, and fighting continued.

1973–1975. Widespread Combat. Fighting continued in many areas despite the cease-fire.

1975, January 1–7. Communists Capture Phuoc Binh. The capital of Phuoc Long fell after a 7-day siege.

1975, January 17. ARVN Counteroffensive in the Mekong Delta. About 2,000 troops attacked along the Cambodian border.

1975, February 24–March 2. Visit of U.S. Congressional Delegation. Eighty-one members went to Indochina to review the situation as a basis for action on request for additional aid.

1975, March 5. Communists Launch Offensive in the Central Highlands. Major fighting by strong North Vietnamese forces resulted in the capture of many towns, cutting of highways, and isolating of garrisons in the Communist advance, and the fall of **Ban Me Thuot** after a fierce battle (March 5–13). In Quang Tri and Thua Thien provinces North Vietnamese forces also made important gains.

1975, March 18–20. ARVN Collapse in the North and West. Government troops withdrew in the face of a growing North Vietnamese offensive. President **Nguyen Van Thieu** announced (March 20) the government's intention to evacuate 2 provinces in the northwest and 9 in the central highlands.

1975, March 25. Hue Falls. The former imperial capital was abandoned, giving the Communists control of Thua Thien Province.

1975, April 1. Danang Falls. Despite plans to hold the city, second-largest in South Vietnam, ARVN troops offered little resistance. The third-largest city, Qui Nhon, was similarly abandoned the following day (April 2).

1975, April 9–22. Battle of Kuon Loc. After fierce fighting in which both sides took heavy casualties and the town changed hands several itmes, it was finally abandoned by the ARVN.

1975, April 21. Thieu Resigns as President. Vice President Tran Van Huong was appointed to replace him.

1975, April 27. Huong Resigns as President. Unacceptable to the Communists for negotiations, he resigned in favor of Lieutenant General **Duong Van Minh.**

1975, April 30. South Vietnam Surrenders to the Communists.

LAOS

Its independence proclaimed (July 19, 1949), and recognized as a neutral nation by the Geneva Conference (see p. 1414), Laos nevertheless became the center of a maelstrom of Communist-inspired outbreaks by the indigenous Pathet Lao, supported by both the U.S.S.R. and Communist China. U.S. support of the Royal Laotian Army (July 9, 1955) was temporarily suspended (October 1960) as the country seethed in a 3-cornered conflict—rightist forces under General **Phoumi Nosavan,** neutralist troops under Premier Prince **Souvanna Phouma,** and the Communist Pathet Lao under Souvanna's half-brother, Prince **Souphanouvong.**

The Plaine des Jarres area in north-central Laos was the arena for most of the fighting, an endless series of inconclusive clashes. Stepped-up assaults by the Pathet Lao brought a concentration of 5,000 U.S. troops into Thailand (May 19, 1962) to protect that nation's border. This force was withdrawn (July 30), its mission accomplished. U.S. military advisers to the Laotian Army were withdrawn by October 7. During 1962, Laotian territory became a convenient communications channel for North Vietnamese troops infiltrating South Vietnam in support of the Viet Cong. On May 17, 1964, the U.S. instituted a continuous aerial reconnaissance sweep of Laos by jet planes. The principal events were:

1953, April 14. Viet Minh Invades Laos. They seized a base abandoned by the French at Samneua. Joining with rebel Laotians, the Viet Minh advanced toward the capital of Luang Prabang, capturing **Xiengkhouang** (April 20). Laos mobilized military forces, and the U.S. rushed military aid. Vietnamese forces began to retreat (early May). French forces retook Xiengkhouang.

1953, October 22. Independence of Laos. France and Laos signed a treaty giving the state full independence and sovereignty within the French Union.

1953–1954. Anti-French Insurgency. This was accompanied by complicated maneuvering by Communist and anti-Communist factions.

1954, July 21. Geneva Accord. (See p. 1414.)

1959, July 30–31. Communist-Led Guerrillas Attack Laotian Army Posts. Communist-led Pathet Lao guerrillas, armed by North Vietnam, attacked Laotian Army posts throughout northern Laos.

1959, September 7. U.N. Investigation. The Security Council voted to inquire into the Laos government's charges of aggression by North Vietnam.

1960, August 9. Coup d'État. The government of Premier **Tiao Samsonith** was overthrown by a military rebellion led by Captain **Kong Le,** a parachute battalion commander. Under Kong Le's sponsorship, neutralist leader Prince **Souvanna Phouma** became premier. Kong Le, soon to become a general, was to play an important, ambiguous, neutralist, but anti-Communist role in subsequent years of the complex civil war.

1961, March 23. U.S. Warnings. President Kennedy announced that the U.S. would not stand idly by and permit Laos to be taken over by advancing externally supported pro-Communist rebel forces. A previous warning had been made without effect by the Eisenhower administration (December 31).

1961, April 3. Cease-Fire between Government and the Pathet Lao.

1962, May 12. U.S. Troop Deployments. As a result of Pathet Lao violation of the 1961 cease-fire agreement and the overrunning of most of northern Laos, President Kennedy ordered a task force of the U.S. Seventh Fleet to move toward the Indochina peninsula. He then ordered 4,000 more U.S. troops to Thailand (where some 1,000 U.S. troops were already stationed, May 15).

1962, July 23. Geneva Agreement on Laos. Fourteen nations guaranteed the neutrality and independence of Laos.

1962, October 5. Withdrawal of U.S. Military Advisers. This was in compliance with the Geneva Agreement. There were approximately 800 U.S. advisers and technicians withdrawn.

1963, April. Renewed Conflict. Major fighting between neutral and Pathet Lao forces stopped after 3 weeks by a cease-fire agreement. Small-scale warfare continued.

1964, May 16–24. Communist Pathet Lao Forces Seize the Plain of Jars. Kong Le's forces were defeated.

1964–1973. Constant Warfare in Laos. The Pathet Lao, supported by strong North Vietnamese forces, ranging in size from 10,000 to 40,000 troops, held the eastern, southern, and northern portions of the country and posed intermittent threats to the government of Prime Minister Souvanna Phouma in Vientiane.

1964, April 19. Coup d'État. Prime Minister Souvanna Phouma was ousted by a right-wing military committee, led by Brigadier General **Kouprasith Abhay,** which seized control of Vientiane. However, after a typical Laotian political ballet, Souvanna Phouma was restored to power, after agreeing to accept the rightist demands for modification of government policy in a merger of rightist and neutralist factions (May 2). This development was denounced by the Pathet Lao.

1965, January 31–February 4. Unsuccessful Army Revolt. Loyal troops cleared rebels from Vientiane, after much fighting. The leader, General **Phoumi Nosavan,** Deputy Prime Minister, fled to Thailand.

1965, March 28–30. Unsuccessful Army Revolt. This was suppressed without bloodshed.

1965, April 16–30. Army Mutiny. Supporters of exiled Deputy Prime Minister Phoumi

Nosavan again attempted a revolt, but were crushed.

1969, April–May. Laotian Troops Recapture Plain of Jars from Communists. The offensive was greatly aided by U.S. close air support. Communist troops still held part of the rim of the plain and were unhindered when they initiated simultaneous operations in southern and central Laos.

1970, February 2. Communists Complete Recapture of Plain of Jars. The region was secured by Pathet Lao troops, supported by North Vietnamese, in an 11-day operation.

1970, April 29–30. North Vietnamese Troops Capture Attopeu.

1972, February 7–March 6. Plain of Jars Battle. Offensive by 4,000 Laotian troops with U.S. air support made some gains. These were all lost to a Communist counteroffensive (February 22).

1973, February 21. Cease-Fire in Laos. The agreement provided for: immediate cessation of hostilities by all Laotian and foreign forces; a new provisional coalition government within 30 days; removal of all foreign troops within 60 days after formation of a government; repatriation of all prisoners within 60 days; supervision by the International Control Commission until a new system could be worked out by the Laotians. The cease-fire was not very effective at first, and at the request of the Vientiane government U.S. B-52 bombers on 2 occasions in the next few days hit Communist positions.

1973, September 14, Renewed Cease-Fire Agreement. This was the most effective cease-fire in Laos in more than 20 years, although sporadic outbreaks of violence continued.

1975, April. Combat Intensity Rises. Pathet Lao forces in Laos began an offensive following major Communist victories in Cambodia and South Vietnam.

CAMBODIA

After its recognition by the Geneva Conference (see p. 1414) as a "neutral" state, Cambodia was in continual friction with her neighbors and with the U.S. as the country veered ever more strongly toward the Communist bloc under its head of state (and ex-king) Prince Norodom

Sihanouk. Sihanouk accepted military aid from the Communist bloc, while placing restrictive conditions upon the reception of U.S. aid. Early (February 18, 1956) he renounced the protection of SEATO. Border clashes with South Vietnam were frequent; free movement through Cambodia of Viet Cong guerrillas was apparently permitted or condoned. Relations with Thailand were equally bad throughout the period.

1950–1954. Widespread Anti-French Insurgency.

1953, June 14. King Norodom Sihanouk into Voluntary Exile. He went to Thailand to promote his fight for complete independence from the French Union.

1963, November 12. Cambodia Refuses U.S. Assistance.

1963, December 12. Cambodia Withdraws Embassy from Washington.

1964–1970. Repeated Border Incidents with Thailand. (See p. 1412.)

1965–1970. Peripheral Involvement in Vietnamese War. Despite the reiterated denials of Chief of State Prince Norodom Sihanouk, practically constant use was made of Cambodian territory by the North Vietnamese, sending reinforcements and supplies to southern South Vietnam via the Ho Chi Minh Trail, and reinforcements by sea through Sihanoukville. The North Vietnamese and the Viet Cong established sanctuary regions and supply depots just inside the Cambodian frontier with South Vietnam. Sihanouk repeatedly denounced U.S. and South Vietnamese pursuit of Communists across the frontier, and occasional U.S. bombings of the sanctuary areas and supply routes.

1969–1970. Guerrilla Warfare against Communist Insurgents. There was increasing violent opposition to the government by Cambodian Communist party **(Khmer Rouge)** guerrillas throughout the country. Collaboration of these antigovernment forces with North Vietnamese and Viet Cong forces was a source of embarrassment as well as a threat to Sihanouk.

1970, March 15. Prime Minister Lon Nol Demands Withdrawal of Communist Forces. In the absence of Prince Sihanouk, the Prime Minister, General **Lon Nol,** responded to anti-Communist demonstrations (which he may have engineered) throughout Cambodia by demanding the immediate withdrawal of North Vietnamese and Viet Cong forces from Cambodian territory. Both refused.

1970, April 30. U.S. Forces Enter Cambodia. (See p. 1330.)

1970, November 1. End of Cambodian Monarchy. Following unanimous legislative action on October 9, Cambodia became the Khmer Republic, ending the 2,000-year-old monarchy.

1972, March 10. Lon Nol Assumes Full Power as Head of State. Sihanouk, who was abroad, went to Peking.

1972, March 21. Communist Bombardment of Phnom Penh. More than 200 rocket and artillery projectiles hit the capital, in the heaviest assault to date.

1972, December. Rocket and Artillery Attacks on Phnom Penh Begin. Shelling by Khmer Rouge forces occurred sporadically, with many civilian casualties, until the end of the war.

1973, March–April. Phnom Penh under Virtual Siege. With all roads to the capital blocked by Communist forces, and the Mekong River supply route also cut by strong forces deployed along the river banks, the capital was in danger of starvation and of running out of necessary military supplies for its hard-pressed garrison. However, strongly protected convoys broke the river blockade (April 8–9), and after the main supply route to the port of Kampong was opened, a large truck convoy reached the city (April 11).

1973, March 17. Lon Nol Declares a State of Emergency. This followed the bombing of the Presidential Palace by a single Cambodian Air Force plane.

1973, August 14. Final U.S. Air Bombardments in Cambodia. (See p. 1333.)

1975, January 1. Rebels Launch Major Drive. In a concerted attack on three fronts around **Phnom Penh,** rebel troops, joined by North Vietnamese and Viet Cong, came within 2 miles of the city's defense line (January 2). All land routes were soon cut. U.S. government-contracted airlift was increased to 10 planes a day.

1975, January 1. Siege of Neak Luong Begins. On the Mekong River, the town commanded the river approach to Phnom Penh.

1975, January 17. Two Convoys Fail to Reach Neak Luong. Both lost ships with ammunition to rebel artillery.

1975, January 23. Ship Convoy Reaches Phnom Penh. Passing through heavy rebel fire, 23 ships arrived from South Vietnam.

1975, February. U.S. Increases Airlift. Flights with arms and ammunition for Phnom Penh were increased to 22–24 a day on February 12 and again 3 days later. Deliveries of food by air were begun on February 27.

1975, February 25. President Ford Requests $222 Million Supplemental Aid. Reportedly less than a month's supply of ammunition remained.

1975, February 28–March 12. Shelling of Phnom Penh's Airport. Intermittent shelling disrupted the airlift, as rebel forces approached within 5 miles of the center of the city.

1975, February 29. U.S. Congressional Delegation Visits Cambodia.

1975, March 11. Military Command Changes. Lieutenant General **Sosthene Fernandez** was removed as commander of the armed forces. Lieutenant General **Saksut Sakhan** replaced him as chief of staff.

1975, April 1. Lon Nol Leaves Cambodia. A collective leadership replaced him.

1975, April 1. Insurgents Capture Neak Luong. The 3-month siege had left the naval base in total ruin. With rebel forces closing in on Phnom Penh and the airport under repeated fire, air traffic was suspended temporarily several times.

1975, April 16. Khmer Rouge Victorious. The government of President Lon Nol (who fled on April 1) surrendered to the Communists, who occupied Phnom Penh.

MALAYSIA

1948, February 1. Federation of Malaya Established. This comprised British colonies on the Malay Peninsula.

1948, February–May. Communist Revolt Begins. This was mainly among the predominantly Chinese element of the population.

1948, June 16. State of Emergency Proclaimed. Guerrilla warfare flared through the Federation. U.K., Australian, and New Zealand troops reinforced the garrison. Terrorism became endemic.

1952, February 7. British Offensive Begins. General Sir **Gerald Templer,** High Commissioner and commander of government forces in the Federation, instituted a concerted, well-planned anti-insurgency campaign. Some 45,000 troops—regulars and special local forces—began warfare against the rebels, combat and psychological.

1954, February 8. Communist High Command Withdraws. British authorities in Kuala Lumpur announced that the Communist party's high command in Malaya had moved to Sumatra. While this was a victory for Britain in their 6-year war, it was also an indication of an attempt to establish a communistic Indonesian front.

1957, August 31. Federation Becomes a Constitutional Monarchy. It remained within the Commonwealth. By this time, the revolt had been suppressed for all practical purposes, though a few pockets of resistance remained in remote jungle areas.

1960, July 31. Emergency Officially Ended. The government announced that the crushing of revolt was completed. Total casualties: Communist rebels, 6,705 killed, 1,286 wounded, 2,696 surrendered; government troops, 2,384 killed, 2,400 wounded.

1962, December 8. Revolt in Borneo. This Indonesian-supported rebellion was quickly suppressed.

1963, September 16. Federation of Malaysia Proclaimed. This included Singapore, Sabah, and Sarawak added to the Federation of Malaya. The new state, a Free World bastion against Communist aggression and encroachments in Southeast Asia, at once became target for attack by Communist-oriented Indonesia (see p. 1422). British military support bolstered the Malaysian defense against interior terrorism and Indonesian raids.

1963–1966. Undeclared War with Indonesia. (See p. 1422.)

1964, May 3. Sukarno Announces Intent to Crush Malaysia.

1964, July 21–23. Communal Rioting in Singapore. The Communists incited Chinese rioting against Malaysia. Disorders were suppressed by police and troops.

1964, July 22. U.S. Pledges Support for Malaysia. This was to bolster the new nation against Indonesian threats.

1965, August 9. Independence for Singapore. By mutual agreement, the city of Singapore withdrew from Malaysia. Initially a British garrison remained responsible for the defense of Singapore.

1971, November 1. ANZUK Force Replaces British Garrison. Units from Australia, New Zealand, and the U.K. made up the new garrison. Singapore began to raise substantial armed forces to share this responsibility in a 5-power defense pact with Malaysia and the other 3 Commonwealth nations.

INDONESIA

1945, August 17. Independence of Republic of Indonesia. This was declared by **Achmed Sukarno** and **Mohammed Hatta** after the collapse of Japan, in an effort to forestall Dutch reoccupation.

1945, September 29. British and Dutch Troops Arrive in Batavia. They began to disarm and repatriate Japanese forces, and to reestablish Dutch control over Netherlands East Indies.

1945, October 14. Hostilities Begin. The Indonesian People's Army declared war against occupying British and Dutch forces.

1945, November 6. Negotiations Rejected. Indonesian republicans rejected the Dutch offer of dominion status and home rule.

1945, November 29. Fall of Surabaya. British troops captured the rebel capital after an intensive battle with Indonesian nationalists.

1946, November 13. Cheribon Agreement. The Dutch recognized the Indonesian Republic (Java, Sumatra, and Madura) and U.S. of Indonesia—to include Borneo, Celebes, Sunda, and Molucca Islands—all under the Netherlands Crown. Clashes with the Dutch continued.

1947, May 4. Nationalists Proclaim Independence of West Java.

1947, July 20. Dutch Offensive on Java. U.N. intervention called for a cease-fire (August 4), but fighting went on despite continuing mediation efforts of the U.N. committee and of U.S. diplomats (1947–1948).

1948, December 19. Dutch Airborne Troops Capture Jogjakarta. This was the capital of the Indonesian rebels. The Dutch soon gained effective control of the entire island of Java (December 25).

1948, December 21. Cease-Fire.

1949, January 28. U.N. Security Council Orders Transfer of Sovereignty. The Netherlands refused, and sporadic hostilities continued.

1949, May 7. Cease-Fire. Dutch troops withdrew from Jogjakarta and Djakarta, new capital of the Indonesian Republic (June 30).

1949, November 2. The Netherlands Grants Full Sovereignty.

1950, August 15. Republic of Indonesia Is Proclaimed.

1950–1961. Endemic Civil War. Unrest, turmoil, and revolt throughout Indonesia, particularly on Sumatra and Celebes.

1955, April 18–27. Bandung Conference. Delegates from 29 Asian and African nations met at Bandung, and announced their aims as elimination of colonialism, independence and self-determination for all peoples, and membership for all nations in the U.N.

1957–1963. Indonesian Harassment against West New Guinea. The Indonesians claimed that this territory (called by them West Irian) should be given to them by the Dutch.

1962, January–August. Sporadic Hostilities. Indonesian torpedo boats off the coast of Dutch New Guinea were attacked by Netherlands forces (January 16). Soon afterward, a guerrilla campaign on Netherlands New Guinea was started by Sukarno (February 20).

1962, August 15. The Netherlands Agrees to Abandon West New Guinea. Formal transfer to Indonesia followed (May 1, 1963).

1962, December. Indonesian-sponsored Revolt in Brunei. The insurgency was suppressed by British troops.

1963, May 18. Sukarno Named President for Life.

1963, September 15. Harassment of Malaysia Begins. Following proclamation of the Federation of Malaysia, Sukarno refused to recognize the new federation, saying "we will fight and destroy it." Continual diplomatic and guerrilla harassment followed, with frequent infiltrations of Indonesian guerrillas into Malaysian territory (see p. 1421).

1965, January 21. Indonesia Withdraws from U.N. This was in protest at Malaysia's being given a seat on the Security Council.

1965, October 1. Communist Coup Effort. The Indonesian Army defeated the Communist effort, and a wave of anti-Communist, anti-Chinese violence swept the islands. The Army, under the leadership of Chief of Staff General **Abdul Haris Nasution** (who became Defense Minister), attempted to break up the Indonesian Communist Party (PKI).

1965, October–December. Massacre of the Communists. The Army and anti-Communist groups killed up to 100,000 PKI members. (Some estimates are much higher.) The result was considerable diminution of the power of President Sukarno, who had been supported by the PKI.

1966, February 21. Sukarno Dismisses Anti-Communist Members of Government. In Sukarno's effort to restore his own power, General Nasution was one of 15 anti-Communist cabinet members dismissed and replaced by left-wingers. There was an immediate violent reaction. Three students, demonstrating outside the Presidential Palace, were killed by police; rioting spread through Djakarta and Java (February 24). Thousands more suspected Communists were killed.

1966, March 12. Military Seizes Control; Sukarno a Figurehead. Leader of the military partial coup d'état was Lieutenant General **Suharto.** The PKI was outlawed; pro-Communists were purged from the government. Sukarno's few remaining powers were later taken by Suharto (February 20, 1967).

1966, June 1. End of Hostilities with Malaysia. This was announced after four days of peace talks in Bangkok, although the status of Sabah and Sarawak on Borneo remained unresolved.

1966, August 11. Treaty With Malaysia. Signed in Djakarta, this formally ended the undeclared hostilities.

1966, September 28. Indonesia Rejoins the U.N. The U.S. resumed economic aid.

1968, March 27. Suharto Named President. This was proclaimed by the Consultative Assembly, formally ending the regime of Sukarno. Five years later Suharto was re-elected (March 1973).

PHILIPPINES

1946, July 4. Republic of the Philippines Established.

1946–1954. Hukbalahap Rebellion. This Communist-led peasant party, dominating central Luzon, conducted civil war against government troops for nearly a decade before being subdued. Primary responsibility for success of the antiguerrilla operations was that of **Ramon Magsaysay,** minister of defense.

1947, March 14. Agreement with U.S. Ninety-nine-year base agreement between the United States and the Philippines.

1952, April 15. Huk Leader Captured. Philippine troops captured **William J. Pomeroy,** leader of the Communist-led Hukbalahaps.

1954. The Philippines Join SEATO. (See p. 1410.)

1962, June 22. The Philippines Claim Sabah Province, North Borneo. This was based upon prior ownership by the Sultan of Sulu before Sabah was seized by Britain in the 19th century. The claim created subsequent tension with Malaysia.

1964–1974. Guerrilla War against Communist "Huks." Recovering from defeat in the early 1950s (see p. 1422), the Philippine Communist party created a new military arm, successor to the Hukbalahap, called Hukbong Magagpalaya Nang Bayan (People's Liberation Army), operating mainly in central Luzon. The tactics of these new "Huks" were similar to those of the Viet Cong, combining terror and assassination with selective measures to gain local good will. Government suppressive activities prevented the Huks from major success but were unable to destroy the guerrillas.

1966, September 16. Agreement Reduces U.S. Base Lease. The term of U.S. leases on bases in the Philippines was reduced from 99 to 25 years.

1966, September 25. Philippine Construction Battalion Reaches Vietnam. The force, eventually reaching 2,000 men, was initially 1,000 strong. It was sent to demonstrate adherence to SEATO policy.

1970–1974. Moslem Insurgency in Southern Islands. Low-level warfare was waged, taking advantage of traditional Moslem-Christian hostility.

1972, September 23. Martial Law in the Philippines. Proclaimed by President **Ferdinand Marcos** as a "last desperate step" to save the islands from Communist-inspired insurgency and chaos, because of the Huk and NPA insurgencies in Luzon and Mindanao. Political opponents claimed this was merely an excuse for dictatorship.

EAST ASIA

Asian and Pacific Council

The Asian and Pacific Council (ASPAC) was established by 9 anti-Communist nations, meeting in Seoul (July 14–16 1966). The members were: Japan, South Korea, the Republic of China (Taiwan), South Vietnam, Thailand, Malaysia, the Philippines, Australia, and New Zealand. At this first meeting the new organization announced its determination to preserve "integrity and sovereignty" in the face of Communist threats, particularly from the People's Republic of China. Other than occasional statements of general support for the efforts of South Vietnam against its Communist enemies, ASPAC has had little influence and has not figured prominently in the policies of any of its members.

China (National Republic)

The civil war between the National Government (of the Kuomintang party) and the Chinese Communist party, which began in 1926, and which was only partially interrupted by the war against Japan, burst into even fiercer flames at the time of the Japanese surrender. **Chiang Kai-shek**'s National Government, decisively defeated, withdrew from the mainland to Formosa (1949), where with U.S. support it defiantly continued to claim to be the legal government of China, now ruled from Peking by **Mao Tse-tung.** The principal events were:

1945, August 14. Treaty with U.S.S.R. This pledged friendship and alliance between Soviet Russia and Chiang Kai-shek's Nationalist Government (see p. 1383). The Manchurian Railway and the port of Dairen were to be held in joint ownership for 30 years. Port Arthur was to become a Soviet-Chinese naval base and the independence of Outer Mongolia recognized.

1945, August. Renewed Chinese Civil War. Chinese Communist forces moved to take over as much of the Japanese-occupied areas as possible, ignoring Chiang's orders to halt. To permit Nationalist compliance with agreed Allied terms of Japanese surrender, and to forestall Communist take-over of all North China, General **Albert C. Wedemeyer,** still Chiang's Chief of Staff, provided American sea and air lift (August–October) to move Nationalist forces to Central and North China. By mid-October about 500,000 Nationalist troops had been so moved.

1945, August 28. American-Sponsored Nationalist-Communist Negotiations. After persuading Chiang to issue an invitation, U.S. Ambassador **Patrick J. Hurley** personally escorted Mao to Chungking for a peace conference. After nearly 2 months, this broke down, when Nationalists discovered that a large Communist force under General **Lin Piao** was quietly moving into southwest Manchuria.

1945, September 30. Arrival of U.S. Marines. To prevent an expected clash of Nationalist and Communist forces, the U.S. 1st Marine Division and other units were landed in eastern Hopei and Shantung. This force soon grew to about 53,000 men. They occupied Peiping, Tientsin, and coastal areas of both provinces.

1945, November 15. Nationalist Offensive in Southwest Manchuria. Nationalist requests to move troops into Manchuria by sea through the Liaotung Peninsula were rejected by the Soviets, who occupied the region under agreed Allied terms for Japanese surrender. Nationalist troops were landed at Chinwangtao, in the area held by U.S. Marines. They attacked across the Great Wall into regions held by Communists, outside the Russian zone of occupation. The well-trained, well-equipped Nationalists pushed aside the Communists and soon held the region as far as Chinchow (November 26).

1945, November 30. Communist Offensive in Shantung. Chen Yi's New Fourth Army occupied much of the province not already held by U.S. Marines.

1945, December 5. Hurley Accuses Foreign Service Officers. Disappointed by failure of his negotiation efforts, Hurley had resigned (November 26). He now charged that his failure had been largely due to obstructive efforts of pro-Communist American Foreign Service officers.

1945, December 14. Marshall as Mediator. U.S. General of the Army George C. Marshall, recently retired as U.S. Army Chief of Staff, was sent to China as personal representative of President Truman, with mission of mediating the dispute.

1946, January 14. Truce in China. Achieved as a result of Marshall's mediation. Despite frequent violations, and nonapplication in Manchuria, this truce remained in effect in most of China for nearly 6 months.

1946, February 25. Nationalist-Communist Accord. National Government and Communist representatives, meeting with General Marshall and other U.S. mediators, agreed to unify the Chinese armed forces into one national army with 50 Nationalist and 10 Communist divisions. This agreement broke down within a few weeks.

Operations in Manchuria, 1946–1948

1946, March 1. Soviet Forces Begin Withdrawal from Manchuria. Chinese Communist troops, scattered about the countryside, moved toward the cities, as Nationalist troops advanced up the main roads and railroads from the southwest. The Soviets had completely dismantled all Japanese-built factories and industrial facilities, and moved the equipment to Siberia. Vast stores of captured Japanese military equipment, however, were left behind by the Russians where they could be seized by the Chinese Communists, enough to equip the entire Chinese Communist Army.

1946, March 10–15. Battle for Mukden (Shenyang). The day following the Russian withdrawal from the city, a battle broke out for control. The Nationalists were successful; they pushed northward.

1946, March 17. First Battle of Szeping. A massive Chinese Communist counterattack drove Nationalist spearheads back. The Communists entrenched this important rail center.

1946, April 14–18. First Battle of Ch'angch'un. A Nationalist contingent of 4,000, airlifted into Ch'angch'un, was driven out by numerically superior Communists.

1946, April 16–May 20. Second Battle of Szeping. The Nationalist New First Army, 70,000 veterans of the Burma Campaigns under **Sun Li-jen,** drove out 110,000 well-entrenched Communists, who claimed they had been attacked by American planes. The Nationalists immediately pushed north toward Ch'angch'un.

1946, April 25–28. Communists Seize Harbin and Tsitsihar. The National Government made no effort to seize these northern cities as the Russians completed their withdrawal from Manchuria (May 3).

1946, May 22. Communists Evacuate Ch'angch'un. Nationalist troops seized the city and continued their northward drive against ineffectual resistance.

1946, June 1. Crossing the Sungari. The Nationalists continued their drive toward Harbin, as Communist resistance stiffened.

1946, June 7–30. Cease-Fire in Manchuria. Brought about by efforts of General Marshall. The Nationalists halted at Shuangcheng. When negotiations broke down, hostilities resumed.

1946, June–December. Stalemate in Manchuria. During the truce the Communists had strengthened their defenses south of Harbin; the Nationalist advance was stalled. The Nationalists, now over 200,000 strong, held the principal centers of southern and central Manchuria, a bridgehead north of the Sungari, and controlled the railroads. The drain of garrisoning these areas precluded assembling and supporting forces large enough to continue the drive toward Harbin. The Communists, who had recruited the disbanded Manchukuan army, had a strength of over 500,000 and held the countryside, but were unable to mount effective attacks against the Nationalists. After cessation of U.S. military assistance (see below), Nationalist forces in Manchuria went completely on the defensive, to conserve supplies and to permit Nationalist offensives elsewhere in China (see p. 1426ff).

1947, January–March. Communist Sungari River Probes. General Lin Piao, commanding Communist forces in Manchuria, launched three offensives across the Sungari, southwest of the Nationalist bridgehead. All were repulsed.

1947, May–June. Sungari River Offensive. Some 270,000 Communists converged on Ch'angch'un, Kirin, and Szeping. All three cities were isolated, and supplied by air. The Nationalists evacuated their bridgehead. Two Nationalist armies were rushed north from Liaotung.

1947, June–July. Third Battle of Szeping. The Communists briefly occupied the rail center (June 16), but were finally repulsed. A lull followed, as both sides prepared for further action (July–August).

1947, September 20. Communists Begin Liaosi Corridor Offensive. The purpose was to cut off Mukden from overland communications to North China. Counter-offensive by Nationalist field commander in Manchuria, **Cheng Tung-kuo,** finally secured the corridor (October 10).

1948, January–February. Renewed Liaosi Corridor Offensive. Chiang Kai-shek flew to Mukden to take personal command. Nationalist counterattacks again secured the corridor. Chiang returned to Nanking.

1948, March–September. Nationalist Erosion. Steady Communist pressure eroded the Nationalist defenses. Nationalists evacuated Kirin to strengthen isolated Ch'angch'un. The defensive attitude and psychology adversely affected Nationalist morale.

1948, September 12. Renewed Liaosi Corridor Offensive. The Communists seized the corridor, repulsing all Nationalist efforts to reopen the line of communications to the south. Chiang flew to Peking to assume command. Finding the situation in Manchuria to be hopeless, he ordered the garrisons to withdraw, fighting their way south. Ch'angch'un was evacuated (October 21).

1948, October 27–30. Battle of Mukden-Chinchow. Retreating Nationalist columns, 3 armies, were struck by a massive Communist counteroffensive. All were killed, captured, or dispersed. The Nationalist commander, General **Liao Yueh-hsiang,** competent Burma veteran, was killed.

1948, November 1. Fall of Mukden. The small remaining Nationalist garrison surrendered. By the end of the year the Communists held all of Manchuria. The Nationalists had lost 300,000 of their best troops.

Operations in North and Central China, 1946–1949

1946, May 1. National Capital to Nanking. The Chinese government officially returned to Nanking from Chungking.

1946, May 5. Hostilities at Hankow. This was one of many breakdowns in the cease-fire established in January (see p. 1424).

July–November. Nationalist North China Offensive. Claiming the provocation of frequent Communist truce violations, Chiang ordered a major offensive to seize North China, hoping to prevent Communists from entrenching themselves. The offensive was highly successful. The Nationalists recovered most of Kiangsu, reopened the Tsinan-Tsingtao railway in Shantung, occupied Jehol and much of Hopeh. The Communists undertook minor counteroffensives, winning temporary successes along the Lung-Hai Railway (Sian-Kaifeng-Hsuchow) and in north Shansi.

1946, July 29. U.S. Halts Military Equipment Assistance. General Marshall, annoyed by the Nationalist offensive, and under strong Communist propaganda attack for U.S. assistance to the Nationalists, ordered an embargo of all U.S. military assistance to both sides. This actually only affected the U.S.-equipped armies of the National Government. Chiang ordered units in Manchuria to go on the defensive, but continued the North China offensive in belief he could win before supplies ran out. This U.S. action had serious psychological as well as practical effects on the National military situation.

1946, September. U.S. Marines Begin Withdrawal. This was interpreted by many Chinese as further evidence of U.S. abandonment of the National Government.

1946, November 8. Chiang Orders Nationalist Cease-fire. He informed General Marshall that he was willing to resume negotiations. Nationalist overtures and U.S. mediation efforts were rejected by the Communists.

1947, January 6. Failure of the Marshall Mission. At Marshall's request, he was recalled by President Truman. Marshall left China, criticizing both sides (January 7). Remaining U.S. Marines in North China (about 12,000) were ordered to withdraw, save for one regiment left in Tientsin under terms of 1901 Boxer Protocol.

1947, January–December. Nationalists on the Defensive. They held the towns and main railroads. Elsewhere the Communists seized the initiative, save for one continuing Nationalist offensive in Shensi.

1947, March 19. Nationalists Capture Yenan. The Nationalist offensive in Shensi captured the Communist capital; Mao Tsetung was forced to flee. Elsewhere the Communists held the initiative, and Mao refused to call back any troops from more important theaters of the war to defend his capital.

1947, October. Communist Offensives. Coordinated with offensives in Manchuria, **Liu Po-ch'eng**'s Central Plains Army and Chen Yi's East China Field Army were active in the area between the Yangtze River and the Lung-Hai Railway and in Shantung. Chen Yi's forces cut the railroad line north of Kaifeng, cutting the main line of communications of Nationalist armies in North China.

1948, March–April. Communist Offensive in Shensi. Troops of General **P'eng Teh-huai** recaptured Yenan (April).

1948, May–September. Communist Offensives in Yellow River Valley. Armies of Chen and Liu steadily reduced Nationalist holdings north of Yellow River. This offensive culminated in the **Battle of Tsinan** (September 14–24), in which 80,000 Nationalist troops defected or were captured.

1948–1949, November–January. Battle of the Hwai Hai. Under overall command of Chen Yi, his army and Liu Po-ch'eng's attacked the Nationalist Seventh and Second Army groups, deployed along the Lung-Hai Railway generally east of Kaifeng. About 500,000 troops were involved on each side. While the East China Field Army pinned down the Seventh Army Group, between Hsuchow and the sea, the Central Plains Field Army smashed into and through the flank of the Second Army Group, west of Hsuchow toward the Hwai River. Efforts of the Seventh and Second Army Groups to retreat to the Hwai were blocked. Much of the Second Army Group broke through, but the Seventh Army Group was destroyed. Total Nationalist casualties exceeded 250,000 men; among those killed were the commanders of both Nationalist Army groups.

1949, January 21. Chiang Resigns. Vice President **Li Tsung-jen** became Acting President.

1949, January 22. Fall of Peking. Nationalist General **Fu-Tso-yi** surrendered to the Communists after a long siege. Mao Tsetung soon thereafter moved the Communist capital to Peking from Yenan.

1949, February. Evacuation of Last U.S. Troop Contingent. Withdrawal of the U.S. 3rd Marine Regiment from Tientsin was considered by both sides to indicate American abandonment of the National Government.

1949, April 1. Nationalist Peace Effort. Li Tsung-jen sent a delegation to Peking to seek Communist agreement to a division of China at the Yangtze. The Communists rejected this, insisting upon Nationalist surrender.

1949, April 20. Communists Cross the Yangtze River. Liu Po-Ch'eng's redesignated Second Field Army and Chen Yi's Third Field Army crossed on a broad front between Nanking and Wuhan. During the crossing two British warships on the Yangtze were attacked and severely damaged by Communist artillery. As Communist troops approached Nanking, the movement of the National Government capital to Canton (begun January 19) was completed.

1949, April 22. Fall of Nanking. This was followed by the capture of Hsuchow (April 26), Wuhan (May 17), Nanchang (May 23), and Shanghai (May 27). Two other important beleaguered cities north of the Yangtze also surrendered: Taiyuan (April 24) and Sian (May 20).

1949, May–December. Nationalist Collapse. Many Nationalist commanders and troop units defected to the Communists. As the Communist armies approached Canton (October), the capital was shifted to Chungking.

1949, August 5. U.S. White Paper. This State Department document, criticizing the National Government, formally announced cutoff of all further military aid.

1949, October 15. Fall of Canton. Chinese Communist troops occupied Canton without opposition. Chiang returned to head the collapsing National Government.

1949, November 30. Fall of Chungking. Chiang established a new capital at Ch'engtu.

1949, December 7. Withdrawal to Formosa. Chiang's government and all his remaining troops successfully completed withdrawal from the mainland as Communist columns approached Ch'engtu. Nationalist troops retained offshore islands of Quemoy, Tachen, and Matsu.

1950. National Government Reforms. Following institution of social and political reforms of the sort promised while the government was still on the mainland, the U.S. resumed economic and military assistance.

1950, June 25. Chiang Offers Military Assistance to U.N. in Korea. Favorably considered by General MacArthur, the offer was turned down by President Truman, who (June 27) ordered the U.S. Seventh Fleet to prevent either Red Chinese attack on Formosa or Nationalist assault against the mainland.

1950, July 24. Intensive Artillery Bombardment of Quemoy.

1951, January 30. U.S. Military Assistance Group Established.

1952, February 1. U.S.S.R. Censured by U.N.

General Assembly. A resolution was approved charging Russia with obstructing the efforts of the National Government of China to retain control of Manchuria following Japan's surrender, and giving military assistance to the Chinese Communists.

1953, February 2. Chiang "Unfettered." President Eisenhower declared that the Seventh Fleet would no longer "serve as a defensive arm of Communist China."

1954, August 17. Communist Threats against Formosa. President Eisenhower said that the Seventh Fleet would go to the defense of Formosa if the Chinese Communists should attempt to invade.

1954, September 3. Heavy Bombardment of Quemoy. Increased activity threatened an invasion of Formosa, and the Seventh Fleet moved to take up positions to defend it.

1954, December 2. Mutual Defense Treaty with the United States. Fear of further involvement of the U.S. in Asian war led to new restrictions on Nationalist China, whose territorial limits were described in the treaty (signed March 3, 1955) as "Formosa and the Pescadores." Communist bombardment of Quemoy and Matsu continued.

1955, January. Tachen Islands Threatened. Intense Chinese Communist pressure against the islands by airplane raids and small-craft raids. The U.S. Seventh Fleet helped to evacuate 25,000 military and 17,000 civilians from the islands (February 6–11). Meanwhile, President Eisenhower asked Congress for emergency powers to permit U.S. armed forces to protect Formosa and the Pescadores islands, and to assist the National Government in defending the islands (January 24).

1955, June 7. "De Facto" Cease-Fire. An uneasy truce settled over Quemoy and Matsu.

1958, August 23. Blockade of Quemoy by Communist Artillery. Continuous bombardment interrupted supply to the islands for both garrison and civilian population. Strenuous effort of Seventh Fleet, convoying supply by water and air, defeated the Red plan (September). The blockade fire gradually died down (October), dwindling to almost nothing (June 1958).

1962, March 24. Chinese Communist Planes over Quemoy. Rumors of pending invasion and heavy mainland troop concentrations produced another tense situation. Once again U.S. sea-power potential asserted itself. Chinese Nationalist planes shot down several Communist aircraft. The U.S. warned the Communists to keep hands off Quemoy and Matsu (January 22). All threat of invasion soon ended.

1962–1974. Civil War Continues. For several years tension between Formosa and the mainland was undiminished, with sporadic guerrilla operations by the Nationalists harassing the Chinese mainland. This tension abated, however, in the later 1960s, with only occasional, prepublicized, shellings of Quemoy reminding both sides that nominally, at least, they were still at war. Severe blows were struck at hopes of Nationalist return to the mainland first by the replacement of the Republic of China in the United Nations by the People's Republic (1972), and second by the United States' recognition of the rival government in Peking (1973).

CHINA (PEOPLE'S REPUBLIC)

1949, September 21. People's Republic Proclaimed at Peking. Mao Tse-tung was named chairman of the Central People's Government; **Chou En-lai,** premier. Immediate recognition was granted by the U.S.S.R. and its satellites, also by India, Burma, and Ceylon. Great Britain soon recognized the new state (January 6, 1950).

1950, February 15. Treaty of Friendship and Alliance with U.S.S.R.

1950, April 23. Communist Conquest of Hainan Island Completed.

1950, October 7. Invasion of Tibet. (See p. 1429.)

1950, October 26. Intervention in Korean War. (See p. 1359.)

1954–1964, September 3. Bombardments of Quemoy. (See p. 1427.)

1957–1965. Rift with U.S.S.R. Increasing ideological differences and mutually conflicting power ambitions by the end of the period widened the crack in the "monolithic" structure of world communism.

1959, August 29. Indian Border Violated by Chinese Troops. (See p. 1405.)

1960, January 28. Sino-Burmese Treaty. (See p. 1411.)

1960, June 3. Anti-Chinese Revolt in Tibet. (See p. 1430.)

1960, June 29. Border Friction with Nepal.

1962, October 20–November 21. India-China Border War. (See p. 1405.) Before the fighting ended, some 3,213 Indian soldiers and 800 Indian civilians had been made prisoner; total casualties are unknown. Prisoners were later returned (April 1963).

1963–1965. Chinese Support of North Vietnamese Aggression. (See p. 1415.) This became another issue exacerbating strained relations between the U.S.S.R. and Communist China.

1964–1974. Intermittent Border Violence along Sino-Soviet Frontiers. This was the result of tensions created by Chinese demands for return to China of vast areas of Soviet East Asia taken by Czarist Russia in the 19th century. Most incidents seem to have occurred at places where border demarcation was unclear.

1964, October 16. China Explodes Its First Atomic Bomb. China became the world's fifth nuclear power.

1966–1969. "Cultural Revolution." Unrest and violence spread throughout much of mainland China, apparently to some extent deliberately inspired by Chairman Mao Tse-tung for the purpose of rejuvenating the revolutionary spirit among members of the Communist party and the people. Initially the Army seems to have supported, or at least endured, excesses of Communist youth, but as the unrest continued unabated, Army leaders seem to have taken the lead in restoring order, apparently with the somewhat reluctant approval of Mao.

1966, October 27. China Reports Firing a Nuclear Missile. This was China's fourth nuclear explosion, the first in which actual weaponry was tested.

1967, June 17. China Explodes Its First Hydrogen Bomb. China's fifth nuclear explosion, like the others, was conducted at the test site at Lop Nor, in Sinkiang.

1969, March 2–15. Undeclared Combat Between Chinese and Soviet Troops. These clashes took place at different points along the Manchurian frontier. Although there had been no previous public announcements, these were mainly intensifications of armed clashes between Chinese and Soviet forces that had been going on for at least 5 years. Clashes occurred frequently in subsequent months, with both sides accusing the other of provocation.

TIBET

1949, November 24. Communist "Liberation" of Tibet Urged. This was a radio appeal from Peking by the **Panchen Lama**—refugee rival of Tibet's nominal ruler, the **Dalai Lama.** The Chinese Communist government soon announced its intention of doing just this (January 1, 1950).

1950, October. Chinese Communist Invasion. A large Chinese force swept across the frontiers despite a Tibetan appeal to the U.N. (November 10), soon overrunning the entire country. The Dalai Lama was permitted to remain as a figurehead ruler in Lhasa. Widespread revolt continued despite fierce Communist repressive measures (1950–1954).

1954, Spring and Summer. Widespread Revolt. This was suppressed by Chinese Communist troops. Most of the 40,000 rebels were killed or executed.

1956–1959. Renewed Unrest. Mass deportations, Chinese infiltrations, and forced Tibetan labor on the military highway connecting Lhasa and Chungking stirred the population.

1959, March 10–27. Rebellion. This was suppressed by Chinese Communist troops. The Dalai Lama fled from Lhasa to India, where political asylum was afforded. He formally accused Communist China of genocide and suppression of human rights, asserting that 65,000 Tibetans had been killed in the revolt, 10,000 young people and children deported to China, and 5 million Chinese moved into Tibet in a resettlement project.

1959–1974. Continued Guerrilla Warfare. Resistance was sporadic, weak, and relatively ineffective.

MONGOLIAN PEOPLE'S REPUBLIC

1945, October 20. Mongolia Votes for Independence from China. This was the overwhelming result of a plebiscite required under the terms of the Sino-Soviet treaty of friendship and alliance (August 14 1945; see p. 1423).

1946, February 27. Treaty with the U.S.S.R. This treaty of friendship, for 10 years, has been regularly renewed.

1947, June 5–8. Mongolian Raid into Sinkiang. Mongolian troops attacked Peitashan to rescue Mongols captured by Chinese in prior border skirmishes.

1945, December 27. Moscow Declaration. The U.S., Soviet, British, and French foreign ministers announced the establishment of a U.S.–Soviet Joint Commission for the purpose of unifying Korea in accordance with the terms of the Cairo Agreement (see p. 1423).

1946–1947. Failure of the U.S.–Soviet Joint Commission. No agreement could be reached on the establishment of an interim government for all Korea.

NORTH KOREA

1948, May 1. Soviets Proclaim North Korean Independence. By establishing the Democratic People's Republic, the U.S.S.R. defied a planned U.N. plebiscite for all Korea. The president was veteran Communist **Kim Il Sung.**

1948, October 19–December 25. Russian Troops Withdraw. A large Soviet training mission remained in North Korea.

1950–1953. Korean War. (See p. 1355.)

1961, July 6. Treaty with the U.S.S.R. Soviet Russia provided assurance of defense protection, plus financial and military equipment assistance.

1968, January 23. *Pueblo* Incident. The The USS *Pueblo,* a Navy electronic intelligence vessel, was attacked and seized by North Korean

gunboats, on the pretext that the vessel had violated the territorial waters of North Korea. U.S. naval forces, including the nuclear-powered carrier USS *Enterprise,* took station off the coast of North Korea, but no reprisals were taken.

1968, December 22. Release of the Crew of the *Pueblo*. This was the result of a complicated agreement with North Korea, whereby the United States representative on the Mixed Armistice Commission in Panmunjon signed a "confession" that the vessel had been engaged in "espionage," then immediately, and with the agreement of the North Koreans, denounced the document as false.

1969, April 15. North Koreans Shoot Down U.S. Reconnaissance Plane. The attack by North Korean aircraft took place about 100 miles off the coast of North Korea. Again U.S. naval forces threatened North Korea, but no action was taken.

SOUTH KOREA

1948, August 15. Proclamation of the Republic of Korea. This was the result of U.N.-supervised elections early in 1948. The first president of the new republic was **Syngman Rhee.**

1948, October 20–27. Communist-Inspired Army Revolt. A Communist cell in a military unit sparked an uprising which briefly controlled the cities of Yosu and Sunchon in southern South Korea. After vicious fighting, in which at least 1,000 were killed on both sides, loyal troops and police suppressed the insurrection. Another revolt in Taegu was quickly subdued (November 3).

1948–1950. Tension in Korea. Border incidents, Communist infiltration across the border, and Communist-inspired disorders throughout South Korea continued and intensified.

1949, June 29. Withdrawal of U.S. Occupation Forces Completed. (See p. 1354.)

1950, June 25. North Korean Invasion of South Korea. (See p. 1355.)

1950–1953. Korean War. (See p. 1355.)

1953–1971. Continued Tension. Deliberate Communist violations of armistice force limitations in North Korea caused the U.S. and South Korea to announce a compensatory build-up of forces and weapons (beginning June 21, 1957). North Korean raids intensified during the period 1967–1971, but no large-scale hostilities occurred.

1960, April 6–27. Korean Violence Forces Resignation of Rhee. Demonstrations and protests by students against the repressive regime of President Syngman Rhee led to widespread riots and disorders. Rhee's efforts to reorganize the government failed to satisfy the protesters, and Rhee resigned, apparently in part as a result of pressure from the United States. Foreign Minister **Huh Chung** became the acting president under the constitution.

1961, May 16. Coup d'État. General **Chung Hee Park** seized power and dismissed the existing government. Later Park was formally elected president (October 15, 1963).

1965, February 25. First South Korean Troops Arrive in South Vietnam. A contingent of 600 troops arrived, scheduled to be increased to 2,000. By early 1966 this force had risen to 21,000, and by late October 1966, to 41,000, including 2 divisions organized as a corps. This was a demonstration of solidarity with the United States.

JAPAN

1945–1951. MacArthur as Supreme Commander for the Allied Powers. Under the firm control and guidance of MacArthur's military government, the Japanese government and nation began recovery from the devastation of the war.

1946, November 3. New Constitution. This became effective May 3, 1947. Among its provisions was a renunciation of the right to wage war.

1951, September 8. Peace Treaty with the Allies. Unable to obtain Soviet agreement to negotiate a peace treaty, the U.S. and 48 other non-Communist nations signed a treaty with Japan (effective April 28, 1952). At the same time, the U.S. signed a bilateral defense agreement with Japan.

1951–1954. Japan Begins Limited Rearmament. Despite apparently sincere devotion to the war-renunciation clause of the constitution, it became apparent to Japan that internal and external security required military forces. With U.S. encouragement, Japan began to develop small "self-defense forces."

1954, March 8. Mutual Defense Agreement with U.S. Under this the U.S. was to give Japan about $100 million in subsidies for production of munitions and food.

1954, July 1. Official Rearmament Approved. After prolonged national and legislative debate, Japan enacted legislation authorizing new armed forces.

1956, October 19. State of War with Russia Terminated. A joint Japanese-Soviet declaration.

1960, January 19. Renewed Mutual Defense Treaty with the U.S.

1972, May 15. Okinawa Returns to Japanese Control. This was a result of a treaty with the United States (June 17, 1971).

1972, September 29. Peace Treaty with Communist China. Signed by Japanese Prime Minister **Kakuei Tanaka** and Chinese People's Republic Prime Minister **Chou En-lai,** this ended the state of war which had technically existed between Japan and China since 1937. Simultaneously Japan severed relations with the Republic of China (Formosa).

AFRICA

Before World War II, Africa, the second largest continent in land area, included but one truly independent nation: tiny, unimportant Liberia. The Union of South Africa, in fact independent, as a dominion was a part of the British Empire. Egypt, nominally independent, was actually under British protection and influence. Ethiopia, which had been truly independent, had recently been conquered by Italy (see p. 1139). In the 20 years after World War II, complete independence was achieved by all nations and regions of Africa, save for a few insignificant Spanish coastal colonies and the large Portuguese colonies of Angola and Mozambique. This achievement of independence, however, did not in the slightest halt the working of the forces of nationalism (and related anticolonialism) in the independent nations or in the few remaining colonial areas. Revolution, new nationalism, new and indigenous imperialism, Communist subversion (of two varieties, one directed from Moscow and the other from Peking), racial antagonisms, and sweeping technological change kept most of Africa in constant turbulence for the entire period. There were internal and external military actions and hostilities of one sort or another in practically every nation and colonial region of the continent. Only the most impor- tant will be noted here.

ORGANIZATION OF AFRICAN UNITY

The Organization of African Unity (OAU) was formed in 1963 through the efforts of the leaders of Nigeria, Ethiopia, and Guinea and was the culmination of earlier efforts to form a broad-based continental organization. All African countries are members except South Africa and Rhodesia. The charter prescribes noninterference in the internal affairs of states, observance of sovereignty and territorial integrity of members, peaceful settlement of disputes, condemna- tion of political assassination and subversive activities, nonalignment with power blocs, and emancipation of the white-ruled African territories. (The OAU Liberation Committee supports the various liberation movements directed against the white-dominated regimes.)

The OAU succeeded in arbitrating the Algerian-Moroccan border war of 1963 and in helping Tanzania replace with African troops the British troops which had quelled its 1964 army mutiny. However, it was ineffective in assisting Zaire (then Democratic Republic of the Congo) during its rebellion in 1964–1965, in resolving the Nigerian-Biafran civil war in 1969, in reconciling various rival liberation movements, or in successfully prosecuting a war of libera- tion in Southern Africa or Rhodesia. Following an allegedly Portuguese-inspired mercenary raid on Guinea (December 1970) consideration was given to establishment of a common African army, but nothing came of that. At a meeting of African leaders at Cairo, July 21, 1964, it was agreed that nuclear weapons would be banned from Africa.

AFRICAN AND MALAGASY COMMON ORGANIZATION

The African and Malagasy Common Organization (OCAM) was formed in 1965 as an outgrowth of earlier attempts at cooperation among French-speaking states, including the African and Malagasy Union and the Regional Council of France, Ivory Coast, Niger, and Dahomey. Aside from political and economic motives, the prime factors were establishment of a common front to meet Ghana's then subversive activities, Chinese Communist infiltration and subversion, and endemic chaos in the former Belgian Congo (now Zaire). OCAM is largely consultative, and no common defense staff or organization exists. Most members still have bilateral defense treaties with France, which ensures their immediate internal and external security when threatened beyond their means to cope. The 14 members are: Cameroon, Central African Republic, Chad, Congo (People's Republic of the Congo), Dahomey, Gabon, Ivory Coast, Malagasy Republic, Niger, Rwanda, Senegal, Togo, Upper Volta, and Zaire.

NORTH AFRICA

Algeria

1945–1954. Autonomy Demanded. The first clash (May 8, 1949) between nationalists and French caused the death of 88 French and more than 1,000 Algerians.

1954–1962. Open Rebellion. The FLN (Front de Libération Nationale, organized 1951) started organized warfare, which was to continue until freedom had been attained. Use of Tunisian bases by the FLN strained French relations with Tunisia (see p. 1435). The insurrection drew nearly one-half of the entire French Army into Algeria, with resulting casualties of 10,200 French soldiers and some 70,000 Algerian insurgents killed.

1958, May 13. French Officer Uprising. Brigadier General **Jacques Massu** established a Committee of Public Safety, protesting against political leadership in the war. This started the chain of political events in France that brought de Gaulle to power (see p. 1377).

1958, June 1. De Gaulle Offers Self-Determination by Referendum. This was opposed by the *pieds noirs* (Algerians of French descent). Rioting and terrorism were manipulated by the "Secret Army" (or OAS, an extremist group organized by *pieds noirs* and French military men). De Gaulle visited Algeria and demanded dissolution of the Committee of Public Safety, restoring French government control in Algeria.

1960, January 22–February 1. Uprising of French Rightists in Algiers. They were opposed to de Gaulle's policy of self-determination for Algeria. This was suppressed by loyal French troops under General **Maurice Challe.**

1961, January 6–8. French Voters Support de Gaulle's Algerian Program. An overwhelming majority approved a referendum to permit Algerian self-determination.

1961, April 22–26. French Military Revolt. A mutiny headed by Generals Challe and **Raoul Salan** was quickly put down (April 25) by loyal French troops on de Gaulle's order (see p. 1377). Salan escaped and directed intensified OAS terrorism in France and Algeria.

1961, May 20. Peace Talks Begin. These took place at Evian-les-Bains in France between representatives of the French government and the rebel Algerian provisional government.

1962, March 7–18. Cease-Fire. Evian negotiations brought a cease-fire between Moslem nationalists (FLN) and French Army. **Ahmed Ben Bella** was chosen as premier.

1962, July 3. Algerian Independence.

1962–1965. Continuing Internal Unrest. There was widespread opposition to Ben Bella's relatively inefficient and dictatorial government, closely aligned with the U.S.S.R. and with the U.A.R. There were several mutinies, revolts, and uprisings around the country.

1963, October 6–12. Berber Revolt. Under the leadership of a dissident Army officer, Colonel **Mohand ou el Hadj,** Berber tribesmen of the Kabylin Mountains region rose against the government. President Ben Bella claimed the insurgency was incited by Morocco, as a result of tense relations over a border dispute (see below). The defeated rebels were forced to flee into the mountains. Colonel Mohand soon ended his revolt and joined in the war against Morocco (October 24).

1963, October 13–30. Border War with Morocco. (See p. 1435.)

1965, June 19. Overthrow of Ben Bella. Control of the nation was seized in a near-bloodless *coup d'état* by the army commander, Communist-trained Colonel **Houari Boume**

dienne, a hero of the revolution against France, who was soon installed as president (July 5).

1967, December 13. Attempted Coup d'État. The uprising, involving army officers, was suppressed.

1967, June. Algeria Sends Contingent to Support Egypt. (See p. 1343.)

Libya

1949–1951. Anticolonial Rioting. This was directed at both British occupation authorities and Italian administration.

1951, December 24. Independence of Libya.

1953, July 29. Treaty with Britain. This gave Britain 20-year rights to maintain military establishments in Libya, in return for which Britain was to pay Libya £1 million a year for 5 years for economic development and £2.75 million a year to aid in balancing the Libyan budget.

1954, September 9. Agreement with the U.S. An agreement was signed in Benghazi, giving the United States use of air bases in Libya in return for payment of $5 million in 1954, $2 million yearly for 20 years.

1969, September 1. Coup d'État. King **Mohommed Idris Al Mahdi as-Sanusi** was overthrown, while out of the country, by a military uprising led by young officers, of whom the leader was Captain **Muammar el Qaddafi.**

1969, December 10. Unsuccessful Coup Attempt. Qaddafi blamed "invisible foreign hands."

1970, January 21. Sale of French Aircraft to Libya. The sale of 100 French Mirage-III supersonic fighter jets to Libya was seen by most of the world as a transparent French evasion of their own embargo on weapons for the warring nations of the Middle East, since it was generally assumed (despite French denials) that these planes would be turned over by Libya to Egypt.

1970, June 11. Departure of Last Americans. As a result of pressure from the government of President Qaddafi, the United States abandoned Wheelus Air Force Base in Libya.

1971, September 1. Loose Union with Egypt and Syria. (See p. 1395.)

1972, August 2. Agreement to Unite with Egypt. This was the result of 3 days of conversations between Presidents Sadat of Egypt and Qaddafi of Libya at Tobruk and Benghazi.

1973, August 29. Reaffirmation of Unification with Egypt. The joint announcement by Sadat and Qaddafi made it clear that the unification would be a slow and gradual process, despite the obvious desire of Qaddafi for immediate union.

1974, January 12–14. Plan for Union with Tunisia. Joint announcement of the merger (January 12) was soon followed by a Tunisian announcement (January 15) that it would be delayed, probably indefinitely.

Morocco

1947–1953. Nationalistic Unrest. Sultan **Mohammed V** gave his support to the nationalist movement, in defiance of the French administration.

1953, August 15–20. French-Inspired Uprising. Tribal leaders in Marrakesh rose against the sultan, under the influence of pro-French leader **Thami Al-Glaoui.** The sultan was deposed and sent into exile by the French.

1953–1955. Increased Unrest. Terrorism and guerrilla operations of nationalists spread throughout Morocco.

1955, August 19–November 5. Intensified Guerrilla Hostilities. French efforts to suppress the risings of Berber tribes and the terrorism in the countryside and in the cities were not very successful.

1955, November 5. France Agrees to Independence. Mohammed V was restored to power.

1956, March 2. Protectorate Status Ceases. France and Morocco by mutual agreement terminated the Treaty of Fez (March 30, 1912). Spain relinquished her protectorate over Spanish Morocco (April 17), and the international status of the Tangier zone later ended (October 29).

1957–1964. Foreign Troops and Bases Withdrawn. Morocco called for evacuation of foreign troops. The last of a large complex of U.S. air bases was returned in 1964; French and Spanish forces had been previously withdrawn. As the period ended, the enclave of Spanish Sahara still remained *in statu quo.*

1957, November–December. Border Clashes at Ifni. Moroccan irregulars were repulsed by Spanish troops after seizing much of the colony. Later Spain ceded Ifni to Morocco (April 1, 1958).

1958, October 21. Pro-Monarchist Revolt. Army units, supporting King Mohammed V in his dispute with the Istiqlal Party, were attacked by loyal troops. Although disowned by the King, they were soon joined by Rif tribesmen in the Tasa area (November).

1959–1960. Revolt in Rif. Antigovernment operations in the Rif continued (see above) despite severe rebel defeats (January 1959) until the region was finally pacified (April 1960). Leader of the government troops was Crown Prince **Montay Hassan.**

1963, October 13–November 4. Border War with Algeria. Large-scale hostilities broke out along a disputed frontier area in the Atlas Mountain-Sahara Desert region after prolonged tension and a number of incidents. A cease-fire was arranged through the mediation of Emperor Haile Selassie of Ethiopia and President **Modibo Keita** of Mali. The Moroccans had the better of the fighting in the Hassi-Beida, Tindouf, Figuig area.

1971, July 10. Attempted Coup. The revolt was suppressed in bloody fighting which began when insurgents attacked the royal palace during a birthday party for King **Hassan II.**

1972, August 17. Attempted Coup. The leader of the revolt, trusted General **Mohommed Offkir,** Minister of Defense since the previous coup attempt, committed suicide when an attempt to shoot down the King's plane failed.

1973, March 2–7. Guerrilla Activity. Guerrillas, apparently from outside the country, disrupted at least 2 towns in southern Morocco. They were defeated and more than 100 were captured, some admitting support from Libya.

Tunisia

1952, March. Violence and Unrest. National self-determination came to a boiling point in disorders and riots directed against French rule.

1955, June 3. Full Internal Autonomy Granted by France. This was effective September 1.

1956, March 17. Independence. All former treaties and conventions were abrogated. **Habib Bourguiba** was chosen premier (March 25). France retained several military bases.

1956, July 25. Republic Proclaimed. Bourguiba became president. Despite disagreements—and one subsequent conflict—with France, Bourguiba kept his nation oriented to the West.

1957, May 26–June 7. Clashes with France. These were the result of activities of French and Tunisian troops in the vicinity of the Algerian border. Similar incidents occurred sporadically until the end of the insurgency in Algeria (see p. 1433).

1958, February–June. Clashes with France. These sporadic border incidents were mostly French punitive action in response to Tunisia-based operations of Algerian nationalists.

1961, July 19–22. Hostilities with France. (See p. 1377.) Tunisian attacks against French military posts brought prompt retaliation. French troops occupied Bizerte. Hostilities were ended by U.N. mediators.

1962, June. French Evacuation Completed. Air-base rights were retained by France.

1962, December 22. Unsuccessful Coup Attempt.

1965, April 28. Tunisia Boycotts Arab League Meetings. This was because the other states refused to consider President Bourgiba's proposals for peace negotiations with Israel.

1966, October 3. Tunisia Breaks Relations with United Arab Republic. This was a result of continued disagreement about peace negotiations with Israel.

1974, January 12–15. Plan for Union with Libya. (See p. 1434.)

EAST AFRICA

Ethiopia

1952, September 11. Union with Eritrea. Moslem Eritrea and Christian Ethiopia were united in a federation.

1954, May 14. Agreement with the U.S. Ethiopia gave the United States 99-year military base rights.

1960, December 13–17. Military Revolt. During the absence of Emperor Haile Selassie, members of the Imperial Guard seized control of Addis Ababa and proclaimed Crown Prince **Asfa-Wossen** the new emperor. Loyal troops suppressed the rebellion. The emperor pardoned his son, who had apparently acted under duress.

1964. Frontier Warfare with Somalia. (See p. 1437.)

1965–1974. Revolt in Eritrea. Eritrea's Assembly had voted unanimously for permanent union with Ethiopia (November 14, 1962). However, nationalist, Communist, and pan-Arab elements had joined forces to compel dissolution of the union. This dissidence, apparently financed by neighboring Arab states (primarily Sudan and Somalia), flared into guerrilla war.

1971–1974. Intensification of Eritrean Revolt. The Eritrean Liberation Front (ELF) stepped up guerrilla-terrorist activities against the government that spread to Addis Ababa.

1974, February 26–28. Army Mutiny. This spread from Asmara to Addis Abada. In an effort to retain control, Emperor Haile Selassie formed a new cabinet and raised Army pay. The nation remained, however, under virtual military control.

1974, June 28. Army Seizes Control. The Emperor became a figurehead, and his remaining power was soon reduced (August 16).

1974, September 12. Army Deposes Haile Selassie. Crown Prince **Asfa Wossen,** in Geneva, said he would serve as a constitutional monarch, as the Army demanded, but he didn't return. General **Aman Michael Andom,** chairman of the Provisional Military Government, became virtual head of state.

1974, November 22–24. Coup d'État. A power struggle among leaders of the Military Council ruling Ethiopia led to the ouster of General Aman (November 22), who was killed the next day during, or executed after, an armed clash of rival forces near his home. This was followed by the execution of about 60 former officials, including 32 military officers as well as cabinet ministers, provincial governors, and members of the nobility, including at least one royal prince. One reason for the power struggle was differences in philosophy and policy with respect to the continuing Eritrean separatist insurrection.

Kenya

1945–1952. Unrest and Violence. This British crown colony became a hotbed of revolution. A secret organization—the Mau Mau—began a campaign of dissidence, which finally erupted in an appalling area-wide blood bath, white colonists and native negroes alike, men, women, and children, being murdered under conditions of terror and treachery.

MAU MAU REVOLT

1952, October 20. Britain Declares a State of Emergency. Britain sent a warship and troops to Kenya to restore order and to suppress the Mau Mau uprising. Guerrilla war spread.

1953, January–May. British Military Measures. Major military and punitive measures were initiated. Leading Kikuyu tribe nationalist leaders, known or suspected to have connections with the Mau Mau Society, were arrested, tried, and convicted, including a 7-year prison sentence for Mau Mau leader **Jomo Kenyatta** (October 20). Central Kenya was sealed off from the rest of the country, and a separate East African command comprising

Kenya, Uganda, and Tanganyika was set up under General Sir **George Erskine.**

1953, June 15. British Victory in the Aberdare Forest. More than 125 Mau Mau were killed, bringing the total of terrorists killed since October to approximately 1,000. Meanwhile, the colonial government undertook measures to improve housing conditions in Nairobi, and soon afterward dropped leaflets over known Mau Mau strongholds promising lenient treatment to all who surrendered and who were not guilty of murder or serious crimes.

1955, February–June. Climactic Campaign Begins. Some 10,000 troops dispersed about 4,000 terrorists in the Mt. Kenya and Aberdare areas.

1955, September 2. Britain Begins to Reduce Forces. Since October 1952, almost 10,000 terrorists had been killed, 1,538 had surrendered, and over 24,000 had been captured or were held as suspects. The campaign against the remaining scattered dissidents continued into early 1956.

1961, August 14. Kenyatta Released from Prison. He immediately became leader of the principal political independence party and began negotiations with the British for independence.

1963, March. Frontier Clashes with Somalia. Somali claims to frontier regions of northern Kenya led to border hostilities. Somalia broke relations with Britain (March 14).

1963, June 1. Kenyatta Prime Minister of Kenya.

1963, December 12. Independence of Kenya. She became an independent state within the British Commonwealth of Nations. Assured by the new status, and under promise of amnesty, Mau Mau adherents began surrendering en masse.

1963, December. Diplomatic Relations Broken with Somalia. Serious border warfare broke out.

1964, January. Unrest and Violence. Communist-inspired violence spread from Zanzibar and Tanganyika.

1964, January 25. British Troops Restore Order in Kenya. British intervention was requested by Kenyatta to suppress Communist-inspired native uprisings. The period ended with the nation threatened by internal turmoil and external war.

1968, January 31. Peace with Somalia. An informal understanding of October 1967 was affirmed by the resumption of diplomatic relations.

Malagasy Republic (Madagascar)

1947–1948. Revolt against France. A nationalist uprising, centering on the east coast, was suppressed by French troops after much bloodshed.

1960, June 25. Independence. The Malagasy Republic elected to remain a member of the French community.

1971, April 1. Rebellion Suppressed. Leftist rebels in Tulear Province, southern Madagascar, were overwhelmed by loyal troops.

1972, May 18. Military Coup d'État. The government of President **Philibultsiranana** was ousted following 4 days of violence. General **Gabriel Pomanantsoa** became head of state.

Somalia

1948–1960. Anticolonial Disorders. These were directed against British occupiers and Italian administrators.

1960, July 1. Independence of Somalia. Italian and British Somaliland were combined as a single state. Somalia almost immediately claimed substantial regions of Ethiopia and Kenya, where Somali populations had been placed by arbitrary colonial frontiers.

1960, August 14. Border Clashes along Ethiopian Frontier.

1961, December 10. Unsuccessful Army Revolt.

1963–1968. Hostilities with Kenya. (See above.)

1964, February 8. Renewed Ethiopian-Somalian Hostilities. Despite a truce resulting from mediation of the Organization of African States, frontier warfare continued.

1968, January 31. Peace with Ethiopia.

1968, September 20. Peace with France. Frontier violence by Somali intruders into the French Territory of the Afars and Issas was ended by agreement.

1969, October 21. Coup d'État. Army and police officers joined to overthrow the regime and to establish a Supreme Revolutionary Council, under Army control.

1970, April 21. Unsuccessful Coup Attempt.

1971, May 25. Unsuccessful Coup Attempt.

Sudan

1953, February 12. Self-Government. Egypt and Great Britain signed an agreement providing for self-government in the Anglo-Egyptian Sudan.

1955, August 16. British and Egyptian Withdrawal Demanded. The parliament of Sudan asked Britain and Egypt to evacuate their troops from the Sudan in 90 days. Britain with 900 troops and Egypt with 500 agreed to be out by November 12.

1955–1972. Revolt in Southern Sudan. The black inhabitants of southern Sudan, in the jungled swamps of the Upper Nile, rebelled against the Arab-Moslem ruling group in Khartoum. Small-scale but bitter fighting continued for 8 years.

1956, January 1. Independent Republic Proclaimed.

1958, November 17. Military Coup. Control of the government of Sudan was seized by Lieutenant General Ibraham Abboud.

1959, November 10. Unsuccessful Coup Attempt.

1964, October–November. Violence Forces Overthrow of Abboud Government. Unable to quell mounting disorders, and faced with a possible military coup, General **Abboud** resigned as president. He was replaced by a council under Premier Sir **el-Khatim el-Khalifa.**

1966, December 28. Unsuccessful Coup Attempt. Loyal army troops subdued insurgents in a training center, after a sharp fight.

1967, June. Sudan Sends Units to Support Egypt in Sinai.

1969, May 25. Coup d'État. The government of Prime Minister **Mohommed Ahmed Mahgoub** was overthrown by a leftist military coup headed by Colonel **Mohommed Gafaar al-Nimeiry.**

1970, March 30–31. Unsuccessful Coup Attempt. In bloody fighting rebel leader Imam **el-Hadi Ahmed el-Mahdi** was killed by troops loyal to President Nimeiry.

1971, July 19–21. Coup and Countercoup. A group of Communist-oriented officers, apparently with covert Soviet support, briefly seized control of the government in a bloodless *coup d'état.* President Nimeiry, however, apparently with support from Egypt and Libya, staged a countercoup and regained power.

1972, March 27. Agreement Ends North-South Civil War. The Arab-Moslem government of northern Sudan reached an accommodation with the black Christians and pagans of the south, ending 17 years of strife.

1973, March 2. Palestinian Terrorists Assassinate American and Belgian Diplomats. The Black September terrorists were convicted of murder (June 1974), but freed by the government and released to the Palestine Liberation Organization (PLO) in Cairo. The U.S. at once recalled its new ambassador.

Tanganyika, Zanzibar, and Tanzania

1961, December 9. Independence of Tanganyika. It remained in the British Commonwealth.

1963, December 10. Independence of Zanzibar. It remained in the British Commonwealth.

1964, January 12. Zanzibar Rebellion. The government was overthrown by African nationalist rebels, some of whom had been trained in Communist China. Nationalist unrest, stirred up by Communists, spread to nearby mainland nations; widespread mutinies resulted.

1964, January 25. British Intervention. At the request of local governments, British troops suppressed mutinies of African troops in Tanganyika, Kenya, and Uganda.

1964, April 26. Establishment of Tanzania. This resulted from the merger of Tanganyika and Zanzibar.

1971, August 24–30. Border Clashes with Uganda. President Amin's pretext for attacks on Tanzanian troops was the accusation that Tanzania was assisting Ugandan rebels (see below).

Uganda

1949–1962. Anticolonial Disorders.

1962, October 9. Independence of Uganda. Freedom was granted by Great Britain to the kingdom, under King **Mutesa II,** Kabaka of Buganda. The new state was initially governed as a dominion, with a loose federal relationship between Uganda proper and the kingdom of Buganda.

1963, October 9. Uganda a Republic. King Mutesa II was elected president (as **Edward Mutesa**) under a new constitution, under which Uganda became a sovereign state within the British Commonwealth. Prime Minister was **Milton Obote,** principal Ugandan politician.

1966, March 3. Obote Deposes President Mutesa. The President fled to Kampala, Buganda, where he was still the Kabaka, and began a separatist movement, demanding withdrawal of Uganda troops by May.

1966, May 23–24. Uganda Suppresses Bugandan Separatist Movement. In 2 days of fighting in Kampala, federal troops seized the Kabaka's palace and restored federal control to Buganda. Mutesa escaped to England. Obote became president.

1971, January 25. Bloody Coup d'État. While Obote was attending a Commonwealth conference in Singapore, a group of officers, headed by Major General **Idi Amin,** seized control of the government and overwhelmed loyal troops. Obote took refuge in Tanzania.

1971, August 24–30. Border Skirmishes with Tanzania. Tense situations between Tanzania and Uganda following Amin's coup resulted in a protracted outburst of hostilities along the border west of Lake Victoria.

CENTRAL AFRICA

Union of Central African States

This loose federation for defense and economic cooperation was established by Zaire (then still known as Congo-Kinshasa), Chad, and the Central African Republic (April 2–3, 1968). It has not played an important role in the policy of its members.

Burundi

1962, July 1. Independence of Burundi. A former German colony and later Belgian mandate and trusteeship, Burundi became an independent monarchy under King **Mwami Mwambutsa,** of the dominant but minority Watusi (Tutsi) tribe. The first years of independence were extremely shaky.

1966, July 8. Coup d'État. Prince **Charles Ndizeye** overthrew his father and proclaimed himself King **Mwami Ntare V.**

1966, November 28. Coup d'État. The new King was overthrown by his appointed Prime Minister, Colonel **Michael Micombero,** who proclaimed himself president. Burundi became a republic.

1970, October 19–20. Unsuccessful Army Mutiny and Coup Attempt. Army officers from the majority, but suppressed, Bahutu (or Hutu) tribe, aided by some Watusi soldiers and police, staged an abortive coup attempt. Bahutu tribesmen revolted and massacred several thousand Watusis. The fierce and bloody Watusi reprisal eliminated all of the Bahutu leaders in and out of the government and army.

1972, April 29–July 31. Civil War and Massacres in Burundi. An effort by Bahutu tribesmen and some Watusi monarchists to free former King Mwami Ntare V failed. Mwami was killed, but armed rebellions spread throughout the country. At least 100,000 people were killed in the first 2 months, largely in reprisal massacres by official and unofficial Watusi organizations. The reprisal killings ended in August, with the government of President Micombero firmly in control.

Central African Republic

1960, August 13. Independence of the Central African Republic. A former French colony, the CAR had been self-governing since 1958. Its President, **David Dacko** established strong ties with Communist China.

1966, January 1. Coup d'État. Military officers, led by Colonel **Jean Bedel Bokassa,** ousted the Dacko government. Bokassa became chief of state and immediately broke relations with Communist China.

1967, November 16. French Troops Support Regime. They were airlifted into the capital, Bangui, to help President Bokassa suppress a threatened uprising.

Republic of Chad

1960, August 11. Independence of Chad. A former French colony, Chad had been autonomous since 1958.

1968–1971. Rebellion of Northern Arabs. This was initiated by a fierce battle in the Tibesti Desert region (March 15 1968). The revolt was suppressed by the central government with the assistance of French troops (see p. 1377). French troops were later withdrawn (June 1971).

1971, August 27. Unsuccessful Coup Attempt. The rebels apparently were supported by Libya. Chad broke diplomatic relations with Libya. Relations were resumed in 1973 when Chad broke relations with Israel.

Malawi

1959–1964. Anti-Colonial Disorders in Nyasaland.

1964, July 6. Independence of Malawi. This was granted by Great Britain to the former self-governing colony of Nyasaland. The new republic joined the British Commonwealth.

Rwanda

1959–1961. Civil War in Belgian Rwanda. Shortly after self-government was granted by Belgium, and following U.N.-supervised elections, the suppressed Bahutu (or Hutu) majority rose in revolt against the minority but dominant Watusi (or Tutsi) tribesmen and seized control after bloody tribal warfare against the giant Watusis.

1962, July 1. Independence of Rwanda. A former German colony and later Belgian mandate and trusteeship, Rwanda became a republic, with the majority Bahutu tribe controlling the country.

Zaire

1949–1960. Widespread Unrest and Disorder in Belgian Congo. Colonial authorities were only partially successful in maintaining order.

1960, June 30. Belgium Grants Independence to the Congo. The first president, **Joseph Kasavubu,** appointed leftist **Patrice Lumumba** as premier. The new republic was unprepared for independence and chaos followed; soldiers and civilians rioted, looted, raped, and murdered. The white population fled the country, taking their expertise with them and adding further to the chaos. Central control disappeared from remote provinces. A few pockets of order were kept by some 10,000 Belgian troops remaining in the Congo, mostly in Katanga, where they were protecting the extensive manufacturing complex created by Belgium to process Katanga's great natural wealth.

1960, July 11. Katanga Proclaims Independence. Moise Tshombe, leader of Katanga, refused Lumumba's demands to submit to central control and to oust Belgian troops; he proclaimed secession of Katanga from the Congo and requested more Belgian military assistance to meet threatened Congo invasion. Lumumba appealed to the U.N. for military assistance in suppressing the Katanga revolt.

1960, July 14. International Crisis. The U.N. Security Council approved establishment of a U.N. security force by Secretary General Dag Hammarskjöld to restore order in the Congo (see p. 1368). The first contingent (from Tunisia) reached Leopoldville the next day, as Khrushchev was threatening military intervention on behalf of Lumumba. Eventual strength of the force was 20,000 men.

1960, July 22. U.N. Demands Belgian Troop Withdrawal. The Belgians complied partially (July 31), save for a few local security detachments and a garrison in Katanga remaining at Tshombe's request.

1960, August 12. Katanga Crisis. Hammarskjöld and 240 Swedish U.N. troops arrived at the Elizabethville airport in Katanga. He repeated earlier demands that all Belgian troops be withdrawn and replaced by U.N. troops. Tshombe refused to permit the U.N. force entry into Elizabethville and threatened to use force if necessary. Erratic Lumumba denounced Hammarskjöld for using white troops, for conniving with Tshombe, and for not placing U.N. forces under his command.

1960, August 24. Revolt in Kasai. Lumumba sent a military force to Kasai to suppress revolt of Buluba chief **Albert Kalonji.** Results were inconclusive.

1960, August 30. Belgium Announces Combat Troop Withdrawal. Hammarskjöld insisted that some still remained on in Katanga.

1960, September 14. Lumumba Overthrown. Colonel **Joseph Mobutu,** Army Chief of Staff, seized virtual control of the government. Kasavubu named **Joseph Ileo** as premier. Lumumba, at first arrested, then fled to east-central Congo.

1960, September 20. U.N. Votes Confidence in Hammarskjöld. Following Soviet representative **Valentin A. Zorin**'s bitter denunciation of Hammarskjöld's policies and action in the Congo, the U.N. overwhelmingly voted its confidence in the Secretary General.

1960, October–November. Widespread Violence and Disorders. These were largely the responsibility of undisciplined Congo troops; there were some clashes with U.N. troops.

1960, December 1. Lumumba Arrested. Government troops seized him and flew him to Leopoldville.

1960, December 14. Stanleyville Revolt. **Antoine Gizenga,** who had been Lumumba's vice-premier, proclaimed himself premier and established a pro-Communist government in Stanleyville. His adherents began to expand their control over much of east-central Congo.

1961, February 9. Lumumba Murdered. Kasavubu ordered Lumumba transferred to a "more secure" prison in Katanga (January 17). Soon afterward, under circumstances not clear, Lumumba was murdered. Tshombe has been accused, probably correctly, of responsibility.

1961, February 21. Katanga Mobilization. This was ordered by Tshombe in response to U.N. threats to force integration of Katanga with the Congo.

1961, February 24. U.N. Action against Stanleyville Requested. Premier Ileo asked for help after Gizenga forces seized Luluabourg, capital of Kasai province.

1961, March 12. Proclamation of a New Congo Federation. This was a reorganization of the government, proclaimed at a meeting in Tananarive, Malagasy, of President Kasavubu, Premier Ileo, and all regional leaders except Gizenga.

1961, April 17. Congo-U.N. Agreement. This document, signed by Congolese President Joseph Kasavubu, authorized the U.N. to use force if necessary to prevent civil war in the Congo.

1961, April 26–June 2. Tshombe Arrested. Following a "unity" conference in Leopoldville, Tshombe was seized and detained by the Congo government. After promising to bring Katanga into the Congo, he was released by

Mobutu. Tshombe thereupon repudiated his agreements made under duress.

1961, August 1. New Government. The reconvened Congo parliament elected **Cyrille Adoula** as premier; Gizenga was named vice-premier.

1961, August 21. U.N. Action in Katanga. After a gradual force build-up, U.N. troops in Katanga seized communications centers in and around Elizabethville to force Tshombe to dismiss white mercenary officers. Reluctantly, he complied, but began measures for the defense of Elizabethville.

1961, September 13–21. Hostilities in Katanga. U.N. forces, attempting to seize control of Elizabethville, were unsuccessful; they lost face while Tshombe and his army gained prestige. Tragically, Hammarskjöld, flying in to bring about a cease-fire, was killed in a plane crash near Ndola, Northern Rhodesia (September 18).

1961, November. Congolese Invasion of Katanga Repulsed. Congolese forces were repelled by the Katanga Army.

1961, December 18. U.N. Capture of Elizabethville. This led to an agreement between Tshombe and Premier Adoula to restore the unity of the Congo (December 21). Tshombe again failed to comply and retained autonomy.

1962, January. Renewed Stanleyville Revolt. Gizenga again attempted an uprising from Stanleyville, but his troops were defeated by Congo Army troops; he was dismissed from the government by Adoula.

1962, December 29–1963, January 15. U.N. Offensive in Katanga. Operations against Elizabethville were begun in response to Katangese provocations, and with the objective of ending Katanga's secessioin from the Congo. Katanga forces were completely defeated, and Tshombe forced to flee. After accepting the integration of his province with the central government (January 15), he went into exile.

1963–1964. Steady Reduction of U.N. Forces. This was largely because of lack of funds to support them. Unrest and revolt continued throughout the country, particularly in the northeastern region.

1964, June 16. American Civilians Fighting as Mercenaries. The U.S. State Department conceded, after several denials, that "some American civilian pilots under contract with the Congolese Government have flown T-28 sorties in the last few days in the eastern part of the Congo." This was part of the central government's effort to suppress a revolt of Bafulero tribesmen in Kivu Province.

1964, June 30. Last U.N. Troops Leave the Congo.

1964, July 9. Moise Tshombe Named Premier. This was obviously a last desperate effort to achieve stability in the Congo. Returning from exile, he immediately began a campaign of combined conciliation and threat of force with the various rebel groups. He also strengthened the Congo Army by bringing in white mercenary troops and officers to help train it. Further training assistance was given by a U.S. military-aid mission. Tshombe, considered a traitor to Africans (even by many moderate African leaders) because of his history in Katanga and his more recent recruiting of white troops, was bitterly denounced throughout Africa, and particularly in Communist-bloc states.

1964, August 30. Congolese Army Retakes Albertville. This had been in rebel hands for 2 months.

1964, September–October. Rebel Gains in East-Central Congo. The alarming increase in strength was largely due to assistance from Communists, and from the U.A.R., through Sudan and Uganda.

1964, November. Congo Army Prepares Offensive against Stanleyville. The rebels then seized some 2,000 white hostages and threatened massacre if Congolese troops approached the rebel capital.

1964, November 25–27. Belgian-U.S. Intervention. A surprise airborne landing by a Belgian paratroop battalion, flown by U.S. air units from Belgium via Ascension Island, seized Stanleyville and rescued some 1,650 white hostages. Violent outcries from the Communist bloc and from anti-Tshombe Africans led to the overly quick withdrawal of the Belgian troops and American planes. As a result, other planned rescue missions were abandoned, and several hundred white hostages were brutally massacred.

1965, October 13. Tshombe Ousted from Office. In a power struggle, the Prime Minister was defeated by President Kasavubu.

1965, November 25. Coup d'État. General Mobutu, Commander in Chief of the Army, ousted Kasavubu in a bloodless coup and named himself president. (He later changed his name to Mobutu Sese Seho.)

1966, May 30. Unsuccessful Coup Attempt. Mobutu gained more strength.

1967, June–November. Revolt by Katangese and Foreign Mercenaries. Belgian and French mercenaries attempted to seize control of Katanga and Kivu provinces. Lack of local support, and prompt counteraction by Mobutu, soon resulted in collapse of the revolt. The mercenaries fled the country.

1967, August 9. Rebels Capture Bukava. The capital of Kivu Province was seized by 1,500 Katangese rebels and 160 mercenaries. The rebels failed to receive expected local support and were soon attacked by forces of the central government.

1967, November 5. Rebels Flee to Rwanda; Revolt Collapses. After recapturing Bukava, Congolese troops pursued some 2,000 rebels and mercenaries across the border of neighboring Rwanda.

1968–1974. Stability Returns to the Congo.

1971, October 27. Congo's Name Changed to Republic of Zaire. This was part of the policy of President Mobutu to replace with African names those brought in by colonists in the 19th and 20th centuries.

Zambia

1952–1964. Anticolonial Disorders in Northern Rhodesia.

1964, January 22. Independence of Zambia. This was granted by Britain to the former self-governing protectorate.

1965–1974. Zambia Becomes a Base for Anticolonial Forces. While scrupulously avoiding incidents with Portuguese Angola and Mozambique, and with Rhodesia, Zambia did allow insurgent forces to operate from its territory. The result was strained relations, and some border clashes with Portuguese and white Rhodesian frontier forces.

WEST AFRICA

Angola

1961, February 4–7. Antigovernment Riots. These were apparently connected with the *Santa Maria* incident (see p. 1381).

1961–1974. Rebellion against Portugal. The efforts of separatist rebels to throw off Portuguese colonial rule were generally unsuccessful. Portuguese military repression eliminated most of the rival groups from Angolan territory. Raids against the Portuguese continued, however, from insurgent bases in Zaire and Zambia.

1974, October 15. Cease-fire Agreed. After the coup in Portugal (see p. 1381), the new Portuguese government moved rapidly to grant autonomy, and the opportunity for independence, to its African colonies. The problem was complicated by the rivalries of different insurgent groups. Unrest and violence continued, mostly between the rival national groups.

Cameroon

1956–1959. Anticolonial Terrorist Activities.

1960, January 1. Independence of East Cameroon. This former French mandate and trusteeship (the eastern part of the former German colony of Cameroons) had been autonomous for 2 years.

1961, October 1. Establishment of the Federal Republic of Cameroon. Following a referendum, most of West Cameroon, a former British mandate and trusteeship (the western part of the old German colony), joined East Cameroon to form a federal union. (The northern part of West Cameroon voted to join Nigeria.)

1962–1971. Insurgent Rebellion. Communist insurgents, based in Congo (Brazzaville), kept up a low-keyed rebellion against the central government until the capture and execution of their leader (1971).

People's Republic of the Congo

1960, August 15. Independence of Congo (Brazzaville). A former French colony, Congo had for 2 years been autonomous. Taking the name of Republic of Congo, the new nation was generally referred to as Congo (Brazzaville), to distinguish it from the Democratic Republic of Congo, known as Congo (Kinshasa). This need to distinguish the two Congos by means of their capitals became unnecessary when Congo (Kinshasa) changed its name to Zaire (see p. 1443).

1963, August 15. Coup d'État. The government of President **Fulbert Youlou** was overthrown by a trade-union coup (see also p. 1377). The new President, **Alphonse Massamba-Debat,** established close relations with Communist China.

1966, June 28–29. Coup d'État.

1968, September 4. Coup d'État. Military officers led by Major **Marien Ngoubai** ousted the former government, after fierce fighting between the Army and Cuban-trained militants. Ngoubai soon became president (January 1 1969). The pro-Peking orientation was not changed. The following year the name of the state was changed to People's Republic of the Congo (January 1970).

Dahomey

1960, August 1. Independence of Dahomey. Self-governing since 1958, Dahomey was granted independence by France.

1963, October 28. Coup d'État. The government of President **Hubert Maga** was overthrown by a group of officers under General **Christophe Soglo,** who installed a civilian government.

1965, December 22. Coup d'État. Again General Soglo led a military coup. This time he retained power as president.

1967, December 17. Coup d'État. A military junta under Major **Maurice Kouandete** overthrew President Soglo.

1969, July 12. Unsuccessful Coup Attempt.

1969, October 21. Unsuccessful Coup Attempt.

1969, December 10. Coup d'État. In the sixth coup or coup attempt since independence, the Army overthrew the government of President **Emile D. Zinsou.**

1972, October 26. Coup d'État. Major **Mattneu Kerekou** became president.

Gabon Republic

1960, August 17. Independence of Gabon. Granted by France, the former colonial power.

1964, February 17–18. Unsuccessful Coup Attempt. The government of President **Leon Mba** was briefly overthrown by the Army, but was restored the next day when French troops flew in and routed the revolutionaries.

The Gambia

1965, February 18. Independence of The Gambia. Granted by Britain. The Gambia changed its government from dominion status to a republic within the Commonwealth (April 1970).

Ghana

1948–1956. Anticolonial Disorders.

1957, March 6. Independence of Ghana. Initially with dominion status, Ghana became a republic within the Commonwealth, with **Kwame Nkrumah** as president (July 1 1960).

1966, February 24. Coup d'État. The armed forces seized power during the absence of Nkrumah on a visit to Communist China. Leader of the revolt was Lieutenant Colonel (soon General) **Joseph Ankrah.**

1966, March–April. Invasion Threat from Guinea. President **Sekou Toure** of Guinea, after welcoming Nkrumah (March 10), threatened to use force to restore his old friend to power. He backed down in the face of mobilization by Ghana and Ivory Coast.

1972, January 13. Coup d'État. Prime Minister **Kafi A. Busia** was deposed by army officers while he was in London for medical treatment. The military men, under Colonel **Ignatius Kutu Acheampong,** suppressed a counter-coup (January 15).

Guinea

1958, October 2. Independence of Guinea. Under the leadership of Prime Minister Sekou Toure, Guinea refused to participate in a French plan for gradual independence of its former African colonies, and proclaimed independence. Toure became president.

1966, March–April. Threat of War with Ivory Coast and Ghana. (See above.)

1970, November 22–24. Portuguese-Supported Invasion Attempt. Dissident Guineans, with some Portuguese support, attempted a seaborne invasion but were repulsed. The Portuguese support was in retaliation for Toure's support of rebels in Portuguese Guinea. Portugal was censured by the United Nations.

Guinea-Bissau

1962–1974. Rebellion against Portugal. Rebels in Portuguese Guinea were the most successful of the African nationalists opposing Portuguese rule. By the time of the coup in Portugal (see p. 1381), the rebel organization—African Independence Party of Guinea and Cape Verde (PAIGC)—held at least 50 percent of the countryside.

1974, July 27. Independence of Guinea-Bissau. Portuguese Guinea was the first of Portugal's dissident African colonies to receive independence.

Ivory Coast

1950–1960. Sporadic Anticolonial Disorders.

1960, August 7. Independence of the Ivory Coast. Granted independence by France after a 2-year period of autonomy, under the leadership of President **Felix Houphouet-Boigny.**

1966, March–April. Threat of War with Guinea. When President Sekou Toure of Guinea threatened to invade Ghana, to reinstate Nkrumah as president (see above), he planned to march through the Ivory Coast. President Houphouet-Boigny refused passage and mobilized his army. Toure backed down.

Liberia

1963, February 5. Unsuccessful Coup Attempt.

Mali (Sudanese Republic)

1959, January 17. The Mali Federation Established. The Sudanese Republic, former French Sudan, elected in 1958 to accept French assistance in reaching independence through progressive autonomy. The region joined with Senegal to establish a new federation.

1960, June 20. Independence of the Sudanese Republic. Despite increasing internal strains, the federation with Senegal continued.

1960, August 20. The Sudanese Republic Withdraws from Federation with Senegal. The new state officially adopted the name Republic of Mali (September 22).

1968, November 19. Coup d'État. The socialist government of President **Mobido Keita** was overthrown by a military group under Lieutenant **Moussa Traore,** who became president (December 6).

1971, April 7. Unsuccessful Coup Attempt.

Mauritania

1957, February 15. Unsuccessful Efforts to Annex Mauritania to Morocco. A so-called

Moroccan Army of Liberation, based in Morocco, although disavowed by the Moroccan government, raided into Mauritania in an unsuccessful attempt to incite rebellion and to join Mauritania to Morocco, which claims sovereignty. The effort was suppressed by French troops.

1960, November 28. Independence of Mauritania. Granted by France.

1962, April 1. Rebel Invasion Attempt from Morocco. The rebels were driven out, again preventing efforts to reunite Morocco and Mauritania.

Niger

1960, August 3. Independence of Niger. Granted by France.

1961–1974. Endemic Unrest. This was exacerbated by famine in the early 1970s.

Nigeria

1960, October 1. Independence of Nigeria. It became a republic within the British Commonwealth. Internal tensions soon appeared, based largely on regional tribal allegiance and jealousies.

1966, January 16. Coup d'État. The government of Prime Minister Sir **Abudakar Tafawa Balewa** was overthrown by dissident army officers. Balewa was killed. Following a power struggle among the officers, the army commander, General **Johnson Aguyi-Ironsi,** took control and established a provisional federal military government.

1966, July 29. Coup d'État. A group of mutinous army officers, mostly from the northern Hausa tribe, kidnapped the Chief of State, General Aguyi-Ironsi, and killed him. During another confused power struggle among officers with regional ties, a group of northern officers led by Colonel **Yakubu Gowon** seized control. Bloodshed continued in the country, with most of the victims members of the industrious southeastern Ibo tribe, predominantly Christians, who had achieved coveted positions in

government or commerce in other regions. Most of the Ibo survivors moved from the Moslem northern regions back to their own area.

1967, May 30. Secession of Biafra from Nigeria. Southeastern Nigeria, inhabited mostly by Ibos and rich in agricultural and mineral resources, primarily petroleum, declared independence from Nigeria, under the leadership of Army Lieutenant Colonel **Chukwuemeka Odumegwu Ojukwu,** who became president of the Republic of Biafra.

CIVIL WAR

1967, June 7–1970, January 12. Nigeria-Biafra Civil War. Colonel Gowon's federal government refused to accept the secession of Biafra and moved to subdue the rebellious region by force. Initial efforts of federal troops—about 12,000 strong—to move into Biafra were repulsed. Biafran forces, hastily recruited, invaded the mid-western region and captured the capital, **Benia** (August 9 1967).

1967, September–1968, May. Federal Invasion of Biafra. With substantial superiority in numbers and quality of equipment, the expanded military forces of the federal government slowly occupied almost half of Biafra, but were halted on a defensive line protecting the Biafran capital of Umuahia. Tentative peace negotiations in London (May) and Kampala (June) were fruitless.

1968, September. Renewed Nigerian Offensive. Federal troops captured **Aba** (September 4) and **Owerri** (September 16). Again the Biafran defenses stiffened and halted the invading forces. Biafra had lost direct access to the sea but maintained contact with the outside world through a tenuous airlift to Fernando Po, a Spanish colony.

1969, February–March. Biafran Counteroffensive. In a surprise counteroffensive, Biafran troops pushed southward in an effort to reopen a seaport. They reached the outskirts of Aba before being halted (March 3). A stalemate followed, during which Nigerian aircraft repeatedly bombed Biafran military targets and civilian centers.

1969, June–December. Final Nigerian Offensive. In a slow but steady offensive, the enlarged, well-equipped Nigerian Army—now about 180,000 strong—overwhelmed the desperate Biafran defense.

1970, January 12. Surrender of Biafra. The last organized defense collapsed. Colonel Ojukwa fled by air to seek refuge in the Ivory Coast. The population of Biafra, which had been about 12 million at the start of the revolt, had suffered approximately 2 million dead (many of these children), mostly from starvation. The region, which had been one of the best-developed in Africa before the war, was devastated.

Senegal

1959, January 17. Senegal and Sudan Form the Mali Federation. These colonial regions, which had been granted autonomy by France in 1958, joined to form a federated republic.

1960, June 20. Independence of Senegal Granted by France. The federal union with the Sudanese Republic continued.

1960, June 20. Collapse of the Mali Federation. The Sudanese Republic adopted the name of Mali (see p. 1445); Senegal continued an independent status.

1962, December 17. Unsuccessful Coup Attempt.

Sierra Leone

1955–1960. Anticolonial Disturbances.

1961, April 27. Independence of Sierra Leone. It assumed dominion status within the British Commonwealth.

1967, March 23. Coup d'État. A military junta seized power without bloodshed, under the leadership of Majors **Charles Blake** and **Andrew T. Juxon-Smith.**

1968, January 25. Coup d'État. Military insurgents seized power.

1968, April 18. Coup d'État. A group of noncommissioned officers overthrew the government of Colonel **Juxon-Smith.** They called

back from exile Colonel **John Bangura** to head a new junta, which installed **Siaka Stevens** as Prime Minister.

1968, November 21–25. Violence Suppressed by Government.

1970, October 5. Unsuccessful Coup Attempt.

1971, January 21. Unsuccessful Coup Attempt.

1971, April 19. Sierra Leone Becomes a Republic. Prime Minister Stevens became president. The nation remained in the Commonwealth.

Togo

1960, April 27. Independence of Togo. Granted by France, following 2 years of autonomy.

1963, January 13. Coup d'État. A military junta of noncommissioned officers seized power after killing President **Sylvanus Olympio.**

1967, January 13. Coup d'État. The government of President **Nicholas Grunitzky** was overthrown without bloodshed by Army Chief of Staff Lieutenant Colonel **Etienne Eyadema.**

1970, August 8. Unsuccessful Coup Attempt.

Upper Volta

1960, August 5. Independence of Upper Volta. Granted by France, following 2 years of autonomy.

1966, January 3. Coup d'État. Army Chief of Staff Lieutenant Colonel **Sangoule Lamizana** bloodlessly overthrew the government of President **Maurice Yameogo.** Lamizana proclaimed himself chief of state.

SOUTHERN AFRICA

Botswana

1966, September 30. Independence of Botswana. The former British colony of Bechuanaland remained in the Commonwealth.

Lesotho

1966, October 4, Independence of Lesotho. The former British protectorate of Basutoland, completely surrounded by the Union of South Africa, remained in the Commonwealth.

1970, January 30. Coup d'État. In a bloodless and apparently reasonably amicable action, the Prime Minister, Chief **Jonathan,** placed King **Moshoeshoe II** in house arrest and seized power. The King left the country for 6 months, returned, took an oath not to get involved in politics, and was returned to the throne.

Mozambique

1962–1974. Rebellion against Portugal. A number of minor separatist movements united under the Frente de Liberatacao de Mocambique (FRELIMO), which established its headquarters in Dar es Salaam. Although other rival groups were organized, FRELIMO carried most of the burden of sporadic guerrilla war against the Portuguese Army and was able to establish control in a number of regions of the country, operating mainly from bases in Tanzania and (to a lesser extent) Zambia.

1974, September 7. Independence from Portugal Agreed. After the coup in Portugal (see p. 1381), the new Portuguese government moved promptly to grant independence to its African colonies. By agreement signed by Portuguese and FRELIMO representatives in Lusaka, Zambia, a cease-fire was initiated. Formal independence followed (June 25, 1975).

Rhodesia

1960, July–October. Rioting in Southern Rhodesia.

1965, November 11. Unilateral Declaration of Independence. The white population (270,000) of the British colony of Southern Rhodesia, fearful of British policies that seemed directed toward placing political power in the hands of the majority black population (5,400,000), declared independence from

Great Britain, under Prime Minister **Ian D. Smith.** Economic sanctions were announced by Great Britain, the U.S., and most African nations.

1968, May 9. United Nations Security Council Orders Trade Embargo on Rhodesia. This action was taken after the failure of repeated British peaceful efforts to get Rhodesia to grant political rights to the black population.

1970, March 2. Rhodesia Proclaims Itself a Republic.

1971–1974. Low-Level Insurgency. Although no immediately serious threat to the stability of the Rhodesian government, incursions of black insurgents, based in neighboring countries, particularly Zambia, aroused fears and tension in the white population.

Republic of South Africa

1949–1962. Sporadic Race Riots. Of these the most significant were at Durban (January 1949; June 1959; January 1960), Kimberley (November 1952), and Sharpeville (March 1960). They were suppressed ruthlessly by government police and troops.

1949, July 13. The United Nations Defied. The Union of South Africa rejected a U.N. Trusteeship Council demand that Southwest Africa become part of the trusteeships system.

1950–1953. Korean War. South Africa provided an air unit to the U.N. command.

1960, March 21. Sharpeville Race Riot. Police opened fire on a crowd of 20,000 blacks attacking a police station; 56 were killed, 162 wounded. Other riots, breaking out throughout the country, were quickly suppressed. There was a wave of international revulsion against South Africa and its racial policy of *apartheid* (apartness).

1961, May 31. South Africa Becomes a Republic. The new republic withdrew from the British Commonwealth.

1966, October 27. United Nations Resolves and End to South African Mandate over South West Africa. This was a League of Nations mandate, after World War I, over the former German colony. South Africa refused to recognize the authority of the U.N. to interfere with its authority over the territory.

1970, July 23. United Nations Security Council Arms Embargo. This was in response to repeated South African defiance of the authority of the U.N. over South West Africa.

1974, November 25. U.N. General Assembly Votes to Suspend South Africa from Membership.

NORTH AMERICA

UNITED STATES

1945, September 14. Pearl Harbor Investigation. A Joint Congressional Committee was appointed to investigate the Pearl Harbor disaster. The investigation opened 2 months later (November 15), and submitted an inconclusive report (July 20, 1946).

1945, December 19–1947, July 26. Military Unification Controversy. President Truman sent a special message to Congress outlining a program of unification for the Army and the Navy and the establishment of the new Air Force in a single Department of Defense. Senior naval officers testifying before the Senate Military Affairs Committee had already expressed vehement opposition to the unification concept. Despite this opposition, President Truman later submitted to Congress a specific plan for the merger of the Army and Navy and a new Air Force in a Department of National Defense (June 15, 1946).

1945–1947. The Marshall Mission to China.

1946, January–March. Crisis in Iran. (See p. 1402.)

1946, March 5. Churchill Speech at Fulton, Mo. In an address at Westminster College, Churchill advocated a fraternal association between the U.S. and Great Britain to deter Soviet aggression. He coined the expression "Iron Curtain."

1946, April 4. Horse Cavalry Abolished. Remaining cavalry units and individuals were merged with armored forces; cavalry disappeared as a separate service.

1946, July 1–25. Bikini Nuclear Tests. First peacetime weapons tests in history were carried out at Bikini Atoll in the Marshall Islands in the Pacific.

1946, December 16. Establishment of Unified Overseas Commands. This was ordered by President Truman on the advice of the Joint Chiefs of Staff.

1947, January 1. Civilian Control of Nuclear Affairs. The U.S. Atomic Energy Commission formally took command of all U.S. atomic-energy affairs.

1947, January 7. General Marshall Appointed Secretary of State. The State Department made public his report on his mission to China, in which he criticized both reactionaries in the Nanking government and the Chinese Communists.

1947, March 12. Truman Doctrine. This proposed economic and military aid to nations threatened by Communist aggression. Specifically, President Truman asked Congress for $400 million to give economic and political aid to Greece and Turkey, both seriously threatened by the possibility of Soviet aggression and both further endangered by Britain's recent decision to withdraw forces from Greece. This was approved by Congress (May 15) and signed by the President (May 22).

1947, March 31. The Selective Service Act Expired.

1947, June 5. Marshall Plan. In a speech at Harvard, Secretary of State Marshall suggested American economic assistance to help Europe recover and to gain the strength necessary to avoid internal subversion and external aggression.

1947, June 5. Satellite Peace Treaties. The Senate ratified peace treaties with Italy, Hungary, Rumania, and Bulgaria.

1947, July 26. The National Security Act of

1947. This was based upon the unification plan submitted a year earlier by President Truman. Three armed services were unified within a National Military Establishment. Secretary of the Navy **James Forrestal** was appointed the first Secretary of Defense.

1947, September 26. Establishment of the U.S. Air Force. Pursuant to the Act of July 26 (see above).

1948, March 27. Key West Agreement. Publication of an agreement between the 3 armed services to resolve disputes regarding their respective roles and missions in the national defense of the U.S.

1948, April 3. Marshall Plan in Operation. President Truman signed the first foreign-aid bill for $6.98 billion.

1948, June–1949, May. Berlin Blockade. (See p. 1378.)

1948, June 11. Vandenberg Resolution. A resolution sponsored by Senator **Arthur H. Vandenberg** proclaimed U.S. policy to give military aid to defensive alliances among the free nations of the world.

1948, June 19. New Selective Service Act. Conscription was reinstated by Congress.

1948, August 23. Newport Agreement; New Roles and Missions. At a meeting at Newport, R.I., Secretary of Defense Forrestal and the Joint Chiefs of Staff agreed on further revision of the roles and missions of the armed forces.

1949, April 23. Construction Stopped on Supercarrier. Secretary of Defense Johnson ordered abandonment of construction on the 65,000-ton USS *United States.*

1949, June 3–August 25. B-36 Controversy. This began with accusations and rumors in the press (by Navy adherents in an interservice controversy) that the B-36 bomber had been chosen as the principal Air Force weapon for personal and political reasons. Secretary of Defense **Louis Johnson** and Secretary of the Air Force **Stuart Symington** were anonymously accused of having personal-gain motives in the decision. After a 2-month investigation, the House Armed Services Committee cleared top-ranking government officials and Air Force officers of "charges and insinuations that collusion, fraud, corruption, influence, or favoritism played any part whatsoever in the procurement of the B-36 bomber."

1949, July 21. North Atlantic Treaty Ratified. This was signed by President Truman (July 25) after Senate ratification. The treaty went into effect after ratification by all 12 signatories (August 24).

1949, August 2. Military Reorganization. Congress approved military recommendations suggested by the Hoover Committee and by Secretary of Defense Forrestal (December 1948). The National Military Establishment was renamed the Department of Defense with increased powers for the Secretary of Defense. The Departments of Army, Navy, and Air Force were reduced from cabinet to departmental rank, and a Chairman was provided for the Joint Chiefs of Staff.

1949, August 5. State Department White Paper on China. (See p. 1427.) Secretary of State **Dean Acheson** blamed Generalissimo Chiang Kāi-shek for the defeats of the National Government. Dr. **Wellington Koo,** Chinese Ambassador to the U.S., acknowledged that China might have made some errors, but insisted that mistakes were not confined to his country (August 7).

1949, October 3–27. "Revolt of the Admirals." This was a continuation of the B-36 controversy. Senior Navy officers publicly and privately charged that the Army and Air Force were trying to destroy naval aviation in order to reduce Navy influence in the military establishment. The Chief of Naval Operations, Admiral **Louis Denfeld,** and Admiral **Arthur Radford,** Commander in Chief of the Pacific Fleet, wrote strong letters criticizing unification of the armed forces. In a subsequent Navy Department investigation, it was ascertained that the instigator of this controversy was Captain **John G. Crommelin,** U.S.N. During a House Armed Services Committee investigation, Admiral Radford attacked the B-36 and the concept of nuclear war as advocated by the Air Force. In subsequent testimony before the House Armed Services Committee, Admiral

Denfeld accused the other services of not accepting the Navy "in full partnership." General **Clifton B. Cates,** Commandant of the Marine Corps, personally attacked Secretary of Defense Johnson and the Chiefs of Staff of the Army and Air Force (October 16). Army Chief of Staff General **Omar Bradley,** in testifying before the committee, declared that the Navy admirals were "fancy Dans" "in open rebellion" against civilian authority. Following an investigation, Navy Secretary **Francis P. Matthews** recommended that Truman relieve Admiral Denfeld as Chief of Naval Operations; this effectively ended the revolt (October 27).

1949, October 29. Controversy on Air Force Build-up. President Truman impounded more than $600 million earmarked by Congress for a 58-group Air Force; he instructed Secretary of Defense Johnson not to build more than 48 groups.

1949, December 7. Secretary of Defense Johnson "Cuts Fat." He announced that by more efficient operation the 1949–1950 budget of $15.7 billion had been cut to $13.0 billion for 1950 without reducing preparedness. Most military men believed preparedness was dangerously impaired, an opinion seemingly corroborated by early results in Korea (see p. 1356).

1950, January 5. Truman Announces Continued Economic Aid to Nationalist China. No more military assistance was contemplated.

1950, May 5. Enactment of Uniform Code of Military Justice.

1950, June 25. Korean War Begins. (See p. 1354.)

1950, June 30. Partial Mobilization; Selective Service Extended. President Truman signed a bill extending the Selective Service Act until July 9, 1951. The measure also authorized the President to call the National Guard and Organized Reserves for 21 months of active service.

1950, September 18. Congress Approves General Marshall as Secretary of Defense. Legislation was necessary, since he had recently been a Regular Army officer.

1950, December 8. Truman Confers with British Prime Minister Attlee. Attlee had come to Washington to convey British hopes that the atomic bomb would not be used first by the United States in the Korean War. President Truman stated that he hoped that world conditions would never call for the use of the atomic bomb. They agreed to support the U.N. in attempts to achieve a free and independent Korea.

1950, December 16. Truman Proclaims State of National Emergency. This was to facilitate prosecution of the war in Korea.

1951, May 1. Publication of Wedemeyer Report. This controversial report on China and Korea had been submitted to President Truman, on September 9, 1947, by retiring General Albert C. Wedemeyer. It had been partially described in the State Department's White Paper on China (August 5, 1949).

1951, May–June. Senate Investigates Relief of General MacArthur. (See p. 1361.)

1951, September 1. Security Treaty with Australia and New Zealand (ANZUS Pact).

1952, March 4. Universal Military Training. The House of Representatives defeated a proposal for universal military training, based upon recommendations of an advisory commission to President Truman (June 1, 1947) and of former Army Chief of Staff Eisenhower (February 15, 1948).

1952, July 13. Military Aid to Yugoslavia Approved. (See also p. 1387.)

1952, August 4. ANZUS Pact. The U.S., Australia, and New Zealand established a Pacific Council. At the first meeting in Washington, the ANZUS Council pledged to guard against the threat of communism and maintain peace in the Pacific (September 9–10, 1953).

1952, November. First Hydrogen Weapon. The U.S. exploded a thermonuclear weapon at Eniwetok. President Truman had broken an AEC controversy earlier by ordering the development (January 31, 1950).

1953, April 3. Military Reorganization Plan. President Eisenhower proposed to give civilian officials in the Defense Department more control. The plan was approved by Congress (June 27).

1953, December 19. Emphasis on Air Power. President Eisenhower and the National Security Council supported plans for the Department of Defense to emphasize air power and continental defenses by increasing Air Force strength and budget, decreasing the Navy and Marine Corps by 15 percent, and decreasing the Army by one-third. The next budget request (January 21, 1954) provided for an Army budget of $10.198 billion, Navy $10.493 billion, and Air Force $16.209 billion. Later General Ridgway, in a magazine article, denied President Eisenhower's statement that the JCS had approved Army cuts (January 1956).

1954, March 8. President Eisenhower's Report on the Mutual Security Act. The U.S. had shipped $7.7 billion of arms and other military equipment to allies since October 1949, and about $3.8 billion in 1953. Almost $6 billion had gone to Western European nations alone, who themselves had spent over $35 billion to build up NATO defenses, of which $11.5 billion had been spent in 1953. Since 1949, military aid to Greece and Turkey totaled $761 million; Far East aid totaled $1.18 billion.

1954, March 16. Presidential Authority under NATO and Rio Treaties. Secretary of State John Foster Dulles said the President had authority to order instant retaliation without consulting Congress in the event of an attack against the U.S., its western European allies, or the Western Hemisphere.

1954, May 7. U.S. Rejects Russian Application to NATO. Britain and France had been consulted prior to the rejection.

1954, September 30. First Atomic-Powered Submarine. The USS *Nautilus* was commissioned.

1955, January 12. "Massive Retaliation" Concept. Secretary of State Dulles announced that the President and the National Security Council had taken a basic decision "to depend primarily upon a great capacity to retaliate instantly [against aggression anywhere] by means and at places of our choosing."

1955, January 17. Defense Budget Controversy. The national defense budget of $34 billion included $15.6 billion for the Air Force, $9.7 billion for the Navy, and $8.85 billion for the Army. General Matthew B. Ridgway, Army Chief of Staff, subsequently testified before the House Armed Services Committee that this budget, and planned cuts in Army armed forces, jeopardized the safety and security of the U.S.

1955, July 11. Opening of U.S. Air Force Academy. Temporarily at Lowry Air Force Base, Denver, Col., it was permanently located at Colorado Springs (September 1958).

1956, May 18. Outbreak of "Colonels' Revolt." Newspaper publication of "leaked" Army and Air Force staff papers revealed bitter interservice rivalry, with Army jealousy and suspicion of the Air Force particularly outspoken in a paper challenging Air Force doctrine that national security lay mainly in air power. Newspapermen, remembering the "Revolt of the Admirals" (see p. 1450), noted that the Army papers had been prepared by a small staff group composed primarily of colonels. Secretary of Defense **Charles E. Wilson** called a special press conference of the Joint Chiefs of Staff to try to demonstrate interservice solidarity (May 21). The Army officers involved were reprimanded and ordered away from the Pentagon.

1956, October 28. Eisenhower Warns Israel. He warned against taking any "forceful initiative in the Middle East." When Israel attacked Egypt the following day (see p. 1341) and Britain and France subsequently became involved at Suez (see p. 1342), Eisenhower forcefully opposed their "aggressions" directly and in the U.N.

1956, November 26. Army Roles and Missions Curtailed. Secretary Wilson gave the Air Force control of all missiles with a range of more than 200 miles. He also severely restricted the Army's planned aviation program.

1956, February 12. Army Opposes Force Levels. Chief of Staff General M. D. Taylor said that 19 authorized divisions were inadequate; 27 or 28 were required to back up U.S. international commitments.

1957, March 9. "Eisenhower Doctrine." The President signed bills authorizing him to use armed forces in Middle East if necessary.

1956, December. End of an Era. The Army announced deactivation of the last mule unit (December 1) and the end of carrier pigeons in the Signal Corps (December 4).

1957, September 24–25. Eisenhower Sends Troops to Little Rock. This ended local defiance of court-ordered school integration.

1958, January 31. First U.S. Satellite. Designed and developed by the U.S. Army, this was "Explorer I." Later, after several more successful launchings, the Army Ballistic Missile Agency was transferred to the newly created National Aeronautics and Space Administration, taking the Army out of space exploration (October 21, 1959).

1958, March 8. Deactivation of Last U.S. Battleship. This was the USS *Wisconsin*.

1958, May 12. NORAD Established. Formal establishment of the Joint Canadian-U.S. North American Air Defense Command confirmed measures already initiated through informal agreement of the two governments (October 1957). NORAD Headquarters was established at Colorado Springs.

1958, May 20. Strategic Army Corps Established. The Pentagon announced that 4 combat-ready divisions, comprising (with supporting troops) 150,000 men, were being combined into a force capable of action at short notice "to meet or reinforce any initial emergency requirements throughout the world."

1958, July 15. U.S. Intervention in Lebanon.

1958, August 6. Department of Defense Reorganization Act. This was the result of a plan submitted to Congress by President Eisenhower (April 16), to (1) stop "unworthy and sometimes costly [interservice] bickering"; (2) assure "clear-cut civilian responsibility, unified strategic planning and direction and completely unified commands"; (3) stop "inefficiency and needless duplication"; and (4) assure "safety and solvency." The Act (a) substantially strengthened the position and authority of the Secretary of Defense in relation to the service departments;

(b) authorized a limited general staff type organization for the Joint Staff, capable of more efficient and more comprehensive service to the Joint Chiefs of Staff; and (c) removed service elements in unified commands from the command jurisdiction of the service secretaries and chiefs of staff.

1959, March 3. Warning of Submarine Menace. Admiral **Arleigh A. Burke,** Chief of Naval Operations, warned of the ever-present danger to U.S. warships and commercial shipping posed by Soviet submarines in international waters.

1959, December 10. Withdrawal of U.S. Troops from Iceland.

1959, December 30. First Operational Polaris Nuclear Submarine. USS *George Washington* was commissioned.

1960, January 19. No Missile Gap. Defense Secretary **Thomas S. Gates** told the Senate Armed Services Committee that previously announced Pentagon estimates of a "missile gap," or "deterrent gap," were based on evaluation of Soviet production potentiality, rather than actual Soviet production. He later (March 16) told the Senate Preparedness Subcommittee that the U.S. had, and would maintain, a nuclear destructive power "several times" greater than that of the U.S.S.R.

1960, May 1. U-2 Reconnaissance Plane Shot Down over Russia. Khrushchev made the announcement to the Supreme Soviet that the plane had been shot down from an altitude of 65,000 feet near Sverdlovsk.

1960, May 10. First Submerged Circumnavigation of the Earth. U.S. nuclear-powered submarine *Triton* went 41,519 miles in 84 days.

1960, May 17. Summit Conference Collapses in Paris. Khrushchev, angrily denouncing American spying by the U-2 plane over Russia, broke up the meeting. President Eisenhower later reported on TV to the people on the U-2 incident and the failure of the summit conference (May 25).

1960, June 13. Soviet Spy Net. Senator **J. William Fulbright** made public a U.S. State Department report that the Soviet bloc maintained a network of 300,000 spies throughout the world.

1960, July 14. U.S. Reaffirms Monroe Doctrine. A response to Khrushchev's threats to retaliate with missiles if the U.S. should intervene militarily in Cuba.

1960, July 20. First Successful Polaris Firing. This was from the submerged nuclear submarine USS *George Washington.*

1960, September 24. Launching of USS *Enterprise*. This was the largest ship ever built and the world's first nuclear-powered aircraft carrier.

1960, November 1. Nuclear Submarine Base Agreement with United Kingdom.

1960, November 14. First Polaris Patrol Mission. The USS *George Washington,* armed with 16 thermonuclear Polaris misiles, sailed from Charleston, S.C.

1961, January. Kennedy Reappraises U.S. Defense Posture. In his first State of the Union Message, he said that he had ordered an appraisal of U.S. strategy, and had directed action to increase U.S. airlift capacity, to step up the Polaris submarine program, and to accelerate the missile program. Shortly thereafter, in a revised budget (March 28), President Kennedy requested $1.954 billion more in defense appropriations than the $41.84-billion budget submitted by Eisenhower.

1961, April 15–20. Bay of Pigs Incident in Cuba. (See p. 1460.)

1961, May 5. First U.S. Manned Space Flight. This was a suborbital flight made by Navy Commander **Alan B. Shepard, Jr.,** launched from Cape Canaveral, Fla., and landed safely in the Atlantic after reaching an altitude of 116.5 miles.

1961, June 3–4. Kennedy-Khrushchev Meeting at Vienna. No agreement was reached, and Kennedy left the meeting apparently with grave doubts as to a peaceful future.

1961, July 10. Berlin Crisis. (See p. 1379.) The United States rejected a Soviet proposal that the U.S., Britain, and France withdraw their forces from West Berlin, to be replaced by a smaller U.N.-supervised force.

1961, September 26. Establishment of the Arms Control and Disarmament Agency.

1962, January 3. Army Increase. President Kennedy announced an increase from 14 to 16 divisions. The Army soon thereafter (February 23) announced an increase in the size of the Strategic Army Corps from 3 to 8 divisions, and its strength had gone from 90,000 to 160,000 men in 6 months since the beginning of the Berlin crisis.

1962, May 6. First Polaris Nuclear Warhead Test. The missile was fired from the nuclear submarine USS *Ethan Allen* and exploded in the Christmas Island testing area.

Cuban Missile Crisis, September–November 1962

1962, September 4. Soviet Military Aid to Cuba. President Kennedy announced that Cuba's military strength had been increased by deliveries of Soviet equipment, but that there was no evidence of significant offensive capability in Cuba. A few days later, despite prodding from members of Congress (particularly Senator **Kenneth Keating**), President Kennedy said that he opposed any invasion of Cuba (September 12). Next day he warned the U.S.S.R. and Cuba against any build-up of offensive strength.

1962, October 22. Crisis Begins. President Kennedy announced to the nation on TV that U.S. surveillance had "established the fact that a series of offensive missile sites is now in preparation" in Cuba that could menace most of the major cities of the Western Hemisphere, and that jet bombers capable of carrying nuclear weapons were being uncrated. He said he had ordered a naval and air quarantine of Cuba that would not be lifted until all offensive weapons were dismantled and removed from Cuba under U.N. supervision. He declared that the launching of any nuclear missile from Cuba against any Western Hemisphere nation would be considered an attack on the U.S. "requiring a full retaliatory response upon the Soviet Union." U.S. forces were placed on alert, and preparations were begun to invade Cuba if necessary.

1962, October 23. Action in U.N. and OAS. A

U.S. demand for dismantling of the bases was lodged in the U.N. Security Council. The Council of the Organization of American States approved a resolution authorizing the use of force to carry out the quarantine.

1962, October 23. Soviet Alert. Alerting its armed forces, the Soviet government challenged the U.S. right to quarantine its shipments to Cuba. U.S. invasion preparations continued.

1962, October 24–29. Secret U.S.-Soviet Negotiations. Prime Minister Khrushchev backed down after an exchange of letters with President Kennedy. He agreed to halt construction of bases in Cuba, to dismantle and remove Soviet missiles there under U.N. supervision. In turn Kennedy agreed to lift the quarantine when the U.N. had taken the necessary measures, and pledged that the U.S. would not invade Cuba, and would withdraw its missiles from Turkey (a step ordered before the crisis, but not publicized).

1962, November 2. Quarantine Lifted. President Kennedy reported to the nation that the Soviet missile bases were being dismantled, and "progress is now being made for the restoration of peace in the Caribbean." The U.S. Defense Department later announced that the U.S.S.R. had begun withdrawal of its jet bombers from Cuba, as pledged by Khrushchev (December 3). In response to congressional criticism, Secretary of Defense **Robert S. McNamara** proved by photographs that offensive weapons had been fully removed from Cuba (February 6 1963). The U.S. and U.S.S.R. later reported to U.N. Secretary General Thant that the crisis was ended (January 7 1963).

1962, December 19. Missile Inventory. The Air Force announced that the U.S. had 200 operational ICBM's: 126 Atlas, 54 Titan, and 20 solid-fuel Minuteman.

1963, May 9. Russians Remain in Cuba. The Defense Department estimated that 17,500 Russians were still in Cuba, including 5,000 combat troops.

1963, August 30. Opening of the "Hot Line." Direct communications were provided between White House and Kremlin.

1963, October 22–24. Operation Big Lift. Fifteen thousand men in a U.S. division were airlifted 5,600 miles in 63 hours and 20 minutes from the U.S. to West Germany.

1964, February 6. Cuba Cuts off Supply of Water to Guantánamo Base. A U.S. distillation plant made the base self-sufficient.

1964, August 7. Gulf of Tonkin Resolution. (See p. 1415.)

1964, September 17. Announcement of Perfection of U.S. Antimissile Defense Systems. President Johnson also announced that U.S. weapons could intercept and destroy hostile armed satellites in orbit. Secretary of Defense McNamara described 2 rocket systems: the Army's 3-stage solid-fueled "Nike Zeus" and the Air Force's liquid-fueled "Thor."

1964, December 12. Defense Secretary McNamara Proposes to Combine Army Reserve and National Guard. After violent public and Congressional protest the plan was dropped.

1965, February 7–1973, January 23. Vietnam War, (See p. 1321.)

1965, April 28. U.S. Intervention in the Dominican Republic. (See p. 1460.)

1965, September 30. Reorganization of Army Reserve and National Guard. Secretary McNamara achieved part of his objective of reserve component reorganization by eliminating a number of Reserve units and increasing the readiness of the remaining Reserve and National Guard units. Additional reorganization to this same end was accomplished 2 years later (June 2 1967).

1967, July 12–17. Race Riots in Newark, N.J. National Guard troops were employed to restore order.

1967, September 18. McNamara Announces Plans for Antiballistic Missile System. "Nike-X" (later called "Sentinel") would be a "light" system employing "Sprint" and "Spartan" missiles, capable of countering any long-range missile that Communist China might deploy in the next 10 years. A system to prevent Soviet missiles from reaching this country would be impossibly expensive. Congress failed to approve.

1968, January 23. *Pueblo* **Incident. (See p. 1430.)**

1968, March 2. C-5A "Galaxy" Displayed. This jet transport was the world's largest aircraft.

1968, March 30. NORAD Renewed. The U.S. and Canada agreed to extend the treaty against long-range bombing attack (see p. 1453) for another 5 years. Canada was not committed to involvement in continental defense against missiles.

1968, March 31. President Johnson Announces Cessation of Bombing of North Vietnam. (See p. 1329.)

1968, May 29. Disappearance of Nuclear Submarine U.S.S. *Scorpion.* This was announced after the vessel was 2 days overdue from a 3-month training exercise. There was a crew of 99 aboard.

1969, March 14. President Nixon Announces the "Safeguard" Antiballistic Missile System. This modification of the "Sentinel" system plan was approved by Congress (November 6).

1969, March 29–1970, March 17. Investigation of Alleged 1968 Massacre of Vietnamese Civilians at My Lai. (See p. 1328.) The allegation was made in a published letter to Secretary of Defense McNamara and several members of Congress by a former soldier, **Ronald L. Ridenhour.** An investigating committee headed by Lieutenant General William R. Peers was appointed by Secretary of the Army **Stanley Resor.** Nearly one year later Resor announced that the Peers investigation report had been submitted to him (March 14), that it established that a massacre had in fact occurred, and that information about the incident had been "kept from being passed up the chain of command" by "several individuals."

1969, July 20. First Men Reach Surface of Moon. Astronauts **Neil Armstrong** and Colonel **Edwin Aldrin, Jr.,** U.S.A.F., were carried to the moon by the Apollo 11 spacecraft.

1969, October 15. Vietnam War "Moratorium." Hundreds of thousands of Americans participated in an antiwar demonstration. In a similar demonstration 250,000 "marched on Washington" a month later (November 15).

1969, November 24. Nuclear Nonproliferation Treaty Signed. Signing by President Richard M. Nixon and Soviet President Nikolai V. Podgorny took place on the same day. The Senate had ratified it March 13.

1970, May 4. Kent State University Incident. Ohio National Guard troops, called out to control disorders at Kent State, fired into demonstrating students with insufficient provocation, killing four.

1970, May 25. Announcement of Deployment of MIRV Missiles. Multiple Individual Reentry Vehicles, or multiple warheads, were fixed on U.S. intercontinental ballistic missiles (ICBMs).

1970, August 28. Successful Test of Safeguard ABM System. A "Spartan" ABM missile launched from Kwajalein intercepted the nose cone of a "Minuteman" ICBM, launched from Vandenburg Air Force Base, 4,200 miles away.

1970, November 23. U.S. Coast Guard Returns Defector to Soviet Control. When a member of a crew of a Soviet fishing trawler jumped from his vessel to the USCGC *Vigilant,* Soviet sailors were allowed to board and seize him. The *Vigilant*'s Commander **Ralph W. Eustis** was subsequently relieved of command, and the First Coast Guard District commander and his chief of staff were retired.

1971, March 29. Lieutenant William H. Calley Convicted by Court-martial of Murder of Civilians at My Lai. This ended more than 4 months of frequently interrupted proceedings and 13 days of deliberation by the military court. Because of discrepancies in testimony, Calley was found guilty of only 22 of the 102 murders with which he had been charged. Four days later he was sentenced to life imprisonment. Although Calley's guilt in the taking of innocent lives was clear, paradoxically the verdict was opposed by as many liberal "dove" opponents of the war (who thought Calley an unfortunate victim of a cover-up by superiors who were responsible for the war and its alleged atrocities) as it was by ardent "hawk" supporters of the war (who thought he had merely been trying to do his duty in obedience to orders as he understood them). The resulting public furor assured lenient treatment for Calley and his eventual release from a few months' confinement in prison 3½ years after his conviction (December 1974). Also contributing to sympathy for Calley was the fact that only 4 people had been tried for the incident and the alleged cover-up, and only Calley had been convicted. (See pp. 1328, 1456.)

1971, April 12. Largest "Conventional" Bomb Announced. This was the 15,000-pound "daisy cutter," being used in Vietnam to clear jungle areas the size of a baseball park, when set to explode 7 feet above ground.

1971, April 19–24. Antiwar Demonstrations. Vietnam veterans were prominent among the demonstrators in Washington and San Francisco.

1971, May 19. Defeat of the Mansfield Amendment. Senator **Mike Mansfield's** proposal for a 50 percent reduction of the 300,000 U.S. combat troops in Europe was defeated 61-36.

1971, June 13. *New York Times* **Publishes "Pentagon Papers."** A highly classified official record of U.S. involvement in Vietnam, prepared in the late 1960s at direction of Secretary of Defense McNamara, was covertly delivered to the *New York Times* by **Daniel Ellsberg,** a former government official and later a research analyst with the Rand Corporation, who was authorized access to the closely guarded documents. Ignoring the official classification stamps on the documents, the *Times* began publishing excerpts from the 47-volume collection until prohibited by government injunction 2 days later. The *Times* appealed the case to the Supreme Court, which ruled in favor of the newspaper. Publication was renewed, and selections were published in book form. Ellsberg was later tried for violation of government security legislation, but the case was dismissed on technical grounds.

1971, June 17. Treaty with Japan on Okinawa. The island was returned to Japan, but U.S. bases were to remain.

1971, November 6. Explosion of 5-Megaton H-Bomb under Amchitka Island. This was the last of some 20 underground tests for a "Spartan" missile warhead. Efforts by environ-

mentalists and antiwar protesters to block the test were rejected by the Supreme Court. There were none of the immediate effects prophesied by the protesters.

1972, May 27. Cessation of Work on "Safeguard" ABM System. This was the result of a "summit" agreement between President Richard M. Nixon and Soviet Chairman Brezhnev (see p. 1371).

1972, June 3. Partial Release of Peers Report on My Lai Massacre. The report of Peers' investigating committee charged that some 300 civilians had been slaughtered in the My Lai incident, and that there was serious misconduct on the part of Americal Division commander, Major General **Samuel W. Koster,** and his assistant division commander, Brigadier General **George H. Young, Jr.,** in their failure to investigate the affair or to take disciplinary action against subordinates who had been obviously guilty of misconduct either in the operation which led to the massacre or in subsequent failure to take corrective or investigative action. General Koster was demoted and reprimanded (May 19, 1971) and allowed to retire; General Young was also reprimanded and allowed to retire. The full report was released 2 years later (November 13; 1974 see above and pp. 1328, 1456.)

1972, September 14. U.S. Senate Approves U.S.-Soviet Offensive Nuclear Missile Agreement. (See p. 1371.) By the "Jackson Amendment," proposed by Senator **Henry M. Jackson,** Senate approval was contingent upon U.S. reservation that the accord should not limit the U.S. "to levels of intercontinental strategic forces inferior to the limits provided for the Soviet Union." The agreement had permitted the U.S.S.R. to have about 1,618 ICBMs, including some 300 massive SS-9s, compared to 1,054 American ICBMs, generally smaller, but mostly with multiple warhead (MIRV) capability, in addition to a substantially larger fleet of U.S. long-range bombers.

1973, January 23. Vietnam Peace Accord. (See p. 1333.)

1973, January 27. End of U.S. Conscription. It was announced by Defense Secretary

Melvin Laird, 6 months earlier than expected, principally because of the end of the Vietnam War. Legislation remained in effect for use in event of an emergency.

1973, May 14–June 22. First Orbiting Space Station. A Navy crew manned the station, "Skylab."

1973, November 7. Congress Limits Presidential Authority to Commit Forces to Hostilities Abroad. Congress passed, over President Nixon's veto, legislation requiring (1) the president to inform Congress within 48 hours of committing any U.S. forces to a foreign conflict, or of "substantially" enlarging the size of a combat force already in a foreign country, and (2) the end of such commitment within 60 days unless its continuation was specifically authorized by Congress.

1974, December 14. U.S. Senate Ratifies Geneva Protocol. After a delay of nearly 40 years the Senate Foreign Relations Committee agreed to U.S. adherence to the Protocol on prohibition of gas and biological warfare (see pp. 1123, 1370).

CANADA

1950–1953. Korean War. (See p. 1354.) Canada contributed 1 brigade of infantry, 1 artillery group, and an armored battalion to U.N. forces.

1953, January 8. Agreement on Radioactive Resources. Agreement between Britain, U.S.,

and Canada to share in uranium ore produced in Australia.

1954, May 13. St. Lawrence Seaway Approved by U.S. President Eisenhower signed legislation authorizing the U.S. to join Canada in constructing the Seaway. This was later dedicated by Queen Elizabeth and President Eisenhower (June 26, 1959).

1954, November 19. Joint Hemisphere Defense. The U.S. and Canada announced plans to construct a Distant Early Warning (DEW) radar line across Arctic Canada.

1958, May 19. NORAD Established. (See p. 1453.)

1963, May 11. U.S. Nuclear Warheads to Canada. Canada accepted U.S. Nuclear warheads for missiles installed on Canadian soil and used by Canadian NATO forces.

1963, August 16. Joint Control of Nuclear Air Defense Weapons. U.S. and Canada signed an agreement under which the U.S. would arm the Canadian Air Defense System.

1968, March 30. NORAD Extended for Five Years. (See p. 1456.)

1969, September 19. Canada Announces Reduction of NATO Force in Europe. Following a previous announcement (April 3, 1969), Prime Minister **Pierre Elliott Trudeau's** government announced that Canadian forces in Europe were to be reduced from 9,800 to 5,000 men in 1970, and that Canada would end its nuclear role in NATO by 1972.

LATIN AMERICA

ORGANIZATION OF AMERICAN STATES

1947, September 2. Treaty of Rio de Janeiro. The Inter-American Defense Treaty, transforming the old Pan American Union into the Organization of American States, was signed by all nations of the Western Hemisphere except Canada. This was ratified by the U.S. Senate (December 8) and became effective when ratified by the 14th nation, Costa Rica (December 3, 1948).

1948, April 26. Charter for the Organization of American States. This was established at a conference in Bogotá.

1954, March 1–28. Caracas Resolution. The Tenth Inter-American Conference of Foreign Ministers, at Caracas, declared that control of the political institutions of any American state by the Communist movement, an extension of the political system of a Continental power outside the Western Hemisphere, would be a threat to the peace of America.

1959, August 12–18. Emergency OAS Ses-

sion. This was in response to Cuban threats of invasion or infiltration in the Caribbean area.

1962, February 14. Cuba Excluded from the Organization of American States.

1964, July 26. OAS Votes Sanctions against Cuba. This action was sparked mainly by Cuban subversion efforts in Venezuela in 1963. OAS members were to sever all diplomatic and consular relations with Cuba; all trade and transportation to Cuba was banned, save for food and medicine.

1964, August 3. Mexico Refuses to Adopt Sanctions against Cuba. (See above.)

1968, May 6. OAS Establishes Peace Force for the Dominican Republic. (See p. 1460.)

MEXICO

1964, August 3. Mexico Refuses to Adopt Sanctions against Cuba. (See above.) Mexico was thus the only state in the hemisphere retaining diplomatic relations and normal trade connection with Cuba.

1968, July 26–October 2. Student Disorders at National University. In an effort to end 7 weeks of sporadic violence, the Mexican Army seized the university (September 18). Student protests erupted into bloody battles, mostly involving students against police (September 19–24). There were several deaths. Two months of unrest climaxed in a major battle between the Army and students, in which about 50 students were killed. The result, however, was an end to the disorders.

CARIBBEAN REGION

Anguilla

1967, July 11. Independence of Anguilla. The tiny island withdrew from confederation with St. Kitts and Nevis, and vainly sought association with Britain, Canada, or the U.S.

1969, March 19. British Occupation. Unrest and economic troubles led Britain to send a mixed force of 100 paratroops, Marines, and

policemen to take control. They were soon withdrawn (September 15), but the island remained virtually a British dependency.

Cuba

1952, March 10. Coup d'État. General **Fulgencio Batista** seized control of the country and established a dictatorship.

1953, July 26–27. Uprising in Santiago and Bayamo Suppressed. An effort by **Fidel Castro** to seize a government armory was defeated; Castro and his brother Raúl were captured and imprisoned.

1956, April 29. Rebellion Suppressed. The uprising occurred at Matanzas.

1956, December 2. Cuba Claims Castro Killed. Castro, who had been released from prison, went to Mexico and led an insurgent group landing in Oriente Province (November 30), was defeated and presumed (erroneously) to have been killed. Actually he and his followers fled to safety in the Sierra Maestra Mountains.

1957–1958. Insurrection. Revolutionaries under Fidel Castro carried out a successful guerrilla campaign from the Sierra Maestra Mountains of Oriente Province, gaining increasing popular support. Castro took the offensive and moved out of the mountains (October 1958).

1959, January 1. Castro Victorious. Batista fled the country as the revolutionaries swept through the country and seized Havana (January 8).

1959, January 7. U.S. Recognizes Castro Government.

1960, May 7–Sept. 25. Cuba Joins Soviet Bloc. This was initiated by resumption of diplomatic relations (May 7), which had been broken in 1952; culminated by Khrushchev's threat to use intercontinental rockets to support Cuba if attacked by U.S. (September 25).

1960, November 1. Castro Rebuffed. President Eisenhower, in response to Castro's threats against Guantánamo, said the U.S. would "take whatever steps are necessary to defend" the base.

1961, January 3. U.S. Breaks Diplomatic Relations.

1961, April 15–20. Bay of Pigs Incident. An attempted invasion of Cuba by approximately 1,400 anti-Castro Cuban revolutionaries clandestinely supported by the U.S. Central Intelligence Agency, who did not receive the air and naval support from U.S. forces which they had been promised, was defeated by Castro forces. This failure was a serious blow to American prestige.

1961, May 1. Castro Proclaims Cuba a Socialist Nation. Secretary of State **Dean Rusk** next day said that Cuba had become a full-fledged member of the Communist bloc.

1962, September 11. Soviet Threatens War in Support of Cuba. The U.S.S.R. accused the U.S. of preparing for an invasion of Cuba, and warned that any U.S. attack on Cuba or on Soviet ships bound for Cuba would mean war.

1962, September–December. Missile Crisis. (See p. 1454.)

1963, March 30. U.S. Bans Exile Raids Against Cuba.

1964, July 26. OAS Sanctions against Cuba.

1965, November 6. Cuba Permits Refugees to Fly to United States. These flights continued until August 1971, during which time 246,000 Cubans flew from Havana to Miami.

1966, February 6. Castro Accuses Communist China of "Betraying" Cuban Revolution. He denounced Chinese propaganda efforts in Cuba and failure of China to provide promised assistance.

Dominican Republic

1949, June 20–21. Revolt Suppressed.

1959, June 23. Cuban Invasion. Cuban-supported invasion of the Dominican Republic by 86 men was crushed by forces of dictator **Leonidas Trujillo.**

1961, May 30. Assassination of Trujillo. This ended the repressive dictatorial regime, which had lasted for 31 years.

1962, January 13. Unsuccessful Coup Attempt.

1963, April 27–30. Dispute with Haiti. This resulted from a raid by Haitian police into the Dominican embassy in Port au Prince. The OAS mediated the dispute.

1963, September 25. Military Coup d'État. The leftist government of President **Juan Bosch** was overthrown by military leaders. Later the government was returned to civilian control (October).

1965, April 24–25. Military Coup d'État. The civilian triumvirate was overthrown. A bloody civil war broke out between leftist and military adherents of former President Bosch and Army and Air Force units under the command of Brigadier General **Elias Wessin y Wessin.** Although Wessin gained control of much of the capital and the country, fighting continued.

1965, April 28. United States Intervention. President Johnson sent in a force of Marines to protect Americans. These were followed by U.S. Army airborne units, as the fighting continued, with the pro-Bosch "rebels" regaining some lost ground, and the adherence of previously uncommitted army units.

1965, May 5. Truce Agreement. Under pressure from the U.S. and the OAS, a cease-fire was agreed upon between the rebel units, under the provisional President, Colonel **Francisco Caamans Deno** and a newly established 3-man military junta led by Brigadier General **Antonio Imbert Barreras.** U.S. forces in the country were 12,439 Army troops and 6,924 Marines. More than 2,000 Dominican deaths were reported in the confused fighting.

1965, May 6. OAS Agrees to Establish Interamerican Peace Force. This would replace the United States force. Participating countries were Venezuela, Brazil, Guatemala, Costa Rica, Honduras, Paraguay, and the U.S.

1965, May 13–19. Renewed Civil War. The junta attacked the rebel forces and drove most of them from the capital. U.S. forces, attempting to hold open an International Security Zone, were accused of aiding both sides. Under U.S. pressure a new truce was agreed.

1965, May 23. OAS Interamerican Armed Force Becomes Operational. The United States began to reduce its 21,000-man force, which had had 19 men killed during the hostilities.

1965, August 31. Provisional Government Approved. Following resignation of the 3-man junta, the rival factions agreed with an OAS Peace Committee to establish a government under Provisional President **Hector Garcia-Godoy.** Sporadic violence continued in the country, however, for several months, becoming more intense in early 1966.

1966, June 3. Election of President Joaquin Balaguer. A centrist, he defeated left-wing candidate Juan Bosch and a right-wing candidate.

1966, June 28. Withdrawal of OAS Force Begins. The first units of the 6,000-man U.S. contingent of the 8,200-man force began to leave the country following an OAS resolution (June 24).

Haiti

1946, January 11. Military Coup d'État. President **Élie Lescot** was overthrown. The military junta selected **Dumarsais Estimé** as president (August 16).

1950, April. Military Coup. Estimé was ousted and replaced by Colonel **Paul Magloire** (October 23).

1956–1958. Chaos in Haiti. Magloire was forced to resign (December 13, 1956) and **François Duvalier** was elected president (September 22, 1957). By ruthless repression he restored order (July 1958).

1959, August 12–18. Emergency session of the Organization of American States was called to ease tensions in the Caribbean area.

1959, August 13. Cuban Invasion. The invading force of 30 armed men was crushed.

1960, November 22. Duvalier Proclaims Martial Law. The dictator took advantage of disorders at the University of Haiti to increase his control over the country.

1962–1963. Border Incidents with Dominican Republic.

1963, August 5–7. Rebel Invasion from the Dominican Republic. A force of Haitian ex-

iles, hoping to overthrow the government of dictator President François Duvalier, landed by sea in northern Haiti, having embarked in the Dominican Republic. The rebels were defeated and fled across the border into the Dominican Republic, which was charged with complicity by Haiti.

1969, June 4. Unidentified Aircraft Attacks Port-au-Prince. Six homemade incendiary bombs dropped from a transport plane caused 3 deaths. Haiti accused Cuba.

Trinidad-Tobago

1962, August 31. Independence Granted by Great Britain.

1970, April 21–25. Riots and Army Mutiny. Racial tension between blacks (47 percent of population) and East Indians (36 percent) flared into violence, which became more serious when perhaps one-fourth of the 800-man army mutinied in support of black-power rioters. Loyal troops suppressed the rebellion after much shooting but negligible bloodshed.

CENTRAL AMERICA

Costa Rica

1948, April 13–20. Revolution. The dictatorial government of President **Teodoro Picado** was overthrown by rebels led by Colonel **José Figueres.**

1948, December 12. Invasion by Armed Rebels from Nicaragua. This was repelled; strained relations with Nicaragua resulted.

1949, April 3. Unsuccessful Coup Attempt.

1955, January 11. Invasion and Rebellion Suppressed. President José Figueres again accused Nicaragua of aggression and asked the OAS Council for aid. An OAS committee set up as a result of this complaint reported (February 17) that the rebels were mostly Costa Ricans who had been based in Nicaragua, and called for conciliation of the dispute between the 2 nations.

1956, January 9. Agreement with Nicara-

gua. It provided for effective cooperation in joint surveillance of borders.

1960, November 9–14. Clash with Nicaraguan Rebels. (See p. 1463.)

El Salvador

1948, December 12. Military Revolt. President **Castaneda Castro** was overthrown, and replaced by a revolutionary junta.

1960, October 26. Coup d'État. The government of President **Jose Maria Lemus** was overthrown by a military junta led by Colonel **Cesar Yanes Urias.**

1969, June 24–28. War with Honduras. Long-smoldering economic and territorial border disputes (1967–1969) were sparked to war by rioting over a series of soccer games. El Salvadoran troops had occupied considerable Honduran territory when both sides accepted a cease-fire arranged by the OAS. Sporadic fighting continued for more than a month, however, with the advantage remaining with El Salvador.

1972, March 25. Attempted Coup d'État. Troops loyal to the president, General **Fidel Sanchez Hernandez,** suppressed a military revolt in brief but bloody fighting; more than 300 were killed on both sides.

Guatemala

1949, July 16–18. Rebellion Suppressed.

1954, June 18–29. Revolution. Anti-Communist forces under Lieutenant Colonel **Carlos Castillo Armas** invaded the country and called on the people to overthrow leftist President **Jacobo Arbenz Guzmán.** Arbenz sought asylum in the Mexican embassy (June 27). A cease-fire was effective after about 100 casualties had been incurred on both sides. Castillo Armas became leader of a junta, and shortly afterward was declared president (July 1).

1955, January 20. Revolt Suppressed. Continuing Communist-inspired insurgency.

1957, October 25. Coup d'État. President **Luis Arturo Gonzalez Lopez** was ousted by a military junta headed by Colonel **Oscar Mendoza.** Colonel **Guillermo Flores Anendaño** was proclaimed president (October 26).

1960, November 16. Threatened Cuban Invasion. U.S. warships protected Guatemala and Nicaragua, threatened by invasion from Cuba.

1963, March 30. Coup d'État. Colonel **Enrique Peralta Azurdia** proclaimed himself president. Low-level, Communist-inspired insurgency continued to cause sporadic violence throughout the country.

1968, August 18. Communist Guerrillas Kill U.S. Ambassador. John Gordon Mein was slain in an apparent kidnap attempt.

1970, April 5. Communist Guerrillas Kill West German Ambassador. Count **Karl von Spreti** was kidnapped, then assassinated.

Honduras

1956, October 21. Military Coup. A junta took control after charges of election fraud against President **Julio Lozano Diaz** caused violence and bloodshed (October 7).

1957, May 2–3. Border Clashes with Nicaragua. (See p. 1463.)

1961, January 27. Invasion Attempt from Nicaragua. Honduran rebels, possibly encouraged by Nicaragua, were repulsed.

1963, October 3, Coup d'État. President **Ramon Villeda Morales** was ousted after sharp fighting between rebel and loyal troops by Colonel **Osvaldo Lopez Arellano,** chief of the armed forces, who proclaimed himself President.

1969, June 24–28. War with El Salvador. (See above.)

Nicaragua

1947, May 26. Coup d'État. General **Anastasio Somoza** overthrew the government of President **Leonardo Argüello.**

1948–1955. Disputes with Costa Rica. (See p. 1461.)

1956, September 21. Assassination of President Somoza. His son, **Luís,** was elected by Congress to complete his term.

1957, May 2–3. Border Clashes with Honduras. Troops of both sides withdrew following OAS mediation.

1959, May 30–June 14. Attempted Insurgent Invasion. Air-transported rebels were defeated by forces of General **Anastasio Somoza Debayle** (brother of President **Luis A. Somoza Debayle**), who accused Cuba of fomenting insurrection.

1960, January–May. Insurgent Border Raids. These were from Honduras (January 2, February 29, and May 14) and Costa Rica (January 9).

1960, November 11–15. Insurgent Invasion from Costa Rica. This was defeated by Nicaraguan troops, aided by Costa Rican troops (November 9–14). Cuba was blamed for supporting the insurgency.

1961, July 26. Sandinista National Liberation Front (FSLN). Nicaraguan insurgents, trained and indoctrinated in Cuba, established the new organization in Honduras, named in honor of the 1920–1930s revolutionary leader, César Augusto Sandino (see p. 1148).

1963. Sandinista Base Established in Nicaragua. The FSLN began guerrilla operations in the northern mountains of Matagalpa, but failed to rally the local peasants to the cause of armed insurrection. Sporadic raids continued.

1967, January 22–23. Attempted Coup d'État. A coalition of civilians and dissatisfied military officers attempted a coup that was suppressed by the National Guard, commanded by Anastasio Somoza Debayle; this youngest son of Anastasio Somoza was elected president (February 5).

1972–1974. Civil Unrest. The increasingly despotic rule of dictator Somoza and his ruthless National Guard alienated much of the population.

Panama

1949, November 26. Coup d'État. Dr. **Arnulfo Arias** was installed as president by the national police.

1951, May 10. Coup d'État. President **Arnulfo Arias** was overthrown in violent fighting. He was succeeded by Vice President **Alicibiades Arosmena.**

1955, January 25. Treaty with the U.S. Yearly payments for the Canal Zone were increased from $430,000 to $1,930,000.

1959, April 24–May 1. Cuban-Based Insurgent Invasion. More than 100 invaders, many Cuban, were defeated, and most killed or captured.

1959, November 3. Anti-U.S. Riots in Panama.

1964, January 9–10. Anti-U.S. Riots. Panama broke diplomatic relations with the U.S. because of riots in the Canal Zone.

1964, April 3. Resumption of Diplomatic Relations with the U.S.

1968, October 11. Coup d'État. Recently inaugurated President Arnulfo Arias was overthrown by the National Guard. A military junta elected 2 of its members, Colonel **Bolivar Urrutia** and Colonel **Jose M. Pinella,** as president and vice president, respectively.

1969, December 15. Coup d'État. Brigadier General **Omar Torrijos,** who had been a principal leader in the October 11 coup against President Arias, ousted the 2 colonels who had outmaneuvered him to gain power after the coup.

SOUTH AMERICA

Argentina

1948, March 4. Agreement with Chile against Britain. The 2 nations agreed on joint defense of their rights in the Antarctic and the Falkland Islands against British claims and occupation.

1951, September 28. Revolt Suppressed. President **Juan Perón** blamed the military revolt, suppressed by loyal troops, on the activities of former U.S. Ambassador **Spruille Braden.** A state of virtual martial law was declared, which gave Perón dictatorial powers.

1955, June–September. Violence, Disorders, Unrest. Minor military and civilian revolts were ruthlessly suppressed by the Perón regime. Perón was excommunicated by the

Catholic Church for suppressing Catholic schools and imprisoning priests.

1955, September 16–19. Perón Overthrown. As disorders spread, Perón declared a state of siege (September 11). The armed forces then rose in a brief revolt. Perón fled. A junta under Major General **Eduardo Leonardi** took control.

1955, November 13. Military Revolt. Leonardi, accused of being "fascist," was overthrown and replaced by Major General **Pedro Aramburu.**

1956, June 10–14. Peronist Revolt Suppressed. Later Peronist plots were discovered and smashed (August 15, November 22).

1960, June 13. Unsuccessful Revolt. An army unit in San Luis surrendered without bloodshed when no other military unit responded to radio appeals for a national uprising.

1962, March 29. Military Coup d'État. President **Arturo Frondizi** was ousted by the armed forces.

1962, August 8–11. Army Mutiny. President **Jose Maria Guido** accepted Army demands for a change in the War Secretary and Army commander in chief.

1962, December 11–12. Military Revolt. This was suppressed by loyal troops.

1963, April 2–5. Military Mutiny and Rebellion. The rebels, mostly naval, were defeated after brief fighting at scattered locations.

1966, June 28. Coup d'État. A military junta ousted President **Arturo Illia** and installed Lieutenant General **Juan Carlos Ongania** as provisional president.

1970–1975. Growth of Terrorist Violence. The country was plagued by a rash of kidnappings and assassinations.

1970, June 8. Coup d'État. A military junta ousted President Ongania and installed Brigadier General **Roberto Marcelo Levingston** as president.

1971, March 23. Coup d'État. A military junta, headed by Army Commander General **Alejandro Augustin Lanusse,** ousted President Levingston.

1971, July 1. Britain and Argentina Agree to Disagree on the Falkland Islands. Although neither side relinquished its claim to sovereignty (Malvinas Islands to Argentina), they agreed to resolve outstanding issues by negotiation.

1973, July 13. Return of Peron. He was elected president, with his second wife, **Isabel,** vice president. She succeeded him upon his death (July 1, 1974).

Bolivia

1946, July 17–21. Popular Revolution. President **Gualberto Villaroel** was killed and succeeded by a liberal government.

1949, May–September. Unrest and Rebellion in the Tin Mines. Suppressed by the army, but left the nation on the verge of bankruptcy.

1950, January 14. Unsuccessful Coup Attempt.

1950, May 18–19. Unsuccessful Coup Attempt.

1951, May 16. Military Coup. A 10-man military junta, led by General **Hugo Ballivián,** seized control in Bolivia.

1952, April 8–11. Revolution. The junta was overthrown by a popular revolt under **Hernán Siles Zuazo.** Doctor **Victor Paz Estenssoro** was proclaimed president (April 16).

1953, November 9. Unsuccessful Revolt.

1958, May 16–22. Unsuccessful Revolt. Uprising in Santa Cruz, suppressed by troops.

1958, October 21. Revolt Suppressed.

1959, April 19. Unsuccessful Revolt in La Paz.

1959, June 26. Unsuccessful Revolt in Santa Cruz.

1960, March 19. National Police Revolt. Crushed by the army.

1964, November 3–4. Military Revolt. President Paz Estenssoro was overthrown by Army General **Alfredo Ovando Candia** and Air Force General **René Barrientos Ortuno.** Ovando soon resigned, leaving popular Barrientos in control.

1966–1967. Cuban-Sponsored Insurgency. Guerrilla activity intensified after arrival of

Ernesto (Che) Guevara from Cuba (November 7, 1966).

1967, October 8–16. Elimination of Che Guevara and His Guerrillas. Military forces commanded by Brigadier General **Juan Jose Torres,** an anti-Communist officer with socialist political views, methodically tracked down, surrounded, then annihilated Che and his small band of followers. Guevara, wounded, was captured (October 8) and died the next day.

1969, April 27. Death of President Barrientos. He was killed in a helicopter crash. He was succeeded by Vice President **Adolfo Siles Salinas.**

1969, September 26. Coup d'État. General Ovando seized control in a bloodless coup and named himself president.

1970, October 6–7. Coup d'État and Countercoup. President Ovando resigned under pressure from right-wing military officers. During the confusion, however, leftist General Torres seized control and proclaimed himself president.

1971, August 19–22. Coup d'État. Right-wing military officers, led by Colonel **Hugo Banzer Saurez,** ousted President Torres after a brief and not very intensive struggle.

Brazil

1954, August 8–25. Civil Disorder. Unable to restore order, President **Getulio Vargas** resigned under military pressure, then committed suicide (August 24).

1955, November 11. Coup d'État. Military officers, led by Lieutenant General **H. B. D. Teixeira Lott,** deposed Acting President **Carlos Coimbra da Luz,** who was accused of fostering a coup to cancel the election of **Juscelino Kubitschek** as president. Kubitschek took office as scheduled (January 31, 1956).

1961, August 26. Military Coup.

1963, September 12. Military Revolt Suppressed.

1964, March 31. Military Revolution. President **Joao Goulart** was deposed and a military dictatorship was established.

1969, August 31. Military Junta Takes Control. Leaders of the 3 armed forces assumed authority after President **Artur da Costa e Silva** was incapacitated by a stroke.

Chile

1973, September 11. Coup d'État. The Chilean tradition of military subordination to civilian political authority ended when a 4-man military junta seized power from President **Salvador Allende Gossens,** an avowed Marxist, after 3 years of increasing chaos under his presidency. (It is not clear to what extent this chaos was the result of Allende's inefficiency, of obstructionism by right-wing civilian and military elements, and of subversion prompted by the U.S. Central Intelligence Agency.) During brief but bitter fighting for the Presidential Palace, Allende was killed, apparently by suicide.

Colombia

1945–1965. Endemic Civil War Continues.

1948, April 9–10. Uprising in Bogotá. This revolt, embarrassing to the government because it occurred during a meeting of the Inter-American Conference, was suppressed.

1948–1958. La Violencia; Low-Scale Guerrilla Insurrection. A variety of local and national issues and problems were exploited by Colombian Communists to stir up a constant state of revolt and insurrection in rural areas of the country. At least 250,000 violent deaths occurred during this decade of violence.

1953, June 13. Military Coup d'État. President **Laureano Gomez** was overthrown by the military under Lieutenant General **Gustavo Rojas Pinilla.**

1957, May 10. Military Coup d'État. Rojas Pinilla overthrown by junta.

Ecuador

1947, March 14. Unsuccessful Coup Attempt.

1947, August 23. Military Coup d'État. President **Jose Maria Velasco Ibarra** was overthrown in a bloodless revolt by Colonel **Carlos Mancheno.** (Velasco Ibarra had also been ousted in 1935.)

1947, September 1–3. Successful Counterrevolution. Mancheno was overthrown and replaced by **Carlos Julio Arosemena Monroy.**

1951–1955. Dispute with Peru. This was due to disputed frontier locations in an area which had been in dispute for over a century. There were intermittent border clashes and arguments before the International Court of Justice.

1961, November 7–9. Coup d'État. Antigovernment riots led to the overthrow of the government of President Jose Maria Velasco Ibarra and clashes between military units supporting rival candidates to replace Velasco. After brief fighting, Army and Air Force leaders agreed to install Vice President **Carlos Julio Arosemena Monroy** as president.

1963, July 11. Coup d'État. Military leaders ousted President Arosemena, who was replaced by a 4-man military junta.

1966, March 29. Antimilitary Coup. The junta resigned as a result of antigovernment riots.

1971, January 18. United States Suspends Arms Aid to Ecuador. This was a result of numerous incidents (28 since 1966) of Ecuadoran seizure of American tuna fishing boats, more than 12 miles off shore but within Ecuador's claimed territorial waters, extending 200 miles from shore. Ecuadoran seizures continued.

1972, February 15. Military Coup d'État. President Jose Maria Velasco Ibarra was ousted by a military junta, which named General **Guillermo Rodriguez Lara** as President. This was the fourth time Velasco had been ousted as president by a military coup.

Guyana

1953, October 9. British Intervention. The left-wing government of Prime Minister **Cheddi Jagan** was ousted by British troops to prevent a Communist takeover of the self-governing colony.

1962–1966. Endemic Violence. Promised independence in 1966, the colony was torn by disorders, riots, and bloodshed, primarily because of racial and political strife between East Indians (49 percent of the population) and blacks (32 percent).

1966, May 26. Independence of Guyana. Despite the continuing explosive internal situation, Britain granted the promised independence, leaving a military garrison in the country. Following national elections Guyana became a republic, but remained in the British Commonwealth (February 23, 1970).

1969, January 2–4. Rebellion Crushed. Rebels from Brazil, apparently supported by large ranch owners and possibly by Venezuela (because of a long-standing border dispute) were driven out; some were killed and captured.

1969, August 19–September 10. Border Clash with Surinam. This resulted from Dutch occupation of territory in a disputed border region. It ended when the Dutch withdrew.

Paraguay

1947, March 30–August 20. Civil War. Efforts by former President **Rafael Franco** to seize control were defeated by the government of President General **Higinio Morínigo.**

1948–1949. Chaos. After the retirement of Morínigo the nation had 5 presidents in 5 months.

1954, May 5. Army Revolt. President **Frederico Chaves** was deposed. General **Alfredo Stroessner,** commander of the armed forces, later was installed as president (August 15).

1959–1960. Rebel Invasions. About 1,000 rebels, based in Argentina, invaded and were crushed (December 1959). President Stroessner blamed the action on Cuba. Six smaller invasions were also crushed (1960).

1963–1974. Stability under Relaxed Dictatorial Control. President Stroessner, allowing limited opposition political activity, was thrice reelected president (1963, 1968, 1973).

Peru

1948, October 27–29. Military Revolt. The government of President **José Bustamante** was overthrown by a military junta under General **Manuel Odría.**

1951–1955. Disputes with Ecuador. (See p. 1466.)

1956, February 16–25. Revolt at Iquilas. The military uprising collapsed when all army units failed to join.

1962, July 18. Military Coup d'État. President **Manuel Prado Ugarteche** was overthrown. A military junta took over, under the leadership of General **Ricardo Perez Godoy.**

1963, March 3. Coup d'État. Chief of State Perez was ousted by the junta following 2 months of internal violence. He was succeeded as chief of state by General **Nicholas Lindley Lopez.**

1964–1965. Communist Revolt. A very minor state of revolt was maintained by Castroite Communist agents in remote Andes regions.

1968, October 3. Coup d'État. The government of President **Fernando Belaunde Terry** was ousted without bloodshed; Army Chief of Staff General **Juan Velasco Alvarado** proclaimed himself president.

Uruguay

1961, January 11–12. Riots and Street Fighting. The government blamed these on Cuban and Soviet agitation.

1967–1973. Era of the Tupamaros. This leftist guerrilla organization, including members of many prominent families and a large number of professionals, terrorized the nation with large-scale robberies and kidnappings their specialty. Victims included U.S. police advisor Dan A. Mitrione, kidnapped, then killed (August 10, 1970), and British Ambassador Geoffrey Jackson, kidnapped and held for 8 months (January 8–September 9, 1971).

1973, February 10. Military Control Granted by President. Under great pressure from rebellious military leaders, President **Juan Maria Bordaberry** ceded practically all governmental authority to military commanders, to permit them to marshal the resources of the nation against the Tupamaro guerrillas.

1973, February–December. Elimination of the Tupamaros. The Army conducted a wholesale campaign, arresting and imprisoning hundreds of members. Some fled to Argentina, where they led an antigovernment group.

Venezuela

1945, October 18. Military Revolt.

1948, November 24. Bloodless Coup d'État. Colonel **Carlos Delgado Chalbaud** seized control. He was later assassinated (November 13, 1950) and replaced by **Germán Suárez Flámerich.**

1951, October 13. Revolt Suppressed.

1952, December 2. Coup d'État. Colonel **Marcos Perez Jimenez,** a member of the junta, seized power "by decision of the armed forces," nullifying the results of a national election 3 days earlier.

1958, January 1–23. Revolt. President **Marcos Pérez Jiménez** suppressed the first revolutionary actions, but was ousted by a military junta in a renewed revolt (January 21–23).

1958, July 23. Counter-revolt Suppressed.

1958, September 7. Unsuccessful Coup Attempt.

1960–1974. Communist Insurgency. Continuous small-scale harassment of the central government through the period.

1961, June 26. Military Uprising Suppressed.

1962, May 4–5, June 4. Military Uprisings Suppressed. President **Romulo Betancourt** became more firmly seated in control.

1963, November. Castro Plot Discovered. Cuban agents and large quantities of arms were captured by government troops.

AUSTRALASIA

Australia

1950–1953. The Korean War. (See p. 1354.) Australia contributed 2 infantry battalions, 1 air fighter squadron, 1 air transport squadron, 1 aircraft carrier, 2 destroyers, and a frigate to the United Nations forces.

1965–1972. The Vietnam War. (See p. 1321.) As a member of SEATO, Australia contributed forces ranging from a battalion to a brigade group and supporting elements (about 7,500 troops) to the allied forces supporting the government of South Vietnam.

1965, June 3. First Australians Arrive in Vietnam. A contingent of more than 100 infantrymen arrived, soon followed by 400 more (June 8).

1966, March 8. Conscription Adopted. This was the first peacetime draft in Australia's history, provoking antigovernment demonstrations. The purpose was to send draftees to an augmented Australian contingent in Vietnam.

1971, November 1. ANZUK Force Established at Singapore. (See p. 1421.)

1972, December 5. End of Conscription. The newly elected government of Prime Minister **Gough Whitlam** carried out its electoral campaign promises of ending the peacetime draft and withdrawing from Vietnam.

1972, December 18. Last Unit Withdraws from Vietnam.

1973, January 21–22. Australia and New Zealand Reaffirm Ties with SEATO and the United States. Announcements were made by Prime Ministers Gough Whitlam and **Norman Kirk** at a meeting in Wellington, N.Z.

NEW ZEALAND

1950–1953. The Korean War. (See p. 1354.) New Zealand contributed an artillery group and 2 frigates to the United Nations forces.

1965–1972. The Vietnam War. As a member of SEATO, New Zealand contributed an artillery battery (later increased to a battalion of 550 men) to the allied forces supporting the government of South Vietnam. (See p. 1321.)

1971, November 1. ANZUK Force Established at Singapore. (See p. 1421.)

1973, January 21–22. New Zealand and Australia Reaffirm Ties with SEATO and the United States. (See above.)

1973, July 21–29. New Zealand Warship Enters French Nuclear Test Zone. As a protest against French nuclear tests near Tahiti (see p. 1377), a New Zealand frigate cruised just outside the 12-mile limit of the territorial waters of Mururoa Atoll, where the French were conducting their tests.

NAURU

1968, January 31. Independendence. An Australian-administered United Nations trust territory (1947–1968), Nauru was a small, nitrate-rich island northeast of the Solomon Island chain.

FIJI

1970, October 10. Independence. This was the 96th anniversary of Fiji becoming a British crown colony (1874).

XXII

THE DOMINANCE OF TECHNOLOGY:

1975–1991

SIGNIFICANT WARS

Iran–Iraq Gulf War

1980, September 9. Iraq Invades Iran. Sensing military weakness in Iran because of the Iranian revolution, Iraq launched a major offensive to resolve the long-standing border dispute (see p. 1403). The Iraqis at first met slight resistance. However, they were very cautious and failed to exploit opportunities for quick victory. The Iranians recovered from the initial defeats and a stalemate ensued along a battle line penetrating into Iran more than 30 miles in some places. The city of Khorramshahr was captured by the Iraqis.

1980–1982. Stalemate. Both sides made repeated efforts to break through the opposing lines. Casualties were heavy, but little was accomplished.

1982, January 23–March 3. Jordanian Volunteers Aid Iraqi Gulf War Efforts. (See p. 1512.)

1982, March 22–30. Iranian Counteroffensive. The Iranians began a determined attempt to drive the Iraqi forces from Iranian soil, and pushed Iraqi forces back as much as 24 miles in some places.

1982, April 30–May 20. Renewed Iranian Offensive. The Iranians again pushed the Iraqis back. The Iranians drove until they approached the Iraqi defenses near the port city of Khorramshahr, which the Iraqis had captured early in the war.

1982, May 22–23. Recapture of Khorramshahr. In the night attack the Iranians encircled the Iraqi force in Khorramshahr, which surrendered. The Iranians captured large quantities of Soviet arms, apparently supplied by Syria. The Iranians, ebullient because of their victory, proclaimed as their war aim and "greatest right" the deposition of Iraqi President Saddam Hussein (May 26).

1982, June 10–20. Iraq Offers Truce and Troop Withdrawal. Iraq proposed a truce and promised to withdraw all Iraqi troops from Iranian soil within 2 weeks of an Iranian acceptance of the truce. Iraq also declared a unilateral cease-fire. Iran responded by reiterating its demand for Hussein's removal from office (June 11, June 20).

1982, July 14–30. Iranian Offensive. Iran launched another offensive. The Iranians made some headway, but suffered heavy casualties for minimal gains.

1982, July 21–August 30. Air War Escalates. Iranian planes attacked Baghdad (July 21). Iraq retaliated with air attacks on Khargh Island, striking Iran's oil shipping facilities and sinking 2 merchant ships (August 18–30).

1982, September–November. New Iranian Offensives. The Iranians regained some territory on the northern front near the border town of Sumar, which Iraq had taken early in the war (October 1). The Iranians again attacked west of Dezful and advanced 3 miles into Iraqi territory near the town of Mandali (November 2). Iraqi troops counterattacked and drove the Iranians back to the border. To the south the Iranians advanced to within artillery range of the stategically important Baghdad-Basra highway (November 17).

1982, December 21. Arab Peace Initiative. The leaders of Algeria, United Arab Emirates, and Saudi Arabia attempted to mediate between the two warring gulf states.

1983, February 2–March 9. Iraqi Air Attacks Cause Massive Oil Spill. Iraqi air attacks on Iranian Persian Gulf oil-producing facilities resulted in the largest oil spill in Persian Gulf history.

1983, February 7–16. Al Amarah Offensive. Iranians advanced toward the Baghdad-Basra road in an attempt to cut that road at Al Amarah. They reached within 30 miles of Al Amarah before being halted and thrown back by an Iraqi counterattack. The Iraqis claimed they knocked out 100 Iranian tanks and took 1,000 POWs.

1983, April–October. Repeated Iranian Offensives. Attacks west of Dezful failed to make significant gains (April 11–14). Next the Iranians launched an offensive into northern Iraq (July 23). Advances were negligible. Iranian troops then tried to break the Iraqi line west of Dezful in a major offensive (July 30). The Iraqis repulsed a number of Iranian thrusts, then counterattacked, but were repulsed in turn (August 6–12). Casualties were severe on both sides. The Iranians launched still another offensive on the northern front and closed a salient that had been opened by Iranian Kurdish rebels (October 20).

1983, July 20. Iraq again Attacks Iranian Oil Centers. Iraqi planes struck Iranian oil industry facilities.

1984, February 11–22. Renewed Iranian Offensive. Iran tried to carry the war deeper into Iraq by resuming the offensive on both the northern and central fronts (February 11). Iraq claimed that on the central front its troops had approached the Baghdad-Basra highway. The Iranians renewed attacks toward Basra in the south (February 22). The Iraqis repulsed the attack, only after severe loss of troops and some territory. Their counterattacks to recapture undeveloped Iraqi oil fields on Majnoon Island also failed. The Iranians claimed that Iraq used poison gas in these battles. International observers confirmed that mustard gas was apparently employed. During this battle Iraqi planes attacked Iranian oil installations near Kargh Island.

1986, July. Capture of Faw. A surprise Iranian offensive, spearheaded by carefully-trained commandos employing infiltration tactics, defeated Iraqi forces on the Faw peninsula and captured the area. This battle signalled the failure of Saddam Hussein's efforts to wage the war on a "limited budget."

1986, Autumn. Shifting Iraqi Strategy. Abandoning the essentially defensive strategy employed since 1982, Saddam Hussein and the Iraqi Ba'athist leadership opted for a more vigorous approach to the war. They expanded the army from 12 divisions to over 40, tapping previously untouched university students as a manpower source for the vastly expanded Republican Guard. The Iraqis also began to employ their air force more aggressively, and with an eye toward closer cooperation with the army in ground operations.

1987, May 17. Attack on the USS *Stark*. Iraqi aircraft, apparently by mistake, attacked the frigate USS *Stark* in the Persian Gulf with French-made Exocet surface-skimmer missiles. The *Stark*'s crew suffered 37 fatalities, and Iraq apologized.

1988, February–August. The War of the Cities. The Iraqis struck at Iranian cities with their Scud missiles, causing considerable damage. The Iranian response, hampered by geography and a lack of missiles, was limited.

1988, April–August. Iraqi Offensives. A series of Iraqi offensives showed the feeble state of the Iranian Army. Four offensives resulted in lopsided victories, with heavy Iranian losses but minimal Iraqi casualties. Faw was recaptured (May), and Iranian forces were left in disarray.

1988, July 3. *Vincennes* Incident. The cruiser USS *Vincennes,* operating in the Persian Gulf, mistook a civilian Iranian airliner for an attacking aircraft, and shot it down, killing all 290 passengers and crew. The airliner had not answered repeated requests for identification, and it was well within missile-launch range of the *Vincennes* when it was shot down. The U.S. government offered to pay compensation to relatives of the dead.

1988, August. Iranian Reorganization. The *Pasdaran* (Revolutionary Guard Corps) was brought under army command, and Speaker of the *majlis* (parliament) Rafsanjani was appointed supreme military commander by Ayatollah Khomeini.

1988, August 20. Cease-fire. Iran announced its acceptance of a U.N.-sponsored cease-fire. Ayatollah Khomeini compared this to "taking poison," but recent defeats coupled with growing Iranian war-weariness compelled the government to end the war.

Britain–Argentina Falkland Islands War

1982, April 2. Argentina Invades the Falklands. A 2,000-man task force landed at Port Stanley. The 84-man Royal Marine garrison surrendered after a 3-hour fight.

1982, April 3. Argentinians Seize South Georgia. A small task force landed at Grytviken, South Georgia, and took control of the island after a 7-hour battle with a 22-man detachment of Royal Marines.

1982, April 3. U.N. Declares Argentina Aggressor. The British took the case to the U.N. Security Council, which voted to declare Ar-

gentina an aggressor and demanded immediate withdrawal (Resolution 502). Argentina rejected the resolution, but offered to negotiate with Britain. The British refused to negotiate until Argentina withdrew from the Falklands.

1982, April 5. Falkland Task Force Leaves Britain. Two light aircraft carriers (HMS *Hermes* and HMS *Invincible*) and 28 other vessels sailed from Portsmouth for the Falklands. The naval task force included an amphibious task force of 2,000 Royal Marine commandos.

The commander of the Falkland Islands Task Force, Rear Admiral **Sir John Woodward,** sailed from Gibraltar with 7 warships and Royal Fleet Auxiliary (RFA) support vessels for Ascension Island where they joined the ships sailing from Britain. Approximately 2,000 additional troops sailed from Southhampton soon after on the passenger cruise ship *Canberra,* refitted to be used as a troop and supply ship (April 9.) By this time the RAF had deployed Hercules and VC-10 Air Transport aircraft at Ascension Island, 3,750 miles northeast of the Falklands. Ascension became a vital supply base for the Falklands Task Force, which would receive tons of stores from Ascension transported by the RAF transport aircraft and by ships of the Royal Fleet Auxiliary (RFA) and Merchant Navy. Ascension also served as a base for RAF bombers and reconnaissance aircraft.

1982, April 8–28. U.S. Attempts Diplomatic Solution. Secretary of State **Alexander Haig** carried on shuttle diplomacy by flying back and forth between Buenos Aires and London in an attempt to reach a peaceful solution to the crisis. Both sides refused to budge from their positions. The U.S. then announced support for Britain (April 30).

1982, April 12. Britain Announces Falklands Quarantine. Britain announced that as of April 28 any Argentine vessel within 200 miles of the Falklands would be considered an aggressor and be treated accordingly.

1982, April 18. British Reconnaissance Units Land. Special Boat Service (SBS) troops (Royal Marines) and Special Air Service (SAS) troops (Army) began secret landings on the Falklands to carry out intelligence operations, which later proved invaluable to the invasion force.

1982, April 25–26. British Recapture South Georgia. British helicopter machine-gun and rocket fire forced an Argentine submarine to run aground at Grytviken Harbor. The next morning a contingent of Royal Marines overcame the Argentine detachment at Grytviken and SAS troops easily recaptured Leith, taking 156 Argentines captive. This concluded operations in South Georgia. The recapture of South Georgia was not only a psychologically important victory, but was of strategic value as well, giving British ships a staging area.

1982, April 28. British Blockade Goes into Effect.

1982, May 1–3. British Air and Naval Successes. A British Vulcan bomber from Ascension Island, and Harrier jump jets from the task force's light carriers, pounded Port Stanley's airfield (May 1). The field was effectively closed. (There were 4 additional Vulcan bomb-

ing attacks in the course of the campaign.) Harriers also bombed a smaller airfield at Goose Green while Sea King helicopters strafed Argentine positions at Darwin. That same day a British Harrier shot down an Argentine Mirage with a Side-winder air-to-air missile. Naval combat intensified when Britain's nuclear-powered submarine HMS *Conqueror* hit the World War II vintage cruiser *General Belgrano* with 2 wire-guided torpedos (May 2). The *General Belgrano* sank in less than an hour; 368 men were lost.

1982, May 4. Argentine Reprisal. An Argentinian Super Etendard fighter-bomber fired an Exocet air-to-sea missile, hitting HMS *Sheffield* 30 miles away. Although the *Sheffield* did not sink, the damage was too extensive to warrant salvaging, and she was scuttled.

1982, May 12. *Queen Elizabeth 2* Sails for Falklands. Refitted for service as a troop transport, the liner left for the Falklands carrying the 5th British Army Brigade, composed of Welsh Guards, Scots Guards, and the 7th Gurkhas.

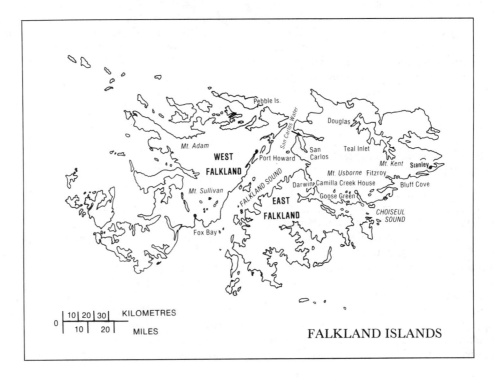

FALKLAND ISLANDS

1982, May 14–15. Pebble Island Raid. In a daring night raid, SAS troops with destroyer naval gunfire support severely damaged the Argentine base at Pebble Island, destroying 11 aircraft, the base radio station, and fuel and ammunition dumps.

1982, May 21. San Carlos Beachhead Established. The 3rd Marine Commando Brigade, reinforced with 2 battalions from the Army's Parachute Regiment, established a beachhead at San Carlos on the west coast of East Falkland. British Harriers downed 15 Argentine planes, while losing one Harrier. However, 3 British ships were hit by Argentine planes and one, the HMS *Ardent,* was lost. By the end of the day 3,000 British troops were ashore.

1982, May 23–25. Argentine Air Attacks and British Buildup. Argentine aircraft continued to mount strong attacks against British ships. The British lost 1 ship on May 23 and 2 on May 25. Nonetheless, the British continued their buildup and by May 24 they had 5,000 men and 5,000 tons of ordnance and supplies ashore. British Harriers and surface-to-air missiles fired from ships and by troops ashore shot down 36 Argentinian planes.

1982, May 26–27. Advance on Port Darwin. The British plan was to strike at Port Stanley as quickly as possible. But first it was determined to secure the southern flank and lines of communication by eliminating Argentine positions at Goose Green and Darwin approximately 20 miles southeast of the San Carlos beachhead. The 2nd Battalion of the Parachute Regiment moved south at night from San Carlos toward Darwin.

1982, May 28. British Take Darwin. At 0200, with supporting naval gunfire from British ships offshore, the attack on Darwin began. The 2nd Battalion also got air support from Harriers. The Argentines offered staunch resistance but were pushed out of their positions into the settlement. They surrendered the following afternoon and the paratroops pushed on to Goose Green (May 28).

1982, May 28–29. Goose Green Seized. The Argentines were well dug in and a bitter fight continued into the night. Early in the morning the last Argentinian position surrendered to the paratroops. The British loss was 17 killed, with 31 wounded. The Argentine loss was 250 killed and 121 wounded. The British took 1,000 prisoners.

COMMENT. *The British victories at Darwin and Goose Green secured the British southern flank, and lines of communication to the beachhead at San Carlos. They also gave the British ground forces a decided psychological advantage over the Argentinians.*

1982, May 27–June 1. Advance on Stanley. While the 2nd Battalion of the Parachute Regiment was advancing on Darwin (see above), the 3rd Marine Commando Brigade's 45th Royal Marine Commando and the Parachute Regiment's 3rd Battalion marched northeast from the San Carlos beachhead to Douglas and thence southeast to Teal Inlet, a 50-mile journey over rough terrain (May 27–30). Major General **Jeremy Moore** assumed direct control of the land operations from San Carlos. Meanwhile, a small task force of SAS troops had been dropped by helicopter on Mount Kent, 10 miles west of Stanley (May 28). It was joined by the 42nd Royal Marine Commando and a supporting light artillery unit (June 1–2).

1982, June 1. 5th Infantry Brigade Lands at Beachhead. The 5th Infantry Brigade (Scots Guards, Welsh Guards, and Gurkha Rifles) began landing at San Carlos. At the same time, the 2nd Battalion of the Parachute Regiment advanced from Goose Green to Fitzroy (18 miles southwest of Stanley), which had been abandoned by the Argentinians.

1982, June 6–8. Advance to Bluff Cove. Elements of the 5th Brigade moved forward to Goose Green and were then transported by sea to Bluff Cove northeast of Fitzroy to take up position for the attack on Port Stanley. Disembarkment operations were hindered by air attacks from Argentine Mirages. Two British troop transport ships were hit and abandoned with heavy losses (June 8). British Harriers downed 4 Mirages. British ships and planes began to soften up Argentinian positions at Port Stanley.

1982, June 11–12. British Take Heights Overlooking Stanley. The 3rd Royal Marine Commando Brigade launched a night attack on 3 key Argentine strong points: Mount Longdon, Two Sisters, and Mount Harriet. Despite surprise, resistance was substantial at Mount Longdon on the north flank and Two Sisters in the center. Mount Harriet, enveloped from the rear, fell quickly.

1982, June 13–14. Final Night Assault. The 2nd Parachute Battalion (victors at Darwin and Goose Green) made a surprise night attack on Wireless Ridge 3 miles east of Mount Longdon. To the south a battalion of Scots Guards took Tumbledown Mountain after a tough fight with Argentine Marines. Gurkhas then passed through the Scots Guards to take Mount William less than 4 miles west of Port Stanley. A

planned third phase of the operation proved unnecessary as Argentinian resistance collapsed.

1982, June 14. Argentine Surrender. The Argentine military governor of the Falklands, General **Mario Benjamin Menendez,** surrendered to General Moore.

1982, June 14. Surrender of South Thule. An Argentine naval detachment that had maintained a research facility at South Thule since 1976 surrendered to a small task force of Royal Marine Commandos without a struggle.

COMMENT. *The British successfully carried out a very difficult operation on short notice over very long lines of communication (over 8,000 miles). The war was an excellent proving ground for modern high technology weaponry. However, the principal lesson was that wars are still won by well-trained, determined soldiers, sailors, and airmen.*

The 1982 War in Lebanon*

1982, June 6. Operation "Peace for Galilee." At 11 A.M., 3 division-size Israeli task forces crossed the border into Lebanon. Their immediate objective was to entrap and destroy PLO forces in southern Lebanon. On the left, Task Force A, with support from a small amphibious force that landed north of Tyre, moved rapidly along the narrow coastal plain bypassing or smashing PLO strongpoints. By nightfall they had surrounded Tyre. In the center, Task Force C captured key Litani River bridgeheads, took Beaufort Castle, secured the Arnoun Heights, and encircled Nabatiye. On the right Task Force H advanced up the Bekaa Valley toward Hasbaiya to secure the eastern flank of Task Force C. Although Task Force H at first experienced no serious resistance, upon entering the Hasbaiya area the Israelis came under fire from Syrian artillery. A fourth task force, Task Force B, composed of naval commandos, paratroops, armor, and artillery, made a surprise amphibious landing north of Sidon at the mouth of the Awali River at 11 P.M. They overran several PLO units and isolated PLO forces in Sidon.

*See map, p. 1506.

1982, June 7. Second Day, Western Sector. While elements of Task Force A moved into Tyre and eliminated PLO resistance south of the Litani River, others approached Sidon from the south. Reserves mopped up areas that had been bypassed. Near Tyre, Task Force A units surrounded and attacked the PLO base at the Rashidiye refugee camp. Meanwhile, Task Force B broke into 3 battle groups. One moved north within 7 kilometers of Damour, south of Beirut; another enlarged the beachhead to the east, forcing the PLO defenders to retreat into the mountains; the third blocked Sidon from the north. Elements of Task Force C took Nabatiye and cleared the Arnoun-Nabatiye plateau of pockets of resistance while other elements pushed northwest toward Sidon to link up with detachments of Task Force A and B to encircle Sidon. Task Force D now entered Lebanon in Task Force C's sector, and by evening approached the Bessri Bridge north of Jezzine to cut off the coastal plain area from the Bekaa Valley. This task force received fire from Syrian artillery as forward elements approached the Bessri River. Although careful to avoid battle with the Syrians, the Israelis returned fire to silence the Syrian artillery.

1982, June 7. Second Day in the Bekaa Valley. Task Force H did not advance, but exchanged fire with PLO and Syrian elements. During the day Task Force H was joined by task forces V and Z to form the Bekaa Forces Group (BFG), about 2 divisions in strength, commanded by Major General **Avigdor Ben Gal.** The mission of BFG was to gain control of the Bekaa Valley and the flanking mountains: the western slopes of the Anti-Lebanon Mountains and Mount Hermon on the east, and the Lebanon Mountains on the west. The BFG would then cut the Beirut-Damascus highway.

1982, June 8. Third Day, Western Sector. Task Force A carried out mop-up operations in Tyre and the Rashidiye refugee camp, and captured Sidon. The Israelis were careful to avoid casualties to Lebanese or Palestinian civilians. Task Force B (reinforced by a battle group from Task Force A) and elements of Task Force C continued to advance slowly on Damour. Opposition was light. Meanwhile, Task Force D reached the Bessri River Bridge where it met and defeated the Syrian 85th Brigade. Toward evening the task force reached Beit el Dine and Ain Zhalta about 10 miles south of the Beirut-Damascus highway.

1982, June 8. Third Day in the Bekaa Valley. Task Force Z attacked the Syrian right flank which, with PLO units, held Jezzine. The combined infantry and armor attack crushed the Syrian defenders, destroying 32 tanks and inflicting heavy casualties. To the east Task Force H captured the village of Mimes and by evening was about 15 kilometers from the main Syrian defense zone in the Rashaiya area.

1982, June 9. Fourth Day on the Coast. Elements of Task Force B attacked Damour which, despite fierce PLO resistance, was taken by evening. Task Force C and elements of Task Force A and remaining units of Task Force B bypassed Damour and drove north reaching the southern suburbs of Beirut by nightfall.

1982, June 9. Fourth Day; Israeli Air Victory. The Israeli Air Force attacked Syrian SAM missile batteries in the Bekaa Valley destroying 17 out of 19 in a 3-hour battle without

loss of a single plane. Twenty-nine Syrian MiGs were shot down.

1982, June 9. Fourth Day in the Bekaa Valley. Task Force H advanced up the Valley and through the Anti-Lebanon Mountains to envelop the Syrian-PLO left flank. Resistance was substantial, but progress was rapid. Task Force Z advanced north from Jezzine toward the Syrian Kara defense zone. Task Force V was committed to combat in the center of the Bekaa Valley and by evening reached the village of Dneibe.

1982, June 10. Fifth Day on the Coast. Task Force A continued to mop up in southern Lebanon, including Sidon and the El Hilweh refugee camp where there was considerable resistance, but proceeded slowly to avoid civilian casualties. Task Force B advanced toward the mountain town of Kfar Matta while Task Force C, reinforced by elements of Task Force A, reached the outskirts of Khalde about 3 miles from the Beirut international airport. Task Force D mopped up in Ain Zhalta, approached Ain Dara and came to within 2 miles of the Beirut-Damascus highway.

1982, June 10. Fifth Day in the Bekaa Valley. The Israelis approached the Syrian main defense area in 3 task forces on a broad front. In the center Task Force H drove up the Valley and cleared out PLO and Syrian units deployed between the Litani River on the west and the Hasbani River to the east. To the west Task Force Z advanced as far as Zalia, within 10 miles of the Beirut-Damascus highway. On the east, in very mountainous terrain, Task Force V also advanced well into the main Syrian defense zone, south of the Beirut-Damascus road. In the air the Israeli Air Force knocked out 2 more SAM batteries.

1982, June 11. Sixth Day; Israelis Close on Beirut and the Beirut-Damascus Highway. Khalde, on the outskirts of Beirut, fell to Task Force C and Israeli troops took up positions on hills overlooking the Beirut Airport. Task Force D approached the Beirut-Damascus highway, which it could control with fire from the dominating hills. To the east in the Bekaa Valley, the Israelis probed the new Syrian defense line just below the Beirut-Damascus highway.

1982, June 11. Sixth Day; Cease-Fire. The fierce fighting of the 2 previous days had cost the Syrians heavy losses in both men and armor. Both they and the Israelis agreed to a cease-fire as of noon. However, fighting with the PLO continued.

COMMENT. *In 6 days of fighting, the Israelis had destroyed the military forces of the PLO in Lebanon and captured enormous quantities of equipment. They had also decisively defeated the Syrians and had artillery in position within range of Damascus.*

1982, June 12–15. Mop-Up; Resistance at El Hilweh Refugee Camp. The Israelis continued to battle remnants of the PLO in the Sidon area, particularly at the El Hilweh refugee camp. The PLO held thousands of Palestinians hostage in their stronghold, hoping at least for a propaganda victory. The Israelis surrounded the camp and advanced slowly, minimizing casualties to civilians and to their own troops. They gradually squeezed the PLO fighters into the center of the camp and finally overwhelmed the survivors (June 15).

1982, June 12–13, Battle of Ain Aanoub. The Israelis, attempting to link up with Christian Lebanese Phalangist forces near Beirut, battled PLO units and Syrian commandos deployed with the PLO. The Syrians and Palestinians were forced to retreat. They lost many tanks and APCs and suffered heavy casualties. The Israelis then linked up with the Phalangists. PLO and Syrian troops in Beirut were cut off from the Syrian forces to the east.

1982, June 14. Israelis Reach Beirut. Advancing to the eastern and southeastern suburbs of Beirut, the Israelis closed on the Syrian and PLO forces trapped in Beirut. They also took control of the southern part of the international airport. The Israelis subsequently deployed along the "green line," separating Christian eastern Beirut from the Moslem western sector of the city (June 18). The Israelis had effectively invested Beirut.

1982, June 22–26. Final Israeli Offensive. The Israelis launched a combined arms offensive toward the Beirut-Damascus highway just east of Beirut. Combat raged for 4 days. By June 25, the Syrians were driven from the outskirts of Beirut. The Israelis controlled the Beirut-Damascus highway from east Beirut to Sofar, 10 miles to the east of the city, and most of the ridges north and south of the highway. Only the Dahr el Baider Ridge separated the Israeli forces in the western sector from the Syrians in the northern Bekaa Valley.

1982, June 26. New Cease Fire. With a new cease-fire in effect major offensive combat operations were complete; however, the Israelis continued to besiege Beirut in order to assure the withdrawal of the PLO and Syrians from the city.

1982, June 26–September 3. Siege of Beirut. Having trapped the PLO and Syrians in west and southwest Beirut, the Israelis determined to coerce the PLO to leave on terms favorable to Israel rather than assault the city. Wishing to minimize civilian casualties the IDF had pilots drop leaflets warning civilians to leave the city along the coastal road controlled by Christian militamen or by the Israeli-controlled Beirut-Damascus highway (June 27). The Israelis then began aerial, naval, and artillery bombardment of the Moslem sectors of the city. The Israelis demanded the PLO surrender their arms and leave Lebanon. While the Israelis were conducting their siege operations, U.S. Special Envoy **Philip Habib,** with the aid of King **Fahd** of Saudi Arabia, conducted multilateral negotiations between the Israelis, the PLO, the government of Lebanon, and the Syrians. The Israelis tightened the noose around West Beirut by turning off essential utilities and prohibiting food and medical supplies from being sent into the besieged sectors of the city (July 3). They later eased these restrictions (July 7). Habib won PLO approval of a withdrawal plan (August 6). The Israeli Cabinet agreed to the Habib plan (August 19) and the PLO and Syrians began withdrawing 2 days later (see below). The Israelis lifted the siege as the last PLO fighters departed the city (September 3).

1982, August 21–28. Deployment of Multinational Peacekeeping Force. With the agreement of the governments of Israel, Syria, and Lebanon, a multinational peacekeeping force with independent contingents of about 1,200 men each from the United States (U.S. Marines), France, and Italy was deployed in

Beirut to ensure safe withdrawal of the PLO and Syrian forces besieged in Beirut by Israeli forces.
1982, September 16–18. Massacre at Sabra and Shatilla Refugee Camps. Phalangist militiamen slaughtered at least 400 men, women, and children while conducting an operation in search of PLO fighters. Israeli troops in the vicinity had permitted the Phalangists to enter the camp to search for PLO soldiers, and apparently did nothing to prevent these massacres.

THE 1982 WAR IN LEBANON (JUNE 6–SEPTEMBER 3): STRENGTHS AND LOSSES

	Strengths	Killed	Wounded	Missing or Prisoners	Tanks Lost	Aircraft Lost
Israeli Armed Forces	65,000	305	1,230	2	40	2
Palestine Liberation Organization	18,000	1,100	2,350	5,100	20	—
Syrian Armed Forces	45,000	1,350	4,800	220	420	92
Total Syrian & PLO Forces	63,000	2,450	7,150	5,320	440	92

1982, September 28. Kahan Commission of Inquiry Established. The Israeli government announced that **Yitzhak Kahan,** president of the Supreme Court, would head a commission which would conduct an official inquiry to determine whether or not the IDF or the government of Prime Minister **Menachem Begin** could be held responsible for the massacre of Palestinian refugees at the Sabra and Shatilla camps. The inquiry began several weeks later (October 16) and was completed 3 months later (January 16).

1983, February 8. Israeli Report on Massacre. An official Israeli government inquiry conducted by the Kahan Commission into the Sabra and Shatilla camp massacres concluded that Phalangist militiamen bore all direct responsibility for the massacres. However, Israeli officials were charged with indirect responsibility for failure to heed information that Lebanese Christian Phalangists were likely to seek revenge on Palestinians for the assassination of Lebanon's President-elect **Bashir Gemayel** (see p. 1505). Israeli Prime Minister Menachem Begin and Defense Minister Ariel Sharon were held to be indirectly responsible for the massacres, as were IDF Chief of Staff Lieutenant General **Rafael Eytan,** Director of Military Intelligence Major General **Yehoshua Saguy** and Brigadier General **Amos Yaron,** commander of the Israeli division deployed in Beirut.

1983, November 24. Israeli-PLO Prisoner Exchange. The Israelis and PLO agreed to exchange 4,500 PLO prisoners of war held by the Israelis for 6 Israelis held by the PLO since their capture more than a year before.

Kuwait (Second Gulf) War

1990, April–July. Iraqi Pressure on Kuwait. Dictator-president Saddam Hussein of Iraq brought pressure against Kuwait to yield on 3 main issues. First, he wanted control of Bubiyan and Warbah Islands to improve Iraqi access to the Persian Gulf from Umm Qasr. Second, he desired concessions on repayment of loans provided by Kuwait during the war with Iran, and on disputed oilfield rights in northern Kuwait. Finally, he wanted a higher OPEC-wide oil price, to help pay for postwar reconstruction in Iraq. He hoped to capitalize on widespread popular Arab dissatisfaction with the West, with conservative Arab governments, and with the slow pace of economic development. Few observers credited his threats to attack Kuwait if his demands were not met.

1990, August 2. Iraqi Invasion of Kuwait. Iraqi troops, some moving by helicopter, invaded and occupied Kuwait. The emir fled the first day; the Iraqis were in full control by August 4. The U.S., the European Community, and Japan swiftly imposed trade sanctions (August 3–4), and Iraqi actions were condemned by the Soviet Union and 14 of 21 Arab League members (August 4).

1990, August 2–6. U.N. Resolutions, U.S. Response. In immediate response to the Iraqi invasion, the U.N. Security Council passed Resolution 660, condemning Iraqi aggression against Kuwait and demanding immediate withdrawal. President Bush ordered U.S. forces to Saudi Arabia (Operation Desert Shield) to protect that country against anticipated Iraqi aggression (August 6). The U.N. Security Council (Resolution 661) imposed economic sanctions on Iraq.

1990, August 8. Iraq Formally Annexes Kuwait. Kuwait became the nineteenth province of Iraq.

1990, August 9. U.S. Forces in Saudi Arabia. A 2,300-man brigade of the 82nd Airborne Division and FB-111 attack aircraft arrived in Saudi Arabia, and B-52 bombers in Diego Garcia, as the first stage of U.S. deployments. Egyptian and Moroccan forces began arriving soon after (August 11). This was the beginning of a major U.S. force deployment to Saudi Arabia, matched by smaller forces from many other U.N. members.

1990, August 13–17. Iraqi "Guest" Policy. With the crisis deepening, Saddam Hussein announced that Westerners in Kuwait and Iraq would be kept in Iraq as "guests," and warned that in case of war they would be used as "human shields" around vital Iraqi installations.

1990, August 22–23. Crisis Deepens. President Bush authorized the activation of 50,000 reservists for U.S. Armed Forces. Iraqi troops surrounded 9 embassies in Kuwait which had refused to close (August 23) following Iraqi annexation of Kuwait.

1990, August 25. U.N. Blockade. In Resolution 665, the Security Council voted 13–0 (Cuba and Yemen abstained) to authorize U.S. and allied warships in the Gulf to use force to impose the sanctions of Resolution 661.

1990, September–1991, February. "Rape of Kuwait." Refugees from Kuwait report massive looting, arson, and random violence by Iraqi troops. The most blatant activity, Iraqi sabotage of Kuwaiti oilwells, left over 500 wellhead fires burning by the war's end. Outcry over Iraqi actions in the West was balanced by approval among many pro-Saddam Arabs, who were jealous of Kuwaiti oil wealth.

1990, September 25. U.N. Air Embargo. In Resolution 670, the Security Council extended the embargo-blockade to cover airborne traffic.

1990, October 29. Iraqi Damage Liability. In Resolution 674, the Security Council affirmed Iraq's liability for damages associated with its invasion and occupation of Kuwait.

1990, November 8. U.S. Force Expansion. U.S. forces in Saudi Arabia and vicinity totaled about 225,000, including 82nd Airborne, 24th Mechanized Infantry, and 101st Airborne (Airmobile) Divisions, and Air Force, Navy, and Marine forces. President Bush announced that another 200,000 troops would be deployed to ensure "an adequate offensive military option."

1990, November 18. Iraqi Response. Saddam Hussein announced that 670,000 Iraqi troops would be deployed to Basra and Kuwait provinces to match the increased U.N.–U.S. deployment. He also announced that, barring hostilities, he would release "guests" at the end of the year.

1990, November 29. U.N. Ultimatum. In Resolution 678, the Security Council authorized members "cooperating with the government of Kuwait . . . to use all means necessary" to compel immediate and unconditional Iraqi withdrawal from Kuwait, if Iraq did not comply with previous resolutions by January 15, 1991.

1990, November 30–December 15. Release of Hostages. Iraq permitted all remaining Americans, Europeans, Japanese, and Soviets to leave Iraq and Kuwait. The U.S. State Department reported that some 500 Americans voluntarily remained in Kuwait and Iraq (December 11). The U.S. closed its embassy in

Kuwait and evacuated the 5 remaining staff members (December 14).

1991, January 17. War Begins. Following the expiration of the deadline imposed by U.N. Resolution 678 (midnight January 15), U.N. air forces begin operations against Iraq at 4 A.M., local time, as part of Operation Desert Storm.

During the first day U.S., French, British, and Saudi aircraft flew some 1,200 sorties, with over 600 combat sorties, striking at Iraqi industrial sites, air defense installations, airfields, and military command and communications facilities.

Comparison of Forces

U.N. COALITION

U.S. forces were: 6 aircraft carrier battle groups, 2 Marine divisions (1st and 2nd), and 7 Army divisions (1st Cavalry, 1st and 3rd Armored, 1st and 24th Mechanized, 82nd Airborne, and 101st Airmobile), and 2nd and 3rd Armored Cavalry Regiments, with some 380,000 ground troops, about 2,200 tanks, 500 combat helicopters, and 1,500 combat aircraft in-theater. U.N. Allied forces were about 110,000 combat troops in 8 divisions, with about 1,200 tanks, 150 helicopters, and 350 combat aircraft. The U.N. Allied forces included contingents from the United Kingdom, France, Italy, Canada, Kuwait, Saudi Arabia, Egypt, Syria, Qatar, Bahrain,

Multiple Launch Rocket System (MLRS) in Saudi Arabia, 1991 (*U.S. Army photo*)

UAE, Oman, Morocco, Czechoslovakia, Pakistan, Bangladesh, Senegal, and Niger; there were also Belgian and German contingents based in Turkey. Commanding the coalition forces was American Army General H. Norman Schwarzkopf.

IRAQ

Iraq had about 550,000 troops in Kuwait and southwestern Iraq, organized in some 42 divisions with 4,200 tanks and 150 helicopters, supported by about 550 combat-ready aircraft.

1991, January 16–February 22. The Air War. Allied air forces range virtually unchallenged over Iraq, striking at military, transport, and communications targets. Western precision-guided and "smart" munitions proved very effective, and gun-camera video footage of impressively precise weapons strikes gained wide circulation in the Western media. By the third week of February, the Iraqi Air Force had lost over 40 planes to 21 lost by the Allies, with 30 Iraqi planes downed against no Allied losses in air-to-air combat. There were more than 70,000 Allied sorties, just over half being combat sorties over enemy territory. About 140 Iraqi planes flew to Iran, where they were interned.

1991, January 19–February 26. The Scud Campaign. Employing a large arsenal of Soviet-made Scud surface-to-surface ballistic missiles, plus Iraqi-made variants, Saddam Hussein struck at both Saudi Arabia and Israel (the latter country not even a belligerent). The Scuds inflicted remarkably few casualties, partly because they were limited to conventional high-explosive warheads, and partly because of the effectiveness of U.S. Patriot surface-to-air missiles (SAMs) in intercepting and destroying incoming Scuds. The U.N. Allied air forces also made great efforts to interdict Scud launching sites, but mobile Scud launchers made the task difficult, and scattered launches continued until the end of the war. There were 35 Scud launches in the first week, 18 in the second week, 8 in the third, and about 20 more toward the end of the war, including a total of 39 fired at Israel. The heaviest loss of life from Scuds occurred near the end of the war, when a Scud struck and destroyed a barracks building near Dharhan housing U.S. troops,

killing 24 National Guard personnel (February 25).

1991, January 29–31. The Battles for Wafrah and Khafji. Apparently hoping to provoke a battle with U.N. ground forces, several Iraqi battalions launched a series of limited attacks against the Saudi border towns of Wafrah and Khafji, beginning in the predawn hours of January 29. After initial confusion (partly caused by Iraqi attempts to feign surrender), Saudi, Qatari, and U.S. Marine units drove the attackers from Wafrah (early on January 30) and then from Khafji (January 31). U.N. casualties were light, but the Iraqis lost at least 30 tanks and other armored vehicles, and several hundred Iraqi soldiers were captured. The U.S. suffered 11 killed and about 35 wounded, including 6 Marines who were killed in "fratricide," when a Maverick missile from a U.S. F-16 destroyed a LAV-25 light APC (January 29).

1991, February 13. Baghdad Claims U.S. Air Attacks on Civilians. U.S. aircraft, probably F-117A "Stealth" fighter-bombers, struck and destroyed a communications bunker in Baghdad. Iraq claimed the bunker was a civilian bomb shelter, and saturated the TV waves with pictures of charred bodies and weeping relatives. The U.N. command repeated that its policy was to avoid strikes on strictly civilian targets. Although civilians (possibly military dependants) were apparently sheltered on the lower level, there was undoubtedly military communications equipment on the upper floor.

1991, February 15–22. Last-Minute Soviet Peace Initiative. The U.S.S.R. made a last-ditch (and ultimately futile) effort to arrange a peaceful settlement. The U.N. Allies, led by the U.S., France, and Great Britain, rejected 2 withdrawal proposals, which ignored or conflicted with various portions of the Security Council resolutions. The Soviets repeated an earlier commitment to support, but not participate in, U.N. military actions.

1991, February 22. President Bush's Ultimatum. President Bush demanded that Iraqi forces withdraw from Kuwait by noon EST, February 23, or face further Allied action. The ultimatum was denounced by Saddam Hussein as "shameful."

1991, February 23. Operation Desert Saber. The Ground War Begins. The Allied ground assault began about 4 A.M., local time. It encompassed 3 main components: (1) a main outflanking move, through southern Iraq, well to the west of Kuwait, by the VII and XVIII Corps to which the British 1st Armored Division and the French 6th Light Armored Division respectively, were attached; (2) a thrust northward along the coast toward Kuwait City

by the U.S. Marines with Saudi and other Pan-Arab forces; and (3) an attack northeastward through southwestern Kuwait, toward Kuwait City mounted by the Pan-Arab Force (largely Saudis, Egyptians, and Syrians). The main effort began as an attack due north toward the Euphrates (February 23). The XVIII Corps, with the French, isolated the operations area from the rest of Iraq and protected the western flank. The VII Corps, with the British 1st Armored Division, turned eastward and drove east, south of the Euphrates River toward Basra and northern Kuwait, while Marine and Pan-Arab forces approached Kuwait City (February 24). U.N. forces reached the outskirts of Kuwait City and drew near Basra (February 25). The vaunted Republican Guard, elite units of the Iraqi Army, were easily overwhelmed by U.S. and British armor (February 25–26). This last defeat marked the end of organized Iraqi resistance.

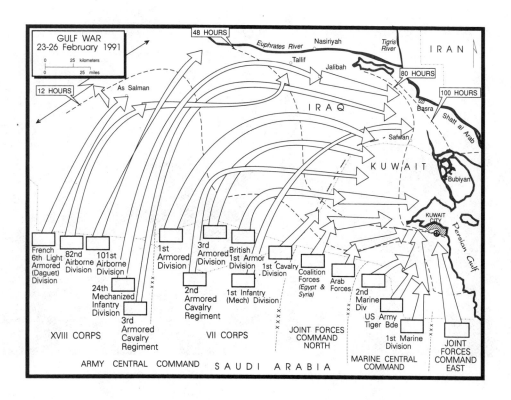

U.N. casualties were astonishingly low: 95 killed, 368 wounded, and about 20 missing during the 100-hour operation. In many instances Iraqi defenders abandoned their positions before U.N. troops reached them, and news reports showed heavy damage inflicted by U.N. aircraft and helicopters on columns of fleeing Iraqi vehicles. Iraqi losses were difficult to assess, but at least 60,000 Iraqis were captured, and as many as 30,000–50,000 were killed, and at least 50,000 wounded in air raids or ground combat between January and the war's end. Most Iraqi heavy equipment in the combat area was destroyed (much of it by U.N. aircraft), or captured.

The poor Iraqi performance, in contrast to even the worst-case Western forecasts, may be attributed to 3 main considerations. First, the U.N. forces had air supremacy, and so not only denied the Iraqis aerial intelligence but were also able to strike at will any Iraqi ground tar-

gets. Second, Iraqi morale, unit cohesion, and combat effectiveness were lower than expected, partly due to the intensive air attacks they suffered throughout the war. Widespread desertions, of both officers and enlisted men, in many units seriously weakened Iraqi capacity to resist on the ground. Only elements of the Republican Guard and a few other scattered units offered anything like organized resistance, and Iraqi actions were hesitant and uncoordinated. The U.N. plan of attack made the most of both Iraqi weaknesses and Allied strengths, particularly enemy reconnaissance failure and shattered command structure. Third, the surprise achieved by coalition forces in the ground campaign completed the demoralization of Iraqi forces.

March 1991. Revolt and Civil Unrest in Iraq. Outside the largely unpopulated regions occupied by Allied troops, the authority of Saddam Hussein's Ba'athist government was

A.H.-64 Apache Helicopter in Saudi Arabia, 1991 (*U.S. Army photo*)

challenged by uncoordinated antigovernment factions among the Iraqi people. In the south and southeast, Iraqi Shi'ites gained control of Basra, An-Nasiriyah, and other cities before Iraqi army forces (including brigades from the 5th "Baghdad" Motorized Division of the Republican Guard) suppressed the uprisings and restored order. In the north, Kurdish rebels at first liberated most of Iraqi Kurdistan from government control. The Kurdish revolt was crushed by Republican Guard troops from Baghdad, plus forces from the Basra region after the Shi'ite uprising was crushed.

INTERNATIONAL PEACEKEEPING

THE UNITED NATIONS

1975–1984. United Nations Force in Cyprus (UNIFCYP). The U.N. Security Council maintained a peacekeeping force in Cyprus.

1975–1984. United Nations Emergency Force (UNEF) in Sinai. The U.N. Security Council continued to extend the mission of the UNEF in the Sinai until faced with the prospect of a Soviet veto. The Security Council voted (July 24, 1979) to eliminate this force, and to rely solely upon observers from the United Nations Truce Supervision Organization (UNTSO). (See pp. 1353.)

1975–1984. Golan Heights U.N. Disengagement Observer Force (UNDOF). The U.N. Security Council continued to deploy a peacekeeping force in the Golan Heights. (See p. 1354.)

1978, March 19. U.N. Peacekeeping in Lebanon. The U.N. Security Council passed a resolution (March 19) calling for Israeli troop withdrawal from southern Lebanon and their replacement with a United Nations Interim Force in Lebanon (UNIFIL). The UNIFIL troops began to replace the Israelis soon after the passage of the resolution (April) and have remained in Lebanon to date (1984). (See p. 1505.)

OTHER PEACEKEEPING

1975, July 30–August 2. Helsinki Security Conference. Representatives of 35 countries met in the Finnish capital to discuss security arrangements for Europe, including the sanctity of national borders, the free passage of citizens across national borders, exchange of information on military maneuvers, and human rights. The Helsinki Accord accepted by the conference was a political triumph for the Soviet Union. Although not a binding agreement, the Accord in effect ratified the existing boundaries of Europe and legitimized Soviet dominance of Eastern Europe. However, the human rights section of the agreement, clearly not being observed in the U.S.S.R., has become a significant political embarrassment to the U.S.S.R.

1975, July 22. OAU Arbitration Effort in Angola. The Organization of African Unity sent appeals to the warring factions in Angola (see p. 1539) calling for a cease-fire and asking them to send representatives to Kampala—where the OAU was holding its twelfth summit meeting. The appeal was unheeded. On the basis of a report by a special OAU commission (October 1–4) President **Idi Amin Dada** of Uganda, OAU chairman, recommended that an OAU peacekeeping force be sent to Angola (November 15). The Popular Movement for the Liberation of Angola (MPLA), the *de facto* government and one of the factions, cited the presence of South African troops in Angola as a basis for rejecting the OAU peacekeeping effort (December 18). Because of deadlock in an emergency meeting (January 10–13, 1976) no OAU actions were taken.

1977, August 5–9. OAU Attempts to Achieve Peace in Ogaden. The OAU's Border Mediation Committee called on Ethiopia and the Somalia-backed Ogaden rebels to stop fighting. There was no response to this by the participants, or to a later mediation effort (February 9, 1978).

1979, March 2. Arab League Mediation of Yemen Conflict. The Arab League accepted a role as a mediator in the border conflicts of Yemen and South Yemen (see p. 1505).

1981, December 17–1982, June 30. OAU Peacekeeping Force in Chad. (See p. 1539.)

1982, August 26. OAU Offers Cease-Fire Plan in Moroccan-Sahara War. The OAU proposed a cease-fire plan which called for a bilateral withdrawal of belligerents and a referendum to decide the future of the Western Sahara. There was no response. (See p. 1548.)

1982, October 4. Proposal for Central America Peace Forum. The United States and 6 other American nations (Costa Rica, Belize, El Salvador, Honduras, Jamaica, and Colombia) agreed to a peace plan calling for (1) a "verifia-

ble and reciprocal" regional treaty, banning arms supplies to warring factions, subversion of national governments, and foreign military advisers in any nation in the region; and (2) development of a "Peace Forum," in which regional states would work to assure democratic elections. The immediate objective of the plan was to end fighting between Honduras and Nicaragua and end internal hostilities in El Salvador and Nicaragua. It was rejected by Nicaragua because of the role of the U.S. in its formulation (November 3). (See p. 1562.)

1983, January 7–9. Contadora Group Peace Initiatives. In an effort to bring peace to Central America, the foreign ministers of 4 Latin American nations (Mexico, Panama, Venezuela, and Colombia) met on the Panamanian island of Contadora. There were no results from this or subsequent meetings (April 10–12 and July 17).

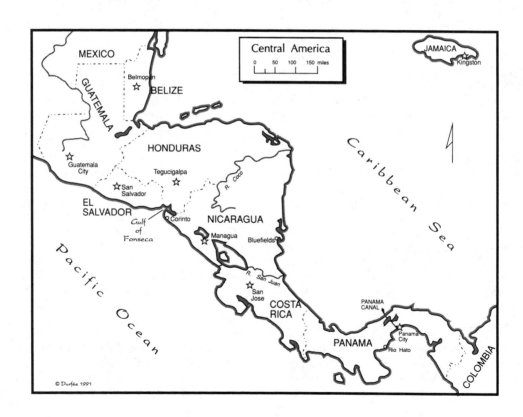

DISARMAMENT ACTIVITIES

1975–1984. The Geneva Conference on Disarmament. The Conference on Disarmament (founded in 1962 as the 18-Nation Disarmament Committee) continued to meet annually; China became the fortieth nation to join (1980). The Conference was attempting to negotiate a comprehensive ban on nuclear testing and on production and use of chemical weapons in warfare, and was negotiating a ban on development or production of radiological fissionable materials. The Soviet Union and the United States each prepared versions of a treaty for ban on chemical weapons (1984).

1975, March 26. Biological Weapons Ban Takes Effect. The 1972 Convention banning the production and stockpiling of biological and toxin weapons went into effect. There were subsequently charges that the Soviet Union and its ally Vietnam had violated this treaty and the earlier (1925) Geneva Protocol (see pp. 1493, 1526).

1979, June 18. SALT II Agreement Signed. U.S. President **Jimmy Carter** and Soviet President **Leonid Brezhnev,** in Vienna, Austria, signed a treaty designed to fill certain loopholes not covered by SALT I: to provide additional verification of compliance; to prohibit the deployment of mobile ICBMs or ALCMs (air-launched cruise missiles); and to prohibit flight testing of mobile ICBMs. The new treaty set limits on the numbers of warheads and on the numbers of launchers. Limits on new systems were also set by allowing only 1 new type of sea-based ICBM and 1 new type of land-based ICBM. Also the treaty prohibited the controversial Soviet "Backfire" bomber from increasing its radius of action, so it could not threaten the U.S. without refueling. American critics of the treaty charged that even with the limitations on warheads, the Soviets would have a sufficient number—when combined with their existing advantage in "throw weight" (explosive power) and improved accuracy—to threaten U.S. landbased missiles in the early 1980s. The critics also charged that the treaty was not veri-

fiable, and that it would place the U.S. in a position of strategic inferiority without achieving strategic stability. Because of these criticisms the U.S. Senate refused to ratify the treaty. However, President Carter pledged to abide by its terms as long as the Soviets did likewise.

1979, December 12. NATO Proposes MNF Negotiations. The foreign and defense ministers of the member states of NATO agreed to deploy United States medium-range nuclear missiles to replace 1,000 obsolete U.S. tactical missiles which would be removed (see p. 1487). They also agreed that negotiations between the United States and the Soviet Union on the reduction of medium-range nuclear forces (MNF) should begin at the earliest possible date.

1981, September 30. Proposal for Nuclear Weapons "Freeze." Scientists from 40 countries at the 31st "Pugwash" meeting agreed that nuclear weapons should be frozen at current levels. This idea was picked up and became the basis for nuclear weapons "freeze" movements in many countries. Legislation was introduced into both houses of the U.S. Congress (May 4 and October 31, 1982).

1981, November 30. Medium-Range Nuclear Force Talks Begin. U.S. and Soviet negotiating teams met in Geneva to discuss reductions of medium-range nuclear forces (MNF) in Europe. The principal issue was the reduction or elimination of Soviet SS-20 missiles in the western U.S.S.R., and of U.S. Pershing II missiles and cruise missiles, which NATO planned to deploy in Western Europe. **Paul Nitze** was the chief U.S. negotiator; his Soviet counterpart was **Yuli A. Kvitsinsky.** In a so-called "zero based option," the U.S. offered not to deploy its planned missiles if the U.S.S.R. destroyed all of its SS-20s. One reason for Soviet rejection was that this would allow the U.K. and France to retain in their strategic deterrent forces some 167 missiles capable of reaching the U.S.S.R.

1982, March 16. Brezhnev Announces Freeze on SS-20 Deployment. Soviet leader Leonid Brezhnev announced that the Soviet Union was freezing MRBM forces at present levels and that some missiles would be dismantled. At that time the U.S.S.R. had about 280 SS-20 missiles deployed in the western U.S.S.R. The missiles to be dismantled were presumably older SS-4 and SS-5 missiles.

1982, May 31. START Talks Set to Begin; U.S. Adheres to SALT Treaties. The Soviet Union and the United States announced that the two superpowers had agreed to begin Strategic Arms Reduction Treaty (START) negotiations. U.S. President **Ronald Reagan** also asserted that the U.S. would continue to abide by the existing strategic arms treaties (SALT I and II) as long as the Soviet Union continued to honor those treaties.

1982, June 29. START Negotiations Begin. The chief of the U.S. negotiating team was retired Lieutenant General **Edward Rowny.** The Soviet team was headed by **Victor Karpov.**

1982, December 21. Andropov's MNF Proposal. New Soviet leader **Yuri Andropov** proposed to reduce the Soviet SS-20 force from 280 to 162 missiles, establishing parity with the missiles of the U.K. and France, in return for a U.S. commitment not to deploy the Pershings or cruise missiles in Europe. The U.S. refused to consider British and French missiles in bilateral talks, particularly since these were national deterrent strategic missiles, and not part of NATO's forces. The U.K. and France announced their refusal to have their missiles considered in bilateral talks. However, British Prime Minister **Margaret Thatcher** (January 18) and Germany's Premier **Helmut Kohl** (March 12)

suggested that the U.S. retreat somewhat from its negotiating position (zero-based option) and make a counterproposal to break the impasse.

1983, March 29–30. U.S. and NATO's Interim MRBM Solution. In response to the urging of NATO allies, President Reagan offered an alternative to the U.S. zero-based option plan on the reduction of medium-range ballistic missiles: he called for equality in warheads between the two superpowers. Reagan pointed out that the Andropov plan (see above), with 162 SS-20 missiles, would still leave the U.S.S.R. with 486 warheads because each SS-20 carried three warheads. The Soviet negotiating team subsequently rejected the Interim Solution (June 24).

1983, November 23. Soviet Walkout from MNF Talks. In protest of the NATO deployment of the first medium-range missiles in West Germany and England (see pp. 1487, 1491) the Soviet negotiators withdrew from the negotiations in Geneva. Also the Soviet negotiators at both the START (Strategic Arms Reduction Treaty) and the MBFR (Mutual Balanced Force Reductions) negotiations used the deployment as a reason for halting negotiations.

1984, January 23. United States Alleges Soviet SALT and Geneva Violations. President Reagan presented to the U.S. Congress charges of Soviet violations of the Interim SALT Agreement and the unratified SALT II Treaty. He also charged violations of the 1925 Geneva Protocol and 1972 Geneva Convention by production and stockpiling of biological weapons. The U.S.S.R. promptly countered with its own list of U.S. violations.

WESTERN EUROPE

WESTERN EUROPEAN DEFENSE ARRANGEMENTS: NATO, 1975–1983

1977, June 9. Soviet SS-20 Missile Deployment Reported to NATO. NATO Secretary

General **Joseph Luns** reported that NATO intelligence had determined that the Soviet Union had deployed SS-20 MRBMs in the western regions of the Soviet Union. The SS-20 had 3 warheads and a 5,000-kilometer range.

The NATO countries of Western Europe were all within the range of the SS-20.

1979, December 11–12. NATO Decides to Deploy MRBMs. Meeting in Brussels the NATO foreign ministers agreed to deploy 108 Pershing II IRBMs and 464 land-based cruise missiles in Western Europe to counter the Soviet deployment of SS-20 MRBMs. At the same time it was agreed that the U.S. would pursue negotiations with the Soviets on the reduction of medium-range nuclear missiles in Europe. This "two track" approach became the basis for subsequent NATO and U.S. negotiating positions regarding the reduction of medium-range nuclear forces in Europe (see p. 1485).

1975–1983. Mutual Balanced Force Reduction (MBFR) Talks. The NATO countries and the Warsaw Pact countries continued MBFR negotiations, but no real progress was made in reaching agreement on the reduction of the size of the conventional forces that each alliance has deployed in Europe. The major problem was the inability of the participants to agree on the strength of ground forces deployed by the Warsaw Pact. NATO estimates of Warsaw Pact troop strength were consistently higher than that acknowledged by the Warsaw Pact.

1981–1983. Western European Disarmament Movement. There were numerous large-scale disarmament protests in England, the Netherlands, and West Germany against the impending installation of Pershing II and cruise missiles in Western Europe.

1983, November 22–28. NATO Begins Deploying MRBMs. The first Pershing IIs and land-based cruise missiles were deployed in NATO member countries England and the Federal Republic of Germany.

1989, April 20. NATO Debates Modernizing Short-Range Missiles. The U.S. and the U.K. had argued that the current missiles needed to be upgraded. In a concession to West German Chancellor **Helmut Kohl,** who was under public pressure to oppose the modernization, it was agreed to defer a decision on the matter.

1990, May 3. Agreements on Reunified Germany. Following reunification, Germany was to retain full membership in NATO. Soviet troops would be allowed to stay in East Germany; no Western troops would be stationed there. In addition, Germany was not allowed to possess its own nuclear, chemical, or biological weapons, and the German armed forces were limited in size.

1990, November 19. NATO and Warsaw Pact Leaders Sign CFE Treaty. The treaty limited each alliance to 20,000 tanks, 20,000 artillery guns, 30,000 armored vehicles, 6,800 combat aircraft, and 2,000 attack helicopters.

UNITED KINGDOM

1975–1984. Continuing Civil War in Northern Ireland. Despite the deployment of substantial British Army forces in the area, violence, bloodshed, and property destruction plagued Northern Ireland, spilling over into the Irish Republic and Great Britain. The hatred and intransigence of both radical Protestants and the outlawed Irish Republican Army (IRA) prevented any compromise. Authorities of the U.K. and the Irish Republic agreed to punish political terrorists on either side of the border between Northern Ireland and the Republic of Ireland. Most notable incidents were the assassinations of the British ambassador to Ireland and his secretary (July 21, 1976) and of Lord Mountbatten of Burma (August 27, 1979).

1975, November 17. "Cod War" with Iceland. Following expiration of a 2-year interim fishing agreement between the United Kingdom and Iceland, harassment of British fishermen was begun by Icelandic Coast Guard vessels within Iceland's declared 200-mile territorial limit. Three British frigates were dispatched to Icelandic waters (November 24). Iceland closed its ports and airports to British traffic. Following mediation by NATO Secretary General Joseph Luns, the British frigates withdrew (January 19, 1976). Continued clashes led the British to send 2 frigates back into the fishing area (February 5). Diplomatic relations were severed (February 19). However, negotiations continued. As a result of

M1A1 Tank in Saudi Arabia, 1991 (*U.S. Army photo*)

agreement on British fishing rights, the British warships withdrew (May 30). The agreement was signed (June 1) and diplomatic relations were restored (June 3, 1976).

1982, January. Dispute with Argentina on Falklands Issue. *La Prensa,* a major Buenos Aires newspaper, reported that Argentine General **Leopoldo Galtieri,** President and leader of the ruling junta, had pledged to return the Falkland Islands—called the Malvinas by the Argentinians—to Argentine control by January 1, 1983. The history of the Falkland/Malvinas dispute between Argentina and Britain can be summarized as follows: In 1829 the first Argentinian settlers arrived at the Falklands but found British settlers already on the islands. Clashes between British and Argentinian settlers led to the establishment of a British administration over the islands in 1833. Argentinian settlers were expelled. Although Argentina did not attempt to establish any more

settlements, it never relinquished its claim. More recently (July 1, 1971) the two countries agreed to settle their dispute by negotiations (see p. 1464). These efforts were unsuccessful (see Argentina, p. 1564).

1982, March 19. South Georgia Incident. A crew of Argentinian workers landed at Leith in South Georgia (part of the Falklands colony) to salvage scrap metal from 3 abandoned whaling stations. Britain sent a diplomatic protest alleging that the crew had illegally entered British territory. Argentina sent 5 warships to the area and issued a communiqué asserting that Argentinian workers did not need permission from the British to be on native soil. The British Foreign Minister informed Parliament that a potentially dangerous situation prevailed in the Falklands (March 30).

1982, April 2–June 14. Falkland Islands War. (See p. 1471.)

1984, April 17. Shooting Incident at the Libyan Embassy. During a demonstration by Libyans opposed to Libyan ruler Muammar Qaddafi, a burst of machine-gun fire from the Libyan embassy in London killed a British policewoman and wounded 10 demonstrators. In response, Britain cut off diplomatic relations with Libya (April 30.)

1984, October 12. IRA Attempts to Assassinate Prime Minister. A hotel in Brighton, site of the annual conference of the Conservative Party was bombed. A member of Parliament and 3 others were killed, 32 were injured. Prime Minister **Margaret Thatcher** had left the building just prior to the explosion.

1985, February 20. Britain Endorses Strategic Defense Initiative. In an address to the U.S. Congress, Prime Minister Thatcher backed the Reagan plan, citing the Soviet return to arms negotiations as a rationale for a continued Western arms build-up. (See p. 1552.)

1985, November 15. Agreement on Northern Ireland. Great Britain and the Republic of Ireland signed an accord that gave the Republic a formal consultative role in the governance of Northern Ireland.

1988, March. Irish-British Violence. A British SAS anti-terrorist team in Gibralter killed 3 unarmed IRA men who had been planning a bomb attack (March 6). At the funeral of the men in Belfast, a Protestant gunman killed 3 and injured dozens (March 16). At the funeral of one of the 3 killed in Belfast, 2 British soldiers drove into the funeral procession by mistake, and were beaten and shot to death (March 19).

1988. IRA Bombings. Six British soldiers were killed in Northern Ireland by a bomb placed in their car (June 15). A bomb in a bus killed 8 British soldiers and injured 28 other civilians in Belfast (August 20).

THE NETHERLANDS

1981, February 28. Dutch Parliament Approves the Deployment of MRBMs. The Dutch Parliament voted to permit deployment of U.S. medium-range cruise missiles and Pershing II missiles on Dutch soil as part of NATO's deterrent.

FRANCE

1975, March 4–December 15. French Military Personnel Unrest. Public protest marches by French military draftees led to a substantial pay increase. Later the French government established a new and relaxed code of discipline (July 16). However, unrest continued. Army units attempted to form unions (November 3 and 4 and December 9). Following an inquiry, ringleaders were arrested (December 15).

1975, April 13. French Intervention in Chad. (See p. 1538.)

1975–1977, French Activity in Afars and Issas. (See p. 1535.)

1976, February 4. French-Somali Troops Clash.

1978, May 19. French Intervention in Zaire.

1978–1983. French Interventions in Chad. (See p. 1538.)

1980, June 26. French Announce Neutron Bomb Test. French President **Valéry Giscard d'Estaing** announced that France had tested enhanced radiation weapons (the so-called "neutron bomb") in the South Pacific. The decision on whether or not to deploy it would be deferred until a later date.

1982–1984. French Peacekeeping Contingent in Beirut. (See pp. 1476, 1505–1509.)

1982, October 7. Nuclear Force Modernization Plan Announced. French Defense Minister **Charles Hernu** said that France intended to give "absolute priority" to the development of improved nuclear capability and that conventional ground forces would be cut 10 percent. This plan subsequently drew fire from French military leaders. General **Jean Delaunay** (Army Chief of Staff) protested that Hernu's plan would reduce the French Army by more than 30,000 men and that reduced spending on conventional forces would result in equipment obsolescence. French Air Force officials also complained that the French Air Force would lose 3,000 badly needed combat aircraft. French Foreign Minister **Claude Cheysson** defended Hernu's plan, asserting that the basis of effective deterrence is nuclear forces. Hernu also answered the critics by saying that conventional capabilities would be increased through a "rapid action and assistance command" able to intervene quickly in any European conflict. He also pointed out that the plan contemplated construction of a nuclear-powered aircraft carrier, a nuclear-powered submarine, and 3 hunter-killer submarines.

1983, May 18. France Supports NATO MNF Deployment. French President **François Mitterand** and German Chancellor Helmut Kohl formally announced support for the scheduled deployment of Pershing and cruise missiles in Western Europe. Mitterand reaffirmed this position at a meeting of NATO's Atlantic Council (June 10), while also reaffirming retention of the French *force de frappe* (independent nuclear deterrence force).

1983, October 23. Suicide Attack on French Peacekeeping Force in Lebanon. (See p. 1508.)

1983, November 8. French Catholic Bishops Support Concept of Deterrence. In a position differing from that of American Catholic bishops (see p. 1552), the French bishops issued a pastoral letter defending nuclear deterrence as being consistent with the Just War doctrine of the Catholic Church. Rejecting pacifism, the French bishops stated that the argument for peace at any price would lead to a situation in which the West would not have the means to defend itself, and would only serve to encourage aggression.

1984, November 10. French Agree to Withdraw Troops from Chad. France later announced (November 16) that its reconnaissance flights over Chad would continue as Libya had failed to withdraw its troops. In addition, 1,200 French troops would remain in Chad.

1985, May 4. French President François Mitterrand Announces Opposition to Strategic Defense Iniatiative Research.

1985, July 10. Sinking of Greenpeace Ship. The ship, the *Rainbow Warrior,* was sunk by French Secret Service agents while at anchor in Auckland, New Zealand. A photographer aboard was killed. The ship was preparing for a protest trip to French nuclear test facilities in the South Pacific. France admitted its complicity in the sinking (September 22). Two French agents were captured by New Zealand police, pled guilty to manslaughter, and were sentenced to 10 years in prison (November 4).

1988, May 5. French Commandos Free Hostages on New Caledonia. Melanesian separatists had held 23 French hostages for 2 weeks; all the hostages were released, 2 commandos and 19 Melanesians were killed.

WEST GERMANY

1979, June 7. Schmidt Warns of Dangers of Soviet Missiles. In a speech at Harvard University's commencement exercises, German Chancellor **Helmut Schmidt** warned that NATO must respond to the Soviet deployment of the SS-20 medium-range ballistic missiles (MRBMs).

1981, May 17. Schmidt Threatens Resignation over MNF Deployments. Defying the opposition of his Social Democratic Party (SPD), Chancellor Schmidt threatened to resign if the party failed to support his approval of the deployment of U.S. medium-range nuclear missiles in Germany as part of the NATO deterrent.

1981, May 26. German Parliament Approves Deployment of Pershing and Cruise Missiles. Despite opposition in his own party, Schmidt won parliamentary approval for the deployment of U.S. cruise missiles and Pershing IIs in West Germany as part of the NATO tactical nuclear force.

1983, April 18. West German Bishops Issue Pastoral Letter on War and Peace. The Roman Catholic bishops of West Germany issued a pastoral letter strongly endorsing nuclear deterrence as a necessary but regrettable means of keeping the peace.

1983, July 4–7. Kohl Negotiates with Soviets about MNF Deployment and Reduc-

tions. German Chancellor **Helmut Kohl** unsuccessfully attempted to persuade Soviet leaders to reconsider their position at the MNF (medium-range nuclear force) reduction talks in Geneva and to accept the NATO interim reduction plan. Later, Soviet Premier **Nikolai Tikhonov** warned that if the West German parliament did not reverse its position on the MNF deployment, the Soviet Union would have to deploy additional missiles and that in the event of war, West Germany would suffer devastation.

1983, November 22, West German Parliament Votes Approval of MRBMs. The West German Bundestag voted to permit the scheduled deployment of U.S. medium-range ballistic missiles on West German soil. This vote assured the Soviet walkout from the Geneva negotiations on MNF reduction. (See p. 1486.)

1985, December 18. Agreement on Strategic Defense Iniatiative. German industry was to participate in the development of technologies for the program.

SWEDEN

1981, October 27–November 6. Soviet Submarine Grounded in Swedish Territorial Waters. (See p. 1493.)

DENMARK

1983, December 1. Danish Parliament Opposes MNF Deployment. The Danish Parliament voted against NATO deployment of U.S. Pershing IIs and ground-launched cruise missiles in Denmark (December 1). Subsequently, Parliament voted to withhold from NATO funds that were to be allocated for deployment of the missiles (December 7).

ITALY

1981–1982, December 17–January 28. American General Held Hostage by Terrorists. Brigadier General **James L. Dozier** was seized in his Verona apartment (December 17) and held by terrorists for 42 days. After an extensive search a special Italian police Anti-

Terrorist Task Force stormed the building where Dozier was being held and rescued him.

1982–1984. Italian Peacekeeping Contingent in Beirut. (See pp. 1476, 1505–1509.)

1983, March 30. Italy Backs Medium-Range Missile Deployment. The Italian government endorsed the U.S.-NATO interim solution on the reduction of medium-range nuclear missile forces in Europe. The Italians later specifically endorsed deployment of Pershing IIs and ground-launched cruise missiles (May 29).

SPAIN

1975. Spanish Withdrawal from Rio de Oro. (See p. 1548.)

1975, November 22. Restoration of Spanish Monarchy. After the death of Generalissimo Francisco Franco, his protégé, **Juan Carlos de Borbón,** was proclaimed king. Almost immediately he began to change Spain into a parliamentary democracy.

1979–1983. Terrorism by Basque Separatists. Terrorism by radical Basque separatists increased in frequency and intensity. Spanish military officers became targets of many attacks. A home-rule bill sponsored by King Juan Carlos was approved by Basque voters in a popular referendum (October 25, 1980). The radicals, however, wanted total independence and continued their violence. The government deployed troops in the Basque region (March 23, 1981). Violence and defiance continued both in the Basque provinces and in other parts of Spain.

1981, February 23. Attempted Coup. Dissatisfaction with Spain's parliamentary democracy led to an attempt by senior military officers to seize power and overthrow the government. One conspirator led civil guards in seizing and holding hostage the cabinet and 350 members of the lower house of the legislature. The commander in Valencia declared martial law, while a principal military adviser to the King, General **Alfonso Armada Comyn,** urged the King to seize control and establish military rule. Juan Carlos refused, and by telephone calls and a radio address persuaded the army to remain loyal to the government. The conspirators surrendered and were arrested (February 24–26).

They were later tried by court-martial and sentenced to prison terms.

1982, May 30. Spain Enters NATO. Following the approval of each NATO member nation, Spain became the sixteenth member of the alliance.

1982, October 2. Coup Foiled. Spanish authorities loyal to the King and the democratic regime discovered plans for a coup to take place just before the national elections scheduled for October 27. Three senior officers were arrested.

1983, February 24. Base Protocol with the United States. Spain and the U.S. signed an agreement which permitted the U.S. to retain its bases in Spain.

1983, June 2. Spain Declines to Sign Medium-Range Missile Deployment Agreement. The Spanish government rejected a NATO agreement permitting future deployment of Pershing II and ground-launched cruise missiles on Spanish soil.

1986, March 12. Spain to Stay in NATO. Voters supported the government in a referendum which stipulated that there would be no nuclear weapons in Spain. Spanish armed forces were to remain outside the NATO unified command structure, and U.S. troop levels in Spain would be reduced below 12,500.

PORTUGAL

1975. Portuguese Mediation in Angola. (See p. 1539.)

1975, March 11. Rightist Coup Fails. Troops loyal to the ruling Armed Forces Movement (MFA) thwarted a coup attempt. Former provisional President General Antonio de Spinola was accused of leading the coup but denied complicity.

1975, November 25–28. Leftist Coup Fails. Following more than a month of political unrest, military and civilian leftists attempted to overthrow the new moderate government of Vice Admiral **José Pinheiro de Azevedo.** Rebellious air force troops seized 4 air bases, the Air Force Command Center, and the national radio and television stations. Army commandos suppressed the revolt in hard fighting.

1982, August 12. Council of Revolution Eliminated. The transition from military to civilian rule was completed when the Portuguese Parliament voted the Revolutionary Council out of existence.

SOVIET UNION

1975–1983. Soviet Military Advisers in Africa, Asia, and Latin America. The Soviet Union gave military and technical aid and sent military advisers to a number of Third World countries (see Cuba, Angola, Ethiopia, South Yemen, Yemen, Afghanistan). In addition to military advisers in client nations, the Soviets were able to project power by greatly upgrading their naval capabilities and obtaining naval and air base rights at a number of strategically located areas around the world including Cuba and Vietnam.

1976, December 26. Soviets Strive for Strategic Nuclear Superiority. The *New York Times* reported that the CIA's latest national intelligence estimate (NIE) was far more pessimistic than in preceding years. The estimates indicated that the U.S.S.R. was trying to achieve strategic superiority.

1977, September 26. Soviet–Warsaw Pact Ground Force Troop Strength Increase Reported. Newspaper reports asserted that in 10 years the U.S.S.R. had increased its ground forces by 30 divisions, to a total strength of 170 divisions.

1979, May 30. Soviet Union Purportedly Seeks First-Strike Capability. United States Secretary of Defense **Harold Brown** speaking at the U.S. Naval Academy in Annapolis, Md., declared that the Soviets were seeking to develop a "first-strike" capability. This would be an ability to destroy all U.S. retaliatory capabilities in one preemptive strike.

1979, December 17. Mission of Soviet Navy Enlarged to Include Global Role. U.S. Navy specialists asserted that the Soviet Union was building a navy with the capability of projecting Soviet power anywhere on the globe. This buildup was being performed under the direction of Admiral **Sergey G. Gorshkov,** chief of the Soviet Navy.

1979, December 25–28. Soviets Invade Afghanistan. (See p. 1519.)

1980, January 3. Soviet Union Rejects MNF Talks with NATO. The Soviets rejected a NATO offer to negotiate the reduction of MRBMs. They claimed that the NATO decision (December 12, 1979) to deploy new Pershing II and ground-launched cruise missiles "destroyed the basis for negotiation" (see p. 1485).

1980, March 20. Soviets Deny Biological Weapons Violation. The Soviets denied that an anthrax epidemic (April 1979) in the city of Sverdlosk was due to an accident that occurred in the manufacture of biological weapons.

1980, April 18. Soviets Test-Launch Killer Satellite. The Soviets launched a killer satellite, Cosmos 1174. The target, Cosmos 1171, had been launched earlier (April 3).

1980, September 18. Reported Improvement in Strategic and ABM Capabilities. London's International Institute for Strategic Studies (I.I.S.S.) reported that Soviet ICBMs were now "significantly more accurate" than earlier generation missiles, thus weakening U.S. deterrence. U.S. intelligence sources had also ascertained that the Soviets had improved their ABM capabilities with a phased-array radar to direct interceptor rockets. The U.S. government did not consider this to be a violation of SALT I because each side was allowed 1 ABM site under the terms of the treaty.

1980, October 2. *Kirov* **Guided Missile Cruiser in Service.** The nuclear-powered guided missile cruiser *Kirov* was reported in active service, giving the Soviet Union substantially increased offensive and defensive strategic naval capability.

1981, January 9. Soviets Launch New Attack Submarine. The new submarine class, called *Oscar* by U.S. Naval Intelligence, was 10,000 tons and could attain speeds up to 40 knots.

1981, September 23. Soviet MNF Superiority Reported. London's I.I.S.S. reported that the U.S.S.R. had more than 3 times as many warheads in their MNF than did NATO.

1981, September 29. Soviet Military Capabilities Enhanced Through Technology Transfers. A U.S. Defense Department booklet on Soviet military capabilities reported that the Soviet Union had used a variety of means, legal and illegal, to acquire Western technology to improve military capabilities.

1981, October 27–November 6. Soviet Submarine Grounded in Swedish Territorial Waters. A Soviet submarine ran aground near the Karlskrona Naval Base. The U.S.S.R. maintained this was due to poor weather and faulty navigational equipment, but this explanation was rejected by Sweden. Swedish investigation discovered radiation near the submarine's torpedo tubes. Sweden pointed out that this was incompatible with Soviet efforts to persuade Sweden to agree that Scandinavia and the Baltic should be a nuclear-free zone. The Soviets insisted there was no radiation in the submarine.

1983, September 1. Soviets Down Korean Commercial Aircraft. Soviet pilots shot down Korean Air Lines plane Flight 007 near Sakhalin Island. The plane bound for Seoul had apparently strayed off course. After tracking the plane for several hours a Soviet aircraft shot down the plane with a missile, killing all 269 passengers and crew.

1983, November 23. Soviets Walk Out from MRBM (MNF) Reduction Talks. (See p. 1486.)

1983, December 8. START Talks Recess without Resumption Date. (See p. 1486.)

1983, December 15. MBFR Talks Adjourn without Setting New Date. (See p. 1486.)

1984, February 9. Death of Andropov. Soviet Communist Party General Secretary **Yuri V. Andropov** died after 15 months in power. **Konstatin U. Chernenko** was chosen by the Politburo to replace him.

1984, March 16. Talks on Reduction of Conventional Forces in Europe Resumed in Vienna.

1984, May 20. U.S.S.R. Reports More Armed Submarines Are Deployed off U.S. Coast. Marshal **Dmitri Ustinov,** Soviet defense minister, said that the deployment was in response to the deployment of U.S. intermediate-range nuclear missiles in Europe.

1985, January 7–8. Agreement on Resumption of Arms Talks. U.S. Secretary of State **George Schultz** and Soviet Foreign Minister **Andrei Gromyko,** meeting in Geneva, agreed to resume negotiations on the reduction of nuclear arms.

1985, March 10. Death of Chernenko. Mikhail Gorbachev was chosen by the Politburo as his successor.

1985, March 12. U.S.-Soviet Arms Talks Begin. The two countries agreed to divide the negotiations into subgroups to deal with strategic nuclear weapons, intermediate-range nuclear weapons, and space weapons.

1985, March 25. Russian Guard Kills U.S. Officer. U.S. Army Major **Arthur Nicholson, Jr.,** was shot while observing Soviet tank facilities near Ludwigslust, East Germany. Soviet claims that he was in a prohibited area were rejected by the U.S. Nicholson was assigned to a military liaison unit in East Germany observing military activities as provided for in post–World War II agreements.

1985, April 7. Gorbachev Proclaims Missile Moratorium. The Soviet leader announced that the U.S.S.R. would cease to deploy intermediate-range nuclear missiles. The U.S. refused to enact a similar freeze, citing the 10:1 Soviet advantage in such missiles.

1986, January 15. Gorbachev Calls for World Ban on Nuclear Weapons. He announced a 3-month extension of the Soviet moratorium on nuclear tests and proposed a worldwide ban on nuclear weapons by the year 2000.

1986, April 25. Explosion at Chernobyl Nuclear Power Plant. The explosion and subsequent fire released massive quantities of radioactive material over the Ukraine, as well as much of the Soviet Union and Europe.

1986, October 11–12. Reagan and Gorbachev Meet in Iceland. Although they came close to major agreements on arms control issues, the opportunity for a comprehensive settlement was lost over Soviet opposition to the U.S. Strategic Defense Initiative (SDI or "Star Wars").

1987, December 8. INF Treaty Signed. The treaty provided for the dismantling of all 1,752 U.S. and 859 Soviet intermediate-range nuclear missiles with a range of 300 to 3,400 miles. The Warsaw Pact endorsed the treaty as well (December 11).

1988, January 7. Soviets Announce Withdrawal from Afghanistan. The withdrawal was to begin when an agreement was reached at U.S.-sponsored talks between the Soviet Union and Afghan rebels. It was later announced (February 8) that the withdrawal was to begin on May 15 and be completed within 10 months, provided that a settlement was reached with Afghan rebels.

1988, February–March. Unrest in Azerbaijan. Christian-Moslem hostilities resulted in the deaths of 31 (Armenian sources claimed that over 300 were killed) in Sumgait, Soviet Republic of Azerbaijan (February 28). Armenian Christians had been demanding that the Nagorno-Karabakh region of Azerbaijan be reunited with the Soviet Republic of Armenia. When the Soviet Presidium rejected Armenian reunification (March 23) there were violent protests in Armenia. Soviet troops and secret police were sent to Yerevan, the Armenian capital, to prevent planned demonstrations.

1988, April 14. U.S.S.R. Signs Agreement on Afghan Pullout. The withdrawal began as scheduled (May 15).

1988, October 1. President Gorbachev. Implementing reforms initiated earlier (June) Gorbachev was named Soviet President by the Supreme Soviet.

1988, November 16. Estonia Challenges Soviet Control. The Supreme Soviet of Estonia asserted the right to veto national laws affecting Estonia. The Presidium of the U.S.S.R. declared the action unconstitutional (November 26). Gorbachev deplored the disastrous rise of nationalism within the U.S.S.R. (November 27).

1988, December 7. Gorbachev Announces Unilateral Conventional Force Reduction. At the U.N. he announced that the Soviet armed forces would be cut by 10 percent, about 500,000 men. Some 10,000 tanks, 8,500 artillery guns, and 800 aircraft would be withdrawn from Eastern Europe.

1989, February 15. Soviets Complete Afghan Pullout. Soviet losses in the war were given as 15,000 killed and 37,000 wounded.

1989, April 9. Soviet Troops Use Poison Gas to Suppress Riot in Georgia. An unknown number of demonstrators in Tbilisi were killed. The Soviet Foreign Ministry claimed that only tear gas had been used, but *Izvestia* reported that a number of people had died from "chemical agents." Georgian toxicologists confirmed that nerve gases had been used.

1989, June 3–10. Ethnic Unrest in Uzbekistan. Bloody clashes erupted between Sunni Moslem Uzbeks and Shiite Moslem Meskhetians. At least 90 were killed and more than 1,000 injured.

1989, August 23. Baltic Republics Denounce 1939 Hitler-Stalin Pact. The Lithuanian Popular Front called for complete independence from the Soviet Union.

1989, September. Nationalist Unrest Continues. Ukrainian nationalists called for the transformation of the Soviet Union into a confederation of autonomous republics (September 8–10). Gorbachev warned leaders of Armenia and Azerbaijan to negotiate an end to their hostilities, and tensions appeared to ease (September 25).

1989, November 27. Denunciation of Lithuanian Independence. The Soviet Politburo condemned Lithuania's move toward greater freedom.

1990, January. Rioting in Azerbaijan. Demonstrators destroyed border stations along the Soviet-Iranian border. President Gorbachev ordered the airlift of 11,000 troops into the region (January 13–14). Entering Baku with tanks and armored personnel carriers, Soviet forces killed at least 93 Azerbaijanis (January 20–25).

1990, February 12. Riots in Dushanbe, Soviet Republic of Tadzhikistan. The death toll was reported as 18.

1990, March 11. Soviet Troops Begin Withdrawal from Hungary. All forces were to be out by July 1991.

1990, March 11. Lithuania Declares Independence. President Gorbachev called the move illegitimate and invalid and warned of possible military action. Soviet paratroops seized the headquarters of the Lithuanian Communist Party in Vilnius (March 25–27). The Soviet government initiated an economic embargo of Lithuania, including cutoff of all oil and natural gas supplies (April 19).

1990, May 4. Latvia Declares Independence. President Gorbachev declared that the Baltic independence movements had no legal basis.

1990, May 27. Further Unrest in Armenia. Soviet troops in Yerevan killed 6 allegedly armed civilians. In the resulting riots, 23 were killed.

1990, June 4–13. Unrest in Kirghizia. Clashes between Uzbeks and Kirghiz left 148 dead. Soviet officials declared a state of emergency.

1990, June 29. Lithuania Places a Moratorium on Independence. The action by the Lithuanian Parliament resulted in the end of the Soviet embargos (July 2).

ALBANIA

1990, June 28. Opposition Clashes with Security Police. Demonstrators sought refuge in foreign embassies. France and Italy evacuated 4,500 refugees after the government gave permission for mass emigrations.

1990, December 9. Antigovernment Protests. Students clashed with police in Tirana. The government announced the legalization of independent political parties (December 11).

BULGARIA

1990, January 5. Ethnic Strife. The government's cancellation of programs designed to force the assimilation of ethnic Turks was denounced by Bulgarian nationalists. Tension was relieved by a compromise that allowed freedom of religion and the use of Turkish or Slavic personal names. Bulgarian was declared the national language and display of foreign (i.e., Turkish) flags was forbidden.

1990, January 15. Repudiation of Communist Domination. The National Assembly unanimously revoked the constitutional provisions guaranteeing the Communists control of politics and government.

1990, August 26. Anti-Communist Riots.
The headquarters of the Socialist Party (formerly Communist Party) in Sofia was set on fire.

CZECHOSLOVAKIA

1989, November–December. Political Upheaval. The Communist Party leadership resigned following 8 days of antigovernment demonstrations (November 24). The new leadership agreed to give up the Communist monopoly on power (November 28). Continuing demonstrations led to the resignation of President **Gustav Husak** (December 10). **Alexander Dubcek,** Czech leader during the "Prague Spring" of 1968, was unanimously elected chairman of Parliament (December 28) and **Vaclav Havel,** a noted dissident, was elected President (December 29).

1990, 26 February. Withdrawal of Soviet Troops. Under terms of an agreement with the Czech government, Soviet troops began a withdrawal from Czechoslovakia.

EAST GERMANY

1989, January–October. Refugees Create Tension. By June, 44,263 East Germans had emigrated to the West, mostly via the newly-opened Hungarian border (see below). The East German government filed a diplomatic protest, but the Soviet Foreign Ministry refused to intervene (September 12). Following mass demonstrations against the government in Leipzig and other East German cities, the East German Communist Party ousted leader Erich Honecker (October 18).

1989, November 7. Fall of the East German Government. Premier **Willi Stoph** and his cabinet resigned under opposition pressure; the following day most of the Politburo also resigned. Virtually all restrictions on travel to the West were removed (November 9). Demands were made by the opposition for the dissolution of the Communist Party and for the reunification of East and West Germany (December 3).

1989, December 22. End of the Berlin Wall.
The dismantling of the wall and other obstacles along the intra-German border began.

1990, September 20. Reunification Treaty Ratified by East and West Germany.

1990, October 3. Reunification of East and West Germany. Berlin was designated as the new capital, although for the time being most government offices remained in Bonn.

HUNGARY

1989, May 2. Hungary Dismantles Border.
With the removal of a 150-mile-long barbed-wire fence, Hungary became the first Eastern European country to open a border with the West.

1989, June 16. Imre Nagy Honored. With government approval, a memorial service was held for Nagy and others executed after the 1956 revolt was crushed.

1989, October 19. Opposition Parties Sanctioned. Parliament voted overwhelmingly to legalize opposition parties, ending one-party Communist rule. Stalinist laws were eliminated from the constitution, and the official name of the country was changed to the Republic of Hungary, deleting the word People's from the title.

1990, March 11. Soviet Troops Begin Withdrawal. By agreement with the Hungarian government, all Soviet troops were to be withdrawn from Hungary.

1990, June 26. Parliament Votes to Withdraw from Warsaw Pact.

POLAND

1980–1981. Labor Unrest in Poland. Months of labor unrest and strikes led to the establishment of a national labor union, Solidarity, reluctantly accepted by the Polish government to avoid increased violence (August 30, 1980). However, unrest continued.

1981, February 9. Military Government Installed. The inability of the civilian government to restore stability led the Communist Party, presumably to forestall Soviet military intervention, to install the defense minister,

General **Wojciech Jaruzelski,** as Prime Minister. Despite strong military leadership and repressive measures, unrest continued throughout the country.

1981, September 4–11. Soviets Conduct War Games near Polish Border. Over 100,000 Soviet troops participated in combat exercises near the Polish border in violation of the Helsinki Accord (see International Peacekeeping, p. 1483) since the Soviets did not reveal the number of troops involved in the games until the U.S. accused them of the violation. The size and timing of the war games, coinciding with the Polish Solidarity national congress, led observers to believe that this was intended to intimidate the Poles.

1981, October 18. Jaruzelski Becomes Head of Polish Communist Party. In an attempt to strengthen the hand of the government, the Polish Communist Party appointed General Jaruzelski as First Secretary of the Polish Communist Party.

1981, December 13. Jaruzelski Imposes Martial Law. Worsening conditions in Poland and the continuing threat of armed Soviet intervention led Jaruzelski to impose martial law. Restrictions on civil and personal liberties were directed particularly against the activities of the Solidarity labor movement.

1982, March 13. Warsaw Pact Military Maneuvers in Northern Poland. The Soviet Union, Poland, and East Germany conducted Warsaw Pact military exercises in northern Poland. These exercises were again seen as an attempt to intimidate the Poles, who were restive under martial law, which had been imposed 3 months earlier (December 13).

1983, July 21–23. End of Martial Law in Poland. General Jaruzelski announced the end of 19 months of martial law (July 21) effective noon the next day. However, living conditions did not improve, since a number of harsh and oppressive laws were left intact.

1989, June 4. Communists Lose Election. The Communists were soundly defeated, and 99 of 100 seats in the Senate went to Solidarity-sponsored candidates. In the lower house, the Sejm, Solidarity won all 161 seats allotted to opposition candidates. Of 299 seats reserved for Communists, voters rejected most of the party-sponsored candidates, voting instead for reformers within the party.

1989, August 14. Communists Unable to Form Government. The Communist Party candidate for premier, General Czeslaw Kiszack, was unable to form a government. An agreement between Jaruzelski and Walesa led to the selection of Solidarity member **Tadeusz Mazowiecki** as premier. The U.S.S.R. announced acceptance of the change in regime, the first time that power had passed from a ruling Communist Party by democratic means in Eastern Europe.

1990, May 27. Local Elections. Solidarity-backed candidates won more than 40 percent of seats, Social Democrats (formerly the United Workers Party, or Communists) won 2 percent.

1990, December 9. Walesa Elected President.

ROMANIA

1989, 20 November. Ceausescu Rejects Reform. The Communist Party President vowed that Romania would never abandon socialism so long as he remained in power. Unenthusiastic crowds were forced by police and armed forces to carry signs supportive of the government while listening to the speech.

1989, December. Fall of Ceausescu. Following the indiscriminate killing of protestors by government security forces in Timisoara (Temesvar) (December 20–22), President Ceausescu fled Bucharest as fighting broke out between security forces and the opposition, which was supported by elements of the army. Ceausescu was captured (December 23), tried, convicted of genocide and gross abuses of power, and executed with his wife (December 25). An interim government was formed under **Ion Iliescu** amidst continuing resistance by security forces (December 25–31).

1990, May 19–20. Ethnic Unrest. Romanians clashed with ethnic Hungarians in Tirgu-Mures, in Transylvania. At least 3 were killed before the interim government sent in 500 troops to restore order.

1990, December 14. Antigovernment Strikes. Timisoara became the center of anti-government activity as strikers closed down the city. The government began talks with the opposition National Liberal party in an effort to conciliate the protestors (December 17).

YUGOSLAVIA

1990, January 22. Renunciation of Communist Domination. The Congress of the League of Communists voted to repudiate the party's constitutionally guaranteed control. Disputes arose between liberal Slovenes and traditionalist Serbs in the government.

1990, January 24. Anti-Serbian Rioting. Ethnic Albanians in Kosovo Province in the Republic of Serbia protested Serbian political control. The official death toll was 19.

1990, July 2. Kosovo Declares Independence from Serbia. The Serbian government, denounced the action as unconstitutional and dissolved the government and parliament of Kosovo, assuming direct control of the province (July 5).

1990, July 3. Slovenia Declares Sovereignty. The republic declared that its laws took precedence over those of the Yugoslavian Federation.

1990, December 22. Slovenia Votes for Independence from Yugoslavia.

GREECE

1980–1983. Strained Relations with NATO Allies. The Socialist government of Greece opposed NATO's planned MNF deployment in Western Europe. The unresolved problem of a partitioned Cyprus with 25,000 Turkish troops deployed in northern Cyprus further strained Greek-NATO relations. Prime Minister **Andreas George Papandreou** repeatedly threatened to force the U.S. to give up military bases in Greece. Papandreou did, however, renew the base agreement with the U.S. for an additional 5 years (August 20, 1983).

TURKEY

1975–1984. Continued Deployment of Turkish Troops in Cyprus. Turkey continued to deploy 25,000 troops in Cyprus. But after the Cypriot Turkish Federal State declared its independence and permanent separation from Cyprus, the Turks made a token withdrawal of 1,500 troops (see p. 1499).

1978, December 26. Martial Law Enacted in Thirteen Provinces. In an effort to bring about a halt to internal religious violence between Sunni and Shi'ite Moslems, the government (with the approval of parliament) imposed martial law in 13 of Turkey's 67 provinces (including Ankara and Istanbul). Later this was extended to 6 additional provinces (April 25, 1979).

1980, September 12. Military Coup d'État. Following months of violence due to political and religious rivalry, the Chief of General Staff, General **Kenan Evren,** established a military dictatorship. He pledged a return to democratic government when public order was restored.

1981, January 15. Constituent Assembly Date Selected. General Evren announced his intention to convene a Constituent Assembly to draft a new democratic constitution. This was done June 30.

1982, July 17. Constituent Assembly Submits Draft Constitution. This was approved by a 90 percent vote in a national referendum (November 7).

1983, December 13. End of Military Rule. Following parliamentary elections (November 6), General Evren and his junta resigned (December 11) and a civilian government took office under Premier **Turgut Ozal** (December 13).

CYPRUS

1975, February 13. Turkish Cypriot Federal State Established. Greek Cypriot President **Archbishop Makarios III** refused to recognize the new state. Only Turkey recognized the legitimacy of the new state. Negotiations between the two Cypriot factions made no progress in solving the partition issue. **Rauf Denktash** was chosen President of Turkish Cyprus (June 20, 1976).

1977, February 12. Partition Negotiations. The Turkish Cypriots accepted territorial concessions in exchange for recognition of their Turkish Cypriot Federated State. However, talks broke down when the Greek Cypriots alleged that the Turkish Cypriots reneged on their part of the bargain.

1983, November 15. Turkish Cyprus Declares Independence. Turkey continued to deploy troops in Cyprus but made the first troop withdrawal since the 1974 invasion.

THE MIDDLE EAST

THE ARAB LEAGUE

1976, June 10. Arab League Establishes Peacekeeping Force in Lebanon. The Arab League voted to establish an Arab League peacekeeping force with troops from Algeria, Libya, Saudi Arabia, and Sudan. Syria, which already had nearly 16,000 troops in Lebanon, was also to participate. The force was dominated by the Syrians, who continued to maintain their strong presence in Lebanon (see p. 1505).

1976, September 6. PLO Granted Full Membership in the Arab League. The Arab League unanimously approved an Egyptian-sponsored motion to grant full membership to the Palestinian Liberation Organization (PLO).

1976, October 17. Arab League Deterrent Force for Lebanon. Since Syria either would not or could not bring a cessation of civil war in Lebanon, the leaders of Lebanon, Egypt, Syria, Saudia Arabia, Kuwait, and PLO chief **Yasir Arafat** met at Riyadh, Saudia Arabia, and agreed to establish an Arab League Deterrent Force in Lebanon. The Deterrent Force would be under the personal leadership of President **Elias Sarkas** of Lebanon. The new Deterrent Force was still largely made up of Syrians. Saudi Arabia, the United Arab Emirates (UAE), Sudan, and North Yemen sent small troop contingents, but withdrew them when it became clear that Syria intended to continue its dominance of the Arab League Force (June 1979). The Syrians continued to maintain large forces in Lebanon nominally under the auspices of the Arab League. Later the Arab League chose not to renew the mandate for its Deterrent Force in Lebanon (July 17, 1982; see p. 1505).

1978, November 1. Arab League Denounces Camp David Accords. The Arab League denounced the Camp David Accords and the agreement of Egypt and Israel to negotiate a treaty (see p. 1502).

1979, March 31. Arab League Imposes Sanctions on Egypt. Because of the Israeli-Egyptian peace treaty, the leaders of 18 of the member nations of the Arab League imposed a boycott of Egyptian goods and agreed to sever diplomatic ties with Cairo. However, 2 member states, Sudan and Oman, retained relations with Egypt and absented themselves from the meeting (see p. 1503).

1989, May 23. Conference in Casablanca. Participation by President Mubarak of Egypt marked the Arab League's implicit acceptance of the Camp David peace treaty with Israel.

1990, August 10. Response to Iraqi Invasion of Kuwait. Twelve of 21 League members met and passed a resolution condemning Iraq's invasion of fellow League member Kuwait. The 9 members abstaining were Jordan, Mauritania, Yemen, Sudan, Libya, Tunisia, Chad, Algeria, and the PLO.

1990, October 31. League Headquarters Returns to Cairo. The Arab League moved its headquarters from Tunis back to Cairo, after an absence of 10 years.

ISRAEL

1976, June 27–28. Entebbe Incident. PLO guerrillas hijacked an Air France commercial aircraft en route to Paris from Athens and diverted it to Entebbe Airport, Uganda. They held the crew and passengers hostage. The hijackers demanded the release of 53 PLO political prisoners in Israel, West Germany, Switzerland, France, and Kenya. The hijackers released 47 passengers but continued to hold hostage 98 passengers and the crew. Uganda's President Idi Amin cooperated with the hijackers, deploying troops to help guard the hostages who were held in the airport terminal. The Israeli government expressed a willingness to negotiate while secretly planning an armed rescue operation.

1976, July 3–4. Entebbe Rescue Mission: Operation "Jonathan." In a daring, meticulously planned and executed rescue operation, Israeli commandos under Lieutenant Colonel **Jonathan Netanyahu** landed at the Entebbe airport, overcame Ugandan and PLO resistance, and freed the hostages. To assure a safe exit the Israelis also destroyed Ugandan air force MiGs based at the airport. The Israelis killed the 7 PLO guerrillas and 20 Ugandan soldiers in the fighting. Three hostages and the Israeli commander, Colonel Netanyahu, were killed; 3 Israeli soldiers were wounded.

1978, March 11. The Coastal Road Massacre. Operating out of Lebanon, 11 seaborne PLO guerrillas landed south of Haifa, killed 1 U.S. civilian on the beach, and attacked a taxi, killing 6 more civilians. The guerrillas commandeered a bus bound for Haifa and made the driver take them toward Tel Aviv. They stopped another northbound bus and opened fire, killing several passengers. They continued south coming within 10 miles of Tel Aviv when they were halted by Israeli troops. In the ensuing fire fight, 9 guerrillas were killed and 2 captured; 35 Israeli civilians were killed and 75 wounded.

1978, March 14–21. The Litani River Operation. In retaliation for the Coastal Road Massacre, the Israeli Defense Force (IDF) launched an invasion of southern Lebanon. The mission was to destroy PLO forces south of the Litani River. Because of the Israeli Cabinet's insistence that casualties be kept to a minimum, the IDF advance was deliberate, enabling most of the PLO fighters to escape beyond the Litani River. Nonetheless, the Israelis were able to secure all of southern Lebanon. They avoided Tyre and the Palestinian refugee camp at Rashidiye.

1978, April 11–June 13. Israeli Withdrawal and the Establishment of a United Nations Force in Lebanon. In accordance with U.N. Security Council Resolution 425, the Israelis withdrew from Lebanon. A United Nations Interim Force in Lebanon (UNIFIL) replaced Israeli troops in southern Lebanon. However, the Israelis insisted that the Lebanese troops and militia commanded by Lebanese Army Major **Saad Haddad** be given control over a strip of land averaging 8 miles in depth just north of the Israeli-Lebanese border. Haddad subsequently named this territory the Independent Republic of Free Lebanon (April 18, 1979). The PLO, however, derisively referred to it as Haddadland.

1978, April–1981, April. PLO Buildup in Lebanon. Following the Litani River Operation, the Israeli withdrawal, and the establishment of the UNIFIL peacekeeping force, the PLO resumed its policy of harassment of Israel. Despite the presence of the UNIFIL forces and Major Haddad's force in southern Lebanon, the PLO made numerous attempts to infiltrate over land and by sea. There were more than 50 artillery and mortar shelling incidents. As a result of these actions 10 Israelis were killed and 57 were wounded. This was a much smaller casualty rate than that before the Litani River Operation, but was considered unacceptable by Israel and the people of northern Galilee. The Israelis carried out numerous air and ground retaliatory attacks upon the PLO in Lebanon, inflicting many casualties. The PLO leaders began reorganizing and changing tactics. They fortified base areas in southern Lebanon. At the same time they began to build up a conventional military force with the assistance of large quantities of arms from the U.S.S.R., including tanks, artillery, and multiple rocket launchers.

1978–1979. Camp David and Peace Treaty with Egypt. (See p. 1504.)

1981, June 7. Israeli Air Attack Destroys

Iraqi Nuclear Reactor. Israeli aircraft bombed and destroyed Iraq's sole nuclear power plant, under construction and nearing completion outside of Baghdad, at Osirak. Prime Minister Menachem Begin justified the preemptive raid because Israel believed that Iraqis intended to produce atomic weapons in the plant and to use them against Israel.

1981, July–1982, May. Uneasy Cease-Fire with PLO. Following a cease-fire arranged by U.S. mediator, Ambassador Philip Habib (July 24), an uneasy truce ensued. The PLO, apparently convinced that any retaliatory Israeli offensive would be limited, like the 1978 Litani River Operation, continued its terrorism against Israeli civilian targets, inside Israel, on the West Bank, and in foreign countries. Israel built up forces just south of the border.

1982, June 3. Israeli Diplomat Wounded. The attempted assassination of Israel's ambassador to England culminated a series of PLO terrorist attacks that prompted a massive Israeli reprisal 3 days later.

1982, June 6. Israeli Invasion of Lebanon. (See p. 1505.)

1983, February 8. Kahan Commission Report. (See p. 1477.)

1983, May 17. Agreement with Lebanon. (See p. 1506.)

1985, February 16. Israeli Withdrawal from Lebanon Begins. The Israeli Defense Forces began to withdraw from Lebanon. The withdrawal, except for a narrow strip of southern Lebanon (Israel's so-called "security zone"), was completed in June.

1985, October 1. Bombing of PLO Headquarters in Tunis. In response to a terrorist attack in Cyprus (September), Israeli Air Force F-15s flew 1,500 miles to bomb PLO headquarters in Tunis. The strike killed 67 Arabs; PLO chairman Yasir Arafat was away. Arab states condemned the attack, but the U.S. stated the attack was a "legitimate response" to terrorism.

1987, December. The *Intifada*. Palestinians began a series of uncoordinated demonstrations and protests against Israeli occupation in the occupied West Bank and Gaza. As it became organized the uprising was dubbed *Intifada* ("uprising" or "popular revolt"), and encompassed economic and cultural countermeasures to Israeli occupation. Many of the protests were violent, involving clashes between stone-throwing youths and Israeli police and army reservists. Israeli efforts to control the revolt, and to destroy the *Intifada*'s organization, were largely ineffective. The revolt gained wide support abroad for the Palestinian cause.

1988, May 2–5. Raid into Lebanon. Israeli ground and air forces, responding to guerrilla raids into northern Israel, mounted a punitive expedition into southern Lebanon (May 2). This culminated in a sharp battle with Shi'ite militiamen, won by the Israelis and their Christian militia allies with tank, helicopter, naval gunfire, and air support (May 4).

1990, October 8. Temple Mount Incident. Amid rising tensions because of the Kuwait crisis, Israeli border guards fired into a rioting Palestinian crowd near the Al-Aqsa mosque on Temple Mount in Jerusalem, killing about 20 and wounding about 150. Israel was condemned in the U.N. Security Council (2 resolutions, on October 12 and 24), supported by the United States.

1991, January 17–February 26. Israel Attacked by Iraqi Scud Missiles. Fulfilling prewar promises, Saddam Hussein launched Scud missiles against Israeli population centers. There were few casualties because of the inaccuracy of these weapons, coupled with the arrival of U.S. Patriot SAM-ATBM units and Iraqi inability to employ chemical warheads. Under U.S. pressure, the Israelis uncharacteristically abstained from striking back, despite considerable public clamor to do so as the sporadic bombardment wore on. (See also p. 1480.)

EGYPT

1975, March 29. Reopening of Suez Canal. Egyptian President **Anwar Sadat** announced that Egypt had reopened the Suez Canal. Israeli ships were barred from using the canal. Sadat also announced that he would renew the mandate of the U.N. Sinai peacekeeping force (see U.N., p. 1483).

Patriot Missile launched in Intercept Test
(*U.S. Army photo*)

1975, July–1976 August. Tense Relations with Libya. Egypt charged Libya with numerous acts of terrorism and plans for terrorist acts in Egypt.

1975, September 1. Egypt and Israel Agree on Sinai Disengagement. After a summer of negotiating, Israel and Egypt reached agreement on Sinai troop withdrawals. The Israelis agreed to withdraw from the Mitla and Gidi passes and return the Abu Rudeis oil fields to Egypt. Egypt agreed to allow Israel to ship nonmilitary goods through the Suez Canal. Both sides agreed to limit the number of troops in the Sinai Peninsula to 8,000. Americans were to operate an early warning system in the passes.

1975, November 30. Israeli Withdrawal. In accordance with the Sinai disengagement agreement, the Israelis withdrew from a 90-mile strip, including the oil fields.

1976, March 14. Sadat Ends Soviet Friendship Treaty. Citing Soviet refusal to rebuild Egyptian military capability after the 1973 Arab-Israeli War and the Soviet unwillingness to reschedule debts, Egyptian President Sadat renounced the 1971 Friendship Treaty with the Soviet Union. Sadat was harshly critical of the Soviets for preventing India from supplying Egypt with MiG parts from a Soviet-licensed MiG plant in India. Sadat later signed an arms deal with China to procure the spare parts for the MiGs (April 21).

1976, July 13–18. Egyptian-Sudanese-Saudi Arabian Cooperation Pacts Signed. Egyptian President Anwar Sadat and Sudanese President **Gaafar al-Nimeiry** agreed to a nonaggression pact to deter Libyan aggression against Sudan. The two leaders flew to Jiddah, where they met Saudi Arabia's King **Khalid** and concluded a trilateral agreement on military, political, and economic cooperation. Sadat and Nimeiry were convinced that Libya's Qaddafi had been behind an aborted coup attempt in Sudan earlier that month (see p. 1535).

1976, December 18–21. Joint Political Command Established between Syria and Egypt. Talks between Egypt's President Anwar Sadat and Syria's President Hafez Assad concluded with an agreement to coordinate the defense and foreign policies of their two countries.

1977, January 16–19. Troops Used to Quell Food Riots. Egyptian troops suppressed riots resulting from government approved rises in food prices.

1977, July 12–19. Border Incidents with Libya. Relations between Libya and Egypt worsened as Egyptian authorities arrested Libyan saboteurs in the border region (July 12). Libyan officials responded by arresting 10 Egyptians (July 16). These incidents culminated in an armed clash.

1977, July 21–24. Egypt–Libya: Border Conflict. Egyptian troops turned back a Libyan combined arms attack across the Egyptian-Libyan border. The Egyptians claimed to have shot down 2 Libyan aircraft and knocked out

Soldiers from the 101st Airborne Division write their thoughts on abandoned SCUD Missile somewhere in Iraq (*U.S. Army photo by SPC Elliot*)

40 tanks. Egyptian losses were light. The Egyptians retaliated by launching a counterattack across the Libyan border and an air attack against a Libyan air base. Satisfied that Libyan dictator **Muammar Qaddafi** had for the time being learned his lesson, Sadat called a cease-fire (July 25). Sadat warned that "that maniac [Qaddafi] is playing with fire."

1977, November 10–18. Sadat Offers to Go to Israel. In a series of dramatic announcements, President Sadat declared he was ready to visit Jerusalem to discuss peace. Prime Minister Begin reciprocated by inviting Sadat.

1977, November 20–22. Sadat Visits Jerusalem. Sadat boldly visited the Israeli capital and addressed the Knesset. Sadat and Begin pledged that they would work for peace and that they would end the enmity between their two countries. A peace process was begun. However, leading Arab nations, especially Libya and Syria, expressed indignation at Sadat's visit and efforts to bring an end to hostilities with Israel. Libya severed diplomatic ties with Egypt.

1977, December 2–5. Arab States Form Anti-Egyptian Front. Representatives of Libya, Syria, Iraq, Algeria, South Yemen, and the PLO met in Tripoli and agreed to form an anti-Egyptian front. Although they condemned Sadat's peace initiative as an act of "high treason," they did not eliminate hopes for eventual rapprochement. The relative mildness of the condemnation prompted a break between Iraq and the other Arab members of the new anti-Egyptian front, causing the Iraqi representative to leave the conference early. Sadat promptly recalled his ambassadors from the five member nations (December 5).

1978, February 23. Egyptian Raid on Cyprus. Arab terrorists claiming to be PLO attacked and killed an Egyptian newspaper editor in Nicosia, Cyprus (February 19), and took 30 hostages. Sadat sent Egyptian commandos on a rescue mission. The commandos killed the terrorists, but then were attacked themselves by a unit of the Cypriot National Guard. Fifteen commandos were killed in the ensuing battle. Others were captured. Egyptian officials were furious and subsequently broke off diplomatic relations with Cyprus.

1978, September 7–17. Camp David Talks and Accords. U.S. President Jimmy Carter arbitrated negotiations between Sadat and Begin for over a week until they agreed on a working plan for a peace treaty. They agreed that the West Bank would have autonomy and that Egypt and Israel would sign a peace treaty within 3 months of the signing of the Camp David Accords.

1979, March 20. U.S. Agrees to Provide Arms to Egypt and Compensation to Israel. As a preliminary to signature of the Egyptian-Israeli Peace Treaty, the U.S. pledged to provide Egypt with warplanes and weapons worth $2 billion and to compensate Israel for its withdrawal from the Sinai with $3 billion.

1979, March 26. Egyptian-Israeli Peace Treaty Signed. After months of difficult negotiations, Egyptian President Sadat and Israeli Prime Minister Begin signed a peace treaty between their two countries in Washington. Both leaders gave U.S. President Jimmy Carter credit for his vital role in the negotiations. Under the terms of the treaty, the Israelis pledged a staged withdrawal of troops and settlers from the Sinai Peninsula, to be completed in 3 years. Israel would use the Suez Canal and pledged to begin negotiations for Palestinian self-rule a month after ratification.

1979, March 27. Israel Starts Withdrawal. Israeli troops began the first stage of withdrawal.

1979, September 8. Egypt Helps Morocco Battle Polisario. Egyptian arms were received by Morocco to assist in the war with Polisario (see p. 1549)

1979, October 6. First U.S. Arms Shipment to Egypt. In a parade commemorating the sixth anniversary of Egypt's crossing of the Suez Canal to begin the October War (1973) Egypt displayed the first new weapons systems received from the United States. The recent Egyptian acquisitions displayed at the parade were F-4 Phantom jet fighter-bombers and M-113 armored personnel carriers (APCs).

1980, January 8. U.S. and Egypt Joint Training Exercises. It was announced that U.S. and Egyptian air forces had recently concluded joint training exercises at Egyptian Air Base at Luxor.

1981, October 6. Sadat Assassinated. President Anwar Sadat was assassinated while he was reviewing a parade commemorating the Egyptian crossing of the Suez Canal at the beginning of the 1973 war with Israel. The assassins, one a soldier, were Moslem fundamentalists opposed to Sadat's peace with Israel and his liberal interpretation of Islamic law. He was replaced by former Vice President Hosni Mubarak.

1981, November 14. Operation Bright Star Begins. Egypt began joint maneuvers with Sudan, Somalia, Oman, and the U.S.

1982, April 25. Israelis Complete Withdrawal from the Sinai.

1983, August 18–September 18. Operation Bright Star 83. Over 5,000 U.S. troops participated with Egyptian troops in joint maneuvers.

1985, October 7–14. The *Achille Lauro* Hijacking. Four members of the Palestinian Liberation Front (PLF) hijacked the Italian cruise liner *Achille Lauro* off the Egyptian coast (October 7), killed an American passenger, and then surrendered to PLO officials after negotiations (October 9). U.S. carrier-borne F-14 fighters intercepted the civilian airliner carrying the 4 terrorists to Tunis (headquarters of the PLO) over international waters, and forced it to land in Sicily, where the hijackers were arrested and charged (October 11). Although the Egyptian government had condemned the original hijacking, it protested the U.S. action as "piracy." U.S. success was marred when Italy allowed PLF chief Abu Abbas to leave, over U.S. protests (October 12).

1990, August 8–1991, February 26. Egyptian

Role in Kuwait Crisis. President Mubarak announced Egyptian willingness to contribute to a multinational force in the gulf (August 8, 1990); Egyptian troops began arriving in Saudi Arabia soon after (August 11). The Egyptians deployed an armored division, a mechanized division, commando units, and support troops in Saudi Arabia (January 1991). These forces took part in the U.N. coalition offensive (February 24–28, 1991; see p. 1476).

YEMEN (NORTH YEMEN)

1977, October 10. Assassinations of President and Vice-President. President **Ibrahim al-Hamdi** and his brother Colonel **Abdullah Mohammed al-Hamdi** were assassinated. The assassins apparently hoped to prevent planned discussions with South Yemen regarding a possible merger of the two countries.

1978, June 24. President Assassinated. The new President, **Ahmed Hussein al-Ghashmi,** was killed when the briefcase of an envoy from South Yemen President **Salem Rubaya Ali** exploded. The South Yemen envoy was also killed in the blast. North Yemen authorities charged that the assassination was planned by Rubaya. Other Arab sources said that the bomb was planted by North Yemen political opponents of Ghashmi. A 3-man military council replaced Ghashmi.

1979, February 24–26. Border Fighting with South Yemen. North Yemen officials alleged that South Yemeni troops had attacked across the border. South Yemen officials denied the charge and claimed they were responding to an invasion of their territory by North Yemen. Both countries agreed to let the Arab League mediate (February 28).

1979, March 17. Cease-Fire. Both North and South Yemen agreed to a cease-fire.

1979, March 29, National Unification Agreement. Following the cease-fire, reconciliation talks were held between North and South Yemen. Despite apparent agreement, the unification did not take place.

1980, May–June. Renewed Conflict with Southern Yemen. (See p. 1514.)

LEBANON

1975–1984. Lebanese Civil War. The establishment of the PLO in Lebanon exacerbated tensions between various Moslem and Christian factions. Full-scale civil war broke out (April 13, 1975). This conflict continued to rage throughout the remainder of the decade and into the 1980s. The PLO, whose objective was the destruction of Israel, used Lebanon as a staging area from which to conduct raids into Israel and to use long-range artillery and rockets to shell Israeli territory.

1976, March. Syrian Intervention. Under the pretext of acting as a peacemaker between warring Lebanese factions, Syria dispatched troops to Lebanon, initially to prevent the PLO and Moslem Lebanese factions from defeating the Christians.

1978, March 14–21. Israeli Invasion of Southern Lebanon: The Litani River Operation. (See p. 1500.)

1978, April 11–June 13. Israeli Withdrawal and the Establishment of United Nations Interim Force in Lebanon. (See p. 1500.)

1978, April–1981, June. PLO Buildup in Lebanon. The PLO built an extensive military/political infrastructure in Lebanon. They received large shipments of arms and equipment from the Soviet Union (see p. 1500).

1982, June–August. Israeli Invasion of Lebanon. (See p. 1475.)

1982, August 21–28. Deployment of Multinational Peacekeeping Force. (See p. 1476.)

1982, August 27–September 3. Evacuation of PLO and Syrian Forces from Beirut. All 15,000 Syrian and PLO forces were evacuated, most by sea; the Syrians and a small Palestine Liberation Army (PLA) contingent went overland through Israeli lines to Damascus. The multinational force contingents withdrew shortly after this (September 10).

1982, September 13. Assassination of President-elect Bashir Gemayel. Lebanese

President-elect Gemayel was assassinated by terrorists allegedly backed by Syria. Israeli troops occupied Beirut for the declared purpose of keeping order and preventing bloodshed.

1982, September 16–19. Sabra and Shatilla Refugee Camp Massacres. (See p. 1477.)

1982, September 20–22. Multinational Peacekeeping Force Redeployed in Beirut. This was in response to world horror because of the Sabra and Shatilla massacres. The U.S. Marine position was just east of the airport. The Italian contingent was further north, closer to Beirut. The French contingent was still further north, on the outskirts of West Beirut. They were later joined by a token British contingent of about 150 men, who took position near the U.S. Marines.

1982–1983. Buildup of PLO in Northern Lebanon. Despite their defeat, the PLO were able to reorganize and began rebuilding their forces in Lebanon.

1983, May 17. Lebanon-Israel Agreement. With the diplomatic aid of the United States, Israel and the Lebanese government of President **Amin Gemayel** reached an agreement which was, in effect, a treaty between the two nations, ending a war which had existed since 1947. Israel was recognized by the Lebanese government and relations between the two countries were at least partially normalized.

1983, September 3. Partial Israeli Withdrawal. Against the wishes of the United States, which wanted the Israelis to remain in the Chouf Mountains to maintain order, the Israelis withdrew their forces (about 10,000 troops) from the Beirut area and the Chouf (Shouf) Mountains to new positions south of the Awali River.

1983, September 4–25. Civil War in the Chouf Mountains. As the Israelis were completing their withdrawal, Phalangist troops and the Lebanese Army began to move into the areas evacuated. The local Druze Moslem militia began immediately attacking both Phalangist and Lebanese army forces. The Druze attacked major strongholds at Aley (Alieh) 10 miles east of Beirut and Bhamdun south of the Beirut-Damascus highway (September 4–6). In subsequent fighting, the Druze militiamen surrounded Deir al Qamar (September 10), and attacked Suk al Gharb, which directly overlooked the President's palace (September 10–11). A Druze armored assault was turned back from Suk al Gharb with the help of U.S. naval gunfire support from offshore (September 19). The warring Lebanese factions agreed to a cease-fire arranged by Saudi mediator Prince **Bandar bin Sultan** (September 25). During this fighting considerable fire from the Druze

guns landed in and near the multinational force positions, mostly near the U.S. Marines. Much of this was fire from stray rounds. Much of it was deliberate. The cease-fire was subsequently broken on a number of occasions.

1983, September 10–October 20. U.S. Navy Supports Marine Peacekeeping Efforts in Beirut. The U.S. government authorized air strikes and gunfire support from the U.S. Navy squadron offshore near Beirut, to help defend the Marine positions near the Beirut Airport. The Navy was also authorized to fire in support of nearby Lebanese Army positions if these were believed to contribute to the security of the Marine positions. The result was occasional fire from U.S. ships against nearby positions of Druze and other antigovernment militia, and against their supporting artillery in the Syrian occupied zone further east. U.S. carrier aircraft began to fly reconnaissance missions over west-central Lebanon.

Carrier USS *Hancock* (*U.S. Navy photo*)

1983, October 23. Suicide Attack on U.S. and French Peacekeeping Force. An Arab terrorist allegedly belonging to a pro-Iranian and pro-Syrian terrorist organization drove a truck loaded with explosives into a combined headquarters and barracks building in the U.S. Marine compound; 241 servicemen were killed. This led to widespread demands in the United States that the Marines be brought home. An official inquiry determined that inadequate security precautions had been taken, and blamed security failures on the local Marine commander, and Navy and Army senior officers in the chain of command. President Ronald Reagan assumed responsibility and no officers were court-martialed. Simultaneously a similar attack on the French peacekeeping force compound killed 58 French soldiers and wounded 15.

1983, November 13–14. Renewed Fighting between Druze Forces and Lebanese Army. After a number of cease-fire violations in weeks following the September 25 cease-fire agreement, fighting erupted at Suk al Gharb in the mountains just east of and overlooking Beirut. The Druze militia began shelling Christian East Beirut. Intense combat ensued between the militias and the Lebanese Army. Again U.S. Marine positions at the airport came under fire. Fighting continued until a new cease-fire was arranged (December 27).

1983, December 4. Two U.S. Planes Shot Down by Syrian SAMs. Following Syrian firing on U.S. reconnaissance planes (December 3) U.S. Navy carrier planes carried out attacks against Syrian air defense positions in the Bekaa Valley. In the attack 2 planes were shot down by Syrian SAMs. One crew was rescued; in the other the pilot was killed, and the navigator, Lieutenant **Robert Goodman,** captured. The Syrians held Goodman as a prisoner of war.

1983, December. Continued Shelling of Marine Position. The U.S. Marine positions were repeatedly shelled. The Marines returned fire.

1983, December. Pro- and Anti-Arafat PLO Forces Battle in Tripoli Area. PLO dissidents, dissatisfied with **Yasir Arafat's** leadership, had been fighting against PLO forces loyal to Arafat for several months. The dissidents, supported by Syria, drove the loyalist contingents back to the port city of Tripoli. A cease-fire was finally arranged between the PLO combatants; Arafat and troops loyal to him were allowed to leave Lebanon by ship. Israeli ships at first barred Arafat's departure, but eventually (after pressure from the U.S.) permitted him to escape (December 20).

1983, December–1984, January 3. Jesse Jackson Wins Goodman's Release. Reverend **Jesse Jackson,** a Democratic Party candidate for President of the United States, appealed personally to Syrian President Hafez Assad in Damascus and won the release of Navy flyer Lieutenant Robert Goodman. The Syrians announced that the release was for humanitarian reasons.

1984, January 15. Death of Haddad. Major Saad Haddad, commander of Lebanese troops and militia in southern Lebanon, died after a lengthy illness (see p. 1500). Haddad, who had been expelled from the Lebanese Army by the pro-Syrian government in 1979, was reinstated to his rank just before his death by the Lebanese State Court, Lebanon's highest judical body. The Court, composed of 1 Christian and 2 Moslems, held that Haddad's dismissal had been unjust and prompted by international pressure. He had, in fact, been continuously paid by the Lebanese Army, and had received regular logistical support from Lebanon, through Haifa. The Israelis praised Haddad for his courage and sense of soldierly duty. Haddad, a Christian, was admired and praised by fellow officers of the Lebanese Army, Moslem and Christian.

1984, January 6–March 5. Druze and Shi'ite Moslems Defeat Lebanese Army. Intense fighting again erupted between Druze and Shi'ite militiamen and Lebanese Army units (January 6). After a brief lull, the fighting intensified (January 13), and the Syrian-backed Shi'ite and Druze forces won a complete victory over the Army east and south of Beirut. At the same time, Shi'ite militiamen drove the Lebanese Army out of West Beirut. President

Gemayel tacitly acknowledged the changed milieu by going to Damascus to conduct talks with Syrian President Assad (February 29–March 1) in order to strike the best deal he could and retain some power in Lebanon. Upon returning to Lebanon at the conclusion of the negotiations (March 5), he renounced the Lebanese agreement with Israel (see p. 1506).

1984, February 7–March 31. Withdrawal of the Multinational Peacekeeping Force from Lebanon. President Reagan ordered the withdrawal of the U.S. Marine peacekeeping force from Beirut. The next day the British contingent of the multinational peacekeeping force was withdrawn and the Italian government likewise announced the withdrawal of its peacekeeping contingent (February 8). The Marines began withdrawing to ships offshore 2 weeks later (February 21). The withdrawal of the U.S. peacekeeping force was complete some 5 weeks later when President Reagan announced that the U.S. Peacekeeping role was at an end (March 31). That same day the French completed the withdrawal of their peacekeeping force.

1984–1988. Continuing Civil Strife. Although major fighting was relatively infrequent, none of the factions involved, including the Syrians, was able to establish order in Lebanon, or even in Beirut itself. Terrorism and minor skirmishing were virtually everyday events, contributing to a slow erosion of Lebanese national life.

1985, February–June. Israeli Withdrawal. Three years after the opening of Ariel Sharon's "Peace for Galilee" operation, the last Israeli troops withdrew from Lebanon north of a narrow Israeli-declared "security zone" in the south.

1988, May 6–12. Fighting Between Amal and Hezbollah. The Shi'ite Amal and Hezbollah (Army of God) militias waged a series of bloody clashes in suburban Beirut, each attempting to assert leadership of Lebanese Shi'ite Moslems.

1989, March 17. "War of Liberation." Following Christian attacks on West Beirut, and a Moslem car-bomb explosion (March 17), Christian General Michel Aoun (appointed commander of the Lebanese Army by Gemayel in late 1988) declared a "War of Liberation" to rid Lebanon of Syrian forces and their allies.

1989, September 6. U.S. Evacuates Embassy. As U.S. relations with Aoun deteriorated, the U.S. evacuated its Beirut embassy, leaving Aoun increasingly isolated diplomatically.

1989, August–October. Ta'if Conference. Meeting in the Saudi Arabian city of Ta'if for security reasons, members of the 1972 National Assembly tried to hammer out a settlement of the Lebanese civil war and crisis. Complete agreement was stymied by General Aoun's intransigence, and by concerns over a timetable for Syrian withdrawal.

1990, January 30. Aoun Attacks Christian Militia. General Aoun's troops attacked the Christian Lebanese Forces militia. Within 3 months the casualties totaled 4,000, including nearly 1,000 dead.

1990, April 3. Lebanese Forces Declare for Hrawi. President **Elias Hrawi** gained an important ally when **Samir Geagea,** leader of the Christian Phalangist Lebanese Forces militia, declared his allegiance to Hrawi.

1990, September 28–October 13. Defeat of Aoun. Syrian and Lebanese Army forces blockaded Aoun's positions in the hills east of Beirut. A Syrian air and ground assault (October 12) compelled Aoun to quit the Presidential Palace and seek political asylum in the French embassy (October 13).

1990, October 25. Militia Withdrawal Agreements. President Hrawi obtained agreement for the withdrawal of all militias from Beirut. He later (November 7) announced the intended dissolution of all militias by March 1991, and the withdrawal of all Syrian forces by September 1992.

SYRIA

1975, June 3–July 8. Increased Tensions and Border Incidents with Iraq. (See p. 1511.)
1975, June 12–August 22. Coordination of Military Policy with Jordan. (See p. 1512.)

1975, December 25–31. Joint Jordanian-Syrian Military Exercises. (See p. 1512.)

1976–1981. Syrian Buildup in Lebanon. Syria moved into Lebanon under the pretext of being a peacemaker between Christian militia on one side and the PLO and its Lebanese Moslem leftist allies on the other (March 1976; see pp. 1499, 1505). The Syrians subsequently received approval from the Arab League to contribute troops to an Arab League Deterrent Force (October 17–18, 1976; see p. 1499). The Syrians gradually increased the size of their forces in Lebanon to about 20,000 troops and shifted their support from the Lebanese Christians to the Moslem factions. The Syrians were deployed in Beirut and south of the Damascus-Beirut highway, along the western slopes of Mount Hermon and in the Bekaa Valley.

1976, June–December. Strained Relations with Iraq. (See p. 1511.)

1977, February 4. Joint Political Command Established between Syria and Egypt. (See p. 1502.)

1977, December 2–5. Arab States Form Anti-Sadat Front at Tripoli Conference. (See p. 1503.)

1978, January 11. Syrian-Soviet Arms Deal. President Hafez Assad made an arms deal with the Soviet Union in which Syria was to receive additional SAMs, tanks, and planes.

1978, October 24–26. Syria and Iraq Agree on Political and Military Coordination. In response to the Israeli-Egyptian Camp David Accords the governments of Syria and Iraq put aside their serious policy differences and pledged to establish a bilateral committee to coordinate military strategy, defense, and foreign policies in order to "serve as the basis for a full military union."

1979, June 19. Joint Political Command with Iraq Established. In accordance with their earlier agreement, Syria and Iraq established a Joint Political Command to coordinate defense, foreign, and economic policies.

1979, August 8. Assad Charged with Supporting Attempted Coup in Iraq. Saddam Hussein, new chairman of Iraq's Revolutionary Command, accused Syrian President Assad of supporting a recent coup attempt in Iraq (see p. 1511). This ended all planning for policy coordination.

1980, September–1988, August. Syrian Support for Iran. Syria was the only Arab state to support Iran during the Iran-Iraq war, largely because of President Assad's hostility to Saddam Hussein and the Iraqi Ba'athist regime.

1982, February 2–24. Syrian Troops Crush Hama Revolt. Efforts by the Syrian Army to suppress the illegal Moslem Brotherhood in Hama led to a full-scale revolt by Sunni Moslem fundamentalists. Fighting was bitter and thousands of Hama residents were killed. Eventually the troops employed armor to destroy whole sections of the city.

1982, March 7. Assad Blames Iraq for Hama Revolt. Syrian President Assad accused Iraq of arming the Moslem Brotherhood, enabling them to revolt against Syrian authority.

1982, June 6–26. War with Israel in Lebanon. (See p. 1474.)

1982, September–December. Syria Rebuilds Armed Forces with Soviet Aid. Syria, with the aid of the Soviet Union, rebuilt the forces shattered in the recent war with Israel. The U.S.S.R. not only replaced the arms and equipment the Syrians had lost in Lebanon but supplied additional weaponry as well.

1983, December 4. Syrian SAMs Shoot Down American Planes. (See p. 1508.)

1984, February 29–March 1. Assad Receives Gemayel. Lebanese President Amin Gemayel traveled to Damascus to meet Syrian President Hafez Assad. Syria had gained virtual domination over all of Lebanon except the southern area occupied by Israel.

1990, April–October 13. Syria Helps Defeat General Aoun. Syrian forces lent strong support, including tanks and aircraft, to efforts by Lebanese President Hrawi to drive out and defeat General Aoun. (See Lebanon, p. 1509.)

1990, August 10. Syria Joins Coalition Against Iraq. At the emergency Arab summit in Cairo, Syria pledges its support against Iraq. This move placed Syria and the U.S. in the same camp. Syria eventually deployed its 9th Armored Division (15,000 men and 300 tanks) to Saudi Arabia (November–December).

IRAQ

1975, June 3–July 8. Strained Relations with Syria. Within weeks of settling a dispute over Syrian control of the flow of water of the Euphrates River (June 3), relations again became strained. Iraq charged Syria with violating Iraqi air space and with armed incursions of Syrian troops into Iraq. Iraq also claimed that Syria was aiding Kurdish rebels in their attempt to win autonomy (see p. 1399). Syria countered by ousting the Iraqi military attaché from Damascus and closing its military attaché's office in Baghdad (July 8).

1976, May 1–30. Kurd Revolt Renewed. Hostilities with Kurd tribesmen in northeast Iraq flared up. The new fighting was caused by the Iraqi policy of forcing Kurds to relocate. The Kurd insurgents were apparently receiving aid from Syria.

1976, June–December Increased Tension with Syria. Syrian intervention in Lebanon further strained relations between Iraq and Syria. Claiming that Syria posed a grave threat to its security Iraq called up its reserves and massed troops on its border with Syria (June 9–12). Syria responded by concentrating forces in the border region. Relations were further strained by Syrian acknowledgment of aid to Kurd rebels (July 4). Syria accused Iraq of closing the common border in order to cover large-scale troop movements (November 2). Relations remained tense as Syrian Foreign Minister **Abdel Halim Khaddam** was wounded in an assassination attempt traced to the radical Palestinian Black June movement, based in Iraq (December 2), while Iraq blamed Syria for attempting to assassinate Saddam Hussein, Deputy Chairman of the Iraqi Revolutionary Command Council (December 14).

1977, July–October. Continued Strained Relations with Syria. (See p. 1510.)

1977, December 2–5. Arab States Form Anti-Sadat Front at Tripoli Conference. (See p. 1503.)

1978, October 24–26. Syria and Iraq Agree on Joint Political-Military Command. (See p. 1510.)

1979, March 1. Kurdish Guerrilla Chieftan Dies. General **Mustafa Barzani,** leader of the Kurdish Democratic Party (KDP) and the Kurdish struggle for autonomy, died in Washington (see pp. 1398–1399). Barzani's sons continued the struggle for autonomy.

1979, June 4–14. Iraqi Counterinsurgency Operations into Iran. Iraqi troops in pursuit of Kurd guerrilla fighters raided Iranian territory, and Iraqi jets strafed Iranian villages (June 4–5). These attacks evoked a warning from the Iranian government (June 14).

1979, June 19, Joint Syrian-Iraqi Command Established. (See p. 1510.)

1979, July 20–August 8. Attempted Coup Foiled; Syrian Involvement Charged. An attempted coup against the new regime of President Saddam Hussein was foiled, and one of the major conspirators allegedly confessed to receiving Syrian support. Syrian President Assad denied the charge (August 6). Twenty-two co-conspirators, including five members of the Revolutionary Command Council, were executed by firing squad (August 8). The allegations against Assad ended any chance for the implementation of the two nations' plans for a joint political and military command (see p. 1510).

1979, September 1. Iran Accuses Iraq of Aiding Kurd Rebels. (See p. 1515.)

1980, September–1988, August 20. Iran–Iraq Gulf War. (See p. 1469.)

1980, October 1. Kurds Escalate Guerrilla Activities. A London-based spokesman for the Kurd Democratic Party said that the guerrillas fighting for Kurd autonomy had escalated their insurgency efforts against the Iraqi government. The guerrillas, led by **Massoud** and **Idris Barzani,** sought to take advantage of the war between Iraq and Iran in order to win autonomy from Iraq.

1981, June 7. Israeli Air Attack Destroys Iraq's Nuclear Reactor. (See pp. 1500–1501.)

1989, January. Attempted Coup. Saddam

Hussein survived an attempted coup mounted by dissident army officers.

1990, March–April. Seizure of Advanced Weapons Bound for Iraq. U.S. and British customs officials, cooperating in an elaborate "sting" operation, seized a shipment of nuclear trigger devices bound for Iraq (March 28). British officials later confiscated lengths of tubing said to be parts for an Iraqi "supergun," reputedly devised by maverick U.S.–Canadian gun designer **Gerald Bull** (March 11).

1990, April 2. Saddam Threatens Chemical Attack on Israel. In a public speech, Saddam Hussein threatened to use chemical weapons against Israel in the event of an Israeli nuclear attack on Iraq or any other Arab nation. The implied chemical capacity caused considerable alarm after the invasion of Kuwait in August (see p. 1472).

1990, April. Air Force Officers Executed. Reports surfaced outside Iraq that 17 air force officers were executed for plotting against Saddam Hussein.

JORDAN

1975, August 22. Joint Jordanian and Syrian Political-Military Command Formed. Jordan's King Hussein and Syrian President Hafez Assad issued a joint communiqué, announcing the establishing of a joint high command to formulate, coordinate, and execute political and military initiatives against Israel.

1975, December 25–31. Joint Military Exercises with Syria. Two Jordanian brigades of about 10,000 troops joined Syrian forces for maneuvers in Syria, simulating defensive operations against a hypothetical Israeli invasion.

1978–1980. Cooling Relations between Syria and Jordan.

1980, November 26–December 11. Border Tension with Syria. Apparently irritated by King Hussein's announcement of support for Iraq in the Iran-Iraq War, pro-Iranian President Assad of Syria ordered a troop buildup on the Jordanian-Syrian border. Hussein placed his forces on alert, and ordered several units to the border area. The two leaders bitterly de-

nounced each other in radio addresses to their nations. Israeli Prime Minister Begin announced that Israel would not tolerate a Syrian invasion of Jordan. Syria issued a list of demands on Jordan, including a call for Jordan to reassert its pledge recognizing the PLO as the only legitimate representative of the Palestinian people (December 2). Hussein rejected the Syrian demands. The crisis was eased by Saudi Arabian mediation. The Syrians withdrew their troops from the border (December 11); next day the Jordanians followed suit.

1982, January 28–March 3. Hussein Reaffirms Jordanian Support of Iraq War Effort; Calls for Volunteers. The Jordanian volunteer contingent was to be called the Yarmuk Force, in honor of the Arab victory over the Byzantines (636 A.D.; see p. 249). In his plea for volunteers Hussein denounced Iran as a threat to the Arab world.

1988, July 31. Jordan Abandons West Bank Claims. In a dramatic announcement, King Hussein abandoned Jordan's claims on, and responsibilities to, the West Bank. This move effectively surrendered leadership in the occupied zones to the PLO and leaders of the *Intifada.*

1990, April 18–20. Riots in Southern Jordan. Protesting widespread price increases, Jordanians rioted in the southern towns of Ma'an, Karak, Tafilah, and Shobek. Eight died before order was restored.

SAUDI ARABIA

1978, February 14. Saudi Arabia Buys F-15 Fighters from the U.S. The U.S. agreed to sell Saudi Arabia 60 F-15 fighter planes for $2.5 billion as part of a larger deal involving the sale of planes to Israel and Egypt (see p. 1504). The sale was later approved by the U.S. Senate (May 15) with the provision that the planes not be deployed near Israel.

1979, March 9. U.S. Sends AWACs to Saudi Arabia. The 2 Airborne Warning and Control reconnaissance planes were intended to help both the U.S. and Saudi Arabia improve intelligence capabilities in the Persian Gulf area, threatened by revolution in Iran. These were American planes, with American crews, but providing information to the Saudi armed forces. The U.S. government believed that this would help the U.S. retain Saudi Arabian friendship. Some observers, particularly in Israel, believed this was naïve.

1980, September 30. More AWACs to Saudis. The U.S. government shipped 4 more U.S. AWAC reconnaissance planes to Saudi Arabia to bolster American interest, influence, and capability in the Persian Gulf area in the aftermath of the Iranian Revolution.

1981, August 8. Saudi Crown Prince Offers Peace Plan for Middle East. Crown Prince Fahd's plan provided for Israeli withdrawal from all occupied territory, the removal of all Israeli West Bank settlements, the establishment of a Palestine State, and the right of all peoples to live in peace. Some Israelis thought the plan offered some hope for peace because of the clause calling for the right of all peoples to live in peace, which was implicit admission of Israeli's right to exist as a sovereign nation. However, the Israeli government rejected the plan, as did the PLO.

1981, October 29. U.S. Senate Approves Sale of AWAC Aircraft to Saudi Arabia. This was achieved only after a bitter parliamentary debate forced by supporters of Israel, which objected strongly to the sale. The Reagan administration apparently hoped that this might influence Saudi Arabia to agree to the establishment of U.S. bases in Saudi Arabia.

1982, January 26. Regional Defense Agreement. Saudi Arabia, Kuwait, Bahrain, Qatar, Oman, and the United Arab Emirates agreed to establish a joint military command and an interlinked air defense system. Prompting this decision were the security challenges posed by the Iranian Revolution, the Gulf War, and the Soviet invasion of Afghanistan.

1987, March. Iranian Pilgrims Riot in Mecca. While on pilgrimage to Mecca (the Moslem holy city), Iranian pilgrims rioted. Saudi security forces reacted swiftly and harshly, provoking a war of words between Iran and Saudi Arabia, and leading to the Saudis limiting Ira-

nian pilgrims to 45,000. Due to Saudi fears of a recurrence, and later evidence that the riots had been planned in advance, the Saudis did not lift the ban until autumn 1990.

1989, March. Saudi-Iraqi Nonaggression Pact.

1990, August–1991, February. Saudi Arabia in the Kuwait Crisis. Fearing invasion after Iraq's conquest of Kuwait, Saudi Arabia played a leading role in mobilizing Arab states against Iraq. Saudi Arabia's request for U.S. aid, and willingness to tolerate U.S. troops on its soil, provoked widespread criticism in the Arab world. The Saudi armed forces, despite lack of experience, performed creditably during the Kuwait War, playing major roles in the Battle of Khafji (January 30–31) and the liberation of Kuwait City (February 24–26). (See p. 1472.)

OMAN (MUSCAT AND OMAN)

1975–1983. Dhofar Separatist Movement Continues Guerrilla Struggle. Oman continued to battle against the Dhofar separatist movement insurgents. The Dhofar rebels, who had been assisted by Southern Yemen, now also received aid from Libya.

1975, February 2. Defense Pact with Iran. Iran agreed to give Oman air and naval support against foreign aggression. Iranian troops also assisted the Omani Army against the Dhofar rebels.

1980, June 4. Assistance and Base Agreement with the United States. Oman agreed to permit U.S. access to selected naval and air bases in exchange for U.S. military and economic aid.

1982, January 26. Regional Defense League Established. (See above.)

SOUTHERN YEMEN (PEOPLE'S DEMOCRATIC REPUBLIC OF YEMEN)

1975, November 30. Cuban Military Advisers in South Yemen. Press reports were subsequently confirmed by the U.S. State Department.

1978, June 25–27. President Deposed and Executed. It is not clear whether President **Salem Rubaya Ali** was overthrown—after a brief battle—because of his reported role in the assassination of President al-Ghashmi in North Yemen (see p. 1505) or because he was suspected by Communist colleagues of possibly reducing his links with the Soviets and improving relations with the U.S. and Saudi Arabia. The Arab League, suspecting the South Yemeni leadership of involvement in Ghashmi's death, suspended economic and cultural ties with South Yemen (July 2).

1980, April 23. President Resigns. Abdel Fattah Ismail was pressured to resign as president of South Yemen because of serious policy differences with Premier **Ali Nasser Mohammed al-Hasani.** Ismail was apparently too eager to bring about the proposed unification with North Yemen.

1980, May 1–June 1. Fighting Intensifies between North and South Yemen. There was a significant rise in armed conflict between North and South Yemen.

1980, June 9. Increase in Soviet-Cuban Presence in South Yemen. The number of Soviet advisers in South Yemen had reportedly risen to more than 1,000 and the number of Cubans to 4,000. The internal security and intelligence systems were allegedly headed by East Germans.

1981, August 19. Treaty with Libya and Ethiopia. The 3 countries pledged reduction of U.S. influence in the region.

1986, January 13–24. Civil War. After gunmen working for President **Ali Nasser Mohammed al-Hasani** killed several of his political opponents at a cabinet meeting (January 13), fighting spread throughout the capital. A week of bloody combat left hundreds dead, much military equipment destroyed, and much of Aden burnt or damaged. Husaini fled the country (January 24).

1989, March. Amnesty for Civil War. Chairman of the Presidium of the Supreme People's Council, **Haidar Abu Bakr al Attar,** announced an amnesty for the losing faction in the 1986 civil war. Former chairman Husaini,

in exile in the Republic of Yemen (North), announced he had no further political ambitions. The amnesty led to improved relations with North Yemen.

1989, June 22–1990, May 22. Yemeni Unification. A series of negotiations led to the selection of a 5-man presidential council, and formal unification of the two Yemens as the Republic of Yemen (May 22, 1990).

1990, August 2–1991, February 26. Yemeni Role in the Kuwait Crisis. The new Republic of Yemen, to the intense displeasure of Saudi Arabia, supported Iraq during the crisis, notably voting against U.N. Security Council Resolution 678 (November 29). Saudi economic sanctions, and their expulsion of Yemeni "guest workers," caused considerable economic hardship in Yemen.

IRAN

1975–1978. Unrest Becomes Rebellion. Reza Shah Pahlavi combined repression with concessions. Arbitrary arrests were made by the secret police (Savak). Opposition political parties were abolished. Nevertheless, opposition to the Shah's rule increased. Political reforms and the release of political prisoners (September–October 1978) failed to satisfy the opposition.

1978, November 5–6. Military Rule Imposed; National Elections Promised. The Shah established martial law while seeking national reconciliation by promising elections no later than by June 1979. Again repression combined with concessions failed to satisfy the National Front opposition; civil unrest continued.

1978, December 29. Shah Names New Premier for Civilian Government. In a last attempt to satisfy his opponents, the Shah named opposition leader **Shahpur Bakhtiar** to head a civilian government. Bakhtiar accepted and established a government (January 6). However, the response of the National Front opposition was to demand the Shah's abdication and to label Bakhtiar a traitor.

1979, January 13. Khomeini Demands Establishment of Islamic Republic and Shah

Departs Iran. In Paris the exiled **Ayatollah Ruhollah Khomeini,** senior Iranian Islamic (Shi'ite) clergyman, demanded that Iran become an Islamic republic. He established a Revolutionary Council.

1979, January 16. Departure of the Shah. Although the Shah did not formally give up his throne, his departure was tantamount to abdication.

1979, January 16–February 11. Bakhtiar-Khomeini Power Struggle. In an effort to restore order and preserve his government, Bakhtiar temporarily closed Iran's airports to prevent Khomeini's return. When the airports were reopened (January 29) Khomeini immediately flew to Teheran from Paris (February 1), proclaimed the establishment of an Islamic republic, and named **Mehdi Bazargan** premier of a provisional government. Armed civilians and rebel soldiers battled government troops in a climactic 3-day struggle, which ended with the overthrow of the Bakhtiar government (February 9–11).

1979, February 12. Khomeini's Islamic Republic Established. Although Khomeini himself had no official cabinet portfolio, he exercised virtually unlimited power.

1979, February 14. First Invasion of U.S. Embassy. Iranian "students" broke into the diplomatic compound of the U.S. embassy and took a U.S. Marine hostage, but later freed him (February 21).

1979, February 26–December 30. Rebellion and Civil Strife in Provinces. A wave of civil unrest, violence, and in some cases outright rebellion swept through several Iranian provinces. Uprisings began in Azerbaijan, where an army barracks was seized (February 22). In response to unrest in Kurdistan, the Kurds were granted limited autonomy (March 25). This did not satisfy them and armed clashes continued throughout the summer.

1979, August–October. Civil War in Kurdistan. Khomeini mobilized the armed forces in order to wipe out the Kurdish rebellion (August 19). The Iranian government also accused Iraq of assisting the Kurdish rebels. A determined offensive by government troops drove the Kurds from their strongholds and across the border into Iraq (September 6). But guerrilla warfare continued. The Kurds regained a key stronghold, Mehabad (October 20), then called for a cease-fire (October 22), which was subsequently granted.

1979, November 3. Seizure of U.S. Embassy. Militant Iranian "students" broke into the U.S. embassy compound and seized some 90 hostages, including 62 (later 66) Americans. Of these, 13 were soon released because they were black and/or women (November 19–20). The 53 remaining Americans were held as prisoners, and were subjected to abuse and mistreatment by their captors. One hostage was eventually released because he was severely ill (July 11, 1980).

1979, December 5. Renewed Hostilities in Kurdistan and Azerbaijan. Following a national plebiscite and a new constitution the Kurds rose again in rebellion. In Azerbaijan the Moslem People's Party took almost complete control of the province, but the central government soon regained authority.

1980–1984. Kurd Revolt Continues. Despite numerous offensives the Iranian Armed Forces were unable to suppress the Kurd rebellion. The Iranian Kurds received aid from Iran's arch enemy Iraq, while Iraq's rebellious Kurds allied themselves with Iran (see p. 1511).

1980, April 24–25. U.S. Efforts to Rescue Hostages: "Desert One." After repeated diplomatic efforts had been rebuffed, an American Joint Service rescue mission was mounted. Helicopters and transport planes with assault troops met in an isolated spot in central Iran ("Desert One"). The breakdown of 3 helicopters forced cancellation of the mission. During withdrawal 8 Americans were killed when a helicopter collided with a transport plane loaded with fuel. An official U.S. investigating board later severely criticized planning and command arrangements for the mission.

COMMENT. *The hostage crisis had already hurt American international prestige; the U.S. government appeared to all the world as being weak and irresolute. The failed rescue mission reinforced these assessments.*

1981, January 20. Release of the U.S. Hostages. The 52 Americans held hostage in Iran were released in return for U.S. agreement to free Iranian assets frozen in the U.S. following seizure of the U.S. embassy in Teheran.

1982, April 7–19. Plots Uncovered: Attempted Coup Foiled. Foreign Minister **Sadegh Ghotbzadeh** and several Iranian army officers were arrested for planning a coup and the assassination of Ayatollah Khomeini. They were subsequently tried and executed.

1989, June 3. Death of Ayatollah Khomeini. Sayyed Ali Khamenei, then serving as president of the republic, succeeded Khomeini as *rabhar* (spiritual guide).

1990, September. Iranian Response to the Kuwait Crisis. The Iranian government firmly opposed the invasion of Kuwait, and supported the U.N. embargoes, but also opposed U.N. military operations against Iraq. Iraqi foreign minister Tariq Aziz visited Teheran to consolidate the peace arrangements (September 9), and these were substantially completed in October. Iran reopened diplomatic relations with Great Britain (September 27), as part of an effort to normalize relations with Europe.

SOUTH ASIA

INDIA

1975, February 25. Sheik Mohammed Abdullah Reinstated as Chief Minister of Kashmir. The Indian government reinstated the "Lion of Kashmir" as the senior official in the Indian sector of the disputed province 22 years after removing him from power (see p. 1404). Pakistan claimed that this abrogated a 1972 agreement to maintain the status quo (see p. 1408).

1975–1979. Tribal Revolts in Northeast India. Government troops carried out counterinsurgency operations against tribal rebels seeking independence for Mizoram and Nagaland. The Nagas ended their 20-year rebellion in exchange for a general amnesty (November 11). Despite losses exceeding 1,000 per year the Mizos continued their guerrilla war against India.

1975–1976. Border Incidents and Strained Relations with Bangladesh. (See p. 1522.)

1975, October 20. Border Clash with China. Both India and China claimed the other side had violated the frontier.

1979, January 5. Resumption of Naga Insurgency. Nagas raided from Nagaland into neighboring Assam state. They demanded that parts of Assam should be ceded to Nagaland.

1979, April 20. Parliamentary Request for Atomic Weapons Rejected. The Indian government turned down a request from its parliament to produce atomic weapons to meet the possible threat of Pakistan's manufacture of these weapons.

1981, April 26. India Threatens Nuclear Weapons Production and Deployment. K. Subrahmanyam, a military adviser to the Indian government, stated that the threat posed by a possible Pakistani deployment of nuclear weapons warranted India's production of these weapons.

1981, November 3–23. India-Pakistan Border Incidents. There was an increase of border incidents as tensions rose because of military aid received by Pakistan from the U.S. and because of Indian fear of Pakistan achieving nuclear capabilities.

1979–1982. Indian Troops Battle Insurgents in States. Indian army units carried out widespread counterinsurgency or riot suppression operations in Assam, Gujerat, Manipur, Mizo, Tripura, and Bombay.

1982, May 17–20. Indian-Sino Border Talks Held. India and China held border talks concerning disputed territories in the Ladakh region of Kashmir, which was under control of the Chinese, and along the border of Bhutan under Indian control.

1983, April 6. Troops Restore Order in Assam. Indian army troops were needed to restore order in Assam, racked by struggles between Assamese and Bengali immigrants.

1983, June 7. Punjab Disturbances Require Paramilitary Troops Intervention. Civil disturbances caused by rival Sikh and Hindu groups required the intervention of central government paramilitary troops to maintain order.

1984, June 6. Storming of the Golden Temple in Amritsar. Sikh guerrillas under **Sant Jinhar Singh Bhindranwale,** seeking increased autonomy from India, occupied this Sikh holy shrine. They were assaulted by the Indian Army in a bloody operation that left over 400 Sikhs dead, including Bhindranwale. This not only made him a martyr for Sikh patriotism, it increased support for the radicals among moderate Sikhs, and led to increased sectarian violence in Punjab.

1984, October 31. Assassination of Indira Ghandi. The prime minister of India was murdered at her residence in New Delhi by two of her Sikh bodyguards, who were afterward killed by police. This may have been a response to the Golden Temple assault. **Rajiv Ghandi,** the prime minister's son, succeeded her in that office.

1987, May. Direct Rule Imposed on Punjab. After 2½ years of sporadic efforts at moderation and reconciliation, continuing guerrilla actions in Punjab caused Prime Minister Rajiv Ghandi to declare direct rule from New Delhi over Punjab. This restored some degree of order, but alienated many Punjabis.

1988, May. Siege of the Golden Temple. Indian security forces besieged Sikh guerrillas at the Golden Temple complex in Amritsar, killing 46. Guerrilla attacks continue.

1988, November. Coup Attempt in Maldives Thwarted. An effort by Tamil guerrillas from Sri Lanka to overthrow the government of the Maldives was blocked by Indian paratroopers.

1990, January–April. Unrest in Kashmir. Demonstrators clashed with police, and government forces imposed curfews, as secessionist sentiment grew in Kashmir among the Moslem majority. Indian security forces fired into a crowd of demonstrators, killing at least 32 (March 1).

1990, November 28. Direct Rule in Assam. Following separatist violence in this northeast Indian state, New Delhi imposed direct rule in an effort to restore order.

PAKISTAN

1975, February 5–28. U.S. Military Aid Strains India-Pakistan Relations. U.S. President **Gerald Ford** lifted a 10-year arms embargo to Pakistan. India protested that this would encourage Pakistan's Prime Minister **Zulfikar Ali Bhutto** in a bellicose anti-Indian policy. India was not placated when subsequently the U.S. lifted its arms embargo against India (February 24).

1975, February 8–March 4. Unrest in Northwest Frontier Province Charged to Afghanistan. Pakistan alleged that Afghanistan was responsible for subversion and collusion with Pakistani dissidents.

1976, February 25. Pakistan Purchases French Nuclear Reprocessing Plant. Foreign arms control analysts interpreted this as a reaction to India's successful nuclear detonation (May 18, 1974). (See p. 1516.)

1976, September 3–10. Kohistani Tribal Revolt Smashed. More than 10,000 troops with armor and air support were necessary to suppress the revolt.

1977, April 21. Martial Law Imposed in Three Cities. Prime Minister **Ali Bhutto,** who was being accused of election corruption, responded to increased violence and unrest by imposing martial law in Karachi, Lahore, and Hyderabad. The martial law decree was later declared unconstitutional by the Supreme Court of Pakistan (June 3) and was lifted 3 days later (June 6).

1977, July 5. Military Coup d'État. Growing unrest, and the election corruption charges, led to a military takeover by Army Chief of Staff General **Mohammed Zia ul-Haq.** Prime Minister Bhutto, as well as opposition party leaders, were jailed; martial law was imposed. Bhutto was soon released. Zia promised to hold elections in October, but soon reneged on this pledge.

1977, September 5. Bhutto Arrested and Charged with Murder. The former Prime Minister was tried and convicted of ordering the murder of a political foe (March 18, 1978).

1979, April 4. Execution of Ali Bhutto. General Zia rejected foreign appeals for clemency.

1979, April 6. U.S. Cuts Military Aid to Pakistan. This was based upon evidence that Pakistan was attempting to develop a nuclear weapons capability. Pakistan denied the charge, asserting only an interest in development of nuclear power for peaceful purposes (April 18). Subsequent talks failed to persuade the U.S. to change its position (October 17–18).

1980, January 15. U.S. Reverses Position on Military Aid. The Soviet Union's invasion of Afghanistan prompted a reversal of the U.S. position on military aid to Pakistan. Despite misgivings about Pakistan's nuclear development intentions, the U.S. offered $400 million in military aid to Pakistan. The U.S. also announced its intent to invoke a 1959 Security Treaty permitting the U.S. to intervene militarily if Pakistan were invaded by the U.S.S.R. Pakistan's President Zia welcomed the pledge but rejected the military aid offer as insignificant.

1980, September 26. Soviet Helicopters Violate Pakistan Border. Pakistani authorities charged that Soviet helicopter gunships violated Pakistani territory and attacked Pakistani troops. Pakistan asserted that this was only one of several such violations since the Soviet invasion of Afghanistan.

1981, March 23–April 21. U.S. Renews Military Aid to Pakistan. The U.S. government offered Pakistan $500 million in military aid to meet the security challenge posed to Pakistan by the Soviet invasion of Afghanistan.

1981, October 16. Pakistan Offers Renunciation of War Treaty with India. India rejected the offer by President Mohammed Zia ul-Haq, dismissing it as a propaganda move (November 11).

1981, November 25. Pakistan Threatens to Develop Nuclear Weapons Capacity. Western press sources reported that Pakistan's President Mohammed Zia ul-Haq told visiting Turkish journalists that Pakistan was striving to develop nuclear weapons because this was the only way to survive militarily and politically in the present international milieu.

1982, January 29–31. Treaty Talks with India. The Indian government reversed its opposition to Pakistan's proposal for a nonaggression treaty and began negotiations with Pakistan in the hopes of arriving at a genuine rapprochement. However, India withdrew from the negotiations because of anti-Indian remarks made by a Pakistani diplomat at the U.N. (February 25).

1988, August 17. Death of President Zia. He and 28 senior officials were killed when their plane crashed under mysterious circumstances. Rumors circulated that this was the work of dissident KHAD (Afghanistan's secret police) or KGB agents.

1988, November 16. Bhutto Wins Elections. Benazir Bhutto, daughter of executed prime minister Zulkifar Ali Bhutto, led the left-populist Pakistan People's Party (PPP) to victory in the first free general elections since 1973, and Bhutto became the first female prime minister of an Islamic country (December 2).

1990, August 6. Bhutto Dismissed as Prime Minister. President Ghulam Ishaq Khan, supported by the armed forces, accused Bhutto of "corruption and nepotism," and dismissed her after 20 months in office. He also declared a state of emergency, and authorized the military to seize the national TV station and telephone system. Bhutto characterized the action as a "constitutional coup" by conservative Islamicists and the military.

AFGHANISTAN

1975, August 30. Economic Assistance Agreement with the U.S.S.R. Under this agreement Afghanistan was to receive economic aid for 30 years.

1978, January 1–March 1. Cooled Relations with Soviets. President Daoud (see p. 1409) became concerned about growing Soviet influence and drastically reduced the number of Soviet advisers from about 1,000 to 200.

1978, April 27–28. Marxist Coup. Daoud was overthrown and killed in a Soviet-backed coup by **Noor Mohammad Taraki,** who began a forced modernization and reeducation program. This program was resented and opposed by most Afghans. Between 8,000 and 12,000 opponents were ruthlessly killed.

1979, February 14. U.S. Ambassador Killed. Ambassador **Adolph Dubs,** taken hostage by Moslem extremists, was killed in a gun battle when Soviets and Afghan authorities stormed the hotel where Dubs was being held. The U.S. State Department, which had urged caution, protested the role the Soviets played in the affair.

1979, March 27. Amin Becomes Premier. President Taraki named **Hafizullah Amin** as Premier. Although the Communist party tightened control over the government, popular resistance grew into full-scale rebellion. Many deserters from the Afghan Army joined rebel groups (called *mujahedeen* freedom fighters or holy warriors), who gained control of 22 of Afghanistan's 28 provinces.

1979, September 16–18. Taraki's Downfall. Amin overthrew Taraki, who was slain in a gun battle. Amin's coup was apparently accomplished without Soviet approval. Friction grew between Amin and his Soviet advisers.

Soviet Invasion of Afghanistan

1979, December 1–16. Soviet Buildup. The Soviets increased the size of their garrisons at the two air bases in Kabul, and began secret preparations for the ouster of Amin. A partial mobilization of Soviet troops just north of Afghanistan was quietly initiated. The Kremlin apparently feared that the mujahedeen would overthrow Amin and end Soviet influence in Afghanistan, and were determined to prevent this at any cost.

1979, December 24. Soviet Seizure of Kabul Airport. This was done by special Soviet forces, directed by First Deputy Minister of Internal Affairs (MVD) Lieutenant General **V. S. Paputin.**

1979, December 25–28. Occupation of Kabul. A massive airlift of Soviet troops began Christmas Day. Three airborne divisions—103rd, 104th, and 105th—were flown in while 4 motorized rifle divisions invaded Afghanistan overland from the north. The 105th Division

occupied Kabul against considerable resistance from elements of the Afghan Army and the local population. Amin and his ministers were isolated in his palace. A Soviet task force attacked the palace and Amin was killed, either in the fighting or by execution. General Paputin was also killed during the battle.

1979, December 28. Karmal Installed As President. Amin's former Vice President, **Babrak Karmal,** now ambassador to Czechoslovakia, was recalled by the Soviets to become President.

COMMENT. *The Soviet invasion of Afghanistan inflamed world public opinion. The Soviet move assured that the U.S. Senate would refuse to ratify the Strategic Arms Limitation Treaty (SALT II Treaty) that had been signed by Presidents Carter and Brezhnev the preceding spring (see p. 1485). President Carter enacted a grain embargo against the Soviet Union and canceled U.S. participation in the 1980 Summer Olympics, which were to be held in Moscow.*

1980, January–February. Popular Resistance Intensifies. Following the initial success of the December coup, opposition to the Soviets became widespread and fierce. Convoys were ambushed and numerous armored vehicles were destroyed. Several towns held by the Soviets were surrounded and besieged by mujahedeen. Soviet activities were constrained by the severe weather conditions that prevailed in Afghanistan in January and February.

1980, February 21–23. Uprising in Kabul. This was crushed by Soviet troops. About 500 Afghans were killed and another 1,200 jailed.

1980, March–April. Soviet Offensive. Soviet objectives were: (1) beseiged towns would be relieved, (2) the mujahedeen were to be driven from the fertile valleys and the strategic roadways that traversed them, (3) a security zone would be established on the border near the Khyber Pass, the route by which the mujahedeen were getting supplies and arms from Pakistan, and (4) all of the bases of resistance in the mountainous regions of Afghanistan were to be eliminated.

1980, March 1. Phase One. The Soviets employed MiG-21s to strafe rebel positions with napalm and rockets. Then Mi-24 Hind helicopters showered mujahedeen positions with more napalm and rocket fire. Next Mi-6 Hook helicopters dropped airborne assault troops on the positions. All Soviet garrisons were soon relieved and the remaining mujahedeen driven into the hills.

1980, April–May. Phases Two and Three. In order to drive the mujahedeen from the valleys and win control of strategic roadways the Soviets used combined arms, air-land attacks, featuring Mi-24s. Although these attacks met with strong resistance and the mujahedeen were able to inflict heavy casualties on the Soviets, they were forced to withdraw to the hills with heavy losses. A security zone was soon established, but the Soviets were unable to prevent the movement of supplies and arms from Pakistan, nor were they able to prevent mujahedeen raids in the strategic valleys.

1980–1985. Continued Afghan Resistance. Gradually the Soviets increased the size of their forces from 85,000 to 105,000, but were unable to break the resistance movement. The rebels continued to attack Soviet convoys and supply lines with considerable success. These attacks enabled the mujahedeen to acquire badly needed arms and ammunition, in addition to supplies from Pakistan. The Soviets mounted numerous offensives, inflicting heavy casualties on the Afghan resistance fighters, but themselves suffered severe casualties from guerrilla attacks and ground operations. The Soviets not only were not able to pacify the mountain areas but were unable to maintain uninterrupted control of the rural valleys. Afghanis alleged that the Soviets used chemical and biological weapons in Afghanistan, in violation of the 1925 Geneva Protocol and the 1972 Geneva ban on the manufacture of biological weapons. The U.S. Department of State also alleged Soviet use of illegal biochemical weapons. The mujahedeen were hampered by religious and tribal rivalries and shortages of arms, equipment, and ammunition. They were unable to create a single political opposition group or a unified command structure. The Soviets, despite setbacks and high casualties (estimates of Soviet casualties by 1984 were about

15,000), showed no signs of giving up in Afghanistan. They expanded the runways at 19 airports and constructed new airports at Kandahar, Herat, Dadadhshan and Mazarree Sharif, as well as a base in the Wakhan corridor near the Chinese border. In 1984, the air bases, in easy range of the Persian Gulf and the Indian Ocean, held an estimated 400 planes, including MiG-23s and Ilyushin-38s.

1985, May. Rebel United Front. Seven major rebel guerrilla groups met in Peshawar, Pakistan, to lay the groundwork for a united front.

1986. Increased U.S. Aid to Rebels. As Soviet offensive operations against the rebels intensified, the U.S. increased its aid, funneling sophisticated weapons like Stinger shoulder-launched surface-to-air missiles to the rebels through Pakistan. The rebels were able to frustrate several Soviet operations, as the fearsome Mi-8 Hip and Mi-24 Hind helicopter gunships were especially vulnerable to Stingers. The Soviets begin to suffer from war-weariness; veterans returning to civilian life reported posttraumatic stress syndrome; their protests of poor treatment echoed the complaints of many U.S. Vietnam veterans.

1986, May 4. Najibullah Replaces Karmal in Coup. Mohammed Najibullah, former head of the KHAD secret police, replaced Babrak Karmal as secretary general of the People's Democratic Party of Afghanistan (PDPA) in a bloodless coup. The following year Najibullah was elected to a 7-year presidential term (November 1987).

1986, October. Najibullah's Reconciliation Policy. After a meeting with Soviet leader Mikhail Gorbachev, Najibullah announced a policy of national reconciliation, offering a unilateral cease-fire and limited government power-sharing to the rebels in an effort to end the civil war. The offer was rejected.

1988, February. Soviet Withdrawal Planned. Contingent up on success of talks between Pakistan and Najibullah by March 15, the Soviets announced they would begin withdrawing all their forces from Afghanistan by May 15, 1988. They expected their withdrawal to take 10 months.

1988, April 14. Geneva Accords between Pakistan and Afghanistan. This followed a meeting between Gorbachev and Najibullah in Tashkent.

1988, May 15–1989, February 15. Soviet Withdrawal. An orderly disengagement was completed in 9 months without serious problems. (See p. 1494.) Soviet advisors remained in Afghanistan, and continued to support the Najibullah regime with some $500 million in supplies and weapons every month.

1989, February 19. Najibullah Declares Martial Law. Najibullah reconstructed his government, increasing the role of the PDPA and sharply favoring the hard-liners.

1989, July. Rebel Repulse at Jalalabad. This reverse showed that common expectations of a postwithdrawal government collapse were misplaced.

1990, October. Rebel Offensive. Tarin Kowt (capital of Oruzgat province) and Qalat (capital of Zabol province) fell in the first weeks. Otherwise the situation was little changed. The rebels controlled the countryside, but lacked both the heavy weapons and the necessary organization and command structure to capture the cities. Despite control of major cities, the government could accomplish little.

SRI LANKA

1977, February 15. State of Emergency Ended. A state of emergency, which had been imposed during the PLF revolt 6 years earlier, was lifted (see p. 1409).

1981–1984. Racial Violence, Civil Disturbance, and Insurgency. Racially motivated violent clashes between the Sinhalese majority and the Tamil minority intensified. A state of emergency was declared (August 17, 1981). However, violence continued, and Tamil insurgents frequently attacked government troops.

1983. Tamil-Sinhalese Violence. Efforts of the Tamil minority, concentrated in the north and east of the country, to gain greater autonomy led to widespread violence. The government again imposed a state of emergency (July). By

year's end several Tamil guerrilla-terrorist groups were active; the major group was the Liberation Tigers of Tamil Eelam (LTTE).

1984–1987. Tamil Violence Continues. Government forces were unable to restore order, and Tamil-Sinhalese sectarian violence continued despite attempted mediation by India.

1987, June. Government Forces Attack Jaffna. After a Tamil terrorist attack in Colombo killed hundreds (April 17), President **Jayawardene** activitated army reserve forces (May) and ordered a major attack on the rebel-controlled city of Jaffna.

1987, July 29. Indian-Brokered Settlement. Under pressure from Tamil groups in India (who were supplying the LTTE), the Indian government pressured Sri Lanka to accept a "settlement" involving withdrawal of Sri Lankan army troops from the Jaffna area, and limited autonomy to the Tamils in exchange for the surrender of arms. The LTTE refused this agreement, and Indian army troops were committed as peacekeeping forces. Radical Sinhalese nationalists began a guerrilla-terrorist campaign against government forces, complaining that the government had yielded too much to the Tamils and India.

1988. Indian Peacekeeping Efforts. In early 1988 Indian forces (about 50,000 men in 4 infantry divisions, plus support units) mounted a series of operations against LTTE forces, including an unsuccessful airborne assault on Jaffna.

1989–1990. Continuing Violence. Despite a cease-fire with the LTTE (May), violence continued. The total death toll for the 6-year civil war reached at least 11,000, including 1,000 Indians (December).

1989, August. India Announces Troop Withdrawal. In response to Sri Lankan pressure, the main withdrawal began in October. Within 6 months the last troops withdrew, effectively leaving northern and eastern Sri Lanka in the hands of Tamil rebels (March 24, 1990).

1990. LTTE-Government Fighting Renewed. The LTTE ended a 13-month truce with attacks on 10 police stations (June 14). Following another brief cease-fire were months of hard fighting with over 2,200 combatants killed. Tamil forces expanded the war by striking at hitherto neutral Moslem communities (August 3, 12). The LTTE announced a unilateral cease-fire (December 31), and indicated willingness to hold peace talks with the government in 1991.

BANGLADESH

1975, August 15. Coup d'État. Minister of Commerce **Khandakar Mushtaque Ahmed,** supported by army officers, led a successful coup which toppled the government of Sheik **Mujibur Rahman,** President and founder of Bangladesh. (See p. 1409.) Mujibur was killed. Martial law was established (August 20).

1975, November 3–6. Military Coup d'État. Major General **Ziaur Rahman** overthrew President Moshtaque. Rahman appointed former Bangladesh Supreme Court Chief Justice **Abu Sadat Mohammed Sayem** president, who appointed a 3-man advisory Council of Armed Forces Chiefs. Continuing internal unrest and border violations were blamed on India, which denied complicity.

1975–1991. Unrest in the Chittagong Hill Tracts. Guerrilla warfare resulted from native resentment of Bengali immigration.

1976, November 29–30. Martial Law Proclaimed. General Ziaur Rahman became virtual dictator.

1977, April 21–May 29. Rahman Consolidates Power. Bangladesh President Sayem resigned, and named General Rahman to succeed him. A national referendum approved Rahman's martial law government by a 99 percent majority (May 29).

1977, October 2. Coup Attempt Fails. Rahman's troops defeated dissident insurgents in a brief but bloody battle. Leaders of the coup were tried and executed (October 27).

1979, April 7. Martial Law Ends. Bangladesh President Ziaur Rahman announced the end of martial law, and promised the return to civilian rule.

1979, April 18. Accord with India Reached. Bangladesh reached an agreement with India on a number of issues that had caused strained relations between the two countries.

1981, May 30–June 2. Rahman Murdered in Coup Attempt. President Ziaur Rahman was killed in Chittagong. However, loyal troops put down the revolt led by General **Manzur Ahmed,** who was killed. Other leading conspirators were tried later. Vice President **Abdus Sattar** took office as President and later won a national presidential election (November 15).

1982, March 25. Coup d'État. Army Chief of Staff General **Hossein Mohammed Ershad** seized power in a bloodless coup. He later assumed the office of President (December 11).

1986, August. Ershad Retires from Military Command. Following his party's electoral victory (May), General Ershad retired from his military posts. He handily won the presidential election (October), securing 84 percent of the vote.

1986, November. End of Martial Law. Following his electoral victory, Ershad restored the constitution.

1989, July–November. Campaign to Force Ershad's Resignation. Following local election violence (February–March), opposition parties demanded Ershad's resignation (July 12). Responding to mounting demonstrations, Ershad imposed a state of emergency (November 27).

1990, December 4. Ershad Resigns. The opposition parties selected Supreme Court Chief Justice **Shahabuddin Ahmed** as interim president (December 5). Ershad and his wife were arrested (December 12).

NEPAL

1989, March. Trade Dispute with India. The expiration of a trade and transit treaty with India triggered a dispute. India closed all but 2 of 15 crossing points. Part of the problem was Nepal's recent purchase of Chinese arms.

1990, April 6–19. Democratic Revolution. Troops and police killed 63 persons during a pro-democracy demonstration in Katmandu (April 6). King **Birendra Bir Bikram Shah**

Deva bowed to political pressure and lifted a 29-year ban on political parties (April 8). Krishna Prasad Bhattarai was appointed as prime minister (April 19) and inaugurated Nepal's first independent government in nearly 3 decades.

SOUTHEAST ASIA

BURMA

The principal threats to Burma during the period 1975–1985 were internal. The guerrilla activities of the Kachin, Karen, and Shan separatist factions threatened the continued existence of the Socialist Republic of the Union of Burma. Communist guerrillas, especially active in the Shan State and supported by the People's Republic of China (PRC), and student unrest threatened the continued dominance of Burma's sole legal political organization, the Socialist Program Party.

1975–1991. Continuous Guerrilla Warfare in Burma. Burmese government troops continued to battle Communist and tribal insurgents (see p. 1411). The People's Republic of China supplied rebels with arms and equipment while maintaining tolerable relations with Burma. Some of the rebel groups funded their activities by growing and selling opium, and the political situation hampered drug control efforts.

1983, October 9. Korean Officials Killed in Terrorist Bombing. A bomb explosion during a state visit by South Korean President Chun Doo Hwan killed 4 members of his cabinet and 13 other South Korean officials. Chun, apparently the target, arrived just after the explosion. A Burmese investigation found North Korea responsible. Burma severed relations with North Korea.

1987, September. Student Riots. Students protesting the one-party regime of General Ne Win rioted following clashes with police.

1988, July. Ne Win Resigns. Under intense popular pressure, he resigned the presidency and chairmanship of the State Council to make way for a democratic civilian government. **Aung San Suu Kyi,** daughter of dead national hero and leader Aung San, played a prominent role in the democratic movement until her arrest (July 1989).

1988, September 18–28. Military Coup. After President Maung Maung's announcement of elections (September 10), the Army and its National Unity Party allies deposed the civilian government (September 18). General **Saw Maung,** an ally of Ne Win, headed a military junta. The official death toll was 263, but foreign diplomats estimated casualties in the thousands.

1989, July. Mass Arrests. Responding to continuing demonstrations for democracy, the Saw Maung regime arrested hundreds of protestors, but promised elections in April or May 1990.

1990, May 27. Elections. The National League for Democracy, led by Aung San Suu Kyi, won a landslide victory in National Assembly elections.

1990, July–December. Government Crackdown. Ruthless suppression of demonstrations and dissent by the Saw Maung regime virtually voided the May election results, since the junta refused to convene the Assembly.

THAILAND

The post–Vietnam War period was marked by coup attempts against the ruling military clique, military operations against Communist guerrillas along the Thailand-Malaysia border, and increasing tensions along the Thailand-Kampuchean border. Vietnamese invasion of Kampuchea and the overthrow of the Pol Pot regime posed the gravest threat to Thailand. The Kampuchean refugee problem strained Thai resources, while armed incursions by Vietnamese troops and their Kampuchean allies resulted in the Thai military confronting a foreign enemy for the first time since the withdrawal of Thai troops from Vietnam. In the aftermath of Vietnam, U.S. aid was limited to arms, munitions, and equipment. The member states of the Association of Southeast Asian Nations (ASEAN) provided Thailand with financial and nonmilitary aid, as well as diplomatic and moral support.

1975–1980. Border Clashes with Laos. There were a number of border clashes with Laos varying in size and intensity.

1975–1978. Border Clashes with Cambodia. (See p. 1527.)

1975–1976. Thai Troops Battle Communist Insurgency. Government troops continued to carry out counterinsurgency operations against Thai guerrillas, mostly in southern Thailand, but also in the northeast. The government claimed that the guerrillas had been trained and were receiving aid from China, Cambodia, and Vietnam.

1975, May 14–19. Thai Protest against American Use of Thai Bases. Thai officials were furious because the U.S. used Thai bases as a staging area for the *Mayaguez* rescue operations (see p. 1527). The U.S. government issued a formal apology, which Thailand accepted (May 19).

1975–1976. U.S. Withdrawal from Thailand. The U.S. began to withdraw planes and troops (June 6–7). All U.S. planes were withdrawn from Thailand before the end of the year (December 19). All troops were withdrawn the following year (June 20).

1976, October 8. Military Coup. The return of exiled former Premier **Thanom Kittikachorn** (September 19) led to demonstrations and riots by university students (October 4–6). To restore order, a military coup was led by Admiral **Sangad Chaloryu. Thanin Kraivichien** (a civilian) was appointed premier, but Admiral Chaloryu retained control as defense minister.

1975–1977. Thai-Malaysian Counterinsurgency Cooperation. Thailand and Malaysia agreed to cooperate in counter-insur-gency efforts against Communist insurgents in the Kra Isthmus (April 11, 1975). After initial cooperative successes, the two countries later agreed that forces of either country could cross their common border to pursue guerrillas and pledged to continue joint counterinsurgency operations (March 4, 1977). Another major success was achieved later in the year in southern Thailand near the Malaysian border (July–August).

1977, March 26. Attempted Coup. General **Chalard Hiranyasiri** failed in his attempt to gain power.

1977, October 20. Military Coup. Defense Minister Admiral Sangad Chaloryu seized control in a bloodless coup.

1978, July 15. Thailand and Cambodia Agree to End Border Hostilities. After neary 4 years of numerous armed border clashes of varying size and intensity, the Thai and Cambodian governments agreed to end their border conflict and exchange envoys.

1979–1984. Vietnamese Incursions into Thailand. Vietnamese troops clashed with Thai troops on a number of occasions while pursuing Cambodian guerrillas. (See p. 1525.)

1980, June 15–July 23. Border Clashes with Laos. Despite earlier agreements to end hostilities, incidents continued, mostly provoked by the Laotians.

1991, February 6. Coup d'État. Citing government corruption (but probably more unhappy with the government's rapprochement with Vietnam, and the declining role of the armed forces in Thai politics), General **Sunthorn Kongsompong** imposed direct military rule.

VIETNAM

The victory of the Socialist Republic of Vietnam (April 1975), and the reunification of the nation, did not bring peace to the Vietnamese people. While invading and occupying Kampuchea, at the same time Vietnam fought off a limited Chinese invasion of the northern provinces of Vietnam. Vietnamese operations against the Khmer Rouge led to violation of the Thai frontier and skirmishes with Thai forces. Ongoing involvement in Kampuchea further isolated Vietnam from its Southeast Asian neighbors, while the unresolved issue of U.S. troops missing in action (MIA) prevented normalization of relations with the U.S.

1975–1977. Border War with Cambodia. Vietnamese troops clashed repeatedly with troops of the Khmer Rouge Communist regime in Cambodia.

1976, July 2. North and South Vietnam Officially Reunited. The North Vietnamese Communist regime completed the reunification of the country, establishing the Socialist Republic of Vietnam.

1977, November–December. Vietnamese Invasion of Cambodia. Vietnam forces crushed a Cambodian division at Snoul (November 18–19), then halted its advance 10 miles inside the Cambodian border (December 3). The Vietnamese soon resumed the offensive and drove 70 miles into southern Cambodia before halting (December 21). At the same time, Vietnam denounced the atrocities the Khmer Rouge had committed against the population of Cambodia (see p. 1527).

1979, January 2–15. Renewed Vietnamese Invasion of Cambodia. In reaction to the Pol Pot regime's war of extermination against the Kampuchean rebels and the Cambodian civilian population, Vietnam invaded Cambodia. The Vietnamese quickly smashed Cambodian forces in 6 provincial capitals and captured Phnom Penh, the Cambodian capital, where they established a puppet regime. Pol Pot's forces were crushed, but maintained several pockets of resistance in the Cambodian countryside.

1979, February 17–March 15. Chinese Invasion of Vietnam. Before dawn, 25,000 Chinese troops and 1,200 tanks crossed the Vietnamese border and penetrated Vietnamese territory from 12 to 20 miles deep. But then the drive slowed. It is, however, unclear whether the Chinese slowed their offensive because of effective Vietnamese resistance or because the Chinese deliberately stopped their advance (as they claimed). No further major actions were fought. The Chinese began to withdraw, proclaiming they had accomplished their mission (March 5). They announced that the withdrawal was completed 10 days later. No reports of losses were given by either side, but Vietnamese losses probably exceeded those of the Chinese.

1979–1989. Vietnamese Counterinsurgency Operations in Cambodia. Vietnamese forces and their Cambodian allies conducted largely successful counterinsurgency operations against the forces of Pol Pot's deposed Khmer Rouge regime, taking a major headquarters and driving many Khmer Rouge troops into Thailand (March 16–31, 1979). Nonetheless, strong pockets of guerrilla resistance remained, both Khmer Rouge guerrilla forces of Pol Pot and forces loyal to Prince Norodom Sihanouk as well (see below and p. 1527).

1979–1991, Strained Relations with Thailand. Because of the Vietnamese invasion of Cambodia, large numbers of Cambodians fled to neighboring Thailand. Vietnam charged Thailand with aiding rebels, while Thai officials charged Vietnam with genocide in Cambodia. Vietnamese incursions into Thailand caused tensions between the two nations (see p. 1524).

1979, April 9–1980, March 6. Chinese-Vietnamese Reconciliation Talks. No real progress was made; relations remained strained.

1979, July–1983, April. Border Incidents with China.

1980–1991, Low-Level Insurgency in South Vietnam. Soldiers of the former Army of the Republic of Vietnam continued to conduct small-scale guerrilla operations against the Communist regime.

1989. Withdrawal from Cambodia. Hoping to reduce military commitments (and strain on their failing economy), and hoping to ease ten- sion with Thailand, the Vietnamese unilaterally withdrew from Cambodia. Vietnamese advisers and technicians remained in Cambodia to aid government efforts against the Khmer Rouge and Sihanouk's forces.

1989, October–November. Partial Soviet Withdrawal from Cam Ranh Bay. The So- viets withdrew their MiG-23 fighters and Tu-16 bombers from the airbase, leaving 6 to 10 mili- tary aircraft.

LAOS

Soon after gaining control of the country, the Pathet Lao-dominated Lao People's Democratic Republic initiated military operations against the Meo, many of whom fled to Thailand and Burma. Firmly controlled by a pro-Vietnamese government, Laos served as a logistical base for Vietnamese operations in Kampuchea.

1975, May–August. Pathet Lao Consoli- dates Power. The Pathet Lao continued to be successful in the long civil war in Laos (see p. 1418). After securing key positions around Vientiane, the Pathet Lao agreed to a cease-fire (May 7). Soon after this, Pathet Lao troops took control of all of southern Laos (May 16–20). Prime Minister Souvanna Phouma ordered government troops to cease resistance to the Pathet Lao (May 23). The Royal Laotian Army and the Pathet Lao were unified (June 17). Two months later the Pathet Lao assumed control in Vientiane Province, giving them complete con- trol of Laos (August 23).

1975–1980. Border Incidents with Thailand. (See p. 1524.)

1975–1977. Insurrection of Meo Tribesmen. Some 5,000 Meo tribesmen, formerly armed and trained by the U.S. Central Intelligence Agency, undertook guerrilla operations against Pathet Lao troops. The Meos inflicted heavy casualties on the Pathet Lao, despite Viet- namese assistance to the Pathet Lao.

1975, December 1–3. Laos Becomes a Social- ist Republic. The ancient monarchy was ended and the rule of the Pathet Lao legit- imatized.

1977, July 18. Friendship and Assistance Treaties with Vietnam. These legitimatized the presence of 40,000 Vietnamese troops al- ready stationed in Laos.

1978, February 10–March 28. Counterin- surgency Operations Against Meo Tribes- men. Laotian troops and some 30,000 Vietnamese troops eliminated most resistance by Meo tribesmen and other dissidents. None- theless, strong pockets of guerrilla resistance remained.

1979, March 7–15. Chinese Border Inci- dents. Laotian authorities reported several in- cursions into Laos by Chinese troops.

1979–1983. Meo Accusations of Laotian and Vietnamese Use of Chemical Weapons. Meo tribesmen who had fled Laos charged that Pathet Lao and Vietnamese troops had used poison gas and biochemical weapons against them (November 3). These reports were re- peated as guerrilla resistance continued.

CAMBODIA (KAMPUCHEA)

The victory of the Khmer Rouge and the establishment of the Pol Pot government marked the beginning of the most tragic episode in Cambodian history. The brutal, repressive, and homicidal policies of the government resulted in mass executions and mass starvation. This situation gave Vietnam a pretext for invasion. The overthrow of the Pol Pot regime and its replacement with the pro-Vietnamese government of **Heng Samrin,** the People's Republic of Kampuchea, however, did not end turmoil or suffering. Khmer Rouge guerrilla units, as well as those of "reactionary" groups, continued to resist the Vietnamese, when not engaged in fighting each other.

1975–1978. Khmer Rouge Massacres. Conducting what was tantamount to a war on the entire population, the Khmer Rouge government of Premier **Pol Pot** was responsible for the deaths of over 2 million people in its attempts to wipe out resistance to its rule.

1975–1978. Border Clashes with Thailand. There were numerous border clashes with Thailand varying in size and intensity. In part these stemmed from the large numbers of refugees seeking to escape the Khmer Rouge massacres. The two countries signed an agreement pledging to pacify their borders (July 19, 1978; see p. 1524).

1975, May 12–14. *Mayaguez* **Incident.** While in international waters some 60 miles off the coast of Cambodia, the American merchant ship S.S. *Mayaguez* was captured and its 39 crewmen taken prisoner by a Cambodian gunboat, then transported to the Cambodian port of Sihanoukville. Failing to win the release of the crew through diplomatic means, President Gerald Ford ordered U.S. Marines to conduct a rescue mission. After being airlifted to Thailand from bases in Okinawa and the Philippines, the Marines conducted a helicopter attack on Tang Island (off Sihanoukville) where the *Mayaguez* was thought to be held. At the same time, carrier aircraft attacked a nearby Cambodian air base and gunboats in Sihanoukville. In the course of the rescue operations, the Cambodians released the *Mayaguez* crewmen in a captured Thai fishing boat. The U.S. losses were 18 killed in action and 50 wounded. The Cambo-

dians lost 17 aircraft and 3 gunboats and an unknown number of personnel.

1975–1979, December. Border War with Vietnam. (See p. 1525.)

1976–1978. Resistance to Khmer Rouge Builds. Defectors from the Cambodian Army, with Vietnamese support, formed a resistance movement called the Movement for Khmer National Liberation (April 21, 1976).

1979–1989. Vietnamese Invasion and Counterinsurgency Operations in Cambodia (See p. 1525.)

1979, October 4. Sihanouk Announces Formation of New Resistance Movement. Prince Norodom Sihanouk, former Cambodian chief of state in exile in Peking, announced the formation of a new resistance movement in opposition to both the Khmer Rouge guerrillas and the Vietnam-backed Kampuchean regime. The resistance movement was to be called the Confederation of the Khmer Nation. Later nominal coordination was established between the Confederation and Khmer Rouge guerrillas (June, 1982).

1989, January–July. Peace Initiaives. Promising negotiations toward a peaceful settlement of Cambodia's decades-long civil war failed in France (July 23–25), largely because of disputes on the role of the dreaded, still-powerful Khmer Rouge. Heavy fighting resumed, and continued inconclusively through 1990.

1989, April–September. Vietnamese Withdrawal. (See p. 1526.)

MALAYSIA

Continued activities of Communist guerrillas along the Thai-Malaysia border provoked numerous counterinsurgency operations, both unilateral and in conjunction with Thai forces, weakening the Malaysian Communist party. By late 1983, many guerrillas had had enough and surrendered to either Thai or Malaysian forces.

1977, January 14–17. Combined Thai-Malaysian Guerrilla Operations. (See p. 1524.)

1977, March 4. Border Pact Signed with Thailand. (See p. 1524.)

1977, July–August. Combined Thai-Malaysian Guerrilla Operations. (See p. 1524.)

1989, December 2. End of Communist Insurgency. After 41 years of guerrilla war, the Communist party of Malaya agreed to end its insurgency. Some 1,200 guerrillas laid down their weapons, ending one of the longest-running guerrilla struggles in history.

SINGAPORE

Singaporean military operations during this period were limited to anti-piracy patrols in the Philip Channel.

INDONESIA

During the late 1970s and early 1980s Indonesia was engaged in a counterinsurgency effort against the *Frente Revolucionario Timorense de Libertacao e Independencia* (FRETILIN) in East Timor. In early 1978, Indonesia faced a new insurgency in West Irian: the Free Papua Movement, which sought independence for western New Guinea.

1975–1984. Guerrilla Warfare in Timor. Indonesian troops battled guerrillas in East Timor after the Revolutionary Front for Independent East Timor declared its independence from Portugal (November 28, 1975).

1975, November 29. Annexation of Timor. The Timor Democratic Union (UDT) declared that Timor had become part of Indonesia.

PHILIPPINES

The administration of President **Ferdinand Marcos** faced 3 serious internal challenges: a Moslem insurgency on Mindanao and in the Sulu Archipelago, led by the Moro National Liberation Front (MNLF) and its military wing, the Bangsa Moro Army; a Communist insurgency in Luzon led by the Communist party of the Philippines and its military wing, the New People's Army (NPA); and the growing opposition of the professional and working classes to both martial law and Marcos' dictatorial rule. Despite some successes against the MNLF and the NPA, and repressive measures against critics of the government, Marcos' control of the Philippines remained at risk.

1975–1991. Moslem Rebellion Continues in Southern Islands. (See p. 1423.) A series of government concessions and negotiated agreements reduced violence (1988–1990), but minor unrest and scattered fighting continued, especially in outlying areas.

1975–1986. New People's Army Guerrilla War. The Maoist New People's Army (NPA), a successor to the Huks (see p. 1422), waged continual guerrilla war against government forces.

1976, February 20. SEATO Disbanded. The Southeast Asia Treaty Organization (SEATO), founded after the 1954 Geneva Conference (see p. 1414) in order to provide collective security for the region, formally came to an end with ceremonies in Manila.

1981, January 17. Marcos Pronounces End to Martial Law. This had been in effect over 9 years (p. 1423).

1983, August 21. Assassination of Aquino. President Marcos's main political foe, **Benigno Aquino,** was assassinated upon his return to Manila from a period of self-exile. Popular indignation flared in the Philippines because of suspicion that Aquino had been killed by soldiers on order of Marcos.

1985, November 3–4. Marcos Announces Election Plans. Under increasing pressure because of the Aquino assassination, Marcos announced a "snap" presidential election for February 7, 1986.

1986, February 7–24. Election and Overthrow of Marcos. Aquino's widow, **Corazon C. Aquino,** ran against Marcos. Fraudulent official election results declared Marcos the winner (February 15), but provoked public outcry and mass anti-Marcos demonstrations (February 16–23). When the army refused to obey Marcos' orders to attack the demonstrators, he fled the country for exile in the U.S. (February 24). Aquino became interim president, and reappointed **Juan Ponce Enrile** as defense minister and Lieutenant General **Fidel Ramos** as chief of staff. Both men had played key roles in deposing Marcos.

1986–1991. Continuing NPA Guerrilla War. To reduce support for the NPA, President Aquino made serious efforts at rural reform and offered repeated amnesty initiatives. By 1989–1990 these efforts, plus army pressure and bloody internal power struggles, had damaged the NPA. It could field only 19,000 armed fighters (down from over 25,000 in the mid-1980s), and turned increasingly to urban terrorism.

1989, May 13. U.S. Airmen Killed by NPA. Their bodies were found near Clark Air Force Base on Luzon. This reflected the NPA's new terrorist tactics, and exploited popular demonstrations against the U.S. bases in the Philippines.

1989, December 1–9. Attempted Coup. This was the most serious of 6 military revolts endured by the Aquino administration. Rebels seized military bases, TV stations, and buildings in the financial district (December 1). In heavy fighting, loyalist forces (supported morally but not physically by U.S. F-4 fighter-bombers from Clark AFB) gradually gained the upper hand. President Aquino declared a state of emergency (December 6), and the last rebels soon surrendered (in Manila December 7, on Cebu December 9). Casualties totaled at least 98 dead and over 500 wounded. The coup's suspected leader, cashiered Army Colonel **Gregorio Honasan** of the Reform the Armed Forces Movement (RAM), who had led a previous coup (November 1987) but escaped custody after his capture, remained at large.

1990, August–September. Terrorist Bombing Campaign. Over 20 bombs exploded over a 2-month period in Manila. These were generally regarded as an effort by dissident military officers in the RAM and YOU (Young Officers' Union) movements to destabilize the government.

1990, September 18–22. U.S.–Philippines Base Negotiations. These took place in Manila, in anticipation of early expiration of the current agreement (September 1991).

1990, October 4–6. Military Revolt. Some 200 soldiers, led by Colonel **Alexander Noble,** captured an outpost on Mindanao and declared the island an independent republic. The Air Force bombed their stronghold (October 5), and the rebels surrendered the next day.

PAPUA, NEW GUINEA

1975, September 16. Independence from Australia.

1975–1976. Bougainville Separatist Movement. Island leader **John Momis** led a secessionist movement, which conducted attacks on Papuan government installations and committed other acts of sabotage before arriving at a negotiated settlement with New Guinea Premier **Michael Somare** (March 26, 1976).

1978–1991. Low-level Insurgency in Western New Guinea (Irian Jaya; see p. 1528). Papua New Guinea is not officially involved (nor does the government back the anti-Indonesia insurgents), but Indonesian actions against the insurgents have strained relations between Indonesia and Papua. In part because of this ongoing security problem, Australia maintains a permanent military presence in Papua.

EAST ASIA

CHINA (PEOPLE'S REPUBLIC)

1976, September 9. Death of Mao Zedong. The death of the original leader of China's Communist party, theoretician of guerrilla warfare, and founder of the People's Republic of China, provided an opportunity for the emerging pragmatic leadership of the Communist party to turn away from idealism to practicality in the affairs of the most populous nation on earth. The most powerful man in the New China soon became **Deng Xiaoping,** whose position of strength was finally consolidated when he forced Mao's immediate successor, **Hua Guofeng,** to resign as Premier (September 4, 1980).

1979, January 1, Diplomatic Relations with the United States. This was a major diplomatic triumph for the P.R.C. The U.S. agreed to withdraw diplomatic recognition from the Nationalist Government of Taiwan, and to abrogate its defense treaty with the N.R.C. The U.S. accepted the status of Taiwan as a province of China, and the P.R.C. agreed that it would not use force to regain control over Taiwan.

1979, February 17. Invasion of Vietnam. (See p. 1525.)

1980, May 18. China Test-Launches ICBM. The Chinese made their first successful test-launch of an ICBM; the missile traveled 6,200 miles.

1987, January 1–5. Widespread Student Demonstrations. Thousands of student in 11 Chinese cities demonstrated peacefully for more democracy and freedom of expression.

1989, April–May. Student Unrest in Beijing. Crowds of students and supporters marched through the city and congregated in T'ienanmen Square, demanding more democratic freedom. Troops were assembled in and near the city, but violence was avoided by both sides. Martial law was declared (May 20).

1989, June 3–4. Bloodshed in T'ienanmen Square. Large bodies of troops, assembled in the outskirts of the city, converged on the square, overwhelming sometimes serious resistance at street barricades with tanks and small arms fire. Estimates of deaths range from a few hundred to more than 1,500. The government's brutal suppression of the demonstration was protested by many nations, including the U.S., which severed all official contact with the Chinese government, short of breaking diplomatic relations.

TAIWAN (NATIONAL REPUBLIC OF CHINA)

1975, April 5. Death of Chiang Kai-shek. He was succeeded as President of the N.R.C. by his son, **Chiang Ching-kuo.**

1979, January 1. Abrogation of Diplomatic Relations and Treaties with United States. The U.S. took this step in order to establish diplomatic relations with the People's Republic of China (see above). But semi-diplomatic relations were retained through liaison offices, and the practical effect of the change was initially negligible.

1982, July 17. U.S. Fighter Plane Co-production Agreement with Taiwan. U.S. President Ronald Reagan signed an agreement with Nationalist China extending the co-production of F-5E fighter planes by the two countries. The People's Republic of China promptly protested that this was a violation of the agreement by which diplomatic relations had been established.

1982, August 18. U.S. to Reduce Arms Supply to Nationalist China. U.S. President Reagan assured the P.R.C. that the U.S. would continue to reduce U.S. arms supplies to the N.R.C. on Taiwan.

1987, July 21. End of Martial Law. This had been imposed 38 years earlier, when the National Government fled to Taiwan from mainland China.

TIBET

1975–1984. Continued Guerrilla Warfare. There was continued low-level insurgency against Chinese rule in Tibet (see p. 1430).

1989, March 5–8. Unrest in Tibet. Troops fired on demonstrators demanding independence. Martial law was declared in Lhasa (March 7).

1990, May 1. Martial Law Lifted in Lhasa.

MONGOLIA

1990, March 14. Mongolia Ends Communist Dictatorship. A new multiparty political system, with free elections, was inaugurated.

NORTH KOREA

1975–1984. Naval Incidents and Continued Tense Relations Along Border with South Korea. (See below.)

1976, August 18. North Korean Assault on South Korean–U.S. Work Crew. North Korean soldiers, armed with metal pikes and axes, assaulted a South Korean–U.S. military personnel work team in the DMZ while they were trimming a tree in order to improve surveillance. Two U.S. Army officers were killed and 5 U.S. enlisted men were wounded. Another South Korean–U.S. work crew returned 3 days later and cut the tree down.

SOUTH KOREA

1975–1984. Clashes with North Korea. There were a number of border incidents and clashes between the North and South Korean navies.

1976, August 18. North Korean Assault on South Korean–U.S. Work Crew. (See above.)

1977, March 9. Planned U.S. Troop Withdrawal. (See p. 1549.)

1978, April 28. South Korean Airliner Attacked over Soviet Territory. Soviet fighter plane fired on and forced a Korean commercial plane to land in Soviet territory. Two passengers were killed and 13 were injured during the incident. The Soviet Union claimed that the airliner was attacked only after the Korean pilot had failed to respond to warning signals. Korea claimed that the Soviets had failed to warn the pilot.

1979, December 12–13. Martial Law Chief Arrested. General **Chung Seung Hwa** was arrested by subordinates for complicity in the slaying of President Park. Chung was subsequently tried and sentenced to 10 years in prison for his role in the assassination (March 13, 1980).

1980, May 18. Martial Law Imposed. Following rioting and antigovernment demonstrations by students and others throughout South Korea, total martial law was imposed by General **Lee Hi Song.**

1981, January 24. Martial Law Decree Lifted. In fulfillment of a promise made upon his accession to the presidency (September 1, 1980) **Chun Doo Hwan** announced the end of martial law in South Korea. South Korea had been under some form of martial law since the assassination of South Korean President Chung Hee Park.

1983, September 1. Soviets Down Korean Commercial Aircraft. (See p. 1493.)

1983, October 9. Korean Officials Killed in Terrorist Bombing. (See p. 1523.)

1987, November. North Korean Terrorists Destroy South Korean Airliner. En route from Abu Dhabi to Singapore the plane went down in the Andaman Sea, near Burma, after a bomb exploded. One of two terrorists later (after extradition to South Korea) confessed that she and her colleague had followed orders from the North Korean government.

1988, May–June. Student Unrest. Students across the country clashed violently with police while demanding reunification of Korea.

JAPAN

1982–1983. Five-Year Military Buildup Plan. The Japanese government announced a 5-year plan of military development with the objective of developing a capability to defend 1,000 miles beyond its seacoast (July 23, 1982). The plan was the first attempt at providing for its own national security since its defeat in the World War II, and the establishment of the National Self-Defense Force (see p. 1431). The Japanese Diet voted funds for this buildup (December 30, 1982; July 12, 1983).

1987, January 23. Japan Increases Levels of Defense Expenditures. Responding to U.S. pressure, the Japanese government's defense budget, for the first time since World War II, exceeded 1 percent of the Gross National Product.

AFRICA

EAST AFRICA

Ethiopia

1975, January 31–July 31. Civil War Escalates in Eritrea. Ethiopian forces continued to battle Eritrean secessionist rebels in Asmara and north of the provincial capital. The rebels were unsuccessful in their attempt to cut the main road between the Ethiopian capital, Addis Ababa, and the deep-water port of Assab on the Red Sea. The rebels also failed to capture Asmara and Keren, northwest of Asmara. Although the rebels failed to capture any key urban center, they did control most areas near Asmara. The two principal rebel factions, the Eritrean Liberation Front-Revolutionary Command (ELF-RC) and the Eritrean Popular Liberation Front (EPLF), joined forces for these operations.

1976, May 16. Ethiopian Peace Offer. Brigadier General **Tafari Banti,** Ethiopian chief of state and nominal head of the Military Council, offered amnesty to Eritrean political prisoners, economic aid to the Eritrean province, and also held out the possibility of regional autonomy for the Eritreans if they would cease fighting.

1976, May 18–22. Ethiopian Peasant Volunteer Offensive. While the Ethiopian government was offering peace, it was raising a volunteer army of as many as 50,000 peasants to launch an offensive against the Eritrean rebels. The drive bogged down almost immediately when the Eritreans blew up a key bridge. The peasant militia was soon disbanded (June 13).

1976, December, Ethiopian Sign Soviet Assistance Pact. Ethiopian leaders signed a long-term military aid treaty with the Soviet Union.

1977, January 5–August 25. Eritrean Successes. The EPLF won an important psychological victory by taking the town of Karora (January 5). Soon after they captured Naqfa, a district capital (March 23). Keren, the second largest city in Eritrea and the key to control of the north and west of the province, was captured (July 3). Taking advantage of Ethiopian involvement in Ogaden, the Eritreans attacked the deep-water port of Massawa on the Red Sea, 45 miles east of Asmara (July 15). Massawa was encircled and almost the entire northern part of the province was in rebel hands (August 25).

1977, January–June. Rebellion in Ogaden Province. A majority of the sparse population of Ethiopia's desert Ogaden Province were Somalis. Undoubtedly encouraged by the Soviet-supported Somalia government, the Western Somalia Liberation Front (WSLF) took up arms to throw off Ethiopian rule and to join Somalia. Ethiopian forces in the province were forced to retreat to fortified mountain bases at Jijiga, Harar, and Diredawa.

1977, February 3. Mengistu Seizes Power. Colonel **Mengistu Haile Mariam** and his followers bested Brigadier General Tafari Banti in a bloody gun battle in Addis Ababa in which Banti was killed. Mengistu ruled virtually unopposed.

OGADEN WAR WITH SOMALIA

1977, July 13. Somalian Invasion of Ogaden. Encouraged by Soviet economic and military assistance, and by Ethiopia's involvement in Eritrea, Somali President **Siad Barre** sent troops into Ogaden Province from northwest Somalia to help the WSLF, with the ultimate objective of annexing Ogaden to Somalia. Barre insisted, however, that the invasion troops were volunteers, and denied any regular Somali troop involvement. Barre's action was embarrassing to the U.S.S.R., which had also begun to provide military assistance to Ethiopia.

1977, August 16–19. Battle of Diredawa. Ethiopia troops repulsed attacks by Somali and WSLF troops to take their fortified positions at Diredawa. The extent to which this Ethiopian success was due to the presence of Soviet advisers and a small contingent of Cuban combat troops is unclear. The Somalis, however, were able to block the railroad from Addis Ababa to Djibouti on the Red Sea.

1977, September 4–14. Battle of Jijiga. Somalis and WSLF defeated Ethiopian forces, destroying or capturing 50 tanks. Casualties were heavy on both sides. The Somalis seized the Kara Marda Pass (September 23–27) and advanced from north and east to blockade Harar. Ethiopian troop buildup, assisted by Soviet advisers and Cuban combat troops, halted further Somali advance, and a stalemate ensued. The Soviets attempted to arrange a settlement between their warring clients.

1977, November 13. Soviets Expelled from Somalia. Infuriated by the fact that his presumed Soviet allies were also assisting his Ethiopian enemies, Siad Barre expelled his Soviet military and technical advisers, and required them to evacuate their naval and submarine bases at Mogadishu and elsewhere on the Indian Ocean coast. Soviet Colonel General **Grigory Borisov,** who commanded the Soviet mission in Somalia, went to Ethiopia.

1977, November–1978; January. Soviet and Cuban Buildup in Ethiopia. Soviet adviser strength increased from a handful to about 1,500; Cuban troop strength increased from about 500 to about 11,000. The Soviet mission was commanded by Army General **Vasily Petrov,** who had been First Deputy Commander of Soviet ground forces. He was assisted by General Borisov after he arrived from Somalia. A combined Ethiopian-Cuban offensive into the Ogaden was planned.

1978, February 6. Relief of Harar. Ethiopian and Cuban troops, assisted by air support provided by Soviet and Cuban pilots, drove the Somalis from their positions around Harar. They were repulsed, however, by Somali and WSLF troops dug in near Babile and the Kara Marda Pass (February 14).

1978, March 2–5. Battle of Diredawa-Jijiga. An Ethiopian-Cuban offensive was spearheaded by a Cuban paratroop battalion airlifted to the vicinity of Diredawa, and routed the Somali forces blocking the railroad. The Ethiopian 10th Division, advancing along the railroad, enveloped the principal Somali positions around Jijiga from the north and east. The Cuban paratroops moved against the rear of the Somali positions, while 3 Cuban infantry regiments and an armored regiment drove southward and eastward through the Kara Marda Pass. The Somalis, virtually surrounded, collapsed after suffering heavy casualties.

1978, March 8. Somali Withdrawal from Ogaden. President Barre asked for a cease-fire, and withdrew his forces. WSLF resistance continued, but was easily controlled by the victorious Ethiopians, who were now able to turn their major attention to the rebellion in Eritrea (see p. 1532).

1978, April 6–30. Ethiopian Buildup in Eritrea. Mengistu was determined to seek a military solution with or without Soviet-Cuban aid in Eritrea, despite Soviet advice to seek a diplomatic settlement. He began massing 100,000 troops along the Eritrean border.

1978, July–November. Ethiopian Offensive in Eritrea. Ethiopian troops advanced into Eritrea between Tigre and Bagemdir (July 26). They drove the Eritreans away from Massawa and advanced toward Keren. They took Keren after a number of bloody battles (November 27). The Eritreans retreated to the northwest corner of the province.

1979–1988. Civil War in the North. Despite modern Soviet weaponry and Soviet, Cuban, and East German advisers, government troops were stalemated in their decades-long (since c. 1962) struggle with guerrilla forces in Tigré province and Eritrea. The well-organized and experienced guerrilla forces were fairly well-equipped. The government's doctrinaire program of Marxist reforms, coupled with blatant oppression of non-Amhara ethnic groups, antagonized many Ethiopians and indirectly aided the rebels.

1988–1989. Reverses in Eritrea and Tigré. A series of major government offensives in Eritrea and Tigré were soundly defeated. Many government troops surrendered, and large quantities of arms and supplies fell into rebel hands. For the first time, the Tigrean Peoples' Liberation Front (TPLF) posed a serious threat to government forces. The initiative passed to the rebels.

1989, May. Military Coup Attempt. Dissident army officers attempted a coup against the government in mid-month while President Mengistu was out of the country. After several days of hard fighting around Addis Ababa and Asmara, they were defeated. Twelve senior officers were later executed (May 1990).

1989, Autumn–1991, Spring. Rebel Victories. War-weary loyalist troops were forced back on all fronts in their war against concentrated efforts by Tigrean and Eritrean insur-gents. Asmara, capital of Eritrea, came under rebel siege, and fighting came within 100 km of Addis Ababa (late 1990).

1991, May. Government Collapse. Mengistu fled the country (May 21), and a few days later the rebels occupied Addis Ababa and set up a new government (May 28).

Seychelles

1976, June 28. Seychelles Granted Independence from Great Britain.

1977, June 5. Coup d'État. The new Socialist government of **France Albert René** established friendly ties with the Soviet Union, granting access to Seychelles naval facilities.

1978, April 29. Coup Attempt Fails. Officials loyal to President France Albert René uncovered a plot to overthrow René and replace him with former President **James Mancham.**

1981, November 25. Coup Attempt Fails. A band of 50 mercenaries, veterans of the Katanga insurrection in the Congo (Zaire, 1961–1967; see pp. 1440–1443) led by soldier of fortune "Mad Mike" Hoare, failed to oust President René's regime by force of arms. Hoare and his band arrived in the Seychelles disguised as rugby players, but fighting broke out when a customs guard noticed that one member had an automatic weapon. A 20-hour gun battle ended with the capture of 5 of the mercenaries, one killed in the course of the action, while Hoare and 43 others made their escape on a comandeered Air India airliner and landed in South Africa. Subsequently, the mercenaries were tried in South African court but received light sentences, fueling the speculation that the South African government had authorized the operation.

1982, August 18–19, Army Revolt. An uprising by dissident army troops was suppressed with the assistance of Tanzanian forces.

Somalia

1975, July 6. Soviet Base in Somalia Confirmed. A U.S. congressional delegation concluded, despite Soviet denials, that the Soviet Union had established a base for naval refueling and storage of weapons at Berbera.

1976, April 5. Soviet Advisers in Somalia. Foreign observers reported the presence of more than 2,000 Soviet military advisers in Somalia, as well as 600 Cubans.

1977, January–June. Involvement in Ogaden. President Siad Barre encouraged ethnic Somali insurgents in Ethiopia's Ogaden Province (see p. 1532).

1977–1978. Ogaden War with Ethiopia. (See p. 1533.)

1978, April 9. Army Revolt Suppressed. Troops loyal to Somali President Siad Barre easily suppressed a revolt of dissident army officers, who blamed the political leadership for the defeat in the Ogaden Desert War.

1980, August 21. U.S.–Somalia Military Cooperation Agreement. Somalia signed an agreement which permitted the U.S. to use Somali naval and air force facilities at Mogadishu on the Indian Ocean and Berbera on the Gulf of Aden. The agreement provided the Somalis with $25 million in military assistance for 1 year (1981) and additional aid in following years.

1988, Summer. Disorders in North. Tribally based unrest in northern Somalia (largely among Issaq Somalis) provoked fierce government repression. Minor guerrilla activity began in the countryside.

1988–1990. Spreading Insurrection. Despite vigorous government counterinsurgency actions, guerrilla forces of the Somali National Movement (SNM) slowly gained ground in northern and central Somalia.

1990, December 27–January 29, 1991. Somalian Revolution. Siad Barre's increasing oppression and guerrilla successes in the countryside sparked popular insurrection in Mogadishu (December 27–28). Fighting spread throughout the country, and the death toll soon reached at least 1,500 (January 14). Despite initially superior government resources, the rebels gradually gained the upper hand. Siad Barre fled the capital (January 27), and fighting ceased in Mogadishu. A provi-

sional government was installed (early February 1991).

Djibouti (French Somaliland)

1975–1977. Unrest in French Somaliland (Territory of Afars and Issas. Internal unrest was compounded by border incidents between French troops and Ethiopian and Somali forces, resulting from hostilities between those countries and insurgency in Ethiopia of separatist Afars and Issas tribesmen.

1977, June 27. Independence of Djibouti. The new nation—renamed Djibouti—began with tension and strained relations with neighboring Ethiopia and Somalia.

Sudan

1975, September 5. Coup Attempt Fails. Troops loyal to President Gaafar al-Nimeiry crushed rebel army officers who had seized a radio station and announced the overthrow of the government.

1976, July 2–3. Coup Attempt Fails. Rebel soldiers attempting to overthrow the Nimeiry government were overwhelmed by loyal troops.

1976, July 13–19. Egyptian, Sudanese, and Saudi Defense Pacts. (See p. 1502.)

1977–1978. Border Skirmishes with Ethiopia. Occasional small-scale fighting occurred due to Sudan's support of the Eritrean separatists.

1977, February 2–6. Separatists Defeated. Elements of the air force, attempting to set up a separate government in the southern portion of the country, were crushed by government troops.

1977, February 27–28. Sudan Enters Joint Command with Egypt and Syria. (See p. 1502.)

1977–1984. Continued Insurgency in the South. Despite earlier agreements (see p. 1438) the dissident Christians and pagans of the south continued their resistance.

1983, May 15–16. Army Mutiny Crushed. Local government troops suppressed a mutiny of disgruntled Sudanese Army enlisted personnel and noncommissioned officers.

1983, September. Rebellion in the South. Under pressure from the Islamic Brotherhood, President Nimeiry decreed legal changes to bring Sudanese law into accord with the *Shari'ah,* and divided the southern region into 3 parts. Colonel **John Garang** became leader of the Sudanese People's Liberation Army (SPLA), combining antigovernment guerrilla forces.

1983–1985. Guerrilla Warfare. Government control in the south was limited to a few larger garrison towns, while most of the countryside was in rebel hands.

1985, April 6. Coup d'État. While President Nimeiry was in Cairo, returning from a visit to the U.S., he was deposed by senior army officers. Defense Minister General **Abdel Rahman Siwar el-Dahab** headed the new government. Guerrilla conflict in the south died away.

1986–1991. Guerrilla Warfare Resumes. In reaction to the Moslem character of the new government, the SPLA resumed guerrilla warfare in the south. Periodic and sporadic famines (1987–1988) both limited military operations and increased civilian suffering. In the northern portions of the southern region, government-sponsored Moslem militias engaged in irregular warfare against neighboring Christian and pagan tribes, reportedly committing atrocities and massacres, and running slaving expeditions.

1989, June 30. Military Coup. The government of Saddiq al-Mahdi was overthrown by army officers, who had been pressing him to affirm a compromise peace arrangement with the SPLA.

1990, March–September. Three Failed Coup Attempts. Government security forces foiled 2 separate coup attempts by dissident army officers (March, April) and noncommissioned officers (September).

1990, August. Southern Rebels Capture Border Garrison Towns. Forces of Garang's SPLA captured the garrison towns of Rath-wulou (on the Zairan border) and Kit Madho (on the Ethiopian border).

Uganda

1976, July 4. Israeli-Entebbe Operation. (See p. 1500.)

1977–1978. Tension with Tanzania. (See p. 1537.)

1978, October 30–November 27. Ugandan Invasion of Tanzania. Ugandan troops penetrated 20 miles into northwestern Tanzania. Tanzania launched counterattacks (November 11). The Ugandans withdrew (November 27).

1979, January–June. Tanzanian Invasion. Tanzanian troops entered Uganda in support of Ugandan exiles attempting to overthrow President Idi Amin Dada. Libyan ruler Colonel Muammar Qaddafi flew troops to aid Amin, but the Libyans and Ugandans were routed by the Tanzanians. A provisional government was formed (April 11–13). Tanzanian forces occupied Kampala and systematically wiped out remaining pockets of resistance (April 18–June 3). Tanzanian troops remained in Uganda for the next 2 years.

1980, May 11. Military Coup d'État. Brigadier General **David Oyite Ojok** deposed the civilian President **Godfrey Binaisa** and established a ruling military junta. Tanzanian troops in Uganda did not intervene. The junta reinstated political parties, then relinquished control of the government (September 17).

1980, October 7–15. Amin Supporters Invade Uganda. Ugandan rebels, loyal to former President Idi Amin, invaded Uganda from Zaire and Sudan. They captured Arua and Koboko, but fled Uganda when Tanzanian troops arrived.

1981. Civil Unrest and Guerrilla Warfare. Following national elections (December 10–11, 1980) civil unrest and guerrilla warfare began in opposition to the government of President **Milton Obote.**

1981, June 30. Tanzanian Troops Withdraw. Tanzanian troops concluded a 2-year peacekeeping mission. President Milton Obote protested the departure of the Tanzanian troops because of guerrilla activity in Uganda.

1985, July 27–29. Milton Obote Deposed. President Obote was deposed by mutinous army forces under the command of Brigadier **Basilio Olara Okello,** and fled to Kenya. Brigadier Okello suspended the constitution, and Lieutenant General Tito Okello (no relation) was named the new ruler (July 29).

1985, December 17. Government Accord with NRA. New Ugandan ruler General Tito Okello concluded an accord with the rebel National Resistance Army (NRA) of **Yoweri Museveni.**

1986, January 15–29. NRA Overthrows Okello. Museveni, frustrated by Okello's failure to control the army, launched an offensive from his base area in southwestern Uganda. NRA troops captured Kampala (January 26) and Jinja (January 27) after heavy fighting. Okello fled to Sudan (January 28). Museveni was sworn in as President on January 29.

1986–1991. Continuing Low-level Insurgency. Low-level guerrilla warfare continued in the north and west of the country. There were many reports of both NRA and guerrilla massacres and atrocities.

1989, January 3–6. Idi Amin Attempts to Return. Former Ugandan dictator Idi Amin traveled to Zaire, equipped with false papers, apparently intending to enter Uganda clandestinely. Zaire refused Uganda's requests for Amin's extradition, and eventually sent him back to exile in Saudi Arabia.

Tanzania

1977–1978. Tension with Uganda. Uganda's President Idi Amin made repeated charges of Tanzanian plans to invade Uganda. President **Julius Nyerere** of Tanzania denied the charges.

1978, October 30–November 27. Ugandan Invasion of Tanzania. (See 1536.)

1979, January–June. Tanzanian Invasion of Uganda. (See 1536.)

1979–1981. Tanzanian Peacekeeping Force in Uganda. (See 1536.)

1982, August 18–19. Tanzanian Intervention in Seychelles. (See p. 1534.)

CENTRAL AFRICA

Burundi

1976, November 1. Coup d'État. The government of President **Michel Micombero** was overthrown in a bloodless military coup led by Lieutenant Colonel **Jean-Baptiste Bagaza.**

1988, August–September. Ethnic Clashes. Following minor violence between the majority Hutu tribe and the dominant but minority Tutsi, the Tutsi-manned army went on a rampage, killing or injuring thousands of virtually unarmed Hutu. Some 50,000 Hutu fled to neighboring countries. Amidst international protests, government reformes encouraged most refugees to return (late 1989).

Rwanda

1990, September 30–October 5. Rwandese Patriotic Front Invasion. A force of 5,000–10,000 exiled Tutsi, based in Uganda and led by Major General **Fred Rwigyenna,** invaded Rwanda, intending to depose the Hutu government and restore Tutsi power (September 30). This mirrored the ethnic conflict in neighboring Burundi in 1988. President **Juvénal Habyarimana** appealed to Belgium for military aid (October 3). The invasion was repulsed with the help of 500 Belgian, 300 French, and 500 Zairan troops. Habyarimana declared a state of emergency (October 8), and later announced a cease-fire and agreed to the return of the exiles (October 18).

Central African Empire (Central African Republic)

1976, December 4. Empire Proclaimed. President Jean-Bedel Bokassa officially renamed the Central African Republic the Central African Empire and proclaimed himself Emperor Bokassa I.

1979, January 20–21. Student Riots Suppressed. A governmental order that university students wear state uniforms resulted in rioting and looting in the capital city of Bangui. Troops from neighboring Zaire helped put down the riots.

1979, September 20. Coup d'État. Emperor Bokassa I was overthrown by former President David Dacko with French help in a bloodless coup d'État. The country was renamed Central African Republic.

1981, September 21. Coup d'État. President Dacko was overthrown in a bloodless coup by the Army Chief of Staff, General **André Lolingoa,** who named himself president.

Chad

1975, April 13–15. Coup d'État. General **Noel Odingar** led a successful coup against the government of Chad's first president, **N'garta Tombalbaye,** who was killed. General **Felix Malloum,** commander of the armed forces, was named head of the Military Council (April 15).

1975, September 27. Malloum Orders French Troops Out of Chad. President Malloum ordered France to withdraw its troops from Chad because France had been negotiating for the release of French nationals held hostage by a Moslem rebel movement, Frolinat, led by **Hissen Habre.**

1975, October 13–October 27. French Withdrawal. This ended 78 years of French military presence in Chad.

1976, April 13. Attempted Coup. Frolinat guerrillas failed in an attempt to assassinate Malloum.

1976–1977. Libyan Intervention. Libyan leader Muammar Qaddafi provided aid for Frolinat rebels in northern Chad and claimed 37,000 square miles of northern Chad (September 9, 1976).

1977, July 18. French Military Assistance. At the request of Chad's government, French transport planes airlifted Chad troops to the battle front in northern Chad. France also sent a few military advisers.

1978, February 1–February 24. Rebel Success. A Frolinat offensive overran nearly 80 percent of the country (February 7). President

Malloum agreed to a cease-fire (February 20), then met with the leaders of Libya, Niger, and Sudan to negotiate a peaceful solution (February 24). But the Frolinat broke the cease-fire by launching a major offensive (April 15).

1978, April 26–June 6. French Intervention. At the request of President Malloum, about 500 troops of the French Foreign Legion intervened on behalf of the government and combined with 1,500 government troops to stop the rebel offensive (June 6). Additional French Foreign Legionnaires soon raised the total of French troops to around 2,500.

1978, August 29. Habre Named Premier. Malloum named guerrilla chieftain Hissen Habre as premier after his faction split with Frolinat.

1979, February 12–17. Attempted Coup. Prime Minister Habre attempted to overthrow President Malloum. French troops helped the government quell the insurrection. Habre fled to the north where he kept his rebellion alive.

1979, March 16–23. National Coalition Government Formed. An agreement was reached between the government, Habre's Frolinat faction, and the rival Frolinat faction headed by **Goukouni Oueddi** to halt civil war and form a coalition provisional government (March 16). It was also agreed that all French troops would be withdrawn. A new government was established with Oueddi at its head (March 23).

1979, April 20. Libyan Invasion Repulsed. Under the pretext of aiding the rebels in northern Chad, Libyan troops invaded. They were repulsed with French assistance. The French began a slow troop reduction.

1980, January 18. Congolese Peacekeepers Arrive. Under the auspices of the Organization of African Unity (OAU), a Congolese peacekeeping force arrived in Chad.

1980, March 22–December 16. Renewed Revolt. Fighting erupted in Ndjamena, the capital city, after Malloum was succeeded as president by Oueddi. The revolt was finally suppressed with the assistance of Libyan troops requested by Oueddi. Habre fled to Cameroon.

1981, January 7. Quaddafi Proclaims Unity of Chad and Libya. This announcement of complete unity was endorsed by Oueddi. Chad was initially occupied by about 6,000 Libyan troops.

1981, October 31. Agreement for Libyan Withdrawal. Oueddi's request for withdrawal was supported by the OAU. Qaddafi reluctantly complied; the Libyan troops were replaced by a smaller OAU peacekeeping force (December 17).

1982, March 21–June 8. Renewed Civil War. After the Libyan departure Habre renewed his rebellion and overthrew Oueddi (June 8).

1983, June–July. Oueddi Renews Civil War. The former President, assisted by Libyan troops, gained control of most of the northern half of the country.

1983, July 30. Government Resurgence. After the rebels seized the Faya-Largeau oasis, Habre directed a counteroffensive to recapture the oasis.

1983, August 3–10. Libyan Intervention. Following a savage air bombardment by Soviet-built Libyan planes, Oueddi's rebels recaptured Faya-Largeau.

1983, August 9–13. France Intervenes. Some 1,000 French paratroopers landed at Ndjamena, then advanced north to Salaland Arada, in the middle of the country. A nonshooting stalemate ensued.

1983, August 17. Cease-Fire. The rival Chad factions agree to a French-arranged cease-fire.

1984, September. France and Libya Negotiate Troop Withdrawal. Mutual withdrawal of troops was agreed. The French troops departed (early 1984), but Libyan troops remained in the Aouzou Strip along the Chad-Libya border.

1986, February–March. Renewed Libyan Incursions. Libyan forces in the Aouzou Strip mounted several major operations to the south, but were repulsed by Chadian forces with French assistance.

1987, March. Chadian Victory. Chadian forces, using pickup trucks equipped with sophisticated Milan and TOW antitank missiles, and employing highly effective hit-and-run tactics, attacked and destroyed the Libyan forces in the Aouzou Strip. The Libyans abandoned much equipment, including artillery,

tanks, air defense weapons, trucks, and a few aircraft. After further border skirmishes, a truce was arranged (September).

1989, April. Coup Attempt Foiled. An attempt to depose President Habré was foiled by loyalist security forces. One coup leader, **Hassan Djamouss,** was killed and the other, General **Idriss Deby** (a former defense minister) fled to Sudan, from which he mounted sporadic guerrilla attacks. Although unsuccessful, the coup marked the disintegration of the northern tribal coalition which had brought Habré to power.

1990, April. Rebel Incursions. Rebel forces, led by exiled former defense minister General Idriss Deby, mounted a series of raids from bases in Sudan.

1990, November 10–December 4. Habré Overthrown. Troops of Deby's Popular Salvation Movement invaded Chad and captured Abéché (November 29). French troops did not intervene. President Hissan Habré fled to Cameroon (December 1), and Deby captured N'Djamena, suspended Parliament, and declared himself president (December 2–4). He promised multiparty democracy, but in light of his ties to Libya, the U.S. organized an airlift to Nigeria for Libyan exiles in Chad (December 6).

Zaire

1989, February–March. Student Unrest. University student unrest was triggered by discovery of a student's body near a military base. Despite rumors that several dozen student demonstrators had been killed by security forces, the government admitted to only 1 death, and closed the universities in response to protests (February 15–March 15).

WEST AFRICA

Angola

1975, January 15. Alvor Agreement. The 3 major opposing factions in Angola met at Alvor, Portugal: Dr. **Agostinho Neto**'s Popular Movement for the Liberation of Angola

(MPLA); **Holden Alvaro Roberto**'s National Front for the Liberation of Angola (FNLA); and **Jonas Savimbi**'s National Union for the Total Independence of Angola (UNITA). They agreed upon a coalition government. Despite this, fighting continued through the spring and summer, punctuated by cease-fires and truces.

1975, March–April. Soviet Support to MPLA. Massive shipments of arms arrived for Neto's faction. Several hundred Cuban military "advisers" also arrived to support Neto.

1975, August 9. Coalition Government Collapses. UNITA and FNLA withdrew from the government, leaving Neto and MPLA in control. Intensified civil war resumed.

1975, August 14. South African Intervention. P. W. Botha, defense minister of the Republic of South Africa, seeking to assure the defeat of Neto's MPLA faction, sent an armored task force into southern Angola. The South Africans twice defeated Cuban troops.

1975, December 12. Battle of Bridge 14. Despite a 3-to-1 advantage in troop strength, Cuban troops were again decisively defeated by the South Africans. By this time there were approximately 7,500 Cuban combat troops in Angola.

1976, February 8. FNLA Crushed. Neto's troops and their Cuban allies defeated FNLA forces and took their major stronghold, Huambo. Roberto and many of his followers fled to neighboring Zambia.

1976, April 30. South Africans Withdraw from Angola. The South African government reached an agreement with MPLA guaranteeing protection of South African economic interests in Angola, provided South African combat troops were withdrawn from Angola. This was a major setback for Savimbi's UNITA.

1976, June–August. Cuban Offensive. Cuban troops launched a combined arms offensive against Savimbi troops hoping to crush UNITA as a rival force in Angolan politics (June 3). The Cubans drove Savimbi's troops from central Angola into the southeast corner of Angola.

1976, September–November. Renewed Cuban Offensive. In an attempt to end all resistance to MPLA rule of Angola, an estimated 15,000 Cuban combat troops launched an offensive against UNITA strongholds in southern Angola. They had only limited success and withdrew.

1976–1990. Savimbi's War of Insurgency. MPLA forces and their Cuban allies were frustrated by UNITA's guerrilla tactics. Savimbi's troops successfully attacked supply routes and raided MPLA camps. Their efforts were aided by South African intervention in southern Angola (1975–1982, 1983–1984; see p. 1543). By the early 1980s, Savimbi's forces had gained effective control of the southeastern third of Angola, and cut the Benguela-Zaire railway several times. Despite these successes, UNITA was unable to win a decisive victory, but neither could the MPLA defeat UNITA, even with Cuban aid.

1989–1990. Tentative Peace in Angola. An Angolan peace settlement, brokered by several African countries, and negotiated at Kinshasha, Zaire (July 1989), broke down amid mutual recriminations between Savimbi and Angolan President **José Eduardo dos Santos.** A new peace effort, supported by both the U.S. and the Soviet Union, met with more success (May–December 1990). Savimbi announced a tentative agreement at a peace conference in Washington (December 14).

1989, January–1991, July. Cuban Withdrawal. As part of efforts toward a peace settlement in Angola, and negotiations relating to Namibian independence (see below), Cuba began to withdraw its 50,000 troops in Angola. Complete withdrawal was due to be completed in July 1991.

People's Republic of Benin (Dahomey)

1975, January 21. Attempted Coup. A brief rebellion led by the minister of labor was put down by government troops.

1975, October 18. Coup Plot Thwarted. Security forces discovered and suppressed a plot to overthrow President Mathieu Kerekou.

1977, January 16. Attempted Coup.

Burkina Faso (Upper Volta)

1980, November 25. Coup d'État. President Sangoule Lamizana was overthrown in a military coup and replaced by Colonel **Saye Zerbo.**

1982, November 17. Coup d'État. Zerbo was overthrown by dissident soldiers led by Major General **Jean-Baptiste Ouedraogo.**

1983, August 5. Coup d'État. Ouedraogo was ousted by former Premier **Thomas Sankara.**

1985, December. Border Clashes with Mali. These sprang from a quarrel over the mineral-rich Agacher Strip, which was also the source of minor conflict in 1974–75. After 5 days of skirmishing, the quarrel was referred to the International Court of Justice at The Hague, and a settlement was eventually reached, dividing the disputed area (1988).

1987, October 15. Coup Topples Sankara. He and 8 others were killed in a coup led by fellow National Revolutionary Council (CNR) member Captain **Blaise Campaoré,** who headed a triumvirate, including fellow CNR members Major **Jean-Baptiste Lingari** and Captain **Henri Zongo,** called the Popular Front (FP).

1989, September–December. Coup Attempts. President Campaoré's fellow FP members Lingari and Zongo were executed by firing squad, allegedly for planning a coup (September 14). The government later announced it had detected and foiled another plot (December 26).

Cameroon

1976, November 16. Border Clash with Gabon. (See below.)

1984, April. Unsuccessful Military Coup. Dissident army officers mounted a coup attempt against President **Paul Biya,** but loyal army forces defeated the rebels.

People's Republic of the Congo (Congo, Brazzaville)

1978, August 14. Attempted Coup. Plans to overthrow the military government of President **Joachim Yombi Opango** were thwarted.

1990, May 22. Political Unrest. After the discovery of the body of opposition leader Joseph Rendjambe, antigovernment protesters rioted in Libreville and Port Gentil. French troops in Gabon (1 Marine infantry regiment) evacuated French citizens, and order returned in early June.

Ivory Coast (Cote D'Ivoire)

1990, May 14–16. Military Unrest. Some 1,000 army conscripts mutinied in Abidjan, demanding higher pay (May 14). Two days later, air force personnel seized the airport, joined by hundreds of soldiers. Most returned to their barracks following government concessions (May 16–17).

Gabon Republic

1976, November 17. Border Clash with Cameroon. Gabon's common border with Cameroon was closed following a Cameroon rocket attack launched on a police post.

1978, June 6–1979, August 14. Peacekeeping Force Sent to Zaire.

Gambia

1981, July 30. Attempted Coup. A coup attempted by paramilitary police and armed civilians enjoyed initial success until Senegalese troops intervened to put down the revolt.

Ghana

1975, December 23. Attempted Coup.

1978, July 5. Coup d'État. Ghana's head of state, General Ignatius Kutu Acheampong, was deposed by his deputy, Lieutenant General **Fred Akuffo.**

1979, May 15. Attempted Coup. A coup attempt by Flight Lieutenant **Jerry Rawlings** was unsuccessful, resulting in Rawling's imprisonment.

1979, June 4. Coup d'État. After being freed from prison by rebels, Rawlings led a successful coup against General Akuffo's government. A civilian government was installed.

1981, December 31. Coup d'État. Rawlings seized complete control from the civilian government, claiming corruption and incompetence.

Guinea

1979, April 17. Troops Sent to Liberia. (See below.)

1980, January 18. Peacekeeping Forces Sent to Chad.

Guinea-Bissau

1975, March 30. Attempted Coup.

1980, November 14. Coup d'État. President **Luis de Almeida Cabral** was ousted by Premier **Joao Bernardo Vieira.**

Liberia

1979, April 14–15. Price Increase Riots. Following the government's proposed increase in the price of rice, demonstrations and riots flared in the capital of Monrovia. Order was restored with the help of 200 troops from neighboring Guinea.

1980, April 12. Coup d'État. Sergeant **Samuel K. Doe,** leading enlisted men of the Liberian Army, overthrew President **William R. Tolbert,** promising an end to corruption in government. Tolbert and 27 others were killed while soldiers went on a looting spree in Monrovia for several days.

1989, December. Coup Attempt Suppressed. Forces loyal to President Doe defeated a coup attempt; afterwards hundreds of refugees fled to the Ivory Coast.

1990, January. Civil War Begins. The National Patriotic Front (NPF) guerrilla army, based in Guinea and led by **Charles Taylor,** invaded Liberia to depose Doe. Taylor was a former finance minister in Doe's government, and his effort posed a real threat to Doe's shaky control. As NPF troops advanced, the war acquired inter-tribal overtones as conflict grew between Doe's Krahn tribe and the rival Gio and Mano tribes.

1990, June 4. U.S. Navy Task Force Arrives. A U.S. Navy force, carrying 2,000 Marines, arrived off Monrovia to conduct an evacuation of U.S. citizens from Monrovia, should such be needed.

1990, July 2. Fighting in Monrovia. Doe's forces controlled little outside of the capital, and the Presidential Palace was besieged. The political situation grew more complex when NPF leader **Prince Yormie Johnson** split with Taylor and began his own faction (July 6). Government troops massacred 600 Gio and Mano refugees at a Lutheran Mission in Monrovia (July 30), inflaming ethnic rivalries.

1990, August 5–8. Western Civilians Evacuated. Following Johnson's threat to take Western hostages (August 4), U.S. Marines mounted a 3-day evacuation effort, extricating U.S. embassy personnel and some 125 civilians. Johnson, meanwhile, captured 22 civilians (August 6), but released them unharmed the next day.

1990, August 24. International Intervention. A 3,000-man force from the Economic Community of West African States (ECOWAS) landed in Monrovia to restore order and help arrange a compromise political settlement. This ECOWAS Monitoring Group (ECOMOG) included contingents from Nigeria, Ghana, Gambia, Guinea, and Sierra Leone, commanded by Ghanian General **Arnold Quainoo.** ECOMOG gained control of much of the capital in scattered fighting (August 25–September 2).

1990, September 9–10. Capture and Death of Doe. While traveling to ECOMOG headquarters to arrange his departure, Doe was intercepted and captured by Johnson's troops (September 9), and later killed (September 10).

1990, September–November 28. Continued Fighting. Factional fighting continued in Liberia after Doe's death between ECOMOG, Johnson's and Taylor's factions of the NPF, and forces loyal to Doe's defense Minister, Brigadier

General **David Nimblay.** ECOWAS finally pressured all parties into accepting a cease-fire negotiated at Bamako, Mali (November 26–28). There was, however, no political settlement, many Liberians were starving, and the threat of renewed civil war remained (early 1991).

Mali

1985, December. Border Clash with Burkina Faso. (See p. 1541.)

Mauritania

1975, December 10. Mauritania Seizes Southern Sahara. While Moroccan troops were seizing the northern ⅔ of the former Spanish colony of Sahara (see p. 1548), Mauritanian troops seized Southern Sahara. They were immediately attacked by indigenous guerrilla forces seeking independence for Sahara: the Popular Front for the Liberation of Saqiat al Hamra and Rio de Oro (Polisario). A fierce 3-way struggle for the control of Sahara ensued, with Polisario guerrillas fighting both Mauritanians and Moroccans.

1978, July 10. Coup d'État. President **Moktar Ould Daddah** was overthrown in a bloodless coup led by Army Chief of Staff Colonel **Mustapha Ould Salek.** Salek began negotiating with the Polisarios for a cease-fire in Sahara. A peace treaty was signed the following year and all Mauritanian troops were withdrawn from the Sahara (August 5, 1979).

1980, January 4. Coup d'État. President **Mohammed Mahmoud Ould Luly** was overthrown by Premier **Mohammed Khouna Ould Haidalla** in a bloodless coup. Luly had been in power for only 6 months following Salek's resignation (June 4, 1979).

1981, March 16. Attempted Coup. A coup attempt by Mauritanian military exiles was put down by the military government. Morocco was blamed for the attempt because of anger over Mauritania's backing out of the Polisario war.

1989, April–August. Border Clash with Senegal. Following a brief trade war with Senegal (January), violence flared after 2 Senegalese peasants were killed by Mauritanian border police (April 9). Ensuing clashes, and widespread atrocities by both sides, left at least 300 dead and perhaps 300,000 refugees (April–August). Diplomatic relations with Senegal were broken (August). Although violence died away, the basic disputes remained a problem into 1991.

Niger

1976, March 15. Attempted Coup.
1983, October 6. Attempted Coup. Easily suppressed by troops loyal to President **Seyni Kountche.**
1989, May. Clashes with Tuareg Separatists. Nomadic Tuareg tribesmen clashed with army forces near Tchintabaraden in the north (early May). Government forces reported 6 soldiers and 31 Tuareg killed, but international news reports claimed army troops had killed hundreds of nomads in reprisals (June–August). The government invited Amnesty International to investigate (September).

Nigeria

1975, July 29. Coup d'État. General **Yakubu Gowon** was deposed in a bloodless military coup, and replaced by Brigadier **Muritala Rufai Mohammed.**
1976, February 13. Attempted Coup. A military coup was suppressed, but rebels assassinated Nigerian head of state General Muritala Rufai Mohammed. Lieutenant General **Olusegun Obsanjo** was chosen as the new chief of the Supreme Military Council.
1983, December 31. Coup d'État. President **Shehu Shagari** was overthrown in a military coup by Major General **Mohammed Buhari,** ending 4 years of civilian rule.
1985, August 27. Coup d'État. General Buhari was deposed by Major General **Ibrahim Babangida.** This was Nigeria's sixth coup in 20 years.

1989, May. Popular Unrest. Government austerity measures led to extensive riots (late May). The government responded with mass arrests, and closed 6 universities for a year.

1990, April 22. Attempted Coup. Disaffected army officers, led by Major **Gideon Orkar** attempted to depose General Babangida. Loyalist troops defeated the rebels and restored order; Orkar and 42 others were convicted and executed by firing squad (July 27). The coup attempt was provoked by the dismissal of Lieutenant General Domkat Yah Bali, a Christian, in a cabinet reshuffle (December 29, 1989), which led to fears of "Moslemization" in the armed forces.

Senegal

1981, July 30. Assistance to Gambia. Senegalese troops assisted Gambia in the suppression of a coup attempt (see p. 1541).

1989, April–August. Border Clashes with Mauritania. See above, under Mauritania.

1990, May 18–22. Border Clash with Guinea-Bissau. A territorial dispute led to 2 major clashes (May 18–19, 22) when Senegalese troops crossed the border. Several soldiers on both sides were killed before both countries agreed to withdraw their forces from the border area.

1990, June 19. Attack by MFDC Guerrillas. Guerrillas of the outlawed *Mouvement des Forces Démocratiques de Casamance* struck at security forces, wounding 9 Senegalese.

Togo

1978, June 19–1979, August 14. Peacekeeping Force in Zaire.

SOUTHERN AFRICA

Mozambique

1975–1991. Guerrilla War. Mozambican troops waged a counterinsurgency struggle against the National Resistance Movement (NMR, or Renamo), which received South African support through 1984–1985. Subsequently many Renamo forces turned to terrorism and brigandage to support themselves. To maintain the railway link through Mozambique to Sofala (Beira), several thousand Zimbabwean troops were deployed in Mozambique after 1980–1981. Mozambique provided support for the military wing of the African National Congress (ANC, outlawed in South Africa). By late 1990, the war had caused an estimated 600,000 war-related deaths, with at least again as many dead from war-caused starvation, and more hundreds of thousands driven from their homes by the fighting.

1975, December 17–19. Coup Attempt. An attempted coup by dissatisfied army officers and policemen was thwarted after 2 days of bloody fighting.

1984, March 16. N'komati Accords with South Africa. Hard-pressed by Renamo, President **Samora Machel** and President Botha of South Africa agreed that in return for Mozambique abandoning support of the ANC in South Africa, South Africa would end support for Renamo. The extent to which South Africa fulfilled the bargain was unclear.

1986, October 19. President Machel Killed in Air Crash. The president and 25 others were killed when their plane crashed in South Africa; an ANC spokesman charged that South Africa had sabotaged the plane. Machel was succeeded by **Joaquim Chissanó.**

1989, May–September. Government Initiatives. Following on guerrilla success in late 1988 and early 1989, the government of President Chissanó embarked on a new policy. While moving steadily away from one-party Marxism and toward a multiparty free-enterprise system, Chissanó also opened a major antiguerrilla offensive in Sofala province.

1990, July–December. Peace Initiatives. Chissanó's policies elicited some response from Renamo's leadership. Government and guerrilla representatives met in Rome (July), and after 3 weeks negotiations concluded a partial cease-fire (December), raising hopes for a settlement in 1991.

Lesotho

1982, December 9. South Africa Raids ANC Base. (See p. 1546.)

Zimbabwe-Rhodesia

1975–1976. Guerrilla Warfare Intensifies. There were two main forces operating in Rhodesia against Prime Minister Ian Smith's minority white government: **Joshua Nkomo**'s Zimbabwe African People's Union (ZAPU) and **Robert Mugabe**'s Zimbabwe African National Union (ZANU). Nkomo negotiated with Smith throughout 1975, but when no progress was made he renewed guerrilla activities, operating out of neighboring Zambia. Continued pressure was also maintained by Mugabe's insurgents operating out of Mozambique. Both Mugabe's and Nkomo's groups received substantial aid from the Soviet Union. Economic sanctions and diplomatic pressures from Western democracies forced Ian Smith to accept in principle the concept of majority rule for Rhodesia, and to schedule elections. Bishop **Abel Muzorewa,** leader of moderate opposition to Smith's minority white government, returned to bolster support before elections (October 1976).

1979, April 17–24. Muzorewa Wins Election. In an election boycotted by both Mugabe and Nkomo, Bishop Muzorewa emerged the clear winner. But ZAPU and ZANU continued guerrilla warfare.

1979, September 10–24. London Conference—British Rule Established. In an effort to defuse the situation in Rhodesia, the government of Prime Minister Margaret Thatcher intervened. Agreement of all parties was reached to return Rhodesia to colony status, and elections were scheduled on terms accepted by Mugabe and Nkomo.

1980, February 27–29. Mugabe Wins Election. Robert Mugabe's ZANU party won a huge majority of votes and seats in the new parliament.

1986. Unrest in Matabeleland. Nkomo's ZAPU party, based largely among the Matabele of southwestern Zimbabwe, was frustrated by ZANU's lock on national politics. Scattered armed clashes between security forces and armed ZAPU supporters occurred throughout the year, threatening on several occasions to erupt in outright civil war. Although fighting had died away by late 1986, the crisis was not finally resolved until ZANU and ZAPU concluded a merger agreement (December 1987).

Republic of South Africa

1975–1982. South African Operations in Angola and Namibia. South African troops carried out several large-scale interventions in the Angolan civil war on behalf of Jonas Savimbi's UNITA forces (see p. 1540), as well as continuing counterinsurgency operations against SWAPO bases in northern Namibia and Angola (see p. 1546).

1976, June 16. Soweto Incident. In Soweto, Southwestern townships outside Johannesburg, security forces opened fire on a crowd of demonstrators protesting government apartheid policies, killing dozens. This incident sparked 3 weeks of riots and protests before order was restored.

1976, June–1991. Violence in the Townships. The black-populated suburban township of South Africa's major cities suffered considerable unrest following the Soweto riots. Scattered antigovernment protests, some violent, continued until the late 1980s. By 1989–1990, liberalization of apartheid statutes reduced black protests against white rule. However, disputes between the Xhosa-based, Marxist-oriented ANC (and sometimes their more moderate United Democratic Front (UDF) allies) and the western-oriented, Zulu-based Inkatha organization (headed by Zulu chief **Mangosuthu Gatsha Buthelezi**), exacerbated by ethnic rivalries, produced increasing violence among black South Africans. Some of these clashes produced local civil wars, especially in Natal, where Xhosa and Zulu live close together. Black-on-black violence had claimed at least 4,000 lives by December 1990, and continued into 1991.

1976–1981. "Independence" of Homelands. Four South African tribal homelands, Transkei (October 26, 1976), Bophuthatswana (December 6, 1977), Venda (September 13, 1979), and Ciskei (December 4, 1981), were granted theoretical independence from South Africa. In practice, they were dependent on South African financial, technical, and defense support. Furthermore, most black leaders, especially in the ANC, regarded them as an effort to "divide and conquer" black opposition. No country besides South Africa has recognized them as sovereign states.

1981–1982. Counterinsurgency Raids into Mozambique. South African commandos staged a number of successful raids on South African National Congress bases in Mozambique.

1982, December 9. Counterinsurgency Raids into Lesotho. South African commandos raided suspected ANC bases in Lesotho. The headquarters was discovered in a residential area, and a number of civilians were killed as well as ANC guerrillas.

1983, May 20–23. Terrorist Attack and Retaliatory Air Strike. A bomb explosion outside the South African Air Force headquarters killed 18 civilians and wounded 200 others (May 20). South Africa retaliated by an air attack on a suspected ANC base in Matola, a suburb of Maputo, Mozambique. Mozambique charged that the attack had hit only civilian businesses and homes. South Africa disputed this claim, asserting that they had hit an ANC guerrilla base.

1983, December 6–1984, January 8. Counterinsurgency Operations in Angola. South Africa troops attacked SWAPO positions in southern Angola, smashing the staging area for a planned guerrilla operation in Namibia. They continued their offensive. At Cuvelai the South Africans defeated a combined Cuban, Angolan, and SWAPO force, killing 324, while suffering only 21 casualties (January 3–5).

1984, January 17–March 16. Pact with Mozambique. Representatives from South Africa and Mozambique met at Komatiport on the Komati River and after extensive negotiations agreed that they cease supporting insurgents in the others' country.

1984, January 31. South Africa Begins Withdrawal from Angola.

1984, February 23. SWAPO Truce Set. The government of South Africa and SWAPO leader **Sam Nujoma** agreed to a truce in Namibia.

1985, July 20–1986, March 7. State of Emergency. In response to growing inability of security forces to control black townships, President Botha declared a nationwide state of emergency, granting the security forces extraordinary powers. The U.S. responded with economic sanctions (July 31, 1985). Nearly 7,000 detainees were released when the state of emergency ended.

1986, May 19. South Africa Attacks ANC Guerrilla Bases. South African air force planes and army special forces units attacked ANC guerrilla bases and strongholds in three neighboring nations—Zimbabwe, Botswana, and Zambia.

1986, June 12. Nationwide State of Emergency. This gave virtually unlimited powers to the security forces to suppress demonstrations and unrest.

1988, February 10. Coup in Bophuthatswana. A group of dissident military personnel ousted President Lucas Mangope, but were defeated when South African forces intervened.

1990, February 2. ANC Unbanned. Along with the South African Communist party (SACP) and 33 other anti-apartheid organizations, the ANC was allowed to resume legal activities. This move by President **F. W. de Klerk** was followed by the release of imprisoned ANC leader **Nelson Mandela** (February 11). These events sparked unrest in 3 of 4 "homelands" as they unbanned antiapartheid groups and moved toward reintegration with South Africa.

1990, April. Air Force Headquarters Raided. South African Air Force headquarters in Pretoria was raided by ultra-rightists, led by **Piet**

Rudolf. The robbers made off with automatic weapons, but Rudolf was arrested later (September).

1990, May 4. Groote Schuur Minute. After extensive deliberation, and prodded by Mandela, the ANC and SACP announced they would make efforts to limit violence and work toward lifting the current state of emergency. This was followed by the Pretoria Minute (August 8), in which the ANC and SACP renounced armed struggle.

1990, July–September. Inkatha "Offensive" in Transvaal. Armed members of Inkatha moved to strengthen their hold over black townships surrounding Pretoria and Johannesburg, leading to some 800 deaths. The ANC protested, and crowds as large as 80,000 gathered around migrant laborer hostels, which were Inkatha bases.

1990, September 15. Government Response. In an effort to halt black-on-black violence, the government began Operation Iron Fist, declaring 15 townships emergency zones. Curfews were imposed (but lifted within weeks), and police vehicles were allowed to mount machine-guns.

Namibia (Southwest Africa)

Britain's League of Nations Mandate over the former German colony of Southwest Africa after World War I was administered by the Union of South Africa, which later became the Republic of South Africa. South Africa continued to administer the region when the former mandates became U.N. trusteeships after World War II. Popular dissatisfaction with South African rule, and doubts whether South Africa would ever grant independence to the region, led to the establishment of the Southwest African People's Organization (SWAPO) by independence-minded local leaders (June 1960). SWAPO began to conduct local guerrilla activities against the South African administration, receiving considerable assistance and training from the U.S.S.R. Despite stern repressive measures, insurgency grew and guerrilla warfare continued during the 1960s and early 1970s.

1975, September 1–12. Turnhalle Conference. As a step toward autonomy, and presumably ultimate independence, the South African government convened a constitutional conference at Turnhalle. Participants were tribal leaders and other local leaders (including representatives of ethnic black groups and of the tiny white community). Multi-ethnic political parties (obviously including SWAPO) were excluded, thus effectively barring most educated and westernized blacks. The results were totally unsatisfactory to SWAPO; guerrilla war continued.

1978, December 4–8. Elections Held. In preparation for independence, South African-supervised elections were held. There were large turnouts, as tribal members did their chiefs' bidding and voted. SWAPO boycotted the elections. The SWAPO boycott assured the electoral triumph of the South African-backed Democratic Turnhalle Alliance.

1978–1989. Insurgency. South Africa continued to operate effectively against the insurgents and administer Namibia in preparation for complete independence. SWAPO guerrillas carried out hit-and-run raids into northern Namibia from bases in southern Angola. This led to repeated South African operations in Angola (see p. 1545).

1989, November 15–December 22. Agreement on Namibian Independence. After extensive negotiations, U.S., South African, Cuban, and Angolan representatives meeting in Geneva agreed on a timetable for Namibian independence and for Cuban withdrawal from Angola (November 15). Final signing of the agreement took place at U.N. headquarters in New York (December 22).

1990, March 21. Namibian Independence. This followed from the Geneva agreement, and a cease-fire and disengagement of forces which took place under U.N. supervision (April 1– November 16, 1989).

NORTH AFRICA

Algeria

1976–1978. Tension with Morocco and Mauritania. Algeria's support of the Polisarios resulted in sporadic border skirmishes with Morocco and Mauritania (see p. 1543).

1988, October–1989, February. Unrest and Reform. Over 500 were killed in antigovernment rioting (October). Subsequently, President **Chadli Benjedid** was reelected for a third 5-year term, and a new, liberal constitution was adopted by popular vote (February). Significant political and economic reforms were effected.

1991, April. Algerian Nuclear Weapons Program Revealed. Speaking off the record, U.S. officials disclosed that Algeria, with Chinese assistance, was building a nuclear reactor for weapons production. This program was believed to be a counterpoise to the Libyan ballistic missile program.

Libya

1977, July 12–24. Border Conflict with Egypt. (See p. 1508.)

1981, August 19. U.S. Planes Down Libyan Planes. During U.S. Sixth Fleet naval exercises in the Gulf of Sidra, in the southern Mediterranean, two U.S. navy F-14s were attacked by two Libyan SU-22s. The Americans shot down both Libyan planes.

1ᶜ81–1983. Libyan Intervention in Chad. (See p. 1539.)

1984, April 30. Great Britain Breaks Diplomatic Relations with Libya.

1986, January 8. U.S. Embargo. After linking Libya to a worldwide terrorist campaign, President Reagan issued an executive order freezing Libyan government assets in the U.S. and in U.S. banks, and prohibiting trade and travel by U.S. citizens to Libya.

1986, April 16. U.S. Warplanes Attack Libya. Launched in retaliation for the Libyan-sponsored terror-bombing of a West Berlin disco frequented by U.S. military personnel that killed 2 and injured 200 (April 5), the raid hit government and military targets at Tripoli and Benghazi, including Libyan leader Qaddafi's residence. Qaddafi was not harmed, but Libyan casualties included several civilians. The U.S. lost 1 F-111 fighter-bomber and 2 airmen.

1987, September. OAU-sponsored Cease-fire Halts Fighting between Libya and Chad. (See p. 1539.)

1988, October 3. End of Libya-Chad War. The two nations agreed to cooperate with OAU in determining the status of the mineral-rich Aouzou Strip in northwestern Chad, which Libya continued to occupy.

1989, January 4. Libyan Fighters Downed by U.S. Warplanes. U.S. Navy F-14 jets shot down 2 Libyan MiG-23 fighters that closed on them with "clear hostile intent" over the Mediterranean, 70 miles north of Tobruk. Meanwhile, U.S.–Libyan tensions escalated when the U.S. accused Libya of building a chemical weapons plant at Rabta, near Tripoli. Libya claimed that the plant was a pharmaceuticals facility.

1990, March 15. Purported Fire at Libyan Chemical Plant. Libya simulated a destructive fire at its Rabta chemical plant. The hoax was apparently designed to avert a U.S. air strike.

1990, December. Libya Backs Overthrow of Habré Regime in Chad. (See p. 1539.)

Morocco

1975, December 11. Morocco Moves into West Sahara. Moroccan troops occupied West Sahara (former Spanish colony of Rio de Oro). The Moroccan government laid claim to the northern ²/₃ of the former colony. Morocco's claims were violently disputed by an indigenous movement, the Popular [Front] for the Liberation of Saqiat al-Hamra and Rio de Oro (Polisario). At the same time Mauritania occupied the southern portion of Sahara (see p. 1543).

1976–1984. Moroccan-Polisario War Continues. The Moroccan-Polisario conflict continued unabated. The Moroccans, despite a commitment of 90,000 troops, were unable to suppress the Polisario, while the Polisario were not able to establish effective rule over the region.

1986, August. King Hassan Abrogates Libyan-Moroccan Treaty of Unity. The 2-year-old treaty fell victim to ideological differences between the two countries' leaders.

1988–1991. Western Sahara Dispute. By 1986, Morocco had consolidated control over approximately ⅔ of the contested Western Sahara, and the fortunes of the Polisario Front were waning. Algerian support for Polisario effectively ended with the formation of the Arab Maghreb Union, a regional organization that included both Morocco and Algeria (February 1989). A year of relative quiet in the war zone was briefly disrupted (September–October 1989) by Polisario attacks launched from Mauritania. U.N.-sponsored peace talks between Morocco and the Polisario Front agreed that a referendum to determine the status of the Western Sahara would be held in 1991.

Tunisia

1985, October 1. Israeli Aircraft Bomb PLO Headquarters in Tunis. (See p. 1501.)

1987, November 7. Bourguiba Removed from Power. The 83-year–old leader, who had been president of Tunisia since 1956, was increasingly infirm and apparently senile. He was replaced by Premier **Zine el-Abidine Ben Ali,** a career army officer.

1988, April 16. Assassination of PLO Military Chief in Tunis. Khalil al-Wazir (Abu Jihad) and 3 others were gunned down by masked attackers. The attack was evidently carried out by Israeli intelligence and special operations forces. Al-Wazir was coordinator of the *Intifada* in the Israeli-occupied territories.

NORTH AMERICA

UNITED STATES

1975, February–June. Middle East Peace Efforts. Secretary of State **Henry Kissinger** made a concentrated effort to work out a Sinai Peninsula troop disengagement scheme between Israel and Egypt while also trying to get the two countries and other Middle East nations to convene a general Middle East peace conference in Geneva. Kissinger made a number of trips to the Middle East and also met with Soviet Foreign Minister **Andrei Gromyko** in an effort to involve the Soviet Union in the planned Middle East Geneva Conference (February 16–17). It was largely through Kissinger's efforts that Israel and Egypt eventually worked out a Sinai Disengagement plan (see p. 1502).

1975, February 16–July 21. Mutual Bal- anced Force Reduction (MBFR) Talks. This was the beginning of an effort to achieve equivalent reductions of NATO and Warsaw Pact forces in Central Europe.

1975, May 12–14. *Mayaguez* **Incident. (See p. 1527.)**

1977–1979. U.S. Efforts Toward Middle East Peace. U.S. President **Jimmy Carter** devoted considerable attention to efforts to bring about peace in the Middle East. High points of these efforts were the Camp David Accords and the subsequent Egypt-Israel Peace Treaty–Treaty of Washington (see p. 1504).

1977, March 9. Planned U.S. Troop Withdrawal from Korea. President Carter announced that all U.S. ground troops (32,000) would be withdrawn from South Korea within 4 or 5 years. Soon after this, Army Chief of Staff General **Bernard W. Rogers** announced that the U.S. had begun removing tactical missiles from South Korea and that South Koreans had begun manning some U.S. air defense missiles deployed in South Korea (April 14). The Carter announcement touched off a controversy involving Major General **John K. Singlaub,** chief of staff of U.S. forces in South Korea, who criticized Carter's plan. Singlaub was recalled to Washington and reassigned (May 27). The planned troop withdrawal began later in the year, but was canceled by President Carter after the Soviet Union invaded Afghanistan (see p. 1519).

1977, June 30. Production of the B-1 Bomber Stopped. President Carter canceled production plans for the proposed B-1 strategic bomber because he thought it was not cost-effective and because he and Secretary of Defense **Harold Brown** thought the cruise missile would be a more effective and cheaper deterrent.

1978, April 7. Carter Defers Production of the Neutron Bomb. Following months of urging NATO allies to declare that they would deploy enhanced radiation nuclear weapons ("neutron bomb"), President Carter announced he would not produce the controversial weapon. Proponents of the weapon argued that its radiation would be very effective in penetrating Soviet tanks and stopping a Soviet combined armed offensive on NATO's central front, while causing little damage to nearby structures. Critics questioned this and claimed that radiation from the weapon would threaten civilians in NATO countries. Later that same year (October 18), Carter partially reversed his decision on the neutron bomb and gave the go-ahead to manufacture some parts of the weapon.

1979, January 22. Carter Budgets MX Missile, Trident Submarine, and Cruise Missile. Under growing criticism of his defense policies from influential critics such as Senator Henry Jackson, President Carter requested $675.4 million for the MX (Missile Experimental) mobile ICBM. The proponents of adding the MX to the U.S. strategic arsenal argued that it would strengthen the strategic triad (land-based ICBMs, nuclear missile armed submarines, and long-range strategic bombers), which was seen as increasingly vulnerable because of qualitative improvements in the Soviet ICBM force (see p. 1480). In addition to the MX, President Carter requested funding for one submarine armed with Trident nuclear missiles and low-altitude cruise missiles and hardening of the silos for the existing ICBM force. Critics faulted Carter for denying navy requests for a new super-carrier and for deciding to phase out the large carrier force. Critics also scorned Carter for considering submarines as a possible basing mode for the MX because greater accuracy could be achieved from the land-based mode. Carter later decided upon the controversial horizontal "race-track" basing mode, in which the missiles would be shuttled from point to point along five oval concourses with a number of the underground silos always remaining vacant. This so-called "shell-game" approach was to complicate targeting for the Soviets.

1980, March 1. Development of a Rapid Deployment Force. The Rapid Deployment Task Force (RDTF) was to be a 100,000-man force composed of units from the Army and Marine Corps, whose mission was to respond and deploy quickly, with naval and air support, to crisis situations around the world.

1980, May 6. MX Basing Mode Changed. The Carter administration abandoned the controversial racetrack basing mode for a straight track basing mode because of the opposition of environmentalists and residents of Utah and Nevada. This did not silence opponents, and the problem of finding the proper basing mode was not resolved.

1980, August 5. Change in U.S. Plans for Nuclear Targeting. In Presidential Directive 59 (PD 59), the emphasis in targeting for the U.S. strategic nuclear force was changed from a countervalue to counterforce strategy. Countervalue calls for targeting industrial facilities and population centers, whereas counterforce is directed against Soviet nuclear missiles.

1980, August 20. Development of "Stealth" Bomber Revealed. The Defense Department revealed planned development of a new, long-range strategic bomber capable of penetrating Soviet defenses by evading radar. The aircraft would not be operative until the 1990s. Republican presidential candidate Ronald Reagan charged that the Carter administration had compromised national security by a premature leak of sensitive information for political gain, and that the announcement was made to offset Reagan's criticism of cancellation of the B-1 bomber (September 4).

1981, August 10. U.S. Reversal on Neutron Bomb. U.S. Secretary of Defense **Caspar Weinberger** announced President Reagan's decision to proceed with full production of enhanced radiation weapons. They could be deployed as warheads for Lance missiles or as 8-inch artillery shells. Reagan, however, decided not to ship it to Europe until a later date, when the Allies would not object to its deployment or when it was needed in a crisis situation.

1981, October 2. New Strategic Missile Deployment Plan. President Reagan announced that he was reversing former President Carter's cancellation of the B-1 bomber. Production of 100 of these strategic bombers would serve to replace the aging B-52 force until the "Stealth" strategic bomber was operational in the 1990s. He also announced a new basing mode for the MX missile (see p. 1550). The first 36 of the 100 planned missiles would be deployed in existing superhardened silos (reinforced with concrete and steel). The decision on how to deploy the remaining missiles would be deferred until completion of further study. The Reagan program called for the development of improved Trident nuclear missile submarines with more accurate and more powerful missiles. Also the plan called for improved strategic defenses and for improved command, control, and communications systems.

1982, June 2–5. Conference on "Military Reform Movement." Widespread criticism of U.S. and NATO doctrine and strategy among

B-52 Bomber (*U.S. Air Force photo*)

members of Congress (who had established an informal "Reform Caucus") and many Defense intellectuals culminated in a 3-day conference at West Point, N.Y. The "Reformers" had been asserting for several years that U.S. doctrine, particularly in the army, was based upon a concept of "attrition warfare," in contrast to the ideas of the Reformers, who favored "maneuver warfare." The Reformers were also opposed to the "Forward Defense" concept in NATO. Although few opinions were changed among the participants, the ferment unquestionably influenced changes in doctrine in the U.S. Army, which appeared about this time, putting emphasis on maneuver, defense in depth, offensive warfare concepts, and long-range strikes by deep interdiction with weapons of increasing range, in a modified doctrinal concept which the army called "Air Land Battle."

1982–1984. U.S. Peacekeeping Contingent in Beirut. (See p. 1475.)

1982, November 27. "Dense Pack" Basing Planned for MX Missiles. This provided for basing the silos of the new ICBMs close together, on the assumption that the attacking Soviet missiles would be rendered ineffective because of the principle of "fratricide." Under this concept, incoming missiles would have to be targeted so close together that the explosion of the first missiles destroys the following missiles before they hit the counterforce target. Advocates of the plan said that only ¼ to ⅓ of the U.S. missiles would be destroyed by a Soviet first-strike attack; this was believed to be a rate of survivability so high as to deter a Soviet attack. The Joint Chiefs of Staff were divided on the dense pack basing mode. But public and Congressional opposition was vociferous. Congress voted to withhold funds (December 20).

1983, January 3. Bipartisan MX Basing Commission Appointed. In an attempt to depoliticize and resolve the MX basing mode controversy, President Reagan appointed a commission, headed by Lieutenant General **Brent Scowcroft,** U.S.A.F. Retired, to study and make recommendations regarding possible basing modes for the controversial missile.

1983, March 23. ABM Research and Devel- opment. In an address to Congress, derided by critics as the "Star Wars Speech," President Reagan announced that the United States would pursue the feasibility of perfecting an antiballistic missile (ABM) defense from outer space. Supporters applauded the speech, saying it was an important step in overcoming the mutual-assured destruction (MAD) mentality that had dominated strategic thinking from the late 1960s to the late 1970s. The project to implement this concept was termed the "Strategic Defense Initiative" (SDI).

1983, April 11. Scowcroft Commission Report. The commission recommended that 100 MX missiles with 10 warheads each be deployed in existing Minuteman silos in Wyoming and Nebraska. The MX would be supplemented by a large, but as yet undetermined, number of smaller single-warhead mobile missiles ("Midgetmen"). The commission also stressed "vigorous research and development of an antimissile defense system." President Reagan approved, finding this consistent with his "Star Wars" concept, and endorsed the commission's recommendations (April 19).

1983, May 3. United States Bishops' Pastoral Letter. The Catholic bishops of the U.S. published a pastoral letter in which they condemned any use of nuclear weapons as immoral, regardless of circumstances, and equally condemned the planned or threatened use of such weapons, thus also proscribing deterrence through threat of employing nuclear weapons.

1983. October 23. Suicide Attack on U.S. Peacekeeping Force in Beirut. (See p. 1508.)

1983. October 25–30. Invasion of Grenada. (See p. 1557.)

1983–1986. U.S. Army Light Divisions Created. The Reagan administration created or reorganized 4 regular army divisions on a new pattern. These included the 6th, 7th, and 25th Light Infantry and the 10th Mountain (Light Infantry) divisions. Later the National Guard's 29th Infantry Division was reorganized as a light division (1985–1986). Small by U.S. standards (10,500 personnel) and lightly armed and equipped, they were highly mobile strategically, requiring fewer aircraft sorties to deploy than either airborne division.

1984–1991. SDI Developments. Against stiff congressional opposition, the Reagan administration's so-called "Star Wars" program (see above) made only limited progress. Many critics maintained that the technical problems were too great to permit success, others attacked the high costs of research and development required, and still others stressed the questionable legality of SDI under the ABM provisions of SALT (see p. 1371). Some critics suggested the systems could be defeated by potential countermeasures, including ICBM decoy warheads, as well as SDI's ineffectiveness against aircraft, cruise missiles, and smuggled bombs. In the face of rising expenses and continuing opposition, administration goals dwindled from a concept of a complete shield against ICBM attack to a limited ABM system, akin in purpose to the abandoned "Safeguard" system of the late 1960s.

1986, October. Goldwater-Nichols Defense Reorganization Act. This wide-ranging defense legislation contained several important reforms. Among those were the designation of the Chairman of the Joint Chiefs of Staff as the senior military adviser to both the President and the Secretary of Defense, and the establishment of the position of Vice-Chairman of JCS and his designation as the second-ranking officer in the military hierarchy. This change meant a considerable increase in responsibility and authority for the Chairman and marked a major step in the evolution of the JCS from a collegial body to a true joint-service general staff with a single chief of staff. The full impact of this unique change in U.S. command structure became apparent during the crisis surrounding the Iraqi invasion of Kuwait and in the ensuing Kuwait War (August 1990–February 1991; see p. 1477).

1986, December–1987, May. Iran-*contra* Affair Revealed. As part of the Reagan administration's efforts to direct covert military support to the *contra* rebels in Central America, U.S. officials, including National Security Advisor **Robert C. McFarlane** and his assistants, Rear Admiral **John M. Poindexter** and Marine Lieutenant Colonel **Oliver L. North,** sold weapons to Iran (including advanced TOW antitank missiles, and spare parts for advanced weapons), funneling the proceeds to the *contras* to make up for congressional refusal to approve funds for military support of the *contras* (see p. 1561). These activities, dating from 1984–1986, were revealed in a series of newspaper reports (December 1986–February 1987). Despite congressional investigations, and criminal trials of McFarlane, Poindexter, and North, the level of President Reagan's or Vice President Bush's direct involvement remained unclear.

1987, December 8. INF Treaty. Signed by Presidents Reagan and Gorbachev in Washington, this agreement governed intermediate nuclear forces (INF), specifically the SS-20 (Soviet) and Tomahawk and Pershing II (U.S.-Nato) missile systems. It provided for the destruction of 1,752 U.S. missiles and 859 Soviet weapons. The actual destruction was carried out at plants in the U.S. and the U.S.S.R. under Soviet and American supervision. This represented the first major nuclear arms agreement since the SALT Treaty (1972).

1988, November 10. F-117A Stealth Fighter Revealed. After years of rumors and

Tomahawk Missile in test flight
(*U.S. Navy photo*)

press speculations, the U.S. Air Force confirmed the existence of 52 Lockheed F-117A Stealth light strike fighter aircraft. The combat debut of the F-117A was in Panama (December 1989; see below), where its performance was equivocal. Its performance against Iraq (January–February 1991) was excellent, and did much to disrupt Iraqi command and control in the opening hours and days of the war.

1989, July 17. First Flight of the B-2 Stealth Bomber. Although scheduled for late autumn 1988, this was delayed for over 7 months. Like the F-117A, the Northrop B-2 used special materials and coatings coupled with a carefully designed shape so as to nearly eliminate radar cross-section, and so allow penetration of heavy enemy air defenses. The B-2's strategic role was to attack Soviet mobile missile sites in the event of general or nuclear war. The radically altered international political situation in

1988–1990, coupled with its extremely high cost, rendered the B-2 controversial in congressional budget deliberations.

1989, December 20–24. Operation JUST CAUSE in Panama. Following the killing of a U.S. Marine by Panama Defense Force soldiers, U.S. forces in the Canal Zone, and others flown directly from the U.S., invaded Panama to depose dictator Manuel Noriega. U.S. forces involved totaled 22,500 marine and army troops, from the 82nd Airborne, 7th Light Infantry, and 5th Mechanized Infantry divisions. U.S. forces swiftly occupied the capital and the countryside, overcoming some determined but scattered resistance. U.S. casualties amounted to 23 killed and 220 wounded; Panamanian military losses included at least 500 killed and wounded, but civilian losses may have been higher. Noriega took shelter in the Papal embassy, where he sheltered for several days be-

Refueling an F 114 Stealth Fighter (*U.S. Air Force photo*)

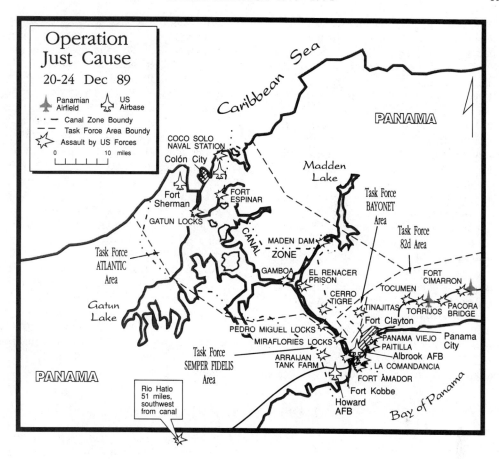

Operation Just Cause
20-24 Dec 89

Panamian Airfield
US Airbase
· - · Canal Zone Boundy
— Task Force Area Boundy
✩ Assault by US Forces

0   10 miles

Caribbean Sea

PANAMA

COCO SOLO NAVAL STATION
Colón City
Madden Lake
Task Force BAYONET Area
Fort Sherman
FORT ESPINAR
GATUN LOCKS
CANAL ZONE
MADEN DAM
Task Force 82d Area
Task Force ATLANTIC Area
GAMBOA
EL RENACER PRISON
FORT CIMARRON
TOCUMEN
Gatun Lake
CERRO TIGRE
TINAJITAS
TORRIJOS
PACORA BRIDGE
Fort Clayton
PEDRO MIGUEL LOCKS
MIRAFLORIES LOCKS
PANAMA VIEJO
PAITILLA
Panama City
Task Force SEMPER FIDELIS Area
ARRAIJAN TANK FARM
Albrook AFB
LA COMANDANCIA
FORT AMADOR
PANAMA
Fort Kobbe
Howard AFB
Bay of Panama

Rio Hatio 51 miles, southwest from canal

---

fore surrendering (January 3, 1990). The U.S. faced heavy criticism abroad, especially from other Latin American nations. A U.N. Security Council resolution condemning the invasion was vetoed by the U.S., France, and Great Britain.

1990, July 8. New Base Agreement with Greece. The U.S. agreed to close 2 bases near Athens but would maintain air and naval facilities at Souda Bay, Crete. Greece would also get credits to buy U.S. weaponry.

1990, August 2–1991, February 26. Response to Iraqi Invasion of Kuwait. (See p. 1477.)

CANADA

1975, May 9. NORAD Pact Renewed. The North American Air Defense treaty, originally signed in 1958, was renewed. The revised treaty gave Canada responsibility for its own air defense for the first time.

1977, July. Armed Forces Enlargement and Modernization Plan Announced. Scheduled to begin in 1978, the plan would add 4,700 men to the 78,000-member Canadian armed forces over a 5-year period, at a $100-million-a-year increase in the yearly budget.

The armed forces equipment was to be completely modernized. West German Leopard tanks would replace the Centurions, while 20 new destroyers were to be purchased to replace some of the aging Canadian ships.

1981, March 11. NORAD Extended for Five Years. (See p. 1555.)

1987. Defense White Paper. Rising defense costs, coupled with Canadian concerns over security for their Arctic domains, led to a major shift in defense policy. Canada abandoned its commitment to Norway in favor of reinforcing the 4th Mechanized Brigade Group at Lahr, West Germany, to form a division in the event of war. Political developments in Europe (1989–1990) led Canada to decide to reduce its 8,000 troops in Europe by 1,400 by late 1991 (September 21, 1990). Canada also announced that its 9-battalion division would be reduced to a large brigade task force of 5 battalions.

1990, July–August. Mohawk Resistance in Quebec. Mohawk tribesmen, angered by plans to convert a sacred tribal site at Oka (60 kilometers west of Montreal) into a golf course, blocked access to the grounds. A raid by the Surête du Quebec (Quebec provincial police) was met by armed Mohawks, and 1 policeman was killed (July 11). The government's use of force mobilized other Mohawks to back the militants at Oka, and Mohawks in the Kahnawake reserve blocked the Mercier Bridge connecting Montreal with the south bank of the St. Lawrence River. Quebec called for federal assistance (August 8), and 3,300 army troops blockaded Mohawk positions at Oka and the Mercier Bridge (August 9–23). Negotiations broke down, and Bourassa ordered the Army to dismantle the barricades (27 August). About 30 Mohawks at Oka, members of the militant Warriors' Society, retreated to a medical clinic on the nearby reserve, but later surrendered (September 26).

1990, August–October. Response to Iraqi Invasion of Kuwait. The Iraqi invasion trapped some 600 Canadians in Kuwait. The government sent 2 destroyers and a supply ship to the Persian Gulf as part of the U.N. blockade force (August 24), and later sent a squadron of 18 CF-18 jet fighters to Saudi Arabia from Germany (October).

LATIN AMERICA

ORGANIZATION OF AMERICAN STATES

1976, October 6. Arbitration of Honduran-Salvadoran Border Disputes. Representatives of the two countries signed a 14-point agreement concerning disputed territory and submitted the dispute to OAS arbitration (see p. 1560).

1977, October 26. OAS Investigates Nicaragua–Costa Rica Border Dispute. A 3-member OAS team arrived in Managua to investigate charges of border violations stemming from attacks by Sandinista rebels across the Costa Rican border (see p. 1555).

MEXICO

1979, January 26. Mexico Supports Guatemala against Insurgents. Mexican troops cooperated with Guatemalan troops in Chiapas, helping them to fight Guatemalan insurgents (see p. 1560).

1982, February 21. Lopez Proposes Central American Peace Plan. Warning against U.S. military intervention in the Central American region, Mexican President **José Lopez Portillo** proposed a plan for ending strife in Central America. This called for: The U.S. to end intervention in Nicaragua; a reduction in the size of Nicaraguan armed forces; and signing of a nonagression pact among Nicaragua, her Central American neighbors, and the U.S. (see Contadora Group, p. 1484).

CARIBBEAN REGION

Cuba

1975–1984. Armed Intervention Abroad and the Proliferation of Revolution. Cuba intervened militarily in a number of countries throughout the period, most notably Angola 1975–1984 (see p. 1540), South Yemen 1978–1984 (see pp. 1513–1514), and Ethiopia 1977–1984 (see p. 1532). Fidel Castro's regime also helped Marxist-Leninist political movements in Grenada, Nicaragua and El Salvador (see pp. 1558–1563).

1983, October 25–30. Cuban Troops Defeated in Grenada. (See below.)

1988, December 27. Cuba Agrees to Withdraw Troops from Angola. After multilateral talks Cuba agreed to a phased withdrawal of its expeditionary force from Angola by July 1991. The force had grown to a peak strength of 50,000 men.

Grenada

1979, March 13. Coup d'État. Maurice Bishop seized power and established a socialist dictatorship on the Caribbean island, which had earlier been granted independence by Britain (February 7, 1973). Bishop sought close ties with Cuba and the Soviet Union. Construction began on a new airport, with a 2-mile runway, supposedly to accommodate tourists, also suitable for military planes. Cuban troops and construction workers as well as Soviet advisers arrived on the island.

1983, October 10–19. Coup d'État. Bishop was suspected by his Marxist colleagues of planning to shift allegiance from the Soviet Union to the United States. He was placed under arrest by his Deputy Prime Minister, **Bernard Coard,** who assumed power. Bishop was temporarily freed by his supporters, but he and many of them were killed by the new regime (October 19). Chaos reigned on the island.

1983, October 23. Caribbean Nations Appeal to U.S. for Help. Noting the chaos on the island, and that the new regime controlled armed forces larger than the combined military strength of all neighboring Caribbean states, the Eastern Caribbean States (a regional grouping including Jamaica, Dominica, Trinidad, Barbados, Tobago, and Belize) formally requested the U.S. to intervene militarily in Grenada to restore order. Since the U.S. government was concerned for the safety of some 1,000 Americans on the island (mostly medical students), President Reagan directed the Defense Department to intervene.

1983, October 25–30. Invasion of Grenada. Hastily planned Operation "Urgent Fury" provided for the landing on Grenada of some 6,000 U.S. troops (U.S. Marines, Army Rangers, and elements of the U.S. Army 82nd Airborne Division), plus token forces (about 500 troops) from the neighboring Caribbean states. Defending the island were approximately 1,000 troops of the Grenadian Army and about 600 Cuban combat engineers. The U.S. forces came ashore at three points in amphibious and airborne assaults. The resistance of the Grenadian troops ranged from slight to negligible; the Cubans fought fiercely. U.S. forces secured the island in about 60 hours. Total U.S. casualties: 18 killed, 83 wounded. Casualties of the defenders: 36 killed in action, 66 wounded in action, 655 captured. Subsequent criticisms of the performance of U.S. forces in this operation appeared in the press.

Haiti

1982, January 9–17. Revolution Thwarted. An attempt to overthrow the government of President **Jean-Claude Duvalier** (son of François Duvalier; see p. 1461) was suppressed.

1986, February 7. Downfall of Duvalier. "President for life" **Jean-Claude Duvalier** ("Baby Doc") fled to exile in France following 3 months of protests, ending 28 years of Duvalier family rule. He was succeeded by Lieutenant General **Henri Namphy,** chief of staff of the Haitian Army, who headed a National Council of Government. Popular reprisals eliminated many of the Tontons Macoutes ("bogeymen"), the dreaded, brutal security forces of Duvalier. Namphy was succeeded by **Leslie Manigat,** an academic, who was elected president (January 1988).

1988, June 19–20. Coup d'État. General Namphy overthrew Manigat, who was exiled; Namphy succeeded to the presidency.

1988, September 17. Coup d'État. Namphy

was ousted by a coup led by Lieutenant General **Prosper Avril.**

1988, September–1990, March. Sporadic Unrest. Avril's government was briefly threatened by 2 successive coup attempts by elite army units loyal to senior officers implicated in the drug trade (April 2 and April 5–10, 1989). The rebels were quickly subdued, and in the aftermath, the army was substantially reduced.

1990, March 12. Avril Resigns. Avril was toppled by continuing unrest. Interim governments gave way eventually to the elected government of leftist Roman Catholic priest, Reverend Jean-Bertrand Aristide (December 16, 1990).

Trinidad and Tobago

1990, July 27–August 1. Abortive Muslim Revolt. The Jamat al Muslimeen sect, led by Imam Yasin Abu Bakr, briefly held Prime Minister Arthur A. M. Robinson and 54 other hostages. Lacking popular support, the rebels surrendered to government forces.

CENTRAL AMERICA

Belize

1981, September 21. Independence from Great Britain. This came despite an ongoing territorial dispute with Guatemala. That dispute, moreover, bars Belize from entry into the Organization of American States (OAS) and protection under the Rio Non-Aggression Pact. Consequently, Britain maintains a 1,600-man garrison in Belize, based on a reinforced infantry battalion.

Costa Rica

1977, October 14–17. Border Dispute with Nicaragua. Costa Rica closed its borders and moved constabulary troops to the Nicaraguan border following the pursuit of Sandinista rebels by Nicaraguan National Guardsmen into Costa Rican territory. Both nations appealed to the OAS, accusing the other of border violations (see p. 1555).

El Salvador

1976. July. Border Clashes with Honduras. (See p. 1560.)

1979, January 20–March 20. Increase in Violence and Guerrilla Activity. Smoldering leftist unrest exploded into violent insurgency as a result of failure of the government to implement a satisfactory land redistribution program. An investigation by the Inter-American Commission of Human Rights resulted in charges that human rights violations were being committed by the El Salvadoran government against its political enemies.

1979, April–October. Continuing Violence. Efforts to suppress the insurgency by President **Carlos Romero** were unsuccessful. The new government of Nicaragua was suspected by the U.S. government of supporting the insurgency, as part of a policy of fomenting revolution and using force of arms to overturn all existing governments in Central America.

1979, October 15. Coup d'État. Disgruntled army officers staged a successful coup, overthrowing Romero.

1979, October–1980, December. Violence and Chaos. Efforts of the new junta to stabilize the country and satisfy the dissidents failed completely, and alienated much of the population. A land reform program was announced (February 11), disavowed (February 20), then a new program was announced (March 6).

1979, March 26. Assassination of Archbishop Romero. Outspoken and much-beloved Archbishop **Oscar Romero** was assassinated by a gang while saying Mass (March 24, 1980). Leftist rebels and right-wing extremists were suspects in the crime.

1980, May 2. Attempted Coup d'État. The effort by former President Romero and disgruntled military officers failed.

1980, December 13. Duarte Named President. The junta appointed moderate Christian Democrat **José Napoleon Duarte** as president, and directed him to reorganize the government.

1981–1991. Continuing Civil War. Despite massive financial and military assistance from the U.S., the Salvadoran Army was unable to suppress the insurgents, who were receiving comparable support from the U.S.S.R. and Cuba, via Nicaragua. Atrocities continued, committed not only by adherents of the leftist guerrillas but also by rightist extremists operating in so-called "death squads" believed to be composed of members of the armed forces, and operating under the direction of military officers. Among the victims were 7 Americans: 4 female missionaries (December 3, 1980), 2 labor leaders advising the Salvadoran government (January 2, 1981), and a naval officer military adviser (May 25, 1983).

Between 1982 and 1988, the death toll amounted to over 50,000, including many victims of the clandestine "death squads."

1982, March 28. Election. Despite a guerrilla boycott and active efforts to frighten people not to vote, there was a large turnout in an election that international observers considered very fair. Duarte's party won a 40 percent plurality of the vote, but a coalition of the right-wing parties, with a majority of right-wing representatives under the leadership of former National Guard officer **Roberto D'Aubuisson,** formed a government. D'Aubuisson, rumored to be one of those responsible for the death squads, failed, however, in his effort to be appointed president. Despite the election, violence and atrocities by both sides continued unabated.

1982, November 12–17. El Salvadoran Counteroffensive. The Salvadoran Army carried out a major counteroffensive in an effort to envelop the guerrilla strongholds and block their escape to Honduras. The Honduran Army also deployed along the common border of the two countries in order to apprehend elements of the guerrilla forces who escaped. The offensive was only partially successful.

1984, May 6. Duarte Elected President. Duarte defeated D'Aubuisson. He promised to seek reconciliation with leftist rebels, without compromising democratic institutions. He also promised to eliminate death squads.

1984, May 24. Military Men Convicted of Death Squad Murders. Largely as a result of pressure from the U.S., the Salvadoran government brought to trial 5 National Guardsmen suspected of murdering the 4 female missionaries (see above). The 5 were tried and convicted of murder.

1985, June 19. Terrorist Attack in San Salvador. FMLN gunmen attacked restaurants and cafés in the capital, killing 4 U.S. Marines, 2 U.S. businessmen, and 7 others; about 15 other people were wounded.

1985, September–October. Duarte's Daughter Kidnapped. President Duarte's daughter, Inés Guadalupe Duarte Duran, and a woman friend were kidnapped by FMLN guerrillas (September 10). After extensive negotiations, the government released 22 guerrillas and allowed another 96 to leave the country for treatment of wounds; in exchange, the FMLN freed 23 kidnapped mayors and some other officials, as well as Duarte Duran and her companion (October 24). President Duarte was criticized by ARENA (among others) for yielding too much to the FMLN.

1988, November 5. FMLN Rejects Cease-Fire. A unilateral government cease-fire was rejected by the FMLN. This occurred against the backdrop of Daniel Ortega's peace initiative (see p. 1563).

1989, November 11–December 15. FMLN Offensive in San Salvador. In a dramatic change in strategy, FMLN forces mounted a major offensive in the capital (November 11). During the first week of fighting, 6 Jesuit priests, their cook, and her daughter were found murdered (November 16); although blamed on right-wing "death squads," army involvement was also suspected (and later proved). Fighting within the capital continued for more than a month. Government forces used artillery and air strikes against FMLN positions in poorer neighborhoods, but there was also combat in wealthier suburbs before battered FMLN forces withdrew into the countryside (mid-December).

1989, May–1990, May. Continuing Death Squad Activities. U.N. human rights observers reported that 3,219 civilians were assassinated during this period.

1990, November–December. FMLN Down Aircraft. FMLN guerrillas, using SA-14 shoulder-launched antiaircraft missiles, downed 2 government aircraft (November 23 and December 4). As a result, the U.S. announced the resumption of military aid to Alfredo Cristiani's ARENA government (December 7). The aid had been suspended earlier (October).

Guatemala

1976–1991. Guerrilla Attacks. A political movement called the Guerrilla Army of the Poor (EGP—*Ejército Guerrillero de los Pobres*) staged a number of raids in various parts of the country, destroying farm equipment and on 1 occasion killing 2 land owners (November 13, 1976). They were especially active in the mountainous El Quiche region north of Guatemala City.

By the late 1980s, strenuous government efforts had substantially reduced guerrilla activities, but right- and left-wing "death squads" continued campaigns of assassination and terror.

1982, March 25. Coup d'État. Brigadier General **Efrain Rios Montt** headed a successful coup. Combining religious fervor with ruthlessness, Rios Montt reduced unrest, but alienated the army.

1983, August 8. Rios Montt Overthrown in Coup. Military officers, headed by Defense Minister General **Oscar Humberto Mehia Vitores,** ousted Rios Montt after a brief struggle.

1985, December 8. Elections. In the first election of a civilian president in 15 years, Christian Democrat Marco Vinicio Cerezo Arevalo won the presidency. Cerezo took office on January 14, 1986, signaling the end of military rule.

1989, May 9. Attempted Coup. Nine civilians and 17 military officers were implicated in a

plot to remove the defense and interior ministers. This threat to the government emerged against a backdrop of increasing violence. Estimates of the monthly death toll from political killings by both right- and left-wing "death squads" ranged from 40 to 200.

1990, December. U.S. Suspends Military Aid. Following the death of a U.S. citizen (June), the U.S. government suspended military aid to Guatemala.

Honduras

1976, July 14–22. Border Clashes with El Salvador. There were a number of armed clashes on the disputed border between El Salvador and Honduras.

1977. Staging Area for Sandinista Forays into Nicaragua. Honduras permitted Nicaraguan Sandinista immigrants of the National Liberation Front (NLF) to conduct operations in Nicaragua from bases in Honduras.

1981–1982. Strained Relations with Nicaragua. Anti-Sandinista guerrilla activity emanating from Honduras led to increased tensions between Nicaragua and Honduras. The Sandinista regime in Nicaragua alleged that the Honduran government had given aid to anti-Sandinista forces within Honduras. Nicaragua increased forces along the border. Honduras received increased military aid from the United States. By early 1982 there were 100 U.S. military advisers in Honduras. Relations deteriorated until both countries withdrew their ambassadors (April 4). Honduran President **Roberto Suazo Cordova,** concerned about Nicaraguan and Cuban subversion in his own country, asked for and received more aid from the U.S. (July 16).

1982, October 4. Latin American Peace Forum. Honduras participated in a peace forum with other South and Central American nations in an attempt to end hostilities in the region (see International Peacekeeping, p. 1483).

1982. November 12. Honduran Troops Aid El Salvadoran Counteroffensive. Honduras supported an El Salvadoran counteroffensive against rebel strongholds by posting 2,000 troops along the Honduran-El Salvadoran border in an attempt to block the rebel passage into Honduras.

1983, August 5. U.S. Begins Honduras Military Maneuvers. The U.S. began large-scale military exercises in Honduras. These were to last as long as 8 months. Ostensible objectives included training Hondurans in counterinsurgency tactics, coordinating with Honduran troops, and training them in amphibious operations and in U.S. field artillery procedures and tactics. The obvious unstated objective was to impose caution on Nicaragua, particularly in its support of the insurgency in El Salvador.

1986, March 21–27. Nicaraguan Incursion. Nicaraguan army forces entered Honduras in pursuit of *contra* guerrillas (March 21–22). After some days of confused fighting (made more confusing by understandable reluctance among Honduran officials to admit presence of *contras* in Honduras), U.S. helicopters were used to insert Honduran troops across the Nicaraguans' main line of retreat, causing them to withdraw hurriedly (March 26–27).

1988, March 16–17. Border Clash with Nicaragua. Nicaraguan army troops entered Honduras to attack *contra* bases there, and clashed with Honduran army troops. The U.S. hurriedly sent 3,200 troops to Honduras, but the Nicaraguans withdrew soon after (March 28–31).

Nicaragua

1975–1976. Successful Counterinsurgency. The National Guard appeared to be winning their counterinsurgency war, and guerrilla activities subsided. However, harsh repressive measures aroused much resentment in the population.

1977. Somoza Under Fire. At home Somoza lost the support of the Catholic Church and much of the middle-class business population. In the international press, the Sandinistas' campaign of violence and terror was ignored, while the Carter administration's citation of

Nicaragua as a violator of human rights was widely applauded.

1977, October. Border Dispute with Costa Rica. (See p. 1558.)

1978, January–September. Violence and Anarchy. Taking advantage of widespread public disobedience and largely effective strikes, the Sandinistas stepped up attacks and also were able to gain many new recruits, including foreigners from East Germany, the PLO, and Cuba.

1978, September 9. Full-Scale Sandinista Offensive Launched. The Sandinistas successfully attacked several large towns and cities. National Guard counteroffensives drove the rebels into the mountains with great difficulty. But the cost was enormous; many civilian casualties were suffered, further alienating the population. The Sandinistas continued their pressure.

1979, May 29–July 17. Sandinista Final Offensive. The rebels resumed the offensive, easily winning control of most of the cities and the countryside in a matter of weeks, and surrounded Managua (July 10). Certain of victory, they named a 5-man ruling junta (July 17). They rejected a belated U.S. request that they share power with more moderate elements of the opposition. At U.S. insistence, Somoza left Managua for Miami.

1979, July 19. Sandinistas Enter Managua in Triumph. The new regime was established with the entry of the Sandinistas into the capital city.

1979–1981. Sandinistas Consolidate Power. At first the government of the Sandinista *commandantes* encouraged a free press and free enterprise, but by late 1980 their policies had swung sharply to the left. They imposed press censorship, and postponed elections until 1985 at the earliest. They also forcibly relocated the Miskito Indians in the northeast, provoking armed resistance. Internationally, they shunned U.S. assistance and drew closer to Cuba and the Soviet Union. By late 1981, there were some 5,000 Cuban, Soviet, and PLO military and technical advisers in Nicaragua.

1981–1990. The *Contra* War. Encouraged by

the Reagan administration in the U.S., Nicaraguan exiles opposed to the Sandinista regime organized a guerrilla army in Honduras and Costa Rica, setting up base areas in both countries (1981–1982). These rebels were called the *contras*, and the Honduras-based forces were largely run by former members of Somoza's National Guard (a fact which did not endear them to most Nicaraguans). Their numbers rose to some 25,000, of whom perhaps 2,500 were active in Nicaragua at any one time (1985–1986). The *contras* enjoyed very modest popular support within Nicaragua, and did not really threaten the regime. Coupled with U.S. economic and diplomatic pressure, however, they eventually produced the end of the Sandinista dictatorship (1990).

1984, April 4–14. CIA "Covert" Mining of Corinto Harbor Revealed. In an operation blatantly contrary to international law, the CIA mounted an undercover effort to mine the port of Corinto. When this program was revealed (April 6–10) in the U.S. the public outcry produced a sharp reaction to government policies in Central America. Congress condemned the action (April 11).

1984, September 1. Two Americans Killed in Nicaragua. Two U.S. citizens, serving as freelance advisers to *contra* forces, were killed when their helicopter was shot down. They were members of Civilian Military Assistance (CMA), a private group with ties to the CIA.

1984, October 14–21. Scandal over CIA "Guerrilla Manual." Public revelation in the U.S. of a CIA manual on guerrilla tactics raised a furor with its suggestion that enemy (in this case, clearly meaning Sandinista) government officials be "neutralized." Some *contra* officials admitted that they had assassinated Sandinista officials.

1984, November 6–12. Soviet Arms Shipment Crisis. The U.S., claiming that a shipment of Soviet military stores bound for Nicaragua included MiG-21 fighters, warned it would not tolerate the delivery of such weapons. Observers noted no jets on the ship when it unloaded (November 7–10). The Sandinistas, ostensibly fearing a U.S. invasion,

moved the country toward a war footing (November 12–13).

1985, April 4–29. *Contra* Funding Dispute. The U.S. Congress initially rebuffed the Reagan administration's requests for aid for the *contras*, despite government pleas not to abandon 15,000 "Freedom Fighters." However, Nicaraguan President Daniel Ortega's trip to Moscow (April 29) provoked reconsideration, and a *contra*-aid bill passed both houses of Congress the same day; additional aid was later approved (June 6–12). Continued popular and congressional opposition to U.S. support of the *contras* eventually induced some within the Reagan administration to resort to illegal and extraconstitutional means to maintain support of the *contras*, actions that produced the "Iran-*contra*" scandal (December 1986–May 1987; see p. 1553). Meanwhile, the U.S. suspended trade with Nicaragua (May 1, 1985).

1986, March 21–27. Incursion into Honduras. (See p. 1561.)

1986, October 5. Plane Downed Carrying Clandestine *Contra* Supplies. The Nicaraguans shot down a C-125K cargo plane and captured a U.S. passenger, **Eugene Hasenfus.** He had been involved in ferrying supplies to *contra* forces inside Nicaragua. While any official connection with the CIA remained murky, the Reagan administration admitted it had ignored the ban on aid to the *contras* (October 19). Hasenfus was tried and convicted by a "people's tribunal," but was later released (December 17, 1987).

1987, November–December. Peace Talks. After Ortega returned from a trip to Moscow (November 15), Nicaragua began a peace initiative. Peace talks broke down twice (December 4, 22) before they were abandoned.

1988, January. *Contra* Supply Planes Downed. An ostensibly private aircraft carrying supplies for the *contras* was shot down, and U.S. citizen **James Denby** captured (January 13). Another plane was downed (January 23), but Denby was freed (January 30).

1988, March 16–18. Nicaraguan Incursion into Honduras. (See above, p. 1561.)

1988, March and May. Abortive *Contra*-

Sandinista Negotiations. Efforts by the Sandinista government and the *contra* rebels to conclude a political settlement to the war were unsuccessful. A 60-day truce was signed (March 24), later extended for 30 days (May 26), but accompanying negotiations were unproductive.

1989, February 14. Tesoro Beach Accords. This regional peace settlement, involving the presidents of Costa Rica, El Salvador, Guatemala, Honduras, and Nicaragua, promised the disbandment of *contra* forces in return for an open election in Nicaragua, scheduled for February 1990. The Sandinistas initiated internal reforms indicating good-faith compliance with the accords (March–April).

1989, August 7. Tela (Honduras) Convention. This agreement among the participants in the Tesoro Beach accord regulated the disbandment of *contra* forces under the auspices of an international commission. Some *contras* refused to participate, but were eventually persuaded to accept the settlement.

1990, February 25. Elections. In a surprise upset, the opposition National Opposition Union (UNO), led by **Violetta Barrios de Chamorro,** won a wide majority in presidential, legislative, and local elections.

1990, October–December. *Contra* **Unrest.** *Contra* veterans, dissatisfied with the postelection settlement, briefly seized control of the town of Waslala (October). Fighting later broke out again (December), leaving 11 dead.

Panama

1977, September 7. Preliminary Draft Canal Treaties Signed. U.S. President Jimmy Carter and Panamanian leader General Omar Torrijos Herrera signed preliminary draft treaties for turning the administration and ownership of the Canal over to Panama. The treaties were ratified by a 2-to-1 majority in a Panamanian plebiscite (October 23).

1978, March 16, April 18. U.S. Senate Ratifies Panama Canal Treaties. The U.S. Senate ratified the two Panama Canal treaties.

President Carter and General Torrijos signed the final draft of the treaties (June 16).

1981, August 1. Torrijos Killed in Plane Crash. His plane crashed in the jungle during an inspection tour.

1983, August. Noriega Becomes Commander of the National Guard. Manuel Antonio Noriega soon reorganized the Guard as the Panamanian Defense Forces (PDF), expanded its size, and increased its political power. He became virtual dictator of Panama.

1988, February 5. Noriega Indicted by U.S. In indictments unsealed in Miami, U.S. prosecutors indicted Noriega on drug-trafficking and money-laundering charges. He denied the accusations, and refused to resign despite public pressure. When President **Eric Arturo Delvalle** tried to dismiss him (February 25), he deposed Delvalle and assumed personal rule (February 26). The U.S. responded with heavy economic pressure and reinforced the Canal Zone garrison (March 1–11), but Noriega remained intransigent and cracked down sharply on internal dissent. He rejected a compromise that would have abandoned U.S. drug charges against him in exchange for his exile (May 25).

1989, May 7. Elections. The result of presidential elections, and accompanying hopes for renewed democracy in Panama, were dashed when Noriega nullified the results. International observers reported numerous instances of blatant fraud, and reports indicated that voters had favored the opposition ticket of **Guillermo Endara** by 3–1. Paramilitary forces, including Noriega's "Dignity battalions," broke up a peaceful protest with baseball bats and metal pipes, injuring hundreds, including both Endara and his running-mate, **Guillermo Ford** (May 10).

1989, October 3. Attempted Coup. Anti-Noriega officers led by Major **Moisés Giroldi Vega** mounted a coup against Noriega. Despite some U.S. support, the rebels were defeated, and 10 officers (including Giroldi) were killed. Noriega conducted a purge of the PDF afterward, and many officers were arrested or fled.

1989, December 20–24. Operation JUST CAUSE. (See p. 1554.)

SOUTH AMERICA

Argentina

1976, January 14. Argentina Places Ambassador to Britain on Indefinite Leave. As a protest against Britain's renewed interest in the Falkland Islands (called Malvinas Islands by Argentinians), the Argentinian government placed their ambassador on indefinite leave. The British reciprocated (January 19).

1976, February 4. Argentinian Ship Fires on British Research Ship. An Argentinian destroyer fired 2 rounds at the British scientific research ship *Shackleton* after requesting that the *Shackleton* be escorted to Tierra del Fuego. The *Shackleton* proceeded to Port Stanley with the Argentinian destroyer giving chase. Six miles from Port Stanley the Argentinian ship fired the 2 warning rounds.

1976, March 24–25. Military Coup d'État. President **Maria Estela "Isabel" Perón,** widow of Juan Perón, was ousted by the Armed Forces Chiefs of Staff, who immediately formed a ruling junta and imposed martial law on Argentina. The junta outlawed 5 extremist parties and initiated a counterinsurgency campaign against guerrilla and political extremists, who had terrorized Argentina for several years.

1976, March 25–1982, July 16. "The Dirty War." Argentina's military junta carried out a relatively successful counterinsurgency war against radical leftist guerrilla forces and political parties, most notably the People's Revolutionary Army (ERP), but in the process carried on a ruthless and indiscriminate campaign of counterterror, committing flagrant violations of the Argentinian citizenry's human rights. During this "Dirty War," as many as 20,000 to 25,000 people "disappeared." Following the defeat of Argentina by Britain in the Falkland Islands War (see p. 1471), the discredited military junta restored civil liberties (July 15) and political parties were allowed to function (July 16).

1977–1978, British and Argentinian Representatives Resume Falkland Negotiations. British and Argentinian diplomats met in New York and carried on negotiations about the future of the Falkland Islands (December 15–18). British and Argentinian negotiators again met in Lima, Peru, to discuss control of the Falkland Islands (February 15–17). The talks took on new meaning because of the discovery of large undersea deposits of oil in the vicinity of the islands.

1979, November 16. Argentina and Britain Resume Exchange of Ambassadors. The British and Argentinian governments sent new ambassadors to each other's capitals. Chargé d'affaires had been the chief officers since Argentina placed its ambassador on indefinite leave (see above).

1981, December 21. Galtieri Takes Power. Lieutenant General **Leopoldo Galtieri** succeeded General **Roberto Eduardo Viola,** who had been compelled to retire by fellow officers after less than 6 months as president.

1982. January–June. Falklands Crisis and War. (See United Kingdom, p. 1471.)

1982, June 17. Galtieri Resigns. Only 3 days after the defeat in the Falklands (Malvinas), Galtieri resigned as both president and Army Chief of Staff. He was succeeded by Major General **Reynaldo Bignone** (July 1), and the navy and air force junta members resigned in protest over his appointment. The junta was reconstituted (September 10), and made progress toward return to civilian rule.

1983, October 30. Elections. In the first elections since Peron's return (1973), **Raul Alfonsin** of the Radical Civic Union was elected president. Alfonsin was inaugurated 6 weeks later (December 10).

1985, October 21. State of Siege. Alfonsin declared a 60-day state of siege during a legal dispute over the arrest of 12 suspected terrorists, including 2 active-duty military officers. Six suspects were released for lack of evidence (October 26).

1986, May–June. Military Unrest. There were several incidents of military unrest and local rebellion following the conviction of 3 former junta members for negligence in prosecution of the Falklands War (May 16).

1987, April 17–22. Army Rebellion. Soldiers

led by Lieutenant Colonel **Aldo Rico** seized the Infantry School at Campo de Mayo near Buenos Aires, protesting the treatment of an officer who had been cashiered for refusing to appear in court on human rights abuse charges (see "Dirty War," p. 1564). The rebellion was suppressed by loyal troops (April 19). There were extensive reports, after the event, that II Corps officers had failed to move promptly against the rebels.

1987, September 27–28. Army Rebellion. Soldiers of the 3rd Infantry Regiment seized their barracks near Buenos Aires, protesting the relief and replacement of Lieutenant Colonel **Dario Fernandez Merguer,** who had refused to act against the Campo de Mayo rebels (April). They surrendered after 4 hours (September 28).

1988, January 16–18. Army Rebellion. Aldo Rico, escaping arrest, took shelter at Monte Caseras with the 4th Infantry Regiment. The regiment's colonel led a mutiny by some 100 soldiers in Rico's support (January 16). Three loyal battalions, totaling 2,000 men, moved against Monte Caseras (January 17). After a sharp fight, they captured the rebel positions and the rebels surrendered (January 18).

1988, December 2–6. Army Rebellion. Led by Colonel **Mohamed Ali Seineldin,** some 500 rebellious soldiers captured the Campo de Mayo Infantry School (December 2). The rebels hoped to win amnesty for officers implicated in human-rights abuses under the military dictatorship of 1976–1983. The commander of a brigade at Cordoba refused to move against the rebels, although the commander insisted the brigade was not joining the mutiny. After some negotiations, Seineldin surrendered (December 6). In all, 855 personnel from 6 units had taken part.

1989, January 25. Attack on La Tablada Barracks. A force of civilian guerrillas attacked the La Tablada Barracks outside Buenos Aires. The armed forces responded swiftly and defeated the guerrillas in a sharp bloody action. Afterward, Alfonsin put a new military funding program before Congress.

1989, June 12. Alfonsin Announces Resignation. Effective June 30, Alfonsin would leave the presidency to President-elect **Carlos Saúl Menem** (elected May 14), ending his regime 6 months early (July 8). Contributing to his decision were food riots (May) and runaway inflation; prices had risen 309,907 percent since he took office (December 1983).

1990, December 3. Abortive Military Rebellion. Dissident army officers Rico and Seineldin, who had led previous rebellions, led several hundred soldiers in an effort to seize Army headquarters in Buenos Aires, and to install Seineldin as Army Chief of Staff. They were defeated by loyalist troops later that day. Casualties included 21 dead. The Menem government later announced it was seeking the death penalty for Rico and Senieldin (December 22).

Bolivia

1976. Guerrilla Activity. The National Liberation Army (ELN), founded by Ché Guevara (see p. 1465), launched a number of raids throughout the year. They were allegedly aided by a Chilean group, the Revolutionary Left Movement.

1978, July 21. Military Annulment of Election Results. Bolivia's first election in 12 years (July 19) was annulled by the candidate of the ruling military government, Air Force General **Juan Pereda Asbun.**

1978, November 24. Military Coup d'État. After 4 months in power, General Pereda was overthrown by fellow junta member Army General **David Padilla Arancibia,** who promised new elections.

1979, July 1–August 6. Return to Civilian Rule. None of the 8 candidates for president achieved a majority in the promised election (July 1). Congress finally selected a provisional president, **Walter Guevara Arce,** bringing a temporary end to military rule.

1979, November 1. Military Coup d'État. Colonel **Alberto Natusch Busch** led a successful coup, overthrowing President Guevara and seizing the presidency. However, he was opposed by other military officers and was forced to resign. The military leaders then selected **Lydia Gueiler Tejada** provisional president (November 16).

1980, July 17. Military Coup d'État. Following a presidential election in which no candidate received a majority, officers under General **Luis García Meza** seized power.

1981, January–August. Continued Civil and Military Unrest. A power struggle within the armed forces led to an armed revolt of senior cadets at the national military academy (March 17). A coup attempt ended in a military stalemate but forced the resignation of García (August 4). A 3-man military junta dominated by former president Natusch took power.

1987, August 28. State of Siege. In response to growing unrest sparked by economic problems, President **Jaime Paz Zamora** declared a 90-day state of siege.

Brazil

1985, January–March. Return of Civilian Rule. Twenty-one years of military rule ended with the election of **Tancredo de Almeida Nevez** (January 15). Nevez fell ill before his inauguration, and Vice President-elect **Jose Sarney** was sworn in (March 15). Sarney became president after Nevez died (April 21).

1990, October 5. Brazilian Nuclear Program Revealed. Minister of Science and Technology **Jose Goldenberg** revealed, in an interview with the *New York Times*, that the Brazilian military had undertaken a secret program to develop nuclear weapons. Following the discovery, President Collor had ordered the program shut down (September 19).

Chile

1973–1980. Junta Rule and Domestic Unrest. The military junta, under the leadership of Army General **Augusto Pinochet Ugarte,** established firm control over the country. There was, however, much internal unrest, and the Revolutionary Left Movement attempted to foment insurrections throughout the period.

1980, September 11. Voters Approve Pinochet's Transition Constitution and Continued Rule. By a better than two-thirds margin the voters approved General Pinochet's constitution, which included his accession to the presidency and the continuation of at least 8 more years of military rule. Pinochet formally assumed the office of president the following year (March 11, 1981).

1983–1984. Civil Unrest. Opposition to Pinochet's rule resulted in a number of riots and some leftist guerrilla activity in rural areas.

1987, July 2–3. Civilian Unrest. Following a general strike, widespread civilian unrest and clashes between protesters and security forces left 8 dead, including a 19-year-old student who was burned to death by security forces.

1987, September 7. Pinochet Survives Assassination Attempt. General Pinochet escaped unharmed from an assassination attempt on his motorcade by the Manuel Rodriguez Patriotic Front.

1988, October 5. Pinochet Loses Plebiscite. Chilean voters overwhelmingly rejected General Pinochet's proposal that he be allowed to continue his military regime. Pinochet publicly acknowledged the result with uncharacteristic grace (October 6). Presidential and congressional elections were scheduled for December.

1990, March 11. Aylwin Sworn in as President of Chile. Following his election (December 14, 1989), Christian Democrat **Patricio Aylwin Azócar** was sworn in as president, succeeding Pinochet, who had deposed the last elected Chilean president, Salvador Allende (1973; see p. 1465).

Colombia

1975–1978. Social Unrest and Insurgency. Colombia suffered from domestic unrest and several insurgency efforts against the government. One of the two major guerrilla movements, the Revolutionary Armed Forces (FARC), attempted to disrupt the national elections (February 26, 1978). Later that year FARC and the second leading revolutionary group, National Liberation Army (ELN), joined forces (August 24).

1979, January 9–November 14. Successful Counterinsurgency Campaign. This effort resulted in the elimination of 2,000 guerrillas and enabled President **Julio Cesar Turbay Ayala** and Defense Minister General **Luis Carlos Camacho Leyva** to pacify the country.

1980–1991. Continuing Guerrilla Activity. Despite setbacks the Colombian insurgents continue to carry on low-level guerrilla activities.

1985, November 6–7. Guerrillas Seize the Palace of Justice. Guerrillas from the notorious M-19 group attacked and seized the Palace of Justice in Bogota (November 6). M-19 had recently broken a truce with the government, claiming that cease-fire promises had not been kept and that reforms were not forthcoming as agreed. The guerrillas took more than 300 people hostage in the palace. Police and army troops used dynamite to breach the palace walls, and stormed the building. The ensuing fire and the battle itself cost 95 lives, including those of 11 Supreme Court justices.

1988, January 25. Attorney General Assassinated. Attorney General **Carlos Mauro Hoyos** was kidnapped and murdered by terrorists in Bogota. The group of Medellín drug lords known as *los Extraditables* (the Extraditables) claimed responsibility. President **Virgilio Barco Vargas** responded by decreeing extended police powers of arrest and detention, and increased penalties for drug trafficking.

1989, August 18. Presidential Candidate Assassinated. Luis Carlos Galán, presidential candidate of the ruling Liberal Party and a strong supporter of a hard-line policy against the Medellín and Cali cocaine cartels, was killed by drug traffickers. This followed the government's crackdown on the drug trade, and in turn led to the declaration of a state of emergency, and the arrest of over 10,000 persons suspected of involvement in the drug trade.

1989, October 28–November 4. Drug Traffickers' Assassination Campaign. The drug lords attempted to maintain their operations and fight legal prosecution and extradition by a campaign of terror. Drug cartel gunmen killed a magistrate, a congressman, a left-wing politician, and 6 policemen in Bogota.

1990, April. Killings in Medellín. During this month, drug-related killings claimed an average of 50 lives per day.

1990, August 7. Drug Lords Proclaim Truce. Following the inauguration of President **César Gaviria Trujillo,** Medellín cartel leader **Pablo Escobar Gaviria** declared a unilateral truce. President Gaviria later offered traffickers immunity from extradition and reduced prison sentences if they surrendered to the authorities.

Ecuador

1975, September 1. Attempted Coup. Armed Forces Chief of Staff General **Raul Gonzalez Alvear** was defeated by troops loyal to President Guillermo Rodriguez Lara.

1976, January 11. Military Coup d'État. The armed forces commanders ousted President Rodriguez in a bloodless coup.

1978, January 16–17. Border Incident with Peru. Brief hostilities broke out along the disputed Marañón River border in the Cordillera del Condor Mountains.

1981, January 28–February 2. Renewed Border Hostilities with Peru. The long-standing border dispute erupted into hostilities when Ecuadoran troops pushed 13 kilometers into Peruvian territory. In Washington the OAS helped negotiate a cease-fire (February 2).

1981, March 6. Truce with Peru. Both sides announced an "immediate peace" along the disputed border and agreed to mediation by the U.S., Chile, Brazil, and Argentina.

Paraguay

1975–1978. Low-Level Insurgency. Army units carried out counterinsurgency operations against Marxist-Leninist guerrillas, and police also conducted operations against leftist guerrillas in Asunción.

1978, May 5. State of Siege Lifted in Three Departments. A state of siege, which had been present since 1947 in the departments of Itapua and Central and Alto Paraná, was lifted. Asunción, the capital, still remained under a state of siege because of the threats posed by Marxist-Leninist guerrillas.

1989, February 3. Coup Deposes Stroessner. General **Andrés Rodriquez** led a successful coup against longtime dictator Alfredo Stroessenr, who had ruled the country since 1954. Scores of people were killed, and Stroessner went into exile in Brazil. Rodriguez's promised free elections followed on May 1, and the Colorado party (PC) won a substantial victory.

Peru

c. 1978–1991. Communist Insurgency. This was the result of efforts by a self-proclaimed Maoist organization with overtones of Inca revivalism, *Sendero Luminoso* (Shining Path), based in the central Peruvian highlands, especially around Ayachuco. *Sendero Luminoso* guerrillas were disciplined and ruthless, and either won over (through promises of land reform) or cowed into submission a sizeable portion of Peru's rural highland population. Although their terrorist bombing campaign (1983) and other efforts to topple the government failed, government efforts to contain their activities met with scant success, and by late 1989 the guerrillas had killed or kidnapped some 15,000 people. In January–July 1990 alone, the guerrilla and counterinsurgency operations cost 1,952 lives, nearly as many as in all of 1989 (1,956). During the late 1980s, *Sendero Luminoso* made considerable headway in efforts to control the Upper Huallaga Valley, a center of coca production, and thereby gained considerable financial resources from "taxes" on coca growers. Government efforts to maintain order were further hampered by the smaller *Movimiento Revolucionario Tupac Amaru* (MRTA) guerrilla-terrorist group, ironically named after a leader of late 18th century Indian resistance to Spanish rule.

1975, August 29. Military Coup d'État. The commander of Peru's 5 military districts issued a joint communiqué removing President Velasco from office and proclaiming General **Francisco Morales Bermúdez Cerrutti** president.

1976, July 9. Attempted Coup d'État. Troops loyal to President Morales crushed a barracks rebellion.

1986, June 17–19. Prison Uprisings. Imprisoned members of *Sendero Luminoso* staged uprisings and took control of 3 prisons (June 17–18). At 2 of these the rebels had obtained arms and constructed fortifications and barricades, which were overcome only when government troops attacked with rocket and armored car support (June 18–19). Official reports (possibly understated) said 156 had died in the uprisings.

Surinam

1980, February 25. Military Coup d'État. Premier **Henck A. E. Arron** was ousted from power in a coup led by 2 army sergeants.

1980, August 13. Military Coup d'État. The military leadership removed President **Johan Ferrier,** who had survived the earlier coup that had ousted Premier Arron.

1982, February 5. Military Coup d'État. The military leadership ousted the civilian government and then imposed military rule.

1989, June 23. Government Truce with Surinamese Liberation Army (SLA). After years of desultory conflict, the government approved an agreement with SLA leader **Ronnie Brunswijk** on conditions for a truce. The SLA drew its support from so-called "Bush Negroes," descendants of escaped slaves who had fled into the interior. Although the truce was ratified by the National Assembly (July 21), the Army refused to accept it. Indian groups also opposed the agreement, fearing for their safety. The insurgency continued.

1990, June–September. Army Offensives. The Army invented a series of offensives against the SLA.

Uruguay

1976, June 12. Coup d'État. President **Juan Maria Bordaberry** was removed by the armed forces in a bloodless coup. Vice President **Alberto Demichelli** temporarily assumed the presidency.

1982, September 28. Military Government Lifts Ten-Year Ban on Political Parties. In preparation for November elections and the restoration of democracy, the government lifted a 10-year ban on political parties.

1985, March–1989, November. Transition to Civilian Rule. Following the military's successful suppression of the Matamoros guerrillas, and bowing to increasing public pressure to restore civilian rule, the ruling military junta supervised a 4-year transitional period.

1989, April 16. Referendum on Military Amnesty. After considerable public debate, a referendum was approved endorsing a law (passed in December 1986) granting amnesty to military personnel for human rights violations committed under military rule (1973–1985). This met the junta's requirement for completion of the transition period and a full return to civilian rule as a result of elections (November 26).

Venezuela

1975–1991. Continued Communist Insurgency. Government troops continued to combat low-level Communist insurgency.

1989, February 27–March 3. Unrest and Riots. Acting to satisfy International Monetary fund requirements, President Carlos Andrés Pérez announced oil and gasoline price increases (February 27). Riots broke out the same day all over the country, but were especially severe in Caracas. Andrés Pérez declared martial law (February 28), and announced wage increases (March 1). Unrest continued for several days until military and police forces were able to restore order. The government reported that 300 had died during the 5 days of riots.

AUSTRALASIA/OCEANIA

Australia

1988. Defense White Paper. Pursuant to its first major reexamination of defense policy since shortly after World War II, the Australian government published its 1988 Defense White Paper. Primary among its issues were 3 principal considerations. First, major conventional war for Australia was judged unlikely, which consequently could safely reduce its defense establishment. Second, security requirements in the Indian Ocean basin demanded a major redeployment of naval forces from the east coast to the west, with major forces based at Perth. Finally, the potential for low-level military actions, including terrorism and potential guerrilla activity, required an increased commitment to defend Australia's thinly populated and vulnerable northeast regions. The White Paper called for greater airmobile capacity and reconnaissance forces to secure the northeastern coast.

New Zealand

1985, November 4. French Agents Sink Greenpeace Vessel in New Zealand. The *Rainbow Warrior,* owned and operated by the international ecology organization Greenpeace, was sunk in Auckland harbor by explosives planted on the vessel's hull. A photographer was killed in the attack, but most of the vessel's crew was ashore at the time and escaped injury. Revelation that the sabotage had been executed by French agents sparked a furor; *Rainbow Warrior* had previously attempted to block French nuclear tests in the Pacific. Although others had planted the charge, a New Zealand court convicted Major **Alain Mafart** and Captain **Dominique Prieur** of manslaughter, since they had directed the effort.

1986, June–August 11. United States Ends ANZUS Relationship with New Zealand. Following the election of a Labor government headed by **David Lange** (July 14, 1984), New

Zealand banned nuclear-armed or nuclear-powered ships from its ports as part of the Nuclear-free New Zealand Act (June 1986). Since this conflicted with U.S. policy "neither to confirm nor deny" the presence or absence of such weapons on its ships, New Zealand applied the law to U.S. vessels. The U.S. decided to suspend transfer of intelligence information to New Zealand, so long as the nuclear-free policy continued (August 11, 1986).

New Caledonia

1984. Ethnic Disputes. Tensions rose between the native Kanaks or Melanesians (45% of the people) and those of European (37%), Polynesian (12%), or Indonesian and Vietnamese descent (6%). The situation was worsened by the French Lemoine statute, which granted full self-government in territorial affairs. The Kanak Socialist National Liberation Front (FLNKS) blocked roads outside the capital of Noumea to isolate it from the countryside. Order was restored when authorities agreed to reformulate plans for independence to secure rights for the Kanaks. In subsequent elections, the FLNKS won a majority of seats outside of Noumea in elections for the territorial assembly (1985).

1986–1988. Continuing Unrest. Despite efforts by the French government to arrange compromise, both Europeans and the FLNKS opposed independence plans, and the FLNKS sponsored strikes, boycotts, demonstrations, and other protests.

1988, April 21–May 5. Hostage Crisis. Clashes between pro-independence Kanak guerrillas and French troops left 7 dead (April 21–May 4). The crisis came to a head when guerrillas kidnapped a number of Europeans, including several *gendarmes*, who were killed by the guerrillas. The surviving hostages were freed by a French commando raid on the guerrilla camp at Ouvéa, leaving 2 commandos and 19 Kanaks dead (May 5). Kanak claims of excessive zeal by the commandos in the so-called "Ouvéa Massacre" were supported by the French De-

fense Ministry, which admitted that some commandos had committed "acts contrary to duty" (May 30).

1989, May. FLNKS Leaders Assassinated. The compromise Matignon Accord (late 1988), which reduced tensions and promised a chance at peaceful settlement, left radical pro-independence members of the FLNKS deeply dissatisfied. The president and vice president of the FLNKS were assassinated by pro-independence radicals. Tensions eased following conciliatory gestures by the government and new elections for the territorial assembly (October).

Fiji

1987, May. Coup d'Etat. Following the election of an Indian-based government (April), elements of the Fijian army, led by Lieutenant Colonel **Sitiveni Rabuka,** deposed the government in a bloodless coup. This reflected deep-seated ethnic tensions, and native Fijian fears of an Indian-dominated government. Rabuka demanded special gurantees of Fijian rights and permanent control of the government. The Governor-General declared a state of emergency, and negotiated a compromise, dependent on new elections and revision of the constitution (August–September).

1987, September. Coup d'État. Frustrated by lack of political progress, Rabuka led another coup and reestablished direct military rule.

Kiribati

1979, July 12. Independence from Great Britain. This nation, with its capital at Bairiki on Tarawa, comprises 33 islands (20 inhabited) in the Gilbert Islands.

Solomon Islands

1978, July 7. Independence from Great Britain. This nation, with its capital at Honiara on Guadalcanal, comprises the double chains of

islands stretching southeast from Rabaul and
New Britain.

Tuvalu

**1979, October 1. Independence from Great
Britain.** This archipelago, with its capital at
Fongafale on Funafuti, was formerly known as
the Ellice Islands.

Vanuatu

**1980, July 30. Independence from France
and Britain.** This nation, with its capital at
Vila on Efate Island, comprises 13 major islands, all former British and French colonies.

BIBLIOGRAPHY*

Note: Those references preceded by an asterisk refer to more than one historical period.

GENERAL**

Albion, Robert G. *Introduction to Military History*. New York: 1929.
Almirante, Gen. D. José. *Bosquejo de la historia militar de España hasta la fin del siglo XVIII España*. 4 vols. Madrid: 1923.
Ardant du Picq, Charles J.J.J. *Battle Studies*. Harrisburg: 1947.
Ballard, George A. *Rulers of the Indian Ocean*. London: 1927.
Belloc, Hilaire. *The Battleground; Syria and Palestine*. Philadelphia: 1936.
Bodart, Gaston. *Losses of Life in Modern Wars*. Oxford: 1916.
———. *Militär-historisches Kriegs-Lexicon, 1618–1905*. Vienna and Leipzig: 1908.
Bretnor, Reginald. *Decisive Warfare*. Harrisburg: 1969.
Brodie, Bernard. *Seapower in the Machine Age*. Princeton, N.J.: 1943.
Brodie, Bernard, and Brodie, Fawn. *From Crossbow to H-Bomb*. New York: 1962.
Cady, John F. *Southeast Asia*. New York: 1964.
Churchill, Winston. *A History of the English-Speaking Peoples*. 4 vols. New York: 1956–1958.
Clausewitz, Carl von. *On War*. Washington: 1956.
Crafts, Alfred, and Buchanan, Percy. *A History of the Far East*. London: 1958.
Craig, Gordon A. *The Politics of the Prussian Army*. New York: 1956.
Creasy, Sir Edward S. *The Fifteen Decisive Battles*. London: 1908.
———. *History of the Ottoman Turks*. 2 vols. London: 1878.
Creswell, John. *Generals and Admirals*. London: 1952.
Delbrueck, Hans. *Geschichte der Kriegskunst im Rahmen der Politischen Geschichte*. 7 vols. Berlin: 1900–1936.
Dodge, Theodore Ayrault. *Great Captains*. Boston: 1895.
Dupuy, R. Ernest. *The Compact History of the United States Army*. New York: 1961.
———. *Men of West Point*. New York: 1951.
———. *Where They Have Trod*. Philadelphia: 1940.
Dupuy, R. Ernest, and Dupuy, Trevor N. *Brave Men and Great Captains*. New York: 1959.
———. *Military Heritage of America*. New York: 1956.†
Dupuy, Trevor N. *Attrition*. Fairfax, Va.: 1990.
———. *The Evolution of Weapons and Warfare*. New York: 1980.
———. *A Genius for War: The German Army and General Staff, 1807–1945*. New York: 1977.
———. *Numbers, Predictions, and War*. New York: 1979.
———. *Understanding Defeat*. New York: 1990.

 * This list does not by any means include all of the works consulted during twenty years of intensive research. It does, however, include most of those that we believe have sufficient importance to be listed as references for further study and reading.
** Apply to three or more historical periods.
† Contains additional bibliography.

_____. *Understanding War*. New York: 1987.

Earle, Edward M., *et al*. *Makers of Modern Strategy*. Princeton, N.J.: 1943.†

Edmonds, James E. *Fighting Fools*. New York: 1938.

Eggenberger, David. *A Dictionary of Battles*. New York: 1967.†

Esposito, Vincent J. (ed.) *The West Point Atlas of American Wars*. New York: 1959.†

Falls, Cyril. *The Art of War from the Age of Napoleon to the Present Day*. New York: 1961.

_____. *A Hundred Years of War, 1850–1950*. New York: 1962.

Fortescue, John W. *A History of the British Army*. 13 vols. New York: 1930.

Fuller, J. F. C. *Armament and History*. New York: 1945.

_____. *The Conduct of War, 1789–1961*. New Brunswick, N.J.: 1961.

_____. *A Military History of the Western World*. 3 vols. New York: 1954.

Ganoe, William A. *The History of the United States Army*. New York: 1942.

Goerlitz, Walter. *History of the German General Staff, 1857–1945*. New York: 1953.

Hall, D. G. E. *A History of Southeast Asia*. London: 1955.

Harbottle, Thomas B. *A Dictionary of Battles*. New York: 1905. (Republished Detroit: 1966.)

Hazard, Harry W. *Atlas of Islamic History*. Princeton: 1954.

Heinl, Robert D., Jr. *Dictionary of Military and Naval Quotations*. Annapolis: 1966.

_____. *Soldiers of the Sea: The United States Marine Corps, 1775–1962*. Annapolis: 1962.

Herrin, Hubert. *A History of Latin America*. New York: 1962.

Historical Evaluation and Research Organization. *Historical Trends Related to Weapon Lethality*. Washington: 1964.

Hitti, Philip K. *The Near East in History*. New York: 1961.

Hittle, James D. *The Military Staff*. Harrisburg: 1949.

Jones, Archer. *The Art of War in the Western World*. Chicago and Urbana, Ill.: 1987.

Kaiser, David. *Politics and War: European Conflict from Philip II to Hitler*. Cambridge, Mass.: 1990.

Keegan, John. *The Face of Battle*. New York: 1976.†

Kennedy, Paul. *The Rise and Fall of the Great Powers*. New York: 1987.†

Langer, William L. *An Encyclopedia of World History*. Boston: 1952.

Larousse Encyclopedia of Modern History. New York: 1964.

Latourette, Kenneth S. *The Chinese: Their History and Culture*. New York: 1934.

Lawford, James, ed. *The Cavalry*. New York: 1976.

Leeb, Wilhelm von. *Defense*. Harrisburg: 1943.

Lewis, Michael. *The Navy of Britain*. London: 1948.

Liddell Hart, Basil H. *The Decisive Wars of History*. London: 1929.

Mahan, Alfred T. *The Influence of Seapower upon History, 1660–1783*. Boston: 1890.

Manucy, Albert. *Artillery Through the Ages*. Washington: 1949.

McNeil, William H. *The Pursuit of Power: Technology, Armed Force, and Society since A.D. 1000*. Oxford: 1982.†

_____. *The Rise of the West*. Chicago: 1963.

Mitchell, William A. *Outlines of the World's Military History*. Washington: 1931.

Montross, Lynn. *War Through the Ages*. New York: 1946.

Morris, Richard B. *Encyclopedia of American History*. New York: 1953.

Morison, Samuel E. *The Oxford History of the American People*. New York: 1965.

Mrazek, James. *The Art of Winning Wars*. New York: 1968.

New Cambridge Modern History. 12 vols. Cambridge, Eng.: 1951.

Oliver, Roland, and Fage, J. D. *A Short History of Africa*. London: 1962.

Palmer, John M. *Washington, Lincoln, Wilson*. New York: 1930.

Paret, Peter, ed. *Makers of Modern Strategy from Machiavelli to the Nuclear Age*. Princeton, N.J.: 1986.†

Payne, L. G. S. *Air Dates*. New York: 1957.

Peterson, Harold L. *A History of Firearms*. New York: 1961.

Phillips, Thomas R. (ed.) *Roots of Strategy*. Harrisburg: 1940.

Potter, E. B., and Nimitz, Chester W. (eds.). *Sea Power*. New York: 1960.

Preston, Richard A.; Wise, Sydney F.; and Werner, Herman O. *Men in Arms*. New York: 1962.

Reischauer, Edwin O., and Fairbank, John. *A History of East Asian Civilization*. Boston: 1960.

Richardson, Lewis Fry. *Statistics of Deadly Quarrels*. Pittsburgh: 1960.

Ropp, Theodore. *War in the Modern World*. Durham: 1959.†

Schlieffen, Alfred von. *Cannae*. Leavenworth: 1931.

Shepherd, William R. *Historical Atlas*. New York: 1929.

Spaulding, Oliver L. *The United States Army in War and Peace*. New York: 1937.

Spaulding, Oliver L.; Nickerson, Hoffman; and Wright, John W. *Warfare: A Study of Military Methods from the Earliest Times*. Washington: 1937.

Spear, Percival. *India*. Ann Arbor: 1961.

Sprout, Harold, and Sprout, Margaret. *The Rise of American Naval Power*. Princeton: 1939.

————. *Toward a New Order of Sea Power*. Princeton: 1940.

Stacy, C. P. *Military History for Canadian Students*. Ottawa: 1953.

Starr, Chester, *et al*. *A History of the World*. Chicago: 1960.

Steele, Matthew F. *American Campaigns*. Washington: 1909.

Todd, Frederick P., and Kredel, Fritz. *Soldiers of the American Army, 1775–1954*. Chicago: 1954.

Toynbee, Arnold. *A Study of History*. Somervell abridgment. 2 vols. New York: 1947–1957.

Turner, Gordon. *A History of Military Affairs in Western Society Since the Eighteenth Century*. New York: 1953.

Upton, Emory. *The Military Policy of the United States*. Washington: 1917.

United States Army. *The Army Almanac*. Washington: 1950.

United States Military Academy. *Summaries of Selected Military Campaigns*. West Point: 1952.

Van Creveld, Martin. *Supplying War: Logistics from Wallenstein to Patton*. Cambridge: 1980.†

Wright, Quincy. *A Study of War*. 2 vols. Chicago: 1942.

ANCIENT WARFARE
(CHAPTERS I–VII, TO A.D. 600)

Adcock, Frank E. *The Greek and Macedonian Art of War*. Berkeley, Calif.: 1957.

————. *The Roman Art of War under the Republic*. Cambridge, Mass.: 1940.

Caesar, Julius. *Commentaries*. (Edited and translated by John Warrington.) London: 1953.

Cambridge Ancient History. 12 vols., with 5 vols. of plates. Cambridge, Eng.: 1923–1939.

Connolly, Peter. *Greece and Rome at War*. London: 1981.

Cottrell, Leonard. *Hannibal, Enemy of Rome*. New York: 1961.

Diodorus Siculus. *Bibliotheca Historica*. (English translation by C. H. Oldfather. Loeb Series.) 12 vols. Cambridge, Mass.: 1933–1957.

Dodge, Theodore Ayrault. *Alexander the Great*. Boston: 1890.

————. *Hannibal*. Boston: 1891.

————. *Julius Caesar*. Boston: 1892.

Engels, Donald W. *Logistics of Alexander the Great and the Macedonian Army*. Berkeley, Calif.: 1978.†

Ferrill, Arther. *The Fall of the Roman Empire*. London: 1986.

Fuller, J. F. C. *The Generalship of Alexander the Great*. New Brunswick: 1958.

————. *Julius Caesar*. New Brunswick: 1965.

*Gibbon, Edward. *Decline and Fall of the Roman Empire*. 3 vols. New York: 1953.

Hanson, Victor D. *The Western Way of War: Infantry Battle in Classical Greece*. New York: 1989.†

Holy Bible, The. Various versions and editions.

Kagan, Donald. *The Archidamian War*. Ithaca, N.Y.: 1974.

_____. *The Fall of the Athenian Empire*. Ithaca, N.Y.: 1981.

_____. *The Outbreak of the Peloponnesian War*. Ithaca, N.Y.: 1969.

_____. *The Peace of Nicias and the Sicilian Expedition*. Ithaca, N.Y.: 1978.

Larousse Encyclopedia of Ancient and Medieval History. New York: 1963.

Liddell Hart, Basil H. *A Greater Than Napoleon—Scipio Africanus*. Edinburgh: 1926.

Livy. *History of Rome*. (English translation by B. O. Foster, E. T. Sage, and A. C. Schlesinger. Loeb Series.) 14 vols. Cambridge, Mass.: 1919–1957.

Mommsen, Theodor. *History of Rome*. 5 vols. New York: 1895.

Plutarchus. *Lives of Themistocles, etc*. New York: 1937.

Polybius. *Histories*. (English translation by W. R. Paton. Loeb Series.) Cambridge, Mass.: 1922–1927.

Robinson, Charles Alexander, Jr. *Alexander the Great*. New York: 1963.

Starr, Chester G. *A History of the Ancient World*. New York: 1965.

Sun Tzu. *The Art of War*. (Translated by Samuel B. Griffith.) New York: 1963.

Tarn, William W. *Alexander the Great*. Cambridge, Eng.: 1948.

Thucydides. *History of the Peloponnesian War*. London: 1954.

Vegetius. *The Military Institutions of the Romans*. In T. R. Phillips (ed.). *Roots of Strategy*. Harrisburg: 1940.

Warry, John. *Warfare in the Classical World*. London: 1980.

Xenophon. *Anabasis*. (English translation of complete works by W. Miller *et al*. Loeb Series.) Cambridge, Mass.: 1914–1925.

_____. *Cyropaedia*. (English translation of complete works by W. Miller *et al*. Loeb Series.) Cambridge, Mass.: 1914–1925.

MEDIEVAL WARFARE
(CHAPTERS VIII–XII, 600–1500)

Allmand, C. T. *Society at War: The Experience of England and France during the Hundred Years' War*. Edinburgh: 1973.†

Beeler, John. *Warfare in Feudal Europe, 730–1200*. Ithaca, N.Y.: 1971.†

Burne, Alfred H. *The Aquincourt War*. London: 1956.

_____. *The Crécy War*. London: 1955.

Cambridge Medieval History. 8 vols. Cambridge, Eng.: 1924–1936.

Charol, Michael (pseud.: Michael Prawdin). *The Mongol Empire*. London: 1952.

Contamine, Philippe. *War in the Middle Ages*. Trans. by Michael Jones. Oxford: 1984.

Costain, Thomas B. *The Last Plantagenets*. New York: 1962.

Glubb, Sir John Bagot. *The Great Arab Conquests*. London: 1963.

Grousset, René. *Conqueror of the World: The Life of Ghingis Khan*. New York: 1966.

_____. *The Epic of the Crusades*. New York: 1971.

Hewitt, H. J. *The Organization of War under Edward III, 1338–1362*. Manchester: 1962.†

Kendall, Paul Murray. *The Yorkist Age*. New York: 1962.

Lamb, Harold. *Charlemagne*. New York: 1954.

_____. *The Crusades*. New York: 1931.

_____. *The Earthshakers*. New York: 1949.

_____. *Genghis Khan, Emperor of All Men*. New York: 1927.

*Mallett, Michael. *Mercenaries and Their Masters: Warfare in Renaissance Italy*. London: 1974.†

*Mallett, Michael, and J. R. Hale. *The Military Organization of a Renaissance State: Venice, c. 1400–1617*. Cambridge: 1984†

Oman, Charles. *A History of the Art of War in the Middle Ages*. London: 1924.

Prescott, William Hickling. *History of the Reign of Ferdinand and Isabella.* 2 vols. New York: 1837.
Smail, R. C. *Crusading Warfare (1097–1193).* Cambridge: 1956†
Vale, M. G. A. *War and Chivalry: Warfare and Aristocratic Culture in England, France, and Burgundy at the End of the Middle Ages.* London: 1981.†
Verbruggen, J. F. *The Art of War in Western Europe during the Middle Ages.* Amsterdam: 1977.
Waley, Arthur (ed.). *The Secret History of the Mongols.* London: 1964.
Wise, Terence. *Medieval Warfare.* New York: 1976.

Early Modern Warfare
(Chapters XIII–XV, 1500–1750)

Camon, Hubert. *Deux grands chefs de guerre du XVII siecle: Condé et Turenne.* Paris: 1899.
Chandler, David. *The Art of Warfare in the Age of Marlborough.* New York: 1976.†
_____. *Marlborough as Military Commander.* New York: 1973.
Dodge, Theodore Ayrault. *Gustavus Adolphus.* Boston: 1890.
*Duffy, Christopher J. *Fire and Stone: The Science of Fortress Warfare, 1660–1860.* Newton Abbot: 1975.
_____. *The Military Experience in the Age of Reason.* New York: 1988.
_____. *Siege Warfare: The Fortress in the Early Modern World, 1494–1660.* London: 1979.
Godley, Eveline. *The Great Condé.* London: 1915.
Guilmartin, J. F., Jr. *Gunpowder and Galleys: Changing Technology and Mediterranean Warfare at Sea in the Sixteenth Century.* Cambridge: 1974.†
Heilmann, Johann Ritter von. *Das Kriegswesen der Kaiserlichen und Schweden zur Zeit des dreissigjaehrigen Krieges.* Leipzig: 1850.
Machiavelli, Niccolò. *The Art of War.* Albany: 1815.
MacMunn, George. *Gustavus Adolphus.* New York: 1931.
Mattingly, Garrett. *The Armada.* New York: 1959.
Oman, Charles. *A History of the Art of War in the Sixteenth Century.* New York: 1937.
Parker, Geoffrey. *The Army of Flanders and the Spanish Road, 1567–1659.* Cambridge: 1972.
_____. *The Military Revolution: Military Innovation and the Rise of the West, 1500–1800.* Cambridge, 1988.†
Prescott, William H. *The History of the Conquest of Mexico.* New York: 1843.
Roberts, Michael. *Gustavus Adolphus: A History of Sweden, 1611–1632.* 2 vols. London: 1953–1958.
Sweden, Armen Generalstaben. Krigshistoriska avdelningen. *Sveriges Krig, 1611–1632.* 6 vols. Stockholm: 1936–1939.
Taylor, Frederick Louis. *The Art of War in Italy, 1494–1529.* Reprint. Westport: 1973.
Thompson, James Westfall. *The Wars of Religion in France.* New York: 1958.
Villermont, Antoine Charles Hennequin, comte de. *Tilly; ou La guerre de trente ans de 1618 à 1632.* 2 vols. Paris: 1860.
Wedgewood, C. V. *The Thirty Years' War.* New York: 1961.

The Century of Revolution
(Chapters XVI and XVII, 1750–1850)

Ballard, Colin. *Napoleon: An Outline.* New York: 1924.
Chandler, David G. *The Campaigns of Napoleon.* New York: 1966.†
Dodge, Theodore Ayrault. *Napoleon.* 4 vols. Boston: 1904.
Duffy, Christopher. *The Army of Frederick the Great.* New York: 1974.†
Dupuy, R. Ernest. *Battle of Hubbardton.* Montpelier: 1960.
Elting, John R. *Swords around the Throne: Napoleon's Grande Armée.* New York: 1988†

Haythornthwaite, Philip J. *The Napoleonic Sourcebook.* New York: 1990.

Jacobs, James R., and Tucker, Glenn. *Compact History of the War of 1812.* New York: 1969.

Johnson, Curtis J. *Battles of the American Revolution.* London: 1975.

Ludwig, Emil. *Napoleon.* New York: 1926.

Mahan, Alfred T. *The Influence of Seapower on the French Revolution.* Boston: 1893.

_____. *Life of Nelson.* 2 vols. Boston: 1900–1907.

Mitchell, Lt. Col. Joseph B. *Decisive Battles of the American Revolution.* New York: 1962.

Napier, Sir Williams. *History of the War in the Peninsula and in the South of France from the Year 1807 to the Year 1814.* London and New York: n.d. (First English edition 1828–1840.)

Parkman, Francis. *Montcalm and Wolfe.* Boston: 1905.

Phipps, Col. Ramsay W. *The Armies of the First French Republic and the Rise of the Marshals of Napoleon.* 5 vols. Cambridge: 1926–1939.

Prucha, Francis Paul. *The Sword of the Republic: the United States Army on the Frontier, 1783–1846.* Bloomington, Ind.: 1977.

Reiners, Ludwig. *Frederick the Great.* New York: 1960.

Roosevelt, Theodore. *The Naval War of 1812.* New York: 1894.

Rothenburg, Gunther E. *The Art of Warfare in the Age of Napoleon.* Bloomington, Ind.: 1980.

Runciman, Sir Walter. *The Tragedy of St. Helena.* New York: 1911.

Scheer, George F., and Rankin, Hugh F. (eds.). *Rebels and Redcoats.* Cleveland: 1957.

Simon, Edith. *The Making of Frederick the Great.* Boston: 1963.

Smith, Justin H. *The War with Mexico.* 2 vols. New York: 1919.

Thaddeus, Victor. *Frederick the Great, the Philosopher King.* New York: 1930.

Trevelyan, George O. *The American Revolution.* 4 vols. London: 1921.

Tucker, Glenn. *Poltroons and Patriots.* 2 vols. Indianapolis: 1954.

Ward, Christopher. *The War of the Revolution.* 2 vols. New York: 1952.

Yorck von Wartenburg, Maximilian. *Napoleon as a General.* 2 vols. London: 1897.

THE EMERGENCE OF THE PROFESSIONAL
(CHAPTER XVIII, 1850–1900)

Ballard, Colin. *The Military Genius of Abraham Lincoln.* Cleveland: 1965.

Bond, Brian (ed.) *Victorian Military Campaigns.* London: 1967.†

Burke, John G. *On the Border with Crook.* New York: 1891.

Catton, Bruce. *The Army of the Potomac.* 3 vols. New York: 1962.

_____. *Centennial History of the Civil War.* 3 vols. New York: 1965.

Churchill, Winston S. *The River War.* London: 1899.

Commager, Henry S. *The Blue and the Gray.* 2 vols. Indianapolis: 1950.

Downey, Fairfax. *Indian-Fighting Army.* New York: 1941.

Dupuy, R. Ernest, and Dupuy, Trevor N. *The Compact History of the Civil War.* New York: 1960.

Farwell, Byron. *Queen Victoria's Little Wars.* New York: 1972.†

Freeman, Douglas S. *Lee's Lieutenants.* 3 vols. New York: 1944.

_____. *R. E. Lee.* 4 vols. New York: 1949.

Fuller, J. F. C. *The Generalship of Ulysses S. Grant.* New York: 1929.

_____. *Grant and Lee.* London: 1933.

Furneaux, Rupert. *The Breakfast War.* New York: 1958.

Gibbs, Peter. *Crimean Blunder.* New York: 1960.

Grant, Ulysses S. *Personal Memoirs.* New York: 1895.

Hattaway, Herman, and Archer Jones. *How the North Won the Civil War.* Urbana, Ill.: 1983.

Hattaway, Herman, Archer Jones, and William S. Still. *Why the South Lost the Civil War.* Athens, Ga.: 1986.

Henderson, George F. R. *Stonewall Jackson and the American Civil War.* London: 1898.

Howard, Michael. *The Franco-Prussian War.* New York: 1962.†

Johnson, Robert Underwood, and Buel, Clarence Clough (eds.). *Battles and Leaders of the Civil War.* 4 vols. New York: 1884–1888.

Livermore, Thomas L. *Numbers and Losses in the Civil War.* Boston: 1901.

McPherson, James B. *Battle Cry of Freedom.* London and New York: 1988.

Mahan, Alfred T. *Lessons of the War with Spain.* Boston: 1899.

Millis, Walter. *The Martial Spirit: A Study of Our War with Spain.* Boston: 1931.

Morris, Donald R. *The Washing of the Spears: The Rise and Fall of the Zulu Nation.* New York: 1965.†

Nye, W. S. *Carbine and Lance.* Norman: 1942.

Pakenham, Thomas. *The Boer War.* London and New York: 1979.†

Porch, Douglas. *The Conquest of Morocco.* New York: 1983.

Sandburg, Carl. *Abraham Lincoln.* New York: 1959.

Utley, Robert M. *Frontier Regulars: The United States Army and the Indian, 1866–1891.* Bloomington, Ind.: 1977.†

Whease, K. C. *Lincoln.* New York: 1948.

Williams, Kenneth P. *Lincoln Finds a General.* 5 vols. New York: 1959.

Williams, T. Harry. *Lincoln and His Generals.* New York: 1952.

Woodham-Smith, Cecily. *The Reason Why.* New York: 1953.

Wyeth, John A. *That Devil Forrest: Life of General Nathan Bedford Forrest.* New York: 1959.

WORLD WAR I AND THE ERA OF TOTAL WAR
(CHAPTER XIX, 1900–1925)

Ayres, Leonard P. *The War with Germany: A Statistical Summary.* Washington: 1919.

Baldwin, Hanson. *World War I.* New York: 1963.

Balfour, Patrick, Baron Kinross. *Ataturk.* New York: 1965.

Barnett, Corelli. *The Swordbearers: Supreme Command in the First World War.* New York: 1964.

Bloch, Ivan S. *The Future of War … Is War Now Impossible?* St. Petersburg, 1898.

Buchan, John. *A History of the Great War.* Boston: 1922.

Churchill, Winston. *The World Crisis.* 6 vols. New York: 1932.

Clubb, O. Edmund. *Twentieth-Century China.* New York: 1964.

*Dornbusch, Charles E. *Histories of American Army Units.* Washington: 1956.

Dupuy, R. Ernest. *Five Days to War.* Harrisburg: 1967.

———. *Perish by the Sword.* Harrisburg: 1939.

Dupuy, R. Ernest, and Baumer, William H. *Little Wars of the U.S.* New York: 1969.

Dupuy, Trevor N., *et al. Military History of World War I.* 12 vols. New York: 1967.

Falls, Cyril. *The Great War, 1914–1918.* New York: 1959.

Fischer, Fritz. *Germany's Aims in the First World War.* New York: 1967.

Fuller, J. F. C. *Machine Warfare.* Washington: 1943.

Hamilton, Ian A. S. *A Staff Officer's Scrap Book.* London: 1905.

Hayes, Grace P. *Compact History of World War I.* New York: 1972.†

Higgins, Trumbull. *Winston Churchill and the Dardanelles.* New York: 1957.

Hindenburg, Paul von. *Out of My Life.* 2 vols. New York: 1921.

Hoffmann, Max. *The War of Lost Opportunities.* New York: 1925.

Horne, Alistair. *The Price of Glory: Verdun 1916.* New York: 1962.

James, Robert R. *Gallipoli.* London: 1965.

*Kennan, George F. *Russia and the West.* Boston: 1962.
_____. *Soviet Foreign Policy, 1917–1941.* Princeton: 1960.
Landor, A. H. S. *China and the Allies.* 2 vols. New York: 1901.
Lawrence, T. E. *Seven Pillars of Wisdom.* New York: 1926.
Liman von Sanders, Otto V. K. *My Five Years in Turkey.* Annapolis: 1927.
*Liu, F. F. *A Military History of Modern China.* Princeton: 1956.
Ludendorff, Erich. *Ludendorff's Own Story.* New York: 1920.
MacArthur, Douglas. *Reminiscences.* New York: 1964.
McEntee, Gerard L. *Military History of the World War.* New York: 1937.
Millis, Walter. *The Road to War.* Boston: 1935.
Moorehead, Alan. *Gallipoli.* New York: 1945.
_____. *The Russian Revolution.* New York: 1958.
Pitt, Barrie. *1918—The Last Act.* New York: 1963.
Sexton, William T. *Soldiers in the Philippines.* Harrisburg: 1939.
Showalter, Dennis E. *Tannenburg: Clash of Empires.* Hamden, Conn.: 1991.†
Stamps, T. Dodson, and Esposito, Vincent J. *A Short History of World War I.* West Point: 1950.
Taylor, Edmond. *The Fall of the Dynasties.* Garden City: 1963.
The Times, London. *The Times Diary and Index of the War, 1914 to 1918.* London: 1921.
Tschuppik, Karl. *Ludendorff: The Tragedy of a Military Mind.* Boston: 1932.
Tuchman, Barbara. *The Guns of August.* New York: 1962.†
U.S. War Department. *Reports of the Chief of Staff.* Washington: 1913–1915, incl.
Warner, Denis and Peggy. *The Tide at Sunrise: A History of the Russo-Japanese War 1904–1905.* New York: 1974.
Wheeler-Bennett, John W. *Nemesis of Power: The German Army in Politics, 1918–1945.* 2nd ed. New York: 1964.
_____. *Wooden Titan: Hindenburg in Twenty Years of German History, 1914–1934.* New York: 1936.
Wavell, Archibald. *Allenby—A Study in Greatness.* New York: 1941.
Young, Brig. Peter, ed. *The Marshall-Cavendish Illustrated Encyclopedia of World War I.* 12 vols. London and New York: 1986.

WORLD WAR II AND THE DAWN OF THE NUCLEAR AGE
(CHAPTER XX, 1925–1945)

Allen, Louis. *The Longest Campaign: Burma, 1942–1945.* London: 1981.
Auphan, Paul, and Mordal, Jacques. *The French Navy in World War II.* Annapolis: 1959.
Baldwin, Hanson W. *Battles Lost and Won: Great Campaigns of World War II.* New York: 1966.
Barnett, Corelli, ed. *Hitler's Generals.* New York: 1989.
Beevor, Antony. *The Spanish Civil War.* London: 1982.†
Blumenson, Martin. *Anzio: The Gamble That Failed.* Philadelphia: 1963.
_____. *The Duel for France.* Boston: 1963.
Bryant, Arthur. *Triumph in the West.* New York: 1959.
_____. *The Turn of the Tide.* New York: 1957.
Bullock, Alan. *Hitler: A Study in Tyranny.* New York: 1962.
Bush, Vannevar. *Modern Arms and Free Men.* New York: 1949.
Butow, J. C. *Japan's Decision to Surrender.* Stanford, Calif.: 1954.
Chew, Allen F. *The White Death: The Epic of the Russo-Finnish Winter War.* Lansing, Mich.: 1972.†
Churchill, Randolph. *Winston S. Churchill.* 2 vols. Boston: 1967.
Churchill, Winston S. *The Second World War.* 6 vols. Boston: 1948–1953.

Clark, Alan. *Barbarossa: The Russian-German Conflict, 1941–45.* New York: 1965.

Coox, Alvin D. and Saburo Hayashi. *Koqun: The Japanese Army in the Pacific War.* Quantico, Va.: 1957.

Cowles, Virginia. *Winston Churchill.* New York: 1953.

Craven, Wesley F., and Cate, James L. (eds.). *The Army Air Forces in World War II.* 6 vols. Chicago: 1948–1955.

De Gaulle, Charles. *War Memoirs.* 3 vols. London: 1959.

Dorn, Frank. *The Sino-Japanese War, 1937–1941.* New York: 1974.

Drea, Edward J. *Nomonhan: Japanese-Soviet Tactical Combat, 1939.* Ft. Leavenworth, Kan.: 1981.

Dupuy, R. Ernest. *St. Vith--Lion in the Way.* Washington: 1949.

*Dupuy, Trevor N. *Military History of the Chinese Civil War.* New York: 1969.

_____. *Military History of World War II.* 19 vols. New York: 1966.

Dupuy, Trevor N., and Martell, Paul. *Great Battles of the Eastern Front.* New York: 1982.

Eisenhower, Dwight D. *Crusade in Europe.* New York: 1948.

Erickson, John. *The Road to Stalingrad.* New York: 1975.

_____. *The Road to Berlin.* Boulder, Colo.: 1983.

Feis, Herbert. *The Atomic Bomb and the End of the War in the Pacific.* Princeton, N.J., 1961.

Frank, Richard B. *Guadalcanal.* New York: 1990.†

Freiden, Seymour, and Richardson, William. *The Fatal Decisions.* New York: 1956.

Fuller, J. F. C. *The Second World War.* New York: 1949.

Gilbert, Felix. *Hitler Directs His War: The Secret Records of His Daily Military Conferences.* New York: 1950.

Glubb, Sir John B. *War in the Desert.* London: 1960.

Griffith, Samuel B. *The Chinese People's Army.* New York: 1967.

Guderian, Heinz. *Panzer Leader.* New York: 1952.

Higgins, Trumbull. *Winston Churchill and the Second Front.* New York: 1957.

History of the Second World War; United Kingdom Military Series. 24 vols. London: 1956–1969.

The Illustrated London News. *Winston Churchill, the Greatest Figure of Our Time.*

Isely, Jeter A., and Crowl, P. A. *The U.S. Marines and Amphibious War.* Princeton, N.J.: 1951.

Jackson, Gabriel. *The Spanish Republic and the Civil War, 1931–1939.* Princeton, N.J.: 1965.†

Jackson, W. G. F. *The Battle for Italy.* New York: 1967.

Liddell Hart, Basil H. *The German Generals Talk.* London: 1968.

_____. *The Liddell Hart Memoirs.* 2 vols. New York: 1966.

Manstein, Eric von. *Lost Victories.* Chicago: 1958.

Marshall, George C.; Arnold, Henry H.; and King, Ernest J. *The War Reports. . . .* Philadelphia: 1947.

Marshall, Samuel L. A. *Men Against Fire.* Washington: 1947.

Moran, Lord. *Churchill from the Diaries of Lord Moran.* Boston: 1966.

Morison, Samuel E. *History of United States Naval Operations in World War II.* 15 vols. Boston: 1962.

_____. *The Two-Ocean War.* Boston: 1963.

Pitt, Barrie (ed.). *History of the Second World War.* London: 1966–1969.

Pogue, Forrest C. *George C. Marshall.* 3 vols. New York: 1969.†

Prange, Gordon W. *At Dawn We Slept: The Untold Story of Pearl Harbor.* New York: 1981.†

Prittie, Terence. *Germans Against Hitler.* Boston: 1964.

Ruge, Friedrich. *Der Zeekrieg: The German Navy's Story, 1939–1945.* Annapolis: 1957.

Salisbury, Harrison E. *The Long March: The Untold Story.* New York: 1985.†

Shirer, William L. *The Rise and Fall of the Third Reich.* New York: 1960.

Slim, William J. *Defeat into Victory.* London: 1956.

Smyth, Henry D. *Atomic Energy for Nuclear Purposes.* Princeton: 1945.

Snyder, Louis L. *Hitler and Nazism.* New York: 1961.

Spears, Edward L. *Assignment to Catastrophe.* 2 vols. New York: 1955.

Speer, Albert. *Inside the Third Reich*. New York: 1970.

Stamps, T. Dodson, and Espostio, Vincent J. *Military History of World War II*. 2 vols. West Point: 1953.

Thomas, Hugh. *The Spanish Civil War*. New York: 1961.†

Trevor-Roper, H. R. *The Last Days of Hitler*. New York: 1947.

Truscott, Lucian. *Command Missions*. New York: 1954.

*Tsou Tang. *America's Failure in China*. Chicago: 1963.

U.S. Army, Office of the Chief of Military History. *The U.S. Army in World War II*. 80 vols. Washington: 1950–1968.

U.S. Strategic Bombing Survey. *Overall Report (European War), etc*. Washington: 1945–1947.

*Wedemeyer, Albert C. *Wedemeyer Reports*. New York: 1958.

*Werth, Alexander. *France, 1940–1955*. New York: 1956.

_____. *Russia at War, 1941–1945*. New York: 1964.

White, Theodore H., and Jacoby, Annalee. *Thunder out of China*. New York: 1946.

*Willoughby, Charles A., and Chamberlain, John. *MacArthur, 1941–1951*. New York: 1954.

Wilmot, Chester. *The Struggle for Europe*. New York: 1952.

Ziemke, Earl. *From Stalingrad to Berlin*. Washington: 1968.

SUPERPOWERS IN THE NUCLEAR AGE
(CHAPTER XXI, 1945–1984)

Badri, Hassan el; Magdoub, Taha el; and Zohdy, Mohommed Dia el Din. *The Ramadan War*. Dunn Loring: 1978.

Blackett, Patrick. *Studies of War, Nuclear and Conventional*. New York: 1962.

Blair, Clay. *The Forgotten War: America in Korea, 1950–1953*. New York: 1987.†

Brodie, Bernard. *Strategy in the Missile Age*. Princeton: 1959.

Chubin, Shahran, and Charles Tripp. *Iran and Iraq at War*. London: 1988.

Cordesman, Anthony H. *The Iran-Iraq War and Western Security, 1984–1987*. London: 1987.

Dunn, Frederick; Brodie, Bernard; *et al*. *The Absolute Weapon*. New York: 1946.

Dupuy, Trevor N. *Elusive Victory: The Arab-Israeli Wars, 1947–1974*. New York: 1980.†

Dupuy, Trevor N., Curt Johnson, David L. Bongard, and Arnold C. Dupuy. *How to Defeat Saddam Hussein*. New York: 1991.†

Fall, Bernard B. *Street Without Joy*. Harrisburg: 1964.†

Fehrenbach, T. R. *This Kind of War*. New York: 1963.

Halberstam, David. *The Best and the Brightest*. New York: 1972.

Hastings, Max, and Simon Jenkins. *The Battle for the Falklands*. New York: 1983.†

Herzog, Haim. *The War of Atonement*. Tel Aviv: 1975.

Horne, Alistair. *A Savage War of Peace: Algeria, 1954–1962*. New York: 1987.†

Hsieh, Alice L. *Communist China's Strategy in the Nuclear Era*. Englewood Cliffs: 1962.

Kahn, Herman. *On Thermonuclear War*. Princeton: 1960.

Kissinger, Henry. *Nuclear Weapons and Foreign Policy*. New York: 1957.

Leckie, Robert. *Conflict: The History of the Korean War*. New York: 1962.

Marshall, S. L. A. *The Military History of the Korean War*. New York: 1963.

_____. *The River and the Gauntlet*. New York: 1953.

_____. *Sinai Victory*. New York: 1958.

Osgood, Robert. *Limited War*. Chicago: 1957.

Pinson, Koppel S. *Modern Germany: Its History and Civilization*. 2nd ed. New York: 1966.

Scott, William F., and Scott, Harrier F. *The Armed Forces of the USSR*. Boulder: 1979.

Sheehan, Neil. *A Bright Shining Lie: John Paul Vann and the American Experience in Vietnam.* New York: 1989.†

Slessor, John. *Strategy for the West.* New York: 1954.

Smith, Jean E. *The Defense of Berlin.* Baltimore: 1963.

Trager, Frank N. *Marxism in Southeast Asia.* Stanford: 1959.

U.S. Army, Office of Chief of Military History. *The U.S. Army in the Korean War.* 5 vols. Washington: 1951–1966.

GENERAL INDEX

Note: Numbers followed by an "i" indicate illustrations

Abbas I the Great, Shah of Persia, 551, 641–42
Abbas II, Shah of Persia, 642
Abbas III, Shah of Persia, 710
Al-Abbas, early Moslem leader, 252
Abbasid caliphate, 252, 286–88, 333, 423
'Abd Allah ibn Sa'd, Moslem leader, 253
Abd el Kader, Emir of Mascara, 855
Abd el Krim, Riff chieftain, 1098–99
Abd er-Rahman I, Governor of Spain, 221–23
Abd er-Rahman II, Emir of Cordova, 280
Abd er-Rahman III, Caliph of Cordova, 280
Abd er-Rahman ibn Mu'awiya, Omyyad emir of Cordova, 227
Abdullah ibn Tahir, Abbasid general, 289
Abdulla ibn Zubayr, Omayyad anti-Caliph, 250
Abdulla ibn Zubayr, Omayyad general, 250
Abd ul-Malik, Omayyad caliph, 250
Abdulmalik-al-Mozaffar, son of Al-Mansour, 324
Abe clan, Japan, 356
Abe, Hiroaki, Japanese admiral, 1259
Abercrombie, James, British general, 772
Abrams, Creighton W., U.S. general, 1316, 1329, 1332
Abrhia, Himyar viceroy, 210
Absalom, son of David, 12
Absalon, Danish soldier-statesman, 316
Abu Abdullah al-Husain, Fatimid leader, 290
Abu Ahmad al-Muwaffak, Turkish leader, 287
abu-Bakr ibn'Umar, Almoravid leader, 333
Abul Bakr Malik al-Adil, Turkish leader, 335
Abu' Bekr, first Mohammedan caliph, 249
Abu Ishak al-Mu'tasim, Abassid caliph, 287
Abu'l Abbas, Abbasid caliph, 252
Abu'l Kasim al-Kiam, Fatimid leader, 289
Abu Muslim, Abbasid revolutionary, 252
Abu Sa'id al-Jannabi, Shi'ite leader, 287
Abu Yazid Makhlad, Kharijite leader, 290

Abu Zora Tarif, Arab general, 218
Abyssinia, 176, 210, 929. *See also* Ethiopia.
Achaean League, 60, 62, 96
Acheampong, Ignatius Kutu, Ghana colonel, 1445
Acheson, Dean, U.S. Secretary of State, 1450
Achille Lauro, 1504
Adachi, Hotaze, Japanese general, 1264
Adams-Onís, Treaty of (February 22, 1819), 880
Adan, Abraham (Bren), Israeli general, 1349
Adarman, Persian general, 209
Adelaide, Queen of Italy, 279
Adenauer, Konrad, German Chancellor, 1379
Adhemar du Puy, French bishop and Crusader, 307, 338
Adherbal, Carthaginian admiral, 67, 101
Aditya I, Chola king, 292
"Admirals Revolt," U.S. Navy, 1450–51
Adoula, Cyrille, Congolese Premier, 1442
Adrian I, Pope, 219
Adrianople, Peace of (1713), 675; Treaty of (September 16, 1829), 847; Truce of (February, 1568), 536
Aduatuca (Tongres), Gaul, 114
Aegidus, ruler of Gaul, 192
Aegina, Greece, 33
Aelfgar, Earl of Mercia, 309
Aemilianus, Emperor of Rome, 154
Aemilianus, Publius Scripio, 99
Aemilius, Quintus, Roman consul, 66
Aethelbald, King of Mercia, 220
Aethelfrith, King of Northumbria, 220
Aethelred of Wessex, 270
Aethelred the Unready, King of England, 276, 308, 309
Aethelstan, King of Wessex, 268, 276
Aethelwold, Saxon leader, 276
Aethelwulf, King of Wessex, 270
Aetius, Roman general, 181, 187–91
Aetolian League, 60, 62
al-Afdal Shahinshah, son of Badr, 335
al-Afdal, son of Saladin, 335
Afonso (Affonso). *See* Alfonso
Afghan Dynasty of North India, 429
Afghanistan, 709, 712, 764, 864–65, 938–39, 1102–03, 1142, 1408–09, 1494, 1495, 1518–21. *See also* specific wars in War Index.

Afonso I of Kongo, West African king, 565
Afrainus, L., Roman general, 119
African and Malagasy Common Organization (OCAM), 1432–33
Afshin, Abbasid general, 287
Aftakin, Turkish general, 288
Agadir Incident, Morocco, 1089, 1098
Agathocles, Tyrant of Syracuse, 42, 64
"Age of Tyrants", 26
Agesilaus, King of Sparta, 47
Agger, 304
Aghlabid Dynasty of North Africa, 289–90
Agis, Spartan king, 35
Agis II, Spartan king, 55
"A-Go," Operation, 1284
Agricola, Gnaeus Julius, Roman general, 141
Agrippa, Marcus Vipsanius, Roman general and admiral, 95, 125, 126
Agrippina, wife of Claudius, 140
Aguinaldo, Emilio, Philippine leader, 950
Aguyi-Ironsi, Johnson, Nigerian general, 1446
Ahab, King of Israel, 12
Ahenobarbus, Lucius Domitius, Roman consul, 119
Ahmad Al-Mu'tamid, Abbasid caliph, 287
Ahmadnagar, India, 644
Ahmed ibn Tulun, Turkish governor, 289
Ahmed ibn Ali al-Thani, Qatar sheikh, 1401
Ahmose II, Egyptian Pharaoh, 23
Aidan, Scottish King, 220
Aigun, Treaty of (1858), 945
Ainsworth, W. L., U.S. admiral, 1268
Air Force
  after World War II, 1320
  British. *See* Royal Air Force.
  helicopters in, 1367, 1482i
  Israeli, 1351
  Japanese, 1260, 1269–70, 1279, 1305
  Soviet, 1366–67
  United States, 1366–67
    in World War II, 1206, 1222–23, 1243–44, 1269–70, 1279, 1292, 1306–7
Airplanes. *See also* specific types, e.g., Spitfire.
  development of, 1006, 1023
  strategy of, 1112–13, 1120–23
  in World War I, 1049, 1062i

1585

INDEX OF WARS

Notes: Numbers followed by the letter "i" indicate illustrations. "A.D." is used only for years 1–99.
If a year does not have "B.C." assume it is "A.D."

INDEX OF BATTLES AND SIEGES

Notes: Numbers followed by the letter "i" indicate illustrations. "A.D." is used only for years 1–99. If a year does not have "B.C." assume it is "A.D."

Aachen (1944), 1216
Aba (1968), 1446
Abensberg (1809), 825
Aberdare Forest (1953), 1437
Aberdeen (1644), 604
Aboukir
  (1799), 753
  (1801), 814
Abu Ageila (1956), 1341
Abydos (989), 286
Acoma Pueblo (1598), 569
Acre
  (1189–1191), 344
  (1291), 418
Acropolis (1825), 849
Actium (31 B.C.), 126–27, 127i
Ad Decimum (533), 202
Adabiya (1973), 1350
Adana (964), 286
Adasa (161 B.C.), 99
Adda (490), 194
Addis Ababa
  (1936), 1139
  (1941), 1173
Aden (1513), 556
Admiralties (1944), 1282
Adrianople
  (323), 166
  (378), 147, 171
  (1205), 412, 419
  (1255), 420
  (1355), 411
  (1913), 1016
Aduatuca (57 B.C.), 114
Aduwa (1896), 929
Adys (256 B.C.), 67
Aegates Islands (241 B.C.), 68
Aegospotami (405 B.C.), 36–37
Afyon (1922), 1087
Agincourt (1415), 445–50, 449i
Agra (1708), 713
Agrigentum (262 B.C.), 66
Ahmadnagar (1803), 860
Ahwaz (1915), 1048
Aiglaesthrep (455), 195
Ain Aanoub (1982), 1476
Aisne
  (57 B.C.), 114
  (1914), 1029
  (1917), 1061
  (1918), 1073
Aitape (1944), 1283
Aix-en-Provence (102 B.C.), 101–2
Ajnadain (634), 249
Akra (721), 251

Akraba (633), 249
Akyab (1943), 1262
Al Kasr al Kebir (1578), 534, 564
Alabama
  Kearsarge vs. (1864), 988
Alam Halfa (1942), 1186
Alamance Creek (1771), 774
Alamo (1836), 882
Alarcos (1195), 325
Albemarle (1864), 988
Albuera (1811), 836
Alcácer do Sol (1217), 408
Alcantara (1580), 534
Aleppo
  (944), 288
  (969), 286
  (1400), 423, 425
Alesia (52 B.C.), 117i, 117–18
Aleutian Islands
  (1942), 1253–54
  (1943), 1272
Alexandretta (1294), 403
Alexandria
  (48 B.C.), 122
  (1798), 752
  (1801), 814
Alexandrovsk (1919), 1102
Alfarrobeira (1449), 469
Alford (1645), 604
Algeciras
  (1344), 407
  (1801), 815
Algiers (1942), 1189
Alhucemas Bay (1925), 1099
Aligarh (1803), 860–61
Aliwal (1846), 863
Aljubarrota (1385), 408
Alkmaar (1572), 529
Allatoona (1864), 985–86
Allia (390 B.C.), 64
Alma (1854), 904–5
Almansa (1707), 685
Almería (1489), 469
Alnwick (1093 and 1174), 315
Alt Breisach (1703), 678
Alte Veste (1632), 589
Amadiye (1973), 1350
Amberg
  (1745), 693
  (1796), 747
Amblève (716), 221
Ambuila (1665), 656
Ambur (1749), 715
Ameixal (1663), 626
Amiens (1918), 1073–74

Amorgos (322 B.C.), 60
Amphipolis (422 B.C.), 34
An Loc (1972), 1332
Anchialus
  (708), 246
  (917), 285
Ancyra (236 B.C.), 61
Andernach
  (876), 277
  (941), 279
Andkhui (1205), 423, 428–29
Andrianople (718), 243
Andros (245 B.C.), 61
Angaur (1944), 1287
Angkor (1430–1431), 483
Angora (1402), 412, 425
Annapolis Royal (1744), 723
Antietam (1862), 962–63, 963i
Antioch
  (218), 152
  (969), 286
  (1084), 330
  (1097–1098), 339–40
  (1119), 342
Antwerp
  (1576), 529
  (1586), 530
  (1830), 843
  (1914), 1029
Anual (1921), 1099
Anzio (1944), 1207–8
Aous (198 B.C.), 95
Appomattox (1865), 991
Aquae Sextae (102 B.C.), 101–2
Aquileia (394), 172
Aquilonia (293 B.C.), 65
Arakan (1943–1944), 1274
Arar (58 B.C.), 113
Arausio (105 B.C.), 101, 107
Araxes (589), 209
Arbela (331 B.C.), 55–56, 56i
Arcadiopolis
  (970), 286
  (1194), 332
Arcis-sur-Aube (1814), 834
Arcola (1796), 750
Arcot (1751), 764–65
Ardennes
  (1914), 1024
  (1944–1945), 1217i, 1217–18
Argaon (1803), 860
Argentan (1944), 1212
Argentaria (378), 170
Argentorate (357), 167–68
Arges River (1916), 1054